AA002406

Proceedings of ASME 2023 International Design Engineering Technical Conferences and Computers and Information in Engineering Conference

(IDETC-CIE2023)

Volume 3B

49th Design Automation Conference (DAC)

August 20-23, 2023
Boston, Massachusetts

Conference Sponsors
Design Engineering Division

Computers and Information
in Engineering Division

THE AMERICAN SOCIETY OF MECHANICAL ENGINEERS
Two Park Avenue * New York, N.Y. 10016

© 2023 The American Society of Mechanical Engineers, 2 Park Avenue, New York, NY 10016, USA
(www.asme.org)

All rights reserved. Printed in the United States of America. Except as permitted under the United States Copyright Act of 1976, no part of this publication may be reproduced or distributed in any form or by any means, or stored in a database or retrieval system, without the prior written permission of the publisher.

INFORMATION CONTAINED IN THIS WORK HAS BEEN OBTAINED BY THE AMERICAN SOCIETY OF MECHANICAL ENGINEERS FROM SOURCES BELIEVED TO BE RELIABLE. HOWEVER, NEITHER ASME NOR ITS AUTHORS OR EDITORS GUARANTEE THE ACCURACY OR COMPLETENESS OF ANY INFORMATION PUBLISHED IN THIS WORK. NEITHER ASME NOR ITS AUTHORS AND EDITORS SHALL BE RESPONSIBLE FOR ANY ERRORS, OMISSIONS, OR DAMAGES ARISING OUT OF THE USE OF THIS INFORMATION. THE WORK IS PUBLISHED WITH THE UNDERSTANDING THAT ASME AND ITS AUTHORS AND EDITORS ARE SUPPLYING INFORMATION BUT ARE NOT ATTEMPTING TO RENDER ENGINEERING OR OTHER PROFESSIONAL SERVICES. IF SUCH ENGINEERING OR PROFESSIONAL SERVICES ARE REQUIRED, THE ASSISTANCE OF AN APPROPRIATE PROFESSIONAL SHOULD BE SOUGHT.

ASME shall not be responsible for statements or opinions advanced in papers or . . . printed in its publications (B7.1.3). Statement from the Bylaws.

For authorization to photocopy material for internal or personal use under those circumstances not falling within the fair use provisions of the Copyright Act, contact the Copyright Clearance Center (CCC), 222 Rosewood Drive, Danvers, MA 01923, tel: 978-750-8400, www.copyright.com.

Requests for special permission or bulk reproduction should be addressed to the ASME Publishing Department, or submitted online at: https://www.asme.org/publications-submissions/journals/information-for-authors/journalguidelines/rights-and-permissions

ISBN: 978-0-7918-8731-8

WELCOME TO ASME IDETC-CIE 2023!

On behalf of the ASME IDETC-CIE 2023 Conference Organizing Committee, it is our great pleasure to welcome you to Boston Park Plaza to participate in this year's conference! The International Design Engineering Technical Conferences & Computers and Information in Engineering Conference (IDETC-CIE) stands proudly as one of the premier ASME conferences. The 2023 ASME-IDETC-CIE marks a significant milestone as the second in-person conference after the challenges of the COVID-19 pandemic.

This year, the response to our call for presentations was beyond our expectations! We appreciate the excitement and commitment shown by our community, and we can't wait to experience the breadth of innovative ideas in the presentations. A special thanks goes to all the sub-conference organizing committees and the technical committees. Without your dedication and hard work, the success of this conference would not have been possible.

This year, we have dedicated efforts to enhance industry participation as we recognize the importance of collaboration between industry and academia. Recognizing the significance of diversity and inclusivity, we have curated activities that support our underserved and underrepresented community members.

Located in the vibrant city of Boston, a central hub for global engagement, this conference is very accessible to participants around the world. We invite you to take full advantage of this opportunity to share knowledge, collaborate, and forge new connections within our community at ASME IDETC-CIE 2023!!

Sincerely,

Sachin Goyal
Associate Professor, Department of Mechanical Engineering
University of California, Merced
Chair, ASME IDETC-CIE 2023 Conference Organizing Committee

Andreas Müller
Professor, Institute of Robotics
Johannes Kepler University Linz
Chair, ASME IDETC-CIE 2023 Conference Organizing Committee

Faez Ahmed
Assistant Professor, Department of Mechanical Engineering
Massachusetts Institute of Technology
Local Chair, ASME IDETC-CIE 2023 Conference Organizing Committee

Beshoy Morkos
Associate Professor, College of Engineering
University of Georgia
Student Activities Chair, ASME IDETC-CIE 2023 Conference Organizing Committee

ASME 25TH INTERNATIONAL CONFERENCE ON ADVANCED VEHICLE TECHNOLOGIES (AVT)

Ole Balling
Conference Chair

Angelo Bonfitto
Program Chair

The Vehicle Design Committee (VDC) promotes innovative analytical, computational, and experimental investigations in the dynamics, control, and design of full vehicle systems, subsystems, and components. With the increasing demands on driving safety and autonomy, the human-vehicle interaction, advanced driver assistance systems, and connected vehicles as well as sustainable propulsion systems and their coupling with the driver/vehicle system are included in the spectrum of topics addressed by VDC. Our members perform fundamental and applied research, and they implement technology for light/heavy vehicle design, modeling, and validation. The VDC is pleased to welcome you to the 24th International Conference on Advanced Vehicle Technologies, held as a part of the 2022 ASME IDETC-CIE. This year the AVT conference will consist of eight symposia for a total of eight sessions in the areas of: Ground Vehicles Dynamics and Controls; Modeling and Testing Tire-Terrain Interaction; Methods for Ground Vehicle Systems Design; Vehicle Electrification and Powertrain Design; Light Vehicles Design; Military and Commercial Ground Vehicle Design; and Intelligent Vehicles. We sincerely appreciate the time and services of these symposium organizers.

This Year the VDC is especially honored to host Dr. Bo Persson from the Peter Grünberg Institute, Jülich, Germany, for the William Milliken Lecture, which is entitled "Rubber friction, tire dynamics and ABS braking simulations". In addition, VDC with support from ASME-DED organizes a Workshop on Autonomous and Connected Vehicles that is open to any ASME member or IDETC attendee.

A Best Paper and a Student Best Paper (for papers authored and submitted by a student as the primary author) are awarded for conference papers that best exemplify the research advances in ground vehicle engineering based on peer reviews and the award committee's ranking.
We truly hope that this year's AVT Conference will provide you with an exciting, enriching, and rewarding experience!

43RD COMPUTERS AND INFORMATION IN ENGINEERING DIVISION CONFERENCE (CIE)

Greetings All Attendees!

The Computers and Information in Engineering Division of ASME welcomes all IDETC-CIE 2023 Conference participants to the 43rd Annual Computers and Information in Engineering Conference (CIE) in Boston, MA (USA).

The CIE conference is a premier venue for the international exchange of technical, scientific, and application knowledge related to the theory and practice of computing to support engineering activities. It provides a forum for researchers, practitioners, educators, and students from academia, industry, and government research labs to share their latest findings and challenges with the broader research community, foster collaborations, and build a sustainable research and education community.

This year we are pleased to report that there will be over 140 technical presentations in the following technical and special topic sessions, organized around the four Technical Committees of the CIE Division, namely: Advanced Modeling and Simulation, Computer-Aided Product and Process Design, Systems Engineering and Information Knowledge Management, and Virtual Environments and Systems.

Advanced Modeling and Simulation (AMS):
- Inverse Problems in Science and Engineering
- Computational Multiphysics Applications
- Uncertainty Quantification in Simulation and Model Verification & Validation
- Simulation in Advanced Manufacturing
- Material Characterization Methods and Applications

Computer-Aided Product and Process Development (CAPPD):
- Human-In-the Loop for Product Design and Automation
- Digital Human Modeling for Design and Manufacturing
- Product and Process Design Automation for Industry 4.0
- Data-Driven Product Design and Fabrication

Systems Engineering Information Knowledge Management (SEIKM):
- Design Informatics
- Systems Engineering and Complex Systems
- Knowledge Capture, Reuse, and Management
- Smart Manufacturing Informatics
- Advanced Manufacturing for Bioeconomy and Circular Economy

Virtual Environments and Systems (VES):
- Designing User Experiences for Virtual Environments
- Virtual Systems for Engineering Applications
- VES Show-and-Tell

AI + ML Approaches for Engineering (General)

Joint Sessions:
- Digital Twin: Advanced Human Modeling and Simulation in Engineering
- Digital Twin Modeling and Analytics for Advanced Manufacturing
- Physics-Informed Machine Learning for Design and Advanced Manufacturing
- Artificial Intelligence and Machine Learning in Design and Manufacturing
- Design, Simulation and Optimization for Additive Manufacturing

In addition to the technical presentations, we will host several specialized events. Accompanying a CIE Keynote Talk, four panels of leading experts from industry, government, and academia will convene to discuss topics related to the future of Computers and Information in Engineering. The Journal of Computing and Information Science in Engineering (JCISE) Spotlight panel session will highlight top articles published over the past year. At the graduate student poster session, select graduate students, each the recipient of an award stipend, will showcase their excellent works.

In addition, we will use the CIE Luncheon to recognize conference best paper awards and the CIE Division awards. We invite you all to join us at the CIE Awards Ceremony Luncheon on Tuesday August 22nd to recognize some of the outstanding research being conducted by peers, colleagues, and students alike. As always, this year's conference would not be possible without the outstanding efforts andcontributions from ASME volunteers. This year's CIE Technical Committee meetings and Division meeting will be held on the evening of Tuesday, August 22nd. It is at these meetings where we acknowledge contributors from the past year while setting the stage for the upcoming year's activities. Please plan to attend and/or join one of these meetings to become further involved in CIE activities.

We would like to thank and recognize the Technical Committee leadership this year for their hard work and contributions:
Advanced Modeling and Simulation (AMS)
- **Piyush Pandita**, Chair
- **Ahn Tran**, Vice Chair
Computer Aided Product and Process Design (CAPPD)
- **Anand Balu Nellippallil**, Chair
- **Jida Huang**, Vice Chair
Systems Engineering and Information Knowledge Management (SEIKM)
- **Douglas Van Bossuyt**, Chair
- **Dazhong Wu**, Vice Chair
Virtual Environments and Systems (VES)
- **Vinayak Krishnamurthy**, Chair
- **Yunbo "WILL" Zhang**, Vice Chair

We would like to use this opportunity to thank our symposium organizers, including Seung-Kyum Choi, Piyush Pandita, Ahn Tran, James Yang, Ashish M. Chaudhari, John Michopoulos, John Steuben, Brian Dennis, Athanasios Iliopoulos, Guanglu Zhang, Zhimin Xi, Chao Hu, Yan Wang, Gaurav Ameta, Bjorn Johansson, Chiradeep Sen, Ehsan Esfahani, Anand Balu Nellippallil, Jida Huang, Tsz Ho Kwok, Giorgio Colombo, Daniele Regazzoni, Satchit Ramnath, Marco Rossoni, Anand Balu Nellippallil, Giovanni Berselli, Weiss Cohen, Jida Huang, Jun Wang, Luis Segura, Yan Lu, Zhuo Yang, Dazgong Wu, Douglas Van Bossuyt, Yaoyao Fiona Zhao, Ying Liu, Zhenghui Sha, Farhad Ameri, Chris Hoyle, Mutahar Safdar, Hyunwoong Ko, Boonserm Kulvatunyou, Evan

Wallace, Vincenzo Ferrero, Senthil Chandrasegaran, Rebecca Friesen, Ronak Mohanty, Vinayak Krishnamurthy, Junfeng Ma, Jinjuan She, Yunbo "WILL" Zhang, Yujiang Xiang, Xianlian Alex Zhou, Dehao Liu, Sheng Yang, Yanglong Lu, Jiarui Xie, Yaoyao Fiona Zhao, Jaehyuk Kim, Fahad Milaat, Jun Wang, Chih-Hsing Chu, Dehao Liu for their efforts and hard work in paper review coordination and recommendation. We would like to thank all reviewers for their time to provide valuable feedback and help maintain high standards and improve the quality of the conference. Last but not the least, we thank all authors for submitting and sharing their latest work to shape the research directions in this community.

Moreover, we thank you for your participation in the various activities of our CIE community. We look forward to seeing you all again next year!

49TH ASME DESIGN
AUTOMATION CONFERENCE (DAC)

Dear Colleagues,

On behalf of the DAC Executive Committee, welcome to the **49th ASME Design Automation Conference (DAC)**!

Following a rigorous review process, this year's DAC technical program consists of 124 accepted papers in 25 active research areas (corresponding approximately to an acceptance rate of 92%). For the first time this year, we also solicited and accepted 23 presentation-only submissions. The technical program will be presented from Monday, August 21 to Wednesday, August 23.

Complementing our technical sessions, we will host a signature event on "**Design for Safe and Reliable Autonomous Systems**", consisting of a panel of top experts from AI modeling, autonomous vehicles, and additive manufacturing, including:
- *Dr. Qi Hommes*, Senior Director, ZooX
- *Mr. Chris Robinson*, Senior Product Manager, Ansys
- *Dr. Heng Huang*, Professor, University of Maryland – College Park
- *Dr. Rajiv Malhotra*, Associate Professor, Rutgers University – New Brunswick

Please join us for the DAC committee meeting on the evening of Tuesday, August 22. During that meeting, we will also present the DAC Young Investigator Award winner and the DAC Best Paper Award winner. We look forward to having our community come together, meet old friends, and make new ones.

From the accepted papers, ten were identified as "Papers of Distinction". These papers are listed below (ordered by paper number and including the assigned session):

- DETC2023-109380: "*Model Consistency for Mechanical Design: Bridging Lumped and Distributed Parameter Models With a Priori Guarantees*", by Randi Wang and Morad Behandish

- DETC2023-110756: "*Mixed-Variable Global Sensitiviy Analysis With Applications to Data-Driven Combinatorial Materials Design*", by Yigitcan Comlek, Liwei Wang, and Wei Chen
- DETC2023-114999: "*Machine Learning-Based Model Bias Correction by Fusing Cae Data With Test Data for Vehicle Crashworthiness*", by Yang Li, Saeed Barbat, Zhenyan Gao, Guosong Li, Ying Zhao, Jice Zeng, and Zhen Hu.
- DETC2023-116586: "*A Reliability-Based Optimization Framework for Planning Operational Profiles for Unmanned Systems*", by Indranil Hazra, Joseph Southgate, Arko Chatterjee, Shapour Azarm, Katrina M. Groth, and Matthew Weiner.
- DETC2023-116622: "*Accounting for Model and Data Uncertainty in Machine Learning Assisted Mechanical Design*", by Xiaoping Du
- DETC2023-116743: "*Characterizing Designs via Isometric Embeddings: Applications to Airfoil Inverse Design*", by Qiuyi Chen and Mark Fuge
- DETC2023-116896: "*Concurrent Probabilistic Control Co-Design and Layout Optimization of Wave Energy Converter Farms Using Surrogate Modeling*", by Saeed Azad and Daniel R. Herber
- DETC2023-116962: "*Advise: AI-Accelerated Design of Evidence Synthesis for Global Development*", by Kristen Edwards, Binyang Song, Jaron Porciello, Carolyn Huang, Faez Ahmed, and Mark Engelbert.
- DETC2023-117013: "*Integrated Sustainable Product Design With Warranty and End-of-Use Considerations*", by Xinyang Liu and Pingfeng Wang
- DETC2023-117400: "*Car Drag Coefficient Prediction With Depth and Normal Renderings*", by Binyang Song, Chenyang Yuan, Faez Ahmed, Nikos Arechiga, and Frank Permenter

Authors from our community will present these and many other excellent papers throughout the conference. We encourage you to support your colleagues by attending their presentations and participating in the discussions.

Finally, organizing the conference requires the generous effort of many individuals. We are particularly grateful to all session organizers and paper review coordinators:

Faez Ahmed, Janet K. Allen, Jesse Austin-Breneman, A. Emrah Bayrak, Morad Behandish, Bill Bernstein, Ramin Bostanabad, Amy Bilton, Wei (Wayne) Chen, Souma Chowdhury, Daniel Cooper, Xiaoping Du, Bryony DuPont, Paul Egan, Ehsan Esfahani, Cong Feng, Yan Fu, Payam Ghassemi, Joshua Hamel, Daniel Herber, Zhen Hu, Horea Ilies, Namwoo Kang, Leifur Leifsson, Mian Li, Xingchen Liu, Yuanzhi Liu, Nordica MacCarty, Ali Mehmani, Nicholas Meisel, Zhenjun Ming, Farrokh Mistree, Seung Ki Moon, Beshoy Morkos, Venkat Nemani, Saigopal Nelaturi, Julián Norato, Philip Odonkor, Herschel Pangborn, Rahul Renu, Daniel Selva, Ada-Rhodes Short, Binyang Song, Eun Suk Suh, Ahn Tran, Zequn Wang, Kate Whitefoot, Natasha Wright, Hongyi Xu, Nita Yodo, Jie Zhang, Zhibo Zhang, Fiona Zhao, Yuqing Zhou

On behalf of the entire DAC community, we welcome you to another enjoyable and thought-provoking Design Automation Conference.

We look forward to seeing you in Boston!

Christopher McComb
Conference Chair

Chao Hu
Program Chair

20TH INTERNATIONAL CONFERENCE ON DESIGN EDUCATION (DEC)

On behalf of the Design Education Committee, we welcome you to the 20th annual International Conference on Design Education. The focus of this conference is on design education among educators, practitioners, and researchers.

This year's DEC Program consists of four technical symposia – (DEC-1) *Implementation, Assessment and Research Methods Across the Curriculum* (DEC-1), *Diversity and Inclusion in Design Education* (DEC-2), *Innovative Practices in Design Education* (DEC-3), and *Demos and Presentation Only* (DEC-4). The Demos and Presentation Only session will include presentations and provide ample opportunity for discussion with the presenters to give feedback on emerging design education research. Refer to the conference Technical Program for the times and locations of the technical sessions. In addition to our technical symposia, we will be continuing our mentorship program for graduate students.

The DEC Best Paper for the 2023 Conference is:

IDETC2023-116688, "Nature Versus Nurture: The Influence of Classroom Creative Climate on Risk-Taking Preferences of Engineering Students," Authors: Aoran Peng, Jessica Menold, Scarlett Miller

We extend special appreciation to our technical session Review Coordinators: Mohammad Fazelpour, Elizabeth Starkey, and Charlotte de Vries. We also give our sincerest thanks to all the reviewers of technical papers; they have ensured the quality of this year's conference.

The DEC technical committee meeting will be posted in the Technical Program. At the meeting we present many of the DEC Awards and plan for next year's conference, which includes the election of new committee leadership members. Everyone is welcome to attend, including new attendees and graduate students. Our meeting is streamlined to respect members' participation in other committees.

Nicholas Meisel
Conference Chair

Rahul Renu
Conference Program Chair

28TH DESIGN FOR MANUFACTURING AND
THE LIFE CYCLE CONFERENCE (DFMLC)

The ASME Design for Manufacturing and the Life Cycle Committee welcomes participants to the 28th Annual Design for Manufacturing and the Life Cycle Conference. The ASME Design for Manufacturing and the Life Cycle Conference is the main international forum for the exchange of technical and scientific information on the theory and practice of Integrated Product and Process Development, Sustainable Design and Manufacturing, Product Lifecycle Management (PLM), and Design for X (DFX) Methods. This conference provides a forum for researchers, practitioners, and educators from academia, government organizations, and industry to share their latest results and challenges with the research community.

We are happy to report that this year's conference continues to feature many new and exciting results and methods to be presented as part of the conference's technical sessions. This year's DFMLC conference includes 21 technical papers and 22 technical presentations across 8 sessions, as follows:

- Session 1: Life Cycle & Human Factors Decision Making
- Session 2: Modeling and Optimization for Sustainable Design and Manufacturing
- Session 3: Design for Supply Chain, End of Life Recovery, and Large Systems
- Session 4: Design for Manufacturing and Assembly
- Session 5: Design for Additive Manufacturing 1
- Session 6: Design for Additive Manufacturing 2
- Session 7: Design of Product-Service and Energy Systems
- Session 8: Special Session: Design Tool & Commercialization Showcase

We would like to thank all the authors for submitting papers, the paper reviewers for sharing their time and expertise, and the session chairs/co-chairs for their participation. Special thanks go to the DFMLC Special Session Chair, Albert Patterson, and the paper review coordinators/co-coordinators for managing the papers through the review process: Hao Zhang, Vincenzo Ferrero, William Bernstein, Bryony Dupont, Yong Hoon Lee, Sara Behdad, Yongxian Zhu, Soonjo Kwon, Satya Peddada, Yaoyao Zhao, Xinyi (Serena) Xiao, Albert Patterson, Amin Mirkouei, Abigail Clarke-Sather, Paul Egan. Your participation and hard work have been vital for the success of the DFMLC conference!

This year, Dr. Gul Kremer, Dean for the University of Dayton School of Engineering, will present the DFMLC keynote lecture. Professor Kremer's research accomplishments focus on applied decision sciences and operations research for product and design systems, and other research interests include sustainability, system complexity, design creativity, and engineering education. There will be a presentation of the 2023 DFMLC Conference Kos Ishii-Toshiba Award for sustained and meritorious contributions to design for manufacturing and the life cycle at the DED luncheon on Monday, August 21st.

The 2023 DFMLC Conference also features a special presentation session. The "Design Tool & Commercialization Showcase" highlights new design tools developed by the members of the ASME Design community in both digital and physical forms.

The DFMLC technical committee meeting will include a review of DFMLC activities during the 2022-2023 cycle. The DFMLC Awards, including the Best Paper Award for the 2023 DFMLC conference, will also be presented in this meeting, and the technical committee will plan for next year's conference. Everyone is welcome to attend.
On behalf of the entire DFMLC community, we welcome you to the 28th Design for Manufacturing and the Life Cycle virtual conference!

Paul Egan
Conference Program Chair

Daniel Cooper
Conference Chair

35TH INTERNATIONAL CONFERENCE
ON DESIGN THEORY AND METHODOLOGY (DTM)

On behalf of the ASME Design Theory and Methodology Committee, we would like to welcome you to the 35th International Conference on Design Theory and Methodology (DTM). Our conference focuses on fundamental design theory and methodologies, and their application in engineering contexts, with contributions provided by both researchers and practitioners.

This 2023 DTM conference includes 54 technical paper presentations and 10 lightning talks. Thematically, the conference includes contributions associated with our four broad foci: Design Theory, Design Methods, Design People, and Design Practice. In addition, this year's conference features a joint session between the Design Education Committee, the Design for Manufacture and Lifecycle Committee, and DTM titled *DfAM Principles and their Education*. This year's conference also features a student poster session where selected Ph.D. students showcase their dissertation proposals.

There were 72 papers submitted and reviewed by an incredible cohort of review coordinators and reviewers. A total of 239 reviews were completed by 155 reviewers. The review coordinators for this year's conference include: Ambrosio Valencia Romero, Astrid Layton, Christine Toh, Hyeonik Song, James Righter, Jinjuan She, Joshua Summers, Kelley Dugan, Kosa Goucher-Lambert, Maha Haji, Mansur M. Arief, Paul Grogan, Rohan Prabhu, Srinivasan Venkataraman, Vivek Rao, Vrushank Phadnis, Youyi Bi, and Zhenghui Sha. It is through the service of these individuals that we are able to maintain the high-quality expectations of the DTM conference.

We are excited to welcome you to this year's conference and hope that you find it engaging, informative, and beneficial.

Conference Chair
Dr. Vimal K. Viswanathan
San Jose State University

Program Chair
Dr. Rahul S. Renu
Francis Marion University

19TH IEEE/ASME INTERNATIONAL CONFERENCE ON MECHATRONICS AND EMBEDDED SYSTEMS AND APPLICATIONS (MESA 2023)

We are pleased to welcome everyone to the 19th IEEE/ASME International Conference on Mechatronics and Embedded Systems and Applications (MESA 2023). The goal of the MESA 2023 is to bring together experts from the fields of mechatronic and embedded systems, disseminate the recent advances in the area, discuss future research directions, and exchange application experience. MESA 2023 will especially bring out and highlight the latest research results and developments in Industry 4.0 and Artificial Intelligence (AI) in the fields of mechatronics and embedded systems. The success of MESA 2023 would be impossible without the tireless effort and dedicated work of the Members of the Organizing Committees. We would like to express our sincere thanks to Symposium Chairs for their wisdom and hard work in coordinating the review of all submitted papers. We are grateful for Members of the International Program Committee and reviewers for their thorough review of the papers. This year the program committee selected about 30 technical presentations following a review process by two or more expert reviewers for each proposed paper. We sincerely hope that MESA 2023 will be a place for excellent discussions that will put forward new ideas advance educational endeavors and promote active research collaborations.

Conference Chair,
Prof. Adriano Mancini
Università Politecnica delle Marche, Ancona, IT

Program Chair,
Prof. Matteo Claudio Palpacelli
Università Politecnica delle Marche, Ancona, IT

17TH INTERNATIONAL CONFERENCE ON MICRO- AND NANOSYSTEMS (MNS)

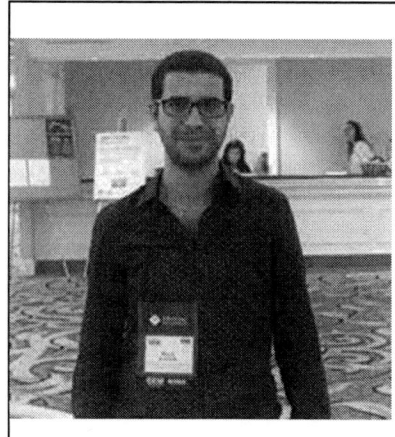

Conference Chair:
Prof. Najib Kacem, Ph.D., FEMTO-ST Institute, France,
najib.kacem@femto-st.fr

Program Chair:
Prof. Jian Zhao, Ph.D., Dalian University of Technology, China,
jzhao@dlut.edu.cn

Welcome to the17th International Conference on Micro- and Nano-systems (MNS) with the topic of "The Next Advances in MEMS", we would like to welcome you and thank you for participating. This conference, sponsored by the Technical Committee of Micro and Nano-systems, an integral part of the ASME Design Engineering Division, will provide researchers in industry, academia, and government a forum to exchange scientific and technical information related to recent developments and emerging issues in the design, mechanics, dynamics, control, and fabrication of micro- (MEMS) and nano-scale (NEMS) systems.

This conference is organized around 5 technical sessions, one of which is jointly offered with the 19th International Conference on Multibody Systems, Nonlinear Dynamics, and Control and the 35th Conference on Mechanical Vibration and Noise:

- Keynote Lecture: Professor Ashwin A. Seshia
- MNS-1: Nonlinear Dynamics and Vibrations of MEMS and NEMS (joint session with MSNDC and VIB)
  - Organizers: Najib Kacem (najib.kacem@femto-st.fr), Hanna Cho (cho.867@osu.edu), Jian Zhao (jzhao@dlut.edu.cn)
- MNS-2: Micro/Nano Bioengineering
  - Organizers: Dumitru Caruntu (Dumitru.Caruntu@utrgv.edu), Brian Jensen (bdjensen@byu.edu), Chu-Yu Huang (tomhuang@nchu.edu.tw)
- MNS-3: Micro/Nano Robotics and Functional Materials
  - Organizers: Irene Fassi (Irene.Fassi@stiima.cnr.it), Yu Liu (yu.liu@vip.163.com), Hoe Joon Kim (joonkim@dgist.ac.kr), Mohammad H. Hasan (hhasan_mohammad@columbusstate.edu), Longquiu Li (longqiuli@hit.edu.cn)
- MNS-4: Micro/Nano IoT, Sensors and Computing
  - Organizers: Muhammad Raziuddin A. Khan (muhammad.khan@navy.mil), Fadi Alsaleem (falsaleem2@unl.edu), Pourkamali Anaraki Siavash (Siavash.Pourkamali@utdallas.edu)
- MNS-5: Micro/Nano Power Sources and Storage
  - Organizers: Oliver M. Barham (oliver@olivermbarham.com), Muhammad Raziuddin A. Khan (muhammad.khan@navy.mil), Marc Litz (marc.s.litz.civ@mail.mil)

This conference provides a forum for researchers, practitioners, educators, and students from industry, academia, and government research labs to share their latest findings and challenges with the broader research community, foster collaborations, and build a sustainable research community.

We are pleased to offer Pr Ashwin A. Seshia as the MNS keynote speaker. Ashwin A. Seshia is a Professor of Microsystems Technology in the Department of Engineering at Cambridge University and a Fellow of Queens' College, Cambridge. He has acted in numerous service and leadership roles for the MEMS, sensors, and frequency control technical communities, and is currently an Editor of the IEEE Journal of Microelectromechanical Systems and a member of the executive committee of the European Frequency and Time Forum.

We would like to thank all the authors for submitting papers and talks and sharing their work in our conference. We would also like to thank the reviewers for providing valuable feedback to help improve the reporting and the quality of the conference, and finally the session chairs and co-chairs that worked on coordinating the paper review process.

We welcome conference participants to become involved with our technical committee. If you are interested in becoming involved in helping to organize our conference, please contact a conference organizer to inquire, and feel free to attend the technical committee meeting which will be held on Tuesday evening, Aug 22nd, from 6-7pm. This meeting is open to all. Room locations are announced in the program. Our community will continue to grow and flourish with your active participation as we work to define our vision for future events.

We welcome you to the17th International Conference on Micro- and Nanosystems (MNS)!
Sincerely,
Najib, Jian, and the entire 2023 MNS Conference team.

47TH MECHANISMS AND ROBOTICS CONFERENCE (MR)

The Mechanisms and Robotics Technical Committee of the ASME Design Engineering Division would like to warmly welcome you to the 47th Mechanisms and Robotics Conference, the premier international forum for the exchange of technical and scientific information on the theory and application of mechanical systems, mechanisms, and robotics.
The first conference, as The Conference on Mechanisms, was held at Purdue University, West Lafayette, Indiana, in 1953. ASME took over the conference and formed the ASME Biennial Mechanisms Conference in 1964. The conference was renamed the ASME Biennial Mechanisms and Robotics Conference in 2000. Starting in 2005, the conference became an annual conference, the ASME Mechanisms and Robotics Conference. Nowadays, the Mechanisms and Robotics Conference is held annually as a part of the ASME International Design Engineering Technical Conferences & Computers and Information in Engineering Conference.

This year we have assembled an exciting conference program and a slate of activities for the attendees, with 100 peer-reviewed technical papers and six technical presentations/posters organized into 9 technical symposia, a keynote speech, an early career invited talk session, a symposium keynote speech, and the Student Mechanisms and Robot Design Competition. Paper topics range throughout areas central to the design of mechanical, mechatronic, and robotic systems, including kinematics, dynamics, design, analysis and validation, compliant mechanisms, origami-based design, metamaterials for mechanisms, novel mechanisms and robots, mobile robots, and various applications. Our Keynote Speech will be given by Prof. Jian Dai, Director of Institute of Robotics at Southern University of Science and Technology and Honorary Chair Professor of King's College London, with his speech entitled: "Reconfiguration that Evolves into Robotics in Arts, Healthcare, and Production".

Submitted papers were eligible for several awards, including the Mechanisms and Robotics Best Paper award, A.T. Yang Memorial award, and Compliant Mechanisms award. The authors of selected papers of the Mechanisms and Robotics Conference are invited to submit enhanced archival versions of their papers to an IDETC Special Issue of the *ASME Journal of Mechanisms and Robotics*. We would like to thank Jian Dai, Chair of the Awards Committee, for coordinating the selection of the awards. Please attend our award session preceding the MR Keynote speech for the presentation of these awards.

The conference and program chairs would like to extend special thanks to all the volunteers who participated in the peer-review process to produce this high-quality program, especially the symposium organizers who coordinated the process:

MR-1: Mechanisms Synthesis & Analysis: Latifah Nurahmi, Kuan-Lun Hsu, Jieyu Wang
MR-2: Theoretical & Computational Kinematics (A.T. Yang Symposium): Nina Robson, Hongliang Shi, Haohan Zhang
MR-3: Compliant Mechanisms: Hongzhe Zhao, Jovana Jovanova, Giovanni Berselli
MR-4: Origami-Based Engineering Design: Shikui Chen, Suyi Li, Jared Butler
MR-5: Motion Planning, Dynamics, and Control of Robots: Damien Chablat, Joo Kim, Andreas Mueller, Jeffrey W. Herrmann
MR-6: Medical and Rehabilitation Robotics: Carl Nelson, Abbas Fattah
MR-7: Novel Mechanisms, Robots, and Applications: Guowu Wei, Reza Fotouhi, Salih Abdelaziz
MR-8: Soft and Continuum Mechanisms: Girish Krishnan, Sree Kalyan Patiballa, Vishesh Vikas
MR-9: Design, Analysis and Fabrication of Architected Materials and Structures: Nilesh D. Mankame, Pablo D. Zavattieri, David Restrepo, Tian "Tim" Chen
MR-10: Student Mechanism and Robot Design Competition: Long Wang, Gaurav Singh, Haiyang Li, Huijuan Feng, Colette Abah
MR-11: Special Early Career Session of Invited Presentations: Mark Plecnik

We extend special thanks to all authors, reviewers, presenters, symposium organizers, session chairs, and other volunteers who have contributed to the overall success of the conference. We trust that you will enjoy the conference and look forward to your continued support to our future Mechanisms and Robotics Conferences.

Conference Chair and Co-Chair:
Dongming Gan, Purdue University
Ketao Zhang, Queens Mary University of London

Program Chair and co-Chairs:
Guangbo Hao, University College Cork
Yu She, Purdue University
Mark Plecnik, University of Notre Dame

19TH INTERNATIONAL CONFERENCE ON MULTIBODY SYSTEMS, NONLINEAR DYNAMICS, AND CONTROL (MSNDC)

On behalf of the ASME Technical Committee on Multibody Systems and Nonlinear Dynamics, we extend a wholehearted welcome to the attendees of the 19th International Conference on Multibody Systems, Nonlinear Dynamics, and Control (MSNDC). Consisting of 14 symposia, the conference features nearly 70 presentations covering traditional and emerging topics in the broad areas of multibody systems and nonlinear dynamics. This event presents a unique opportunity for researchers, practitioners, educators, and students to report their accomplishments, exchange ideas, and become familiar with emerging trends in the field. The conference is organizing the MSNDC Best Paper and Best Student Paper competitions.

This year, we are honored to recognize Professor Gábor Stépán as the recipient of the Lyapunov Award for his seminal contributions in numerical methods and cutting-edge applications such as machine tool vibrations, balancing, wheel shimmy, vehicle traction, breaking, stability of robot control, human balancing and traffic control. Established in 2003, the Lyapunov Award recognizes lifelong contributions to the field of nonlinear dynamics.

We are honored to host two keynote lectures by Professor Arvind Raman and Professor Aki Mikkola. Professor Arvind Raman is the Robert V. Adams Professor in Mechanical Engineering at Purdue University. His research focuses on exploiting nonlinear dynamics for innovations in diverse interdisciplinary areas such as nanotechnology, biomechanics, and appropriate technologies for sustainable development. He is an ASME fellow, an ASME Gustus Larson Memorial Award recipient, Keeley fellow (Oxford), College of Engineering outstanding young investigator awardee, and NSF CAREER awardee.

Professor Aki Mikkola is a Professor in the Department of Mechanical Engineering at LUT University, Lappeenranta, Finland. Currently, he leads the research team of the Laboratory of Machine Design. He has been awarded five patents, has contributed to more than 150 peer-reviewed journal papers, and has presented more than 100 conference articles. His major research activities are related to flexible multibody dynamics, rotating structures, and biomechanics. He is currently Editor-in-Chief of the Journal of Multibody System Dynamics (Springer).

Last but not least, we would like to acknowledge the all-important effort and contribution made by the symposium organizers as well as manuscript reviewers – thank you very much. Your help has been essential. We would also like to thank all contributors for choosing this conference as the venue for sharing the outcomes of their intellectual pursuits.

Conference Co-Chairs:
Hiroyuki Sugiyama
University of Iowa
Pierpaolo Belardinelli
University of Marche

Program Co-Chairs:
Kiran X. D'Souza
Ohio State University
Grzegorz Orzechowski
LUT University

2023 POWER TRANSMISSION AND GEARING CONFERENCE (PTG)

On behalf of the ASME Technical Committee on Power Transmission and Gearing (PTG), we would like to extend a wholehearted welcome to the attendees of the 2023 International Conference on Power Transmission and Gearing. We thank all of the authors for choosing this forum to share their latest research findings, and all of those who chose to attend this conference. The contributions of the leading researchers and practitioners from around the world make this conference an ideal forum and opportunity for enhancing the technology of power transmission and gearing and exchanging ideas. We hope you will take advantage of this unique opportunity to learn about the latest research work, become familiar with emerging trends in the field and network as well.

PTG 2023 features outstanding full research papers and presentations covering a wide range of topics on power transmission and gearing, which include:

- Gear Geometry
- Gear Analysis
- Materials, Fatigue
- Gear Dynamics and Noise
- Gearbox Design, Reliability, and Diagnostics
- Gear Manufacturing
- Lubrication and Efficiency
- Bearings, Clutches, Couplings, and Splines
-

We acknowledge and thank all the reviewers for their support and assistance and all the following members of the PTG Committee for their dedicated service and efforts in organizing this conference:

Paris Altidis, Flexco Steel Lacing Company
Christopher Cooley, Oakland University
Richard Dippery, Kettering University
Brian Dykas, US Army Research Laboratory
Qi Fan, The Gleason Works
Alfonso Fuentes, Rochester Institute of Technology
Robert Giachetti, Exponent
Robert Handschuh, NASA Glenn Research Center
Adrian Hood, US Army Research Laboratory
Mohammad Hotait, General Motors
Don Houser, The Ohio State University
Murat Inalpolat, University of Massachusetts Lowell
Ahmet Kahraman, The Ohio State University
Mark Klein, Honda Motor Company
Mohsen Kolivand, Meritor
Timothy Krantz, NASA Glenn Research Center
Sheng Li, Wright State University
Teik C. Lim, New Jersey Institute of Technology
Kenneth Nowaczyk, Ford Motor Company
Robert Parker, University of Utah
Alfred Pettinger, Engineering Systems Inc.
Steve Siegert, Borg Warner
Avinash Singh, General Motors
Jeremy Wagner, John Deere Product Engineering Center
Yawen Wang, University of Texas, Arlington
Jon Williams, Hilliard Corporation
Brian Wilson, Advanced Drivetrain Engineering
Carlos Wink, Eaton Vehicle Group

We look forward to a successful conference and hope that you enjoy all the events and your stay in Boston as well.

David Talbot
Ohio State University
Conference Chair

Hai Xu
General Motors
Program Chair

35TH CONFERENCE ON MECHANICAL VIBRATION AND NOISE (VIB)

On behalf of the Technical Committee on Vibration and Sound (TCVS), we cordially welcome you to the 35th Conference on Vibration and Noise (VIB). This conference covers a broad spectrum of topics related to vibratory systems including those at emerging frontiers of science and engineering as well as traditional fields where mechanical vibrations are essential. VIB provides a setting for dissemination and discussion of the state of the art of modeling, analysis, and experimentation in all aspects of vibration and noise research. This year's conference includes close collaborations with other IDETC tracks to bring together researchers with similar interests, enhance the technical program, and improve the attendee experience. The following symposia make up this year's VIB:

VIB-1	Dynamics and Waves in Structures and Metamaterials
VIB-2	Vibration and Stability of Mechanical Systems
VIB-3	Energy Harvesting
VIB-4 and MR-4	Origami-Inspired Engineering: Design, Dynamics, and Everything in Between
VIB-5, MNS-1 and MSNDC-13	Nonlinear Dynamics and Vibrations of MEMS and NEMS
VIB-6 and MSNDC-3	Contact Dynamics and Jointed Structures
VIB-7 and MSNDC-14	Industrial Applications of Vibration, Acoustics and Dynamics
VIB-8 and MSNDC-12	Rotating Systems and Rotor Dynamics
VIB-9 and MSNDC-4	Nonlinear Systems & Phenomena
VIB-10 and MSNDC-6	Machine Learning Applications in Vibrations and Dynamics
VIB-11 and MSNDC-7	Time-delay, Time-varying and Discontinuous Dynamical Systems
VIB-12 and MSNDC-11	Dynamics and Control of Smart Structures and Systems
VIB-13	Dynamics of Biological, Bio-Inspired and Biomimetic Systems
VIB-14 and PTG-3	Gear Dynamics and Noise
VIB-15	Vibration Measurement, Signal Processing, and Structural Damage Detection

VIB is highlighted by keynote lectures celebrating top honors given by TCVS. Professor Steven Shaw from the Florida Institute of Technology is the recipient of the J. P. Den Hartog Award for lifetime contributions to the teaching and practice of vibration engineering. His keynote talk is titled "Centrifugal Pendulum Vibration Absorbers – from Den Hartog to Now". Professor Bogdan Epureanu from the University of Michigan will present a keynote talk titled "Physics-Informed Data-Driven Methods for Reduced Order Modeling in Dynamics." He is the recipient of the N. O. Myklestad Award in recognition of his innovate contributions to vibration engineering in the area of multi-physical systems. Professor Kathryn Matlack from the University of Illinois Urbana-Champaign is the recipient of the C. D. Mote, Jr. Early Career Award for research excellence in the field of vibration and acoustics. Her keynote talk is titled "Manipulating Vibrations with Phononic Materials".

As part of VIB, TCVS graciously sponsors a student paper competition and student travel support program. This year's conference includes a new undergraduate research symposium on dynamics, vibration and acoustics.

We gratefully acknowledge the efforts of the VIB symposium organizers, reviewers, and authors. It is your efforts that make this conference vibrant.

Christopher G. Cooley   Mark Jankauski
Oakland University   Montana State University
Conference Chair   Technical Program Chair

2023 IDETC-CIE CONFERENCE ORGANIZING COMMITTEE

Sachin Goyal
University of California
Conference Co-Chair

Andreas Müller
Johannes Kepler University Linz
Conference Co-Chair

Beshoy Morkos
University of Georgia
Student Activities Chair

Faez Ahmed
Massachusetts Institute of Technology
Local Chair

Computers & Information in Engineering Division Executive Committee (CIE)
Past Chair/Awards Chair: Mahesh Mani
Chair: Paul Witherell
Vice Chair & Conference Chair: Caterina Rizzi
Program Chair: Robert Wendrich
Secretary: Krishananand Kaipa
Member at Large: John Steuben
Member at Large: Daniela Faas
Liaison: Marc Halpern

Design Engineering Division Executive Committee (DED)
Past Chair: Dane Quinn
Chair/Vice Chair: Micky Caruntu
Technical Committee Liaison: Scott Ferguson
Conference Executive: Scarlet R. Miller
Secretary: Mary Frecker
Honors & Awards: Stefano Lenci
Publications: Michael Kokkolaras

CONFERENCE ORGANIZERS

25th International Conference on Advanced Vehicle Technologies (AVT)

Costin Untaroiu
Virginia Tech University
Conference Chair

Ole Balling
Aarhus University, Denmark
Program Chair

43rd Computers and Information in Engineering Division Conference (CIE)

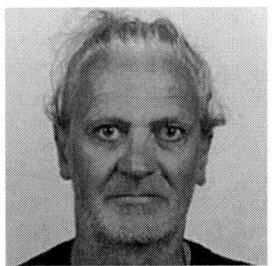

Caterina Rizzi
University of Bergamo
Conference Chair

Robert E. Wendrich
Rawshaping Technology RBSO
Program Chair

49th ASME Design Automation Conference (DAC)

Chris McComb
Penn State University
Conference Chair

Chao Hu
University of Connecticut
Program Chair

20th International Conference on Design Education (DEC)

Nicholas Alexander Meisel
Penn State University
Conference Co-Chair

Rahul Renu
Francis Marion University
Conference Co-Chair

28th Design for Manufacturing and the Life Cycle Conference (DFMLC)

Daniel Cooper
University of Michigan
Conference Chair

Paul Egan
University of Michigan
Texas Tech University

35th International Conference on Design Theory and Methodology (DTM)

Vimal K. Viswanathan
San Jose State University
Conference Chair

Rahul Renu
Francis Marion University
Program Chair

19th IEEE/ASME International Conference on Mechatronic and Embedded Systems and Applications (MESA)

Adriano Mancini
Università Politecnica delle Marche
Conference Chair

Matteo-Claudio Palpacelli
Università Politecnica delle Marche
Program Chair

47th Mechanisms and Robotics Conference (MR)

Dongming Gan
Purdue University
Conference Chair

Ketao Zhang
*Queen Mary
University of London*
Conference Co-Chair

Dongming Gan
Guangbo Hao
Program Chair

Yu She
Purdue University
Program Co-Chair

Mark Plecnik
*University of Notre
Dame*
Program Co-Chair

17th International Conference on Micro- and Nano-Systems (MNS)

Najib Kasem
*Univ. Bourgogne
Franche-Comté,*
Program Co-Chair

Jian Zhao
*Dalian University of
Technology*
Program Co-Chair

19th International Conference on Multibody Systems, Nonlinear Dynamics, and Control (MSNDC)

Pierpaolo Belardinelli
Polytechnic University of Marche
Conference Chair

Hiroyuki Sugiyama
University of Iowa
Conference Co-Chair

Kiran X. D'Souza
Ohio State University
Program Co-Chair

Grzegorz Orzechowski
LUT University
Program Co-Chair

2023 International Power Transmission and Gearing Conference (PTG)

David Talbot
The Ohio State University
Program Chair

Hai Xu
Oakland University
Program Co-Chair

35th Conference on Mechanical Vibration and Noise (VIB)

Chris Cooley
Oakland University
Conference Chair

Mark Jankauski
Montana State University
Program Chair

REVIEWERS

Mohammad A. Al-Shudeifat

Colette Abah

Salih Abdelaziz

Abdessattar Abdelkefi

Lanie Abi

Gizem Acar

Gabriele Maria Achilli

Hamid Afshari

Fatemeh Afzali

Alice Agogino

Milton Aguirre

Malena Agyemang

Faez Ahmed

Muhammad Hassaan Ahmed

Christopher Aksland

Louay Al Roomi

Oreoluwa Alabi

Michael Alexander

Khaled Alhazza

Hessein Ali

Sami Alkharabsheh

Douglas Allaire

Janet K. Allen

James Allison

Ali Almandeel

Mohammad Alsager Alzayed

Fadi Alsaleem

Abdulrahman Al-Shanoon

Farhad Ameri

Gaurav Ameta

Ali Amoozandeh

Di An

Luis Angel

Anusha Anisetti

Joel Anstrom

Nicole Apetre

K. Arabian

Vigen Arakelyan

Andrea Arena

Mojtaba Arezoomand

Mansur Arief

Keisuke Arikawa

Ryan Arlitt

Jessica Armstrong

Mauricio Arredondo

Alessio Artoni

Manan Arya

Keivan Asadi

Alireza Asadpoure

Doris Aschenbrenner

Simone Asci

Omar Ashour

Vanessa Audrey

Daniel Aukes

Jesse Austin-Breneman

Alkim Avsar

Saeed Azad

Enrico Babilio

Erin E. Bachynski-Polić

Noah Bagazinski

Duygu Bagci Das

XianXu Bai

Youdun Bai

Nikhil Bajaj

Albin Bajrami

Firooz Bakhtiari-Nejad

Ole Balling

Federico Ballo

Hyunseung Bang

Qifang Bao

Oumar Barry

Vito Basile

Alain Batailly

Mattia Battarra

Alparslan Emrah Bayrak

Sara Behdad

Amir Behjat

MohammadMahdi Behzadi

Pierpaolo Belardinelli

Pinhas Ben-Tzvi

William Bernstein

Giovanni Berselli

Harish Bezawada

Anindya Bhaduri

Kiran Bhole

Anudeep Bhoopalam

Pranav Bhounsule

Youyi Bi

Michele Bici

Dustin Bielecki

Amy Bilton

Andrew Birnbaum

Thijs Blad

Lucienne Blessing

Angelo Bonfitto

Joran Booth

Monica Bordegoni

Yuri Borgianni

Ramin Bostanabad

Armin Bosten

Joe Bradley

Adam R. Brink

Levent Burak Kara

Ravi Burla

Grace Burleson

Alex Burnap

Jared Butler

Asad Butt

Srikumar C. Gopalakrishnan

Jonathan Cagan

Runze Cai

Massimo Callegari

Benjamin Calmé

Chris Cameron

Jake Campbell

Stephen Canfield

Antonio Caputi

Luca Carbonari

Biagio Carboni

Stephane Caro

Julia Carroll

Marina Carulli

Dumitru Caruntu

Giandomenico Caruso

Arnaldo Casalotti

Damien Chablat

James Chagdes

Surya Chakrabarti

Nilanjan Chakraborty

Joel Chan

Kuei-Yuan Chan

Senthil Chandrasegaran

Ching-Yuan Chang

Tse-Shao Chang

Abheek Chatterjee

Ashish Chaudhari

Prathamesh Chaudhari

Brian Chell

Shikui Chen

Yu-Hsun Chen

Liuqing Chen

Jiangce Chen

Shuping Chen

Wei Chen

Cheng Chen

Genliang Chen

Zhen Chen

Hongrui Chen

Chong Chen

Zheyuan Chen

Yangquan Chen

Yao Cheng

Christine Chevallereau

Heng Chi

Shital Chiddarwar

Hyunkyoo Cho

Hanna Cho

Seung-Kyum Choi

Leah Chong

Souma Chowdhury

Sanjib Chowdhury

Chih-Hsing Chu

Ching-Wei Chuang

Erik Chumacero-Polanco

Hyun-Joon Chung

Wu Chunnong

Ender Cigeroglu

Abigail Clarke-Sather

Timothy Cleary

François Cluzel

Bryan Cochran

Peter Coffin

David Cohen

Courtney Cole

Arianne Collopy

Giorgio Colombo

Justin Conzola

Christopher Cooley

Daniele Costa

Ronald Couch

Tonghui Cui

Yaxin Cui

Xu Cui

Ranting Cui

Daicong Da

Xiang Dai

Shanna Daly

Nicole Damen

Revanth Damerla

Karthik Dantu

Francesco Danzi

Oguzhan Das

Madhurima Das

Tim Davenport

Michael Dawson

Jan De Jong

Charlotte De Vries

Ebru Demir

Onan Demirel

Shiguang Deng

Kshiteej Deshmukh

Shrinath Deshpande

Harish Devaraj

Gaurang Dharap

Somayajulu L. Dhulipala

Ahmet Dindar

Joseph Distefano

Donald Docimo

Zoltan Dombovari

Bin Dong

Jiayuan Dong

Guoying Dong

Andy Dong

Ata Donmez

Daniel Dopico

Greg Dorgant

Alberto Doria

Arinan Dourado

Kiran D'souza

Xiaoping Du

Xianping Du

Ping Du

Daniel Duecker

Kelley Dugan

Shammo Dutta

Salvador Echeveste

Kristen Edwards

Paul Egan

Mohammed El Kihal

Richard Ellingham

Aliakbar Eranpurwala

Jose Escalona

Ehsan T Esfahani

Lorenzo Failla

Huashuai Fan

Qi Fan

Hongbin Fang

Lezheng Fang

Irene Fassi

Abbas Fattah

Claudio Favi

Mohammad Fazelpour

Brian Feeny

Zhang Feihong

Cong Feng

Huijuan Feng

Zhipeng Feng

Yanbiao Feng

Scott Ferguson

Javier Fernández Aceituno

Vincenzo Ferrero

Francesco Ferrise

Brett Fiedler

Matteo Forlini

Reza Fotouhi

Mary Frecker

Rebecca Friesen

Matthew Fronk

Katherine Fu

Jiaming Fu

Yan Fu

Kazuko Fuchi

Alfonso Fuentes Aznar

Mark Fuge

Lawrence Funke

Alessandro Galdelli

Juan Galvis

Dongming Gan

Anthony Garland

Diego Garzon-Alvarado

Joseph Gattas

John Gero

Johannes Gerstmayr

Masood Ghasemi

Payam Ghassemi

Bogdan-George Gherman

Anna Ghidotti

Seyede Fatemeh Ghoreishi

Sayan Ghosh

Amaninder Gill

Andrew Gillman

Daniel Giraldo-Guzman

Massimiliano Gobbi

Francisco Gonzalez

Alex Gorodetsky

Kosa Goucher-Lambert

Marc Gouttefarde

Benjamin Graber

Paul Graham

Daniele Grandi

Paul Grogan

Magdalena Grohman

James Guest

David Guirguis

Yi Guo

Tinghao Guo

Xin Guo

Aakash Gupta

Joshua Gyory

Karl Haapala

Mahdi Haghshenas-Jaryani

Ichiro Hagiwara

David Hajdu

Maha Haji

Amal Z. Hajjaj

John Hall

Luke Hallum

Denise Halverson

Josh Hamel

Ji Han

Buddhika Hapuwatte

Brianne Hargrove

Megan Harris

Md Nahid Hasan

Mohammad Hasan

George Hazelrigg

Haiyang He

Shawky Hegazy

Seyed Mohammadreza Heidari

Daniel Herber

Just Herder

Jeffrey Herrmann

Nathan Hertlein

Amin Heyrani Nobari

Ethan Hilton

Dion Hogervorst

Derek Hollenbeck

Katja Holtta-Otto

Takatoshi Hondo

Yao Hong

Isaac Hong

Jonathan Hopkins

Imre Horvath

Yehia Hossam El Din

Mohammad Hotait

Larry Howell

Christopher Hoyle

Kuan-Lun Hsu

Chao Hu

Lingnan Hu

Weiwei Hu

Zhen Hu

Fu Hu

Chien-Ming Huang

Jida Huang

Chu-Yu Huang

Shih-Chun Huang

Aihua Huang

Shyh-Chour Huang

Jianzhe Huang

Guanyu Huang

Shen Huipijng

Daniel Hulse

Christine Human

Dongwook Hwang

Edoardo Ida

Horea Ilies

Athanasos Iliopoulos

Farhad Imani

Murat Inalpolat

Giovanni Incerti

Nizar Jaber

Prakhar Jaiswal

Ankur Jaiswal

Sagil James

Anand Jammulamadaka

Laurent Jay

Paramsothy Jayakumar

Brian Jensen

David Jensen

Nand Jha

Hao Ji

Chen Jiang

Long Jiang

Zhujin Jiang

Jiefeng Jiang

Yi Jin

Jian Jin

Yan Jin

Enze Jin

Bjorn Johansson

Albert Jones

Cole Joslyn

Jovana Jovanova

Sri Sadhan Jujjavarapu

Rama Krishna K.

Najib Kacem

Ibrahim Falih Kadhim

Ahmet Kahraman

Vinayak J Kalas

Amir Mohamad Kamalirad

Elisabeth Kames

Namwoo Kang

Sean Kelly

Pramod Khadilkar

Qasim Khadim

Mostafa Khalil

Muhammad Khan

Firas Khasawneh

Raj Pradip Khawale

Jivtesh Khurana

Namhun Kim

Hoe Joon Kim

Harrison Kim

Joo H. Kim

Jaehyuk Kim

Euiyoung Kim

Yves Klett

Adam Klodowski

Hyunwoong Ko

Eric Kolb

Ali Kolivand

Xianwen Kong

Lingyu Kong

Rasool Koosha

Stijn Koppen

Maulik C. Kotecha

Jozsef Kovecses

Julia Kramer

Timothy Krantz

Vinayak Raman Krishnamurthy

Prajit Krisshnakumar

Sai Aditya Raman Kuchibhatla

Boonserm Kulvatunyou

Ajeet Kumar

Rajesh Kumar

Shivesh Kumar

Jyoti Kumar

Chandan Kumar Sahu

Chin-Hsing Kuo

Eric Kurstak

Michael Kutzer

Tsz Ho Kwok

Elisa Kwon

Soonjo Kwon

Kai Lan

Brandon Lane

Daniel Lanzoni

Carlye Lauff

Astrid Layton

Michael Leamy

Yong Hoon Lee

Ikjin Lee

Jyh-jone Lee

Mathias Legrand

David Lehotzky

Leifur Leifsson

Matthew Leineweber

Stefano Lenci

Nikita Letov

Honglin Li

Suyi Li

Sheng Li

Jinghao Li

Hewenxuan LI

Haiyang Li

Yan Li

Xinyu Li

Zhuoxuan Li

Mian Li

Bowen Li

Longqiu Li

shiyao Li

Xiaofan Li

Yuxiang Li

Yuhua Li

Wei Li

Yan Liang

Haiguang Liao

Hao-Yu Liao

Jiankan Liao

Ting Liao

Po Ting Lin

Jianing Lin

Julie S. Linsey

Marc Litz

Xin Liu

Dehao Liu

Zuolin Liu

Zhao Liu

Yu Liu

Tianchen Liu

Yuanzhi Liu

Chang Liu

Xingchen Liu

Chao Liu

Feng Liu

Alvaro Lopez- Varela

Mostaan Lotfalian Saremi

Karen Lozano

Yanglong Lu

Yaodong Lu

Yan Lu

Lele Luan

Jose Lugo

Urbano Lugris

Chuan Luo

Jianxi Luo

Craig Lusk

Liye Lv

Matthew Lynch

Shengnan Lyu

Junfeng Ma

Nordica Maccarty

Ameneh Maghsoodi

Spencer Magleby

Gargi Majumder

Richard Malak Jr.

Adriano Mancini

Charles Manion

Hemanth Manjunatha

Peter Manzl

Jessica Gissella Maradey Lazaro

Edoardo Marconi

Marco Marconi

Carianne Martinez

Valeriy Martynyuk

Matteo Massaro

Ion Matei

Pawandeep Matharu

Jayant Mathur

Christopher Mattson

Christopher Mccomb

Lachlan McKenzie

John Mcphee

Lior Medina

Ali Mehmani

Nicholas Meisel

Giovanni Meneghetti

Jessica Menold

Diwesh Meshram

John Michopoulos

Aki Mikkola

Fahad Milaat

Jelena Milisavljevic-Syed

Scarlett R. Miller

Zhenjun Ming

Amin Mirkouei

Farrokh Mistree

Kentaro Miura

Ronak Mohanty

Tushar Mollik

Tamas Molnar

Ryan Monroe

Youngjin Moon

Marco Morandini

Beshoy Morkos

Federico Morosi

Daniel Morris

Simir Moschini

Zissimos Mourelatos

Huda Mousavi

Chandan Mozumder

Andreas Mueller

Andreas Muller

Luis Munoz

Alexander Murphy

Andrew Murray

Rosemarie Murray

David Myszka

Frank Naets

Jacquelyn Nagel

David Najera-Flores

Joel Najmon

Ananya Nandy

Austin Nash

Paromita Nath	Jitesh Panchal	Rohan Prabhu
Anand Balu Nellippallil	Piyush Pandita	Phanisri Pratapa
Todd Nelson	Herschel Pangborn	Christopher G. Pretty
Carl Nelson	Meghashyam Panyam	Anurag Purwar
Venkat Nemani	Dimitrios Papadimitriou	Feng Qian
Federico Neri	Kijung Park	Xiaoping Qian
Paul Nesline	Hyongju Park	Jian Qin
Rodrigo Nicoletti	Yeun Park	Lifang Qiu
Dean Nieusma	Robert G. Parker	Justin Quan
Joep Nijssen	Matthew Parkinson	D. Dane Quinn
Yutaka Nomaguchi	Victor Parque	Eliott Radcliffe
Julian Norato	Apurva Patel	Sajjad Raeisi
Mostafa Nouh	Parth Patel	Madhu Raghavan
Kenneth Nowaczyk	Sree Kalyan Patiballa	Hossam Ragheb
Konstantina Ntarladima	Albert Patterson	Jubeyer Rahman
Latifah Nurahmi	Steve Paul	Ayush Raina
Zachary Ochitwa	Satya Peddada	Vishal Ramadoss
Manaswin Oddiraju	William Peng	Venkat Ramakrishnan
Philip Odonkor	Tao Peng	Adhiti Raman
Eduardo Okabe	Giuseppe Pennisi	Devarajan Ramanujan
Alison Olechowski	Alfred Pettinger	Satchit Ramnath
Andrew Olewnik	James Pflumm	Vivek Rao
Grzegorz Orzechowski	Vrushank Phadnis	Sandipp Krishnan Ravi
Kevin Otto	Cyril Picard	Antonio Recuero
Emilio Ottonello	Bharath Pidaparthi	Daniele Regazzoni
Hassen Ouakad	Rocco Pietrini	Lyle Regenwetter
Enes Timur Ozdemir	Cecil Piya	William Regli
Alberto Padovan	Mark Plecnik	Tahira Reid
Safvan Palathingal	Alex Pletta	Yi Ren
Giacomo Palmieri	Jonisha Pollard	Rahul Sharan Renu
Matteo C. Palpacelli	Mirco Polonara	James Righter
Tan Pan	Siavash Pourkamali	David Rios-Zapata

Caterina Rizzi

Nina Robson

Steven Rodriguez

Bao Rong

David Rosen

Marco Rossoni

Clark Roubicek

Sipu Ruan

Davide Russo

Lokaditya Ryali

Jana Saadi

Mostafa Sabbaghi

Omid Saber

Mutahar Safdar

Alex Sahar

Tarik Sahin

Anshuman Kumar Sahu

Michael Saidani

Akira Saito

Alejandro Salado

Andrea Salvatore

Corina Sandu

Brandon Sargent

Bahadir Sarikaya

Soumalya Sarkar

Prabir Sarkar

Sree Shankar Satheesh Babu

Anupam Saxena

Mark Schenk

Ryan Schkoda

James Schmiedeler

Adam Schroeder

Christopher Sebastian

Carolyn Seepersad

Hullas Sehgal

Robert Seifried

Daniel Selva

Chiradeep Sen

Radu Serban

Valeria Settimi

Thurston Sexton

Zhenghui Sha

Divya Shah

Devanshi Shah

V.R. Shanmukhasundaram

Ashu Sharma

Saurav Sharma

Shashank Sharma

Mohammad Shavezipur

Jinjuan She

Yu She

Tripp Shealy

Gulai Shen

i.y. Shen

Chengzhi Shi

Hongliang Shi

Zhenghong Shi

Yong Shi

Yi-Pei Shih

Shailesh Shirguppikar

Geng Shixiong

Ada-Rhodes Short

Timothy Simpson

Siddharth Singh

Yogesh Singh

Vishal Singh

Shubhendu Kumar Singh

Alok Sinha

Kevin Skenes

Brian Slaboch

Gim Song Soh

Hyeonik Song

Xueguan Song

Binyang Song

Rujun Song

Carl Sorensen

Nicolas F. Soria Zurita

Christian Spreafico

Saketh Sridhara

Tino Stankovic

Elizabeth Starkey

Giulia Stefani

John Steuben

Daniel Suarez

Hiroyuki Sugiyama

Joshua Summers

Xiaoguang Sun

Sijie Sun

Krishnan Suresh

Sita Syal

Henrik Tamás Sykora

Brian Sylcott

Wei-Che Tai

Shun Takai

David Talbot

Tsz Ling Elaine Tang

Yunlong Tang

Alessandro Tasora

Ayse Tekes

Yishen Tian

Meng-Hsuan Tien

Christine Toh

Serife Tol

Daniel Torres

Andres Tovar

Shahrzad Towfighian

Alex Towse

Anh Tran

Cameron Turner

Pedro Urda

Kedar Vaidya

Nima Valadbeigi

Ambrosio Valencia-Romero

Homero Valladares

Anton van Beek

Douglas Van Bossuyt

Noe Vargas Hernandez

Soumya Vasisht

Pavan Tejaswi Velivela

Srinivasan Venkataraman

Swaminath Venkataswaran

Christopher Vermillion

Matteo Verotti

Adwait Verulkar

Vishesh Vikas

Jairo Viola

Antoni Viros I Martin

Vimal Viswanathan

Andrea Vitali

Nguyen Vu Linh

Evan Wallace

Shu Wan

Zequn Wang

Pingfeng Wang

Jing Wang

Pai Wang

Ruyue Wang

Fengxia Wang

Tingwei Wang

Haoqi Wang

Jieyu Wang

Jun Wang

Randi Wang

Kai Wang

Yan Wang

Wei Wang

Zihan Wang

Liwei Wang

Shoufei Wang

Gou-jen Wang

Wang Wang

Sibao Wang

Jonathan Weaver-Rosen

Guowu Wei

Xinqi Wei

Kate Whitefoot

Richard Wiebe

Gloria Wiens

Justin Wilbanks

Jon Williams

Matthew Williams

Carlos H. Wink

Amos Winter

Paul Witherell

Andrew Wodehouse

Kristin Wood

Natasha Wright

Dazhong Wu

Guangqiang Wu

Hao Wu

Di Wu

Zhimin Xi

Guanghui Xia

Songtao Xia

Yiwei Xia

Yujiang Xiang

Jiarui Xie

Siyuan Xing

Cenbo Xiong

Yi Xiong

Haohua Xiu

Hongyi Xu

Hai Xu

Qingsong Xu

Leidong Xu

Dong Xu

Mostafa Yacoub

Darshan Yadav

Hiroki Yamashita

Peng Yan

Zhipei Yan

Wei Yan

Haosen Yang

Zhuo Yang

Sheng Yang

Jingyi Yang

Hang Yang

Xiaonan Yang

Wenhao Yang

James Yang

Ruoyu Yang

Maria C. Yang

Xiaoou Yang

Chao-Lung Yang

Yuan Yao

T.-J. Yeh

Karthik Yerrapragada

Juan Yi

Zhong You

Behrooz Yousefzadeh

Liangyao Yu

Xinxin Yu

Xiangping Yu

Yongguang Yu

Sichen Yuan

Andrea Zanoni

Chen Zeng

Jice Zeng

Yuxin Zhai

Jie Zhang

Yunbo Zhang

Zhen Zhang

Haifeng Zhang

Ying Zhang

Peng Zhang

Hao Zhang

Dong Zhang

Qi Zhang

Siqi Zhang

Zhibo Zhang

Xiaoxu Zhang

Yang Zhang

Shengli Zhang

Guanglu Zhang

Xiaojia Shelly Zhang

Siyuan Zhang

Haohan Zhang

He Zhang

Shanglong Zhang

Yaoyao Zhao

Donghua Zhao

Hongzhe Zhao

Ping Zhao

Jian Zhao

Liqian Zhao

Yinjun Zhao

Linchuan Zhao

Zhen Zhao

Bujingda Zheng

Yi Zheng

Kevin Zheng

Kai Zhou

Yuqing Zhou

Hong Zhou

Xianlian Zhou

Jianhua Zhou

Xiang Zhou

Shengxi Zhou

Weidong Zhu

Jiaxiang Zhu

Damijan Zorko

Hongxiang Zou

Chengzhe Zou

Jianyong Zuo

Andreas Zwölfer

Marcel Zydeck

Wei Yan

Haosen Yang

Zhuo Yang

Sheng Yang

Jingyi Yang

Hang Yang

Xiaonan Yang

Wenhao Yang

James Yang

Ruoyu Yang

Maria C. Yang

Xiaoou Yang

Chao-Lung Yang

Yuan Yao

T.-J. Yeh

Karthik Yerrapragada

Juan Yi

Zhong You

Behrooz Yousefzadeh

Liangyao Yu

Xinxin Yu

Xiangping Yu

Yongguang Yu

Sichen Yuan

Andrea Zanoni

Chen Zeng

Jice Zeng

Yuxin Zhai

Jie Zhang

Yunbo Zhang

Zhen Zhang

Haifeng Zhang

Ying Zhang

Peng Zhang

Hao Zhang

Dong Zhang

Qi Zhang

Siqi Zhang

Zhibo Zhang

Xiaoxu Zhang

Yang Zhang

Shengli Zhang

Guanglu Zhang

Xiaojia Shelly Zhang

Siyuan Zhang

Haohan Zhang

He Zhang

Shanglong Zhang

Yaoyao Zhao

Donghua Zhao

Hongzhe Zhao

Ping Zhao

Jian Zhao

Liqian Zhao

Yinjun Zhao

Linchuan Zhao

Zhen Zhao

Bujingda Zheng

Yi Zheng

Kevin Zheng

Kai Zhou

Yuqing Zhou

Hong Zhou

Xianlian Zhou

Jianhua Zhou

Xiang Zhou

Shengxi Zhou

Weidong Zhu

Jiaxiang Zhu

Damijan Zorko

Hongxiang Zou

Chengzhe Zou

Jianyong Zuo

Andreas Zwölfer

Marcel Zydeck

PROCEEDINGS OF ASME 2023 INTERNATIONAL DESIGN ENGINEERING TECHNICAL CONFERENCES AND COMPUTERS AND INFORMATION IN ENGINEERING CONFERENCE (IDETC-CIE2023)

49th Design Automation Conference (DAC) Part Two
Table of Contents

Design of Complex Systems

Heuristics for Solver-Aware Systems Architecting (SASA): A Reinforcement Learning Approach — DETC2023-115030
Vikranth S. Gadi, Taylan G. Topcu, Zoe Szajnfarber, and Jitesh H. Panchal

Improving Change Management by Quantifying the Relationships Between Margins and Design Variable Flexibility — DETC2023-115312
Lindsey Jacobson and Scott Ferguson

Electronics Design and Verification for Robots With Actuation and Sensing Requirements — DETC2023-115313
Dongsheng Chen, Zonghao Huang, and Cynthia Sung

Knowledge Transfer in Self-Organizing Systems and Impact of Social Ability — DETC2023-116392
Bingling Huang and Yan Jin

Framing Wicked Problems Through Evidentiary and Interpretative Analysis — DETC2023-117285
Mayank J. Bhalerao, Wesley T. Honeycutt, Ashok K. Das, Janet K. Allen, and Farrokh Mistree

Design of Engineering Materials and Structures

Orientation Optimization With Topological Derivative for Anisotropic Materials With Rotational Symmetry — DETC2023-110477
Masaki Noda, Kei Matsushima, and Takayuki Yamada

Mixed-Variable Global Sensitivity Analysis With Applications to Data-Driven Combinatorial Materials Design — DETC2023-110756
Yigitcan Comlek, Liwei Wang, and Wei Chen

Tackling an Exact Maximum Stress Minimization Problem With Gradient-Free Topology Optimization Incorporating a Deep Generative Model — DETC2023-111265
Misato Kato, Taisei Kii, Kentaro Yaji, and Kikuo Fujita

An Improved Shape Annealing Algorithm for the Generation of Coated DNA Origami Nanostructures — DETC2023-113633
Bolutito Babatunde, Jonathan Cagan, and Rebecca E. Taylor

Mitigating the Effects of Source-Dependent Bias and Noise on Multi-Source Bayesian Optimization: Application to Materials Design — DETC2023-114414
Zahra Zanjani Foumani, Amin Yousefpour, Mehdi Shishehbor, and Ramin Bostanabad

Design of Mixed-Category Stochastic Microstructures: A Comparison of Curvature Functional-Based and Deep Generative Model-Based Methods
Leidong Xu, Kiarash Naghavi Khanghah, and Hongyi Xu

DETC2023-114601

High-Frequency Band Gap Design of Porous Phononic Crystals by Topology Optimization
Naoki Murai, Yuki Noguchi, and Takayuki Yamada

DETC2023-114619

Selective Amplification and Suppression of Strain in a Multi-Axis Force Sensor Using Topology Optimization
Myung Kyun Sung, Soobum Lee, Devin E. Burns, and Jude Thaddeus Persia

DETC2023-114920

Data-Driven Design of High Electron Mobility Transistor Devices Using Physics-Informed Gaussian Process Modeling
Anabel Renteria, Yanwen Xu, Bayan Hamdan, Zhou Li, Sergio Cordero, Debbie G. Senesky, and Pingfeng Wang

DETC2023-117200

Comparing Derivatives of Neural Networks for Regression
Joel C. Najmon and Andres Tovar

DETC2023-117571

Engineering for Global Development

Design and Evaluation of an Automatic Scheduling-Manual Operation Tool to Bring Precision Irrigation to Resource-Constrained Farmers
Georgia D. Van de Zande, Carolyn Sheline, Susan Amrose, Jeffrey Costello, Aditya Ghodgaonkar, Fiona Grant, and Amos G. Winter, V

DETC2023-112470

Lab-to-Market Design of an Electrodialysis-Based Home-Scale Water Desalination System
Marie Floryan, Quinn Bowers, Zachary Sternberg, Sahas Gembali, Akshita Goyal, Jonathan Bessette, Soraya Honarparvar, and Amos Winter

DETC2023-112625

Feasibility of Small-Scale, Off-Grid Desalination in Navajo Nation
Melissa Brei

DETC2023-113479

Feeling the Distance: Exploring Novice Designers' Perceptions of the Psychological Distance Towards and Empathy Induced by Problem Variations
Jenna Herzog, Rebekah Fodale, Mohammad Alsager Alzayed, Elizabeth M. Starkey, and Rohan Prabhu

DETC2023-114540

Reducing the Barriers to Designing 3D-Printable Prosthetics In Resource-Constrained Environments
Junghun Lee, Andrew Chesang, Michael Gichane, Moise Busogi, Jean Byiringiro, and Conrad Tucker

DETC2023-116399

Achieving High Performance and Low Cost: Development of a High-Performing Passive Prosthetic Knee for Emerging Markets
Madison Reddie, Saloni Bedi, Manasi Vaidya, Amari Griffin, Nina T. Petelina, and Amos G. Winter

DETC2023-116478

Quantifying Resilience Trade-Offs for Small-Scale Farms: A System Optimization Study in Uganda
DETC2023-116657
Jesse Austin-Breneman, Praneet Nallan Chakravarthula, Alvin B. Kimbowa, Peter Ozaveshe Oviroh, Samuel Boahen, Emmanuel Wokulira Miyingo, and Panos Y. Papalambros

Fifty-Five Prompt Questions for Identifying Social Impacts of Engineered Products
DETC2023-116725
Christopher A. Mattson, Thomas B. Geilman, Joshua F. Cook-Wright, Christopher S. Mabey, Eric Dahlin, and John L. Salmon

The Giving Garden: Realizing Community and Fostering a Connection to the Land Through Co-Design in Duluth, Minnesota
DETC2023-116756
Austin Konrath, Abigail R. Clarke-Sather, Regina Laroche, Morgan Bliss, and Rumbidzai Masawi

Design Interviews Conducted by Intra- and Intercultural Teams: A Case Study on Dialysis in Zimbabwe
DETC2023-116953
Micki Grover, Carlye A. Lauff, Chiratidzo Ndhlovu, and Natasha C. Wright

AI-Accelerated Design of Evidence Synthesis for Global Development
DETC2023-116962
Kristen M. Edwards, Binyang Song, Jaron Porciello, Mark Engelbert, Carolyn Huang, and Faez Ahmed

Geometric Modeling and Algorithms for Design and Manufacturing

Model Consistency for Mechanical Design: Bridging Lumped and Distributed Parameter Models With A Priori Guarantees
DETC2023-109380
Randi Wang and Morad Behandish

Methods for Creating Additive Printing Paths on Nonplanar Surfaces
DETC2023-110757
Liam Rudd, Zahra Faghihrasoul, and Matthew I. Campbell

Multi-Material Topology Optimization Considering the Bounding Box Dimension Constraint and Assemblability Based on the Extended Level Set Method in Two Dimensions
DETC2023-111214
Yukun Feng, Yuki Noguchi, and Takayuki Yamada

Finding Chain Nets of Solids for 3D Printability
DETC2023-114669
Matthew Lawrence, Scott Tomlinson, and Bashir Khoda

An Empirical, Deterministic Design Theory for Compact Drip Emitter Labyrinths
DETC2023-116552
Aditya Ghodgaonkar, Emily Welsh, Benjamin Judge, Michael Bono, and Amos G. Winter, V

How to Encode Microstructure in Machine Learning: A Comparison Study
DETC2023-116704
Yulun Wu and Yumeng Li

Metamodel-Based Design Optimization (MBDO)

Constrained Bayesian Optimization Methods Using Regression and Classification Gaussian Processes As Constraints — DETC2023-109993
Cole Jetton, Chengda Li, and Christopher Hoyle

Topology Optimization With Quantum Approximate Bayesian Optimization Algorithm — DETC2023-116549
Jungin E. Kim and Yan Wang

Concurrent Probabilistic Control Co-Design and Layout Optimization of Wave Energy Converter Farms Using Surrogate Modeling — DETC2023-116896
Saeed Azad and Daniel R. Herber

Multi-Task Multi-Fidelity Machine Learning for Reliability-Based Design With Partially Observed Information — DETC2023-117032
Yanwen Xu, Hao Wu, Zheng Liu, and Pingfeng Wang

Multidisciplinary Design Optimization, Multiobjective Optimization, and Sensitivity Analysis

Data-Driven Multifidelity Topology Design With a Latent Crossover Operation — DETC2023-111079
Taisei Kii, Kentaro Yaji, Kikuo Fujita, Zhenghui Sha, and Carolyn C. Seepersad

Topology Optimization of Rarefied Gas Devices With Discrete Velocity Method — DETC2023-111436
Kaiwen Guan, Kei Matsushima, and Takayuki Yamada

A GPU-Based Parallel Bound-and-Classify Method for Continuous Constraint Satisfaction Problems — DETC2023-112414
Wangchuan Feng, Guanglu Zhang, and Jonathan Cagan

Multi-Material and Multi-Joint Topology Optimization Considering Multiple Design Spaces — DETC2023-113451
Il Yong Kim, Yuhao Huang, and Luke Crispo

Robust Topology Optimization of Synchronous Reluctance Motors Using Cardinal Basis Function Based Level Set Method — DETC2023-115068
Jiawei Tian, David Torrey, Fang Luo, Jon Longtin, and Shikui Chen

Optimization of 3D Printing While Traveling En Route to Extend Range of UAS for Multi-Location Mission Scenarios — DETC2023-116618
Tevin Dickerson, John L. Salmon, and Christopher A. Mattson

Efficient Robust Design Space Visualization and Exploration for Many-Objective Problems – A Vehicular Crashworthiness Example — DETC2023-117199
Niharika Balaji, Mathew Baby, Gehendra Sharma, Rashmi Rama Sushil, Palaniappan Ramu, and Anand Balu Nellippallil

Design and Decision Making Under Uncertainty

Satisficing Strategy in Engineering Design   **DETC2023-109302**
    Lin Guo and Suhao Chen

Quantification Model Uncertainty of Label-Free Machine Learning for   **DETC2023-112948**
Multidisciplinary Systems Analysis
    Huiru Li, Jitesh H. Panchal, and Xiaoping Du

Efficient Airfoil Geometric Uncertainty Quantification Using Neural Network   **DETC2023-114954**
Models and Sequential Sampling
    Pavankumar Koratikere and Leifur Leifsson

Multi-Agent Bayesian Optimization for Unknown Design Space Exploration   **DETC2023-115112**
    Siyu Chen, Alparslan Emrah Bayrak, and Zhenghui Sha

A Comparison of Bayesian Acquisition Functions for Use In Surrogate Multi-   **DETC2023-116320**
Objective Feasibility Robust Optimization With Interval Uncertainty
    Randall J. Kania and Shapour Azarm

Robust Design for Product Adaptation Considering Changes in Configurations   **DETC2023-116614**
and Parameters
    Reza Deabae and Deyi Xue

Integrated Sustainable Product Design With Warranty and End-of-Use   **DETC2023-117013**
Considerations
    Xinyang Liu and Pingfeng Wang

Uncertainty Quantification on Mechanical Behavior of Corroded Plate With   **DETC2023-117050**
Statistical Shape Modeling
    Hao Wu, Parth Bansal, Zheng Liu, Yumeng Li, and Pingfeng Wang

An Efficient Surrogate Modeling Method for Reliability-Based Global Path   **DETC2023-117348**
Planning of Off-Road Autonomous Ground Vehicles
    *Jianhua Yin, Zhen Hu, Zissimos P. Mourelatos, David Gorsich, Amandeep
        Singh, and Seth Tau*

Computational Design for Biomedical Applications

Pareto Optimization of Tissue and Blood Vessel Growth in 3D Printed Bone   **DETC2023-115147**
Scaffolds
    Amit M. E. Arefin and Paul F. Egan

Designing Programmable Ferromagnetic Soft Metastructures for Minimally   **DETC2023-116342**
Invasive Endovascular Therapy
    *Ran Zhuang, Jiawei Tian, Apostolos Tassiopoulos, Chandramouli
        Sadasivan, Xianfeng David Gu, and Shikui Chen*

Human-Artificial Intelligence Collaboration in Engineering System Design

How Does Agency Impact Human-AI Collaborative Design Space Exploration? A Case Study on Ship Design With Deep Generative Models
Shahroz Khan, Panagiotis Kaklis, and Kosa Goucher-Lambert

DETC2023-112570

Adaptation and Challenges in Human-AI Partnership for the Design of Complex Engineering Systems
Zeda Xu, Chloe Hong, Nicolás F. Soria Zurita, Joshua T. Gyory, Gary Stump, Hannah Nolte, Jonathan Cagan, and Christopher McComb

DETC2023-115176

Let's Chat If You Are Unhappy – The Effect of Emotions on Interaction Experience and Trust Toward Empathetic Chatbots
Ting Liao and Bei Yan

DETC2023-115318

A Multi-Objective Bayesian Optimized Human Assessed Multi-Target Generated Spectral Recommender System for Rapid Pareto Discoveries of Material Properties
Arpan Biswas, Yongtao Liu, Maxim Ziatdinov, Yu-Chen Liu, Stephen Jesse, Jan-Chi Yang, Sergei Kalinin, and Rama Vasudevan

DETC2023-116956

Understanding the Relation Between Designer Search Strategies and Designer Learning During Design Space Exploration
Hyeonik Song and Daniel Selva

DETC2023-116984

Design of Autonomous Systems

Large-Scale Path Planning in Complex Environments Based on Genetic Algorithm
Chuanhui Hu and Yan Jin

DETC2023-116340

An Enhanced Timed Elastic Band Method for Autonomous Navigation and its Collision Avoidance Reliability Analysis
Zhimin Xi

DETC2023-116695

On How a Self-Organizing System Produces Collective Behavior
Jinhui Cao, Zhenjun Ming, Janet K. Allen, and Farrokh Mistree

DETC2023-116875

Evolving Cyber-Physical-Social Systems

Chat Generative Pretrained Transformer: Extinction of the Designer or Rise of an Augmented Designer
Amaninder Singh Gill

DETC2023-116971

Special Session with DFMLC: Modeling and Optimization for Sustainable Design and Manufacturing

An Approach for Predicting Social, Environmental, and Economic Product Impacts and Navigating the Associated Impact Trade-Space in Engineering Design **DETC2023-116719**
    Christopher S. Mabey, Tevin J. Dickerson, John L. Salmon, and Christopher A. Mattson

The Automation of Artistic and Creative Design

The Generation of Novel Art Using Collaborative ML Models **DETC2023-116825**
    Ada-Rhodes Short

State of the Art: A Review of AI Art Generation Methods for Rigorous Design **DETC2023-116833**
    Lauren Bertelsen and Ada-Rhodes Short

UQ of ML Models for Data-Driven Design

Machine Learning-Based Model Bias Correction by Fusing CAE Data With Test Data for Vehicle Crashworthiness **DETC2023-114999**
    Jice Zeng, Ying Zhao, Guosong Li, Zhenyan Gao, Yang Li, Saeed Barbat, and Zhen Hu

Accounting for Model Uncertainty in Machine Learning Assisted Mechanical Design **DETC2023-116622**
    Xiaoping Du

AI-Driven Design Innovation

An Integrated Approach to Designing Robust Turbocompressors on Gas Bearings Through Surrogate Modeling and Constrained Multi-Objective Optimization **DETC2023-115201**
    Soheyl Massoudi, Cyril Picard, and Jürg Schiffmann

Teaching AI to Design From Humans: a Comparison of Behavioral Cloning Architectures **DETC2023-115280**
    Ghazal Bozorgmehry Boozarjomehry and Joseph Thekinen

Transfer Reinforcement Learning: Feature Transferability in Ship Collision Avoidance **DETC2023-116709**
    Xinrui Wang and Yan Jin

Toward Artificial Empathy for Human-Centered Design: A Framework **DETC2023-117266**
    Qihao Zhu and Jianxi Luo

TABLE OF CONTENTS

Heuristics for Solver-Aware Systems Architecting (SASA): A Reinforcement Learning Approach.................... 1
Vikranth S. Gadi, Taylan G. Topcu, Zoe Szajnfarber, Jitesh H. Panchal

Improving Change Management by Quantifying the Relationships Between Margins and Design
Variable Flexibility 13
Lindsey Jacobson, Scott Ferguson

Electronics Design and Verification for Robots With Actuation and Sensing Requirements 27
Dongsheng Chen, Zonghao Huang, Cynthia Sung

Knowledge Transfer in Self-Organizing Systems and Impact of Social Ability................................ 37
Bingling Huang, Yan Jin

Framing Wicked Problems Through Evidentiary and Interpretative Analysis 47
Mayank J. Bhalerao, Wesley T. Honeycutt, Ashok K. Das, Janet K. Allen, Farrokh Mistree

Orientation Optimization With Topological Derivative for Anisotropic Materials With Rotational
Symmetry 65
Masaki Noda, Kei Matsushima, Takayuki Yamada

Mixed-Variable Global Sensitivity Analysis With Applications to Data-Driven Combinatorial
Materials Design.......................... 71
Yigitcan Comlek, Liwei Wang, Wei Chen

Tackling an Exact Maximum Stress Minimization Problem With Gradient-Free Topology
Optimization Incorporating a Deep Generative Model 81
Misato Kato, Taisei Kii, Kentaro Yaji, Kikuo Fujita

An Improved Shape Annealing Algorithm for the Generation of Coated DNA Origami
Nanostructures.......................... 91
Bolutito Babatunde, Jonathan Cagan, Rebecca E. Taylor

Mitigating the Effects of Source-Dependent Bias and Noise on Multi-Source Bayesian
Optimization: Application to Materials Design.......................... 103
Zahra Zanjani Foumani, Amin Yousefpour, Mehdi Shishehbor, Ramin Bostanabad

Design of Mixed-Category Stochastic Microstructures: A Comparison of Curvature Functional-
Based and Deep Generative Model-Based Methods 113
Leidong Xu, Kiarash Naghavi Khanghah, Hongyi Xu

High-Frequency Band Gap Design of Porous Phononic Crystals by Topology Optimization 124
Naoki Murai, Yuki Noguchi, Takayuki Yamada

Selective Amplification and Suppression of Strain in a Multi-Axis Force Sensor Using Topology
Optimization.......................... 130
Myung Kyun Sung, Soobum Lee, Devin E. Burns, Jude Thaddeus Persia

Data-Driven Design of High Electron Mobility Transistor Devices Using Physics-Informed
Gaussian Process Modeling.......................... 137
*Anabel Renteria, Yanwen Xu, Bayan Hamdan, Zhou Li, Sergio Cordero, Debbie G. Senesky,
Pingfeng Wang*

Comparing Derivatives of Neural Networks for Regression 144
    Joel C. Najmon, Andres Tovar

Design and Evaluation of an Automatic Scheduling-Manual Operation Tool to Bring Precision
Irrigation to Resource-Constrained Farmers 151
    Georgia D. Van de Zande, Carolyn Sheline, Susan Amrose, Jeffrey Costello, Aditya
    Ghodgaonkar, Fiona Grant, Amos G. Winter V

Lab-to-Market Design of an Electrodialysis-Based Home-Scale Water Desalination System 162
    Marie Floryan, Quinn Bowers, Zachary Sternberg, Sahas Gembali, Akshita Goyal, Jonathan
    Bessette, Soraya Honarparvar, Amos Winter

Feasibility of Small-Scale, Off-Grid Desalination in Navajo Nation 173
    Melissa Brei

Feeling the Distance: Exploring Novice Designers' Perceptions of the Psychological Distance
Towards and Empathy Induced by Problem Variations 180
    Jenna Herzog, Rebekah Fodale, Mohammad Alsager Alzayed, Elizabeth M. Starkey, Rohan
    Prabhu

Reducing the Barriers to Designing 3D-Printable Prosthetics In Resource-Constrained
Environments 190
    Junghun Lee, Andrew Chesang, Michael Gichane, Moise Busogi, Jean Byiringiro, Conrad
    Tucker

Achieving High Performance and Low Cost: Development of a High-Performing Passive
Prosthetic Knee for Emerging Markets 200
    Madison Reddie, Saloni Bedi, Manasi Vaidya, Amari Griffin, Nina T. Petelina, Amos G.
    Winter

Quantifying Resilience Trade-Offs for Small-Scale Farms: A System Optimization Study in Uganda 209
    Jesse Austin-Breneman, Praneet Nallan Chakravarthula, Alvin B. Kimbowa, Peter Ozaveshe
    Oviroh, Samuel Boahen, Emmanuel Wokulira Miyingo, Panos Y. Papalambros

Fifty-Five Prompt Questions for Identifying Social Impacts of Engineered Products 219
    Christopher A. Mattson, Thomas B. Geilman, Joshua F. Cook-Wright, Christopher S. Mabey,
    Eric Dahlin, John L. Salmon

The Giving Garden: Realizing Community and Fostering a Connection to the Land Through Co-
Design in Duluth, Minnesota 232
    Austin Konrath, Abigail R. Clarke-Sather, Regina Laroche, Morgan Bliss, Rumbidzai Masawi

Design Interviews Conducted by Intra- and Intercultural Teams: A Case Study on Dialysis in
Zimbabwe 242
    Micki Grover, Carlye A. Lauff, Chiratidzo Ndhlovu, Natasha C. Wright

AI-Accelerated Design of Evidence Synthesis for Global Development 254
    Kristen M. Edwards, Binyang Song, Jaron Porciello, Mark Engelbert, Carolyn Huang, Faez
    Ahmed

Model Consistency for Mechanical Design: Bridging Lumped and Distributed Parameter Models
With A Priori Guarantees 268
    Randi Wang, Morad Behandish

Methods for Creating Additive Printing Paths on Nonplanar Surfaces 278
    Liam Rudd, Zahra Faghihrasoul, Matthew I. Campbell

Multi-Material Topology Optimization Considering the Bounding Box Dimension Constraint and Assemblability Based on the Extended Level Set Method in Two Dimensions 293
Yukun Feng, Yuki Noguchi, Takayuki Yamada

Finding Chain Nets of Solids for 3D Printability 300
Matthew Lawrence, Scott Tomlinson, Bashir Khoda

An Empirical, Deterministic Design Theory for Compact Drip Emitter Labyrinths 307
Aditya Ghodgaonkar, Emily Welsh, Benjamin Judge, Michael Bono, Amos G. Winter V

How to Encode Microstructure in Machine Learning: A Comparison Study 320
Yulun Wu, Yumeng Li

Constrained Bayesian Optimization Methods Using Regression and Classification Gaussian Processes As Constraints 328
Cole Jetton, Chengda Li, Christopher Hoyle

Topology Optimization With Quantum Approximate Bayesian Optimization Algorithm 340
Jungin E. Kim, Yan Wang

Concurrent Probabilistic Control Co-Design and Layout Optimization of Wave Energy Converter Farms Using Surrogate Modeling 351
Saeed Azad, Daniel R. Herber

Multi-Task Multi-Fidelity Machine Learning for Reliability-Based Design With Partially Observed Information 365
Yanwen Xu, Hao Wu, Zheng Liu, Pingfeng Wang

Data-Driven Multifidelity Topology Design With a Latent Crossover Operation 374
Taisei Kii, Kentaro Yaji, Kikuo Fujita, Zhenghui Sha, Carolyn C. Seepersad

Topology Optimization of Rarefied Gas Devices With Discrete Velocity Method 385
Kaiwen Guan, Kei Matsushima, Takayuki Yamada

A GPU-Based Parallel Bound-and-Classify Method for Continuous Constraint Satisfaction Problems 391
Wangchuan Feng, Guanglu Zhang, Jonathan Cagan

Multi-Material and Multi-Joint Topology Optimization Considering Multiple Design Spaces 402
Il Yong Kim, Yuhao Huang, Luke Crispo

Robust Topology Optimization of Synchronous Reluctance Motors Using Cardinal Basis Function Based Level Set Method 409
Jiawei Tian, David Torrey, Fang Luo, Jon Longtin, Shikui Chen

Optimization of 3D Printing While Traveling En Route to Extend Range of UAS for Multi-Location Mission Scenarios 419
Tevin Dickerson, John L. Salmon, Christopher A. Mattson

Efficient Robust Design Space Visualization and Exploration for Many-Objective Problems - A Vehicular Crashworthiness Example 432
Niharika Balaji, Mathew Baby, Gehendra Sharma, Rashmi Rama Sushil, Palaniappan Ramu, Anand Balu Nellippallil

Satisficing Strategy in Engineering Design 447
Lin Guo, Suhao Chen

Quantification Model Uncertainty of Label-Free Machine Learning for Multidisciplinary Systems Analysis 460
    Huiru Li, Jitesh H. Panchal, Xiaoping Du

Efficient Airfoil Geometric Uncertainty Quantification Using Neural Network Models and Sequential Sampling 470
    Pavankumar Koratikere, Leifur Leifsson

Multi-Agent Bayesian Optimization for Unknown Design Space Exploration 480
    Siyu Chen, Alparslan Emrah Bayrak, Zhenghui Sha

A Comparison of Bayesian Acquisition Functions for Use In Surrogate Multi-Objective Feasibility Robust Optimization With Interval Uncertainty 490
    Randall J. Kania, Shapour Azarm

Robust Design for Product Adaptation Considering Changes in Configurations and Parameters 500
    Reza Deabae, Deyi Xue

Integrated Sustainable Product Design With Warranty and End-of-Use Considerations 510
    Xinyang Liu, Pingfeng Wang

Uncertainty Quantification on Mechanical Behavior of Corroded Plate With Statistical Shape Modeling 520
    Hao Wu, Parth Bansal, Zheng Liu, Yumeng Li, Pingfeng Wang

An Efficient Surrogate Modeling Method for Reliability-Based Global Path Planning of Off-Road Autonomous Ground Vehicles 527
    Jianhua Yin, Zhen Hu, Zissimos P. Mourelatos, David Gorsich, Amandeep Singh, Seth Tau

Pareto Optimization of Tissue and Blood Vessel Growth in 3D Printed Bone Scaffolds 535
    Amit M. E. Arefin, Paul F. Egan

Designing Programmable Ferromagnetic Soft Metastructures for Minimally Invasive Endovascular Therapy 546
    Ran Zhuang, Jiawei Tian, Apostolos Tassiopoulos, Chandramouli Sadasivan, Xianfeng David Gu, Shikui Chen

How Does Agency Impact Human-AI Collaborative Design Space Exploration? A Case Study on Ship Design With Deep Generative Models 557
    Shahroz Khan, Panagiotis Kaklis, Kosa Goucher-Lambert

Adaptation and Challenges in Human-AI Partnership for the Design of Complex Engineering Systems 570
    Zeda Xu, Chloe Hong, Nicolas F. Soria Zurita, Joshua T. Gyory, Gary Stump, Hannah Nolte, Jonathan Cagan, Christopher McComb

Let's Chat If You Are Unhappy - The Effect of Emotions on Interaction Experience and Trust Toward Empathetic Chatbots 583
    Ting Liao, Bei Yan

A Multi-Objective Bayesian Optimized Human Assessed Multi-Target Generated Spectral Recommender System for Rapid Pareto Discoveries of Material Properties 594
    Arpan Biswas, Yongtao Liu, Maxim Ziatdinov, Yu-Chen Liu, Stephen Jesse, Jan-Chi Yang, Sergei Kalinin, Rama Vasudevan

Understanding the Relation Between Designer Search Strategies and Designer Learning During Design Space Exploration 607
Hyeonik Song, Daniel Selva

Large-Scale Path Planning in Complex Environments Based on Genetic Algorithm 617
Chuanhui Hu, Yan Jin

An Enhanced Timed Elastic Band Method for Autonomous Navigation and its Collision Avoidance Reliability Analysis 631
Zhimin Xi

On How a Self-Organizing System Produces Collective Behavior 641
Jinhui Cao, Zhenjun Ming, Janet K. Allen, Farrokh Mistree

Chat Generative Pretrained Transformer: Extinction of the Designer or Rise of an Augmented Designer 657
Amaninder Singh Gill

An Approach for Predicting Social, Environmental, and Economic Product Impacts and Navigating the Associated Impact Trade-Space in Engineering Design 662
Christopher S. Mabey, Tevin J. Dickerson, John L. Salmon, Christopher A. Mattson

The Generation of Novel Art Using Collaborative ML Models 675
Ada-Rhodes Short

State of the Art: A Review of AI Art Generation Methods for Rigorous Design 717
Lauren Bertelsen, Ada-Rhodes Short

Machine Learning-Based Model Bias Correction by Fusing CAE Data With Test Data for Vehicle Crashworthiness 727
Jice Zeng, Ying Zhao, Guosong Li, Zhenyan Gao, Yang Li, Saeed Barbat, Zhen Hu

Accounting for Model Uncertainty in Machine Learning Assisted Mechanical Design 738
Xiaoping Du

An Integrated Approach to Designing Robust Turbocompressors on Gas Bearings Through Surrogate Modeling and Constrained Multi-Objective Optimization 746
Soheyl Massoudi, Cyril Picard, Jurg Schiffmann

Teaching AI to Design From Humans: a Comparison of Behavioral Cloning Architectures 761
Ghazal Bozorgmehry Boozarjomehry, Joseph Thekinen

Transfer Reinforcement Learning: Feature Transferability in Ship Collision Avoidance 770
Xinrui Wang, Yan Jin

Toward Artificial Empathy for Human-Centered Design: A Framework 784
Qihao Zhu, Jianxi Luo

Author Index

Proceedings of the ASME 2023
International Design Engineering Technical Conferences and
Computers and Information in Engineering Conference
IDETC-CIE2023
August 20-23, 2023, Boston, Massachusetts

DETC2023-115030

HEURISTICS FOR SOLVER-AWARE SYSTEMS ARCHITECTING (SASA): A REINFORCEMENT LEARNING APPROACH

Vikranth S. Gadi
School of Mechanical Engineering
Purdue University
West Lafayette, Indiana 47907

Taylan G. Topcu
Grado Department of Industrial and Systems Engineering
Virginia Tech
Blacksburg VA 24061

Zoe Szajnfarber
Engineering Management and Systems Engineering
George Washington University
Washington DC 20052

Jitesh H. Panchal
School of Mechanical Engineering
Purdue University
West Lafayette, Indiana 47907

ABSTRACT

The crowdsourcing literature has shown that domain experts are not always the best solvers for complex system design problems. Novices and specialists in adjacent domains can, under certain conditions, provide novel solutions at lower costs. Additionally, the best types of solvers for different sub-problems are dependent on the architecture of complex systems. The assignment of solvers based on the architecture, referred to as Solver-Aware System Architecting (SASA), expands traditional system architecting practices by considering solver characteristics and contractual incentive mechanisms in the design process and aims to improve complex system design and innovation by leveraging the strengths of domain experts, crowds, and specialists for different parts of the problem. Given the complexity of system design problems and the variety of solvers available, it is desirable to have heuristics to guide solver assignment to different problems. Developing effective heuristics for solver assignments in complex system design is challenging due to a large number of possible combinations of problem-solver pairs. To address this challenge, this paper presents a computational approach using a multi-armed bandit

(MAB) formulation to generate heuristics for solver assignment. The approach is demonstrated using a simple and idealized problem of golf, which has characteristics similar to design problems, including how the problem is decomposed into sub-problems, and solved by different solvers. The results show that the proposed approach is effective in deriving a rich set of heuristics for the golf problem, and can be extended in the future to more complex systems design problems.

Keywords: Design heuristics, decision making, reinforcement learning, multi-armed bandit, systems engineering, systems architecture, solver-aware system architecting.

1 Introduction

Traditional complex system design practices primary rely on experts with domain-specific knowledge [1, 2]. Such experts are generally in short supply and they can be costly to employ [3]. Additionally, exclusive dependence on them can lead to "innovation blindness" [4]. On the other hand, there is a growing understanding of using non-experts for complex system design problems [5, 6]. For example, crowdsourcing has been used for innovative design and problem-solving in engineering system

Copyright © 2023 by ASME

design [7, 8]. However, system architecting traditionally focuses on the best structure for the operational context/environment, without considering who is solving different parts of the problem [9, 10]. The past literature ignores the potential of architectural innovation and using different kinds of solvers, including the capabilities of the solvers themselves [11, 12]. Recent literature has shown the best system architecture depends on *who* is solving the problem, and *how* they are engaging in the process [13].

The approach of jointly considering both problem decomposition and solving capacity is referred to as Solver-Aware System Architecting (SASA) [13]. SASA allows organizations to benefit from the unique strengths of domain experts, crowds, and specialists to enhance complex system design and innovation. The SASA framework expands on traditional architecting practices by including solver characteristics and contractual incentive mechanisms in the design process. Szajnfarber and co-authors [13] demonstrate the benefits of concurrently considering problem architecture and solver characteristics for improving solution efficacy using an abstract simulation model [14, 15]. Solver assignment refers to the process of pairing solvers (e.g. experts, crowds, specialists) with a specific problem or task.

The best solution for a complex system design problem depends on the design preferences which are specific criteria, values, or attributes about the design of a complex system [16, 17]. Consequently, the SASA framework brings forth an additional dimension to consider in exploration of the design space: the solver-assignment decisions. By incorporating design preferences into the SASA process, designers can more effectively evaluate alternatives based on how well they meet the desired design objectives. The right combination of technical architectures and solvers can lead to improved quality of the design solution, along with reduction of development costs and reliance on a single type of solver, such as experts.

With the increasing complexity of systems [18], the number of possible solver assignments for a given problem can be overwhelming, making it challenging to choose the most appropriate one. The complexity of the solver assignment process can be reduced by developing heuristics that guide designers in choosing the best solvers based on the problem characteristics. A heuristic is a problem-solving strategy that uses a practical, rule-of-thumb approach to finding a solution. Heuristics are used when finding an optimal solution is not feasible or when a quick, satisfactory solution is needed [19]. Heuristics take into consideration the characteristics of the problem and use past experience to arrive at satisficing solutions.

In their earlier work on SASA, Szajnfarber and co-authors [13] identified two heuristics for guiding solver assignment through pair-wise comparisons. These heuristics are (i) isolate subproblems that match an external specialty, and (ii) leverage crowdsourcing tournaments to tackle problems with a low-entry barrier and high variability. The heuristics strike a balance between the experts' tendency to gravitate towards dominant solutions and the crowd's ability to fully explore the design space. The authors of [13] acknowledged the difficulty in fully exploring the space of potential heuristics due to the limitations of traditional statistical methods to disentangle overlapping and sequential assignments, and the associated computational complexity. This restricts the extensibility of the SASA framework to more realistic system design problems.

To develop the best heuristics for solver assignment we would like to explore the design space as we make decisions about which solver to assign to each sub-problem. In complex design spaces, sequential decision making using reinforcement learning (RL) can help in managing this exploration-exploitation trade-off [20]. In this paper we formulate the solver assignment process in the abstract simulation model developed for SASA framework as a one-state Markov Decision Process (MDP), and make use of Muli-armed bandits (MAB) [20], which is a reinforcement learning technique, to solve the one-state MDP.

The approach results in *inclusionary* and *exclusionary* heuristics for the solver-assignment problem. An inclusionary heuristic suggests that a particular solver type should be used for a given sub-problem because that solver's performance is highly likely to be better than the performance of the other solvers. In other words, inclusionary heuristics are used to identify the best solver for a specific task. On the other hand, an exclusionary heuristic suggests that a particular solver should not be used for a given sub-problem because it has consistently performed poorly in similar situations. Exclusionary heuristics are used to identify the solvers that should be avoided for a specific task.

The results demonstrate that reinforcement learning is a viable method for finding exclusionary and inclusionary heuristics for SASA. Our findings illustrate how the optimal solver-assignment strategy can be adapted based on the designer's relative preference between design performance and cost. The paper is organized as follows. We present a review of the literature on solver asssignment in system architecting and heuristics in engineering design in Section 2. Section 3 provides an overview of the approach followed in this paper. Section 4 presents the reference golf model, the formulation of the reinforcement learning problem, and the resulting inclusionary and exclusionary heuristics for the golf problem. Finally, closing comments are presented in Section 5.

2 Background

2.1 Solver Assignment in Systems Engineering and System Architecting

Systems engineering is a discipline that focuses on designing complex systems in a way that allows subtasks to be completed efficiently in parallel and eventually integrated to form a system that delivers value over its lifecycle [21].

Copyright © 2023 by ASME

Crawley et al. define a system architecture as "the embodiment of concept, the allocation of physical/informational function to the elements of form, and the definition of relationships among the elements and with the surrounding context" [22]. Fixson et al. [23] further explore the relationship between product architecture, innovation, and industry structure and highlight the trend towards increased modularity in products and how this can impact competition and industry structure. A research study assessed fractionated architectures using a custom-built spacecraft model and found that they can have a stronger value proposition by attaining a lesser life cycle cost and longer mission lifetime relative to a comparable monolithic spacecraft for high-resolution earth imaging mission [24].

To summarize, several studies have shown how architectural choices influence the value of the design artifacts [25–29]; however, these predominantly focus on the uncertainties associated with the operational environment and disregard the influence of the actors who will execute the design process. In this conundrum, SASA enhances the typical architecture screening process by incorporating solver allocation as an integral aspect of the development process. This new step involves characterizing different solver capabilities in relation to various subfunctions, and then evaluating architectures for their performance under different solving configurations.

2.2 Heuristics in Engineering Design

Heuristics play a significant role in the engineering design process and have been extensively studied in the design literature. Fu et al. [30] define heuristics as context-sensitive guidelines, based on intuition, experience, or implicit knowledge, which guide the design process towards a satisfactory, yet not necessarily optimal solution. Heuristics are used at all stages of the design process, including the creation of new design concepts and decision making during the later stages. A protocol study by Yilmaz et al. [31] found that engineers from various domains regularly utilize heuristics in the conceptual design phase. Yilmaz and Seifart [32] also demonstrate that the use of heuristics by experts during the early stages of design can result in innovative and creative solutions. Fillingim et al. [33] conducted interviews with ten experts at the Jet Propulsion Laboratory and identified 101 heuristics used in space mission design. Heuristics are also employed in the later stages of design, such as optimization, to simplify the process of finding optimal designs. However, Deshmukh et al. [19] caution that while these heuristics can be useful in managing optimization tasks, they can also lead to unintended constraints and cognitive biases. To conclude, while heuristics play an important role in engineering design and are useful in various ways, they could also lead to undesirable consequences. Thus, it is necessary to carefully evaluate the potential implications of using heuristics in design.

In summary, the systems design literature extensively studied how technical decisions should be made and identified

heuristics for practically addressing these concerns. However, with SASA, there is a pressing need to expand design heuristics to system architecting and solver assignment decisions. While machine learning methods have been used to explore design spaces [34], such techniques have not been explored for developing heuristics for systems architecting and solver assignment decisions. We address this gap by presenting a reinforcement learning based approach to extract heuristics for the solver assignment problem. The approach is discussed in the following section.

3 Approach

We aim to extract heuristics for solver assignment in SASA while maximizing the solution quality and reducing the total cost associated with that solution. Our objective is to determine the design preferences under which solvers outside domain-experts, who can be specialists in other domains or amateurs that can be engaged through a crowd-sourcing contest, can provide high-quality solutions to complex problems that are traditionally considered to be the domain of experts. To achieve this, we will utilize a simulation model to analyze the effects of different problem decompositions and solver assignments on the solution performance and the cost of generating the solutions and describe it in Section 3.1. In Section 3.2 we model formulate the solver assignment process as a Markov Decision Process (MDP), and in Section 3.3 we describe an approach to solve this MDP and to extract heuristics.

3.1 A Model of the Problem Decomposition and Solver Assignment Process

Problem: Considering design as a problem solving process, we represent the problem of a systems engineer mathematically using a design space $\mathcal{X}$ and a multidimensional solution performance space $\mathcal{Y}$. The design space represents the set of variables, parameters, and constraints that define the design possibilities which are the solutions of the problem. It also involves a preference function $V(Y)$ where $Y \in \mathcal{Y}$ that maps the points in a multi-dimemsional performance space to a scalar value for comparing different solutions.

Problem Decomposition: The problem architecture is a set of sub-problems and interdependencies among them. The sub-problems and their architectures can be represented as a network or a design structure matrix (DSM) [35, 36]. The DSM is a matrix, where the rows and columns represent the element of the system, which may be components or functions, and the cells of the matrix represent the interactions or dependencies between the elements. The architecture may also contain design rules which dictate how the sub-problems are coordinated. For example, in the DSM shown in Figure 1, we decompose the problem into two sub-problems based on the dependencies between the sub-problems. The coordination rules dictate how the different sub-problems are coordinated (e.g., how the output

Copyright © 2023 by ASME

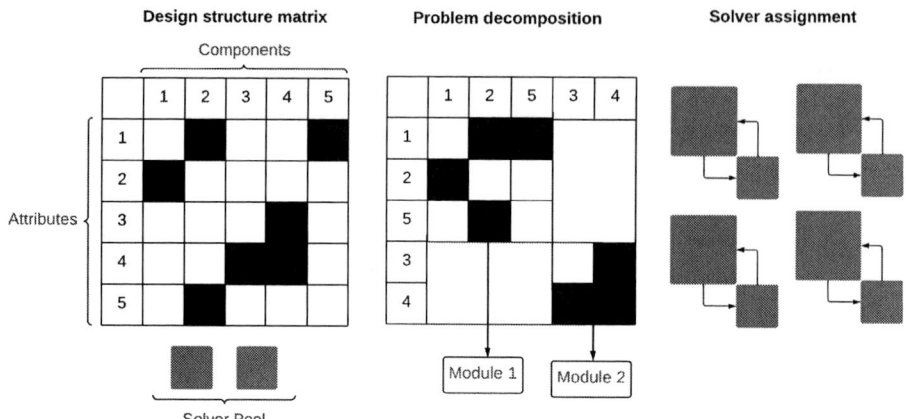

FIGURE 1: Illustration of the Problem Decomposition and Solver Assignment Process

of a module is used as an input to the other module) to achieve the overall system objectives. The design and coordination rules together describe how the decomposition is organized and managed.

Solver Assignment: The solver assignment space contains all possible assignments of solvers to sub-problems in the problem architecture. A solver from the solver pool is assigned to each sub-problem in the problem architecture, resulting in an assignment of solvers to sub-problems. Each solver has a probability distribution describing how well he/she achieves the objective for that sub-problem. As each problem architecture might contain a different number of sub-problems, the size of the set of solver assignment varies depending on the problem architecture. Alternative system architectures can be operationalized as task structures, which means they are represented in a way that can be practically executed by assigning solvers to different modules within the architecture. Multiple problem task structures, which serve as operational representations of different architectures, are defined to solve the problem.

Alternate solver types are assigned to tasks in each task structure, represented with blue and green colors in Figure 1, and problem-solving can be simulated for each combination, with solvers executing their assigned tasks in the context of the overall task structure.

Solving Process: To represent the output of the solving process, we use an output vector $Y \in \mathscr{Y}$ which captures the performance and cost metrics resulting from the solver assignment process and can be used for comparison between different assignments of solvers to sub-problems. Each component of the output vector corresponds to a different performance or cost metric, depending on the specific problem being considered.

3.2 Solver Assignment as a Markov Decision Process

Solver assignment can be modeled as a single-step Markov decision process (MDP). In this model, the solver assignment process at each step is a solver-assignment decision. The state of the system at each step includes the current assignment of solvers to sub-problems, as well as any relevant information about the problem architecture and the solver pool. The agent receives a reward based on the quality of the resulting solution generated from the problem solving process. The reward also depends on the cost associated with the solution. Both the quality and cost are components of the output vector Y.

Mathematically, we can represent the single-step MDP of solver assignment as a tuple $\langle S, A, T, R \rangle$, where:

1. S is the set of possible states of the system, which includes all possible assignments of solvers to sub-problems.
2. A is the set of possible actions that the agent can take, which includes all possible assignments of solvers to sub-problems inside a problem. By choosing a specific action, the agent assigns a particular set of solvers to sub-problems in a problem.
3. T is the transition function that describes the probability of transitioning from one state to another, given an action. In this case, the transition function is deterministic, as the new state is entirely determined by the chosen action.
4. R is the reward function that describes the reward received by the agent for a given state and action. The reward function is based on the quality of the resulting assignment, as well as any costs associated with the assignment process. It is defined as a linear combination of performance and cost of the simulation output,

$$R = q - w_c * c \qquad (1)$$

Copyright © 2023 by ASME

where w_c serves as the weight of the cost.

The solver-assignment decision is made by an agent whose goal is to find the optimal policy π, which maps each state to the optimal action to take in that state, in order to maximize the expected total reward over time. This can be expressed mathematically as $\pi(s) = \arg\max_{a \in A} Q(s,a)$, where $Q(s,a)$ is the optimal action-value function, which represents the expected reward for taking action a in state s and following the optimal policy thereafter.

We make the following assumptions about the solver assignment process.

A1: The set of actions does not change over time. The set of possible actions, i.e., assigning solvers to sub-problems in a problem, remains constant throughout the entire process. Once a solver is assigned to a sub-problem, it remains assigned to that sub-problem until the end of the solving process.

A2: Each action is independent. The assignment of solvers to problems is independent of other assignments. The assignment of a solver to one sub-problem does not affect the assignment of solvers to other sub-problems.

A3: The rewards are drawn randomly from a probability distribution that is specific to each action. The rewards obtained from assigning a solver to a sub-problem are based on the solver's expertise and effort required to solve the sub-problem, which are specific to each assignment. The rewards for assigning a solver to one sub-problem are independent of the rewards obtained from assigning solvers to other sub-problems.

A4: The environment is stationary. The reward distribution of actions does not change over time. The expertise and effort required to solve a sub-problem do not change over time, and the expertise of the solvers also remains constant. Therefore, the rewards obtained from assigning solvers to sub-problems do not change over time.

These assumptions allow us to formulate the single-state MDP as a multi-armed bandit problem. The assumptions mentioned above simplify the solver assignment process by reducing it to a single-state MDP, where the action taken by the solver in the current state will have an immediate impact on the reward received. Since the environment is stationary (Assumption 4), the reward distribution of actions does not change over time, which means that the history of the solver's actions and rewards is not important for determining the reward of the current action.

Under these assumptions, the solver assignment process can be seen as a multi-armed bandit (MAB) problem, where each assignment of solvers to a problem is an arm, and the solving process is a pull of the arm. The goal is to maximize the cumulative reward received over a finite number of pulls. The rewards of each arm (sub-problem) are independent and randomly drawn from a probability distribution specific to each arm (Assumptions 2 and 3). Therefore, the multi-armed bandit problem formulation allows us to efficiently explore the space of assignments and exploit the best assignments based on the observed rewards.

3.3 Reinforcement Learning for Identifying Heuristics

In the multi-armed bandit problem, the agent has the choice of selecting one of k different actions, referred to as "arms". After making a selection, denoted by A_t at time step t, the agent receives a reward R_t. The objective of the agent is to maximize the cumulative reward by choosing optimal actions.

Typically, each action has an associated value, also known as an expected reward. For any action a in the action space $\mathscr{A}$ which consists of n_a actions, $q_*(a)$ which is the true expected reward for action a, the expected reward can be represented as follows:

$$q_*(a) = \mathbb{E}[R_t | A_t = a]. \tag{2}$$

As the actual values of the actions are not known for certain, an estimate of the value of action a at time step t can be computed as $Q_t(a)$. In MAB problems, it is often assumed that $Q_t(a)$ is close to the true expected reward $q_*(a)$ and these estimates are used to determine the action to be taken.

In this paper, we use the epsilon-greedy algorithm (ε-greedy), which is a representative algorithm of the MAB.

$$a = \begin{cases} \arg\max_{a \in \mathscr{A}} q_*(a) & \text{with probability } (1 - \varepsilon) \\ \widehat{a} & \text{with probability } \varepsilon \end{cases} \tag{3}$$

In the ε-greedy, action is derived by the following Equations 1 and 2, where $\widehat{a}$ and ε represent an action selected randomly and the probability of selecting a random action, respectively. The ε-greedy algorithm balances the exploration of new actions (by randomly selecting an action with probability ε) and exploitation of the current best action (by selecting the action with the highest expected reward with probability $1 - \varepsilon$).

We set ε to 0.05, which means that 5 percent of the time a random action is selected, while the other 95 percent of the time the action with the highest expected reward is selected. This value of ε is a commonly used default in MAB problems as it strikes a reasonable balance between exploration and exploitation. However, the specific value of ε can be adjusted based on the problem at hand, as well as the trade-off between exploration and exploitation desired by the user.

We evaluate the probability of the reward obtained by assigning a specific solver type to a module (R_{solver}) to be greater than the reward obtained by any other solver ($R_{\neg solver}$), i.e., $P(R_{solver} > R_{\neg solver})$. If this probability exceeds a

Copyright © 2023 by ASME

certain threshold τ, we consider it as an inclusionary heuristic. Similarly, if $P(R_{solver} < R_{\neg solver})$ exceeds the threshold τ, we consider it as an exclusionary heuristic, i.e., that type of solver should not be assigned.

4 Demonstration of the Approach using an Idealized Problem

In this section, we describe a model of golf played by different players as a simplified representation of SASA. The model is developed by Szajnfarber and coauthors, and described in detail in [13].

4.1 Golf as an Illustrative Problem for SASA

The golf model is composed of four main elements: a reference problem, its alternative task structures (or modularizations), solver assignment, and the solving process.

The Reference Problem: A nine-hole, one-dimensional golf game is modeled, where the goal is to move the ball 700 yards from the tee to the pin using a sequence of generic golf strokes [14]. The simulation implemented three stroke types that are representative of different tasks associated with the design process: (i) *driving* the ball off the tee, where the goal is to move the ball as far as possible, (ii) *approach*, where the goal is to get the ball as close to the hole as possible; and finally (iii) *putting*, where the goal is to accurately sink the ball in the hole. Each model run starts with a drive off the tee, ends with the ball in the hole, the next stroke type is determined based on the position of the ball on the field. The outcomes are evaluated based on performance and cost to complete the nine holes.

Alternative Architectures: The baseline single hole golf game (H) is decomposed into three modules based on functional similarities and the stroke types associated with them [37, 38]: the Tee (T), where the objective is to maximize distance, the Fairway (F), where the goal is both distance and accuracy, and the Green (G), where the goal is to accurately putt the ball into the hole and overshoots are penalized. These three modules were organized into three alternative task structures of the reference problem through introduction of interface rules [39, 40] (i.e., pick the ball that is furthest down the field): TS, LG, and TFG. Here, the L module represents the long game that is essentially a combination of T and F; where the objective is to reach the green from the tee in the least number of strokes. Whereas, S module refers to the short game, a combination of F and G, where the goal is to sink the ball in as few strokes as possible.

Solver Types and Assignment: Three player types are defined using probability distributions for each stroke type to map their relative goodness into the simulated model. These player types are analogous to different solver types in engineering design with different expertise. The first is domain-experts that are representative of cross-trained multi-disciplinary experts or concurrent engineering design teams (e.g., NASA Team X). These are represented as *Professional* golfers in the simulation; they are better performing on the reference problem H level and are good in all aspects of the game. The second player type is *Specialists*, who are representative of experts in a neighbouring that shares one common function with the multi-disciplinary design task, such as structural design or electronics. In the golf model, these are represented as long-drivers who are excellent only at driving off the Tee and mediocre in every other task. The third solver archetype is *Amateurs*, which are representative of independent solvers in the crowd that can be engaged through an open innovation contest. Compared to Professionals, Amateurs are modeled to have significantly lower average capabilities and higher variability in all aspects of golf. Specialists are assumed to behave identically to Amateurs in all tasks except driving. Relative solver capabilities were calibrated by using real-world golf statistics and are represented by normal distributions.

Assignment of these solver archetypes to alternative task structures is represented by modeling contractual mechanisms. For Professionals, it is assumed that they are engaged with a regular employment contract where the solving effort is incentivized by a fixed price (similar to a salary). Specialists are assumed to be engaged via a bidding process with three participants, where the best performance is awarded the task and the runner-ups are only compensated a fraction of their efforts for participation. Finally, Amateurs are assumed to be contracted through an open innovation contest, a winner takes all crowd-sourcing competition with a fixed-prize pot [41], similar to an amateur golf tournament. It is assumed that a hundred amateurs participate in each crowd-sourcing contest and the best performance is selected ex-post.

Solving Process: The solving processes are simulated by using the alternative task structures {H, TS, LG, TFG}, the contract mechanisms, solver assignment to a given task, and the abilities of the solver archetypes to finish a nine-hole game of golf. Below, we elaborate how performance and cost are accounted for.

Performance (P) is measured by the total number of strokes taken to complete the game. Let s be the number of strokes, j be different tasks in the assignment, and i be the number of holes, performance is computed as:

$$P = \Sigma_1^i \Sigma_1^j s_j. \tag{4}$$

Cost (C) is calculated in three main categories: architecting, execution, and integration. The architecting stage involves planning and task allocation, the execution stage is when the work is done by assigned groups, and the integration stage is when the outputs of the tasks are combined. In the baseline model, the wages for professionals (w_P), amateurs (w_A), and specialists (w_S) are set at 10, 1, and 12 cost units per stroke respectively. The cost consists of three components: (a)

Copyright © 2023 by ASME

architecting cost, (b) execution cost, and (c) integration cost.

The architecting cost is assumed negligible for the undecomposed problem H and constant for all modularized architectures. As it is typically done in the industry, we assume that architecting is performed by an expert thus architecting costs are equal to W_P.

The execution costs are determined by the solver assignment and their associated cost function. Each solver is assumed to be assigned to a task by a specific contractual structure. For Professionals, it is assumed that they followed a conventional employment contract and paid a wage (W_P) for each stroke (S_P) they take. Thus, the cost for professional is $C_{WP} = (W_P S_P)$. It is assumed that specialists work as technical contractors, where the cost of execution involved a bidding phase with participation costs plus the winning specialist's wage for the task. It is assumed that the number of bidders (m) for specialists is always three, and the bidding cost (b) is one-tenth of the specialist's wage (w_S). Hence, the cost for specialist, $C_{WS} = b(m-1) + w_S S_S$. For amateurs, it is assumed that they participate in a winner-takes-all crowdsourcing contest and the cost for the firm is fixed prize purse (p). Thus, the cost for amateur is $C_{WA} = p$, one tenth of the cost of W_P.

Finally, integration involves coordinating the hand-off rules between modules, and it is assumed that one low-skilled ball tracker is required per participant to keep track of the number of strokes taken by each ball. Hence, the cost of integration is equal to (nW_{TC}), where n is the number of participants (e.g., 100 for an amateur tournament) and W_{TC} is the wage for each tournament coordinator.

4.2 Multi-armed Bandit Formulation for the Golf Problem

In this subsection, we describe the Multi-Armed bandit based reinforcement learning model (action, and reward) used for decision making for solver assignment in the golf model and compare its performance with random sampling. For each alternative task structure in the golf example (H, LG, TFG, TS), we can assign combinations of available solvers (Professional, Amateur, Specialist). This constitutes 48 possible scenarios in which solver assignments can take place. These combinations serve as "arms" for the MAB problem, and they constitute the action space, $\mathcal{A}$ ($n_a = 48$). We use the reward (R) definition as described in Equation (1), where q is the performance P, and c is the cost C. If the optimal reward is known, regret is defined as the difference between the optimal reward and the reward for the MAB. We visualize the average regret using the ε-greedy policy described in Section 3.3 and random sampling for 25 runs in Figure 2(a). We observe that the regret starts converging for the ε-greedy policy after around 100 episodes while it continues to fluctuate in the case of random sampling. In the context of the SASA problem, the best arm corresponds to the optimal solver-assignment. Therefore, the MAB approach

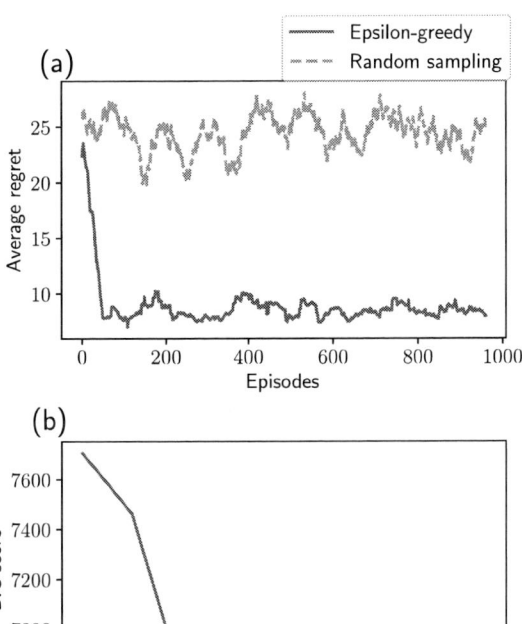

FIGURE 2: RL model for Golf. **a**: Comparison of ε - greedy policy with random sampling. **b**: BIC score is minimum when GMM has 3 components.

can identify the optimal solver-assignment more quickly and with less regret than random sampling. This is because the MAB algorithm learns from previous assignments and focuses on the most promising solvers, whereas random sampling does not take into account the history of assignments and may waste computational resources on sub-optimal solvers.

To identify heuristics, we compare the reward distribution across various solver assignments in that sub-problem and for that, we chose to fit the data using an the Gaussian mixture model (GMM). The GMM is a probabilistic model that assumes all the data points are generated from a mixture of a finite number of Gaussian distributions with unknown parameters. We fit the probability density functions of the reward distributions for the corresponding solver assignments using a GMM. To find the number of components for GMM we use the Bayesian information criterion (BIC). BIC rewards goodness of fit (as assessed by the likelihood function), it includes a penalty that is an increasing function of the number of estimated parameters. The penalty discourages over-fitting, which is desired because increasing the number of parameters in the model almost always improves the goodness of the fit. We calculate the average BIC

Copyright © 2023 by ASME

score for all the distributions to find the number of components for the GMM and visualize it in Figure 2(b). We observe that the BIC score is minimum when GMM has three components.

4.3 Identifying Heuristics for the Golf Problem

We use 3-component GMM to fit the reward distributions and visualize them for module T in Figure 3(a). We use the fitted GMM models to generate heuristics. For example, we compute $P(R_{solver} - R_{\neg solver} > 0)$ for Amateurs and Specialists and visualize it in Figure 3(b) and Figure 3(c), respectively. We observe that in approximately 86.5 percent of the cases, Amateurs performed worse than Professionals and Specialists, and in approximately 89.2 percent of the cases Specialists performed better than Professionals and Amateurs. Hence for this module, we can conclude that on using a threshold $\tau = 0.8$, assignment of Amateurs is an exclusionary heuristic, and assignment of Specialists is an inclusionary heuristic.

4.4 Results: Extraction of Heuristics

For all simulations, we used $\varepsilon = 0.05$, $t = 500$, and ran the model for 1000 runs (50 samples of 20 runs). We varied the weight of cost in the reward function (w_c), and the threshold (τ) used to identify the heuristics. We present the identified heuristics for $\tau = 0.5$ in Table 1 and $\tau = 0.8$ in Table 2. Short labels "Pro, Am, and Spec" are used for Professional, Amateur, and Specialist solvers, respectively. The inclusionary heuristics are shown in normal text, exclusionary heuristics in strikethrough text, and "-" indicates cases when no heuristic was identified. For example, Table 1 shows that for $\tau = 0.5$ whenever there is a T, assign a Spec to it and whenever there is a T-F, assign a Spec to T, and Pro to F. The ranges for the weight of cost (w_c) were selected based on the reward signals, such that there is a balance between improving performance and lowering cost within the range. The results of the study show that the selection of solver-assignment heuristics significantly affects the reward. In particular, we observed that the inclusionary and exclusionary heuristics varied depending on the specific module, w_c, and τ.

Threshold $\tau = 0.5$ suggests that a particular assignment is better than the other available assignment in 50 percent of the cases. We choose to present the results for $\tau = 0.5$ and $\tau = 0.8$ to show the effect of setting a relatively high and low thresholds for a heuristic. A lower threshold allows us to generate a greater number of heuristics, this helps us to understand the overall trends of solver assignments for sub-problems and changing w_c. We can be more certain about the heuristics generated using a greater threshold but it does not results in many heuristics for solver assignment. As the value of w_c increases, the optimal choices tend to shift towards the lower-cost options. This is because the weight given to cost in the objective function is higher, making it more important to minimize cost even at the expense of some performance. On comparing Table 1 and Table 2, we observe that the total number of heuristics

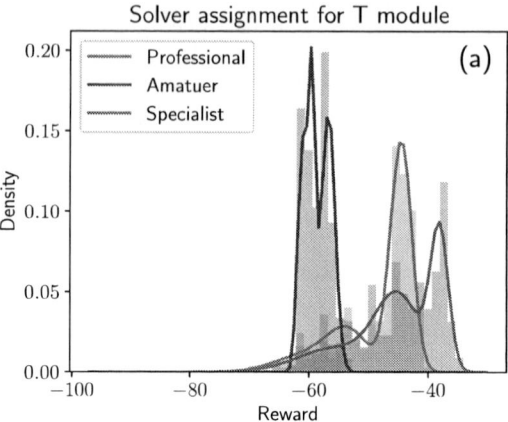

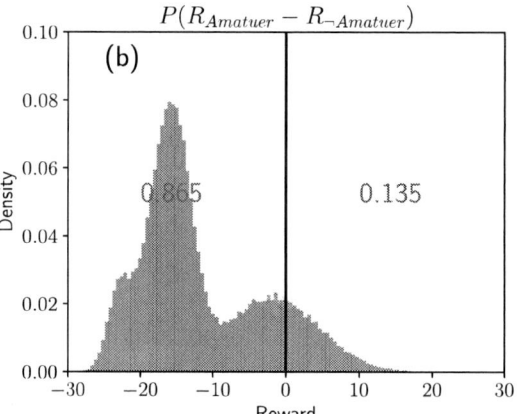

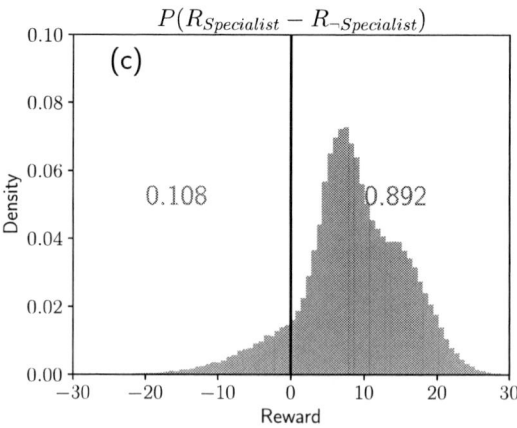

FIGURE 3: (a) Probability density for reward corresponding to solver assignment in module T; (b) Amateur solver assignment heuristic for module T; (c) Specialist solver assignment heuristic for module T.

Copyright © 2023 by ASME

TABLE 1: Heuristics identified for various modules and w_c values given $\tau = 0.5$. The inclusionary heuristics are shown in normal text, exclusionary heuristics in strikethrough text, and "-" indicates cases when no heuristic was identified.

w_c	0.0	0.25	0.5	0.75	1.0
T	Spec; ~~Am~~	Spec; ~~Am~~	Spec; ~~Am~~	Spec; ~~Am~~	Spec; ~~Am~~
F	Pro; ~~Spec~~	Pro; ~~Spec~~	Pro; ~~Spec~~	~~Spec~~	Am; ~~Spec~~
G	Pro; ~~Am~~	Pro; ~~Am~~	~~Spec~~	~~Spec~~	Pro; ~~Am~~
L	Pro; ~~Am~~	Pro; ~~Am~~	Pro; ~~Am~~	Pro; ~~Spec~~	Am; ~~Spec~~
S	Pro; ~~Spec~~	Pro; ~~Spec~~	Am; ~~Spec~~	Am; ~~Spec~~	Am; ~~Spec~~
H	Pro; ~~Am~~	Pro; ~~Spec~~	Am; ~~Spec~~	Am; ~~Spec~~	Am; ~~Spec~~
T-F	Spec-Pro	Spec-Pro; ~~Am-Spec~~	Spec-Pro; ~~Am-Spec~~	Spec-Pro; ~~Am-Spec~~	Spec-Pro; ~~Am-Spec~~
F-G	-	Pro-Pro	Pro-Pro	Pro-Pro	Am-Am
L-G	Pro-Pro	Pro-Pro; ~~Am-Spec~~	Pro-Pro; ~~Am-Spec~~	Pro-Pro; ~~Spec-Spec~~	Am-Am; ~~Spec-Spec~~
T-S	Spec-Pro	Spec-Pro; ~~Am-Spec~~	Spec-Pro; ~~Am-Spec~~	Spec-Pro; ~~Am-Spec~~	Spec-Pro; ~~Am-Spec~~
T-F-G	-	Spec-Pro-Pro; ~~Am-Spec-Spec~~	~~Am-Spec-Spec~~	-	-

TABLE 2: Heuristics identified for various modules and w_c values given $\tau = 0.8$. The inclusionary heuristics are shown in normal text, exclusionary heuristics in strikethrough text, and "-" indicates cases when no heuristic was identified.

w_c	0.0	0.25	0.5	0.75	1.0
T	Spec	-	-	-	-
F	-	Pro ~~Spec~~	Pro ~~Spec~~	~~Spec~~	~~Spec~~
G	-	-	-	-	-
L	~~Am~~	~~Am~~	-	-	
S	~~Spec~~	~~Spec~~	Am; ~~Spec~~	~~Spec~~	Am; ~~Spec~~
H	Pro; ~~Am~~	Pro; ~~Spec~~	Am; ~~Spec~~	Am; ~~Spec~~	Am; ~~Spec~~
T-F	-	Spec-Pro	Spec-Pro	-	-
F-G	-	-	-	-	-
L-G	-	Pro-Pro; ~~Am-Spec~~	-	Pro-Pro	-
T-S	-	Spec-Pro	Spec-Pro	Spec-Pro	Spec-Pro
T-F-G	-	Spec-Pro-Pro; ~~Am-Spec-Spec~~	~~Am-Spec-Spec~~	-	-

identified reduces as we increase the threshold but we can be more confident about them. Therefore, heuristics identified for $\tau = 0.8$ also apply for $\tau = 0.5$, but not vice versa.

In Table 1 we identify the solver-assignment heuristics which always hold for the w_c values considered. For module T, assignment of Spec is always an inclusionary heuristic, while the assignment of Am is always an exclusionary heuristic. For modules F and S, the assigning the solver Spec is always an exclusionary heuristic. For modules T-F and T-S, the assigning of solvers Spec-Pro is always an inclusionary heuristic. We further identify the heuristics that appeared in most but not all cases. The solver-assignment of Pro is an inclusionary heuristic for module G. The solver-assignment of Am-Spec heuristic is an exclusionary heuristic for module T-S. The solver-assignment of Pro-Pro is an inclusionary heuristic for module L-G. Finally, we identify the solver-assignment pairs that change as the weight of cost increases. We observe that as the value of w_c increases, there is a general trend towards the exclusion of specialist

Copyright © 2023 by ASME

and professional solvers in favor of amateur solvers. For modules F, S, L, and H, the inclusionary heuristic changes from solver-assignment of Pro to Am as the weight of cost increases. For problem types L and H, the exclusionary heuristic changes from solver-assignment of Am to Spec. The inclusionary heuristic changes from solver-assignment of Pro-Pro to Am-Am for modules F-G and L-G as w_c increases.

5 Discussion

In this paper, a computational approach to generate heuristics for solver assignment in complex system design problems using a multi-armed bandit (MAB) formulation is presented. The approach is demonstrated using an abstract golf problem and the results show that the proposed approach is effective in deriving a set of solver-assignment heuristics for a reference design problem in way that can be adjusted with respect to designer preferences. Inclusionary and exclusionary heuristics varied depending on the specific module, w_c, and τ; allowing the designer to pick the parameters w_c and τ based on their preferences for cost sensitivity and certainty of performance.

In the prior work, two heuristics were identified for guiding solver assignment through pair-wise comparisons: to isolate subproblems that match an external specialty and to leverage tournaments to fully explore highly variable problem spaces [13]. Although the potential value of combinatorial solver-assignment heuristics was recognized, their identification was left out of scope due the complexity of the solution space. To that end, the RL approach illustrated in this paper provides a richer exploration of the design space, accompanied by computational savings, and the ability to identify solver assignment heuristics for each of the specific sub-problems and all of their possible combinations. Furthermore, we not only identified the same set of solver-assignment heuristics that were documented in previous work, but are also able to explore the overall heuristic space, building confidence in the transferability of the approach to other complex design problems.

We have limited this study to using three kinds of solvers and the ε-greedy algorithm. Future research opportunities include using other MAB algorithms such as Thompson Sampling and Upper Confidence Bound, including more types of solvers, contractual structures, interface rules, and more realistic design problems beyond golf. Additionally, future research could investigate including decision making to generate architectures within the MDP, which can be solved with multi-step RL formulations such as tabular methods and policy gradients.

ACKNOWLEDGMENT

We gratefully acknowledge the financial support from the National Science Foundation through NSF CMMI Grants 2129539 (Purdue) and 2129574 (the George Washington University & Virginia Tech). Any opinions, findings, and conclusions or recommendations expressed in this material are those of the authors and do not necessarily reflect the views of the National Science Foundation.

REFERENCES

[1] Carlile, P. R., 2004. "Transferring, translating, and transforming: An integrative framework for managing knowledge across boundaries". *Organization science, 15*(5), pp. 555–568.

[2] Vincenti, W. G., et al., 1990. *What engineers know and how they know it*, Vol. 141. Baltimore: Johns Hopkins University Press.

[3] Cappelli, P. H., 2015. "Skill gaps, skill shortages, and skill mismatches: Evidence and arguments for the united states". *ILR review, 68*(2), pp. 251–290.

[4] Leonardi, P. M., 2011. "Innovation blindness: Culture, frames, and cross-boundary problem construction in the development of new technology concepts". *Organization Science, 22*(2), pp. 347–369.

[5] Chesbrough, H. W., 2003. "A Better Way to Innovate". *Harvard Business Review, 81*(7), July, pp. 12–13. Number: 7 Publisher: Harvard Business School Publication Corp.

[6] Gambardella, A., Raasch, C., and von Hippel, E., 2016. "The User Innovation Paradigm: Impacts on Markets and Welfare". *Management Science, 63*(5), Apr., pp. 1450–1468. Number: 5 Publisher: INFORMS.

[7] Panchal, J. H., et al., 2015. "Using crowds in engineering design–towards a holistic framework". In DS 80-8 Proceedings of the 20th International Conference on Engineering Design (ICED 15) Vol 8: Innovation and Creativity, Milan, Italy, 27-30.07. 15, pp. 041–050.

[8] Chaudhari, A. M., Sha, Z., and Panchal, J. H., 2018. "Analyzing participant behaviors in design crowdsourcing contests using causal inference on field data". *Journal of Mechanical Design, 140*(9), p. 091401.

[9] Szajnfarber, Z., and Weigel, A. L., 2013. "A process model of technology innovation in governmental agencies: Insights from NASA's science directorate". *Acta Astronautica, 84*, Mar., pp. 56–68.

[10] Vrolijk, A., and Szajnfarber, Z., 2015. "When Policy Structures Technology: Balancing upfront decomposition and in-process coordination in Europe's decentralized space technology ecosystem". *Acta Astronautica, 106*, Jan., pp. 33–46.

[11] Maier, M. W., and Rechtin, E., 2009. *The art of systems architecting*. CRC press.

[12] Crawley, E., Cameron, B., and Selva, D., 2015. *System architecture: Strategy and product development for complex systems*. Prentice Hall Press.

[13] Szajnfarber, Z., Topcu, T. G., and Lifshitz-Assaf, H., 2022. "Towards a solver-aware systems architecting framework:

Copyright © 2023 by ASME

leveraging experts, specialists and the crowd to design innovative complex systems". *Design Science, 8*, p. e10.

[14] Szajnfarber, Z., Grogan, P. T., Panchal, J. H., and Gralla, E. L., 2020. "A call for consensus on the use of representative model worlds in systems engineering and design". *Systems Engineering, 23*(4), pp. 436–442. Number: 4 _eprint: https://onlinelibrary.wiley.com/doi/pdf/10.1002/sys.21536.

[15] Chaudhari, A. M., Gralla, E. L., Szajnfarber, Z., Grogan, P. T., and Panchal, J. H., 2020. "Designing Representative Model Worlds to Study Socio-Technical Phenomena: A Case Study of Communication Patterns in Engineering Systems Design". *Journal of Mechanical Design, 142*(12), Sept. Number: 12.

[16] Hazelrigg, G., 1998. "A Framework for Decision-Based Engineering Design". *Journal of Mechanical Design, 120*(4), Dec., pp. 653–658. Number: 4.

[17] Collopy, P. D., and Hollingsworth, P. M., 2011. "Value-Driven Design". *Journal of Aircraft, 48*(3), May, pp. 749–759. Number: 3.

[18] Hennig, A., Topcu, T. G., and Szajnfarber, Z., 2021. "So You Think Your System Is Complex?: Why and How Existing Complexity Measures Rarely Agree". *Journal of Mechanical Design, 144*(4), Nov. Number: 4.

[19] Deshmukh, A. P., Thurston, D. L., and Allison, J. T., 2016. "Heuristics for formulating design optimization models: Their uses and pitfalls". In 5th Internatios Engineering Systems Symposium, CESUN 2016.

[20] Sutton, R. S., and Barto, A. G., 2018. *Reinforcement learning: An introduction.* MIT press.

[21] Haskins, C., Forsberg, K., and Krueger, M., 2011. "Systems engineering handbook: A guide for system life cycle processes and activities". Incose San Diego, CA (US).

[22] Crawley, E., Cameron, B., and Selva, D., 2015. *System architecture: strategy and product development for complex systems.* Prentice Hall Press.

[23] Fixson, S. K., and Park, J.-K., 2008. "The power of integrality: Linkages between product architecture, innovation, and industry structure". *Research Policy, 37*(8), pp. 1296–1316.

[24] O'Neill, M. G., and Weigel, A. L., 2011. "Assessing fractionated spacecraft value propositions for earth imaging space missions". *Journal of Spacecraft and Rockets, 48*(6), pp. 974–986.

[25] Boas, R., Cameron, B. G., and Crawley, E. F., 2013. "Divergence and lifecycle offsets in product families with commonality". *Systems Engineering, 16*(2), pp. 175–192. Number: 2 _eprint: https://onlinelibrary.wiley.com/doi/pdf/10.1002/sys.21223.

[26] Ross, A. M., Rhodes, D. H., and Hastings, D. E., 2008. "Defining changeability: Reconciling flexibility, adaptability, scalability, modifiability, and robustness for maintaining system lifecycle value". *Systems Engineering, 11*(3), pp. 246–262. Number: 3.

[27] Mosleh, M., Ludlow, P., and Heydari, B., 2016. "Distributed Resource Management in Systems of Systems: An Architecture Perspective". *Systems Engineering, 19*(4), pp. 362–374. Number: 4 _eprint: https://onlinelibrary.wiley.com/doi/pdf/10.1002/sys.21342.

[28] Mosleh, M., Dalili, K., and Heydari, B., 2018. "Distributed or Monolithic? A Computational Architecture Decision Framework". *IEEE Systems Journal, 12*(1), Mar., pp. 125–136. Number: 1 Conference Name: IEEE Systems Journal.

[29] Topcu, T. G., and Mesmer, B. L., 2018. "Incorporating end-user models and associated uncertainties to investigate multiple stakeholder preferences in system design". *Research in Engineering Design, 29*(3), July, pp. 411–431. Number: 3.

[30] Fu, K. K., Yang, M. C., and Wood, K. L., 2016. "Design principles: Literature review, analysis, and future directions". *Journal of Mechanical Design, 138*(10), Aug.

[31] Yilmaz, S., Daly, S. R., Seifert, C. M., and Gonzalez, R., 2015. "How do designers generate new ideas? design heuristics across two disciplines". *Design Science, 1*, Nov.

[32] Yilmaz, S., and Seifert, C. M., 2011. "Creativity through design heuristics: A case study of expert product design". *Design Studies, 32*(4), July, pp. 384–415.

[33] Fillingim, K. B., Nwaeri, R. O., Borja, F., Fu, K., and Paredis, C. J. J., 2019. "Design heuristics: Extraction and classification methods with jet propulsion laboratory's architecture team". *Journal of Mechanical Design*, June, p. 1.

[34] Ororbia, M. E., and Warn, G. P., 2023. "Design synthesis of structural systems as a markov decision process solved with deep reinforcement learning". *Journal of Mechanical Design, 145*(6), p. 061701.

[35] Steward, D. V., 1981. "The design structure system: A method for managing the design of complex systems". *IEEE Transactions on Engineering Management, EM-28*(3), Aug., pp. 71–74. Number: 3 Conference Name: IEEE Transactions on Engineering Management.

[36] Browning, T. R., 2016. "Design Structure Matrix Extensions and Innovations: A Survey and New Opportunities". *IEEE Transactions on Engineering Management, 63*(1), Feb., pp. 27–52. Number: 1 Conference Name: IEEE Transactions on Engineering Management.

[37] Parnas, D. L., 1972. "On the criteria to be used in decomposing systems into modules". *Pioneers and Their Contributions to Software Engineering, 15*(12), pp. 1053–1058. Number: 12.

[38] Ulrich, K. T., 2003. *Product design and development.* Tata

Copyright © 2023 by ASME

McGraw-Hill Education.

[39] Baldwin, C. Y., and Clark, K. B., 2000. *Design rules: The power of modularity*, Vol. 1. MIT press, Cambridge, MA.

[40] Topcu, T. G., Mukherjee, S., Hennig, A. I., and Szajnfarber, Z., 2021. "The Dark Side of Modularity: How Decomposing Problems can Increase System Complexity". *Journal of Mechanical Design,* **144**(3), p. 031403. Number: 3.

[41] Taylor, C. R., 1995. "Digging for golden carrots: An analysis of research tournaments". *The American Economic Review*, pp. 872–890. Publisher: JSTOR.

Proceedings of the ASME 2023
International Design Engineering Technical Conferences and
Computers and Information in Engineering Conference
IDETC-CIE2023
August 20-23, 2023, Boston, Massachusetts

DETC2023-115312

IMPROVING CHANGE MANAGEMENT BY QUANTIFYING THE RELATIONSHIPS BETWEEN MARGINS AND DESIGN VARIABLE FLEXIBILITY

Lindsey Jacobson
Graduate Research Assistant
North Carolina State University
Raleigh, NC 27695, USA
ltjacob3@ncsu.edu

Scott Ferguson
Associate Professor
North Carolina State University
Raleigh, NC 27695, USA
scott_ferguson@ncsu.edu

ABSTRACT

Managing system changes is one of the most challenging and important tasks facing today's engineering designers. Beyond the numerous design revisions that occur during the design process, accelerating rates of technological and societal growth strain existing systems, and those that will produce the greatest value are the systems that can be adapted and evolved. The complexity of modern engineered systems requires that significant resources be invested so that changes can be managed throughout the design and redesign processes. Implementing even relatively small modifications can require extensive design effort in systems with many coupled and entangled relationships. There is significant evidence that design margins can effectively limit change propagation, yet the relationship between margins and change propagation has not been fully studied. In this paper, we take a step toward quantifying how margins create flexibility at a component level in an engineered system. We define a decision variable's (DV) Complex Flexibility Range (CFR) as the extent by which a variable can change while maintaining feasibility with respect to all requirements and constraints. A high-powered rocket's recovery system is modeled, and the CFRs are calculated for multiple DVs. We use the quantified CFRs to demonstrate how margins create flexibility at the level of DVs and how changing a decision variable impacts margins and DV flexibility. Several applications for CFRs are identified, including 1) predicting change propagation, 2) identifying change absorbers, 3) communicating changes in margin/flexibility, and 4) negotiating DV selection and requirements definition. This work provides a foundation for further study of how margins can be allocated and for studying how systems can be efficiently, and effectively, adapted and evolved.

Keywords: Design margins, change propagation, system adaptability, engineering change management

1. INTRODUCTION

Our society is built around engineered systems that operate in an increasingly dynamic world. Because these systems serve foundational societal needs, we place great demands on them and find reduced value in legacy systems as requirements and objectives change. However, replacing legacy systems is often too costly and modifying existing systems can be challenging if they are not designed in ways that make them amenable to change. We believe that adaptability is a pillar of holistic perspectives of sustainability, given that systems unable to adapt will inevitably lose value and will eventually require expensive modification or replacement.

Yet, before engineered systems even reach the field, designers often struggle through the process of modifying a system throughout the design process. Modern systems are complex and entangled. Though proposed changes may seem simple at first consideration, implementing a change may require considering a significant number of interfaces, requirements, and constraints. Changes can also propagate through a network of system relationships. Specifically, it has been observed that change propagates through dependencies within and across four domains: requirements, functions, components, and design process [1]. Change propagation has also been modelled more simply, by relating direct parameters that can be adjusted directly by designers and target parameters that represent the design specifications that should be satisfied [2].

There has been increased interest in studying the role that margins play in helping engineering designers plan for change and mitigate change propagation. Eckert et al. note that

Copyright © 2023 by ASME

engineers make reference to the idea of margins by using phrases like "keeping something in reserve" or "having a buffer" [3]. Yet, El Fassi et al. [4] describe the lack of approaches and tools for tracking margins and the lack of documentation around the assumptions made by engineers when those margins are incorporated. Our goal in this paper is to take a step toward quantifying how margins create flexibility at a component level in an engineered system. This work provides a foundation to study strategic allocation of margin to enable system adaptability.

1.1 The Link Between System Change, Change Propagation, and Margins

There is great interest in predicting how easily systems may be changed and the extent an initial change may propagate. The Change Prediction Method (CPM) was developed as a tool to give engineers insight into the risk of change propagation occurring as a function of estimated likelihoods and impacts of changes [5]. Important scholarly work has explored the application and extension of CPM. As a single example, CPM has been applied to model how changes may impact product requirements [6].

The calculation of risk in CPM is often driven by subjective estimates of likelihood and impact. Researchers have proposed reducing subjectivity by using interface data to inform impact and likelihood metrics [7]. However, these values become a function of estimated probability distributions, which may still require some subjectivity to define. It has been suggested that change propagation methods that account for the true values of parameters and constraints would likely be more accurate than existing probabilistic models [8].

The subjectivity of impact and likelihood metrics is of greater concern in the design of complex systems where designers may lack sufficient system knowledge to develop these metrics. Researchers have studied the ability of engineers to develop mappings of system relationships and have noted that it is unlikely that a single person could map all connections in a complex system, even if they have many years of experience working with it [9]. While this principle makes subjective elicitation for CPM more difficult, it also demonstrates the need for design tools that identify impacts of changes. Such tools could help prevent unintentional change propagation.

There is evidence that margins limit change propagation by acting as change absorbers [8,10]. Margin has been defined as the "the extent to which a parameter value exceeds what it needs to meet its functional requirements regardless of the motivation for which the margin was included" [10]. Tolerance margin has specifically been described as the extent to which a component can change before causing further redesign [11]. Prior work by our research team explored one approach for linking margins and CPM [12]. Subsequent calculations of risk were performed by modeling the consumption of margins as an increase in the likelihood of change for a given element in the system architecture after a design modification occurred. We could then trace the rate at which the number of propagating changes multiplied as the design went under continued refinement.

In a recent literature review on change propagation, the authors suggest that further research is needed to determine the relationships between margin and change propagation [13]. Specifically, there is interest in studying how margins influence whether a change will be absorbed or propagated [13].

1.2 Modeling and Quantifying Margins and the Link to Flexibility

Researchers have been investigating how to model the relationship between margin and change propagation, as proposed in [13]. Lebjioui discusses using CPM as a method to elicit and analyze margins, given that if components can absorb change, they should have some margin [14]. Long presents a foundation for relating excess and change propagation to assess a system's lifecycle value [15]. Yet, few methods exist that quantify the relationship between margins and change propagation. The Margin Value Method was developed to manage margin with respect to changeability and proposes a quantification of how margins contribute to change absorption potential [11]. In this study, we are also interested in investigating how margins are used to absorb change at a component level. However, our work is distinguished from MVM in that we do not assume there is a structured design process and we do not assume that we can define target thresholds at the component level. Rather, we consider system requirements and constraints as points of reference for defining margins and derive the flexibility of design variables with respect to system-level margins. We believe that modelling margin with respect to system requirements may be beneficial when there is little to no structure in a design process and when it is difficult to pass down target requirements for components.

In addition, there are limitations in how margins are tracked throughout the design process. Margins may be added by multiple designers or stakeholders, and these margins often are not documented or communicated [14]. In collaborative design settings, individual designers may add their own margins without communicating them, which can cause margin stacking and overdesign [16]. Clearances are a type of margin that have been identified as important in managing change propagation, but they often exist at the intersection of subsystems and design teams, which makes negotiation over the use of clearances difficult [3]. If multiple teams decide to reduce a clearance simultaneously, their decisions may be incompatible, resulting in infeasibility and change propagation. Our modelling approach addresses these challenges by providing designers with perspective on how their local design changes, implemented at a component level, will impact performance, margins, and flexibility of other variables throughout the system. By showing how margins afford flexibility to design variables and how the modification of design variables impact system feasibility and flexibility, our model can be used as a communication and negotiation tool in redesign tasks involving large design teams.

Copyright © 2023 by ASME

2. QUANTIFYING THE RELATIONSHIP BETWEEN MARGINS AND DV FLEXIBILITY

The aim of our investigation is demonstrating the quantified relationship between parameter margins and decision variable flexibility. We will show how 1) the margin associated with a parameter constrains the amount that a decision variable can be modified, and 2) how changes in a decision variable—as the result of a design modification—impact the amount of margin associated with each parameter. As stated in Section 1, while the scholarly work around margins and change propagation have resulted in substantial advancements, the 1) quantification of margins, 2) tracing of margins, and 3) graphical representation of margins are active areas of research. The approach presented in this section addresses these issues so that we can gain insight about how the presentation of margins might help designers better understand how their local/individualized design decisions impact the design solution space.

The approach that we use is presented in Figure 1. This process consists of three major activities: 1) System identification and requirement mapping, 2) Mapping the system architecture, and 3) Quantifying parameter margins and each decision variable's Complete Flexibility Range.

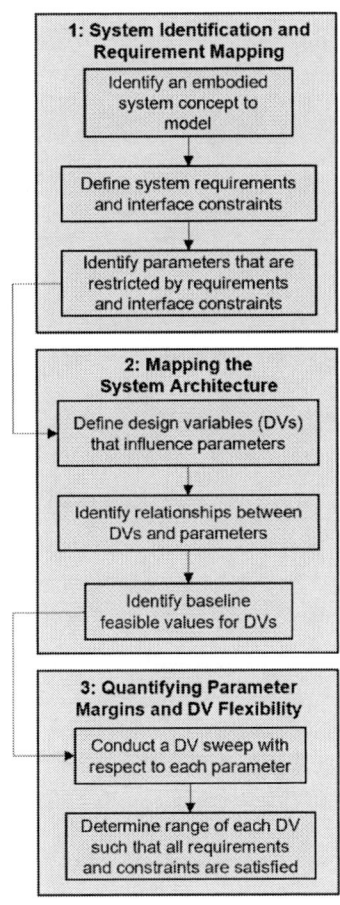

FIGURE 1: APPROACH TO RELATE MARGINS AND DV FLEXIBILITY.

Step 1: System identification and requirement mapping

We begin this process by identifying the embodied system concept that will be modeled. We start with an embodied system concept because a system architecture exists and (baseline) decision variable specifications have been established. If the system being modeled has not reached the embodiment phase of design, the system architecture may be understood but the sizing of variables and components may not be specified. Quantifying margins requires information about variable specifications.

Next, we define the requirements and interface constraints for the embodied system concept. Requirements and constraints govern solution feasibility and are used in quantifying margin. As described in Section 1, margin has been defined as the difference between a parameter and the requirement on that parameter. In this study, we identify parameters that are restricted by system requirements and interface constraints. System requirements and constraints serve as a consistent point of reference for defining margins in any design process. This representation also allows us to relate margins and system feasibility. Together, requirements and parameters define feasibility (does the current parameter satisfy the requirement?) and margins (given that the requirement is met, what is the quantified distance between the parameter and requirement?).

Step 2: Mapping the system architecture

As governed by the solution architecture, parameters are a function of one or more decision variables. Throughout the design process, engineers must manage relationships between variables they control and parameters that they care about (those that must meet requirements and satisfy constraints). There are many design tools for modeling and managing relationships between variables and parameters, including HD-DSMs [17], MDMs [18], interface control documents [19], and RDSP-DV trees [20]. Decision variables (DV), as defined last year in [20] when describing a rocket problem, are:

"design choices that can be made directly by the design team. The decision variables discussed in this mapping are the specifications describing the size and characteristics of each existing component. As an example, the decision variables for a motor could be: diameter, length, fuel type, assembly time, initial mass, burnout mass, mass burn rate, thrust curve, and location within the vehicle."

Baseline values for DVs are identified from the embodied system concept, and a sweep of each DV is conducted with respect to each parameter. In this process, each DV is modified independently to determine the effect on each parameter. Each DV is swept within a "reasonable" range, defined by the designer, around its baseline value. While a single DV is varied, other DVs maintain their baseline feasible value. Sweeping a DV parallels sweeping a parameter in a sensitivity study.

Copyright © 2023 by ASME

Step 3: Quantifying parameter margins and a decision variable's Complete Flexibility Range

The results of the DV sweeps are then used for quantifying the feasible range of each DV where all requirements and constraints are satisfied. We define this range as a decision variable's Complete Flexibility Range (CFR). To calculate the CFR for a DV x_j, inequalities are developed to relate the sweep of x_j with respect to each parameter, $P_i(x_j)$, and the required value of the parameter, denoted as R_i. Solving this inequality produces a set of x_j for which P_i meets its requirement, otherwise known as the flexibility of x_j with respect to parameter i. This set is referred to as L_{ij}. Then, the CFR is calculated as the union of x_j's flexibilities with respect to each parameter. The equations used to determine a CFR for the jth DV in a system with n parameters are shown below.

$$P_i(x_j) \le R_i \quad for \ i = 1{:}n \tag{1}$$

$$CFR_j = \cup_{i=1{:}n} L_{ij} \tag{2}$$

As a simple example of quantifying a DV's Complete Flexibility Range, consider a sweep of DV x_1 over a range of $[0,1]$ with respect to parameter P_1. Consider further that P_1 must be less than R_1. Changing x_1 changes P_1, and if x_1 is reduced the margin on P_1 is consumed. Too great of a reduction in x_1 may result in P_1 no longer meeting its requirement, and an infeasible design is the outcome of such a modification. Designers must then change other DVs to make the solution feasible again, with the possibility of even more changes propagating.

We define DV flexibility with respect to a parameter as the range over which a DV value can independently change while 1) maintaining the parameter's feasibility, and 2) preventing change propagation. These concepts are visualized in Fig. 2, with the red dashed line representing the requirement (R_1) on P_1, the blue curve representing the value of P_1 over the DV sweep of x_1, and the margin and flexibility range shown as labelled. R_1 is used to label the limit on the flexibility range of x_1 with respect to P_1.

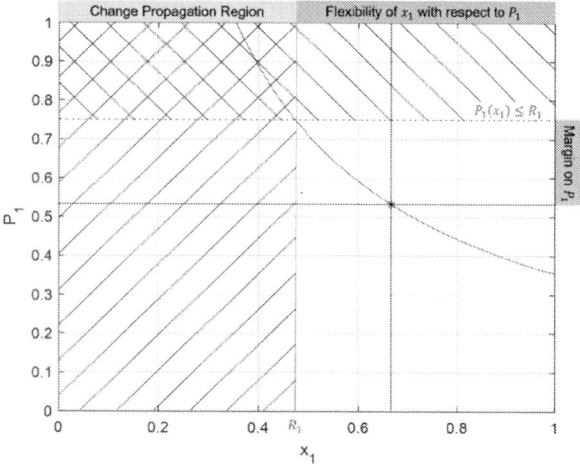

FIGURE 2: FLEXIBILITY OF x_1 WITH RESPECT TO P_1.

A decision variable's Complete Flexibility Range is the union of the DV's flexibility ranges with respect to all parameters. A designer who is only concerned with satisfying R_1, or is unaware of x_1's relationship to other requirements, may change the value of x_1 within its flexibility range with respect to R_1. However, x_1 is likely related to many other parameters and margins. Without awareness of these relationships, a designer may change x_1 in a way that drastically reduces margins, contributes to overdesign, and/or causes significant change propagation. Therefore, it is appropriate for designers to consider a DV's flexibility range with respect to the set of all system parameters as it gives complete insight into the system-level ramifications of local design decisions.

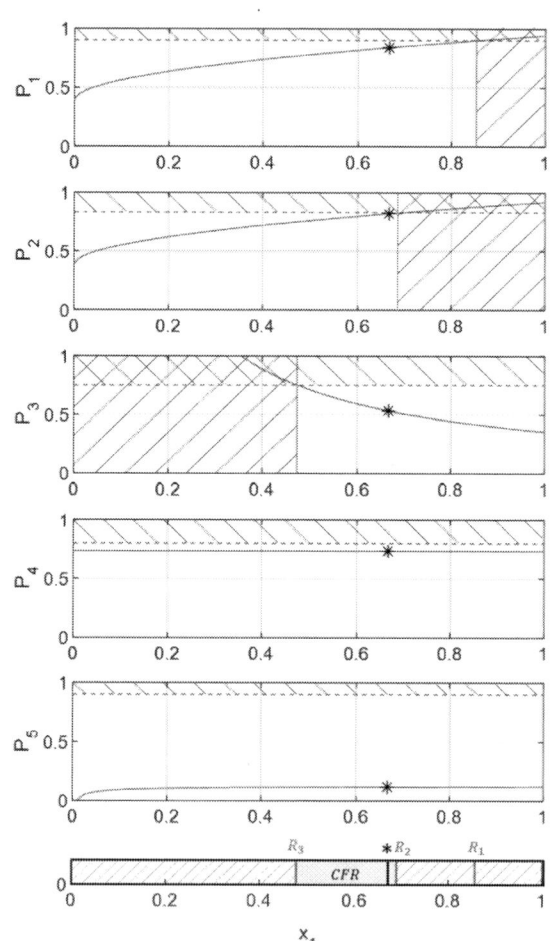

FIGURE 3: COMPLETE FLEXIBILITY RANGE (CFR) FOR x_1.

A DV's CFR is identified as the union of multiple flexibility ranges, as illustrated in Figure 3. The flexibility ranges of x_1 with respect to P_1, P_2, and P_3 are bounded given that there are values of x_1 in the sample range that cause P_1, P_2, and P_3 to become infeasible However, the flexibility ranges of x_1 with respect to P_4 and P_5 are unbounded given that there are no values of x_1 in the sample range that cause P_4 or P_5 to become infeasible. As depicted, the CFR of a DV is the range over which the DV can

Copyright © 2023 by ASME

independently change without making the solution infeasible with respect to any requirement or constraint.

The CFR of a DV is found within a one-dimensional slice of the feasible region in the direction of that DV such that the slice intersects with the set of DVs associated with the embodied system concept. A visualization of this slice, shown at the bottom of Figure 3, provides perspective on how margins create opportunities for absorbing change at the DV level.

This figure also illustrates how changing values of DVs can reduce or expand various system margins. We believe that computing and reporting Complete Flexibility Ranges can support margin tracking through the design process and help communicate how changing a DV will impact performance, margins, and the flexibility of other DVs.

3. APPLICATION OF APPROACH USING A HIGH-POWERED ROCKET EXAMPLE
3.1 Calculating CFRs in the Recovery System

In this section we quantify the CFRs for a set of decision variables using the exemplar of a recovery system for a high-powered rocket. We then discuss and demonstrate how the CFRs provide insights about how changing a DV affects system margins and the ability to modify other DVs.

Step 1: System identification and requirement mapping

Table 1: REQUIREMENTS FOR ROCKET RECOVERY SYSTEM.

	Requirement
R_1	The descent time (P_1) from apogee to landing must be less than 90 seconds.
R_2	The vehicle's drift distance from the launch pad (P_2) must be less than 2500 ft.
R_3	The landing kinetic energy (P_3) of each vehicle section must be less than 75 ft-lb.
R_4	Vehicle descent velocity under drogue parachute (P_4) must be less than 120 ft/s.
R_5	The stress of the main parachute opening shock on the nosecone bulkhead (P_5) must be less than 4500 psi.
R_6	The stress of the main parachute opening shock on the AV bay bulkhead (P_6) must be less than 4500 psi.
R_7	The drogue parachute deployment delay (P_7) must be no more than 2 seconds.
R_8	The main parachute deployment altitude (P_8) must be at least 500 ft above ground.

Our paper at the 2022 IDETC explored the system architecture and yearly evolution of a high-powered rocket designed by students participating in the NASA Student Launch (SL) competition [21]. We mapped the rocket's architecture and discussed the coupling of DVs associated with the recovery system. We revisit this example in this paper because of the

challenge that coupling creates when managing margins and managing change.

Our embodied recovery system concept is based on a standard dual-deployment recovery architecture. That is, a drogue parachute (DP) deploys at a specified time after apogee to initially slow the vehicle down. After the vehicle has descended to a specified altitude, the main parachute (MP) deploys and slows the vehicle until it reaches its landing speed.

The recovery system must meet multiple requirements, with some of the most important being the vehicle's kinetic energy upon landing, descent time, and drift distance. These requirements are established by the competition sponsor. Some teams derive additional requirements to account for system interfaces and lessons learned from previous launches. In this study, we consider requirements provided by the competition sponsor [22] and requirements commonly imposed by a student design team [23,24]. These requirements are shown in Table 1. Each requirement specifies a value that a parameter must reach, as described in the approach. Parameters are denoted as P.

Step 2: Mapping the system architecture

Multiple DVs influence the recovery system parameters identified in Table 1. In [20], we developed RDSP-DV (Requirement-Driven System Parameter-DV) trees for each parameter so that we could define the set of DVs that are relevant in this investigation. An RDSP-DV tree for drift distance (P_2) is shown in Figure 4. We select this RDSP-DV tree as the relationships between drogue descent velocity (P_4), descent time (P_1), and DVs are also defined within it.

Engineering designers define recovery system parameters by selecting the radius and coefficient of drag of each parachute, the drogue deployment delay time, the main parachute deployment altitude, the mass and location of each element in the system, vehicle separation points, and vehicle apogee. In a deeper study, the mass and location of each component and the vehicle section separation points could be included as DVs, but as a simplification, we model burnout section masses as DVs. Further, DVs that influence vehicle apogee could be included, but we treat vehicle apogee as a constant in this study to maintain a reasonable scope. All DVs of interest in this study are listed in Table 2.

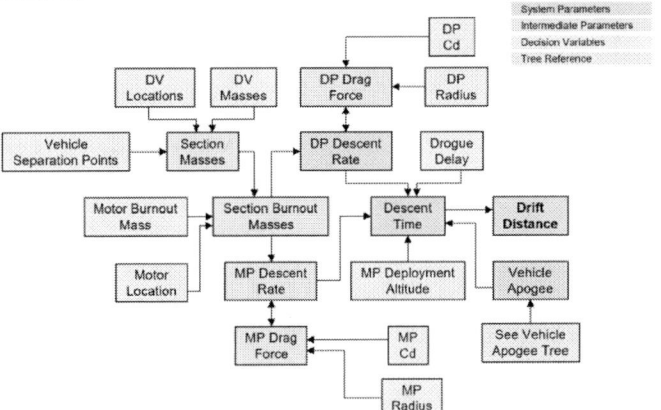

FIGURE 4: DRIFT DISTANCE RDSP-DV TREE.

Copyright © 2023 by ASME

Table 2: DVS IN HIGH-POWERED ROCKET RECOVERY SYSTEM.

	Design Variable
x_1	Drogue parachute coefficient of drag (C_d)
x_2	Drogue parachute radius
x_3	Drogue parachute deployment delay
x_4	Main parachute coefficient of drag (C_d)
x_5	Main parachute radius
x_6	Main parachute deployment altitude
x_7	Burnout mass of vehicle section 1
x_8	Burnout mass of vehicle section 2
x_9	Burnout mass of vehicle section 3

After identifying general relationships in the RDSP-DV trees, mathematical relationships between parameters and DVs were derived or approximated using principles of physics. A matrix illustrating the relationships and couplings between parameters and DVs is shown in Figure 5. The vehicle section masses (x_7, x_8, and x_9) are the most coupled, with each influencing six of the parameters. The drogue parachute characteristics (x_1 and x_2) are also highly coupled, with each influencing five parameters.

	x_1	x_2	x_3	x_4	x_5	x_6	x_7	x_8	x_9
P_1									
P_2									
P_3									
P_4									
P_5									
P_6									
P_7									
P_8									

FIGURE 5: DRIFT DISTANCE RDSP-DV TREE.

Step 3: Quantifying parameter margins and a decision variable's Complete Flexibility Range

Baseline DV values for the vehicle's recovery system were determined by referencing previous high-powered rocketry design reports. A DV sweep was then conducted with respect to each parameter. Each DV was varied independently over a sample range containing the baseline feasible value and 100 uniformly spaced points were sampled within each range. Sample ranges were estimated based on what DV values are available commercially and what design teams have reported in the past [23,24]. Values for the baseline and the samples used in our study are reported in Table 3. DV sweeps associated with the baseline configuration are shown in Figure 8 in the appendix. The DV sweeps with respect to P_7 and P_8 are not shown in Figure 8 because they only relate to one DV.

Finally, parameter margins were determined, and CFRs were developed for each DV in the baseline system configuration using the DV sweep results. These CFRs are shown in Figure 6.

Table 3: BASELINE SYSTEM DV VALUES AND SAMPLE RANGES.

DV	Baseline System Values	Sample Range
x_1	1.3	$0 - 2.5$
x_2	0.75 feet	$0 - 2$ feet
x_3	0 seconds	$0 - 4$ seconds
x_4	2	$0 - 3$
x_5	5 feet	$0 - 8$ feet
x_6	600 feet	$0 - 1000$ feet
x_7	0.207 slugs	$0 - 0.8$ slugs
x_8	0.220 slugs	$0 - 0.8$ slugs
x_9	0.601 slugs	$0 - 0.8$ slugs

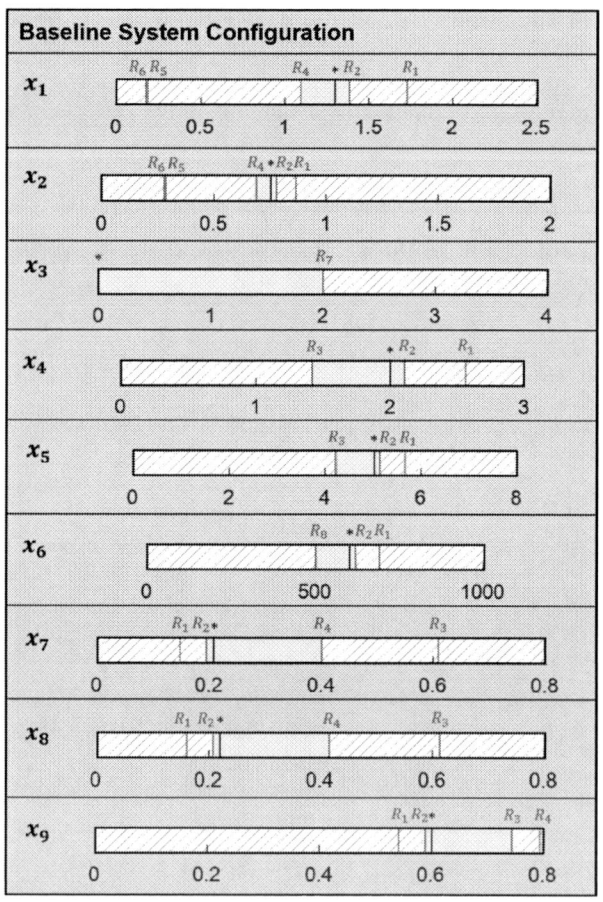

FIGURE 6: BASELINE COMPLETE FLEXIBILITY RANGES.

3.2 Calculating CFRs in the Recovery System of a High-Powered Rocket

We demonstrate the impact of a change on the system by modifying a single DV, x_5. The main parachute radius impacts the rocket's descent time, drift distance, landing kinetic energy,

Copyright © 2023 by ASME

and stresses due to opening shock. A designer may change the parachute radius because some payloads may have a higher probability of success if the drift distance is lower (drones and transmitters), while other payloads may have a higher probability of success if the landing kinetic energy is lower (landers). When design teams modify their payloads or launch entirely new payloads, they may develop new recovery objectives, prompting them to modify DVs such as x_5. The modification to the system that we consider is summarized in Table 4. DV sweeps for each DV were conducted after this modification and are shown in Figure 9 in the appendix.

Table 4: MODIFICATIONS TO BASELINE SYSTEM CONFIGURATION.

DV	Baseline System Values	Modified System Values	Sample Range
x_5	5 feet	4.5 feet	$0 - 8$ feet

The set of CFRs for each system configuration are shown side by side in Figure 7, with the baseline system CFRs on the left and the modified system CFRs on the right. The upper and lower bounds of each CFR associated with each configuration are shown in Table 5. The percentage of the sample range that each CFR covers is also included in Table 5. In addition, we report the percentage of each DV sample range that is feasible with respect to each requirement in two tables in the appendix. These percentages are termed feasible range percentages (FRPs). The FRPs associated with the baseline system configuration are in Table 7 in the appendix, and those associated with the modified configuration are in Table 8 in the appendix.

The parameter margins associated with each of the system configurations are shown in Table 6. The margin as a percentage of the parameter requirement is also provided in Table 6 and is referred to as normalized margin. The calculation of normalized margin on the ith parameter is shown below where M_{N_i} represents normalized margin, M_i represents parameter margin, and R_i represents the requirement of that parameter.

$$M_{N_i} = \frac{M_i}{R_i} \qquad (3)$$

In the baseline system and modified system CFRs, parameters with greater normalized margin generally produced higher FRPs with respect to the sample range of related DVs. For example, Table 6 shows that there is high normalized margin on the stress in each bulkhead and the drogue deployment delay (P_5, P_6, and P_7) in each configuration. The margin on these parameters corresponds to FRPs ranging from 85% to 100%, as shown in Table 7 and Table 8 in the appendix, with most FRPs being 100%. In other words, the requirements on bulkhead stress and drogue deployment delay restrict flexibility on only up to 15% of the sample range of any DV. These margins contribute to significant flexibility, allowing designers to modify the main parachute characteristics, deployment settings, and section burnout masses without any concern that the vehicle's bulkheads would need to be redesigned. Designers would only need to

consider redesign if the drogue parachute characteristics were reduced significantly.

In contrast, Table 6 shows that there is low normalized margin on the vehicle's descent time, drift distance, and descent velocity under drogue parachute (P_1, P_2, and P_4) in each configuration. These margins correspond to FRPs ranging from 26% to 100%, as shown in Table 7 and Table 8 in the appendix. This means that the requirements on descent time, drift distance, and drogue descent speed may restrict flexibility on up to 74% of a DV's sample range. In Figure 7, R_1, R_2, and R_4 are often shown to be limiting DV flexibility. The drift distance requirement is most restricting. Designers cannot reduce section masses or increase any parachute characteristics much without making the drift distance too high. Though it is not shown here, this may have further implications. If the section masses have a minimum limit, this may inhibit the ability to reach a higher target apogee.

As discussed previously, the CFRs show how much a DV could change independently while maintaining solution feasibility. In previous literature, it has been said that margins support system flexibility by limiting change propagation [8,10]. The CFRs generated in this study demonstrate this principle. The extensive margins on P_5, P_6, and P_7 allow designers to launch a range of payloads and pursue unique recovery objectives without concern that change will propagate through these parameters. The narrow margins on P_1, P_2, and P_4 prevent designers from efficiently accommodating these changes.

The derivation of CFRs in Section 2 implies that CFRs are heavily dependent on the current configuration of the system, including the system architecture, requirements, and the baseline values of DVs. The difference in CFRs between the baseline system and the modified system demonstrates that CFRs are highly sensitive to even a small change in a single DV. The modified system features a 10% change to the main parachute radius. This modification caused the vehicle's drift distance, descent time, landing kinetic energy, and bulkhead stresses to change. Accordingly, the margins on each of these parameters changed, resulting in consumption and expansion of the flexibility of all DVs aside from the main parachute radius. The margins on drift distance and landing kinetic energy changed most drastically, with the margin on drift distance quintupling and the margin on landing kinetic energy being reduced to under half of its original value.

Before the modification, P_3 corresponded to FRPs ranging from 47% to 93%, as shown in Table 7. After the modification, P_3 corresponded to FRPs ranging from 42% to 82%, as shown in Table 8. The reduction in FRPs is caused by the reduction in margin on P_3. The flexibility ranges of section masses are most affected by this reduction. With less margin on the landing kinetic energy, designers are more limited in how much they can increase vehicle section masses before change propagates, especially in section three of the vehicle, which often houses the payload. This may be a major concern if there is uncertainty about the payload mass or if the team plans to use the vehicle to launch a wide variety of payloads in the future.

Copyright © 2023 by ASME

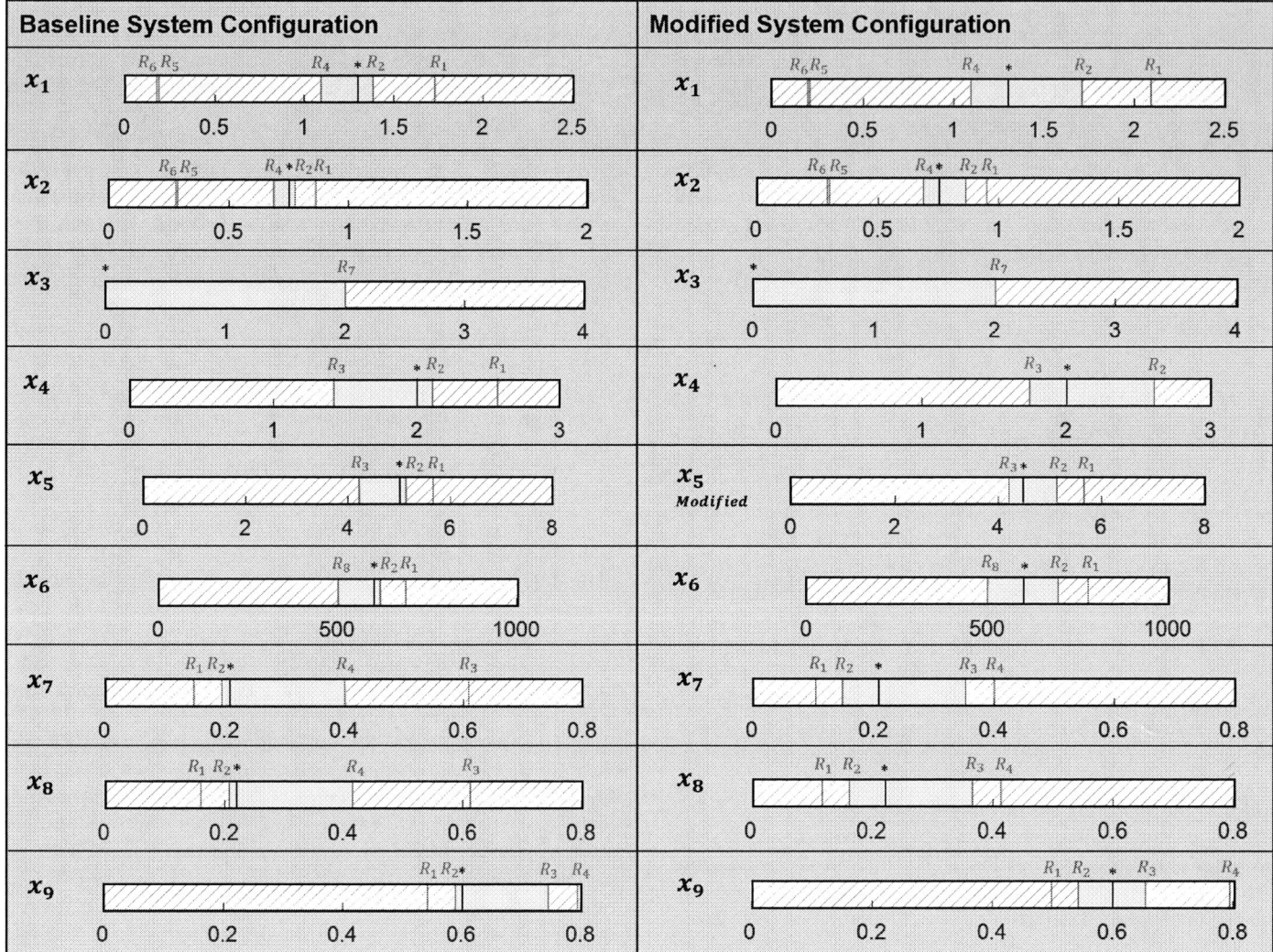

FIGURE 7: VISUALIZATION OF BASELINE SYSTEM AND MODIFIED SYSTEM COMPLETE FLEXIBILITY RANGES (CFRS).

Table 5: COMPARISON OF BASELINE AND MODIFIED SYSTEM COMPLETE FLEXIBILITY RANGES (CFRS).

DV	Baseline System Configuration			Modified System Configuration		
	CFR Lower Bound	CFR Upper Bound	CFR Percentage of Sample Range	CFR Lower Bound	CFR Upper Bound	CFR Percentage of Sample Range
x_1	1.09	1.38	12%	1.09	1.71	25%
x_2	0.69	0.77	4%	0.69	0.86	9%
x_3	0	2	50%	0	2	50%
x_4	1.42	2.11	23%	1.75	2.60	28%
x_5	4.21	5.13	12%	4.21	5.13	12%
x_6	500	618	12%	500	697	20%
x_7	0.15	0.40	31%	0.15	0.35	25%
x_8	0.16	0.41	31%	0.15	0.36	26%
x_9	0.54	0.74	25%	0.54	0.65	14%

Copyright © 2023 by ASME

Table 6: COMPARISON OF BASELINE AND MODIFIED SYSTEM PARAMETER MARGINS.

Parameter	Requirement	Baseline System Configuration		Modified System Configuration	
		Margin	Normalized Margin	Margin	Normalized Margin
P_1	< 90 seconds	5.95 seconds	7%	10.46 seconds	12%
P_2	< 2500 feet	34.53 feet	1%	166.7 feet	7%
P_3	< 75 ft-lb	21.74 ft-lb	29%	9.25 ft-lb	12%
P_4	< 120 ft/s	9.92 ft/s	8%	9.92 ft/s	8%
P_5	< 4500 psi	3903 psi	87%	3847 psi	85%
P_6	< 4500 psi	3939 psi	88%	3886 psi	86%
P_7	< 2 seconds	2 seconds	100%	2 seconds	100%
P_8	< 500 feet	100 feet	20%	100 feet	20%

While the flexibility ranges on the masses decrease, other flexibilities increase. Before the modification, P_2 corresponded to FRPs ranging from 26% to 100%, as shown in Table 7. After the modification, P_2 corresponded to FRPs ranging from 32% to 100%, as shown in Table 8. The increase in FRPs is caused by the increase in margin on P_2. The flexibilities of the drogue parachute characteristics are significantly impacted by this increase in margin. Designers can increase the drogue parachute characteristics more without concern that change will propagate. This may be beneficial if designers seek to descend more slowly under drogue parachute, perhaps to take clear pictures of the launch field below during descent. The CFRs calculated for this system illustrate how a change to a parameter margin relates to changes in the flexibility of the DVs related to that parameter.

Ultimately, changing a DV can cause changes to multiple parameters, parameter margins, and DV flexibilities. Though it is not shown, CFRs could also demonstrate how changing a requirement can change multiple parameter margins and DV flexibilities. If these changes are substantial enough, the system may become infeasible, causing change propagation. If the system does not become infeasible, changes to DVs and to requirements can consume or expand parameter margins. Communicating these changes in margins is important given that they impact the flexibility of other DVs. If these changes are not properly communicated, designers may consume margin in conflicting ways, resulting in system infeasibility, or designers may simultaneously expand margins, resulting in overdesign. Additionally, forecasting changes in margin and changes in DV flexibility could inform decision making with respect to change, thereby supporting change management. In the next section, we discuss how CFRs could be used as a communication and forecasting tool in the design process.

4. DISCUSSION

CFRs quantify the range over which a DV can change while maintaining solution feasibility with respect to all requirements. These ranges provide perspective about how requirements relate to DVs and how requirements are actively constraining the solution space. This information can be used by designers to manage change in four major applications: 1) determining whether a DV change will propagate, 2) identifying DVs as immediate change absorbers, 3) communicating reduction and expansion of margins, and 4) negotiating DV selection and the specification placed on the requirement. These applications are discussed in detail in this section.

4.1 Predicting Change Propagation

Concept: By definition, CFRs quantify the range over which DVs can be modified before change propagates. If designers know the CFRs associated with their system configuration, they can quickly assess whether proposed modifications will result in change propagation and what parameters would be impacted. Developing a common understanding of change pathways with respect to DV modifications is especially important when designers are working independently. If designers lack system perspective, that is if they lack understanding of the many ways in which a DV contributes to parameters and system feasibility, they may inadvertently propagate change through their local design decisions.

In context of the high-powered rocket example: Consider that the payload team requests a decrease in the vehicle's drift distance to yield a greater probability that their transmission device will be successful. The recovery team may immediately consider altering x_2, the drogue parachute radius, given that it has a strong impact on drift distance. Reasonably, they may propose changing x_2 from 0.75 feet to 0.5 feet. However, modifying x_2 in such a way would propagate change through R_4, as shown in x_2's CFR in Figure 6. It is possible that R_4 was not an immediate concern of theirs, that they had forgotten about R_4, or that they were unaware of R_4 because the requirement was imposed by another design group. However, in analyzing the CFR for x_2, the recovery team would quickly realize that this decision would propagate change. If designers analyze CFRs prior to making a change, they should have a more complete understanding of how their local changes impact the system, and they should never propagate change unintentionally.

4.2 Identifying Change Absorbers

Concept: The CFRs also reveal what DVs might be capable of absorbing changes. The CFRs show how each DV can be modified without causing further change propagation. If there is a change in a requirement or a change in objectives, referring to

Copyright © 2023 by ASME

the CFRs could reveal favorable change pathways. Referring to the DV sweeps with respect to each parameter, shown in Figure 8 and Figure 9 of the appendix, could reveal precisely how single variables can be tuned to achieve some desired effect on the system parameter.

In context of the high-powered rocket example: Consider again the payload team's request to reduce drift distance. Designers may refer to the DV sweeps associated with drift distance, P_2, for the baseline system. These sweeps are displayed in the second row of Figure 8 in the appendix. Designers would quickly see that they could change any of the nine DVs to reduce drift distance. Using the parameter curves with respect to each DV, they could determine how much each DV would need to be modified to achieve the desired reduction in drift distance. Then, these DV modifications could be assessed with the CFRs to determine whether any DVs could be modified to achieve the objective while maintaining feasibility.

We observe that the set of flexibility ranges with respect to a parameter and the set of CFRs on DVs could be evaluated to determine whether changes could be absorbed in a single modification. In future work, we plan to investigate how flexibility ranges and CFRs could support identification of change pathways in scenarios where no single modification of a DV would produce a feasible solution. We also plan to investigate how designers could use flexibility ranges to guide the selection of an adaptation strategy, considering tradeoffs between change propagation, performance, and preservation of flexibility.

4.3 Communicating Changes in Margin/Flexibility

Concept: Even when DVs are modified in such a way that they absorb change immediately, these modifications change parameters, margins, and the flexibility of other DVs. If designers are unaware how a DV modification has impacted system margins, they may make consecutive changes that consume or expand the same margins. This could result in change propagation or overdesign. Such outcomes may occur with a higher risk when designers work independently. Designers may independently make modifications that are feasible and further their own objectives, but they may not know when to communicate those changes to others. However, these modifications should be communicated because they affect parameter margins and ultimately impact how other designers can modify their DVs before change propagates. Likewise, if designers make local decisions that expand margins, this should also be communicated to prevent overdesign and to reveal increases in flexibility that can be used by other designers

In context of the high-powered rocket example: Consider the modification of x_5 discussed in Section 3.2. x_5 is modified within its CFR, so the system remains feasible after the modification and change does not propagate through DVs. However, in the new system state, the CFRs have changed as shown in Figure 7. Most notably, there was a significant reduction in the CFR of x_9, the mass of section 3 of the vehicle. Prior to the change, the CFR for x_9 extended over 25% of the sample range, but after the change, the CFR for x_9 extended over only 14% of the sample range. Given that there may be greater uncertainty about the mass of the section of the vehicle, or that there may be great motivation to alter the mass of a section to accommodate a new payload, this reduction in the CFR of x_9 may be significant to design objectives and to avoiding change propagation in the future. Perhaps the recovery team finds the reduction in x_9's flexibility to be acceptable, but the payload team finds it unacceptable. In discussing the reduction of x_9's flexibility, designers may reconsider the change to x_5, seeking a different change pathway to achieve their desired goal. We observe that quantifying CFRs for a potential system modification could provide meaningful insights about the ability to make future changes.

4.4 Negotiating DV Selection and Requirement Definition

Concept: Finally, the CFRs provide perspective on how requirements are actively constraining the selection of DVs. The representation of a CFR shows how much a DV selection can be modified before certain requirements become unsatisfied.

In context of the high-powered rocket example: As discussed in section 3.2, the requirements on parameters P_1, P_2, and P_4 significantly constrain DV flexibility in comparison to the requirements on P_5, P_6, and P_7, which collectively do little to constrain DV flexibility. While we discussed this with respect to FRPs, this can also be seen visually in the CFRs by comparing where requirements impose bounds on the DVs. In observing how requirements constrain DVs, designers may reconsider some of their DV selections to take advantage of existing margins or to reduce unnecessary margins. Designers may also consider relaxing requirements definitions to provide greater DV flexibility.

Consider the requirements R_5 and R_6. These are interface requirements defined by the structures team based on the design of their bulkheads. The properties of the bulkheads could be modeled as DVs, but they were omitted from consideration because they belong to a different subsystem. However, in noticing the DV flexibility afforded by the bulkheads, the recovery team may discover through investigation that they never expect to approach the requirement, no matter how they change DVs. At least in the current configuration, the CFRs suggest that so long as there is a reasonably sized drogue parachute deployed, R_5 and R_6 should remain satisfied. These margins provide the entire design team with reassurance that changes to the recovery system should not propagate back to the vehicle's structure, suggesting that the two systems should be able to be modified independently. However, these margins may also present an opportunity. Since the recovery team does not anticipate hitting R_5 and R_6, it is possible that the bulkheads may not need to be as strong and could be redesigned to have less mass. This change could provide performance benefits with respect to many vehicle parameters. This provides an opportunity for the design team to discuss whether the existing margin has been incorporated because of uncertainties, or whether the margins are the property of an unrefined design. CFRs can reveal areas in which DVs may be overdesigned and

Copyright © 2023 by ASME

contribute towards unnecessary margin. Designers could also use CFRs to identify areas in which requirements could be relaxed to provide valuable DV flexibility.

4.5 Limitations

The limitation of this approach is that it assumes that DVs are changed independently and does not provide direct insight about how manipulation of multiple variables at once may impact the system. In cases where DVs are coupled in commercial off-the-shelf (COTS) components or where DVs are coupled in a component due to physical principles (ex: mass and volume), the baseline system CFRs cannot be used to predict change propagation or identify the DVs that serve as change absorbers. Regardless of coupling, the CFRs could still be used to review requirements, and the new system's CFRs could still be modeled to communicate the impact of changes. Future research is needed that explores how coupled DVs can be represented in the form of a CFR.

We also acknowledge the possible computational cost associated with generating CFRs. First, quantifying CFRs requires mathematical models of system relationships. Mathematical models may not always be known, and while approximations can be used, it is uncertain how approximations may limit the accuracy of the tool. Future research that investigates the quantification of flexibility ranges with other means of modeling and the implications of using model approximations. Brahma and Wynn propose a method for integrating FEA data into MVM, and it is possible that similar finite element methods could be integrated in modeling CFRs [25]. Furthermore, it may be computationally expensive to generate CFRs for a system every time its state changes, especially when the system has many DVs and when those DVs are modified frequently.

5. CONCLUSION

In this study, we propose an approach for modeling, communicating, managing, and operationalizing DV flexibility in an engineered system. We represent DV flexibilities as CFRs, which are defined as the range over which a DV can be independently modified while maintaining solution feasibility with respect to all requirements. In other words, CFRs represent how much a DV can be modified before change propagates. The derivation of CFRs implies that CFRs are dependent on the system architecture, defined requirements, and the current values of DVs. We provide evidence that DV flexibility is related to parameter margins and propose a heuristic that increasing a parameter margin generally increases the flexibility of DVs with respect to that parameter.

CFRs are first quantified for a baseline configuration of a high-powered rocketry recovery system. The impact of a design modification is demonstrated to exemplify how CFRs allow for the management of design changes, margins, and DV flexibility. We discuss how designers could refer to CFRs to prevent unintended and otherwise unforeseen change propagation. The CFRs can also provide designers with perspective on a DV's ability to absorb change in the current configuration. In these applications, the CFRs are limited in that they can only forecast change propagation and change absorption for uncoupled DVs. Future work should investigate how CFRs could be adapted to predict change propagation with respect to coupled DVs.

We also discuss how CFRs can be used to predict and communicate the impacts of local design changes on flexibility across a system. Designers may consider how their proposed modification would impact other flexibilities before committing to the change. In addition, communicating changes in flexibility after a modification could prevent independent designers from expanding margin unnecessarily or from consuming margin in conflicting ways, which may cause inadvertent and unforeseen change propagation. Finally, there is evidence that CFRs could be used to negotiate DV selections and requirement definitions. CFRs provide perspective on how requirements are actively constraining the flexibility of DVs. In reviewing CFRs, designers may discover that some requirements are overly constraining DV flexibility and that there is significant value in relaxing them. Conversely, in reviewing CFRs, designers may find that there is a superfluous amount of margin on some parameters, suggesting that some DVs specifications may be contributing to overdesign.

Quantifying CFRs requires mathematical models of system relationships, though these models may be approximations. Additional research is needed that investigates the implications of using model approximations when quantifying CFRs. Further, advances in digital engineering tools can be explored for their impact in reducing the modeling and computational burden of quantifying CFRs.

The work presented in this paper provides a foundation for future study of system adaptability. By modeling flexibility at a DV level, CFRs could provide insights about how margin could be allocated to enable system adaptability and how designers could strategically manage change and redesign. Future work could also investigate the extent to which coupling of DVs in COTS components drives change propagation. CFRs may reveal significant flexibility in a single DV, but designers may be unable to modify that DV independently unless they design a component in house. Finally, it is possible that CFRs could have applications in set-based design, as the collection of CFRs for a current system configuration represents a feasible set of design solutions that only require a single change.

ACKNOWLEDGEMENTS

We gratefully acknowledge funding from the NC State University Provost's Doctoral Fellowship.

REFERENCES

[1] Ahmad, N., Wynn, D. C., and Clarkson, P. J., 2013, "Change Impact on a Product and Its Redesign Process: A Tool for Knowledge Capture and Reuse," Res Eng Des, **24**(3), pp. 219–244.

[2] Yang, F., and Duan, G. J., 2012, "Developing a Parameter Linkage-Based Method for Searching Change Propagation Paths," Res Eng Des, **23**(4), pp. 353–372.

Copyright © 2023 by ASME

[3] Eckert, C., Isaksson, O., Lebjioui, S., Earl, C. F., and Edlund, S., 2020, "Design Margins in Industrial Practice," Design Science, pp. 1–37.

[4] El Fassi, S., Guenov, M. D., and Riaz, A., 2020, "An Assumption Network-Based Approach to Support Margin Allocation and Management," *Proceedings of the Design Society: DESIGN Conference*, Cambridge University Press, pp. 2275–2284.

[5] Clarkson, P. J., Simons, C., and Eckert, C., 2004, "Predicting Change Propagation in Complex Design," Journal of Mechanical Design, Transactions of the ASME, 126(5), pp. 788–797.

[6] Koh, E. C. Y., Caldwell, N. H. M., and Clarkson, P. J., 2012, "A Method to Assess the Effects of Engineering Change Propagation," Res Eng Des, 23(4), pp. 329–351.

[7] Hamraz, B., Hisarciklilar, O., Rahmani, K., Wynn, D. C., Thomson, V., and Clarkson, P. J., 2013, "Change Prediction Using Interface Data," Concurr Eng Res Appl, 21(2), pp. 141–154.

[8] Brahma, A., and Wynn, D. C., 2021, "A Study on the Mechanisms of Change Propagation in Mechanical Design," Journal of Mechanical Design, Transactions of the ASME, 143(12).

[9] Jarratt, T., Eckert, C., and Clarkson, P. J., 2004, "Development of a Product Model to Support Engineering Change Management," *Proceedings of the TMCE*.

[10] Eckert, C., Isaksson, O., and Earl, C., 2019, "Design Margins: A Hidden Issue in Industry," Design Science, 5, pp. 1–25.

[11] Li, R., Yi, H., and Cao, H., 2022, "Towards Understanding Dynamic Design Change Propagation in Complex Product Development via Complex Network Approach," Int J Prod Res, 60(9), pp. 2733–2752.

[12] Long, D., and Ferguson, S., 2020, "Studying Dynamic Change Probabilities and Their Role in Change Propagation," Journal of Mechanical Design, Transactions of the ASME, 142(10), pp. 1–14.

[13] Brahma, A., and Wynn, D. C., 2022, "Concepts of Change Propagation Analysis in Engineering Design," Res Eng Des.

[14] Lebjioui, S., "Investigating and Managing Design Margins throughout the Product Development Process."

[15] Long, D., and Ferguson, S., 2019, *An Excess Based Approach to Change Propagation*.

[16] Eckert, C., and Isaksson, O., 2017, "Safety Margins and Design Margins: A Differentiation between Interconnected Concepts," *Procedia CIRP*, Elsevier B.V., pp. 267–272.

[17] Tilstra, A. H., Seepersad, C. C., and Wood, K. L., 2012, "A High-Definition Design Structure Matrix (HDDSM) for the Quantitative Assessment of Product Architecture," Journal of Engineering Design, 23(10–11), pp. 767–789.

[18] Browning, T. R., 2016, "Design Structure Matrix Extensions and Innovations: A Survey and New Opportunities," IEEE Trans Eng Manag, 63(1), pp. 27–52.

[19] Weiss, S. I., 2013, "Interface Definition and Management," *Product and Systems Development*, John Wiley & Sons, Inc., pp. 47–54.

[20] Jacobson, L., and Ferguson, S., 2022, "Requirements Mapping of a High-Powered Rocket System to Explain Solution Similarities Across Generations," Proceedings of the ASME 2022 International Design Engineering Technical Conference & Computers and Information in Engineering Conference.

[21] Adams, J., 2023, "NASA Student Launch" [Online]. Available: https://www.nasa.gov/stem/studentlaunch/home/index.html. [Accessed: 11-Mar-2023].

[22] National Aeronautics and Space Administration, 2023, "2023 NASA Student Launch Handbook and Request for Proposal," pp. 1–135 [Online]. Available: www.nasa.gov.

[23] High-Powered Rocketry at NC State University, 2022, *Tacho Lycos 2022 NASA Student Launch Critical Design Review*.

[24] High-Powered Rocketry at NC State University, 2023, *Tacho Lycos 2023 NASA Student Launch Critical Design Review*.

[25] Brahma, A., and Wynn, D. C., 2020, "Calculating Target Thresholds for the Margin Value Method Using Computational Tools," *Proceedings of the Design Society: DESIGN Conference*, Cambridge University Press, pp. 111–120.

6. APPENDIX

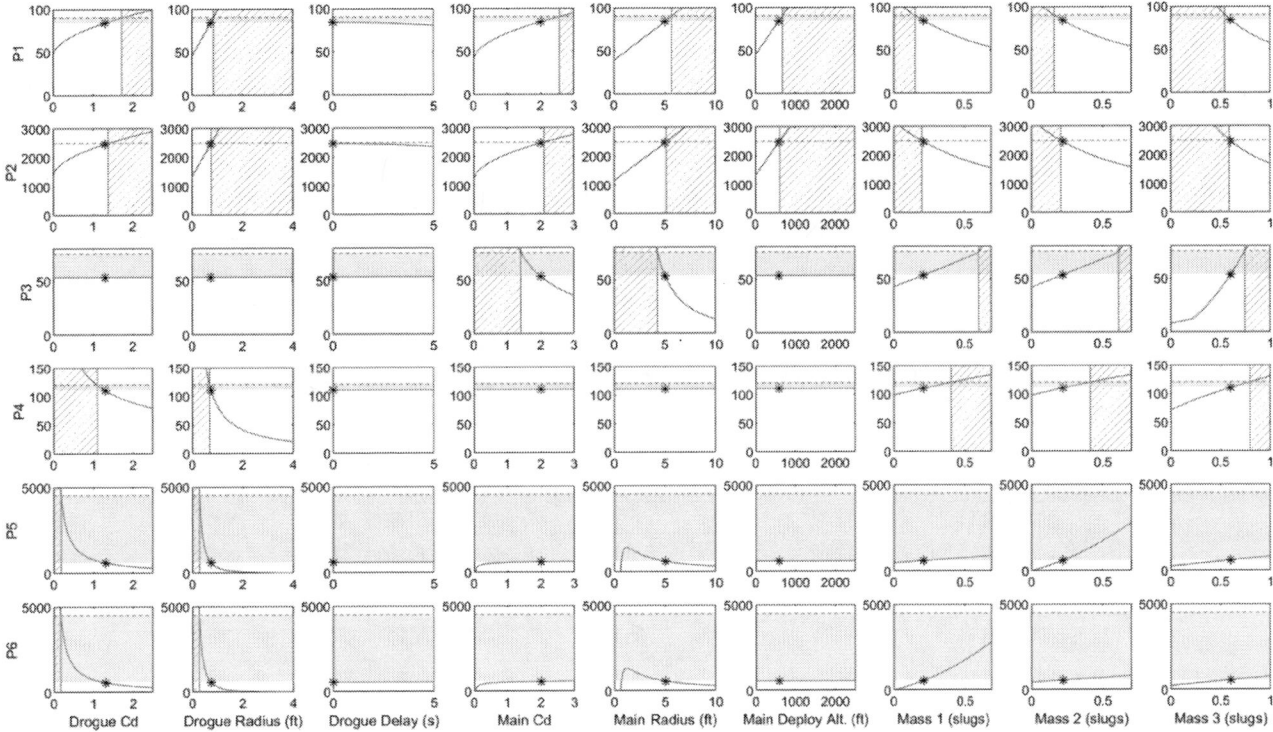

FIGURE 8: BASELINE SYSTEM DV SWEEPS.

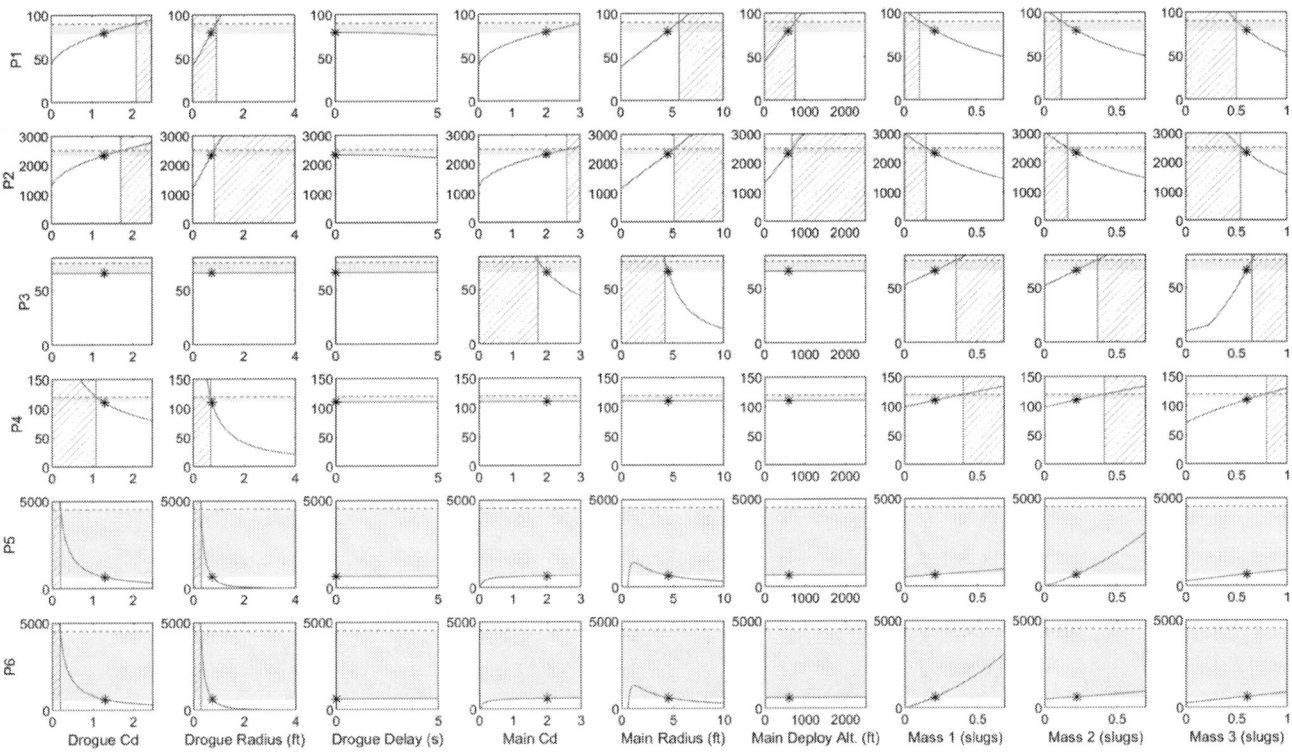

FIGURE 9: MODIFIED SYSTEM DV SWEEPS.

Copyright © 2023 by ASME

Table 7: BASELINE SYSTEM FEASIBLE RANGE PERCENTAGES (FRPS)

	x_1	x_2	x_3	x_4	x_5	x_6	x_7	x_8	x_9
P_1	69%	43%	100%	85%	71%	69%	82%	80%	26%
P_2	55%	39%	100%	70%	64%	62%	76%	74%	26%
P_3				53%	47%		76%	77%	93%
P_4	56%	66%					50%	52%	99%
P_5	92%	86%		100%	100%		100%	100%	100%
P_6	92%	86%		100%	100%		100%	100%	100%
P_7			100%						
P_8						50%			

Table 8: BASELINE SYSTEM FEASIBLE RANGE PERCENTAGES (FRPS)

	x_1	x_2	x_3	x_4	x_5	x_6	x_7	x_8	x_9
P_1	84%	48%	100%	100%	71%	78%	87%	85%	48%
P_2	68%	43%	100%	87%	64%	70%	82%	81%	32%
P_3				42%	47%		46%	46%	82%
P_4	56%	66%					50%	52%	99%
P	92%	85%		100%	100%		100%	100%	100%
P_6	93%	85%		100%	100%		100%	100%	100%
P_7			100%						
P_8						50%			

Copyright © 2023 by ASME

Proceedings of the ASME 2023
International Design Engineering Technical Conferences and
Computers and Information in Engineering Conference
IDETC-CIE2023
August 20-23, 2023, Boston, Massachusetts

DETC2023-115313

ELECTRONICS DESIGN AND VERIFICATION FOR ROBOTS WITH ACTUATION AND SENSING REQUIREMENTS

Dongsheng Chen[1],*, Zonghao Huang[1], Cynthia Sung[1],

[1]University of Pennsylvania, Philadelphia, PA

ABSTRACT

Robot design is a challenging problem involving a balance between the robot's mechanical design, kinematic structure, and actuation and sensing capabilities. Recent work in computational robot design has focused on mechanical design while assuming that the given actuators are sufficient for the task. At the same time, existing electronics design tools ignore the physical requirements of the actuators and sensors in the circuit. In this paper, we present the first system that closes the loop between the two, incorporating a robot's mechanical requirements into its circuit design process. We show that the problem can be solved using an iterative search consisting of two parts. First, a dynamic simulator converts the mechanical design and the given task into concrete actuation and sensing requirements. Second, a circuit generator executes a branch-and-bound search to convert the design requirements into a feasible electronic design. The system iterates through both of these steps, a process that is sometimes required since the electronics components add mass that may affect the robot's design requirements. We demonstrate this approach on two examples – a manipulator and a quadruped – showing in both cases that the system is able to generate a valid electronics design.

Keywords: Electronics design, computational tool, generative design

1. INTRODUCTION

Robot design is a difficult problem. Robots include multiple subsystems: a mechanical layer, sensing and actuation, and computation for planning and control. A valid robot design must account for design requirements within each of these subsystems and interdependencies between them. Consider, for example, a wheeled robot that must carry a load at a certain speed. If the load is large, the actuators are likely larger in size as well. This increases the load on the base, which affects its dimensions and material. In addition, larger actuators often draw larger

currents, which affects the circuit components and power source. In recent decades, cheaper electronics and easily programmable controllers like Arduino[1] and Raspberry Pi[2] have increased access to the robotic hardware required to build these systems. However, resolving the constraints between them is still a time-consuming iterative design process. Thus, the engineering knowledge to design a functional system remains a barrier to casual designers.

Efforts to simplify electromechanical design can be roughly divided into three domains: those that focus on mechanical design, those focusing on electrical design, and those focusing on electromechanical interactions.

Systems focusing on mechanical design mostly study a geometry's effect on robot performance or how to design complicated geometric objects. For example, [1] uses a differentiable simulator to determine the fastest, stable, and efficient shape of a soft fish. In legged robot design [2–4], users can focus on morphology design and the system will automatically optimize and complete the fabrication plan with guaranteed stable motions. Drag-and-drop interfaces such as those in [4, 5] allow users to simultaneously explore the geometry and motion design space, seeing how changes in the robot's mechanical structure affect its trajectory. However, all of these systems ignore the actuation requirements for the system and assume that the user's specified motion is always possible to implement.

For electronics, there exist more granular tools to help user to design circuits. Online resources for component selection[3] and auto-completion[4] can simplify the design process but still require users to understand what components to use. Instead, abstracting lower level part selections and wire connections into higher level block diagram connections helps both in circuit correctness and less effort input [6], which can save users time for more creative design ideas. Systems like Firtzing [7] adopt

---

*Corresponding author: dschen@seas.upenn.edu

[1]Arduino. https://www.arduino.cc
[2]Raspberry Pi. https://www.raspberrypi.org/
[3]EDAsolver. http://edasolver.com/
[4]circuito.io. https://www.circuito.io/

Copyright © 2023 by ASME

this idea by providing a drag-and-drop interface for users with an auto-complete feature [8] to suggest candidate components based on a circuits database. Chen et al [9] take a step further by directly converting natural language oral instructions into a functional circuit. Trigger-Action [10] further converts abstract events (e.g., when the temperature is higher than) and actions (e.g., make a buzz) into a detailed circuit plan. Some systems rely on a detailed hardware description language (HDL), which allows users to specify electronic circuits precisely. Like software programming, this makes users more productive by freeing them from worrying about lower level details of circuit designs. Ramesh et al [11] propose a circuit synthesis method where users specify electronics and annotate control logic between them through code, while the rest of the circuit is generated automatically. Lin et al [12] build on this work and propose a formal HDL that can accommodate multiple levels of abstraction of circuit design. Such HDLs can be combined with visual editors for high-level block diagrams [13] or circuit geometric layouts [14], which subsequent user studies have shown can help users in the design process [13]. Finally, [15, 16] enable users to extend these ideas past traditional rigid PCB design to design electronic modules on curved surfaces such as gloves or masks with more appropriate placements for interactive prototypes. However, similarly to robot design systems, most of these approaches only focus on electronics design itself and ignore physical requirements on the selected components and thus are not able to handle a complex robot co-design problem.

Systems that address not only the challenges of mechanical design and electrical design separately, but also the conflicts between them are few. Optimization-based approaches such as [17, 18] discuss how to cast these problems as a co-design problem, but still leave it up to the user to define the problem. Work in [19] provides a scripting interface for users to define robots, including their mechanical and electrical components, but assumes that the list of components the user specifies is always correct and provides no verification of the design. The closest work that attempts to complete eletromechanical co-design is [20], which adds verification for customized PCBs of electromechanical designs. Similarly, [21] can design the layout of electronics components and their corresponding fixtures and enclosures, given all components and a geometry. However, these systems still require users to know what components to use and where to place them. These systems all ignore an important component of robot design – the actuation and sensing – instead assuming that whatever actuator is available is sufficient for the robot's task. As a robot design becomes larger and more complex, this assumption is less likely to hold. Actuation and power often occupy a large portion of the robot's mass and volume [22–24], and designers spend a significant amount of time ensuring the actuation systems are sufficient.

We propose that a new design paradigm that incorporates structural requirements into electronics design is required in order to computationally generate feasible robots. The contributions of this paper are: 1) A framework that integrates actuation and functional constraints into the circuit generation process; 2) an algorithm for auto-generation of a complete circuit design, including pin connections between components; and

3) experimental validation of the system on two example robots: a quadruped and a manipulator arm. Our system uses off-the-shelf components that user can easily acquire from vendors instead of custom PCBs. Note that we assume the mechanical design (including kinematics and materials) is specified and focus on the feasibility of the electronics and actuation system. More functionality such as concurrent mechanical design optimization and electronics placement can be incorporated in future work.

The remainder of this paper is structured as follows. Sec. 2 introduces our robot and parts representation and defines problems to be solved. Sec. 3 gives an overview of the system. Sec. 4 describes how a robot's dynamic requirements are extracted from the simulation. Sec. 5 details the circuit generation procedure. Sec. 6 describes experimental results, wherein the system was applied to a manipulator arm and a quadruped. Sec. 7 concludes with limitations of this work and directions for future work.

2. ROBOT AND PARTS REPRESENTATION

Computing over a robot design requires a formal representation of the robot, the task, and the components in the database. We outline each of these elements.

2.1 Robot Representation

Mechanical Design. A mechanical design is a linkage consisting of joints and links. This design can be represented as a graph $\mathcal{M} = (\mathcal{L}, \mathcal{J})$, where vertices are the n_L links $\mathcal{L} = \{\mathcal{L}_1, \cdots, \mathcal{L}_{n_L}\}$ and edges are the n_J joints $\mathcal{J} = \{\mathcal{J}_1, \cdots, \mathcal{J}_{n_J}\}$. Because the robot is a physical object, each of these joints and links additionally contains information about physical quantities. In particular, a joint $\mathcal{J}_i$ has two links $\mathcal{L}_i, \mathcal{L}_{i+1}$ that it connects, a type (R or P) indicating whether it is a revolute or prismatic joint, a state q_i indicating its current position, and a range of motion $range_i^q$. A link $\mathcal{L}_j$ has x, y, z bounding box dimensions, a mass m_j, and an inertia I_j.

We store this information in the URDF[5] format since many existing tools like [5] can generate URDF representations of kinematic structures, making it easy for users to use our system with their robot designs.

Electronics Design. Similarly to the mechanical design, the electronics design for a robot is a graph $\mathcal{E} = (\mathcal{C}, \mathcal{N})$ where the n_C vertices $\mathcal{C} = \{\mathcal{C}_1, \cdots, \mathcal{C}_{n_C}\}$ are the individual components and the n_N edges $\mathcal{N} = \{\mathcal{N}_1, \cdots, \mathcal{N}_{n_N}\}$ are pin connections. These objects are expanded in further detail in Sec. 2.3.

Robot. A robot design $\mathcal{R}$ is defined as the combination of a mechanical design and an electronics design, i.e., $\mathcal{R} = (\mathcal{M}, \mathcal{E})$.

Task. The purpose of designing a robot is to complete a particular task. For the purposes of this work, we divide tasks into two types: actuation and sensing. An actuation task $\mathcal{T}_k^a$ is a motion sequence for the robot. That is, the task is a mapping $\mathcal{T}_k^a : \mathbb{R} \to \mathbb{R}^{n_J}$ from a time $t \in \mathbb{R}$ to a vector of joint states $[q_1, \cdots, q_{n_J}]$. A sensing task $\mathcal{T}_k^s$ is a pair $\mathcal{T}_k^s = (\mathbf{p}_k, f_k)$, where

---

[5]URDF. http://wiki.ros.org/urdf

Copyright © 2023 by ASME

$\mathbf{p}_k$ is a position on the robot body (typically the location of a joint or the center of a link) and f_k is one of the possible functions listed in Table 1. Then the task specification for a robot is a set of tasks $\mathcal{T} = \{\mathcal{T}_1^a, \cdots, \mathcal{T}_1^s, \cdots\}$ that the robot must accomplish.

Electomechanical Design Problem The problem to be solved is defined as follows.

Problem 1 *Given a mechanical design $\mathcal{M}$ and tasks $\mathcal{T}$, find an electronics design $\mathcal{E}$ such that the designed robot $\mathcal{R} = (\mathcal{M}, \mathcal{E})$ is able to complete the tasks in $\mathcal{T}$.*

2.2 Design Requirements

In order to solve this problem, we must first convert the mechanical design and tasks into a concrete set of design requirements. For actuation tasks, we focus on two types of requirements. Each actuation task $\mathcal{T}_k^a$ requires that the joints on the mechanical design achieve particular velocity and force/torque ranges. Combining them together yields the full design requirement that the actuators at each of the joints $\mathcal{J}_i$ must be able to achieve $range_i^\tau = [\tau_i^{min}, \tau_i^{max}]$ forces/torques and $range_i^v = [v_i^{min}, v_i^{max}]$ velocities. For a sensing task $\mathcal{T}_k^s$, we map the function f_k onto a component type as indicated in Table 1. The design requirement is then that at least one component of this type must be included in the electronics design for each such task. Understanding the mapping of more complex functions onto component types would fit into this framework and we reserve it for future work.

Converting the mechanical design and task into these requirements allows us to split Problem 1 into two subproblems.

Problem 2 *Given a mechanical design $\mathcal{M}$ and tasks $\mathcal{T}$, find the associated design requirements consisting of force/torque and velocity actuator ranges and required sensing components.*

Problem 3 *Given force/torque and velocity actuator ranges and required sensing components, find an electronics design $\mathcal{E}$ that satisfies these design requirements.*

Note that although we have written two distinct problem statements, they are actually related since the weight of the electronics will affect force/torque ranges, and the resulting force/torque ranges affect the electronics design.

2.3 Electronics Design

Components The electronics components $\mathcal{C}_i$ that we consider in this system are, at the most basic level, collections of pins $\mathcal{P}_i$ with functional relationships $\mathcal{A}_i$ between the input and output voltages $U_{i,in/out}$ and currents $I_{i,in/out}$. The number of pins in groups can vary with different versions of components. We use

Function	Sensor
Velocity/Position Control	Encoder
Force Control	Force sensor
Remote Control	Bluetooth
Line Tracking	Camera

TABLE 1: FUNCTION-SENSOR MAPPING

Property	Possible Values
Class	Power, Function, Both
Function type	Analog, Digital, PWM, I²C, UART, SPI
Physical type	Normal pin, USB, JST
Dependent pins	Pin 1, Pin 2, ..., Pin n
IO type	Input, Output, Bidirection
Connection type	One-to-one, One-to-many
Voltage range	$[U_{min}, U_{max}]$
Current range	$[I_{min}, I_{max}]$

TABLE 2: PROPERTIES USED TO DEFINE A PIN TYPE, FUNCTION, AND POSSIBLE CONNECTIONS.

$$\underbrace{\begin{bmatrix} -1 & 1 & 0 & 0 & 0 & V \\ 0 & 0 & k_t & -1 & 0 & 0 \\ \frac{1}{k_e} & -\frac{1}{k_e} & -\frac{R}{k_e} & 0 & -1 & 0 \end{bmatrix}}_{\mathcal{A}_{DC}} \begin{bmatrix} P_1 \\ P_2 \\ I \\ \tau \\ \omega \\ 1 \end{bmatrix} \geq 0$$

FIGURE 1: A SAMPLE DC MOTOR MODEL.

a template-driven approach to model each type of component so that new components can be added easily by following the template structure.

We first model pins as shown in Table 2. Each pin $p_i^j \in \mathcal{P}_i$ has a set of properties that define the pin's function, type and possible connections. For example, the pin Class $class_i^j \in \{power, function, both\}$ indicates the type of pin. The pin's $class_i^j$, function $function_i^j$, physical shape $physical_i^j$, connection type $connection_i^j$, voltage range $range_i^{V,j}$ and current range $range_i^{I,j}$ dictate what other pins p_i^k can be connected to. The dependencies list $\mathbf{p}_i^D = \{p_q, p_r, \cdots\}$ includes all pins that must be connected in order for p_i^j to work. For example, in order to output voltage from an H-bridge, the power supply and ground must be connected. With this information, a pin's properties are determined: $p_i^j = (class_i^j, function_i^j, physical_i^j, \mathbf{p}_i^D, io_i^j, connection_i^j, range_i^{V,j}, range_i^{I,j})$.

In addition to basic pin relationships, some components can perform specific functions. For example, a motor outputs torque and velocity when a voltage is applied in a given load condition. These components are represented as extensions (practically, subclasses) of our base template. Our current database contains 10 types of off-the-shelf components that include typical actuators and sensors as well as other supporting components such as motor drivers. They are: {DC Motor(6), Servo Motor(4), Encoder(2), Force Sensor(3), Camera(2), Bluetooth(2), H-Bridge(4), Voltage Regulator(2), Micro-Controller(3), Battery(6)}. The number in parentheses indicates the number of components of each type. Although the current number of components is small, in the future, the database could be enhanced through existing databases such as Fritzing or through systems such as [25], which automatically parses datasheets for component specifications.

The functional relations between pins and other outputs for a component $\mathcal{C}_i$ are represented as linear equality/inequality constraints. We choose a linear model because it can be

Copyright © 2023 by ASME

represented as matrix of coefficients $\mathcal{A}_i$ and linear feasibility problems are faster to solve. If the true model of a component has nonlinearities, we store the linearized model, where coefficients in $\mathcal{A}_i$ may contain terms from the component parameters. An example of DC motor model $\mathcal{A}_{DC}$ is shown in Figure 1. The 3×6 matrix is the model. The column vector stores decision variables of the model x, which will be discussed more in Sec. 5. The variables P_1, P_2 are two motor pin voltages, I is current across the motor, τ is the torque the motor can generate, ω is the corresponding angular velocity. Inside the model, V is the rated voltage, k_e, k_t are motor constants and R is the internal resistance. These are specific parameters for each component and are pulled from the component's datasheet. Multiplying the first row of the matrix with the column vector gives an expression which constrains that the voltage difference between the two leads of the motor must not exceed the rated voltage. Similarly, multiplying the second and third rows of the matrix with the column vector produces constraints on output torque and velocity of the motor. Other components include other types of information, for example, communication frequency for sensors, and general information such as size, weight and cost. See the Appendix for a full description of all of the components in our database.

Pin Connections Given this representation, a pin connection $\mathcal{N}_j \in \mathcal{N}$ is represented as follows: $\mathcal{N}_j = (p_{i_1}^{j_1}, p_{i_2}^{j_2})$. That is, the connection directly connects individual pins within the components making up the electronics design.

3. SYSTEM OVERVIEW

The purpose of our system is to automatically generate an actuation, sensing, and electronics plan for an input robot design with a specified task requirement. Doing this requires a sequence of steps, including interpreting the robot and task specification, generating a list of candidate electronic components, generating a candidate circuit design, and verifying the design for functionality and physical feasibility. The workflow is shown in Figure 2. Given the robot and task specification, a dynamic simulator addresses Problem 2, converting them into design requirements. A circuit generator then addresses Problem 3, converting these design requirements into a functional electronics design using a branch-and-bound search. The design can be optimized toward being simple, low-cost, and light-weight, etc.

Note that in most robotic systems, actuation and power occupy a substantial portion of the robot's total mass [22, 23]. Therefore, simply generating the electronics subsystem is insufficient. Extra mass introduced by newly added motors and batteries must be passed back through the simulator to ensure that the robot still works with the updated weight. If the design is still valid and the proposed actuators can still perform the given task, then a successful design process is accomplished. If it fails, the circuit generator must try again with the updated design requirements. Thus, our system applies an iterative approach, repeatedly simulating and updating the circuit design until it converges to a feasible solution or terminates because the problem is infeasible with the available components.

Algorithm 1: CIRCUITGENERATION($\{range_i^\tau\}$, $\{range_i^v\}$, $\mathcal{T}^s$, COST)

Input: force/torque requirements $\{range_i^\tau\}$, velocity requirements $\{range_i^v\}$, sensing tasks $\mathcal{T}^2$, cost metric cost
Output: electronics design $\mathcal{C}$
// generate lists of possible components

1 **for each** $\mathcal{J}_i \in \mathcal{J}$, $\mathcal{C}_{i,prop} \leftarrow$ {actuators satisfying $range_i^\tau$ and $range_i^v$};
2 **for each** $\mathcal{T}_k^s \in \mathcal{T}^s$, $\mathcal{C}_{k,prop} \leftarrow$ {sensors satisfying task $\mathcal{T}_k^s$};
3 $mincost \leftarrow \infty$; $\mathcal{C} \leftarrow \{\}$; $\mathcal{N} \leftarrow \{\}$; // initialize search
4 **foreach** *permutation* $\mathcal{C}'$ in ($\mathcal{C}_{i,prop}$, $\mathcal{C}_{k,prop}$) **do**
5     $\{mincost_{new}, \mathcal{C}_{new}, \mathcal{N}_{new}\} \leftarrow$ BranchAndBound($\mathcal{C}'$, $mincost$, cost);
6   **if** $mincost_{new} < mincost$ **then**
7        $mincost = mincost_{new}$; $\mathcal{C} = \mathcal{C}_{new}$; $\mathcal{N} = \mathcal{N}_{new}$;
8   **end**
9 **end**
10 $\mathcal{C} \leftarrow (\mathcal{C}, \mathcal{N})$;

---

4. ACTUATION REQUIREMENTS VIA DYNAMIC SIMULATION

In order to extract design requirements, the given mechanical design $\mathcal{M}$ is simulated forward in time for each of the actuation tasks $\mathcal{T}_k^a$. As the design executes the joint trajectories required, individual joint velocities and forces/torques are measured over time and the upper and lower bounds are recorded, yielding a required force/torque range $range_{k,i}^\tau = [\tau_{k,i}^{min}, \tau_{k,i}^{max}]$ and a required velocity range $range_{k,i}^v = [v_{k,i}^{min}, v_{k,i}^{max}]$ for each joint $\mathcal{J}_i$. For each joint, then, the actuation design requirement is that the actuator must be able to achieve $range_i^\tau = [\min_k \tau_{k,i}^{min}, \max_k \tau_{k,i}^{max}]$, and similarly for the velocity.

We use Project Chrono [26] to simulate mechanical designs since it enables us to directly import URDFs and programmatically modify mass and inertia values. The mechanical designs are driven to follow the desired joint trajectories using PI control over the position ($k_P = 1$, $k_I = 0.2$) and velocity ($k_P = 1$, $k_I = 0.3$) with a simulation step size of 0.005 s.

5. CIRCUIT GENERATION

The circuit generator converts design requirements into a valid electronics design (Alg. 1). The system starts by selecting actuators based on the actuator requirements and sensors from the required component types and then infers the rest of the necessary components. The inference and search process is similar to the classic circuit satisfiability problem, which is NP-complete. We choose to use a branch-and-bound method for this step (ref. Alg. 2). During branching, required components are inferred, connected and verified. During bounding, less optimal designs are pruned to accelerate the search.

5.1 Branch

In branching stage, new candidate components are connected to the electronics design through a sequence of inference, connection and verification steps. For each requirement, a

Copyright © 2023 by ASME

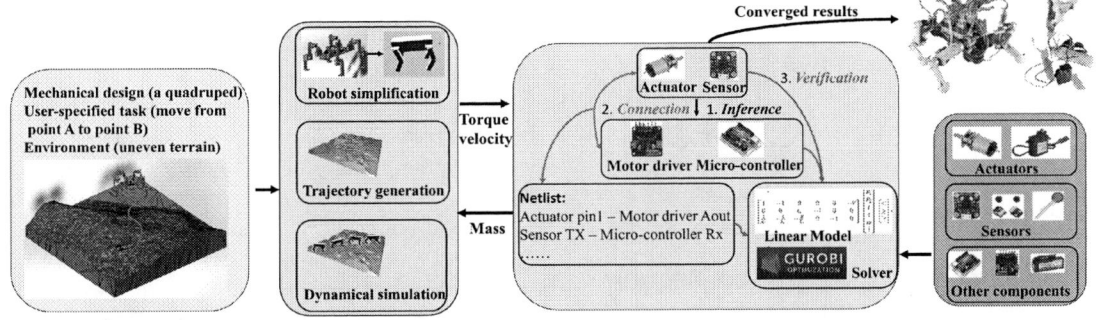

FIGURE 2: SYSTEM OVERVIEW

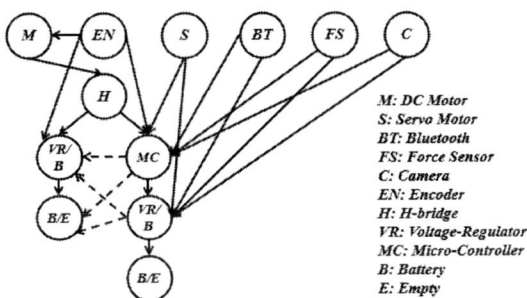

M: DC Motor
S: Servo Motor
BT: Bluetooth
FS: Force Sensor
C: Camera
EN: Encoder
H: H-bridge
VR: Voltage-Regulator
MC: Micro-Controller
B: Battery
E: Empty

FIGURE 3: INFERENCE TEMPLATE

list of potential components is generated from the database and candidate components are selected. Pin connections are generated and the sub-circuit is verified using an optimization that also generates voltage and current requirements for additional components needed to support the existing design. The process continues until a full circuit is completed.

5.1.1 Infer. The inference step involves determining whether and which new components must be added to the electronics design for it to be functional. Initially, this decision is driven by the actuator and sensor requirements. A list of potential actuators for each joint is selected from the database based on torque and velocity requirements. A list of potential sensors for each required function is also selected. In later stages, this decision is driven by the existing components in the electronics design, following a template (Figure 3), which we have developed through our own design experience. The inference procedure follows arrows in the template to generate all dependent components. For example, introducing a DC motor (M) to the system requires the addition of an H-bridge (H), which in turn requires a power source (VR/B) and a micro-controller (MC).

To generate a full circuit, the system begins at the root node and sequentially moves down the layers of the template to identify candidate components. Choosing candidate components requires two decisions: the version and the number of it. To determine possible versions, the system matches two layers' component voltage and current ranges. If the output voltage range of a

---

Algorithm 2: BRANCHANDBOUND($\mathscr{C}_{req}$, $mincost$, COST)

Input: required components $\mathscr{C}_{req}$, cost $mincost$ of $\mathscr{C}_{req}$, cost metric cost

Output: total cost $mincost_{new}$ of completed circuit; generated components $\mathscr{C}_{new}$; pin connections $\mathscr{N}_{new}$

1   $\mathscr{C}_{new} \leftarrow \mathscr{C}_{req}$;
    $\{mincost_{new}, \mathscr{N}_{new}\} \leftarrow$ OptimizeConnections($\mathscr{C}_{req}$);
    // Infer: Exit if no new components are required

2   **for each** $\mathscr{C}_i \in \mathscr{C}$, $\mathscr{C}_{i,prop} \leftarrow$ {candidate next-layer components from Fig. 3};

3   **if** *all* $\mathscr{C}_{i,prop}$ *empty* **then** return ;

4   **foreach** *permutation* $\mathscr{C}'$ in ($\mathscr{C}_{i,prop}$ **do**   // Branch

5     $\{mincost', \mathscr{N}'\} \leftarrow$ OptimizeConnections($\mathscr{C}_{req} \cup \mathscr{C}'$);
       // Connect/Verify

6   **if** ($\mathscr{C}'$, $\mathscr{N}'$) *is valid* **then**

7   **if** $mincost' > mincost$ **then return** ;   // Bound

8        $\{mincost_{new}, \mathscr{C}_{new}, \mathscr{N}_{new}\} \leftarrow$ BranchAndBound($\mathscr{C}_{req}$

9        $\cup \mathscr{C}'$, $mincost$, cost)

10   **end**

11 **end**

---

component intersects with input voltage range of its dependency ($range_i^{Vout} \cap range_j^{Vin} \neq \emptyset$), where component $\mathscr{C}_j$ depends on $\mathscr{C}_i$, and the maximum output current is greater than the required input current ($I_{max,i}^{out} \geq I_j^{in}$), then these components are candidates. The number of components is decided as the minimum number required to have enough output pins for the dependency and to deliver enough output current.

In cases where candidate components of the same type are identified, these components are combined to reduce the complexity of the circuit. These possible overlaps are indicated by the dashed lines in the template. First, the system checks to see which existing components in the circuit are the same type as the new layer. For example, in the first layer, if the motors can be controlled using the same H-bridge, then they are treated as one set of constraints. To make this decision, a graph is constructed with nodes being the existing components and edges indicating components whose voltage and current ranges overlap. A minimum clique cover on the graph is performed to identify the required smallest number of different components.

Copyright © 2023 by ASME

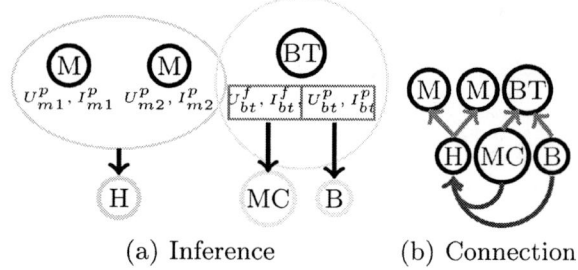

| | (a) Inference | (b) Connection |

FIGURE 4: AN EXAMPLE OF INFERENCE STEP AND CONNECTION STEP

H.AOUT1 – M1.+	H.AOUT2 – M1.-
H.BOUT2 – M2.-	H.BOUT1 – M2.+
MC.4 – H.AIN1	MC.5 – H.AIN2
MC.6 – H.BIN1	MC.7 – H.BIN2
MC.50 – BT.MISO	MC.51 – BT.MOSI
MC.52 – BT.SCK	MC.53 – BT.CS
B.+ – BT.VIN	B.+ – H.VIN
B.+ – MC.Vin	B.- – BT.GND
B.- – H.GND	B.- – MC.GND

FIGURE 5: EXAMPLE OUTPUT FOR A REMOTE-CONTROL TWO-WHEELED ROBOT.

5.1.2 Connect. Once components are selected, the system connects them into a sub-circuit. The relations between components are generated using a pin-matching procedure similar to [20], allowing only pins with same class and compatible I/O relations to connect. The pin matching algorithm is run between the previous component's connected pins' dependencies list and all of the new component's pins. Valid connections are created by checking the availability of the pin, intersection of voltage ranges, and compatibility of function. The connection order follows the template in Figure 3, i.e., starting with actuators and sensors and moving through new components as they are added.

5.1.3 Verify. Although the sub-circuit satisfies the connection constraints between two layers of components, it is yet unknown whether it satisfies the constraints imposed by all components, including ones that are not considered during the connection step. Therefore, once the sub-circuit is connected, it is verified through an optimization. Consider a sub-circuit that consists of components $\mathcal{C}$ and pin connections $\mathcal{N}$. To check feasibility of the sub-circuit, individual models $\mathcal{A}_i$ for each of the components $\mathcal{C}_i \in \mathcal{C}$ are included as constraints in the optimization problem. In addition, for each connection $(p_{i_1}^{j_1}, p_{i_2}^{j_2}) \in \mathcal{N}$, there is a constraint that the voltages of the two pins must match $V_{i_1}^{j_1} = V_{i_2}^{j_2}$. The optimization problem is solved using an iterative Newton-Raphson search. Within each iteration, the linearized problem is solved using the Gurobi solver[6], and the component models $\mathcal{A}_i$ are updated accordingly. If the problem has valid solutions (no conflicts between pin voltages and currents within components), then the sub-circuit is verified as functional and can be used to infer later components.

---

[6]Gurobi. https://www.gurobi.com/

5.2 Bound

For any given design requirements, there are many electronics designs that satisfy them. We could terminate the search at the first valid circuit found and the resulting robot would be feasible. However, the first valid option may also be unnecessarily complex. We select 6 common metrics to optimize the circuit design towards being simple, low-cost and lightweight: number of components, number of component pins, number of connections, monetary cost, total weight, and power consumption. The metric used to rank electronics designs is a weighted average of these 6 metrics. Since all of these metrics increase as components are added and, through the inference step, components are only ever added or merged into the existing set, this weighted cost function can be used to prune the search for a more optimal electronics design. If at any stage, the cost of the current sub-circuit exceeds that of the as-yet best solution, then search along that branch is terminated. Once users are satisfied with the current design, even if the circuit is still not globally optimal, they can stop the process at any time.

5.3 Example Walkthrough

As an example, we walk through the design of a remote-controlled two-wheeled robot. The flow of this process is shown in Figure 4. For simplicity, assume actuators and sensors are already specified from the database – 2×Pololu 4797 (DC motor) and 1×Bluefruit SPI (Bluetooth) – although the real system will do a search over all possible motors and remote control options. Let m_1 and m_2 denote the two motor components and bt denote the Bluetooth component. Based on the inference template, the DC motors require H-bridges and the Bluetooth module requires micro-controllers and batteries. The first step is to run the optimization to determine the voltage and current ranges of pins for these components. The result for this layer is $V_{m_1}^p = V_{m_2}^p = [3.78V, 6V]$, $V_{bt}^f = [0V, 3.3V]$, $V_{bt}^p = [3.3V, 16V]$ and $I_{m_1}^p = I_{m_2}^p = 0.03A$, $I_{bt}^f = 10^{-5}A$, $I_{bt}^p = 0.03A$, where superscripts p and f indicate the power and function pins, respectively. Based on these voltage ranges, a minimum clique cover determines that the motors can share the same H-bridge. Then, the data used to infer specific components are: $\{V_m^p = [3.78V, 6V], \quad I_m^p = 0.06\}$ for the H-bridge; $\{V_{bt}^f = [0V, 3.3V], \quad I_{bt}^f = 10^{-5}\}$ for the micro-controller; $\{V_{bt}^p = [3.3V, 16V], \quad I_{bt}^p = 0.03\}$ for the battery. With these constraints, one possible combination for next level components is: h =DRV8835 (a dual-channel H-bridge), mc =Arduino Mega, and b =6LR61 (9V battery). Since DRV8835 is a dual-channel H-bridge, one suffices to support both motors. The Bluetooth only requires one set of SPI pins which Arduino can provide and a 9V battery is also sufficient to power all components. The last step is to connect them together using the pin matching algorithm, resulting in the design shown in Figure 5. Components connected by a red line are connected first.

6. EXPERIMENTS

We implemented this system in C++ on a Windows 10 computer with an Intel i7-8700 @ 3.4GHz CPU and 16GB memory. The dynamical simulation uses the Project Chrono [26] and the optimization package for circuit generation is Gurobi[6].

Copyright © 2023 by ASME

TABLE 3: CIRCUIT GENERATION WITHOUT DYNAMIC CONSTRAINTS

Test	# Inputs	Circuit Size	# Searched	# Pruned	# Total
1	1	4	59	45	136
2	1	4	46	34	105
3	4	8	272	247	968
4	21	29	447	407	777

Inputs: 1) 1 DC Motor. 2) 1 Camera. 3) 2 DC Motors, 2 Encoders.
4) 8 DC Motors, 8 Encoders, 4 Servos, 1 Camera.

All components in the database are stored in text files using Protobuf[7].

6.1 Circuit Generation without Robot Constraints

To check the validity of our branch-and-bound search, we first evaluate the performance of the circuit generation algorithm in the absence of task constraints. Table 3 reports the size of the input and resulting circuits. We include how many circuits were searched and of those, how many branches were pruned, as compared to the total number of possible circuit candidates. The results show that our system is able to deal with a large range of circuit complexity, from the smallest circuit with a single actuator or sensor to a much more complex example with 12 actuators and 9 sensors. In general, the first valid circuit is found very quickly: Even the most complex design takes only 40 ms. The pruning works effectively to accelerate the search process by reducing the number of sub-circuits explored to between 28% and 57% of that when no pruning is applied.

6.2 Circuit Generation for Robots and Hardware Verification

We then test the full workflow of our system from design to manufacturing on two robots, a line-following quadruped and a 2-joint manipulator with force-sensing gripper. Figure 6 shows the input mechanical design, the final physical prototype. Table 4a and 4b show the mass and dimension details. The system is able to generate a valid electronics design for both robots. When building the physical prototypes, all components were connected by following the pin netlist generated by the system. The quadruped is able to walk and use its camera. The manipulator is able to pick up objects and sense grip force.

6.2.1 Circuit Generation and Fabrication of Quadruped Robot.
The quadruped example contains 4 legs, each having 3 joints, and demonstrates our system's ability to process complicated mechanical designs. The input sensing task is to follow a white line on the uneven terrain in Fig. 2. The gait of the robot is designed manually, resulting in three individual actuation tasks enabling the robot to walk forward, turn left, and turn right. The design objective for the quadruped is minimum weight. To match the massless leg assumption of the dynamic simulator, we include in the component library only the motors with small size and weight (max mass of 10 g). To simplify the fabrication process and to achieve a symmetrical design, we also constrain the system to use same motors for the same joints of each leg. These additional constraints are inputted

[7]Protocol buffers. https://developers.google.com/protocol-buffers/

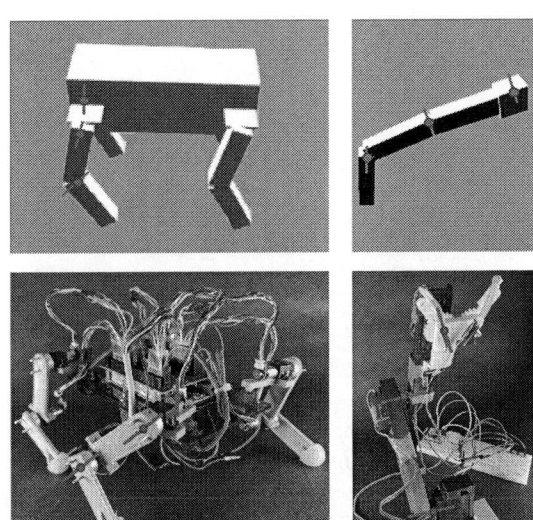

FIGURE 6: FIRST ROW: FROM LEFT TO RIGHT: INPUT QUADRUPED MODEL; INPUT MANIPULATOR MODEL. SECOND ROW, FROM LEFT TO RIGHT: FABRICATED QUADRUPED; FABRICATED MANIPULATOR. RED DOTS AND LINES INDICATE JOINT CENTERS AND AXES OF ROTATION.

manually and demonstrate the ability of our system to deal with more complex design requirements than just those outlined in our problem statement.

The resulting electronics design includes 8 DC motors (Pololu 3050) with 8 encoders (Pololu 4761), 4 servo motors (LS-009AF), 4 H-bridges (DRV 8835), 1 camera (MU Vision Sensor) and 1 Arduino Mega that has enough I/O pins to accommodate all the above components. The system also chose one 4.8V battery for servomotors and H-bridges and one 9V for the Arduino. The whole electronics design took 10 s to generate. The fabricated quadruped prototype is able to use this exact circuit to walk and operate the camera to see the environment.

When testing the quadruped, we noticed that its dynamic performance did not entirely match that of the simulation and

TABLE 4: WEIGHT AND DIMENSION OF ROBOT PARTS

Component	Mass (g)	Dimensions (mm)
Quadruped Base	105.1	$180 \times 80 \times 21$
Quadruped Link1	4.4	$40 \times 30 \times 5$
Quadruped Link2	16.8	$80 \times 40 \times 15$
Quadruped Link3	9.5	$70 \times 18 \times 18$
DC Motor	9.5	$10 \times 12 \times 26$
Servo Motor	9	$22.3 \times 11.8 \times 26.3$
Other Electronics	135.4	-

(a) Quadruped

Component	Mass (g)	Dimensions (mm)
Arm Link1	30.2	$150 \times 28 \times 18$
Arm Link2	27.4	$150 \times 28 \times 18$
Gripper	24.2	$102.3 \times 39.1 \times 21.2$
Servo Motor	40	$40.8 \times 20.1 \times 38$

(b) Manipulator

Copyright © 2023 by ASME

TABLE 5: SYSTEM RUNNING TIME FOR MANIPULATOR

Test Mass	Process	DST (s)	CGT (s)	CVT (s)
200 g	Generation 1	15.043	0.035	-
	Verification 1	14.87	-	0.002
242 g	Generation 1	21.275	0.05	-
	Generation 2	21.867	0.026	-
	Verification 2	21.833	-	0.003
640 g	Generation 1	18.91	0.028	-
	Generation 2	18.86	0.02	-
5000 g	Generation 1	29.396	0.025	-

DST: dynamic simulation time. CGT: circuit generation time. CVT: circuit verification time.

it tended to lose balance on steep terrain. We believe that the cause is that the wires, connectors, and mounting hardware were not included in the simulation, resulting in inaccurate mass and inertia estimates. This unmodeled extra weight could be added by estimating the additional mass from the number of connections.

6.2.2 Circuit Generation and Fabrication of Manipulator. The manipulator example demonstrates the benefits of iterative simulation and circuit generation in our system. The mechanical design of the manipulator contains 2 joints and a gripper. The inputted actuation task for this system is picking up objects with different mass. The sensing task is that the gripper measures the force with which it picks up the object. We use minimum number of components and minimum number of pins with equal weights as the objective. The system's output circuit actuates both joints and the gripper with the same servomotor (which varies for the different masses). Two force sensors (Pololu 1695) are selected for the gripper. One Arduino Mini and a 6V battery form the remaining components.

The computation time required for each test mass is shown in Table 5. Complete electronics generation involves simulating twice: first to generate circuit and second to verify the dynamical system's validity with added mass. If verification fails, either a new circuit is generated or it is infeasible and the algorithm terminates. When the weight is small (200 g), the system is able to successfully find a design in one iteration. When the weight is large (5000 g), the system quickly determines that no solution exists. Between these two extremes, the iterate is needed. For a 242 g mass, the system's first proposal fails the verification step, causing it to compute a revised circuit that ultimately passes. Conversely, the 640 g causes the system to determine design infeasibility on the second iteration.

Our results demonstrate the importance of accounting for electronics in a physical robot design. Depending on the inputs, the mass and torque limits of the actuators have a large effect on the ability to generate a valid circuit. Further, it may take multiple iterations in design process to determine that the robot is not feasible at all given the available actuators.

Anecdotally, we note that the servomotors ultimately used for the 200 g physical prototype are not the ones in the initial circuit generated by the system. When fabricating, we realized that we ran out of the servos selected by the system. Upon updating the database, the system generated a valid design that could be fabricated with our available components, although it was suboptimal compared to the initial design. That is to say, the database can be customized to the user's specific inventory, and the system still functions correctly.

7. CONCLUSIONS

In summary, we present a closed-loop system that assists in robot design for a desired task. The system iteratively determines design requirements through a dynamical simulation and uses these requirements to generate an electronics design through a template-based branch-and-bound search. Experiments on a quadruped and a manipulator arm show that the generated electronics design is capable of providing the needed torque and velocity to drive both robots to achieve their tasks. The results show the importance of incorporating the electronics design into the robot's dynamical requirements and vice versa.

In future work, we expect to extend this system to other properties of the electronics design that can affect the robot's performance such as the dimensions and locations of electronics. In addition, the current algorithm outputs pin connections without a proper circuit layout or assembly plan. As a result, the real constructed robots used many wires and are suboptimal. We expect that our design framework could be extended to support variations of this type.

In addition, scalability is an intrinsically difficult problem due to the non-existence of design standards like mechanical parts. Every manufacturer has their own design rules, which makes it hard to model those components with an absolute generality since one can easily invalidate a general model by introducing a brand-new function into an Integrated Circuit (IC) component. Our approach parameterizes the electronic component into a pin-based data file in an attempt to make it easy in the future to create new virtual components. The word "easy" in this case means that the designer still needs to copy the voltage, current and other information from the datasheet and arrange them in the given pattern order. Future work may incorporate strategies such as [25] that can automatically parse datasheets information to alleviate manual data transfer burden.

Finally, for the future goal of making full robot design simpler and more intuitive for non-engineers, we aim to include additional metrics such as battery life estimation and more complex mappings between a task and a needed sensor or actuator. It remains to be determined whether these additional metrics or design requirements can be incorporated into the existing simulation framework and the branch-and-bound search.

ACKNOWLEDGEMENTS

Support for this project has been provided in part by NSF Grant No. 1845339 and 1138847.

REFERENCES

[1] Ma, Pingchuan, Du, Tao, Zhang, John Z., Wu, Kui, Spielberg, Andrew, Katzschmann, Robert K. and Matusik, Wojciech. "DiffAqua." *ACM Transactions on Graphics* Vol. 40 No. 4 (2021): pp. 1–14.

[2] Megaro, Vittorio, Thomaszewski, Bernhard, Nitti, Maurizio, Hilliges, Otmar, Gross, Markus and Coros, Stelian. "Interactive Design of 3D-Printable Robotic Creatures." *ACM Trans. Graph.* Vol. 34 No. 6 (2015).

Copyright © 2023 by ASME

[3] Desai, Ruta, Yuan, Ye and Coros, Stelian. "Computational abstractions for interactive design of robotic devices." *Intl. Conf. on Robotics and Automation*: pp. 1196–1203. 2017.

[4] Ha, Sehoon, Coros, Stelian, Alspach, Alexander, Kim, Joohyung and Yamane, Katsu. "Computational co-optimization of design parameters and motion trajectories for robotic systems." *Intl. Journal of Robotics Research* Vol. 37 No. 13-14 (2018): pp. 1521–1536.

[5] Schulz, Adriana, Sung, Cynthia, Spielberg, Andrew, Zhao, Wei, Cheng, Robin, Grinspun, Eitan, Rus, Daniela and Matusik, Wojciech. "Interactive robogami: An end-to-end system for design of robots with ground locomotion." *Intl. Journal of Robotics Research* Vol. 36 No. 10 (2017): pp. 1131–1147.

[6] Lin, Richard, Ramesh, Rohit, Iannopollo, Antonio, Sangiovanni Vincentelli, Alberto, Dutta, Prabal, Alon, Elad and Hartmann, Björn. "Beyond Schematic Capture: Meaningful Abstractions for Better Electronics Design Tools." *Conf. on Human Factors in Computing Systems*: p. 283. 2019.

[7] Knörig, André, Wettach, Reto and Cohen, Jonathan. "Fritzing: a tool for advancing electronic prototyping for designers." *Intl. Conf. on Tangible and Embedded Interaction*: pp. 351–358. 2009.

[8] Lo, Jo-Yu, Huang, Da-Yuan, Kuo, Tzu-Sheng, Sun, Chen-Kuo, Gong, Jun, Seyed, Teddy, Yang, Xing-Dong and Chen, Bing-Yu. "AutoFritz: Autocomplete for Prototyping Virtual Breadboard Circuits." *Conf. on Human Factors in Computing Systems*: p. 403. 2019.

[9] Chen, Taizhou, Xu, Lantian and Zhu, Kening. "FritzBot: A data-driven conversational agent for physical-computing system design." *Intl. Journal of Human-Computer Studies* Vol. 155 (2021): p. 102699.

[10] Anderson, Fraser, Grossman, Tovi and Fitzmaurice, George. "Trigger-Action-Circuits: Leveraging Generative Design to Enable Novices to Design and Build Circuitry." *ACM Symp. on User Interface Software and Technology*: pp. 331–342. 2017.

[11] Ramesh, Rohit, Lin, Richard, Iannopollo, Antonio, Sangiovanni-Vincentelli, Alberto, Hartmann, Björn and Dutta, Prabal. "Turning coders into makers: the promise of embedded design generation." *ACM Symp. on Computational Fabrication*: p. 4. 2017.

[12] Lin, Richard, Ramesh, Rohit, Chi, Connie, Jain, Nikhil, Nuqui, Ryan, Dutta, Prabal and Hartmann, Björn. *Polymorphic Blocks: Unifying High-Level Specification and Low-Level Control for Circuit Board Design.* ACM (2020): p. 529–540.

[13] Lin, Richard, Ramesh, Rohit, Jain, Nikhil, Koe, Josephine, Nuqui, Ryan, Dutta, Prabal and Hartmann, Bjoern. "Weaving Schematics and Code: Interactive Visual Editing for Hardware Description Languages." *The 34th Annual ACM Symposium on User Interface Software and Technology*: p. 1039–1049. 2021. Association for Computing Machinery, New York, NY, USA.

[14] McElroy, Leo, Bolsée, Quentin, Peek, Nadya and Gershenfeld, Neil. "SVG-PCB: A Web-Based Bidirectional Electronics Board Editor." *Proceedings of the 7th Annual ACM Symposium on Computational Fabrication*. 2022. Association for Computing Machinery, New York, NY, USA.

[15] Hodges, Steve, Villar, Nicolas, Chen, Nicholas, Chugh, Tushar, Qi, Jie, Nowacka, Diana and Kawahara, Yoshihiro. "Circuit Stickers: Peel-and-Stick Construction of Interactive Electronic Prototypes." *Proceedings of the SIGCHI Conference on Human Factors in Computing Systems*: p. 1743–1746. 2014. Association for Computing Machinery, New York, NY, USA.

[16] Zhu, Junyi, Zhu, Yunyi, Cui, Jiaming, Cheng, Leon, Snowden, Jackson, Chounlakone, Mark, Wessely, Michael and Mueller, Stefanie. "MorphSensor: A 3D Electronic Design Tool for Reforming Sensor Modules." *Proceedings of the 33rd Annual ACM Symposium on User Interface Software and Technology*: p. 541–553. 2020. Association for Computing Machinery, New York, NY, USA.

[17] Censi, Andrea. "A class of co-design problems with cyclic constraints and their solution." *IEEE Robotics and Automation Letters* Vol. 2 No. 1 (2016): pp. 96–103.

[18] Dawson, Charles and Fan, Chuchu. "Certifiable Robot Design Optimization using Differentiable Programming." (2022).

[19] Mehta, Ankur, DelPreto, Joseph, Shaya, Benjamin and Rus, Daniela. "Cogeneration of mechanical, electrical, and software designs for printable robots from structural specifications." *Intl. Conf. on Intelligent Robots and Systems* (2014): pp. 2892–2897.

[20] Bezzo, Nicola, Gebhard, Peter, Lee, Insup, Piccoli, Matthew, Kumar, Vijay and Yim, Mark. "Rapid co-design of electro-mechanical specifications for robotic systems." *Intl. Design Engineering Technical Conf. and Computers and Information in Engineering Conf.*, Vol. 57199: p. V009T07A009. 2015.

[21] Desai, Ruta, McCann, James and Coros, Stelian. "Assembly-aware Design of Printable Electromechanical Devices." *ACM Symp. on User Interface Software and Technology*: pp. 457–472. 2018.

[22] Ma, Kevin Y., Chirarattananon, Pakpong, Fuller, Sawyer B. and Wood, Robert J. "Controlled Flight of a Biologically Inspired, Insect-Scale Robot." *Science* Vol. 340 No. 6132 (2013): pp. 603–607.

[23] Bledt, Gerardo, Powell, Matthew J., Katz, Benjamin, Di Carlo, Jared, Wensing, Patrick M. and Kim, Sangbae. "MIT Cheetah 3: Design and Control of a Robust, Dynamic Quadruped Robot." *Intl. Conf. on Intelligent Robots and Systems*: pp. 2245–2252. 2018.

[24] Davey, Jay, Kwok, Ngai and Yim, Mark. "Emulating self-reconfigurable robots-design of the SMORES system." *Intl. Conf. on Intelligent Robots and Systems*: pp. 4464–4469. 2012.

[25] Hsiao, Luke, Wu, Sen, Chiang, Nicholas, Ré, Christopher and Levis, Philip. "Creating Hardware Component Knowledge Bases with Training Data Generation and Multi-Task Learning." *ACM Trans. Embed. Comput. Syst.* Vol. 19 No. 6 (2020).

Copyright © 2023 by ASME

[26] Tasora, Alessandro, Serban, Radu, Mazhar, Hammad, Pazouki, Arman, Melanz, Daniel, Fleischmann, Jonathan, Taylor, Michael, Sugiyama, Hiroyuki and Negrut, Dan. "Chrono: An open source multi-physics dynamics engine." *Intl. Conf. on High Performance Computing in Science and Engineering*: pp. 19–49. 2015.

APPENDIX A. COMPONENTS LIBRARY MODELS

Every component's model includes variables which are unknowns to be solved and parameters which are model-specific constants, The variables and parameters are related through a linear matrix to formulate equality/inequality constraints. In the coefficient matrix, parameters V_i^u, V_i^l represent the upper and lower voltage bounds for a pin.

A.1 Battery model

The model has 2 variables + and −, which are output voltages of two leads of the battery, and one parameter V, which is the nominal voltage of the battery.

$$\begin{bmatrix} 1 & 0 & -V \\ 0 & 1 & 0 \end{bmatrix} \begin{bmatrix} + \\ - \\ 1 \end{bmatrix} = 0$$

A.2 Bluetooth model

The model has 7 variables: VIN, GND are input voltages for powering pins. $SCK, MISO, MOSI, CS$ are voltages for SPI pins. OT is the voltage of additional pins.

$$\begin{bmatrix} 1 & 0 & 0 & 0 & 0 & 0 & 0 & -V_1^u \\ 0 & 1 & 0 & 0 & 0 & 0 & 0 & -V_2^u \\ 0 & 0 & 1 & 0 & 0 & 0 & 0 & -V_3^u \\ 0 & 0 & 0 & 1 & 0 & 0 & 0 & -V_4^u \\ 0 & 0 & 0 & 0 & 1 & 0 & 0 & -V_5^u \\ 0 & 0 & 0 & 0 & 0 & 1 & 0 & -V_6^u \\ 0 & 0 & 0 & 0 & 0 & 0 & 1 & -V_7^u \end{bmatrix} \begin{bmatrix} VIN \\ GND \\ SCK \\ MISO \\ MOSI \\ CS \\ OT \\ 1 \end{bmatrix} \leq 0$$

A.3 Camera model

The model has 5 variables: $5V, GND$ are input voltages for powering pins. TX, RX are voltages for Serial pins. OT is the voltage of additional pins.

$$\begin{bmatrix} 1 & 0 & 0 & 0 & 0 & -V_1^u \\ 0 & 1 & 0 & 0 & 0 & -V_2^u \\ 0 & 0 & 1 & 0 & 0 & -V_3^u \\ 0 & 0 & 0 & 1 & 0 & -V_4^u \\ 0 & 0 & 0 & 0 & 1 & -V_5^u \end{bmatrix} \begin{bmatrix} 5V \\ GND \\ TX \\ RX \\ OT \\ 1 \end{bmatrix} \leq 0$$

A.4 Encoder model

The model has 6 variables: VCC, GND are the input voltages for powering pins. $M1, M2$ are voltages for motor pins. $OUTA, OUTB$ are voltages of two signal pins.

$$\begin{bmatrix} 1 & 0 & 0 & 0 & 0 & 0 & -V_1^u \\ 0 & 1 & 0 & 0 & 0 & 0 & -V_2^u \\ 0 & 0 & 1 & 0 & 0 & 0 & -V_3^u \\ 0 & 0 & 0 & 1 & 0 & 0 & -V_4^u \\ 0 & 0 & 0 & 0 & 1 & 0 & -V_5^u \\ 0 & 0 & 0 & 0 & 0 & 1 & -V_6^u \end{bmatrix} \begin{bmatrix} VCC \\ GND \\ M1 \\ M2 \\ OUTA \\ OUTB \\ 1 \end{bmatrix} \leq 0$$

A.5 Force sensor model

The model has 2 variables: $PIN1, PIN2$ are voltages of two ends of a force sensor.

$$\begin{bmatrix} 1 & 0 & -V_1^u \\ 0 & 1 & -V_2^u \end{bmatrix} \begin{bmatrix} PIN1 \\ PIN2 \\ 1 \end{bmatrix} \leq 0$$

A.6 H-bridge model

The model has 9 variables and is a linearized model. VCC is the input voltage of the motor pin. $5V, GND$ are the logic input voltages for powering pins. $IN1, IN2$ are voltages for two input pins for one channel. $OUT1, OUT2$ are voltages for two outputs pins for one channel. $DUTY1, DUTY2$ are corresponding duty cycle values for two input pins. Its parameter V_{log} is the logic high voltage value.

$$\begin{bmatrix} 1 & 0 & 0 & 0 & 0 & 0 & 0 & 0 & -V_1^l \\ 0 & 1 & 0 & 0 & 0 & 0 & 0 & 0 & -V_2^l \\ 0 & 0 & 1 & 0 & 0 & 0 & 0 & 0 & -V_{log} \\ 0 & 0 & 0 & 1 & 0 & 0 & -V_{log} & 0 & 0 \\ 0 & 0 & 0 & 0 & 1 & 0 & 0 & -V_{log} & 0 \\ -1 & 0 & 0 & 0 & 0 & 1 & 0 & -V_1^l & 0 & V_1^l \\ -1 & 0 & 0 & 0 & 0 & 0 & 1 & 0 & -V_1^l & V_1^l \end{bmatrix} \begin{bmatrix} VCC \\ 5V \\ GND \\ IN1 \\ IN2 \\ OUT1 \\ OUT2 \\ DUTY1 \\ DUTY2 \\ 1 \end{bmatrix} \begin{bmatrix} \geq \\ \geq \\ \geq \\ \geq \\ \geq \\ = \\ = \end{bmatrix} 0$$

A.7 Micro-controller model

Because a Micro-controller has too many pins, we use one pin to represent one category of pins here. For all possible pins, only pins that are used in the connection stage will appear in the model. For the Micro-controller model, VCC, GND are input voltages for powering pins. DIG is the voltage for all digital pins. ANA is the voltage for all analog pins. $DUTY$ is corresponding duty cycle values for digital pins.

$$\begin{bmatrix} 1 & 0 & 0 & 0 & 0 & -V_1^l \\ 0 & 1 & 0 & 0 & 0 & -V_2^l \\ 0 & 0 & 1 & 0 & -V_3^l & 0 \\ 0 & 0 & 0 & 1 & 0 & -V_4^l \end{bmatrix} \begin{bmatrix} VCC \\ GND \\ DIG \\ ANA \\ DUTY \\ 1 \end{bmatrix} \geq 0$$

A.8 Servo model

The model is almost identical as the DC motor model. The only difference is that the servo model has one extra signal pin, which needs to satisfy one more voltage inequality constraint.

A.9 Voltage regulator model

The model has 4 variables. $VIN+, VIN-$ represent voltages for input pins. $VOUT+, VOUT-$ represent voltages for output pins. Its parameter V_{in} is the upper bound voltage for VIN. V_{out} is the upper bound voltage for $VOUT$. V_{diff} is the minimal voltage difference between VIN and $VOUT$.

$$\begin{bmatrix} 1 & 0 & 0 & 0 & -V_{in} \\ 0 & 1 & 0 & 0 & -V_{out} \\ -1 & 1 & 0 & 0 & V_{diff} \\ 0 & 0 & 1 & 0 & 0 \\ 0 & 0 & 0 & 1 & 0 \end{bmatrix} \begin{bmatrix} VIN+ \\ VOUT+ \\ VIN- \\ VOUT- \\ 1 \end{bmatrix} \leq 0$$

Copyright © 2023 by ASME

Proceedings of the ASME 2023
International Design Engineering Technical Conferences and
Computers and Information in Engineering Conference
IDETC-CIE2023
August 20-23, 2023, Boston, Massachusetts

DETC2023-116392

KNOWLEDGE TRANSFER IN SELF-ORGANIZING SYSTEMS AND IMPACT OF SOCIAL ABILITY

Bingling Huang
Dept. of Aerospace & Mechanical Engineering
University of Southern California
Los Angeles, USA
binglinh@usc.edu

Yan Jin*
Dept. of Aerospace & Mechanical Engineering
University of Southern California
Los Angeles, USA
yjin@usc.edu
(*corresponding author)

ABSTRACT

Multiagent systems have evolved in both research and practice domains over the past decades. Application domains have also broadened into robotic and vehicular systems, among others. A self-organizing system comprises multiple agents and can adapt to changing operational environments. Early work on the self-organizing system has tackled the system design problem by introducing information fields and logical rules. As tasks become more complex, the difficulty of devising the fields and the rules has become harder. To overcome this limitation, a multiagent reinforcement learning based approach has been explored, aiming to let agents acquire the task knowledge by themselves. The research results thus far have verified the effectiveness of the learning-based approach. However, the training process of reinforcement learning takes a long time and a vast amount of computational resources. The question arises: can the previously trained team knowledge, embedded in neural networks, be reused so that the new training time can be significantly reduced? This paper addresses this overall question by investigating the effectiveness of transferring the knowledge of neural networks learned by a certain agent team to another team of the same domain but with different team sizes. Furthermore, the learning team has social abilities: they not only focus on the task states of the environment while learning but also observe the behavior of other agents. The results of this study have demonstrated the potential benefits as well as limitations of knowledge transfer between teams of the same domain but with different team sizes.

Keywords: Self-organizing systems, multiagent reinforcement learning, social learning, knowledge transfer.

1 INTRODUCTION

Muti-agent systems (MAS) have been applied to wide engineering fields, such as production lines, warehouse robotics, air force, and navy systems, etc. As the task environments become more unpredictable and complicated, the demands of a high level of automation and intelligence arise, leading to higher requirements on the MAS design. Therefore, constructing a MAS with high robustness, adaptability, resilience, and flexibility has drawn great attention from designers and researchers.

Self-organization is a concept that origins from biological sciences [1]. It refers to the process of organization to an arising overall order from members' local interactions [2] and an ability of a class of self-organizing systems (SOSs) to change their internal structure and functions in response to changing external circumstances [3]. Due to its advanced features, such as high-level adaptability and decentralized control [4], it has become one of the promising strategies for solving multiagent system design problems in the engineering fields. For example, swarm robots [5] can be organized in a self-organizing system to perform repetitive or dangerous tasks like package delivery. Robot teams behave cooperatively and spontaneously without global control or detailed decentralized design. Even though an individual may have limited sensing and movement abilities, teams have the potential to achieve incredible team accomplishments.

There are many methods to assist in achieving self-organization. Zouein has developed a DNA-based Cellular Formation Representation framework (cFORE) [6] to address issues of adaptive systems design. However, the adaptability of the framework is limited by the pre-defined design information

Copyright © 2023 by ASME

stored in the dDNA. Chen [7] has proposed a biology-inspired system representation named Behavior-based design DNA (B-dDNA) [7]. Chiang [8] has further detailed the design modeling for the interaction among agents. Khani [9] has considered both "task" and "social" aspects in one self-organizing system and proposed a two-field mechanism to enable CSO systems [9]. However, those works suffered the issue of field definition and description. Especially as the task complexity increases, describing fields in proper mathematical forms may become impossible.

To solve the issue, Ji [3] has utilized multiagent reinforcement learning (MARL) to design self-organizing systems and carried out physical structure assembly simulations. The method can be employed where multiple agents learn how to accomplish the shared task collaboratively by maximizing a shared reward function. It excels at encouraging agents to learn through the trial-end-error process.

However, training agents from novices have a high cost. Learning from thousands of trials makes this method consume a substantial amount of time and computational sources. Especially when the environments are complex, and designers hold limited prior task knowledge, agent teams need to explore the possible strategies from sparse reward fields [3]. Therefore, it is desired that well-trained knowledge can be accumulated, re-utilized, and even inherited among different teams to avoid repetitive training. Transferring knowledge between tasks can not only reduce computational costs but also enhance the capability of agents in handling complex tasks. Many researchers have contributed to assist the agent in learning fast and becoming experts.

In [10], the authors propose REPrepresentation and REPAINT algorithms for knowledge transfer in deep reinforcement learning to accelerate learning processes. The algorithm not only employs the pre-trained policy but also utilizes an advantage-based experience selection method. The idea of selecting samples based on their relatedness help an agent to choose useful samples and conduct more effective learning. In [11] authors propose Actor-Mimic, which is a transfer reinforcement learning approach to mimic expert decisions for multi-task learning. In [12], the authors have investigated knowledge transfer from simple tasks to relatively complex ones and utilized the acquired knowledge as a "teacher" to guide a novice agent towards enhanced capabilities. It is found that 'copy expert' can help increase the learning efficiency for high similarity tasks but is not efficient for more complex tasks and with low similarity.

However, previous studies have primarily focus on knowledge transfer in single agent learning scenarios. As more agents are involved in a task, the situations of knowledge transfer become more complicated. In multiagent reinforcement learning, the knowledge is embedded in a cluster of value functions or neural networks. An individual's policy is not only determined by the task environment but also influenced by the actions of other agents.

In [13], the authors focus on knowledge transfer in MARL and propose a new algorithm to let novice agents to learning from the environment and teachers parallelly. The method is flexible on the network structure. However, it assumes that well-trained policy has high quality. In complex engineering tasks, it is hard to have a good "teacher" that already masters the task domain knowledge well.

In [14], authors adopt game theory-based MARL and conduct value function transferring in the learning process. However, the method utilized the agents with sing-agent knowledge (e.g., local value function).

Some research has found that task features are a crucial factor in the effect of learning success [15], and two tasks can be considered to be similar when they share the same state-action space [16] [17]. Glatt et al. found that the similarity between tasks plays an important role in the success of knowledge transfer in applications [18]. It can improve the learning speed for a new task.

However, the complicated process of transferring knowledge among teams in different contexts and the teamwork involved in the joint policies pose a big challenge to achieving successful knowledge transfer. Thus, it is an urgent demand for an in-depth understanding of knowledge transfer mechanisms among different teams for complex tasks.

Drawing from the previous discussion, the primary objective of this paper is to investigate the transfer of knowledge between teams of varying sizes and the impact of knowledge quality on the knowledge transfer process in the same task domains. In addition, this paper will investigate the effect of social abilities as well. Social abilities can bring individuals closer to each other, but this may result in a loss of flexibility. Therefore, the effect of trained teams' social ability will be examined. It should be noted that in this paper, social ability is limited to observing the behavior of other agents rather than active information exchange. The following three research questions will be addressed. 1) *What are the effects of knowledge transfer between teams of different sizes?* 2) *What is the impact of knowledge quality on the transformation?* 3) *Whether the effect may change when the trained teams have the social ability?* To answer the questions, we conduct the simulations of 'L'-assembly tasks involving collision avoidance in *pygame* [19] and utilize Multiagent Deep Reinforcement Learning (MADRL) methods to train agent teams.

It is hoped that well-trained knowledge still holds the possibility of being re-deployed by different teams without further training. Additionally, the findings are expected to assist the next steps in knowledge transfer in different task domains.

The remaining sections of this paper are structured as follows. Section 2 provides a detailed description of the methodology employed in this study. In Section 3, we present the case study design and its implementation. Section 4 reports the experimental results and provides a discussion of their implications. Finally, in Section 5, we summarize the findings and highlight potential future research.

Copyright © 2023 by ASME

2 METHODS

2.1 Multiagent Reinforcement Learning

Since many applications naturally involve more than one collaborative learner in a task [20], multiagent reinforcement learning (MARL) has been broadly explored for multiple agents acting, interacting, and influencing the environment simultaneously.

In the cooperative MARL, the centralized training and decentralized implementation (CTDE) [20] addresses training agents using centralized information but executes in a decentralized manner [21]. However, it is hard for agents to select their local actions based on a global view [20]. Besides, a common policy of the agent team suffers risks in coordinating to achieve a long-run reward. Therefore, we adopt the paradigm of multiple individual learners, where each agent of the system performs its single-agent reinforcement learning algorithm [3] and holds its own deep neural network to learn from the observed information.

In MARL, several agents learn from trials in the process of maximizing the reward function based on a finite Markov Decision Process (MDP), which is a mathematical framework used to model decision-making problems. The cooperative MARL is modeled in a tuple of $< S, A, P, R, \gamma, N, t >$ to describe a multiagent system in a sequence of discrete-time steps [22]. At each time step t, N represents as the number of agents; $S_t \in S$ stands the state of the environment; $\alpha_t \in A$ is the representation of the joint actions of all agents and a_t^i stands the individual's action choice at step t. $P(s_{t+1}|s_t, a_t): S \times A \times S \rightarrow [0,1]$ is the transition function for modeling the environment and determining the environmental states. The system actions are defined as a joint set $A = A^1 \times \cdots \times A^N$ [23]. The joint policy depends on all agents' learned policies, presented in $^S\Pi \sim {}^S\Pi^1 \times {}^S\Pi^2 \times ... \times {}^S\Pi^N$. In cooperative tasks, all agents share a team reward function for a common goal since they emphasize the system interests rather than separate local awards. R presents the reward function that tells numerical scores to assess the team performance, and γ is a discount factor that is used for calculating return values. The problem of solving an MDP is to find a policy π that can maximize the accumulated reward [24].

Q-learning [25] is a popular reinforcement learning algorithm that updates the cumulative rewards of actions based on the temporal difference of value estimations, shown in equation (1).

$$Q(S_t, A_t) \leftarrow Q(S_t, A_t) \\ + \alpha[R_{t+1}\gamma max_\alpha Q(S_{t+1}, \alpha) \\ - Q(S_t - A_t)] \quad (1)$$

Deep Q-learning [26] [27] has been developed in recent years to replace the Q-table with a Q-network with weights θ_i to process complex situations in an end-to-end way. The Q-network updates its weights θ_i at each iteration i by minimizing the loss function shown as:

$$L_i(\theta_i) = E[(R_{t+1} + \gamma \, max_\alpha \, Q(S_{t+1}, \alpha) - Q(S_t, A_t))^2] \quad (2)$$

In individual learning, each agent views others as part of the environment and conducts training through the Deep Q-learning algorithm [27] to process complex situations in an end-to-end way [3] [28]. The loss function is shown in equation (3) in ith agent's Q-network updates its weight θ_j^i at each iteration j.

$$L_j(\theta_j^i) = E\left[(r_{t+1} + \gamma \, max_{a^i} \, Q(s_{t+1}, a_{t+1}^i \mid {}^-\theta_j^i) \\ - Q(s_t, a_t^i \mid \theta_j^i))^2\right] \quad (3)$$

where θ_j^i is the parameters of the target networks and θ_j^i is the parameters of the online network.

The cooperative task domains are considered to maximize the accumulative rewards resulting from the joint actions of all agents over time. Their separate neural networks are trained independently and simultaneously to capture the cooperative behavior codes in a dynamic environment.

2.2 Integrate Social Ability in Multiagent Systems

Many experimental studies have shown the limitation of individual learning [29] [30]. Instinctively, ignoring other members' existence may limit teamwork. Therefore, in order to boost the capability of an agent team, we integrate social ability into the multiagent systems. In this paper, the social ability is referred to as "being aware of the existence and statuses of other teammates." It is hoped to simulate real practices when robots can send and receive signals via sensors to broadcast and communicate [31].

Agents conduct learning based on the information they receive [32]. In independent learning, agents only have task-relevant information, such as the environment features, task progress, and objects, as shown in Figure 1. However, the information is usually deficient for conducting complex tasks or coordinating a large team. Thus, *social views* are introduced, which indicate the interactional information among team members received by individual agent. As shown in Figure 1, such interactional information can be the agent's ID, name, and policy in term of Q-tables, neural networks, action-state trajectories and etc.

In real world scenarios, robots can utilize sensors to send and receive signals which allow them to hold broad views. Thus, by providing both task-related and social-related information to agents during the simulation, we can enable them to exhibit social abilities, which refers to the team's ability to synchronize information among team members at each step. The task-related and social-related information are obtained from a robot's *task views* and *social views* respectively. By holding social views, agents are able to conduct social learning and are the potential to output better teamwork.

Copyright © 2023 by ASME

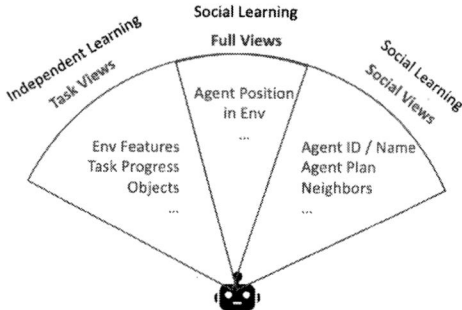

Figure 1: Task Views and Social Views

2.3 Multiagent Reinforcement Learning with Social Ability

In MARL, the information that each agent receives is represented in the *state*. Each agent trans its own neural networks to master how to behave based on the received information, as shown in equation (3). Therefore, a social state ^{K}S is introduced to represent social information. It is defined based on the robot's social ability. The task-relevant information is saved in the environmental state ^{E}S. They are appended together as ^{S}S to work on the training and iteration.

$$^{S}S = {}^{E}S \frown {}^{K}S \tag{4}$$

Therefore, the loss function for *ith* agent is updated with the new *state* shown as below:

$$L_t^i\left({}^{S}\theta_t^i\right) = E_{s,a,r,\,{}^{S}s'}\left[\left(r_t^i + \gamma \max_{a'} Q\left({}^{S}s_t^{i'},\, a'\middle|\, \theta_{t-1}^i\right)\right.\right.$$
$$\left.\left. - Q\left({}^{S}s_t^i,\, a\middle|\, \theta_t^i\right)\right)^2\right] \tag{5}$$

2.4 Knowledge Transfer between Different Team Sizes

Reinforcement learning can be applied to capture domain knowledge for various tasks [18]. However, the cost is high for training repetitively under similar circumstances. It is desired that earned knowledge can be reused and transferred among different teams with high knowledge quality. In that way, the knowledge can be applied more efficiently and costly.

In MADRL, the task knowledge is embedded in well-trained neural networks (NNs). The parameters and structure of the NNs save information on how to behave in a certain task. Therefore, we will explore the mechanism and performance of well-trained NNs to be transferred to other teams with different team sizes in performing the same task. The results are expected to provide hints on the mechanism of knowledge transfer in the same task domain.

In this paper, the teams that are trained from the novice are named Trained Teams (Tr), while the teams that deploy the Trained Teams' knowledge are named Test Teams (Ts). Correspondingly, the size of a Trained Team is written as #Tr, while the size of a Test Team is represented as #Ts. For

convenience, the knowledge transferred from Tr to Ts in a task is represented as $NNs_{\#Tr \to \#Ts}^{task}$.

Based on the cases, there are three possible circumstances may occur.

- C1: #Tr == #Ts
- C2: #Tr > #Ts
- C3: #Tr < #Ts

In C1, the two team sizes are identical. Thus, the knowledge is transferred correspondingly from one agent in Tr to the other in Ts. In C2 and C3, the sizes of the two teams are different. When #Tr > #Ts, Test Teams randomly select $NNs_{\#Tr \to \#Ts}^{task}$ to apply [33]. When #Tr > #Ts, Test Teams repeatedly deployed $NNs_{\#Tr \to \#Ts}^{task}$ [33].

3 EXPERIMENT DESIGN

3.1 "L"-Shape Assembly Task involving Collision Avoidance

Precision assembly plays an important role in engineering. For instance, one product needs to be assembled from parts. During the process, it is desired to avoid any collisions to ensure safety [34]. Therefore, we will focus on an "L"-shape assembly involving collision avoidance task to simulate some scenarios in a factory.

As shown in Figure 2, the "L-shape" assembly task involving collision avoidance requires an agent team to transport a Dynamic Box and assemble it with a Target Box to form an "L" shape. During the task process, the Dynamic Box needs to avoid any collisions with surrounding walls or obstacles in the field.

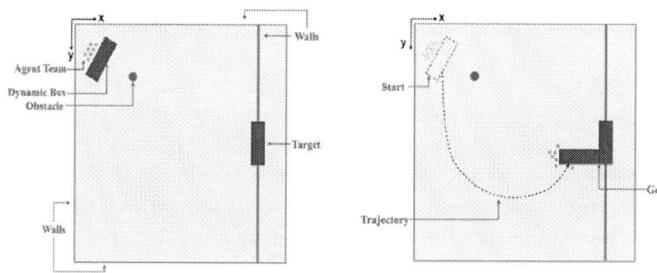

Figure 2: Task Description

The task environment is simulated by *pygame* [19] as a 1000×1000 pixels field, including a Dynamic Box, a static Target Box, obstacles, and four surrounding walls. The Dynamic Box (the upper-left brown rectangle) can be pushed by agent teams, and its mass is defined as 1 kg. The Target Box sits in the middle of the right wall and stays static during the task implementation process. The sizes of the dynamic and target boxes are 180×60 pixels. The agent team represented as green squares need to spontaneously organize themselves and find a suitable trajectory to assemble two boxes in the "L" configuration. At the same time, they are required to guarantee to avoid collisions of the dynamic box on the walls or obstacles. Figure 2 illustrates the task start and goal status. In this paper, we will focus on solving

Copyright © 2023 by ASME

two tasks with one static obstacle and three obstacles, respectively. As shown in Figure. 3, the task with one obstacle in the field is named as Task A, while another one contains three obstacles is named as Task B. The details of the settings are provided in Table 1.

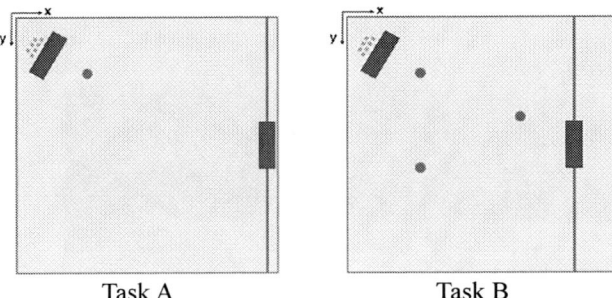

Figure 3: Task A and B Illustrations

In assembly tasks, how to achieve high precision of final configuration is critical for a successful task. That put forwards the high requirements on the strategies and optimal policy learned by the agent teams. In this task, they not only need to cooperatively control the Dynamic Box's orientation and displacement but also to look for a suitable trajectory avoiding obstacles on their way. Physically, the box's movement depends on the joint impulses of all agents at each step. Thus, the well-trained knowledge becomes valuable to be deployed for further steps and the high efficiency of using knowledge are crucial.

Table 1: Task Settings

Task Name	Task A	Task B
Field size	1000 * 1000	1000 * 1000
Box size	180 * 60	180 * 60
Target box size	180 * 60	180 * 60
Box start center coordination	(150, 180)	(150, 180)
Target box center coordination	(950, 500)	(950, 500)
Obstacle center coordination	(350, 170)	(350, 170)
		(350, 370)
		(500, 270)
Box mass (kg)	1	1
Push impulse ($1\ N \cdot s$)	1	1

Previous experiments have demonstrated that Task A can be accomplished by individual learning teams, while Task B requires agent teams with social ability, as the three obstacles in Task B leave little room for the dynamic box to pass through. Therefore, Task A is devised to address the first research question about the impact of team size on the knowledge transfer process, while Task B is developed to investigate the other two research questions about the effects of knowledge quality and the team's social ability. This paper aims to investigate the knowledge transfer mechanism within the same task domain by addressing the research questions. The goal is to enhance group intelligence in order to successfully accomplish increasingly complex tasks.

3.2 State Space and Action Space

The self-organizing system is trained by Deep Q-learning, and an ε–$greedy$ strategy is applied to explore optimal policy [35]. The hyperparameters of the algorithm are shown below in Table 2.

Table 2: Hyperparameters of training

Training episodes	20,000
Discount factor	0.99
Memory buffer size	1000
Mini-batch size	32
Target network update frequency	200
Learning rate	0.001
Neural network size	(63, 64, 128, 6)
epsilon	1 → 0.01

Action Space: In assembly tasks, agent teams are capable of transporting the Dynamic Box through pushing. Therefore, the surrounding area of the dynamic box is divided into six regions, shown in Figure 4. During the simulation, an individual agent can select one of six positions to push the box with 1 N·s impulse. As each agent is relatively small, there can be multiple agents in the same region at each step.

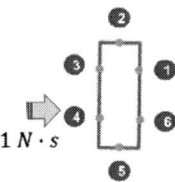

Figure 4: Action Space

Therefore, the action space is defined as:

$$A = <a_1, a_2, a_3, a_4, a_5, a_6> \qquad (6)$$

where a_i means push i position with impulse of $1\ N \cdot s$

State Space: The *environmental state* ^ES is set as a 63-digit tuple, covering the information of the vicinity situation around the Dynamic Box, and Dynamic Box's velocity in x-direction, y-direction, and angular velocity. Thus, the *environmental state* can be defined as:

$$^ES = <Vicinity\ Situation, v_X, v_Y, \omega> \qquad (7)$$

where *Vicinity Situation* is represented in a tuple with a binary number ("0" or "1") to show if there exist obstacles/target boxes around the Dynamic Box. It can be sensed by sensors on the Dynamic Box in the range of 200-pixel range.

The *social state* ^SS is represented in a tuple to show social information. As known from the action space setting, an agent stands in one of six regions to give impulse at each step. Therefore, it can be defined that agents with social abilities are able to observe if there exist other team members in all other regions. And use one indicator I to show the existence. When I equals 1, it means the region is pushed by some team members. When I equals 0, it means that no one has pushed in this region. Thus, the *social state* is shown in equation (8).

Copyright © 2023 by ASME

$$^{S}S_i = < I_{left2}, I_{left1}, I_{opposite}, I_{right1}, I_{right2}, > \qquad (8)$$

where I_{left1} and I_{right1} are indicators to show if the left and right adjacent areas of the *ith* agent are pushed by someone. I_{left2} and I_{right2} point to the secondary adjacent areas. $I_{opposite}$ points to the opposite area of the *ith* agent. It is worth mentioning that the *left* and *right* is the relevant position from the views of the *ith* agent.

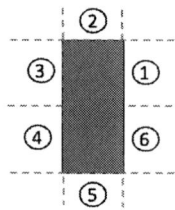

Figure 5: Region Division around the Dynamic Box

3.3 Knowledge Transfer between Teams

In the experiments, our focus is on the knowledge transfer mechanism between teams of varying sizes, as well as the impact of social ability on the transfer process. Therefore, #Tr is fixed as 10 to be trained from novice to master Task A and Task B separately. In order to successfully complete Task B, the 10-agent teams can hold the social ability and use both *environmental* and *social* states for training. #Ts is set as 2, 4, 6, 8, 12, 14, 16, and 18 for each group of study.

3.4 Knowledge Quality

The knowledge quality is assessed based on how well the task is performed by the Test Teams and the difference between the task performances of the Train Team and its corresponding Test Team.

In the "L"-shape assembly task, the task performance will be evaluated through the final configurations. The final configurations can be represented in two angles [28], which are self-angle α and relevant angle β. They both can be defined as the relevant positions of two boxes and sensed by the sensors on the Dynamic Box, shown in the Figure. 6.

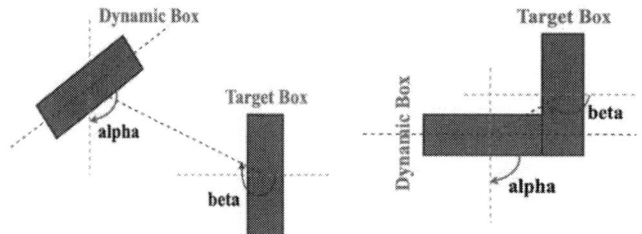

Figure 6: Angles for Defining Final Configurations

A perfect "L-shape" means $\alpha = 90°$ and $\beta = 135°$, represented by a red dot shown in Figure 6.

$$\alpha_{goal} = 90° \qquad (9)$$
$$\beta_{goal} = 135° \qquad (10)$$

We can plot them in a scatter plot [28] as shown in Figure 7. The task results of learning teams are presented in a coordinate system whose x-axis is angle α, the y-axis is angle β, and one dot presents one learning team's task result.

The center red dot means the goal. Other dots are the task results of the trained agent teams. The closer to the red dot the results are, the better quality of the final "L-shape" is. In each group, we conduct 20 cases with different random seeds 1-20.

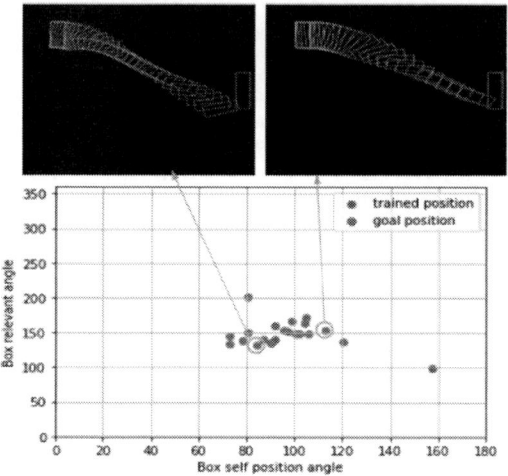

Figure 7: Task Performance

Furthermore, the statistical aspects can be assessed by calculating the Euclidean Distance between a learning team's performance and the goal in the space of $< \alpha, \beta >$, which can be expressed as:

$$D_i = \sqrt{(\alpha_i - \alpha_{goal})^2 + (\beta_i - \beta_{goal})^2} \qquad (11)$$

Where α_i and β_i are obtained from the ith learning team's final configuration. and α_{goal} and β_{goal} are known from equations (9) and (10). The Euclidean distance measure assists in understanding the knowledge quality and further processing the results. Smaller distance values indicate better final configurations, while zero distance means a perfect "L-shape."

4 RESULTS AND DISCUSSION

4.1 Effect of knowledge transfer among teams with different sizes

As introduced, well-trained knowledge is embedded in neural networks and can be transferred among different teams. In this study, the effect of transferring between different team sizes is explored.

Figure. 8 presents the 10-Agent teams' training results in Task A as a baseline, while Figure. 9 shows the results of transferring $NN^{TaskA}_{\#Tr=10}$ (knowledge of 10-Agent Teams in Task A) into Test Teams with the size of 2, 4, 6, 8, 12, 14, 16, and 18, respectively. It can be observed that the green dots are distributed densely around the red dot (goal), meaning that the quality of

Copyright © 2023 by ASME

$NN_{\#Tr=10}^{TaskA}$ is pretty high, and the Training Teams can implement a very good "L"-assembly.

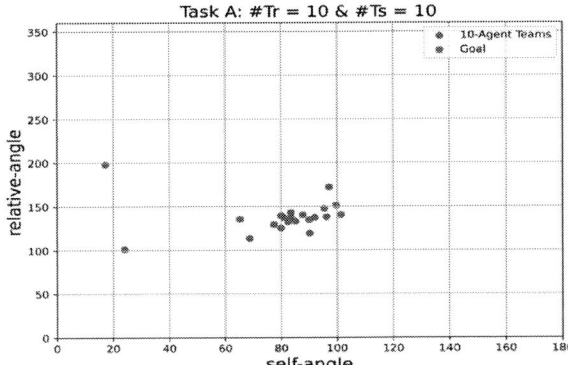

Figure 8: Final Configurations of 10-Agent Teams in Task A.

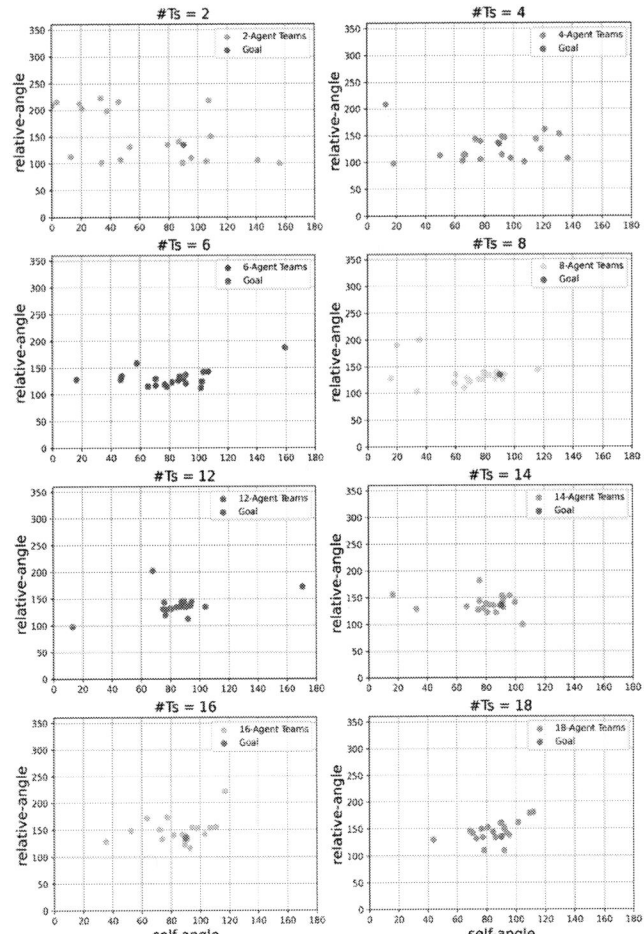

Figure 9: Final Configurations of transferring $NNs_{\#Tr=10}^{TaskA}$ to teams with size of 2, 4, 6, 8, 12, 14, 16, 18 in Task A.

The sub-figures in Figure. 9 show the cases utilizing $NNs_{\#Tr=10}^{TaskA}$ in teams of #Ts = 2 to #Ts=18. The first four sub-

figures show the C2 situation (#Tr > #Ts). It can be seen that the distribution of experimental dots becomes sparser as the #Ts decreases. Especially in the case of (#Tr = 2), nearly no Test Team can output a good performance. The last four sub-figures present the C3 situation (#Tr < #Ts). In the C3, the distributions remained as dense as the Training Team's performance. Even in the case of (#Ts = 18), it still shows a high quality of knowledge.

The results show the trend that the knowledge quality is largely hurt when transferred to smaller teams, but it is easy to remain or only slightly decrease when transferred to larger teams.

The trends are demonstrated clearly when the results are processed quantitively in the *Euclidean distance*. In Figure. 10, the x-axis shows the #Ts from 2 to 18, and the y-axis shows the *Euclidean distance* value. Results are plotted in boxplots, where the box area shows the data range from the first quartile (Q1) to the third quartile (Q3). The flier points shown in Fig. 10 are those out of the 1.5 * IQR (Q3-Q1). It can be observed that cases in the C3 (#Ts = 12, 14, 16, and 18) keep relatively similar results compared with the Training Team (#Tr = #Ts = 10). However, the training results are worse with larger *medians* and *deviations* in the C2 (#Ts = 2, 4, 6, and 8).

The tendency provides hints on the mechanism of knowledge transfer between different teams. Members of Training Teams master the behavior codes in respect of their own specialization. In that way, the teams can cooperatively complete the tasks.

When the knowledge is transferred to larger teams, the labor division can be relatively flexible, which increases the possibility of success. The Test Teams can still function even if partial agents cannot perform well. However, knowledge transfer becomes challenging when Test Teams have fewer team members. The $NN_{\#Tr=10}^{TaskA}$ holds a certain amount of knowledge of a specific role in a team. However, a smaller Test Team size means that some knowledge cannot be inherited. It requires members in smaller Test Teams to be more capable in the task performed. That increases the difficulty greatly and influences the transfer quality.

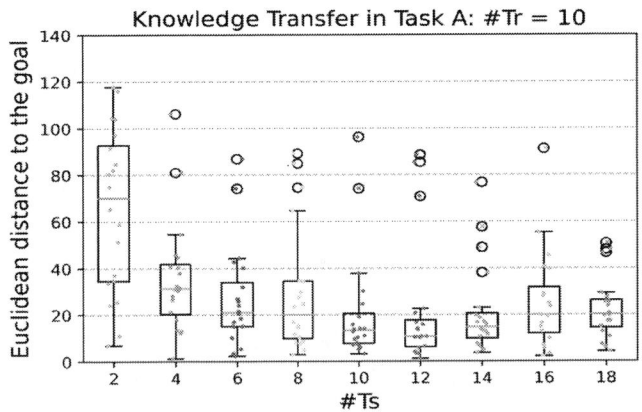

Figure 10: Boxplot of Euclidean Distance of transferring $NNs_{\#Tr=10}^{TaskA}$ to teams with size of 2, 4, 6, 8, 10, 12, 14, 16, 18 in Task A.

Copyright © 2023 by ASME

4.2 Impact of knowledge quality in transforming

In Section 4.1, the 10-Agent Training Teams can well master how to perform Task A and hold a great knowledge quality. In this study, we will look into the impact of knowledge quality on the transforming process. Thus, the Training Teams are trained to perform the more complicated task, Task B.

As introduced in Section 3.1, Task B is difficult to complete since three obstacles only leave very narrow space for the Dynamic Box to pass through. Agent teams are supposed to have high-quality cooperation to avoid collisions and assembly of an "L"-shape at the same time.

The training and transferred results in Task B are shown in the Figure. 11, where the fifth bar presents the performance of Training Teams and others represent the transferred cases. Comparing the teams' performance in Task A in Figure. 10, the performance of 10-Agent Training Teams in Task B has a higher *median*, larger *deviation*, and worse *outliers*. It indicates that the trained knowledge quality is lower.

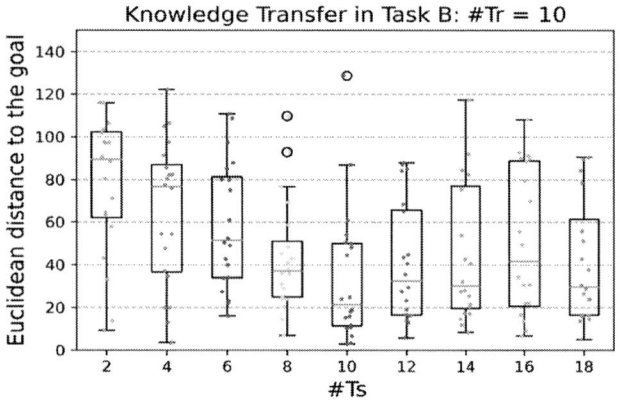

Figure 11: Boxplot of Euclidean Distance of transferring $NNs^{TaskB}_{\#Tr=10}$ to teams with size of 2, 4, 6, 8, 10, 12, 14, 16, & 18 in Task B.

Under this circumstance of lower quality of trained knowledge, it can be seen that the transferred knowledge qualities are all not satisfied in both C2 and C3 conditions in Figure. 11. The dots are very sparse and have large *deviations*.

It indicates that the trained knowledge quality is crucial for knowledge transfer. When the trained knowledge quality is low, the Test Teams' performances become worse regardless of Test Team size.

4.3 Impact of social ability on knowledge transfer

In order to complete Task B, we add the social ability in the agent teams via the method in Section 3.2. Figure. 12 shows the performance of 10-Agent Training Teams with social ability in Task B. It can be seen that the teams' performance has improved greatly by broadening the social views. The dots distribute much denser, which indicates the trained knowledge quality is good.

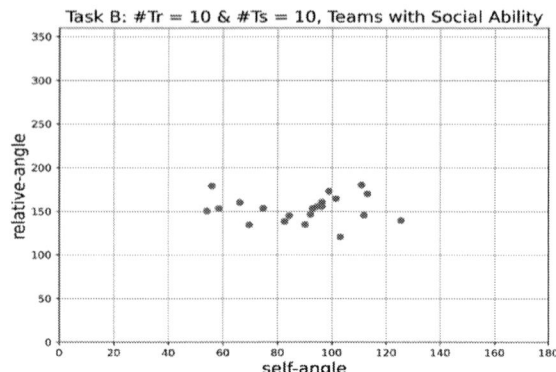

Figure 12: Final Configurations of 10-Agent Teams with Social Abilities in Task B.

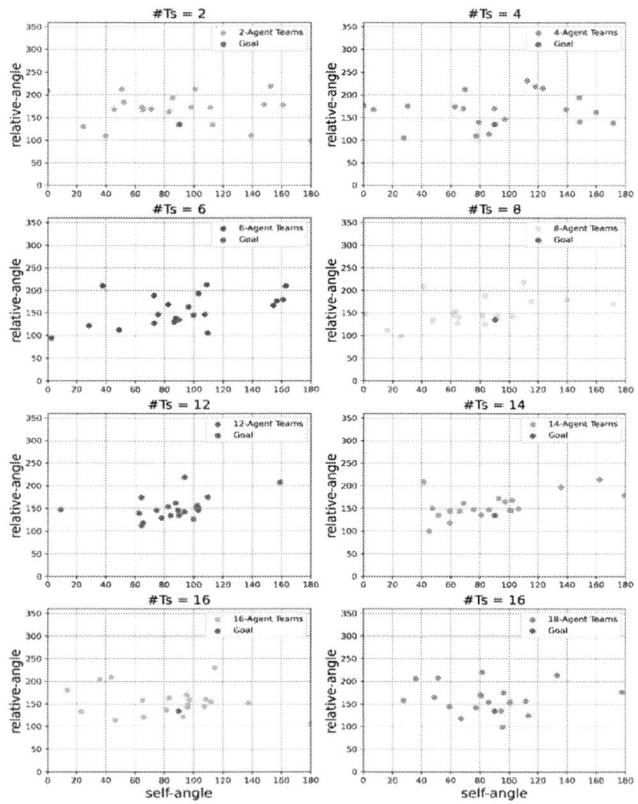

Figure 13: Final Configurations of transferring $NNs^{TaskB}_{\#Tr=10}$ to teams with social abilities with the size of 2, 4, 6, 8, 10, 12, 14, 16, and 18 in Task B.

Similarly, sub-figures in Figure. 13 shows the output of transferring $NN^{TaskB}_{\#Tr=10}$ in Test Teams in sizes of 2, 4, 6, 8, 12, 14, 16, and 18, respectively. The results show that only (#Tr=12) and (#Tr=14) groups have good outputs, keeping dense distributions, while most of the Test Teams (#Tr = 2, 4, 6, 8, 16, and 18) all suffer the failure in utilizing the $NN^{TaskB}_{\#Tr=10}$. The trend can be clearly shown in Figure. 14 as well.

Copyright © 2023 by ASME

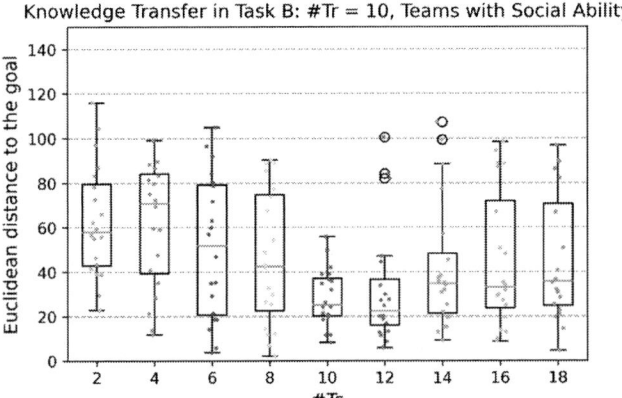

Figure 14: Boxplot of Euclidean Distance of transferring $NNs_{\#Tr=10}^{TaskB}$ to teams with social abilities with the size of 2, 4, 6, 8, 10, 12, 14, 16, and 18 in Task B.

The results indicate that social ability does have influences on knowledge transfer. Even though the trained knowledge quality is high, the transferring difficulty remains high when Trained Teams have the social abilities.

Social ability does help the Training Teams to be more capable of complex tasks; however, the learned strategies are complicated to process *environmental* and *social* information. Most likely, the learned policy relies more on teamwork and cooperation, leading the knowledge becomes more sensitive to the team size. That explains why the knowledge trained from a team with social ability is only easily transferred to similar team sizes.

5 CONCLUSION

This paper has discussed the knowledge transfer among the different sizes of teams in the context of "L"-shape assembly tasks involving collision avoidance in self-organizing systems. What's more, it also examines how the mechanism is affected when the training teams possess social abilities. Based on the findings presented in this paper, we can draw several conclusions that address the research questions outlined in the first section:

- The well-trained knowledge is easier to be transferred from trained teams to larger test teams compared to smaller teams. When transferring learned knowledge to smaller teams, the knowledge quality is hurt largely.
- The trained knowledge quality is crucial for knowledge transfer. When the trained knowledge quality is low, it is hard to be transferred with satisfactory quality.
- The presence of social ability significantly influences the knowledge transfer mechanism. The transfer process is highly sensitive to the team size, highlighting the need for careful consideration of these factors in the design and implementation of self-organizing systems.

The findings of this paper provide a comprehensive insight of the mechanism of knowledge transfer between different teams in the same problem domains. The results demonstrate that well-trained knowledge can be deployed by different teams, and the transfer success is contingent upon the knowledge quality and the training team's social ability. Future research will expand to explore knowledge transfer across different task domains.

ACKNOWLEDGEMENTS

This paper is based on the work supported by the Autonomous Ship Consortium (ASC) with members of BEMAC Corporation, ClassNK, MTI Co. Ltd., Nihon Shipyard Co. (NSY), Tokyo KEIKI Inc., and National Maritime Research Institute of Japan. The authors are grateful for their support and collaboration on this research.

REFERENCES

[1] Crommelinck, Marc, Bernard Feltz, and Philippe Goujon, eds. Self-organization and emergence in life sciences. Dordrecht, The Netherlands:: Springer, 2006.

[2] Klir, George J., and W. Ross Ashby. "Requisite variety and its implications for the control of complex systems." Facets of systems science (1991): 405-417.

[3] Ji, Hao, and Yan Jin. "Designing Self-Organizing Systems With Deep Multi-Agent Reinforcement Learning." International Design Engineering Technical Conferences and Computers and Information in Engineering Conference. Vol. 59278. American Society of Mechanical Engineers, 2019.

[4] Jin, Yan, and Chang Chen. "Cellular self-organizing systems: A field-based behavior regulation approach." AI EDAM 28.2 (2014): 115-128.

[5] Brambilla, Manuele, et al. "Swarm robotics: a review from the swarm engineering perspective." Swarm Intelligence 7 (2013): 1-41.

[6] Zouein, George, Chang Chen, and Yan Jin. "Create adaptive systems through "DNA" guided cellular formation." Design Creativity 2010. Springer London, 2011.

[7] Chen, Chang, and Yan Jin. "A behavior based approach to cellular self-organizing systems design." International Design Engineering Technical Conferences and Computers and Information in Engineering Conference. Vol. 54860. 2011.

[8] Chiang, Winston, and Yan Jin. "Toward a meta-model of behavioral interaction for designing complex adaptive systems." International Design Engineering Technical Conferences and Computers and Information in Engineering Conference. Vol. 54792. 2011.

[9] Khani, Newsha, James Humann, and Yan Jin. "Effect of social structuring in self-organizing systems." Journal of Mechanical Design 138.4 (2016): 041101.

[10] Tao, Yunzhe, et al. "Repaint: Knowledge transfer in deep reinforcement learning." International Conference on Machine Learning. PMLR, 2021.

Copyright © 2023 by ASME

[11] Parisotto, Emilio, Jimmy Lei Ba, and Ruslan Salakhutdinov. "Actor-mimic: Deep multitask and transfer reinforcement learning." arXiv preprint arXiv:1511.06342 (2015).

[12] Liu, Xiongqing, and Yan Jin. "Reinforcement learning-based collision avoidance: impact of reward function and knowledge transfer." AI EDAM 34.2 (2020): 207-222.

[13] Liu, Wenzhang, et al. "Knowledge transfer in multi-agent reinforcement learning with incremental number of agents." Journal of systems engineering and electronics 33.2 (2022): 447-460.

[14] Hu, Yujing, Yang Gao, and Bo An. "Learning in Multi-agent Systems with Sparse Interactions by Knowledge Transfer and Game Abstraction." AAMAS. 2015.

[15] Wang, Xinrui, and Yan Jin. "Work Process Transfer Reinforcement Learning: Feature Extraction and Finetuning in Ship Collision Avoidance." International Design Engineering Technical Conferences and Computers and Information in Engineering Conference. Vol. 86212. American Society of Mechanical Engineers, 2022.

[16] Taylor, Matthew E., and Peter Stone. "Transfer learning for reinforcement learning domains: A survey." Journal of Machine Learning Research 10.7 (2009).

[17] Phillips, Caitlin. Knowledge transfer in Markov decision processes. Technical report, McGill University, School of Computer Science, 2006. URL http://www. cs. mcgill. ca/cphill/CDMP/summary. pdf, 2006.

[18] Glatt, Ruben, Felipe Leno Da Silva, and Anna Helena Reali Costa. "Towards knowledge transfer in deep reinforcement learning." 2016 5th Brazilian Conference on Intelligent Systems (BRACIS). IEEE, 2016.

[19] P. Shinners, "Pygame-Python Game Development. Retrieved from http://www.pygame.org,".

[20] Chen, Gang. "A New Framework for Multi-Agent Reinforcement Learning--Centralized Training and Exploration with Decentralized Execution via Policy Distillation." arXiv preprint arXiv:1910.09152 (2019).

[21] Lyu, Xueguang, et al. "Contrasting centralized and decentralized critics in multi-agent reinforcement learning." arXiv preprint arXiv:2102.04402 (2021).

[22] Buşoniu, Lucian, Robert Babuška, and Bart De Schutter. "Multi-agent reinforcement learning: An overview." Innovations in multi-agent systems and applications-1 (2010): 183-221.

[23] Wang, Ying, and Clarence W. De Silva. "Multi-robot box-pushing: Single-agent q-learning vs. team q-learning." 2006 IEEE/RSJ international conference on intelligent robots and systems. IEEE, 2006.

[24] Devlin, Sam, Daniel Kudenko, and Marek Grześ. "An empirical study of potential-based reward shaping and advice in complex, multi-agent systems." Advances in Complex Systems 14.02 (2011): 251-278.

[25] Watkins, Christopher John Cornish Hellaby. "Learning from delayed rewards." (1989).

[26] Mnih, Volodymyr, et al. "Playing atari with deep reinforcement learning." arXiv preprint arXiv:1312.5602 (2013).

[27] Mnih, Volodymyr, et al. "Human-level control through deep reinforcement learning." nature 518.7540 (2015): 529-533.

[28] Huang, Bingling, and Yan Jin. "Reward shaping in multiagent reinforcement learning for self-organizing systems in assembly tasks." Advanced Engineering Informatics 54 (2022): 101800.

[29] Du, Yali, et al. "Liir: Learning individual intrinsic reward in multi-agent reinforcement learning." Advances in Neural Information Processing Systems 32 (2019).

[30] Wang, Li, et al. "Individual Reward Assisted Multi-Agent Reinforcement Learning." International Conference on Machine Learning. PMLR, 2022.

[31] V Klingspor, Volker, John Demiris, and Michael Kaiser. "Human-robot communication and machine learning." Applied Artificial Intelligence 11.7 (1997): 719-746.

[32] Ndousse, Kamal K., et al. "Emergent social learning via multi-agent reinforcement learning." International Conference on Machine Learning. PMLR, 2021.

[33] Ji, Hao, and Yan Jin. "Evaluating the learning and performance characteristics of self-organizing systems with different task features." AI EDAM 35.4 (2021): 404-422.

[34] Ji, Sanghoon, et al. "Learning-based automation of robotic assembly for smart manufacturing." Proceedings of the IEEE 109.4 (2021): 423-440.

[35] Wunder, Michael, Michael L. Littman, and Monica Babes. "Classes of multiagent q-learning dynamics with epsilon-greedy exploration." Proceedings of the 27th International Conference on Machine Learning (ICML-10). 2010.

Proceedings of the ASME 2023
International Design Engineering Technical Conferences and
Computers and Information in Engineering Conference
IDETC-CIE2023
August 20-23, 2023, Boston, Massachusetts

DETC2023-117285

FRAMING WICKED PROBLEMS THROUGH EVIDENTIARY AND INTERPRETATIVE ANALYSIS

Mayank J. Bhalerao
Graduate Research Assistant
The Systems Realization Laboratory @ OU
University of Oklahoma, Norman, OK, USA

Wesley T. Honeycutt
Research Associate
OU GeoCarb
University of Oklahoma, Norman, OK, USA

Ashok K. Das
Founder, CEO
SunMoksha Power Pvt. Ltd.
Bengaluru, INDIA

Janet K. Allen[1]
John and Mary Moore Chair and Professor
The Systems Realization Laboratory @ OU
University of Oklahoma, Norman, OK, USA

Farrokh Mistree
L.A. Comp Chair and Professor
The Systems Realization Laboratory @ OU
University of Oklahoma, Norman, OK, USA

ABSTRACT

Wicked problems are characterized by incomplete and conflicting information. To frame a wicked problem, it is necessary to analyze the interaction between variables and thence identify a reduced set of variables that are key to designing a socio-economic-technical system.

In this paper we propose using a combination of interpretative and evidentiary analysis through the application of Dilemma Triangle Method and System Dynamics, respectively. We propose a computational framework that allows a designer to convert heuristics into insights by using System Dynamics modelling, thus allowing a designer to analyze the interaction between variables. Further, our framework is based on the notion of involving human-in-the-loop, wherein wicked problems are framed through synergistic actions between a human -and a computer. The benefits of using this framework are

- *Converting heuristics into insights,*
- *Understanding the interaction between variables by analyzing the behavior of the system,*

- *Identifying the correct size of the problem by eliminating the variables that cannot be used to design a socio-economic-technical system.*

To demonstrate the efficacy of the framework we use data pertaining to Kantashol village in Jharkhand, India. The data was provided by SunMoksha Power Pvt. Ltd. Our focus in this paper is on describing the framework rather than the results on the ground in India.

Keywords: Wicked Problems, Evidentiary Analysis, Interpretative Analysis, Heuristics, Human-in-the-loop, System Dynamics, Dilemma Triangle Method

GLOSSARY

Wicked Problem: A class of problems which are ill formulated, where the information is confusing and conflicting, where there are many clients and decision makers with conflicting values, and where the ramifications in the whole system are thoroughly confusing [1].

Decision Maker: An individual who can affect, through his/her decisions, the achievement of objectives for an organization.

---

[1] Corresponding author: janet.allen@ou.edu

Copyright © 2023 by ASME

Stakeholder: An individual who can affect or is affected by the achievement of the objectives for an organization.

Framing: Identifying the problem correctly before solving a problem to ensure that the problem are correctly addressed.

Interpretative Analysis: An approach for analyzing qualitative data that involves exploring and interpreting the meaning of data from the perspectives of individuals/actors involved [2].

Evidentiary Analysis: Analysis that involves the systematic process of collecting, examining, and evaluating data and analyzing it with rigorous research methods, in order to provide decision support through that evidence [3].

Heuristics: Heuristics are the assumptions, experiences, domain expertise, that are applied in a way to hasten the process of approaching a solution.

Thematic Area: An area or category in which issues related to the same subject are considered.

Human-in-the-loop: Humans act as an embedded component in the system where their intent, emotions, cognition, etc. are intrinsic part of the computational system [4,5].

1. FRAME OF REFERENCE

1.1 Wicked Problems – Definition, Characteristics and Broader Impacts

Horst Rittel defines wicked problems as 'a class of social system problems which are ill formulated, where the information is confusing, where there are many clients and decision makers with conflicting values, and where the ramifications in the whole system are thoroughly confusing', which is considered as one of the earliest definitions of wicked problem; see editorial by Churchman, [1]. Rittel and Weber in their seminal paper emphasize the notion of focusing on the nexus of goal formulation, problem definition and equity issues. Social processes are seen as links connecting the open systems into large, interconnected networks which follow continuity of input-output relations. Rittel and Weber enunciate the importance of correctly identifying and framing the wicked problem by stating *"In that structural framework it has become less apparent where problem centres lie, and less apparent where and how we should intervene even if we do happen to know what aims we seek"* [6]. Further, they state that describing and locating the problem is one of the most challenging and intractable difficulties to address. The ten characteristics of a wicked problem recognized by Rittel and Weber are:

i. *There is no definitive formulation of wicked problems. They are difficult to frame.*[2]
ii. *Wicked problems have no stopping rule.*
iii. *Solutions to wicked problems are not true or false, but good or bad.*
iv. *There is no immediate and no ultimate test of a good solution to a wicked problem.*
v. Every solution to a wicked problem is a "one-shot operation"; because there is no opportunity to learn by trial-and-error, every attempt counts significantly.
vi. Wicked problems do not have an enumerable (or an exhaustively describable) set of potential solutions, nor is there a well-described set of permissible operations that may be incorporated into the plan.
vii. Every wicked problem is essentially unique.
viii. Every wicked problem can be a symptom of another problem.
ix. *There are numerous explanations for a wicked problem.*
x. *The planner has no right to be wrong.*

The characteristics that we address in this paper are italicized. Rittel and Weber contend that a systems approach is appropriate to frame wicked problems. They argue that for wicked problems one cannot understand the problem without knowing about its context and that the systems approach of the 'first generation' is futile to deal with wicked problems. Accordingly, a designer might be overwhelmed and feel paralyzed about addressing wicked problems. We believe that wicked problems can be addressed by identifying the correct size of the problem, instead of getting overwhelmed by its notion. Paralysis occurs when one acts too reflexively and considers wicked problems so overwhelming that it discourages them from doing anything about it [7]. Termeer et al. emphasize the significance of small wins to tackle wicked problems and its value in bringing in transformative change [8]. We agree with the notion that it is important for a designer to take small steps to address the "wickedness" embodied wicked problems. This is reflected in our proposed approach for framing a wicked problem.

The United Nations General Assembly adopted the 2030 Agenda for Sustainable Development [9] where the major focus is on "transform[ing] the world to better meet human needs" and "leave no one behind and create a world of dignity". "We need to tackle root causes and do more to integrate the economic, social and environmental dimensions of sustainable development." In a subsequent editorial published in Nature in 2020, the writers contend that the world is almost set to miss all the goals except two, namely, "eliminating preventable deaths among newborns and under-fives," and "getting children into primary schools", which are the closest to being achieved [10]. Eden et. al. argue that irrespective of Covid 19 pandemic, the agenda to achieve the goals was inevitable due to the fact that the United Nations are addressing issues which are wicked problems [11]. Instead of trying to solve wicked problems, designers/ policymakers should focus on managing or coping with the wicked problems [12]. We believe that in order to manage or cope with wicked problems it is important to correctly frame the problem at the start of a design process. This has broader impacts in terms of correctly identifying the problem and going to the core of the wickedness of the problems in order to solve them. In this paper we consider an example of wicked problem in a village

---

[2] We have italicized the characteristics that we have considered in the proposed framework.

Copyright © 2023 by ASME

in India, namely, Kantashol. The three thematic areas that we consider in framing the Kantashol wicked problem are water, forestry, and agriculture. These thematic areas are anchored in demographics, culture and socioeconomics associated with this village. We require a comprehensive understanding when we deal with wicked problems with the goal of sustainable development. We observe that water, forestry, and agriculture are interdependent areas with challenges that are intertwined, however, they are approached separately in silos [13]. We suggest that wickedness of the problem lies in modeling the synergy between the different thematic areas of consideration that makes it essential to initially frame the wicked problem and correctly identify the variables of consequence and its size. With the goal to provide a framework to frame wicked problems as well as anchor with sustainable development goals we select three Drivers[3] for our problem namely, People, Planet and Prosperity[4] [14].

1.2 Interpretative Analysis and Evidentiary Analysis

Various authors have commented on the role of interpretative and evidentiary analysis in approaching wicked problems. Evidence-based analysis for public policy analysis is mired in debates in terms of its utility. Several authors stress the need for an orderly approach and explore the evidentiary analysis to aid policy making [15]. Daviter, argues that the main aim of evidence-based analysis, that is evidentiary analysis, is to provide design options for conflicting interpretations by enabling analytical tasks to be more objectifiable [16]. Various authors in the past argue that the problems that are ill structured (wicked problems) are not open to analytical methods. Strong evidence seldom contributes in the analysis of problems when the boundaries are not well defined [17]. Authors who contest evidentiary analysis assert that the evidence is often value-laden which suggests that the evidence is more likely based on biased conclusions [18]. Authors argue that, with this notion, evidence-based analysis for wicked problems like public policy are arrived through an order of ranking of technological method rather than consensus between various stakeholders and actors involved in the process [19]. However, the notion of evidentiary analysis for wicked problems is widely bolstered by academicians, administrators, and politicians [20]. Daviter [12] argues that when we deal with wicked problems, the knowledge base we have is often 'fragmented' and 'contested' due to the notion of available evidence being 'incomplete', 'inconclusive' and 'incommensurable'.

In this paper we account for the issues cited in the preceding paragraph by proposing a framework to frame a wicked problem that embodies the integration of interpretative and evidentiary analysis through Dilemma Triangle Method and System Dynamics, respectively. We suggest that framing of any wicked problem and identifying the core of it allows a designer to understand the problem and proceed in a structured way to address the wickedness of any wicked problem. Through the inclusion of interpretative and evidentiary analysis we enable humans in the loop to account for human cognition, mental capabilities, and socio-cultural elements [21]. We suggest that by including a human-in-the-loop we facilitate the efficient framing of wicked problems by maximizing the synergy between human abilities and computational capabilities.

1.3. Dilemma Triangle Method and System Dynamics
1.3.1. Dilemma Triangle Method (DTM)

The Dilemma Triangle Method is used to identify the dilemmas embodied in wicked problems. A Dilemma is defined as follows: Dilemma: *"A dilemma is a difficult choice between two options, each of which is unacceptable or unfavorable"* [22].

A schematic of the Dilemma Triangle Method is shown in Figure 1. We use the following key words in the Dilemma Triangle Method:

Driver: *These are the thematic areas that are key to framing a wicked problem and thence used in identifying an appropriate solution or way forward. There is no limit on the number of Drivers that can be considered.*

Focus: *A single statement used to define the goal that must be achieved for the Driver. There can be several Foci for each Driver.*

Issues: *Issues must be addressed to satisfy the Focus that must be achieved for the Driver.*

In the Dilemma Triangle Method, we select the thematic areas that contribute to the wicked problem and identify the Drivers for each thematic area; see Figure 1. As shown in Block 1, Figure 1, we identify three Drivers for each thematic area. Further, we define the Focus using experience and judgment each Driver that enables us to establish the boundaries This includes taking into account the perspectives of multiple stakeholders involved. Once we have the Focus for each Driver, we list the Issues as shown in Block 2; Figure 1, that are key to achieving the specified Focus. Further the two important stages in the Dilemma Triangle Method which help in managing the dilemmas are creating Tension Matrix and identifying the Dilemmas.

A) Tension Matrix (Block 3; Figure 1)

A Tension Matrix is created to identify the relation between two Issues. This matrix is a foundational step in identifying Dilemmas. There are four relations between Issues which can be identified through the Tension Matrix:

---

[3] Key words associated with the Dilemma Triangle Method are shown in Courier font.

[4] *Progress* used in our earlier publications has been replaced by *prosperity* to conform to the definition adopted for sustainable development at COP 26 in Glasgow.

Copyright © 2023 by ASME

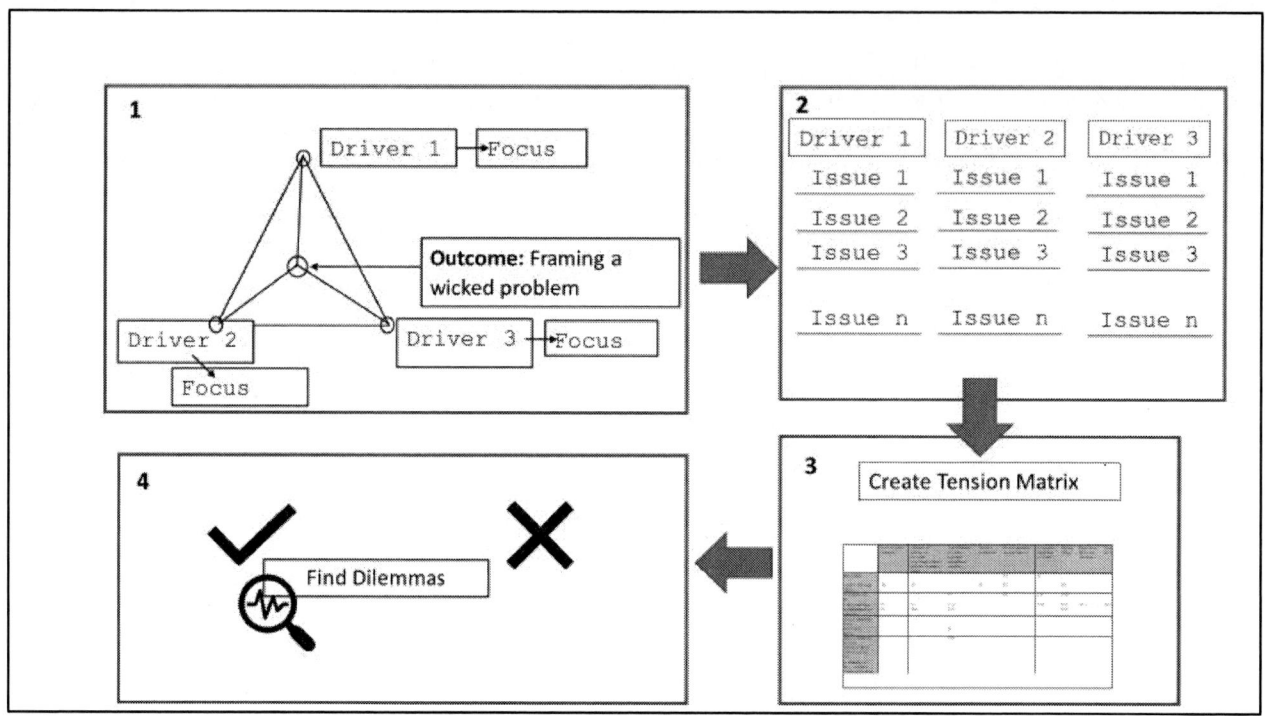

FIGURE 1: GENERALIZED DILEMMA TRIANGLE METHOD

a) *Tension:* A tension results when an `Issue` associated with one `Driver` negatively impacts an `Issue` associated with another `Driver`.

b) *Dependent:* A dependent arises when the `Issue` associated with one `Driver` positively impacts an `Issue` associated with another `Driver`.

c) *Inter-Tension:* When one `Issue` negatively impacts an `Issue` of a different thematic area.

d) *Inter-dependent:* When one `Issue` positively impacts an `Issue` of different thematic area.

B) Identifying Dilemmas (Block 4; Figure 1)

`Dilemmas` are identified based on the `Tension Matrix` constructed. `Dilemmas`, when correctly identified through combination of evidentiary and interpretive analysis help us to frame a wicked problem, the framework for which is described in this paper.

In several agricultural villages, relevant to this effort, the income of villages depends on the forest. If the villagers stop cutting trees and accessing the forest utilities their income will reduce. Thus, this happens to be a dilemma. Further, villagers who practice agriculture have over exploited water for their personal and agricultural use. They practice excessive tillage in farms which has other detrimental impacts on the planet. Thus, this is again a dilemma and precludes sustainable development. The three `Drivers` we select to promote sustainable development in such a village are People, Planet and Prosperity. This is anchored in the test problem used in this paper. The use of the Dilemma Triangle Method to manage dilemmas in one

thematic area for sustainable rural development of India is documented in Reference [22]. Further, the Dilemma Triangle Method is expanded to three thematic areas to provide a method for social entrepreneurs to develop value propositions [24]. The application of Dilemma Triangle Method along with systems dynamics to propose policies and value propositions is presented in [23–24] In our earlier papers we focused on:

- Karkaria et al. [23] use the combination of Dilemma Triangle Method and System Dynamics to determine policies. Their main objective is to propose policies to ensure sustainable development. In this paper, we focus on framing wicked problems through iteration by incorporating evidentiary and interpretative analysis while maximizing synergy between computational capabilities and human abilities.

- Kamala et al. [24] focus on using System Dynamics to create value propositions for social entrepreneurs. Their objective is to aid social entrepreneurs to provide decision support to choose right value proposition required for the intervention and evaluate its pre-impact. Whereas through the framework proposed in this paper, a designer can identify the wickedness of the problem and frame wicked problems through interpretive and evidentiary analysis.

1.3.2.System Dynamics Modelling

We use System Dynamics to model the system and enable a designer to simulate and analyze the behavior of the system to gain insights that support decision making. Through simulation of system using System Dynamics, we understand the effect of variables on each other and gain insight on the interaction between variables and their impact on system model. Systems Dynamics necessitates constructing a causal loop diagram and a stock and flow diagram.

a) *Causal Loop Diagram*
Causal loop diagrams are an effective way of mapping the relationship between variables. These allow a designer to link the variables with one another and understand the interconnections of variables in a system. Further, causal loop diagrams help a designer understand the system as a whole and provides an opportunity to enhance the system structure. A causal loop diagram is an effective tool for story telling in order to communicate the understanding of the elements of system and system as whole.

b) *Stock and Flow Diagram*
The creation of stock and flow diagrams allows a designer to simulate the system and get insights on the interaction between variables. Through the simulations created through stock and flow diagram, a designer gain insights of the systems behavior by simulating the system which acts as an important tool for decision support when complex systems are involved. The two important elements namely stock and flow are defined as follows:
Stock: Stock is the accumulation of a quantity at any state of the system.
Flow: A flow is entity which increases or decreases the magnitude of stock.

2. FRAMEWORK FOR FRAMING A WICKED PROBLEM

In this section we describe a framework that can be used by a designer to frame a wicked problem and identify the variables that can be used to design a socio-economic-technical system. We enable a designer to convert the early-stage heuristics into insights to frame a wicked problem through evidentiary and interpretative analysis with a human-in-the-loop.

2.1 Approach for Framing a Wicked Problem
In Figure 2, we illustrate our approach for framing a wicked problem. Given a wicked problem, in the initial stages a designer has heuristics anchored in past experience. However, to frame a wicked problem a designer needs to generate evidence-based insight to augment what is currently known to him/her. Thus, based on the heuristics, a designer invokes the Dilemma Triangle Method. Using the information generated and deductive speculation a designer then constructs a Systems Dynamics model to model the system and thereby gain insight into the behavior of the system. With these insights a designer modifies the input to the Dilemma Triangle Method by considering the evidence-based insights gained through exercising the System Dynamics model. This process is repeated until a designer is satisfied with the outcome. In summary, a designer carries out interpretative analysis through the Dilemma Triangle Method and evidentiary analysis through System Dynamics. This process allows a designer to synthesize the heuristics and experiences into insights through deductive speculation to frame a wicked problem.

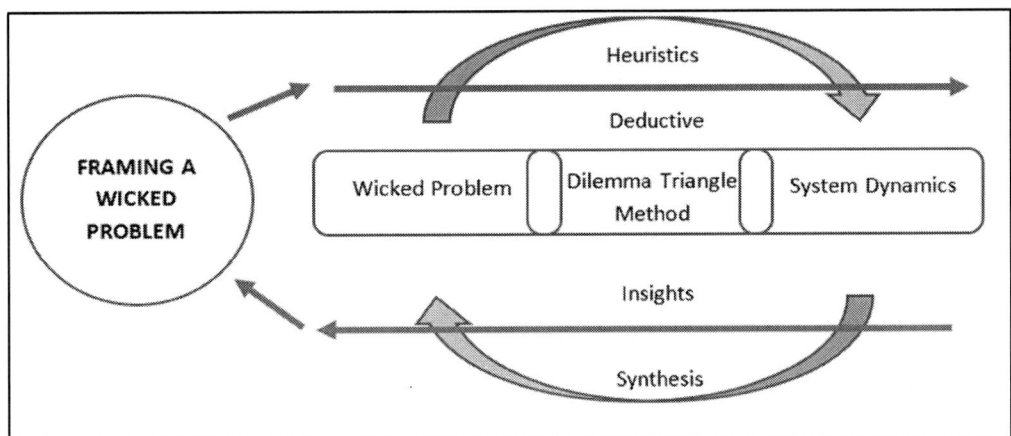

FIGURE 2: OVERVIEW OF APPROACH

The rest of the paper is organized as follows: In Section 2 a description of the proposed framework is presented. In Section 3, the test problem and the approach used while exercising the proposed framework is described. In Section 4, the results and the efficacy of the proposed framework are discussed. A discussion of the results is included in Section 5. In Section 7 appropriate closing remarks are documented.

2.2 Features of the Proposed Framework
We propose a framework to frame the wicked problems through a structured process. The features of the proposed framework are summarized in Figure 3.
Conversion of heuristics into insights: While dealing with wicked problems, it is evident that a designer have incomplete information which is often confusing and

Copyright © 2023 by ASME

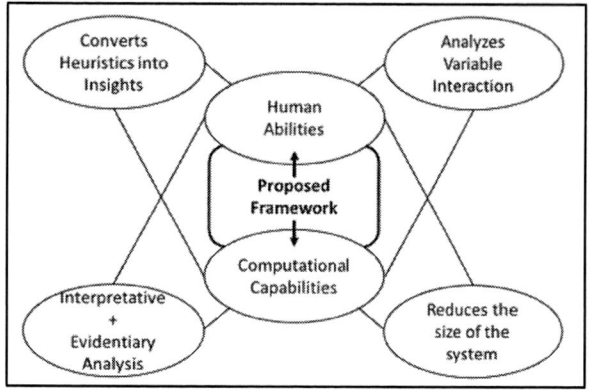

FIGURE 3: FEATURES OF THE PROPOSED FRAMEWORK

conflicting. The value of this feature of the framework is to attain insights from the initial heuristics. As shown in Figure 2, the initial invocation of the Dilemma Triangle Method is based on the heuristics a designer have. Through the advancement from Dilemma Triangle Method to System Dynamics model, we create a foundation for having insights based on heuristics. The system behavior analysis as shown by Block 2A in Figure 4 is the stage where the heuristics are converted into insights.

Analyze interaction between variables and identify the correct size of the problem: When we consider a wicked problem, we have a plethora of variables that are of interest to a designer. However, some of these might not be significant. It is therefore important for a designer to understand the interaction between these variables and their overall effect on the wicked problem. This allows a designer to identify the variables that are significant which further emphasizes the third peculiar feature of the proposed framework, through which we enable a designer to identify the correct size of the problem. This anchors in Block 2 and Block 2A of Figure 4.

Maximizes synergy between human capabilities and computational abilities: Due to the characteristics of wicked problems, it can seldom be modelled alone with computational abilities. Human cognitive characteristics play a very important role in addressing wicked problems through judgements, perspectives, and experiences of humans. The value added through this feature is to maximize the synergy between human abilities and computational capabilities. Through the Dilemma Triangle Method a designer brings to bear his/her judgement (qualitative information) that is anchored in experience. By exercising the Systems Dynamics model is able to transform judgment (qualitative information); see Blocks 1, 2 and 3 in Figure 4.

Integration of Interpretative and Evidentiary analysis: Interpretative and evidentiary analysis play a very important role and have their own significance for addressing various types of problems. Through the proposed framework,

authors provide an opportunity for a designer to have both; interpretative analysis through the Dilemma Triangle Method and evidentiary analysis by simulating the system by virtue of System Dynamics model.

2.3 Description of the Framework
The conceptual design of the framework is divided into three main building blocks is shown in Figure 4. Further, in Figure 5 we present the detailed framework. In this section we discuss three steps in the context of their utility in the framework.

Step 1: Dilemma Triangle Method
Identify the `Focus` and list the `Issues` that are key to attaining the `Focus`. This is based on heuristics and observations gained

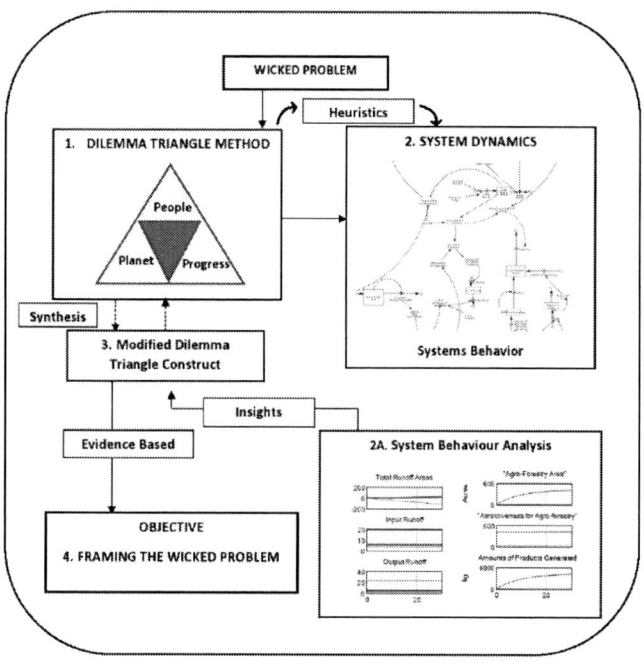

FIGURE 4: MENTAL MODEL FOR THE PROPOSED FRAMEWORK

by the knowledge of the wicked problem we have. At this stage we do not create the `Tension Matrix` to arrive at `Dilemmas`. Through Dilemma Triangle Method we carry out interpretive analysis which allows us to incorporate the behavioral, cognitive, and social elements in the analysis to frame wicked problems. This interpretive analysis further helps a designer to analyze the qualitative data involving interpreting the data from the perspectives of actors involved. The information used is based on heuristics anchored in what the designer knows at this time.

Copyright © 2023 by ASME

Step 2: Systems Dynamics
Create a Systems Dynamics model to gain insights for framing a wicked problem. A designer develops causal loop diagrams

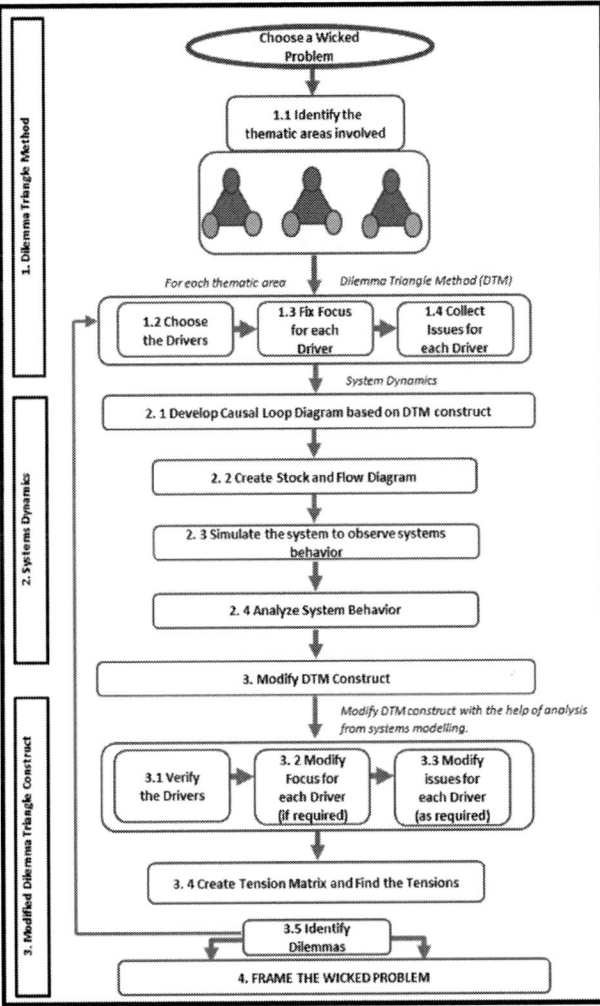

FIGURE 5: PROPOSED FRAMEWORK TO FRAME WICKED PROBLEMS

based on the Dilemma Triangle construct from Step 1. Further, designer creates stock and flow diagrams and simulate the system. Through this structured process a designer is able to simulate and observe the behavior of the system. A designer analyzes the effects of interaction between variables and gains insight on the important variables that affect/govern the behavior of the simulated system and also those variables that a designer is unable to control. With these insights a designer eliminates the variables that do not impact the outcome and those that are not under a designer's control. This enables a designer to isolate the "wickedness" associated with the wicked problem and deal with it appropriately.

Thus, through System Dynamics a designer is able to carry out the evidentiary analysis by gathering insights and evidence by simulating the system and observing the interaction

between variables. Through the framework, a designer have the opportunity to collect, evaluate and examine data by incorporating the interpretive analysis as well to provide decision support to frame wicked problems.

Step 3: Modified Dilemma Triangle Method
As discussed in Step 2, at this stage a designer has gained insights from the System Dynamics model through the simulation of the system. The designer leverages the insights gleaned and modifies the Dilemma Triangle created in Step 1. This enables a designer to modify the `Tensions` and `Dilemmas` and add observations based on insights which impact the system. The `Dilemmas` constitute the framing of the wicked problem. Each `Dilemma` needs to be resolved. The resolution will typically involve some combination of technical, regulatory, policy, financial and social consideration.

Through this framework, we utilise the interpretative analysis through the Dilemma Triangle Method compounded with the evidence-based analysis through System Dynamics to frame wicked problems. Thus, a designer who uses this framework, can convert the heuristics into insights and frame the wicked problem through a structured process to ensure correct identification of the problem and a way forward.

3 TEST PROBLEM OF KANTASHOL VILLAGE TO DEMONSTRATE THE EFFICACY OF THE FRAMEWORK

To illustrate the efficacy of the framework we use data for Kantashol village in Jharkhand, India provided by Dr. Ashok Das and his colleagues at SunMoksha Power Pvt. Ltd. The average rainfall is around 60-65 days in a year with an annual count of 800-1310 mm, due to which villagers use excessive amount of ground water which happen to be one of the reason of overexploitation of the ground water. Villagers are over dependent on the forestry for their livelihood and the practice of agriculture is limited due to various reasons. The average temperature in the village is around 40-45 degree Celsius. The village has marginal road transport. Since the income of villages depends on forestry, if the villagers stop cutting trees and accessing the forest utilities their income will be reduced. Thus, this happens to be a dilemma. Further, villagers over exploiting the water for their personal and agricultural use exacerbates challenges. The local practice of excessive tillage in farms which has various detrimental impacts on the planet including erosion which, in time, reduces the acreage of available land to the villagers, affecting the planet adversely. The situation is a wicked problem due to the incomplete, conflicting, and confusing information.

Copyright © 2023 by ASME

Approach

Having identified this as a wicked problem based on insights from various stakeholders, we begin to assess the situation borrowing heavily from the expertise from the SunMoksha team. SunMoksha is an international partnership between scholars and local industry in India that strives to develop and field-deploy clean and sustainable technology solutions and provides consulting services for rural development and urban sustainability. The SunMoksha team consists of experts and professional with a passion for sustainable development, and decades of experience in technology, engineering and management across Asia, Africa and the USA including team members working on the ground which grants us real-time insights on problems faced at individual and community level in terms of people, planet, and prosperity.

The framework illustrated in Figure 5 is systematically exercised for the Kantashol village problem. A discussion on the efficacy of the framework through the test problem is presented in Section 4. The following steps are explained with respect to the test problem and are tied with the steps illustrated in Figure 5.

Step 1 Figure 5: Dilemma Triangle Method
Step 1.1: Identify the thematic areas involved

Initially we select the important thematic areas involved based on the lifestyle and situations of the village. The thematic areas we select are water, forestry, and agriculture. However, to illustrate the framework we combine all three thematic areas in the Dilemma Triangle Method. The system dynamics model, however, has all three thematic areas involved to ensure we get accurate insights.

Step 1.2: Define the Drivers

We select the three Drivers with the goal of improving the progress of the people and at the same to ensure sustainable development. The three Drivers we select are People, Planet and Prosperity.

Step 1.3: Fix Focus for each Driver

Based on the situation in Kantashol we define the Focus of each Driver. This is based on taking perspectives of various stakeholders and the data collected from the SunMoksha team. We take into consideration various factors affecting the livelihood and progress of the people of the village at an individual-level as well as community-level alongside ensuring that the development is sustainable.

Step 2 Figure 5: Systems Dynamics Modelling
Step 2.2: Create Causal Loop Diagram

This first step in System Dynamics modeling anchors with creating causal loop diagram. These causal loops help designers to map relationship between different variables. The interpretive analysis carried during the Dilemma Triangle Method in Step 1 forms the foundation for constructing causal loop diagrams.

Step 2.2: Create Stock and Flow diagram

Once we have mapped the relationship and dependencies amongst variables, we create the stock and flow diagram through which we simulate the system. To create the stock and flow diagram we specify the relationship between variables by inserting the mathematical equation and values, the data for which we acquire from the SunMoksha team. Based on the qualitative and quantitative data and survey we classify variables as objective variables and decision variables. This allows us to categorize the variables in to simulate the system.

The information that cannot be quantified is incorporated in the framework through Dilemma Triangle Method. While applying Dilemma Triangle Method we use heuristics that are judgements anchored in experiences. This is where the interpretative analysis is executed. The information rather than being made up is anchored with deductive reasoning followed through the steps in the framework. Dilemma Triangle Method forms the foundation for the information that is fed in stock and flow diagram.

Step 2.3: Simulate the system to observe systems behavior

At this stage we simulate the system by conducting various experiments. For example, we vary the amount of tillage and multi cropping in order to understand its effect on the profits incurred and the runoff areas. This step helps us to simulate the system and understand the behavior of the system by analyzing the effects of interaction between variables and on the system.

At this stage we have the interpretations based on the initial Dilemma Triangle construct and evidence based on the system dynamics modelling. We simulate the Kantashol village based on the system dynamics model and critically analyze the behavior. Through this analysis we understand the variables which do not affect the outcome as well as attain insights to identify the problem correctly.

Step 3 Figure 5: Modified Dilemma Triangle Construct
Based on the insights and evidence attained through Step 1 and Step 2, we modify the Dilemma Triangle. This allows us to identify the wicked problem based on interpretations that are based on expert views, and judgements, by inclusion of all the stakeholders as well as on the evidence gathered based on computational simulations. The interpretative analysis is compounded through evidentiary analysis by the aid of System Dynamics.

Accordingly, we have the opportunity to modify the Focus and Issues through Steps 3.1, 3.2 and 3.3. Further, we construct the Tension Matrix for the Issues that ties with the Step 3.4. The Tension Matrix allows us to find the Dilemmas thereby help us frame the wicked problem.

Step 4 Figure 5: Frame the Wicked Problem
After creating the tension matrix, we identify the dilemmas which enable us to frame the wicked problem which is our main objective in this paper.

Copyright © 2023 by ASME

With the application of framework, we have a perfect reconciliation of interpretative analysis and evidence-based analysis thus ensuring the fidelity of the framing of the wicked problem through human abilities compounded through computational capabilities.

4 RESULTS AND DEMONSTRATION OF THE EFFICACY OF THE PROPOSED FRAMEWORK

In this section we comment on the efficacy of the proposed framework. We discuss the results in three parts, namely, Initial Dilemma Triangle construct, System Dynamics Modelling, and Modified Dilemma Triangle construct.

4.1 Dilemma Triangle Method (Step 1; Figure 5)

a. People

FOCUS: To improve the quality of life of people by providing them with adequate nutritious food, water for various purposes, and promote sustainable agroforestry for the sustainable development of the village.

ISSUES

1. Lack of agriculture and crop diversification
2. Absence of policies to promote agroforestry and strict government policies to access the forests.
3. Unavailability of water due to lack of facilities, excessive runoff, less rainfall, etc.

b. Planet

FOCUS: To preserve forest and its biodiversity, prevent runoff, preserve fertility of the agricultural land, and utilize water resources wisely.

ISSUES

1. Excessive tillage for agriculture
2. Overdependence on agroforestry and lack of sustainable agricultural practices.
3. Excessive depletion of water and lack of awareness to maintain the quality of the water.

c. Prosperity

FOCUS: To enhance the sources of income for the farmers, to ensure progress in the income of villagers, and to provide reliable and feasible sources of water for varied purposes.

ISSUES

1. Excessive runoff due to high tillage resulting in large barren lands in long term.
2. Unawareness of appropriate agricultural practices resulting in high reliance on agroforestry.
3. Unavailability of water due to limited access and an unstable power supply accompanied by unknown wastage of water.

For the initial Dilemma Triangle construct, we define the Focus and list the Issues encountered to achieve the Focus for each Driver. This is based on the information from various sources

and includes data, experiences, and judgements. Based on the Issues and Focus for each Driver, we create a System Dynamics model and simulate the system to help us understand the interaction between variables.

In this paper we use two types of variables, namely, objective variables and decision variables. We define decision variables as the variables which a designer can assign a value or a set of values to achieve a goal or assess its effect on desired outcome, whereas the variables that represent the objective or goal and measure the effectiveness of the solution are defined as objective variables.

We assess the effect of decision variables on objective variables through System Dynamics.

Based on the information gained, we recognize that for the development to take place we need to increase the disposable income of the villagers without adversely affecting the planet. Thus, in order to ensure this, we demonstrate the utility of the proposed framework by selecting the following objective variables which are of significance to the villagers:

1. Overall Profit
2. Total Runoff Areas

We use following decision variables to assess their effects on above mentioned objective variables:

1. Amount of tillage
2. Multi-cropping

The results presented are specific to the case of the village selected due to the intricacies of various social factors. The main objective of the authors is to demonstrate the efficacy of the proposed framework in order to provide decision support for a designer to identify variables that impact the outcome of a wicked problem which has a plethora of variables under consideration.

This helps us to identify the core of the problem and frame the wicked problems with maximum fidelity. The levels of decision variables are kept as per the judgements and experiences of the authors and may vary for every case. We have simulated the system multiple times to understand the system's behavior and interaction between variables in order to make decisions on the levels of decision variables to assess profit and total runoff areas.

4.2 System Dynamics – Results and Analysis (Step 2; Figure 5)

Values of Input Variables to System Dynamics Model (Data Obtained from SunMoksha Pvt. Ltd.) are presented in Table 1. The stock and flow diagram of the System Dynamics model is shown in Figure 6. We simulate the system in order to analyze the interaction between variables. We change the magnitude/level of the decision variables in order to understand the effects of each decision variable on the objective variables. We keep the total area available as the highest value considering that we have the entire area for utilization.

Copyright © 2023 by ASME

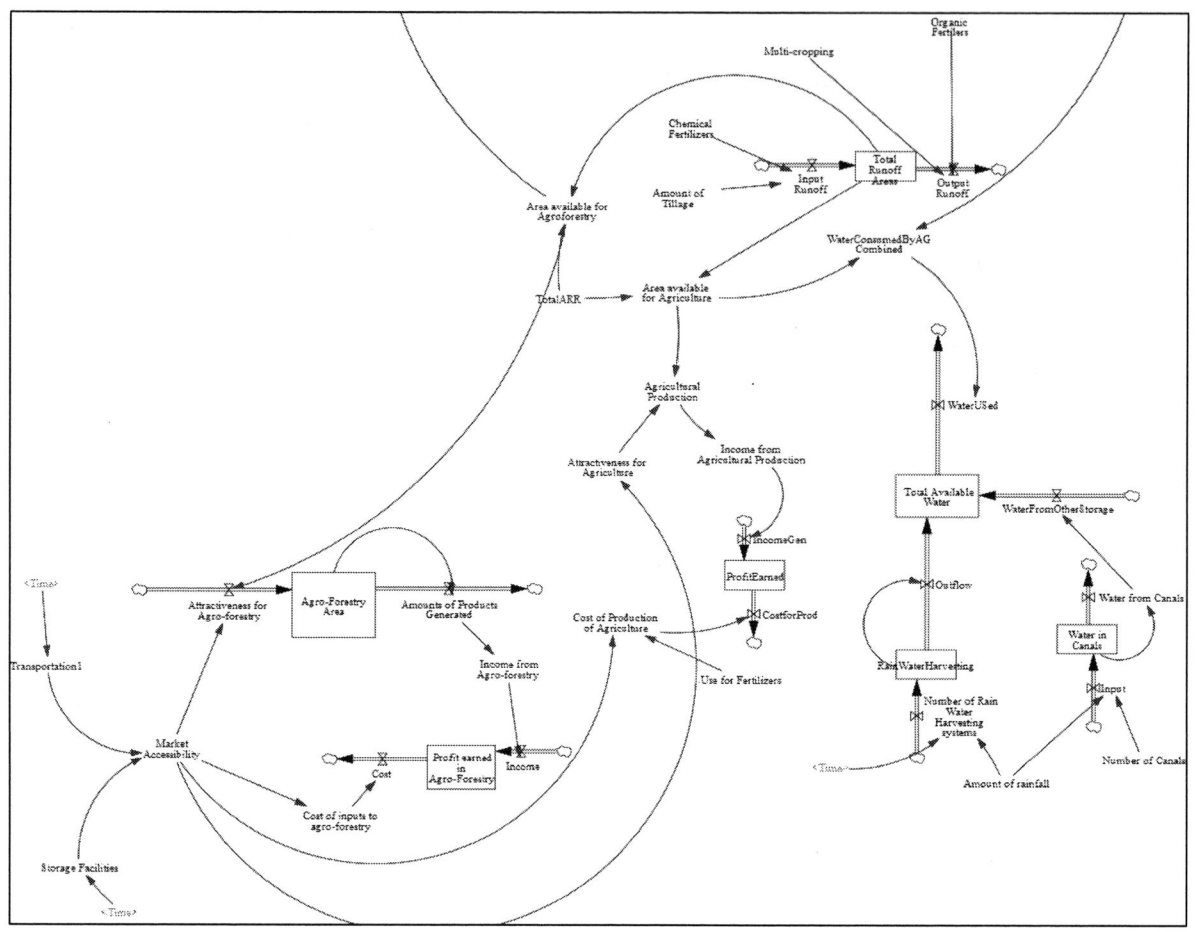

FIGURE 6: STOCK AND FLOW DIAGRAM OF THE KANTASHOL VILLAGE

The control volume is defined by the thematic areas which a designer selects during the Step 1 of the framework. With the thematic areas defined (in our case Forestry, Agriculture, and Water) we define the external boundaries of the control system. The drivers for these thematic areas enable a designer to mark the internal boundaries of the control volume. The consensus on this decision anchors with the experiences and judgements taken during interpretive analysis through Dilemma Triangle Method. While framing wicked problems we do not model the uncertainties, instead observe, and analyze them through simulation of systems. This analysis is then used to make modifications by iterating through the Dilemma Triangle Method and System Dynamics model to frame wicked problems. The insights gleaned through the simulations can then be utilized to synthesize information and mitigate uncertainties to identify the correct size of the problem, concentrate on core of the wickedness subsequently aiding in framing the wicked problems.

4.2.1 *Results of System Dynamics Modelling*

We initially present the effect of amount of tillage on profit earned and the total runoff areas in Figure 7 and Figure 8. The effect of multi cropping on profit earned and total runoff areas is presented in Figure 9 and Figure 10. The results presented are with low tillage, medium tillage, and high tillage. We keep the value of multi-cropping and the rest of the variables at medium level.

While considering the multi cropping decision variable, we consider the maximum tillage being done in order to be sure that the effect of multi cropping overcomes tillage. We consider two independent variables in the System Dynamics model. The first being amount of tillage and the second being the multi cropping. We assess the effects of these variables on two objective variables, namely, the total profit earned and total runoff areas.

Copyright © 2023 by ASME

TABLE 1: VALUES FOR INPUT VARIABLES OF THE SYSTEM DYNAMICS MODEL

Input Parameter	Value
Storage Parameter	6 facilities
Transportation	50 vehicles
Amount of Tillage	5000 acres
Cost of chemical fertilizer	310 rupees per acre
Cost of organic fertilizers	150 rupees per acre
Multi-cropping	Normal – 2.5 crops/year – (Variable)
Number of post processing equipment	50 machines
Irrigation cost	5000 rupees per acre
Animal Labor Cost	33 rupees per acre
Cost of Seeds	2000 rupees per acre
Manure cost	150 rupees per acre
Human Labor Cost	5000 rupees per acre
Electricity cost	181 rupees per acre
Diesel Cost	520 rupees per acre
Total Area	20000 acres
Number of canals	5 canals
Amount of rainfall per month	1800 mm per annum
Number of Rain Water Harvesting systems	10 systems

Initially we simulate the system by changing the level of amount of tillage. We consider three levels/cases. One when no tillage is done, another when we simulate moderate tillage and the third with high levels of tillage. As seen through Figure 7, the amount of tillage does not drastically affect the overall profits. The profits remain relatively equal for all the levels of tillage. Thus, the progress of the village does not hamper due to the amount of tillage. However, when we assess the effects of the amount of tillage on the runoff areas, we find that the higher the amount of tillage the larger are the runoff areas.

Runoff is a deleterious factor for the productivity of the agricultural land in the long term. Moreover, the soil loses its fertility which affects Planet. Larger run-off areas results in the decrease in the level of water absorption levels in the soil. To avoid any bias in the results, we maintain other factors at their moderate level during the simulation to analyze the system behavior with respect to change in the quantity of tillage. With the results as shown in Figure 7 and Figure 8, it is evident that

the amount of tillage has a significant impact on the total run off areas. Further, in order to assess the impact of multi-cropping on the entire system model, we simulate the system by changing the intensity of the levels of multi-cropping. We assess the impact of change in multi cropping on two factors, namely, the 'overall profit' and the 'total runoff'. The results are presented in Figure 9 and Figure 10.

Analyzing the behavior of the simulated system, we observed that with the increase in multi-cropping there is an increase in the overall profit. Further, on comparing Figure 7 and Figure 9 we observe that the magnitude of decrease in profit due to high tillage is less than that of the magnitude of increase in profits by increasing the levels of multi cropping.

The effect of multi-cropping on runoff areas is observed through the results presented in Figure 10. With increase in the multi cropping, the runoff areas decreases. While simulating the system for multi cropping decision variable we keep the levels of amount of tillage at the highest levels in order to confirm the efficacy of multi cropping for increase in profits and decrease in runoff areas. Through the behavior of the simulated systems, we come to the following conclusions with respect to the decision variables and the objective variables as shown in Table 2. The effect of interaction between decision variables and objective variables considered to demonstrate the framework is summarized in Table 2.

TABLE 2: EFFECT OF DECISION VARIABLES ON OBJECTIVE VARIABLE

Decision Variables	Objective Variables	
	Overall Profit	Total Run off areas
Amount of Tillage (↑)	Slightly decreases	Increases considerably
Multi Cropping (↑)	Increases	Decreases considerably

With the results obtained by simulating the system and the explanation provided we identify that the multi-cropping decision variable impacts the outcome. This helps us to identify the core of the problem by eliminating the amount of tillage decision variable. With the results through systems dynamics and the justification provided above we assert that multi-cropping is a significant decision variable while amount of tillage, though of interest to us is not of significance to the model and, thus we eliminate it. Through this we identify the correct size of the problem by identifying the variables which have significant impact on the model/outcome. The utility of this framework comes along with the framing of the wicked problem through evidence based structured process by converting heuristics into insights.

Framing the problem has great significance in order to get robust solutions especially when dealing with wicked socio-economic-technical problems that involve many variables (quantitative and qualitative) that need to be dealt with. Problem framing requires correctly identifying the problem by simulating

Copyright © 2023 by ASME

the system in order to understand its behavior with respect to the variables of interest. In order to model a system and arrive at robust solutions it becomes important to initially frame the wicked problems with high fidelity. The efficacy of the solutions proposed for wicked problems depends on how well the problem is framed.

To achieve this, it becomes important to gain insight on variables which affect the problem to identify the core of the problem where wickedness lies so that it is modelled to provide decision support to the decision makers. This allows us to redefine the Focus and Issues in the Dilemma Triangle Method in order to understand the exact problem and frame it by revisiting the initial Dilemma Triangle construct (Step 1; Figure 5) that we developed in the initial stages of the framework. Further, we modify the Dilemma Triangle constructed in the initial phase in order to enhance it with the insights gained on the interaction between variables through the behavior of the simulated system.

4.3 Modified Dilemma Triangle (Step 3; Figure 5)

After gaining insights of behavior of the simulated system through System Dynamics as shown in Step 2 of the framework presented in Figure 5, we utilize those insights to modify the Dilemma Triangle in order to frame the wicked problem. These insights are discussed in the Section 4.2.1

The Drivers of the Dilemma Triangle remain the same as before: People, Planet and Prosperity. We modify the Focus (if required) and Issues (as required) based on the results of system dynamics model. We convert heuristics into insights to frame a wicked problem through the integration of evidentiary and interpretative analysis.

a. People

FOCUS: To improve the quality of life of people by providing them adequate nutritious food, water for various purposes, and promote sustainable agroforestry for the sustainable development of the village.

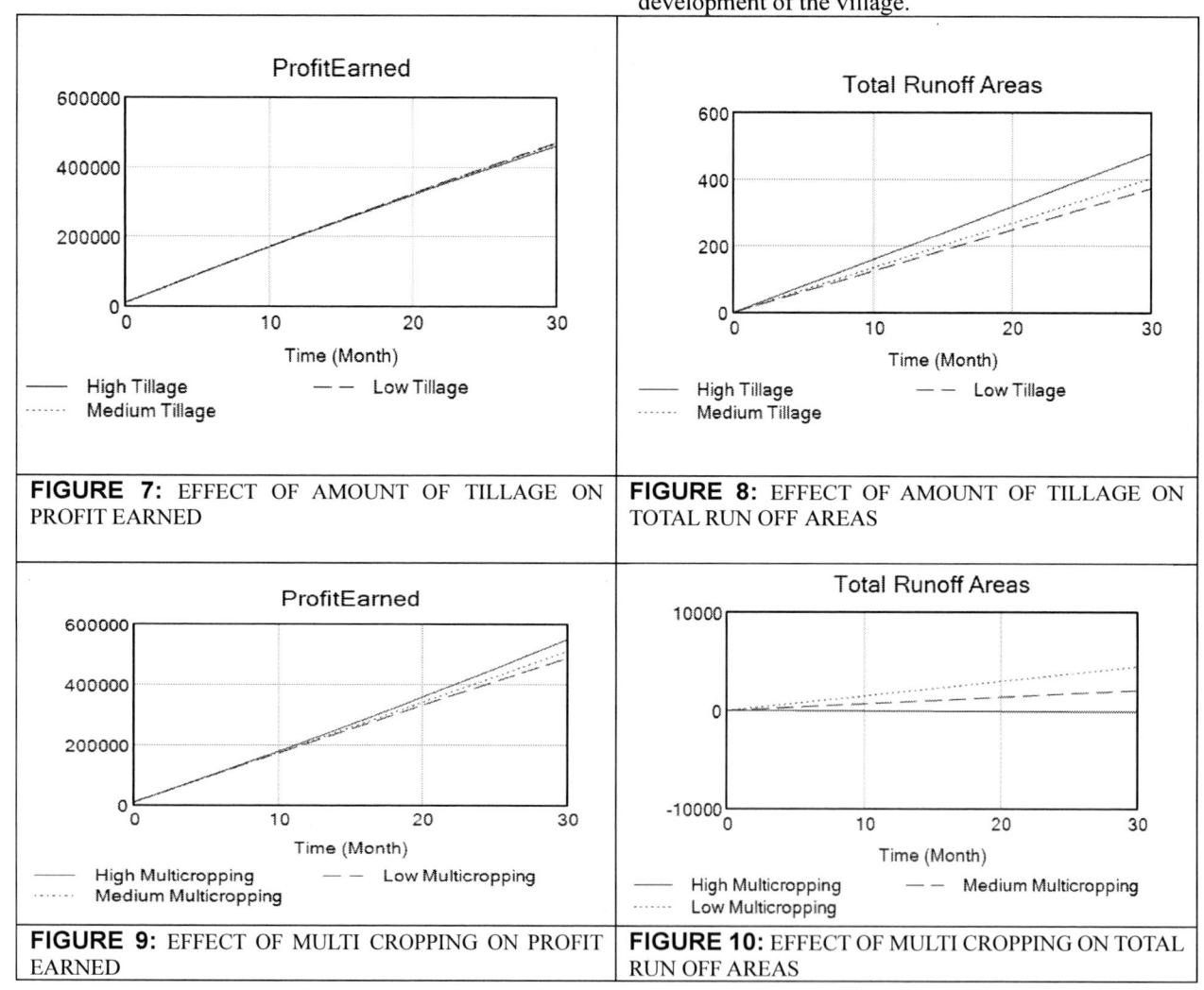

FIGURE 7: EFFECT OF AMOUNT OF TILLAGE ON PROFIT EARNED

FIGURE 8: EFFECT OF AMOUNT OF TILLAGE ON TOTAL RUN OFF AREAS

FIGURE 9: EFFECT OF MULTI CROPPING ON PROFIT EARNED

FIGURE 10: EFFECT OF MULTI CROPPING ON TOTAL RUN OFF AREAS

Copyright © 2023 by ASME

ISSUES
1. Lack of crop diversification and multi-cropping practices in agriculture leading to limited production of nutritious food.
2. Absence of policies to promote agroforestry and strict government policies to access the forests.
3. Unavailability of water due to lack of facilities and improper management in agricultural practices.

b. Planet
FOCUS: To preserve forest and its biodiversity, prevent runoff, preserve fertility of the agricultural land, and utilize water resources wisely.
ISSUES
1. Lack of multi-cropping practices in agriculture contributes to the soil losing its fertility and restricting water to be soaked in the ground leading to the unavailability and depletion of ground water in a village.
2. Overdependence on agroforestry and lack of sustainable agricultural practices.
3. Excessive depletion of water due to mono cropping in agriculture requiring high tillage which enables the soil to lose its fertility.

c. Prosperity
FOCUS: To enhance the disposal income for the farmers, to ensure progress in the income of villagers, and to provide reliable and feasible sources of water for varied purposes.
ISSUES
1. Lack of multi-cropping practices in agriculture. Monocropping results in higher tillage of land which affects the water holding capacity of the soil as well as degrading its fertility.
2. Excessive use of chemical fertilizers to enhance the production in mono cropping resulting in degradation of land in long terms and thus affecting prosperity in long term.
3. Unavailability of water due to limited access and unstable power supply accompanied by prevailing monocropping practices leading to affect the quality of soil and water holding capacity of soil.

With the modified Focus and Issues in the dilemma triangle we now proceed with the further steps of dilemma triangle method. The Tension Matrix is shown in Table 3. In the context of Tension Matrix we enumerate Tensions (which are denoted by 'T' in Table 3) to demonstrate the framework and then proceed to find dilemmas which helps a designer frame the problem through the proposed framework.

The efficacy of the framework is observed from the Dilemmas identified. The Dilemmas are presented in Table 4. We identify the correct problem and the associated Dilemmas using insights gained from system behavior through system dynamics. For example, instead of concentrating on the amount

of tillage to prevent runoff areas which was initially considered by most of the stakeholders, we could reframe it through identifying the correct alternatives by studying its interaction with other variables. Agriculture requires tillage and thus taking steps to stop it affects the agriculture adversely.

We know that excessive tillage causes increases runoff areas which hampers the water absorbing capacity of the soil Through the framework, we understand that it is the agricultural practices which are more important to Focus than emphasizing directly on stopping tillage which is neither a viable solution knowing its necessity in agriculture. The dilemmas which anchor with framing the wicked problem are presented in Table 4. Further, knowing its interaction with other variables we also eliminate the variable in modelling the system further to identify the dilemmas/frame the problems. This demonstrates the efficacy of the framework provided. Thus, through the proposed framework, a designer can:
1. Convert heuristics into insights to frame a wicked problem through the integration of evidentiary and interpretative analysis.
2. Understand the interaction between variables through behavior of the simulated system by the virtue of System Dynamics.
3. Identify the correct size of the problem by eliminating the variables which do not impact the outcome/model and are not relevant to the wicked problem.

These three aspects together help in efficient framing of the wicked problem.

In this paper we account for the key issues that are italicized in the ten characteristics of wicked problem recognized by Rittel and Weber [6] which is discussed in Section 1. We summarize how we address these key issues by the application of our framework in Table 5.

5. DISCUSSION

In this section we comment on the bottlenecks, verification of the framework as well as list a few limitations observed while applying it to frame a wicked problem.

The bottlenecks in the framework include undertaking interpretive analysis through Dilemma Triangle Method. This requires judgement and heuristics to be incorporated at the initial stage of the analysis. This includes exploring and interpreting data from the perspective of individuals which requires a designer to have experience with Dilemma Triangle Method. Further, the iteration from System Dynamics model back to Dilemma Triangle Method requires analysis of interaction of variables to modify the initial exercise of the Dilemma Triangle Method. This requires a designer to move back and forth and iterate in order to identify the wickedness of the problem.

Copyright © 2023 by ASME

TABLE 3: TENSION MATRIX *(T – Tension, D – Dependent)*

	Lack of multi cropping	Absence of policies to promote agroforestry/Strict policies to access forest	Unavailability of water – improper agricultural practices	Limited production	Overdependence on agroforestry	Excessive depletion of water	Higher tillage	Excessive chemical fertilisers	Unavailability of water
Higher tillage					T1	D			
Excessive chemical fertilisers	T2	T3		T4	T5		T6		
Unavailability of water			T7		T8	T9	T10		
Limited production	D	D	T 11			T12	T13	T14	T15
Overdependence on agroforestry	D	T16	T17				T18		
Excessive depletion of water									
Lack of multi cropping			D						
Absence of policies to promote agroforestry/Strict policies to access forest			T19						
Unavailability of water – improper agricultural practices									

TABLE 4: DESCRIPTION OF DILEMMAS

Sr. No.	Dilemma- *Framing Wicked Problems*	Dilemma arrived from	Justification
1	How can we promote multi-cropping by reducing the use of chemical fertilizers and promoting organic fertilizers by ensuring decent production levels?	Tensions 2 and 4	Through the systems behavior it is evident that multi-cropping has a significant role to play for the planet and prosperity of the village. Further, through simulations we could realize that organic fertilizers play an important role in the development of planet. Monocropping comes with excessive use of chemical fertilizers (to increase production and gain profits) to match with the nutrients that soil has lost. Thus, it is important to promote multi-cropping to reduce the use of chemical fertilizers whilst ensuring that the production through agriculture is decent.
2	How can we ensure higher production and progress of the village by promoting eco-friendly methods of agriculture without exploiting the available water?	Tensions 4 and 11	The use of fertilizers in mono cropping not only affects the overall progress of the village but also degrades the planet. However, villagers are unaware about efficient and proper methods of agriculture. With the introduction of new methods there are high chances of exploitation of natural resources like water. Thus, it is very important to ensure that we preserve the planet and also help people and ensure prosperity by adopting eco-friendly methods of agriculture.
3	How can we reduce the overdependence of farmers on agroforestry and enable them to get decent production and sources of production through agricultural practices without making them overly relied on chemical fertilisers?	T1, T5, T15, T17	Reducing overdependence of agroforestry comes with excessive agriculture. Increased participation in agriculture results in exploitation of resources and degradation of planet due to excessive monocropping, high tillage, improper use of water, etc. Excessive monocropping and high tillage results in loosening of the soils water holding capacity which results in waste of available water. Thus, it is important to reduce the overdependence of farmers whilst also ensuring that their participation in agriculture does not harms the planet.

Copyright © 2023 by ASME

Table 5: SUMMARY OF KEY ISSUES ADDRESSED

Wicked Problem Characteristics	Contextualization
There is no definitive formulation of wicked problems. They are difficult to frame.	The framework provides a holistic approach to frame and identify correct size of wicked problems.
Wicked problems have no stopping rule.	We provide an opportunity to iterate through the framework. This augments the combination of interpretive and evidentiary analysis to understand the nature of the problem. This allows a designer to do better with every iteration until reaching a certain point. By iterating, a designer can identify the finer details and relation between variables which result in appropriate framing of wicked problem.
Solutions to wicked problems are not true or false, but good or bad.	Through the combination of Dilemma Triangle Method and System Dynamics we approach towards getting better insights for every iteration we have, enabling us to frame problems correctly which is the first step to approach better solutions.
There is no immediate and no ultimate test of a good solution to a wicked problem.	Through the framework we provide an opportunity for a designer to involve human intelligence and cognitive abilities which augment computational capabilities in order to enhance the fidelity of framing wicked problems. Through this we give a designer a frame to enhance test-ability.
There are numerous explanations for a wicked problem.	With this framework the problem becomes concrete for the a designer to identify the focus without making foregone conclusions about prior explanations.
The planner has no right to be wrong.	Through the framework we provide an opportunity for the a designer to analyze the dilemmas and the causal relationships between different elements (variables). Through the framework a designer can carry analysis to interpret and compare a variety of conflicting scenarios resulting in improvement of the characteristics and 'corroboration' of results with the wickedness of the problem. Thus, with this framework we provide an opportunity to 'improve' rather than aiming to find the truth.

Verification of the Proposed Framework: In Steps 1, 2, and 3 as shown in Figure 5, and further discussed in detail in Section 4, we gain input from the SunMoksha team of experts. After demonstrating the efficacy of the framework for the test problem, we confirm the results obtained through the proposed framework with the experts in SunMoksha Team as well as the villagers. We are informed that more such insights can be gained by modifying the thematic areas and modifying the System Dynamics model. The verification of the proposed framework comes through multiple trials and experimental simulations run by the authors. These experimental trials are carried out with multiple scenarios and verified accordingly in accordance with the SunMoksha team and the authors. The validation of the framework is yet to be carried out.

Limitations of the Proposed Framework: The framework is based on the assumption that the Dilemma Triangle Method is executed rationally without any bias. Further, we assume that all the stakeholders act rationally which ideally should be the case, however, it might not happen. Further, the evolution of variables is not considered in this work, and we assume that variables are proportional with respect to time. This brings us with the following limitations:

i. The efficacy of the framework depends on the expertise, judgement, and interpretations of humans. Inclusion of bias by human cognitive capabilities make the results obtained through the framework futile and it negatively impacts the framing of wicked problems.

ii. System Dynamics used to simulate the system through *predefined data.*

What is the impact of a predefined set of rules and data? When we create a model (even at an initial level) we are explicitly defining the relations between different variables. For example, in stock and flow diagram we specify the causal relations between variables and define it through equations, which is essentially training the model on how it should behave. This implies that we are training the variables on how they should behave with any change which is done through simulations or changing the equations.

Copyright © 2023 by ASME

6. CLOSING REMARKS

"How can a designer frame a wicked problem and identify the variables that can be used to design a socio-economic-technical system?"

In response to the question, in this paper, we propose a framework to frame wicked problems through interpretative analysis by utilizing Dilemma Triangle Method and evidentiary analysis through the System Dynamics modelling. With the proposed framework, we provide an opportunity to a designer to frame any wicked problem. The framework is designed to convert heuristics that a designer has in the initial phase, to insights by the analysis of interaction between variables through systems simulation by using System Dynamics. The framework consists of three stages, the first being the Dilemma Triangle Method, (Step 1; Figure 5) which is used to identify the Drivers and define the Focus of these Drivers. At this stage we do not identify the dilemmas. The second stage (Step 2; Figure 5) includes analyzing the system behavior and the variable interaction through System Dynamics. Through this stage we get insights on behavior of the simulated system and the interaction between variables. This analysis enables us to identify variables which impact wicked problems and thus identify the correct size of the problem by identifying its core. Further, the third stage (Step 3; Figure 5) is revisiting the Dilemma Triangle constructed in Step 1 and modifying it with the insights gained through System Dynamics. This allows us to modify the tensions based on insights gained through the behavior of the simulated system and then identify dilemmas to frame wicked problems.

To illustrate the efficacy of the framework we use an example of Kantashol village, a socio-economic-technical system, which is in Jharkhand, India. In Kantashol, villagers are overdependent on forestry for their livelihood and the practice of agriculture is limited due to multiple reasons including lack of water, insufficient yields due to improper methods of agriculture etc. Villagers have overexploited the ground water, and there is excessive runoff due to excessive tillage for agriculture. We see that there are three thematic areas that need to be considered, namely; Forestry, Agriculture, and Water. We classify this as a wicked problem due to the incomplete and conflicting information available, conflicting perspectives of the villagers in Kantashol, multiple explanations of the existing problems in the village, and the existence of multiple tensions between the Drivers. Initially we attain heuristics from the expertise of the SunMoksha team and experiences of the villagers. We define Drivers, Focus for each Driver, and Issues for each Drivers through the heuristics attained. Further, we create a System Dynamics model to convert the heuristics into insight through the qualitative and quantitative data from SunMoksha team. The results we get from System Dynamics model help us to gather insight on the interaction between variables. This is demonstrated in Section 5 by eliminating the 'amount of tillage' decision variable by showing its interaction with 'multi-cropping' decision variable. We demonstrate through the System Dynamics modelling that 'multi-cropping' is a significant variable that affects the problem, and which suppresses the negative effects caused by 'amount of tillage'. Moreover, we demonstrate that its effect on the objective variables, namely, 'total runoff areas' and 'overall profit' is insignificant in comparison to 'multi-cropping' variable. Thus, we identify the variables which impact the core of the problem leading us to gain insights on the correct size of the problem. Thus, through the proposed framework, a designer can:

i. Convert heuristics into insights to frame a wicked problem through the integration of evidentiary and interpretative analysis.

ii. Understand the interaction between variables through behavior of the simulated system by the virtue of System Dynamics.

iii. Identify the correct size of the problem by eliminating the variables which do not impact the outcome/model, are not relevant to the wicked problem, or not under a designer's control.

The utility of the proposed framework for framing a wicked problem are:

i. Enhancing the synergy between human-computer interaction by allowing human-in-the-loop to enhance framing of the wicked problem through computational capabilities and human abilities.

ii. Enables a designer to convert the heuristics into insights through a structured process.

iii. Perfect integration of interpretative and evidentiary analysis to frame the wicked problems which forms the fundamental step of modelling a public policy.

The proposed framework can be extended in various domains to frame wicked problems. In the following discussion we expand the possibility of application of the proposed framework in different research areas for varied problems. Problem (i) and (ii) are presented by NSF-NASA in a workshop with an objective to provide the context to a designer on wicked problems and extreme design problems whereas Problem (iii) is the one that authors are working on.

i. *Revitalizing Rural Communities that Depend on One Industry*

   Many regions in industrialized countries:
   - Remain isolated and lack access to goods, services, and resources that are vital to thriving.
   - Often suffer from single-industry economic dependencies that limit growth opportunities and upward, both individually and regionally.

   Goal: For rural communities to thrive and become resilient outside of single industry infrastructure.

ii. *Democratizing Medical Supply Delivery*
   - Current medical supply transport is plagued by losses, including a relatively high temperature excursion rate.

- Delivery includes a diverse supply chain.
- Access to medical supplies is limited in some communities in the U.S. and abroad.
- Current U.S. regulatory and liability frameworks do not account for medical transport by non-traditional vehicles such as a drone.

iii. *Environmental Justice in Oklahoma City*

Urban atmospheric pollution is driven by policy decisions negotiated by competing interests including local and regional governments, industry, and citizen's groups. Policy goals may greatly impact exposure to pollutants harmful to health and wellbeing, exemplified in extrema by historic redlining of minority groups in dense clusters near industrial emitters, generating urban canyon effects which further trap already significant pollutants with known health impacts. As the environmental sensor revolution quietly takes place in our urban centers, an opportunity arises to inject cyber system infrastructure data into extant social decision-making frameworks which shape our physical future by helping planners make better informed design decisions for uncertain policy futures. We have identified a sample region in Oklahoma City which meets the historic context above and presents a growth opportunity for a prototype framework to integrate with current policy-Drivers, including growth of sensing infrastructure in the region.

Wicked problems exist in every research area. The preceding are some examples in which the framework for framing wicked problems using evidentiary and interpretive analysis may be used by a designer. The authors have provided an opportunity to a designer to frame wicked problems through evidentiary and interpretative analysis by involving human-in-the-loop and identifying correct size of the problem and the variables which impact the wickedness of wicked problems.

ACKNOWLEDGEMENTS

We acknowledge the support the SunMoksha Pvt. Ltd. team on the ground for their assistance with the data, expertise, and shared opinions. We acknowledge Ayushi Sharma, Vispi Karkaria and Abhishek Yadav for sharing their expertise with us. Janet K. Allen and Farrokh Mistree gratefully acknowledge the John and Mary Moore Chair and the L.A. Comp Chair at the University of Oklahoma.

REFERENCES

[1] Churchman, C. W., 1967, "Guest Editorial: Wicked Problems," Management Science, **14**(4,), pp. B141–B142.

[2] Charmaz, K., 2006, "Constructing Grounded Theory: A Practical Guide Through Qualitative Analysis," Sage Publications Ltd., London.

[3] Kirschke, S., Avellán, T., Benavides, L., Caucci, S., Hahn, A., Müller, A., and Rubio Giraldo, C. B., 2023, "Results-Based Management of Wicked Problems? Indicators and Comparative Evidence from Latin America," Environmental Policy and Governance, **33**(1), pp. 3–16.

[4] Munir, S., Stankovic, J. A., Liang, C.-J. M., and Lin, S., 2013, "Cyber Physical System Challenges for Human-in-the-Loop Control," Proceedings of 8th International Workshop on Feedback Computing (Feedback Computing 13), pp. 1–4.

[5] Schirner, G., Erdogmus, D., Chowdhury, K., and Padir, T., 2013, "The Future of Human-in-the-Loop Cyber-Physical Systems," IEEE Computer, **46**(1), pp. 36–45.

[6] Rittel, H. W. J., and Webber, M. M., 1973, "Dilemmas in a General Theory of Planning," Policy Sci, **4**(2), pp. 155–169.

[7] Termeer, C. J. A. M., Dewulf, A., Breeman, G., and Stiller, S. J., 2015, "Governance Capabilities for Dealing Wisely With Wicked Problems," Administration & Society, **47**(6), pp. 680–710.

[8] Termeer, C. J. A. M., and Dewulf, A., 2019, "A Small Wins Framework to Overcome the Evaluation Paradox of Governing Wicked Problems," Policy and Society, **38**(2), pp. 298–314.

[9] "United Nations. 2015b. Transforming Our World: The 2030 Agenda for Sustainable Development. Resolution Adopted by the General Assembly on 25 September 2015. A/RES/70/1. UN General Assembly, Seventieth Session. Agenda Items 15 and 116. New York: United Nations."

[10] Nature, 2020, "Editorial: Get the Sustainable Development Goals Back on Track," Nature, **577**(7788), pp. 7–8.

[11] Eden, L., and Wagstaff, M. F., 2021, "Evidence-Based Policymaking and the Wicked Problem of SDG 5 Gender Equality," J Int Bus Policy, **4**(1), pp. 28–57.

[12] Daviter, F., 2017, "Coping, Taming or Solving: Alternative Approaches to the Governance of Wicked Problems," Policy Studies, **38**(6), pp. 571–588.

[13] "Principal Sustainability Components: Empirical Analysis of Synergies between the Three Pillars of Sustainability: International Journal of Sustainable Development & World Ecology: Vol 19, No 5" [Online]. Available: https://www.tandfonline.com/doi/abs/10.1080/13504509. 2012.696220. [Accessed: 09-Mar-2023].

[14] Flint, R. W., 2013, *Practice of Sustainable Community Development: A Participatory Framework for Change*, Springer, New York, NY.

[15] Burton, P., 2006, "Modernising the Policy Process," Policy Studies, **27**(3), pp. 173–195.

[16] Daviter, F., 2019, "Policy Analysis in the Face of Complexity: What Kind of Knowledge to Tackle Wicked Problems?," Public Policy and Administration, **34**(1), pp. 62–83.

[17] Mitroff, I. I., 1974, *The Subjective Side of Science : A Philosophical Inquiry into the Psychology of the Apollo*

Moon Scientists, Amsterdam : Elsevier Scientific Pub. Co. ; New York : American Elsevier Pub. Co.

[18] Hammersley, M., 2005, "Is the Evidence-Based Practice Movement Doing More Good than Harm? Reflections on Iain Chalmers' Case for Research-Based Policy Making and Practice," Evidence and Policy, **1**(1), pp. 85–100.

[19] Biesta, G., 2007, "Why 'What Works' Won't Work: Evidence-Based Practice and the Democratic Deficit in Educational Research," Educational Theory, **57**(1), pp. 1–22.

[20] Newman, J., and Head, B. W., 2017, "Wicked Tendencies in Policy Problems: Rethinking the Distinction between Social and Technical Problems," Policy and Society, **36**(3), pp. 414–429.

[21] Tyworth, M., Giacobe, N. A., Mancuso, V. F., McNeese, M. D., and Hall, D. L., 2013, "A Human-in-the-Loop Approach to Understanding Situation Awareness in Cyber Defence Analysis," SESA, ICST, EAI Endorsed Transactions on Security and Safety, **vol.1 13(2):e6**(2).

[22] Yadav, A., Das, A. K., Roy, R. B., Chatterjee, A., Allen, J. K., and Mistree, F., 2017, "Identifying and Managing Dilemmas for Sustainable Development of Rural India," ASME International Conference on Design Theory and Methodology, Paper Number: DETC2017-67592.

[23] Karkaria, V., Das, A. K., Yadav, A., Sharma, A., Allen, J. K., and Mistree, F., 2021, "A Computational Framework for Social Entrepreneurs to Determine Policies for Sustainable Development," ASME 47th Design Automation Conference, Paper Number DETC2021-70827.

[24] Kamala, V. V. R., Das, A. K., Sharma, A., Allen, J. K., and Mistree, F., 2022, "A Method for Social Entrepreneurs to Develop Value Propositions for Sustainable Development," International Journal for Sustainable Development and Planning, **17**(8), pp. 2347–2356.

[25] Kamala, V. V. R., 2020, "A Computational Framework to Foster Sustainable Rural Development in Indian Off-Grid Villages," MS Thesis Industrial and Systems Engineering, University of Oklahoma, Norman, OK.

**Proceedings of the ASME 2023
International Design Engineering Technical Conferences and
Computers and Information in Engineering Conference
IDETC-CIE2023
August 20-23, 2023, Boston, Massachusetts**

DETC2023-110477

ORIENTATION OPTIMIZATION WITH TOPOLOGICAL DERIVATIVE
FOR ANISOTROPIC MATERIALS WITH ROTATIONAL SYMMETRY

Masaki Noda, Kei Matsushima, Takayuki Yamada*

The University of Tokyo, Tokyo, Japan

ABSTRACT

This paper proposes a method for optimizing the orientation of multi-directionally reinforced materials that exhibit rotational symmetry, such as woven CFRP sheets or laminated CFRP sheets. The orientation is represented in a "relaxed" Cartesian format, and symmetry modifications are employed to prevent overlapping design variables for identical material constants. The optimal orientation is estimated using topological derivatives for anisotropic materials, and the orientation distribution is subsequently optimized based on this estimate. The proposed method is validated using a range of numerical examples of elasticity problems.

Keywords: Orientation optimization, Topological derivative, Topology optimization, Rotational symmetric property, Plain weave carbon fiber sheet

1. INTRODUCTION

The weight reduction of mechanical structures is crucial in industries like aerospace and automotive for achieving lower fuel consumption and longer cruising ranges, which ultimately contribute to a more sustainable society. To achieve lightweight structures while maintaining other mechanical performance factors, such as stiffness and durability, composite materials are increasingly being used [1]. Composite materials, including Carbon Fiber Reinforced Plastics (CFRP), are anisotropic, meaning that the overall mechanical performance depends on the orientation, i.e., in which direction the composite material is placed. Therefore, careful attention must be paid to the orientation of the composites for optimum performance of overall the machine structure.

To optimize the orientation of anisotropic materials, significant research has been conducted [2–6]. An exhaustive review of the literature can be found in [7]. To optimize orientation, it is essential to parameterize the orientation. One simple method to parameterize orientation is to use the orientation direction θ.

However, parameterizing orientation by the angle of rotation introduces an ambiguity in the expression for rotation. Specifically, the material constants before and after rotation are the same when the material is rotated by a certain angle, leading to multiple values of parameter θ indicating the optimal orientation. In two-dimensional problems, any material will have the same material constant due to the 2π rotation. This problem is called 2π-ambiguity. This ambiguity results in problems such as multiple optimal solutions or a discontinuity in the distribution of optimal orientation angles within the structure. While ensuring the uniqueness of the optimal solution can be achieved by assuming that the final optimal solution is replaced by a material constant [8] or by bounding the design variable [9], discontinuities in the solution distribution may still pose a troublesome problem [10]. In order to regularize the optimization problem, it is often necessary to guarantee the smoothness of the solution. As a result, a discontinuous distribution of solutions cannot be directly represented, and instead must be replaced by a smoothly transitioning distribution of solutions. This may lead to optimal solutions with orientations that were not originally intended for the structure.

To avoid the discontinuities by the 2π-ambiguity, Nomura et al. [5] utilize the Cartesian representation of the orientation angle θ.

For materials without chiral symmetries, the π-ambiguity should also be considered. To address such issues, Chu et al. [10] use the Cartesian representation for 2θ in the smoothing step of orientation, rather than θ, in order to avoid the π-ambiguity. In addition, Allaire et al. [8] consider the singularity of the orientation field by using a submanifold.

Thus, several previous studies have addressed the issue of the π-ambiguity.

However, materials that have $2N$-fold rotational symmetry are also important from an engineering standpoint. For example, woven CFRP and laminated CFRP with multiple sheets, which are bidirectionally or multi-directionally reinforced CFRP, have rotationally symmetric material constants [11–13]. Additionally,

---

*Corresponding author: t.yamada@mech.t.u-tokyo.ac.jp

Copyright © 2023 by ASME

periodic microstructures often have rotational symmetry [14–16], and graphene, a one-atom-thick allotrope of carbon, has unusual two-dimensional Dirac-like electronic excitations due to the rotational symmetry of its honeycomb structure [17].

Optimizing the orientation of $2N$-fold rotational symmetric materials requires consideration of the π/N-ambiguity. Therefore, in this study, we utilize a Cartesian representation for $2N\theta$ to solve the π/N-ambiguity.

The rest of this paper is organized as follows. Section 2 presents the mathematical formulation of the proposed method. The section starts by defining design variables that represent orientations of materials with symmetric properties. The update scheme for the orientation using the estimated optimal orientation is then introduced, followed by a method to estimate the optimal orientation. In Sec. 3, the validity and usefulness of the proposed method are confirmed by applying it to a linear elasticity problem. Finally, Sec. 4 presents the conclusion of this study.

2. FORMULATION OF ORIENTATION OPTIMIZATION METHOD

2.1 Orientation representation

We consider a two-dimensional structure in which anisotropic materials are spatially distributed. The anisotropy is described by a scalar field (orientation) θ inside the anisotropic materials, which determines the governing equations of the structure. The case of linear elasticity is shown in Section 3. When the material constant has $2N$-fold rotational symmetry, in the proposed method, the orientation θ is represented using auxiliary variables ξ and η as follows:

$$\cos(2N\theta) = \xi,$$
$$\sin(2N\theta) = \eta,$$
$$\sqrt{\xi^2 + \eta^2} = 1. \tag{1}$$

Using these auxiliary variables, we avoid the multivaluedness of θ, which is periodic in π/N.

To facilitate numerical processing in optimization, following the method proposed by Nomura et al. [5], the Cartesian representation is "relaxed" as follows:

$$\cos(2N\theta) = \frac{\xi}{\sqrt{\xi^2 + \eta^2}},$$
$$\sin(2N\theta) = \frac{\eta}{\sqrt{\xi^2 + \eta^2}},$$
$$\sqrt{\xi^2 + \eta^2} \le 1. \tag{2}$$

Figure 1 shows the orientation representation using the auxiliary variables ξ and η for the case of $N = 2$.

2.2 Update scheme

We seek to find the optimal distribution of θ that maximizes (or minimizes) a given objective functional J, such as compliance. To obtain the optimal orientation, the auxiliary variables ξ and η are updated as follows:

$$\xi + \tau\nabla^2\xi = K\cos(2N\theta^{\text{opt}}) + (1-K)\xi^{\text{pre}},$$
$$\eta + \tau\nabla^2\eta = K\sin(2N\theta^{\text{opt}}) + (1-K)\eta^{\text{pre}}, \tag{3}$$

where ξ^{pre} and η^{pre} are the values of ξ and η at the previous step, respectively. K is the step size, and τ is a regularization coefficient that ensures the smooth distribution of the variables in space. If K is small, the algorithm tends to be stable, whereas a large value of K leads to a rapid convergence of the iterative scheme. A very small or zero value of τ may lead to a discontinuous distribution of ξ and η, while a large value of τ may cause the values of ξ and η to be constant throughout the calculation domain. More details on these parameters can be found in [18].

The value of θ^{opt} represents the estimated optimal orientation, which will be discussed in the next subsection. By updating the auxiliary variables ξ and η using Eq. (3), the orientation approaches the optimal orientation, avoiding local optima.

2.3 Estimation of optimal orientation

In some cases, the optimal orientation of a material can be computed exactly, as reported in the literature [2]. However, in general, it is a challenging task to calculate the optimal orientation exactly. Therefore, in this study, we estimate the optimal orientation as the orientation that minimizes a topological derivative, which is an explicit function in terms of orientation θ. The topological derivative is the sensitivity to the substitution of one material for another within an infinitesimal small circular domain Ω_ε centered at x and of radius ε. We assume that the θ-oriented domain is replaced with a material having a different orientation $\tilde{\theta}$. A topological derivative $D^{\theta\to\tilde{\theta}}J$ of object functional J is defined as follows:

$$D^{\theta\to\tilde{\theta}}J(x) = \lim_{\varepsilon\to+0}\frac{J_\varepsilon - J}{\text{meas}(\Omega_\varepsilon)}, \tag{4}$$

where J and J_ε are the objective function before and after replacing materials, $\text{meas}(\Omega_\varepsilon)$ is the measure of the domain Ω_ε.

Then, the estimated optimal orientation can be written as follows:

$$\theta^{\text{opt}}(x) = \underset{\tilde{\theta}\in[0,\pi/N)}{\arg\min}\, D^{\theta\to\tilde{\theta}}J(x). \tag{5}$$

3. NUMERICAL EXAMPLE

To verify the validity of the proposed method, several numerical examples are provided.

First, the proposed method is applied to an orientation optimization problem for a linear elasticity problem, which is formulated as follows:

$$\min_{\xi,\eta} J = \int_{\Gamma^t} t \cdot u(x)\mathrm{d}x, \tag{6}$$

$$\text{subject to} \int_D \mathscr{E}(u(x)) : C(x) : \mathscr{E}(v(x))\mathrm{d}x$$

$$= \int_{\Gamma^t} t \cdot v(x)\mathrm{d}x, \tag{7}$$

$$u(x) = 0 \quad x \in \Gamma^u, \tag{8}$$

where $\mathscr{E}(u) := \frac{1}{2}(\nabla u + \nabla u^T)$ is the strain tensor, u is a displacement field, v is a test function, the vector t is a traction force imposed on a boundary Γ^t, boundary Γ^u is a fixed boundary, D is

Copyright © 2023 by ASME

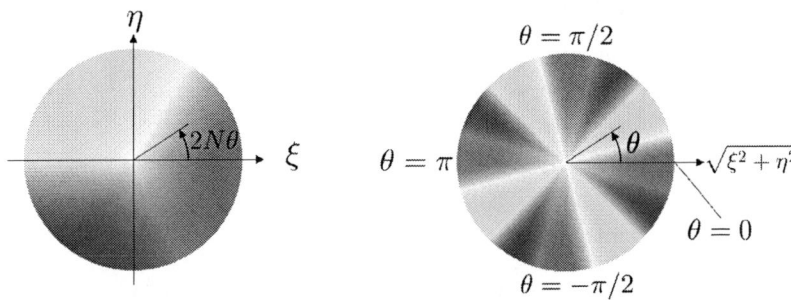

FIGURE 1: ORIENTATION REPRESENTATION. ON THE LEFT IS CARTESIAN COORDINATE SPACE AND ON THE RIGHT IS THE POLAR CO-ORDINATE SPACE, WITH EACH MATERIAL CONSTANT SET FOR EACH COLOR POINT.

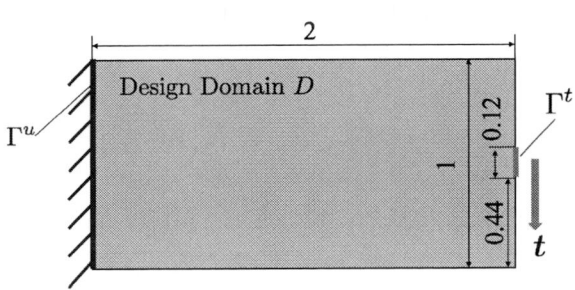

FIGURE 2: FIXED DESIGN DOMAIN AND BOUNDARY CONDITIONS

a design domain, and $C(x)$ is an elastic modulus tensor defined as follows:

$$C(x) = C^F\Big(\theta\big(\xi(x),\eta(x)\big)\Big), \qquad (9)$$

$$C^F_{ijkl}(\theta) = \overline{C^F}_{mnpq} R_{im}(\theta) R_{jn}(\theta) R_{kp}(\theta) R_{lq}(\theta), \qquad (10)$$

where $C^F(\theta)$ is the elastic modulus tensor oriented in θ of anisotropic material, whose components are defined in Eq. (10), $\overline{C^F}_{mnpq}$ are the components of a given anisotropic elastic modulus tensor $\overline{C^F}$, and $R_{ij}(\theta)$ are the components of a rotation tensor $R(\theta)$ defined as follows:

$$R(\theta) = \begin{pmatrix} \cos\theta & \sin\theta \\ -\sin\theta & \cos\theta \end{pmatrix}. \qquad (11)$$

Note that $\overline{C^F}$ should have $2N$-fold rotation symmetry, i.e., $\overline{C^F} = C^F(\frac{n}{N}\pi)$ $(n \in \{1,\ldots,2N\})$.

In this situation, the topological derivative is calculated as follows:

$$\mathrm{D}^{\theta\to\bar\theta} J = \mathscr{E}(u) : A^{\theta\to\bar\theta} : \mathscr{E}(u), \qquad (12)$$

where $A^{\theta\to\bar\theta}$ is the elastic moment tensor [19, 20].

Figure 2 shows the design domain D and boundaries. The traction force is set to $t = (0,-1)^T$. The anisotropic elastic modulus tensor $\overline{C^F}$ is set as follows:

$$E = 100,$$
$$G = 10,$$
$$\nu = 0.3,$$

$$\overline{C^F}_{ijkl} = \begin{cases} \frac{E}{1-\nu^2} & i=j=k=l \\ \frac{E\nu}{2(1-\nu^2)} & i=j, k=l, i \neq k \\ \frac{G}{2(1+\nu)} & i \neq j, k \neq l, i=k \\ 0 & \text{otherwise} \end{cases}. \qquad (13)$$

Figures 3 show the optimization results. As shown in Figs. 3, the orientation changes drastically throughout the optimization process. Figure 3d shows the principal stress direction for the final design. By comparing Figs. 3c and 3d, we can observe that the orientation and principal stress direction mostly coincide, indicating that a reasonable optimal result has been successfully obtained.

Next, the proposed method is applied to the simultaneous optimization of orientation and topology. Specifically, we consider an anisotropic material denoted by F, an isotropic material denoted by I, and a void denoted by V.

The optimization problem we aim to solve is formulated as follows:

$$\text{design variables:} \Omega_I, \Omega_V, \Omega_F, \xi, \eta, \qquad (14)$$

$$\min J = \int_{\Gamma^t} t \cdot u(x)\mathrm{d}x, \qquad (15)$$

$$\text{subject to } \int_D \mathscr{E}\big(u(x)\big) : C^M(x) : \mathscr{E}\big(v(x)\big)\mathrm{d}x$$
$$= \int_{\Gamma^t} t \cdot v(x)\mathrm{d}x, \qquad (16)$$

$$u(x) = 0 \quad x \in \Gamma^u, \qquad (17)$$

$$\Omega_I \cup \Omega_V \cup \Omega_F = D, \qquad (18)$$

$$\Omega_a \cap \Omega_b = \emptyset \text{ for } b \neq a, \qquad (19)$$

$$\int_{\Omega_I} \rho_I \mathrm{d}x + \int_{\Omega_F} \rho_F \mathrm{d}x \leq \hat{W}, \qquad (20)$$

where Ω_I, Ω_V, and Ω_F denote the domains of materials I, V, and F, respectively. Additionally, ρ_I and ρ_F represent the mass density of materials I and F, respectively. Furthermore, $\hat{W}$ denotes

Copyright © 2023 by ASME

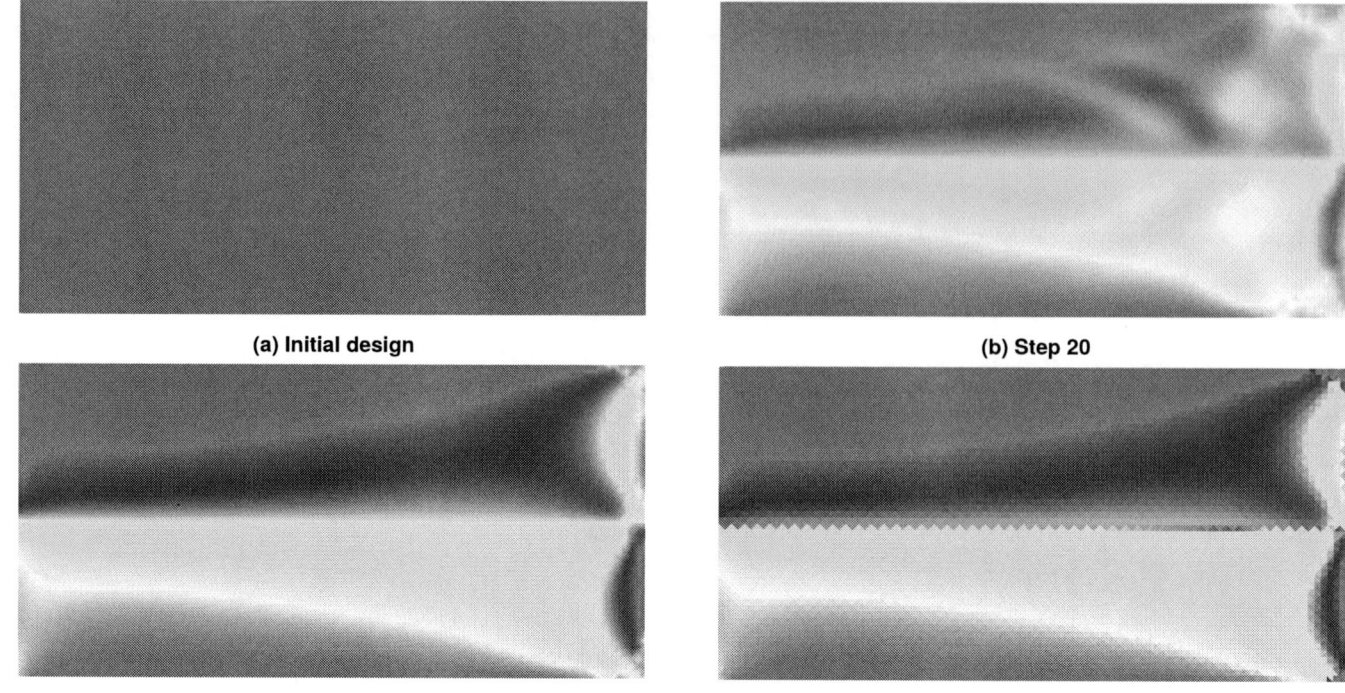

(a) Initial design

(b) Step 20

(c) Final design

(d) Principal stress direction for final design

FIGURE 3: RESULT FOR ORIENTATION-ONLY OPTIMIZATION. THE ORIENTATION AND PRINCIPAL STRESS DIRECTIONS ARE REPRESENTED BY COLORS, AS SHOWN IN FIG. 1.

the maximum weight for the weight constraint, and $C^M(x)$ is a multi-material elastic modulus tensor defined as follows:

$$C^M(x) = \begin{cases} C^I & x \in \Omega_I \\ C^V & x \in \Omega_V \\ C^F(\theta) & x \in \Omega_V \end{cases}, \qquad (21)$$

where C^I is an isotropic elastic modulus tensor. To represent the void domain, we adopt the Ersatz material approach by assigning a small isotropic elastic modulus tensor, denoted by C^V.

The proposed method is utilized to optimize the orientation. Domains Ω_I, Ω_V, and Ω_F, for each material I, V, and F, respectively, are optimized using an extended level set-based multi-material topology optimization method, proposed in [21].

The maximum weight is set to $\hat{W} = 1.2$. The mass densities are set to $\rho_I = 2$ and $\rho_F = 1$. The design domain, boundary conditions, and anisotropic elastic modulus $\overline{C^F}$ are set the same as in the anisotropic-only case, as shown in Fig. 2 and Eq. (13). The elastic modulus tensors for isotropic material C^I and void

C^V are set as follows:

$$E^I = 200,$$
$$E^V = 0.01,$$
$$v^I = 0.3,$$
$$v^V = 0.3,$$

$$C_{ijkl}^I = \begin{cases} \frac{E^I}{1-(v^I)^2} & i=j=k=l \\ \frac{E^I v^I}{2(1-(v^I)^2)} & i=j, k=l, i \neq k \\ \frac{E^I}{2(1+v^I)} & i \neq j, k \neq l, i=k \\ 0 & \text{otherwise} \end{cases},$$

$$C_{ijkl}^V = \frac{E^V}{E^I} C_{ijkl}^I. \qquad (22)$$

Figures 4 show the optimization result.

In Fig. 4e, one of the reinforced directions is in accordance with the principal stress direction as shown in Fig. 4f. Isotropic materials with a large Young's modulus are placed near the root and load boundary where stress is concentrated. Therefore, the obtained structure is a reasonable optimal structure.

4. CONCLUSION

This study proposes a method to optimize the orientation of multi-directionally reinforced anisotropic materials. The proposed method is then applied to solve the mean compliance minimization problem for anisotropic materials exhibiting four-fold rotational symmetry, resulting in reasonable and optimal solutions.

Copyright © 2023 by ASME

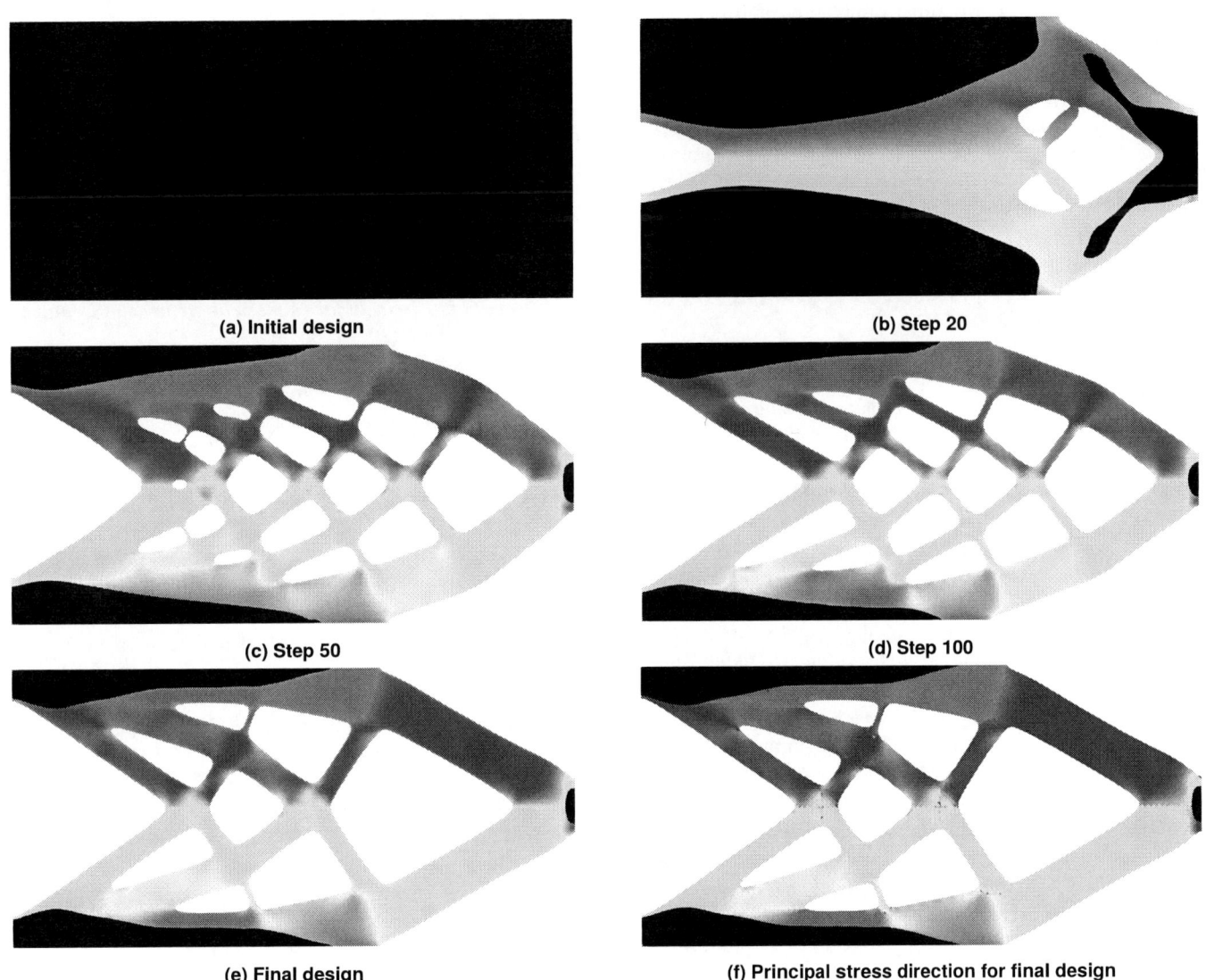

(a) Initial design

(b) Step 20

(c) Step 50

(d) Step 100

(e) Final design

(f) Principal stress direction for final design

FIGURE 4: RESULT FOR ORIENTATION AND TOPOLOGY OPTIMIZATION. THE ORIENTATION AND PRINCIPAL STRESS DIRECTIONS ARE REPRESENTED BY COLORS, AS SHOWN IN FIG. 1. BLACK AND WHITE DOMAINS INDICATE ISOTROPIC MATERIAL AREAS AND CAVITIES, RESPECTIVELY.

Copyright © 2023 by ASME

ACKNOWLEDGMENTS

This work was supported by JST FOREST Program (Grant Number JPMJFR202J, Japan).

REFERENCES

[1] Soutis, Costas. "Carbon fiber reinforced plastics in aircraft construction." *Materials Science and Engineering: A* Vol. 412 No. 1-2 (2005): pp. 171–176.

[2] Pedersen, Pauli. "On optimal orientation of orthotropic materials." *Structural optimization* Vol. 1 (1989): pp. 101–106.

[3] Bendsoe, Martin P, Guedes, JM, Haber, Robert B, Pedersen, P and Taylor, JE. "An analytical model to predict optimal material properties in the context of optimal structural design." *Journal of Applied Mechanics* Vol. 61 (1994): pp. 930–937.

[4] Stegmann, Jan and Lund, Erik. "Discrete material optimization of general composite shell structures." *International Journal for Numerical Methods in Engineering* Vol. 62 No. 14 (2005): pp. 2009–2027.

[5] Nomura, Tsuyoshi, Dede, Ercan M, Lee, Jaewook, Yamasaki, Shintaro, Matsumori, Tadayoshi, Kawamoto, Atsushi and Kikuchi, Noboru. "General topology optimization method with continuous and discrete orientation design using isoparametric projection." *International Journal for Numerical Methods in Engineering* Vol. 101 No. 8 (2015): pp. 571–605.

[6] Desai, Akshay, Mogra, Mihir, Sridhara, Saketh, Kumar, Kiran, Sesha, Gundavarapu and Ananthasuresh, GK. "Topological-derivative-based design of stiff fiber-reinforced structures with optimally oriented continuous fibers." *Structural and Multidisciplinary Optimization* Vol. 63 (2021): pp. 703–720.

[7] Nikbakt, S, Kamarian, S and Shakeri, M. "A review on optimization of composite structures Part I: Laminated composites." *Composite Structures* Vol. 195 (2018): pp. 158–185.

[8] Allaire, Grégoire, Geoffroy-Donders, Perle and Pantz, Olivier. "Topology optimization of modulated and oriented periodic microstructures by the homogenization method." *Computers & Mathematics with Applications* Vol. 78 No. 7 (2019): pp. 2197–2229.

[9] Xia, Qi and Shi, Tielin. "A cascadic multilevel optimization algorithm for the design of composite structures with curvilinear fiber based on Shepard interpolation." *Composite Structures* Vol. 188 (2018): pp. 209–219.

[10] Chu, Sheng, Xiao, Mi, Gao, Liang, Zhang, Yan and Zhang, Jinhao. "Robust topology optimization for fiber-reinforced composite structures under loading uncertainty." *Computer Methods in Applied Mechanics and Engineering* Vol. 384 (2021): p. 113935.

[11] Fujita, Akihiro, Hamada, Hiroyuki and Maekawa, Zenichiro. "Tensile properties of carbon fiber triaxial woven fabric composites." *Journal of composite materials* Vol. 27 No. 15 (1993): pp. 1428–1442.

[12] Hochard, Ch, Aubourg, P-A and Charles, J-P. "Modelling of the mechanical behaviour of woven-fabric CFRP laminates up to failure." *Composites Science and Technology* Vol. 61 No. 2 (2001): pp. 221–230.

[13] Tyler, T. "Developments in triaxial woven fabrics." *Specialist yarn and fabric structures: developments and applications* Vol. 141 (2011): pp. 141–163.

[14] Sigmund, Ole and Torquato, Salvatore. "Composites with extremal thermal expansion coefficients." *Applied Physics Letters* Vol. 69 No. 21 (1996): pp. 3203–3205.

[15] Neves, Miguel M, Sigmund, Ole and Bendsøe, Martin P. "Topology optimization of periodic microstructures with a penalization of highly localized buckling modes." *International Journal for Numerical Methods in Engineering* Vol. 54 No. 6 (2002): pp. 809–834.

[16] Andreassen, Erik and Jensen, Jakob Søndergaard. "Topology optimization of periodic microstructures for enhanced dynamic properties of viscoelastic composite materials." *Structural and Multidisciplinary Optimization* Vol. 49 No. 5 (2014): pp. 695–705.

[17] Neto, AH Castro, Guinea, Francisco, Peres, Nuno MR, Novoselov, Kostya S and Geim, Andre K. "The electronic properties of graphene." *Reviews of modern physics* Vol. 81 No. 1 (2009): p. 109.

[18] Yamada, Takayuki, Izui, Kazuhiro, Nishiwaki, Shinji and Takezawa, Akihiro. "A topology optimization method based on the level set method incorporating a fictitious interface energy." *Computer Methods in Applied Mechanics and Engineering* Vol. 199 No. 45-48 (2010): pp. 2876–2891.

[19] Bonnet, Marc and Delgado, Gabriel. "The topological derivative in anisotropic elasticity." *Quarterly Journal of Mechanics and Applied Mathematics* Vol. 66 No. 4 (2013): pp. 557–586.

[20] Delgado, Gabriel and Bonnet, Marc. "The topological derivative of stress-based cost functionals in anisotropic elasticity." *Computers & Mathematics with Applications* Vol. 69 No. 10 (2015): pp. 1144–1166.

[21] Noda, Masaki, Noguchi, Yuki and Yamada, Takayuki. "Extended level set method: A multiphase representation with perfect symmetric property, and its application to multi-material topology optimization." *Computer Methods in Applied Mechanics and Engineering* Vol. 393 (2022): p. 114742.

Copyright © 2023 by ASME

**Proceedings of the ASME 2023
International Design Engineering Technical Conferences and
Computers and Information in Engineering Conference
IDETC-CIE2023
August 20-23, 2023, Boston, Massachusetts**

DETC2023-110756

MIXED-VARIABLE GLOBAL SENSITIVITY ANALYSIS WITH APPLICATIONS TO DATA-DRIVEN COMBINATORIAL MATERIALS DESIGN

Yigitcan Comlek[1], Liwei Wang[1], Wei Chen[1],*

[1]Northwestern University
Evanston, IL

***Corresponding Author:** Dr. Wei Chen (weichen@northwestern.edu)

ABSTRACT

Global Sensitivity Analysis (GSA) is the study of the influence of any given inputs on the outputs of a model. In the context of engineering design, GSA has been widely used to understand both individual and collective contributions of design variables on the design objectives. So far, global sensitivity studies have often been limited to design spaces with only quantitative (numerical) design variables. However, many engineering systems also contain, if not only, qualitative (categorical) design variables in addition to quantitative design variables. In this paper, we integrate the novel Latent Variable Gaussian Process (LVGP) with Sobol' analysis to develop the first metamodel-based mixed-variable GSA method. Through two analytical case studies, we first validate and demonstrate the effectiveness of our proposed method for mixed-variable problems. Furthermore, while the new metamodel-based mixed-variable GSA method can benefit various engineering design applications, we employ our method with multi-objective Bayesian optimization (BO) to accelerate the Pareto front design exploration in many-level combinatorial design spaces. Specifically, we implement a sensitivity-aware design framework for metal-organic framework (MOF) materials that are constructed only from qualitative design variables and show the benefits of our method for expediting the exploration of novel MOF candidates from a many-level large combinatorial design space.

Keywords: Global Sensitivity Analysis, Metamodels, Latent Variable Gaussian Process, Mixed-Variable Design Spaces, Bayesian Optimization

NOMENCLATURE

LVGP	Latent Variable Gaussian Process
GSA	Global Sensitivity Analysis
MV	Mixed-Variable
BO	Bayesian Optimization

1. INTRODUCTION

In the context of engineering design research, sensitivity analysis (SA) can be defined as the study of how the input design variables, independently or interactively, influence the output, i.e., the design objective of interest [1]. SA can help designers build design models, whether by identifying design factors that contribute to achieving goals, calibrating model parameters, observing input and output uncertainties, reducing design space and verifying whether the general trend of a model matches with the real system.

SA can be considered on two levels: local and global [2]. Local sensitivity analysis focuses on understanding the impact of local perturbations of design variables on the design objective [3]. On the other hand, global sensitivity analysis (GSA) aims to apportion the design variables' variations to objective variation over the entire range. Among various types of approaches for GSA, the variance-based GSA is the most commonly used [4-6]. Variance-based methods aim to analyze how much of the overall variability of the response can be related to the variability of the design variables [7]. Within the variance-based methods, Sobol' indices are commonly used sensitivity measures that explain the contribution of each design variable on the variability of the response through two sensitivity metrics, i.e., Main Sensitivity Index (MSI) and Total Sensitivity Index (TSI) [8]. The MSI aims to quantify how each design variable contributes to the response individually, whereas the TSI quantifies the interaction effect within design variables on the response. Evaluations of Sobol' metrics typically require thousands of samples for statistical analysis, which is infeasible if the design response is computationally expensive to obtain [9, 10]. However, with the recent advancements in metamodeling and machine learning, the high computational cost of evaluating responses has been replaced by accurate metamodels, including but not limited to Kriging models, Neural Networks, and Gaussian Process models. Within this context, Chen et al. have developed

Copyright © 2023 by ASME

analytical sensitivity analysis using metamodels to accelerate the calculations of Sobol' sensitivity metrics [11].

Thus far, to the best of the authors' knowledge, the GSA techniques are limited to studies with only quantitative (numerical) design variables, while qualitative (categorical) design variables are ubiquitous in many engineering applications such as materials design and topology optimization [12, 13]. There is a clear need for incorporating qualitative design variables into GSA studies to understand their influences on the design objectives. The key challenge hindering the application of GSA for qualitative variables is the lack of a corresponding sampling technique and accurate surrogate modeling for mixed-variable design problems. The aim of this research is thus to create a mixed-variable GSA method by integrating Latent Variable Gaussian Process (LVGP) and a new mixed-variable sampling methodology for metamodel-based Sobol' indices calculations. The LVGP allows the incorporation of qualitative variables into Gaussian Process (GP) models by implicitly mapping each qualitative variable into a quantitative space through low-dimensional latent variables [14]. With this approach, the qualitative variables can be directly implemented into the GSA studies through metamodel-based Sobol' indices calculations. Furthermore, this mixed-variable GSA approach is integrated with Bayesian optimization to accelerate the search in many-level combinatorial materials design.

The outline of the paper is as follows. In Section 2, the LVGP and metamodel-based Sobol' sensitivity indices are explained in detail. Next, the mixed-variable GSA method is introduced. In Section 3, the method is tested on two mathematical functions commonly used in literature for GSA studies. Finally, a design framework consisting of mixed-variable GSA and Bayesian optimization is demonstrated to accelerate the novel metal-organic framework materials. The mixed-variable GSA method has been shown to accelerate the exploration of optimum solutions in a large combinatorial design space.

2. METAMODEL-BASED MIXED-VARIABLE GLOBAL SENSITIVITY ANALYSIS METHOD

With the integration of LVGP and metamodel-based Sobol' indices, we are proposing a metamodel-based mixed-variable GSA method to incorporate qualitative variables into GSA applications.

2.1 Latent Variable Gaussian Process (LVGP)

Gaussian processes (GPs) are well-known interpolation-based metamodels that have been used in many engineering applications [15]. However, due to the nature of the GP correlation functions, qualitative variables cannot be implemented directly without a well-defined distance metric. The key idea behind the LVGP modeling is that for every qualitative variable, there exists an underlying, potentially high-dimensional, quantitative space that explains the qualitative variable's influence on the response of interest. Using this knowledge, the latent variable approach aims to construct an

implicit mapping of qualitative variables into a low-dimensional quantitative latent space, approximating the effects of underlying variables.

Consider a design space with mixed-variable input, $w = [x^T, t^T]^T$ where $x = [x_1, x_2, \ldots, x_q]^T \in R^q$ are quantitative variables and $t = [t_1, t_2, \ldots, t_m]^T$ are qualitative design variables. Each qualitative design variable t_j has l_j design options (levels), i.e., $t_j \in \{1, 2, \ldots, l_j\}$ for $j = 1, 2, \ldots m$. For real physical models, there are quantitative variables $v(t_j) = [v_1(t_j), v_2(t_j) \ldots, v_n(t_j)] \in R^n$ underlying each qualitative variable, which could be unknown, undiscovered, or extremely high-dimensional. The key idea of LVGP is to learn a low-dimensional latent space to approximate the original underlying qualitative space via statistical inference. Although the dimensions of the learned latent variable vector $z \in R^k$ can be freely chosen, a two-dimensional (2D) latent vector, $k = 2$, is usually sufficient in most engineering designs [14], which is also adopted in this study. Therefore, each level of a qualitative variable t_j is represented by a 2D latent vector $z(t_j) = [z_{j,1}, z_{j,2}]^T$. Then, the transformed design space becomes $h = [x^T, z(t)^T]^T \in R^{(q+m \times 2)}$ with $z(t) = [z_{1,1}, z_{1,2}, \ldots, z_{j,1}, z_{j,2}, \ldots, z_{m,1}, z_{m,2}]^T$. An illustration of the implicit mapping from qualitative variables, represented by different shapes, to two-dimensional latent variables is shown in Figure 1.

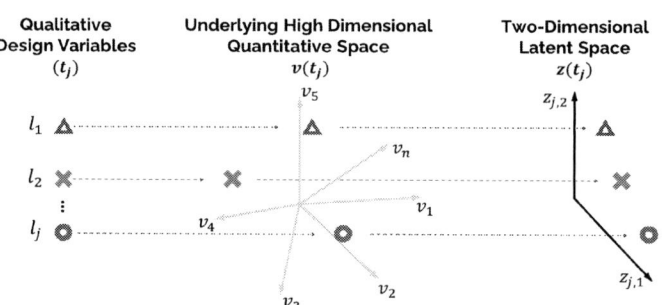

FIGURE 1: THE MAPPING OF QUALITATIVE VARIABLES ONTO 2-DIMENSIONAL LATENT SPACE.

Now, consider a single response GP model with a prior constant mean μ to describe the mean response at any given point in the transformed design space h. A zero-mean Gaussian Process is used to capture the variance of the response, described by a covariance function $K(h, h')$. The covariance function $K(h, h') = \sigma^2 \cdot c(h, h')$ describes the relationship or the correlation of responses at any pairs of input points h and h', where σ^2 represents the prior variance of the GP model and $c(h, h')$ is the correlation function. LVGP extends the commonly used Gaussian correlation function to include latent variables,

$$c(h, h') = \exp\left(\sum_{i=1}^{q} \phi_i (x_i - x_i')^2 - \sum_{j=1}^{m} \left\| z_{j,1} - z_{j,1}' \right\|_2^2 + \left\| z_{j,2}(t_j) - z_{j,2}' \right\|_2^2 \right), \quad (1)$$

where ϕ_i is a scaling parameter that will be estimated for each quantitative variable x_i. Note that the scaling parameters of latent variables are a unity vector since these scaling factors are absorbed into the estimated latent variable values $z(t)$. The rationale behind this correlation function is that points closed in the design space h should also exhibit similar output patterns. For a size-n training sample set with input $W_s = \left[w^{(1)}, w^{(2)}, \dots, w^{(n)} \right]^T$ and output $y_s = \left[y^{(1)}, y^{(2)}, \dots, y^{(n)} \right]^T$, the parameters, μ, σ, and ϕ, along with the 2D mapped latent variables, $z(t)$, are estimated through Maximum Likelihood Estimation (MLE), i.e., finding parameters to maximize the log-likelihood function,

$$l(\mu, \sigma, \phi, z) = -\frac{n}{2} \ln(\sigma^2) - \frac{1}{2} ln|C(z, \phi)| - \frac{1}{2\sigma^2} (y_s - \mu \mathbf{1})^T C(z, \phi)^{-1} (y_s - \mu \mathbf{1}), (2)$$

where C is the $n \times n$ correlation matrix with $C_{ij} = c(h^{(i)}, h^{(j)})$ for $i, j = 1, 2, 3, \dots, n$, $h^{(i)}$ is the transformed input from the sample $w^{(i)}$, $\mathbf{1}$ is a vector of ones with dimensions of $n \times 1$. Once the latent variables $z(t)$ and the parameters u, σ, and ϕ are estimated, the fitted model will be used for metamodel-based GSA. Readers interested further in LVGP are referred to the original paper [14].

2.2 Metamodel-Based Sobol' Indices Calculations for Global Sensitivity Analysis

Sobol' sensitivity indices are variation-based GSA metrics, with the Main Sensitivity Index (MSI) and Total Sensitivity Index (TSI) indicating the contribution of each design variable and their interactions on the response's variation, respectively. For a design system with q continuous design variables, the Sobol' indices can be calculated as below:

$$S_i = \frac{Var_x \left[E_{x_1, \dots, x_q} y \middle| x_i \right]}{Var_{x_1, \dots, x_q}[y]} = \frac{V_i}{V}, (3)$$
$$S_i^t = S_i + S_{i,\sim i}, (4)$$
$$for\ i = 1, \dots, q$$

where y is the response of interest, V_i is the variance of the response with respect to the changes in design variable x_i, V is the variance of the response, S_i is the MSI, S_i^t is the TSI of the design variables, and $S_{i,\sim i}$ is the higher order Sobol' sensitivity indices between variable x_i and remaining variables $x_{\sim i}$. To evaluate the indices, a Monte Carlo sampling is usually adopted, requiring a large number of response evaluations that are unaffordable for design applications with expensive simulation costs. To overcome this challenge, Chen et al. developed metamodel-based Sobol' indices calculations through tensor product formulation [11]. This methodology enabled the integration of metamodels to provide fast and accurate response evaluation for GSA studies. Herein, we use LVGP as our mixed-variable metamodel and extend the formulation in [11] to be

$$S_i = \frac{Var_{w_i} E_{w, \dots, w_d} \left[E \left[y_{LVGP|W_s, y_s}(w) \right] \middle| w_i \right]}{Var_{w_i, \dots, w_d} E \left[y_{LVGP|W_s, y_s}(w) \right]}, (5)$$
$$for\ i = 1, \dots, d$$

Here, the $w = [x^T, t^T]^T \in \mathrm{R}^d$, where $d = q + m$, corresponds to the mixed-variable design space with quantitative (x_q) and qualitative (t_m) variables, and $y_{LVGP|W_s, y_s}$ is the predicted response using the LVGP model trained on the sample set W_s, and response y_s. It should be noted that the 2D latent variables have been shown to be sufficient and are employed for response prediction only. Sobol' indices are calculated by directly sampling mixed-variable design space, using the original qualitative variable t instead of its 2D latent variables. More details on the sampling are described in the next section. With the integration of metamodel-based GSA calculations and LVGP, we can then incorporate qualitative variables into GSA applications. Although there are other metamodel options that can handle mixed-variable inputs, such as Random Forest (RF), the prediction accuracy of LVGP was shown to be better in most cases when using the same sample size [14]. Therefore, we chose LVGP as our metamodel in this study.

2.3 Metamodel-Based Mixed-Variable Global Sensitivity Analysis Method

Given a mixed-variable design space, the GSA is done through the method shown in Figure 2. First, the mixed-variable samples are fed into the LVGP for model fitting. Once the trained model is ready, the Sobol' analysis starts with sampling the qualitative and quantitative variables from quasi-random Sobol sets, filling the given design space range in a highly uniform manner. In particular, to achieve discrete sampling for qualitative variables, once the Sobol set is created, the sample set is sliced into l_j number of sections for each qualitative variable t_j. In this way, each level is sampled equally in the quasi-random set. Next, the LVGP model is employed to make fast and accurate predictions on the sample space. Finally, using the tensor product formulation given in Equation (5), the Sobol' sensitivity metrics, MSI and TSI are obtained. Although the demonstrated framework is named for mixed-variable design spaces, the framework is also applicable for design spaces with only qualitative variables.

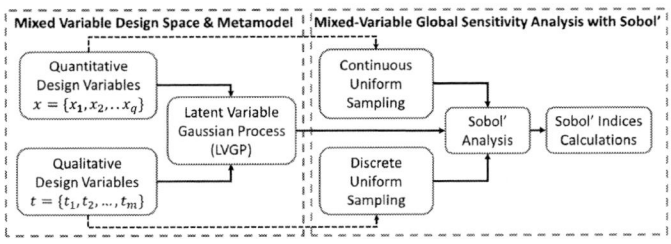

FIGURE 2: THE METAMODEL BASED MIXED-VARIABLE GLOBAL SENSITIVITY ANALYSIS METHOD

Copyright © 2023 by ASME

3. VALIDATION WITH NUMERICAL EXAMPLES

Before applying our method to real engineering design applications, we validated it through well-known testing functions for GSA studies [16]. To achieve this, we first converted some design variables into qualitative variables by assigning some predetermined discrete values to the variables. This allows us to convert quantitative design spaces into mixed-variable design spaces as test cases. As the design spaces are converted to mixed-variable sets, we expect that the influence of both the existing quantitative variables and the newly generated qualitative variables will change from their original effects. Therefore, we implemented a two-stage verification mechanism to validate our Mixed-Variable GSA method. First, we compared the Mixed-Variable (MV) Sobol' indices obtained from our method with the Ground-Truth Mixed-Variable (True-MV) Sobol' indices. Here the True-MV indices correspond to the ground-truth value of Sobol' metrics when the design space is mixed-variable. Since the equations are known, the true values of the metrics can be obtained through the sampling scheme defined in Section 2.3. The validation at this stage checks whether the mixed-variable metrics obtained from our method are matching well with the true mixed-variable metrics. In the second stage, we compared the MV Sobol' indices with the ground-truth Sobol' values obtained from the GSA with only quantitative design variables. The key idea behind this validation is that as the number of levels, i.e., design choices, increase, the Sobol' indices values must approach the true Sobol' metrics with only quantitative variables. This is due to the fact that the continuous sampling space can be considered as a discrete space with an infinite number of levels. As a result, we expected to see the convergence of MV Sobol' indices towards the ground-truth Sobol' indices values. With the defined two validation criteria, we validated our method on two mathematical functions commonly used in GSA studies: Ishigami and Hartman 6D functions.

3.1 Validation with Ishigami Function

The Ishigami function is a highly non-linear testing function with three variables that have been extensively used in GSA studies [17]. The formulation of the Ishigami function, as well as its ground-truth Sobol' sensitivity indices, are given in Equation (6) and Table 1, respectively.

$$f(x) = \sin(x_1) + 7\sin^2(x_2) + 0.1x_3^4\sin(x_1) \quad (6)$$
$$for \; x = [x_1, x_2, x_3] \, \epsilon \, [-\pi, \pi]$$

TABLE 1. SOBOL' SENSITIVITY INDICES OF THE ISHIGAMI FUNCTION WITH QUANTITATIVE INPUTS

Design Variable	Main Sensitivity Index (MSI)	Total Sensitivity Index (TSI)
x_1	0.3138	0.5575
x_2	0.4413	0.4424
x_3	0	0.2436

To implement the Ishigami function as a mixed-variable problem, we converted the design variables x_1 and x_3 to qualitative variables by assigning different discrete values (levels) between the given ranges in Equation (6). For ease of illustration, we renamed x_1 as t_1 and x_3 as t_3. For instance, in the case of 3 levels, the variables t_1 and t_3 can only take the discrete values of $(-\pi, 0, \pi)$. To test each validation stage, we looked at cases with a number of levels ranging from [2, 20]. For each level instance, we trained an LVGP model to learn the response surface of the Ishigami function. Next, we used our proposed method to calculate the MSI and TSI for all instances. Finally, we obtained the ground-truth discrete and quantitative (True) Sobol' metrics from the Ishigami function and compared the results. Figure 3 shows both the MSI and TSI values obtained for validation. The first column shows the MSI and the second column shows the TSI for all three design variables.

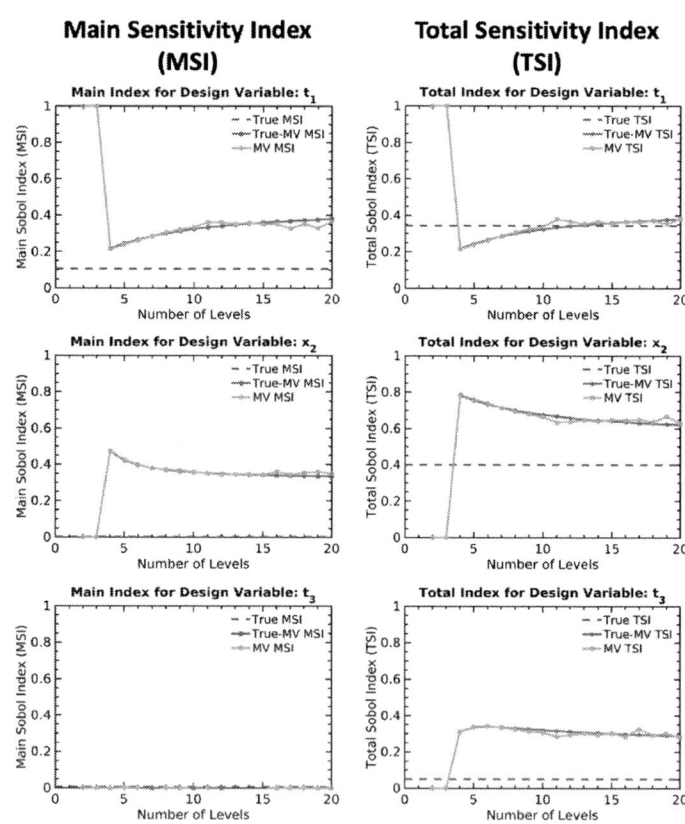

FIGURE 3: SOBOL' SENSITIVITY INDICES OF THE ISHIGAMI FUNCTION WITH MIXED-VARIABLE INPUTS.

For the cases with qualitative variables, t_1 and t_3, we observe a great match between the true discrete Sobol' values (True-MV MSI & TSI) and the values obtained from our method (MV MSI & TSI). Specifically, the method was able to identify not only the zero individual contribution provided by t_3, but also its total interaction contribution with other design variables.

Copyright © 2023 by ASME

Furthermore, although x_2 was considered as a quantitative variable, we expected to observe a notable change in its Sobol' values as the number of levels associated with the other two qualitative variables vary. Therefore, we examined the changes in Sobol' values of x_2 as a function of the number of levels of t_1 and t_2. Our method was able to capture this change in Sobol' values very well for both the MSI and TSI. Finally, the convergence of Sobol' metrics to the true quantitative Sobol' sensitivity values (True TSI & MSI) demonstrated and validated the correct implementation of our method.

3.2 Validation with Hartmann 6D Function

We continued to validate our method with another well-known mathematical function for GSA studies, the Hartmann 6-Dimensional (6D) function [18]. As the name suggests, this function contains six design variables and is formulated as Equation 7.

$$f(\pmb{x}) = \sum_{i=1}^{4} \alpha_i \exp\left(\sum_{j=1}^{6} A_{ij}\left(x_j - P_{ij}\right)^2\right) \quad (7)$$
$$for\ \pmb{x} = [x_1, x_2, x_3, x_4, x_5, x_6] \in [0,1]$$

The variables $\pmb{\alpha}$, A, and P are denoted as constants and their corresponding values are given in [18]. Implementing a similar transformation as the Ishigami function, we converted two of the design variables, x_1 and x_6 to qualitative variable, t_1 and t_2, by assigning different discrete values. The number of discrete options ranged from [2,14]. The main reason behind choosing these two variables is because their continuous TSI values, shown in Table 2, contribute to the response the most compared to the remaining four design variables. As a result, the goal is to observe if our method can capture the same contribution. For demonstration purposes, the sensitivity values of the two qualitative variables are shown only. We implemented our method once again through the two-stage validation process.

TABLE 2. SOBOL' SENSITIVITY INDICES OF TWO QUANTIATIVE DESIGN VARIABLES FROM THE HARTMAN 6D FUNCTION

Design Variable	Main Sensitivity Index (MSI)	Total Sensitivity Index (TSI)
x_2	0.0025	0.3992
x_6	0.0086	0.4812

Figure 4 below shows the well-matching Sobol' sensitivity indices between the true discrete indices (True-MV MSI & TSI) and indices obtained from our method (MV MSI & TSI) for the two qualitative variables. Specifically, the method was able to capture the high contribution of TSI for the two qualitative variables, which was the main goal of this study. Furthermore, the convergence towards ground-truth Sobol' values (True TSI & MSI) demonstrates the effectiveness of our method once again. Although not shown here, the same matching trend is also

observed for the remaining four quantitative variables (x_1, x_2, x_3, x_4).

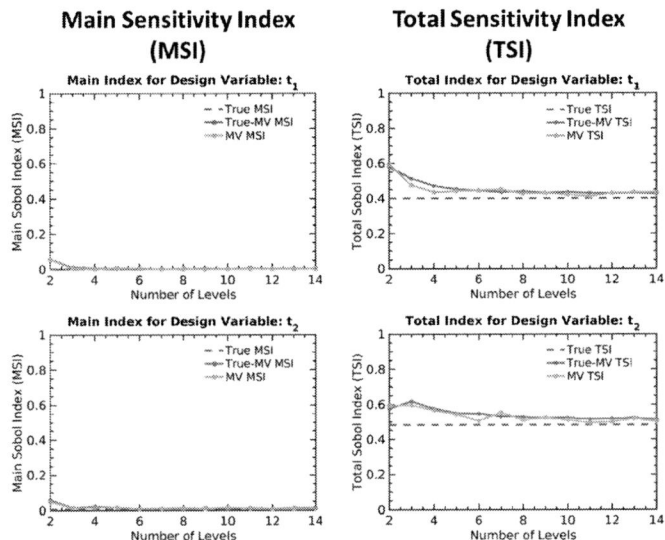

FIGURE 4: SOBOL SENSITIVITY INDICES VALUES OF THE HARTMAN 6D FUNCTION.

4. ACCELERATING COMBINATORIAL MATERIALS DESIGN BY INTEGRATING BAYESIAN OPTIMIZATION

In this section, we show how our method can be utilized in engineering applications, using the design of metal-organic frameworks (MOFs) as a demonstrative example. Specifically, in Section 4.1 we applied our GSA method to the qualitative space of MOFs for two gas absorption properties and explained the physical meaning behind the obtained sensitivity indices. Finally, in Section 4.2 we demonstrated how the sensitivity information obtained from our method can benefit the design of MOFs for accelerating the exploration of novel candidates from a large combinatorial design space.

4.1 Global Sensitivity Analysis for the Design of Metal-Organic Frameworks for Gas Absorption Properties

MOFs are a class of porous crystalline materials that are formed by arranging organic linkers, inorganic nodes, and functional groups in various topologies. Due to their highly tunable nature, MOFs have been promising solutions for numerous applications including gas storage, gas separation, and catalysis [19-21]. Recently, we have implemented LVGP for large combinatorial design spaces of MOFs [22]. In this work, we focused on the design of MOFs for carbon dioxide (CO_2) absorption properties. Specifically, we designed MOFs for their carbon capture capabilities, *CO_2 working capacity* property, and separation from nitrogen (N_2), *CO_2/N_2 selectivity* property, as these properties are involved in many CO_2 applications, such as carbon capture from flue gas emissions and natural gas sweetening [23].

Copyright © 2023 by ASME

MOFs can be represented by a vector of qualitative variables, [A-B-C-D], where each element in the vector takes an integer number to represent the qualitative design choice of that variable. Starting from the element "A", each element in the vector corresponds to a qualitative building block design variable, specifically known as Nodular Building Block 1 (N-BB1), Nodular Building Block 2 (N-BB2), Connecting Building Block 1 (C-BB1), and Connecting Building Block 2 (C-BB2), respectively. These variables represent the core building blocks that make up the material. Each choice of building block contains different of design options represented by an integer.

In our previous work, we considered a design space of 1001 MOFs constructed from combinations of 7, 4, 6, and 6 unique qualitative design variables of N-BB1, N-BB2, C-BB1, and C-BB2, respectively. The design variables are given in Figure 5.

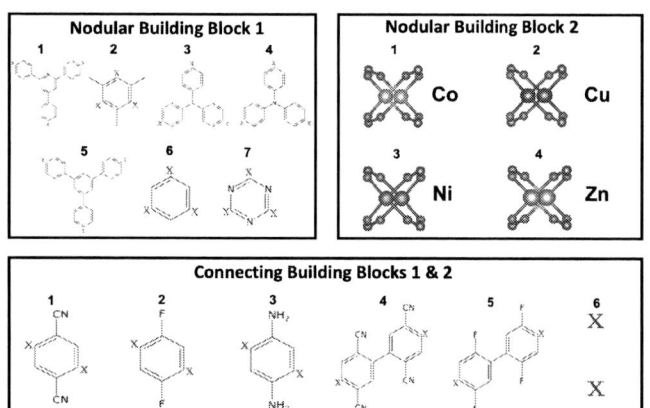

FIGURE 5: THE DESIGN CHOICES FOR THE METAL-ORGANIC FRAMEWORK MATERIAL CONSTRUCTION.

Here, we used the same dataset and fitted an LVGP model on the entire design space to capture the relationships between the design variables and the two gas absorption properties. Next, we implemented the proposed mixed-variable global sensitivity analysis method to quantify the contribution of qualitative design variables to the variability in the properties. The obtained Sobol' indices, MSI and TSI, for the two properties are displayed in Figure 6.

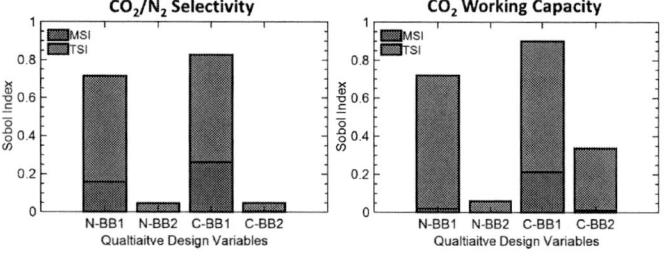

FIGURE 6: SOBOL SENSITIVITY INDICES VALUES OF THE FOUR QUALITATIVE DESIGN VARIABLES.

The obtained results show that the variables N-BB1 and C-BB1 influence the CO_2/N_2 selectivity significantly. Specifically, their interaction effect influences the property as the TSI values of both variables are much higher than others. Similar results are also obtained for CO_2 working capacity property. These insights obtained from the GSA method can actually be explained and justified by the chemical mechanism behind it. Based on the knowledge from material science, the two properties are highly dependent on the size of the MOF structure, which could be defined by the MOF's largest cavity diameter (LCD) [22]. As shown in Figure 7, the LCD and both properties are negatively influenced by each other. In other words, MOFs with smaller LCD values tend to have high CO_2 gas absorption properties and vice versa. Meanwhile, the LCD of the MOF is controlled by the building blocks that surround the material. Therefore, we can use LCD as an intermediate descriptor to unravel the relationship between the structural design variables and properties to justify the proposed mixed-variable GSA.

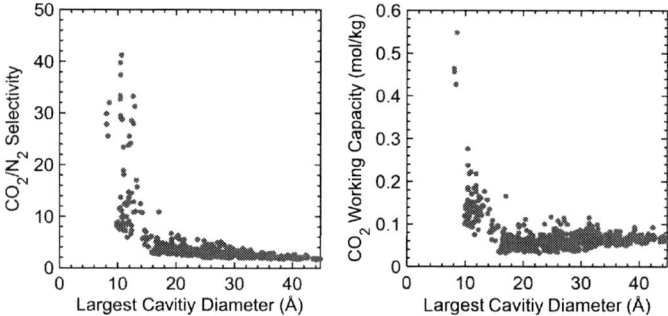

FIGURE 7: LARGEST CAVITY DIAMETER (LCD) COMPARED WITH THE TWO CO_2 GAS ABSORPTION PROPERTIES.

Specifically, the N-BB1 and C-BB1 determine the overall shape of the material. As seen in Figure 5, the N-BB2 is a choice of metal node, which does not influence the shape significantly. Similarly, the C-BB2 connects the N-BB2 to the MOF, which does not contribute to the overall shape of the structure. Figure 8 further shows the variations of LCD with respect to design choices within each design variable. The upper and lower tails of each box demonstrate the maximum and minimum, the vertical edges of the boxes show the 1st and 3rd quantile, and the red line shows the median of the LCD values for that specific design choice (level). It is immediately apparent that the choice design variables for N-BB1 and C-BB1 significantly change the LCD. On the other hand, the choice of N-BB2 or C-BB2, does not influence the LCD crucially. Thus, the LCD distributions of the variables can explain the sensitivity values obtained from our method.

The qualitative GSA enabled us to conclude that Nodular BB1 and Connecting BB2 are the design variables that contribute to the two absorption properties the most. In the next section, we used this important knowledge to accelerate the discovery and design of novel MOFs.

Copyright © 2023 by ASME

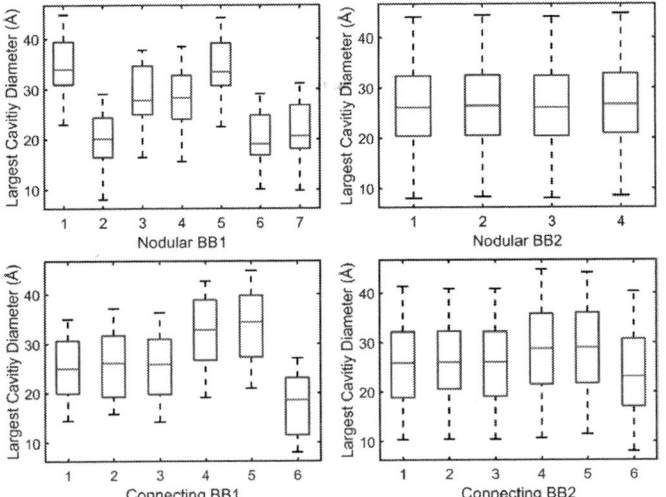

FIGURE 8: INFLUENCE OF DESIGN CHOICES ON THE LARGEST CAVITY DIAMETER

4.2 The Sensitivity Aware Multi-Objective Design Framework

In this section, we implemented a new sensitivity-aware design framework using the knowledge obtained from the GSA study for the design of MOFs for the two gas absorption properties, CO_2 working capacity and CO_2/N_2 selectivity. Our design optimization goal is to identify MOF designs that lie on the Pareto front of the two conflicting gas absorption properties. Here, the Pareto front corresponds to MOF designs that possess properties higher than the rest of the design space but cannot be improved without sacrificing the other property of interest [24].

First, we explain the new sensitivity-aware framework, demonstrate its results, and finally compare the new framework with the "vanilla" design framework without using the sensitivity analysis [22].

The new sensitivity-aware design framework consists of two stages as shown in Algorithm 1. For ease of notation, the four design variables are represented using the vector notation, [A, B, C, D] and their respective two properties, CO_2 working capacity (Y_1) and CO_2/N_2 selectivity (Y_2), are described with [Y] = [Y_1, Y_2]. In the first stage, the framework searches for optimum design candidates only in the Nodular BB1 (A) and Connecting BB1 (C) space. This is because, based on the conclusions obtained from our GSA framework, they are the most important factors for gas absorption properties, while the remaining two factors have little influence. By focusing only on those important factors, we can significantly reduce the search space to facilitate the design optimization.

To initiate the first stage, a design of experiment (DOE) with all four variables is created from sliced-optimum Latin hypercube sampling [25]. In order to facilitate notation, this initial design space, along with its corresponding properties, is

denoted as D_{ABCD}. Two LVGP models are trained on the D_{ABCD} for each property with all four design variables as inputs, denoted as $LVGP_{ABCD}$. To simplify notation, the $LVGP_{ABCD}$ contains two LVGP models trained on the two properties. We use these two models to make predictions on all the MOF design candidates for each property. The predicted properties for both objectives are represented with $\hat{Y}_{ABCD}$. Next, we create a new dataset, D_{AC}, from the initial DOE with only the unique [A, C] combinations as the input. For each unique [A, C], we collect the maximum property values predicted by the $LVGP_{ABCD}$ models for all combinations of [B, D] as the outputs for the new dataset. In other words, the outputs represent the best properties we are expected to get for a given [A, C]. A second set of two LVGP models, $LVGP_{AC}$, is then trained on this new dataset with only unique [A, C] combinations as the input and their respective predicted maximum property values as outputs.

ALGORITHM 1: THE SENSITIVITY-AWARE DESIGN FRAMEWORK

1:	*Start* First Stage
2:	$D_{ABCD} \leftarrow [(A, B, C, D), Y]$
3:	**for** *iteration* = 1, 2, ..., *n* **do**
4:	Train $LVGP_{ABCD}$ with D_{ABCD}
5:	$D_{AC} \leftarrow [(A, C), \max_{(B, D)} (\hat{Y}_{ABCD} \mid A, C)]$
6:	Train $LVGP_{AC}$ with D_{AC}
7:	Find A_{can}, C_{can} with BO and $LVGP_{AC}$
8:	**for** each B, D given A_{can}, C_{can} **do**
9:	Find B_{can}, D_{can} with BO & $LVGP_{ABCD}$
10:	**end for**
11:	Add [(A_{can}, B_{can}, C_{can}, D_{can}), Y_{can}] to D_{ABCD}
12:	**end for**
13:	Obtain (A_{opt}, C_{opt}) from D_{AC}
14:	*End* First Stage
15:	*Start* Second Stage
16:	**while** (A_{PF}, B_{PF}, C_{PF}, D_{PF}) $\notin$ D_{ABCD} **do**
17:	**for** each X_B, X_D given (A_{opt}, C_{opt}) **do**
18:	Find A_{can}, B_{can}, C_{can}, D_{can} with BO & $LVGP_{ABCD}$
19:	**end for**
20:	Add [(A_{can}, B_{can}, C_{can}, D_{can}), Y_{can}] $\rightarrow$ D_{ABCD}
21:	Train $LVGP_{ABCD}$ with D_{ABCD}
22:	**end while**
23:	Obtain (A_{PF}, B_{PF}, C_{PF}, D_{PF}) from D_{ABCD}
24:	*End* Second Stage

A multi-objective Bayesian optimization (BO) is then implemented in this first stage to search for the most optimal [A, C] combination using $LVGP_{AC}$ to obtain the acquisition function, i.e., Expected Improvement (EI) [26]. For the multi-objective scenarios of this study, BO aims to identify the Pareto front solutions. This multi-objective BO identifies the [A, C] candidate, denoted as [A_{can}, C_{can}] that is expected to improve the current known Pareto front designs in the [A, C] space. However, a full MOF representation vector that also includes Nodular BB2

Copyright © 2023 by ASME

(B) and Connecting BB2 (C) is needed to evaluate the true gas absorption values. To mitigate this, a second multi-objective BO is run with fixed $[A_{can}, C_{can}]$ to identify the best $[B, D]$ candidate, denoted as $[B_{can}, D_{can}]$, using $LVGP_{ABCD}$ to obtain the EI values. After that, a new MOF candidate with full representation, i.e., $[A_{can}, B_{can} C_{can}, D_{can}]$, is identified and passed to property evaluations to complete one iteration of the first stage. This candidate is added to the list of explored MOF candidates and the $LVGP_{ABCD}$ model is updated accordingly for the next iterations of the first stage of the framework. The first stage is terminated after reaching the predefined maximum number of iterations, with all the optimum candidates $[A_{opt}, C_{opt}]$ on the Pareto front (PF) identified.

In the second stage, the original design space has been reduced to a much smaller space, in which $[A, C]$ can only be selected from $[A_{opt}, C_{opt}]$. Again, we use BO to search for the Pareto front set in this reduced design space with $LVGP_{ABCD}$ as the surrogate model. At the end of the second stage, the optimum candidates representing the Pareto front $[A_{PF}, B_{PF}, C_{PF}, D_{PF}]$ are evaluated and the design process is concluded.

The GSA results obtained from our framework enabled us to direct the initial focus of the exploration toward novel Nodular BB1 and Connecting BB2 candidates.

4.3 Results

In this section, we demonstrate the capability of our sensitivity-aware design framework on the MOF design space and compare it with a "vanilla" design framework. Without the GSA information, the vanilla framework directly uses four-variable LVGP models in the multi-objective BO to search for the Pareto front in the original full design space. This large MOF design space contains 47,470 MOF candidates constructed from combinations of 4, 7, 41, and 42 combinations of Nodular Building Block 1 (N-BB1), Nodular Building Block 2 (N-BB2), Connecting Building Block 1 (C-BB1), and Connecting Building Block 2 (C-BB2), respectively. We have previously simulated the properties of all 47,740 candidates using Grand Canonical Monte Carlo (GCMC) simulations and identified 7 Pareto Front candidates to be explored by both frameworks. The details of the GCMC simulations can be found in [22]. The property space, along with the Pareto front points of two properties is given in Figure 9, which will be used as the ground truth for later validation.

To account for at least one of the design options (levels) of the C-BB2, the vanilla framework started optimization with a DOE that contains 42 MOFs obtained from Sliced-OLHS. After continuing for 190 iterations of multi-objective BO, the vanilla framework identified all the Pareto front designs, exploring 232 MOFs in total. The number of explored MOFs corresponded to 0.49% of the entire design space, which demonstrates the effectiveness of LVGP on large combinatorial design spaces once again. This result was used as a benchmark for our sensitivity-aware design framework.

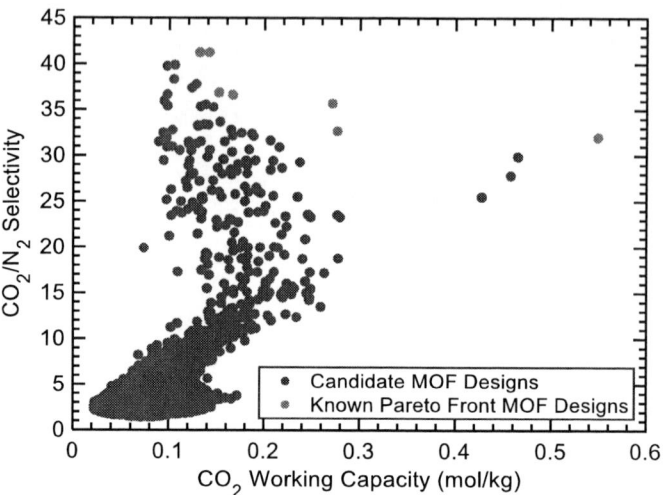

FIGURE 9: THE KNOWN PROPERTY SPACE OF CO_2 WORKING CAPACITY AND CO_2/N_2 SELECTIVITY

Starting with the same DOE used in the vanilla framework, we ran the first stage of the sensitivity-aware design framework until 100 MOFs are explored, including the 42 initial MOFs, to identify the optimal $[A_{opt}, C_{opt}]$ design variables. At the end of this 58-iteration optimization, two optimal $[A_{opt}, C_{opt}]$ candidates are identified and passed on to the second stage. The optimal candidates contained two different design options for the Nodular BB1 (A) and only one design variable for the Connecting BB1 (C). In the second stage, the third BO framework is run to identify the optimal candidates from a much smaller design space, as the design space searched reduced from 47,740 candidates to a design space of 336 candidates that are formed by combinations of 2 N-BB1, 4 N-BB2, 1 C-BB1, and 42 C-BB2. Using the previous $LVGP_{ABCD}$ model, the second stage is initiated to identify the true Pareto Front designs, $[A_{PF}, B_{PF}, C_{PF}, D_{PF}]$. After 113 iterations of the second stage, all the Pareto front MOF designs were identified. Including the 100 MOFs identified from the first stage, a total of 213 MOFs were explored to find the true Pareto front MOF designs, corresponding to 0.45% of the design space. Figure 10 and Figure 11 show the optimization history search and the evolution of the Pareto front for both frameworks, respectively. As seen from Figure 10, the sensitivity-aware design framework was able to identify all Pareto designs at a faster rate compared to the vanilla framework. Similarly, the Pareto evolution shown in Figure 11 demonstrates that the sensitivity-aware design framework can swiftly expand the Pareto front, specifically at the earlier stages of optimization, while the vanilla framework takes longer to do so. This can be very advantageous for optimization applications with limited resources.

Copyright © 2023 by ASME

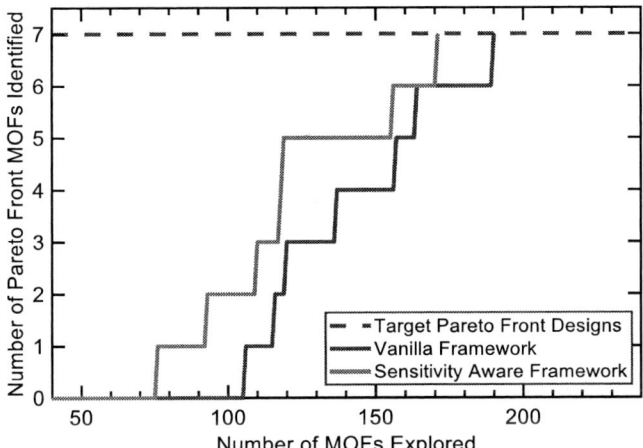

FIGURE 10: PARETO FRONT OPTIMIZATION HISTORY FOR TWO DESIGN FRAMEWORKS

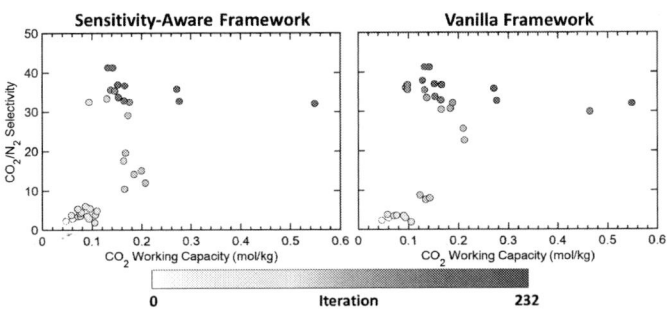

FIGURE 11: THE PARETO FRONT EVOLUTION HISTORY FOR TWO DESIGN FRAMEWORKS

The optimization results obtained from the sensitivity-aware design framework show an 8.2% improvement compared to the number of MOFs explored with the vanilla design framework. Although the improvement percentage is marginal, the results are significant compared to the already well-performing vanilla framework, which only explored 0.49% of the design space. Furthermore, since the evaluation of properties is expensive, the 8% improvement provides significant time and cost savings for designers. It can be expected that the level of improvement will become much higher for other design applications with more variables and larger design spaces.

The obtained results show that our metamodel-based mixed-variable GSA method can play an important role in the design of MOFs. The method not only provides physical insights regarding the design choices but also can assist in better design optimization by directing the focus of optimization towards better solutions to significantly reduce the search space.

5. CONCLUSION

Sensitivity analysis is a very powerful technique to evaluate design models in multiple aspects. However, current sensitivity efforts under global sensitivity analysis techniques fail to consider mixed-variable design spaces. In this paper, we developed a metamodel-based global sensitivity analysis method that can accommodate models with mixed variables or only qualitative variables. We validated the effectiveness of our method on two well-known mathematical equations used in sensitivity studies. This validation concluded that our method is able to capture the relationships between and within mixed-variable design spaces on the response of interest effectively. Next, we demonstrated the benefit of incorporating our method into combinatorial and many-level engineering materials design applications. Specifically, the newly defined sensitivity-aware design framework accelerated the exploration of novel material candidates, surpassing the existing highly efficient design framework, for the design of metal-organic framework materials used in gas absorption applications. We envision that our new metamodel-based mixed-variable global sensitivity analysis method can further assist design engineers in building more accurate mixed-variable design models by identifying the most influential design variables, observing input and output uncertainties, and calibrating model parameters.

ACKNOWLEDGEMENTS

We acknowledge the supports from the Advanced Research Projects Agency-Energy (ARPA-E), U.S. Department of Energy, under Award Number DE-AR0001209, and National Science Foundation (NSF) Grant 1729743. Finally, we acknowledge Professor Randall Q. Snurr, and Thang Duc Pham from Department of Chemical and Biological Engineering at Northwestern University for generating the metal-organic framework data, and Yi-Ping Chen from the Department of Mechanical Engineering at Northwestern University for helping with the manuscript figures.

REFERENCES

[1] Saltelli, A., Tarantola, S., Campolongo, F., and Ratto, M., 2004, Sensitivity analysis in practice: a guide to assessing scientific models, Wiley Online Library.

[2] Morio, J., 2011, "Global and local sensitivity analysis methods for a physical system," European Journal of Physics, 32(6), p. 1577.

[3] Iooss, B., and Lemaître, P., 2015, "A Review on Global Sensitivity Analysis Methods," Uncertainty Management in Simulation-Optimization of Complex Systems: Algorithms and Applications, G. Dellino, and C. Meloni, eds., Springer US, Boston, MA, pp. 101-122.

[4] Borgonovo, E., and Plischke, E., 2016, "Sensitivity analysis: A review of recent advances," European Journal of Operational Research, 248(3), pp. 869-887.

[5] Lo Piano, S., Ferretti, F., Puy, A., Albrecht, D., and Saltelli, A., 2021, "Variance-based sensitivity analysis: The quest for better estimators and designs between explorativity and economy," Reliability Engineering & System Safety, 206, p. 107300.

Copyright © 2023 by ASME

[6] Chan, K., Saltelli, A., and Tarantola, S., "Sensitivity Analysis Of Model Output: Variance-based Methods Make The Difference," Proc. Winter Simulation Conference Proceedings, pp. 261-268.

[7] Norton, J., 2015, "An introduction to sensitivity assessment of simulation models," Environmental Modelling & Software, 69, pp. 166-174.

[8] Sobol´, I. M., 2001, "Global sensitivity indices for nonlinear mathematical models and their Monte Carlo estimates," Mathematics and Computers in Simulation, 55(1), pp. 271-280.

[9] Marrel, A., Iooss, B., Laurent, B., and Roustant, O., 2009, "Calculations of Sobol indices for the Gaussian process metamodel," Reliability Engineering & System Safety, 94(3), pp. 742-751.

[10] Saltelli, A., Annoni, P., Azzini, I., Campolongo, F., Ratto, M., and Tarantola, S., 2010, "Variance based sensitivity analysis of model output. Design and estimator for the total sensitivity index," Comput. Phys. Commun., 181(2), pp. 259-270.

[11] Chen, W., Jin, R., and Sudjianto, A., 2004, "Analytical Variance-Based Global Sensitivity Analysis in Simulation-Based Design Under Uncertainty," J. Mech. Des., 127(5), pp. 875-886.

[12] Zhang, Y., Apley, D. W., and Chen, W., 2020, "Bayesian Optimization for Materials Design with Mixed Quantitative and Qualitative Variables," Sci. Rep., 10(1), p. 4924.

[13] Wang, Y., Iyer, A., Chen, W., and Rondinelli, J. M., 2020, "Featureless adaptive optimization accelerates functional electronic materials design," Appl. Phys. Rev., 7(4), p. 041403.

[14] Zhang, Y., Tao, S., Chen, W., and Apley, D. W., 2020, "A Latent Variable Approach to Gaussian Process Modeling with Qualitative and Quantitative Factors," Technometrics, 62(3), pp. 291-302.

[15] Rasmussen, C. E., and Williams, C. K. I., 2005, "Gaussian Processes for Machine Learning," The MIT Press.

[16] Azzini, I., and Rosati, R., 2022, "A function dataset for benchmarking in sensitivity analysis," Data Brief, 42, p. 108071.

[17] Ishigami, T., and Homma, T., "An importance quantification technique in uncertainty analysis for computer models," Proc. [1990] Proceedings. First International Symposium on Uncertainty Modeling and Analysis, pp. 398-403.

[18] Caflisch, R. E., 1998, "Monte Carlo and quasi-Monte Carlo methods," Acta Numerica, 7, pp. 1-49.

[19] Li, H., Wang, K., Sun, Y., Lollar, C. T., Li, J., and Zhou, H.-C., 2018, "Recent advances in gas storage and separation using metal–organic frameworks," Mater. Today, 21(2), pp. 108-121.

[20] Shah, M., McCarthy, M. C., Sachdeva, S., Lee, A. K., and Jeong, H.-K., 2012, "Current Status of Metal–Organic Framework Membranes for Gas Separations: Promises and Challenges," Ind. Eng. Chem. Res., 51(5), pp. 2179-2199.

[21] Freund, R., Zaremba, O., Arnauts, G., Ameloot, R., Skorupskii, G., Dincă, M., Bavykina, A., Gascon, J., Ejsmont, A., Goscianska, J., Kalmutzki, M., Lächelt, U., Ploetz, E., Diercks, C. S., and Wuttke, S., 2021, "The Current Status of MOF and COF Applications," Angew. Chem. Int. Ed., 60(45), pp. 23975-24001.

[22] Comlek, Y., Pham, T. D., Snurr, R., and Chen, W., 2023, "Rapid Design of Top-Performing Metal-Organic Frameworks with Qualitative Representations of Building Blocks," arXiv preprint arXiv:2302.09184.

[23] Qian, Q., Asinger, P. A., Lee, M. J., Han, G., Mizrahi Rodriguez, K., Lin, S., Benedetti, F. M., Wu, A. X., Chi, W. S., and Smith, Z. P., 2020, "MOF-Based Membranes for Gas Separations," Chem. Rev., 120(16), pp. 8161-8266.

[24] Censor, Y., 1977, "Pareto optimality in multiobjective problems," Appl. Math. Optim., 4(1), pp. 41-59.

[25] Ba, S., Myers, W. R., and Brenneman, W. A., 2015, "Optimal Sliced Latin Hypercube Designs," Technometrics, 57(4), pp. 479-487.

[26] Jones, D. R., 2001, "A Taxonomy of Global Optimization Methods Based on Response Surfaces," J. Global Optim., 21(4), pp. 345-383.

Copyright © 2023 by ASME

Proceedings of the ASME 2023
International Design Engineering Technical Conferences and
Computers and Information in Engineering Conference
IDETC-CIE2023
August 20-23, 2023, Boston, Massachusetts

DETC2023-111265

TACKLING AN EXACT MAXIMUM STRESS MINIMIZATION PROBLEM WITH GRADIENT-FREE TOPOLOGY OPTIMIZATION INCORPORATING A DEEP GENERATIVE MODEL

Misato Kato[1], Taisei Kii[1], Kentaro Yaji[1,*], Kikuo Fujita[1]

[1]Department of Mechanical Engineering, Osaka University, Osaka 565-0871, Japan

ABSTRACT

Maximum stress minimization is an active research topic in topology optimization. In general, gradient-based topology optimization has been used to solve such a problem, but it fundamentally faces problematic aspects. For example, singularity problems require relaxation treatments. They transform an exact minimax problem, such as the maximum stress minimization problem, into any differentiable pseudo-problem. Therefore, gradient-based topology optimization can be regarded as a low-fidelity optimization that deals with pseudo-models. Our interest is that there is room for improvement regarding the exact maximum stress in the optimized designs of the conventional method. To clarify, we focus on data-driven multifidelity topology design (MFTD) that optimizes topology without gradient information, even under a high degree of design freedom. Its basic idea is based on evolutionary algorithms (EAs). In the optimization process, the initial design candidates are generated by solving low-fidelity optimization problems. Then, they are iteratively updated a deep generative model by conducting high-fidelity forward analyses. This paper tackles a bi-objective problem of the exact maximum stress and volume minimization by data-driven MFTD incorporating initial solutions composed of the optimized designs derived by solving the gradient-based topology optimization using the p-norm stress measure. The numerical results indicate that the optimized designs by data-driven MFTD completely dominate the initial solutions and achieve a volume reduction of up to 10% under the same maximum stress value.

Keywords: Topology optimization, Data-driven design, Maximum stress minimization

1. INTRODUCTION

In recent years, weight reduction has been emphasized in the industrial sector to reduce resources and improve fuel efficiency, while designs must ensure strength and rigidity for safety and performance. In particular, stress is one of the most critical factors

in structural design. Moreover, the resolution of stress concentration often requires time-consuming tasks for trial and error in geometry adjustments. Therefore, optimal design considering stress is expected to enhance the design quality and reduce the design time.

Topology optimization is a practical design methodology for effectively designing lightweight structures. It allows maximum mass reduction while retaining the necessary parts to ensure structural performance [1, 2]. However, maximum stress minimization is a problematic issue in topology optimization. Therefore, research on such an issue has attracted attention and has been an active field, while the mean compliance minimization problem—maximum stiffness structural design—has been well established. This gap is because conventional topology optimization methods concerning maximum stress minimization face three issues, as follows [3].

Singularity The singularity problem occurs when using density-based topology optimization in general. As the design variables approach zero to generate void regions, there are persistent non-zero stresses that do not disappear [4–8]. These regions represent holes without material and should ideally have zero stress. However, due to residual strains, they can generate unexpectedly high stress values. Therefore, the density-based topology optimization must avoid such singularity by using relaxed stress values which are not exact [9–11].

Strong nonlinearity Stress concentration occurs easily in re-entrant corners such as L-bracket, thin structures, and edges and is very sensitive to changes in structure. Additionally, it is susceptible to the initial guess and is greatly affected by parameters and formulations.

Stress localization Stress values at all elements are critical in finding their maximum, but the computational burden is enormous. Therefore, alternative ways to evaluate the stress, e.g., *p*-norm and Kresselmeier-Steinhauser functions, have

---

**Corresponding author: yaji@mech.eng.osaka-u.ac.jp

Copyright © 2023 by ASME

been proposed to approximate the local stress values as a single differentiable function [12, 13]. These relaxation techniques significantly reduce the computation burden, but cannot fully capture the exact stress behavior.

As mentioned above, solving maximum stress minimization problems requires many relaxation treatments, even though only pseudo-models are solvable. The original root of these difficulties relies on the existence of the intermediate state of the design variable field, i.e., grayscale, which is essential to use a gradient-based optimizer. Furthermore, as various patterns of optimized designs exist as local optima due to the multimodality of the solution space, it can be concluded that gradient-based topology optimization is not the best option for solving exact maximum stress minimization problems.

To radically overcome the above issues, gradient-free optimization is a promising option to deal with the 0/1 design variable field. It means the design variable field is defined as the characteristic function used in the original concept of topology optimization [1]. Evolutionary algorithms (EAs) are the representative gradient-free optimization [14] and are based on the evolutionary mechanisms of living organisms. Many works [15] demonstrated that EAs could be applied to topology optimization problems. However, EA-based topology optimization typically requires a large number of function calls of the forward analysis along with exponentially increasing the computational burden, namely, the *curse of dimensionality*. To overcome this limitation, Yaji et al. [16] recently proposed a gradient-free topology optimization framework called *data-driven multifidelity topology design (MFTD)*, which can deal with a relatively large number of design variables. This framework is based on the combination of multifidelity design guided by topology optimization [17] and data-driven topology design [18], incorporating a deep generative model as a crossover-like operator in EAs. The idea of data-driven MFTD is that design candidates, generated through solving low-fidelity topology optimization problems, are updated based on an EA strategy with high-fidelity evaluation. The framework was applied to topology optimization problems that are hard to solve directly with conventional gradient-based methods, e.g., minimax and turbulent flow problems.

In this paper, we focus on whether the data-driven MFTD can solve an exact maximum stress minimization problem without any relaxation treatment used in the conventional gradient-based topology optimization. We hypothesize that there is room for improvement in terms of the exact maximum stress in the optimized designs of the conventional gradient-based topology optimization. Here, we define "fidelity" in terms of the exactness of the problem. To clarify, a low-fidelity optimization is the gradient-based topology optimization using pseudo-density and stress relaxation techniques, while high-fidelity evaluations use 0/1 density and exact stress values. In this way, we investigate whether the optimized designs given by solving gradient-based topology optimization using the *p*-norm could be updated on the exact maximum stress minimization problem via data-driven MFTD.

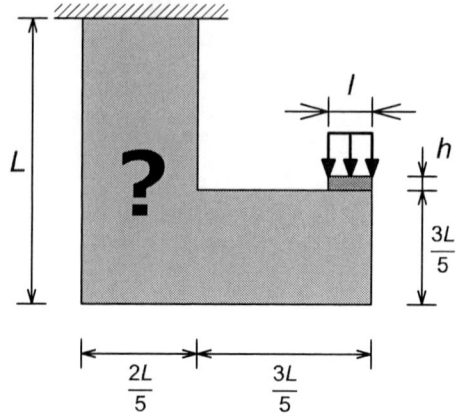

FIGURE 1: PROBLEM DEFINITION

2. PROBLEM SETTINGS

This study deals with the bi-objective optimization of maximum stress minimization and volume minimization. Note that although topology optimization problems are typically formulated as a single objective with constraints due to the use of gradient-based optimizers, many design optimization problems require such a multiobjective formulation [19].

The L-bracket, widely used as a benchmark for maximum stress minimization problems, is used as the analysis model in this study. Here, stress is concentrated at the re-entrant corner, where the material is actively removed during optimization to reduce the maximum stress. The dimensions, loads, and boundary conditions are shown in Fig. 1. The distributed load is applied on the top of the non-design domain.

Let us consider the discrete form of a maximum stress minimization problem. A formulation of the bi-objective optimization problem can be written as follows:

$$
\begin{aligned}
\text{find} \quad & \boldsymbol{\rho} = \rho_i \ (i = 1, 2, ..., n), \\
\text{that minimize} \quad & \sigma_{\max} = \max(\sigma_{vm,i}), \\
& V = \sum_{i=1}^{n} v_i \rho_i, \\
\text{subject to} \quad & \rho_i \in \{0, 1\}.
\end{aligned}
\tag{1}
$$

Herein, $\boldsymbol{\rho}$ is the design variable vector with the binary values, the so-called characteristic function, n is the number of elements, σ_{vm} is the von Mises stress, and v_i is the elemental volume. Note that the optimization problem defined by Eq. (1) is formulated as the exact maximum stress minimization problem with the characteristic function used in the original concept of topology optimization [1]. Since such an ideal formulation cannot be solved directly by gradient-based topology optimization in general, relaxation techniques are essential to derive a reasonable optimized structure.

3. CONVENTIONAL APPROACH

We here discuss the standard density-based topology optimization method for solving maximum stress minimization problems. In this study, the analysis domain is discretized using the finite element method, in which the pseudo-density corresponding

Copyright © 2023 by ASME

to the design variables, $\rho_i \in [0, 1]$, are defined at each element on the basis of the solid isotropic material with penalization (SIMP) [5, 6, 9, 10].

3.1 Density Filter

Directly using ρ_i as the design variables may cause an ill-posed problem, e.g., a checkerboard pattern, even if ρ_i is continuous. Hence, we introduce the material density $\tilde{\rho}_i$ defined using the density filter [20], as follows:

$$\tilde{\rho}_i = \frac{\sum_{j \in \Omega_i} w_j \rho_i}{\sum_{j \in \Omega_i} w_j},$$
$$w_j = \frac{r_0 - r_j}{r_0}, \tag{2}$$

where w_j is the weight factor, r_0 is the filter radius, and r_j is a center point at each element in the effective area of the filter radius, Ω_i. The density filter provides spatial smoothness to the design variables ρ_i, preventing checkerboard and avoiding stress concentration due to sharp structures.

3.2 Equilibrium Equation

Under the assumption of linear elasticity and static behavior, the discrete form of the equilibrium equation can be represented as

$$\boldsymbol{K u} = \boldsymbol{F}, \tag{3}$$

where $\boldsymbol{K}$ is the global stiffness matrix, $\boldsymbol{u}$ is the vector of global nodal displacements, and $\boldsymbol{F}$ is a vector of the known external loads, respectively. $\boldsymbol{K}$ can be adequately built by using the stiffness matrices of the elements $\boldsymbol{K}_i$ and the modified Young's modulus E_{SIMP}, given by:

$$\boldsymbol{K} = \sum_{i=1}^{n} E_{\text{SIMP}}(\tilde{\rho}_i) \boldsymbol{K}_i$$
$$E_{\text{SIMP}}(\tilde{\rho}_i) = E_{\min} + \tilde{\rho}_i^p (E_0 - E_{\min}), \tag{4}$$

where E_0 and $E_{\min}$ are Young's modulus of the solid and void phases. p is the penalization coefficient to promote binarization to achieve a crisp design finally. Typically, $p = 3$ has been used in topology optimization problems concerning the maximum stress minimization/constraint [3, 12, 21, 22].

3.3 Stress Relaxation

It is widely known that maximum stress minimization problems face a singularity problem, in which a stress value increases when the material density is decreased at a local point. Hence, stress relaxation is necessary to avoid singularity and stabilize the optimization process. There are various studies on stress relaxation methods, e.g., the ε-relaxation and smooth envelope functions [10, 23]. In this study, we use the qp-parametrization [3, 21, 22], one of the popular relaxation methods.

The vector of stress at evaluation point i can be written in Voigt notation as

$$\boldsymbol{\sigma}_i = (\sigma_{ix}\, \sigma_{iy}\, \sigma_{iz}\, \tau_{ixy}\, \tau_{iyz}\, \tau_{izx})^{\mathsf{T}}. \tag{5}$$

The penalized and relaxed stress measure $\hat{\sigma}_i$ is given as

$$\hat{\sigma}_i(\tilde{\rho}_i) = \eta(\tilde{\rho}_i)\sigma_i,$$
$$\eta(\tilde{\rho}_i) = \tilde{\rho}_i^q, \tag{6}$$

where q is the penalty coefficient, typically $q = 0.5$ is used. This relaxation technique has the effect of penalizing intermediate values of the material density and promoting binarization. In addition, the following holds for both extreme values:

$$\hat{\sigma}_i(\tilde{\rho}_i = 1) = \sigma_i, \tag{7}$$
$$\hat{\sigma}_i(\tilde{\rho}_i = 0) = \lim_{\tilde{\rho}_i \to 0} \hat{\sigma}_i(\tilde{\rho}_i) = 0. \tag{8}$$

Herein, Eq. (8) justifies that the qp-parametrization can avoid the singularity problem [24].

3.4 *P-norm Stress Measure*

Let us consider that the stress index is the relaxed von Mises stress, defined as

$$\hat{\sigma}_{vm,i} = (\hat{\sigma}_{ix}^2 + \hat{\sigma}_{iy}^2 + \hat{\sigma}_{iz}^2 - \hat{\sigma}_{ix}\hat{\sigma}_{iy}$$
$$- \hat{\sigma}_{iy}\hat{\sigma}_{iz} - \hat{\sigma}_{iz}\hat{\sigma}_{ix} + 3\hat{\tau}_{ixy}^2 + 3\hat{\tau}_{iyz}^2 + 3\hat{\tau}_{izx}^2)^{\frac{1}{2}}. \tag{9}$$

In the optimization problem (1), we replace the objective function expressed as the maximum stress with $\sigma_{\max} = \max(\hat{\sigma}_{vm,i})$.

The maximum stress is not differentiable and typically needs to be approximated using a global function. In this study, we use p-norm stress measure given by

$$\sigma_{PN} = \left(\frac{1}{n} \sum_{i=1}^{n} \hat{\sigma}_{vm,i}^P \right)^{1/P}. \tag{10}$$

Herein, σ_{PN} approaches $\sigma_{\max}$ to when $P \to \infty$, but it is impossible because $P \gg 1$ causes numerical instability. Therefore, the penalty factor P significantly impacts the search performance. In previous works, although $P = 8$ was often chosen from the viewpoint of numerical stability during optimization [3], it cannot generally capture the maximum stress. To realize a large P setting, we employ the continuation method [2]. It allows numerical stability by increasing the penalty coefficient P at every pre-determined step during optimization.

3.5 Problem Formulation

The original optimization problem has two objectives: maximum stress minimization and volume minimization. To solve this bi-objective problem using the conventional gradient-based topology optimization methods, we use the ε-constraint approach [25, 26] to replace the problem with a single objective of minimizing the maximum stress under the various volume constraint values. Consequently, the optimization problem can be formulated as follows:

$$\text{find} \quad \rho = \rho_i \ (i = 1, 2, ..., n)$$

$$\text{that minimize} \quad \sigma_{PN} = \left(\frac{1}{n} \sum_{i=1}^{n} \hat{\sigma}_{vm,i}^P \right)^{1/P}$$

$$\text{subject to} \quad V = \sum_{i=1}^{n} v_i \rho_i \le V_{\max}$$

$$\rho_i \in [0, 1]. \tag{11}$$

Copyright © 2023 by ASME

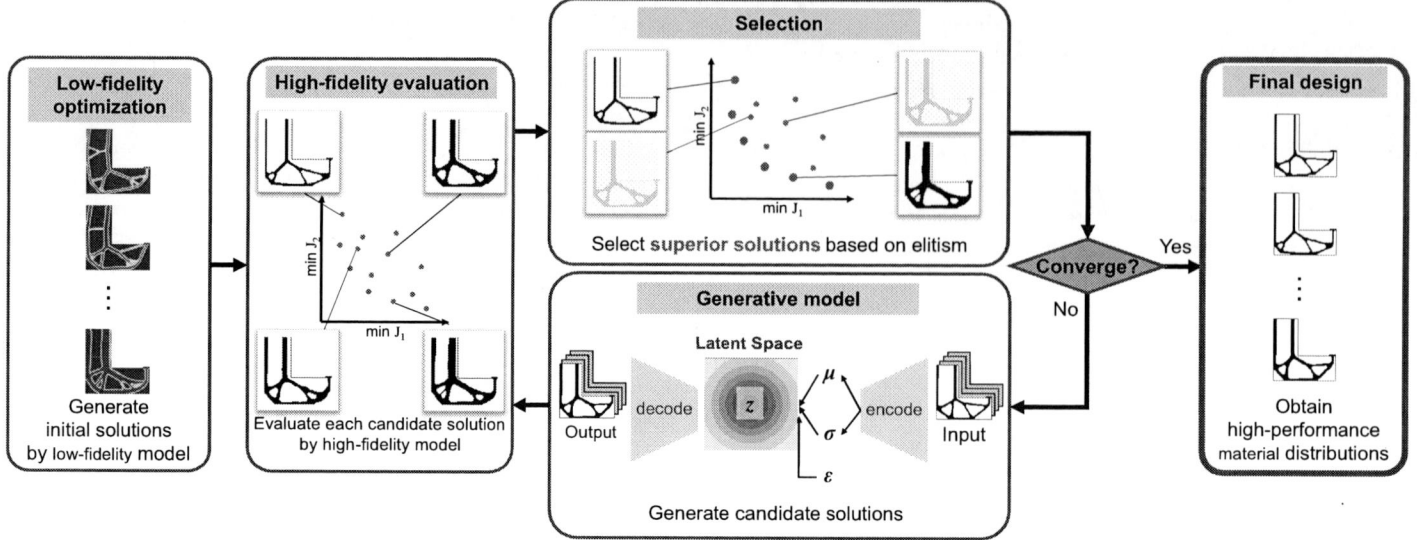

FIGURE 2: DATA PROCESS FLOW OF DATA-DRIVEN MULTIFIDELITY TOPOLOGY DESIGN

We solve this optimization problem by using the method of moving asymptotes (MMA) [27], a popular gradient-based optimizer in the research community of topology optimization.

4. PROPOSED APPROACH

This study investigates whether optimized designs obtained by solving the optimization problem (11) can be improved on the original topology optimization problem (1). To clarify, we tackle solving the original minimax problem defined as Eq. (1) by data-driven multifidelity topology design (MFTD) [16] that is a gradient-free topology optimization framework under a high degree of design freedom. The procedures of data-driven MFTD are shown in Fig. 2. Each step is briefly described below. Note that this paper omits a mutation-like operation proposed in the original framework for simplicity. The detailed procedures can be found in the original paper [16].

4.1 Low-fidelity Optimization

A pseudo-optimization problem is solved in which a low-fidelity model incorporating design parameters called seeding parameters is used to generate various design candidates. In this study, we use the relaxed problem formulation in Eq. (11) as the low-fidelity optimization problem in which the maximum volume V_{max} is varied on the basis of ε-constraint method to generate various patterns of initial designs for the framework.

4.2 High-fidelity Evaluation

In this step, all the candidates are evaluated on the original objective function space in Eq. (1), namely, the exact maximum von Mises stress σ_{max} and the volume V. It should be emphasized that the framework only requires the forward analysis of the original high-fidelity model without any gradient information on the objective and constraint functions.

4.3 Selection

Based on the results of the high-fidelity evaluation, superior candidates are selected using an elite strategy of EAs. In

this study, we use the nondominated sorting genetic algorithm II (NSGA-II) as a selection algorithm to rank and select candidates based on the Pareto dominance relation in the objective function space [28].

4.4 Generative Model

Variational autoencoder (VAE) [29] is used for generating new candidates from the input data composed of the selected elite solutions. VAE consists of two neural networks, namely, an encoder and a decoder. As shown in Fig. 2, it has the capability to extract the information that constitutes the structure from the high-dimensional input data and to reduce it to a lower-dimensional manifold called latent space. In the standard VAE, the latent variable z is defined as

$$z = \mu + \sigma \circ \varepsilon, \tag{12}$$

where μ is the mean, σ is the variance, $\circ$ is the element-wise product, and ε is a random vector following the standard normal distribution.

With this architecture, VAE uses the same dataset for inputs and outputs to perform unsupervised learning to construct the latent space. The following loss function L_{VAE} is used for learning [30]:

$$L_{VAE} = L_{recon} + L_{KL}, \tag{13}$$

where L_{recon} is the reconstruction loss measured by the mean squared error, and L_{KL} is the Kullback-Leibler (KL) divergence given by

$$L_{KL} = -\frac{\varrho}{2} \sum_{i=1}^{N_{lt}} \left(1 + \log(\sigma_i^2) - \mu_i^2 - \sigma_i^2 \right), \tag{14}$$

where N_{lt} is the dimension of the latent space, μ_i and σ_i are the ith components of μ and σ, respectively. And, ϱ is the weight parameter that controls the influence of the KL divergence that works to regularize the latent field to be $N(0, 1)$.

Copyright © 2023 by ASME

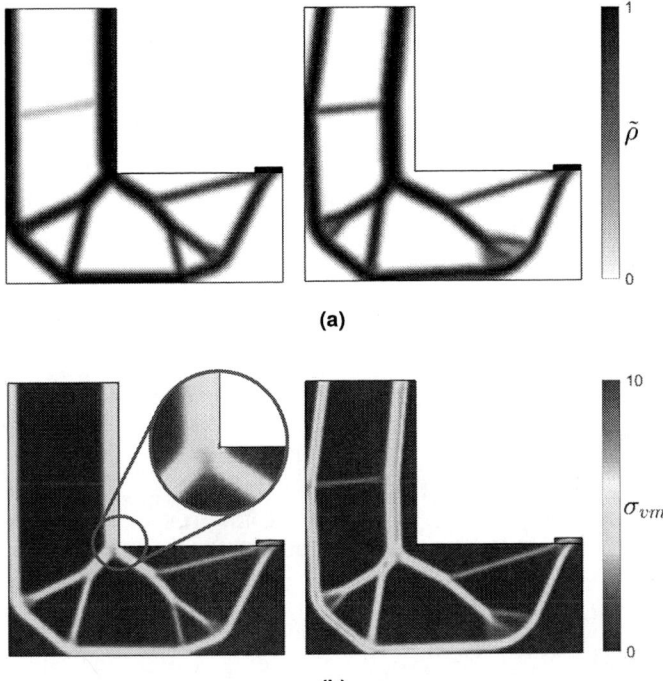

(a)

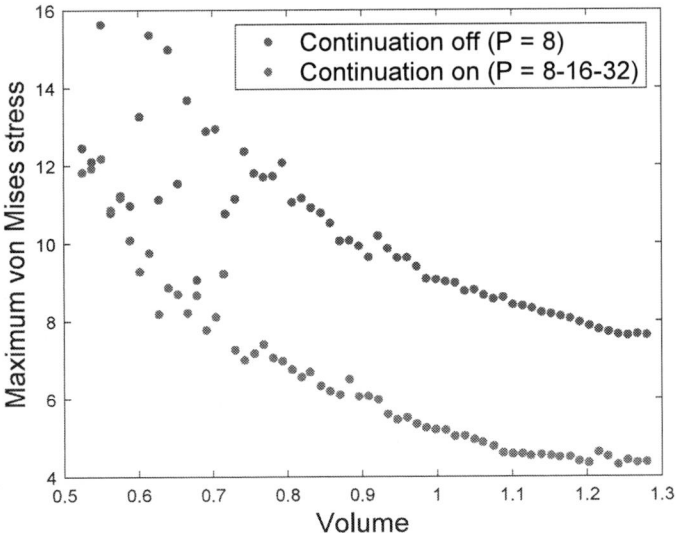

(b)

FIGURE 3: OPTIMIZED DESIGNS USING THE FIXED PENALTY $p = 8$ (LEFT) AND THE CONTINUATION METHOD $p = 8\text{-}16\text{-}32$ (RIGHT) UNDER THE SAME VOLUME CONSTRAINT: (A) MATERIAL DISTRIBUTION, (B) VON MISES STRESS

FIGURE 4: EFFECT OF THE CONTINUATION METHOD ON THE MAXIMUM VON MISES STRESS

Since the dimensionality is significantly reduced from the input and output layers to the lower dimensional latent space, it is expected that important features of the training data can be extracted in this space. Based on this assumption, the material distribution is generated by random sampling in the latent space. The sampling vector is composed of uniformly distributed random numbers in [-4, 4], which covers 99.7% of the normal

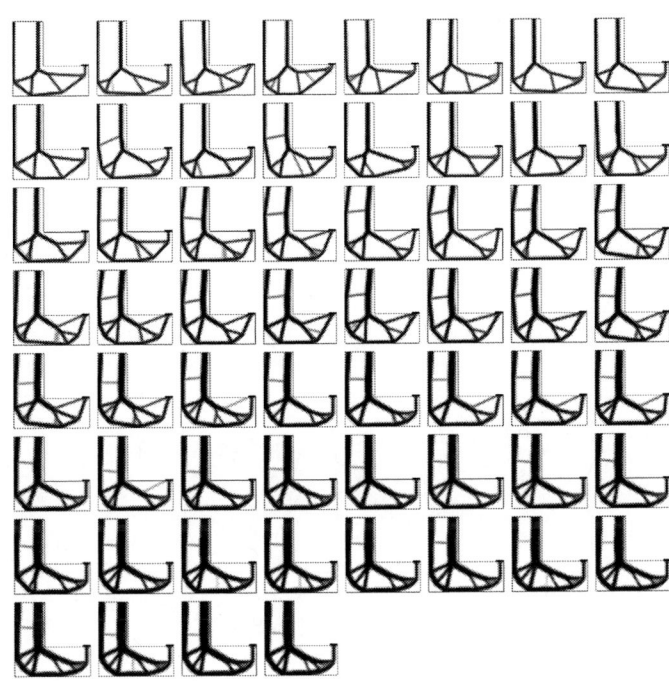

FIGURE 5: INITIAL SOLUTIONS BY THE LOW-FIDELITY OPTIMIZATION

distribution within $\pm 4\sigma$ for each latent variable, and the material distribution is output from the sampling vector using the learned VAE. In this way, a variety of solutions are generated that inherit essential features of the training data. The solutions are evaluated using the high-fidelity forward analysis, and the solution updating is repeated until converging the Pareto front.

5. RESULTS AND DISCUSSION

We now introduce numerical examples of a maximum stress minimization problem and demonstrate the efficacy of the data-driven MFTD to find room for improvement of the optimized designs in the conventional gradient-based topology optimization.

5.1 Design Settings

We deal with the L-bracket shown in Fig. 1, where all the constants are dimensionless values, and $L = 2$, $l = 0.2$, $h = 0.04$. The Young's modulus is set as $E_0 = 1$ for the solid and $E_{\min} = 10^{-9}$ for the void. Poisson's ratio is $\nu = 0.3$. The load is $F = 1$ applied to the normal direction of the top boundary of the non-design domain. The filter radius is $r = 0.05$, and the number of finite elements is 6400. Although body-fitted meshes are necessary to accurately evaluate stress values, we used square meshes for the first step in this paper.

5.2 Generation of Initial Solutions

This study uses the optimized designs generated by solving the gradient-based topology optimization method formulated by Eq. (11), as the initial solutions. The sensitivity analysis for deriving the gradient information is performed using the adjoint method, and we use the MMA as the gradient-based optimizer

Copyright © 2023 by ASME

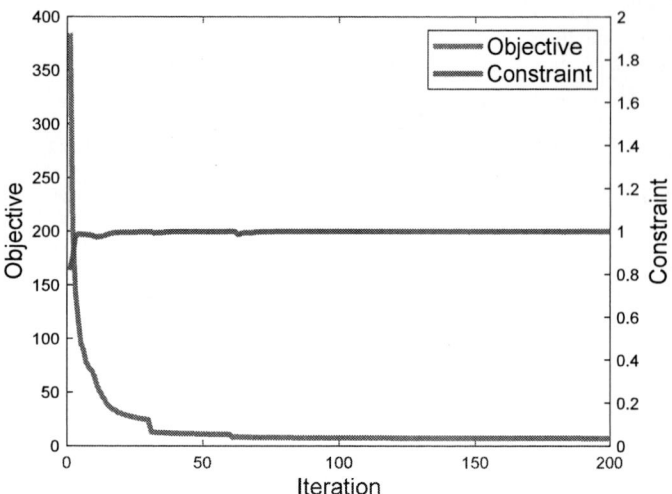

FIGURE 6: CONVERGENCE HISTORIES OF THE OBJECTIVE FUNCTION σ_{PN} AND THE NORMALIZED VOLUME CONSTRAINT V/V_{max} WITH $V_{max} = 35\%$, IN THE LOW-FIDELITY TOPOLOGY OPTIMIZATION WITH THE CONTINUATION METHOD

[27]. The convergence criterion is simply set to the maximum iteration of 200. The move limit of the MMA is set to 0.05. The volume constraint is adjusted from 20% to 50% in 0.5% increments for generating various design candidates.

This study aims to verify whether the performance of the optimized designs by the gradient-based topology optimization method can be improved by further optimization with data-driven MFTD and to demonstrate the usefulness of the gradient-free method for the maximum stress minimization problem. Therefore, it is necessary to prepare the best initial solutions as much as possible. We employ the continuation method for the penalty parameter P in the p-norm to accurately capture the maximum stress during optimization. Although many previous works claim $P = 8$ is a suitable value of the p-norm, this value affects numerical stability and may fail to capture the maximum stress. Fig. 3 shows the material and stress distributions in 35% volume under $P = 8$ and the continuation method, in which P is increased every 30 steps as 8-16-32. As shown in Fig. 3, the re-entrant corner filled with the material causes stress concentration, while the continuation method avoids such a singularity problem. Fig. 4 shows the effect of the continuation method on the maximum von Mises stress. The results indicate that the continuation method achieves a better Pareto front compared with the $P = 8$ fixed case. Hence, we use the continuation method in low-fidelity optimization.

Fig. 5 shows the optimized designs generated by the gradient-based topology optimization using the continuation method. Here, let us focus on the structure of the obtained results. As shown in Fig. 5, various structures appear even when the volumes are close to each other. At first glance, there is no regularity between the change in structure and the increase in volume, and the number of members also differs. On the other hand, very similar structures are observed even if the volumes are different.

Fig. 6 shows the convergence history of the objective function defined as the p-norm stress measure defined as Eq. (10) under the 35% volume constraint with the continuation method.

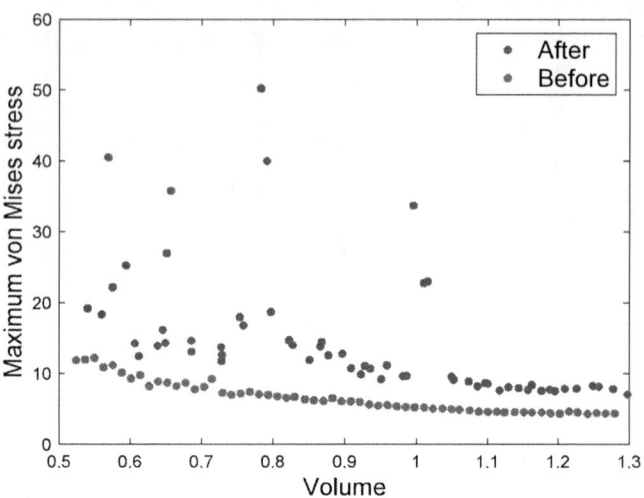

FIGURE 7: EFFECT OF BINARIZATION FOR THE DESIGN VARIABLES ON THE MAXIMUM VON MISES STRESS

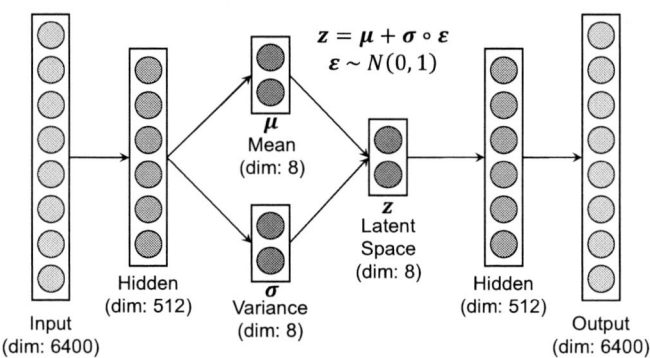

FIGURE 8: ARCHITECTURE OF VAE

The objective and constraint functions converge successfully, and it can be said that an optimized structure is obtained under the relaxation treatments for the gradient-based topology optimization method at least.

Fig. 7 shows the effect of binarization in which $\rho_i \in [0, 1]$ is replaced with $\rho_i \in \{0, 1\}$ for tackling the original optimization problem (1) and indicates that the maximum stress changes significantly. It should be emphasized that this is a serious problem since crisp structures are typically required in actual design. In this study, we use the binarized solutions of low-fidelity optimization as the initial solutions for data-driven MFTD.

5.3 Improvement of Pareto Solutions

The initial solution obtained in Fig. 5 is input to the data-driven MFTD to produce the Pareto solution. Then, we directly solve the original optimization problem (1) concerning the exact maximum stress minimization. The convergence criterion is simply but sufficiently set to the maximum iteration number of 5000. Here, 100 elite solutions survive every optimization step. Fig. 8 shows the architecture of VAE. It uses a simple neural network based on the multilayer perceptron.

Fig. 9 shows the final results of the structures. Interestingly, it can be confirmed that the optimized structures yield a similar

Copyright © 2023 by ASME

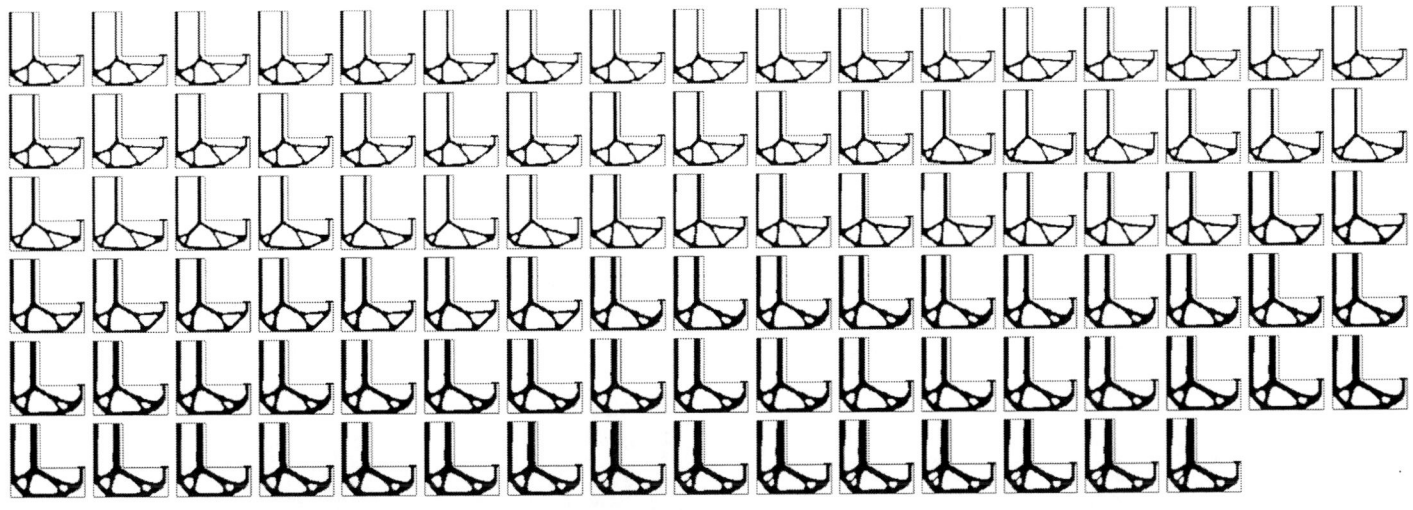

FIGURE 9: MATERIAL DISTRIBUTIONS OF THE ELITE DATA AT ITERATION 5000

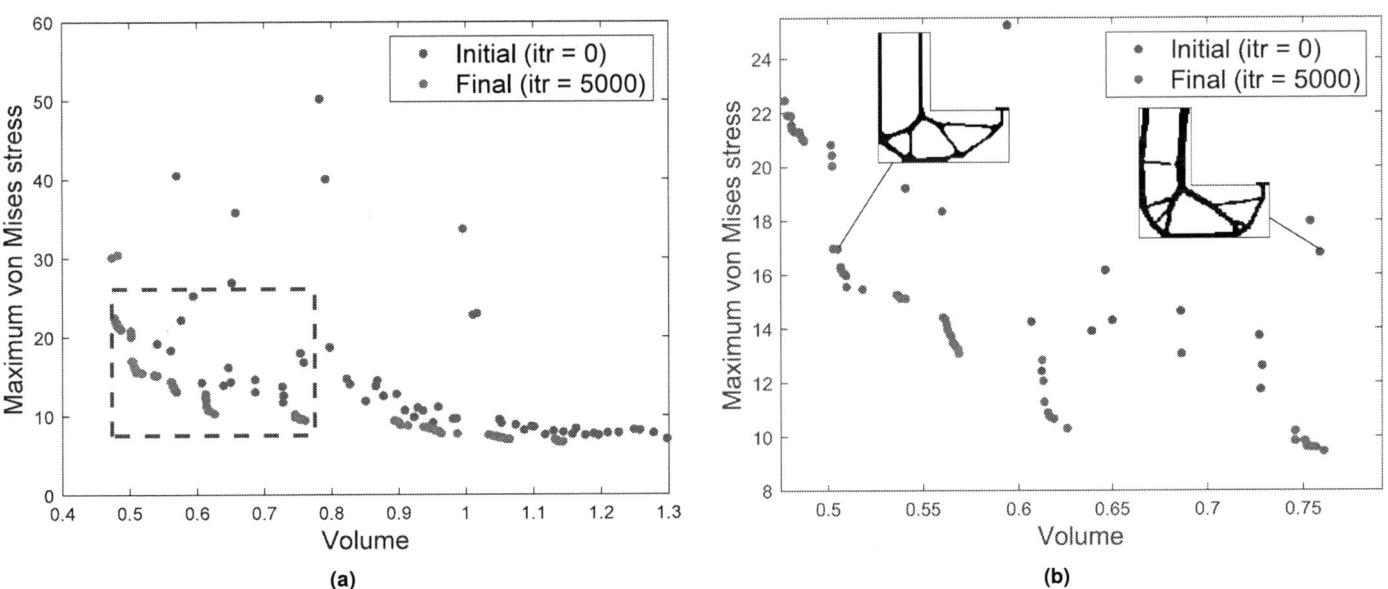

FIGURE 10: OBJECTIVE SPACE: : (A) WHOLE OBJECTIVE SPACE, (B) ZOOMING IN THE DASHED BOX WITH MATERIAL DISTRIBUTIONS

topology while the initial solutions are scattered by volume. This result indicates that although we cannot prove the obtained Pareto solution is the global optimum, it can be expected that data-driven MFTD achieves a unique Pareto solution in the sense of the global search.

Fig. 10 shows the objective function space, in which it can be confirmed that the final result completely dominates the initial solutions. It means that the data-driven MFTD overtakes the performance of conventional gradient-based topology optimization. In concrete, it achieves up to 10% volume reduction under the same stress values. Fig. 10b shows an example of the initial and final solutions, where the objective function values are $\sigma_{max} = 16.81$ with $V = 0.7589$, and $\sigma_{max} = 16.97$ with $V = 0.5052$, respectively. Fig. 11 shows the convergence history of the hypervolume indicator which is a measure of the convergence performance of multi-objective optimization. In the case

of two objective functions, it is represented by the area formed by the reference point and the Pareto front in the objective space. As shown in Fig. 11, the hypervolume gradually improves up to 5000 iterations. Note that this is a relatively large number of iterations compared with other optimization problems investigated in the previous work [16]. Fig. 12 shows the material and stress distributions. Focusing on material distributions, the initial solution is linear and angular including the holes, whereas the holes of the final solution are round through the data-driven MFTD. This difference is thought to have alleviated the stress concentration in the corners. As for the overall structure in the initial solutions, each bar has the same thickness, which causes corners to appear in the connection parts and makes stress concentration more likely. On the other hand, in the final results, there is no limit on the thickness of the bars, and the framework of the members is distributed in a straight or radial pattern. Additionally, it is noticeable that the

Copyright © 2023 by ASME

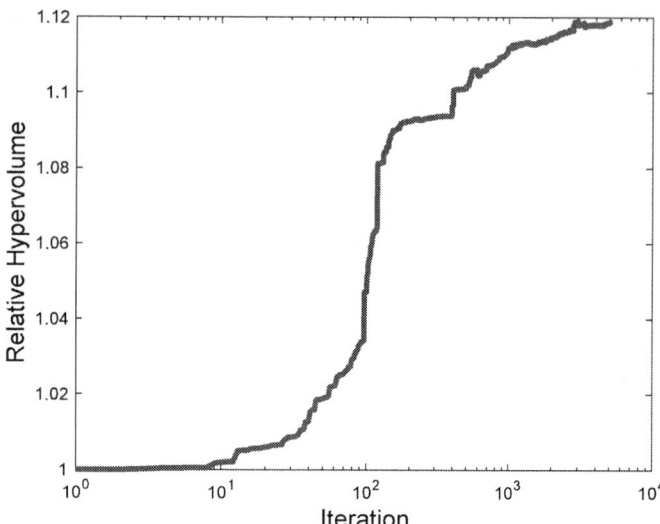

FIGURE 11: CONVERGENCE HISTORY OF THE HYPERVOLUME IN-DICATOR

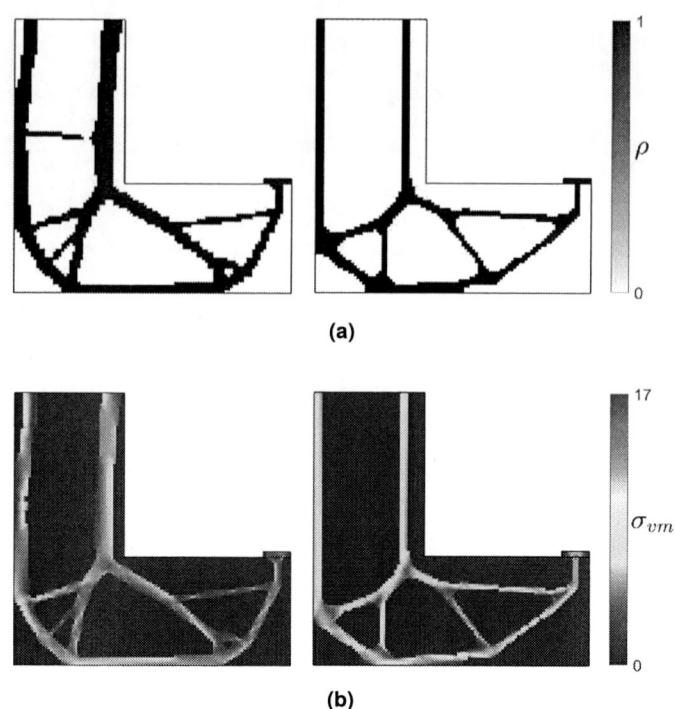

FIGURE 12: AN INITIAL SOLUTION (LEFT) AND A FINAL SOLUTION (RIGHT) UNDER THE SIMILAR MAXIMUM VON MISES STRESS: (A) MATERIAL DISTRIBUTION, (B) VON MISES STRESS

surface and connection parts draw smooth curves. In addition, as shown in Fig. 12, a disconnected bar exists in the initial solution due to the use of the binarized design variables with respect to the relaxed design variables, and it is clear that there is room for further volume reduction in the gradient-based topology optimization method. On the other hand, since such an unreasonable structure does not survive in the optimization process of data-driven MFTD, the final set of optimized designs is composed of reasonable structures in terms of the original objectives as shown in Fig. 9.

6. CONCLUSION

In this paper, we investigated whether there is room for improvement in optimized designs of a maximum stress minimization problem, which is solved by the standard gradient-based topology optimization method. For this, we first discussed the limitations of the conventional methods due to the use of gradient-based optimizers and then tackled solving the maximum stress minimization problem by a data-driven MFTD framework, a gradient-free topology optimization method. We demonstrated the efficacy of the proposed framework through a numerical example, in which initial solutions for the data-driven MFTD are derived by solving the gradient-based topology optimization using the p-norm stress measure, and they are updated using VAE based on the manner of EAs. We found that the optimized designs by data-driven MFTD completely dominate those of the initial solutions. This result justifies that data-driven MFTD could find room for the improvement of conventional gradient-based topology optimization for the maximum stress minimization problem under the investigated condition at least.

As for future work, we consider that our framework can be applied to further complex and practical optimization problems, such as considering geometrical nonlinearity under the exact maximum stress minimization/constraint.

ACKNOWLEDGMENTS

This work was supported by JSPS KAKENHI Grant Numbers 20KK0329 and 20H02054.

REFERENCES

[1] Bendsøe, M. P. and Kikuchi, N. "Generating Optimal Topologies in Structural Design Using a Homogenization Method." *Computer Methods in Applied Mechanics and Engineering* Vol. 71 No. 2 (1988): pp. 197–224. DOI 10.1016/0045-7825(88)90086-2.

[2] Bendsøe, M. P. and Sigmund, O. *Topology Optimization:Theory, Methods, and Applications.* Springer, Berlin, Heidelberg (2004). DOI 10.1007/978-3-662-05086-6.

[3] Le, C., Norato, J., Bruns, T., Ha, C. and Tortorelli, D. "Stress-based Topology Optimization for Continua." *Structural and Multidisciplinary Optimization* Vol. 41 No. 4 (2010): pp. 605–620. DOI 10.1007/s00158-009-0440-y.

[4] Kirsch, U. "On Singular Topologies in Optimum Structural Design." *Structural Optimization* Vol. 2 No. 3 (1990): pp. 133–142. DOI 10.1007/BF01836562.

[5] Rozvany, G. I. N. and Birker, T. "On Singular Topologies in Exact Layout Optimization." *Structural Optimization* Vol. 8 No. 4 (1994): pp. 228–235. DOI 10.1007/BF01742707.

[6] Cheng, G. D. and Jiang, Z. "Study on Topology Optimization with Stress Constraints." *Engineering Optimization* Vol. 20 No. 2 (1992): pp. 129–148. DOI 10.1080/03052159208941276.

[7] Norato, J., Smith, H., Deaton, J. and Kolonay, R. "A maximum-rectifier-function approach to stress-constrained

Copyright © 2023 by ASME

topology optimization." *Structural and Multidisciplinary Optimization* Vol. 65 No. 10. DOI 10.1007/s00158-022-03357-z.

[8] Verbart, A., Langelaar, M. and Keulen, F. van. "Damage approach: A new method for topology optimization with local stress constraints." *Structural and Multidisciplinary Optimization* Vol. 53 No. 5 (2016): pp. 1081–1098. DOI 10.1007/s00158-015-1318-9.

[9] Rozvany, G. I. N. "On Design-dependent Constraints and Singular Topologies." *Structural and Multidisciplinary Optimization* Vol. 21 No. 2 (2001): pp. 164–172. DOI 10.1007/s001580050181.

[10] Cheng, G. D. and Guo, X. "ϵ-relaxed Approach in Structural Topology Optimization." *Structural Optimization* Vol. 13 No. 4 (1997): pp. 258–266. DOI 10.1007/BF01197454.

[11] Moon, Seung Jae and Yoon, Gil Ho. "A newly developed qp-relaxation method for element connectivity parameterization to achieve stress-based topology optimization for geometrically nonlinear structures." *Computer Methods in Applied Mechanics and Engineering* Vol. 265 (2013): pp. 226–241. DOI 10.1016/j.cma.2013.07.001.

[12] París, J., Navarrina, F., Colominas, I. and Casteleiro, M. "Topology Optimization of Continuum Structures with Local Stress Constraints." *Structural and Multidisciplinary Optimization* Vol. 39 No. 4 (2009): pp. 419–437. DOI 10.1007/s00158-008-0336-2.

[13] Yang, R. J. and Chen, C. J. "Stress-based Topology Optimization." *Structural Optimization* Vol. 12 No. 2 (1996): pp. 98–105. DOI 10.1007/BF01196941.

[14] Coello, C. A. C., Lamont, G. B. and Veldhuizen, D. A. V. *Evolutionary Algorithms for Solving Multi-objective Problems.* Springer, New York, NY (2007). DOI 10.1007/978-0-387-36797-2.

[15] Guirguis, D., Aulig, N., Picelli, R., Zhu, B., Zhou, Y., Vicente, W., Iorio, F., Olhofer, M., Matusik, W., Coello Coello, C. A. and Saitou, K. "Evolutionary Black-Box Topology Optimization: Challenges and Promises." *IEEE Transactions on Evolutionary Computation* Vol. 24 No. 4 (2020): pp. 613–633. DOI 10.1109/TEVC.2019.2954411.

[16] Yaji, K., Yamasaki, S. and Fujita, K. "Data-driven Multifidelity Topology Design Using a Deep Generative Model: Application to Forced Convection Heat Transfer Problems." *Computer Methods in Applied Mechanics and Engineering* Vol. 388 (2022): p. 114284. DOI 10.1016/j.cma.2021.114284.

[17] Yaji, K., Yamasaki, S. and Fujita, K. "Multifidelity Design Guided by Topology Optimization." *Structural and Multidisciplinary Optimization* Vol. 61 No. 3 (2020): pp. 1071–1085. DOI 10.1007/s00158-019-02406-4.

[18] Yamasaki, S., Yaji, K. and Fujita, K. "Data-driven Topology Design Using a Deep Generative Model." *Structural and Multidisciplinary Optimization* Vol. 64 No. 3 (2021): pp. 1401–1420. DOI 10.1007/s00158-021-02926-y.

[19] Xia, M., Zhou, Q., Sykulski, J., Yang, S. and Ma, Y. "A Multi-Objective Topology Optimization Methodology Based on Pareto Optimal Min-Cut." *IEEE Transactions on Magnetics* Vol. 56 No. 3 (2020): pp. 1–5. DOI 10.1109/TMAG.2019.2955386.

[20] Bruns, T. E. and Tortorelli, D. A. "Topology Optimization of Non-linear Elastic Structures and Compliant Mechanisms." *Computer Methods in Applied Mechanics and Engineering* Vol. 190 No. 26 (2001): pp. 3443–3459. DOI 10.1016/S0045-7825(00)00278-4.

[21] Bruggi, M. "On an Alternative Approach to Stress Constraints Relaxation in Topology Optimization." *Structural and Multidisciplinary Optimization* Vol. 36 No. 2 (2008): pp. 125–141. DOI 10.1007/s00158-007-0203-6.

[22] Holmberg, E., Torstenfelt, B. and Klarbring, A. "Stress Constrained Topology Optimization." *Structural and Multidisciplinary Optimization* Vol. 48 No. 1 (2013): pp. 33–47. DOI 10.1007/s00158-012-0880-7.

[23] Rozvany, G. I. N. and Sobieszczanski-Sobieski, J. "New Optimality Criteria Methods: Forcing Uniqueness of the Adjoint Strains by Corner-rounding at Constraint Intersections." *Structural Optimization* Vol. 4 No. 3 (1992): pp. 244–246. DOI 10.1007/BF01742752.

[24] Kočvara, M. and Stingl, M. "Solving Stress Constrained Problems in Topology and Material Optimization." *Structural and Multidisciplinary Optimization* Vol. 46 No. 1 (2012): pp. 1–15. DOI 10.1007/s00158-012-0762-z.

[25] Marler, R. T. and Arora, J. S. "Survey of Multi-objective Optimization Methods for Engineering." *Structural and Multidisciplinary Optimization* Vol. 26 No. 6 (2004): pp. 369–395. DOI 10.1007/s00158-003-0368-6.

[26] Miettinen, K. *Nonlinear Multiobjective Optimization.* Springer, New York, NY (1998). DOI 10.1007/978-1-4615-5563-6.

[27] Svanberg, K. "The Method of Moving Asymptotes—a New Method for Structural Optimization." *International Journal for Numerical Methods in Engineering* Vol. 24 No. 2 (1987): pp. 359–373. DOI 10.1002/nme.1620240207.

[28] Deb, K., Pratap, A., Agarwal, S. and Meyarivan, T. "A Fast and Elitist Multiobjective Genetic Algorithm: NSGA-II." *IEEE Transactions on Evolutionary Computation* Vol. 6 No. 2 (2002): pp. 182–197. DOI 10.1109/4235.996017.

[29] Kingma, D. P. and Welling, M. "Auto-Encoding Variational Bayes." arXiv preprint (2013). DOI 10.48550/arXiv.1312.6114.

[30] Kingma, D. P. and Ba, J. "Adam: A Method for Stochastic Optimization." arXiv preprint (2014). DOI 10.48550/arXiv.1412.6980.

[31] Sigmund, O. and Petersson, J. "Numerical Instabilities in Topology Optimization: A survey on Procedures Dealing with Checkerboards, Mesh-dependencies and Local Minima." *Structural Optimization* Vol. 16 No. 1 (1998): pp. 68–75. DOI 10.1007/BF01214002.

[32] Svanberg, K. "Mma and Gcmma, Versions September 2007." 2007.

[33] Foster, David. *Generative Deep Learning: Teaching Machines to Paint, Write, Compose, and Play.* O'Reilly Media (2019).

Copyright © 2023 by ASME

[34] Madeira, J. F. A., Rodrigues, H. C. and Pina, H. L. "Multi-objective Optimization of Structures Topology by Genetic Algorithms." *Advances in Engineering Software* Vol. 36 No. 1 (2005): pp. 21–28. DOI 10.1016/j.advengsoft.2003.07.001.

[35] Balamurugan, R., Ramakrishnan, C. V. and Singh, N. "Performance Evaluation of a Two Stage Adaptive Genetic Algorithm (TSAGA) in Structural Topology Optimization." *Applied Soft Computing* Vol. 8 No. 4 (2008): pp. 1568–4946. DOI 10.1016/j.asoc.2007.10.022.

[36] Madeira, J. F. A., Pina, H. L. and Rodrigues, H. C. "GA Topology Optimization Using Random Keys for Tree Encoding of Structures." *Structural and Multidisciplinary Optimization* Vol. 40 No. 1 (2010): pp. 227–240. DOI 10.1007/s00158-008-0353-1.

[37] Bendsøe, M. P. "Optimal Shape Design as a Material Distribution Problem." *Structural Optimization* Vol. 1 No. 4 (1989): pp. 193–202. DOI 10.1007/BF01650949.

Proceedings of the ASME 2023
International Design Engineering Technical Conferences and
Computers and Information in Engineering Conference
IDETC-CIE2023
August 20-23, 2023, Boston, Massachusetts

DETC2023-113633

AN IMPROVED SHAPE ANNEALING ALGORITHM FOR THE GENERATION OF COATED DNA ORIGAMI NANOSTRUCTURES

Bolutito Babatunde
Department of
Mechanical Engineering
Carnegie Mellon
University
Pittsburgh, PA

Jonathan Cagan
Department of
Mechanical Engineering
Carnegie Mellon
University
Pittsburgh, PA

Rebecca E. Taylor
Department of
Mechanical Engineering
Carnegie Mellon
University
Pittsburgh, PA

ABSTRACT

In recent years, the field of structural DNA nanotechnology has advanced rapidly due to transformative design tools. Although these tools have been revolutionary, they still bear one overall limitation of requiring users to fully conceptualize their designs before designing. Recently, a simple computational casting technique was developed using generative optimization strategies to automate the design of DNA origami nanostructures. This approach employs a shape annealing algorithm, which creates a formal language of honeycomb DNA origami nanostructures with shape grammars and drives designs from the language towards a desired configuration using simulated annealing. This initial demonstration of the approach can generate novel scaffold routing schemes for creating solid or hollow structures constrained by boundaries of polyhedral meshes. The results from the initial approach, particularly from the hollow structures, reveal a challenging design space. This simple technique generates novel scaffold routing schemes that do not replicate the overall polyhedral mesh shape and is limited in its ability of controlling scaffold path exploration in the design space. This paper demonstrates an approach for varying effective wall thickness and improving quality of polyhedral mesh coverage for hollow structures that can be tuned and optimized by introducing a more refined computational casting technique. We achieve these improvements through changes in the simulated annealing algorithm by adding a Hustin move set algorithm that dynamically adjusts the performance of the overall design and redefining how these hollow designs are articulated. The results in this work illustrate how the shape annealing algorithm can navigate a challenging design space to generate effective hollow designs.

Keywords: design optimization, generative design, micro and nano systems design and synthesis of, biomaterials, design automation, structural DNA nanotechnology, and DNA origami

NOMENCLATURE

bp	base pair
f(x)	objective function
M	mutations or iterations
L	limit
T	temperature
T_p	previous temperature
T_i	initial temperature
α	alpha (reduction factor)
r	shape rule
Q_r	quality factory for the rth shape rule
n_r	number of times the rth shape rule is called at T_p
s	total number of shape rules
$P_{i,r}$	initial probability if the rth shape rule
P_u	probability multiplier
$prob_i$	initial probability for the Hustin move set
Q_{total}	total quality factor over all shape rules
x	minimum perpendicular Euclidean distance from the outer wall of input mesh to the center of the 7 bp scaffold section
d	set desired distance from the outer wall of input mesh

Copyright © 2023 by ASME

1. INTRODUCTION

In the nanoworld, there has been an exponential increase in the development of tools for the design of structures and machines using DNA nanotechnology [1–10]. These approaches are based on Watson-Crick base pairing in DNA, wherein adenine (A) and guanine (G) pair with thymine (T) and cytosine (C) via hydrogen bonds [11–13]. Figure 1 shows a 3D molecular representation of a 7-base pair (bp) single helix section of a DNA double helix with simplifications where a single helix can be represented using ball-and-stick models or cylindrical building blocks. By harnessing the programmable binding of nucleobases, the users can control the assembly of biomechanical nanostructures [14]. The most common method for forming complex DNA nanostructures is an approach called DNA origami. As shown in Figure 2a, this self-assembly method utilizes a long single-stranded DNA (ssDNA), also known as a "scaffold", that is folded and then cinched together with hundreds of oligonucleotides, also known as "staples", to form a desired configuration [12,15]. A common scaffold used in the field of DNA origami is M13mp18 (M13), which is obtained from a bacteriophage and is 7249 bases in length [16]. The DNA origami designs can be represented in 2D schematics and 3D renderings in which the scaffold and staples are simplified to cylinders (Figure 2b). The DNA origami folding method can create multilayer or wireframe architectures with a myriad of applications including nanosensors [17], nanolithography [18], nanomachines [19], and nanocasting [20].

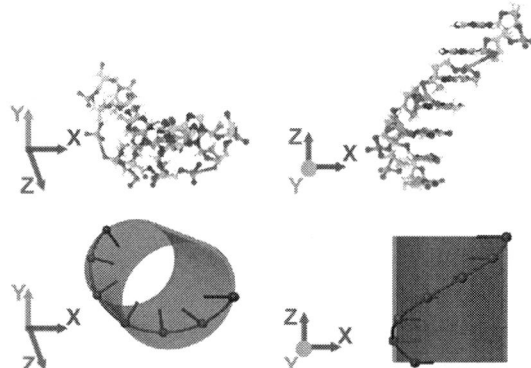

FIGURE 1: REPRESENTATION OF 7 BP SCAFFOLD SEGMENT: (TOP) HELIX REPRESENTATION. (BOTTOM) SIMPLIFIED REPRESENTATION.

Computational casting is a new approach that uses the shape annealing generative algorithm [21] to automate the routing of the scaffold strand around or within an input polyhedral mesh, resulting in coated-type or solid designs, respectively [22]. The shape annealing algorithm uses shape grammars [23] to create a language of DNA nanostructures or shape rules that defines the path through space, and simulated annealing [24] drives the language towards an optimally preferred configuration. This technique paves the way for novel design tools in which humans are not required to define the scaffold routing for irregular multilayer DNA origami nanostructures. By using a complex polyhedral mesh as a design boundary, this approach adds a multilayer functionality that complements existing automated tools for generating 2D and 3D wireframe DNA origami nanostructures [3,5–10]. Here, polyhedral mesh refers to a simple geometric object. This method is particularly groundbreaking in the nanoworld because the generated scaffold routes are capable of being optimized for structure and function. It is important to note that the polyhedral mesh is merely a boundary that is removed after the scaffold is generated.

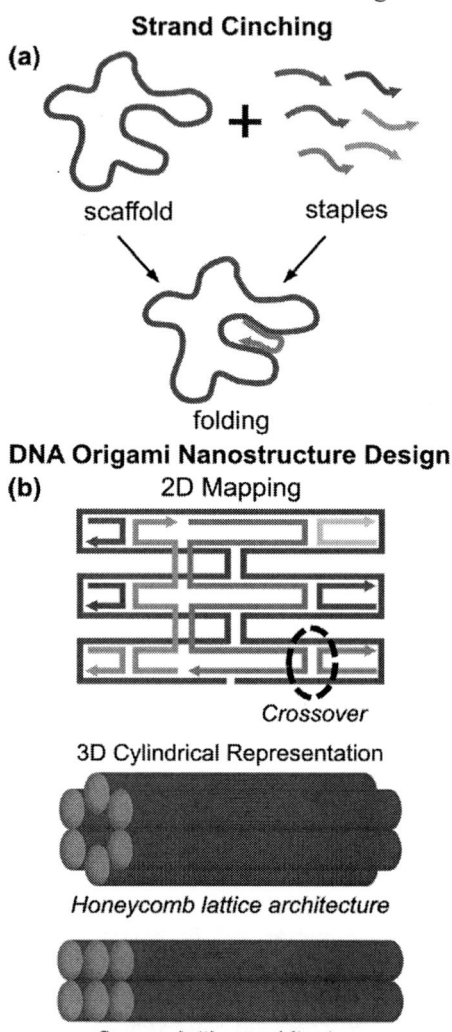

FIGURE 2: DNA ORIGAMI SELF-ASSEMBLY METHOD: (A) SCAFFOLD FOLDING (B) DNA ORIGAMI NANOSTRUCTURE DESIGN VISUALIZATION.

In this work we address the quality and thickness of generated designs using a refined version of our computational casting technique to route the scaffold around a tetrahedral mesh (Figure 3). This has not been demonstrated in prior work. The refined technique demonstrates that the objective function from the simulated annealing algorithm can tune the mean effective thickness of the coated-type designs. Therefore, this approach can achieve desired coatings with a variety of scaffold lengths.

Copyright © 2023 by ASME

This feature is of the utmost importance to the DNA nanotechnology community because the scaffold material length is often a key constraint for design, limiting the size of what can be built.

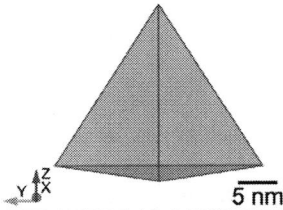

FIGURE 3: INPUT TETRAHEDRAL MESH FOR REFINED AND SIMPLE TECHNIQUE WITH AXIS ALIGNED BOUNDING BOX (L, W, H) DIMENSIONS OF (21.7, 25.0 ,20.4) NM.

This work also investigates the density within the walls for various thicknesses, determining that internal porosity or quality can also be tuned by the objective function. Such an approach which has not been addressed with prior algorithms shows how a design algorithm applied to this new problem can have a profound impact in the output quality. This paper addresses performance limitations in previous work for coated-type designs which was apparent in large gaps within the walls and low wall density where the overall structure did not replicate the shape of the tetrahedral mesh (Figure 4) [22]. Figure 4 displays an example of the cross-sectional view generated with the simple algorithm from previous work. In the figure, the rectangle to the left highlights the location of the cross-sectional view on the overall structure. This same cross-sectional representation is maintained for the refined technique. Such limitations show that the design problem of coating a scaffold around irregular geometries creates a challenging design space that cannot be tackled with simple optimization strategies. To tackle the challenging design space and further improve the quality of the coated designs, this work appends the simulated annealing algorithm with the Hustin move set [25]. This technique dynamically adjusts the probability of the shape rules to improve the performance of the algorithm, tune the wall thickness, and decrease the large gaps observed in previously generated designs.

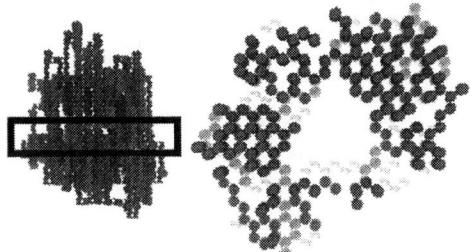

FIGURE 4: EXAMPLE OF GENERATED DNA ORIGAMI DESIGN USING SIMPLE TECHNIQUE WITH TOTAL SCAFFOLD LENGTH OF 12,901 BP.

2. MATERIALS AND METHODS

This section provides a brief description of the generative methods in the computational casting approach from previous

work and details the refinements made to tune the mean effective thickness of generated coated-type DNA origami designs. The same three shape grammar rules are used, and a refined simulated annealing algorithm is then used to control the formation of these structures to meet the well-defined coating criteria. In the simulated annealing algorithm, coating behavior is rearticulated by replacing the objective function from previous work. The Hustin move set is then introduced as a modification to the simulated annealing algorithm to iteratively increase the probability of selected shape rules that help boost the algorithm's coating performance.

2.1 Shape Grammars

In shape annealing, a shape grammar is a set of 2D or 3D shape rules that are applied sequentially from a starting geometric shape to generate a set of designs or language [23], while simulated annealing controls the evolution of the starting shape. In this case, the geometric shape takes on key features of a single-stranded DNA (ssDNA) in 3D form with a helical diameter of 2 nm and an axial rise per base of 0.34 nm or roughly 34.3° turn per base. The shape grammar rules in this work mirror the rules from the previous computational casting technique [22], which adopt the honeycomb lattice architecture from the popular manual computer-aided design (CAD) tool, caDNAno [2]. Since the honeycomb lattice permits a union (or crossover) between neighboring helices at multiples of 7 bases at roughly 240°, the design language divides the scaffold into 7 bp sections ensuring continuous scaffold growth through the addition of a new section.

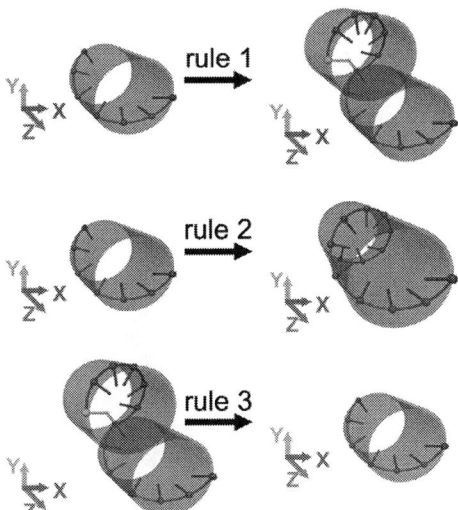

FIGURE 5: SHAPE GRAMMAR

Figure 5 illustrates simple examples of the three-rule shape grammar used in the previous work. The blue single helix shows helix growth in the +z-direction while the red single helix shows helix growth in the -z direction. The first rule is the crossover rule which generates a 7 bp increment of scaffold on the nearest of the three possible neighboring helices at angles 0°, 120°, and 240°. Figure 5 shows the crossover at 120°. The second rule is

93

Copyright © 2023 by ASME

the extension rule which extends the scaffold by a 7 bp section along the z-axis of its current helix. Figure 5 shows extension of the blue single helix. The third rule is the reversal rule which undoes a previous call of the first and second rule to stop the scaffold from being stuck and encourages scaffold exploration within the given design space. Figure 5 shows the undoing of Rule 1. The rules create a language or set of instructions that drive the path of a continuous scaffold strand along a caDNAno honeycomb lattice architecture.

2.2 Optimization with Simulated Annealing

Shape annealing is a version of simulated annealing which is a stochastic optimization technique, where each state is calculated using a randomly generated shape rule as a move set. In previous work, the simulated annealing algorithm has an outer and inner loop. Within the inner loop of M mutations or iterations, each randomly generated feasible state is assessed using an objective function, $f(x)$. For objective minimization, if $\Delta f(x) \leq 0$, where the evaluated current and previous states are compared, the current state is accepted. However, if the condition is false, a probability of adding the state is calculated:

$$P_{add} = e^{-\frac{\Delta f(x)}{T}}, \qquad (1)$$

where $\Delta f(x)$ is the change in objective function and T is the temperature. The current state can only be added if a normalized random number is less than the calculated probability. If a set limit (L) of added states has been reached, the algorithm jumps out of the inner loop. In the outer loop, if no states have been added then the algorithm has reached equilibrium and can be terminated. If the condition is false, the temperature is lowered with an exponential function:

$$T = \alpha T_p, \qquad (2)$$

where T_p is the previous temperature value and α, which is between 0 and 1, is the reduction factor. The algorithm runs until either equilibrium is reached or the temperature reaches zero.

2.2.1 Hustin Move Set

In this paper, we apply an additional step within the outer loop of the simulated annealing algorithm called the Hustin move set, which adjusts the probability of selecting a given rule based on the performance of the algorithm at any time [25]. The rationale for dynamically modifying the probabilities using the Hustin move set is that the refined technique will increase the probability of selecting the appropriate shape rule for an increased coverage and wall density. This therefore decreases the number of poor shape rule acceptances made during the anneal. This common technique has been used to address the problem of optimal nonorthogonal routing of components in a chemical plant [26]. This problem is similar to the DNA scaffold routing problem in this work where the Hustin move set is used to modify the probability of selecting each shape rule. Previously, the computational casting technique set an equal probability of selecting each shape rule throughout an anneal. The generated results from this fixed shape rule selection probability display

sparse scaffold coverage of input polyhedral mesh and low wall density in the overall structure (Figure 4). Based on analysis, we hypothesize these performance limitations to originate from a lower call in the extension shape rule (rule 2) due to a higher ratio of neighboring helical sections of only 7 bp in length (rule 1) within regions of sparse scaffold coverage. Therefore, modification of shape probability selection is necessary.

In the modified simulated annealing algorithm, the probabilities of implementing moves are updated after each temperature reduction as a function of the success of the moves (or shape rules) in accomplishing the objective at the previous temperature. The measure of success per shape rule (r) is determined by a quality factor,

$$Q_r = \frac{1}{n_r} \sum_{\substack{rules \\ accepted}} |\Delta f(x)|, \qquad r = 1, \ldots s, \qquad (5)$$

where n_r is the number of times the rth shape rule was called at the previous temperature, $\Delta f(x)$ is the change in objective due to an accepted shape rule from objective minimization, and s is the total number of shape rules. If a shape rule is not called at a specific temperature value, then its quality factor is set to zero. At the start of an anneal, each shape rule probability is initialized to ensure that each rule is likely to be called. While this initial probability could be skewed with an initial bias or uniform for all shape rules, the sum of all initial probabilities must equal 1:

$$\sum_{\substack{intial \\ probabilities}} P_{i,r} = 1, \qquad (6)$$

where $P_{i,r}$ is the initial probability of the rth shape rule. Here, the initial probability for the extension rule is much higher than the crossover and reversal rules. Once the quality factor is calculated, the probability of selecting the rth shape rule is updated with the following equation:

$$P_r = P_{i,r} P_u + (1 - P_u)\frac{Q_r}{Q_{total}}, \qquad r = 1, \ldots s, \qquad (7)$$

where P_u, which is between 0 and 1, is a multiplier that guarantees the relevance of the initial bias per temperature reduction and Q_{total} is the total quality factor over all shape rules. A shape rule with a larger quality factor will have a higher call probability at the next temperature.

2.3 Shape Annealing

Shape annealing uses simulated annealing to control when a randomly selected shape rule (or scaffold section) is accepted to the overall continuous scaffold routing pattern at a given state. Of note, prior uses of shape annealing in design focused on the design of truss structures [27]. One contribution of this work is the extension and demonstration of the algorithm applied at the nano scale to design DNA structures.

Figure 6 illustrates a flow chart of the revised shape annealing casting algorithm including the extension to include the Hustin move set. In the flowchart, there are three key inputs: starting shape, specification, and objective and constraints. The starting shape, which is a match from the previous technique, is a blue scaffold 7 bp scaffold section. The specifications are the same key shape annealing parameters from previous work with an appended modification: initial temperature (T_i), limit (L), mutations (M), reduction factor (α), and initial probability ($prob_i$) for Hustin move set. The constraints and objectives mirror the previous technique, where the constraint given is the tetrahedral mesh as a boundary and the objective is for scaffold to wall distance minimization. The algorithm focuses on shape rule transformation where each shape rule is evaluated based on the refined simulated annealing algorithm. In the flow chart the crossover rule (rule 1) is selected and applied to the starting shape which consists of a single blue segment which is transformed to a blue and red 14 bp long scaffold. The shape rule probabilities are updated per temperature reduction and the image for this step is an example where the cylinders in red are the accepted shape rules used to calculate the new probabilities. Once the overall shape has reached convergence, the shape annealing algorithm will stop and the output will be the final design.

2.4 Structural Validation with Coarse-Grained Simulations

Validating the shape and stability of these generated complex DNA origami designs is essential for ensuring structure formation during the time-intensive physical experimentation. There are a variety of established structural validation tools, which model the traits of DNA origami through computer simulations that provide detailed predictions of the behavior of the DNA nanostructures comparable to the results from physical experimentation [15,28–35]. OxDNA is a coarse-grained DNA model that is used as a tool to capture the structural, mechanical, and thermodynamic characteristics of DNA at the nucleotide resolution level over timescales ranging from, but not limited to, nanoseconds to microseconds that are relevant for studies of nanostructure behavior [32–34]. OxDNA provides invaluable information about the local and global dynamic properties of these DNA nanostructures such as the response of the structure due to internal stress [35], yielding under tension [36], and actuation [37].

This work uses oxDNA to model the generated DNA origami designs following a common relaxation process [31] in which overlapping nucleotide volumes and stretched backbone bonds are corrected. The general goal of the relaxation process is to reach a steady state through energetic minimization of the system. The relaxation process is two-part, where a minimization algorithm runs for 2000 steps and a relaxation molecular dynamics simulation run for 10^7 steps. After the DNA origami nanostructure is well relaxed, another molecular dynamics simulation step is subsequently run for 10^7 steps with the absence of added forces or modified potentials used for root mean-square fluctuations (RMSF) calculations. After running simulations, these structures are analyzed using a standardized set of tools [38].

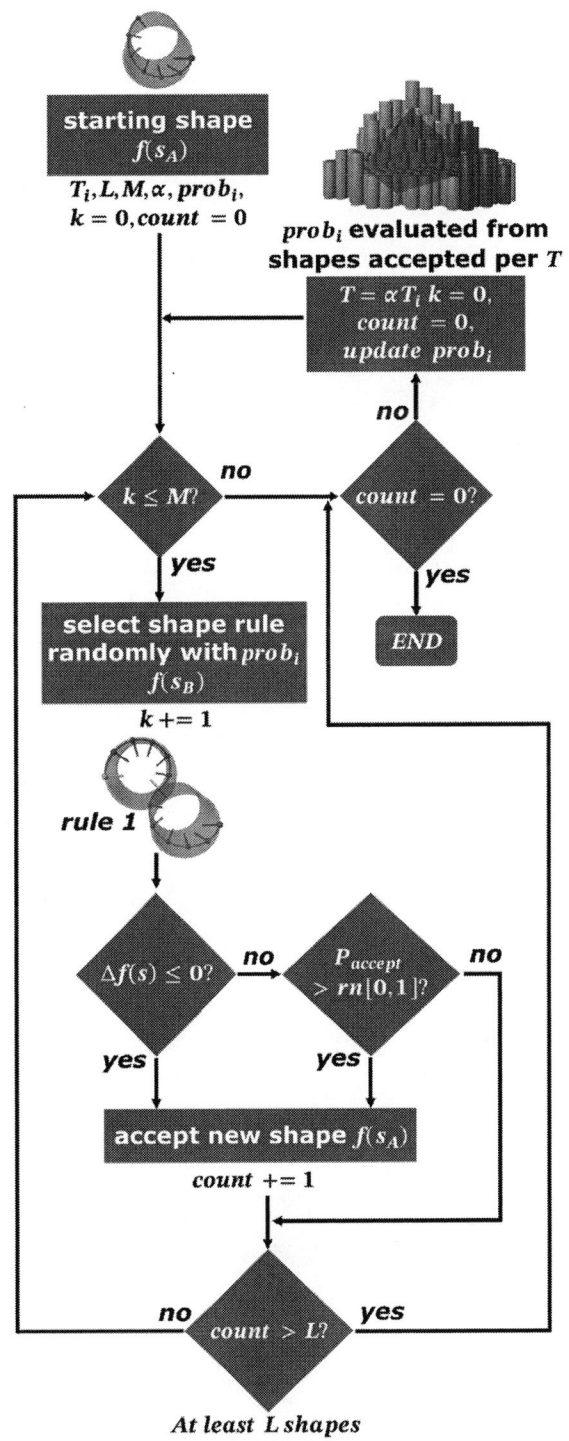

FIGURE 6: SHAPE ANNEALING ALGORITHM FLOWCHART

RMSF is one of the common structural analyses that measures the fluctuation of the DNA origami nanostructure during simulation. One way to calculate RMSF is by averaging the configurations generated over the entire simulation trajectory

Copyright © 2023 by ASME

[38]. It is important to note that all simulations in this work are performed at 295 K and a 15.15 fs time step (0.005 simulation units). All additional parameters used are detailed in literature by Doye et al. [31].

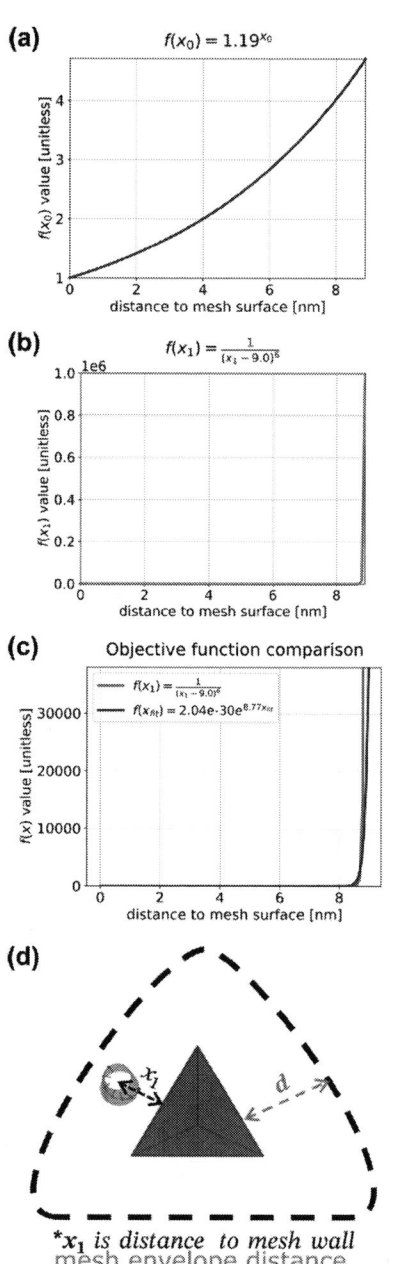

FIGURE 7: $F(X)$ REARTICULATION (A) PREVIOUS $F(X_0)$ (B) CURRENT $F(X_1)$ (C) EXPONENTIAL VS. HYPERBOLIC $F(X)$ COMPARISON (D) ILLUSTRATION OF $F(X)$ VARIABLES.

2.5 Rearticulation of Coating Behavior

As stated previously, the objective function defines the desired characteristic of coating and is used to assess whether a shape rule is accepted or rejected. In previous work, coating behavior is achieved by minimizing the distance from the center of generated helical 7 bp scaffold sections towards the outer wall of the input triangular mesh. The distance is then scaled along the following objective function to articulate coating [22]:

$$f(x) = 1.19^x, \qquad (8)$$

where x is the minimum perpendicular Euclidean distance from the outer wall of the input mesh to the 7 bp scaffold section. The constant value in the previous objective function is purely arbitrary and is selected from a range of 1.00 to 2.00 where values approaching 1.00 generates designs that are loosely coated with higher coverage while values approaching 2.00 generates designs that are tightly coated with limited coverage of the polyhedral mesh. The rationale for selecting an exponential function is because the shape rules generated at a closer distance to the mesh have lower $f(x)$ values and therefore minimize $\Delta f(x)$ while the shape rules generated at a further distance from the mesh would have a higher likelihood of rejection due to much higher $f(x)$ values. In the previous technique, internal porosity or quality is traded for input mesh coverage. This is shown in the plot from Figure 7(a). Although this objective function demonstrates the capability of coating, the coated designs do not replicate the overall shape of the input polyhedral mesh as shown in Figure 4. This is due to the slow ramp of the exponential curve in Figure 7(a) where the shape rules generated at a closer distance to the mesh surface do not fully minimize $\Delta f(x)$ and the shape rules generated at a further distance from the mesh do not have comparably higher $f(x)$ values. Evidently, a change in objective function is necessary.

To increase quality and internal porosity of overall coated designs, the generated scaffold sections are limited to set desired distances measured from the outer wall of the desired mesh to a location on the outside of the mesh and the objective function is replaced with a hyperbolic function following a more step-like behavior:

$$f(x) = \frac{1}{(d-x)^6}, \qquad (9)$$

where x is the minimum perpendicular Euclidean distance from the outer wall of the input mesh to the 7 bp scaffold section and d is the set desired distance from outer wall of input mesh in which scaffold sections can be accepted to control effective wall thickness. The step-like behavior allows for more acceptance of feasible scaffold sections that tightly coat the outer wall of the input mesh without compromising porosity since the $f(x)$ values are all close to zero before the set desired distance. As the hyperbolic function approaches the set-desired distance, there is an abrupt increase in $f(x)$ values, which lowers the likelihood of shape rule acceptance according to the shape annealing algorithm. This is shown in Figure 7(b). This hyperbolic function also allows for physically relevant parameters as opposed to the arbitrary parameters of the exponential function shown in Figure 7(c). The exponent on the refined objective function is selected from a range of 2 to 9 where values approaching 2 show a more gradual increase in function values with distances close to d while values approaching 9 show a sharper increase in function values that become too computationally intensive with distances closer to d. 6 is a more appropriate exponent value since it

Copyright © 2023 by ASME

creates function values with distances closer to d that are computationally feasible to calculate and still maintain a sharp increase in function values. Figure 7(d) paints a picture of a 7 bp scaffold section accepted within the set desired distance.

2.6 Hustin Move Set Initial Probability

The Hustin move set initial probability used for all designs generated with the refined shape annealing algorithm in order of shape rules are 0.18 for the crossover rule, 0.44 for the extension rule, and 0.38 for the reversal rule. An initial probability search was implemented before changing the objective function with set desired distances by varying each shape rule from (0,1) to find structures that have the largest number of scaffold sections within an arbitrary distance of 12.0 nm from the mesh surface. This arbitrary distance was selected because the resulting scaffold routes showed sparse searches of the vast design space beyond $d = 12.0$ nm. The rationale for setting a bound for measuring the total scaffold sections is to find the shape rule probabilities with a higher porous layer close to the mesh surface. Since regions with large gaps from the previous algorithm occur due to a reduction in the application of the extension rule, the refined algorithm has a lower probability for the crossover rule and the highest probability for the extension rule. The reversal rule has a reasonably high probability to ensure that the algorithm does not converge on a local minimum, or the scaffold is not stuck during an anneal and can fully explore the design space. The probabilities are then dynamically adjusted during run time as described in Section 2.2.1.

3. RESULTS

By using a 3D tetrahedral mesh as input, the refined shape annealing algorithm can generate coated-type designs with a variety of effective wall thicknesses. After generating DNA origami designs of varying thicknesses, the designs are converted to the caDNAno JSON file format using a customized scadnano Python script [39]. The designs are then auto-stapled using the caDNAno auto-stapling function [2] with additional manual editing due to the non-traditional scaffold routings for effective stapling. After stapling, the JSON files are converted using the TacoxDNA platform [40] to PDB files, which are visualized using ChimeraX [41,42]. The JSON files are also converted to oxDNA topology and configuration files which are used as input files to run simulations. The output files after running simulations are used as input for the RMSF analysis scripts. The results from the analysis are then visualized using oxView [38] and ChimeraX.

Since the refined shape annealing algorithm uses a probabilistic technique, the generated continuous scaffold can get stuck, because the algorithm converges to a local minimum. This is typically avoided by selecting the best generated solution from several algorithm runs [43]. In this work, the algorithm is run in 10 batches of 10 (100 times in total), where the best generated solution is selected per batch. This is done per d from 8.0 to 12.0 nm. Since the metric of success is a tightly packed generation of scaffold within the set desired distance, the 10

DNA origami designs with the highest scaffold length are selected. An exhaustive search is performed to select the best shape annealing parameters per d that follow the metric of success stated by varying the limit (L) and temperature (T), which were found to be more sensitive parameters during an anneal. Table 1 shows the shape annealing parameters per set desired distance using the tetrahedral mesh as input. Table 2 shows the mean and standard deviation of 10 best generated solutions per 10 batches using the tetrahedral input mesh. To demonstrate the capability of the refined shape annealing algorithm in generating DNA origami designs with varying effective wall thicknesses, Figure 8 shows the cross-sectional view of the top three out of 10 designs with $d = [8.0, 9.0, 12.0]$ nm. Figure 9 shows the fully relaxed configurations of the structures generated in Figure 8 after structural validation through oxDNA simulations. Figure 10 shows the root mean-square fluctuations (RMSF) of the structures generated from oxDNA. Figure 12 illustrates the average effective wall thickness analysis of the top 10 generated DNA origami designs per batches of 10.

To ensure the method of analysis in Figure 12 is not mesh specific, the results from a cube input mesh with an edge length of 21.2 nm using the refined technique with $d = 9.0$ nm are also analyzed under the same method. Figure 11 shows an example of a fully relaxed configuration and RMSF from the 10 best generated DNA origami designs using the cube input mesh. In Figure 13, the same method of analysis of the top 10 generated DNA origami designs per batches of 10 for the cube input mesh is used to illustrate effective wall thickness. Tables 3 & 4 show the shape annealing parameters for $d = 9.0$ nm and the mean and standard deviation of the 10 best generated solutions per batch, respectively, using the input cube mesh.

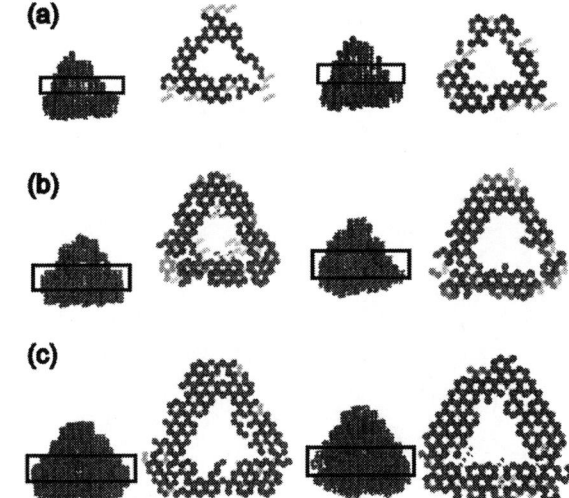

FIGURE 8: TWO OF BEST GENERATED DNA ORIGAMI DESIGNS FOR TETRAHEDRAL MESH USING REFINED TECHNIQUE WITH TOTAL SCAFFOLD LENGTHS OF: (A) 5,306 BP AND 5,257 BP FOR $D = 8.0$ NM, (B) 7,273 BP AND 6,979 BP FOR $D = 9.0$ NM, AND (C) 12,264 BP AND 12,586 BP FOR $D = 12.0$ NM.

Copyright © 2023 by ASME

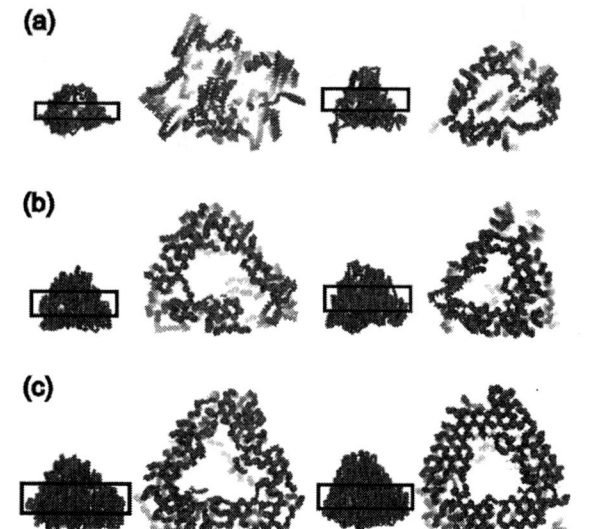

FIGURE 9: FULLY RELAXED CONFIGRUATIONS FROM OXDNA SIMULATIONS OF STRUCTURES IN FIGURE 8 FOR: (A) $D = 8.0$ NM, (B) $D = 9.0$ NM, AND (C) $D = 12.0$ NM.

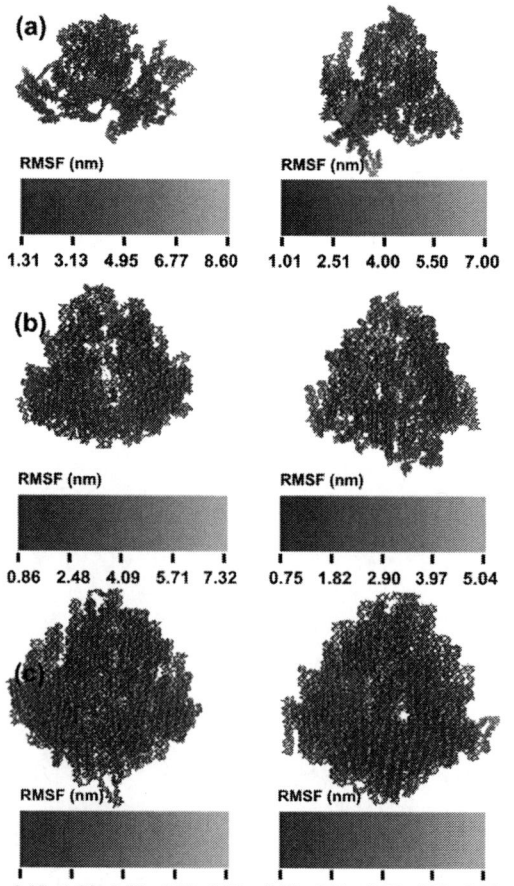

FIGURE 10: RMSF OF STRUCTURES IN FIGURE 8 FOR: (A) $D = 8.0$ NM, (B) $D = 9.0$ NM, AND (C) $D = 12.0$ NM. RMSF PATTERNS USE A COLORMAP FROM VIOLET TO YELLOW.

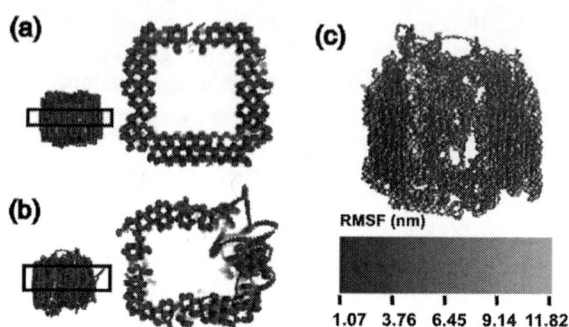

FIGURE 11: ONE OF BEST GENERATED DNA ORIGAMI DESIGNS USING REFINED TECHNIQUE AND CUBE MESH FOR $D = 9.0$ NM (A) WITH TOTAL SCAFFOLD LENGTH OF 13,937 NM (B) FULLY RELAXED CONFIGURATION (C) RMSF PATTERN USING COLORMAP FROM VIOLET TO YELLOW.

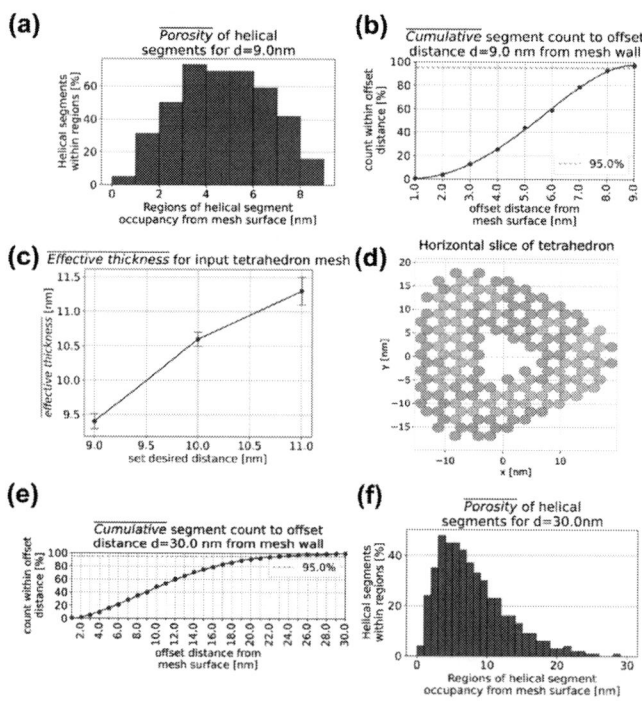

FIGURE 12: EFFECTIVE THICKNESS ANALYSIS FOR TETRAHEDRAL MESH USING REFINED TECHNIQUE: (A) AVG POROSITY ANALYSIS FOR $D = 9.0$ NM (B) AVG EFFECTIVE THICKNESS ANALYSIS FOR $D = 9.0$ NM (C) AVG EFFECTIVE THICKNESS COMPARISON (D) GRID FILLING VISUALIZATION AT MID-HEIGHT OF A STRUCTURE SHOWING UNOCCUPIED GRIDS IN RED. USING SIMPLE TECHNIQUE: (E) AVG EFFECTIVE THICKNESS ANALYSIS (F) AVG POROSITY ANALYSIS

Copyright © 2023 by ASME

(a)

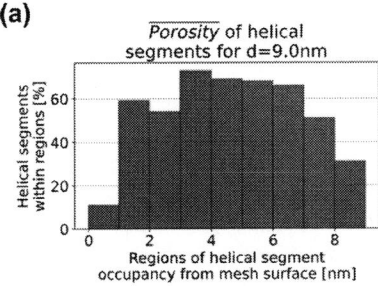

(b)

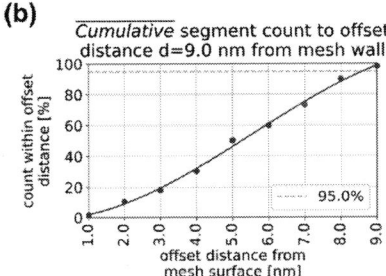

(c)

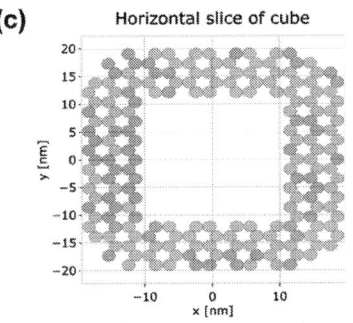

FIGURE 13: EFFECTIVE THICKNESS ANALYSIS FOR CUBE MESH USING REFINED TECHNIQUE FOR $D = 9.0$ NM (A) AVG POROSITY ANALYSIS (B) AVG EFFECTIVE WALL THICKNESS ANALYSIS (C) GRID FILLING VISUALIZATION AT MID-HEIGHT OF A STRUCTURE SHOWING UNOCCUPIED GRIDS IN RED.

TABLE 1 REFINED SHAPE ANNEALING OPTIMIZATION PARAMETERS USING TETRAHEDRAL MESH AS INPUT

d [nm]	T	L	α	M
8.0	956	280	0.98	510
9.0	941	300	0.98	510
10.0	980	260	0.98	510
11.0	888	340	0.98	510
12.0	912	260	0.98	510

TABLE 2 AVG AND STD DEV OF TOTAL SCAFFOLD LENGTH (sl)OF TOP 10 GENERATED DESIGNS PER 10 BATCHES WITH INPUT TETRAHEDRAL MESH USING REFINED TECHNIQUE

d [nm]	s̄l	σ
8.0	5187	30
9.0	6755	38
10.0	8456	34
11.0	10549	69
12.0	12138	40

TABLE 3 REFINED SHAPE ANNEALING OPTIMIZATION PARAMETERS USING CUBE MESH AS INPUT

d [nm]	T	L	α	M
9.0	975	360	0.98	510

TABLE 4 AVG AND STD DEV OF TOTAL SCAFFOLD LENGTH (sl)OF TOP 10 GENERATED DESIGNS PER 10 BATCHES WITH INPUT CUBE MESH USING REFINED TECHNIQUE

d [nm]	s̄l	σ
9.0	13818	49

4. DISCUSSION

Figure 8-10 illustrate the capability of the refined shape annealing algorithm in controlling the quality of the effective wall thickness of coated-type designs using a complex tetrahedral boundary. Figure 8 illustrates the tight coating of scaffold sections which is highlighted in the maintenance of the triangular shape toward the edges of the generated structures. This is unlike the structure generated in Figure 4 using the simple shape annealing algorithm where the outer walls of the structure form an arbitrary shape. Figures 9 and 10 illustrate the simulation results for the structures in Figure 8 with $d = [8.0, 9.0, 12.0]$ nm. Although the fully relaxed configurations and RMSF calculations in Figure 9 and 10 show major deviations for $d = 8.0$ nm from the intended final configuration, there are only slight deviations for $d = 9.0$ nm and higher. Figure 10 illustrates a decrease in regions with high fluctuations from increasing the set desired distance values. The results from the RMSF calculations in Figure 10 show some regions of high RMSF values where the crossover rule (rule 1) is accepted recursively with lack of variety in the acceptance of the crossover rule and extension rule (rule 2). These regions where rule 1 is recursively accepted are subjected to high fluctuations, which should be decreased before physical characterization. A post-processing step would be a prime solution to address such limitations by applying the new extension rule to those regions. Instead of scaffold sections that are only 7 bp in length, the scaffold sections would then be at least 14 bp in length which would create more effective stapling and reductions in the number of high fluctuations. In the future, it would be interesting to explore how the generated scaffold route is affected by the incorporation of RMSF data.

Figure 12 illustrates verification of increasing internal porosity and quality in changing effective wall thickness values of the generated scaffold sections of the top 10 designs. The distance of each helical segment in Figure 12 is calculated from its center. The set desired distances of 9.0 nm to 11.0 nm were selected for analysis since values lower than 9.0 nm are unable to generate structures maintaining structural integrity while values higher than 11.0 nm generate structures with scaffold lengths orders of magnitude longer than the M13 scaffold. Figure 12 (a) illustrates the average porosity within the walls of the top 10 generated designs using the refined technique by calculating the percent of helical segments generated out of the theoretical maximum helical segments within set regions from the mesh

Copyright © 2023 by ASME

surface to d. The plotted region starts from $[0.0,1.0)$ nm and increases by 1.0 nm. Although Figure 12 (a) only shows the porosity analysis for $d = 9.0$ nm, the analyses for higher distances show similar trends. While Figure 12 (a) shows a trend that is approximately uniform with a higher percentage of helical segments within the set regions, Figure 12 (f) shows a trend that is approximately exponential with a lower percentage of helical segments within set regions which therefore demonstrates an increase in effective thickness quality and porosity. Improvement in uniformity can also be quantitatively measured with a coefficient of variation (CV) formula [44] which shows the diversity of a dataset with respect to the mean:

$$CV = \frac{\sigma}{\mu}, \qquad (10)$$

where σ is the standard deviation and μ is the mean. The CV ratio is directly proportional to the variability within the data set in relation to the mean [44]. The histogram of the original algorithm in Figure 12 (f) has a CV of 0.96 while the histogram of the updated algorithm shows improvements with a CV of 0.49.

Using the refined technique, an effective wall thickness value is calculated from the cumulative percent of helical segments within an offset distance from the mesh surface to d in Figure 12 (b). The offset distance with 95.0% of total scaffold sections plus 1.0 nm to account for the radius of the scaffold is the effective wall thickness value. Although Figure 12(b) only shows the effective wall thickness analysis for $d = 9.0$ nm, the analyses for higher distances show similar trends. Figure 12 (c) shows an increase in effective wall thickness value that approximates to the respective set desired distance using the refined technique. Figure 12 (e), which uses the simple technique, has a slower ramp to the total number of scaffold sections than in Figure 12 (b), demonstrating a steady decrease in generated scaffold segments towards to the edges of the structure. This steady decrease demonstrates the lack of generated structures with tight coatings unlike the plot illustrated in Figure 12 (b). Figure 12 (d) provides a visualization of the filling within the walls at the middle of the last generated structure with $d = 9.0$ nm from Figure 8 (a) using the refined technique with 45% of grids filled, which certainly shows a tighter packing than the previous technique in Figure 4.

Using the refined technique, a simple cube mesh is used as an input with $d = 9.0$ nm to demonstrate the generality of this analysis method. Figure 11 (a) illustrates the versatility of the refined shape annealing algorithm in utilizing arbitrary input mesh files to generate unique scaffold routes. Figure 11 (b) shows a slight deviation from the intended final configuration. The same conclusions can be made in Figure 10 for Figure 11 (c) where regions of high RMSF values are due to lack of variety in sequentially accepted shape rules. Using the same analysis method in Figure 12 (a), Figure 13(a) illustrates the average porosity with the cube as an input mesh that also follows an approximately uniform trend. However, Figure 13(a) has a CV of 0.36 thus showing more uniformity within the wall of the structure with a cube input mesh than a tetrahedron input mesh. Using the same analysis method in Figure 12 (b), the average effective thickness value calculated is 9.0 nm. Interestingly, in

Figure 13 (c) 69% of grids are filled, which is much higher than the number of grids filled with the tetrahedral mesh as input. This leads to the hypothesis that prism-like meshes have a higher porosity due to the alignment of the helical axis to the mesh wall. In the future, it would be interesting to investigate the relationship between porosity and scaffold axis orientation.

Although the refined algorithm shows major improvements in the articulation of coating, Figure 8 and 11 show small gaps within the walls of the generated designs. This is attributed to the scaffold being unable to fully search the complex design space during an anneal. A post-processing step can also be used to address the lack of total design space exploration in the future. Expanding our simple shape grammar rules would be an interesting way of solving the gap problem. This step will only target regions with small gaps to ensure adequate filling. The same shape annealing algorithm would run, with only one differing property where a new shape grammar rule would replace the current extension rule (rule 2) where the scaffold could be extended by 14 bp instead of 7 bp.

In addition to the application of a post-processing step, future work will focus on automating the stapling process for full design automation. Another desired next step would be developing an algorithm to unite the start and end location of the generated scaffold pattern to accommodate both linear and circular scaffolds.

5. CONCLUSION

The results show that shape annealing is apt for automating varying effective wall thicknesses to generate novel DNA origami designs with high dense layering within the walls. By applying design methods, the refined shape annealing algorithm can navigate the challenging physical space. Such design methods include adding the Hustin move set to favor rules that follow the objective function and redefining coating to show that desired characteristics can be well coded into DNA origami designs. The results, which introduce new capabilities to enable the generation of an array of effective quality solutions, are an important step for creating a tool for the DNA origami design community. This opens doors to many desirable traits to address the long-standing challenges in DNA origami design that could be coded into the structures such as increasing nanostructure yield [45] and closing the design-iteration loop due to added knowledge of design-property relationship [46].

ACKNOWLEDGEMENTS

Sources of support include the Department of Defense (DoD) National Defense Science and Engineering Graduate (NDSEG) Fellowship and the National Science Foundation (NSF) Award CMMI-2113301. The authors would like to thank D. Sebastian Arias and Mitch Fogelson for support in fleshing out intricacies in the algorithm.

Copyright © 2023 by ASME

REFERENCES

[1] Williams, S., Lund, K., Lin, C., Wonka, P., Lindsay, S., and Yan, H., 2009, "Tiamat: A Three-Dimensional Editing Tool for Complex DNA Structures," *DNA Computing*, A. Goel, F.C. Simmel, and P. Sosík, eds., Springer Berlin Heidelberg, Berlin, Heidelberg, pp. 90–101.

[2] Douglas, S. M., Marblestone, A. H., Teerapittayanon, S., Vazquez, A., Church, G. M., and Shih, W. M., 2009, "Rapid Prototyping of 3D DNA-Origami Shapes with CaDNAno," Nucleic Acids Res, **37**(15), pp. 5001–5006.

[3] Jun, H., Zhang, F., Shepherd, T., Ratanalert, S., Qi, X., Yan, H., and Bathe, M., 2019, "Autonomously Designed Free-Form 2D DNA Origami," Science Advances, **5**(1), p. eaav0655.

[4] Huang, C.-M., Kucinic, A., Johnson, J. A., Su, H.-J., and Castro, C. E., 2020, *Integrating Computer-Aided Engineering and Computer-Aided Design for DNA Assemblies*, Bioengineering.

[5] Benson, E., Mohammed, A., Gardell, J., Masich, S., Czeizler, E., Orponen, P., and Högberg, B., 2015, "DNA Rendering of Polyhedral Meshes at the Nanoscale," Nature, **523**(7561), pp. 441–444.

[6] Veneziano, R., Ratanalert, S., Zhang, K., Zhang, F., Yan, H., Chiu, W., and Bathe, M., 2016, "Designer Nanoscale DNA Assemblies Programmed from the Top Down," Science, **352**(6293), pp. 1534–1534.

[7] Jun, H., Shepherd, T. R., Zhang, K., Bricker, W. P., Li, S., Chiu, W., and Bathe, M., 2019, "Automated Sequence Design of 3D Polyhedral Wireframe DNA Origami with Honeycomb Edges," ACS Nano, **13**(2), pp. 2083–2093.

[8] Jun, H., Wang, X., Bricker, W. P., Jackson, S., and Bathe, M., 2020, "Rapid Prototyping of Wireframe Scaffolded DNA Origami Using ATHENA," p. 2020.02.09.940320.

[9] 2021, "Lcbb/Athena."

[10] "METIS – Three-Dimensional, Algorithmically-Generated Library of DNA Origami Shapes" [Online]. Available: https://metis-dna-origami.org/. [Accessed: 06-Feb-2022].

[11] Seeman, N. C., 1982, "Nucleic Acid Junctions and Lattices," Journal of Theoretical Biology, **99**(2), pp. 237–247.

[12] Rothemund, P. W. K., 2006, "Folding DNA to Create Nanoscale Shapes and Patterns," Nature, **440**(7082), pp. 297–302.

[13] Seeman, N. C., and Sleiman, H. F., 2017, "DNA Nanotechnology," Nature Reviews Materials, **3**(1), pp. 1–23.

[14] Tørring, T., and Gothelf, K. V., 2013, "DNA Nanotechnology: A Curiosity or a Promising Technology?," F1000Prime Rep, **5**.

[15] Castro, C. E., Kilchherr, F., Kim, D.-N., Shiao, E. L., Wauer, T., Wortmann, P., Bathe, M., and Dietz, H., 2011, "A Primer to Scaffolded DNA Origami," Nat Methods, **8**(3), pp. 221–229.

[16] Said, H., Schüller, V. J., Eber, F. J., Wege, C., Liedl, T., and Richert, C., 2012, "M1.3 – a Small Scaffold for DNA Origami," Nanoscale, **5**(1), pp. 284–290.

[17] Sun, W., Boulais, E., Hakobyan, Y., Wang, W. L., Guan, A., Bathe, M., and Yin, P., 2014, "Casting Inorganic Structures with DNA Molds," Science, **346**(6210), pp. 1258361–1258361.

[18] Du, K., Park, M., Ding, J., Hu, H., and Zhang, Z., 2017, "Sub-10 Nm Patterning with DNA Nanostructures: A Short Perspective," Nanotechnology, **28**(44), p. 442501.

[19] Douglas, S. M., Bachelet, I., and Church, G. M., 2012, "A Logic-Gated Nanorobot for Targeted Transport of Molecular Payloads," Science, **335**(6070), pp. 831–834.

[20] Selnihhin, D., Sparvath, S. M., Preus, S., Birkedal, V., and Andersen, E. S., 2018, "Multifluorophore DNA Origami Beacon as a Biosensing Platform," ACS Nano, **12**(6), pp. 5699–5708.

[21] Cagan, J., and Mitchell, W. J., 1993, "Optimally Directed Shape Generation by Shape Annealing," Environ. Plann. B, **20**(1), pp. 5–12.

[22] Babatunde, B., Arias, D. S., Cagan, J., and Taylor, R. E., 2021, "Generating DNA Origami Nanostructures through Shape Annealing," Applied Sciences, **11**(7), p. 2950.

[23] Stiny, G., 1980, "Introduction to Shape and Shape Grammars," Environ. Plann. B, **7**(3), pp. 343–351.

[24] Kirkpatrick, S., Gelatt, C. D., and Vecchi, M. P., 1983, "Optimization by Simulated Annealing," Science, **220**(4598), pp. 671–680.

[25] Hustin, S., and Sangiovanni-Vincentelli, A., 1987, "TIM, a New Standard Cell Placement Program Based on the Simulated Annealing Algorithm," Hilton Head, SC.

[26] Szykman, S., and Cagan, J., 1996, "Synthesis of Optimal Nonorthogonal Routes," Journal of Mechanical Design, **118**(3), pp. 419–424.

[27] Shea, K., and Cagan, J., 1999, "Languages and Semantics of Grammatical Discrete Structures," AIEDAM, **13**(4), pp. 241–251.

[28] Kim, D.-N., Kilchherr, F., Dietz, H., and Bathe, M., 2012, "Quantitative Prediction of 3D Solution Shape and Flexibility of Nucleic Acid Nanostructures," Nucleic Acids Res, **40**(7), pp. 2862–2868.

[29] Maffeo, C., and Aksimentiev, A., 2020, "MrDNA: A Multi-Resolution Model for Predicting the Structure and Dynamics of DNA Systems," Nucleic Acids Research, **48**(9), pp. 5135–5146.

[30] Yoo, J., and Aksimentiev, A., 2013, "In Situ Structure and Dynamics of DNA Origami Determined through Molecular Dynamics Simulations," Proceedings of the National Academy of Sciences of the United States of America, **110**(50), pp. 20099–20104.

[31] Doye, J. P. K., Fowler, H., Prešern, D., Bohlin, J., Rovigatti, L., Romano, F., Šulc, P., Wong, C. K., Louis, A. A., Schreck, J. S., Engel, M. C., Matthies, M., Benson, E., Poppleton, E., and Snodin, B. E. K., 2020,

Copyright © 2023 by ASME

"The OxDNA Coarse-Grained Model as a Tool to Simulate DNA Origami," arXiv:2004.05052 [cond-mat].

[32] Ouldridge, T. E., Louis, A. A., and Doye, J. P. K., 2011, "Structural, Mechanical, and Thermodynamic Properties of a Coarse-Grained DNA Model," J. Chem. Phys., **134**(8), p. 085101.

[33] Šulc, P., Romano, F., Ouldridge, T. E., Rovigatti, L., Doye, J. P. K., and Louis, A. A., 2012, "Sequence-Dependent Thermodynamics of a Coarse-Grained DNA Model," The Journal of Chemical Physics, **137**(13), p. 135101.

[34] Snodin, B. E. K., Randisi, F., Mosayebi, M., Sulc, P., Schreck, J. S., Romano, F., Ouldridge, T. E., Tsukanov, R., Nir, E., Louis, A. A., and Doye, J. P. K., 2015, "Introducing Improved Structural Properties and Salt Dependence into a Coarse-Grained Model of DNA," The Journal of Chemical Physics, **142**(23), p. 234901.

[35] Snodin, B. E. K., Schreck, J. S., Romano, F., Louis, A. A., and Doye, J. P. K., 2019, "Coarse-Grained Modelling of the Structural Properties of DNA Origami," Nucleic Acids Research, **47**(3), pp. 1585–1597.

[36] Engel, M. C., Smith, D. M., Jobst, M. A., Sajfutdinow, M., Liedl, T., Romano, F., Rovigatti, L., Louis, A. A., and Doye, J. P. K., 2018, "Force-Induced Unravelling of DNA Origami," ACS Nano, **12**(7), pp. 6734–6747.

[37] Shi, Z., and Arya, G., 2020, "Free Energy Landscape of Salt-Actuated Reconfigurable DNA Nanodevices," Nucleic Acids Research, **48**(2), pp. 548–560.

[38] Poppleton, E., Bohlin, J., Matthies, M., Sharma, S., Zhang, F., and Šulc, P., 2020, "Design, Optimization and Analysis of Large DNA and RNA Nanostructures through Interactive Visualization, Editing and Molecular Simulation," Nucleic Acids Res, **48**(12), p. e72.

[39] Doty, D., Lee, B. L., and Stérin, T., 2020, "Scadnano: A Browser-Based, Scriptable Tool for Designing DNA Nanostructures," arXiv:2005.11841 [cs, q-bio].

[40] Suma, A., Poppleton, E., Matthies, M., Šulc, P., Romano, F., Louis, A. A., Doye, J. P. K., Micheletti, C., and Rovigatti, L., 2019, "TacoxDNA: A User-Friendly Web Server for Simulations of Complex DNA Structures, from Single Strands to Origami," Journal of Computational Chemistry, **40**(29), pp. 2586–2595.

[41] Goddard, T. D., Huang, C. C., Meng, E. C., Pettersen, E. F., Couch, G. S., Morris, J. H., and Ferrin, T. E., 2018, "UCSF ChimeraX: Meeting Modern Challenges in Visualization and Analysis," Protein Science, **27**(1), pp. 14–25.

[42] Pettersen, E. F., Goddard, T. D., Huang, C. C., Meng, E. C., Couch, G. S., Croll, T. I., Morris, J. H., and Ferrin, T. E., 2021, "UCSF ChimeraX: Structure Visualization for Researchers, Educators, and Developers," Protein Science, **30**(1), pp. 70–82.

[43] Rutenbar, R. A., 1989, "Simulated Annealing Algorithms: An Overview," IEEE Circuits and Devices Magazine, **5**(1), pp. 19–26.

[44] Everitt, B. S., 2002, *The Cambridge Dictionary of Statistics*, Cambridge University Press, Cambridge, UNITED KINGDOM.

[45] Ke, Y., Bellot, G., Voigt, N. V., Fradkov, E., and Shih, W. M., 2012, "Two Design Strategies for Enhancement of Multilayer–DNA-Origami Folding: Underwinding for Specific Intercalator Rescue and Staple-Break Positioning," Chem. Sci., **3**(8), pp. 2587–2597.

[46] Majikes, J. M., and Liddle, J. A., 2020, "DNA Origami Design: A How-To Tutorial," J. RES. NATL. INST. STAN., **126**, p. 126001.

Proceedings of the ASME 2023
International Design Engineering Technical Conferences and
Computers and Information in Engineering Conference
IDETC-CIE2023
August 20-23, 2023, Boston, Massachusetts

DETC2023-114414

MITIGATING THE EFFECTS OF SOURCE-DEPENDENT BIAS AND NOISE ON MULTI-SOURCE BAYESIAN OPTIMIZATION: APPLICATION TO MATERIALS DESIGN

Zahra Zanjani Foumani, Amin Yousefpour, Mehdi Shishehbor, Ramin Bostanabad[1]
Department of Mechanical and Aerospace Engineering, University of California Irvine,
Irvine, CA, USA

ABSTRACT

Bayesian optimization (BO) is a sequential optimization strategy that is increasingly employed in a wide range of areas including materials design. In real world applications, acquiring high-fidelity (HF) data through physical experiments or HF simulations is the major cost component of BO. To alleviate this bottleneck, multi-fidelity (MF) methods are increasingly used to forgo the sole reliance on the expensive HF data and reduce the sampling costs by querying inexpensive low-fidelity (LF) sources whose data are correlated with HF samples. Existing multi-fidelity BO (MFBO) methods operate under the following two assumptions: (1) Leveraging global (rather than local) correlation between HF and LF sources, and (2) Associating all the data sources with the same noise process. These assumptions dramatically reduce the performance of MFBO when LF sources are only locally correlated with the HF source or when the noise variance varies across the data sources. To dispense with these incorrect assumptions, we propose an MF emulation method that (1) learns a noise model for each data source, and (2) enables BO to leverage highly biased LF sources which are only locally correlated with the HF source. We illustrate the performance of our method through analytical examples and engineering problems on materials design.

Keywords: Bayesian optimization; multi-fidelity modeling; emulation; heterogenous noise modeling; interval score.

1. INTRODUCTION

Bayesian optimization (BO) is a sequential and sample-efficient global optimization technique that is increasingly used in the optimization of expensive-to-evaluate (and typically black-box) functions [1]. BO has two main ingredients: an emulator which is typically a Gaussian process (GP) and an acquisition function (AF) [2]. The first step in BO is to train an emulator on some initial data. Then, an auxiliary optimization is

solved to determine the new sample that should be added to the training data. The objective function of this auxiliary optimization is the AF whose evaluation relies on the emulator. Given the new sample, the training data is updated and the entire emulation-sampling process is repeated until the convergence conditions are met [3].

Although BO is a highly efficient technique, the total cost of optimization can be substantial if it solely relies on the accurate but expensive high-fidelity (HF) data source. To mitigate this issue, multi-fidelity (MF) techniques are widely adopted [4-6] where one uses multiple data sources of varying levels of accuracy and cost in BO. The fundamental principle behind MF techniques is to exploit the correlation between low-fidelity (LF) and HF data to decrease the overall sampling costs [7, 8].

Over the past two decades many multi-fidelity BO (MFBO) strategies have been proposed which primarily differ in terms of their emulator and AF. Most existing strategies rely on the Co-Kriging method [9], Kennedy and O'Hagan's bi-fidelity approach [10], and the BoTorch package [11]. These MFBO methods have some major drawbacks such as inability to simultaneously leverage multiple LF sources, sensitivity to the sampling costs (where highly inexpensive LF sources cause numerical and convergence issues), and presuming simple bias forms for the LF sources.

Some of these limitations are recently addressed in [12] where the authors propose to (1) use latent map Gaussian processes (LMGPs) for emulation, and (2) quantify the information value of LF and HF samples differently. Their AF is cost-aware in that it considers the sampling cost in quantifying the value of HF and LF data points. Henceforth, we refer to this method as $MFBO_\alpha$.

While $MFBO_\alpha$ performs much better than competing MF approaches, it has two main limitations which are demonstrated

---

[1] Corresponding author. Email: Raminb@uci.edu

Copyright © 2023 by ASME

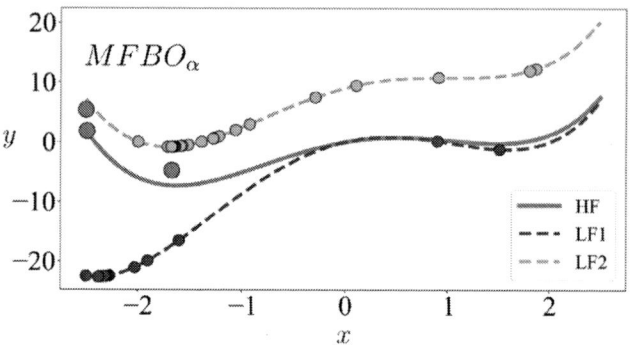 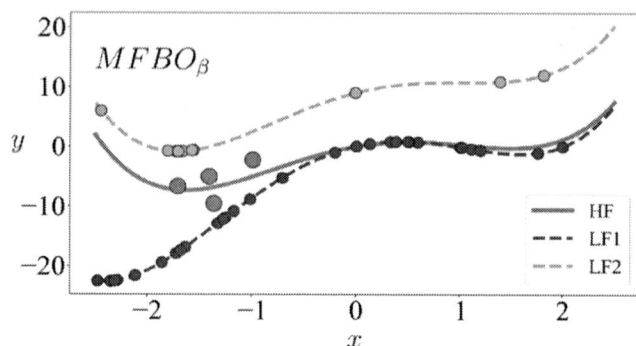

FIGURE 1 EFFECT OF HETEROGENOUS NOISE AND MODEL FIDELITY ON MFBO: HF data are noisy and expensive while the LF data are deterministic and cheap. In this example, LF1 is more correlated with the HF source for $x > 0$ while LF2 correlates better with HF for $x < 0$. The sampling cost of the HF and two LF sources are 1000/100/100, respectively. $MFBO_\alpha$ uses both LF sources and learns a single noise process for the three sources. $MFBO_\beta$ is the approach we propose in this paper to increase sampling efficiency and solution accuracy. $MFBO_\beta$ more effectively explores the space (as it samples more points in $x < 0$) and better leverages LF1 in $x > 0$. As for LF2, $MFBO_\beta$ mostly samples from $x < 0$, since this region includes the optimum of LF2 and is more correlated with the HF. Initial data are not shown in these figures.

with a simple 1D example in FIGURE 1. Firstly, $MFBO_\alpha$ excludes highly biased LF sources from BO with the rationale that they can steer the search process in the wrong direction. This exclusion is done before BO starts since the decision is made based on the fidelity manifold of LMGP that is trained on the *initial* data. However, this manifold only quantifies global accuracy of LF sources with respect to the HF source. That is, if an LF source is only correlated with the HF data in a small region, $MFBO_\alpha$ discards it. We argue that such an early exclusion is suboptimal since that small region may contain the global optimum of the HF source.

The second limitation of $MFBO_\alpha$ is that it assumes all sources are corrupted with the same noise process (with unknown noise variance). However, MF datasets typically have different levels of noise especially if some sources represent deterministic computer simulations while others are physical experiments [13, 14]. In such applications, $MFBO_\alpha$ overestimates the uncertainties which, in turn, reduces the performance of MFBO.

To address these two limitations, we introduce $MFBO_\beta$ for multi-fidelity cost-aware Bayesian optimization. $MFBO_\beta$ has the same AFs as $MFBO_\alpha$ and can leverage an arbitrary number of LF sources in optimizing an HF source. Unlike $MFBO_\alpha$, $MFBO_\beta$ never discards any LF sources (regardless of its bias with respect to the HF source) and estimates a noise process for each data source. We argue that $MFBO_\beta$ quantifies the uncertainties more accurately than $MFBO_\alpha$ and thus achieves a higher performance in MFBO. FIGURE 1 schematically demonstrates the advantages of $MFBO_\beta$ over $MFBO_\alpha$ in a 1D example where there are one HF and two LF sources.

The rest of the paper is organized as follows. We provide the methodological details in Section 2 and then evaluate the performance of $MFBO_\beta$ via multiple ablation studies in Section 3. We conclude the paper in Section 4 by summarizing our contributions and providing future research directions.

2. METHODS

In this section, we first provide some background on LMGP and MF modeling with LMGP in Section 2.1 and Section 2.2, respectively. We then propose our efficient mechanism for inversely learning a noise process for each data source in Section 2.3 . Next, we introduce the cost-aware AF of $MFBO_\beta$ in Section 2.4. Finally, in Section 2.5 we elaborate on our idea that improves the uncertainty quantification (UQ) capabilities of LMGPs and, in turn, benefits MFBO.

2.1 Latent Map Gaussian Process (LMGP)

Gaussian processes (GPs) are emulators which assume the training data come from a multivariate normal distribution with parametric mean and covariance functions. Following this assumption, the training data can be modeled as:

$$y(\boldsymbol{x}) = \beta + \xi(\boldsymbol{x}) \tag{1}$$

where $\boldsymbol{x} = [x_1, x_2, \ldots, x_{dx}]^T$ is the input vector, $y(\boldsymbol{x})$ is the output, β is an unknown coefficient, and $\xi(\boldsymbol{x})$ is a zero-mean GP with the covariance function:

$$cov\big(\xi(\boldsymbol{x}), \xi(\boldsymbol{x}')\big) = c(\boldsymbol{x}, \boldsymbol{x}') = \sigma^2 r(\boldsymbol{x}, \boldsymbol{x}') \tag{2}$$

where σ^2 is the variance of the process and $r(\cdot, \cdot)$ is the parametric correlation function. In this paper we use the Gaussian correlation function defined as:

$$r(\boldsymbol{x}, \boldsymbol{x}') = \exp\left\{ -\sum_{i=1}^{dx} 10^{\omega_i} \, (x_i - x_i')^2 \right\} \tag{3}$$

where $\boldsymbol{\omega} = [\omega_1, \omega_2, \ldots, \omega_{dx}]^T$ are the scale parameters. GP modeling highly depends on the choice of the correlation function which measures the distance between any two points.

Copyright © 2023 by ASME

To directly use GPs in MF modeling, we follow [15] who convert MF modeling to a manifold learning problem via LMGPs which are extensions of GPs that can handle categorical data [16] while providing a visualizable manifold that can be used to interpret the correlation among data sources.

Denoting the categorical inputs by $t = [t_1, t_2, ..., t_{dt}]^T$ where variable t_i has l_i distinct levels, LMGP maps each combination of the categorical levels to a point in a learned quantitative manifold. To this end, LMGP assigns a unique vector to each combination of the categorical variables and then learns a linear transformation that maps these unique vectors into a compact manifold with dimensionality dz:

$$z(t) = \zeta(t)A \tag{4}$$

where t denotes a specific combination of the categorical variables, $z(t)$ is the $1 \times dz$ posterior latent representation of t, $\zeta(t)$ is a unique prior vector representation of t, and A is a rectangular matrix that maps $\zeta(t)$ to $z(t)$. In this paper, grouped one-hot encoding is used to generate the prior vectors and hence the dimensionality of $\zeta(t)$ and A are $1 \times \sum_{i=1}^{dt} l_i$ and $\sum_{i=1}^{dt} l_i \times dz$, respectively. These mapped points can now be directly embedded in the correlation function as:

$$
\begin{aligned}
r(u, u') = &\exp\left\{-\sum_{i=1}^{dx} 10^{\omega_i} (x_i - x_i')^2\right\} \\
&\times \exp\left\{-\sum_{i=1}^{dz} \left(z_i(t) - z_i(t')\right)^2\right\}
\end{aligned} \tag{5}
$$

where $u = [x; t]$ and $z(t) = [z_1(t), z_2(t), ..., z_{dz}(t)]$ is the location in the learned latent space corresponding to the specific combination of the categorical variables denoted by t.

LMGP estimates the hyperparameters $(\beta, A, \omega, \sigma^2)$ via maximum a posteriori (MAP) which, assuming $dz = 2$, provides point-estimates for $dx + 2 \times \sum_{i=1}^{dt} l_i$ variables. Upon parameter estimation, LMGP uses the conditional distribution formulas to predict the response distribution at the arbitrary point u with the following mean and variance:

$$\mathbb{E}[y(u)] = \mu(u) = \hat{\beta} + r^T(u) R^{-1}(y - \mathbf{1}_{n \times 1}\hat{\beta}) \tag{6}$$

$$
\begin{aligned}
c(y(u), y(u)) &= \sigma^2(u) \\
&= \hat{\sigma}^2(1 - r^T(u)R^{-1}r(u) \\
&+ (g(u))^2(\mathbf{1}_{1 \times n}R^{-1}\mathbf{1}_{n \times 1})^{-1})
\end{aligned} \tag{7}
$$

where n is number of training samples, $\mathbb{E}$ denotes expectation, $\mathbf{1}_{a \times b}$ is an $a \times b$ matrix of ones, $r(u)$ is an $n \times 1$ vector with the ith element $r(u^i, u)$, R is an $n \times n$ matrix with $R_{ij} = r(u^i, u^j)$, and $g(u) = 1 - \mathbf{1}_{1 \times n}R^{-1}r(u)$.

2.2 Multi-fidelity Emulation via LMGP

The first step to MF emulation with LMGP is to augment the inputs with the additional categorical variable s that indicates the source of a sample, i.e., $s = \{'1', '2', ..., 'ds'\}$ where the jth element corresponds to source j for $j = 1, ..., ds$. Subsequently, the training data from all sources are concatenated and used in LMGP to build an MF emulator. Upon training, to predict the objective value of a point x from source j, x is concatenated with the categorical variable s that corresponds to source j and fed into the trained LMGP. We refer the readers to [17] for more detail but note here that in case the input variables already contain some categorical features (see Section 3.2 for an example), we endow LMGP with two manifolds where one encodes the fidelity variable s while the other manifold encodes the rest of the categorical variables.

It has been recently shown [18] that LMGPs have the following primary advantages over other MF emulators: (1) they provide a more flexible and accurate mechanism to build MF emulators since they learn the relations between the sources in a nonlinear manifold, (2) they learn all the sources quite accurately rather than just emulating the HF source, and (3) they provide a visualizable global metric for comparing the relative discrepancies/similarities among the data sources.

2.3 Source-dependent Noise Modeling

The presence of noise significantly affects the performance of BO, and incorrectly modeling it can cause over-exploration or under-exploration of the search space. To mitigate the effects of noise in BO, we reformulate LMGPs to independently model a noise process for each data source. This reformulation can improve the accuracy of the model in noisy regions and, in turn, guide the search toward the global optimum when the modeled is deployed in MFBO.

To model noise in GPs, the nugget or jitter parameter, δ, is used [19] to replace R with $R_\delta = R + \delta I$ where I is an $n \times n$ identity matrix. With this approach, the estimated stationary noise variance in the data is $\delta\sigma^2$ and the mean and variance formulations in Eq. (6) and Eq. (7) are modified by using R_δ instead of R.

Although incorporating this modification in the correlation matrix can enhance the performance of the emulator and BO in single-fidelity (SF) problems, it does not yield the same benefits in MF optimization. This is likely because of the dissimilar nature of the data sources and their corresponding noises. When dealing with multiple sources of data, each source may suffer from different levels and types of noise. Consider a bi-fidelity dataset where the HF data comes from an experimental setup and is subject to measurement noise, while the LF data is generated by a deterministic computer code which has a systematic bias due to missing physics. In this case, using only one nugget parameter in LMGP for MF emulation is obviously not an optimum choice.

To address this issue effectively, we propose to use multiple nugget parameters in the emulator. Specifically, we define the nugget vector $\delta = [\delta_1, \delta_2, ..., \delta_{ds}]$ and update the correlation matrix as follows:

$$R_\delta = R + N_\delta \tag{8}$$

Copyright © 2023 by ASME

where N_δ denotes an $n \times n$ diagonal matrix whose $(i,i)^{th}$ element is the nugget element corresponding to the data source of the ith sample. For instance, suppose the ith sample (u^i) is generated by source ds. Then, $(i,i)^{th}$ element of N_δ is δ_{ds}. Then, we use Eq. (8) to build the correlation matrix of LMGP and jointly estimate all the parameters via MAP as:

$$
\begin{aligned}
[\hat{\beta}, \hat{\sigma}, \widehat{\omega}, \widehat{A}, \widehat{\delta}] &= \underset{\beta,\sigma,\omega,A,\delta}{\text{argmin}} \; L_{MAP} = \\
&\underset{\beta,\sigma,\omega,A,\delta}{\text{argmin}} \left(\frac{n}{2} \log(\sigma^2) + \frac{1}{2} \log(|R_\delta|) \right. \\
&+ \frac{1}{2\sigma^2} (y - \mathbf{1}_{n\times 1}\beta)^T R_\delta^{-1} (y \\
&\left. - \mathbf{1}_{n\times 1}\beta) + \log\left(\underset{\beta,\sigma,\omega,A,\delta}{p(\cdot)} \right) \right)
\end{aligned}
\tag{9}
$$

where $p(\cdot)$ is the prior of the hyperparameters. We define independent priors for each parameter where $\omega^i \sim N(-3,3), \beta \sim N(0,1), A^{ij} \sim N(0,3), \sigma \sim LN(0,3)^2$, and $\delta^i \sim LHS(0,0.01)^3$ [20]. This approach increases the number of hyperparameters to $dx + 2 \times \sum_{i=1}^{dt} l_i + ds$.

2.4 Multi-source Cost-aware Acquisition Function

The choice of AF is crucial in MFBO since it must consider the biases of LF data and source-dependent sampling costs in addition to balancing exploration and exploitation. To capture these goals, separate AFs are defined in [17] for LF and HF sources with a focus on exploration and exploitation, respectively.

Following the idea of proposing an AF with a focus on exploration for the LF sources, the AF of the jth LF source ($j \neq l$, l denotes the HF source) is defined as the exploration part of expected improvement (EI) in $MFBO_\alpha$:

$$
\gamma_{LF}(u; j) = \sigma_j(u)\phi\left(\frac{y_j^* - \mu_j(u)}{\sigma_j(u)}\right)
\tag{10}
$$

where y_j^* is the best function value obtained so far from source j and $\phi(\cdot)$ denotes the probability density function (PDF) of the standard normal variable. $\sigma_j(u)$ and $\mu_j(u)$ are the standard deviation and mean, respectively, of point u from source j which we estimate via:

$$
\mu(u) = \hat{\beta} + r^T(u) R_\delta^{-1}(y - \mathbf{1}_{n\times 1}\hat{\beta})
\tag{11}
$$

$$
\begin{aligned}
c(y(u), y(u)) &= \sigma^2(u) \\
&= \hat{\sigma}^2 \left(1 - r^T(u)R_\delta^{-1}r(u) \right. \\
&\left. + (g(u))^2 (\mathbf{1}_{1\times n}R_\delta^{-1}\mathbf{1}_{n\times 1})^{-1} \right) \\
&+ \hat{\delta}_j
\end{aligned}
\tag{12}
$$

where $g(u) = 1 - \mathbf{1}_{1\times n}R_\delta^{-1}r(u)$ and $\hat{\delta}_j$ is the estimated nugget parameter for source j.

As probability of improvement (PI) is computationally efficient and emphasizes exploitation, $MFBO_\alpha$ utilizes it as the AF for the HF data source. Accordingly, $MFBO_\beta$ uses PI for the HF source (source l) with the new standard deviation and mean calculated based on the Eq. (11) and Eq. (12):

$$
\gamma_{HF}(u; l) = \psi\left(\frac{\mu_l(u) - y_l^*}{\sigma_l(u)}\right)
\tag{13}
$$

where $\psi(\cdot)$ is the cumulative density function (CDF) of the standard normal distribution.

In each iteration of BO, we first use the mentioned AFs to solve ds auxiliary optimizations to find the candidate points with the highest acquisition value from each source. We then scale these values by the corresponding sampling costs to obtain the following composite AF:

$$
\gamma_{MFBO_\beta}(u; j) = \begin{cases} \dfrac{\gamma_{LF}(u; j)}{O(j)} & j = 1, \ldots, ds \;\; and \;\; j \neq l \\[3mm] \dfrac{\gamma_{HF}(u; l)}{O(l)} & j = l \end{cases}
\tag{14}
$$

where $O(j)$ is the cost of acquiring one sample from source j. We determine the final candidate point (and the source that it should be sampled from) via:

$$
[u^{k+1}, j^{k+1}] = \underset{u,j}{\text{argmax}} \; \gamma_{MFBO_\beta}(u; j)
\tag{15}
$$

2.5 Emulation for Exploration

The composite AF in Eq. (14) quantifies the information value of LF samples via Eq. (10) whose value scales with the prediction uncertainties, i.e., $\sigma(u)$. The source-dependent noise modeling of Section 2.3 improves LMGP's ability in learning the uncertainty by introducing a few more hyperparameters. However, the added hyperparameters may result into overfitting and, in turn, deteriorate the predicted uncertainties [21, 22].

A related issue is the effect of large *local* biases of LF sources which can inflate the uncertainty quite substantially and, as a result, increase $\gamma_{LF}(u; j)$. This increase causes MFBO to repeatedly sample from the biased LF sources (see FIGURE 1 where $MFBO_\alpha$ takes many samples from LF1 in the $x < 0$ region while LF1 is quite biased for $x < 0$). Such repeated samplings reduce the efficiency of MFBO and may cause numerical issues or even convergence to a suboptimal solution.

To address the above issues simultaneously, we argue that the training process of the emulator should increase the importance of UQ which directly affects the exploration part of MFBO. To this end, we leverage strictly proper scoring rules while training LMGPs.

---

[2] Log-Normal
[3] Log-Half-Horseshoe with zero lower bound and scale parameter 0.01.

Scoring rules [23, 24] evaluate a *probabilistic* prediction by assigning a numerical score to it. The scoring rule of an emulator is (strictly) proper if matching the predicted distribution with the underlying sample distribution (uniquely) maximizes the expected score for any sample [25]. The probabilistic nature of LMGP's prediction motivates us to use the negatively oriented interval score (hereafter denoted by IS) to evaluate the UQ capabilities of LMGPs. We choose IS since it is robust to outliers, rewards narrow prediction intervals, and is flexible in the choice of desired coverage levels [26, 27].

IS is a special case of quantile prediction that penalizes the model for each observation that is not inside the $(1 - v) * 100\%$ prediction interval. The lower ($\mathcal{L}^i$) and upper ($\mathcal{U}^i$) endpoints of this prediction interval for the ith observation are their predictive quantiles at levels $\frac{v}{2}$ and $1 - \frac{v}{2}$, respectively. So, we calculate the IS as:

$$IS_v = \qquad\qquad (16)$$

$$\frac{1}{n}\sum_{i=1}^{n}(\mathcal{U}^i - \mathcal{L}^i) + \frac{2}{v}(\mathcal{L}^i - y(\boldsymbol{u}^i))\,\mathbb{1}\{y(\boldsymbol{u}^i) < \mathcal{L}^i\}$$
$$+ \frac{2}{v}(y(\boldsymbol{u}^i) - \mathcal{U}^i)\,\mathbb{1}\{y(\boldsymbol{u}^i) > \mathcal{U}^i\}$$

where $\mathbb{1}\{\cdot\}$ is an indicator function which is 1 if its condition holds and zero otherwise [28, 29]. We use $v = 0.05$ (95% prediction interval), so $\mathcal{U}^i = \mu(\boldsymbol{u}^i) + 1.96\,\sigma(\boldsymbol{u}^i)$ and $\mathcal{L}^i = \mu(\boldsymbol{u}^i) - 1.96\,\sigma(\boldsymbol{u}^i)$.

Having defined the IS, we now formulate the new objective function for training LMGPs where $IS_{0.05}$ is used as a penalty term during hyperparameter estimation to increase the focus on UQ. Since the effectiveness of this penalization mechanism depends on the value of the posterior, we introduce an adaptive coefficient whose magnitude depends on the posterior value. With this penalty term, we estimate the hyperparameters of LMGP via:

$$[\hat{\beta}, \hat{\sigma}, \widehat{\boldsymbol{\omega}}, \widehat{\boldsymbol{A}}, \widehat{\boldsymbol{\delta}}] = \underset{\beta,\sigma,\omega,A,\delta}{\text{argmin}}\ L_{MAP}$$
$$+ \varepsilon |L_{MAP}| \times IS_{0.05} \qquad (17)$$

where $|\cdot|$ denotes the absolute function and ε is a user-defined scaling parameter. In this paper, we use $\varepsilon = 0.08$.

3. RESULTS AND DISCUSSION

We demonstrate the performance of $MFBO_\beta$ on two analytic examples (details in TABLE 1) and two real-world problems. In each case, we compare the results against those of $MFBO_\alpha$ and single-fidelity BO (*SFBO*). While *SFBO* uses EI as its AF, $MFBO_\beta$ and $MFBO_\alpha$ use the AFs introduced in Section 2.4.

We assume that the cost of querying any of the data sources is much higher than the computational costs of BO (i.e., fitting LMGP and solving the auxiliary optimization problem). Therefore, we compare the methods based on their capability to identify the global optimum of the HF source and the overall data collection cost. By comparing these methods, we aim to demonstrate: (1) the advantages of estimating noise process for each data source, (2) that using IS improves the prediction which also enhances the convergence of BO (our defined AFs highly rely on the quality of the prediction), and (3) that deploying IS eliminates the need for excluding highly biased fidelity sources.

We use the same stop conditions across the three methods to clearly demonstrate the benefits of our two contributions. In particular, the optimization is stopped when either of the following happens: (1) the overall sampling cost exceeds a pre-determined maximum budget, or (2) the best HF sample does not change over 50 iterations. The maximum budget for the analytical examples is 40000 units, while it is 1000 and 1800 for the two real-world examples as their data collection cost is much lower than that of analytical ones.

3.1 Analytical Examples

We consider two analytical examples (*Wing, Borehole*) with the input dimensionality of 10 and 8, respectively. To challenge the convergence and better illustrate the power of separate noise estimation, we only add noise to the HF data (the noise variance is defined based on the range of each function and is shown in TABLE 1). Both examples are single response and details regarding their formulation, initialization, and sampling cost is presented in TABLE 1. To assess the robustness of the results and quantify the effect of random initial data, we repeat the optimization process 20 times for each example with each of the three methods (all initial data are generated via Sobol sequence).

In each example, the relative root mean squared error (RRMSE) is calculated between LF sources and their corresponding HF source based on 1000 samples to show the relative accuracy of the LF sources (presented in TABLE 1). Based on these numbers, in *Borehole*, unlike *Wing*, the source ID, true fidelity level (based on the RRMSEs), and sampling cost are not related. For instance, although the first LF source is the most expensive one, it has the least accuracy.

$MFBO_\alpha$ excludes the highly biased LF sources from BO before any new samples are obtained. This exclusion is done based on the latent map of the LMGP model that is trained on the *initial* data. FIGURE 2 shows the latent maps of *Wing* and *Borehole* examples where Source 1 represents the HF source, and the rest of the sources represent LF ones. As shown in FIGURE 2, while all the fidelity sources of *Wing* are beneficial (since the points encoding the LF sources are very close to the HF point), the first two LF sources of *Borehole* are not correlated enough with the HF (their latent positions are distant from that of the HF) and hence are excluded in $MFBO_\alpha$. However, $MFBO_\beta$ does not require this exclusion because it leverages the biased LF sources merely in the regions that they are correlated with the HF source. In this paper, we do not exclude the biased sources in $MFBO_\alpha$ to better compare it with our proposed method.

FIGURE 3 summarizes the convergence history of each example by depicting the best HF sample found by each method versus its accumulated sampling cost. As we expect, MF methods ($MFBO_\beta$ and $MFBO_\alpha$) outperform *SFBO* in *Wing*

Copyright © 2023 by ASME

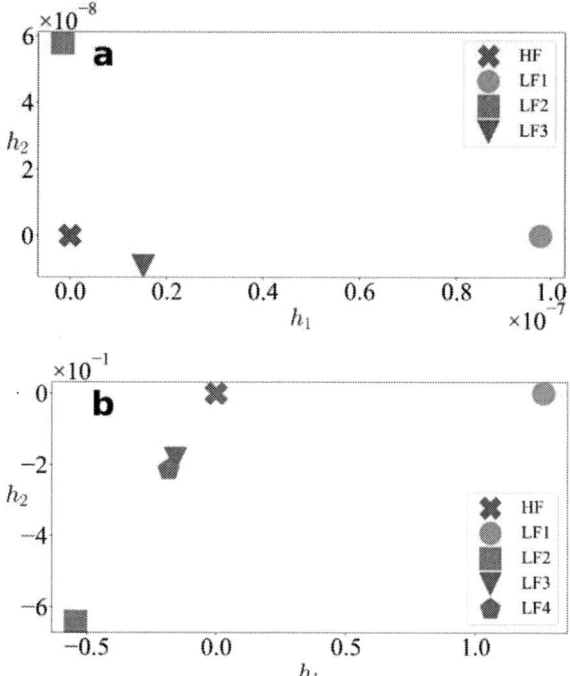

FIGURE 2 FIDELITY MANIFOLDS OF ANALYTIC EXAMPLES: The plots in **(a)** and **(b)** are obtained by fitting an LMGP to the initial data in the *Wing* and *Borehole* examples, respectively. Due to the consistency across the 20 repetitions, the plots are randomly chosen among them. In **(b)**, the HF source is encoded far from LF1 and LF2 which indicates that these two sources have large biases with respect to the HF source. $MFBO_\alpha$ excludes these two sources from the BO while $MFBO_\beta$ does not.

(FIGURE 3a) by leveraging the inexpensive LF sources that are globally correlated with the HF source. However, the superior performance of $MFBO_\beta$ is more obvious in *Borehole* with biased data sources.

In *Borehole* (FIGURE 3b), all the thin red curves ($MFBO_\alpha$) are straight lines, except for two curves. This means that for 18 repetitions, the optimization process fails to improve. The reason behind this lack of improvement is that $MFBO_\alpha$ fails to handle *local* correlation of the LF sources and samples points that steer the optimization in the wrong direction. Consequently, $MFBO_\alpha$ cannot find any efficient HF sample with large enough information value to compensate for its high sampling cost which results in the lack of improvement. Conversely, all the thin green curves ($MFBO_\beta$) converge to a value very close to the ground truth. In addition, $MFBO_\beta$ yields almost the same convergence value as *SFBO*, but with lower computational cost. This instance further demonstrates the effectiveness of our proposed AFs, since *SFBO* is very accurate due to only sampling from HF and not dealing with local biases.

FIGURE 4 illustrates the statistics of the convergence cost and value for each repetition. The pink boxes show the variations in convergence values through 20 repetitions and the blue ones demonstrate the convergence cost. The dashed pink line is the

ground truth we aim to find. As also shown in FIGURE 3, in all the examples, $MFBO_\beta$ outperforms other methods considering convergence value and cost. As demonstrated in FIGURE 4a, unlike MF methods, the increase in dimensionality (8 in *Borehole* to 10 in *Wing*) adversely affects the performance of *SFBO*, as it cannot find the ground truth of *Wing* despite merely sampling from the HF source. In addition, utilizing beneficial information provided by the inexpensive unbiased LF sources, causes the better performance of $MFBO_\beta$ and $MFBO_\alpha$ compared to *SFBO*. The lower variations in the convergence values of $MFBO_\beta$ through 20 repetitions (smaller pink boxes and whiskers) and lower convergence costs compared to $MFBO_\alpha$, further show the superiority of our proposed method; $MFBO_\beta$ outperforms $MFBO_\alpha$ even in the absence of biased sources.

The rationale behind the lower cost of $MFBO_\alpha$ in *Borehole* (FIGURE 4b) can be attributed to the observation made for FIGURE 3b; in 18 repetitions of *Borehole*, the optimization is not improved at all. So, in all those repetitions, the optimization meets the second stop condition and is terminated in the 50th iteration while it is not converged. This fact is also obvious in the long whiskers of the pink boxplots of $MFBO_\alpha$ (high variation on the convergence values in each repetition) and the median which is far from the ground truth.

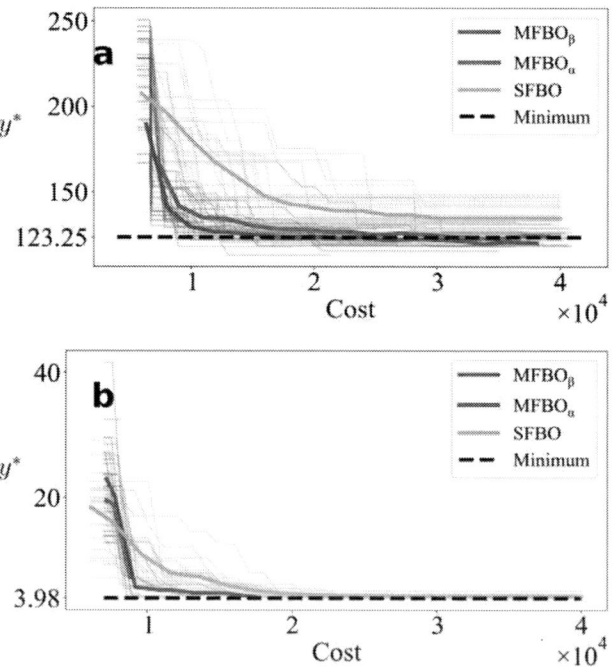

FIGURE 3 CONVERGENCE HISTORIES: The plots depict the best HF sample found by each approach versus their sampling costs accumulated during the BO iterations (the cost of initial data is included). **(a)** and **(b)** refer to *Wing* and *Borehole*, respectively. The thin curves show the convergence history of each repetition, and the solid thick ones indicate the average behavior across the 20 repetitions. In all the examples, $MFBO_\beta$ outperforms $MFBO_\alpha$ in terms of both convergence value and cost. In both examples, *SFBO* performs the worst. The ground truth is represented by the black dashed line.

Copyright © 2023 by ASME

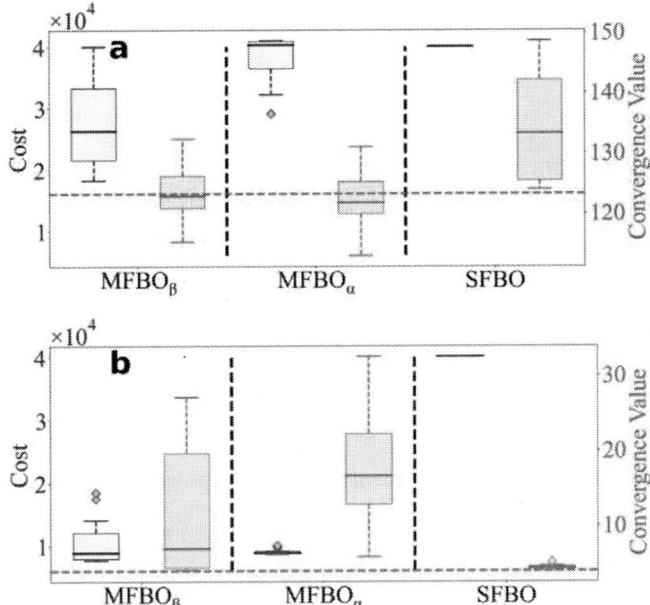

FIGURE 4 ACCUMULATED COSTS AND CONVERGENCE VALUES: The blue boxplots illustrate the accumulated costs of each repetition, and the pink ones demonstrate their convergence values for the three methods. The pink dashed line indicates the ground truth. **(a)** and **(b)** refer to *Wing* and *Borehole*, respectively. In both examples, the median of the convergence box of $MFBO_\beta$ is very closer to the ground truth than that of the $MFBO_\alpha$, while its cost is much lower which proves the superiority of our proposed approach.

3.2 Real-world Datasets

In this section, we study two materials design problems where the aim is to find the composition that best optimizes the property of interest. We do not add noise to these two examples as they are noisy themselves. Both examples have categorical inputs and the HF and LF data are obtained via simulations (based on the density functional theory) with different fidelity levels.

The first problem is bi-fidelity where the goal is to find the member of the nanolaminate ternary alloy (*NTA*) family with the largest bulk modulus [30]. The HF and LF datasets are 10-dimensional (7 quantitative and 3 categorical where the latter have 10, 12, and 2 levels), single response with 224 samples each. We define the cost ratio of 10/1 for the fidelity sources and initialize the BO with 20 HF and 10 LF samples (the composition with the largest bulk modulus is excluded from the HF dataset). To quantify the robustness of the proposed method to the random initial data, we repeat this process 20 times for each BO method.

The second problem is hybrid organic–inorganic perovskite (*HOIP*) crystals that aim to find the compound with the smallest inter-molecular binding energy [31]. There are 3 fidelity sources for this example, 1 HF and 2 LFs, with the same dimensionality (1 output and 3 categorical inputs with 10, 3, and 16 levels) but different sizes. The HF dataset has 480 samples, while the first

and second LF sources have 179 and 240 samples, respectively. We assign the cost ratio of 15/10/5 to the fidelity sources and initialize the BO with (15, 20, 15) samples for the HF and LF sources, respectively (the best compound is excluded). We repeat the BO process 20 times to assess the robustness.

As mentioned before, the first step in $MFBO_\alpha$ is to train an LMGP to the initial data in each problem to exclude the highly biased sources. As illustrated in FIGURE 5, the latent points of the fidelity sources of *NTA* are very close in the learned fidelity manifold which demonstrates a high correlation between its fidelity sources. However, both latent points of LF sources in *HOIP* are far from the HF one, so they both should be excluded. By excluding both LF sources, the MF problem converts to the SF one, so $MFBO_\alpha$ is not applicable to it anymore. Therefore, we do not exclude the biased LF sources from *HOIP* in this paper to be able to compare the performance of $MFBO_\beta$ with $MFBO_\alpha$.

A summary of the convergence history of *NTA* and *HOIP* is depicted in FIGURE 6 by showing the best HF sample found by each method versus its accumulated sampling cost. In *NTA* (FIGURE 6a), the LF sources are globally correlated with the HF; consequently, MF methods perform better than *SFBO* by using inexpensive and informative LF sources. Additionally, higher prediction accuracy of the emulator of $MFBO_\beta$ results in a more efficient sampling and faster convergence of BO in $MFBO_\beta$ compared to $MFBO_\alpha$. Regarding the spike in the

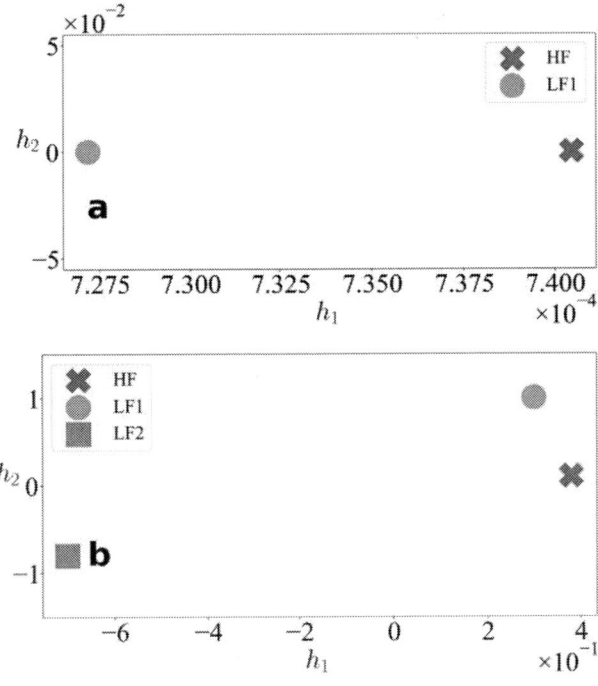

FIGURE 5 FIDELITY MANIFOLDS OF REAL-WORLD EXAMPLES: The plots in **(a)** and **(b)** are obtained by fitting an LMGP to the initial data in the *NTA* and *HOIP* examples, respectively. Due to the consistency across the 20 repetitions, the plots are randomly chosen among them. In **(b)**, $MFBO_\alpha$ excludes LF1 and LF2 as they are encoded far from the HF.

Copyright © 2023 by ASME

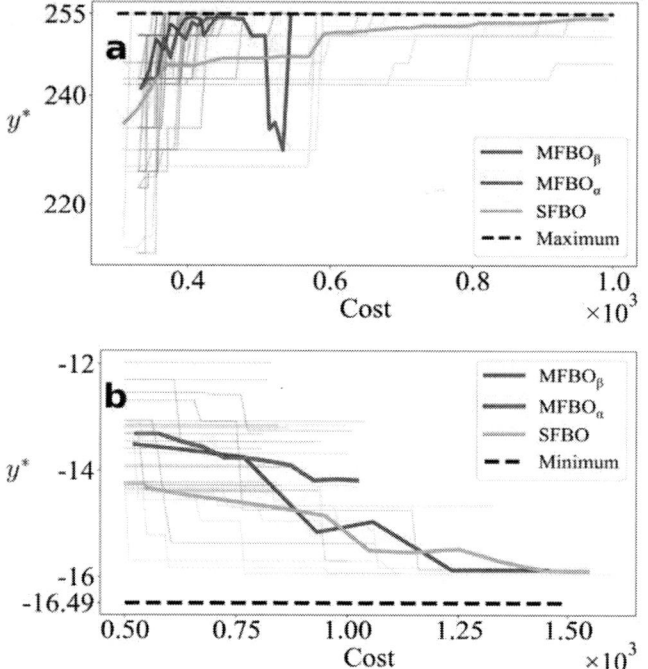

FIGURE 6 CONVERGENCE HISTORIES: The plots depict the best HF sample found by each approach versus their sampling costs accumulated during the BO iterations (the cost of initial data is included). **(a)** and **(b)** refer to *NTA* and *HOIP*, respectively. The solid thick curves indicate the average behavior across the 20 repetitions. $MFBO_\beta$ performs better than $MFBO_\alpha$ in both examples in case of convergence value and cost.

convergence plot of $MFBO_\alpha$, we note that 18 repetitions converge at costs below 500. Consequently, the thick red line (which is the average across the 20 repetitions) becomes highly sensitive to the convergence values after cost exceeds 500 since it is an average of only 2 values. Specifically, in one of those two repetitions, the maximum bulk modulus found is stabilized at 237 for many iterations, but it converges to the ground truth value of 255 when the cost reaches 544, resulting in a sudden jump.

The superiority of $MFBO_\beta$ is more obvious in *HOIP* (FIGURE 6b) with biased sources. In *HOIP* example, as we expect, $MFBO_\alpha$ converges to a sub-optimal compound since both LF sources are only locally correlated with the HF source. So, the AFs fail to sample valuable points to improve the optimization as they cannot find the region where the LF sources are beneficial and informative. Additionally, each data source is obtained from a distinct process, so it suffers from different types and levels of noise. Therefore, estimating a single noise for all the data sources in $MFBO_\alpha$ results in a poor emulation and further exacerbates the performance of AFs. $MFBO_\beta$ overcomes these issues by focusing more on UQ and estimating separate noise processes, resulting in better performance, and outperforming $MFBO_\alpha$.

Similar to Section 3.1, we also show the details of the accumulated convergence cost and values of each repetition in FIGURE 7. In this regard, convergence of all methods to the ground truth in all the repetitions in *NTA* (FIGURE 7a), makes the pink boxplots become straight lines (zero variation). However, the two outliers in *SFBO* show its disability in finding the ground truth in two repetitions despite merely sampling from the accurate HF source. Additionally, the faster convergence of $MFBO_\beta$ compared to other methods associates it with much lower convergence costs.

In *HOIP* (FIGURE 7b), as shown in FIGURE 6b, $MFBO_\alpha$ fails to manage the noise and biased sources, and converges to a sub-optimal compound. However, $MFBO_\beta$ leverages the biased LF sources only in the regions that they provide effective information by taking advantage of the penalized objective function which significantly improves the prediction. Furthermore, the large variation in the convergence values of *SFBO* and its much higher cost, further highlights the superiority of $MFBO_\beta$ in efficiency and robustness, though the median of the convergence boxplot of *SFBO* is closer to the ground truth.

4. CONCLUSION

In this paper, we develop a novel method to improve the performance of multi-fidelity cost-aware BO techniques. Our method enhances the accuracy and convergence rate of MFBO through two main contributions. Firstly, we enable the emulator to estimate separate noise processes for each source of data. This feature increases the accuracy of the trained model since

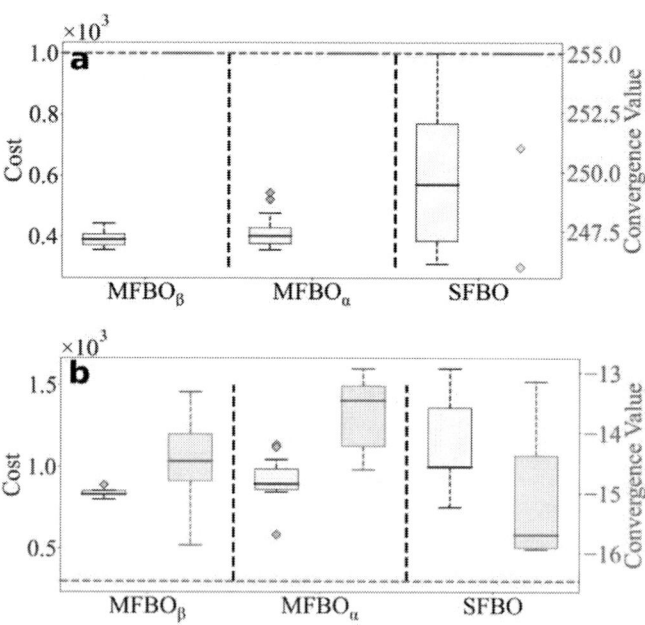

FIGURE 7 ACCUMULATED COSTS AND CONVERGENCE VALUES: The blue boxplots illustrate the accumulated costs of each repetition, and the pink ones demonstrate their convergence values. The pink dashed line indicates the ground truth. **(a)** and **(b)** refer to *NTA* and *HOIP*, respectively. **(a)** All the methods found the ground truth in all iterations, so there is no variation.

Copyright © 2023 by ASME

different data sources may exhibit different types and levels of noise. Secondly, we define a new objective function penalized by IS to (1) further improve the accuracy of the prediction, (2) increase the focus on UQ, and (3) forgo the need to exclude biased data sources. The main advantages of our method are its efficient and superior performance in the presence of highly biased data sources, and no required prior knowledge about the accuracy of the data sources. We illustrate these advantages through analytic and real-world examples.

We propose to use two fixed AFs in each iteration. However, one can also customize the choice of AFs for different iterations using adaptive approaches. The examples presented in this paper are limited to single-objective problems. In addition, in all the

examples we report the noisy data as the result, but one can increase the accuracy by excluding the noise effects on the convergence value. We intend to explore these avenues further in future research, aiming to extend our proposed method to handle multi-objective problems with adaptive AFs.

ACKNOWLEDGEMENTS

We appreciate the support from National Science Foundation (award numbers OAC-2211908 and OAC-2103708), the Early Career Faculty grant from NASA's Space Technology Research Grants Program (award number 80NSSC21K1809), and the UC National Laboratory Fees Research Program of the University of California (Grant Number L22CR4520).

APPENDIX

A Table of Numerical Examples

TABLE 1 LIST OF ANALYTIC FUNCTIONS: n denotes the number of initial samples. The relative root mean squared error (RRMSE) of an LF source is calculated by comparing its output to that of the HF source at 10000 random points. The cost column is the cost of obtaining a sample from the corresponding source. The last column denotes the variance of the added noise to the HF source.

Name	Source ID	Formulation	n	RRMSE	Cost	Noise variance
Borehole	HF	$\dfrac{2\pi T_u \left(H_u - H_l\right)}{\ln\left(\frac{r}{r_w}\right)\left(1 + \frac{2\,LT_u}{\ln\left(\frac{r}{r_w}\right)r_w^2 k_w} + \frac{T_u}{T_l}\right)}$	5	-	1000	16
	LF1	$\dfrac{2\pi T_u \left(H_u - 0.8\,H_l\right)}{\ln\left(\frac{r}{r_w}\right)\left(1 + \frac{1\,LT_u}{\ln\left(\frac{r}{r_w}\right)r_w^2 k_w} + \frac{T_u}{T_l}\right)}$	5	4.40	100	-
	LF2	$\dfrac{2\pi T_u \left(H_u + 3H_l\right)}{\ln\left(\frac{r}{r_w}\right)\left(1 + \frac{8\,LT_u}{\ln\left(\frac{r}{r_w}\right)r_w^2 k_w} + 0.75\,\frac{T_u}{T_l}\right)}$	50	1.54	10	-
	LF3	$\dfrac{2\pi T_u \left(1.1\,H_u - H_l\right)}{\ln\left(\frac{4r}{r_w}\right)\left(1 + \frac{3\,LT_u}{\ln\left(\frac{r}{r_w}\right)r_w^2 k_w} + \frac{T_u}{T_l}\right)}$	5	1.3	100	-
	LF4	$\dfrac{2\pi T_u \left(1.05\,H_u - H_l\right)}{\ln\left(\frac{2r}{r_w}\right)\left(1 + \frac{2\,LT_u}{\ln\left(\frac{r}{r_w}\right)r_w^2 k_w} + \frac{T_u}{T_l}\right)}$	50	1.3	10	-
Wing	HF	$0.36 s_w^{0.758} w_{fw}^{0.0035}\left(\frac{A}{cos^2(\Lambda)}\right)^{0.6} q^{0.006}\lambda^{0.04}\left(\frac{100 t_c}{\cos(\Lambda)}\right)^{-0.3}\left(N_z\,W_{dg}\right)^{0.49}$ $+\; s_w w_p$	5	-	1000	9
	LF1	$0.36 s_w^{0.758} w_{fw}^{0.0035}\left(\frac{A}{cos^2(\Lambda)}\right)^{0.6} q^{0.006}\lambda^{0.04}\left(\frac{100 t_c}{\cos(\Lambda)}\right)^{-0.3}\left(N_z\,W_{dg}\right)^{0.49}$ $+\; w_p$	5	0.19	100	-
	LF2	$0.36 s_w^{0.8} w_{fw}^{0.0035}\left(\frac{A}{cos^2(\Lambda)}\right)^{0.6} q^{0.006}\lambda^{0.04}\left(\frac{100 t_c}{\cos(\Lambda)}\right)^{-0.3}\left(N_z\,W_{dg}\right)^{0.49}$ $+\; w_p$	10	1.14	10	-
	LF3	$0.36 s_w^{0.9} w_{fw}^{0.0035}\left(\frac{A}{cos^2(\Lambda)}\right)^{0.6} q^{0.006}\lambda^{0.04}\left(\frac{100 t_c}{\cos(\Lambda)}\right)^{-0.3}\left(N_z\,W_{dg}\right)^{0.49}$	50	5.75	1	-

Copyright © 2023 by ASME

References

1. Shahriari, B., et al., *Taking the human out of the loop: A review of Bayesian optimization.* Proceedings of the IEEE, 2015. **104**(1): p. 148-175.

2. Frazier, P.I.J.a.p.a., *A tutorial on Bayesian optimization.* 2018.

3. Li, S. *Multi-Fidelity Bayesian Optimization via Deep Neural Networks.* in *Neural Information Processing Systems.* 2020. Vancouver.

4. Song, J., Y. Chen, and Y. Yue. *A general framework for multi-fidelity bayesian optimization with gaussian processes.* in *The 22nd International Conference on Artificial Intelligence and Statistics.* 2019. PMLR.

5. Takeno, S., et al. *Multi-fidelity Bayesian optimization with max-value entropy search and its parallelization.* in *International Conference on Machine Learning.* 2020. PMLR.

6. Zhang, S., et al. *An efficient multi-fidelity bayesian optimization approach for analog circuit synthesis.* in *Proceedings of the 56th Annual Design Automation Conference 2019.* 2019.

7. Shu, L., P. Jiang, and Y. Wang, *A multi-fidelity Bayesian optimization approach based on the expected further improvement.* Structural and Multidisciplinary Optimization, 2021. **63**: p. 1709-1719.

8. Tran, A., T. Wildey, and S. McCann, *sMF-BO-2CoGP: A sequential multi-fidelity constrained Bayesian optimization framework for design applications.* Journal of Computing and Information Science in Engineering, 2020. **20**(3).

9. Xiao, M., et al., *Extended Co-Kriging interpolation method based on multi-fidelity data.* 2018. **323**: p. 120-131.

10. Kennedy, *Bayesian calibration of computer models.* Journal of the Royal Statistical Society: Series B (Statistical Methodology),, 2001.

11. Gardner, *Blackbox matrix-matrix gaussian process inference with gpu acceleration.* Advances in neural information processing systems, 2018.

12. Foumani, Z.Z., et al., *Multi-fidelity cost-aware Bayesian optimization.* 2023. **407**: p. 115937.

13. Escamilla-Ambrosio, P.J. and N. Mort. *Hybrid Kalman filter-fuzzy logic adaptive multisensor data fusion architectures.* in *42nd IEEE International Conference on Decision and Control (IEEE Cat. No. 03CH37475).* 2003. IEEE.

14. Kreibich, O., J. Neuzil, and R.J.I.T.o.I.E. Smid, *Quality-based multiple-sensor fusion in an industrial wireless sensor network for MCM.* 2013. **61**(9): p. 4903-4911.

15. Eweis-Labolle, J.T., N. Oune, and R.J.J.o.M.D. Bostanabad, *Data Fusion With Latent Map Gaussian Processes.* 2022. **144**(9): p. 091703.

16. Oune, N., *Latent map Gaussian processes for mixed variable metamodeling.* Computer Methods in Applied Mechanics and Engineering 2021.

17. Zanjani Foumani, Z., *Multi-Fidelity Cost-Aware Bayesian Optimization.* arXiv preprint arXiv:2211.02732, 2022.

18. Oune, N., R.J.C.M.i.A.M. Bostanabad, and Engineering, *Latent map Gaussian processes for mixed variable metamodeling.* 2021. **387**: p. 114128.

19. Bostanabad, R., et al., *Leveraging the nugget parameter for efficient Gaussian process modeling.* International journal for numerical methods in engineering, 2018. **114**(5): p. 501-516.

20. Carvalho, C.M., N.G. Polson, and J.G.J.B. Scott, *The horseshoe estimator for sparse signals.* 2010. **97**(2): p. 465-480.

21. Gal, Y., M. Van Der Wilk, and C.E.J.A.i.n.i.p.s. Rasmussen, *Distributed variational inference in sparse Gaussian process regression and latent variable models.* 2014. **27**.

22. Mohammed, R.O. and G.C. Cawley. *Over-fitting in model selection with Gaussian process regression.* in *Machine Learning and Data Mining in Pattern Recognition: 13th International Conference, MLDM 2017, New York, NY, USA, July 15-20, 2017, Proceedings 13.* 2017. Springer.

23. Lindley, D.V.J.I.S.R.R.I.d.S., *Scoring rules and the inevitability of probability.* 1982: p. 1-11.

24. Winkler, R.L., et al., *Scoring rules and the evaluation of probabilities.* 1996. **5**: p. 1-60.

25. Gneiting, T. and A.E.J.J.o.t.A.s.A. Raftery, *Strictly proper scoring rules, prediction, and estimation.* 2007. **102**(477): p. 359-378.

26. Bracher, J., et al., *Evaluating epidemic forecasts in an interval format.* PLoS computational biology, 2021. **17**(2): p. e1008618.

27. Mitchell, K. and C. Ferro, *Proper scoring rules for interval probabilistic forecasts.* Quarterly Journal of the Royal Meteorological Society, 2017. **143**(704): p. 1597-1607.

28. Gneiting, T. and A.E. Raftery, *Strictly proper scoring rules, prediction, and estimation.* Journal of the American statistical Association, 2007. **102**(477): p. 359-378.

29. Mora, C., et al., *Probabilistic Neural Data Fusion for Learning from an Arbitrary Number of Multi-fidelity Data Sets.* arXiv preprint arXiv:2301.13271, 2023.

30. Yeom, C.-U., *Performance evaluation of automobile fuel consumption us-.* Symmetry, 2019.

31. Herbol, H.C., et al., *Efficient search of compositional space for hybrid organic–inorganic perovskites via Bayesian optimization.* npj Computational Materials, 2018. **4**(1): p. 51.

Copyright © 2023 by ASME

Proceedings of the ASME 2023
International Design Engineering Technical Conferences and
Computers and Information in Engineering Conference
IDETC-CIE2023
August 20-23, 2023, Boston, Massachusetts

DETC2023-114601

DESIGN OF MIXED-CATEGORY STOCHASTIC MICROSTRUCTURES: A COMPARISON OF CURVATURE FUNCTIONAL-BASED AND DEEP GENERATIVE MODEL-BASED METHODS

Leidong Xu, Kiarash Naghavi Khanghah, Hongyi Xu*

Mechanical Engineering, University of Connecticut, Storrs, CT 06269

* Email: hongyi.3.xu@uconn.edu

ABSTRACT

Bridging the gaps among various categories of stochastic microstructures remains a challenge in the design representation of microstructural materials. Each microstructure category requires certain unique mathematical and statistical methods to define the design space (design representation). The design representation methods are usually incompatible between two different categories of stochastic microstructures. The common practice of pre-selecting the microstructure category and the associated design representation method before conducting rigorous computational design limits the design freedom and reduces the possibility of obtaining innovative microstructure designs. To overcome this issue, this paper proposes and compares two methods, the deep generative modeling-based method and the curvature functional-based method, to understand their pros and cons in designing mixed-category stochastic microstructures for desired properties. For the deep generative modeling-based method, the Variational Autoencoder is employed to generate an unstructured latent space as the design space. For the curvature functional-based method, the microstructure geometry is represented by curvature functionals, of which the functional parameters are employed as the microstructure design variables. Regressors of the microstructure design variables-property relationship are trained for microstructure design optimization. A comparative study is conducted to understand the relative merits of these two methods in terms of computational cost, continuous transition, design scalability, design diversity, dimensionality of the design space, interpretability of the statistical equivalency, and design performance.

Keywords: Stochastic microstructures; Microstructure design; Deep generative model; Curvature functional; Design representation.

1. INTRODUCTION

By designing the microstructures of architected materials, a wide spectrum of properties, such as strength [1-3], ductility [4], energy density [5, 6], and thermal conductivity [1, 7, 8], can be achieved to meet engineering requirements. Here we focus on stochastic microstructures, of which the statistical variations in structural characteristics are induced by uncertainties in the manufacturing processes [9-11], defects or porosities [12], or the inherent randomness at the micro- or nano-scale [13, 14]. In the field of engineered architected metamaterials, designers have looked into stochastic structure designs to achieve higher energy absorption [6, 15, 16], compatibility with traditional manufacturing techniques [17, 18], and robustness against defects [19].

In the literature, a variety of statistical characterization and stochastic reconstruction-based approaches have been proposed for designing stochastic microstructures. Statistical characterization is a process that generates statistical descriptors and functions of the stochastic microstructure features observed from digital images (e.g., microscopic images). Stochastic reconstruction is a process that re-generates statistically equivalent microstructures based on the input statistical descriptors and functions. One simple and straightforward way is to characterize microstructures with physically meaningful parametric descriptors such as volume fraction, particle/pore size, fiber length, fiber orientation, etc. In addition, high dimensional statistical functions including N-point correlation functions [20-23], spectrum density function [24, 25], and random fields [26, 27] have also been applied to describe the complex stochastic microstructure morphologies. One major limitation of these methods is that each stochastic microstructure category requires some unique mathematical and statistical representations that are incompatible with other categories. For

Copyright © 2023 by ASME

example, random fiber composites require fiber orientation tensor [10, 28], random particle composites require the statistical distribution of particle diameters [29, 30], granular alloy microstructures require both grain orientation and crystal orientation [31], and spinodal-like structures can be described with spectrum density function [25]. Therefore, a designer needs to decide the microstructure category before defining the design space and conducting computational design. The step of pre-selecting the microstructure category limits the design freedom and reduces the possibility of obtaining innovative microstructure designs.

In recent years, deep generative models, such as Variational Autoencoders (VAEs), generative adversarial networks (GANs), and their variations, have been employed in stochastic microstructure reconstruction and design [16, 32-38]. However, the aforementioned works only consider a limited number of microstructure categories [39] and do not focus on bridging the gaps among various categories. In our previous work [40], we established a deep generative modeling framework that learns a unified microstructure design space based on multiple categories of stochastic microstructures (random fibers, random particles, random ellipses, random node-edge networks, and random amorphous microstructures) and deterministic, periodic microstructures (e.g., cellular metamaterials). This framework enables a smooth transition between stochastic and deterministic structural patterns in the property-driven microstructure design. However, this framework only handles 2D microstructure images and is demanding on training data and computational resources, so its application to 3D microstructure design is limited by the curse of dimensionality.

To address the aforementioned challenges, here we investigate two approaches that have the capability of generating a unified design space that embodies various categories of stochastic microstructures:

(i) A data-driven approach based on the deep generative model;

(ii) A mathematics-based approach that is established upon the curvature functionals.

As shown in **Figure 1**, these two methods are employed in design representation to create a parametric design space for stochastic microstructure design. With the obtained design space, Design of Experiments (DOE), supervised learning of the microstructure-property relationship, and property-driven design will be conducted to generate new microstructure designs. A comparative study will be presented to discuss the pros and cons of the two methods.

The remainder of the paper is organized as follows. Section 2 introduces a deep generative model-based design methodology. Section 3 introduces the curvature functional-based design methods. In section 4, a microstructure design case is presented to compare the two methods. Section 5 presents a comprehensive discussion of the comparison of the two methods. Section 6 concludes this paper.

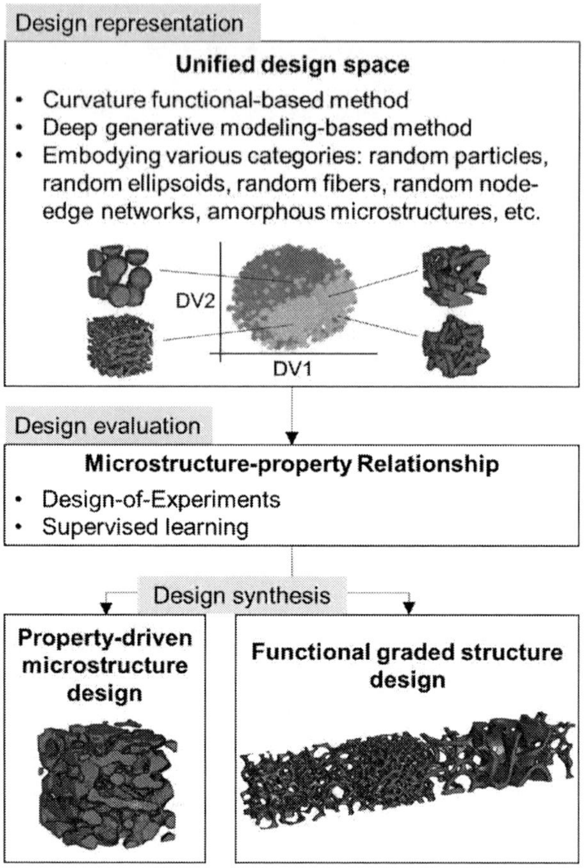

Figure 1: Design of mixed-category stochastic microstructures. The focus of this paper is to compare the curvature functional-based and the deep generative model-based methods in the design representation. Both methods will be employed to create a unified design space that embodies various categories of stochastic microstructures for the property-driven microstructure design.

2. DEEP GENERATIVE MODEL-BASED METHOD

One way to bridge the gap among different microstructure categories is to leverage the data-driven approach, e.g., deep feature learning, to learn a unified design space based on a large and diverse microstructure database that embodies various categories of microstructures. We first established a 3D stochastic microstructure database by leveraging the stochastic reconstruction methods proposed in our previous works, including the statistical descriptor-based method [10, 30, 41, 42], the space tessellation-based method [9], the spectrum density function (SDF)-based random field method [14], etc. This database consists of 40,000 microstructural images with a resolution of 64×64×64, and the microstructure samples can be classified into five categories: random particles, random fibers, random ellipsoids, random node-edge networks, and amorphous microstructures. Samples from each category are shown in **Figure 2**. The dataset is divided into a training set and a test set in a ratio of 9:1.

Copyright © 2023 by ASME

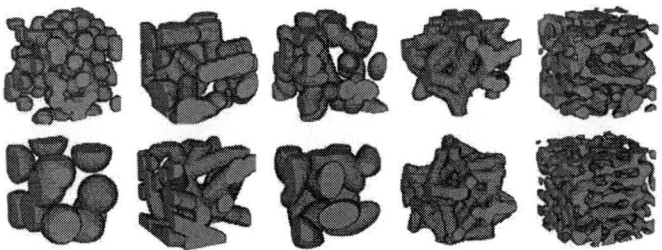

Figure 2: Examples of microstructure samples in the database for deep generative modeling. From left to right: random particles, random fibers, random ellipsoids, random node-edge networks, and amorphous microstructures.

2.1. Microstructure representation by VAE

VAE is a deep generative model that consists of two major components: an encoder network and a decoder network. The encoder network maps the input data to a Gaussian distribution in the latent space, which allows for the generation of novel data samples through sampling from the learned distribution. The decoder network takes the latent representation as the input and reconstructs the original data. The key feature of VAE is the introduction of a probabilistic approach to encode the input data into the latent space. Rather than mapping the input data to a single point in the latent space, the VAE maps the input data to a probability distribution over the latent space. Compared to other generative models, e.g., GAN and diffusion model, VAE provides an interpretable latent space, which can be used as a low-dimensional design space. The similarity of structural features can be measured by the distance in the latent space of VAE. Moreover, GAN models suffer from diminished gradient, model collapse, and other training instability issues that limit their application to complex datasets, such as mix-category microstructure datasets. Therefore, VAE is selected in this study. A general loss function of a vanilla VAE is expressed as:

$$L_i(\boldsymbol{\theta}, \boldsymbol{\phi}) = -E_{\boldsymbol{z} \sim q_{\boldsymbol{\theta}}(\boldsymbol{z}|\boldsymbol{x}_i)}[log\, p_{\boldsymbol{\phi}}(\boldsymbol{x}_i \mid \boldsymbol{z})] + \\ D_{KL}(q_{\boldsymbol{\theta}}(\boldsymbol{z} \mid \boldsymbol{x}_i)|p(\boldsymbol{z})) \quad (1)$$

where $\boldsymbol{\theta}$ and $\boldsymbol{\phi}$ are the parameters of the decoder and encoder, respectively, and $\boldsymbol{x}_i$ is input microstructure image data for our case, and $\boldsymbol{z}$ denotes the latent vectors. The first term, $-E_{\boldsymbol{z} \sim q_{\boldsymbol{\theta}}(\boldsymbol{z}|\boldsymbol{x}_i)}[log\, p_{\boldsymbol{\phi}}(\boldsymbol{x}_i \mid \boldsymbol{z})]$, is the reconstruction loss that measures the pixel-level error between the input and reconstruction. The second term, $D_{KL}(q_{\boldsymbol{\theta}}(\boldsymbol{z} \mid \boldsymbol{x}_i)|p(\boldsymbol{z}))$,

denotes the KL loss and ensures that the learned distribution q follows the true prior distribution p. Practically, including the KL term in the loss function can avoid overfitting and also regularize the latent space to reduce discontinuities in the latent space.

Figure 3 shows our implementation of the VAE to generate a parametric latent space representation of the stochastic microstructures as the design space. The encoder follows a VGG-style architecture, in which the convolution layer blocks are followed by the fully connected layers. The dimension of latent vectors is set at 256 based on the results of trials, in order to balance the reconstruction quality and the time efficiency of conducting optimal microstructure search in the latent space.

We also explored other variants of VAE in this work. Literature and our previous work suggest that including a style loss term in the loss function typically enhances reconstruction quality significantly [40, 43]. However, the small improvement in quality comes at the cost of a substantial increase in computational complexity due to the tensor permutation process on each image. We also tested an architecture that incorporates the style loss [43], but did not observe an improvement in the reconstruction quality. Furthermore, we experimented with a Gaussian-mixture VAE [44], but did not observe any significant benefits either. After a thorough exploration of these options, we decided to employ a vanilla VAE for its computational efficiency.

2.2 Property-driven microstructure design and generation of functionally graded structure designs by VAE

As discussed in Section 1, we adopt the surrogate model-based optimization approach to design microstructures for desired properties. The latent variables are considered as microstructure design variables. DOE is conducted in latent space to generate a dataset for training the microstructure-property surrogate models. Multi-response Gaussian Process (GP) regression models are employed to establish the relationship between the latent variables and the mechanical properties.

As the computational cost of design evaluation (by surrogate model) during the optimization process is not a concern here, we select the Genetic Algorithm (GA) to solve the design problem. GA, and other evolutionary algorithms, have the advantage of avoiding local minima. For multi-objective optimization problems, Non-dominated Sorting Genetic Algorithm II (NSGA-II) [45] is employed as the optimizer.

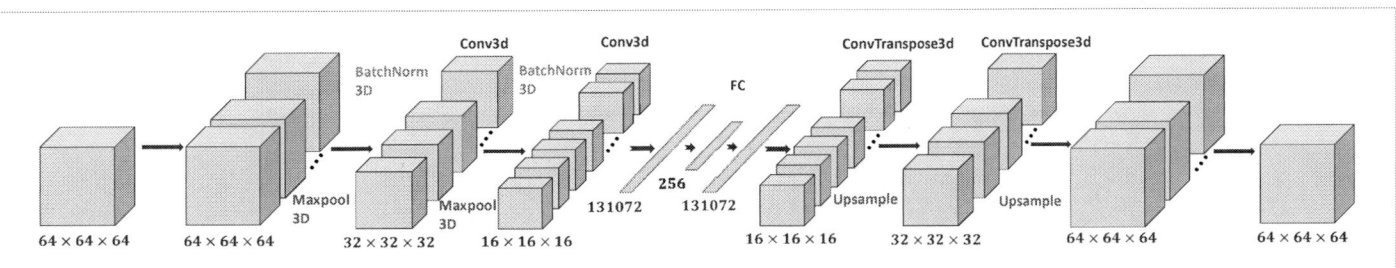

Figure 3: Architecture of the Variational Autoencoder. The reduced dimensional latent space is employed as the design space.

Copyright © 2023 by ASME

The optimal designs are first obtained in the format of latent vector, and the corresponding microstructure images are reconstructed by the decoder. The properties of the optimal microstructure designs are verified by simulations, as there always exist discrepancies between the surrogate model-predicted properties and the true values.

In addition to designing microstructure units, we also investigate the VAE model's capability of generating functionally graded structure designs. A functionally graded structure is characterized by the variation in structure gradually over volume, resulting in corresponding continuous changes in the properties. A series of microstructure units are generated by conducting spherical linear interpolation [46] between two microstructure unit samples in the latent space. A gradual change in the microstructure features can be observed in this series of designs. A functionally graded structure can be generated by assembling those microstructure units sequentially (**Figure 4**). Due to the discrete nature of the microstructure interpolation, one outstanding shortcoming is the lack of continuity at the interface between two adjacent microstructure units. The presence of discontinuities at the interface can lead to local stress concentrations that may weaken the overall strength of the structure and even cause it to failure. Non-smooth transitions in the interfaces can be observed, as shown in the side views in **Figure 4 (d)**.

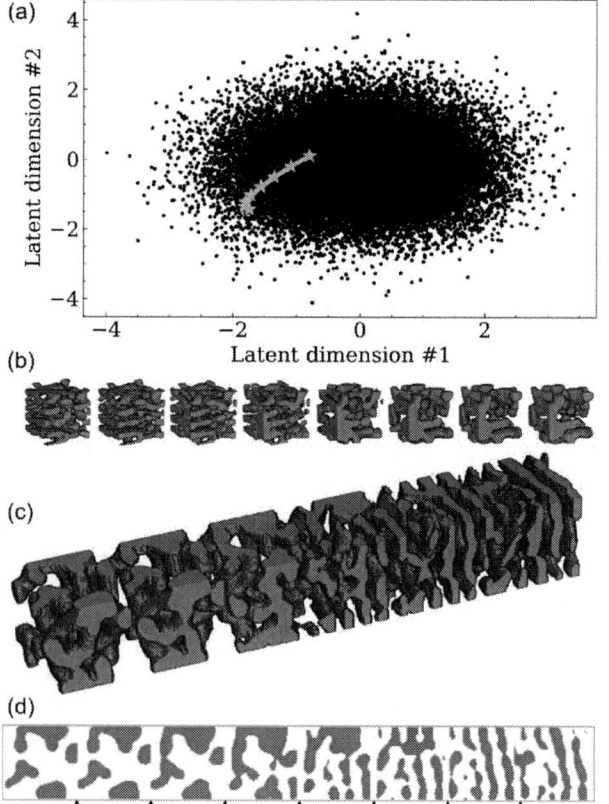

Figure 4: Functional graded structure design by the deep generative modeling-based method. (a) A series of designs are

generated along a certain path in two selected dimensions of the 256-dimensional latent space. (b) Each star in the path is decoded into a microstructure unit. (c) A functionally graded structure design is created by assembling the microstructure units. (d) Side view of the 3D functionally graded structure. The interfaces among adjacent units are marked by triangles. Due to the discrete nature of the sampling process, non-smooth transitions can be observed at the interfaces among microstructure units.

3. CURVATUAL FUNCTIONAL-BASED METHOD

3.1. Microstructure representation by curvature functionals

Curvature functionals are capable of generating a variety of complex shapes and have been demonstrated as a powerful tool for designing bio-mimetic scaffold [47]. Curvature functionals employ a phase-field formulation to diffuse an approximation of a vast range of shape textures. The resulting approximation is used as a loss function, in conjunction with modern automatic differentiation optimizers, to generate geometries from a random field initialization. When compared to the phase-field [48, 49] and statistical functional approaches [25], such as spinodal microstructures generated by Gaussian random field (GRF) [2, 50], curvature functionals have the ability to generate a broader range of topologies. These include laminar, spherical, pearly thin wall, and tube shapes, and are governed by seven generation parameters $\boldsymbol{a} = [a_{2,0}, a_{0,2}, a_{1,1}, a_{1,0}, a_{0,1}, a_{0,0}]$ and m_0. However, the mathematical meaning of the generation parameters is yet fully explored which limits the capability in directly using this method for inverse design. To address this limitation, we utilize the supervised learning method to establish the relation between generation parameters and properties to enable the property-driven microstructure design.

Gaussian curvature is a differential geometry measure of the curvature of a surface at a given point, which is defined as the production of the principal curvatures κ_1, κ_2 by

$$K = \kappa_1 \kappa_2 \qquad (2)$$

The complex microstructure surface under constant volume is modelled as a curvature functional

$$\boldsymbol{F}(S) = \int_S p(\kappa_1, \kappa_2) dA \qquad (3)$$

where p is the second order polynomial of the principal curvatures of the entire surface S. p is restricted at the degree of 2, as it is efficient to generate topological features. The curvature functionals can be expanded as

$$\boldsymbol{F}(S) = \int_S \left(a_{2,0}\kappa_1^2 + a_{1,1}\kappa_1\kappa_2 + a_{0,2}\kappa_2^2 + a_{1,0}\kappa_1 + a_{0,1}\kappa_2 + a_{0,0}\right)dA = \int_S \left(\sum_{|\alpha| \leq 2} a_\alpha (\kappa_1\kappa_2)^\alpha\right)dA \qquad (4)$$

Generally, it is convenient to refine this kind of 2D surface functionals to scalar fields u in 3D volume by diffusion approximation. And the matrix field $\mathcal{M}_u^\epsilon$ is introduced as:

$$\mathcal{M}_u^\epsilon = -\epsilon \operatorname{Hess} u + \frac{w'(u)}{\epsilon} n_u \otimes n_u \qquad (5)$$

whose trace is equal to

$$\operatorname{Tr}\mathcal{M}_u^\epsilon = -\epsilon \Delta u + \frac{w'(u)}{\epsilon} \qquad (6)$$

Copyright © 2023 by ASME

116

Applied phase-field approximation and further simplification, the final representation of the phase-field $\mathcal{F}_\epsilon(u)$ can be written as

$$\mathcal{F}_\epsilon(u) = \int_\Omega \left[\frac{a_{2,0}+a_{0,2}-a_{1,1}}{2\epsilon} \|\mathcal{M}_u^\epsilon\|^2 + \frac{a_{1,1}}{2\epsilon}(\mathrm{Tr}\mathcal{M}_u^\epsilon)^2 + \frac{a_{2,0}-a_{0,2}}{2\epsilon}\mathrm{Tr}\mathcal{M}_u^\epsilon\sqrt{(2\|\mathcal{M}_u^\epsilon\|^2 - (\mathrm{Tr}\mathcal{M}_u^\epsilon)^2)^+} + \frac{a_{1,0}+a_{0,1}}{2}|\nabla u|\mathrm{Tr}\mathcal{M}_u^\epsilon + \frac{a_{1,0}-a_{0,1}}{2}|\nabla u|\sqrt{(2\|\mathcal{M}_u^\epsilon\|^2 - (\mathrm{Tr}\mathcal{M}_u^\epsilon)^2)^+} + a_{0,0}\epsilon|\nabla u|^2 \right] dx \quad (7)$$

To implement the phase-field $\mathcal{F}_\epsilon(u)$ to generate microstructure geometries given a random initialization, a mass-preserving flow can be defined as

$$\dot{u} = \Delta \frac{\partial \mathcal{F}_\epsilon}{\partial u} \quad (8)$$

This form can also be repressed as

$$u = \nabla \cdot A + m_0 \quad (9)$$

where $A: \Omega \to \mathbb{R}^3$ is a periodic vector field, and $m_0 \in \mathbb{R}$ is the desired value of the average $\bar{u}$ which also approximates the volume fraction by $\frac{m_0+1}{2}$. Finally, an energy function is defined as:

$$G_\epsilon(A) = \mathcal{F}_\epsilon(\nabla \cdot A + m_0) \quad (10)$$

with a gradient of

$$\frac{\partial G_\epsilon}{\partial A}(A) = -\nabla \frac{\partial \mathcal{F}_\epsilon}{\partial u}(u) \quad (11)$$

This energy function is used as the loss function with an auto-differentiation tool that iteratively optimizes u to evolve a random vector field A_0 until the energy function meets the convergence criterion. Empirically, A_0 can be drawn from a uniform distribution. Random initialization of the structure image in the curvature functional method results in diverse yet statistically equivalent stochastic reconstructions of microstructures that share the same input generation parameters a and m_0. Therefore, the generation variables can be considered as a quantitative representation of an infinite set of random but statistically equivalent microstructures, which makes this method suitable for generating random but statistically equivalent stochastic microstructure designs. Several examples of statistically equivalent microstructure samples generated from the same a vector are shown in **Figure 5 (a~d)**.

3.2 Property-driven microstructure design and generation of graded functional structure designs by curvature functionals

Following the flowchart in **Figure 1**, we propose a surrogate model-based optimization approach for microstructure design. The surrogate model of the relationship between the generation parameters a and material property is established using GP regression. It is to be noted that random but statistically equivalent microstructures will be generated for a given set of design variables. Therefore, we generated ten samples from ten fixed random initializations (A) for the same design variable vector, and then simulated the mechanical properties of all ten samples. We generated a total of 20,000 samples using 2,000 sets of generation parameters. Similar to the method presented in

Section 2, we adopt GA and NSGA-II as the optimizer to solve the property-driven design problem. In the last step, the digital images of the microstructure designs are reconstructed based on the design variable vector a.

Here we also investigated the curvature functional-based method's capability of generating functional graded structure designs. One advantage of the curvature functional method is that a smooth transition between different categories of microstructures can be easily obtained by varying the values of the generation parameters continuously. **Figure 5(e)** shows a functional graded design generated based on continuous functions of the generation parameters a along the longitudinal direction.

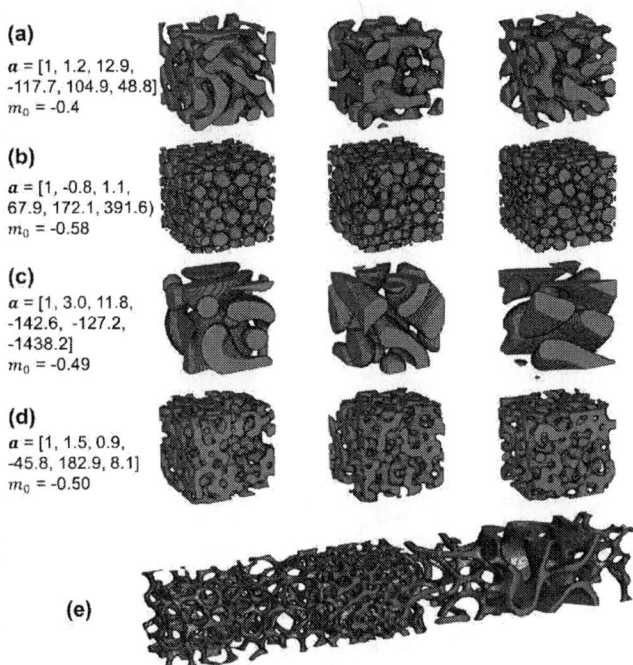

Figure 5: (a)~(d) Design variable vectors and the corresponding statistically equivalent microstructure samples. Each row shows three stochastic samples of the same microstructure design and the corresponding generation parameters. (e) A functionally graded structure obtained by the curvature functional-based method. It is created from continuous functions of the generation parameters a.

4. A COMPARATIVE STUDY WITH A DESIGN FOR STIFFNESS PROBLEM

In this section, we present a design case to compare the deep generative model-based and the curvature functional-based design representation methods in two aspects: the accuracy of the microstructure-property regressor and the performance of the optimal designs obtained with each method.

Here we define a multi-objective microstructure design problem that maximizes the Young's moduli along X-, Y-, and Z- directions. Design constraints are defined to guarantee close-to-isotropic designs, i.e., the differences between the maximum/minimum modulus and the median modulus of the

Copyright © 2023 by ASME

three directions are within 3%. Therefore, the optimization problem can be formulated as

$$\max E_i(\mathbf{z}) \text{ or } \max E_i(\mathbf{a}, m_0), i = X, Y, Z \quad (12)$$

subject to:

$$\frac{|E_{\text{highest}} - E_{\text{medium}}|}{E_{\text{medium}}} < 3\% \quad (13)$$

$$\frac{|E_{\text{lowest}} - E_{\text{medium}}|}{E_{\text{medium}}} < 3\% \quad (14)$$

If using the curvature functional-based method, additional constraints are needed to guarantee the convergence of microstructure image reconstruction:

$$\max(u) > 0.1 \quad (14)$$
$$\min(u) < -0.1 \quad (15)$$
$$\text{discrepancy}(u) < 0.75 \quad (16)$$

where the discrepancy is a measurement of how much the scalar fields u deviate from a tanh profile phase field function [47]. As this research focuses on investigating the influence of microstructure morphology on the properties, the volume fraction is set as a constant (0.4).

As a preparation for exploring the relationship between microstructure and the property of interest, in this case, elasticity, we performed finite element simulations on all microstructure samples by ABAQUS. The 0-1 matrices that represent the binary microstructure images are transformed into hexahedral meshes. The elastic modulus and Poisson's ratio of the 1 phase in the microstructure are $E_{Boron} = 379300$ MPa and $\gamma_{Boron} = 0.1$, whereas $E_{Aluminum} = 68300$ Mpa, $\gamma_{Aluminum} = 0.3$ for the 0 phase. The Young's moduli (E_x, E_y, E_z) in the X-, Y-, and Z-direction are calculated from the compliance matrix. The infinitesimal displacement boundary conditions are shown in **Figure 6**.

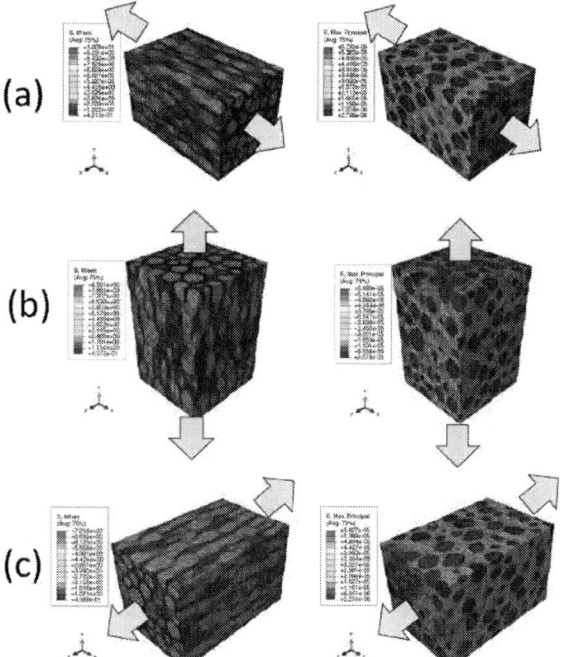

Figure 6: Elasticity property analysis on a microstructure for the maximum in-plane strain and the maximum von Mises stress in (a) X-direction, (b) Y-direction, and (c) Z-direction.

The dimensionality of the design space has a strong impact on the predictability of the GP regressors. The design space generated by VAE has a dimensionality of 256. By contrast, the design space of the curvature functional-based method is only 7. More input variables indicate a potentially better capability to capture complex microstructure features, but practically, a high dimensional input space poses a significant challenge to establishing the design variable-property relationship by surrogate modeling because a lot more training data points are required to fully cover the input space. In **Table 1**, we present a comparison of three GP models: VAE latent space-based GP model with a dataset of 40000 samples, VAE latent space-based GP model with a dataset of 20000 samples, and curvature functional-based GP model with a dataset of 20000 samples. In each training, the dataset is split into a training set (90%) and a test set (10%). The model accuracy, R^2, is evaluated based on the test set. The curvature functional-based GP model has a higher accuracy, even when comparing with the VAE-based GP model that uses twice as many training data points.

Table 1: Prediction accuracies of the GP regression models with the design spaces generated by the VAE-based method and the curvature functional-based method.

Model (size of the dataset)	R^2 score		
	E_x	E_y	E_z
GP w/VAE (40000)	0.743	0.681	0.746
GP w/VAE (20000)	0.686	0.620	0.688
GP w/ curvature functional (20000)	0.811	0.803	0.775

Another point worth noting is that some combination of generation parameters in the curvature functional method may generate ill-posed geometric which may have zero level set and floating fragments, where such fragments can lead to unrealistic microstructures in composite material and porous material from both design and manufacturing perspectives. Therefore, three criteria, $\max(u)$, $\min(u)$, and discrepancy ratio, are required to identify ill-posed phase-field u during the optimization process. These three criteria must be included as inequality constraints in optimization to ensure successful reconstructions of the final microstructure designs. Experimentally, we observe that these three constraint functions limit the number of feasible designs significantly.

The Pareto frontiers obtained with the two methods are compared in **Figure 7**. The performances of the design points in these plots are the true values obtained from verification simulations. Due to the predicted errors of the microstructure-property model, some of the optimal designs violate the design constraints on isotropicity. For the VAE-based method, only 10% of the optimal designs in the Pareto frontier satisfy the design constraints. Among the feasible designs, we can hardly find designs that rank in the top 10% compared to the samples in

Copyright © 2023 by ASME

the microstructure database, with respect to the properties of interest.

On the other hand, more than 70% percent of optimal designs found by the curvature functional approach are isotropic, according to the results of verification simulations. Furthermore, almost all of the feasible solution rank in the 10% compared to the samples in the microstructure database. **Figure 8 (a)~(d)** show several examples of the optimal designs obtained by the curvature functional-based method, and **Figure 8 (e) and (f)** show the optimal designs obtained by the VAE-based method.

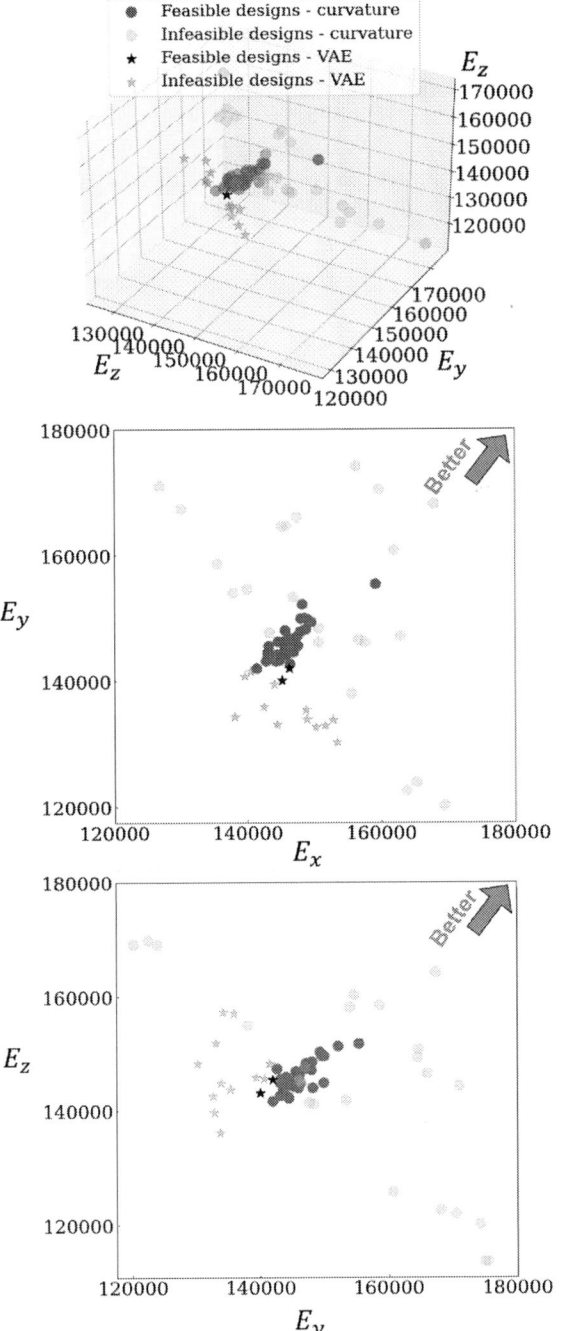

Figure 7: Pareto frontiers obtained by both design approaches. As there are three design objectives, one 3D view and two 2D views of the performance space are provided. The design objective is to maximize E_x, E_y, and E_z. The feasible design points are in dark colors and the infeasible design points are in light colors.

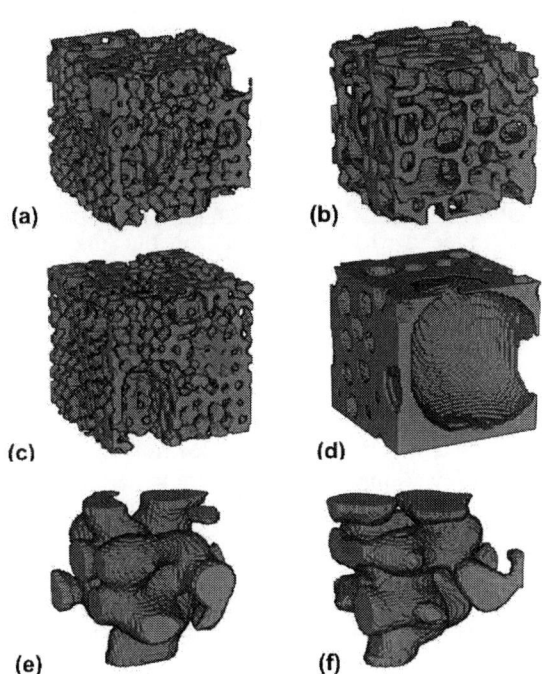

Figure 8: (a)~(d) Optimal designs from the curvature functional-based optimization approach. (a) a = [1, 4.0, 0.3, 75, 200, -2060] and m_0 = -0.26. (b) a = [1, 3.93, 3.79, 36, 194, 2998] and m_0 = -0.43. (c) a = [1, 3.99, 0.34, 45, 197, 1422] and m_0 = -0.19. (d) a = [1, 3.98, 0.22, 40, 198, 1431] and m_0 = -0.30. (e) and (f) Two optimal designs from the VAE-based design approach.

5. UNDERSTANDING THE PROS AND CONS OF THE TWO DESIGN REPRESENTATION METHODS

As summarized in **Table 2**, the pros and cons of the deep generative modeling-based method and the curvature functional-based method are discussed in terms of seven criteria: computational cost, continuous transition in functionally graded structure design, scalability of the microstructure design, design diversity, dimensionality of the design space, and design performance.

<u>Computational cost</u>: To obtain a design space that embodies various categories of microstructures, the deep generative modeling-based approach requires significant computing resources for data generating and model training. On the other hand, the curvature functional-based method incurs minimal costs in defining the design space, while computing the viability constraints (Equation 14~16) during the optimization process is relatively computationally expensive.

Copyright © 2023 by ASME

Continuous transition in functionally graded structure design: When creating functionally graded structure designs, the curvature functional-based method can guarantee a smooth transition among various microstructure patterns. With the deep generative model-based method, the functionally graded structure design is created by assembling a series of microstructure units, which correspond to discrete points in the latent space. Therefore, a smooth transition between microstructure units cannot be guaranteed. This issue could potentially be mitigated (but not resolved) by applying circular spatial padding to the transposed convolutional layer in the deep generative model [51], but the impacts on reconstruction quality and computational complexity need further investigation.

Scalability of the microstructure design: The deep generative models, which are trained on the images directly, cannot generate images with a wide range of sizes and resolutions. By contrast, the curvature functional-based method can easily map the design variables to an arbitrary domain size (**Figure 9**).

Design diversity: The deep generative models have the advantage over the curvature functionals. Theoretically, the deep generative models can be extended to embody any type of microstructure (e.g., microstructures with triangular inclusions) as long as the training data are available. The curvature functionals can only generate microstructures with curved surfaces.

Dimensionality of the design space: The curvature functional-based method has the advantage in generating a low dimensional design space. Although we can also set the dimensionality of the VAE latent space to a very low value (e.g. 8, the same as the design space of the curvature functional method) by modifying the fully connected layers in encoder, in practice, it will lead to a much poorer reconstruction accuracy. The high dimensionality of the VAE latent space poses a significant challenge to establishing the microstructure-property relationship, as well as searching for the optimal microstructure designs in the design space.

Interpretability of statistical equivalency among stochastic microstructure designs: It is a unique requirement for stochastic microstructure design. From the perspective of statistical characterization and stochastic reconstruction, one "design" actually represents an infinite number of microstructure samples that are random but statistically equivalent. The design representation by curvature functional parameters can provide this capability. By contrast, in the latent space learned by the deep generative model, each point corresponds to one specific, unique microstructure image. The distance between the points is a measurement of the pixel-to-pixel similarity of the two images, instead of the similarity in the statistical sense. As shown in **Figure 10**, two statistically equivalent random particle microstructure samples are far apart in terms of the Euclidean distance in latent space, while the random particle microstructure #1 is closer to the quasi-random microstructure. Therefore, it is not possible to define statistical equivalency purely based on the distance in the latent space. We acknowledge the possibility of generating random but statistically equivalent microstructures by

introducing empirical statistical descriptors into the loss function of deep generative models (e.g., GAN) [52], but then again, it loops back to our original research question: how to select proper descriptors for describing stochastic microstructures without compromising the design freedom.

Design performance: The performances of the optimal designs are influenced by two factors: the accuracy of the microstructure-property surrogate model, and the effectiveness of design exploration/searching in the design spaces generated by each method. Although the curvature functional-based method demonstrates better performances in the presented case study, we should be cautious to make a conclusion. In our previous work [40] and literature [53], it has been demonstrated that training the VAE and the latent variable-property regressor simultaneously can improve the property prediction accuracy. This paper focuses on the capability of learning a unified design space, so the simultaneously training of the latent space and the property repressor is out of scope and not included.

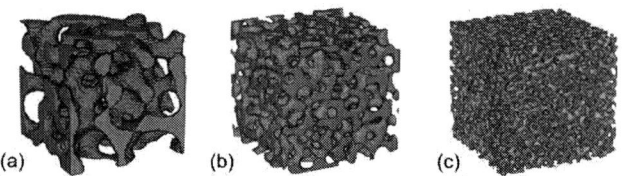

Figure 9: Scalability of the curvature functional-based method: microstructure designs generated from the same design variable vector $\boldsymbol{a} = [1, 2.8, 2, -10, -10, 25]$ and $m_0 = -0.25$ with sizes of (a) 64^3 (b) 128^3 (c) 256^3 voxels by the curvature functional-based method.

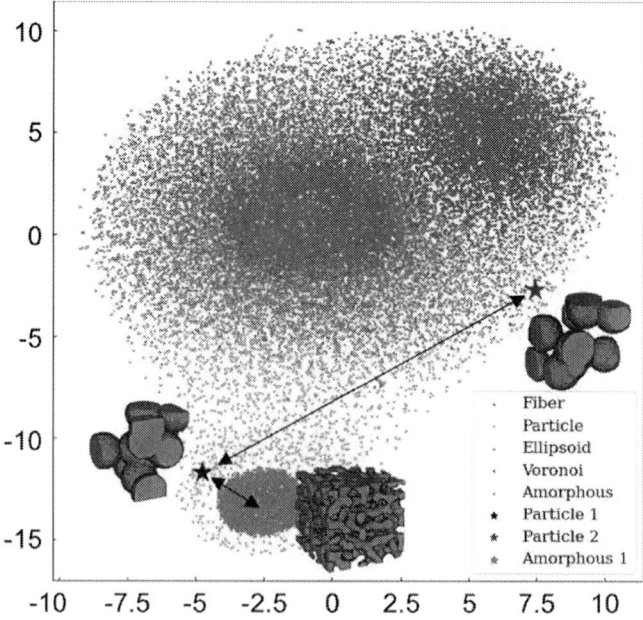

Figure 10: 2D t-SNE representations of VAE latent space. It is observed that the distance between two statistically equivalent

Copyright © 2023 by ASME

random particle microstructures is larger than that between a random particle microstructure and an amorphous microstructure. Therefore, the Euclidean distance in the latent space cannot be used to identify statistically equivalent microstructures.

Table 2: Overview of the comparison between the deep generative model-based and curvature functional-based methods in the design representation. The criterion with * is only valid for the methods and case study presented in this paper and further investigations are needed for other cases. "+" means better, "-" means worse.

Criteria	Deep generative model-based method	Curvature Functional-based method
Computational Cost	−	+
Continuous transition	−	+
Design scalability	−	+
Design diversity	+	−
Dimensionality of the design space	−	+
Interpretability of statistical equivalency	−	+
Design performance*	−	+

6. CONCLUSION AND FUTURE WORK

In this paper, we proposed and compared two methods for generating a unified design space that embodies various categories of stochastic microstructures: the deep generative model-based method and the curvature functional-based method. For the deep generative model-based method, the latent space learned from a highly diversified microstructure database is employed as the microstructure design space. For the curvature functional-based method, the generation parameters in the functionals are used as microstructure design variables. We established surrogate models to predict the relationship between microstructure design variables and the properties of interest, and conducted surrogate model-based optimization to design microstructures for desired properties. Furthermore, we applied the two methods to generate functionally graded structure designs. We present a comprehensive discussion and comparison of each method, outlining their respective advantages and drawbacks. This discussion serves to inform the design process for architecture and composite materials, aiding in the selection of an appropriate method based on the desired outcomes.

In our future work, we plan to test both methods on more engineering case studies to deepen our understanding of the strengths of each method. We are also aiming to extend both

methods to the design of 3D mixed-stochasticity microstructural systems that embody both deterministic and stochastic microstructure units.

ACKNOWLEDGEMENT

The authors gratefully acknowledge financial support from the National Science Foundation (CAREER Award CMMI-2142290). L.X. sincerely thanks the insightful discussions with Anna Song at Imperial College London.

REFERENCES

[1] P.P. Meyer, C. Bonatti, T. Tancogne-Dejean, D. Mohr, Graph-based metamaterials: Deep learning of structure-property relations, Materials & Design 223 (2022) 111175.
[2] S. Kumar, S. Tan, L. Zheng, D.M. Kochmann, Inverse-designed spinodoid metamaterials, npj Computational Materials 6(1) (2020) 73.
[3] Z. Wang, W. Xian, M.R. Baccouche, H. Lanzerath, Y. Li, H. Xu, Design of Phononic Bandgap Metamaterials Based on Gaussian Mixture Beta Variational Autoencoder and Iterative Model Updating, Journal of Mechanical Design 144(4) (2022) 041705.
[4] Y. Kim, H.K. Park, J. Jung, P. Asghari-Rad, S. Lee, J.Y. Kim, H.G. Jung, H.S. Kim, Exploration of optimal microstructure and mechanical properties in continuous microstructure space using a variational autoencoder, Materials & Design 202 (2021) 109544.
[5] M. Guo, J. Jiang, Z. Shen, Y. Lin, C.-W. Nan, Y. Shen, High-energy-density ferroelectric polymer nanocomposites for capacitive energy storage: enhanced breakdown strength and improved discharge efficiency, Materials Today 29 (2019) 49-67.
[6] Q. Ye, Y. Liu, H. Lin, M. Li, H. Yang, Multi-band metamaterial absorber made of multi-gap SRRs structure, Applied Physics A 107 (2012) 155-160.
[7] K. Yaji, S. Yamasaki, K. Fujita, Data-driven multifidelity topology design using a deep generative model: Application to forced convection heat transfer problems, Computer Methods in Applied Mechanics and Engineering 388 (2022) 114284.
[8] P. Du, A. Zebrowski, J. Zola, B. Ganapathysubramanian, O. Wodo, Microstructure design using graphs, npj Computational Materials 4(1) (2018) 50.
[9] Y. Li, Z. Chen, L. Su, W. Chen, X. Jin, H. Xu, Stochastic reconstruction and microstructure modeling of SMC chopped fiber composites, Composite Structures 200 (2018) 153-164.
[10] Z. Chen, T. Huang, Y. Shao, Y. Li, H. Xu, K. Avery, D. Zeng, W. Chen, X. Su, Multiscale finite element modeling of sheet molding compound (SMC) composite structure based on stochastic mesostructure reconstruction, Composite Structures 188 (2018) 25-38.
[11] W.M. Tucho, V.H. Lysne, H. Austbø, A. Sjolyst-Kverneland, V. Hansen, Investigation of effects of process parameters on microstructure and hardness of SLM manufactured SS316L, Journal of Alloys and Compounds 740 (2018) 910-925.

Copyright © 2023 by ASME

[12] S. Wang, J. Ning, L. Zhu, Z. Yang, W. Yan, Y. Dun, P. Xue, P. Xu, S. Bose, A. Bandyopadhyay, Role of porosity defects in metal 3D printing: Formation mechanisms, impacts on properties and mitigation strategies, Materials Today (2022).

[13] T. Tran‐Phu, R. Daiyan, X.M.C. Ta, R. Amal, A. Tricoli, From Stochastic Self‐Assembly of Nanoparticles to Nanostructured (Photo) Electrocatalysts for Renewable Power‐to‐X Applications via Scalable Flame Synthesis, Advanced Functional Materials 32(13) (2022) 2110020.

[14] H. Xu, J. Zhu, D.P. Finegan, H. Zhao, X. Lu, W. Li, N. Hoffman, A. Bertei, P. Shearing, M.Z. Bazant, Guiding the Design of Heterogeneous Electrode Microstructures for Li‐Ion Batteries: Microscopic Imaging, Predictive Modeling, and Machine Learning, Advanced Energy Materials 11(19) (2021) 2003908.

[15] A. Guell Izard, J. Bauer, C. Crook, V. Turlo, L. Valdevit, Ultrahigh energy absorption multifunctional spinodal nanoarchitectures, Small 15(45) (2019) 1903834.

[16] Z. Yang, X. Li, L. Catherine Brinson, A.N. Choudhary, W. Chen, A. Agrawal, Microstructural materials design via deep adversarial learning methodology, Journal of Mechanical Design 140(11) (2018).

[17] C.M. Portela, A. Vidyasagar, S. Krödel, T. Weissenbach, D.W. Yee, J.R. Greer, D.M. Kochmann, Extreme mechanical resilience of self-assembled nanolabyrinthine materials, Proceedings of the National Academy of Sciences 117(11) (2020) 5686-5693.

[18] Y. Li, Z. Chen, H. Xu, J. Dahl, D. Zeng, M. Mirdamadi, X. Su, Modeling and simulation of compression molding process for sheet molding compound (SMC) of chopped carbon fiber composites, SAE International Journal of Materials and Manufacturing 10(2) (2017) 130-137.

[19] Q.T. Do, C.H.P. Nguyen, Y. Choi, Homogenization-based optimum design of additively manufactured Voronoi cellular structures, Additive Manufacturing 45 (2021) 102057.

[20] Y. Jiao, F. Stillinger, S. Torquato, Modeling heterogeneous materials via two-point correlation functions: Basic principles, Physical Review E 76(3) (2007) 031110.

[21] Y. Jiao, F.H. Stillinger, S. Torquato, Modeling heterogeneous materials via two-point correlation functions. II. Algorithmic details and applications, Phys Rev E 77(3) (2008).

[22] S. Torquato, Optimal design of heterogeneous materials, Annu Rev Mater Res 40 (2010) 101-129.

[23] Y. Zhang, M. Yan, Y. Wan, Z. Jiao, Y. Chen, F. Chen, C. Xia, M. Ni, High-throughput 3D reconstruction of stochastic heterogeneous microstructures in energy storage materials, npj Computational Materials 5(1) (2019) 1-8.

[24] S. Yu, C. Wang, Y. Zhang, B. Dong, Z. Jiang, X. Chen, W. Chen, C. Sun, Design of non-deterministic quasi-random nanophotonic structures using Fourier space representations, Scientific reports 7(1) (2017) 1-10.

[25] A. Iyer, R. Dulal, Y. Zhang, U.F. Ghumman, T. Chien, G. Balasubramanian, W. Chen, Designing anisotropic microstructures with spectral density function, Comp Mater Sci 179 (2020) 109559.

[26] M. Grigoriu, Random field models for two-phase microstructures, J Appl Phys 94(6) (2003) 3762-3770.

[27] E. Levina, P.J. Bickel, Texture synthesis and nonparametric resampling of random fields, The Annals of Statistics 34(4) (2006) 1751-1773.

[28] S.G. Advani, C.L. Tucker III, The use of tensors to describe and predict fiber orientation in short fiber composites, Journal of rheology 31(8) (1987) 751-784.

[29] H. Xu, R. Liu, A. Choudhary, W. Chen, A machine learning-based design representation method for designing heterogeneous microstructures, Journal of Mechanical Design 137(5) (2015) 051403.

[30] H. Xu, D.A. Dikin, C. Burkhart, W. Chen, Descriptor-based methodology for statistical characterization and 3D reconstruction of microstructural materials, Comp Mater Sci 85 (2014) 206-216.

[31] Y. Li, H. Xu, W.-J. Lai, Z. Li, X. Su, A Multiscale Material Modeling Approach to Predict the Mechanical Properties of Powder Bed Fusion (PBF) Metal with Consideration of Microstructure Uncertainties, Fourth ASTM Symposium on Structural Integrity of Additive Manufactured Materials & Parts, Oxon Hill, Maryland, USA, 2019.

[32] R.K. Tan, N.L. Zhang, W. Ye, A deep learning–based method for the design of microstructural materials, Structural and Multidisciplinary Optimization 61 (2020) 1417-1438.

[33] R. Bostanabad, Reconstruction of 3d microstructures from 2d images via transfer learning, Computer-Aided Design 128 (2020) 102906.

[34] A. Dahari, S. Kench, I. Squires, S.J. Cooper, Fusion of complementary 2D and 3D mesostructural datasets using generative adversarial networks, Advanced Energy Materials 13(2) (2023) 2202407.

[35] S. Noguchi, J. Inoue, Stochastic characterization and reconstruction of material microstructures for establishment of process-structure-property linkage using the deep generative model, Physical Review E 104(2) (2021) 025302.

[36] S. Kench, S.J. Cooper, Generating three-dimensional structures from a two-dimensional slice with generative adversarial network-based dimensionality expansion, Nature Machine Intelligence 3(4) (2021) 299-305.

[37] R. Cang, Y. Xu, S. Chen, Y. Liu, Y. Jiao, M. Yi Ren, Microstructure representation and reconstruction of heterogeneous materials via deep belief network for computational material design, Journal of Mechanical Design 139(7) (2017).

[38] J. Jung, J.I. Yoon, H.K. Park, H. Jo, H.S. Kim, Microstructure design using machine learning generated low dimensional and continuous design space, Materialia 11 (2020) 100690.

[39] S. Deng, C. Mora, D. Apelian, R. Bostanabad, Data-Driven Calibration of Multifidelity Multiscale Fracture Models Via Latent Map Gaussian Process, Journal of Mechanical Design 145(1) (2023) 011705.

[40] L. Xu, N. Hoffman, Z. Wang, H. Xu, Harnessing structural stochasticity in the computational discovery and design of microstructures, Materials & Design 223 (2022) 111223.

[41] Y. Li, W. Chen, H. Xu, X. Jin, 3D representative volume element reconstruction of fiber composites via orientation tensor and substructure features, 31st Annual Technical Conference of the American Society for Composites, American Society for Composites, 2016.

[42] N. Hoffman, J. Lee, W. Li, J. Zhu, H. Xu, A Stochastic Microstructure Reconstruction-Based Mechanical and Transport Modeling Approachfor Learning the Microstructure-Property Relationship of Li-Ion Battery Graphite Anodes, 239th ECS meeting, Digital Meeting, 2021.

[43] R. Cang, H. Li, H. Yao, Y. Jiao, Y. Ren, Improving direct physical properties prediction of heterogeneous materials from imaging data via convolutional neural network and a morphology-aware generative model, Comp Mater Sci 150 (2018) 212-221.

[44] N. Dilokthanakul, P.A. Mediano, M. Garnelo, M.C. Lee, H. Salimbeni, K. Arulkumaran, M. Shanahan, Deep unsupervised clustering with gaussian mixture variational autoencoders, arXiv preprint arXiv:1611.02648 (2016).

[45] K. Deb, A. Pratap, S. Agarwal, T. Meyarivan, A fast and elitist multiobjective genetic algorithm: NSGA-II, IEEE transactions on evolutionary computation 6(2) (2002) 182-197.

[46] K. Shoemake, Animating rotation with quaternion curves, Proceedings of the 12th annual conference on Computer graphics and interactive techniques, 1985, pp. 245-254.

[47] A. Song, Generation of tubular and membranous shape textures with curvature functionals, Journal of Mathematical Imaging and Vision 64(1) (2022) 17-40.

[48] P.-A. Geslin, I. McCue, B. Gaskey, J. Erlebacher, A. Karma, Topology-generating interfacial pattern formation during liquid metal dealloying, Nature communications 6(1) (2015) 8887.

[49] D. Montes de Oca Zapiain, J.A. Stewart, R. Dingreville, Accelerating phase-field-based microstructure evolution predictions via surrogate models trained by machine learning methods, npj Computational Materials 7(1) (2021) 3.

[50] F.V. Senhora, E.D. Sanders, G.H. Paulino, Optimally‐Tailored Spinodal Architected Materials for Multiscale Design and Manufacturing, Advanced Materials 34(26) (2022) 2109304.

[51] A. Gayon-Lombardo, L. Mosser, N.P. Brandon, S.J. Cooper, Pores for thought: generative adversarial networks for stochastic reconstruction of 3D multi-phase electrode microstructures with periodic boundaries, npj Computational Materials 6(1) (2020) 1-11.

[52] F. Zhang, Q. Teng, H. Chen, X. He, X. Dong, Slice-to-voxel stochastic reconstructions on porous media with hybrid deep generative model, Comp Mater Sci 186 (2021) 110018.

[53] L. Wang, Y.-C. Chan, F. Ahmed, Z. Liu, P. Zhu, W. Chen, Deep generative modeling for mechanistic-based learning and design of metamaterial systems, Comput Method Appl M 372 (2020) 113377.

Proceedings of the ASME 2023
International Design Engineering Technical Conferences and
Computers and Information in Engineering Conference
IDETC-CIE2023
August 20-23, 2023, Boston, Massachusetts

DETC2023-114619

HIGH-FREQUENCY BAND GAP DESIGN OF POROUS PHONONIC CRYSTALS BY TOPOLOGY OPTIMIZATION

Naoki Murai[1], Yuki Noguchi[1,2], Takayuki Yamada[1,2,*]

[1]Department of Engineering, Graduate School of Engineering, The Universtiy of Tokyo, Tokyo 113-8656, Japan
[2]Department of Stragetec Studies, Institute of Engineering Innovation, Graduate School of Engineering, The University of Tokyo, Tokyo 113-8656, Japan

ABSTRACT

Phononic crystals can control the propagation of elastic waves by their unique dispersion properties such as band gaps. In recent years, the application of phononic crystals for controlling high-frequency elastic waves, which is important for managing thermal conductivity in the nanoscale and for wireless communication, has been attracting attention. This research proposes a design method of phononic crystals which exhibit band gaps at high band order for the applications with high-frequency elastic waves. To consider manufacturability in nanoscale, phononic crystals made of a single material are assumed. A level set-based topology optimization is used for the design of phononic crystals. An objective function is proposed to achieve a band gap within an aimed range of frequency. As a numerical example, 2-dimensional silicon phononic crystal with hexagonal lattice is optimized. Our approach successfully generates a phononic crystal with a band gap at high band order.

Keywords: Phononic crystals, Topology optimization, Eigenvalue problems, Band gaps, High-frequency

1. INTRODUCTION

Phononic crystals are periodic composite structures which exhibit unique dispersion properties. One of the outstanding properties of phononic crystals is band gap. Band gap is a frequency band where the propagation of elastic waves is prohibited. There are many applications that utilize band gap, vibration reduction[1], energy harvetsting[2], and heat management in microscale[3].

The width of band gaps is important for the application of phononic crystals. A lot of previous studies proposed design methods for phononic crystals which exhibit a wide band gap. Band gaps of phononic crystals are mainly caused by two phenomena: scattering of the waves at inclusions arranged with the

periodicity corresponding to the wavelength, and resonance of the waves in the unit cell structure. Therefore, the dispersion properties depend on the shape, topology, size, and material properties of inclusions in the phononic crystals. The structural design of the inclusions is very important to design phononic crystals. However, it is difficult to apply a trial and error approach to design phononic crystals that exhibit desired dispersion properties such as wide band gaps.

Topology optimization methods are powerful tools for designing phononic crystals. The methods optimize the distribution of materials within a predefined region, which yields the highest degree of design freedom in structural optimization methods. The use of topology optimization was first introduced in the design of phononic crystals by Sigmund and Jensen [4]. Their study found that the phononic crystals composed of two materials with the high-contrast in their material properties (density and Young's modulus) could exhibit a wider band gap than the low-contrast cases. Since this pioneering study, many researchers have investigated how to maximize band gaps using topology optimization. Most of them optimized two-phase phononic crystals with high-contrast material properties. However, from the perspective of manufacturability at nanoscale, phononic crystals should be constructed from a single material. The phononic crystals of this type are referred to as porous phononic crystals. The structure of porous phononic crystals is characterized by a matrix of solid material with void regions filled with air. As a result of the high-contrast in material properties between the solid and the air, they exhibit wide band gaps at low frequencies. Structural designs of porous phononic crystals have been studied by several studies[5, 6], but they have treated on band gaps in the low-frequency range with unit cell sizes of similar orders to the wavelength of elastic waves. The size of the phononic crystals is important and should be large from the viewpoint of manufacturability, especially for applications at nanoscale.

*Corresponding author: t.yamada@mech.t.u-tokyo.ac.jp

Copyright © 2023 by ASME

To address this problem, we focus on band gaps at high band order which can control high frequency range of the elastic waves. By optimizing the dispersion properties in high order, the designed phononic crystals are larger in size compared to the wavelength of the elastic waves. However, a few studies [7] focused on band gaps at high band order. In this research, we design unit cell structure of 2D porous phononic crystals with a band gap at high band order by a level set-based topology optimization[8].

In this study, 2D porous phononic crystals which exhibit band gap at high order is designed by topology optimization. First, the problem of propagation of elastic waves in periodic structures is described. The derivation of the eigenvalue problem is explained by introducing the elastodynamic equation. Next, level set-based topology optimization is explained and the optimization problem is formulated based on an objective function that we propose. Then, the numerical implementation of the optimization algorithm is shown. As a numerical example, optimization of a porous phononic crystal which exhibits band gap at high band order including aimed frequency is demonstrated. The validity of the proposed method is verified through the examples.

2. ELASTODYNAMIC PROBLEMS IN PHONONIC CRYSTALS

We consider a 2-dimensional periodic structure in a hexagonal lattice made of an elastic material and void. Figure 1(a) shows a primitive cell with the lattice vector $\boldsymbol{R} = (\boldsymbol{r}_1, \boldsymbol{r}_2)$. a is the lattice constant. D denote entire of unit cell and Ω is occupied by a material. Ω_{void} is air and is excluded in the analysis of the elastic wave propagation. Γ_{per} is the outer boundary of Ω on which the periodic boundary conditions are applied. We consider a hexagonal lattice with sixfold symmetry (Fig. 1(c)) according to a previous research[9], which reports that phononic crystals of hexagonal lattice could exhibit wider band gaps than the square lattice. The set of right-angled triangles divided by the folding lines are denoted by D_i ($i = 1, 2, \cdots 12$). The gray-colored zone in Fig. 1(d) represents the irreducible Brillouin zone.

On the assumption that the elastic waves are time harmonic with the angular frequency ω, the propagation of the waves in the structure Ω is governed by the elastodynamic equation:

$$-\text{div}(\boldsymbol{\sigma}) = \rho(\boldsymbol{r})\omega^2 \boldsymbol{u}, \qquad (1)$$

where $\boldsymbol{\sigma}$ is the stress tensor and $\boldsymbol{u}$ is the displacement vector. ρ is the mass density of the elastic material. $\boldsymbol{r}$ is the position vector. ω is the angular frequency. We assume plane strain conditions in this study. The stress is governed by Hooke's law,

$$\sigma_{ij} = \lambda(\boldsymbol{r})\varepsilon_{kk}\delta_{ij} + 2\mu(\boldsymbol{r})\varepsilon_{ij}, \qquad (2)$$

$$\varepsilon_{ij} = \frac{1}{2}\left(\frac{\partial u_i}{\partial x_j} + \frac{\partial u_j}{\partial x_i}\right). \qquad (3)$$

ε_{ij} is the strain tensor. λ and μ are the Lame's coefficients.

According to the Bloch–Floquet theorem, we can obtain the dispersion properties by solving (1) in a unit cell with periodic boundary conditions. The displacement vector $\boldsymbol{u}$ is rewritten as:

$$\boldsymbol{u}(\boldsymbol{r}, t) = \hat{\boldsymbol{u}}(\boldsymbol{r})\exp(i\boldsymbol{k} \cdot \boldsymbol{r}). \qquad (4)$$

$\boldsymbol{k}$ is the Bloch wave vector. $\hat{\boldsymbol{u}}$ is periodic in the unit cell D. That is, the following relationship holds:

$$\hat{\boldsymbol{u}}(\boldsymbol{r} + \boldsymbol{R}) = \hat{\boldsymbol{u}}(\boldsymbol{r}). \qquad (5)$$

Substituting (4) to (1), we obtain the eigenvalue problem. The eigenvalue problem is solved by the FEM (Finite Element Method) and can be written as:

$$(\mathbf{K}(k) - \omega^2 \mathbf{M}(r))\mathbf{u} = \mathbf{0}. \qquad (6)$$

$\mathbf{K}$ and $\mathbf{M}$ are the stiffness matrix and the mass matrix. $\mathbf{u} = \hat{\boldsymbol{u}}$ is the eigenvector. To examine the dispersion curves which show the relation between wavevector $\boldsymbol{k}$ and eigenvalues ω, the eigenvalue problem is solved with the wave vector $\boldsymbol{k}$ sweeping across the boundaries of the irreducible Brillouin zone ($\Gamma - M - X - \Gamma$). Corresponding to $\boldsymbol{k}$, the discretized wavevector $\boldsymbol{k}_m$ is used when estimating the dispersion relation of the phononic crystals by the FEM.

3. FORMULATION OF AN OPTIMIZATION PROBLEM

3.1 Level set-based topology optimization

We design unit cell structure of phononic crystals with a band gap at high band order based on a level set-based topology optimization method[8]. The object function $J(\omega_\phi)$ expresses the performance of phononic crystals. Here, we denoted the eigenvalue as ω_ϕ to emphasize that ω implicitly depends on ϕ. Configuration of the material in the unit cell D is represented by a level set function, as follows:

$$\begin{cases} -1 \le \phi(\boldsymbol{x}) < 0 & \text{if } \boldsymbol{x} \in \Omega_{\text{void}}, \\ \phi(\boldsymbol{x}) = 0 & \text{if } \boldsymbol{x} \in \partial\Omega_{\text{void}}, \\ 0 < \phi(\boldsymbol{x}) \le 1 & \text{if } \boldsymbol{x} \in \Omega. \end{cases} \qquad (7)$$

In the optimization process, ϕ is updated to minimize J and optimized configuration can be obtained. The optimization problem is formulated as:

$$\min_\phi J(\omega_\phi) \qquad (8)$$

$$\text{subject to the eigenvalue problem (Eq. 6).} \qquad (9)$$

ϕ is updated by solving a reaction-diffusion equation[8] under periodic boundary conditions on Γ_{per},

$$\frac{\partial\phi}{\partial t} = -K(J' - \tau\nabla^2\phi) \text{ in } D. \qquad (10)$$

t is a fictitious time. $K(> 0)$ is a constant, and J' is a design sensitivity of J. τ is a regularization parameter, which we set to a small positive value.

3.2 Objective function

Here, we define the objective function J. Most of the previous studies define objective function as the ratio of the width of band gaps and the middle of the gap. For example, the following objective function has been used:

$$\begin{aligned} F &= \frac{\Delta\omega}{\omega_{\text{mid}}} \\ &= \frac{\min_m(\omega_{n+1}(\boldsymbol{k}_m)) - \max_m(\omega_n(\boldsymbol{k}_m))}{(\min_m(\omega_{n+1}(\boldsymbol{k}_m)) + \max_m(\omega_n(\boldsymbol{k}_m)))/2}. \end{aligned} \qquad (11)$$

Copyright © 2023 by ASME

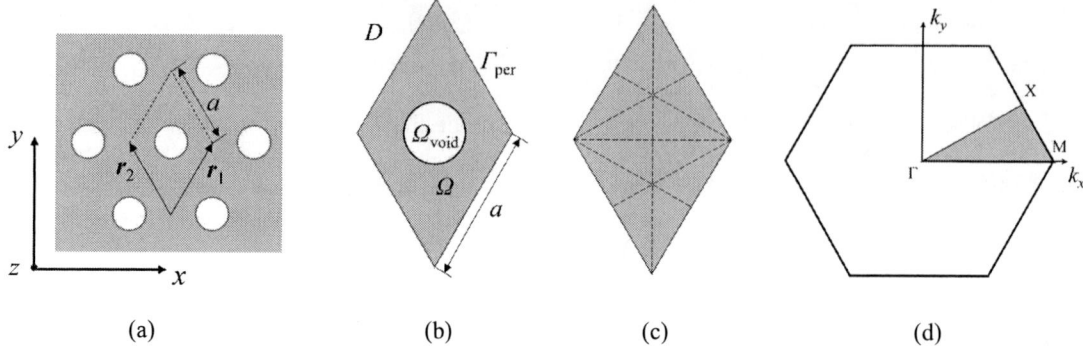

FIGURE 1: HEXAGONAL LATTICE (A) LATTICE VECTOR (B) PRIMITIVE CELL (C) SIXFOLD SYMMETRY (D) IRREDUCIBLE BRILLOUIN ZONE

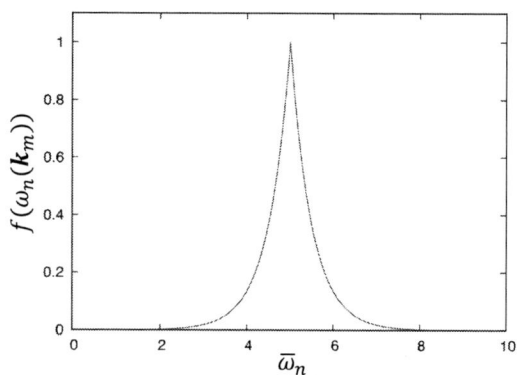

FIGURE 2: PLOT OF $f(\omega_n(k_m))$ (C=10, $\bar{\omega}_{ref}$ = 5.0).

$\omega_n(k_m)$ is the eigenfrequency of n-th band. By minimizing F using topology optimization methods, a large band gap in desired order between n-th and $(n + 1)$-th bands is obtained. However, considering highorder bands, this definition encounters difficulties in selectiong n and it does not allow for the selection of the frequency range where the band gap is desired to be. In cases where multiple bands intersect within the aimed frequency range, the optimization process is expected to be unstable, as F only evaluates two eigenfrequencies and their associated wavevectors

To overcome these problems, we define an objective function J as:

$$J = \sum_{n,m} f(\omega_n(k_m)), \qquad (12)$$

$$f(\omega_n(k_m)) = \exp(-c|\frac{\bar{\omega}_n(k_m)}{\bar{\omega}_{ref}} - 1|). \qquad (13)$$

In Eq. (13), we introduced a normalized frequency $\bar{\omega} = \omega a/2\pi C_t$ to prevent the effect of the magnitude of ω to the value of the objective function (Eq. (12)). $C_t = \sqrt{\frac{\mu}{\rho}}$ is the speed of the transverse elastic wave in the elastic material. $\bar{\omega}_{ref} > 0$ is a reference normalized frequency where a band gap is desired to be exhibited. c is a positive constant that weights eigenvalues corresponding to the gap with the reference frequency, $|\frac{\bar{\omega}_n}{\bar{\omega}_{ref}} - 1|$. Figure 2 shows the plot of $f(\omega_n(k_m))$ in Eq. (13). If c is set to

a smaller value, the weight of bands far from ω_{ref} becomes more significant. The proposed objective function takes into account all bands theoretically, and is suitable for creating a new band gap in the frequency range where three or more bands exist before the optimization.

3.3 Topological derivative

In this study, the topological derivative of the eigenfrequency ω_n, $D_T\omega_n$ is used as the design sensitivity J' in Eq. (10). Considering the case that a small cicular hole (Ω_ε) of the radius ε is created in Ω, $D_T\omega_n$ is defined as:

$$D_T\omega_n = \lim_{\varepsilon \to 0} \frac{\omega_n(\Omega \setminus \overline{\Omega}_\varepsilon) - \omega_n(\Omega)}{V(\varepsilon)}. \qquad (14)$$

$V(\varepsilon)$ is defined so that the limit of the right hand term exists. Ammari et al. [10] derived the topological derivative for the eigenvalue problem. Based on this work, J' is expressed as follows:

$$J' = \sum_{n,m} \frac{\partial f}{\partial \omega_n(k_m)} D_T\omega_n(k_m), \qquad (15)$$

$$\frac{\partial J}{\partial \omega_n(k_m)} = \begin{cases} \frac{c}{\bar{\omega}_{ref}}\exp(-c(\frac{\bar{\omega}_n(k_m)}{\bar{\omega}_{ref}} - 1)) & \text{if } \omega_n(k_m) > \omega_{ref} \\ -\frac{c}{\bar{\omega}_{ref}}\exp(-c(1 - \frac{\bar{\omega}_n(k_m)}{\bar{\omega}_{ref}})) & \text{if } \omega_n(k_m) > \omega_{ref} \end{cases}$$
$$(16)$$

$$D_T\omega_n(k_m) = A_{ijkl}\varepsilon_{ij}\varepsilon_{kl} - (\omega_n(k_m))^2\rho u_i u_i. \qquad (17)$$

We remark that J is not differentiable if $\omega_n(k_m) = \omega_{ref}$ and $\frac{\partial J}{\partial \omega_n}$ is not defined in the case. However, this does not make an issue in the optimization calculation by setting the parameter c to a small enough value such that the contribution of f in Eq. (13) to J at that point can be ignored. A_{ijkl} in Eq. (17) is defined as:

$$A_{ijkl} = \frac{3(1-v)}{2(1+v)(7-5v)}\left\{\frac{-(1-14v+15v^2)E}{(1-2v)^2}\delta_{ij}\delta_{kl} \\ +5E(\delta_{ik}\delta_{jl} + \delta_{il}\delta_{jk})\right\}, \quad (18)$$

where E and v are Young's modulus and Poisson's ratio of the elastic material occupying Ω, respectively.

Copyright © 2023 by ASME

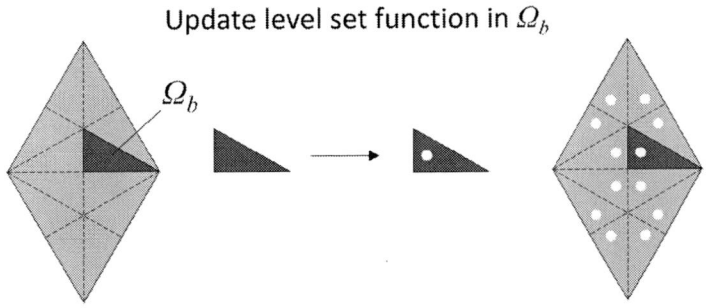

Update level set function in Ω_b

FIGURE 3: UPDATE SCHEME OF THE LEVEL SET FUNCTION.

4. IMPLEMENTATION OF OPTIMIZATION PROBLEM

The optimization algorithm is summarized below.

step **1**: Define initial configuration of unit cell and parameters.

step **2**: Solve the eigenvalue problem (6) by FEM.

step **3**: Compute the objective function (12).

step **4**: Compute the sensitivity (15).

step **5**: Update the level set functions ϕ by solving reaction-diffusion equation (10) by FEM.

step **6**: Remesh unit cell based on the updated level set function ϕ.

step **7**: Back to *step* **2**.

The eigenvalue problem (Eq. 6) and the reaction–diffusion equation (Eq. 10) are solved by the open-source FEM software FreeFem++[11]. To maintain the sixfold symmetry of the unit cell, the reaction–diffusion equation (10) is solved within one of the triangles D_i divided by the folding lines. Then, the updated *phi* is projected to the entire unit cell D according to the symmetry. We choose one of the triangles and denote D_b (the dark gray-colored region in Fig. 3, for example). Then, J'_{average} is defined as the average of J' in D_i. Finally, the design sensitivity in D_b is written as:

$$J'(D_b) = wJ'_{\text{average}} + (1 - w)J'_{\text{pre}}, \tag{19}$$

where, J'_{pre} is the sensitivity in D_b in the previous step of the optimization (set to 0 for the first step), and w is the weight to the sensitivity at the current step and the previous step. This weight is introduced to prevent instability of the optimization computation.

After updating the level set function ϕ, the mesh of computational domain D is refined to obtain mesh over Ω. The refined meshes over D and Ω are used to solve the reaction-diffusion equation and the eigenvalue problems, respectively. This remeshing procedure is performed by the open-source platform Mmg[12].

5. RESULT

We choose silicon as the elastic material. The density is set as $\rho = 2331 [\text{kg/m}^3]$, and Lame's coefficients are $\lambda = 6.390 [\text{GPa}]$ and $\mu = 7.962 [\text{GPa}]$. The lattice constant is $a = 1.0 [\mu\text{m}]$. The initial configuration of the unit cell and the dispersion curves are shown in Figure 4. The eigenfrequency is normalized as $\omega a / 2\pi C_t$ where $C_t = 1848 [\text{m/s}]$ is the speed of the transverse elastic wave in silicon.

Based on these settings, the configuration of the unit cell is optimized with the reference normalized frequency $\bar{\omega}_{\text{ref}} = 5$. The constant c of the objective function and the weight w are set to $c = 10$ and $w = 0.5$. There are four bands that cross with the line $\bar{\omega} = 5$ with the initial configuration, as shown in Fig. 4(a). If the objective function is defined by the width of band gaps like Eq. (11), only two bands will be optimized, and the other two bands may remain in the aimed frequency range, which will disturb to the realization of the band gap. On the other hand, the proposed objective function considers all of them and creates a band gap even if three or more bands exist in the aimed range. Figure 5 shows optimized unit cell and dispersion curves of the optimized structure. The optimized shape has thin beams and disconnected resions, as shown in Fig. 5(b). There is a band gap around the reference frequency $\bar{\omega}_{\text{ref}} = 5$ which the initial unit cell does not exhibit. The gray-colored zone in Fig. 5(d) highlights the band gap that we aimed for in the optimization. Therefore, our approach has successfully designed phononic crystals with a band gap at high order within the aimed frequency range.

6. CONCLUSION

A design method is presented for porous phononic crystals that exhibit a band gap at high band order using level set-based topology optimization. We proposed the objective function to create a new band gap within aimed frequency bands where multiple bands exist and intersect. This approach is effective in the design of phononic crystals with band gaps, particulary within the high-frequency range. As a numerical example, we show an optimized configuration of the unit cell of a porous phononic crystal that achieves a band gap at high band order within the aimed frequency range.

As a future work, study on initial configuration is important. Since the optimization result is sensitive to the initial configuration, it is important to use appropriate settings of the initial configuration. Previous studies[13, 14] reported that the number of the inclusions in the unit cell is related to the order of band where the band gap exists. As another problem, the structure could be divided into several parts in the optimization process. In our setting of analysis, the situation does not make sense since

Copyright © 2023 by ASME

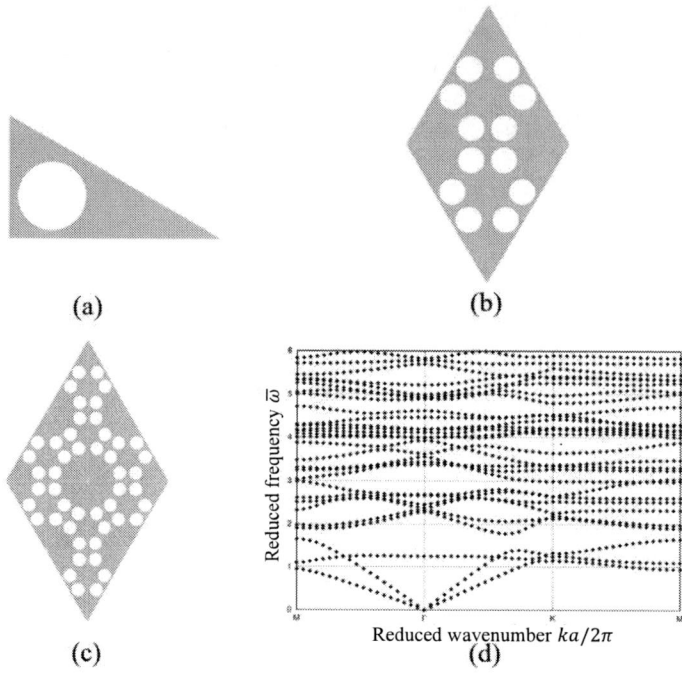

FIGURE 4: INITIAL CONFIGURATION OF UNIT CELL: (A) Ω_b (B) UNIT CELL Ω (C) STRUCTURE CONSISTING OF 4 UNIT CELLS (D) DISPERSION CURVES.

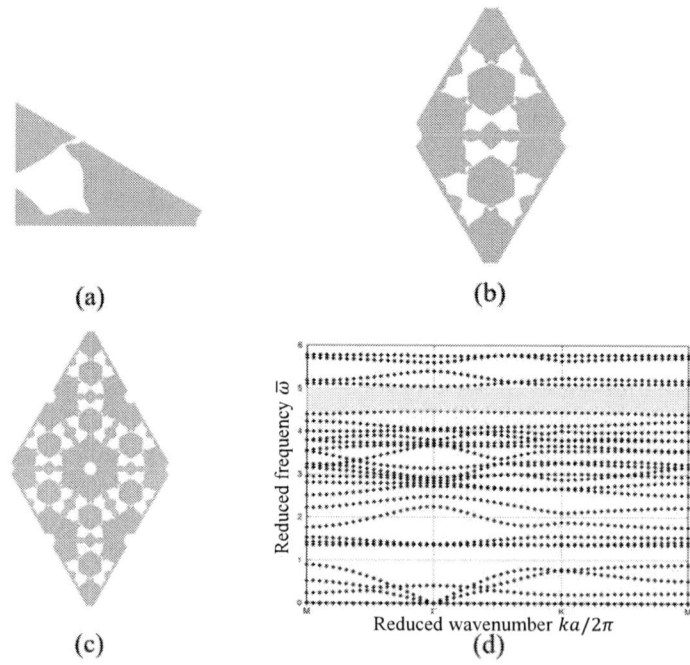

FIGURE 5: OPTIMIZED CONFIGURATION OF UNIT CELL: (A) Ω_b (B) UNIT CELL Ω (C) STRUCTURE CONSISTING OF 4 UNIT CELLS (D) DISPERSION CURVES.

the disconnected regions affect each other not at all. We need to introduce a constraint to connect the structures with the entire part of the phononic crystals.

ACKNOWLEDGMENTS
Place any acknowledgments here.

REFERENCES

[1] Yu, Dianlong, Wen, Jihong, Zhao, Honggang, Liu, Yaozong and Wen, Xisen. "Vibration reduction by using the idea of phononic crystals in a pipe-conveying fluid." *Journal of Sound and vibration* Vol. 318 No. 1-2 (2008): pp. 193–205.

[2] Lv, Hangyuan, Tian, Xiaoyong, Wang, Michael Yu and Li, Dichen. "Vibration energy harvesting using a phononic crystal with point defect states." *Applied Physics Letters* Vol. 102 No. 3 (2013): p. 034103.

[3] Hopkins, Patrick E, Reinke, Charles M, Su, Mehmet F, Olsson III, Roy H, Shaner, Eric A, Leseman, Zayd C, Serrano, Justin R, Phinney, Leslie M and El-Kady, Ihab. "Reduction in the thermal conductivity of single crystalline silicon by phononic crystal patterning." *Nano letters* Vol. 11 No. 1 (2011): pp. 107–112.

[4] Sigmund, Ole and Søndergaard Jensen, Jakob. "Systematic design of phononic band–gap materials and structures by topology optimization." *Philosophical Transactions of the Royal Society of London. Series A: Mathematical, Physical and Engineering Sciences* Vol. 361 No. 1806 (2003): pp. 1001–1019.

[5] Dong, Hao-Wen, Su, Xiao-Xing and Wang, Yue-Sheng. "Multi-objective optimization of two-dimensional porous phononic crystals." *Journal of Physics D: Applied Physics* Vol. 47 No. 15 (2014): p. 155301.

[6] Hedayatrasa, Saeid, Abhary, Kazem, Uddin, Mohammad and Ng, Ching-Tai. "Optimum design of phononic crystal perforated plate structures for widest bandgap of fundamental guided wave modes and maximized in-plane stiffness."
Journal of the Mechanics and Physics of Solids Vol. 89 (2016): pp. 31–58.

[7] Li, Yang Fan, Meng, Fei, Li, Shuo, Jia, Baohua, Zhou, Shiwei and Huang, Xiaodong. "Designing broad phononic band gaps for in-plane modes." *Physics Letters A* Vol. 382 No. 10 (2018): pp. 679–684. URL https://www.sciencedirect.com/science/article/pii/S0375960118300021.

[8] Yamada, Takayuki, Izui, Kazuhiro, Nishiwaki, Shinji and Takezawa, Akihiro. "A topology optimization method based on the level set method incorporating a fictitious interface energy." *Computer Methods in Applied Mechanics and Engineering* Vol. 199 No. 45-48 (2010): pp. 2876–2891.

[9] Zhang, Zhaoxuan, Li, Yang Fan, Meng, Fei and Huang, Xiaodong. "Topological design of phononic band gap crystals with sixfold symmetric hexagonal lattice." *Computational Materials Science* Vol. 139 (2017): pp. 97–105.

[10] Ammari, Habib, Calmon, Pierre and Iakovleva, Ekaterina. "Direct elastic imaging of a small inclusion." *SIAM Journal on Imaging Sciences* Vol. 1 No. 2 (2008): pp. 169–187.

[11] Hecht, F. "New development in FreeFem++." *J. Numer. Math.* Vol. 20 No. 3-4 (2012): pp. 251–265. URL https://freefem.org/.

[12] Dapogny, Charles, Dobrzynski, Cécile and Frey, Pascal. "Three-dimensional adaptive domain remeshing, implicit domain meshing, and applications to free and moving boundary problems." *Journal of computational physics* Vol. 262 (2014): pp. 358–378.

[13] Sigmund, Ole and Hougaard, Kristian. "Geometric properties of optimal photonic crystals." *Physical review letters* Vol. 100 No. 15 (2008): p. 153904.

[14] Li, Yang Fan, Meng, Fei, Li, Shuo, Jia, Baohua, Zhou, Shiwei and Huang, Xiaodong. "Designing broad phononic band gaps for in-plane modes." *Physics Letters A* Vol. 382 No. 10 (2018): pp. 679–684.

Copyright © 2023 by ASME

Proceedings of the ASME 2023
International Design Engineering Technical Conferences and
Computers and Information in Engineering Conference
IDETC-CIE2023
August 20-23, 2023, Boston, Massachusetts

DETC2023-114920

SELECTIVE AMPLIFICATION AND SUPPRESSION OF STRAIN IN A MULTI-AXIS FORCE SENSOR USING TOPOLOGY OPTIMIZATION

Myung Kyun Sung	**Soobum Lee**	**Devin E. Burns**	**Jude Thaddeus Persia**
University of Maryland Baltimore County Baltimore, MD	University of Maryland Baltimore County Baltimore, MD	NASA Langley Research Center Langley, VA	University of Maryland Baltimore County Baltimore, MD

ABSTRACT

This paper proposes an innovative conceptual design of a wind tunnel balance axial section using topology optimization. A wind tunnel balance is a sensor that measures six force/moment components from a wind tunnel model. It is also a structural link between the wind tunnel model and supporting hardware. An axial section, one of the six measurement sections in the balance, is difficult to design because it is often required to resolve an axial force which is much smaller than other force components. Topology optimization is used in this paper to obtain a non-intuitive conceptual design of an axial section. To realize the design requirements, a new top-down symmetric design formulation is suggested to amplify the gauge reading under a small axial loading and to suppress the gauge reading under nonaxial loadings. The formulation assumes the use of a conventional full Wheatstone bridge circuit. The projection method is extensively used to consider manufacturing uncertainties. Then, a postprocessing strategy is used to generate a manufacturable geometry. The postprocessing leverages the Multi-Objective Genetic Algorithm function in ANSYS Workbench to ensure the design requirements of the balance structure are met while maintaining a smooth geometry profile. Satisfactory sensing performance is verified from the postprocessed model using commercial FEM software.

Keywords: topology optimization, wind tunnel balance, six-axis sensor, Wheatstone bridge

NOMENCLATURE

$\mathbf{F}_a$	axial force
$\mathbf{F}_n$	normal force
$\mathbf{M}_n$	counter-acting moment
$\mathbf{M}_p$	pitching moment
σ_{si}	stress at sensor i
σ_{max}	maximum von Mises stress

1. INTRODUCTION

Six-component force and moment sensors are widely used in both industrial and research settings. Many of these sensors utilize an elastic deformation to measure force and moment components precisely [1,2]. A wind tunnel balance is a six-axis sensor that measures six force/moment components (F_x, F_y, F_z, M_x, M_y, M_z: axial/normal/side force, roll/yaw/pitching moment, respectively) from a test model in a wind tunnel. An internal balance is mounted in the test model, and provides the structural link between the wind tunnel model and ground [3]. Using a conventional balance design established in the mid-1950s that utilizes multiple Electrical Discharge Machining (EDM) processes, balances require an average of one year to manufacture and cost an average 120,000 U.S. dollars [4]. To reduce the time and cost, additive manufacturing is recently introduced [5,6].

The balance needs to be designed to measure forces and moments accurately when force components are applied simultaneously [7]. The balance is divided into six measurement sections to measure forces and moments separately, and the design of the axial section is the most challenging because the axial force is substantially smaller than other force and moment components.

There have been several studies to reduce the coupling error using shape optimization [7], signal processing [8], and machine learning [9-11]. Additional studies have sought to accurately measure the axial force using shape optimization of the vertical hinge [12-13]. Another study modified the Wheatstone bridge equation for interference error reduction [14].

This material is declared a work of the U.S. Government and is not subject to Copyright protection in the United States.

Copyright © 2023 by The United States Government

Consideration of high force ratio is essential in case the drag force is significantly low. However, previous studies based on the conventional force balance geometry with minor geometry changes [15,16] were only demonstrated for a low axial-to-normal force ratio −1:4 [12, 17], 1:6.25 [14] – unlike the high force ratio (1:13+) attempted in this study. In addition, the conventional geometry requires multiple cutting processes using plunge electrical discharge machining (EDM) due to its complex geometry and it makes the manufacturing process very costly and time consuming with approximately 50% of the manufacturing time is spent on the axial section [4].

This paper develops a new conceptual design of a force balance using robust topology optimization that can detect a very small axial force. In detail, we develop a new design of the axial section for the wind tunnel balance that has adequate stress from a small axial load and negligible stress from a large normal load and pitching moment (more than ten times larger than the axial load). Structural safety is also contemplated in the design formulation by handling the maximum von Mises stress. Two different design studies – symmetric and asymmetric sensor layouts – are conducted and the performances are compared. Additional post-processing is performed for easy manufacturing.

2. DESIGN METHODOLOGY

This section describes the design methodology of a wind tunnel balance (axial section) using topology optimization. The balance structure needs to be designed so that a full Wheatstone bridge circuit can detect changes of resistance from four strain gauges when deformation occurs in the structure. Fig. 1 (a) shows two strain gauge layouts to be studied in this paper (details are explained later in this section). One resistor pair (R_1 and R_2) are placed on the top and the other pair (R_3 and R_4) is attached symmetrically on the bottom of the front face. When the resistors are connected to the full Wheatstone bridge circuit (Fig. 1 (b)), the output to input ratio of the Wheatstone bridge circuit is written as:

$$\frac{V_{out}}{V_{in}} = \frac{1}{4}\left(\frac{\Delta R_1}{R_1} - \frac{\Delta R_2}{R_2} + \frac{\Delta R_3}{R_3} - \frac{\Delta R_4}{R_4}\right) \quad (1)$$

where V_{out} and V_{in} are output and input voltage of the Wheatstone bridge circuit, and i is the gauge index. The resistances can be replaced with the strains, or $\Delta R_i/R_i = K\varepsilon_i$, as:

$$\frac{V_{out}}{V_{in}} = \frac{K}{4}(\varepsilon_1 - \varepsilon_2 + \varepsilon_3 - \varepsilon_4) \quad (2)$$

where K is the gauge factor.

For the design of the axial section, a two dimensional geometry is considered (147×38 mm^2) as shown in Fig. 2 that is extruded into the z direction to form a three dimensional geometry. The figure also indicates the design domain (grey) where topology optimization is conducted. We use a 4-noded squared quadratic element (size = 0.0625mm^2) for finite element analysis. 89376 elements are used for the entire domain, and 39520 elements are used in the design domain. The left end is fixed, and the loadings – the axial, normal loading and pitching moment ($\mathbf{F}_a$, $\mathbf{F}_n+\mathbf{M}_n$, $\mathbf{M}_p$) – are applied at the center of the right end. $\mathbf{M}_n$ is a counter-acting moment accompanied by $\mathbf{F}_n$ to generate a pure normal force at moment center (which is near the sensor location)

In this study a top-down symmetric gauge layout is investigated as shown in Fig. 1. Strain gauges are placed on the y-z plane and y directional (vertical) strains are measured. The design domain has the horizontal symmetric line through which the geometry and the sensor layout are mirrored to the lower half. When an axial force applied, we expect a bending beam between the sensors (e.g., between 1 and 2; and between 3 and 4) that generates different signs of strain but the same amplitude, and the appropriate connection to the bridge circuit can generate a satisfactory gauge reading.

Compared to our previous study [18] where a moderate sensor response requirement was utilized (σ_0 = 15 and 30MPa), this paper considers a more practical sensor response requirement (σ_0 =75MPa).

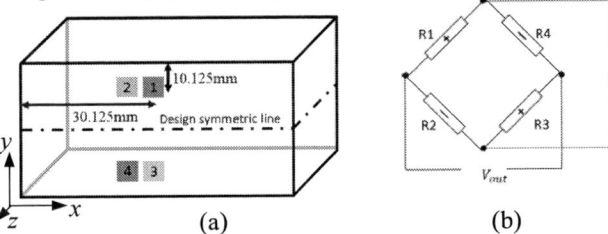

FIGURE 1: SENSOR LOCATIONS IN A DESIGN DOMAIN AND THE CORRESPONDING WHEATSTONE BRIDGE CONFIGURATION

2.1 Design formulation

The topology optimization formulation is explained in this section. Three loading conditions – axial force, normal force, and pitching moment – are considered. The normal force is accompanied with the counter-acting moment ($\mathbf{M}_n$) to apply a pure normal force at the defined moment center of the balance (near the sensor location). The normal force is around 13 times larger than the axial force. Out of plane loadings (side force, rolling, and yawing moments) are not considered in this paper for two reasons. First, the axial section geometry is uniform throughout the thickness direction (z) and robust to the out of plane forces. Second, there are cage sections that measure out of plane forces located forward and/or aft of the axial section [19]. The loadings are defined as: axial force ($\mathbf{F}_a$) = 8.9 N, normal force ($\mathbf{F}_n$) = 116.8 N, counter-acting moment ($\mathbf{M}_n$) = 8.43 Nm, pitching moment ($\mathbf{M}_p$) = 7.43 Nm.

As explained in the introduction, robust topology optimization can consider manufacturing uncertainties by utilizing multiple filtering methods [20,21]. Sensitivity filtering [22] and density filtering [23] are two popular filtering methods widely adapted in topology optimization, but can generate a "grey" area that makes it hard to determine a clear geometry. One well known remedy to eliminate the grey area is the projection method [24-26] that was expanded to consider manufacturing uncertainties, known as the robust approach [27]. In this study,

Copyright © 2023 by The United States Government

the robust approach is investigated to consider manufacturing uncertainty. The robust approach generates three different designs (dilated, intermediate, and eroded) based on three different density threshold points (η_d, η_i, and η_e) that have different thicknesses. The intermediate design is a representative design with over-edging (dilated) and under-edging (eroded) uncertainties. More details can be found in [25].

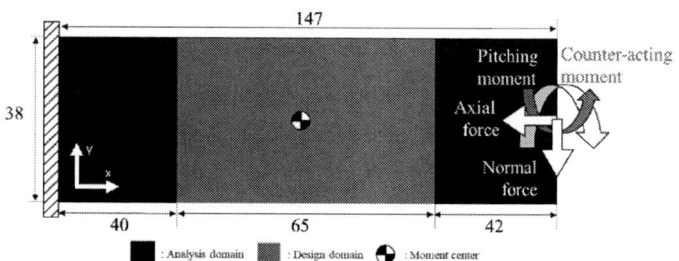

FIGURE 2: DOMAIN SETUP AND LOADING/BOUNDARY CONDITIONS

We consider the following design formulation:

$$max\, V(\bar{x}) \tag{3}$$

$$s.t.: \sigma_{max} = c^{(d,i,e)}\sigma_{pnorm}^{(d,i,e)} \leq \sigma_{safe}\ by\ \Sigma\mathbf{F} \tag{4}$$

$$\sigma_s = \left|\frac{\sigma_{s1}(\bar{x}^{(d,i,e)})+\sigma_{s3}(\bar{x}^{(d,i,e)})}{2}\right| \geq \sigma_0 \quad by\mathbf{F}_a \tag{5}$$

where V is a volume of the design, and $\bar{x}$ is a physical design variable from the projection method [24]. The objective function in Eq. (3) maximizes the volume for both studies, which minimizes material to cut using conventional manufacturing. The constraint (Eq. (4)) is the safety constraint under the combined loadings ($\Sigma\mathbf{F} = \mathbf{F}_a$ (the axial force) + $\mathbf{F}_n$ (normal force) + $\mathbf{M}_n$ (Counter-acting moment) + $\mathbf{M}_p$ (pitching moment)), to have the maximum von Mises stress (σ_{max}) less than its threshold (σ_{safe}). c is a normalization coefficient to convert P-norm to σ_{max}. Superscripts d, i, and e represent the robust density that are dilated, intermediate and eroded, respectively. The normalized P-norm stress is used which is formulated as [28]:

$$\sigma_{pnorm} = \left[\sum_{i=1}^{n} v_i \tilde{\sigma}_i^{Pn}\right]^{\frac{1}{Pn}} \tag{6}$$

where i is element index, n is the number of elements in the design domain, and v_i is elemental volume. The P-norm stress with the infinity norm parameter ($Pn = \infty$) represents the maximum von Mises stress of the structure [29]. In this study, the norm parameter is found by trial and error as 2.5 ($Pn = 2.5$). The tilde on top of σ_i indicates a relaxed stress at element i. The relaxation method is applied to the design variables [28] to generate a smooth stress function and overcome the singularity issue. Handling stress for topology optimization is not a focus of this paper, but more details can be found in [28, 30-32]. Regarding sensor performance, Eq. (5) is considered where subscripts indicate the sensor index. To satisfy the axial force sensing requirement, the application of $\mathbf{F}_a$ must yield a bridge

strain reading higher than σ_0. We do not need to have a constraint to suppress sensor readings by the bending force components (normal force and pitching moment) due to the design symmetry – Sensor 1 and 3 will have the exact same magnitude but different sign under normal load or pitching moments, and this results in cancelled strain (zero output) in the Wheatstone bridge circuit.

Isotropic 300 maraging steel is considered for manufacturing (Young's modulus (E_0) = 189 GPa; Poisson's ratio (γ) = 0.3). The maximum von Mises stress of the structure (σ_{safe} in Eq. (4)) is set as 897 MPa. At the sensors, it is required to generate a sufficient directional stress larger than $\sigma_0 = 75$ MPa under the axial force and a restricted directional stress (less than 10% of axial loading case) under the normal force or pitching moment. Vertical (y) directional stress is used for evaluating the sensor performance. Use of horizontal (x) directional stress is not considered because it is hard to design a flexible structure aligned to the direction of the axial load. Table 1 summarizes the other optimization parameters.

Table 1 Optimization parameters

Description	Parameter	Value
Young's modulus	E_0	189 GPa
Poisson's ratio	γ	0.3
Maximum von Mises stress allowance	σ_{safe}	897 MPa
Directional stress requirement for sensor	σ_0	75 MPa
Restriction coefficient for sensor performance	α	10%
Norm parameter for P-norm	Pn	2.5
relaxation power [28]	r	0.5
Penalization power	P	3.0
Density threshold points for the robust approach [26]	$\eta^{(d,i,e)}$	0.4, 0.5, 0.6

These optimization problems are solved using the in-house code in MATLAB, with Method of Moving Asymptotes (MMA) algorithm [33]. The optimization is assumed to converge when density change ratio becomes less than 1%.

3. RESULTS AND DISCUSSION

3.1 Summary of the design results
This section describes the balance design results. Fig. 3 shows the design results (intermediate for the robust topology optimization). The arrows indicate sensor locations (sensor 1 and 3) and the red circle indicates the maximum von Mises stress location. Table 2 shows the maximum von Mises stress and sensor performance under each loading condition for the dilated, intermediate, and eroded designs.

All the constraints are satisfied except for σ_{max} (eroded) and the sensor performance (dilated) as underlined in Table 2. Due to the design symmetry, Sensor 1 and 3 readings have the same sign and magnitude under the axial load but different sign under

Copyright © 2023 by The United States Government

normal load or pitching moments. The latter case results in signal cancellation in the Wheatstone bridge circuit that always shows zero output.

FIGURE 3: TOPOLOGY DESIGN RESULT

Table 2 Summary of topology study results

Unit: MPa	dilated	intermediate	eroded
σ_{max} by ΣF	551.297	696.106	**1105.833**
σ_{s1} by F_a	**30.684**	75.000	75.000
σ_{s3} by F_a	**30.684**	75.000	75.000
σ_{s1} by $F_n + M_n$	-89.897	-158.282	-22.948
σ_{s3} by $F_n + M_n$	89.897	158.282	22.948
σ_{s1} by M_p	-15.828	-20.797	-4.438
σ_{s3} by M_p	15.828	20.797	4.438

3.1 Analysis on stress and sensing mechanism

This section performs a detailed kinematic analysis and stress analysis for each study. The design has seven bodies (1 ~ 7) and eleven joints (a ~ k) in the upper half domain as shown in Fig. 4 which displays the deformed shapes under the axial load. In this figure, the deformation directions are indicated using arrows. The deformation is exaggerated (x200) for better understanding.

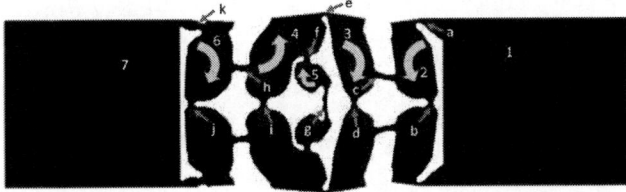

FIGURE 4: DEFORMED SHAPE UNDER AXIAL LOAD WITH BODY AND JOINT NUMBERING (EXAGGERATED BY 200 TIMES)

Bodies 1 and 7 transfer compressed force to Bodies 2, 3, 4, and 6 and their rotating deformation extends the sensor location to satisfy the sensor requirement (75 MPa) as reviewed in Fig. 5(a). In this figure, the high directional (y) stress is found at the sensor under the axial load. The stresses at Sensors 1 and 3 have the identical magnitude and sign. On the other hand, when the normal load or pitching moment is applied, the sensors have different signs of stress (the same magnitude) as shown in Fig. 5 (b), (c). The same effect occurs from Sensors 2 and 4 and these stress values are canceled through the Wheatstone bridge circuit. The von Mises stress contour subject to all combined load is shown in Fig. 5 (d), which confirms the maximum von Mises stress (696 MPa) is less than the threshold ($\sigma_{safe} = 897$ MPa).

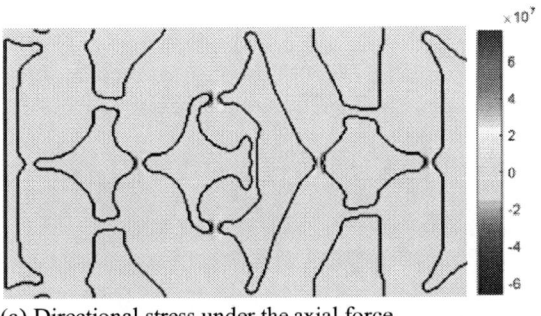

(a) Directional stress under the axial force

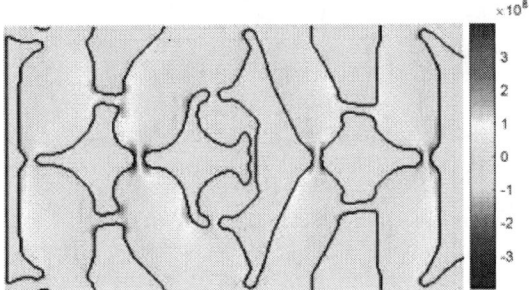

(b) Directional stress under normal load

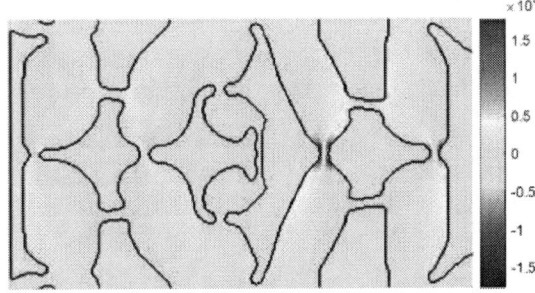

(c) Directional stress under pitching moment

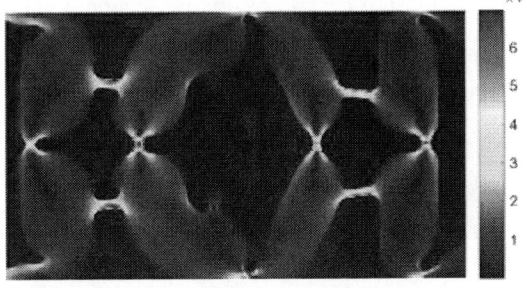

(d) von mises stress under the combined load
FIGURE 5: STRESS CONTOURS

4. POSTPROCESSING AND PERFORMANCE VERIFICATION

This section explains postprocessing for accurate boundary representation, ease of manufacturing, and experimental design verification using the manufactured prototype.

4.1 Postprocessing for topology optimized design

The topology optimization results are converted to geometric CAD models using ANSYS workbench. The flexure joints are represented as rectangular corner-filleted hinges with

Copyright © 2023 by The United States Government

length and width parameters. To make use of the half-bridge and ensure the space for Sensor 2, a minor shape change has been made to the sensor joint (f). Fig. 6 illustrates the recreation of each geometric model, as well as the parameterized dimensions of individual joints – lengths and widths.

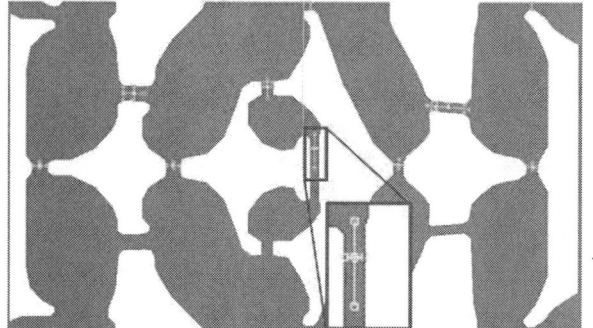

FIGURE 6: PARAMETERIZED DIMENSIONS AND CAD RECREATION

Finite element meshes are generated as shown in Fig. 7 (a). We generated a fine mapped mesh in the joint region (element size 0.18 mm), such that the element count along the length of the joint is retained during parametric changes of the geometry. This is to evaluate the stress value more accurately and minimize the stress variation by free mesh generation.

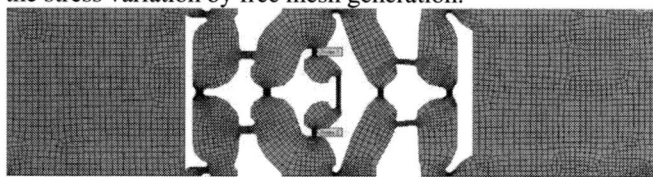

FIGURE 7: MESH GENERATION FOR POSTPROCESSING (28272 ELEMENTS)

This controlled mesh may result in different stress values relative to the topology optimization results in Section 3. The optimization formulation in the postprocessing is more focused on improvement of stress values. ANSYS Workbench is used for the postprocessing optimization of both studies, and is formulated as:

$$\min \sigma_{max} \qquad (7)$$
$$s.t.: |\sigma_{s1}| \geq \sigma_0 \, by \, F_a \qquad (8)$$
$$|\sigma_{s3}| \geq \sigma_0 \, by \, F_a \qquad (9)$$

The objective function in Eq. (7) is chosen to minimize the maximum von Mises stress of the structure, expecting a more reduced maximum von Mises stress. Each constraint in Eqs. (8 ~ 9) acts to maintain the sensor performance relative to the topology optimization.

In the MOGA optimization process, joint widths are permitted to change in a range of ±25% of the original values. The postprocessing optimization results are obtained after 1,000 iterations. The parameter (Table 3) and performance (Table 4) changes are summarized where the parameters with more than 15% change are bolded in Table 3. The corresponding stress

contours are shown in Fig. 8 and the final bridge strain readings are summarized in Table 5.

Table 3 Postprocessed results – width change (in mm)

Joint index	Graphical estimates	Results from optimization
a	0.790	0.778
b	1.185	1.185
c	1.160	1.094
d	1.280	1.096
e	**0.691**	**0.863**
f	1.170	1.210
g	0.889	0.905
h	**1.420**	**1.696**
i	**1.284**	**1.602**
j	1.679	1.700

Table 4 Postprocessed results on stress (in MPa)

Stresses	Values
σ_{max} by $\Sigma\mathbf{F}$	770.07
σ_{s1} by $\mathbf{F}_a$	75.326
σ_{s3} by $\mathbf{F}_a$	75.322
σ_{s1} by $\mathbf{F}_n + \mathbf{M}_n$	-50.343
σ_{s3} by $\mathbf{F}_n + \mathbf{M}_n$	50.378
σ_{s1} by $\mathbf{M}_p$	-86.721
σ_{s3} by $\mathbf{M}_p$	86.699

The most notable changes are increased widths of the joints directly around the sensor location from both studies. The joints are optimized to reduce the maximum von Mises stress – we observe more than a 15% increase in the widths of joints e, h, and i to help reduce stress levels. However, the thicknesses are balanced such that the sensor readings remain satisfied. Joint d is the only one to decrease in width to compensate for required flexibility. To summarize, the results generate sensor performance close to the original topology results while also producing negligible readings under non-axial loads.

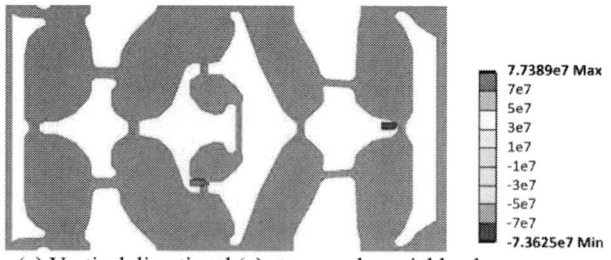

(a) Vertical directional (y) stress under axial load

Copyright © 2023 by The United States Government

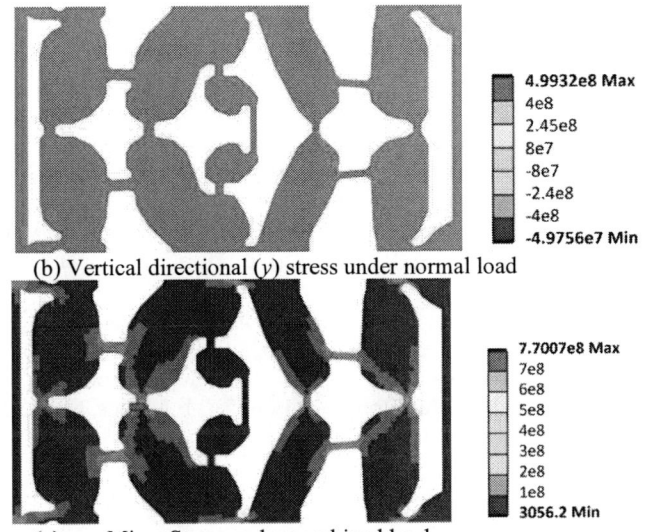

(b) Vertical directional (y) stress under normal load

4.9932e8 Max
4e8
2.45e8
8e7
-8e7
-2.4e8
-4e8
-4.9756e7 Min

7.7007e8 Max
7e8
6e8
5e8
4e8
3e8
2e8
1e8
3056.2 Min

(c) von Mises Stress under combined load

FIGURE 8: STRESS CONTOURS AFTER JOINT PARAMETERS OPTIMIZATION (IN PA)

Table 5 Bridge strain reading by FEM (Unit: μm/m)

Loading	Sensor index				Full bridge strain reading
	1	2	3	4	
Axial, F_a	399	-369	399	-369	384
Normal, $F_n + M_n$	-460	474	459	-474	0
Pitching, M_p	-267	278	267	-278	0

5. CONCLUSION

In this paper, we suggest new internal wind tunnel balance designs based on two different design studies – symmetric and asymmetric design. The optimization problem was formulated to obtain substantial strain at the gauge locations with a small axial load applied and restricted strain with large normal load and pitching moment applied. This problem is different from a compliant mechanism design that is focused on displacement performance. In addition, minimization of von Mises stress under all combined load was considered for structural safety. The use of robust topology optimization generated conceptual designs. Topologically optimized designs were postprocessed to obtain a clear geometry profile for manufacturing, and we could obtain the bridge strain readings that effectively amplify the axial response and cancel the nonaxial response.

ACKNOWLEDGEMENTS

The authors gratefully acknowledge NASA Langley Research Center and the National Institute of Aerospace for their support from Prime Contract No. NNL13AA08B, Subcontract No. T19-601049-UMBC.

REFERENCES

[1] S.A. Liu, H.L. Tzo, A novel six-component force sensor of good measurement isotropy and sensitivities, Sensors Actuators A Phys. 100 (2002) 223–230. https://doi.org/10.1016/S0924-4247(02)00135-8.

[2] J.O. Templeman, B.B. Sheil, T. Sun, Multi-axis force sensors: A state-of-the-art review, Sensors Actuators A Phys. 304 (2020) 111772. https://doi.org/https://doi.org/10.1016/j.sna.2019.111772.

[3] K. Hufnagel, G. Schewe, Force and Moment Measurement, in: C. Tropea, A.L. Yarin, J.F. Foss (Eds.), Springer Handb. Exp. Fluid Mech., Springer Berlin Heidelberg, Berlin, Heidelberg, 2007: pp. 563–616. https://doi.org/10.1007/978-3-540-30299-5_8.

[4] D. Cahill, F. Steinle, S. Richardson, Evaluation of wind tunnel internal force balances from seven vendors, AIAA Pap. (2004) 10438–10447.

[5] D.E. Burns, A. Kudzal, B. McWilliams, J. Manjarres, D. Hedges, P.A. Parker, Investigating Additively Manufactured 17-4 PH for Structural Applications, J. Mater. Eng. Perform. 28 (2019) 4943–4951. https://doi.org/10.1007/s11665-019-04206-9.

[6] D.E. Burns, P.A. Parker, Additively Manufactured Wind-Tunnel Balance, J. Aircr. (2020) 1–6. https://doi.org/10.2514/1.C035889.

[7] M.K. Kang, S. Lee, J.H. Kim, Shape optimization of a mechanically decoupled six-axis force/torque sensor, Sensors Actuators, A Phys. (2014). https://doi.org/10.1016/j.sna.2014.01.001.

[8] G.A. Kebede, A.R. Ahmad, S.-C. Lee, C.-Y. Lin, Decoupled Six-Axis Force–Moment Sensor with a Novel Strain Gauge Arrangement and Error Reduction Techniques, Sensors. 19 (2019). https://doi.org/10.3390/s19133012.

[9] Q. Liang, J. Long, G. Coppola, D. Zhang, W. Sun, Novel decoupling algorithm based on parallel voltage extreme learning machine (PV-ELM) for six-axis F/M sensors, Robot. Comput. Integr. Manuf. (2019). https://doi.org/10.1016/j.rcim.2018.12.014.

[10] H. Lee, H.J. Lim, A. Chattopadhyay, Data-driven system health monitoring technique using autoencoder for the safety management of commercial aircraft, Neural Comput. Appl. (2020). https://doi.org/10.1007/s00521-020-05186-x.

[11] H. Lee, G. Li, A. Rai, A. Chattopadhyay, Real-time anomaly detection framework using a support vector regression for the safety monitoring of commercial aircraft, Adv. Eng. Informatics. 44 (2020) 101071. https://doi.org/https://doi.org/10.1016/j.aei.2020.101071.

[12] L. Swapna, B. Katta, B.S. Suresh, Shape Optimization of a Drag Force Element of a Force Transducer for Wind Tunnel Measurements, 5 (2015) 33–38. https://doi.org/10.5923/c.jmea.201502.07.

[13] S. Skube, H. Bennett, Wind tunnel balance and method of use, US 10,267,708 B2, 2019.

[14] S. Zhang, X. Li, H. Ma, H. Wen, Mechanical analysis of normal force interference on axial force measurement for internal sting balance, Aerosp. Sci. Technol. 58 (2016) 351–357. https://doi.org/10.1016/j.ast.2016.08.028.

Copyright © 2023 by The United States Government

[15] R.M. Hansen, Mechanical Design and fabrication of strain-gage balance, 1956.

[16] M. Dubois, Six-component strain-gage balances for large wind tunnels, Exp. Mech. 21 (1981) 401–407. https://doi.org/10.1007/BF02327141.

[17] M. Samardžic, D. Marinkovski, Z. Anastasijević, D. Curčić, Z. Rajić, An elastic element of the forced oscillation apparatus for dynamic wind tunnel measurements, Aerosp. Sci. Technol. 50 (2016) 272–280. https://doi.org/10.1016/j.ast.2016.01.011.

[18] M.K. Sung, S. Lee, D.E. Burns, Robust topology optimization of a flexural structure considering multi-stress performance for force sensing and structural safety, Struct. Multidiscip. Optim. 65 (2021) 6. https://doi.org/10.1007/s00158-021-03088-7.

[19] D.E. Burns, P.A. Parker, B.D. Phillips, T.L. Webb III, D. Landman, Wind Tunnel Balance Design: A NASA Langley Perspective, 2020. https://ntrs.nasa.gov/citations/20200002944.

[20] O. Sigmund, J. Petersson, Numerical instabilities in topology optimization: A survey on procedures dealing with checkerboards, mesh-dependencies and local minima, Struct. Optim. 16 (1998) 68–75. https://doi.org/10.1007/BF01214002.

[21] G.-W. Jang, J.H. Jeong, Y.Y. Kim, D. Sheen, C. Park, M.-N. Kim, Checkerboard-free topology optimization using non-conforming finite elements, Int. J. Numer. Methods Eng. 57 (2003) 1717–1735. https://doi.org/10.1002/nme.738.

[22] O. Sigmund, On the Design of Compliant Mechanisms Using Topology Optimization, Mech. Struct. Mach. 25 (1997) 493–524. https://doi.org/10.1080/08905459708945415.

[23] T.E. Bruns, D.A. Tortorelli, Topology optimization of non-linear elastic structures and compliant mechanisms, Comput. Methods Appl. Mech. Eng. 190 (2001) 3443–3459. https://doi.org/10.1016/S0045-7825(00)00278-4.

[24] J.K. Guest, J.H. Prévost, T. Belytschko, Achieving minimum length scale in topology optimization using nodal design variables and projection functions, Int. J. Numer. Methods Eng. 61 (2004) 238–254. https://doi.org/10.1002/nme.1064.

[25] O. Sigmund, Morphology-based black and white filters for topology optimization, Struct. Multidiscip. Optim. 33 (2007) 401–424. https://doi.org/10.1007/s00158-006-0087-x.

[26] F. Wang, B.S. Lazarov, O. Sigmund, On projection methods, convergence and robust formulations in topology optimization, Struct. Multidiscip. Optim. 43 (2011) 767–784. https://doi.org/10.1007/s00158-010-0602-y.

[27] O. Sigmund, Manufacturing tolerant topology optimization, Acta Mech. Sin. 25 (2009) 227–239. https://doi.org/10.1007/s10409-009-0240-z.

[28] C. Le, J. Norato, T. Bruns, C. Ha, D. Tortorelli, Stress-based topology optimization for continua, Struct. Multidiscip. Optim. 41 (2010) 605–620. https://doi.org/10.1007/s00158-009-0440-y.

[29] D.M. De Leon, J. Alexandersen, J.S. Jun, O. Sigmund, Stress-constrained topology optimization for compliant mechanism design, Struct. Multidiscip. Optim. 52 (2015) 929–943. https://doi.org/10.1007/s00158-015-1279-z.

[30] M. Bruggi, On an alternative approach to stress constraints relaxation in topology optimization, Struct. Multidiscip. Optim. 36 (2008) 125–141. https://doi.org/10.1007/s00158-007-0203-6.

[31] M.P. Bendsøe, Optimal shape design as a material distribution problem, Struct. Optim. 1 (1989) 193–202. https://doi.org/10.1007/BF01650949.

[32] M.K. Sung, S.B. Lee, D.E. Burns, Design of a Wind Tunnel Balance using Topology Optimization Considering Multifunctional Stress Performance, in: AIAA Scitech 2021 Forum, American Institute of Aeronautics and Astronautics, 2021. https://doi.org/doi:10.2514/6.2021-1687.

[33] K. Svanberg, The method of moving asymptotes - a new method for structural optimization, Int. J. Numer. Methods Eng. 24 (1987) 359–373.

Copyright © 2023 by The United States Government

Proceedings of the ASME 2023
International Design Engineering Technical Conferences and
Computers and Information in Engineering Conference
IDETC-CIE2023
August 20-23, 2023, Boston, Massachusetts

DETC2023-117200

DATA-DRIVEN DESIGN OF HIGH ELECTRON MOBILITY TRANSISTOR DEVICES USING PHYSICS-INFORMED GAUSSIAN PROCESS MODELING

Anabel Renteria[1], Yanwen Xu[1], Bayan Hamdan[1], Zhou Li[2], Sergio Cordero[2],
Debbie Senesky[2], and Pingfeng Wang[1,*]

[1] Industrial and Enterprise Systems Engineering, University of Illinois at Urbana-Champaign
Urbana, IL 61801, United States
[2] Aeronautics and Astronautics, Stanford University, Stanford, CA 94305, United States.

ABSTRACT

High electron-mobility transistors (HEMTs) have emerged as an attractive alternative for high-efficiency power systems, due to their good material properties to perform at high voltages, temperatures, and frequencies. For that reason, design optimization of HEMTs becomes imperative to ensure the quality and capability of the device when in service. There have been models derived from experimentation to guide the design of HEMTs. Nonetheless, due to its expensive manufacturing process, the relationship of the temperature channel with respect to the design parameters has not been investigated thoroughly. This paper presents a multiphysics finite element (FE) simulation to predict the HEMT's device maximum channel temperature when varying different design parameters. Furthermore, Gaussian Process (GP) based surrogate model was developed using the simulation results as the training database with adaptive sampling techniques for the optimization process. The proposed high-fidelity surrogate model effectively predicts the channel temperature of the HEMT device and enables an optimum search over the design space.

Keywords: HEMT device, design optimization, multiphysics FE simulation, surrogate model.

1. INTRODUCTION

High-efficiency power devices are a growing market for several industrial applications including voltage converters, electric vehicles, cellphones chargers, communication services, among others [1]. Transistors are divided into three types, bipolar junction transistors (BJT), field-effect transistors (FET), and insulated-gate bipolar transistors (IGBT) [2–5]. High electron-mobility transistors (HEMTs) are classified as FET and they are fabricated with two semiconductors materials with different band gaps that provides better capabilities to operate at higher frequencies, temperature, and pressure conditions than ordinary transistors [6]. Aluminum gallium nitrate (Al_xGaN_{1-x}) / gallium nitrate (GaN) HEMTs have received a lot of attention due to their capability to leverage a two-dimensional electron gas (2DEG), which enhances the performance of these transistors. The 2DEG is produced as a result of spontaneous piezoelectric polarization of the semiconductor materials, which is influenced by externals mechanical strains and temperature conditions thus affecting the performance of the transistor [7]. In addition, the DC drain current characteristics of HETMs change depending on the temperature effects because it alters the energy band offset [8].

Self-heating of HEMTs is a common observed issue that is produced due to Joule heating, which is dependent on power conditions [9]. Additionally, temperature accelerated kinetics becomes a critical factor that affects the overall reliability of the

---

*Corresponding Author. Email: pingfeng@illinois.edu

Copyright © 2023 by ASME

transistor as it leads undesired hotspots and to potential failure modes such as thermal degradation. High levels of temperature are a sign of degradation on the HEMT device that is mainly affected by high voltage conditions produced by gate leakage [10].

It is also well known that the performance of AlGaN/GaN transistors such as parameters of HEMTs and their output performance. For example, Russo and Di Carlo studied the effect of the source to gate distance with respect to the current-voltage characteristics at the drain contact. Results showed that downscaling the source to gate distance improves the device performance by enhancing the output current at the drain [11]. Bordoloi et al. studied the reliability effect on different geometrical shapes for the gate on HEMTs devices. It was concluded that different geometrical shapes of gates affected the threshold voltage due to variations on capacity, however, the fluctuations were considered as minimally significant [12]. Even though previous studies have shown some interesting results into how the design parameters affect the overall performance of HEMTs, there are not studies that provide a detailed understanding of the relationship between design of AlGaN/GaN HEMTs and their effect on the channel temperature.

Gaussian Process (GP) surrogate-based models have demonstrated to be flexible to capture complex nonlinear responses and have shown great performance due to their capability of estimating the model uncertainty [13]. In addition, GP models incorporate multifidelity information to improve the accuracy of the prediction [14]. Due to all the advantages of GP compared with other surrogate models, GP has increased its role for diverse engineering applications. However, some of the challenges to construct the GP model for HEMTs device optimization could be obtaining enough high-fidelity data. In addition, the accuracy estimation of the GP surrogate model highly depends on the quantity of training sample points, which require large amounts of data to provide an accurate prediction. For that reason, surrogate models could result in an expensive and time-consuming process, which can be mitigated by using adaptive sampling strategies to improve the model efficiency while minimizing the computational effort. Thus, the incorporation of GP model with adaptive sampling strategies offers a less expensive computational cost associated with data acquisition.

In this study, an AlGaN/GaN HEMT device is modeled by FE simulation to obtain some insight about the performance given different design parameters. The transistor performance is further analyzed using the proposed physics informed GP (MPI-GP) model assisted with adaptive sampling techniques. The performance of the transistor can be improved using the optimized geometrical design that minimizes the channel temperature. The proposed methodology makes it possible to obtain more physically consistent results to maximize the performance of HEMT devices while increasing the reliability and overall performance of the device.

2. METHODOLOGY

In this section, the methodology used for this study is presented. Section 2.1 describes the details of the multiphysics simulation model. Section 2.2 presents the GP-based surrogate modeling approach and Section 2.3 shows the adaptive sampling strategy.

2.1 Multiphysics Simulation Model

This section introduces the FE multiphysics simulation model used to represent the mechanism of a HEMT device and obtain the channel temperature produced at the 2DEG. First, the representation of the total 2DEG charge density ns, is expressed as a function of temperature and strain as [7]

$$ns(\varepsilon, T) = \frac{q_p^\alpha(\varepsilon) - q_p^\gamma(\varepsilon)}{e} - \frac{k^\alpha}{e^2 t^\alpha} \\ * [e\phi_b - \Delta E_C + (E_F - E_C^\gamma)](\varepsilon, T) \quad (1)$$

where $q_p^\alpha(\varepsilon)$ represents the spontaneous piezoelectric polarization of the AlGaN layer as a function of strain, $q_p^\gamma(\varepsilon)$ is the spontaneous piezoelectric polarization of the GaN layer as a function of strain, e is the electron charge constant, k^α is the dielectric permittivity and t^α is the thickness of the AlGaN layer.

The influence of the Schottky contact is represented with the term $e\phi_b$ is expressed as a temperature and strain dependent function as

$$e\phi_b = \phi_{Ni} + \zeta_{Ni} T - X^\alpha + \frac{q_p^\alpha}{e^2 D_{it}} \quad (2)$$

where ϕ_{Ni} and ζ_{Ni} are the gate metal function and temperature coefficient. X^α represents the electron affinity and D_{it} is the constant density of AlGaN interface state.

The band offset ΔE_C represents the differences in the band gap energies of the semiconductor materials, which is influenced by temperature and strain conditions as

$$\Delta E_C = 0.7(E_G^\alpha(\varepsilon, T) - E_G^\gamma(\varepsilon, T)) \quad (3)$$

where E_G^α and E_G^γ represent the band gap given the temperature conditions for AlGaN and GaN layer respectively. Lastly, $(E_F - E_C^\gamma)$ showed in Equation (1) is solved using the Fermi-Dirac semiconductor statistics.

The 2DEG channel mobility usually depends on the quality of the semiconductor material, however, the mobility is also dependent on the temperature conditions, which can be expressed as

$$\mu(T) = \mu_0 \left(\frac{T}{300}\right)^{-\theta_T} \quad (4)$$

where μ_0 is the mobility of the 2DEG at room temperature and θ_T represents the temperature coefficient of mobility and thus the

Copyright © 2023 by ASME

2DEG sheet conductivity can be analytically calculated as a function of strain and temperature. Once the charge in the 2DEG is known, then the threshold voltage (V_{th}) required to turn off the HEMT device can be defined as a function of strain and temperature conditions as

$$V_{th} = -\frac{n_s(\varepsilon, T)et^\alpha}{k^\alpha} \qquad (5)$$

Additionally, the thermal conduction equation used to represent the self-heating of HEMTs is represented as

$$\rho c \frac{\partial T}{\partial t} = \nabla \cdot K\nabla T + j_n \cdot E \qquad (6)$$

where ρ represents the material's mass density, c is the heat capacity, K the thermal conductivity, E represents the electric field, while $j_n \cdot E$ refers to the joule heat generated in the device by the electron current.

2.2 GP-based Surrogate Modeling

GP based surrogate models [15–19] were combined with the FE multiphysics model to estimate the AlGaN/GaN HEMT performance (i.e., the maximum channel temperature measured at the 2DEG) given the inputs of the design geometry parameters. The training set used to construct the GP surrogate model was built using sample points extracted from the multiphysics simulations results. The GP based surrogate model could be expressed as

$$G_K(x) = T(x) + S(x) \qquad (7)$$

where $G_K(x)$ represents the prediction of the performance results, expressed as a function at point x, $T(x)$ represents the trend function, and it is a polynomial term of x that interpolates the input sample points. $S(x)$ is a Gaussian stochastic process with zero mean and variance σ^2. In addition, the polynomial term $T(x)$ is the product of regression basis function f and the regression coefficients β,

$$T(x) = f^T(x)\beta \qquad (8)$$

The co-variance function between two arbitrary input points x_i and x_j can be defined as

$$Cov[S(x_i), S(x_j)] = \sigma^2 R(x_i, x_j) \qquad (9)$$

where $R(x_i, x_j)$ represents the automatic relevance determination – squared exponential (ARD-SE) correlation function matrix. The kernel function defines the covariance structure of the GP model, and it is also responsible for capturing the underlying physics of the problem. For highly non-linear problems, the ARD-SE kernel is an appropriate approach, and it can be expresses as

$$R(x_i, x_j) = \exp\left[-\sum_{p=1}^{N} a_p |x_i^p - x_j^p|^{b_p}\right] \qquad (10)$$

where a_p and b_p are parameters of the GP based surrogate model. With n pairs of observations $O = [X, Y]$, where X is the input from the dataset and Y is the input performance, then the log-likelihood function of the Kriging model can be expressed as

$$\ln L = -\frac{1}{2}\left[\begin{array}{c} n\ln(2\pi) + n\ln\sigma^2 + \ln|R| \\ +\frac{1}{2\sigma^2}(Y - F^T\beta)^T R^{-1}(Y - F^T\beta) \end{array}\right] \qquad (11)$$

where F represents the regression basis function at X and $F_{ij} = f_j(x_i)$. Then β and σ^2 can be solved by maximizing the likelihood function respectively from

$$\beta = [F^T R^{-1} F]^{-1} F^T R^{-1} Y \qquad (12)$$

$$\sigma^2 = \frac{(Y - F^T\beta)^T R^{-1}(Y - F^T\beta)}{n} \qquad (13)$$

where n represents the number of samples. Consequently, with the GP model, the response for any given new point x' can be estimated as

$$G_K(x') = F^T\beta + r^T R^{-1}(Y - F^T\beta) \qquad (14)$$

where r is the correlation vector between x' and the sampled points X. In addition, the mean square error $e(x')$ can be computed by

$$e(x') = \sigma^2\left[1 - r^T R^{-1} r + \frac{(1 - F^T R^{-1} r)^2}{F^T R^{-1} F}\right] \qquad (15)$$

Thus, the prediction of the response value at the new sample point x' using the GP model can be considered as a random variable that follows a normal distribution with mean $G_K(x')$ and variance $e(x')$.

2.3 Adaptive Sampling Strategy

Adaptive sampling strategies are usually used to reduce the computational effort of surrogate models and are used as an alternative to improve the model efficiency. Adaptive sampling techniques help to select new training sample points at each iteration based on the uncertainty quantification of the previous sample points around the design framework. Initially, the GP surrogate model was trained with a small data set from the simulation model and started the optimization process considering different parameter design ranges. Then, candidate sample points that show a large amount of uncertainty compared with the initial design point were identified and tested by the FE model. The outputs obtained from the candidate samples are added to the training data set of the GP model, then the

Copyright © 2023 by ASME

optimization process moves to the next design point. This iterative process was used to update the GP model prediction until the accuracy of the model satisfies the conditions set to minimize the uncertainties, thus leading to selecting the optimum design for the HEMT device.

3. RESULTS AND DISCUSSION

This section introduces a detailed explanation of the results obtained in this optimization approach. First, section 3.1 describes first how the FE multiphysics model was built and the results observed for channel temperature measured at the 2DEG of the HEMT device. Section 3.2 shows the parametric design optimization approach and the GP surface predictive model, with some guidance into the best design to minimize the channel temperature.

3.1 FE Multiphysics Simulation

The cross-section of an AlGaN/GaN HEMT device considered for the FE simulation model is shown in Figure 1. The HEMT device considered for this study consists of a GaN layer and $Al_{0.25}GaN_{0.75}$ layer as the semiconductor materials, a substrate of silicon carbide (SiC), an aluminum oxide (Al_2O_3) passivation layer, standard ohmic contacts (i.e., source and drain) and a Schottky contact (i.e., gate). The device total length was fixed at 10 μm and the working temperature was evaluated as 300K and 500K.

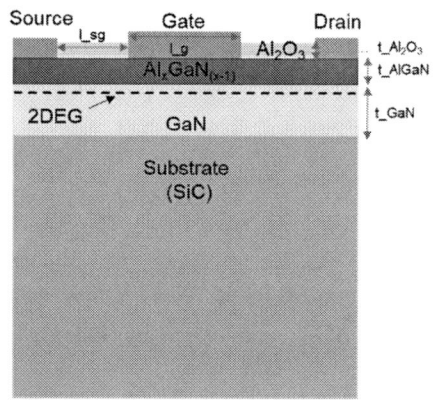

FIGURE 1: CROSS-SECTIONAL DESIGN OF HEMT DEVICE

The thermal behavior of the HEMT device was represented in a 2D model in COMSOL multiphysics software. The materials used were considered as isotropic, which means that it was assumed to have uniform material properties in all directions. The bottom of the structure was defined as a perfect heat sink with isothermal boundary conditions, while the remaining sides of the HEMT structure were set with adiabatic boundary conditions as shown in Figure 2. The material properties used for the thermal behavior of HEMT are summarized in Table 1, which were taken from [20–23]. Additionally, the mechanical properties used to build the simulation model are shown in Table 2. The values used to construct the elasticity matrix and the coupling matrix were taken from [24], which helps to represent

the piezoelectric properties of the semiconductor materials. Then the thermal model is also coupled with the electro model to represent the heat source boundary condition produced in the AlGaN/GaN due to electron mobility. The HEMT device is simulated with voltage source to gate (Vs) set to 0 V, voltage on the drain (Vd) was tested from 5 V and voltage on the gate (Vg) was tested from -1 V.

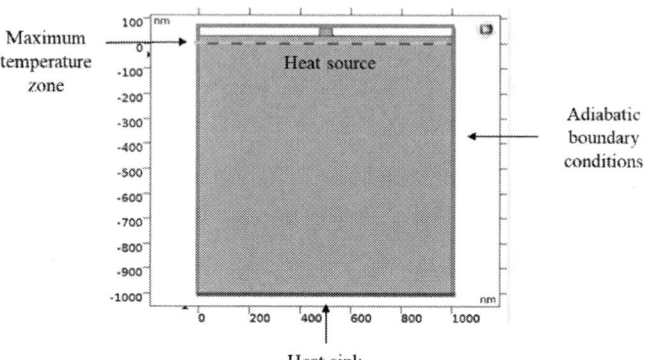

FIGURE 2: BOUNDARY CONDITIONS OF HEMT DEVICE

TABLE 1. MATERIAL PROPERTIES FOR THE FE SIMULATION MODEL

Material	Thermal Conductivity, W/(mK)	Heat Capacity, J/(KgK)	Density, Kg/m³
GaN	$160 \left(\frac{T}{300}\right)^{-1.4}$	415	6095
AlGaN	$27 \left(\frac{T}{300}\right)^{-1.15}$	492.62	5386.25
Al_2O_3	$49 \left(\frac{T}{300}\right)^{-1}$	730	3965
SiC	$387 \left(\frac{T}{293}\right)^{-1.49}$	750	3100

TABLE 2. MECHANICAL MATERIAL PROPERTIES

Material property	GaN	AlN
ϵ_r	9.7	9.375
C_{11} (GPa)	367	396
C_{12} (GPa)	135	137
C_{13} (GPa)	103	108
C_{33} (GPa)	405	373
C_{44} (GPa)	95	116
e_{31} (C/m²)	-0.34	-0.53
e_{33} (C/m²)	0.67	1.5

Figure 3 shows the results on the channel temperature distribution at working temperature of T0 = 300 K and 500 K, where it can be observed that the maximum temperature is reached by the drain side of the gate side of the transistor.

Copyright © 2023 by ASME

Similarly, Figure 4 shows the surface temperature distribution across the HETM device, which shows the hotspot area due to self-heating. In addition, it can be observed in Figure 3 that a higher channel temperature is expected when increasing the voltage applied to the gate (Vg) of the transistor. Since the heat generated in the HEMT device is highly concentrated at a specific zone, then the area for heat exchange with air becomes really small, and thus resulting as a negligible heat dissipation from the top of the device [25]. Similar results for temperature profile along the interface of HEMTs have been found in previous studies [26,27].

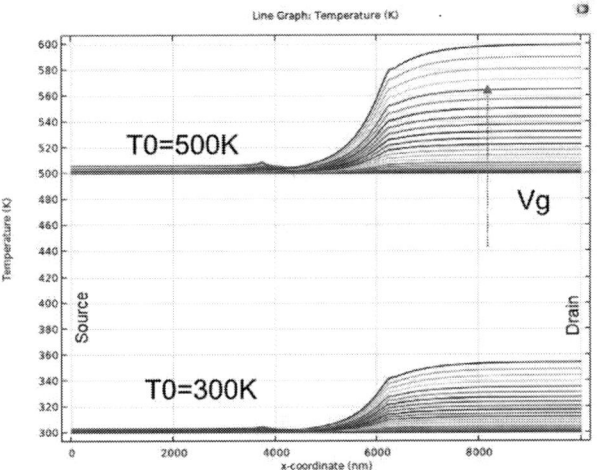

FIGURE 3: CHANNEL TEMPERATURE DISTRIBUTION ALONG THE 2DEG OF THE HEMT DEVICE

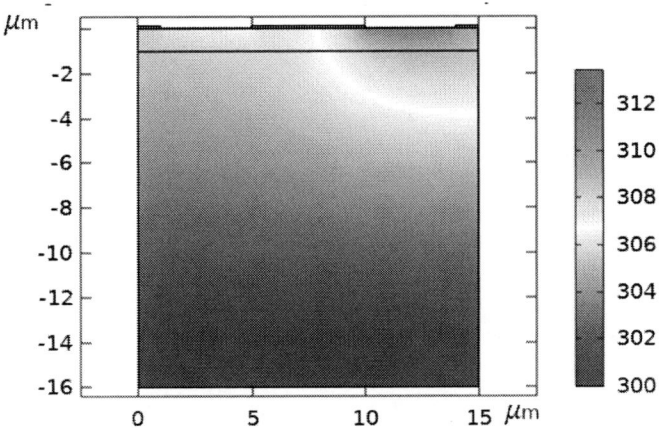

FIGURE 4: SURFACE TEMPERATURE DISTRIBUTION ALONG THE HEMT DEVICE

3.2 Parametric Design Optimization

Five design variables were considered for this study for the HEMT device, including the thickness of GaN layer, thickness of the AlGaN layer, thickness of the passivation layer, source to gate distance and length of the gate. Those design variables were selected as it is usually a controllable parameter in the manufacturing process of GaN transistors and it might have an influence in the system output performance. The controllable design parameters and range studied for optimization of the AlGaN/GaN HEMT device were determined based on manufacturing capabilities and are summarized in Table 3. Our goal is to find the best geometry design for the HEMT device, which minimizes the channel temperature performances at a working temperature condition of $T0 = 300$ K in the design space.

TABLE 3. DESIGN PARAMETERS OF HEMT DEVICE

Parameter	Range	Description
T_{GaN}	1-2 μm	Thickness GaN
T_{AlGaN}	10-30 nm	Thickness AlGaN
T_{Al2O3}	70-120 nm	Thickness Al_2O_3
L_{SG}	2-4 μm	Source to gate distance
L_G	1-4 μm	Length gate

The adaptive sampling technique used in this optimization approach works by iteratively selecting the input points based on the surrogate model's predicted accuracy. By focusing on the input points that are most likely to improve the accuracy of the model, the technique reduces the number of high-fidelity simulations or experiments required to train the surrogate model.

However, there are trade-offs between the accuracy of the model and the reduction in computational effort. If the adaptive sampling technique is too aggressive in selecting input points, it may miss important features of the problem, leading to a less accurate model. On the other hand, if the technique is too conservative, it may require too many high-fidelity simulations or experiments, leading to a more computationally expensive model. Therefore, the choice of the termination criterion and the selection of suitable metrics to assess the accuracy of the model are critical in achieving an appropriate balance between accuracy and computational effort.

The relationship between design parameters and channel temperature obtained from the GP model is shown in Figure 5. It can be observed from the constructed GP model that smaller gate sizes contribute to decrease the channel temperature. Similarly, the distance from the gate to the source contact seems to have a small influence on the maximum surface temperature. These results are in good agreement with literature that showed that the gate size does have an impact on temperature performance of HEMTs [28–30]. On the other hand, it is also observed that thickness of the AlGaN material has a high influence on the maximum channel temperature. This result was expected as the thickness of AlGaN directly impact the 2DEG charge density [31,32].

Optimal design strategies can be applied according to the channel temperature requirements via the proposed methodology. Table 4 shows a comparison between the optimal designs that aim to minimize the channel temperature of the transistor and the designs that lead to higher channel temperature

Copyright © 2023 by ASME

performance at the 2DEG. The design that minimizes the surface temperature considering manufacturing capabilities is T_{GaN} of 1 µm, T_{AlGaN} of 10 nm, T_{Al2O3} of 120 nm, L_{SG} of 3 µm, and L_G of 1 µm with a maximum surface temperature of 304.82 K. Furthermore, it can be observed that the design that leads to the highest channel temperature of 340.54 K is by using T_{GaN} of 2 µm, T_{AlGaN} of 30 nm, T_{Al2O3} of 70 nm, L_{SG} of 4 µm, and L_G of 4 µm. Therefore, using the proposed strategy, the temperature of the hotspot can be reduced by 10.49%.

TABLE 4. HEMT DESIGN OPTIMIZATION RESULTS

Design parameters (T_{GaN}, T_{AlGaN}, T_{Al2O3}, L_{SG}, L_G)	Surface Temperature (K)
(1,10,120,3,1)	304.82
(2,20,120,2,1)	308.59
(1.5,30,70,2,1)	333.71
(2,30,70,4,4)	340.54

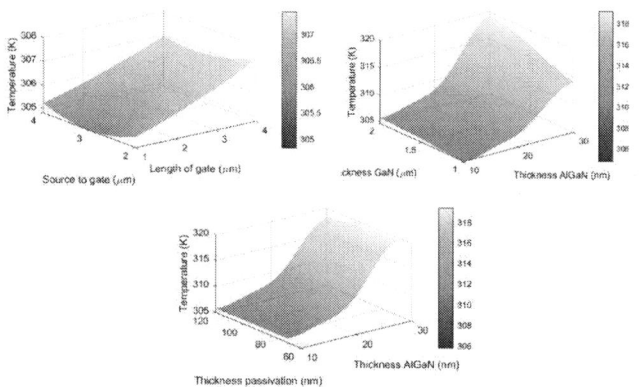

FIGURE 5: MAXIMUM CHANNEL TEMPERATURE PREDICTIVE MODEL ALONG THE 2DEG OF THE HEMT DEVICE

4. CONCLUSION

This paper proposed a FE multiphysics GP model approach to investigate the relationship between different parameter designs of the HEMT device with respect to the channel temperature. The FE simulation model was employed to represent the HEMT device mechanism, where the channel temperature was the output performance metric of interest. Then, the GP model was constructed with the FE simulation with adaptive sampling methods contributed to minimize the computational effort while being able to search through the whole design space. The predicted surface response provided valuable information and guidance for the HEMT design.

ACKNOWLEDGEMENTS

This research is partially supported by the National Science Foundation (NSF) the Engineering Research Center for Power Optimization of Electro-Thermal Systems (POETS) with cooperative agreement EEC-1449548, and the Alfred P. Sloan Foundation through the Energy and Environmental Sensors program with grant # G-2020-12455.

REFERENCES

[1] Mishra, U. K., Parikh, P., and Wu, Y.-F., 2002, "AlGaN/GaN HEMTs-an Overview of Device Operation and Applications," Proceedings of the IEEE, **90**(6), pp. 1022–1031.

[2] Sinha, S. K., and Chaudhury, S., 2013, "Impact of Oxide Thickness on Gate Capacitance—A Comprehensive Analysis on MOSFET, Nanowire FET, and CNTFET Devices," IEEE Transactions on Nanotechnology, **12**(6), pp. 958–964.

[3] Gordon, R., and Bejger, A., 2022, "Effect of Temperature Change on Acoustic Emission Signal in IGBT Transistors of Marine Propulsion System Converters," Energies, **15**(12), p. 4276.

[4] Sadighbayan, D., Hasanzadeh, M., and Ghafar-Zadeh, E., 2020, "Biosensing Based on Field-Effect Transistors (FET): Recent Progress and Challenges," TrAC Trends in Analytical Chemistry, **133**, p. 116067.

[5] Gao, Y., Huang, A. Q., Krishnaswami, S., Richmond, J., and Agarwal, A. K., 2008, "Comparison of Static and Switching Characteristics of 1200 V 4H-SiC BJT and 1200 V Si-IGBT," IEEE Transactions on Industry Applications, **44**(3), pp. 887–893.

[6] Maier, D., Alomari, M., Grandjean, N., Carlin, J.-F., Diforte-Poisson, M.-A., Dua, C., Delage, S., and Kohn, E., 2012, "InAlN/GaN HEMTs for Operation in the 1000 C Regime: A First Experiment," IEEE Electron Device Letters, **33**(7), pp. 985–987.

[7] Yalamarthy, A. S., and Senesky, D. G., 2016, "Strain- and Temperature-Induced Effects in AlGaN/GaN High Electron Mobility Transistors," Semicond. Sci. Technol., **31**(3), p. 035024.

[8] Huque, M. A., Eliza, S. A., Rahman, T., Huq, H. F., and Islam, S. K., 2009, "Temperature Dependent Analytical Model for Current–Voltage Characteristics of AlGaN/GaN Power HEMT," Solid-State Electronics, **53**(3), pp. 341–348.

[9] Durna, Y., Kocer, H., Aras, Y. E., Soydan, M. C., Butun, B., and Ozbay, E., 2022, "Correlation-Based Study of FEA and IR Thermography to Reveal the 2DEG Temperature of a Multi-Fingered High-Power GaN HEMT," Journal of Applied Physics, **131**(8), p. 085107.

[10] Marcon, D., Kauerauf, T., Medjdoub, F., Das, J., Van Hove, M., Srivastava, P., Cheng, K., Leys, M., Mertens, R., Decoutere, S., Meneghesso, G., Zanoni, E., and Borghs, G., 2010, "A Comprehensive Reliability Investigation of the Voltage-, Temperature- and Device Geometry-Dependence of the Gate Degradation on State-of-the-Art GaN-on-Si HEMTs," *2010 International Electron Devices Meeting*, p. 20.3.1-20.3.4.

[11] Russo, S., and Di Carlo, A., 2007, "Influence of the Source–Gate Distance on the AlGaN/GaN HEMT Performance,"

Copyright © 2023 by ASME

IEEE Transactions on Electron Devices, **54**(5), pp. 1071–1075.

[12] Bordoloi, S., Ray, A., and Trivedi, G., 2021, "Introspection Into Reliability Aspects in AlGaN/GaN HEMTs With Gate Geometry Modification," IEEE Access, **9**, pp. 99828–99841.

[13] Satria Palar, P., Rizki Zuhal, L., and Shimoyama, K., 2020, "Gaussian Process Surrogate Model with Composite Kernel Learning for Engineering Design," AIAA Journal, **58**(4), pp. 1864–1880.

[14] Giselle Fernández-Godino, M., Park, C., Kim, N. H., and Haftka, R. T., 2019, "Issues in Deciding Whether to Use Multifidelity Surrogates," AIAA Journal, **57**(5), pp. 2039–2054.

[15] Xu, Y., Renteria, A., and Wang, P., 2022, "Adaptive Surrogate Models with Partially Observed Information," Reliability Engineering & System Safety, **225**, p. 108566.

[16] Wu, H., Zhu, Z., and Du, X., 2020, "System Reliability Analysis With Autocorrelated Kriging Predictions," Journal of Mechanical Design, **142**(10).

[17] Wu, H., Hu, Z., and Du, X., 2020, "Time-Dependent System Reliability Analysis With Second-Order Reliability Method," Journal of Mechanical Design, **143**(3).

[18] Wu, H., and Du, X., 2022, "Envelope Method for Time- and Space-Dependent Reliability Prediction," ASCE-ASME J Risk and Uncert in Engrg Sys Part B Mech Engrg, **8**(4).

[19] Xu, Y., Lalwani, A. V., Arora, K., Zheng, Z., Renteria, A., Senesky, D. G., and Wang, P., 2022, "Hall-Effect Sensor Design with Physics-Informed Gaussian Process Modeling," IEEE Sensors Journal, pp. 1–1.

[20] Stevens, L. E., "Thermo-Piezo-Electro-Mechanical Simulation of AlGaN (Aluminum Gallium Nitride) / GaN (Gallium Nitride) High Electron Mobility Transistors," M.S., Utah State University.

[21] Azarifar, M., and Donmezer, N., 2016, "A Roadmap for Building Thermal Models for AlGaN/GaN HEMTs: Simplifications and Beyond," American Society of Mechanical Engineers Digital Collection.

[22] Huseynov, E. M., Naghiyev, T. G., and Aliyeva, U. S., 2020, "Thermal Parameters Investigation of Neutron-Irradiated Nanocrystalline Silicon Carbide (3C–SiC) Using DTA, TGA and DTG Methods," Physica B: Condensed Matter, **577**, p. 411788.

[23] Nagaral, M., Auradi, V., Parashivamurthy, K. I., Kori, S. A., and Shivananda, B. K., 2018, "Synthesis and Characterization of Al6061-SiC-Graphite Composites Fabricated by Liquid Metallurgy," Materials Today: Proceedings, **5**(1, Part 3), pp. 2836–2843.

[24] Ambacher, O., Majewski, J., Miskys, C., Link, A., Hermann, M., Eickhoff, M., Stutzmann, M., Bernardini, F., Fiorentini, V., Tilak, V., Schaff, B., and Eastman, L. F., 2002, "Pyroelectric Properties of Al(In)GaN/GaN Hetero- and Quantum Well Structures," J. Phys.: Condens. Matter, **14**(13), p. 3399.

[25] Wang, A., Tadjer, M. J., and Calle, F., 2013, "Simulation of Thermal Management in AlGaN/GaN HEMTs with Integrated Diamond Heat Spreaders," Semicond. Sci. Technol., **28**(5), p. 055010.

[26] Helou, A. E., Tadjer, M. J., Hobart, K. D., and Raad, P. E., 2018, "Full 3D Thermal Simulation of GaN HEMT Using Ultra-Fast Self-Adaptive Computations Driven by Experimentally Determined Thermal Maps," *2018 24rd International Workshop on Thermal Investigations of ICs and Systems (THERMINIC)*, pp. 1–3.

[27] Heller, E., Choi, S., Dorsey, D., Vetury, R., and Graham, S., 2013, "Electrical and Structural Dependence of Operating Temperature of AlGaN/GaN HEMTs," Microelectronics Reliability, **53**(6), pp. 872–877.

[28] Gu, Y., Huang, W., Zhang, Y., Sui, J., Wang, Y., Guo, H., Zhou, J., Chen, B., and Zou, X., 2022, "Temperature-Dependent Dynamic Performance of p-GaN Gate HEMT on Si," IEEE Transactions on Electron Devices, **69**(6), pp. 3302–3309.

[29] Chen, X., Boumaiza, S., and Wei, L., 2019, "Self-Heating and Equivalent Channel Temperature in Short Gate Length GaN HEMTs," IEEE Transactions on Electron Devices, **66**(9), pp. 3748–3755.

[30] Saadaoui, S., Fathallah, O., and Maaref, H., 2022, "Effects of Gate Length on GaN HEMT Performance at Room Temperature," Journal of Physics and Chemistry of Solids, **161**, p. 110418.

[31] Khan, Md. A. K., Alim, M. A., and Gaquiere, C., 2021, "2DEG Transport Properties over Temperature for AlGaN/GaN HEMT and AlGaN/InGaN/GaN PHEMT," Microelectronic Engineering, **238**, p. 111508.

[32] Wang, Y.-H., Liang, Y. C., Samudra, G. S., Chang, T.-F., Huang, C.-F., Yuan, L., and Lo, G.-Q., 2013, "Modelling Temperature Dependence on AlGaN/GaN Power HEMT Device Characteristics," Semicond. Sci. Technol., **28**(12), p. 125010.

Copyright © 2023 by ASME

Proceedings of the ASME 2023
International Design Engineering Technical Conferences and
Computers and Information in Engineering Conference
IDETC-CIE2023
August 20-23, 2023, Boston, Massachusetts

DETC2023-117571

COMPARING DERIVATIVES OF NEURAL NETWORKS FOR REGRESSION

Joel C. Najmon[1], Andres Tovar[2],*

[1]Purdue University, West Lafayette, IN
[2]Indiana Univ.-Purdue Univ., Indianapolis, IN

ABSTRACT

In the past decades, neural networks have rapidly grown in popularity as a way to model complex non-linear relationships. The computational efficiently and flexibility of neural networks has made them popular for machine learning-based optimization methods. As such the derivative of a neural network's output is required for gradient-based optimization algorithms. Recently, there have been several works towards improving derivatives of neural network targets, however there is yet to be done a comparative study on the different derivation methods for the derivative of a neural network's targets with respect to its input features. Consequently, this paper's objective is to implement and compare common methods for obtaining or approximating the derivative of neural network targets with respect to their inputs. The methods studied include analytical derivatives, finite differences, complex step approximation, and automatic differentiation. The methods are tested by training deep multilayer perceptrons for regression with several analytical functions. The derivatives of the neural network-derived methods are evaluated against the exact derivative of the test functions. Results show that all of the derivation methods provide the same derivative approximation to near working precision of the computer. Implementation of the study is done using the TensorFlow library in a provided Python code.

Keywords: neural networks, derivative approximation, regression, TensorFlow, automatic differentiation

1. INTRODUCTION

In the past decades, neural networks have rapidly grown in popularity as a way to model complex non-linear relationships. Neural networks are well-suited to this task as they can be trained with only data (i.e., without an explicit function) and can efficiently produce model predictions once trained. A common application of neural networks is for regression problems [1] where the network is trained to predict the response of an expensive black-box function. Due to these benefits, neural networks for regression can be paired with optimization methods to create surrogate-based [2] or machine learning-based [3] optimization methods. A limitation of these methods is that gradient-based optimization algorithms can be difficult to implement as sensitivity coefficients of the black-box functions can be difficult or impractical to obtain.

Recently, there have been several works towards improving derivatives of neural network targets and their accompanying frameworks. Kissel and Diepold [4] proposed the use of least-squares approximated derivatives to train a Sobolev norm neural network for functions where target derivatives are not directly available. Meanwhile, Avrutskiy [5] showed that if several orders of target's derivatives are known, then feedforward neural networks can be trained with increase precision. This is done by incorporating deviations of the target's derivatives from the known derivatives into an extended cost function. Later Kiran and Naik [6] applied complex step derivative approximation to a feedforward neural network in order to accurately obtain derivatives of a regression task. Whereas Ledesma et al. [7] used the chain rule to take an analytical derivative of a multilayer perceptron neural network in order to derive a differential neural network. The differential neural network is based off of the original network and consequently does not need to be trained. Similarly, Rodini [8] derived and proposed a simple recursive algorithm for computing the first- and second-order derivatives of a deep neural network via analytical derivatives.

While a variety of derivative methods have been proposed and studied for neural networks, a comparative study has not been done to compare and contrast the performance of each method. As such the objective of this paper is to implement and compare common methods for obtaining or approximating the derivative of neural network targets with respect to their inputs. The methods studied include analytical derivatives, central finite difference approximation, complex step derivative approximation, and automatic differentiation. The methods are tested by training deep

---

*Corresponding author: tovara@iupui.edu

Copyright © 2023 by ASME

multilayer perceptron neural networks for regression with several analytical functions, which each have an exact analytical derivative. The training of the networks and testing of the methods is done using the TensorFlow library in Python. The code is made available to the reader in the Appendix. The derivatives of the neural network-derived methods are evaluated against the exact derivative of the test functions.

The paper is organized as follows: the four studied derivative methods are presented in Sec. 2. Next the methods are compared for deep multilayer perceptrons, corresponding to analytical test functions, in Sec. 3. The conclusions of the study are given in Sec. 4 and the associated Python code is provided in Appendix A.

2. NEURAL NETWORKS-DERIVED DERIVATIVE METHODS

Neural networks or artificial neural networks are a type of machine learning model that can be trained to recognize patterns or make predictions. Neural networks consist of interconnected neurons that perform weighted operations (i.e., additions and multiplications) passed through non-linear activation functions [9]. The weights and bias variables of each neuron can be optimized to minimize the loss or objective function of the model (e.g., mean squared error between the target and the network's predicted output). Multiple layers of these neurons can be connected to form a deep neural network architecture. The simplest neural network architecture is found in a feedforward neural network where input data is only passed in the forward direction to the next layer. For the neural network-derived derivative study of this work, a fully connected class of feedforward neural networks known as multilayer perceptrons (MLPs) is utilized.

Consider a general MLP that consists of an input layer, an output layer, and N_L hidden layers (Fig. 1). The input and output layers have neurons that are equal to the input and output dimensions of the dataset to be trained, while each hidden layer has N_N neurons. The weight matrices $\mathbf{W}_n$ and bias vectors $\mathbf{B}_n$ of layer n modify the corresponding input $\mathbf{y}_{n-1}$ of the layer before the layer's activation function F_n is applied to calculate the output of the layer $\mathbf{y}_n$. This is given by

$$\mathbf{y}_n = F_n(\mathbf{W}_n \mathbf{y}_{n-1} + \mathbf{B}_n) \tag{1}$$

where n ranges from 1 to $N_L + 2$. The total number of layers is $N_L + 2$ due to the input and output layers. For the input layer ($n = 1$), the input feature set $\mathbf{x}$ is equated to $\mathbf{y}_0$. Similarly the output of the MLP is $\mathbf{y}_{N_L+2}$.

2.1 Analytical Derivative

Since the output of a network is computed through a series of nested activation functions, an analytical derivative of the network's outputs can be computed with respect to its inputs by applying the chain rule. Consider the condensed output of the MLP given by

$$\mathbf{y}_{N_L+2} = F_{N_L+2}\left(F_{N_L+1}\left(\ \cdots\ F_2\left(F_1\left(\mathbf{x}\right)\right)\right)\right). \tag{2}$$

Taking the derivative of (2) with respect to the input $\mathbf{x}$ is done via the chain rule to give

$$\frac{d\mathbf{y}_{N_L+2}}{d\mathbf{x}} = \frac{d\mathbf{y}_{N_L+2}}{d\mathbf{y}_{N_L+1}} \frac{d\mathbf{y}_{N_L+1}}{d\mathbf{y}_{N_L}} \cdots \frac{d\mathbf{y}_2}{d\mathbf{y}_1} \frac{d\mathbf{y}_1}{d\mathbf{x}}. \tag{3}$$

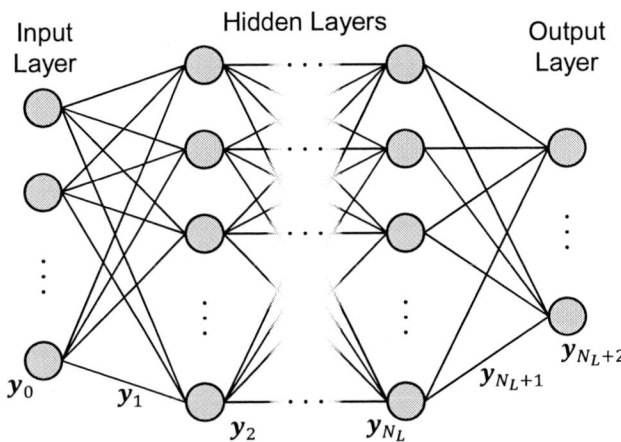

FIGURE 1: ARCHITECTURE OF A MULTILAYER PERCEPTRON FEEDFORWARD NEURAL NETWORK.

Since $\mathbf{x} = \mathbf{y}_0$, (3) can be expressed in Lagrange notation as

$$\frac{d\mathbf{y}_{N_L+2}}{d\mathbf{x}} = \prod_{k=1}^{N_L+2} \frac{d\mathbf{y}_k}{d\mathbf{y}_{k-1}}. \tag{4}$$

When evaluating Eq. (4), additional $\mathbf{W}_n$ factors will accumulate inside of the product operator as a result of applying the chain rule. It is important to note that these additional $\mathbf{W}_n$ may appear to be a simple matter of matrix multiplication, however this is not always the case. Due to the nested activation functions of the neural network, these additional $\mathbf{W}_n$ matrices should be multiplied with element-wise multiplication. See Appendix A for an implementation of this calculation in Python.

2.2 Central Finite Difference Approximation

Due to the inexpensive function evaluations afforded by neural networks, the central finite difference approximation (CFDA) method can be efficiently applied to approximate derivatives. As such the CFDA of the MLP with respect to input j is given by

$$\frac{d\mathbf{y}_{N_L+2}}{dx_j} \approx \frac{\mathbf{y}_{N_L+2}\left(\mathbf{x} + h\delta_j\right) - \mathbf{y}_{N_L+2}\left(\mathbf{x} - h\delta_j\right)}{2h} \tag{5}$$

where h is a small real number (e.g., $h = 10^{-6}$) and $\delta_j = [0, \ldots, 0, 1, 0, \ldots, 0]^\mathsf{T}$ with the number 1 located at the jth row.

2.3 Complex Step Derivative Approximation

If the activation functions employed in the MLP are holomorphic [10], then the complex step derivative approximation (CSDA) method [11] can also be applied. The CSDA of the MLP with respect to input j is given by

$$\frac{d\mathbf{y}_{N_L+2}}{dx_j} \approx \frac{\text{Im}\left[\mathbf{y}_{N_L+2}\left(\mathbf{x} + \eta\delta_j\right)\right]}{\eta} \tag{6}$$

where η is a small real number (e.g., $\eta = 10^{-12}$). Since the CSDA does not involve a difference operation it is not subjected to subtractive cancellation errors. This constitutes a significant advantage over finite difference-based approximations [12].

Copyright © 2023 by ASME

2.4 Automatic Differentiation

Automatic differentiation is a set of techniques to evaluate the derivative of a function defined by a computer program [13]. Automatic differentiation works by overloading standard elementary operators and functions with a derivative rule in addition to their function value. Similar to the previously derived analytical derivative, the chain rule can be repeatedly applied to these operations allowing for derivatives of an arbitrary order to be computed automatically to working precision. The downside to automatic differentiation is that it requires careful implementation into a software package [14], which means that its availability can be limited. TensorFlow natively supports automatic differentiation [15] which makes its implementation straightforward.

3. COMPARISON OF NEURAL NETWORK DERIVATIVES

In order to compare the accuracy of the various derivative methods, several analytical test functions are selected for training of the deep MLPs for regression. Table 1 presents the four analytical test functions $y^{(r)}$ and their exact derivatives $\frac{dy^{(r)}(x)}{dx}$ where r is the function's identity number.

TABLE 1: ANALYTICAL TEST FUNCTIONS USED TO TRAIN THE DEEP MLPS AND THEIR CORRESPONDING EXACT DERIVATIVES.

Function	Exact Derivative
$y^{(1)}(x) = x^2$	$\frac{dy^{(1)}(x)}{dx} = 2x$
$y^{(2)}(x) = \sin(x)$	$\frac{dy^{(2)}(x)}{dx} = \cos(x)$
$y^{(3)}(x) = e^{-x}$	$\frac{dy^{(3)}(x)}{dx} = -e^{-x}$
$y^{(4)}(x) = x^3 - 2x^2 + 0.75x + 0.5$	$\frac{dy^{(4)}(x)}{dx} = 3x^2 - 4x + 0.75$

When it comes to selecting a deep neural network model parameters there is an vast amount options that can be selected in regards to the architecture, activation functions, optimization algorithms, and other parameters [1]. This study tunes the model parameters by individually varying each parameter until the model's mean squared error (MSE) stops decreasing. Every time a parameter is changed, three models are trained to decrease the likelihood of having a poor initialization conditions. The best performing models are used to progress further tuning. Following this process, the model parameters are set to $N_L = 2$ hidden layer, $N_N = 10$ neurons per hidden layer, and using the sigmoid activation function. The output layer uses a linear activation function. The optimization algorithm is set to the Adaptive Moment Estimation (Adam) method [16]. The models are trained with a training dataset of size $N_r = 5000$ for 10000 epochs. A testing dataset of size $N_t = 2143$ is selected so that the testing dataset is 30% and the training dataset 70% of the now total dataset size of $N_T = 7143$.

Once the models had been trained and tested, an interesting phenomenon became immediately evident. Regardless of the function tested or the derivative method considered, all of the four derivative methods produced identical $\frac{dy(x)}{dx}$ values. In other words, all four of the methods are capable of solving for a neural network's derivative to near working precision of the computer. The perturbation values for the CFDA and CSDA methods were set to $h = 10^{-6}$ and $\eta = 10^{-12}$. If these values

were made sufficiently small or large enough, then the derivative values would start to differ for the CFDA and CSDA methods, however this is due to these approximation methods becoming unstable. The analytical derivative through the neural network and automatic differentiation where found to always be the same. Figures 2 to 5 compare the four analytical test functions and its derivatives to the results found using the MLP's prediction and the neural network-derived derivative methods. Only *one* line is plotted for the neural network-derived derivative methods since they all produced the same derivative value for a given input.

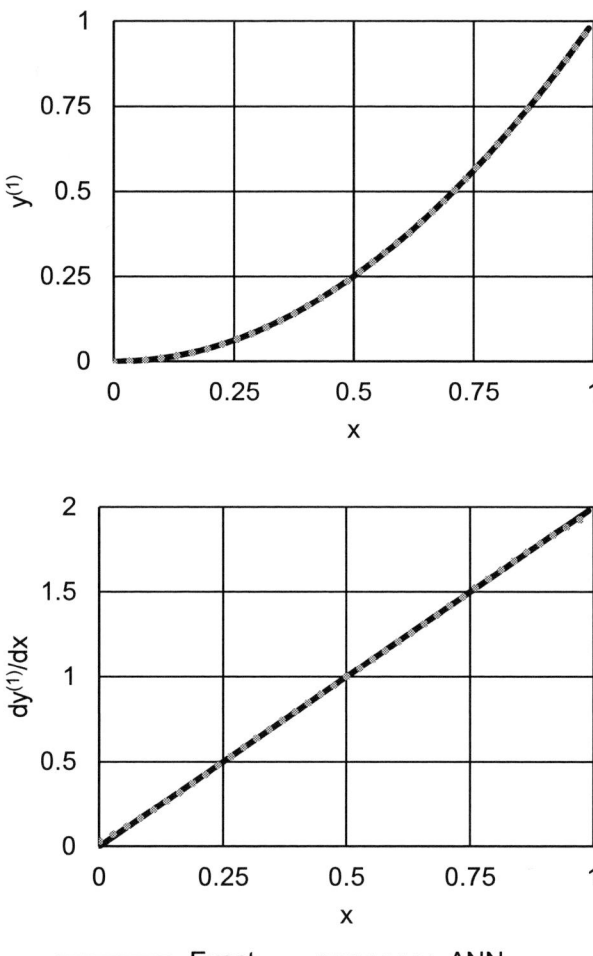

FIGURE 2: THE FIRST TEST FUNCTION AND ITS DERIVATIVE FOR THE EXACT FUNCTION AND ITS NEURAL NETWORK MODEL.

The function and its derivative approximated through the neural network match closely to the exact function and its analytical derivative for all of the functions. The neural network-derived derivative value differ the most near the bounds of the training datasets. Under the training circumstances, this was most clearly seen in function 2 (Fig. 3). Table 2 presents the mean squared error for the testing dataset between the test function and its derivative. From this table we see a trend that the better a neural network is at predicting a function value, the better it will be at predicting the function's derivative.

Copyright © 2023 by ASME

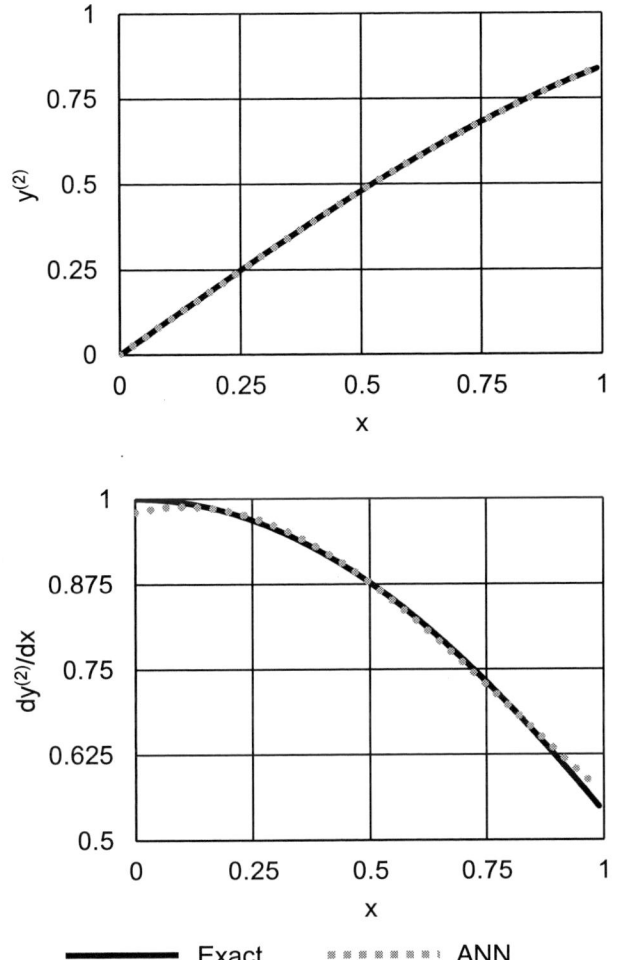

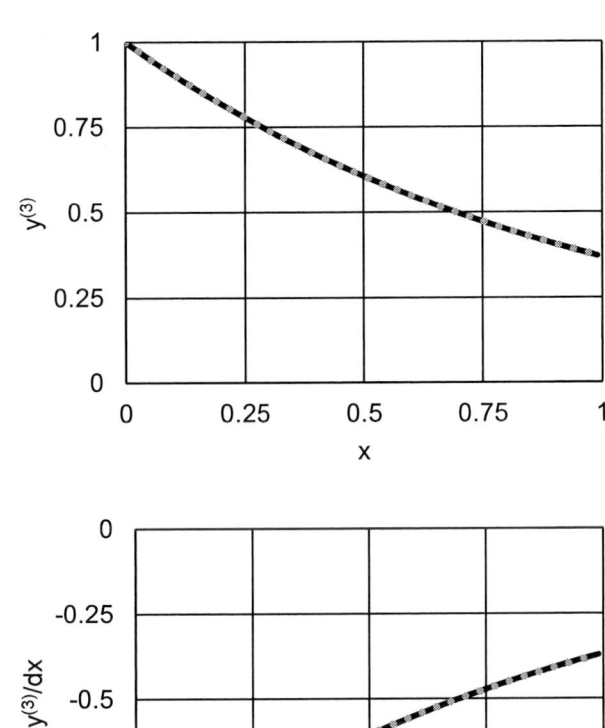

FIGURE 3: THE SECOND TEST FUNCTION AND ITS DERIVATIVE FOR THE EXACT FUNCTION AND ITS NEURAL NETWORK MODEL.

FIGURE 4: THE THIRD TEST FUNCTION AND ITS DERIVATIVE FOR THE EXACT FUNCTION AND ITS NEURAL NETWORK MODEL.

4. CONCLUSION

This paper presents a study on several methods for the calculation of neural network derivatives. The derivation methods studied in this paper include analytical derivatives, central finite differences, complex step, and automatic differentiation. In order to compare the accuracy of the various derivative methods, several analytical test functions are selected for training of the deep MLPs for regression. Interestingly, the four derivation methods produced identical results for the derivatives of the deep MLPs with respect to their input. As such, all of the methods presented in this paper are capable of solving for a neural network's derivative to near working precision of the computer.

All of the neural networks were able to successfully predict their corresponding test function. It was seen that the neural network-derived derivatives deviated more from the true derivative near the bounds of the training dataset, however this behavior was only witnessed in some of the test functions. As such neural network-derived derivatives should always be verified with derivatives derived from the exact function, especially for samples near the boundary of the design domain. In the common

case that the true analytical derivative of the function is unknown, then the CFDA method or the CSDA method (if applicable) were found to be sufficiently suitable for verification. Furthermore, the mean squared error between the true and neural network-derived functions and derivatives showed that the better a deep MLP can predict the function value, then the better its derivative prediction will be.

This study is limited to a one dimensional function with a single variable input. Current work is underway to extend the study to multivariate input and output functions. The computational cost of each neural network-derived derivative method should also be compared for a more comprehensive study.

ACKNOWLEDGEMENTS

This manuscript is based upon work supported by the National Science Foundation under Grant No. 1842164, Candent Technologies, and the Naval Air Warfare Center under Contract No. N68335-22-C-0448. The authors gratefully acknowledge their support. Any opinions, findings, conclusions, or recommendations expressed in this manuscript are those of the authors and do not necessarily reflect the views of our supporters.

Copyright © 2023 by ASME

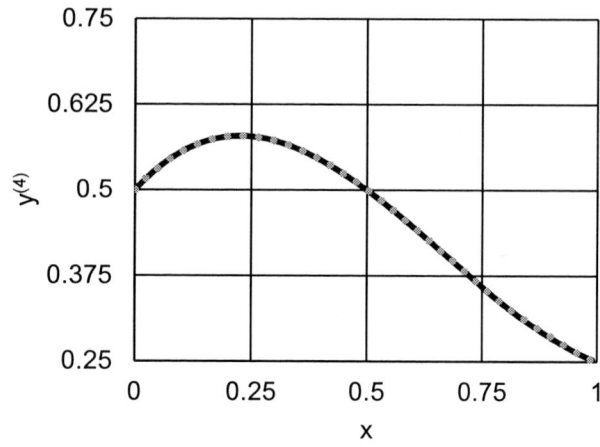

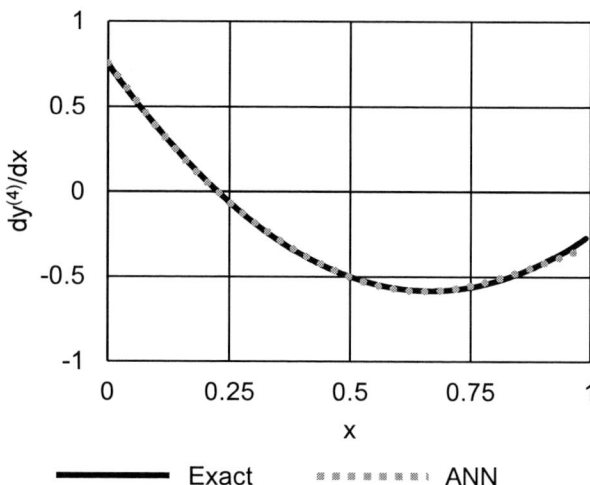

FIGURE 5: THE FOURTH TEST FUNCTION AND ITS DERIVATIVE FOR THE EXACT FUNCTION AND ITS NEURAL NETWORK MODEL.

TABLE 2: MEAN SQUARED ERRORS OF THE FOUR TEST FUNCTIONS AND THEIR DERIVATIVES.

Function	MSE of $y(x)$	MSE of $\frac{dy(x)}{dx}$
1	1.71×10^{-7}	4.41×10^{-5}
2	2.06×10^{-7}	5.80×10^{-5}
3	1.16×10^{-8}	2.28×10^{-6}
4	4.95×10^{-7}	1.30×10^{-4}

REFERENCES

[1] Das, Laya, Sivaram, Abhishek and Venkatasubramanian, Venkat. "Hidden representations in deep neural networks: Part 2. Regression problems." *Computers Chemical Engineering* Vol. 139 (2020): p. 106895.

[2] Forrester, Alexander IJ and Keane, Andy J. "Recent advances in surrogate-based optimization." *Progress in aerospace sciences* Vol. 45 No. 1-3 (2009): pp. 50–79.

[3] Owoyele, Opeoluwa and Pal, Pinaki. "A novel machine learning-based optimization algorithm (ActivO) for accelerating simulation-driven engine design." *Applied Energy* Vol. 285 (2021): p. 116455.

[4] Kissel, M. and Diepold, K. "Sobolev Training with Approximated Derivatives for Black-Box Function Regression with Neural Networks." *Machine Learning and Knowledge Discovery in Databases. European Conference, ECML PKDD 2019, 16-20 Sept. 2019*, Vol. pt.II. Conference proceedings: pp. 399–414. 2020. Springer International Publishing. DOI 10.1007/978-3-030-46147-8_24.

[5] Avrutskiy, V. I. "Enhancing Function Approximation Abilities of Neural Networks by Training Derivatives." *IEEE Transactions on Neural Networks and Learning Systems* Vol. 32 No. 2 (2021): pp. 916–24. DOI 10.1109/TNNLS.2020.2979706.

[6] Kiran, Ravi and Naik, Dayakar L. "Novel sensitivity method for evaluating the first derivative of the feed-forward neural network outputs." *Journal of Big Data* Vol. 8 No. 1 (2021): pp. 1–13.

[7] Ledesma, Sergio, Almanza-Ojeda, Dora-Luz, Ibarra-Manzano, Mario-Alberto, Yepez, Eduardo Cabal, Avina-Cervantes, Juan Gabriel and Fallavollita, Pascal. "Differential neural networks (DNN)." *IEEE Access* Vol. 8 (2020): pp. 156530–156538. DOI 10.1109/ACCESS.2020.3019307.

[8] Rodini, Simone. "Analytical derivatives of neural networks." *Computer Physics Communications* Vol. 270 (2022): p. 108169.

[9] McClure, Nick. *TensorFlow machine learning cookbook.* PACKT publishing Ltd (2017).

[10] Kaup, Ludger and Kaup, Burchard. *Holomorphic functions of several variables: an introduction to the fundamental theory.* Vol. 3. Walter de Gruyter (2011).

[11] Squire, William and Trapp, George. "Using complex variables to estimate derivatives of real functions." *SIAM review* Vol. 40 No. 1 (1998): pp. 110–112.

[12] Martins, Joaquim RRA, Sturdza, Peter and Alonso, Juan J. "The complex-step derivative approximation." *ACM Transactions on Mathematical Software (TOMS)* Vol. 29 No. 3 (2003): pp. 245–262.

[13] Neidinger, R. D. "Introduction to Automatic Differentiation and MATLAB Object-Oriented Programming." *Siam Review* Vol. 52 No. 3 (2010): pp. 545–563. DOI 10.1137/080743627.

[14] Margossian, Charles C. "A review of automatic differentiation and its efficient implementation." *Wiley interdisciplinary reviews: data mining and knowledge discovery* Vol. 9 No. 4 (2019): p. e1305.

[15] Abadi, Martín, Barham, Paul, Chen, Jianmin, Chen, Zhifeng, Davis, Andy, Dean, Jeffrey, Devin, Matthieu, Ghemawat, Sanjay, Irving, Geoffrey and Isard, Michael. "TensorFlow: A System for Large-Scale Machine Learning." *12th USENIX symposium on operating systems design and implementation (OSDI 16)*. Conference proceedings: pp. 265–283. 2016.

[16] Ruder, Sebastian. "An overview of gradient descent optimization algorithms." *arXiv preprint arXiv:1609.04747* (2016).

Copyright © 2023 by ASME

APPENDIX A. PYTHON CODE

```python
1   # NN Derivatives
2   # by Joel C. Najmon
3   # Python 3.8
4   # IMPORT PACKAGES
5   import numpy as np  # version 1.22.2
6   from scipy import stats as stats  # version 1.8.0
7   import tensorflow as tf  # version 2.2.0
8   precision = tf.float64
9
10  #%% DEFINE TEST FUNCTIONS
11  ydim = 1  # y dimension
12  xdim = 1  # x dimension
13  x_lb = 0
14  x_ub = 1
15  y_lb = 0
16  y_ub = 1
17
18  # Function 1
19  # yx = lambda x:  x**2
20  # dyx = lambda x: 2*x
21
22  # Function 2
23  yx = lambda x:  np.sin(x)
24  dyx = lambda x: np.cos(x)
25
26  # Function 3
27  # yx = lambda x:  np.exp(-x)
28  # dyx = lambda x: -np.exp(-x)
29
30  # Function 4
31  # yx = lambda x:  x**3 - 2*x**2 + 0.75*x + 0.5
32  # dyx = lambda x: 3*x**2 - 4*x + 0.75
33
34  #%% TRAINING FEATURE SETS
35  # number of training feature sets
36  Nr = int(5e3)
37  LHS = stats.qmc.LatinHypercube(xdim)
38  xtrain = LHS.random(Nr) * (x_ub - x_lb) + x_lb
39  ytrain = yx(xtrain)
40  nxtrain = (xtrain - x_lb)/(x_ub - x_lb)
41  nytrain = (ytrain - y_lb)/(y_ub - y_lb)
42  #%% TESTING FEATURE SETS
43  # number of testing feature sets
44  Nt = int(np.round((Nr / 0.7) - Nr, decimals=0))
45  xtest = LHS.random(Nt) * (x_ub - x_lb) + x_lb
46  nxtest = tf.Variable((xtest - x_lb) /
47                       (x_ub - x_lb), dtype=precision)
48
49  #%% TRAIN NN
50  NL = int(2)  # number of hidden layers
51  NN = int(10)  # number of neurons per hidden layer
52  kernel = 'none'
53  hidden_activation = 'sigmoid'
54  output_activation = 'linear'
55
56  tf.keras.backend.set_floatx(precision.name)
57
58  inputs = tf.keras.Input(shape=(xdim, 1))
59  outputs = tf.keras.layers.Dense(1)(inputs)
60  # create ann that has sequential layers
```

```python
61  nn_model = tf.keras.models.Sequential()
62  nn_model.add(tf.keras.Input(shape=(xdim, 1)))
63  # loss = tf.keras.losses.MeanAbsoluteError()
64  loss = tf.keras.losses.MeanSquaredError()
65  optimizer =\
66      tf.keras.optimizers.Adam(learning_rate=0.001)
67
68  for i in range(NL):
69      nn_model.add(tf.keras.layers.Dense(
70          NN, hidden_activation, kernel))
71  nn_model.add(tf.keras.layers.Dense(
72      ydim, output_activation, kernel))
73  nn_model.compile(loss=loss,
74                   optimizer=optimizer,
75                   metrics=['mse'])
76
77  nn_model.fit(nxtrain, nytrain, batch_size=256,
78               epochs=100, verbose=1)
79
80  #%% METHODS FOR y(x)
81  # 1) true function
82  y1 = yx(xtest)
83
84  # 2) NN prediction
85  y2 = nn_model.predict(nxtest.numpy())\
86      .reshape(Nt, ydim)
87
88  # 3) Manual NN prediction
89  # y3 calculated below in Method 1 for dy(x)/dx
90
91  #%% METHODS FOR dy(x)/dx
92  # 0) Analytical derivative of true function
93  dydx0 = np.zeros((Nt, ydim, xdim))
94  for d in np.arange(0, xdim):
95      dydx0[:, :, d] = dyx(nxtest.numpy())\
96          .reshape(Nt, ydim) * (x_ub - x_lb) + x_lb
97
98  # 1) Analytical derivative (chain rule)
99  Wn = []  # initialize lists
100 Bn = []
101 Yn = []
102 dYn = []
103 Wn.append([])  # store NN weights
104 Bn.append([])  # store NN biases
105 for L in np.arange(0, NL+1):
106     Wn.append(nn_model.layers[L]
107               .weights[0].numpy().T)
108     Bn.append(nn_model.layers[L]
109               .bias.numpy().T.reshape(-1, 1))
110 Yn.append(nxtest.numpy().T)  # store input
111 # store an "initial dYn" term for convenience
112 dYn.append(Wn[1].T.reshape(-1, 1, xdim))
113
114 # if else statement to assign lambda function for
115 # derivative of activation function in hidden
116 # layers. Create custom functions if not here.
117 if hidden_activation == 'sigmoid':
118     # derivative of sigmoid activation function
119     dhidden_dx = lambda x: np.exp(-x)\
120                  / (1 + np.exp(-x))**2
121 elif hidden_activation == 'relu':
```

```python
122     # derivative of relu activation function
123     dhidden_dx = lambda x: (tf.nn.relu(x)
124                                 / x).numpy()
125
126 # if else statement to assign lambda function for
127 # derivative of activation function in output
128 # layers.
129 if output_activation == 'linear':
130     # derivative of linear activation function
131     doutput_dx = lambda x: np.ones(x.shape())
132
133 # manually loop through hidden layers to
134 # calculate analytical derivative with chain rule
135 for L in np.arange(1, NL+1):
136     Yn.append(getattr(tf.nn,
137     hidden_activation)(np.matmul(Wn[L], Yn[L-1])
138                         + Bn[L]).numpy())
139     dYn.append(np.matmul(Wn[L + 1]
140     .reshape(1, -1, NN),
141     (dYn[L - 1].reshape(NN, -1, xdim)
142     * dhidden_dx(np.matmul(Wn[L], Yn[L - 1])
143     + Bn[L]).reshape(NN, Nt, xdim))
144                         .reshape(xdim, NN, Nt)))
145 # calculate output of NN
146 Yn.append(np.matmul(Wn[-1], Yn[-1]) + Bn[-1])
147 y3 = Yn[-1].reshape(Nt, ydim)
148 # calculate derivative of NN
149 dYn.append(dYn[-1])
150 dydx1 = dYn[-1].reshape(Nt, ydim, xdim)\
151        * (x_ub - x_lb) + x_lb
152
153 # 2) CFDA
154 h_cfd = 1e-6
155 dydx2 = np.zeros((Nt, ydim, xdim))
156 for d in np.arange(0, xdim):
157     h_mat = np.zeros((Nt, xdim))
158     h_mat[:, [d]] = np.ones((Nt, 1))*h_cfd
159     dydx2[:, :, d] =\
160     ((nn_model.predict(nxtest.numpy() + h_mat) -
161       nn_model.predict(nxtest.numpy() - h_mat)) /
162      (2 * h_cfd)).reshape(Nt, ydim)\
163     * (x_ub - x_lb) + x_lb
164
165 # 3) CSDA
166 h_csm = 1e-12
167 dydx3 = np.zeros((Nt, ydim, xdim))
168 for d in np.arange(0, xdim):
169     h_mat = 0j*np.zeros((xdim, Nt))
170     h_mat[[d], :] = 1j*np.ones((1, Nt))*h_csm
171     Yn_csm = [nxtest.numpy().T + h_mat]
172     # manually loop through hidden layers to
173     # calculate NN prediction with complex input
174     for L in np.arange(1, NL+1):
175         if hidden_activation == 'relu':
176             # since the inequalities of relu do not
177             # support complex numbers, simply just
178             # pass through all values.
179             Yn_csm.append(np.matmul(Wn[L] + 0j,
180                             Yn_csm[L-1]) + Bn[L])
181         else:
182             Yn_csm.append(getattr(tf.nn,
183             hidden_activation)(np.matmul(Wn[L] +
184             0j, Yn_csm[L - 1]) + Bn[L]).numpy())
185     Yn_csm.append(np.matmul(Wn[-1] + 0j,
186                     Yn_csm[-1]) + Bn[-1])
187     dydx3[:, :, d] = (np.imag(Yn_csm[-1]) /
188         h_csm).reshape(Nt, ydim) *\
189                     (x_ub - x_lb) + x_lb
190
191 # 4) Automatic Differentiation
192 dydx4 = np.zeros((Nt, ydim, xdim))
193 with tf.GradientTape(persistent = True) as tape:
194     y_pred = nxtest
195     for layer in nn_model.layers:
196         y_pred = layer(y_pred)
197 dydx4[:, :, 0] = tape.gradient(y_pred,
198 nxtest).numpy() * (x_ub - x_lb) + x_lb
199
200 # ERROR CALCULATIONS
201 mse =\
202 tf.metrics.mean_squared_error(
203 y_true=y1.squeeze(), y_pred=y2.squeeze()).numpy()
204 dydx1_er =\
205     tf.metrics.mean_squared_error(
206 y_true=dydx0.squeeze(), y_pred=dydx1.squeeze())\
207         .numpy()
208 dydx2_er =\
209     tf.metrics.mean_squared_error(
210 y_true=dydx0.squeeze(), y_pred=dydx2.squeeze())\
211         .numpy()
212 dydx3_er =\
213     tf.metrics.mean_squared_error(
214 y_true=dydx0.squeeze(), y_pred=dydx3.squeeze())\
215         .numpy()
216 dydx4_er =\
217     tf.metrics.mean_squared_error(
218 y_true=dydx0.squeeze(), y_pred=dydx4.squeeze())\
219         .numpy()
220
221 print(' ')
222 print('FUNCTION ERRORS:')
223 print('Mean Absolute Error: ', mae)
224 print('Mean Squared Error:  ', mse)
225 print(' ')
226 print('DERIVATIVE ERRORS:')
227 print('Analytical (Chain Rule):    ', dydx1_er)
228 print('Central Finite Difference: ', dydx2_er)
229 print('Complex Step Method:       ', dydx3_er)
230 print('Automatic Differentiation: ', dydx4_er)
```

Proceedings of the ASME 2023
International Design Engineering Technical Conferences and
Computers and Information in Engineering Conference
IDETC-CIE2023
August 20-23, 2023, Boston, Massachusetts

DETC2023-112470

DESIGN AND EVALUATION OF AN AUTOMATIC SCHEDULING-MANUAL OPERATION TOOL TO BRING PRECISION IRRIGATION TO RESOURCE-CONSTRAINED FARMERS

Georgia D. Van de Zande *
Department of Mechanical Engineering
Massachusetts Institute of Technology
Cambridge, Massachusetts 02139
Email: gdvdz@mit.edu

Carolyn Sheline
Department of Mechanical Engineering
Massachusetts Institute of Technology
Cambridge, Massachusetts 02139
Email: csheline@mit.edu

Susan Amrose
Department of Mechanical Engineering
Massachusetts Institute of Technology
Cambridge, Massachusetts 02139
Email: samrose@mit.edu

Jeffrey Costello
Department of Mechanical Engineering
Massachusetts Institute of Technology
Cambridge, Massachusetts 02139
Email: jcostllo@mit.edu

Aditya Ghodgaonkar
Department of Mechanical Engineering
Massachusetts Institute of Technology
Cambridge, Massachusetts 02139
Email: adighod@mit.edu

Fiona Grant
Department of Mechanical Engineering
Massachusetts Institute of Technology
Cambridge, Massachusetts 02139
Email: fionag@mit.edu

Amos G. Winter, V
Department of Mechanical Engineering
Massachusetts Institute of Technology
Cambridge, Massachusetts 02139
Email: awinter@mit.edu

ABSTRACT

As populations increase and freshwater supplies decrease, adopting water- and energy-efficient irrigation practices is crucial, particularly in resource-constrained regions. Here, farmers

are often unable to purchase the equipment used in precision irrigation, a practice that implements the automatic scheduling of irrigation events to achieve high efficiency. Currently, no irrigation methods exist that combine the automatic scheduling of irrigation events with the manual operation of valves, a common practice on low-income farms. This work introduces a de-

---

*Address all correspondence related to ASME style format and figures to this author.

Copyright © 2023 by ASME

sign concept for an automatic scheduling and manual operation (AS-MO) tool that addresses the efficiency needs of resource-constrained farms and integrates into current manual practices. However, it is unknown how farmers would value such a tool. Through interviews and focus groups facilitated by a series of storyboards and a physical prototype, the proposed concept was evaluated by farmers and key market stakeholders in Kenya, Jordan, and Morocco. Results showed that farmers in Kenya and Jordan in particular valued the proposed AS-MO concept because they want increased efficiency on their farms but did not want to install automatic valves for cost and complexity concerns. A possible market was also found in Morocco, but a majority of interviewed farms preferred automatic valve operation due to large farm sizes. Interviewees provided feedback on how to improve the tool's design in future iterations. If adopted at scale, this AS-MO tool could increase efficiency on farms that otherwise cannot afford current precision irrigation technology, improving sustainable agriculture worldwide.

1 INTRODUCTION

The aim of this work is to evaluate a means of bringing many of the water and energy efficiency benefits of precision irrigation to resource-constrained regions without the high equipment costs and complexity of existing methods.

The second Sustainable Development Goal (SDG 2) calls for the achievement of food security by 2030 [1]. This aim is particularly imperative in low- and middle-income regions such as East Africa (EA) and the Middle East and North Africa (MENA), where over 33% and 10% of the population, respectively, is projected to be undernourished in 2030 [2]. Numerous studies have shown that increasing access to irrigation is an effective path to achieve food security in these regions [3–5]; however, irrigation is a water- and energy-intensive process, counter to the additional aim of SDG 2 to promote sustainable agriculture. The high water use of irrigation is particularly challenging in arid and semi-arid regions like MENA and EA, respectively. Prior work has suggested that water- and energy-saving technologies could particularly benefit medium-scale farms in EA and small- to medium-scale farms in MENA, many of which have access to capital to pay for some of this technology [6]. In EA, these farms, generally sized between five and 15 acres, rely on hired manual labor to feed the growing city centers [7]. In MENA, the farm size scale is country-dependent, with the small-scale farms generally ranging from five to 25 acres and the medium-scale farms generally ranging from 50–120 acres. Both small- and medium-scale farms typically rely on hired manual labor, but medium-scale farms may also have specialized labor such as a farm manager or agronomist. The growing number of small- and medium-scale farms has the promise to increase food security in EA and MENA, but doing so sustainably remains a challenge [8,9].

Solar-powered drip irrigation has been proposed for regions with high solar irradiance as a means to increase irrigation with minimal water and fossil fuel use [10–12]. Solar is especially applicable in rural EA where access to grid electricity can be uncommon [13]. Drip irrigation uses a network of pipes and emitters to deliver water directly to crops' roots, saving up to 50% of water compared to flood irrigation, a commonly-used method [14]. On farms with deep boreholes, this method can also be energy-efficient because less water used means less energy needed to operate a pump. In off-grid irrigation systems, the amount of energy used is critical because this value dictates the capital cost of solar panels, one of the largest system costs [6].

Further, the water- and energy-saving ability of solar-powered drip irrigation systems depends on how farmers operate the systems on a daily basis [15]. Savings can be realized when precision irrigation methods are introduced to optimally operate the system and deliver the exact amount of water needed. Many precision irrigation technologies measure farm and weather conditions, calculate ideal irrigation schedules, and use automated valves to carry out these schedules [16, 17]. Precision irrigation systems often rely on arrays of sensors, solenoid valves, and proprietary hardware and software [18–21]. These technologies increase the efficiency of irrigation systems but can cost up to tens of thousands of dollars to equip an entire farm [22].

Unfortunately, economic constraints in EA and MENA make it difficult for medium-scale farmers to adopt existing precision irrigation equipment. In contrast, these farms often employ local laborers to both monitor and carry out irrigation tasks using manual valves [6]. These laborers use inexpensive but time-consuming and often imprecise manual methods for determining when to irrigate, like "stick" and "ball" tests [22]. In a stick test, a laborer inserts a stick 10 cm into the soil. If it comes out with dirt attached, the soil is moist enough. In a ball test, a farmer forms a handful of dirt into a ball. If the ball crumbles when let go, the soil is too dry. The irrigation experience of hired laborers varies widely, so farms cannot rely on these binary tests to deliver the most water- and energy-efficient irrigation. Human laborers also typically rely on observations of current and past weather and crop conditions, lacking the ability to make accurate forecasts such as those used in precision irrigation. In addition, relying on past conditions alone does not account for changes in climactic conditions as global temperatures rise [23,24]. Inaccurate forecasting of weather conditions can negatively impact the reliability of solar-powered irrigation systems on cloudy days if farms have not properly planned for future weather events.

Some existing products attempt to bridge the gap between fully automated precision irrigation and fully manual heuristic methods. However, these products are timer-based and largely fall short of delivering the efficiency and prediction benefits of precision irrigation. As two examples, the Pro-C irrigation controller (Hunter Industries, California) and the SST1200OUT irrigation timer (Rain Bird Corporation, California) are relatively low-cost products—in the $100–300 range—that control a series of solenoid valves to carry out predetermined irrigation schedules. While these products are affordable to many farms, they still rely on the farmer to determine and input the irrigation schedule. Even for the most experienced farmers, it is extremely challenging to determine an irrigation schedule that has been optimized for both water- and energy-savings. In addition, these devices cannot deliver the computationally intensive optimization benefits of conventional precision irrigation.

To identify potential opportunities to realize some of the key benefits of precision irrigation with minimal complexity and cost, Fig. 1 characterizes two of the critical actions of irrigation system control. The first looks at determining a schedule of irrigation events (e.g., scheduling), and the second at operating valves in a hydraulic network (e.g., operation). Each of these actions can be done either manually by a farmer or automatically by the system, resulting in four distinct design spaces. Fully automated precision irrigation systems are in the lower right quadrant, while fully manual methods, like stick or ball tests paired manual valves, are in the upper left. Existing irrigation timers fall into the manual scheduling and automatic operation quadrant. To the best of the authors' knowledge, no commercial technologies exist that can deliver the automatic scheduling benefits of precision irrigation to farms that primarily rely on the manual operation of valves, such as the resource-constrained farms in EA and MENA. The lower left quadrant of Fig. 1 highlights this gap in the design space.

We hypothesize that a technology in the automatic scheduling and manual operation (AS-MO) design space is well-suited for the small- to medium-scale farmers typically found in EA and MENA. Automatic scheduling that relies on low-cost sensors and cloud computing could provide farmers with irrigation schedules that have been optimized for water and energy efficiency, enabling them to access several of precision irrigation's key benefits. Further, with an AS-MO tool, these cost-constrained farms could continue to rely on manual operation, leveraging the manual labor available in these regions. This approach could minimize farmers' costs while easing adoption by simplifying the installation of new equipment.

Implementing an AS-MO irrigation control strategy has several challenges. First, a fully optimized, auto-generated schedule might change every day or even every minute as the system integrates new inputs of current weather and farm conditions. Second, the generated schedule might be highly complex. Prior work has proposed that irrigation systems strategically turn on and off different sections at different times of day to make the best use of available solar power [15]. As humans prefer easy-to-use tools, farmers might decide the frustrations of a frequently-changing schedule are not worth the efficiency benefits of precision irrigation. The perceived desire of farmers to adopt an AS-MO tool in this context is unknown but critical to its potential to create an impact in EA and MENA markets. Filling this knowledge gap requires understanding how to design an AS-MO tool that provides

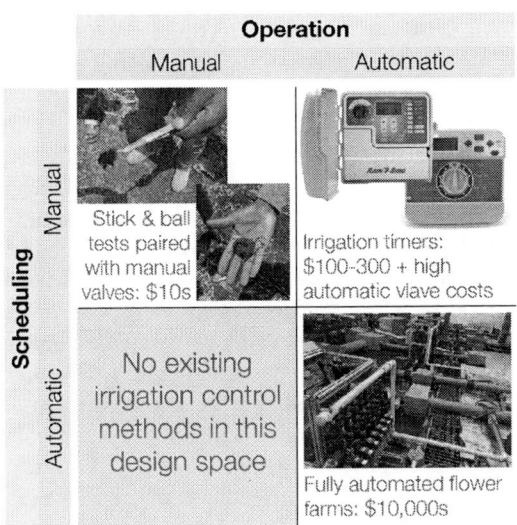

FIGURE 1: Visualization of the design space of irrigation system control methods with regard to two key elements: scheduling and operating. Existing methods typically fill three of the four design spaces. This work proposed a tool to fill the gap in the automatic scheduling and manual operation space. This work evaluates this design concept's fitness for medium-scale farmers in EA and MENA against existing solutions that use other control methods.

access to precision irrigation's efficiency benefits while aligning with the current practices of EA and MENA farms.

To evaluate the potential viability of an AS-MO tool in the EA and MENA marketplace and to better understand how farmers might value and interact with such a tool in practice, this paper addresses the following research aims:

1. Characterize an AS-MO tool architecture that effectively transmits key benefits of precision irrigation while integrating into the current practices and capabilities of target farms, informed by prior market analysis and recent innovations.
2. Substantiate the value of an AS-MO tool among potential users in EA and MENA markets and assess their desire to adopt a tool with this architecture, relying on storyboard-based interviews and focus groups.
3. Assess target farmers' satisfaction with a proposed user interface for an AS-MO tool and identify avenues for improvement, based on interactions with a medium-fidelity prototype of the tool within interviews and focus groups.

2 THE PROPOSED AS-MO TOOL

To introduce automatic scheduling and manual operation on EA and MENA farms, a tool was sought that (1) requires a low infrastructure investment and (2) does not require complex maintenance. To facilitate automatic scheduling, multiple precision irrigation algorithms, including the ones mentioned in Section 1, were evaluated for their fitness against these user needs. A scheduling theory being developed by the Global Engineering and Research (GEAR) Lab was chosen.

The left-hand side of Fig. 2 shows how an AS-MO tool implementing this theory could meet the needs of EA and MENA farmers. This theory leverages cloud computing to build an optimal irrigation schedule and characterize soil moisture without the use of soil moisture sensors, which are expensive and complex to calibrate. It does this using soil water balance calculations and several inputs from the farm [14, 25]. Farm inputs include readings from several simple weather sensors, solar panel power readings, and user inputs regarding system component specifications and agronomy details, such as solar array capacity, pump operating points, irrigation block areas, crop types, and soil texture. By relying on cloud computing and machine learning, this theory enables cost-constrained farmers to gain access to the forecasting benefits of precision irrigation. Doing so with few inexpensive sensors meets farmers' need for minimal infrastructure and maintenance.

Further, the GEAR Lab theory strategically coordinates irrigation events throughout the farm (top left of Fig. 2), which increases system reliability on cloudy days and reduces the capital cost of solar-powered systems. By predicting and then matching the pumping energy needed to meet crop water demand (light blue boxes in the power-time plot) with the forecasted available power (dark blue line), the tool can efficiently schedule irrigation events. For example, the tool might schedule one, two, or three blocks to be open during periods of forecasted low, medium, and high solar irradiance, respectively. Generating schedules in this way has the potential to reduce power system costs by up to 30% for medium-scale Kenyan farms [6, 22].

Scheduling algorithms, including the GEAR Lab theory, often rely on frequently-updated irrigation events to most efficiently meet irrigation demands. In an AS-MO tool, these schedules must be communicated to farmers in a way that is easy to follow. To accomplish this, the proposed tool was designed to send messages to farmers' cell phones, products which are increasingly more common in low-resource countries [26]. Researchers have shown that Short Message Service (SMS) reminders can improve health outcomes in Kenya. These frequent, timely reminders have improved immunization timeliness and adherence to antiretroviral treatment [27, 28]. This work hypothesizes that the idea of frequent notifications can effectively address the challenge of implementing complex irrigation schedules on resource-constrained farms.

The tool sends notifications with schedule information to farmers throughout the day (right-hand side of Fig. 2). At the beginning of each day, the tool determines an irrigation schedule and presents it to the farmer. The farmer has the option to accept or slightly modify this preliminary schedule. Once the accepted schedule begins, the tool sends additional messages to the farmer's phone, reminding them to manually open or close valves according to the schedule (lower right of Fig. 2). The farmer would then manually open or close valves as directed and then confirms the action was complete.

At the end of an irrigation event, the farmer has the option to add 10 additional minutes of irrigation time if they notice insufficient water delivery. In a preliminary evaluation of this AS-MO concept conducted in October 2021, the ability for farmers to slightly adjust the irrigation schedule during the day was found to be important [22]. This time-adding feature was integrated into the design concept to meet this user need. The order and duration of irrigation events were still automatically scheduled and communicated to farmers to enable manual valve operation. These interactions are repeated throughout the day, according to the predetermined irrigation schedule.

3 DESIGN OF AN INTERACTION PROTOTYPE

A physical prototype of the AS-MO tool that simulated a farmer's daily interaction with it was designed. Prototypes are known to increase the quality of feedback given by interview participants because they allow a potential user to imagine interacting with the proposed device [29, 30]. This mechanism was used to evaluate how farmers and stakeholders respond to the basic elements of an AS-MO tool. The prototype itself consisted of three components: a mobile phone, a control box, and a weather station (Fig. 3).

The phone was equipped with Telegram, a common messaging app (Telegram FZ-LLC, 2023). Telegram users can have conversations with bots that deliver pre-programmed messages, and these bots can ask users short answer questions that determine the messaging path the bot takes next. For this study, a Telegram bot was created to walk participants through the following set of simulated AS-MO tool interactions:

- Provide farmers with a sample daily irrigation schedule, simulating the first message a farmer would receive each morning;
- Ask farmers if they approved of that day's irrigation schedule;
- Send a message prompting the farmer to manually open or close a valve when an irrigation event started or ended, respectively;
- Give farmers the ability to add an additional 10 minutes of irrigation time when an irrigation block is scheduled to end, and then update the schedule based on this choice; and
- Give farmers the ability to skip a block before irrigation starts, and then update the schedule based on this choice.

These interactions aimed to allow the research team to elicit feedback on these core design decisions.

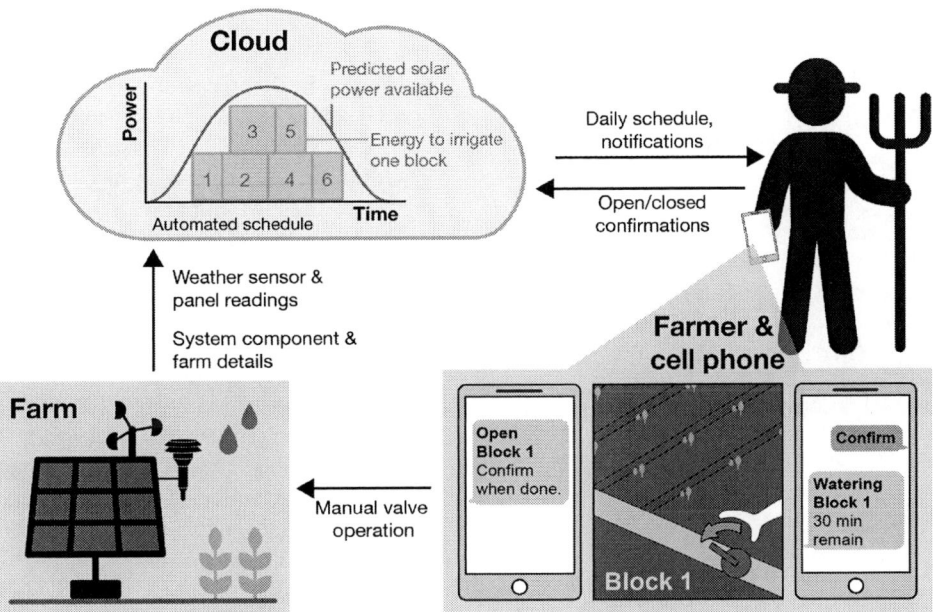

FIGURE 2: A depiction of the proposed AS-MO tool and the system on which it relies. Details about the farm and irrigation system are fed into a cloud-based algorithm that automatically generates an efficient irrigation schedule. This schedule is then communicated to a farmer's phone via notifications that are sent at the beginning of the day and at the start and end of each irrigation event. These messages direct a farmer to carry out the generated schedule by manually operating valves. When farmers confirm that actions have been completed, it informs the algorithm how closely the schedule was followed so it can adjust the next day's schedule accordingly.

The prototype control box consisted of an e-Ink screen mounted on a black box of a similar size anticipated for the controller (approximately 230x150x70 mm). Inside the box was a battery and a Raspberry Pi that carried out the Telegram bot's script. The box did not have any physical modes of interaction (e.g., buttons or dials), but it was designed to:

– Display the open/closed status of irrigation blocks based on confirmations a participant made in Telegram;
– Display a countdown telling the user when the next irrigation event was scheduled to occur; and
– Demonstrate to participants the anticipated size of a permanently-mounted control box.

The prototype weather station included the number and type of weather sensors that would be required to generate an optimized irrigation schedule, including wind speed, wind direction, ambient light, solar irradiance, precipitation, temperature, and humidity. This allowed the research team to elicit feedback on the weather information that participants found most valuable.

4 PROTOTYPE-BASED INTERVIEW METHODS

The physical prototype was designed to help participants describe what would be most valuable and most frustrating about using the AS-MO tool. To reach these aims, interviews and focus groups were conducted with potential users and market stakeholders in Kenya, Jordan, and Morocco, an approach inspired by Lean Startup methodologies [31].

During interviews and focus groups, participants were first introduced to the tool design concept with a set of storyboards using a protocol designed for a preliminary study to evaluate the concept [22]. After the storyboard introduction, participants were given the physical prototype designed to help them answer questions relating to the value and daily use of the proposed tool. Specifically, (1) What is the most useful information they think the tool could provide? (2) How do farmers think they would or would not use the tool daily? and (3) What drawbacks do they think they would encounter when using the tool? Specific interview questions targeted these broader research questions, but the semi-structured nature of the interviews and focus groups meant that not all participants were asked the same specific questions.

During the study, it was made clear to participants that interacting with the prototype alone would not open or close valves, as the valves would not be automatic. Rather, the user would manually perform these actions in the field and then use Telegram on the phone to confirm once complete.

As the prototype was intended to assess user interactions rather than the efficacy of the automatic schedule determination, a mock irrigation schedule was presented to the user. The dura-

Copyright © 2023 by ASME

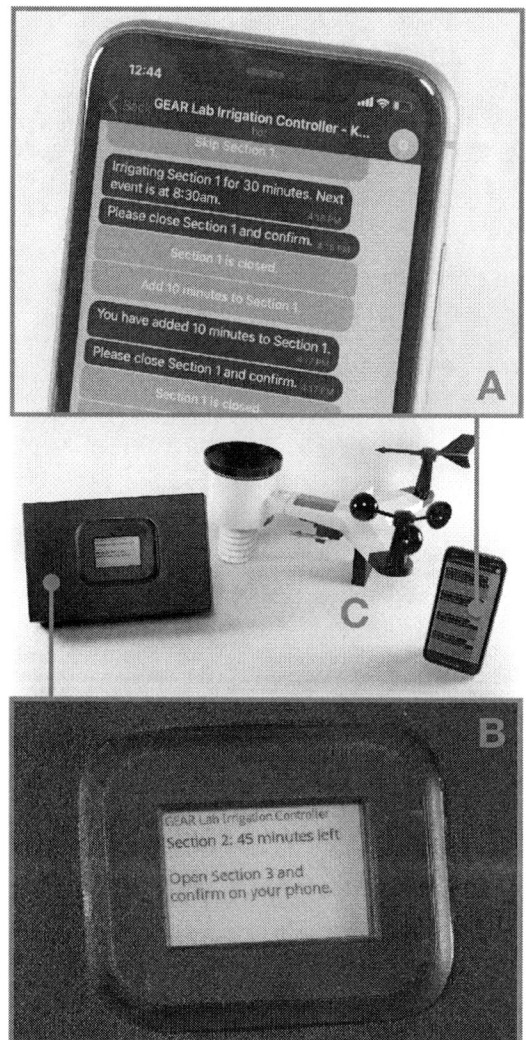

FIGURE 3: The three components of the physical prototype used to facilitate interviews and focus groups. The phone (A) was equipped with a Telegram bot that stepped farmers through a key set of interactions with the tool. The control box (B) displayed the status of these interactions and directed farmers to interact on the phone. The low-cost weather station (C) showed farmers what data the tool might collect: wind speed, wind direction, ambient light, solar irradiance, precipitation, temperature, and humidity.

tions of irrigation events were also shortened for the study, and participants were made aware of these adjustments.

In total, 22 prototype-based interviews and focus groups with farmers were conducted (seven in Kenya, five in Morocco, and 10 in Jordan), involving a total of 40 farmers (13 farmers in Kenya, 11 in Morocco, and 16 in Jordan). These farmers were associated with 22 farms, ranging from 3–10 acres in Kenya, 5–120 acres in Morocco, and 4–120 acres in Jordan. These farm size ranges in all three countries were representative of the ranges in each country for which solar-powered drip irrigation would be most feasible [6]. Eight Kenyan farmers had previously participated in the preliminary set of interviews and focus groups, so they were already familiar with the design concept [22]. Unfortunately, due to travel complications, three interviews in Morocco were conducted without the physical prototype. These protocols involved only the storyboards.

The prototyped-based interviews were also conducted with 30 stakeholders (five in Kenya, five in Jordan, and 20 in Morocco) who were broadly familiar with the EA irrigation market were also recruited for interviews. Stakeholders included irrigation engineers, managers of irrigation equipment distributors, borehole drillers, agronomists, and government officials. These stakeholders represented professional viewpoints of different sectors of the irrigation and agriculture markets. They have collectively helped thousands of farmers improve their farms, so they could provide perspectives on a large population of farmers in ways that individual farmers could not. Interviews with stakeholders followed a similar protocol as interviews with farmers and sought to assess the tool's potential as a viable product in EA and MENA markets. All interviews and focus groups took place in March 2022, and all protocols were approved by the Massachusetts Institute of Technology Institutional Review Board (protocol E-4098).

5 RESULTS

5.1 Substantiation of the tool's value

In 23 out of 36 interviews (nine in Kenya, seven in Morocco, and seven in Jordan), farmers asserted that the AS-MO tool would likely be adopted by farmers in the target user group, a result consistent with prior work [22].

The most valuable benefits of the tool according to participants were alleviating water scarcity concerns and preventing over-irrigation. Farmers and stakeholders alike noted that climate change has altered seasonal rains such that they are no longer predictable. Farmers can no longer reliably anticipate water availability based on historical trends. Participants claimed that an automatic scheduling tool could aid them as they plan irrigation events.

Farmers in particular also noted that the tool could save them effort, money, and time. In contrast, three stakeholders and two farmers were concerned that using the tool could potentially increase the amount of time that a laborer was needed on the farm. This discrepancy suggests the need to explore whether the tool saves or increases labor and time when used over long periods.

Copyright © 2023 by ASME

5.2 Farmer scheduling and operation preferences

Figure 4 summarizes the scheduling and operation preferences noted from the 36 farmer and stakeholder interviews and focus groups. Operation preferences are broken down by country.

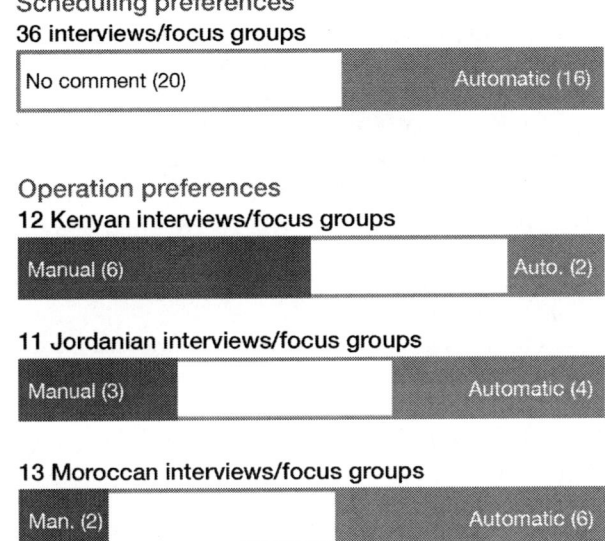

FIGURE 4: A summary of both farmer and stakeholder preferences for scheduling and operation. Automatic scheduling was preferred over manual scheduling by all participants who had a preference. Preference for manual operation over automatic operation differed by country. Not all participants mentioned a preference, so they are visualized by the white space.

In 13 of 22 interviews, farmers noted that they particularly appreciated the automatic scheduling aspect of the AS-MO tool. This result suggests that this is an important feature for Jordanian and Moroccan farmers in addition to Kenyan farmers. Farmers noted that an automatically-determined schedule specific to their farm and weather conditions could improve their yields.

There was disagreement among farmers on their preference for manual versus automatic operation of valves. In 12 interviews (two in Kenya, four in Jordan, and six in Morocco), farmer or stakeholder participants preferred automatic valve operation, while in 11 interviews (six in Kenya, three in Jordan, and two in Morocco), manual valve operation was preferred. The preference for automatic operation was particularly driven by MENA participants who operated or served on larger farms. On larger farms, participants claimed that automatic operation was worth the investment because laborers would otherwise need to walk long distances to manually operate valves, wasting time and potentially increasing labor costs. Several of the larger farms had already installed automated solenoid valves and asked if the tool could be adapted to operate those valves.

Kenyan farmers in particular favored manual valve operation over automatic operation, with only two of seven Kenyan farmers claiming a preference for automatic valves. Here, manual valves were heavily preferred over solenoid valves due to their low cost. Study participants also noted that the reliability and familiarity with manual valves in the region could benefit Kenyan farmers more than solenoid valves. Several participants in Jordan also had a preference for manual valves, suggesting that an AS-MO tool could have promise in these markets.

In all three countries, the majority of farmers liked the ability to add more time or change the schedule slightly, suggesting that they value retaining some degree of manual control.

Participants in all three countries commented on the importance of demonstrating the tool to farmers before they would be likely to adopt the technology, a result consistent with literature about farmers in Tanzania, South Africa, and Morocco [32–34]. Nine farmers claimed they would need to closely monitor the tool on their own farm for a period of time before trusting that the automatic schedule determination was sufficient. This result stresses the importance of demonstrating the tool before farmers can realize its full benefits.

5.3 Features to consider adding to the AS-MO tool

Study participants suggested several features that they would like to see in future iterations of the AS-MO tool design. Both farmers and stakeholders expressed a preference for using a custom app to communicate with the AS-MO tool as opposed to using a messaging app like Telegram. Participants claimed that a custom app would provide more functionality, citing several key benefits.

First, participants noted that inputting the farm details needed for the automatic scheduling aspects of the tool could be easier with a custom app. Farmers and agronomists agreed that they would accept the need to update farm details when they change crops as long as it was easy. Several farmers reported changing their crop selections every few weeks, while others remained more consistent. Participants noted that the process of entering and updating farm details could be cumbersome if not designed well. A custom app would allow for the greatest flexibility when inputting these key details.

Second, a custom app would allow different users to visualize their farm data in different ways, reflecting differences in the types of information that various stakeholders reported finding the most valuable. Farm managers and farm employees reported that detailed data on crop irrigation needs and weather forecasts would be most valuable. Conversely, farm owners reported that they would be less concerned with their farm's daily operational status and more concerned with the overall status. Distributors noted that they could use system operating data to monitor the

Copyright © 2023 by ASME

equipment that they had sold that might still be under warranty. These results demonstrate that a variety of interfaces highlighting different information might be needed to account for the diversity of user roles, which a custom app could provide.

Finally, several participants were concerned that a messaging-based interaction could be difficult for illiterate laborers to use. An app would allow for the use of more symbols, or even voiced instructions, making the tool more accessible.

In addition to a custom app, another key feature was mentioned by study participants as being potentially useful. While most farmers preferred for the main interaction to be through their phones, 11 participants suggested that farmers should have the ability to interact with the control box without a phone. Numerous reasons were cited as to why a phone might not be available. For example, the phone could be broken, the battery could be dead, someone else could be using the phone, or the cellular service could be poor. Seven participants in Jordan and Morocco claimed that a well-designed app would be sufficient and that they would not need any interaction with the control box. However, these participants had larger farms with potentially more access to capital and did not report having the phone and service problems reported more frequently on smaller farms. These results suggest that critical interactions with the AS-MO tool should be integrated into a control box design so that farmers who need it have consistent access.

6 DISCUSSION

This work demonstrated that the proposed AS-MO tool has the potential to bring the efficiency benefits of precision irrigation to medium-scale farms in Kenya and small- and medium-scale farms in Jordan and Morocco. It could do this by bridging the gap between existing, expensive precision irrigation technologies and affordable, easy-to-adopt irrigation methods.

Data from the study validated the assumptions made in Section 1 about the potential benefits of an AS-MO irrigation control method over the other methods in Fig. 1. First, compared to both manual scheduling methods (top half of Fig. 1), an AS-MO tool was hypothesized to address problems that are hard for humans to solve alone, such as creating efficient, reliable irrigation schedules. Discussions with farmers confirmed that doing so was difficult, time-consuming, and sometimes not possible without the use of sensors and calculations. The increase in efficiency and reliability provided by automatic scheduling was found valuable by most farmers, confirming initial hypotheses.

Second, compared to automatic control and automatic operation (bottom right of Fig. 1), AS-MO was predicted to deliver value to farmers for its familiarity and affordability. Some farmers preferred manual valves over automatic ones because they were concerned about the reliability of solenoid valves, a technology with which they had little familiarity. Several farmers also valued the ability to continue visually inspecting each block

after each irrigation event. Farmers' preferences to continue certain practices that are currently a part of many farms' operations suggest equipment familiarity is a priority. Farmers, particularly farm owners, also expressed interest in the AS-MO tool because it was lower cost than a fully-automated system, suggesting that the tool's affordability is also a priority for the targeted farms.

Results from Kenya, Jordan, and Morocco are anticipated to be applicable to the larger regions of EA and MENA, respectively, so differences in farmer preferences between the three countries could also predict differences in the two regions. One key difference between the regions was that it appeared that several interviewed Jordanian and Moroccan farmers were more familiar with current precision irrigation techniques than farmers in Kenya were. They were more excited about a fully automated system because they knew and trusted automated valves. On the other hand, Kenyan farmers and stakeholders more frequently expressed skepticism about automated valves, claiming they might break frequently.

A second difference between the regions was that there were mixed preferences for manual valve operation over automatic in Jordan and Morocco compared to a strong preference for manual operation in Kenya. While these results showed a slight preference for full automation in the Jordanian and Moroccan markets, it does not necessarily mean that an AS-MO tool could not provide value in the MENA region. Wider ranges of farm sizes were interviewed in Jordan and Morocco than in Kenya, and the larger farms were particularly interested in automatic valves. These large farms appeared to have more access to capital than the other studied farms, suggesting that the AS-MO tool concept might not be applicable to farms that fit this profile. However, there was strong interest in manual valves among the smaller farms in Jordan and Morocco which appeared to have less access to capital, suggesting there is likely a MENA market sector that is interested in an AS-MO tool in the way the Kenyan farmers were. Future exploration of the EA and MENA markets could confirm if the differences seen in Kenya, Jordan, and Morocco reflect the differences between EA and MENA as whole regions.

The proposed AS-MO tool could potentially be a good segue product for farmers who are transitioning from fully manual to fully automated. Several study participants pointed out that it would be beneficial for the tool to be adapted to include automatic valve operation, especially on larger or wealthier farms. This result suggests the participants saw the potential for the AS-MO tool to be "upgraded" from a semi-manual/semi-automatic tool to a fully automatic tool according to users' needs. There are likely cases where a farm first sees a need to address the challenge of automating irrigation schedules, so they adopt the AS-MO tool. Once that farm grows to the point where manual valve operation also becomes challenging, the farm could install solenoid valves and a new control box to operate them. At this point, the farm could continue using the same automatic scheduling methods as the AS-MO tool used, so the irrigation schedules

Copyright © 2023 by ASME

are familiar and trusted. In the app, the farmer could input that the farm is now fully automated, and the tool could control the solenoid valves rather than sending instructions to laborers' cell phones. If this tool could ease the transition from fully manual to fully automatic, it could help farmers adopt further benefits of conventional precision irrigation, like automatic operation.

This work demonstrates the successful use of a methodology in which the research team identified opportunities to automate complex tasks while designing ways for users to complete these tasks in simpler, manual ways. The goal of this approach was to gain some benefits of automation while also realizing other benefits of manual work in order to lower overall product costs. Interviewees suggested this semi-automatic/semi-manual product architecture could be valuable if applied to fertigation, suggesting that this approach could have implications past the specific example of irrigation in the MENA and EA markets. Additional opportunities could include home gardening or landscaping. To apply a semi-automatic/semi-manual architecture to a new area, it is helpful for researchers and designers to break down a problem into the necessary actions (e.g., scheduling and operation, in this case). They can then understand which actions are simpler to perform manually and which would be more difficult. For the difficult actions only, researchers and designers would then identify ways in which technology could improve those actions. New technology may need to be invented to communicate complex operations to users who are carrying out manual actions. This strategy of pairing automated actions with manual actions could open new areas for innovation while serving users' needs best.

Several limitations existed in this study. The small number of farmer interviews does not necessarily give a generalized opinion of all potential users in EA and MENA. To attempt to mitigate this limitation, lead users, early adopters, and market stakeholders were recruited for the study. However, because these participants were more familiar with advanced technology, they might have a higher preference for automation than the general population would. This may have led to more disinterest in the AS-MO tool than is potentially accurate in a group of target users.

A second limitation is that users did not interact with a fully-functioning prototype for an extended period of time. The prototype performed basic interactions, not in-frequent or edge-case interactions like inputting details of a farm or managing a failure in the system. The prototype also did not calculate an irrigation schedule specific to a farm but instead used a preprogrammed schedule. Had farmers seen a higher fidelity prototype, they might have had a stronger critique of the automatic scheduling aspect of the tool, especially if it calculated a schedule drastically different than they expect.

7 CONCLUSIONS AND FUTURE WORK

The objective of this work was to evaluate a potential means of bringing the water and energy efficiency benefits of precision irrigation to resource-constrained regions like EA and MENA. To do this, a design concept for an AS-MO tool that could communicate complex but efficient irrigation events to farmers was characterized. To evaluate this concept a set of storyboards and a physical prototype of the tool were used in Kenya, Jordan, and Morocco to facilitate further interviews and focus groups with farmers and stakeholders.

The results demonstrated that the proposed AS-MO tool has the potential to enable target farmers to realize the energy- and water-saving benefits of precision irrigation. The majority of all interviewed farmers were interested in the automatic scheduling aspect of the AS-MO tool. They recognized how implementing water- and energy-efficient schedules could save them time, effort, and money on their farms. Kenyan farmers and small-scale farmers in Jordan and Morocco also liked the manual valve operation that an AS-MO tool affords. They felt more confident in adopting low-cost, familiar hardware like manual valves over solenoid valves.

Interviews with farmers and stakeholders also provided insights on how farmers might best interact with the AS-MO tool. Results suggested that a smartphone app should be designed in order to enable key user interactions with the tool. Results showed that it was valuable to give farmers the flexibility to change the predetermined schedule, even slightly. Farmers liked the ability to add time to each irrigation event in case they thought the tool delivered an insufficient amount. They also liked they could shorten, pause, or cancel an event if needed. An app-based interaction should include different data visualizations for various user profiles, such as managers, owners, and laborers. Further, a limited set of critical interactions should be made possible on the permanently-mounted control box for when phones are unavailable. A screen that shows the status and several buttons or a dial could meet this user need.

Stakeholders and farm owners in a position to buy such a tool suggested the tool has the potential to become a viable commercial product in the studied countries. Several stakeholders claimed it could benefit the growing number of solar-powered drip irrigation users.

To bring the AS-MO tool design concept to fruition, further research is needed to learn how farmers interact with a functioning AS-MO tool for an extended period of time. This study only addressed the core functions of the proposed AS-MO tool. Other functions need to be prototyped and tested. It is also necessary to study the interactions farmers have with the AS-MO tool over the course of a season to understand how to improve it for future users. This tool must be demonstrated under these conditions to gain further user feedback. The study also assumed that the perspectives of Kenyan farmers and Jordanian and Moroccan farmers would represent the perspectives of EA and MENA farmers, respectively. Future work should expand regional coverage to confirm or deny this assumption. With these next steps, future development on an AS-MO tool could help bring water-

Copyright © 2023 by ASME

and energy-efficient irrigation to resource-constrained regions like EA and MENA.

ACKNOWLEDGMENT

The authors would like to thank our regional partners—Davis & Shirtliff, MIRRA Jordan, ICARDA Morocco, and INRA Morocco—for connecting us with interview participants and the participants themselves for sharing their perspectives. Thank you to Professors Glen Urban and Maria Yang for providing guidance on this work. Final acknowledgments go to the Julia Burke Foundation, USAID (Cooperative Agreement Number AID-OAA-A-16-00058), and the MIT Department of Mechanical Engineering for funding this work.

REFERENCES

[1] United Nations, *The Sustainable Development Goals Report*. United Nations, 2020.

[2] FAO, IFAD, UNICEF, WFP and WHO, *The State of Food Security and Nutrition in the World 2020*. Rome: FAO, IFAD, UNICEF, WFP and WHO, 2020.

[3] T. Amede, "Technical and institutional attributes constraining the performance of small-scale irrigation in Ethiopia," *Water Resources and Rural Development*, vol. 6, pp. 78–91, Nov. 2015.

[4] T. Shah, S. Verma, and P. Pavelic, "Understanding smallholder irrigation in Sub-Saharan Africa: results of a sample survey from nine countries," *Water International*, vol. 38, pp. 809–826, Oct. 2013.

[5] S. N. Ngigi, J. N. Thome, D. W. Waweru, and H. G. Blank, "Low-cost irrigation for poverty reduction: an evaluation of low-head drip irrigation technologies in Kenya," annual report, International Water Management Institute (IWMI), Colombo, Sri Lanka, 2001.

[6] G. D. Van de Zande, S. Amrose, E. Donlon, P. Shamshery, and A. G. Winter V, "Identifying opportunities for irrigation systems to meet the specic needs of farmers in East Africa," *Submitted to Irrigation Science*, 2022.

[7] T. Jayne, J. Chamberlin, L. Traub, N. Sitko, M. Muyanga, F. K. Yeboah, W. Anseeuw, A. Chapoto, A. Wineman, C. Nkonde, and R. Kachule, "Africa's changing farm size distribution patterns: the rise of mediumâscale farms," *Agricultural Economics*, vol. 47, pp. 197–214, Nov. 2016.

[8] G. Jobbins, J. Kalpakian, A. Chriyaa, A. Legrouri, and E. H. El Mzouri, "To what end? Drip irrigation and the water-energy-food nexus in Morocco," *International Journal of Water Resources Development*, vol. 31, pp. 393–406, July 2015.

[9] F. A. Ward and M. Pulido-Velazquez, "Water conservation in irrigation can increase water use," *Proceedings of the National Academy of Sciences*, vol. 105, pp. 18215–18220, Nov. 2008.

[10] P. Schmitter, K. S. Kibret, N. Lefore, and J. Barron, "Suitability mapping framework for solar photovoltaic pumps for smallholder farmers in sub-Saharan Africa," *Applied Geography*, vol. 94, pp. 41–57, May 2018.

[11] E. S. Hrayshat and M. S. Al-Soud, "Potential of solar energy development for water pumping in Jordan," *Renewable Energy*, vol. 29, pp. 1393–1399, July 2004.

[12] M. Aliyu, G. Hassan, S. A. Said, M. U. Siddiqui, A. T. Alawami, and I. M. Elamin, "A review of solar-powered water pumping systems," *Renewable and Sustainable Energy Reviews*, vol. 87, pp. 61–76, May 2018.

[13] M. P. Blimpo and M. Cosgrove-Davies, "Electricity Access in Sub-Saharan Africa," p. 167, 2019.

[14] R. Allen, L. S. Pereira, D. Raes, and M. Smith, "FAO Irrigation and Drainage Paper No. 56," tech. rep., Rome, 1998.

[15] F. Grant, C. Sheline, J. Sokol, S. Amrose, E. Brownell, V. Nangia, and A. G. Winter, "Creating a Solar-Powered Drip Irrigation Optimal Performance model (SDrOP) to lower the cost of drip irrigation systems for smallholder farmers," *Applied Energy*, vol. 323, p. 119563, Oct. 2022.

[16] E. A. Abioye, M. S. Z. Abidin, M. S. A. Mahmud, S. Buyamin, M. H. I. Ishak, M. K. I. A. Rahman, A. O. Otuoze, P. Onotu, and M. S. A. Ramli, "A review on monitoring and advanced control strategies for precision irrigation," *Computers and Electronics in Agriculture*, vol. 173, p. 105441, June 2020.

[17] A. Srinivasan, ed., *Handbook of Precision Agriculture: Principles and Applications*. Binghamton, NY: The Haworth Press, Inc., 2006.

[18] I. Yahyaoui, F. Tadeo, and M. V. Segatto, "Energy and water management for drip-irrigation of tomatoes in a semi- arid district," *Agricultural Water Management*, vol. 183, pp. 4–15, Mar. 2017.

[19] A. Merida Garcia, I. Fernandez Garcia, E. Camacho Poyato, P. Montesinos Barrios, and J. Rodriguez Diaz, "Coupling irrigation scheduling with solar energy production in a smart irrigation management system," *Journal of Cleaner Production*, vol. 175, pp. 670–682, Feb. 2018.

[20] V. Zavala, R. LÃ³pez-Luque, J. Reca, J. MartÃnez, and M. Lao, "Optimal management of a multisector standalone direct pumping photovoltaic irrigation system," *Applied Energy*, vol. 260, p. 114261, Feb. 2020.

[21] O. Adeyemi, I. Grove, S. Peets, and T. Norton, "Advanced Monitoring and Management Systems for Improving Sustainability in Precision Irrigation," *Sustainability*, vol. 9, p. 353, Feb. 2017.

[22] G. D. Van de Zande, C. Sheline, and A. G. Winter, "Evaluating the Potential for a Novel Irrigation System Controller to Be Adopted by Medium-Scale Contract Farmers in East Africa," in *Volume 6: 34th International Conference on De-*

Copyright © 2023 by ASME

sign Theory and Methodology (DTM), (St. Louis, Missouri, USA), p. V006T06A037, American Society of Mechanical Engineers, Aug. 2022.

[23] A. Sheshadri, M. Borrus, M. Yoder, and T. Robinson, "Midlatitude Error Growth in Atmospheric GCMs: The Role of Eddy Growth Rate," *Geophysical Research Letters*, vol. 48, Dec. 2021.

[24] J. Woetzel, D. Pinner, H. Samandari, H. Engel, M. Krishnan, R. McCullough, T. Melzer, and S. Boettiger, "How will African farmers adjust to changing patterns of precipitation?," tech. rep., McKinsey Global Institute, 2020.

[25] J. Doorenbos and A. Kassam, "FAO Irrigation and Drainage Paper 33: Yield Response to Water," tech. rep., FAO, Rome, 1979.

[26] Deloitte, "Sub-Saharan Africa Mobile Observatory," tech. rep., 2012.

[27] D. G. Gibson, B. Ochieng, E. W. Kagucia, J. Were, K. Hayford, L. H. Moulton, O. S. Levine, F. Odhiambo, K. L. O'Brien, and D. R. Feikin, "Mobile phone-delivered reminders and incentives to improve childhood immunisation coverage and timeliness in Kenya (M-SIMU): a cluster randomised controlled trial," *The Lancet Global Health*, vol. 5, pp. e428–e438, Apr. 2017.

[28] C. Pop-Eleches, H. Thirumurthy, J. P. Habyarimana, J. G. Zivin, M. P. Goldstein, D. de Walque, L. MacKeen, J. Haberer, S. Kimaiyo, J. Sidle, D. Ngare, and D. R. Bangsberg, "Mobile phone technologies improve adherence to antiretroviral treatment in a resource-limited setting: a randomized controlled trial of text message reminders," *AIDS*, vol. 25, pp. 825–834, Mar. 2011.

[29] M. J. Coulentianos, I. Rodriguez-Calero, S. R. Daly, and K. H. Sienko, "Global health front-end medical device design: The use of prototypes to engage stakeholders," *Development Engineering*, vol. 5, p. 100055, 2020.

[30] C. A. Lauff, D. Knight, D. Kotys-Schwartz, and M. E. Rentschler, "The role of prototypes in communication between stakeholders," *Design Studies*, vol. 66, pp. 1–34, Jan. 2020.

[31] D. S. Silva, A. Ghezzi, R. B. d. Aguiar, M. N. Cortimiglia, and C. S. ten Caten, "Lean Startup, Agile Methodologies and Customer Development for business model innovation: A systematic review and research agenda," *International Journal of Entrepreneurial Behavior & Research*, vol. 26, pp. 595–628, May 2020.

[32] B. G. Mgendi, S. Mao, and F. Qiao, "Does agricultural training and demonstration matter in technology adoption? The empirical evidence from small rice farmers in Tanzania," *Technology in Society*, vol. 70, p. 102024, Aug. 2022.

[33] K. Thinda, A. Ogundeji, J. Belle, and T. Ojo, "Understanding the adoption of climate change adaptation strategies among smallholder farmers: Evidence from land reform beneficiaries in South Africa," *Land Use Policy*, vol. 99,

p. 104858, Dec. 2020.

[34] M. Benouniche, M. Kuper, A. Hammani, and H. Boesveld, "Making the user visible: analysing irrigation practices and farmersâ logic to explain actual drip irrigation performance," *Irrigation Science*, vol. 32, pp. 405–420, Nov. 2014.

Copyright © 2023 by ASME

**Proceedings of the ASME 2023
International Design Engineering Technical Conferences and
Computers and Information in Engineering Conference
IDETC-CIE2023
August 20-23, 2023, Boston, Massachusetts**

DETC2023-112625

LAB-TO-MARKET DESIGN OF AN ELECTRODIALYSIS-BASED HOME-SCALE WATER DESALINATION SYSTEM

Marie Floryan[1] , Quinn Bowers[1], Zachary Sternberg [1], Sahas Gembali [1], Akshita Goyal [1], Jonathan Bessette [1], Soraya Honarparvar [1], Amos Winter[1],*

[1]Massachusetts Institute of Technology, Cambridge, MA

ABSTRACT

Water scarcity is increasingly becoming a problem in India, so much so that a recently proposed piece of legislation would mandate home-scale water desalination systems to have a 60% recovery ratio [1], significantly higher than existing systems. We developed an electrodialysis-based home-scale desalination system for obtaining drinking water that will meet and surpass this new proposed recovery ratio. This project builds heavily on previous and current work by the MIT Global Engineering and Research (GEAR) Lab and was conducted with Eureka Forbes Limited (EFL), a major stakeholder in India's home-scale desalination industry. Two critical design modules within the desalinator are explored - the system integration and fluid reversal mechanism and the electrodialysis stack end cap analysis. We developed a model framework that will output the validity of end cap designs based on established design requirements, designed fluid reversal mechanisms for distinct markets, and fabricated a prototype desalination cabinet that fits within the EFL's design constraints. We also discuss the details of our collaboration with both academic and industrial partners, extract specific lessons learned about the lab-to-market design process, and discuss some of the specific challenges faced by a design group in our circumstances and the ways we resolved them.

Keywords: desalination, electrodialysis, India

1. INTRODUCTION

Drinking water is the most basic global health need to be addressed as per the World Health Organization (WHO) [2]. Access to clean drinking water is also one of the UN Sustainable Development Goals [3].

Even though the local municipal bodies in India treat water for physical and biological contaminants before supplying it to households (with varying degrees efficacy depending on location, temperature, season [4, 5]), high total dissolved salts (TDS) levels present in tap water across much of India make it unfit for drinking (typical ranges up to 2000 ppm). The upper limit of

TDS levels in tap water in India is 500 ppm [6], while the WHO rates water with TDS greater than 500 ppm as "unacceptable" and TDS between 375-500 ppm as "poor", and 125-250 ppm as "good" [7]. Water desalination techniques such as Reverse Osmosis (RO) can help reduce the TDS of water making it healthy for drinking when used in combination with a sediment filter and UV disinfection treatment [8]. RO can remove up to 99%+ of the dissolved salts (ions), particles, colloids, organics, bacteria, and pyrogens from the feed water through semi-permeable membranes at the industrial scale [9]. However, in-home RO systems have a recovery rate of 25% to 50% [10].

Only 2% of 22,000 survey participants in a LocalCircles survey get drinkable quality water from their local body and do not require additional purification steps, while 65% use some kind of at-home filtration mechanism to obtain drinkable water [11]. There are also many households in India that have bore wells dug in their houses to directly access groundwater that is untreated and often needs purification before drinking [11].

Tap water salinity has increased the adoption rate of standalone home-scale water purification systems over the years. Approximately 12.94 million RO units are sold in India every year [12]. These units can hold 5-15 L of water depending on the size and are designed to be mounted on walls or fit inside kitchen cabinets.

The taste of water from RO-based systems has become an indicator of water quality for users over the years. Consumers associate the taste of 100 ppm TDS with good, drinking water, even though 375 ppm is fit for drinking [7]. Consequently, many users employ domestic water treatment systems on water already suitable for drinking due to taste preference, leading to excess waste water due to the low recovery of RO systems. On average 75L of water is required to obtain just 15L of drinking water (average per day consumption per family of four) from an RO system which is 20% recovery ratio (a measure of the water usage efficiency of the system) [13]. This amount of water wastage, in addition to the water usage from crop irrigation, has caused increased stress on the groundwater table in India, leading to an upcoming legislative decision of mandating at least 60% water

---

*Corresponding author: awinter@mit.edu
Documentation for `asmeconf.cls`: Version 1.34, June 5, 2023.

recovery rate in any new home-scale desalination systems [1].

Electrodialysis (ED)-based membrane desalination technology (currently used in industrial desalination) has a water recovery rate of up to 90% [10], making it a potentially more sustainable alternative to the conventional RO system for low TDS water. But, ED-based systems are not currently designed for home-scale desalination. MIT GEAR lab has successfully created a lab proof of concept for home-scale desalination [14–16]. This paper discusses the product design journey of a lab-scale setup to a home-scale desalination unit designed for manufacturing, assembly, and service.

1.1 Understanding the Indian Market

The demand for drinking water is expected to grow to 73 billion (Bn) cubic meters in 2025 and 102 Bn cubic meters by 2050 in India [17], thus making it even more essential for membrane technology in home scale setup to be more efficient. The Indian market for water purification cabinets was worth USD 754.2 million in 2020 and is projected to reach USD 1.9375 Bn by the year 2027, growing at a cumulative annual growth rate (CAGR) of 14.3% during the forecast period (2020-2027) [17]. This growth can be attributed to the expanding middle class and subsequent increase in health concerns.

Our project was conducted in partnership with the leading manufacturer and supplier of these RO-based water purification units in India- Eureka Forbes Limited (EFL). EFL owns 56% market share of water purifiers in India and has become a household name since their first filtration product called 'Aquaguard'[13].

The current products from EFL are a combination of RO and UV filtration technology and are priced between USD 180 - USD 460 [18]. Since India is a cost-sensitive market, a sizeable departure from their premium market segment would lower the percentage adoption of any new technology restricting us to work within market price constraints.

2. BACKGROUND

2.1 Lab Version of Home Scale Electrodialysis Desalination

GEAR Lab at MIT has been working on creating a home-scale version of the electrodialysis technology that can provide up to 75% water recovery based on the salinity of groundwater in India, far exceeding the 60% policy mandate[1]. The electrodialysis membranes used in such an electrodialysis system are shown in Figure 1 A. 21 membrane pairs are alternately sandwiched between two platinum electrodes. Corrosive acid is typically produced in the anode compartment, and so platinum is typically used for the electrodes due to its highly non-corrosive and non-reactive nature, though carbon electrodes are also emerging. The stack of membranes is then compressed inside two solid end caps to close the system and prevent leakage. Water passes through the ED stack where the electrodes separate salt ions and heavy metal ions into cations and anions (Figure 1 B), directing them into the concentrated water stream for disposal, thus leaving desalinated water for the diluate water stream. This process is discussed in more detail in other GEAR lab publications [10] [15] [14].

The overall system required to achieve clean drinking water from the electrodialysis technology is shown in Figure 2. Components like sediment filter, antiscalant, copper/zinc post carbon

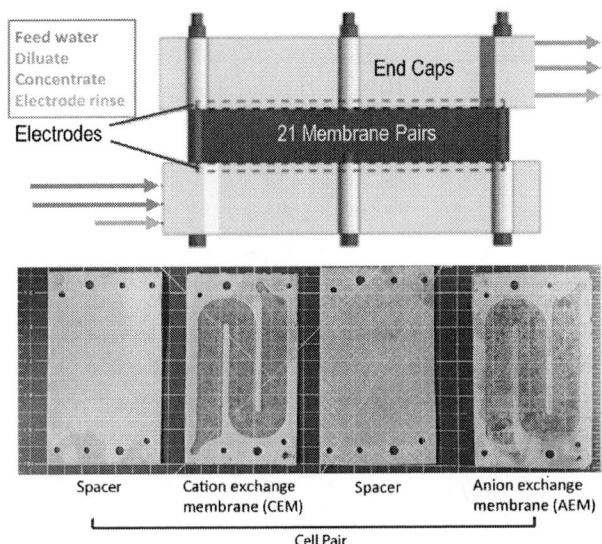

FIGURE 1: ED STACK. (A) WHOLE STACK WITH CRITICAL COMPONENTS AND WATER STREAMS LABELLED. (B) MEMBRANE PAIRS.

TABLE 1: MAINTENANCE COST COMPARISON BETWEEN A TYPICAL RO CABINET AND THE PROPOSED ED CABINET

Maintenance	RO Cabinet	ED Cabinet
Maintenance type	RO cartridge replacement	Acid tank refill
Frequency	Annual	Annual
Annual cost	50 USD	10 USD
5-year lifetime maintenance cost	250 USD	50 USD

treatment, UV filter listed in the high-level system architecture already exist in Eureka Forbes' RO line of products, but changes to the fluid controls and pressure pumps were required to replace the RO system parts with ED parts. Although ED membranes have an initial high capital cost, they have the potential to last as long as 7 years under proper use conditions [19]. The cost of a full ED stack (membranes and end caps both) is USD $85, much higher than that of RO membrane cartridges, which must be replaced annually. To operate at the required price points of the Indian Market, and maximize the lifetime of the ED membranes, components of the RO fluid control system were removed or modified and and the introduction of an acid dosing module was added. The increased system efficiency may lower overall cost of operation over a five year period, by reducing maintenance costs (Table 1).

2.2 Challenges with the Lab-Scale Setup

The GEAR Lab's lab-scale setup (Figure 5), uses industrial components, both large in volume and non-scalable for manufacturing. In order to create a viable product, a redesign of system components and fittings with allocated space within a cabinet is

Copyright © 2023 by ASME

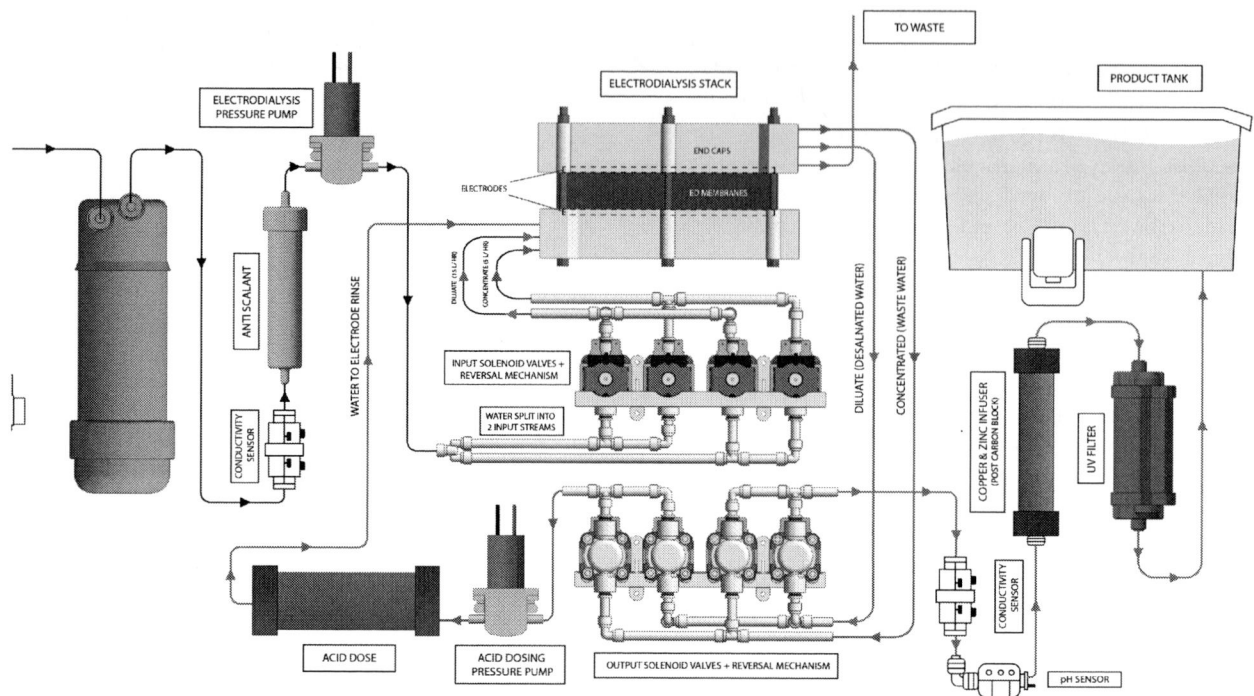

FIGURE 2: LAYOUT OF ALL THE COMPONENTS REQUIRED FOR WATER DESALINATION IN THE ED SYSTEM. LINES WITH ARROWS INDICATE WATER FLOW PATHS, WHERE BLACK IS FEED WATER, PURPLE IS ACID DOSING STREAM, BLUE IS DILUATE, AND RED IS CONCENTRATE.

required along with a detailed design for manufacturing (DFM), analysis and plan.

We elected to work on three modules of the ED cabinet, to bring the product closer to launch and commercial success. We evaluated the fluid control mechanism, which swaps which channel in the ED stack is concentrate and which is diluate. We also integrated the system, to fit all the system components into a cabinet while minimizing points of failure. Finally, we redesigned the ED stack end caps, to optimize for size, cost, and effectiveness. Sections 3 and 4 discuss in detail the design explorations, system analysis, and decision-making for the critical modules.

3. SYSTEM INTEGRATION AND FLUID CONTROLS
3.1 Background
RO systems force a single feed water stream through a filter to reduce salinity, whereas ED systems split the feed into two channels, the diluate stream and the concentrate stream, prior to entry to the stack. The charge across the electrodes moves the salt ions out of the diluate stream and into the concentrate stream; the concentrate cycles through to rinse the electrodes and is then disposed of, while the diluate continues to the final UV disinfection stage before being deposited into the product tank.

Over time, the channel for the concentrate stream will experience scaling and buildup; therefore, critical to the operational lifetime of the ED stack, the GEAR Lab has implemented a reversal network module designed to flush both channels, as well as swap which channel contains the concentrate and which contains

the diluate [20]. The process shown in Figure 3 requires a matrix of solenoids and T/Y-Splits in order to flip the input streams for the diluate and concentrate, then restore the streams for the output. Two conditions define which is the diluate and which is concentrate: (i) the flow rate, 5L/hr for concentrate , and 15L/hr for diluate, and (ii) the polarity of the electrodes which defines the direction in which the ions are moving. The concentrate flow rate influences recovery ratio and concentrate TDS. The reversal ideally occurs every 7L of clean water produced, with a brief flush to disposal in between each cycle. Additional systems to reduce scale buildup include sediment filtration and antiscalant stages prior to entering the ED stack and acid dosing system after the stack, prior to the electrode rinse.

In order to manage these dynamic fluid streams, this section will discuss the process and challenges presented when designing and implementing a reversal network module into the system, inclusive of the volumetric constraints and integration of new components into the current system architecture.

3.2 Methods
Four ED fluid flow architectures were considered to optimize the reversal module architecture. The selected architecture (Figure 3) features one upstream flow regulator and two sets of four solenoid valves, one upstream and one downstream of the ED stack. The alternative options considered using various combinations of one or two pressure regulators, T/Y-split valves, and standard solenoid valves. The final architecture was chosen based

Copyright © 2023 by ASME

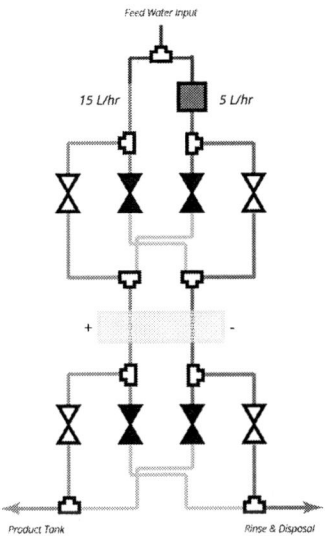

FIGURE 3: SELECTED FLUID CIRCUIT ARCHITECTURE. PURPLE IS FEED WATER, BLUE IS DILUATE, AND RED IS CONCENTRATE. SNAPSHOT IN ONE STATE, ALTERNATE STATE WOULD SEE OPEN SOLENOIDS BECOME CLOSED AND VICE-VERSA.

on cost and component availability requirements.

Since the electronic pressure regulator and multichannel solenoids are cost-prohibitive and not common off-the-shelf components, methods were developed to determine if the novel system would fit and how it could be integrated into EFL's existing product architecture.

Three methods were performed in order to understand the volume distribution in the cabinet: (i) product tear down (ii) in-depth CAD study and modeling, and (iii) physical prototyping The product tear down of the EFL *Superb* and *Enhance* models showed the different ways in which EFL configures their products, what frameworks and components are consistent across products, and which manufacturing methods may be available in the system design. The team used a CAD model of another EFL cabinet, the *Astor*, that then enabled us to (i) identify an available envelope for our ED Stack and reversal modules (ii) explore positioning for the critical modules (iii) model plumbing architectures linking components and (iv) estimate the volumetric distribution of different modules in the system. Lastly, the physical prototype would validate the design and integration of new components and accessories.

3.3 Results

Overall, we confirmed configuration Option #1 can indeed fit within the confines of the EFL *Astor* cabinet when RO-specific components are removed and have performed a volumetric study shown in Figure 4 of both the RO model and our proposed ED model.

Within the *Astor* cabinet's volume, the 7-Liter product tank occupies approximately 25.72%. About half of that space by-volume (14.45%) is occupied by the ED Stack with its maximum thickness end caps (9.17%) and the reversal solenoids (5.28%). The *Astor* employs a 4 bar of pressure, 1.1 L/min flow rate feed

pump which occupies 5.39% of the cabinet volume; our team has identified a pump offering 2.4 bar of pressure and 2 L/min flow rate which meets the requirements to support ED and only occupied 1.04% of the cabinet volume, or one-fifth the size of the original pump (Figure 4 E). In addition to the stack and reversal network, an antiscalant tube and a 100 mL acid dosing module have been introduced into the ED cabinet, while the Pre-Ag filter and RO filter have both been removed. Additionally, the sediment filter, UV filter, post-carbon filter, DC adapter have all carried over based on the system requirements from EFL and the GEAR Lab. All tubing and interior connections have been planned and modeled in CAD in order to account for all viable remaining space. Thus, the remaining free space has been broken into two categories: usable and unusable. Usable represents rectilinear voids where additional components currently not accounted for (i.e. control PCB) could be integrated into future iterations of the design.

The remaining space in the cabinet has been deemed unusable since design features, curvatures, and protrusions inhibit the introduction of new components. While the amount of unusable space exhibits negligible change in the two configurations, the ED model has just under half the amount of usable free space remaining.

3.4 Discussion

The main challenge is inserting the ED Stack and reversal network in the cabinet, as compact as possible, while accommodating for both the components carried over from the RO unit, as well as the additional new support components (Figure 4 A). The weight of the stack and solenoids can provide a stable center of gravity, so we prioritized central placement. The GEAR Lab advised the stack should stand vertically to minimize leakage; additionally, the stack has been placed in the rear of the unit to prevent tampering by the customer.

Custom brackets have been designed to support both the ED stack and the solenoid blocks and attach them to the center support wall. The design of the brackets has adhered to DFM considerations for injection-molding with consistent wall thickness, zero undercuts, and minimized material where possible; the brackets were fabricated using 3D printing for testing in the physical prototype.

The ED stack bracket features a "center-out" design with a raised center section based on the 22-24 mm thickness of the pressure-applied membranes, while keeping the outer edges open to allow for continued experimentation of the end cap thickness, which will be discussed in Section 4. Each unit uses two brackets, a top and bottom, to compress the stack in place; however, due to the weight of the stack, the bottom bracket rests on the inset in the enclosure for the feed and disposal valves.

The solenoid brackets feature a custom-fit bed to block four solenoids together. While the solenoids are oriented to pass water through vertically, 90° angle press-fits, T/Y-Splits, and short lengths of tube are utilized to bus solenoids together horizontally, effectively creating an ad hoc "multi-channel" solenoid system in as efficient a space as possible. By flipping the input block and the output block 180° (Figure 4 B-D), we save approximately 20 mm of vertical footprint layering the tubing. If the press-fits and

Copyright © 2023 by ASME

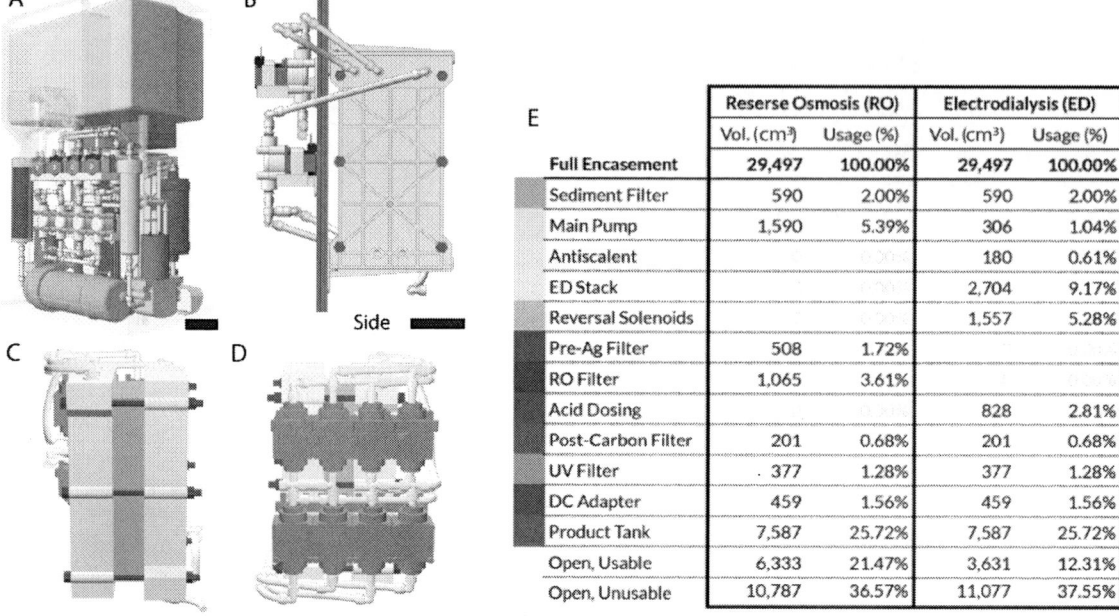

	Reserse Osmosis (RO)		Electrodialysis (ED)	
	Vol. (cm³)	Usage (%)	Vol. (cm³)	Usage (%)
Full Encasement	**29,497**	**100.00%**	**29,497**	**100.00%**
Sediment Filter	590	2.00%	590	2.00%
Main Pump	1,590	5.39%	306	1.04%
Antiscalent			180	0.61%
ED Stack			2,704	9.17%
Reversal Solenoids			1,557	5.28%
Pre-Ag Filter	508	1.72%		
RO Filter	1,065	3.61%		
Acid Dosing			828	2.81%
Post-Carbon Filter	201	0.68%	201	0.68%
UV Filter	377	1.28%	377	1.28%
DC Adapter	459	1.56%	459	1.56%
Product Tank	7,587	25.72%	7,587	25.72%
Open, Usable	6,333	21.47%	3,631	12.31%
Open, Unusable	10,787	36.57%	11,077	37.55%

FIGURE 4: ED CABINET MOCK UP. (A) 3D LAYOUT OF ALL REQUIRED ED CABINET COMPONENTS. (B) SIDE, (C) BACK, AND (D) FRONT VIEW OF THE VALVE AND STACK MODULE. (E) COMPARISON OF A VOLUMETRIC BREAKDOWN OF COMPONENTS BETWEEN AN RO CABINET AND A THE PROPOSED ED CABINET. SCALE BAR IS 4 CM.

tubing could be consolidated into custom injection-molded components, part counts, assembly time, and points of failure could be further reduced, however this is a costly procedure and would likely not be implemented in a minimum viable product. Furthermore, the exact valves will have to be chosen by EFL as they will also have to conform to their maintenance and component failure practices.

Throughout this process, it was discovered that most EFL cabinets, regardless of range, occupy an estimated 300 mm x 280 mm x 500 mm rectilinear envelope, however, each product's encasement has been spatially optimized based on its interior components. In the *Astor*, concave curves cut out the back corners of the enclosure; if more space is required, squaring off these corners could not only add interior volume, but also improve the ratio of usable space to unusable space (currently approx. 1:2).

The findings presented here can be applied to other home-scale ED systems, but generally this design process was specific to EFL. Future steps for the system integration and fluid control include:

- Identifying the optimal envelope for control PCB Investigating custom injection-molded fluid bus to reduce the quantity of press-fit joints

- Confirming pump, solenoid models and availabilities with EFL's third-party vendors

- Performing design review with EFL engineering and R&D teams to refine design choices

- Integrating all components into physical model to perform system validation testing

- Further optimizing module layout

4. ELECTRODIALYSIS STACK DESIGN

4.1 Background

The end caps are critical to preventing leakage from the stack (Figure 1). The end caps must be stiff enough to provide sufficient compression to the stack membranes, while also minimizing the amount of material to minimize end cap weight, volume, and cost. We developed a framework that evaluates the capability of end caps to meet these design requirements and allows for end cap design iteration and informs end cap designs for different membrane geometries. We also discuss non-structural end cap design changes that minimize manufacturing costs and simplify assembly.

4.2 Stiffness Design

We want our end caps to be thin and light to reduce material cost, but we concurrently needed to ensure sufficient stiffness for appropriate load distributions and sealing. There are two pressures to consider when designing end caps: (i) there's a distributed load from pressure due to water flow through the stack and (ii) that there are reaction forces applied via the end cap bolts through preload. These two pressures tend to push the end caps apart, causing them to bow in the center. Furthermore, while the preload pressure works to compress and seal the membranes, the water pressure will work to push the membranes apart, causing leakage. The leakage will occur either at the outer perimeter of the stack, or inside of the stack in between membranes, which would cause the fluid circuit to short circuit, reducing the water flow path length and thus reducing the salt removal efficiency of

Copyright © 2023 by ASME

the stack. We were unable to find guidelines to determine the preload pressure necessary to ensure there is no leakage between membranes so we developed an experimental model coupled with a simulation to estimate the minimum preload for varying end cap geometries.

First we devised a scalar salt removal effectiveness metric,

$$effectiveness = \frac{concentrate\ TDS - diluate\ TDS}{feed\ TDS} \quad (1)$$

which is based on measuring the TDS of the in-flowing feed water and the out-flowing concentrate and diluate streams. At baseline, when the water is running through the stack without the desalination function turned on, the TDS of the concentrate and diluate streams will each be equal to the feed TDS, giving an effectiveness of 0. The maximum possible effectiveness is 2, when all salt from the diluate is removed to the concentrate (assuming equivalent flow rates or a recovery ratio of 50%; testing was done at steady state and with the same flow rates in each stream, so the results are self-consistent). Assuming no TDS losses,

$$concentrate\ TDS + diluate\ TDS = 2 * feed\ TDS \quad (2)$$

. When there is internal stack leakage, we expect the effectiveness of the system to decrease. We want to ensure our end caps maximize this effectiveness value, without changing any of the other operational settings of the ED system. When salt removal operational settings are held constant, the only variable that influences the effectiveness is the time the water spends within the stack, and so effectiveness can be used as a measure of internal stack leakage. Effectiveness, as we have defined it, is a universal parameter that can be applied to other membrane geometries of similar channel width and height, and is independent of the end cap geometry.

Our goal was to develop a framework that can be used to design end caps for different membrane geometries without having to physically prototype the end caps on each iteration to know if the design will prevent leakage between the membranes, and thus maximize the effectiveness. Furthermore, the framework can also be used to design end caps with less material while ensuring they meet their stiffness requirement to prevent leakage. To develop this framework, we first used an experimental method to determine the plateau effectiveness of the bench-top prototype and extract membrane stack material properties. We then built a CAD and FEA model of the membrane stack and end cap to identify the minimum membrane deflection necessary to prevent leakage. We then iterated through several end cap geometries and plotted a phase diagram that identifies a end cap design region that will satisfy operational requirements. Finally, this process was validated with an experiment using this framework-driven end cap design methodology, and then the process was applied to other membrane geometries.

4.3 Experiment Procedure

Our experimental setup is depicted in Figure 5. We created a feed water supply of roughly 500 ppm TDS, then used a conductivity sensor to find the actual salinity of our feed water.

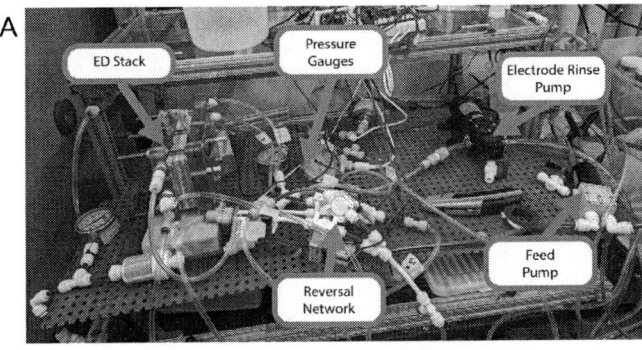

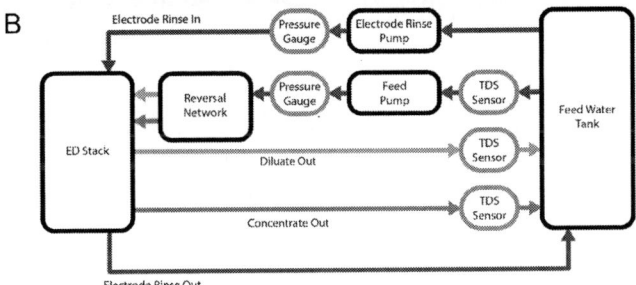

FIGURE 5: EXPERIMENTAL SETUP. (A) BENCH TOP SETUP AND (B) DIAGRAM OF THE SETUP.

We created end caps with a known geometry and attached them to the stack. We then used a torque wrench to set the six bolts connecting the end caps to a known preload. We measured the distance between the end caps i.e. the thickness of the membrane stack. Finally, the system was run for several minutes and the feed pressure, electrode rinse pressure, and the TDS of the concentrate, diluate, and feed channels were measured (Figure 6 A). All these measurements were taken two times each for a range of bolt preloads between 1 and 4 Nm. The user-controlled desalination parameters were held constant among all experiments. The system was run at 16 V and about 600 mA, running on the order of tens of watts. RO systems typically run on the order of 100s of watts.

4.4 Analysis

Bolt preload was measured as a torque applied to the bolt, but for further analysis was converted to a force, as

$$Bolt\ force = \frac{bolt\ torque}{bolt\ diameter * k} \quad (3)$$

where k = 0.2 [21]. This k is the bolt torque constant, and is determined by a variety of factors including the materials of the nuts and bolts and the alignment of the bolt holes. It captures friction losses on the bolt, and can only be measured experimentally. We did not do this, and instead used the generally accepted value of 0.2.

We also needed to estimate the bulk stiffness of the membrane stack. We did this by estimating the material property Young's Modulus, E. Young's modulus is defined as

$$E = \frac{\sigma}{\epsilon} \quad (4)$$

Copyright © 2023 by ASME

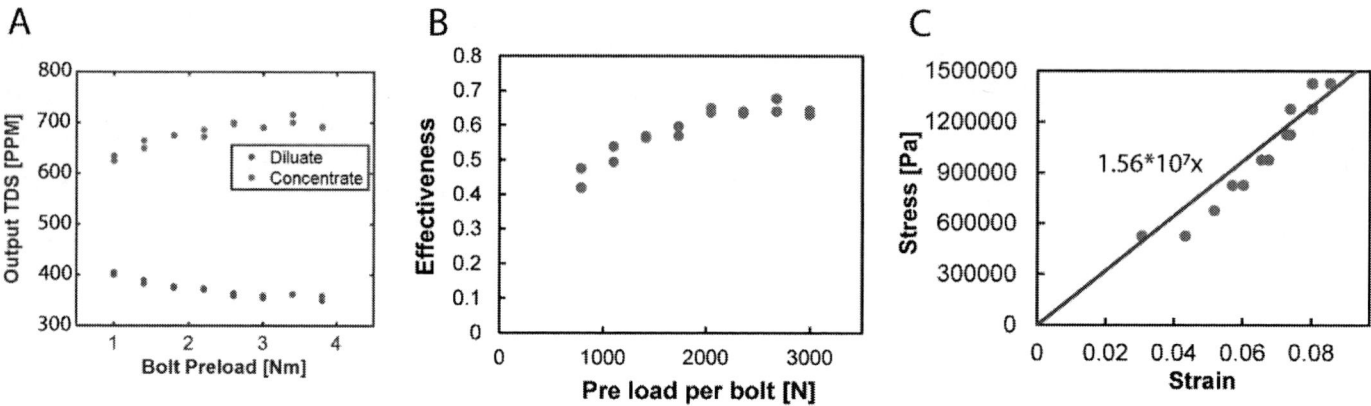

FIGURE 6: EXPERIMENTAL OUTPUTS. (A) TDS OF DILUATE AND CONCENTRATE AND (B) EFFECTIVENESS FOR INCREASING BOLT PRELOAD. (C) STRESS-STRAIN PLOT FOR MEMBRANE STACK IN COMPRESSION.

where strain, ϵ is the fractional compression of the membrane stack from rest, and σ is the applied stress. The thickness of the membrane stack was measured at each preload, and was used to compute ϵ. The stress, σ, was approximated as the total bolt force over the area of the membrane stack. We plotted these stress and strain values and calculated the slope to estimate the Young's modulus of the membrane stack to be approximately 15.5 MPa.

We note that plateau effectiveness is 0.65, and so we chose the minimum effectiveness to be 0.585, 90% of the plateau effectiveness of 0.65 (Figure 6). We expect this is the peak effectiveness of the system at its current settings, and that the effectiveness is lower at lower preloads due entirely to internal leakage short-cutting the tortuous path. We then used SolidWorks to model the interaction between the end caps and the membrane stuck and run finite element analysis (FEA) to find the minimum compression of the membrane stack for each condition. The model of the end cap makes use of known geometries and material properties, and the membrane stack model makes use of known geometries and computed Young's modulus (Figure 6). Due to symmetry between the two end caps, half of the stack was modelled to reduce computational effort. The bottom of the membrane stack model (center plane in the whole stack), was fixed, and the end cap was constrained to make contact with the perimeter of the membrane stack. Bolts were modelled and a downward force was applied to model the bolt preload. The internal water pressure acting on the end cap from below was modelled as a uniform pressure acting upwards onto the end cap. Each experimental preload condition was simulated and the minimum membrane deflection that corresponds to the previously identified minimum bolt preload force of 1400 N was found to be 0.23 mm (Figure 6).

4.5 Geometric Design Process

The one inch end caps used in the experiment provided the necessary stiffness to compress the membranes to prevent leakage and ensure effective operation. Motivated by reducing material consumption and lowering weight, end caps of lower thicknesses were simulated to find an optimal end cap that balances both the high stiffness and the lower material usage requirements. The simulation results are summarized in a phase plot that plots the max % end cap deformation (which occurs in the center of the

end cap), against the applied bolt preload (Figure 7). Three regions were identified on this plot. The leak region, in blue, represents conditions where membrane deflection is not sufficient to prevent leakage. The high stress region, in orange, identifies conditions where the simulated maximum von Mises stress is large compared to the material tensile and flexural strength. The feasible conditions fall into the remaining unshaded region. Here, the membranes are compressed enough to prevent leakage, and the stresses in the end cap fall within an allowable range.

The analysis above suggested that a 0.75 inch end cap with 2362 N bolt preload should have the same deformation and effectiveness, as the one inch end caps used for the initial experiment with 1732 N bolt preload. To validate our analysis, we manufactured end caps 0.75 inches thick, and tested them using the same experimental setup as previously mentioned.

Based on the FEA, the 0.75 inch end caps were expected to be less effective below bolt torques of 2000 N. From the experimental validation results (Figure 7), the validation effectiveness approaches the same plateau effectiveness of 0.65, but it reaches this plateau at a higher preload than the initial experiment. This agrees with the FEA model, which predicted a higher preload required to fully compress the membrane stack. The final validation was to plot the minimum membrane compression as a function of preload (Figure 7). The plot shows that a higher preload must be applied onto the 0.75 inch end caps to ensure they meet the minimum 0.23 mm membrane compression requirement.

A similar simulation analysis was conducted on a different membrane and end cap geometry (Figure 7 C). Here, the aspect ratio of the membrane and end caps is longer. A ribbed geometry for the end caps was also tested in effort to further reduce material. It was assumed that the material properties of the membrane stack remained the same. We then used FEA to iterate through possible end cap geometries, all solid blocks of ABS and between 0.5 and 1 inch thick. The results are shown in a phase plot (Figure 7). Only a few geometries were able to satisfy both the stress and deformation requirements.

This FEA analysis, now validated, was used to design end caps for new membrane geometries. We found that with acrylonitrile butadiene styrene (ABS), the von Mises stress region

Copyright © 2023 by ASME

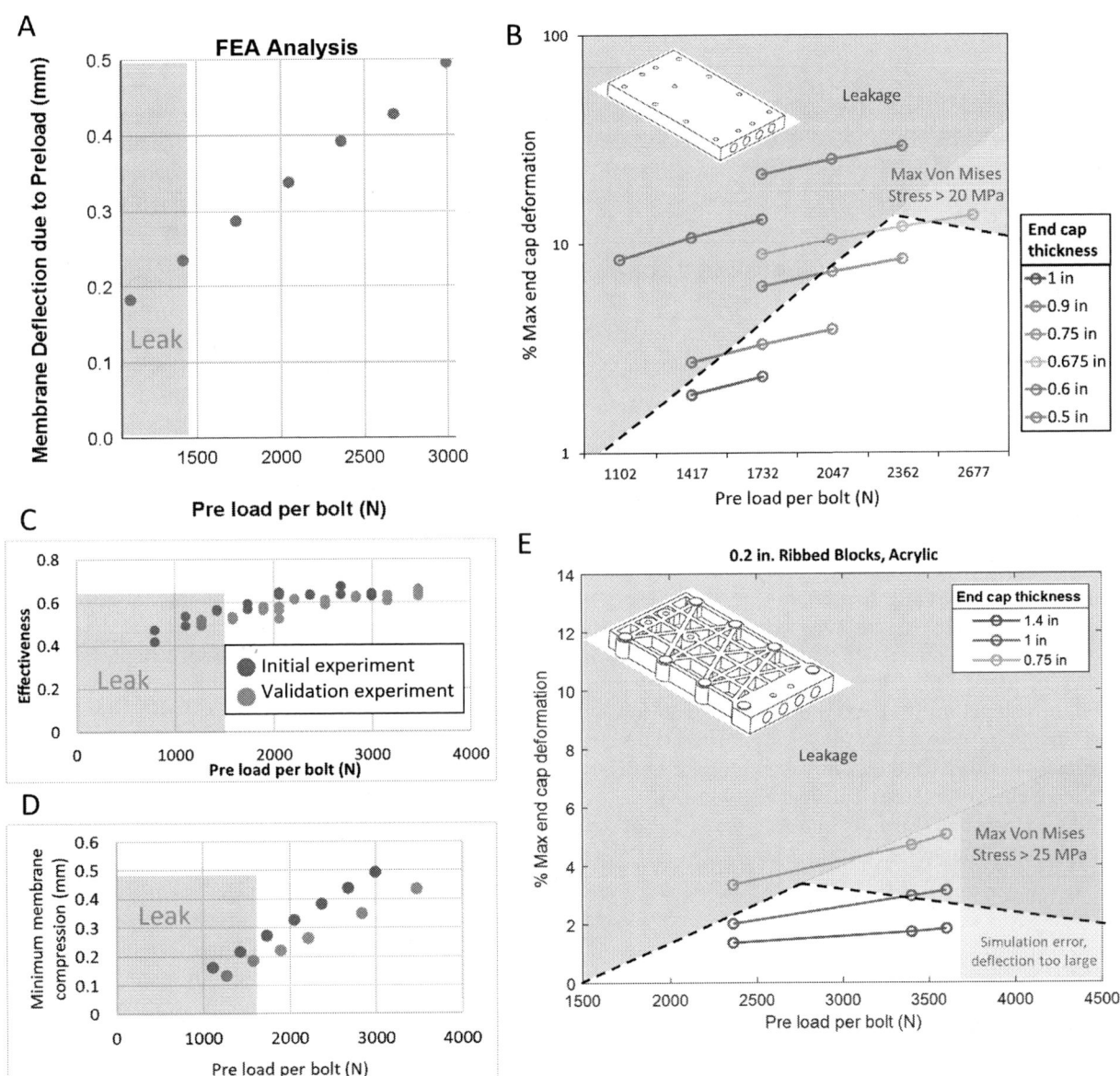

FIGURE 7: SIMULATION OUTPUTS. (A) FEA ANALYSIS FOR IDENTIFYING THE MINIMUM MEMBRANE DEFLECTION FOR LEAKAGE. (B) PHASE PLOT FOR EXPERIMENTAL SOLID END CAP DESIGN, WHERE BLUE REGION REPRESENTS LEAKAGE, ORANGE AREA REPRESENTS HIGH STRESS OPERATING REGIME, AND THE WHITE AREA REPRESENTS ADEQUATE OPERATION. COMPARISON OF EFFECTIVENESS (C) AND MINIMUM MEMBRANE COMPRESSION BEFORE LEAKAGE (D) BETWEEN INITIAL AND VALIDATION EXPERIMENTS. (E) PHASE PLOT FOR THEORETICAL RIBBED END CAP DESIGN WITH 0.2 IN WIDE RIBBING. COLORED REGIONS SYMBOLIZE SAME OPERATING CONDITIONS AS IN (B).

Copyright © 2023 by ASME

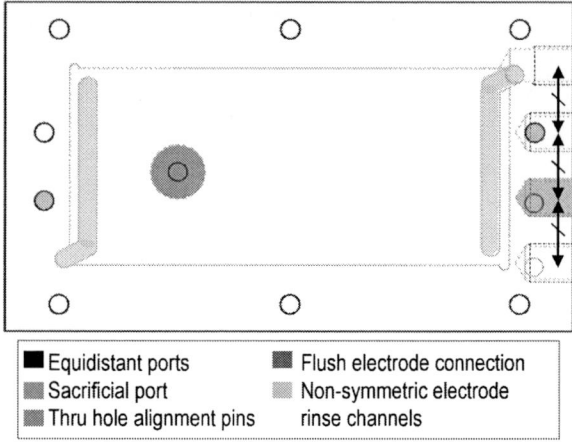

Equidistant ports
Sacrificial port
Thru hole alignment pins
Flush electrode connection
Non-symmetric electrode rinse channels

FIGURE 8: DESIGN FOR MANUFACTURING AND ASSEMBLY MODIFICATIONS TO THE END CAP DESIGN.

left too little workable design space. To have more freedom in the geometry of the end cap, the material was changed to acrylic, a stronger plastic with a higher EI. This change, in conjunction with the longer aspect ratio of the membrane geometry used and internal fluid pressure, allowed us to use ribbed end caps to meet the design requirements. End caps of constant rib thickness and varying heights are plotted in Figure 7 E.

4.6 Design for Manufacturing and Assembly

We determined early on that the end caps would be injection molded due to the volume and material of parts produced. Moreover, our industry partners, EFL, already injection mold other components and the cabinet itself. Capital costs for further injection molding should be low, thanks to existing relationships with manufacturers. To minimize costs further, we prefer both end caps to be identical, so both can be produced with only a single mold. This was accomplished by ensuring the end caps have rotational symmetry (Figure 8):

- Two through holes for alignment pins were added to have symmetry.

- Adding a sacrificial valve port and ensuring all water inlet and outlet valves on the end cap are equidistant. This sacrificial valve port will be plugged during the stack assembly process.

- The internal geometries of the electrode rinse channel were modified to have rotational symmetry.

- The electrode connected was modified so it sits flush with the electrode surface. This was done to minimize leakage at the electrode connection.

4.7 Discussion

The salt cut in these experiments was not to drinkable level because the stack tested was not optimized and the voltage and current applied were lower than required, and future work exploring the optimal stack design is currently being conducted. However, the results are useful in indicating required bolt preloads

and in crafting a design framework for stack fastening. Due to the increments used in increasing bolt torque, as well as the simplistic conversion of preloaded bolt torque to bolt force, the border between an end cap geometry that will leak and one that will not is not precisely defined. The phase plots in their current state serve more to inform regions of operation that are likely to meet the design requirements. Higher resolution FEA analyses coupled with long term aging experiments must be performed to accurately understand the behavior of the stack, and how preload and end cap geometry affect the effectiveness of the desalination process. More experiments must be performed to determine if there exists a region where membrane compression is too large and impedes the water flow within the channels of the membrane stack. If such a case exists, then this region will exist in the bottom right corner of the phase plot, where the preload is high and a stiffer end cap is used. There were some discrepancies between the initial experiment and the FEA validation experiment. The discrepancy in Young's moduli can be attributed to experimental error in measuring the thickness of the membrane stack, difference in initial membrane stack thickness, and differences in stack setup that may have caused uneven loading between the two experiments. A ribbed end cap is desired because it will allow for uniform, thin wall thickness everywhere along the end cap, making it suitable for injection molding and most cost effective for large production volumes. The FEA pipeline shows that an injection moldable, ribbed end cap can meet the design requirements if stronger plastics are used. An optimization study must be done with the end caps to determine the optimal geometry in terms of cost, ease and quality of injection molding, and compatibility with surrounding components within the cabinet.

5. CONCLUSIONS

After working on the critical modules of the ED desalination system- materials, components, design, and manufacturing processes have been identified that meet EFL's cost goals and make the team's proposed solution technically feasible and desirable by the Indian market. A fluid reversal system was identified for this initial product demonstration stage, cabinet modifications were identified to ensure all ED and desalination components can fit within the standard EFL cabinet shell volume, and a process for evaluating end cap effectiveness for preventing leakage and compressing the membranes and spacers was developed. The final steps of the project requires testing in different terrains across India for varying input TDS levels. The product user testing should provide evidence that the system is able to desalinate water effectively at varying TDS levels with at least a 70% recovery ratio. The product also needs to be tested for shipping and service training to ensure the high quality of the product and service.

The interdisciplinary learning from the team and systems thinking approach for complex problem-solving attributed to the success of this project. Lab-to-market productization requires a deep understanding of the technology, the user, manufacturing, supply chain, and market volatility, to make decisions that may not be the most desired solution options(in an ideal state) but are the ones that can be scaled for impact and profitability. As a team with skills in mechanical engineering, industrial design, human-centered design, business, and strategy, our solutions were

Copyright © 2023 by ASME

vetted from every lens and geared towards an understanding of developing markets and product commercialization.

An average Indian household of 4 members on average consumes 15 L of drinking water per day. Reverse Osmosis technology, operating at 20% efficiency, wastes 60L of water per household per day. When replaced with electrodialysis technology, the same family would waste only 5L of water per day. Given 1.2 million RO units sold per year in India [17], 66 million L of water per day and 2.4 trillion L per year can be saved by adopting this technology in India alone. Home-scale ED desalination based water purification has the potential to disrupt the water desalination industry across the globe where access to clean drinking water is a challenge, including saline groundwater within the United States.

ACKNOWLEDGMENTS

This project was funded by EFL and MIT Department of Mechanical Engineering. We would like to acknowledge the EFL team that worked with us and guided us throughout this process. In particular, we would like to acknowledge Dr. Satish Kumar and Suresh Redhu. This project was done as part of the 2.76 Global Engineering class at MIT.

REFERENCES

[1] PTI. "NGT asks govt to ban RO-purifiers that demineralises water." The Hindu (2020). Accessed March 4, 2023, URL https://www.thehindu.com/news/national/ngt-asks-govt-to-ban-ro-purifiers-that-demineralises-water/article32075468.ece.

[2] WHO (ed.). "Drinking-water quality guidelines." World Health Organization (2023). Accessed March 4, 2023, URL https://www.who.int/teams/environment-climate-change-and-health/water-sanitation-and-health/water-safety-and-quality/drinking-water-quality-guidelines.

[3] of Economic, UN Department and Affairs, Social (eds.). "6. Ensure availability and sustainable management of water and sanitation for all." United Nations: Department of Economic and Social Affairs (2023). Accessed March 4, 2023, URL https://sdgs.un.org/goals/goal6.

[4] Ramteke, PW, Bhattacharjee, JW, Pathak, SP and Kalra, N. "Evaluation of coliforms as indicators of water quality in India." *Journal of applied microbiology* Vol. 72 No. 4 (1992): pp. 352–356.

[5] Kumpel, Emily and Nelson, Kara L. "Comparing microbial water quality in an intermittent and continuous piped water supply." *Water research* Vol. 47 No. 14 (2013): pp. 5176–5188.

[6] Systems, Kent Ro. "What are Total Dissolved Solids (TDS) How to Reduce Them?" Kent Healthcare Products (2019). Accessed May 12, 2023, URL https://www.kent.co.in/blog/what-are-total-dissolved-solids-tds-how-to-reduce-them/#:~:text=According%20to%20the%20Bureau%20of,%2C%20however%2C%20is%20300%20ppm.

[7] "Total dissolved solids in Drinking-water." World Health Organization (2003). Accessed May 12, 2023, URL https://cdn.who.int/media/docs/default-source/wash-documents/wash-chemicals/tds.pdf?sfvrsn=3e6d651e_4.

[8] Publishing, IWA (ed.). "Reverse Osmosis and Removal of Minerals from Drinking Water." The International Water Association (2016). Accessed March 4, 2023, URL https://www.iwapublishing.com/news/reverse-osmosis-and-removal-minerals-drinking-water.

[9] industrial water, PURETEC (ed.). "What is Reverse Osmosis?" PURETEC industrial water (2022). Accessed March 4, 2023, URL https://puretecwater.com/reverse-osmosis/what-is-reverse-osmosis.

[10] Nayar, Kishor G., Sundararaman, Prithiviraj, O'Connor, Catherine L., Schacherl, Jeffrey D., Heath, Michael L., Gabriel, Mario Orozco, Shah, Sahil R., Wright, Natasha C. and Winter, V, Amos G. "Feasibility study of an electrodialysis system for in-home water desalination in urban India." *Development Engineering* Vol. 2 (2017): pp. 38–46. DOI https://doi.org/10.1016/j.deveng.2016.12.001. URL https://www.sciencedirect.com/science/article/pii/S2352728516300045.

[11] Circles, Local. "Only 35% Indian households surveyed rate the quality of piped water they get from local body as good." Local Circles (2023). Accessed March 4, 2023, URL https://www.localcircles.com/a/press/page/drinking-water-survey.

[12] Analyst, Chem (ed.). "India Reverse Osmosis Membrane Market Analysis: Plant Caapcity, Production, Operating Efficiency, Process, DEmand Supply, End Use, Application, Sales Channel, Region, Competition, Trade, Market Analysis, 2015-2030." Chem Analyst (2021). Accessed March 4, 2023, URL https://www.chemanalyst.com/industry-report/india-reverse-osmosis-membrane-market-482#:~:text=%5BOnline%20Quarterly%20Update%5D%20India%20Reverse,CAGR%20of%208.95%25%20until%20FY2030.

[13] "Interview with Dr. Satish Kumar." personal communication (2022). General Manager of Water Category, Eureka Forbes Limited.

[14] Wright, Natasha C., Shah, Sahil R., Amrose, Susan E. and Winter, Amos G. "A robust model of brackish water electrodialysis desalination with experimental comparison at different size scales." *Desalination* Vol. 443 (2018): pp. 27–43. DOI https://doi.org/10.1016/j.desal.2018.04.018. URL https://www.sciencedirect.com/science/article/pii/S0011916417325262.

[15] Shah, Sahil R., Walter, Sandra L. and Winter, Amos G. "Using feed-forward voltage-control to increase the ion removal rate during batch electrodialysis desalination of brackish water." *Desalination* Vol. 457 (2019): pp. 62–74. DOI https://doi.org/10.1016/j.desal.2019.01.022. URL https://www.sciencedirect.com/science/article/pii/S0011916418317776.

[16] Shah, Sahil R., Wright, Natasha C., Nepsky, Patrick A. and Winter, Amos G. "Cost-optimal design of a batch electrodialysis system for domestic desalination of brackish groundwater." *Desalination* Vol. 443 (2018): pp. 198–211. DOI https://doi.org/10.1016/j.desal.2018.05.010.

Copyright © 2023 by ASME

URL https://www.sciencedirect.com/science/article/pii/S0011916417325729.

[17] Consulting, Blue Weave (ed.). "India Water Purifiers Market, By Technology (RO Water Purifiers, UV Water Purifiers, Gravity Based Water Purifiers, Sediment Filters), By Distribution Channel (Retail Distributors, Online Suppliers, Direct-to-Consumers), By End-User (Industrial, Commercial, Household), By Region (North, South, East, West) Trend Analysis, Competitive Market Share Forecast, 2017-2027." Blue Weave Consulting (2021). Accessed March 4, 2023, URL https://www.blueweaveconsulting.com/report/india-water-purifier-market-1857.

[18] Limited, Eureka Forbes (ed.). "Paani ka doctor." Eureka Forbes Limited (2023). Accessed March 4, 2023, URL https://www.eurekaforbes.com/paani-ka-doctor.

[19] Lenntech (ed.). "Electrodialysis reversal (EDR)." Lenntech (2023). Accessed March 4, 2023, URL https://www.lenntech.com/processes/electrodialysis-reversal-edr-.htm#:~:text=The%20average%20life%2Dspan%20of,between%

205%20and%207%20years.

[20] Varner, Hannah. "Architecture and unit design of a capital cost optimized, household electrodialysis desalination device with continuous flow." Master's Thesis, Massachusetts Institute of Technology, Cambridge, MA. 2020. URL https://dspace.mit.edu/handle/1721.1/143619.

[21] INSTITUTE, NDUSTRIAL FASTENERS (ed.). "hat is the Proper Torque to Use on a Given Bolt?" Brighton Best (2021). Accessed March 4, 2023, URL https://www.brightonbest.com/pfconline/PDF/WhatIsProperTorque.pdf.

[22] Tiseo, Ian. "Average water consumption per person across India in 2021, by age group." Statista (2023). Accessed March 4, 2023, URL https://www.statista.com/statistics/1137274/india-average-water-consumption-per-person-by-age-group/.

[23] Winter, Amos. "Home-use desalination." GEAR Lab, MIT (2023). Accessed March 4, 2023, URL https://www.gear.mit.edu/home-use-desalination.

Proceedings of the ASME 2023
International Design Engineering Technical Conferences and
Computers and Information in Engineering Conference
IDETC-CIE2023
August 20-23, 2023, Boston, Massachusetts

DETC2023-113479

FEASIBILITY OF SMALL-SCALE, OFF-GRID DESALINATION IN NAVAJO NATION

Melissa Brei*

Massachusetts Institute of Technology, Cambridge, MA

ABSTRACT

Water scarcity is affecting billions of people around the world and the utilization of brackish water, defined as having total dissolved salt content between 500-10,000 mg/L, could provide a potential solution. This paper focuses on the Navajo Nation in the southwest of the United States, an extremely water stressed region, due to many homes lacking piped water and electricity. Previous work points to membrane desalination being a suitable off-grid technology for providing potable water. By analyzing overlaps of various maps, feasible regions for desalination were identified due to high potential for positive impact and opportunity for desalination to succeed. This geographic analysis revealed the southern region as the focal point for this study, which was corroborated by various partners. Water haulers and stakeholders in diverse organizations affiliated with water management were interviewed to elucidate design requirements for a highly-adoptable desalination system. Engaging with stakeholders at the beginning of the design process validated the proposal of desalination as a technical solution. Additionally, the interviews revealed sensitive design requirements and unique challenges that will aid the designers in the following stages of the design process. Future work will include a robust evaluation of membrane desalination technologies to select the best suited for this application.

Keywords: Collaborative Design, Design Methodology, Design Process

1. INTRODUCTION

Water scarcity is a rising problem around the world. 17 countries, home to a quarter of the global population, face "extremely high" water stress [1]. This problem is being exacerbated by rising sea levels causing saltwater intrusion in groundwater sources that 2.5 billion people worldwide rely on for drinking water [2, 3]. Considering the low availability of freshwater, utilizing brackish water, defined as having 500-10,000 mg/L of total dissolved salts (TDS), could double the available supply [4, 5]. The Navajo Nation, located in the southwest of the United States, is one region that would benefit from brackish water utilization.

It has high levels of water stress, exacerbated by a recent severe drought, and plentiful brackish water and will therefore be the focus of this study.

The tribe's utility service, Navajo Tribal Utility Authority (NTUA), reported that 30% of households lack piped water which the U.S. Indian Health Service (IHS) estimated to be 9,650 homes [6]. Many of these people haul water weekly, sometimes daily for livestock, from windmill-powered wells located around the Navajo Nation. During summer months, some wells run dry causing early morning long lines and further trips to find a running water point. Due to lost time, fuel costs, and vehicle repair costs, Navajo water haulers pay significantly more for water. Studies completed by the Navajo Nation Department of Water Resources estimate they pay $0.012-$0.035 per liter compared to $0.0004-$0.0011 in nearby suburban areas [7, 8].

For many of these families, piping infrastructure is not feasible due to sparsely distributed and remote houses [9]. Not only would it incur high costs to build and maintain, but it could also lead to health consequences from stagnant water. Many of the same households also lack electricity and the NTUA estimates it will cost $40,000/home to connect them [10]. This lack of electricity limits the implementation of at-home or regional solutions that can provide clean water. The other issue is the distributed contaminants in the groundwater. Arsenic and salt naturally occur in the groundwater while uranium has infected water in the western area of Navajo Nation due to mismanaged United States mining operations [11]. The presence of these harmful contaminants limits water access in several areas.

A potential solution for increasing water access in the area is membrane desalination. Depending on the technology, desalination can extract many different types of contaminants, including the ones listed above, and can be supplemented with pre- and post-processing filtration [12, 13]. These technologies have also been successfully demonstrated and implemented for small-scale, off-grid applications [12, 14–17]. The objective of this study is to identify populations in the Navajo Nation that are best positioned to benefit from desalination and elucidate their design requirements.

---

*Corresponding author: mbrei@mit.edu

Copyright © 2023 by ASME

2. FEASIBLE REGIONS FOR DESALINATION

To identify feasible regions for desalination, various maps of the area were analyzed to locate overlaps with high potential for impact and opportunity.

2.1 Impact

Desalination has the potential to provide a significant positive impact in regions with high water stress where many homes also lack piped water. The Aqueduct Water Risk Atlas, developed by the World Resources Institute, was used to illustrate how the water stress varies in the region in Fig. 1a [18]. The number of homes without piped water also varies from region to region. This map, shown in Fig. 1b, was formed by the Water Access Coordination Group during the COVID-19 pandemic [6]. Analyzing the overlap of these two maps, the south, west, and center show potential for desalination to provide a positive impact if it can supply clean water. In the center of the Navajo Nation, the NTUA is expanding their piped network, so it will not be considered a high priority for this study.

2.2 Opportunity

Even if desalination has potential to be impactful in a region, certain conditions must be met to ensure it can operate successfully. There needs to be ample brackish water and for an off-grid system, a renewable energy source must be present. A benefit of this region's arid climate is the abundance of solar irradiance as seen in Fig. 2a. Much of the Navajo Nation is also located on the Colorado Plateau so there is plentiful wind that can be harnessed for energy as well. While everywhere in the Navajo Nation has strong opportunity for renewable energy, not all regions have brackish water. Figure 2b was made by interpolating well data provided by the United States Geological Survey [19]. This data does not cover the entire Navajo Nation and is also outdated but can be used to identify regions in the target range. The south and west have an opportunity for off-grid desalination in addition to impact due to the prevalence of brackish water and renewable energy sources. The west is home to several abandoned uranium mines that have affected the groundwater in the area [11]. While desalination technology has the capability of removing uranium from drinking water, the brine will contain hazardous levels of uranium. Transportation of this brine to the proper facility will be prohibitively expensive for an off-grid application. Therefore, the primary focus of this study is on the southern region of the Navajo Nation.

3. DESIGN REQUIREMENT FORMULATION

There are several stakeholders involved with water management in the Navajo Nation. It is a sovereign nation that works closely with many departments in the U.S. federal and state governments. Figure 3 shows a stakeholder map that illustrates how each group would interact with a desalination device. There are a few nuances to address:

- NTUA oversees wells that are connected to their piped network. These wells typically need minimal treatment, such as chlorine to kill biological contaminants. The Department of Water Resources (DWR) oversees livestock wells which are typically brackish water.

- Certain chapters are able to make their own business decisions, like funding a desalination unit, without going through the Navajo Nation federal government.

- Due to limited personnel capacity, there is a potential need for a third party organization, non-profit or for-profit, to operate and conduct maintenance on the desalination unit.

3.1 Methods

To elucidate the design requirements for a desalination system, a variety of stakeholders were interviewed:

- Water Haulers: 23 participants located in Indian Wells, Birdsprings, or Leupp

- Chapter Officials: President and Secretary of Indian Wells, Chapter Manager and Secretary of Birdsprings, Chapter Manager of Leupp, and Community Services Coordinator and Council Delegate of Round Rock (see Fig. 1b)

- DWR: Director and head of Technical Construction and Operations

- NTUA: Deputy General Manager and manager of a wastewater facility

- Navajo Nation EPA: Department Manager of Surface and Groundwater Protection

- U.S. Bureau of Reclamation: Program Development Division Manager

Interviews with water haulers were conducted after the study was approved by the Navajo Nation Human Research Review Board (NNHRRB) and received exempt approval from MIT Committee on the Use of Humans as Experimental Subjects. Each household was compensated $20 and the one hour, semi-structured interviews touched on the following topics:

- Typical water hauling trip

- Cost of water hauling

- Willingness to pay for more accessible drinking water

- Perception of a reliable water source

- Wastewater management

The other stakeholder interviews began with general questions to learn their perspective on the state of accessible drinking water and an ideal solution. From there, questions relevant to the stakeholder's expertise, such as brine management for the EPA and demand trends for the NTUA, were asked. Detailed notes and audio, if consented, were recorded during each stakeholder interview and organized into several topics. Patterns were identified that represent participant's perceptions and knowledge regarding technological solutions for producing drinking water. Design requirements were then generated in three main categories: production volume, cost, and water quality.

Copyright © 2023 by ASME

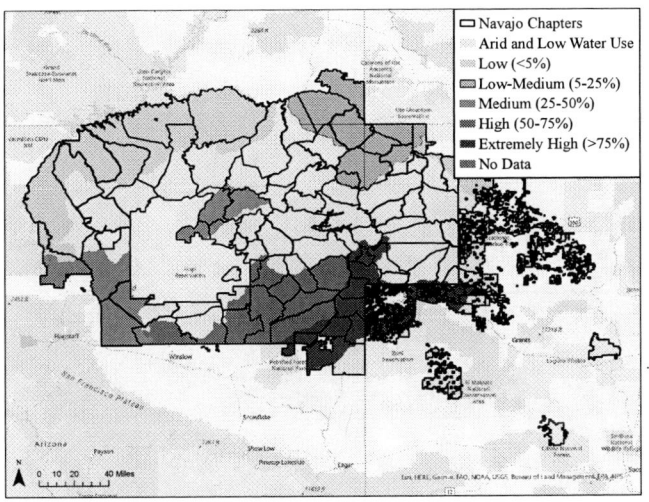

(a) Water scarcity in the Navajo Nation provided by the Aqueduct Water Risk Atlas [18]

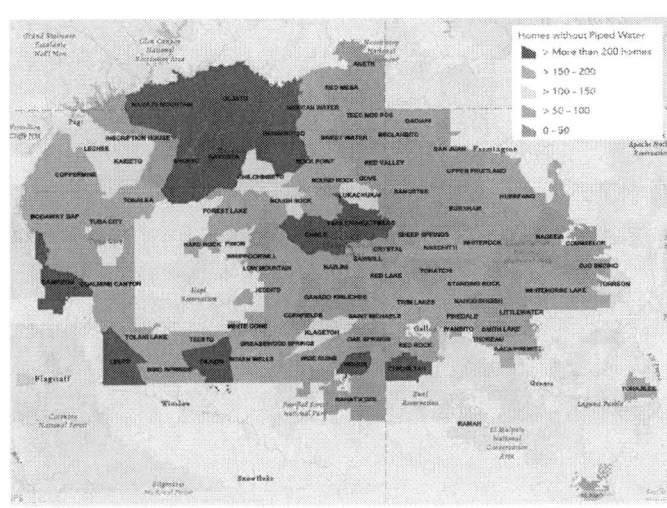

(b) Map of homes without piped water compiled by the Water Access Coordination Group [6].

FIGURE 1: DESALINATION HAS THE POTENTIAL TO PROVIDE A SIGNIFICANT POSITIVE IMPACT IN REGIONS WITH HIGH WATER STRESS AND DENSITY OF HOMES LACKING PIPED WATER.

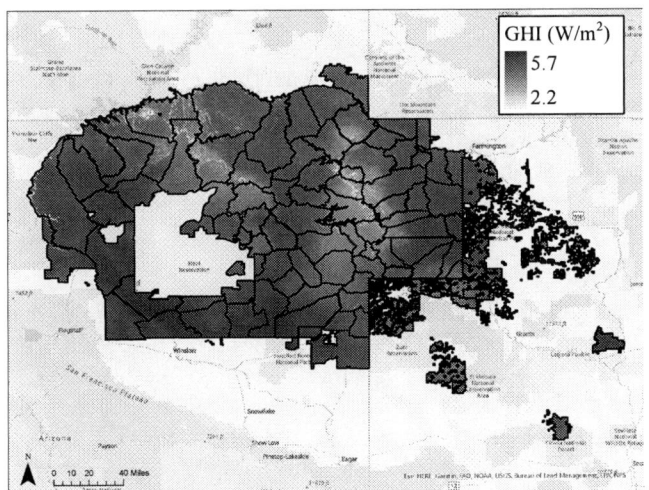

(a) Solar irradiance, represented by global horizontal irradiation (GHI), in the Navajo Nation provided by the Global Solar Atlas 2.0 [20].

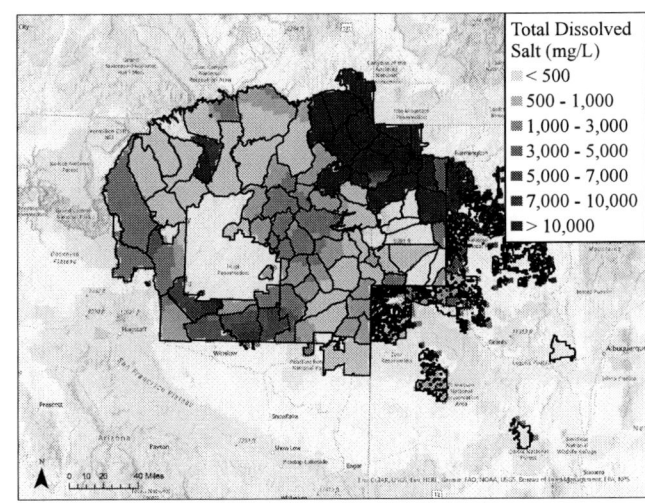

(b) Map of TDS levels around the Navajo Nation using interpolated well data from the United States Geological Survey [19].

FIGURE 2: DUE TO THE PREVALENCE OF SOLAR IRRADIANCE AND BRACKISH GROUNDWATER IN SEVERAL REGIONS, OFF-GRID DESALINATION HAS A STRONG OPPORTUNITY TO BE SUCCESSFUL.

Copyright © 2023 by ASME

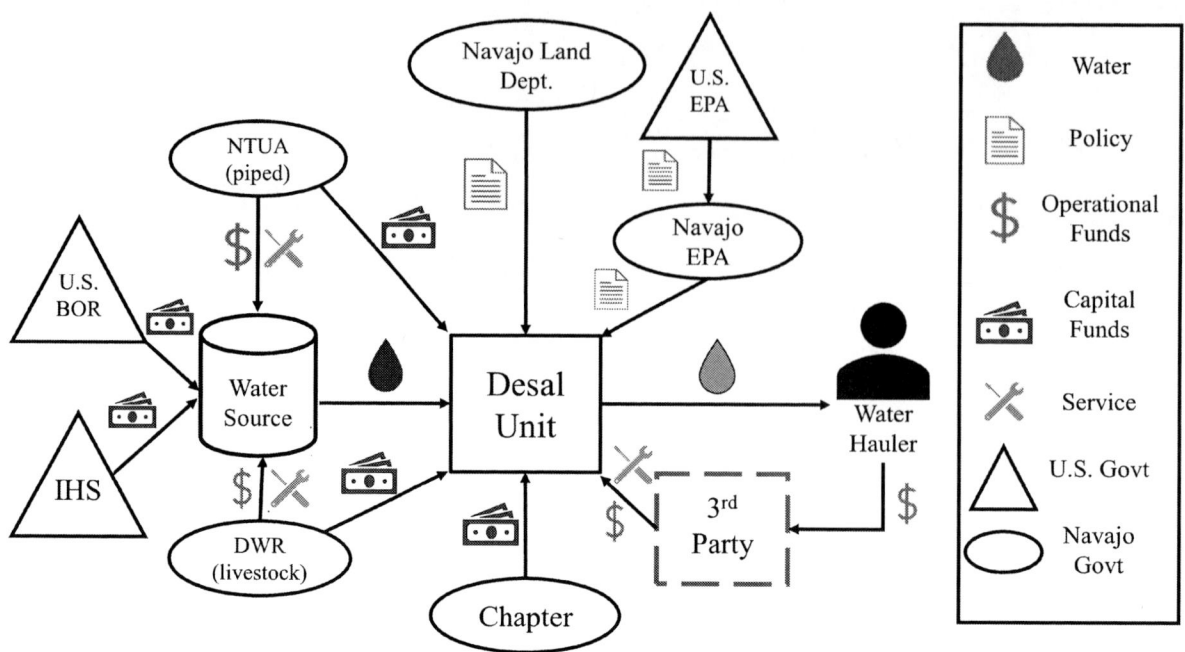

FIGURE 3: STAKEHOLDERS ASSOCIATED WITH A DESALINATION DEVICE IN THE NAVAJO NATION AND THEIR RELATIONSHIPS. NAVAJO TRIBAL UTILITY AUTHORITY (NTUA) AND DEPT. OF WATER RESOURCES (DWR) ARE THE PRIMARY NAVAJO ENTITIES IN CHARGE OF WATER INFRASTRUCTURE. THE U.S. BUREAU OF RECLAMATION (BOR) AND INDIAN HEALTH SERVICES (IHS) FUND PROJECTS FOR INCREASING WATER ACCESS IN THE AREA.

3.2 Production Volume

Interviews with stakeholders presented two promising solution schemes: home-scale and community-scale. Both schemes would allow for people to haul water from a closer source and reduce time, vehicle wear, and gas money.

For home-scale, the user would haul from a brackish well, which are more prevalent in the area, then desalinate with a system at home. Many people live in a cluster of homes with extended family members, so the production volume would depend on the amount of people in the cluster. From the water haulers interviewed, a household cluster could have 1-15 people living there. The WHO recommends 19L/day/person which is a conservative estimate since Navajo people are water conscious. With this estimate, the home-scale production volume requirement is 19-285 L/day. Desalination will produce brine that is more concentrated than the hauled water fed into the system. Interviews with water haulers revealed many dispose of their grey water by tossing it outside. The recovery ratio, the amount of drinking water produced to water used, should be maximized while not negatively impacting the environment due to improper disposal of highly concentrated brine. Research is currently being conducted into safe salinity levels for the small volumes of wastewater that home systems could produce by the Wright Group at University of Minnesota. The minimum recovery ratio of 50% was chosen for this study since more drinking water should be produced than wasted.

For community-scale, the desalination system would be located at a brackish well rather than a home. Since homes are widely dispersed in the Navajo Nation, a community watering point could service a wide range of people. The lower range, 20 people, represents two average household clusters while the upper range represents an entire chapter, 1,000 people. While people typically haul weekly or biweekly, the average daily production, using 19L/day/person, is the design requirement. Some days may produce more or less but should average to produce 380 - 19,000 L/day depending on the community size. A community-scale system can take advantage of traditional brine management techniques, like an evaporation pond, so it can safely produce more concentrated brine than home-scale systems. Since this area is water stressed, the amount of waste sent to an evaporation pond should be minimized, so the recovery ratio must be $\geq 70\%$.

3.3 Cost

The cost of a desalination system will be a major motivating factor in the adoption. Three cost metrics must be considered: capital cost, operating cost, and levelized cost of water (LCOW). Capital cost is a lower priority than the other two since there exists several avenues to apply for one-time funding. The system must be able to compete with commercialized, on-grid systems of the same scale. Operating cost was another metric that every stakeholder considered important. Due to limited personnel and budgets, funds for operating costs are harder to consistently obtain. Thus, minimizing this cost is critical for the long-term implementation. Finally, LCOW, cost to produce a cubic meter of water amortized over the lifetime of the system, is a common performance metric used to compare different technologies. The system should cost the same, ideally less, than the current system of hauling water. Interviews with chapter officials revealed the average price of delivering water is $0.013/liter and the low-

Copyright © 2023 by ASME

est estimate for hauling is $0.012/liter [7]. Many water haulers purchase bottled water from the grocery store and reported spending approximately $4 for a pack of forty 0.5 L bottles, which is $0.2/liter. To be conservative, the lowest value, $0.012/liter, will be the design requirement for both size scales.

3.4 Water Quality

Water quality consists of three primary categories: biological, chemical, and physical. The design chosen must be able to meet government regulations and user preference. Biological contaminants are critical to extract since they cause infectious diseases which the WHO reports as the most common risk associated with drinking water [21]. Interviews with the EPA and DWR revealed that the most common chemical contaminants in southern Navajo Nation are arsenic, iron, manganese, and salt. The chosen design must be able to extract these contaminants to produce drinking water that meets EPA standards [22]. While physical water quality has limited health consequences compared to biological and chemical, it can cause the user to reject the water. Since many people buy bottled water, the TDS should be lower than 500 mg/L to ensure the drinking water is accepted by the community [23].

4. DISCUSSION AND FUTURE WORK

By engaging with stakeholders at the beginning of the design process, this study was able to identify the core issues, understand sensitive requirements, and reveal unique challenges pertaining to the availability of drinking water in the Navajo Nation. From speaking with water haulers, many expressed their ideal solution to be piped water or a private well. However, government agencies claimed a piped network is infeasible for many homes due to the low population density. For private wells, high quality water is not available at every home and wells are very expensive to drill in the region. Based off these interviews, accessibility to quality drinking water was identified as a core issue. Additionally, many water haulers expressed frustration at the reliability of well infrastructure. The remote location of the wells, lack of funding, and minimal personnel, results in delayed maintenance. By speaking with a variety of stakeholders, a holistic view of the problem was achieved and the core issues were identified as accessibility and reliability. Accessibility gives merit to desalination as a technical solution since it could treat the water nearby and enable more usable sources in the area. In the next stages of the design process, reliability will be a priority for the designers.

The stakeholder interviews also increased understanding of the community's relative sensitivity to all the design requirements. For an economically-constrained region, it was critical to learn that operating costs are much more sensitive than capital costs. In remote areas with poor road conditions, it can be difficult to provide frequent technical service, especially if less funding is available for long-term operations. Interviews with water haulers also revealed a preference for bottled water, imposing a more strict requirement on product salinity. Even though more energy is required to desalinate to a lower salinity, people will no drink, let alone purchase, water that is undesirable. Speaking with stakeholders early allows the designer to not only identify design requirements but also their relative importance to community members.

Finally, conducting semi-structured interviews enabled stakeholders to reveal unique challenges that otherwise would not be uncovered until field testing. Several water haulers mentioned that horses prefer high quality water and livestock are an important facet of the Navajo culture. Since horses drink 2-10 times the amount of water as people in the Navajo Nation, this could have a significant impact on the future design, both home-scale and community-scale. For home-scale, another potential issue is people hauling from a different source because their closest well ran dry or the well pump malfunctioned. If a design assumes a certain water chemistry, it may not be flexible enough to produce quality water with a different feedwater make-up. Additionally, desalination produces brine which will be logistically difficult to manage for a dispersed, home-scale-focused operation. Brine and feedwater can be controlled more easily with community-scale designs but there exists a new set of challenges. Every stakeholder mentioned vandalism of public water sources and was unable to point to a specific cause. There was also little to no identifiable trend for the time of day or week that people haul water. This random demand can significantly affect a future design that prioritizes reliability. These insights would not have been obtained without speaking with a diverse group of stakeholders. By engaging at the beginning of the design process, the designers ensured that desalination is a worthy fit for the problem and know what requirements to prioritize in the next steps.

There are three potential membrane desalination technologies that both meet the design requirements elucidated in this study and have demonstrated past academic and commercial success in similar applications: reverse osmosis (RO), electrodialysis (ED), and nanofiltration (NF). Other studies have shown that NF is less effective at removing monovalent ions, like salt, from brackish water so it will only be considered in conjunction with another technology. RO and ED have both been demonstrated at a range of size scales, feedwater salinities, product salinities, and energy sources [12, 24]. Despite studies showing that ED is more cost-effective than RO for feedwater salinities ≤ 5000 ppm, RO accounts for 80% of the brackish water desalination market [25]. At this range, ED uses less energy to create one unit of water, so for an off-grid system, ED will have reduced capital cost due to smaller energy demands [24–26]. ED is also able to achieve higher recovery ratios and longer membrane lifetime, increasing its suitability for this application. RO's dominant market share is due to numerous reasons including a more developed supply chain, less complexity, and more contaminant removal [12, 24]. Future work involves a robust analysis of the optimal desalination system design that will be benchmarked against current commercial systems.

5. CONCLUSION

This work defines design requirements for a remote, small-scale desalination system based in the Navajo Nation. It identified the southern region of the Navajo Nation as the most likely to benefit from membrane desalination due to the prevalence of brackish water and solar irradiance. Additionally, the high water stress and density of homes without piped water illustrate the

Copyright © 2023 by ASME

TABLE 1: SUMMARY OF DESIGN REQUIREMENTS

Design Requirement	Home-scale	Community-scale
Production Volume	19-285 L/day	380-19,000 L/day
Recovery Ratio	$\geq 50\%$	$\geq 70\%$
Capital Cost	$\leq$ current commercial systems	$\leq$ current commercial systems
Operating Cost	minimized	minimized
LCOW	$\leq \$0.012$/liter	$\leq \$0.012$/liter
Feedwater Salinity	500-5,000 mg/L	500-5,000 mg/L
Product Salinity	200-500 mg/L	200-500 mg/L

potential for desalination to make a positive impact in the area. Interviews with various stakeholder groups were used to generate design requirements for a desalination system operating in this region, summarized in Table 1. The important insights gained from the stakeholder interviews emphasize the importance of engaging the community from the beginning of the design process. Additionally, the interactions with a diverse set of stakeholders resulted in a holistic image of the current eco-system and areas for improvement. Future work will include an analysis of different desalination technologies and subsequent optimal designs to ensure the design requirements are met.

6. ACKNOWLEDGEMENTS

This work is funded by the United States Bureau of Reclamation and the Julia Burke Foundation. Thank you to Jeff Costello, Ben Judge, Ian Manning, Dr. Natasha Wright, and Jimmy Tran for assisting in conducting and analyzing interviews.

REFERENCES

[1] "Updated Global Water Risk Atlas Reveals Top Water-Stressed Countries and States." (2019). Accessed 2022-10-10, URL https://www.wri.org/news/release-updated-global-water-risk-atlas-reveals-top-water-stressed-countries-and-states.

[2] Jasechko, Scott, Perrone, Debra, Befus, Kevin M., Bayani Cardenas, M., Ferguson, Grant, Gleeson, Tom, Luijendijk, Elco, McDonnell, Jeffrey J., Taylor, Richard G., Wada, Yoshihide and Kirchner, James W. "Global Aquifers Dominated by Fossil Groundwaters but Wells Vulnerable to Modern Contamination." Vol. 10 No. 6 : pp. 425–429. DOI 10.1038/ngeo2943. URL https://doi.org/10.1038/ngeo2943.

[3] Abu-alnaeem, Madhat Farouk, Yusoff, Ismail, Ng, Tham Fatt, Alias, Yatimah and Raksmey, May. "Assessment of Groundwater Salinity and Quality in Gaza Coastal Aquifer, Gaza Strip, Palestine: An Integrated Statistical, Geostatistical and Hydrogeochemical Approaches Study." Vol. 615 : pp. 972–989. DOI 10.1016/j.scitotenv.2017.09.320. URL https://www.sciencedirect.com/science/article/pii/S0048969717326712.

[4] Zektser, I. S. and Everett, Lorne G. *Groundwater Resources of the World and Their Use*. UNESCO.

[5] Gleick, Peter H. *Water in Crisis: A Guide to the World's Fresh Water Resources*. Oxford University Press.

Accessed 2022-10-10, URL http://catdir.loc.gov/catdir/enhancements/fy0604/92030061-t.html.

[6] "Navajo Safe Water: Protecting You and Your Family's Health." ArcGIS StoryMaps. Accessed 2022-10-14, URL https://storymaps.arcgis.com/stories/1b4dc0d978c74d97a559e615730d4cd4.

[7] "Navajo Nation Drought Contingency Plan 2003." Docslib. Accessed 2022-10-14, URL https://docslib.org/doc/5884781/navajo-nation-drought-contingency-plan-2003.

[8] "DWR2011 Water Resource Development Strategy for the Navajo Nation.Pdf."

[9] Calkins, Mark A. "Sanitation Deficiency System (SDS): A Guide for Reporting Sanitation Deficiencies for American Indian and Alaska Native Homes and Communities." : p. 82.

[10] "Lighting the Navajo Nation | American Public Power Association." Accessed 2022-10-14, URL https://www.publicpower.org/periodical/article/lighting-navajo-nation.

[11] Ingram, Jani C., Jones, Lindsey, Credo, Jonathan and Rock, Tommy. "Uranium and Arsenic Unregulated Water Issues on Navajo Lands." Vol. 38 No. 3 : p. 031003. DOI 10.1116/1.5142283. Accessed 2022-10-14, URL 32226218, Accessed 2022-10-14, URL https://www.ncbi.nlm.nih.gov/pmc/articles/PMC7083651/.

[12] Peter-Varbanets, Maryna, Zurbrügg, Chris, Swartz, Chris and Pronk, Wouter. "Decentralized Systems for Potable Water and the Potential of Membrane Technology." Vol. 43 No. 2 : pp. 245–265. DOI 10.1016/j.watres.2008.10.030. Accessed 2022-10-03, URL https://linkinghub.elsevier.com/retrieve/pii/S0043135408004983.

[13] Baker, Richard W. *Membrane Technology and Applications*. McGraw-Hill Professional Engineering, McGraw-Hill.

[14] Sanna, Anas, Buchspies, Benedikt, Ernst, Mathias and Kaltschmitt, Martin. "Decentralized Brackish Water Reverse Osmosis Desalination Plant Based on PV and Pumped Storage - Technical Analysis." Vol. 516 : p. 115232. DOI 10.1016/j.desal.2021.115232. Accessed 2022-10-03, URL https://linkinghub.elsevier.com/retrieve/pii/S0011916421003039.

[15] Malek, P., Ortiz, J.M. and Schulte-Herbrüggen, H.M.A. "Decentralized Desalination of Brackish Water Using an Electrodialysis System Directly Powered by Wind Energy." Vol. 377 : pp. 54–64.

Copyright © 2023 by ASME

DOI 10.1016/j.desal.2015.08.023. Accessed 2022-10-03, URL https://linkinghub.elsevier.com/retrieve/pii/S001191641530059X.

[16] Sharon, H. and Reddy, K.S. "A Review of Solar Energy Driven Desalination Technologies." Vol. 41 : pp. 1080–1118. DOI 10.1016/j.rser.2014.09.002. Accessed 2022-10-03, URL https://linkinghub.elsevier.com/retrieve/pii/S1364032114007758.

[17] Ghazi, Zeyad Moustafa, Rizvi, Syeda Warisha Fatima, Shahid, Wafa Mohammad, Abdulhameed, Adil Muhammad, Saleem, Haleema and Zaidi, Syed Javaid. "An Overview of Water Desalination Systems Integrated with Renewable Energy Sources." Vol. 542 : p. 116063. DOI 10.1016/j.desal.2022.116063. Accessed 2022-10-17, URL https://www.sciencedirect.com/science/article/pii/S0011916422005185.

[18] Hofste, Rutger Willem, Kuzma, Samantha, Walker, Sara, Sutanudjaja, Edwin H., Bierkens, Marc F. P., Kuijper, Marijn J. M., Sanchez, Marta Faneca, Beek, Rens Van, Wada, Yoshihide, Rodríguez, Sandra Galvis and Reig, Paul. "Aqueduct 3.0: Updated Decision-Relevant Global Water Risk Indicators." Accessed 2022-11-07, URL https://www.wri.org/research/aqueduct-30-updated-decision-relevant-global-water-risk-indicators.

[19] Qi, Sharon L. and Harris, Alta C. "Geochemical Database for the National Brackish Groundwater Assessment of the United States." DOI 10.5066/F72F7KK1. Accessed 2022-11-08, URL https://www.sciencebase.gov/catalog/item/583dfd9ee4b088b77f520d07.

[20] "Global Solar Atlas." Accessed 2022-11-08, URL https://globalsolaratlas.info/support/terms-of-use.

[21] Organization, World Health and WHO. *Guidelines for Drinking-water Quality*. World Health Organization.

[22] "Navajo Nation Surface Water Quality Standards 2015."

[23] Bruvold, William H. "Mineral Taste and the Potability of Domestic Water." Vol. 4 No. 5 : pp. 331–340. DOI 10.1016/0043-1354(70)90074-6. Accessed 2022-11-08, URL https://www.sciencedirect.com/science/article/pii/0043135470900746.

[24] Patel, Sohum K., Biesheuvel, P. Maarten and Elimelech, Menachem. "Energy Consumption of Brackish Water Desalination: Identifying the Sweet Spots for Electrodialysis and Reverse Osmosis." Vol. 1 No. 5 : pp. 851–864. DOI 10.1021/acsestengg.0c00192. Accessed 2022-09-08, URL https://doi.org/10.1021/acsestengg.0c00192.

[25] Wright, Natasha C. and Winter, Amos G. "Justification for Community-Scale Photovoltaic-Powered Electrodialysis Desalination Systems for Inland Rural Villages in India." Vol. 352 : pp. 82–91. DOI 10.1016/j.desal.2014.07.035. Accessed 2022-09-12, URL https://linkinghub.elsevier.com/retrieve/pii/S0011916414004160.

[26] He, Wei, Le Henaff, Anne-Claire, Amrose, Susan, Buonassisi, Tonio, Peters, Ian Marius and Winter, Amos G. "Voltage- and Flow-Controlled Electrodialysis Batch Operation: Flexible and Optimized Brackish Water Desalination." Vol. 500 : p. 114837. DOI 10.1016/j.desal.2020.114837. Accessed 2022-10-17, URL https://linkinghub.elsevier.com/retrieve/pii/S0011916420315150.

Copyright © 2023 by ASME

Proceedings of the ASME 2023
International Design Engineering Technical Conferences and
Computers and Information in Engineering Conference
IDETC-CIE2023
August 20-23, 2023, Boston, Massachusetts

DETC2023-114540

FEELING THE DISTANCE: EXPLORING NOVICE DESIGNERS' PERCEPTIONS OF THE PSYCHOLOGICAL DISTANCE TOWARDS AND EMPATHY INDUCED BY PROBLEM VARIATIONS

Jenna Herzog
Dept. of Chemical Engineering
Lafayette College
Easton, PA, USA
herzogj@lafayette.edu

Rebekah Fodale
Dept. of Mechanical Engineering
Lafayette College
Easton, PA, USA
fodaler@lafayette.edu

Mohammad Alsager Alzayed
Dept. of Industrial and Management Systems Engineering
Kuwait University
Kuwait City, Kuwait
mohammad.alsageralzayed@ku.edu.kw

Elizabeth M. Starkey
School of Engineering Design and Innovation
The Pennsylvania State University
University Park, PA, USA
ems413@psu.edu

Rohan Prabhu
Dept. of Mechanical Engineering
Lafayette College
Easton, PA, USA
prabhur@lafayette.edu
Corresponding author

ABSTRACT

The accelerating depletion of natural resources has necessitated the design of environmentally sustainable engineering solutions. To meet this need for sustainable solutions, designers must actively incorporate considerations of environmental impact in their design decisions. Prior research suggests that the effects of climate change are often perceived to be psychologically distant, and this distance could inhibit individuals from actively engaging in environmentally sustainable behavior. Little research has investigated the impact of problem framing based on designers' psychological distance on design performance. Furthermore, research suggests that empathy development could be an effective mechanism for bridging psychological distance. However, little research has assessed the utility of empathy-invoking problem framing in sustainable engineering design practice and education. Our aim in this study is to explore this research gap by comparing student designers' perceptions of different problem formulations. Specifically, we tested the effects of variations in (1) the socio-spatial context and (2) the empathy focus of a similar design problem. The effects of these variations on the perceived psychological distance and empathy-invoking nature were tested through a 2x2 between-subjects experiment. From the results, we see that the variations in the problem

formulation did not relate to either the perceived psychological distance or the perceived empathy-invoking nature. These findings suggest that issues related to environmental sustainability tend to be perceived similarly, despite differences in their socio-spatial context and empathy-invoking nature.

Keywords: *environmental sustainability; problem formulation; psychological distance; trait empathy.*

1. INTRODUCTION

Engineering has a ubiquitous presence in modern life, with most products and services around us being outcomes of some form of engineering design [1]. The growing reliance on engineered and manufactured products has resulted in the overconsumption of natural resources, necessitating the need to consider environmental impact in the early stages of engineering design [2]. Problem finding and framing are typically the first and arguably the most important stages of design as designers often set the foundation for their design requirements in these stages [3,4]. Design requirements are critical as they can impact the outcomes of concept generation and selection decisions, thereby determining the effectiveness of the design process as a whole [5]. Therefore, designers must sufficiently emphasize environmental considerations when framing design problems.

Copyright © 2023 by ASME

Design problems have numerous characteristics such as complexity [6], constrainedness [7], descriptiveness [8], and social-spatial context [9], each of which has been shown to have varying effects on design outcomes. One such characteristic of design problems is the psychological distance of the design context from the designer. Psychological distance is defined as the separation between the self and external concepts [10] and has been posited to compose four dimensions: (1) temporal, (2) spatial, (3) social, and (4) hypothetical [11]. Prior research suggests that the more psychologically distant an object or event is from an individual, the less accessible that object is to that individual's consciousness [10]. Investigating the role of designers' psychological distance is particularly important in sustainable design tasks since prior research has found that an individual's response to environmental issues depends on their geographical, social, and political backgrounds [12]. Moreover, the effects of climate change are typically perceived to be psychologically distant [13], thus preventing individuals from actively engaging in environmentally sustainable behavior [14]. Therefore, problem-framing informed by designers' psychological distance could be a useful strategy to encourage the active consideration of environmental impact in design. Such emphasis on psychological distance in problem framing could also work towards leveraging designers' motivation to solve the problem at hand [15].

Prior research suggests that perspective-taking – a form of invoking empathy – can help bridge psychological distance [16]. For example, personas, or fictional characters that portray archetypes of real users [17], can help designers understand the main characteristics of the end-user. Moreover, empathy-invoking activities have been shown to result in better teamwork, problem contextualization, and design inspiration [18]. While these findings support empathy-driven problem formulation in design, limited research has studied the use of empathy-invoking problem formulation as a mechanism to bridge psychological distance. The use of empathy development remains particularly unexplored in the context of environmentally sustainable design.

Motivated by this research gap, we aim to study novice designers' perceptions of different problem formulations in a sustainability-focused design task. Specifically, we conducted a survey-based between-subjects study asking participants to report their perceived psychological distance towards and empathy invoked by one of four problem variants. The participants' responses were statistically analyzed to investigate whether the four problem variants were perceived to be different by novice designers. Before introducing the details of our study, we review prior work in Section 2, with the research questions (RQs) and our corresponding hypotheses presented in Section 3. Our experimental methods are discussed in detail in Section 4, the results of the experiment are discussed in Section 5, and the implications of these results are discussed in Section 6. Finally, we provide concluding remarks in Section 7 and discuss the limitations of our study with potential directions for future research in Section 8.

2. RELATED WORK

Our aim in this paper is to understand novice designers' perceptions of different problem formulations in sustainability-focused design tasks. Toward this aim, we reviewed prior work on the role of problem formulations in engineering design as discussed next. We also reviewed prior research on psychological distance towards environmental sustainability and the use of empathy invoking problems in engineering design, as discussed next.

2.1. The Role of Problem Formulation in Engineering Design Practice and Education

Design is often referred to as the coevolution of the problem and solution spaces [3,4], making problem framing a critical aspect of design. This importance of problem framing is also highlighted in the Componential Model of Creativity, in which Amabile [19] posits problem identification to be the first stage of creative cognition.

Given the importance of problem framing in design, several researchers have studied the impact of different aspects of design problems on design performance. For example, Onarheim [7] studies the effect of the constrainedness of design problems on designers' information search strategies in design tasks. The author finds that while under-constrained design problems fail to provide sufficient direction for information search, over-constrained design problems limit designers from seeking novel ways of solving the problem. Similarly, Prabhu et al. [20] compare the creativity and manufacturability of additively manufactured solutions generated for different problems related to the COVID-19 pandemic. The problems compared by the authors varied in their urgency and technical complexity. From their results, the authors find that solutions generated for problems of high urgency and technical complexity were least manufacturable with additive manufacturing. Researchers have also tested the impact of the focus of evaluation in design problems on design creativity. For example, giving specific goals for creative production has been found to encourage more creative responses [8]. This use of explicit goals has also been shown to improve the effectiveness of providing designers with ideation tools such as brainstorming [21,22]. These findings suggest that the framing and definition of problems could have important impacts on designers' performance when solving these design problems.

In addition to these studies into the role of problem framing in design practice, researchers have also studied the impact of problem formulation on the effectiveness of project- and problem-based design education. For example, Prabhu et al. [6] compare the use of problems of different complexities in design for additive manufacturing education. The authors find that complex problems with explicit constraints are more effective in engaging students to learn about additive manufacturing's design freedoms. Moreover, they find that these complex problems encourage students to leverage these design freedoms to generate more unique solutions. Similarly, Alsager Alzayed et al. [9] compare the effects of different social-spatial contexts (i.e., perceived developing and developed countries) on empathy

Copyright © 2023 by ASME

development in introductory design courses. The authors find that the problem context did not have significant impacts on empathy development. This result echoes the findings of Burleson et al. [23] who find that despite the importance of socio-political context in design, these factors are not frequently taken into account by designers. These findings call for the need for more effective design educational interventions that encourage designers to consider the social context of design problems throughout the design process. Towards this need, Svilha et al. [24] present a framework for framing socio-technical problems in engineering education. Their proposed framework suggests that educators account for four aspects of design problems: (1) the relevance to students' lives, (2) the complexity of the sociotechnical context, (3) the accessibility to the problem, and (4) open-endedness with sufficient direction. Through an analysis of previously used design problems, the authors posit that problems that meet these four criteria are more effective in design education.

From these studies, we see that problem framing has important implications for the outcomes of design practice. While characteristics of design problems such as complexity and rewards have been extensively studied, little attention is given to the socio-spatial context of design problems. Moreover, little research has investigated designers' perceptions of different problem contexts and formulations, especially in sustainable design education. Our aim in this research is to explore this research gap through an experimental study. Specifically, we aim to study how students perceive different problem contexts and formulations in terms of their psychological distance and empathy-invoking nature. Before doing so, prior work on the role of psychological distance in environmental sustainability is reviewed, as discussed next.

2.2. Psychological Distance in the Context of Environmental Sustainability

Liberman and Trope [10] introduce the idea of psychological distance as the separation between the self and external concepts through time. Additionally, they suggest that psychological distance is informed by personal experiences about certain objects or points in time [10]. According to more recent definitions in the Construal Level Theory, "the construct of psychological distance is a subjective experience that something is close or far away from the self, here, and now" ([25], pg. 2). Moreover, Construal Level Theory suggests that the more psychologically distant an object or event is from an individual the less accessible that object is to that individual's consciousness [10]. Individuals have also been observed to think of psychologically distant objects or events in less detailed and more uncertain terms [26].

Additionally, researchers posit that psychological distance manifests along four dimensions: (1) temporal, (2) spatial, (3) social, and (4) hypothetical [11]. Temporal distance refers to the difference between the moment in the past or future of the event occurring, and the time of the individual evaluating or experiencing the event. The distance the object or event is from the individual in space is known as spatial distance. Social

psychological distance is between the individual and other individuals, groups, or communities. Finally, hypothetical distance defines the certainty or likelihood with which an individual perceives an event will occur.

Given the important role of psychological distance in determining individuals' behavior, its impact has been studied in several contexts. For example, psychological proximity to scientific findings has been shown to impact its perception as being relevant for the local community (i.e., spatial) and for the present time (i.e., temporal) [25]. In addition, proximity to scientific findings has also been shown to influence their perceptions as being tangible, in terms of it having practical implications and tangible effects on the world (i.e., hypothetical), as well as it is conducted by individuals that are approachable and similar to oneself (i.e., social proximity) [25].

Similar findings have been made in the context of climate change and environmental sustainability [12]. Understanding public perceptions of climate change is crucial to obtain the desired societal transformation needed to preserve our planet [12]. Although global warming impacts are occurring worldwide, individuals are more or less responsive to these factors depending on their geographical, social, and political backgrounds [12]. Overall when evaluating society's attention to different environmental challenges, climate change is often perceived as distant on several dimensions [27]. This psychological distance from climate change could limit individuals from acting toward addressing these challenges. This lack of action is particularly important considering how vast and overwhelming tackling environmental issues can seem to humans. The typical response, when faced with distant or global problems, is to feel helpless and less in control [13]. Studies have shown that reducing the socio-spatial distance of climate change makes the issue seem more relevant and promotes behavioral engagement [28]. Furthermore, studies have even suggested that highlighting the negative impacts of distant and abstract challenges could better motivate individuals to take action, especially in the context of sustainable design problems [29].

The impacts of climate change may not only feel socio-spatially distant on a global scale but also feel temporally distant. For instance, it is a common belief that the effects of global warming are not currently being felt and that the impacts of any mitigation efforts will only benefit future generations [13]. That large of a time frame feels abstract and unreliable to individuals, and therefore, environmental crises are not perceived as urgent. This psychological distance and lack of urgency could inhibit active engagement in environmentally sustainable behaviors at the individual and organizational levels [14].

As discussed in Section 2.1, problem framing plays an important role in determining the outcomes of engineering design practice and education. When designing for environmental sustainability, designers' psychological distance from the context could impact their design outcomes. Bardwell [15] suggests that cognitive psychology and conflict management are key components of effective problem-framing. Effective problem-framing creates an environment that encourages participation and results in more positive outcomes

Copyright © 2023 by ASME

when approaching these pressing environmental problems [15]. Furthermore, Bardwell explains how environmental choices reflect political and social values beyond scientific values and designers tend to have "a bias towards the familiar" when it comes to problem-solving. That is, individuals are inclined to use their previous personal experiences to solve problems which leads to psychologically distant problems feeling more obscure.

Therefore, problem-framing informed by designers' psychological distance could be a useful strategy to encourage designers to better understand the issue and generate environmentally sustainable solutions. Such emphasis on psychological distance in problem framing could also work towards leveraging designers' motivation to solve the problem at hand [15]. Despite the critical role that psychological distance plays in driving environmentally sustainable action, little research has investigated its role in problem framing in engineering design. Our aim in this research is to take a first step in this direction by studying student designers' perceived psychological distance from different problem formulations. Before doing so, we review prior work on the role of empathy in problem formulation, as discussed next.

2.3. Empathy-Driven Problem Formulation in Engineering Design

Empathy, or the reactions of one individual to the observed experiences of another [30], has been considered a core tenet of engineering design due to its ability to help designers relate to the end-user and understand the meaning of different experiences to these users [31,32]. Through a qualitative study, researchers found that invoking empathy in the engineering classroom was related to effective teamwork, problem contextualization, and design inspiration [18]. Similarly, empathy-invoking design experiences have been shown to encourage novice designers to generate ideas of high creativity [33].

Motivated by this importance of empathy, design researchers have tested various human-centered design tools that invoke designers' empathy in the problem formulation stage. Some examples of such tools include empathy maps [34], user journey maps [35], and extreme user simulations [36]. Of these various empathy-invoking tools, one of the most common tools used in design practice is the persona: a fictional character that portrays an archetype of a real user with a description of this user's interests, needs, and other key characteristics [17,37,38]. In design practice and research, information such as name, image, occupation, and age are typically included to represent a persona [39]. Personas are often represented in textual form accompanied by an image representing the user [40].

These empathy-invoking tools, especially personas, are effective in helping designers better understand the users' needs. As discussed in Section 2.2, empathy development, particularly through perspective-taking, could be an effective mechanism for bridging psychological distance. Despite the growing body of research on the utility of empathy-driven problem formulation, little research has investigated whether these empathy-driven tools are suitable for psychologically distant design contexts such as environmental sustainability. Such an investigation can help understand the utility of empathy development in encouraging designers to consider environmental sustainability in the design process. Motivated by this research gap, our aim in this paper is to study student designers' perceptions of different problem formulations. Specifically, we aim to compare design problems that vary in (1) their socio-spatial distance and (2) their empathy-invoking nature. Towards this aim, we seek to answer the RQs and test the corresponding hypotheses as discussed next.

3. RESEARCH QUESTIONS

In light of this prior work, our aim in this paper is to investigate student designers' perceptions of the psychological distance and empathy-invoking nature of sustainable design problems. Toward this aim, we seek answers to the following research questions (see Figure 1):

- RQ1: How do the variations in the formulation of sustainable design problems relate to their perceived psychological distance?
- RQ2: How do the variations in the formulation of sustainable design problems relate to their perceived empathy-invoking nature?

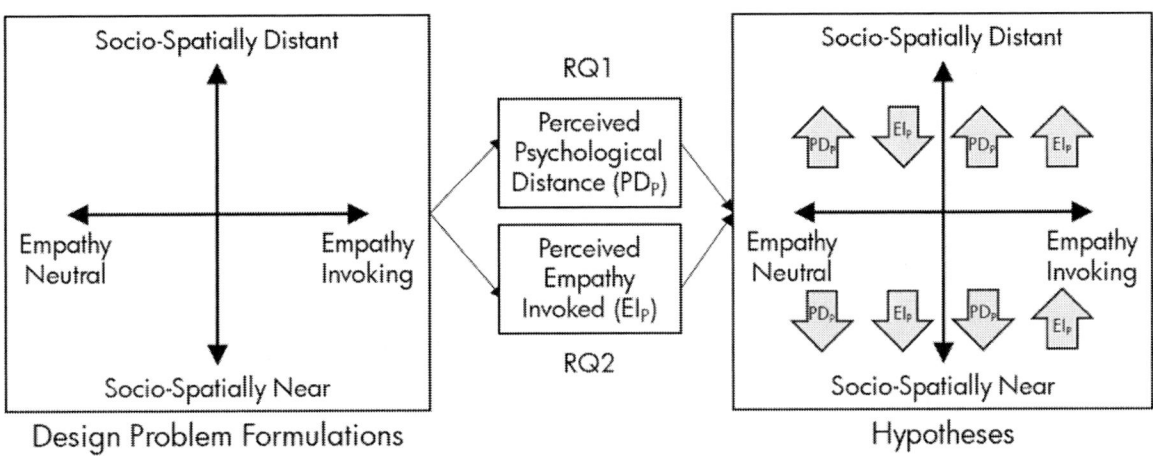

Figure 1. Overview of our study design, research questions, and corresponding hypotheses in this paper

Copyright © 2023 by ASME

We hypothesize that participants who are given the design problem with the detailed persona would report higher scores on the perceived empathy invoked by the design problem. This hypothesis is based on prior work suggesting the use of personas and detailed narratives to invoke empathy toward the end user [41]. We also hypothesize that participants who are given the socio-spatially near problem would report lower scores on the perceived psychological distance. This hypothesis is based on prior work suggesting the importance of socio-spatial distance in determining the perceived psychological distance from certain objects or events [28].

4. EXPERIMENTAL METHODS

Our aim in this paper is to investigate student designers' perceptions of the psychological distance and empathy-invoking nature of sustainable design problems. Towards this research aim and to answer the RQs presented in Section 3, we conducted a survey-based experiment with novice engineering design students. The experiment was reviewed and approved by the university's Institutional Review Board before it was conducted. The details of the experiment along with the various problem formulations tested are discussed next.

4.1. Participants

The participants were recruited from a first-year introductory course on engineering design at a large research university in the northeastern United States. Of the 138 participants consenting to participate in the experiment, 88 self-identified as male, 49 self-identified as female, with the remaining not reporting their gender. Additionally, 115 participants were in their first year of study, 20 in their second year of study, and one in their third year of study, with the remaining participants not reporting their year of study. This demographic information was collected to understand the distribution of the sample and was not used in our analyses. However, these demographic identifiers could impact our results, therefore, calling for future work to test these effects.

4.2. Developing the Problem Variants

Since our main aim is to understand how different problem formulations are perceived by novice engineering designers, we formulated four variants of the similarly contextualized problem. These problem formulations varied based on two factors: (1) the socio-spatial distance and (2) the empathy-invoking nature. All four problem variants focused on providing access to clean water and sanitation and were presented in relation to Sustainable Development Goal #6: Clean Water and Sanitation [42]. Two variations in the socio-spatial distance were created: (1) a socio-spatially problem situated in Pennsylvania, USA, and (2) a socio-spatially far problem situated in Northern India. Similarly, two variations of the empathy focus of the design problem were created: (1) an empathy-neutral problem without an explicit persona, and (2) an empathy-invoking problem comprising an explicit persona. These variations resulted in a 2x2 study, and the four problem formulations are presented in Table 1.

Table 1. The four problem variants compared in our experiment

	Socio-Spatially Near	Socio-Spatially Far
Empathy Neutral	A family in Pennsylvania is facing the negative impacts of climate change. The rising temperatures limit their agricultural production due to the lack of precipitation and clean water. Design a solution so that this family can have access to clean water for their land.	A family in Northern India is facing the negative impacts of climate change. The rising temperatures limit their agricultural production due to the lack of precipitation and clean water. Design a solution so that this family can have access to clean water for their land.
Empathy Invoking	The Williams family in Pennsylvania has lived on the same farm for five generations. They are dependent on the farm as their main source of income and food for their children. The lack of rainfall due to climate change has devastated their farm and caused a drastic decrease in their crop yield. As a result, their children are malnourished, and they have lost their income. Moreover, they are heartbroken by the prospect of losing their family farming business. Design a solution so that this family can have access to clean water for their land.	The Singh family of Northern India has lived on the same farm for five generations. They are dependent on the farm as their main source of income and food for their children. The lack of rainfall due to climate change has devastated their farm and caused a drastic decrease in their crop yield. As a result, their children are malnourished, and they have lost their income. Moreover, they are heartbroken by the prospect of losing their family farming business. Design a solution so that this family can have access to clean water for their land.

4.3. Survey Procedure and Instruments

Our aim in this paper is to test student designers' perceptions of four variants of a sustainability-focused design problem. To collect data about students' perceptions, we conducted an electronic survey sent via email. Participation in the study was voluntary, and informed consent was gathered through Microsoft Forms. Notably, participants were offered extra credit for completing the survey by their course instructors.

The survey comprised two scales: (1) perceived psychological distance and (2) perceived empathy invoked, with seven items each. Each participant was randomly assigned one of the four problem variants presented in Table 1 and was asked to respond to a fourteen-item survey on a Likert-type scale of 1 = Strongly Disagree to 6 = Strongly Agree [43]. The details of each scale and the corresponding scale items are discussed next.

Copyright © 2023 by ASME

4.3.1. Perceived Psychological Distance

Psychological distance is defined as the extent to which certain concepts are present in one's reality [26]. Prior research shows that psychological proximity includes an individual's perception of the problem as (1) relevant to the community, (2) occurring in the present time, and (3) being tangible in its implications and effects on the world [25]. Psychological distance has been measured based on four key components: (1) the distance between the target event in the past or the future and the individual in time (temporal distance), (2) how far the individual is from the target event in space (spatial distance), (3) the social affinity or similarity between the individual and the occurrence (social distance), and (4) the perceived probability that an event will happen (uncertainty) [11].

Based on these four theorized components of psychological distance, we designed a survey to evaluate novice designers' perceptions of their psychological distance from the sustainability-focused design problem presented to them (see Section 4.2). Specifically, the seven-item scale presented in Table 2 was used to evaluate participants' perceived psychological distance across the four components. Specifically, we included two items each for temporal, social, and hypothetical distance, and one item for spatial distance. These items were generated through common themes and scale items used in prior studies such as [25]. Participants were asked to respond to each item in the context of the presented design problem on a scale of 1 = Strongly Disagree to 6 = Strongly Agree. We used a six-point scale to minimize the reliance on neutral responses [44].

The internal consistency of the scale was assessed by computing Cronbach's α [45]. The scale item scores showed a moderate internal consistency through an observed Cronbach's α = 0.43, 95% CI [0.28, 0.57]. An average of the participants' scores on the seven scale items was computed to obtain an overall perceived psychological distance score. These overall scores were used as the data to answer the first RQ.

Table 2. Scale items used to measure the perceived psychological distance from the design problem

Item #	Scale Item
PD1	I have never met a family like the one mentioned in the problem statement.
PD2	The seriousness of climate change effects is not as severe as presented in this problem statement.
PD3	When I think about problems such as this, I usually think of it occurring in countries that are far away.
PD4	Situations similar to this problem are immediate threats affecting people right now.
PD5	I don't think these kinds of problems will significantly impact people I know.
PD6	I am uncertain that problems similar to the one above is really happening around the world.
PD7	I feel a sense of urgency to change my behavior in order to solve problems such as this one.

4.3.2. Empathy Invoking Nature of the Design Problem:

Empathy is defined as the reactions of one individual to the observed experiences of another [30]. Prior research in design highlights the importance of empathy in the problem definition and formulation stages of the design process [46]. Given the importance of empathy in design, researchers have used various measures for assessing designers' empathy, ranging from surveys to task-based assessments [47].

Using previously proposed measures as a starting point, a seven-item scale was developed to assess the empathy-invoking nature of the design problem. This scale was based on Hess et al.'s [48] measure of empathy in design. Participants were asked to respond with their agreement to the statements summarized in Table 3 on a Likert-type scale with 1 = Strongly Disagree and 6 = Strongly Agree. The internal consistency of the participants' responses was assessed by computing Cronbach's α [45]. The scale scores showed a moderate internal consistency as established through an observed Cronbach's α = 0.65, 95% CI [0.55, 0.74]. An average of the scores on the seven scale items was computed to obtain an overall empathy-invoking score. These overall mean scores were used as the data to answer the second research question.

Table 3. Scale items that were used to measure the perceived empathy invoked by the problem

Item #	Scale Item
EM1	I can imagine the everyday activities of the user mentioned in the problem statement above.
EM2	I can imagine how a user feels when they experience a similar problem as above.
EM3	I can relate to the challenges that users experience in their everyday life.
EM4	I feel sorry for users who experience similar problems as above.
EM5	I feel concern for users who face similar challenges to those presented in the problem statement.
EM6	I feel a desire to identify ways to improve the experiences of the users mentioned in the problem statement.
EM7	I feel guilty if I am unable to understand the perspectives of the users mentioned in the problem statement.

5. DATA ANALYSIS AND RESULTS

To answer the RQs presented in Section 3, the data collected from the experiment were analyzed using quantitative methods. The details of the analyses and the corresponding results are discussed in this section.

5.1. RQ1: How do the variations in the formulation of sustainable design problems relate to their perceived psychological distance?

To answer the first research question, we performed a two-way ANOVA with the overall perceived psychological distance score as the dependent variable. We used the two components of problem formulation, i.e., socio-spatial distance and empathy-focus, as the independent variables. Before performing the

Copyright © 2023 by ASME

analysis, we first tested whether our data were normally distributed. A Shapiro-Wilk test [49] revealed that the data were normally distributed ($p > 0.05$). Next, we established the homogeneity of variances across the four problem types through Levene's test ($p > 0.05$) [50]. Finally, no significant outliers were observed based on a visual inspection of the boxplots.

From the results of the ANOVA, we see that first, there was no significant interaction between the two independent variables ($F (1, 130) = 0.04$, $p = 0.84$, SSE = 0.02). Therefore, we tested the main effects of the two problem formulation components. From the results, we see that neither the socio-spatial distance nor the empathy focus of the problem formulation had significant effects on the perceived psychological distance ($F (1, 130) = 0.003$, $p = 0.95$, SSE = 0.001, and $F (1, 130) = 1.41$, $p = 0.24$, SSE = 0.56, respectively). We present a comparison of the perceived psychological distance scores between the four problem types in Figure 2. A discussion of the implications of the results is outlined in Section 6.

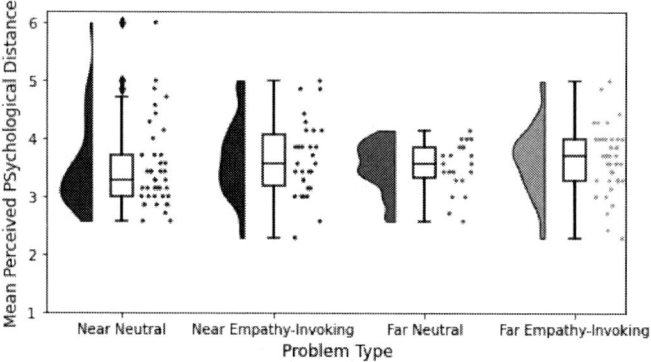

Figure 2. Comparing perceived psychological distance across the four problem variations (lower scores correspond to lower psychological distance)

5.2. RQ2: How do the variations in the formulation of sustainable design problems relate to their perceived empathy-invoking nature?

To answer the second RQ, we performed a two-way ANOVA with the overall perceived psychological distance score as the dependent variable and the problem type as the independent variable. Before performing the analysis, we first tested whether our data were normally distributed. A Shapiro-Wilk test revealed that the data were not normally distributed ($p = 0.03$). Next, we established the homogeneity of variances across the four problem types through Levene's test ($p > 0.05$). Finally, no significant outliers were observed based on a visual inspection of the boxplots. Despite the violation of the assumption of normal distribution, we performed the ANOVA given the robustness of the analysis to violations of normality.

From the results of the ANOVA, we see that first, there was no significant interaction between the two independent variables ($F (1, 130) = 0.007$, $p = 0.93$, SSE = 0.003). Therefore, we tested the main effects of the two problem formulation components.

From the results, we see that neither the socio-spatial distance nor the empathy focus of the problem formulation had significant effects on the perceived psychological distance ($F (1, 130) = 0.14$, $p = 0.71$, SSE = 0.06, and $F (1, 130) = 0.006$, $p = 0.94$, SSE = 0.002, respectively). We present a comparison of the perceived psychological distance scores between the four problem types in Figure 3. A discussion of the implications of the results can be found in Section 6.

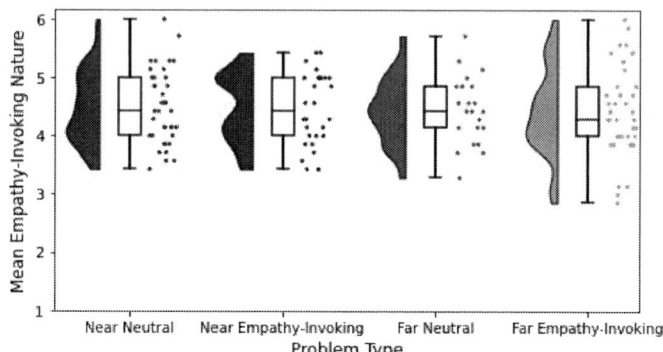

Figure 3. Comparing the perceived empathy-invoking nature of the four problem formulations (lower scores indicate lower perceived empathy invoked)

6. DISCUSSION

Our aim in this research was to investigate student designers' perceptions of the psychological distance and empathy-invoking nature of different sustainable design problem formulations. Towards this aim, we conducted a survey-based experiment with novice engineering design students and two key findings were observed:

- Variations in the problem formulation did not relate to participants' perceived psychological distance.
- Variations in the problem formulation did not relate to the perceived empathy-invoking nature of the design problems.

The first key finding from this study is that the variations in the socio-spatial context and the empathy focus of a design problem did not relate to participants' perceived psychological distance. Specifically, the average perceived psychological distance of all problems was similar for all four problem formulations (Near-Neutral = 3.49, Near-Empathy = 3.65, Far-Neutral = 3.63, Far Empathy = 3.58), and these scores were close to the scale center (i.e., 3.5). While we chose problem statements that were designed to be socio-spatially near or far, the content of the problem (access to clean water) may be distant for all our participants, therefore obfuscating the differences in psychological distance due to the manipulation. Prior research suggests that individuals associate certain environmental challenges (e.g., rising water levels) more strongly with climate change compared to others (e.g., rising temperatures) [51]. This finding, therefore, calls for a deeper look into students' perceived psychological distance towards different problem contexts besides access to clean water and sanitation.

Copyright © 2023 by ASME

Secondly, this finding could be attributed to the use of an aggregated psychological distance score. That is, in our study, we aggregated psychological distance along the four dimensions (i.e., social, spatial, temporal, and hypotheticality) into a single average score. Despite the high internal reliability in the scale, individuals could demonstrate differences in the degree of psychological distance based on the dimension under investigation. Using an aggregate score could have obscured differences present in the scores for each dimension, as alluded to in prior work [26]. Future research must, therefore, investigate differences between the problem formulations along each dimension of psychological distance.

The second key finding is that the variations in the problem formulation did not relate to the perceived empathy-invoking nature of the design problems. Specifically, we see that the average empathy-invoking scores of participants were high for all four problem statements (Near-Neutral = 4.48, Near-Empathy = 4.48, Far-Neutral = 4.44, Far Empathy = 4.43). These results suggest that sustainable design tasks are perceived to be empathy-invoking regardless of the socio-spatial distance from the user or the inclusion of a persona in the problem formulation.

Taken together, our findings suggest that the design problem formulations used in our study are perceived similarly by novice designers. However, one caveat in this observation is that all four problem variations focused on the same environmental challenge: clean water access. Therefore, to validate these findings, future research should examine if, and how, different design tasks, beyond those related to sustainability, impact student designers' perceptions. Moreover, our study was designed to assess student designers' *perceptions* of design problems. These perceptions might not always translate into effective design outcomes such as problem needs identification or ideation effectiveness. Thus, future research must investigate whether the variations in the socio-spatial context and the empathy focus of a design problem impact design outcomes such as problem needs identification and ideation effectiveness.

7. CONCLUDING REMARKS

Our goal in this study was to explore student designers' perceptions of the psychological distance and empathy-invoking nature of different sustainable design problem formulations. Specifically, we tested the effects of variations in (1) the socio-spatial context and (2) the empathy focus of a similar design problem. All four problem variants focused on providing access to clean water and sanitation and were presented in relation to Sustainable Development Goal #6: Clean Water and Sanitation. From the results of our survey-based study, we see that the variations in the problem formulations did not relate to either the perceived psychological distance or the perceived empathy-invoking nature of the design problem. This finding suggests that issues related to environmental sustainability tend to be perceived similarly, despite differences in their socio-spatial context and empathy-invoking nature. Furthermore, this finding calls for a deeper investigation into the role of empathy development tools beyond personas in psychologically distant contexts such as environmental sustainability.

8. LIMITATIONS AND DIRECTIONS FOR FUTURE WORK

The findings from this study revealed some insights into novice designers' perceptions of different problem formulations in sustainability-focused design tasks. However, certain limitations exist that present avenues for future research. First, this study was conducted with engineering students from a large northeastern university in the United States of America. Sustainable design tasks are complex, and students might not be yet equipped to understand the details of those issues. First-year students, while motivated to solve climate change, can often not grapple with how large of a scale this problem is. Thus, future research is warranted to extend these findings to individuals from different cultural backgrounds and levels of expertise. Second, all four problem variations used in the experimental design of this study focused on sustainable design issues. Future iterations of this research should examine if, and how, different problem statements, not related to sustainability, impact student designers' perceptions of the psychological distance and empathy-invoking nature of the design tasks. Third, the results from this study are based on self-reported surveys that capture students' perceptions. The variations in socio-spatial context and the empathy focus of a design problem might impact design outcomes such a problem needs identification and ideation effectiveness, which calls for future research. Fourth, the empathy-invoking scenario used in this study was text-based which might potentially be limiting in invoking participants' empathy. Thus, future iterations of this research could include videos or photos of the personas. Finally, the components of psychological distance were used as an aggregate in this study. Researchers such as Trope and Lieberman [26] found that the measurements of the four components of psychological distance are not necessarily proportional to one another, and hence an aggregate measure might be misleading. Thus, future iterations of this study should explore the different components of psychological distance.

ACKNOWLEDGEMENTS

Jenna Herzog and Rebekah Fodale were funded by the Claire Boothe Luce Research Scholarship and the Engineering Division at Lafayette College.

REFERENCES

[1] Thursby, M., 2014, "The Importance of Engineering: Education, Employment, and Innovation," The Bridge, **44**.

[2] Bragança, L., Vieira, S. M., and Andrade, J. B., 2014, "Early Stage Design Decisions: The Way to Achieve Sustainable Buildings at Lower Costs," The Scientific World Journal, **2014**, pp. 1–8.

[3] Dorst, K., 2006, "Design Problems and Design Paradoxes," Design Issues, **22**(3), pp. 4–17.

[4] Dorst, K., and Cross, N., 2001, "Creativity in the Design Process: Co-Evolution of Problem–Solution," Design Studies, **22**(5), pp. 425–437.

[5] Suwa, M., Gero, J., and Purcell, T., 2000, "Unexpected Discoveries and S-Invention of Design Requirements:

Important Vehicles for a Design Process," Design Studies, **21**(6), pp. 539–567.

[6] Prabhu, R., Miller, S. R., Simpson, T. W., and Meisel, N. A., 2020, "Complex Solutions for Complex Problems? Exploring the Role of Design Task Choice on Learning, Design for Additive Manufacturing Use, and Creativity," Journal of Mechanical Design, **142**(3), p. 031121.

[7] Onarheim, B., 2012, "Creativity from Constraints in Engineering Design: Lessons Learned at Coloplast," Journal of Engineering Design, **23**(4), pp. 323–336.

[8] Shalley, C. E., 1991, "Effects of Productivity Goals, Creativity Goals, and Personal Discretion on Individual Creativity.," Journal of Applied Psychology, **76**(2), pp. 179–185.

[9] Alsager Alzayed, M., McComb, C., Menold, J., Huff, J., and Miller, S. R., 2021, "Are You Feeling Me? An Exploration of Empathy Development in Engineering Design Education," Journal of Mechanical Design, **143**(11), pp. 1–57.

[10] Trope, Y., and Liberman, N., 2003, "Temporal Construal.," Psychological Review, **110**(3), pp. 403–421.

[11] Liu, Q., Zhang, X., Huang, S., Zhang, L., and Zhao, Y., 2020, "Exploring Consumers’ Buying Behavior in a Large Online Promotion Activity: The Role of Psychological Distance and Involvement," J. theor. appl. electron. commer. res., **15**(1), pp. 0–0.

[12] Semenza, J. C., Hall, D. E., Wilson, D. J., Bontempo, B. D., Sailor, D. J., and George, L. A., 2008, "Public Perception of Climate Change," American Journal of Preventive Medicine, **35**(5), pp. 479–487.

[13] Busse, M., and Menzel, S., 2014, "The Role of Perceived Socio-Spatial Distance in Adolescents' Willingness to Engage in pro-Environmental Behavior," Journal of Environmental Psychology, **40**, pp. 412–420.

[14] V. Boivin, D., and Boiral, O., 2022, "So Close, Yet So Far Away: Exploring the Role of Psychological Distance from Climate Change on Corporate Sustainability," Sustainability, **14**(18), p. 11576.

[15] Bardwell, L. V., 1991, "Problem-Framing: A Perspective on Environmental Problem-Solving," Environmental Management, **15**(5), pp. 603–612.

[16] Loy, L. S., and Spence, A., 2020, "Reducing, and Bridging, the Psychological Distance of Climate Change," Journal of Environmental Psychology, **67**, p. 101388.

[17] Grudin, J., and Pruitt, J., 2002, "Personas, Participatory Design and Product Development: An Infrastructure for Engagement," *PDC 02 Proceedings of the Participatory Design Conference*, Malmo, Sweden.

[18] Fila, N., and Hess, J., 2016, "In Their Shoes: Student Perspectives on the Connection between Empathy and Engineering," *2016 ASEE Annual Conference & Exposition Proceedings*, ASEE Conferences, New Orleans, Louisiana, p. 25640.

[19] Amabile, T., 1996, *Creativity in Context: Update to the Social Psychology of Creativity*, Westview Press.

[20] Prabhu, R., Berthel, J. T., Masia, J. S., Meisel, N. A., and Simpson, T. W., 2022, "Rapid Response! Investigating the Effects of Problem Definition on the Characteristics of Additively Manufactured Solutions for COVID-19," Journal of Mechanical Design, **144**(5), p. 054502.

[21] Litchfield, R. C., 2009, "Brainstorming Rules as Assigned Goals: Does Brainstorming Really Improve Idea Quantity?," Motiv Emot, **33**(1), pp. 25–31.

[22] Litchfield, R. C., Fan, J., and Brown, V. R., 2011, "Directing Idea Generation Using Brainstorming with Specific Novelty Goals," Motiv Emot, **35**(2), pp. 135–143.

[23] Burleson, G., Herrera, S. V. S., Toyama, K., and Sienko, K. H., 2023, "Incorporating Contextual Factors Into Engineering Design Processes: An Analysis of Novice Practice," Journal of Mechanical Design, **145**(2), p. 021401.

[24] Svihla, V., Wilson-Fetrow, M., Chen, Y., Chi, E., Datye, A. K., Han, S. M., Gomez, J., and Olewnik, A., 2021, "The Educative Design Problem Framework: Relevance, Sociotechnical Complexity, Accessibility, and Nondeterministic High Ceilings," p. 17.

[25] Većkalov, B., Zarzeczna, N., McPhetres, J., van Harreveld, F., and Rutjens, B. T., 2022, "Psychological Distance to Science as a Predictor of Science Skepticism Across Domains," Pers Soc Psychol Bull, p. 014616722211181.

[26] Trope, Y., and Liberman, N., 2010, "Construal-Level Theory of Psychological Distance," Psychol Rev, **117**(2), pp. 440–463.

[27] Spence, A., Poortinga, W., and Pidgeon, N., 2012, "The Psychological Distance of Climate Change: Psychological Distance of Climate Change," Risk Analysis, **32**(6), pp. 957–972.

[28] Loy, L. S., and Spence, A., 2020, "Reducing, and Bridging, the Psychological Distance of Climate Change," Journal of Environmental Psychology, **67**, p. 101388.

[29] Chu, H., 2022, "Construing Climate Change: Psychological Distance, Individual Difference, and Construal Level of Climate Change," Environmental Communication, **16**(7), pp. 883–899.

[30] Davis, M. H., 1983, "Measuring Individual Differences in Empathy: Evidence for a Multidimensional Approach.," Journal of Personality and Social Psychology, **44**(1), pp. 113–126.

[31] McGinley, C., and Dong, H., 2011, "Designing with Information and Empathy: Delivering Human Information to Designers," The Design Journal, **14**(2), pp. 187–206.

[32] Battarbee, K., 2004, *Co-Experience: Understanding User Experiences in Social Interaction*, University of Art and Design in Helsinki, Helsinki.

[33] Johnson, D. G., Genco, N., Saunders, M. N., Williams, P., Seepersad, C. C., and Hölttä-Otto, K., 2014, "An Experimental Investigation of the Effectiveness of Empathic Experience Design for Innovative Concept Generation," Journal of Mechanical Design, **136**(5), p. 051009.

[34] Melo, Á. H. da S., Rivero, L., Santos, J. S. dos, and Barreto, R. da S., 2020, "EmpathyAut: An Empathy Map for People

Copyright © 2023 by ASME

with Autism," *Proceedings of the 19th Brazilian Symposium on Human Factors in Computing Systems*, ACM, Diamantina Brazil, pp. 1–6.

[35] Endmann, A., and Keßner, D., 2016, "User Journey Mapping – A Method in User Experience Design," i-com, **15**(1), pp. 105–110.

[36] Raviselvam, S., Höltta-Otto, K., and Wood, K. L., 2016, "User Extreme Conditions to Enhance Designer Empathy and Creativity: Applications Using Visual Impairment," *Volume 7: 28th International Conference on Design Theory and Methodology*, American Society of Mechanical Engineers, Charlotte, North Carolina, USA, p. V007T06A005.

[37] Chang, Y., Lim, Y., and Stolterman, E., 2008, "Personas: From Theory to Practices," *Proceedings of the 5th Nordic Conference on Human-Computer Interaction: Building Bridges*, ACM, Lund Sweden, pp. 439–442.

[38] LeRouge, C., Ma, J., Sneha, S., and Tolle, K., 2013, "User Profiles and Personas in the Design and Development of Consumer Health Technologies," International Journal of Medical Informatics, **82**(11), pp. e251–e268.

[39] Cooper, A., 2004, *The Inmates Are Running the Asylum*, Sams, Indianapolis, IN.

[40] Ferreira, B., Silva, W., Oliveira, E., and Conte, T., 2015, "Designing Personas with Empathy Map," pp. 501–505.

[41] Ohm, F., Vogel, D., Sehner, S., Wijnen-Meijer, M., and Harendza, S., 2013, "Details Acquired from Medical History and Patients' Experience of Empathy – Two Sides of the Same Coin," BMC Med Educ, **13**(1), p. 67.

[42] Lee, B. X., Kjaerulf, F., Turner, S., Cohen, L., Donnelly, P. D., Muggah, R., Davis, R., Realini, A., Kieselbach, B., MacGregor, L. S., Waller, I., Gordon, R., Moloney-Kitts, M., Lee, G., and Gilligan, J., 2016, "Transforming Our World: Implementing the 2030 Agenda Through Sustainable Development Goal Indicators," J Public Health Pol, **37**(S1), pp. 13–31.

[43] Arnold, W. E., McCroskey, J. C., and Prichard, S. V. O., 1967, "The Likert-type Scale," Today's Speech, **15**(2), pp. 31–33.

[44] Guy, R. F., and Norvell, M., 1977, "The Neutral Point on a Likert Scale," The Journal of Psychology, **95**(2), pp. 199–204.

[45] Cronbach, L. J., 1951, "Coefficient Alpha and the Internal Structure of Tests," Psychometrika, **16**(3), p. 38.

[46] Li, J., and Holtta-Otto, K., 2022, "Inconstant Empathy – Interpersonal Factors That Influence the Incompleteness of User Understanding," Journal of Mechanical Design, pp. 1–38.

[47] Surma-aho, A., and Höltta-Otto, K., 2022, "Conceptualization and Operationalization of Empathy in Design Research," Design Studies, **78**, p. 101075.

[48] Hess, J., Sanders, E., and Fila, N. D., 2022, "Measuring and Promoting Empathic Formation in a Multidisciplinary Engineering Design Course," *ASEE 2022 Annual Conference*, Minneapolis, Minnesota, p. 26.

[49] Shapiro, S. S., and Wilk, M. B., 1965, "An Analysis of Variance Test for Normality (Complete Samples)," Biometrika, **52**(3/4), p. 22.

[50] Gastwirth, J. L., Gel, Y. R., and Miao, W., 2009, "The Impact of Levene's Test of Equality of Variances on Statistical Theory and Practice," Statist. Sci., **24**(3).

[51] Leviston, Z., Price, J., and Bishop, B., 2014, "Imagining Climate Change: The Role of Implicit Associations and Affective Psychological Distancing in Climate Change Responses: Implicit Associations with Climate Change," Eur. J. Soc. Psychol., **44**(5), pp. 441–454.

Copyright © 2023 by ASME

Proceedings of the ASME 2023
International Design Engineering Technical Conferences and
Computers and Information in Engineering Conference
IDETC-CIE2023
August 20-23, 2023, Boston, Massachusetts

DETC2023-116399

REDUCING THE BARRIERS TO DESIGNING 3D-PRINTABLE PROSTHETICS IN RESOURCE-CONSTRAINED ENVIRONMENTS

Junghun Lee[1], Andrew Chesang [2], Michael Gichane[2], Moise Busogi [3], Jean Byiringiro [2]
Conrad Tucker [1,3]

[1] Carnegie Mellon University, USA
[2] Dedan Kimathi University of Technology, Kenya
[3] Carnegie Mellon University Africa, Rwanda

ABSTRACT

Prosthetic devices remain unaffordable to many patients in resource-constrained environments. The design process of personalized prosthetic devices is sophisticated and requires specialized equipment and trained professionals who may have limited availability. The situation is further exacerbated by the fact that the resources are frequently situated away from underserved communities, compounding the issue of accessibility. Lowering the barriers to accessing required equipment and minimizing the involvement of trained professionals in designing and manufacturing prosthetic devices can potentially minimize the required visits to the prosthetists and lower the associated costs. Traditional prosthetic and orthotic device creation methods can take up to several months. There are attempts to leverage advanced technologies to enhance the design process, but they require special equipment such as 3D scanners and MRI equipment in addition to the active involvement of specialists.

This work proposes a computer vision method for creating a virtual model of the affected human limb. We also propose an automated approach for generating the socket from the virtual model and scaling other parts of the prosthetic device. The prosthetic device designed through the proposed process is realized via a 3D printer. We present a case study demonstrating the feasibility of our approach, using a mobile phone and open-source software to reconstruct a virtual representation of an upper limb and customize a prosthetic hand with minimal input from a specialist.

Keywords: Virtual model, Computational Design, Prosthetics, Computer Vision

1. INTRODUCTION

Many patients suffer from damage to their limbs. The loss of limbs severely affects a patient's everyday life in terms of functionality and social interaction. Prosthetics and orthotics are used to support, realign, or redistribute pressure across part of a person's musculoskeletal system [1,2]. However, for these devices to be effective, they must be customized to each person's unique anatomy and functional needs [3-7]. This is especially critical for children facing significant challenges due to physical growth and psychological development [8,9].

Despite the necessity of prosthetic devices, they are inaccessible to many patients, especially in resource-constrained environments [10,11]. According to the World Health Organization (WHO), only one in ten people in need can access assistive products, including prostheses and orthoses due to their high cost, lack of awareness, unavailability of trained personnel, policy, and financing issues [12]. The situation can be exacerbated in conflict areas due to the high risk of upper limb loss during warfare [13].

A major limitation of traditional methods for producing customized prosthetic devices is the labor-intensive hand-crafting procedure [11]. The traditional process is based on the artisan's skills that requires some level of subjectivity. In this method, prosthetists manually cast the patient's body with physical molds and produce a customized assistive device [14,15]. Following this process, the patients must visit the hospital multiple times to cast their bodies and adjust prosthetic devices. The traditional method of creating a prosthetic and orthotic device can take up to several months [16]. There are attempts to leverage advanced technologies to enhance the design process, but they require special equipment such as 3D scanners and MRI equipment, in addition to the active involvement of specialists [5-7]. To address these challenges, this paper introduces a method that streamlines the process from prosthetic device design to fabrication without requiring specialized equipment and with minimal reliance on scarce trained professionals. The paper is structured as follows: Section 2 introduces related work, and Section 3 describes the proposed approach to construct the virtual model and customize the prosthetic device. In section 4, we present a case study for a prosthetic hand to demonstrate the effectiveness of the proposed method. Finally, we discuss the conclusions and future plans in Section 5.

Copyright © 2023 by ASME

2. RELATED WORK

The typical production process of prosthetic devices requires prosthetists to manually cast the patient's body and produce a customized assistive device [14,15]. Numerous studies have attempted to explore the benefits of 3D printing in the production of prosthetics and orthotics [17]. For instance, researchers have developed body-powered 3D-printed prosthetic hand designs [13-15]. Some authors have mainly focused on developing materials, methods, and equipment for the 3D printing of prosthetic devices [18-21]. One notable movement towards an accessible prosthetic hand is e-NABLE Forum. e-NABLE Forum is a volunteer movement using 3D printing to create open-source 3D-printed upper limb assistive devices [22]. They provide various open-source designs and production guidelines for many different needs of prosthetic hands. Although it is widely used, designs in e-NABLE Forum are only provided with a limited variety of sizes. Therefore, much work is still needed to customize the prosthetic hands to perfectly meet an individual's requirements. There are attempts to provide scalable prosthetic device designs in e-NABLE Forum, but they only adopt limited parameters or require a skillful modeler as a volunteer.

Existing research recognizes the critical role of reducing the cost and labor of prosthetics and orthoses using virtual models and CAD tools [23-26]. These processes can provide quicker and cheaper customized designs, compared to traditional approaches. However, it requires special equipment and frequent involvement of specialists. For example, Moreman et al. created a virtual model of the damaged foot using MRI equipment. They offset the surface of the virtual model to generate a socket for the prosthetic device [7]. Górski et al. created a virtual model of the patient's residual arm with a 3D scanner and automated the design process using scripts and macros in the CAD Software [24]. Rai et al. generated a prosthetic device socket to cover the surface of the residual body. The authors employed topology optimization techniques to optimize the lattice structure from the virtual amputated foot, acquired using MRI equipment [25]. Steer et al. proposed a method to generate prosthetic device sockets with a parametric design and genetic algorithm from a virtual model [26].

This research aims to contribute to the efforts of making prosthetic devices more affordable and accessible in remote communities and resource-constrained environments by expediting the design process and utilizing more cost-effective techniques. With this goal in mind, a virtual model of the residual body is created to customize a prosthetic device. With the virtual model of the residual body, a customized prosthetic device can be created with minimum patient participation. We use photogrammetry to virtually reconstruct the residual body from images captured by a mobile phone. Photogrammetry refers to the process of deriving metric information about an object through measurements made on the photographs of the object [27]. Photogrammetry has the advantage that it only requires images of the target scene to reconstruct the virtual model. In this process, patients only need to create the virtual representation of the residual body by capturing images of the residual body and sharing them. That, in turn, can potentially reduce the number of visits needed in the early stage of the design and minimize the need for costly facilities. In addition to the affordability and reduction of turn-around time, research has proven the effectiveness of photogrammetry by using images taken by mobile phones [28-30]. We also partly automate the customization process of the prosthetic device so that the process can be done with the minimum involvement of a specialist. This can potentially reduce the cost and labor in the prosthetic device design process.

3. METHOD

The proposed method for the design process consists of two steps: i) virtual reconstruction and ii) prosthetic device customization and fabrication (i.e., 3D printing). The first step is virtual reconstructing from images captured from a mobile phone. There have been several attempts to substitute the traditional plaster casting process of designing prosthetics and orthotics with a virtual model [23-26]. However, these attempts still require special equipment, such as 3D scanners and MRI equipment, and the active involvement of specialists. We aim to overcome this limitation by using only a mobile phone to reconstruct the virtual model. Mobile phones are highly accessible and user-friendly devices. According to The World Bank, the rate of Mobile cellular subscriptions in sub-Saharan Africa stood at 93% in 2021 [31]. Although it is essential to recognize that not all cellular subscriptions guarantee access to a phone camera, it is worth noting that the number of smartphone connections is expected to exceed 700 million by 2025, resulting in a projected adoption rate of 66% in the region [32].

The second step is the customization and fabrication of the prosthetic device. In this step, we generate a socket for the prosthetic device from a virtual model of the residual body created in the previous step and scale the other parts of the prosthetic device according to it. Once all components of the prosthetic device have been customized, the device is fabricated using 3D printing technology. The prosthetic customization process using the virtual model is automated using scriptable, open-source CAD software. This reduces the high cost and intensive labor required in the traditional customized prosthetic creation process [11]. The proposed process for creating a 3D printable prosthetic device requires patients only to provide photographs of their residual bodies, which must be taken following proper guidelines. An overview of the proposed prosthetic device design process is shown in Figure 1.

Copyright © 2023 by ASME

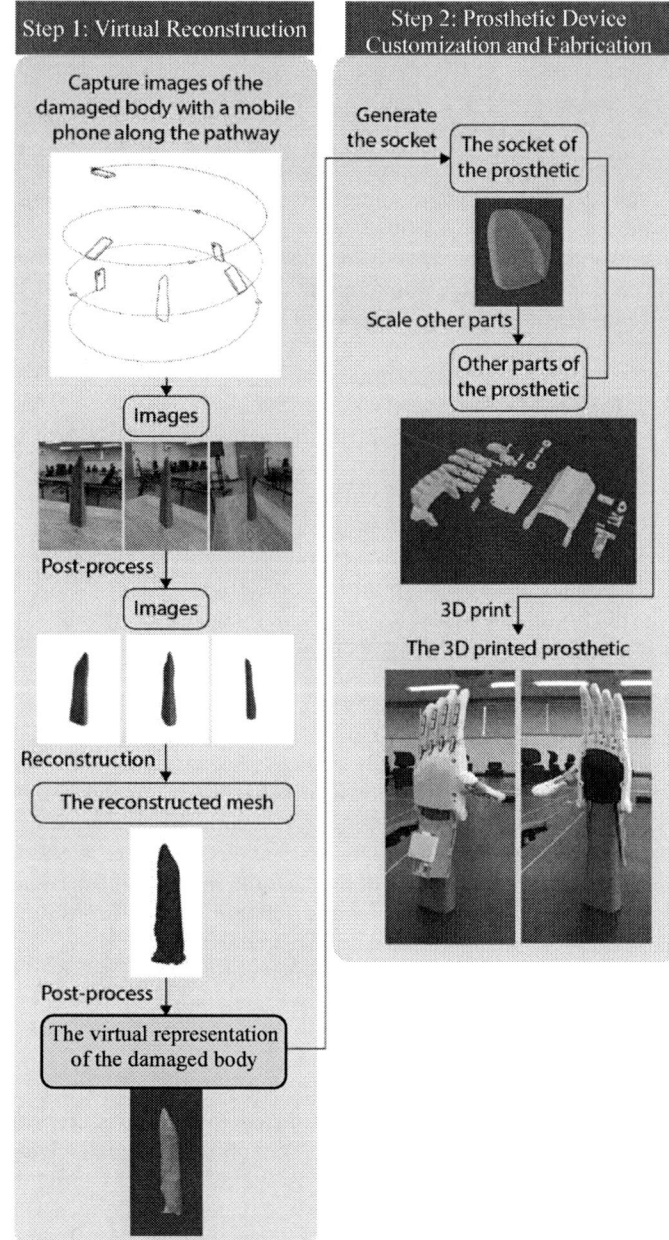

FIGURE 1: Overview of the method

3.1 Virtual Model Reconstruction

In this section, we explain how we reconstruct the virtual model of the residual body from images captured from a mobile phone. We use photogrammetry to reconstruct the virtual model of the patient's residual body. A sequence of images with a minimum side overlap of 60% and a frontal overlap of 80% maximum from diverse views is required to achieve the best result in photography [33]. However, it is challenging to consider such a condition every time a picture is taken. Therefore, Munhoz et al. proposed a pathway to ensure an appropriate and diverse view of photographs, as shown in Figure 2 [34]. The

proposed pathway successfully created a virtual representation of the infant hip model by photographing along the proposed pathway. To facilitate real-world adoption and reduce entry barriers, it is advisable to implement a mobile application that provides guidance and semi-automates the process. This approach would ensure that the captured images meet the necessary standards in terms of both quality and quantity for photogrammetric purposes.

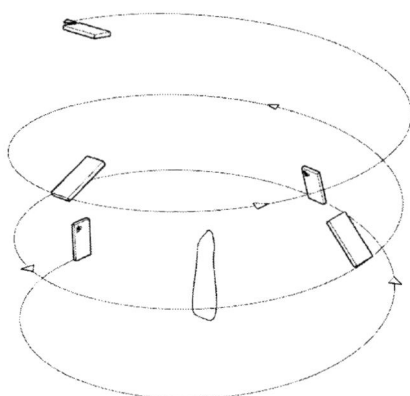

FIGURE 2: The pathway designed by Munhoz et al. [34]

The challenge in using photogrammetry to reconstruct the virtual model is identifying the patient's residual body inside the whole scene. To address this problem, we post-process the image data for photogrammetry and mesh reconstructed from photogrammetry. We remove the background of captured images before photogrammetry and remove the vestige of the unremoved background in ˊ the mesh created from photogrammetry. These processes can be done automatically to reduce the related specialist's involvement.

The quality of photogrammetry is affected by the camera and environment determined by the following factors: scene size, material properties, quality of the textures, shooting time, amount of light, varying light or objects, camera device's quality, and setting [35]. Therefore, camera settings should be adjusted, and the right environment should be selected to gain sharp images without motion blur and depth blur. A mobile application or a guideline can help users unfamiliar with capturing good-quality images.

The concrete process of reconstructing the virtual model consists of three steps. First, take pictures of the residual body with a mobile phone and send them. The patient will have to be assisted by a secondary person who will capture the images of the residual body. This step is done on the patient's side without the involvement of trained specialists. Second, post-process them for photogrammetry. Finally, photogrammetry is used to reconstruct the virtual model from post-processed images, and the reconstructed mesh is post-processed. These two steps can be done without patient involvement and can be automated. The reconstruction process proposed in the research is described in Figure 3.

Copyright © 2023 by ASME

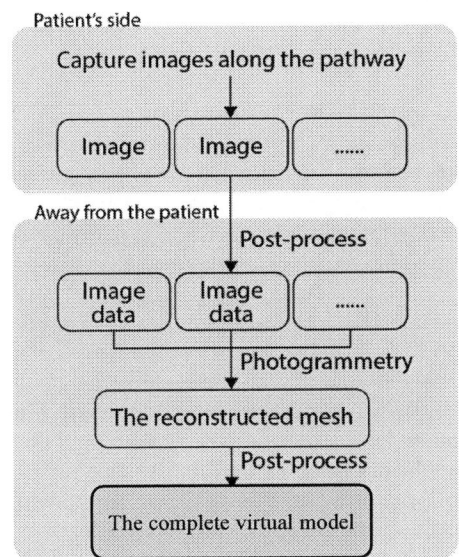

FIGURE 3: Overview of the virtual model reconstruction process

We select capturing multiple images to gather information on the residual body. Capturing multiple images has three major advantages over taking videos. First, videos can include blurry or out-focused frames. Meanwhile, many mobile phones have functions to prevent this when capturing images. Also, people can immediately check the captured images to prevent bad-quality images. The bad quality of images significantly harms the quality of the photogrammetry result. Second, videos do not include meta-data, while photographs include such information. Metadata is camera data attached to the image file, which includes the F-stop, ISO, focal length, and exposure time. These data are important in the reconstruction process in photogrammetry. Lastly, taking a video can be burdensome because one must hold the camera in an appropriate view for an extended period of time.

After capturing the image of the residual body, we post-process the image data. To avoid the task of identifying the target object in the virtual model of the whole scene, the background of the image data is removed so that only the virtual model of the target object is reconstructed. We reconstruct the virtual model using photogrammetry after the images are post-processed.

After the mesh is reconstructed from photogrammetry, it undergoes three post-processing steps. First, we identify the residual body and clear other parts of the mesh, as there may be leftover artifacts from the background removal process due to the imperfection of these techniques. Second, we smooth the mesh since surfaces generated by photogrammetry may be rigid due to imperfections in the reconstruction process. We can approximate the human body's organic shape by smoothing the surface. Last, we must position and orient the virtual model in the correct position and angle to customize the prosthetic device from it.

3.2 Prosthetic Device Customization and Fabrication

In this section, we explain how we customize and fabricate the prosthetic device using the virtual model created in the previous stage. The whole process consists of importing the virtual model, generating the socket from the virtual model, and scaling other parts of the prosthetic device. These processes are automated with scriptable CAD tools to minimize the involvement of specialists. The customized prosthetic design is fabricated via 3D-printing techniques. The main challenge of the customization process is to provide a customized socket. The socket is a part of the prosthetic device where the user's residual body interfaces with the prosthetic device. The role of the socket in this research is to comfortably fill the vacant space between the patient's residual body and the prosthetic device. The challenge is to generate a socket customized to the residual bodies of patients with all different shapes. Two problems must be addressed to generate a customized socket. First, the socket must fit the residual body of the patient. To generate a socket that fits the patient's residual body, we remove the shape of the patient's residual body from an initial socket using a boolean difference operation. Afterward, tolerance is given between the patient's body and the socket for the convenience of wearing. In this research, the initial socket is designed in the shape of the prosthetic hand and is filled with no vacant area. The virtual model created in the previous stage represents the shape of the patient's residual body. Examples of the initial socket and operated socket are shown in Figure 4.

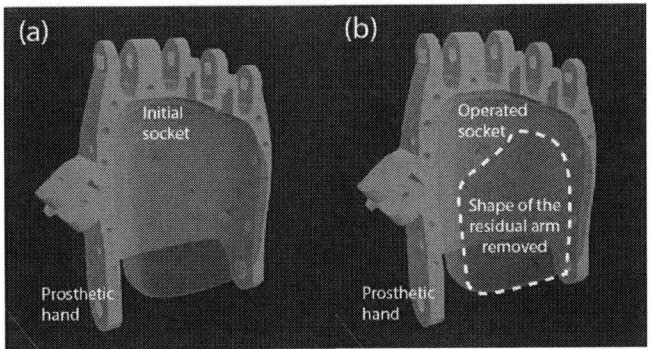

FIGURE 4: (a) An example of the initial socket (b) An example of the operated socket

Second, the socket must be the correct size. It must be big enough to cover the patient's residual body but as small as possible to be light, cheap, and convenient. We use the operated socket's width, length, and height to size the socket. When the initial socket is bigger than the virtual model of the residual body, the dimensions of the operated socket change according to the scale of the initial socket. However, when the initial socket is smaller than the virtual model of the residual body, the dimensions of the operated socket are affected by the shape of the residual body, as shown in Figure 5 (c), because the socket is damaged by the boolean difference operation with the virtual model of the residual body. Therefore, we can find the right size of the socket by observing how the dimensions of the operated

Copyright © 2023 by ASME

socket change. The operated socket's dimensions start changing differently from the scale of the initial socket if the operated socket is being affected by the virtual model of the residual body. We can scale the socket by finding the initial socket's minimum scale covering the residual body. The scaling process proposed in the research is described in Figure 6. After finding the minimum scale of the initial socket, we increase the scale to give minimum thickness and fix the scale. The minimum scale depends on the material and printer used for 3D printing. After the scale is fixed, we remove the shape of the residual body from the initial socket. Shapes of the operated sockets according to the scale of the initial socket are shown in Figure 5. The scales shown in the images are relative to the prosthetic hand's initial modeling, which has no relationship with the real-world scale. After finding the right socket size for the patient, we adjust the size of other parts of the prosthetic device to match the socket. After the size of every part is adjusted, the prosthetic device is ready to be 3D-printed.

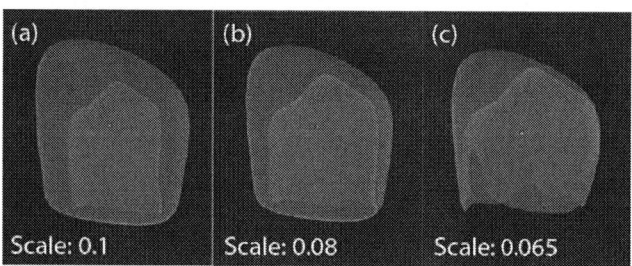

FIGURE 5: The shape of operated sockets for prosthetic hands according to their initial size.

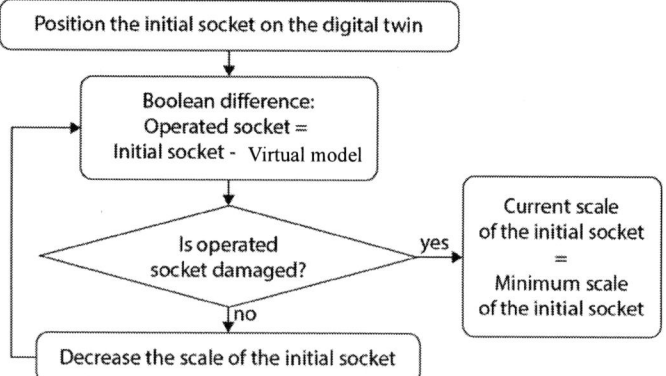

FIGURE 6: Overview of the socket scaling process

4. CASE STUDY

In this section, a case study is introduced demonstrating the proposed design process of the 3D-printable prosthetic hand. This case study represents a scenario where a trans-carpal-damaged patient creates a customized prosthetic hand design through the proposed design process. This case study is based on prosthetic hands for trans-carpal damage, one of the most common forms of upper limb loss [36]. But the scope of the proposed design process can be expanded to prosthetic devices for all types of limb damage. Also, the 3D-printed generic model

is used throughout the case study to represent the residual arm as a proof of concept before official clinical trials.

4.1 Virtual model Reconstruction

This section presents how we follow the process proposed in the method section to reconstruct the virtual model of the 3D-printed generic hand model. First, we capture images of the target object with a mobile phone in a pathway suggested by Munhoz et al. [34]. We use Samsung Galaxy A50 with three rear cameras of 25MP wide-angle camera, an 8MP ultra wide-angle camera, and a 5MP depth sensor camera. The phone used in the case study has a function for optimizing camera settings according to the scene. The case study is conducted in an environment with several fluorescent lamps of white light and windows open, letting the sunlight in. Although there are chairs and desks in the background, they don't cause any problems. The number of pictures taken is 228. To post-process the photographs, we remove the image's background using the Python "remBG" module [38]. Meta-data of the image is removed together when we use "remBG" to remove the background. Therefore, we use "PIL" to restore the meta-data of each image [39]. The whole process of post-processing captured images is done automatically using Python.

Then, we use Meshroom to reconstruct the virtual model from the post-processed images. Meshroom is a free, open-source 3D reconstruction software based on AliceVision, a photogrammetry framework [40]. Meshroom is scriptable using Python, which enables the automation of the whole reconstruction process. Computer resources of i7-8700 CPU, 32GB Ram, and NVIDIA GeForce RTX 2070 are used to run Meshroom. CUDA-enabled GPU and 32GB of RAM are enough to achieve the best result of Meshroom [40]. 228 images are provided to Meshroom, and Meshroom selects 207 images for reconstruction. Meshroom automatically sorts out low-quality images and uses only the qualifying data for the final photogrammetry. The following attributes are used for the Meshroom: Describer type=sift, akaze, Describer density=high, Describer quality=high, Depth map down scale=2, guided matching, disabled CPU, and enabled mesh filter for finding the biggest surface.

After the photogrammetry, we post-process the reconstructed mesh. Blender is used for this process. Blender is an open-source computer graphic tool supported by Blender Foundation which aims to provide access to an entirely free/open-source 3D creation method [41]. Blender has two significant advantages in this project. First, Blender is a free software with an extensive community of users and contributors. This feature makes Blender more accessible than other computer graphics tools and lowers the barrier to contribution and modification to the project. Second, Blender is scriptable using Python. This feature enables automation of the computational design process. In this case study, the post-processing of the mesh and customization of the prosthetic hand is scripted using Blender and Python, which enabled the process without specialists' involvement.

The post-processing stage consists of three steps. First, clear the mesh to remove the vestige of the unremoved background.

Copyright © 2023 by ASME

The *Loose* operation in Blender splits the mesh, removing all residual meshes and leaving the biggest volume. Second, *Smooth* operation in Blender is used multiple times to smooth the mesh. Lastly, the mesh needs to be placed in an upright position for the next process. This involves rotating the mesh in all possible angles to find the angle with the longest vertical height. The final virtual models compared to the target object are shown in Figure 7. Although the final virtual model resembles the target object, several limitations exist. It still has rigid textures; some parts were inaccurate and/or missing. Incomplete parts of the virtual model are due to a lack of high-quality image data of the object from specific angles.

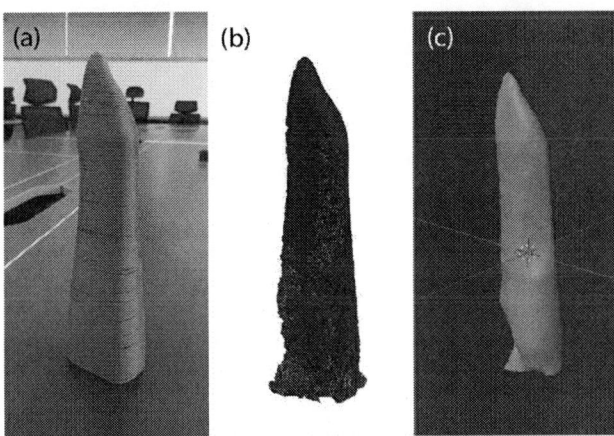

FIGURE 7: (a) Target object, (b) Mesh reconstructed by photogrammetry, (c) The final virtual model after post-processing.

4.2 Prosthetic Device Customization and Fabrication

This section presents how the prosthetic device is customized and fabricated in the case study. Due to the imperfection of the virtual model, the case study of customization and fabrication is conducted based on the digital data used to 3D-print the generic residual arm. The open-source software Blender is used for the customization. From the generic model of the residual arm shown in Figure 8(a), we generate the socket shown in Figure 8(b) without the involvement of a specialist. The process is achieved by scripting the prosthetic device customization process explained in the method section with Blender and Python. First, the initial socket is positioned according to the residual arm. Then the socket is scaled by observing the scale change of the operated socket. The minimum scale of the initial socket in this case study is 0.055. The scale used inside Blender is not related to the real-world scale. Finally, we provide space for the residual arm to the initial socket by a boolean difference operation with the residual arm. The operated socket is used as the final socket for the prosthetic.

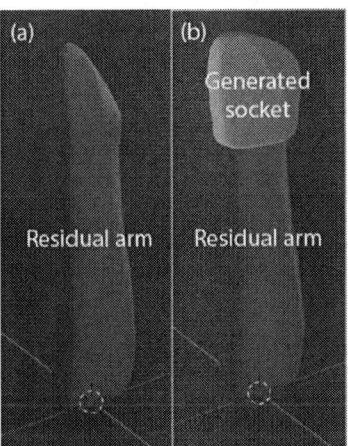

FIGURE 8: (a) The generic model of the residual arm (b) The generated socket from the generic model of the residual arm

After generating the socket, we scale the remaining parts of the prosthetic device according to the scale of the generated socket. The goal is to match the palm size, which determines the prosthetic hand's inner side to fit exactly the outside of the socket. Other prosthetic parts, such as fingers, are adjusted accordingly to fit the palm. In this case study, we modified and used the Pheonix V2 from the e-NABLE Forum as a prosthetic device, which is a body-powered prosthetic designed for patients with trans-carpal damage [37]. We augment the gauntlet of the prosthetic so that it doesn't need thermoforming for the fabrication and substitute ready-made joints with 3D-printed joints. Also, we add 3D-printable sockets to the prosthetic hand to prevent it from dangling. The adopted design lets the patient move their fingers by bending their wrists. The prosthetic hand design used in the case study is shown in Figure 9. The overview of the parts of the final prosthetic hand is shown in Figure 10.

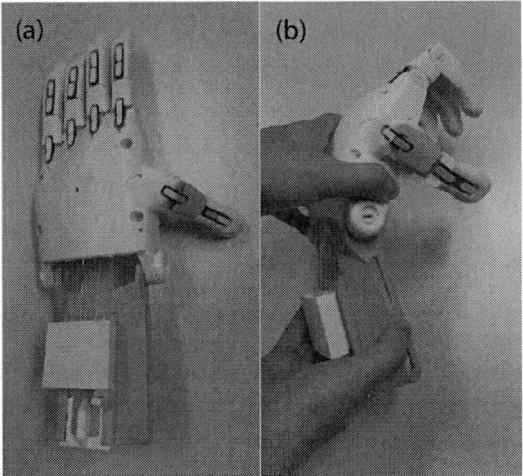

FIGURE 9: (a) Modified Pheonix V2, which is used for this case study, (b) Modified Pheoix V2 with finger bending

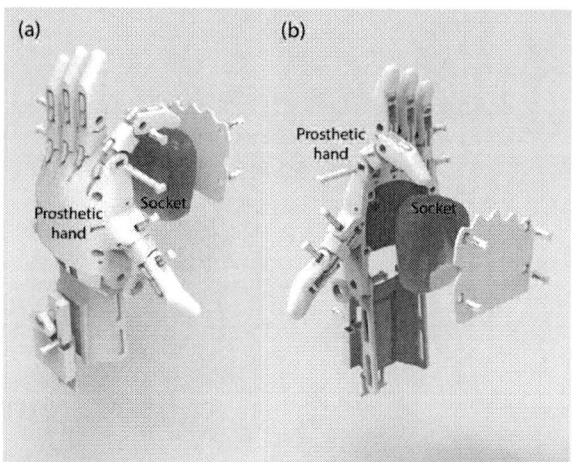

FIGURE 10: (a) Overview of the final prosthetic hand assembly from the back view (b) Overview of the final prosthetic hand assembly from the front view

After customizing the prosthetic hand, we 3D print the prosthetic hand and the socket. Figure 11 shows the final results. Figure 11 shows how the prosthetic hand design customized by our process is applied to the 3D-printed generic arm model. The arm model and the prosthetic device are printed by the 3D printer named Flash Forge 3, and 1.75mm PLA filament was used with the FDM method. They are printed on the following setting: height layer=0.18mm, fill density=15%, outer cell layer=2. The socket is printed using a 3D printer named Figure 4, and Rubber 65A BLK is used as the material with the SLA method. The socket is printed on the following setting: layer thickness=50μ.

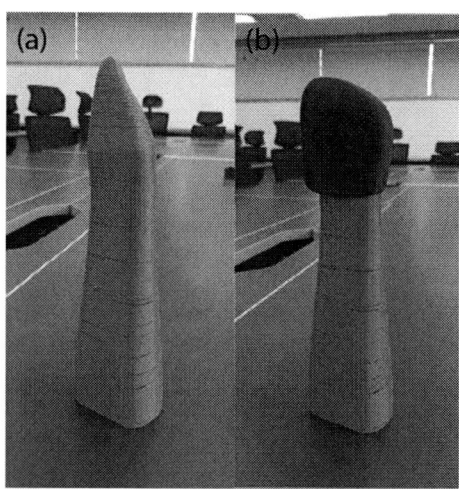

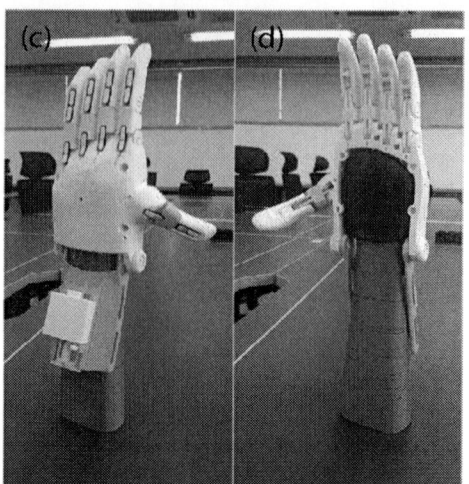

Figure 11: (a) The 3D-printed generic hand model, (b) The hand model with the printed socket, (c) The hand model with the socket and the prosthetic hand, (d) The front side of the assembled hand model before closing the palm

The prosthetic hand designed through the proposed process demonstrates the potential of the process. Regarding costs incurred during the case study, the primary cost involved is in 3D printing, and the primary equipment required are a mobile phone, 3D printers, and a computer. The time spent on each step in the case study is summarized in Table 1. Most of the time is spent on 3D-printing the prosthetic device. The time for 3D printing can be significantly reduced by operating multiple 3D printers to fabricate many parts simultaneously. The required time for creating prostheses and orthoses through the traditional method varies depending on several factors: artisan's skill, environment, the shape of the residual body, etc., but the required duration is approximately one week to several months [14-16].

Table 1: Approximate time spent on each step in the proposed design process

Task		Time
Capture the target object		3 min
Post-process images		8 min
Photogrammetry		3 h 21 min
Post-process the mesh		2 min
Customize the Prosthetic		1 min
	Socket	11 h

Copyright © 2023 by ASME

3D printing	Prosthetic	48 h 55 min
Total		**63 h 30 min**

5. CONCLUSION AND DISCUSSION

In this study, we propose an approach to reduce the time and cost of the prosthetic device design process to increase accessibility for patients in resource-constrained environments. We also present a case study that demonstrates the proposed design process. The result of the case study shows that the prosthetic device created through the suggested design process fits the generic hand model used to represent the patient's residual body in the case study. The suggested design process reconstructs the virtual model with only a mobile phone which is highly accessible in resource-constrained environments [31,32]. This can reduce the barrier caused by the inaccessibility of special equipment and specialists in these areas. Moreover, capturing images of the residual body can rely on the effort of the patient's side. This can reduce the number of visits required to the prosthetic specialist, which can be situated away in underserved communities [12]. As presented in the case study, post-processing images and meshes and scaling the prosthetic is automated using open-source software. Other processes in the suggested framework, such as photogrammetry and 3d printing, also have the potential to be automated. These can significantly reduce the involvement of related specialists, reducing costs and barriers. The time spent on the presented case study is 63.5 hours. The required time can significantly decrease if one can operate multiple 3D printers to print the prosthetic. The time required for the suggested design process is much faster than traditional methods for creating prosthetics which is about one week to several months [14-16]. This can lower the cost and barrier to creating and maintaining prosthetic devices. This especially benefits children whose bodies physically grow very fast [8,9].

However, several limitations exist along the process. To begin with, researchers captured images of the target object in the presented case study. This process originally belongs to the patient's side. Further research is needed on whether this process can solely rely on the effort of the patient's side. Developing a mobile application or guideline to assist people in taking enough high-quality photos and setting up the environment might be needed. The desired image quality and environment on photogrammetry should also be explored to approach this topic. So far, post-processing of the images only included removing the background, and the results are imperfect. Future studies will achieve better photogrammetry results by improving the quality of removing background and post-processing the captured images more actively. For example, sharpening or increasing the contrast of images. In the presented case study, researchers manually adjusted the parameters for the photogrammetry. However, the tool we used for photogrammetry in the case study is scriptable. In the future, we will find the parameters that generally work well for reconstructing the residual body or

suggest an algorithm that recommends optimal parameters for each case to automate photogrammetry. We merely smoothed the surface of the reconstructed mesh to manage the rigid surface. In the future, we will find a scientific way to approximate the organic shape of the human body. In the presented case study, the scale of the virtual model is imperfect. Although photogrammetry includes the scaling process, it is imperfect, especially after removing the background. To address this problem, we can adopt technologies that measure the length of the target object through a mobile phone which is already a pre-installed application in some smartphones [42-45]. CCTags can also be used to precisely scale the virtual model. Meshroom, the photogrammetry tool used in the case study, provides a library for CCTags to enable subpixel precision while being largely robust to challenging shooting conditions [46]. We only use the surface information of the residual body to create the prosthetic. However, we can generate a more sophisticated design by adopting the anatomy data of the residual body or other body parts. In the case study, we simply change the scale of the prosthetic hand. This can have several shortcomings. The surface of the prosthetic device can get too thick or thin. There can be a limitation to customizing. For example, we cannot change the length of the finger. Parts of the prosthetic become incompatible between patients because every patient has a different body size. Therefore, in the future, we must regenerate the design for each patient every time beyond merely changing the scale of the prosthetic device. We may also adopt other metrics of the body, such as the finger length of the other hand, for this process. Last, some figures used along the process are based on the rules of thumb. For example, the tolerance given on the socket, the minimum thickness of the socket, and the number of smooth functions applied to the virtual model. We will find more scientific figures in the future.

There are also topics to improve the socket design. The socket design in this research is merely a vacant space between the patient's residual body and the prosthetic device. This design has limitations regarding weight, cost, comfort, and ventilation. In future studies, we will suggest a more sophisticated socket design by considering ergonomics and implementing lattice structure and topology optimization in the socket generation process. The current socket design can cause discomfort to patients because the weight distribution is not considered. The vulnerability of the socket and the prosthetic device to fatigue caused by consistent use is also not considered. To achieve a more sophisticated socket design, we must include finite element analysis and physical simulation in the suggested design process, which can also be automated. By using 3D printers as means of manufacturing, we can increase the complexity of design for the sake of the patient without increasing the required time and cost. Moreover, printable elastic materials, including the material used in this project, may cause an allergic reaction in some patients [47]. Therefore, we must suggest a socket design that can be manufactured with material verified to be safe in contact with the skin.

The presented case study has several limitations in representing the proposed design process. It is based on a 3D-printed generic arm model, researchers did the work that actually belongs to the patient's side, and the case study was not conducted in a resource-constrained environment. In the future, we must go through the suggested design process and clinical trials with real patients and users in a resource-constrained area to verify the concept. We will also compare costs for the suggested design process and conventional methods for creating prosthetics in resource-constrained environments. Furthermore, we will expand the study's scope beyond trans-carpal damage.

ACKNOWLEDGEMENTS

This work was partly funded by the CMU Africa African Engineering and Technology Network (AFRETEC) seed grant. Any opinions, findings, or conclusions in this paper are those of the authors and do not necessarily reflect the sponsors' views.

REFERENCES

[1] Jochan, M. C., and Ravikumar, K., 2020, "A Review on Prosthetics and Orthotics for Amputees and Disabled," Journal of Critical Reviews, 7(15), pp. 2175–2189.

[2] Ikeda, A. J., Grabowski, A. M., Lindsley, A., Sadeghi-Demneh, E., and Reisinger, K. D., 2014, "A Scoping Literature Review of the Provision of Orthoses and Prostheses in Resource-Limited Environments 2000–2010. Part Two: Research and Outcomes," Prosthetics and orthotics international, 38(5), pp. 343–362.

[3] Hawke, F., Burns, J., Radford, J. A., and Du Toit, V., 2008, "Custom-made Foot Orthoses for the Treatment of Foot Pain," Cochrane Database of Systematic Reviews, (3).

[4] Trotter, L. C., and Pierrynowski, M. R., 2008, "Ability of Foot Care Professionals to Cast Feet Using the Nonweightbearing Plaster and the Gait-Referenced Foam Casting Techniques," Journal of the American Podiatric Medical Association, 98(1), pp. 14–18.

[5] Amrutsagar, L., Parit, G., Ghyar, R., and Bhallamudi, R., 2020, "Parametric Design and Hybrid Fabrication of Above-Knee Prosthesis," Indian journal of orthopaedics, 54, pp. 381–390.

[6] Mavroidis, C., Ranky, R. G., Sivak, M. L., Patritti, B. L., DiPisa, J., Caddle, A., Gilhooly, K., Govoni, L., Sivak, S., and Lancia, M., 2011, "Patient Specific Ankle-Foot Orthoses Using Rapid Prototyping," Journal of neuroengineering and rehabilitation, 8(1), pp. 1–11.

[7] Vujaklija, I., & Farina, D. (2018). 3D printed upper limb prosthetics. Expert review of medical devices, 15(7), 505-512.

[8] Burn, M. B., Ta, A., and Gogola, G. R., 2016, "Three-Dimensional Printing of Prosthetic Hands for Children," The Journal of hand surgery, 41(5), pp. e103–e109.

[9] Krebs, D. E., Edelstein, J. E., and Thornby, M. A., 1991, "Prosthetic Management of Children with Limb Deficiencies," Physical therapy, 71(12), pp. 920–934.

[10] Marino, M., Pattni, S., Greenberg, M., Miller, A., Hocker, E., Ritter, S., and Mehta, K., 2015, "Access to Prosthetic Devices in Developing Countries: Pathways and Challenges," *2015 IEEE Global Humanitarian Technology Conference (GHTC)*, IEEE, pp. 45–51.

[11] Strait, E., 2006, "Prosthetics in Developing Countries," Prosthetic Resident, 1, pp. 1–3.

[12] Organization, W. H., 2017, "WHO Standards for Prosthetics and Orthotics."

[13] Cabibihan, J.-J., Alkhatib, F., Mudassir, M., Lambert, L. A., Al-Kwifi, O. S., Diab, K., and Mahdi, E., 2021, "Suitability of the Openly Accessible 3D Printed Prosthetic Hands for War-Wounded Children," Frontiers in Robotics and AI, 7, p. 594196.

[14] Chen, R. K., Jin, Y., Wensman, J., and Shih, A., 2016, "Additive Manufacturing of Custom Orthoses and Prostheses—A Review," Additive manufacturing, 12, pp. 77–89.

[15] Wang, Y., Tan, Q., Pu, F., Boone, D., and Zhang, M., 2020, "A Review of the Application of Additive Manufacturing in Prosthetic and Orthotic Clinics from a Biomechanical Perspective," Engineering, 6(11), pp. 1258–1266.

[16] Uustal, H., 2006, "Prosthetics and Orthotics," Essential Physical Medicine and Rehabilitation, pp. 101–118.

[17] Banga, H. K., Kalra, P., Belokar, R. M., and Kumar, R., 2020, "Design and Fabrication of Prosthetic and Orthotic Product by 3D Printing," *Prosthetics and Orthotics*, IntechOpen.

[18] Barrios-Muriel, J., Romero-Sánchez, F., Alonso-Sánchez, F. J., and Salgado, D. R., 2020, "Advances in Orthotic and Prosthetic Manufacturing: A Technology Review," Materials, 13(2), p. 295.

[19] Saleh, J. M., and Dalgarno, K. W., 2009, "Cost and Benefit Analysis of Fused Deposition Modelling (FDM) Technique and Selective Laser Sintering (SLS) for Fabrication of Customised Foot Orthoses," *Innovative Developments in Design and Manufacturing*, CRC Press, pp. 723–728.

[20] Faustini, M. C., Neptune, R. R., Crawford, R. H., and Stanhope, S. J., 2008, "Manufacture of Passive Dynamic Ankle–Foot Orthoses Using Selective Laser Sintering," IEEE transactions on biomedical engineering, 55(2), pp. 784–790.

[21] Harper, N. G., Russell, E. M., Wilken, J. M., and Neptune, R. R., 2014, "Selective Laser Sintered versus Carbon Fiber Passive-Dynamic Ankle-Foot Orthoses: A Comparison of Patient Walking Performance," Journal of biomechanical engineering, 136(9).

[22] 2019, "e-NABLE Forum," e-NABLE [Online]. Available: https://hub.e-nable.org/s/e-nable-forum/wiki/Welcome+to+e-NABLE%21. [Accessed: 13-Mar-2023].

[23] Shih, A., Park, D. W., Yang, Y.-Y. D., Chisena, R., and Wu, D., 2017, "Cloud-Based Design and Additive Manufacturing of Custom Orthoses," Procedia Cirp, 63, pp. 156–160.

[24] Górski, F., Wichniarek, R., Kuczko, W., and Żukowska, M., 2021, "Study on Properties of Automatically Designed 3d-Printed Customized Prosthetic Sockets," Materials, 14(18), p. 5240.

[25] Rai, P., Jankiraman, V., Teacher, M., Velu, R., Kumar, S. A., Binedell, T., and Subburaj, K., 2022, "Design and Optimization of a 3D Printed Prosthetic Socket for Transtibial Amputees," Materials Today: Proceedings, 70, pp. 454–464.

[26] Steer, J. W., Grudniewski, P. A., Browne, M., Worsley, P. R., Sobey, A. J., and Dickinson, A. S., 2020, "Predictive Prosthetic Socket Design: Part 2—Generating Person-Specific Candidate Designs Using Multi-Objective Genetic Algorithms,"

Copyright © 2023 by ASME

Biomechanics and modeling in mechanobiology, **19**(4), pp. 1347–1360.

[27] Mikhail, E. M., Bethel, J. S., and McGlone, J. C., 2001, *Introduction to Modern Photogrammetry*, John Wiley & Sons.

[28] Fawzy, H. E.-D., 2015, "The Accuracy of Mobile Phone Camera Instead of High Resolution Camera in Digital Close Range Photogrammetry," International Journal of Civil Engineering & Technology (IJCIET), **6**(1), pp. 76–85.

[29] Chikatsu, H., and Takahashi, Y., 2009, "Comparative Evaluation of Consumer Grade Cameras and Mobile Phone Cameras for Close Range Photogrammetry," *Videometrics, Range Imaging, and Applications X*, SPIE, pp. 130–141.

[30] Ortiz-Sanz, J., Gil-Docampo, M., Rego-Sanmartín, T., Arza-García, M., and Tucci, G., 2021, "A PBeL for Training Non-Experts in Mobile-Based Photogrammetry and Accurate 3-D Recording of Small-Size/Non-Complex Objects," Measurement, **178**, p. 109338.

[31] "Mobile Cellular Subscriptions (per 100 People) - Sub-Saharan Africa | Data" [Online]. Available: https://data.worldbank.org/indicator/IT.CEL.SETS.P2?locations =ZG. [Accessed: 13-Mar-2023].

[32] Global System for Mobile Communications Association. The mobile economy: Sub-Saharan Africa 2019, 2019

[33] "Capturing — Meshroom V2021.0.1 Documentation" [Online]. Available: https://meshroom-manual.readthedocs.io/en/latest/capturing/capturing.html. [Accessed: 13-Mar-2023].

[34] Munhoz, R., Moraes, C. A. da C., Tanaka, H., and Kunkel, M. E., 2016, "A Digital Approach for Design and Fabrication by Rapid Prototyping of Orthosis for Developmental Dysplasia of the Hip," Research on Biomedical Engineering, **32**, pp. 63–73.

[35] "Tutorial: Meshroom for Beginners — Meshroom V2021.0.1 Documentation" [Online]. Available: https://meshroom-manual.readthedocs.io/en/latest/tutorials/sketchfab/sketchfab.html. [Accessed: 13-Mar-2023].

[36] Cordella, F., Ciancio, A. L., Sacchetti, R., Davalli, A., Cutti, A. G., Guglielmelli, E., and Zollo, L., 2016, "Literature Review on Needs of Upper Limb Prosthesis Users," Frontiers in neuroscience, **10**, p. 209.

[37] 2021, "Phoenix v2 Hand," e-NABLE [Online]. Available: https://hub.e-nable.org/p/devices?p=Phoenix+v2+Hand. [Accessed: 13-Mar-2023].

[38] singhaman092, 2023, "Rembg."

[39] Clark, A. and others, 2015, "Pillow (Pil Fork) Documentation," readthedocs.

[40] "AliceVision | Meshroom - 3D Reconstruction Software" [Online]. Available: https://alicevision.org/#meshroom. [Accessed: 13-Mar-2023].

[41] Foundation, B., "Blender.Org - Home of the Blender Project - Free and Open 3D Creation Software," blender.org.

[42] Valockỳ, F., Drahoš, P., and Haffner, O., 2020, "Measure Distance between Camera and Object Using Camera Sensor," *2020 Cybernetics & Informatics (K&I)*, IEEE, pp. 1–4.

[43] Holzmann, C., and Hochgatterer, M., 2012, "Measuring Distance with Mobile Phones Using Single-Camera Stereo Vision," *2012 32nd International Conference on Distributed Computing Systems Workshops*, IEEE, pp. 88–93.

[44] Laotrakunchai, S., Wongkaew, A., and Patanukhom, K., 2013, "Measurement of Size and Distance of Objects Using Mobile Devices," *2013 International Conference on Signal-Image Technology & Internet-Based Systems*, IEEE, pp. 156–161.

[45] "Using Quick Measure on My Samsung Phone | Samsung Australia" [Online]. Available: https://www.samsung.com/au/support/mobile-devices/using-quick-measure/. [Accessed: 28-Apr-2023]

[46] "Related Projects — Meshroom V2021.0.1 Documentation" [Online]. Available: https://meshroom-manual.readthedocs.io/en/latest/more/related-projects/related-projects.html. [Accessed: 05-May-2023].

[47] 2020, "Figure 4 RUBBER-65A BLK," 3D Systems [Online]. Available: https://www.3dsystems.com/materials/figure-4-rubber-65a-blk. [Accessed: 13-Mar-2023].

Proceedings of the ASME 2023
International Design Engineering Technical Conferences and
Computers and Information in Engineering Conference
IDETC-CIE2023
August 20-23, 2023, Boston, Massachusetts

DETC2023-116478

ACHIEVING HIGH PERFORMANCE AND LOW COST: DEVELOPMENT OF A HIGH-PERFORMING PASSIVE PROSTHETIC KNEE FOR EMERGING MARKETS

Madison Reddie[1,†], Saloni Bedi[1,†], Manasi Vaidya[1,†], Amari Griffin[2,†], Nina T. Petelina[1,*] Amos G. Winter[1]

[1]Massachusetts Institute of Technology, Cambridge, MA
[2]Harvard University, Cambridge, MA

ABSTRACT

There is significant need for low-cost, high-performance prosthetic knees in low- and middle-income countries (LMICs) due to a large number of amputees and particularly challenging socioeconomic and environmental conditions. Prostheses are important for maintaining one's participation in society, culture, and the economy, but many are either prohibitively expensive or do not provide near-able-bodied kinematics. Poor performing prosthetic knees cause discomfort and draw unwanted attention to transfemoral amputees. In this study, we refine the design of a high-performing, single-axis, passive prosthetic knee developed with a focus on the Indian market in order to reduce cost, weight, and part count; enhance manufacturability; and improve aesthetics. The load paths and functional componentry were critically analyzed to identify opportunities to streamline the design while maintaining strength and the near-able-bodied kinematics offered by the original design. The part count was reduced almost four-fold, and the mass of the prosthesis was reduced three-fold. An enclosure was also designed to encase the functional componentry in an aesthetically acceptable package. The changes made to the design are believed to significantly advance the usability and commercial viability of the prosthetic knee. This study may serve as an example of how products developed for emerging markets may achieve affordability without sacrificing performance.

Keywords: prosthesis, design for manufacturing, emerging markets

1. INTRODUCTION

USAID reports that 100 million people around the world need prosthetic or orthotic devices, but only 10% actually have access to such devices, with much of the unmet need found in low- and middle-income countries (LMICs) [1, 2]. For those who need

[†]Joint first authors
[*]Corresponding author: petelina@mit.edu
Documentation for `asmeconf.cls`: Version 1.34, May 12, 2023.

them, these devices are critical to quality of life; independence; and the ability to participate in society, culture, and the economy [1]. The intricate structure and function of the knee make knee prostheses among the most challenging assistive devices to engineer, leaving many of the estimated nine million transfemoral amputees around the world underserved [3]. India alone is home to 300,000 transfemoral amputees, representing a huge need and potential for socioeconomic impact [4].

There are generally two segments in the prosthetic knee market: knees that are affordable but have poor biomechanical performance, and knees with high biomechanical performance that are prohibitively expensive for the average Indian amputee. Cheaper knees often lack the stability, durability, and provision of natural gait that is demanded by all prosthesis users. Challenges unique to amputees in LMICs, such as chronic unemployment, lack of proximity to orthopedists, stigma, and frequent ambulation over uneven terrain, make performance on these metrics even more important [2, 5, 6]. This market gap has motivated the development of a high-performance prosthetic knee for India through a project whose progress has been reported in [2, 4, 7, 8].

This prosthetic knee is a passive, single-axis joint that has been parametrically designed drawing on both gait biomechanics and perspectives of Indian transfemoral amputees [2]. It demonstrates high biomechanical performance through modules focused on stance stability and swing damping. However, the design to date is heavy, bulky, complex, and unenclosed, making the knee expensive to manufacture, difficult to use, and aesthetically unacceptable. Thus, the objective of this study is to revise the knee design to reduce the joint weight, volume, part count, and cost and to encase the functional componentry in a cosmetically satisfactory enclosure.

In this paper, we first present the previous knee design and further specify the redesign objectives. Second, we discuss our approach to redesigning the functional componentry of the knee, part by part. Third, we present the design of an enclosure for

Copyright © 2023 by ASME

the knee joint. Fourth, we present the results of the mechanical redesign and enclosure design and compare the new structure to the previous knee. We discuss how our approach can be used as a model for achieving both high performance and low cost in products for emerging markets. Finally, we present limitations and future work.

2. MATERIALS AND METHODS

2.1 Previous Design

The modularly designed prosthetic knee joint (Fig. 1) provides near-able-bodied kinematics through two primary modules: one controlling knee flexion and stability and one that provides damping during leg swing. Further detail regarding gait biomechanics, stance phases, and the development of these two modules can be found in [2, 4, 7, 8]. Labeled parts and the names used to reference them appear in Fig. 2 and Table 1 (damper not pictured in Fig. 2). The flexion and stability module uses a four bar-linkage latch (parts 8 and 9 in Fig. 2) that locks the knee, preventing flexion and buckling during early and mid-stance, and is disengaged as the ground reaction force (GRF) crosses a virtual axis created by the linkages (part 9). The virtual axis was deliberately placed such that the moment on the latch generated by the GRF reverses direction in late stance, when the GRF crosses the axis. With the latch unlocked, the knee piece (part 3) can rotate about the axle (part 2), and the user can flex the knee.

After toe-off, as the user swings the limb forward, a rotary hydraulic damper (round part in the back of Fig. 1) is engaged via a one-way roller clutch paired to the axis about which the knee rotates to prevent over-swing while facilitating ground clearance. The damper shears a viscous silicone oil between sets of concentric fixed and rotating plates, which generates a damping coefficient based on the user's body mass. A bias spring (part 7, held by part 5) restores the latch to the locked position, and a hard stop (part 6) prevents hyperextension by over-rotation of the knee piece (part 3). While the stability and damping modules together offer near-able-bodied kinematics passively, the mechanical designs of the stability module and integrated knee have yet to be optimized for weight, manufacturability, cost, and aesthetics.

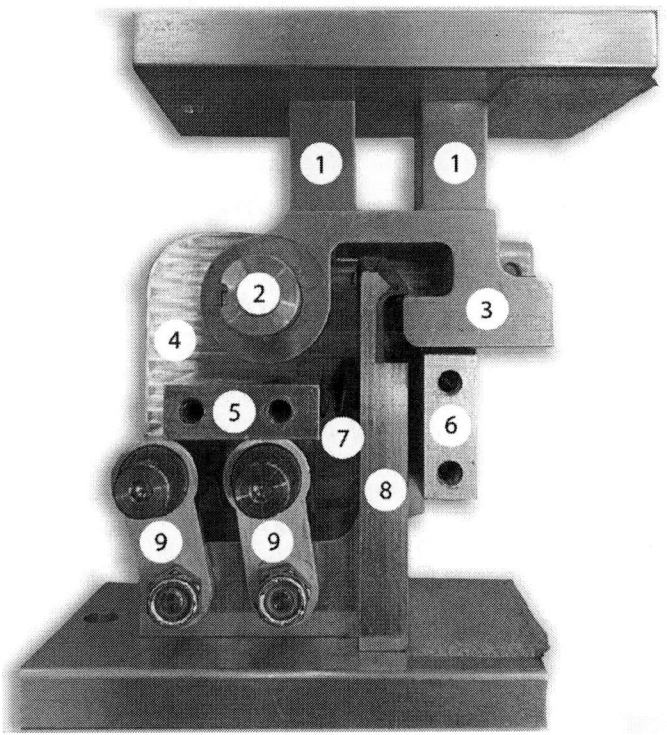

FIGURE 2: PREVIOUS KNEE DESIGN WITH ONE SIDE PLATE AND DAMPER REMOVED. PART NAMES APPEAR IN TABLE 1.

TABLE 1: FUNCTIONAL COMPONENTS IN PREVIOUS KNEE DESIGN

Part no.	Part name
1	columns
2	axle
3	knee piece
4	side plate
5	spring holder
6	hard stop
7	bias spring
8	latch
9	linkages

FIGURE 1: PREVIOUS KNEE PROTOTYPE

Copyright © 2023 by ASME

Most components were made from Aluminum 6061, with the latch part (8) being made from Aluminum 7075 for extra strength. The axle system (2) includes a key, two oil-infused brass bushings, two roller bearings, a clutch, and four washers. The spring holder (5) and hard stop (6) each use four screws and washers. The linkages (9) include four screws, four nuts, four oil-infused brass bushings, four ball bearings, and four washers. Finally, there are six spacers between the side plates (4) and other components. The cost of the parts for this one-off prototype reached approximately 1000USD. The prototype is shown disassembled in Fig. 3.

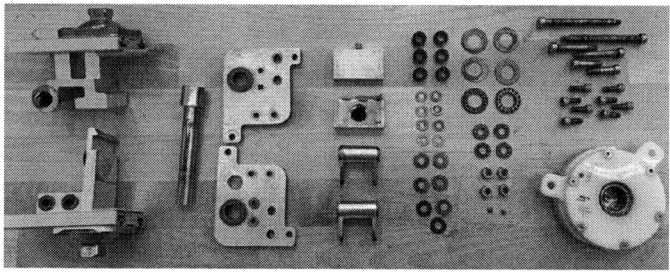

FIGURE 3: PREVIOUS KNEE PROTOTYPE DISASSEMBLED

Given prior research and interactions with Indian users in prototype tests [4], along with the specifications of knees currently popular in emerging markets [5], we sought to maintain the knee's kinematics and support users within the 95th percentile body mass in India (76 kg) [9] while reducing the weight of the knee joint to under 1 kg, reducing its manufacturing cost to less than 100 USD, and ensuring a natural and discreet appearance under pants, which entails making the joint small enough to be encased by an inconspicuous enclosure. We also assigned a minimum safety factor of 2.4 to increase the robustness of the knee. The mechanical design of the functional componentry was refined first, and an enclosure was later developed around the revised structure.

2.2 Mechanical Design

Loading. The prosthetic knee transmits the user's body mass and GRFs between the upper and lower prosthesis. The body mass load travels down the two columns (part 1 in Fig. 2) through the knee piece (part 3) to the hard stop (6) and axle (2) and then through the side plates (4). The side plates transmit the load to the linkages (9), through which the load travels to the bottom the of the latch (8) and the lower prosthesis. It is unknown how much of the load is transmitted through the hard stop (6) compared to the axle (2). During particular phases of the stance cycle, the latching mechanism (the interface of parts 3 and 8) is also loaded. To be conservative, it was assumed that each possible load path bears the maximum possible load. Peduzzi de Castro et al. [10] experimentally measured GRF in unilateral transfemoral amputees wearing passive prosthetic knees and found a maximum GRF at heel strike of 101.6 +/- 5.7% of the user's body mass. Using +2 standard deviations, 76 kg body mass produces a GRF of 840 N.

Four-bar Linkage. As the integrity of the latch is critical to preventing buckling of the knee, the latch (part 8) strength was an initial focus. Finite element analysis (FEA) in Fusion 360 was used to understand the baseline strength of the starting latch design as well as the stresses in the part generated by an 840 N upward force on the latching face. Forces were modeled as uniformly distributed over entire faces in the analyses in this study. At their originally designed widths of 38 mm in their initially specified material (Aluminum 7075), the latch lever and tip provided safety factors near seven. FEA showed a maximum stress of 61 MPa at the edge where the latch tip face meets the latch lever. As Aluminum 7075 is unnecessarily strong for this load case, we opted to use Aluminum 6061, the common aluminum alloy used in the rest of the joint, for the latch. Some engineering plastics could handle this loading, but the higher strength of aluminum allows for greater part size reduction, enabling a more compact enclosure. We assumed a yield strength of Aluminum 6061-T6 of 276 MPa. To achieve a safety factor of 2.4, maximum stress should not exceed 115 MPa, approximately double the FEA maximum stress output.

FEA demonstrated that the enlargement of the cross-section of the latch lever (part 8) below the knee piece (part 3) did not significantly affect the lever's strength or deflection under load, so the additional material (accounting for 25% of the latch's volume) was removed. The latch was further narrowed to 25mm. The latch tip, besides its width being narrowed, was left intact to maintain a substantial contact surface with the knee piece for stable latching.

The two linkages (part 9) need only move a small amount (approximately 5 mm, just enough to eliminate the interference at the latching mechanism) under a modest force from the GRF (about 27 N), making them a reasonable candidate for replacement by a compliant mechanism. The feasibility of a compliant mechanism in place of the linkages and bias spring (7) was therefore investigated. However, because the body mass and GRF loads are transmitted via the linkages between the knee piece (3)/upper prosthesis and lower prosthesis, a compliant mechanism could not satisfy the latch translation requirements while also transmitting body mass and GRF loads without buckling.

Using only a single set of linkages, as opposed to the symmetrical structure in the previous design, was also considered, but this adds manufacturing complexity to the latch (8), which could otherwise be CNC machined from a single direction. Thus, the basic structure of the original four-bar (parts 8 and 9) was retained. As the large vertical forces on the linkages (9) are balanced, making stresses in the linkages relatively small, the linkages were switched from aluminum to an engineering plastic (Nylon 6/6 or Delrin, depending on manufacturer preference). Plastic linkages allow near-frictionless rotation about their fasteners, so the fasteners, bushings, bearings, and washers were replaced with off-the-shelf press-fit steel pins, significantly reducing part count. The width of the four-bar system was reduced to match the new latch width, and the required pin size was calculated based on the Von Mises equivalent stress from shear and bending stress under the 840 N load, plus the safety factor. While the side plates (4) are intended to be in direct contact with the linkages, since friction is not a concern with plastic linkages, a conservative spacing of 1 mm between the plates (4) and linkages (9) was allowed in the pin stress calculations to account for any imperfections in manufacturing and assembly. The pin diameter

Copyright © 2023 by ASME

was rounded up to the nearest standard size to reduce part cost and increase ease of part sourcing, and the linkages were then resized to accommodate the necessary pins with minimal material.

With smaller linkages (9), the latch lever (8) could also be shortened while maintaining clearance between the linkages and knee piece (3). The tail of the latch was also made smaller around the smaller pins.

Axle and Knee Piece. The required size of the axle (2) where it interfaces with the one-way roller clutch in the damper was determined in a prior study [8]. The cross-section of the axle decreases as it emerges from the clutch and keys into the knee piece (3). We avoided further reducing the axle cross-section as a large reduction in cross-sectional area would significantly weaken the axle.

The width of the knee piece (3) was reduced to match the new width of the latch (8) plus two linkages (9), and the depth of the latching face of the knee piece was minimized based on the latch tip geometry. The cross-sectional area of the knee piece structure surrounding the latch tip was also minimized using FEA for the maximum possible latch loading.

The set of columns (1) atop the knee piece serve to transmit load from the upper prosthesis down to the knee piece (3). Structural buckling formulae were used to determine the minimum cross-sectional dimensions for a column for this purpose at the length required by the updated latch (8) and knee piece (3) designs. This area was less than the cross-sectional area of a standard pyramid adapter (about 1 cm square). Thus, we used a single column, and since the column must lead to a male pyramid adapter for compatibility with various thigh sockets (upper prosthesis), the thickness was set to the approximate length of a pyramid adapter. The width was set to the same as the rest of the knee piece (3) so that the part forms a 2D shape that may be milled from a single direction. An adapter can also be milled into the top of the column by one additional CNC operation, in which case the column could also be narrowed to the width of the adapter.

Alternative materials for the latch (8) and knee piece (3), such as steel and titanium, were considered, but aluminum offers favorable machinability, strength-to-weight ratio, and cost and is sufficiently strong for the loads experienced by the knee. Shrinking the parts beyond the dimensions required for aluminum does not provide significant marginal benefit for compactness.

Other Components. As done with the linkage fasteners, an off-the-shelf press-fit steel pin was substituted for the hard stop (6), which previously consisted of a block of aluminum and four screws and washers. Although the knee piece (3) is intended to extend to the side plates (4), a conservative spacing between each side plate and each end of the knee piece of 1 mm was used in analysis as with the linkage (9) pins. A minimum pin diameter was then found from the analytically calculated Von Mises equivalent stress from the shear and bending loads.

The similar aluminum block and four fasteners used to place the bias spring (part 5) were another target for design simplification and part reduction. The addition of an enclosure around the knee mechanism presented an opportunity to build a feature into the inside of the enclosure to place the spring (7). To streamline this mechanism, the spring action was moved to the posterior-most surface of the latch (8), behind the linkages. A concentric cylindrical feature was then to be incorporated into the inside of the enclosure to hold the compression spring to this surface.

Based on the revised knee structure, the side plates (4) were reconfigured to join the axle (2), hard stop (6), and top linkage (9) pins with as little material as possible. They were determined to be made of 3 mm thick (half of their prior thickness) Nylon 6/6 or Delrin, as forces on the side plates are always balanced, putting them under low stress, and the vertically oriented plate shape is strong.

The new knee joint was 3D printed to ensure that the knee dynamics remain intact. The damper design was parametrically optimized in [8] and was thus not a focus of this study but will be revisited in future work given the significant changes to the rest of the mechanical design.

2.3 Enclosure Design

Enclosing the knee is important to product aesthetics and market acceptance as well as safety (preventing user exposure to pinch points and sharp edges). Previously collected user feedback emphasized the need for the prosthesis to resemble a physiological knee under clothing. Prostheses in high-income markets are often designed to appear sleek and high-tech, whereas popular prostheses in LMIC markets are more commonly designed to resemble physiological limbs [2, 5, 6]. Thus, the enclosure for this knee was designed in a horizontal cylindrical shape for discreetness under clothing. The enclosure will not be load bearing and can therefore be manufactured from a durable plastic. The enclosure details were designed specifically for ease of injection molding to lower manufacturing costs.

The enclosure consists of four parts that snap together and attach to the mechanical structure of the knee via the axle (part 2 in Fig. 2). The top and bottom sections of the enclosure are each more than half of a cylinder in order to anchor to the axle, making them challenging to eject from injection molds in single pieces. Separating the enclosure into four parts also eases assembly of the enclosed mechanical knee.

The top two parts of the enclosure rotate with the knee piece about the axle. The enclosure fits tightly around the damper for compactness, and adapters protrude through openings at the top and bottom for compatibility with existing upper and lower prostheses. Internal ribbing strengthens the parts in case of impact. Prototypes were iteratively 3D printed to test the pivoting of the parts, required spacing and tolerances, and smooth rotation. Prototypes were made with ABS and sanded smooth.

3. RESULTS
3.1 Mechanical Design

We found the four-bar latch (part 8), knee piece (3), axle (2), hard stop (6), bias spring (5 and 7), and side plate (4) systems of the knee in Figs. 1 and 2 to be over-engineered and inefficient. The improved design is pictured in Fig. 4.

Aluminum 6061-T6 was substituted for the Aluminum 7075 used in the latch (8), and the latch lever and tip were reduced from 38 mm wide and 70 mm tall to 20 mm wide and 35 mm tall. The extension of the cross-section below the latch face was

Copyright © 2023 by ASME

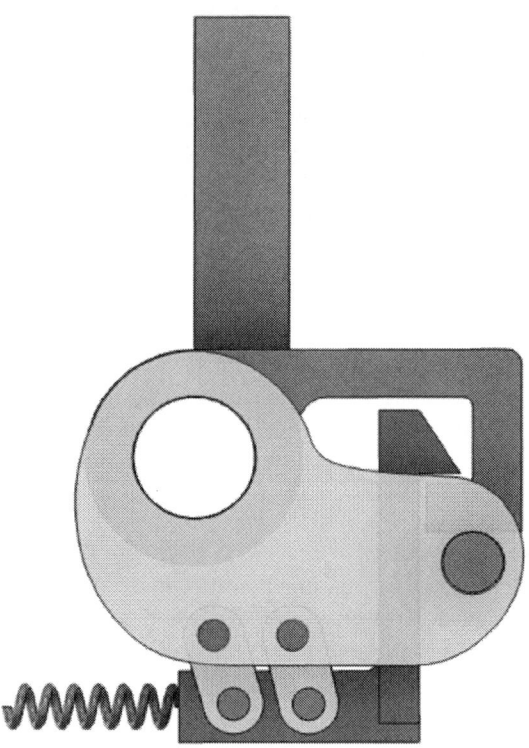

FIGURE 4: CAD MODEL OF UPDATED DESIGN OF FUNCTIONAL COMPONENTRY OF THE KNEE

FIGURE 5: FEA-CALCULATED STRESS DISTRIBUTION IN THE UPDATED LATCH PART DESIGN

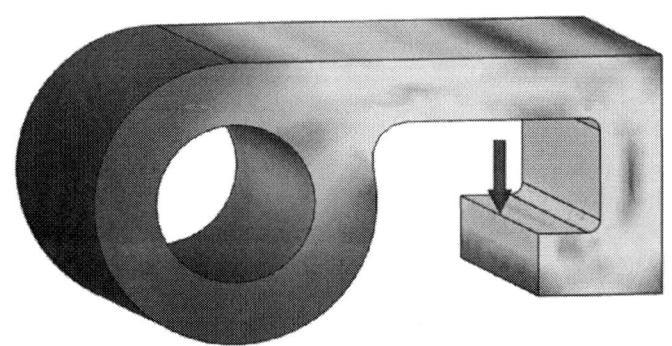

FIGURE 6: FEA-CALCULATED STRESS DISTRIBUTION IN THE UPDATED KNEE PART DESIGN

also removed. The updated latch design thus reduced the volume and weight of the latch by 76% (versus the prior design). FEA demonstrated a maximum stress of 99.2MPa and corresponding minimum safety factor of 2.8 for Aluminum 6061, which exceeds the target safety factor of 2.4 (Fig. 5).

The knee piece (3) was condensed around the latch (8), with its thickness almost halved to 5 mm, while the previous axle (2) size was maintained due to requirements imposed by the damper's roller clutch. The cross-section of the knee piece was thus reduced from 38 mm wide and almost 10 mm thick to about 32 mm wide and 5 mm thick, and the two columns (1) were replaced with one, effectively reducing the part volume by 47%. FEA for the updated knee piece showed a maximum stress of 93.8MPa and minimum safety factor of 2.9 (Fig. 6). The latch and knee piece can each be CNC machined in a single operation from one side. Die casting was considered for these parts but ruled out due to the potential for material weakening from porosity.

The linkage (9) fasteners were replaced with off-the-shelf steel pins of the minimum required diameter per the loading (rounded up to the nearest standard size), and the linkages were condensed around the pins and switched to an engineering plastic. The plastic linkages can be injection molded and mitigate friction between themselves and the pins, allowing the elimination of the oil-infused brass bushings, ball bearings, and washers.

The hard stop piece (6) and the four fasteners anchoring it to the side plates (4) were likewise replaced with a simple, off-the-shelf press-fit steel pin. The aluminum block and four associated fasteners (5) used to place the bias spring (7) were replaced with

an injection molded feature on the inside of the enclosure, which adds negligible marginal cost to the enclosure. Each of these substitutions also eliminated washers from the assembly. Finally, the side plates (4) were reconfigured to transmit loads between parts using minimal material. They were also made plastic, which enables other parts to move while in direct contact with them with minimal friction. Like the linkages, they can be injection molded. These changes result in a new mechanical design with a total part volume 76% lower than that of the previous design and a mass approximately 67% lower than that of the previous design. Further comparisons are shown in Table 2. A side-by-side of to-scale prototypes of the previous and new designs is pictured in Fig. 7.

3.2 Enclosure

CAD models of the enclosure design are pictured in Figs. 9 and 10, and an exploded view of the functional componentry and enclosure is shown in Fig. 11. The enclosure is 96 mm

Copyright © 2023 by ASME

TABLE 2: COMPARISON OF THE PREVIOUS AND NEW MECHANICAL DESIGNS

Part(s)	Previous	New	% change
Latch (8) vol	$32.6cm^3$	$7.8cm^3$	-76%
Knee pc (3) vol	$37.8cm^3$	$20cm^3$	-47%
Linkages (9) vol	$10.6cm^3$	$0.8cm^3$	-93%
Hard stop (6) vol	$9.8cm^3$	$1.2cm^3$	-88%
Spring holder (5) vol	$9.8cm^3$	0	-100%
Side plate (4) vol	$52.1cm^3$	$5.9cm^3$	-89%
Vol of above parts	$152.7cm^3$	$35.7cm^3$	-77%
Mass of above parts	1500g	500g	-67%
# fasteners	14	0	-100%
# internal parts	64	17	-73%

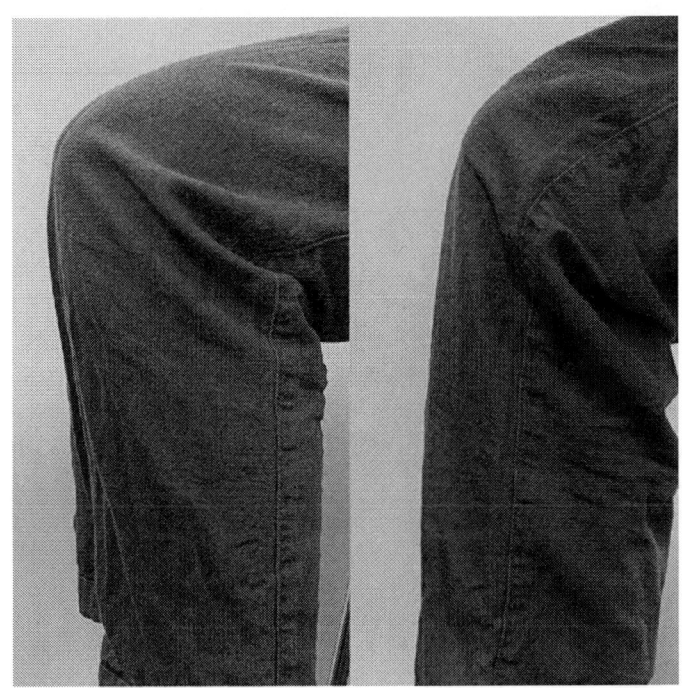

FIGURE 8: A PHYSIOLOGICAL KNEE (LEFT) UNDER PANTS, AND A PROTOTYPED PROSTHETIC KNEE ENCLOSURE (FIG. 12) UNDER THE SAME PAIR OF PANTS (RIGHT).

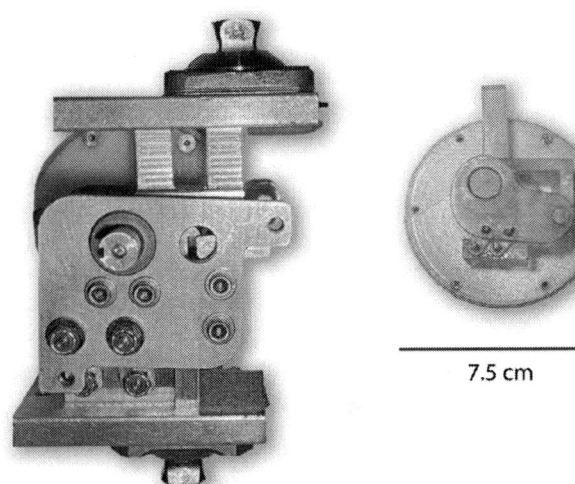

FIGURE 7: TO-SCALE PROTOTYPES OF THE PREVIOUS (LEFT, ALUMINUM) AND NEW (RIGHT, 3D PRINTED) MECHANICAL DESIGNS. FOR SCALE, THE DAMPERS BOTH HAVE A DIAMETER OF APPROXIMATELY 75 MM.

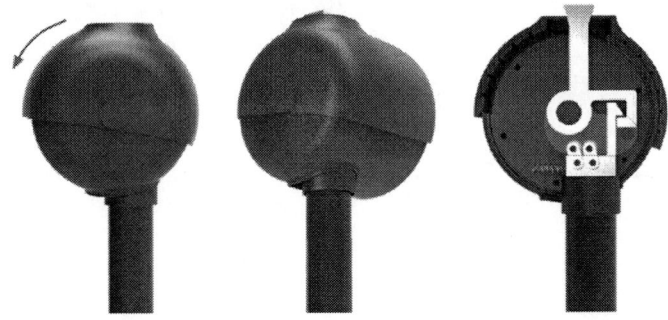

FIGURE 9: SIDE, ANGLED, AND SECTION VIEWS OF CAD MODEL OF THE ENCLOSURE

in diameter and approximately 85 mm wide, putting it within the range of physiological knees. The size and shape give it an appearance not unlike that of physiological knees under pants, as pictured in Fig. 8, thus satisfying the basic aesthetic requirement of discreetness.

Furthermore, the enclosure seals off pinch points and sharp edges in the functional componentry, improving safety and the user experience. Radial rib lines on the inside of the top left and bottom right parts (Fig. 9) add strength and rigidity to the parts, which makes the enclosure less likely to permanently deform or break due to impacts, such as a user falling on the knee. The four-piece structure lowers the cost of injection molding, as less-than-half cylinders are easier than more-than-half cylinders to eject from molds, and affords the assembly process greater flexibility. A 3D printed prototype (Fig. 12) attests to the ease of assembly of the snap fit enclosure.

4. DISCUSSION

In this study, we performed a design iteration on the prosthetic knee joint designed in [4, 7, 8] to provide stable, natural gait. We sought to maintain the exceptional performance achieved in stability and swing damping while reducing the size, weight, and cost of the mechanical design and enclosing it in a compact, aesthetically acceptable package. The loads on the knee were analyzed, and load-bearing components were reduced to the minimum sizes necessary to support maximum loads with a minimum safety factor of 2.4. The functionalities of each component were analyzed, and simplified or combined mechanisms were substituted where possible.

The knee was simplified from over 60 distinct components to 17 (plus four for the enclosure) and reduced from approximately 1.5 kg to 0.5 kg. The volume of the major internal componentry was reduced by 76%, largely by condensing the latching

Copyright © 2023 by ASME

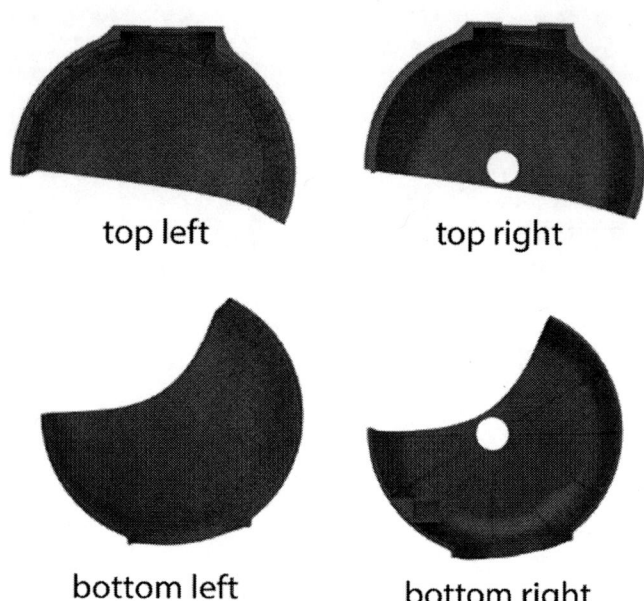

FIGURE 10: CAD MODELS OF THE FOUR PARTS OF THE ENCLO-SURE

top left top right

bottom left bottom right

FIGURE 11: EXPLODED VIEW CAD MODEL OF THE NEW MECHAN-ICAL AND ENCLOSURE DESIGNS

mechanism (parts 3 and 8 in Fig. 2). The hard stop (part 6) was miniaturized, and the two linkages (part 9) were condensed around pins and switched from metal to plastic. The side plates (4) were reduced in size based on this new joint structure and also switched to plastic. The spring holder and its fasteners were removed entirely. An enclosure for the knee that resembles a physiological knee under clothes was also designed. Based on prior user feedback and benchmarking of popular prosthetic knees in LMICs, we expect these changes to significantly enhance the usability and commercial viability of the knee. Importantly, a more natural appearance under clothes can reduce the discrimination experienced by prosthesis users in many cultures. Eliminating all 14 fasteners may also contribute to the longevity of the knee compared to other prosthetic knees, as a trial of the Red Cross knee in Tanzania found that loosening of fasteners was a common cause of instability [6].

Based on preliminary discussions about the design with manufacturers Alfaset and FURPO, manufacturing cost is projected to approach our cost target. The manufacturers encouraged our use of standard-sized components and a four-piece injection molded enclosure (as opposed to two pieces) for their contributions to cost reduction.

FIGURE 12: FRONT (LEFT) AND SIDE (RIGHT) VIEWS OF A PROTOTYPE OF THE ENCLOSURE

For nine million transfemoral amputees, prosthetic knees are critical to social and economic life. Unfortunately, high-performance prosthetic knee design has proved to be an engineering challenge, and most prosthetic knees tend to be either low-performance and inexpensive or high-performance and unaffordable for most users living in LMICs. The prosthetic knee market is a microcosm of a larger trend in global product development. Many projects taking aim at LMIC markets fall into the "trap" of sacrificing performance or features to reduce cost, often based on the misconception that LMIC markets do not demand performance on par with that seen in high-income markets [11]. In many cases, LMIC contexts may in fact demand higher performance. For example, a clinical study of two above-knee prostheses in Tanzania found that participants used their prostheses for a median of 14 hours per day [6], indicating that prostheses for this population should be exceptionally durable and comfortable to accommodate such heavy use. In India, a discreet appearance under clothing is more important than in the U.S. due to cultural factors.

Although earlier design iterations of this high-performance knee were expensive, in this study, we have demonstrated how thoughtful engineering analysis and design for manufacturing can be applied to to develop a simultaneously high-performance and low-cost product that meets the needs of an LMIC market. By conducting an additional design iteration, spending extra time and effort analyzing and simplifying structures and condensing functionalities, we were able to dramatically reduce the complexity of a knee prosthesis that offers near-able-bodied gait. Throughout this process, we did not strip down or sacrifice any functionality, with the exception of reducing factors of safety to no less than 2.4.

4.1 Limitations and Future Work

The structural components of the knee were redesigned to reduce part volume and weight using hand calculations and simple finite element analyses in this study. In future work, topology optimization may be utilized to further minimize the knee size and weight. While the redesigned knee has been CAD modeled and finite element analyses have been performed for all load-bearing component redesigns, a complete prototype of the updated knee has yet to be fabricated from the intended knee materials. Fatigue life has also not yet been evaluated. ISO standards require a life of

Copyright © 2023 by ASME

3 million cycles (approximately 3 years of use) for prostheses [12]. Load cells placed on a new, high-fidelity knee prototype may help elucidate more precisely the distribution of loads between different load paths, which will be useful for fatigue analysis. In future work, innovations on the damper described in [8] will continue.

New prototypes integrating design changes will be constructed from the selected materials in order to confirm that the knee kinematics are intact and for fatigue testing. Updated prototypes will also be tested with transfemoral amputees locally and in India, and feedback will be collected regarding the knee's appearance and usability. Of particular concern is whether the asymmetry of the enclosed knee about the upper and lower prosthesis caused by the damper is bothersome to users. Finally, FEA only accounts for predicted use cases, so longitudinal studies of prosthesis performance in real-world conditions should also be conducted. Depending on prosthetic knee user falling patterns, impact testing of the enclosed knee may be warranted. The infiltration of dust and other elements into the enclosure may affect prosthesis function and longevity and should also be studied. Seals may be added around the openings of the enclosure in future work based on the findings.

5. CONCLUSIONS

In this study, we redesigned the mechanical structure of a passive, single-axis prosthetic knee that provides near-able-bodied kinematics to increase usability and commercial viability in low- and middle-income markets. Engineering analyses facilitated the reduction of the knee's part count from over 60 components to 17, with the sum of part volumes and masses reduced by 76% and 67%, respectively. An enclosure has also been designed to encase the mechanism and create a discreet profile. This study demonstrates how engineering analysis and design for manufacturing can be employed to reduce product cost without sacrificing performance.

ACKNOWLEDGMENTS

This work was funded by the MIT Department of Mechanical Engineering and was conducted as a part of the 2.76 Global Engineering class in fall 2022. We would like to thank our collaborator at Jaipur Foot (BMVSS), Dr. Pooja Mukul, for her continued guidance of this project. Thank you to Collin Goldbach and Dr. Amy Carleton for providing technical and communication support to the project.

REFERENCES

[1] Sexton, Sandra. "Rehabilitation of People with Physical Disabilities in Developing Countries." Technical report no. USAID. 2016. URL https://www.ispoint.org/wp-content/uploads/2022/02/rehabilitation_of_people_wit-1.pdf.

[2] Narang, Yashraj S. "Identification of Design Requirements for a High-performance, Low-cost, Passive Prosthetic Knee through User Analysis and Dynamic Simulation." Master's Thesis, Massachusetts Institute of Technology, Cambridge, MA. 2013. URL https://core.ac.uk/download/pdf/18321468.pdf.

[3] McDonald, Cody L., Westcott-McCoy, Sarah, Weaver, Marcia R, Haagsma, Juanita and Kartin, Deborah. "Global Prevalence of Traumatic Non-fatal Limb Amputation." *Prosthetics and Orthotics International* Vol. 45 No. 2 (2021): pp. 105–114. DOI 10.1177/0309364620972258.

[4] Arelekatti, V. N. Murthy, Petelina, Nina T., Johnson, W. Brett, Major, Matthew J. and Amos G. Winter, V. "Design of a Four-bar Latch Mechanism and a Shear-based Rotary Viscous Damper for Single-axis Prosthetic Knees." *ASME Journal of Mechanisms and Robotics* Vol. 14 No. 3 (2022). DOI 10.1115/1.4052804.

[5] Hamner, Samuel R., Narayan, Vinesh G. and Donaldson, Krista M. "Designing for Scale: Development of the Re-Motion Knee for Global Emerging Markets." *Annals of Biomedical Engineering* Vol. 41 No. 9 (2013): pp. 1851–1859. DOI 10.1007/s10439-013-0792-8.

[6] Jensen, J. S. and Raab, W. "Clinical Field Testing of Trans-femoral Prosthetic Technologies: Resin-wood and ICRC-polypropylene." *Prosthetics and Orthotics International* Vol. 28 (2004): pp. 141–151. DOI 10.1080/03093640408726699.

[7] Berringer, Molly A., Boehmcke, Paige J., Fischman, Jason Z., Huang, Athena Y., Joh, Youngjun, Warner, J. Cali, Arelekatti, V. N. Murthy, Major, Matthew J. and Winter, Amos G. "Modular Design of a Passive, Low-cost Prosthetic Knee Mechanism to Enable Able-bodied Kinematics for Users with Transfemoral Amputation." *Proceedings of the ASME 2017 International Design Engineering Technical Conferences.* DETC2017-68278. Cleveland, USA, August 6–9, 2017. DOI 10.1115/DETC2017-68278.

[8] Arelekatti, V. N. Murthy, Petelina, Nina T., Johnson, W. Brett, Winter, Amos G. and Major, Matthew J. "Design of a Passive, Shear-based Rotary Hydraulic Damper for Single-axis Prosthetic Knees." *Proceedings of the ASME 2018 International Design Engineering Technical Conferences.* DETC2018-85962. Quebec City. Canada, August 26–29, 2018. DOI 10.1115/DETC2018-85962.

[9] "National Family Health Survey, India." Technical Report No. 3. International Institute for Population Sciences, Ministry of Health and Family Welfare, Government of India. 2006.

[10] de Castro, Marcelo Peduzzi, Soares, Denise, Mendes, Emília and Machado, Leandro. "Plantar Pressures and Ground Reaction Forces during Walking of Individuals with Unilateral Transfemoral Amputation." *PM & R: The Journal of Injury, Function, and Rehabilitation* Vol. 6 No. 8 (2014): pp. 698–707. DOI 10.1016/j.pmrj.2014.01.019.

[11] Winter, Amos and Govindarajan, Vijay. "Engineering Reverse Innovations." *Harvard Business Review* Vol. July–August 2015 (2015). URL https://hbr.org/2015/07/engineering-reverse-innovations.

[12] "Prosthetics — Structural Testing of Lower-limb Prostheses — Requirements and Test Methods." Technical Report No. 10328:2016. ISO. 2016. URL https://www.iso.org/standard/70205.html.

[13] Stevens, Thomas T. "Stochastic Fields and Their Digital

Simulation." *Stochastic Methods*. Martimius Publishers, Dordrecht (1999): pp. 22–36.

Copyright © 2023 by ASME

Proceedings of the ASME 2023
International Design Engineering Technical Conferences and
Computers and Information in Engineering Conference
IDETC-CIE2023
August 20-23, 2023, Boston, Massachusetts

DETC2023-116657

QUANTIFYING RESILIENCE TRADE-OFFS FOR SMALL-SCALE FARMS: A SYSTEM OPTIMIZATION STUDY IN UGANDA

Jesse Austin-Breneman*
Global Design Laboratory
University of Michigan
Ann Arbor, MI 48109
Email: jausbren@umich.edu

Praneet Nallan Chakravarthula
Dept. of Mechanical Engineering
University of Michigan
Ann Arbor, MI 48109
Email: praneet@umich.edu

Alvin B. Kimbowa
Dept. of Electrical and Computer Engin.
Makerere University
Kampala, Uganda
Email: alvin.kimbowa@students.mak.ac.ug

Peter Ozaveshe Oviroh
Dept. of Mechanical Engineering Science
University of Johannesburg,
Johannesburg, South Africa
Email: poviroh@uj.ac.za

Samuel Boahen
Mechanical Engineering Department
Kwame Nkrumah Univ. of Science and Technology
PMB, Kumasi, Ghana
Email: s.boahen@knust.edu.gh

Emmanuel Wokulira Miyingo
Dept. of Electrical and Computer Engin.
Makerere University
Kampala, Uganda
Email: emmanuel.miyingo@mak.ac.ug

Panos Y. Papalambros
Optimal Design Laboratory
University of Michigan
Ann Arbor, MI 48109
Email: pyp@umich.edu

ABSTRACT

Resilience is broadly understood as the capacity of a system to absorb, adapt, and transform in response to significant changes in its external environment. Improving resilience requires design trade-offs, for example, increases in equipment capacity and capital costs. This paper explores resilience trade-offs

for food security and crop production of agricultural systems representing the small-scale farms dominant in low-income regions of the world. In this study, resilience assessment metrics are incorporated into a water-energy-food nexus optimization method to quantify resilience trade-offs. As an illustrative case, the previously developed system modeling and optimization framework INRCD-OPT is applied to the case study of small-scale farming in the Kampala region of Uganda to examine trade-offs between

---

*Address all correspondence to this author.

Copyright © 2023 by ASME

investments in irrigation driven by renewable energy microgrids and farm resilience during drought years. Optimal farm designs for baseline climate parameters show that a tailored microgrid capacity and irrigation strategy is needed to maximize profit under different climate conditions. Results optimizing for average profit across varying climate conditions show that a "resilient" farm design balances a smaller microgrid with a more aggressive irrigation strategy, navigating the trade-off between capital investment, operating cost, and crop yields under drought conditions. This type of analysis can inform long-term planning for resilience and the value of investments to achieve increased food security and crop production stability.

1 Introduction

Small-scale farms of less than two hectares make up 84% of global farms, providing food and economic activity for the majority of the low-income population of the world [1]. Due to their importance in the global food system, improving outcomes for these farms is a key part of any plan to progress towards the United Nations Sustainable Development Goals [2]. Increases in profitability and sustainability for small farms would result in significant reductions in global poverty and hunger. Although there are a number of obstacles to realizing this goal, many researchers have identified a lack of resilience as a critical challenge facing small farmers [3]. With limited access to financial resources and dedicated infrastructure, small-scale farms are especially susceptible to shocks, such as extreme weather events, climate change, or marketplace disruptions [4].

Resilience is a widely-studied concept in the sustainability literature [5]. Definitions vary by discipline and application, but resilience is generally defined as the capacity of a system to absorb, adapt, and transform in response to an external shock [6]. Prior work by the authors provides an overview of resilience research as applied to farming systems [7]. This systematic literature review categorized resilience research on agriculture by geographic area, focus, and findings. Results showed that research in different geographic regions tended to focus on different dimensions of resilience. For example, a higher percentage of studies examined transformative resilience capabilities in Europe than in Africa. There were also differences by region in the types of shocks studied. Shocks induced by climate change, such as droughts and heat waves, pose a serious threat to food security, particularly for farming communities in developing countries and their resilience [8].

This article focuses on resilience of small-scale farms in Uganda with respect to drought conditions. The previously-developed modeling and system optimization tool, INRCD-OPT [9], allows exploration of crop production and profit gains under different scenarios. Using this modeling framework, we explore the resilience of typical small farms when faced with drought years rather than years with average rainfall. In particular, we examine the tradeoffs between investing in sustainable irrigation to support crop yields, especially in drought years, and the installation and operation costs of irrigation. Here, sustainable irrigation refers to irrigation systems with pumping operated by a microgrid that uses renewable energy such as photovoltaic cells. The Uganda location was selected due to the composition of the research team and access to the local community.

In the remainder of the article we present related work and the research gap addressed in this paper; the methodology we followed including background for the case study, the system model description, and the simulation and optimization framework; computational results and discussion; and a conclusion on research limitations and future work.

2 Related Work

This study integrates research from several fields including engineering design, sustainability, agriculture, and system engineering. We briefly review the concepts and past work relevant to the present case study. Further details on past work can be found in the cited references.

2.1 Engineering Systems Resilience

Engineering systems are mostly developed with the objective to operate at an optimal design point in a steady equilibrium state. Resilience of engineering systems focuses on the rate of return of the system to the designed equilibrium state when there is a disturbance [6]. As such, the resilience of engineering systems entails the return of performance variables such as the "system function" to its designed operating point or restoring the state of the system to its designed point after a disturbance. These disturbances may be caused by disruptions in system operations as a result of accidents or external shocks, changes to new permanent operating states, stress or disasters.

The ability of the system to absorb, resist, withstand, react, and adapt to these disruptions, or change its state or function to operate at its designed equilibrium state in the presence of disruptions is characteristic of its resilience [10]. Alternatively, resilience of engineering systems can be perceived as the ability of the system to respond to disturbances by moving the optimal design state of the system to another favorable state or form [11]. Quantification of resilience of an engineering system requires evaluation of the performance metrics that characterize its optimal design state, the disturbances occurring, the response of the system to the disturbances, and the new state of the system after the response.

2.2 Resilience in Farming Systems

Resilience of farming systems is necessary for food security, sustainability of farmlands, and economic benefits to farmers. Here we define a farming system as the Water-Food-Energy

Copyright © 2023 by ASME

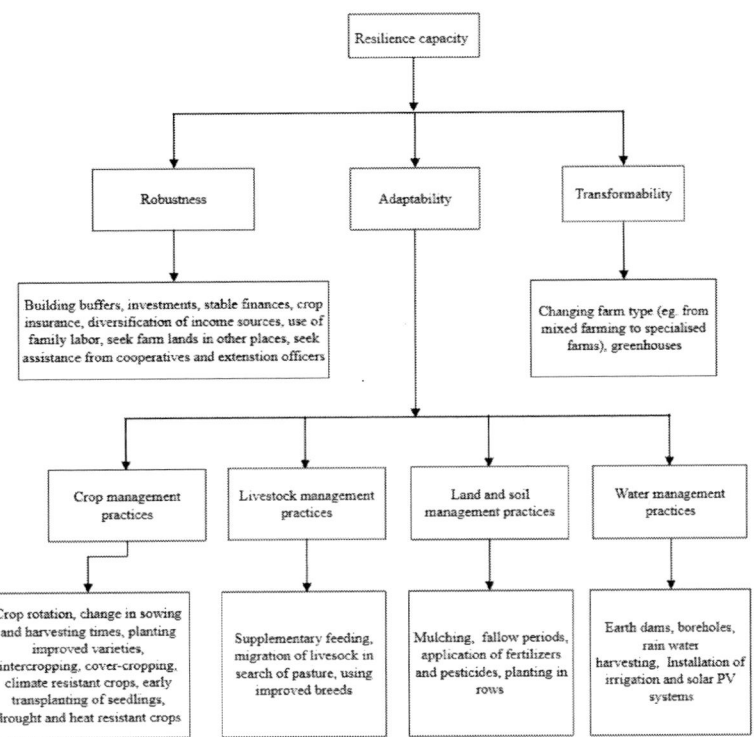

FIGURE 1: Resilience capacity of farming systems

(WEF) Nexus of a farm or collection of farms comprising the crops, irrigation, and energy subsystems [9], [12]. A farm is said to be resilient when it is able to produce the expected yield and profit in the face of shocks that directly affect the operations of the farm, such as drought, flooding, heat waves, pest infestations, and indirectly affect the supply chain such as price volatility and pandemics [13]. In the presence of these shocks, a resilient farming system has the ability to anticipate, adjust and recover, in order to produce the desired yield and profit. This ability denotes the resilience capacity of the farming system.

Generally, resilience capacity of farming systems can be classified into robustness, adaptability and transformability [5]. A robust farming system continues to operate in the face of a shock without changing its output. For such farms, farmers diversify their income sources, build buffer stocks, insure their crops, join farm cooperatives to receive knowledge on resilient farming practices, or change their planting and harvesting periods to ensure their resilience [14]. Robustness is mostly practiced among farmers in Europe and North America since most of these farmers have the knowledge, capacity, support from government, and cooperative organizations to ensure robustness of their farms [15] [16].

Most farmers across the globe practice adaptability to ensure the resilience of their farms [14]. To gain adaptability,

farmers change their farm inputs and production in response to a shock without compromising on the basic structures of the farm. Adaptable resilience methods employed by farmers include crop management practices such as planting high-quality seeds, planting drought-resistant and flood-resistant crops; livestock management practices such as decreasing livestock stock density; land and soil management practices such as mulching, inter-cropping, allowing fallow periods, application of fertilizers and pesticides; and water management practices such as building irrigation systems [17]. Transformability deals with changing the whole farming system or structure into an entirely new farming system [18]. This approach is quite expensive and is, therefore, not common among farmers, particularly in the developing world.

Figure 1 summarizes resilience capacity and the corresponding practices as used in farming systems. Quantification of the resilience of a farming system requires an assessment of the shocks affecting the farming system, the resilience method employed, and how the applied resilience method affects yield and profit in the face of a shock.

2.3 INRCD projects

The integration of conservation and development principles to ensure sustainability and environmental protection during eco-

Copyright © 2023 by ASME

nomic wealth creation led to the concept of Integrated Natural Resource Conservation and Development (INRCD) projects that evolved from early efforts by the World-Wide Fund for Nature in the 1980s [19]. INRCD projects are global initiatives to promote economic growth in local communities while conserving natural resources [20]. At its core, INRCD combines socioeconomic investments with integrated conservation and developmental projects using system design optimization tools to ensure sustainability and profit at the community level. The INRCD-OPT tool can assist stakeholders in visualizing trade-offs between conflicting objectives, comprehending the significance of cross-sectoral interactions, and selecting the best system for their scenario [9].

2.4 Research Gap

Previous research based on this concept of INRCD projects [9, 20–22] has not addressed the issue of resilience of such projects in the face of shocks that disrupt their operation and cause insecurity in income and food supply.

Small-scale farmers with farms of less than two hectares, especially in low-income settings, frequently operate under constraints that hinder making decisions regarding technology investments. The farmers may not have enough time, information, or analytical capabilities to evaluate potential alternatives properly and with the appropriate time horizon. These farms are also particularly vulnerable to shocks, such as extreme droughts or commodity crop price crashes. Thus, small-scale farmers can realize significant potential benefits from improvements to the resilience of their farms and a clearer understanding of the trade-offs involved when seeking such resilience and the associated investments.

Prior work in resilience assessment tools and other similar decision-support tools for general agricultural systems can help farmers understand trade-offs and improve their decision-making and resulting outcomes. However, there are two key gaps in this body of work: 1) existing tools have not been tailored to either small-scale farms or the low-income settings of interest, and 2) analytical tools often consider only one subsystem and may miss important interactions between different subsystems.

The case study presented in this paper examines the design of a farming system for a small Ugandan farm of one hectare. The Ugandan agricultural sector is dominated by these types of farms, with an average farm size of 0.97 hectares [23]. The term 'system' may appear somewhat verbose for such a humble implementation, yet it is an accurate technical term for the problem addressed. Moreover, when considering the large number of such farms, the system perspective becomes much more pertinent. Using the INRCD-OPT tool, the crop selection, irrigation subsystem design, and energy subsystem design are optimized for different resilience metrics and conditions.

3 Methodology

This section reviews the motivation and background for the particular case study, and presents the farming system model, followed by a description of the simulation and optimization framework used in the study.

3.1 Uganda Case Study Background

Based on prior work and interactions with collaborating farmers in Uganda, the farm is modeled as a system comprised of water (W), energy (E), and food (F) subsystems, referred to also as a WEF Nexus system [12]. In order to develop a model capable of estimating crop and profit outcomes, a number of assumptions and simplifications were made and are further elaborated in the following sections. We consider a region in Uganda with small landholder farms, particularly the agricultural settlements in Kampala. We assumme a 4-hectare farm with uninterrupted access to water and no electrical grid access. Electricity is provided through solar photovoltaic (PV) panels, batteries, or a backup diesel generator.

Agriculture in Uganda is seriously threatened by drought, which causes significant output losses [24–26]. This threat is an effect of climate change [27, 28] and needs urgent attention as it is resulting to hunger and loss of livelihoods. Most of the farmers in Uganda entirely depend on nature to grow crops, i.e., rain-fed crop production [28] They tend to plant according to traditional seasons for seasonal crops such as maize, beans, or groundnuts. They usually plant at the beginning of the rain seasons, i.e., early March and mid-August or beginning of September [29, 30]. However, the traditional rain seasons in Uganda have become quite irregular and hard to predict [31]

The government of Uganda is currently (2023) working to support small-scale farmers by contributing up to 75 percent of installation costs to farmer-specific irrigation systems in various parts of Uganda, part of the World Bank's "Uganda - Irrigation for Climate Resilience Project (ICRP)" [32]. This is intended to motivate small-scale farmers to move from full dependence on seasonal rains for the growth of their crops in order to modernize their farming activities and stabilize their incomes [30]. Several farmers have been assessed and positively evaluated for eventual co-funding under Uganda Government's micro-irrigation initiative for small-scale farmers [33]. A number of them already have some infrastructure, i.e., water source and garden(s) to be irrigated. Owing to the intermittent drought events in Uganda, we have incorporated such drought scenarios in the system modeling to account for different weather conditions, and also to offer a broader perspective of the functionality of the INRCD simulation and optimization framework.

3.2 Farming System Model

System-level modeling of the general Agriculture-Energy (Ag-En) INRCD system is depicted in Figure 4. Weather (in-

Copyright © 2023 by ASME

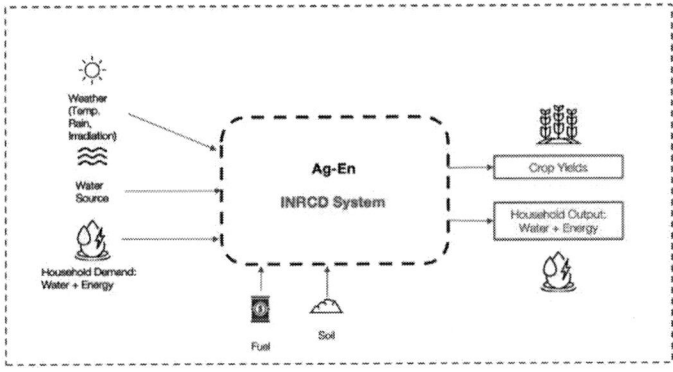

FIGURE 2: Overall Ag-En INRCD system model representation

cluding temperature, precipitation, and irradiance), residential energy and water consumption, soil type and related properties, and fuel for any supplementary non-renewable energy source (in this case, a diesel auxiliary generator) are system inputs. Crop yields, water requirements, and energy production for household use are outputs of the system. Crop selection and irrigation strategy, piping dimensions, pump selection, microgrid system characteristics (e.g., battery selection, solar panel selection, auxiliary generator characteristics), and the energy supply strategy (the microgrid 'dispatch' strategy)are decision variables. The system model consists of three coupled subsystems described below. More details on the subsystems and their interactions in the model can be found in an earlier publication [9].

3.2.1 Subsystem 1: Food (Crops)

The food subsystem calculates the estimated crop yield and daily water requirement based on the crop selection, irrigation method, and climate data. Based on interactions with local farmers, available crop options include maize, soybeans, dry beans, paddy rice, or nothing at all. The climate data for both the baseline and drought conditions is created using the LARS-WG stochastic weather generator [34]. Assuming a certain system architecture and a particular set of environmental variables, the food subsystem also calculates a subsystem cost. The Deficit Irrigation Toolbox (DIT) [35] is used to simulate crop growth over a year using the climate data and thereby calculate yield [9, 35]. The water demand to achieve the crop yields and associated income is passed on to the irrigation subsystem.

3.2.2 Subsystem 2: Water (Irrigation)

An adequate time-dependent irrigation schedule is necessary, and it needs to take into account the location, irrigation system, and crop selection in question. This schedule, or irrigation strategy, is generated essentially by solving an optimal control problem for water supply over time with control objective the maximization of crop yield. As mentioned, this subsystem model is executed by

DIT. The irrigation subsystem calculates the pumping system's energy consumption depending on the irrigation system's parameters and the demand for irrigation. In this study, the irrigation type is drip irrigation. Design variables are the diameters of the main pipe, submain pipe, lateral pipes and others. The pipe head loss, groundwater elevation differential, and water emitter operating pressure are used to calculate total system pressure. More details can be found in previously reported work [9].

3.2.3 Subsystem 3: Energy (Microgrid)

We assume no main grid access to electricity is available at rural locations around Kampala. The usual power source in this setting is only a diesel generator, but aiming at sustainable development we consider a combination of renewable energy resources and a diesel generator backup. This subsystem is configured with solar photovoltaic (PV) panels, batteries, and auxiliary diesel generator to provide the energy needed by the irrigation system and farm household. The subsystem model has three design variables, number of PV panels, number of batteries, and diesel generator size. For a given set of design variables, an inner control optimization problem is solved to determine the microgrid's dispatch (control) strategy for selecting the supply source of energy. At each point in time along the simulation, the dispatch strategy seeks to minimize the overall cost of electricity production required to meet pumping and household power demands. The energy model takes as input pump energy demand, temperature, irradiation, and the characteristics of the different power sources, and calculates the cost incurred in meeting the pump and household energy demand.

3.2.4 Weather Model

We consider different weather conditions as inputs to the model for drought and normal or baseline years. The LARS-WG stochastic weather generator uses regional climate models to generate local-scale climate scenarios. For the presented case study, regional-level historical climate data from Kampala, Uganda, was obtained from the NASA Langley Research Center Prediction of Worldwide Energy Resource (POWER) project [36]. This information was used to create daily time series of maximum and minimum temperatures, precipitation, and solar radiation. Two climate time series were generated, a baseline climate and a drought climate. The drought time series was created with similar temperatures, but significantly less precipitation. The total annual precipitation for the baseline case is 1135mm and 906mm for the drought case. With greater solar irradiance and less precipitation, the percentage of evapotranspiration, which is the percentage of water that does not enter the soil, is significantly higher for the drought time series. Figure 3 shows the daily evapotranspiration for both the baseline and drought conditions.

Copyright © 2023 by ASME

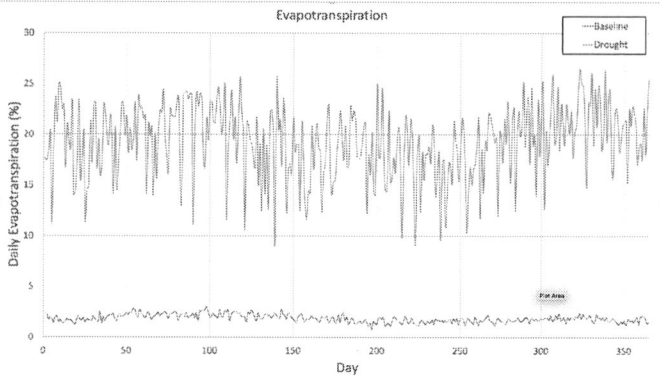

FIGURE 3: Daily Evapotranspiration Percentage

3.3 INRCD-OPT Simulations and Optimization

In order to capture trade-offs between the different subsystems with regards to resilience objectives, the INRCD-OPT model was optimized for three cases: maximize annual profit with baseline weather/climate, maximize annual profit with drought weather/climate, and maximize a 'combined' objective. For the first two cases, the INRCD-OPT model optimized the subsystem design variables to maximize profit for a single year simulated using their respective LARS-WG time series. For the third case, the subsystem design variables are optimized to maximize the average profit over two years, where the first year uses the baseline time series and the second year uses the drought time series.

For each scenario, baseline, drought, and combined, the INRCD-OPT model was optimized using a two-phase process, with a global search as the first phase followed by a local search as the second phase. In Phase 1, all of the design variables, including the microgrid variables, are considered continuous. In order to search the design space broadly, the non-gradient pattern search function from the MATLAB global optimization toolbox was used to optimize the farm system starting from an arbitrary feasible point. A number of other gradient-free optimization algorithms were tested and rejected as convergence time and solutions were inferior to those from pattern search. Previous testing of INRCD-OPT also led to the selection of Generalized Pattern Search as the search algorithm and OrthoMADS as the polling algorithm. The best design points identified in Phase 1 were then used as initial starting points for the Phase 2 local search. It should be noted that results from pattern search meet only heuristic termination criteria and not satisfaction of optimality conditions.

In Phase 2, the gradient-based Sequential Quadratic Programming (SQP) algorithm fmincon in the MATLAB optimization toolbox was used to identify the closest optimal point to the termination point from Phase 1. SQP was selected due to its fast local convergence and satisfaction of optimality conditions. Because the number of PV panels and number of batteries in the microgrid are discrete variables, all possible combinations of the nearest bounding discrete values for these variables were generated. For each combination, a separate instance of SQP was run with these variables fixed at their discrete values as parameters, resulting in a unique optimal point. The objective function values for all identified points were compared, and the largest one was selected as the system's optimum. Again, it should be noted that in general global optimality cannot be claimed for any of these results and the results presented here are the best we obtained. Further details of the optimization model and solution strategy for INRCD-OPT can be found in prior work by the authors [9].

4 Results

In this section, results from the optimization of the INRCD-OPT model are presented. Using the weather files discussed in Section 3.2.4 as inputs, three instances were simulated. Figure 4 shows the results from the two-phase optimization process for a farm in Kampala under the three climatic scenarios - baseline weather only (on the left), drought weather only (center), and combined (right). The combined scenario uses the two-year timeline where the baseline climate (Year 2000) in Kampala is followed by the simulated drought climate.

From the results, the irrigation strategy during a drought year is to water more frequently and possibly water greater amounts per day during the farming seasons. Due to significantly lower precipitation, higher average temperatures, and low evapotranspiration, there is also an increase in the number of photovoltaic panels during a drought year and a slightly lower reliance on the diesel generator to meet energy demands. Consequently, worse-than-ideal weather conditions and a bigger microgrid resulted in a lower crop yield and profit, respectively, during the drought year. Table 1 shows the farm system designs with optimal values for each subsystem and the resulting profit under baseline, drought, and combined scenarios.

5 Discussion

The most notable result in the case study is that the INRCD-OPT model highlights that trade-offs between the microgrid selection and the irrigation strategy are important to maximizing the profit across both scenarios. The INRCD-OPT model found similar results across all three optimization scenarios for several of the design variables. For example, the optimal crop allocation was the same, with 100% of the field area dedicated to either maize or rice depending on the growing season. This is likely due to maize and rice having the highest potential revenue given the combinations of possible yield and unit price and the use of profit as the sole objective. Profit is sensitive to changes in crop allocation. At the optimal point, a change of 5% in crop area

Copyright © 2023 by ASME

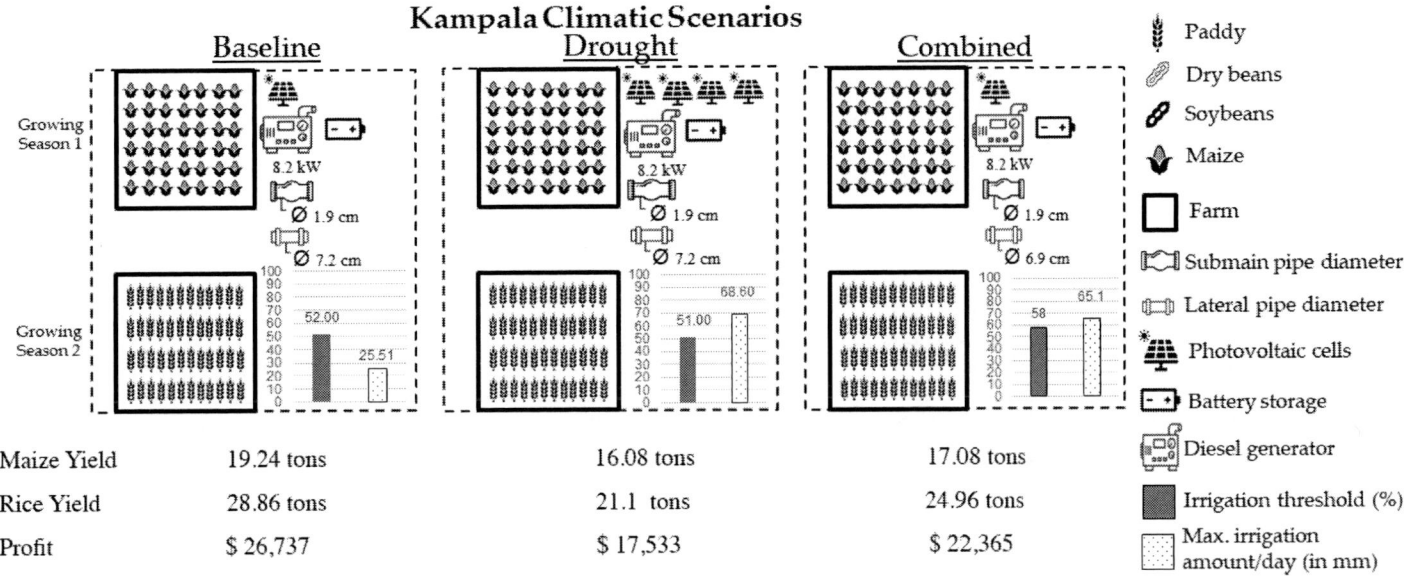

FIGURE 4: Optimal design and outcomes for maximizing profitability for the farm in Kampala under three climatic scenarios: baseline climate only (left), drought climate only (center), and combined climate (right). Outcomes such as profit and crop yield are indicated below each figure

from maize and rice to soy and dry beans results in a loss of almost $1000 in profit. This selection of a single crop for each growing season does not necessarily reflect the situation on the ground, as farmers in this area may have other objectives, such as a diversity of produce for their own consumption. It was not obvious prior to the use of the tool that a mono-crop would be the optimal choice across both climate scenarios. Similarly, the tool selected the lower bounds of the pipe diameter sizes for all cases. The cost of the irrigation system, and therefore the total profit, was sensitive to small changes in pipe diameter.

These results indicate that for this location resilience of profit across both scenarios depends on tailoring the micro-grid selection and irrigation strategy. In the drought scenario, the higher average temperatures, irradiance, and evapotranspiration lead to a lower overall crop yield. Perhaps to mitigate this effect, the optimal farming system design has an increased daily irrigation allowance and increased PV cells. This increase in microgrid capacity is likely due to the increase in required pumping time which grows from 61 hours in the baseline case to 101 hours in the drought. The corresponding energy requirement also increases from 933 kW to 1262 kW. In short, when faced with drought conditions, the optimal design responds to the greater irrigation and energy needs by moving to a greater solar capacity for the microgrid. Although this results in higher capital expenditures, it reduces operating costs to a higher degree. By contrast, the baseline scenario uses less energy as there are fewer dry days requiring irrigation and therefore relies on the diesel generator more often. These results are consistent with common sense.

While not a validation per se, this consistency encourages further exploration of the INRCD-OPT tool for supporting decision making.

In the combined scenario, the model calculates the average profit across both scenarios for the same farm system design. The identified optimal system has the smaller capacity microgrid associated with the baseline scenario but an irrigation allowance similar to that of the drought scenario. The irrigation threshold is also increased which means that, for the combined case, the irrigation strategy is to water more frequently and at a potentially greater amount. However, because the irrigation allowance is an irrigation control scheme parameter that provides an upper bound for the daily amount of water, this limit is likely never reached under the baseline conditions. Under the drought conditions, the optimal combined farm has a greater energy load, 1604 kW, compared the the farm optimized for drought conditions, 1262 kW. This is an increase due to increased reliance on the diesel generator. This result is notable because the "resilient" farm design is similar to the baseline farm. The tool suggests for this location the farmer does not need to invest in a different microgrid to prepare for a potential drought.

In sum, the system optimization model reinforces insights into system behavior when designing for resilience is simple scenarios. The objective function used here was a simple average across potential climate scenarios, but a similar approach can be used for more complicated time-dependent resilience metrics.

Copyright © 2023 by ASME

TABLE 1: Optimal Farm System Designs

Variable	Baseline	Drought	Combined
Crop Allocation Variables			
Maize (%)	100	100	100
Dry Beans (%)	0	0	0
Rice (%)	100	100	100
Soy Beans (%)	0	0	0
Irrigation System Variables			
Lateral pipe diameter (m)	0.019	0.019	0.019
Submain pipe diameter (m)	0.072	0.072	0.069
Irrigation threshold (%)	52	51	58
Maximum irrigation allowable per day (mm)	25.5	68.6	65.1
Microgrid Variables			
PV cells (units)	1	4	1
Batteries (units)	1	1	1
Generator power (kW)	8.2	8	8.2
Profit (USD)	26,737	17,533	22,365

6 Conclusion

We presented a modeling and optimization study for designing a small-scale agricultural system with increased resilience to drought conditions in the in rural Kampala region of Uganda. The INRCD-OPT framework can holistically examine tradeoffs between the crop, irrigation, and microgrid subsystems. In the study, the optimal farm for baseline and drought scenarios was determined along with a resilient system design that produced the highest average profit across all three scenarios. The INRCD-OPT model also indicated that the crop allocation did not change depending on the climate conditions, but that drought conditions required a greater irrigation strategy and corresponding capacity increase for the solar energy microgrid. The combined scenario demonstrated that a resilient farm should have a greater irrigation allowance, but no change in the solar capacity of the microgrid.

There are several limitations in the presented work. First, the resilience metric was a simple average across the potential scenarios. There is a rich body of work in resilience assessment, and future work should incorporate these assessment models into the objective functions of the optimization model. Second, the study was limited to parameters associated with Kampala, Uganda as this was the site of prior work by the authors and with the great-est amount of information available. Future work should include other locations already experiencing drought events to explore other potential implications to resilience optimization for agricultural systems. Finally, the study considered only two climate scenarios, baseline and drought, and did not account for sequencing of multiple year events. Future work should include a richer set of climate change scenarios and the associated modeling and optimization results for resilience over a longer time horizon and more varied conditions.

A broader limitation is the assumption of non-rain water being available in the quantities needed in the different scenarios. Water over-consumption is an increasing problem is East Africa and an important consideration for both sustainability and resilience which is not addressed in this work.

REFERENCES

[1] FAO, 2020. *The State of Food and Agriculture 2020. Overcoming water challenges in agriculture.* Food and Agriculture Organization of the United Nations, Rome, Italy.

[2] Abraham, M., and Pingali, P., 2020. "Transforming smallholder agriculture to achieve the SDGs". *The role of smallholder farms in food and nutrition security*, pp. 173–209.

[3] Czekaj, M., Adamsone-Fiskovica, A., Tyran, E., and Kilis, E., 2020. "Small farms' resilience strategies to face economic, social, and environmental disturbances in selected regions in Poland and Latvia". *Global Food Security*, **26**, p. 100416. DOI: 10.1016/j.gfs.2020.100416.

[4] Mizik, T., 2021. "Climate-smart agriculture on small-scale farms: A systematic literature review". *Agronomy*, **11**(6), p. 1096. DOI: 10.3390/agronomy11061096.

[5] Meuwissen, M. P., Feindt, P. H., Spiegel, A., Termeer, C. J., Mathijs, E., De Mey, Y., Finger, R., Balmann, A., Wauters, E., Urquhart, J., et al., 2019. "A framework to assess the resilience of farming systems". *Agricultural Systems*, **176**, p. 102656. DOI: 10.1016/j.agsy.2019.102656.

[6] Hoiling, C., 1973. "Resilience and sustainability of ecological systems". *Annual Review of Ecology, Evolution, and Systematics*, **4**, pp. 1–23.

[7] Boahen, S., Oviroh, P. O., Austin-Breneman, J., Miyingo, Emmanuel, W., and Papalambros, P. Y., 2023. "Understanding resilience of agricultural systems: A systematic literature review". *Proceedings of the Design Society, International Conference in Engineering Design 2023. To appear.*

[8] EU, 2020. A farm to fork strategy for a fair, healthy and environmentally-friendly food system. Available at: https://eur-lex.europa.eu/legal-content/EN/TXT/?uri=CELEX:52020DC0381 (accessed 2022-10-22).

[9] Oviroh, P. O., Austin-Breneman, J., Chien, C.-C., Chakravarthula, P. N., Harikumar, V., Shiva, P., Kim-

Copyright © 2023 by ASME

bowa, A. B., Luntz, J., Miyingo, E. W., and Papalambros, P. Y., 2023. "Micro Water-Energy-Food (MicroWEF) Nexus: A system design optimization framework for Integrated Natural Resource Conservation and Development (INRCD) projects at community scale". *Applied Energy,* **333**, p. 120583. DOI: 10.1016/j.apenergy.2022.120583.

[10] Wied, M., Oehmen, J., and Welo, T., 2020. "Conceptualizing resilience in engineering systems: An analysis of the literature". *Systems Engineering,* **23**, 1, pp. 3–13. DOI: 10.1002/sys.21491.

[11] Mayar, K., Carmichael, D. G., and Shen, X., 2022. "Resilience and Systems—A Review". *Sustainability (Switzerland),* **14**, 7. DOI: 10.3390/su14148327.

[12] Food, and of the United Nations, A. O., 2014. *The Water-Energy-Food Nexus - A new approach in support of food security and sustainable agriculture.* Food and Agriculture Organization of the United Nations, Rome, Italy.

[13] Spiegel, A., Slijper, T., de Mey, Y., Meuwissen, M. P., Poortvliet, P. M., Rommel, J., Hansson, H., Vigani, M., Soriano, B., Wauters, E., Appel, F., Antonioli, F., Gavrilescu, C., Gradziuk, P., Finger, R., and Feindt, P. H. "Resilience capacities as perceived by european farmers". *Agricultural Systems,* 10. DOI:10.1016/j.agsy.2021.103224.

[14] Benabderrazik, K., Kopainsky, B., Monastyrnaya, E., Thompson, W., Tazi, L., Joerin, J., and Six, J., 2022. "Climate resilience and the human-water dynamics. The case of tomato production in Morocco". *Science of the Total Environment,* **849**, 11. DOI: 10.1016/j.scitotenv.2022.157597.

[15] Doran, E. M., Zia, A., Hurley, S. E., Tsai, Y., Koliba, C., Adair, C., Schattman, R. E., Rizzo, D. M., and Méndez, V. E., 2020. "Social-psychological determinants of farmer intention to adopt nutrient best management practices: Implications for resilient adaptation to climate change". *Journal of Environmental Management,* **276**, 12. DOI: 10.1016/j.jenvman.2020.111304.

[16] Herrera, H., Schütz, L., Paas, W., Reidsma, P., and Kopainsky, B., 2022. "Understanding resilience of farming systems: Insights from system dynamics modelling for an arable farming system in the Netherlands". *Ecological Modelling,* **464**, 2. DOI: 10.1016/j.ecolmodel.2021.109848.

[17] Onyeneke, R. U., Amadi, M. U., Njoku, C. L., and Osuji, E. E., 2021. "Climate change perception and uptake of climate-smart agriculture in rice production in Ebonyi state, Nigeria". *Atmosphere,* **12**, 11. DOI: 10.3390/atmos12111503.

[18] Slijper, T., Urquhart, J., Poortvliet, P. M., Soriano, B., and Meuwissen, M. P., 2022. "Exploring how social capital and learning are related to the resilience of Dutch arable farmers". *Agricultural Systems,* **198**, 4. DOI: 10.1016/j.agsy.2022.103385.

[19] Hughes, R., and Flintan, F., 2001. *Integrating conservation and development experience: a review and bibliography of the ICDP literature.* London: International Institute for Environment and Development.

[20] Barlow, T., Biddanda, M., Mendke, S., Miyingo, E., Sicko, A., Papalambros, P. Y., Chien, C.-C., and O'Neal, W., 2021. "A system design optimization model for integrated natural resource conservation and development in an agricultural community". *Proceedings of the Design Society,* **1**, pp. 273–282.

[21] Rajski, P. V., and Papalambros, P. Y., 2021. "Integrated natural resource and conservation development project: A review of success factors from a systems perspective". *Proceedings of the Design Society,* **1**, pp. 1867–1876. DOI:10.1017/pds.2021.448.

[22] Rajski, P., Sicko, A., and Papalambros, P., 2022. "Modeling social benefits in system design optimization of integrated natural resources conservation and development (inrcd) projects: Identification and quantification of design attributes from extant literature". *Proceedings of the Design Society,* **2**, pp. 1099–1108. DOI: 10.1017/pds.2022.112.

[23] Priegnitz, U., Lommen, W. J., Onakuse, S., and Struik, P. C., 2019. "A farm typology for adoption of innovations in potato production in southwestern uganda". *Frontiers in Sustainable Food Systems,* **3**, p. 68. DOI: 10.3389/fsufs.2019.00068.

[24] Nansamba, M., Sibiya, J., Tumuhimbise, R., Ocimati, W., Kikulwe, E., Karamura, D., and Karamura, E., 2022. "Assessing drought effects on banana production and on-farm coping strategies by farmers—A study in the cattle corridor of Uganda". *Climatic Change,* **173**(3-4), p. 21. DOI: 10.1007/s10584-022-03408-w.

[25] Mfitumukiza, D., Barasa, B., Kiggundu, N., Nyarwaya, A., and Muzei, J. P., 2020. "Smallholder farmers' perceived evaluation of agricultural drought adaptation technologies used in Uganda: Constraints and opportunities". *Journal of Arid Environments,* **177**, p. 104137. DOI: 10.1016/j.jaridenv.2020.104137.

[26] Akwango, D., Obaa, B. B., Turyahabwe, N., Baguma, Y., and Egeru, A., 2017. "Effect of drought early warning system on household food security in Karamoja subregion, Uganda". *Agriculture & Food Security,* **6**(1), pp. 1–12. DOI: 10.1186/s40066-017-0120-x.

[27] Magesa, B. A., Mohan, G., Matsuda, H., Melts, I., Kefi, M., and Fukushi, K., 2023. "Understanding the farmers' choices and adoption of adaptation strategies, and plans to climate change impact in africa: A systematic review". *Climate Services,* **30**, p. 100362. DOI: 10.1016/j.cliser.2023.100362.

[28] Wichern, J., Hammond, J., van Wijk, M. T., Giller, K. E., and Descheemaeker, K., 2023. "Production variability and adaptation strategies of Ugandan smallholders in the face of climate variability and market shocks". *Climate Risk Man-*

agement, p. 100490. DOI: 10.1016/j.crm.2023.100490.

[29] Wanyama, J., Ssegane, H., Kisekka, I., Komakech, A. J., Banadda, N., Zziwa, A., Ebong, T. O., Mutumba, C., Kiggundu, N., Kayizi, R. K., et al., 2017. "Irrigation development in Uganda: constraints, lessons learned, and future perspectives". *Journal of Irrigation and Drainage Engineering,* **143**(5), p. 04017003. DOI: 10.1061/(ASCE)IR.1943-4774.0001159.

[30] Micro Scale Irrigation Program – Ministry of Agriculture, Animal Industry and Fisheries. https://www.agriculture.go.ug/micro-scale-irrigation-program/. accessed on 2023-03-13.

[31] Mulinde, C., Majaliwa, J., Twinomuhangi, R., Mfitumukiza, D., Komutunga, E., Ampaire, E., Asiimwe, J., Van Asten, P., and Jassogne, L., 2019. "Perceived climate risks and adaptation drivers in diverse coffee landscapes of Uganda". *NJAS-Wageningen Journal of Life Sciences,* **88**, pp. 31–44. DOI:10.1016/j.njas.2018.12.002.

[32] Harriet Nattabi, G. I., 2020. Uganda-irrigation for climate resilience project (icrp). https://projects.worldbank.org/en/projects-operations/project-detail/P163836;. accessed 2023-04-04.

[33] Magoum, I., 2023. Uganda: 160 farmers to be provided with irrigation systems in kapelebyonge. https://www.afrik21.africa/en/uganda-160-farmers-to-be-provided-with-irrigation-systems-in-kapelebyong/. accessed 2023-04-04.

[34] Semenov, M. A., Barrow, E. M., and Lars-Wg, A., 2002. "A stochastic weather generator for use in climate impact studies". *User Man Herts UK*, pp. 1–27.

[35] Schuetze, N., and Mialyk, O., 2019. "Deficit irrigation toolbox: A new tool to improve crop water productivity and food security under limited water resources.". In Geophysical Research Abstracts, Vol. 21.

[36] Jordan, T. L., Hutchinson, M. A., Watkins, V. E., Langford, W. M., Cagle, C. M., and Logan, M. J., 2007. "National Aeronautics and Space Administration Langley Research Center's Design Criteria for Small Unmanned Aerial Vehicle Development".

Proceedings of the ASME 2023
International Design Engineering Technical Conferences and
Computers and Information in Engineering Conference
IDETC-CIE2023
August 20-23, 2023, Boston, Massachusetts

DETC2023-116725

FIFTY-FIVE PROMPT QUESTIONS FOR IDENTIFYING SOCIAL IMPACTS OF ENGINEERED PRODUCTS

Christopher A. Mattson[1,*], Thomas B. Geilman[1], Joshua F. Cook-Wright[1],
Christopher S. Mabey[1], Eric Dahlin[2], John L. Salmon[1]

[1] Department of Mechanical Engineering, Brigham Young University, Provo, UT
[2] Department of Sociology, Brigham Young University, Provo, UT

ABSTRACT

This paper introduces 55 prompt questions that can be used by design teams to consider the social impacts of the engineered products they develop. These 55 questions were developed by a team of engineers and social scientists to help design teams consider the wide range of social impacts that can result from their design decisions. After their development, these 55 questions were tested in a controlled experiment involving 12 design teams. Given a one-hour period of time, 6 control teams were asked to identify many social impacts within each of the 11 social impact categories identified by Rainock et al. [1], while 6 treatment groups were asked to do the same while using the 55 questions as prompts to the ideation session. Considering all 1271 social impacts identified by the teams combined and using 99% confidence intervals, the analysis of the data shows that the 55 questions cause teams to more evenly identify high quality, high variety, high novelty, impacts across all 11 social impact categories during an ideation session, as opposed to focusing too heavily on a subset of impact categories. The questions (treatment) do this without reducing the quantity, quality, or novelty of impacts identified, compared to the control group.

Keywords: Social Impact, Sustainable Development, Design Method Validation

1. INTRODUCTION

Most engineered products have an impact on society [2]. Those social impacts affect sustainable development, as do environmental and economic impacts [3]. Thoughtful engineering decisions made in consideration of all three impact areas are most likely to support the United Nations (UN) Sustainable Development Goals (SDGs) and address the global challenges of our day, including climate change [4, 5]. This paper is focused solely on

improving the design team's ability to meaningfully consider the social dimension of sustainable design.

As a way of decomposing the challenging task of considering and evaluating the social impact of engineered products, Rainock et al., carried out a substantial literature survey of 121 papers from 72 different journal sources in numerous disciplines [1]. The goal of the survey was to identify the various ways engineered products impact society. Rainock et al., identified 11 social impact categories. They are:

- Impacts on Health & Safety
- Impacts on Education
- Impacts on Paid Work
- Impacts on Conflict & Crime
- Impacts on Family
- Impacts on Gender
- Impacts on Human Rights
- Impacts on Stratification
- Impacts on Social Networks & Communication
- Impacts on Population Change
- Impacts on Cultural Identity & Heritage

Ottosson et al., asked a multidisciplinary team consisting of eight engineering and social scientists to map these social impacts onto 150 products designed for social good. With this mapping, they identified the conditional and joint probability of impacts in these categories being co-present in any one product [6]. They found that in no case did a product have impact in only one of these categories, thus pointing to the value of considering various social impacts for any one product.

Pack et al., mapped these social impacts to industry practice by interviewing 46 individuals at 34 companies in search of the

---

*Corresponding author: mattson@byu.edu
Version 1.34, May 15, 2023.

Copyright © 2023 by ASME

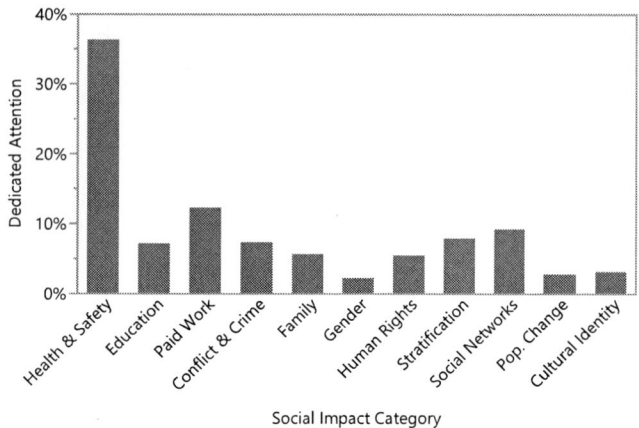

FIGURE 1: RESULTS OF PACK ET AL.'S INDUSTRY SURVEY ON ATTENTION GIVEN TO SOCIAL IMPACTS BASED ON CATEGORY [7].

breadth to which social impacts are considered by design and engineering professionals [7]. The study found that social impacts were considered by these professionals, but not in a way that was as comprehensive or as holistic when compared to the social impact categories identified by Rainock et al. [1]. Pack et al., found that industry professionals were significantly more focused on health and safety impacts, seemingly at the expense of the other social impact categories. Figure 1 illustrates the main finding from Pack et al., that is pertinent to the present paper [7].

Also discovered by Pack et al., was the reality that design teams have very few tools at their disposal for considering the social impacts of their design and engineering decisions [7].

Various tools have since been developed with hopes of facilitating a design team's consideration of social impacts during the development process. Those include social FMEA [8], social impact modeling [9], use of social impact sensors [10, 11], and more. In recent years there has been an increase in the development of complex systems-based models, including those that rely heavily on adoption models [12], agent-based models [13], and optimization techniques. Although meaningful, there is still a shortage of simple, generalized, and quick-to-use tools that can positively guide design teams to more fully consider the social dimension of their work.

This paper introduces a simple set of thought questions, designed by a set of engineers and social scientists, to be used as prompts when considering social impacts for a given product. This set of questions was tested in a controlled experiment and shown with 99% confidence to help teams more evenly identify high quality, high variety, high novelty, impacts across all 11 social impact categories during an ideation session, as opposed to focusing too heavily on a subset of impact categories. Importantly, these questions support the team in this way without reducing the quantity of impacts identified.

To present the questions, experiment, and findings, the remainder of the paper is organized as follows: Section 2 presents a historical perspective on social impact consideration as well as a review of pertinent literature. Section 3 presents the 55 questions

and their development, followed by Section 4, which describes the test used to validate the questions. Section 5 presents results. Finally in Section 6 concluding remarks and limitations are presented.

2. HISTORICAL PERSPECTIVE AND LITERATURE SURVEY

The social impact of technology has been a topic discussed in the archival literature since the 1940s, beginning with Marcuse who described how technological solutions strongly influence human behavior [2]. His vivid examples include highway design with its careful placement of vehicle parking near scenic vistas, refueling locations, and signage regarding leisure and refreshment. Implied from this example are social impacts on cultural identity and heritage, paid work, networks and communication, and population change. He provides other examples that extend from technology's influence on *behavior* to its influence on *thoughts and priorities* when he reports "the average man hardly cares for any living being with the intensity and persistence he shows for his automobile." This implies the technology's social impact on gender, family, and possibly more. Marcuse's thesis is powerful; the engineer – through technology development – is a social leader. While a few specific social impacts can be *implied* from his examples, Marcuse provides little guidance for the engineer to grasp or plan for the social impacts he or she has.

In 1947, with growing acknowledgment of technology's positive and negative role in society, Bartlett examined the social impacts of the era's most impactful technology: the radio [14]. His study used measures of listener audience size and location coupled with a variety of social surveys to draw conclusions about the radio's social impact. Bartlett discusses the radio's influence on voters and reelection, farmers, and society's confidence in news reporting. Bartlett reports various details including that farm-family cohesiveness increased with the adoption of the radio. He concludes that the radio "made life more appealing", and that its wide and rapid diffusion was due to its universality. Though more explicit than Marcuse [2] in articulating specific social impacts, Barlett's review of the radio's social impacts does not attempt to establish social impact theories or methodology.

By the 1960s however, theoretical underpinning still present in modern sustainable development emerge as researchers focused on what is now considered traditional socio-economics (impacts on health, education, and income). Centering on health and safety, Starr introduced a basic utility-risk framework for evaluating social impacts [15]. To identify the social impacts needed to carry out his framework, Starr indicates that readily available historical data on accidents and health are "stepping stones" for design teams to find social benefits and costs of technology. While true, his guidance is minimal, leaving design teams with complex socio-technical systems that are difficult to decompose from a social impact perspective. The challenge is worsened by accelerated technology diffusion, which Starr identifies as "engineering developments involving new technology... become deeply integrated into the system of society before their impact is evident or measurable" [15].

In the 1980s, quantitative measures of social impact become more present in the literature, centered primarily on technology safety [16, 17]. Kenney [16] introduces a von Neumann-

Copyright © 2023 by ASME

Morgenstern utility function [18], which is executed using a hierarchy of *individual* and *societal* level impacts – all based on fatalities caused by technology. While valuable for its important step toward quantifying social impact, the impacts are so narrowly focused on fatality that the broader meanings of social impact are lost.

Also focused on fatalities, Slovic, et al. widen Kenny's evaluation of social impact by considering pain, suffering, and economic hardship of victims and their family and friends, as well as public distress and economic turmoil that can result from larger scale technological accidents [17]. Their study draws on various empirical studies involving drugs, transportation, weapons, and more to capture societal perceptions of hazard and risk.

In the early 2000s, as the Millennium Development Goals gained popularity [19], social sustainability research began capturing the social impacts of *business practice* in a deeper way (with some extensions to engineering) [20–24]. In their seminal work, Labuschagne and Brent directly address the question "what social criteria must a social impact assessment method consider and measure?" While never explicitly answered due to the unique nature of each enterprise, they review 31 frameworks and guidelines related to Social Impact Assessment (SIA) [20], ranging from the United Nations Commission on Sustainable Development to sustainability metrics proposed by the Institution of Chemical Engineers, to the Dow Jones Sustainability World Indexes Guide. Their review maps frameworks and guidelines to 18 social criteria influenced by business structure and practice. These include the criteria of (i) economic welfare and employment, (ii) community involvement of company, and (iii) fair labor practices. Only one of their identified social criteria definitively relates to engineered products: product responsibility. Only 4 of the 31 frameworks and guidelines mapped to product responsibility.

With these studies in the early 2000s [20–24], an important pattern emerges: the identification of broad social impact categories (e.g., equity), more focused sub-categories (e.g., gender equality), and measurable social indicators (e.g., ratio of average female wage and male wage).

Following these developments to introduce social sustainability measures into business practice and structure, more specific social impact considerations become noticeable in the engineering literature. For example, Rojanomon et al., when choosing run-of-river hydro-power sites in Thailand, explicitly consider the hydro-power project's impact on community member health, education, employment, quality of life, household changes, community changes, use of nearby forest, community perception/attitude toward project, and community support [25]. With this and other project studies [26], we see social impacts as applied to specific engineering projects, but we do not see generalized methods in the early 2000s to guide design teams in identifying social impacts pertinent to their specific project.

In recent years, however, more generalized engineering-centric methods for considering the social impact of engineered products appear in the literature [27], including methods such as *design justice* [28, 29], with its emphasis on equity. Considering 374 engineering papers published between 2012 and 2022, Armstrong et al., observe various trends related to Rainock's 11

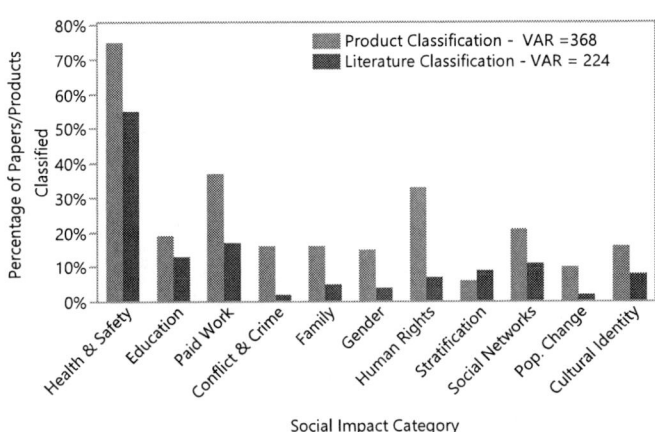

FIGURE 2: QUANTITY OF PAPERS/PRODUCTS FOR EACH SOCIAL IMPACT CATEGORY FROM [27]. NOTICE WIDE VARIATION ACROSS IMPACT CATEGORIES, AND LARGE DIFFERENCES BETWEEN PRODUCT AND LITERATURE CLASSIFICATIONS.

social impact cateogries [1], including large variation in literature attention across the 11 social impact categories, as well as noticeable mismatch between impacts engineered products have and the attention given to those categories in the literature. These observations are illustrated in Figure 2.

Of the 374 papers reviewed by Armstrong et al., 134 of them were coded as including heuristics and/or frameworks meant to guide engineering design teams. Notable among these is the work by Ottosson et al. [6], where design teams are guided to non-obvious impact categories through correlation tables derived by joint and conditional probabilities of various impacts being co-present. While these tables can be used very quickly by any design team, the study is limited to simply guiding teams to one or more of the 11 social impact categories and does not explicitly help teams identify specific social impacts within those categories.

Stevenson et al. [9] provide a more in-depth approach, which allows teams to identify pertinent and specific social impact indicators within the 11 social impact categories. A significant drawback to this approach, however, is that it requires dozens of hours of design work to explore social impact categories, converge upon pertinent ones, and extract the necessary data needed to carry out this approach.

In review, we see the early recognition (1940s) of the engineers' role as technology designer, and the influence of engineered products on society [2, 14]. By the 1960s we see a coalescence around impacts related to socio-economics (heath, education, and paid work) around which basic utility-risk frameworks emerge [15]. More focused mathematical models for social impact, and consequently less comprehensive socially, gain traction in the literature in the 1980s [16, 17], but not without deep criticism regarding their accuracy. A wider perspective on social sustainability appears in the literature in the early 2000s with the explicit articulation of the broad social impact categories and specific social impact indicators popularized with the Millennium Development Goals [20–24], though these were heavily centered on business practice, not engineering. From 2012-2022

Copyright © 2023 by ASME

we see the emergence of numerous engineering studies including social impacts, though we continue to see a noticeable imbalance in how broadly social impacts are considered in the engineering literature [27].

3. DEVELOPMENT OF THE 55 QUESTIONS

Having conducted the study presented in Section 2, we believe there is an opportunity to help design teams identify a wider range of social impact than is facilitated by the current literature. While there are meaningful models and methods for assessing [30] or predicting [13] the social impact of engineered products, these are often resource-intensive to create. Additionally, these models and methods require insights and outlooks not often associated with engineering, such as those related to a stakeholder's cultural identity and heritage and how that might be enhanced or compromised by engineering decisions. These shortcomings often prevent engineers from considering social impact more fully in the engineering process.

With the goal to remove some of these shortcomings, we sought the development of a simple design activity that could be completed in one hour with little training, yet provide meaningful insights through broad consideration of social impact. The activity format we designed is a series of thought questions about the potential social impacts of a product across 11 social impact categories. Because there are 55 such questions, we simply refer to them as *The 55 Questions* or *The 55 Prompt Questions*.

Within an interdisciplinary team of engineers and social scientists, initial brainstorming led to a large set of questions that would take a design team more than one hour to consider. The questions were refined and reduced to five questions for each social impact. The first four questions in each category are detailed and thought-provoking, while the final question in each category encourages thought about other potential impacts not identified by the first questions. This fifth question is essential to illustrate to those using the questions that they do not encompass all possible social impacts.

The questions are designed for teams to spend one minute reading and internalizing each question, and identifying 2 to 3 potential impacts as prompted by the question. The 55 questions are listed below, by social impact category.

Impacts on Health & Safety

1. In what ways could the product improve/change the health of users or aid users in healthy practices?
2. In what ways could the product (unintentional or not) harm users, or have long term or addictive effects?
3. In what ways could the product protect/prevent users from harm or safety hazards?
4. In what ways could the product affect mental and/or emotional health?
5. In what other ways could the product impact health and safety (positive or negative)?

Impacts on Education

6. In what ways could the product provide formal or informal education or skill training to users?

7. In what ways could the product require or provide specialized education/training to use?
8. In what ways could the product be used in the creation, discovery, or sharing of new knowledge?
9. In what ways could the product change access to education by gender, socioeconomic status, age, or race?
10. In what other ways could the product impact education (positive or negative)?

Impacts on Paid Work

11. In what ways could the product change the output, efficiency, or ability to produce a good or service?
12. In what ways could the product create jobs or skilled labor? Could it replace or eliminate jobs?
13. In what ways could the product affect the safety/well-being of employees or protect worker rights?
14. In what ways could the product facilitate the creation, management, or growth of businesses?
15. In what other ways could the product impact paid work (positive or negative)?

Impacts on Conflict & Crime

16. In what ways could the product help detect/prevent/prosecute crime or help ensure fair legal process?
17. In what ways could the product be used for crime such as violence, theft, sexual abuse, substance abuse, or fraud?
18. In what ways could the product expose or protect personal information/privacy?
19. In what ways could the product increase interpersonal conflict/contention (road rage, arguments, litigation)?
20. In what other ways could the product impact conflict and crime (positive or negative)?

Impacts on Family

21. In what ways could the product alter the way family members interact with each other?
22. In what ways could the product strengthen or weaken family ties, including spending time together?
23. In what ways could the product be used simultaneously by or shared between family members?
24. In what ways could the product change family roles (household work, child-rearing, income earning, etc.)?
25. In what other ways could the product impact family (positive or negative)?

Impacts on Gender

26. In what ways could the product amplify gender specific issues (health, sanitation, or gender norms etc.)?
27. In what ways is the product's usability or ergonomics affected by the user's gender?
28. In what ways could the product be sold in gender specific product lines, or marketed to specific genders?
29. In what ways could the product maintain/uphold/challenge gender roles and norms (cultural expectations)?
30. In what other ways could the product impact gender (positive or negative)?

Copyright © 2023 by ASME

Impacts on Human Rights

31. In what ways could the product provide/extend the most basic human rights (water, energy, etc.) to users?
32. In what ways could the product affect access to public services or democratic processes for all people?
33. In what ways could the product affect personal freedoms (religion, assembly, speech)?
34. In what ways could the product influence how human rights are protected/ violations reported/prosecuted?
35. In what other ways could the product impact human rights (positive or negative)?

Impacts on Stratification

36. In what ways could the product be used to distinguish between social or economic groups?
37. In what ways could the product be accessible to all people, or could it decrease access/accessibility?
38. In what ways could the product provide access to goods/services to those who were previously excluded?
39. In what ways could the product be used to improve or degrade one's socioeconomic status?
40. In what other ways could the product impact stratification (positive or negative)?

Impacts on Social Networks & Communication

41. In what ways could the product improve or impair the ability of users to communicate?
42. In what ways could the product change the way people communicate or the content of communication?
43. In what ways could the product facilitate/sustain the creation of new relationships and communities?
44. In what ways could the product provide equitable opportunities for communication and connection?
45. In what other ways could the product impact social networks and communication (positive or negative)?

Impacts on Population Change

46. In what ways could the product generate/produce population change (immigration, move-ins, travel, etc.)?
47. In what ways could the product affect birth rate/death rate?
48. In what ways could the product affect living conditions in an area that would encourage population change?
49. In what ways could the product allow populations to move from place to place seasonally or otherwise?
50. In what other ways could the product impact population change (positive or negative)?

Impacts on Cultural Identity & Heritage

51. In what ways could the product be used to express someone's cultural values, norms, and beliefs?
52. In what ways could the product be in conflict with any cultural norms or religious practices?
53. In what ways could the product move behaviors away from traditional practices?
54. In what ways could the product create/alter/protect culture?
55. In what other ways could the product impact cultural identity and heritage (positive or negative)?

4. METHOD USED TO TEST THE 55 QUESTIONS

In this section, a team-based experiment designed to test the effectiveness of the 55 prompt questions is described. The data evaluation methods used to code and score the output of the team in preparation for statistical significance testing are also described.

Frey and Dym [31] provide a meaningful and convincing argument regarding the necessity of testing design methods to validate claims made by method developers. Consistent with their argument, we designed a controlled experiment to test the effectiveness of the 55 questions. This experiment produced data from which statistical analysis could reveal the influence of the treatment (use of 55 questions), if any.

4.1 Experiment Description

To test the 55 prompt questions, an experiment was conducted that compared the output of 12 randomly-formed design teams. Six teams were randomly chosen to be included in treatment or control groups. During the experiment, teams did not know if they were part of the treatment or control. Treatment groups were instructed to complete a specific activity with the aid of the 55 questions presented in Section 3, while the control groups were asked to complete the same activity without the questions. All teams were given 55 minutes to identify many specific potential social impacts for a specific product evaluated by all teams.

Immediately before the activity began, all teams were given a basic presentation introducing Rainock et. al's 11 social impact categories [1], with examples. After the briefing, teams moved into individual team spaces, which consisted of a whiteboard, a sealed activity packet, and approximately 100 square feet of space to work separated from other teams. The treatment and control groups were in the same large space but were visually separated to prevent ideas or expectations from being shared between the treatment and control.

The sealed activity packet consisted of (i) a short written set of instructions specific to treatment and control groups, (ii) a brief description of the 11 social impact categories for reference during the activity, (iii) the written design brief, which described the product they were to evaluate, and (iv) for the treatment group only, the 55 prompt questions.

At the beginning of the activity, teams were asked to open the activity packet and follow the written instructions. The instructions asked the 6 control teams to identify many specific social impacts of the Global Village Shelter (see Figure 3) using Rainock et al.'s 11 social impact categories as a guide [1]. The 6 treatment teams were asked to do the same, but were given the 55 prompt questions to guide their ideation.

All groups were given 55 minutes to ideate and record possible social impacts of the Global Village Shelter. The control teams were told to spend about five minutes on each of the 11 social impact categories while the treatment groups were told to spend about one minute per question (which is equivalent to five minutes per category). Both groups were encouraged to produce two to three ideas per minute. Ideas were recorded on sticky notes during the experiment and later recorded digitally in a spreadsheet. Importantly, teams acted independently after

Copyright © 2023 by ASME

Design Brief:

Global Village Shelters, made from biodegradable laminated material, are low-cost temporary emergency shelters that can last up to eighteen months. Prefabricated, shipped flat, and requiring no tools to assemble, they are easy to deploy. The first prototypes were sent to Afghanistan and Grenada, and later used in tsunami-hit countries in Asia; Pakistan's Azad Kashmir Province, which was devastated by an earthquake, and to Gulfport, Mississippi; after Hurricane Katrina.

FIGURE 3: DESIGN BRIEF GIVEN TO ALL GROUPS DETAILING THE PRODUCT FOR WHICH SOCIAL IMPACTS WERE TO BE IDENTIFIED.

the instruction to open the design packet was given. While the organizers were present to make observations, they did not interact with the teams and did not answer questions in order to avoid giving any one team additional information or any unfair advantage.

4.2 Product Considered in Experiment

The product selected for this experiment was a modular plastic housing unit called the Global Village Shelter, designed by Ferrara Design inc. with Architecture for Humanity. This product is designed to provide temporary housing for displaced persons such as political refugees or victims of natural disasters.

To introduce participants to this product a simple design brief was included in the sealed activity packet. The design brief is shown in Figure 3. This product was chosen because it can be simply and quickly understood by the participants, who were unfamiliar with this product.

4.3 Participants

The participants in this experiment were undergraduate students at Brigham Young University in Provo, UT, USA, with demographics as shown in Table 1. As shown in Table 2, participants were also asked to comment on their previous experience with design and social impact. Brigham Young University's Institutional Review Board (IRB) approved the role of the participants in the experiment. Participants were recruited before the day of the event using printed fliers and classroom announcements in various engineering courses. No announcements were made by course instructors, nor were participants made to feel that grades

in their courses would be influenced by their participation in the experiment. Participants were monetarily compensated for their time, and provided a meal before the experiment.

TABLE 1: PARTICIPANT DEMOGRAPHICS

Participant Demographic	Treatment	Control
Field of Study (Major)		
Total Participants	20	18
Mechanical Engineering (ME)	16	14
Applied Math (ME Focus)	0	1
Finance	0	1
Chemical Engineering	0	1
Experience Design	0	1
Pre-Mechanical Engineering	1	0
Open	1	0
Manufacturing	1	0
Biology	1	0
Year of University Study		
Year 1 (Freshman)	4	3
Year 2 (Sophomore)	7	6
Year 3 (Junior)	4	4
Years 4 or 5 (Senior)	5	5
Gender		
Female	5	3
Male	15	15

4.4 Method for Evaluating Team Output

We used content analysis [32] initially to code the social impacts identified by the teams participating in the study. Then we rated the quality and novelty of each impact. This was a multi-part process involving:

- Part 1: Sorting impacts by social impact category

- Part 2: Clustering similar impacts together within each team's set of responses

- Part 3: Rating the quality of each identified impact

- Part 4: Rating the novelty of each identified impact

4.4.1 Sorting impacts by social impact category (Part 1). During the test, 1,134 sticky notes were produced by the teams. These impacts were digitally listed in a spreadsheet. To minimize potential sources of bias, this list was disconnected from any identifying information (team number, treatment/control, etc.) and randomized before being given to the sorting team. Following conventional procedures for coding qualitative data, first, two research assistants working individually sorted each impact into one of the 11 social impact categories. These groupings were then compared and conflicts were discussed by a group of two highly experienced researchers in the social impact space, and at least one of the research assistants who did the initial sorting. 122 impacts were removed from consideration because they could not

Copyright © 2023 by ASME

TABLE 2: EXPERIENCE WITH SOCIAL IMPACT AND DESIGN OF STUDY PARTICIPANTS

Survey Question Administered Pre-Ideation	T	C
Design is something I do as a hobby	6	4
I have taken design courses	9	9
My major is design centered	8	9
I have done design at an internship or job	1	3
I do design research	1	0
I know what design is, but have no experience	5	5
I have no knowledge about design	2	0
Social Impacts are a personal interest or hobby	1	3
I have taken classes on Social Impact	2	0
My major is Social Impact centered	1	1
I have done Social Impact work at an internship, job, or volunteer position (outside of church service)	2	2
I know what social impacts are but have no personal experience	9	7
I have no knowledge about Social Impact	8	7

be considered social impacts; rather, they were solely environmental impacts, enterprise-level economic impacts, or they were otherwise unintelligible statements. Other identified social impacts spanned multiple social impact categories and were thus assigned to all applicable categories as separate impacts, thus resulting in a total of 1271 identified impacts, and 1149 intelligible social impacts.

4.4.2 Clustering similar impacts together within each team's set of responses (Part 2). Part 1 of the sorting and evaluation process was completed before Parts 2-4. To remove bias, Part 2 was done independently of Parts 3 and 4. This was done because Part 2 exposes which teams identified which impacts, though it was not disclosed which teams were part of treatment or control groups. Raters who participated in Part 2 did not participate in Parts 3 or 4.

After the results of Part 1 were sorted, two research assistants independently clustered team-identified impacts into supersets that captured the same basic impact, while being blind to which teams were treatment and which were control. For example, the impacts "protection from UV" and "protection from wind" were grouped together into a superset related to "protection from exposure to nature." These were clustered together because the research assistants deemed these impacts to be insufficiently different. While a team identifying 30 potential impacts in a single category may appear impressive, it is ultimately less useful if those impacts are the same or very similar.

4.4.3 Rating the quality of each identified impact (Part 3). Each of the 1271 impacts identified in the experiment were randomized and then rated for their quality in two dimensions: the quality of the impact articulation, and the potential of the identified impact to influence design decision making. All impact quality ratings were made by a single expert reviewer with significant experience in social impact modeling. A single reviewer was used to better ensure that any observed differences

between the control and treatment groups were not artificially induced by differences in reviewer perspectives. Both quality ratings were given based on a scale from 1-5 with 1 being a low quality score and a 5 being a high quality score. Table 3 shows the full rubric used for assigning quality articulation scores while Table 4 shows the rubric for scoring quality of potential influence on product decisions.

4.4.4 Rating the novelty of each identified impact (Part 4). Like Part 3, the novelty of each of the 1271 impacts was rated in two dimensions: the novelty of the impact in-domain, and novelty out-of-domain. In-domain novelty is a measure of how often the rater believes the social impact would be identified within the specified problem domain (in this case, temporary housing). Out-of-domain novelty is a measure of how often the rater believes the impact would be identified outside of the specified problem domain. For each novelty evaluation, a rating of 1-3 was given with 1 being common and 3 being novel. If an identified social impact was rated as common within the specified domain, it was automatically rated as common outside of the specified domain. Table 5 shows the rubric used by the rater during the evaluation. Novelty was rated by the same expert from Part 3, but not at the same time Part 3 was being completed.

5. TEST RESULTS

The data resulting from the tests described in Section 4 were analyzed using common statistical methods. Two general tests were conducted: a test evaluating *differences in mean*, and another evaluating *differences in variation* across control and treatment groups. Because it is possible that the 55 questions cause teams to more evenly identify impacts across all social impact categories (as opposed to focusing on only a few), differences in variance may be present. Therefore, we applied the two-sample Welch's T-test, which allows for unequal variances between control and treatment groups.

The null hypothesis for all T-tests was that the mean values from the control group (C) and treatment group (T) are indistinguishable. For each individual T-test – if found to be true – the null hypothesis means that the treatment had no impact on team performance relative to the design prompt for that test. The null hypothesis for the one-tailed F-Tests carried out in this study was that the variance from the treatment group (T) is not lower than the variance from the control group (C). For each individual F-test – if found to be true – the null hypothesis means that the treatment had no impact on team performance relative to the design prompt for that test. We performed each test at 90%, 95%, and 99% confidence intervals (C.I.). If the p value for the test is greater than 0.10, 0.05, or 0.01, respectively, the null hypothesis cannot be rejected.

Summary statistics and results of the statistical tests are shown in Table 6. Note the bolded and highlighted text in the table, pointing to statistically significant results. As described more fully below, we conclude the following from the statistical testing:

1. The 55 questions (treatment) cause teams to more evenly identify high quality, high variety, high novelty impacts across all 11 social impact categories during an ideation

Copyright © 2023 by ASME

TABLE 3: RUBRIC USED TO RATE QUALITY ARTICULATION OF IDENTIFIED SOCIAL IMPACTS

Score	Description	Example
5	Stated as a viable social impact or states viable impact with identified social impact category (Not necessarily verbatim)	"degradation in health due to lack of sanitation facilities"
4	Stated as a product concept/feature with an obvious and viable social impact or obviously related to a social impact but is not stated as a product concept/feature	"Could create sense of worthlessness as shelters fall apart"
3	Stated as a product concept/feature without an obvious impact or possible inferred secondary impact	"More parties with more people living close together"
2	Poor result in the ideation activity (Gaming the ideation process) or not obviously connected to product	"Family"
1	Stated without enough information to understand or not at all related to social impact	"72 hour kit with shelter" or "Find lighter material"

TABLE 4: RUBRIC USED TO RATE QUALITY OF EXPECTED INFLUENCE OF IDENTIFIED SOCIAL IMPACTS

Score	Description	Example
5	Identified impact will definitely influence product or system decision making	"Degradation in health due to lack of sanitation facilities"
4	Identified impact will probably influence product or system decision making	"Different shelter sizes/combine them for bigger families"
3	Identified impact will possibly influence product or system decision making	"Include volunteer groups to help connect/find people with dispatch teams"
2	Identified impact will probably not influence product or system decision making	"Could detract from rebuilding efforts"
1	Identified impact will definitely not influence product or system decision making	"Boredom"

TABLE 5: RUBRIC USED FOR ASSIGNING VARIETY SCORES

Domain	Score	Frequency of Impact being referenced inside or outside domain
In Domain	3	Never seen (Less than 1% of the time)
	2	Rarely seen (Less than 5% of the time)
	1	Common (More than 5% of the time)
Out of Domain	3	Never seen (Less than 1% of the time)
	2	Rarely seen (Less than 5% of the time)
	1	Common (More than 5% of the time)

session, as opposed to focusing too heavily on a subset of impact categories.

2. The 55 questions do this without reducing the total quantity of impacts identified. In other words, using the 55 questions does not reduce the quantity of high quality, high novelty, high variety impacts identified.

These conclusions about the 55 questions are meaningful since they indicate that the questions promote team consideration of social impacts more evenly than is (i) found in current industry practice [7], (ii) observed in commercial products [6], and (iii) treated in the sustainability literature [27]. Importantly, the 55 questions are shown to do this without compromising team productivity when ideating potential social impacts for a given product.

In the remainder of this section, each of the data sets shown in Table 6 are described in greater detail.

5.1 First Test: Total Quantity of Impacts

The first column of data in Table 6 is labeled *Total Quantity of Impacts*. This data is total quantity of impacts identified in each category by the control groups summed (C), and in each category by the treatment groups summed (T). A bar chart of that data is shown in Figure 4. The summary statistics indicate that there is no statistically significant difference between the control and treatment groups when considering total quantity. Notice that while it can be observed from Table 6 that the control group consistently identified a larger total number of impacts across all three tests, these observations are not statistically significant as demonstrated by the T tests. This is an important finding because it indicates that although the treatment group was required to read, internalize, and use as inspiration the 55 prompt questions, this did not observably reduce team output.

Also, it can be seen for *Total Quantity of Impacts* that the

TABLE 6: DATA FROM EXPERIMENT WITH STATISTICAL TEST RESULTS.

	Total Quantity of Impacts		Quantity of High Quality and High Variety Impacts		Quantity of High Quality, High Variety, and High Novelty Impacts	
	C[2]	T[3]	C	T	C	T
No Impact (removed[1])	62	60	2	2	0	0
Heath & Safety	131	95	44	27	1	4
Education	42	40	16	22	4	6
Paid Work	49	59	17	20	2	2
Conflict & Crime	67	48	35	28	5	5
Family	46	63	23	29	3	3
Gender	21	31	15	15	3	4
Human Rights	25	31	9	9	1	1
Stratification	57	49	26	17	7	3
Social Networks & Communication	63	69	17	19	9	3
Population Change	29	49	11	14	4	3
Cultural Heritage & Identity	49	36	24	16	17	6
Total	641	630	239	218	56	40
Total (*No Impact* removed)	579	570	237	216	56	40
Mean	52.64	51.82	21.55	19.64	5.09	3.64
Variance	898.05	362.36	109.67	40.45	21.49	2.45
T Value	0.076		0.517		0.986	
T Critical (90% C.I.)	1.337		1.337		1.337	
T Critical (95% C.I.)	1.746		1.746		1.746	
T Critical (99% C.I.)	2.583		2.583		2.583	
p value (T test)	0.940		0.306		0.172	
Conclusion (T test)	Treatment has no negative influence on mean		Treatment has no negative influence on mean		Treatment has no negative influence on mean	
F Statistic	2.48		2.711		8.756	
F Critical (90% C.I.)	2.32		2.32		2.32	
F Critical (95% C.I.)	2.98		2.98		2.98	
F Critical (99% C.I.)	4.85		4.85		4.85	
p value (F test)	**0.092**		**0.066**		**0.001**	
Conclusion (F test)	Treatment significantly **reduces variation**		Treatment significantly **reduces variation**		Treatment significantly **reduces variation**	

1: Deemed purely an environmental impact, or enterprise economic impact. 2: C represents control. 3: T represents treatment.

F test indicates no statistically significant difference between the control and treatments groups, when using a 95 or 99% C.I.. For those C.I.s, this means that by *total quantity*, the 55 questions did not produce a more even consideration of social impacts. The table does indicate, however, that when using a 90% C.I., the 55 questions do produce a statistically significant difference in the treatment group output.

5.2 Second Test: Quantity of High Quality, High Variety Impacts

The second column of data in Table 6 is labeled *Quantity of High-Quality, High-Variety Impacts*. This column is a filtered subset of the data from the first column. It was filtered before the statistical tests were performed and is based on this rationale: for any ideation activity, it is common to expect a large number of low-quality ideas. It is also common to expect repeated ideas

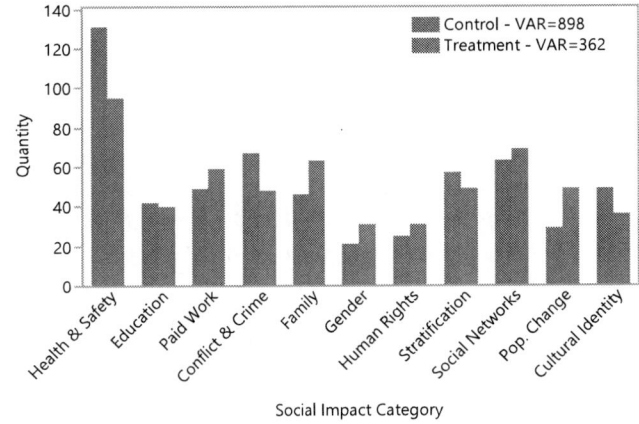

FIGURE 4: TOTAL QUANTITY OF IMPACTS

Copyright © 2023 by ASME

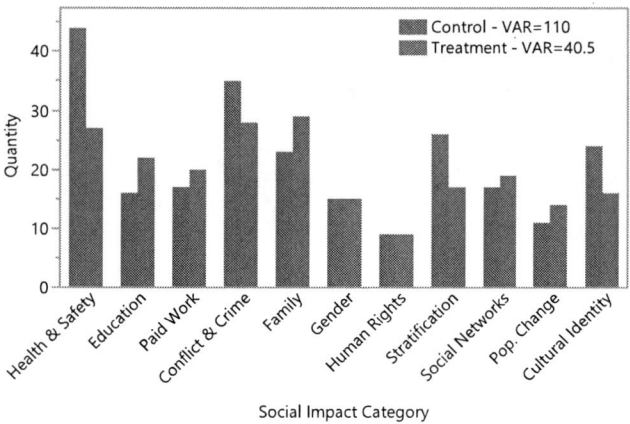

FIGURE 5: QUANTITY OF HIGH-QUALITY AND HIGH-VARIETY IMPACTS

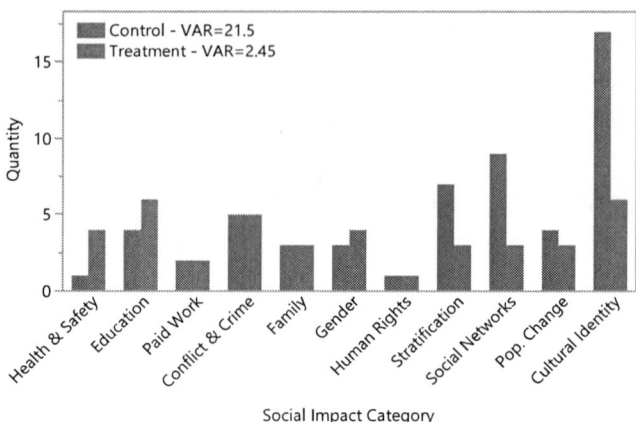

FIGURE 6: QUANTITY OF HIGH-QUALITY, HIGH-VARIETY, HIGH-NOVELTY IMPACTS

to appear. The filtering simply removed low-quality ideas and duplicate ideas. Therefore, the remaining set is considered to be high quality and high variety. A bar chart of this filtered set is shown in Figure 5.

An aggregate quality score was determined by simply using the minimum quality rating between *quality of articulation* and *influence on decision making* ratings made by experts. Such an aggregation prevented well articulated impacts that have little influence on decision making from receiving a high-quality rating. Low-quality ideas were determined by removing any idea with an aggregate quality score below 3.

The variety was determined using the method described in Section 4 and by simply counting the number of unique impacts after duplicates were clustered together.

The T test indicates that there is no difference in the number of remaining impacts (after filtering) between the control and treatment groups. But the F test indicates that the 55 question treatment produced less variation in the number of high quality, high variety ideas across the 11 social impact categories – when considering a 90% C.I.. Thus, with 90% confidence, we conclude that the 55 question treatment causes teams to identify high quality, high variety impacts more evenly across the 11 social impact categories.

5.3 Third Test: Total Quantity High Quality, High Variety, High Novelty Impacts

The third column of data in Table 6 is labeled *Quantity of High Quality, High Variety, High Novelty Impacts*. This column is another layer of filtering compared to the second column. Here, low novelty impacts are filtered out. Thus the remaining impacts are considered high-quality, high-variety, and high-novelty. A bar chart of this full filtered set is shown in Figure 6.

Novelty was rated by an expert reviewer using a 3-point scale as described in Section 4. All impacts receiving a rating of 2 or 3 in the out-of-domain criteria were retained. We chose this particular filtering because it kept only those impacts that truly novel both in and out of domain. As with the second test, we see no difference in means, but we do see a difference in variation

between control and treatment groups. Therefore we conclude, with 99% confidence, that the 55 question treatment causes teams to identify high quality, high variety, high novelty impacts more evenly across the 11 social impact categories when compared to the control group.

6. DISCUSSION AND CONCLUSIONS

In this paper, we have introduced 55 prompt questions for design teams to use when trying to identify social impacts for a given products. We created these 55 prompt questions with the intent to help design teams spend one hour identifying meaningful social impacts across a wide range of social impact categories. With at least 90% confidence and up to 99% confidence, the summary statistics show that compared to a set of control groups that did not have the 55 questions, the 55 questions are effective at helping teams consider social impact more evenly across Rainock et al.'s [1] 11 social impact categories, as opposed to focusing on only a subset of impacts. Although the 55 questions require additional time to read, internalize, and use as prompts during the ideation, this additional requirement does not change the quantity of team output in a statistically significant way. Ultimately this means the 55 questions help the team more deeply consider the wide range of impacts from Rainock et al. [1] without negatively affecting team performance relative to the three tests shown in Table 6.

To put these results into visual perspective, and as an anecdote, consider the output of 4 teams among the 12 who participated in this paper's experiment. A single control team and a single treatment team are compared in Figures 7-8. The comparison shown in Figure 7 shows a more even consideration of impacts by the treatment group, compared to the control group which has focused more heavily on Health & Safety, Conflict & Crime, and Stratification compared to the treatment. To be fair, Figure 8 represents a different potential result. Here the control and the treatment are statistically no different in their means nor their variances. While this is a possible outcome, it is important to observe that using the 55 prompt questions did not negatively affect team performance.

Copyright © 2023 by ASME

228

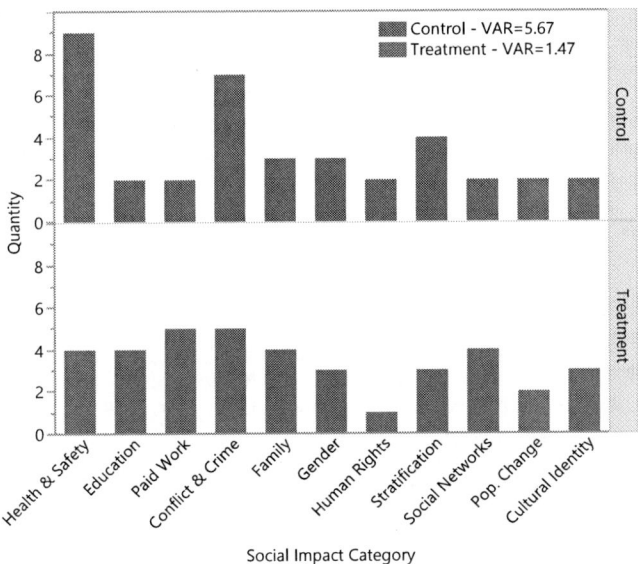

FIGURE 7: OUTPUT OF MEANINGFUL IMPACTS FROM TEAM 10 (CONTROL) COMPARED TO THAT OF TEAM 5 (TREATMENT)

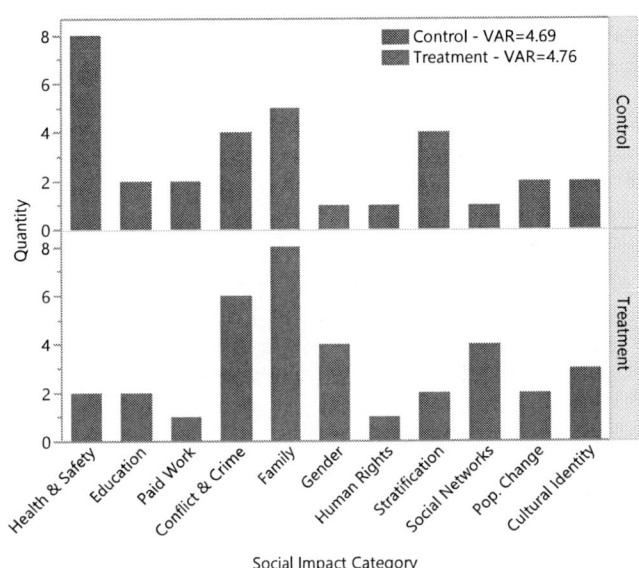

FIGURE 8: OUTPUT OF MEANINGFUL IMPACTS FROM TEAM 1 (CONTROL) COMPARED TO THAT OF TEAM 2 (TREATMENT)

Statistically speaking, and when considering all of the control group data together and all of the treatment group data together, neither the control nor the treatment produced more concepts across categories (difference in means test), but the treatment group produced less variation in quantity of impacts identified across the impact categories (difference in variance test).

Aside from the summary statistics, the participant survey issued directly after the ideation portion of the experiment, and completed by each participant, indicates that the control and treatment groups are indistinguishable in all ways except three:

1. In the time allotted, what percentage of all potential Social Impacts of the Product do you think your team identified? Control Groups: 62%. Treatment Groups: 71%.

2. To what extent of all the design actions your team generated are feasible? Control Groups: 80.5%. Treatment Groups: 83.9%.

3. If you were to implement all design actions you generated, how much would they improve the social impact of the product? Control Groups: 72.2%. Treatment Groups: 79.4%.

The 55 questions and the tests used to validate their effectiveness are not without limitations. The questions themselves are limited to 4 specific questions per social impact category and one generic question per category. This small set of questions may unduly limit the social impacts considered to primarily those related to the questions themselves. Additionally, questions were only developed and tested for Rainock et al.'s [1] 11 social impact categories. Other categories not part of those 11 may be important for a given project. In the test performed, the effectiveness of each question was not evaluated. Such testing could lead to question refinement that may have a measurable effect.

Regarding the testing performed, the results are potentially limited to those who match the demographic of the test subject – undergraduate engineering students. A different result may be observed if tested with experienced professionals. The test results are also limited to the scenario tested, which is that social impacts were identified for a project the test subjects were not designing themselves. Different observations may be made if the test subjects were more intimately involved in the design of the product under consideration.

Based on the findings presented in this paper, coupled with our experience in social impact modeling for engineering projects, we recommend teams spend one hour using the 55 prompt questions to identify a set of potential social impacts for the product with which they are involved. We believe that using these questions will help teams avoid fixation on a subset of impacts and thus consider social impact more holistically.

ACKNOWLEDGMENTS

The authors gratefully acknowledge the contributions of Dr. Phillip Stevenson and engineer Andrew Armstrong for their early-stage contributions to 55 prompt questions. Likewise we acknowledge social scientists Johnny Cope and Rachel Samsion for their careful review of the 55 prompt questions before they were used in testing. We also acknowledge the generous funding of Crocker Ventures.

REFERENCES

[1] Rainock, Meagan, Everett, Dallin, Pack, Andrew, Dahlin, Eric C and Mattson, Christopher A. "The social impacts of products: a review." *Impact assessment and project appraIsal* Vol. 36 No. 3 (2018): pp. 230–241.

[2] Marcuse, Herbert. "Some social implications of modern technology." *Zeitschrift für Sozialforschung* Vol. 9 No. 3 (1941): pp. 414–439.

[3] Mattson, Christopher A, Pack, Andrew T, Lofthouse, Vicky and Bhamra, Tracy. "Using a product's sustainability space

Copyright © 2023 by ASME

as a design exploration tool." *Design Science* Vol. 5 (2019): p. e1.

[4] Cf, ODDS. "Transforming our world: the 2030 Agenda for Sustainable Development." *United Nations: New York, NY, USA* (2015).

[5] Burleson, Grace, Lajoie, Jason, Mabey, Christopher, Sours, Patrick, Ventrella, Jennifer, Peiffer, Erin, Stine, Emma, Stettler Kleine, Marie, MacDonald, Laura, Austin-Breneman, Jesse, Javernick-Will, Amy, Winter, Amos, Lucena, Juan, Knight, David, Daniel, Scott, Thomas, Evan, Mattson, Christopher and Aranda, Iana. "Advancing sustainable development: Emerging factors and futures for the engineering field." *Sustainability* Vol. 15 No. 20 (2023): p. 7869. DOI 10.3390/su15107869.

[6] Ottosson, Hans J, Mattson, Christopher A and Dahlin, Eric C. "Analysis of perceived social impacts of existing products designed for the developing world, with implications for new product development." *Journal of Mechanical Design* Vol. 142 No. 5 (2020).

[7] Pack, Andrew T, Rose Phipps, Emma, Mattson, Christopher A and Dahlin, Eric C. "Social impact in product design, an exploration of current industry practices." *Journal of Mechanical Design* Vol. 142 No. 7 (2020).

[8] Armstrong, Andrew G, Mattson, Christopher A, Salmon, John L and Dahlin, Eric C. "FMEA-Inspired Analysis for Social Impact of Engineered Products." *International Design Engineering Technical Conferences and Computers and Information in Engineering Conference*, Vol. 85390: p. V03BT03A017. 2021. American Society of Mechanical Engineers.

[9] Stevenson, Phillip D, Mattson, Christopher A and Dahlin, Eric C. "A method for creating product social impact models of engineered products." *Journal of Mechanical Design* Vol. 142 No. 4 (2020).

[10] Thomas, Evan, Wilson, Daniel, Kathuni, Styvers, Libey, Anna, Chintalapati, Pranav and Coyle, Jeremy. "A contribution to drought resilience in East Africa through groundwater pump monitoring informed by in-situ instrumentation, remote sensing and ensemble machine learning." *Science of The Total Environment* Vol. 780 (2021): p. 146486.

[11] Stringham, Bryan J and Mattson, Christopher A. "Design of remote data collection devices for social impact indicators of products in developing countries." *Development Engineering* Vol. 6 (2021): p. 100062.

[12] Kiesling, Elmar, Günther, Markus, Stummer, Christian and Wakolbinger, Lea M. "Agent-based simulation of innovation diffusion: a review." *Central European Journal of Operations Research* Vol. 20 (2012): pp. 183–230.

[13] Mabey, Christopher S, Armstrong, Andrew G, Mattson, Christopher A, Salmon, John L, Hatch, Nile W and Dahlin, Eric C. "A computational simulation-based framework for estimating potential product impact during product design." *Design Science* Vol. 7 (2021): p. e15.

[14] Bartlett, Kenneth G. "Social impact of the radio." *The Annals of the American Academy of Political and Social Science* Vol. 250 No. 1 (1947): pp. 89–97.

[15] Starr, Chauncey. "Social benefit versus technological risk: what is our society willing to pay for safety?" *Science* Vol. 165 No. 3899 (1969): pp. 1232–1238.

[16] Keeney, Ralph L. "Evaluating alternatives involving potential fatalities." *Operations Research* Vol. 28 No. 1 (1980): pp. 188–205.

[17] Slovic, Paul, Lichtenstein, Sarah and Fischhoff, Baruch. "Modeling the societal impact of fatal accidents." *Management Science* Vol. 30 No. 4 (1984): pp. 464–474.

[18] Jv, Neumann, Morgenstern, Oskar et al. "Theory of games and economic behavior." (1944).

[19] Sachs, Jeffrey D. "From millennium development goals to sustainable development goals." *The lancet* Vol. 379 No. 9832 (2012): pp. 2206–2211.

[20] Labuschagne, Carin and Brent, Alan. "Social indicators for sustainable project and technology life cycle management in the process industry (13 pp+ 4)." *The International Journal of Life Cycle Assessment* Vol. 11 (2006): pp. 3–15.

[21] Labuschagne, Carin, Brent, Alan C and Van Erck, Ron PG. "Assessing the sustainability performances of industries." *Journal of cleaner production* Vol. 13 No. 4 (2005): pp. 373–385.

[22] Labuschagne, Carin, Brent, Alan C and Claasen, Schalk J. "Environmental and social impact considerations for sustainable project life cycle management in the process industry." *Corporate Social Responsibility and Environmental Management* Vol. 12 No. 1 (2005): pp. 38–54.

[23] Hutchins, Margot J and Sutherland, John W. "An exploration of measures of social sustainability and their application to supply chain decisions." *Journal of cleaner production* Vol. 16 No. 15 (2008): pp. 1688–1698.

[24] Bai, Chunguang and Sarkis, Joseph. "Integrating sustainability into supplier selection with grey system and rough set methodologies." *International Journal of Production Economics* Vol. 124 No. 1 (2010): pp. 252–264.

[25] Rojanamon, Pannathat, Chaisomphob, Taweep and Bureekul, Thawilwadee. "Application of geographical information system to site selection of small run-of-river hydropower project by considering engineering/economic/environmental criteria and social impact." *Renewable and Sustainable Energy Reviews* Vol. 13 No. 9 (2009): pp. 2336–2348.

[26] Sabini, Luca, Muzio, Daniel and Alderman, Neil. "25 years of 'sustainable projects'. What we know and what the literature says." *International Journal of Project Management* Vol. 37 No. 6 (2019): pp. 820–838.

[27] Armstrong, Andrew G, Suk, Hailie, Mabey, Christopher S, Mattson, Christopher A, Hall, John and Salmon, John L. "Systematic Review and Classification of the Engineering for Global Development Literature Based on Design Tools and Methods for Social Impact Consideration." *Journal of Mechanical Design* Vol. 145 No. 3 (2023): p. 030801.

[28] Costanza-Chock, Sasha. *Design justice: Community-led practices to build the worlds we need.* The MIT Press (2020).

[29] Das, Madhurima, Roeder, Gillian, Ostrowski, Anastasia K, Yang, Maria C and Verma, Aditi. "What Do We Mean

Copyright © 2023 by ASME

When We Write About Ethics, Equity, and Justice in Engineering Design?" *International Design Engineering Technical Conferences and Computers and Information in Engineering Conference*, Vol. 86267: p. V006T06A036. 2022. American Society of Mechanical Engineers.

[30] Petti, Luigia, Serreli, Monica and Di Cesare, Silvia. "Systematic literature review in social life cycle assessment." *The International Journal of Life Cycle Assessment* Vol. 23 (2018): pp. 422–431.

[31] Frey, Daniel D and Dym, Clive L. "Validation of design methods: lessons from medicine." *Research in Engineering Design* Vol. 17 (2006): pp. 45–57.

[32] Krippendorff, Klaus. *Content analysis: An introduction to its methodology*. Sage publications (2018).

[33] Sewell, William H et al. "Construction and Standardization of a Scale for the Measurement of the Socio-Economic Status of Oklahoma Farm Families." (1940).

[34] Sarkis, Joseph, Helms, Marilyn Michelle and Hervani, Aref A. "Reverse logistics and social sustainability." *Corporate social responsibility and environmental management* Vol. 17 No. 6 (2010): pp. 337–354.

[35] Armstrong, Andrew G, Mattson, Christopher A and Lewis, Randy S. "Factors leading to sustainable social impact on the affected communities of engineering service learning projects." *Development Engineering* Vol. 6 (2021): p. 100066.

Proceedings of the ASME 2023
International Design Engineering Technical Conferences and
Computers and Information in Engineering Conference
IDETC-CIE2023
August 20-23, 2023, Boston, Massachusetts

DETC2023-116756

THE GIVING GARDEN: REALIZING COMMUNITY AND FOSTERING A CONNECTION TO THE LAND THROUGH CO-DESIGN IN DULUTH, MINNESOTA

Austin Konrath
University of Minnesota Duluth
Department of Civil Engineering
Duluth, Minnesota, USA

Abigail R. Clarke-Sather*
University of Minnesota Duluth
Department of Mechanical and
Industrial Engineering
Duluth, Minnesota, USA,
abbie@d.umn.edu*

Regina Laroche
St. Mark Giving Garden
Co- Founder
Duluth, Minnesota, USA

Morgan Bliss
St. Mark Giving Garden Committee Member
Duluth, Minnesota, USA

Rumbidzai Masawi
St. Mark Giving Garden Lead Gardener
Duluth, Minnesota, USA

ABSTRACT

Communities with Healthy Food Priority Areas (HFPA) across the U.S. decreased the quality of life and well-being of individuals. Duluth, Minnesota suffers from several HFPAs. In response, a local organization, "the St. Mark Giving Garden", promotes healthy living, centering African American resiliency and traditional identity through the communal production of culturally relevant foods. The organization started a community garden in an existing garden site to remedy the food disparities in the Central and East Hillside neighborhoods of Duluth.

After success in the 2022 growing season, the organization looked to expand into a new, larger site. The St. Mark Giving Garden utilized co-design to develop this future community garden site. A leadership committee was organized including members from multiple community stakeholders and partner organizations. This committee defined the goals of the project and then began holding community design sessions to begin the development process. Co-design places the end-user at the foundation of a design project. In this community garden project, the Central and East Hillside neighborhoods became an integral part of the design process; community members presented concerns, desires, and additions to elevate the project and completely benefit the community.

Co-design provides a powerful opportunity for generating an effective product that holistically meets all of the end-users needs. This project compared a community garden design draft produced without co-design and a design produced with co-design. The co-designed product is superior in several ways relative to the preliminary design: it maximized the type of space by decreasing the area of communal space by 30% and

increasing the area of exercise and performance space by 8000% and 600% respectively; minimized the number of unnecessary elements, such as eliminating more than 50% of fencing; and increased the number of connections between elements by 200%. These improvements are more reflective of the values and needs of the community and assist in better serving the community.

Key Words: Community gardens, inclusive design, co-design, engineering for global development (EGD), disenfranchised communities.

1. INTRODUCTION

Ensuring the longevity and well-being of any population begins with providing access to quality food. In 2012, the USDA published a study highlighting the more than 6,500 Healthy Food Priority Areas (HFPA), (poor neighborhoods with restricted access to affordable groceries), around the US and their negative impacts on the people living in them [1], [2]. Living in an HFPA often results in a diet composed mostly of processed foods, absent of healthy produce. This diet then results in higher incidences of obesity, cardiovascular disease, diabetes, and other weight-related conditions [1]. Healthy Food Priority Areas are caused by factors such as redlining, economic fragmentation, and the worsening economics of operating a small business [3]; the lack of healthy food access therefore disproportionately affects minoritized and poor communities further contributing to societal inequity [1], [3].

While being an urban area with a population of over 80,000 people, the city of Duluth offers a very rural feel

Copyright © 2023 by ASME

relative to many comparably sized cities. Duluth sits in northeastern Minnesota, USA at the edge of many pristine wildernesses such as the Boundary Waters Canoe Area and the Superior National Forest. Furthermore, due to its northern latitude and proximity to Lake Superior, Duluth experiences very mild, short summers and harsh, long winters. The short growing season makes local agriculture a very difficult endeavor.

Duluth is no exception when it comes to food disparities. 17.3 percent of households in Duluth sit below the poverty line, nearly double the Minnesota state average percentage of 9.5% [4]. This level of poverty consequently creates a large incidence of disenfranchised communities. These communities struggle with limited access to healthy foods and other economic barriers which detract from quality of life [3].

The residents of the Central and East Hillside communities, in particular, are subject to several disparities relative to other Duluthians. Due to generations of systemic social neglect, this community's relationship with the land, healthy food, and neighbors has been severed, leading to physical, mental, and spiritual health crises in these communities. The ultimate culmination of these crises is an 11-year shorter life expectancy for individuals in these communities compared to surrounding neighborhoods that have access to healthy resources [5].

To rectify this disparity and rejuvenate the Central and East Hillside communities, the St. Mark African Methodist Episcopal (AME) Church broke ground on a community garden named, "The Giving Garden" in 2020. In the 2022 growing season, the 800-square-foot space produced more than 2000 pounds of produce, which it donated to members of the community in need of healthy food [5]. Furthermore, the Giving Garden cultivates as much human connection as it does nutritional mass. The soil of the garden absorbs sunlight and nutrients, but it also absorbs the music, stories, gratitude, and voices of the youth and elders who cherish the space [6]. From this soil, the seeds of a cross-cultural community are planted. Giving Garden programming attempts to accomplish this with community events such as Juneteenth Celebrations, Garden Closing Thanksgiving ceremonies, anti-racism training for volunteers, and youth programs such as Hiking in Harmony (a youth camp meant to cultivate a connection to nature and to the other campers) and Youth Employment Services (YES) events.

However, this space only yields a fraction of the healing that the Hillside communities require. The Giving Garden is therefore looking to expand into an adjacent neglected city park to maximize its positive contributions to the community. The land of the Hillside Sport Court Park has long stood dormant and no longer benefits the community fully. Expanding into this space will not only increase the available growing space by over a factor of 10, but will also rejuvenate the land and strengthen community kinship.

The community knows what necessities they are lacking and what they would like in their community spaces; the members of the community are therefore foundational to the design process. Co-design is the integration of the stakeholder in the design process to create highly usable and accessible spaces. It acknowledges the diversity of the parties involved and focuses on opening forums where collaboration may occur [7], [8]. Under co-design, the designer is simply an illustrator who brings the community's ideas to life. This technique allows those not well-versed in design techniques and approaches to collaborate constructively in the design process [7].

Co-designed projects more closely reflect the expectations and requirements of end-users, ultimately leading to increased use of the end product [9]. Co-design provides the opportunity for allowing advocacy of the community's needs and bringing awareness to problems that plague communities. The designer is then empowered to incorporate solutions to those issues into the design to create a safe and effective design [8]. Furthermore, co-design requires regular interaction with the community, resulting in the emergence of empathy between the user and designer [7]. This empathy allows their relationship to transcend a designer-user relationship and become a mutually beneficial partnership. This connection then gives birth to an appreciation for the diversity of the community that the designer can harness to integrate the community's identity into the final product [7].

Previous community garden co-design projects laid the groundwork for the planning process of this project. The first step in co-designed community garden projects in Italy [8], [9], Chile [10], [11], and Ethiopia [12] was to define the goals of the project for presenting to the community. Visioning events and interviews were then held to begin identifying means of benefiting the community and barriers the project might encounter [11]. In each project, leaders and members of the community were interviewed to receive their input on the project. Additionally, each stakeholder group was consulted to ensure equal representation in the design process [10], [12]. Specific focus was put on reaching every involved party in the community through many community-visioning events at a variety of venues. Co-design must honor the intrinsic diversity of each stakeholder group in order to produce a design that equally benefits all members of the community [12].

After the visioning stage comes the design and revision stage. At this stage, stakeholders are presented with a visual/physical representation of the design for their input. It is imperative that during this portion of co-design, elements that benefit the community are concretely implemented into the design, and significant research is conducted to turn the barriers of the project into benefits [11]. Often barriers are transformed into positives through a multidisciplinary approach which can be achieved through the assistance of the community's local knowledge and expertise [8]. A powerful outcome of this multidisciplinary approach is the redefining of the mission of the project. When tackling different barriers in the co-design process, the community constantly redefines and augments the goal of the project to bring about the greatest good in the community. The evolution of the goal of the project results in a superior product compared to the initial vision and is expected as stakeholders' ideas morph over time [9].

A common theme among these projects was the dedication of a significant amount of time to planning and community

Copyright © 2023 by ASME

design events. Co-design requires an enormous time investment in order to reap its benefits [12]. The reviewed projects had great difficulty generating a significant amount of community participation; many events had to be held to collect feedback from a representative sample of the community. Furthermore, the designers found it extremely tedious to work with the community. The communities the projects were serving were not well versed in community garden design so significant effort had to be made to make the community design events accessible to everyone [8]. While this is cumbersome, the projects concluded that this diligence in creating accessibility for all helped every voice be heard within the design process [9].

Co-design applies to a wide variety of projects. Many designers often neglect to incorporate the end user, but by reuniting the end user with the design process, a higher-quality product may be created. This study seeks to provide an example of the implementation of co-design principles in the design of a community garden for a disenfranchised neighborhood. This project aims to compare a design produced with community input and a design produced without community input to quantify the benefits of co-design.

2. METHODS

The first author partnered with the St. Mark Giving Garden to carry out and evaluate the co-design process in their community garden expansion project. To achieve proper co-design practices, research on implementing co-design and previous co-design projects was initially conducted using the Kathryn A. Martin Library and Google Scholar. The keywords, "co-design", "community garden", and "user-centered design" were primarily used.

Prior to this study, the Giving Garden committee produced a design utilizing informal input from only Giving Garden committee members with the second author. This provided an opportunity to compare a limited initial draft with a final design produced through co-design.

2.1 Giving Garden Co-Design Process

Integrating co-design began with the leadership of the Giving Garden. The Giving Garden committee, consisting of leaders in the community, students, and representatives for the project's partners, met weekly online over Zoom to openly discuss goals, guiding principles, and requirements from September 1st - December 15th before opening the forum to the community. Through these meetings, the Giving Garden established the ultimate goal of the project to (1) expand the outreach of the Giving Garden; (2) increase space for intergenerational and community connection; and (3) revive the underutilized land (an overgrown baseball field and soccer pitch). This allowed the project to gain direction, enabling all involved parties to be on the same page as the project progressed.

The Giving Garden is not an isolated project: it could not exist without its partnerships with numerous regional organizations (Figure 1). In total, the Giving Garden benefits

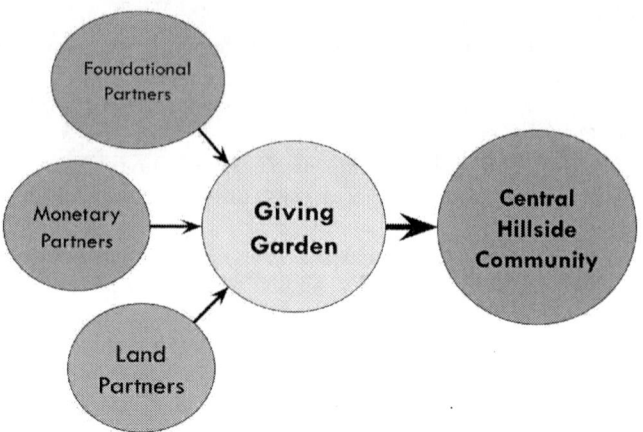

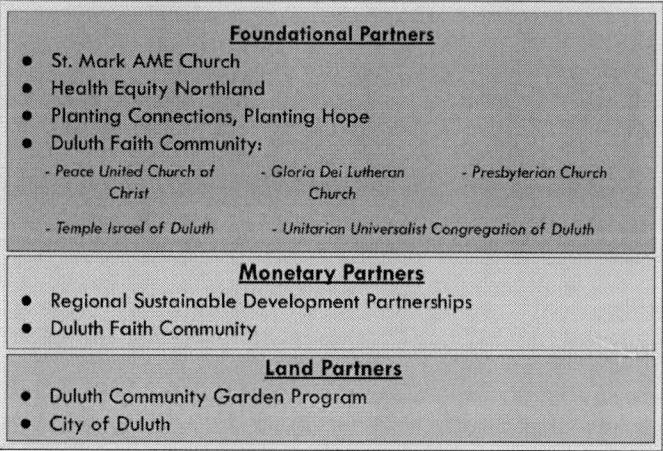

FIGURE 1: GIVING GARDEN STAKEHOLDER FRAMEWORK

from the contributions of more than 30 entities in the Duluth and Minnesota area. Each contributes to a different aspect of the Giving Garden mission: some offer monetary support through grants and donations, some provide personnel and event support through volunteers and programming, and others supply the land and space for the Giving Garden to accomplish its goals. A diagram detailing the web of the five primary stakeholders in the Giving Garden can be found in Figure 1.

Once the goals of this project were established, a community and stakeholder listening session was conducted utilizing the "World Cafe Method" - a simple, yet effective, means of hosting a large group of people in open discussion [13]. In a world cafe, individuals are split into small groups, typically no more than 5 people, to create a welcoming environment where everyone can contribute freely. A host welcomes the community members, presents a brief introduction of the project, and explicitly states the overall purpose of the event. Each table has a host who poses a guiding question to their group for discussion. Participants are encouraged to take notes, draw, use color, or express their ideas by any means they deem appropriate. Once the discussion is complete, participants move to another table to discuss a

Copyright © 2023 by ASME

different element of the project [13]. Collection of visual ideas is a necessity; graphic recording makes ideas accessible for all participants while also providing the opportunity for further expansion of those ideas [13], [14].

The world cafe event was held on September 24, 2022, in the current Giving Garden. Four Giving Garden youth leaders led 15 individuals in a design discussion. In addition, several representatives from foundational and monetary partners were present at the event to provide their thoughts and concerns on the project. This event provided valuable insight into the expectations of community members and stakeholders; these insights allowed the drafting of a new community garden design to begin.

Much of the design took place utilizing computer-aided design software, specifically Solidworks version 2022-2023 and AutoCAD. Creating a visual representation that collects all of the ideas generated from the community assists in significantly progressing the project. This is an essential checkpoint in the design process as the community is able to compare the design with their expectations/vision of the project. From here, the designer and the community can collaborate to identify elements of the design that match expectations or need improvement.

Once the new draft was completed it was presented to 12 Giving Garden committee members for feedback in early December 2022 over the course of 3 weekly meetings. Feedback was given verbally and through email.

From this, the project was refined according to the vision of the committee, including ideas on various means of deepening the impact of this project. The updated design was then presented to the Central and East Hillside community and invested organizations at St. Mark AME church on February 5th, 2023 to bring increased awareness to the project and solicit constructive suggestions. Twenty-five individuals were present in person and several stakeholders were present via a livestream. Concerns and feedback among community members dealt with accessibility, bathroom access, and possible winter programming. These concerns were then addressed in a revised draft.

Furthermore, a one-page flier was developed with a description of the design. The flier was hand distributed to 400 households in the Central and East Hillside neighborhoods and sent to stakeholders to spread awareness and generate support.

Community support is essential for any update to community space; however, community support is insignificant if approval from the governing body, in this case, the city, is not acquired. The Giving Garden submitted its draft to the City of Duluth for approval at the end of October 2022. Importantly, for major projects such as this, the city of Duluth hosts its own community assessment events to gauge interest and assistance in designing the project. This will allow the design to further evolve as the city of Duluth is able to invite a much wider variety of individuals to these discussions and encourage the sharing of their opinions more than a Giving Garden-led discussion could.

2.2 Design Assessment Method

Finally, to measure the benefits of co-design, two methods were used, (1) an anonymous satisfaction survey was conducted to measure community satisfaction with each design and (2) a feature-based evaluation technique was modified from [15] to compare the preliminary design with the current design. This technique specifically focuses on comparing the topology and number of features of each draft and measuring the change in geometry of each element.

The topology of elements plays a large role in construction and significantly influences the interaction of product elements [15]. Our project, therefore, defined topology goals to be maximizing the connections between elements and eliminating unnecessary barriers that restrict the flow of movement through the park. The topology of each design was evaluated through a network diagram that visually represented the evolution of the design and how each element was connected. This diagram enabled easy analysis of the number of connections and identification of barriers present in the designs.

Optimizing the number of each feature is essential in maximizing the efficiency of a product [15]. To compare the features of each design, the number and type of features of each design were tallied and recorded. This optimization required determining the community's opinion on the number of each element and determining whether to increase or decrease the number of each feature.

Lastly, comparing the geometry of each design was conducted similarly to the feature analysis. The designs were compared via the measurement of the total area of each design and the percentage of the total area each type of area made up. This is crucial because the geometry of a product plays an important role in determining functionality [15]. Allocating an appropriate amount of space to each type of area enables the design to attend to all of the community's needs.

The survey that was administered to the Central and East Hillside community and stakeholders was expected to be utilized to validate the changes to the second design. The survey contained an image of the preliminary design and an image of the current design. Potential respondents were asked to rate how completely each design satisfied the requirements of the community on a Likert scale of one to seven where 1 is extremely dissatisfied and 7 is extremely satisfied. This attempted to measure how a design produced with co-design influenced community satisfaction in the end product. Respondents were also asked what elements they would like added to the current design and what elements they wanted to be removed from the current design to further improve the design. Unfortunately, this survey only received one response despite being sent to all community members who subscribed to the Giving Garden mailing list. To accommodate the lack of survey data, improvements on the second design were validated using comments and feedback received from community leaders and members during co-design workshops, in Giving Garden meetings, and while tabling in the area.

Copyright © 2023 by ASME

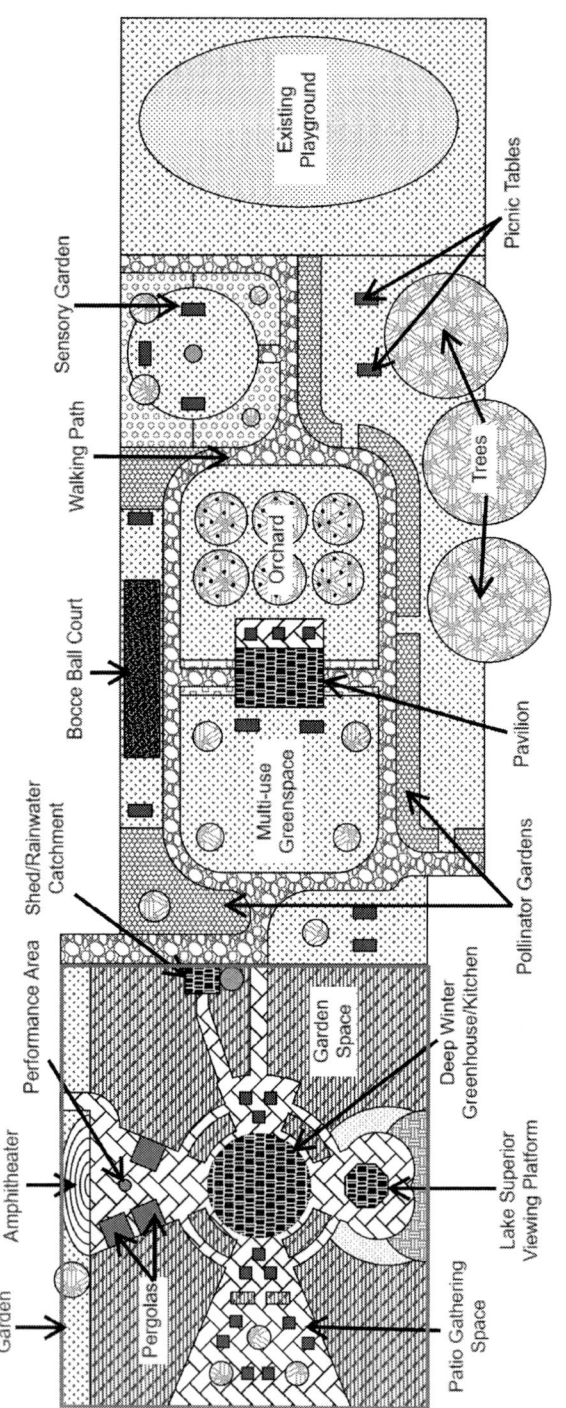

FIGURE 2. CURRENT DESIGN PRODUCED WITH CO-DESIGN

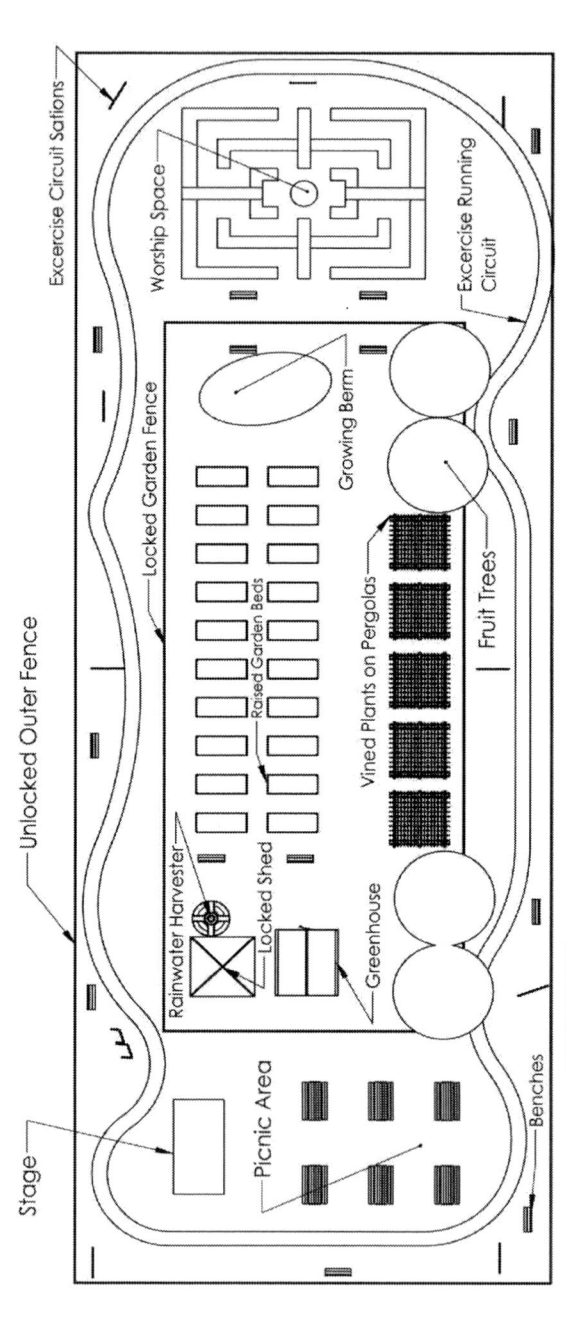

FIGURE 3. PRELIMINARY DESIGN PRODUCED WITHOUT CO-DESIGN

Copyright © 2023 by ASME

3. RESULTS

Prior to the co-design workshops and numerous Giving Garden meetings, a design was created utilizing initial considerations from a small group of individuals from the Giving Garden committee. A short list of desired elements, rough dimensions, and 5 hours was all that contributed to the preliminary design in Figure 3.

Because the community wasn't included in this design process, this draft has several drawbacks. It is detrimentally basic, does not effectively utilize the Sport Court space, nor satisfies all of the needs of the community. It merely meets the arbitrary requirements set forth by the committee.

Rectification was thus required. Co-design workshops were organized and held on September 24th, October 15th, and February 5th. Additionally, various leaders in the community, such as the St. Mark AME pastor, the local YMCA director, and Community Action Duluth representatives, were invited to weekly Giving Garden meetings for their contributions. A total of five meetings were attended by these leaders.

Utilizing this input from the community we reconsidered and defined our main goals for the space to best serve the community. We determined that the goals of this project were to (1) create an environment that generates community connection, (2) cultivate a relationship with the land, and (3) reconnect the community with its cultural heritage. Applying these goals, integrating the elements requested from our co-design practices, and investing more than 5 months of time - from the beginning of October to the beginning of March - enabled our original draft to evolve into the design in Figure 2.

3.1 Description of Second Design

The new design satisfies all three of our goals: through the garden, we can foster a relationship with the land; the numerous gathering spaces encourage community interaction and relationship building; and because the Central Hillside community has a substantial Black population for Duluth, African cultural elements were implemented into the space to assist in reconnecting the community to their African heritage.

According to [16], African communities are typically arranged "around a central area", creating a "garden for everyone". Individual's homes are then oriented around this central area meaning a courtyard behaves as an "outdoor room" for each private dwelling. This is significant because according to many African communities/cultures, the individual mind prefers a communal setting relative to an isolated one. Courtyards in these African communities, therefore, enable communal interaction and satisfy individuals' desire for social connection [16].

The garden in our design thus attempts to emulate the African courtyard. As seen in Figure 2, there is a large communal center around the greenhouse with a Lake Superior viewing platform, seating, and an amphitheater built into the adjacent hillside branching off from the center. These elements offer a variety of settings for community gatherings and activities - providing the opportunity for cultivating community relationships. However, surrounding all of these elements in the design is garden space. When people come to enjoy the courtyard and its surrounding amenities, they will consequently be immersed in the garden. Coming to the courtyard will not only provide physical and psychological benefits, but it will also cultivate a connection to the land through the garden.

The rest of the park (to the right of the garden) offers several amenities which should encourage community use and improve quality of life. These amenities include a pavilion, a bocce ball court, an orchard, a sensory garden, and a playground. This versatility was welcomed by the community and many were excited about the opportunity to renovate the space.

3.2 Analysis of Second Design Improvements

Utilizing co-design resulted in an incredible transformation of the current design. Two major concerns arose during our co-design programming which were not addressed by the first design. The growing season in Duluth is very short - maybe 5 months with a little luck. This limits Giving Garden programming into that short window; limiting the amount of nutritional mass, connection to the land, and community relationships the Giving Garden can cultivate. The obvious solution to bolster these efforts is to expand into the winter months.

Extending the growing season is accomplished in our design through a deep winter greenhouse which will be able to grow produce year-round. It will also provide an indoor gathering space in the winter that is in the presence of food production. With this greenhouse, the Giving Garden gains the opportunity to enhance the health and quality of life of Hillside community members no matter the season. Multiple members of the community expressed great excitement over the possibility of both the deep winter greenhouse and the opportunity for winter activities during the February 5th community meeting.

The other concern regarded the lack of culinary abilities within the community. Because of the Central Hillside community's generational limited access to fresh produce, the community's culinary knowledge of cooking with fresh produce has been greatly reduced. To rectify this issue, the gathering space in the greenhouse will be equipped with a kitchen capable of preparing the produce of the garden in a multitude of ways. Cooking sessions can then be held to restore culinary skills in the community, allowing individuals to better appreciate and utilize fresh fruits and vegetables. This should assist in reducing dependency on processed foods and benefit the health of the community.

Figures 4 and 5 highlight the evolution of the topology between the two designs due to co-design. The topology diagrams in Figures 4 and 5 contrast both the number of each type of element (i.e. structure - orange, gathering space - red, individual space - blue, greenery - green, and miscellaneous - yellow) and how each element contributes to the overall Giving Garden experience. Ergonomically arranging features and connecting them is an essential aspect of product design [15]. The current design contains 9 more elements compared to the

Copyright © 2023 by ASME

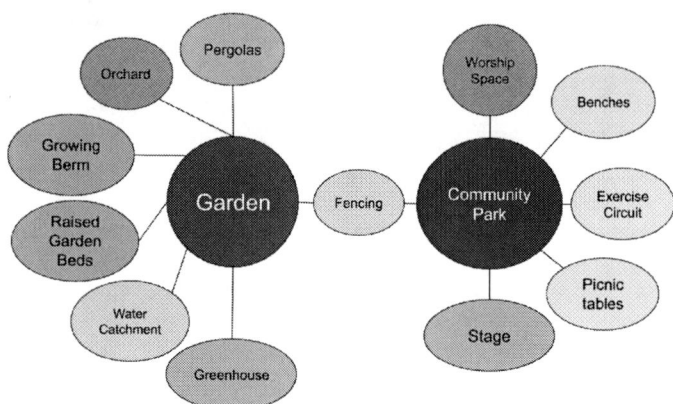

FIGURE 4: NETWORK DIAGRAM - PRELIMINARY DESIGN

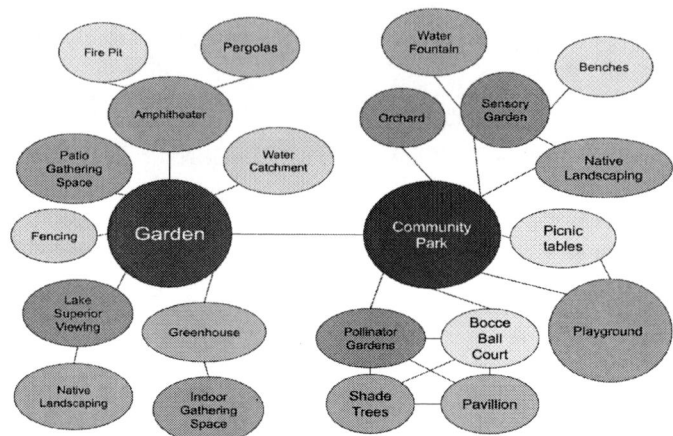

FIGURE 5: NETWORK DIAGRAM - CURRENT DESIGN

preliminary design. Four of these new elements are community gathering spaces, which will help increase the time spent in the park and cultivate an appreciation for the garden. Also, the current design doubled the total number of connections between elements from 13 in the preliminary design to 27 in the current design. These connections exemplify an improved flow and usability of the space.

There was significant input from the community expressing satisfaction with the improved interconnectedness of the second design. Community members were particularly satisfied with the "multi-generational approach" where kids could play on the playground, teens can use the multi-use greenspace, elders could enjoy the pavilion or sensory garden, and their parents could work in the garden while keeping an eye on the children. This sentiment was expressed by a community member who emailed the following comments after the February 5th meeting:

> Love the multi-generational approach! As a parent of a small child, I love the idea of a playground next to a garden.

This evolution of interconnectedness was a direct result of co-design as the community was able to advocate for the connection of specific elements in the park.

Figure 6 and Table 1 calculate the geometric differences between the preliminary design and the current design. There were several calls from the community to increase the scope of the project because the original design was unable to meet all of the requirements of the community - such as the lack of greenhouse and kitchen space. Additionally, members of the Giving Garden community also began to express desires to host a wider variety of events requiring more room to store materials and accommodate individuals. The total area of the project therefore more than doubled to fully utilize the Hillside Sport Court Park. Subsequently, all types of spaces increased in area relative to the preliminary design. However, each type of space did not scale equally. Figure 6 visually represents the change in

percentages of each type of space. This chart is particularly useful in highlighting the redistribution of each type of area to better suit the community's needs and desires. For example, communal space, which had the greatest percentage of area in the preliminary design, decreased in percentage of the total area in the current design by 27.3%. While this may initially appear to detract from the park, this area was reallocated to exercise, performance, landscaping, and greenhouse space - spaces the preliminary design was lacking according to the community. A follow-up email was sent by a community member expressing that this increase in size was greatly appreciated, "Wow from 800 sq ft to the whole sports court!" By decreasing the total amount of communal space, the design was able to address a wider variety of community needs.

TABLE 1: CURRENT (RED) AND PRELIMINARY (BLUE) GEOMETRY COMPARISON

Type of Space	Area (ft²)		Change	Percent of Total Area		Change
Garden/ Food Production	4500	9675	+5175	18.2	17.42	-0.78
Communal	16053	20937	+4884	65.0	37.71	-27.29
Individual	1724	3136	+1412	7.0	5.65	-1.35
Exercise	127	10720	+10593	0.5	19.31	+18.81
Performance	200	1347	+1147	0.8	2.43	+1.63
Landscaping	0	3911	+3911	0	7.04	+7.0
Walkway	1922	4798	+2876	7.8	8.64	+0.8
Greenhouse	174	1000	+826	0.7	1.80	+1.1
Total	24700	55524	+30824	100	100	

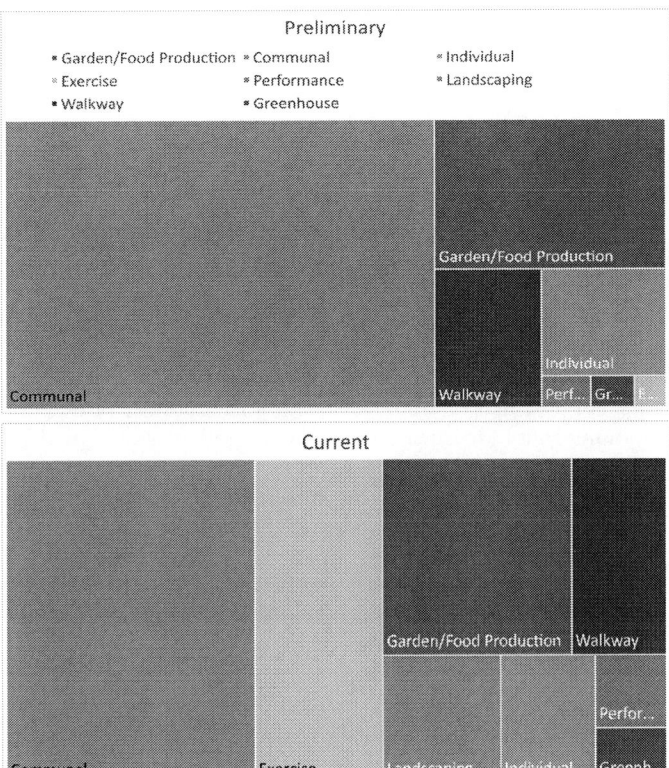

FIGURE 6: TREE MAP COMPARING THE PERCENTAGE OF TOTAL AREA OF EACH TYPE OF SPACE

Community feedback supported this redistribution of area. Many community members and several community leaders particularly appreciated the addition of a deep winter greenhouse to the space to enable winter activities. Several community leaders were also pleased with the multifunctionality of the space, allowing individuals of all circumstances to come to enjoy the garden. A community member emailed the following:

> I like the playground idea and oval solar greenhouse. Maybe having some sports game areas. I like the garden and African Heritage gardening practices. I like the amphitheater and arts.

This person emphasizes the importance of shared communal space for the garden.

Additionally, Table 2 reveals the differences in the number of specific elements of each design. This table highlights the change in what each design values. For example, according to the community, the preliminary design lacked an adequate number of shared communal infrastructure such as picnic tables/benches and pavilions for lounging in the park and trees to provide shade on hot, summer days. The current design, therefore, increased the number of each element.

Furthermore, according to the community, the preliminary design was also investing too many resources in pergolas,

TABLE 2: CURRENT (RED) AND PRELIMINARY (BLUE) DESIGN FEATURE COMPARISON

Element	Quantity		Change
Benches/Picnic Tables	14	26	+12
Pergolas	5	3	-2
Pavilions	0	2	+2
Trees	4	21	+17
Raised Garden Beds	20	10	-10
Fencing (ft.)	1130	550	-580

raised garden beds, and fencing, requiring the current design to reduce these elements. Fencing, in particular, decreased by 51%. While fencing is a necessity in a garden to keep deer and other hungry animals from eating the produce, too much fencing limits the usability of a space and discourages interactions between people and the land. At a weekly meeting, a monetary partner for the project commented that they liked the multifunctionality of the design. The current design thus removed unessential fencing to allow for increased ease of movement through the space to encourage interactions and thus community building.

4. DISCUSSION

It was hoped to survey the community to gauge their satisfaction with each design. This would provide quantitative data on how the community experienced co-design. A survey was sent out via email to stakeholders and community members but over the course of a month, the survey only received one response. This is an example of one of the many difficulties of co-design; it can be very challenging to collect community input from a representative sample passively. Rather an active approach is required to encourage community participation in visioning and designing.

To overcome the poor results from the survey, our World Cafe and St. Mark AME church presentation provided the necessary feedback on our design. Relative to the reviewed co-design projects in Italy [8], [9], Chile [10], [11], and Ethiopia [12], the Giving Garden project received a significant amount of community engagement. The Giving Garden utilized a wide variety of events (e.g. Juneteenth Celebration, Closing Ceremony, and the World Cafe event) to collect extensive feedback from each demographic in the community (e.g. the youth, elderly, families, ethnicities, etc.). For the youth, garden visioning and events held in partnership with schools were the most successful. Feedback was best received from the elderly

Copyright © 2023 by ASME

and families at St. Mark AME church events and through tabling events throughout the community. Our wide demographic net considering youth, families, and the elderly contrasts with many of the projects reviewed prior to this study where only one or two demographics were consulted [8], [9]. This is a result of co-design's lack of a shared, replicable framework. Instead, co-design is applied to each context individually for the best results; however, best practices across contexts can be proposed and tested. In this case, the community park will be a place for cross-generational connection meaning the Giving Garden had to invest lots of resources to acknowledge and incorporate the age diversity of the community.

Furthermore, the Giving Garden's investment in acknowledging the diversity of the community also contributed to the development of the interconnectedness between community groups. At the February 5th workshop, different races, genders, and age groups intermingled and collaborated in discussing the Phase II design, nurturing significant connections among the community. This nurturing of connection is a foundational pillar of the Giving Garden.

4.1 Implications of this Research

Co-design's general purpose is to meet end-user needs through design. In a community, end users and their needs are diverse and divergent. Rarely has connecting end users or members of a community together or facilitating interactions between diverse user groups been the goal of co-design processes; rather the goal focuses on connecting end users to the process of design and the designs they are utilizing. However, building community is achieved by encouraging and facilitating community interactions and connections.

This paper developed a novel technique for measuring connectivity between elements using network diagrams to measure the ability of a design to increase interactions between different end users towards the aim of building community. How connected different functions or aspects of a design are important when designing for communities and communal use. The idea behind this metric is that the more connected the features of the design the more opportunity and connections that will grow between different community members utilizing the design. Using topology comparisons between different designs allows visualization and quantification of connectivity between design elements, which can be critical to community-based co-design. This case study on a garden could have been carried out for any public space, structure, or shared communal resource.

5. CONCLUSIONS

Because of the increased community contributions from our co-design techniques, the second design more effectively addresses the health, cultural, and social needs of a community compared to the first design. There is much more variety in the type of space in the current design than in the preliminary design and there are double the number of connections between elements in the current design. These improvements will help

contribute to an elevated park experience. Therefore, co-design is a key asset in the designer's toolbox; it creates a highly functional and satisfying product that can benefit end-users.

The process of co-design is not straightforward and contradicts many typical design norms. Co-design is extremely slow which can be frustrating for designers accustomed to speeding projects to completion. Community input is needed after each design phase which requires organizing open forums for discussion, collecting community input, and then implementing that feedback into the design. Consequently, it can take several iterations before a satisfactory design is achieved using co-design. However, co-design can greatly assist in developing a superior product. Future co-design projects must therefore allocate a significant portion of time for community input to ensure the final design holistically benefits each stakeholder group.

In addition, it can be difficult to collaborate with community members, particularly if they are not well-versed in the design process. Despite this, active engagement must be encouraged through a wide variety of venues to reach different demographics and engage more people in the community. Each individual's perspective and input on the design are invaluable in ensuring that the product benefits every user equally. Co-design, therefore, assists in promoting equitable design to the highest degree.

Lastly, the collection of community feedback can not be done passively. The satisfaction survey in this project was sent out via email and received only one response because of its passive nature. This project's active feedback collection, such as the World Cafe event and the St. Mark AME presentation, were much more effective. However, these types of events take time to plan which lengthen the timeline of a project. While this meticulousness can be cumbersome, it is the reason why co-design results in a superior product - every aspect of the project is labored over to ensure that each element adds value to the product. This design is subject to change based on evolving feedback from the community. This is an ongoing project and the design is fluid and adaptable to future inputs.

Populations will continue to grow, putting increased stress on our resources, expanding inequalities, and requiring additional space to accommodate people. Thus, it will be society's responsibility to construct spaces that address the individual challenges that face each community. Co-design is a powerful way to develop those spaces to ensure that they are designed equitably and holistically benefit each unique community and encourage interactions between community members with diverse interests and needs.

ACKNOWLEDGMENTS

Great thanks to all Giving Garden members - Janet Kennedy, Meg Litts, Elyse Carter Vosen, Carla Hamilton, and the rest of the Giving Garden Committee for their support and contributions. Thank you to Kathy Bogen, the Kako Foundation, and Mentor North for bringing community children and families into the Giving Garden. Special thanks to Pastor Anthony Galloway, the St. Mark AME congregation,

Copyright © 2023 by ASME

Health Equity Northland, Echoes of Peace, Planting Connections, Planting Hope, and the Central Hillside community for allowing our dreams to become a reality.

REFERENCES

[1] P. Dutko, M. Ver Ploeg, and T. Farrigan. "Characteristics and Influential Factors of Food Deserts," ERR-140, U.S. Department of Agriculture, Economic Research Service, August 2012.

[2] M. Vitcenda, "Extension, UMD research sparks community action on Duluth food desert," *extension.umn.edu*, 2012. [Online]. Available: https://extension.umn.edu/vital-connections/extension-umd-research-sparks-community-action-duluth-food-desert. [Accessed Jan. 11, 2023].

[3] A. Pinea and J. Bennett, "Food access and food deserts: the diverse methods that residents of a neighborhood in Duluth, Minnesota use to provision themselves," *Community Development*, vol. 45, no. 4, pp. 317-336. Jul. 2014.

[4] Duluth Parks and Recreation, Duluth, Minnesota, USA. "Duluth Parks, Recreation, Open Space & Trails Plan," 2022. Accessed: Jan. 11, 2023. [Online]. Available: https://duluthmn.gov/media/14188/10-10-22-essential-spaces-plan-for-council.pdf

[5] J. Kennedy, "Giving garden," *Health Equity Northland*. [Online]. Available: https://healthequitynorthland.org/giving-garden. [Accessed: 11-Jan-2023].

[6] R. LaRoche, "Giving Garden Newsletter," *St. Mark African Methodist Episcopal*, Jan. 1, 2021. [Accessed Jan. 11, 2023].

[7] "What is user centered design?," *The Interaction Design Foundation*. [Online]. Available: https://www.interaction-design.org/literature/topics/user-centered-design. [Accessed: 11-Jan-2023].

[8] N. Fumagalli, E. Fermani, G. Senes, M. Boffi, L. Pola, and P. Inghilleri, "Sustainable Co-Design with Older People: The Case of a Public Restorative Garden in Milan (Italy)," Sustainability, vol. 12, no. 8, p. 3166, Apr. 2020, doi: 10.3390/su12083166.

[9] M. Boffi, L. Pola, N. Fumagalli, E. Fermani, G. Senes, and P. Inghilleri, "Nature experiences of older people for active ageing: An interdisciplinary approach to the co-design of Community Gardens," Frontiers, 27-Aug-2021. [Online].

[10] M. Gaete Cruz , A. Ersoy, D. Czischke and E. van Bueren, "A Framework for Co-Design Processes and Visual Collaborative Methods: An Action Research Through Design in Chile," Urban Planning, vol. 7, no. 3, pp. 363-378, Sept. 2022.

[11] E. Peker and Ataöv Anlı, Governance of climate responsive cities: Exploring cross-scale dynamics. Cham, Switzerland: Springer, 2021.

[12] E. Blynn, E. Harris, M. Wendland, C. Chang, D. Kasungami, M. Ashok, and M. Ayenekulu, "Integrating human-centered design to Advance Global Health: Lessons from 3 programs," Global health, science and practice, 29-Nov-2021. [Online]. Available: https://www.ncbi.nlm.nih.gov/pmc/articles/PMC8628497/. [Accessed: 14-Feb-2023].

[13] J. Brown and D. Isaacs, *The world café: Shaping our futures through conversations that matter*. San Francisco, CA: Berrett-Koehler Publishers Inc., 2006.

[14] "World Cafe Method," *The World Cafe*, 25-Nov-2019. [Online]. Available: https://theworldcafe.com/key-concepts-resources/world-cafe-method. [Accessed: 11-Jan-2023].

[15] V. Cicirello and W. C. Regli, "Machining Feature-based Comparisons of Mechanical Parts ." Carnegie Melon University, Philadelphia, 2001.

[16] G. Steyn, "African courtyard architecture: Typography, art, science and relevance," *Acta Structilia*, vol. 12, no. 2, pp. 106-129, 2005.

Proceedings of the ASME 2023
International Design Engineering Technical Conferences and
Computers and Information in Engineering Conference
IDETC-CIE2023
August 20-23, 2023, Boston, Massachusetts

DETC2023-116953

DESIGN INTERVIEWS CONDUCTED BY INTRA- AND INTERCULTURAL TEAMS:
A CASE STUDY ON DIALYSIS IN ZIMBABWE

Micki Grover[1], Carlye A. Lauff[1], Chiratidzo Ndhlovu[3], and Natasha C. Wright[2]
[1,2]University of Minnesota, Minneapolis, USA
[1]School of Product Design, College of Design, [2]Department of Mechanical Engineering
[3]University of Zimbabwe, Harare, ZW

ABSTRACT

In global development engineering, semi-structured, direct-dialogue interviews are often recommended in order to generate a deep understanding of stakeholders' needs and to create products that meet those needs. In this study, interviews were used to explore the existing dialysis treatment program for end-stage kidney disease in Zimbabwe. This study has two aims: (i) to understand the dialysis service model and limits to its expansion, and (ii) to examine the impact of the cultural background of the interviewing team on interview outcomes.

Virtual training on exploratory interviewing was developed and administered to 12 undergraduate students living in the United States, Uganda, and Zimbabwe. Six teams, each having either an intercultural or intracultural composition conducted field interviews (n=18) with Zimbabwean dialysis professionals to better understand the existing service model associated with hemodialysis (HD) and peritoneal dialysis (PD) treatment modalities. Interviews were coded in NVivo to develop an overall service model map including relevant people, props, and processes. Key limitations to expanding PD programs include: lack of clean water sources, no in-country dialysate production, insufficient financial resources, limited nursing staff, and difficulty in tracking medical information during home-based treatment, among others.

The service model map was additionally used to quantify the number of codes uncovered in individual interviews. Intercultural pairs produced a higher proportion of top scoring interviews than did the intracultural pairs. The small sample size, however, results in only an early indication of potentially replicable findings. The work represents a potential methodology for further research in this space.

Keywords: international development, design ethnography, intercultural interviews, dialysis, global development

1.0 INTRODUCTION

Ethnographic design is an interdisciplinary practice that combines the culturally-conscious methodologies of ethnography with the design process [1], [2]. While the rapid nature of product and service design programs often do not allow enough time for fully immersive anthropological practices, less intensive versions of those practices can be implemented. For instance, building rapport in anthropology may mean living amongst a group of people to establish trust and understanding. In design ethnography, some of the benefits of building rapport can be achieved with the tailored approach of taking time at the beginning of an exploratory interview to connect on a personal level. Another topic of ethnographic research that can be applied to the design process is reflexivity. Reflexivity is the critical consideration of the effect the researcher has on the group being researched during the study [3]–[5]. Design projects that utilize exploratory interviewing require particular emphasis on reflexivity during cross-cultural exchanges. Student interviewers in design projects often have a limited scope of experience with cross-cultural research, and may unknowingly allow their background to influence their results.

The impact of the cultural background of the interviewer has been examined in anthropological research, but has not been thoroughly explored in design ethnography. Exploratory interviews in design research are often conducted in pairs so that one interviewer can lead the discussion while another primarily takes notes. The note-taker is an additional presence in the interview who participates in the initial rapport building and is welcome to contribute to the question-asking portion as well. This study explores the impact of intercultural and intracultural groups of interviewers on interview outcomes. This is one variable that may affect the quality of an interview and could be reported along with reflexively as a way to discuss researcher impact. A sister paper explores a method of

Copyright © 2023 by ASME

measuring interview quality for additional methodological standardization: "Towards quantifying interviews: comparing techniques to evaluate the quality of design interviews" by Grover, Wright, and Lauff.

The design research project to which this methodology study was applied sought to understand the current service model for peritoneal dialysis (PD) in Zimbabwe and limitations to its expansion. Using the data synthesized from the exploratory interviews conducted, a service model map was created and discussed. The interview findings, including the service model map, aim to answer the first research question: *RQ1: What does the current dialysis program look like in Zimbabwe and what is limiting its expansion?*

The methodology study focused on the effect of the cultural background of interviewer pairs and aims to serve as a pilot study on the second research question: *RQ2: How does inter- versus intracultural team composition affect the output of the interview?*

2.0 BACKGROUND
2.1 End-Stage Kidney Disease and Dialysis

Limited access to dialysis, the life saving therapy for end-stage kidney disease (ESKD) patients, may result in up to seven million deaths worldwide every year, more than the deaths from HIV and Tuberculosis combined. Only 16% of ESKD patients in Africa receive the treatment they need [6]. In a 2017 systematic review of ESKD outcomes in African adults and children, 96% of adults and 95% of children who could not access dialysis died or were presumed dead [7].

There are two primary dialysis options for ESKD patients, having equivalent long term survival rates - hemodialysis (HD) and peritoneal dialysis (PD). HD is the most common modality globally; it removes excess fluid and solutes by passing the patient's blood through an external filter (the dialyzer). HD currently accounts for ~95% of dialysis treatment in Zimbabwe. Patients on HD are usually required to visit a dialysis center three times weekly for ~4 hour treatment sessions. The added time and cost associated with transportation to the center is a barrier to consistent treatment. Additionally, HD is unsuitable for many pediatric patients whose blood volume may be too small for an external filter; PD is used instead. PD utilizes a membrane within the abdomen, the peritoneum, as the filter (Figure 1). The peritoneal space is filled with PD fluid through an abdominal catheter; osmotic and concentration gradients between the blood and PD fluid lead to the removal of excess body fluid and solutes, respectively. Patients or their caregivers can often perform this procedure within the comfort of their home. The increase in convenience compared to HD is a major advantage of PD. Studies have also shown that patients undergoing PD require fewer blood transfusions than those undergoing HD [8]. These features typically make PD less expensive in high-income countries and in 2019, the International Society for Nephrology recommended a PD first approach for expanding dialysis access [9].

Over the past three years, the Wright Lab at the University of Minnesota has been developing technologies with the aim of increasing access to PD in low- and middle-income countries. Their focus has been in Nigeria, where PD is currently not available for ESKD patients. In the present study, existing HD and PD service models are explored in hopes of gathering insight into how these programs are implemented in the Zimbabwean context.

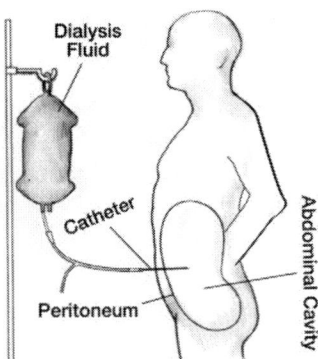

FIGURE 1: PERITONEAL DIALYSIS [10]

2.2 Cross-Cultural Research and Interviewing

In cross-cultural research, extra care must be given to select research methods appropriate to the cultural context [11]. One such choice is that of the interviewers who will conduct the exploratory design interviews for the design project. The cultural differences between the interviewer and the interviewee can impact the interview outcomes. Examples of this in literature are the interviewee being inclined to take on the role of group representative instead of speaking from their own viewpoint [12], the interviewee considering some topics "unknowable" to the interviewer [12], or the interviewee being reluctant to divulge information due to trust concerns [3], [13]. Additionally, the interviewer can miss opportunities for learning by not considering the effect of their own cultural norms and subjectivity of interpretation when exploring unfamiliar topics with the interviewee [4], [14], [15]. However, there are potential benefits to performing design interviews cross-culturally. Interviewers who are less familiar with the culture of their interview subjects may be less complacent than those who are [16]. This could lead those cross-cultural interviewers to ask for details on topics that interviewers familiar with that culture may take for granted. Tacit knowledge within a culture often goes unarticulated, and cross-cultural interviewers have an outsider perspective that may be able to better elicit and capture such findings.

The effect of cultural backgrounds cannot be removed from interactions, but researchers can use their understanding of such effects strategically [17]. There could be a benefit to intercultural pairs, where one member is of a similar cultural background as the interviewee and the other is of a different cultural background. If the benefit of understanding and access is granted to the interviewer with the similar cultural background and the openness to new information is added by the interviewer with a very different cultural background, then a mixed pair may help produce more successful interviews.

Copyright © 2023 by ASME

Additionally, intercultural partners may impact each other by encouraging an overall greater breadth of information to be elicited during interviews. These ideas are explored in this study where groups of interviewers were organized so that intercultural and intracultural pairs were represented. The groups were compared based on the volume of information they gained from each interview. A greater amount of information gained is an indication of an interviewee who felt more comfortable and inclined to share in depth [13]. To measure the volume of information, interview transcripts were coded by items pertinent to building a service model map and those codes were summed to compare between groups.

This study used exploratory semi-structured interviews to help understand the current state of dialysis services in Zimbabwe. A common qualitative research tool, semi-structured interviews have been used by multiple studies aimed to understand different aspects of PD in other parts of the world [18]–[20]. The information gained from this study will be used in the End-Stage Kidney Disease Design Project by the Wright Research Lab. The service model map was solicited as a way to better understand the landscape, resources, barriers, and opportunities present in Zimbabwe. Differing styles of service model maps have been developed in other medical sector studies to better understand and improve processes [21]–[23]. This understanding will be used to shape the implementation plan of the technology under development.

3.0 METHODS

This research was conducted with approval from the Institutional Review Board (IRB) at University of Minnesota under STUDY00013159 and also through the Medical Research Council of Zimbabwe (MRCZ). Written consent was obtained from all student interviewers (n=12); verbal consent was obtained from all interview participants (n=18).

3.1 Interview Training

Twelve undergraduate students were recruited from three different countries to be a part of this study. The locations of recruitment were the United States (Minneapolis), Uganda (Kampala and Mbarara), and Zimbabwe (Harare). Six students from the United States (US), four from Uganda (UG), and two from Zimbabwe (ZW) joined the study. The students were recruited as paid research assistants, performing ethnographic design interviews, as well as study subjects, giving consent for their interviews to be analyzed via the methodology described here. The students' practice and field interviews were also in the development of a method for measuring interview quality based on interviewer skills [24].

It was important that all students had a similar amount of interviewing experience to maintain a skill level as uniform as possible. The students chosen were identified as having minimal interviewer experience, yet the potential to gain those skills. This potential was determined through an application process, a preliminary interview with a research assistant, and an informational meeting with one of the study PIs and research assistant.

The students went through a training program of five virtual workshops aimed at defining and applying design ethnography skills, with a focus in exploratory interviewing. For more details on this program, please refer to [25]. As part of the training program, students were put into groups of two to perform practice interviews. Each student led three practice interviews while their partner took notes and assisted with questions as needed. Original groups were divided into six same-city pairs: US-US, US-US, US-US, UG-UG, UG-UG, ZW-ZW. For the third session of practice interviews, one of the US groups was mixed with one of the UG groups. The final group pairings, which stayed together for all field interviews included these six same-city and inter-city pairs: US-US, US-US, US-UG, US-UG, UG-UG, ZW-ZW. For the remainder of this work, the terms intercultural and intracultural will be used to denote inter-city and same-city pairs, respectively. It is important to note that city of residence is just one of many factors that may define an individual's cultural identity.

3.2 Data Collection

For the ESKD Design Project, the student pairs were tasked with gathering contacts and conducting interviews with dialysis professionals located in Zimbabwe. Students were instructed to seek out differing viewpoints and to go in depth on a wide variety of topics. In total, the students gathered insights from dialysis professionals in the following roles: doctor, pediatrician, dialysis center director, dialysis clinic manager, peritoneal dialysis nurse, ICU nurse, engineer, technician, and pharmacist. The purpose of these interviews was to understand the state of dialysis in Zimbabwe, and to identify the barriers and opportunities for the technology developed by the Wright Lab to achieve its goal of improving peritoneal dialysis (PD) accessibility. Note that per the approved IRB, patients were excluded from the study.

In total, eighteen field interviews were conducted. Each group finished 2-4 interviews. The number of interviews a group contributed to the study depended on their ability to recruit interview participants, schedule their interviews, and complete their interviews and deliverables. The deliverables for each interview were a signed verbal consent form, a recording of the interview, an interview summary (3-5 sentences), a description of primary themes and takeaways, and the notes written during the interview. Interviews were all intended to be over the video conferencing program Zoom to utilize its video recording and automated transcription applications. The groups were also tasked with editing their transcripts for any errors produced by the auto-transcription software. The transcripts were further reviewed and corrected by the graduate research assistant performing the data analysis portion of this study. One member of the group of two Zimbabwean students elected to withdraw from the study due to personal time constraints, so their second interview contribution was performed by only one student. That interview was also the only one conducted in-person. This methodology change was allowed as a solution to difficulties the student had in accessing free and reliable wireless internet for using Zoom.

Copyright © 2023 by ASME

At the end of the study, each student was given the opportunity to have a final meeting with one of the study PIs to give and receive feedback. In these meetings, the students shared comments related to working cross-culturally. Eight students engaged in these feedback sessions; their comments were organized and evaluated to extract insights.

3.3 Data Analysis

3.3.1 Service Model Maps

Findings were identified and organized using the qualitative analysis software NVivo (QSR International). NVivo's platform allows for a video and time stamped transcript to be reviewed side-by-side as sections of text are highlighted and categorized. The process of highlighting and labeling sections of a transcript into categories is called coding. All transcripts were coded in NVivo to identify people, processes, and props. In this context, the "people" are stakeholders in Zimbabwe's dialysis service model, the "processes" are actions that connect the people, and the "props" are items interacted with or exchanged during a process. For example, if an interview participant who is a technician shares that part of maintaining the water plant at the hospital is sending water samples to the lab for testing, the people codes identified would be "technician" and "lab staff", the process code would be "water testing", and the prop would be "water sample."

To illustrate the connections between codes, a service model map was created using the data gathered from all 18 field interviews. The final service model map with all gathered codes and connections is shown in Figure 2. The red boxes in the service model map are the people, the yellow diamonds are the processes, the blue cylinders are the props, and the arrows are the connections between those codes. White shapes are people, processes, or props that are identified as "needed." For example, the interview participants identified transplant surgeons as a resource currently not available in Zimbabwe, so that code is represented by a white box. Clean water was mentioned as difficult to find in the PD portion of the service model, so clean water is represented with a white cylinder. The direction of the arrows show the direction of activity. Dietitians inform and prescribe a diet to the patient, so the arrow goes from the dietitian to the "Inform/Prescribe Diet" process and another arrow from that process to the patient.

There were 149 total codes extracted from the 18 interviews and used to build the service model map. To organize this information, the map was split into the following categories: Pharmacy, Finances, Medical Services, Peritoneal Dialysis Program, Equipment, and Supply Chain. A service model map for each interview was built from the final map by marking mentioned codes with a red star. The title of the interviewee was indicated in the upper left hand corner and their main topic area(s) of focus indicated by highlighting that section of the map green. For instance, an interview with a Pediatrician has the section Medical Services highlighted green while an interview with a Clinic Manager has Finances and Supply Chain highlighted.

3.3.2 Radar Charts and Bar Graphs

The quantity of codes uncovered in each interview were then used to evaluate the effectiveness of each group. Radar charts (Figure 4) were the data visualization method chosen for their ability to encapsulate multi-axis data for side-by-side comparison. Each axis on the radar charts represents a different section of codes: pharmacy, finances, medical services, peritoneal dialysis program, equipment, and supply chain. The value along each axis accounts for the number of codes identified per section. Bar graphs (Figure 5) were generated with the total number of codes per interview for further comparison. The graphs include count average lines to show which groups performed stronger on average.

4.0 RESULTS & DISCUSSION

4.1 Service Model Map

The service model map (Figure 2) is the primary deliverable for the ESKD Design Project, contributing to its technology translation plan. The map was constructed using the information gathered from the 18 field interviews with dialysis stakeholders in Zimbabwe. The final service model map was used as a template to create maps corresponding to each interview that mark the codes identified with a red star, as described in Section 3.3.1. Two examples of the individual interview service model maps are shown in Figures 2 and 3. The final service model map (Figure 2) can be studied to address Research Question 1 regarding understanding the current peritoneal dialysis system in Zimbabwe and its barriers to expansion. The results included are from the perspectives of the individuals interviewed in this study only. The authors are aware that dialysis programs are implemented in different ways in different countries, and that the service model as described here may not be a complete representation of the program available in Zimbabwe.

4.1.1 Peritoneal Dialysis Program

The Peritoneal Dialysis (PD) Program section of the service model map (Figure 2) has 42 codes. PD is a primarily patient-run process, which is reflected in 15 PD processes directly connected to the patient on the map. There are many items needed for PD, which are represented by a group of props connected to the process of "PD Treatment." The props identified as being available were dialysate, PD fluid bags, caps, tubing, gauze, strapping, couch, drip stand, and aseptic solution. The interviewees also named resources that were either not available or difficult to secure. The needed resources are: funding, treatment data, clean water, temperature controlled and ventilated storage for supplies, automated PD with a PD cycler, PD for acute dialysis, and in-hospital PD.

According to the interview participants involved in this study, Parirenyatwa is the only location with a PD program. Bulawayo used to have one such program, but this changed once Cimas became the primary supplier for private dialysis clinics there. Cimas did not support PD, which caused a reduction in the demand for PD dialysate from Datlabs, the local dialysate manufacturer in Bulawayo. Datlabs stopped producing dialysate due to the drop in demand, which resulted in the local hospitals discontinuing PD programs. Datlabs had also been where local nurses were trained in PD practices. This ceased when dialysate was no longer produced there.

The PD Program at Parirenyatwa includes nurses and nephrologists along with a few PD-specific roles. The PD Coordinator and PD Counselor help with evaluating a patient's circumstances through a home visit to approve them for PD. When the patient first starts their treatment, the PD Counselor provides education and a PD Starter Pack, which includes supplies and a diet sheet for guidance on their new dietary requirements. The waste fluid is flushed and the bag waste may be burned through a City Council run incineration program or brought to the hospital for disposal. The PD Counselor continues to be available for the patient to contact by phone and to answer questions and address issues. The most common challenge associated with PD mentioned by the interviewees was infection (e.g. peritonitis). The role of the PD Counselor and PD Coordinator may be taken on by one individual called a PD Nurse.

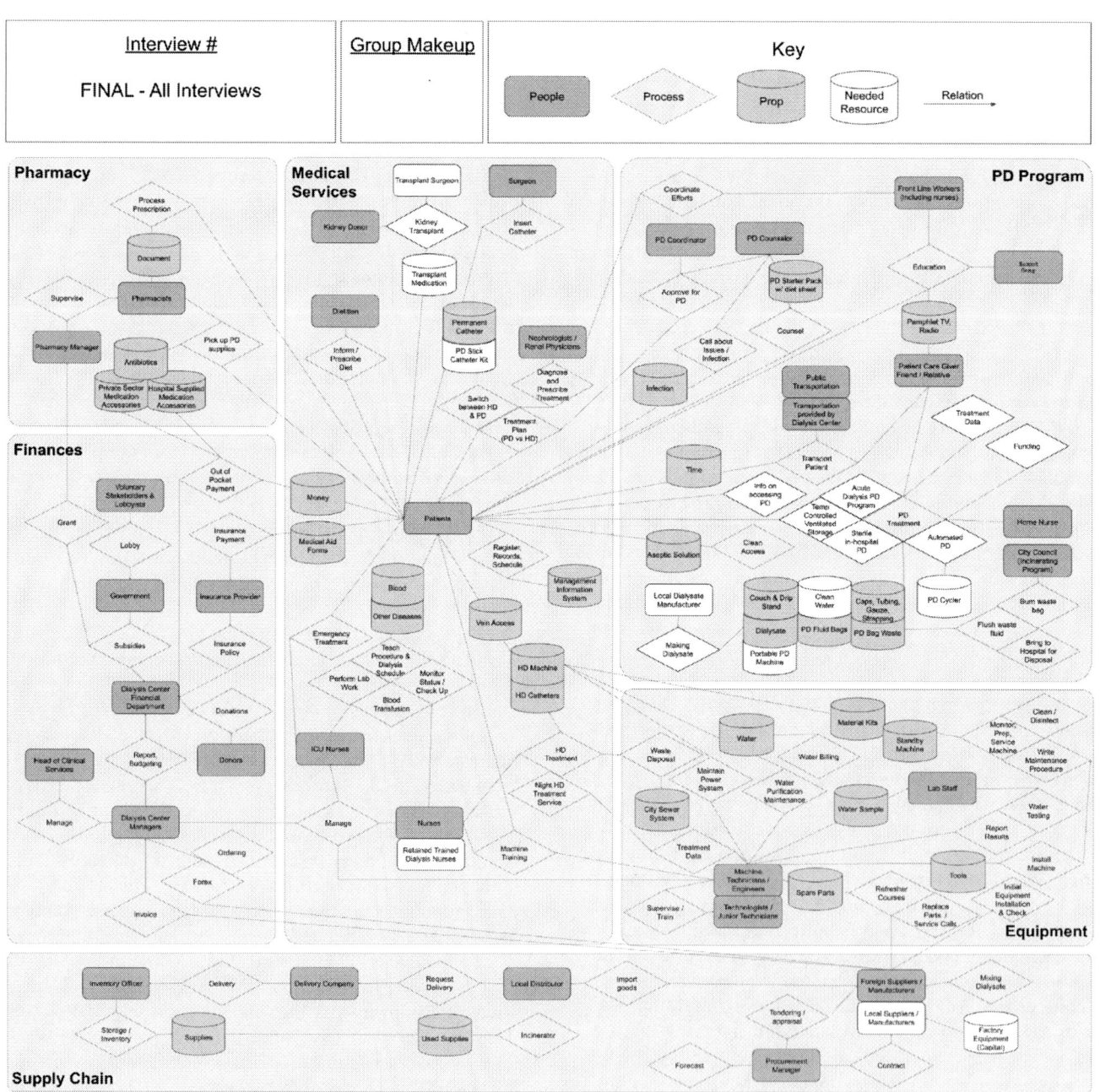

FIGURE 2: FINAL SERVICE MODEL MAP, COMPILED FROM ALL INTERVIEW

Copyright © 2023 by ASME

The interviewees expressed the difficulty some patients have in adjusting to the lifestyle changes required for PD. Interviewees shared stories of patients who did not adhere to hygiene, dietary, and scheduling requirements and the adverse effects this had on their health. For this reason, one of the requirements for a patient to be approved for PD is a reliable support system. The service model map has the code "Patient Care Giver, Friend/Relative" to represent such a support system. Friends and relatives help the patient with their PD treatment and with transportation to the hospital for scheduled check-ups or supply pick-ups. The Pharmacy has PD supplies, dialysate and accessories, which the patient must organize a way to pick up regularly. The transportation process costs the patient time, especially patients that live far from the pharmacy and hospital. Public transportation is an option for some patients and there was a mention of transportation provided by a dialysis center. One interviewee talked about ways to educate people on PD and dialysis, mentioning pamphlets, TV, and Radio as currently utilized methods. There are also support groups for people in dialysis programs so that they can share information and build a community.

Though there are resources in place to support PD in Zimbabwe, interviewees named resources needed to improve the program. A lack of proper funding was brought up as a barrier to PD program expansion. Another difficulty with home-based PD is that it does not generate the same amount and specificity of treatment data as in-center hemodialysis (HD). When complications occur with an HD patient, the logs at the hospital and the diagnostics on the HD machine can be used to troubleshoot and find solutions. When complications occur with PD, the front line workers can only speak with the patient to try and determine the cause. Clean water was important to the PD process but difficult for many potential patients to access. Lack of access to clean water and other environmental concerns, such as not having temperature controlled and ventilated storage for supplies, sometimes led to PD not being prescribed as a viable option for patients. Interviewees also mentioned types of PD that are not currently available in Zimbabwe: automated PD with a PD Cycler, PD for use in treating acute dialysis, and a designated sterile area for PD set up in-hospital. These needed resources were identified by interviewees as potential avenues for improving the current PD program.

4.1.2 Medical Services

The Medical Services section of the service model map has many stakeholders with direction-patient contact, so the patient has been centered in this section of the map. This section includes the majority of codes related to the HD services. The people in this section are nephrologists, surgeons, transplant surgeons, dietitians, and nurses.

Nephrologists are connected to the PD Program section as well as they diagnose, prescribe treatment, and help coordinate efforts with the PD-specific staff. The nephrologist, with the help of the PD Coordinator and PD Counselor, may determine that a patient is a candidate for PD based on their ability to self-administer treatment, their environment, and their support system. If a patient will have difficulty traveling to the hospital for regular HD treatment, the nephrologist is more likely to consider them for the PD program. Determining a patient is appropriate for PD is a multi-step process that includes the home visit of the PD Coordinator and/or PD Counselor.

Many patients are put on HD because it is the quicker option, especially for patients who arrive at the hospital in a state of emergency. There is a process for switching patients from HD to PD, but many patients do not want to go through an additional surgery for PD catheter placement and/or cannot afford the procedure. Surgeons are responsible for inserting catheters. According to one interviewee, there used to be a "PD Stick Catheter Kit" which could be administered by doctors for emergency PD, but the hospital no longer supplies this. Zimbabwe does not currently have a kidney transplant program in place, so the Transplant Surgeon, Kidney Transplant, and Transplant Medication codes are all indicated as "needed" on the service model map.

Nurses are in charge of many processes connected directly to the patient. This is reflected in the service model map where 7 processes connect nurses to patients. The nurse's responsibilities are to teach the patient the procedure and schedule for their dialysis, monitor the status of the patient through check-ups, perform lab work, make blood transfusions, and administer HD Treatment. Some private dialysis clinics offer HD treatments at night, which is important to some patients. If an emergent dialysis patient comes into a hospital that does not regularly offer night sessions, the ICU nurses, who may not be specifically trained in dialysis, will need to administer treatment. One of the resources that is lacking in the current service model is an adequate number of trained dialysis nurses. When nurses are trained in dialysis, new opportunities are opened in other countries, such as South Africa, where there are higher wages, more resources, and better patient outcomes. The dietitian's job is to prescribe the patient a diet conducive to dialysis. In hospitals without dietitians, this role is taken on by nephrologists and nurses.

4.1.3 Pharmacy

The Pharmacy section on the service model map is the smallest section with only 9 codes. Pharmacists and Pharmacy Managers are the people codes identified. The Pharmacist processes the prescription of the patient so that they may receive their medication and accessories (e.g. tubing, gauze, strapping). If the patient has an infection, they will pick up antibiotics from the pharmacy as well. There are both private and government hospital-affiliated public pharmacies. Though some medications and accessories are subsidized in public pharmacies, an interviewee shared that it is not enough to cover all of the patients they need to support. Though the pharmacy may be able to start the year covering a patient's expenses with the subsidies, they will begin charging the patient towards the end of the year when the subsidy runs out. To show this on the map, out-of-pocket payment is connected to both the private sector and hospital supplies. Pharmacy managers supervise

Copyright © 2023 by ASME

pharmacists and manage the grant process with the government to procure the subsidies.

4.1.4 Finances

The concern over funding exists in many portions of the service model map. Dialysis programs are constrained due to lack of funding: the pharmacy runs out of funds for medications, and the government hospitals have reduced hemodialysis for patients to two times per week instead of the typically-recommended three times per week to try and spread money and resources across more patients. If patients in Zimbabwe can afford to go to private clinics, they are often able to receive the recommended three times a week treatment, but many patients cannot. Multiple interviewees expressed remorse for this state of affairs.

The Finances section of the service model map depicts what is happening behind the scenes for institutions to fund their dialysis programs. Patients submit medical aid forms to their insurance provider, who then makes a payment to the dialysis center. The dialysis center is also funded through subsidies from the Zimbabwean government. There is a government policy in place for public hospitals to provide free dialysis services but, according to the interview participants, the amount of funds provided to hospitals has not been adequate for the number of patients. Stakeholders lobby the government for more money for dialysis programs. Dialysis centers are also supported by donations. The people with the highest number of processes identified in this section of the service model are the dialysis center managers. The managers process invoices and place orders from the suppliers, which often involves the foreign currency exchange market. The managers report to the head of clinical services and send budgets to the financial department.

4.1.5 Equipment and Supply Chain

Interview participants also shared details about the supply chain involved for dialysis supplies and equipment. The Inventory Officer code is located in the supply chain section of the service model map, though they would work within the hospital or pharmacy. The inventory officer manages the storage and keeps inventory for supplies. They send used supplies to the incinerator. Goods are imported from suppliers to a local distributor who arranges the delivery of the supplies to the inventory officer. The interview participants always referred to suppliers as foreign entities: Fresenius, Baxter, Mediwise, and Gambro. Local suppliers were identified as a needed resource, but a lack of capital equipment was noted as a barrier to entry for new businesses. Procurement managers handle the tender and bidding process with the suppliers to establish a contract. The procurement managers also forecast the needs of the dialysis clinic to keep supplies and equipment at the needed levels.

The other stakeholder group connected to manufacturers are technicians, who were placed in the Equipment section of the service model map. The manufacturer assists the technician with the initial HD machine installation and verification. The manufacturer trains the technicians on how the equipment operates and the technicians then train the nurses. The manufacturers offer refresher courses for the technicians so that they stay up to date on equipment use and management. Technicians contact the manufacturers for service calls or to request replacement parts as needed. Technicians support the equipment by monitoring, prepping, servicing, cleaning, and writing maintenance procedures for it. They manage many props: spare parts, tools, material kits, and standby machines ready to replace a machine that has gone down so that treatment can resume as quickly as possible. The power system for the hospital and the water purification plant are also managed by technicians. They collect and send water samples to the lab for testing. A report is sent back to the technician and, if anything is wrong, the technician will service the water plant accordingly. Stakeholders in this role may also be called engineers. Some of these processes could alternatively be delegated to the technologists who are supervised and trained by a technician. Most of the process codes mentioned for technicians by interviewees related to in-hospital treatment and HD machines. No codes were directly connected between the PD Program section and the Equipment section of the service model map.

These insights were synthesized from the 18 interviews with stakeholders performed by the student interviewers. The information gathered will factor into the ESKD Design Project for the implementation of their developing technology. By better understanding the current state of the peritoneal dialysis program and dialysis service model as a whole, the Wright Lab can make informed decisions regarding how to best establish their technology to help improve peritoneal dialysis accessibility.

4.1.6 Key Barriers to Dialysis Access

Insufficient funding and challenges with transportation lead patients to receive fewer HD treatment sessions than recommended and to struggle with out-of-pocket payments. While a transition to PD services may help overcome these barriers to HD access, the following key barriers to PD expansion in Zimbabwe were also identified:

- There is currently only one home-based PD program and no in-hospital PD;
- Automated PD is not available;
- There is no local production of dialysate, which increases the price of PD supplies;
- There is a reliance on surgeons for catheter insertion and an inadequate number of available surgeons;
- HD catheters are easier to insert, causing HD to become the default choice in emergency situations;
- Once nurses become trained in dialysis they are often recruited by institutions outside Zimbabwe;
- Not all patients have an adequate home environment for PD supply storage;
- Not all patients have access to clean water, which is required for PD;

Copyright © 2023 by ASME

248

- Transportation for check-up appointments and supply pick-ups is inaccessible to some patients;
- Patients have difficulty adjusting to the PD lifestyle of schedule, diet, and hygiene requirements which can lead to complications, including infection; and
- The available PD equipment does not collect the data needed to guide decisions when complications arise.

These barriers are suggestive of potential policy, education, service, and technology interventions that may be necessary to enable the expansion of PD in Zimbabwe.

4.2 Information Gathered as a Function of Team Composition

The final service model map (Figure 2) was duplicated and used to depict which codes appeared in each interview indicated with red stars (e.g. Figures 3 and 4). The total number of codes per interview was used as a quantitative indicator of interviewer groups' success. Each group is identified by the countries affiliated with each partner. There were four intracultural groups having students from the same country, two groups from the United States, one group from Uganda, and one group from Zimbabwe; and two intercultural groups with one Ugandan student and one US student each. The Zimbabwean group is assumed to have the closest cultural background to the Zimbabwean interviewees. The Ugandan interviewers have a different background than the interviewees, yet closer than the interviewers from the United States.

Figure 3 shows a service model map for an interview with a technician. Many of the codes discovered are in Equipment, the area highlighted as the main focus for the technician. This shows that the interview team went into depth with the interview participant in the area they are most knowledgeable on, which resulted in many connections being made. Figure 4 shows another example of a service model map from an interview with a technician, however this map has far fewer codes and connections in the Equipment section, demonstrating a potential missed opportunity to dig deeper with that participant. Figure 3 is an interview by an intercultural UG-US group while Figure 4 is from an interview by an intracultural UG-UG group. Though these figures show an intercultural group outperforming an intracultural group, it is just one example. To look more broadly at the output of the different groups, bar graphs (Figure 5) and radar charts (Figure 6) were created.

Figure 5 shows the total number of codes identified in each interview, organized chronologically (left to right) within each interviewer group. The average number of codes collected per group is shown with horizontal lines. The averages aid in comparing the intercultural groups (UG-US, color-coded green) from the intercultural groups (UG-UG orange, US-US blue, ZW-ZW yellow). The top performing group based on the average number of codes identified was intercultural (Group 1), however two intracultural groups had similar average output (Groups 3 and 4). The intercultural groups had more high code-producing interviews. Of the top half of interviews, the majority (6 of 9) were conducted by intercultural groups. This is of

particular note as only two of the total six groups were intercultural.

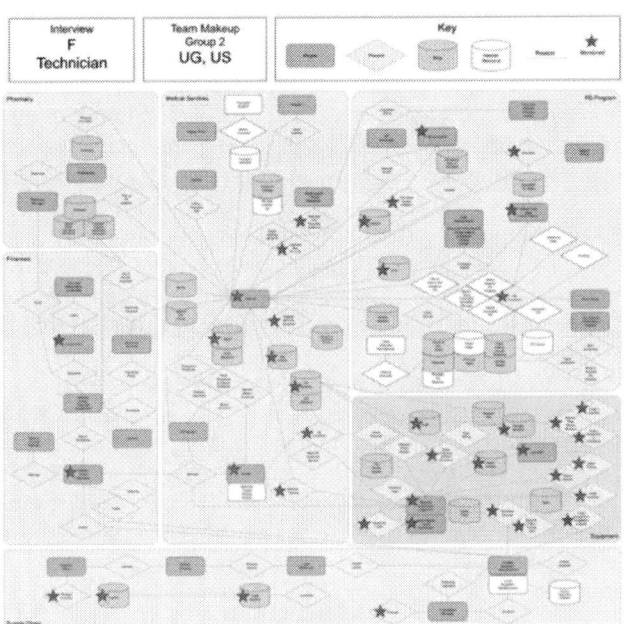

FIGURE 3: SERVICE MODEL MAP FROM FIGURE 2 WITH RED STARS DENOTING CODES GATHERED FROM INTERVIEWEE "F"

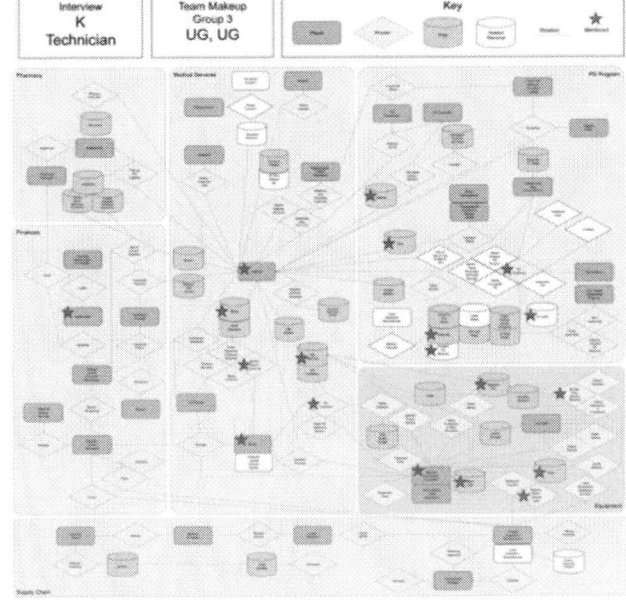

FIGURE 4: SERVICE MODEL MAP FROM FIGURE 2 WITH RED STARS DENOTING CODES GATHERED FROM INTERVIEWEE "K"

The interviews were organized in succession to see potential patterns in performance as groups gained interview experience. Group 1 and 2 produced a reduced number of total codes in each subsequent interview, with Group 1 ranging from 55 to 39 codes and Group 2 ranging from 41 to 28 codes.

Copyright © 2023 by ASME

Groups 3 and 4 did not consistently increase or decrease the number of codes gathered over time. Group 3's third interview yielded 57 codes and their fourth interview yielded 19 codes, the largest range of codes of all groups. Group 5 had an increase between their first interview (19 codes) and their second (30 codes). Group 6 stayed almost constant between their two interviews (18 and 16 codes). Though the intercultural groups had a pattern of decreasing codes in subsequent interviews, the inconsistency of performance across all groups shows that other factors, such as the breadth of knowledge of the interviewee, may have a large impact on the number of codes gathered. Some interviewees may have been uniquely knowledgeable or particularly reluctant to share information. In order to reduce the impact of the individual interviewees, a larger dataset would need to be examined.

One area of interest is how group makeup may affect the breadth of information gathered. The groups were tasked with seeking interview contacts from a wide variety of perspectives to build the most holistic service model map possible. Radar charts (Figure 6) display the number of codes each group uncovered by topic area. These radar charts show how successful each group was in covering a wide breadth of information and which, if any, topic areas were favored.

Each radar chart represents a different group. The color-coding remains consistent with that of the bar graph (Figure 5). The individual interviews are divided to help compare the groups which conducted differing numbers of interviews. Note that groups with more interviews had more opportunities to expand the overall area of their radar chart.

Each angular axis of the radar chart represents a different topic area related to the dialysis service model map (Figure 2). The values on the angular axes are the number of codes identified in that topic area. Radar charts with a greater colored area represent interviews that identified more codes. The direction(s) the radar charts skew show in which topics each group had greater success identifying codes. The average number of codes in each topic area from all 18 interviews was also calculated and added to the charts as a black outline.

Using the black outline on the radar charts, the number of codes each group gathered per topic area can be compared to the group average. Of particular interest are how the intercultural groups compare to the average. The group that gathered the greatest variety of information was Group 1, an intercultural group. Group 1's radar chart extends beyond the group average in all directions. Group 1's greatest focus area, 21 codes gathered for the topic of PD Program in a single interview, far exceeded the average of 10 codes. The other intercultural group, Group 2, exceeded the average in all topic areas except for Pharmacy where they met the average of 2 codes in a single interview. Group 2 only exceeded the averages for the topics of PD Program and Medical Services by a narrow margin (2 and 1 codes respectively). While both intercultural groups performed well, Group 1 contributed the greatest variety of codes while Group 2 performed more closely to the average.

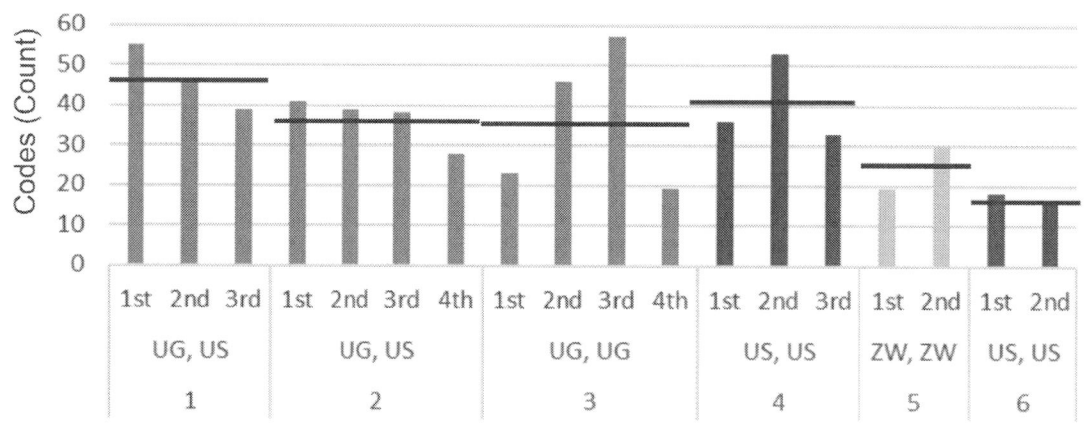

FIGURE 5: GROUP MAKEUP AND INFORMATION UNCOVERED PER INTERVIEW (COUNT OF CODES)

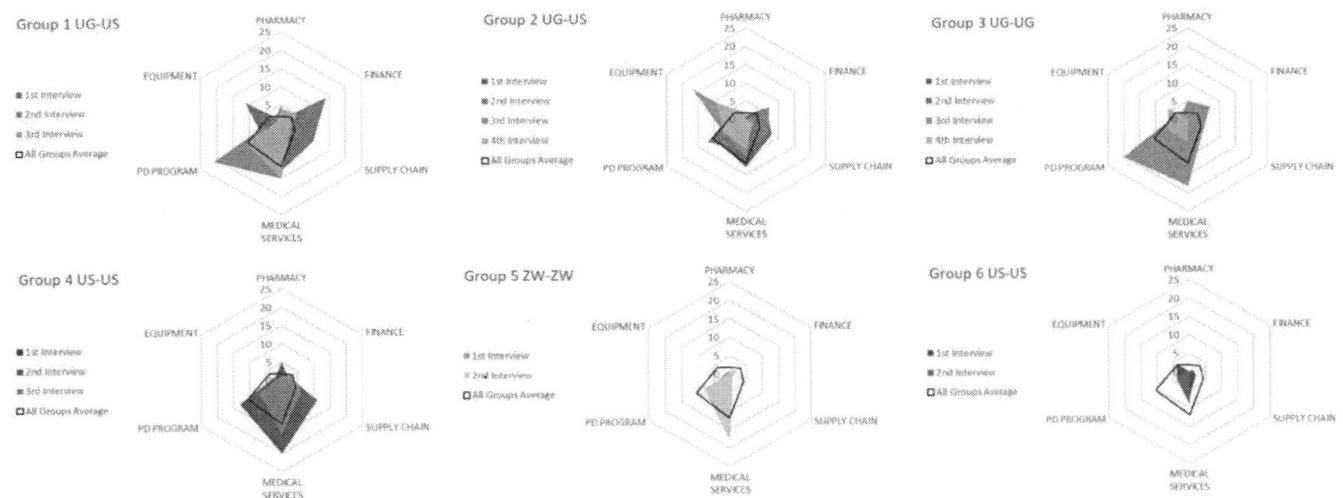

FIGURE 6: INFORMATION GATHERED (INDICATED BY NUMBER OF CODES) PER GROUP, COMPARED TO GROUP AVERAGE

4.3 Themes in Post-Program Conversations

The PIs for this research conducted post-program interviews with 8 of the students. The objective of these interviews was to learn about ways to improve the training program in the future, and to understand more about the group dynamics during the field interviews. There were several themes that emerged.

In general, students enjoyed the opportunity to work cross-culturally. The differences between cultures generated curiosity and empathy, two key components for design interviewing. The students were motivated by the purpose of the End Stage Kidney Disease research project and wanted to contribute to helping the dialysis communities in Zimbabwe. One student from the United States noted that interacting with their peer interviewers from Zimbabwe was helpful in preparing to conduct interviews with people in Zimbabwe. Increasing cross-cultural interactions between students during the training program could further prepare students before field interviews.

The interview training included three rounds of practice interviews which included cross-cultural interactions. The students found the practice interviews to be particularly helpful for skill development. One student noted that it was hard to transition from the practice interview topic of grocery shopping preferences to dialysis systems. The practice session prompt could be changed to a medical focus which would cover topics similar to those for medical device development.

Students found interviewing to be uncomfortable at first. The unfamiliarity of interviewing caused nervousness, but confidence increased with subsequent interviews. Some students wished they had a greater depth of knowledge around the medical device being developed for the project. The students received an overview of the device, but had difficulty with answering interviewee questions about the project. The training could be improved by adding an exercise to allow students to observe and/or practice answering such questions. Time was a large factor in retention of information, both skills and background knowledge on the project. Due to a delay in IRB processing and approval, there was a multi-month time gap between the training program and the field interviews which reduced readiness. Additionally, many students experienced difficulty in securing interview contacts, which resulted in long periods of time (weeks, months) between consecutive interviews, also impacting readiness. Students suggested refresher courses and weekly check-in meetings during field interviews as solutions to this issue. The addition of a training segment on best practices for finding and securing interview contacts is also recommended, as many students were unfamiliar with such methods.

One of the primary difficulties conducting virtual interviews internationally is the technological concerns. Students experienced poor internet connectivity, unclear video and audio, Zoom crashing, and Google Docs not consistently working. Working virtually also required coordination across time zones, which made finding shared availability difficult. Students in intercultural groups needed to find the best way to contact each other with Whatsapp emerging as the most convenient resource for both. Virtual interviewing eliminates travel concerns, but comes with specific technological challenges.

5.0 LIMITATIONS

The NVivo coding was performed by one research assistant and was therefore subject to the limitations of that assistant's perceptions. The summaries and notes submitted by the student groups were also utilized to help ensure all output was extracted from each interview. The service model maps were first constructed using the summaries and notes to incorporate the perspectives of the interviewer groups. Any additional people, processes, and props identified through the NVivo coding process were then added to each map. These multiple perspectives help to achieve a more thorough interpretation of the findings, but are still limited by the perceptual skills of those involved.

Copyright © 2023 by ASME

The practice of making a service model map is inherently variable. There are many different ways to represent and organize the same information. Some codes could have been combined, while others could have been broken down into more detail. Some information lent itself better to being included on a service model map than other pieces of information. The "needed" codes were also a product of interpretation if they existed in Zimbabwe but in inadequate numbers in the opinion of some interview participants.

The number of codes uncovered in an individual interview depends on more variables than interviewer group makeup. One major variable in any interview is the interview participant themselves. Some interview participants are much more knowledgeable about the study area than others, some are more forthcoming, and some are more skilled at being interviewed. The way to reduce the effect of the individual interviewees on the data is to conduct more interviews. However, this study had a limited number of interviews (2-4) performed per group, which makes the results only an early indication of potentially replicable findings.

Another way to increase the strength of the data would be to incorporate more student interviewer groups. This would help to reduce the effect of the skill and commitment level of each interviewer. This study attempted to control skill level by starting with all novice interviewers and having them participate in the same training program. However, the students who emerged from the training program demonstrated unequal skill levels. This is evident by some groups having more success in completing their task of recruiting, organizing, and completing interviews than others. The groups who conducted the lowest number of interviews, groups 5 and 6, also had the lowest average amounts of information in their interviews. This demonstrates that the results of the output are also skewed by the participation level of the groups due to the small number of participants.

This study involved six groups in total, two of which were intercultural. Only one group had students from the same country as the interview participants, and it was an intracultural group. Another area of future exploration could involve mixed groups where one partner has the same cultural background as the interviewees.

Though this study focused on the cultural variable of country of origin, the niche topic of dialysis may have cast all interviewers as outsiders. None of the students were dialysis practitioners, which resulted in many topics being unfamiliar even to the Zimbabwean interviewers.

6.0 CONCLUSION

This study involved 12 student interviewers from three different countries. The students participated in a training program to gain interviewing skills and then conducted interviews as part of the End Stage Kidney Disease Design Project for the Wright Lab. The groups were intentionally organized to compare intercultural and intracultural group makeup on interview output. Here "culture" was defined by the city of the student's undergraduate education. Findings could

be used by teams performing intercultural research to guide interviewer recruitment decisions.

A service model map was created depicting the state of dialysis in Zimbabwe to answer Research Question 1: *What does the current dialysis program look like in Zimbabwe and what is limiting its expansion?* HD is the primary modality for dialysis in Zimbabwe, but HD treatments are not accessible to many patients due to transportation challenges. There is only one home-based PD program for patients who cannot access the HD modality. The barriers to PD program expansion identified in this study include: no local production of dialysate, shortages of trained PD professionals, difficulties in accessing clean water for PD in the home environment, difficulties in tracking medical information in the home environment, and challenges with adjusting to the PD lifestyle, among others.

This study demonstrated methods for answering Research Question 2: *How does team composition (inter- versus intracultural) affect the output of the interview?* The majority of top scoring interviews were conducted by intercultural groups. Though these results were inconclusive largely due to the limited dataset, the techniques described could be utilized in future studies to further explore the research questions proposed. The study measured interview output as the number codes identified. The people, processes, and props were delineated using NVivo and the connections were illustrated on the service model maps created with Google Slides. Radar charts and bar graphs were used to visualize the differences between the groups. Future studies on this topic could employ similar practices to build upon these preliminary findings.

ACKNOWLEDGEMENTS

The authors acknowledge the contribution of the 12 undergraduate students who served as both interviewers and research subjects in this study, and of Jennifer Hoody, who recruited the students for this study and coordinated their training in exploratory design interviewing.

This work was supported by the Center for Global Health and Social Responsibility (CGHSR), and the Kusske Design Initiative (KDI) through the College of Design, both at the University of Minnesota, USA.

REFERENCES

[1]   S. Pink, V. Fors, D. Lanzeni, M. Duque, S. Sumartojo, and Y. Strengers, *Design Ethnography: Research, Responsibilities, and Futures*. London: Routledge, 2022. doi: 10.4324/9781003083665.

[2]   S. Reeves, J. Peller, J. Goldman, and S. Kitto, "Ethnography in qualitative educational research: AMEE Guide No. 80," *Med. Teach.*, vol. 35, no. 8, pp. e1365–e1379, 2013, doi: 10.3109/0142159X.2013.804977.

[3]   A. Kaliber, "Reflecting on the reflectivist approach to qualitative interviewing," *Azimuth*, vol. 8, no. 2, pp. 339–357, 2019.

[4]   K. K. Tsang, D. Liu, and Y. Hong, "Introduction: Qualitative Research Methods in Empirical Social

Sciences Studies—Young Scholars' Perspectives and Experiences," in *Challenges and Opportunities in Qualitative Research: Sharing Young Scholars' Experiences*, K. K. Tsang, D. Liu, and Y. Hong, Eds., Singapore: Springer, 2019, pp. 1–5. doi: 10.1007/978-981-13-5811-1_1.

[5] P.-C. Hsiung, "Teaching Reflexivity in Qualitative Interviewing," *Teach. Sociol.*, vol. 36, no. 3, pp. 211–226, 2008.

[6] C. Huff, "How artificial kidneys and miniaturized dialysis could save millions of lives," *Nature*, vol. 579, no. 7798, pp. 186–188, Mar. 2020, doi: 10.1038/d41586-020-00671-8. ·

[7] G. Ashuntantang *et al.*, "Outcomes in adults and children with end-stage kidney disease requiring dialysis in sub-Saharan Africa: a systematic review," *Lancet Glob. Health*, vol. 5, no. 4, pp. e408–e417, Apr. 2017, doi: 10.1016/S2214-109X(17)30057-8.

[8] A. A. House, B. Pham, and D. E. Pagé, "Transfusion and recombinant human erythropoietin requirements differ between dialysis modalities.," *Nephrol. Dial. Transplant.*, vol. 13, no. 7, pp. 1763–1769, Jul. 1998, doi: 10.1093/ndt/13.7.1763.

[9] A. Bello *et al.*, "Global Kidney Health Atlas: A Report by the International Society of Nephrology on the Global Burden of End-Stage Kidney Disease," Brussels, 2019. Accessed: Mar. 12, 2023. [Online]. Available: https://www.theisn.org/initiatives/global-kidney-health-atlas/

[10] National Institute of Diabetes and Digestive and Kidney Diseases, National Institutes of Health, *Man receiving peritoneal dialysis*. Accessed: Mar. 12, 2023. [Online]. Available: https://www.niddk.nih.gov/news/media-library/18241

[11] R. Y. Bayeck, "The Intersection of Cultural Context and Research Encounter: Focus on Interviewing in Qualitative Research," *Int. J. Qual. Methods*, vol. 20, p. 1609406921995696, Jan. 2021, doi: 10.1177/1609406921995696.

[12] S. Shah, "The researcher/interviewer in intercultural context: a social intruder!," *Br. Educ. Res. J.*, vol. 30, no. 4, pp. 549–575, Aug. 2004, doi: 10.1080/0141192042000237239.

[13] A. Au, "Thinking about Cross-Cultural Differences in Qualitative Interviewing: Practices for More Responsive and Trusting Encounters," *Qual. Rep.*, vol. 24, no. 1, pp. 58–77, Jan. 2019, doi: 10.46743/2160-3715/2019.3403.

[14] P. Liamputtong, "Doing Research in a Cross-Cultural Context: Methodological and Ethical Challenges," in *Doing Cross-Cultural Research: Ethical and Methodological Perspectives*, P. Liamputtong, Ed., in Social Indicators Research Series. Dordrecht: Springer Netherlands, 2008, pp. 3–20. doi: 10.1007/978-1-4020-8567-3_1.

[15] D. T. Peters and L. A. Giacumo, "Ethical and Responsible Cross-Cultural Interviewing: Theory to Practice Guidance for Human Performance and Workplace Learning Professionals," *Perform. Improv.*, vol. 59, no. 1, pp. 26–34, 2020, doi: 10.1002/pfi.21906.

[16] F. Irvine, G. Roberts, and C. Bradbury-Jones, "The Researcher as Insider Versus the Researcher as Outsider: Enhancing Rigour Through Language and Cultural Sensitivity," in *Doing Cross-Cultural Research: Ethical and Methodological Perspectives*, P. Liamputtong, Ed., in Social Indicators Research Series. Dordrecht: Springer Netherlands, 2008, pp. 35–48. doi: 10.1007/978-1-4020-8567-3_3.

[17] V. Reyes, "Ethnographic toolkit: Strategic positionality and researchers' visible and invisible tools in field research," *Ethnography*, vol. 21, no. 2, pp. 220–240, Jun. 2020, doi: 10.1177/1466138118805121.

[18] R. C. Walker, A. Tong, K. Howard, and S. C. Palmer, "Clinicians' experiences with remote patient monitoring in peritoneal dialysis: A semi-structured interview study," *Perit. Dial. Int.*, vol. 40, no. 2, pp. 202–208, Mar. 2020, doi: 10.1177/0896860819887638.

[19] D. J. Campbell, J. C. Craig, D. W. Mudge, F. G. Brown, G. Wong, and A. Tong, "Patients' Perspectives on the Prevention and Treatment of Peritonitis in Peritoneal Dialysis: A Semi-Structured Interview Study," *Perit. Dial. Int.*, vol. 36, no. 6, pp. 631–639, Nov. 2016, doi: 10.3747/pdi.2016.00075.

[20] S. Mougel, H. Tabibi, and M. Rosier, "Narrative interviews to assess quality of life in Peritoneal Dialysis," *Bull. Dial. À Domic.*, vol. 4, no. 3, Art. no. 3, Jul. 2021, doi: 10.25796/bdd.v4i3.62223.

[21] N. S. Lymperopoulos, R. Jeevan, L. Godwin, D. Wilkinson, K. Shokrollahi, and M. I. James, "The Introduction of Standard Operating Procedures to Improve Burn Care in the United Kingdom," *J. Burn Care Res.*, vol. 36, no. 5, pp. 565–573, Sep. 2015, doi: 10.1097/BCR.0000000000000210.

[22] M. Gilchrist, B. D. Franklin, and J. P. Patel, "An outpatient parenteral antibiotic therapy (OPAT) map to identify risks associated with an OPAT service," *J. Antimicrob. Chemother.*, vol. 62, no. 1, pp. 177–183, Jul. 2008, doi: 10.1093/jac/dkn152.

[23] G. T. Jun, C. Morrison, and P. J. Clarkson, "Articulating current service development practices: a qualitative analysis of eleven mental health projects," *BMC Health Serv. Res.*, vol. 14, no. 1, p. 20, Jan. 2014, doi: 10.1186/1472-6963-14-20.

[24] M. Grover, C. Lauff, and N. Wright, "Towards quantifying interviews: comparing techniques to evaluate the quality of design interviews," In Review.

[25] M. Grover, N. Wright, J. Hoody, and C. Lauff, "Developing design ethnography interviewing competencies for novices," presented at the 2022 ASEE Annual Conference & Exposition, Aug. 2022. Accessed: Mar. 04, 2023. [Online]. Available: https://peer.asee.org/developing-design-ethnography-interviewing-competencies-for-novices

Copyright © 2023 by ASME

Proceedings of the ASME 2023
International Design Engineering Technical Conferences and
Computers and Information in Engineering Conference
IDETC-CIE2023
August 20-23, 2023, Boston, Massachusetts

DETC2023-116962

AI-ACCELERATED DESIGN OF EVIDENCE SYNTHESIS FOR GLOBAL DEVELOPMENT

Kristen M. Edwards[1,†,*], Binyang Song[1,†], Jaron Porciello [2], Mark Engelbert[3], Carolyn Huang[3], Faez Ahmed[1],

[1]Massachusetts Institute of Technology, Cambridge, MA
[2]University of Notre Dame, South Bend, IN
[3]International Initiative for Impact Evaluation, Inc.

ABSTRACT

When designing evidence-based policies and programs, decision-makers must distill key information from a vast and rapidly growing literature base. Identifying relevant literature from raw search results is time and resource intensive, and is often done by manual screening. In this study, we develop an AI agent based on a bidirectional encoder representations from transformers (BERT) model and incorporate it into a human team designing an evidence synthesis product for global development. We explore the effectiveness of the human-AI hybrid team in accelerating the evidence synthesis process. To further improve team efficiency, we enhance the human-AI hybrid team through active learning (AL). Specifically, we explore different sampling strategies, including random sampling, least confidence (LC) sampling, and highest priority (HP) sampling, to study their influence on the collaborative screening process. Results show that incorporating the BERT-based AI agent into the human team can reduce the human screening effort by 68.5% compared to the case of no AI assistance and by 16.8% compared to the industry standard of using an n-gram language model to encode texts and a support vector machine (SVM)-based classifier for identifying 80% of all relevant documents. When we apply the HP sampling strategy for AL, the human screening effort can be reduced even more to 78.3% for identifying 80% of all relevant documents compared to no AI assistance. We apply the AL-enhanced human-AI hybrid teaming workflow in the design process of three evidence gap maps (EGMs) for USAID and find it to be highly effective. These findings demonstrate how AI can accelerate the development of evidence synthesis products and promote timely evidence-based decision making in global development in a human-AI hybrid teaming context.

Keywords: AI in design, Natural Language Processing,

[†]Joint first authors
[*]Corresponding author

Global Development, Evidence Synthesis

1. INTRODUCTION

In 2011 the U.S. Agency for International Development (US-AID) released *Evaluation Policy*, and in doing so made an ambitious commitment to rigorously evaluating evidence in order to make evidence-based policy [1]. Evidence-based policy refers to public policy that is based on, or informed by, evaluated and objective evidence. To emphasize the importance of evidence-based policy within USAID and the U.S. government, the Foundations for Evidence-based Policymaking Act of 2018 required all agencies under the Act to "affirm the agency's commitment to conducting rigorous, relevant, evaluations and to using evidence from evaluations to inform policy and practice" [2]. It is imperative in part because these policies dictate the expenditure of billions of dollars. For example, in 2017, USAID spent $1.01 billion on foreign agricultural assistance alone [3].

However, evaluating all available evidence has been made burdensome by the current information explosion. In 2018 alone, global research output in science and engineering was 2.6 million articles, which grew at a rate of 4% annually from 2008-2018 [4]. A person's capacity to understand all available research is limited. Policy-makers have thus turned to evidence synthesis to understand the growing corpus of research available and make informed decisions. Evidence synthesis refers to the process of compiling information and knowledge from many sources and disciplines to inform decisions [5, 6]. However, creating evidence synthesis products like evidence gap maps (EGMs) requires extensive time and effort from human experts. EGMs, as described in the Related Works section, visualize interventions and their associated outcomes [7], and have been shown to provide incredible value to decision-makers in fields ranging from agriculture to public health [5]. For example, Figure 1 represents a portion of an EGM available from 3ie [1]. We can see there is a research gap

---

[1]https://developmentevidence.3ieimpact.org/egm/food-systems-and-nutrition-

Copyright © 2023 by ASME

between the interventions of "water access & management" and "improved seeds" and the outcomes regarding "profit". Policymakers can plan future investments and research accordingly.

Our goal is to accelerate the design of EGMs in the global development space and alleviate the burden of information filtering. The International Initiative for Impact Evaluation (3ie) is one of the global leaders in generating EGMs for decision-making. 3ie's current evidence synthesis process includes significant expert screening of documents and moderate use of machine intelligence, often taking nearly six months to complete [6]. Natural language processing (NLP), a form of artificial intelligence (AI), has long been used for text comprehension. Recently, the rule-based NLP models have attracted some attention and been explored to promote evidence-based decision making in the medical, legal, and global development fields [8–10]. The work that has successfully done so may be improved upon by incorporating the latest transformer- and transfer learning-based NLP models.

1.1 Contributions

Title and abstract (TA) screening is one of the most time-consuming steps in the EGM design process, typically involving comprehending the titles and abstracts of tens or hundreds of thousands of papers for screening. Through collaborating with 3ie, we make the following contributions:

1. We develop a BERT-based AI agent to accelerate the TA screening portion of the EGM design process, and incorporate it into a human team to explore the efficiency gains made through human-AI teaming. With the best combination, our AI agent reduces human effort by 78.3% when identifying 80% of all eligible documents, as compared to no AI assistance.

2. We compare our BERT-based AI agent against the industry standard SVM-based model, and find that the BERT-based model outperforms SVM in both model performance (12% average increase in accuracy for the three EGMs) and saved effort (17% reduction in required effort in the simulated case, and a 46% average reduction in effort for the three deployed EGMs).

3. We identify the optimal training size (5,000 documents) for both model performance and saved effort.

4. We compare active learning strategies and find that by using HP or LC we can decrease human effort by an additional 30% (compared to BERT with no AL) for identifying 80% of all included documents.

5. We support the development of three EGMs: Agriculture, Nutrition, and Resilience.

2. RELATED WORK

In the following sections we describe related work in the fields of evidence gap maps, natural language processing, and active learning, particularly in the context of human-AI teams.

2.1 Evidence Gap Maps

EGMs are one form of evidence synthesis - the process of compiling information and knowledge from many sources and

---
evidence-gap-map

disciplines to inform decisions [5, 6]. Evidence synthesis provides more reliable information about a topic than a single study by systematically collecting, categorizing, and analyzing a broad range of studies [11]. Evidence synthesis for decision making was largely popularized by the biomedical field, but it provides clear benefits for decision-makers in any field [9, 12, 13]. Thus, evidence synthesis is an incredibly valuable tool for decision-makers in global development seeking to design policies and fund research [5].

The International Initiative for Impact Evaluation (3ie) has pioneered the use of EGMs, which present a visual overview of completed and ongoing impact evaluations and systematic reviews in a specific sector [14]. 3ie creates these EGMs via the "thematic [collection] of information about impact evaluation and systematic reviews that measure the effects of international development policies and programmes" [7]. The final product is a matrix, organized by "intervention" categories on the vertical axis and "outcome" categories on the horizontal axis. Interventions are the action taken in the study, and outcomes are the result of the action. Each cell of the matrix contains studies that rigorously evaluate the impact of a specific intervention on a specific outcome.

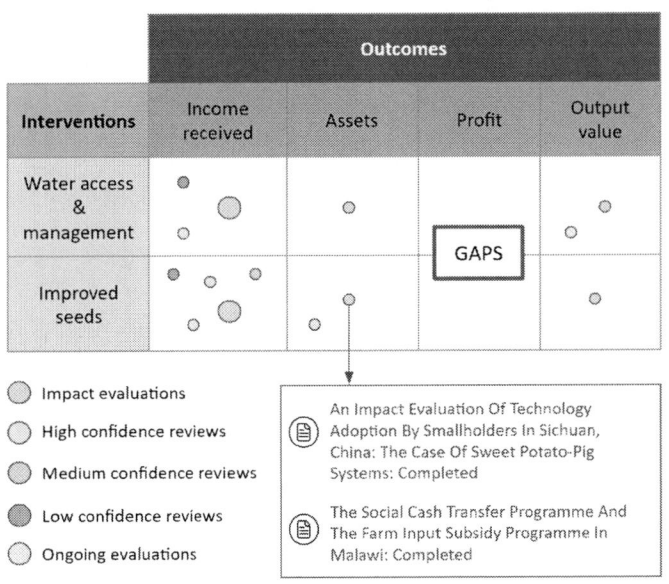

FIGURE 1: A REPRESENTATION OF A PORTION OF A 3IE EGM SHOWING TWO INTERVENTIONS AND FOUR OUTCOMES. RESEARCH GAPS EXIST BETWEEN THE TWO INTERVENTIONS AND THE OUTCOME "PROFIT". DOTS OF DIFFERENT COLORS REPRESENT DIFFERENT EVIDENCE TYPES. DOT SIZES INDICATE HOW MANY DOCUMENTS EXIST IN EACH GROUP.

3ie sets the global standard for EGMs, and the mapping method has been adapted by organizations including the Campbell Collaboration, the World Bank Independent Evaluation Group, and USAID [14]. Like other forms of evidence synthesis, EGMs begin with an expansive and systematic search of scholarly databases and "grey literature" sources (such as repositories of government documents or websites of think-tanks) to identify potentially relevant studies. EGM teams then screen these search

Copyright © 2023 by ASME

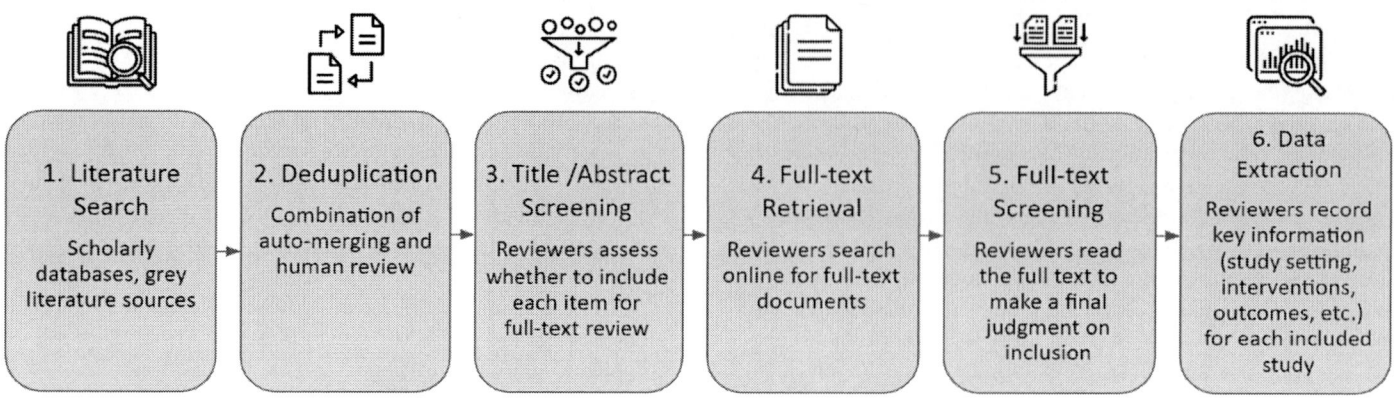

FIGURE 2: A HIGH LEVEL VIEW OF THE CURRENT EGM CREATION PROCESS.

results to identify studies that meet the EGM's criteria for interventions evaluated, outcomes measured, implementation setting, and study design. Once eligible studies are identified, the EGM team extracts information on interventions, outcomes, and other key characteristics of each study to determine its placement in the EGM matrix and to allow for analysis of trends in the literature.

3ie uses a software called EPPI-Reviewer which aids in the creation of EGMs. While EPPI-Reviewer has some machine learning functions that can accelerate screening [15], most EGM tasks are still performed manually. Thus, each EGM requires significant human effort and expertise, with many EGMs requiring nearly six months to complete [6]. Given that one of the main barriers to evidence use among policymakers is the lack of timely research outputs [16], there is a critical need to reduce the time and effort needed to complete the EGM design and development process.

The high level steps of designing an EGM are shown in Figure 2. Our work focuses on step three, in which reviewers screen documents for inclusion in an EGM based on their title and abstract. Selected documents will move on to full-text review. We create three transformer-based NLP models that automatically classify documents for inclusion at this step.

2.2 Natural Language Processing in Evidence Synthesis

NLP is a field of machine learning in which computational machines are trained to understand text and spoken language. Historically in NLP words are represented as vectors where similar words are located near each other in continuous space [17–19]. In the medical field, the development of an NLP-based model for automating evidence synthesis, called BioMedICUS, improved the scalability and performance of text analysis and processing of biomedical and clinical reports [8]. The success of NLP in the medical field has led to its use in other fields, with models like LexNLP, which automatically extracts information from legal text [9].

There are several industry-standard NLP tools used to aid human experts when designing evidence synthesis products like EGMs. The most common tools include EPPI-Reviewer, Rayyan, and RobotReviewer [12, 13], and all of these utilize the frequency based n-gram language model to encode texts and a support vector machine(SVM) classifier or an ensemble of two SVM classifiers

as their primary ML model [20–22]. Their first classifier encodes texts as individual words, pairs of words, and triplets of words (uni-, bi-, and tri-grams). The second classifier uses a unigram model (i.e., each word is considered individually).

Furthermore, rule-based NLP models have been explored to promote evidence-based decision making in multiple fields [10]. However, the rule-based models are often case-specific. It requires significant effort to adapt a rule-based model from one EGM to another. Moreover, it is challenging to capture all the subjective criteria used by humans and embed them into the defined rules comprehensively.

Modern NLP models have been largely shaped by the introduction of the transformer in 2017, which allowed text inputs to be fed in parallel and achieved state-of-the-art results over SVM and other models in many NLP tasks [23]. Bidirectional Encoder Representations from Transformers (BERT) is among the most well-known transformer-based models and has been extensively explored in NLP tasks such as language translation and question answering [24, 25]. Other such models include GPT and models based off of it [26].

Our work explores BERT-based NLP models as a tool for human-AI teams designing EGMs for global development, which involves more unstructured studies and broader domains than other fields. Recent research that is perhaps most similar to our work is srBERT [27], which explores fine-tuning a BERT model with topic-specific articles in order to accelerate the screening process for a systematic review about "moxibustion for improving cognitive impairment" [27]. Our work, on the other hand, works with much larger and broader datasets in order to create EGMs. Additionally, we exhibit the effectiveness of our NLP tool in a real human-AI team and ultimately create three deployed EGMs in Agriculture, Nutrition, and Resilience. We utilize the experience of creating deployed EGMs to explore the nuances of human-AI teaming in this design process.

2.3 Active Learning and Human-AI Teams

In many AI tasks, obtaining labeled training data is expensive and time-consuming [28]. We are motivated to explore avenues to decrease the size of training data needed by using active learning (AL). AL is the concept that an ML algorithm can perform better with less training data if it is allowed to choose

Copyright © 2023 by ASME

the data from which it learns. AL has been applied to deep learning problems such as image classification [29, 30], speech recognition [31], data exfiltration detection [32], and many NLP tasks [33]. There are three main problem setups, or scenarios, in which a learner may be able to ask queries: membership query synthesis [34], stream-based selective sampling [35], and pool-based sampling[36]. The most commonly used pool-based sampling strategies evaluate and rank the entire unlabeled pool in terms of informativeness and then select the best queries [37]. There are also different *query strategies* for choosing which un-labeled instances to query. The most commonly used query strategy is uncertainty sampling [36]. In this strategy, the learner queries the instances for which the learner is least certain how to label [37]. Within the uncertainty sampling category, there are three primary measures that evaluate how uncertain the learner is about each instance: least confidence [38], margin sampling [39], and entropy [40].

During the training, AL tries to optimize the information flow from humans to AI to improve AI performance with less training data. In this study, AI is working collaboratively with humans instead of alone. Accordingly, both the information flow from humans to AI and that from AI to humans are important to the performance of human-AI hybrid teams. Therefore, AL in such a context should consider the bi-directional information flows.

In fact, common barriers stopping human screeners from in-corporating AI into their EGM design process include a mismatch in existing workflows, and a steep learning curve [13]. Research shows that human-AI teams using active learning for a real life task can lose the human agent's trust if the AI agent makes irrel-evant suggestions or predictions during the training process [32]. Therefore, we explore and discuss the effective integration of AI tools into the existing EGM design process. We also use AL-based approaches to maximize the accuracy of the AI classifier, while minimizing the workload put on the human screeners.

3. METHODOLOGY

In this work, we utilize a BERT-based NLP model to accel-erate the design process of evidence gap maps. We utilize the NLP tool in the workflow of a real human-AI team with members of 3ie. We supported the development of three deployed EGMs in the topics of Agriculture, Nutrition, and Resilience.

We analyze how different ML methods and data training sizes affect the human-AI team performance, focusing on the trade-offs between model accuracy and human effort. We compare our fine-tuned BERT model against the industry standard SVM-based NLP tools. In our study, we employed the ensemble of two SVM models described in [22] as the baseline SVM model. Further, we explore the effect of active learning with various query strategies on model accuracy and human effort.

Our work is comprised of two case studies:

1. **Deployed EGM design**: Actively designing and creat-ing three EGMs regarding Agriculture, Nutrition, and Re-silience.

2. **Simulated EGM design**: Using a pre-existing, fully labeled dataset to retrospectively study the most effective classifi-

cation algorithms (SVM vs. BERT) and active learning strategies for EGM creation.

These two case studies present different challenges and prior-ities. In the simulated EGM design, we have the benefit of a fully labeled dataset, which we can practice multiple techniques on. In the deployed EGM design, we are in the real-world situation of creating an EGM from scratch using a human-AI team. There-fore, we only have labels for the documents that we specifically choose to screen.

Further, in the deployed EGM design, we are motivated to design the most comprehensive and informative EGM while ef-ficiently utilizing human resources. Consequently, we want to minimize time that human experts spend screening irrelevant documents, and screen only the relevant documents. This con-trasts the strategy in classical active learning to query or screen the documents we are most uncertain of.

3.1 Dataset Description

For the simulated EGM design, our data is provided by 3ie and is derived from manually labeled documents from 3ie's De-velopment Evidence Portal (DEP) [2] [41], an expansive repository of impact evaluations and systematic reviews in global develop-ment across a wide range of sectors. We utilize a dataset of 68,539 documents screened for inclusion in 3ie's DEP to de-velop and evaluate our classification model. Table 1 shows the key attributes of the dataset, such as title, abstract, and inclusion decisions.

Attribute	Description
Title	Title of the paper.
Abstract	Abstract of the paper.
Keywords	Keywords of the paper.
Year	Publication year.
Publication type	Journal, conference proceeding, re-port, etc.
Source	The source of the paper, e.g., jour-nals or conferences.
Inclusion decision	Whether the paper is included as a relevant study. If not, what is the exclusion criterion.

TABLE 1: THE KEY ATTRIBUTES IN THE DEVELOPMENT EVI-DENCE PORTAL DATASET.

In this study, the title of each paper is integrated into the abstract as a sentence at the beginning. The BERT classification model takes the integrated texts as the input. The label of "in-cluded" or "excluded" is derived from the inclusion decision. To train the binary classification model, the "0" class corresponds to the "excluded" papers, and the "1" class comprises the "in-cluded" papers. This dataset is highly imbalanced, containing 5,281 included papers and 63,258 excluded papers. The crite-ria for excluding the papers are also extracted for training the criterion-specific classification models.

For our deployed EGM design, we are actively designing three EGMs. As indicated in Figure 2, and per 3ie's EGM work-

---

[2]https://developmentevidence.3ieimpact.org/

Copyright © 2023 by ASME

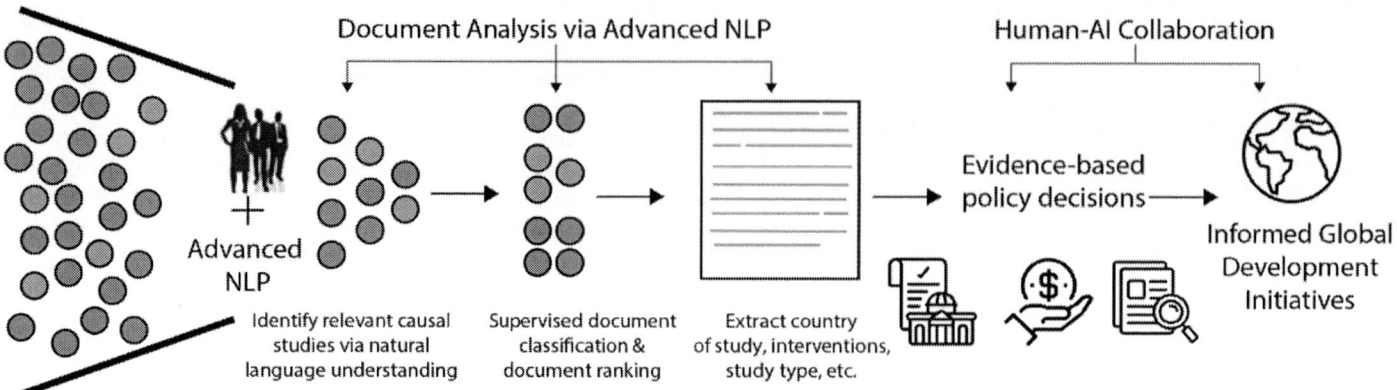

FIGURE 3: PROPOSED UTILIZATION OF NLP TOOLS IN A HUMAN-AI TEAM TO SCREEN, UNDERSTAND, AND CLASSIFY DOCUMENTS (REPRESENTED BY CIRCLES) IN ORDER TO INFORM EVIDENCE-BASED POLICY DECISIONS. OUR GOAL IS TO ACCELERATE THE DESIGN PROCESS FOR EGM PRODUCTS IN THE GLOBAL DEVELOPMENT FIELD.

flow, we gather our initial dataset via a literature search through scholarly databases and grey literature sources. For the three EGMs, their initial dataset sizes are as follows: Agriculture 221k, Nutrition 117k, and Resilience 60k.

3.2 Data Pre-processing

The raw documents are pre-processed to remove noise. Two types of noise are removed in this step. The first is non-English texts. A portion of the papers provide titles and abstracts in multiple languages. Since our models only take texts in English as input, the sentences in languages other than English are noise to the models and should be removed. The second type comprises English text content that is irrelevant to the scope of the document, such as a copyright statement. The pre-processing consists of five steps. (1) Each document is parsed into sentences. (2) A language detection model is used to identify sentences written in non-English languages. (3) We manually label the sentences from 500 documents with the "relevant" and "irrelevant" labels. (4) A BERT classification model is trained on the labeled data to predict the labels of the other sentences. The accuracy of the model is higher than 0.99. (5) Once the irrelevant sentences are removed, the remaining relevant sentences are integrated back into the original documents.

3.3 Priority Score

In this study, the AI agent is operationalized by a BERT binary classification model, which employs a 12-layer pre-trained uncased BERT embedding module with a hidden size of 768. The BERT embedding module is followed by a dropout layer with a drop rate = 0.1 and a linear layer that outputs a 2-dimensional (2D) vector as the final classification prediction. As described above, the AI agent needs to sample or prioritize the unlabeled papers according to the probabilities of being relevant, as predicted by the classification model. This probability is named the *"priority score"* (PS) in definition 1.

Definition 1. *Priority score* is the probability that a paper is a relevant paper predicted by the AI agent, which is calculated by

$PS(p) = softmax(Pred(p))[1]$, where $Pred(p)$ is the prediction output from the classification model for a paper p, which is a 2D vector. The "1" in the equation indicates that PS(p) is the probability of the paper being classified to the "1" class. Following this definition, higher screening priority scores are assigned to the papers with higher predicted probabilities of being relevant.

3.4 Sampling Strategies

According to the predicted PSs, we apply three different query strategies to sample papers from the unscreened subset, which will be labeled and added to the training set in the next iteration.

1. **Least confidence**: The *least confidence (LC)* query strategy is one of the commonly used strategies for active learning, which samples papers that the model is least certain how to classify [37], as shown by $x_{LC}^* = \underset{x}{argmax}\ U(x)$. The classification uncertainty $U(p)$ of the paper p is derived from the classification model output $Pred(p)$ through $U(p) = 1 - max(softmax((Pred(p)))$.

2. **Highest priority**: The *highest priority (HP)* query strategy samples papers with the highest PSs, given by $x_{HP}^* = \underset{x}{argmax}\ PS(x)$. This query strategy is adapted from the uncertainty sampling strategies [36]. For evidence synthesis, all relevant papers need to be verified by a human agent, so the papers most likely to be relevant are first sampled.

3. **Random**: The *random* query strategy randomly samples papers from the unlabeled list without using any informativeness measure. In this case, no AL is applied.

3.5 Human-AI Hybrid Team Workflow

We assume the human-AI hybrid team is tasked with screening a set of papers to identify papers satisfying a given scope. The human agents start the TA screening process by specifying the screening criteria. Then, the AI agent randomly samples a

Copyright © 2023 by ASME

subset of papers, which are screened by the human agents as the initial training set. On this basis, our model is trained to learn the screening criteria from the training. With the learned knowledge, the AI agent predicts the PSs of the unscreened papers. According to the predicted PSs, the AI agent needs to check whether the screen-train-predict-sample loop should stop. If not, it employs a certain strategy to sample a set of papers to be screened for the next iteration.

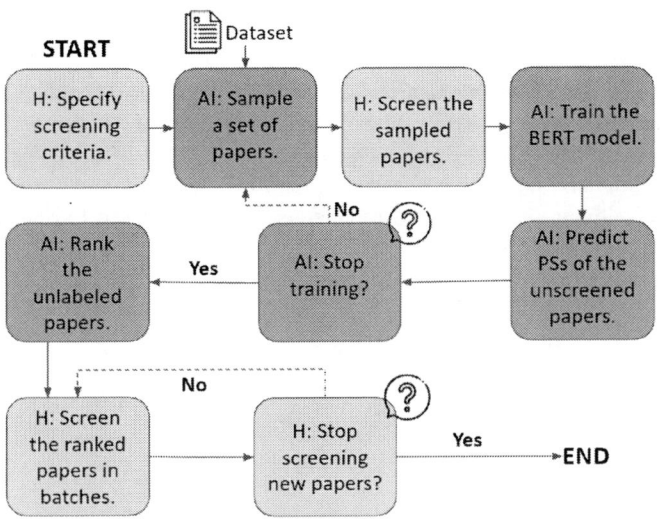

FIGURE 4: THE WORKFLOW OF THE HUMAN-AI HYBRID TEAM. H REPRESENTS HUMAN AGENTS.

Once the AI agent decides to stop training the model after a few iterations, it prioritizes all the unscreened papers according to their predicted PSs. Then, the human agents screen the prioritized papers in batches and decide at which batch the screening process should be ended. In alignment with the typical EGM workflow, each paper must be reviewed by two human screeners before being included for the next stage of screening.

Since the dataset is imbalanced, a random over-sampling method is applied to the training set to make the numbers of samples from both classes equal. In this study, the AI agent samples a batch of 1,000 papers each time. The batch number is selected because it balances the gain from and the cost of updating the model.

3.6 Evaluation Metrics

In the ML domain, accuracy and F1 score are commonly used to evaluate the performance of classification models. However, these metrics alone are not informative enough to assess the performance and efficiency of the human-AI hybrid teams. In this study, we evaluate the performance of the human-AI hybrid teams in terms of *human effort* in definition 2 needed for achieving an *inclusion rate* in definition 3. The computational cost reflects team efficiency from another perspective, which is not discussed in this study.

Definition 2. *Human effort* is defined as the ratio (HE) between the number of papers that need to be screened manually ($n_{screened}$) for identifying a given amount of relevant papers and the total number of papers ($n = 68,539$ in the simulated

EGM design case study) in the dataset, which is calculated by: $HE = n_{screened}/n$.

Definition 3. *Inclusion rate* is the ratio (IR) between the number of included papers being identified ($n_{identified}$) and the total number of included paper ($n_{included} = 5,281$ in the simulated EGM design case study) in the dataset, calculated by: $IR = n_{identified}/n_{included}$. With limited resources, a higher inclusion rate is preferred.

Given a set of scientific papers and the screening criteria, an efficient human-AI hybrid team should minimize the human effort and computational cost for achieving a satisfying inclusion rate or maximize the inclusion rate with available human effort and computational resources. Additionally, the F1 score of the corresponding classification model in each case is also reported for assessing the performance of the AI agent.

4. EXPERIMENTAL SETUP

In this section, we discuss the experimental setups for the two case studies: deployed EGM design and simulated EGM design. The aspects specific to each case study are described in 4.1 and 4.2, while their shared components like baseline techniques and implementation details are described in sections 4.3 and 4.4.

4.1 Deployed EGM Design Experiments

The human-AI team is tasked with designing three EGMs for deployment: an Agriculture, Nutrition, and Resilience EGM. For each one, the human-AI interaction and, therefore, our experiments are targeted at the title and abstract screening process.

The three EGMs are created independently and following the same process. They start with the following size of datasets: Agriculture 221k, Nutrition 117k, and Resilience 60k. The human-AI workflow used for the title and abstract screening of each EGM is shown in Figure 4 and described above in section 3.5. Once the process is complete for each EGM, the human-AI team has labeled a small subset of each dataset. We utilize these labeled datasets for the following experiments.

4.1.1 Model Performance Experiments. We explore the effectiveness of the industry standard SVM methods and our proposed BERT method in classifying documents as relevant or irrelevant for each of the three EGMs. Due to the real-world nature of this case study, we only have labels for those documents which we choose to screen. For each of these labeled datasets, we perform a 85% - 15% train-test-split and determine the classification accuracy.

4.1.2 Human Effort Experiments. We also compared the trained BERT and SVM models in terms of human effort to assess hybrid team efficiency. Specifically, we suppose the documents in the test set would be screened in descending order of priority scores predicted by the BERT and SVM models respectively. Human effort is defined as the percentage of documents that humans need to screen for getting a specific inclusion rate. The hybrid team is more efficient if fewer documents must be screened to obtain the same number of included documents. That is, less human effort is needed.

Copyright © 2023 by ASME

4.2 Simulated EGM Design Experiments

In this case study assume the human-AI hybrid team is tasked with screening a set of 68,539 documents to identify documents satisfying a given scope. This dataset is fully labeled, and we can therefore test the efficacy of different training sizes and active learning sampling strategies.

4.2.1 Training Size Experiments.
For ML model training, a larger training set often benefits model performance but needs more human effort to label the data. In the human-AI hybrid team, the trade-off between the model performance and the required human effort for labeling should be balanced carefully to achieve high team efficiency. We conduct experiments to investigate how the training size affects hybrid team efficiency - both in terms of the model performance (F1 score) and human effort required. Specifically, we start with an initial training set of 1,000 papers; to expand the training set, we randomly sample 1,000 papers, label them, and add them to the training set in each iteration from 1,000 to 6,000. During training, we use 85% of the papers in the training set to train our model and 15% as a validation set. All the other papers compose the testing set.

4.2.2 Active Learning Experiments.
When the AI agent samples new papers to be screened, the query strategy used affects the informativeness of the sampled papers, which further influences the ML model performance and hybrid team efficiency. We compare two different sampling strategies, LC and HP, with random sampling through experiments. For each sampling strategy, we start with the same initial training set of 1,000 papers with the random sampling case. After that, we sample 1,000 new papers using the LC or HP strategy to expand the training in each iteration. We experiment with training sizes ranging from 1,000 to 7,000 for the two sampling strategies.

4.3 Baseline

To answer RQ1 - *how much human effort can be saved when the AI agent is trained on an optimal data size?*, we compare the best case from the experiments with different training sizes with the baseline cases. In the first case, the human team works alone on the same task without any AI assistance. That is, the human agents randomly screen papers from the dataset. The second case employs a support vector machine (SVM)-based classifier, which is developed for retrieving randomized controlled trials and available in the EPPI-Reviewer software [22]. For the second baseline, we also experiment with four different training sizes ranging from 1,000 to 7,000, from which the best model is used as the baseline.

To answer RQ2 - *how much human effort can be further saved by enhancing the hybrid team through active learning?*, the best model from the experiments with different training sizes is used as the baseline, where all the sampled papers are randomly selected. We compare the best models from the experiments with the LC sampling strategy and the HP sampling strategy to the baseline model, respectively.

4.4 Implementation Details

In this study, our models are trained with a learning rate of 1×10^{-5}. There is a warm-up phase at the beginning of the training process, which lasts for one epoch. The experiments were performed on Intel(R) Xeon(R) W-2295 CPU @ 3.00GHz 3.00 GHz, with 18 cores and 256 GB of RAM. Model training and predicting were conducted on Nvidia RTX A5000 GPUs (single GPU per run). Each experiment is repeated five times. When the predicted uncertainties and PSs are needed to sample new papers with the LC and HP strategies, we use the mean values of the predictions from the five runs to improve the repeatability of the results.

5. RESULTS

In the following sections we present the results of our experiments in both case studies: the deployed EGM design, and the simulated EGM design. We compare different ML models, training sizes, and active learning sampling strategies and report their effects on model performance and human effort. Further, we go on to discuss the limitations of our work and the future use of human-AI teams in EGM design.

5.1 Deployed EGM Design Results

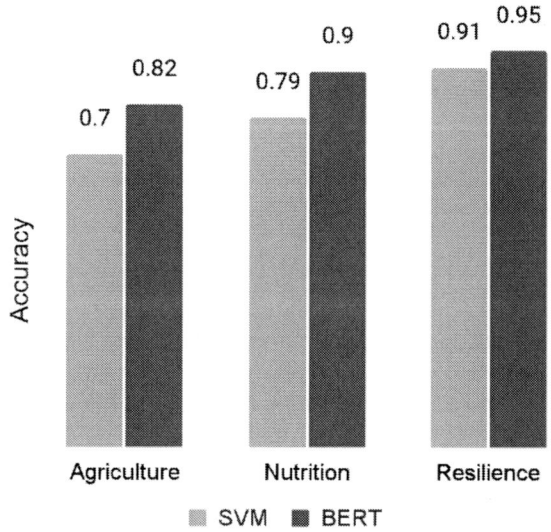

FIGURE 5: THE CLASSIFICATION ACCURACY OF BERT AND SVM MODELS FOR THE THREE EGMS CREATED: AGRICULTURE, NUTRITION, AND RESILIENCE.

In this section we compare inclusion classification done by BERT and SVM models. We compare model performance across two different metrics: overall accuracy, and saved effort. The BERT model is our proposed approach, whereas the SVM model is what tools like EPPI Reviewer utilize to classify documents, and therefore represents the industry standard.

5.1.1 Model Accuracy.
Figure 5 shows the accuracy of the BERT and the SVM classifcation models for each of the three EGMs we created. The results show that for all three EGMs, the BERT model resulted in higher accuracy than the SVM model. The most common NLP tools used to aid EGM creation today are based on an SVM model, so our results suggest that utilizing BERT for classification has benefits over the industry standard EGM-creation tools.

Copyright © 2023 by ASME

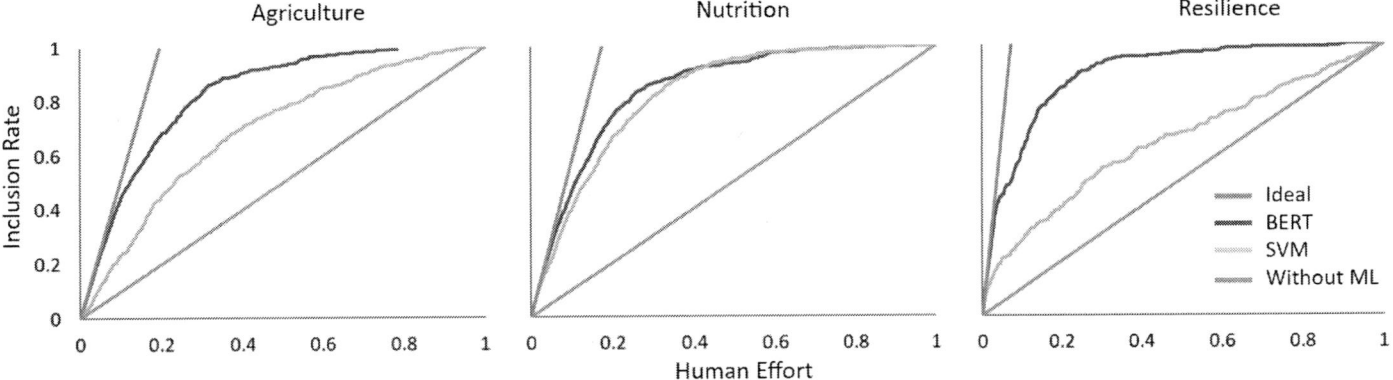

FIGURE 6: HOW INCLUSION RATE VARIES WITH HUMAN EFFORT FOR THE THREE DEPLOYED EGMS. A HIGHER INCLUSION RATE AT A LOWER HUMAN EFFORT IS PREFERRED. BERT OUTPERFORMS SVM IN ALL THREE EGMS.

EGM	SVM	BERT	Percent Effort Saved by BERT
Agriculture	53%	28%	**47%**
Nutrition	29%	24%	**17%**
Resilience	68%	17%	**75%**

TABLE 2: HUMAN EFFORT REQUIRED TO REACH AN 80% INCLUSION RATE FOR EACH OF THE THREE EGMS AND THE TWO ML MODELS, SVM AND BERT. THE PERCENT EFFORT SAVED BY BERT IS CALCULATED AS THE PERCENT DIFFERENCE BETWEEN HUMAN EFFORT FOR BERT AND SVM.

5.1.2 Saved Effort. Figure 6 shows how inclusion rate varies with human effort. The orange "Ideal" line indicates a perfect inclusion rate, where only relevant documents are screened and therefore all documents seen are included. The grey "Without ML" line indicates a case in which human experts must screen all documents at random in order to find all of the included documents. We compare two ML strategies, BERT and SVM, and find that BERT outperforms SVM for all three EGMs.

We carried out a set of experiments with the screened papers of the three EGMs for the comparison. Aiming at an inclusion rate of 80% for the screened papers, we found that the human raters needed 47% less human effort for Agriculture, 17% less human effort for Nutrition, and 75% less human effort for Resilience when working with the BERT-based models rather than with the baseline EPPI-Reviewer's SVM. The raw values of human effort for BERT and SVM for the three EGMs are shown in Table 2. The effort-saving capabilities of the BERT models are further amplified in the real screening process, in which the models are updated multiple times as new labeled documents come in as training data. In this case, the model improves iteratively over time. As its classification accuracy increases, the model can suggest only the most relevant documents to the human raters. This type of active learning is explored in the simulated dataset and described in section 5.2.2.

5.2 Simulated EGM Design Results

5.2.1 Saved Effort. The performance of the human-AI hybrid team is assessed through inclusion rate (IR) and human effort

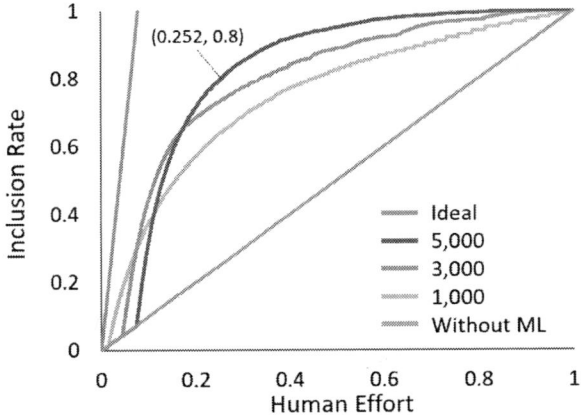

FIGURE 7: HOW INCLUSION RATE VARIES WITH HUMAN EFFORT WHEN OUR BERT-BASED MODEL IS TRAINED WITH DIFFERENT TRAINING SIZES. A TRAINING SIZE OF 5,000 PERFORMS BEST, AS INDICATED BY REACHING AN INCLUSION RATE OF 0.8 WITH THE LOWEST HUMAN EFFORT.

(HE). Figure 7 shows the variation of IR with HE when our model is trained with different training sizes. The grey "Without ML" line in the figure corresponds to the condition where the human agents work alone without any AI agent. Since the human agents randomly screen papers form the dataset, IR is equal to HE in this case. The orange "Ideal" line close to the y-axis denotes the ideal case, in which each screened document is an included document, and no excluded documents are screened. The slope of this line is 5,281 (the total number of included papers) / 68,539 (the total number of papers in the dataset). The other curves in the figure describe how the IR changes as the human agents invest more screening efforts when the BERT-based AI agent is trained on the datasets with different sizes.

Each curve consists of two parts. The first straight line part indicates the process that the human agents label papers from the dataset to prepare the training set. Since the screened papers are randomly selected, the IR is equal to HE. Once the training set is ready, our model is trained on it to predict the PSs of the unlabeled papers. The curved part following the straight line corresponds to

Copyright © 2023 by ASME

the process during which the human agents screen the unlabeled papers sequentially according to the predicted PSs. Since the unlabeled papers are prioritized for screening, the curves are much steeper in this second portion than in the first, which has the same slope as the "Without ML" line. In the curved portion, the initial slopes are close to the slope of the ideal line, then gradually decrease later on. This trend suggests that the papers with higher PSs are more likely to be identified as included papers than the papers with lower PSs, implying the effectiveness of the AI agent in prioritizing the unlabeled papers for screening.

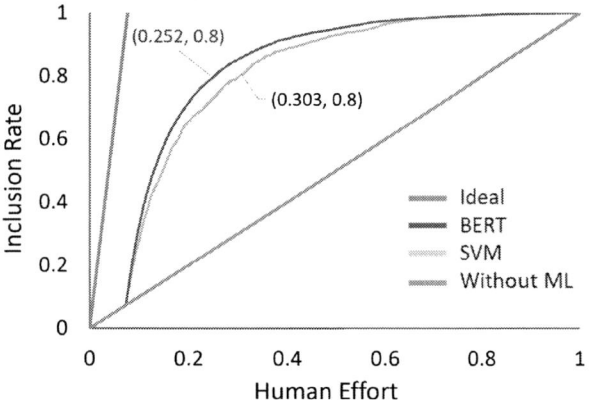

FIGURE 8: HOW INCLUSION RATE VARIES WITH HUMAN EFFORT FOR THE DIFFERENT ML MODELS: OUR MODEL (BERT), THE INDUSTRY STANDARD (SVM), AND "IDEAL" AND "WITHOUT ML" BASELINES.

Since a high-performing human-AI hybrid team can achieve a higher IR with a lower HE, its initial slope should appear closer to the "Ideal" line in 7. As the training size increases, the curve gets steeper, indicating improved model performance. This is in line with the increasing F1 scores shown in Figure 9. However, because a larger training size needs more human labeling effort (i.e., a longer straight line in the first part along the diagonal line), it may also impair the efficiency of the human-AI hybrid team. Given a target IR of 80%, the curves show that the hybrid team gets the highest efficiency when the training size is 5,000. Under this condition, the human agents only need to screen 25.2% of the papers to get an IR of 80%, while they need to screen 80% of the papers to get the same IR in the case without the AI guidance. Therefore, when the BERT-based AI agent is incorporated into the human team, it can save 54.8% human screening effort for getting the IR of 80%.

We also compare the BERT-based model with the SVM model used in the EPPI-Reviewer software in terms of their effectiveness as the AI agent. Similarly, we train the SVM model with different training sizes (1,000, 3,000, 5,000, 7,000), among which the training size of 5,000 needs the least human effort for getting the IR of 80%. Figure 8 compares the best BERT-based model (5,000) and the best SVM model (5,000), suggesting that the BERT-based model enables the human agents to save more screening efforts compared to the SVM model for getting any IR. Specifically, the human agents can save 5.1% more screening efforts when working with the BERT-based AI agent than working

with the SVM-based AI agent for getting the IR of 80%. Therefore, our BERT-based model is more effective in acting as the AI agent.

5.2.2 The Effect of Active Learning. In the following section, we discuss how the strategies for sampling new data to expand the training size affect the performance of the AI agent and the efficiency of the human-AI hybrid team for the TA screening task.

Active Learning, Training Size, and Model Performance

Here we report the results of the experiments with different training sizes and different sampling strategies to demonstrate the effect of incorporating the AI agent into the human team, answering RQ1. Following the protocol of the classification problems with the imbalanced dataset, we use the F1 score computed at the default threshold of 0.5 as the classification metric. In these experiments, the sampled papers are randomly selected. The black curve in Figure 9 shows the variation of the F1 score with the training size. As the training size increases, the F1 score improves with diminishing marginal effect, especially when the training size is larger than 5,000.

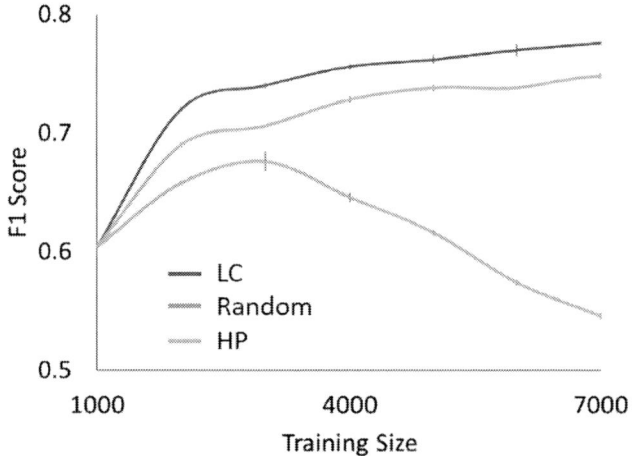

FIGURE 9: HOW MODEL PERFORMANCE, AS SHOWN BY THE F1 SCORE, VARIES WITH TRAINING SIZE AND ACTIVE LEARNING SAMPLING STRATEGY. THE BARS INDICATE ONE STANDARD ERROR. WE FIND THAT THE LEAST CONFIDENCE (LC) STRATEGY PERFORMS THE BEST.

Similar to the random sampling case, the training size affects the performance of our classification model. As shown in Figure 9, a larger training size improves the F1 score when the LC strategy is applied. If we employ the HP strategy, a moderate training size (e.g., 2,000) benefits the F1 score most, and a larger training set impairs the F1 score when its size surpasses a certain value (e.g., 2,000). Overall, sampling new papers using the LC strategy leads to better classification models than randomly sampling new papers, as indicated by the higher F1 score; however, the HP sampling strategy results in worse classification models than random sampling, indicated by the lower F1 scores.

Active Learning, Training Size, and Human Effort

The selected AL sampling strategy and the training size also affect the human-AI team efficiency. Under the random sampling

Copyright © 2023 by ASME

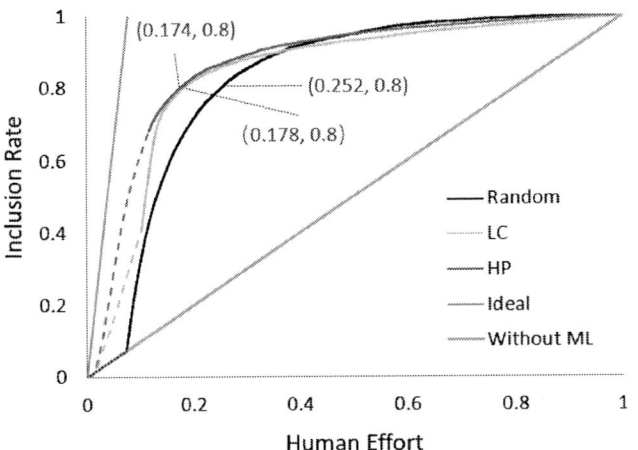

FIGURE 10: HOW DIFFERENT AL SAMPLING STRATEGIES AFFECT THE HUMAN EFFORT AND INCLUSION RATE RELATIONSHIP. THE DOTTED LINE PORTION OF EACH CURVE REPRESENTS THE SCREENING-UPDATING-PREDICTING-SAMPLING ITERATIONS, WHILE THE SOLID LINE PART CORRESPONDS TO THE PROCESS WHEN THE HUMAN AGENTS SCREEN THE PRIORITIZED PAPERS.

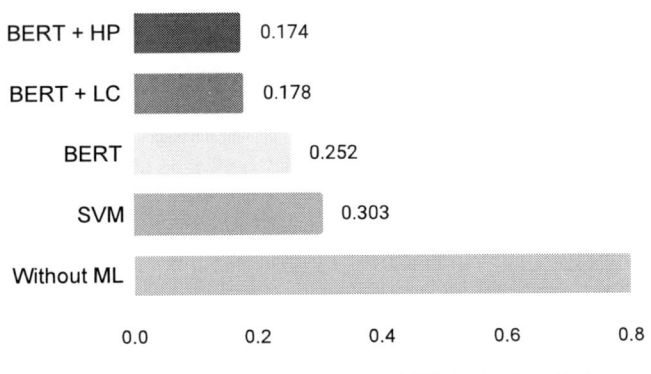

Human Effort to Reach an 80% Inclusion Rate

FIGURE 11: THE HUMAN EFFORT REQUIRED TO REACH AN 80% INCLUSION RATE FOR VARIOUS MODELS. LOWER HUMAN EFFORT IS PREFERRED. WITH NO ML, IT TAKES 80% HUMAN EFFORT TO REACH AN 80% INCLUSION RATE.

condition, a moderate training size can well balance the trade-off between higher model performance and more labeling effort for creating the training data, leading to the highest team efficiency. We observe similar trends for AL. To get an IR of 80%, the human-AI hybrid team achieves the highest team efficiency with a training size of 7,000 when the LC and HP sampling strategies are applied, respectively.

Figure 10 compares the team efficiency among different sampling conditions, including random sampling, LC sampling, and HP sampling. We can see that the efficiency of the human-AI hybrid team is improved substantially with active learning. When the LC and HP strategies are applied, the human agents can respectively save 7.4% and 7.8% screening effort for getting the

IR of 80%. Specifically, the dotted line portion of each curve represents the screen-update-predict-sample iterations (i.e., the "AI: Stop training?" loop in Figure 4), while the solid line part corresponds to the process when the human agents screen the prioritized papers according to the predictions from the finalized AI model.

We can see that the dotted line portions of the LC and the HP curves are much steeper than the dotted line portion of the random sampling curve. The trends suggest that with both the LC and HP sampling strategies, a larger portion of the sampled papers are included papers compared to the random sampling strategy. That is, the LC and the HP sampling strategies, especially HP, improve team screening efficiency substantially during the screening-updating-predicting-sampling iterations. This can be explained by the sampling strategies themselves. The LC strategy samples the papers with the highest classification uncertainties. Given the highly imbalanced dataset, our model is less confident in classifying the papers from the minor class, i.e., the included papers from the "1" class, leading to more papers being sampled from the minor class. The HP strategy samples the papers with the highest predicted PSs, which are more likely to be included papers by the definition of PS.

Moreover, by sampling the papers with the highest classification uncertainties, the LC sampling strategy also enables our model to learn more efficiently from human labeling compared to the other sampling strategies. This is evidenced by the observation that the solid curve part of the blue curve is steeper at the early phase than the solid curve parts of the red and black curves in Figure 10 and the highest F1 scores for LC in Figure 9.

6. DISCUSSION

Figure 11 provides a comprehensive view of the various ML methods we compared, and their effect on human effort. The figure shows the human effort required to reach an 80% inclusion rate for no ML assistance, an SVM-based model, a BERT-based model, and a BERT-based model with the LC or HP AL sampling strategies. We observe that the BERT-based model outperforms SVM, with a 16.8% relative reduction in human effort. The results show that the AL strategies reduce human effort even further, by about 30% compared to BERT without AL. Since our motivation is to accelerate the EGM design process and decrease the resource and time intensity of the process, this result is of great significance.

Within the hybrid team, effective interactions and mutual learning between the human agents and the AI agent can improve team performance significantly. When the LC sampling strategy is applied, both the information flow from the human agents to the AI agent (i.e., human knowledge conveyed in the labeled papers) and the information flow from the AI agent to the human agents (i.e., the AI predictions conveyed in the sampled or prioritized papers) play a role in improving the efficiency of the human-AI hybrid teams. In contrast, when the HP sampling strategy is applied, the information flow from the AI agent to the human agents plays a major role in benefiting hybrid team efficiency, especially during the screen-update-predict-sample iterations. However, in HP, the information flow from the human agents to the AI agent is not as beneficial for improving model performance.

Copyright © 2023 by ASME

In a practical screening process, we only know the labels of a part of the papers in a dataset, which means the actual IR, as well as its overall changing trend, is unknown. In such a scenario, it is difficult to determine when to stop expanding the training set and updating the AI agent and when to stop screening the prioritized papers. The changing scale of the predicted rankings of the unlabeled papers and the growth rate of IR can inform us about the stopping. Small changes in the paper rankings and a low growth rate of IR may suggest we stop updating the AI agent and stop screening the prioritized papers, respectively.

From the records of twelve human screeners working on an agriculture development EGM, we learn that a human screener can screen 38.6 ($SE = 1.00$) papers per hour on average. On this basis, the AL-enhanced AI agent can save human screeners $(80\% - 17.4\%) \times 68,539/38.6 = 1,111.5$ hours for TA screening compared to the case without the AI agent. Compared to the case using EPPT-Reviewer SVM as the AI agent, $(30.3\% - 17.4\%) \times 68,539/38.6 = 229.1$ hours can be saved by our model.

6.1 Discussion on AI-assisted Design of EGMs

The nature of our work creating EGMs for deployment and use by USAID meant that our team faced many real-world challenges. In this section, we discuss the unique challenges and limitations that arise when designing EGMs with AI-assistance.

The Cost of Communication The cost of communicating in a human-AI team can be significant but difficult to quantify as it involves multiple factors. One such factor is the time and effort required to exchange information via email, which can lead to delays and potential miscommunication. Additionally, updating document labels and merging datasets can be a complex and time-consuming task that requires oversight and project management to ensure accuracy. These activities can also be a source of errors that can negatively impact the performance of AI models. Finally, time lags between humans labeling documents, the AI agent receiving the documents and updating the model, and then the AI agent sending back newly ranked documents for human screening means that one team may be operating with incomplete data. Therefore, optimizing communication channels and implementing efficient communication protocols can help reduce the costs associated with human-AI team collaboration.

Trust in AI A major challenge that many AI recommendation systems face is the "cold start" problem. The AI agent must provide some prediction about the documents in the first iteration, but at this point the model knows nothing about the new domain. In our case, we pre-trained our model on documents in the global development space, but this cannot ensure that it would perform well in classifying documents for, say, an Agriculture-specific EGM without any additional training data. This challenge, while common, can lead to distrust in AI from the human team, if they find the initial rankings to be incorrect. Additionally, the training data for our models are labels from people, which can be noisy. The AI model's performance is constrained by the quality of its training data, and therefore to have meaningful and accurate model results, we must begin with consistent high-quality training data.

"When to Stop": A Business Decision Another challenge in the deployed EGM design case study was determining when to stop the screening process, the second question shown in Figure 4. Our human-AI team faced a trade-off here between screening more papers in order to improve model performance, or stopping screening in order to move onto the next step in the EGM process (Figure 2). This is ultimately a business decision in which the team must weigh the resource cost of improving the model, and identifying the most "true positive" documents. We experimented with two techniques for determining "when to stop." The first of these techniques was calculating the similarity of the rankings of the documents when ordered based on the priority score between two consecutive iterations. If the similarity of the two rankings was above a certain value after a screening iteration, we could stop updating the BERT model. The second technique was to terminate the human screening process at a specific real-time inclusion rate, e.g., the number of relevant documents identified from screening 1,000 documents in the current iteration. If the number of relevant documents is lower than a given threshold, the human screening team could stop the screening process. Future work could specifically address the question of when to stop screening, as it is a highly relevant decision for the human-AI team.

Automation of Full-text Screening The EGM design process, as depicted in Figure 2, includes both title and abstract screening, and full-text screening. A natural continuation of our work would be to use NLP to assist in the full-text screening step. This step, however, presents the logistical challenge of obtaining the full-text documents. While many institutions have subscription-based access to scholarly article databases, copyright issues make downloading and using full-text documents a challenge when working among and between different institutions. This ultimately dictated that our project scope remains in the title and abstract screening process alone.

Additionally, to perform full-text screening, one would need to train another BERT model to classify full-text documents for inclusion. This means human screeners would need to generate a training dataset for this task, which would require significant human effort. Large language models (LLMs) may assist in this challenge. LLMs which are trained on billions of documents [42] have a broad understanding of language, and future work can explore whether they can classify full-text documents without domain specific training data.

Limitations and Risks

Here, we discuss the potential of missing studies that should be included during the TA screening process. The percentage of studies that may be missed are influenced by both the accuracy and the percentage of papers being screened. Although the BERT models exhibit significantly higher accuracies (82% - 95%) than the SVM models (70% - 91%) for the three deployed EGMs, their accuracies are still not 100%. Accordingly, if the human raters screen the papers in the descending order of priority scores, the inclusion rate (i.e., the percentage of relevant papers identified by screening a given number of papers) decreases as the priority score drops, according to the characteristics of the classification models. This raises the cost of identifying included papers from the low priority score ranges. To balance the time cost and gain from screening the papers sequentially, the screening team made the decision to stop the screening process when the inclusion

Copyright © 2023 by ASME

rate becomes lower than a certain value. With this setting, it is inevitable that a part of the relevant papers may be missing from the final list.

To estimate the percentage of missing papers, we conducted a set of experiments with a fully labeled dataset (i.e., the DEP dataset) which is similar to the Agriculture, Nutrition, and Resilience datasets. The results show that we can get 80% of the relevant papers by screening 17% of the papers returned from the search process and 90% of the relevant papers by screening 31% of the papers returned from the search process.

Counterfactual Analysis Our team faced the challenge of accurately comparing different document screening techniques - such as using a BERT-based model, an SVM-based model, and no ML model within the real-world setup. It was infeasible for the human raters to create each EGM three separate times in order to compare the entire process for each technique. Therefore, we standardized our comparisons by performing retrospective experiments after the human-AI team had labeled a subset of the data. We present the results using this labeled subset. We further aimed to address this limitation by including the second case study - the simulated EGM design. In this case study, we utilized a fully labeled dataset of 68,539 documents in order to experiment with the various ML models and active learning sampling strategies. However, this challenge means we do not have true counterfactual analyses of how the EGM process would have proceeded without any AI assistance. Future work could further address this limitation by creating each EGM multiple times for each different technique.

6.2 Future Use of Human-AI Teams for EGM Design

The AI sub-field of NLP is experiencing rapid growth. Large language models (LLMs) like OpenAI's ChatGPT [42], and Meta's Galactica [43] are changing the way the world perceives, exploits, and interacts with pre-trained language models. These models were released after we had concluded our EGM creation; however, we predict that their capabilities will shift the way that NLP is utilized in EGM design.

We have performed a number of exploratory experiments to understand LLMs' capability in EGM design. We explored ChatGPT's understanding of the relationship between certain interventions and outcomes by asking it "How can agriculture transformation change poverty, migration, and food security?" The LLM captured the general qualitative relationships between the intervention (agriculture transformation) and the outcomes (poverty, migration, and food security), but did not output any quantitative implications, potential information sources, or indications of how well the relationships have been studied. This suggests to us that deep generative models may be able to capture the intervention-outcome relationships in a coarse resolution, but cannot provide all the detailed information that a human team or human-AI hybrid team can capture.

In a small-scale study, we explored whether LLMs could identify the interventions and outcomes of ten of the documents shown in the EGM in Figure 1. We fed ChatGPT the relevant abstracts and asked it for the intervention and outcome of each abstract. We found that the LLM was able to generate relevant interventions and outcomes. One such response was "The inter-

vention is an agriculture transformation program, which is not further specified in the paper. The outcome variables evaluated in the study are poverty, migration, food security, and agricultural revenue." This generative capability can be powerful in the early stage of EGM design, during which human experts determine the intervention and outcome categories that frame the EGM scope (shown as the column and row headers in Figure 1). Generative text can also be powerful for creating brief summaries of many documents, which falls under the overall goal of evidence synthesis.

7. CONCLUSION

In this paper, we have studied (1) how incorporating the BERT-based AI agent into the human team affects team efficiency in the EGM design process and (2) how enhancing the hybrid team through active learning can improve hybrid team efficiency. We propose a human-AI hybrid teaming workflow during TA screening portion of the EGM design process. We a) design and deploy three EGMs for global development in the areas of Agriculture, Nutrition, and Resilience, and b) conduct simulated experiments with a fully labeled dataset to answer the research questions described above. Our results show that the data size for training the AI agent influences hybrid team efficiency. When the training size is optimized, the incorporation of the BERT-based AI agent can reduce human effort by 68.5% compared to the case without AI assistance and by 16.8% compared to the case using an SVM-based agent for getting to an inclusion rate of 80%. Moreover, enhancing the hybrid team through active learning can further reduce human effort by 30% compared to BERT with no active learning. The proposed human-AI hybrid teaming workflow has been validated in the practical construction process of three EGMs. Therefore, the AL-enhanced human-AI hybrid team can accelerate evidence gap map design, and decision making in the global development field significantly.

ACKNOWLEDGMENTS

This publication was made possible through support provided by the USAID Bureau for Resilience and Food Security, U.S. Agency for International Development. The funders had no role in study design, data collection and analysis, decision to publish, or preparation of the manuscript. The opinions expressed herein are those of the authors and do not necessarily reflect the views of the U.S. Agency for International Development.

REFERENCES

[1] USAID. *Strengthening Evidence-Based Development*. USAID (2016).

[2] 115th Congress, U.S. "Foundations for Evidence-Based Policymaking Act of 2018." (2018).

[3] Kraybill, David and Mercier, Stephanie. "How the United States Benefits from Agricultural and Food Security Investments in Developing Countries." (2019). URL https://www.usaid.gov/sites/default/files/documents/1867/BRIEF_-_US_Benefits_Overview.pdf.

[4] White, Karen. "Publications Output: U.S. Trends and International Comparisons." https://ncses.nsf.gov/pubs/nsb20206/ (2019). Accessed: 14-January-2022.

Copyright © 2023 by ASME

[5] Donnelly, Christl A., Boyd, Ian, Campbell, Philip, Craig, Claire, Vallance, Patrick, Walport, Mark, Whitty, Christopher J. M., Woods, Emma and Wormald, Chris. "Four principles to make evidence synthesis more useful for policy." *Nature* Vol. 558 (2018): pp. 361–364.

[6] Snilstveit, Birte, Vojtkova, Martina, Bhavsar, Ami, Stevenson, Jennifer and Gaarder, Marie. "Evidence & Gap Maps: A tool for promoting evidence informed policy and strategic research agendas." *Journal of Clinical Epidemiology* Vol. 79 (2016): pp. 120–129. DOI 10.1016/j.jclinepi.2016.05.015. Accessed 2019-07-06, URL https://www.jclinepi.com/article/S0895-4356(16)30190-1/abstract.

[7] 3ie. "Evidence Gap Maps." Available at https://www.3ieimpact.org/evidence-hub/evidence-gap-maps (2021/08/04) (2021).

[8] NLPIE. "BioMedICUS." https://nlpie.github.io/biomedicus/ (2019).

[9] Bommarito, Michael J, Katz, Daniel Martin and Detterman, Eric M. "LexNLP: Natural language processing and information extraction for legal and regulatory texts." (2018). URL 1806.03688.

[10] Porciello, Jaron, Ivanina, Maryia, Islam, Maidul, Einarson, Stefan and Hirsh, Haym. "Accelerating evidence-informed decision-making for the Sustainable Development Goals using machine learning." *Nature Machine Intelligence* Vol. 2 (2020): pp. 559–565.

[11] Briner, Rob B. and Denyer, David. "Systematic review and evidence synthesis as a practice and scholarship tool." *Handbook of evidence-based management: Companies, classrooms and research*. New York University Press (2012): pp. 112–129.

[12] Blaizot, Aymeric, Veettil, Sajesh K., Saidoung, Pantakarn, Moreno-Garcia, Carlos Francisco, Wiratunga, Nirmalie, Aceves-Martins, Magaly, Lai, Nai Ming and Chaiyakunapruk, Nathorn. "Using artificial intelligence methods for systematic review in health sciences: A systematic review." *Research Synthesis Methods* Vol. 13 No. 3 (2022): pp. 353–362. DOI https://doi.org/10.1002/jrsm.1553. URL https://onlinelibrary.wiley.com/doi/pdf/10.1002/jrsm.1553, URL https://onlinelibrary.wiley.com/doi/abs/10.1002/jrsm.1553.

[13] Altena, Allard, Spijker, René and Olabarriaga, Silvia. "Usage of Automation Tools in Systematic Reviews." *Research Synthesis Methods* Vol. 10 (2018). DOI 10.1002/jrsm.1335.

[14] 3ie. "Evidence Mapping." Available at https://www.3ieimpact.org/evidence-hub/evidence-gap-maps (2021/08/04) (2021).

[15] O'Mara-Eves, Alison, Thomas, James, McNaught, John, Miwa, Makoto and Ananiadou, Sophia. "Using text mining for study identification in systematic reviews: a systematic review of current approaches." *Systematic reviews* Vol. 4 No. 1 (2015): p. 1. Accessed 2016-03-11, URL http://systematicreviewsjournal.biomedcentral.com/articles/10.1186/2046-4053-4-5. 00000.

[16] Oliver, Kathryn, Innvar, Simon, Lorenc, Theo, Woodman, Jenny and Thomas, James. "A systematic review of barriers to and facilitators of the use of evidence by policymakers." *BMC Health Services Research* Vol. 14 (2014): p. 2. DOI 10/gbfqbx. Accessed 2016-07-30, URL http://dx.doi.org/10.1186/1472-6963-14-2. 00000.

[17] Mikolov, Tomas, Sutskever, Ilya, Chen, Kai, Corrado, Greg and Dean, Jeffrey. "Distributed Representations of Words and Phrases and their Compositionality." (2013). URL 1310.4546.

[18] Mikolov, Tomas, Chen, Kai, Corrado, Greg and Dean, Jeffrey. "Efficient Estimation of Word Representations in Vector Space." (2013). URL 1301.3781.

[19] Pennington, Jeffrey, Socher, Richard and Manning, Christopher D. "Glove: Global Vectors for Word Representation." *EMNLP*, Vol. 14: pp. 1532–1543. 2014.

[20] Ouzzani, Mourad, Hammady, Hossam, Fedorowicz, Zbys and Elmagarmid, Ahmed. "Rayyan—a web and mobile app for systematic reviews." *Systematic Reviews* Vol. 5 No. 210 (2016). DOI https://doi.org/10.1186/s13643-016-0384-4. URL https://link.springer.com/article/10.1186/s13643-016-0384-4#citeas.

[21] Marshall, Iain J, Kuiper, Joel, Banner, Edwards and Wallace, Byron C. "Automating Biomedical Evidence Synthesis: RobotReviewer." 2017. DOI 10.18653/v1/P17-4002.

[22] Thomas, James, McDonald, Steve, Noel-Storr, Anna, Shemilt, Ian, Elliott, Julian, Mavergames, Chris and Marshall, Iain J. "Machine learning reduced workload with minimal risk of missing studies: development and evaluation of a randomized controlled trial classifier for Cochrane Reviews." *Journal of Clinical Epidemiology* Vol. 133 (2021): pp. 140–151. DOI https://doi.org/10.1016/j.jclinepi.2020.11.003. URL https://www.sciencedirect.com/science/article/pii/S0895435620311720.

[23] Vaswani, Ashish, Shazeer, Noam, Parmar, Niki, Uszkoreit, Jakob, Jones, Llion, Gomez, Aidan N., Kaiser, Lukasz and Polosukhin, Illia. "Attention Is All You Need." (2017). URL 1706.03762.

[24] Devlin, Jacob, Chang, Ming-Wei, Lee, Kenton and Toutanova, Kristina. "BERT: Pre-training of Deep Bidirectional Transformers for Language Understanding." (2019). URL 1810.04805.

[25] Liu, Yinhan, Ott, Myle, Goyal, Naman, Du, Jingfei, Joshi, Mandar, Chen, Danqi, Levy, Omer, Lewis, Mike, Zettlemoyer, Luke and Stoyanov, Veselin. "RoBERTa: A Robustly Optimized BERT Pretraining Approach." (2019). URL 1907.11692.

[26] Radford, Alec, Wu, Jeffrey, Child, Rewon, Luan, David, Amodei, Dario and Sutskever, Ilya. "Language Models are Unsupervised Multitask Learners." *OpenAI Blog* (2019).

[27] Aum, Sungmin and Choe, Seon. "srBERT: automatic article classification model for systematic review using BERT." *Systematic Reviews* Vol. 10 No. 285 (2021). DOI https://doi.org/10.1186/s13643-021-01763-w10.1002/jrsm.1335.

[28] Ein-Dor, Liat, Halfon, Alon, Gera, Ariel, Shnarch, Eyal, Dankin, Lena, Choshen, Leshem, Danilevsky, Marina,

Aharonov, Ranit, Katz, Yoav and Slonim, Noam. "Active Learning for BERT: An Empirical Study." *EMNLP*. 2020.

[29] Aggarwal, Umang, Popescu, Adrian and Hudelot, Celine. "Minority Class Oriented Active Learning for Imbalanced Datasets." *2020 25th International Conference on Pattern Recognition (ICPR)*. 2021. IEEE. DOI 10.1109/icpr48806.2021.9412182. URL https://doi.org/10.1109%2Ficpr48806.2021.9412182.

[30] Gal, Yarin, Islam, Riashat and Ghahramani, Zoubin. "Deep Bayesian Active Learning with Image Data." (2017). DOI 10.48550/ARXIV.1703.02910. URL https://arxiv.org/abs/1703.02910.

[31] Tur, Gokhan, Hakkani-Tur, Dilek and Schapire, Robert E. "Combining active and semi-supervised learning for spoken language understanding." *Speech Communication* Vol. 45 (2005): pp. 171–186.

[32] Chung, Mu-Huan, Chignell, Mark, Wang, Lu, Jovicic, Alexandra and Raman, Abhay. "Interactive Machine Learning for Data Exfiltration Detection: Active Learning with Human Expertise." *Proceedings of the IEEE International Conference on Systems, Man, and Cybernetics (SMC)*: pp. 280–287. 2020. DOI 10.1109/SMC42975.2020.9282831.

[33] Olsson, Fredrik. "A literature survey of active machine learning in the context of natural language processing." Technical report no. Swedish Institute of Computer Science. 2009.

[34] Angluin, Dana. "Queries and Concept Learning." *Machine Learning* Vol. 2 (1998): pp. 319–342.

[35] Atlas, Les, Cohn, David and Ladner, Richard. "Training Connectionist Networks with Queries and Selective Sampling." Touretzky, D. (ed.). *Advances in Neural In-formation Processing Systems*, Vol. 2. 1989. Morgan-Kaufmann. URL https://proceedings.neurips.cc/paper/1989/file/b1a59b315fc9a3002ce38bbe070ec3f5-Paper.pdf.

[36] Lewis, David D. and Gale, William A. "A Sequential Algorithm for Training Text Classifiers." *Proceedings of the 17th Annual International ACM SIGIR Conference on Research and Development in Information Retrieval*: p. 3–12. 1994. Springer-Verlag, Berlin, Heidelberg.

[37] Settles, Burr. "Active Learning Literature Survey." Computer Sciences Technical Report 1648. University of Wisconsin–Madison. 2009.

[38] Culotta, Aron and Andrew, McCallum. "Reducing Labeling Effort for Structured Prediction Tasks." *AAAI*: p. 746–751. 2005.

[39] Scheffer, Tobias, Decomain, Christian and Wrobel, Stefan. "Active Hidden Markov Models for Information Extraction." *IDA*. 2001.

[40] Shannon, C. E. "A mathematical theory of communication." *The Bell System Technical Journal* Vol. 27 No. 4 (1948): pp. 623–656. DOI 10.1002/j.1538-7305.1948.tb00917.x.

[41] 3ie. "Development Evidence Portal." Available at https://developmentevidence.3ieimpact.org (2021/08/04) (2021).

[42] OpenAI, TB. "Chatgpt: Optimizing language models for dialogue." *OpenAI* (2022).

[43] Taylor, Ross, Kardas, Marcin, Cucurull, Guillem, Scialom, Thomas, Hartshorn, Anthony, Saravia, Elvis, Poulton, Andrew, Kerkez, Viktor and Stojnic, Robert. "Galactica: A Large Language Model for Science." (2022). DOI 10.48550/ARXIV.2211.09085. URL https://arxiv.org/abs/2211.09085.

Proceedings of the ASME 2023
International Design Engineering Technical Conferences and
Computers and Information in Engineering Conference
IDETC-CIE2023
August 20-23, 2023, Boston, Massachusetts

DETC2023-109380

MODEL CONSISTENCY FOR MECHANICAL DESIGN:
BRIDGING LUMPED AND DISTRIBUTED PARAMETER MODELS WITH A PRIORI GUARANTEES

Randi Wang, Morad Behandish*

Palo Alto Research Center (PARC), Palo Alto, CA

ABSTRACT

Engineering design often involves representation in at least two levels of abstraction: the system-level, represented by lumped parameter models (LPMs), and the geometric-level, represented by distributed parameter models (DPMs). Functional design innovation commonly occurs at the system-level, followed by a geometric-level realization of functional LPM components. However, comparing these two levels in terms of behavioral outcomes can be challenging and time-consuming, leading to delays in design translations between system and mechanical engineers. In this paper, we propose a simulation-free scheme that compares LPMs and spatially-discretized DPMs based on their model specifications and behaviors of interest, regardless of modeling languages and numerical methods. We adopt a model order reduction (MOR) technique that a priori guarantees accuracy, stability, and convergence to improve the computational efficiency of large-scale models. Our approach is demonstrated through the model consistency analysis of several mechanical designs, showing its validity, efficiency, and generality. Our method provides a systematic way to compare system-level and geometric-level designs, improving reliability and facilitating design translation.

Keywords: Model Consistency, System Design, Geometric Design, Model Order Reduction, A Priori Error Analysis

1. INTRODUCTION

In the field of engineering design, the initial step towards creating innovative designs often involves system design, which entails conceptualizing and developing functional structures. Once a system is designed, the subsequent step involves developing its 3D geometry for manufacturing purposes. The objective of the system-based geometric design process is to identify at least one geometric realization that matches the system design. In this paper, we restrict our attention to mechanical systems whose representation at the system-level can be given by mass-spring-damper networks.

1.1 Motivation

The DPMs (e.g., 3D geometric assemblies) and LPMs (e.g., mechanical mass-spring-damper networks) that describe the same physical system at different levels of abstraction are represented using different languages and semantics that cannot be directly translated into one another. For instance, Dassault Systèmes offers two commercial software, Dymola for system modeling and SolidWorks for geometric modeling. The models created in these programs are incompatible and cannot be automatically translated into each other owing to disparities in model types and representations [1–4]. This gap presents a significant challenge for ensuring consistency between the system models and computer-aided design/engineering (CAD/CAE) models. This paper concentrates on the crucial technical issue of systematically verifying the consistency between the two models.

LPMs use a network of lumped components to represent the structure of engineering systems, such as masses, springs, and dampers, which are governed by a system of ordinary differential equations (ODEs) in terms of variables that vary with time [5, 6]. On the other hand, DPMs explicitly consider the geometric and material properties of engineering systems, which are governed by a system of partial differential equations (PDEs) that take into account variables varying with both time and space [7]. Numerical discretization methods can be used to approximate these PDEs by large systems of ODEs upon approximating the spatial continuum in a finite basis [8].

Figure 1 depicts the process of designing a 3D suspension mechanism based on a system design. After creating an LPM in Modelica [9], mechanical parts are used to realize the lumped components and obtain the expected behaviors. The stiffness and damping of the absorber are modeled by a spring-damper pair, without considering the precise geometric realization. Two options for the geometric realization of the absorber are presented. A designer must verify that the selected option behaves as intended by the lumped component before the assembly process (①) in the figure) and on the geometric assembly level (② in the figure) to obtain a qualified design. Currently, the only reliable

---
*Corresponding author: moradbeh@parc.com

Copyright © 2023 by ASME

way to compare design behaviors is to simulate both the LPM and DPM and compare their differential equation solutions a posteriori, which is computationally prohibitive for large-scale models, not to mention the additional challenges related to selecting appropriate time-steps, stability, and convergence. The goal of this paper is to propose a systematic method to check consistency between the system models and CAD/CAE models.

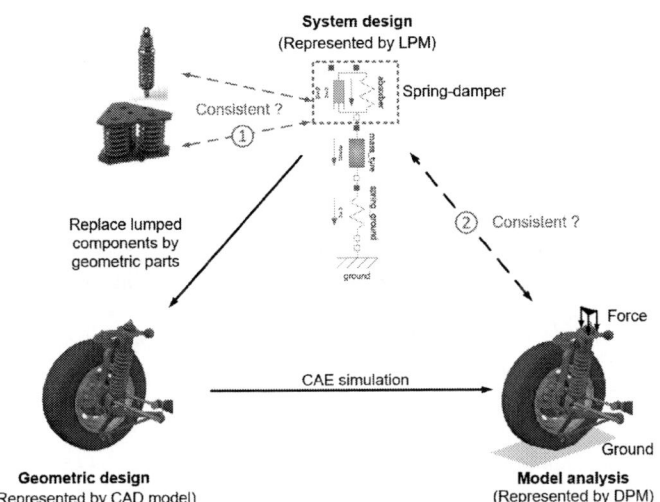

FIGURE 1: THE ROLE OF MODEL CONSISTENCY ANALYSIS IN THE PROCESS OF SYSTEM-BASED GEOMETRIC DESIGN

1.2 Contributions
Our main technical contributions are:

1. We propose a generalizable definition of *consistency* between mechanical LPMs and DPMs that considers both model *specifications* and *solutions*, taking into account factors such as mass, initial and boundary conditions (ICs/BCs), and the behavior of interest (BoI).

2. We develop a *simulation-free* scheme for checking the consistency between LPMs and DPMs based on the definition proposed above. This idea is to compute a priori error bounds between the solutions of both model types by comparing their parameters, circumventing the costly process of solving differential equations.

3. We adopt a MOR technique that a priori guarantees accuracy, stability, and convergence in the simulation-free scheme to improve the efficiency of consistency analysis for large-scale models. The MOR enables the analysis of large-scale designs in a computationally efficient manner.

1.3 Outline
In Section 2, we review the prior works on system-based geometric design, physical model solution comparison, and MOR. In Section 3, we define the problem of model consistency analysis. In Section 4, we present a simulation-free scheme to compare LPMs and DPMs. We illustrate the application of our scheme to various mechanical problems in Section 5. Finally, in Section 6, we summarize the implications of our proposed scheme and suggest potential future research directions.

2. RELATED WORK
2.1 System-Based Geometric Design
The system-to-geometry design strategy has been shown by Ulrich [10] to reduce complexity and clarify the problem-solving process. Most system-based geometric design approaches rely on a one-to-one correspondence between the lumped component in the LPM and the geometric part in a solid model repository. For instance, Finger and Rinderle proposed a component database that contains correspondence between ports of bond graphs and geometric parts for a specific class of mechanical designs [11]. Engelson developed an integrated environment that combines geometric design and system modeling tools to assist engineers in constructing and verifying large, moving rigid-body assemblies [12]. However, these approaches may result in unrealistic designs by ignoring *function-sharing*. Ulrich's work stands out as an exception [10], as it demonstrates a systematic approach to merge multiple functions, abstracted by different system-level components, in fewer geometric parts. Despite the numerous approaches proposed for system-based geometric design [10, 12–16], none of them formally introduce the concept of consistency between system and geometric designs, nor do they provide a systematic approach for assessing the validity of the geometric design with respect to the target system behavior.

2.2 Solution Comparison between LPM and DPM
Comparing the solutions of LPMs and DPMs is a common requirement in model conversion problems, where the two models must be converted to each other with tolerable differences in their respective solutions. In computer graphics, for instance, deformable objects like cloth fabrics and soft tissues are often converted to lumped mass-spring models for faster simulations due to the simplicity and efficiency of LPMs [17–19]. Gelder [20] developed a lumped-spring element based on the geometric angle and length information of 2D linear triangular finite elements, while Vincent et al. [21] extended the method to rectangular finite elements. In both cases, the solutions of the converted LPM and the original finite element model can be directly compared. Suriya et al. in [22] proposed a method for converting 3D deformable objects from finite element models to LPMs by minimizing the difference of the stiffness matrices of the two models. However, the dimension of the LPM solution is often much smaller than that of the DPM after spatial discretization, making direct comparison challenging. In other words, a one-to-one correspondence between the solution dimensions of the two models is typically not guaranteed. This limits the applicability of existing methods to the model solution comparison problem addressed in this paper.

2.3 Model Order Reduction
The field of MOR has seen the development of various techniques [23–27] to approximate a given "full-order" model (FOM) in a numerically efficient and stable manner while preserving certain desired properties. The balanced truncation (BT) method, for instance, removes weakly controllable and observable states from the FOM [28], but can be computationally expensive for large-scale models [29]. The rational Krylov subspace (RKS) method, on the other hand, approximates the FOM transfer function by

Copyright © 2023 by ASME

matching a few significant terms of its Taylor series expansion [30], but does not have a standard rule for choosing the frequency, which is the frequency around which the Taylor series expansion is made. To address this limitation, the iterative rational Krylov algorithm (IRKA) was developed, which iteratively updates the expansion frequency using the reflection of poles of the updated ROM about the imaginary axis at each iteration step until the difference between the transfer functions of FOM and ROM is minimized [31].

However, the IRKA algorithm cannot ensure the error converges to a local minimum [32]. To overcome this limitation, the CUmulative REduction (CURE) scheme was proposed, which adaptively chooses the expansion frequency and incrementally increases the scale of the ROM by monotonically decreasing the norm of the error transfer function to zero through an accumulation process [33]. Furthermore, a stability-preserving, adaptive rational Krylov (SPARK) algorithm [33] was developed to maintain model stability and is usually embedded in the CURE scheme to generate a family of stable ROMs whose orders are increased by sequential accumulation in a single MOR process. This embedded CURE scheme with SPARK algorithm (hereafter abbreviated by SPARK+CURE) was used in [34] to generate a family of physically-interpretable multi-fidelity surrogate LPMs for physical systems governed by PDEs.

The SPARK+CURE method has several key advantages; namely: 1) automatic search for proper expansion frequencies; 2) preserved model stability; 3) guaranteed error convergence; and 4) a priori error bound. This method is particularly suitable for large-scale mechanical systems, as it can significantly improve the time efficiency of model consistency analysis. In Table 1, we compare the important properties of the BT, RKS, IRKA, and SPARK+CURE methods, highlighting the advantages of the latter. For our proposed simulation-free scheme (Section 4), we adopt the SPARK+CURE method to ensure accuracy, stability, and convergence in our model consistency analysis.

TABLE 1: COMPARISON OF FOUR POPULAR MOR METHODS

	BT	RKS	IRKA	SPARK+CURE
Numerical Efficiency	LOW	HIGH	HIGH	HIGH
A priori error bound	YES	NO	NO	YES
Auto order decision	NO	NO	NO	YES
Maintain stability	YES	NO	YES	YES
Guarantee convergence	YES	NO	NO	YES

3. PROBLEM FORMULATION

This section outlines the problem formulation for the model consistency analysis procedure, which involves two model types: the LPM, which provides a system-level description of physical systems, and the DPM, which incorporates the spatiotemporal distribution of materials. The definition of these models and the relationship between their model specifications, behaviors, and

ICs/BCs are crucial for formulating the problem, and will be discussed below.

3.1 Definitions

We focus on linear time-invariant (LTI) mechanical LPMs that can be modeled as networks of interconnected lumped components such as masses, springs, and dampers. To uniquely determine the system behavior, initial mass displacements, velocities, and source terms must be specified, which give rise to a system of ODEs in the time domain. We assume that any physically meaningful mass-spring-damper network can be realized by at least one DPM, which has a continuous material distribution and geometry embedded in 3D space. The DPM can be divided into pieces such that there is a one-to-one correspondence between the spatial integration of mass density and lumped masses. The effective stiffness and damping of the DPM correspond to those of the mass-spring-damper network, and the spatial integration of ICs/BCs and body effects of different pieces correspond to the ICs and source terms of lumped masses. The DPM behavior is described by PDEs defined over a region of spacetime. To compute the effective stiffness and damping, solving PDEs is usually unavoidable. The solution generally involves spatial discretization and numerical methods such as finite difference, element, and volume analysis [8].

Based on the above definitions, we can define *consistency* between an LPM and a DPM as the satisfaction of the following three conditions:

(C1) The total lumped mass value of the LPM matches the spatial integration of the mass density of the DPM.

(C2) The spatial integration of the ICs/BCs, as well as the body effects of the DPM, match the ICs and source terms of the LPM.

(C3) The BoI of the DPM matches the BoI of the LPM in terms of spatial integration.

In other words, for an LPM and a DPM to be consistent, the mass, ICs, source terms, and BoI of the LPM must be equivalent to the mass distribution, initial and boundary conditions, and BoI of the DPM, respectively.

3.2 A mechanical example

Figure 2 depicts an example of an LPM and its corresponding DPM realization, where the geometry of a 3D suspension system is considered. The LPM is composed of a mass, two springs, and a damper, and is subjected to a time-varying external force $\mathbf{f}(t)$. The network is connected to the ground via a spring. On the other hand, the DPM is an assembly of several solid parts that have linear-elastic material distributed in 3D space. The top surface of the assembly is subjected to time-varying pressure $\mathbf{p}(\mathbf{x}, t)$, where $\mathbf{x}$ is the spatial 3D coordinates. The entire assembly is fixed to the ground. To ensure consistency between the LPM and DPM, we require that the following conditions are satisfied: the mass of the DPM matches the total mass of the LPM, i.e., $m = \int \rho(\mathbf{x}) dV$ where $\rho(\mathbf{x})$ is the material density and dV represents an infinitesimal volume element; the external load on the LPM

Copyright © 2023 by ASME

matches the total force on the DPM, i.e., $\mathbf{f}(t) = \int \mathbf{p}(\mathbf{x}, t)dS$ where dS represents an infinitesimal surface element; the pre-specified displacement $\bar{\mathbf{u}}(\mathbf{x}, t)$ of the LPM (which is fixed to the ground) matches that of the DPM, i.e., $\bar{\mathbf{w}}(t) = \frac{1}{S_c} \int \bar{\mathbf{u}}(\mathbf{x}, t)dS = \mathbf{0}$, where S_c is the contact surface area between the tire and the ground; and the solved displacement of the LPM matches that of the DPM, i.e., $\mathbf{w}(t) = \frac{1}{S_p} \int \mathbf{u}(\mathbf{x}, t)dS$, where S_p is the top surface area of the DPM assembly.

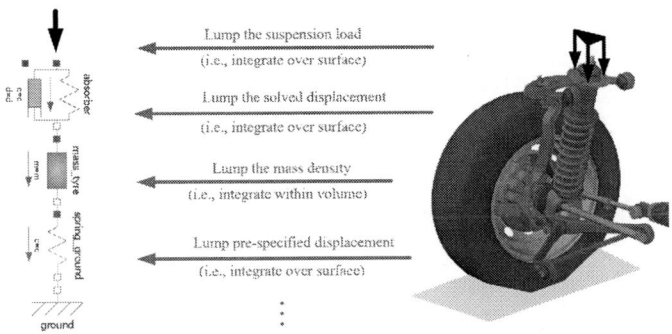

FIGURE 2: AN LPM, ITS DPM REALIZATION, AND THE MODEL SPECIFICATION MATCH

3.3 Formulation

To analyze consistency between LPM and DPM, one has to verify if the conditions (C1) through (C3) are satisfied.

For practical purposes, we introduce an upper-bound (i.e., inequality constraint) to relax the strict equality constraint of BoI in (C3). Noting that an LPM is a ROM of a DPM, its solution can be at best an approximation of the DPM solution, and hence, it is not reasonable to expect a strict equivalence between the solutions of LPM and DPM.

Given an LPM and a DPM with matched mass properties, ICs/BCs, and source terms, our objective is to check if the overall error between pairs of LPM BoI quantities $\mathbf{w}_i(t)$ and the corresponding DPM BoI quantities $\mathbf{u}_i(t)$ for $i = 1, 2, \ldots, n$ is less than an acceptable tolerance $\epsilon > 0$. Symbolically, this can be represented as $\sum_{i=1}^{n} \|\mathbf{u}_i(t) - \mathbf{w}_i(t)\| \leq \epsilon$. Any commonly used vector norm such as L^1, L^2, and L^∞ can be used as the norm $\|\cdot\|$. However, for the sake of specificity, we use L^∞ to provide a measure of the maximum deviation between the DPM and LPM BoI quantities.

Below, we present a simulation-free scheme that compares the BoIs of a given pair of LPM and DPM in terms of the L^∞ norm, for the purpose of model consistency analysis.

4. SIMULATION-FREE SCHEME

Assuming that an LPM and a DPM are provided, along with their matching mass properties, ICs/BCs, and source terms, as well as the BoI quantities' matching form (i.e., spatial integration over the boundary surface), a scheme is presented below for calculating the a priori error between the corresponding BoI quantities of the two models, without requiring explicit numerical solution of differential equations.

We start by deriving the governing equations for both the LPM and the DPM, which are respectively described by a sys-

tem of second-order ODEs and PDEs, respectively. To enable a comparison between the two models, we discretize the PDEs of the DPM using spatial discretization methods [35], thereby converting them to a system of second-order ODEs. We then transform the given ODEs of the LPM and the resulting ODEs (semi-discretized PDEs) of the DPM into their respective state-space forms through linear transformation [36], as illustrated below:

- LPM ODEs in state-space form:

$$
\begin{aligned}
\mathbf{E}_l \mathbf{x}_l(t) &= \mathbf{A}_l \mathbf{x}_l(t) + \mathbf{B}_l h_l(t) \\
\mathbf{y}_l(t) &= \mathbf{C}_l \mathbf{x}_l(t)
\end{aligned} \tag{1}
$$

- DPM ODEs (semi-discretized PDEs) in state-space form:

$$
\begin{aligned}
\mathbf{E}_d \mathbf{x}_d(t) &= \mathbf{A}_d \mathbf{x}_d(t) + \mathbf{B}_d h_d(t) \\
\mathbf{y}_d(t) &= \mathbf{C}_d \mathbf{x}_d(t)
\end{aligned} \tag{2}
$$

A human-readable LPM ODE system (1) can have tens to hundreds of state variables, while the DPM ODE system (2) can have tens or hundreds of thousands, if not millions, of state variables, as a result of semi-discretizing PDEs on a fine mesh. In both equations, $\mathbf{x}(t)$ represents a vector of state variables, $\mathbf{y}(t)$ represents a vector of BoIs, and $h(t)$ represents the input signal of external forces. The descriptor matrix $\mathbf{E}$, dynamic matrix $\mathbf{A}$, input matrix $\mathbf{B}$, and output matrix $\mathbf{C}$ parameterize the LTI systems. The user determines the matrix $\mathbf{C}$ based on the BoI. The matrices $\mathbf{A}$ through $\mathbf{E}$ have the following forms:

$$
\mathbf{A} = \begin{bmatrix} \mathbf{0} & \mathbf{I} \\ -\mathbf{K} & -\mathbf{R} \end{bmatrix}, \mathbf{B} = \begin{bmatrix} \mathbf{0} \\ \mathbf{F} \end{bmatrix}, \mathbf{C} = \mathbf{C}, \mathbf{E} = \begin{bmatrix} \mathbf{I} & \mathbf{0} \\ \mathbf{0} & \mathbf{M} \end{bmatrix}, \tag{3}
$$

where the matrix $\mathbf{I}$ is the identity matrix, $\mathbf{M}$, $\mathbf{K}$, and $\mathbf{R}$ are the mass, stiffness, and damping matrices, respectively, and $\mathbf{F}$ is a vector that contains the external forces. The matrices $\mathbf{M}$, $\mathbf{K}$, and $\mathbf{R}$ used in $\mathbf{A}_l$ through $\mathbf{E}_l$ in (1) are constructed based on the constitutive relationship of lumped components and their adjacency relations. On the other hand, the matrices $\mathbf{M}$, $\mathbf{K}$, and $\mathbf{R}$ used in $\mathbf{A}_d$ through $\mathbf{E}_d$ in (2) are generated from spatial discretization methods [8].

Since both systems of ODEs are LTI, we can match them using linear projections:

(a) The source terms are related by $\mathbf{B}_l(t)h_l(t) = \Gamma_n \mathbf{B}_d(t)h_d(t)$, where Γ_n represents a projection matrix that maps the discrete form of forces of DPM to the lumped forces of LPM.

(b) The initial displacement and velocity terms are related by $\Gamma_I \mathbf{x}_d(0) = \mathbf{x}_l(0)$ and $\Gamma_I \dot{\mathbf{x}}_d(0) = \dot{\mathbf{x}}_l(0)$, where Γ_I denotes the projection matrix that maps the spatially discretized ICs of the DPM to the ICs of the LPM.

(c) The BoI terms are related by $\mathbf{C}_d = \Gamma_f$, where Γ_f is a projection matrix that maps the spatially discretized BoI of DPM to the BoI of LPM.

By substituting the projection in (a) into (1), we obtain a revised state-space form:

Copyright © 2023 by ASME

$$\mathbf{E}_l\mathbf{x}_l(t) = \mathbf{A}_l\mathbf{x}_l(t) + \mathbf{B}'_l h_d(t)$$
$$\mathbf{y}_l(t) = \mathbf{C}_l\mathbf{x}_l(t) \tag{4}$$

This new form indicates that the source term of the LPM can be replaced with the source term of the DPM. This substitution is crucial for computing an upper-bound for the L^∞ error between $\mathbf{y}_l(t)$ and $\mathbf{y}_d(t)$, as demonstrated below.

Given that we are working with LTI systems, we can solve (4) and (2) using Laplace transforms [37]. The solutions take the following forms:

$$\mathbf{y}_l(s) = \underbrace{\left(\mathbf{C}_l(s\mathbf{E}_l - \mathbf{A}_l)^{-1}\mathbf{B}'_l\right)}_{\mathbf{G}_l(s)} h_d(s) \tag{5a}$$

$$\mathbf{y}_d(s) = \underbrace{\left(\mathbf{C}_d(s\mathbf{E}_d - \mathbf{A}_d)^{-1}\mathbf{B}_d\right)}_{\mathbf{G}_d(s)} h_d(s), \tag{5b}$$

where s represents the complex frequency variable, while $\mathbf{G}_l(s)$ and $\mathbf{G}_d(s)$ correspond to the transfer functions of the LPM and semi-discretized DPM, respectively.

When two models with the same input $h_d(t)$ are compared, the maximum difference between their outputs $\mathbf{y}_d(t)$ and $\mathbf{y}_l(t)$ is upper-bounded [38] as follows:

$$\max_{t\in[0,\infty)} \|\mathbf{y}_d(t) - \mathbf{y}_l(t)\|_\infty \le \|\mathbf{G}_d(s) - \mathbf{G}_l(s)\|_{\mathscr{H}_2} \cdot \sqrt{\int_0^\infty \|h_d(t)\|_2 dt}, \tag{6}$$

as long as $h_d(t)$ is a finite energy input (i.e., $\int_0^\infty \|h_d(t)\|_2^2 dt < \infty$), which is the case for most engineering problems. The formula to compute $\|\mathbf{G}_d(s) - \mathbf{G}_l(s)\|_{\mathscr{H}_2}$ is given in 7, where j is the imaginary unit and ω is the frequency in radians per unit time.

$$\|\mathbf{G}_d(s) - \mathbf{G}_l(s)\|_{\mathscr{H}_2} = \left(\frac{1}{2\pi} \int_{-\infty}^\infty |\mathbf{G}_d(j\omega) - \mathbf{G}_l(j\omega)| \, d\omega\right)^{1/2} \tag{7}$$

Notably, the integration term in 6 is constant for a given source term $h_d(t)$, as the BC of DPM is time-invariant. This means that as the norm $\|\mathbf{G}_d(s) - \mathbf{G}_l(s)\|_{\mathscr{H}_2}$ approaches zero, the maximum difference between $\mathbf{y}_d(t)$ and $\mathbf{y}_l(t)$ also approaches zero.

In essence, the deviation between the BoIs of LPM and DPM in the time domain can be approximated by the deviation between $\mathbf{G}_d(s)$ and $\mathbf{G}_l(s)$ in the frequency domain using the $\mathscr{H}_2$ norm. An important observation is that the computation of $\|\mathbf{G}_d(s) - \mathbf{G}_l(s)\|_{\mathscr{H}_2}$ involves solving algebraic equations instead of ODEs, as discussed in [39]. This means that we can avoid solving ODEs altogether. To speed up the computation of these algebraic equations, we use the ROM of the DPM, as shown in the equation below.

$$\mathbf{E}_r\mathbf{x}_r(t) = \mathbf{A}_r\mathbf{x}_r(t) + \mathbf{B}_r h_d(t)$$
$$\mathbf{y}_r(t) = \mathbf{C}_r\mathbf{x}_r(t) \tag{8}$$

In order to enhance the computational efficiency of solving large-scale algebraic equations arising from DPMs, a MOR

method is employed. Specifically, we adopt the SPARK+CURE method from [33], which generates a small-scale surrogate model for the DPM. Due to space constraints, please refer to Panzer's original Ph.D. thesis [33] for more details about this method. This surrogate model has the same source terms $\mathbf{u}_d(t)$ as the DPM in (2), and its transfer function is denoted by $\mathbf{G}_r(s)$. The SPARK+CURE method guarantees an upper-bound $\|\mathbf{G}_d(s) - \mathbf{G}_l(s)\|_{\mathscr{H}_2} \le \bar{\varepsilon}_1$ between the transfer functions of the DPM and the surrogate model. Unlike the spatial-discretized DPM, we assume that the given LPM in (1) has a small scale, so we do not apply MOR to it.

Because both the given LPM and the surrogate model of the semi-discretized DPM have a small state space, computing the value $\bar{\varepsilon}_2$ of $\|\mathbf{G}_r(s) - \mathbf{G}_l(s)\|_{\mathscr{H}_2}$ is time-efficient. With these two values, we can use a triangular inequality to obtain an upper-bound $\bar{\varepsilon}$ for the error between the transfer functions $\mathbf{G}_d(s)$ of the DPM and $\mathbf{G}_l(s)$ of the LPM as follows:

$$\begin{aligned}
\|\mathbf{G}_d(s) - \mathbf{G}_l(s)\|_{\mathscr{H}_2} &= \|\mathbf{G}_d(s) - \mathbf{G}_r(s) + \mathbf{G}_r(s) - \mathbf{G}_l(s)\|_{\mathscr{H}_2} \\
&\le \|\mathbf{G}_d(s) - \mathbf{G}_r(s)\|_{\mathscr{H}_2} + \|\mathbf{G}_r(s) - \mathbf{G}_l(s)\|_{\mathscr{H}_2} \\
&\le \bar{\varepsilon}_1 + \bar{\varepsilon}_2 = \bar{\varepsilon}
\end{aligned} \tag{9}$$

If the upper-bound $\bar{\varepsilon}$ is within the given tolerance, we can view the BoI of DPM and the BoI of LPM as similar.

The equation above provides an upper-bound for the absolute error. To obtain an upper-bound for the relative error, we can use the equation below, where the transfer function $\mathbf{G}_l(s)$ of the LPM is selected as the reference.

$$\frac{\|\mathbf{G}_d(s) - \mathbf{G}_l(s)\|_{\mathscr{H}_2}}{\|\mathbf{G}_l(s)\|_{\mathscr{H}_2}} \le \frac{\bar{\varepsilon}_1 + \bar{\varepsilon}_2}{\|\mathbf{G}_l(s)\|_{\mathscr{H}_2}} = \bar{\varepsilon}_{rel} \tag{10}$$

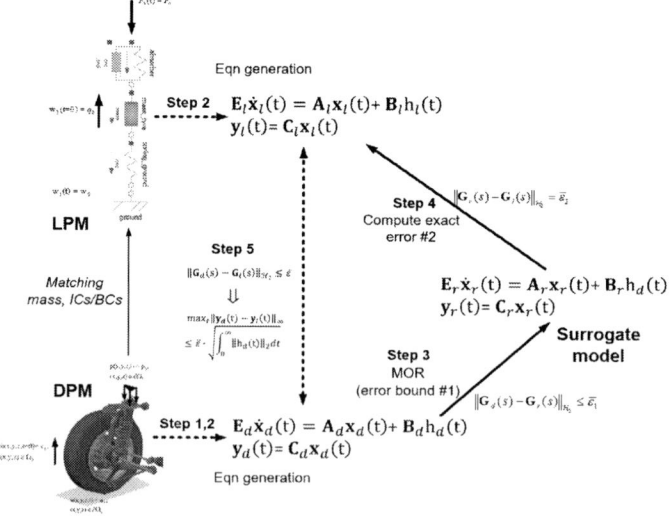

FIGURE 3: SIMULATION-FREE SCHEME TO COMPUTE THE ERROR BOUND BETWEEN THE BOIS OF A PAIR OF LPM AND DPM

Figure 3 provides a visual representation of the five-step simulation-free scheme proposed in this study, where an LPM with masses, springs, and dampers and a DPM with 3D suspension are used as examples. The approach is outlined step-by-step as follows. Given the ODEs of the LPM and the PDEs of the DPM:

1. Use a spatial discretization (e.g. finite element [35]) method to convert the PDEs into a system of ODEs.

2. Convert the ODEs of both models to state-space form.

3. Use the SPARK+CURE method to reduce the scale of the DPM and ensure a priori guaranteed $\mathcal{H}_2$ error $\bar{\varepsilon}_1$ between the DPM and its surrogate model.

4. Calculate the exact $\mathcal{H}_2$ error $\bar{\varepsilon}_2$ between the surrogate DPM and the LPM.

5. Estimate the upper-bound of the $\mathcal{H}_2$ error $\bar{\varepsilon}$ between the transfer functions of the given LPM and semi-discretized DPM using the triangular inequality and verify whether it falls within the acceptable tolerance.

5. APPLICATIONS

In this section, we demonstrate the practical application of the proposed simulation-free scheme by using it to compare the solutions of the LPM and the DPM for two mechanical designs: a bracket (Figure 4b) and a frame (Figure 4c). Both designs have linear isotropic material properties, and their mass properties, ICs/BCs, and the form of the BoI quantities are pre-specified to ensure consistency between the two models.

5.1 Model Specifications

The same LPM topology is utilized for both mechanical designs, as shown in Figure 4a, but with different parameters for the lumped components. The values are provided in Table 2, using SI base units. Both lumped masses have zero initial displacements and velocities.

Two mechanical parts, namely a bracket and a frame, have been designed to implement the LPM instances. The DPMs for both designs have zero initial displacements and velocities. The top surfaces of the bracket and the frame are subjected to pressures that vary with time. The time-varying characteristics of the pressures are depicted in Figures 6a and 6b for the bracket and the frame, respectively.

In order to compare the LPM and DPM models of the two designs, it is necessary to ensure that their mass properties, ICs/BCs, and BoI quantities match. The requirements for this matching are explained in (C1)-(C3) in Section 4, and an example of the matching process is provided in Section 3.2. To maintain consistency, we will follow the same matching procedure as the example in Section 3.2.

5.2 Surrogate DPM of the Bracket

Here, we demonstrate the effectiveness of our simulation-free approach in enhancing the efficiency of model comparison for large-scale models. We used tetrahedral finite elements to discretize the bracket's geometry at three different resolutions (only one sketch is shown in Figure 7a), resulting in three sets of state-space equations with a state variable count ranging from approximately 3,000 to 6,000. We then applied the SPARK+CURE

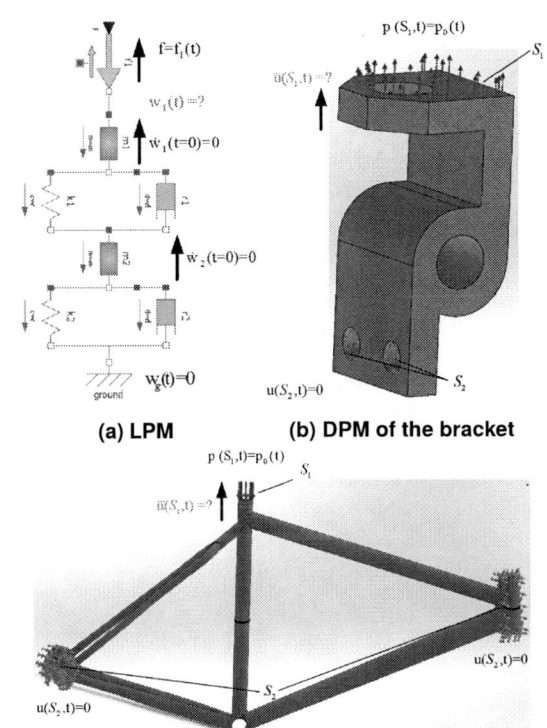

(a) LPM   (b) DPM of the bracket

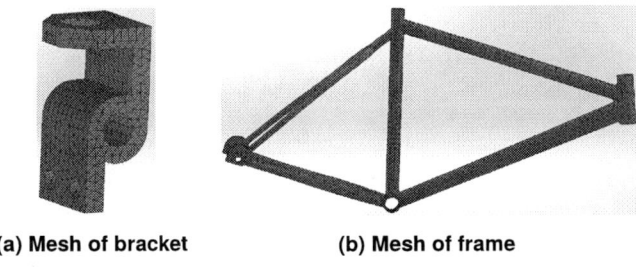

(c) DPM of the frame

FIGURE 4: LPM AND DPMS OF BRACKET AND FRAME

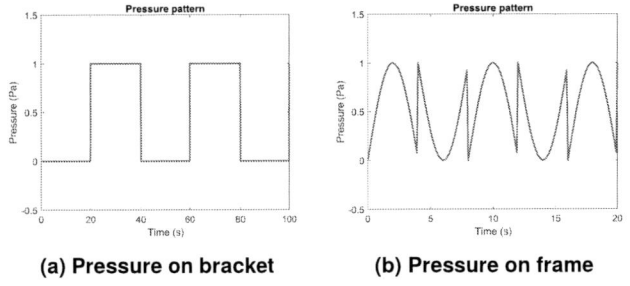

(a) Mesh of bracket   (b) Mesh of frame

FIGURE 5: MESHES OF BRACKET AND FRAME

(a) Pressure on bracket   (b) Pressure on frame

FIGURE 6: PRESSURES ON BRACKET AND FRAME

method to generate three sets of ROMs that have a priori relative $\mathcal{H}_2$ error bounds just below 1%.

To assess the quality of the ROMs, we used the backward Euler method [7] to simulate the DPM and all its surrogate models, using the same solver and time steps. The simulation results are presented in Figures 7a through 7c, where the abscissa denotes

Copyright © 2023 by ASME

TABLE 2: PARAMETERS OF THE LPM USED FOR THE BRACKET AND THE FRAME DESIGNS

	m_1	m_2	k_1	k_2	r_1	r_2	f_1
LPM for Bracket	3.8465×10^5	3.512×10^3	3.316×10^4	4.688×10^3	1.4697×10^5	2.9052×10^3	0.005
LPM for Frame	7.997×10^5	6.9139×10^4	4.8561×10^8	2.2308×10^8	2.8102×10^7	1.5075×10^5	2186.56

computation time and the ordinate denotes the average displacement of the bracket's top surface. Due to limited space, only a few ROM curves are labeled. The solver employed in the simulations was the "sparse state space (sss)" toolbox in Matlab [40], which can analyze high-dimensional dynamical systems with state-space dimensions of $O(10^4)$ or more. This solver preserves system matrix sparsity, allowing it to use computationally efficient operations, such as sparse LU decompositions [41], that would otherwise be infeasible or time-consuming. It is important to note that the simulations were only used to illustrate the accuracy of the surrogate models and were not necessary for our proposed model comparison method.

5.3 Error Analysis

As the order of ROM increased, which corresponds to an increase in the number of state variables of ROM, the simulated response of the ROM more closely approached that of the DPM for all cases. To quantify the error introduced by the MOR process, we calculated two different measures, as shown in Figure 8: the relative $\mathcal{H}_2$ error of the transfer functions and the root mean square error (RMSE) of the simulated response, whose formulas can be found in [42]. The SPARK+CURE method guaranteed the strictly monotonic decay of the bound of the a priori relative $\mathcal{H}_2$ error, as shown in Figures 8d ~ 8f. However, it did not provide a rigorous a priori bound on RMSE, as shown in Figures 8a ~ 8c, although an overall reduction pattern was commonly observed.

For all three cases, the RMSE was around $O(10^{-9})$, two orders of magnitude smaller than the DPM output of $O(10^{-7})$. The a priori relative $\mathcal{H}_2$ error bound showed that if the order of ROM was larger than 36, 38, and 34 in the three cases, respectively, the exact relative $\mathcal{H}_2$ error would be less than $10^{-2.1506} \approx 0.707\%$, $10^{-2.1859} \approx 0.652\%$ and $10^{-2.1797} \approx 0.661\%$. By selecting the ROM of order 34 generated from case 3 as the surrogate DPM for the bracket, we computed the a priori relative $\mathcal{H}_2$ error bound between the discretized DPM and the LPM using (10) as follows:

$$\frac{\|\mathbf{G}_d(s) - \mathbf{G}_l(s)\|_{\mathcal{H}_2}}{\|\mathbf{G}_l(s)\|_{\mathcal{H}_2}} \leq \frac{\bar{\varepsilon}_1 + \bar{\varepsilon}_2}{\|\mathbf{G}_l(s)\|_{\mathcal{H}_2}} = 0.0463 = \bar{\varepsilon}_{rel} \quad (11)$$

Given that the pre-conditions ensure the matches of masses, ICs, and BCs, the LPM and DPM would be considered consistent if the upper bound of error $\bar{\varepsilon} = 0.0463$ is within the user-defined tolerance. To compare the numerical solutions between the LPM and DPM, we plot the errors in Figure 10b, while the numerical solutions are compared in Figure 10a. The maximum difference between the model solutions over the entire time domain is less than 4.8197×10^{-8}, which is one order of magnitude smaller than the maximum model output $O(10^{-6})$.

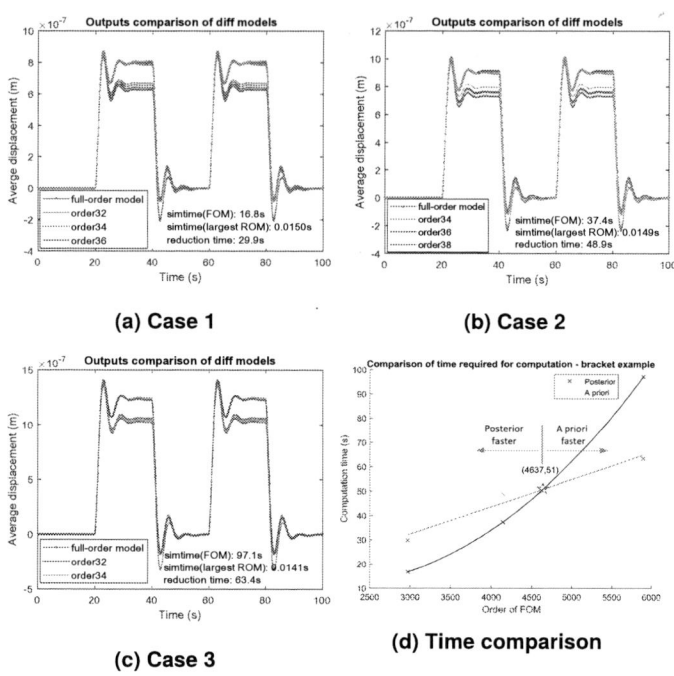

(a) Case 1

(b) Case 2

(c) Case 3

(d) Time comparison

FIGURE 7: COMPARISON OF THE SIMULATION RESULTS AND COMPUTATION TIME BETWEEN DPM AND ROMS - THE BRACKET EXAMPLE

5.4 Time Efficiency Analysis

Through simulations of the DPM and all ROMs for three cases of the bracket, we have obtained two polynomial curves (Figure 7d) that fit the relation between computation time and the order of DPM for both a posterior solution comparison based on direct ODE simulations and our simulation-free scheme. The curves intersect at a certain point, indicating that if the order of the discretized DPM is lower than 4,637, comparing solutions through simulations is more time-efficient. However, if the order is higher than 4,637, our simulation-free scheme will be more time-efficient, with an accuracy loss of at most 1%. It should be noted that if a smaller user-selected relative $\mathcal{H}_2$ error bound is used (i.e., <1%), a new cross-over point that slightly shifts towards a larger order of DPM will be found.

5.5 Surrogate DPM of the Frame and Error Analysis

In a similar manner, we apply the SPARK+CURE approach to the frame design and compare the results of both DPM and ROMs (Figure 9a). The computed relative $\mathcal{H}_2$−error bound is presented in Figure 9b. The figure indicates that if the ROM order is greater than 236, the relative $\mathcal{H}_2$ error between the DPM and the surrogate DPM would be less than $10^{-2.0168} \approx 0.962\%$. Based on this, we can compute the a priori $\mathcal{H}_2$ relative error bound between the discretized DPM and the LPM as follows:

Copyright © 2023 by ASME

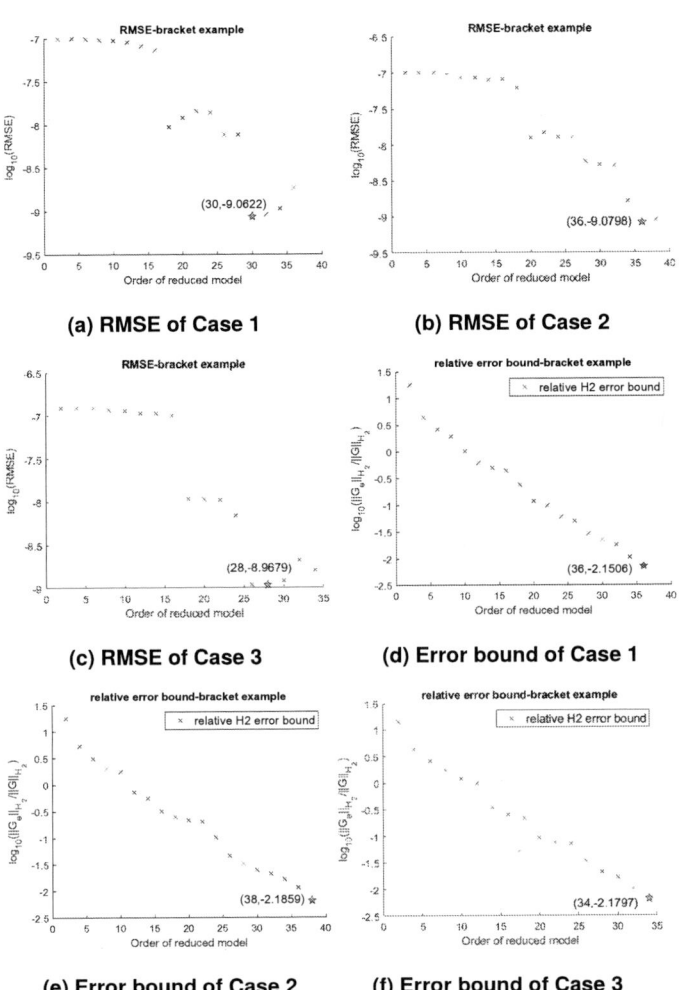

(a) RMSE of Case 1

(b) RMSE of Case 2

(c) RMSE of Case 3

(d) Error bound of Case 1

(e) Error bound of Case 2

(f) Error bound of Case 3

FIGURE 8: RMSE AND A PRIORI RELATIVE $\mathcal{H}_2$ ERROR BOUND - THE BRACKET EXAMPLE

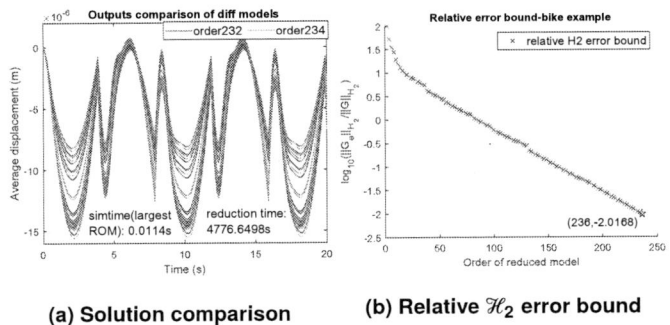

(a) Solution comparison

(b) Relative $\mathcal{H}_2$ error bound

FIGURE 9: COMPARISON OF THE SIMULATION RESULTS BETWEEN DPM AND ROM AND A PRIORI RELATIVE $\mathcal{H}_2$ ERROR BOUND - FRAME EXAMPLE

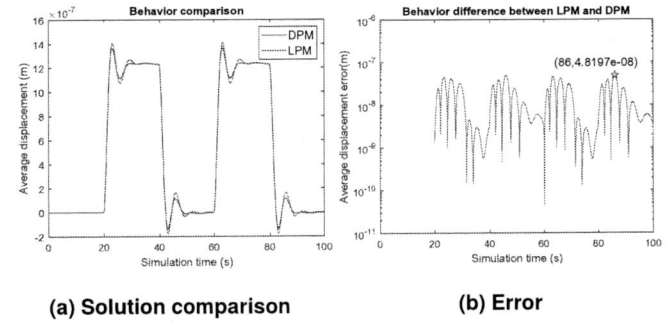

(a) Solution comparison

(b) Error

FIGURE 10: NUMERICAL SOLUTION COMPARISON BETWEEN THE LPM AND THE DPM - BRACKET EXAMPLE

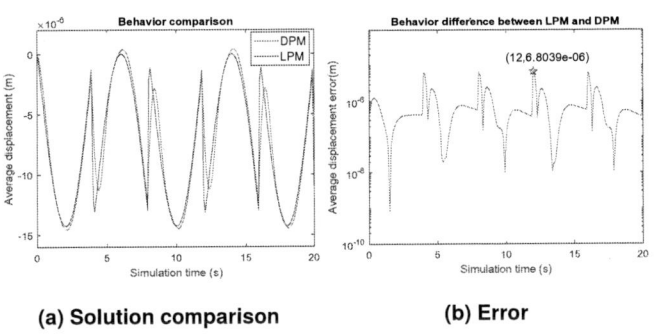

(a) Solution comparison

(b) Error

FIGURE 11: NUMERICAL SOLUTION COMPARISON BETWEEN LPM AND DPM - FRAME EXAMPLE

$$\frac{\|\mathbf{G}_d(s) - \mathbf{G}_l(s)\|_{\mathcal{H}_2}}{\|\mathbf{G}_l(s)\|_{\mathcal{H}_2}} \leq \frac{\bar{\varepsilon}_1 + \bar{\varepsilon}_2}{\|\mathbf{G}_l(s)\|_{\mathcal{H}_2}} = 0.8229 = \bar{\varepsilon}_{rel} \quad (12)$$

The agreement between the LPM and DPM can be determined by the user-defined threshold for the relative error. When the threshold is greater than 0.8229, the models are in agreement. If the threshold is smaller than 0.8229, the models are not in agreement. In Figure 11a, we compare the simulation results of these two models, and their numerical solutions difference is depicted in Figure 11b. It can be observed that the difference is significantly smaller than the displacement of design models in most of the time steps.

6. CONCLUSION

The paper introduces a model consistency concept and presents a simulation-free scheme for comparing the behavior of the LPM and the DPM. By leveraging the SPARK+CURE MOR method, the proposed approach can address the computational challenges associated with large-scale models. Our analysis of two mechanical models demonstrates that our simulation-free scheme provides relatively tight $\mathcal{H}_2$ a priori error bounds for behavior comparison. Furthermore, we show that our proposed scheme is more time-efficient than direct numerical simulation if the order of the discretized DPM is above a certain threshold, with increasing efficiency as the DPM order increases. The proposed approach can be applied to various engineering applica-

Copyright © 2023 by ASME

tions to enhance the system-based geometric design and facilitate integration between system-level and geometric designs.

However, we acknowledge that our framework has some limitations. One limitation is that the scheme to compare model solutions does not account for errors caused by spatial discretization of DPMs, and users must select an appropriate spatial discretization method and resolution. Additionally, the CURE scheme can only be applied to reduce the order of dissipative models, and the a priori guarantees provided by SPARK+CURE are only valid for LTI models. There are two ways to extend our proposed scheme to nonlinear models. The first approach is to divide the nonlinear MOR models into piecewise linear MOR models, and then apply the SPARK+CURE method, similar to the Trajectory PieceWise Linear (TPWL) Method [43]. The second approach is to use MOR methods designed specifically for nonlinear models, such as the Symplectic MOR methods [44], the MOR methods based on the proper orthogonal decomposition [45], and the Gaussian process regression [46], etc. However, many of these methods do not offer rigorous a priori error guarantees.

ACKNOWLEDGMENTS

This material is based upon work supported by the Defense Advanced Research Projects Agency (DARPA) under Agreements No. HR00111990029 and HR00112090065. We are truly grateful to Professor Vadim Shapiro from the University of Wisconsin-Madison for his invaluable contributions to this research. His guidance and insightful discussions have significantly improved the quality of our paper. The responsibility for errors and omissions lies solely with the authors.

REFERENCES

[1] Systemes, Dassault. "Dymola user manual." *Dassault Systemes* (2018).

[2] Chen, Jiangce, Ilies, Horea T and Ding, Caiwen. "Graph-Based Shape Analysis for Heterogeneous Geometric Datasets: Similarity, Retrieval and Substructure Matching." *Computer-Aided Design* Vol. 143 (2022): p. 103125.

[3] Almattar, Tayseer. *Learn SOLIDWORKS 2020: A hands-on guide to becoming an accomplished SOLIDWORKS Associate and Professional.* Packt Publishing Ltd (2019).

[4] Chen, Jiangce and Ilieş, Horea T. "Maximal disjoint ball decompositions for shape modeling and analysis." *Computer-Aided Design* Vol. 126 (2020): p. 102850.

[5] Karnopp, Dean C, Margolis, Donald L and Rosenberg, Ronald C. *System dynamics: modeling, simulation, and control of mechatronic systems.* John Wiley & Sons (2012).

[6] Wang, Randi and Shapiro, Vadim. "Topological semantics for lumped parameter systems modeling." *Advanced Engineering Informatics* Vol. 42 (2019): p. 100958.

[7] Mazumder, Sandip. *Numerical methods for partial differential equations: finite difference and finite volume methods.* Academic Press (2015).

[8] Mattiussi, Claudio. "The finite volume, finite element, and finite difference methods as numerical methods for physical field problems." Vol. 113 (2000): pp. 1–146.

[9] Fritzson, Peter. *Introduction to modeling and simulation of technical and physical systems with Modelica.* John Wiley & Sons (2011).

[10] Ulrich, Karl T. *Product design and development.* Tata McGraw-Hill Education (2003).

[11] Finger, Susan, Rinderle, James et al. *A transformational approach to mechanical design using a bond graph grammar.* [Carnegie Mellon University], Engineering Design Research Center (1990).

[12] Engelson, Vadim, Bunus, Peter, Popescu, Lucian and Fritzson, Peter. "Mechanical CAD with multibody dynamic analysis based on Modelica simulation." *Proceedings of the 44th Scandinavian Conference on Simulation and Modeling*: pp. 18–19. 2003.

[13] Prabhu, DR and Taylor, DL. "Synthesis of systems from specifications containing orientations and positions associated with flow variables." *Proceedings from 1989 ASME Design Automation Conference*: pp. 273–280. 1989.

[14] Greer, James LaMonte. "Effort flow analysis: a methodology for directed product evolution using rigid body and compliant mechanisms." Ph.D. Thesis, The University of Texas at Austin. 2002.

[15] Kota, Sridhar and Ananthasuresh, GK. "Designing compliant mechanisms." *Mechanical Engineering-CIME* Vol. 117 No. 11 (1995): pp. 93–97.

[16] Greer, James L, Jensen, Daniel D and Wood, Kristin L. "Effort flow analysis: a methodology for directed product evolution." *Design Studies* Vol. 25 No. 2 (2004): pp. 193–214.

[17] Mollemans, Wouter, Schutyser, Filip, Van Cleynenbreugel, Johan and Suetens, Paul. "Fast soft tissue deformation with tetrahedral mass spring model for maxillofacial surgery planning systems." *International Conference on Medical Image Computing and Computer-Assisted Intervention*: pp. 371–379. 2004. Springer.

[18] Baraff, David and Witkin, Andrew. "Large steps in cloth simulation." *Proceedings of the 25th annual conference on Computer graphics and interactive techniques*: pp. 43–54. 1998.

[19] Kähler, Kolja, Haber, Jörg and Seidel, Hans-Peter. "Geometry-based muscle modeling for facial animation." *Graphics interface*, Vol. 2001: pp. 37–46. 2001.

[20] Gelder, Allen Van. "Approximate simulation of elastic membranes by triangulated spring meshes." *Journal of graphics tools* Vol. 3 No. 2 (1998): pp. 21–41.

[21] Baudet, Vincent, Beuve, Michaël, Jaillet, Fabrice, Shariat, Behzad and Zara, Florence. "New mass-spring system integrating elasticity parameters in 2D." (2007).

[22] Natsupakpong, Suriya and Çavuşoğlu, M Cenk. "Determination of elasticity parameters in lumped element (mass-spring) models of deformable objects." *Graphical Models* Vol. 72 No. 6 (2010): pp. 61–73.

[23] Baur, Ulrike, Benner, Peter and Feng, Lihong. "Model order reduction for linear and nonlinear systems: a system-theoretic perspective." *Archives of Computational Methods in Engineering* Vol. 21 No. 4 (2014): pp. 331–358.

Copyright © 2023 by ASME

[24] Fang, Dehong, Huang, Zhenwei, Zhang, Jinsong, Hu, Zanao and Tan, Jifu. "Flow pattern investigation of bionic fish by immersed boundary–lattice Boltzmann method and dynamic mode decomposition." *Ocean Engineering* Vol. 248 (2022): p. 110823.

[25] Qu, Zu-Qing. *Model Order Reduction Techniques with Applications in Finite Element Analysis: With Applications in Finite Element Analysis.* Springer Science & Business Media (2004).

[26] Huang, Bin and Wang, Jianhui. "Applications of physics-informed neural networks in power systems-a review." *IEEE Transactions on Power Systems* (2022).

[27] Fang, Dehong, Zhang, Jinsong and Huang, Zhenwei. "Modal analysis on mechanism of bionic fish swimming by dynamic mode decomposition." *Ocean Engineering* Vol. 273 (2023): p. 113897.

[28] Chahlaoui, Younes, Gallivan, Kyle A, Vandendorpe, Antoine and Van Dooren, Paul. "Model reduction of second-order systems." *Dimension Reduction of Large-Scale Systems.* Springer (2005): pp. 149–172.

[29] Besselink, Bart, Tabak, Umut, Lutowska, Agnieszka, Van de Wouw, Nathan, Nijmeijer, H, Rixen, Daniel J, Hochstenbach, ME and Schilders, WHA. "A comparison of model reduction techniques from structural dynamics, numerical mathematics and systems and control." *Journal of Sound and Vibration* Vol. 332 No. 19 (2013): pp. 4403–4422.

[30] Lohmann, Boris and Salimbahrami, Behnam. "Introduction to Krylov subspace methods in model order reduction." *Methods and applications in automation* (2000): pp. 1–13.

[31] Gugercin, Serkan, Antoulas, Athanasios C and Beattie, Cchristopher. "$\mathcal{H}_2$ model reduction for large-scale linear dynamical systems." *SIAM journal on matrix analysis and applications* Vol. 30 No. 2 (2008): pp. 609–638.

[32] Beattie, Christopher A and Gugercin, Serkan. "A trust region method for optimal h 2 model reduction." *Proceedings of the 48h IEEE Conference on Decision and Control (CDC) held jointly with 2009 28th Chinese Control Conference*: pp. 5370–5375. 2009. IEEE.

[33] Panzer, Heiko KF. "Model order reduction by Krylov subspace methods with global error bounds and automatic choice of parameters." Ph.D. Thesis, Technische Universität München. 2014.

[34] Wang, Randi and Behandish, Morad. "Surrogate Modeling for Physical Systems with Preserved Properties and Adjustable Tradeoffs." *arXiv preprint arXiv:2202.01139* (2022).

[35] Bathe, Klaus-Jürgen. *Finite element procedures.* Klaus-Jurgen Bathe (2006).

[36] Atkinson, Kendall, Han, Weimin and Stewart, David E. *Numerical solution of ordinary differential equations.* Vol. 108. John Wiley & Sons (2011).

[37] Callier, Frank M and Desoer, Charles A. *Linear system theory.* Springer Science & Business Media (2012).

[38] Antoulas, Athanasios C, Beattie, Christopher A and Gugercin, Serkan. "Interpolatory model reduction of large-scale dynamical systems." *Efficient modeling and control of large-scale systems.* Springer (2010): pp. 3–58.

[39] Peeters, Jelle and Michiels, Wim. "Computing the H2 norm of large-scale time-delay systems." *IFAC Proceedings Volumes* Vol. 46 No. 3 (2013): pp. 114–119.

[40] Castagnotto, Alessandro, Varona, Maria Cruz, Jeschek, Lisa and Lohmann, Boris. "sss & sssMOR: Analysis and reduction of large-scale dynamic systems in MATLAB." *at-Automatisierungstechnik* Vol. 65 No. 2 (2017): pp. 134–150.

[41] Trefethen, Lloyd N and Bau, David. *Numerical linear algebra.* Vol. 181. Siam (2022).

[42] Wang, Randi. *Consistency Analysis Between Lumped and Distributed Parameter Models.* The University of Wisconsin-Madison (2021).

[43] Rewieński, Michał and White, Jacob. "Model order reduction for nonlinear dynamical systems based on trajectory piecewise-linear approximations." *Linear algebra and its applications* Vol. 415 No. 2-3 (2006): pp. 426–454.

[44] Peng, Liqian and Mohseni, Kamran. "Symplectic model reduction of Hamiltonian systems." *SIAM Journal on Scientific Computing* Vol. 38 No. 1 (2016): pp. A1–A27.

[45] Chaturantabut, Saifon and Sorensen, Danny C. "Nonlinear model reduction via discrete empirical interpolation." *SIAM Journal on Scientific Computing* Vol. 32 No. 5 (2010): pp. 2737–2764.

[46] Marzouk, Youssef M and Najm, Habib N. "Dimensionality reduction and polynomial chaos acceleration of Bayesian inference in inverse problems." *Journal of Computational Physics* Vol. 228 No. 6 (2009): pp. 1862–1902.

Copyright © 2023 by ASME

Proceedings of the ASME 2023
International Design Engineering Technical Conferences and
Computers and Information in Engineering Conference
IDETC-CIE2023
August 20-23, 2023, Boston, Massachusetts

DETC2023-110757

Methods for Creating Additive Printing Paths on Nonplanar Surfaces

Liam Rudd, Zahra Faghihrasoul, Matthew I. Campbell[1]
School of Mechanical, Industrial, and Manufacturing Engineering
Oregon State University
Corvallis, OR, USA

ABSTRACT

One of the ways that additive manufacturing technology continues to improve is the addition of a fourth or fifth degree-of-freedom. Adding rotational axes to the printer head or to the print bed can lead to improvement in the time, cost of printing as well as the quality of the print. We have witnessed similar improvements in subtractive CNC machining as well. The added axes require new computational methods to automatically define toolpaths. In additive, slicer software creates planar cross-sections through the part and automatically defines these toolpaths. With added rotational axes, toolpaths on nonplanar surfaces will be required to fully utilize the advantages of multi-axis printing. This paper reviews current work in nonplanar toolpaths and introduces three new concepts that advance the field. These three concepts are presented in three separate sections – each with their own related work and results. Firstly, we present methods for projecting or unwrapping a nonplanar surface to 2D polygons so that paths can be defined by conventional means. Then, the paths are projected or wrapped back onto the 3D surfaces. Next, two approaches are presented for creating perimeter paths and infill paths that do not transform the slices to 2D planes. These iterative methods simulate the toolpath as particles reacting to virtual forces that guide them to ideal positions. The third builds on an algorithm for creating biologically inspired undulating paths or membranes called differential line growth. These methods can be useful in defining paths for surfaces with high curvature since converting to 2D and back would likely produce too much distortion and paths would not maintain near-equal geodesic distances.

1 INTRODUCTION

As additive manufacturing (AM) machines continue to advance in capabilities and performance – higher precision and quicker runtimes – some systems are finding benefit in adding a fourth and fifth axis of control [1–4]. This parallels the development of CNC mills that also find new applications with additional degrees of freedom [5]. These added movements come in the form of rotational joints that have the benefit of breaking free from traditional planar printing. Doing so means that parts can be printed: 1) with few support structures, 2) with better surface quality, and 3) in less time. However, the slicing and subsequent path planning needed for nonplanar slices is still very much a research question. In this paper, we present various techniques for creating the printed paths on nonplanar surfaces and introduce preliminary research towards new methods for solving this problem.

Generally, there are two types of paths in additive manufacturing: perimeters and infills. Perimeter paths follow the contours of the desired shape to assure a smooth and more accurate geometric shape, while infill operations fill the interior of a given shape with varying degrees of density. Often low densities of infill, such as 20% are used to reduce the material and elapsed time in creating the part. Fully dense, or 100% infill, is common when part strength is key or when post machining is required. With laser powder bed processes, it may be difficult to remove unadulterated power from the voids of parts, thus mandating that 100% infill is used. At any rate, perimeter and infill paths are nearly always defined from two-dimensional polygonal operations. Perimeter paths require offsetting curves, which often leads to self-intersections and

---

[1] Contact author: Matt.Campbell@oregonstate.edu

Copyright © 2023 by ASME

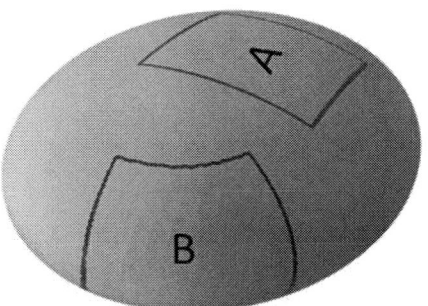

FIGURE 1: AN ELLIPSOID DEMONSTRATES THE ISSUE OF USING 2D METHODS FOR ARBITRARY NONPLANAR PATH PLANNING.

topological path changes that are only reliably solved by approximating paths as polygons. Common infill patterns are also defined with 2D polygon operations as well, and there is no evidence in the literature or commercial tools that this is done by other means. Consider the simple ellipsoid in Figure 1. The square projected at 'A' suffers only minor distortion since this region of the ellipsoid is fairly flat. However, a square projected at 'B' is quite significantly distorted. Printing paths defined from current 2D simplifications will not maintain consistent distances apart and thus the final product of the print will have filament overlap or too much separation within the layer. This leads to the crux of the paper, which is to investigate and define perimeter and infill approaches when the slice to be built is no longer planar. We answer this with three separate sections – each with their own related work, methods, and results subsections. These are summarized in Table 1.

Section 2 looks at a lossless method transforming 3D curves onto a 2D plane. From this transformation, traditional 2D polygonal operations can be used; these 2D paths are then transformed back to the 3D surface. In order for this to function correctly, the 3D shape must be a developable surface; this means that it can be unwrapped to 2D without dilation or compression. Sections 3 and 4 present novel solutions to perimeter and infill challenges respectively – again with their own related work, method, and results subsections. Since these methods do not require the two-dimensional transformations found in existing methods, they may be used on any surface (developable or not). The paper concludes with a discussion of current challenges, conclusions from this paper, and future work involving the methods discussed.

TABLE 1: MAP OF METHODS PRESENTED IN THIS PAPER

	Developable Solids	Non-Developable Solids
Perimeter		Section 3
Infill	Section 2	Section 4

2 PERIMETER AND INFILL PATTERNS FROM THE 2D SIMPLIFICATION

To date, most efforts at defining nonplanar printing paths require the transformation to and from 2D polygons. Essentially this can be accomplished by two methods: conformal transformations or unwrapping. The former is the current state-of-the-art and the main achievement of conformal printing is described in the next section (Section 2.1). Following this, we present a simple concept to unwrap a surface to a 2D plane. As mentioned, this unwrapping is only possible for *developable* surfaces. Here developable is a mathematical term [6] that describes surfaces of this type. The most common developable surfaces are cylinders and cones (see Figure 2), but any prismatic surface can also be classified as developable. When possible, unwrapping should be preferred to conformal printing since no distortions are created in the resulting paths. As multi-axis 3D printing is pursued, one can easily create these cylindrical and conic surfaces with a rotational axis below the workpiece bed.

2.1 Related Work

Recent research in nonplanar 3D printing focuses on creating smooth surface finishes and optimizing mechanical strength through the use of conformal 3D printing. Conformal printing may also be referred to as curved layered fused filament fabrication. It is the procedure of adding a continuous nonplanar layer of filament that conforms to the outer surface of a structure. Alsharhan et al. explored the mechanical advantages of using nonplanar printing for thin-walled structures [7]. The study included comparing failure modes of dome-like specimen printed either with or without an outer conformal layer and without a conformal layer. Pérez-Castillo et al. reviewed the use of curved layers in extrusion additive manufacturing including the characterization of mechanical properties, and various algorithms used to create curved toolpaths [8]. Nayyeri et al. reviewed the benefits of nonplanar slicing over traditional slicing [9]. The improvements are less print time, better surface quality, and higher loading capacities. Furthermore, the authors state that nonplanar slicing and tool pathing algorithms are needed in order to gain the numerous advantages associated with conformal printing.

FIGURE 2: THE TOP ROW IN BLUE SHOWS COMMON DEVELOPABLE SURFACES. THE BOTTOM ROW IN RED SHOWS NONDEVELOPABLE SURFACES.

Copyright © 2023 by ASME

Commonly, nonplanar tool pathing strategies use projection from planar toolpaths to the nonplanar work surface. Llewellyn-Jones uses this strategy for nonplanar conformal printing with a delta printer [10]. Shembekar et al. tested a tool pathing algorithm for nonplanar conformal printing with a 6-DOF robotic manipulator using the same methodology [11]. Bhatt et al. also used the projection method to create toolpaths for conformal wire additive manufacturing. Ahlers et al. proposes the same projection method with an added collision avoidance algorithm [12]. Rodriguez-Padilla used a similar projection method to create paths on a tessellated surface [13]. Although this method is common and works effectively for controlled surfaces, it is highly dependent on the slope of the printing surface as stated by Feng et al. [14]. Therefore, alternative methods are required for creating a more robust method for nonplanar tool pathing. Jin et al. created infill toolpaths by connecting points making up the offset surfaces of each layer [15]. Feng et al. proposed the usage of a discrete geodesic distance field to control the distance between paths [14]. Shan et al. experimented with creating nonplanar layers using isothermal properties on the print surface to create paths [16].

Much previous research has been done for nonplanar additive tool pathing; however, most applications utilize a projection method that works well with a curved surface but will fail with more complex geometries. The algorithms presented in this section have been developed to improve the accuracy in transforming nonplanar surfaces to and from 2D planes for creating more accurate additive manufacturing toolpaths.

2.2 Wrapping Method

The following algorithm makes use of fundamental geometric principles of developable surfaces in the generation of nonplanar toolpaths on surfaces. Unique geometric equations can be created to unwrap developable surfaces into a flat plane, where a developable surface is a surface that can be formed by bending or rolling a planar surface without stretching or tearing [6]. Consider a thin slice created by finding the intersection between the offset of a developable surface and the part to be printed. The unwrapping equations can be applied to the slice

to create a 2D representation of the layer. Then, standard 2D tool pathing algorithms will create layer paths similar to the methods found in 3D printing slicer software (e.g., PrusaSlicer, Cura, etc.). The 2D paths can then be wrapped back onto the original surface through similar equations.

This wrapping method ensures that the dimensions of the part are preserved; however, the method is limited to developable surfaces. Figure 3, shows the entire process of this new wrapping method using an example part geometry constrained to a cylindrical surface. After slicing the part with offset layers of the printing surface, Figure 3a, the outlines of each layer need to be unwrapped from the cylinder using Equations 1 and 2 which create 2D coordinates. Where r_{layer} is the radius at which the current slice is from the center of the cylinder.

$$X_{flat} = X_{wrapped} \tag{1}$$

$$Y_{flat} = r_{layer} * \tan^{-1}\left(\frac{Z_{wrapped}}{Y_{wrapped}}\right) \tag{2}$$

Once unwrapped, Figure 3b, traditional 3D printing perimeters and infill can be created using a common methodology, Figure 3c. The flattened paths can then be re-wrapped onto the cylinder as shown in Figure 3d to create the final toolpaths. Equations 3-5 are used to transform the flat 2D coordinates back into 3D coordinates on the cylinder. Using a cylinder significantly simplifies the equations used for the unwrapping and wrapping processes, but any surface that can be unwrapped without tearing can be solved in a similar manner.

$$X_{wrapped} = X_{flat} \tag{3}$$

$$Y_{wrapped} = r_{layer} * \cos\left(\frac{Y_{flat}}{r_{layer}}\right) \tag{4}$$

$$Z_{wrapped} = r_{layer} * \sin\left(\frac{Y_{flat}}{r_{layer}}\right) \tag{5}$$

While formulating wrapping and unwrapping equations is not trivial, the proposed wrapping method is efficient for constraining 3D paths to developable geometries such as ship

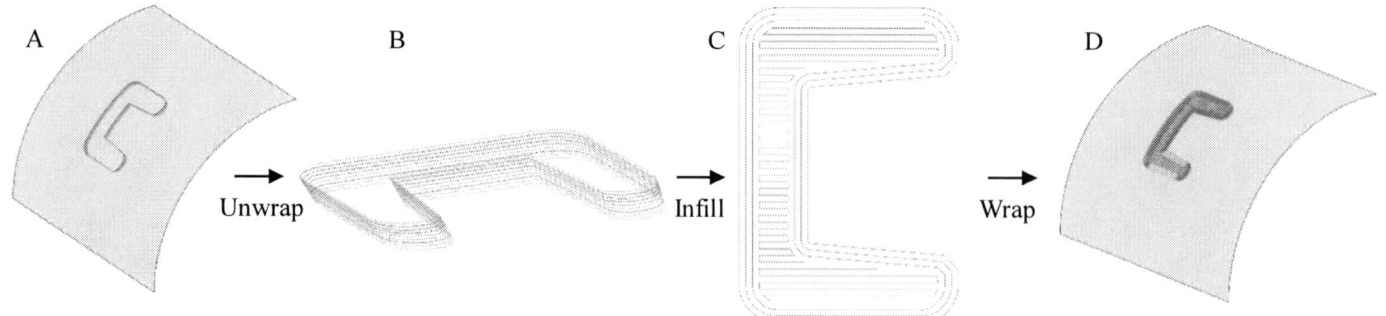

FIGURE 3: A VISUAL REPRESENTATION OF THE ENTIRE PROPOSED WRAPPING METHOD MOVING FROM A) THE INITIAL SLICES, TO B) THE UNWRAPPED OUTLINES, TO C) THE FLAT PERIMETERS AND INFILL, TO D) THE WRAPPED TOOLPATHS ON THE PRINTING SURFACE.

Copyright © 2023 by ASME

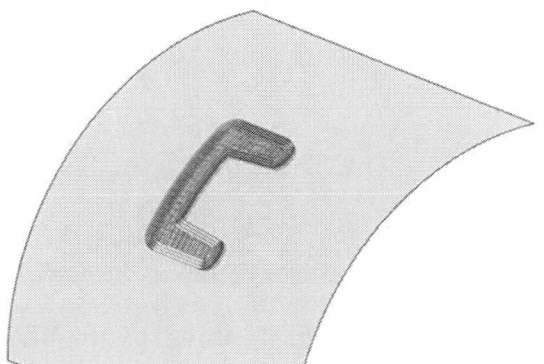

FIGURE 4: TOOL PATHS FOR A C-SHAPED TEST PART GENERATED THROUGH THE PROPOSED WRAPPING METHOD.

hulls and aircraft wings but will break down for non-developable surfaces like a sphere.

2.3 Results

The wrapping method results in toolpaths that are constrained perfectly on the test surface, in our case, a cylinder. Figure 4 shows the simulated toolpaths generated by the wrapping algorithm; these paths were transformed into a single g-code file that was printed using a custom-made multi-axis 3D printer. The final printed part on the cylindrical test surface is shown in Figure 5a. Additional print geometries were experimented with including a rectangular part and a circular prism, Figure 5b and Figure 5c respectively. All toolpaths created using the wrapping method were successfully printed by the custom printer. Notable errors in the printed parts are caused by inaccurate surface calibration techniques and an unoptimized printing profile which are unrelated to the presented method.

3 INFLATION METHOD FOR NONPLANAR PERIMETERS

Perimeters are tool path features used in additive manufacturing to increase the overall quality of the final part. They are created by offsetting the outline inward a distance of an extruded line width to ensure strong bonding mechanics between the outer surface and the infill. The number of perimeters can vary based on the desired stiffness, but in general, one to three perimeters are used in common printing applications. While perimeters can be excluded, they provide important rigidity and surface smoothness that functional parts often require.

3.1 Related Work

Since perimeters are offset lines of 3D boundaries that define the slice, 3D offsetting on surfaces was explored. Holla, et al. developed a method to offset a curve on a tessellation by using the bisecting vector of line segments at each point as the offset direction and applying a modified version of the common planar self-intersecting checks [17]. Similarly, Chen, et al. proposed a method for directly offsetting curves on triangulated mesh surfaces [18]. These algorithms are advantageous since a transformation to another space is not required; however, the accuracy of the methods begins to reduce as the offset distance becomes larger. Detecting and removing self-intersecting sections can become computationally expensive and the robustness of this approach is questionable. Xu, et al. used a mapping-based approach to remove self-intersection of offset paths on mesh surfaces for CNC machining [19]. While revisiting the same point on a tool path is not a significant concern for subtractive methods, self-intersecting toolpaths in additive methods can cause material build up that will decrease the quality of the part and could cause printing failure.

Often Euclidean distances are used when creating 3D curve offsets, but Euclidean distances do not properly describe the total distance along a surface which can lead to gaps or overlapping print lines. Xin, et al. developed a method for

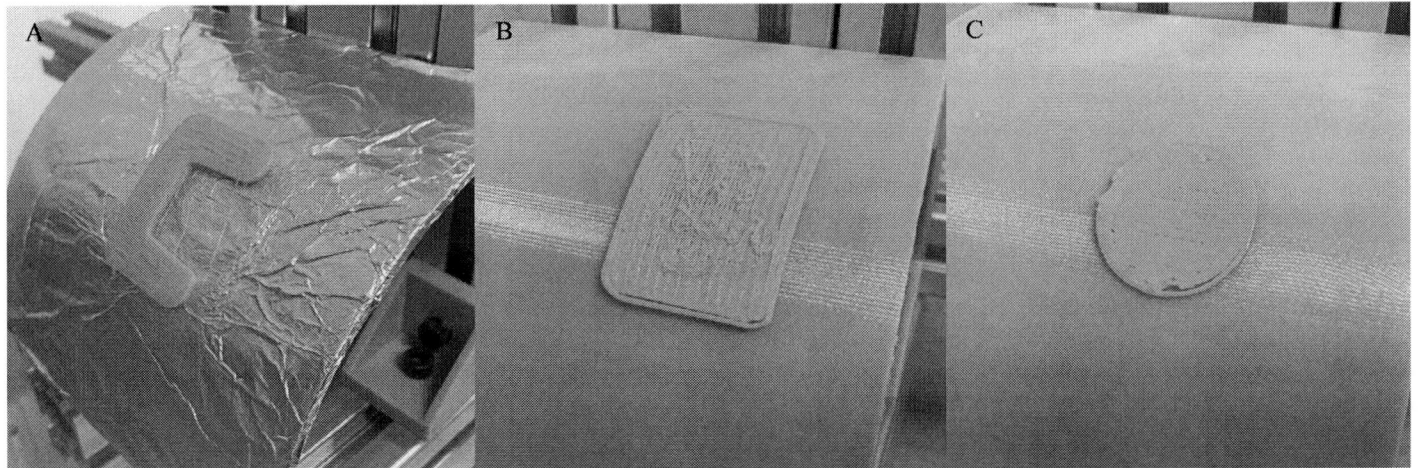

FIGURE 5: TEST PRINTS ON A CYLINDER WITH A KNOWN RADIUS CREATED FROM TOOL PATHS GENERATED THROUGH THE PROPOSED WRAPPING METHOD: A) A C-SHAPED PART, B) A RECTANGULAR PART, AND C) A CIRCULAR PRISM.

Copyright © 2023 by ASME

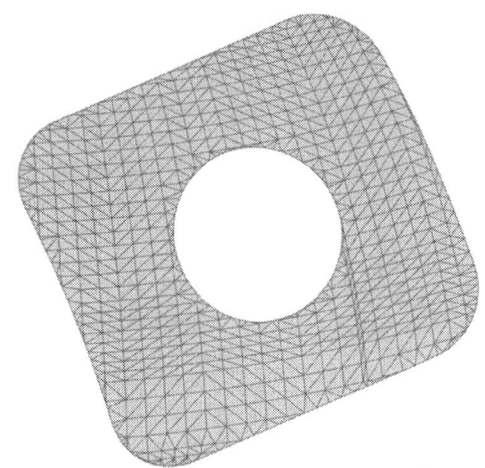

FIGURE 6: THE TESSELLATED FIRST LAYER OF THE SLICED TEST PART.

computing the geodesic offsets of curves on a triangulated mesh [20]. Using the geodesic distance along a surface more accurately represents the distance between lines of filament which would result in a better bond between the perimeter and the outline. Elber and Kim proposed a Euclidean offset of curves on a freeform surface by creating a circle in the plane of the bisecting line at each point on the curve and finding the point of intersections between the circle and the surface [21]. Xu, et al. developed an offsetting method for polyhedral surfaces by using mesh flattening [22]. Hu et al. proposed a region-based path planning method that divided the surface into smaller approximately planar regions [23]. Since these regions are approximately planar, traditional 2D offsetting could be used to generate perimeters.

3.2 Point-to-Face Orthogonal Projection Method

In contrast to the wrapping method proposed in Section 2.2, the following algorithm was developed for constraining curves to general tessellated surfaces. Current projection methods suffer significant distortion effects when projecting 2D toolpaths to highly complex surfaces; this method was developed in an effort to reduce those distortions. Instead of projecting the toolpaths along a static chosen direction, each point is projected to the surface along the normal vector of the closest triangle to the point. While this procedure was developed to work in parallel with the main inflation method described in the next subsection, it can be standalone for accurately transforming conventional 2D toolpaths.

A tessellated slice, Figure 6, is created for each layer by slicing a sample part with offsets of the print surface. The layers are saved as a triangulation surface consisting of an array of vertices and a connectivity array, where the ith row of the connectivity array contains three indices corresponding to three rows in the vertex array which are the coordinates of the ith triangle.

Consider the point, p_0, near the surface of a tessellation, T, with N triangles (Figure 7). A normal vector, $\vec{n}$, is created by

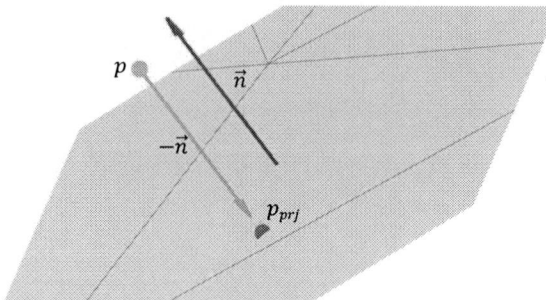

FIGURE 7: A VISUAL REPRESENTATION OF THE METHOD USED TO PROJECT POINT, P, ONTO THE TRIANGULATION USING THE NORMAL VECTORS OF THE TRIANGULAR FACES.

taking the cross product of any two edges for each triangle on T. Equation 6 is used to project p_0 to each plane defined by the triangles in T, as shown in Figure 5 resulting in N new points, p_{prj}. Where, p_0 is the initial point, $\vec{c}$ is the center of the triangle, and $\vec{n}$ is the normal of the triangle.

$$\overrightarrow{p_{prj}} = \overrightarrow{p_0} + \left((\vec{c} - \overrightarrow{p_0}) \cdot \vec{n} \right) \vec{n} \qquad (6)$$

Then, barycentric coordinates (u, v, w) are determined for each point in p_{prj}, each set of coordinates saved into a barycentric array. Each value in this array is altered by applying Equation 7. Where b is a stand-in that represents an individual barycentric coordinate. All values between zero and one in the barycentric array are set equal to zero and all other points are made positive. This process is used to account for how far outside of the triangle the projected point is. The modified barycentric array is then summed along its rows to determine the error of each projected point. Figure 8 shows how any projected point that lies inside the corresponding triangle will have an error of 0, while points that lie outside will have some error. The projected point in p_{prj} with the smallest error is returned as the new point.

$$b = \begin{cases} 0, & 0 \le b \le 1 \\ |b|, & otherwise \end{cases} \qquad (7)$$

While it seems like there should be one and only one triangle with zero error, there are times when the curvature of the surface is so significant that the point falls between the

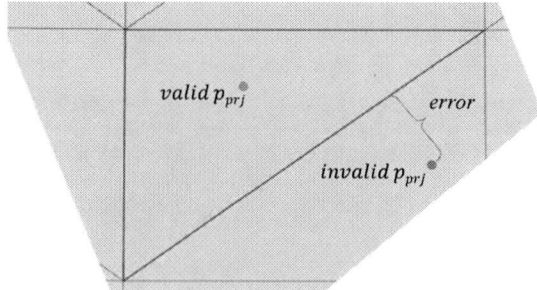

FIGURE 8: AN EXAMPLE OF WHICH POINTS ARE VALID BASED ON THE LOCATION OF THE POINT IN RELATION TO THE TRIANGLE.

Copyright © 2023 by ASME

projected volumes of two adjacent triangles. In these cases, the point with the smallest error is returned. In a case where more than one triangle is valid (i.e., b is zero for multiple triangles), the triangle with the shortest projection distance is chosen for projection. This method is best used for constraining a point to a surface as it is pushed along the same surface, where the fixed projection method would fail as the surface becomes too complex.

3.3 Inflation Method

Efficient methods have been developed for polygonal planar offsetting, and solutions for offsetting nonplanar curves have been implemented into some CAD software tools. However, many of these implementations have not been integrated with slicing software to create a general nonplanar additive tool pathing algorithms and are not robust enough when offsetting on complex surfaces with large offsetting distances. So, the research team was motivated to develop a simpler approach for creating perimeters on nonplanar surfaces that did not require removing self-intersecting regions and other post-offsetting tasks associated with traditional offsetting methods. This section describes a novel method for generating nonplanar offset curves on a general surface for the benefit of creating nonplanar printing toolpaths for multi-axis additive manufacturing. Similar to how a gas fills the volume containing it, the inflation method iteratively expands a single polygon to the edges of a discrete surface. The pseudocode of the proposed method is shown in Figure 9.

It should be noted that the boundary curve, B, is on the triangulated surface, T, and, if necessary, is connected using lines drawn over T to remove any rings (i.e., holes) in the surface so that the entire boundary can be described as one polygon. A volume, V, shown in Figure 10, is defined along the edges to constrain the expanding polygon from growing beyond a set offset distance. The volume is created by sweeping a circular profile, of radius d_{offset}, around the curve creating a tubular path along the shape. This tube is defined as alternating spheres at the vertices and cylinders along the edges. This simplification streamlines the intersection calculations used later in the algorithm. One might think that this boundary can be simply intersected with the triangles to get the perimeter path. While the mathematics of calculating the intersecting points on the between edges of the triangles and the volume is straightforward, connecting these points into a continuous line is not, especially if the topology of the curve is different than the initial curve.

Consider a polygon, S, which is created by randomly choosing a triangle on the surface that has no points within the blue tubular boundary volume. A logic array, L, stores a Boolean value for each point in the polygon. Where, *True* means that the point is fixed, and *False* means that the point is free to move. At the start of each iteration midpoints are added on the edges of S where the length is greater than a maximum distance, d_{max}. Adding new points allows the algorithm to start from a simple triangle and grow an arbitrary number of vertices to meet the necessary complexity of the perimeter. The value of d_{max} should be set so that the resolution of the eventual perimeters is high enough that no edges significantly leave the surface, and that the eventual G-code meets the target tolerances. As a small bookkeeping matter, new values must be added to L at the same indices that points were added to S.

Inputs: S = Initial Polygon on Surface; B = Boundary Curves; T = Triangulated Surface; d_{max} = Max Distance Between Points; d_{offset} = Offset Distance

1. L = *False* array with length = len(S)
2. V = Pipe-shaped offset volume along B with radius d_{max}
3. While not all L == *True*:
4. S, L = Add midpoints and logic rows between lines that have lengths > d_{max}
5. L = Update freeze logic where points are within or on V
6. F = Create pressure array for each point in S
7.   For i from 1 to length(S):
8.   If L_i = *False*:
9. p_{new} = S_i + F_i
10.   If p_{new} is within V:
11. p_{new} = intersection between F_i and V
12. L_i = *True*
13.   End
14.   end
15. p_{new} = Project p_{new} to T
16. S_i = p_{new}
17.   end
18. end

FIGURE 9: PSEUDOCODE FOR INFLATION METHOD

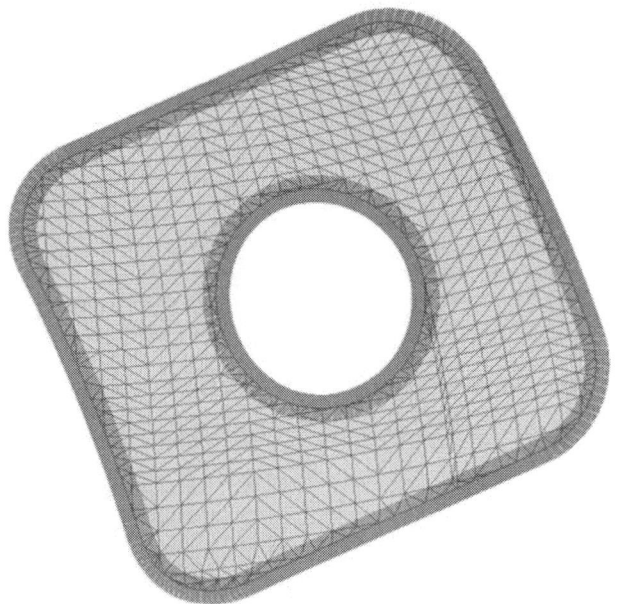

FIGURE 10: AN EXAMPLE OF THE PIPE-SHAPED OFFSET VOLUME USED TO CONSTRAIN THE INFLATING POLYGON A SET DISTANCE FROM BOUNDARY.

Copyright © 2023 by ASME

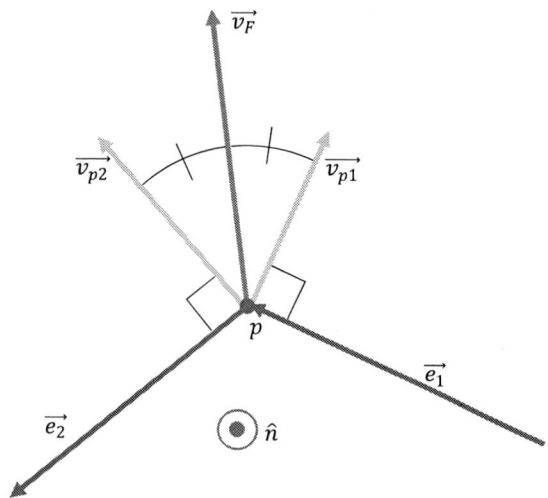

FIGURE 11: THE PROCESS OF GENERATING A PUSHING FORCE FOR EACH POINT IN THE INFLATION ALGORITHM. THIS IS A 2D VIEW OF THE PROCESS FOR SIMPLICITY. THE ENTIRE PROCESS TAKES PLACE ON THE PLANE OF THE CLOSEST TRIANGLE TO THE WORKING POINT.

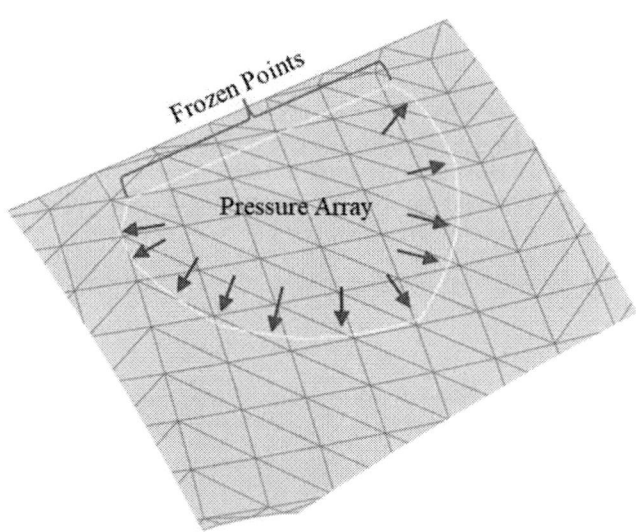

FIGURE 12: A ZOOMED VIEW OF A STEP IN THE INFLATION ALGORITHM. THE ARROWS SHOW THE PUSH VECTORS THAT ARE APPLIED TO EACH UNFROZEN POINT, WHILE THE POINTS AT THE OFFSET EDGE ARE FROZEN AND HAVE NO VECTOR APPLIED TO THEM.

Inflation is achieved by applying a pushing vector at each point on the polygon in every iteration. This moves the points away from one another simulating the inflation effect, which is shown in Figure 11. The pushing force, $\overrightarrow{v_F}$, for each point is defined as the angular bisector between the perpendicular vectors of the two edges connected to that point. Figure 12 shows a top-down view of the vector geometry used to calculate this bisecting vector. Since a discrete definition of the surface is used, the closest triangle to the current point, p, is found and its

normal, $\vec{n}$, is stored in association with p. Perpendicular vectors, $\overrightarrow{v_{p1}}$ and $\overrightarrow{v_{p2}}$, are determined by taking the cross product between the corresponding edges and the normal vector (Equation 8) where the subscript n refers to either edge 1 or edge 2. An angular bisector is determined using these perpendicular vectors, as shown in Equation 9. This angular bisector is calculated for each point on the polygon and used to create a pushing force, $\overrightarrow{v_F}$, for each corresponding point on the growing polygon.

$$\overrightarrow{v_{pi}} = \overrightarrow{e_i} \times \hat{n} \tag{8}$$

$$\overrightarrow{v_{Fi}} = \overrightarrow{v_{pi}} + \overrightarrow{v_{pi+1}} \tag{9}$$

New points are created by adding these push vectors to the corresponding points in S. A query is run on each new point, p_{new}, to check if it falls within the tubular boundary volume. Equations 10 and 11 check if the point falls within one of the boundary's spheres or cylinders, respectively, and Equation 12 ensures that a point falls within a finite cylinder defined by B_n and B_{n+1}. Where B_n is a point on boundary curve that defines the center of the spheres and the first end of the cylinder's axes, and B_{n+1} is an adjacent point to B_n on the boundary curve that defines the second end of the cylinder's axes. The query is run through two checks; first, if Equation 10 returns *True*, then if **both** Equations 11 and 12 return *True*. If either check returns *True* then the intersection between the push vector and the boundary volume must be determined.

$$(p_{new} - B_n) \cdot (S_i - B_n) \leq d_{offset}^2 \tag{10}$$

$$\left\| \frac{(p_{new} - B_n) \times (B_{n+1} - B_n)}{\|B_{n+1} - B_n\|} \right\| \leq d_{offset} \tag{11}$$

$$0 \leq \frac{(p_{new} - B_n) \cdot \frac{B_{n+1} - B_n}{\|B_{n+1} - B_n\|}}{\|B_{n+1} - B_n\|} \leq 1 \tag{12}$$

The intersection between the push vector and the volume is found by solving for t in both Equations 13 and 14. The boundaries of the spheres and cylinders are found where Equations 13 and 14 result in a value of zero, respectively. When the new point has been determined to be inside the volume, there must be an intersection between the pushing line and the volume. Equation 15 defines this pushing line that originates from the previous un-pushed point, S_i, is in the direction of the push vector, $\overrightarrow{v_{Fi}}$, all with respect to t, which is the distance along the line. Where the result is the distance along the pushing line that intersections occur. All values of t not between *zero* and *one* are ignored since these correspond to points outside the segment between the un-pushed point and the new point. The remaining t values are transformed into points on the pushing line using Equation 15. A final check is run to verify that any cylinder intersections satisfy Equation 12 by being within the corresponding finite cylinders. The minimum t value of the remaining t's is found and the corresponding point on the line is used as the new point on the inflating shape. The

Copyright © 2023 by ASME

corresponding row in the logic array is set to *True* and the point is frozen in place.

$$Sph(t) = (p_{new}(t) - B_n) \cdot (p_{new}(t) - B_n) - (d_{offset})^2 \quad (13)$$

$$Cyl(t) = \left\| \frac{(p_{new}(t) - B_n) \times (B_{n+1} - B_n)}{\|B_{n+1} - B_n\|} \right\| - d_{offset} \quad (14)$$

$$p_{new}(t) = S_i + t * \vec{v_{Fi}} \quad (15)$$

After the new point has been calculated, it is projected to the surface using the method described in Section 2.3. The entire method will repeat until all of *L* is *True*, resulting in a nonplanar offset of the input boundary constrained to the discrete surface of *T*. This proposed method could be an effective method for creating offsets of nonplanar curves on general discrete surfaces.

3.4 Results

The point-to-face orthogonal projection method in section 3.2 was developed alongside the iterative pathing approaches discussed in sections 3.3 and 4.2, so the projection of points onto the triangulated surface occurs in every iteration of those growth algorithms. The paths shown in this section were generated using point-to-face orthogonal projection and the inflation algorithm. Error is introduced into this projection method through the discrete approximation of the surface resulting in discrepancies between the simulation and the real surface; however, a relatively fine meshing quality can reduce the error to be negligible.

The current MATLAB implementation of the inflation method is capable of creating offsets for nonplanar curves on surfaces defined as triangulations. Figure 13 shows the final iteration of a single offset produced through this method. Sequential growth stages throughout the algorithm are shown in Figure 14, starting from the left, and ending at the right. The current version of the algorithm takes about *33 seconds* to complete an offset for one layer. Because each point on the inflating shape needs a boundary intersection check ($O(n^2)$) run at each step, as well as a surface projection ($O(n)$), the time complexity of the algorithm is $O(n^2 + n)$ Continued discussion of the proposed inflation method can be found in Section 5.

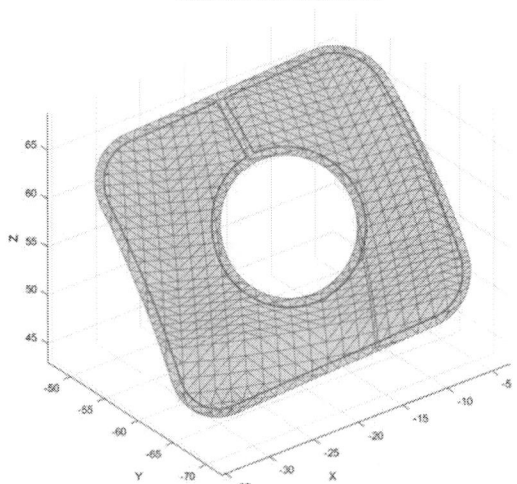

FIGURE 13: FINAL RESULT OF A SINGLE OFFSET ON A DISCRETE REPRESENTATION OF A LAYER USING THE EDGES OF THE LAYER AS THE BOUNDARY SURFACE.

4 CREATING INFILL

The Differential Line Growth algorithm is a simulation of the growth process of a biological system in which an initial curve evolves into a sinuous or highly-branched curve to fill a specific region of space. Differential Line Growth consists of nodes in space, connected by line segments to form paths. A set of rules repeatedly apply to points and update their position to form the curve. These rules are as follows: each node wants to be close to its connected nodes, and simultaneously maintain a minimum distance from all other nodes. In cases where an edge's length exceeds a prescribed magnitude, a new node is injected. Because this algorithm strives to keep points and lines at a prescribed distance, it appears to be a suitable candidate as an infilling pattern tool. One of the advantages of using Differential Line Growth to fill a 2D region is in cases with complex boundaries. Unlike traditional patterns like zigzag and Hilbert Curves, it is flexible and easily adjusts itself to the boundaries.

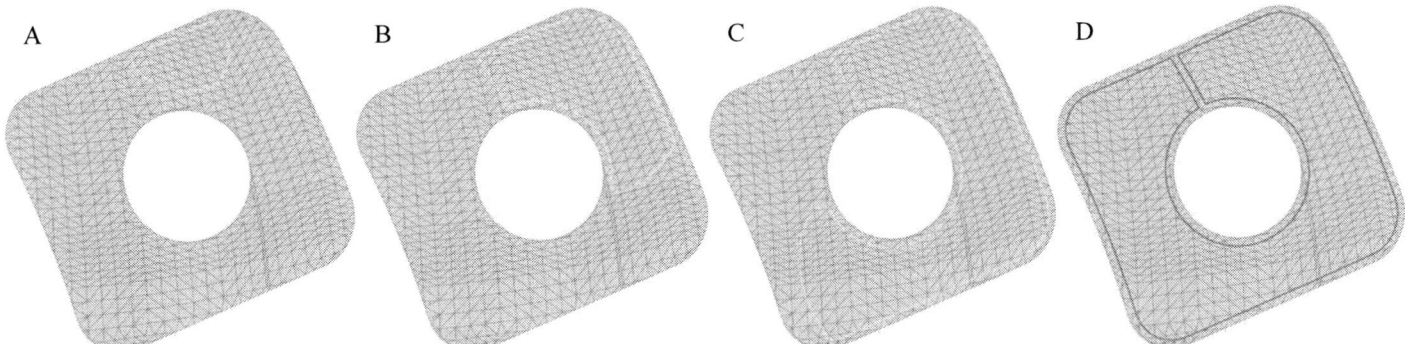

FIGURE 14: THE SEQUENTIAL STAGES OF GROWING THE OFFSET GOING FROM A TO D IN ORDER. A YELLOW OUTLINE SIGNIFIES THAT ALGORITHM IS STILL RUNNING AND THE GREEN CURVE SIGNIFIES THE FINAL RESULT OF THE ALGORITHM.

Copyright © 2023 by ASME

4.1 Related Work

There are numerous papers that describe unique infill approaches for additive manufacturing. In this section, we present a subset of techniques that have been crucial to the development of our nonplanar approach.

Pedersen and Singh synthesized labyrinthine structures to aesthetically fill a given area [24]. By following a texture map, the authors develop a stochastic algorithm where an initial curve evolves into a branching shape effected by spatially varying parameters. The algorithm defines counteracting forces, which are adopted in our approach to repeatedly move each node of the curve. The consequences of these forces result in a smooth and continuous curve. The authors present this method as a means for creating art; our approach further developed their idea for an additive manufacturing context.

In terms of novel infill patterns, Zhao and Huang introduce Connected Fermat Spirals (CFS) pattern to continuously fill a 2D region [25]. Their method consists of four stages: decomposing the region into spiral-able subregions, generating Fermat spirals for the respective area separately, connecting the curves based on spiral-contour tree, and a post-process optimization with the aim of improving the path to avoid sharp turns and to maintain parallel sections. The result of the method is a single continuous path that fills a region. However, this approach may negatively affect mechanical quality, since interval time for printing two adjacent lines is long and the previous adjacent paths have had enough time to cool down. In our Differential Line Growth algorithm, the sinuous shape preserves locality and increases interlocking structures, which results in high stiffness and mechanical quality. In this way, our results are most similar to the Hilbert curve infill patterns. However, the Hilbert curve approach is not effective for arbitrary regions, especially hollow structures with narrow sections [26]. Zhao and Huang provide one of the fewest recent papers that discusses new infill patterns, but – like traditional zigzag or contour-parallel infill methods – this method does not provide a means to transform into arbitrary 3D surfaces.

Kaplan and Bosch provide another approach to create a curve to fill a region. They took advantage of the traveling salesman problem (TSP) to construct a continuous line drawing based on gradients from a photograph or image [26]. Their algorithm converts pixels into a number of points proportional to the darkness of the grayscale. These points are treated as the cities for the TSP. The concept of rephrasing the infill as a traveling salesman problem was also considered by the research team, but on seeing the results in this paper, we realized that path-to-path distances were not maintained, and their method requires an immense calculation cost.

Other than the paper by Pedersen and Singh, there is various documentation about creating differential line growth algorithms for the purpose of mimicking biology or creating art. However technical papers are sparse on the topic, but various resources can be found in blog posts and open-source software. Webb presents how different ways of injecting points into the path affect the resulting curve. In the article, he provides nine case studies [27]. Hoff illustrates attraction and repulsion force during growth of the curve in several short, animated videos [28]. Yu et al. utilize tangent-point energy as a repulsion force to avoid intersection between points that are extremely close and thus enable them to push themselves away from each other. This energy depends on the distance between the point along the curve, and radius of the sphere that includes the points. At the end, they use a complex computation to optimize the energy to keep the curve from self-intersecting [29]. Toncean utilizes a grid method to avoid intersection, where the space is divided by a grid with equal buckets, each node through the grid is assigned with a bucked [30]. But perhaps the most famous example of this method is by Bader et al., in which differential growth is used to create wearable clothes and jewelry [31].

4.2 Differential Line Growth Method

The differential line growth algorithm is initiated from a triangle in the surface and develops it into a complex, ramified curve which fills a defined area. Traditionally, the curve is a simple polygon in that it is a series of line-segments which connect points in space to construct a single closed non-intersecting path. As the algorithm ensues, the position of these points is defined by a series of forces applied by other points. This is essentially a particle physics simulation where forces define accelerations; accelerations define velocities; and velocities define changes in position. Let the number of points be defined by N and the position of each point by P_i, where $1 \leq i \leq N$.

Equation 16 presents a force-balanced equation where five distinct effects lead to a new acceleration for each point. These forces are described in detail in the remainder of this subsection. This vector sum of forces is assigned to the acceleration, a_i, of P_i.

$$a_i = W_b B_i + W_a A_i + W_c C_i + W_f F_i + D_i \qquad (16)$$

This effectively treats all nodes as having one mass unit but – given the user-defined weights from the equation – there is no real loss of generality or exactness from this simplification. Equations 17-19 describe how the node position, P_i, is updated at each iteration based on Δx_i, which is updated from velocity, V_i, and acceleration with help of a given time-step, Δt.

$$V_i = V_{previous} + a_i \Delta t \qquad (17)$$
$$\Delta x_i = V_i \Delta t + \frac{1}{2}(a_i \Delta t^2) \qquad (18)$$
$$P_i = P_{previous} + \Delta x_i \qquad (19)$$

Here, the time step along with the inertia (effectively stored in the velocity update equation) provides added robustness in the algorithm. In fact, the time-step is adjusted dynamically during the process. If the Δx's are large and the polygon self-intersects, then Δt is reduced and the dynamics of the iteration are attempted again. Conversely, if the Δx's are too small, then Δt is increased to arrive at a solution more efficiently.

Copyright © 2023 by ASME

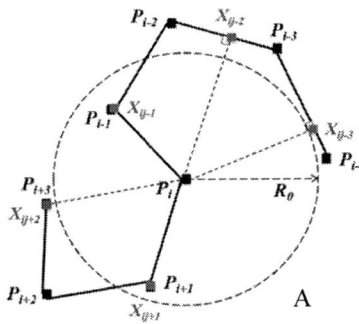

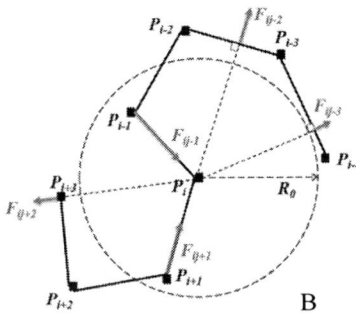

FIGURE 15: A) A VISUAL REPRESENTATION OF XIJ, B) A VISUAL REPRESENTATION OF DIRECTION AND MAGNITUDE OF F_{ij}.

Attractions-Repulsion force: As mentioned before, each node is inclined to be close to its connected nodes through a mutual attraction force. Also, each node maintains a minimum distance from all other nodes. If two nodes become extremely close together, they will exhibit a strong repulsive force. This distance plays a crucial role in forming the shape of the curve. A user-provided input, R_0, defines the target distance between printing paths. The direction and magnitude of force that a segment exerts on a node depends on the distance, r, from the node to its closest point on each nearby segment. At long distances, nodes should exert a weak attraction force. As nodes get closer, the force gets stronger; and if they exceed a prescribed radius, then the force will reverse – repelling nodes that are too close. Thankfully, the Lennard-Jones Potential [32] characterizes these features as seen from Equation 20 and is plotted in Figure 15. The Lennard-Jones constant, σ_{LJ}, in this problem determines the distance that the direction of the force is reversed and is set equal to R_0, the target distance between printer paths. The exponent, L, is nominally 6 when modeling van der Walls forces in nature, but for our purposes, we have reduced it to 3 for better numerical stability.

$$w(r) = \left(\frac{\sigma_{LJ}}{r}\right)^{2L} - \left(\frac{\sigma_{LJ}}{r}\right)^{L} \tag{20}$$

$$f_{ij} = \frac{p_i - x_{ij}}{|p_i - x_{ij}|} \cdot w\left(\left|p_i - x_{ij}\right|\right) \tag{21}$$

$$A_i = \sum_j^N f_{ij} \tag{22}$$

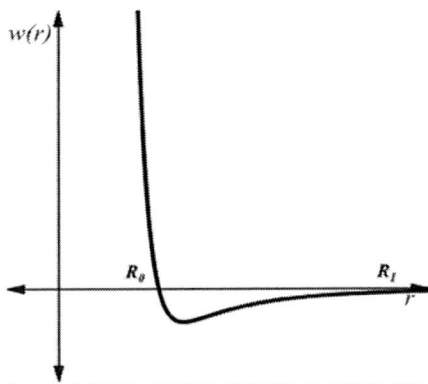

FIGURE 16: LENNARD-JONES CURVE BASED ON DISTANCE

Figure 16 shows geometrical representation of how the attraction-repulsion forces are determined. Note how the red dots in this figure correspond to X_{ij} and f_{ij} in the equations. Here, we are illustrating how the force is found for P_i. The first part of Equation 21 simply establishes this unit direction and the second part determines the vector magnitude from the Lennard-Jones characteristic equation. These forces – as shown in the bottom of Figure 16 as blue arrows – are summed in Equation (22) to determine the node's overall attraction-repulsion force, A_i.

Brownian Force: The next force is called the Brownian motion force, which simply represents a vector with random direction and magnitude. Z_i is a random number with normal distribution in range of [0 1]. v_i is a normal vector with a random direction. $D_{average}$ is the average distance between nodes. This Brownian force adds random fluctuation to the nodes as defined in Equation 23. This force helps to avoid premature stasis in the shape.

$$B_i = Z_i v_i D_{average} \tag{23}$$

Damping Force: When two nodes are extremely close, the repulsive forces become exponentially strong so that nodes can be flung far away. Furthermore, oscillation can happen in the nodes, so a damping force, Equation 25, is included to improve the stability of the dynamics. Damping is simply proportional to the velocity.

$$D_i = -D_{mf} V_{i_{previous}} \tag{24}$$

Fairing Force: Figure 17 demonstrates the fairing force which plays a role in the structure of the curve. Fairing force, F_i, represents the tendency of a node in the curve to align itself in a straight line with the adjacent connected nodes. Consequently, the magnitude of the force has a reversed correlation with the size of the angle between the two segments. This force controls the smoothness of the curve. Equation 25 shows the formula for fairing force.

$$F_i = \frac{(P_{i+1} - P_{i-1}) - 2P_i}{\|(P_{i+1} - P_{i-1}) - 2P_i\|} \cdot (1 - \alpha/\pi) \tag{25}$$

Copyright © 2023 by ASME

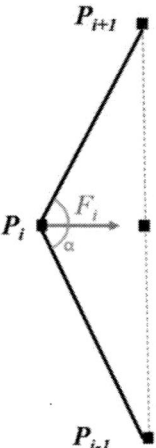

FIGURE 17: THE FAIRING FORCE PULLS TOWARDS THE MIDPOINT OF THE LINE CONNECTING ADJACENT VERTICES.

Here, α represents the internal angle, so the force increases as the angle becomes sharper. The first part of the equation simply establishes a unit vector from the point to the midpoint of the previous and subsequent points.

Boundary Attraction-Repulsion: The last force from Equation 16 is C_i which is applied from the fixed boundaries. To enclose the curve in a specific region, the boundary exerts force on the nodes when they are too close. The nature of this force is the same as A_i where the Lennard-Jones potential is used to regulate the force's magnitude.

After the position of all the nodes are updated, the algorithm then searches throughout the entire curve for edges with lengths larger than maximum allowed distance (d_{max}). If an edge's length exceeds d_{max}, then the edge is split in half. This splitting of edges is where the growth is modeled, and where complexity is built up by the system.

Figure 18 is the differential line growth algorithm pseudocode which shows the steps that the forces are applied and how new nodes are added. There is another function that checks the self-intersection of the curve referenced on line 17. Before starting the next iteration, the algorithm checks geometric validation of the curve, searching for self-intersections and boundary-intersections of the curve. This is a standard function in polygon processing libraries. If an intersection is detected, then Δt is reduced and the process starts again from last valid shape.

4.3 Test Cases

The aforementioned pseudocode is initially implemented in MATLAB and later in C#. Unlike most previous work, the use of a bounding polygon is necessary to create meaningful toolpaths. To create such paths, the team took slices through common tessellated models. Figure 19 shows a common 3D model used for testing. The red slice plane produces a polygonal outline, which is shown in detail in Figure 20. The differential line growth algorithm starting with a triangle placed inside progresses quickly through numerous iterations (Figure 21),

---

Inputs: $G_{initial}$ = Initial Polygon;
$G_{bounding}$ = Boundary- Curves
Outputs: G_{dif} =Differential-Growth Curve
Knobs: R_0= Repulsion-length; R_1= Length-Attraction; D_{max} = Maximum-distance; W_f =Fairing-Weight;
W_b = Brownian-Weight;
W_a =Repulsion-Attraction-Weight; D_{mf} =Damper

0.	Initialize: $G_{dif} = G_{initial}$
1.	Do until converged
2.	For P_i (each point in G_{dif})
3.	Calculate Brownian-force, B_i
4.	Calculate Fairing-Force, F_i
5.	Calculate Repulsion-Attraction force, A_i
6.	For each line segment in G_{dif}
7.	Calculate $X_{segline}$ (closest point in a line segment to P_i)
8.	if $X_{segline} < R_1$
9.	then $A_i = F_{LN}(X_{segline})$,
10.	else $A_i = 0$
11.	Calculate Boundary Collision Force, C_i
12.	Calculate damping force, $D_i = -D_{mf}V_{i,prev}$
13.	$a_i = W_f F_i + W_b B_i + W_a(A_i + C_i) + D_i$
14.	$V_i = V_{i,previous} + a_i \Delta t$
15.	$\Delta x_i = V_i \Delta t + 0.5(a_i \Delta t^2)$
16.	$G_{new} = G_{dif} + \Delta x$
17.	if G_{new} intersect itself or $G_{bounding}$ return to step (1)
18.	for each line segment calculate length, D_{seg}
19.	if $D_{seg} > D_{max}$ add new point
20.	$G_{dif} = G_{new}$

FIGURE 18: PSEUDOCODE FOR THE DIFFERENTIAL LINE GROWTH

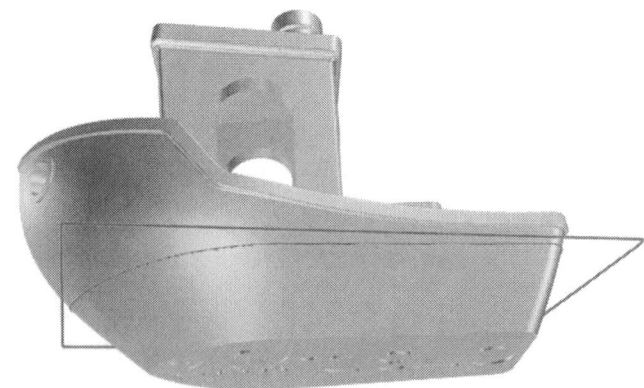

FIGURE 19: THE 3DBENCHY MODEL, A COMMON TEST CASE FOR ADDITIVE PRINTING IS SLICED TO PRODUCE AN EXAMPLE POLYGON FOR THE METHOD.

Copyright © 2023 by ASME

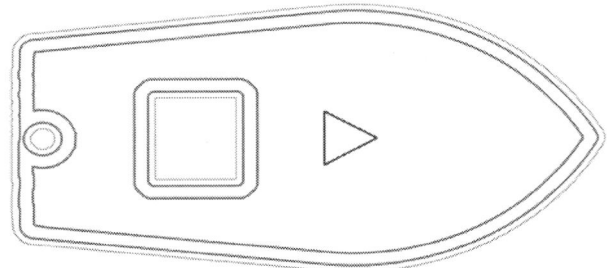

FIGURE 20: POLYGONAL SLICE, SHOWN IN GRAY, IS COMPRISED OF ONE POSITIVE CONTOUR AND TWO NEGATIVE CONTOURS (OR HOLES). TWO PERIMETERS, SHOWN IN RED, ARE DEFINED BY A POLYGONAL OFFSET FUNCTION. THE BLUE REPRESENTS THE STARTING SHAPE FOR THE PRESENTED METHOD.

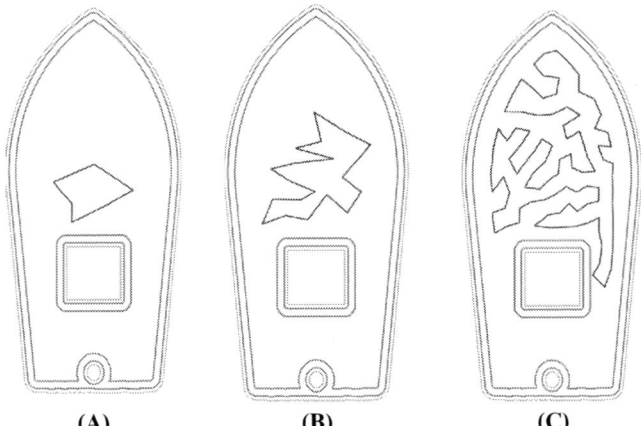

(A)   **(B)**   **(C)**

FIGURE 21: PROGRESSION OF INFILL POLYGON GROWTH: A) 6 EDGES, B) 22 EDGES), C) 98 EDGES

until it converges to the shape shown in Figure 22. This is accomplished in 9.366 seconds. Due to randomness in the algorithm, new shapes are produced with each run.

In the "Stanford Bunny" model (Figure 23a), similar results are generated. The sliced plane is processed as a single polygon with 282 edges. Once again, we first create two offset curves before initiating the differential line growth algorithm from an internal triangle. The process completes in 4.94 seconds and produces the blue polygon in Figure 23b.

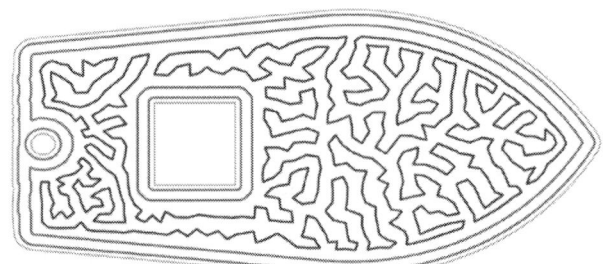

FIGURE 22: THE FINAL RESULT FOR AN INFILL PATH IS ONE CONTINUOUS POLYGON WITH 655 EDGES.

Following this, two additional post-processing steps are implemented. Because the infill is a single polygon, one can open the contour and connect it to the perimeter curves (Figure 23c) which produces a single path for created the entire layer. Figure 23d shows the zoomed-in section where the curves are connected. The second post-processing step is a smoothing operation that is done to improve the movement and surface smoothness of the layer. The final path contains 4,054 edges. In fact, this process could be connected to the previous and next sliced levels of the model to produce longer chains of uninterpreted printing, which in turn means that the nozzle is not turned on and off. This continuous printing is sometimes referred to as vase mode and results in smooth surfaces due to the lack of stringing that happens when cycling the nozzle. It also has the effect of reducing time by eliminating non-productive movements between additive operations. These reasons justify the use of differential line growth for infill operation in conventional planar 3D printing, but the real focus is on extending this to nonplanar surfaces.

An initial test using Blender and a python script shows how differential line growth can be extended to these nonplanar surfaces (Figure 24). The result takes significant time (~2 minutes) to complete but demonstrates the power of the method in creating printer paths that maintain a near uniform thickness. While some of the implementation details are completed in packaged Blender routines, the team will re-implement these within source-code. The triangle membership method presented in Section 2.3 will be used to create an attraction force that continuously pulls the 3D polygon to the given surface. Taken

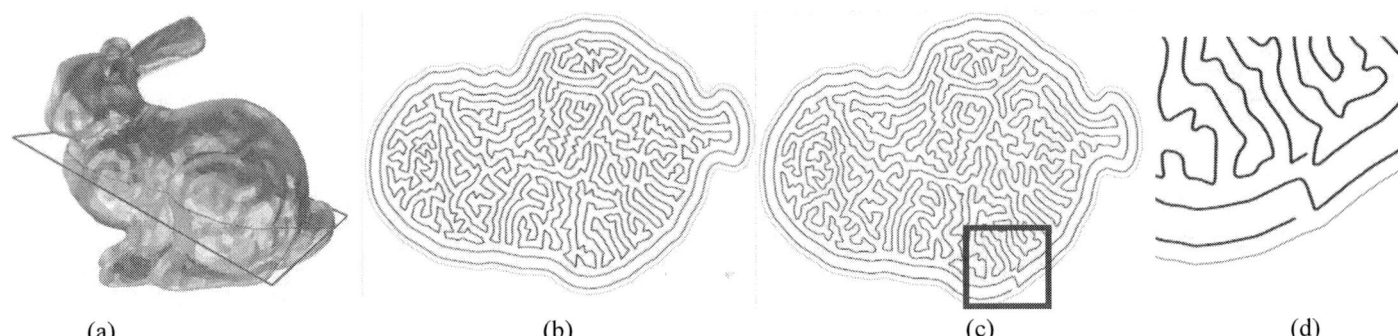

(a)   (b)   (c)   (d)

FIGURE 23: A) THE STANFORD BUNNY IS SLICED TO PRODUCE A POLYGONAL PATCH. B) THE RESULTING PERIMETER AND INFILL CURVES ARE C) JOINED AND SMOOTHED TO PRODUCE A SINGLE CONTINUOUS PATH. B) A ZOOMED-IN SECTION OF THE PATH FROM C.

Copyright © 2023 by ASME

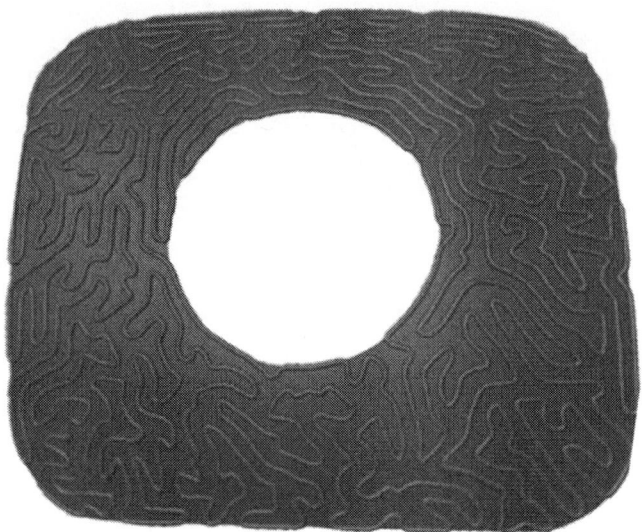

FIGURE 24: THE CURVED SURFACE PRESENTED EARLIER IS USED AS A SIMPLE TEST CASE FOR DIFFERENTIAL LINE GROWTH ON A NONPLANAR SURFACE.

with the perimeter defining method, shown in Section 3.2, continuous paths for arbitrary 3D surfaces will be possible.

5 DISCUSSION

This research is motivated by a desire to create additive toolpaths that are accurate for nonplanar slices without the need to transform to and from 2D planes. The common method for achieving nonplanar toolpaths is to project 2D paths onto a 3D surface. This transformation has previously worked well for many researchers since their slices have large near flat curvatures. However, it is likely that multi-axis additive printing will become more prevalent, and more robust toolpath methods will be required to achieve quality results for increasingly complex parts and surfaces.

Multiple methods have been proposed in this paper relating to the generation of toolpaths for nonplanar additive manufacturing. So, the discussion will focus on the combination of the algorithms discussed in the earlier sections as well as the advantages and disadvantages of each algorithm individually and in combination with one another.

Firstly, in Subsection 2.2, the wrapping method utilizes geometric attributes of developable surfaces to unwrap the nonplanar slices into planar representations. This method still requires the transformation to and from 2D, but the generated toolpaths have proper geodesic distances along the surface. This presented algorithm is a simple and easy to understand method that uses no approximations. However, the simplicity of this method comes at the cost of constraining the user to developable surfaces. While many structures can be closely approximated as a developable surface, this algorithm does not work in a general case. For the appropriate surfaces, the method provides undistorted toolpaths that perfectly conform to the printing surface. Additionally, since this method utilizes

traditional 2D tool pathing strategies, it is not combined with the remaining methods.

Subsection 3.2 discusses a variation of the projection method that can be applied to a general continuous surface. While current projection methods utilize a single direction for projection, this method varies the projection line based upon the location of the point being projected, such that the point will always be projected to the triangle closest to it along that triangle's normal. When applied to a planar representation of the tool path, the distortion effects are reduced. Additionally, integrating this method with a growth-like algorithm – where the projection occurs at every step of the algorithm – actively constrains a growing tool path to a general surface, resulting in no distortion.

Subsection 3.3 describes the first of the two growth algorithms discussed in this paper. This inflation method grows a polygonal offset along a surface by expanding an initial polygon until all the points on the polygon are near the original polygon. Combining this algorithm with the proposed projection method led to promising results which show that the algorithm works well for continuous surfaces as well as polygons that have hole features. This method could also be applied to 2D tool pathing, but the primary motivation for developing this method was for a 3D tool pathing application that could be iteratively constrained to a surface. The simplicity of this algorithm is an advantage over traditional offsetting algorithms that require self-intersection checks, which becomes increasingly complex in three dimensions. However, additional complexities do arise when considering the iterative nature of the algorithm such as deciding on stopping conditions and determining an ideal starting polygon. Some cases exist where the inflation method cannot squeeze through thin gaps between boundaries, and therefore cannot fully describe the offset. In these cases, an additional inflation needs to be run starting in the unfinished regions.

Next, we presented an approach to infill patterns based on differential line growth [24]. Like the inflation method, this is an iterative method that imagines forces acting on the toolpath to form it into an optimal shape where equilibrium in the forces is achieved at the optimum. The organic looking toolpath maintains a fairly equal distance between adjacent paths, but this will need to be validated in future studies. The end results appear to be One benefit of this approach is that a single path is defined; and, as is shown in Figure 23, the infill path can be connected to the perimeter so that an entire layer is printed without changing (i.e., stopping and starting) the deposition rate. This was the main motivation for the Fermat spirals paper [25] and is beneficial in avoiding stringing and reducing non-productive times. The chaotic paths created may also be beneficial in avoiding anisotropic weaknesses in the material. Typical infill patterns portray uniform directions with a given layer but often alter that uniformity in consecutive layers to avoid directional effects that would make the part weak under shearing or bending loads. Our serpentine paths are similar to gyroidal paths in which there is significant in-layer fluctuations. While this might lead to concerns with the printer's machine

Copyright © 2023 by ASME

dynamics, the added strengths may counteract those problems. Further study is necessary to understand these effects.

Currently, our differential line growth approach does not address how the next layers should be changed to best bond to the current one. One could either create new random growth paths or start from the previous one, and let the forces morph it into a new equilibrium. Our future work will examine how the paths of adjacent layers relate and examine the bridging limits that one path creates for subsequent layers. Similarly like the inflation algorithm, cases exist where the method cannot enter a thin region and would require an additional lines to be created to fill unfinished space.

Computationally, the inflation and differential line growth infill approaches take advantage of a unique approach to solving the toolpath problem for nonplanar additive. It is possible that more direct algorithms could be created for such problems, but such approaches would likely contain nonlinear equations that need to be continuously solved through their own iterations. Here the iteration is defined outright, and the interplay between forces and position changes gives the method flexibility and robustness in handling any arbitrary surfaces.

In the future, the speed, strength, time, and material usage of this new method will be compared to traditional infill methods like grid, cubic, hexagonal, or gyroid patterns.

6   CONCLUSION

The methods proposed here offer novel approaches to the geometrical limitations and new capabilities of multi-axis manufacturing; while the methods were intended for use in additive manufacturing, they could be utilized in other applications such as subtractive machining. Further development in optimizing and refining the individual algorithms is needed; however, the current versions of each method provide a promising proof-of-concept in many aspects. The wrapping method achieves accurate surface conformation. The point-to-face orthogonal projection method is an important aspect of achieving iterative surface constraints. And both the inflation and differential growth methods provide novel approaches to creating both 2D and 3D additive toolpaths.

Further work is required to fully integrate the proposed methods into a full tool path algorithm using the combination of the inflation perimeters, differential growth infill, and the point-to-face orthogonal projection methods. While all of the methods proposed work well in isolation of one another, further testing is needed once they are applied in a single tool path generation algorithm. Once an integrated algorithm is created, testing the generated tool path with a capable machine can be achieved.

ACKNOWLEDGEMENTS

We would like to thank Oregon Manufacturing Innovation Center for providing funding for this research.

REFERENCES

[1] Jetton, C., Rudd, L., and Campbell, M. I., 2022, "Systemic Generation of 5-Axis Manufacturing Machines," ASME International Design Engineering Technical Conferences & Computers and Information in Engineering Conference.

[2] Urhal, P., Weightman, A., Diver, C., and Bartolo, P., 2019, "Robot Assisted Additive Manufacturing: A Review," Robot Comput Integr Manuf, **59**, pp. 335–345.

[3] Bhatt, P. M., Kulkarni, A., Malhan, R. K., Shah, B. C., Yoon, Y. J., and Gupta, S. K., 2022, "Automated Planning for Robotic Multi-Resolution Additive Manufacturing," J Comput Inf Sci Eng, **22**(2).

[4] Bhatt, P. M., Malhan, R. K., Shembekar, A. V., Yoon, Y. J., and Gupta, S. K., 2020, "Expanding Capabilities of Additive Manufacturing through Use of Robotics Technologies: A Survey," Addit Manuf, **31**, p. 100933.

[5] ZHU, Z., TANG, X., CHEN, C., PENG, F., YAN, R., ZHOU, L., LI, Z., and WU, J., 2022, "High Precision and Efficiency Robotic Milling of Complex Parts: Challenges, Approaches and Trends," Chinese Journal of Aeronautics, **35**(2), pp. 22–46.

[6] "9.7.1 Differential Geometry of Developable Surfaces" [Online]. Available: https://web.mit.edu/hyperbook/Patrikalakis-Maekawa-Cho/node190.html. [Accessed: 22-Feb-2023].

[7] Alsharhan, A. T., Centea, T., and Gupta, S. K., 2017, "Enhancing Mechanical Properties of Thin-Walled Structures Using Non-Planar Extrusion Based Additive Manufacturing," ASME 2017 12th International Manufacturing Science and Engineering Conference, MSEC 2017 collocated with the JSME/ASME 2017 6th International Conference on Materials and Processing, **2**.

[8] Pérez-Castillo, J. L., Cuan-Urquizo, E., Roman-Flores, A., Olvera-Silva, O., Romero-Muñoz, V., Gómez-Espinosa, A., and Ahmad, R., 2021, "Curved Layered Fused Filament Fabrication: An Overview," Addit Manuf, **47**, p. 102354.

[9] Nayyeri, P., Kourosh Zareinia, ·, and Bougherara, H., "Planar and Nonplanar Slicing Algorithms for Fused Deposition Modeling Technology: A Critical Review," The International Journal of Advanced Manufacturing Technology, **1**, p. 3.

[10] Llewellyn-Jones, T., Allen, R., and Trask, R., 2016, "Curved Layer Fused Filament Fabrication Using Automated Toolpath Generation," 3D Print Addit Manuf, **3**(4), pp. 236–243.

[11] Shembekar, A. V., Yoon, Y. J., Kanyuck, A., and Gupta, S. K., 2018, "Trajectory Planning for Conformal 3D Printing Using Non-Planar Layers," International Design Engineering Technical Conferences and Computers and Information in Engineering Conference.

Copyright © 2023 by ASME

[12] Ahlers, D., Wasserfall, F., Hendrich, N., and Zhang, J., 2019, "3D Printing of Nonplanar Layers for Smooth Surface Generation," IEEE International Conference on Automation Science and Engineering, **2019-August**, pp. 1737–1743.

[13] Rodriguez-Padilla, C., Cuan-Urquizo, E., Roman-Flores, A., Gordillo, J. L., and Vázquez-Hurtado, C., 2021, "Algorithm for the Conformal 3D Printing on Non-Planar Tessellated Surfaces: Applicability in Patterns and Lattices," Applied Sciences 2021, Vol. 11, Page 7509, **11**(16), p. 7509.

[14] Feng, X., Cui, B., Liu, Y., Li, L., Shi, X., and Zhang, X., 2021, "Curved-Layered Material Extrusion Modeling for Thin-Walled Parts by a 5-Axis Machine," Rapid Prototyp J, **27**(7), pp. 1378–1387.

[15] Jin, Y., Du, J., He, Y., and Fu, G., 2017, "Modeling and Process Planning for Curved Layer Fused Deposition," The International Journal of Advanced Manufacturing Technology, **91**(1), pp. 273–285.

[16] Shan, Y., Shui, Y., Hua, J., and Mao, H., 2023, "Additive Manufacturing of Non-Planar Layers Using Isothermal Surface Slicing," J Manuf Process, **86**, pp. 326–335.

[17] Holla, V. D., Shastry, K. G., and Prakash, B. G., 2003, "Offset of Curves on Tessellated Surfaces," Computer-Aided Design, **35**(12), pp. 1099–1108.

[18] Chen, Z. M., Chen, Y. G., and Liang, F. J., 2006, "A New Offset Approach for Curves on Triangle Mesh Surfaces," International Technology and Innovation Conference, pp. 202–208.

[19] Xu, J., Sun, Y., and Zhang, L., 2015, "A Mapping-Based Approach to Eliminating Self-Intersection of Offset Paths on Mesh Surfaces for CNC Machining," Computer-Aided Design, **62**, pp. 131–142.

[20] Xin, S. Q., Ying, X., and He, Y., 2011, "Efficiently Computing Geodesic Offsets on Triangle Meshes by the Extended Xin–Wang Algorithm," Computer-Aided Design, **43**(11), pp. 1468–1476.

[21] Elber, G., and Kim, M. S., 2020, "Euclidean Offset and Bisector Approximations of Curves over Freeform Surfaces," Comput Aided Geom Des, **80**, p. 101850.

[22] Xu, J., Sun, Y., and Wang, S., 2013, "Tool Path Generation by Offsetting Curves on Polyhedral Surfaces Based on Mesh Flattening," International Journal of Advanced Manufacturing Technology, **64**(9–12), pp. 1201–1212.

[23] Hu, Z., Hua, L., Qin, X., Ni, M., Liu, Z., and Liang, C., 2022, "Region-Based Path Planning Method with All Horizontal Welding Position for Robotic Curved Layer Wire and Arc Additive Manufacturing," Robot Comput Integr Manuf, **74**, p. 102286.

[24] Pedersen, H., and Singh, K., 2006, "Organic Labyrinths and Mazes," NPAR Symposium on Non-Photorealistic Animation and Rendering, **2006**, pp. 79–86.

[25] Zhao, H., Gu, F., Huang, Q. X., Garcia, J., Chen, Y., Tu, C., Benes, B., Zhang, H., Cohen-Or, D., and Chen, B., 2016, "Connected Fermat Spirals for Layered Fabrication," ACM Transactions on Graphics (TOG), **35**(4).

[26] Alba, J. C. G., Nunez, D. A., Mauledoux, M., and Aviles, O. F., 2022, "Deposition Toolpath Pattern Comparison: Contour-Parallel and Hilbert Curve Application," International Journal of Mechanical Engineering and Robotics Research, **11**(7), pp. 542–548.

[27] Karplan, C. S., and Bosch, R., 2005, "TSP Art," *Renaissance Banff: Mathematics, Music, Art, Culture*, Banff, Alberta, CA, pp. 301–308.

[28] "Exploring 2D Differential Growth with JavaScript | by Jason Webb | Medium" [Online]. Available: https://medium.com/@jason.webb/2d-differential-growth-in-js-1843fd51b0ce. [Accessed: 09-Mar-2023].

[29] "On Generative Algorithms: Differential Line Inconvergent" [Online]. Available: https://inconvergent.net/generative/differential-line/. [Accessed: 11-Mar-2023].

[30] Yu, C., Schumacher, H., and Crane, K., 2021, "Repulsive Curves," ACM Trans Graph, **40**(2).

[31] "Real-Time Differential Growth in JavaScript" [Online]. Available: http://adrianton3.github.io/blog/art/differential-growth/differential-growth.html. [Accessed: 11-Mar-2023].

[32] "Wanderers on Behance" [Online]. Available: https://www.behance.net/gallery/21605971/Neri-Oxman-Wanderers. [Accessed: 09-Mar-2023].

[33] Lennard-Jones, J. E., 1931, "Cohesion," Proceedings of the Physical Society, **43**(5), p. 461.

Proceedings of the ASME 2023
International Design Engineering Technical Conferences and
Computers and Information in Engineering Conference
IDETC-CIE2023
August 20-23, 2023, Boston, Massachusetts

DETC2023-111214

MULTI-MATERIAL TOPOLOGY OPTIMIZATION CONSIDERING THE BOUNDING BOX DIMENSION CONSTRAINT AND ASSEMBLABILITY BASED ON THE EXTENDED LEVEL SET METHOD IN TWO DIMENSIONS

Yukun Feng[1], Yuki Noguchi[1,2], Takayuki Yamada[1,2,*]

[1]Department of Mechanical Engineering, Graduate School of Engineering, The University of Tokyo, Tokyo 113-8656, Japan.
[2]Department of Strategic Studies, Institute of Engineering Innovation, Graduate School of Engineering, The University of Tokyo, Tokyo 113-8656, Japan.

ABSTRACT

This paper proposes a multi-material topology optimization method considering the bounding box dimension constraint and assemblability of the optimized structures in two dimensions. To handle multi-material topology optimization problems, we first introduce the concept of the extended level set method and the topological derivative. Second, we introduce the dimensional constraint of the two-dimensional bounding box and the assembly constraint for assemblability of the multi-material structure. We also elaborate their mathematical models. We then compute the design sensitivities of the two constraints based on the topological derivative and adjoint method. Third, we formulate the problem of multi-material topology optimization with constraints and explain the algorithm flow based on the finite element method. Finally, we verify our proposed algorithm on two numerical examples in two dimensions. The proposed method with two-dimensional geometry constraints on the optimized structures is applicable to machining methods. A representative application is milling, in which the largest and second-largest dimensions of the workpieces are limited by the working area of the milling machine. In addition, our method is helpful for the assembly structure composed of multiple components. Traditional topology optimization method does not consider the assemblability, which makes the structure difficult to be assembled and decomposed in practical engineering. Our work will assist the manufacturability and assemblability of topology-optimized multi-material structures.

Keywords: Topology optimization, Multi-material structure, Geometry constraint, Design for manufacturing

1. INTRODUCTION

Topology optimization is a commonly used structural optimization method with a higher degree of design freedom than other structural optimization methods. Therefore, it can greatly meet engineering needs. Since its conceptualization by Bendsøe and Kikuchi [1], the topology optimization problem has been applied to single-material cases. However, industrial development has inspired interest in multi-material structures. Owing to their excellent and unique properties, such structures have been applied in various fields such as bio-inspired structures, electromagnetic fields, and optical fields [2, 3]. Therefore, topology optimization of multi-material structures has been extensively researched [4, 5]. Most of the multi-material topology optimization methods are extensions of the topology optimization methods for single-material cases, such as the solid isotropic material with penalty method [6], bidirectional evolutionary structural optimization [7], and the level set method [8]. The extended level set method, recently proposed by Noda et al. [9], is the most general multi-material model among the level set methods for multi-material topology optimization problems.

In practical engineering applications, multi-material structures are often assembled from multiple components. To ensure their manufacturability and assemblability, manufacturing and multi-material structures must be constrained during the topology optimization process. First, the manufacturing process is often restricted by the dimensions of the target objects to be processed. For example, milling is limited by the size of the workpieces. In particular, the largest and second-largest dimensions of the workpieces should be within the working area of the milling machine. Otherwise, the workpieces cannot be manufactured simultaneously and the integrity of the structure is lost [10].

To avoid these problems, a structure assembled from multiple components requires dimensional constraints on each material component. These constraints must consider the maximum dimension of the component. Here we propose a dimensional constraint in two dimensions for manufacturing methods such as

---

*Corresponding author
Documentation for `asmeconf.cls`: Version 1.34, May 9, 2023.

Copyright © 2023 by ASME

milling. By limiting the length and width of the two-dimensional bounding box, we can limit each component to the maximum working area of the machining tool. To ensure easy decomposition and reassembly of the optimized assembly structure, we should also constrain the assemblability [11]. In this work, the dimensional constraints and assembly constraints are applied simultaneously during the optimization process. While ensuring that the structure can be mechanically processed successfully, it also ensures that it can be easily and successfully assembled or decomposed as assemblies. By considering both the manufacturing process and the assembly process, our proposed method will help the real application of the topology-optimized structures.

Based on the concept of the extended level set method, this paper proposes a multi-material topology optimization algorithm combining the dimensional and assembly constraints. The remainder of the paper is organized as follows. First, the extended level set method and the concept of the topological derivative are introduced. Second, the dimensional and assembly constraints are mathematically modeled. Third, a topology optimization problem considering the dimension and assembly constraints is formulated. The corresponding algorithm flow is then constructed. Finally, the algorithm is verified on two numerical examples in two dimensions.

2. EXTENDED LEVEL SET METHOD

This section introduces the extended level set method for multi-material problems. Topology optimization is a design method that optimizes the distribution of materials in the design domain D to obtain the optimal structure. In the single-material level set method proposed by Yamada et al. [12], the solid material phase i in the design domain D is represented by the following characteristic function χ:

$$\chi(x) = \begin{cases} 1 & \text{if} \quad \phi(x) \geq 0 \\ 0 & \text{if} \quad \phi(x) < 0, \end{cases} \quad (1)$$

where x is a point in D and $\phi(x)$ is the level set function, defined as

$$\begin{cases} 0 < \phi(x) \leq 1 & \text{if} \quad x \in \Omega \\ \phi(x) = 0 & \text{if} \quad x \in \partial\Omega \\ -1 \leq \phi(x) < 0 & \text{if} \quad x \in D \setminus \Omega, \end{cases} \quad (2)$$

where Ω is the solid material domain.

In the extended level set method, a problem with M material phases requires $M \times M$ level set functions. For example, the level set functions $\phi_{li}(l = 0, 1, i - 1, i + 1, ..., M - 1)$ remain positive in the domain of material-phase domain i and negative in other phases. Therefore, the characteristic function $\chi^{(i)}$ of material phase i in the extended level set method can be expressed as

$$\chi^{(i)}(x) = \prod_{l \neq i} H(\phi_{li}), \quad (3)$$

where $H(g)$ is the Heaviside function defined as follows:

$$H(g) = \begin{cases} 1 & \text{if} \quad g \geq 0 \\ 0 & \text{if} \quad g < 0, \end{cases} \quad (4)$$

In Noda et al. [9], the level set functions ϕ_{ij} of the material phases i and j are updated using the following reaction-diffusion equation:

$$\frac{\partial \phi_{ij}}{\partial t} = \frac{-D_{ij}J - C^{all}\Sigma_k \lambda_k D_{ij}G_k}{C_{ij}} + \tau_{ij}L^2\nabla^2\phi_{ij}, \quad (5)$$

where t is a fictitious time step, and J is the objective function. $D_{ij}J$ is the topological derivative of the objective function, which is detailed below. C^{all} and C_{ij} are normalizing parameters. λ_k is the Lagrangian multiplier for constraint functions and $D_{ij}G_k$ is the topological derivative of constraint functions. τ_{ij} is used as the regularization parameter. L is the characteristic length of the whole design domain. The term $\nabla^2\phi_{ij}$ smooths the level set functions.

In the extended level set method, the objective function is decreased by placing new material phases. This process is realized through the topological derivative $D_{i \to j}^{\top}J$, which measures the change in the objective function when a small inclusion Ω_ϵ with material j is created in the region of a material phase i. It is computed as

$$D_{i \to j}^{\top}J := \lim_{\epsilon \to 0} \frac{(J_{i \to j} + \delta J_{i \to j}) - J}{meas(\Omega_i \setminus \Omega_\epsilon) - meas(\Omega_i)}, \quad (6)$$

where $\delta J_{i \to j}$ is the change in the objective function after creating the small inclusion. The topological derivative $D_{ij}J$ in a system with materials i and j is expressed as follows:

$$D_{ij}J = D_{i \to j}^{\top}J - D_{j \to i}^{\top}J. \quad (7)$$

3. MODELING OF DIMENSIONAL AND ASSEMBLY CONSTRAINTS

The dimensional constraint limits the dimension of each material component to the maximum working area of the manufacturing tools. Meanwhile, the assembly constraint permits the assemblability of the optimized structure. This section introduces the mathematical models and computes the topological derivatives of the two constraints.

3.1 Dimensional constraints

Based on a previous work [13], the dimensional constraint in this paper is the bounding box, which is often used for object location and detection in computer graphics[14, 15]. Here we propose bounding box constraints for each material component, as shown in Fig. 1. Materials 1,2, and 3 are surrounded by rectangular bounding boxes, which are tightly closed to their boundaries. Then, we can easily restrict the dimension of each material component by limiting the size of its bounding box.

First, we determine the orientations of the bounding boxes by extracting the principal directions of each material component. We define the weighted covariance matrix $A^{(i)}$ for material component i as

$$A^{(i)} = \frac{\int_D \chi^{(i)} x x^{\top} \, d\Omega}{A_w^{(i)}}, \quad (8)$$

where $A_w^{(i)} = \int_D \chi^{(i)} \, d\Omega$. In a two-dimensional case, we compute the two principal directions of the bounding box $u_1^{(i)}$ and $u_2^{(i)}$ by

Copyright © 2023 by ASME

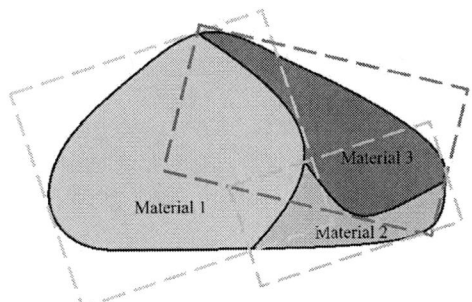

FIGURE 1: THE BOUNDING BOXES OF MATERIAL COMPONENTS IN A MULTI-MATERIAL STRUCTURE.

decomposing $A^{(i)}$. $u_1^{(i)}$ and $u_2^{(i)}$ represent the length and width directions of the bounding box, respectively. We then obtain the centroid $c_p^{(i)}$ along each principal direction p:

$$c_p^{(i)} = \frac{\int_D \chi^{(i)} z_p^{(i)} \, d\Omega}{A_w^{(i)}} \tag{9}$$

where $z_p^{(i)}$ is the projection of the design point along the direction p, $z_p^{(i)} = u_p^{(i)\top} x$.

Approximating the bounding box dimensional constraint using the P-norm function [16], we obtain

$$G_l^{(i)} = C_f \left\{ \int_D \chi^{(i)} [(z_1^{(i)} - c_1^{(i)})^2]^{P_n} \, d\Omega \right\}^{\frac{1}{P_n}} - \left(\frac{L_{max}^{(i)}}{2} \right)^2 \le 0,$$

$$G_w^{(i)} = C_f \left\{ \int_D \chi^{(i)} [(z_2^{(i)} - c_2^{(i)})^2]^{P_n} \, d\Omega \right\}^{\frac{1}{P_n}} - \left(\frac{W_{max}^{(i)}}{2} \right)^2 \le 0, \tag{10}$$

where C_f is an adaptive correction parameter and P_n is the smoothing factor in the P-norm function. It should be noted that the increase of P_n will lead to the increase of topological derivative, which will cause the instability of the optimization problem. Therefore, in this paper, the value of P_n is appropriately set to 1. $G_l^{(i)}$ and $G_w^{(i)}$ are the constraint functions along the length and width directions respectively. $L_{max}^{(i)}$ and $W_{max}^{(i)}$ are the maximum allowable dimensions of the manufacturing tool along the length and width principal directions, respectively.

Consider a small inclusion of material j into a region of material i. The topological derivative of the bounding box constraint is then derived as follows:

$$D_{i \to j}^T G_p^{(i)}$$

$$= K_1 [(z_p^{(i)} - c_p^{(i)})^{2P_n} + \int_D \chi^{(i)} 2P_n (z_p^{(i)} - c_p^{(i)})^{2P_n - 1} \left(\frac{-u_p^{(i)\top} x}{A_w^{(i)}} \right.$$

$$+ \mu \cdot (\delta A) u_p^{(i)}) \, d\Omega], \tag{11}$$

where $K_1 = \frac{C_f}{P_n} [\int_D \chi^{(i)} (z_p^{(i)} - c_p^{(i)})^{2P_n} \, d\Omega]^{\frac{1}{P_n} - 1}$. Here, $p = 1$ and $p = 2$ represent the length and width directions of the bound-

ing box, respectively. Satisfies the following adjoint equation:

$$A_w^{(i)} (AA^\top)^{(i)\top} \mu - A_w^{(i)} w_p^{(i)} \mu = -A_w^{(i)} x - \int_D \chi^{(i)} x \, d\Omega - 2A_w^{(i)} \sigma u_p^{(i)}, \tag{12}$$

where $w_p^{(i)}$ is the eigenvalue of the covariance matrix, and $\sigma = -\frac{u_p^{(i)\top} x}{2} - \frac{u_p^{(i)\top} \int_D \chi^{(i)} x \, d\Omega}{2A_w^{(i)}}$. The topological derivative can be computed after obtaining the adjoint variable μ.

3.2 Assembly constraints

To ensure its easy decomposition and assembly, a multi-material structure must contain no undercuts or interior voids. In this paper, both shapes are prevented by imposing assembly constraints. Note that assembly can be regarded as the reverse process of decomposition. Based on a previous work [17], the present example and constraints take decomposition as an object. The details will be demonstrated in a simple example. Fig. 2 shows an undercut shape of material j stuck in the groove of material i, which prevents easy disassembly. Meanwhile, an interior void of material j is totally inserted in the region of material i and cannot be removed along the prescribed disassembly direction. Under correct assembly constraints, both shapes can be eliminated from the optimized structure.

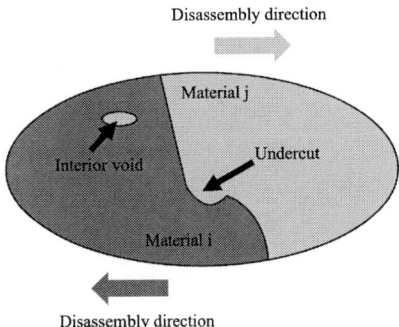

FIGURE 2: SIMPLE EXAMPLE OF UNDERCUT AND INTERIOR VOID.

The proposed assembly constraints are based on a fictitious physical model. Assume that a fictitious heat flow is generated in the region of material i (Fig. 3). The heat flows from the right to the left boundary of the design domain along the prescribed disassembly direction $d^{(i)}$. We then obtain a heat field $\psi^{(i)}(x)$, where $\psi^{(i)} = 1$ along the downstream of the material i and $\psi^{(i)} = 0$ elsewhere in the design domain. The area in which $\psi^{(i)} = 1$ is highlighted in panel (a) of Fig. 3. To obtain the undercut shape and interior void, we extract the region belonging to material j from the heat field in which $\psi^{(i)} = 1$. The extracted undercut and interior void are represented as colored regions in panel (b) of Fig. 3. Through the above process, the undercut and interior void shapes of material j in the region of material i can be extracted under two conditions: (1) $\chi^{(j)} = 1$, (2) $\psi^{(i)} = 1$. Therefore, when decomposing the material component i, we can eliminate the undercut and interior void shapes by minimizing

Copyright © 2023 by ASME

295

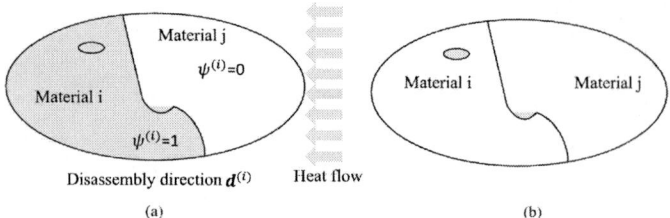

FIGURE 3: (A): THE FICTITIOUS HEAT FIELD. (B): THE EXTRACTED UNDERCUT AND AN INTERIOR VOID.

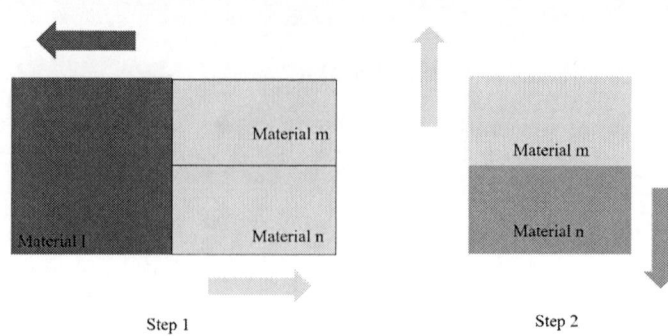

Step 1   Step 2

FIGURE 4: DECOMPOSITION STEP OF MATERIALS l, m, AND n.

the following objective function:

$$F^{(j)} = \int_D \chi^{(j)} \psi^{(i)} \, d\Omega. \qquad (13)$$

Note that the superscript of the objective function j in Eq. (13) represents a constraint on material component j. Considering the symmetry of materials i and j, we define the objective function of the assembly constraint as follows:

$$G_a^{(ij)} = F^{(j)} \cdot F^{(i)} = \int_D \chi^{(j)} \psi^{(i)} \, d\Omega \cdot \int_D \chi^{(i)} \psi^{(j)} \, d\Omega. \qquad (14)$$

The heat field $\psi^{(i)}(x)$ is obtained by solving the following steady advection-diffusion equation:

$$\begin{cases} -L^2 div(E^{(i)} \cdot \nabla\psi^{(i)}) + LV d^{(i)} \cdot \nabla\psi^{(i)} = \beta\chi^{(i)}(1-\psi^{(i)}) & \text{in } D \\ \psi^{(i)} = 0 & \text{on } \Gamma_{in} \\ n^{(i)} \cdot \nabla\psi^{(i)} = 0 & \text{on } \partial D \backslash \Gamma_{(in)}, \end{cases} \qquad (15)$$

where L is the characteristic length. $E^{(i)}$ is the diffusion coefficient tensor, which determines the direction of the heat flow. V is the velocity of the heat flow, and β determines the degree of the heat flow. Here, the diffusion coefficient tensor $E^{(i)}$ is computed as

$$E^{(i)} = d^{(i)} \otimes d^{(i)} + \epsilon \sum_{n=1}^{N-1} e_n^{(i)} \otimes e_n^{(i)}, \qquad (16)$$

where $\otimes$ is the Kronecker product, $e_n^{(i)}$ is the unit vector orthogonal to $d^{(i)}$, and ϵ is a small value (set to 0.01 in this paper [17]).

In a structure composed of more than two materials, the order of the disassembly should be considered. For example, consider a structure of three materials l, m, and n. We first consider materials m and n as a whole component located to the right of material l. We then decompose materials m and n. The whole decomposition process is shown in Fig. 4. Using Eq. (14), the objective function of the overall process can be expressed as follows:

$$\begin{aligned} G_a &= F^{(l)} \cdot F^{(mn)} + F^{(m)} \cdot F^{(n)} \\ &= \int_D \chi^{(l)} \psi^{(mn)} \, d\Omega \cdot \int_D \chi^{(mn)} \psi^{(l)} \, d\Omega \\ &\quad + \int_D \chi^{(m)} \psi^{(n)} \, d\Omega \cdot \int_D \chi^{(n)} \psi^{(m)} \, d\Omega. \end{aligned} \qquad (17)$$

In the following section, we adopt G_a as the objective function of the assembly constraint, which depends on the order and direction of decomposition. Now consider a small inclusion of material j in the region of material i. The topological derivative of the assembly constraint in Eq. (14) becomes

$$\begin{aligned} D_{i \to j}^T G_a^{(i)} &= \{\beta(1-\psi^{(i)})\widetilde{\psi}^{(i)} + \psi^{(i)}\} \int_D \chi^{(i)} \psi^{(j)} \, d\Omega \\ &\quad + \{\beta(1-\psi^{(j)})\widetilde{\psi}^{(j)} + \psi^{(j)}\} \int_D \chi^{(j)} \psi^{(i)} \, d\Omega, \end{aligned} \qquad (18)$$

where $\widetilde{\psi}^{(j)}$ is obtained by solving the following adjoint equation [17]:

$$-L^2 div(E^{(i)} \cdot \nabla\widetilde{\psi}^{(i)}) + LV d^{(i)} \cdot \nabla\widetilde{\psi}^{(i)} = \beta\chi^{(i)}(1-\widetilde{\psi}^{(i)}). \qquad (19)$$

4. FORMULATION OF TOPOLOGY OPTIMIZATION WITH DIMENSIONAL AND ASSEMBLY CONSTRAINTS

The topology optimization problem is implemented as a minimum-means compliance problem. Assume the u represents the displacement field and Γ^u is a fixed boundary. A traction t is applied on boundary Γ^t. The problem can be formulated as

$$\inf_{\chi^{(i)}} J = \int_{\Gamma^t} t \cdot u \, d\Gamma + \gamma G_a,$$

subject to

$$-div(C(\chi^{(i)}) : \varepsilon(u)) = 0 \qquad \text{in } D,$$
$$u = 0 \qquad \text{on } \Gamma^u,$$
$$(C(\chi^{(i)}) : \varepsilon(u)) \cdot n = t \qquad \text{on } \Gamma^t$$
$$G_v^{(i)} = \frac{\int_D \chi^{(i)} \, d\Omega}{\int_D \, d\Omega} - V_{max}^{(i)} \leq 0,$$
$$G_l^{(i)}(u_1^{(i)}) = C_f \left\{ \int_D \chi^{(i)} [(z_1^{(i)} - c_1^{(i)})^2]^{P_n} \, d\Omega \right\}^{\frac{1}{P_n}} - (\frac{L_{max}^{(i)}}{2})^2 \leq 0.$$
$$G_w^{(i)}(u_2^{(i)}) = C_f \left\{ \int_D \chi^{(i)} [(z_2^{(i)} - c_2^{(i)})^2]^{P_n} \, d\Omega \right\}^{\frac{1}{P_n}} - (\frac{W_{max}^{(i)}}{2})^2 \leq 0$$

$u_p^{(i)}$ subject to the eigenvalue problem of matrix $A^{(i)}$, with $i = 1, 2, ..., M$.

$$(20)$$

Copyright © 2023 by ASME

Here, γ is the weight coefficient of the assembly constraint, ε is the infinitesimal strain tensor, $G_v^{(i)}$ is the volume constraint and $V_{max}^{(i)}$ is the maximum allowable volume.

5. OPTIMIZATION ALGORITHM

The optimization algorithm of the proposed algorithm is summarized as follows.

Step 1. Initialize the level set functions ϕ_{ij}.

Step 2. Solve the governing equations by the finite element method.

Step 3. Check the convergence conditions of the objective function and constraints. Terminate the algorithm if all convergence conditions are met.

Step 4. If the algorithm has not converged, calculate the design sensitivity of the problem.

Step 5. Update the level set functions by solving the reaction—diffusion equation in Eq. (5) using the finite element method. Return to Step 2.

6. NUMERICAL EXAMPLES

In this section, our topology optimization algorithm is validated on two numerical examples. In case 1, the test object is a $2[m] \times 1[m]$ cantilever beam. The left boundary of the structure is fixed and a force t is applied on the centroid of the right boundary (See Fig. 5 for details).

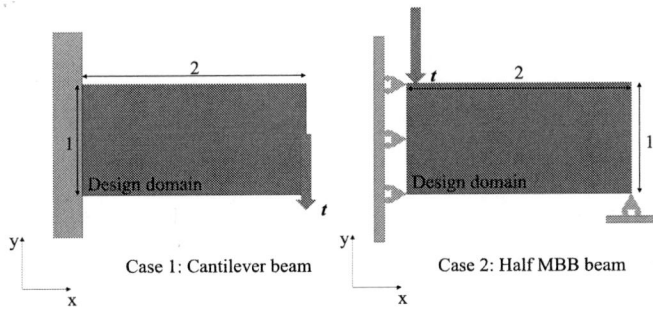

FIGURE 5: CASE 1: TWO-DIMENSIONAL CANTILEVER BEAM SETTING. CASE 2: TWO-DIMENSIONAL HALF MBB BEAM SETTING.

Case 1 includes one void material and two solid materials. Accordingly, we set the material number $M = 3$. The Young's modulus, Poisson's ratio, and maximum volume ratio of (Void material, material 1, and material 2) are set as $(0.1, 200, 150)[GPa]$, $(0.3, 0.3, 0.3)$, and $(100, 30, 30)\%$, respectively. The regularization parameter τ_{ij} and the characteristic length L are set to 1.0×10^{-3} and 1, respectively. The maximum dimensions of the dimensional constraint on each material component, $L_{max}^{(i)}$ and $W_{max}^{(i)}$ are set as $1.3[m]$ and $1[m]$, respectively. For the assembly constraint, the prescribed disassembly directions $\boldsymbol{d}^{(i)}$ of materials 1 and 2 are set to left and right, respectively. The detailed settings are as follows:

$$\boldsymbol{d}^{(1)} = \begin{bmatrix} -1 \\ 0 \end{bmatrix}, \qquad \boldsymbol{d}^{(2)} = \begin{bmatrix} 1 \\ 0 \end{bmatrix} \tag{21}$$

In addition, the flow velocity V is set as $100[m/s]$ and the weighting coefficient γ is set to 1.5. ϵ in the diffusion coefficient tensor $E^{(i)}$ is set to 0.01. The detailed settings of both constraints are shown in Fig. 6. Three different optimized results are shown in Fig. 7. Panel (a) displays the optimized structure under the volume constraint alone. As the dimensional and assembly constraints are excluded, the length dimensions of materials 1 and 2 exceed their prescribed maximum dimensions. In addition, materials 1 and 2 cannot be decomposed along our prescribed direction. The optimized result under the volume and assembly constraints is shown in Fig. 7 (b). In this case, the optimized structure can be decomposed in the left and right directions but the dimensions of each material component fail to meet our requirements. Fig. 7 (c) shows the final optimized design under the volume, dimensional, and assembly constraints, where its decomposition structure is shown in Fig. 7 (d). The dimensions of each material are limited to the bounding boxes, which represent the prescribed maximum dimensions. Moreover, materials 1 and 2 can be decomposed along the prescribed direction.

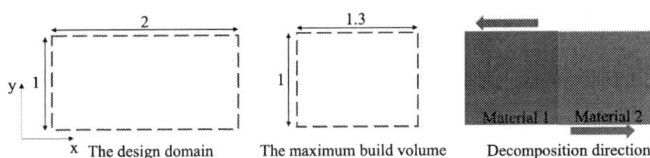

FIGURE 6: DETAILED SETTING OF THE BOUNDING BOX AND THE DISASSEMBLY DIRECTION IN CASE 1.

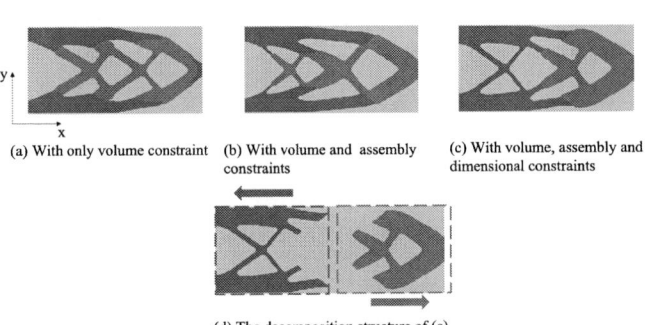

FIGURE 7: OPTIMIZED STRUCTURE UNDER (A) VOLUME CONSTRAINT ONLY, (B) VOLUME AND ASSEMBLY CONSTRAINTS, AND (C) VOLUME, ASSEMBLY, AND DIMENSIONAL CONSTRAINTS. (D) DECOMPOSITION STRUCTURE OF CASE 1. (RED: MATERIAL 1; BLUE: MATERIAL 2;

Next, we verify our algorithm on a $2[m] \times 1[m]$ half MBB beam. The detailed setting of case 2 is shown in Fig. 5. To represent one void material and three solid materials, the material number M is set to 4. The Young's modulus, Poisson's ratio, and maximum volume ratio of (Void material, material 1, material 2, and material 3) are set as $(0.1, 200, 150, 120)[GPa]$, $(0.3, 0.3, 0.3, 0.3)$, and $(100, 13.3, 13.3, 13.3)\%$, respectively. The regularization parameter τ_{ij} and the characteristic length L are set to 1.0×10^{-3} and 1, respectively. The maximum dimensions of the dimensional constraint on each material component, $L_{max}^{(i)}$

Copyright © 2023 by ASME

297

and $W_{max}^{(i)}$ are set as 1.4[m] and 0.75[m], respectively. The prescribed decomposition procedure for the assembly constraint is separated into two steps. First, Materials 2 and 3 are regarded as a whole component and are decomposed from Material 1 along the right—left direction. Next, Materials 2 and 3 are decomposed along the up—down direction. The $d^{(i)}$ are set as follows:

$$d^{(1)} = \begin{bmatrix} -1 \\ 0 \end{bmatrix}, \ d^{(2)} = \begin{bmatrix} 0 \\ 1 \end{bmatrix}, \ d^{(3)} = \begin{bmatrix} 0 \\ -1 \end{bmatrix}, \ d^{(23)} = \begin{bmatrix} 1 \\ 0 \end{bmatrix}$$

(22)

where $d^{(23)}$ is the direction of the heat flow when Materials 2 and 3 are regarded as a whole component. The flow velocity V is set as 100[m/s] and γ is set to 1.5. The detailed settings of the two constraints are shown in Fig. 8. Three different optimized results are shown in Fig. 9. Panel (a) displays the optimized structure under the volume constraint alone. As the dimensional and assembly constraints are excluded, the dimensions of materials 1, 2, and 3 all exceed their prescribed maximum dimensions. In addition, the whole structure cannot be decomposed along our prescribed direction. After applying the volume and assembly constraints, the optimized structure can be decomposed following our prescribed directions and steps, but the dimensions of each material component fail to meet our requirements (Fig. 9 (b)). Fig. 9 (c) shows the final optimized design after applying the volume, dimensional, and assembly constraints (the decomposition structure is shown in Fig. 9 (d)). The prescribed steps decompose Materials 1, 2, and 3 along the prescribed directions. Moreover, the dimensions of each material are limited to the bounding boxes, which represent the prescribed maximum dimensions.

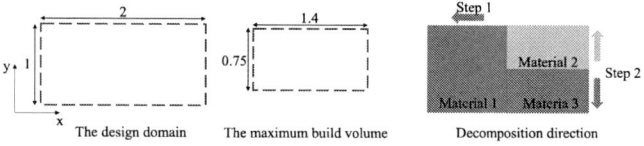

FIGURE 8: DETAILED SETTING OF THE BOUNDING BOX AND THE DISASSEMBLY DIRECTION FOR CASE 2.

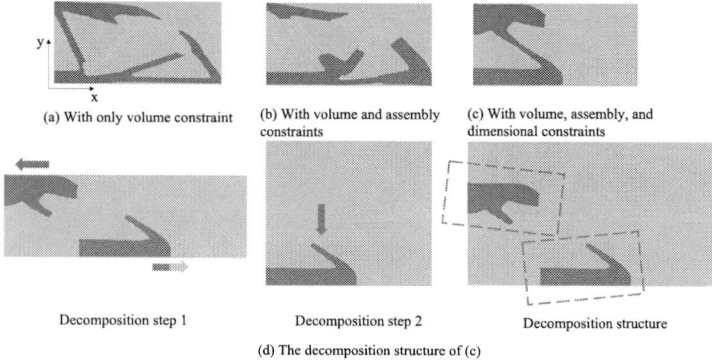

FIGURE 9: OPTIMIZED STRUCTURE UNDER (A) VOLUME CONSTRAINT ONLY, (B) VOLUME AND ASSEMBLY CONSTRAINTS, AND (C) VOLUME, ASSEMBLY, AND DIMENSIONAL CONSTRAINTS. (D) DECOMPOSITION STRUCTURE OF CASE 2. (ORANGE: MATERIAL 1; YELLOW: MATERIAL 2; BLUE: MATERIAL 3)

In the above two cases, the dimensional and assembly constraints were applied during the optimization process. The dimensions of each material component in both optimized results satisfied the prescribed maximum dimensions and the materials were decomposed or assembled in our prescribed direction. Therefore, our proposed algorithm is well verified in the two cases.

7. CONCLUSION

We proposed a multi-material topology optimization method with dimensional and assembly constraints for manufacturing in two dimensions. The contents of this paper are summarized below.

(1) The concept of the extended level set method and topological derivative are briefly introduced.

(2) Mathematical models of the dimensional and assembly constraints are elaborated and their design sensitivities are derived based on the topological derivative and adjoint method.

(3) A multi-material topology optimization problem combining dimensional and assembly constraints is formulated. The algorithm development is based on the finite element method. The algorithm flow is also explained.

(4) Two numerical examples based on the minimum compliance problem are presented in two dimensions. The dimensional and assembly constraints are perfectly satisfied in both cases.

By applying dimensional and assembly constraints, our work will assist the manufacturability and assemblability of topologically optimized multi-material structures for manufacturing methods such as milling. The limitation of our work is the computation efficiency, the combination of the two constraints will increase the computation cost. In our future work, the acceleration method of the constraints will be considered. Besides, three-dimensional examples have not been conducted in this work. In the future, three-dimensional and real cases will be applied to further illustrate the validity of our proposed method.

REFERENCES

[1] Bendsøe, Martin Philip and Kikuchi, Noboru. "Generating optimal topologies in structural design using a homogenization method." *Computer methods in applied mechanics and engineering* Vol. 71 No. 2 (1988): pp. 197–224.

[2] Wang, Rui, Gu, Dongdong, Lin, Kaijie, Chen, Caiyan, Ge, Qing and Li, Deli. "Multi-material additive manufacturing of a bio-inspired layered ceramic/metal structure: Formation mechanisms and mechanical properties." *International Journal of Machine Tools and Manufacture* Vol. 175 (2022): p. 103872.

[3] Han, Daehoon and Lee, Howon. "Recent advances in multi-material additive manufacturing: methods and applications." *Current Opinion in Chemical Engineering* Vol. 28 (2020): pp. 158–166.

[4] Li, Yu, Lai, Yaping, Lu, Gan, Yan, Fucheng, Wei, Peng and Xie, Yi Min. "Innovative design of long-span steel—concrete composite bridge using multi-material topology

Copyright © 2023 by ASME

optimization." *Engineering Structures* Vol. 269 (2022): p. 114838.

[5] Liu, Chih-Hsing, Chen, Yang and Yang, Sy-Yeu. "Topology optimization and prototype of a multimaterial-like compliant finger by varying the infill density in 3D printing." *Soft Robotics* Vol. 9 No. 5 (2022): pp. 837–849.

[6] Xu, Shuzhi, Liu, Jikai, Zou, Bin, Li, Quhao and Ma, Yongsheng. "Stress constrained multi-material topology optimization with the ordered SIMP method." *Computer Methods in Applied Mechanics and Engineering* Vol. 373 (2021): p. 113453.

[7] Gan, Ning and Wang, Qianxuan. "Topology optimization of multiphase materials with dynamic and static characteristics by BESO method." *Advances in Engineering Software* Vol. 151 (2021): p. 102928.

[8] Bai, Jiantao and Zuo, Wenjie. "Multi-material topology optimization of coated structures using level set method." *Composite Structures* Vol. 300 (2022): p. 116074.

[9] Noda, Masaki, Noguchi, Yuki and Yamada, Takayuki. "Extended level set method: A multiphase representation with perfect symmetric property, and its application to multi-material topology optimization." *Computer Methods in Applied Mechanics and Engineering* Vol. 393 (2022): p. 114742.

[10] Yi, Bing and Saitou, Kazuhiro. "Multicomponent topology optimization of functionally graded lattice structures with bulk solid interfaces." *International Journal for Numerical Methods in Engineering* Vol. 122 No. 16 (2021): pp. 4219–4249.

[11] Sato, Yuki, Yamada, Takayuki, Izui, Kazuhiro and Nishiwaki, Shinji. "Manufacturability evaluation for molded parts using fictitious physical models, and its application in topology optimization." *The International Journal of Advanced Manufacturing Technology* Vol. 92 (2017): pp. 1391–1409.

[12] Yamada, Takayuki, Izui, Kazuhiro, Nishiwaki, Shinji and Takezawa, Akihiro. "A topology optimization method based on the level set method incorporating a fictitious interface energy." *Computer Methods in Applied Mechanics and Engineering* Vol. 199 No. 45-48 (2010): pp. 2876–2891.

[13] Feng, Yukun, Noda, Masaki, Noguchi, Yuki, Matsushima, Kei and Yamada, Takayuki. "Multi-Material Topology Optimization for Additive Manufacturing Considering Dimensional Constraints." *Available at SSRN 4332503* .

[14] Tang, Zhi-Ri, Hu, Ruihan, Chen, Yanhua, Sun, Zhao-Hui and Li, Ming. "Multi-expert learning for fusion of pedestrian detection bounding box." *Knowledge-Based Systems* Vol. 241 (2022): p. 108254.

[15] Xia, Chang, Zhang, Bin, Wang, Haijun, Qiao, Si and Zhang, Anqi. "A minimum-volume oriented bounding box strategy for improving the performance of urban cellular automata based on vectorization and parallel computing technology." *GIScience & Remote Sensing* Vol. 57 No. 1 (2020): pp. 91–106.

[16] Moon, Seung Jae and Yoon, Gil Ho. "A newly developed qp-relaxation method for element connectivity parameterization to achieve stress-based topology optimization for geometrically nonlinear structures." *Computer Methods in Applied Mechanics and Engineering* Vol. 265 (2013): pp. 226–241.

[17] Hirosawa, Ryoma, Noda, Masaki, Matsushima, Kei, Noguchi, Yuki and Yamada, Takayuki. "Multicomponent topology optimization method considering assemblability using a fictitious physical model." .

Proceedings of the ASME 2023
International Design Engineering Technical Conferences and
Computers and Information in Engineering Conference
IDETC-CIE2023
August 20-23, 2023, Boston, Massachusetts

DETC2023-114669

FINDING CHAIN NETS OF SOLIDS FOR 3D PRINTABILITY

Matthew Lawrence*[1][2], Scott Tomlinson[2], Bashir Khoda[1]

[1]Department of Mechanical Engineering, University of Maine, Orono, ME
[2]Advanced Structures and Composites Center, University of Maine, Orono, ME

ABSTRACT

3D-printing or additive manufacturing (AM) can be a size-constrained and time-consuming manufacturing method compared to those traditionally used. Printing larger structures as a net could be a possible solution to this limitation. The intent behind this work is to establish a methodology for creating 3D structures with discretized planner segments using smaller printer envelopes. Nets of solids are capable of folding into an enclosed body from coplanar shapes. Nets can be utilized to create larger objects not otherwise possible due to space limitations in manufacturing methods. Chain nets are particularly useful for AM, which can avoid collision between the segmented part net and printer topology. To prevent interference with the printer, the shape must be unfolded in a chain net. Unfolding algorithms of this type can be simplified to a Hamiltonian pathfinding problem. This method can be particularly useful when used in conjunction with printing on a flexible substrate, i.e., fabric. Larger solids can be created by printing each side separately and using the fabric as a joint between the faces. When the sides of the solids are printed, the print bed no longer limits the size of solids that can be manufactured. This would also allow collaborative printing with multiple printer platforms to print different faces of the net simultaneously.

Keywords: Algorithms, Computer-Aided Design

1. INTRODUCTION

The unfolding of 3D objects can have various realistic applications. The applications intended for this work are in additive manufacturing. This paper aims to present a methodology for finding chain nets of solids for the purpose of additive manufacturing onto a flexible substrate such as fabric. 3D printing allows for optimization of panels to better suit the design requirements. Parts can be modified as needed for optimized structural support and multifunctionality.

The methods presented are intended to compensate for printer geometries in the unfolding of polytopes. Segmenting an enclosed polytope into a net will allow for each polytope face to be scaled to the size of the print bed, creating larger structures than otherwise possible with traditional additive manufacturing methods.

Previously printed segments may interfere with the structure of the printer used. The polytope will need to be printed with each face only touching at most two others, leading to no branching in the unfolded shape. An example of the process is shown in Figure 1.

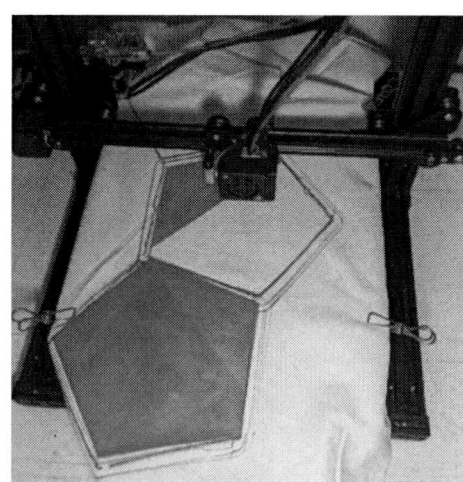

FIGURE 1: THE CHAIN NET OF A POLYTOPE BEING PRINTED [1]

The concept of a net was first introduced in 1525 by Albrecht Dürer with details about methods used [2]. When a

Approved for Public Release – DEVCOM SC PR2023_94367

Copyright © 2023 by ASME

solid in $\mathbb{R}^3$ with polygonal faces is unfolded, the polyhedron is cut along certain edges to be laid flat in a plane until all lie parallel and coincident to the plane. A polyhedron net can be shown in $\mathbb{R}^2$. Figure 2A shows the components of a polytope.

The connectivity of the polyhedron faces can be visualized using a graph. The graph of a polyhedron shows the connectivity of the vertices. The dual graph of a polyhedron shows the connectivity of the faces. The graph for the example polytope is shown in Figure 2B. The dual graph for a polytope is shown in Figure 2C.

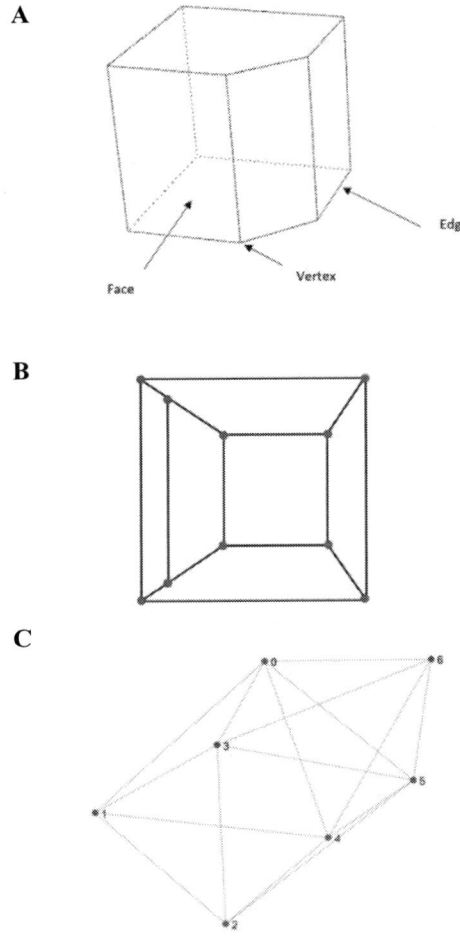

FIGURE 2: (A) THE COMPONENTS OF AN EXAMPLE POLYTOPE (B) THE GRAPH OF THE POLYTOPE (C) THE DUAL GRAPH OF THE POLYTOPE

There are two general ways of storing a graph. Both adjacency lists and adjacency matrices are the preferred methods. A graph is a set of nodes connected by either directed or undirected paths. An adjacency matrix can be utilized to store the connectivity of a graph in matrix form. An adjacency matrix is square with the length being the number of nodes. A value of 1 denotes a connection between 2 vertices in an unweighted graph. A value of 0 denotes no connection. An example of a dual graph of a cube is shown in Figure 3. In a cube, each face is adjacent to every other face, with an opposite face that is not connected. The corresponding adjacency matrix is shown in Equation 1.

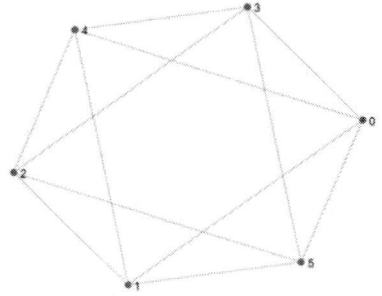

FIGURE 3: THE DUAL GRAPH OF A CUBE

$$
g = \begin{array}{c} \begin{array}{cccccc} V_0 & V_1 & V_2 & V_3 & V_4 & V_5 \end{array} \\ \begin{bmatrix} 0 & 1 & 0 & 1 & 1 & 1 \\ 1 & 0 & 1 & 0 & 1 & 1 \\ 0 & 1 & 0 & 1 & 1 & 1 \\ 1 & 0 & 1 & 0 & 1 & 1 \\ 1 & 1 & 1 & 1 & 0 & 0 \\ 1 & 1 & 1 & 1 & 0 & 0 \end{bmatrix} \end{array} \qquad (1)
$$

A chain-net is a special net where the unfolded faces of the solid border only two others in the unfolded configuration [3]. Not every polyhedron is capable of being unfolded into a chain. An example of a net that is not a chain net is shown in Figure 4A. This is not a chain net since the net contains branches. A chain net of a cube is shown in Figure 4B.

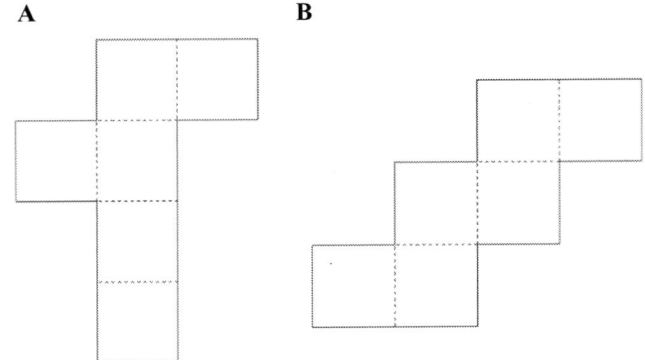

FIGURE 4: (A) A NET OF A CUBE (B) A CHAIN NET OF A CUBE

Chain nets are specifically important since any nets containing branches will not be possible to 3D-print when scaled to the size of the bed without interference with printer topology. The Platonic solids all contain single-chain nets. The amount of

Copyright © 2023 by ASME

single chain nets for each are shown below in Table 1 with flipped and repeated unfolding paths being excluded.

TABLE 1: THE NUMBER OF DISTINCT CHAIN NETS FOR EACH PLATONIC SOLID [4]

Platonic Solid	Unique Chain Nets
Tetrahedron	1
Cube	4
Octahedron	3
Dodecahedron	340
Icosahedron	18

2. METHODOLOGY

Finding a chain net can be simplified to a Hamiltonian pathfinding problem on the dual graph of a polytope [4]. The dual graph includes nodes placed on the centroid of each polytope face. A Hamiltonian path that starts and ends with the same node is called a Hamiltonian cycle. Finding a Hamiltonian cycle is an NP-complete problem [5]. So long as a polytope graph is cut along a Hamiltonian path and the surface angle is less than 180 degrees, the unfolding will not overlap [2].

2.1 Polytope Analysis

The unfolding of a shape first starts with the creation or importation of a solid into Rhino CAD software. Rhino was selected since the built-in script functions can easily extract the parameters of a part. An example polytope was created with its faces labeled via Rhino and is shown in Figure 5.

The adjacency of the faces will be extracted using a visual basic based Rhino script to then be used for pathfinding purposes. An adjacency matrix is ideal for storing the connectivity of the sides. A RhinoScript is utilized to label the faces starting with face 0 and ending with face n-1, where n is the total number of sides on the polytope.

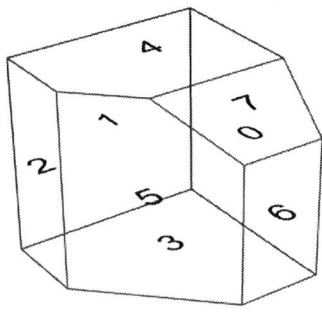

FIGURE 5: A POLYTOPE WITH LABELED FACES IN RHINO

The corresponding adjacency data from the polytope shown in Figure 5 was extracted. The data will be stored in an adjacency list for further processing. A text file is an effective

format for the transfer of the data into MATLAB. This data represents the dual graph of the polytope. An adjacency matrix was then created with MATLAB from the adjacency list. The adjacency matrix is shown in Equation 2. The dual graph corresponding to the adjacency matrix is presented in Figure 6.

$$g = \begin{bmatrix} 0 & 1 & 0 & 1 & 1 & 0 & 1 & 1 \\ 1 & 0 & 1 & 1 & 1 & 0 & 0 & 0 \\ 0 & 1 & 0 & 1 & 1 & 1 & 1 & 1 \\ 1 & 1 & 1 & 0 & 0 & 1 & 1 & 0 \\ 1 & 1 & 1 & 0 & 0 & 1 & 0 & 1 \\ 0 & 0 & 1 & 1 & 1 & 0 & 1 & 1 \\ 1 & 0 & 1 & 1 & 0 & 1 & 0 & 1 \\ 1 & 0 & 1 & 0 & 1 & 1 & 1 & 0 \end{bmatrix} \qquad (2)$$

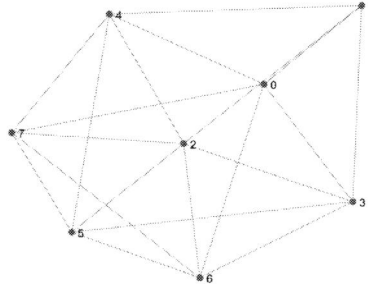

FIGURE 6: THE CONNECTIVITY OF THE POLYTOPE FACES REPRESENTED WITH A DUAL GRAPH

2.2 Pathfinding Methods

A cycle of a graph is a set of edges that creates a path with the starting and ending node being the same [6]. A Hamiltonian cycle is a special type of cycle that passes through each node a single time [7]. It is not possible to determine if a graph has a Hamiltonian cycle without using an exhaustive search method.

Since the unfolding of a shape can be thought of as finding a Hamiltonian Path, conventional pathfinding solvers can be utilized.

Methods for solving the Hamiltonian Path with a brute force or exhaustive approach become problematic when the number of sides is too large. The matrix of possible unfolding patterns will have a width equal to the number of sides and a height equal to the factorial of the number of sides. With a double data type using 8 bytes [8], the storage required will grow immensely. The calculation of the storage used to store the possible unfolding paths for a polytope using double precision is shown in Equation 3 [9] with n being the number of polytope faces. A plot of the storage requirement for a given number of faces is shown in Figure 7.

$$RAM\ used = 8\ bytes * n * n! * 10^{-9} \frac{GB}{byte} \qquad (3)$$

Copyright © 2023 by ASME

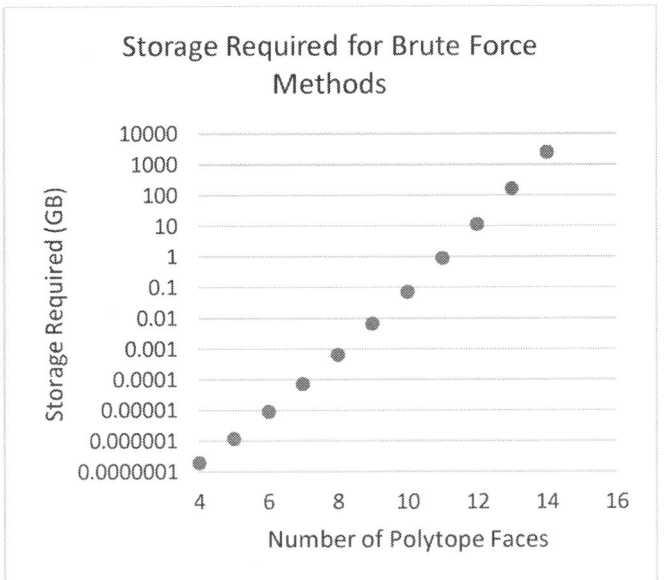

FIGURE 7: THE RAM USED FOR STORING ALL POTENTIAL UNFOLDING PATHS IN GIGABYTES FOR A GIVEN NUMBER OF POLYTOPE FACES

The preferred method of finding a Hamiltonian Path is a heuristic method. This method involves choosing a specific starting and ending node on the graph. The starting node is not particularly important, since the solution will be a cycle with the starting node and ending node being the same.

From the starting node, the algorithm will determine another node to move to in sequential order. If the nodes are not connected or have already been visited, a function will return a false value and choose another node to attempt to move to. This process will be repeated until all nodes have been visited and the path returns to the starting node.

2. RESULTS AND DISCUSSION

Using MATLAB, a cycle in the graph can be found. To proceed with a visual representation, a starting face should be chosen. Face 0 will be used to remain consistent with another viewpoint presented previously. The solution found for starting with face 0 is shown in Figure 8.

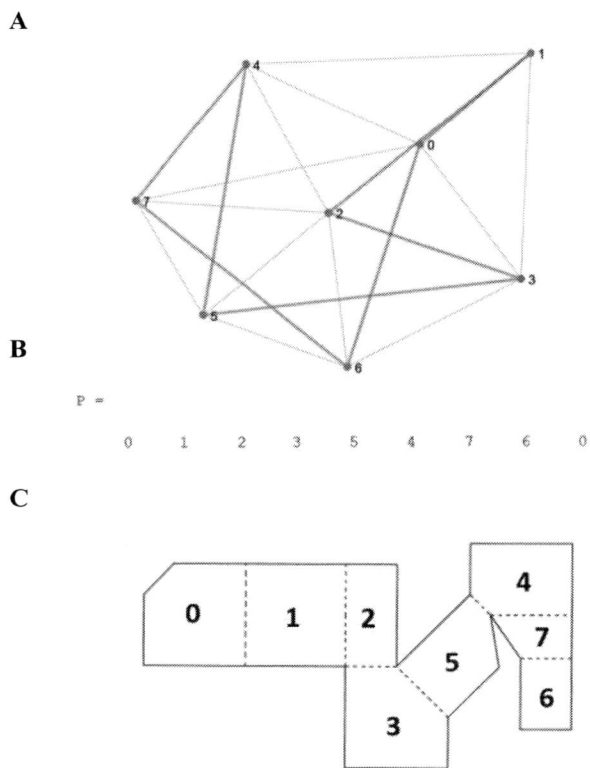

FIGURE 8: (A) A HAMILTONIAN CYCLE FOR THE GRAPH IN RED WITH THE ORIGINAL GRAPH IN BLUE (B) A HAMILTONIAN CYCLE OF THE GRAPH IS FOUND (C) THE NET CORRESPONDING TO THE HAMILTOINIAN PATH

Since heuristic methods are being used, some polytopes can be unfolded that are not possible with a traditional exhaustive search due to storage limitations. An example of a truncated pyramid with a total of 14 faces is shown in Figure 9A. The solution to the pathfinding problem is shown Figure 9B.

Copyright © 2023 by ASME

303

A

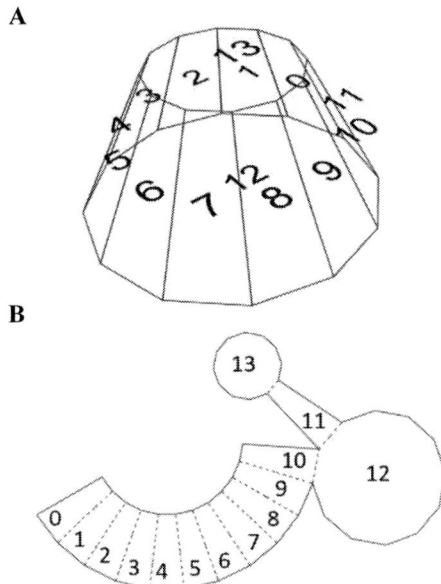

B

FIGURE 9: (A) A 14-SIDED TRUNCATED PYRAMID (B) A SOLUTION TO THE UNFOLDING OF THE 14-SIDED TRUNCATED PYRAMID

The chain net of a polytope can be created and extruded in a CAD software and used as a model for 3D printing. In this example, the chain net of a cube was replicated. The fabric will be used as a flexible hinge between the faces of the polytope. The chain net of the cube is shown in Figure 10A. A final print is shown in Figure 10B. net is shown folded into a cube in Figure 10C.

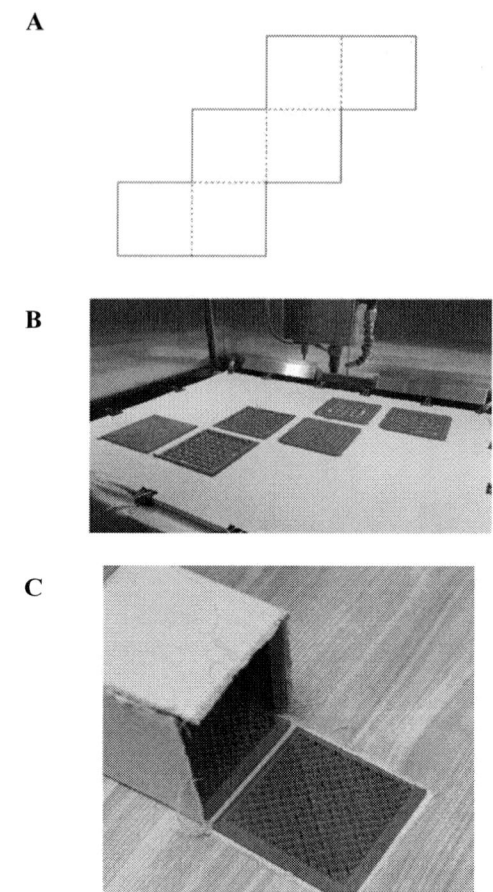

FIGURE 10: (A) A CHAIN NET OF THE CUBE (B) A CHAIN NET OF A CUBE PRINTED ON FABRIC USING A LARGE AREA JUGGERBOT PRINTER (C) AN ASSEMBLED CUBE

Another example of a created and manufactured chain net is a small-scale pyramid shown in Figure 11A. The base has a side length of approximately 4in, with the triangle faces being equilateral. The printed net of the pyramid is shown in Figure 11B. The printed object is shown in Figure 11C.

Copyright © 2023 by ASME

A

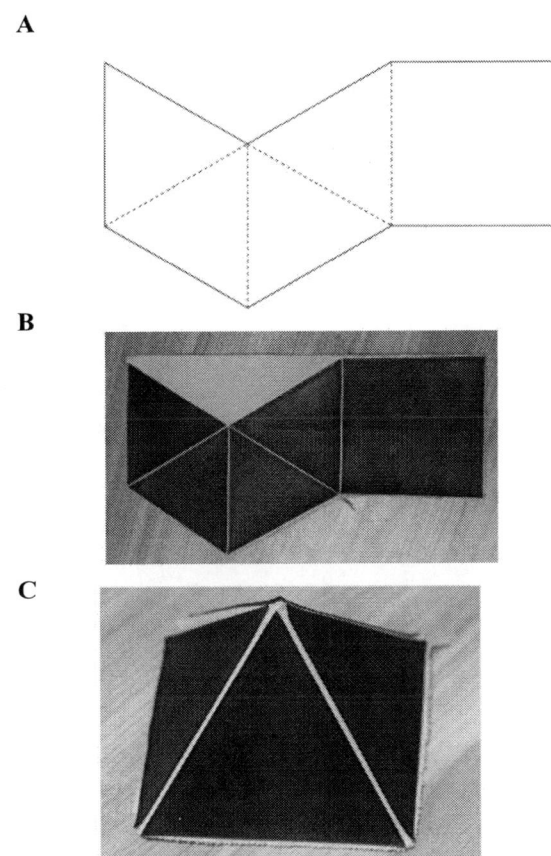

B

C

FIGURE 11: (A) A CHAIN NET OF A PYRAMID (B) THE PRINTED CHAIN NET FOR THE PYRAMID (C) AN ASSEMBLED CUBE

Another application of printing chain nets is the practice of multiple 3D-printers printing simultaneously or collaborative printing. If the polytope faces are printed onto a flexible substrate, the segmented faces need to be printed in the correct locations. Measurements of the flexible substrate can ensure the faces are in the correct location and orientation.

Since the part is segmented, different faces of a polytope can be printed with different printers concurrently. Printing a chain net prevents interference with printer topology since the chain can span between multiple printers without interference from branching of the net. An example of such a printing setup is shown in Figure 12.

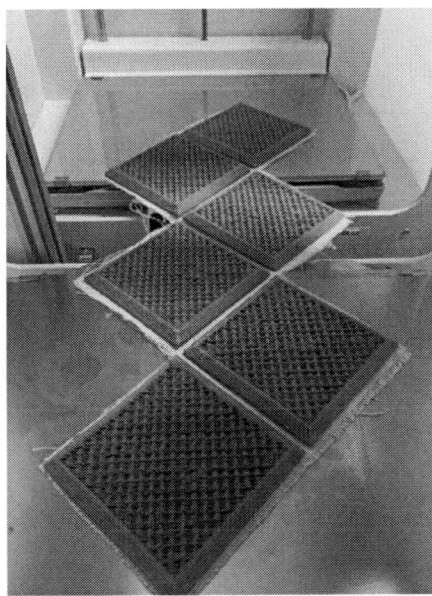

FIGURE 12: A COLLABORATIVE PRINTING SETUP TO PRINT THE NET OF A CUBE SIMULATNIOUSLY ONTO A FABRIC SUBSTRATE

3. CONCLUSION

Hamiltonian Paths can be utilized to find chain nets of polytopes for the purposes of manufacturing. Chain nets are useful to prevent interference with printer topology when printing using the methods presented in this work. The methods presented in this work are applicable for convex polytopes with a Hamiltonian cycle being present in the dual graph. There is no way to check if a graph contains a cycle without an exhaustive search. An exhaustive search is impractical for larger polytopes due to storage limitation as shown in Equation 3. Manufacturing polytopes using a chain net allow for larger structures to be created, bypassing the limitations of a single printer envelope.

ACKNOWLEDGEMENTS

This material is based upon work supported by the U.S. Army Combat Capabilities and Development Command – Soldier Center (DEVCOM SC) under Contract Nos. W911QY-18-C-0101 and W911QY-20-C-0053. Any opinions, findings and conclusions or recommendations expressed in this material are those of the author(s) and do not necessarily reflect the views of the DEVCOM SC.

REFERENCES

[1] M. Blais, S. Tomlinson, and B. Khoda, "Thermoplastics 3D Printing Using Fused Deposition Modeling on Fabrics," in *Volume 3: Advanced Materials: Design, Processing, Characterization, and Applications*, Virtual, Online: American Society of Mechanical Engineers, Nov. 2021, p. V003T03A004. doi: 10.1115/IMECE2021-69695.

[2] "Nets of Polyhedra." Accessed: Feb. 13, 2023. [Online]. Available:

Copyright © 2023 by ASME

https://citeseerx.ist.psu.edu/document?repid=rep1&type=pdf&doi=9fa2c63fd7cf9bf749c9713e0f4855a65402720d

[3] "NETS OF POLYHEDRA." Accessed: Feb. 13, 2023. [Online]. Available: http://faculty.washington.edu/moishe/branko/BG183a.Nets%201.pdf

[4] "NETS OF POLYHEDRA II." Accessed: Feb. 13, 2023. [Online]. Available: http://faculty.washington.edu/moishe/branko/BG183b.Nets%202.pdf

[5] E. W. Weisstein, "Hamiltonian Cycle." https://mathworld.wolfram.com/ (accessed Mar. 13, 2023).

[6] E. W. Weisstein, "Graph Cycle." https://mathworld.wolfram.com/ (accessed May 09, 2023).

[7] E. W. Weisstein, "Hamiltonian Cycle." https://mathworld.wolfram.com/ (accessed May 09, 2023).

[8] "Strategies for Efficient Use of Memory - MATLAB & Simulink." https://www.mathworks.com/help/matlab/matlab_prog/strategies-for-efficient-use-of-memory.html (accessed Feb. 28, 2023).

[9] "How MATLAB Allocates Memory - MATLAB & Simulink." https://www.mathworks.com/help/matlab/matlab_prog/memory-allocation.html (accessed Feb. 28, 2023).

Proceedings of the ASME 2023
International Design Engineering Technical Conferences and
Computers and Information in Engineering Conference
IDETC-CIE2023
August 20-23, 2023, Boston, Massachusetts

DETC2023-116552

AN EMPIRICAL, DETERMINISTIC DESIGN THEORY FOR COMPACT DRIP EMITTER LABYRINTHS

Aditya Ghodgaonkar *
Department of Mechanical Engineering
Massachusetts Institute of Technology
Cambridge, Massachusetts 02139
Email: adighod@mit.edu

Emily Welsh
Department of Mechanical Engineering
Massachusetts Institute of Technology
Cambridge, Massachusetts 02139
Email: ewelsh@mit.edu

Benjamin Judge
Department of Mechanical Engineering
Massachusetts Institute of Technology
Cambridge, Massachusetts 02139
Email: bjudge@mit.edu

Michael Bono
Department of Mechanical Engineering
Massachusetts Institute of Technology
Cambridge, Massachusetts 02139
Email: mikebono@mit.edu

Amos G. Winter, V
Department of Mechanical Engineering
Massachusetts Institute of Technology
Cambridge, Massachusetts 02139
Email: awinter@mit.edu

ABSTRACT

Growing food demand, climate change, and constrained natural resources create the need for large-scale, sustainable agricultural intensification. Despite drip irrigation's ability to be more water efficient than traditional irrigation technologies, its adoption and retention is limited to due to its high hydraulic equipment costs, particularly in low/middle-income countries. As a commodity product, drip emitters contribute directly to raw material costs and additionally dictate tube thickness and related material consumption. This work introduces a new empirical, deterministic design theory for creating compact, low-cost labyrinths, which are otherwise a volume-intensive component of drip irrigation emitters. To simplify design analysis a review of current commercial art, manufacturing process constraints and symmetry-based geometric relationships was conducted, resulting in the labyrinth's tooth tip gap being selected as a key design variable. The tip gap is correlated with the hydraulic performance of a test labyrinth geometry via a Design of Experiments approach. The experiments shed light on two distinct fluid dynamic regimes in the labyrinth based on the tip gap

size and provide an empirical expression between the two. This work demonstrates that simultaneous consideration of symmetry, manufacturing process and design goals enables rapid synthesis of labyrinths that are 43.77% shorter than comparable commercial designs.

1 INTRODUCTION

There is an urgent need for widespread, sustainable agricultural intensification to meet the unmet and growing global food demand against the backdrop of climate change and constrained freshwater resources. Although drip irrigation systems can be highly water efficient (up to 60% compared to flood or furrow irrigation [1–5]), their adoption and retention is constrained by high equipment costs [6, 7], which includes material costs. This work focuses on addressing the raw material cost of manufacturing drip emitters and the tubing to which they are bonded. These two elements alone can contribute up to 50% of the lifetime costs per hectare of a drip irrigation system [8]. One of the barriers against creation of more affordable, durable drip irrigation tubing is the complexity associated with design and development of

---

*Corresponding Author.

Copyright © 2023 by ASME

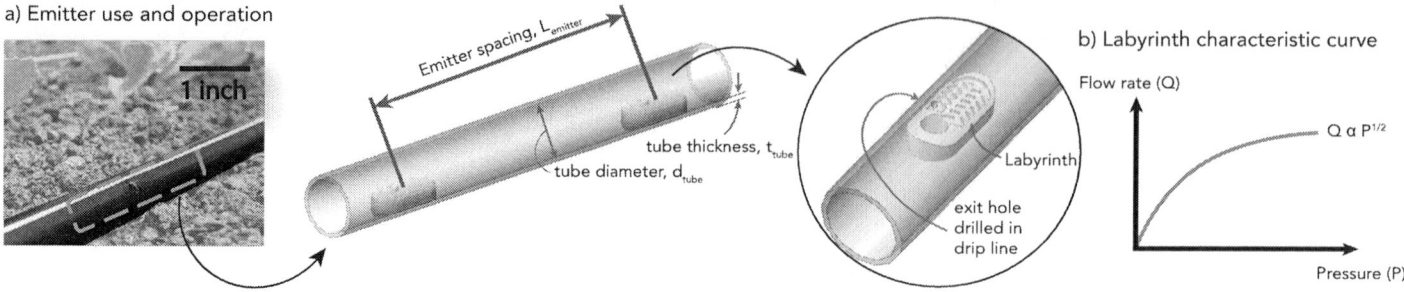

FIGURE 1: a) Inline emitters are injection molded and bonded to the inside of the tube during tube extrusion at precisely set spacings, ($L_{spacing}$). The tube is typically made from High-Density Polyethylene (HDPE) and has an inner diameter d_{tube} and thickness t_{tube}. The labyrinth passage is directly molded into the emitter body and often uses the tube wall as a sealing surface. b) Typical flow rate vs. pressure trend of a labyrinth is quadratic.

emitter labyrinths. These are a volume-intensive hydraulic resistance feature ubiquitous to all emitters (see figure 1). To that end, the objective of this work is to provide a design theory that will enable creation of compact emitter labyrinths. The tool will take the form of an empirically derived mathematical relationship between an emitter labyrinth's performance and selected geometric parameters, subject to manufacturing topology constraints and the goal of reducing emitter volume.

Drip irrigation delivers water directly to the root zone via a network of pressurized tubing and emitters, avoiding water losses associated with conveyance and evaporation. Emitters are flow metering devices that ensure targeted and regulated watering of crops (figure 1). This work focuses on *labyrinths* in *inline* emitters. Inline emitters are molded and inserted into the tube at set spacings and are widely used for growing horticultural crops. The labyrinth is a zig-zag flow feature designed to create pressure loss through turbulent energy dissipation. This hydraulic resistance serves as a load for the pump, ensuring that water is delivered across the field and not only at the head. Functionally, labyrinths perform akin to orifices, i.e., flow rate $\propto$ pressure$^{1/2}$. This relationship allows labyrinth hydraulic performance to be defined by the constant turbulent hydraulic resistance, $K_{lab} = (\Delta P/Q^2)$, where ΔP is the pressure drop across the labyrinth in Pa. , Q is the flow rate in m^3/s, and K_{lab} is presented in kg/m^7.

To illustrate how emitters and tubing can be a significant contributor to upfront equipment costs, figure 2 demonstrates the case study of a Tomato farm in Morocco using inline Pressure-Compensating (PC) emitters [9]. PC emitters use compliant features in addition to the rigid labyrinth to maintain a fixed flow rate above a minimum inlet pressure. The hydraulic components (tubing and emitters) alone are estimated to contribute ~49% of the lifetime costs per hectare [10] (figure 2). Further, for a typical 1000 ft. reel of inline pressure-compensating drip emitter tubing with 1 ft. emitter spacing, 27.5% of the cost can be driven by

material and manufacturing costs of emitters, with tube material contributing up to 27.5% more costs in cases where tube walls are thick ($\geq$ 1 mm) [11]. Crucially, emitters are commodity products – a 100 ft. × 100 ft. field section could have up to 10,000 emitters across it. Reducing the overall volume of a single emitter can therefore have a magnified impact on equipment costs. Since labyrinths contribute significantly to the material used in the emitter body, designing compact labyrinths could contribute to the objective of designing low-cost/volume emitter bodies.

Tube wall thickness and associated cost can be driven by emitter size. The tube wall needs to be thick enough to resist rupture due to kinking near the ends of the emitter body. As a result drip tubing in developed markets is often around 1mm thick [11–14]. In contrast, in scenarios where affordability constrains the amount of material that can be used, emitters are bonded to tube walls as thin as 0.1 mm [15]. As seen in figure 3, the resulting equipment often tears near the emitter edge

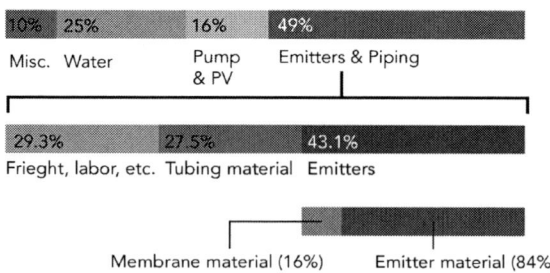

FIGURE 2: Demonstrative case study of a solar-powered drip irrigation farm in Morocco using of Pressure-Compensating (PC) drip emitters for growing tomatoes [9]. The lifetime costs per hectare shows a large contribution from the hydraulics. The cost breakdown of Toro BlueLine PC dripline is used as reference for evaluating contributions from emitters and tubing. The emitter itself has associated material costs [11].

Copyright © 2023 by ASME

and needs to be replaced resulting in replacement costs. Therefore, designing compact emitters could lead to the holistic development of affordable, robust driplines, by allowing for saved emitter raw material to be redirected toward thicker tubes.

The challenge associated with creating compact emitter labyrinths arises from their operating physics (which is driven by turbulent flows) and the large number of geometric design parameters defining them. This makes their design space vast and difficult to efficiently explore. Current industrial labyrinth design processes can be iterative, computationally intensive, and time-consuming [11, 16]. Under the constraints of a competitive market, this hampers the creation of emitters that are economically profitable for the manufacturer and beneficial to the farmer. Academic work on the labyrinth design is often done in the context of studying Non-Pressure-Compensating (NPC) emitters where the labyrinth is the sole flow passage. Several studies have computationally modeled the turbulent flow of existing commercial labyrinths and their reliance on a single or select group of geometric parameters [17, 18], shedding light on the fundamental operation of labyrinths. Some works further conducted physical imaging studies [19–21]. The use of CFD tools to create regression models of Non-Pressure Compensating emitters' hydraulic or clogging performance as a function of select variables has been carried out as well [22, 23]. The utility of data-driven design methods is then demonstrated when these models are combined with single-objective-optimizations or Machine Learning models to determine an optimal variant of the commercial design explored. In most studies, design variables of interest are *heuristically* selected based on the author's intuition of whether they affect hydraulic and/or clog performance or not.

This creates an opportunity to contribute a labyrinth design framework that explores all geometric parameters subject to manufacturing topology constraints while simultaneously considering both material volume and hydraulic resistance. This

could be used by drip irrigation designers for rapidly synthesizing labyrinth geometries that meet a desired flow rate-pressure performance. Broadly, such a design theory could be useful for researchers and engineers engaged in the creation of microfluidic flow devices, particularly passive devices having tortuous geometries akin to the one explored here. These devices have been studied for the control and manipulation of fluid mixing [24], chemical reactions [25, 26], particle focusing and separation [27, 28], and heat exchangers [29]. These applications are rich in fluid phenomena and can be tuned for high efficiency, with Reynolds numbers often in the range $Re \sim \mathscr{O}(0.1-100)$. As an avenue to possibly increase the throughput and performance of these devices, this work offers an alternate tortuous geometry (the 'labyrinth') and its hydraulic design principles. By ensuring fully developed turbulence, the geometry functions consistently over a wider range of Reynolds numbers ($Re > 1000$), while being commercially manufacturable and tunable for flow rate using just 3 parameters.

We present an empirically derived mathematical model that relates labyrinth hydraulic performance/resistance to selected geometric parameters. To facilitate the most meaningful geometry-hydraulic relationship, Sec. 2 develops geometric parameterization and relationships and uses manufacturing constraints to select the most relevant design levers which are the number of tooth pairs, labyrinth depth, and tooth tip gap. A Design of Experiments (DoE) is proposed to evaluate the relationship between tip gap and hydraulic resistance. To ensure both physical and commercial relevance, the DoE is informed by both a review of commercial art and CFD simulations detailed in Sec. 2 as well. To further motivate the value of reducing emitter volume, Sec. 3 uses a simple material volume model to showcase how raw material saved on emitter production can lead to a significant increase in tube wall thickness and potential durability. Sec. 4 details the DoE, prototyping process, and test procedure. Sec. 5 presents the results of the DoE, demonstrating that there exist two distinct flow regimes based on the value of the tip gap. Finally, sec. 6 evaluates how well the design objectives were met, yielding a labyrinth 43.77% shorter than comparable commercial products without any significant loss in performance.

2 LABYRINTHS: GEOMETRY, OPERATION, AND MANUFACTURING

A labyrinth consists of a sequence of staggered teeth-like structures. In this work, we choose to focus on a geometry that has two passages/legs connected by a 'U-turn' passage, as seen in figure 4a.

The fluid flow that passes through the labyrinth needs to adjust to flow around or past these teeth. This creates recirculation zones between teeth which are an integral feature of labyrinths. The onset of these vortices or recirculation zones occurs at low Reynolds numbers (Re $\sim$ 100 − 700) relative to the oft-cited

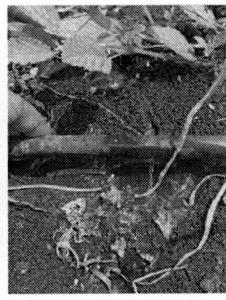

FIGURE 3: Driplines getting damaged due to low wall thickness in Kenya. Left: Kinking and rupture near emitter due to awkward bending of the tube during storage, right: rupture in the tube due to suspected user damage (farmer stepped on the tube during installation). Low tube wall thickness arises due to cost constraints on the product put in place to make it affordable to farmers.

Copyright © 2023 by ASME

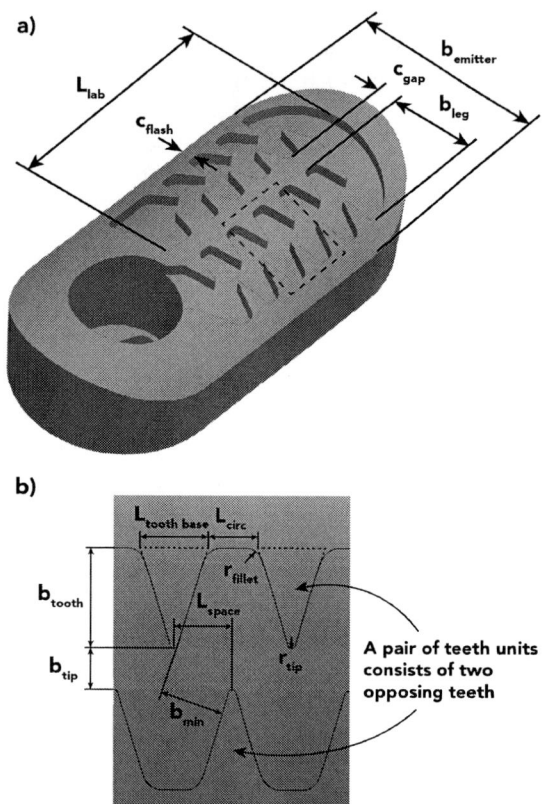

FIGURE 4: Geometric parameterization of a labyrinth. Several parameters are needed to fully define the geometry on a) a macroscopic level, and b) a tooth-level geometry. Not shown in this figure is the labyrinth depth h_{lab}.

threshold for turbulent flow in a circular pipe (2300) [8, 18]. Al-Muhammad et al. [18] conducted detailed computational simulations of a commercial labyrinth geometry and investigated the sources of pressure drop. Their analysis reveals that between three apparent sources of pressure energy dissipation - mean turbulent dissipation, fluctuating turbulent dissipation, and wall friction - mean turbulent dissipation is the primary cause of pressure drop in labyrinths. The geometry of the teeth and spacing between them in turn determines the flow structures and therefore the hydraulic resistance of the labyrinth. A description of a common labyrinth geometry is provided in figure 4.

2.1 Manufacturing considerations

Several geometric parameters introduced in figure 4 are bounded or limited by manufacturing process constraints. Emitters are produced on a commercial scale using injection molding from Polyethylene. In NPC emitters (see figure 1) which are functionally just a labyrinth, the labyrinth is molded onto one side of the emitter body (see figure 1) with the inlet on the other side. The labyrinth is then sealed against the drip tubing during the insertion stage. In the case of Pressure-Compensating (PC) emitters, all the hydraulic features including the labyrinth are molded into the emitter body which is then welded to a sealing cap with an inlet.

Design goals and constraints are introduced in Table 1. The constraints arise from injection molding (packing limits, thermal warp-related limits) and thermoplastic welding (flash clearances, sealing walls, alignment pins needs). For the latter, a large number of considerations arise from the need for an 'energy director' molded onto the emitter cover that helps guide weld energy between the two parts and becomes the site of the weld [11, 30]. A commonly used welding process for plastics is ultrasonic welding (another being laser welding). To ensure that designs developed in this study are viable for both NPC and PC emitters, the more stringent welding-related constraints of PC emitter production were considered. The process of designing the labyrinth must respect these limits to yield a design that not just meets design goals, but can also be produced on a commercial scale.

2.2 Geometric relationships

Labyrinth geometries are often laid out in a manner that allows for simple geometric relationships between labyrinth parameters to be derived on the basis of symmetry. These relationships allow us to identify variables that are redundant. Such variables can be expressed as an algebraic sum of design levers and pre-set constraints.

Width-wise geometric analysis: The overall emitter width $b_{emitter}$ can be considered to be a sum of labyrinth leg width b_{leg} and flash/energy director spacing required for cover-body welding. The emitter width itself is typically limited to ≤ 11 mm to ensure that the emitter can fit within at least 16 mm tube without creating excessive flow resistance (see Table 2). Further breaking down the components of the labyrinth leg width and grouping the constrained terms (Table 1) helps us identify a simple relation between tooth triangle height b_{tooth} and tip gap b_{tip}.

$$b_{emitter} = 2 \times (b_{leg} + c_{flash}) + c_{gap}, \text{ and} \qquad (1)$$
$$b_{leg} = 2 \times b_{tooth} + b_{tip}, \qquad (2)$$
$$\implies b_{emitter} = 2 \times (2b_{tooth} + b_{tip} + c_{flash}) + c_{gap}. \qquad (3)$$
$$\implies b_{tip} = 2b_{tooth} + (\text{constrained terms}). \qquad (4)$$

Length-wise geometric analysis: The length of the labyrinth L_{lab} can be expressed as the product of the number of teeth pairs N_{units} and the space between the teeth L_{space}. Since we constrain L_{space} for prototyping ease, N_{units} can be directly related to L_{lab}

$$L_{lab} = 2 \times N_{units} \times L_{space}. \qquad (5)$$

Copyright © 2023 by ASME

Design feature	Requirement type	Quantification	Justification [?, 30]
Emitter/labyrinth body dimensions			
Length, $L_{emitter}$	Design goal	Minimize	See Sec. 3.1
Manufacturing considerations			
Leg gap, c_{gap}	Constraint	> 1.27 mm	Clearance for energy directors in ultrasonic welding.
Flash clearance, c_{flash}	Constraint	> 1.016 mm	To accommodate energy director and weld flash.
Tooth tip radius, r_{tip}	Constraint	≥ 0.064 mm	Based on mold packing limits.
Base filet, r_{fillet}	Constraint	= 0.012 in.	Test designs were milled (see Sec. 4.1). Constraint was introduced
Tooth spacing, $L_{spacing}$	Constraint	≥ 0.024 in.	to ensure the use of a 0.024-inch end-mill for all prototypes.

TABLE 1: A list of major injection molding and plastic welding constraints that guide the quantification of select emitter dimensions.

Additionally, L_{space} is the sum of the tooth triangle base length and circulation space length

$$L_{space} = \frac{1}{2}\left(L_{circ} + L_{toothbase}\right). \tag{6}$$

In the commercial manufacture of emitters, it is likely that the constraining element is $L_{toothbase}$ (by injection molding packing limits). Additionally, the minimum limit of L_{space} may also be driven by clogging considerations to avoid particle jamming. The end effect is that L_{space} can be evaluated or constrained by manufacturing or performance. This creates a trade-off between the number of tooth units and the labyrinth length via Eqn. 5.

2.3 Downselecting design parameters

The combination of manufacturing constraints and geometric relationships helps reduce the number of design parameters to a select few.

– **Tooth pairs** (N_{units}): Sokol et al. [8] and Narain and Winter [31] have shown that the labyrinth hydraulic resistance $K_{lab} \propto N_{units}$. This simple linear proportionality helps tune hydraulic resistance and hence flow-pressure performance by simply adding or removing teeth pairs, with the likely caveat that L_{circ} and $L_{toothbase}$ are constant. Since $Q(P) = \sqrt{(P/K_{lab})}$, designers will often set N_{units} to meet required flow rate performance.

– **Labyrinth depth** (h_{lab}): This is considered to be a key design variable that could be used to regulate the hydraulic performance of the labyrinth. It is expected that with increasing labyrinth depth (h_{lab}), the resistance decreases monotonically, and therefore study of depth effects is left out of the current work's scope.

– **Tip gap** (b_{tip}): Tip gap represents the component of distance between teeth tips in the width-wise direction. Tip gap is chosen for study instead of tooth triangle height (L_{tooth}) since it has a more intuitive impact on flow structures in the labyrinth and hence resistance. A negative value implies an overlap of teeth (see figure 5).

The current study is limited to exploring the effect of tip gap on hydraulic resistance only. Before developing a model correlating geometry and hydraulic resistance, it is beneficial to review commercial emitters and evaluate the range and distribution of select parameters. This review examined the tip gap, depth, number of teeth pairs, as well as the bounding dimensions of the labyrinth and present them in Table 2.

The physical significance of varying b_{tip}: In the review of commercial labyrinths, tip gaps varied between -90 μm and $+300$ μm. While a clear trend emerges between flow rate/hydraulic resistance and labyrinth depth (deeper depth creates lower resistance and hence higher flow rates), there appears to be a lack of consensus around the tip gap and its impact on flow resistance. To shed more light on the tip gap-resistance relationship, CFD simulations of two identical labyrinths with 11 pairs of teeth and 1 mm depth were conducted at 1 bar inlet pressure (for details, see Appendix A). The two labyrinths had tip gaps $+150$ μm gap (case A, figure 5) and -150 μm 'overlap' (case B, figure 5) respectively. To visualize flow fields, velocity vector plots were generated. Case A showcases how a positive b_{tip} allows for an easier, faster flow of water, leading to high-speed core flow. It also leads to the formation of larger, fast-moving circulation zones between the teeth than in Case B. Case B shows that for a negative b_{tip}, the flow needs to continuously redirect itself leading to slower mean speed and smaller circulation zones. Since turbulent energy dissipation rates are correlated to hydraulic resistance [18], this field was plotted as well. Between Case A and B, a clear shift is observed in dissi-

Copyright © 2023 by ASME

Emitter	$h_{lab} \pm 0.0254$ mm	$b_{tip} \pm 0.01$ mm	N_{units}	$L_{lab} \pm 0.0254$ mm	$b_{emitter} \pm 0.0254$ mm
Jain Turbo Excel NPC (0.8 L/hr)	0.4	0.1016	14.5	16.891	5.9436
Jain Turbo Excel NPC (1.2 L/hr)	0.62	0.0762	12	16.891	5.9436
Jain Turbo Excel NPC (1.6 L/hr)	0.68	0.127	10	16.891	5.9436
Jain Turbo Excel NPC (4 L/hr)	1.05	0.3048	7.5	16.891	5.9436
Jain Turbocascade 1.1 L/hr PC	0.7366	0.1016	21.5	24.003	10.414
Jain Turbocascade 1.6 L/hr PC	0.9144	0.0762	19.5	24.003	10.414
Jain Turbocascade 2.0 L/hr PC	1.0414	0.1016	17.5	24.003	10.414
Toro BlueLine PC 1.0 L/hr	0.5334	-0.08636	15	17.78	11.049
Toro BlueLine PC 2.0 L/hr	1.0668	-0.0254	15	17.78	11.049
Kenyan NPC 1.6 L/hr (locally sourced)	0.7112	0.1778	33.5	29.845	8.5344
Kenyan NPC 3.5 L/hr (locally sourced)	0.6604	0.29464	19.5	29.845	8.5344

TABLE 2: Review of commercial labyrinth designs. NPC flow rates are measured at 10 m of head (1 bar). Fractional tooth pair (N_{units}) implies that the labyrinth had at least one unpaired tooth.

pation hot spots, with Case B having large, localized dissipation hot spots adjacent to the tooth wall. Case B has a larger hydraulic resistance of $K_{lab} = 1.875 \times 10^{17}$ kg/m^7, while Case A has $K_{lab} = 1.113 \times 10^{17}$ kg/m^7. This represents an increase of nearly 69% from Case A to Case B and is marked by a reduction in flow rate from 3.4 L/hr to 2.6 L/hr (at 1 bar). Clearly, changing b_{tip} has a significant impact on the fluid mechanics of the labyrinth, and hence a wide range that respects but goes beyond commercial product dimensions must be considered for the subsequent study.

The significance of this analysis is that we now have a potential avenue towards reducing emitter volume without compromising on the labyrinth hydraulic performance.

3 DESIGN OBJECTIVE: REDUCING LABYRINTH LENGTH

The goal of this work is to provide a design theory for creating compact labyrinths for low-cost, durable drip irrigation. The volume of the emitter body is sensitive to the geometry of the molded flow features which are complex and driven by hydraulic performance, clogging, and manufacturing considerations. To simplify the analysis, the volume of the emitter body is abstracted as the volume of the bounding box for the emitter body. The emitter width and height do not have objectives associated with them in this study. This is further elaborated below.

Emitter width: The emitter body width is constrained to ensure that the emitter body fits within the dripline tube which typically starts from an inner diameter of 16 mm. A review of some commercial emitter bodies (Table 2) indicated that body widths are typically $8.5 - 11$ mm for multi-pass labyrinths (where the labyrinth takes 2-3 passes between inlet and outlet). Therefore the width is considered to be fixed at ~ 11 mm or ~ 0.433 inches.

Emitter height: Designers will typically seek to minimize the height of the emitter body to minimize obstruction to flow past the emitter body. Therefore emitter body height is often the sum of the deepest hydraulic features and the minimum web section that can be injection molded without thermal warp. Since the only feature of interest in the current context is the labyrinth, the emitter body height is considered to be a floating value that is automatically set by the labyrinth depth. Therefore we do not assign a numerical objective to this parameter.

The design goal is therefore to reduce the length of the labyrinth L_{lab}. The shortest labyrinths reviewed in Table 2 are the Jain NPC emitters. Among these, the 0.8 L/hr emitter needs to have the highest hydraulic resistance within the smallest footprint. The question posed to the DoE is then: **could we achieve the same flow rate as a current commercial, compact labyrinth, but with a significantly lesser material by only tuning the tip gap b_{tip}?**

3.1 Implications of reducing labyrinth length

Reducing emitter volume can directly reduce the raw material expenses borne by the manufacturer and potentially the sale price to the farmer. This study offers an alternate way of viewing the benefit of reduced emitter volume - increasing tube thickness and hence durability. Long emitter bodies require a sufficiently

Copyright © 2023 by ASME

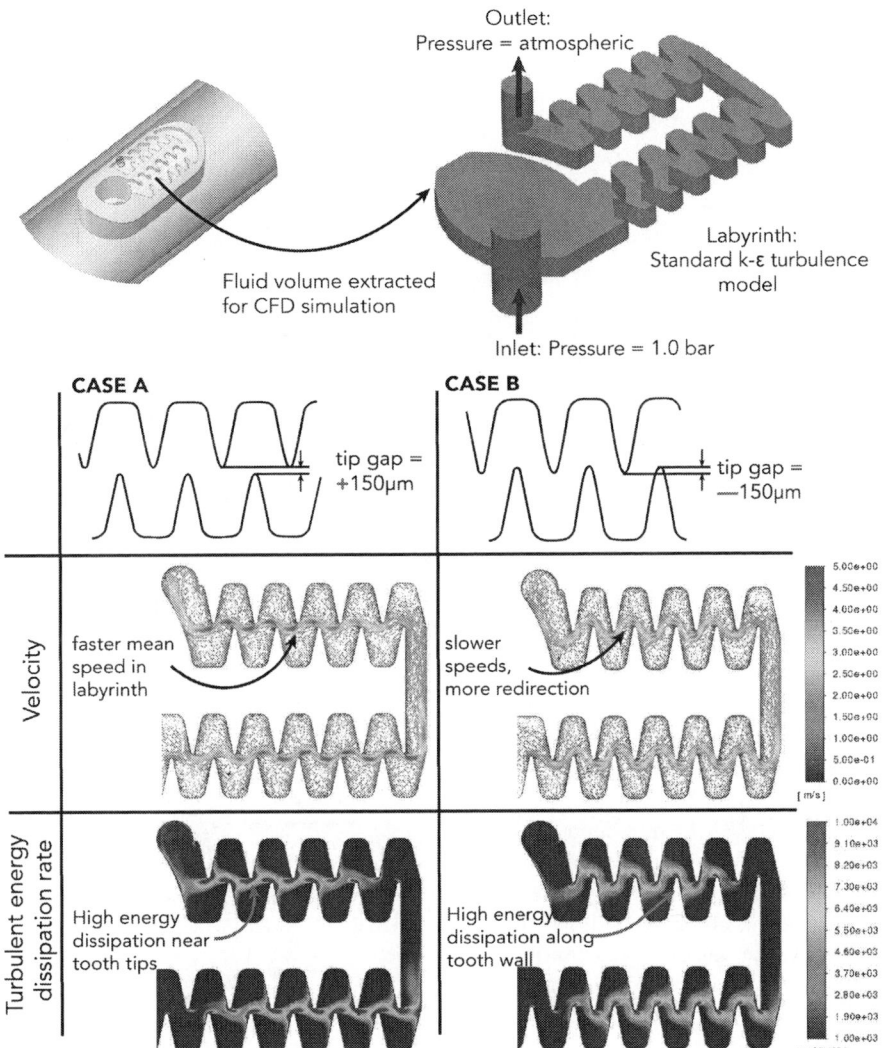

FIGURE 5: Computational simulations of two identical labyrinths ($h_{lab} = 1$ mm) with one geometric variation in tip gap (b_{tip}). Case A (left) has $b_{tip} = +150\ \mu$m, permitting a larger mean velocity and flow rate (3.412 L/hr at 1 bar). Case B (right) features overlapping teeth with $b_{tip} = -150\ \mu$m. This creates a need for continuous redirection of flow, resulting in lower average velocity and high localized energy dissipation along the tooth wall, resulting in a low flow rate (2.629 L/hr at 1 bar). This comparative study showcases how minute changes in certain geometric features can create large changes in flow structures and subsequently hydraulic performance.

thick tube wall to be securely bonded to. If that does not happen (due to cost or affordability constraints), the tube will have a tendency to kink and rupture near the emitter body's edge (figure 3). We propose that reducing the raw material used for molding emitter bodies allows the saved material to be redirected towards thickening or reinforcing the tube wall since both the emitter body and tubing are made from polyethylene. A simple material volume model is constructed to elaborate this argument. The tube material volume for a single dripline is

$$\mathcal{V}_{drip} = \frac{\pi L_{drip}}{4}\left(D_o^2 - D_i^2\right) = \frac{\pi L_{drip} t_{tube}}{2}\left(D_o + D_i\right), \quad (7)$$

where D_o and D_i are the outer and inner diameters respectively, L_{drip} is the length of the dripline, and t_{tube} is the tube thickness. Most of these dimensions are set by agronomy, operational and performance considerations such as tube flow frictional loss and size of the field section. Irrigation engineers, distributors, and

Copyright © 2023 by ASME

manufacturers have various rules of thumb that they use to select the right lateral tubing for a given field [15, 32–34]. Next, the volume contributed by the emitters is

$$\mathcal{V}_{em,total} \approx L_{emitter} \times b_{emitter} \times h_{emitter} \times N_{emitters}, \qquad (8)$$

where $(L \times b \times h)_{emitter}$ is the volume of the bounding box for an emitter body and $N_{emitters}$ is the number of emitters in a dripline. The latter is given by $L_{drip}/L_{spacing}$, where $L_{spacing}$ is the emitter spacing in the tube. In the current context, the emitter length $L_{emitter}$ is used interchangeably with the labyrinth length L_{lab} (see Sec. 2.2 for details). Assuming that the total dripline volume $\mathcal{V}_{tot} = \mathcal{V}_{drip} + \mathcal{V}_{em,total}$ is a constant, it can be differentiated with respect to $L_{emitter}$ and re-integrated to yield the following expression for the increase in tube wall thickness as a function of a fractional reduction in emitter length

$$\Delta t_{tube} = t_{tube} - t_{tube,0}$$
$$= \frac{b_{emitter} h_{emitter} L_{emitter,0}}{\pi D_i L_{spacing}} \times \left(1 - \frac{L_{emitter}}{L_{emitter,0}}\right). \qquad (9)$$

Here the suffix 0 represents a nominal case with respect to which the benefit of reducing emitter length is now demonstrated. We pick the nominal case of an emitter with dimensions $L_{emitter,0} = 50.8$ mm, $b_{emitter} = 5.84$ mm, and $h_{emitter} = 2.54$ mm bonded to a $t_{tube,0} = 0.1$ mm thick tube at $L_{spacing} = 12$ inches. Following the model's underlying assumption of emitter material reduced being equivalent to tube material gained at no added cost, figure 6 shows a linear relationship between tube thickness and emitter length. For the highest emitter count ($L_{spacing} = 12$ in.), reducing emitter length by 50% can enable $\sim 27\%$ increase in nominal tube wall thickness. This could potentially lead to more durable drip line products in cost-constrained markets, enabling longer lifespans and lower replacement costs. In developed markets, it represents a reduction in the cost of manufacturing/sale. Regardless of the economic backdrop, reducing the amount of plastic material used by these commodity products is more sustainable for the environment.

4 DESIGN OF EXPERIMENTS: PROTOTYPING AND PROCEDURE

A comprehensive evaluation of manufacturing constraints, geometry, and commercial art has led to the selection of tip gap b_{tip} as a design variable of interest for tuning hydraulic resistance in compact labyrinths. The impact of the number of units (N_{units}) on resistance has already been characterized [8, 31], and $N_{units} = 11$ was set initially as a starting guess. One tooth pair had to be removed on the second leg to accommodate the exit of the labyrinth. The labyrinth depth is set to 0.5 mm. Five test

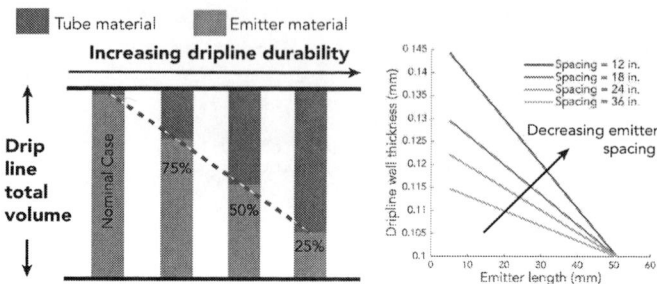

FIGURE 6: Outcome of the material model (Eqn. 9) using the nominal case of a 50.8 mm (2-inches) long emitter body bonded to a 0.1 mm thick drip line. a) Reducing emitter length leads to a proportional increase in tube material, and b) this effect is magnified as emitter spacing is reduced, i.e., emitters per unit length are increased.

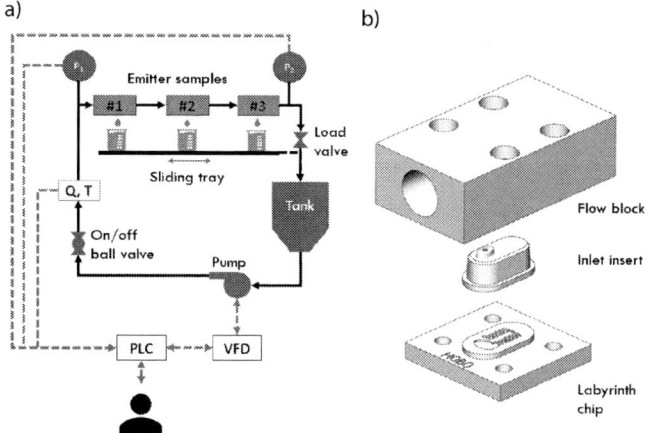

FIGURE 7: a) Schematic of the experimental setup, and b) prototyping assembly used for testing. The labyrinth is milled into the emitter chip.

labyrinths with $b_{tip} = -150, -75, 0, +75, +150$ μm were prepared. This range and step size were selected to cover a significant portion of the commercial range (-90 to 300μm) while also capturing the fluid dynamics that could be driving the resistance mechanism (based on CFD analysis in figure 5). Flow rate-pressure measurements were collected and processed to evaluate hydraulic resistance $K_{lab} = P/Q(P)^2$.

4.1 Prototyping process

The objective of the prototyping procedure is to be able to meet the tight dimensional tolerances expected from emitter dimensions while being flexible to multiple designs, which the commercial production process is not. Additionally, we aim to recreate the hydrodynamics of an emitter inserted inside a tube

Copyright © 2023 by ASME

as closely as possible. Figure 7b shows how the emitter prototype consists of three parts: a flow block, an inlet insert, and the emitter 'chip'.

- **Emitter/labyrinth chip:** The flow features are milled into the flow chip which is made from aluminum. The choice of material is driven by prototype ruggedness, machinability, and cleanliness of parts. We elect to make 2.5D labyrinths (a 2D profile that is milled downwards to a set depth). This prototyping process is based on that used by Sokol et al. [8].
- **Flow block:** The flow block enables the emitter to be placed 'in-line' in the flow tube. It has a through bore and is fitted with barb fittings that are selected to provide maximum diametric continuity for the water as it flows from the tube to the flow block and into the emitter chip. The flow block is milled out of Delrin (Polyoxymethylene or POM). This material is selected for its lightness.
- **Inlet insert:** The emitter chip and the flow block interface via a 3D printed inlet insert. The purpose of this insert is to serve as an emitter-shaped bluff body in the flow while having a realistic inlet that guides water to the emitter chip underneath. Due to its ability to print fine features with high precision, Direct Light Synthesis (Carbon, Inc.) 3D printing procedure was selected for making this part. The inlet used in this study is a 'tower' style inlet that is derived from measured dimensions on a Toro BlueLine PC emitter.

4.2 Testing setup and procedure

figure 7a presents the experimental test setup for flow-pressure measurement. It consists of a flow loop that starts with a plastic storage tank (4-gallon capacity). The tank feeds a centrifugal pump (AMT 368C-98) that circulates water through the flow loop. Approximately 7 liters of tap water were circulated through the system. The emitter samples were placed 'in-line' using 16mm ID, 1 mm thick dripline tubes. Care was taken to ensure at least 12 inches (30.48 cm) of spacing to prevent the fluid wake of one emitter from affecting the performance of the next; this spacing is also a common emitter spacing in commercial products [11–14]. The emitter segment is mounted above a tray that drains into the tank. A sliding platform is used to slide flow measurement cups under the emitters during measurement.

The pump motor is operated using a Variable Frequency Drive (WEG CFW100 series) that was controlled by a Programmable Logic Controller (ClickPLC C0-12DD2E-2-D) via an RS485 communication module. This enables control and monitoring of the pump from a computer. Sensing elements in the system include a flow-temperature transducer (ProSense FMM50-1002, ± 0.010 GPM flow rate, $\pm 0.28°C$ temperature) and pressure transducers (Cynergy3 IPSU-G0075-5, $\leq \pm 0.1875$ psi) at the start and exit of the emitter segment. A Proportional-Integral (PI) control loop was implemented on the PLC, allowing the user to enter a pressure setpoint between $0 - 30$ psi ($0 - 2.08$

bar) at the entrance of the emitter segment (Pressure-1 in figure 7a). The loop then sets, monitors, and adjusts the motor frequency, ensuring steady pressure during readings. For the highest pressure setpoints (> 1 bar), an additional needle valve was used downstream of the emitters to create a load on the pump.

The test process begins by recording the temperature of the water and setting the pressure to 1 bar (14.5 psi). After approximately $1 - 2$ min of conditioning time, the emitter flow is collected in measurement cups for 120 seconds (using a mobile stopwatch, ± 1 sec of operator error). The water collected is then weighed on a scale (USS-DBS61-50, ± 0.01g) and recorded. This process is then repeated for the desired setpoints. The pressure setpoints were logarithmically spaced (8 points/decade) to ensure that multiple readings are made at low pressures and that 1 bar (as a nominal data point) is captured. The focus on low-pressure readings ensures that we sufficiently capture the quadratic relationship between flow rate and pressure. At the end of the test procedure, the end temperature was recorded. In general, we noted an increase of $10 - 15°C$ over the course of a full experiment. At each setpoint, the pressure typically varied between $\pm 10\%$ of the setpoint. This variability in pressure could be reduced by better tuning the control loop in future work. To capture statistical variance stemming from assembly errors and environmental changes (temperature), three complete sets of measurements were carried out with each labyrinth.

5 RESULTS

The five labyrinth geometries that were prototyped are identical to the geometry analyzed in figure 5, and only varied in tip gap. While major dimensions were set by manufacturing constraints (Table 1), it is pertinent to mention that all have the same number of teeth units (11), the same depth (1 mm), and most importantly have a length of only $L_{lab} = 9.5$ mm. This means that the labyrinths presented in the subsequent discussion are already 43.77% shorter than the Jain Turbo Excel NPC labyrinths, which were the shortest commercially available emitters reviewed.

Figure 8a presents the primary observations, in the form of flow rate (L/hr) against pressure (bar). The error bars represent pressure fluctuations (10% of setpoint) and the standard deviation in flow rate across three measurements. To begin with, it is clear that the labyrinths indeed possess a turbulent flow-pressure correlation, i.e., $Q \propto P^{1/2}$. This is further highlighted by how hydraulic resistance is nearly constant with pressure and Reynolds number (figure 8b-c), which aligns with prior research in the field [8, 18, 31]. The Reynolds number is defined as

$$Re = \frac{\rho U D_h}{\mu}, D_h = \frac{4A}{\mathcal{P}}, \qquad (10)$$

where D_h is the hydraulic diameter of the labyrinth, $A = h_{lab} \times b_{min}$ is the cross-sectional area of flow (see figure 4 for b_{min}),

Copyright © 2023 by ASME

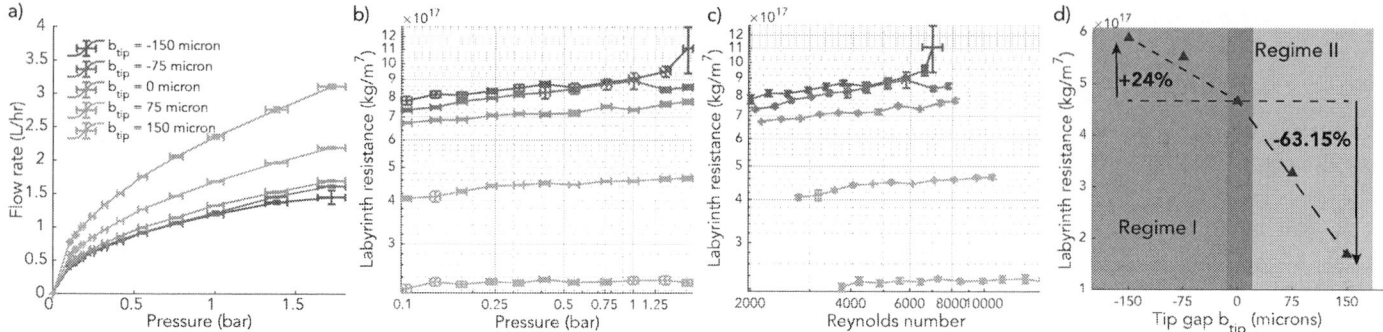

FIGURE 8: Results of the Design of Experiments. a) Flow rate-pressure trends follow the expected square root relationship. The hydraulic resistance at each pressure point was evaluated as $K_{lab} = P/Q(P)^2$ and plotted against b) pressure and c) Reynolds number. The pressure-averaged hydraulic resistance can then be plotted against the design parameter b_{tip}, indicating the existence of two regimes where the effect of changing b_{tip} varies significantly.

$\mathscr{P} = 2 \times (b_{min} + h_{lab})$ is the perimeter, and $U = Q/A$ is the average flow speed. With the lowest evaluated Reynolds number being ~ 2000, no 'onset' regime is observed in the study, i.e., the flow appears to be turbulent throughout the Reynolds number range examined.

Based on preliminary CFD analysis it was expected that labyrinths with $b_{tip} \leq 0$ would possess a more considerable hydraulic resistance than those with $b_{tip} > 0$. Observations in figure 8a validate this hypothesis. Labyrinths with tip $b_{tip} \leq 0$ consistently have a lower flow rate than those with $b_{tip} > 0$. However, the decrease in flow rate with b_{tip} is bounded and we observe a leveling effect. Between $b_{tip} = -150$ to $0\mu m$, flow-pressure contours seemingly collapse onto one another. This is contrasted by the flow-pressure trends for $b_{tip} = +75, +150\mu m$, which differ significantly. At 1 bar, the flow rate increases by 37% as b_{tip} increases from 75 to 150 μm. This is a powerful takeaway from a design perspective because it offers the designer a 'point of diminishing return' when it comes to using the tip gap to tune the hydraulic resistance for desired performance. Indeed b_{tip} appears to be a strong design lever when positive but may have reduced significantly after it becomes negative, i.e., once teeth overlap. This is better visualized using figure 8d which shows how the rate of increase in K_{lab} rapidly declines once $b_{tip} \approx 0$ or < 0 mm.

6 DISCUSSION

This work demonstrates that a careful analysis of geometry and manufacturing process considerations can help identify effective design levers that can subsequently be tuned to design compact labyrinths/emitters that perform on par with the current state of the art. The labyrinths designed by setting constrained geometric parameters to their limiting values (Table 1) and fitting $N_{units} = 11$ pairs within $L_{lab} = 9.5$ mm, behave as expected

($Q \propto P^{1/2}$) and perform hydraulically on par with commercial products.

It is expected that fitting a high resistance/low flow rate labyrinth into a compact footprint is challenging. For the subsequent discussion, the 0.8 L/hr Jain Turbo Excel NPC emitter is used as a benchmark. With the current configuration, the results indicate that simply tuning b_{tip} will not yield a labyrinth that provides a 0.8 L/hr flow rate as the Jain labyrinth does. The Jain labyrinth likely has resistance $K_{lab} = P/Q^2 = 2.025 \times 10^{18}$ kg/m^7. With $b_{tip} \leq 0$ mm, the flow rate at 1 bar is 1.38-1.52 L/hr. The highest pressure-averaged resistance observed in this study was $K_{lab}(b_{tip} = -150\mu m) = 5.887 \times 10^{17}$ kg/m^7. Although results indicate limited upside in further reducing b_{tip}, there is still room to tune N_{units} or h_{lab}. While it is known that $K_{lab} \propto N_{units}$, achieving the resistance of the 0.8 L/hr Jain NPC requires a more than threefold increase N_{units}, which is potentially infeasible in the current architecture. However, the depth of the Jain labyrinth is 0.4 mm, which is shallower than all the proposed labyrinths and likely has a large impact on hydraulic resistance. Therefore there is additional value in also exploring the synergistic effect of labyrinth depth, tip gap, and the number of repeating units. That will be a direction for the continuation of this study.

Nevertheless, this study has directly led to the creation of potential commercially viable compact labyrinths and by extension, non-pressure-compensating emitters. They are 43.77% shorter than the shortest comparable emitter at an equivalent flow rate (1.6 L/hr Jain NPC). Adopting such an emitter could potentially enable a significant increase in dripline wall thickness at no added raw material consumption. Direct wall thickness improvement could aid in resisting damage by animals and birds, as well as accidental user damage.

Secondly, the observations also shed light on the need for tight tolerances in the commercial manufacture of emitters. A change of b_{tip} from 75 μm to 150 μm ($\sim$0.003 in.) can cut

Copyright © 2023 by ASME

resistance by nearly half. For both NPC as well as PC emitters this would represent a significant change in flow-pressure performance and an extremely poor coefficient of manufacturing variability. As a result, the tolerance of this feature is likely to be <0.003 inches. This should serve as a useful guidepost for future academic work in this field.

Third, figure 8d highlights the existence of two possible flow regimes in the labyrinth on either side of $b_{tip} = 0$. For positive tip gaps, the relationship between K_{lab} and b_{tip} is marked by a strong negative slope, while that for negative tip gaps is marked by a much weaker trend. This observation is supported by the CFD analysis (figure 5), where dramatically different flow fields in the same labyrinth were observed when tip gap is reduced from $+150 \, \mu m$ to $-150 \, \mu m$. We propose the existence of two distinct regimes. In Regime I ($b_{tip} < 0$), the flow is forced to continually redirect itself. This leads to a shift in the turbulent energy dissipation hotspots towards the tooth wall and a reduction in the size and velocity of recirculation zones. In Regime II ($b_{tip} > 0$), the clear gap between teeth allows for a large amount of the flow to pass straight through the labyrinth with relatively lesser redirection between the teeth. This leads to the formation of relatively large, high-velocity recirculation zones. Linear fits yield the following piecewise expression (in kg/m^7)

$$K_{lab} = \begin{cases} (-8.127 b_{tip} \times 10^{-3} + 4.668) \times 10^{17}, & b_{tip} < 0 \\ (-19.78 b_{tip} \times 10^{-3} + 4.668) \times 10^{17}, & b_{tip} > 0. \end{cases} \quad (11)$$

7 CONCLUSIONS AND FUTURE WORK

The objective of this work is to contribute towards and establish the value of creating a deterministic design tool for emitter labyrinths. Such a tool could assist with the more rapid industrial design of emitters that are less expensive in terms of raw material expenses and potentially more viable for cost-constrained markets. This in turn could be a step towards wider adoption of drip irrigation. Further, we reveal the richness in flow phenomena and variability in hydraulic performance seen by varying only one commercially and physically significant design parameter (b_{tip}).

The experimental observations indicate that two distinct mechanistic regimes of hydraulic resistance exist with respect to tip gap and that only one (Regime II) may offer large performance tunability at the cost of tighter imposed tolerances. Most critically, this study shows that commercially viable emitters can be rapidly synthesized in a reliable, deterministic fashion by evaluating and integrating manufacturing topology constraints early in the design process. The labyrinths created in this study are up to 43.77% shorter than comparable commercial labyrinths.

Nevertheless, the designs proposed need not represent 'optimal' variants. Here the labyrinth length L_{lab} value was chosen intuitively, and this decision could be further analyzed. However, the review of design goals, geometry, and manufacturing

constraints leads to the downselection of just three design variables and is crucial since it reduces the order of the design space from more than 10 to just 3. This conclusion could help guide a more efficient study and could lead to better-informed data-driven generative models for emitters in the future.

Beyond the tip gap, the labyrinth depth and the number of tooth pair units are significant design levers as well. The effect of changing tooth unit count has already been explored by Sokol et al. [8] and Narain and Winter [31]. However, it could be valuable to next explore the effect of simultaneously changing two or more parameters to tune performance since the parametric effects may not be mutually exclusive.

Finally, the current study focuses on the relationship between selected geometry and hydraulic performance. However, emitter clogging represents a significant challenge for farmers that needs further exploration. A possible direction for the future could be to downselect commercially viable labyrinths developed using the proposed framework and then investigate clogging in them to understand better how geometry affects clogging. This will help manufacturers reduce raw plastic material consumption and costs, and also potentially improve clog resistance, helping create a better farmer experience.

ACKNOWLEDGMENT

The authors would like to acknowledge funding and technical support from the The Toro Company. A special thank you to Luis Niquet for valuable technical guidance and information on emitter design. Thank you to Dave Laybourn, Dr. Charles Schmid, and Jeff Vildibil for technical feedback and guidance on commercial emitter design and manufacturing. We would also like to thank Jeffery Costello for valuable assistance with setting up and tuning the experimental test stand, and Georgia Van de Zande for helpful guidance and perspectives.

REFERENCES

[1] S. Postel, P. Polak, F. Gonzales, and J. Keller, "Drip irrigation for small farmers: A new initiative to alleviate hunger and poverty," *Water International*, vol. 26, no. 1, pp. 3–13, 2001.

[2] A. Vidal, *Case studies on water conservation in the Mediterranean region.* No. 4, Food & Agriculture Org., 2001.

[3] N. Ibragimov, S. R. Evett, Y. Esanbekov, B. S. Kamilov, L. Mirzaev, and J. P. Lamers, "Water use efficiency of irrigated cotton in uzbekistan under drip and furrow irrigation," *Agricultural water management*, vol. 90, no. 1-2, pp. 112–120, 2007.

[4] N. Maisiri, A. Senzanje, J. Rockstrom, and S. Twomlow, "On farm evaluation of the effect of low cost drip irrigation on water and crop productivity compared to conven-

Copyright © 2023 by ASME

tional surface irrigation system," *Physics and Chemistry of the Earth, parts A/B/C*, vol. 30, no. 11-16, pp. 783–791, 2005.

[5] O. Cetin and L. Bilgel, "Effects of different irrigation methods on shedding and yield of cotton," *Agricultural Water Management*, vol. 54, no. 1, pp. 1–15, 2002.

[6] R. E. Namara, R. Nagar, and B. Upadhyay, "Economics, adoption determinants, and impacts of micro-irrigation technologies: empirical results from india," *Irrigation science*, vol. 25, no. 3, pp. 283–297, 2007.

[7] S. T. Hornum and S. Bolwig, "The growth of small-scale irrigation in kenya: The role of private firms in technology diffusion," 2020.

[8] J. Sokol, J. Narain, J. Costello, T. McLaurin, D. Kumar, and A. G. Winter, "Analytical model for predicting activation pressure and flow rate of pressure-compensating inline drip emitters and its use in low-pressure emitter design," *Irrigation Science*, vol. 40, no. 2, pp. 217–237, 2022.

[9] J. A. Sokol, *Parametric design and performance validation of low-cost, low-pressure drip emitters and irrigation systems*. PhD thesis, Massachusetts Institute of Technology, 2020.

[10] J. Sokol, S. Amrose, V. Nangia, S. Talozi, E. Brownell, G. Montanaro, K. Abu Naser, K. Bany Mustafa, A. Bahri, B. Bouazzama, *et al.*, "Energy reduction and uniformity of low-pressure online drip irrigation emitters in field tests," *Water*, vol. 11, no. 6, p. 1195, 2019.

[11] A. Ghodgaonkar. Private communication with Toro Ag, 2020–2022.

[12] "Aries npc technical documentation," tech. rep., Netafim, 2022 [Online].

[13] "Uniram pc technical documentation," tech. rep., Netafim USA, 2022 [Online].

[14] "Turbo cascade pc, pcas, and pcnl technical documentation," tech. rep., Jain, 2022 [Online].

[15] A. Ghodgaonkar and G. Van de Zande. Direct Farmer & Irrigation engineer Interactions in Kenya, Mar. 2022.

[16] H. K. Celik, D. Karayel, N. Caglayan, A. E. Rennie, and I. Akinci, "Rapid prototyping and flow simulation applications in design of agricultural irrigation equipment: case study for a sample in-line drip emitter: the paper is to study cfd and rp application samples on the design issues associated with agricultural irrigation equipment," *Virtual and Physical Prototyping*, vol. 6, no. 1, pp. 47–56, 2011.

[17] J. Feng, Y. Li, W. Wang, and S. Xue, "Effect of optimization forms of flow path on emitter hydraulic and anti-clogging performance in drip irrigation system," *Irrigation science*, vol. 36, pp. 37–47, 2018.

[18] J. Al-Muhammad, S. Tomas, and F. Anselmet, "Modeling a weak turbulent flow in a narrow and wavy channel: case of micro-irrigation," *Irrigation science*, vol. 34, no. 5, pp. 361–377, 2016.

[19] Y. Li, P. Yang, T. Xu, S. Ren, X. Lin, R. Wei, and H. Xu, "Cfd and digital particle tracking to assess flow characteristics in the labyrinth flow path of a drip irrigation emitter," *Irrigation Science*, vol. 26, pp. 427–438, 2008.

[20] L. Yu, N. Li, J. Long, X. Liu, and Q. Yang, "The mechanism of emitter clogging analyzed by cfd–dem simulation and ptv experiment," *Advances in Mechanical Engineering*, vol. 10, no. 1, p. 1687814017743025, 2018.

[21] J. Al-Muhammad, S. Tomas, N. Ait-Mouheb, M. Amielh, and F. Anselmet, "Experimental and numerical characterization of the vortex zones along a labyrinth milli-channel used in drip irrigation," *International Journal of Heat and Fluid Flow*, vol. 80, p. 108500, 2019.

[22] J. Zhang, W. Zhao, Y. Tang, and B. Lu, "Anti-clogging performance evaluation and parameterized design of emitters with labyrinth channels," *Computers and Electronics in Agriculture*, vol. 74, no. 1, pp. 59–65, 2010.

[23] J. Zhang, W. Zhao, and B. Lu, "Rapid prediction of hydraulic performance for emitters with labyrinth channels," *Journal of irrigation and drainage engineering*, vol. 139, no. 5, pp. 414–418, 2013.

[24] S. Movahedirad *et al.*, "Fluid micro-mixing in a passive microchannel: comparison of 2d and 3d numerical simulations," *International Journal of Heat and Mass Transfer*, vol. 139, pp. 907–916, 2019.

[25] A. Haghighinia and S. Movahedirad, "A tri-fluid tortuous microfluidic chip for green synthesis of nanoparticles and inactivation of a model gram-negative bacteria: Intracellular components evaluation," *Journal of Flow Chemistry*, vol. 12, no. 3, pp. 337–352, 2022.

[26] S. Hardt, K. Drese, V. Hessel, and F. Schö̈nfeld, "Passive micro mixers for applications in the micro reactor and μtas field," in *International conference on nanochannels, microchannels, and minichannels*, vol. 41642, pp. 45–55, 2004.

[27] J. Zhang, W. Li, M. Li, G. Alici, and N.-T. Nguyen, "Particle inertial focusing and its mechanism in a serpentine microchannel," *Microfluidics and nanofluidics*, vol. 17, pp. 305–316, 2014.

[28] D. Di Carlo, D. Irimia, R. G. Tompkins, and M. Toner, "Continuous inertial focusing, ordering, and separation of particles in microchannels," *Proceedings of the National Academy of Sciences*, vol. 104, no. 48, pp. 18892–18897, 2007.

[29] Z. Dai, D. F. Fletcher, and B. S. Haynes, "Impact of tortuous geometry on laminar flow heat transfer in microchannels," *International Journal of Heat and Mass Transfer*, vol. 83, pp. 382–398, 2015.

[30] S. K. Bhudolia, G. Gohel, K. F. Leong, and A. Islam, "Advances in ultrasonic welding of thermoplastic composites: a review," *Materials*, vol. 13, no. 6, p. 1284, 2020.

[31] J. Narain and A. G. Winter V, "A hybrid computational and

Copyright © 2023 by ASME

analytical model of inline drip emitters," *Journal of mechanical design*, vol. 141, no. 7, 2019.

[32] Ministry of Agriculture and Natural Resources, Ethiopia, *Guideline for Irrigation Agronomy*, April 2018.

[33] The Toro Company, *Toro Micro-Irrigation Owner's Manual*, 2011.

[34] Netafim, *Drip Irrigation handbook*, 2015.

APPENDIX A

CFD simulation in figure 5 was run on ANSYS Fluent 2022. The fluid volume of the labyrinth was extracted and meshed using tetrahedral elements. The minimum mesh size was set by considering the smallest desired feature resolution. The smallest feature was the tooth tip which had a radius of 0.064 mm. The mesh size was set to ensure at least five cells along the tip.

CFD consideration	Value
Surface mesh minimum size	3.8×10^{-6}m
Surface mesh maximum size	1.56×10^{-5}m
Maximum body mesh size	7.8×10^{-5}m
Mesh type	Tet
Boundary layers	5 (auto-generated)
Turbulence model	Standard $k - \varepsilon$
Boundary conditions	inlet pressure = 1 bar outlet pressure = atm.

Copyright © 2023 by ASME

Proceedings of the ASME 2023
International Design Engineering Technical Conferences and
Computers and Information in Engineering Conference
IDETC-CIE2023
August 20-23, 2023, Boston, Massachusetts

DETC2023-116704

HOW TO ENCODE MICROSTRUCTURE IN MACHINE LEARNING: A COMPARISON STUDY

Yulun Wu[1], Yumeng Li[1,*]

[1]University of Illinois at Urbana-Champaign, Champaign, IL

ABSTRACT

Accurately predicting the response of materials under different loading conditions is crucial for designing and developing new materials with desired properties. However, this process can be computationally expensive and challenging, especially for heterogeneous materials with complex microstructures. Recently, machine learning has been widely used to address the challenge for developing predictive models for various material systems with reduced reliance on extensive experimental testings and repetitive expensive physics simulations. The microstructure of a material plays a critical role in determining its properties, making it a key factor that needs to be accounted for in predictive modeling. Heterogeneous materials, specifically, often have complex microstructures with numerous features like pores, inclusions, and grain boundaries, which need to be accurately captured but is hard to be quantified for developing machine learning based predictive models. Therefore, accurate encoding of microstructural features is essential for making reliable predictions. Nevertheless, how to effectively and efficiently capture the complex microstructural features in developing machine learning based predictive models largely remains an open question for researchers in the field of materials science. In this paper, we present a comparison study of different encoding methods for microstructures in machine learning models. Specifically, we investigate pre-defined encoding methods and automatic encoding methods for a synthetic heterogeneous material system. the performance of each machine learning model is evaluated by predicting material responses such as strain energy. Our results show that convolutional neural networks (CNNs) have the ability to auto-encode the microstructure information of material and make promising prediction, especially when good pre-defined descriptors are not available. Overall, this study provides valuable insights into the performance of different encoding methods for microstructures in machine learning models, and can inform the development of more accurate and efficient models for materials science applications.

*Corresponding author: yumengl@illinois.edu
Documentation for asmeconf.cls: Version 1.32, May 14, 2023.

Keywords: material response, microstructure encoding, convolution neural networks, machine learning

1. INTRODUCTION

Machine learning (ML) models has gained widespread attention recently in the field of material science and engineering due to their ability to accurately and efficiently predict material properties [1–5]. Machine learning based predict models have shown great potential to accelerate materials discovery and design [6–9]. However, the effectiveness of machine learning models depends heavily on the quality and quantity of the data used for training, as well as the ability of the model to accurately represent the underlying physics and chemistry of the material [10]. In the context of materials science, one of the main challenges in developing high fidelity machine learning based predictive models is how to encode microstructure in a way that captures the critical and relevant features [11] for training. This challenge is particularly profound for heterogeneous materials [12]. Heterogeneous materials often have complex microstructures with numerous features, such as pores, inclusions, and grain boundaries, which needs to be captured and quantified to develop the machine learning models [13]. Capturing such complex features requires careful consideration of the modeling and encoding techniques used. Furthermore, the choice of the encoding method used often depends on the specific material being studied and the particular problem being addressed.

To address these challenges, various encoding techniques have been proposed, including descriptor-based methods that quantitatively describe the microstructure using pre-defined descriptors such as volume fraction and two-point correlation functions, and image-based techniques that directly use 2D or 3D images of microstructures as the input for machine learning model [14, 15].

Descriptor-based techniques leverage prior knowledge of microstructure of specific material systems to identify representative features that can impact the material behaviors. Among them, some descriptors are proposed through solid mechanics. Xu et al. proposed a descriptor-based methodology for designing and de-

Copyright © 2023 by ASME

veloping heterogeneous microstructural materials systems, which effectively guides the development of materials systems with desired properties [16]. Liu et al. presented a statistical descriptor-based volume-integral micromechanics model that can predict the effective mechanical properties of heterogeneous materials with arbitrary inclusion shapes [17]. Anand et al. used topological features such as persistent Ricci curvature and persistent homology to describe the crystal structure of halide perovskites improved the accuracy of property prediction models [18]. While others are utilizing the image-based features. Sarkar et al. presented an ensemble method using texture-based features and statistical moments to classify microstructures [19]. Naik et al. proposed a texture-based approach using co-occurrence matrices and statistical features to classify microstructures [20]. Hu et al. suggested a feature extraction method that combined gray-level co-occurrence matrices (GLCM) and SVM to classify different types of microstructures [21].

Descriptor-based method can be very effective at identifying important microstructural characteristics, however, it highly depends on prior knowledge of the material systems that may be not always available. As a result, convolutional neural networks (CNNs) are popular recently to serve as auto feature generators. For instance, Wang et al. proposed a deep learning framework that used a CNN to extract microstructural features and an ANN to predict the mechanical properties of composite materials [22]. Lambard et al. built a generative adversarial network (GAN) that could generate realistic microstructure samples for steel alloys [23]. Seyedebrahimi et al. developed a method that uses Fully Convolutional Neural Network (FCNN) accompanied by a max-voting scheme to achieve to classify different low carbon steel[24]. wu et al. introduced a deep learning-based method that uses a hybrid convolutional and recurrent neural network (CRNN) to extract features from microstructure images and predicts microstructural evolution during spinodal decomposition [25].

Despite the progress made, how to effeictively and accurately capture microstructure information in training dataset for developing machine learning based predictive models remains an open question, and there is no universal method that works for all types of materials. In this paper, we present a comparative study of different encoding methods that have been developed for representing microstructures in machine learning based models. Specifically, we adopted the Material MNIST dataset with hand-writing digit images [26] as representative volume element to represent heterogeneous material systems with an inclusion in the matrix material where the inclusion has a complex geometry. Support vector regression, CNN and FNN are adopted as the machine learning models to work with the predefined descriptor or auto-generated features. Machine learning models are developed to predict strain energy under specific loading conditions. The paper is organized as the following. In Section 2, two types of encoders, i.e. predefined descriptors and automatic encoder, in machine learning models for capturing micorstructure characteristics are illustrated. In Section 3, training dataset, training models and implementation details are introduced. In Section 4, machine learning models with different encodes are compared in terms of the training and testing errors, model size and com-

putational cost. In addition, Spearman correlation are used to evaluate the effectiveness of the features of microstructures used in each model. Our results show that CNNs can auto-encode the microstructure information of materials and make promising predictions. Overall, this study provides valuable insights into the performance of different encoding methods for microstructures in machine learning models and can inform the development of more accurate and efficient models for materials science applications.

2. MICROSTRUCTURE ENCODERS

The general workflow for predicting material responses using ML model is depicted in Fig. 1. An encoder is needed to digitalize the microstructure prior to training the ML model. Therefore, this section introduces two different encoding approaches. Section 2.1 covers the use of pre-defined descriptors as encoders, while Section 2.2 focuses on the use of image-based encoder for presenting the microstructure for training.

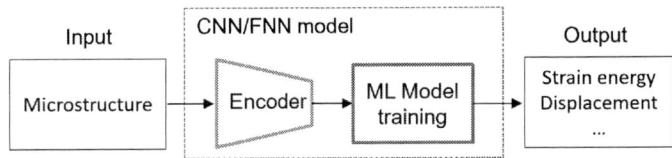

FIGURE 1: WORKFLOW OF PREDICTING MATERIAL'S RESPONSES USING MICROSTRUCTURE INFORMATION.

2.1 Pre-defined descriptor

In the field of materials science, accurately representing microstructure information is critical for developing predictive models, especially for heterogeneous materials with complex microstructures. One commonly used approach is to represent the microstructure using pre-defined descriptors to capture microstructural characteristics that are believed to impact the material properties based on pre-existing knowledge. These descriptors are used as input features for training machine learning models to develop predictive models of the structure-property relationships.

Heterogeneous materials contain multiple-phase materials as Fig. 2, where the weight rations of different phases, the dispersion and shape of each phase are some essential features to capture in developing constitutive material models. Volume fractions are usually used to represent the ration of each inclusion phase to the base material. For the dispersion, two-point correlation functions are used to quantify how well the inclusions are mixed in another material. For poly-crystal materials, grain size and orientation are critical microstructural characteristics, which are usually represented using statistical model like grain size distribution. In this study, we adopt two types of machine learning models, i.e. Support Vector Regression (SVR) and FNN, to work with pre-defined descriptors in order to evaluate how effective these pre-defined descriptors in representing microstructural characteristics in developing data-driven predictive models.

Copyright © 2023 by ASME

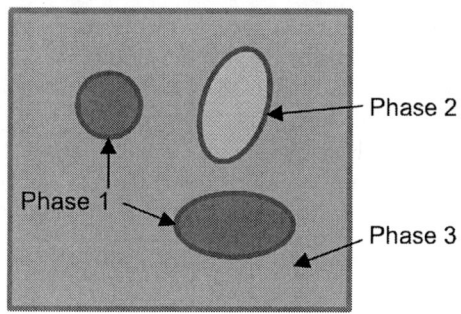

FIGURE 2: ILLUSTRATION OF HETEROGENEOUS MATERIALS.

2.2 Automatic encoder

Convolutional layers play a crucial role in the success of convolutional neural networks (CNNs) for the application of material analysis and design as they provide another promising route to generate feature maps for characterizing microstructures. The CNN can automatically generate features with the microstructure images as the inputs with minimum humane intervention, without relying on pre-defined descriptors. Each convolutional layer takes as input a set of feature maps from the previous layer and applies a convolution operation to produce a set of output feature maps. Mathematically, the output of a convolutional layer can be expressed as:

$$a_{i,j}^{(l)} = f\left(\sum_{i=-\infty}^{\infty} \sum_{j=-\infty}^{\infty} a_{i+u,j+v}^{(l-1)} \cdot k_{rot}(l)_j \cdot \chi(i,j) + b^{(l)} \right), \quad (1)$$

$$\chi(i,j) = \begin{cases} 1, & 0 \le i,j \le n \\ 0, & \text{others} \end{cases} \quad (1a)$$

where $a_{i,j}^{(l)}$ represents the output of the lth convolutional layer, k is an $n \times n$ convolutional kernel, b is the bias, and f is the activation function. The function $\chi(i,j)$ is a binary indicator function that is 1 for input pixels within the image boundary and 0 for those outside. Each output feature map can be considered as a set of weighted input feature maps from the previous layer, and thus serves as a set of learned features that capture relevant information for predicting the output.

The CNN thus acts as an automatic feature generator, extracting features that are highly relevant to the material's properties which are identified by training the weights and biases in the convolutional kernels. It is efficient but one critical drawback is the lack of physical meanings of the generated features based on the microstructure images.

In addition to CNN, FNN can capture microstructural features based on microstructure images. The input layer of FNN can read in the microstructure information represented in the microstructure images without reduction. The output of each fully connected layer can be mathematically represented as shown in Equation 2:

$$a^{(l)} = W^{(l)} \cdot a^{(l-1)} + b^{(l)} \quad (2)$$

Where $a^{(l)}$ represents the output of the lth fully connected layer, W is the weight matrix, and b is the bias term. During the training process, the weights and biases are adjusted to obtain appropriate features in reduced dimension, which can be used for predicting material properties. However, the microstructure is often stretched out into an array when using FNN as the encoder, causing the spatial information to be lost. This may result in a loss of important information that is relevant to the topology and can impact the accuracy of the model.

In this study, both FNN and CNN are used as automatic encoders and compared with pre-defined encoders.

3. DATASET, MODELS AND IMPLEMENTATION DETAILS

In this section, the Material MNIST dataset is introduced in section 3.1 which is used as training data. In section 3.2, all models are introduced. In section 3.2, the implementation details are shown.

3.1 Dataset introduction

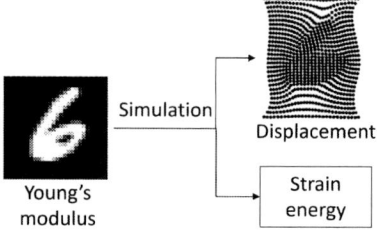

FIGURE 3: ILLUSTRATION OF DATASET FORMATION.

As shown in Fig. 3, to form the Material MNIST dataset as in [26], the hand-written digit images were used to present complex microstructure of heterogeneous materials and were converted to corresponding heterogeneous material blocks using material properties. The Young's modulus of the heterogeneous material block is calculated based on its grayscale bitmap, and the relationship is:

$$E = \frac{b}{255.0}(100.0 - 1.0) + 1.0 \quad (3)$$

Where the E stands for Young's modulus matrix and the b stands for the grayscale bitmap. The range for the grayscale is 0 - 255 so that the range for the transformed Young's modulus is 1 - 100 MPa.

The inputs mapping the heterogeneous material block using corresponding Young's modulus can contain both microstructure details as well as individual constituent's material properties. The compressible Neo-Hookean material model is used for all material blocks to calculate the strain energy and displacement fields under certain boundary conditions as in [27].

$$\psi = \frac{1}{2}\mu[\mathbf{F} : \mathbf{F} - 3 - 2\ln(\det bfF)] + \\ \frac{1}{2}\lambda\left[\frac{1}{2}\left((\det F)^2 - 1\right) - \ln(\det F)\right] \quad (4)$$

$$E = \frac{\mu(3\lambda + 2\mu)}{\lambda + \mu} \quad (5)$$

where ψ is strain energy, F is the deformation gradient, and μ and λ are Lamé parameters for elastic material properties. All strain

Copyright © 2023 by ASME

energies are obtained through finite element simulation using the FEniCS computing platform [28, 29].

The original MNIST dataset contains 10 kinds of hand-written digit images from 0-9. Each image containing the same digit looks also very different. As a result, this dataset includes plenty of different material structures in the training dataset and provides an excellent case for us to test the generality of the CNN model. One task for this dataset is to predict strain energy based on material microstructures. This task is adapted to verify the efficiency of our model. There are four different loading cases in the dataset and a dataset of uniaxial extension load case are used in this research. Several different strains, from 0.00004 to 0.5, are applied to the top edge of all samples for simulation, and the results of the largest strain, 0.5, are used in this research. It has 60000 samples in the training set and 10000 samples in the testing set for the uniaxial extension load case. Those data are sufficient for model training.

3.2 Pre-defined descriptors definition

As mentioned earlier, a hand-written digit comprises materials with varying Young's modulus, ranging from 1 to 100 MPa, and can be approximated as a three-phase material. These phases include the inside area, the edge area, and the background. The inside area has a Young's modulus close to 100 MPa, while the background has a Young's modulus close to 1 MPa. The Young's modulus of the edge area lies in between. The areas of these regions are calculated by counting the total number of pixels in the original image. Young's modulus of the edge area are varying compared with inside area and background. The total Young's modulus for edge area is also calculated as a pre-defined descriptor. The inside and edge areas are combined to obtain the total area of the digit. To capture finer shape information, the minimum, maximum, mean, and standard deviation of each row and column of the digital area are calculated, along with the width and height of the digit. These shape descriptors provide a comprehensive characterization of the geometric properties of the heterogeneous material represented by the hand-written digit image. For each sample, a total of 14 pre-defined descriptors are extracted from the initial Young's modulus matrix as shown in Table. 1, providing information about the microstructure. These descriptors serve as input features to train machine learning models.

TABLE 1: SUMMARY OF ALL 14 PREDEFINED DISCRIPTORS

Category	Index	Features
Area	1	Inside area,
	2	Edge area,
	3	Background area,
	4	Edge total Young's Module
Shape	5	Height,
	6	Width,
	7-10	Maximum/ minimum/ average/ standard deviation of rows
	11-14	Maximum/ minimum/ average/ standard deviation of columns

3.3 Models design

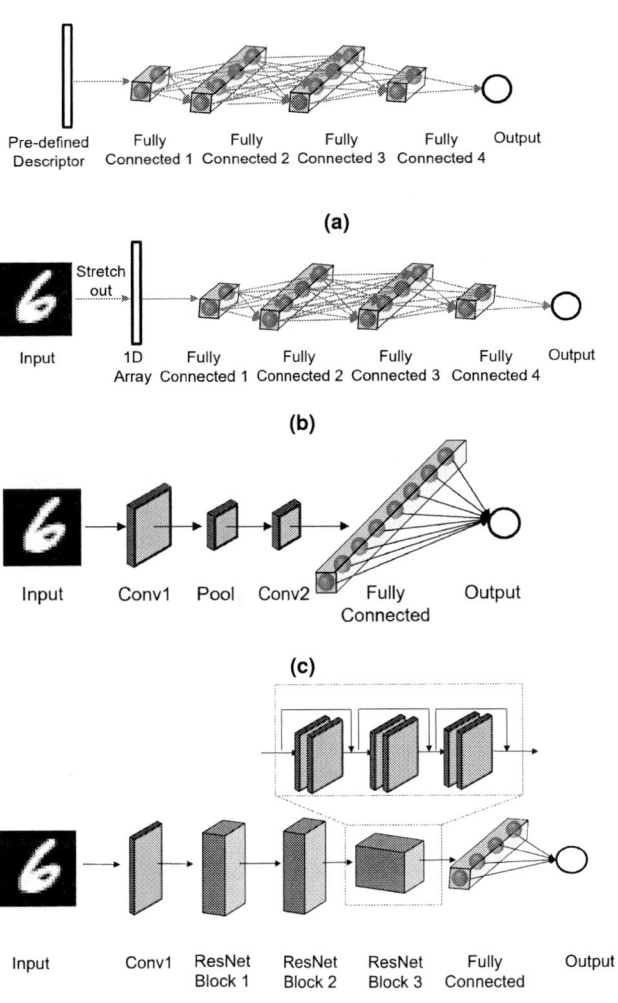

FIGURE 4: STRUCTURES OF (A) FNN USING PRE-DEFINED DE-SCRIPTOR, (B) FNN USING ORIGINAL MICROSTRUCTURE, (C) CNN MODEL AND (D) RESNET20 (DEEPER CNN).

Figure 4 illustrates the architectures of the FNN and CNN models to be used with pre-defined microstructural descriptors and automatic microstructure encoders in this study. For FNN with pre-defined descriptors, features in Table 1 are used as the input. For FNN and CNN using automatic encoders, the material microstructure property matrix as described in section 3.1 serves as the input, with the strain energy being the output. The CNN and FNN architectures are directly adopted from [26], while the ResNet20 is a CNN with deeper layers. The shortcut structure used in the ResNet blocks is borrowed from [30] to prevent the CNN with deep layers from suffering from the gradient vanishing issue. Each layer is represented by a different color. It should be noted that the input is stretched out before being fed into the FNN, resulting in the loss of spatial information regarding the microstructure.

To train all CNN and FNN models, a 24 GB NVIDIA RTX A5000 is used. The Pytorch package is used to implement all CNN and FNN models. The proposed CNN has 2.0 Mb train-

Copyright © 2023 by ASME

able variables and is trained for 100 epochs during each training process. The training batch size is set to 64, while the testing batch size is 1000. A fixed learning rate of 0.0005 is applied to all models, and they are trained from scratch. The activating function for each layer is the leaky ReLU function. The batch normalization layer is used after each convolutional layer. In all case studies, the original training and testing splitter is followed, with 50000 training samples and 10000 testing samples in the Material MNIST dataset.

4. RESULTS AND DISCUSSION

In this section, machine learning models are trained to compare different encoding methods. In section 4.1, pre-defined encoder is adapted, and SVR and FNN are used to demonstrate the pre-defined encoder. In section 4.2, FNN and CNN are trained to demonstrate the automatic encoders. In section 4.3, model size and computational cost will be further compared and discussed.

4.1 Predicting material's response using pre-defined encoder

TABLE 2: PREDICTING STRAIN ENERGY RESULTS OF BENCH-MARK CNN AND FNN MODELS

	Features	Train MAE	Test MAE
SVR	Area	8.9	7.6
	Area + Shape	8.5	7.6
FNN	Area	7.3	7.5
	Area + Shape	7.5	6.9

Two ML models, i.e. SVR and FNN, are adapted to evaluate the predictive accuracy of strain energy using pre-defined descriptors. The results are presented in Table 2. From the table, two observations can be drawn. Firstly, relative to SVR, the FNN model demonstrates a superior ability to establish a relationship between pre-defined descriptors and the final total strain energy of the heterogeneous material. This is evident from its lower Mean Absolute Error (MAE) for both training and testing errors, regardless of the features used, when compared to the SVR model. The FNN model attains the lowest testing MAE of 6.9 when utilizing all pre-defined descriptors as input features. Secondly, the study highlights the difficulty in identifying a relationship between shape features and strain energy. The addition of shape features in the SVR model results in only a minor improvement in error. Even with the FNN model, the enhancement in testing MAE is not significant. Therefore, it is recommended to employ more powerful pre-defined descriptors to enhance the accuracy of both SVR and FNN models. The correlation of all pre-defined descriptors also agrees with the conclusion in Fig.5. All absolute value of Spearman correlation values between pre-defined descriptors and strain energy are less than 0.75.

4.2 Predicting material's response using automatic encoder

The architectures of FNN and CNN are adopted from [26], which are shown in Fig.4b and 4c. Based on the results presented in Table 3, three important conclusions can be drawn. Firstly,

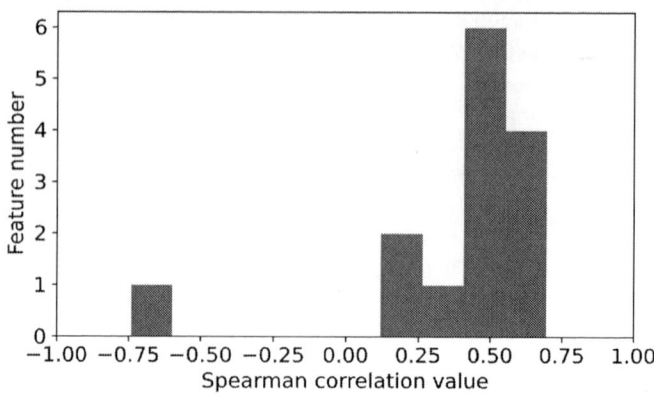

FIGURE 5: HISTOGRAM OF SPEARMAN CORRELATION COEFFICIENTS BETWEEN PRE-DEFINED DESCRIPTORS AND STRAIN ENERGY.

TABLE 3: COMPARISON USING PREDEFINED DESCRIPTORS AND AUTOMATIC ENCODER

	Features	Train MAE	Test MAE
FNN	Area + Shape	7.5	6.9
FNN	Microstructure	1.1	2.1
CNN	Microstructure	1.0	1.3
ResNet20	Microstructure	0.5	1.0

the fact that using FNN as an automatic encoder yields significantly lower MAE than using pre-defined descriptors indicates that microstructure images provide much more valuable information than pre-defined descriptors alone. Secondly, utilizing CNN as an automatic encoder can further reduce the testing MAE by nearly 40%, from 2.1 to 1.3, suggesting that spatial information is a critical factor in material response prediction. As discussed in the introduction, the primary difference between these two automatic encoders is their ability to maintain spatial information. Finally, the accuracy of the predictions can be further improved by tuning the CNN model structure. The ResNet structure is utilized to create a deeper model without the vanishing gradient problem. By increasing the layer from 2 to 20, the ResNet20 model, as shown in Fig.4d, can further reduce the testing MAE to 1.0 and achieve the lowest testing error among all models. The Spearman correlation coefficients between the automatic features generated by the ResNet20 model, which performed the best, and strain energy are also calculated as shown in Fig.6. In the fully connected layer, 64 features are generated, with more than half of them showing a high correlation to the final strain energy. This provides further evidence that the automatic encoder is capable of extracting useful information from the microstructure.

4.3 Model size, computational cost and convergency

Table 4 presents the computational costs and space occupation of the CNN models, which vary depending on their model sizes and structures. Among the models evaluated, ResNet20 achieved the best performance, with a training time of approximately 1 hour, including testing time. The test time for all samples in the testing set was around 10 seconds, and the model's size is

Copyright © 2023 by ASME

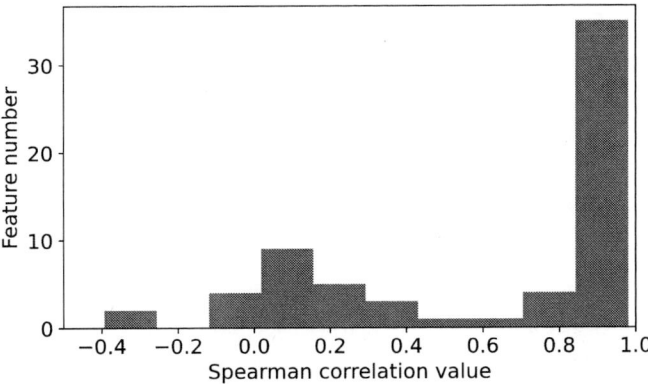

FIGURE 6: HISTOGRAM OF SPEARMAN CORRELATION CO-EFFICIENTS BETWEEN AUTOMATIC GENERATE FEATURES BY RESNET20 AND STRAIN ENERGY.

TABLE 4: MODEL SIZES AND COMPUTATIONAL COSTS FOR DIFFERENT AUTOMATIC ENCODER MODELS

Model	Train time	Test time (10000 samples)	Model size
CNN	20 mins	8 s	1.7 Mb
FNN	20 mins	8 s	64 Mb
ResNet20	1 hour	10 s	1.1 Mb

relatively small compared to other CNN models, making it highly accessible for low computational resources. Despite being more computationally expensive than the original CNN models during the training process, the proposed ResNet20 model can be conveniently trained with fewer resources, indicating its excellent accessibility. Additionally, the testing time for all models is almost the same. When testing a batch of 1000 samples simultaneously, the testing time for one sample could be much less than 1 second. This suggests that the CNN can predict strain energy in real-time performance once trained. Figure 7 displays the learning curves of the three models. The plot demonstrates that all models have reached convergence after being trained.

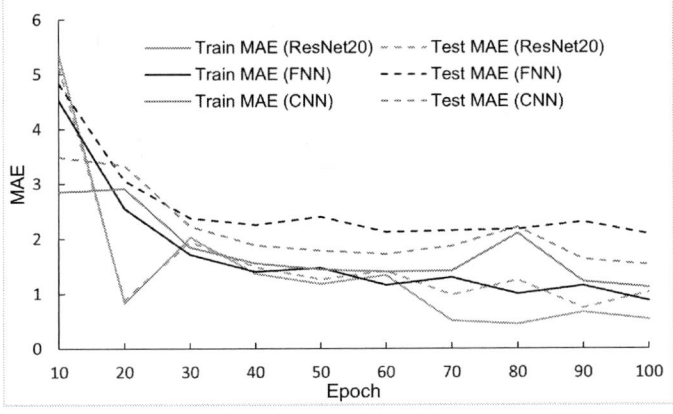

FIGURE 7: TRAINING AND TESTING MAE FOR THREE MODELS.

5. CONCLUSION

In this study, we have conducted a comparative analysis of multiple encoding methods for microstructures using the Mechanical MNIST dataset as a case study. The hand written images in MNIST dataset are used as representative volumne element to represent the microstructures of a heteregenous material system, which contains a inclusion with a complex geometry in the matrix materials. Based on our findings, we have drawn several key conclusions. Firstly, we found that area and shape are two critical aspects that must be considered when predicting material response. While area is relatively easy to quantify, defining a good shape descriptor can be challenging. Secondly, in cases where no suitable pre-defined descriptors are available, we recommend using an automatic encoder to encode microstructure information in developing machine learning based predictive models. This approach has proven to be more effective in capturing essential features of microstructures to leverage machine learning in establishing structure-property relationships. Lastly, we have found that spatial information is also vital for predicting material responses, and the model's structure must be carefully designed to account for this information. Therefore, we recommend that researchers give careful consideration to the structure of their models when encoding microstructure information.

Leveraging the advance in 3D printing techniques, hybrid material systems created through combining various material constituents have the great potential to enable the design of novel materials with multiple functions and optimal properties. To enable the design, the structure-property relationships are much-needed to be developed and represented for the hybrid material systems. Machine learning models are quite powerful to recognize the underlying relationships like the constitutive material models of the hybrid material systems. However, it is critical to adopt an effective method to represent the microstructures of hybrid materials in machine learning based models. This can be quite powerful to enable bio-inspired material design where complex microstructure can be observed. In addition, such topic is also important to use machine learning models to investigate the material behaviors of sophisticated biomaterials. This work is expected to provide some insights on how microstructure can be informed in machine learning models. In the future study, more cases with different microstructures should be involved in the comparative study.

ACKNOWLEDGMENTS

The authors thank the support of University of Illinois at Urbana-Champaign.

REFERENCES

[1] Butler, Keith T, Davies, Daniel W, Cartwright, Hugh, Isayev, Olexandr and Walsh, Aron. "Machine learning for molecular and materials science." *Nature* Vol. 559 No. 7715 (2018): pp. 547–555.

[2] Raccuglia, Paul, Elbert, Katherine C, Adler, Philip D F, Falk, Casey, Wenny, Malia B, Mollo, Aurelio, Zeller, Matthias, Friedler, Sorelle A, Schrier, Joshua, Norquist, Alexander J et al. "Machine-learning-assisted materials

Copyright © 2023 by ASME

discovery using failed experiments." *Nature* Vol. 533 No. 7601 (2016): pp. 73–76.

[3] Chen, Xin, Zhou, Haofei and Li, Yumeng. "Effective design space exploration of gradient nanostructured materials using active learning based surrogate models." *Materials & Design* Vol. 183 (2019): p. 108085.

[4] Yong, Sihan, Zheng, Zhuoyuan, Wang, Pingfeng and Li, Yumeng. "Machine learning assisted design for active cathode materials." *ASME International Mechanical Engineering Congress and Exposition*, Vol. 84508: p. V003T03A007. 2020. American Society of Mechanical Engineers.

[5] Bansal, Parth, Zheng, Zhuoyuan, Shao, Chenhui, Li, Jingjing, Banu, Mihaela, Carlson, Blair E and Li, Yumeng. "Physics-informed machine learning assisted uncertainty quantification for the corrosion of dissimilar material joints." *Reliability Engineering & System Safety* Vol. 227 (2022): p. 108711.

[6] Ramprasad, Rampi, Batra, Rohit, Pilania, Ghanshyam, Mannodi-Kanakkithodi, Arun and Kim, Chiho. "Machine learning in materials informatics: recent applications and prospects." *npj Computational Materials* Vol. 3 No. 1 (2017): p. 54.

[7] Singh, Akash and Li, Yumeng. "Machine Learning Potentials for Graphene." *ASME International Mechanical Engineering Congress and Exposition*, Vol. 86656: p. V003T03A036. 2022. American Society of Mechanical Engineers.

[8] Li, Yumeng, Xiao, Weirong and Wang, Pingfeng. "Uncertainty quantification of artificial neural network based machine learning potentials." *ASME International Mechanical Engineering Congress and Exposition*, Vol. 52170: p. V012T11A030. 2018. American Society of Mechanical Engineers.

[9] Bansal, Parth, Zheng, Zhuoyuan and Li, Yumeng. "Uncertainty Quantification on Galvanic Corrosion Based on Adaptive Surrogate Modeling." *ASME International Mechanical Engineering Congress and Exposition*, Vol. 86717: p. V009T12A005. 2022. American Society of Mechanical Engineers.

[10] Mai, Haoxin, Le, Tu C, Chen, Dehong, Winkler, David A and Caruso, Rachel A. "Machine learning for electrocatalyst and photocatalyst design and discovery." *Chemical Reviews* Vol. 122 No. 16 (2022): pp. 13478–13515.

[11] Ward, Logan, Agrawal, Ankit, Choudhary, Alok and Wolverton, Christopher. "A general-purpose machine learning framework for predicting properties of inorganic materials." *npj Computational Materials* Vol. 2 No. 1 (2016): pp. 1–7.

[12] You, Huaiqian, Zhang, Quinn, Ross, Colton J, Lee, Chung-Hao and Yu, Yue. "Learning deep implicit fourier neural operators (IFNOs) with applications to heterogeneous material modeling." *Computer Methods in Applied Mechanics and Engineering* Vol. 398 (2022): p. 115296.

[13] Fu, Jinlong, Cui, Shaoqing, Cen, Song and Li, Chenfeng. "Statistical characterization and reconstruction of heterogeneous microstructures using deep neural network." *Computer Methods in Applied Mechanics and Engineering* Vol. 373 (2021): p. 113516.

[14] Ghosh, Srimoyee, Bhowmik, Arghya and Kumbhakar, Pathik. "Machine learning in materials informatics: Recent applications and prospects." *Journal of Materials Research* Vol. 34 No. 20 (2019): pp. 3451–3465.

[15] Xue, Dezhen, Balachandran, Prasanna V., Hogden, John and Theiler, James. "Machine learning and materials informatics: Recent progress and prospects." *Advanced Materials* Vol. 32 No. 48 (2020): p. 2004339.

[16] Xu, Hongyi, Li, Yang, Brinson, Catherine and Chen, Wei. "A descriptor-based design methodology for developing heterogeneous microstructural materials system." *Journal of Mechanical Design* Vol. 136 No. 5 (2014): p. 051007.

[17] Liu, Zeliang, Moore, John A, Aldousari, Saad M, Hedia, Hassan S, Asiri, Saeed A and Liu, Wing Kam. "A statistical descriptor based volume-integral micromechanics model of heterogeneous material with arbitrary inclusion shape." *Computational Mechanics* Vol. 55 (2015): pp. 963–981.

[18] Anand, D Vijay, Xu, Qiang, Wee, JunJie, Xia, Kelin and Sum, Tze Chien. "Topological feature engineering for machine learning based halide perovskite materials design." *npj Computational Materials* Vol. 8 No. 1 (2022): p. 203.

[19] Sarkar, Shib Sankar, Sheikh, Khalid Hassan, Mahanty, Arpan, Mali, Kalyani, Ghosh, Aniruddha and Sarkar, Ram. "A harmony search-based wrapper-filter feature selection approach for microstructural image classification." *Integrating Materials and Manufacturing Innovation* Vol. 10 (2021): pp. 1–19.

[20] Naik, Dayakar L, Sajid, Hizb Ullah and Kiran, Ravi. "Texture-based metallurgical phase identification in structural steels: a supervised machine learning approach." *Metals* Vol. 9 No. 5 (2019): p. 546.

[21] Hu, FK, Zhu, ZJ, Wang, K, Zhu, B and Zhang, YS. "Identification of hot stamping fully martenstic microstructure SEM photograph with support vector machine." *Advanced High Strength Steel and Press Hardening: Proceedings of the 4th International Conference on Advanced High Strength Steel and Press Hardening (ICHSU2018)*: pp. 272–278. 2019. World Scientific.

[22] Wang, Ting, Chen, Chao and Yang, Liu. "Deep learning for microstructure analysis and materials design: A review." *Advanced Engineering Materials* Vol. 20 No. 11 (2018): p. 1800255.

[23] Lambard, Guillaume, Yamazaki, Kazuhiko and Demura, Masahiko. "Generation of highly realistic microstructural images of alloys from limited data with a style-based generative adversarial network." *Scientific Reports* Vol. 13 No. 1 (2023): p. 566.

[24] Azimi, Seyed Majid, Britz, Dominik, Engstler, Michael, Fritz, Mario and Mücklich, Frank. "Advanced steel microstructural classification by deep learning methods." *Scientific reports* Vol. 8 No. 1 (2018): p. 2128.

[25] Wu, Peichen, Iquebal, Ashif Sikandar and Ankit, Kumar. "Emulating microstructural evolution during spinodal decomposition using a tensor decomposed convolutional and

recurrent neural network." *Computational Materials Science* Vol. 224 (2023): p. 112187.

[26] Lejeune, Emma. "Mechanical MNIST: A benchmark dataset for mechanical metamodels." *Extreme Mechanics Letters* Vol. 36 (2020): p. 100659.

[27] Kirby, R. C. "Algorithm 839: FIAT, a New Paradigm for Computing Finite Element Basis Functions." *ACM Transactions on Mathematical Software* Vol. 30 (2004): pp. 502–516. DOI 10.1145/1039813.1039820.

[28] Alnæs, Martin, Blechta, Jan, Hake, Johan, Johansson, Anders, Kehlet, Benjamin, Logg, Anders, Richardson, Chris, Ring, Johannes, Rognes, Marie E. and Wells, Garth N. "The FEniCS project version 1.5." *Archive of Numerical Software* Vol. 3 No. 100 (2015): pp. 9–23.

[29] Logg, Anders, Mardal, Kent-Andre and Wells, Garth N. *Automated Solution of Differential Equations by the Finite Element Method: The FEniCS Book.* Vol. 84 of *Lecture Notes in Computational Science and Engineering.* Springer (2012).

[30] Targ, Sasha, Almeida, Diogo and Lyman, Kevin. "Resnet in resnet: Generalizing residual architectures." *arXiv preprint arXiv:1603.08029* (2016).

Copyright © 2023 by ASME

Proceedings of the ASME 2023
International Design Engineering Technical Conferences and
Computers and Information in Engineering Conference
IDETC-CIE2023
August 20-23, 2023, Boston, Massachusetts

DETC2023-109993

CONSTRAINED BAYESIAN OPTIMIZATION METHODS USING REGRESSION AND CLASSIFICATION GAUSSIAN PROCESSES AS CONSTRAINTS

Cole Jetton, Chengda Li, Christopher Hoyle
Design Engineering Lab
Oregon State University
Corvallis, Oregon, 97331

ABSTRACT

In Bayesian optimization, which is used for optimizing computationally expensive black-box problems, constraints need to be considered to arrive at solutions that are both optimal and feasible. Techniques to deal with black-box constraints have been studied extensively, often exploiting the constraint model to sample points that have a high probability of being feasible. These constraints are usually modeled using regression models if the output is a known value on a continuous real scale. However, if the constraint or set of constraints is not modeled explicitly, but rather it is only known whether a design vector is either feasible or infeasible, then we treat the constraints as a categorical constraint. Because of this, these constraints should be modeled using classification models rather than regression methods, which have also been studied. Because of the variety of approaches to handling constraints, there is a need to compare methods for handling both classification constraints as well as continuous constraints modeled with individual regression models. This paper explores and compares four main methods, with two additional ones specifically for classification constraints; these methods handle black-box constraints in Bayesian optimization by modeling the constraints with both regression and classification Gaussian processes. We found that the problem type and number of constraints influence the effectiveness of different approaches with statistical differences in convergence. Regression models can be advantageous in terms of model fit time; however, this is also a function of the total number of constraints. Generally, regression constraints outperformed classification surrogates in terms of minimizing computational time, but the latter was still effective. Overall, this study provides valuable insights into the performance and implementation of different constrained Bayesian optimization techniques. This can help inform engineers on which method is most suitable for their problem, what issues they may encounter during implementation, and give a general understanding of the differences between using regression and classification Gaussian processes as constraints.

Keywords: Algorithm Comparison, Bayesian Optimization, Constraint Handling, Gaussian Process Classification, Gaussian Process Regression, Optimization.

NOMENCLATURE

$\mathcal{C}$	Class of Data
$f(x)$	Objective Function
f^*	Minimum of the Objective Function
f_{min}	Current Minimum of Objective Function
$g(x)$	Inequality Constraint in Negative Null Form
$\mathcal{GP}$	Gaussian Process (GP)
$k_{SE}(x,x')$	Squared Exponential Covariance Function for a GP
$m(x)$	Mean Function for a GP
$u_s(x)$	Surrogate Model of Function u (subscript s)
$\hat{u}(x)$	Mean of Surrogate Model of Function u
x	Independent Variable
x^*	Independent Variable Leading to f^*
$\sigma_u(x)$	Standard Deviation of Surrogate Function $u_s(x)$
ρ	Penalty Value for Penalty Function
$EI(x)$	Expected Improvement Function
$PF(x)$	Probability of Feasibility Function
$p(z\|x)$	Probability of outcome z given x

1 INTRODUCTION

Bayesian optimization has grown in popularity in recent years as an effective and efficient approach to global optimization of arbitrary functions. Especially within the field of engineering, optimization helps minimize the mass of structures, reduce costs, and ensure safe designs [1]. This is generally written in negative null form as described below in Equation 1. The goal is to minimize some objective function, $f(x)$, while satisfying a set of constraints, $c_i(x)$, to arrive at a feasible solution.

Copyright © 2023 by ASME

$$min\ f(x)$$
$$w.r.t.\ \ x \tag{1}$$
$$s.t.$$
$$c_i(x) \leq 0\ for\ i = 1,2,..n$$

If the objective and constraint functions are analytic, or quick to compute, it is relatively easy to use gradient-based methods, either directly or via estimations, to arrive at the optimum. However, as engineering problems have become increasingly complex and computationally intensive, researchers and practitioners have explored Bayesian Optimization as a viable approach to these complex problems [2].

Bayesian optimization is a helpful tool for optimizing so-called "black-box" functions, or functions that are only known in terms of their input and output, since it simultaneously constructs and optimizes a surrogate model by using an acquisition function to find new, and potentially better, sample points [3,4]. Bayesian optimization is commonly used in engineering research and has been effectively used in material design, fluid mechanics, and improving machine learning algorithms [5–8]. For Bayesian optimization to work within the engineering domain, it must *incorporate constraints* to ensure that the solution is feasible. If the constraint is also a black-box function, that must be considered when implementing Bayesian optimization by creating another surrogate model (of the constraint) and using that to guide the optimization process. There are many methods for dealing with these types of constraints: many of them focus on finding sample points that have a high probability of the constraints being satisfied [9–15].

Generally, these constraints have continuous values, such as stress, cost, or temperature, which have their respective limit states based on material or economic constraints. However, sometimes the only information known is whether the constraint is met/not met, and thus it is categorical. These could include feedback from external sources regarding machinability, cost quotas, or possible simulation failures. In the absence of a continuous value, a classification model will best represent these constraints, and they have been extensively studied for their application in engineering design [16]. This research focuses on *binary classification* models as they can represent if a constraint is met/not met; this works well in Bayesian optimization and has been applied to engineering problems [17,18].

These differences in constraint response lead to two types of black-box constraints: regression models when the value is continuous, and classification models when there is a binary response. However, there needs to be a clearer understanding of how each model behaves throughout the optimization process to conclude if one is truly better than the other. This research does not aim to compare all known methods. Rather, it aims to help inform researchers on what to expect when implementing constrained Bayesian optimization, when to use certain types of surrogates for their constraints, and how the shape of the underlying objective and constraint functions could affect the outcome.

2   BACKGROUND

This section introduces the Gaussian Process surrogate models used in this research along with the Bayesian optimization. It highlights the general methods for handling constraints, while Section 3 discusses the specific methods explored in this research.

2.1   Gaussian Process Surrogate Models

A key component of Bayesian optimization is the reliance on a surrogate model, which is a mathematical representation of an outcome of a function or process [2]. This can be modeled using both parametric and nonparametric models. A parametric model assumes the shape of a function and minimizes the error between the training points (x', y') and continuous variables (x, y) by altering a set of parameters to find that predefined shape. For example, a linear regression model, shown in Equation 2, assumes a slope and intercept and changes the parameter β and minimizes the error ϵ for each independent variable to fit the outcome to a hyperplane.

$$y = x^T \beta + \epsilon \tag{2}$$

This contrasts which non-parametric models, such as a Gaussian Process, which do not assume the shape of the response data. A Gaussian Process, often shortened to GP, is a general non-parametric model which can be applied to both regression and classification models [19]. This is represented in Equation 3 below, where the surrogate model $f_s(x)$ is distributed according to a Gaussian Process model. This gives it both a predicted mean value, $\hat{f}(x)$, and a standard deviation, $\sigma_{f(x)}$, as a function of a continuous independent variable x. Since it can predict both the mean and error at any point it is an excellent surrogate model for Bayesian optimization, with both regression and classification surrogates being applied to engineering problems as both objective functions and constraints [5,17,18].

$$f_s(x) \sim \mathcal{GP}\big(m(x), k(x, x')\big) \tag{3}$$

Two key components of Gaussian Processes are the use of a mean function $m(x)$ and covariance function $k(x, x')$, which defines the mean of the GP and the relationship between the training data x' and the new continuous value x. There are many mean and covariance functions to choose from, many of which can be combined, and can be chosen based on an underlying knowledge of what the function should look like [19,20]. A common mean function is the constant mean function, shown in Equation 4, where the mean of the model is a constant instead of an expected slope. The squared exponential covariance function shown in Equation 5, is a common covariance function that describes the relationship between the training points and new independent variable via their squared difference.

$$m(x) = C \tag{4}$$

Copyright © 2023 by ASME

$$k_{SE}(x, x') = \sigma_f exp\left(-\sum_{i=1}^{n} \frac{(x_i - x_i')^2}{2l_i^2}\right) \qquad (5)$$

Each has its own hyperparameters that need to be learned to fit the model to the data, the most common of which is to maximize the marginal likelihood that the model fits the data points [19]. The process involves fitting a latent model before transforming it to find a posterior model to predict the values along a continuous response surface based on the independent variables as in the case of all regression models.

Surrogate models can be further broken down into either regression or classification models [19]. A regression model means the output is a continuous variable as a function of its independent variables, such as stress or temperature. A classification model means the output belongs to a category, such as a constraint being met or not met. In this case, a classification Gaussian Process can be used. These rely on the same process of regression models but are expanded to predict the *probability* that the independent variable belongs to a class [19,21]. It transforms the latent model to predict the probability that the outcome belongs to a class $\mathcal{C}$ given a continuous independent variable x by passing the surrogate model through a sigmoid function S, shown in Equation 6. This is done with either a logistic or Gaussian cumulative distribution function, respectively, known as the logit and probit models [22].

$$p(y = \mathcal{C}|x) = S\big(f_s(x)\big) \qquad (6)$$

In the case of binary classification, or when there are only two classes (such as yes or no, often trained on $x \in \{-1,1\}$) the training data is generally used to predict $p(y = 1|x)$. There are many ways to implement Classification GPs, each with its own advantages and disadvantages, with a big disadvantage being computation time which makes them difficult to implement, especially on large-scale data sets [21].

2.2 Bayesian Optimization

Bayesian optimization is a black-box optimization algorithm for finding the optimum of the function whose expression is unknown [2,4]. These types of functions are typically computationally expensive and Bayesian optimization allows for fewer function computations by building a surrogate model of the black-box function to minimize instead.

The first step of Bayesian optimization is to calculate the output of the objective function with sample points or design variables of interest to the user. These samples are used to construct a surrogate model of the function. Next, an acquisition function selects the next sample point, also called the optimizer candidate, which has the best chance of approaching the optimum of the objective function according to the current observations. The acquisition function considers both the model mean and error to balance finding a more optimal point while improving the model accuracy [4]. The new observation will be used to iteratively build a better surrogate model. The algorithm

stops when the stopping condition is met, including but not limited to local convergence, maximum number of iterations, or model accuracy. The summary is shown in Algorithm 1 with a visual representation in Figure 1. The dashed black line represents the original function with the surrogate model in blue with both the predicted mean and error. Note how the accuracy of the model increases with each new sample before local convergence.

Algorithm 1: General Bayesian Optimization

1. Evaluate objective function $f(x)$ at initial points.
   1.1. Evaluate constraint function(s) $c^i(x)$.
2. **While not converged:**
3. Fit objective function surrogate model $f_s(x)$
   3.1. Fit constraint function(s) surrogate model(s) c_s^i.
4. Choose new sample x_n based on acquisition function.
5. Evaluate objective at the new sample point $f_n = f(x_n)$.
   5.1. Evaluate constraint function(s) at $c_n^i = c^i(x_n)$.
6. Save or update necessary values.
   6.1. Save f^* and x^* from $argmin_x\big(f(x)\big)$.
   6.2. Update penalty, barrier, or related constants.
7. **End while**
8. Return f^* and x^*

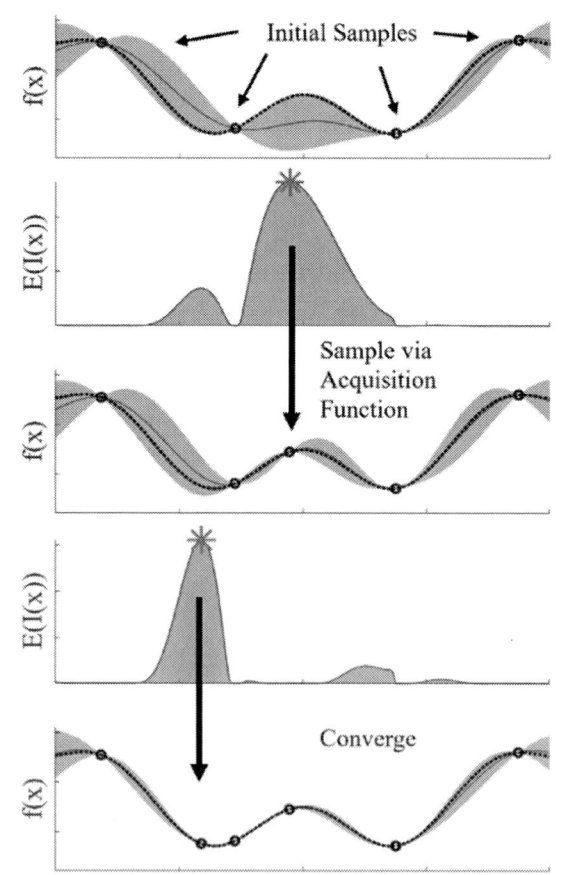

FIGURE 1: BAYESIAN OPTIMIZATION VIA EXPECTED IMPROVEMENT EXAMPLE

Copyright © 2023 by ASME

2.2.1 Acquisition Functions

In Bayesian optimization, the main problem is figuring out how to select the sample points for the next evaluation based on existing observations. The algorithm needs to make a trade-off between exploring regions of uncertainty (exploration) and focusing on regions known to have better objective values (exploitation), so an acquisition function is utilized to do the exploring or exploiting in the design space. In the exploration criteria, acquisition functions tend to select the candidates which are away from the current sample points, so the candidates will be more evenly selected from infill points. On the other hand, in the exploitation criteria, the acquisition functions focus on regions around the current best point; therefore, the optimization might end up at a local optimum.

The Expected Improvement (EI) function, defined in Equations 6 and 7, is the method that balances the exploration and exploitation strategies [23]. The EI function measures each infill point's potential improvement relative to the current best point and is used to balance the exploration-exploitation trade-off. High EI values correspond to points that have a high probability of being better than the current best candidate.

$$EI(x) = \left(f_{min} - \hat{f}(x)\right)\Phi(z) + \sigma_{f(x)}\phi(z) \qquad (6)$$

Where:

$$z = \frac{f_{min} - \hat{f}(x)}{\sigma_{f(x)}} \qquad (7)$$

Note that, in practice, this is often treated as an if-then statement, where $EI(x) = 0$ if the standard deviation is zero to avoid issues with programs attempting to divide by zero.

2.2.2 Constraint Methods

There are many ways to handle constraints in Bayesian optimization which can be divided into two main categories: those that integrate the constraints into the objective function and those that exploit an acquisition function [2]. If the constraints are analytical then those can be used to ensure that all sample points are in the feasible domain.

The methods that integrate the constraint into the objective all follow the same principle of penalty methods and rely on unconstrained optimization methods. In this version of constrained Bayesian optimization, the GP model is constructed based on the output of the objective function and the value of the constraint functions, including some necessary penalty terms, making infeasible values less likely to be sampled [24]. This same principle works on similar methods in Bayesian optimization such as barrier functions, the augmented Lagrangian method, and the alternating direction method of multipliers [2,12,14].

By exploiting the acquisition function, the algorithm can be modified to select points that should be feasible. There are different methods to quantify the likelihood that a proposed candidate will satisfy all constraints of the problem, all of which can be integrated with an acquisition function to determine new sample points [9–11,13]. However, some methods have been known to struggle at constraint boundaries, especially if the optimum lies at the intersection of two or more constraints [15].

3 METHODOLOGY

This section outlines the six acquisition function methods, how they work for both regression and classification surrogate constraints, the four test cases, and the comparison metrics. All programming was done in MATLAB using the Gaussian Process for Machine Learning (GPML) Toolbox, a powerful tool for both regression and classification Gaussian processes [25].

3.1 Constraint Handling Methods

The research used six total acquisition functions, four of which were tested on both regression and classification models with two exclusively used on classification models. Each are in Table 1 with keys representing if the method was applied ($\checkmark$),not applied (X), or modified but still applied (~). The modified version was only done on the penalty function using a classification model and is detailed later. The asterisk on the multiple constraint classification methods highlights how this research treated two constraints as one. This means there was a single binary response on if both constraints were met or not. We next describe the six methods used.

Method 1, the penalty method, is commonly used to handle constraints when optimizing constrained problems [24] and is relatively simple to implement with Bayesian optimization [2,24]. With a penalty method, the acquisition function used is given by eq. 6 to minimize the penalized objective function.

TABLE 1: CONSTRAINT HANDLING METHODS TESTED AND WHERE THEY WERE APPLIED

	Constraint Method	Regression Gaussian Processes		Classification Gaussian Processes	
		Single Constraint	Multiple Constraints	Single Constraint	Multiple Constraints*
1)	Penalty Function	$\checkmark$	$\checkmark$	~	~
2)	$PF(x) \cdot EI(x)$	$\checkmark$	$\checkmark$	$\checkmark$	$\checkmark$
3)	$min\left(1, 2 \cdot PF(x)\right) \cdot EI(x)$	$\checkmark$	$\checkmark$	$\checkmark$	$\checkmark$
4)	$EI(x)\ s.t.\ PF(x) \geq 0.5$	$\checkmark$	$\checkmark$	$\checkmark$	$\checkmark$
5)	$p(y = 1|x) \cdot EI(x)$	X	X	$\checkmark$	$\checkmark$
6)	$EI(x)\ s.t.\ p(y = 1|x) \geq 0.5$	X	X	$\checkmark$	$\checkmark$

Copyright © 2023 by ASME

Several penalty functions have been introduced for constraint handling and the most popular one, the quadratic loss penalty function, is used for the study, defined in Eq. 8.

$$f_k'(x) = f(x) + \rho_k \sum_{i=1}^{n} max\big(0, g_i(x)\big)^2 \qquad (8)$$

Here, $f_k'(x)$ is the transformed objective function, $f(x)$ is the original objective function, $g_i(x)$ are the constraint functions, with ρ_k being the positive penalty parameter that controls the strength of the penalty term at the kth update [1]. This penalty value needs to be updated multiple times, with this research starting the value at 1 and increasing it ten-fold if the acquisition function could not identify an improved point after five tries. This allows the algorithm to build an accurate model by sampling points that are infeasible at higher penalty values but less so at lower ones.

The rest of the methods rely on modifying the acquisition function to select points that have a high probability of not violating any constraints. If the constraint is also a black-box function, then the GP model can be used to find the probability the constraints will be satisfied using Eq. 9 [9]. If the constraints are independent, the probability of violating any of them is a joint probability shown in the new acquisition function of Eq. 10, which is the Expected Improvement times the probability the constraints will be violated, used for Method 2 [2,11]. Here, g_0 is the limit-state value of a constraint function (zero if in negative null form), with $\hat{g}(x)$ and $\sigma_g(x)$ being the predicted mean and standard deviation, respectively.

$$PF(x) = \Phi\left(\frac{g_0 - \hat{g}(x)}{\sigma_{g(x)}}\right) = \frac{1}{2} + \frac{1}{2}\,\mathrm{erf}\left(\frac{g_0 - \hat{g}(x)}{\sigma_{g(x)}}\right) \qquad (9)$$

$$Aq_{Method\,2}(x) = EI(x) \prod_{i=1}^{n} PF_i(x) \qquad (10)$$

One difficulty with this acquisition function is that it is less likely to explore the region close to the constraint boundary since it typically has a lower probability of feasibility. This is especially apparent when two or more constraints intersect. To deal with this, the following acquisition function in Eq. 11 (Method 3) was also tested [15]. This makes the probability of feasibility equal 1 if the original probability was 50% or higher. For an understanding of what this looks like compared to the original $PF(x)$, see Figure 2.

$$Aq_{Method\,3}(x) = EI(x) \prod_{i=1}^{n} min\big(1, 2PF_i(x)\big) \qquad (11)$$

The final main method, Method 4, utilizes the probability of feasibility but treats it as a constraint; there is a probability of feasibility threshold that can be customized by the user. In this method for Bayesian optimization, the acquisition function

maximizes the expected improvement only in areas where the probability of feasibility is above the specified threshold [26]. This is demonstrated below in Eq. 12 and considers all individual constraints. We note that the original paper exploring this method used Support Vector Machines instead of Gaussian Processes for their constraint but still uses the same 50% probability cutoff.

$$Aq_{Method\,4}(x) = EI(x)\ s.t.\ PF_i(x) \geq 0.5\ for\ i\ = 1\ to\ n \quad (12)$$

Each of these modifiers will affect the acquisition function differently, possibly leading to different sample points on each iteration, even with the same information. The second graph of Figure 2 shows how all three methods would behave on some arbitrary expected improvement function. In this example, each selects a different point for the next sample. Note how Method 4 behaves similarly to Method 3 but cannot select any points to the left of the constraint boundary. Over iterations, this could compound leading to different performances of the respective algorithms.

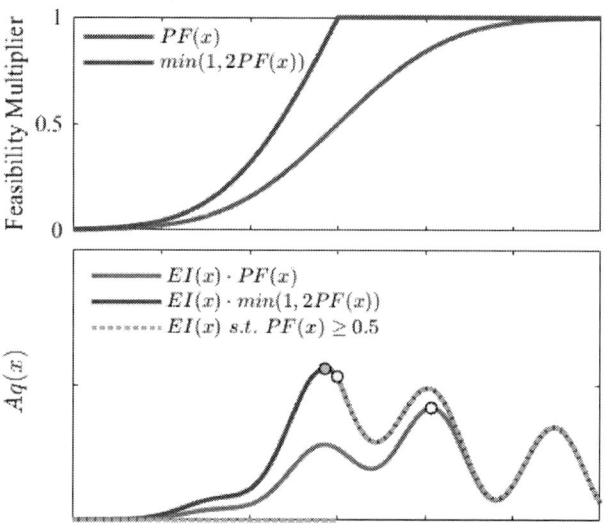

FIGURE 2: VISUALIZATION OF THE FEASIBILITY MULTIPLIER (TOP) AND ACQUISITION FUNCTIONS. (BOTTOM) FOR METHODS 2, 3, AND 4

3.1.1 Regression Constraint Specifics

The last two methods, which are discussed next, do not apply to the regression constraints since they only apply to classification models. All regression constraints were modeled using the same regression Gaussian processes discussed earlier with a squared exponential covariance function and a constant mean function as in Equations 4 and 5. This is because we are assuming no prior knowledge of either the constraint or the objective function.

3.1.2 Classification Constraint Specifics

The classification constraints required a few modifications to the above regression constraints and included two methods based on Methods 2 and 4. To reflect the categorical response of

Copyright © 2023 by ASME

a classification model, a single surrogate was used to represent if all constraints were met. This means the response was a "yes" if and only if all constraints were met. Second, the penalty function is only considered if the constraint was met, meaning a penalty value is used instead of taking the squared value of the response. Finally, the classification constraints were placed in negative-null form to get them to work with the same methods shown for regression constraints. This means the models find $p(g = -1|x)$ where -1 is the response leading to the constraints being satisfied.

This research uses both the direct probability and the latent model for methods traditionally used for regression-based constraints. Both have been used in engineering optimization but need to be compared against each other [17,18]. This leads to the final two methods, shown in Eqs. 13 and 14, which are based on previous research using direct probability to select the next point [17,26].

$$Aq_{Method\,5}(x) = EI(x) \cdot p(g = -1|x) \qquad (13)$$

$$Aq_{Method\,6}(x) = EI(x) \; s.t. \; p(g = -1|x) \geq 0.5 \qquad (14)$$

To model the classification surrogate, this research uses Expectation Propagation, which is an iterative algorithm for determining the probability that an output belongs to a specific class based on a continuous independent variable [27]. The specifics are outside of the scope of this paper, but it is one of many methods and is highlighted for its "practical accuracy" compared to others [21]. This makes it ideal for this experiment as it required multiple runs on the same problem.

3.2   Test Problems

These methods were tested on four different constrained optimization problems. We chose each problem for a variety of reasons but kept them all in two dimensions to allow for quicker computation time and easy visualization. This visualization also allows us to observe the final surrogate models. Each brings something unique, either based on the type of constraint, the shape of the underlying objective function, or the application.

The equations are in Table 2 with graphs of the problems in Figure 4 on the next page.

Except for Problem D, which is an engineering-focused example, all are mathematical examples that are commonly used in algorithm comparison. Problem A is a sphere function with a single linear constraint designed to show that the algorithms work. Problem B had a multimodal problem to explore how the different methods would behave. Problem C introduced two constraints with a Rosenbrock function.

Finally, Problem D is an engineering-related example designed to explore how these methods may behave on problems that engineers may encounter. Here, the goal is to determine the optimal height, H, and the diameter of the bar, d, of a two part truss to minimize its weight subject to stress and buckling constraints under a vertical load as shown in Figure 3 [28]. The members are two steel hollow cylinders with a wall thickness of $t = 0.1$ inches and are pinned together at the top and the ground at a $2B$ distance away where $B = 30$ inches. The magnitude of the load $P = 33$ kips while the bars have a yield strength $\sigma_a = 100$ ksi, a Young's modulus $E = 30 * 10^6$ psi, and a density $D = 0.3$ psi. The solution at the intersection of these two constraints making it ideal for this research.

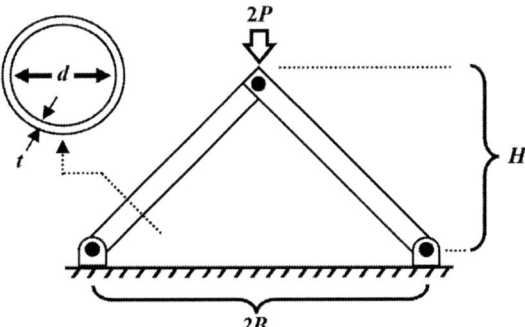

FIGURE 3: TWO-BAR TRUSS PROBLEM [28]

TABLE 2: LIST OF ALL FOUR TEST PROBLEMS INCLUDING THEIR RANGE AND UNIQUE PROPERTIES

	Objective Function	Constraints (All ≤ 0)	Range	Comments
A)	$f_A(x_1, x_2) = x_1^2 + x_2^2$	$g_A(x_1, x_2) = x_1 + x_2 + 0.5$	$x_1 \in (-1,1)$ $x_2 \in (-1,1)$	Simple Objective and Constraint
B)	$f_B(x_1, x_2) = cos(2x_1)cos(x_2) + sin(x_1)$	$g_B(x_1, x_2) = \frac{1}{4}(x_1 + 5)^2 + \frac{1}{100}x_2^2 - 1$	$x_1 \in (-5,0)$ $x_2 \in (-5,5)$	Multi-Modal Objective
C)	$f_C = (1 - x_1)^2 + 10(x_2 - x_1^2)^2$	$g_{C1}(x_1, x_2) = 3x_1 - x_2 - 2$ $g_{C2}(x_1, x_2) = \frac{1}{3}x_1 + x_2 - \frac{4}{3}$	$x_1 \in (-2,2)$ $x_2 \in (-2,2)$	f^* at Constraint Intersection
D)	$f_D(d, H) = 2D\pi d \cdot t \cdot \sqrt{B^2 + H^2}$	$g_{D1}(d, H) = \left(P/(\sigma_a \pi t)\right)\left(\sqrt{B^2 + H^2}/(H \cdot d)\right) - 1$ $g_{D2}(d, H) = \frac{(8P(B^2 + H^2)^{1.5})}{\left(\pi^3 E \cdot t \cdot H \cdot (d^2 + t^2)\right)} - 1$	$d \in (0,10)$ $H \in (0,30)$	Engineering Focused, f^* at Constraint Intersection

Copyright © 2023 by ASME

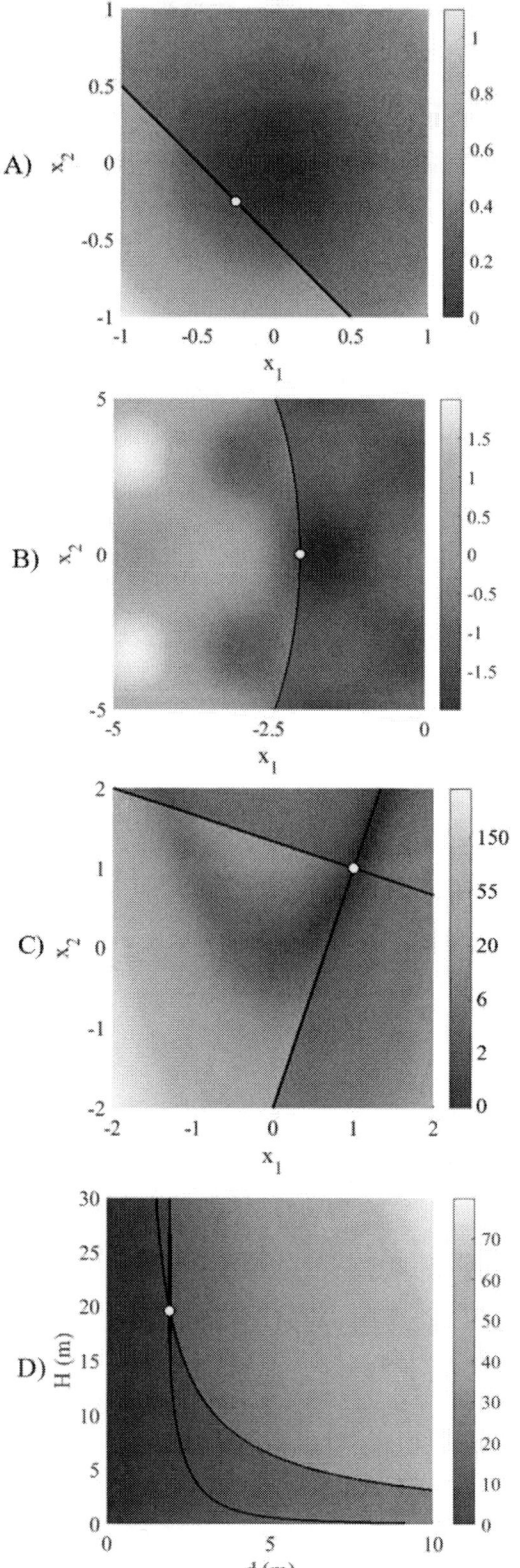

FIGURE 4: VISUALIZATION OF THE FOUR TESTED OBJECTIVE FUNCTIONS WITH THEIR CONSTRAINTS (SHADED AREA) AND THE MINIMUM (YELLOW DOT)

3.3   Test Setup

Overall, 40 tests were run since there are 10 different methods (4 for regression constraints and 6 for classification constraints) on four different test problems. For a statistically significant sample size, each method was run on each test problem 30 times. There were cases where the GPML toolbox errored out and code exceptions were made to discount that case and ensure that there were 30 successful runs. We note that this is something that should be considered when implementing Bayesian optimization.

For the initial sampling, the random number generator required to create the Latin Hypercube Sampling was reset at the beginning of each test, meaning that no method could have the advantage of better initial samples. To ensure that timing tests were fair, all runs were done on the same Dell XPS 15 9500 Laptop with an Intel Core i9-10885H CPU, an eight-core 32 GB of RAM, and an x64-based processor. We note that timing can be affected by a variety of factors and fit time could change by using a different computer, adding in a parallel process, or using different mean and covariance functions. This is why relative timing was used instead of the fitting time itself.

Other details of the algorithm include the method for maximization of the acquisition function and the convergence criteria. The acquisition function was maximized using an exhaustive search under the assumption of some tolerance based on the given problem, such as using subdivisions of 0.01. This was so it would always choose the global maximum since the acquisition function is often multimodal.

The algorithm converged or stopped under one of two conditions. The first was ending at 50 iterations. The second was ending after the standard deviation of the model approached zero, at which the code would select a final minimum value based on the mean of the objective function if that point was feasible, which is shown in Equation 15.

$$\underset{x}{arg\min} \hat{f}(x) \; s.t. \; g_i(x) \leq 0 \tag{15}$$

This follows the common convergence criteria of stopping and has been used previously as an acquisition on its own [2,29]. Additionally, it allows us to see if that method is an effective stopping criterion for these types of problems and, just as importantly, saves time since it reduces iterations. The different methods for convergence need to be considered when setting up Bayesian optimization and we can observe if they are effective.

4   RESULTS

After testing, we compared the results using both quantitative analysis on the metrics of convergence and timing, and qualitative reflection on how they were implemented. This section discusses which methods tended to outperform others, how problem formulation (such as shape and number of constraints) could play a factor, issues with implementation, and general trends across all comparisons.

Copyright © 2023 by ASME

4.1 Convergence Comparison

The first test was to look at the final mean values for each problem and compare them against each other to see if any trends can be observed. Figure 5 below shows the distance between the mean final value of each method to the mean final value of the method that performed the best. It breaks up results based on the test problem and if it was the regression or classification constraint. For visual comparison, this was transformed into a heat map, with blue being closer and red being further away from the lowest value.

Problem	A		B		C		D	
Type	R	C	R	C	R	C	R	C
1)	0.037	0.132	0.964	0.860	1E-03	0.043	5.221	3.167
2)	0.000	0.174	0.000	0.027	0.000	0.000	2.830	5.201
3)	2E-04	0.193	0.011	0.025	0.000	0.000	0.000	6.069
4)	2E-03	0.191	0.070	0.109	3E-06	0.056	4.420	5.040
5)		0.232		0.041		0.000		6.573
6)		0.191		1.425		0.056		5.040
Min	0.1259		-1.0373		0		13.8694	

Method (row label for methods 1–6)

FIGURE 5: HEAT MAP OF DISTANCE FROM BEST AVERAGE MINIMUM FOR EACH PROBLEM

We can observe how, in general, the use of a regression model as the constraint generally had a lower mean than the classification models regardless of the type of problem. It also shows how the penalty methods were less effective in most cases. It also shows how the use of a direct probability model for classification (Methods 5 and 6) either performed the same or slightly worse than the latent model methods.

4.1.1 Statistical Comparison of Means

For a further comparison, we looked at statistical differences between the final values for different methods using Welch's t-tests of means, which is a generalization of the student's t-test where the variances are not assumed to be the same [30]. This was broken down into five tests on the four separate problems to keep them separate and observe how the problem type could influence the outcome. The first test (T1) compares penalty and non-penalty methods. The second test (T2) compares the regression and classification penalty methods. The third test (T3) compares regression and classification Probability of Feasibility methods. The fourth test (T4) compares $PF(x)$ (Method 2) and $p(y = 1|x)$ (Method 5) for the classification models. The final test (T5) compares those same models but with the probability cut-off of 50% (Methods 4 versus 6).

The resulting p-values are shown below in Table 3 as well as which one was better, or if there is there was no statistical difference under the $p = 0.05$ cutoff. We also included which mean performed better, but we note that the lack of a statistically significant p-value should be considered when comparing them. For brevity we use 'CN' for cannot reject the null hypothesis that they are the same, 'S' if the means were the same, 'P' for penalty

methods, 'NP' for non-penalty methods, 'R' for regression constraints, and 'PF' for Probability of Feasibility methods.

TABLE 3: RESULTING *P*-VALUES AND CONCLUSION FROM A TEST OF MEANS BETWEEN TYPES OF CONSTRAINT HANDLING METHODS

	A	B	C	D
T1	$p = 0.543$ CN/P	$p = 1e - 3$ NP	$p = 0.074$ CN/NP	$p = 0.711$ CN/NP
T2	$p = 5e - 7$ R	$p = 8e - 4$ R	$p = 7e - 5$ R	$p = 0.051$ CN/R
T3	$p = 5e - 22$ R	$p = 0.23$ CN	$p = 0.035$ R	$p = 0.002$ R
T4	$p = 0.290$ CN/PF	$p = 0.509$ CN/PF	$p = 1$ CN/S	$p = 0.269$ CN/PF
T5	$p = 1$ CN/S	$p = 3e - 34$ PF	$p = 1$ CN/S	$p = 1$ CN/S

For more than half of the tests, there was no statistical difference in how they performed. This makes it hard to highlight one method over the rest. However, T1 suggests that the non-penalty method may be advantageous to those that use penalties, especially since the penalty methods failed with the multimodal problem. Additionally, for tests T2 and T3, the regression constraint outperformed the classification constraints, which reinforces the conclusions from looking at just the means earlier.

Notably, the switch between using the latent model and the direct probability with the classification surrogate was not significant, as shown in T4 and T5. Most of them arrived at the same final mean values for those problems. When comparing $PF(x)EI(x)$ to $p(g = -1|x)EI(x)$ there were no statistical differences between the means, however, the probability of feasibility method tended to have a lower mean than the direct probability method. The only outstanding difference was for Problem B, or the multimodal problem, where using the 50% Probability of Feasibility cutoff worked significantly better on the latent model. This is most likely due to Method 4's ability to incorporate the error into its model making it easier to explore the objective function and constraint before converging.

4.1.2 Convergence Over Iterations

Beyond the final mean value and their statistical difference, we should examine their convergence trends over iterations. Since there are forty test cases, they were separated by test problems and are shown in Figure 8 in the Appendix. Each subplot compares related methods tracking the current best f^* at each iteration. The first shows the penalty method between regression and classification constraints (Method 1). The second shows the expected improvement times the probability (Methods 2 and 4). The third compares Method 3 for the regression and classification constraints. The final set of graphs shows the probability cut-off methods (Methods 4 and 6).

Comparison 1 shows how the regression penalty method outperformed the classification penalty. This is most likely due

Copyright © 2023 by ASME

to the use of values for the constraints instead of a single penalty value. The only exception to this is the Rosenbrock function (Problem C) but as shown all methods performed well there. Methods 2 and 4 for the classification constraint perform similarly to each other on all four problems, while the regression constraint still has a lower minimum. The method in the Comparison 3 graphs also do not share any discernible trend, except for regression constraints generally converging quicker. Finally, the main trend for the different probability cutoff methods is that for three out of the four cases the direct probability performed the same as the classification probability of feasibility method. The two graphs overlap with each other in the figure. The only exception is the multimodal test problem, suggesting there is a slight advantage of using the latent model.

The graphs also show how the shape of the underlying objective function, as well as how many constraints it has, affects the convergence. The unimodal problems generally show the same behavior between all the methods. The multimodal objective function in Problem B struggles with the penalty methods. This, along with the convergence in graphs in Comparison 4, shows that more exploratory methods are more likely to converge. The more exploratory methods do take more iterations but make up for that with a more accurate model. The engineering test problem is quite interesting as there is no general downward trend. Instead, the current model goes until the point where the error on the objective function is zero and stops convergence early. This is why there is a high variance of the minimum little-to-know statistical difference between the final values.

4.2   Relative Timing

It was also important to look at average fit time per iteration for each model to gain insight into the potential tradeoffs between model types. Figure 6 shows a heat map of the average fit time per iteration for each model, which is adjusted by the relative timing instead of the pure value. This is influenced by a variety of factors such as the shape of the objective function, which type of Gaussian process was used, and if there were multiple constraints.

Problem	A		B		C		D	
Type	R	C	R	C	R	C	R	C
1)	1.079	1.000	1.165	1.000	1.138	1.000	1.000	1.100
2)	1.167	6.968	1.516	11.47	2.532	2.942	2.676	3.597
3)	1.430	5.364	1.940	11.52	2.601	2.952	2.470	3.589
4)	1.524	3.602	1.935	11.47	2.640	3.815	2.607	3.433
5)		6.250		11.79		2.922		3.582
6)		4.165		4.770		3.741		3.411
Min (s)	0.1103		0.0778		0.2212		0.1646	

FIGURE 6: HEAT MAP OF RELATIVE TIMING ADJUSTED FOR EACH PROBLEM'S MINIMUM FITTING TIME

The classification penalty was usually faster, which makes sense because it is a single regression model with a penalizing term. The use of Expectation Propagation in the classification methods drastically increases the computation time. This could be alleviated by using a different method, however that risks losing accuracy [21]. This tradeoff may be negligible if the computation time of the objective function is of greater concern. Problems C and D highlight how using multiple constraint surrogates will directly affect the computation time when using regression models since each constraint needs to be modeled separately. However, since the classification treats the constraints as a single model this trend is not observed. Using two regression constraints is still faster than a single classification one, and there may be a point where it is computationally efficient to model constraints as a single binary classification constraint instead of multiple regression constraints.

The last observation is that the shape of the constraint affected the fit time. Comparing Problems A and B, where the first had a linear constraint while the second had a quadratic one, it is clear that the shape affected how long it took. While the total timing is also affected by the shape of the objective function, the drastic difference between the two classification constrained methods showed that the curve of the Problem B constraint affected the total fit time.

4.3   Qualitative Reflection

Finally, it is important to discuss the implementation of the different methods and other trends that we observed that are unquantifiable. We also wanted to discuss how regression and classification constraints played a role in the performance of the algorithms beyond the convergence values discussed earlier.

Implementing the penalty method for handling the constraints is straightforward. Converting the problem to be unconstrained and can be implemented in Bayesian optimization with minimal modifications. The main difficulty is knowing when to increase the penalty value. This research used five tries without improvement before increasing the, but it could be better to use a different value, a local convergence criterion, or model accuracy instead. This additional knob needs to be considered when implementing penalty methods as they can lead to drastically different outcomes.

On the other hand, those methods that relied on the Probability of Feasibility were harder to implement compared to the penalty method since the second or tertiary models needed to be calculated. The extra probability evaluations involved are less intuitive than setting it up as an unconstrained optimization problem. This included rewriting the code multiple times when adding in an additional constraint compared to just once, but this could be tackled using different programming languages or methods.

We should also observe how the two types of constraint models, regression and classification, model the constraint and affect sample points. To demonstrate this, we show Figure 7 which shows the probability of feasibility after convergence on Problem C. This should be compared to Figure 4 earlier to see

Copyright © 2023 by ASME

their resemblance. Lighter values show more feasible points while darker means less feasible, with red dots being the sample points. With these two linear constraints, the regression surrogate models them perfectly while the classification model includes more error but still mimics the overall shape. We can also see how the regression model led to more exploration before finding the minimum. Essentially, it explored the constraints until those were modeled perfectly. Once it figured out the two constraints it could immediately converge. The classification model was more exploitative, continuously looking near the optimal point before converging.

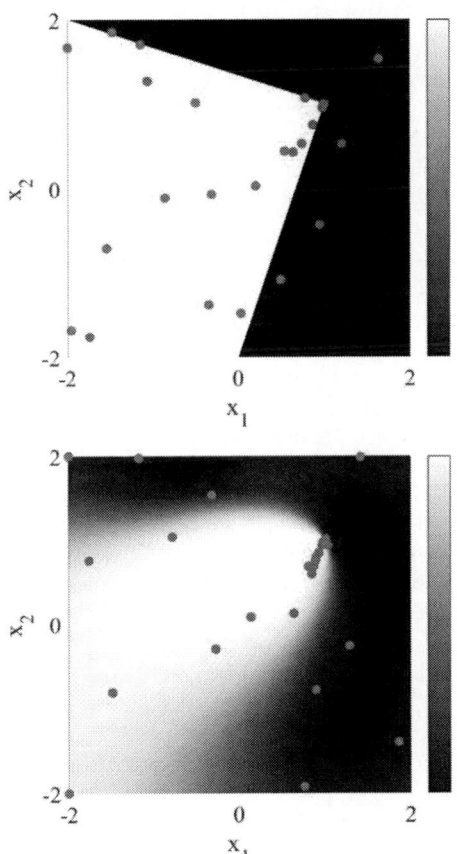

FIGURE 7: COMPARISON OF REGRESSION (TOP) AND CLASSIFICATION (BOTTOM) PROBABILITY OF FEASIBILITY SURROGATES FOR PROBLEM C

5   CONCLUSIONS

In general, if researchers have access to a numerical value for their constraints, they should go with a regression model. This is the case if they plan on using a penalty method or acquisition function based methods. Classification constraints should only be used as a last resort or in situations where feedback is binary since they do not accurately represent the constraint function. Even with multiple constraints, it is quicker to computer two Gaussian process regression models than using Expectation Propagation for classification. This will change with more constraints, and it is possible that at a certain number of constraints, the classification model is faster to fit but that needs to be balanced with other concerns such as the computational cost of evaluating the models.

Researchers should be aware of issues that they can encounter when implementing their own Bayesian optimization algorithms. We discussed how it is easier to switch acquisition functions and make the process converge early if the model accuracy is high. This is especially helpful with functions that may have a high computation cost, but the resulting surrogate model ends up being simpler than it seemed. However, this can lead to poor final values if the constraint model is not as accurate. This suggests the need to incorporate the accuracy of the constraint, essentially modifying Equation 15 to include both.

One drawback of this comparison is that there are more methods than those compared here. As cited earlier, the methods here were either used or directly inspired by those used in the literature. The literature on constrained Bayesian optimization presents a wide range of methods that explore different ways of quantifying constraint feasibility, treating the problem as unconstrained optimization, or even incorporating external knowledge to help build the constraint before starting the optimization process [10,11,13–15,31]. For this, the authors encourage other researchers to find methods that best fit their problem. This involves seeing what type of test problems they used. If a paper demonstrates that the method works well on a computational fluid dynamics problem, then that should be the starting point for implementation.

Another area of research is to explore how these different methods apply to more complex engineering problems, specifically those that use black-box functions instead of analytical problems that are treated as such. That would be computationally expensive since it needs to be run enough times to check for statistical significance but could provide valuable insight into where different methods perform best.

Overall, this study provides valuable insight into different methods for handling black-box constraints, the differences in using regression and classification Gaussian processes as constraints, and potential issues that could be encountered when implementing Bayesian optimization. By understanding these methods, engineers and researchers can make informed decisions on when to implement them based on their problem type.

REFERENCES

[1]   Arora, J. S., 2017, *Introduction to Optimum Design*, Elsevier Inc.

[2]   Forrester, A., Sobester, A., and Keane, A., 2008, *Engineering Design via Surrogate Modelling: A Practical Guide*, John Wiley & Sons Ltd.

[3]   Shahriari, B., Swersky, K., Wang, Z., Adams, R. P., and de Freitas, N., 2016, "Taking the Human out of the Loop: A Review of Bayesian Optimization," Proceedings of the IEEE, **104**(1), pp. 148–175.

Copyright © 2023 by ASME

[4] Frazier, P. I., 2018, "A Tutorial on Bayesian Optimization," (Section 5), pp. 1–22.

[5] Morita, Y., Rezaeiravesh, S., Tabatabaei, N., Vinuesa, R., Fukagata, K., and Schlatter, P., 2022, "Applying Bayesian Optimization with Gaussian Process Regression to Computational Fluid Dynamics Problems," J Comput Phys, **449**.

[6] Frazier, P. I., and Wang, J., 2016, "Bayesian Optimization for Materials Design," Information Science for Materials Discovery and Design, **225**, pp. 45–75.

[7] Kotthoff, L., Wahab, H., and Johnson, P., 2021, "Bayesian Optimization in Materials Science: A Survey," pp. 1–15.

[8] Snoek, J., Larochelle, H., and Adams, R. P., 2012, "Practical Bayesian Optimization of Machine Learning Algorithms," Adv Neural Inf Process Syst, **25**.

[9] Schonlau, M., 1997, "Computer Experiments and Global Optimizat Ion."

[10] Gramacy, R. B., Lee, H. K. H., Holmes, C., and Osborne, M., 2012, "Optimization Under Unknown Constraints," Bayesian Statistics 9, **9780199694**, pp. 1–19.

[11] Gelbart, M. A., Snoek, J., and Adams, R. P., 2014, "Bayesian Optimization with Unknown Constraints," Uncertainty in Artificial Intelligence - Proceedings of the 30th Conference, UAI 2014, pp. 250–259.

[12] Ariafar, S., Coll-Font, J., Brooks, D., and Dy, J., 2019, *ADMMBO: Bayesian Optimization with Unknown Constraints Using ADMM.*

[13] Gardner, J. R., Kusner, M. J., Xu, Z., Weinberge, K. Q., and Cunningham, J. P., 2014, "Bayesian Optimization with Inequality Constraints," Proceedings of the 31 st International Conference on Machine Learning, pp. 937–945.

[14] Gramacy, R. B., Gray, G. A., le Digabel, S., Lee, H. K. H., Ranjan, P., Wells, G., and Wild, S. M., 2016, "Modeling an Augmented Lagrangian for Blackbox Constrained Optimization," Technometrics, **58**(1), pp. 1–11.

[15] Bagheri, S., Branke, J., Konen, W., Deb, K., Allmendinger, R., Fieldsend, J., Quagliarella, D., and Sindhya, K., 2017, "Constraint Handling in Efficient Global Optimization," *GECCO 2017 - Proceedings of the 2017 Genetic and Evolutionary Computation Conference*, Association for Computing Machinery, Inc, pp. 673–680.

[16] Sharpe, C., Wiest, T., Wang, P., and Seepersad, C. C., 2019, "A Comparative Evaluation of Supervised Machine Learning Classification Techniques for Engineering Design Applications," Journal of Mechanical Design, Transactions of the ASME, **141**(12).

[17] Tran, A., Sun, J., Furlan, J. M., Pagalthivarthi, K. V., Visintainer, R. J., and Wang, Y., 2019, "PBO-2GP-3B: A Batch Parallel Known/Unknown Constrained Bayesian Optimization with Feasibility Classification and Its Applications in Computational Fluid Dynamics," Comput Methods Appl Mech Eng, **347**, pp. 827–852.

[18] Valladares, H., and Tovar, A., 2022, "Multi-Objective Bayesian Optimization Supported By Gaussian Process Classifiers and Conditional Probabilities," International Design Engineering Technical Conferences and Computers and Information in Engineering Conference.

[19] Rasmussen, C. E., and Williams, C. K. I., 2006, *Gaussian Processes for Machine Learning*, MIT Press Books, Cambridge, Massachusetts.

[20] Duvenaud, D. K., 2014, "Automatic Model Construction with Gaussian Processes."

[21] Nickisch, H., and Rasmussen, C. E., 2008, "Approximations for Binary Gaussian Process Classification," Journal of Machine Learning Research, **9**, pp. 2035–2078.

[22] Aldrich, J. H., and Nelson, F. D., 1984, *Linear Probability, Logit, and Probit Models*, Sage.

[23] Zhan, D., and Xing, H., 2020, "Expected Improvement for Expensive Optimization: A Review," Journal of Global Optimization, **78**(3), pp. 507–544.

[24] Dhaene, T., Sasena, M. J., Papalambros, P., and Goovaerts, P., 2002, *Exploration of Metamodeling Sampling Criteria for Constrained Global Optimization Related Papers Inverse Surrogat e Modeling: Out Put Performance Space Sampling EXPLORATION OF METAMODELING SAMPLING CRITERIA FOR CONSTRAINED GLOBAL OPTIMIZATION.*

[25] Rasmussen, C. E., and Nickisch, H., 2013, *The GPML Toolbox Version 4.2.*

[26] Basudhar, A., Dribusch, C., Lacaze, S., and Missoum, S., 2012, "Constrained Efficient Global Optimization with Support Vector Machines," Structural and Multidisciplinary Optimization, **46**(2), pp. 201–221.

[27] Minka, T. P., 2001, "Expectation Propagation for Approximate Bayesian Inference," Proceedings of the 17th Conference in Uncertainty in Artificial Intelligence, pp. 362–369.

[28] Belegundu, A. D., and Chandrupatla, T. R., 2019, *Optimization Concepts and Applications in Engineering*, Cambridge University Press.

[29] Zhang, Y., Li, M., Zhang, J., and Li, G., 2016, "Robust Optimization With Parameter and Model Uncertainties Using Gaussian Processes," Journal of Mechanical Design, **138**(11).

[30] WELCH, B. L., 1947, "The Generalization of 'Student's' Problem When Several Different Population Variances Are Involved," Biometrika, **34**(1–2), pp. 28–35.

[31] Tao, T., Zhao, G., and Ren, S., 2020, "An Efficient Kriging-Based Constrained Optimization Algorithm by Global and Local Sampling in Feasible Region," Journal of Mechanical Design, Transactions of the ASME, **142**(5).

Copyright © 2023 by ASME

APPENDIX

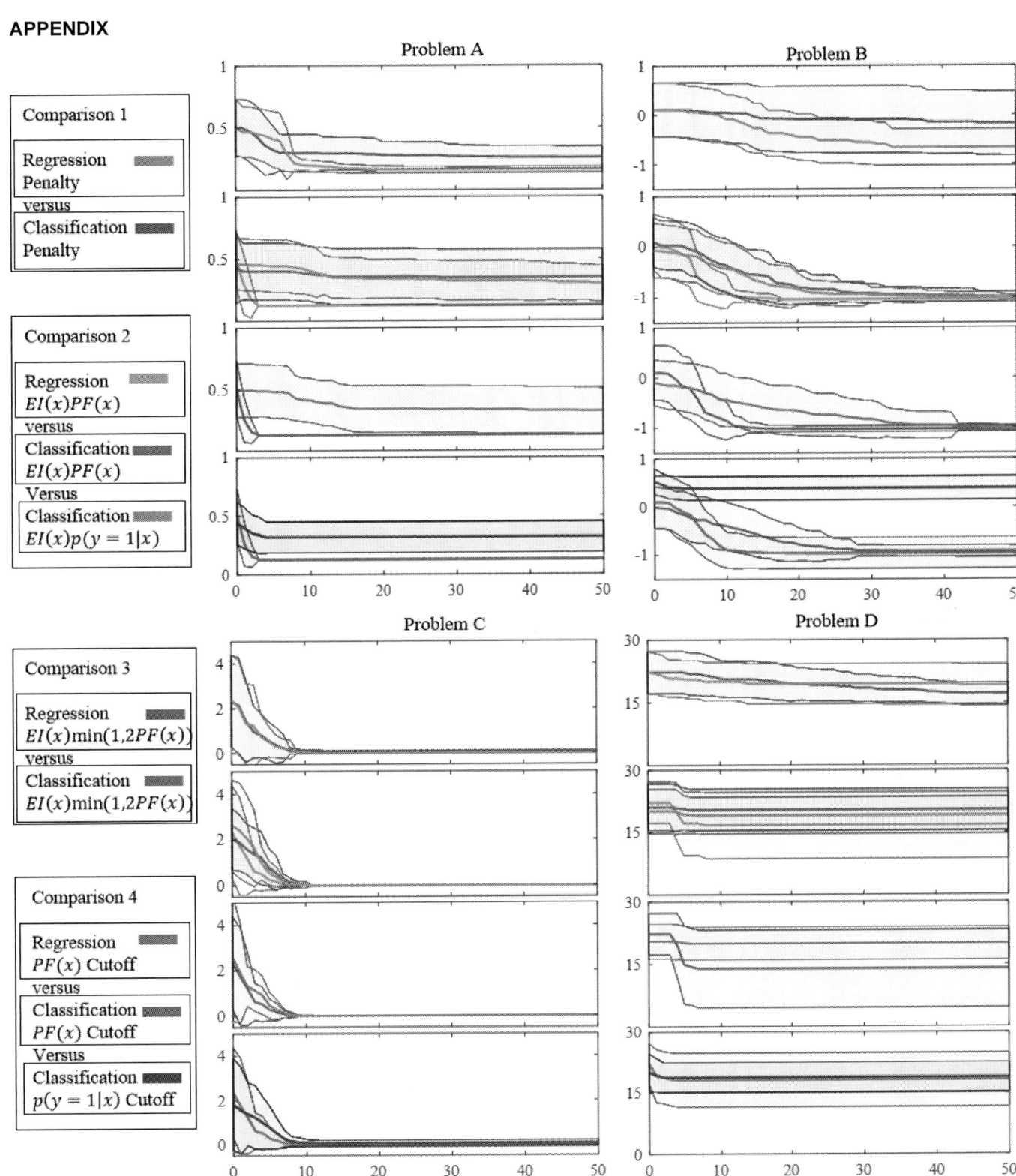

FIGURE 8: CONVERGENCE GRAPHS OF EACH TEST PROBLEM SHOWING THE MEAN AND STANDARD DEVIATION OF THE CURRENT BEST MINIMUM OVER THE ITERATIONS. SEPARATED INTO FOUR DIFFERENT TESTS TO COMPARE RELATED METHODS

Copyright © 2023 by ASME

Proceedings of the ASME 2023
International Design Engineering Technical Conferences and
Computers and Information in Engineering Conference
IDETC-CIE2023
August 20-23, 2023, Boston, Massachusetts

DETC2023-116549

TOPOLOGY OPTIMIZATION WITH QUANTUM APPROXIMATE BAYESIAN OPTIMIZATION ALGORITHM

Jungin E. Kim, Yan Wang
Woodruff School of Mechanical Engineering
Georgia Institute of Technology
Atlanta, GA 30332, USA

ABSTRACT

Despite its wide scope of applications, topology optimization faces a major challenge, which is its high computational expenses caused by the large number of design variables and the underlying physics. Because the number of possible material layouts increases exponentially, the search for the optimum becomes computationally intractable. Recently, quantum computing emerged as an alternative paradigm for solving optimization problems with potentially much higher efficiency than classical computing. The quantum superposition phenomenon allows quantum computer to perform a parallel search of many possible solutions. In this work, a two-mixer quantum approximate Bayesian optimization algorithm (TM-QABOA) is developed as a hybrid quantum-classical optimization algorithm, where Pauli-X mixers and generalized Grover mixers are applied in an alternating fashion. TM-QABOA enables a dynamic exploration-exploitation balance to improve searching efficiency. The exploration is done with the Pauli-X mixers, whereas the exploitation is enhanced by the generalized Grover mixers through amplitude amplification. Surrogate-based Bayesian optimization is used to optimize the hyperparameters of the quantum circuit, i.e., rotation angles. The algorithm can also be used to solve mixed-integer optimization problems. The feasibility of TM-QABOA to solve optimization problems is demonstrated with two simple examples. One is two-dimensional truss design, and the other is metamaterials design.

Keywords: Quantum computing; quantum approximation optimization algorithm; Bayesian optimization; topology optimization; metamodeling; metamaterials

1. INTRODUCTION

Topology optimization (TO) is the process of choosing the material phases in a spatial domain to achieve the desirable mechanical, thermal, acoustic, optic, or other properties. Despite its wide scope of applications, TO faces a major challenge, which

is its high computational expenses from solving the underlying physics problems when a large number of binary design variables for material selections are used. Furthermore, the number of possible material layouts increases exponentially with the number of design variables. Although current TO methods are capable of solving problems with many design variables, it is desirable to further increase the efficiency of finding the optimal solution with fewer iterations.

Some methods have been developed to improve the efficiency of solving TO problems. Continuous density variables have been widely used to replace the original discrete variables so that optimization algorithms for continuous problems can be applied instead of solving combinatorial optimization problems. Nevertheless, the curse-of-dimensionality issue still exists. A common approach to improve efficiency is parallel computing [1], where the original domain is decomposed into smaller sub-domains. Each processor performs computations on one sub-domain. The parallelization can be based on graphical processing unit (GPU) systems [2,3]. Another common approach is dimensionality reduction where the number of design variables is decreased. A variety of dimensionality reduction strategies have been developed, including reduced basis models for approximate solutions [4], reduced series expansion of the material field [5], reduction of the degrees of freedom in finite element formulation [6], and optimizing topology based on geometric primitives [7].

Recently, quantum computing (QC) emerged as an alternative paradigm for solving optimization problems with potentially much higher efficiency than classical computing. The key to improving optimization efficiency is the quantum parallelism, by which multiple solutions are searched simultaneously based on the quantum superposition phenomenon. In QC, the solutions are encoded with quantum bits (or qubits). In contrast to a classical binary bit where the state must be either 0 or 1, a qubit exists in both states simultaneously. Quantum superposition allows quantum computer to perform a parallel search of many possible solutions.

Copyright © 2023 by ASME

One of the most recently developed quantum optimization algorithms is the quantum approximate optimization algorithm (QAOA) [8]. QAOA uses a quantum circuit, or sequence of unitary rotation gates, to alter the basis state amplitudes of the system's superposition. The circuit is an alternating sequence of phase-separating and mixer Hamiltonian operators which encode the objective quantity and perturb the system, respectively. The objective of QAOA is to classically optimize quantum gate rotation angles to increase the amplitude of the most optimal basis state. With a larger amplitude, the optimal basis state has a greater probability of being observed. QAOA at a circuit depth of one is able to achieve higher approximation ratios than the classical guessing approach for NP-difficult problems such as MaxCut and MaxSat [8, 9].

Recently, a quantum approximation Bayesian optimization algorithm (QABOA) [10] was proposed, where surrogate-based Bayesian optimization is used as the outer loop of QAOA. In addition to quantum walk mixers in the circuit, generalized Grover operators are introduced to perform amplitude amplification. In this work, QABOA is applied to solve TO problems. QABOA is also extended to use Pauli-X mixers and generalized Grover mixers in an alternating fashion. The new architecture is named as TM-QABOA. TM-QABOA enables a dynamic exploration-exploitation balance to further improve searching efficiency. The exploration is done with the Pauli-X mixers, whereas the exploitation is enhanced by the generalized Grover mixers through amplitude amplification. QABOA is more efficient than QAOA because the generalized Grover mixers further increase the amplitudes of improved solutions. In addition, QABOA can reduce the number of quantum circuit calculations because the optimization of rotation angles is guided with a surrogate. Furthermore, QABOA can solve mixed-integer problems efficiently. While the discrete variables are optimized with the quantum circuit, the optimization of the continuous variables is done with the surrogate-based Bayesian optimization. By using quantum superposition to perform a parallel search of multiple solutions, quantum optimization is potentially much more efficient in acquiring the optimal solution than classical optimization.

The quantum computer technology is evolving quickly. The practicality of QC in solving optimization problems, however, is currently limited to small problems where only small numbers of qubits are needed. Currently available gate-based quantum computers can only have dozens of qubits. Quantum annealers that are dedicated to the quantum annealing algorithm can have up to 2,000 qubits. In this work, a quantum computer simulator, instead of a physical quantum computer, is used to solve two small TO problems. The first one is a ground-structure truss design problem with eight qubits, and the second one is a mixed-integer metamaterials design problem with six qubits. The purpose of this work is to demonstrate the feasibility of quantum optimization algorithms for solving TO problems. The quantum speedup over the classical TO methods for larger problems need to be studied in the future when more qubits become available. It should be noted that it is straightforward to formulate TO problems with binary variables in QC. It is not necessary to represent material phases as continuous variables as in density-based methods.

The remainder of this paper is structured as follows. A review of existing methods to improve computational efficiency of TO is given in Section 2. An introduction of quantum optimization is presented in Section 3. The proposed TM-QABOA framework is described in Section 4. In Section 5, the TM-QABOA is used to solve two small TO problems. In the first problem, the objective is to minimize the displacement of a targeted node in a two-dimensional truss structure which involves eight discrete variables. In the second problem, the objective is to maximize the strength-weight ratio of metamaterials which involves six discrete and eleven continuous variables. The results demonstrate the usefulness of TM-QABOA in solving TO problems efficiently. Challenges of quantum optimization for TO and future extensions of TM-QABOA are discussed in Section 6.

2. EXISTING METHODS TO IMPROVE COMPUTATIONAL EFFICIENCY OF TOPOLOGY OPTIMIZATION

The standard TO approach is to divide a domain into smaller elements. This approach is computationally expensive because a large number of underlying physics problems must be solved when a large number of design variables are used to describe the material configurations. For this reason, several types of approaches have been developed to improve the efficiency of TO.

One common approach is parallel computing. In general, parallel computing reduces the computational cost of TO by performing computations in multiple spatial domains simultaneously with multiple processors. Parallel computing is applied in several ways to solve TO problems. Mahdavi et al. [11] used the domain decomposition to perform multiple finite-element analyses and sensitivity calculations simultaneously for two-dimensional compliance minimization problems. Borrvall et al. [1] combined the domain decomposition and the conjugate gradient method to solve multiple equilibrium equations for three-dimensional TO problems. To solve structural optimization problems with stress constraints, Paris et al. [12] parallelized first-order sensitivity analyses of stress constraints and search direction calculations by sequential linear programming. Kim et al. [13] performed multiple sensitivity analyses and design variable updates in parallel to solve eigenvalue TO problems. Liu et al. [14] performed a parametrized level set method where the whole computation process is parallelized. This method is also used to solve TO problems with unstructured meshes [15]. Several TO coding algorithms also implement parallel computing including PETSc [16], TopADD which solves problems involving arbitrary domains [17], and PolyTop++ which efficiently manages memory [18]. A notable parallel computing strategy is the use of multi-GPU architectures which use task-level parallelism to evaluate multiple simulation models. For large-scale robust TO problems, multi-GPU systems

Copyright © 2023 by ASME

demonstrate computational speedup over multi-central processing unit systems [2, 3].

Another common approach is dimensionality reduction, where the number of design variables is reduced to improve the search efficiency in TO. Several dimensionality reduction strategies have been devised. Gogu [4] used reduced-order basis models to approximate solutions to equilibrium equations at a low computational cost. Luo and Bao [5] approximated the material field with a series expansion which not only reduces dimensionality, but also avoids the issues of checkerboard patterns and mesh dependency. Zheng et al. [6] decreased the degrees of freedom in the finite-element formulations using nodal displacements while also accelerating convergence with grey-scale suppression to force material density values to be 0 or 1. Koh et al. [19] proposed the multi-frequency quasi-static Ritz vector method which reduces the cost of computing dynamic responses and sensitivity values of dynamic multi-substructure systems. Guo et al. [7] used the geometry of a moving morphable bar to derive a local coordinate system with a smaller number of design variables. Guest et al. [20] used Heavyside projections so that a single design variable represents multiple elements. Kazemi et al. [21] used geometric components such as bar elements as an alternative way to represent the material density. Amir et al. [22] used the multigrid method which solves a linear system of equations given a coarse mesh and interpolates the solutions for a finer mesh. Lu and Wang [23] used compact geometric and topological descriptors so that the topology of metamaterials can be optimized efficiently.

Several methods use machine learning to predict optimal material distributions based on previous TO computations. For example, convolutional neural networks were used to predict optimal designs of permanent magnet motors [24], two-dimensional metamaterials based on the homogenization method [25], and three-dimensional structures given loading conditions and volume fractions [26]. Qiu et al. [27] proposed a deep learning model which combines convolutional and recurrent neural networks to predict material configurations with minimum compliance and deformation. Other types of neural networks have been used. Sun et al. [28] combined a deep neural network with the material-field series expansion approach to reduce the number of design variables which allows for sufficient exploration of the reduced search space. Li et al. [29] used a generative adversarial network (GAN) to predict a near-optimal conductive heat transfer structure with lower resolution and a super-resolution GAN to predict the refined structure. Li et al. [30] combined the Kriging model with lower-dimensional latent variable representations to predict periodic structures with maximized in-plane frequency bandgap. Kallioras et al. [31] used a deep belief network to reduce the number of iterations in the solid isotropic penalization method.

The above methods have demonstrated improvements in computational efficiency of TO. Recently, QC was introduced as an alternative paradigm for solving difficult optimization problems. QC is promising since it can find an optimal solution more efficiently than classical computing. In this work, we propose a new quantum optimization algorithm and use it to solve TO problems.

3. A BRIEF OVERVIEW OF QUANTUM OPTIMIZATION

Quantum optimization algorithms can potentially speed up the search for optima by taking advantage of quantum superposition. In an n-qubit quantum system, the possible solutions for optimization are encoded as 2^n combinations of qubits. Each n-qubit combination $|q_i\rangle$ ($i = 0, 1, ..., 2^n - 1$), also known as computational basis, is a 2^n-dimensional unit vector with the ith entry set to 1. For example, when $n = 1$, $|0\rangle = \begin{bmatrix} 1 \\ 0 \end{bmatrix}$ and $|1\rangle = \begin{bmatrix} 0 \\ 1 \end{bmatrix}$. A complex number a_i is the amplitude associated with $|q_i\rangle$. The amplitudes satisfy the normalization constraint defined as

$$\sum_{i=0}^{2^n-1} a_i \bar{a}_i = 1 \qquad (1)$$

where $\bar{a}_i$ is the complex conjugate of a_i. The state of an n-qubit quantum system, $|s\rangle$, is then defined as

$$|s\rangle = \sum_{i=0}^{2^n-1} a_i |q_i\rangle. \qquad (2)$$

Mathematically, $|s\rangle$ is a normalized vector of amplitudes. When the quantum system is observed, $|s\rangle$ collapses into one of the $|q_i\rangle$ with a probability equivalent to $a_i \bar{a}_i$.

In quantum optimization, the goal is to observe the quantum system at its ground state where the quantum system's energy is minimized. There are several types of quantum optimization algorithms. The first type is adiabatic quantum computing (AQC), also called quantum annealing (QA), which is based on the adiabatic theorem to perform a continuous-time evolution of a Hamiltonian that describes the energy landscape of the quantum system [32]. According to the adiabatic theorem, if the quantum system is initialized at the ground state of an initial Hamiltonian and gradual changes to the system are performed, then the system will end up in the ground state of the target Hamiltonian. Compared to simulated annealing which uses a hill climbing approach, QA performs quantum tunneling which allows the search process to escape local minima more quickly. With the transverse Ising model, QA can achieve the ground-state energy with higher probability than simulated annealing given the same annealing schedule [33].

The second type is hybrid quantum-classical optimization where a quantum circuit and a classical computer are used together. A quantum circuit is a sequence of unitary operators that alter the amplitudes of the system's state. Examples of hybrid quantum-classical optimization algorithms are the variational quantum eigensolver (VQE) [34] and the quantum approximate optimization algorithm (QAOA) [8]. In these algorithms, classical optimizers search for rotation angles which increase the amplitude of the optimal basis state. VQE is dedicated to materials design problems, whereas QAOA is for generic optimization problems. In QAOA, a mixer Hamiltonian is designed to explore the space of computational basis states.

Copyright © 2023 by ASME

The third type of quantum optimization is the adaptive Grover search [35, 36] based on Grover's algorithm [37]. In Grover's algorithm, a target basis state is found by performing amplitude amplification. The amplitude amplification is achieved with a sequence of Grover rotations. Each Grover rotation consists of two reflections. The first reflection negates the amplitudes of the targeted basis states. The second reflection flips all amplitudes about the average. In the adaptive Grover search, the target basis states are selected based on the current best objective value as the threshold. Iteratively, Grover's algorithm is applied to find solutions that are better than the threshold.

Quantum optimization algorithms recently started being applied to solve design optimization problems. QA was applied to optimize the radiative cooling ability of a metamaterial structure, where qubits are used to represent the types of constituent materials [38]. Similarly, QA [39] and VQE [40] are used to predict a protein's spatial structure. QA was also used to determine equilibrium ensembles of dense polymer mixtures [41].

Because of their potential advantage in quantum speedup, quantum optimization algorithms can yield optimal designs which are otherwise difficult to acquire using classical optimization methods. However, the potentials of quantum optimization to solve TO problems has not been fully explored.

4. PROPOSED QUANTUM APPROXIMATE BAYESIAN OPTIMIZATION ALGORITHM WITH TWO MIXERS

In the proposed TM-QABOA, the quantum circuit is combined with Bayesian optimization (BO). The algorithm can be used to solve mixed-integer optimization problems. Fig. 1 shows the architecture of the TM-QABOA. The quantum circuit consists of a series of unitary operators, including phase-separating Hamiltonian operators U_C's, mixer Hamiltonian operators U_B's, and generalized Grover mixers U_G's. The three operators are applied in an alternating fashion. There is a rotation angle associated with each operator. Specifically, $U_C(C(\mathbf{x}_c), \gamma)$ is associated with the phase-separating Hamiltonian C, as a function of continuous design variables $\mathbf{x}_c$, and rotation angle γ. $U_B(B, \beta)$ is associated with the mixer Hamiltonian B and rotation angle β. $U_G(\theta)$ is associated with rotation angle θ. The rotation angles are the hyperparameters that are optimized with BO in the outer loop. BO is also used to optimize continuous design variables in mixed-integer problems.

In this section, the components of the TM-QABOA are described. In Section 4.1, the quantum circuit architecture as well as alternative types of mixers are introduced. In Section 4.2, BO, which is used to optimize the hyperparameters of the quantum circuit, is described. In Section 4.3, the overall TM-QABOA procedure is summarized.

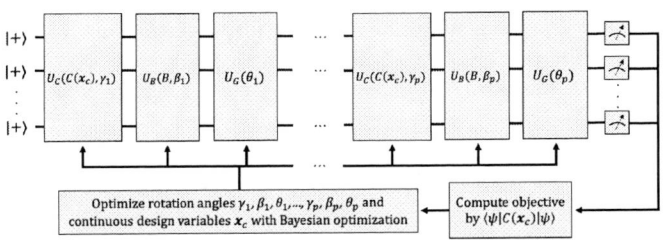

FIGURE 1: The architecture of the proposed TM-QABOA.

4.1 QUANTUM CIRCUIT ARCHITECTURE

The evolution of an n-qubit system performed by the TM-QABOA quantum circuit with a depth of p is defined as

$$|\psi\rangle = U_G(\theta_p)U_B(B, \beta_p)U_C(C(\mathbf{x}_c), \gamma_p) \cdots$$
$$U_G(\theta_1)U_B(B, \beta_1)U_C(C(\mathbf{x}_c), \gamma_1)|+\rangle^{\otimes n} \quad (3)$$

where $|+\rangle$ is the initial state of a qubit with uniform superposition, and $|+\rangle^{\otimes n}$ is the initial state for the n-qubit system. $\otimes$ denotes the tensor product. To prepare the initial state, all qubits are first initialized as $|0\rangle$, then the Hadamard gate is applied to each qubit. This operation results in a uniform superposition with each basis state having an amplitude of $2^{-n/2}$.

The quantum circuit for TM-QABOA is an alternating sequence of three types of gates. The first gate is the phase-separating Hamiltonian operator U_C which encodes the objective quantity being optimized. Given C and γ, U_C is a unitary operator defined as

$$U_C(C(\mathbf{x}_c), \gamma) = e^{-j\gamma C}. \quad (4)$$

Given an objective function $f(q_1, \ldots, q_n)$ where $q_1, \ldots, q_n$ are binary variables, $C(\mathbf{x}_c)$ is computed with the transformation [9]

$$q_i \to \frac{1}{2}(1 - Z_i) \quad (i = 1, \ldots, n) \quad (5)$$

where Z_i is the Pauli-Z gate performed on the ith qubit.

The second gate is the mixer Hamiltonian operator U_B which perturbs the system to change from one state to another. Similar to U_C, U_B is a unitary operator defined as

$$U_B(B, \beta) = e^{-j\beta B}. \quad (6)$$

In the original formulation of QAOA, B is a Pauli-X mixer defined as the sum of n Pauli-X gates.

The third gate is the generalized Grover mixer U_G which performs amplitude amplification on solutions with improved objective values. U_G is a sequence of two reflection operators, defined as

$$U_G(\theta) = U_R U_S(\theta). \quad (7)$$

Copyright © 2023 by ASME

343

Suppose that the state of the system prior to performing $U_G(\theta)$ is $|\phi\rangle$. First, $U_S(\theta)$ is a selective phase shift operator which shifts the phase of target basis states by a rotation angle θ, defined as

$$U_S(\theta) = I - (1 - e^{j\theta}) \sum_{u \in S} |u\rangle\langle u| \tag{8}$$

where I is the identity matrix and S is the set of target basis states. Second, U_R is a reflection operator which reflects $U_S(\theta)|\phi\rangle$ about the average amplitude. U_R is defined as

$$U_R = U_H(I - 2|0\rangle\langle 0|)U_H \tag{9}$$

where U_H are Hadamard gates applied to all qubits. By performing the reflections defined in Eqs. (8) and (9), the amplitudes of the most promising basis states increase.

After the alternating operations in Eq. (3), the measurement on all n qubits collapse $|\psi\rangle$ into a single basis state $|\psi_f\rangle$. The system's energy is then computed by the expectation of C defined as

$$f(\psi_f) = \langle \psi_f | C(x_c) | \psi_f \rangle. \tag{10}$$

The TM-QABOA quantum circuit is constructed with two types of mixers to improve searching efficiency with a dynamic exploration-exploitation balance. The Pauli-X mixer induces exploration by perturbing the system and modifying all basis state amplitudes. As a result, the search is global and all basis states can be reached. The generalized Grover mixer facilitates exploitation of potential optimal solutions through amplitude amplification. The simultaneous applications of both mixers help improve the efficiency of finding the optimal solution. Without the mixer of exploitation, over-exploration may delay the time to obtain the global optimum. Without the mixer of exploration, over-exploitation could increase the chance of being trapped in local optimum.

Similar to QAOA, the most critical component to the optimization performance of TM-QABOA is the mixer operator U_B. Because the choice of U_B determines how the amplitudes are modified, various mixers have been proposed to improve searching efficiency. For instance, the XY mixer has been proposed [42]-[43]. The XY mixer is applied to explore states that are topologically connected as graphs. This topological constraint reduces the searching space so that the computational efficiency can be improved for certain problems. Similarly, the continuous-time quantum walk mixer [44] was introduced to explore the graphs as constrained optimization. A combination of continuous-time quantum walk and generalized Grover operators [10] was recently proposed to enhance both exploration and exploitation. The Grover operator facilitates exploitation and ensures fair sampling of the solutions [45]. Other examples of mixers include a free-axis mixer which performs qubit rotations about any axis in the XY-plane [46], an entangled gate mixer which induces wider exploration of the search space [47], and a mixer defined with adaptive bias fields

to improve runtime based on measured outcomes from previous iterations [48].

Currently, most mixers either reduce the size of the search space or increase exploitation. How to improve the exploration-exploitation balance has not been studied. Current studies of mixers focus more on exploration. Understanding the benefits of exploitation is helpful to design more efficient algorithms.

4.2 BAYESIAN OPTIMIZATION

In the proposed TM-QABOA, BO is used to optimize the rotation angles and continuous design variables in mixed-integer problems. Given a collection of sample points, a surrogate model is constructed and updated iteratively to guide the search process towards the optimal solution. The most widely used surrogate is Gaussian process regression (GPR), which also captures the uncertainty of the surrogate predictions. The surrogate can be very efficiently evaluated to determine the next sample point.

First, an initial collection of sample points is generated through either a pre-existing dataset or random sampling from the search space. The surrogate is then trained on the initial sample points. At sample location x in the search space, the GPR model represents the objective function value $f(x)$ as a Gaussian distribution, defined as

$$f(x) \sim GP(m(x), K(x, x')) \tag{11}$$

where $m(x)$ is the mean function and $K(x, x')$ is a covariance kernel function corresponding to samples x and x'. One type of kernel is the Matérn kernel, defined as

$$K_M(x, x') = \frac{1}{\Gamma(\nu)2^{\nu-1}} \left(\frac{\sqrt{2\nu}}{l} d(x, x') \right)^{\nu} K_{\nu}(\frac{\sqrt{2\nu}}{l} d(x, x')) \tag{12}$$

where ν is the smoothness hyperparameter, l is the length scale, $\Gamma(\nu)$ is the gamma function, $d(x, x')$ is the Euclidean distance between x and x', and K_{ν} is the modified Bessel function of the second kind. With a small value of ν, $K_M(x, x')$ allows the GPR to reliably model significant changes in the objective function.

During each optimization iteration, the next sample is selected by maximizing an acquisition function. The acquisition function $A(x)$ represents how useful x is in finding the optimal solution. $A(x)$ is defined based on the surrogate so that its optimization can be done efficiently. One example of an acquisition function is upper confidence bound (UCB). For minimization, UCB is defined as

$$A_{UCB}(x) = \alpha\sigma(x) - \mu(x) \tag{13}$$

where $\alpha > 0$ is a tradeoff parameter, $\sigma(x)$ is the standard deviation of the objective value, and $\mu(x)$ is the average objective value. The value of α determines the level of balance

Copyright © 2023 by ASME

between exploration and exploitation. When α is small, exploitation is favored where the search process finds sample points with small objective values. When α is large, exploration is favored where the search process finds sample points with large levels of uncertainty. After the next sample is determined, the objective function is evaluated and the surrogate model is updated with the new sample. The iterations continue until the convergence criteria are met.

4.3 TM-QABOA PROCEDURE

In TM-QABOA, the quantum circuit is used to obtain the optimal solution in a discrete subspace whereas BO optimizes the continuous design variables and the hyperparameters of the rotation angles in the circuit. Table 1 outlines the TM-QABOA procedure. Initial samples of the rotation angles $\gamma_i, \beta_i, \theta_i$ ($i = 1, \ldots, p$) and continuous design variables x_c are randomly sampled from the search space. For each initial sample, the quantum circuit in Eq. (3) is run and $|\psi_f\rangle$ is measured. The objective function value f is then computed with $|\psi_f\rangle$ and x_c. The GPR is trained on the collection of initial sample points. During each optimization iteration, the next sample $x^\dagger$ of rotation angles and continuous design variables is obtained from maximizing $A(x)$. The quantum circuit in Eq. (3) is run with $x^\dagger$ and $|\psi_f\rangle$ is measured. The objective function value $f^\dagger$ is computed with $|\psi_f\rangle$. The current best objective function value f^* is updated to $f^\dagger$ if $f^\dagger$ is better than f^*. The GPR is then updated with $\{x^\dagger, f^\dagger\}$. The optimization process continues until the maximum number of iterations is complete or convergence criteria are met. TM-QABOA is implemented and integrated with the Qiskit Aer simulator.

5. TWO EXAMPLES

To demonstrate the feasibility of QC in solving TO problems, TM-QABOA is used to solve two examples. The first problem, discussed in Section 5.1, is a two-dimensional truss design problem. The objective is to minimize the displacement of a targeted node in a two-dimensional truss structure. The second problem, discussed in Section 5.2, is a metamaterial design problem. The objective is to maximize the strength-weight ratio of an aluminum metamaterial given geometrical parameters. The problems are designed with small numbers of design variables to account for the number of qubits available in the simulator.

5.1 TWO-DIMENSIONAL TRUSS DESIGN

Fig. 2 shows a two-dimensional truss structure. The structure contains pinned supports at nodes 1, 2, 5, 6, and 11. The structure is also subjected to a downwards force of $F = 300$ kips at node 4. The solid lines represent members always present in the structure, and the dashed lines represent members that may or may not be present. For the purpose of discussion, the dashed-line members are called optional members. All members are

made of A36 structural steel and have cross-sectional areas of 10 in^2.

TABLE 1: TM-QABOA procedure.

INPUT: Num_init_samples, Num_BO_iter
OUTPUT: $x^*, f(x^*)$

(1)	$i = 0$, $D = \{\}$
(2)	WHILE $i <$ Num_init_samples:
(3)	Sample $x = (\gamma_1, \beta_1, \theta_1, \ldots, \gamma_p, \beta_p, \theta_p, x_c)$ from search space
(4)	Run quantum circuit in Eq. (3) with x
(5)	Measure $|\psi_f\rangle$ from the quantum circuit
(6)	Compute $f = f(x_c, |\psi_f\rangle)$
(7)	Append $\{x, f\}$ to D
(8)	$i = i + 1$
(9)	END WHILE
(10)	Train GPR on D
(11)	$i = 0$
(12)	WHILE $i <$ Num_BO_iter and convergence criteria are not met:
(13)	Find $x^\dagger = \underset{x}{\arg\max}\, A(x)$
(14)	Run quantum circuit in Eq. (3) with $x^\dagger$
(15)	Measure $|\psi_f\rangle$ from the quantum circuit
(16)	Compute $f^\dagger = f(x_c^\dagger, |\psi_f\rangle)$
(17)	IF $f^\dagger < f^*$:
(18)	$x^* = x^\dagger$ $f^* = f^\dagger$
(19)	END IF
(20)	Append $\{x^\dagger, f^\dagger\}$ to D
(21)	Update GPR on D
(22)	$i = i + 1$
(23)	END WHILE

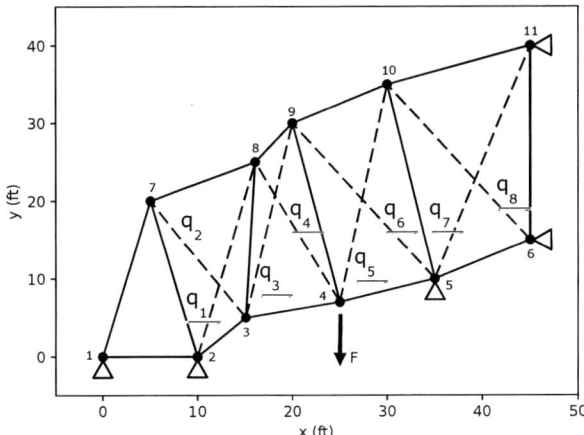

FIGURE 2: Two-dimensional truss structure with solid lines representing members always present in the structure and dashed lines representing members optionally present in the structure.

The objective is to minimize the downward displacement δ at node 4. This is done by determining which of the eight optional members are present in the truss. The presence of an optional member is denoted by a binary variable q_i $(i = 1, ..., 8)$, where $q_i = 1$ if the member is present, or $q_i = 0$ if absent. The downward displacement at node 4 is determined from a weighted sum of displacement values for all possible truss layouts. This sum is defined as

$$f(q_1, ..., q_8) = \sum_{i=0}^{2^8-1} w_i \delta(e_i) \qquad (14)$$

where

$$w_i = \begin{cases} 1 & ((q_1, ..., q_8) = e_i) \\ 0 & ((q_1, ..., q_8) \neq e_i) \end{cases}. \qquad (15)$$

Here, e_i represents a specific binary string value of $(q_1, ..., q_8)$. In Eq. (15), $w_i = 1$ when $(q_1, ..., q_8)$ takes the value of e_i, and $w_i = 0$ otherwise. Each displacement value $\delta(e_i)$ is computed from the linear system $KU = F$ where F is the force vector, K is the stiffness matrix, and U is the displacement vector.

In this problem, three optimization algorithms are performed, including the discrete BO method [49] where a discrete kernel is utilized, the X-QABOA where the quantum walk mixer in the original QABOA [10] is replaced with the Pauli-X mixer, and the proposed TM-QABOA. All three algorithms involve BO where the GPR surrogates guide the search process towards the optimal solution. Each algorithm is repeated for three runs to determine average convergence behaviors. The search space for discrete BO consists of the eight binary variables q_i $(i = 1, ..., 8)$. The search space for X-QABOA consists of six rotation angles γ_i and β_i $(i = 1, 2, 3)$. The search space for TM-QABOA consists of six rotation angles $\gamma_i, \beta_i,$ and θ_i $(i = 1, 2)$. All rotation angles range from 0 to 2π. For X-QABOA and TM-QABOA, the quantum circuit is performed on eight qubits which each represent the presence of an optional member.

The initial dataset is constructed with five sample points randomly acquired from the search space. The same five points are used to initialize the GPR for all runs of all algorithms. Twenty-five BO iterations are perfomed. For discrete BO, the GPR is defined with the squared-exponential kernel. The squared-exponential kernel is defined with the Hamming distance function. For both X-QABOA and TM-QABOA, the GPR is defined with the Matérn kernel set to $\nu = 0.5$ to model significant displacement changes with small rotation angle changes. For all algorithms, the UCB acquisition function with $\alpha = 1$ is used to select the next sample point. Simulated annealing is used to sample 20,000 points from the search space, each of which is used to evaluate the UCB function.

Fig. 3 shows the average convergences of the three optimization algorithms. It is observed that TM-QABOA on average results in a better objective value than X-QABOA. This suggests that a dynamic exploration-exploitation balance can help improve the efficiency of finding the optimal solution. The amplitude amplification performed by the generalized Grover mixers increases the amplitudes of better solutions and

suppresses the ones of the less promising solutions. It is also observed that discrete BO performs slightly better than TM-QABOA in terms of the number of iterations. However, because the difference between both average objective values is small, TM-QABOA can sometimes result in a better solution than discrete BO. Furthermore, from iterations 20 to 23, the average objective values in TM-QABOA are slightly lower than those in discrete BO.

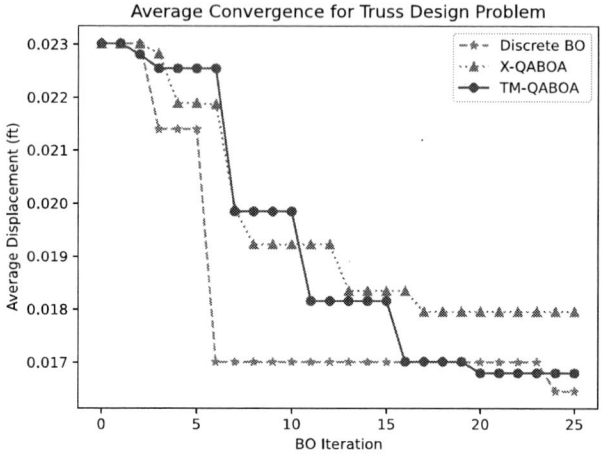

FIGURE 3: Average convergence behaviors of discrete BO, X-QABOA, and TM-QABOA for the two-dimensional truss design problem.

5.2 METAMATERIALS DESIGN

To reduce the searching space in metamaterials design, the periodic surface modeling [50] was recently applied for shape and topology optimization of metamaterials [23]. The interface of metamaterials is described by the periodic surface model, defined as [50]

$$\phi(r) = \sum_{k=1}^{K} \mu_k \cos [2\pi \kappa_k (p_k \cdot r)] \qquad (16)$$

where $r \in \mathbb{R}^4$ is a position vector in homogeneous coordinates, μ_k is a periodic moment, κ_k is a scale factor, $p_k \in \mathbb{R}^4$ is a basis vector, and K is the total number of basis vectors. In this example, $\kappa_k = 1$.

Since Eq. (16) describes an implicit surface, intersection operations are needed to construct a solid metamaterial structure. First, two implicit surfaces ϕ and ϕ_c are generated by the intersection operation defined as

$$\phi_s = \max (\phi, -\phi_c). \qquad (17)$$

ϕ_c is defined as

$$\phi_c = \phi - C \qquad (18)$$

Copyright © 2023 by ASME

where C is the offset between the surfaces. Next, the solid metamaterial structure is constructed with the intersection operation defined as

$$\phi_a = \max(\phi_b, \phi_c) \tag{19}$$

where ϕ_b is a cube with side length of 0.03 m. The resulting structure after Eq. (19) is a representative volume element (RVE) of the metamaterial.

The objective of this example is to maximize the strength-weight ratio of a metamaterial structure. The material of this structure is aluminium with an elastic modulus of 68 GPa and Poisson's ratio of 0.32. Maximizing the strength-weight ratio is equivalent to minimizing

$$SV(\phi_a) = \alpha\left(\frac{V_e(\phi_a)}{V_0}\right) + (1 - \alpha)\left(\frac{S_{max}(\phi_a)}{S_0}\right) \tag{20}$$

where $V_e(\phi_a)$ is the volume of the RVE, V_0 is a reference volume of 10^{-5} m^3, $S_{max}(\phi_a)$ is the maximum von Mises stress, S_0 is the reference maximum von Mises stress equal to 5×10^7 Pa, and α is a weight parameter set to 0.5. SV is computed by performing finite-element simulations on the model. The finite-element simulations are conducted with a maximum surface Delaunay ball radius of 0.0012 m, a maximum facet distance of 0.00017 m, and a maximum circumference edge of 6.

The design variables which define ϕ_a include basis vectors, periodic moments, and the offset between the two periodic surfaces. A collection of ten basis vectors is defined as

$$P = \{P_1, P_2, P_3, P_4, Q_1, Q_2, Q_3, Q_4, Q_5, Q_6\} =$$

$$= \left\{ \begin{bmatrix} 1 \\ 0 \\ 0 \\ 0 \end{bmatrix}, \begin{bmatrix} 0 \\ 1 \\ 0 \\ 0 \end{bmatrix}, \begin{bmatrix} 0 \\ 0 \\ 1 \\ 0 \end{bmatrix}, \begin{bmatrix} 0 \\ 0 \\ 0 \\ 0 \end{bmatrix}, \begin{bmatrix} 1 \\ 1 \\ 0 \\ 0 \end{bmatrix}, \begin{bmatrix} -1 \\ 1 \\ 0 \\ 0 \end{bmatrix}, \begin{bmatrix} 1 \\ 0 \\ 1 \\ 0 \end{bmatrix}, \begin{bmatrix} -1 \\ 0 \\ 1 \\ 0 \end{bmatrix}, \begin{bmatrix} 0 \\ 1 \\ 1 \\ 0 \end{bmatrix}, \begin{bmatrix} 0 \\ -1 \\ 1 \\ 0 \end{bmatrix} \right\} \tag{21}$$

where P_i ($i = 1, ..., 4$) is always present in the metamaterial structure and Q_i ($i = 1, ..., 6$) is optionally present in the structure. In other words, there are six binary design variables q_i ($i = 1, ..., 6$) representing whether Q_i is present in the structure. P_i ($i = 1, ..., 4$) remains present in the structure to ensure a porous geometry. Next are ten periodic moments $\tilde{\mu}_{P_1}, ..., \tilde{\mu}_{P_4}, \tilde{\mu}_{Q_1}, ..., \tilde{\mu}_{Q_6}$ which describe the magnitudes of the lattice directions. The periodic moments are ten continuous design variables ranging from 0 to 300. Last is the offset C between the implicit surfaces ϕ and ϕ_c. This is one continuous variable ranging from 150 to 300.

Even with the ten basis vectors defined in Eq. (21), a wide variety of topologies can be obtained. Two examples in Fig. 4 are used to illustrate. The structure in Fig. 4(a) is defined with the basis vectors P_1, P_2, P_3, and P_4, whereas the structure in Fig. 4(b) is defined with the basis vectors P_1, P_2, P_3, P_4, and Q_6. For both structures, the periodic moments of all basis vectors are 100 and the offset between implicit surfaces is 250.

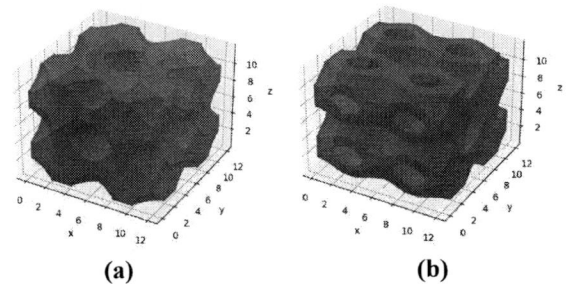

FIGURE 4: Metamaterial structures defined with PS models. The basis vectors include **(a)** P_1, P_2, P_3, P_4, and **(b)** P_1, P_2, P_3, P_4, Q_6.

In total, there are seventeen design variables, six of which are binary and the other eleven are continuous. Eq. (20) is expressed as a weighted sum defined as

$$f(q_1, ..., q_6, \tilde{\mu}, C) = \sum_{i=0}^{2^6-1} w_i SV(e_i, \tilde{\mu}, C) \tag{22}$$

where

$$w_i = \begin{cases} 1 & ((q_1, ..., q_6) = e_i) \\ 0 & ((q_1, ..., q_6) \neq e_i) \end{cases}, \tag{23}$$

and $\tilde{\mu} = (\tilde{\mu}_{P_1}, ..., \tilde{\mu}_{P_4}, \tilde{\mu}_{Q_1}, ..., \tilde{\mu}_{Q_6})$. Here, e_i represents a specific binary string value of $(q_1, ..., q_6)$.

Discrete BO, X-QABOA, and TM-QABOA are each repeated for three runs to determine average convergence behaviors. The search space for discrete BO consists of the seventeen design variables q_i ($i = 1, ..., 6$), $\tilde{\mu}$, and C. The BO search space for X-QABOA consists of q_i ($i = 1, ..., 6$), $\tilde{\mu}$, and six rotation angles γ_i and β_i ($j = 1, 2, 3$). The BO search space for TM-QABOA consists of q_i ($i = 1, ..., 6$), $\tilde{\mu}$, and six rotation angles denoted as γ_j, β_j, and θ_j ($j = 1, 2$). All rotation angles range from 0 to 2π.

Each run is conducted with the same initial dataset of five random sample points and twenty BO iterations. For discrete BO, the GPR is defined with the squared-exponential kernel. The squared-exponential kernel is defined with the Hamming distance function. For X-QABOA and TM-QABOA, the GPR is defined with the Matérn kernel with $\nu = 0.5$ to model significant SV value changes with small changes in design variables. The UCB acquisition function with $\alpha = 1$ is maximized to determine the next sample point from 20,000 points sampled from simulated annealing.

The average convergence performances for the three algorithms are shown in Fig. 5. Both X-QABOA and TM-QABOA result in faster convergence than discrete BO. The two quantum optimization algorithms perform very similarly. X-QABOA results in a slightly better average objective value after twenty iterations. From iterations 6 to 13, TM-QABOA exhibits lower average objective values than X-QABOA. This suggests

Copyright © 2023 by ASME

that a dynamic exploration-exploitation balance is useful in increasing searching efficiency.

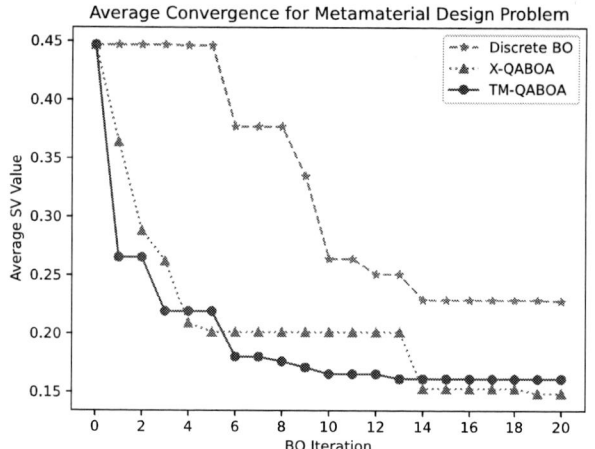

FIGURE 5: Average convergence behaviors of discrete BO, X-QABOA, and TM-QABOA for the aluminum metamaterial design problem.

6. CONCLUDING REMARKS

The main challenge of TO for engineering applications is the high computational cost when a large number of design variables are involved. Improving the computational efficiency of TO is an important research task. In this paper, the feasibility of using QC algorithms to solve the combinatorial problems in TO is demonstrated. A hybrid quantum-classical algorithm, TM-QABOA, is proposed to use the strengths of quantum optimization and Bayesian optimization simultaneously. Quantum optimization can search discrete combinations efficiently, whereas surrogate-based Baysian optimization improves the efficiency of global search for continuous variables. The proposed TM-QABOA can improve searching efficiency of quantum circuits with two mixers in an alternating fashion. The two mixers enable a dynamic balance between exploration and exploitation. This balance can improve the searching efficiency. In comparison with QAOA, QABOA utilizes a surrogate to guide the optimization. Surrogate-based optimization can also improve searching efficiency by reducing the number of objective function evaluations, which can also reduce the number of quantum circuit runs. The numerical results of two TO problems show that TM-QABOA can solve small TO problems efficiently. Therefore, the exploration-exploitation balance is important in designing mixers for QAOA-type algorithms.

This work is the first known attempt of applying QAOA-type algorithms to solve TO problems. Although QC currently has limited practicality due to the small number of available qubits, it is observed in both problems that QC is useful in increasing the searching efficiency. As quantum computer technology continues to evolve, quantum computers with larger numbers of qubits will be available to solve larger-scale TO problems. In addition, the depth of the quantum circuit is currently limited by the decoherence in quantum computers. That is, the quantum states cannot be kept for longer time to run more unitary operators before they collapse because of the interactions with the external environment. The two limitations need to be overcome in order to solve real-world TO problems.

In future work, the scalability of QABOA in terms of the number of qubits and the number of design variables should be studied to fully understand its performance in solving TO problems. Another extension is to perform sensitivity studies on the convergence of TM-QABOA with different quantum circuit depths, which correspond to different numbers of rotation angles. The convergence is also affected by the choices of acquisition function in BO and the kernels in the GPR. The proposed TM-QABOA is capable of handling mixed-integer optimization problems. The continuous design variables are optimized with BO. Another extension is to instead use the quantum side of the algorithm to optimize continuous design variables. Using a quantum computer to compute continuous variables itself is not well-developed and requires major research efforts.

REFERENCES

1. Borrvall, T., & Petersson, J. (2001). Large-scale topology optimization in 3D using parallel computing. *Computer Methods in Applied Mechanics and Engineering*, **190** (46-47), 6201-6229.
2. Martínez-Frutos, J., & Herrero-Pérez, D. (2016). Large-scale robust topology optimization using multi-GPU systems. *Computer Methods in Applied Mechanics and Engineering*, **311**, 393-414.
3. Martínez-Frutos, J., Martínez-Castejón, P. J., & Herrero-Pérez, D. (2017). Efficient topology optimization using GPU computing with multilevel granularity. *Advances in Engineering Software*, **106**, 47-62.
4. Gogu, C. (2015). Improving the efficiency of large scale topology optimization through on-the-fly reduced order model construction. *International Journal for Numerical Methods in Engineering*, **101** (4), 281-304.
5. Luo, Y., & Bao, J. (2019). A material-field series-expansion method for topology optimization of continuum structures. *Computers & Structures*, **225**, 106122.
6. Zheng, W., Wang, Y., Zheng, Y., & Da, D. (2020). Efficient topology optimization based on DOF reduction and convergence acceleration methods. *Advances in Engineering Software*, **149**, 102890.
7. Guo, G., Zhao, Y., & Zuo, W. (2022). Explicit and efficient topology optimization for three-dimensional structures considering geometrical nonlinearity. *Advances in Engineering Software*, **173**, 103238.
8. Farhi, E., Goldstone, J., & Gutmann, S. (2014). A quantum approximate optimization algorithm. *arXiv preprint arXiv:1411.4028.*
9. Wang, Z., Hadfield, S., Jiang, Z., & Rieffel, E. G. (2018). Quantum approximate optimization algorithm for MaxCut: A fermionic view. *Physical Review A*, **97** (2), 022304.
10. Wang, Y. (2021). A quantum approximate Bayesian optimization algorithm for continuous problems. In *IISE*

Copyright © 2023 by ASME

Annual Conference. Proceedings (pp. 235-240). Institute of Industrial and Systems Engineers (IISE).

11. Mahdavi, A., Balaji, R., Frecker, M., & Mockensturm, E. M. (2006). Topology optimization of 2D continua for minimum compliance using parallel computing. *Structural and Multidisciplinary Optimization*, **32**, 121-132.

12. Paris, J., Colominas, I., Navarrina, F., & Casteleiro, M. (2013). Parallel computing in topology optimization of structures with stress constraints. *Computers & Structures*, **125**, 62-73.

13. Kim, T. S., Kim, J. E., & Kim, Y. Y. (2004). Parallelized structural topology optimization for eigenvalue problems. *International Journal of Solids and Structures*, **41** (9-10), 2623-2641.

14. Liu, H., Tian, Y., Zong, H., Ma, Q., Wang, M. Y., & Zhang, L. (2019). Fully parallel level set method for large-scale structural topology optimization. *Computers & Structures*, **221**, 13-27.

15. Lin, H., Liu, H., & Wei, P. (2022). A parallel parameterized level set topology optimization framework for large-scale structures with unstructured meshes. *Computer Methods in Applied Mechanics and Engineering*, **397**, 115112.

16. Aage, N., Andreassen, E., & Lazarov, B. S. (2015). Topology optimization using PETSc: An easy-to-use, fully parallel, open source topology optimization framework. *Structural and Multidisciplinary Optimization*, **51**, 565-572.

17. Zhang, Z. D., Ibhadode, O., Bonakdar, A., & Toyserkani, E. (2021). TopADD: a 2D/3D integrated topology optimization parallel-computing framework for arbitrary design domains. *Structural and Multidisciplinary Optimization*, **64** (3), 1701-1723.

18. Duarte, L. S., Celes, W., Pereira, A., M. Menezes, I. F., & Paulino, G. H. (2015). PolyTop++: an efficient alternative for serial and parallel topology optimization on CPUs & GPUs. *Structural and Multidisciplinary Optimization*, **52**, 845-859.

19. Koh, H. S., Kim, J. H., & Yoon, G. H. (2020). Efficient topology optimization of multicomponent structure using substructuring-based model order reduction method. *Computers & Structures*, **228**, 106146.

20. Guest, J. K., & Smith Genut, L. C. (2010). Reducing dimensionality in topology optimization using adaptive design variable fields. *International Journal for Numerical Methods in Engineering*, **81** (8), 1019-1045.

21. Kazemi, H., Vaziri, A., & Norato, J. A. (2018). Topology optimization of structures made of discrete geometric components with different materials. *Journal of Mechanical Design*, **140** (11), 111401.

22. Amir, O., Aage, N., & Lazarov, B. S. (2014). On multigrid-CG for efficient topology optimization. *Structural and Multidisciplinary Optimization*, **49**, 815-829.

23. Lu, Y., & Wang, Y. (2022). Structural optimization of metamaterials based on periodic surface modeling. *Computer Methods in Applied Mechanics and Engineering*, **395**, 115057.

24. Sasaki, H., & Igarashi, H. (2019). Topology optimization accelerated by deep learning. *IEEE Transactions on Magnetics*, **55** (6), 1-5.

25. Kollmann, H. T., Abueidda, D. W., Koric, S., Guleryuz, E., & Sobh, N. A. (2020). Deep learning for topology optimization of 2D metamaterials. *Materials & Design*, **196**, 109098.

26. Xiang, C., Wang, D., Pan, Y., Chen, A., Zhou, X., & Zhang, Y. (2022). Accelerated topology optimization design of 3D structures based on deep learning. *Structural and Multidisciplinary Optimization*, **65** (3), 99.

27. Qiu, C., Du, S., & Yang, J. (2021). A deep learning approach for efficient topology optimization based on the element removal strategy. *Materials & Design*, **212**, 110179.

28. Sun, Z., Wang, Y., Liu, P., & Luo, Y. (2022). Topological dimensionality reduction-based machine learning for efficient gradient-free 3D topology optimization. *Materials & Design*, **220**, 110885.

29. Li, B., Huang, C., Li, X., Zheng, S., & Hong, J. (2019). Non-iterative structural topology optimization using deep learning. *Computer-Aided Design*, **115**, 172-180.

30. Li, M., Cheng, Z., Jia, G., & Shi, Z. (2019). Dimension reduction and surrogate based topology optimization of periodic structures. *Composite Structures*, **229**, 111385.

31. Kallioras, N. A., Kazakis, G., & Lagaros, N. D. (2020). Accelerated topology optimization by means of deep learning. *Structural and Multidisciplinary Optimization*, **62** (3), 1185-1212.

32. Farhi, E., Goldstone, J., Gutmann, S., & Sipser, M. (2000). Quantum computation by adiabatic evolution. *arXiv preprint quant-ph/0001106*.

33. Kadowaki, T., & Nishimori, H. (1998). Quantum annealing in the transverse Ising model. *Physical Review E*, **58** (5), 5355.

34. Peruzzo, A., McClean, J., Shadbolt, P., Yung, M. H., Zhou, X. Q., Love, P. J., Aspuru-Guzik, A. & O'Brien, J. L. (2014). A variational eigenvalue solver on a photonic quantum processor. *Nature Communications*, **5** (1), 4213.

35. Baritompa, W. P., Bulger, D. W., & Wood, G. R. (2005). Grover's quantum algorithm applied to global optimization. *SIAM Journal on Optimization*, **15** (4), 1170-1184.

36. Wang, Y. (2014, August). Global optimization with quantum walk enhanced Grover search. In *International Design Engineering Technical Conferences and Computers and Information in Engineering Conference* (Vol. 46322, p. V02BT03A027). American Society of Mechanical Engineers.

37. Grover, L. K. (1996, July). A fast quantum mechanical algorithm for database search. In *Proceedings of the Twenty-Eighth Annual ACM Symposium on Theory of Computing* (pp. 212-219).

38. Kitai, K., Guo, J., Ju, S., Tanaka, S., Tsuda, K., Shiomi, J., & Tamura, R. (2020). Designing metamaterials with

Copyright © 2023 by ASME

quantum annealing and factorization machines. *Physical Review Research*, **2** (1), 013319.

39. Perdomo-Ortiz, A., Dickson, N., Drew-Brook, M., Rose, G., & Aspuru-Guzik, A. (2012). Finding low-energy conformations of lattice protein models by quantum annealing. *Scientific Reports*, **2** (1), 1-7.

40. Robert, A., Barkoutsos, P. K., Woerner, S., & Tavernelli, I. (2021). Resource-efficient quantum algorithm for protein folding. *npj Quantum Information*, **7** (1), 38.

41. Micheletti, C., Hauke, P., & Faccioli, P. (2021). Polymer physics by quantum computing. *Physical Review Letters*, **127** (8), 080501.

42. Hadfield, S., Wang, Z., O'gorman, B., Rieffel, E. G., Venturelli, D., & Biswas, R. (2019). From the quantum approximate optimization algorithm to a quantum alternating operator ansatz. *Algorithms*, **12** (2), 34.

43. Wang, Z., Rubin, N. C., Dominy, J. M., & Rieffel, E. G. (2020). X y mixers: Analytical and numerical results for the quantum alternating operator ansatz. *Physical Review A*, **101** (1), 012320.

44. Marsh, S., & Wang, J. B. (2019). A quantum walk-assisted approximate algorithm for bounded NP optimisation problems. *Quantum Information Processing*, **18**, 1-18.

45. Bärtschi, A., & Eidenbenz, S. (2020, October). Grover mixers for QAOA: Shifting complexity from mixer design to state preparation. In *2020 IEEE International Conference on Quantum Computing and Engineering (QCE)* (pp. 72-82). IEEE.

46. Govia, L. C. G., Poole, C., Saffman, M., & Krovi, H. K. (2021). Freedom of the mixer rotation axis improves performance in the quantum approximate optimization algorithm. *Physical Review A*, **104** (6), 062428.

47. Chen, Y., Zhu, L., Mayhall, N. J., Barnes, E., & Economou, S. E. (2022, June). How much entanglement do quantum optimization algorithms require?. In *Quantum 2.0* (pp. QM4A-2). Optica Publishing Group.

48. Yu, Y., Cao, C., Dewey, C., Wang, X. B., Shannon, N., & Joynt, R. (2022). Quantum approximate optimization algorithm with adaptive bias fields. *Physical Review Research*, **4** (2), 023249.

49. Wang, Y. (2021). Design of trustworthy cyber–physical–social systems with discrete Bayesian optimization. *Journal of Mechanical Design*, **143** (7).

50. Wang, Y. (2007). Periodic surface modeling for computer aided nano design. *Computer-Aided Design*, **39** (3), 179-189.

Proceedings of the ASME 2023
International Design Engineering Technical Conferences and
Computers and Information in Engineering Conference
IDETC-CIE2023
August 20-23, 2023, Boston, Massachusetts

DETC2023-116896

CONCURRENT PROBABILISTIC CONTROL CO-DESIGN AND LAYOUT OPTIMIZATION OF WAVE ENERGY CONVERTER FARMS USING SURROGATE MODELING

Saeed Azad[*]
Postdoctoral Fellow
Department of Systems Engineering
Colorado State University
Fort Collins, CO 80523
Email: saeed.azad@colostate.edu

Daniel R. Herber
Assistant Professor
Department of Systems Engineering
Colorado State University
Fort Collins, CO 80523
Email: daniel.herber@colostate.edu

ABSTRACT

Wave energy converters (WECs) are a promising candidate for meeting the increasing energy demands of today's society. It is known that the sizing and power take-off (PTO) control of WEC devices have a major impact on their performance. In addition, to improve power generation, WECs must be optimally deployed within a farm. While such individual aspects have been investigated for various WECs, potential improvements may be attained by leveraging an integrated, system-level design approach that considers all of these aspects. However, the computational complexity of estimating the hydrodynamic interaction effects significantly increases for large numbers of WECs. In this article, we undertake this challenge by developing data-driven surrogate models using artificial neural networks and the principles of many-body expansion. The effectiveness of this approach is demonstrated by solving a concurrent plant (i.e., sizing), control (i.e., PTO parameters), and layout optimization of heaving cylinder WEC devices. WEC dynamics were modeled in the frequency domain, subject to probabilistic incident waves with farms of 3, 5, 7, and 10 WECs. The results indicate promising directions toward a practical framework for array design investigations with more tractable computational demands.

Keywords: surrogate modeling; control co-design; layout optimization; wave energy converter farms; energy systems

1 INTRODUCTION

Wave energy, with its temporal and spatial availability, low variability, and high predictability, is a promising source of renewable energy [1]. However, its technology readiness level (TRL), which is often used to classify the development maturity of a new technology, is still low compared to wind and solar technologies [2], indicating that more investment and research are required to improve their techno-economic performance [3].

Among the many approaches undertaken as a part of this effort, the primary focus has been on the sizing of the device (i.e., plant) and/or its power take-off (PTO) (i.e., control) [4–8]. Since the economic viability of wave energy converter (WEC) devices depends strongly on the energy generated, WECs must be carefully deployed in an array. While WEC arrays reduce the installation, maintenance, and operation costs [9], they result in a complex hydrodynamic interaction effect that appears mainly due to the presence of multiple WECs in close proximity [10]. This interaction effect can be constructive or destructive [11]. Therefore, the spatial configuration of WECs within an array must be carefully selected to ensure a constructive effect. This exploration is generally done through array or layout optimization of WECs [9, 12, 13].

As emphasized in Ref. [14], potential improvements in WEC array performance may be realized by leveraging a system-level design framework that considers plant, control, and layout concurrently. However, accurate estimation of the hydrodynamic

---

[*]Corresponding author, saeed.azad@colostate.edu

interaction effect in array optimization in general, and this integrated design framework in particular, is computationally expensive [15]. For example, a single call to the boundary element method (BEM) solver Nemoh for a two- and three-WEC farm using axisymmetric meshes takes about 384 s, and 1305 s, respectively. Because of this cost, the design of WEC farms using accurate numerical solutions has been generally limited in the complexity of the array (fixed array geometries such as square and triangular [11]), and the total number of WECs in the farm [15].

In this article, we estimate the complex hydrodynamic interaction effect, up to second-order, by constructing data-driven surrogate models using artificial neural networks (ANNs), along with a hierarchical interaction decomposition approach inspired by concepts from many-body expansion (MBE) [16, 17]. Since BEM solvers are known to provide the accurate hydrodynamics [18], the ANN models were trained on data generated from an open-source, BEM solver Nemoh [19–21], for one- and two-WEC studies, leading to a second-order approximation of the complex hydrodynamic interaction effect. These surrogate models are then utilized within the optimization framework to implement a surrogate-assisted optimization approach that has the potential to efficiently address the issue of accurate estimation of hydrodynamic coefficients within large arrays. Through this, we aim to develop a practical framework that enables the design of more complex WEC arrays compared to other similar works in the literature with fewer assumptions [15].

This article presents some preliminary results towards the concurrent optimization of plant (uniform across the farm), control, and layout for heaving cylinder WEC devices within a farm. This framework naturally combines an stochastic in expectation uncertain control co-design (SE-UCCD) [22] formulation with the layout optimization problem. By enabling the simultaneous implementation of an integrated design approach, known as control co-design (CCD) [8, 23–26] with layout optimization for half-submerged cylindrical, heaving WEC devices in the presence of uncertainties from incident waves, this article attempts to pave the way for more complex investigations of WEC farms.

The heaving WEC is characterized as a point absorber, i.e., having a relatively small dimension with respect to the prevailing wavelength [1]. This device is modeled in the frequency domain, with a PTO system that exerts a load force on the oscillating body while storing energy (thereby providing a bi-directional power flow through a reactive control strategy) [1]. To calculate the absorbed power, the probability distribution of waves is constructed based on historical data collected from Pacific Islands in close proximity to the Hawaiian Islands over the lifetime of the device. This data set is openly accessible from Refs. [27, 28]. The optimization problem is formulated with the objective function of maximizing the expected value of power over the lifetime of the farm per unit volume of the device.

The remainder of this article is organized as follows: Sec. 2

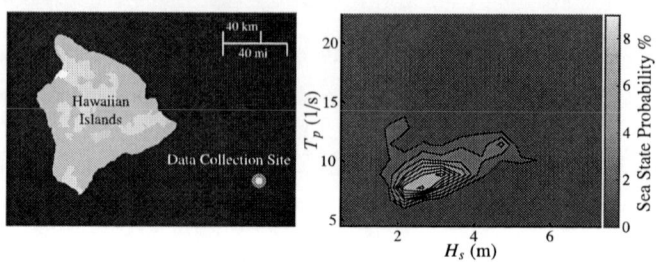

FIGURE 1: Data collection site and the wave scatter diagram created using the joint probability distribution of significant wave heights and wave periods.

presents the methodological discussion, including probabilistic wave modeling, wave-structure interactions, dynamics and equations of motions for WECs, array geometry and considerations, surrogate modeling with an emphasis on ANN, and a brief introduction into principles of MBE; Sec. 3 describes the construction and validation of surrogate models using ANNs and the BEM solver Nemoh; Sec. 4 starts with introducing the concurrent UCCD and layout optimization problem and then presents results associated with multiple array investigations; Finally, Sec. 5 presents conclusions and limitations of the current study, as well as potential future work.

2 METHODS

In this section, we start by describing the wave climate and its modeling considerations. Calculation of hydrodynamic coefficients, along with the basics of potential flow theory and boundary element method in Nemoh are presented next. Then equations of motion for a heaving cylinder WEC are introduced in the frequency domain, and some considerations regarding layout geometry are presented. The section concludes with a discussion on surrogate models and many-body expansion principles.

2.1 Wave Climate and Modeling

In wave energy applications, wind-generated gravity surface waves are of interest. These are waves resulting from the wind blowing over the ocean surface, dominated by gravity and inertial forces [1]. The site of interest is located in the Pacific Islands, off the coast of the Hawaiian Islands. 30 years of historical data, collected from 1976 through 2005, is used to estimate the joint probability distribution of waves for various significant wave heights H_s and wave periods T_p. Since the resulting data form a nonparametric representation of the probability density function, a kernel distribution characterized by a smoothing function and a bandwidth value was utilized. Using a Gaussian quadrature approach with n_{gq} points in each dimension, we first obtained the Legendre-Gauss nodes and weights. The probability distribution of waves was then constructed using these points in Matlab's *ks-*

density function, resulting in the annual probability matrix. Figure 1 presents the location of the site, along with the wave contour diagram for year 1 of the study.

The sea state is described using the JOint North Sea WAve Project (JONSWAP) spectrum, defined as:

$$S_{JS}(H_s, T_p, \omega) = \alpha_s \omega^{-5} \exp\left[-\beta_s \omega^{-4}\right] \tag{1}$$

where ω is the angular frequency, and α_s and β_s are parameters of the spectrum defined as:

$$\alpha_s = \frac{\beta_s}{4} H_s^2 C(\gamma)\gamma^r \tag{2}$$

$$\beta_s = \frac{5}{4}\omega_p^4 \tag{3}$$

where ω_p is the peak angular frequency, and $C(\cdot)$ is a normalizing factor calculated as:

$$C(\gamma) = 1 - 0.287\ln(\gamma) \tag{4}$$

In these equations, γ is defined as:

$$\gamma = \begin{cases} 5 & \text{for } \frac{T_p}{\sqrt{H_s}} \leq 3.6 \\ \exp(5.75 - 1.15\frac{T_p}{\sqrt{H_s}}) & \text{for } 3.6 \leq \frac{T_p}{\sqrt{H_s}} \leq 5 \\ 1 & \text{for } \frac{T_p}{\sqrt{H_s}} > 5 \end{cases} \tag{5}$$

Finally, r is defined as:

$$r = \exp\left[\frac{-1}{2\sigma^2}\left(\frac{\omega}{\omega_p} - 1\right)^2\right] \quad \text{where } \sigma = \begin{cases} 0.07 \text{ for } \omega \leq \omega_p \\ 0.09 \text{ for } \omega > \omega_p \end{cases} \tag{6}$$

For more details on this spectrum and its associated parameters, the readers are referred to Ref. [1, pp. 71].

Using this spectrum, the incident wave field can be modeled by irregular waves constructed through the superposition of n_r regular waves. Assuming that the angle of wave direction β_w is 0, the irregular incident wave can be approximated as:

$$\eta(x,t) = \sum_{i=1}^{n_r} \frac{H_i}{2} \cos(k_i x - \omega_i t + \phi_i) \tag{7}$$

where H_i is the wave amplitude, θ is the randomly generated wave phase, and k is the wave number satisfying the dispersion relation:

$$\omega^2 = gk\tanh kh \approx \begin{cases} gk & \text{as } kh \to \infty \\ gk^2h & \text{as } kh \to 0 \end{cases} \tag{8}$$

where g is the gravitational acceleration, and h is the water depth. Figure (2) shows the JONSWAP spectrum and a wave signal generated for an arbitrary H_s and T_p associated with the site of study.

2.2 Wave-Structure Interactions

Before discussing the design and dynamics of WEC devices, it is necessary to understand the ocean waves and wave-structure interactions. This section describes some of the basic principles without going into too much of the mathematical description of such concepts. Interested readers may refer to Refs. [1, 29, 30]

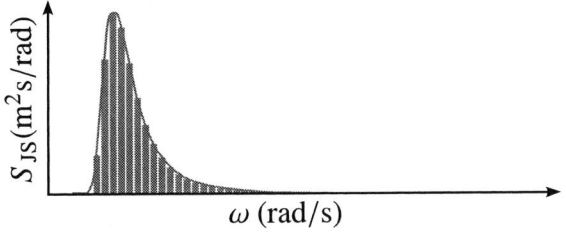

(a) JONSWAP spectrum.

(b) Irregular wave signal resulting from superposition of harmonic waves.

FIGURE 2: JONSWAP spectrum and an irregular wave signal created using 200 harmonic waves.

for more details.

2.2.1 Linear Potential Flow Theory.
In their most fundamental form, the ocean waves may be described through Navier-Stokes equations, which along with continuity equation and incompressibility constraint, result in a system of nonlinear partial differential equations [1]. The numerical solution of this system of equations through computational fluid dynamics (CFD) is computationally expensive. However, by making simplifying assumptions, such as incompressible, inviscid (negligible viscosity), and irrotational flow, one can utilize potential flow theory, which is characterized by the Laplace equation, and boundary constraints at the free surface and any rigid boundary. By further assuming non-steep waves (i.e., small wave height in relation to wave length), the free-surface boundary constraints can be linearized, resulting in linear potential theory, also known as Airy wave theory [1].

2.2.2 Hydrodynamic and Hydrostatic Forces.
In linear potential flow theory, the fluid velocity potential ϕ is divided into potentials corresponding to incident ϕ_i, scattered ϕ_s, and radiated waves ϕ_r:

$$\phi = \phi_i + \phi_s + \phi_r \tag{9}$$

Incident waves represent the propagation of the wave in the absence of any structure. Scattered waves appear as a result of the interaction of the waves and the motionless structures, while radiated waves result from the motion of the structure.

Copyright © 2023 by ASME

The force and force moment on a structure is separated into forces resulting from the incident and scattered waves, also known as the excitation force and the radiation force. The excitation force is created by the action of incident waves on motionless structures. Therefore, the excitation force captures two effects: the motion of the incident wave, also known as the Froude-Krylov force, as well as the scattering effect. The radiation force characterizes the forces applied to the structure as a result of its own oscillatory motion in the absence of an incident wave field. The real part of the radiation force constitutes what is known as the added mass, while the imaginary part is the damping coefficients.

The forces created by scattered and radiated waves are of significant importance in the layout optimization of WEC devices because, as opposed to the incident wave that travels only in one direction, the scattered and radiated waves propagate along all directions, thereby affecting all WEC devices, regardless of their location. Therefore, they redistribute part of the incident energy in all directions [18]. Because of this phenomenon, the design, control, and layout optimization of WECs have strong interactions and must be approached concurrently. Hydrostatic forces result from the change in the hydrostatic pressure on the wetted surface of the body as it moves from its equilibrium position. Analytical and numerical solutions exist to calculate these forces. In the following section, we briefly describe one such numerical method known as the boundary element method or BEM.

2.2.3 Boundary Element Method in Nemoh.
In BEM, systems of partial differential equations from linear potential theory are formulated into the boundary integral form and transformed into a problem of source distribution on the body surface using Green's theorem [31]. While many BEM solvers are available, in this study we use Nemoh, an open-source frequency-domain solver. To use `Nemoh`, the user must provide the mesh for the desired body. Using the `Matlab` wrapper provided in the software package, a symmetric mesh was generated for the cylindrical WEC device based on the specified radius and draft dimensions and used in `Nemoh`.

2.3 Dynamics and Control of WECs
Despite high nonlinearities, the hydrodynamic interactions between a WEC device and ocean waves can be simplified when the device exhibits small amplitudes in its oscillatory motions. This, along with a linear representation of forces involved in WEC dynamics, can be utilized to carry out the modeling in the frequency domain. These frequency-domain models are perhaps the first design space that engineers thoroughly investigate to gain early-stage insights before moving on to the more expensive time-domain models. Due to their computational efficiency, these models are suitable for the preliminary assessment of a system-level design framework in which farm layout design considerations are accompanied by control co-design efforts. In

this section, we briefly present the relevant formulations for the dynamics of WEC arrays in the frequency domain.

Using linear potential flow theory, and considering regular waves with radial frequency of ω and unit amplitude as an input, the equation of motion for n_{wec} buoys in the frequency domain can be described as:

$$-\omega^2 \mathbf{M}\hat{\xi}(\omega) = \hat{\mathbf{F}}_{\text{FK}}(\omega) + \hat{\mathbf{F}}_s(\omega) + \hat{\mathbf{F}}_r(\omega) + \hat{\mathbf{F}}_{\text{hs}}(\omega) + \hat{\mathbf{F}}_{\text{pto}}(\omega) \quad (10)$$

where $\hat{\ast}$ is the complex amplitude of $\ast$, $\hat{\xi}(\cdot) \in \mathbb{R}^{n_{\text{wec}} \times 1}$ is the displacement vector, $\hat{\mathbf{F}}_{\text{FK}}(\cdot)$ is the Froude-Krylov force, $\hat{\mathbf{F}}_s(\cdot)$ is the scattering force vector, $\hat{\mathbf{F}}_r(\cdot)$ is the radiation force, $\hat{\mathbf{F}}_{\text{hs}}(\cdot)$ is the hydrostatic force, $\hat{\mathbf{F}}_{\text{pto}}(\cdot)$ is the power-take-off (PTO) force, all defined in $\in \mathbb{R}^{n_{\text{wec}} \times 1}$. In addition, $\mathbf{M} \in \mathbb{R}^{n_{\text{wec}} \times n_{\text{wec}}}$ is the diagonal mass matrix. The excitation force is defined as the sum of Froude-Krylov and scattering forces:

$$\hat{\mathbf{F}}_e(\omega) = \hat{\mathbf{F}}_{\text{FK}}(\omega) + \hat{\mathbf{F}}_s(\omega) \quad (11)$$

The radiation force is calculated as a function of hydrodynamic damping coefficient matrix $\mathbf{B}(\cdot)$, and added mass coefficient matrix $\mathbf{A}(\cdot)$ as:

$$\hat{\mathbf{F}}_r(\omega) = -i\omega \mathbf{B}(\omega)\hat{\xi}(\omega) + \omega^2 \mathbf{A}(\omega)\hat{\xi}(\omega) \quad (12)$$

where $\mathbf{B}(\cdot) \in \mathbb{R}^{n_{\text{wec}} \times n_{\text{wec}}}$ captures the dissipated energy transmitted from WEC motions to the water (propagating away from the body dissipative effect), and $\mathbf{A}(\cdot) \in \mathbb{R}^{n_{\text{wec}} \times n_{\text{wec}}}$ represents the inertial increase due to water displacement as a result of the WEC motion (reactive effect) [29]. Considering n_{wec} WEC devices in a fixed array with only heave motion, $\mathbf{A}(\omega)$ and $\mathbf{B}(\omega)$ matrices have the following form [15]:

$$\mathbf{A}(\omega) = \begin{bmatrix} a_{11}(\omega) & a_{12}(\omega) & a_{13}(\omega) & \cdots & a_{1n_{\text{wec}}}(\omega) \\ a_{21}(\omega) & a_{22}(\omega) & a_{23}(\omega) & \cdots & a_{2n_{\text{wec}}}(\omega) \\ \vdots & \vdots & \vdots & \vdots & \vdots \\ a_{n_{\text{wec}}1}(\omega) & \cdots & \cdots & \cdots & a_{n_{\text{wec}}n_{\text{wec}}}(\omega) \end{bmatrix} \quad (13)$$

$$\mathbf{B}(\omega) = \begin{bmatrix} b_{11}(\omega) & b_{12}(\omega) & b_{13}(\omega) & \cdots & b_{1n_{\text{wec}}}(\omega) \\ b_{21}(\omega) & b_{22}(\omega) & b_{23}(\omega) & \cdots & b_{2n_{\text{wec}}}(\omega) \\ \vdots & \vdots & \vdots & \vdots & \vdots \\ b_{n_{\text{wec}}1}(\omega) & \cdots & \cdots & \cdots & b_{n_{\text{wec}}n_{\text{wec}}}(\omega) \end{bmatrix} \quad (14)$$

The hydrostatic force, which results from the balance between buoyancy and gravity is calculated as:

$$\hat{\mathbf{F}}_{\text{hs}}(\omega) = -G\hat{\xi}(\omega) \quad (15)$$

where G is the hydrostatic coefficient, calculate as $G = \rho g S$, with S being the cross-sectional area at the undisturbed sea level calculated as a function of the WEC radius: πR_{wec}^2 for a heaving cylinder.

At the core of WEC device functionality is a PTO system that converts mechanical motion into electricity. PTO design has significant implications on the sizing and economic performance of WECs [3, 32]. Therefore it is necessary to investigate the PTO system from the early stages of the design process. It is well known within the WEC literature that incorporating active control strategies significantly increases the energy gener-

Copyright © 2023 by ASME

ated through WECs [6, 33, 34]. Reactive control is one of the earliest control strategies developed for WEC devices. It enables a bidirectional power flow between the PTO spring and the buoy [1]. In its linear form, the PTO force is composed of two contributions:

$$\hat{\mathbf{F}}_{\text{pto}}(\omega) = -i\omega\mathbf{B}_{\text{pto}}\hat{\boldsymbol{\xi}}(\omega) - \mathbf{K}_{\text{pto}}\hat{\boldsymbol{\xi}}(\omega) \qquad (16)$$

where $\mathbf{K}_{\text{pto}} \in \mathbb{R}^{n_{\text{wec}} \times n_{\text{wec}}}$ and $\mathbf{B}_{\text{pto}} \in \mathbb{R}^{n_{\text{wec}} \times n_{\text{wec}}}$ are diagonal matrices of stiffness and damping of PTO systems for WEC devices, respectively. The complex amplitude of the motions of all WECs for a regular wave of frequency ω and unit amplitude can now be described as a transfer function matrix:

$$\hat{\boldsymbol{\xi}}(\omega) = \mathbf{H}(\omega)\hat{\mathbf{F}}_e(\omega) \qquad (17)$$

$$\mathbf{H}(\omega) = \left[[\omega^2(\mathbf{M}+\mathbf{A}(\omega))+G+\mathbf{K}_{\text{pto}}] + i\omega(\mathbf{B}(\omega)+\mathbf{B}_{\text{pto}})\right]^{-1} \qquad (18)$$

From here, captor velocity and acceleration can be calculated as:

$$\hat{\dot{\boldsymbol{\xi}}} = i\omega\hat{\boldsymbol{\xi}} \qquad (19)$$

$$\hat{\ddot{\boldsymbol{\xi}}} = -\omega^2\hat{\boldsymbol{\xi}} \qquad (20)$$

These complex amplitudes can be used to represent the displacement, velocity, and acceleration in time through the following equations:

$$\boldsymbol{\xi}(t) = \text{Re}\{\hat{\boldsymbol{\xi}}(\omega)\exp(i\omega t)\} \qquad (21)$$

$$\dot{\boldsymbol{\xi}}(t) = \text{Re}\{i\omega\hat{\boldsymbol{\xi}}(\omega)\exp(i\omega t)\} \qquad (22)$$

$$\ddot{\boldsymbol{\xi}}(t) = \text{Re}\{-\omega^2\hat{\boldsymbol{\xi}}(\omega)\exp(i\omega t)\} \qquad (23)$$

The time-averaged absorbed mechanical power for a sea state with significant wave height of H_s and peak period of T_p can then be described as:

$$\mathbf{p}_m(H_s, T_p, \omega) = \frac{1}{2}\omega^2\hat{\boldsymbol{\xi}}^T\mathbf{B}_{\text{pto}}\hat{\boldsymbol{\xi}} \qquad (24)$$

To calculate the absorbed power for the site of interest for year y, it is necessary to estimate the power production of the wave farm in each desired sea state by integrating the product of the wave spectrum with the time-averaged power of Eq. (24) over all frequencies [11, 12]:

$$\mathbf{p}_i(H_s, T_p, y) = \int_0^\infty 2S_{JS}(H_s, T_p, \omega)\mathbf{p}_m(H_s, T_p, \omega)d\omega \qquad (25)$$

where $\mathbf{p}_i(H_s, T_p, y)$ is the mechanical power matrix. Considering all sea states (which are now discretized by n_{gq} Gauss quadrature points), this equation can be estimated using [13]:

$$\mathbf{p}_i(H_s, T_p, y) = \sum_{k=0}^{n_w} 2\Delta\omega_k S_{JS}(H_s, T_p, \omega_k)\mathbf{p}_m(H_s, T_p, \omega_k) \qquad (26)$$

where n_w is the number of frequencies in the discretized form. Now, considering the number of years in the study, where $y = 1, 2, \ldots n_{yr}$ and their associated probability matrices, the average power can be calculated as [35]:

$$p_a = \eta_{\text{pcc}}\eta_{\text{oa}}\eta_t \sum_{y=1}^{n_{yr}} \mathbf{p}_i(H_s, T_p, y)\mathbf{p}_r(H_s, T_p, y) \qquad (27)$$

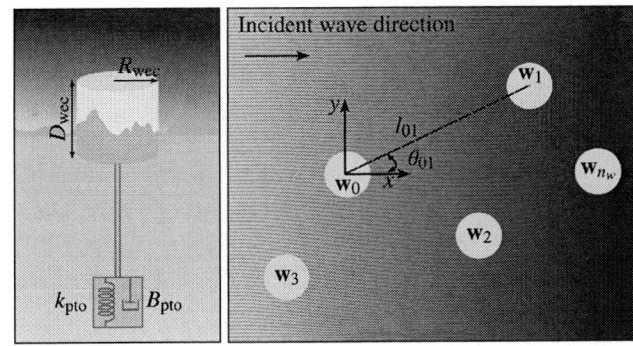

FIGURE 3: Illustration of WEC and its layout considerations.

where p_a is the average power. The joint probability distribution of the wave climate in the yth year of the study is described by $\mathbf{p}_r(H_s, T_p, y)$. In this equation, η_{pcc} is the efficiency of the power conversion chain, η_{oa} is the operational availability, and η_t is the transmission efficiency. Ideally, maximizing the ratio between the average power and the total cost is desired, including development, infrastructure, PTO, maintenance, etc. In this article, however, we use a much simpler objective function that considers the average power per unit volume of the wave-absorbing body, which is expressed as [30]:

$$p_v = \frac{p_a}{\pi R_{\text{wec}}^2 D_{\text{wec}}} \qquad (28)$$

where p_v is the average power per unit volume of the device. R_{wec} and D_{wec} are the radius and draft of the heaving cylinder WEC device, respectively.

2.4 Array Considerations

In this study, we consider an array with a total of n_{wec} WEC devices that are fully characterized by the $2-$by$-n_{\text{wec}}$ dimensional layout matrix $\mathbf{w} = [\mathbf{w}_1, \mathbf{w}_2, \cdots, \mathbf{w}_{n_{\text{wec}}}]$. Each element of n_{wec} is a vector composed of the center coordinates of each body in the Cartesian coordinate system, such that the pth body is characterized by $\mathbf{w}_p = [x_p, y_p]^T$. These bodies are subject to an incident wave propagating along the direction with angle β_w relative to the x-axis. Since the coordinate system may be rotated for these axisymmetric bodies, the incident wave angle may be set to 0 without any loss of generality. We further characterize the relative distance between two WECs, namely pth and qth bodies, and their relative angle with respect to the positive direction of the x-axis as l_{pq} and θ_{pq}, respectively. Figure 3 illustrates WECs in an array.

2.5 Surrogate Models for Hydrodynamic Interactions

Surrogate models provide a valuable tool for a simplified approximation of input/output relationships for computationally expensive tasks. The usage of surrogate models in general, and ANNs in particular in the WEC literature, is not new. Anderlini et al.

Copyright © 2023 by ASME

355

used ANNs to map the significant wave height, wave energy period, PTO damping, and its stiffness to the mean absorbed power in Ref. [36]. A shallow neural network machine learning algorithm was used in Ref. [37] to find the optimal latching times for the latching control of WECs. An ANN was developed in Ref. [38] for identification and control of a fully-submerged WEC. Reference [17] developed surrogate models to capture the hydrodynamic interactions effect within a WEC farm using Gaussian process regression. In this work, we follow the path of Ref. [17] to develop surrogate models that can estimate the hydrodynamic interaction effect using hierarchical interaction decomposition and MBE principles with ANNs.

2.5.1 Artificial Neural Networks.

ANN is a model comprised of an interconnected network of nodes or neurons in a layered structure. Each neuron uses a transfer function (such as the logistic function) to output a value between 0 and 1 depending on the weighted sum of its inputs. These neurons are characterized by a weight term that defines the input/output function of the network [39]. These weights, along with the network structure, may be adjusted to ensure that ANN performs reasonably well in predicting the outputs. These weights are found by fitting or learning on the training data through an algorithm that computes the gradient vector to minimize a prediction error. A performance measure, such as the mean-squared error, characterizes the error. For more information about ANN, the readers are referred to Ref. [40].

In this article, we utilize feedforward shallow neural networks. In such networks, the neurons in each hidden layer use the outputs of all nodes from the previous layer as input. This structure allows the network to capture complex nonlinear relationships in the data. Calibrating the ANN may take some effort; however, the resulting surrogate models can predict the desirable outputs with high accuracy and, thus, facilitate the effective implementation of layout optimization for WECs.

2.5.2 Many-body Expansion.

Estimating the interaction effect among a large number of bodies is a problem relevant to a multitude of disciplines, including quantum mechanics [41] and molecular dynamics [16]. Many-body expansion method (MBE) has been developed to estimate the total interaction effect among n_{wec} bodies as the summation of effects corresponding to a finite number of clusters. These clusters are systematically selected to capture the effects of a single, two-, three-, and m-bodies [42]. Representing the desired hydrodynamic interaction effect with $\psi(\mathbf{w})$, MBE up to m clusters can be estimated as [17]:

$$\psi(\mathbf{w}) \approx \sum_{i=1}^{n_{\text{wec}}} \psi(\mathbf{w}_i) + \sum_{i=1}^{n_{\text{wec}}-1} \sum_{j>i}^{n_{\text{wec}}} \Delta\psi(\mathbf{w}_i, w_j) + \dots \quad (29)$$

$$+ \sum_{i=1}^{n_{\text{wec}}-2} \sum_{j>i}^{n_{\text{wec}}-1} \sum_{k>j}^{n_{\text{wec}}} \Delta\psi(\mathbf{w}_i, \mathbf{w}_j, \mathbf{w}_k) + \dots$$

$$+ \sum_{i=1}^{n_{\text{wec}}-m} \dots \sum_{k>j}^{n_{\text{wec}}} \Delta\psi(\mathbf{w}_i, \dots, \mathbf{w}_k)$$

Using an alternative notation $\{\mathbf{w}\}_l^m$ to represent the lth distinct m-body cluster, this formulation can be written as:

$$\psi(\mathbf{w}) \approx \sum_{l=1}^{n_{\text{wec}}} \psi(\{\mathbf{w}\}_l^1) + \sum_{l=1}^{n_{\text{wec}}!/[2!(n_{\text{wec}}-2)!]} \Delta\psi(\{\mathbf{w}\}_l^2) + \dots \quad (30)$$

$$+ \sum_{l=1}^{n_{\text{wec}}!/[3!(n_{\text{wec}}-3)!]} \Delta\psi(\{\mathbf{w}\}_l^3) + \dots$$

$$+ \sum_{l=1}^{n_{\text{wec}}!/[m!(n_{\text{wec}}-m)!]} \Delta\psi(\{\mathbf{w}\}_l^m)$$

where $\Delta\psi(\cdot)$ is the additive interaction effect, calculated as:

$$\Delta\psi(\{\mathbf{w}\}_l^m) = \psi(\{\mathbf{w}\}_l^m) - \left[\sum_{r=1}^{m} \psi(\{\{\mathbf{w}\}_l^m\}_r^1) + \dots \right. \quad (31)$$

$$\left. + \sum_{r=1}^{m!/[2!(m-2)!]} \Delta\psi(\{\{\mathbf{w}\}_l^m\}_r^2) + \dots + \sum_{r=1}^{m} \Delta\psi(\{\{\mathbf{w}\}_l^m\}_r^{m-1}) \right]$$

In this study, we only consider interaction effects of two-body clusters. Therefore, for each pair of p and q WECs, the additive effect can be written as:

$$\Delta\psi(\mathbf{w}_p, \mathbf{w}_q) = \psi(\mathbf{w}_p, \mathbf{w}_q) - \psi(\mathbf{w}_p) - \psi(\mathbf{w}_q) \quad (32)$$

3 CONSTRUCTING SURROGATE MODELS

In this section, we discuss some considerations in developing ANNs for hydrodynamic interaction effects and demonstrate their capability to estimate the hydrodynamic coefficients with reasonable accuracy.

3.1 Developing Surrogate Models

In order to generate the training data for ANNs, the first step is to identify inputs and outputs. In this article, we are interested in estimating radiation and excitation forces exerted on the WEC for all one- and two-WEC clusters. This results in an output structure with components associated with the added mass, damping coefficient, and the real and imaginary parts of the excitation force. These quantities of interest (QoI) constitute the output vector of our ANNs and are normalized according to the following relationships [6, 17, 43]:

$$\bar{\mathbf{F}}_e = \hat{\mathbf{F}}_e/(\rho g \pi R_{\text{wec}}^2 D_{\text{wec}}) \quad (33)$$

$$\bar{\mathbf{A}} = \mathbf{A}/(\rho \pi R_{\text{wec}}^2 D_{\text{wec}}) \quad (34)$$

$$\bar{\mathbf{B}} = \mathbf{B}/(\omega \rho \pi R_{\text{wec}}^2 D_{\text{wec}}). \quad (35)$$

Although it is theoretically possible to develop a single ANN with multiple inputs and outputs, we use a single ANN for each of these outputs in this work. This separation helps with simplifying the task of tuning network parameters.

For the calculation of quantities of interest in the single-body

Copyright © 2023 by ASME

cluster, a sufficient number of samples for different WEC radii and draft dimensions must be provided only at the center of the coordinate system, i.e., $\mathbf{w}_0 = [0,0]^T$. This is because the radiation output from these surrogate models is independent of the location of the WEC (i.e., $\mathbf{A}(\mathbf{w}) = \mathbf{A}([0,0]^T)$ and $\mathbf{B}(\mathbf{w}) = \mathbf{B}([0,0]^T)$), however, it depends on plant specifications, such as WEC radius and draft. The excitation force output is invariant with respect to translations along the y-axis and is symmetric with respect to the wave propagation direction (i.e., x-axis); however, for translation by L along the x direction, a phase shift is created by $\exp(-ikL)$, where k is the wave number and $i = \sqrt{-1}$.

The training set for developing ANNs should adequately represent the space of WEC dimensions, including WEC radius and draft. Since the BEM solution is created in the frequency domain, radial frequency is also an input. These inputs are normalized according to the following relationships:

$$\tilde{R}_{\text{wec}} = R_{\text{wec}}/\bar{R}_{\text{wec}} \tag{36}$$

$$\tilde{D}_{\text{wec}} = D_{\text{wec}}/\bar{D}_{\text{wec}} \tag{37}$$

$$\tilde{\omega} = \omega/\bar{\omega} \tag{38}$$

where $\bar{R}_{\text{wec}}$, $\bar{D}_{\text{wec}}$, and $\bar{\omega}$ are maximum WEC radius, draft, and radial frequencies, respectively. The input vector for the single-body cluster is then defined as $\tilde{v}_1 = [\tilde{R}_{\text{wec}}, \tilde{D}_{\text{wec}}, \tilde{\omega}]^T$. The QoI for the 1-body cluster constitutes the normalized added mass $\tilde{a}$, damping coefficient $\tilde{b}$, and the real and imaginary parts of the excitation force $\tilde{f}_e$. These quantities are represented as $\tilde{y}_1 = [\tilde{a}, \tilde{b}, \text{Re}\{\tilde{f}_e\}, \text{Im}\{\tilde{f}_e\}]^T$. The resulting surrogate models for 1-body clusters are then defined as:

$$\tilde{y}_1 = f_1(\tilde{v}_1) \tag{39}$$

where f_1 is the vector of resulting ANN functions for the single-body cluster, composed of $f_1 = [f_1^a, f_1^b, f_1^{fr}, f_1^{fim}]^T$.

The interaction effect among 2-body clusters is characterized by an input vector that, in addition to WEC radius, draft, and radial frequency, includes the normalized relative distance l_{pq} and relative angle θ_{pq} between the two bodies:

$$\tilde{l}_{pq} = l_{pq}/\bar{l}_{pq} \tag{40}$$

$$\tilde{\theta}_{pq} = \theta_{pq}/\bar{\theta}_{pq} \tag{41}$$

where $\bar{l}_{pq}$, and $\bar{\theta}_{pq}$ are the maximum distance and angle considered in producing Nemoh results. The input vector is then described as $v_2 = [\tilde{R}_{\text{wec}}, \tilde{D}_{\text{wec}}, \tilde{l}_{pq}, \tilde{\theta}_{pq}, \tilde{\omega}]^T$. The radiation output entails elements corresponding to the additive effect described in Eqs. (31)–(32). However, due to numerical reasons, it is more straightforward to structure the two-body surrogate models with direct outputs from Nemoh. Any required transformation, such as the ones described in Eqs. (31)–(32), can be performed once the models are developed. The 2-body system radiation effect depends strictly on the relative distance of the bodies, while the excitation force needs only be investigated in the range of $[0, \pi]$ because any other angle can be mapped to that solution [17]. These outputs are defined as $\tilde{y}_2 = [\tilde{a}_{11}, \tilde{a}_{12}, \tilde{b}_{11}, \tilde{b}_{12}, \text{Re}\{\tilde{f}_{e_{11}}\}, \text{Im}\{\tilde{f}_{e_{11}}\}]^T$,

where the indices $*_{11}$ and $*_{12}$ correspond to the interaction effect between elements of the mass matrix in Eq. (13) with an array size of $n_{\text{wec}} = 2$. The 2-body cluster surrogate models can be defined as:

$$\tilde{y}_2 = f_2(\tilde{v}_2) \tag{42}$$

where f_2 is the vector of resulting ANN functions for the two-body cluster, composed of $f_2 = [f_2^{a_{11}}, f_2^{a_{12}}, f_2^{b_{11}}, f_2^{b_{12}}, f_2^{fr}, f_2^{fim}]^T$. From here, the additive effect from the 2-body cluster associated with the radiation interaction can be defined as:

$$\Delta \tilde{a}_{11} = \tilde{a}_{11} - \tilde{a} = f_2^{a_{11}}(\tilde{v}_2) - f_1^a(\tilde{v}_1) \tag{43}$$

$$\Delta \tilde{a}_{12} = f_2^{a_{12}}(\tilde{v}_2) \tag{44}$$

$$\Delta \tilde{b}_{11} = \tilde{b}_{11} - \tilde{b} = f_2^{b_{11}}(\tilde{v}_2) - f_1^b(\tilde{v}_1) \tag{45}$$

$$\Delta \tilde{b}_{12} = f_2^{b_{12}}(\tilde{v}_2) \tag{46}$$

For excitation force, the additive effect is captured as:

$$\Delta \tilde{f}_{e11} = (\tilde{f}_e - \tilde{f}_{e_{11}}) \exp(ikL) \tag{47}$$

$$= \left([f_{fr}^1(\tilde{v}_1) + if_{fim}^1(\tilde{v}_1)] - [f_{fr}^2(\tilde{v}_2) + if_{fim}^2(\tilde{v}_2)] \right) \exp(ikL)$$

Other quantities of interest, such as those corresponding to the additive effect of the first body on the second one, may be simply calculated by swapping the order of the bodies.

3.2 Data Processing

In addition to the normalization scheme discussed in Sec. 3.1, further processing of data is necessary in order to improve the performance of ANNs.

Design-informed considerations are utilized in order to generate appropriate inputs for developing ANNs. While preventing us from producing impractical solutions, such considerations improve the training performance by limiting the range of outputs in the training set. Here, two practical design decisions are considered when generating the training data. First, extreme and unreasonable design combinations are avoided by only considering cases where the radius and draft ratio are within an acceptable range. This limitation prevents us from generating solutions that have little physical viability. Second, as will be described in Sec. 4.1 Eq. (48b), a safety distance, proportional to the radius of the WEC, is necessary for the reliable maintenance of WEC devices. This safety distance is also considered when generating data for the training of ANNs.

Despite the provisions described above, it was occasionally observed that the data set entailed few extremely high and/or low data points. This issue might be associated with irregular frequencies in Nemoh solution, which arise due to a fundamental error in BEM formulation [31]. These data points were identified and replaced by the mean of their neighboring data points. More advanced methods for removal of irregular frequencies from Nemoh solutions must be investigated in future work [44].

Since the BEM was implemented for a wide range of WEC radii, drafts, distances, and angles, even after normalization, the

Copyright © 2023 by ASME

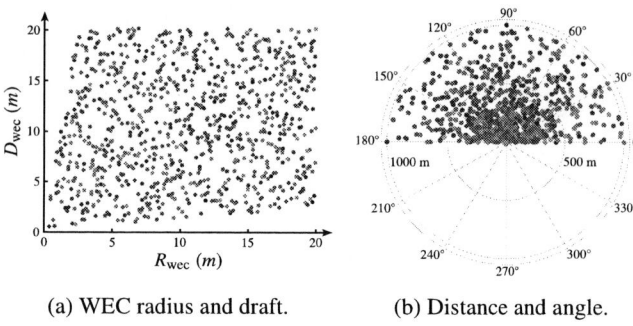

(a) WEC radius and draft. (b) Distance and angle.

FIGURE 4: Sampling the design space using LHS for two-body interactions.

TABLE 1: Settings for creating training data using Nemoh.

Symbol	Definition	Value	Unit
ω	frequency	[0.1 7]	rad/s
h	water depth	50	m
β_w	wave angle	0	rad
R_{wec}	WEC radius	[0.5, 20]	m
D_{wec}	WEC draft	[0.5, 20]	m
l_{pq}	Relative distance	$[2R_{\text{wec}} + s_d, 1000]$	m
θ_{pq}	Relative angle	$[0, \pi]$	rad
n_{s_1}	# 1-body samples	225	-
n_{s_2}	# 2-body samples	1000	-

range of the QoI was extremely high among some of the solution sets. After some trial and error with various normalization schemes, it was decided that transforming each solution set to the range of $[-1, 1]$ through a linear transformation results in the most satisfactory performance. However, to do this, additional ANNs need to be developed in order to estimate the range and offset of these linear transformations. Through this approach, we ensure that ANNs preserve the shape or profile of the QoI. This approach resulted in a total of 30 surrogate models.

3.3 Validation of Surrogate Models
The BEM solver Nemoh was utilized to create training samples for the development of surrogate models. A total of 225 and 1000 Nemoh calls were made to generate data for the 1-body- and 2-body-related surrogate models, respectively. To improve computational efficiency, customized code was developed to implement Nemoh in parallel, without manipulating Nemoh's source code and executable files. The resulting package is openly accessible on GitHub upon [45].

Uniform spacing was used in creating training data for the 1-body ANNs. However, for the 2-body ANNs, a Latin-hypercube sampling (LHS) approach was implemented. For the distance dimension, this LHS approach was performed using logarithmically-spaced samples. The LHS for the two-body ANNs is shown in Fig. 4, where each unique color corresponds to an individual data point in the four-dimensional space.

The hydrodynamic coefficients were obtained at the constant water depth of $h = 50$ m for the range of radial frequencies from $\omega_0 = 0.1$ rad/s to $\omega_f = 7$ rad/s, with a total of 50 evenly-spaced values. The mean squared error was consistently used for performance assessment of the ANNs, and the training data was divided such that 70% was used for training, 15% for validation, and 15% for tests. All the ANNs were trained for up to 30000 epochs, with early stopping of 10000 maximum failures in the validation set using Matlab. General information about generating the training set is presented in Tab. 1.

The resulting ANNs are first validated against arbitrary de-

sign specifications using Nemoh to ensure reasonable performance. Using $R_{\text{wec}} = 8$ m and $D_{\text{wec}} = 4$ m, the validation plots for added mass, damping coefficient, and real and imaginary parts of the excitation force for the 1-body cluster are presented in Fig. 5, where it is clear that the ANNs are in reasonable agreement with Nemoh. Similarly, for the two-body interaction effect, the developed ANNs were compared with Nemoh results for an arbitrary case of $R_{\text{wec}} = 15$ m, $D_{\text{wec}} = 8$ m, $l_{pq} = 200$ m, and $\theta_{pq} = 0.078$ radians. The comparison plots for the added mass and damping coefficients and the real and imaginary parts of the excitation forces are shown in Fig. 6.

4 CONCURRENT UCCD AND LAYOUT OPTIMIZATION
Optimal spacing of various WEC technologies within an array and optimal sizing and control of WEC devices have been popular topics in the literature, often investigated individually [46–48]. The significant impact of these elements on WEC performance has convinced more researchers to concurrently investigate such attributes of these complex systems. For example, Ref. [15] carries out a concurrent sizing and layout optimization of WECs by taking advantage of the simplified q-factor equation with the point-absorber approximation. Various optimization methods have been used for the layout optimization of WECs, including sequential quadratic programming [49], genetic algorithm (GA) [50], and heuristics [51]. This section presents the simultaneous formulation and some preliminary results for the concurrent plant, control, and layout optimization of WEC farms.

4.1 Problem Formulation
Since manufacturing costs associated with WECs of different dimensions become prohibitive, it is reasonable to assume that WEC dimensions are uniform across the farm. However, to maximize energy generation, control parameters associated with the PTO system of each WEC, along with their location (thus, the configuration of the farm), may be varied. Expressing the objective function as the average power per unit volume of the device,

Copyright © 2023 by ASME

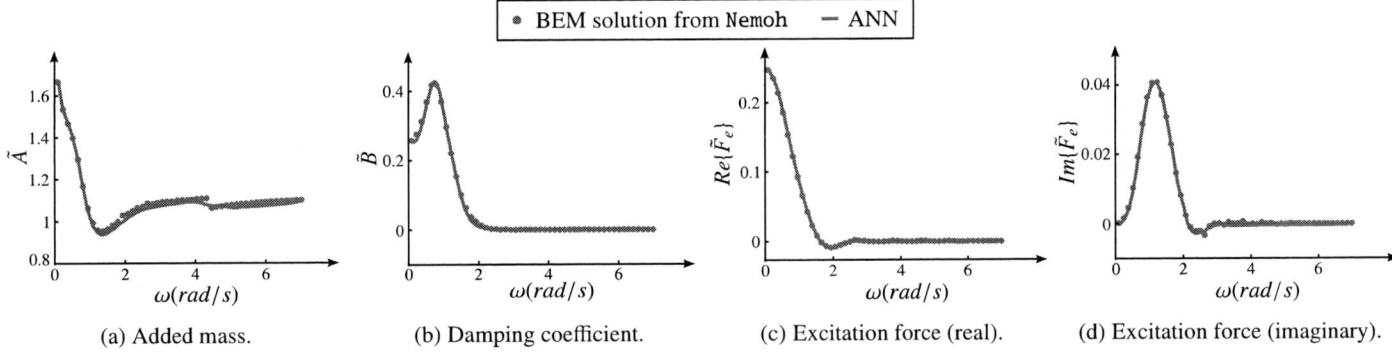

(a) Added mass.   (b) Damping coefficient.   (c) Excitation force (real).   (d) Excitation force (imaginary).

FIGURE 5: Validation of ANNs with BEM solution from Nemoh using $R_{\text{wec}} = 8$ m and $D_{\text{wec}} = 4$ m.

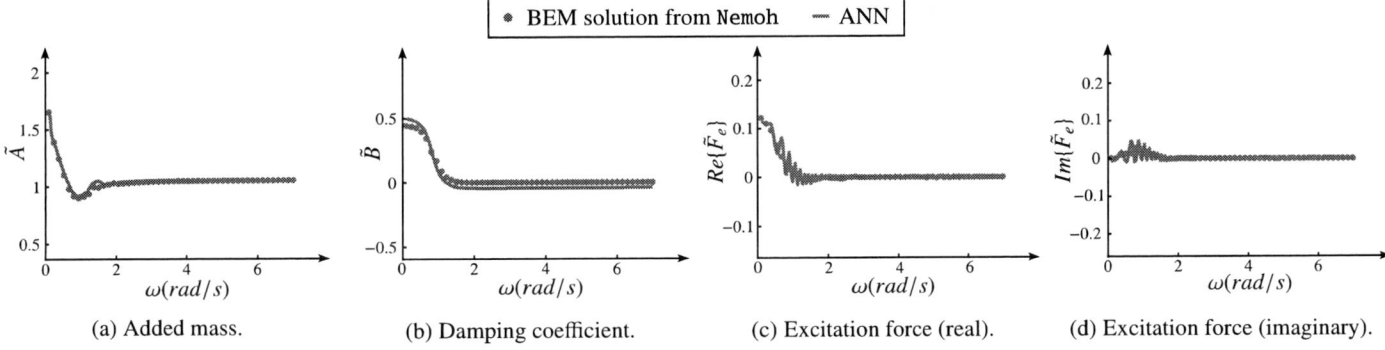

(a) Added mass.   (b) Damping coefficient.   (c) Excitation force (real).   (d) Excitation force (imaginary).

FIGURE 6: Validation of select ANNs with BEM solution from Nemoh using $R_{\text{wec}} = 8$ m, $D_{\text{wec}} = 15$ m, $l_{pq} = 200$ m, and $\theta_{pq} = 0.078$ radians.

the concurrent UCCD and layout optimization problem can be formulated as:

$$\underset{p,u,\mathbf{w}}{\text{minimize:}} \quad -p_v(\boldsymbol{p}, \boldsymbol{u}, \mathbf{w}) \tag{48a}$$

$$\text{subject to:} \quad \begin{aligned} 2R_{\text{wec}} + s_d - \boldsymbol{L}_{pq} &\leq 0 \\ \forall \ p, q &= 1, 2, \ldots, n_{\text{wec}} \quad p \neq q \end{aligned} \tag{48b}$$

$$\underline{\boldsymbol{p}} \leq \boldsymbol{p} \leq \bar{\boldsymbol{p}} \tag{48c}$$

$$\underline{\boldsymbol{u}} \leq \boldsymbol{u} \leq \bar{\boldsymbol{u}} \tag{48d}$$

$$\underline{\mathbf{w}} \leq \mathbf{w} \leq \bar{\mathbf{w}} \tag{48e}$$

$$\text{where:} \quad \boldsymbol{p} = [R_{\text{wec}}, D_{\text{wec}}]^T \in \mathbb{R}^2$$

$$\boldsymbol{u} = [\mathbf{K}_{\text{pto}}, \mathbf{B}_{\text{pto}}]^T \in \mathbb{R}^{2n_{\text{wec}}}$$

$$\mathbf{w} = [\boldsymbol{x}, \boldsymbol{y}] \in \mathbb{R}^{2(n_{\text{wec}}-1)}$$

where $\boldsymbol{p}$ is the vector of plant optimization variables composed of WEC radius R_{wec} and draft D_{wec} (uniform across the farm). $\boldsymbol{u}$ is the vector of time-independent control parameters for each of the WEC devices, composed of PTO spring stiffness $\mathbf{K}_{\text{pto}}$ and damping $\mathbf{B}_{\text{pto}}$. The layout design vector $\mathbf{w}$ is composed of $(n_{\text{wec}} - 1)$ coordinates, corresponding to the x- and y-axis locations of WEC devices. Note that here, the first WEC is always assumed to be

positioned at the center of the coordinate system and, thus, is not included in $\mathbf{w}$. Equation (48b) ensures that the minimum distance between each pair of WECs is larger than the WEC diameter plus an additional safe distance s_d, which is required to allow maintenance ships to pass safely [12]. Here, we calculate the safe distance as a function of R_{wec}, such that $s_d = (R_{\text{wec}}/5) \times 50$ m. All of the optimization variables are confined within their associated bounds, as described by Eqs. (48c)–(48e). The farm area is restricted to a box with dimensions of $\pm 0.5 \times \sqrt{20000 n_{\text{wec}}}$ m in both x and y axes [12]. These details, along with some additional problem data, are presented in Table 2.

4.2 Results and Discussion

The concurrent plant, control, and layout optimization problem was implemented for four case studies with $n_{\text{wec}} = 3, 5, 7$, and 10. Due to the complex nature of the problem, global optimization is required. Therefore, Matlab's *surrogateopt* solver, specifically developed for the global optimization of time-consuming objective functions, was used for these case studies and ran for a total of 300 function evaluations using parallel computation. The computer architecture used for these case studies is a desktop workstation with an AMD Ryzen Threadripper 3970X 32-core

Copyright © 2023 by ASME

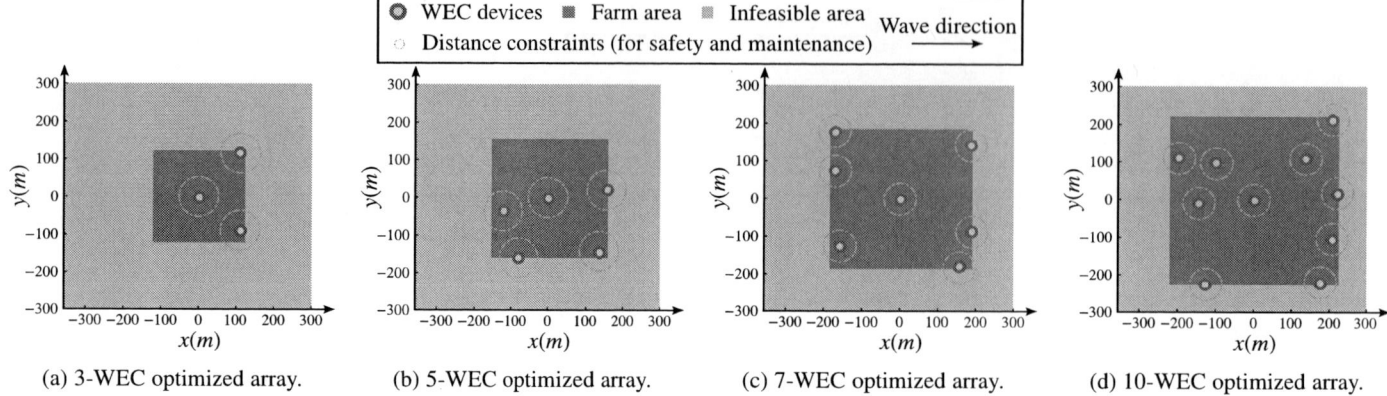

FIGURE 7: Optimized array layout with farm area and distant constraints.

TABLE 2: Problem parameters.

Option	Value	Option	Value
$\underline{R}_{wec}$	0.5 m	$\bar{R}_{wec}$	10 m
$\underline{D}_{wec}$	0.5 m	$\bar{D}_{wec}$	10 m
$\underline{k}_{pto}$	-3×10^8 N/m	$\bar{k}_{pto}$	3×10^8 N/m
$\underline{B}_{pto}$	0 Ns/m	$\bar{B}_{pto}$	3×10^8 Ns/m
$\underline{x}$	$-0.5\sqrt{2n_{wec} \times 10^4}$ m	$\bar{x}$	$0.5\sqrt{2n_{wec} \times 10^4}$ m
$\underline{y}$	$-0.5\sqrt{2n_{wec} \times 10^4}$ m	$\bar{y}$	$0.5\sqrt{2n_{wec} \times 10^4}$ m
ρ	1025 kg/m^3	g	9.81 m/s^2
s_d	$50 \times R_{wec}/5$ m	n_{wec}	[3, 5, 7, 10]
n_{yr}	30 years	n_r	200
n_{gq}	20	η_{pcc}	0.8
η_{oa}	0.95	η_t	0.98

processor at 3.69 GHz, 128 GB of RAM, 64-bit windows 10 Enterprise LTSC version 1809, and Matlab R2022a. The results are presented in Table 3.

According to this table, the proposed framework enables the concurrent control co-design and layout optimization of WEC farms with significant improvements in computational efficiency, making it possible to investigate large arrays within a reasonable amount of time. This improvement in efficiency is signified by the fact that with the proposed approach, a 10-WEC array optimization study can be performed much faster than a single Nemoh call to solve a 3-WEC hydrodynamic problem. An expected tradeoff is the accuracy of the estimations, which is mainly affected by two contributions, one from overlooking 3 and higher-body interactions in developing MBE models and the other from inherent errors associated with the development of surrogate models. Upon further improvement of the accuracy of these estimations, the proposed method may offer more practical, early-stage insights into the interactions between the design, control, and layout of these complex devices.

The optimized layout configurations associated with all of the studies are described in Table 3 and visualized along with farm area and distance constraints in Fig. 7. The table shows that the radius of WEC devices is relatively high (about 8.83 m) for 3-WEC farm, while it decreases in size for the other studies. This result seems to be consistent with some of the results from the literature, where it is shown that multiple smaller devices result in a higher power [52–54]. However, complete validation of such literature results requires the optimizer to determine the number of WEC devices, which can be a future direction of this research. The draft dimensions in all of these studies remain very close to the lower bound of 0.5 m. This result is also consistent with the results from the literature, where it is shown that the draft size does not have a significant impact on power generation [55].

The variations in $\mathbf{B}_{pto}$ and $\mathbf{K}_{pto}$ are much more diverse, indicating that individual control of WECs is necessary for improved device performance. Here, we have allowed the spring stiffness to take negative values. These stiffness results enable a reactive phase control strategy [56] that is often used to extend the range of resonance conditions in WEC design [57, 58]. Despite the limitations of this approach (as stated in Ref. [56]), it is shown that phase control with a negative spring stiffness increases the absorbed power [57]. In each case study, the PTO damping is selected by the optimizer in order to maximize the energy production per volume of the WEC device over the lifetime of the farm. This approach considers the availability of wave resources when designing the PTO damping coefficient.

The optimized layout configurations and their associated distance constraints are presented in Fig. 7. While Ref. [51] has shown that, for the most part, the best layout arrays are symmetric with respect to the wave direction, this symmetry only occurs for the 3-WEC study in Fig. 7a. This outcome might be mainly associated with (*i*) the need to further improve the estimation of the hydrodynamic interaction effect by accounting

Copyright © 2023 by ASME

TABLE 3: Concurrent UCCD and layout optimization solutions.

Case Study	R_{wec} [m]	D_{wec} [m]	$\mathbf{B}_{pto}$ [Ns/m]	$\mathbf{K}_{pto}$ [N/m]	w [m]	p_v [MW/m^3]	Time [s]
3-WEC	8.83	0.54	1.7×10^8	-2.97×10^8	[0,0]	175.58	97
			1.65×10^8	2.53×10^8	[109, -89.52]		
			1.1×10^8	-4.21×10^6	[114.07, 115.89]		
5-WEC	7.12	0.5	2.64×10^8	1.44×10^8	[0,0]	566.02	143
			1.67×10^8	5.26×10^6	[-138.76, -119.09]		
			1.67×10^8	2.56×10^8	[-24.07, -154.43]		
			2.94×10^8	-2.95×10^8	[-41.54, 129.69]		
			2.72×10^8	2.73×10^8	[-49.97, -70.61]		
7-WEC	6.71	0.62	8.86×10^7	-2.95×10^8	[0,0]	2.21×10^3	219
			1.94×10^8	-3.24×10^7	[154, -181.38]		
			1.16×10^8	1.32×10^8	[-169.51, 73.9]		
			6.66×10^7	-8.66×10^6	[-171.96, 178.79]		
			2.84×10^8	2.32×10^8	[185.47, 142.23]		
			2.67×10^8	1.69×10^8	[187.08, -86.12]		
			2.67×10^8	4.4×10^7	[-156.86, -124.98]		
10-WEC	7.32	0.5	2.24×10^8	-1.73×10^8	[0,0]	4.12×10^3	379
			2.99×10^8	-2.56×10^7	[-100.66, 98.45]		
			1.29×10^8	-2.83×10^8	[-127.97, -223.61]		
			6.9×10^7	1.82×10^8	[-145.75, 8.36]		
			2.41×10^8	-2.08×10^7	[219.94, 18.84]		
			2.23×10^8	-6.74×10^7	[-198.02, 112.55]		
			1.89×10^8	2.14×10^7	[206.39, 214.89]		
			2.68×10^7	1.45×10^8	[204.82, -102.91]		
			1.81×10^8	-2.27×10^8	[174.45, -218.4]		
			1.66×10^8	1.49×10^7	[135.33, 109.09]		

for higher-order terms in MBE, *(ii)* improving upon the accuracy of the lower-order surrogate models, and *(iii)* allowing the optimizer to run for a higher number of function evaluations. Additionally, it is also possible that the influence of the separating distance between WECs is limited due to proper tuning of the PTO damping coefficient [11].

The generated power per unit volume of the device is also shown in the table. It is evident that the power generation per unit volume of the device, as well as the overall power generation (not shown here), increases as we increase the number of devices in the farm. The average amount of power generated by an individual device per year in each layout also increases as we increase the number of devices in the farm.

While the results presented here offer some insights into the effectiveness of the proposed approach in terms of computational expense by utilizing MBE and surrogate modeling, conclusions regarding trends in array configurations and the amount of generated power are avoided at this stage. This decision is because the accuracy tradeoff in developing surrogate models must be fur-

ther investigated, and higher-order terms must be included in the MBE approach in order to capture the complex interactions in the farm better. In addition, this study considers no limits on PTO force and buoy displacement. This assumption implies that the device can generate power from waves with very high amplitudes without considering any of the physical, structural, and operational constraints. For these reasons, we avoid making any recommendations regarding optimal layout configurations but emphasize that the proposed approach and methodology have the potential to address the computational bottleneck in the concurrent plant, control, and layout optimization of large arrays.

5 CONCLUSION

Due to nonlinear and complex dynamics, the sizing, control, and array layout optimization of wave energy converters (WECs) are coupled disciplines and must be approached concurrently from the early stages of the design process. However, estimating the hydrodynamic interaction effect for this integrated design study

Copyright © 2023 by ASME

is computationally complex. Therefore, in this article, we developed data-driven artificial neural networks (ANNs) and utilized the principles of many-body expansion (MBE) to efficiently calculate the hydrodynamic interaction effect up to the second-order terms. The heaving WEC model was developed in the frequency domain, with WEC radius and draft as plant optimization variables that were assumed to be uniform across the farm. The power take-off control damping, stiffness parameters, and array layout were optimized for each single WEC device using a global optimization algorithm. The results indicated significant computational efficiency compared to the direct usage of boundary element method solvers in the optimizer loop.

While the current study shows promising directions for efficiently estimating the hydrodynamic interaction effect, larger array studies necessitate using higher-order terms in the MBE equation. In addition, the variations in water depth and geographical location of the farm must be considered to make general recommendations. More complex objective functions that consider techno-economic-related metrics involving costs for manufacturing, operation, PTO, maintenance, etc., must be investigated. Finally, force saturation limits are necessary to ensure practical infrastructure requirements and must be considered as additional constraints in the optimization problem for future work.

ACKNOWLEDGMENTS

The authors gratefully acknowledge the financial support from National Science Foundation Engineering Design and Systems Engineering Program, USA under grant number CMMI-2034040. We would also like to thank Gaofeng Jia and Akshat Chulahwat for their contributions.

REFERENCES

[1] Ning, D., and Ding, B., 2022. *Modelling and Optimization of Wave Energy Converters*. CRC Press. doi: 10.1201/9781003198956

[2] Straub, J., 2015. "In search of technology readiness level (TRL) 10". *Aerosp. Sci. Technol.,* **46**, pp. 312–320. doi: 10.1016/j.ast.2015.07.007

[3] Tan, J., Polinder, H., Laguna, A. J., Wellens, P., and Miedema, S. A., 2021. "The influence of sizing of wave energy converters on the techno-economic performance". *J. Mar. Sci. Eng.,* **9**(1), p. 52. doi: 10.3390/jmse9010052

[4] McCabe, R., Murphy, O., and Haji, M., 2022. "Multidisciplinary optimization to reduce cost and power variation of a wave energy converter". In International Design Engineering Technical Conferences, p. V03AT03A023. doi: 10.1115/DETC2022-90227

[5] Neshat, M., Sergiienko, N. Y., Amini, E., Majidi Nezhad, M., Astiaso Garcia, D., Alexander, B., and Wagner, M., 2020. "A new bi-level optimisation framework for optimising a multi-mode wave energy converter design: A case study for the Marettimo Island, Mediterranean Sea". *Energies,* **13**(20), p. 5498. doi: 10.3390/en13205498

[6] Herber, D. R., and Allison, J. T., 2013. "Wave energy extraction maximization in irregular ocean waves using pseudospectral methods". In International Design Engineering Technical Conferences, p. V03AT03A018. doi: 10.1115/DETC2013-12600

[7] Tan, J., Polinder, H., Laguna, A. J., and Miedema, S., 2022. "The application of the spectral domain modeling to the power take-off sizing of heaving wave energy converters". *Appl. Ocean Res.,* **122**, p. 103110. doi: 10.1016/j.apor.2022.103110

[8] Coe, R. G., Bacelli, G., Olson, S., Neary, V. S., and Topper, M. B., 2020. "Initial conceptual demonstration of control co-design for WEC optimization". *J. Ocean Eng. Mar. Energy,* **6**(4), pp. 441–449. doi: 10.1007/s40722-020-00181-9

[9] Abdulkadir, H., and Abdelkhalik, O., 2023. "Optimization of heterogeneous arrays of wave energy converters". *Ocean Eng.,* **272**, p. 113818. doi: 10.1016/j.oceaneng.2023.113818

[10] Falnes, J., 1980. "Radiation impedance matrix and optimum power absorption for interacting oscillators in surface waves". *Appl. Ocean Res.,* **2**(2), pp. 75–80. doi: 10.1016/0141-1187(80)90032-2

[11] Borgarino, B., Babarit, A., and Ferrant, P., 2012. "Impact of wave interactions effects on energy absorption in large arrays of wave energy converters". *Ocean Eng.,* **41**, pp. 79–88. doi: 10.1016/j.oceaneng.2011.12.025

[12] Neshat, M., Mirjalili, S., Sergiienko, N. Y., Esmaeilzadeh, S., Amini, E., Heydari, A., and Garcia, D. A., 2022. "Layout optimisation of offshore wave energy converters using a novel multi-swarm cooperative algorithm with backtracking strategy: A case study from coasts of Australia". *Energy,* **239**, p. 122463. doi: 10.1016/j.energy.2021.122463

[13] Mercadé Ruiz, P., Nava, V., Topper, M. B., Ruiz Minguela, P., Ferri, F., and Kofoed, J. P., 2017. "Layout optimisation of wave energy converter arrays". *Energies,* **10**(9), p. 1262. doi: 10.3390/en10091262

[14] Ringwood, J. V., Zhan, S., and Faedo, N., 2023. "Empowering wave energy with control technology: Possibilities and pitfalls". *Annu. Rev. Control.* doi: 10.1016/j.arcontrol.2023.04.004

[15] Lyu, J., Abdelkhalik, O., and Gauchia, L., 2019. "Optimization of dimensions and layout of an array of wave energy converters". *Ocean Eng.,* **192**, p. 106543. doi: 10.1016/j.oceaneng.2019.106543

[16] Gordon, M. S., Fedorov, D. G., Pruitt, S. R., and Slipchenko, L. V., 2012. "Fragmentation methods: A route to accurate calculations on large systems". *Chem. Rev.,* **112**(1), pp. 632–672. doi: 10.1021/cr200093j

[17] Zhang, J., Taflanidis, A. A., and Scruggs, J. T., 2020. "Surrogate modeling of hydrodynamic forces between multiple floating bodies through a hierarchical interaction decomposition". *J. Comput. Phys.,* **408**, p. 109298. doi: 10.1016/j.jcp.2020.109298

[18] Babarit, A., 2013. "On the park effect in arrays of oscillating wave energy converters". *Renew. Energ.,* **58**, pp. 68–78. doi: 10.1016/j.renene.2013.03.008

[19] Babarit, A., and Delhommeau, G., 2015. "Theoretical and numerical aspects of the open source BEM solver NEMOH". In European Wave and Tidal Energy Conference.

[20] Babarit, A., and Delhommeau, G., 2022. "Computation of second-order wave loads on floating offshore wind turbine platforms in bichromatic bi-directional waves using open-source potential flow solver NEMOH". In 18émes Journées de l'Hydrodynamique.

Copyright © 2023 by ASME

doi: 10.5281/zenodo.7418379

[21] Kurnia, R., Ducrozet, G., and Gilloteaux, J.-C., 2022. "Second-order difference and sum-frequency wave loads in the open-source potential flow solver NEMOH". In International Conference on Offshore Mechanics and Arctic Engineering, p. V05AT06A019. doi: 10.1115/OMAE2022-79163

[22] Azad, S., and Herber, D. R., 2022. "Control co-design under uncertainties: formulations". In International Design Engineering Technical Conferences, p. V03AT03A008. doi: 10.1115/DETC2022-89507

[23] Garcia-Sanz, M., 2019. "Control co-design: an engineering game changer". *Adv. Control Appl.: Engineering and Industrial Systems, 1*(1), Dec., p. e18. doi: 10.1002/adc2.18

[24] Ströfer, C. A. M., Gaebele, D. T., Coe, R. G., and Bacelli, G., 2023. "Control co-design of power take-off systems for wave energy converters using WecOptTool". *IEEE Trans. Sustain. Energy.* doi: 10.1109/TSTE.2023.3272868

[25] O'Sullivan, A. C., and Lightbody, G., 2017. "Co-design of a wave energy converter using constrained predictive control". *Renew. Energ., 102*, pp. 142–156. doi: 10.1016/j.renene.2016.10.034

[26] Peña-Sanchez, Y., García-Violini, D., and Ringwood, J. V., 2022. "Control co-design of power take-off parameters for wave energy systems". *IFAC-PapersOnLine, 55*(27), pp. 311–316. doi: 10.1016/j.ifacol.2022.10.531

[27] Storlazzi, C. D., Shope, J. B., Erikson, L. H., Hegermiller, C. A., and Barnard, P. L., 2015. "Future wave and wind projections for United States and United States-affiliated Pacific Islands". *US Geological Survey Open-File Report, 1001*, p. 426. doi: 10.3133/ofr20151001

[28] Erikson, L. H., Hegermiller, C. E., Barnard, P. L., and Storlazzi, C. D., 2016. "Wave projections for United States mainland coasts". *US Geological Survey Pamphlet to Accompany Data Release, 585*. doi: 10.5066/F7D798GR

[29] Folley, M., 2016. *Numerical Modelling of Wave Energy Converters*. Academic Press. doi: 10.1016/C2014-0-04006-3

[30] Falnes, J., and Kurniawan, A., 2020. *Ocean Waves and Oscillating Systems*, Vol. 8. Cambridge University Press. doi: 10.1017/CBO9780511754630

[31] Penalba, M., Kelly, T., and Ringwood, J., 2017. "Using NEMOH for modelling wave energy converters: A comparative study with WAMIT". In European Wave and Tidal Energy Conference.

[32] Tan, J., Polinder, H., Wellens, P., and Miedema, S., 2020. "A feasibility study on downsizing of power take off system of wave energy converters". In *Developments in Renewable Energies Offshore*. CRC Press, pp. 140–148. doi: 10.1201/9781003134572-18

[33] Tedeschi, E., Molinas, M., Carraro, M., and Mattavelli, P., 2010. "Analysis of power extraction from irregular waves by all-electric power take off". In Energy Conversion Congress and Exposition, pp. 2370–2377. doi: 10.1109/ECCE.2010.5617893

[34] Clément, A., and Babarit, A., 2012. "Discrete control of resonant wave energy devices". *Philos. Trans. Royal Soc. A, 370*(1959), pp. 288–314. doi: 10.1098/rsta.2011.0132

[35] Neary, V. S., Lawson, M., Previsic, M., Copping, A., Hallett, K. C., Labonte, A., Rieks, J., Murray, D., et al., 2014. Methodology for design and economic analysis of marine energy conversion (MEC) technologies. Tech. Report SAND2014-9040, Sandia National Laboratories, Mar.

[36] Anderlini, E., Forehand, D., Bannon, E., and Abusara, M., 2017. "Reactive control of a wave energy converter using artificial neural networks". *Int. J. Mar. Energy, 19*, pp. 207–220. doi: 10.1016/j.ijome.2017.08.001

[37] Thomas, S., Eriksson, M., Göteman, M., Hann, M., Isberg, J., and Engström, J., 2018. "Experimental and numerical collaborative latching control of wave energy converter arrays". *Energies, 11*(11), p. 3036. doi: 10.3390/en11113036

[38] Valério, D., Mendes, M. J., Beirão, P., and da Costa, J. S., 2008. "Identification and control of the AWS using neural network models". *Appl. Ocean Res., 30*(3), pp. 178–188. doi: 10.1016/j.apor.2008.11.002

[39] LeCun, Y., Bengio, Y., and Hinton, G., 2015. "Deep learning". *Nature, 521*(7553), pp. 436–444. doi: 10.1038/nature14539

[40] Haykin, S., 1999. *Neural Networks: A Comprehensive Foundation*, 2nd ed. Prentice Hall.

[41] Lee, T., and Yang, C., 1957. "Many-body problem in quantum mechanics and quantum statistical mechanics". *Phys. Rev., 105*(3), p. 1119. doi: 10.1103/PhysRev.105.1119

[42] Suarez, E., Diaz, N., and Suarez, D., 2009. "Thermochemical fragment energy method for biomolecules: Application to a collagen model peptide". *J. Chem. Theory Comput., 5*(6), pp. 1667–1679. doi: 10.1021/ct8005002

[43] Mavrakos, S., and Koumoutsakos, P., 1987. "Hydrodynamic interaction among vertical axisymmetric bodies restrained in waves". *Appl. Ocean Res., 9*(3), pp. 128–140. doi: 10.1016/0141-1187(87)90017-4

[44] Kelly, T., Zabala, I., Pena-Sanchez, Y., Ringwood, J., Henriques, J., and Blanco, J. M., 2022. "A post-processing technique for removing 'irregular frequencies' and other issues in the results from BEM solvers". *Int. J. Mar. Energy, 5*(1), pp. 123–131. doi: 10.36688/imej.5.123-131

[45] https://github.com/AzadSaeed/Parallel-Nemoh.

[46] Sergiienko, N. Y., Cocho, M., Cazzolato, B. S., and Pichard, A., 2021. "Effect of a model predictive control on the design of a power take-off system for wave energy converters". *Appl. Ocean Res., 115*, p. 102836. doi: 10.1016/j.apor.2021.102836

[47] Esmaeilzadeh, S., and Alam, M.-R., 2019. "Shape optimization of wave energy converters for broadband directional incident waves". *Ocean Eng., 174*, pp. 186–200. doi: 10.1016/j.oceaneng.2019.01.029

[48] McGuinness, J. P., and Thomas, G., 2017. "The constrained optimisation of small linear arrays of heaving point absorbers. part i: The influence of spacing". *Int. J. Mar. Energy, 20*, pp. 33–44. doi: 10.1016/j.ijome.2017.07.005

[49] Fitzgerald, C., and Thomas, G., 2007. "A preliminary study on the optimal formation of an array of wave power devices". In European Wave and Tidal Energy Conference, pp. 11–14.

[50] Tay, Z. Y., and Venugopal, V., 2017. "Optimization of spacing for oscillating wave surge converter arrays using genetic algorithm". *J. Waterw. Port, Coast. Ocean Eng., 143*(2), p. 04016019. doi: 10.1061/(ASCE)WW.1943-5460.0000368

[51] Moarefdoost, M. M., Snyder, L. V., and Alnajjab, B., 2017. "Layouts for ocean wave energy farms: Models, properties, and optimization". *Omega, 66*, pp. 185–194.

Copyright © 2023 by ASME

doi: 10.1016/j.omega.2016.06.004

[52] O'connor, M., Lewis, T., and Dalton, G., 2013. "Techno-economic performance of the Pelamis P1 and Wavestar at different ratings and various locations in Europe". *Renew. Energ., 50*, pp. 889–900. doi: 10.1016/j.renene.2012.08.009

[53] De Andres, A., Guanche, R., Vidal, C., and Losada, I., 2015. "Adaptability of a generic wave energy converter to different climate conditions". *Renew. Energ., 78*, pp. 322–333. doi: 10.1016/j.renene.2015.01.020

[54] De Andres, A., Maillet, J., Hals Todalshaug, J., Möller, P., Bould, D., and Jeffrey, H., 2016. "Techno-economic related metrics for a wave energy converters feasibility assessment". *Sustainability, 8*(11), p. 1109. doi: 10.3390/su8111109

[55] Khojasteh, D., and Kamali, R., 2016. "Evaluation of wave energy absorption by heaving point absorbers at various hot spots in Iran seas". *Energy, 109*, pp. 629–640. doi: 10.1016/j.energy.2016.05.054

[56] Antonio, F., 2010. "Wave energy utilization: A review of the technologies". *Renew. Sust. Energ. Rev., 14*(3), pp. 899–918. doi: 10.1016/j.rser.2009.11.003

[57] Todalshaug, J. H., Ásgeirsson, G. S., Hjálmarsson, E., Maillet, J., Möller, P., Pires, P., Guérinel, M., and Lopes, M., 2016. "Tank testing of an inherently phase-controlled wave energy converter". *Int. J. Mar. Energy, 15*, pp. 68–84. doi: 10.1016/j.ijome.2016.04.007

[58] Peretta, S., Ruol, P., Martinelli, L., Tetu, A., and Kofoed, J. P., 2015. "Effect of a negative stiffness mechanism on the performance of the WEPTOS rotors". In VI International Conference on Computational Methods in Marine Engineering, pp. 58–72.

Nomenclature

Acronyms

ANN	artificial neural network
BEM	boundary element method
CCD	control co-design
CFD	computational fluid dynamics
LHS	Latin hypercube sampling
MBE	many body expansion
PTO	power take-off
QoI	quantities of interest
SE-UCCD	stochastic in expectation uncertain control co-design
TRL	technology readiness level
WEC	wave energy converter

Select Variables

$\mathbf{A}$	added mass matrix
$\mathbf{B}$	damping coefficient matrix
B_{pto}	PTO damping
D_{wec}	WEC draft
$\hat{\mathbf{F}}_{\text{FK}}$	Froude-Krylov force
$\hat{\mathbf{F}}_s$	scattering force
$\hat{\mathbf{F}}_r$	radiation force
$\hat{\mathbf{F}}_{\text{hs}}$	hydrostatic force
$\hat{\mathbf{F}}_{\text{pto}}$	PTO force
G	hydrostatic coefficient
$\hat{\mathbf{F}}_e$	excitation force
H_s	significant wave height
k_{pto}	PTO stiffness
$\mathbf{M}$	mass matrix
R_{wec}	WEC radius
S	cross-sectional area at undisturbed sea level
$S_{JS}(\cdot)$	JONSWAP spectrum
T_p	wave period
$\mathbf{W}$	layout matrix
f_1	first-order ANN function vector
f_2	second-order ANN function vector
h	water depth
l_{pq}	relative distance between pth and qth WEC
m	order of MBE
n_{gq}	number of Gauss quadrature points
n_{yr}	life of the farm
p_v	power per unit volume
p_m	time-averaged absorbed mechanical power
$\tilde{v}_1$	ANN input for 1-body cluster
$\tilde{v}_2$	ANN input for 2-body cluster
w_i	layout vector for ith WEC
$\tilde{y}_1$	ANN output for 1-body cluster
$\tilde{y}_2$	ANN output for 2-body cluster
α	spectrum parameter
β	spectrum parameter
β_w	wave angle
$\Delta\psi$	additive hydrodynamic effect
θ_{pq}	relative angle between pth and qth WEC
$\hat{\xi}$	displacement matrix
ρ	water density
ψ	hydrodynamic effect
ω	radial frequency

Copyright © 2023 by ASME

Proceedings of the ASME 2023
International Design Engineering Technical Conferences and
Computers and Information in Engineering Conference
IDETC-CIE2023
August 20-23, 2023, Boston, Massachusetts

DETC2023-117032

MULTI-TASK MULTI-FIDELITY MACHINE LEARNING FOR RELIABILITY-BASED DESIGN WITH PARTIALLY OBSERVED INFORMATION

Yanwen Xu, Hao Wu, Zheng Liu and Pingfeng Wang*

Department of Industrial and Enterprise Systems Engineering
University of Illinois at Urbana-Champaign
Urbana, IL 61801, United States.

ABSTRACT

In complex engineering systems, assessing system performance and underlying failure mechanisms with respect to uncertain variables requires repeated testing, which is often limited by test capacity and computational budget and fails to accurately capture the complex system's high-dimensional nature. A method that can efficiently use information that is partially available from various sources is thus urgently needed for complex system design. This paper presents a multi-fidelity surrogate modeling strategy that efficiently utilizes partially observed information (POI) from various sources, including data with different fidelity and dimensionality. Additionally, in reliability analysis and design optimization tasks, multiple constraints must be evaluated concurrently for each design point. However, as the complexity of systems increases, the number of constraints grows, resulting in a rapid increase in computational effort. Therefore, a multi-fidelity multi-task surrogate modeling framework with POI was proposed to aid in the development of surrogate models, which increases the effectiveness of reliability analysis. The proposed multi-fidelity multi-task machine learning (MFMT-ML) model utilizes a Bayesian framework, which significantly improves the predictive model's performance and provides uncertainty quantification of the prediction. It also offers premium features such as using multi-fidelity sources of data points and POI, allowing simultaneous evaluation of multiple constraints through a single test, and offering a highly accurate and efficient reliability-based design optimization framework through knowledge sharing. By incorporating partially observed information from various sources, our approach offers a promising avenue for improving system performance prediction accuracy and efficiency while reducing the cost and complexity of complex system design.

Keywords: Machine Learning, Multi-fidelity Model, Multi-task Learning, Reliability Analysis, Design Optimization.

---

*Corresponding author: pingfeng@illinois.edu

NOMENCLATURE

b	Global covariance structure
d	Design variables
$\mathcal{F}$	Variational lower bound
g	Performance limit
G	Performance measure
I	Unite vector
k	Local covaruiance structure
K	Covariance function
K^f	Semipositive definite matrix
KL	Kullback-Leibler divergence
L	Lower bound
POI	Partially observed information
q	Variational distribution
$\mathbf{R}$	Correlation function
S_i	Diagonal matrix
σ^2	Variance
μ	mean value
U	Upper bound
x^*	Test dataset
$\hat{y}$	Predicted system response/performance
Y	Observed system response/performance
Z^O	Fully observed datasets
Z^U	Partially observed datasets
$\otimes$	Kronecker product

1. INTRODUCTION

The performance of the product and system with respect to design variables must be determined repeatedly through testing in complex system analysis, uncertainty quantification, and reliability-based design tasks in order to support and enhance the reliability of product designs. In many complex real-world systems, there are usually many design variables involved, which entails high design costs. Numerous physical experiments and/or simulations are typically needed to study system performance,

Copyright © 2023 by ASME

and they are frequently prohibitively expensive, especially for such iterative tasks. Therefore, reducing the computational cost is a key current focus in many engineering applications. Surrogate modeling [1–3], also known as metamodeling or response surface modeling, is a powerful tool for approximating complex and computationally expensive models with simpler and faster models. The goal is to generate a model that accurately represents the behavior of the original system with a minimum number of evaluations. This is accomplished by using a variety of regression techniques, including radial basis functions, artificial neural networks [4, 5], and Gaussian Process (GP) models [6–8], to map a set of input parameters to the corresponding output responses. Many different applications, including engineering design, optimization, and uncertainty analysis, frequently use surrogate modeling. The GP model, a nonparametric approach that makes no assumptions about the functional form and provides a full probability distribution over the outputs, exhibits superior performance among surrogate models with distinguishing features, allowing quantification of the uncertainty in the predictions.

The GP model is data-driven, which imposes requirements that the dataset must be sufficiently large[9, 10], particularly for high-dimensional problems [11, 12]. Multi-source datasets and partially observed datasets could be useful in order to lessen the burden of data collection[13]. Multisource data is characterized by information collected from diverse sources, including various experiments, simulations, or measurements. Partially Observed Information (POI) describes instances where complete and pertinent information is not accessible. The information acquired from different data sources or derived from models with varying levels of simplification typically exhibits multiple fidelity and dimensionality levels, resulting in a partially observed information dataset.

Prior research has primarily concentrated on utilizing fully observed information, commonly depicted by a series of system input-output sample points, to build a surrogate model. Nevertheless, the presence of multi-source and partially observed information presents difficulties for conventional surrogate modeling techniques. For instance, a standard GP surrogate model is unable to deal with POI directly and necessitates fully observed data. GP-based multi-fidelity models, like the Co-Kriging model, enable the utilization of multi-fidelity information but also necessitate fully observed data with comparable dimensions. In order to manage the POI dataset that comes with sample points collected from various sources with varying dimensions and fidelity, latent variable models are frequently utilized. These models aim to restore the missing value from the corresponding observations. Our previous research [14–16] put forward an adaptable surrogate modeling technique that handles partially observed information and leverages all available data to enhance the model's capabilities through the Bayesian Gaussian process latent variable model (GPLVM) [17, 18].

In addition, complex engineering systems often require the use of multiple system performance functions to define the system condition or failure, which can lead to costly and repetitive function evaluations, especially for reliability analysis and reliability-based design optimization (RBDO) tasks. In RBDO tasks, reliability analysis [19–22] with multiple limit state func-

tions is essential for identifying critical components and guiding the design process towards more reliable systems. The objective of RBDO tasks is to find a design that meets multiple constraints and a reliability constraint specifying the required probability of failure. To achieve this, multiple constraints must be assessed simultaneously, corresponding to each design and step in the reliability estimation and optimization process. As the number of constraints increases, the number of performance evaluations grows, resulting in a significant increase in computational effort. Reliability-based design optimization with multiple constraints is important for engineering design, quality control, and risk assessment, among other applications, as the optimization results can guide the design process towards more reliable systems and identify optimal designs that meet multiple requirements while ensuring a specified level of reliability.

In order to efficiently assess multiple performance functions with limited data sets, a multitask modeling strategy can be integrated into the construction of a surrogate model. Multitask learning, which was initially introduced by Caruana [23], is a machine learning method that aims to improve the performance of several related tasks by taking advantage of the information and knowledge that is shared across them. In Schwaighofer et al. [24], GP-based versions of multi-task models were introduced, and an Expectation-Maximization (EM) algorithm was proposed for learning. Shi et al. [25] also explored similar techniques. Additionally, Yu et al. [26] conducted an extensive study on the relationships between the linear model and GPs to develop a multi-task GP formulation. In addition, Bonilla et al. [27] introduced the idea of two matrices to model covariance between inputs and tasks respectively, the term "multi-task Gaussian process" has mostly referred to the choice of covariance structure. Some further developments have been discussed by Hayashi et al. [28], Rakitsch et al. [29], and Zhu and Sun [30]. Nguyen and Bonilla [31] proposed a sparse approximation for multi-task GP inference, which is known as an example of linear models of coregionalization (LMC) in geostatistics literature. Alvarez and Lawrence [32] provides a unified view on the topic as well as an efficient strategy for constructing computationally efficient approximations. It is important to note that this paper defines a different multi-task paradigm for GPs, focusing on sharing information through the mean function rather than the covariance structure. Swersky et al. [33] worked on Bayesian hyper-parameter optimization in LMC models. Overall, instead of training separate models for each task, multitask learning enables the model to learn multiple tasks concurrently, resulting in more effective use of data and improved generalization performance. Shared features, shared representations, or shared parameters can be used to represent the shared information between tasks. The objective of multitask learning is to find a shared representation that is beneficial for all tasks, while still allowing each task to have its own specialized representation if necessary [34].

In this study, we proposed a multi-task multi-fidelity machine learning method to boost the effectiveness of the surrogate model when face with challenges related to data scarcity. By creating a mechanism for information sharing between various tasks, the suggested method can handle several related tasks at once. Additionally, the multi-fidelity framework permits the use

Copyright © 2023 by ASME

of multi-source, multi-fidelity POI datasets to further increase the effectiveness of data usage. The main contributions of this work are twofold: Firstly, the proposed method effectively utilizes partially available information from multiple sources with varying dimensions and fidelity. Secondly, the approach allows the model to exploit the shared information and knowledge across tasks, resulting in improved performance for all tasks, while still enabling each task to have its own specialized representation. The proposed framework outperforms the conventional approach of training separate Gaussian process (GP)-based models for each system response or constraint in both surrogate modeling and reliability-based design optimization tasks.

The remainder of this paper is organized as follows. The proposed method is discussed in Section 2. A vehicle deisgn example is provided in Section 3, followed by conclusions in Section 4.

2. METHODOLOGY

2.1 Formulation of a Standard Gaussian Process

The GP-based surrogate models have become one of the most popular methods to develop computationally efficient surrogate models in many engineering design applications, including simulation-based design optimization and uncertainty analysis. First, for the standard GP formulation, all variables are fully observed. Assume a dataset with input $X \in R^{N \times Q}$ and observed target $Y \in R^{ND}$ are stored. A standard GP model can be generally expressed as

$$Y_{GP}(\mathbf{x}) = f(\mathbf{x}) + S(\mathbf{x}) \tag{1}$$

where $S(\mathbf{x})$ is a Gaussian stochastic process with zero mean and variance σ^2. GP priors are placed on the mapping f, so that the function follows a Gaussian distribution where the covariance function K between arbitrary two input values x_i and x_j can be defined as

$$K = \text{Cov}\left[S(\mathbf{x}_i), S(\mathbf{x}_j)\right] = \sigma^2 \mathbf{R}(\mathbf{x}_i, \mathbf{x}_j) \tag{2}$$

where $\mathbf{R}(\mathbf{x}_i, \mathbf{x}_j)$ denotes the correlation function matrix, and can be defined as

$$\mathbf{R}(\mathbf{x}_i, \mathbf{x}_j) = \text{Corr}(\mathbf{x}_i, \mathbf{x}_j) = \exp\left[-\sum_{p=1}^{N} a_p \left|x_i^p - x_j^p\right|^{b_p}\right] \tag{3}$$

where N is the number of observations; a_p and b_p, $p = 1, \cdots, N$, are parameters of the GP model. Then, the model likelihood $p(Y|X)$ is:

$$\int_f p(Y \mid f) p(f \mid X) df = \prod N\left(Y \mid 0, K + \sigma^2 \mathbf{I}\right) \tag{4}$$

The traditional Gaussian process (GP) model has limitations in its ability to handle multi-fidelity or partially observed information. It can only use the fully observed sample points, which makes it difficult to accurately model complex systems. This limitation results in high model development costs and a lack of flexibility in the GP model.

2.2 Gaussian Process Latent Variable Model for POI

Obtaining high-fidelity and high-dimensional fully observed data for complex systems can be challenging and expensive due to limitations in test capacity and computational resources. However, partially observed data from various sources can be utilized to help accurately capture the complex engineering systems. This approach can enable a comprehensive utilization of all available multi-fidelity information, including data from different sources, instead of relying solely on fully observed information that may be difficult to obtain or derive.

In the absence of fully observed inputs, the input X need to be recovered from the output Y through maximum likelihood. Since X is a latent variable, it can be assigned to a prior density given by the standard normal density. The prior for X is:

$$p(X) = \prod_{i=1}^{N} N\left(x_i \mid 0, I_Q\right) \tag{5}$$

where each x_i is the i-th row of X. I_Q is the unite vector with length Q. The Bayesian GP-LVM is employed to integrate X out by constructing a variational lower bound,

$$\mathcal{F} \leq \log \int_X p(Y) = \log \int_X p(Y \mid X) p(X) \tag{6}$$

with a variational distribution

$$q(X) = \prod_{i=1}^{N} N\left(x_i \mid \mu_i, S_i\right) \tag{7}$$

where S_i is a diagonal matrix. The formula for variational bound is derived as

$$\mathcal{F} = \langle \log p(Y \mid X) \rangle_{q(X)} - KL(q(X) \| p(X)) \tag{8}$$

where $<> q(X)$ denotes the expectation with respect to $q(X)$. KL term represents the Kullback-Leibler divergence with

$$KL(q(X) \| p(X)) = \int q(X) \log \frac{q(X)}{p(X)} dX \tag{9}$$

When inputs are partially observed, existing missing values in the dataset are divided into fully and partially observed subsets as $Z = (Z^O, Z^U)$, where O and U denote fully and partially observed datasets respectively. The variational distribution becomes

$$q(X \mid Z) = q(X^o \mid Z^o) q\left(X^U \mid Z^U\right) \tag{10}$$

And $log(Y|Z)$ is approximated with the variational lower bound

$$\mathcal{F} = \langle \log p(Y \mid X) \rangle_{q(X|Z)} - KL(q(X \mid Z) \| p(X)) \tag{11}$$

The variational approach aims to approximate the true posterior with a variational distribution. The maximization of the variational lower bound provides a Bayesian training procedure that is robust to overfitting. Moreover, the likelihood function can be computed in closed form for exponential square kernels [17].

Copyright © 2023 by ASME

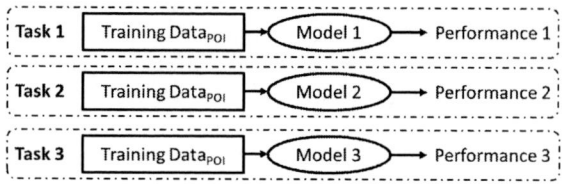

Multi-fidelity Single Task Learning with POI

(a) Single task learning framework

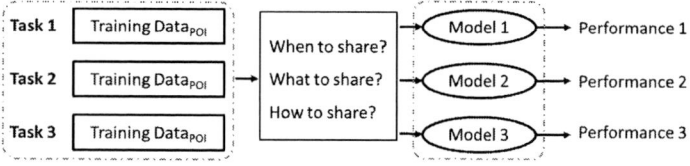

Multi-fidelity Multi-Task Learning with POI

(b) Multi-task learning framework

FIGURE 1: THE COMPARISON BETWEEN MULTI-FIDELITY SINGLE TASK AND MULTI-FIDELITY MULTI-TASK LEARNING FRAMEWORK

The integration of partially observed information using a multi-fidelity surrogate modeling strategy can greatly reduce the amount of required expensive fully observed data and improve prediction accuracy and efficiency. Thus, partially observed data can offer a promising avenue for improving system performance prediction accuracy and efficiency, while also reducing the cost and complexity of complex system design.

2.3 Multi-fidelity Multi-task Learning Framework

In engineering applications, numerous design constraints and physics prior information are frequently imposed on the system for a variety of reasons, including manufacturing policies, diverse failure modes, safety standards, and customer preferences. As the number of constraints grows, the number of required training data points and performance evaluations increases quickly during testing, leading to a substantial rise in computational efforts. To address the challenge of efficiently evaluating multiple performance functions with limited data sources during reliability analysis or design optimization tasks, a multitask modeling strategy can be introduced into surrogate model construction.

Multi-task learning aims to learn m learning tasks that are related by using the knowledge contained in all or some of the other tasks to improve the learning of a model for each task [35]. Fig. 1 shows a comparison between single and multi-task learning with POI. In multi-fidelity single-task learning (as shown in Fig. 1a), the training models for different tasks are independent; in contrast, for multi-fidelity multi-task learning (as shown in Fig. 1b), the training models for different tasks are shared. Multitask learning generally involves three aspects (as shown in Fig. 1b): when to share, what to share, and how to share.

The "when to share" problem in multi-task learning involves deciding whether to use a single-task or multi-task model for a given task. Currently, this decision is often made by human experts and are few studies on the topic. One solution is to formulate the decision as a model selection problem and use techniques such as cross-validation, but this can be computationally heavy

and require more training data. An advanced solution is to use multi-task models that can reduce to single-task models under certain conditions, allowing the training data to determine the appropriate form.

Determining "what to share" involves deciding how knowledge can be shared among all tasks. Feature-based multi-task learning seeks to learn common features across different tasks in order to share knowledge, while parameter-based multi-task learning uses model parameters from one task to aid in learning model parameters for other tasks, often through regularization techniques. After choosing "what to share," "how to share" details specific ways to transfer knowledge between tasks.

Therefore, in this study, the development of multi-fidelity multi-task models requires a covariance function which can simultaneously describe the correlation between different input parameters for a single task and the correlation between different input parameters for multiple tasks at the same time [36]. Therefore, this covariance function should satisfy two requirements: (1) to be able to simulate the local covariance structure of single-task learning; (2) to be able to simulate the global cross-covariance structure of multiple tasks [37]. To accomplish the above purpose, Alvarez et al. [38] proposed that the covariance function could be composed of the product of the local covariance structure $k(x, x')$ and the global covariance structure $b(x, x')$. $k(x, x')$ is used to describe the intra-correlation of a single task and $b(x, x')$ is used for describing the intercorrelation of different tasks.

$$k_{MTGP}(x, x') = k(x, x') b(x, x') \tag{12}$$

The structure of MTGP can be further expanded into a linear weighted form of different independent covariance functions. If there are Q separable structure, it can be expressed as:

$$k_{MTGP}(x, x') = \sum_{i=1}^{Q} k_i(x, x') b_i(x, x') \tag{13}$$

Based on the above property, Bonilla [27] proposed to connect the tasks in series to develop a MTGP structure. In terms of the m-dimensional input $(x_1, x_2, \cdots, x_m)^T$, if the corresponding multi-task output is $y = (y_1, y_2, \cdots, y_n)$, then:

$$\bar{f}_l(x_*) = \left(K_l^f \otimes K_*^x \right)^T \sum{}^{-1} y \tag{14}$$

$$\sum = K^f \otimes K^x + D \otimes I \tag{15}$$

where $\bar{f}_l(x)$ is the average predicted values of x; $\otimes$ is Kronecker product; K^f is a semipositive definite matrix; K_l^f represents the l-th column of K^f; K_x^* is the covariance matrix between test dataset x^* and the training dataset; and K_x is the covariance matrix of the training dataset. D is an $m \times m$ diagonal matrix.

Multitask learning strategy enhances the power of the model by jointly learning multiple tasks, the model can capture shared features between tasks, leading to improved generalization performance on each task. This strategy enables us to find a shared representation that is beneficial for all tasks, while still allowing each task to have its own specialized representation if necessary.

Copyright © 2023 by ASME

Furthermore, multitask learning strategy can improve the efficiency of the learning process by reducing the amount of data needed for each individual task and can be useful in scenarios where there are limited resources for training and deploying separate models for each task, as a single model can be used to perform multiple related tasks.

2.4 Multi-task Multi-fidelity Machine Learning Algorithm

In this subsection, we provide a comprehensive overview of the proposed method and present the detailed steps involved. The algorithmic implementation of the proposed method is provided in Algorithm. 1. The proposed method begins with the collection of data from multiple sources with varying levels of fidelity and dimensionality. Next, a multi-fidelity multi-task Gaussian process model is trained to capture the underlying relationships between the input and output variables of the system. The model is trained using a Bayesian framework, which not only improves the model's predictive performance but also provides uncertainty quantification of the prediction. The proposed method focuses on sharing information through multiple data sources and tasks

via the latent variable and MTGP structure respectively, which allows for the efficient use of partially observed and multi-fidelity data for multiple tasks. Finally, the trained model is used to perform system modeling and reliability-based design optimization tasks in section 3.

3. VEHICLE DESIGN EXAMPLE
3.1 Modeling Vehicle Performances

The first example is a vehicle design problem (shown in Fig.2a) described by a set of design variables $(x_1, x_2, \cdots, x_8)$. According to safety regulation of vehicle side impact, vehicle design must meet internal and regulated side impact requirements. The objective of the design is to reduce the total weight while maintaining or improving the performance of vehicle in side impact. The response surfaces for three performance measures (shown in Fig.2b) are constructed from a vehicle side impact model as $G_j \leq g_j$, $j = 1, 2, 3$, where the performance limits g_j form a vector $g = [1, 32, 32]^T$. The response surfaces of the vehicle weight f and the performance measures G_i are defined in Eq (16).

$$
\begin{aligned}
f(weight) &= 1.98 + 4.90d_1 + 6.67d_2 + 6.98d_3 + 4.01d_4 + 1.78d_5 + 2.73d_7 \\
G_1 &= 1.16 - 0.3717X_2X_4 - 0.00931X_2X_8 - 0.484X_3X_7 + 0.01343X_6X_8; \\
G_2 &= 28.98 + 3.818X_3 - 4.2X_1X_2 + 0.0207X_5X_8 + 6.63X_6X_7 - 7.7X_4X_5 + 0.32X_7X_8; \\
G_3 &= 46.36 - 9.9X_2 - 12.9X_1X_8 + 0.1107X_3X_8.
\end{aligned}
\tag{16}
$$

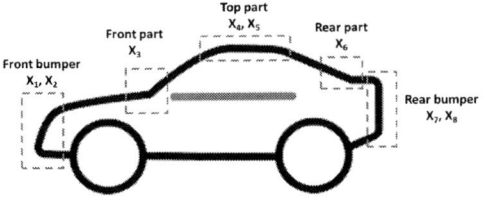

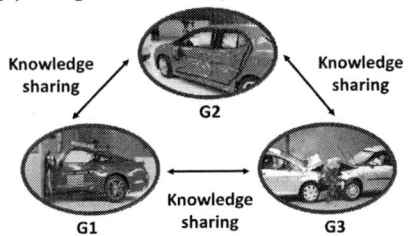

(a) Design variables $x_1 - x_8$ of the vehicle

(b) Knowledge sharing mechanism between three performance functions

FIGURE 2: VEHICLE DESIGN PARAMETERS AND PERFORMANCE FUNCTIONS

For training the surrogate model (as shown in Fig. 3), we used 30 fully observed sample points (10 points per constraint) from source 1, and 60 partially observed sample points (20 points per constraint) from sources 2 to 5, as presented in Table 1. Three models were used for comparison: a single-task Gaussian process (GP) model, a single-task Gaussian process latent variable model

TABLE 1: MISSING PATTERN FOR PARTIALLY OBSERVED INFORMATION

Variable	X_1	X_2	X_3	X_4	X_5	X_6	X_7	X_8
Source 1	o	o	o	o	o	o	o	o
Source 2	o	o		o	o	o	o	o
Source 3	o	o	o			o	o	o
Source 4	o	o	o	o	o		o	o
Source 5	o	o	o	o	o	o		

(GPLVM), and the proposed multi-fidelity multi-task GPLVM (MM-GPLVM). The traditional single-task GP model could only use the 30 fully observed samples for training, while the single-task GPLVM and MM-GPLVM were trained using both 30 fully observed and 60 partially observed samples. We evaluated the accuracy of the surrogate models using the mean square error with 10,000 testing sample points. Mean Square Error (MSE) was used as evaluation metric for measuring the accuracy of the models. It calculates the average squared difference between the predicted and actual values of the target variable. The formula for MSE is:

$$
MSE = \frac{1}{n} * \sum_{i=1}^{n} (y_i - \hat{y}_i)^2
\tag{17}
$$

where n is the number of samples in the dataset, y_i is the actual value of the target variable for the i-th sample, and $\hat{y}_i$ is the predicted value of the target variable for the i-th sample. The MSE value is always non-negative and a smaller value indicates

Copyright © 2023 by ASME

Algorithm 1 MULTI-TASK MULTI-FIDELITY MACHINE LEARNING WITH PARTIALLY OBSERVED INFORMATION

1: Given: Multi-fidelity training datasets, including fully observed data (Z^o, Y^o) and partially observed data (Z^u, Y^u)
2: Define a small value, e.g. $\epsilon = 10^{-9}$
3: Initialize $q(X^o) = \prod_{i=1}^{n} N(X_{i,:}^o | Z_{i,:}^o, \epsilon I)$
4: *Fix $q(X^o)$ in the optimizer*
5: Train a variational GP-LVM model M^o given the above $q(X^o)$ and Y^o
6: **for** $i = 1, \ldots, |Y^u|$ **do**,
7:      Build multi-task learning covariance function and optimize the marginal likelihood function as Eq.(14)
8:      Reconstruct missing value and location with $p(\hat{X}_{i,:}^u | y_i^u, M^o) \approx q(\hat{X}_{i,:}^u) = N(\hat{X}_{i,:}^u | \mu_{i,:}^u, \hat{S}_i^u)$
9:      Initialize $q(X_{i,:}^u) = N(X_{i,:}^u | \mu_{i,:}^u, S_i^u)$ as follows
10:      **for** j=1,..., q **do**,
11:          **if** $Z_{i,j}^u$ is observed **then**
12:              $\mu_{i,j}^u = Z_{i,j}^u$ and $(S_i^u)_{j,j} = \epsilon$
13:              Fix $\mu_{i,j}^u$, $(S_i^u)_{j,j}$ in the optimizer
14:          **else**
15:              $\mu_{i,j}^u = \hat{\mu}_{i,j}^u$ and $(S_i^u)_{j,j} = (\hat{S}_i^u)_{j,j}$
16: Incorporate MTGP structure to the marginal likelihood optimization process as Eq.(14)
17: Train a multi-task multi-fidelity variational GP-LVM model $M^{o,u}$ using the initial $q(X^o)$ and $q(X^u)$ defined above and data Y^o, Y^u
18: Predictions can be made using model $M^{o,u}$ for all tasks.

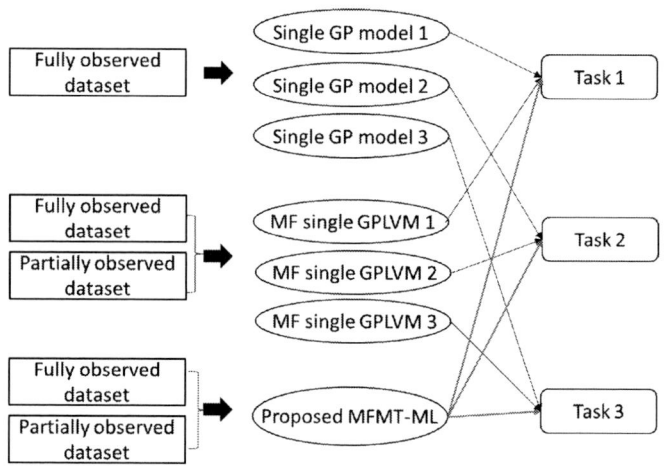

FIGURE 3: MODEL TRAINING FRAMEWORK OF SINGLE GP MODEL, MULTI-FIDELITY SINGLE TASK, AND MULTI-FIDELITY MULTI-TASK MODEL

a better performance of the model.

The results presented in Tab.2 show that the traditional single-task GP model has poor accuracy and high mean square error (MSE) due to the limited fully observed training samples. The GPLVM model performs better by utilizing partially observed samples, but the proposed MM-GPLVM model further improves the performance and prediction accuracy by using a multi-task strategy. This approach is particularly beneficial for product design and optimization applications as the additional available data can enhance the model to provide more accurate system response and reliability estimations.

The information sharing and utilization mechanism of MTL provides several advantages besides the data augmentation. For example, certain features may be simpler to learn for one task but harder for another, perhaps because one task interacts with the

features in a more complicated manner or because other features are obstructing the model's ability to learn those features. By using MTL, the model can "eavesdrop" and learn those features from another task where they are easier to learn.

3.2 Vehicle Design Optimization

Next, we present the formulation of a reliability-based design optimization (RBDO) model for the vehicle design example. The objective of the RBDO model is to generate designs that minimize the system design costs while ensuring a target level of reliability. In the RBDO model, the mean values of the random design parameters are commonly used as design variables, allowing the system design to be optimized while satisfying all reliability constraints. As a result, the design solution obtained by solving the RBDO model provides a higher level of reliability than that obtained by the deterministic design model. The mathematical formulation of the RBDO model is described in [39, 40].

$$
\begin{aligned}
\text{Minimize} \quad & f(\mathbf{d}) \\
\text{Subject to} \quad & \Pr\left(G_j(\mathbf{X}; \mathbf{d}) \le g_j\right) \ge 99\%, j = 1, \ldots, 3 \quad (18)\\
& d_i^L \le d_i \le d_i^U, i = 1, \ldots, 8
\end{aligned}
$$

where d is a vector of design variables, $f(d)$ is the objective function. Superscripts 'L' and 'U' represent the lower bound and upper bound, respectively. The primary challenge in solving the RBDO model as shown in Eq. (18) lies in the accurate and efficient evaluation of the probabilistic constraints during the iterative design optimization process, since considerable computational or experimental efforts are normally required.

Efficient design optimization for a vehicle can be achieved with the aid of MM-GPLVM in repeated function evaluations. The proposed MM-GPLVM surrogate model identifies the optimal design combination as [0.5, 1.3, 0.5, 0.5, 0.5, 0.93, 0.19, 0.34], which minimizes the total vehicle weight to 20. The resulting optimization satisfies

Copyright © 2023 by ASME

TABLE 2: COMPARISON OF THE MODEL ACCURACY BETWEEN SINGLE AND MULTI-TASK GP-BASED SURROGATE MODELS

Prediction Error (MSE)		G_1	G_2	G_3
	Single GP model	0.219	0.213	0.056
Surrogate models	Multi-fidelity single GPLVM	0.062	0.142	0.016
	Multi-fidelity multitask GPLVM	0.026	0.037	0.012

TABLE 3: THE NUMBER OF FUNCTION EVALUATION DURING THE RBDO PROCESS

Number of function evaluations		G_1	G_2	G_3
	Single GP model	34	34	34
RBDO with	Multi-fidelity single GPLVM	18	18	18
	Multi-fidelity multitask GPLVM		14	

the reliability requirements with 99% confidence, accounting for design uncertainties. The proposed approach exhibits premium features, including (1) utilizing multi-fidelity sources of data points and POI, (2) enabling simultaneous evaluation of multiple constraints through a single test, and (3) providing a highly accurate and efficient reliability-based design optimization framework through knowledge sharing. The specifics of the number of function evaluations during the RBDO process are summarized in Tab. 3.

4. CONCLUSION

In conclusion, this paper presents a novel multi-fidelity multi-task machine learning model for system design with partially observed information. The proposed approach effectively leverages all available multi-fidelity information, including data from other tasks and data points from different sources, to reduce reliance on expensive fully observed information that may be difficult to obtain or derive. By utilizing a multi-fidelity approach, the proposed method can improve prediction accuracy and efficiency while reducing the amount of required fully observed data for multiple tasks.

The numerical example presented in this study demonstrates that the proposed multi-fidelity multi-task machine learning strategy yields significant improvements over traditional methods in terms of prediction accuracy and efficiency. Furthermore, the multi-task learning mechanism utilized in this study enhances the efficiency of the reliability-based design optimization process by concurrent evaluations of multiple tasks, which is crucial for complex system design.

Overall, this paper provides a promising approach for improving system performance prediction accuracy and efficiency while reducing the cost and complexity of model development. The proposed method has the potential to greatly impact the field of system design by reducing the reliance on expensive fully observed data and enabling the efficient utilization of multi-fidelity multi-task information. Future research can explore the extension of this approach to more complex engineering systems.

ACKNOWLEDGMENTS

This research is partially supported by the National Science Foundation (NSF) the Engineering Research Center for Power Optimization of Electro-Thermal Systems (POETS) with cooperative agreement EEC-1449548, and the Alfred P. Sloan Foundation through the Energy and Environmental Sensors program with grant # G-2020-12455.

REFERENCES

[1] Kohtz, Sara, Xu, Yanwen, Zheng, Zhuoyuan and Wang, Pingfeng. "Physics-informed machine learning model for battery state of health prognostics using partial charging segments." *Mechanical Systems and Signal Processing* Vol. 172 (2022): p. 109002.

[2] Bansal, Parth, Zheng, Zhuoyuan, Shao, Chenhui, Li, Jingjing, Banu, Mihaela, Carlson, Blair E and Li, Yumeng. "Physics-informed machine learning assisted uncertainty quantification for the corrosion of dissimilar material joints." *Reliability Engineering & System Safety* Vol. 227 (2022): p. 108711.

[3] Chen, Xin, Zhou, Haofei and Li, Yumeng. "Effective design space exploration of gradient nanostructured materials using active learning based surrogate models." *Materials & Design* Vol. 183 (2019): p. 108085.

[4] Li, Yumeng, Xiao, Weirong and Wang, Pingfeng. "Uncertainty quantification of artificial neural network based machine learning potentials." *ASME International Mechanical Engineering Congress and Exposition*, Vol. 52170: p. V012T11A030. 2018. American Society of Mechanical Engineers.

[5] Liu, Dehao and Wang, Yan. "Multi-fidelity physics-constrained neural network and its application in materials modeling." *Journal of Mechanical Design* Vol. 141 No. 12 (2019).

[6] Zheng, Zhuoyuan, Chen, Bo, Xu, Yanwen, Fritz, Nathan, Gurumukhi, Yashraj, Cook, John, Ates, Mehmet N, Miljkovic, Nenad, Braun, Paul V and Wang, Pingfeng. "A Gaussian process-based crack pattern modeling approach for battery anode materials design." *Journal of Electrochemical Energy Conversion and Storage* Vol. 18 No. 1 (2021): p. 011011.

[7] Xu, Yanwen, Lalwani, Anand Vikas, Arora, Kanika, Zheng, Zhuoyuan, Renteria, Anabel, Senesky, Debbie G and

Copyright © 2023 by ASME

Wang, Pingfeng. "Hall-Effect Sensor Design With Physics-Informed Gaussian Process Modeling." *IEEE Sensors Journal* Vol. 22 No. 23 (2022): pp. 22519–22528.

[8] Zheng, Zhuoyuan, Xu, Yanwen and Wang, Pingfeng. "Uncertainty quantification analysis on mechanical properties of the structured silicon anode via surrogate models." *Journal of The Electrochemical Society* Vol. 168 No. 4 (2021): p. 040508.

[9] Xu, Yanwen and Wang, Pingfeng. "A comparison of numerical optimizers in developing high dimensional surrogate models." *International Design Engineering Technical Conferences and Computers and Information in Engineering Conference*, Vol. 59193: p. V02BT03A037. 2019. American Society of Mechanical Engineers.

[10] Xu, Yanwen and Wang, Pingfeng. "Rare event estimation of high dimensional problems with confidence intervals." *IIE Annual Conference. Proceedings*: pp. 31A–36A. 2020. Institute of Industrial and Systems Engineers (IISE).

[11] Xu, Yanwen and Wang, Pingfeng. "Sequential Sampling Based Reliability Analysis for High Dimensional Rare Events With Confidence Intervals." *International Design Engineering Technical Conferences and Computers and Information in Engineering Conference*, Vol. 84010: p. V11BT11A039. 2020. American Society of Mechanical Engineers.

[12] Xu, Yanwen and Wang, Pingfeng. "An Enhanced Squared Exponential Kernel With Manhattan Similarity Measure for High Dimensional Gaussian Process Models." *International Design Engineering Technical Conferences and Computers and Information in Engineering Conference*, Vol. 85390: p. V03BT03A025. 2021. American Society of Mechanical Engineers.

[13] Xu, Yanwen, Kohtz, Sara, Boakye, Jessica, Gardoni, Paolo and Wang, Pingfeng. "Physics-informed machine learning for reliability and systems safety applications: State of the art and challenges." *Reliability Engineering & System Safety* (2022): p. 108900.

[14] Xu, Yanwen, Renteria, Anabel and Wang, Pingfeng. "Adaptive surrogate models with partially observed information." *Reliability Engineering & System Safety* Vol. 225 (2022): p. 108566.

[15] Xu, Yanwen and Wang, Pingfeng. "Adaptive Surrogate Models for Uncertainty Quantification with Partially Observed Information." *AIAA SCITECH 2022 Forum*: p. 1439. 2022.

[16] Xu, Yanwen and Wang, Pingfeng. "Hierarchical Surrogate Modeling With Multiple Order Partially Observed Information." *International Design Engineering Technical Conferences and Computers and Information in Engineering Conference*, Vol. 86236: p. V03BT03A025. 2022. American Society of Mechanical Engineers.

[17] Damianou, Andreas C, Titsias, Michalis K and Lawrence, Neil. "Variational inference for latent variables and uncertain inputs in Gaussian processes." (2016).

[18] Titsias, Michalis and Lawrence, Neil D. "Bayesian Gaussian process latent variable model." *Proceedings of the thirteenth international conference on artificial intelligence and statistics*: pp. 844–851. 2010. JMLR Workshop and Conference Proceedings.

[19] Xu, Yanwen and Wang, Pingfeng. "Reliability Analysis with Partially Observed Information." *2022 Annual Reliability and Maintainability Symposium (RAMS)*: pp. 1–6. 2022. IEEE.

[20] Wu, Hao, Zhu, Zhifu and Du, Xiaoping. "System reliability analysis with autocorrelated kriging predictions." *Journal of Mechanical Design* Vol. 142 No. 10 (2020).

[21] Wu, Hao, Hu, Zhangli and Du, Xiaoping. "Time-dependent system reliability analysis with second-order reliability method." *Journal of Mechanical Design* Vol. 143 No. 3 (2021).

[22] Wu, Hao and Du, Xiaoping. "Envelope Method for Time- and Space-Dependent Reliability Prediction." *ASCE-ASME Journal of Risk and Uncertainty in Engineering Systems, Part B: Mechanical Engineering* Vol. 8 No. 4 (2022): p. 041201.

[23] Caruana, Rich. *Multitask learning*. Springer (1998).

[24] Schwaighofer, Anton, Tresp, Volker and Yu, Kai. "Learning Gaussian process kernels via hierarchical Bayes." *Advances in neural information processing systems* Vol. 17 (2004).

[25] Shi, Jian Qing, Murray-Smith, Roderick and Titterington, D Michael. "Hierarchical Gaussian process mixtures for regression." *Statistics and computing* Vol. 15 (2005): pp. 31–41.

[26] Yu, Kai, Tresp, Volker and Schwaighofer, Anton. "Learning Gaussian processes from multiple tasks." *Proceedings of the 22nd international conference on Machine learning*: pp. 1012–1019. 2005.

[27] Bonilla, Edwin V, Chai, Kian and Williams, Christopher. "Multi-task Gaussian process prediction." *Advances in neural information processing systems* Vol. 20 (2007).

[28] Hayashi, Kohei, Takenouchi, Takashi, Tomioka, Ryota and Kashima, Hisashi. "Self-measuring similarity for multi-task gaussian process." *Proceedings of ICML Workshop on Unsupervised and Transfer Learning*: pp. 145–153. 2012. JMLR Workshop and Conference Proceedings.

[29] Rakitsch, Barbara, Lippert, Christoph, Borgwardt, Karsten and Stegle, Oliver. "It is all in the noise: Efficient multi-task Gaussian process inference with structured residuals." *Advances in neural information processing systems* Vol. 26 (2013).

[30] Zhu, Jiang and Sun, Shiliang. "Multi-task sparse Gaussian processes with improved multi-task sparsity regularization." *Pattern Recognition: 6th Chinese Conference, CCPR 2014, Changsha, China, November 17-19, 2014. Proceedings, Part I 6*: pp. 54–62. 2014. Springer.

[31] Nguyen, Trung V, Bonilla, Edwin V et al. "Collaborative Multi-output Gaussian Processes." *UAI*: pp. 643–652. 2014.

[32] Alvarez, Mauricio A and Lawrence, Neil D. "Computationally efficient convolved multiple output Gaussian processes." *The Journal of Machine Learning Research* Vol. 12 (2011): pp. 1459–1500.

Copyright © 2023 by ASME

[33] Swersky, Kevin, Snoek, Jasper and Adams, Ryan P. "Multi-task bayesian optimization." *Advances in neural information processing systems* Vol. 26 (2013).

[34] Tan, Zhongfu, De, Gejirifu, Li, Menglu, Lin, Hongyu, Yang, Shenbo, Huang, Liling and Tan, Qinkun. "Combined electricity-heat-cooling-gas load forecasting model for integrated energy system based on multi-task learning and least square support vector machine." *Journal of cleaner production* Vol. 248 (2020): p. 119252.

[35] Zhang, Yu and Yang, Qiang. "A survey on multi-task learning." *IEEE Transactions on Knowledge and Data Engineering* Vol. 34 No. 12 (2021): pp. 5586–5609.

[36] Liu, Haitao, Ong, Yew-Soon, Shen, Xiaobo and Cai, Jianfei. "When Gaussian process meets big data: A review of scalable GPs." *IEEE transactions on neural networks and learning systems* Vol. 31 No. 11 (2020): pp. 4405–4423.

[37] Genton, Marc G and Kleiber, William. "Cross-covariance functions for multivariate geostatistics." (2015).

[38] Alvarez, Mauricio A, Rosasco, Lorenzo, Lawrence, Neil D et al. "Kernels for vector-valued functions: A review." *Foundations and Trends in Machine Learning* Vol. 4 No. 3 (2012): pp. 195–266.

[39] Xu, Yanwen and Wang, Pingfeng. "CVaR Formulation of Reliability-Based Design Problems Considering the Risk of Extreme Failure Events." *2021 Annual Reliability and Maintainability Symposium (RAMS)*: pp. 1–5. 2021. IEEE.

[40] Zhao, Zilong, Xu, Yanwen, Lin, Yu-Feng, Wang, Xinlei and Wang, Pingfeng. "Probabilistic modeling and reliability-based design optimization of a ground source heat pump system." *Applied Thermal Engineering* Vol. 197 (2021): p. 117341.

[41] Dempster, Arthur P, Laird, Nan M and Rubin, Donald B. "Maximum likelihood from incomplete data via the EM algorithm." *Journal of the royal statistical society: series B (methodological)* Vol. 39 No. 1 (1977): pp. 1–22.

[42] Williams, Christopher KI and Rasmussen, Carl Edward. *Gaussian processes for machine learning.* Vol. 2. MIT press Cambridge, MA (2006).

Copyright © 2023 by ASME

Proceedings of the ASME 2023
International Design Engineering Technical Conferences and
Computers and Information in Engineering Conference
IDETC-CIE2023
August 20-23, 2023, Boston, Massachusetts

DETC2023-111079

DATA-DRIVEN MULTIFIDELITY TOPOLOGY DESIGN WITH A LATENT CROSSOVER OPERATION

Taisei Kii[1], Kentaro Yaji[1,*], Kikuo Fujita[1], Zhenghui Sha[2], Carolyn C. Seepersad[2]

[1]Department of Mechanical Engineering, Osaka University, Osaka 565-0871, Japan
[2]Walker Department of Mechanical Engineering, University of Texas Austin, Austin, TX 78712

ABSTRACT

Topology optimization is one of the most flexible structural optimization methodologies. However, in exchange for its high degree of design freedom, typical topology optimization cannot avoid multimodality, where multiple local optima exist. This study focuses on developing a gradient-free topology optimization framework to avoid being trapped in bad local optima. Its core is a data-driven multifidelity topology design (MFTD) method, in which design candidates generated by solving low-fidelity topology optimization problems are updated based on evolutionary algorithms (EAs) through high-fidelity evaluation. The key component of the data-driven MFTD is a deep generative model that compresses the dimension of the original data into a low-dimensional manifold, i.e., the latent space. In the original framework, convergence variability and premature convergence problems arise as the generative process is performed randomly in the latent space. Inspired by a popular crossover operation, we propose a data-driven MFTD framework incorporating a new crossover operation called latent crossover. We apply the proposed method to a maximum stress minimization problem in 2D structural mechanics. The results demonstrate that the latent crossover improves convergence stability compared to the original method. Furthermore, the optimized designs exhibit performance comparable to or better than that in conventional gradient-based topology optimization using the P-norm measure.

Keywords: Topology optimization; Deep generative model; Maximum stress minimization; Latent crossover

1. INTRODUCTION

Topology optimization, first proposed by Bendsøe and Kikuchi [1], enables the determination of an optimized material distribution for a structural optimization problem and offers a high degree of design freedom [2]. While this attractive feature makes it applicable to various structural design problems, topology op-

timization faces challenges with multimodality, where multiple local optima exist in the solution space. That is, gradient-based optimizers used in conventional topology optimization methods may fall into low-performance local optima. This intractable characteristic is often seen in a strongly nonlinear problem, e.g., minimax problems; thus it is challenging to obtain structures that exhibit high levels of performance.

One of the standard ways to overcome the problem of multimodality is evolutionary algorithms (EAs) since they are gradient-free [3]. An EA, such as the genetic algorithm, mimics the evolutionary mechanisms of living organisms, and solutions are represented as strings of genes. The solution search is performed by applying three basic genetic operations: selection, crossover, and mutation, to a population of individuals. Each iteration of these genetic operations is referred to as a generation. The selection is an operation that retains individuals with relatively better objective function values in the population for the next generation. The crossover is an operation that partially exchanges genes between selected individuals to generate new individuals (offspring) that inherit traits from old ones (parents). However, if some individuals in the population have significantly higher fitness than others in the early stages of the search, they may weed out others by selection and crossover, leading to a loss of diversity and a high probability of premature convergence [4]. The mutation is an operation that introduces new genes into the population by changing a portion of the genes of selected individuals, which helps maintain diversity in the population. Several methods [5–8] have been proposed to solve topology optimization problems using EAs, taking advantage of their gradient-free nature. While they can perform a global search for strongly nonlinear problems, Sigmund [9] has pointed out issues with EA-based topology optimization. That is, topology optimization problems often require a large number of design variables, and the computational cost of the EA increases exponentially with the number of design variables due to the so-called *curse of dimensionality*.

As a potentially promising way to avoid the curse of dimen-

---

*Corresponding author: yaji@mech.eng.osaka-u.ac.jp

Copyright © 2023 by ASME

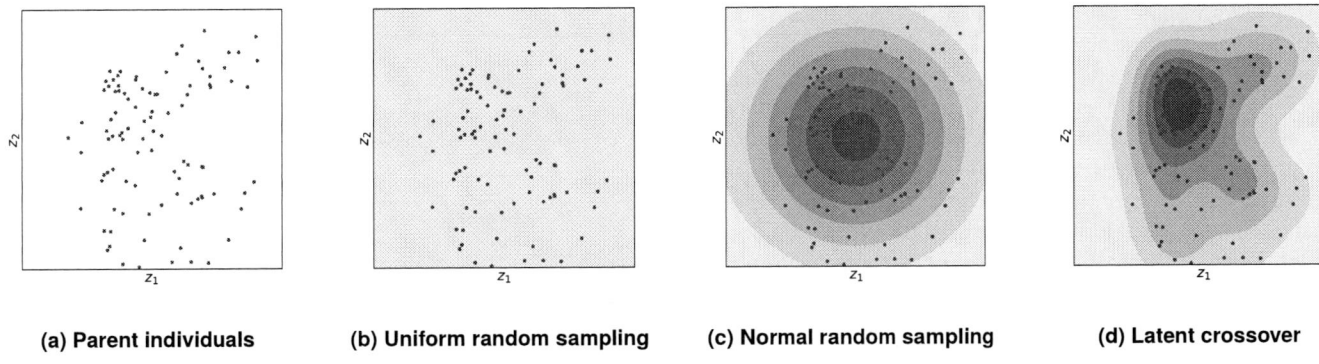

(a) Parent individuals   (b) Uniform random sampling   (c) Normal random sampling   (d) Latent crossover

FIGURE 1: THE PROBABILITY DISTRIBUTION FOR GENERATING OFFSPRING IN 2D LATENT SPACE

sionality, some deep generative models can dramatically reduce the dimensionality of the topology optimization problem. In deep generative models such as variational autoencoders (VAEs) [10] and generative adversarial networks (GANs) [11], an encoder is built to compress high-dimensional data into a low-dimensional manifold, called latent space. In addition, a decoder reconstructs high-dimensional data from the latent space. As a review paper [12] mentioned, relevant studies on deep generative models for engineering design problems have increased dramatically in recent years. As pioneering work, Guo et al. [13] proposed a data-driven indirect design representation for high-dimensional design problems, which iteratively optimizes the latent space of a VAE as the design variable field. Oh et al. [14] proposed a design framework that iteratively trains a GAN to generate a variety of designs. Kazemi et al. [15] proposed a method to generate conceptual designs using a GAN for multi-physics topology optimization problems.

Based on combining EAs and deep generative models, Yaji et al. [16] proposed a data-driven multifidelity topology design (MFTD) method that enables gradient-free topology optimization under a high degree of design freedom. The basic idea of data-driven MFTD is that design candidates, generated by solving low-fidelity topology optimization problems, are iteratively updated using an EA that guides queries to a high-fidelity analysis model. The key to this framework builds upon data-driven topology design [17], incorporating a VAE as a crossover-like operation for each optimization step. The effectiveness of the framework was demonstrated for topology optimization problems that are hard to solve directly with conventional methods, such as minimax and turbulent flow problems. However, since the generative process in a VAE is based on a uniform random sampling in the latent space, it is expected that the effectiveness of the approach can be improved if the crossover operation is adopted based on EAs.

This paper proposes a data-driven MFTD framework incorporating a particular crossover operation based on EAs, called *latent crossover*. Specifically, simplex crossover (SPX) [18]— a crossover operator of real-coded genetic algorithms (RC-GAs) [19]—is used for latent crossover. We apply the proposed method to a maximum stress minimization problem of an L-bracket and verify the effectiveness of latent crossover, compar-

ing it with the original data-driven MFTD. We also discuss its usefulness by comparing the results of the proposed method with those of gradient-based topology optimization using the P-norm measure for the maximum stress minimization problem.

2. LATENT CROSSOVER

In data-driven MFTD [16], whose details are descrived in Section 3, the high-dimensional material distribution data of the design candidates are encoded by a VAE into low-dimensional real-valued latent variables that correspond to EA genes, making the framework similar to RCGAs among EAs. Its high representation flexibility makes crossover more important in the RCGA than in the binary GA, and it has been the subject of various studies. For example, Kita and Yamamura [20] proposed a theory called the function specialization hypothesis concerning the selection and crossover operators in RCGAs, which includes the following ideas:

- The selection operator eliminates individuals with low fitness and, meanwhile, selects and replicates those with high fitness. Therefore, it is designed to narrow the population distribution gradually.

- The crossover operator transforms the distribution by combining parent individuals to generate offspring and is designed to retain the ability to generate new offspring for a finite population, but not to change the population distribution.

The following design guideline [21–23] for RCGA crossover operators was proposed focusing on the perspective of statistics to concretize the above theory. That is, the crossover operator should be designed to inherit statistics such as the mean vector and variance/covariance matrix of the population.

In data-driven MFTD, candidate solutions are generated through random sampling from the latent space of a VAE, so in terms of the genetic distribution and statistics of the population, we consider the probability distribution of the generated offspring. Fig. 1 shows an example of the probability distribution for generating offspring in a two-dimensional latent space. The darker areas have a higher probability of generating offspring. Assuming that the distribution of the parent population, as shown

Copyright © 2023 by ASME

in Fig. 1a, is given, data-driven MFTD performs sampling by uniform random numbers in the latent space, regardless of the distribution of the parent population. The resulting probability distribution of the generated offspring becomes the one shown in Fig. 1b. It cannot be said that the statistics of the parent population are inherited. Although the use of a VAE as a deep generative model enables a crossover-like operation in data-driven MFTD, it remains only as an operation similar to crossover and cannot be considered strictly performing crossover because of random sampling. Since the input data follows a normal distribution in the latent space due to the nature of VAEs, generating offspring through sampling based on a normal distribution rather than a uniform distribution is also possible. However, as shown in Fig. 1c, the probability of generated offspring does not follow the distribution of the parent population; therefore, the statistics of the parent population are not inherited in this case. Based on EA's concept, preserving the diversity of the population helps prevent premature convergence, but crossover-like sampling from the latent space using random sampling can lead to an early loss of diversity in the population. This results in fluctuation in convergence and, in the worst case, failures to perform a global search, leading to the possibility of getting stuck in local optima.

As mentioned above, it is impossible to strictly inherit the statistical characteristics of the parent population through random sampling. According to its nature, a crossover operation generates offspring by targeting small areas for parents who are close together and large areas for those who are far apart [24]. Thus, applying latent crossover to the parent population in Fig. 1a, the probability distribution of generated offspring is expected to become the one shown in Fig. 1d. Therefore, it can be said that a crossover operation in the latent space, i.e., the latent crossover, is promising.

3. FRAMEWORK

3.1 Data-Driven MFTD with Latent Crossover

Data-driven MFTD focuses on solving the following general multi-objective topology optimization problem:

$$
\begin{aligned}
\underset{\boldsymbol{\gamma}}{\text{minimize}} \quad & [J_1(\boldsymbol{\gamma}), J_2(\boldsymbol{\gamma}), \ldots, J_{r_0}(\boldsymbol{\gamma})] \\
\text{subject to} \quad & G_j(\boldsymbol{\gamma}) \le 0, \\
& \gamma_e \in \{0, 1\}, \; e = 1, 2, \ldots, N.
\end{aligned}
\tag{1}
$$

Here, J_i ($i = 1, 2, \ldots, r_o$) and G_j ($j = 1, 2, \ldots, r_c$) are the objective and constraint functions, respectively. The optimization problem defined by Eq. (1) is a 0-1 optimization problem with $\boldsymbol{\gamma}$ composed of N design variables. Since such a problem is a nonlinear mathematical optimization problem with a massive number of design variables, we adopt the concept of multifidelity topology design (MFTD) [25] and divide the problem of Eq. (1) into two procedures: low-fidelity optimization and high-fidelity evaluation, to solve the problem.

Using the MFTD approach and a deep generative model, data-driven MFTD iteratively updates solution candidates in a gradient-free manner similar to EAs. Note that the latent space is updated at every optimization step. The schematic flowchart of the proposed data-driven MFTD with latent crossover is shown in Fig. 2, and the details of each step are explained below.

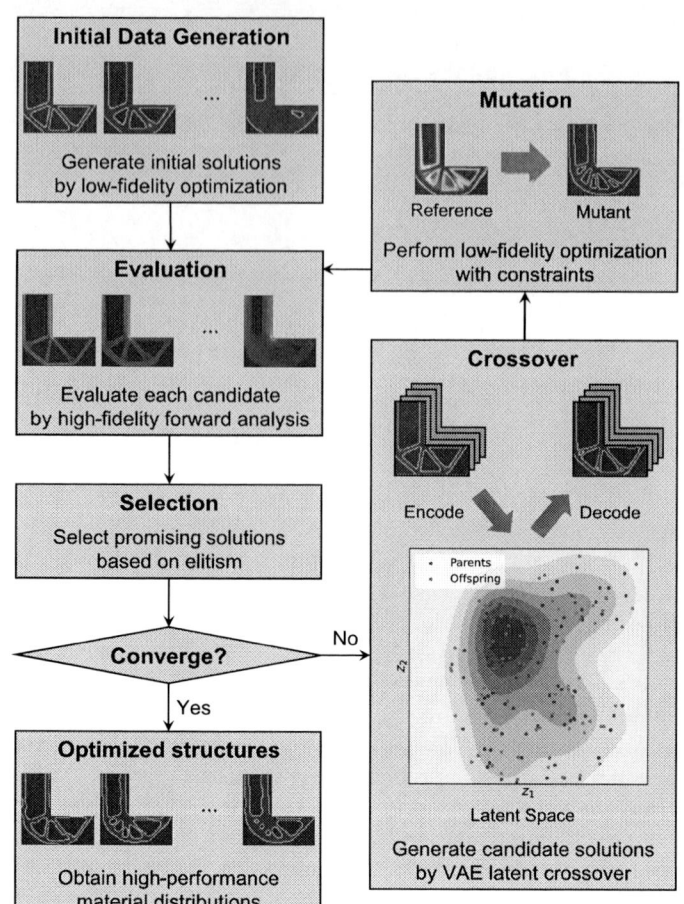

FIGURE 2: SCHEMATIC FLOWCHART OF DATA-DRIVEN MFTD WITH LATENT CROSSOVER

Initial Data Generation For the original optimization problem of Eq. (1), we solve a low-fidelity optimization problem formulated as follows, which can be easily solved as a simple pseudo-problem:

$$
\begin{aligned}
\underset{\boldsymbol{\gamma}^{(k)}}{\text{minimize}} \quad & \widetilde{J}_i(\boldsymbol{\gamma}^{(k)}) \\
\text{subject to} \quad & \widetilde{G}_j(\boldsymbol{\gamma}^{(k)}, \mathbf{s}^{(k)}) \le 0, \\
& \gamma_e^{(k)} \in [0, 1], \; e = 1, 2, \ldots, N \\
\text{for given} \quad & \mathbf{s}^{(k)}, \; k = 1, 2, \ldots, K.
\end{aligned}
\tag{2}
$$

Here, $\widetilde{J}_i$ and $\widetilde{G}_j$ are the objective and constraint functions for the low-fidelity optimization problem, respectively, which can be easily computed by pseudo-functions. Additionally, $\mathbf{s} = [s_1, s_2, \ldots, s_{N_{sd}}]$ represents the set of N_{sd} types of artificial design parameters called seeding parameters, and $\mathbf{s}^{(k)}$ is the sample point of $\mathbf{s}$. For instance, the seeding parameters are defined as a maximum limit of a constraint and optimization parameters such as a filter radius. By solving the relaxed low-fidelity optimization problem of Eq. (2) under various seeding parameter settings, where $\gamma_e^{(k)}$ is relaxed to $[0, 1]$, K kinds of promising and diverse material distributions are prepared as initial solutions.

Copyright © 2023 by ASME

Evaluation The performance of candidate solutions is evaluated using a high-fidelity analysis model, which is used to compute the original multiple objective functions J_i and G_j in Eq. (1).

Selection As mentioned in Section 2, the selection is a critical genetic operation in RCGAs. For problems as in Eq. (1), it is necessary to evaluate solutions using multiple objective functions and select those to be preserved in the next generation. This paper uses the nondominated sorting genetic algorithm II (NSGA-II) [26] strategy as a selection algorithm, which selects candidates in a multi-objective manner by ranking them based on the Pareto dominance relation using distances in the objective function space. The nondominated candidate solutions, which are not dominated by any other solutions, are selected from the population, and then a set of Pareto solutions is constructed.

Crossover A VAE is trained with the Pareto solution set as input to construct a latent space, where high-dimensional material distributions are encoded into low-dimensional latent variables. The latent crossover is performed using these latent variables to generate offspring in the latent space. Decoding the offspring generated by latent crossover yields new material distributions that inherit the characteristics of the input data, and candidate solutions are generated. The details of the VAE and the latent crossover operation are described in Sections 3.2 and 3.3, respectively.

Mutation The latent space of the VAE is constructed using the Pareto solution set of the current generation and corresponds to a subspace in which the solutions are distributed. Even if the mutation method of RCGAs, such as the nonuniform mutation operator [27], is applied, its outcome is limited to a specific subspace against the whole solution space. This limitation is because such a mutation only performs a local search in the subspace around the solutions distributed in the whole solution space. Thus, it cannot be expected to maintain the diversity of the population and prevent premature convergence, as discussed in Section 1.

Therefore, under the following constraint function, the low-fidelity optimization problem is solved using the same method as when generating initial data:

$$\widetilde{G}_{\text{mut}}(\boldsymbol{\gamma}^{(m)}) = \sum_{e=1}^{N} v_e \gamma_e^{(m)} \gamma_e^{\text{ref}(m)} \le \widetilde{G}_{\text{mut}}^{\max} |D|, \qquad (3)$$

where $m = 1, 2, \ldots, N_{\text{mut}}$ is the number of mutants, v_e is the elemental volume, $\bar{G}_{\text{mut}}^{\max}$ is a parameter that controls the degree of overlap between the reference material distribution $\gamma^{\text{ref}(m)}$ and the design variable $\gamma^{(m)}$, and $|D| = \sum_{e=1}^{N} v_e$ is the volume of D. In brief, the role of the constraint of Eq. (3) is to generate a different material distribution from $\gamma_e^{\text{ref}(m)}$.

This paper uses the average value of material distributions in a given generation as a reference structure. This average distribution can be considered to be representative of the material distributions of the population. By solving the low-fidelity optimization problem with the constraint function

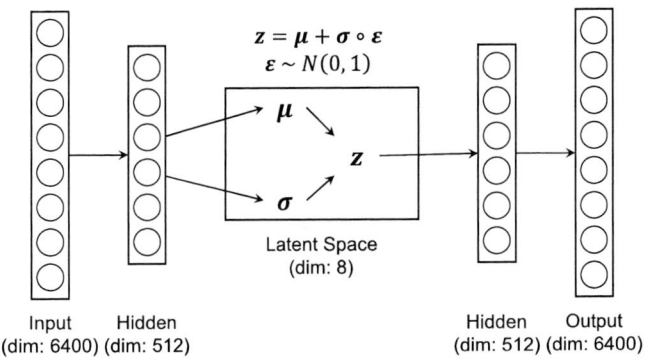

FIGURE 3: ARCHITECTURE OF VAE

of Eq. (3) and the reference structure, promising candidate solutions can be generated with unique features that are not present in the population. This approach enables a mutation-like operation, similar to the mutation in EAs, to maintain diversity and prevent premature convergence. It should be noted that the mutants added to the population through this operation are still limited to a specific subspace and may not search the whole solution space comprehensively.

3.2 Variational Autoencoder

Fig. 3 shows the architecture of the VAE used in the numerical examples in Section 4. 6400 input/output elements are combined into two 8-dimensional layers, $\boldsymbol{\mu}$ and $\boldsymbol{\sigma}$, through a hidden layer of 512 dimensions. $\boldsymbol{\mu}$ is the mean vector, and $\boldsymbol{\sigma}$ is the variance vector of the latent variables $\mathbf{z}$. The following equation defines the latent variable vector $\mathbf{z}$:

$$\mathbf{z} = \boldsymbol{\mu} + \boldsymbol{\sigma} \circ \boldsymbol{\varepsilon}, \qquad (4)$$

where $\circ$ is the operator that calculates the element-wise product, and $\boldsymbol{\varepsilon}$ is a vector of random numbers from the standard normal distribution. In VAEs, unsupervised learning is performed using the same dataset for both input and output, constructing the latent space. The following loss function L_{VAE} is used for the training:

$$L_{\text{VAE}} := L_{\text{recon}} + \varrho L_{\text{KL}}, \qquad (5)$$

$$L_{\text{KL}} = -\frac{1}{2} \sum_{i=1}^{N_{\text{lt}}} \left(1 + \log(\sigma_i^2) - \mu_i^2 - \sigma_i^2\right), \qquad (6)$$

where N_{lt} is the dimension of the latent space, μ_i and σ_i are the i-th elements of $\boldsymbol{\mu}$ and $\boldsymbol{\sigma}$. L_{recon} is a reconstruction loss using mean squared error, and L_{KL} is known as the Kullback-Leibler (KL) divergence. ϱ is the weight parameter that controls the influence of the KL divergence to regularize the latent space to the standard normal distribution.

Compared to simple dimensionality reduction using autoencoders (AEs), VAEs are trained by incorporating probabilistic variation through $\boldsymbol{\varepsilon}$, allowing for estimation of the given dataset distribution, and can be used as a deep generative model for continuous data generation. When using material distributions as a dataset for topology optimization, essential features within the dataset are extracted by compressing them into dramatically

Copyright © 2023 by ASME

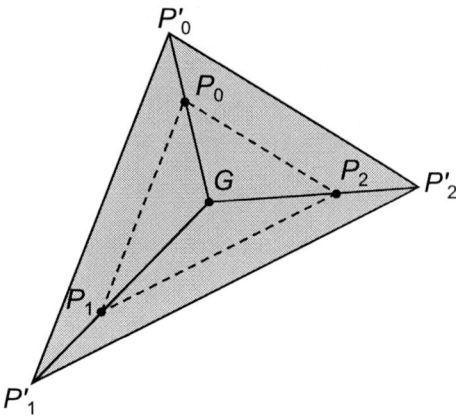

FIGURE 4: SPX OFFSPRING GENERATION AREA FOR 2D

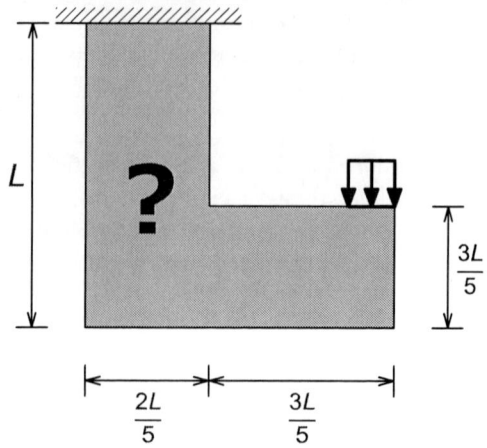

FIGURE 5: DESIGN PROBLEM OF L-BRACKET

smaller latent variables. According to the standard normal distribution, latent variables do not take extremely large or small values. To represent all material distributions without excessive randomness, original data-driven MFTD generates offspring by sampling uniform random numbers in $[-4, 4]$, which covers 99.7% of the data within $\pm 4\sigma$, for each latent variable. However, as mentioned in Section 2, the generation of probability distribution shown in Fig. 1 can be problematic. In this paper, we perform latent crossover using the crossover operator explained in Section 3.3.

3.3 Simplex Crossover

Due to the high degree of freedom of representing genes as real-valued vectors, the RCGA has limited offspring that can be generated from selected parent individuals using crossover operators, such as the single-point crossover commonly used in binary evolutionary algorithms. Several crossover operators for RCGAs [18, 28, 29] have been proposed to address this issue. This paper uses the simplex crossover (SPX) [18] for a latent crossover operator. SPX is one of the multi-parent crossover operators for RCGAs that generates offspring using three or more parent individuals and is consistent with the crossover design guidelines [21–23] as it inherits the average value and covariance matrix of the population.

When the search space is defined as the real n-dimensional space $\mathbb{R}^n$, where individuals are represented as vectors of real numbers, the algorithm for SPX is as follows.

(1) Randomly select $(n + 1)$ parent individuals $P_0, P_1, \ldots, P_n$ from the population.

(2) Calculate the centroid G of the parent individuals as follows:

$$G = \sum_{i=0}^{n} P_i. \tag{7}$$

(3) Calculate x_k and C_k for $k = 0, 1, \ldots, n$ as follows:

$$x_k = G + \varepsilon(P_k - G), \tag{8}$$

$$C_k = \begin{cases} \mathbf{0} & (k = 0) \\ r_{k-1}(x_{k-1} - x_k + C_{k-1}) & (k = 1, \ldots, n). \end{cases} \tag{9}$$

Here, ε is a parameter called the expansion rate, and $\sqrt{n+2}$ is the recommended value for inheriting population statistics [18]. r_k is obtained by transforming a uniform random number $u(0, 1)$ in the interval $[0, 1]$ as follows:

$$r_k = \begin{cases} 0 & (k < 0) \\ u(0, 1)^{\frac{1}{k+1}} & (k = 1, \ldots, n - 1) \\ 1 & (k \geq 1). \end{cases} \tag{10}$$

(4) Generate a child individual C as follows:

$$C = x_n + C_n. \tag{11}$$

With these procedures, SPX generates offspring uniformly within the enclosed space of the ε-extended polytope $P'_0, P'_1, \ldots, P'_n$ centered at the centroid of the parent individuals $P_0, P_1, \ldots, P_n$, as shown in Fig. 4. Therefore, SPX is a crossover operator that achieves a balance between exploration and exploitation [30].

4. NUMERICAL EXAMPLES
4.1 Problem Setting

This study applies the proposed method to the design problem of a two-dimensional L-bracket. It is widely used as a benchmark for stress-based topology design [31–34] and is a minimax problem with its high nonlinearities caused by the stress singularity, called re-entrant corner, at the inner corner. It can be formulated as the following multi-objective optimization problem:

$$\underset{\gamma}{\text{minimize}} \quad J_1 = \max\left(\sigma_{\text{vM}}\right),$$

$$J_2 = \sum_{e=1}^{N} v_e \gamma_e \tag{12}$$

$$\text{subject to} \quad \gamma_e \in \{0, 1\}, \quad e = 1, 2, \ldots, N.$$

Here, σ_{vM} is the von Mises stress, the maximum of which is an objective function, and the volume is the other objective function.

Copyright © 2023 by ASME

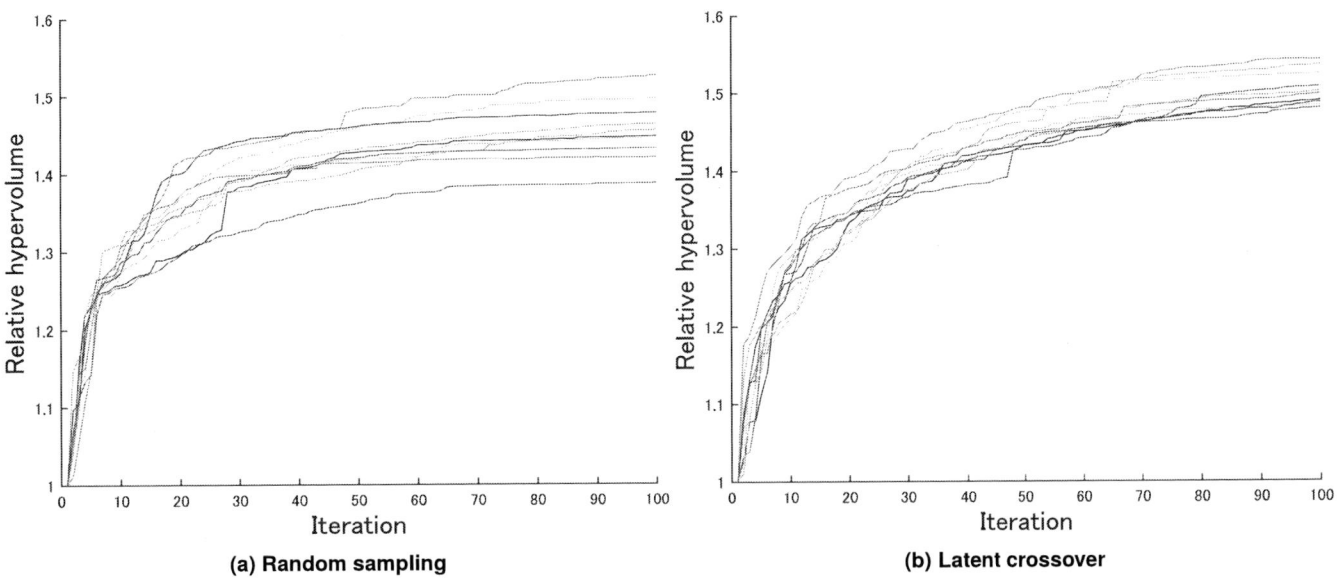

(a) Random sampling **(b) Latent crossover**

FIGURE 6: HYPERVOLUME FOR TEN TRIALS

Note that the design variables are defined as discrete values, 0 or 1, to deal with the ideal topology optimization problem with high-fidelity evaluation.

The design domain and boundary conditions for the L-bracket, as shown in Fig. 5, include fixing the upper end and applying a vertical downward distributed load at the top corner to avoid stress concentration. The length of the bracket is set to $L = 2$, and the design domain is divided into 6400 square elements ($N = 6400$). Young's modulus of the structural material is set to 1, one of the voids is set to 1×10^{-9} instead of 0 to avoid the singular stiffness matrix, and Poisson's ratio is set to 0.3.

Under the assumption that a promising solution can be obtained even with stiffness maximization [32], we formulate the minimum compliance problem and use it as a low-fidelity optimization problem:

$$
\begin{aligned}
&\underset{\boldsymbol{\gamma}^{(k)}}{\text{minimize}} && \widetilde{J}_1 = \mathbf{f}^\mathsf{T} \mathbf{u} \\
&\text{subject to} && \widetilde{J}_2 = \sum_{e=1}^{N} v_e \gamma_e^{(k)} \le s^{(k)}, \\
& && \gamma_e^{(k)} \in [0,1], \ e = 1, 2, \ldots, N \\
&\text{for given} && s^{(k)}.
\end{aligned}
\tag{13}
$$

Here, $\mathbf{f}$ and $\mathbf{u}$ are vectors in the equilibrium equation, namely, $\mathbf{Ku} = \mathbf{f}$, with the global stiffness matrix $\mathbf{K}$. In Eq. (13), the volume is converted from an objective function to a constraint function based on the ε-constraint method for the original optimization problem of Eq. (12), and since $\gamma_e^{(k)}$ is relaxed to $[0,1]$, this problem can be easily solved using the density-based method [2]. Note that a design variable filter [35] is applied to ensure the smoothness of $\boldsymbol{\gamma}$ in D.

As for the parameters related to the overall procedure, the number of initial data and Pareto solutions from the selection operation are both set to 100. Regarding the parameters related to the mutation operation, N_{mut} is set to 16 and $\widetilde{G}_{\text{mut}}$ is set to 0.01.

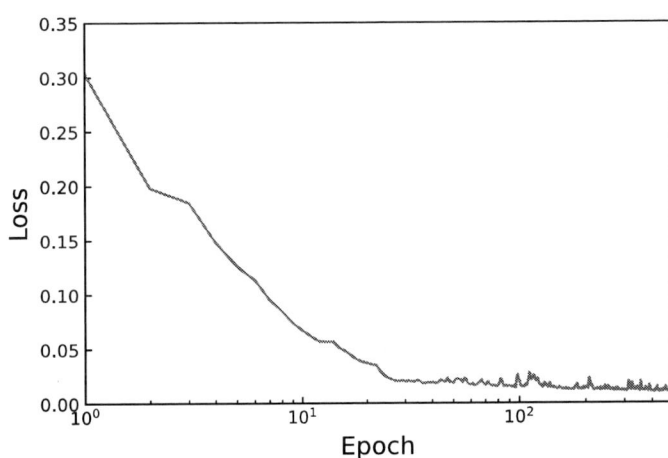

FIGURE 7: LEARNING HISTORY AT ITERATION 0 FOR THE ARCHITECTURE IN FIGURE 3, THE LOSS FUNCTION IN EQUATION 5, AND THE TRAINING DATA DISCRIVED IN SECTION 4.4

During the latent crossover, 9 parent individuals are used by the SPX method because the dimension of the VAE latent space is 8.

4.2 Verification of VAE Model

First, we verify the VAE model and parameters, which plays a central role in the data-driven MFTD. After preliminary studies on the hyperparameters, we establish the VAE architecture as shown in Fig. 3. The VAE is trained with 100 material distribution samples with 500 epochs, a batch size of 20, and a learning rate of 0.001. The training is terminated if the loss function L_{VAE} of Eq. (5) is not improved in every iteration for a total of 50 iterations.

Fig. 7 shows the history of the loss function in Eq. (5) during training using the material distribution data at iteration 0 descrived in Section 4.4 as an example. The loss function converges smoothly, indicating that the VAE is appropriately trained

Copyright © 2023 by ASME

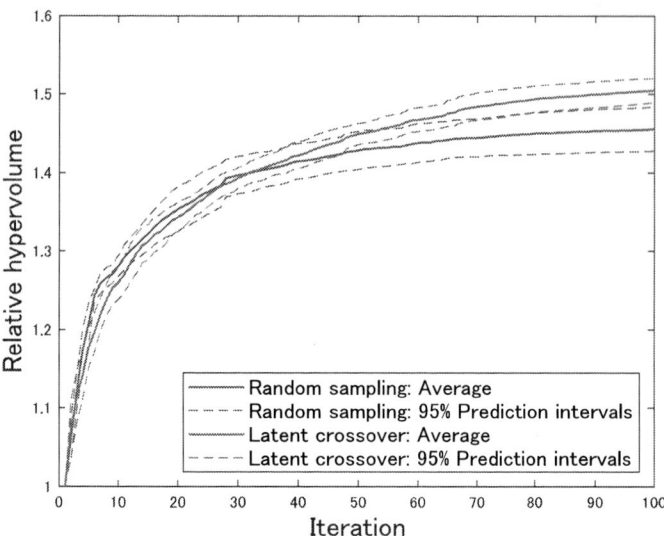

FIGURE 8: COMPARISON OF HYPERVOLUME

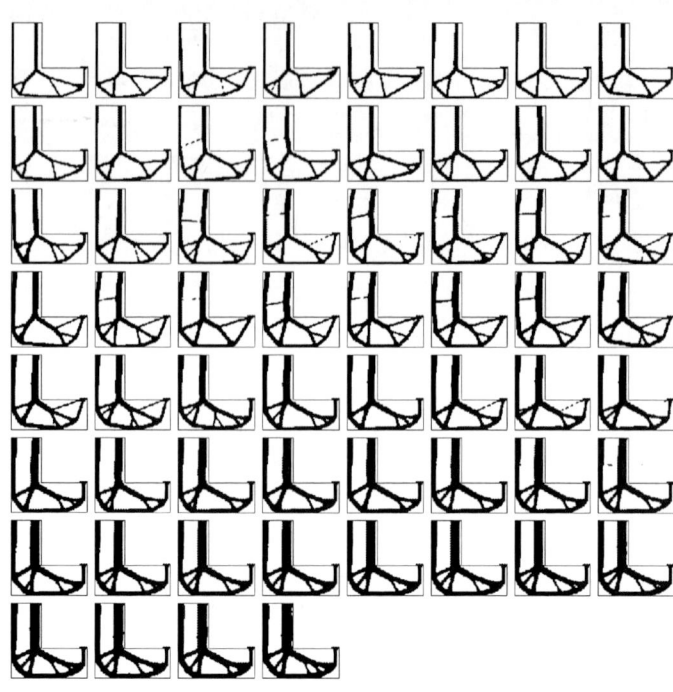

FIGURE 9: OPTIMIZED STRUCTURES BY GRADIENT-BASED TOPOLOGY OPTIMIZATION USING THE *P*-NORM MEASURE

under the investigated condition.

4.3 Verification of Latent Crossover Effect

For the problem set up in Section 4.1, we compare the original and proposed data-driven MFTD frameworks. Since both methods involve random effects, we evaluate and compare them using the hypervolume indicator [36] over ten trials, which is normalized using the initial one. The hypervolume is a measure for the convergence performance of multi-objective optimization. In the case of two objectives, it is represented by the area formed by the reference point and the Pareto front in the objective space. Although mutation is usually performed at regular intervals of iterations, we confirmed that in the case of this design problem, the mutants are selected only once at the beginning, and no mutants are selected as elite solutions thereafter. Therefore, we used the initial data composed of the mutants and initial solutions to compare them with the search performance by crossover without mutation.

Fig. 6 shows the iteration history of the hypervolume indicator over ten trials. In terms of the value at 100 iterations, random sampling in Fig. 6a shows a considerable variation in the range from 1.38 to 1.52, while the latent crossover in Fig. 6b remains stable in the range from 1.48 to 1.54. The average values of each hypervolume indicator in the ten trials are plotted in Fig. 8. Up to iteration 30, the value of random sampling is higher than that of latent crossover. However, after iteration 30, this relationship is reversed, and at iteration 100, the average value of random sampling is 1.45, while that of latent crossover is 1.50, indicating a difference of 5%. In addition, at iteration 100, the lower limit of the 95% prediction intervals for the latent crossover case exceeds the upper limit for the random sampling case. A t-test was performed on the hypervolume values at iteration 100, and the p-value was 0.00180, which is less than 0.05. Therefore, it can be considered statistically significant that the latent crossover outperforms the random sampling.

The SPX operator used as the latent crossover operator gradually changes the population distribution while inheriting the

statistics, so the increase in hypervolume is slower in the early stages of the search (up to iteration 30) compared to the random sampling. However, this approach maintains diversity and prevents premature convergence, which leads to a more advanced Pareto front in the final iteration (at iteration 100). This improvement can be explained based on the theory that the balance between exploration and exploitation [30], i.e., expanding the Pareto front and advancing it, respectively, is significant in EAs. From these results and discussions, it can be concluded that data-driven MFTD achieved stable and high search performance with the latent crossover based on the theory of RCGAs.

4.4 Validity of Optimized Structure

Next, we compare structures obtained through data-driven MFTD with structures obtained through direct optimization using a gradient-based approach without relying on MFTD principles. However, deriving sensitivity analysis for the optimization problem defined by Eq. (12) is impossible because J_1 is the maximum value of the von Mises stress and γ_e is a discrete value $\{0, 1\}$. Therefore, we use the *P*-norm measure [37, 38], commonly used in stress-based topology optimization [32, 33], and relax γ_e to $[0, 1]$, as follows:

$$
\begin{aligned}
\underset{\gamma}{\text{minimize}} \quad & J = \left(\frac{1}{N} \sum_{e=1}^{N} (\sigma_{\text{vM}})^P \right)^{\frac{1}{P}} \\
\text{subject to} \quad & G = \sum_{e=1}^{N} v_e \gamma_e \leq V_{\max} |D|, \\
& \gamma_e \in [0, 1], \ e = 1, 2, \ldots, N.
\end{aligned}
\tag{14}
$$

Here, P is the stress norm parameter, and J is called P-norm stress. For the multi-objective problem formulated in Eq. (12), the

Copyright © 2023 by ASME

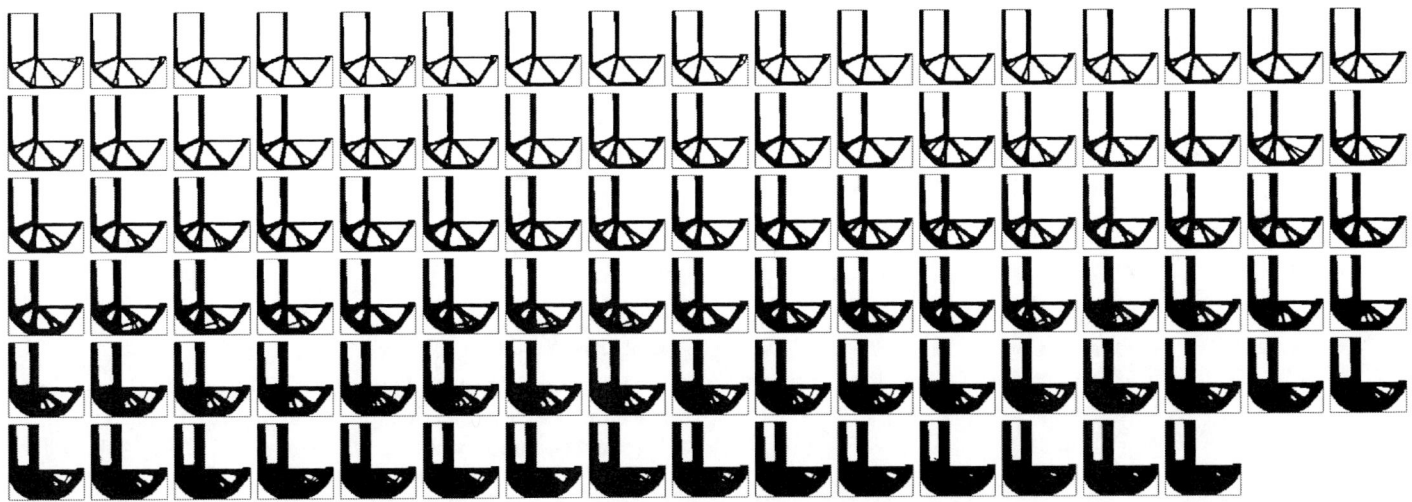

FIGURE 10: INITIAL DATA GENERATED BY SOLVING A MEAN COMPLIANCE MINIMIZATION PROBLEM UNDER VARIOUS VOLUME CONSTRAINT SETTINGS AS THE LOW-FIDELITY TOPOLOGY OPTIMIZATION PROBLEM

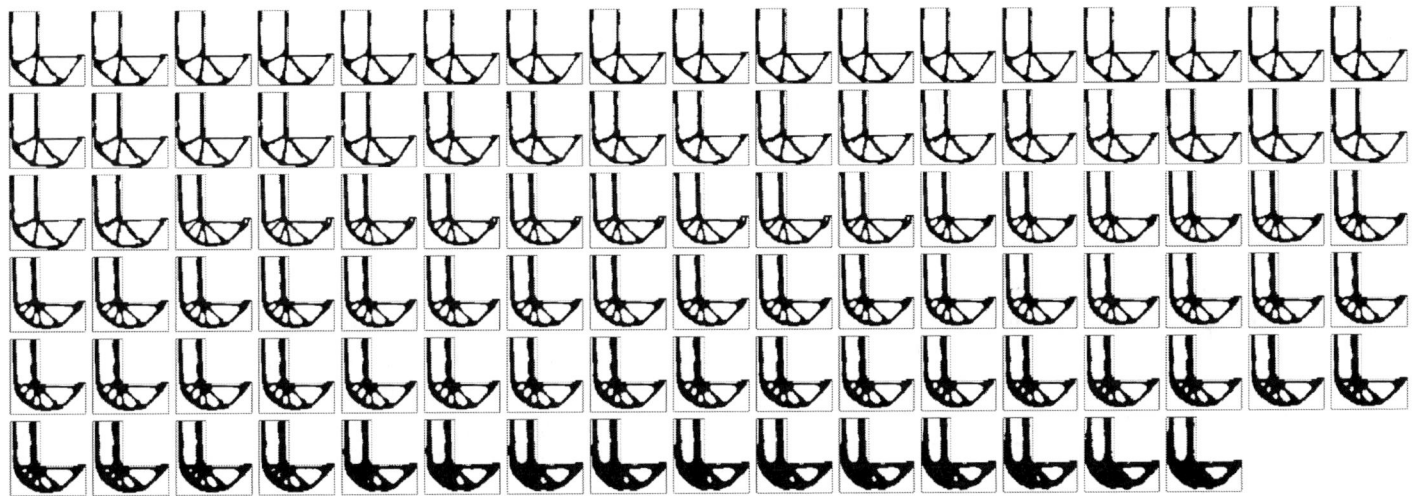

FIGURE 11: OPTIMIZED STRUCTURES BY DATA-DRIVEN MFTD

volume is set as the constraint function based on the ε-constraint method. When the stress norm parameter $P \to \infty$, the P-norm stress approaches the maximum stress value $\max(\sigma_{vM})$, but the smoothness is lost. On the other hand, when $P = 1$, the smoothness is maintained, but it approaches the average stress value, resulting in an optimized structure closer to the compliance minimum design. Previous studies [32, 34] have shown that $P = 8$ yields the most reasonable design, but in order to set an objective function closer to the maximum stress value, this paper uses a method that iteratively increases P from 8 to 16 and 32 using the continuation method [39]. This operation enables us to use a more rigorous approximation function while stably solving the optimization problem. We set the P-norm stress based on the result of the continuous approach as the objective function and use the method of moving asymptotes (MMA) [40] as the optimization method. Fig. 9 shows 60 directly solved designs while changing V_{max} from 20% to 50% in 0.5% increments.

Fig. 10 shows the initial dataset obtained by solving the low-

fidelity optimization problem in Eq. (13), and Fig. 11 shows the optimized structures obtained by data-driven MFTD with 300 iterations. The initial dataset, which consists of compliance minimization designs, has structures that cause stress concentration at their re-entrant corners, whereas the structures obtained by data-driven MFTD have rounded shapes with their re-entrant corners smoothed out. The improved performance and reduced volume can be seen by comparing the plots of iteration 0 and iteration 300 in the objective function space shown in Fig. 12.

When comparing the performance of Pareto solutions obtained by data-driven MFTD and the solutions obtained by direct optimization in the objective space shown in Fig. 12, it can be confirmed that equivalent performance solutions are obtained in general. In particular, data-driven MFTD outperforms direct optimization for solutions in the volume range of 0.7 to 1.0. In addition, while the solutions for direct optimization are sparsely distributed in the objective space due to the instability of the objective function, the solutions for data-driven MFTD form an

Copyright © 2023 by ASME

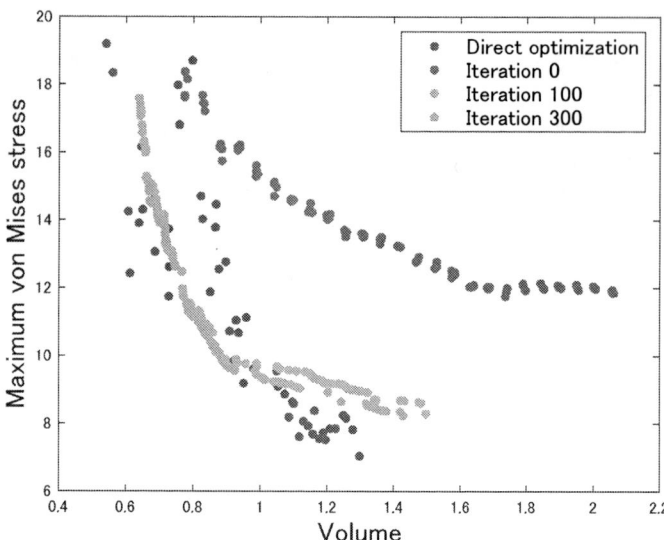

FIGURE 12: OBJECTIVE SPACE ON THE VOLUME AND MAXIMUM VON MISES STRESS

orderly Pareto front.

The designs shown in Fig. 9 obtained by direct optimization are structures that appear to be composed of straight members and often have triangular or rectangular voids. One advantage of data-driven MFTD is that material distributions are represented as vectors and updated using a VAE, eliminating the need for sensitivity analysis. Therefore, as in Eq. (12), the maximum stress can be used directly as the objective function. This feature leads to generally curved structures with rounded appearances, as shown in Fig. 11, suggesting that stress concentration is avoided. In addition, it is found that various patterns are obtained due to the multimodality caused by the strong nonlinearity of the P-norm measure, while the final results of the proposed method in Fig. 11 relatively tend to have a similar feature in terms of their topology. Although we cannot prove that the obtained Pareto solutions are the global optima respectively, the result indicates that the data-driven MFTD method is likely to yield a unique set of Pareto solutions through an extensive search process.

5. CONCLUSION

This paper proposed a data-driven multifidelity topology design (MFTD) framework incorporating latent crossover that performs crossover in the latent space of the variational autoencoder (VAE). Since the latent space is constructed as continuous real numbers, this paper employed the simplex crossover (SPX) as a latent crossover operator based on the theoretical aspects of crossover in real-coded genetic algorithms (RCGAs). The results showed that the proposed method improves the search performance compared to the original method, which performs random sampling in the latent space. As an interesting aspect, this paper confirms that the proposed method achieves almost the same performance as that of gradient-based topology optimization using the P-norm measure for the maximum stress minimization problem, despite only solving mean compliance minimization as the low-fidelity topology optimization problem. Furthermore, it was found that the final results of the proposed method tend to

achieve a similar topology, while the optimized results of the gradient-based method are various patterns due to the multimodality caused by the strong nonlinearity of the P-norm measure. Hence, the data-driven MFTD approach is expected to yield a unique set of Pareto solutions through gradient-free searching.

To verify the efficacy of the proposed framework on different optimization problems, we plan to apply it to other problems involving strongly nonlinear physical phenomena, such as turbulence and geometric nonlinearity.

ACKNOWLEDGMENTS

The authors would like to thank Misato Kato, a graduate student at Osaka University, for providing the data used in Fig. 9. The second author was supported by JSPS KAKENHI Grant Numbers 20KK0329 and 20H02054.

REFERENCES

[1] Bendsøe, M. P. and Kikuchi, N. "Generating Optimal Topologies in Structural Design Using a Homogenization Method." *Computer Methods in Applied Mechanics and Engineering* Vol. 71 No. 2 (1988): pp. 197–224. DOI 10.1016/0045-7825(88)90086-2.

[2] Bendsøe, M. P. and Sigmund, O. *Topology Optimization: Theory, Methods, and Applications.* Springer Science & Business Media (2003).

[3] Mitchell, M. and Taylor, C. E. "Evolutionary Computation: An Overview." *Annual Review of Ecology and Systematics* Vol. 30 No. 1 (1999): pp. 593–616. DOI 10.1146/annurev.ecolsys.30.1.593.

[4] Goldberg, D. E. *Genetic Algorithms in Search, Optimization, and Machine Learning.* Addison Wesley (1989).

[5] Wang, S. Y. and Tai, K. "Structural Topology Design Optimization Using Genetic Algorithms with a Bit-Array Representation." *Computer Methods in Applied Mechanics and Engineering* Vol. 194 No. 36-38 (2005): pp. 3749–3770. DOI 10.1016/j.cma.2004.09.003.

[6] Madeira, J. F. A., Pina, H. L. and Rodrigues, H. C. "GA Topology Optimization Using Random Keys for Tree Encoding of Structures." *Structural and Multidisciplinary Optimization* Vol. 40 (2010): pp. 227–240. DOI 10.1007/s00158-008-0353-1.

[7] Zhou, H. "Topology Optimization of Compliant Mechanisms Using Hybrid Discretization Model." *Journal of Mechanical Design* Vol. 132 No. 11 (2010): p. 111003. DOI 10.1115/1.4002663.

[8] Balamurugan, R., Ramakrishnan, C. V. and Swaminathan, N. "A Two Phase Approach Based on Skeleton Convergence and Geometric Variables for Topology Optimization Using Genetic Algorithm." *Structural and Multidisciplinary Optimization* Vol. 43 (2011): pp. 381–404. DOI 10.1007/s00158-010-0560-4.

[9] Sigmund, O. "On the Usefulness of Non-Gradient Approaches in Topology Optimization." *Structural and Multidisciplinary Optimization* Vol. 43 (2011): pp. 589–596. DOI 10.1007/s00158-011-0638-7.

Copyright © 2023 by ASME

[10] Kingma, D. P. and Welling, M. "Auto-Encoding Variational Bayes." arXiv preprint (2013). DOI 10.48550/arXiv.1312.6114.

[11] Goodfellow, I. J., Pouget-Abadie, J., Mirza, M., Xu, B., Warde-Farley, D., Ozair, S., Courville, A. and Bengio, Y. "Generative Adversarial Networks." arXiv preprint (2014). DOI 10.48550/arXiv.1406.2661.

[12] Regenwetter, L., Nobari, A. H. and Ahmed, F. "Deep Generative Models in Engineering Design: A Review." *Journal of Mechanical Design* Vol. 144 No. 7 (2022): p. 071704. DOI 10.1115/1.4053859.

[13] Guo, T., Lohan, D. J., Cang, R., Ren, M. Y. and Allison, J. T. "An Indirect Design Representation for Topology Optimization Using Variational Autoencoder and Style Transfer." *AIAA/ASCE/AHS/ASC Structures, Structural Dynamics, and Materials Conference* (2018): p. 0804. DOI 10.2514/6.2018-0804.

[14] Oh, S., Jung, Y., Kim, S., Lee, I. and Kang, N. "Deep Generative Design: Integration of Topology Optimization and Generative Models." *Journal of Mechanical Design* Vol. 141 No. 11 (2019): p. 111405. DOI 10.1115/1.4044229.

[15] Kazemi, H., Seepersad, C. C. and Kim, H. A. "Multiphysics Design Optimization via Generative Adversarial Networks." *Journal of Mechanical Design* Vol. 144 No. 12 (2022): p. 121702. DOI 10.1115/1.4055377.

[16] Yaji, K., Yamasaki, S. and Fujita, K. "Data-Driven Multifidelity Topology Design Using a Deep Generative Model: Application to Forced Convection Heat Transfer Problems." *Computer Methods in Applied Mechanics and Engineering* Vol. 388 (2022): p. 114284. DOI 10.1016/j.cma.2021.114284.

[17] Yamasaki, S., Yaji, K. and Fujita, K. "Data-Driven Topology Design Using a Deep Generative Model." *Structural and Multidisciplinary Optimization* Vol. 64 No. 3 (2021): pp. 1401–1420. DOI 10.1007/s00158-021-02926-y.

[18] Tsutsui, S., Yamamura, M. and Higuchi, T. "Multi-Parent Recombination with Simplex Crossover in Real Coded Genetic Algorithms." *Proceedings of the 1st Annual Conference on Genetic and Evolutionary Computation* Vol. 1 (1999): pp. 657–664. URL https://dl.acm.org/doi/pdf/10.5555/2933923.2933986.

[19] Herrera, F., Lozano, M. and Verdegay, J. L. "Tackling Real-Coded Genetic Algorithms: Operators and Tools for Behavioural Analysis." *Artificial Intelligence Review* Vol. 12 (1998): pp. 265–319. DOI 10.1023/A:1006504901164.

[20] Kita, H. and Yamamura, M. "A Functional Specialization Hypothesis for Designing Genetic Algorithms." *IEEE International Conference on Systems, Man, and Cybernetics* Vol. 3 (1999): pp. 579–584. DOI 10.1109/ICSMC.1999.823277.

[21] Kita, H. and Yamamura, M. "Design Guidelines for Genetic Algorithms based on Function Specialization Hypothesis." *Journal of the Society of Instrument and Control Engineers* Vol. 38 No. 10 (1999): pp. 612–617. DOI 10.11499/sicejl1962.38.612.

[22] Kita, H., Ono, I. and Kobayasi, S. "Multi-Parental Extension of the Unimodal Normal Distribution Crossover for Real-Coded Genetic Algorithms." *Transactions of the Society of Instrument and Control Engineers* Vol. 36 No. 10 (2000): pp. 875–883. DOI 10.9746/sicetr1965.36.875.

[23] Beyer, H. G. and Deb, K. "On Aelf-Adaptive Features in Real-Parameter Evolutionary Algorithms." *IEEE Transactions on Evolutionary Computation* Vol. 5 No. 3 (2001): pp. 250–270. DOI 10.1109/4235.930314.

[24] Herrera, F., Lozano, M. and Sánchez, A. M. "A Taxonomy for the Crossover Operator for Real-Coded Genetic Algorithms: An Experimental Study." *International Journal of Intelligent Systems* Vol. 18 No. 3 (2003): pp. 309–338. DOI 10.1002/int.10091.

[25] Yaji, K., Yamasaki, S. and Fujita, K. "Multifidelity Design Guided by Topology Optimization." *Structural and Multidisciplinary Optimization* Vol. 61 (2020): pp. 1071–1085. DOI 10.1007/s00158-019-02406-4.

[26] Deb, K., Pratap, A., Agarwal, S. and Meyarivan, T. "A Fast and Elitist Multiobjective Genetic Algorithm: NSGA-II." *IEEE Transactions on Evolutionary Computation* Vol. 6 No. 2 (2002): pp. 182–197. DOI 10.1109/4235.996017.

[27] Michalewicz, Z. *Genetic Algorithms + Data Structures = Evolution Programs*. Springer Berlin, Heidelberg (1996).

[28] Eshelman, L. J., Mathias, K. E. and Schaffer, J. D. "Crossover Operator Biases: Exploiting the Population Distribution." *Proceedings of the 7th International Conference on Genetic Algorithms* (1997): pp. 354–361.

[29] Ono, I., Kita, H. and Kobayashi, S. "A Real-Coded Genetic Algorithm Using the Unimodal Normal Distribution Crossover." *Advances in Evolutionary Computing: Theory and Applications* (2003): pp. 213–237.

[30] Črepinšek, M., Liu, S. H. and Mernik, M. "Exploration and Exploitation in Evolutionary Algorithms: A Survey." *ACM Computing Surveys* Vol. 45 No. 3 (2013). DOI 10.1145/2480741.2480752.

[31] Duysinx, P. and Bendsøe, M. P. "Topology Optimization of Continuum Structures with Local Stress Constraints." *International Journal for Numerical Methods in Engineering* Vol. 43 No. 8 (1998): pp. 1453–1478. DOI 10.1002/(SICI)1097-0207(19981230)43:8<1453::AID-NME480>3.0.CO;2-2.

[32] Le, C., Norato, J. A., Bruns, T., Ha, C. and Tortorelli, D. "Stress-Based Topology Optimization for Continua." *Structural and Multidisciplinary Optimization* Vol. 41 (2010): pp. 605–620. DOI 10.1007/s00158-009-0440-y.

[33] Holmberg, E., Torstenfelt, B. and Klarbring, A. "Stress Constrained Topology Optimization." *Structural and Multidisciplinary Optimization* Vol. 48 (2013): pp. 33–47. DOI 10.1007/s00158-012-0880-7.

[34] Norato, J. A., Smith, H. A., Deaton, J. D. and Kolonay, R. M. "A Maximum-Rectifier-Function Approach to Stress-Constrained Topology Optimization." *Structural and Multidisciplinary Optimization* Vol. 65 No. 10 (2022): p. 286. DOI 10.1007/s00158-022-03357-z.

[35] Bourdin, B. "Filters in Topology Optimization." *International Journal for Numerical Methods in Engineering* Vol. 50 No. 9 (2001): pp. 2143–2158. DOI 10.1002/nme.116.

[36] Shang, K., Ishibuchi, H., He, L. and Pang, L. M. "A Survey on the Hypervolume Indicator in Evolutionary Multi-objective Optimization." *IEEE Transactions on Evolutionary Computation* Vol. 25 No. 1 (2021): pp. 1–20. DOI 10.1109/TEVC.2020.3013290.

[37] Yang, R. J. and Chen, C. J. "Stress-Based Topology Optimization." *Structural Optimization* Vol. 12 (1996): pp. 98–105. DOI 10.1007/BF01196941.

[38] Duysinx, P. and Sigmund, O. "New Developments in Handling Stress Constraints in Optimal Material Distribution." *Proceedings of the 7th AIAA/USAF/NASA/ISSMO Symposium on Multidisciplinary Analysis and Optimization* (1998)DOI 10.2514/6.1998-4906.

[39] Li, L. and Khandelwal, K. "Volume Preserving Projection Filters and Continuation Methods in Topology Optimization." *Engineering Structures* Vol. 85 (2015): pp. 144–161. DOI 10.1016/j.engstruct.2014.10.052.

[40] Svanberg, K. "The method of moving asymptotes—a new method for structural optimization." *International Journal for Numerical Methods in Engineering* Vol. 24 No. 2 (1987): pp. 359–373. DOI 10.1002/nme.1620240207.

Copyright © 2023 by ASME

Proceedings of the ASME 2023
International Design Engineering Technical Conferences and
Computers and Information in Engineering Conference
IDETC-CIE2023
August 20-23, 2023, Boston, Massachusetts

DETC2023-111436

TOPOLOGY OPTIMIZATION OF RAREFIED GAS DEVICES WITH DISCRETE VELOCITY METHOD

Guan Kaiwen	Kei Matsushima	Yamada Takayuki *
The University of Tokyo	The University of Tokyo	The University of Tokyo
Tokyo, Japan	Tokyo, Japan	Tokyo, Japan

ABSTRACT

This paper presents a topology optimization method for rarefied gas devices. Based on the discrete velocity method, which is a conventional iterative scheme to solve the Boltzmann equation, we propose an extension to include a pseudo design density that characterizes distribution of materials in the design domain. Using a discrete version of the optimization problem, we obtain a system of discrete adjoint equations, which can be solved by recording necessary information during the discrete velocity method calculation. Design sensitivity to maximize (or minimize) an objective function can be calculated from the solution of the adjoint equations. Validity of the optimization method is demonstrated through optimization of a bent pipe.

Keywords: Topology optimization, Rarefied gas, Discrete velocity method, Adjoint equations, Sensitivity analysis

1. INTRODUCTION

Rarefied gas flows are the flows where the mean free path of gas molecules are not negligible compared to the characteristic length of the flow field. As the degree of rarefication increases, collisions between gas molecules lose their dominance to reflection at surrounding solid boundaries. As a result, velocity distribution of gas molecules can be far away from the equilibrium state described by the Maxwellian distribution, and constitutive relation in classical fluid dynamics, as is described through the Navier–Stokes–Fourier equations, may no longer hold. The more fundamental Boltzmann equation is used to describe rarefied gas behavior. During the last few decades, several numerical methods such as direct simulation Monte Carlo (DSMC) [1], discrete velocity method (DVM) [2], and lattice Boltzmann method [3], have been developed and used for analysis and design of devices in rarefied gas flows.

To improve the design of rarefied gas devices, topology optimization was applied to the system governed by Boltzmann equation in a previous study [4]. Topology optimization allows direct manipulation of material distribution, thus provides the maximum degree of freedom. In [4], density method was used with DSMC, where a design variable indicates the presence of solid or fluid in the design domain. During simulation, molecule velocities are modified by local pseudo density value, in order to resemble reflection at solid boundaries. However, as simulation molecules are still allowed to enter solid region, discrepancy between flow fields obtained through an explicitly expressed reflection boundary and one expressed through pseudo density still occurs. On top of that, the method proposed can only deal with diffuse reflection boundary conditions, and statistical noise is always present, which can only be suppressed by increasing sampling range.

In this study, we apply topology optimization of rarefied gas flows using the discrete velocity method (DVM). The strategy is similar to that is followed in [4]. We shall demonstrate that with DVM, the pseudo design model will be more accurate, fully recovering the flow field as is calculated in a computational domain confined by explicitly stated solid boundaries. Besides that, different reflection boundary conditions can be applied, and smooth, deterministic results can be obtained with higher efficiency.

The structure of this paper is as follows. We first introduce the traditional DVM as our basic tool for solving the Boltzmann equation. Then we propose our extension of DVM to incorporate the pseudo design density. After that, we formulate the discrete version of optimization problem, derive the discrete adjoint equations, and obtain the expression for design sensitivity. Finally, numerical example for optimization design of a bent pipe is provided.

1.1 Discrete Velocity Method

The governing equation for single-component rarefied gas flows is the Boltzmann equation, which is given by

$$\frac{\partial f}{\partial t} + \vec{v} \cdot \nabla_x f + \vec{F} \cdot \nabla_v f = C(f). \tag{1}$$

$f(\vec{v}, \vec{x}, t)$ is the velocity distribution function (VDF) for a single gas molecule that describes the probability of one specific gas molecule having velocity $\vec{v}$ at position $\vec{x}$ and time t. $\vec{F}$ is the

---

*Corresponding author: t.yamada@mech.t.u-tokyo.ac.jp
Documentation for `asmeconf.cls`: Version 1.34, May 12, 2023.

Copyright © 2023 by ASME

external force per unit mass, and $C(f)$ is the collision operator on f describing the rate of change in f due to binary collisions of gas molecules. To avoid evaluating the complicated collision integral, Shakhov model [5] is often used, in which $C(f)$ is given by

$$C^s(f) = \frac{f^s(\vec{v}, \vec{x}, t) - f(\vec{v}, \vec{x}, t)}{\tau(\vec{x}, t)}. \tag{2}$$

Here $\tau = \mu/p$ is the relaxation time, μ is viscosity and p is pressure. The target state f^s is given through Maxwellian distribution f^w:

$$f^s(\vec{v}, \vec{x}, t) = f^w \left[1 + (1 - \text{Pr}) \frac{\vec{Q} \cdot \vec{c}}{5pRT} \left(\frac{c^2}{RT} - 5 \right) \right], \tag{3}$$

where $\vec{c} = \vec{v} - \vec{U}$ is the peculiar velocity of the molecule, $\vec{Q}$ is heat flux vector, R is specific gas constant, and T is temperature. The Maxwellian distribution is given by

$$f^w = \frac{\rho}{(2\pi RT)^{3/2}} \exp\left(-\frac{c^2}{2RT}\right), \tag{4}$$

where ρ is density of gas. Neglecting the external force term, Eq. (1) is simplified to

$$\frac{\partial f}{\partial t} + \vec{v} \cdot \nabla_x f = \frac{1}{\tau}(f^s - f). \tag{5}$$

Substituting f by its value at a discrete set of velocity points $\{\vec{v}_1, \vec{v}_2, \ldots, \vec{v}_n\}$, and writing $f_n(\vec{x}, t) = f(\vec{v}_n, \vec{x}, t)$, Eq. (5) is broken down into a system of n equations for each velocity component,

$$\frac{\partial f_n}{\partial t} + \vec{v}_n \cdot \nabla_x f_n = \frac{1}{\tau}(f_n^s - f_n). \tag{6}$$

As is suggested in [6], we use forward Euler scheme for the time derivative, treat convection term $\vec{v}_n$ and f_n in Shakhov model implicitly, and keep f^s explicitly, thus Eq. (6) can be discretized in time as

$$\frac{1}{\Delta t}(f_n^{k+1} - f_n^k) + \vec{v}_n \cdot \nabla_x f_n^{k+1} = \frac{1}{\tau^k}(f_n^{s,k} - f_n^{k+1}). \tag{7}$$

Implicit treatment allows us to use very large time steps so that steady state can be found quickly.

The spatial gradient is solved through finite volume method (FVM). For simplicity, we restrict ourselves to two dimensional cases. Assuming the computational domain is discretized by a structured grid, where each cell can be indexed by (i, j), the FVM version of Eq. (7) is obtained by integrating both sides over a computational cell (with subscript n omitted for readability):

$$\left(\frac{1}{\Delta t} + \frac{1}{\tau_{i,j}^k}\right) \Omega_{i,j} \Delta f_{i,j}^k + \sum_m \vec{S}_m \cdot \vec{v} \Delta f_m^k$$
$$= \frac{\Omega_{i,j}}{\tau_{i,j}^k}(f_{i,j}^{s,k} - f_{i,j}^k) - \sum_m \vec{S}_m \cdot \vec{v} f_m^k, \tag{8}$$

where $\Omega_{i,j}$ is volume of cell (i, j), $\vec{S}$ is the out-pointing normal vector of the cell's faces, whose magnitude is equal to face area, m is index of cell faces, and Δf^k represents the difference between

step $k + 1$ and k, namely, $\Delta f^k = f^{k+1} - f^k$. f_m is the value of f at cell faces, which can be constructed via first-order upwind scheme, or other higher-order schemes.

Now we can apply Lower-Upper Symmetric Gauss-Seidel (LU-SGS) method to solve the implicit system by two sweeps [6]. Let RHS denote the right hand side of Eq. (8), we obtain $\Delta f_{i,j}$ by two sweeps of the structured grid.

Forward :
$$\mathbf{D}_{i,j} \Delta f_{i,j}^* + \mathbf{L}_{i,j}^x \Delta f_{i-1,j}^* + \mathbf{L}_{i,j}^y \Delta f_{i,j-1}^* = \text{RHS}_{i,j}, \tag{9}$$

Backward :
$$\Delta f_{i,j} = \Delta f_{i,j}^* - (\mathbf{U}_{i,j}^x \Delta f_{i+1,j}^* - \mathbf{U}_{i,j}^y \Delta f_{i,j+1}^*)\mathbf{D}_{i,j}^{-1}. \tag{10}$$

The coefficients are defined by

$$\mathbf{D}_{i,j} = \frac{\Omega_{i,j}}{\Delta t} + \frac{\Omega_{i,j}}{\tau_{i,j}} + |\vec{S}_x \cdot \vec{v}| + |\vec{S}_y \cdot \vec{v}|, \tag{11}$$

$$\mathbf{L}_{i,j}^x = \frac{1}{2}\vec{S}_x \cdot \vec{v} \left[1 - \text{sign}(\vec{S}_x \cdot \vec{v}) \right], \tag{12}$$

$$\mathbf{L}_{i,j}^y = \frac{1}{2}\vec{S}_y \cdot \vec{v} \left[1 - \text{sign}(\vec{S}_y \cdot \vec{v}) \right], \tag{13}$$

$$\mathbf{U}_{i,j}^x = -\frac{1}{2}\vec{S}_x \cdot \vec{v} \left[1 + \text{sign}(\vec{S}_x \cdot \vec{v}) \right], \tag{14}$$

$$\mathbf{U}_{i,j}^y = -\frac{1}{2}\vec{S}_y \cdot \vec{v} \left[1 + \text{sign}(\vec{S}_y \cdot \vec{v}) \right], \tag{15}$$

where $\vec{S}_x = \frac{1}{2}(\vec{S}_{i-1/2, j} + \vec{S}_{i+1/2, j})$, $\vec{S}_y = \frac{1}{2}(\vec{S}_{i, j-1/2} + \vec{S}_{i, j+1/2})$, and $\text{sign}(\cdot)$ is the sign function which returns 1 for positive input, and 0 otherwise.

After Δf^k is calculated, the discretized DVM is updated by $f_{i,j}^{k+1} = f_{i,j}^k + \Delta f_{i,j}^k$ for every velocity component $\vec{v}_n$.

2. FORMULATING OPTIMIZATION PROBLEM

2.1 Density method

One fundamental principle of topology optimization is to replace the structural optimization problem with a material distribution problem. In the design domain $\mathcal{D}$, for optimization of rarefied gas devices, we assume a characteristic function χ to indicate presence of solid or fluid. For instance, letting fluid domain be $\mathcal{F}$, and solid domain be $\mathcal{S} = \mathcal{D} \backslash \mathcal{F}$, we have

$$\chi(\vec{x}) = \begin{cases} 1 & \text{for } \vec{x} \in \mathcal{F}, \\ 0 & \text{for } \vec{x} \in \mathcal{S}. \end{cases} \tag{16}$$

However, as χ can take either 1 or 0 at arbitrary points, infinitely small structures are not prohibited, resulting in the optimization problem being ill-posed. An alternative is to use a normalized density $\alpha \in \mathcal{L}^\infty(\mathcal{D}; [0, 1])$ with increased regularity as the design variable [7]. However, introducing α means we allow some intermediate state between solid ($\alpha = 0$) and fluid ($\alpha = 1$). The conventional DVM needs to be extended to incorporate α in such a way that authentically reflects material distribution.

2.2 The extended DVM

Using pseudo design density α for topology optimization requires the numerical method for Boltzmann equation to accurately simulate rarefied gas flows under various material distribution conditions. Under the FVM framework for solving Eq. (8),

Copyright © 2023 by ASME

we consider the practical case where α is represented by its values at the center of a finite number of computational cells, which can be written as vector $\vec{A} = \{\alpha_{1,1}, \alpha_{1,2}, \ldots, \alpha_{i,j}, \ldots\}$. As baselines, we would like the following properties to be satisfied for the extended DVM:

- The resulting steady state of flow field changes continuously with $\vec{A}$.

- For a given $\vec{A}$ whose elements are either 0 or 1, the resulting flow field in flow region obtained through extended DVM should be the same compared to flow field obtained through conventional DVM, where the computational domain is set as the union of all the FVM cells where $\alpha = 1$, and proper boundary conditions are applied.

In order to satisfy the criterion above, the following scheme is proposed.

2.2.1 Macroscopic quantities. In conventional DVM, macroscopic flow properties including density ρ, velocity $\vec{U}$, temperature T, pressure p, and heat flux $\vec{Q}$ are calculated from the discrete VDF f, either as outputs when the flow field reaches steady state, or as arguments to be supplied to the Shakhov model in order to calculate the collision term. In the extended DVM, we suggest $\vec{U}, T, p, \vec{Q}$ to be modified by the local value of α. Letting a prime denote modified values, we write

$$\vec{U}'_{i,j} = \alpha_{i,j}\vec{U}_{i,j}, \tag{17}$$

$$T'_{i,j} = \alpha_{i,j}T_{i,j} + (1 - \alpha_{i,j})T^b_{i,j}, \tag{18}$$

$$p'_{i,j} = \rho_{i,j}RT'_{i,j}, \tag{19}$$

$$\vec{Q}'_{i,j} = \alpha_{i,j}\vec{Q}_{i,j}, \tag{20}$$

where T^b is the background temperature, which is decided before DVM starts, and controls the temperature of solid, if present, in the design domain.

In the extended DVM, these modified macroscopic quantities are used in Shakhov model and as outputs at steady state. It is clear that when $\alpha_{i,j} = 1$, no change is applied, effectively. But when $\alpha_{i,j} = 0$, the result is some stationary gas with fixed temperature and zero internal heat flux, which indicates existence of solid.

2.2.2 Convection and reflection. Apart from having fixed macroscopic quantities, another distinction that separates solid and fluid is the absence of mass exchange. In conventional DVM, convection changes are calculated under the FVM framework, by evaluating the surface fluxes across all the boundaries of each computational cell. Let ϕ represent the mass flux across the shared surface of two adjacent FVM cells, whose detailed expression depends on the FVM scheme one chooses. This elemental flux will result in a loss in mass for the upwind cell, and a gain of equal amount for the downwind cell, thus ensuring conservation.

In the extended DVM, the above model no longer holds. If the upwind cell has $\alpha = 0$, it should represent solid region that does not emit mass into fluid region. In addition, if the downwind cell has $\alpha = 0$, mass flux from upwind cell should be reflected, instead of entering. Therefore, we propose the following correction for convection terms by the pseudo density.

- The loss for the upwind cell, denoted by ϕ^u, is given by $\phi^u = \alpha^u\phi$, where α^u is the pseudo density in the upwind cell.

- The gain for the downwind cell, denoted by ϕ^d, is given by $\phi^d = \alpha^d\phi^u$, where α^d is the pseudo density in the downwind cell.

- The difference $\phi^r = \phi^u - \phi^d$ is the amount to be reflected.

Note that, after the above correction, there will be zero exchange of mass between computational cells if either one has $\alpha = 0$. In the mean time, conservation of mass is preserved by considering a reflection of amount ϕ^r at every cell surface. Also note that, this extension does not depend on the actual reflection model one chooses. In practice, any reflection model can be used depending on the setting of the optimization problem.

2.2.3 Solution of the implicit system via LU-SGS. Since the macroscopic quantities and the convection terms have changed in the extended DVM, the implicit LU-SGS solver should also be corrected. In the RHS of Eq. (8), the relaxation state $f^{s,k}_{i,j}$ should be calculated with the corrected macroscopic quantities. Flow resulting from reflection fluxes ϕ^r should be accounted for in $\sum_m \vec{S}_m \cdot \vec{v} f^k_m$ as an explicit source term.

For the coefficients of the sweeping process, note that after correction by α, the loss for upwind cell is α^u times compared to conventional DVM. And the mass exchange between adjacent cells is $\alpha^u\alpha^d$ times the original value. Therefore, the coefficients are corrected by

$$\mathbf{D}_{i,j} = \frac{\Omega_{i,j}}{\Delta t} + \frac{\Omega_{i,j}}{\tau_{i,j}} + \left(|\vec{S}_x \cdot \vec{v}| + |\vec{S}_y \cdot \vec{v}|\right)\alpha_{i,j}, \tag{21}$$

$$\mathbf{L}^x_{i,j} = \frac{1}{2}\vec{S}_x \cdot \vec{v}\left[1 - \text{sign}(\vec{S}_x \cdot \vec{v})\right]\alpha_{i,j}\alpha_{i-1,j}, \tag{22}$$

$$\mathbf{L}^y_{i,j} = \frac{1}{2}\vec{S}_y \cdot \vec{v}\left[1 - \text{sign}(\vec{S}_y \cdot \vec{v})\right]\alpha_{i,j}\alpha_{i,j-1}, \tag{23}$$

$$\mathbf{U}^x_{i,j} = -\frac{1}{2}\vec{S}_x \cdot \vec{v}\left[1 + \text{sign}(\vec{S}_x \cdot \vec{v})\right]\alpha_{i,j}\alpha_{i+1,j}, \tag{24}$$

$$\mathbf{U}^y_{i,j} = -\frac{1}{2}\vec{S}_y \cdot \vec{v}\left[1 + \text{sign}(\vec{S}_y \cdot \vec{v})\right]\alpha_{i,j}\alpha_{i,j+1}. \tag{25}$$

2.2.4 Summary of the extended DVM. It is clear from the above derivation process, that the extended DVM satisfies the requirements stated in section 2.2. α is only involved in simple algebraic operations, which ensures continuity. And when $\alpha = 1$, the result will be identical compared to conventional DVM. Identical reflection boundary conditions are ensured by the construction of convection fluxes ϕ^u, ϕ^d, and ϕ^r.

A summary of the extended DVM is presented in Algorithm 1.

2.3 The discrete optimization problem

Based on the extended DVM, we can formally write down the discrete optimization problem. Let $\vec{f}$ be the vector of all the discrete values of f at every velocity point in every computational cell. The extended DVM can be viewed as an operator acting on $\vec{f}$, which updates $\vec{f}^k$ to $\vec{f}^{k+1}$ according to Algorithm

Copyright © 2023 by ASME

Algorithm 1 EXTENDED DVM
___
1: Initialize the discrete VDF.
2: **while** Flow field has not converged **do**
3:     Calculate macroscopic quantities.
4:     Calculate collision term through Shakhov model.
5:     Calculate convection changes.
6:     Update VDF by LU-SGS.
7: **end while**
___

1. Symbolically, we write $\vec{f}^{k+1} = L(\vec{f}^k; \vec{A})$. Let the objective to be maximized be a real-valued function r defined on $\vec{f}$. In most cases, values of $\vec{f}$ at specific points are of little interest, and the focus is on macroscopic quantities of the flow field. However, as macroscopic quantities are derived as moments of the VDF, we still write r as a function defined over $\vec{f}$ for consistency. For sufficiently large integer K such that the flow field reaches steady state after K DVM iterations, we consider the following optimization problem:

$$\sup_{\vec{A}} \quad r = r(\vec{f}^K) \tag{26}$$

$$\text{Subject to}: \quad \vec{f}^K = L(\vec{f}^{K-1}; \vec{A}) \tag{27}$$

$$\vdots$$

$$\vec{f}^1 = L(\vec{f}^0; \vec{A}) \tag{28}$$

$$\sum_{i,j} \alpha_{i,j} \Omega_{i,j} \leq V_{\max} \tag{29}$$

where $\vec{f}^0$ in Eq. (28) is the initial condition for VDF, and $V_{\max}$ in Eq. (29) is a volume constraint on fluid region.

3. NUMERICAL IMPLEMENTATION

3.1 Sensitivity analysis

Optimization of the objective functional under volume constraint is realized through the augmented Lagrangian multiplier method. Based on the optimization problem stated in the previous section, we consider the following Lagrangian J defined by

$$J = r(\vec{f}^K) - \sum_{k=1}^{K} \vec{\Lambda}^k \cdot \left[\vec{f}^k - L(\vec{f}^{k-1}; \vec{A})\right]$$
$$- \zeta\left(\vec{A} \cdot \vec{\Omega} - V_{\max}\right) + \frac{\sigma}{2}\left(\vec{A} \cdot \vec{\Omega} - V_{\max}\right)^2. \tag{30}$$

$\vec{\Lambda}^k$ are vectors of Lagrangian multipliers that have the same size as $\vec{f}$, and they correspond to each iterative equation from Eq. (27) to Eq. (28). $\vec{\Omega}$ is the vector containing all cell volumes $\Omega_{i,j}$, ζ is the Lagrangian multiplier for inequality constraint, and $\sigma > 0$ is a penalty factor.

First-order optimum condition requires that the Fréchet derivative of J with respect to $\vec{f}^k$ are all zero, which gives a

system of K equations for $\vec{\Lambda}$

$$\nabla_f r(\vec{f}^K) - \vec{\Lambda}^K = 0, \tag{31}$$

$$\vec{\Lambda}^K \left[\nabla_f L(\vec{f}^{k-1}; \vec{A})\right]^T - \vec{\Lambda}^{k-1} = 0, \tag{32}$$

$$\vdots$$

$$\vec{\Lambda}^2 \left[\nabla_f L(\vec{f}^1; \vec{A})\right]^T - \vec{\Lambda}^1 = 0, \tag{33}$$

where a T in the superscript denotes matrix transposition. Note that, these equations form a final value problem. Equation (31) offers the final condition for $\vec{\Lambda}^K$ via the gradient of the objective functional, and Eq. (32) to Eq. (33) describes the iterative relation between $\vec{\Lambda}^{k+1}$ and $\vec{\Lambda}^k$. Since the extended DVM process is a deterministic one, the gradients $\nabla_f L$ can be explicitly evaluated, therefore all $\vec{\Lambda}$ can be explicitly evaluated.

Once the adjoint variables are obtained, we calculate design sensitivity $\vec{W}$ by evaluating the Fréchet derivative of J with respect to $\vec{A}$.

$$\vec{W} = \sum_{k=1}^{K} \vec{\Lambda}^k \cdot \left[\nabla_A L(\vec{f}^{k-1}; \vec{A})\right]^T - \zeta\vec{\Omega} + \sigma(\vec{A} \cdot \vec{\Omega} - V_{\max})\vec{\Omega}. \tag{34}$$

Again, since the DMV process is deterministic, the gradients $\nabla_A L$ can be explicitly evaluated.

3.2 Optimization algorithm

Sensitivity $\vec{W}$ gives us a measure of how the objective functional will change with respect to design variable α under the inequality volume constraint. We now give a brief outline of the steepest gradient algorithm to find the optimal $\vec{A}$ in Algorithm 2. The step length τ is a fixed positive number decided before optimization. As $\vec{A}$ changes throughout optimization loop, Lagrangian multiplier ζ is updated accordingly. After each update, we cast $\vec{A}$ into the range $[0, 1]$, where α is initially defined on. According to the Karush–Kuhn–Tucker condition, ζ will be either zero (for an inactive constraint) or positive (for an active constraint). A properly decided $\zeta_{\max} > 0$, together with normalizing $\vec{W}$ can help convergence. Finally, we decide that $\vec{A}$ has converged when the relative change between loops is less than $\epsilon = 10^{-3}$.

Algorithm 2 OPTIMIZATION ALGORITHM
___
1: Initialize $\vec{A}$ and ζ.
2: **while** $\vec{A}$ has not converged **do**
3:     Perform DVM calculation in Algorithm 1.
4:     Solve the adjoint equations and obtain $\vec{W}$.
5:     Update ζ by $\zeta \leftarrow \zeta + \sigma(\vec{A} \cdot \vec{\Omega} - V_{\max})$.
6:     Set $\zeta \leftarrow \max(\min(\zeta, \zeta_{\max}), 0)$.
7:     Set $\vec{A} \leftarrow \tau\vec{W}/|\vec{W}|$, where $|\cdot|$ is the Euclidean norm.
8:     Set $\vec{A} \leftarrow \max(\min(\vec{A}, 1), 0)$.
9: **end while**
___

Copyright © 2023 by ASME

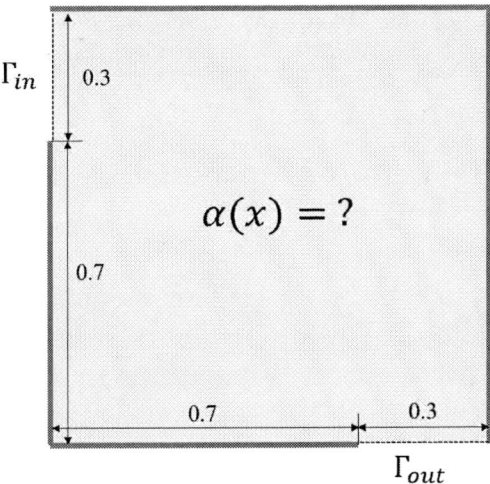

FIGURE 1: GEOMETRY SETTING OF THE BENT PIPE OPTIMIZATION PROBLEM.

FIGURE 2: OPTIMIZED DESIGN FOR THE BENT PIPE. WHITE REGION REPRESENTS FLUID, AND BLACK REGION REPRESENTS SOLID.

4. NUMERICAL EXAMPLE

Here, we present the optimized result for a two-dimensional bent pipe design. The working gas is chosen as Argon, and solid background temperature is set to $T^b = 273$ K. The initial pressure $p_0 > 0$ of gas is decided so that the Knudsen number at 273 K is 10.0. The computational domain is a 1 m by 1 m square, which is enclosed by diffuse solid walls at T^b, as shown in Fig. 1, with an open inlet Γ_{in} at the left boundary and an outlet Γ_{out} at the bottom boundary. Both the inlet and the outlet have width 0.3 m. At Γ_{in}, pressure is set to $1.001p_0$, temperature is fixed at 273 K, and we require that macroscopic velocity in the y-direction is zero. For Γ_{out} we only set pressure to be $0.999p_0$. Other macroscopic conditions are decided by local values during DVM calculation according to [8]. The computational domain is discretized by a uniform 50×50 square mesh, and the velocity domain in DVM is discretized by 100×100 uniform velocity points in the range $[-4c_0, 4c_0]$, where $c_0 = \sqrt{2RT^b}$.

The objective function to be maximized is chosen as the average flow rate at the inlet. Since pressure and temperature are fixed at the inlet, we only consider velocity in the x-direction and define r as

$$r(\vec{f}) = \sum_{(i,j)\text{at inlet}} U_{i,j}, \qquad (35)$$

where $U_{i,j} = \left(\sum_n f_{n,i,j}\vec{v}_{n,x}\right) / \left(\sum_n f_{n,i,j}\right)$ is local macroscopic velocity in the x-direction, and $\vec{v}_{n,x}$ is the x-component of the discrete velocity $\vec{v}_n$.

The volume constraint in the design domain is set to $V_{max} = 0.4$, and all solid surfaces are treated as diffuse reflectors. The optimal design obtained is shown in Fig. 2. All the calculations are programmed using FORTRAN, and executed on a Windows system with an Intel CPU.

5. CONCLUSION

In this paper, a topology optimization method for rarefied gas devices is presented. The governing Boltzmann equation is solved via the deterministic DVM. A pseudo density model is constructed based on conventional DVM to characterize material distribution. Topology optimization problem is formulated in discrete form by viewing DVM process as operator acting on a discrete distribution function values. Design sensitivity is obtained using augmented Lagrangian multiplier method, and evaluated by solving discrete adjoint equations. The optimal design is obtained by steepest gradient method. Numerical example for bent pipe optimization is presented to demonstrate the validity of the proposed method.

REFERENCES

[1] Bird, G. A. *Molecular Gas Dynamics And The Direct Simulation Of Gas Flows* (1994).

[2] Aristov, Vladimir. *Direct Methods for Solving the Boltzmann Equation and Study of Nonequilibrium Flows*. Vol. 60 (2001). DOI 10.1007/978-94-010-0866-2.

[3] Chen, Shiyi and Doolen, Gary D. "Lattice Boltzmann Method for Fluid Flows." *Annual Review of Fluid Mechanics* Vol. 30 No. 1 (1998): pp. 329–364. DOI 10.1146/annurev.fluid.30.1.329. URL https://doi.org/10.1146/annurev.fluid.30.1.329, URL https://doi.org/10.1146/annurev.fluid.30.1.329.

[4] Guan, Kaiwen, Matsushima, Kei, Noguchi, Yuki and Yamada, Takayuki. "Topology optimization for rarefied gas flow problems using density method and adjoint IP-DSMC." *Journal of Computational Physics* Vol. 474 (2023): p. 111788. DOI https://doi.org/10.1016/j.jcp.2022.111788. URL https://www.sciencedirect.com/science/article/pii/S0021999122008518.

[5] Shakhov, E. M. "Generalization of the Krook kinetic relaxation equation." *Fluid Dynamics* Vol. 3 (1968). DOI - 10.1007/BF01029546. URL -https://doi.org/10.1007/BF01029546.

[6] Zhu, Lianhua, Pi, Xingcai, Su, Wei, Li, Zhi-Hui, Zhang, Yonghao and Wu, Lei. "General synthetic iterative scheme for nonlinear gas kinetic simulation of multi-scale rarefied gas flows." *Journal of Computational Physics* Vol. 430 (2021):

Copyright © 2023 by ASME

p. 110091. DOI https://doi.org/10.1016/j.jcp.2020.110091. URL https://www.sciencedirect.com/science/article/pii/S0021999120308652.

[7] Bendsøe, M P and Sigmund, O. "Material interpolation schemes in topology optimization." *Archive of Applied Mechanics* Vol. 69 No. 9 (1999): pp. 635–654.

[8] Fang, Yichuan and Liou, William W. "Computations of the Flow and Heat Transfer in Microdevices Using DSMC With Implicit Boundary Conditions ." *Journal of Heat Transfer* Vol. 124 No. 2 (2001): pp. 338–345. DOI 10.1115/1.1447933. URL https://asmedigitalcollection.asme.org/heattransfer/article-pdf/124/2/338/5749463/338_1.pdf, URL https://doi.org/10.1115/1.1447933.

Copyright © 2023 by ASME

Proceedings of the ASME 2023
International Design Engineering Technical Conferences and
Computers and Information in Engineering Conference
IDETC-CIE2023
August 20-23, 2023, Boston, Massachusetts

DETC2023-112414

A GPU-BASED PARALLEL BOUND-AND-CLASSIFY METHOD FOR CONTINUOUS CONSTRAINT SATISFACTION PROBLEMS

Wangchuan Feng, Guanglu Zhang, Jonathan Cagan

Department of Mechanical Engineering, Carnegie Mellon University, Pittsburgh, PA

ABSTRACT

Continuous constraint satisfaction is prevalent in many science and engineering fields. When solving continuous constraint satisfaction problems, it is more advantageous for practitioners to derive all feasible regions (i.e., the solution space) rather than a limited number of solution points, since these feasible regions facilitate innovative design concept generation and design tradeoff evaluation. Several CPU-based branch-and-prune methods and geometric approximation methods have been proposed in prior research to derive feasible regions for continuous constraint satisfaction problems. However, these methods have not been extensively adopted in practice, mainly because of their high computational expense. To overcome the computational bottleneck of extant CPU-based methods, this paper introduces a GPU-based parallel bound-and-classify method to derive feasible regions for continuous constraint satisfaction problems in a reasonable computational time. Using interval arithmetic, coupled with the computational power of GPU, this method iteratively partitions the design space into many subregions and classifies these subregions as feasible, infeasible, and indeterminate regions. To visualize these classified regions in the design space, a planar visualization approach that projects all classified regions into one figure is also proposed in this paper. The GPU-based parallel bound-and-classify method and the planar visualization approach are validated through a case study of welded beam design. The case study shows that the method and the approach can solve the welded beam design problem and visualize the result effectively and efficiently. A four-step procedure for implementing the GPU-based parallel bound-and-classify method and the planar visualization approach in practice is also outlined.

Keywords: constraint satisfaction, design optimization, engineering design, GPU programming, interval arithmetic, parallel computing

1. INTRODUCTION

A continuous constraint satisfaction problem (CCSP), also known as a numerical constraint satisfaction problem (NCSP), is defined by sets of variable values that must satisfy a finite number of constraints, where each variable ranges over a continuous domain. Continuous constraint satisfaction has significant applications in many science and engineering fields, such as material science, chemical engineering, and mechanical engineering [1-4]. For example, material scientists need to assess the manufacturing feasibility of a material based on constraints determined by the material composition and its processing conditions (e.g., pressure and temperature) [5]; mechanical engineers are required to meet several constraints related to system and component performances (e.g., speed and payload) when they design an unmanned aerial vehicle [6].

For a continuous constraint satisfaction problem, the allowable ranges of variables define a design space. Each solution of the problem (i.e., a set of variable values that satisfy all specified constraints) corresponds to a point in the design space, referred to as a solution point. In practice, sampling-based methods [7, 8] have been commonly employed to derive multiple solution points for continuous constraint satisfaction problems. However, in contrast to solution points, when the closed-form expression of each constraint is available, it is more advantageous for practitioners to derive all feasible regions, also known as solution regions or solution space, in the design space, where any solution point within a feasible region satisfies all specified constraints. The feasible regions derived from continuous constraint satisfaction allow practitioners to generate innovative design concepts and evaluate design tradeoffs [9, 10]. In addition, these feasible regions also facilitate the negotiation process during multi-party collaboration [11].

To derive the feasible regions for continuous constraint satisfaction problems, several branch-and-prune methods [12-15] and geometric approximation methods [9, 16-18] have been proposed in prior research, but these methods have not been extensively adopted in practice, mainly because of their high computational expense. Although the recent advances in the graphics processing unit (GPU) offer new opportunities to overcome the computational bottleneck [19], extant branch-

Copyright © 2023 by ASME

and-prune methods and geometric approximation methods cannot utilize the computational power of GPU because these methods are designed for central processing unit (CPU) sequential computing.

Moreover, extant branch-and-prune methods and geometric approximation methods output sets of numerical values that represent the location of each feasible region in the design space. In contrast to these numerical values, a visual representation of all feasible regions in the design space is more helpful for practitioners to identify solutions of the problem and make informed decisions accordingly. However, when a continuous constraint satisfaction problem includes more than three variables (i.e., $n > 3$), the feasible regions derived from these extant methods are usually visualized through multiple 2-dimensional or 3-dimensional figures. Practitioners have to jump back and forth between these figures and cannot visually observe all n-dimensional feasible regions in one figure.

In this paper, a GPU-based parallel bound-and-classify method is introduced to derive the feasible regions for continuous constraint satisfaction problems. The method applies to a finite number of inequality constraints that have closed-form expressions. Using interval arithmetic, coupled with the computational power of GPU, this method iteratively partitions the design space into many subregions and classifies these subregions as feasible, infeasible, and indeterminate regions, where any point within an infeasible region cannot satisfy one or more specified constraints. Specifically, at each iteration step, the indeterminate regions classified in the last iteration step are partitioned into many subregions based on the compute capability of the practitioner's GPU, and these subregions are evaluated using interval arithmetic and classified based on the specified constraints in parallel on the GPU. Besides the GPU-based parallel bound-and-classify method, a planar visualization approach that projects all n-dimensional feasible regions in the design space into one figure is also proposed to enable more effective solution identification for continuous constraint satisfaction problems.

This paper begins with a brief review of extant methods to solve continuous constraint satisfaction problems and the background knowledge of interval arithmetic in Section 2. Section 3 presents the GPU-based parallel bound-and-classify method and the planar visualization approach for continuous constraint satisfaction problems. Four steps for implementing the GPU-based parallel bound-and-classify method and the planar visualization approach in practice are provided in Section 4. The application of the method and the approach is demonstrated through a case study of welded beam design in Section 5. The paper concludes with a discussion of the contribution of the work and future research directions.

2. BACKGROUND AND RELATED WORK

This section briefly reviews extant methods for continuous constraint satisfaction, with a focus on the engineering design context. A thorough review of constraint satisfaction, where variables are defined in continuous and discrete domains, can be found in books and review papers written or edited by Apt [20], Tsang [21], Rossi et al. [22], Kumar [23], and Brailsford et al. [24], among others. The background knowledge of interval arithmetic is also provided in this section.

2.1 Extant Methods for Continuous Constraint Satisfaction

The goal of continuous constraint satisfaction is to find sets of variable values that satisfy all specified constraints. Each set of variable values that satisfy all specified constraints corresponds to a solution point in the design space. Sampling-based methods, also known as designs for computational experiments [25-27], such as maximum entropy designs, mean squared-error designs, Latin hypercubes, and randomized orthogonal arrays, are commonly employed to derive multiple solution points for continuous constraint satisfaction problems [7, 8]. Using these sampling-based methods, a finite number of sample points are first chosen from the design space. The corresponding variable values at each sample point are then plugged into the specified constraints to filter out the points at which one or multiple constraints cannot be satisfied, and the remaining points are the solution points. These sampling-based methods provide a straightforward way to explore the design space, but these methods are not able to delineate feasible regions from the design space, where any solution point within a feasible region satisfies all specified constraints. Without the location of each feasible region in the design space, practitioners are limited to considering factors that cannot be formally expressed (e.g., aesthetics and socio-economic factors) at a finite number of solution points during their design process [10]. Moreover, practitioners cannot comprehensively evaluate design tradeoffs (e.g., the impact of adding a new constraint or removing an existing constraint) based on the solution points derived by sampling-based methods [9].

When the closed-form expression of each constraint is available, several branch-and-prune methods [12-15] and geometric approximation methods [9, 16-18] have been proposed to delineate feasible regions from the design space for continuous constraint satisfaction problems. These branch-and-prune methods aim to generate a set of n-dimensional hyperrectangular regions, also known as boxes, that enclose all feasible regions in the design space. A typical branch-and-prune method relies on an iterative process that consists of many branching steps and pruning steps, where a branching step partitions a chosen region into two or more subregions in a specific dimension in the design space (i.e., only partition the range of one specific variable in the chosen region), and a pruning step reduces the size of a chosen region by eliminating infeasible subregions within the region [12, 13]. Commonly used techniques in the pruning step include but are not limited to hull consistency, box consistency, constraint inversion, and dichotomous search [12, 28, 29]. The iterative process is halted when a user-specified region size tolerance is reached. In contrast, extant geometric approximation methods endeavor to represent feasible regions in the design space using many three-dimensional geometric elements, such as orthogonal polyhedra [17], parallelepipeds [18], and polytopes [9]. For example, a polytope

Copyright © 2023 by ASME

representation method introduced by Devanathan and Ramani [9] includes the following four steps:

1. Formulate the continuous constraint satisfaction problem.
2. Transform all constraints into ternary constraints.
3. Create the feasible regions for each ternary constraint separately.
4. Prune these feasible regions using consistency to obtain the feasible regions for all ternary constraints.

As demonstrated by the four-step procedure above, extant geometric approximation methods are only able to address ternary constraints, where each constraint includes three variables. For constraints involving more than three variables, practitioners must introduce new variables (i.e., auxiliary variables) and rewrite each of these constraints as multiple ternary constraints [16, 30].

Despite several successful applications, these branch-and-prune methods and geometric approximation methods have not been extensively adopted in practice, especially in engineering design. Sampling-based methods are still commonly used in the engineering design process [6, 31]. A major reason is that these branch-and-prune methods and geometric approximation methods are significantly more computationally expensive than sampling-based methods. Specifically, extant branch-and-prune methods usually take many iteration steps to reach the user-specified region size tolerance because each branching step only partitions the chosen region in one dimension (e.g., bisect the range of one variable and keep intact the ranges of other variables in the chosen region) [12]; extant geometric approximation methods often require very high resolution (i.e., the size of each geometric element must be very small) to avoid erroneous approximation of the feasible regions for all ternary constraints [9]. Notably, the recent advances in GPU offer new opportunities to overcome the computational bottleneck [19]. With the falling price of GPU, many personal laptops and desktops are now equipped with a dedicated GPU. However, since GPU is designed for *parallel computing*, extant branch-and-prune methods and geometric approximation methods based on CPU *sequential computing* cannot utilize the computational power of GPU.

In addition, the result outputted from extant branch-and-prune methods and geometric approximation methods are sets of numerical values that represent the location of each feasible region. In practice, a visual representation of all feasible regions in the design space is helpful for practitioners to identify solutions of the problem and make informed decisions accordingly. However, when a continuous constraint satisfaction problem includes more than three variables (i.e., $n > 3$), the feasible regions derived from these extant methods are usually visualized through multiple 2-dimensional or 3-dimensional figures. For example, a total of ten 3-dimensional figures are necessary to visualize feasible regions when a constraint satisfaction problem includes five variables [15]. As a result, practitioners have to jump back and forth between these figures when they generate or evaluate design concepts based on feasible regions in the n-dimensional design space.

2.2 Interval Arithmetic

To overcome the computational bottleneck discussed in Section 2.1, a GPU-based parallel bound-and-classify method is introduced in this paper for continuous constraint satisfaction problems based on interval arithmetic. Interval arithmetic, also known as interval analysis or interval computation, is a mathematical technique designed to automatically provide rigorous bounds on rounding errors, approximation errors, and propagated uncertainties in mathematical computation [32, 33]. Unlike floating-point computations where each value is represented as a number (e.g., 2.01), interval arithmetic represents each value as a range (e.g., [1.95, 2.11]). A list of interval arithmetic operations has been specified in IEEE 1788 standard [34], and four basic interval arithmetic operations (i.e., addition, subtraction, multiplication, and division) are provided in Eqs. (1) - (4) for demonstration purposes [35], where an interval $[a, b]$ is represented in square brackets, a is the lower bound, and b is the upper bound of the interval:

$$[x_1, x_2] + [y_1, y_2] = [x_1 + y_1, x_2 + y_2], \tag{1}$$

$$[x_1, x_2] - [y_1, y_2] = [x_1 - y_2, x_2 - y_1], \tag{2}$$

$$[x_1, x_2] \cdot [y_1, y_2] = \begin{bmatrix} \min\{x_1 y_1, x_1 y_2, x_2 y_1, x_2 y_2\}, \\ \max\{x_1 y_1, x_1 y_2, x_2 y_1, x_2 y_2\} \end{bmatrix}, \tag{3}$$

$$\frac{[x_1, x_2]}{[y_1, y_2]} = [x_1, x_2] \cdot \frac{1}{[y_1, y_2]}. \tag{4}$$

In practice, since each numerical value is approximated to its closest binary machine number and stored in computers [36], outward rounding is implemented for every interval arithmetic operation to provide rigorous bounds for the solution [37]. The outwardly rounded lower bound (left endpoint) is the closest machine number less than or equal to the exact lower bound, and the outwardly rounded upper bound (right endpoint) is the closest machine number greater than or equal to the exact upper bound [32]. For example, the interval solution [1.9986, 2.0014] could be outwardly rounded as [1.998, 2.002] with four significant digits.

Although the outwardly rounded upper and lower bounds derived by interval arithmetic operations have been proved to create an interval that contains all possible solutions (i.e., rigorous bounds for the solution) [32, 33], many interval computations including multiple interval arithmetic operations suffer the dependence problem, also known as interval dependency. The dependence problem results in overestimated bounds for the solution (i.e., an interval with unwanted excess width). For example, for the function $f(x) = x - x^2$, where $x \in [0, 1]$, the exact range of the function value is [0, 0.25]; the interval arithmetic operations shown in Eq. (2) and Eq. (3) give [0, 1] - [0, 1] · [0, 1] = [-1, 1], where the result [-1, 1] includes the exact range [0, 0.25], but it has significant excess width. Such overestimated bounds are produced because each occurrence of the variable x is treated as a different variable in the interval computation. Notably, if every variable occurs only once in a function, the interval evaluation of the function gives the exact range of the function value. However, when any variable occurs more than once in a function, the dependence problem often appears. In

Copyright © 2023 by ASME

practice, an interval computation including multiple interval arithmetic operations is often expressed as a sequence of operations, commonly known as a *code list*, to facilitate computer programming [32]. The dependence problem could be identified by counting the number of times each variable occurs in a code list. For example, a function is defined as

$$g(x_1, x_2) = e^{x_1 + x_2^2} - x_1 \cdot x_2, \qquad (5)$$

the interval evaluation of the function defined by Eq. (5) is expressed as a code list:

$$T_1 = x_2^2 \qquad (6)$$

$$T_2 = x_1 + T_1, \qquad (7)$$

$$T_3 = e^{T_2} \qquad (8)$$

$$T_4 = x_1 \cdot x_2 \qquad (9)$$

$$g(x_1, x_2) = T_3 - T_4. \qquad (10)$$

Since both variables x_1 and x_2 occur more than once in the code list, the dependence problem appears when the code list represented by Eqs. (6) - (10) is employed to compute the bounds of the function value based on interval arithmetic.

The dependence problem is a major factor that impedes the application of interval arithmetic. Several approaches have been introduced by researchers to produce exact bounds or at least sharper bounds for interval computation [28, 32, 38-40]. One major approach is to rewrite the function in another form [28, 32], such as a form where every variable only occurs once (if possible), the centered form, the mean value form, or a Taylor series. In the first example, the function could be rewritten as $f(x) = x - x^2 = - (x - 0.5)^2 + 0.25$, and the interval evaluation of the rewritten function leads to the exact range [0, 0.25]. Another approach, known as splitting, is to divide the range of each variable into d subintervals and then take the union of the d interval evaluations over the elements of the subdivision. It has been proved that the union of the d interval evaluations converges to the exact range of the function value when d approaches infinity [32]. In the first example, when the range $x \in$ [0, 1] is uniformly divided into 1,000 subintervals, the union of the 1,000 interval evaluations of the function $f(x) = x - x^2$ in each subinterval gives [-0.001, 0.251], which is close to the exact range [0, 0.25] [32].

3. REGION CLASSIFICATION AND VISUALIZATION FOR CONTINUOUS CONSTRAINT SATISFACTION

Based on the research gap discussed in Section 2.1, a GPU-based parallel bound-and-classify method is introduced in this section to derive feasible regions in the design space for continuous constraint satisfaction problems. Unlike extant CPU-based branch-and-prune methods and geometric approximation methods that only evaluate one region in the design space at a time, at each iteration step, the method introduced in this section partitions indeterminate regions in the design space into many subregions based on the compute capability of the practitioner's GPU and evaluates these subregions in parallel on the GPU. Such GPU-based parallel computing significantly improves problem-solving efficiency. Moreover, the method is

able to solve the continuous constraint satisfaction problem involving more than three variables, and practitioners do not need to transform all constraints into ternary constraints when they implement the method. Besides the GPU-based parallel bound-and-classify method, a planar visualization approach is also introduced in this section to project all n-dimensional feasible regions in the design space into one figure for continuous constraint satisfaction problems. Practitioners can observe the location of each feasible region in the design space from the figure and make informed decisions accordingly.

3.1 The GPU-based Parallel Bound-and-Classify Method

For a continuous constraint satisfaction problem with a finite set of inequality constraints $C = \{c_1, c_2, ..., c_m\}$, these inequality constraints are defined as

$$\{c_j : g_j(x_1, x_2, ..., x_n) \leq 0\}, \qquad (11)$$

where $j \in \{1, 2, ..., m\}$. There are a set of n variables $X = \{x_1, x_2, ..., x_n\}$, and these variables range over continuous domains $D = \{d_1, d_2, ..., d_n\}$ defined as

$$\{d_i : x_i \in [x_i^{lb}, x_i^{ub}]\}, \qquad (12)$$

where $i \in \{1, 2, ..., n\}$, and x_i^{lb} and x_i^{ub} are the lower and upper bounds of variable x_i, respectively. The objective is to find sets of n variable values that satisfy all the constraints $C = \{c_1, c_2, ..., c_m\}$. Notably, $g_j(x_1, x_2, ..., x_n)$ in Eq. (11) represents the closed-form expression of an inequality constraint function. Solving the continuous constraint satisfaction problems where constraints do not have closed-form expressions is beyond the scope of this paper. Equality constraints are also not included in the continuous constraint satisfaction problem defined by Eq. (11) and Eq. (12). If feasible, each equality constraint could be employed to eliminate a variable from a continuous constraint satisfaction problem in practice. An equality constraint also could be replaced by two inequality constraints that are close to each other when such an approximation is acceptable [16]. For example, an equality constraint $f(X) = 0$ may be replaced by two inequality constraints $f(X) - \varepsilon \leq 0$ and $- f(X) - \varepsilon \leq 0$ (i.e., $-\varepsilon \leq f(X) \leq \varepsilon$), where ε is a user-specified small positive value.

To solve the continuous constraint satisfaction problem, the n-dimensional design space established by Eq. (12) is iteratively partitioned into many subregions, and these subregions are classified as feasible, infeasible, and indeterminate regions, where these three types of regions are defined as:

Feasible region: if any point within the region satisfies all the constraints $C = \{c_1, c_2, ..., c_m\}$.

Infeasible region: if any point within the region cannot satisfy one or multiple constraints in $C = \{c_1, c_2, ..., c_m\}$.

Indeterminate region: if points within the region may or may not satisfy all the constraints $C = \{c_1, c_2, ..., c_m\}$.

The whole design space is first evaluated using interval arithmetic. Each constraint function $g_j(X)$ is expressed as a code list, as demonstrated in Section 2.2. The allowable ranges of n variables defined by Eq. (12) are then plugged into each code list, and the sequence of interval arithmetic operations in each code list yields $[DLB_j, DUB_j]$, where DLB_j and DUB_j are the

Copyright © 2023 by ASME

lower and upper bounds for the value of the constraint function $g_j(X)$ within the design space, respectively. Notably, the interval bounds derived by each interval arithmetic operation are outwardly rounded to the user-specified digits or significant digits (e.g., eight significant digits), and the interval $[DLB_j, DUB_j]$ is therefore guaranteed to include all possible values of $g_j(X)$ within the design space. If any variable occurs more than once in a code list, the dependency problem needs to be considered, and the function rewriting approach and/or the splitting approach discussed in Section 2.2 could be used to derive the exact or sharper bounds for the value of the constraint function $g_j(X)$ within the design space.

Based on the lower and upper bounds for the value of each constraint function, DLB_j and DUB_j, and the inequality defined by Eq. (11), the whole design space is classified as a feasible region, an infeasible region, or an indeterminate region. Specifically, if $DUB_j \leq 0$ for all $j \in \{1, 2, \ldots, m\}$, the design space is a feasible region; if $DLB_j > 0$ for any $j \in \{1, 2, \ldots, m\}$, the design space is an infeasible region; otherwise, the design space is an indeterminate region. When the whole design space is classified as a feasible region or an infeasible region, practitioners need to check each constraint and consider tightening or relaxing certain constraint(s) if necessary.

When the whole design space is classified as an indeterminate region, the design space is iteratively partitioned into subregions. At each iteration step, a GPU-based parallel bound-and-classify process is carried out to classify these subregions as feasible regions, infeasible regions, and indeterminate regions. Only the indeterminate regions are further partitioned in the next iteration step. The iteration process is halted when the user-specified stopping criteria are satisfied. The partition strategy, the GPU-based parallel bound-and-classify process, and the stopping criteria are discussed as follows.

At the beginning of each iteration step, the indeterminate region(s) classified in the last iteration step are partitioned into many subregions. Each indeterminate region is partitioned in multiple dimensions at the same time using a user-specified partition method, such as uniform partition and golden ratio partition. Different scales and coordinate systems also could be employed if necessary. For example, a logarithmic scale could be applied to a variable if the variable is defined over a large range (e.g., $[10^{-6}, 10^6]$). The total number of subregions that results from the partition is chosen based on the compute capability of the practitioner's GPU. For example, at each iteration step, each of the indeterminate regions could be partitioned in all its n dimensions at the same time. In each dimension, the interval of the corresponding variable is uniformly partitioned into k subintervals, where the integer k is estimated through:

$$k \approx \sqrt[n]{\frac{N_{GPU}}{N_{idt}}} \text{ when } N_{idt} \cdot 2^n < N_{GPU}, \tag{13}$$

$$k = 2 \text{ when } N_{idt} \cdot 2^n \geq N_{GPU} \tag{14}$$

where N_{GPU} is the number of shading units in the practitioner's GPU, and N_{idt} is the number of indeterminate regions that need to be partitioned at the iteration step. Notably, the partition strategy defined by Eq. (13) and Eq. (14) is presented here for demonstration purposes, and practitioners are recommended to customize the partition strategy based on their specific problem when they implement the method. The partition strategy defined by Eq. (13) and Eq. (14) avoids GPU underutilization in the first few iteration steps when N_{idt} has a small value (e.g., $N_{idt} = 1$ in the first iteration step), but this partition strategy can lead to a significant number of subregions when many indeterminate regions need to be partitioned in the iteration step (e.g., $N_{idt} > 10^6$) or the continuous constraint satisfaction problem includes many variables (e.g., $n > 20$).

The subregions that result from the partition are then evaluated and classified as feasible, infeasible, and indeterminate regions in parallel on the GPU during each iteration step (i.e., a GPU-based parallel bound-and-classify process during an iteration step). Based on the code list of each constraint function $g_j(X)$, the interval evaluation of each constraint function within each subregion gives $[LB_j^{(s)}, UB_j^{(s)}]$. Here, $LB_j^{(s)}$ and $UB_j^{(s)}$ are the outwardly rounded lower and upper bounds for the value of the constraint function $g_j(X)$, respectively, s is the number of the subregion, and $s \in \{1, 2, \ldots, N_s\}$, where N_s is the total number of subregions that need to be evaluated at the iteration step. The subregion s is classified as a feasible region if $UB_j^{(s)} \leq 0$ for all $j \in \{1, 2, \ldots, m\}$. If $LB_j^{(s)} > 0$ for any $j \in \{1, 2, \ldots, m\}$, the subregion s is classified as an infeasible region. If the subregion s is not a feasible or infeasible region, it is classified as an indeterminate region. Notably, when any variable occurs more than once in a code list, practitioners are recommended to use the function rewriting approach and/or the splitting approach discussed in Section 2.2 to derive the exact or sharper bounds for the value of the constraint function $g_j(X)$ within each subregion (i.e., reduce excess width for $[LB_j^{(s)}, UB_j^{(s)}]$) at least during the first few iteration steps. Using the exact or sharper bounds, more subregions could be classified as feasible and infeasible regions in an iteration step, and thus less indeterminate regions need to be partitioned in the next iteration step.

At the end of each iteration step, if the user-specified stopping criteria are satisfied, the iteration process is halted. Otherwise, a new iteration step begins, and the indeterminate region(s) classified in the last iteration step are further partitioned. Practitioners can specify one or multiple stopping criteria based on computational power and n-dimensional region volume when they implement the method. The stopping criteria related to computational power include but are not limited to the computer run time, the maximum number of iteration steps, the maximum number of subintervals over the range of each variable, and the region size tolerance (e.g., the smallest region width in any specific dimension or all dimensions). The stopping criteria related to n-dimensional region volume could be defined using the increment of the n-dimensional volume of all classified feasible regions between two adjacent iteration steps or the n-dimensional volume of all classified indeterminate regions at the end of an iteration step. Notably, since a user-specified stopping criterion related to n-dimensional region volume may take an unaffordable computer

Copyright © 2023 by ASME

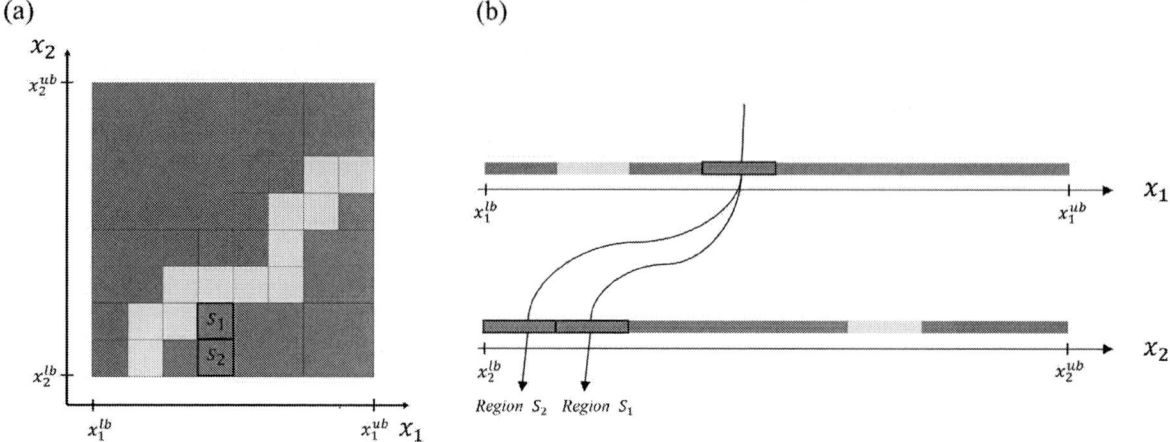

FIGURE 1: AN ILLUSTRATION OF THE PLANAR VISUALIZATION APPROACH (A) N-DIMENSIONAL INFEASIBLE (RED), INDETERMINATE (YELLOW), AND FEASIBLE (GREEN) REGIONS AND (B) PROJECTION FROM N-DIMENSIONAL REGIONS TO ONE FIGURE WITH N HORIZONTAL AXES

run time to be satisfied (or even may not be satisfied), at least one stopping criterion related to computational power should be employed in practice to halt the iteration process in due time. For example, practitioners could specify the maximum number of iteration steps and a limit value for the n-dimensional volume of indeterminate regions as two stopping criteria when they implement the method, and the iteration process is halted if any of these two stopping criteria is satisfied.

When the iteration process is halted, all feasible, infeasible, and indeterminate regions classified during the iteration process are outputted as sets of intervals, where each interval set includes n intervals that correspond to one classified region in the design space. The union of all feasible regions creates a *sound* representation of the solution space for the continuous constraint satisfaction problem defined by Eq. (11) and Eq. (12) since any point within these feasible regions satisfies all specified constraints. Meanwhile, the union of all feasible and indeterminate regions creates a *complete* representation of the solution space for the problem since all solution points are guaranteed to be included in these feasible and indeterminate regions.

3.2 The Planar Visualization Approach

All n-dimensional feasible, infeasible, and indeterminate regions classified during the iteration process are projected into one figure using a planar visualization approach illustrated in Figure 1. The n-dimensional regions classified by the method are shown in Figure 1(a), where red, yellow, and green regions represent infeasible, indeterminate, and feasible regions, respectively. There are two variables in Figure 1 (i.e., $n = 2$) for a clear illustration of the projection process, and the planar visualization approach is applicable to high-dimensional problems (e.g., $n > 3$) as demonstrated in the case study in Section 5. The classified regions shown in Figure 1(a) are projected into one figure with n horizontal axes for planar visualization, as shown in Figure 1(b). Each horizontal axis in Figure 1(b) corresponds to one variable. The allowable range of each variable defined by Eq. (12) is paved with red, yellow, and green blocks in Fig-

ure 1(b). The interpretations of these three types of blocks are presented as follows.

The length of a red block in the horizontal axis defines a specific range for the corresponding variable, where any variable value that belongs to the specific range cannot satisfy one or multiple constraints defined by Eq. (11) *regardless of* the values of other variables.

The length of a yellow block in the horizontal axis defines a specific range for the corresponding variable, where any variable value that belongs to the specific range may or may not satisfy all the constraints defined by Eq. (11) *regardless of* the values of other variables.

The length of a green block in the horizontal axis defines a specific range for the corresponding variable, where any variable value that belongs to the specific range satisfies all the constraints defined by Eq. (11) *only if* the values of other variables belong to certain ranges.

The planar visualization approach projects all classified n-dimensional regions into one figure, and practitioners can observe the ranges for variables where the solution for the continuous constraint satisfaction problem could exist from the green blocks in the figure. In addition, once practitioners specify a specific range for a variable within a green block, the ranges of other variables that lead to the feasible region(s) in the design space could be highlighted and connected by arrows in the planar visualization figure, as illustrated by region S_1 and region S_2 in Figure 1(b).

4. FOUR STEPS TO SOLVE CONTINUOUS CONSTRAINT SATISFACTION PROBLEMS

A GPU-based parallel bound-and-classify method is introduced in Section 3 for continuous constraint satisfaction problems. The method iteratively partitions the design space into many subregions and classifies these subregions as feasible, infeasible, and indeterminate regions. The classified regions are then projected into one figure using a planar visualization approach. In this section, the procedure to implement the GPU-

Copyright © 2023 by ASME

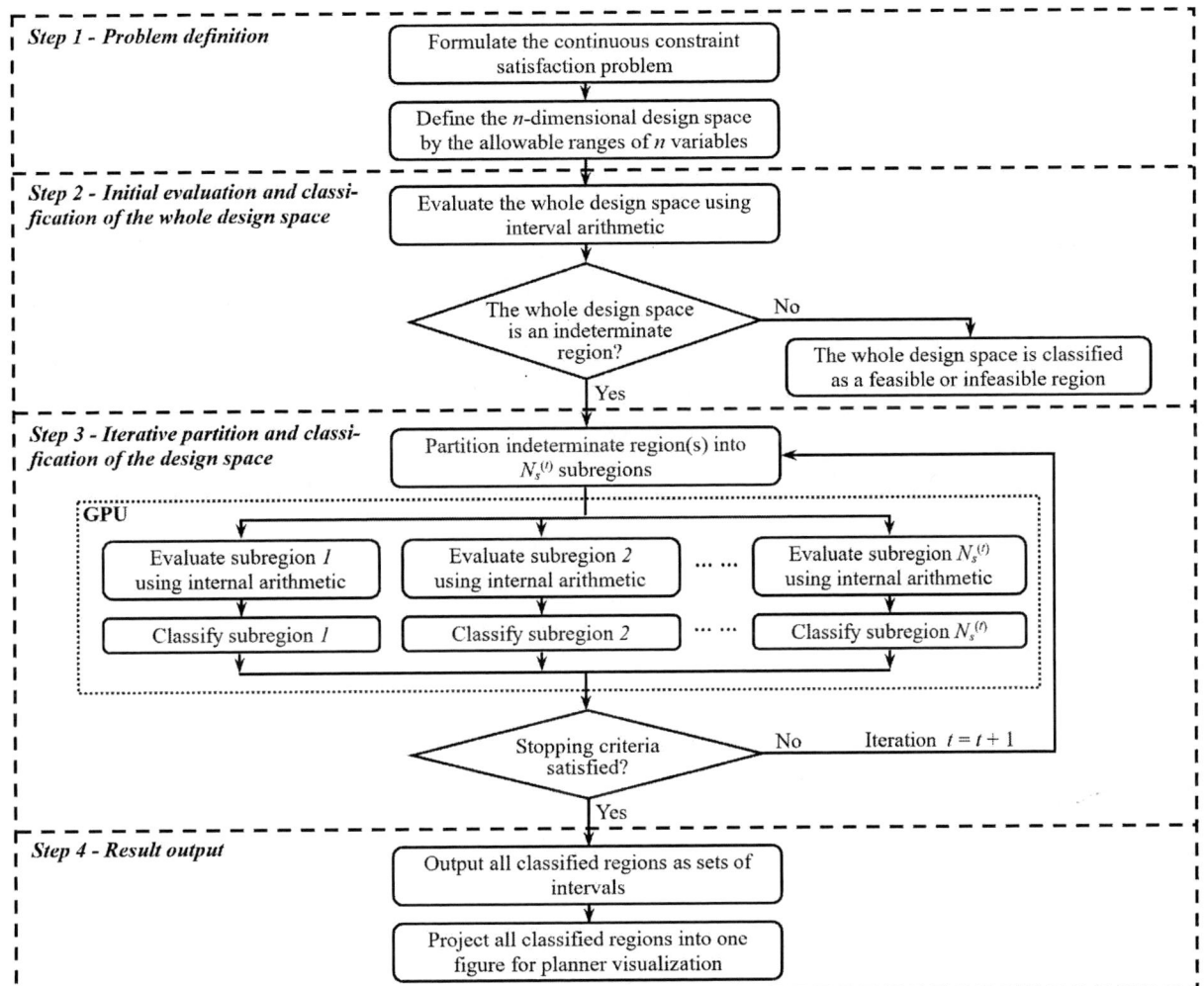

FIGURE 2: FOUR STEPS TO APPLY THE GPU-BASED PARALLEL BOUND-AND-CLASSIFY METHOD AND THE PLANAR VISUALIZATION APPROACH

based parallel bound-and-classify method and the planar visualization approach is summarized in four steps. These four steps are illustrated in Figure 2. Practitioners can follow these steps to solve continuous constraint satisfaction problems and visualize their results.

Step 1 - Problem definition: A continuous constraint satisfaction problem is defined and formulated based on Eq. (11) and Eq. (12). The allowable ranges of n variables define an n-dimensional design space. All constraints are formulated as inequality constraints with closed-form expressions. If equality constraint(s) exist, each equality constraint could be employed to eliminate a variable from the problem or replaced by two inequality constraints that are close to each other when such an approximation is acceptable.

Step 2 - Initial evaluation and classification of the whole design space: The constraint functions are evaluated within the whole design space using interval arithmetic. Based on the interval evaluations of all constraint functions, the whole design space is classified as a feasible, infeasible, or indeterminate region. If the whole design space is classified as a feasible or

infeasible region, practitioners need to check each constraint and consider tightening or relaxing certain constraint(s) if necessary.

Step 3 - Iterative partition and classification of the design space: If the whole design space is classified as an indeterminate region, the design space is iteratively partitioned into many subregions, and these subregions are classified as feasible regions, infeasible regions, and indeterminate regions. Specifically, at each iteration step, the indeterminate region(s) classified in the last iteration step are partitioned into many subregions using a user-specified partition strategy, as exemplified by Eqs. (13) and (14). The constraint functions are then evaluated within these subregions using interval arithmetic in parallel on a GPU. Each subregion is classified as a feasible, infeasible, or indeterminate region based on the interval evaluations of all constraint functions within the subregion. The iteration process is halted when the user-specified stopping criteria are satisfied.

Step 4 - Result output: When the iteration process is halted, all feasible, infeasible, and indeterminate regions classified during the iteration process are outputted as sets of intervals,

Copyright © 2023 by ASME

where each interval set corresponds to one classified region in the design space. Besides these sets of intervals, all classified regions are also projected into one figure for planar visualization. The green blocks in the figure indicate the ranges for variables where the solution for the continuous constraint satisfaction problem could exist.

5. CASE STUDY OF THE WELDED BEAM DESIGN

To demonstrate the four-step procedure outlined in Section 4, a continuous constraint satisfaction problem for welded beam design is employed as a case study in this section. The results of the case study validate the effectiveness and efficiency of the GPU-based parallel bound-and-classify method and the planar visualization approach. Using the method and the approach, the four-dimensional feasible regions of the problem are derived successfully as sets of intervals, and these feasible regions are also projected into one figure for planar visualization. The results outputted from the method and the approach facilitate the welded beam design concept generation and evaluation. The source code to implement the GPU-based parallel bound-and-classify method for this case study is included in the supplementary material.

The welded beam design is a benchmark engineering design problem [41, 42], where a rectangular beam is designed to be welded onto a primary structure, as illustrated in Figure 3. The original problem is adapted as a continuous constraint satisfaction problem including four design variables, $X = \{x_1, x_2, x_3, x_4\}$, where x_1 is the weld height, x_2 is the length of the beam that is welded onto the primary structure, x_3 is the height of the beam, and x_4 is the thickness of the beam. The allowable ranges of these four variables, $D = \{d_1, d_2, d_3, d_4\}$, are defined in the unit of inch as

$$\{d_1 : x_1 \in [0.001, 1]\}, \tag{15}$$

$$\{d_2 : x_2 \in [0.001, 8]\}, \tag{16}$$

$$\{d_3 : x_3 \in [5, 30]\}, \tag{17}$$

$$\{d_4 : x_4 \in [0.001, 1]\}. \tag{18}$$

Six constants involved in the welded beam design problem are provided in Table 1.

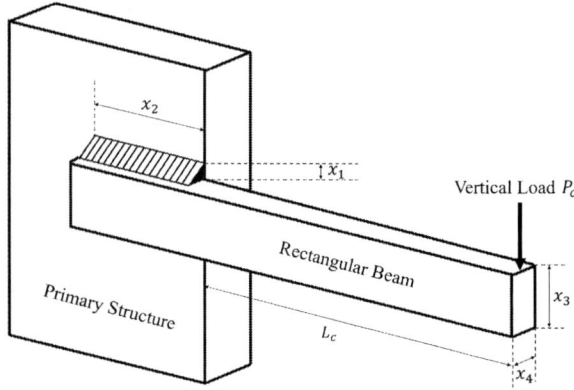

FIGURE 3: WELDED BEAM DESIGN PROBLEM

TABLE 1: SIX CONSTANTS INVOLVED IN THE WELDED BEAM DESIGN PROBLEM

	Definition	Value
P_c	The vertical load applied on the beam	6,000 pounds
L_c	Distance from load to the point of support	14 inches
E_c	Young's modulus of the beam material	30×10^6 psi
G_c	Shear modulus of the beam material	12×10^6 psi
S_1	Weld material cost per unit volume	0.10471 dollar/in^3
S_2	Beam material cost per unit volume	0.04811 dollar/in^3

Eight design constraints, $C = \{c_1, c_2, ..., c_8\}$, are defined by
1. The weld shear stress is less than or equal to 13,600 psi.

$$\{c_1 : g_1(X) = \tau(X) - 13600 \le 0\}, \tag{19}$$

$$\tau(X) = \sqrt{(\tau')^2 + 2\tau'\tau'' \frac{x_2}{2R} + (\tau'')^2}, \tag{20}$$

where τ' is the primary stress acting over the weld throat area defined by Eq. (21), and τ'' is the secondary torsional stress defined by Eqs. (22) - (25).

$$\tau' = \frac{P_c}{\sqrt{2}x_1x_2}, \tag{21}$$

$$\tau'' = \frac{MR}{J}, \tag{22}$$

$$M = P_c \left(L_c + \frac{x_2}{2}\right), \tag{23}$$

$$R = \sqrt{\frac{x_2^2}{4} + \left(\frac{x_1 + x_3}{2}\right)^2}, \tag{24}$$

$$J = 2\sqrt{2}x_1x_2 \left[\frac{x_2^2}{12} + \left(\frac{x_1 + x_3}{2}\right)^2\right]. \tag{25}$$

2. The normal stress in the beam is less than or equal to 30,000 psi.

$$\{c_2 : g_2(X) = \sigma(X) - 30000 \le 0\}, \tag{26}$$

$$\sigma(X) = \frac{6P_cL_c}{x_4x_3^2}. \tag{27}$$

3. The weld height cannot exceed beam thickness.

$$\{c_3 : g_3(X) = x_1 - x_4 \le 0\}. \tag{28}$$

4. The overall material cost cannot exceed $5.

$$\{c_4 : g_4(X) = S_1x_1^2x_2 + S_2x_3x_4(L_c + x_2) - 5 \le 0\}. \tag{29}$$

5. The weld height should be larger than or equal to 0.125 inches.

$$\{c_5 : g_5(X) = 0.125 - x_1 \le 0\}. \tag{30}$$

6. The maximum allowable beam-end deflection is 0.25 inches.

$$\{c_6 : g_6(X) = \delta(X) - 0.25 \le 0\}, \tag{31}$$

$$\delta(X) = \frac{4P_cL_c^3}{E_cx_3^3x_4}. \tag{32}$$

7. The 6,000 lb vertical load does not cause the beam to buckle.

$$\{c_7 : g_7(X) = 6000 - B(X) \le 0\}, \tag{33}$$

Copyright © 2023 by ASME

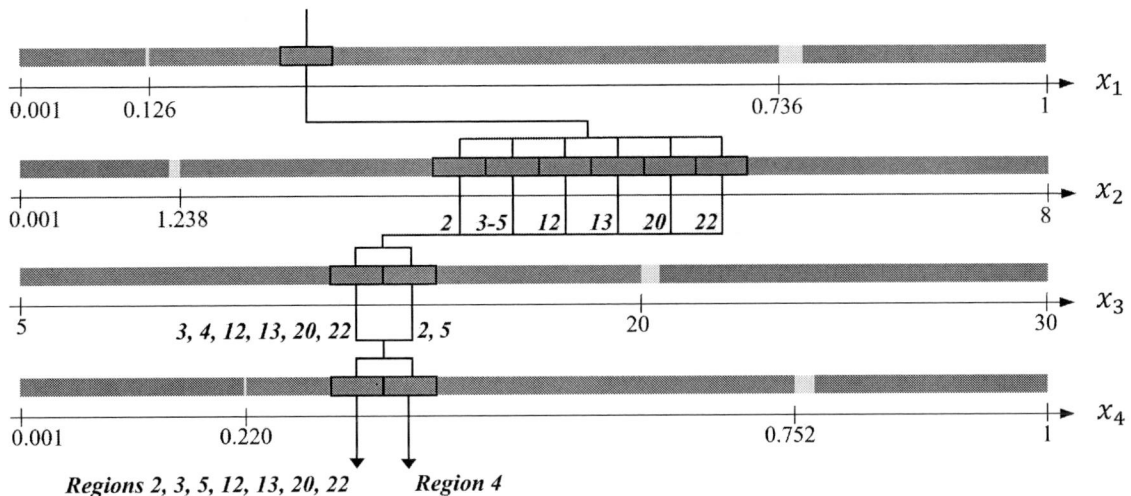

FIGURE 4: PLANAR VISUALIZATION FOR THE RESULT OF THE WELDED BEAM DESIGN CASE STUDY

where the beam buckling load, $B(X)$, is derived by

$$B(X) = \frac{4.013\sqrt{E_c G_c \frac{x_3^2 x_4^6}{36}}}{L_c^2} \left(1 - \frac{x_3}{2L_c}\sqrt{\frac{E_c}{4G_c}}\right). \quad (34)$$

8. The overall fabrication cost cannot exceed \$3.5.

$$\{c_8 : g_8(X) = F(X) - 3.5 \le 0\}, \quad (35)$$

$$F(X) = (1 + S_1)x_1^2 x_2 + S_2 x_3 x_4 (L_c + x_2). \quad (36)$$

To derive the feasible regions of the continuous constraint satisfaction problem defined by Eqs. (15) - (36), the GPU-based parallel bound-and-classify method presented in Section 3 and Section 4 is employed. The allowable ranges of the four variables given by Eqs. (15) - (18) define a four-dimensional design space for the problem. The whole design space is first evaluated using interval arithmetic. The code lists of the eight constraint functions, $g_j(X)$, are generated and provided in the supplementary material. The sequence of interval arithmetic operations in each code list yields the outwardly rounded lower and upper bounds for the value of the constraint function $g_j(X)$ within the design space, $[DLB_j, DUB_j]$, where $j \in \{1, 2, \ldots, 8\}$, and the whole design space is classified as an indeterminate region accordingly. The design space is then iteratively partitioned into many subregions, and these subregions are evaluated and classified as feasible regions, infeasible regions, and indeterminate regions in parallel on a GPU at each iteration step. The partition strategy defined by Eq. (13) and Eq. (14) in Section 3 is employed in this case study for demonstration purposes. At each iteration step, the splitting approach discussed in Section 2.2 is used to derive the sharper bounds for the value of the constraint function $g_j(X)$ within each subregion. For demonstration purposes, the stopping criterion is set as the smallest region width in all dimensions is less than 0.05 inches. The stopping criterion is satisfied after seven iteration steps, and a total of 192,065,215 classified regions are outputted as sets of intervals. Among these classified regions, there are 43,072,368 feasible regions, 59,961,905 infeasible regions, and 89,030,942 indeterminate regions. Notably, when the iterative process is halted,

the four-dimensional volume of the 89,030,942 indeterminate regions is less than 0.1% of the volume of the whole design space. The 43,072,368 sets of intervals for the feasible regions are provided in the supplementary material.

All the 192,065,215 four-dimensional feasible, infeasible, and indeterminate regions are also projected into one figure using the planar visualization approach presented in Section 3 and Section 4, as shown in Figure 4. Practitioners can observe the ranges for the four variables within which all eight constraints could be satisfied from the green blocks in the figure (i.e., $x_1 \in [0.126, 0.736]$, $x_2 \in [1.238, 8]$, $x_3 \in [5, 20]$, and $x_4 \in [0.220, 0.752]$ in Figure 4). In addition, once a specific range for a variable within a green block is specified by practitioners (e.g., $x_1 \in [0.251, 0.301]$ in Figure 4), the ranges of other variables that lead to the feasible regions derived from the GPU-based parallel bound-and-classify method (i.e., any of the 43,072,368 feasible regions) are connected by arrows, and practitioners can generate or evaluate welded beam design concepts accordingly. The eight sets of intervals corresponding to the feasible regions 2, 3-5, 12, 13, 20, and 22 in Figure 4 could be found in the supplementary material.

The effectiveness and efficiency of the GPU-based parallel bound-and-classify method could be compared with that of two GPU-based exhaustive search methods using the same computing resources. Here, the performance of the GPU-based parallel bound-and-classify method is not compared with that of extant CPU-based methods since a fair comparison cannot be made by using different computing resources (i.e., CPU vs. GPU in this case) and a GPU-based method is usually significantly faster than its CPU-based counterparts [19]. The sampling-based exhaustive search method covers the whole design space with sample points, and the distance between two adjacent sample points is 0.05 inches. The values of the eight constraint functions are calculated at these sample points in parallel on a GPU to check whether each sample point is a solution point. Notably, the sampling-based exhaustive search method is not able to determine whether the region enclosed by several sample points is a feasible region. In contrast, the region-based

Copyright © 2023 by ASME

TABLE 2: COMPUTATIONAL TIMES OF THREE GPU-BASED METHODS

Device	Sampling-based exhaustive search method	Region-based exhaustive search method	Parallel bound-and-classify method
Laptop with one GPU	5,032 seconds	~2,768,000 seconds	3,883 seconds
Server with one GPU	2,439 seconds	~446,000 seconds	431 seconds

exhaustive search method partitions the design space as four-dimensional hypercubes with an edge length of 0.05 inches. The regions enclosed by these hypercubes are evaluated and classified as feasible, infeasible, and indeterminate regions using interval arithmetic in parallel on a GPU. Importantly, the GPU-based parallel bound-and-classify method introduced in Section 3 is a region-based iterative search method. The GPU-based parallel bound-and-classify method iteratively partitions the design space into many subregions. These subregions are evaluated and classified as feasible, infeasible, and indeterminate regions in parallel on the GPU at each iteration step, and the iteration process is halted when the smallest region width in all dimensions is less than 0.05 inches. Using these three methods to solve the welded beam design problem defined by Eqs. (15) - (36), the results indicate that the GPU-based parallel bound-and-classify method derives the same feasible regions as the region-based exhaustive search method. In addition, using the same computing resource (a typical laptop with one GPU and a typical server with one GPU, respectively), the GPU-based parallel bound-and-classify method consumes significantly less computational time than both the sampling-based exhaustive search method and the region-based exhaustive search method. Specifically, as shown in Table 2, when the same server is employed as the computing resource, the GPU-based parallel bound-and-classify method consumes three orders less computational time than the region-based exhaustive search method but derives the same feasible regions for the problem; when compared to the sampling-based exhaustive search method, the GPU-based parallel bound-and-classify method only consumes 18% computational time but can derive feasible regions for the problem rather than only check whether the vertexes of each four-dimensional hypercube are solution points.

6. CONCLUSIONS AND FUTURE WORK

A GPU-based parallel bound-and-classify method is introduced to solve continuous constraint satisfaction problems. The novelty of this method is that the method iteratively partitions the design space into many subregions based on the compute capability of the practitioner's GPU; at each iteration step, the method evaluates and classifies many subregions in parallel on the GPU. A planar visualization approach that projects all classified regions in the design space into one figure is also proposed for continuous constraint satisfaction problems. The application of the GPU-based parallel bound-and-classify method

and the planar visualization approach is demonstrated through a case study of welded beam design.

Using the GPU-based parallel bound-and-classify method, practitioners can overcome the computational bottleneck of extant CPU-based methods and derive the feasible regions as sets of intervals for continuous constraint satisfaction problems in a reasonable computational time. All these feasible regions are also projected into one figure by using the planar visualization approach, and the figure facilitates practitioners' design concept generation and evaluation.

This work has limitations that offer opportunities for future research. As stated in Section 3, the GPU-based parallel bound-and-classify method introduced in this paper only applies to continuous constraint satisfaction problems with a finite set of inequality constraints, where each inequality constraint has a closed-form expression. Future work could extend the method to continuous constraint satisfaction problems with both equality and inequality constraints, where each constraint may not have a closed-form expression (e.g., differential equations could be involved in one or multiple constraints). In addition, the partition strategy defined by Eq. (13) and Eq. (14) is employed in the case study that involves four design variables (i.e., $n = 4$). Notably, the partition strategy defined by Eq. (13) and Eq. (14) can lead to a significant number of subregions that need to be evaluated and classified at each iteration step for high-dimensional continuous constraint satisfaction problems (e.g., $n > 20$). In future research, the partition strategy could be modified based on the compute capability of the practitioner's GPU for high-dimensional problems. For example, at each iteration step, each indeterminate region could be partitioned in l out of n dimensions rather than all its n dimensions, where $l < n$. The modified partition strategy could improve the efficiency of the GPU-based parallel bound-and-classify method when solving high-dimensional continuous constraint satisfaction problems.

ACKNOWLEDGMENTS

This material is partially supported by the Air Force Office of Scientific Research through grant FA9550-18-0088. Any opinions, findings, conclusions, or recommendations expressed in this material are those of the authors and do not necessarily reflect the views of the sponsor.

SUPPLEMENTARY MATERIAL

The code lists, the sets of intervals for the feasible regions, and the Python source code to implement the GPU-based parallel bound-and-classify method for the welded beam design case study are freely available for scientific research through the GitHub repository: https://github.com/CMU-Integrated-Design-Innovation-Group/PITSA.

REFERENCES

[1] Arróyave, R., Gibbons, S., Galvan, E., and Malak, R., 2016, "The Inverse Phase Stability Problem as a Constraint Satisfaction Problem: Application to Materials Design," JOM, 68(5), pp. 1385-1395.

Copyright © 2023 by ASME

[2] Swaney, R. E., and Grossmann, I. E., 1985, "An Index for Operational Flexibility in Chemical Process Design. Part I: Formulation and Theory," AIChE J., 31(4), pp. 621-630.

[3] Malan, S., Milanese, M., and Taragna, M., 1997, "Robust Analysis and Design of Control Systems Using Interval Arithmetic," Automatica, 33(7), pp. 1363-1372.

[4] Yvars, P. A., 2008, "Using Constraint Satisfaction for Designing Mechanical Systems," IJIDeM, 2(3), pp. 161-167.

[5] Galvan, E., Malak, R. J., Gibbons, S., and Arroyave, R., 2017, "A Constraint Satisfaction Algorithm for the Generalized Inverse Phase Stability Problem," J. Mech. Des., 139(1), p. 011401.

[6] Larson, B. J., and Mattson, C. A., 2012, "Design Space Exploration for Quantifying a System Model's Feasible Domain," J. Mech. Des., 134(4), p. 041010.

[7] Simpson, T. W., Poplinski, J., Koch, P. N., and Allen, J. K., 2001, "Metamodels for Computer-based Engineering Design: Survey and Recommendations," Eng. Comput., 17(2), pp. 129-150.

[8] Wang, G. G., and Shan, S., 2007, "Review of Metamodeling Techniques in Support of Engineering Design Optimization," J. Mech. Des., 129(4), pp. 370-380.

[9] Devanathan, S., and Ramani, K., 2010, "Creating Polytope Representations of Design Spaces for Visual Exploration Using Consistency Techniques," J. Mech. Des., 132(8), p. 081011.

[10] Gelle, E., Faltings, B. V., Clément, D. E., and Smith, I. F., 2000, "Constraint Satisfaction Methods for Applications in Engineering," Eng. Comput., 16(2), pp. 81-95.

[11] Lottaz, C., Clément, D. E., Faltings, B. V., and Smith, I. F., 1999, "Constraint-based Support for Collaboration in Design and Construction," J. Comput. Civ. Eng., 13(1), pp. 23-35.

[12] Granvilliers, L., and Benhamou, F., 2006, "Algorithm 852: Realpaver: An Interval Solver Using Constraint Satisfaction Techniques," ACM TOMS, 32(1), pp. 138-156.

[13] Vu, X.-H., 2005, "Rigorous Solution Techniques for Numerical Constraint Satisfaction Problems," Ph.D., EPFL, Lausanne, CH.

[14] Ratschan, S., 2006, "Efficient Solving of Quantified Inequality Constraints Over the Real Numbers," ACM TOCL, 7(4), pp. 723-748.

[15] Hu, J., Aminzadeh, M., and Wang, Y., 2014, "Searching Feasible Design Space by Solving Quantified Constraint Satisfaction Problems," J. Mech. Des., 136(3), p. 031002.

[16] Sam-Haroud, D., and Faltings, B., 1996, "Consistency Techniques for Continuous Constraints," Constraints, 1(1), pp. 85-118.

[17] Vu, X.-H., Sam-Haroud, D., and Silaghi, M.-C., 2002, "Numerical Constraint Satisfaction Problems with Non-isolated Solutions," Proc. International Workshop on Global Optimization and Constraint Satisfaction, pp. 194-210.

[18] Goldsztejn, A., and Granvilliers, L., 2010, "A New Framework for Sharp and Efficient Resolution of NCSP with Manifolds of Solutions," Constraints, 15(2), pp. 190-212.

[19] Owens, J. D., Houston, M., Luebke, D., Green, S., Stone, J. E., and Phillips, J. C., 2008, "GPU Computing," Proc. IEEE, 96(5), pp. 879-899.

[20] Apt, K., 2003, *Principles of Constraint Programming*, Cambridge University Press, Cambridge, UK.

[21] Tsang, E., 1993, *Foundations of Constraint Satisfaction*, Academic Press, London, UK.

[22] Rossi, F., Van Beek, P., and Walsh, T., 2006, *Handbook of Constraint Programming*, Elsevier, Amsterdam, NL.

[23] Kumar, V., 1992, "Algorithms for Constraint-Satisfaction Problems: A Survey," AI Mag., 13(1), pp. 32-32.

[24] Brailsford, S. C., Potts, C. N., and Smith, B. M., 1999, "Constraint Satisfaction Problems: Algorithms and Applications," Eur. J. Oper. Res., 119(3), pp. 557-581.

[25] Chaloner, K., and Verdinelli, I., 1995, "Bayesian Experimental Design: A Review," Stat. Sci., pp. 273-304.

[26] Morris, M. D., and Mitchell, T. J., 1995, "Exploratory Designs for Computational Experiments," J. Stat. Plan. Inference, 43(3), pp. 381-402.

[27] Pronzato, L., and Müller, W. G., 2012, "Design of Computer Experiments: Space Filling and Beyond," Stat. Comput., 22(3), pp. 681-701.

[28] Hansen, E., and Walster, G. W., 2004, *Global Optimization Using Interval Analysis*, Marcel Dekker, Inc., New York.

[29] Benhamou, F., Goualard, F., Granvilliers, L., and Puget, J., 1999, "Revising Hull and Box Consistency," Proc. International Conference on Logic Programming, pp. 230-244.

[30] Sam, J., 1995, "Constraint Consistency Techniques for Continuous Domains," Ph.D., EPFL, Lausanne, CH.

[31] Han, H., Chang, S., and Kim, H., 2017, "A Systematic Approach to Identifying a Set of Feasible Designs," ASME DETC2017-68003.

[32] Moore, R. E., Kearfott, R. B., and Cloud, M. J., 2009, *Introduction to Interval Analysis*, SIAM, Phila., PA.

[33] Jaulin, L., Kieffer, M., Didrit, O., and Walter, E., 2001, *Applied Interval Analysis*, Springer, London, UK.

[34] IEEE 1788-2015 Standard for Interval Arithmetic.

[35] Moore, R. E., 1962, "Interval Arithmetic and Automatic Error Analysis in Digital Computing," Ph.D., Stanford University, CA.

[36] IEEE 754-2019 Standard for Floating-Point Arithmetic.

[37] Kulisch, U. W., and Miranker, W. L., 1981, *Computer Arithmetic in Theory and Practice*, Academic Press, New York.

[38] Ratschek, H., and Rokne, J., 1984, *Computer Methods for the Range of Functions*, Halsted Press, New York.

[39] Rokne, J. G., 1986, "Low Complexity k-dimensional Centered Forms," Computing, 37(3), pp. 247-253.

[40] Neumaier, A., 1990, *Interval Methods for Systems of Equations*, Cambridge University Press, Cambridge, UK.

[41] Ragsdell, K., and Phillips, D., 1976, "Optimal Design of a Class of Welded Structures Using Geometric Programming," J. Eng. Ind., 98(3), pp. 1021-1025.

[42] Rao, S. S., 2009, *Engineering Optimization: Theory and Practice*, John Wiley & Sons, Hoboken, NJ.

Copyright © 2023 by ASME

Proceedings of the ASME 2023
International Design Engineering Technical Conferences and
Computers and Information in Engineering Conference
IDETC-CIE2023
August 20-23, 2023, Boston, Massachusetts

DETC2023-113451

MULTI-MATERIAL AND MULTI-JOINT TOPOLOGY OPTIMIZATION CONSIDERING MULTIPLE DESIGN SPACES

Il Yong Kim[1*], Yuhao Huang[1], Luke Crispo[1]
[1]Queen's University, Kingston, ON, Canada
*kimiy@queensu.ca

ABSTRACT

Today, automakers are focusing on cost reduction and lightweighting, by utilizing a combination of different materials in the vehicle design. Topology optimization is a numerical tool that provides more design freedom than other methods, such as size optimization and shape optimization. Multi-material topology optimization can optimize both material layout and material distribution to improve structural performance. However, these methods assume that dissimilar materials are perfectly bonded, which limits the manufacturability of the design. This work presents a multi-material and multi-joint topology optimization methodology that considers additional design variables for joints in the material interface region. The mechanical properties of joints are included in the analysis, which affects the overall structural behavior and the optimized result. This paper firstly introduces topology optimization methods and material interpolation functions for multiple design spaces. Then, the material interface region detection method is explained. After that, sensitivity analysis for different responses is conducted. Lastly, the results of some example models demonstrate this methodology can be used to control mass and joining cost in multiple design spaces within a structure.

Keywords: Multi-material topology optimization; multi-joint topology optimization; joint design; structural optimization

1. INTRODUCTION

Topology optimization (TO) has become an active research area in both academic and industry since the introduction by Bendsøe and Kikuchi in 1988 [1]. TO is a numerical tool used to determine the (local) optimal material distribution within the design space for certain problem statements. In the density approach, the design variables are the pseudo density of each element and ranged between 0 and 1, where 0 indicates a void

element and 1 indicates a solid element. The solid isotropic material with penalization (SIMP) method is used to interpolate material properties from void to solid [2]. Single material topology optimization (SMTO) can be extended to multi-material topology optimization (MMTO) to improve design freedom, and therefore yield better solutions. However, MMTO assumes that the interface between dissimilar materials is perfectly bonded together, which is not realistic from a manufacturing perspective. Shah et al. added a cost constraint for interface regions, but the mechanical properties of joints were not simulated [3]. The first work that considered joint mechanical properties was done by Woischwill and Kim, where the optimization was decomposed into two subproblems, MMTO and multi-joint topology optimization (MJTO) [4]. However, this method was limited to 2D geometry. Florea et al. extended this method to 3D geometry and considered tooling accessibility constraints [5]. However, the methods used by Woischwill and Kim, and Florea et al. had two limitations. Firstly, a specialized finite element mesh was required. Secondly, the optimization process was decoupled into MMTO and MJTO, which was not a simultaneous multi-material and multi-joint topology optimization (MM-MJ-TO). Theoretically, simultaneous optimization would converge to a better result, compared to decoupled optimization. After that, Florea et al. integrated spatial gradient into SIMP formulation and developed simultaneous MM-MJ-TO [6]. In this method, a specialized mesh was not needed. Also, Florea et al. demonstrated a strong dependence between optimal joint design and optimal structural material design. Crispo et al. proposed a new spatial gradient computation method and studied the effect of joint costs on the structural material distribution [7].

The previous MM-MJ-TO framework was limited to single design space. However, industry models often contain multiple design spaces. This paper extends the MM-MJ-TO framework to

Copyright © 2023 by ASME

multiple design spaces, which allows users to control the distribution of the material and joint cost fractions in different locations within the structure.

A design space is defined as a set of elements specified by the user (i.e., a list of non-ordered unique element IDs). Each design space can have its own material candidates and manufacturing methods. For simplicity, in this paper, all design spaces use same material candidates with MM-MJ-TO as interpolation scheme. Certain responses, such mass fraction or cost fraction, can reference a specific design space. Compliance response must reference all elements. It should be noted that elements may be shared between design spaces.

2. MATERIALS AND METHODS
2.1 Problem Statement
The problem statement for MM-MJ-TO is shown in (1). The objective is to minimize total compliance of all elements $C(x)$, and the problem is subjected to a set of mass fraction constraints $g_i^1(x)$ and joint cost fraction constraints $g_i^2(x)$. The design variables will control the layout and distribution of two structural materials and two joint materials. Where material 1 and material 2 are the stiffer and weaker structural material candidates, respectively; and material 3 and material 4 are the stiffer and weaker joint material candidates respectively. The geometry is discretized into a finite number of elements. In the designable regions, each element has four design variables, where x_j^1 and x_j^2 determine the existence and selection of structural materials, respectively; x_j^3 and x_j^4 determine the existence and selection of joint materials, respectively.

$$\underset{x}{\text{minimize}} \quad C(x) = u^T K u$$

$$\text{subject to} \quad K u = f$$

$$g_i^1(x) = \frac{1}{\sum_{j \in N_i} v_j \rho^1} \sum_{j \in N_i} \left(v_j \sum_{k=1}^{4} w_j^k \rho^k \right) \leq \bar{M}_i$$

$$g_i^2(x) = \frac{1}{\sum_{j} v_j} \sum_{j \in N_i} \left(v_j \sum_{k=3}^{4} w_j^k q^k \right) \leq \bar{Q}_i$$

$$x_j^k \in (0,1], \quad k = 1, 2, 3, 4$$

$$i = 1, \dots, M$$

The model also needs to satisfy a linear static equation, where K is the global stiffness matrix, u is the displacement vector, and f is the applied force vector. Both K and u depend on all design variables. There are total $2M$ constraints, where each constraint is imposed on a design space set N_i, with $\bar{M}_i$ as the upper bound for i-th mass fraction constraint, $\bar{Q}_i$ as the upper bound for the i-th joint cost fraction constraint, v_j is the volume of j-th element, and ρ^k is the density of the k-th material candidate. For the joining cost fraction constraint, q^k defines the joint cost of the k-th material candidate, and is only valid for the joint materials. The mass weighting factor w_j^k represents the amount of each k-th material candidate in each j-th element, with

values ranging from 0 (void) to 1 (solid), and is interpolated from the design variables.

2.2 Material Interpolation
The interpolated Young's modulus E_j can be expressed as the weighted sum of Young's modulus of four material candidates in (2), where ω_j^k is the stiffness weighted factor for j-th element and k-th material, and E^k is the stiffness of the k-th material. In the converged result, each element should only contain one material, which means one weighting factor is 1 while the other three are 0. Intermediate densities are discouraged through a penalty factor p with values between 3 and 5. The interface existence variable n_j depends on the structural material distribution.

$$E_j(x_j^k) = \sum_{k=1}^{4} \omega_j^k E^k$$

$$\omega_j^1 = (1 - n_j)(x_j^1)^p (x_j^2)^p$$

$$\omega_j^2 = (1 - n_j)(x_j^1)^p (1 - (x_j^2)^p)$$

$$\omega_j^3 = n_j (x_j^3)^p (x_j^4)^p$$

$$\omega_j^4 = n_j (x_j^3)^p (1 - (x_j^4)^p)$$

Similarly, (3) shows the interpolation for density and the expression for weighting factor w_j^k. Penalization is not needed here.

$$\rho_j(x_j^k) = \sum_{k=1}^{4} w_j^k \rho^k$$

$$w_j^1 = (1 - n_j) x_j^1 x_j^2$$

$$w_j^2 = (1 - n_j) x_j^1 (1 - x_j^2)$$

$$w_j^3 = n_j x_j^3 x_j^4$$

$$w_j^4 = n_j x_j^3 (1 - x_j^4)$$

If all materials are isotropic and are assumed to have the same Poisson's ratio, then the interpolated element stiffness matrix $K_j(x_j^k)$ can be calculated using (4), where K_j^k is the element stiffness matrix for solid candidate material. If the Poisson's ratios are slightly different, then the average Poisson ratio could be used as an approximation. For anisotropic materials or very different Poisson's ratios, discrete material optimization (DMO) is recommended [8].

$$K_j(x_j^k) = E_j \frac{K_j^1}{E^1} = E_j \frac{K_j^2}{E^2} = E_j \frac{K_j^3}{E^3} = E_j \frac{K_j^4}{E^4}$$

The element stiffness matrix K_j is resized into global dimension by inserting empty columns and rows at correct degrees of freedom. The global stiffness matrix K is the

Copyright © 2023 by ASME

summation of all resized matrices, $\underset{\sim}{K}_j^R$, including both designable and non-designable elements, N.

$$\underset{\sim}{K} = \sum_{j \in N} \underset{\sim}{K}_j^R \tag{5}$$

2.3 Interface Detection

Interfaces between materials are identified through a spatial gradient calculation of the material selection design variable field $\underset{\sim}{x}^2$ using an unstructured approach [9]. The spatial gradient in the m-th spatial direction $\nabla_m x_j^2$ is calculated using (6), where h_{ji} is a distance term between the j-th and i-th element centroids and θ_{mji} is the angle between the m-th direction and a vector from the j-th to the i-th element centroid. The summation is calculated over the N_j neighbouring elements within a search radius r. The magnitude of the spatial gradient is then calculated.

$$\nabla_m x_j^2 = \frac{1}{\sum\limits_{i \in N_j} h_{ji} v_i} \sum_{i \in N_j} h_{ji} v_i \left(x_i^2\right)^p \cos\theta_{mji}$$

$$h_{ji} = \max\left(0, r - \text{dist}(j,i)\right) \tag{6}$$

$$\left\|\nabla_m x_j^2\right\| = \sqrt{\sum_{m=1}^{3} \nabla_m x_j^2}$$

The spatial gradient magnitude is thinned and projected in (7) to calculate the final interface term used in the material interpolation schemes. The β projection parameter controls the slope and η controls the horizonal shift of the Heaviside function. This work uses projection parameters of $\beta = 70$ and $\eta = 0.02$. After the optimization, all interface values of $n_j > 0.8$ are considered to be joining elements.

$$n_j = \frac{1}{1 + e^{-2\beta\left[\left(1 - x_j^2\right)\left\|\nabla_m x_j^2\right\| - \eta\right]}} \tag{7}$$

In this paper, only the interface between different structural materials will be identified for joint placement. The boundary between design spaces will not be considered to require joints.

2.1 Sensitivity Analysis

Sensitivity for the total compliance including all designable and non-designable elements is shown in (8).

$$\frac{\partial C}{\partial x_j} = -\underset{\sim}{u}^T \frac{\partial \underset{\sim}{K}}{\partial x_j} \underset{\sim}{u} \tag{8}$$

The expression for the stiffness matrix can be derived as shown in (9).

$$\frac{\partial \underset{\sim}{K}}{\partial x_j} = \frac{K_j}{E_j} \frac{\partial E_j}{\partial x_j} \tag{9}$$

The sensitivity of the mass fraction and joint cost fraction constraints g_i^1 and g_i^2 is shown in (10).

$$\frac{\partial g_i^1}{\partial x_j} = \frac{1}{\sum\limits_{l \in N_i} v_l \rho^1} \sum_{l \in N_i} \left(v_l \sum_{k=1}^{4} \frac{\partial w_l^k}{\partial x_j} \rho^k \delta_{lj}\right)$$

$$\frac{\partial g_i^2}{\partial x_j} = \frac{1}{\sum\limits_{j \in N_i} v_j} \sum_{l \in N_i} \left(v_l \sum_{k=3}^{4} \frac{\partial w_l^k}{\partial x_j} q^k \delta_{lj}\right) \tag{10}$$

$$\delta_{lj} = \begin{cases} 1 & \text{if } l = j \\ 0 & \text{if } l \neq j \end{cases}$$

The sensitivities of the material, mass, and cost interpolation schemes, and for interface detection calculations are explained in detail in [7] and are not presented in this work.

3. RESULTS AND DISCUSSION

The MM-MJ-TO is implemented through in-house code using the MMA optimizer [10], with the FEA analysis performed in Altair OptiStruct.

3.1 Cantilever Beam

The 2D cantilever beam model is shown in Figure 1. The left edge is constrained in all degrees of freedom, and a force of 1000 N is applied on the middle node of the right edge in downwards direction. The geometry is discretized into 40 × 200 CQUAD4 elements. The left half of the model is design space 1 (DS 1) and the right half of the model is design space 2 (DS 2).

FIGURE 1: 2D CANTILEVER DESIGN SPACES AND BOUNDARY CONDITIONS

The two structural materials are titanium and aluminum, while the two joint materials are weld and adhesive. The material properties are shown in Table 1. The Poisson's ratio for all materials is set to be 0.3. In real life models, the joint interface regions are thinner than the element size, therefore homogenization is needed to approximate joint properties. Since the purpose of this paper is to demonstrate the methodology, the joints are assumed to be isotropic and with arbitrarily defined properties. A methodology for joint material modelling using numerical homogenization can be found in [11].

Four case studies have been conducted with various mass fraction constraints without using any joint cost fraction constraints. The optimized results are shown in Figure 2. The amount of each material in the optimized geometry is reported as a mass fraction (MF) in percentage value calculated relative to a design space of solid titanium.

Copyright © 2023 by ASME

TABLE 1: MATERIAL PROPERTIES USED FOR CANTILEVER BEAM PROBLEM

Material Property	Titanium	Aluminum	Weld	Adhesive
Stiffness E^k (GPa)	110	70	50	30
Density ρ^k (g/cm³)	4.5	2.7	3.6	2.4

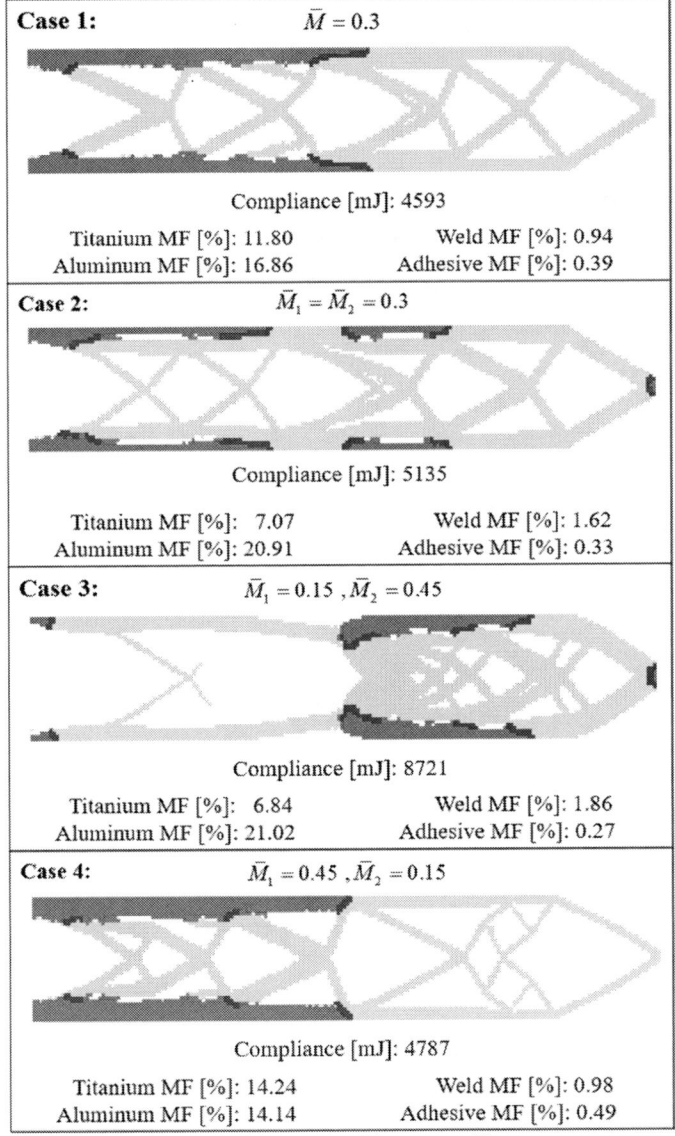

Titanium ⬛ **Aluminum** ☐ **Weld** ⬛ **Adhesive** ⬛

FIGURE 2: CANTILEVER BEAM CASE STUDIES DEMONSTRATING THE EFFECT OF VARYING MASS FRACTION (MF) CONSTRAINT LIMITS

In case 1, the problem is solved with a single design space with a 30% mass fraction constraint on all elements. In case 2, both DS 1 and DS 2 are subjected to a 30% mass fraction constraint, resulting in an overall equivalent mass, but with an even distribution between the design spaces. In case 3 and 4, DS 1 and 2 are subjected to two combinations of 15% and 45% mass fraction constraints, resulting in an overall resultant mass fraction of 30%.

Titanium, the stronger material, is placed on the top and bottom sides of the cantilever beam (and the load applied location in some cases), because these regions have high compliance and need to be reinforced. All joints are located between titanium and aluminum, which is expected.

Case 1 has 11.8% better objective (compliance) than case 2 because the additional mass fraction constraint in case 2 reduces the design freedom. However, case 2 has a more uniform distribution of material. Case 3 has the worst objective because less material is placed on the left where the bending stress is highest and should be strengthened. Case 4 has a similar objective value and topology to case 1.

Design space mass fraction constraints influence the material distribution and the usage of joints within the structure. For example, Case 2 and Case 3 require more joints because there are five distinct titanium regions present in the solution (compared to two distinct regions in Case 1), even though the mass of titanium is smaller in Case 2 and Case 3.

It should be noted that some aluminum members are disconnected in Case 3. This is because the figures present thresholded material types instead of the intermediate design variables values. The disconnected members in Case 3 were below the threshold value and therefore thresholded to void.

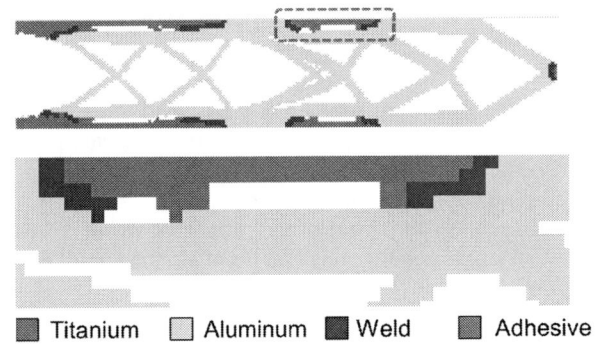

⬛ Titanium ☐ Aluminum ⬛ Weld ⬛ Adhesive

FIGURE 3: PLACEMENT OF JOINT MATERIALS IN MULTI-MATERIAL CONNECTION UNDER TENSILE LOAD

Figure 3 shows a closeup view of a tension member from Case 2 result. The optimizer uses the joint selection design variable to place most of weld (the stiffer joint) near the end points of the titanium member, and adhesive joints closer to the middle. This configuration is efficient for transferring tension load between titanium and aluminum members.

3.2 Truck Chassis Frame

The proposed method is applied to a truck chassis frame model to demonstrate its effectiveness on practical problems with complex loading and geometry. In addition, joining cost fraction constraints are introduced to this problem. The model

Copyright © 2023 by ASME

shown in Figure 4 is meshed using approximately 180,000 hexahedral elements with side lengths of about 30mm divided into three design spaces (DS). DS 1, DS 2, and DS 3 represent the frame, cab, and bed of the truck, respectively, which are typically manufactured and assembled individually before a final assembly stage. This division of the geometry allows for the overall structure to be optimized while providing more control over the mass and joining cost in specific regions. It is important to note this model does not have a small enough element size to generate any small-scale features of the truck assembly and should instead be used as a first step in the conceptual design process to identify material distribution, followed by further stages of optimization on individual components with a refined mesh.

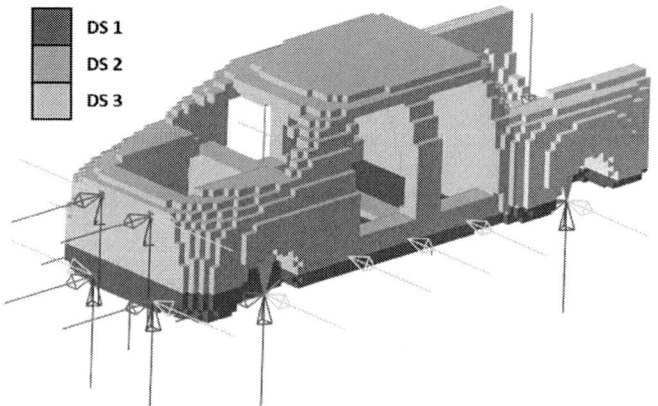

FIGURE 4: FINITE ELEMENT MODEL FOR TRUCK CHASSIS FRAME WITH THREE DESIGN SPACES (DS)

A unique load is applied in 35 different load steps, each with an arbitrary magnitude of 1000 N distributed using RBE3 elements. The compliance objective function is calculated as a weighted sum of the compliance of each load step. The four points representing the wheels are connected using an RBE3 element to a single central node that is constrained in all degrees of freedom.

The truck chassis frame problem is solved using steel and aluminum candidate materials, as well as weld and adhesive joint candidate materials. The associated material properties are summarized in Table 2. As with the cantilever beam problem, the Poisson's ratio for all materials is set to 0.3 and approximate joint properties are determined assuming homogenized material properties along interface elements. Joint cost values are arbitrarily defined with a value of 2 for welds and 1 for adhesives. These values represent a relative magnitude of joining cost between candidate joint materials, and therefore the exact values of joining cost are not physically meaningful.

As shown in Case 1 of Figure 5, the truck model was first solved without joint cost constraints using mass fraction values of 10% for DS 1 and DS 3 and 30% for DS 2. This solution placed steel around all load application points due to the high compliance in these regions. In addition, steel regions connect the two bottom steel members beneath the middle of the

passenger compartment and reinforce from the front to back of the truck chassis through the A and C pillars. In this problem statement without cost, welds are placed at all interfaces between the steel and aluminum components.

TABLE 2: MATERIAL PROPERTIES USED FOR TRUCK CHASSIS FRAME PROBLEM

Material Property	Steel	Aluminum	Weld	Adhesive
Stiffness E^k (GPa)	210	70	112	42
Density ρ^k (g/cm³)	7.85	2.7	5.3	4.2
Joint Cost q^k ($/cm³)	-	-	2	1

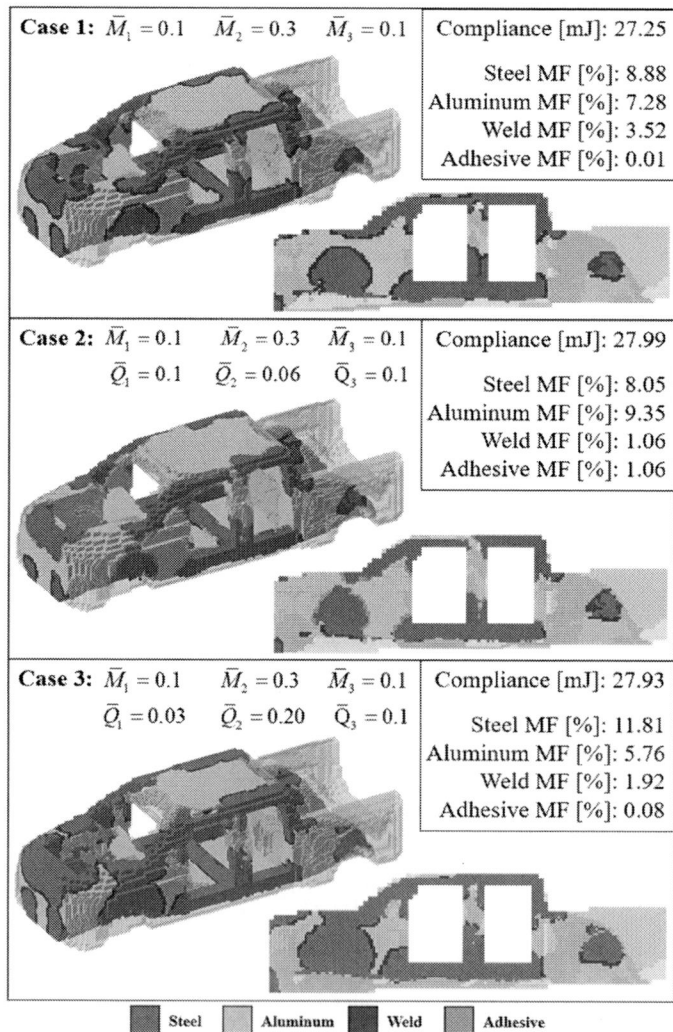

FIGURE 5: OPTIMIZATION RESULTS FOR TRUCK CHASSIS FRAME STUDY WITH VARYING JOINT COST FRACTION CONSTRAINT LIMITS

Copyright © 2023 by ASME

In Case 2, a loose joint cost fraction constraint of 10% is introduced for DS 1 and DS 3 and a tight joint cost fraction constraint is assigned to DS 2. The joint cost fraction constraint functions as a volume fraction constraint applied only to joint materials, with different cost weighting factors applied as per Table 2. The optimized result in Case 2 has a similar structure to Case 1 in DS 1 and DS 3, as the joint cost constraints are not limiting the design ($g_1^2 = 0.046$, $g_3^2 = 0.048$). In DS 2, the distribution of steel and aluminum is similar to Case 1, while the joint distribution is changed drastically to meet the cost fraction constraint. Joint elements are removed from many of the interface regions (indicating no connection to be used in that location) and adhesives instead of welds are used to connect between structural materials. The additional constraint results in a 3% increase in compliance compared to Case 1. Looking at the mass fractions of these designs, Case 2 uses 70% less weld material and increases the adhesive use to satisfy the joining cost constraint. The structural material distribution also changes with a 28% increase in aluminum in Case 2.

The final problem statement in Case 3 uses a strict joint cost constraint of 3% in DS 1 and looser joint cost constraints in DS 2 and DS 3. In this case, the optimization converges to a different distribution of structural material, with more aluminum material added beneath the engine area. There is also 30% more steel used compared to Case 1, with the majority added to DS 1. These changes occur because the optimization converges to a different local optimum solution.

The changes to DS 1 are highlighted in Figure 6. Note that the floating elements shown in Case 3 are connected to material from the design spaces hidden from this image. The results are not symmetric due to small rounding errors that build up over the course of the optimization. This problem is more prevalent in the MM-MJ-TO approach due to the steep Heaviside thresholding apparent in interface elements, which amplify small numerical differences.

Case 1:

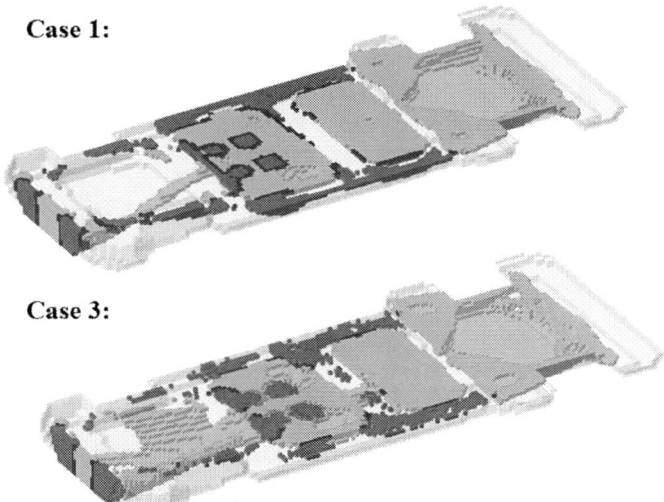

Case 3:

FIGURE 6: COMPARISON OF TRUCK FRAME STRUCTURE IN DS 1 FOR CASE 1 AND CASE 3 RESULTS

4. CONCLUSION

This work presents a multi-material and multi-joint topology optimization methodology with the capability to restrict mass and cost fraction across multiple design spaces. The approach was applied on an academic cantilever beam problem to show the implications of using more than one design space in MM-MJ-TO problems. The truck chassis frame model demonstrates the effectiveness of this approach to reduce the joining material in specific regions of the geometry.

In terms of future work, the interface between the design space and non-design space should be considered. Adding the capability to consider manufacturing constraints, such as extrusion, symmetry, or casting, would further improve the manufacturability of the design. Additionally, supporting more optimization problem statements, with the inclusion of displacement or stress constraints, would expand the scope of the tool.

REFERENCES

[1] Bendsøe, Martin Philip and Kikuchi, Noboru. "Generating optimal topologies in structural design using a homogenization method." *Computer Methods in Applied Mechanics and Engineering* Vol. 71 No. 2 (1988): pp. 197–224. DOI 10.1016/0045-7825(88)90086-2.

[2] Bendsøe, Martin Philip. "Optimal shape design as a material distribution problem." *Structural Optimization*, Vol. 1 No. 4, (1989): pp. 193–202. DOI 10.1007/BF01650949.

[3] Shah, Vishrut, Pamwar, Manish, Sangha, Balbir and Kim, Il Yong. "Material interface control in multi-material topology optimization using pseudo-cost domain method." *International Journal for Numerical Methods in Engineering* Vol. 122 No. 2 (2020): pp. 455–482. DOI 10.1002/nme.6545.

[4] Woischwill, Christopher and Kim, Il Yong. "Multimaterial multijoint topology optimization." *International Journal for Numerical Methods in Engineering* Vol. 115 No. 13 (2018): pp. 1552–1579. DOI 10.1002/nme.5908.

[5] Florea, Vlad, Pamwar, Manish, Sangha, Balbir and Kim, Il Yong. "3D multi-material and multi-joint topology optimization with tooling accessibility constraints." *Structural and Multidisciplinary Optimization* Vol. 60 No. 6 (2019): pp. 2531–2558. DOI 10.1007/s00158-019-02344-1.

[6] Florea, Vlad, Pamwar, Manish, Sangha, Balbir and Kim, Il Yong. "Simultaneous single-loop multimaterial and multijoint topology optimization." *International Journal for Numerical Methods in Engineering* Vol. 121 No. 7, (2020): pp. 1558–1594.

[7] Crispo, Luke, Roper, Stephen William Knox, Bohrer, Rubens, Morin, Rosalie and Kim, Il Yong. "Multi-material and multi-joint topology optimization for lightweight and cost-effective design." *Proceedings of the ASME IDETC/CIE.* DETC2021-67317. Virtual, Online, August 17–19, 2021. DOI 10.1115/DETC2021-67317.

[8] Li, Daozhong and Kim, Il Yong. "Modified element stacking method for multi-material topology optimization with anisotropic materials." *Structural and Multidisciplinary*

Copyright © 2023 by ASME

Optimization Vol. 61 No. 2 (2020): pp. 525-541. DOI 10.1007/s00158-019-02372-x.

[9] Crispo, Luke, Bohrer, Rubens, Roper, Stephen William Knox and Kim, Il Yong. "Spatial gradient interface detection in topology optimization for an unstructured mesh." *Structural and Multidisciplinary Optimization* Vol. 63 No. 1 (2021): pp. 515–522. DOI 10.1007/s00158-020-02688-z.

[10] Svanberg, Krister. "The method of moving asymptotes–a new method for structural optimization." *International Journal for Numerical Methods in Engineering* Vol. 24 No. 2 (1987): pp. 359–373. DOI 10.1002/nme.1620240207.

[11] Sirola, Tim. "Multi-Joint Topology Optimization for Stiffness Constrained Design Problems." MASc Thesis. Queen's University, Kingston, Canada. 2022. URL hdl.handle.net/1974/31332.

Proceedings of the ASME 2023
International Design Engineering Technical Conferences and
Computers and Information in Engineering Conference
IDETC-CIE2023
August 20-23, 2023, Boston, Massachusetts

DETC2023-115068

ROBUST TOPOLOGY OPTIMIZATION OF SYNCHRONOUS RELUCTANCE MOTORS USING CARDINAL BASIS FUNCTION BASED LEVEL SET METHOD

Jiawei Tian[1], David Torrey[2], Fang Luo[3], Jon Longtin[1], Shikui Chen[1,*]

Department of Mechanical Engineering[1]
Department of Electrical and Computer Engineering[3]
State University of New York at Stony Brook, Stony Brook, New York, USA, 11794
GE Research[2], Niskayuna, New York, USA, 12309
Email: {Jiawei.Tian, Fang.Luo, Jon.Longtin, Shikui.Chen}@stonybrook.edu;
torrey@ge.com

ABSTRACT

Synchronous reluctance motors (SynRMs) have gained considerable attention in the field of electric vehicles as they reduce the need for permanent magnets in the rotor, resulting in less material and manufacturing costs. However, their lower average torque and torque ripple vibrations have been identified as key issues that require resolution. In this study, we present a SynRM design framework employing the cardinal basis functions (CBF)-based parametric level set method. The SynRms design problem is recast as a variational problem constrained by Maxwell's equations which describe the behavior of electric and magnetic fields in the SynRM. A continuum shape sensitivity analysis is carried out using the material derivative and adjoint method. A distance regularization energy function is employed to maintain the level set function as a signed distance function during the optimization. The parametric topology optimization problem is computationally solved using the Method of Moving Asymptotes (MMA). To demonstrate the effectiveness of our approach, we present a numerical example that compares the torque characteristics of the optimal design with those of a reference design. Preliminary results show that the optimized SynRM has a 30.30% increase in average torque, along with a slight increase in torque ripple, compared to the reference model.

1 Introduction

The recent focus in motor design has shifted towards increasing efficiency, reducing cost, and improving environmental friendliness [1]. Having no permanent magnets in the rotor, synchronous reluctance motors (SynRMs) can offer lower material and manufacturing costs [2]. In view of this, synchronous reluctance motors are an attractive option for industrial applications such as pumps, fans, traction, and electric vehicles.

In contrast to the permanent magnet (PM) machines [3], synchronous reluctance motors generate torque utilizing magnetic reluctance variation [4]. While SynRMs avoids the usage of PM in its structures, their limitations can not be ignored. One major disadvantage of SynRMs is their relatively lower average torque compared to other types of motors, including induction motor (IM) [5], permanent magnet synchronous motor (PMSM) [6]. Additionally, SynRMs suffer from torque ripple due to the interaction between the spatial harmonics of magnetomotive force (MMF) and the rotor geometry [7]. Theoretically, the torque performance of SynRM depends on the excitation current and the motor design itself [8]. Consequently, there is a need to identify the optimal layout of the rotor to improve the torque characteristics. Traditionally, design optimization of electric machines is based on the parameterized studies [9], which focus on finding the optimized shape and size, providing limited flexibility.

Because gradient-based topology optimization (TO) pro-

---

*Address all correspondence to this author.

Copyright © 2023 by ASME and GE Research

vides higher flexibility, several studies have been conducted on topology optimization for SynRMs. Notably, Lee et al. [10, 11] implemented a density-based topology optimization framework to determine an optimized layout for the iron webs and bridges of SynRMs. This work considers multiple aspects, including torque performance, manufacturability, and structural safety, as simultaneous optimization objectives. Additionally, Lee et al. [8] efficiently incorporated a design-dependent current phase angle into the topology optimization process for SynRMs. Okamoto et al. [12] employed an MMA-based TO method to optimize the rotor core layout of SynRMs, with a focus on enhancing the torque performance of and investigating the impact of the rotor bridges. Yamashita et al. [13] applied the level set method to designing SynRMs for reduced iron loss and improved torque characteristics. Park et al. [14, 15] employed the level set method to improve the average torque by maximizing the magnetic energy between two rotor positions. In [16], the rotor of SynRMs was optimized using the ON/OFF method, and the iron loss, including eddy current loss and hysteresis loss, were newly considered in the optimization. Besides SynRMs, topology optimization for permanent magnet machines has been implemented using a number of methods, including bidirectional evolutionary structural optimization (BESO) [17, 18], level set method [19, 20] and density-based methods [21, 22].

Topology optimization for electric machine design is currently in an active research area, and it entails a number of challenges that need to be addressed. The optimized result using the density-based method usually accompanies by the intermediate value and blurring boundaries, and the On-Off approach usually generates the checkerboard pattern [23]. These drawbacks will bring about troubles for further manufacturing. Although the filter projection can partly mitigate this issue and improve the optimized design's manufacturability, the selection of appropriate filtering radius directly affects the patterns of the optimization results [24]. A common way is conducting the parametric study to investigate the effects of various filtering radii on the optimized result, which is time-consuming [8]. Since the level set method directly evolves the design boundary [25], it can ensure a clear boundary in the final design and lessen the need for post-processing. However, for the conventional level set approach, to guarantee the boundary evolves one grid interval length per time step, the step size needs to be sufficiently small. This will cause an excessive number of iterations before the objective function converges. Additionally, the conventional level set method requires the design velocity field extension to the whole design domain, which results in additional computational costs.

In view of this, we propose a cardinal basis function (CBF) based level set method to design SynRMs. By parameterization of level set function with cardinal basis function (CBF), we can transform the original Hamilton-Jacobi PDE into a system of ODE to lessen the computational burden [26, 27]. In addition, the reinitialization scheme in the conventional level set method

is replaced by the minimization of a distance regularization energy function along with the objective function [28]. This avoids the undesirable periodical suspension in the optimization process. With the parameterization of the level set function, the advanced optimization solver, such as Method of Moving Asymptotes (MMA) [29], can be employed to find the optimal design. The convergence speed can remarkably improve with these pros brought by the CBF-based level set method.

The rest of the paper is organized as follows: Section.2 introduces the modeling of synchronous reluctance motors (SynRMs). Section.3 presents the details on topology optimization of SynRMs, including cardinal basis function (CBF) based level set method, problem formulation, and shape sensitivity analysis, followed by one numerical example given in Section.4. Section.5 concludes the paper and outlines perspective work.

2 Modeling of Synchronous Reluctance Motors (SynRMs)

The SynRM consists of a laminated steel rotor and a laminated stator excited by a poly-phase winding. The winding is typically an integral slot winding, though a fractional slot winding could be applied. This study uses a four-pole SynRM as the numerical design model for topology optimization. The 2D geometry of the quarter SynRM model is illustrated in Fig.1, which is driven by balanced three-phase currents $i_u(t)$, $i_v(t)$ and $i_w(t)$:

$$
\begin{aligned}
i_u(t) &= I_m \sin\left(\omega_c t - \theta\right) \\
i_v(t) &= I_m \sin\left(\omega_c t - \frac{2\pi}{3} - \theta\right) \\
i_w(t) &= I_m \sin\left(\omega_c t + \frac{2\pi}{3} - \theta\right) ,
\end{aligned}
\tag{1}
$$

where I_m is the amplitude of the current set to 120 A. The current phase angle θ is set as $\pi/3$ during the optimization process. The rotor rotates with a rotational velocity 1200 r/min. The active length of this SynRM is 80 mm. The radial airgap length is 5 mm. For the material properties, a linear material with relative permeability μ_r is set 5000 for the rotor iron, where the design domain is placed. Although local magnetic saturation brought by the nonlinear constitutive relation between magnetic flux density **B** and magnetic field strength **H** will impact the magnetic performance, a linear relation is adopted at the current stage. This is because incorporating the nonlinearity into the current optimization model will overly complicate the shape sensitivity analysis, especially for solving the adjoint equation. To evaluate the effect of relative permeability, it is advisable to consider different relative permeability values. For the non-design area, the nonlinear soft iron is chosen for the stator. A copper electrical conductivity of 6×10^7 S/m is used for the coils.

Copyright © 2023 by ASME and GE Research

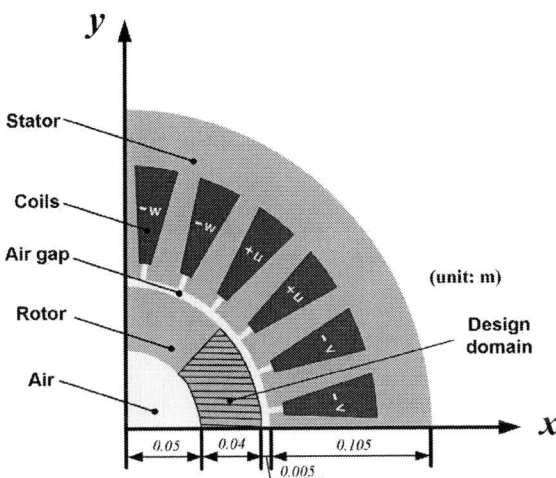

FIGURE 1: CAD model of synchronous reluctance motors (Syn-RMs).

Rotor rotation is modeled using the embedded physics interface for rotating machinery in COMSOL Multiphysics, where the sliding mesh interface facilitates the separation of the rotor and the stator. In addition, an anti-periodic boundary condition is applied to both sides of the geometry, as depicted in Fig. 1.

3 Topology Optimization of Synchronous Reluctance Motors (SynRMs)

3.1 Conventional Level Set Method

Conventionally, the level set function Φ is a Lipschitz continuous real-valued function defined in $\mathbb{R}^2$ or $\mathbb{R}^3$ [30]. With the level set method, the structure boundary $\partial\Omega$, highlighted as the red curve, is implicitly represented by the zero isosurface of the level set function with one-higher dimension, as illustrated in Fig. 2. According to the sign of the level set function, the design domain D can be divided into three parts, indicating the material, the interface, and the void, respectively. The level-set representation can be formulated as Eq. 2:

$$
\begin{cases}
\Phi(x,t) > 0, & x \in \Omega, & \text{material} \\
\Phi(x,t) = 0, & x \in \partial\Omega, & \text{boundary} \\
\Phi(x,t) < 0, & x \in D/\Omega, & \text{void}
\end{cases}, \quad (2)
$$

where x is the coordinates of an arbitrary point in the design domain and t is a pseudo time for the dynamic shape optimization process. The motion of the material interface is governed by the Hamilton-Jacobi equation:

$$
\frac{\partial\Phi(x,t)}{\partial t} - V_n|\nabla\Phi(x,t)| = 0, \quad (3)
$$

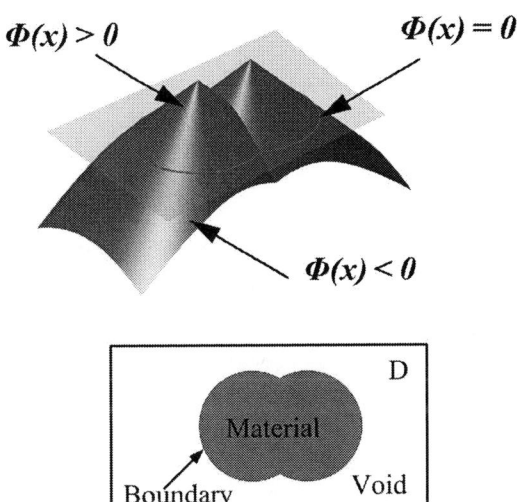

FIGURE 2: A schematic of level set representation.

where V_n is the normal velocity field contributing to the shape variation.

3.2 CBF-Based Parametric Level Set Method

In this section, a cardinal basis function (CBF) is constructed based on the radial basis function partition of unity (RBF-POU) collocation method to parameterize the level set function [31]. For the derivations of CBF construction using RBF, see [27] and the references cited therein. Given n different nodes $x_1, x_2, ..., x_n \in \mathbb{R}^2$ or $\mathbb{R}^3$, a level set function can be generally interpolated with cardinal basis functions (CBFs) in the following form,

$$
\Phi(x,t) = \sum_{j=1}^{n} \Psi_j(x)\mu_j(t), \quad (4)
$$

where $\mu_j(t)$ is actually the value of level set function at j_{th} node. $\Psi_j(x)$ is the constructed CBF, which is equal to 1 at the center node and 0 at other nodes. The CBF with this Kronecker delta properties can be expressed as follows:

$$
\Psi_j(x_i) = \begin{cases} 1, & i = j \\ 0, & i \neq j \end{cases}, \quad (5)
$$

After the parameterization of the level set function using CBF, the design variable at each node owns obvious physical meaning, which is, namely, the value of the level set function. By substi-

Copyright © 2023 by ASME and GE Research

tuting the Eq. 4 into the Hamilton-Jacobi Eq. 3, the original PDE is converted to an ODE in the following form,

$$\sum_{j=1}^{n} \dot{\mu}_j(t)\Psi_j(\boldsymbol{x}) - V_n|\nabla\Phi| = 0. \tag{6}$$

From the above equation, the normal velocity field V_n can be obtained as follows:

$$V_n = \frac{1}{|\nabla\Phi|}\sum_{j=1}^{n} \dot{\mu}_j(t)\Psi_j(\boldsymbol{x}). \tag{7}$$

3.3 Problem Formulation

The design target is to improve the average torque of SynRM and reduce its torque ripple simultaneously. In this study, the torque ripple can be measured as variance Var, and the average torque and the torque ripple can be expressed as follows:

$$E(T_i) = \frac{1}{n}\sum_{j=1}^{n} T_i \quad, \tag{8}$$

$$Var(T_i) = \frac{1}{n}\sum_{j=1}^{n} (T_i - E(T_i))^2 \quad, \tag{9}$$

where T_i is the torque at the i_{th} rotor position. Thus, the optimization objective can be formulated to maximize the mean of the torque and minimize its variance, which can be expressed as:

$$\begin{aligned} \textit{Maximize:} \quad & F = E(T_i) - Var(T_i), \\ \textit{Subject to:} \quad & a(\boldsymbol{A},\overline{\boldsymbol{A}}) = l(\overline{\boldsymbol{A}}), \quad \forall \overline{\boldsymbol{A}} \in U \\ & V(\Omega) = V^* \quad, \end{aligned} \tag{10}$$

where $V(\Omega)$ is the volume ratio of iron and V^* is the target volume ratio. The energy bilinear form $a(\boldsymbol{A},\overline{\boldsymbol{A}})$, the source linear form $l(\overline{\boldsymbol{A}})$ of magnetostatic system without permanent magnet and the volume of iron $V(\Omega)$ are described by:

$$a(\boldsymbol{A},\overline{\boldsymbol{A}}) = \int_{\Omega} v\boldsymbol{B}(\boldsymbol{A}) \cdot \boldsymbol{B}(\overline{\boldsymbol{A}})\mathrm{d}\Omega \quad, \tag{11a}$$

$$l(\overline{\boldsymbol{A}}) = \int_{\Omega} \boldsymbol{J} \cdot \overline{\boldsymbol{A}}\mathrm{d}\Omega \quad, \tag{11b}$$

$$V(\Omega) = \int_{\Omega} H(\Phi)\mathrm{d}\Omega \quad, \tag{11c}$$

where $H(\Phi)$ represents the Heaviside function and v, $\boldsymbol{B}$ and $\boldsymbol{J}$ represent the magnetic reluctivity, magnetic flux density, and current density, respectively. A is the magnetic vector potential, and the arbitrary virtual vector potential $\overline{A}$ belongs to the space of admissible vector potential U:

$$U = \left\{ \overline{\boldsymbol{A}} \in \left[H^1(\Omega) \right] \mid \overline{\boldsymbol{A}} = 0 \text{ on } \boldsymbol{x} \in \Gamma \right\} \quad, \tag{12}$$

where Γ denotes the Dirichlet essential boundary and $H^1(\Omega)$ represents the Sobolev space of first-order [32].

For accurate interpolation of material properties and effective avoidance of numerical instability during the optimization process [26, 27], a distance regularization energy function is introduced here. This function needs to be minimized along the objective function to maintain the distance-regularized level set function. The distance regularization energy function R is given in the following form:

$$R = \int_{\Omega} P(|\nabla\Phi|)\mathrm{d}\Omega \quad, \tag{13}$$

where $P(|\nabla\Phi|)$ is the regularization energy potential density, which is proposed by Li et al. [28] and formulated as follows:

$$P = \begin{cases} \dfrac{1}{(2\pi)^2}\left(1 - \cos\left(2\pi|\nabla\Phi|\right)\right), & |\nabla\Phi| < 1 \quad, \\ \dfrac{1}{2}\left(|\nabla\Phi| - 1\right)^2, & |\nabla\Phi| > 1 \quad. \end{cases} \tag{14}$$

3.4 Shape Sensitivity Analysis

This section details how to conduct the shape sensitivity using the material time derivative. For the derivation, a general magneto-static system is considered, and the whole magneto-static system can be divided into two sub-domains Ω_1 and Ω_2 with the interface of γ, shown in Fig. 3. The domain Ω_1 and Ω_2 have a distribution of v_1, J_1, M_1 and v_2, J_2, M_2, respectively.

For derivation convenience, we consider a general objective function that is defined inside a region Ω_2 in Fig. 3 as

$$F = \int_{\Omega_2} f(\boldsymbol{B}(A_2))\,k\,\mathrm{d}\Omega \quad, \tag{15}$$

where k is the localizing factor, which is used to select the integral domain for the objective function, where it is equal to 1. Except in this area, the localizing factor k is zero.

Firstly, the objective function is coupled with the magnetostatic governing equation using the Lagrange multiplier method as follows:

$$L(\boldsymbol{A},\overline{\boldsymbol{A}}) = F + l(\overline{\boldsymbol{A}}) - a(\boldsymbol{A},\overline{\boldsymbol{A}}) \quad. \tag{16}$$

Copyright © 2023 by ASME and GE Research

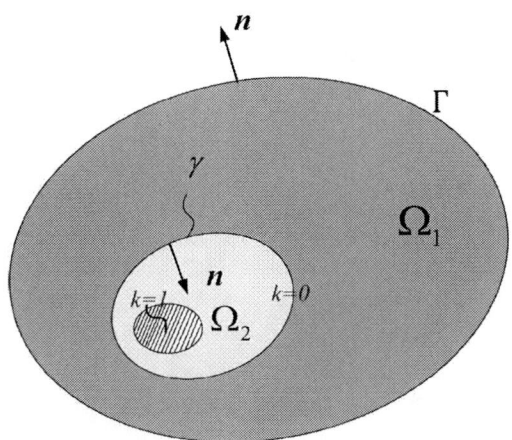

FIGURE 3: A schematic of the magnetostatic system interface.

Next, the material time derivative is utilized to derive the shape sensitivity [30, 33, 34]:

$$\frac{DL(A,\overline{A})}{Dt} = \frac{DF}{Dt} + \frac{Dl(\overline{A})}{Dt} - \frac{Da(A,\overline{A})}{Dt} \quad . \quad (17)$$

For conciseness, the derivative of the Lagrangian is directly presented as follows:

$$\frac{DL}{Dt} = \int_{\gamma} \frac{1}{\mu_0}\left(\frac{1}{\mu_r}-1\right)B(A_1)\cdot B(\overline{A}_2)V_n\mathrm{d}s \quad , \quad (18)$$

where γ is the interface between the iron Ω_1 and the air Ω_2. The adjoint variable $\overline{A}_2$ can be obtained by solving the following equation with the corresponding boundary condition.

$$\int_{\Omega_1} \frac{1}{\mu_1}B(A_1)\cdot B(\overline{A}_1)\mathrm{d}\Omega + \int_{\Omega_2} \frac{1}{\mu_2}B(A_2)\cdot B(\overline{A}_2)\mathrm{d}\Omega$$
$$= \int_{\Omega_2} \frac{\partial f}{\partial \mathbf{B}_2}\cdot B(\dot{A}_2)k\mathrm{d}\Omega \quad , \quad (19)$$

where $\frac{\partial f}{\partial \mathbf{B}_2} = \left(\frac{\partial f}{\partial B_{2x}}, \frac{\partial f}{\partial B_{2y}}\right)^T$. By plugging Equation (7) into Equation (18), the material derivative of Lagrangian can be assembled as

$$\frac{DL}{Dt} = \sum_{j=1}^{n} \dot{\mu}_j(t)\int_{\gamma}\left[\frac{1}{\mu_0}\left(\frac{1}{\mu_r}-1\right)B(A_1)\cdot B(\overline{A}_2)\right]\frac{1}{|\nabla\Phi|}\Psi_j(\mathbf{x})\mathrm{d}s \quad . \quad (20)$$

Similarly, the material derivative of the volume constraint can be

formulated as

$$\frac{DV}{Dt} = \int_{\gamma} V_n\mathrm{d}s = \int_{\gamma}\frac{1}{|\nabla\Phi|}\sum_{j=1}^{n}\dot{\mu}_j(t)\Psi_j(\mathbf{x})\mathrm{d}s \quad . \quad (21)$$

With the chain rule, the material derivatives of Lagrangian and volume constraint can also be expressed as

$$\frac{DL}{Dt} = \frac{\partial L}{\partial \mu(t)}\frac{\partial \mu(t)}{\partial t} = \frac{\partial L}{\partial \mu(t)}\dot{\mu}(t) \quad , \quad (22a)$$

$$\frac{DV}{Dt} = \frac{\partial V}{\partial \mu(t)}\frac{\partial \mu(t)}{\partial t} = \frac{\partial V}{\partial \mu(t)}\dot{\mu}(t) \quad . \quad (22b)$$

To solve the optimization problem (10), the advanced gradient-based optimizer, method of moving asymptotes (MMA) [35, 29], is implemented in this study. By comparing the corresponding parts of Eqs. (20), (21) and (22), the sensitivity of the objective function F and volume constraint V can be formulated as follows:

$$\frac{\partial F}{\partial \mu_j(t)} = \int_{\gamma}\left[\frac{1}{\mu_0}\left(\frac{1}{\mu_r}-1\right)B(A_1)\cdot B(\overline{A}_2)\right]\frac{1}{|\nabla\Phi|}\Psi_j(\mathbf{x})\mathrm{d}s \quad , \quad (23a)$$

$$\frac{\partial V}{\partial \mu_j(t)} = \int_{\gamma}\frac{1}{|\nabla\Phi|}\Psi_j(\mathbf{x})\mathrm{d}s \quad . \quad (23b)$$

It is noted that the boundary integration in the above equation can be converted into a domain integration by using the Dirac delta function δ as:

$$\mathrm{d}s = \delta(\Phi)|\nabla\Phi|\mathrm{d}\Omega. \quad (24)$$

In addition, the sensitivity of the distance regularization energy functional R can be derived as follows:

$$\frac{\partial R}{\partial \mu_j(t)} = -\nabla\cdot(d_p(|\nabla\Phi|)\nabla\Phi)\cdot\Psi_j(\mathbf{x}) \quad , \quad (25)$$

where d_p is defined as [28]

$$d_p(s) \triangleq \frac{p'(s)}{s}, \quad (26)$$

where $p'(s)$ is the first derivative of the regularization energy potential density defined in Eq. (14). Finally, the shape sensitivity

Copyright © 2023 by ASME and GE Research

413

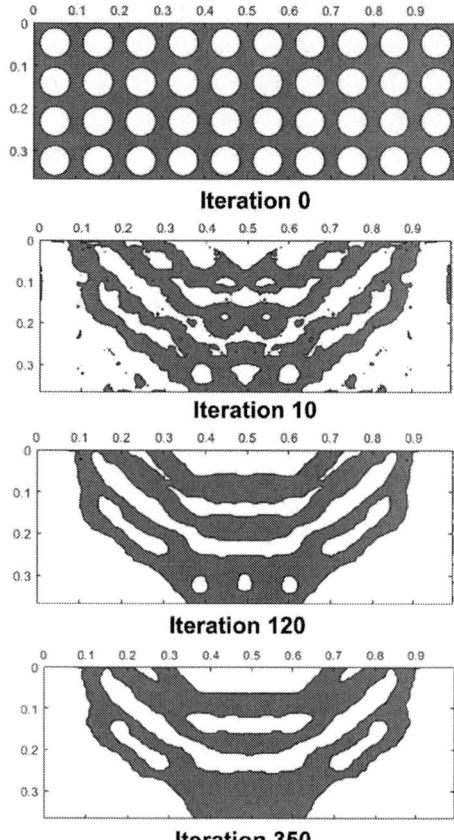

FIGURE 4: The optimization design evolution for SynRM design.

of the total objective function J coupling the distance regularization energy function R with the original objective function F can be expressed in the form of domain integration as:

$$\frac{\partial J}{\partial \mu_j(t)} = \int_\Omega \left[\frac{1}{\mu_0} \left(\frac{1}{\mu_r} - 1 \right) \boldsymbol{B}(\boldsymbol{A}_1) \cdot \boldsymbol{B}(\overline{\boldsymbol{A}}_2) \right] \delta(\Phi) \Psi_j(\boldsymbol{x}) \mathrm{d}\Omega$$
$$+ w \frac{\partial R}{\partial \mu_j(t)} \quad , \tag{27}$$

where w is the weighting factor. Similarly, the shape sensitivity of the volume constraint expressed in the form of domain integration is also given here:

$$\frac{\partial V}{\partial \mu_j(t)} = \int_\Omega \delta(\Phi) \Psi_j(\boldsymbol{x}) \mathrm{d}\Omega \quad . \tag{28}$$

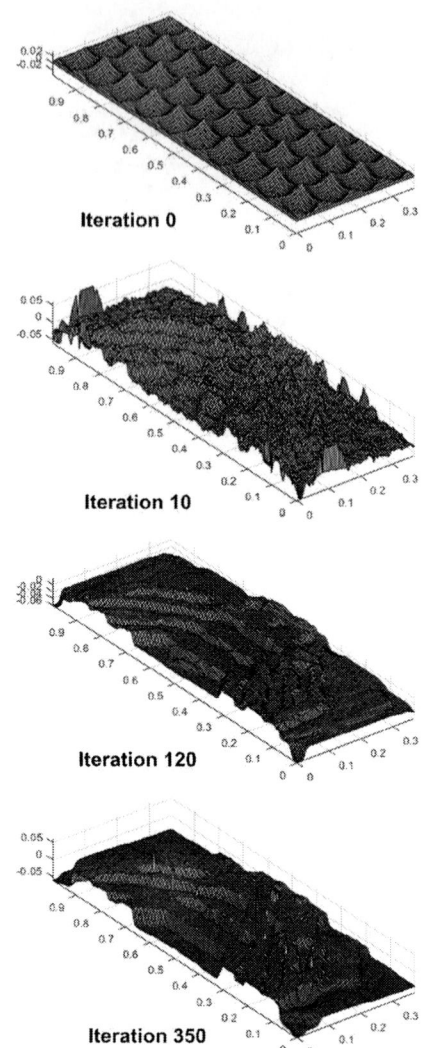

FIGURE 5: The evolution of level set function Φ.

4 Design Example

The proposed design optimization method is applied to a four-pole Synchronous reluctance motor in Sec. 2. To obtain a symmetric design, only the left half domain is studied, and the design variable is symmetrically assigned along the axis of 45 degree, as shown in Fig. 1. The relative permeability of the material is interpolated using the Heaviside function of the level set function Φ. The initial volume fraction of iron in the rotor is 0.6078, and the target volume fraction target is set to 0.4.

From Sec. 3.4, it is noted that the calculation of gradient and divergence is vital in the shape sensitivity analysis. In general, structured quadrilateral meshes have several advantages over triangle meshes when it comes to computing gradients and divergences. The gradient of a scalar function on a structured

Copyright © 2023 by ASME and GE Research

quadrilateral mesh can be approximated using a bilinear interpolation scheme, which is simpler and more accurate than the linear interpolation scheme used for triangle meshes. In addition, the divergence of a vector field can be calculated using a simple finite difference scheme that takes advantage of the structured quadrilateral mesh. According to this, the conformal mapping theory [36, 37] is employed to parameterize the 2D triangle meshed irregular design domain onto a structured quadrilateral meshed rectangular domain, where the level set function is defined. Then, the proposed extended level set method is applied to the design in this study. For more details, readers are referred to [38, 39, 40, 41]. In the implementation process, the design domain is meshed with 11363 triangular elements before conformally mapped to a $0.3708\ m \times 1\ m$ rectangular domain, where the level set function is defined and discretized with 65×174 grids.

In this study, a total of 7 rotor positions (every 10 degrees from 0 to $60°$) were investigated. Due to the rotational nature of the design domain, the meshes employed to discretize the said domain must undergo rotational movement. The rotation-moving mesh is expressed as

$$\begin{bmatrix} x_r \\ y_r \end{bmatrix} = \begin{bmatrix} \cos\theta & -\sin\theta \\ \sin\theta & \cos\theta \end{bmatrix} \begin{bmatrix} x - X_p \\ y - Y_p \end{bmatrix} + \begin{bmatrix} X_p \\ Y_p \end{bmatrix}, \quad (29)$$

where x_r and y_r represent the mesh coordinates after rotation relative to a random point (X_p, P_p). x and y are the mesh coordinates in the original design domain and θ represents the mechanical rotation angle. Figure 4 shows the design evolution on the rectangular domain, and its corresponding level set function Φ is given in Fig. 5. The design evolution in the rotor of SynRM after conformal mapping is illustrated in Fig. 6. The optimization history, including the average torque, torque variance, volume ratio, and distance regularization energy, is given in Fig. 7. The volume of the permanent magnet is 40.01% when the optimization ends. Since there is no least reluctance position in the initial design, the average reluctance torque is 0. After the optimization ends, the average torque increased to $14.71\ N \cdot m$, and the torque variance converged at 1.21. In addition, the distance regularization energy R remains at a relatively low level, which guarantees the distance-regularized level set evolution during the optimization process.

To verify the effectiveness of the proposed method, we also investigated the torque performance of a reference synchronous reluctance motor with a multi-layer flux barrier, shown in Fig. 8. The same boundary conditions are applied to the reference model, and its calculated torque characteristics are compared with the optimized SynRM in Fig. 9. Our optimized model can generate an average torque at $14.88\ N$ while the average torque of the reference model is $11.42\ N$, so it turns out that the average torque increased by 30.30%. The torque variance of the

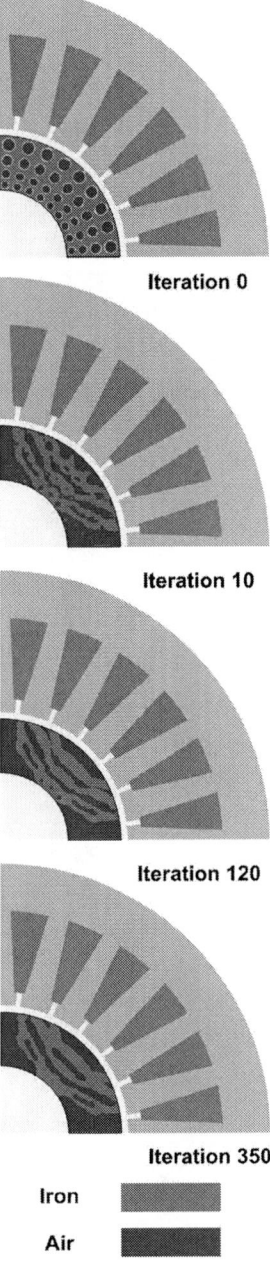

FIGURE 6: The design evolution in the rotor of SynRM.

optimized design is 0.228425, while the reference model only has 0.11 torque variance. One reason leading to this deficiency can attribute to the few rotor positions considered. In addition, an appropriate weight should be chosen to balance the average torque and torque variance in the objective function formulation.

Finally, the full 2.5D optimized rotor was extruded and assembled with the rotor bridge, shaft, and stator for unity, as

Copyright © 2023 by ASME and GE Research

shown in Fig. 10.

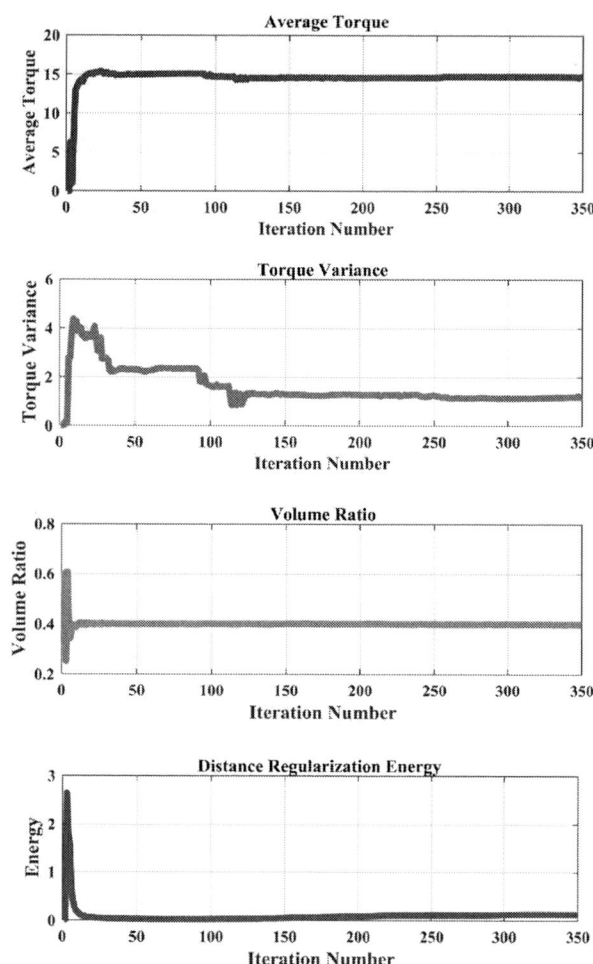

FIGURE 7: The iteration history for SynRM design.

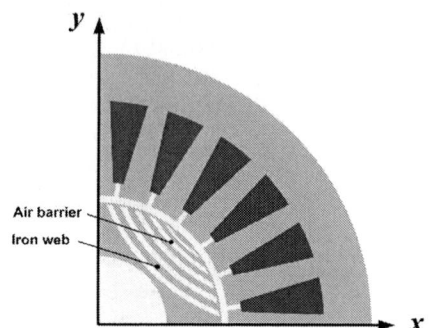

FIGURE 8: The CAD model of a reference SynRM.

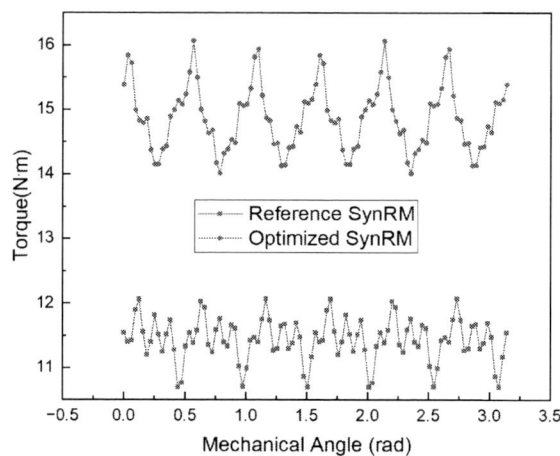

FIGURE 9: Torque performance comparison between reference and optimized SynRM.

5 Conclusion and Future Work

The rotor structure of the four-pole synchronous reluctance motor (SynRM) was designed using the cardinal basis function (CBF) based level set method. Although the solver for the SynRM is a time-dependent study, a magnetostatic field analysis is carried out at every rotor position in this study. The design problem is reformulated to balance average torque improvement and ripple reduction. One numerical example was carried out to verify the proposed method. The torque performance of the optimized SynRM is compared with that of a reference model. Pre-

liminary results show that the optimized SynRM has a 30.30% increase in average torque, along with a slight increase in torque ripple, compared to the reference model. Based on these, there is still much room for improvement in future studies. One of our efforts will incorporate the nonlinear B-H curve into the topology optimization of the rotor structure to accurately investigate the influence of magnetic saturation on torque performance. Besides, to further reduce the torque ripple, an increasing number of rotor positions should be considered in the objective function formulation, with attention to the issue of computational cost.

Acknowledgement

We would like to acknowledge the financial support provided by the National Science Foundation (CMMI-1762287; PFI-RP-2213852), GE Renewable Energy (Awards No.: 95031, 97223), the Advanced Energy Research and Technology Center (AERTC), and the Center for Integrated Electric Energy Systems

Copyright © 2023 by ASME and GE Research

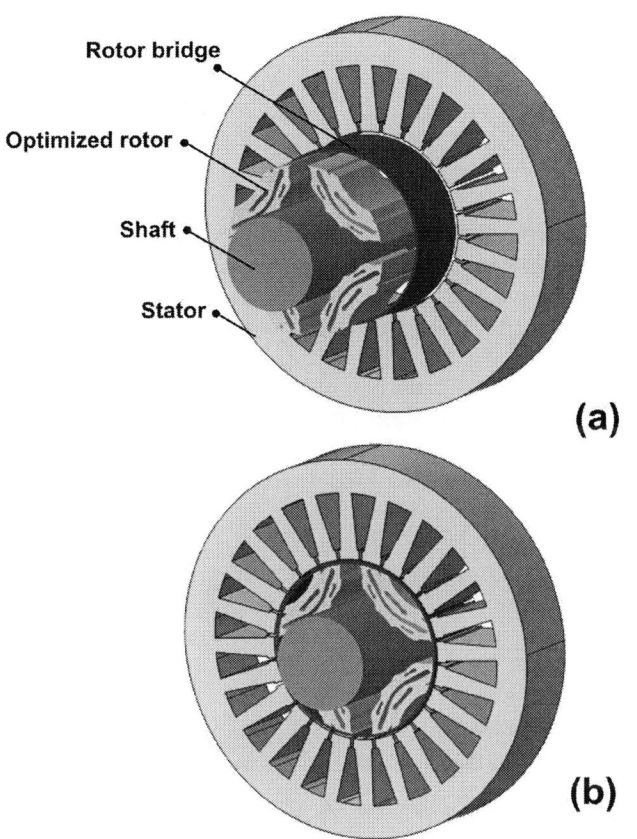

FIGURE 10: (a) The 3D components of optimized SynRM. (b) 3D assembly of optimized SynRM.

(CIEES) at the State University of New York (SUNY) at Stony Brook (SBU). We also extend our gratitude to Peter Donnelly, David Hamilton, Michael Antonucci, and Prathamesh Hajirnis at the SBU Office of Economic Development for their invaluable support throughout this project.

REFERENCES

[1] Pham, T., Kwon, P., and Foster, S. "Additive manufacturing and topology optimization of magnetic materials for electrical machines—a review". *Energies, 14*(2), p. 283.

[2] Betz, R. E., Lagerquist, R., Jovanovic, M., Miller, T. J., and Middleton, R. H. "Control of synchronous reluctance machines". *IEEE Transactions on Industry Applications, 29*(6), pp. 1110–1122.

[3] Bonthu, S. S. R., Arafat, A., and Choi, S. "Comparisons of rare-earth and rare-earth-free external rotor permanent magnet assisted synchronous reluctance motors". *IEEE Transactions on Industrial Electronics, 64*(12), pp. 9729–9738.

[4] Heidari, H., Rassõlkin, A., Kallaste, A., Vaimann, T., Andriushchenko, E., Belahcen, A., and Lukichev, D. V. "A review of synchronous reluctance motor-drive advancements". *Sustainability, 13*(2), p. 729.

[5] Bocker, J., and Mathapati, S. "State of the art of induction motor control". In 2007 IEEE International Electric Machines & Drives Conference, Vol. 2, IEEE, pp. 1459–1464.

[6] Fei, W., and Luk, P. C.-K. "Torque ripple reduction of a direct-drive permanent-magnet synchronous machine by material-efficient axial pole pairing". *IEEE Transactions on Industrial Electronics, 59*(6), pp. 2601–2611.

[7] Park, J.-M., Kim, S. I., Hong, J. P., and Lee, J. H. "Rotor design on torque ripple reduction for a synchronous reluctance motor with concentrated winding using response surface methodology". *IEEE Transactions on Magnetics, 42*(10), pp. 3479–3481.

[8] Lee, C., and Jang, I. G. "Multi-material topology optimization for the PMSMs under the consideration of the MTPA control". *Structural and Multidisciplinary Optimization, 65*(9), p. 263.

[9] Nishanth, F., Johnson, M., and Severson, E. L. "A review of thermal analysis and management of power dense electric machines". In 2021 IEEE International Electric Machines & Drives Conference (IEMDC), IEEE, pp. 1–8.

[10] Lee, C., Lee, J., and Jang, I. G. "Topology optimization for the manufacturable and structurally safe synchronous reluctance motors with multiple iron webs and bridges". *IEEE Transactions on Industrial Electronics, 70*(1), pp. 678–687.

[11] Lee, C., and Jang, I. G. "Topology optimization of multiple-barrier synchronous reluctance motors with initial random hollow circles". *Structural and Multidisciplinary Optimization, 64*, pp. 2213–2224.

[12] Okamoto, Y., Hoshino, R., Wakao, S., and Tsuburaya, T. "Improvement of torque characteristics for a synchronous reluctance motor using MMA-based topology optimization method". *IEEE transactions on magnetics, 54*(3), pp. 1–4.

[13] Yamashita, Y., and Okamoto, Y. "Design optimization of synchronous reluctance motor for reducing iron loss and improving torque characteristics using topology optimization based on the level-set method". *IEEE Transactions on Magnetics, 56*(3), pp. 1–4.

[14] Heo, S. H., Baek, M. K., Lee, K. H., Hong, S. G., and Park, I. H. "Shape and topology optimization of rotor in synchronous reluctance motor using continuum sensitivity and adaptive level set method". In 2013 International Conference on Electrical Machines and Systems (ICEMS), IEEE, pp. 129–133.

[15] Kim, Y. S., and Park, I. H. "Topology optimization of rotor in synchronous reluctance motor using level set method and shape design sensitivity". *IEEE Transactions on Applied Superconductivity, 20*(3), pp. 1093–1096.

Copyright © 2023 by ASME and GE Research

[16] Sato, S., Sato, T., and Igarashi, H. "Topology optimization of synchronous reluctance motor using normalized Gaussian network". *IEEE transactions on magnetics, 51*(3), pp. 1–4.

[17] Takahashi, N., Yamada, T., and Miyagi, D. "Examination of optimal design of IPM motor using ON/OFF method". *IEEE Transactions on Magnetics, 46*(8), pp. 3149–3152.

[18] Okamoto, Y., Tominaga, Y., Wakao, S., and Sato, S. "Topology optimization of rotor core combined with identification of current phase angle in IPM motor using multistep genetic algorithm". *IEEE Transactions on Magnetics, 50*(2), pp. 725–728.

[19] Choi, J. S., Izui, K., Nishiwaki, S., Kawamoto, A., and Nomura, T. "Topology optimization of the stator for minimizing cogging torque of IPM motors". *IEEE Transactions on Magnetics, 47*(10), pp. 3024–3027.

[20] Tian, J., Zhuang, R., Cilia, J., Rangarajan, A., Luo, F., Longtin, J., and Chen, S. "Topology Optimization of Permanent Magnets for Generators Using Level Set Methods". In International Design Engineering Technical Conferences and Computers and Information in Engineering Conference, Vol. 86236, American Society of Mechanical Engineers, p. V03BT03A037.

[21] Hermann, A. N. A., Mijatovic, N., and Henriksen, M. L. "Topology optimisation of PMSM rotor for pump application". In 2016 XXII International Conference on Electrical Machines (ICEM), IEEE, pp. 2119–2125.

[22] Guo, F., Tang, N., and Brown, I. P. "Magneto-structural combined dimensional and topology optimization of interior permanent magnet synchronous machine rotors". *IEEE Transactions on Industry Applications, 58*(6), pp. 7241–7250.

[23] Mohamodhosen, B. S. B. "Topology optimisation of electromagnetic devices". PhD thesis, Ecole Centrale de Lille.

[24] Lazarov, B. S., and Sigmund, O. "Filters in topology optimization based on Helmholtz-type differential equations". *International Journal for Numerical Methods in Engineering, 86*(6), pp. 765–781.

[25] Wang, M. Y., Wang, X., and Guo, D. "A level set method for structural topology optimization". *Computer methods in applied mechanics and engineering, 192*(1-2), pp. 227–246.

[26] Jiang, L., and Chen, S. "Parametric structural shape & topology optimization with a variational distance-regularized level set method". *Computer Methods in Applied Mechanics and Engineering, 321*, pp. 316–336.

[27] Jiang, L., Chen, S., and Jiao, X. "Parametric shape and topology optimization: A new level set approach based on cardinal basis functions". *International Journal for Numerical Methods in Engineering, 114*(1), pp. 66–87.

[28] Li, C., Xu, C., Gui, C., and Fox, M. D. "Distance regularized level set evolution and its application to image segmentation". *IEEE transactions on image processing, 19*(12), pp. 3243–3254.

[29] Svanberg, K. "MMA and GCMMA-two methods for nonlinear optimization". *vol, 1*, pp. 1–15.

[30] Allaire, G., Jouve, F., and Toader, A.-M. "Structural optimization using sensitivity analysis and a level-set method". *Journal of computational physics, 194*(1), pp. 363–393.

[31] Safdari-Vaighani, A., Heryudono, A., and Larsson, E. "A radial basis function partition of unity collocation method for convection–diffusion equations arising in financial applications". *Journal of Scientific Computing, 64*(2), pp. 341–367.

[32] Adams, R. A., and Fournier, J. J., 2003. *Sobolev spaces.* Elsevier.

[33] Choi, K. K., and Kim, N.-H., 2006. *Structural sensitivity analysis and optimization 1: linear systems.* Springer Science & Business Media.

[34] Park, I. H., 2019. *Design Sensitivity Analysis and Optimization of Electromagnetic Systems.* Springer.

[35] Svanberg, K. "The method of moving asymptotes—a new method for structural optimization". *International journal for numerical methods in engineering, 24*(2), pp. 359–373.

[36] Lui, L. M., Gu, X., Chan, T. F., Yau, S.-T., et al. "Variational method on riemann surfaces using conformal parameterization and its applications to image processing". *Methods and Applications of Analysis, 15*(4), pp. 513–538.

[37] Gu, X., Wang, Y., Chan, T. F., Thompson, P. M., and Yau, S.-T. "Genus zero surface conformal mapping and its application to brain surface mapping". *IEEE transactions on medical imaging, 23*(8), pp. 949–958.

[38] Ye, Q., Guo, Y., Chen, S., Lei, N., and Gu, X. D. "Topology optimization of conformal structures on manifolds using extended level set methods (X-LSM) and conformal geometry theory". *Computer Methods in Applied Mechanics and Engineering, 344*, pp. 164–185.

[39] Tian, J., Zhao, X., Gu, X. D., and Chen, S. "Designing ferromagnetic soft robots (FerroSoRo) with level-set-based multiphysics topology optimization". In 2020 IEEE International Conference on Robotics and Automation (ICRA), IEEE, pp. 10067–10074.

[40] Tian, J., Li, M., Han, Z., Chen, Y., Gu, X. D., Ge, Q., and Chen, S. "Conformal topology optimization of multimaterial ferromagnetic soft active structures using an extended level set method". *Computer Methods in Applied Mechanics and Engineering, 389*, p. 114394.

[41] Xu, X., Gu, X. D., and Chen, S. "Topology optimization of thermal cloaks in euclidean spaces and manifolds using an extended level set method". *International Journal of Heat and Mass Transfer, 202*, p. 123720.

Copyright © 2023 by ASME and GE Research

Proceedings of the ASME 2023
International Design Engineering Technical Conferences and
Computers and Information in Engineering Conference
IDETC-CIE2023
August 20-23, 2023, Boston, Massachusetts

DETC2023-116618

OPTIMIZATION OF 3D PRINTING WHILE TRAVELING EN ROUTE TO EXTEND RANGE OF UAS FOR MULTI-LOCATION MISSION SCENARIOS

Tevin Dickerson [1], John L. Salmon[1,*], Christopher A. Mattson[1]

[1]Brigham Young University, Provo, UT

ABSTRACT

The nexus of two relatively recent technologies, additive manufacturing and unmanned aircraft systems (UAS), has enabled new and unique capabilities that have only started to be realized in integrated systems. This paper explores and quantifies the impact of 3D printing parts for UAS, or entire UAS systems, on an agent platform while it travels to multiple locations as part of a mission objective. The printed or enhanced UAS can then be released at launch points farther away from the goal locations. This, in turn, can accelerate mission completion times and reduce travel costs. The methodology and optimization problem are developed by making use of Apollonius circles extensively with interesting examples presented of the integration of these two technologies. Results indicate that based on the print capability and agent travel speed, using the sequence provided by a Traveling Salesman's problem solution is not always optimal. Thousands of scenarios are optimized across the design space with some interesting phenomena discussed in the analysis.

Keywords: additive manufacturing, unmanned aircraft systems, 3D printing, constrained optimization, traveling salesman problem, Apollonius circles

1. INTRODUCTION

Unmanned aircraft systems (UAS) have become increasingly prevalent in the 21st century. These systems can range from recreational, to commercial, and combat use. From the US-Afghanistan war in 2001 to the Russia-Ukraine war starting in 2022, UAS systems have seen an explosion of applications, from Search and Rescue (SAR) [1], product delivery [2], reconnaissance missions [3], and even to Loitering Munitions [4]. With their expanded use and adoption into vast and vital roles, production and use of these systems are implementing optimizing techniques that could further UAS applications and reduce costs, increase effectiveness, or both.

*Corresponding author: johnsalmon@byu.edu

FIGURE 1: CONCEPT ILLUSTRATION OF A 3D PRINTER IN THE BACK OF A TRUCK THAT COULD PRINT PARTS WHILE TRAVELING

Concurrently, advances in additive manufacturing are making it possible to build complex and reliable parts with the use of a single machine [5, 6]. Material options for additive manufacturing are also expanding with various choices among plastics, metals, and composites [5–7]. The use of additive manufacturing for UAS systems is also becoming more advantageous [6]. Light-weight and inexpensive parts can be produced, prototyped, and implemented in a short duration of time [6]. The symbiosis of these two technologies, UAS and 3D printing, together suggest that the fabrication of unmanned aircraft systems in an operating environment could lead to saving time, reducing cost, or improving the performance of the various assets within the system.

3D printing systems are becoming smaller and more viable to transport. They could be placed in mobile platforms which could include trucks, navel vessels, aircraft, or even backpacks (see Figure 1).

A recent example of this implementation is the USS Bataan which has multiple 3D printers aboard able to print plastics and metal parts [8] while the ship is in motion. Thus, a mobile agent,

Copyright © 2023 by ASME

such as a truck or naval ship, having the capability to manufacture a UAS while moving or 'en route,' could adapt the UAS design to meet the needs of any situation by manufacturing a specific part, attachment, or component in the direct operating environment. Since many UAS are designed for a specific role in mind, requirement changes or adjustments to a mission or purpose can result in product overdesign, extra cost, or delays. For SAR applications a stable efficient airframe is desirable to extend the loitering time and to provide a stable platform for cameras and sensors [9]. On the other hand, a Loitering Munitions UAS needs to be agile, quick, and avoid detection. As expected, certain aspects of these systems will be widely different however the components can have many similarities. A simple, basic, or default UAS design with a standard frame, structure, and propulsive plant could have components altered or swapped out to achieve different overall system capabilities. These parts could be 3D printed in situ, on-demand, and while an agent is traveling to the launch location of the particular UAS mission. The new parts are then integrated into the basic UAS design to extend its default capabilities.

Consider, for example, a scenario where a truck is delivering medicine and pharmaceuticals to a set of tribal villages that cannot be reached due to terrain, forest, or other barriers similar to the company "Zipline"[10]. The truck drives as close as possible to the villages and then uses a UAS to deliver and fly the last few kilometers. Since the UAS carry only a few hundred grams of tablets or pills for any given flight, the UASs are cheap, as light as possible, and fly only one direction (i.e. they are attritable and do not return to the agent[10]). On any given day due to rain, flooding, or other events, the UAS may be over-designed or under-designed. It would be ideal to have a default UAS that can fly, say 1 km, but can be extended to 2 km or more, by 3D printing parts that can increase its range while the truck is in the operating environment. Instead of carrying a variety of UAS sizes and ranges, 3D printing en route can allow for the adaptation of the environmental conditions at the time of use.

A similar situation can be established for military units defending an area that could change from being a passive to an active situation at any time [11]. An attritable UAS designed to surveil certain locations, released from a remote point could have its range extended through 3D printing, to keep the warfighter safer and further from a potentially hostile location. Or, the default UAS could be switched to a loitering munition UAS without having to bring in specifically designed loitering munitions components. Although the most common use for UAS in the military is for Intelligence, Surveillance, and Reconnaissance (ISR) [12] there are also other additional uses that include communications, resupply, and additional combative support [12, 13] similar to the medicine delivery mission described above.

As stated previously, regardless of the application domain, the agent carrying the 3D printer in these scenarios could be, *i)* a human (with a backpack) printing and sending out UAS to cut down on the walking distance to locations that may be radioactive, *ii)* a boat exploring islands from a distance to ascertain depth or harbor status, *iii)* a truck as described above, or *iv)* even an airship releasing smaller drones to record atmospheric data and other tasks at specific locations. These examples are clearly not

currently adopted and would require additional support structures for full operational implementation. For example, with humans carrying 3D printers to better access radioactive zones, it may be that two or three humans must carry the power sources (i.e. batteries), the filament, and the printer itself as a team. However, their average travel speed is still pertinent such that they may only print at night, (i.e. while not moving) for an effective or average print speed that is closed to "halved" because of the traveling that may be conducted during the day.

The objective of this paper is to quantify the impact of 3D printing UAS, or parts of UAS, en route by an agent needing to 'visit' more than one location leveraging a UAS for the final kilometers to reduce the time or distance necessary.

2. BACKGROUND

Similar ideas for additive manufacturing UAS systems have been adapted by General Atomics [14]. General Atomics has developed a small UAS aircraft called the Sparrowhawk. It is designed to be a launch and retrieve system from current General Atomics platforms carrying additional sensors or payloads [15]. General Atomics claims that a sample Sparrowhawk was built with 3D printing techniques for its integrated structure design [14]. They also claim that the manufacturing and assembly took less than 2 days with assembly processes taking approximately 20 minutes [14]. This example of additive manufacturing shows how 3D printing en route could become feasible in the near future. This also shows that materials are not only limited to plastic material limits.

For a fixed-wing UAS the fundamental components to consider printing en route are the wings or parts of the wings. Different operational conditions could result in different wing structure requirements [9], but at a very high level, an increased aspect ratio would increase the lift-to-drag ratio and likely increase the range of the UAS [16]. Thus, the wings could be completely built en route or added to the wing structure of a default UAS design in sections (to increase the span of the wings). Another example of adding new sections to a wing could be winglets used for their ability to reduce induced drag [17]. An addition of winglets could decrease the amount of induced drag by 5-6% under cruise conditions [17] and thus increase the range. A similar analysis for implementing winglets on then MALE UAV found increases in the lift and drag ratios that resulted in a flight time increase of 10% [16]. Although pre-printing winglets and carrying those is a reasonable alternative, similar arguments for over- or under-designing the UAS are still valid. But more importantly, printing *in situ* responds to the operational environment more directly if weather, terrain, or other mission objectives demand adjustments.

Other options could be available for increasing the flight time. Additional storage space could be made for more batteries and similar design options could be implemented for multirotor configurations such as adding additional motors or batteries as needed to increase the flight time.

For the purposes of this paper, we will assume that there is a relationship such that the UAS flight time increases from 3D printing a part while the agent is driving, walking, or traveling to a location. Since most printers can be assumed to print at a constant rate, we likewise assume the rate at which the flight

Copyright © 2023 by ASME

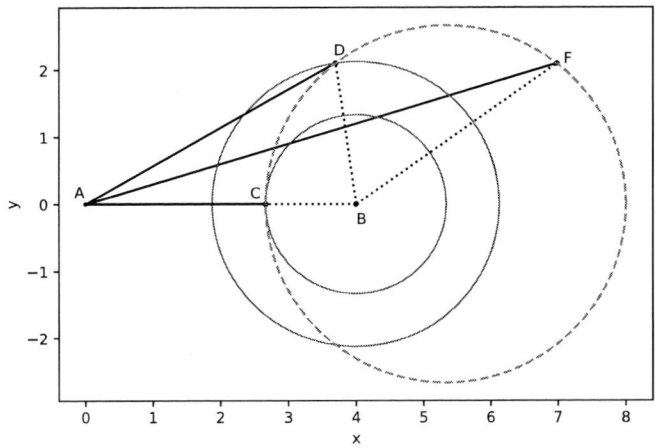

FIGURE 2: EXAMPLE APOLLONIUS CIRCLE AND SOLUTIONS OF LAUNCH POINTS FOR VARIOUS INCIDENT DIRECTIONS FOR A 1-LOCATION SCENARIO, WITH $k = 0.5, r_0 = 0$.

time increases is a constant and thus a linear relationship with the print time available or the agent travel time (the actual relationship will be non-linear but the linear assumption is invoked herein for the limited scope of this paper). For example, if an agent is flying or driving 40 km, at 40km/h, and the onboard 3D printer can print a part that extends the UAS range by 2 km during that same time, the printer's 'speed' is 2 km/h. This speed ratio, i.e. 2/40, is a key parameter described and explored in the following section. As mentioned above, we recognize that there are nonlinear relationships at play between flight range increases and printer speeds. However, for purposes of this initial exploration and analysis, a linear approximation of these relationships and the setup (to be described next) can still be used with some modifications to permit nonlinear cases in future research.

3. METHODOLOGY

For this paper we will assume that the domain that the agent moves in is void of obstacles and flat terrain. This assumption could apply to aviation, naval, or in some human travel domains in which straight, unobstructed paths, and constant velocity movements are likely to apply.

3.1 Apollonius Circles

To model the effect of fabricating a UAS on a moving agent the properties of an Apollonius circle will be used [18]. An Apollonius circle, as shown by the red dashed circle in Figure 2 is defined by any two unique points, say A, B, and that a third point, C, exists such that the ratio between the distances $\|\overrightarrow{BC}\|$ and $\|\overrightarrow{AC}\|$ remains a constant.

$$\frac{\|\overrightarrow{BC}\|}{\|\overrightarrow{AC}\|} = k \tag{1}$$

The locus of points in which point C satisfies the above equation defines the Apollonius circle. In Figure 2, point C satisfies this equality when $k = 0.5$, which means, as an example,

$\|\overrightarrow{BC}\| = 1.33$ and $\|\overrightarrow{AC}\| = 2.67$. Likewise, $\|\overrightarrow{BF}\|$ is half the length of $\|\overrightarrow{AF}\|$, and $\|\overrightarrow{BD}\|$ is half the length of $\|\overrightarrow{AD}\|$.

The radius of the Apollonius circle will change depending upon the magnitude of k. As k becomes smaller and approaches 0, the Apollonius circle becomes correspondingly smaller and its center point moves towards point B, in the limit.

For our current purposes, we will use the Apollonius circle to define possible launch points of a UAS to fly the remaining distance, such that the agent never needs to visit, in person, a particular location demanded as part of the overall mission. Continuing the discussion above, if an agent is printing while traveling from point A to point B and the print rate is high enough with respect to the agent's speed, such that for every 1 km the agent traveled the UAS could fly 0.5 km (i.e., $k = 0.5$), the UAS can be released at point C and the agent's mission is complete. Of course, the agent can travel to point D (from A) which takes longer, but then there is more time to 3D print en route and as a result, the UAS has an even larger range and can fly to the location B from the launch point of point D. This analysis holds for all points and locations around the Apollonius circle. The shortest mission complete time is a direct approach towards B but other angles will be necessary when more than one location is part of the scenario discussed later.

In the proceeding discussion, as the agent is moving towards a launch point and begins manufacturing the UAS, we assume that it is completely printed and assembled just in time for release. It is also assumed that the agent will travel at constant velocity V in a straight direction toward the launch point.

As described, as the manufacturing time increases, the maximum flight time or range of the UAS will increase. In order for the UAS to reach the location, it must be released at a point such that the UAS can reach the location, (i.e., within its range). This creates a circle of radius $r(t)$ around point B. For example, if the agent is printing while traveling from A to D, but chooses not to release the UAS at D it still must release that UAS at a point on a circle defined by the radius of $\|\overrightarrow{BD}\|$ (or closer) represented with the larger blue circle in Figure 2.

This radius represents all possible launch positions such that the flight duration of the UAS can make it to the location with a direct approach. The launch position space will be represented as S.

$$S \equiv B + r(t) \begin{bmatrix} \cos\theta \\ \sin\theta \end{bmatrix} \tag{2}$$

The radius $r(t)$ will increase as the agent travels because the range of the UAS is increasing due to the 3D printing en route. The rate at which $r(t)$ increases is represented by U and is the rate at which the flight range increases with respect to the manufacturing time. In other words, U is the 'speed' of the printer and in connection with the speed of the agent, defined as V, this results in $k = U/V$. The angle θ is the angle between $\overrightarrow{BC}$ and the positive x-axis. Mathematically, the radius $r(t)$ grows according to:

$$r(t) = Ut \tag{3}$$

Copyright © 2023 by ASME

For a single location, a direct approach to that location will result in the shortest path solution. This path will produce a point C at which the UAS will be released. At point C, the Apollonius circle and the smaller blue circle will intersect as shown in Figure 2. Since the time t_1 it takes for the agent to reach this launch location C and the time it takes for the radius to increase to this launch location are the same, the distances and speeds can be related as shown in equation 4.

$$t_1 = \frac{\|\overrightarrow{AC}\|}{V} = \frac{r(t_1)}{U} \qquad (4)$$

Rearranging equation 4, a ratio of the distances and speeds is created and defines the constant k value, described previously, for an Apollonius circle as defined in equation 5.

$$\frac{\|\overrightarrow{BC}\|}{\|\overrightarrow{AC}\|} = \frac{r(t_1)}{\|\overrightarrow{AC}\|} = \frac{U}{V} = k \qquad (5)$$

If the agent were to take a path that was not directed toward B as shown in Figure 2, such as towards D, it would result in a new solution with a new launch space S that has a larger radius $r(t)$. Since k, or the ratio of the speeds, is constant for a given scenario, we can use the same Apollonius circle as before to define all of the launch positions at any intercept that is not in line with a direct path to B. This means that the radius $r(t)$ will be known for any value of θ given a k, or $r(\theta, k)$. In practice, a feasible angle θ must be constrained within the interval $[\alpha, -\alpha]$, where α is the angle when the path is tangent to the Apollonius circle.

The same conclusion will arise if there is an initial starting radius r_0 or $r(t)|_{t=0} > 0$. This non-zero radius means that *without* 3D printing en route, the UAS can reach the location from that radius. This is represented in Figure 3 as the small black circle, with $r_0 = 1$. If the agent traveled to this r_0 and then released a UAS with a default range of 1 km, it would be able to fly the remainder from H to B. Ideally, the agent could print an enhancement for the UAS while traveling to H and consequently increase the range for this default UAS, thereby permitting the launch point to be earlier (such as at point G). The UAS could now fly the full distance from G to B now that it has been enhanced with the part that was made while traveling to point G.

The initial circle (shown in black, with radius r_0 around point B in Figure 2) will grow at the same rate U over time to become the smaller of the two blue circles. For a given k ratio, the Apollonius circle, now using points A and H, is shown with a red dashed line. The intersection of $r(t)$ for some t and this Apollonius circle is the point G at which an enhanced UAS, now with more than 1 km range, would be able to fly all the way to B.

$$\Delta r = r(t) - r_0 \qquad (6)$$

$$t_1 = \frac{d}{V} = \frac{\Delta r}{U} \qquad (7)$$

$$\frac{\|\overrightarrow{GH}\|}{\|\overrightarrow{AG}\|} = \frac{\Delta r}{\|\overrightarrow{AG}\|} = \frac{U}{V} \qquad (8)$$

Performing similar logic as before with r_0, we arrive at the same conclusion that the Apollonius circle can be used to model

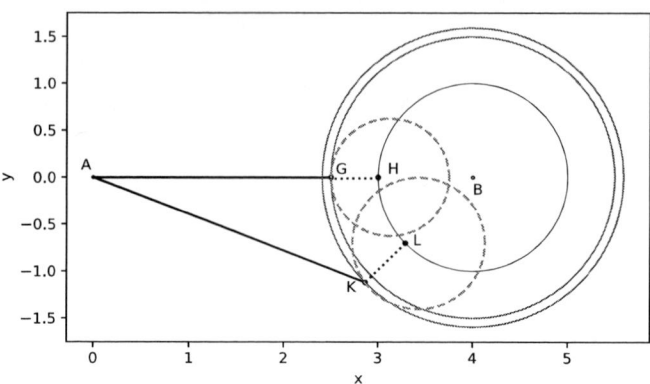

FIGURE 3: EXAMPLE SOLUTIONS OF LAUNCH POINTS, APOLLONIUS CIRCLES, AND INCIDENT DIRECTIONS FOR A 1-LOCATION SCENARIO, WITH $k = 0.2$, $r_0 = 1$ TO SHOW DETAIL.

the intercept location in any feasible direction. An example of this is presented at point K in Figure 3 where the radius has expanded to intersect with the Apollonius circle, but this time the Apollonius circle will use A and L for the two points.

In summary, when $r_0 = 0$ a UAS would need to be entirely manufactured while en route. Any modification would change the radius value by equation 3. For a single location scenario, the shortest route traveled by the agent will be in line with the foci (two points) of the Apollonius circle. As the ratio k approaches infinity, the total distance needed to travel by the agent will decrease compared to this default condition. These 1-Location scenarios with different k ratios are shown in Figure 4 on the left column. The lower left example of Figure 4 shows a value when $r_0 = 0$ and $k = 0$, suggesting that the agent themself needs to travel from the origin (called L_0) all the way to location 1 (called L_1). The second from the bottom has $r_0 = 1$ and $k = 0$ suggesting that no printing is performed en route but the default UAS can be released at the edge of the blue circle. The third from the bottom has $r_0 = 1$ and $k = 0.1$, indicative of printing en route and then enhancing the UAS with that component such that it can be released a little outside the back radius (i.e. $\Delta r > 0$). Finally, the top left shows the same situation but now $k = 0.5$. The resultant launch point is even further back from the black default radius r_0.

The conditions become more complex with more than one location. Conditions could change depending on launch points, the number of manufacturing assets (i.e. printers), and the order or sequence of locations visited. For manufacturing processes, a combination of printers and/or complex production lines could be used to increase the number of UASs (or enhancement parts) developed at one time. However, for this paper, it will be assumed that only one UAS can be developed/enhanced at a time by the onboard resources of the agent. This means that each segment along the path, d_i, defined as the distance from p_{i-1} to p_i (for example, $d_1 = \|\overrightarrow{p_0 p_1}\|$) where p_i is the launch point associated with each ith location L_i is the path during which printing can occur. The length of d_i will limit the growth of each corresponding radius Δr_i since the agent is assumed to travel along each d_i at the constant speed of V.

For a 2-Location scenario, the longest path would result if

Copyright © 2023 by ASME

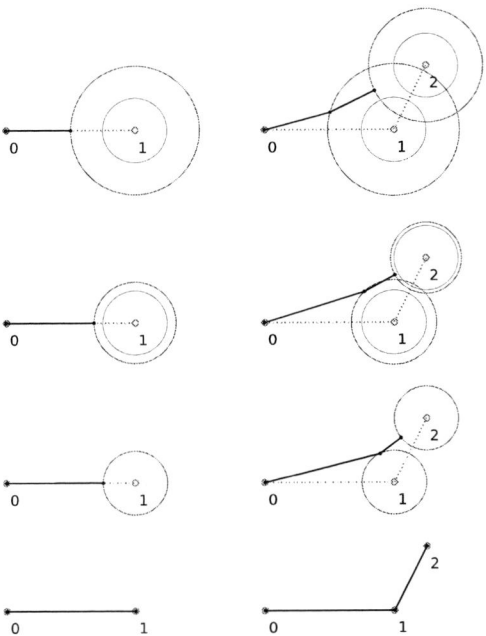

FIGURE 4: EXAMPLE 1-LOCATION (LEFT COLUMN) AND 2-LOCATION (RIGHT COLUMN) SCENARIOS. FROM BOTTOM TO TOP, RESPECTIVELY: ($r_0 = 0$ AND $k = 0$), ($r_0 = 1$ AND $k = 0$), ($r_0 = 1$ AND $k = 0.1$), ($r_0 = 1$ AND $k = 0.5$)

the agent had to travel to each location without the assistance of a UAS. This would occur when U approaches 0 and when no default UAS are available to launch at radius r_0. Thus, as the ratio k increases, the path length becomes continually shorter until it reaches 0 when k is infinity in the limit. If U were set to zero the problem revert to the traditional traveling salesman problem (TSP) (or the traveling salesman problem with neighborhoods (TSPN), if $r_0 > 0$). TSPN is similar to TSP however each *area* needs to be visited instead of each location [19].

Likewise, in a 2-Location scenario, the sequential Apollonius circles begin to become dependent upon each other. The right column of figure 4 shows the same conditions as the left side but now with two locations. The bottom two right figures show the TSP and TSPN solutions, with $r_0 = 0$, and $r_0 = 1$ respectively. The top right two figures show differing values of k with $r_0 = 1$ in both cases. As k increases we can see that not only does r_1 and r_2 change but θ_1 changes drastically. Under different relative locations for L_1 and L_2, both θ_1 and θ_2 values can change. However, the L_2 location in the right column examples of Figure 4 will behave similarly to a 1-Location scenario (left side of Figure 4) since it is the final segment along the total path. This is observed because the last location will not have any influence on downstream locations since there are none.

3.2 Objective Function

It is desirable to decrease the total time required for the agent to visit all locations and complete the mission through the use of UAS. Given that the agent's speed is assumed constant the total time will be proportional to the total distance traveled. Thus

minimizing the total distance or path length and still visiting each location, via UAS, becomes the objective. Likewise, in order to minimize the total path length D the launch points p_i need to be optimized such that:

$$D = \sum_{i=1}^{n} d_i(\theta_i, r_i) \tag{9}$$

$$= \sum_{i=1}^{n} \|p_i(\theta_i, r_i) - p_{i-1}(\theta_{i-1}, r_{i-1})\|_2 \tag{10}$$

where d_i is the euclidean distance between launch points p_{i-1} and p_i, which are functions of θ_i and r_i. All launch points with the exception of p_0 (which is the origin) need to be constrained by their Apollonius circle based on the points p_{i-1} and p_i, or more compactly their distance d_i, such that:

$$h_i(\theta_i) \equiv \frac{\Delta r_i}{d_i} - \frac{U}{V} = 0 \tag{11}$$

Under certain circumstances, a default UAS may be considered over-designed for a mission. It should be possible, therefore, that an agent could choose to launch a UAS from a point that falls inside a maximum possible Apollonius circle size constrained by equation 11. Any larger Apollonius circle would require a larger k or larger U value. In certain scenarios, an optimal solution may exist that permits launching a UAS from inside its Apollonius circle (or even inside r_0) resulting in a shorter travel path for the agent. To account for this possibility, the equality constraint above is reconfigured to an inequality constraint:

$$g_i(\theta_i, r_i) \equiv \frac{\Delta r_i}{d_i} - \frac{U}{V} \leq 0 \tag{12}$$

Since V and U are constants, they define the maximum Apollonius circle. Any launch point outside Δr_i would violate this constraint (i.e., required Δr_i becomes too large). On the other hand, in order to prevent optimization from exploiting negative Δr_i values, the boundary constraint of $r(t_i) \geq 0$ is also established. In other words, if $r(t_i) = 0$, the solution becomes trivial as any value of θ_i will satisfy the optimization problem.

3.3 Permutations and Traveling Salesman Problem

Given that equation 10 requires a sequence of locations for a full scenario, this problem must also be optimized in terms of the best sequence of locations. Therefore, this problem also includes similarities between the traveling salesman problem (TSP) or the traveling salesman problem with neighborhoods (TSPN). Solving the TSP part of the current problem could be implemented with many different strategies such as 2Opt or 3Opt heuristics discussed in [20].

However, if the sequence of locations was chosen that corresponds with the optimal sequence for TSP, this would not be the same solution for the current domain with 3D printing en route. Under certain values of k, a more optimal solution (i.e., a shorter total path length) will require traveling first to a launch point associated with the *farthest* location from the origin. This strategy essentially allows for a larger Δr_i value bringing the launch point closer to its neighboring locations. In order to account for

Copyright © 2023 by ASME

423

these scenarios, all the sequence permutations of an *N*-Location scenario will be analyzed in this paper, since the TSP solution is not a guaranteed solution when *k* and *r* can vary, as will be demonstrated.

3.4 Example 2-Location Scenarios and Launch Point Sets

Figure 5 shows three different locations for L_2 designated as 2a, 2b, or 2c. The agent will start flying (or traveling) and begin printing parts for the UAS which will be released at the first point in the path. We assume that location 1 needs to be visited before location 2 for this section.

With $k = 0.4$, the 1 km L_1 circle will grow exactly 40% of the distance the agent traveled to arrive at the point of L_1 release. At the moment of UAS release, we assume it immediately starts printing the part for the UAS for the next launch point and travels head-on to the second and final location. Of course, the L_2 circle will grow accordingly during that second segment (and only that second segment) at the same 40% rate and thus the L_2 release point will occur farther than 1 km away from the L_2 location.

Based on the relative locations of any 2-Location scenario, the total path length will be longer or shorter than this example but it will always be less than the total distance if the agent traveled all the way to the center of L_1 and L_2 locations. The path that cuts through the 1 km radius of L_1, on its way to L_{2c}, is of particular interest that will be discussed later in this paper. One can correctly assume that a default range UAS could be released at any time on or within that circle. Note that the agent's mission is considered complete once the UAS is released at the final release point. Return trips to the origin are not included in the path length optimization which is also discussed in later sections of this paper.

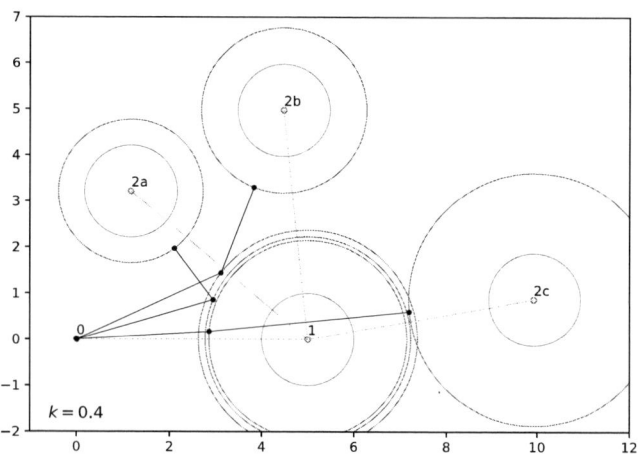

FIGURE 5: THREE POTENTIAL LOCATIONS FOR L_2 WITH OPTI-MAL PATH AND LAUNCH POINTS FOR L_1, L_{2a}, L_{2b}, AND L_{2c}

Figure 6 includes seven different possible positions for L_2. This results in seven points (colored black) where the agent would launch the UAS to visit L_1 and then from that point, seven more points (colored green) where it would travel to before releasing another UAS to visit L_2. These two collections of points form two loci of points (one black and one green), which define the respective optimal launching points for UAS.

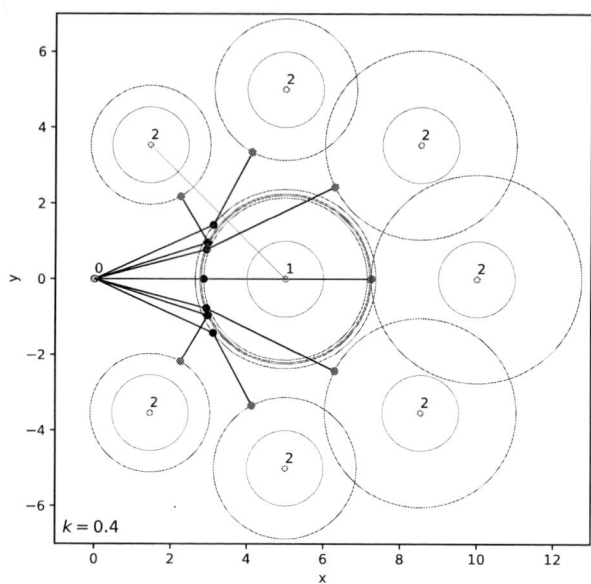

FIGURE 6: SEVEN EVENLY SPREAD (AROUND L_1) POTENTIAL LOCATIONS FOR L_2 WITH OPTIMAL PATH AND RELEASE OR LAUNCH POINTS FOR L_1 IN BLACK AND L_2 IN GREEN

To obtain Figure 7, we add 363 other potential positions for L_2, (spread around in a circle from L_1, for the full 360 degrees), resulting in the full locus of launch points in black for L_1 and in green for L_2. The top right example in Figure 7 has $k = 0.5$ which is close to the $k = 0.4$ in Figure 6. For the scenario presented, the L_1 locus of points (black) will usually assume a curved-C shape while the L_2 locus of points (green) will be more similar to an obloidal shape.

The other three sub-figures show the same two loci of points for the two launch points with different print speeds. Interestingly, for the distances shown, the L_2 locus of points will fold in on itself as U becomes larger than 1. This is because if L_2 is back in the direction of the origin (after visiting L_1) the distance to fly back to the launch point will be quite small.

3.5 Optimization Problem Formulation

For this paper, a comparison between the TSP, TSPN, and the new Apollonius problem will be conducted. The number of locations will range from 1-5. The permutation of the given number of locations will be conducted. A 15 x 15 square km area will be used to simulate an operating environment. The agent will start at position L_0 (which is also $p_0 = (0,0)$) for all scenarios. Locations are randomly placed in the environment, all with the same initial radius r_0. The agent speed, V, will be set at 1, and U will vary in order to change the k ratio value within a range between 0 and 1.

The optimization problem is formally stated as follows:

Copyright © 2023 by ASME

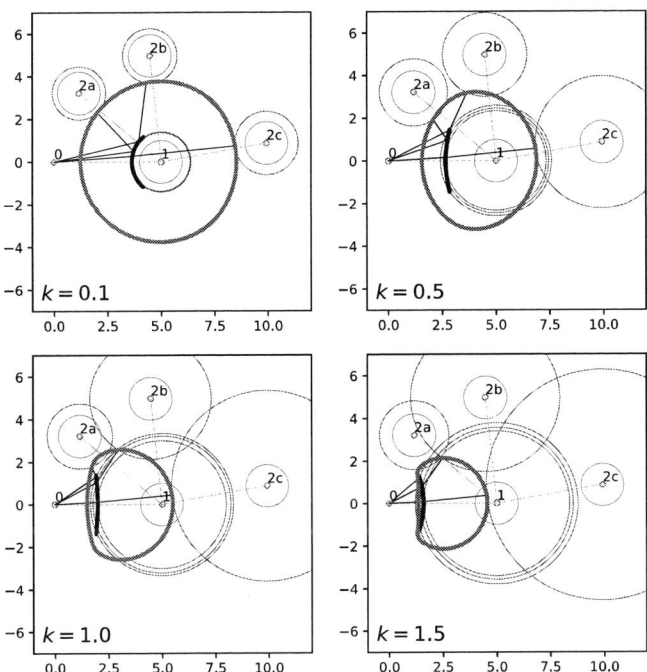

FIGURE 7: DEFINING THE LOCUS OF POINTS FOR THE RELEASE OR LAUNCH POINT FOR L_1 AND L_2 UNDER FOUR DIFFERENT PRINT SPEED CONDITIONS.

$$\min_{r,\theta} \quad D = \sum_{i=1}^{n} d_i(\theta_i, r_i)$$

$$\text{subject to:} \quad g_i(\theta_i, r_i) = \frac{\Delta r_i}{d_i} - \frac{U}{V} \leq 0$$

$$r_i \geq 0 \quad \forall\, i$$

Where the objective function was discussed in equation 10 and the launch positions will be constrained by equation 12. This will bind the launch positions to be on or within their respective Apollonius circle. Once the agent has reached the launch point the UAS is assumed attritable and will not need to return to the agent. Given this, the mission will be assumed complete when the agent has dropped off the last drone. The python optimization package scipy.minimize was used for the analysis presented in the next section.

4. RESULTS AND ANALYSIS

For a specified sequence of locations, the optimizer will minimize the path from the L_0 origin, $(0,0)$, to all the locations, in order, which satisfies the constraint equations specified above. Even if the print speed is set at zero, $k = 0$, the agent can still make use of the UAS to fly the final 1 km to the location. The optimizer will define the UAS launch or release points that lie on those 1 km circles constraints as shown in the top left of Figure 8. The optimal path, as in previous examples, is defined by the black points and sold black line segments.

When 3D printing is applied en route those launch points will move away from their respective location in a concentric circle, because the range of the UAS will be extended beyond 1 km with the 3D printed parts that were fabricated while the agent was traveling to that location. The top right example in Figure 8 has the print speed set at $k = 0.25$ and a shorter optimal distance. With faster print speeds, the launch points will be even further from their locations, (bottom left of Figure 8). Finally, under the condition of very fast print speeds, the optimal sequence may change as well to further minimize the distance. This may have an interesting effect in that some individual line segments will be longer or shorter but the summation of those segments will result in an overall shorter path length D. The following section will explore in more detail the potential changes in optimal sequence and indirectly the minimum path length.

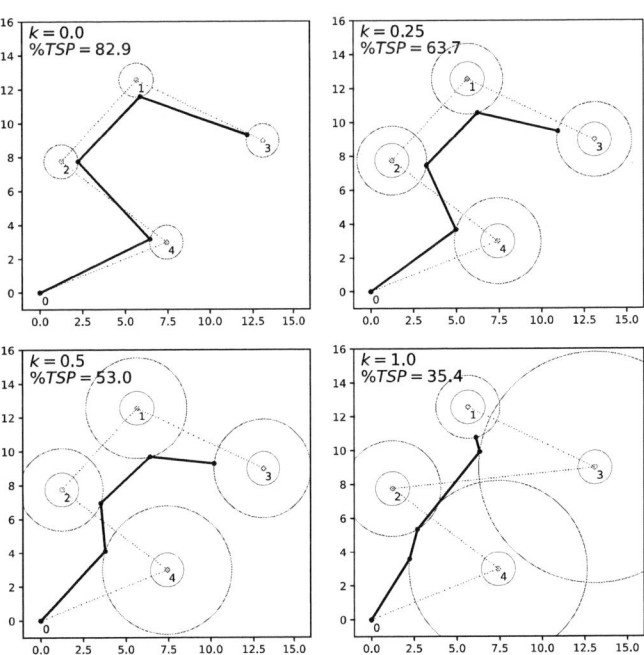

FIGURE 8: MINIMUM PATH LENGTH (BLACK LINE) FOR THE SAME 4-LOCATION SCENARIO UNDER DIFFERENT PRINT SPEED CONDITIONS. THE PERCENTAGE OF OPTIMAL PATH TO THE TSP SOLUTION (RED DASHED LINE) IS ALSO PROVIDED

4.1 Comparison to TSP Distance

For any particular scenario, there exists n-factorial number of sequences any of which could be the shortest path to visiting all the locations. For example, in the 4-Location scenario of Figure 8, calculating all 24 different sequences is necessary to identify the shortest path. This is commonly known as TSP, introduced before, and is known to be a NP-complete problem[21–23].

In Figure 9, small 4-segment links represent the 24 different sequences for the same 4-Location scenario as above. The shortest path or TSP distance is found at the top left sequence in Figure 9. This sequence corresponds to 3 out of the 4 conditions in 8. Interestingly, under different conditions, such as when the 3D print speed is large enough, the shortest path uses a sequence that is not the TSP sequence (e.g. the bottom right of Figure 8).

Copyright © 2023 by ASME

This particular 4-segment link is found on the top row, second from the left, in Figure 9. Without 3D printing, this sequence would be significantly longer than the TSP distance but the interrelationships between the specific sequence and the impacts of 3D printing for extending the range of unmanned systems make it optimal.

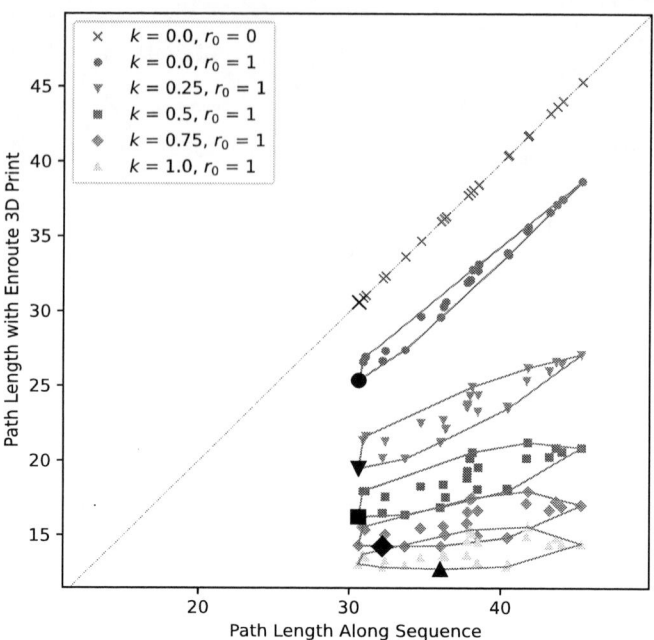

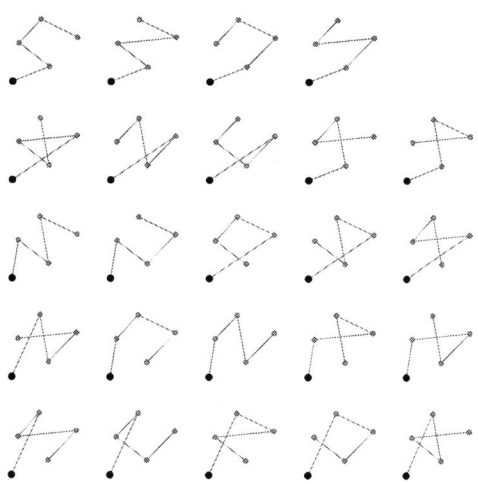

FIGURE 9: REPRESENTATIONS OF THE 24 POSSIBLE SEQUENCES THAT AN AGENT CAN USE TO VISIT THE LOCATIONS IN A 4-LOCATION SCENARIO

What this suggests is that *when an agent can print en route, using the TSP sequence is not always optimal* in terms of total required distance to travel.

In order to explore this phenomenon, we plot the path lengths when the agent must visit each location within each of the 24 possible sequences along the 45-degree line in Figure 10. Each of these is designated with an "x" and the minimum path length across all sequences (i.e., the TSP distance ≈ 31) is the smallest of this set designated with a larger black "x." This condition assumes no 3D printing (i.e., $k = 0$) and no usage of UAS for the final kilometer, (i.e., $r = 0$).

In the next analysis, we assume that the conditions are such that the agent can travel within 1 km of each location and then release or launch a UAS that can fly the remainder of the way in order to complete the mission objective (e.g., delivery, reconnaissance, surveillance, etc.). This condition is represented in the top left figure of Figure 8 and does not require 3D printing (under the assumption that the default range for the small UAS is 1 km). Since there are 24 sequences for this condition of $k = 0$ and $r = 1$, the 24 optimal distances are also plotted in Figure 10 as blue circles. However, since these distances are shorter in general than the $r = 0$ condition, the group of 24 circles are plotted below the 45-degree line. Therefore, the x-values represent the corresponding distance if $r = 0$ (as before) and the y-values represent the distance with the condition $r = 1$ and $k = 0$ resulting in a value lower or below the 45-degree line. The convex hull is added as an outline to this group of points to better indicate its extent.

This pattern is repeated for the next condition of $k = 0.2$

FIGURE 10: PLOT OF THE DISTANCES (Y-VALUES) FOR ALL 24 SEQUENCES UNDER DIFFERENT CONDITIONS. THE X-VALUES ARE ALWAYS THE PATH DISTANCES CORRESPONDING TO THE ORIGINAL SCENARIO WITHOUT 3D PRINTING OR OFFSET UAS RANGE

and $r = 1$ with inverted red triangles. Since this condition now implements 3D printing en route, the total distance traveled by the agent is further reduced and the set of points is moved further down compared to the previous group. The same process is repeated for the additional conditions presented in the legend of Figure 10. In each condition, the group's minimum distance is shown with a larger black point with the corresponding symbol. Each point will have a corresponding point in the other groups based on the same sequence.

The interesting condition occurs when the print speed reaches a large enough value such as $k = 0.8$ or $k = 1.0$ in Figure 10. The minimum distance for these conditions now no longer corresponds with the TSP distance of 31 km. The black triangle and black plus symbols correspond with a sequence that was around 36 km originally but then became only 13 km when 3D printing en route was used. Thus, a deviation from the TSP sequence is more optimal under these conditions. There is still a 'red plus symbol,' (x,y)=(31,13.5), corresponding to that TSP distance but it is now longer than (or above) the 'black plus symbol,' (x,y)=(35,13).

The fascinating discovery is that sticking to the TSP sequence would have resulted in a sub-optimal path plan for this scenario if the print speed was above 0.8. Other scenarios will experience a similar sequence change for the minimum path length based on the value of U and the relative geometry of the actual location positions.

This phenomenon is observed more dramatically with more locations in a scenario. Figure 11 represents a scenario with 7 locations and therefore 5040 sequences. The shortest distance (TSP distance) is again represented with a black "x" and sits on

Copyright © 2023 by ASME

the 45-degree line. The other conditions are again represented with 5040x6 optimizations of sequences and conditions. Each condition has a minimum that is designated with a larger black symbol of the same type. Interestingly, and similar to the example above, the sequence that results in the optimal path for each condition changes as U changes. The seven black symbols correspond to different sequences but with a decreasing path length as the print speed increases.

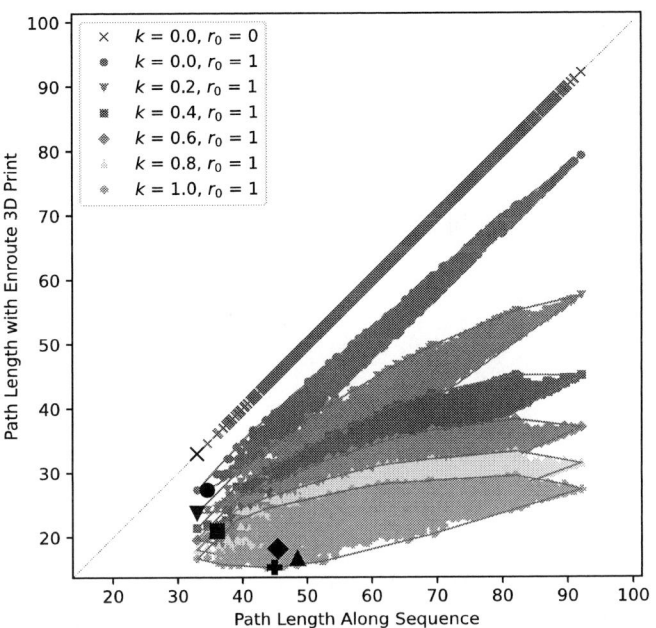

FIGURE 11: PLOT OF THE DISTANCES FOR ALL 5040 SEQUENCES UNDER DIFFERENT CONDITIONS FOR A SCENARIO WITH 7 LOCATIONS.

We can take these seven points and plot these distances against the associated print speeds U along the x-axis as shown in Figure 12. The TSP sequence is [0,5,3,1,7,4,2,6] and has a length of 33 km. The second condition also has $k = 0$ but now $r = 1$ and therefore is plotted directly below the TSP point. This second condition, with a black circle, has a minimum path length of 27 km, and a sequence of [0,5,3,1,7,2,4,6] (note the reversal of L_2 with L_4 compared to the TSP sequence). The optimal sequence changes back when $k = 0.2$ to the TSP sequence. Even more interesting, the optimal sequence continues to change 4 more times at every increase of U up through the last condition when $k = 1.0$ The specific sequence is listed off as a row vector for each point in Figure 12. The changes are sometimes minor with simple reversals of two locations, but some of the changes are more significant. This particular 7-Location example is chosen because of its interesting property of multiple changes and doesn't happen for all scenarios. However, sequence changes are observed quite often and especially with larger values of U.

The inset scenario, bordered in green, of Figure 12 shows the sequence for the minimum path length when $k = 0.4$ with sequence [0,5,3,1,6,2,4,7]. This sequence follows the TSP sequence for the first three locations but then deviates quite substantially in the next four locations.

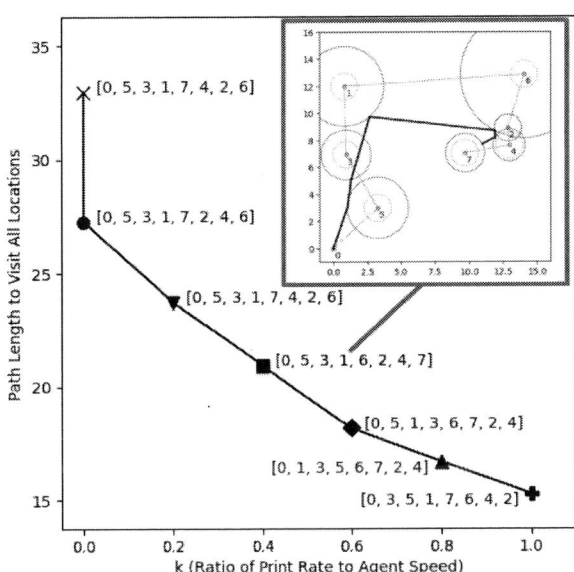

FIGURE 12: REDUCTION IN TOTAL DISTANCE WITH DIFFERENT U/V SPEED RATIOS

4.2 Optimization across the Full Scenario Space

Finally, the process described in the previous sub-section is repeated for 1000s of scenarios across the virtual environment and then analyzed to characterize the reduction in path length when 3D printing can be performed on demand and en route to extend the range of the default UAS.

Initially, 1000 single-location scenarios are randomly selected throughout the 15 x 15 km square area, defined by the boundaries: $0 <= x_i <= 15$, and $0 <= y_i <= 15$, where $i = 1, ..., 1000$. In each of these scenarios, the agent starts at L_0 or $(0,0)$ and flies directly towards the L_1 location. The optimal solution is trivial for these scenarios and can be calculated outside of the optimizer. Essentially, for a single location scenario, the TSP distance is the simple euclidean distance of L_1 from L_0. Without 3D printing en route, that distance is reduced by 1 km if using the prefabricated or default UAS. Likewise, that distance is reduced further if printing the supplementary parts is enabled for extending the UAS range en route. Finally, the faster the print speed (U) the shorter the path because the longer the UAS range is extended. These can all be calculated outside the optimizer because all paths are co-linear with the original TSP path.

For the 1000 2-Location scenarios, 1000 3-Location scenarios, and so on, the optimizer is required to account for the non-trivial solutions that can occur under interesting location positions among the various scenarios. In each of the N-Location scenarios, with N>1, the optimizer will calculate the minimal path, D^*, for the agent for all the sequences for that particular scenario. Thus, for a two-location scenario, it will calculate sequence [0,1,2] and [0,2,1], for all U/V ratios of interest. Likewise, for the 3-Location scenarios, it will perform this opti-

Copyright © 2023 by ASME

mization six times for sequences: [0,1,2,3], [0,1,3,2], [0,2,1,3], [0,2,3,1], [0,3,1,2], [0,3,2,1]. Table 1 summarized the total number of optimizations required to explore the design space up to and including scenarios with five locations. Each scenario will have a corresponding number of sequences to analyze. Further, each scenario and sequence is optimized for five different values of the 3D print speed, U. The final column sums the total number of optimizations within a location set and across the entire design space.

The limit to include up to 5-Location scenarios is mostly a computational expense constraint. The more than 3/4 of a million optimizations (i.e., 765,000) required just under 12 hours to complete. Due to the requirement and guarantee of finding the TSP distance in order to make comparisons, the full factorial set of all sequences for every scenario is necessary as listed in the second column of Table 1. As a result, exploring the 6-Location scenarios would push the compute time to more than a week. Explorations into higher location scenarios were performed but with much smaller sample sizes (i.e. much less than 1000).

TABLE 1: OPTIMIZATION EXPERIMENTS ACROSS THE DESIGN SPACE

N	Sequences	Scenarios	k settings	Optimizations
1	1	1000	5	5000
2	2	1000	5	10000
3	6	1000	5	30000
4	24	1000	5	120000
5	120	1000	5	600000
Total				765000

The results of the 765,000 optimizations are presented in Figure 13. Along the x-axis are the different locations in a particular sequence or scenario. For each particular scenario size (i.e. number of locations), six bar plots are presented for the six respective conditions under which the optimization was performed for the scenarios. Each box plot represents the distribution of only the optimal distance (and sequence) of 1000 scenarios for the condition specified in the legend. The red box-plots assume no default UAS range ($r = 0$) and no 3D printing en route. In other words, the red box-plots represent the simple path required if the agent was to visit all locations by himself. This is confirmed in the 1-Location scenarios where the maximum distance from the origin would be around 21 km, and the furthest one can be from the agent's starting location in a 15 x 15 km area. A similar argument can be made for the mean distance value which is near 12 km shown in Figure 13.

The blue box-plots show the conditions where the agent uses the default UAS range but does not print en route and therefore $r = 1$, and $k = 0$ for these conditions. This pattern continues for the other four box plots, all with $r = 1$, and with $k = 0.25$, $k = 0.5$, $k = 0.75$, and $k = 1.0$ respectively.

As expected the average agent path distance decreases as the print speed increases for the scenarios in Figure 13. Comparisons between box-plots *within* a Number of locations category are feasible because the locations for the 1000 scenarios are all the same, but the conditions (i.e. r and k) are not.

Connecting the mean values for a particular condition across

the different number of location groups would show an increasing trend line but with a diminishing in steepness with increased location number (e.g., connecting the mean values for the yellow box-plots series). This trend line is seen for each condition but with differing amounts of diminishing.

A different trend line could be observed by connecting the mean values within a category (i.e. the six box-plots for each Number of Locations group). This trend line continually decreases but at a slower rate as k increases. This suggests an asymptote towards the theoretical limit of a distance of zero, but also points to the selection of a reasonable print speed based on the law of diminishing returns, where the marginal gains may be insufficient to justify printing any faster. True, printing faster will always save time (via saving distance), but at decreasing amounts as U increases.

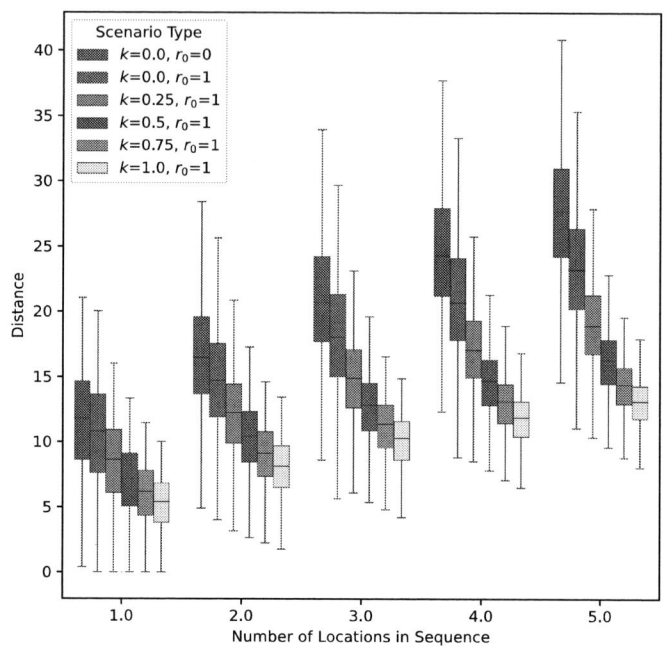

FIGURE 13: SUMMARY RESULTS OF 765000 OPTIMIZATION RESULTS DEMONSTRATING DISTRIBUTIONS OF REDUCTION IN TOTAL DISTANCE. BOX PLOTS ARE SEPARATED BY DIFFERENT U/V SPEED RATIOS AND DIFFERENT SCENARIOS

In order to make generalized comparisons between the expected improvement from 3D printing en route across the design space, we divide each optimal path length by the corresponding TSP path length for the same scenarios. In other words, each of the distances from the 1000 scenarios and for all conditions is divided by the TSP distance for that corresponding scenario. Therefore, the ratio is equal to 1 (or 100%) when we divide the TSP distance by itself which devolves to a "zero thickness" red box-plot at the top of Figure 14. All other conditions, and by extension box-plots, are compared to this maximum distance for each of the 5000 scenarios represented, with the five different conditions.

Although the box-plots are actually distributions, the overall average relationship can be confirmed by comparing the mean of the yellow box-plot (≈ 5), to the red boxplot (mean of 12) in

Copyright © 2023 by ASME

Figure 13 and then observing the mean of the yellow box-plot for N=1, ($\approx$ 45%) in Figure 13 which is close to the 5/12 estimate.

The interesting observation in Figure 13 is the relatively flat trend lines, after connecting mean values for a particular condition across groups. For example, the mean values for the yellow box-plots, when $r = 1$, and $k = 1.0$ seem to all reach a steady state value of around 45%. Across a randomly selected scenario and location, one could therefore expect this reduction with a U/V ratio of 1, regardless of the number of locations in the scenario. Understandably, this value only holds for the particular size of the area and relative speeds under investigation but a steady state value in other environments would be expected.

On the other extreme, with the blue box-plots, representing the conditions when $r = 1$, and $k = 0.0$, the trend line does seem to decrease with every addition of another location in the scenario. But this trend would flatten out as well as the density of locations in the area increases. Thus, a type of saturation is reached where further reduction, in terms of percentage, becomes almost flat.

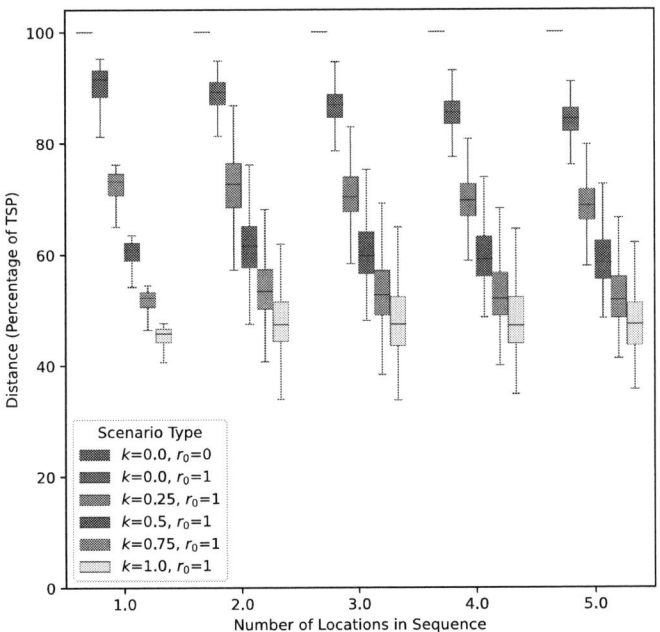

FIGURE 14: DISTRIBUTIONS OF 765000 OPTIMIZATION RESULTS DEMONSTRATING THE REDUCTION IN TOTAL PATH LENGTH (COMPARED TO THE TSP DISTANCE). BOX PLOTS ARE SEPARATED BY THE SCENARIO TYPE AND DIFFERENT U/V SPEED RATIOS.

Although the foregoing figure (Figure 14) demonstrates the expected reduction in the total path length compared to the TSP distance, certain scenarios represented in this summary figure have interesting non-intuitive results that validate the optimization process within the stated domain. For example, in Figure 15, the sequence of locations for UAS launch points is [0,3,5,4,2,1]. In this figure, Location 5 is designated with a green circle indicating that the path cuts across the 1-km radius (i.e., closest approach to L_5 < 1 km). As before, the red dotted line indicates the required path if the agent were to visit each location in the

designated sequence itself. This suggests that printing for L_3 starts at the origin and continues until the launch point for L_3 at which time both UASs for L_5 and L_3 are released concurrently because of the coincidence of those launch points. This represents the intuitive optimal solution (with $D^* = 18.802$) because no printing for L_5 is needed due to the default UAS range of 1 km.

However, the true optimal path can improve slightly upon this solution as can be seen in Figure 16 with $D^* = 18.797$. To obtain the true minimal path length for this scenario, the agent should print the addition for the L_5 UAS first, release it accordingly, and then start printing the L_3 UAS. Because of the interdependence with the downstream locations, the path length (black lines) for sequence [0,5,3,4,2,1] is shorter than sequence [0,3,5,4,2,1]. The trade-off options are essentially printing two smaller parts or one larger part. In this scenario, the two smaller parts strategy is optimal although counter-intuitive. Many of these occurrences happen and are extant in the summary data presented above. Most often they will occur when a general alignment in direction of two locations is found to be similar along the particular sequence (such as in L_5 and L_3 in this scenario). These situations also occur when the radius overlap is possible for some sequences of scenarios and thus the particular circles and therefore release points can intersect.

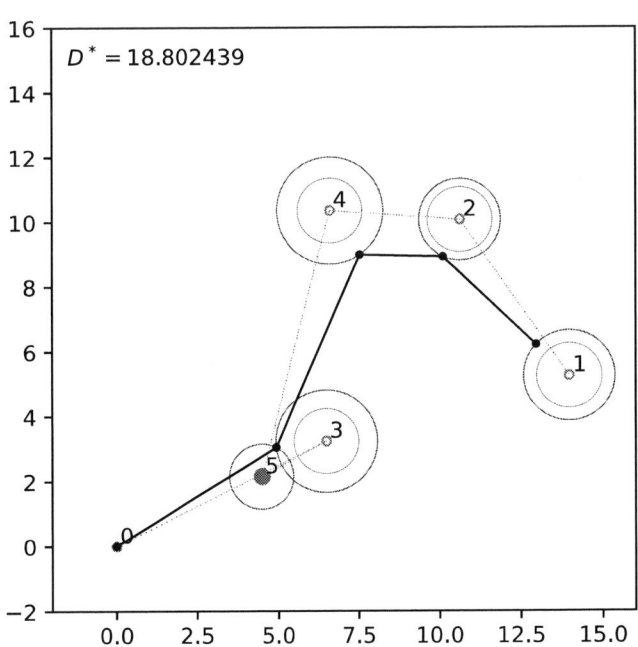

FIGURE 15: EXAMPLE SCENARIO WITH NON-OPTIMAL SEQUENCE, WITH K=0.1 AND SEQUENCE: [0,3,5,4,2,1]

5. FUTURE WORK, LIMITATIONS, AND OTHER TYPES OF OPTIMIZATION

This analysis was performed using a brute force approach exploring all the permutations to determine the optimum solution. This, however, is expensive computationally and prohibitively so with scenarios with numbers of locations greater than 5. Future

Copyright © 2023 by ASME

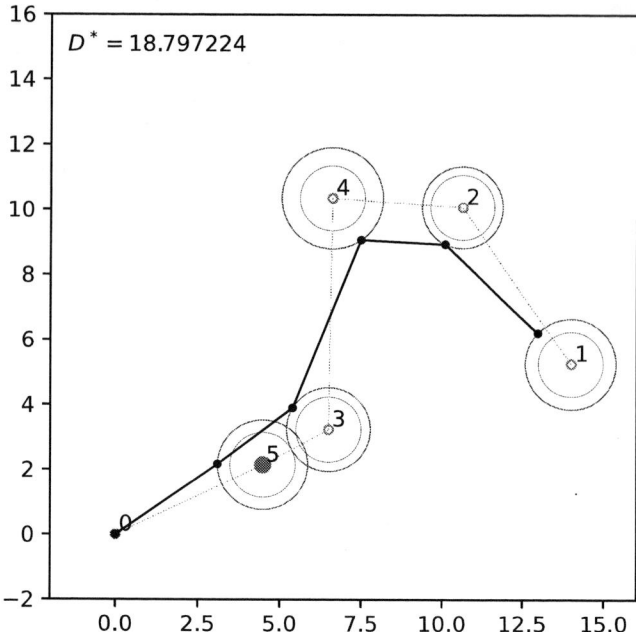

FIGURE 16: EXAMPLE SCENARIO WITH NON-INTUITIVE OPTIMAL SEQUENCE, WITH K=0.1 AND SEQUENCE: [0,5,3,4,2,1]

$$\sum_{r=1}^{n} \frac{n!}{(n-r)!} \qquad (13)$$

The foregoing analysis only explored $n!$ scenarios which included all locations. It is possible a sub-set sequence would still satisfy the mission requirements because a path segment might, by chance, cut through the radius of an out-of-set location. What's more, the reason any sub-set permutation of a series needs to be considered is based on the location for which the agent is printing, since the printing sequence could be different than the launching sequence. So far it has been assumed that the agent only prints for the location to which it is traveling toward. There are some scenarios where a more optimal solution may exist where printing for a future location and *then* the current location would result in a shorter path. This is left for future work.

Straight unobstructed paths are also assumed in this analysis which may not apply to some real-world cases. Future work could include obstacle avoidance schemes like vector fields, random trees structures, or path constraints and scores. This would cause interesting changes to the Apollonius Circle concept where the effective speed in a straight line direction would be reduced or, on the other hand, warping of the circle could be applied based on terrain constraints.

Lastly, the number of locations in every scenario presented in this paper allows for exhaustive testing and optimization. With scenarios with more than eight locations, the computational expense becomes a major factor in a full factorial or brute force approach. In future work, algorithms to approach the optimal path, although never guaranteeing it, will be explored. Of course, many of these may initially be based on the traditional TSP algorithms that make use of the 2opt or 3opt processes or other gradient-free methods, but as described, the TSP sequence *does not* always provide the shortest path under certain values of k. These methods to explore larger scenarios while reducing the computational expense are left for future research.

work will compare a permutation analysis to different types of heuristics to help limit the number of computations needed to find an acceptable solution if not the global optimum.

This analysis assumes that the agent does not return to its initial starting position after the end of every mission. It is likely that in many situations an agent will be asked to return to its starting position after mission completion. For inexpensive reconnaissance UAS and loitering munitions the assumption that the agent doesn't return is appropriate, however, some UAS systems will need to be returned. Any returning to the origin or UAS returning to the agent is left for future work and is beyond the scope of what's included in this paper. It is guaranteed that many of the solutions would change as the last location will now influence the objective function since one more segment is added to the total path distance, and more specifically, the distance from p_n to p_0.

This analysis also assumed that the UAS are attritable. This is clearly wasteful in some examples presented previously and the agent will understandably want to retrieve the UAS after the UAS has delivered a product, surveilled, or otherwise completed its mission. Future work will include having the agent return to the agent after each UAS mission is complete, with additional expected constraints on time and pick-up points.

We recognize that there are some scenarios where permutations will not provide the global optimum. In order to account for these situations and to find the optimal solution, the permutation and any sub-set permutation of the series need to be tested. This causes the number of sequences that needed to be evaluated to be:

6. CONCLUSION

The integration of additive manufacturing while traveling en route and UAS technology could enable the reduction in time of completing missions. Apollonius circles provide a feasible technique into solving the unique problem of visiting multiple locations using UAS in a mission scenario. Additive manufacturing en route could provide benefit of lowering the time and distance cost in an operational environment for a variety of locations where agents carrying 3D printing technology have print rates sufficiently high to extend or enhance the range of attritable UAS for certain missions. Optimizing for minimum distance in these multiple location scenarios does not always align with the intuitive expectation of the traveling salesman sequence solution because of the interdependence of the location positions, the 3D print rate, and the agent's speed. Further advances in this field could encourage system-design decisions that could reduce costs and provide increased flexibility in operational missions.

Copyright © 2023 by ASME

REFERENCES

[1] Oh, Donggeun and Han, Junghee. "Smart search system of autonomous flight uavs for disaster rescue." *Sensors* Vol. 21 No. 20 (2021): pp. 1–19. DOI 10.3390/s21206810.

[2] Sacramento, David, Pisinger, David and Ropke, Stefan. "An adaptive large neighborhood search metaheuristic for the vehicle routing problem with drones." *Transportation Research Part C: Emerging Technologies* Vol. 102 No. March (2019): pp. 289–315. DOI 10.1016/j.trc.2019.02.018.

[3] Christensen, Carsten and Salmon, John. "Principles for small-unit sUAS tactical deployment from a combat-simulating agent-based model analysis." *Expert Systems with Applications* Vol. 190 (2022): p. 116156. DOI 10.1016/j.eswa.2021.116156.

[4] Ling, Haifeng, Luo, Hongchuan, Bai, Linyuan, Zhu, Tao, Wang, Qing and Yu, Lidong. "Planning with Nonlinear Seeker Measurement." Vol. 2020 (2020).

[5] Hapon, Viacheslav, Teplyk, Yevgenii and Tuz, Mykola. "Additive Technology In Aircraft Manufacturing." *Proceedings of the National Aviation University* Vol. 84 No. 3 (2020): pp. 38–43. DOI 10.18372/2306-1472.84.14951.

[6] Goh, G. D., Agarwala, S., Goh, G. L., Dikshit, V., Sing, S. L. and Yeong, W. Y. "Additive manufacturing in unmanned aerial vehicles (UAVs): Challenges and potential." *Aerospace Science and Technology* Vol. 63 (2017): pp. 140–151. DOI 10.1016/j.ast.2016.12.019.

[7] Goh, G. D., Toh, William, Yap, Y. L., Ng, T. Y. and Yeong, W. Y. "Additively manufactured continuous carbon fiber-reinforced thermoplastic for topology optimized unmanned aerial vehicle structures." *Composites Part B: Engineering* Vol. 216 No. February (2021). DOI 10.1016/j.compositesb.2021.108840.

[8] Surface, Naval and Atlantic, Force. "Metal 3D Printer Installed on USS Bataan." No. November (2022): pp. 1–3.

[9] Pecho, Pavol, Ažaltovič, Viliam, Kandera, Branislav and Bugaj, Martin. "Introduction study of design and layout of UAVs 3D printed wings in relation to optimal lightweight and load distribution." *Transportation Research Procedia* Vol. 40 (2019): pp. 861–868. DOI 10.1016/j.trpro.2019.07.121.

[10] McCall, Becky. "Sub-Saharan Africa leads the way in medical drones." *Lancet (London, England)* Vol. 393 No. 10166 (2019): pp. 17–18. DOI 10.1016/S0140-6736(18)33253-7.

[11] U.S. Army. "Offense and Defense ADP 3-90." No. August 2018 (2019): pp. 2018–2020.

[12] Ciolponea, Constantin-Adrian. "The Integration of Unmanned Aircraft System (UAS) in Current Combat Operations." *Land Forces Academy Review* Vol. 27 No. 4 (2022): pp. 333–347. DOI 10.2478/raft-2022-0042.

[13] Christensen, Carsten Douglas. "An Agent-based Decision Support Framework for sUAS Deployment in Small Infantry Units." *ProQuest Dissertations and Theses* (2020).

[14] Atomics, General. "GA-ASI Partners with Divergent Technologies, Inc." General Atomics (2023).

[15] Atomics, General. "GA-ASI Conducts Sparrowhawk sUAS Flight Tests." General Atomics (2020).

[16] Panagiotou, P., Kaparos, P. and Yakinthos, K. "Winglet design and optimization for a MALE UAV using CFD." *Aerospace Science and Technology* Vol. 39 (2014): pp. 190–205. DOI 10.1016/j.ast.2014.09.006.

[17] Rajendran, Saravanan. "Design of Parametric Winglets and Wing tip devices – A Conceptual Design Approach." (2012): p. 71.

[18] Liberman, Edward J. "Air force institute of technology." No. March (2015).

[19] Mennell, William Kenneth. "Heuristics for solving three routing problems: Close-enough traveling salesman problem, close-enough vehicle routing problem, and sequence-dependent team orienteering problem." *ProQuest Dissertations and Theses* (2009): p. 2900.

[20] Alatartsev, Sergey, Augustine, Marcus and Ortmeier, Frank. "Constricting insertion heuristic for traveling salesman problem with neighborhoods." *ICAPS 2013 - Proceedings of the 23rd International Conference on Automated Planning and Scheduling* (2013): pp. 2–10.

[21] De Berg, Mark, Gudmundsson, Joachim, Katz, Matthew J., Levcopoulos, Christos, Overmars, Mark H. and Van Der Stappen, A. Frank. "TSP with neighborhoods of varying size." *Journal of Algorithms* Vol. 57 No. 1 (2005): pp. 22–36. DOI 10.1016/j.jalgor.2005.01.010.

[22] Arkin, Esther M. and Hassin, Refael. "Approximation algorithms for the geometric covering salesman problem." *Discrete Applied Mathematics* Vol. 55 No. 3 (1994): pp. 197–218. DOI 10.1016/0166-218X(94)90008-6.

[23] Dumitrescu, Adrian and Mitchell, Joseph S.B. "Approximation algorithms for TSP with neighborhoods in the plane." *Journal of Algorithms* Vol. 48 No. 1 (2003): pp. 135–159. DOI 10.1016/S0196-6774(03)00047-6.

Proceedings of the ASME 2023
International Design Engineering Technical Conferences and
Computers and Information in Engineering Conference
IDETC-CIE2023
August 20-23, 2023, Boston, Massachusetts

DETC2023-117199

EFFICIENT ROBUST DESIGN SPACE VISUALIZATION AND EXPLORATION FOR MANY-OBJECTIVE PROBLEMS – A VEHICULAR CRASHWORTHINESS EXAMPLE

Niharika Balaji
M.S Student
The Systems Realization Laboratory @ FIT
Florida Institute of Technology, Melbourne, FL, USA

Mathew Baby
Doctoral Candidate
The Systems Realization Laboratory @ FIT
Florida Institute of Technology, Melbourne, FL, USA

Gehendra Sharma
Research Engineer
Center for Advanced Vehicular Systems
Mississippi State University
Starkville, MS, USA

Rashmi Rama Sushil
M.S Student
ADOPT Laboratory,
Department of Engineering Design,
Indian Institute of Technology Madras, TN, India

Palaniappan Ramu
Associate Professor
ADOPT Laboratory,
Department of Engineering Design,
Indian Institute of Technology Madras, TN, India

Anand Balu Nellippallil[1]
Assistant Professor
The Systems Realization Laboratory @ FIT
Florida Institute of Technology, Melbourne, FL, USA

ABSTRACT

The design of vehicular components is often complicated by the presence of many conflicting goals, such as improved energy absorption for safety, mass reduction for increased fuel economy, and so on. The uncertainties associated with the manufacturing processes involved in realizing the components impact vehicular performance. Hence, the design of vehicular components needs to account for many conflicting goals and associated uncertainties to ensure performance. We look at managing the uncertainties involved by exploring the design space and seeking a ranged set of 'satisficing robust solutions' that are relatively insensitive to uncertainties while still meeting the requirements of the designers.

In this paper, we present a robust design exploration framework that combines the robust concept exploration framework with a machine learning-based visualization technique called interpretable Self-Organizing Map (iSOM) to support the design exploration and visualization for problems with many (more than three) conflicting goals. In the framework presented, we examine uncertainty associated with the design variables that define the many conflicting goals. A robust goal

formulation using the Design Capability Index (DCI) construct is employed in the framework to account for the design variable uncertainties. The use of iSOM for visualization of design spaces allows designers to overcome the limitation of conventional ternary plot-based visualization, where designers are limited to visualizing a maximum of 3 goals at a time. The framework presented supports designers in efficiently visualizing and exploring high-dimensional robust design spaces and identifying satisficing robust solutions, which was not possible previously.
Keywords: Surrogate modeling, Robust design, High dimensional design space exploration, iSOM, Satisficing robust solutions.

1. FRAME OF REFERENCE

In the automotive sector, there always exists a push toward designing vehicles and vehicular components with improved performance [1]. Safety, efficiency, and economy are key focus areas for improvements in the design of vehicles and vehicular components. It has been proven that fuel consumption can be cut down by 6% to 8% by reducing vehicle mass by 10% [2]. In addition, lightweight designs also bring considerable benefits in

---

[1] Corresponding author, Email: anellippallil@fit.edu

terms of improved vehicle performance and reduced emissions. Given these benefits and the need to satisfy tighter emission norms, weight/mass reduction is always a focus in vehicular design. Designers look at the use of lighter materials to achieve lightweight designs. But the focus on lightweight designs to achieve improved efficiency and economy often comes at the cost of safety. The aspect of safety is accounted for during design through the consideration of the 'crashworthiness' of vehicular/vehicular component designs. The ability of a vehicle to prevent injury and fatality of the occupants during collisions is known as 'crashworthiness' [3]. With more stringent statutory safety requirements, manufacturers are required to develop safer cars that can protect occupants from fatal injuries during crashes. Hence, the study of vehicle crashworthiness to assess the safety aspect of the design of the vehicle and its components has become vital. The improvement of the crashworthiness of the vehicle body or automobile structures and components has received substantial attention from researchers during the past few decades. The designer's focus in designing for crashworthiness is on realizing sturdy, crush-proof passenger-survival components that can absorb the maximum amount of energy and dissipate it in a stable and regulated manner in the event of a crash. This focus on maximizing energy absorption could typically result in component designs with increased thickness and subsequently increased mass. Hence, it becomes challenging for designers to make decisions that successfully maintain a balance between crashworthiness and light-weighting [4]. Therefore, the need is to support the design of vehicles and vehicular components that meets the conflicting goals – safety and lightweight design. The most difficult task for designers is to enhance the crashworthiness of automobiles with the rapid growth of technology [5]. A trade-off must be made to retain the structural integrity of the vehicle components and allow the vehicle's structure to absorb the maximum amount of energy through structural deformation for occupant safety in the event of an accident [5]. Vehicular experimental crash testing is one method employed to ensure the safety of lightweight vehicular and component designs. But, the experimental crash testing approach is expensive and time-consuming and, therefore, should only be considered during the final stage of the design [4]. An alternative approach is to perform computer simulations. While simulation-based approaches are relatively less expensive, they have limitations arising from: a) lack of computational resources, b) computational complexity, and c) design challenges such as the need to account for multiple conflicting goals, uncertainties in design variables, and visualize and explore the high-dimensional design spaces. Different approaches are discussed in the literature to improve the crashworthiness of lightweight vehicle designs and obtain optimal solutions. For simple problems, single objective optimization formulations have been used, but real-world vehicular design problems typically involve several conflicting goals and require trade-offs to be made between the conflicting goals. Hence, multi-objective formulations have been widely discussed in vehicle crashworthiness design literature. For instance, a multi-objective problem is considered for designing composite absorbers for the

crashworthiness of vehicles by using the radial basis function in combination with genetic algorithms [6]. Parrish and co-authors [7] present an approach that combines the two different fidelity models, i.e., the one-step solver and incremental step solver, as a correction function with an artificial bee colony algorithm for sheet metal forming example. A combination of the Kriging model and NSGA II (non-dominated sorting genetic algorithm) is used for designing vehicle occupant restraint systems to overcome the limitation of conventional methods [6, 8, 9]. Acar and co-authors [10] present a comparison of polynomial response surfaces, radial basis, and Kriging models for designing thin-walled energy absorption tubes in crashworthiness. A comparison of multi-objective optimization and single-objective optimization from a Pareto perspective is presented in [11] to highlight the importance of multi-objective optimization. The Kriging approach is used in conjunction with a multi-objective genetic algorithm (GA) to build crash-worthy vehicles for foam-filled bitubal structures [12]. A gradient algorithm is used in topological optimization for vehicle front rail structure [13]. For vehicle crashworthiness, a framework is proposed for simulation-based design considering an efficient global optimization technique [14]. Jin and coauthors [15] carry out a comparative study of four meta-modeling techniques and discuss their use for multiple objective design problems. The approaches discussed above are deterministic approaches of optimization in the crashworthiness design of vehicles and do not consider any uncertainties that affect the design process.

Simulation-based design approaches for the design of vehicles and vehicular components are subjected to uncertainties arising from the random noises, variability in design variables (associated with manufacturing variations), and the models used in the simulations. It is vital to account for these uncertainties as they significantly impact the performance of the designs, especially when safety is the focus. Several approaches that consider uncertainty in simulation-based design have been discussed in the literature. Gu and co-authors [16] compare the deterministic approach and reliability-based approach optimization for vehicle crashworthiness under multiple impact collisions considering sampling techniques and reliability analysis. In [17], a reliability-based optimization approach is presented to account for the uncertainty involved in designing vehicles for a side crash. Sun and co-authors [18] considered a parametric uncertainty in the design of foam-filled thin-walled structures and used a robust design method to solve the problem. A discussion of the various types of uncertainties and the utility of robust design approaches in comparison with other approaches is presented in [19]. A sequential optimization and reliability assessment technique is used to consider uncertainties in design variables and improve the efficiency of probabilistic design for a vehicle crashworthiness example during a side collision [20]. A multi-objective robust optimization technique is proposed for full front vehicle impact to increase energy absorption and reduce the structural weight considering parametric uncertainties [21]. In [22], the adaptive robust design optimization method is employed to consider the variations in the manufacturing processes during metal forming. An efficient

Copyright © 2023 by ASME

adaptive response surface strategy method is used to minimize the computational simulations and consider data uncertainty in vehicle crashworthiness [23]. Hou and coauthors [24] use the response surface method for optimizing the design of tapered circular tubes, and a comparison is made with three different configurations. The above optimization-based approaches are aimed at helping designers identify unique single-point solutions by employing computationally intensive iterative optimization loops. This results in these approaches being unsuitable for early-stage design exploration, where the designer's focus is on quickly identifying a ranged set of solutions that meets their requirements.

Uncertainties in the manufacturing process and in the materials' mechanical properties impact vehicular components' crashworthiness performance. The uncertainties are categorized into different types: i) Natural Uncertainty: unpredictable uncertainty due to natural randomness; ii) Model Parameter Uncertainty: uncertainty that occurs as a result of inaccurate or insufficient input data; iii) Model Structure Uncertainty: uncertainty that occurs due to the assumptions made in developing the models used; and iv) Propagated Uncertainty: uncertainty that is a combination of the above three uncertainties which is propagated from one level to another during design. These uncertainties in the system will lead to undesirable variations in system performance. While some types can significantly impact system performance, others might be almost insignificant. Hence there is a need to manage the impact of the uncertainties during simulation-based design to ensure performance. This can be achieved by considering robust design methods that help designers identify solutions that are relatively insensitive to uncertainties. In robust design, the focus is not on eliminating the sources of uncertainties but rather on managing their impact [25, 26] on performance. There are three types of information that are considered during the robust design of processes or products. They are uncertainty due to: i) noise factors, ii) control factors, and iii) responses. Noise factors

cannot be controlled, and control factors are the design variables controllable by the designer (see Figure 1). Responses are measures that are indicative of how well the product or the process is performing. Robust design is classified into three types: i) Type I robust design, where the focus is on identifying design variables that meet the performance requirements despite variability in noise factors, ii) Type II robust design, where the focus is on identifying design variables that meet the performance requirements despite variability in the design variables themselves, and iii) Type III robust design, where the focus is on identifying design variables that meet the performance requirements given the variability in the model used.

From a systems design perspective, we view design as a simulation-aided, decision-based process. We, therefore, follow the Decision-Based Design (DBD) paradigm wherein designers make a series of decisions given the information available [27]. Foundational to our work is the Decision Support Problem Technique in DBD [28, 29], anchored in the notion of bounded rationality proposed by Herbert A. Simon [30]. Given that the models employed in simulations are incomplete, inaccurate, of different fidelity, and are approximations of reality, we seek 'satisficing solutions' for the design problem at hand by exploring the solution space. A satisficing solution [31] is one that 'satisfy' and 'suffice' the designers' requirements for the conflicting goals present. The compromise Decision Support Problem (cDSP) [32] is a well-established DSP construct in the literature that is used to explore satisficing solutions for multiple conflicting goals during the early stages of design.

Robust design constructs like Design Capability Index (DCI) [33] and Error Margin Index (EMI) [34] in conjunction with the cDSP construct are employed to help generate satisficing robust solutions that are relatively insensitive to uncertainties while still meeting the multiple conflicting design requirements. DCI is employed for Type I and II robust designs, whereas EMI is employed for Type III robust designs.

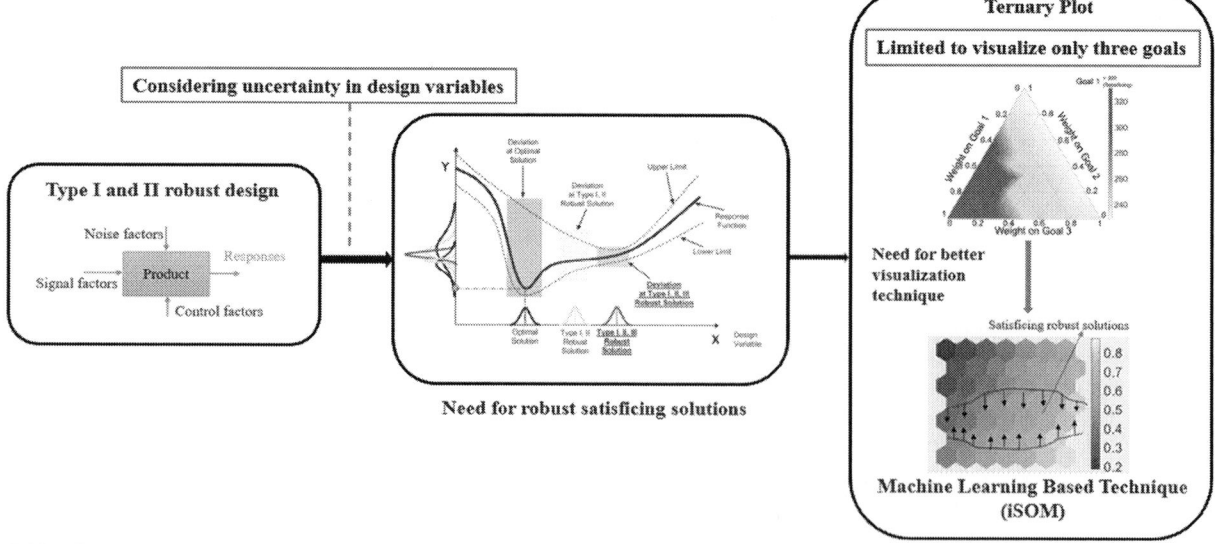

FIGURE 1. Uncertainties during design and the requirement for effective visualization and robust methods.

Copyright © 2023 by ASME

The robust design/solution space generated needs to be visualized to aid designers in making informed tradeoff decisions among the many conflicting goals. Different types of visualization techniques are employed to facilitate design space exploration. Some of them include ternary plots, coordinated plots, nested axis plots, and many others [35]. Sobester and coauthors discuss the use of nested axis plots and tile plots in design space visualization [36]. Tile plots are a combination of matrices, with each tile having a defined width, height, and area which are proportional to input values. The use of nested axis plots for visualization requires some variables to be fixed to capture the variation of the responses with respect to other variables. The use of the above methods assumes the independence of the variables and cannot be carried out for all the dimensions. It becomes challenging to analyze the data without compromising their correctness as the amount and dimension of the data increase. Dimensionality reduction techniques like principal component analysis, singular value decomposition, and independent component analysis have been proposed to deal with the above limitation. These techniques and their appropriate applications are discussed in [37]. The use of trade space exploration tools for visualizing multidimensional data considering a satellite design test example is presented in [38]. In order to carry out design space exploration, other researchers employ methods like ternary plots and scatter plots. However, these are restricted to three or four dimensions only [39]. Given the limitation in the current set of visualization approaches, there is a need for improved methods that enable designers to quickly visualize the high-dimensional design spaces to understand the tradeoffs and relations and thereby make informed design decisions during the early stages of design. The machine learning-based visualization technique, interpretable Self-Organizing Map (iSOM), is a visualization approach that aids designers in making informed decisions by helping visualize the high-dimensional design spaces in 2 dimensions. In this paper, we present a robust design exploration framework to support designers in systematically generating, visualizing, and exploring high-dimensional robust design spaces to identify satisficing robust solutions for many conflicting goals. In the framework, we combine the DCI and cDSP constructs with iSOM based visualization technique, to support the identification of satisficing robust solutions. We showcase the efficacy of the framework using the car design problem for crashworthiness which involves many conflicting goals and uncertainties in the design variables.

The outline of the rest of the paper is as follows. We describe the problem in Section 2. The robust design exploration framework is presented and discussed in detail in Section 3. In Section 4, the utility of the framework is demonstrated using a test problem that involves vehicular design for crashworthiness. In Section 5, we present the visualization technique iSOM (interpretable Self-Organizing Map) and discuss the results of the implementation of the framework for the test problem. We close the paper with our remarks in Section 6.

2. PROBLEM DESCRIPTION

One of the leading causes of loss of human life is the critical injuries sustained during vehicular collisions/crashes [40]. Hence, there exists a need to ensure safety during the design of vehicles to minimize injuries to human occupants and human fatalities. Vehicular crash tests are typically conducted to ensure that the vehicle's design meets mandated standards for crashworthiness. Energy absorption during the crash is one key parameter used to evaluate a vehicle's safety performance. When a collision occurs, the energy generated is partly absorbed by the vehicle's structure and components, and the remaining energy is transferred to the components of the vehicle. The transfer of energy results in the vehicle bouncing back after the crash and can result in severe injuries or fatalities to the human occupants. The greater the energy absorbed by the vehicle structure and components during the initial stages of a collision, the lower the harm done to its occupants. In this paper, we consider a vehicular side crash scenario. The forces that affect the occupants during a side crash are directly related to the weight and size of the vehicle and components, and the intensity of these forces is correlated with the probability of injury [10]. In addition to the need for safety, designers are also required to consider the need to ensure reduced mass/weight for improved efficiency and emissions. In this paper, we address the design of safe lightweight vehicles during a side crash scenario. There could be negative effects on the vehicle's safety in terms of certain safety parameters when the mass of the vehicle is reduced. Given the complex interactions and generally conflictive nature of lightweight and crash performance requirements, the design of lightweight and safe vehicles requires the simultaneous exploration of weight reduction and energy absorption goals. The energy absorbed upon collision is significantly influenced by the design of the energy-absorbing structure. A 1996 Dodge Neon car model is used in this paper. The energy-absorbing structures identified include five components, as shown in Figure 2 and listed in Table 1.

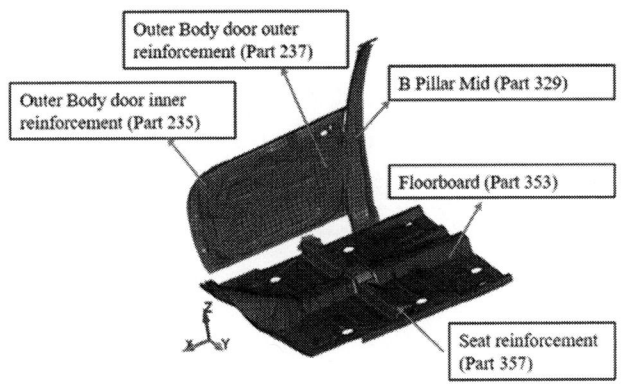

FIGURE 2. Five components considered for side crash scenario

Copyright © 2023 by ASME

TABLE 1. Description of the selected components

Part No.	Part Description
235	OB-DOOR-FT-I-R (Outer body door inner reinforcement)
237	OB-DOOR-FT-O-R (Outer body door outer reinforcement)
329	CH-B-PILLAR-MID-R (B Pillar Mid)
353	CH-CBN-FLOORBRD-FT (Floorboard)
357	CH-CBN-SEAT-REINF-FT (Seat Reinforcement)

The goals of the design problem considered are to maximize energy absorbed individually by three components, namely Parts 235, 237 and 353 (the first three goals) and minimize the combined mass of all 5 components (fourth goal). The thicknesses of five selected components are the design variables. The choice of the 3 components is based on two criteria: i) the potential for highest mass reduction, and ii) components that has potential for higher energy absorption. The above assessment was carried out based on the data from a trial-side crash simulation run, where the mass and total energy absorbed by car components were recorded. The components that had higher mass or absorbed greater energy or met both criteria were selected because, with these parts selected, there is a greater potential to improve the vehicle design in terms of the goals of the problem considered. In Table 1, details of the five components selected for vehicular side crash problem is provided. The process of designing a safe lightweight vehicle is complex owing to i) the large number of design variables and their complex interactions, and ii) many conflicting goals. Due to the above characteristics, there are potentially thousands of designs to consider, especially during the early stages of design. Hence, it is impossible to neither rely on the experience of the design engineer to select a design that is safer nor carry out expensive and time-consuming physical tests. Therefore, we look at the use of simulations to come up with candidate designs/solutions and then explore the design/solution spaces to identify 'satisficing solutions' for the conflicting goals present. The design of safe lightweight vehicles is also subject to uncertainties like random noises that are attributed to random variations in material properties and characteristics, uncertainties in design variables that are attributed to the variations in manufacturing/processing, and uncertainties in the models used that are attributed to the surrogates that are approximations of the true relations between design variables and responses. Hence, there is also the need to account for these uncertainties in the simulation-based design of safe lightweight vehicles.

We look at 'robust design', where the focus is on managing the uncertainties involved by identifying solutions that are relatively insensitive to the uncertainties present. Hence, our focus in this paper is on identifying 'satisficing robust solutions,' that 'satisfice' the designers' requirements for the conflicting goals present while still being relatively insensitive to the uncertainties involved. The current approaches that support the robust design exploration are limited by their ability to consider a maximum of 3 goals for visualization and exploration. The design of complex systems, including the design of safe lightweight vehicles, is typically characterized by the presence of more than 3 goals. Hence, there is a need to support the robust design exploration of design problems with many (more than 3) goals.

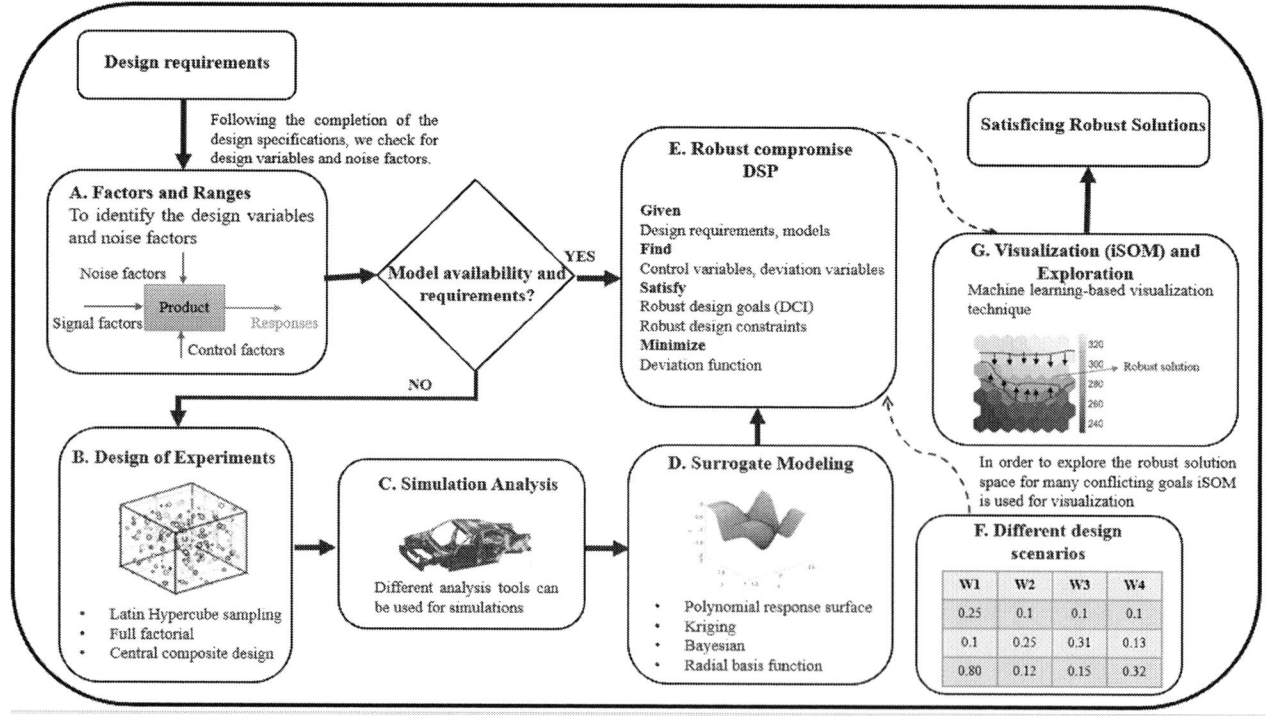

FIGURE 3. Robust design exploration framework

Copyright © 2023 by ASME

We, therefore, present a robust design exploration framework for many goal problems that allows designers to systematically: i) generate robust solution space for the problem at hand, and ii) visualize and explore the solution space to identify satisficing robust solutions for the many goals.

3. ROBUST DESIGN EXPLORATION FRAMEWORK

The proposed robust design exploration framework, is discussed in this section and shown in Figure 3. The proposed framework is based on the Robust Concept Exploration Method (RCEM) presented by Chen and coauthors [41]. The framework is described using Steps A to G below.

Step A: In this step, problem-specific information in terms of design variables, their bounds, and constraints are identified by the designer for the given design requirements. Following this, a check is carried out to identify if mathematical models that relate the design variable to the goals are available explicitly. If models are available, the designer proceeds to *Step E*. If models are not available, then the designer proceeds to *Step B*.

Step B: If mathematical models that capture the problem-specific relations are not available, the designer seeks to develop data-driven surrogate models that capture the relations using simulations. A Design of Experiments (DOE) is carried out to determine the design points at which simulations need to be carried out to generate the data required for surrogate modeling. Designers can choose from various DOE techniques available in the literature to span the design space effectively. For the design problem discussed in Section 2, we use the Latin Hypercube Sampling (LHS). The benefit of using an LHS sampling is that it can span the design space efficiently without having to use many points for problems with many dimensions.

Step C: Next, Finite Element (FE) vehicular crashworthiness simulations are carried out by the designer at the design points identified in Step B, to generate the simulation data for developing surrogate models. For the problem presented, the FE simulations are carried out using a 1996 Dodge Neon model.

Step D: In Step D, surrogate modeling is carried out. The surrogate models so developed are the approximate representation of the actual underlying function model and are utilized by the designer to map the design space to response/performance space based on the data generated from the FE simulations in Step C. In this paper, we make use of response surface methodology to develop second-order polynomial response models for the goals of the problem presented in Section 2.

Step E: In this step, the designer makes use of the surrogate models developed in Step D and the models identified in Step A and formulates the decision support problem. The designer uses the compromise Decision Support Problem (cDSP) with DCI constructs incorporated into the formulation to account for uncertainties in design variables. The cDSP allows designers to model problems with many conflicting goals. The cDSP is a hybrid of mathematical and goal programming. The problem-specific information is captured in the cDSP using the four keywords - *Given, Find, Satisfy,* and *Minimize*. Using the cDSP,

the designer seeks to minimize the weighted sum of deviations of the goal values achieved from their targets. The designer can generate multiple design solutions by assigning different weights to the different goal deviations in the cDSP.

Step F: After the robust cDSP is formulated in Step E, the designer establishes different design scenarios to exercise the cDSP formulation. Each design scenario corresponds to a specific combination of weights assigned to the deviation of the goal values from their targets (weights add up to 1). The weights are indicative of the preferences of the designer for the different conflicting goals present. The cDSP is then exercised for the design scenarios by the designer, and the results of the same are reported.

Step G: In Step G, the designer visualizes the design and/or solution spaces based on the solutions obtained from exercising the robust cDSP formulation for the different design scenarios. The visualization assists the designer in systematically exploring the design space to identify satisficing robust solutions for the many goals problem. In this paper, we make use of the machine learning-based visualization technique called interpretable Self-Organizing Maps (iSOM) to visualize the solution space for the many goals.

4. DEMONSTRATION OF THE UTILITY OF THE ROBUST DESIGN EXPLORATION FRAMEWORK USING VEHICULAR CAR CRASH DESIGN PROBLEM

The framework's utility is demonstrated using the vehicular car crash design problem. The designer follows the steps presented in Section 3 (see Figure 3) to identify robust satisficing solutions for the many (4) goals of the problem while accounting for the uncertainties in the design variables (thickness of the 5 chosen components, see Table 1).

Step A: Determination of design variables

The designer starts by identifying the design requirements. For the design problem considered, the designer aims at maximizing energy absorption for a side impact scenario while minimizing the mass by controlling the thickness of the five important components identified (see Table 1). It is also required to ensure that the design is robust to uncertainties in the thicknesses of the components. These uncertainties occur due to faults in the manufacturing process, geometric intolerances, and human errors. The four goals of the car side crash problem considered are to maximize energy absorption for five components and minimize the total combined weight of all five components. The thicknesses of the five selected components are the design variables. The designer next checks to see if models that relate the responses and the input design variables are available explicitly. For the problem considered, mathematical models are not directly available, and therefore, we proceed to step B.

Step B: Design of Experiments

Since models that capture the relations are not available for the 4 goals of the problem, we develop data-driven surrogate models. An LHS DoE is used to generate a set of points for carrying out FE vehicular crash simulations. In this paper, 44

Copyright © 2023 by ASME

437

LHS design points are generated for the car side crash problem, and these 44 points are the different combinations of the values of the five thickness variables within their upper and lower bounds.

Step C: Finite element simulation for vehicular side crash

Next, FE simulations are carried out to generate data required for surrogate modeling, as described in Step C of the framework, see Figure 3. The FE model used in the side crash simulation is a vehicle model developed by the United States National Crash Analysis Center [42]. Later this model was modified by researchers at the Center for Advanced Vehicular Systems (CAVS) at Mississippi State University [43, 44]. In Figure 4, we depict the Dodge Neon FE simulation setup for the side-crash scenario.

FIGURE 4. LS Dyna FE car crash simulation

The dashboard, inside door paneling, steering wheel/column assembly, driver's seat, and under-the-hood parts, are all included in the interior of the vehicle model. There are a total of 221,049 elements and 433,287 nodes in this modified FE car model. The moving deformable barrier (MDB) developed by Fang and co-authors [44] serves as the impacting vehicle model in the FE simulation. In this paper, we consider side crash scenarios for the vehicle model. The software LS-Dyna is used to simulate every single car crash instance.

Step D: Surrogate modeling

Response surface methodology is employed to develop second-order polynomial response models for the four goals in terms of the design variables. The FE simulations discussed in Step C are run for the LHS DOE points identified in Step B2. Polynomial response surface models of different orders (first order, second order, and third order) are created for all 4 goals. Based on the coefficient of determination (R^2) and Cross-Validation Mean Absolute Error (CV-MAE) values of the different surrogate models, the second order is found to have the best fit when compared to other orders. The second-order polynomial response model takes the form shown in Equation 1.

$$\hat{y} = \beta_0 + \beta_1 t_1 + \beta_2 t_2 + \beta_3 t_3 + \beta_4 t_4 + \beta_5 t_5 + \beta_{11} t_1^2 + \beta_{12} t_1 t_2 + \beta_{13} t_1 t_3 + \beta_{14} t_1 t_4 + \beta_{15} t_1 t_5 + \beta_{22} t_2^2 + \beta_{23} t_2 t_3 + \beta_{24} t_2 t_4 + \beta_{25} t^2 t_5 + \beta_{34} t_3 t_4 + \beta_{35} t_3 t_5 + \beta_{44} t_4^2 + \beta_{45} t_4 t_5 + \beta_{55} t_5^2 \tag{1}$$

The thickness of the five components namely, Part 235, Part 237, Part 329, Part 353, and Part 357 are given by t_1, t_2, t_3, t_4, and t_5, respectively. In Table 2 we show the R^2 values for the second-order polynomial response surface models developed. The surrogate models developed are included in Table A1 of Appendix A.

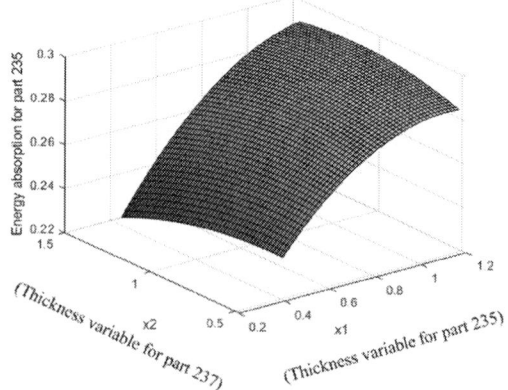

5a. Energy absorption for Part 235 (value× 10^6 in N-mm)

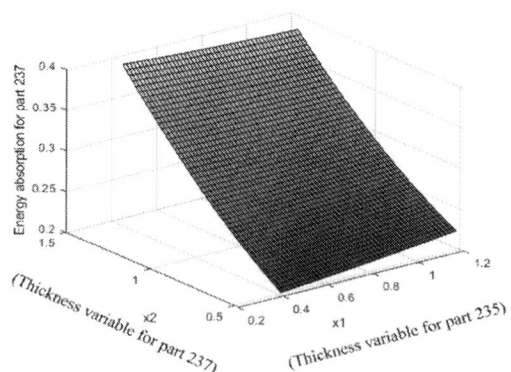

5b. Energy absorption for Part 237 (value× 10^6 in N-mm)

Copyright © 2023 by ASME

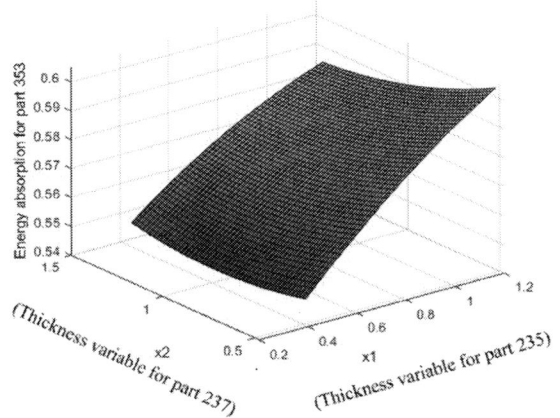

5c. Energy absorption for Part 353 (value× 10^6 in N-mm)

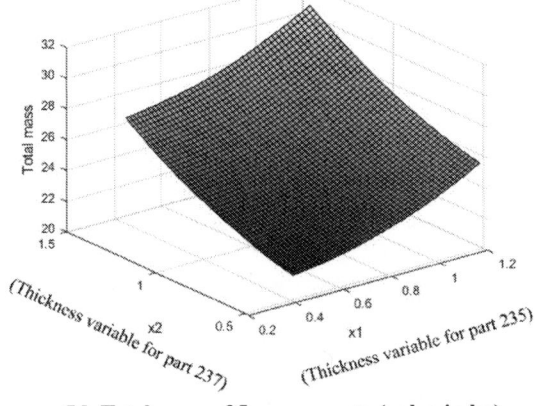

5d. Total mass of 5 components (value in kg)

FIGURE 5. Mean response models for the 4 goals

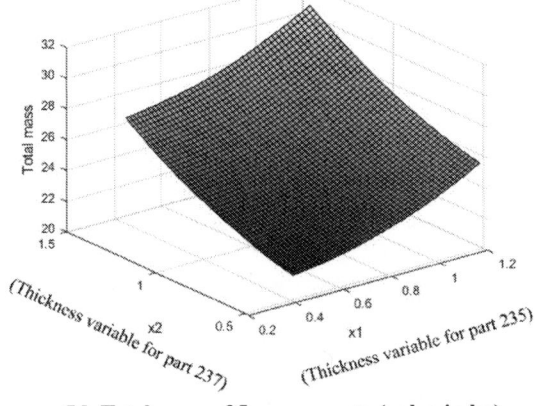

FIGURE 6. DCI construct and calculations

In Figure 5, mean response models for the four goals considered in this paper are plotted, keeping three thickness variables fixed.

TABLE 2. R² values

Components	R² Value
Energy absorption for Part 235 (E_{235})	0.9200
Energy absorption for Part 237 (E_{237})	0.9955
Energy absorption for Part 329 (E_{329})	0.9985
Energy absorption for Part 353 (E_{353})	0.8609
Energy absorption for Part 357 (E_{357})	0.9821
Total Mass of all Parts (Mass)	0.9813

Step E: Robust cDSP Formulation – With DCI Construct

After the surrogate models are created for the 4 goals in Step D, the robust cDSP for the problem is formulated. Compromise decision support problem is used to formulate the problem. The four important keywords: Given, Find, Satisfy, and Minimize, capture the problem-specific information. Formulating and solving the given problem for robust design as cDSP construct is explained in [31]. The robust design formulation is achieved using two metrics: i) Design Capability Index (DCI), and ii) Error Margin Index (EMI). DCI is incorporated while considering uncertainty in design variables, and EMI is considered when there is uncertainty in the model. In this paper,

we use the cDSP in conjunction with the DCI construct to manage the uncertainty in the thickness design variables. DCI is a metric developed for evaluating how well a variety of design criteria may be satisfied by a variety of design specifications considering uncertainty in the design variables themselves. DCI is employed as goal formulation in the cDSP for the car design problem, which is represented in Table 3. DCIs are used as performance and robustness indices for systems. There are three different cases of DCI, as shown in Figure 6, based on the nature of the goal. The mean system performance should always be within LRL (lower requirement limit) and URL (upper requirement limit). The measure of safety against failure caused by uncertain parameters increases with the DCI value. When the requirements are provided as a range of values rather than a target value due to the uncertainty in the design variables, it is crucial to understand the extent of values that satisfy the criteria.

The steps to formulate the problem as DCI construct is given below.

Step 1: Estimate the response variation caused by changes in the design variable using a first-order Taylor series expansion. The response variations are given by Equation 2.

$$\Delta Y = \sum_{i=1}^{n} |\frac{\partial f}{\partial x_i}| . \Delta x_i \qquad (2)$$

where, ΔY represents function model equation, $\frac{\partial f}{\partial x_i}$ represents the differentiation of the function equation, and Δx_i represents the variance allowed in design variable.

Step 2: Next the DCI is calculated as per Equation 3 given below.

$$DCI = (\mu_y - LRL) / \Delta Y \text{ (Smaller is better case)} \qquad (3)$$

where, μ_y is the mean response function and LRL is the lower requirement limit for smaller is better scenario.

Step F: Exercising cDSP for different scenarios

The cDSP presented in Table 3 is exercised for different design scenarios to generate different solutions. 44 design

Copyright © 2023 by ASME

scenarios are considered to explore the design space and to obtain robust solutions for the given test problem. The different weight scenarios represent different preferences for the 4 goals. Selected design scenarios and corresponding weights assigned to goals are shown in Table 4. When designers focus on maximizing just one goal, Scenarios 1 through 4 are used, see Table 4. For instance, when we want to maximize goal 3 then scenario 3 is considered where the full weightage is given to the third goal. Scenario 5 represents a situation where all the goals are given equal priority.

TABLE 3. cDSP with DCI construct

Given

Design requirements for the problem
- Maximize energy absorption for three components, E_{235}, E_{237}, E_{353} (3 goals)
- Minimize the mass of the components, Mass (fourth goal)

Requirements for cDSP goals
- To maximize the DCI formulation for all the four goals. A DCI Target value of 20 is defined for all goals.

Find

μ_x (mean of system variables): thickness values of parts selected - Part 235 (t_1), Part 237 (t_2), Part 329 (t_3), Part 353 (t_4), Part 357 (t_5)
d_i^+, d_i^- (deviation variables)

Satisfy

System constraints
- Constraint for mass
  Mass $\geq$ 30 (kg)
- Energy Absorption constraint
  $E_{235} \geq 0.23 \times 10^6$ (N-mm)
  $E_{237} \geq 0.22 \times 10^6$ (N-mm)
  $E_{353} \geq 0.49 \times 10^6$ (N-mm)

System goals
Goal 1
- Maximize DCI for Part 235 Energy Absorption
  DCI E_{235} (x) / DCI $E_{235, \text{Target}}$ $+d_1^- - d_1^+ = 1$

Goal 2
- Maximize DCI for Part 237 Energy Absorption
  DCI E_{237} (x) / DCI $E_{237, \text{Target}}$ $+d_2^- - d_2^+ = 1$

Goal 3
- Maximize DCI for Part 353 Energy Absorption
  DCI E_{353} (x) / DCI $E_{353, \text{Target}}$ $+d_3^- - d_3^+ = 1$

Goal 4
- Maximize DCI for Total Mass of all Parts
  DCI Mass(x) / DCI Mass, $_{\text{Target}}$ $+d_4^- - d_4^+ = 1$

Minimize

To minimize deviation function

$$Z = \sum_{i=1}^{4} W_i(d_i^- + d_i^+); \quad \sum_{i=1}^{4} W_i = 1$$

TABLE 4. Weight scenarios for the four goals

Scenarios	Weight 1	Weight 2	Weight 3	Weight 4
1	1	0	0	0
2	0	1	0	0
3	0	0	1	0
4	0	0	0	1
5	0.25	0.25	0.25	0.25
6	0.46	0.23	0.23	0.08
7	0.24	0.17	0.28	0.31
-	-	-	-	-
15	0.3	0.26	0.19	0.25
16	0.26	0.02	0.07	0.65
-	-	-	-	-
43	0.12	0.14	0.39	0.35
44	0.03	0.48	0.35	0.14

Step G: Visualization and exploration of robust design space

Next, the robust solution space of the problem is visualized by the designer using iSOM. A brief discussion of iSOM and its application to the problem is discussed below. iSOM visualization helps the designer systematically explore and visualize the design space to identify satisficing robust solutions.

5. INTERPRETABLE SELF-ORGANIZING MAPS (iSOM)

Many goals design problem often involve the idea of tradeoff between these goals due to their potential conflicting nature. The designers interest lies in improving the design and understanding the tradeoffs involved in such design problem [45]. The flexibility to explore and visualize the design space allows the designer to make better design decisions. The most popular neural network model is Kohonen's self-organizing map (SOM) [46]. SOM converts the higher dimensional data into lower dimensional data for easy interpretation of the solutions. SOM consists of two layers: input and output layer as depicted in Figure 7.

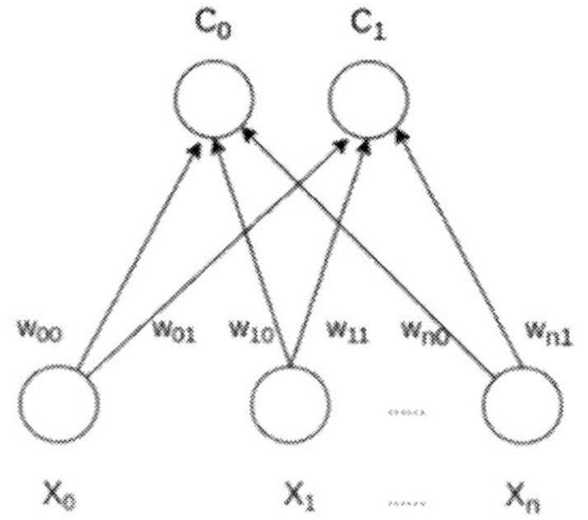

FIGURE 7. Vector mapping during training in SOM

In SOM, first, the nodes are initialized by assigning random values. Next, the Euclidean distance is calculated from input vectors (X_0, X_1, ..., X_n) to clusters (C_0 and C_1), and the one with the minimum Euclidean distance is considered the winning vector. The weight rule is updated, and the SOM network is trained, and these trained vectors are used for new clusters. In Conventional SOM the node with the smallest distance from the input vector is considered as best matching unit. The process is continued till the error reduction metrics are met and trained SOM weights are obtained as the output.

Copyright © 2023 by ASME

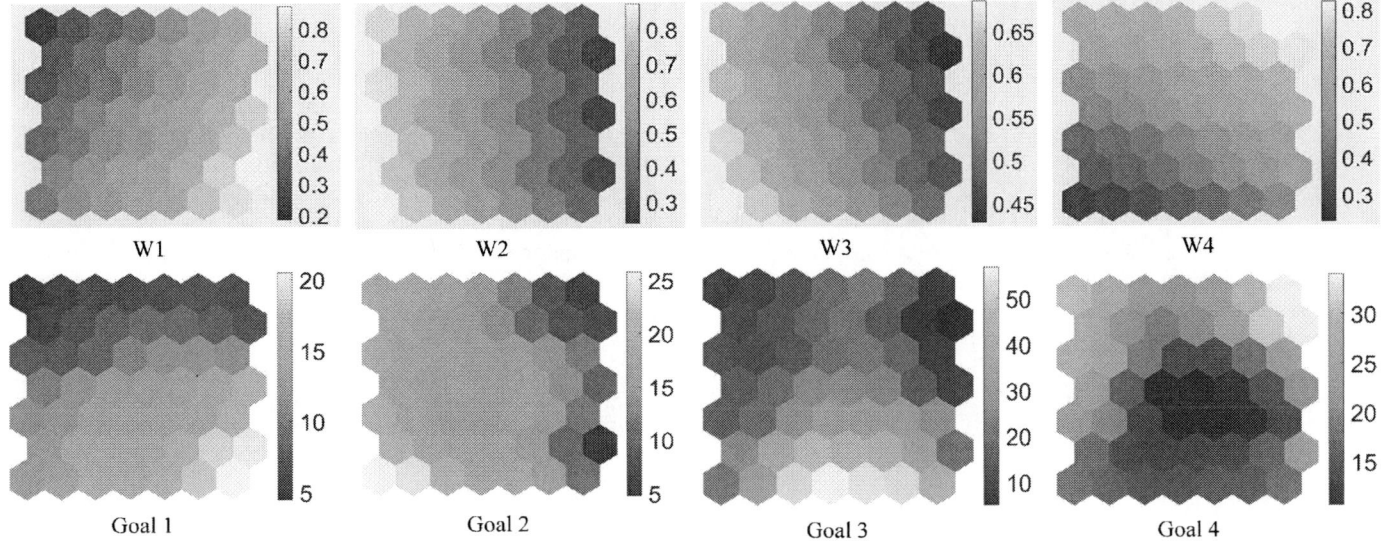

FIGURE 8. Input and Output component plots for car side crash

The conventional self-organizing maps (cSOM) have a limitation of self-folding that affects the visualization of the results [35]. To overcome this limitation of cSOM, interpretable self-organizing maps (iSOM) were developed where results are easily interpretable and avoid self-folding [35]. In iSOM, the selection of the best matching unit (BMU) is different and only the input data is considered, and the update step simply uses the response value and leaves out the input variables. These modifications prevent the iSOM grid from self-intersecting or folding [35].

To visualize robust solution space generated at the end of Step G, we make use of the machine learning-based visualization technique called iSOM. In Figure 8, we represent the 2D component planes of iSOM for the weights (inputs) and goals (outputs) for the car side crash problem. Weight 1 (see W1 in Figure 8) refers to the weightage assigned to Goal 1 (maximize DCI value for component 235), Weight 2 (W2 in Figure 8) refers to the weightage assigned to Goal 2 (maximize DCI value for component 237), Weight 3 (W3 in Figure 8) refers to the weightage assigned to Goal 3 (maximize DCI value for component 353), and Weight 4 (W4 in Figure 8) refers to the weightage assigned to for Goal 4 (maximize DCI value for total mass). In Figure 8, the scales for the goal plots represent the achieved DCI values. A high DCI value indicates that the mean value of the goal is away from the LRL (lower requirement limit) or URL (upper requirement limit) identified and has minimum variance about its mean. In order to identify satisficing robust solutions for all 4 goals, a systematic approach is employed. From Figure 8, the designer can capture the trends of the goals with varying weights on the goals. It is seen that high values of DCI for Goal 1 are achieved when W1 value is high. Similarly, each DCI values of goals increase with an increase in respective weights. On analyzing the iSOM plots in Figure 8, the conflicting nature of the goals of the car side crash problem becomes evident as regions that satisfy the DCI values of one goal conflict with the high DCI regions for another goal. The use of iSOM plots

can make the exploration and interpretation of solutions easier when designers are interested in maximizing or minimizing certain goals.

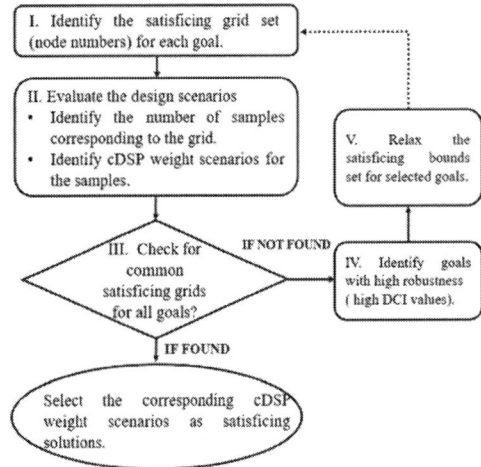

FIGURE 9. Steps involved in the systematic identification of common satisficing solutions for all goals using iSOM

We employ a systematic approach to explore the robust design space using the output component iSOM plots to identify satisficing robust solutions for the many goals. The steps of the systematic approach followed are shown in Figure 9. Based on the satisficing limits set by the designer for the various goals, we start with Step I (see Figure 9) to identify the satisficing grid set for each goal. This involves finding the node numbers of the iSOM grid points that fall within the satisficing limits of each goal. The node numbers of the iSOM grid points (49 grid points) are shown in Figure 10.

Copyright © 2023 by ASME

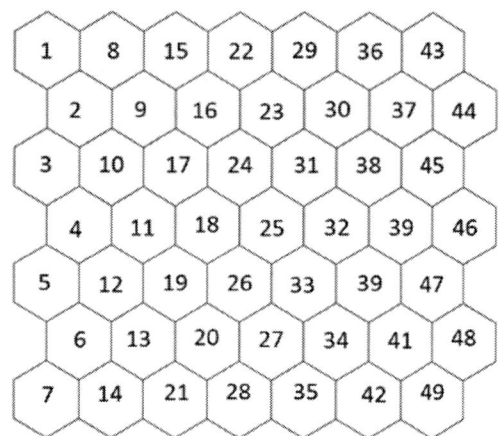

Figure 10. iSOM Grid point location numbers

Next, in Step II, we evaluate the design scenario corresponding to the identified satisficing grid set for each goal and identify the design scenarios that are mapped to the grids. We then move to Step III, where we look for common satisficing grid points. If any are found, we go to the last step and select the cDSP weight scenarios corresponding to the common satisficing grid points as the common robust satisficing solutions. When no common satisficing grid points are identified, we proceed to Step IV to determine goals with high robustness (high DCI values). Next, in Step V, we relax the satisficing limits set for selected goals (goals with high DCI values identified in Step V). We then repeat Steps I through V in a loop until common satisficing grid point/points are identified. Following the systematic approach described above, we choose the Regions of Interest (RoI) based on the design requirements identified for the car crash problem. These RoIs are represented by the highlighted hexagons in Figure 11a.

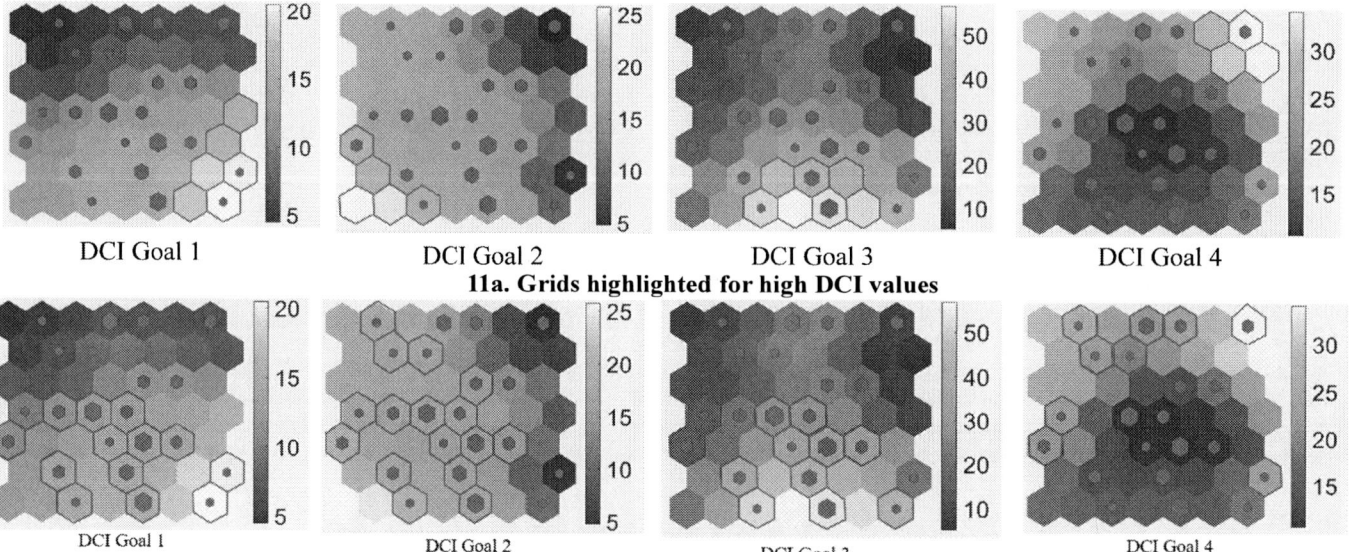

11a. Grids highlighted for high DCI values

11b. Grids highlighted after relaxing satisficing goal limits

FIGURE 11. iSOM plots with highlighted grids for satisficing goals

TABLE 5. Weight scenarios and design variable values considering high DCI values

Goals	Grids	No of scenarios	Weight scenario	Design Variables (Thickness of Parts) in mm				
				Part 235 (t_1)	Part 237 (t_2)	Part 329 (t_3)	Part 353 (t_4)	Part 357 (t_5)
Goal 1	49	1	24	0.794759	0.785839	0.397125	0.913072	0.383625
	48	1	1	0.999904	0.578394	0.377797	0.5714	0.407019
Goal 2	5	2	8	0.88638	0.822095	0.465883	0.735803	0.468546
			2	0.9725	0.826967	0.5295	0.600292	0.5115
	21	1	30	0.953159	0.776908	0.579186	1.01582	0.343664
Goal 3	27	3	36	0.988117	0.776135	0.582469	1.01565	0.343664
			33	0.957301	0.83545	0.3621080	0.73001	0.517434
			14	0.792591	0.785477	0.397125	0.913072	0.383625
	35	4	6	0.747891	0.800108	0.397125	0.823183	0.383625
			18	0.794759	0.785839	0.397125	0.913072	0.383625
			20	0.633364	0.804622	0.364031	0.808268	0.573625
			32	0.792591	0.785477	0.397125	0.913072	0.383625
	21	1	30	0.692604	0.835467	0.360528	0.563781	0.343664
Goal 4	43	2	4	0.583963	0.604725	0.355451	0.53114	0.437419
			6	0.632911	0.626524	0.357554	0.602653	0.344142

Copyright © 2023 by ASME

TABLE 6. Weight scenarios and design variable values after relaxing satisficing limits

Goals	Grids	No of scenarios	Weight scenario	Design Variables (Thickness of Parts) in mm				
				Part 235 (t_1)	Part 237 (t_2)	Part 329 (t_3)	Part 353 (t_4)	Part 357 (t_5)
Goals 1, 2, 3 and 4	4	1	17	0.36525	0.25654	0.65894	0.589546	0.345879
	5	2	8	0.324589	0.52451	0.23145	0.364217	0.42514
			2	0.59785	0.345698	0.612785	0.84573	0.32456

To demonstrate this, we first consider the maximum DCI values for all goals as the satisficing limit. This corresponds to the following DCI values

(i) Goal 1 DCI values ≥ 20;
(ii) Goal 2 DCI values ≥ 25;
(iii) Goal 3 DCI values ≥ 50; and
(iv) Goal 4 DCI values ≥ 30

On analyzing the iSOM plots that meets this requirement in Figure 11a, we see that grid points 49 and 48 are mapped to two scenarios that meet Goal 1. Grids 5 and 21 are mapped to 3 scenarios that meet Goal 2. Similarly, for Goal 3, grids 27, 35, and 21 are mapped to 8 scenarios, and for Goal 4, grid 43 is mapped to two scenarios. In Table 5, we list the design scenarios and design variable values for chosen satisficing grids for all the 4 goals of the test problem. When considering high values of DCI, all the design variable values that are in Table 5 are satisficing. Designer can select of the design variable values for each goal based on the requirements. For example, considering Goal 3, any of the identified design variable value sets can be chosen from the weight scenarios mapped to grids 27, 35, and 21. However, from the grids selected for all the 4 goals, it is observed that there are no common regions for the given problem. To arrive at common satisficing regions for all the goals, the satisficing limits of the goals are relaxed by following the systematic approach presented. For this, we check for the achieved DCI values of all goals. From the solution space generated, it is clear that all the solutions identified are having DCI values much greater than one thus making the solutions relatively insensitive to uncertainties. Thus, the potential to relax the satisficing limits of all four goals is possible. Based on this, the satisficing DCI limits for all four goals are relaxed and the following values are identified as acceptable:

(i) Goal 1 DCI values ≥ 15;
(ii) Goal 2 DCI values ≥ 15;
(iii) Goal 3 DCI values ≥ 20; and
(iv) Goal 4 DCI values ≥ 20

The highlighted hexagons after relaxing the satisficing goal limits are depicted in Figure 11b. On analyzing the iSOM plots for the relaxed satisficing limits, we notice that the Grid points 4 and 5 are common to all goals and satisfy the relaxed design requirements. In Table 6, we list the design scenarios and design variable values corresponding to the common satisficing grids points identified. The weight scenarios corresponding to these grids are identified as the satisficing robust solutions for the many conflicting vehicular crash problem.

Using iSOM combined with the robust cDSP, the designers are able to easily visualize, interpret, and explore the solutions based on the design requirements of the many-objective problem and identify robust solutions sets that satisfice the designer's requirements.

6. CLOSING REMARKS

In this paper, we propose a generic framework that integrates robust compromise decision support problem construct with interpretable self-organizing maps for efficient visualization and exploration of problems with many conflicting goals. The functionalities offered by the framework for a designer include: i) capability to manage the uncertainty in design variables for problems with many conflicting goals, (ii) capability to explore and visualize the solution space simultaneously for many conflicting goals and understand and interpret the correlations between the goals, between the goals and design variables, and between the design variables, (iii) capability to systematically identify common satisficing robust solutions sets for many objectives. The utility of the framework is demonstrated for a vehicular crash problem with 5 variables and 4 goals. The presented framework is generic and supports designers in efficiently visualizing and exploring design and solution spaces for the simulation-based realization of complex engineered systems characterized by the presence of many conflicting objectives and different sources of uncertainty.

ACKNOWLEDGEMENTS

NB, MB, and ABN thank the Department of Mechanical and Civil Engineering, Florida Institute of Technology, for the support.

REFERENCES

1. Fang, H., Solanki, K., and Horstemeyer, M. F., 2005, "Numerical Simulations of Multiple Vehicle Crashes and Multidisciplinary Crashworthiness Optimization," *International Journal of Crashworthiness*, vol.10, no.2, pp. 161-172.
2. Acar, E., Guler, M. A., Gerçeker, B., Cerit, M. E., and Bayram, B., 2011, "Multi-Objective Crashworthiness Optimization of Tapered Thin-Walled Tubes with Axisymmetric Indentations," *Thin-Walled Structures*, vol.49, no.1, pp. 94-105.
3. Duan, S., Tao, Y., Han, X., Yang, X., Hou, S., and Hu, Z., 2014, "Investigation on Structure Optimization of Crashworthiness of Fiber Reinforced Polymers Materials," *Composites Part B: Engineering*, vol. 60, pp. 471-478.
4. Fang, J., Sun, G., Qiu, N., Kim, N. H., and Li, Q., 2017, "On Design Optimization for Structural Crashworthiness and its State of the Art," *Structural and Multidisciplinary Optimization*, vol.55, pp. 1091-1119.

Copyright © 2023 by ASME

5. Avalle, M., Chiandussi, G., and Belingardi, G., 2002, "Design Optimization by Response Surface Methodology: Application to Crashworthiness Design of Vehicle Structures," *Structural and Multidisciplinary Optimization*, vol.24, no. 4, pp. 325-332.
6. Lanzi, L., Castelletti, L., and Anghileri, M., 2004, "Multi-objective Optimisation of Composite Absorber Shape Under Crashworthiness Requirements," *Composite structures*, vol.65, no. 3-4, pp. 433-441.
7. Parrish, A., Rais-Rohani, M., and Najafi, A., 2012, "Crashworthiness Optimisation of Vehicle Structures with Magnesium Alloy Parts," *International Journal of Crashworthiness*, vol.17, no.3, pp. 259-281.
8. Gu, X., Sun, G., Li, G., Huang, X., Li, Y., and Li, Q., 2013, "Multiobjective Optimization Design for Vehicle Occupant Restraint System Under Frontal Impact," *Structural and Multidisciplinary Optimization*, vol.47, pp. 465-477.
9. Coello, C. A. C., Lamont, G. B., and Van Veldhuizen, D. A., 2007, Evolutionary Algorithms for Solving Multi-Objective Problems, Springer.
10. Acar, E., Guler, M., Gerceker, B., Cerit, M., and Bayram, B., 2011, "Multi-Objective Crashworthiness Optimization of Tapered Thin-Walled Tubes with Axisymmetric Indentations," *Thin-Walled Structures*, vol.49, no.1, pp. 94-105.
11. Hou, S., Li, Q., Long, S., Yang, X., and Li, W., 2009, "Crashworthiness Design for Foam Filled Thin-Wall Structures," *Materials & design*, vol.30, no.6, pp. 2024-2032.
12. Zhang, Y., Sun, G., Li, G., Luo, Z., and Li, Q., 2012, "Optimization of Foam-Filled Bitubal Structures for Crashworthiness Criteria," *Materials & Design*, vol.38, pp. 99-109.
13. Soto, C. A., 2004, "Structural Topology Optimization for Crashworthiness," *International Journal of Crashworthiness*, vol. 9,no.3, pp. 277-283.
14. Hamza, K., and Shalaby, M., 2014, "A Framework for Parallelized Efficient Global Optimization with Application to Vehicle Crashworthiness Optimization," *Engineering Optimization*, vol.46, no.9, pp. 1200-1221.
15. Jin, R., Chen, W., and Simpson, T. W., 2001, "Comparative Studies of Metamodelling Techniques under Multiple Modelling Criteria," *Structural and multidisciplinary optimization*, vol. 23, pp. 1-13.
16. Gu, X., Dai, J., Huang, X., and Li, G., 2017, "Reliable Optimisation Design of Vehicle Structure Crashworthiness under Multiple Impact Cases," *International Journal of Crashworthiness*, vol. 22,no.1, pp. 26-37.
17. Youn, B. D., Choi, K., Yang, R.-J., and Gu, L., 2004, "Reliability-Based Design Optimization for Crashworthiness of Vehicle Side Impact," *Structural and Multidisciplinary Optimization*, vol. 26, pp. 272-283.
18. Sun, G., Song, X., Baek, S., and Li, Q., 2014, "Robust Optimization of Foam-Filled Thin-Walled Structure Based on Sequential Kriging Metamodel," *Structural and Multidisciplinary Optimization*, vol. 49, pp. 897-913.
19. Aspenberg, D., Jergeus, J., and Nilsson, L., 2013, "Robust Optimization of Front Members in a Full Frontal Car Impact," *Engineering Optimization*, vol.45, no.3, pp. 245-264.
20. Du, X., and Chen, W., 2004, "Sequential Optimization and Reliability Assessment Method for Efficient Probabilistic Design," *J. Mech. Des.*, vol. 126,no.2, pp. 225-233.
21. Sun, G., Li, G., Zhou, S., Li, H., Hou, S., and Li, Q., 2011, "Crashworthiness Design of Vehicle by Using Multiobjective Robust Optimization," *Structural and Multidisciplinary Optimization*, vol. 44, no.1, pp. 99-110.
22. Tang, Y., and Chen, J., 2009, "Robust Design of Sheet Metal Forming Process Based on Adaptive Importance Sampling," *Structural and Multidisciplinary Optimization*, vol. 39, pp. 531-544.
23. Shi, L., Yang, R.-J., and Zhu, P., 2013, "An Adaptive Response Surface Method for Crashworthiness Optimization," *Engineering Optimization*, vol. 45, no.11, pp. 1365-1377.
24. Hou, S., Han, X., Sun, G., Long, S., Li, W., Yang, X., and Li, Q., 2011, "Multiobjective Optimization for Tapered Circular Tubes," *Thin-Walled Structures*, vol. 49, no. 7, pp. 855-863.
25. Muster, D., and Mistree, F., 1988, "The Decision Support Problem Technique in Engineering Design," *International Journal of Applied Engineering Education*, vol.4, no.1, pp. 23-33.
26. Allen, J. K., Seepersad, C., Choi, H., and Mistree, F., 2006, "Robust Design for Multiscale and Multidisciplinary Applications."
27. Mistree, F., 1988, "The Decision-Support Problem Technique in Engineering Design," *The International Journal of Applied Engineering Education*, vol.4, pp. 23.
28. Mistree, F., and Allen, J. K., 1997, "Position Paper Optimization in Decision-Based Design," *Optimization in Industry, Palm Coast, FL, Mar*, pp. 23-27.
29. Simon, H. A., 2013, Administrative Behavior, Simon and Schuster.
30. Simon, H. A., 1956, "Rational Choice and the Structure of the Environment," *Psychological Review*, vol.63, no.2, pp. 129-138.
31. Hughes, O. F., 1993, "Compromise Decision Support Problem and the Adaptive Linear Programming Algorithm," *Progress In Astronautics and Aeronautics: Structural Optimization: Status and Promise*, vol.150, pp. 251.
32. Chen, W., Allen, J. K., Tsui, K.-L., and Mistree, F., 1996, "A Procedure for Robust Design: Minimizing Variations Caused by Noise Factors and Control Factors."
33. Chen, W., Simpson, T. W., Allen, J. K., and Mistree, F., 1999, "Satisfying Ranged Sets of Design Requirements using Design Capability Indices as Metrics," *Engineering Optimization*, vol. 31,no.5, pp. 615-619.
34. Choi, H.-J., Austin, R., Allen, J. K., Mcdowell, D. L., Mistree, F., and Benson, D. J., 2005, "An Approach for Robust Design of Reactive Power Metal Mixtures Based on Non-Deterministic Micro-Scale Shock Simulation," *Journal of Computer-Aided Materials Design*, vol. 12, pp. 57-85.

35. Sushil, R. R., Baby, M., Sharma, G., Balu Nellippallil, A., and Ramu, P., "Data Driven Integrated Design Space Exploration Using iSOM," *Proc. International Design Engineering Technical Conferences and Computers and Information in Engineering Conference*, American Society of Mechanical Engineers, p. V03AT03A014.

36. Sobester, A., Forrester, A., and Keane, A., 2008, Engineering Design via Surrogate Modeling: A Practical Guide, John Wiley & Sons.

37. Ayesha, S., Hanif, M. K., and Talib, R., 2020, "Overview and Comparative Study of Dimensionality Reduction Techniques for High Dimensional Data," *Information Fusion*, vol.59, pp. 44-58.

38. Simpson, T. W., Miller, S., Tibor, E. B., Yukish, M. A., Stump, G., Kannan, H., Mesmer, B., Winer, E. H., and Bloebaum, C. L., 2017, "Adding Value to Trade Space Exploration when Designing Complex Engineered Systems," *Systems Engineering*, vol.20, no.2, pp. 131-146.

39. Santi, F., Dickson, M. M., Espa, G., and Giuliani, D., 2022, "Plot3logit: Ternary Plots for Interpreting Trinomial Regression Models," *Journal of Statistical Software, Code Snippets*, vol.103, no.1, pp. 1 - 27.

40. Fang, H., Solanki, K., and Horstemeyer, M., 2005, "Numerical Simulations of Multiple Vehicle Crashes and Multidisciplinary Crashworthiness Optimization," *International Journal of Crashworthiness*, vol.10, pp. 161-172.

41. Wei, C., Allen, J. K., and Mistree, F., 1997, "A Robust Concept Exploration Method for Enhancing Productivity in Concurrent Systems Design," *Concurrent Engineering*, vol. 5, no.3, pp. 203-217.

42. Zaouk, A. K., Marzougui, D., and Bedewi, N. E., 2000, "Development of a Detailed Vehicle Finite Element Model Part I: Methodology," *International Journal of Crashworthiness*, vol. 5, no.1, pp. 25-36.

43. Horstemeyer, M. F., Ren, X. C., Fang, H., Acar, E., and Wang, P. T., 2009, "A Comparative Study of Design Optimisation Methodologies for Side-Impact Crashworthiness, using Injury-Based Versus Energy-Based Criterion," *International Journal of Crashworthiness*, vol.14, no.2, pp. 125-138.

44. Fang, H., Rais-Rohani, M., Liu, Z., and Horstemeyer, M. F., 2005, "A Comparative Study of Metamodeling Methods for Multiobjective Crashworthiness Optimization," *Computers & Structures*, vol.83, no.25, pp. 2121-2136.

45. Nagar, D., Ramu, P., and Deb, K., "Interpretable Self-Organizing Maps (iSOM) for Visualization of Pareto Front in Multiple Objective Optimization," *Proc. Evolutionary Multi-Criterion Optimization*, H. Ishibuchi, Q. Zhang, R. Cheng, K. Li, H. Li, H. Wang, and A. Zhou, eds., Springer International Publishing, pp. 645-655.

46. Yin, H., 2008, "The Self-Organizing Maps: Background, Theories, Extensions and Applications," *Computational Intelligence: A Compendium*, J. Fulcher, and L. C. Jain, eds., Springer Berlin Heidelberg, Berlin, Heidelberg, pp. 715-762.

APPENDIX A - Surrogate models developed

Table A1. Response surface models developed

Function	Response surface models
Energy Absorption for Part 235 (N-mm)	$E_{235} = (0.19739 + 0.14887 \times t_1 + 0.02670 \times t_2 - 0.05124 \times t_3 - 0.08292 \times t_4 + 0.01850 \times t_5 - 0.08935 \times t_1^2 + 0.02951 \times t_1 \times t_2 - 0.00579 \times t_1 \times t_3 + 0.07720 \times t_1 \times t_4 - 0.02256 \times t_1 \times t_5 - 0.02076 \times t_2^2 - 0.01874 \times t_2 \times t_3 - 0.00806 \times t_2 \times t_4 + 0.00479 \times t_2 \times t_5 + 0.03565 \times t_3^2 + 0.04727 \times t_3 \times t_4 - 0.01389 \times t_3 \times t_5 + 0.06025 \times t_4^2 - 0.0511 \times t_4 \times t_5 + 0.04532 \times t_5^2) \times 10^6$
Energy Absorption for Part 237 (N-mm)	$E_{237} = (0.13666 + 0.00607 \times t_1 + 0.09939 \times t_2 + 0.03214 \times t_3 + 0.02319 \times t_4 - 0.02941 \times t_5 + 0.008624 \times t_1^2 - 0.04545 \times t_1 \times t_2 + 0.21449 \times t_1 \times t_3 - 0.0618 \times t_1 \times t_4 + 0.02878 \times t_1 \times t_5 + 0.06637 \times t_2^2 - 0.0321 \times t_2 \times t_3 + 0.031315 \times t_2 \times t_4 - 0.00037 \times t_2 \times t_5 - 0.01601 \times t_3^2 + 0.001301 \times t_3 \times t_4 - 0.00013 \times t_3 \times t_5 - 0.01339 \times t_4^2 + 0.01777 \times t_4 \times t_5 + 0.003009 \times t_5^2) \times 10^6$
Energy Absorption for Part 329 (N-mm)	$E_{329} = (0.015484 - 0.02342 \times t_1 + 0.024195 \times t_2 + 0.238752 \times t_3 - 0.04171 \times t_4 + 0.011807 \times t_5 + 0.005085 \times t_1^2 - 0.013 \times t_1 \times t_2 - 0.0312 \times t_1 \times t_3 + 0.0513 \times t_1 \times t_4 + 0.01696 \times t_1 \times t_5 + 0.001905 \times t_2^2 - 0.03205 \times t_2 \times t_3 + 0.011237 \times t_2 \times t_4 + 0.000675 \times t_2 \times t_5 - 0.03904 \times t_3^2 + 0.080816 \times t_3 \times t_4 + 0.041151 \times t_3 \times t_5 + 0.000641 \times t_4^2 - 0.06095 \times t_4 \times t_5 - 0.00502 \times t_5^2) \times 10^6$
Energy Absorption for Part 353 (N-mm)	$E_{353} = (0.080989 - 0.05628 \times t_1 - 0.03778 \times t_2 + 0.09809 \times t_3 + 1.05099 \times t_4 + 0.114237 \times t_5 - 0.02441 \times t_1^2 - 0.02385 \times t_1 \times t_2 - 0.087018 \times t_1 \times t_3 + 0.0509 \times t_1 \times t_4 + 0.116148 \times t_1 \times t_5 + 0.015803 \times t_2^2 + 0.060604 \times t_2 \times t_3 - 0.0333 \times t_2 \times t_4 - 0.00434 \times t_2 \times t_5 - 0.11714 \times t_3^2 - 0.08324 \times t_3 \times t_4 - 0.02029 \times t_3 \times t_5 - 0.48092 \times t_4^2 - 0.14211 \times t_4 \times t_5 - 0.7409 \times t_5^2) \times 10^6$
Energy Absorption for Part 357 (N-mm)	$E_{357} = (0.124979 - 017563 \times t_1 - 0.01787 \times t_2 - 0.0144 \times t_3 - 0.06605 \times t_4 + 0.2426 \times t_5 - 0.00679 \times t_1^2 + 0.05797 \times t_1 \times t_2 + 0.012006 \times t_1 \times t_3 + 0.140194 \times t_1 \times t_4 + 0.071128 \times t_1 \times t_5 + 0.012219 \times t_2^2 - 0.00968 \times t_2 \times t_3 - 0.0443 \times t_2 \times t_4 - 0.04542 \times t_2 \times t_5 + 0.0568 \times t_3^2 + 0.1267 \times t_3 \times t_4 + 0.055 \times$

	$t_3 \times t_5 + 0.0384 \times t_4{}^2 \quad -0.1261 \times t_4 \times t_5 - 0.0072 \times t_5{}^2) \times 10^6$
Total Mass of all Parts (kg)	Mass $= 6.90989 - 3.708636 \times t_1 - 1.850079 \times t_2 + 6.84746 \times t_3 + 13.566 \times t_4 + 1.251 \times t_5 \quad +4.57118 \times t_1{}^2 + 0.007921 \times t_1 \times t_2 + 5.538308 \times t_1 \times t_3 - 1.3646 \times t_1 \times t_4 - 1.6955 \times t_1 \times t_5 + 3.1740 \times t_2{}^2 \quad +1.0138 \times t_2 \times t_3 + 2.582 \times t_2 \times t_4 - 1.7904 \times t_2 \times t_5 - 6.903 \times t_3{}^2 - 7.156 \times t_3 \times t_4 + 6.10069 \times t_3 \times t_5 + 2.253031 \times t_4{}^2 \quad +7.952 \times t_4 \times t_5 - 4.5043 \times t_5{}^2$

Proceedings of the ASME 2023
International Design Engineering Technical Conferences and
Computers and Information in Engineering Conference
IDETC-CIE2023
August 20-23, 2023, Boston, Massachusetts

DETC2023-109302

SATISFICING STRATEGY IN ENGINEERING DESIGN

Lin Guo[1]
Industrial Engineering
South Dakota School of Mines and Technology,
Rapid City, SD, USA

Suhao Chen
Industrial Engineering
South Dakota School of Mines and Technology,
Rapid City, SD, USA

ABSTRACT

In engineering-design problems, usually, there are multiple goals (or objectives), continuous and discrete variables, nonlinear equations, nonconvex equations, goals with various units, and coupled decisions that are required. Ideally, the target of the multiple goals needs to be reached simultaneously within the feasible space bounded by constraints and bounds. However, the optimal solution to a problem that incorporates one or more of the complexities above may not be available or obtained by designers within the available resource. To deal with all those complexities, a modeling strategy named "satisficing" was proposed in the 1980's and applied to various design problems in the past three decades. The satisficing strategy allows designers to find "good enough" solutions that "make things work," yet the solutions may not be optimal. Moreover, with the satisficing strategy, designers may explore the solution space and gain knowledge on the performance through the design space. To carry out the satisficing strategy, a model formulation framework, the compromise Decision Support Problem (cDSP), was once proposed for modeling engineering-design problems, and a solution algorithm, the Adaptive Linear Programming algorithm (ALP), was created for linearizing and solving a nonlinear cDSP and output satisficing solutions, and a solver, the Decision Support in the Design of Engineering Systems (DSIDES), was developed to realize cDSP formulation and solving. In the past 30 years, satisficing strategy has been improved and applied to a wide variety of engineering-design problems. Derived methods, concepts, and platforms using the satisficing strategy are developed. This paper is a review of the representative publications regarding utilizing the satisficing strategy in managing engineering design. Among the publications, different topics are reviewed and summarized, including methods, theories, and frameworks to facilitate the robust design, multiscale and microstructure design, multistage design and concurrent design, coupled decision-making in

designs, exploration of the solution space and design space, multigoal and multidisciplinary design, and knowledge-based design and platformation. We generalize the specialties, advantages, and scope of applications of the methods in the satisficing strategy. We expect this paper provides useful information on when and how designers may or should attach to satisficing as a strategy for their problems.

KEYWORDS

Satisficing; engineering-design problems; robust design; knowledge-based design; multigoal design; multidisciplinary design; microstructure design; multistage design.

1 SPECIAL REQUIREMENTS FOR METHODS IN ENGINEERING DESIGN

Engineering design is a task that involves a group of designers with different interests and knowledge to make decisions that comply with their mutual requirements [1]. As decision-making requires designers to select the desired solution among several alternatives regarding the output performance, one or more utility functions are usually consciously or subconsciously applied by the designers to measure the output performance. Designers can define a utility function through different methods, for example, through game theory [2], analytic hierarchy process (AHP) [3], Pareto comparisons [4], etc. However, regardless of the seeming rationale that those methods present, the use of utility theory may result in poor decision-making and may not be appropriate for engineering design [5]. In addition, it is difficult for designers to identify and maintain the optimal solution using optimization methods, given that there are various complexities and uncertainties underlying an engineering-design problem. Moreover, the mission of engineering design is more than obtaining optimal solutions. There are other tasks and foci, such as acquiring knowledge on the output performance in a wide

---

[1] Corresponding author: lin.guo@sdsmt.edu

Copyright © 2023 by ASME

range of solution space (not only the near-optimal area), improving the robustness of the decision model, coupling the continuous and discrete decisions, connecting sub-models with different levels of refinement or fidelity, and so on.

Designers accept problematic assumptions when applying utility theory. Optimization is one method based on utility theory-based methods that enable designers to select the optimal solution among alternatives. Nevertheless, not all design problems can be solved using optimization. Even if the optimal solution to the decision model is acquired, it may not be optimal or even feasible for the physical problem, as the optimality conditions can be easily broken [6]. This often results from the incapability of the decision model to capture all information and requirements [7] of the design. One source of the incompleteness and inaccuracy of the model is that by modeling the objective function using utility theory, designers naturally accept the assumption that all decision-makers are rational and every detail in the decision model perfectly maps the physical world to the model abstraction. Such an assumption is often wrong, especially in engineering design.

Complexities incorporated in engineering design cause convergence failure. Engineering-design problems encounter multiple complexities [8]; therefore, the decision models may have nonlinear, nonconvex equations, multiple objectives with different units and scales, and objectives with different levels of achievability. These complexities sometimes lead to immature convergence or no convergence.

Multiple sources of variation bring uncertainties. How do uncertainties take place? From the view of the source of uncertainties, there are variations i) in environmental or other noise-noise factors [9], ii) in design (or decision) variables – control factors [10], iii) brought by modeling methods [11, 12], and iv) brought by the process of managing the previous three types of variation [13]. How do uncertainties affect the design? From the perspective of optimality conditions, uncertainties that break the equilibrium of any Karush-Kuhn-Tucker (KKT) conditions destroy the optimality of a solution, thereby making an optimal solution infeasible or useless [14].

The tasks of engineering design include way more than finding the optimal solution. The demand for case-by-case, general, or reusable knowledge and insight through the designing process entails design methods to enable learning model behaviors [15], exploring the solution space to satisfy a variety of design scenarios [16, 17], determining the refinement of a model and its sub-models [18, 19], and choose the proper set of solutions as the input of the previous stage of design [20-22]. Although optimization methods sometimes work efficiently in returning optimal solutions to mathematical models, they do not support knowledge attainment.

Therefore, we need a design strategy to overcome the shortcomings in optimization methods, including avoiding the assumption in utility theory-based functions, managing the typical complexities and uncertainties in engineering-design problems, and supporting knowledge learning through post-solution analysis.

In Section 2, we briefly introduce the satisficing strategy as an alternative to optimization – how is satisficing strategy different, and why is it sometimes fitter for engineering design? We describe a packet of methods that realize the satisficing strategy. In Section 3, we summarize the research topics, derived work (new features and modules, tools, methods, theories, concepts, frameworks, and platforms), applications, and contributions to satisficing strategy. In Section 4, we anticipate the way forward of satisficing strategy regarding the ideas, functions, and applications. In Section 5, we conclude the contributions and limitations of this paper.

2 SATISFICING STRATEGY AND WHY
Modeling methods for engineering-design problems fall into two categories, i) formulating a complex problem exactly and searching for the optimal solution or near-optimal solutions, and ii) approximating a complex problem and finding a set of solutions that output acceptable performance. We name the first strategy "optimizing" and the second one "satisficing." The major differences between the two strategies are represented in three aspects: model formulation, solution algorithms, and post-solution analysis [7, 14]. In this section, we describe the typical features of the satisficing strategy regarding those three aspects.

The concept of satisficing solution or satisficing searching is proposed by Herbert Simon [23-25] as "good enough" solutions that make the design output acceptable but maybe not optimal performances. Later, the idea of satisficing is put into practice by several researchers. For decades, Mistree and his research team [7, 26-28] realize satisficing strategy in engineering design using the compromise Decision Support Problem (cDSP) as the formulation construct, the Adaptive Linear Programming algorithm (ALP) as the solution algorithm, and the Decision Support in the Design of Engineering Systems (DSIDES) as a platform and solver to formulate and solve a cDSP with the ALP. Besides, there are a few other scholars [29-37] who anticipate or apply satisficing as a practical or affordable solution strategy in their design. Although they reach satisficing using various modeling and measurement methods, they all agree on the idea that "good enough" but not necessarily optimal solutions well balance the performance of their designs and the workload of finding the solutions. For decades, the cDSP-ALP-DSDIES community has systematically developed and derived a packet of design methods along the satisficing strategy, especially for engineering designs. In this paper, we discuss satisficing strategy mainly through reviewing this community's representative works.

2.1 The Specialty in the Formulation
Examples of the formulations using the optimizing strategy include mathematical optimization and its variants, such as goal

Copyright © 2023 by ASME

programming. Solutions are usually obtained by identifying the Pareto front consisting of nondominated or near-optimal solutions using optimization solution algorithms. The formulation of design problems using satisficing strategy, namely, the compromise Decision Support Problem (cDSP), has the key features that allow designers to identify satisficing solutions that meet the necessary KKT condition but not the sufficient condition.

The format of a nonlinear optimization problem is like this: For a given objective function $f(x)$, Euler and Lagrange develop the Euler-Lagrange equation forming the second-order ordinary differential equations $\nabla_{xx}^2 f(x)$ to facilitate finding the stationary solutions. The value of the variables that maximize $f(x)$ within the feasible set $\mathcal{F}$ is the solution to the optimization problem, where $\mathcal{F}$ is the set bounded by constraints and bounds. The format of an optimization problem $\mathbb{O}$ can be represented as follows. x is the vector of decision variables as real numbers. $g_i(x)$ is the ith inequality constraint. $h_j(x)$ is the ith equality constraint. Any point x that is a local extrema of the set mapped by multiplying active equations with a nonnegative vector, is a local optimum of $\mathbb{O}$ [38], denoted as x^*. The elements of such a non-negative vector are Lagrange multipliers, μ and λ.

The format of an optimization problem $\mathbb{O}$:
Given
$$f: \mathbb{R}^n \to \mathbb{R}, \mathcal{F} \subseteq \mathbb{R}^n$$
$$\mathcal{F} = \left\{ x \in \mathbb{R}^n | g_i(x) \geq 0, i = 1, \dots, m, h_j(x) = 0, j = 1, \dots, \ell \right\}$$
Find
$$x^*: f(x^*) \succcurlyeq f(x), \forall x \in \mathcal{F}$$

One variant of optimization is goal programming. The format of a goal programming problem $\mathbb{O}^{goal}$ is represented as follows. A target value T is predefined for the objective function $f(x)$ as the right-hand side value, so the objective becomes an equation, and we call it a goal. d^- and d^+ are deviation variable measuring the underachievement and over-achievement of the goal towards its target. The problem is solved by minimizing the deviation variables, which is minimizing the difference between $f(x)$ and T. In other words, goal programming is aimed at finding T's closest projection on $\mathcal{F}$.

The format of a goal programming problem $\mathbb{O}^{goal}$:
Given
$$f: \mathbb{R}^n \to \mathbb{R}, \mathcal{F} \subseteq \mathbb{R}^n$$
$$\mathcal{F} = \left\{ x \in \mathbb{R}^n | g_i(x) \geq 0, i = 1, \dots, m, h_j(x) = 0, j = 1, \dots, \ell, d^- \cdot d^+ = 0, 0 \leq d^{\mp} \leq 1, Goal: f(x) + d^- - d^+ = T \right\}$$
Find
$$x^*: \mathbb{P}_{x \in \mathcal{F}}(Goal: f(x) = T)$$

In the cDSP, elements of mathematical programming and goal programming are combined. A cDSP $\mathbb{C}$ is represented as follows. For a nonlinear cDSP, we first linearize the nonlinear equations,

including nonlinear constraints and nonlinear goal. Therefore, the nonlinear cDSP first becomes a linear problem with a linear goal $Goal^{linear}$, a linear feasible space $\mathcal{F}^{li}$ bounded by linear constraints $g(x)^{li} \geq 0$ and $h(x)^{li} = 0$. Thus, using a cDSP, we seek the closest projection from the linear goal set onto a linear feasible set. We define the solution as a satisficing solution, and we use x^s to denote it.

The format of a cDSP $\mathbb{C}$:
Given
$$f: \mathbb{R}^n \to \mathbb{R}, \mathcal{F} \subseteq \mathbb{R}^n$$
$$\mathcal{F}^{li} = \left\{ x \in \mathbb{R}^n | g_i(x)^{li} \geq 0, i = 1, \dots, m, h_j(x)^{li} = 0, j = 1, \dots, \ell, d^- \cdot d^+ = 0, 0 \leq d^{\mp} \leq 1, Goal^{li}: \frac{f(x)^{li}}{T} + d^- - d^+ = 1 \right\}$$
Find
$$x^s: \mathbb{P}_{x \in \mathcal{F}^{li}}(Goal^{li}: f(x)^{li} = T)$$

The difference between x^* and x^s is that x^* conforms both the necessary (first-order) and sufficient (second-order) KKT conditions, whereas x^s conforms the necessary KKT condition but may not conforms the sufficient KKT condition. This is because the second derivative of the linear equations, $Goal^{li}$, $g_i(x)^{li} \geq 0$, and $h_j(x)^{li} = 0$, degenerates − as a result, no uncertainty may affect the feasibility of x^s since no uncertainty gets a chance to break the equilibrium of the second-order Lagrange equation. In addition, when the convexity of $\mathcal{F}$ is greater than the convexity of the $f(x)$, x^* may not be identified as the second-order Lagrange equation has no solution, but x^s is obtainable due to its irrelevancy with the second-order KKT condition. The details of the mathematical demonstration and illustration using example problems are in Chapter 2 of [6].

2.2 The Specialty in the Solution Algorithm
There are different ways of categorizing the solution algorithms. Sharma and Kumar [39] propose that all solution algorithms fall into two categories, classical methods and meta-heuristics algorithms. Classical methods are based on vertex searching or solving equation set to meet the KKT conditions, (e.g., Lagrange's multiplier method, Branching, etc.) Metaheuristics algorithms include deterministic searching methods (Tabu search and local search) and stochastic searching-based methods (Simulated Annealing, Hill Climbing, Ant Colony algorithm, evolutionary algorithm, etc.). Abualigah et al. [40] classify solution algorithms into two types − metaheuristics algorithms and heuristics algorithms. The former contains three sub-categories, local-based algorithms, evolutionary algorithms, and swarm-based algorithms. The latter specifically refers to vertex searching algorithms such as the Simplex algorithm. Daound et al. [41] refine the definition by identifying four sub-categories in the metaheuristic algorithms: physics-based, population-based, human-based, and evolutionary-based algorithms; they define the local searching algorithms as heuristic algorithms.

Copyright © 2023 by ASME

In this paper, not only do we care about the way of solving a problem, but we also discuss whether the solution is optimal and whether a set of nonoptimal but good enough solutions may provide designers with more options and relevant knowledge in the design space. We use a matrix to categorize the solutions algorithms, as Table 1 shows. The two dimensions of the matrix are the optimality of the converged solutions and the searching conditions. No interior-point searching algorithms guarantee the optimality unless KKT conditions are applied to facilitate the examination of the optimality. Such an examination is a posteriori supplementary method; therefore, we define all interior-point searching algorithms in the category of converging near-optimal solutions.

The ALP algorithm performs both interior-point searching and vertex searching to maintain a balance between exploitation and exploration. With the module "XPLORE" in DSIDES, a wide-spread interior-point random search is performed to select a good starting point (see Figure 1), then the problem is linearized around the starting point using the Second-order Sequential linear programming algorithm (SLIP2), and then the linear problem is solved using dual simplex. The linearization can go through a number of iterations to make sure the nonlinear problem is linearized sufficiently and accurately at different local optimums. As there are often more constraints than variables in the engineering-design problems, using the dual simplex is easier than using the simplex. In this way, the solution is on the vertex of the linearized problem and meets the necessary KKT condition to ensure near optimality. The difference between a satisficing solution, a near-optimal solution, and the optimum is illustrated in Figure 2.

Table 1 Classifying solution algorithms regarding the optimality of solutions and the searching conditions

	Converge the optimal solution	Type of converged solutions		Strength	Weakness
		Near-optimal	Satisficing		
Vertex search	Simplex, Dual Simplex, Lagrange's multipliers	SLP, SQL	ALP	Requiring relatively low computing power; may give model formulation insight	May not work effective for nonconvex problems
Interior-point search		Meta heuristic interior-point searching algorithms (NSGA, GBO, etc.)		Easy to deal with nonlinear, high-dimension problems	Requiring large computing power; no insight of formulation modification

With the satisficing solution algorithm – the ALP, why the nonconvex problem can be dealt with, and why is the linearization relatively accurate? Two mechanisms in the ALP make it possible to linearize the nonconvex – using the second-order sequential linearization (SLIP2) algorithm and the accumulation of the linear constraints. The SLIP2 first uses the parabola (the one that passes through the starting point and the two intersection points of the surface and x1-x2 plane) to approximate the nonlinear surface, then linearizes the parabola into a plane. And the linearized planes from multiple iterations are accumulated as the linear constraints to replace the nonlinear constraint. In this way, the nonconvex equations can be linearized, and the linearized constraints are more accurate. The detailed mathematical descriptions and illustrations are in [42].

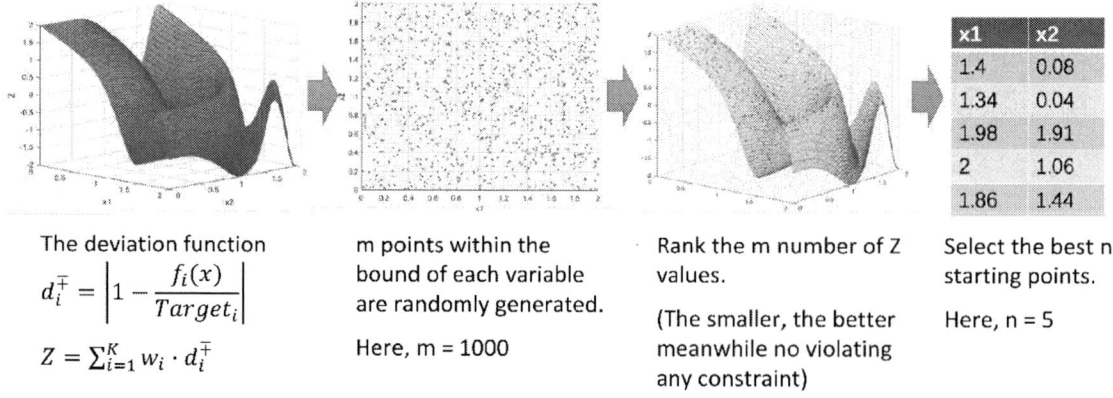

The deviation function
$$d_i^{\mp} = \left| 1 - \frac{f_i(x)}{Target_i} \right|$$

$$Z = \sum_{i=1}^{K} w_i \cdot d_i^{\mp}$$

m points within the bound of each variable are randomly generated.

Here, m = 1000

Rank the m number of Z values.

(The smaller, the better meanwhile no violating any constraint)

Select the best n starting points.

Here, n = 5

x1	x2
1.4	0.08
1.34	0.04
1.98	1.91
2	1.06
1.86	1.44

Figure 1 Using the "XPLORE" module in DSIDES to identify "n" best starting points

Copyright © 2023 by ASME

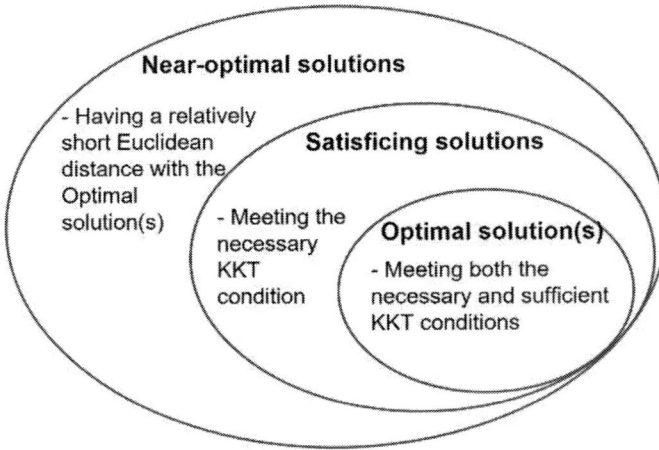

Figure 2 Relation between the optimal, satisficing, and near-optimal solutions

2.3 Features that Support Knowledge Attainment and Coupled Decision-Making

Managing engineering-design problems requires designers to find a set of satisficing solutions and way more than that – gain knowledge about the design space, make coupled decisions, and tailor the methods to engineering designs in different fields. The satisficing strategy provides a good construct to complete those tasks. For example, DSIDES provides the formulation construct and solvers for compromise DSP (cDSP), selection DSP (sDSP), and any sequences or combinations of the compromise and selection DSPs, which support the mixture of continuous and discrete decision-making.

We explain how the satisficing strategy realized using cDSP-ALP-DSIDES support various tasks in engineering design by guiding readers through the relevant publications in the past three decades in Section 3.

2.4 Scope of Applications and Limitations

Satisficing strategy facilitates designers to deal with engineering problems with multiple goals, goals with various units and achievability, nonlinear features, continuous and discrete variables, evolving design preferences, and coupled decision-making. However, for problems that do not contain the features above, the satisficing strategy may not outperform optimization methods.

In addition, to formulate a problem into a cDSP, a target for each goal should be defined. For an objective that has no target value, or when designers have no idea about the target value, the satisficing strategy may not be the best option.

3 Applications and Evolution of Satisficing Strategy

There are seven major contributions (subjects) in the satisficing engineering designs in the past three decades: i) robust design methods, ii) multiscale and microstructure design, iii) multistage design and concurrent design, iv) coupled decision-making, v)

exploration of the solution space and design space, vi) multigoal (multiobjective) or multidisciplinary design, and vii) knowledge-based design and platformation. There are overlaps among those subjects, as some applications in one subject are the verifications of the methods in other subjects. In this section, we select the representative work in each subject based on the utility, popularity, and extendibility of the publications and summarize the typical features, contributions, and scope of applications.

3.1 Robust Design Managing Four Types of Uncertainty

The idea of robust design using the satisficing strategy is to make the design relatively insensitive to different types of uncertainty [11, 13]. Uncertainties are classified into four types based on their sources [11]: Type I – the uncertainty brought by the variation in environmental noise or other noise factors [9], Type II – the uncertainty brought by the variation in design variables of control factors [10], Type III – the uncertainty introduced by modeling methods [11, 12], and Type IV – the uncertainty introduced during the management of the previous three types of uncertainty [13]. There are four types of robust design using the satisficing strategy, which facilitate the designers to manger the four types of uncertainty.

Taguchi develops the "parameter design concept" to use a two-part orthogonal array for experimental designs using the "signal-to-noise-ratio" as an optimization criterion. This is later defined by Chen et al. [10]. With the identification of one limitation of Taguchi's method – no accurate solution can be yielded for highly nonlinear problems, and given the method is criticized by the statistical community [43], Chen et al. propose a variation to Taguchi's method by integrating the Response Surface Methodology (RSM) with the cDSP to manage both Type I and Type II uncertainty. Meanwhile, Simpson et al. [44] and Chen et al. [45, 46] propose the Robust Concept Exploration Method (RCEM) and use the principles to determine a range set of top-level design specifications to realize the robust design. Later, Choi et al. [47] use RCEM with Error Margin Index (RCEM-EMI) to realize the robust design of materials by employing the EMI to indicate the mean and spread of system performance considering the variability in design variables and models. The authors [47] also propose RCEM with Design Capability (RCEM-DCI) to determine whether a ranged design specification is capable of satisfying a ranged set of design requirements. Choi et al. [13, 48] propose an Inductive Design Exploration Method (IDEM) for designing materials and products concurrently and systematically to manage Type IV uncertainty by sequentially identifying a ranged set of feasible specifications and searching the feasible spaces regarding the top-level design requirements.

Given the foundation of four types of robust design – the RCEM and IDEM with indices, the satisficing community later expands the method, derive RCEM methods with new indices, and applies them in different fields of engineering design. Wang et al. [49] propose a design exploration method for adaptive design

Copyright © 2023 by ASME

451

systems that include using local regression and inverse IDEM using an example problem in the design of a photonic crystal coupler and waveguide. Sinha et al. [50-52] propose to use the Inductive Discrete Constraints Evaluation (IDCE), which is to sequentially identify feasible regions in design space using a metric called Hyper Dimensional-Error Margin Index (HD-EMI) indicating the degree of reliability of the model. Kulkarni et al. [53] and Gautham et al. [54] later apply HD-EMI in the multistage, inverse design of the heat treatment operation. Samadiani et al. [55] and Panchal et al. [56] integrate the Proper Orthogonal Decomposition-based (POD-based), multiscale model with cDSP to realize the robust design of datacenter cell in thermal design. Messer [57] and Rippel [58] apply and enrich the RCEM using the early stage of design of a pressure vessel by presenting three alternative methods, namely, second derivative, multiple derivatives, and multiple point method, to measure the robustness through estimating the variance. Goh et al. [59-61] propose the Integrated Multiscale Robust Design (IMRD) for traversing the Integrated Computational Materials Engineering (ICME) in multiscale heterogeneous internal structure evolution.

In summary, the four types of robust design using satisficing strategy are using the cDSP and ALP as the basic modeling construct and solution algorithm, applying indices to measure and control variations from multiple sources, and sometimes integrating other modeling techniques to identify and work on a range of feasible solutions to ensure the design is insensitive to uncertainties.

3.2 Multiscale and Microstructure Design in Material Design

Material design is a field that often encounters nonlinear equations, variations from multiple sources, multiscale or hierarchical structures, etc., so it requires satisficing as a design strategy. For this topic, the contributions are mainly from the applications of the robust design method in material design.

Seepersad et al. [62] present the Type I, II, and III robust design method for designing materials in mesoscopic scales by topologically and parametric tailoring them to achieve desired properties. Allen et al. [11] use the first three types of the robust design methods on multidisciplinary, multiscale design with multiple sources of uncertainty. Thompson et al. [63] consider the material properties as uncertain variables and apply the robust design method to provide ranges of acceptable material properties for framing subsequent material design or selection for designing blast-resistant panels. McDowell et al. [64] give an example for designing a plasticity-related microstructure – a four-phase reactive power metal-metal oxide mixture for initiation of exothermic reactions under shock-wave loading. Later McDowell et al. [65] integrate the design of multiscale material, multifunctional material, and product design, and the proposed integrated method also works for other systems-based design of materials. Seepersad et al. [66] propose a two-stage topology design approach that requires customized multifunctional properties for designing cellular materials.

Samadiani et al. [55] use a test problem of datacenter cell internal design to demonstrate the efficacy of the proposed method with the integration of POD-based, multiscale modeling with the cDSP. The method can be used for other simulation-based, multiscale designs. Shukla et al. [16, 67] deal with a ladle refining and continuous casting problem by using the satisficing strategy to facilitate producing new grades of steel to meet stringent sets of property specifications by exploring the solution space. Goh et al. [61] select the satisficing strategy for it allows to find feasible ranged sets of solutions and avails the ICME horizontal Process-Structure-Property-Performance (PSPP) in multiscale heterogeneous internal structure evolution. Beemaraj et al. [68] propose a method for designing a robust composite structure subjected to different loading conditions.

Why do designers choose the satisficing strategy in material design, especially when designing multiscale, microstructures, or needing to cope with the material properties with other requirements in a process or product? Typical reasons include i) the product is expected to be robust to the variation in the material property, ii) the material attributes can be designed to be robust against uncertainties due to random variation of microstructure, iii) designers only need to determine the value of decision variables that satisfy the rigid requirements (constraints) and achieve the desired properties (goals) as closely as possible, which means they do not have to squeeze to the optimal properties, iv) material design for a product is often a concurrent design that comprises material design or selection and the dimension or geometry design of the product, v) robust solutions can be founded to reduce the variability in response for simulation-based design or metamodel-based design, and vi) the solution space can be explored relatively sufficiently not only to output a solution but to observe and predict model behaviors and in the wide design space, which can be useful information for new material experiment and innovation.

3.3 Multistage Design and Concurrent Design

Multistage design and concurrent design are also applications of robust design methods, specifically RCEM. The applications include engine design, aircraft design, gear transmission, multifunctional material design, process chain design, product and product family design, and steel production.

Concurrent design: Rangarajan et al. [69] adopt RCEM as a framework to address tribological considerations in the concurrent design of automobile engine lubricated components. Chen et al. [70] use RCEM to evaluate design alternatives and develop top-level specifications for High-Speed Civil Transport (HSCT) aircraft. Simpson et al. [71] propose a product variety tradeoff evaluation method for assessing alternative product platform concepts with varying levels of commonality – the challenge lies in balancing the commonality and performance of products in a product family. The authors verify the method using gear transmission and general aviation aircraft design. McDowell et al. [65] perform concurrent design by integrating multiscale, multifunctional, and product design at the same time.

Copyright © 2023 by ASME

Inverse design: Nellippallil and coauthors deal with a series of multistage, inverse design problems in steel manufacturing. Nellippallil et al. [72] present an inverse design method based on empirical models and response surface models to support the integration that facilitates information flow between stages of the hot rod process chain. The stages are designed by passing the information obtained after exercising the end-stage cDSP to the earlier-stage cDSPs. Then, the authors improve the inverse design method [22] to achieve the integrated design exploration of materials, products, and manufacturing processes through the vertical and horizontal integration of models. They demonstrate the efficacy of the method using a hot rod rolling and cooling process chain problem by exploring the processing paths and microstructure in an inverse manner. Nellippallil and the coauthors further improve the method into a goal-oriented Inverse Design method [73] to empower the capability to carry out a microstructure-mediated design satisficing specific processing performance of a product. They apply the method to design the thermos-mechanical processing of a steel rod. The authors then summarize their work as an approach [74] that facilitates co-design across the materials, products, and manufacturing processes. With this goal-oriented, inverse co-design approach, Fonville et al. [20] design two components of an American football helmet, the composite shell and foam liner. The method can be used in other fields when the multistage, concurrent design of multiple elements (material, product, manufacturing, etc.) is needed and when only the end-stage performance or requirements are available.

Designers choose the robust design method under the satisficing strategy for concurrent, multistage, and inverse design projects because they need to manage the uncertainties brought by connecting the stages, concurrent decision-making, and the lack of knowledge on the association between the early-stage decision variable and the end performance of a product or a system.

3.4 Coupled Decision-Making
Coupled-decision-making problems often have common features with the multistage, concurrent design problems – designers need to determine two things while interdependent on each other in a system, for example, the material and the product specification. In this subsection, we focus on the problems that require hierarchical ways of coupling the selection decisions and compromise decisions. Another type of coupled decision-making is making decisions by different teams with multiple backgrounds or fulfilling multiple disciplines and managing the interactions. We categorize those problems in multidisciplinary design and illustrate them in Section 3.5.

Besides the works on steel manufacturing described in Section 3.3, which require the design of materials and products at the same time, some applications and derived methods in steel casting and gear design also require material selection and geometric design simultaneously. Kumar et al. [15] develop an integrated design framework, Platform for the Realization of Engineered Materials and Products (PREMAP), based on metamodels and cDSP. The authors utilize the framework to assist in making decisions when an existing configuration for continuous casting is unable to meet the requirements. The approach can be adopted for integrating the host of operations for materials development with specific properties and the coupled design of products and materials. Gautham et al. [54] envisage PREMAP as a platform for the purpose of integrating models, knowledge, and data for designing both the material and the product. Kulkarni et al. [75] use PREMAP as a platform for the realization of engineered materials and products for gear design by exploiting the synergy between component design, material design, and manufacturing. Later, Kulkarni et al. [76] propose a method, the concept exploration method, based on the cDSP, which is demonstrated for gear design which requires the simultaneous exploration of geometry, material, and manufacturing spaces to exploit synergies. Sharma et al. [77] present a method for robust design using coupled decisions to identify design scenarios that are relatively insensitive to uncertainties. Those design scenarios are modeled as coupled decisions to account for influence among decisions. The method is tested using a gearbox design problem – selecting gear material and determining gearbox geometry. Based on this work, Sharma et al. [78] later propose a decision classification scheme, the Multilevel Decision Scenario Matrix (MDSM), to develop design processes involving decision interactions.

As DSIDES can couple the cDSP and sDSP in different levels, hierarchies, and sequences, when the selection and compromise decisions need to be made at the same time, designers tend to adopt cDSP-ALP-DSIDES and derived design frameworks or methods in coupled decision-making.

3.5 Multigoal (Multiobjective) or Multidisciplinary design through Exploring the Solution Space
Engineering-design problems often have multiple goals (or objectives). Most test problems referred to in Sections 3.1-3.4 are multigoal problems. However, in this subsection, we only introduce the publications that focus on the methods of the exploration of the design preferences, tradeoffs among goals, and interrelationship among goals, especially when there are three or more goals and the design encompass multiple disciplines. Since multigoal problems are usually managed through exploring the solution space (consists of vectors of decision variables) and design space (consists of scenarios of formulating and compromising multiple goals), we combine the reviewing of the work in two categories – "v) exploration of the solution space and design space" and "vi) multigoal (multiobjective) or multidisciplinary design" – here in one section.

Xiao et al. [79] propose the Collaborative Multidisciplinary Decision-making Methodology (CMDM) and use it to design a robot arm with three goals – deformation towards 0.5 mm, von Mises stress around 6 Mpa, and weight around 3.5 g. under working loads around. The CMDM enables designers to integrate the cDSP with Game Theory and apply DCI to obtain robust solutions. The cDSP is used to implement and exchange design scenarios while making coupling decisions. Game Theory is adopted for managing interactions between teams' decisions. DCI is used for maintaining design freedom to accommodate downstream changes. Wang et al. [49] propose the Design Exploration Method for Adaptive Design Systems (DEM-ADS) to design a photonic crystal coupler and waveguide using the mean response function from simulations as the objective and setting the EMIs between the objective and its upper and lower bounds as goals. The method is a derived method of the IDEM in the robust design. Ahmed et al. [80] design a multigoal gear blank without prior knowledge of preferences among goals. They also explore different scenarios in the two ways (Archimedean and Preemptive) of goal formulation and gain knowledge on design scenarios associated with the design preferences. Smith et al. [81] design a thermal plant based on the Rankine cycle with six goals and demonstrate the exploration of the solution space to compromise the goas using both the Archimedean (weighted combination) and Preemptive (Lexicographic) methods. Gautham et al. [82] and use the IDEM to design a heat treatment process. The five material properties become goals in the decision model. They study the tradeoffs among the properties and find ranged sets of specifications by exploring the design preferences. Anapagaddi et al. [83] and Sabeghi et al. [84, 85] explore the solution space of multigoal design problems in continuous casting and ladle refining by evolving the weights of the goals so as to produce new grades of steels to meet stringent sets of property specifications. Goh et al. [60], Sabeghi et al. [85], Nellippallil et al. [21, 22, 72-74], and Guo et al. [42, 86] apply the Ternary plot to visualize the desired weight range for three-goal problems using applications in steel manufacturing, material design, and supply chain design. Guo et al. [8] propose a domain-independent algorithm, the Adaptive Leveling-Weighting-Clustering algorithm (ALWC), to explore the design scenarios for multigoal and many-goal (when there are more than three goals) problems. Using ALWC, designers may explore combinations and priorities of the goals based on their interrelationships.

Engineering design problems in nature have multiple goals, and often those goals are formulated using knowledge and requirements from various disciplines. In the early stage of design, designers rely on their domain expertise or intuition to compromise the goals with certain design scenarios and meet different design preferences. However, a robust method should output desired performance without too much domain expertise or assumptions. The methods of the exploration of the solution space and design space can reduce designers' dependence on prior domain knowledge and attain knowledge along with the exploration. The design methods in satisficing strategy well support solution space exploration as sets of satisficing solutions under multiple design scenarios can be identified, and model behaviors associated with those solutions and design scenarios can be recorded and analyzed.

3.6 Knowledge-based Design and Platformation

Theoretically, the awareness, discovery, reconstruction, storage, and reuse of knowledge can take place in any design problem. Not only the satisficing design strategy supports the knowledge-based design, but all kinds of methods should do. However, because the engineering designs using satisficing strategy, especially the RCEM, enforce rich post-solution analyses, designers can obtain relatively more knowledge about the robust design, solution space and design space performance, and model behaviors. The beauty of knowledge recognized in design is that it can be reused and guide the next design project, and it can be generalized into rules and packaged into templates or platforms that ease future designs in various fields.

Knowledge and templates/tools about design problems: Chen et al. [70] propose the RCEM not only as a method for facilitating robust design but also a method enhancing design productivity by increasing design knowledge in the early stages of designs. "Design knowledge is increased by implementing sophisticated integrated systems analysis in the early stages of design and making decisions based on better information." Zha et al. [87] integrate the cDSP with the Fuzzy Synthetic Decision model (FSD) as a knowledge-intensive collaboration paradigm to support hybrid decisions. Given that collaborative design decisions have objective and subjective features, the proposed method hybrid the selection and compromise decisions and archive knowledge on collaborations. Given the demand for a reusable design process and tool to support interactions in designs, Panchal et al. [88] derive a method, the modular decision-centric approach. The authors develop generic computational templates to instantiate four foundations to support concurrent design. The four foundations include hierarchical systems, separate information, decision-centric activities, and model interactions. Pederson et al. [89] propose a domain-independent approach for realizing hierarchical product platforms. The authors focus on how to synthesize numerical taxonomy and technology diffusion in a systematic multiobjective decision-making method.

Knowledge and platforms about design methods: Wang et al. propose an ontology-based uncertainty management approach [90] in designing robust decision workflows and a knowledge representation approach [91] for designing complex engineered systems. In their two publications [90, 91], they propose a decision-centric design process representation scheme, the Phase-Event-Information X (PEI-X) diagram, for designing workflows. Later, Wang et al. [92] propose a knowledge-based design guidance system (KBDGS) to support Cyber-Physical-Product-Service Systems (CPPSS) design. Ming et al. [93, 94] use an ontology to represent knowledge of decision interactions and realize it as a cloud-based platform [95]. The authors architecture the cloud-based platform to support a human-cyber-physical view of systems realization ecosystem. The decision support platforms, KBDGS and PDSIDES, can be extended as cloud-based platforms, which lead to possibilities in AI-based

Copyright © 2023 by ASME

design and self-organizing systems design in the age of Industry 4.0.

4 WAY FORWARD IN INDUSTRY 4.0

Industry 4.0, a transformative industrial revolution, with new technologies and applications like cloud-based design, cloud-based manufacturing, and the Internet of Things, makes it possible for all participants in the whole process chain to be networked, self-organized, and collaborate on decentralized decision-making [96]. The question is, why satisficing strategy?

The answer is: A networked, self-organized system that requires decision-making may have all the features that engineering-design problems have – nonlinear, nonconvex mathematical relations among factors, a number of conflicting goals, variations in controllable and uncontrollable factors, coupled decisions required, multistage and inverse decisions required, knowledge is created and can be generalized and reused anywhere all the time, etc. Above all, robust design is still the urged demand in Industry 4.0. We have used the previous three sections to demonstrate that satisficing strategy allows decision-makers to manage the features above relatively well. Some instructive concepts and ideas have already been proposed or foreseen by the satisficing community.

Nellippallil et al. [96] present the architecture and functionalities of a cloud-based computational platform to support mass collaboration and open innovation. By reviewing the evolution of engineering design in connected products, end-to-end digital integration, mass customization and personalization, data-driven design, digital twins, etc., Jiao et al. [97] envision a human-cyber-physical view of the systems realization ecosystems. Milisavljevic-Syed et al. [98] propose to leverage the advantages of digitization and Artificial Intelligence to address smart manufacturing in a networked manufacturing system. Ming et al. [99] identify the requirements of the decision support platform for design Engineering 4.0 and propose an architecture to resolve the challenges. Later they propose a framework for a Cloud-Based Platform for Decision Support in the Design of Engineered Systems (CB-PDSIDES) [100].

In summary, future research on engineering design is mainly about i) how to graft new technologies and ideas into the satisficing methodology, ii) how to utilize new technologies and ideas to better derive and innovate design methods under the satisficing strategy, and iii) make the knowledge recognition and knowledge-based design available for all participants in the cyber-physical-social system.

5 CLOSING REMARKS

In this paper, we first discuss the motivation for developing and employing the satisficing strategy in engineering designs. Optimization, as a popular method and strategy, is based on the utility theory. However, decision-makers, just like all other human beings, cannot be completely rational and knowledgeable to develop a utility function to map a problem in the physical world as a model in the mathematical world; therefore, the optimal solution to an optimization problem may not exist or be accessible. Even if the optimal solution is obtained, it may not be useful when being implemented in the physical world. In this sense, we need another modeling strategy that facilitates finding useful, "good enough" (but not necessarily optimal) solutions, which we define as satisficing solutions.

In addition, for engineering-design problems, besides solving the problem and obtaining a solution, there are other tasks, such as acquiring knowledge and insight about the output performance in certain ranges of solution space, improving the robustness of the design, coupling selection decisions with compromise decisions, integrating multiple models with different types or levels of fidelity, from multiple stages, etc. Therefore, a design strategy that enables designers to avoid the drawbacks of the optimizing strategy and work on the tasks above is needed. And that is satisficing strategy. There are different methods that give satisficing solutions and attain knowledge through post-solution analysis. In this paper, we focus on the methods based on the cDSP-ALP-DSIDES due to the consistency, productivity, and richness of their relevant work.

Then, we analyze the specialty of the formulation construct (cDSP), the solution algorithm (ALP), and the solver (DSIDES) that supports yielding satisficing solutions and realizing the features that support the design. Those features include the usefulness of the solutions, insensitivity of the solutions to uncertainties from various sources, and relatively easy and cheap to obtain.

By reviewing the publications in the cDSP-ALP-DSIDES community in the past 30 years, we classify the contributions in seven subjects: i) robust design methods, ii) multiscale and microstructure design, iii) multistage design and concurrent design, iv) coupled decision-making, v) exploration of the solution space and design space, vi) multigoal (multiobjective) or multidisciplinary design, and vii) knowledge-based design and platformation. We review the representative work in each subject, summarize their contributions and applied field, and discuss why the authors choose satisficing methods as the foundation to derive their methods.

Finally, we envision the value and development of satisficing methodologies in the age of Industry 4.0. Given the complexity of design in the digital era, satisficing strategy has advantages in producing robust design. We hope this paper can provide useful information on when and why designers may choose the satisficing strategy, concepts, methods, and tools to make their designs robust, smart, and valuable.

ACKNOWLEDGMENTS

Lin Guo acknowledges the financial support from the Pietz Professorship and Start-Up Fund from the Office of the Vice

Copyright © 2023 by ASME

President for Research and Partnerships and the Office of the Provost at the South Dakota School of Mines and Technology.

REFERENCES

[1] Reich, D., Green, R. E., Kircher, M., Krause, J., Patterson, N., Durand, E. Y., Viola, B., Briggs, A. W., Stenzel, U., and Johnson, P. L., 2010, "Genetic history of an archaic hominin group from Denisova Cave in Siberia," Nature, 468(7327), pp. 1053-1060.

[2] Nash Jr, J. F., 1950, "The bargaining problem," Econometrica: Journal of the econometric society, pp. 155-162.

[3] Saaty, T. L., 1979, "Optimization by the analytic hierarchy process," AIR FORCE OFFICE OF SCIENTIFIC RESEARCHBOLLING AFB DC.

[4] Sen, A., 2018, Collective choice and social welfare, Harvard University Press.

[5] Hazelrigg, G. A., 2010, "The Pugh controlled convergence method: model-based evaluation and implications for design theory," Research in Engineering Design, 21, pp. 143-144.

[6] Guo, L., 2021, "Model Evolution for the Realization of Complex Systems," Doctor of Philosophy Dissertation, University of Oklahoma, Norman, OK.

[7] Mistree, F., Hughes, Owen F, Bras, Bert, 1993, "Compromise decision support problem and the adaptive linear programming algorithm," Progress In Astronautics and Aeronautics: Structural Optimization: Status and Promise, 150, p. 251.

[8] Guo, L., Milisavljevic-Syed, J., Wang, R., Huang, Y., Allen, J. K., and Mistree, F., 2022, "Managing multi-goal design problems using adaptive leveling-weighting-clustering algorithm," Research in Engineering Design, pp. 1-22.

[9] Taguchi, G., 1985, "Quality engineering in Japan," Communications in Statistics-Theory and Methods, 14(11), pp. 2785-2801.

[10] Chen, W., Allen, J. K., Tsui, K.-L., and Mistree, F., 1996, "A Procedure for Robust Design: Minimizing Variations Caused by Noise Factors and Control Factors."

[11] Allen, J. K., Seepersad, C., Choi, H., and Mistree, F., 2006, "Robust Design for Multiscale and Multidisciplinary Applications."

[12] Allen, J. K., Seepersad, C. C., Mistree, F., Savannah, G., and Savannah, G., 2006, "A Survey of Robust Design with Applications to Multidisciplinary and Multiscale Systems," Journal of Mechanical Design, 128(4), pp. 832-843.

[13] Choi, H.-J., Allen, J. K., Rosen, D., McDowell, D. L., and Mistree, F., "An inductive design exploration method for the integrated design of multi-scale materials and products," Proc. International design engineering technical conferences and computers and information in engineering conference, pp. 859-870.

[14] Guo, L., 2021, Model evolution for the realization of complex systems, The University of Oklahoma.

[15] Kumar, P., Goyal, S., Singh, A. K., Allen, J. K., Panchal, J. H., and Mistree, F., "PREMΛP: exploring the design space for continuous casting of steel," Proc. ICoRD'13: Global Product Development, Springer, pp. 759-772.

[16] Shukla, R., Kulkarni, N. H., Gautham, B., Singh, A. K., Mistree, F., Allen, J. K., and Panchal, J. H., 2015, "Design exploration of engineered materials, products, and associated manufacturing processes," JOM, 67(1), pp. 94-107.

[17] Shukla, R., Goyal, S., Singh, A. K., Panchal, J. H., Allen, J. K., and Mistree, F., "An approach to robust process design for continuous casting of slab," Proc. International Design Engineering Technical Conferences and Computers and Information in Engineering Conference, American Society of Mechanical Engineers, p. V02BT03A005.

[18] Panchal, J. H., Paredis, C. J., Allen, J. K., and Mistree, F., 2008, "A value-of-information based approach to simulation model refinement," Engineering Optimization, 40(3), pp. 223-251.

[19] Messer, M., Panchal, J., Allen, J., Mistree, F., Krishnamurthy, V., Klein, B., and Yoder, P., "Designing embodiment design processes using a value-of-information-based approach with applications for integrated product and materials design," Proc. International Design Engineering Technical Conferences and Computers and Information in Engineering Conference, pp. 823-840.

[20] Fonville, T. R., Nellippallil, A. B., Horstemeyer, M., Allen, J. K., and Mistree, F., "A goal-oriented, inverse decision-based method for an American Football Helmet," Proc. International Design Engineering Technical Conferences and Computers and Information in Engineering Conference, American Society of Mechanical Engineers, p. V02BT03A026.

[21] Nellippallil, A. B., Mohan, P., Allen, J. K., and Mistree, F., 2020, "An inverse, decision-based design method for robust concept exploration," Journal of Mechanical Design, 142(8), p. 081703.

[22] Nellippallil, A. B., Rangaraj, V., Gautham, B., Singh, A. K., Allen, J. K., and Mistree, F., 2018, "An inverse, decision-based design method for integrated design exploration of materials, products, and manufacturing processes," Journal of Mechanical Design, 140(11).

[23] Simon, H. A., 1956, "Rational choice and the structure of the environment," Psychological review, 63(2), p. 129.

[24] Simon, H. A., 1973, "Organization man: Rational or self-actualizing?," Public Administration Review, 33(4), pp. 346-353.

[25] Simon, H. A., 1976, "From Substantive to Procedural Rationality," 25 years of economic theory, Springer, pp. 65-86.

[26] Mistree, F., Hughes, O., and Phuoc, H., 1981, "An optimization method for the design of large, highly constrained complex systems," Engineering Optimization, 5(3), pp. 179-197.

[27] Mistree, F., Smith, W., Bras, B., Allen, J., and Muster, D., 1990, "Decision-based design: a contemporary paradigm for ship design," Transactions, Society of Naval Architects and Marine Engineers, 98(1990), pp. 565-597.

[28] Mistree, F., and Kamal, S., 1985, DSIDES: Decision support in the design of engineering systems, University of Houston.

[29] Salado, A., and Nilchiani, R., 2016, "Reducing excess requirements through orthogonal categorizations during problem formulation: Results of a factorial experiment," IEEE

Copyright © 2023 by ASME

Transactions on Systems, Man, and Cybernetics: Systems, 47(3), pp. 405-415.

[30] Ball, L. J., Evans, J. S. B., and Dennis, I., 1994, "Cognitive processes in engineering design: A longitudinal study," Ergonomics, 37(11), pp. 1753-1786.

[31] Ruiz-Lopez, T., Rodriguez-Dominguez, C., Noguera, M., Rodríguez, M. J., Benghazi, K., and Garrido, J. L., 2013, "Applying model-driven engineering to a method for systematic treatment of NFRs in AmI systems," Journal of Ambient Intelligence and Smart Environments, 5(3), pp. 287-310.

[32] Loch, C., Mihm, J., and Huchzermeier, A., 2003, "Concurrent engineering and design oscillations in complex engineering projects," Concurrent Engineering, 11(3), pp. 187-199.

[33] Utomo, C., and Rahmawati, Y., 2020, "Agreement options for negotiation on material location decision of housing development," Construction Innovation, 20(2), pp. 209-222.

[34] Arjmandzadeh, Z., Nazemi, A., and Safi, M., 2019, "Solving multiobjective random interval programming problems by a capable neural network framework," Applied Intelligence, 49, pp. 1566-1579.

[35] Tchangani, A. P., 2010, "Considering bipolarity of attributes with regards to objectives in decisions evaluation," Engineering Economics, 21(5).

[36] Chester, M. V., and Allenby, B., 2019, "Infrastructure as a wicked complex process," Elementa: Science of the Anthropocene, 7.

[37] Herman, J. D., Reed, P. M., Zeff, H. B., and Characklis, G. W., 2015, "How should robustness be defined for water systems planning under change?," Journal of Water Resources Planning and Management, 141(10), p. 04015012.

[38] Courant, R., and Hilbert, D., 1953, "Methods of Mathematical Physics, vol. I, 1953," New York: Interscience.

[39] Sharma, S., and Kumar, V., 2022, "A Comprehensive Review on Multi-objective Optimization Techniques: Past, Present and Future," Archives of Computational Methods in Engineering, 29(7), pp. 5605-5633.

[40] Abualigah, L., Shehab, M., Alshinwan, M., and Alabool, H., 2020, "Salp swarm algorithm: a comprehensive survey," Neural Computing and Applications, 32, pp. 11195-11215.

[41] Daoud, M. S., Shehab, M., Al-Mimi, H. M., Abualigah, L., Zitar, R. A., and Shambour, M. K. Y., 2022, "Gradient-Based Optimizer (GBO): A Review, Theory, Variants, and Applications," Archives of Computational Methods in Engineering, pp. 1-19.

[42] Guo, L., Balu Nellippallil, A., Smith, W. F., Allen, J. K., and Mistree, F., "Adaptive Linear Programming Algorithm With Parameter Learning for Managing Engineering-Design Problems," Proc. International Design Engineering Technical Conferences and Computers and Information in Engineering Conference, American Society of Mechanical Engineers, p. V11BT11A029.

[43] Box, G., 1988, "Signal-to-noise ratios, performance criteria, and transformations," Technometrics, 30(1), pp. 1-17.

[44] Simpson, T., Allen, J., Chen, W., and Mistree, F., "Conceptual design of a family of products through the use of

the Robust Concept Extrapolation Method," Proc. 6th Symposium on Multidisciplinary Analysis and Optimization, p. 4161.

[45] Chen, W., Allen, J. K., Mavris, D. N., and Mistree, F., 1996, "A concept exploration method for determining robust top-level specifications," Engineering Optimization+ A35, 26(2), pp. 137-158.

[46] Chen, W., Allen, J. K., and Mistree, F., 1997, "A robust concept exploration method for enhancing productivity in concurrent systems design," Concurrent Engineering, 5(3), pp. 203-217.

[47] Choi, H.-J., Austin, R., Allen, J. K., Mcdowell, D. L., Mistree, F., and Benson, D. J., 2005, "An approach for robust design of reactive power metal mixtures based on non-deterministic micro-scale shock simulation," Journal of Computer-Aided Materials Design, 12, pp. 57-85.

[48] Choi, H.-J., 2005, A robust design method for model and propagated uncertainty, Georgia Institute of Technology.

[49] Wang, C., Krishnamurthy, V., Klein, B., Choi, S.-K., Allen, J. K., and Mistree, F., "A Design Exploration Method for Adaptive Design Systems," Proc. International Design Engineering Technical Conferences and Computers and Information in Engineering Conference, pp. 455-466.

[50] Sinha, A., Bera, N., Allen, J. K., Panchal, J. H., and Mistree, F., 2013, "Uncertainty management in the design of multiscale systems," Journal of Mechanical Design, 135(1).

[51] Sinha, A., Chakraborty, M., Ghosh, S., Kumar, C., Panchal, J. H., Allen, J. K., McDowell, D. L., and Mistree, F., "Microstructure-Mediated Integration of Material and Product Design: Undersea Submersible," Proc. International Design Engineering Technical Conferences and Computers and Information in Engineering Conference, pp. 467-478.

[52] Sinha, A., Panchal, J. H., Allen, J. K., and Mistree, F., "Managing uncertainty in multiscale systems via simulation model refinement," Proc. International Design Engineering Technical Conferences and Computers and Information in Engineering Conference, pp. 471-483.

[53] Kulkarni, N., Gupta, R., Khan, D., Gautham, B., Allen, J. K., Panchal, J., and Mistree, F., "Inverse design of manufacturing process chains," Proc. International Design Engineering Technical Conferences and Computers and Information in Engineering Conference, American Society of Mechanical Engineers, p. V02BT03A004.

[54] Gautham, B., Singh, A. K., Ghaisas, S. S., Reddy, S. S., and Mistree, F., "PREMAP: a platform for the realization of engineered materials and products," Proc. ICoRD'13: Global Product Development, Springer, pp. 1301-1313.

[55] Samadiani, E., Joshi, Y., Allen, J. K., and Mistree, F., 2010, "Adaptable robust design of multi-scale convective systems applied to energy efficient data centers," Numerical Heat Transfer, Part A: Applications, 57(2), pp. 69-100.

[56] Panchal, J. H., Choi, H.-J., Shephard, J., Allen, J. K., McDowell, D. L., and Mistree, F., "A strategy for simulation-based multiscale, multi-functional products and associated design processes," Proc. International Design Engineering

Copyright © 2023 by ASME

Technical Conferences and Computers and Information in Engineering Conference, pp. 845-857.

[57] Messer, M., Pedersen, K., Allen, J. K., & Mistree, F., 2006, "Domain independent approach to designing hierarchical platforms on multiple levels of abstraction and scales," International Design Engineering Technical Conferences and Computers and Information in Engineering Conference, pp. 367-377.

[58] Ripppel, M., Choi, S., Mistree, F., and Allen, J., "Alternatives to Taylor series approximation for the variance estimation in robust design," Proc. 13th AIAA/ISSMO Multidisciplinary Analysis Optimization Conference, p. 9083.

[59] Goh, C.-H., Ahmed, S., Dachowicz, A. P., Allen, J. K., and Mistree, F., "Integrated multiscale robust design considering microstructure evolution and material properties in the hot rolling process," Proc. International Design Engineering Technical Conferences and Computers and Information in Engineering Conference, American Society of Mechanical Engineers, p. V02BT03A001.

[60] Goh, C.-H., Dachowicz, A. P., Allen, J. K., and Mistree, F., "Exploring the performance-property-structure solution space in friction stir welding," Proc. Proceedings of the 3rd World Congress on Integrated Computational Materials Engineering (ICME 2015), Springer, pp. 347-354.

[61] Goh, C. H., Dachowicz, A. P., Allen, J. K., and Mistree, F., 2018, "A Computational Method for the Design of Materials Accounting for the Process–Structure–Property–Performance (PSPP) Relationship," Integrated Computational Materials Engineering (ICME) for Metals: Concepts and Case Studies, pp. 539-572.

[62] Seepersad, C. C., Allen, J. K., McDowell, D. L., and Mistree, F., "Robust design of cellular materials with topological and dimensional imperfections," Proc. International Design Engineering Technical Conferences and Computers and Information in Engineering Conference, pp. 807-821.

[63] Thompson, S., Muchnick, H., Choi, H., McDowell, D., Allen, J., & Mistree, F., 2006, "Robust materials design of blast resistant panels," 11th AIAA/ISSMO Multidisciplinary Analysis and Optimization Conference.

[64] McDowell, D. L., Choi, H. J., Panchal, J., Austin, R., Allen, J., & Mistree, F., 2007, "Plasticity-related microstructure-property relations for materials design," Key Engineering Materials, 340, pp. 21-30.

[65] McDowell, D. L., Panchal, J., Choi, H.-J., Seepersad, C., Allen, J., and Mistree, F., 2009, Integrated design of multiscale, multifunctional materials and products, Butterworth-Heinemann.

[66] Seepersad, C. C., Allen, J. K., McDowell, D. L., and Mistree, F., 2008, "Multifunctional topology design of cellular material structures," Journal of Mechanical Design, 130(3).

[67] Shukla, R., Anapagaddi, R., Singh, A. K., Panchal, J. H., Mistree, F., and Allen, J. K., 2016, "Design Exploration to Determine Process Parameters of Ladle Refining for an Industrial Application," steel research international, 87(10), pp. 1333-1343.

[68] Beemaraj, S. B., Pathan, R. K., Salvi, A. G., Sharma, G., Mistree, F., and Allen, J. K., "Inverse Multi-Scale Robust Design of Composite Structures Using Design Capability Indices," Proc. International Design Engineering Technical Conferences and Computers and Information in Engineering Conference, American Society of Mechanical Engineers, p. V009T009A017.

[69] Rangarajan, B., Mistree, F., and Sorab, J., "Incorporating Tribological Considerations in the Robust Concurrent Design of Automobile Engine Lubricated Components," Proc. International Design Engineering Technical Conferences and Computers and Information in Engineering Conference, American Society of Mechanical Engineers, p. V002T002A031.

[70] Chen, W., Allen, J. K., and Mistree, F., 2000, "Design Knowledge Development for Productivity Enhancement in Concurrent Systems Design," Knowledge-Based Systems, Elsevier, pp. 1037-1060.

[71] Simpson, T. W., Seepersad, C. C., and Mistree, F., 2001, "Balancing commonality and performance within the concurrent design of multiple products in a product family," Concurrent Engineering, 9(3), pp. 177-190.

[72] Nellippallil, A. B., Song, K. N., Goh, C.-H., Zagade, P., Gautham, B., Allen, J. K., and Mistree, F., 2017, "A goal-oriented, sequential, inverse design method for the horizontal integration of a multistage hot rod rolling system," Journal of Mechanical Design, 139(3), p. 031403.

[73] Nellippallil, A. B., Mohan, P., Allen, J. K., and Mistree, F., "Inverse thermo-mechanical processing (ITMP) design of a steel rod during hot rolling process," Proc. International Design Engineering Technical Conferences and Computers and Information in Engineering Conference, American Society of Mechanical Engineers, p. V02AT03A053.

[74] Nellippallil, A. B., Allen, J. K., Gautham, B., Singh, A. K., and Mistree, F., 2020, Architecting robust co-design of materials, products, and manufacturing processes, Springer.

[75] Kulkarni, N., Zagade, P. R., Gautham, B., Panchal, J. H., Allen, J. K., and Mistree, F., "PREMAP: Exploring the Design and Materials Space for Gears," Proc. ICoRD'13: Global Product Development, Springer, pp. 745-757.

[76] Kulkarni, N., Gautham, B., Zagade, P., Panchal, J., Allen, J. K., and Mistree, F., 2015, "Exploring the geometry and material space in gear design," Engineering Optimization, 47(4), pp. 561-577.

[77] Sharma, G., Allen, J. K., and Mistree, F., 2021, "A method for robust design in a coupled decision environment," Design Science, 7, p. e23.

[78] Sharma, G., Allen, J. K., and Mistree, F., 2022, "Designing concurrently and hierarchically coupled engineered systems," Engineering Optimization, pp. 1-21.

[79] Xiao, A., Seepersad, C. C., Allen, J. K., Rosen, D. W., & Mistree, F., 2007, "Design for manufacturing: application of collaborative multidisciplinary decision-making methodology," Engineering Optimization, 39(4), p. 22.

[80] Ahmed, S., Goh, C.-H., Allen, J. K., Mistree, F., Zagade, P., and Gautham, B., "Hot forging of automobile steel gear blanks: an exploration of the solution space," Proc. International Design Engineering Technical Conferences and Computers and Information in Engineering Conference, American Society of Mechanical Engineers, p. V02BT03A003.

Copyright © 2023 by ASME

[81] Smith, W. F., Milisavljevic, J., Sabeghi, M., Allen, J. K., and Mistree, F., "The realization of engineered systems with considerations of complexity," Proc. International Design Engineering Technical Conferences and Computers and Information in Engineering Conference, American Society of Mechanical Engineers, p. V007T006A019.

[82] Gautham, B., Kulkarni, N., Zagade, P., Allen, J. K., Mistree, F., and Panchal, J., "ICME for the Integrated Design of an Automotive Gear Considering Uncertainty," Proc. Proceedings of the 3rd World Congress on Integrated Computational Materials Engineering (ICME 2015), Wiley Online Library, pp. 323-330.

[83] Anapagaddi, R., Shukla, R., Goyal, S., Singh, A. K., Allen, J. K., Panchal, J. H., and Mistree, F., "Exploration of the design space in continuous casting Tundish," Proc. International Design Engineering Technical Conferences and Computers and Information in Engineering Conference, American Society of Mechanical Engineers, p. V02BT03A006.

[84] Sabeghi, M., Shukla, R., Allen, J. K., and Mistree, F., "Solution space exploration of the process design for continuous casting of steel," Proc. International Design Engineering Technical Conferences and Computers and Information in Engineering Conference, American Society of Mechanical Engineers, p. V02BT03A005.

[85] Sabeghi, M., Smith, W., Allen, J. K., and Mistree, F., "Solution space exploration in model-based realization of engineered systems," Proc. International Design Engineering Technical Conferences and Computers and Information in Engineering Conference, American Society of Mechanical Engineers, p. V02AT03A015.

[86] Guo, L., Chen, S., Allen, J. K., and Mistree, F., 2021, "A Framework for Designing the Customer-Order Decoupling Point to Facilitate Mass Customization," Journal of Mechanical Design, 143(2).

[87] Zha, X. F., Sriram, R. D., Fernandez, M. G., and Mistree, F., 2008, "Knowledge-intensive collaborative decision support for design processes: A hybrid decision support model and agent," Computers in Industry, 59(9), pp. 905-922.

[88] Panchal, J. H., Gero Fernández, M., Paredis, C. J., Allen, J. K., and Mistree, F., 2009, "A modular decision-centric approach for reusable design processes," Concurrent Engineering, 17(1), pp. 5-19.

[89] Pedersen, K., Messer, M., Allen, J. K., and Mistree, F., 2013, "Hierarchical product platform design: a domain-independent approach," Ships and Offshore Structures, 8(3-4), pp. 367-382.

[90] Wang, R., Nellippallil, A. B., Wang, G., Yan, Y., Allen, J. K., and Mistree, F., 2019, "Ontology-based uncertainty management approach in designing of robust decision workflows," Journal of Engineering Design, 30(10-12), pp. 726-757.

[91] Wang, R., Nellippallil, A. B., Wang, G., Yan, Y., Allen, J. K., and Mistree, F., 2021, "A process knowledge representation approach for decision support in design of complex engineered systems," Advanced Engineering Informatics, 48, p. 101257.

[92] Wang, R., Milisavljevic-Syed, J., Guo, L., Huang, Y., and Wang, G., 2021, "Knowledge-based Design Guidance System for Cloud-based Decision Support in the Design of Complex Engineered Systems," Journal of Mechanical Design, 143(7).

[93] Ming, Z., Sharma, G., Allen, J. K., and Mistree, F., 2020, "An ontology for representing knowledge of decision interactions in decision-based design," Computers in Industry, 114, p. 103145.

[94] Ming, Z., Wang, G., Yan, Y., Panchal, J. H., Goh, C. H., Allen, J. K., and Mistree, F., 2018, "Ontology-based representation of design decision hierarchies," Journal of Computing and Information Science in Engineering, 18(1).

[95] Ming, Z., Nellippallil, A. B., Yan, Y., Wang, G., Goh, C. H., Allen, J. K., and Mistree, F., 2018, "PDSIDES—A knowledge-based platform for decision support in the design of engineering systems," Journal of Computing and Information Science in Engineering, 18(4).

[96] Nellippallil, A. B., Ming, Z., Allen, J. K., and Mistree, F., 2019, "Cloud-based materials and product realization—fostering ICME via industry 4.0," Integrating Materials and Manufacturing Innovation, 8, pp. 107-121.

[97] Jiao, R., Communri, S., Panchal, J., Milisavljevic-Syed, J., Allen, J. K., Mistree, F., and Schaefer, D., 2021, "Design engineering in the age of industry 4.0," Journal of Mechanical Design, 143(7).

[98] Milisavljevic-Syed, J., Allen, J. K., Communri, S., Mistree, F., Milisavljevic-Syed, J., Allen, J. K., Communri, S., and Mistree, F., 2020, "Decision-Based Design of Networked Manufacturing Systems (NMS)," Architecting Networked Engineered Systems: Manufacturing Systems Design for Industry 4.0, pp. 41-70.

[99] Ming, Z., Nellippallil, A. B., Wang, R., Allen, J. K., Wang, G., Yan, Y., and Mistree, F., 2022, "Requirements and architecture of the decision support platform for design engineering 4.0," Architecting a Knowledge-Based Platform for Design Engineering 4.0, Springer, pp. 1-22.

[100] Ming, Z., Nellippallil, A. B., Wang, R., Allen, J. K., Wang, G., Yan, Y., and Mistree, F., 2022, "Extending PDSIDES to CB-PDSIDES: New Opportunities in Design Engineering 4.0," Architecting A Knowledge-Based Platform for Design Engineering 4.0, Springer, pp. 213-237.

Copyright © 2023 by ASME

Proceedings of the ASME 2023
International Design Engineering Technical Conferences and
Computers and Information in Engineering Conference
IDETC-CIE2023
August 20-23, 2023, Boston, Massachusetts

DETC2023-112948

QUANTIFICATION MODEL UNCERTAINTY OF LABEL-FREE MACHINE LEARNING FOR MULTIDISCIPLINARY SYSTEMS ANALYSIS

Huiru Li[1, 2]
[1]**Department of Mechanical and Energy Engineer**
Indiana University-Purdue University
Indianapolis
Indianapolis, IN
[2]**School of Mechanical Engineering**
Purdue University
West Lafayette

Jitesh H. Panchal
School of Mechanical Engineering
Purdue University
West Lafayette, IN

Xiaoping Du
Department of Mechanical and Energy Engineer
Indiana University-Purdue University Indianapolis
Indianapolis, IN

ABSTRACT

Many engineering systems involve multiple interacting disciplines or subsystems. For a design or analysis task, unknown linking variables, which are those variables that are outputs of some disciplines and inputs of other disciplines, are obtained by solving the system of implicit interdisciplinary compatibility equations for a given set of system inputs. This study creates surrogate models for linking variables using label-free training with neural networks. The compatibility equations are embedded in the cost function of the model training. They are calculated and are not solved for given input training variables, thereby avoiding label acquisition. To quantify the prediction errors of the surrogate models, we build their error models with Gaussian Process regression, which uses the existing training points and the derivatives of the compatibility equations at the training points. The error models are then used to compensate for the errors of neural network surrogate models of the linking variables, producing more accurate predictions of linking variables with quantified model uncertainty for predicting system responses. The linking variables with quantified model uncertainty are then used to predict the system responses and associated prediction errors. We demonstrate the effectiveness of the proposed method by the application to a propane combustion problem.

Keywords: Multidisciplinary systems, uncertainty quantification, machine learning, neural network, Gaussian process

1. INTRODUCTION

The complexity of engineering systems arises from their involvement of multiple disciplines that interact with one another. This interaction in multidisciplinary systems is common in various fields, including aerospace [1], marine applications [2, 3], automobile engineering [4], and renewable energy [5, 6]. For instance, a system with coupled fluid and structure disciplines [7] and another with coupled aerodynamic and structure disciplines [8] represent some of the many multidisciplinary systems encountered in engineering.

Computational models, derived from domain physics, are commonly used in multidisciplinary systems analysis and design. Computational models are usually expensive to run. To this end, advanced statistical methods and scientific machine learning (SciML) have been increasingly used to create surrogates for computational models. Although surrogate models are much more efficient, we need to account for their prediction error, which is the discrepancy between the prediction from a surrogate model and the original and validated computational model. The model error may be large, especially in the design

Copyright © 2023 by ASME

space that has not been well covered by training points. The error may cause safety concerns [9], for example, the collision of a self-driving car [10].

We never know the model error unless we call the original computational model. But we can estimate the prediction error by quantifying the model uncertainty of the surrogate model. This type of uncertainty is known as epistemic uncertainty and arises due to a lack of knowledge, including ignorance, assumptions, and simplifications in model formulations [11-14]. The present study is concerned with examining epistemic uncertainty in surrogate models.

The other type of uncertainty is aleatory uncertainty, which stems from the random nature of factors, such as user environment and manufacturing. Variables that exhibit aleatory uncertainty are typically modeled using probability distributions. If a machine learning model is used for prediction and design, input variables may be random variables. They are therefore random variables with aleatory uncertainty. While accommodating aleatory uncertainty is not the focus of this study, the methodology developed herein can be readily applied to address both model and aleatory uncertainty in engineering analysis and design.

Recent research has placed greater emphasis on studying epistemic uncertainty in deep learning [14-18]. The majority of these studies have focused on supervised learning. A proof-of-concept study was performed to quantify the prediction uncertainty of a response whose labels are unavailable during training, using a physics-based label-free regression approach [19]. A neural network model is first built by including a system of implicit equations from which the responses are solved for. The equations are satisfied automatically during the process of model training, therefore eliminating the need for labels. Then the Gaussian Process (GP) regression [20-24] is used to build an error model for the response. Although GP regression cannot be label free, the model errors are estimated by approximating the system of implicit equations at the training points of the input variables. This method can be applied to prediction and design tasks that involve both model and aleatory uncertainty.

This study builds upon the above method [19] and extends it to multidisciplinary analysis (MDA). The linking variables, which maintain subsystem compatibility, are obtained by solving a system of implicit compatibility equations, a major task of MDA. The label-free method is then used to construct models for the linking variables. Once these models are available, they are used to predict system responses, which are functions of the system inputs and linking variables. Prediction can be accomplished much more quickly and easily than with traditional MDA methods. Additionally, the prediction uncertainty can be quantified by using the model uncertainty of the linking variables. A demonstration is presented using the example of propane combustion.

The remaining sections of the paper are organized as follows: Section 2 provides a review of multidisciplinary systems. Section 3 describes the proposed method in detail followed by an example and the corresponding results in Section

4. Section 5 concludes the paper and highlights potential future work.

2. MULTIDISCIPLINARY SYSTEMS

Figure 1 illustrates a multidisciplinary system consisting of three interrelated disciplines or subsystems. The system's variables are explained below.

Input variables: They include shared input variables by all subsystems and local input variables.

- x_s: shared input random variables for all disciplines
- x_i: local input random variables of subsystem i

Linking variables: They are internal variables which maintain copmatability among subsystems. They are invisible from outside of the system.

- y_{ij}, $i, j = 1,2, \ldots, m, i \neq j$: linking variables, which are outputs of subsystem i and inputs to subsystem j
- $y_i = \{y_{ij} | i, j = 1,2, \ldots, m, i \neq j\}$: the set of linking variables generated as outputs from subsystem i and are taken as inputs to the other subsystems
- $y^i = \{y_1, y_2, \ldots, y_{i-1}, y_{i+1}, \ldots, y_m\}$: the set of linking variables taken as inputs to subsystem i

System responses (output variables): They are functions of input variables and linking variables.

- $z_i, i = 1,2, \ldots, v$

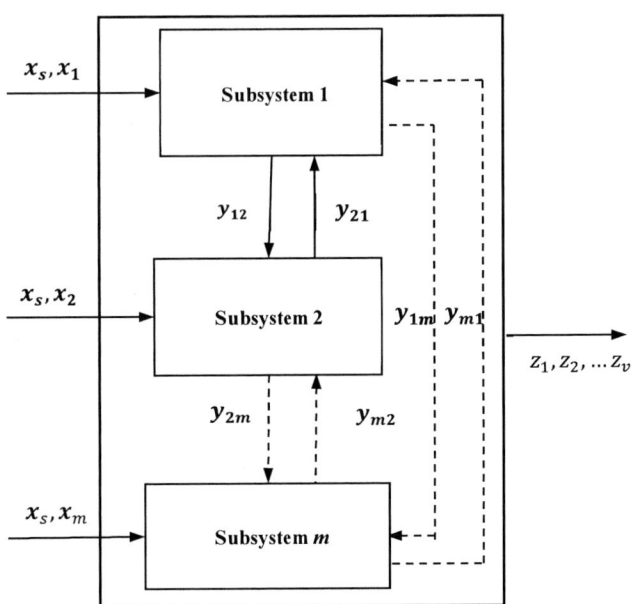

FIGURE 1: A MULTIDISCIPLINARY SYSTEM

For subsystem i, output linking variables y_i are given as

$$y_i = f_{y_i}(x_s, x_i, y^i) = f_{y_i}(x_s, x_i, y_1, \ldots, y_{i-1}, y_{i+1}, \ldots, y_m) \ (1)$$

where $f_{y_i}(\cdot)$ are disciplinary models of subsystem i.

Copyright © 2023 by ASME

During MDA, all linking variables are obtained by solving the system of compatibility equations given in Eq. (1).

Then system responses $\boldsymbol{z}$ are obtained by

$$z = f_z(x_s, x_1, \ldots, x_n, y_1, \cdots, y_m) \qquad (2)$$

3. METHOD

The computation of compatibility equations $\boldsymbol{f}_{y_i}(\cdot)$ for linking variables is typically expensive, which can slow down the MDA process. This is because MDA involves an iterative solution approach, where the linking variables are solved for a given set of input variables. This iterative process needs many rounds of evaluations of disciplinary models $\boldsymbol{f}_{y_i}(\cdot)$. It also requires communication between disciplines, since each discipline has its own set of models that depend on the outputs of models in other disciplines.

It is therefore desirable to construct surrogate models for the disciplinary models. In this section, we discuss how to extend the label-free neural network method [19] to build surrogate models without the need to solve compatibility equations. We will also discuss how the Gaussian Process (GP) method can be employed to quantify model uncertainty.

3.1 Label-free neural network learning

We extend the label-free neural network method [19] for MDA, whose task is to predict the system responses given a set of input variables. To use the label-free method, we combine all the input variables, including shared and disciplinary input variables, into $\boldsymbol{x} = (\boldsymbol{x}_s^{\mathsf{T}}, \boldsymbol{x}_1^{\mathsf{T}}, \boldsymbol{x}_2^{\mathsf{T}}, \ldots)^{\mathsf{T}}$ and all linking variables into $\boldsymbol{y} = (\boldsymbol{y}_1, \boldsymbol{y}_2, \ldots, \boldsymbol{y}_m)$. The compatibility equations are then rewritten as

$$h_i(x, y) = y_i - f_{y_i}(x, y^i) = 0, i = 1,2,\ldots,m \qquad (3)$$

The system responses are then given by

$$z = f_z(x, y) \qquad (4)$$

Directly employing the neural network method requires the labels of linking variables to be solved first for a given set of input variables $\boldsymbol{x}$. However, as previously discussed, solving for these labels can be computationally expensive and involves significant communication between disciplines. To circumvent this issue, we embed the system of compatibility equations in the training process of the neural network, rather than solving them before training. This approach allows us to build surrogate models without the need to generate labels, resulting in significant computational savings. Let the surrogate models be denoted as follows:

$$y_i = f_{y_i}^{NN}(x), \qquad i = 1,2,\ldots,m \qquad (5)$$

They are built with training data $\mathbb{D} = (\boldsymbol{x}^{(1)}, \boldsymbol{x}^{(2)}, \ldots, \boldsymbol{x}^{(d)})$, where d is the number of training points. The hyperparameters $\boldsymbol{w}$ of the neural network, including weights and biases, are determined by minimizing the residuals of $h_i(\boldsymbol{x}^{(j)}, \boldsymbol{f}_y^{NN}(\boldsymbol{x}^{(j)}; \boldsymbol{w})) = 0$ in the following loss function:

$$L(w) = \sum_{j=1}^{d} \sum_{i=1}^{m} h_i^2\left(x^{(j)}, f_{y_i}^{NN}(x^{(j)}; w)\right) \qquad (6)$$

After the surrogate models $y_i = f_{y_i}^{NN}(x), i = 1,2,\ldots,m$, are built, system responses can then be predicted at a new design point $\boldsymbol{x}$.

$$z = f_z(x, y) = f_z\left(x, f_y^{NN}(x)\right) \qquad (7)$$

3.2 Model uncertainty quantification

Since there are prediction errors in the neural network models $\boldsymbol{f}_y^{NN}(\boldsymbol{x})$, the errors will propagate through Eq. (7) to system predictions, impacting the accuracy of system predictions. It is therefore essential to quantify the errors associated with the system response predictions for critical applications. To achieve this, we employ the label-free method described in [19] to quantify the model uncertainty in $\boldsymbol{f}_y^{NN}(\boldsymbol{x})$ and derive a method to quantify the model uncertainty in the predictions of $\boldsymbol{z}$.

The error models of the linking variables are built by the GP method and are denoted by $\varepsilon_{y_i} = \varepsilon_{y_i}^{GP}(\boldsymbol{x})$, $i = 1,2,\ldots,m$, where ε_{y_i} is the model error of the i-th surrogate model $f_{y_i}^{NN}(\boldsymbol{x})$.

The model error is defined by

$$\varepsilon_{y_i}(x) = y_i(x) - f_{y_i}^{NN}(x) \qquad (8)$$

Since no labels of the model errors are available, they are estimated. The residual of $h_i(\boldsymbol{x})$ at training point $\boldsymbol{x}^{(j)}$ is given by $h_i(\boldsymbol{x}^{(j)}, \boldsymbol{y}^{(j)}) \approx h_i(\boldsymbol{x}^{(j)}, \boldsymbol{y}^{NN}(\boldsymbol{x}^{(j)}))$, in which

$$y^{NN}\left(x^{(j)}\right) = (f_1^{NN}\left(x^{(j)}\right), f_2^{NN}\left(x^{(j)}\right), \ldots, f_m^{NN}\left(x^{(j)}\right)^{\mathsf{T}} \qquad (9)$$

The true linking variable is

$$y_i\left(x^{(j)}\right) = f_i^{NN}\left(x^{(j)}\right) + \varepsilon_{y_i}^{GP}\left(x^{(j)}\right) \qquad (10)$$

The compatibility equations are rewritten as

$$h_i\left(x^{(j)}, y^{(j)}\right) = h_i\left(x^{(j)}, y^{NN}\left(x^{(j)}\right)\right) + \varepsilon_{y_i}^{GP}\left(x^{(j)}\right) = 0 \quad (11)$$

where

$$\varepsilon_y^{GP}\left(x^{(j)}\right) = \left(\varepsilon_{y_1}\left(x^{(j)}\right), \varepsilon_{y_2}\left(x^{(j)}\right), \ldots, \varepsilon_{y_m}\left(x^{(j)}\right)\right)^{\mathsf{T}} \qquad (12)$$

The first order Taylor expansion series at $\left(\boldsymbol{x}^{(j)}, \boldsymbol{y}^{NN}(\boldsymbol{x}^{(j)})\right)$ yields

$$h_i\left(x^{(j)}, y^{(j)}\right) \approx h_i\left(x^{(j)}, y^{NN}\left(x^{(j)}\right)\right) + \nabla_{h_i}^{\mathsf{T}} \varepsilon_y^{GP}\left(x^{(j)}\right) = 0 \quad (13)$$

where

$$\nabla_{h_i} = \left(\frac{\partial h_i}{\partial y_1}, \frac{\partial h_i}{\partial y_2}, \cdots \frac{\partial h_i}{\partial y_m}\right)^{\mathsf{T}} \Big|_{x^{(j)}} = 0 \qquad (14)$$

and ∇_{h_i} is a gradient that contains the derivatives of h_i with respect to the outputs of the surrogate models. We then have a system of equations for the model errors.

Copyright © 2023 by ASME

$$h_i\big(\boldsymbol{x}^{(j)}, y^{NN}(\boldsymbol{x}^{(j)})\big) + \nabla_{h_i}^{\top}\,\boldsymbol{\varepsilon}_y^{GP}(\boldsymbol{x}^{(j)}) = 0,\, i = 1,2,\ldots,m \quad (15)$$

The equations are rewritten as

$$\mathbf{A}^{(j)}\boldsymbol{\varepsilon}^{(j)} + \mathbf{B}^{(j)} = 0 \quad (16)$$

where

$$\mathbf{A}^{(j)} = \begin{pmatrix} \dfrac{\partial h_1}{\partial y_1} & \dfrac{\partial h_1}{\partial y_2} & \cdots & \dfrac{\partial h_1}{\partial y_m} \\ \dfrac{\partial h_2}{\partial y_1} & \dfrac{\partial h_2}{\partial y_2} & \cdots & \dfrac{\partial h_2}{\partial y_m} \\ \vdots & \vdots & \ddots & \vdots \\ \dfrac{\partial h_m}{\partial y_1} & \dfrac{\partial h_m}{\partial y_2} & \cdots & \dfrac{\partial h_m}{\partial y_m} \end{pmatrix}\Bigg|_{\boldsymbol{x}^{(j)}} \quad (17)$$

$$\mathbf{B}^{(j)} = \begin{pmatrix} h_1\big(\boldsymbol{x}^{(j)}, y^{NN}(\boldsymbol{x}^{(j)})\big) \\ h_2\big(\boldsymbol{x}^{(j)}, y^{NN}(\boldsymbol{x}^{(j)})\big) \\ \cdots \cdots \\ h_m\big(\boldsymbol{x}^{(j)}, y^{NN}(\boldsymbol{x}^{(j)})\big) \end{pmatrix} \quad (18)$$

The model errors $\boldsymbol{\varepsilon}_y^{(j)} = \boldsymbol{\varepsilon}_y^{GP}(\boldsymbol{x}^{(j)})$ at training points $\boldsymbol{x}^{(j)}$ are then estimated by

$$\boldsymbol{\varepsilon}_{y_i}^{(j)} = \big(\mathbf{A}^{(j)}\big)^{-1}\mathbf{B}^{(j)} \quad (19)$$

Dataset of $\big(\boldsymbol{x}^{(j)}, \boldsymbol{\varepsilon}_{y_i}^{(j)}\big)$, $j = 1,2\ldots d$, $i = 1,2\ldots m$, is then used to create GP models $\varepsilon_{y_i}^{GP}(\boldsymbol{x})$ for model errors ε_{y_i}. The linking variables are predicted by

$$y_i = f_i^{NN}(\boldsymbol{x}) + \varepsilon_{y_i}^{GP}(\boldsymbol{x}) \quad (20)$$

The prediction of a linking variables is a random variable since $\varepsilon_{y_i}^{GP}$ follows a normal distribution $N\big(\mu_{\varepsilon_{y_i}}^{GP}, \sigma_{\varepsilon_{y_i}}^{GP}\big)$, where $\mu_{\varepsilon_{y_i}}^{GP}$ and $\sigma_{\varepsilon_{y_i}}^{GP}$ are the mean and standard deviation of the model error, which are the outputs of the GP error model. Hence the prediction also follows a normal distribution with the mean and standard deviation given below.

$$\mu_{y_i} = f_{y_i}^{NN}(\boldsymbol{x}) + \mu_{\varepsilon_{y_i}}^{GP}(\boldsymbol{x}) \quad (21)$$

and the standard deviation is given by

$$\sigma_{y_i} = \sigma_{\varepsilon_{y_i}}^{GP}(\boldsymbol{x}) \quad (22)$$

Both $\mu_{\varepsilon_{y_i}}^{GP}(\boldsymbol{x})$ and $\sigma_{\varepsilon_{y_i}}^{GP}(\boldsymbol{x})$ are outputs of the GP model.

After obtaining the model errors of the linking variables, we now discuss the quantification of model uncertainty in system responses. With the distribution of the predicted linking variables $\boldsymbol{y}$ available, we can find the distribution of the system response $z = f_z(\boldsymbol{x}, \boldsymbol{y})$. The distribution will then provide the best prediction or the mean of the system response and the associated uncertainty (standard deviation).

There are many ways to find the distribution of z, such as Monte Carlo simulation and advanced reliability methods [23, 25, 26]. Herein we use the most efficient method or the First Order Second Moment (FOSM) Method [27, 28]. Using FOSM,

we employ the first order Taylor expansion at the means of the projected linking variables $\boldsymbol{\mu}_y$ in Eq. (21) at a new design test point $\boldsymbol{x}_{\text{new}}$. The approximation is given by

$$z \approx f_z(\boldsymbol{x}, \boldsymbol{\mu}_y) + \nabla^{\top} f_z(\boldsymbol{x}_{\text{new}})(\boldsymbol{x} - \boldsymbol{x}_{\text{new}})$$
$$+ \nabla^{\top} f_z(\boldsymbol{y}(\boldsymbol{x}_{\text{new}}))(\boldsymbol{y} - \boldsymbol{y}(\boldsymbol{x}_{\text{new}})) \quad (23)$$

where $\nabla f_z(\boldsymbol{x}_{\text{new}})$ and $\nabla^{\top} f_z(\boldsymbol{y}(\boldsymbol{x}_{\text{new}}))$ are the gradients of $f_z(\cdot)$ at $\boldsymbol{y}(\boldsymbol{x}_{\text{new}})$ and are given by

$$\nabla f_z(\boldsymbol{x}_{\text{new}}) = \left(\frac{\partial f_z}{\partial x_i}\bigg|_{\text{new}}\right)_{i=1}^{m} \quad (24)$$

$$\nabla f_z(\boldsymbol{y}(\boldsymbol{x}_{\text{new}})) = \left(\frac{\partial f_z}{\partial y_i}\bigg|_{\text{new}}\right)_{i=1}^{m} \quad (25)$$

The model error of z at $\boldsymbol{x}_{\text{new}}$ is then given by

$$\varepsilon_z \approx z - f_z(\boldsymbol{x}, \boldsymbol{\mu}_y) = \nabla^{\top} f_z(\boldsymbol{y}(\boldsymbol{x}_{\text{new}}))\boldsymbol{\varepsilon}_y \quad (26)$$

ε_z follows a Gaussian distribution $N(\mu_{\varepsilon_z}, \sigma_{\varepsilon_z})$ since it is a linear combination of $\boldsymbol{\varepsilon}_y$ whose elements follow Gaussian distributions. The mean and standard deviation of ε_z are given by

$$\mu_{\varepsilon_z} \approx \nabla^{\top} f_z(\boldsymbol{y}(\boldsymbol{x}_{\text{new}}))\boldsymbol{\mu}_{\varepsilon_y} \quad (27)$$

$$\sigma_{\varepsilon_z} \approx \left[\sum_{i=1}^{m}\left(\frac{\partial f_z}{\partial y_i}\bigg|_{\text{new}}\right)^2 \left(\sigma_{\varepsilon_{y_i}}^{GP}(\boldsymbol{x}_{\text{new}})\right)^2\right]^{1/2} \quad (28)$$

Eq. (28) holds if the GP error models of the linking variables are built independently; or in other words, the model errors of the linking variables are independent.

If the GP error models are trained simultaneously by methods such as Co-Kriging [29, 30], the model errors of the linking variables are dependent, and we should use the following equation.

$$\sigma_{\varepsilon_z}^2 \approx \sum_{i=1}^{m}\sum_{j=1}^{m} \frac{\partial f_z}{\partial y_i}\bigg|_{\boldsymbol{x}_{\text{new}}} \frac{\partial f_z}{\partial y_j}\bigg|_{\boldsymbol{x}_{\text{new}}} Cov\big(\varepsilon_{y_i}, \varepsilon_{y_j}\big) \quad (29)$$

where $Cov\big(\varepsilon_{y_i}, \varepsilon_{y_j}\big)$ is a covariance of ε_{y_i} and ε_{y_j}. It is available after the models are trained. Eq. (29) can be rewritten as

$$\sigma_{\varepsilon_z}^2 \approx \sum_{i=1}^{m}\left(\frac{\partial f_z}{\partial y_i}\bigg|_{\boldsymbol{x}_{\text{new}}}\right)^2 \sigma_{y_i}^2 +$$
$$2\sum_{i=1}^{m}\sum_{j>i}^{m} \frac{\partial f_z}{\partial y_i}\bigg|_{\boldsymbol{x}_{\text{new}}} \frac{\partial f_z}{\partial y_j}\bigg|_{\boldsymbol{x}_{\text{new}}} Cov\big(\varepsilon_{y_i}, \varepsilon_{y_j}\big) \quad (30)$$

The derivatives of z with respect to linking variables are needed for the error estimate as shown in Eqs. (29) and (30).

Copyright © 2023 by ASME

However, not all derivatives with respect all linking variables are required since only part of all linking variables will appear in the model of z. This can reduce the computational cost.

If predictions at more than two test points are required for a design task, we can also obtain the joint distribution of the model prediction errors. Next, we use u design points $X_{\text{new}} = \left(x_{1,\text{new}}^\top, x_{2,\text{new}}^\top, \dots, x_{u,\text{new}}^\top\right)^\top$ and the models of linking variables, which are independently built, as an example. Let the predictions at X_{new} are $z(X_{\text{new}}) = \left(z_{(1)}, z_{(2)}, \dots, z_{(u)}\right)^\top$, where $z_{(i)}$ is the system response at $x_{i,\text{new}}, i = 1,2,\dots,u$. They follow a multivariate Gaussian distribution with the mean vector μ_{ε_Z} and covariance matric Σ_{ε_Z}. μ_{ε_Z} is given by

$$\mu_{\varepsilon_Z} = \left[\mu_{\varepsilon_Z}(x_{1,\text{new}}), \mu_{\varepsilon_Z}(x_{2,\text{new}}), \dots, \mu_{\varepsilon_Z}(x_{u,\text{new}})\right]^\top \approx C\mu_{\varepsilon_y} \quad (31)$$

where C can be calculated as $\nabla^\top f_z\big(y(x_{i,\text{new}})\big)$. Eq. (31) can be expressed as

$$\mu_{\varepsilon_Z}(x_{i,\text{new}}) \approx \nabla^\top f_z\big(y(x_{i,\text{new}})\big)\mu_{\varepsilon_y}(x_{i,\text{new}}) \quad (32)$$

Σ_{ε_Z} is given by

$$\Sigma_{\varepsilon_Z}(X_{\text{new}}) = \begin{pmatrix} \sigma_{\varepsilon_Z}^2(x_{1,\text{new}}) & Cov_{12} & \cdots & Cov_{1m} \\ Cov_{21} & \sigma_{\varepsilon_Z}^2(x_{2,\text{new}}) & \cdots & Cov_{2v} \\ \cdots & \cdots & \cdots & \cdots \\ Cov_{v1} & Cov_{v2} & \cdots & \sigma_{\varepsilon_Z}^2(x_{v,\text{new}}) \end{pmatrix} \quad (33)$$

where the covariance $Cov_{ij}, i,j = 1,2,\dots,u, i \neq j$, are given by

$$Cov_{ij} = Cov\left(\varepsilon_z(x_{i,\text{new}}), \varepsilon_z(x_{j,\text{new}})\right)$$
$$= \frac{\partial f_z}{\partial y_i}\bigg|_{x_{i,\text{new}}} \frac{\partial f_z}{\partial y_i}\bigg|_{x_{i,\text{new}}} Cov\left(\varepsilon_{y_i}, \varepsilon_{y_j}\right) \quad (34)$$

4. RESULTS AND DISCUSSION

In this section, we explore how to apply the proposed method to a combustion of propane problem [31-35], which is a widely used benchmark problem for multidisciplinary systems analysis and design. To demonstrate the effectiveness of our approach, we present a modified version of this problem.

Linking variables used in this example are $y = (y_1, y_2, \dots, y_5)^\top$ and all input variables are $x = (x_1, x_2, \dots, x_4)^\top$. The system of compatibility equations for the linking variables are given below.

$$h_1 = K_5 y_1 (3 - x_1) - x_1 y_3 \quad (35)$$

$$h_2 = K_8 x_1 (3 - x_1) y_4 P \quad (36)$$

$$h_3 = K_{10} x_1^2 - (3 - x_1)^2 x_{10} P \quad (37)$$

$$h_4 = 2y_1 + 2y_5 + x_3 + x_4 - 8 \quad (38)$$

$$h_5 = y_5 - \sum_{j=1}^{4} x_j - \sum_{i=1}^{4} y_i - (3 - x_1) - (4R - 2x_3) \quad (39)$$

The constants are $K_5, K_6, K_7, K_9 = 1$; $K_8, K_{10} = 0.1$; $R = 10$, and $P = 40$.

System responses are given by

$$z_1 = K_6 y_1^{\frac{1}{2}}(3 - x_1)^{\frac{1}{2}} - x_1^{\frac{1}{2}}\left(\frac{P}{y_5}\right)^{\frac{1}{2}} \quad (40)$$

$$z_2 = K_7 x_1^{\frac{1}{2}} y_1^{\frac{1}{2}} - (3 - x_1)^{\frac{1}{2}} x_4 \left(\frac{P}{y_5}\right)^{\frac{1}{2}} \quad (41)$$

$$z_3 = K_9 x_1 x_2^{\frac{1}{2}} - (3 - x_1)^{\frac{1}{2}}(4R - 2x_3)\left(\frac{P}{y_5}\right)^{1/2} \quad (42)$$

We first build neural network (NN) models $y_i = f_{y_i}^{NN}(x), i = 1,2,\dots,5$, for link variables, using 150 training points. The network consists of six layers, including an input feature layer and a self-defined output regression layer. The neural network structure and parameters are listed in Table 1.

Table 1 NETWORK STRUCTURE AND PARAMETERS

Layer	Number of Neurons
Input layer	4
Batch Normalized layer 1	-
Activation function	Sigmoid
Fully connected layer 1	20
Batch Normalized layer 2	-
Activation function	Sigmoid
Fully connected layer 2	20
Batch Normalized layer 3	-
Activation function	Sigmoid
Fully connected layer 3	20
Batch Normalized layer 4	-
Activation function	Sigmoid
Fully connected layer 4	20
Regression output layer	5

To evaluate the accuracy of label-free learning, we perform an actual MDA at the training points to obtain the true labels of the linking variables. We then compare these true labels with the predictions generated by the neural network (NN) models. To save space, we display scatter plots of three linking variables in Figs. 2 to 4. Although the predictions by the NN models exhibit a high degree of accuracy, there are still some errors.

To evaluate the accuracy of the NN models, we use 400 test points. Figs. 5 to 7 show scatter plots of the true labels and predictions by the NN models at the test points. The figures indicate larger errors at the testing points than those at the training points in general. To compensate and quantify these errors, we proceed to create GP error models $\varepsilon_i^{GP}(x,t), i = 1,2,\dots,5$, for the model errors.

Copyright © 2023 by ASME

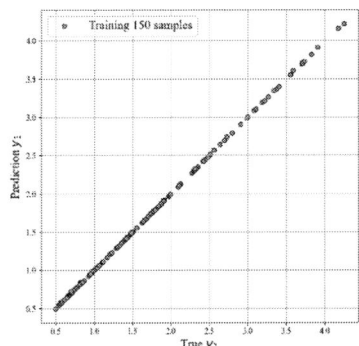

FIGURE 2: y_1 AT TRAINING POINTS

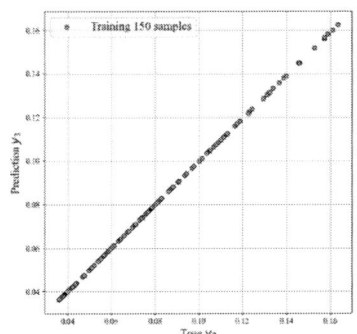

FIGURE 3: y_3 AT TRAINING POINTS

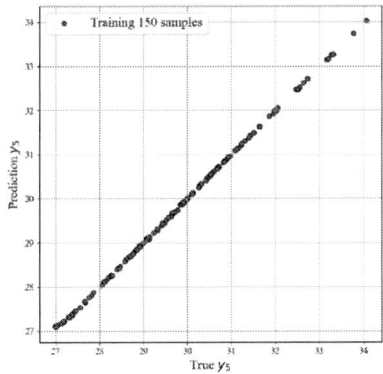

FIGURE 4: y_5 AT TRAINING POINTS

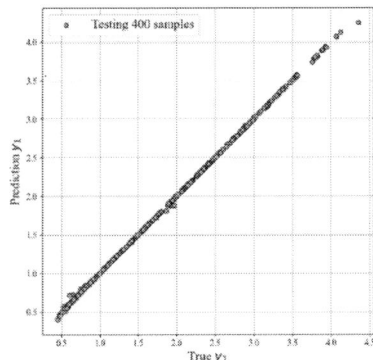

FIGURE 5: y_1 AT TEST POINTS

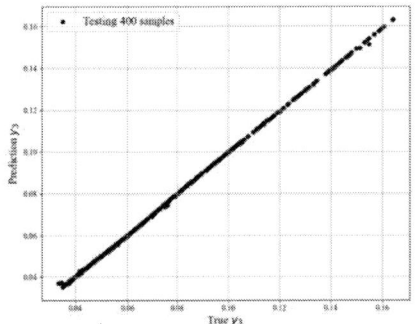

FIGURE 6: y_3 AT TEST POINTS

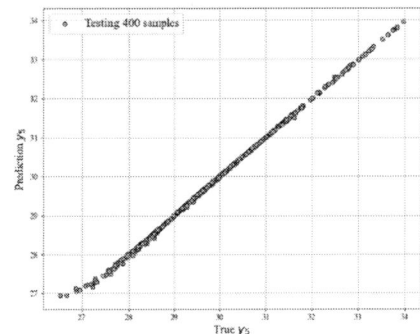

FIGURE 7: y_5 AT TEST POINTS

We now build models of prediction errors at the training points using our proposed methods. To obtain the estimated labels of the model errors, we generate approximated model prediction errors at the training points using the proposed methods. Take y_1 and y_5 as an example. Figs. 8 and 9. illustrate that when these approximated errors are used to correct the original errors at the training points, the accuracy of the NN models improves. If we use the approximated model errors as labels to build GP error models subsequently, we can not only reduce prediction errors but also quantify their uncertainty. This enables us to predict the linking variables and system responses more accurately.

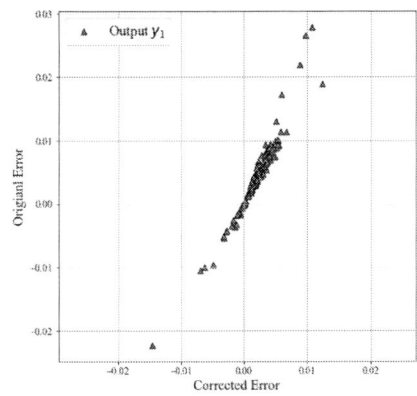

FIGURE 8: NN MODEL ERRORS AND CORRECTED ERRORS IN y_1

Copyright © 2023 by ASME

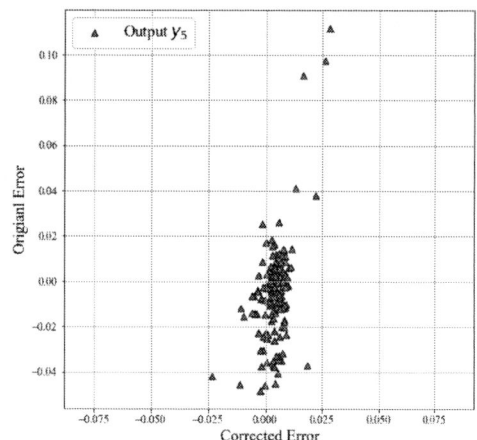

FIGURE 9: NN MODEL ERRORS AND CORRECTED ERRORS IN y_5

After the GP error models are built, we test their effectiveness on the over predictions of linking variables. The scatter plot of the mean prediction error for y_1 and y_5 at the 400 test points in Figs. 10 to 11 indicate the higher accuracy of the mean predictions after the prediction errors are compensated for the mean error predictions from the GP models.

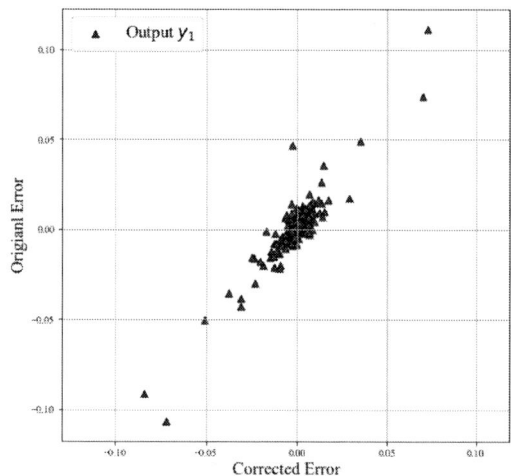

FIGURE 10: NN MODEL ERROR of y_1 AND CORRECTED ERRORS BY GP MODELS

After the model uncertainty of the linking variables is quantified, we now quantify the model uncertainty of system responses. Table 2 provides the mean predictions and the associated standard deviations of the three responses at a test $x = (1.5010, 18.5151, 1.0536, 0.8345)^\top$. This test point is in the region which is well covers by the training. As a result, the prediction uncertainty is small as indicated by the small values of the three standard deviations.

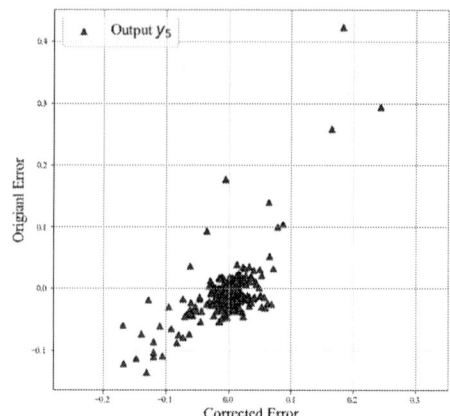

FIGURE 11: NN MODEL ERROR of y_5 AND CORRECTED ERRORS BY GP MODELS

Table 2 PREDICTION OF RESPONSES AT TEST POINT IN COVERED REGION

	μ_{z_i}	σ_{z_i}
Z_1	0.1107	2.2221×10^{-4}
Z_2	0.3453	2.1845×10^{-4}
Z_3	-59.5602	3.4900×10^{-3}

Table 3 provides the mean predictions and the associated standard deviations of the three responses at another test $x = (0.5010, 17.0251, 0.8022, 0.5026)^\top$. This test point far away from the region which is well covers by the training. As expected, the prediction uncertainty is large. This is indicated by the larger values of the three standard deviations.

Table 3 PREDICTION OF RESPONSES AT TEST POINT IN UNCOVERED REGION

	μ_{z_i}	σ_{z_i}
Z_1	-0.2756	1.0744×10^{-1}
Z_2	-1.3195	4.9006×10^{-2}
Z_3	-111.0342	4.3558×10^{-1}

To illustrate the prediction uncertainty graphically, we plot the predictions of the three responses with respect to x_1 by fixing other input variables. As a result, we can show curves of the mean predictions and the 95 confidence bounds. Two cases are reported here. Case 1 is for predictions in the region ($x_1 \in [1,2]$) that is well covered by training points while Case 2 for predictions in the region ($x_1 < 1$ or $x_1 > 2$) that is not adequately covered by training points.

Case 1 is shown in Figs. 12 to 14. The prediction uncertainty is very small, and the mean prediction curves are almost identical to the bounds of the 95 confidence curves.

Copyright © 2023 by ASME

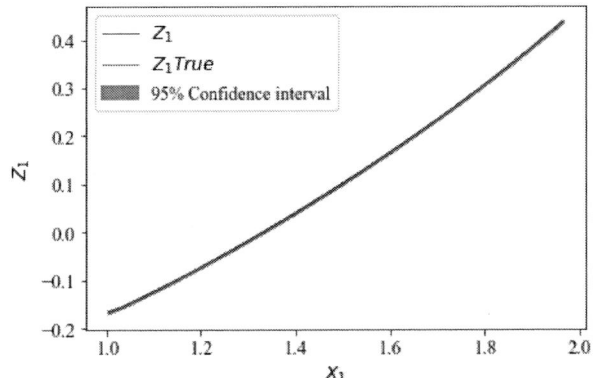

FIGURE 12: PREDICTION OF z_1 IN REGION WELL COVERED BY TRAINING POINTS

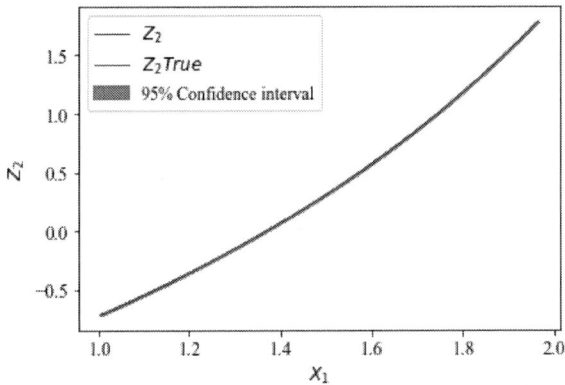

FIGURE 13: PREDICTION OF z_2 IN REGION WELL COVERED BY TRAINING POINTS

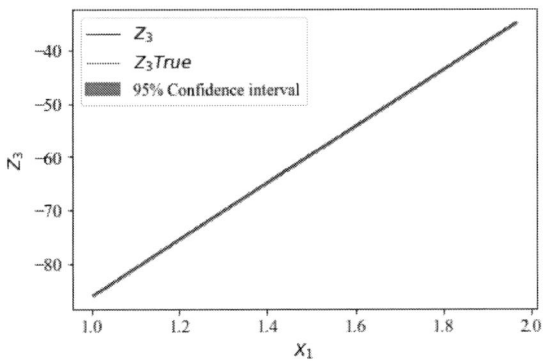

FIGURE 14: PREDICTION OF z_3 IN REGION WELL COVERED BY TRAINING POINTS

Case 2 is shown in Fig. 15 to 17. The prediction uncertainty is large in the region ($x_1 < 1$ or $x_1 > 2$), which is not adequately covered by training points. The 95 percent confidence intervals are much wider than those in Case 1. This is an indication of large prediction error due to inadequate training in the region. If a design is identified in this region, sufficient conservativeness should be imbedded into the design.

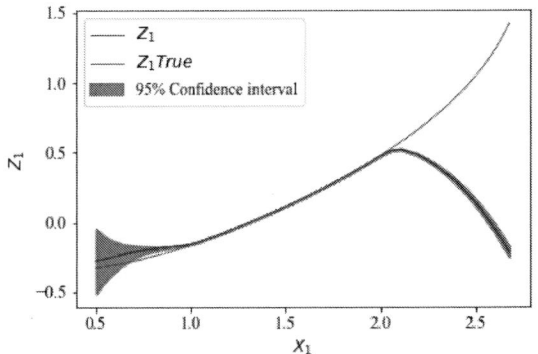

FIGURE 15: PREDICTION OF z_1 IN REGION ($x_1 < 1$ or $x_1 > 2$) NOT ADEQUATELY COVERED BY TRAINING POINTS

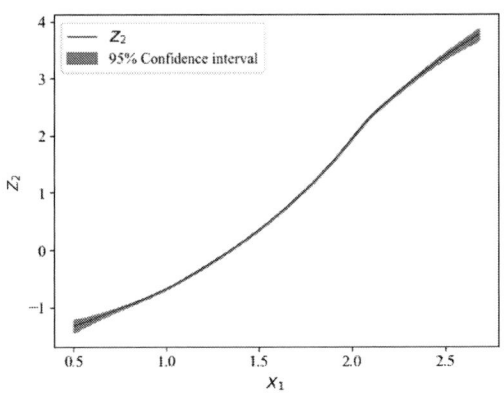

FIGURE 16: PREDICTION OF z_2 IN REGION ($x_1 < 1$ or $x_1 > 2$) NOT ADEQUATELY COVERED BY TRAINING POINTS

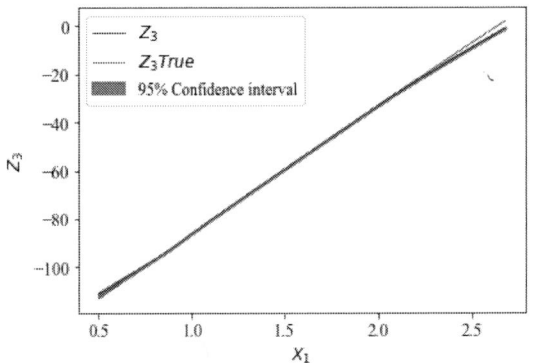

FIGURE 17: PREDICTION OF z_3 IN REGION ($x_1 < 1$ or $x_1 > 2$) NOT ADEQUATELY COVERED BY TRAINING POINTS

5. CONCLUSION

This study demonstrates the feasibility of using label-free machine learning in multidisciplinary systems analysis (MDA)

Copyright © 2023 by ASME

through a proof-of-concept investigation. The embedding of system compatibility equations in the training of neural network models allows for label-free training, eliminating the need for expensive linking variable solving for label generation. The use of Gaussian Process regression provides quantification of the model uncertainty of the neural network models. With these models, system responses can be predicted with estimated errors.

Further research is needed to evaluate the proposed method with more real-world problems and its application in multidisciplinary design optimization (MDO). The computational cost or the efficiency of the proposed method will be evaluated by the comparison of using the traditional MDA methods. Additionally, the integration of aleatory (data) uncertainty and model (epistemic) uncertainty in MDA and MDO will be explored in future research. Overall, this study provides a promising approach to perform label-free machine learning for MDA, and there is potential for further advancements in MDO.

ACKNOWLEDGEMENTS

The support from National Science Foundation under Grant No 1923799 and AI Institute at IUPUI is acknowledged.

REFERENCES

[1] Maute, K., and Allen, M., 2004, "Conceptual design of aeroelastic structures by topology optimization," Structural and Multidisciplinary Optimization, 27(1-2), pp. 27-42.

[2] Peri, D., and Campana, E. F., 2003, "Multidisciplinary design optimization of a naval surface combatant," Journal of Ship Research, 47(1), pp. 1-12.

[3] Hart, C. G., and Vlahopoulos, N., 2010, "An integrated multidisciplinary particle swarm optimization approach to conceptual ship design," Structural and Multidisciplinary Optimization, 41(3), pp. 481-494.

[4] Kodiyalam, S., Yang, R., Gu, L., and Tho, C.-H., 2004, "Multidisciplinary design optimization of a vehicle system in a scalable, high performance computing environment," Structural and Multidisciplinary Optimization, 26(3-4), pp. 256-263.

[5] Grujicic, M., Arakere, G., Pandurangan, B., Sellappan, V., Vallejo, A., and Ozen, M., 2010, "Multidisciplinary design optimization for glass-fiber epoxy-matrix composite 5 MW horizontal-axis wind-turbine blades," Journal of materials engineering and performance, 19(8), pp. 1116-1127.

[6] Hu, Z., and Du, X., 2012, "Reliability analysis for hydrokinetic turbine blades," Renewable Energy, 48, pp. 251-262.

[7] Lund, E., Møller, H., and Jakobsen, L. A., 2003, "Shape design optimization of stationary fluid-structure interaction problems with large displacements and turbulence," Structural and Multidisciplinary Optimization, 25(5-6), pp. 383-392.

[8] Guo, J., and Du, X., 2010, "Reliability analysis for multidisciplinary systems with random and interval variables," AIAA journal, 48(1), pp. 82-91.

[9] Varshney, K. R., "Engineering safety in machine learning," Proc. 2016 Information Theory and Applications Workshop (ITA), IEEE, pp. 1-5.

[10] Hong, J.-W., Wang, Y., and Lanz, P., 2020, "Why is artificial intelligence blamed more? Analysis of faulting artificial intelligence for self-driving car accidents in experimental settings," International Journal of Human–Computer Interaction, 36(18), pp. 1768-1774.

[11] Liang, C., Mahadevan, S., and Sankararaman, S., 2015, "Stochastic Multidisciplinary Analysis Under Epistemic Uncertainty," Journal of Mechanical Design, 137(2).

[12] Li, H., and Du, X., "A Bayesian Approach to Recovering Missing Component Dependence for System Reliability Prediction via Synergy Between Physics and Data," Proc. ASME 2021 International Design Engineering Technical Conferences and Computers and Information in Engineering Conference V03BT03A009.

[13] Li, H., and Du, X., 2021, "Recovering Missing Component Dependence for System Reliability Prediction via Synergy Between Physics and Data," Journal of Mechanical Design, 144(4).

[14] Islam, M. M., Li, H., Zhang, X., Du, X., and Yu, H., "Supervised Surrogate Modeling for Hagen-Poiseuille and Womersley flows," Proc. APS Division of Fluid Dynamics Meeting Abstracts, p. N01. 099.

[15] Hüllermeier, E., and Waegeman, W., 2021, "Aleatory and epistemic uncertainty in machine learning: An introduction to concepts and methods," Machine Learning, 110(3), pp. 457-506.

[16] Senge, R., Bösner, S., Dembczyński, K., Haasenritter, J., Hirsch, O., Donner-Banzhoff, N., and Hüllermeier, E., 2014, "Reliable classification: Learning classifiers that distinguish aleatory and epistemic uncertainty," Information Sciences, 255, pp. 16-29.

[17] Shaker, M. H., and Hüllermeier, E., "Aleatory and epistemic uncertainty with random forests," Proc. International Symposium on Intelligent Data Analysis, Springer, pp. 444-456.

[18] Li, H., Yin, J., and Du, X., "Label Free Uncertainty Quantification," Proc. AIAA SCITECH 2022 Forum, p. 1097.

[19] Li, H., Yin, J., and Du, X., "Uncertainty Quantification of Physics-Based Label-Free Deep Learning and Probabilistic Prediction of Extreme Events," Proc. International Design Engineering Technical Conferences and Computers and Information in Engineering Conference, American Society of Mechanical Engineers, p. V03BT03A001.

[20] Pandita, P., Tsilifis, P., Awalgaonkar, N. M., Bilionis, I., and Panchal, J., 2021, "Surrogate-based sequential Bayesian experimental design using non-stationary Gaussian Processes," Computer Methods in Applied Mechanics and Engineering, 385, p. 114007.

[21] Wu, H., Zhu, Z., and Du, X., 2020, "System Reliability Analysis With Autocorrelated Kriging Predictions," Journal of Mechanical Design, 142(10).

[22] Wu, H., and Du, X., 2023, "Time- and Space-Dependent Reliability-Based Design With Envelope Method," Journal of Mechanical Design, 145(3).

Copyright © 2023 by ASME

[23] Wu, H., and Du, X., 2022, "Envelope Method for Time- and Space-Dependent Reliability Prediction," ASCE-ASME J Risk and Uncert in Engrg Sys Part B Mech Engrg, 8(4).

[24] Li, H., Islam, M., Yu, H., and Du, X., "Physics-Based Regression vs. CFD for Hagen-Poiseuille and Womersley Flows and Uncertainty Quantification," Proc. Eleventh International Conference on Computational Fluid Dynamics (ICCFD11).

[25] Yin, J., and Du, X., 2021, "High-Dimensional Reliability Method Accounting for Important and Unimportant Input Variables," Journal of Mechanical Design, 144(4).

[26] Yin, J., and Du, X., 2022, "Active learning with generalized sliced inverse regression for high-dimensional reliability analysis," Structural Safety, 94, p. 102151.

[27] Mallor, C., Calvo, S., Núñez, J. L., Rodríguez-Barrachina, R., and Landaberea, A., 2020, "Full second-order approach for expected value and variance prediction of probabilistic fatigue crack growth life," International Journal of Fatigue, 133, p. 105454.

[28] Huang, B., and Du, X., 2008, "Probabilistic uncertainty analysis by mean-value first order Saddlepoint Approximation," Reliability Engineering & System Safety, 93(2), pp. 325-336.

[29] Stein, A., and Corsten, L., 1991, "Universal kriging and cokriging as a regression procedure," Biometrics, pp. 575-587.

[30] Marcotte, D., 1991, "Cokriging with MATLAB," Computers & Geosciences, 17(9), pp. 1265-1280.

[31] Tedford, N. P., and Martins, J. R., 2010, "Benchmarking multidisciplinary design optimization algorithms," Optimization and Engineering, 11, pp. 159-183.

[32] Tedford, N., 2007, Comparison of MDO architectures within a universal framework.

[33] Wang, D., Wang, G. G., and Naterer, G. F., 2007, "Collaboration pursuing method for multidisciplinary design optimization problems," AIAA journal, 45(5), pp. 1091-1103.

[34] Yuan, W., Liu, Y., Wang, H., and Ye, X., 2017, "A serialization-based partial decoupling approach for multidisciplinary design optimization of complex systems," Proceedings of the Institution of Mechanical Engineers, Part B: Journal of Engineering Manufacture, 231(14), pp. 2608-2621.

[35] Kodiyalam, S., 1998, Evaluation of methods for multidisciplinary design optimization (MDO), Phase I, Citeseer.

Copyright © 2023 by ASME

Proceedings of the ASME 2023
International Design Engineering Technical Conferences and
Computers and Information in Engineering Conference
IDETC-CIE2023
August 20-23, 2023, Boston, Massachusetts

DETC2023-114954

EFFICIENT AIRFOIL GEOMETRIC UNCERTAINTY QUANTIFICATION USING NEURAL NETWORK MODELS AND SEQUENTIAL SAMPLING

Pavankumar Koratikere
School of Aeronautics and Astronautics
Purdue University
West Lafayette, Indiana 47907
Email: pkoratik@purdue.edu

Leifur Leifsson[*]
School of Aeronautics and Astronautics
Purdue University
West Lafayette, Indiana 47907
Email: leifur@purdue.edu

ABSTRACT

In this paper, a new and unique surrogate-based forward propagation algorithm for aerodynamic geometric uncertainty quantification (UQ) is proposed. The proposed algorithm extends the recent efficient global optimization with neural network (NN)-based prediction and uncertainty (EGONN) algorithm which was created for unconstrained optimization problems. The proposed extended EGONN algorithm for UQ (uqEGONN) constructs a global surrogate model of the aerodynamics characteristics using a NN based on data sampled from physics-based computational fluid dynamics (CFD) simulations. The NN model is adaptively enhanced using a NN model of the prediction uncertainty. In each sampling cycle, the prediction NN is used to compute the summary statistics with Monte Carlo simulations (MCS). The algorithm terminates when the absolute relative change in the summary statistics reach a specified tolerance, or when a specified maximum number of samples is reached. The algorithm is demonstrated on the UQ of the RAE 2822 airfoil at Mach 0.734, angle of attack 2.89 deg, and Reynolds number of 6.5 million. The geometric uncertainty is represented using twelve normally distributed parameters. The results show that the proposed algorithm yields comparable results as those obtained from direct CFD-based MCS with less than 100 CFD samples. The mean and standard deviation of the airfoil drag coefficient are within 0.45 drag counts, for the lift coefficient it is within 0.04 lift counts, and for the pitching moment coefficient it is within 0.11 ($\times 10^{-2}$).

INTRODUCTION

Quantifying the impacts of uncertainties on engineered system's performance is important. Examples of uncertainties include manufacturing process deviations from specifications, varying loading conditions, and inherent variability in some system parameters. These uncertainties can lead to a direct impact on the system performance. The aerodynamic surfaces, such as airfoils and wings, are no different and are subject to many uncertainties, thus there is a need to perform uncertainty quantification [1, 2, 3, 4, 5, 6, 7]. In this work, the effects of geometric uncertainty on the performance of an airfoil in transonic flow is investigated.

Uncertainty quantification (UQ) of a system can be broadly divided into three major steps: (1) identifying the types of uncertainties, (2) modeling the input parameter uncertainties, and (3) sampling the input uncertainties and propagating through the system process to yield the output probability distribution of the quantities of interest (QoIs), and, subsequently, computing the summary statistics [8]. The focus of this work is on step three. Specifically, the goal of this work is to create a nonintrusive method to efficiently propagate the aerodynamic surface geometric uncertainties through computational fluid dynamics (CFD) models. The challenges are (1) a large number of uncertain parameters, (2) expensive simulations, and (3) many samples needed for converged statistics.

Nonintrusive forward propagation methods can be broadly classified as perturbation methods [9], direct quadrature [10, 11], polynomial chaos [12, 13], and Monte Carlo simulation (MCS)

---

[*]Corresponding author.

Copyright © 2023 by ASME

[14, 15]. Perturbation methods use a local Taylor series expansion of the functional output. These methods are limited to local modeling and need at least first-order derivative information. Direct quadrature uses numerical quadrature to evaluate the statistics. This method is limited to low-dimensional problems, although sparse grids partially alleviate this issue. Polynomial chaos represent uncertain parameters as a sum of orthogonal basis functions and can yield the statistics and the output distributions. This method is, however, limited to small number of dimensions. Monte Carlo methods approximate the statistics and output distributions using random sampling. These methods are easy to use and are independent of the problem dimension. However, a major weakness is their inefficiency, i.e., many samples are required to obtain converged values of the statistics.

Expensive (time-consuming) simulations are replaced in surrogate methods with fast approximation model [16]. In this way, the computational cost is shifted over to the creation of a surrogate model. In the context of simulation-based UQ [17], the surrogate needs to represent the uncertain output response of the simulation model in terms of the global uncertain input parameter space. If that is possible, then the summary statistics and output distributions can be estimated using the aforementioned forward propagation methods.

Kriging is a widely used surrogate modeling method capable of approximating nonlinear responses [18]. The advantage of kriging is that it comes with a prediction uncertainty, which enables sequential sampling of the parameter space to enhance the kriging prediction. A weakness of kriging is that it is limited to small data sets, however the use of graphical processing units (GPUs) have partially relieved this issue [19]. On other hand, neural networks (NNs) [20] scale more efficiently for large data sets [21, 22] while still being capable of handling nonlinear responses. A major limitation of NNs, however, is that uncertainty estimates are not readily available for a single prediction [21]. Consequently, it is necessary to make use of an ensemble of NNs with a range of predictions [23, 24, 25, 26] or use dropout to represent model uncertainty [27]. Those algorithms are computationally very intensive.

The efficient global optimization (EGO) algorithm is a widely used approach for sequential sampling [28]. Typically, EGO is used with kriging since it comes with a prediction uncertainty model. In a recent work by the authors [29], EGO was combined with NNs, in an algorithm called EGONN, to address some of the challenges of using kriging with EGO. Specifically, within the new and unique EGONN algorithm, the underlying function response is modelled using one NN and its prediction uncertainty is modelled with another NN. This enables an efficient sequential sampling of the parameter space to adaptively enhance the NN prediction model by maximizing the expected improvement infill criterion. The EGONN algorithm proposed in [29] was created to solve unconstrained optimization problems.

In this paper, a novel extension of the EGONN algorithm to handle UQ problems for aerodynamic surfaces in external flow is proposed. The goal is create an accurate global surrogate model of the true function, in this case the aerodynamic responses, that can be sampled quickly using MCS to compute the desired statistical information for the purpose of UQ. The proposed extension to EGONN is to model the spatial error of the prediction NN in order to construct the prediction uncertainty model. By maximizing the prediction error NN model, a new sample point is determined and appended to the current data set for training the prediction NN model. Since the prediction is being updated in each sequential sampling cycle it allows for termination of the UQ process based on convergence of the summary statistics, which is another new and unique feature of the proposed algorithm. This is achieved in the following way. In each iteration of the sequential sampling the current prediction NN is used to perform a MCS to yield the summary statistics and the change in the predicted mean and standard deviation with respect to the previous iteration is calculated. The algorithm terminates if the change in the statistics is below a pre-specified tolerance or a pre-specified maximum number of function evaluations has been reached. The proposed EGONN for UQ (uqEGONN) is demonstrated on geometric UQ of an airfoil shape in transonic flow and compared against direct MCS of the true function. It is shown that the proposed uqEGONN algorithm is able to yield accurate summary statistics as well as output distributions at a low computational cost.

The next section describes the proposed uqEGONN algorithm. The following section presents the numerical results of the airfoil UQ problem. Concluding remarks and possible next steps in this work are presented in the last section.

METHODS

In this section, the proposed uqEGONN algorithm is presented and described in detail. The functionality of the algorithm is validated by applying to an analytical example and comparing against the results of direct MCS.

Proposed uqEGONN algorithm

A surrogate-based forward propagation approach for simulation-based UQ is proposed. The proposed uqEGONN algorithm is given in Algorithm 1. Initially, two separate data sets, $(\mathbf{X}, \mathbf{Y})$ and $(\mathbf{X}, \mathbf{Y})_u$, are generated using design of experiments, such as space-filling Latin hypercube sampling (LHS) [30]. $(\mathbf{X}, \mathbf{Y})$ is for training a prediction model NN_y and $(\mathbf{X}, \mathbf{Y})_u$ is for training a prediction uncertainty model NN_u. The next steps in the algorithm comprise the sequential sampling. In the first step, the NN_y is fit to the current training data set $(\mathbf{X}, \mathbf{Y})$. In the second step, NN_y is evaluated at $\mathbf{X}$ and $\mathbf{X}_u$ to yield $\widehat{\mathbf{Y}}$ and $\widehat{\mathbf{Y}}_u$, respectively. This data is used to compute the spatial prediction errors

Copyright © 2023 by ASME

Algorithm 1 UQ with EGONN (uqEGONN)

Require: initial data sets $(\mathbf{X}, \mathbf{Y})$ and $(\mathbf{X}, \mathbf{Y})_u$
  repeat
    fit NN_y to data $(\mathbf{X}, \mathbf{Y})$
    use NN_y to get $\widehat{\mathbf{Y}}$ at $\mathbf{X}$ and $\widehat{\mathbf{Y}}_u$ at $\mathbf{X}_u$
    compute errors: $\mathbf{S} \leftarrow \sqrt{(\mathbf{Y} - \widehat{\mathbf{Y}})^2}$ and $\mathbf{S}_u \leftarrow \sqrt{(\mathbf{Y}_u - \widehat{\mathbf{Y}}_u)^2}$
    combine data: $\widetilde{\mathbf{X}} \leftarrow \mathbf{X} \cup \mathbf{X}_u, \widetilde{\mathbf{S}} \leftarrow \mathbf{S} \cup \mathbf{S}_u$
    fit NN_u to data $(\widetilde{\mathbf{X}}, \widetilde{\mathbf{S}})$
    $\mathbf{P} \leftarrow \arg\max \widehat{\mathbf{S}}(\mathbf{x})$
    $\mathbf{X} \leftarrow \mathbf{X} \cup \mathbf{P}$
    $\mathbf{Y} \leftarrow \mathbf{Y} \cup y(\mathbf{P})$
    $\mu, \sigma \leftarrow$ Monte Carlo simulation using NN_y
  until convergence

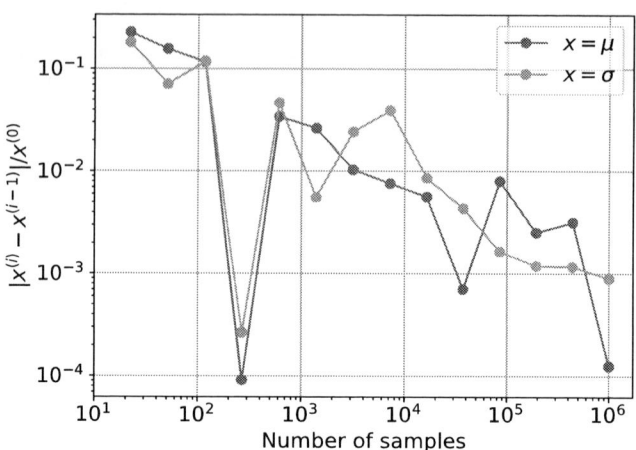

FIGURE 1: Monte Carlo simulations on the ishigami function: convergence of the mean and standard deviation of Y

$\mathbf{S} = \sqrt{(\mathbf{Y} - \widehat{\mathbf{Y}})^2}$ and $\mathbf{S}_u = \sqrt{(\mathbf{Y}_u - \widehat{\mathbf{Y}}_u)^2}$. The data is then appended as $\widetilde{\mathbf{X}} = \mathbf{X} \cup \mathbf{X}_u$, and $\widetilde{\mathbf{S}} = \mathbf{S} \cup \mathbf{S}_u$. The prediction uncertainty model NN_u is fit to the data set $(\widetilde{\mathbf{X}}, \widetilde{\mathbf{S}})$. A new sampling point $\mathbf{P}$ is now found by maximizing the prediction uncertainty model $\widehat{\mathbf{S}}(\mathbf{x}) = NN_u(\mathbf{x})$. The new sample point $\mathbf{P}$ is appended to $\mathbf{X}$ and the corresponding function value $y(\mathbf{P})$ is appended to $\mathbf{Y}$.

In the last step, the prediction model NN_y is used in MCS to yield the summary statistics, the mean μ and standard deviation σ, in each sampling cycle. This allows for terminating the algorithm based on the convergence of the summary statistics. In particular, the algorithm terminates if the absolute relative change in the mean, calculated as $|\mu^{(i)} - \mu^{(i-1)}|/|\mu^{(0)}|$, where i is the number of the sampling cycle, is less than a predefined tolerance, τ_μ, and the absolute relative change in the standard deviation, calculated as $|\sigma^{(i)} - \sigma^{(i-1)}|/|\sigma^{(0)}|$, is less than a predefined tolerance, τ_σ, or if the number of sequential sample cycles exceeds a predefined maximum number, N_{max}.

Validation example

The proposed uqEGONN algorithm is validated using the Ishigami function [31], which is written as

$$Y = f(x_1, x_2, x_3) = \sin(x_1) + a\sin^2(x_2) + bx_3^4 \sin(x_1), \quad (1)$$

where $a = 7$, $b = 0.1$, and $x_i \sim \mathcal{U}[-\pi, \pi] \; \forall \; i = 1, 2, 3$. The convergence of MCS on the true analytical function (1) is shown in Fig. 1. It can be seen that MCS needs around one million samples to reach a converged mean $\mu = 3.50$ and standard deviation $\sigma = 3.72$. Figure 2 shows the true output distribution obtained by MCS.

The output of uqEGONN is compared with the values obtained from MCS. The two initial data-sets $(\mathbf{X}, \mathbf{Y})$ and $(\mathbf{X}, \mathbf{Y})_u$,

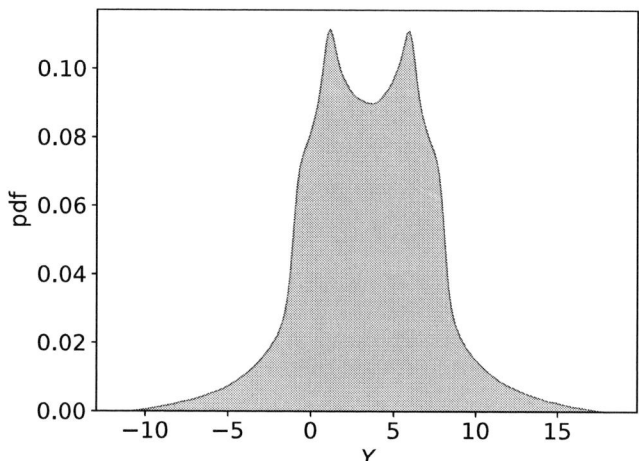

FIGURE 2: Monte Carlo simulation output distribution for ishigami function

consisting of 100 and 50 points, respectively, are generated using LHS sampling [30] within the bounds described in (1). For this problem, the architecture of NN_y and NN_u is kept the same. Both NN consist of 2 hidden layers, each with 16 neurons, and the hyperbolic tan function is used for activation. The Adam optimizer [32] is used for training both the NNs with a learning rate of 0.001 and 10,000 epochs. The NNs are implemented using Tensorflow [33] and hyperparameters are tuned in such a manner that both the NNs slightly over-fit so that fitting overall improves when new samples are added. The values of tolerances are set to $\tau_\mu = 0.001$ and $\tau_\sigma = 0.001$, and the maximum number of sequential samples (N_{max}) is set to 100.

Copyright © 2023 by ASME

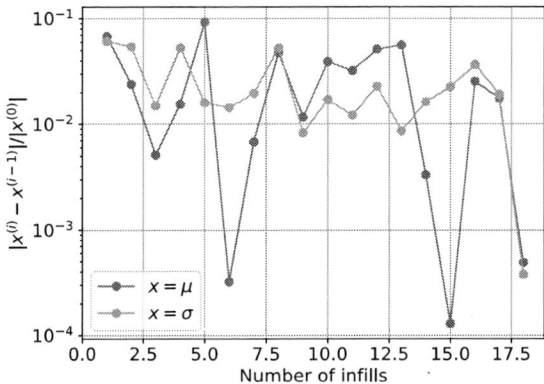

FIGURE 3: Convergence of uqEGONN for the ishigami function

TABLE 1: Comparison of statistics computed using true function and uqEGONN

Parameter	Statistic	True	uqEGONN	Δ
Y	μ	3.50	3.49	-0.01
	σ	3.72	3.74	-0.02

Figure 3 shows the convergence of the uqEGONN algorithm for the Ishigami function. The process is terminated after 18 infill points since the absolute relative difference is below the set tolerance value. The upEGONN, with only 118 samples, yields $\mu = 3.49$ and $\sigma = 3.74$ (shown in Table 1) which is very close to the true values obtained using 1 million samples.

Figure 4 compares the distribution of Y obtained from uqE-GONN with true distribution. Initially, the distribution does not match with the true distribution, but as new samples are added, the distribution improves and finally reaches close to the true distribution. There are still small inconsistencies between the distributions which can be improved by reducing the tolerance but that will require more infill samples.

NUMERICAL EXPERIMENTS

This section presents the numerical results of applying the proposed uqEGONN algorithm for the geometric UQ of an airfoil in transonic flow. The results are compared with those obtained using MCS directly on the computational simulation model.

Problem description

The uncertainties in the lift and drag coefficients of the RAE 2822 airfoil due to surface variability are quantified. The freestream Mach number is set to 0.734, the angle of attack to 2.89 deg, and the Reynolds number to 6.5×10^6. The shape of the air-

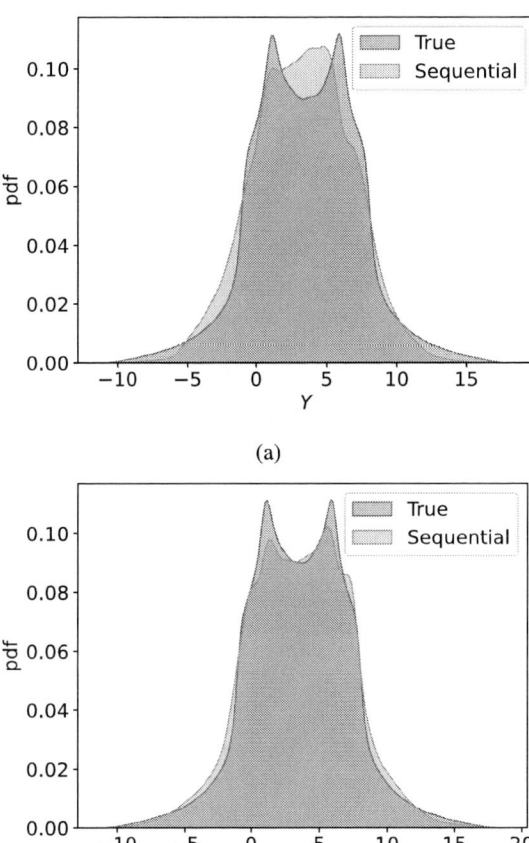

(a)

(b)

FIGURE 4: Ishigami function output distribution from MCS of True function and uqEGONN (Sequential): (a) initial, (b) final.

foil is parameterized using class shape transformation (CST) [34] and implemented within pyGeo [35]. A total of twelve CST coefficients are used for the parameterization, six for upper surface and six for lower surface. To represent the geometric uncertainty in the shape, each CST coefficient is assumed to be normally distributed with the mean being the value of the coefficient and the standard deviation is such that 99% of all the perturbations remain within $\pm 10\%$ of the coefficient value. Figure 5 shows the RAE 2822 airfoil with the region of uncertainty around the shape.

CFD simulations

The flow around the airfoil is simulated using the compressible Reynolds-averaged Navier-Stokes (RANS) equations, coupled with the Spalart-Allmaras [36] turbulence model. ADflow [37] is used to solve these equations which is a finite-volume structured multiblock mesh solver. The approximate Newton Krylov

Copyright © 2023 by ASME

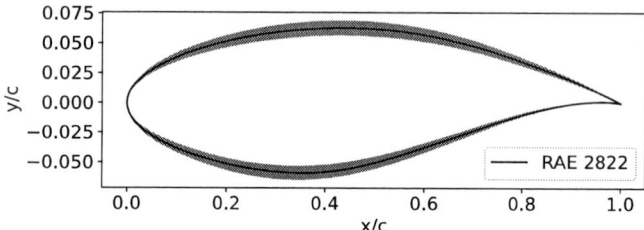

FIGURE 5: The RAE 2822 airfoil and the geometric uncertain regions of the upper and lower surfaced (shown in blue).

method [38] initializes the solver and reduces the relative residual norm to 10^{-6}. After this, the Newton Krylov method is used to reduce the relative residual norm to 10^{-14}.

A structured O-shape computational mesh around the airfoil is created for ADflow using pyHyp [39]. A grid convergence study is performed to find an accurate and computationally efficient mesh. Table 2 outlines the result of the study. The mesh is extended until 100c and the y^+ value is maintained under 1 for all the levels. In this work, L1 level mesh (shown in Fig. 6) is used for performing all the computations. Figure 7 shows the Mach contour plot of the flow near the airfoil and the corresponding coefficient of pressure (C_p) distribution for the analysis performed at given conditions.

TABLE 2: Grid convergence study for the RAE 2822 at $M_\infty = 0.734$, $Re = 6.5 \times 10^6$, and $C_l = 82.4$ lift counts.

Level	Number of cells	C_d (d.c.)	$C_m(\times 10^2)$	α (deg.)
L0	512,000	195.58	-9.6	2.828
L1	128,000	200.55	-9.4	2.891
L2	32,000	213.26	-9.1	3.043
L3	8,000	235.05	-8.6	3.278

Setup of uqEGONN

In this work, the **X** vector denotes shape of the airfoil and consists of 12 features, each representing a CST coefficient. LHS is used to generate 50 **X** and 25 $\mathbf{X}_u$ samples within the bounds of $\pm 10\%$ of the CST coefficients. Then, 75 CFD simulations are performed in total based on sampled shapes to compute **Y** and $\mathbf{Y}_u$ which consists of the lift coefficient (C_l), drag coefficient (C_d) and pitching moment coefficient (C_m) values. For this problem, the NN_y consists of 2 hidden layers, with 3 and 2 neurons, respectively. The NN_u also consists of 2 hidden layers, with 5 and

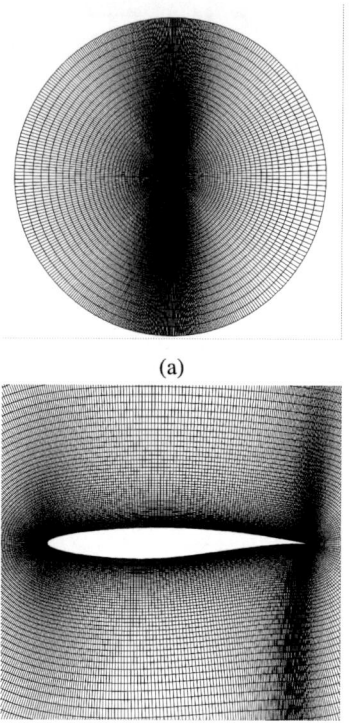

(a)

(b)

FIGURE 6: L1 computational mesh for the airfoil flow simulation: (a) the far-field, and (b) a zoom in near the airfoil surface.

4 neurons, respectively. The activation function used in both the NNs is the hyperbolic tanh. The Adam [32] optimizer is used for training both the NNs with a learning rate of 0.001 and 10,000 epochs. The NNs are implemented within Tensorflow [33] and hyperparameters are tuned in such a manner that both the NNs slightly over-fit so that fitting overall improves when new samples are added. The values of tolerances are set to $\tau_\mu = 0.005$ and $\tau_\sigma = 0.005$, and the maximum number of sequential samples (N_{max}) is set to 50.

Results

The summary statistics and output distributions of the airfoil geometric UQ are obtained for the true function by performing direct MCS of the CFD model using 1,000 random samples of the twelve uncertain CST coefficients. Ideally, more samples should be used to compute the statistics but the computational cost of the CFD model is the limiting factor for this problem. Figures 8, 9 and 10 show the convergence of the MCS for the coefficients of lift (C_l), drag (C_d), and pitching moment (C_m), respectively. Figures 11, 12 and 13 show the output distributions of C_l, C_d and C_m. Table 3 lists the converged values of the statistics for these 1,000 CFD samples. Here, the definitions of lift counts (l.c.) and

Copyright © 2023 by ASME

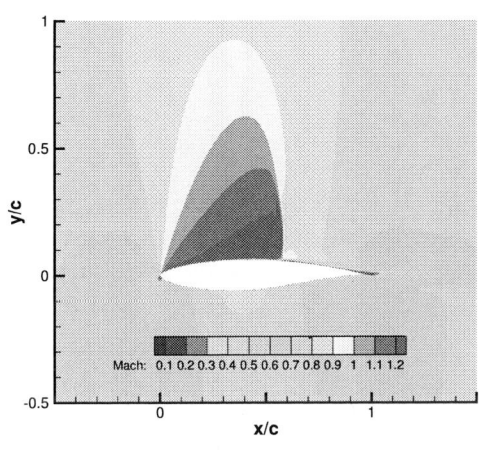

(a)

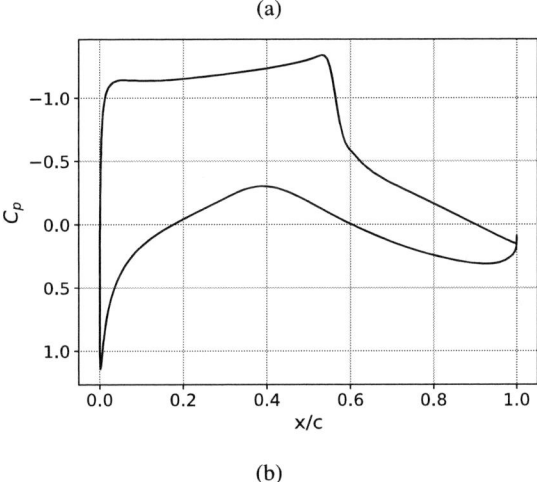

(b)

FIGURE 7: Flow field of the RAE2822 airfoil at given conditions: (a) Mach contours, and (b) surface pressure distribution.

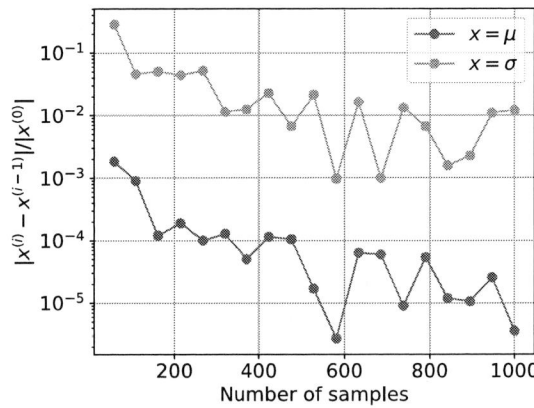

FIGURE 8: Monte Carlo simulations on the CFD model: convergence of the mean and standard deviation of the lift coefficient.

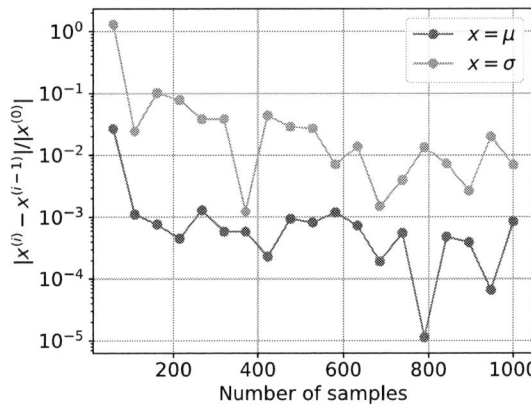

FIGURE 9: Monte Carlo simulations on the CFD model: convergence of the mean and standard deviation of the drag coefficient.

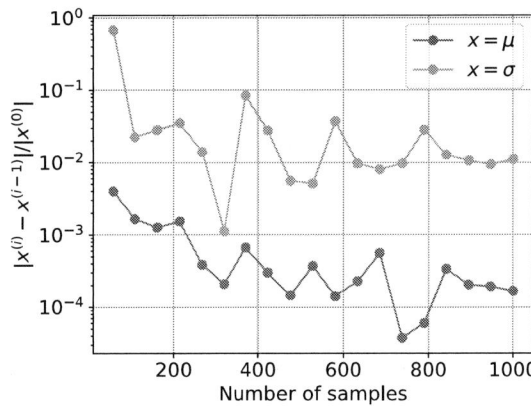

FIGURE 10: Monte Carlo simulations on the CFD model: convergence of the mean and standard deviation of the pitching moment coefficient.

drag counts (d.c.) are $\Delta C_l = 10^{-2}$ and $\Delta C_d = 10^{-4}$, respectively. The variations in C_l and C_m are not large, but C_d varies significantly, which is evident from the large value of σ. Any variation in the geometry of the airfoil influences the location and strength of the shock (shown in Fig. 7), which will have a strong effect on drag coefficient. The uncertainty in the values of C_l, C_d and C_m warrants the need for performing UQ due to uncertainties in airfoil shape.

Figures 14, 15, and 16 show the convergence of the uqE-GONN algorithm for the coefficients of lift (C_l), drag (C_d), and pitching moment (C_m), respectively. For each output, the uqE-GONN process is repeated since NN_y can only handle scalar output. It can be seen that algorithm terminates on the convergence criteria for C_l after 22 infills, for C_d after 18 infills, and for C_m after 14 infills. These infill sampling points are in addition to the

Copyright © 2023 by ASME

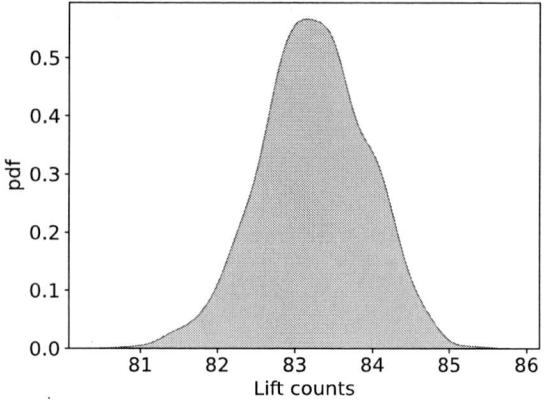

FIGURE 11: Monte Carlo simulation output distribution for the lift coefficient.

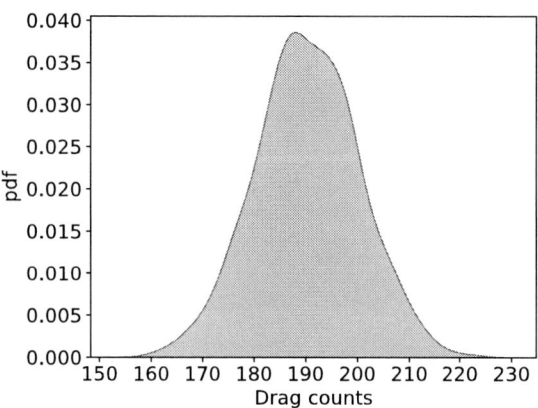

FIGURE 12: Monte Carlo simulation output distribution for the drag coefficient.

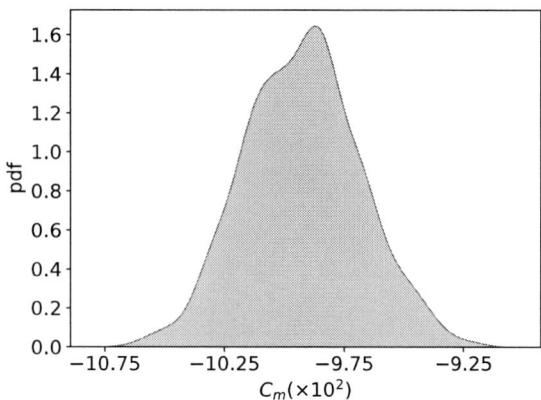

FIGURE 13: Monte Carlo simulation output distribution for the pitching moment coefficient.

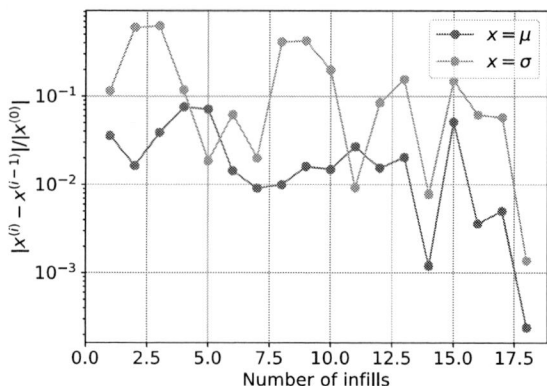

FIGURE 14: Convergence of uqEGONN for the drag coefficient.

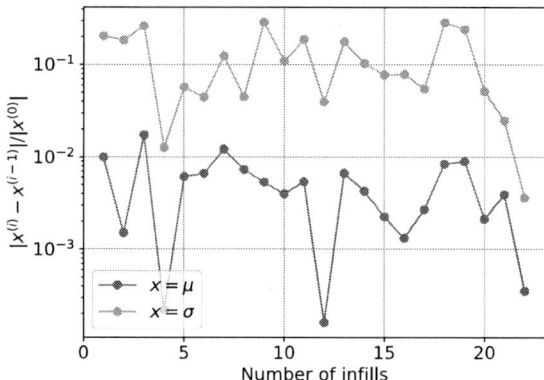

FIGURE 15: Convergence of uqEGONN for the lift coefficient.

75 initial samples.

Table 3 lists the converged values of the statistics for uqE-GONN. It can be seen that μ and σ for C_d is within 0.45 d.c. of the true values, for C_l it is within 0.05 l.c., and for C_m it is within 0.11 ($\times 10^{-2}$). Figures 17, 18 and 19 shows the output distributions of C_l, C_d and C_m obtained from the uqEGONN algorithm (Sequential) and from the direct MCS of the CFD model (True). It can be seen that the output distributions from uqEGONN based on only initial samples are significantly different than those of the true function (cf. Figs. 17(a), 18(a) and 19(a)), but with the additional infill samples the output distributions align quite well (cf. Figs. 17(b), 18(b) and 19(b)).

CONCLUSION

A novel surrogate-based forward propagation algorithm for simulation-based aerodynamic geometric uncertainty quantification (UQ) is proposed. The distinct features of the algorithm are the sequential sampling and updating of a neural network (NN)

Copyright © 2023 by ASME

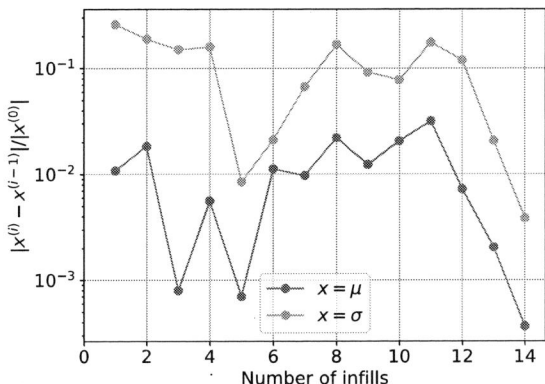

FIGURE 16: Convergence of uqEGONN for the pitching moment coefficient.

TABLE 3: Comparison of the statistics from direct Monte Carlo simulations on the CFD model and uqEGONN.

Parameter	Statistic	CFD	uqEGONN	Δ
C_d (d.c.)	μ	190.28	190.73	0.45
	σ	9.98	10.14	0.16
C_l (l.c.)	μ	83.24	83.20	-0.04
	σ	0.69	0.68	-0.01
C_m ($\times 10^2$)	μ	-9.93	-9.82	0.11
	σ	0.24	0.26	0.02

predictions of the nonlinear aerodynamic responses and the associated prediction uncertainties, automated sequential sampling termination criteria based on the absolute relative changes in the summary statistics, and the elimination of validation testing data sets and the arbitrary choices of convergence metrics, such as those based on the root mean squared error.

A numerical example of geometric UQ for an airfoil in transonic flow shows that comparable summary statistics and output probability density functions are obtained for the aerodynamic coefficients by the proposed algorithm as those obtained by directly embedding the computational fluid dynamics model in the Monte Carlo simulations (MCS). The proposed algorithm obtains the characteristics at a low computational cost.

Future work includes characterizing the proposed algorithm on higher dimensional simulation-based aerodynamic problems, such as wings and turbine blades, investigating the effects of reducing the number of initial samples, and automating the hyperparameter tuning of the NN architecture in every sampling cycle due to the increasing number of samples.

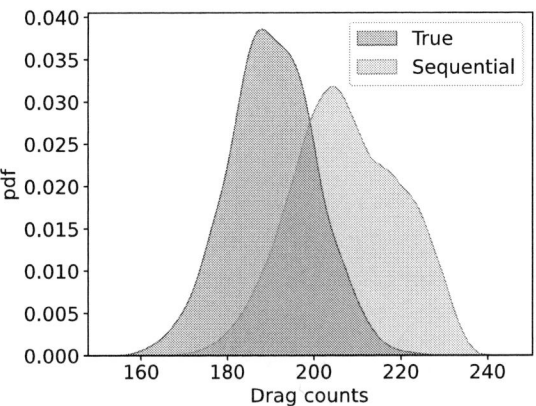

(a)

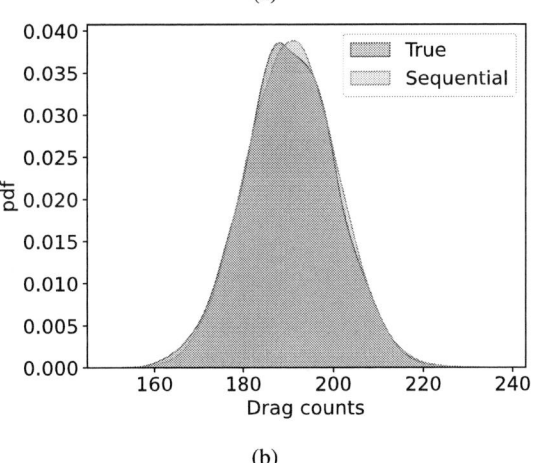

(b)

FIGURE 17: Drag coefficient output distribution from MCS of CFD (True) and uqEGONN (Sequential): (a) initial, (b) final.

REFERENCES

[1] Wu, X., Zhang, W., and Song, S., 2017. "Uncertainty quantification and sensitivity analysis of transonic aerodynamics with geometric uncertainty". *International Journal of Aerospace Engineering,* **2017**.

[2] DeGennaro, A. M., Rowley, C. W., and Martinelli, L., 2015. "Uncertainty quantification for airfoil icing using polynomial chaos expansions". *Journal of Aircraft,* **52**(5), pp. 1404–1411.

[3] Liu, D., Litvinenko, A., Schillings, C., and Schulz, V., 2017. "Quantification of airfoil geometry-induced aerodynamic uncertainties—comparison of approaches". *SIAM/ASA Journal on Uncertainty Quantification,* **5**(1), pp. 334–352.

[4] Marepally, K., Jung, Y. S., Baeder, J., and Vijayakumar, G., 2022. "Uncertainty quantification of wind turbine airfoil aerodynamics with geometric uncertainty". In Journal

Copyright © 2023 by ASME

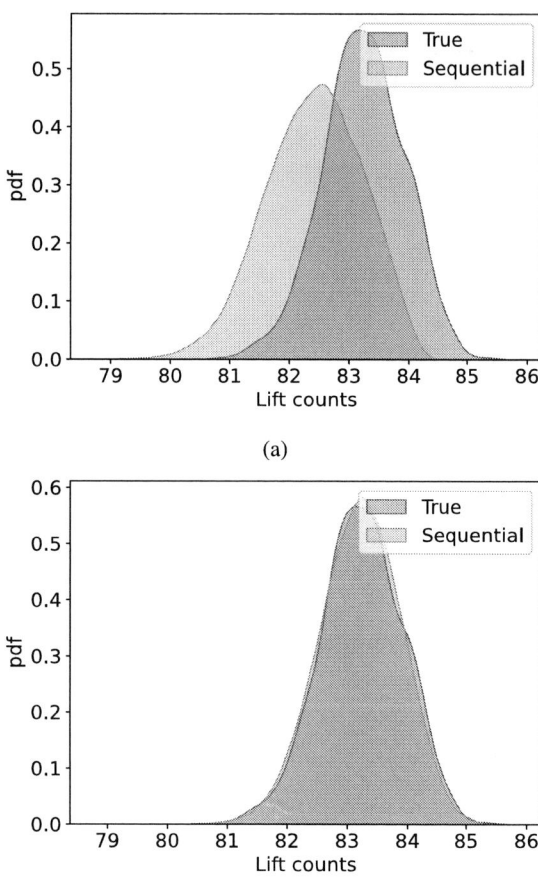

(a)

(b)

FIGURE 18: Lift coefficient output distribution from MCS of CFD (True) and uqEGONN (Sequential): (a) initial, (b) final.

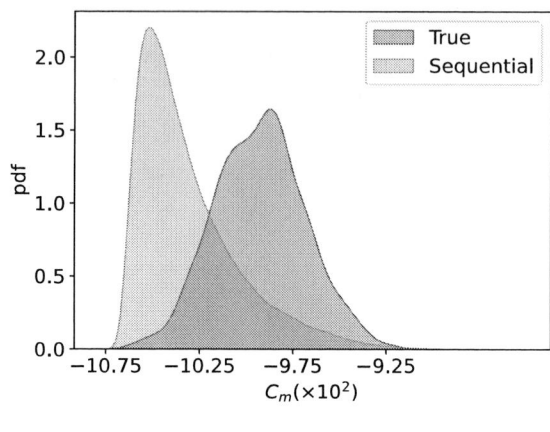

(a)

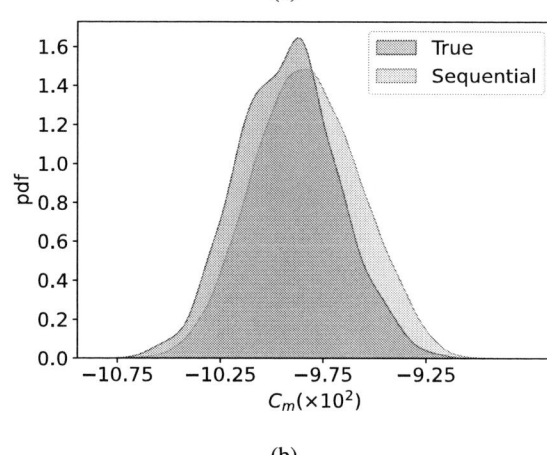

(b)

FIGURE 19: Pitching moment coefficient output distribution from MCS of CFD (True) and uqEGONN (Sequential): (a) initial, (b) final.

of Physics: Conference Series, Vol. 2265, IOP Publishing, p. 042041.

[5] Kawai, S., Yamazaki, W., and Oyama, A., 2021. "Uncertainty quantification of supersonic biplane airfoil by multielement polynomial chaos based on edge detection". In AIAA Scitech 2021 Forum, p. 0179.

[6] Hosder, S., Walters, R., and Balch, M., 2008. "Efficient uncertainty quantification applied to the aeroelastic analysis of a transonic wing". In 46th AIAA Aerospace Sciences Meeting and Exhibit, p. 729.

[7] Chaudhuri, A., Haftka, R. T., Ifju, P., Chang, K., Tyler, C., and Schmitz, T., 2015. "Experimental flapping wing optimization and uncertainty quantification using limited samples". *Structural and Multidisciplinary Optimization,* *51*, pp. 957–970.

[8] Yao, W., Chen, X., Luo, W., Van Tooren, M., and Guo, J., 2011. "Review of uncertainty-based multidisci-

plinary design optimization methods for aerospace vehicles". *Progress in Aerospace Sciences,* *47*(6), pp. 450–479.

[9] Wong, F. S., 1985. "First-order, second-moment methods". *Computers & structures,* *20*(4), pp. 779–791.

[10] Attar, P. J., and Vedula, P., 2013. "On convergence of moments in uncertainty quantification based on direct quadrature". *Reliability Engineering & System Safety,* *111*, pp. 119–125.

[11] Monahan, J. F., 2011. *Numerical methods of statistics.* Cambridge University Press.

[12] Knio, O. M., and Le Maitre, O., 2006. "Uncertainty propagation in cfd using polynomial chaos decomposition". *Fluid dynamics research,* *38*(9), p. 616.

[13] Najm, H. N., 2008. "Uncertainty propagation and polynomial chaos techniques in cfd.". *Annual Review of Fluid Mechanics*(SAND2008-1167J).

Copyright © 2023 by ASME

[14] Helton, J. C., Johnson, J. D., Sallaberry, C. J., and Storlie, C. B., 2006. "Survey of sampling-based methods for uncertainty and sensitivity analysis". *Reliability Engineering & System Safety, 91*(10-11), pp. 1175–1209.

[15] Landau, D., and Binder, K., 2021. *A guide to Monte Carlo simulations in statistical physics*. Cambridge university press.

[16] Queipo, N. V., Haftka, R. T., Shyy, W., Goel, T., Vaidyanathan, R., and Tucker, P. K., 2005. "Surrogate-based analysis and optimization". *Progress in Aerospace Sciences, 21*(1), pp. 1–28.

[17] Shields, M. D., Au, S.-K., and Sudret, B., 2019. Advances in simulation-based uncertainty quantification and reliability analysis.

[18] Forrester, A. I. J., and Keane, A. J., 2009. "Recent advances in surrogate-based optimization". *Progress in Aerospace Sciences, 41*(1–3), pp. 50–79.

[19] Wang, K., Pleiss, G., Gardner, J., Tyree, S., Weinberger, K. Q., and Wilson, A. G., 2019. "Exact gaussian processes on a million data points". *Advances in Neural Information Processing Systems, 32*.

[20] Goodfellow, I., Bengio, Y., and Courville, A., 2016. *Deep Learning*. The MIT Press, Cambridge, MA.

[21] Lim, Y.-F., Ng, C. K., Vaitesswar, U. S., and Hippalgaonkar, K., 2021. "Extrapolative Bayesian Optimization with Gaussain Process and Neural Network Ensemble Surrogate Models". *Advanced Intelligent Sytems, 3*, p. 2100101.

[22] Snoek, J., Rippel, O., Swersky, K., Kiros, R., Satish, N., Sundaram, N., Patwary, M. A., and Adams, R. P., 2015. "Scalable bayesian optimization using deep neural networks". In Proceedings of the 32nd International Conference on Machine Learning, pp. 2171–2180.

[23] Lampinen, J., and Vehtari, A., 2001. "Bayesian approach for neural networks - review and case studies". *Neural Networks, 14*(3), pp. 257–274.

[24] Titterington, D. M., 2004. "Bayeisan methods for neural networks and related models". *Statistical Science, 19*(1), pp. 128–139.

[25] Goan, E., and Fookes, C., 2020. "Bayesian neural networks: An introduction and survey". In Mengersen, K., Pudlo, P., Robert, C. (eds) Case Studies in Applied Bayesian Data Science. Lecture Notes in Mathematics, pp. 45–87.

[26] Lakshminarayanan, B., Pritzel, A., and Blundell, C., 2017. "Simple and scalable predictive uncertainty estimation using deep ensembles". *Advances in Neural Information Processing Systems, 30*.

[27] Gal, Y., and Ghahramani, Z., 2016. "Dropout as a bayesian approximation: Representing model uncertainty in deep learning". In Proceedings of the 33rd International Conference on Machine Learning, pp. 1050–1059.

[28] Jones, D. R., Schonlau, M., and Welch, W. J., 1998. "Efficient Global Optimization of Expensive Black-Box Functions". *Journal of Global Optimization, 13*(4), pp. 455–492.

[29] Koratikere, P., Leifsson, L. T., Barnet, L., and Bryden, K., 2023. "Efficient global optimization algorithm using neural network-based prediction and uncertainty". In AIAA SCITECH 2023 Forum, p. 2683.

[30] McKay, M., Conover, W., and Beckman, R., 1979. "A comparison of three methods for selecting values of input variables in the analysis of output from a computer code". *Technometrics, 21*, pp. 239–245.

[31] Ishigami, T., and Homma, T., 1990. "An importance quantification technique in uncertainty analysis for computer models". In [1990] Proceedings. First international symposium on uncertainty modeling and analysis, IEEE, pp. 398–403.

[32] Kingma, D. P., and Ba, J., 2014. Adam: A method for stochastic optimization. arXiv:1412.6980.

[33] Abadi, M., and et al., 2015. TensorFlow: Large-scale machine learning on heterogeneous systems. Software available from tensorflow.org.

[34] Kulfan, B. M., 2008. "Universal parametric geometry representation method". *Journal of aircraft, 45*(1), pp. 142–158.

[35] Kenway, G., Kennedy, G., and Martins, J. R., 2010. "A CAD-free approach to high-fidelity aerostructural optimization". In 13th AIAA/ISSMO multidisciplinary analysis optimization conference.

[36] Spalart, P. R., and Allmaras, S. R., Reno, NV, January 6-9, 1992. "A one equation turbulence model for aerodynamic flows". In 38th AIAA Aerospace Sciences Meeting and Exhibit, Vol. 92-0439.

[37] Mader, C. A., Kenway, G. K., Yildirim, A., and Martins, J. R., 2020. "ADflow: An open-source computational fluid dynamics solver for aerodynamic and multidisciplinary optimization". *Journal of Aerospace Information Systems, 17*(9), pp. 508–527.

[38] Yildirim, A., Kenway, G. K., Mader, C. A., and Martins, J. R., 2019. "A jacobian-free approximate newton–krylov startup strategy for rans simulations". *Journal of Computational Physics, 397*, p. 108741.

[39] Secco, N. R., Kenway, G. K., He, P., Mader, C., and Martins, J. R., 2021. "Efficient mesh generation and deformation for aerodynamic shape optimization". *AIAA Journal, 59*(4), pp. 1151–1168.

Copyright © 2023 by ASME

Proceedings of the ASME 2023
International Design Engineering Technical Conferences and
Computers and Information in Engineering Conference
IDETC-CIE2023
August 20-23, 2023, Boston, Massachusetts

DETC2023-115112

MULTI-AGENT BAYESIAN OPTIMIZATION FOR UNKNOWN DESIGN SPACE EXPLORATION

Siyu Chen[1], Alparslan Emrah Bayrak [2], Zhenghui Sha[1,*],

[1] Walker Department of Mechanical Engineering, The University of Texas at Austin, Austin, TX
[2] School of Systems and Enterprises, Stevens Institute of Technology, Hoboken, NJ

ABSTRACT

This paper proposes a multi-agent Bayesian optimization (MABO) framework as a reference model for rational design teams to study the effects of information exchange on a team's search performance in finding global optimum of complex objective functions with many local optima. The core idea of the framework has three main steps. First, the design space is divided into regions based on the number of agents involved in the search. In each region, only one agent works on the part of the objective function. Second, a global-local communication strategy is developed to allow agents in local searches to share their sampled design points with a global evaluator. The global evaluator computes the posterior mean and variance based on all sampled points from local agents and evaluates the acquisition function (e.g., the expected improvement) to recommend the next sampling decisions for local agents. Third, when making the decision about where to sample next, each local agent only has access to the expected improvement evaluated in its local region and chooses the design that yields the largest value locally. To evaluate how the information exchange between agents and between local and global impact the search results, our framework is compared with a multi-agent model that does not allow information sharing and global-local interaction. Furthermore, we evaluated the performance of the model based on benchmark functions with varying complexities and also investigated the impact of the number of agents on search performance. We observe that when information sharing is allowed and global-local interaction is enabled in all scenarios, there is a significant improvement in convergence speed as well as the success rate of convergence.

Keywords: Multi-agent System (MAS), Bayesian Optimization (BO), Design Team, Design Space Exploration

---

*Corresponding author: zsha@austin.utexas.edu

1. INTRODUCTION

Design space exploration is the process of finding the best design solution that meets all requirements and constraints by exploring and evaluating different design alternatives available for a given problem [1]. It is typically a sequential process where knowledge of the design space is acquired through a series of design assessments, rather than starting with a full understanding of the design space of evaluation metrics at the beginning. Furthermore, a design process, particularly for complex design problems, usually involves more than just one person making decisions. It is rather a team effort where decisions made by other members influence the decision of each member.

Coordinating the decisions of team members in such a process is an essential problem that could significantly influence the effectiveness of design teams [2]. The impact of mechanisms and the frequency of interactions between design team members on design outcomes have been studied in the literature [3, 4]. As an example, the work by McComb et al. finds that if design team members all work on the same configuration problem, the optimal number of interactions between team members should be zero [5]. However, in practice, it is not quite common to see teams where all members work on the same problem, but there is rather a division of labor where each member works on a different sub-problem within a system-level design problem. Such a case deserves a new investigation since the coordination of subproblems requires some information sharing between team members [6]. In this paper, we computationally study the impact of communication between team members (i.e., computational design agents) on design exploration performance when each agent works on a different portion of the design space. We use rational agents that act to maximize their own utility, meanwhile can share their respective design samples with a global evaluator for optimal decision-making, in order to study the impact of information sharing independent of confounding human factors.

Design space exploration can be formulated as a black-box optimization problem when the space or function form is un-

Copyright © 2023 by ASME

known to designers. Bayesian optimization (BO) is knowledge-based reasoning to explore unknown design spaces informed by past experience, or data [7], which provides rational design recommendations with future sampling decisions. BO typically estimates the model using a Gaussian process, which takes into account the uncertainty during the search and adopts an acquisition function that determines where to sample next in the process. This approach effectively balances exploration and exploitation to achieve the optimum [8].

The idea of BO has been used for decades in the literature, including for well-known design optimization algorithms such as Efficient Global Optimization [9]. Other examples include the work in [10], which proposes a BO approach that can effectively handle variable-size design space problems, with results demonstrating superior performance compared to other optimization methods. BO has also been applied to practical design application contexts as well. The study in [11] argues that BO is an efficient and effective method of exploring the design space of hardware accelerators, which can significantly reduce the exploration time while still achieving high-quality designs. Another example in [12] introduces a data-driven approach to design space exploration and exploitation based on BO for additive manufacturing.

In spite of those successful applications, however, for problems with a large design space and complex non-linear functions, where many local optima exist, it is difficult for a single BO to efficiently determine the global optimal value within the design space. For complex design problems, the Multidisciplinary Design Optimization (MDO) literature promotes the use of decomposition-based approaches where a system-level problem is partitioned into smaller subproblems to be solved in a coordinated fashion [13]. For instance, Bayrak et al. show that partitioning a high-dimensional problem into smaller subproblems enables the finding of effective solutions with multiple agents compared to a case where the entire problem is addressed by a single agent [14].

This paper presents a multi-agent BO (MABO) framework to tackle the optimization of complex design problems, enabling a team of agents to learn from their local design space and collaborate by sharing global information with the other agents in the team. We use this framework as a reference model to computationally study the value of interactions and information sharing within design teams where multiple rational agents combine their information to solve local design problems and find the global optimal solution. To the best of our knowledge, the use of MABO as a model of rational design teams to study the impact of information sharing accounting for uncertainty in design space is new in the literature. Such a study can provide upper bounds on team performance with an appropriate information-sharing mechanism.

The MABO framework comprises three main steps: 1) *Design space partitioning*: The design space is divided into regions based on the number of agents involved in the search. Each region is searched by a single agent, which helps avoid redundant search efforts and reduce computational cost; 2) *Global-local communication strategy*: The MABO framework incorporates a global-local interaction that enables agents in local searches to share their sampled design points with a global evaluator. The global evaluator evaluates the acquisition function based on the sampled points from all agents to recommend the next sampling decisions for the agents. This communication strategy allows for effective information sharing and the utilization of global knowledge, which helps avoid getting trapped in local optima. 3) *Local decision-making*: Each agent only has access to the expected improvement evaluated in its local region and chooses the design that yields the largest value locally. This allows for an effective exploration of the design space while prioritizing local optima. The paper evaluates the performance of the proposed MABO framework in different information-sharing scenarios using benchmark functions of varying complexity and number of agents. The results demonstrate that the MABO framework significantly improves the convergence speed compared to a multi-agent model that does not allow information sharing and global-local interaction.

This paper is structured as follows. Section 2 provides technical background on BO. Next, Section 3 delves into the technical details of the proposed MABO framework, outlining the problem formulation and the solution approach with multiple agents. Sections 4 and 5 present the experimental settings and results, comparing the proposed MABO method with a global-local interaction and a method without a global evaluator. Section 6 further discusses the research results and draws insights into the impact of information sharing on design team performance. Finally, Section 7 concludes the paper with a summary of findings and limitations that lead to future work.

2. BACKGROUND ON BAYESIAN OPTIMIZATION

We use BO to model the decision-making of each agent in a team to find the optimum of an unknown objective function in its local region defined by a design space partitioning. This section provides a brief introduction to the preliminaries of BO. A typical BO process comprises two major components, 1) a statistical inference method, typically Gaussian process (GP) regression, to model the unknown objective function value with uncertainty and 2) an acquisition function to decide where to sample in the design space [15]. In order to find the global optimum of a black-box objective function $f(\mathbf{x})$, where $\mathbf{x} \in A$; A is a d-dimensional design space domain $A \subseteq \mathbb{R}^d$, BO updates the posterior mean and variance of a Gaussian process given the prior data, and an acquisition function selects the next best guess for the optimum based on the updated posterior probability distribution. Note that this process assumes that agents always act rationally to maximize their own utility represented by the acquisition function and does not include confounding human factors in the analysis.

2.1 Gaussian process

Gaussian process is a commonly used statistical inference model, which defines a distribution over possible unknown functions [16]. BO realizes the reasoning about $f(\mathbf{x})$ by choosing an appropriate Gaussian process prior:

$$\mathbf{f}(\mathbf{x}_{1:k}) \sim \mathcal{GP}(\mu_0(\mathbf{x}_{1:k}), \Sigma_0(\mathbf{x}_{1:k}, \mathbf{x}_{1:k})), \quad (1)$$

where the set of observations is $\mathcal{D} = (\mathbf{x}_{1:k}, \mathbf{f}(\mathbf{x}_{1:k}))$, $\mathbf{x}_{1:k} = [\mathbf{x}_1, \ldots, \mathbf{x}_k]$, $\mathbf{f}(\mathbf{x}_{1:k}) = [f(\mathbf{x}_1), \ldots, f(\mathbf{x}_k)]$,

Copyright © 2023 by ASME

$\mu_0(\mathbf{x}_{1:k}) = [\mu_0(\mathbf{x}_1),\ldots,\mu_0(\mathbf{x}_k)]$ is the mean vector by evaluating a mean function μ_0 at each $\mathbf{x}_1,\ldots,\mathbf{x}_k$, and $\Sigma_0(\mathbf{x}_{1:k},\mathbf{x}_{1:k}) = [\Sigma_0(\mathbf{x}_1,\mathbf{x}_1),\ldots,\Sigma_0(\mathbf{x}_1,\mathbf{x}_k);\ldots;\Sigma_0(\mathbf{x}_k,\mathbf{x}_1),\ldots,\Sigma_0(\mathbf{x}_k,\mathbf{x}_k)]$ is constructed by covariance $\Sigma_0(\cdot,\cdot)$ between each observation. Given the observation data $\mathscr{D}$, the posterior probability distribution is defined as [15]:

$$f(\mathbf{x}) \mid \mathbf{f}(\mathbf{x}_{1:k}) \sim \mathscr{GP}\left(\mu(\mathbf{x}),\sigma^2(\mathbf{x})\right)$$
$$\mu(\mathbf{x}) = \Sigma_0(\mathbf{x},\mathbf{x}_{1:k})\,\Sigma_0(\mathbf{x}_{1:k},\mathbf{x}_{1:k})^{-1}\left(\mathbf{f}(\mathbf{x}_{1:k}) - \mu_0(\mathbf{x}_{1:k})\right) + \mu_0(\mathbf{x})$$
$$\sigma^2(\mathbf{x}) = \Sigma_0(\mathbf{x},\mathbf{x}) - \Sigma_0(\mathbf{x},\mathbf{x}_{1:k})\,\Sigma_0(\mathbf{x}_{1:k},\mathbf{x}_{1:k})^{-1}\Sigma_0(\mathbf{x}_{1:k},\mathbf{x}),$$
$$(2)$$

where $\mu(\mathbf{x})$ denotes the posterior mean and and $\sigma^2(\mathbf{x})$ denotes the posterior variance.

2.2 Acquisition function

An acquisition function is a heuristic used to determine the next point to sample in the search space. This function takes the probabilistic surrogate model, introduced in Section 2.1 that approximates the objective function as input. The next observation (sampling point) for the search is selected by optimizing the acquisition function [15] while balancing the search strategy between exploration and exploitation. Several acquisition functions are widely used in BO, such as Probability of Improvement (PI) [17], Expected Improvement (EI) [7], Lower Confidence Bound (LCB) [18] and Thompson Sampling (TS) [19]. In this study, we adopt EI for the acquisition function due to its high sensitivity to improvements and fast convergence speed to the optimum [9].

Assuming that the design problem is formulated as a minimization problem, the corresponding utility function can be defined as follows:

$$u(\mathbf{x}) = max(0, f^*(\mathbf{x}) - f(\mathbf{x})), \qquad (3)$$

where $f^*(\cdot)$ is the minimum value of $f(\cdot)$ observed so far. Using this utility, the acquisition function with EI can be formulated as follows:

$$
\begin{aligned}
a_{\mathrm{EI}}(\mathbf{x}) &= \mathbb{E}[u(\mathbf{x}) \mid \mathbf{x}, \mathscr{D}]\\
&= \int_{-\infty}^{f^*} (f^* - f)\,\mathcal{N}(f^*; \mu(\mathbf{x}), \sigma^2(\mathbf{x}))\mathrm{d}f\\
&= (f^* - \mu(\mathbf{x}))\,\Phi\left(f^*; \mu(\mathbf{x}), \sigma^2(\mathbf{x})\right)\\
&\quad + \sigma^2(\mathbf{x})\mathcal{N}\left(f^*; \mu(\mathbf{x}), \sigma^2(\mathbf{x})\right),
\end{aligned}
\qquad (4)
$$

where μ and σ^2 are mean and variance functions of the posterior probability distribution for f given by Eq. (2).

With the acquisition function a_{EI} shown in Eq. (4), the next sampling point $\mathbf{x}$ is selected as the one that maximizes EI. The two terms in Eq. (4) can be viewed as a trade-off between exploiting the information from evaluating the points with low mean values and exploring the points with high uncertainty.

3. MULTI-AGENT BAYESIAN OPTIMIZATION
3.1 Problem setup

In this section, we show the problem formulation to find the minimum of a black-box function in a d-dimensional design space $A \subseteq \mathbb{R}^d$ with N agents in a team. The goal of agent i, where $i \in \{1, 2, ..., N\}$ is to find the location of global minimum $\mathbf{x}^*$

$$\mathbf{x}^* = argmin_{\mathbf{x} \in A} f(\mathbf{x}), \qquad (5)$$

where $f(\cdot)$ is a black-box objective function and $\mathbf{x} = (x_1, x_2, ..., x_d) \in A$.

We assume that the design task is divided into N regions and that each agent is assigned to one unique region in the design space $A_i \subseteq A$ to divide the labor among the agents in a team. We assume that the division of labor is done at the beginning before the design search starts. In this formulation, none of the agents is allowed to search beyond their assigned region but only communicate by sharing the local points they sampled in the past. The search stops either after a predefined number of steps or when an agent arrives sufficiently close to the local minimum (e.g. when the smallest convergence rate among N agents is lower than a predefined threshold). Figure 1a shows an example to illustrate this idea. In this example, the objective function is an unknown function of an MAS consisting of three agents. The design space has been partitioned into three local regions depicted in Figure 1b. In each region, only one agent is assigned to search the space locally to find the global minimum (marked as a red star), whereas none of them knows where the global minimum is. We do not allow multiple agents in one region because that does not fundamentally change the problem to be solved and only influences the convergence speed [5] because that simply increases the number of sampling points in each search.

3.2 Solution process with MABO

To computationally study the solution to the problem presented in the preceding section using a team of agents, we develop a multi-agent Bayesian Optimization (MABO) process that allows all agents to work collaboratively to find the optimal solution in an unknown design space. The algorithm for our proposed MABO process is described in Algorithm 1. In this MAS, an important feature is that it has a global evaluator and the agents are allowed to contribute their locally sampled points to the global; meanwhile, the information processed at global will be shared back to each individual for their decision-making on future moves. We use this global-local interaction mechanism to study the value of interactions among team members (i.e., agents) in solving a complex problem (with multiple local optima) when there is a division of labor among team members. We perform this analysis in comparison to a case where the agents do not use this global evaluator, i.e., they do not communicate with each other.

Specifically in this MAS, when agents are allowed to communicate, a Gaussian process, as introduced in Section 2.1, uses Bayesian inference to estimate the probability distribution of potential values for $f(\mathbf{x})$ at a candidate point $\mathbf{x}$. The posterior distribution is iteratively updated by the global evaluator collecting observation data contributed by all agents. The acquisition function is located at the global level that evaluates the value of the EI at a new point $\mathbf{x}$ based on the current posterior distribution over f. After global evaluation, each agent can only access the EI information defined in its local region A_i and will choose the next sampling point that produces the local maximum of EI.

Copyright © 2023 by ASME

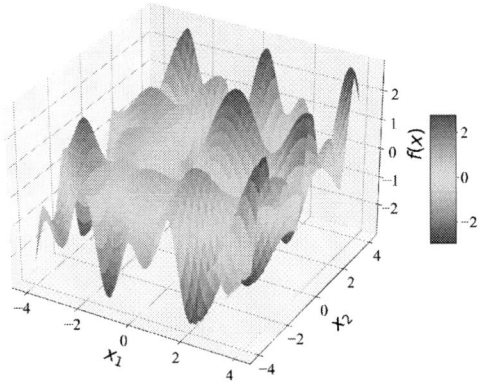

(a) Objective function

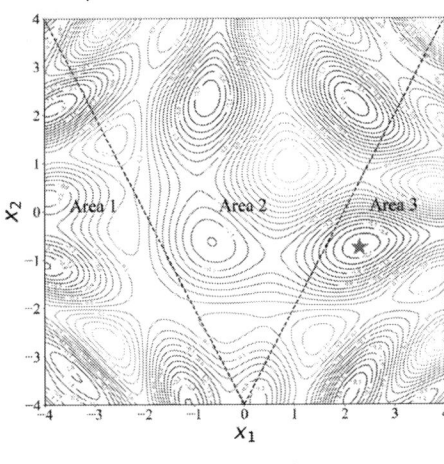

(b) Contour plot

FIGURE 1: AN EXAMPLE OF THE OBJECTIVE FUNCTION IN 2D DESIGN SPACE.

With the objective function shown in Figure 1a, one snapshot of an iteration in MABO using Algorithm 1 is illustrated in Figure 2. Each agent picks one new point in each iteration in its local region divided by the dashed line. The global evaluator collects all sampled points (the black points) from each agent to evaluate the posterior means and variance shown in the subfigures of the first and second columns. Based on this evaluation, the acquisition function value is updated, as shown in the subfigures in the third column. To move forward, the agents select the location of the maximum value of the acquisition function in their respective local region (the red points) as the next sampling point. In Figure 3, the sampled points of each agent are marked with different colors, where the numbers on the points indicate the search in each iteration step.

This approach (when agents share design samples with each other through a global evaluator) can effectively find the global optimum with high convergence speed (albeit without any theoretical guarantees) by involving multiple agents working on their local regions in each iteration. The global evaluator can acquire multiple observations within a single iteration, thereby enabling a comprehensive evaluation of the objective function across all local regions at each iteration.

Algorithm 1 MULTI-AGENT BAYESIAN OPTIMIZATION

Set local region A_i for each agent in an unknown design space A

Place a Gaussian process prior $\mathcal{D}_i = (\mathbf{x}_{1:k}, \mathbf{f}(\mathbf{x}_{1:k}))$ in A_i for each agent

Observe f at the initial step in design space A

for $Step = 1$ to $MAX - Step$ **do**

    Update the posterior probability distribution with all available data on each local region A_i as Eq. (2)

    Update the acquisition function Eq. (4)

    for agent 1 to N **do**

        Let $\mathbf{x}_i$ be a optimizer of acquisition function Eq. (4) in the local region A_i

        Observe $f(\mathbf{x}_i)$.

    end for

    Collect all data $(\mathbf{x}_i, f(\mathbf{x}_i))$ from agents

end for

Return a solution: the point $\mathbf{x}$ evaluated with the global optimum $f^*(\mathbf{x})$.

---

4. EXPERIMENT SETUP

To investigate the impact of information exchange and interactions between local agents (through the global evaluator), we created another MABO process that does not have a global evaluator and each agent solves a BO in its own region as a baseline model. Therefore, we compare the following two MABO processes in terms of their convergence speeds.

- *Method 1: the MABO process without a global evaluator;*

- *Method 2: the proposed MABO proposed with a global evaluator enabled.*

We test these two processes on two different benchmark functions with varying complexities, as shown in Table 1 and displayed in Figure 4: 1) the Cosines function and 2) the Eggholder function. These two functions are widely recognized within the global optimization literature, as they have been frequently utilized as benchmarks in the context of BO. This is evidenced by their previous use in research studies such as [20, 21]. Compared to the former function, the latter is more complex as it contains many more local minima and maxima in the search space.

To study the scalability of the findings and evaluate the impact of the number of agents on search performance, we performed computational studies with three and five agents. As a result, we create three different experimental scenarios: 1) MABO of the Cosines function with an MAS of three agents, 2) MABO of the Eggholder function with an MAS of three agents, and 3) MABO of the Eggholder function with an MAS of five agents. To keep a fair comparison, the number of maximum iterations for sampling is set to 50 for all experiments, and the initial number of samples for the Gaussian process prior $\mathcal{D}_i = (\mathbf{x}_{1:k}, \mathbf{f}(\mathbf{x}_{1:k}))$ in A_i is set to $k = 5$ in each method and scenario in Algorithm 1.

5. RESULTS

Scenario 1: MABO of the Cosines function with an MAS of three agents. Each agent is assigned to its local region defined

Copyright © 2023 by ASME

TABLE 1: TWO BLACK-BOX FUNCTIONS

Name	Formula	Global minimum	Global domain A
Cosines	$f(\mathbf{x}) = 1 - (x_1^2 + x_2^2 - 0.3cos(3\pi x_1) - 0.3cos(3\pi x_2))$	$f(0.314, 0.303) = -1.596$	$\{\mathbf{x}\| -1 \leq \mathbf{x} \leq 1\}$
Eggholder	$f(\mathbf{x}) = -(x_2 + 47)\sin\left(\sqrt{\|x_2 + \frac{x_1}{2} + 47\|}\right)$ $-x_1 \sin\left(\sqrt{\|x_1 - (x_2 + 47)\|}\right)$	$f(512, 404.232) = -959.641$	$\{\mathbf{x}\| -520 \leq \mathbf{x} \leq 520\}$

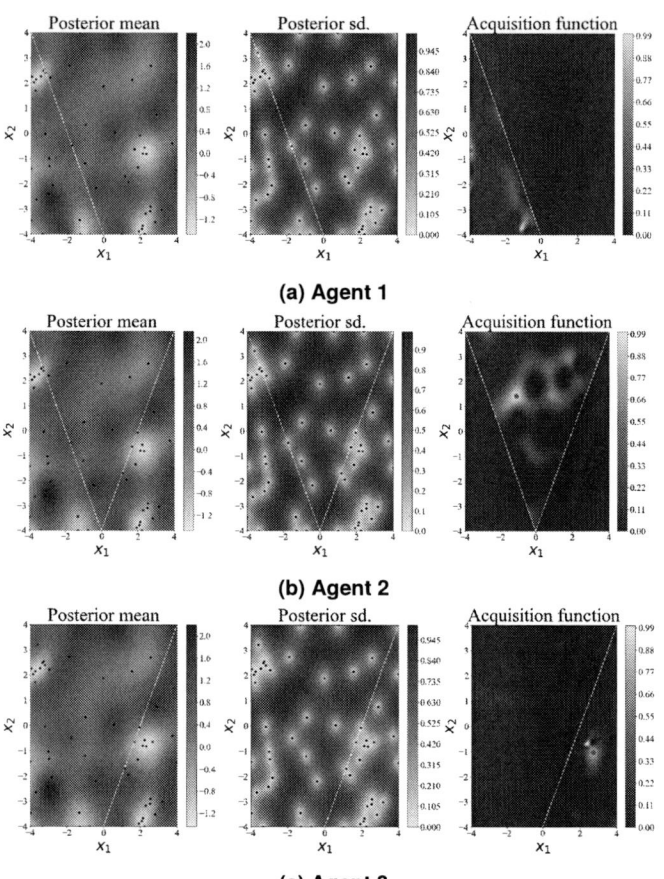

(a) Agent 1

(b) Agent 2

(c) Agent 3

FIGURE 2: THE GP MODEL AND ACQUISITION FUNCTION WITH A GLOBAL EVALUATOR ARE ENABLED FOR AGENTS 1, 2, AND 3.

in Table 2. The sampling process and the search trajectory (indicated by the number index) of each agent in its own region are displayed in Figure 5. The best $f(\mathbf{x})$ (i.e., f^* in Eq. (4)) observed so far in each step shown in Figure 6 describes the convergence speed for an MAS. According to Figure 6a, Agent 2 in Method 1 (i.e., the MABO without a global evaluator) was able to successfully find the global minimum $f(0.314, 0.303) = -1.596$ in 31 steps. Agent 1 and Agent 3 reached their local minima in 20 and 22 steps, respectively. With Method 2 (i.e., the MABO with a global evaluator), Agent 2 reached the global minimum in only six steps, as shown in Figure 6b. Agent 1 and Agent 3 found their local minima in 9 and 18 steps, respectively.

Scenario 2: MABO of the Eggholder function with an MAS of three agents. The local region for each agent is defined in Table 3 and the global minimum of this Eggholder function is located in area 3, $f(512, 404.232) = -959.641$. Due to the increased complexity of the Eggholder function, finding the global mini-

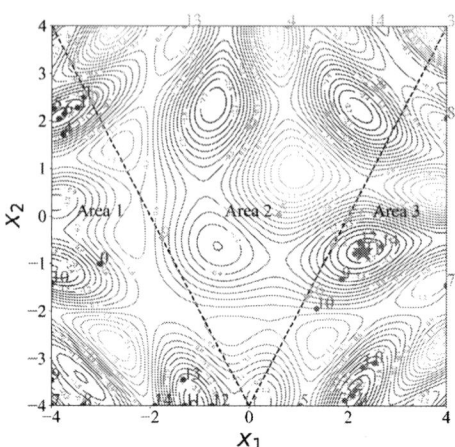

FIGURE 3: SAMPLING PROCESS FOR THREE AGENTS. THE RED STAR IS THE TRUE OPTIMUM. THE NUMBERS LABELED ON THE POINTS ARE THE SEARCH IN EACH ITERATION STEP.

TABLE 2: COSINES FUNCTION. LOCAL DESIGN SPACE DOMAIN FOR THREE AGENTS.

	Local region A_i
Agent 1	$\{\mathbf{x}\| -2x_1 + x_2 + 2 \leq 0\}$
Agent 2	$\{\mathbf{x}\| -2x_1 - x_2 - 2 \leq 0, 2x_1 - x_2 - 2 \leq 0\}$
Agent 3	$\{\mathbf{x}\| -2x_1 + x_2 + 2 \leq 0\}$

mum using MABO without global-local information exchange is a challenging task. As shown in Figure 8a, Agent 3 cannot reach the global minimum even after 50 steps. Actually, at Step 49, the best performance achieved is from Agent 2. In Method 2 with a global evaluator, MAS accomplished the search in 32 steps, as illustrated in Figure 8b.

Scenario 3: MABO of the Eggholder function with an MAS of five agents. In this scenario, the MABO methods were experimented with the same function, i.e., the Eggholder function, but the MAS in each method was expanded to five agents. The local region of each agent is defined in Table 4. Figure 9 demonstrates the sampling points and the corresponding trajectories for both methods. It is observed that many sampling points are clustered in

TABLE 3: EGGHOLDER FUNCTION. LOCAL DESIGN SPACE DOMAIN FOR THREE AGENTS.

	Local region A_i
Agent 1	$\{\mathbf{x}\| -2x_1 + x_2 + 520 \leq 0\}$
Agent 2	$\{\mathbf{x}\| -2x_1 - x_2 - 520 \leq 0, 2x_1 - x_2 - 520 \leq 0\}$
Agent 3	$\{\mathbf{x}\| -2x_1 + x_2 + 520 \leq 0\}$

Copyright © 2023 by ASME

local minima. This indicates that agents in both methods tend to continue to exploit a location region if the previous sampling point in that region yields the best $f(\mathbf{x})$ observed so far. The agents start exploring other unknown spaces if no further improvement can be made since the last-found best $f(\mathbf{x})$ in their current local search regions. As shown in Figure 10a, Agent 5 in Method 1 reaches the global minimum $f(512, 404.232) = -959.641$ in 43 steps, while Agent 5 in Method 2 finds the minimum in just seven steps, as illustrated in Figure 10b.

6. DISCUSSION

The findings regarding the convergence speed of the methods in different scenarios are summarized in Table 5. The comparison between Scenario 1 and Scenario 2 indicates that, as the complexity of the function increases, more iterations are needed during the search for convergence. However, using the global

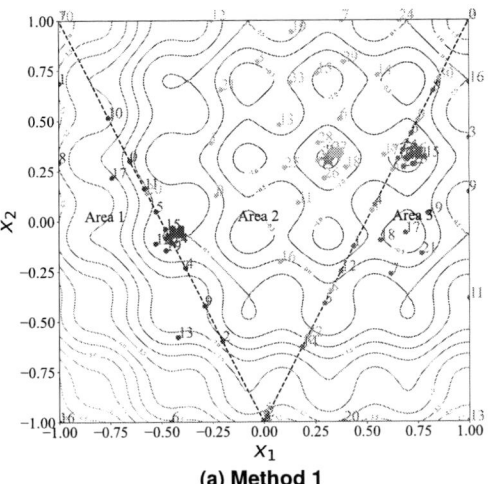

(a) Method 1

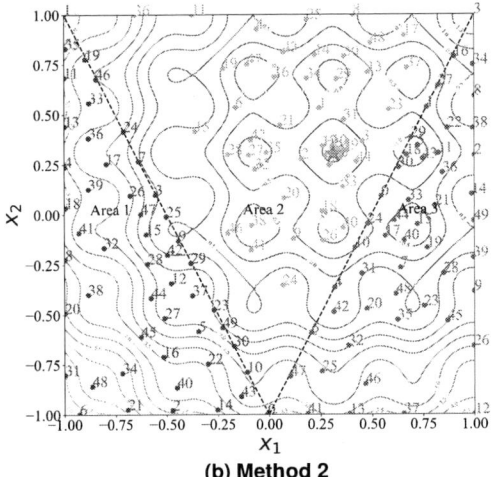

(b) Method 2

FIGURE 5: DESIGN SPACE EXPLORATION IN SCENARIO 1

evaluator (and interactions among agents) is found to be more robust against the increased complexity because Method 2 successfully finds the global minimum within 50 steps, yet Method 1 failed in Scenario 2. Comparing the results of Scenario 2 and Scenario 3 indicates that both MABO methods become more efficient when more agents are available in a team. In particular, the convergence speed of our proposed MABO method (i.e., Method 2) with five agents is about 4.5 times faster than that with three agents when dealing with the same Eggholder function. In all scenarios, the results indicate that the MABO with global-local information exchange outperforms the one without such information exchange.

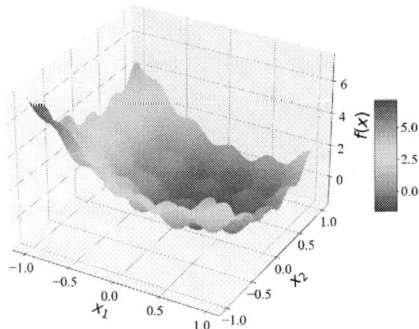

(a) Cosines function

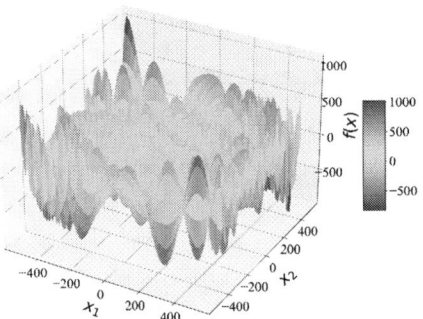

(b) Eggholder function

FIGURE 4: TWO BENCHMARK FUNCTIONS. THE EGGHOLDER FUNCTION IS MUCH MORE COMPLEX THAN THE COSINE FUNCTION, WITH MUCH MORE LOCAL OPTIMA THAN THE LATTER.

TABLE 4: EGGHOLDER FUNCTION. LOCAL DESIGN SPACE DOMAIN FOR FIVE AGENTS.

	Local region A_i
Agent 1	$\{\mathbf{x} \| 3x_1 + x_2 + 1040 \le 0\}$
Agent 2	$\{\mathbf{x} \| -3x_1 - x_2 - 1040 \le 0\}$
Agent 3	$\{\mathbf{x} \| 6x_1 + x_2 - 520 \le 0, -6x_1 + x_2 - 520 \le 0\}$
Agent 4	$\{\mathbf{x} \| 3x_1 - x_2 - 1040 \le 0, -6x_1 - x_2 + 520 \le 0\}$
Agent 5	$\{\mathbf{x} \| -3x_1 + x_2 + 1040 \le 0\}$

TABLE 5: CONVERGENCE SPEED FOR EACH SCENARIO

Obj. Func.	MAS with three agents		MAS with five agents	
	Method 1	Method 2	Method 1	Method 2
Cosines	31	6	N/A	N/A
Eggholder	>200	32	43	7

Copyright © 2023 by ASME

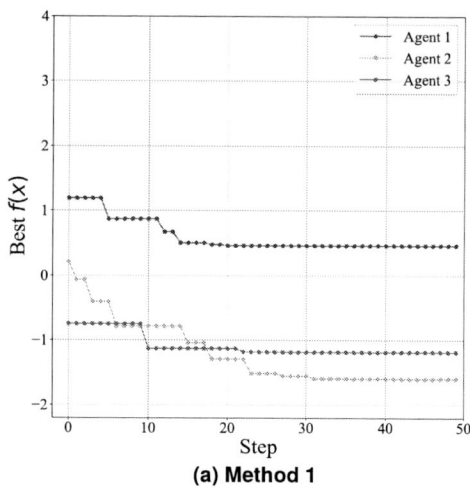

(a) Method 1

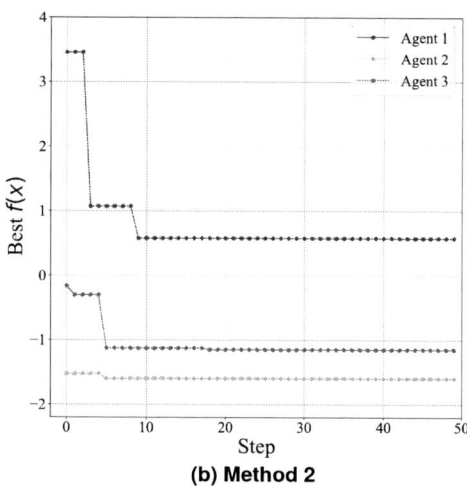

(b) Method 2

FIGURE 6: CONVERGENCE SPEED IN SCENARIO 1

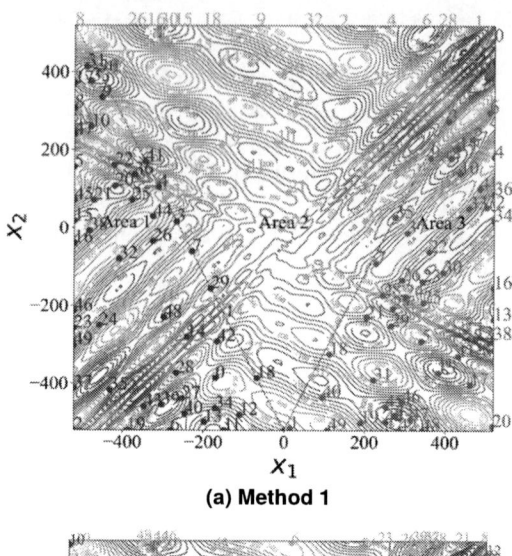

(a) Method 1

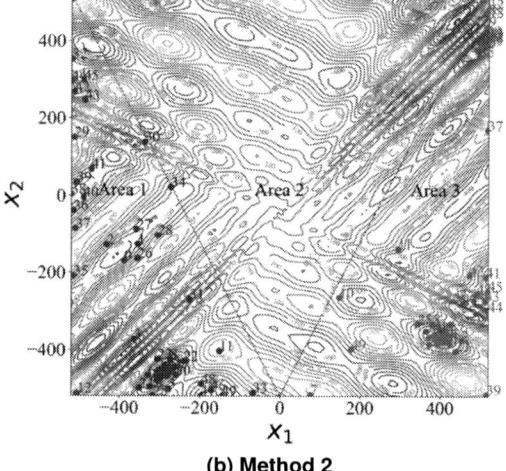

(b) Method 2

FIGURE 7: DESIGN SPACE EXPLORATION IN SCENARIO 2

Although these results are intuitive, they lead to some useful take-aways for design teams in practice. First, they provide a different conclusion about interactions between team members from the computational findings in the literature. Recall that the study by McComb et al. suggests that team members should not interact with each other when they all work on the same problem without a division of labor [5] whereas the existence of a division of labor in a team benefits from interactions for effective solutions based on our findings. Note that this result does not contradict this literature since the study conditions in terms of task allocations are different. Combining our results with those from [5], we can see when team communication is beneficial and when it is not. In our study, even though the agents are responsible for their own regions, information from other regions in the design space leads these agents to find solutions in their own regions faster than in the case where there is no interaction among agents. The convergence results also support the argument that teams are as good as the most vital link in the team [22], i.e., the performance of a team is determined by the best agent in the team. Note that

these results do not include any human factors that might add some adverse effects due to communication problems, trust and cognitive biases.

7. CONCLUSION AND FUTURE WORK

The study is motivated to answer the following question: What is the impact of information exchange among agents in a multi-agent system (MAS) on their performance in searching a complex unknown design space collaboratively for the global optimum? To this end, the paper presents a multi-agent Bayesian optimization (MABO) framework that addresses the challenges of finding the global optimum of complex objective functions with many local optima. The framework involves dividing the design space into local regions and allowing agents in local searches to share their sampled design points with a global evaluator. The global evaluator computes the posterior mean and variance and evaluates the acquisition function to recommend the next sampling decisions for local agents. To answer the motivation question, the proposed method was compared with a MABO method that does not allow information exchange. The results show that

Copyright © 2023 by ASME

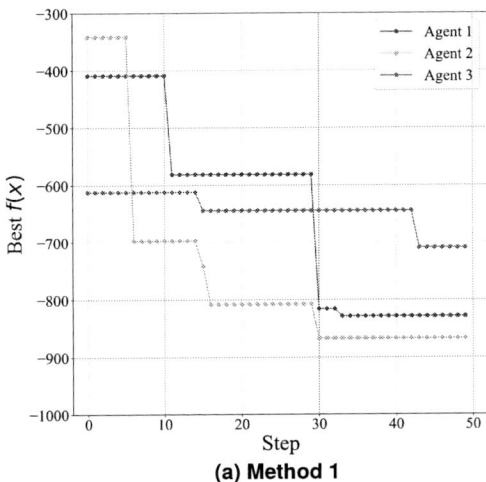

(a) Method 1

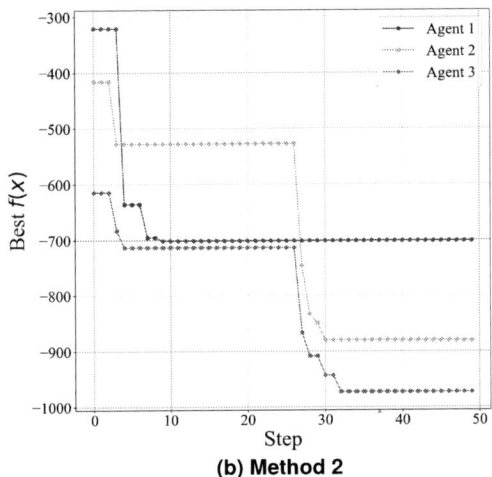

(b) Method 2

FIGURE 8: CONVERGENCE SPEED IN SCENARIO 2

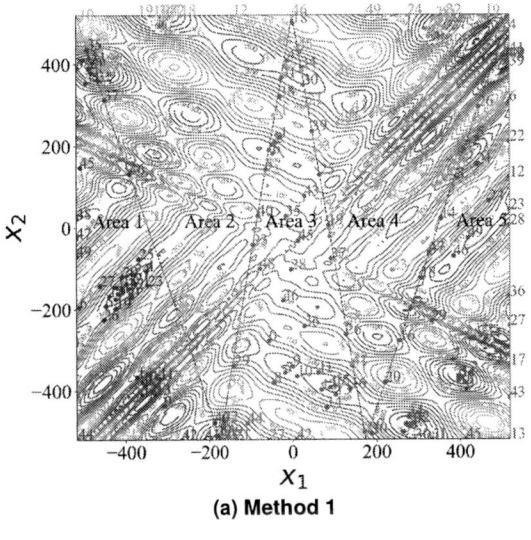

(a) Method 1

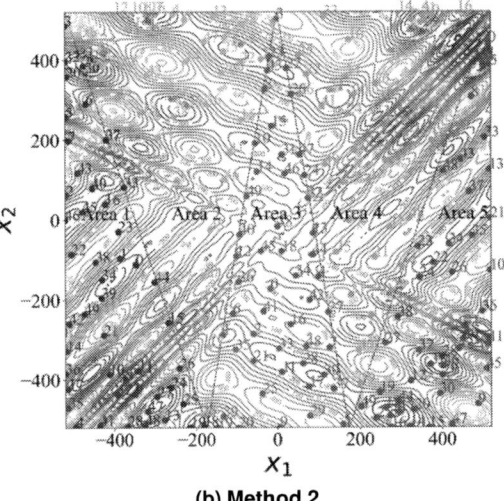

(b) Method 2

FIGURE 9: DESIGN SPACE EXPLORATION IN SCENARIO 3

allowing information exchange and global-local interaction significantly improves convergence speed (more than five times on average), regardless of function complexity and MAS team size. Therefore, we conclude that the proposed MABO framework is effective in exploring complex design spaces and finding the optimal solution.

Although promising, there are a number of limitations in our current study, which lead to several future directions. First, we did not consider the cost in the current framework. However, in reality, the cost could be associated with search and information sharing. In our future study, in addition to evaluating search performance quantified by convergence speed, cost shall be the other dimension in performance evaluation. Second, an unknown space must involve unknown constants. In our current study, agents are assumed to have access anywhere in the designated region. In future work, this assumption can be relaxed by introducing "infeasible area" in the local search regions, so agents are not allowed to sample candidate points in the infeasible areas. Moreover, we can investigate how the location of the infeasible area would influence the agents' search performance.

Third, in the current study, we conducted the experiment with the MABO framework using one particular acquisition function and a special kernel setting of the Gaussian process. In future studies, experimental results will be collected with more comprehensive hyperparameter settings, so a complete picture of the agents' search performance can be obtained. Fourth, this paper only presents the MABO solution to 2-dimensional design space exploration. In higher-dimensional cases, more complex division strategies could be required to partition the space. Lastly, since information exchange and global-local interaction are enabled in our framework, this opens many research questions for the community. For example, when the amount of information and frequency can be controlled, how could the variations influence the agents' search performance? In our future work, we are motivated to answer these questions based on the work presented.

ACKNOWLEDGMENTS

The authors greatly acknowledge the financial support provided by the Walker Department of Mechanical Engineering at

Copyright © 2023 by ASME

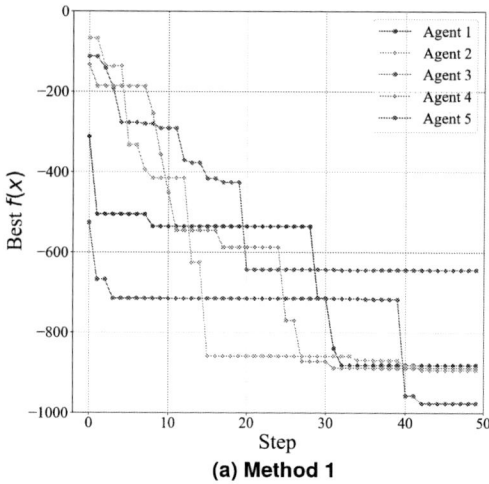

(a) Method 1

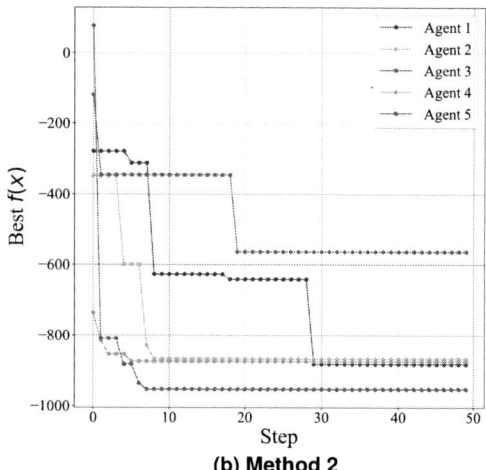

(b) Method 2

FIGURE 10: CONVERGENCE SPEED IN SCENARIO 3

the University of Texas at Austin.

REFERENCES

[1] Nardi, Luigi, Koeplinger, David and Olukotun, Kunle. "Practical design space exploration." *2019 IEEE 27th International Symposium on Modeling, Analysis, and Simulation of Computer and Telecommunication Systems (MASCOTS)*: pp. 347–358. 2019. IEEE.

[2] Gyory, Joshua T, Soria Zurita, Nicolás F, Cagan, Jonathan and Mccomb, Christopher. "Independence or Interaction? Understanding the Benefits and Limitations of Nominally-Inspired and Interacting Sub-Structured Teams in a Virtual and Interdisciplinary Engineering Design Task." *Journal of Mechanical Design* : pp. 1–42.

[3] Gyory, Joshua T, Kotovsky, Kenneth and Cagan, Jonathan. "The influence of process management: uncovering the impact of real-time managerial interventions via a topic modeling approach." *Journal of Mechanical Design* Vol. 143 No. 11 (2021).

[4] McComb, Christopher, Cagan, Jonathan and Kotovsky, Kenneth. "Lifting the Veil: Drawing insights about design teams from a cognitively-inspired computational model." *Design Studies* Vol. 40 (2015): pp. 119–142.

[5] McComb, Christopher, Cagan, Jonathan and Kotovsky, Kenneth. "Optimizing design teams based on problem properties: computational team simulations and an applied empirical test." *Journal of Mechanical Design* Vol. 139 No. 4 (2017): p. 041101.

[6] Allison, James T., Kokkolaras, Michael and Papalambros, Panos Y. "Optimal Partitioning and Coordination Decisions in Decomposition-Based Design Optimization." *Journal of Mechanical Design* Vol. 131 No. 8 (2009). 081008.

[7] Snoek, Jasper, Larochelle, Hugo and Adams, Ryan P. "Practical bayesian optimization of machine learning algorithms." *Advances in neural information processing systems* Vol. 25 (2012).

[8] Greenhill, Stewart, Rana, Santu, Gupta, Sunil, Vellanki, Pratibha and Venkatesh, Svetha. "Bayesian optimization for adaptive experimental design: A review." *IEEE access* Vol. 8 (2020): pp. 13937–13948.

[9] Jones, Donald R, Schonlau, Matthias and Welch, William J. "Efficient global optimization of expensive black-box functions." *Journal of Global optimization* Vol. 13 No. 4 (1998): p. 455.

[10] Pelamatti, Julien, Brevault, Loïc, Balesdent, Mathieu, Talbi, El-Ghazali and Guerin, Yannick. "Bayesian optimization of variable-size design space problems." *Optimization and Engineering* Vol. 22 (2021): pp. 387–447.

[11] Reagen, Brandon, Hernández-Lobato, José Miguel, Adolf, Robert, Gelbart, Michael, Whatmough, Paul, Wei, Gu-Yeon and Brooks, David. "A case for efficient accelerator design space exploration via bayesian optimization." *2017 IEEE/ACM International Symposium on Low Power Electronics and Design (ISLPED)*: pp. 1–6. 2017. IEEE.

[12] Xiong, Yi, Duong, Pham Luu Trung, Wang, Dong, Park, Sang-In, Ge, Qi, Raghavan, Nagarajan and Rosen, David W. "Data-driven design space exploration and exploitation for design for additive manufacturing." *Journal of Mechanical Design* Vol. 141 No. 10 (2019).

[13] Tosserams, S, Hofkamp, AT, Etman, LFP and Rooda, JE. "A specification language for problem partitioning in decomposition-based design optimization." *Structural and Multidisciplinary Optimization* Vol. 42 (2010): pp. 707–723.

[14] Bayrak, Alparslan Emrah, McComb, Christopher, Cagan, Jonathan and Kotovsky, Kenneth. "A strategic decision-making architecture toward hybrid teams for dynamic competitive problems." *Decision Support Systems* Vol. 144 (2021): p. 113490.

[15] Frazier, Peter I. "A tutorial on Bayesian optimization." *arXiv preprint arXiv:1807.02811* (2018).

[16] Rasmussen, Carl Edward. "Gaussian processes in machine learning." *Summer school on machine learning*: pp. 63–71. 2003. Springer.

[17] Kushner, H. J. "A New Method of Locating the Maximum Point of an Arbitrary Multipeak Curve in the Presence of

Copyright © 2023 by ASME

Noise." *Journal of Basic Engineering* Vol. 86 No. 1 (1964): pp. 97–106.

[18] Srinivas, Niranjan, Krause, Andreas, Kakade, Sham M and Seeger, Matthias. "Gaussian process optimization in the bandit setting: No regret and experimental design." *arXiv preprint arXiv:0912.3995* (2009).

[19] Russo, Daniel J, Van Roy, Benjamin, Kazerouni, Abbas, Osband, Ian, Wen, Zheng et al. "A tutorial on thompson sampling." *Foundations and Trends® in Machine Learning* Vol. 11 No. 1 (2018): pp. 1–96.

[20] González, Javier, Dai, Zhenwen, Hennig, Philipp and Lawrence, Neil. "Batch Bayesian optimization via local penalization." *Artificial intelligence and statistics*: pp. 648–657. 2016. PMLR.

[21] De Palma, Alessandro, Mendler-Dünner, Celestine, Parnell, Thomas, Anghel, Andreea and Pozidis, Haralampos. "Sampling acquisition functions for batch Bayesian optimization." *arXiv preprint arXiv:1903.09434* (2019).

[22] Brownell, Ethan, Cagan, Jonathan and Kotovsky, Kenneth. "Only as strong as the strongest link: The relative contribution of individual team member proficiency in configuration design." *Journal of Mechanical Design* Vol. 143 No. 8 (2021).

Copyright © 2023 by ASME

Proceedings of the ASME 2023
International Design Engineering Technical Conferences and
Computers and Information in Engineering Conference
IDETC-CIE2023
August 20-23, 2023, Boston, Massachusetts

DETC2023-116320

A COMPARISON OF BAYESIAN ACQUISITION FUNCTIONS FOR USE IN SURROGATE MULTI-OBJECTIVE FEASIBILITY ROBUST OPTIMIZATION WITH INTERVAL UNCERTAINTY

Randall J. Kania[1,3] and Shapour Azarm[2,3]
[1]Assistant Clinical Professor, First-year Innovation and Research Experience
[2]Professor, Department of Mechanical Engineering
[3]University of Maryland, College Park, MD 20742

ABSTRACT

Engineering design optimization problems are commonly multi-objective, involve computationally expensive simulations, and have uncertainty. In these problems, consideration of uncertainty can produce feasibly robust optimized design solutions, i.e., optimized and feasible under uncertainty. In general, methods for solving multi-objective robust optimization problems can be computationally expensive. To ease the computational burden and make these methods more accessible, a surrogate-based bi-objective Bayesian feasibility robust optimization method from the literature has been extended in this paper to higher numbers of objectives. A key component of the proposed method is a new acquisition function. The acquisition function proposed in this paper uses a Monte Carlo approach to approximate the expected centroidal improvement of the solution set. The proposed acquisition function is compared against exact (closed form) acquisition function-based approaches and other methods from the literature. The experimental results from this study suggest the multi-objective robust optimization framework is insensitive to the selection of acquisition function.

Keywords: Robust Optimization, Multi-objective, Bayesian, Acquisition Function, Surrogate Modeling

NOMENCLATURE

a	constant
d	dimension of design vector
E	expectation operator
f_i	objective function i
$\bar{f}_i$	centroid of region for objective function i
$\hat{f}_i$	surrogate model of objective function i
$F(G)$	total set of sampled objective (constraint) values for training models
f_i^{ND}	component of current non-dominated solutions for the ith objective function
$f_i^{ND,c}$	component of non-dominated solution closest to candidate point for the ith objective function
f_{min}	current best sample for single objective Bayesian optimization
g_j	constraint function j
$\hat{g}_j$	surrogate model of constraint function j
$i(I)$	index of (total number of) objective functions
Imp	improvement over current candidate solution
$j(J)$	index of (total number of) constraint functions
$n(N)$	index of (total number of) design variables
P	probability operator
$\hat{s}$	standard deviation of Gaussian process
u	vector of uncertain parameters
$u_l(u_u)$	lower (upper) bounds of uncertain parameters
x	vector of design variables
X	set of design vectors for model training
x_c	candidate design vector
$x_l(x_u)$	lower (upper) bounds of design variables
μ	mean function of Gaussian process
ν	number of non-dominated solutions

1. INTRODUCTION

In formulating engineering design optimization problems with computationally expensive simulations, assumptions are often made to reduce computational costs. These assumptions can have the drawback of removing key factors from consideration. Two such factors are the existence of multiple competing objectives and the presence of uncertainty. When competing objectives are considered, their consideration is most often limited to two objectives, again due to simplification and computational cost. However, by considering fewer objectives, a portion of the design space is removed from the view of the designer. In this work, the efficacy of methods for considering higher numbers of (more than two) objective

Copyright © 2023 by ASME

functions while considering uncertainty in a design optimization setting are studied.

Multi-objective optimization problems have been solved by classical methods such as objective weighting and epsilon constraints which convert the problem to a single objective [1]. Solving these iteratively by varying the weights and epsilon values, provides a set of optimal solutions which are called: non-dominated (Pareto) solutions, hereafter also referred to as non-dominated solution frontier. Another class of approaches for solving multi-objective optimization problems has been to use population based metaheuristics like the genetic algorithm to produce a non-dominated set of solutions [2]. Population based methods can be well suited to this task because they are already performing operations on multiple solutions. Methods such as NSGA-II allow for a non-dominated set to be found in a single solver run by incorporating metrics such as population diversity along with individual fitness [3].

Although these population-based methods arrive at the non-dominated solutions via a single run of the optimizer, the number of times that objective and constraint functions need to be evaluated can be high, due to the population size and number of generations needed for convergence. In cases where function evaluations are costly, the population-based methods can become prohibitively expensive. One approach for alleviating the cost is to use surrogate models with Bayesian optimization. In Bayesian optimization, the acquisition function is optimized in each iteration to select the next candidate solution to sample. The choice of function affects the balance between exploration of design space and exploitation of the current model. An acquisition function for two objectives was developed in [4] and is often called either the centroidal or Euclidean expected improvement method. The method was further popularized in [5], largely due to the provided MATLAB implementation of their proposed method. It is based on the expected improvement criteria [6] which seeks to balance exploration of the design space with exploitation of the current model estimate. Although the closed form solution presented in [4] is only applicable to two objectives, the method was later extended to multiple objectives [7]. Another method for using Bayesian optimization for bi-objective problems was presented in [8] using the lower confidence bound acquisition function. A number of other methods for Bayesian optimization of higher numbers of objective functions have been developed in recent years [9], [10]. The majority of these methods involve characterizing improvement of the non-dominated solution set as a whole into a single scalar value and solving it repeatedly, somewhat analogous to the classical multi-objective techniques [4], [8], [11]–[15]. These multi-objective Bayesian optimization methods make it much more affordable for a decision maker or designer to create a set of non-dominated solutions balancing their multiple competing objectives. Another challenge faced by the designer is the presence of uncertainty.

Uncertainty handling is a topic that has been addressed in the field of robust optimization. It has been well studied for single objective problems ranging from linear programming [16] to more general non-convex optimization [17]. Robust optimization for multiple objective functions has also been explored [18]–[20]. Among these works, the use of surrogate modelling has been shown previously to reduce the computational costs of expensive function evaluations. For two objective functions, it has been further shown how Bayesian optimization techniques can bolster robust optimization [21]. The method in [21] can use a number of different acquisition functions in the robust optimization methodology. The acquisition function used in that work was the closed-form bi-objective expected improvement function from [4]. In the current work, the method is expanded using a number of different acquisition functions to extend it beyond two objective functions [22].

The contributions of this work are as follows. This paper contributes to a comparison study of several multi-objective Bayesian optimization acquisition functions in problems containing uncertainty. Specifically, first, a new Monte-Carlo based acquisition function is presented. This acquisition function can circumvent the computational costs of other methods by omitting exact calculations of expected improvement. Second, a method for combining acquisition functions for greater than two objectives with the non-dominated relaxation method for multi-objective Bayesian feasibility robust optimization under interval uncertainty is provided. Two scalable examples and an engineering example are used to compare the efficacies of the existing and proposed multi-objective acquisition functions when combined with the non-dominated relaxation technique for multi-objective Bayesian robust optimization. These methods are also compared against two population-based methods from the literature, one with surrogate modelling and one without.

The rest of the paper is organized as follows: Section 2 contains a description of the problem being addressed by the proposed methodology as well as the challenges this problem presents. Next, Section 3 presents background information on acquisition functions used in multi-objective Bayesian optimization. Section 4 then details the proposed multi-objective Bayesian robust optimization methodology. Section 5 discusses the example problems used to compare the proposed method with other methods from the literature along with the results and some analysis. Finally, Section 6 contains the conclusions drawn from the study.

2. PROBLEM STATEMENT AND CHALLENGES
The multi-objective robust optimization problem under interval uncertainty has the form seen in Eq. (1).

$$\begin{aligned} \min_{x} \; & f_i(x) & i \in 1, \dots, I \\ s.t. \; & g_j(x, u) \leq 0 & j \in 1, \dots, J \\ & \forall u \in [u_l, u_u] \\ & x \in [x_l, x_u] \end{aligned} \quad (1)$$

Here, the I objective functions to be minimized are given by f_i, the design variables are contained in the vector x which is bounded below and above by x_l and x_u respectively, the J

Copyright © 2023 by ASME

inequality constraints are g_j, and the uncertain parameters present in the constraints are given by $\boldsymbol{u}$ which is bounded below and above respectively by $\boldsymbol{u}_l$ and $\boldsymbol{u}_u$.

There are two main challenges for solving the problem in Eq. (1), particularly when using an acquisition function with surrogate models. The first challenge is that as the number of objective functions increases from two to three and beyond, the geometry dimension and complexity of the non-dominated solution frontier are increased from a curve (or piecewise linear) to a nonlinear hyper-surface (or hyper-plane). In Figure 1(a), a simple example of a solution set to a bi-objective optimization problem is shown. The non-dominated solution frontier in this case is piecewise linear and is along the black piecewise line. Moving to a three-objective optimization problem in Figure 1(b), the non-dominated solution frontier is represented by the grey surface. As such, the number of samples needed to approximate these geometries goes up exponentially as the number of objective functions is increased. Keeping the number of function evaluations low for such problems becomes even more imperative.

The second challenge that is faced in problems with more than two objective functions is the lack of ordering or arrangement of the solutions. In a bi-objective optimization problem, sorting the non-dominated solutions by one objective automatically sorts them by the second objective in reverse

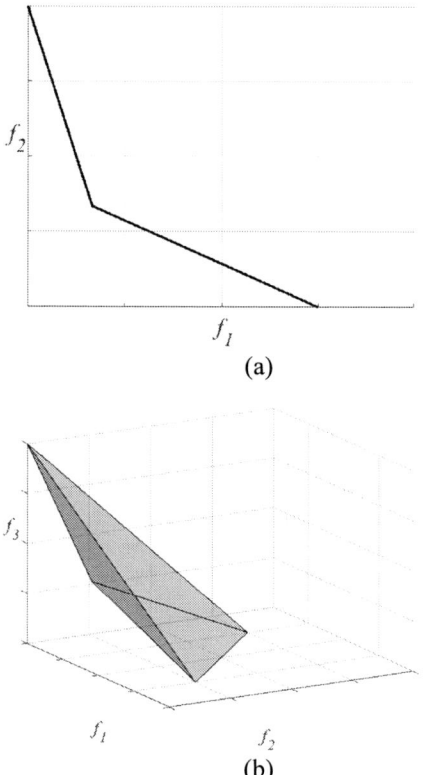

Figure 1: A set of optimal solutions along (a) a curve in a two-objective space and (b) a surface in a three-objective space.

order. Beyond two objectives, this property no longer holds and integration of acquisition functions over the non-dominated space requires a more sophisticated programmatic approach.

The next section will discuss two current acquisition functions for finding the next location to sample in a Bayesian optimization scheme.

3. ACQUISITION FUNCTIONS: BACKGROUND

3.1. Multi-Objective Centroidal Acquisition Functions

The non-dominated relaxation algorithm, as presented in [21], is applicable to robust optimization problems of any number of objective functions. However, the acquisition function used, the centroidal method as adapted from [4], only applies to problems with two objectives. Variations of the centroidal method and other acquisition functions are thus sought. Expansion of the centroidal method to higher numbers of objectives has been discussed before [12] and shown [7] but the formulations are not readily available. These existing methods require a partitioning of the non-dominated space, as seen in Figure 2, which can become intractable at higher dimensions.

The most straightforward method partitions the space into $(v + 1)^I$ cells over which the improvement metrics are calculated. Here, v is the number of non-dominated solutions, and I is the number of objectives [23]. This expression for the number of cells comes from dividing the non-dominated space by every combination of objective function values from the current non-dominated set. A demonstration in a two-objective space is shown in Figure 2 for a minimization problem. Here, there are six non-dominated solutions shown as black dots and two reference points B and G shown as empty circles. These are the "bad" point and the "good" point, respectively. The bad point should be chosen to be dominated by every solution, and the good point should dominate every solution. These points can generally be taken to be $(-\infty, \dots, -\infty)$ for a good point, and $(+\infty, \dots, +\infty)$ for a bad point, for a multi-objective minimization problem, as in Eq. (1). Each non-dominated solution defines the lower bound for a region in each dimension and the upper bound for a region in each dimension. The number of upper bounds on partitioned regions is then equal to the number of non-dominated solutions plus one for the bad point or $v + 1$. This is true for each objective function. The partitioned regions (or cells) are then the intersections of the regions bounded by the non-dominated solutions and the good and bad point. For I objective functions, this gives us $(v + 1)^I$. For the six non-dominated solutions in Figure 2, with two objective functions, there are $(6 + 1)^2 = 49$ regions. Note that this process partitions the entire objective space. It must then be established which of the cells are in the dominated and non-dominated regions. Faster partitioning methods have been developed to reduce the number of partitions needed [7], [24]–[26] as well as an approximation that ignores less impactful partitions [23].

Copyright © 2023 by ASME

One method to avoid dependency on the partitioning is with a Monte Carlo sampling. A method for doing so is discussed in this article.

3.2. Multi-Objective Expected Hypervolume Improvement

Another acquisition function which has received much attention is the expected hypervolume improvement. The Walking Fish Group (WFG) method was originally developed as a means of efficiently calculating the hypervolume difference [25], but has also been adapted for partitioning the non-dominated region [7]. It utilizes a branch-and-bound procedure based on the principle that cells which dominate a non-dominated cell will also be non-dominated. This allows for cells to be grouped to form fewer, larger cells without exploring each one.

4. METHODOLOGY

The proposed methodology, the Centroidal Monte Carlo (CMC) method, adopts the robust optimization scheme with non-dominated relaxation from [21] and extends it to multi-objective optimization problems. The non-dominated relaxation method is combined with a general Monte Carlo version of the centroidal expected improvement acquisition function, termed Centroidal Monte Carlo (CMC), and is presented here for multiple objectives. Instead of integrating the probability distribution function over the entire non-dominated domain with exponentially increasing numbers of cells, a finite set of samples can be taken using CMC as described here. Figure 3 shows an example sampling stage for CMC with two objectives. Accordingly, the steps in the CMC method are presented below. For clarity and simplification, constraint considerations are omitted for the time being.

1. Train Gaussian Process (GP) model on given training set X of designs x and the corresponding set of objective function values F for each of the objective functions.
2. Generate a candidate design x_c to be evaluated.

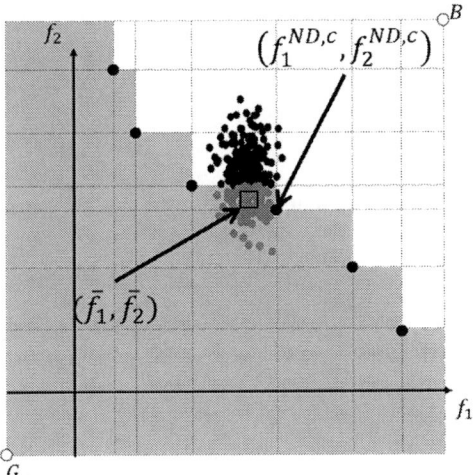

Figure 3: Demonstration of the Centroidal Monte Carlo method for multi-objective expected improvement

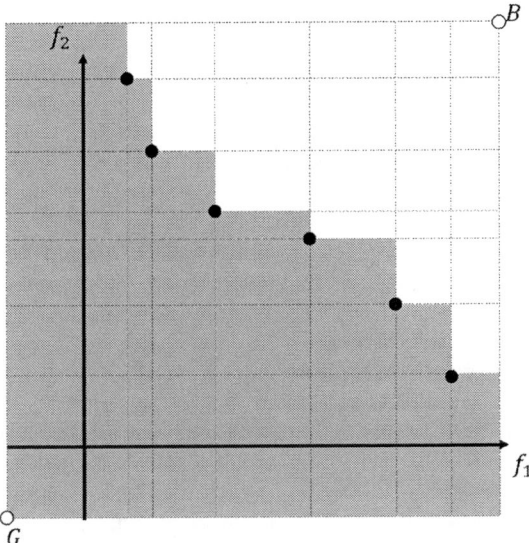

Figure 2: An example of a partitioning of the objective space based on the current non-dominated solutions and a "good" point G and a "bad" point B. The number of partitions is $(v + 1)^I = (6 + 1)^2 = 49$

3. Draw samples $f_{candidate}$ from the multi-variate normal distribution ($Normal$) described by the predicted output of the surrogate models of the objective functions at point x_c as in Eq. (2).

$$f_{candidate} \sim Normal\big(\boldsymbol{\mu}(x_c), \boldsymbol{\sigma}(x_c)\big) \qquad (2)$$

Here, $\boldsymbol{\mu}(x)$ and $\boldsymbol{\sigma}(x)$ are I dimensional vectors of the mean prediction and error (or standard deviation) prediction of the surrogates for the I objective functions at a given configuration of designs x. These random samples are demonstrated in Figure 3 (for illustration of a two-objective case) as the small points (both black and red).

4. Filter samples according to their relationship to the current non-dominated frontier as either "non-dominated" or "dominated". In Figure 3, the non-dominated random samples are shown in red.
5. Integrate the positions of the non-dominated samples to find the centroid of the improvement region $\bar{f}$. This is indicated by the black square labelled $\big(\bar{f}_1, \bar{f}_2\big)$ in Figure 3.
6. Calculate the probability of improvement (probability that the predicted mean value will be on the non-dominated frontier) as the fraction of $f_{candidate}$ samples that are non-dominated.
7. Calculate the expected improvement as in Eq. (3) as the product of Euclidean distance between the centroid $\bar{f}$ and the closest existing non-dominated solution, $f^{ND,c}$ in Figure 3, and the probability of improvement $P[Imp(x)]$.

Copyright © 2023 by ASME

$$E[Imp(\pmb{x})] = P[Imp(\pmb{x})] \sqrt{\sum_i \left(\bar{f_i} - f_i^{ND,c}\right)^2} \qquad (3)$$

8. Maximize the expected improvement in Eq. (3) to identify the next design to evaluate. Add this new design to $\pmb{X}$ and its corresponding objective values to $\pmb{F}$.
9. Return to 1. and repeat until computational budget is exhausted.

This acquisition function can then be combined with the non-dominated relaxation method [21], to find robust multi-objective solutions by replacing the previous expected improvement calculations. The non-dominated relaxation method iteratively trains surrogate models, solves the robust optimization problem for the known scenarios, updates models and estimates of feasible points, and then repeats. The resulting method uses Gaussian process surrogates, trained on the previously gathered data, to approximate the objective and constraint function values at a candidate design.

For higher numbers of objective functions, the process is more difficult to visualize. Instead of the rectangular cells visible in Figure 3, the cells become volumes or hypervolumes for more objective functions. While the numbers of cells to integrate over for analytical acquisition functions increases rapidly, the CMC follows the same procedure by only checking for non-domination.

Constraints can now be added back in. GP models are constructed in the same manner for the constraint functions using the sampled constraint values in $\pmb{G}$. They are handled using the strategy in Eq. (4)

$$\hat{g}(\pmb{x}, \pmb{u}) - a\hat{s}(\pmb{x}, \pmb{u}) \le 0 \qquad (4)$$

relaxing the mean constraint value in proportion to the prediction uncertainty of the GP by a factor of a. For this work, the proportionality constant a was taken to be 1. Optimizing the CMC acquisition function, a point which will be feasible for all known scenarios and is expected to improve the non-dominated set the most is sampled. The approximated worst-case constraint violation is then solved for, producing a design and scenario pair with the optimal design and its worst-case scenario. The objective and constraint GP models are retrained with the new sample point (the design and scenario pair) and the scenario is added to the list of worst-case scenarios. The updated models are then used to estimate which of the current feasible points are not robust and expected to dominate a robust point. These are removed for the current iteration. The expected centroidal improvement over the remaining non-dominated points can then be calculated, starting the procedure over.

Several examples and their results for the above method and for some methods from the literature are provided in the next section.

5. EXAMPLES

In this section several examples are provided for optimization problems with multiple objectives and parameters with interval uncertainty in the constraint functions. The first two problems are examples from [27], which can be tested at varying numbers of design variables and/or objectives and constraints. Here, both problems are taken to have three objective functions. The third problem is a modification of the engineering example used in [21], the optimization of a three-bar truss. The methods being compared are the robust optimization loop described in [21] with the proposed CMC acquisition function, as proposed in this work, an exact calculation of the centroidal method, or Exact Centroid (EC) [4], [5] for multiple objective functions [7], and the expected hypervolume improvement (EHVI) using the Walking Fish Group (WFG) algorithm [25]. Additionally, Approximation Assisted – Multi-Objective Robust Optimization (AA-MORO) [18] and the deterministic Multi-Objective Robust Optimization (MORO) [18] are used to solve the robust optimization problems.

Solutions of problems with three or more objectives are difficult to compare by visual inspection, so instead several quality metrics are used to describe the results. The metrics used here are the same as those described in [21], namely Hypervolume Difference (HD) to measure convergence based on the size of the non-dominated region with smaller values being better [28], Overall Spread (OS) to measure spread as the hypervolume contained between the furthest end points with larger values being better [28], and Spacing (SP) to measure the regularity of the solutions as the standard deviation of the distances between subsequent non-dominated solutions with smaller values being better [29]. Additionally, the number of non-dominated robust solutions is reported for each method. All values are reported with their mean and standard deviation over the five independent runs of each method. The mean and standard deviation were then used to perform a two-sample z-test with a significance level of 0.05 to test if the differences in performance between the metrics for the proposed CMC and other tested methods were statistically significant. Except where otherwise stated, the differences were statistically significant.

The deterministic MORO [18] method used MATLAB's 'gamultiobj' [30] function with all default values to solve the robust optimization problem. All four surrogate methods used an initial sample size equal to: $\frac{(2N+1)(2N+2)}{2} + 1$, where N is the total number of design variables and uncertain parameters. Each method was run for five iterations, adding five new sample points each iteration. Verification was also performed for all results by testing the constraints for feasibility against 10,000 random scenarios. Results from the verification study are omitted because all solutions for all methods tested were found to be fully feasibly robust, i.e., feasible under all scenarios.

Computational expense was measured in function calls or evaluations. This is not reported in the results because the surrogate-based methods are all allotted the same number of function calls and the non-surrogate-based method (MORO) always used several orders of magnitude more function calls. For instance, in Example 1 with 5 variables all of the surrogate-

Copyright © 2023 by ASME

based methods used 282 function calls while MORO averaged substantially more with 4,949,870 function calls.

One thing to note in the results is that the OS values are greater than 1 for some examples. This is not typical since it is generally defined as the ratio of the differences between the extreme non-dominated solution frontier and the differences between finite ideal and bad points [28]. To create a fair comparison, values for these points were set in advance. The extreme solutions for all of the methods exceeded the pre-determined ideal and bad points which led to the above results lying outside of the 0 to 1 interval. This indicates that the initial choice of the ideal and bad point could have been better selected to suit the problem. In any case, the larger the OS value, the better it is.

5.1. Example 1: DTLZ8

The DTLZ8 problem is the first of two constrained variable-size test problems introduced in [27] which can be solved for different numbers of objective functions and design variables. There are I objective functions, I constraints, N variables, and N uncertain parameters. It has been modified here to have uncertainty in each of the design variables. The formulation is provided in Eq. (5) . The constant c is the uncertainty in the design variable. Here the value of c is 0.1.

$$
\begin{aligned}
&\min_{x} \ f_i(\boldsymbol{x}) && i \in 1, \dots, I \\
&s.t. \ \ g_i(\boldsymbol{x}, \boldsymbol{u}) \le 0 && i \in 1, \dots, I \\
&\quad u_n \in [-c, c] \\
&\quad x_n \in [0 + c, 1 - c] \\
&\quad \bar{x}_n = x_n + u_n && n \in 1, \dots, N
\end{aligned}
$$

where:

$$
f_i(\boldsymbol{x}) = \frac{1}{\left\lfloor \frac{d}{I} \right\rfloor} \sum_{n = \left\lfloor (i-1)\frac{d}{I} \right\rfloor}^{\left\lfloor i\frac{d}{I} \right\rfloor} x_n \tag{5}
$$

$$
g_i(\boldsymbol{x}, \boldsymbol{u}) = 1 - f_I(\overline{\boldsymbol{x}}) - 4 f_i(\overline{\boldsymbol{x}}), \ i = 1, \dots, (I - 1)
$$

$$
g_I(\boldsymbol{x}, \boldsymbol{u}) = 1 - 2 f_I(\overline{\boldsymbol{x}}) - \min_{\substack{n \ne i \\ n, i \in 1, \dots, I-1}} [f_n(\overline{\boldsymbol{x}}) + f_i(\overline{\boldsymbol{x}})]
$$

The DTLZ8 problem was solved for three objective functions and three constraints. Tests were performed for the case with five variables and five uncertain parameters (uncertain variables) and the case with ten variables and ten uncertain parameters (uncertain variables). The results are given in Table and Table 4 respectively.

Recalling that for HD lower values are better, for OS higher values are better, for SP lower values are better, and more solutions are better, the proposed CMC approach combined with the non-dominated relaxation algorithm had the worst HD value and second worst values for the other quality

metrics. The second method tested, EHVI combined with the non-dominated relaxation algorithm performed similarly but with better HD. The two methods from the literature, AA-MORO and MORO, generally performed the best.

For the ten-variable case, we see two general trends. First, differences in performance for all methods decreased for all parameters except OS. The second general trend is that OS values for all of the methods are an order of magnitude smaller, suggesting a clustering of the solutions for all methods. Other than the number of solutions and the spacing metric, the proposed CMC method again performed worse in comparison. One possible explanation of this performance could be due to DTLZ8 being the least complex of the provided examples. The emphasis on balancing exploration and exploitation in non-dominated relaxation algorithm may not be paying off in a region that is easier to exploit.

5.2. Example 2: DTLZ9

The DTLZ9 problem is the second of the constrained, variable-sized test problems from [27]. A version of this problem was previously modified to include uncertainty in [17], but it was used in a single-objective optimization setting. The form used here, in Eq. (6), has I objectives, $I - 1$ constraints, N design variables, and N uncertain parameters. The uncertainty is added to the design variables. The amount of uncertainty c is set to 0.1. Results for the three objective, five variable case are shown in Table , and for the three objective, ten variable case are shown in Table 4.

Table 1: Quantitative results for Example 1 as mean (standard deviation) for 5 variables

Approach	# Solutions	HD	OS	SP
CMC	52	0.73	2.39	0.09
	(3.35)	(0.01)	(0.10)	(0.01)
EHVI [25]	34.4	0.69	2.05	0.14
	(4.45)	(0.03)	(0.36)	(0.01)
AA-MORO [18]	68.6	0.70	2.56	0.08
	(2.3)	(0.01)	(0.13)	(0.00)
MORO [18]	67.8	0.65	2.41	0.07
	(1.72)	(0.01)	(0.15)	(0.01)

Table 2: Quantitative results for Example 1 as mean (standard deviation) for 10 variables

Approach	# Solutions	HD	OS	SP
CMC	74.8	0.61	0.56	0.05
	(7.93)	(0.01)	(0.04)	(0.01)
EHVI [25]	44	0.61	0.47	0.06
	(6.4)	(0.01)	(0.05)	(0.01)
AA-MORO [18]	68.8	0.58	0.76	0.05
	(1.85)	(0.03)	(0.12)	(0.01)
MORO [18]	69.6	0.57	0.83	0.05
	(0.8)	(0.01)	(0.01)	(0.00)

Copyright © 2023 by ASME

$$\min_{x} f_i(\boldsymbol{x}) \qquad i \in 1, \dots, I$$
$$s.t. \ g_i(\boldsymbol{x}, \boldsymbol{u}) \le 0 \qquad i \in 1, \dots, I-1$$
$$\forall u_n \in [-c, c],$$
$$x_n \in [0+c, 1-c]$$
$$\bar{x}_n = x_n + u_n \qquad n \in 1, \dots, N$$

where: $\hspace{6cm}$ (6)

$$f_i(\boldsymbol{x}) = \sum_{n=\left\lfloor (i-1)\frac{d}{I} \right\rfloor}^{\left\lfloor i\frac{d}{I} \right\rfloor} x_n^{0.1}$$

$$g_i(\boldsymbol{x}, \boldsymbol{u}) = 1 - f_i^2(\bar{\boldsymbol{x}}) - f_i^2(\bar{\boldsymbol{x}})$$

Compared to DTLZ8, the DTLZ9 problem is more complicated due to its nonconvex constraints. In this case, the HD, OS, and SP metrics were all very close. However, the three acquisition functions used with the non-dominated relaxation algorithm all produced low numbers of solutions compared to MORO and AA-MORO.

Moving to the ten-variable case, the EC outperformed the others in terms of HD. The proposed approach shows an advantage in OS but the improvement over EHVI is not statistically significant. MORO had the best SP value while EC had the worst. The number of solutions is again low for the methods used with the non-dominated relaxation algorithm, but the CMC method performed better than the other two such methods.

Next, the DTLZ9 problem is expanded to four objectives and 10 variables, and the results are shown in Table 5. The HD values show CMC on par with AA-MORO and MORO performing slightly better. The proposed CMC performed the best for OS and only trailed behind AA-MORO at a level that was not statistically significant. Compared to the previous, smaller, cases the relative performance of CMC is noticeably improved. Similar to the bi-objective case discussed in [21], this starts to hint at advantages in problems of increased size or complexity and needs to be investigated with more problems of larger sizes.

Table 3: Quantitative results for Example 2 as mean (standard deviation) for 5 variables

Approach	# Solutions	HD	OS	SP
CMC	26.4	0.92	3.37	0.03
	(3.35)	(0.01)	(0.36)	(0.01)
EC [7]	18.4	0.93	3.86	0.06
	(5.00)	(0.00)	(0.04)	(0.02)
EHVI [25]	20	0.91	3.57	0.04
	(1.67)	(0.00)	(0.07)	(0.01)
AA-MORO [18]	70	0.93	3.78	0.02
	(0)	(0.01)	(0.08)	(0.01)
MORO [18]	70	0.93	3.22	0.01
	(0)	(0.00)	(0.39)	(0.00)

Table 4: Quantitative results for Example 2 as mean (standard deviation) for 10 variables

Approach	# Solutions	HD	OS	SP
CMC	41.67	0.81	2.24	0.03
	(11.79)	(0.00)	(0.03)	(0.01)
EC [7]	15.5	0.77	1.9	0.14
	(4.03)	(0.01)	(0.01)	(0.03)
EHVI [25]	25.8	0.81	2.22	0.06
	(5.9)	(0.00)	(0.03)	(0.01)
AA-MORO [18]	70	0.79	2.00	0.02
	(0)	(0.01)	(0.04)	(0.01)
MORO [18]	70	0.79	1.97	0.01
	(0)	(0.00)	(0.02)	(0.00)

Table 5: Quantitative results for Example 2 as mean (standard deviation) for 4 objectives and 10 variables

Approach	# Solutions	HD	OS	SP
CMC	112.5	0.89	0.98	0.04
	(17.89)	(0.00)	(0.01)	(0.01)
AA-MORO [18]	71.4	0.89	0.96	0.04
	(2.8)	(0.00)	(0.02)	(0.00)
MORO [18]	70	0.88	0.91	0.05
	(0)	(0.01)	(0.01)	(0.01)

5.3. Example 3: Three Bar Truss

The three-bar truss problem is used here as an engineering example. It was originally formulated in [31] for a single objective. The formulation was modified to include uncertainty in [32], and a second objective was added in [33]. It has been further modified here to allow for a third objective to be optimized. In addition to minimizing the horizontal tip deflection u and the volume V of members 1 and 3, the stress in the first member σ_1 should also be minimized. The corresponding constraint on the stress in the first member is also removed. The new formulation is provided in Eq. (7) with three objective, four constraints, two design variables, and two uncertain parameters.

$$\min_{A_1, \theta} f_1 = u(A_1, \theta)$$
$$\min_{A_1, \theta} f_2 = V(A_1, \theta)$$
$$\min_{A_1, \theta} f_3 = \sigma_1(A_1, \theta)$$
$$s.t. \ g_1 = \sigma_2 \le \sigma_{yield}$$
$$g_2 = v \le v_{max}$$
$$g_3 = |u(A_1, \theta) - u(A_1 + \Delta A_1, \theta + \Delta \theta)| \quad (7)$$
$$\le \Delta f_1$$
$$g_4 = |V(A_1, \theta) - V(A_1 + \Delta A_1, \theta + \Delta \theta)|$$
$$\le \Delta f_2$$
$$2 \le A_1 \le 10 \ cm^2, \quad 30 \le \theta \le 40°$$
$$-0.1 \le \Delta A_1 \le 0.1 \ cm^2, -5° \le \Delta \theta \le 5°$$

Additionally, in the context of this problem, v is the vertical tip deflection, σ_2 is the stress in the second member, A_1, A_2 are the areas of the first and second members, θ is the angle of the

Copyright © 2023 by ASME

applied load P, and l is the height of the truss. The definitions of all calculated values are given in Eq. (8) with constants $l = 1\,m, P = 100\,kN\,GPa, E = 70\,GPa$, and $A_2 = 2\,cm^2$.

$$u = \frac{\sqrt{2}lP\cos(\theta)}{A_1 E}$$
$$v = \frac{\sqrt{2}lP\sin(\theta)}{(A_1 + \sqrt{2}A_2)E}$$
$$\sigma_1 = \frac{P}{\sqrt{2}}\left[\frac{\cos(\theta)}{A_1} + \frac{\sin(\theta)}{A_1 + \sqrt{2}A_2}\right] \qquad (8)$$
$$\sigma_2 = \frac{\sqrt{2}P\sin(\theta)}{(A_1 + \sqrt{2}A_2)}$$
$$V = 2\sqrt{2}A_1 l + A_2 l$$

The results for the three objective three-bar truss problem are provided in Table 6. The CMC and EHVI methods produced the greatest number of solutions and had the greatest OS values. The EHVI acquisition function combined with the non-dominated relaxation algorithm also had the best average HD value. Other than more irregular spacing of the solutions, EHVI appears to have generally performed the best for this example.

6. CONCLUSION

This paper contributes to the existing literature along the following directions. Firstly, a new Centroidal Monte Carlo (CMC) method for calculating the centroidal improvement acquisition function was constructed and tested. The proposed CMC approach does not rely on a partitioning of the non-dominated domain. This is beneficial because the partitioning alone can be computationally expensive, and cumbersome, as the number of objective functions increases.

Secondly, a comparison of multiple different acquisition functions in the context of surrogate multi-objective feasibility robust optimization was conducted. The acquisition functions considered for the comparison were the proposed CMC method, an exact centroidal method, and the expected hypervolume improvement. Each of these was inserted into the non-dominated relaxation algorithm and applied to several example problems. These methods were also compared with two methods from the literature, AA-MORO, which did not use an acquisition function, and MORO, which did not use surrogates. Insight is thus gained into their applicability in a robust optimization setting.

The results from the study show that there is little difference in performance between the methods which use an acquisition function, i.e., the proposed CMC versus EC, and EHVI, for most cases. For the simplest of the examples, DTLZ8, the population-based methods from the literature, AA-MORO and MORO, performed the best. For DTLZ9 and the three-bar truss engineering problem, the three methods based on the non-dominated relaxation algorithm started to show relative

Table 6: Quantitative results for Example 3 as mean (standard deviation)

Approach	# Solutions	HD	OS	SP
CMC	77.6	0.11	2.66	0.05
	(1.85)	(0.01)	(0.01)	(0.01)
EHVI [25]	74.5	0.004	2.66	0.09
	(2.5)	(0.03)	(0.01)	(0.02)
AA-MORO [18]	70	0.11	2.45	0.05
	(0)	(0.03)	(0.04)	(0.01)
MORO [18]	70	0.10	2.61	0.05
	(0)	(0.02)	(0.04)	(0.00)

improvement over MORO and AA-MORO, particularly with respect to the HD metric. Based on that trend, and the results of [21], it seems likely that the proposed methods would show improved performance for problems of greater dimension (more variables, parameters, objectives, and constraints) and complexity (greater nonlinearities in the functions). This is because the population-based methods seen here are more exploitation focused and thus do very well in simpler problems. The more balanced approach of the Bayesian optimization methods will commit more resources to exploration which may be less beneficial for such problems. It can also be inferred that the non-dominated relaxation algorithm is not particularly sensitive to the choice of acquisition function.

This work would benefit from future studies on wall clock time, relating the number of samples each iteration to the number of objectives, larger problem sizes, and more diverse acquisition functions. It is worth noting that all of the quality metrics can only give insight into the specific values that they measure. Different decision makers and applications may value some metrics more over others.

ACKNOWLEDGEMENTS

This study was based on the dissertation work [22] of the first author, Dr. Randall Kania, while he was a Ph.D. student with the Department of Mechanical Engineering at the University of Maryland, College Park. The work presented in this paper was supported in part by a gift from Dr. Alex Mehr (PhD '03) through the Design Decision Support Laboratory Research and Education Fund. The authors also acknowledge the University of Maryland supercomputing resources (http://hpcc.umd.edu) made available for conducting the research reported in this paper.

REFERENCES

[1] V. Chankong and Y. Y. Haimes, *Multiobjective Decision Making: Theory and Methodology*. in North Holland series in system science and engineering, no. 8. New York: North Holland, 1983.

[2] K. Deb, *Multi-Objective Optimization using Evolutionary Algorithms*. Chichester, England: John Wiley & Sons, 2001.

[3] K. Deb, A. Pratap, S. Agarwal, and T. Meyarivan, "A Fast and Elitist Multiobjective Genetic Algorithm: NSGA-II,"

Copyright © 2023 by ASME

IEEE Trans. Evol. Comput., vol. 6, no. 2, pp. 182–197, Apr. 2002, doi: 10.1109/4235.996017.

[4] A. J. Keane, "Statistical Improvement Criteria for Use in Multiobjective Design Optimization," *AIAA J.*, vol. 44, no. 4, pp. 879–891, Apr. 2006, doi: 10.2514/1.16875.

[5] A. Forrester, A. Sobester, and A. Keane, *Engineering Design Via Surrogate Modelling: A Practical Guide.* John Wiley & Sons, 2008.

[6] D. R. Jones, M. Schonlau, and W. J. Welch, "Efficient Global Optimization of Expensive Black-Box Functions," *J. Glob. Optim.*, vol. 13, no. 4, pp. 455–492, 1998, doi: 10.1023/A:1008306431147.

[7] I. Couckuyt, D. Deschrijver, and T. Dhaene, "Fast Calculation of Multiobjective Probability of Improvement and Expected Improvement Criteria for Pareto Optimization," *J. Glob. Optim.*, vol. 60, no. 3, pp. 575–594, Nov. 2014, doi: 10.1007/s10898-013-0118-2.

[8] L. Shu, P. Jiang, X. Shao, and Y. Wang, "A New Multi-Objective Bayesian Optimization Formulation with the Acquisition Function for Convergence and Diversity," *J. Mech. Des.*, vol. 142, no. 9, p. 091703, Sep. 2020, doi: 10.1115/1.4046508.

[9] D. Zhan and H. Xing, "Expected Improvement for Expensive Optimization: A Review," *J. Glob. Optim.*, vol. 78, no. 3, pp. 507–544, Nov. 2020, doi: 10.1007/s10898-020-00923-x.

[10] S. Rojas-Gonzalez and I. Van Nieuwenhuyse, "A Survey on Kriging-Based Infill Algorithms for Multiobjective Simulation Optimization," *Comput. Oper. Res.*, vol. 116, p. 104869, Apr. 2020, doi: 10.1016/j.cor.2019.104869.

[11] J. Knowles, "ParEGO A Hybrid Algorithm with Online Landscape Approximation for Expensive Multiobjective Optimization Problems," *IEEE Trans. Evol. Comput.*, vol. 10, no. 1, pp. 50–66, 2006.

[12] M. T. M. Emmerich, A. H. Deutz, and J. W. Klinkenberg, "Hypervolume-Based Expected Improvement: Monotonicity Properties and Exact Computation," in *2011 IEEE Congress of Evolutionary Computation (CEC)*, New Orleans, LA, USA: IEEE, Jun. 2011, pp. 2147–2154. doi: 10.1109/CEC.2011.5949880.

[13] T. Wagner, M. Emmerich, A. Deutz, and W. Ponweiser, "On Expected-Improvement Criteria for Model-based Multi-objective Optimization," in *Parallel Problem Solving from Nature, PPSN XI*, R. Schaefer, C. Cotta, J. Kołodziej, and G. Rudolph, Eds., Berlin, Heidelberg: Springer Berlin Heidelberg, 2010, pp. 718–727. doi: 10.1007/978-3-642-15844-5_72.

[14] W. Ponweiser, T. Wagner, D. Biermann, and M. Vincze, "Multiobjective Optimization on a Limited Budget of Evaluations Using Model-Assisted S-Metric Selection," p. 11.

[15] W. Liu, Q. Zhang, E. Tsang, C. Liu, and B. Virginas, "On the Performance of Metamodel Assisted MOEA/D," in *Advances in Computation and Intelligence*, L. Kang, Y. Liu, and S. Zeng, Eds., in Lecture Notes in Computer Science, vol. 4683. Berlin, Heidelberg: Springer Berlin

Heidelberg, 2007, pp. 547–557. doi: 10.1007/978-3-540-74581-5_60.

[16] A. Ben-Tal, L. El Ghaoui, and A. Nemirovski, "Robust Optimization." Princeton University Press, Princeton, NJ, 2009. doi: 10.1515/9781400831050.

[17] E. Rudnick-Cohen, J. W. Herrmann, and S. Azarm, "Non-Convex Feasibility Robust Optimization Via Scenario Generation and Local Refinement," *J. Mech. Des.*, vol. 142, no. 5, p. 051703, May 2020, doi: 10.1115/1.4044918.

[18] W. Hu, M. Li, S. Azarm, and A. Almansoori, "Multi-Objective Robust Optimization under Interval Uncertainty Using Online Approximation and Constraint Cuts," *J. Mech. Des.*, vol. 133, no. 6, pp. 061002–061002, Jun. 2011, doi: 10.1115/1.4003918.

[19] S. Gunawan and S. Azarm, "Multi-Objective Robust Optimization Using a Sensitivity Region Concept," *Struct. Multidiscip. Optim.*, vol. 29, no. 1, pp. 50–60, Aug. 2004, doi: 10.1007/s00158-004-0450-8.

[20] M. Li, S. Azarm, and A. Boyars, "A New Deterministic Approach Using Sensitivity Region Measures for Multi-Objective Robust and Feasibility Robust Design Optimization," *J. Mech. Des.*, vol. 128, no. 4, pp. 874–883, Jul. 2006, doi: 10.1115/1.2202884.

[21] R. J. Kania and S. Azarm, "Bi-Objective Surrogate Feasibility Robust Design Optimization Utilizing Expected Non-Dominated Improvement With Relaxation," *J. Mech. Des.*, vol. 145, no. 3, p. 031703, Mar. 2023, doi: 10.1115/1.4055738.

[22] R. J. Kania, "Single- and Multi-Objective Feasibility Robust Optimization under Interval Uncertainty with Surrogate Modeling," Ph.D. Dissertation, University of Maryland, College Park, MD, 2022.

[23] I. Couckuyt, D. Deschrijver, and T. Dhaene, "Towards Efficient Multiobjective Optimization: Multiobjective statistical criterions," in *2012 IEEE Congress on Evolutionary Computation*, Brisbane, Australia: IEEE, Jun. 2012, pp. 1–8. doi: 10.1109/CEC.2012.6256586.

[24] R. Lacour, K. Klamroth, and C. M. Fonseca, "A Box Decomposition Algorithm to Compute the Hypervolume Indicator," *Comput. Oper. Res.*, vol. 79, pp. 347–360, Mar. 2017, doi: 10.1016/j.cor.2016.06.021.

[25] L. While, L. Bradstreet, and L. Barone, "A Fast Way of Calculating Exact Hypervolumes," *IEEE Trans. Evol. Comput.*, vol. 16, no. 1, pp. 86–95, Feb. 2012, doi: 10.1109/TEVC.2010.2077298.

[26] K. Dächert, K. Klamroth, R. Lacour, and D. Vanderpooten, "Efficient Computation of the Search Region in Multi-Objective Optimization," *Eur. J. Oper. Res.*, vol. 260, no. 3, pp. 841–855, Aug. 2017, doi: 10.1016/j.ejor.2016.05.029.

[27] K. Deb, L. Thiele, M. Laumanns, and E. Zitzler, "Scalable Test Problems for Evolutionary Multiobjective Optimization," in *Evolutionary Multiobjective Optimization: Theoretical Advances and Applications*, A. Abraham, L. C. Jain, and R. Goldberg, Eds., in Advanced

Copyright © 2023 by ASME

information and knowledge processing. New York: Springer, 2005, pp. 105–145.

[28] J. Wu and S. Azarm, "Metrics for Quality Assessment of a Multiobjective Design Optimization Solution Set," *J. Mech. Des.*, vol. 123, no. 1, p. 18, 2001, doi: 10.1115/1.1329875.

[29] J. R. Schott, "Fault Tolerant Design Using Single and Multicriteria Genetic Algorithm Optimization," Ph.D. Dissertation, Massachusetts Institute of Technology, Cambridge, Mass, 1995.

[30] "MATLAB (R2021b)." The Mathworks Inc., Natick, Massachusetts, 2022.

[31] L. A. Schmit, "Structural Design by Systematic Synthesis," in *Proceeds of the Second National Conference on Electronic Computation*, 1960.

[32] S. Gunawan, "Parameter Sensitivity Measures for Single Objective, Multi-Objective, and Feasibility Robust Design Optimization," Ph.D. Dissertation, University of Maryland, College Park, MD, 2004. Accessed: Aug. 18, 2015. [Online]. Available: http://drum.lib.umd.edu/handle/1903/1542

[33] T. Xie *et al.*, "Advanced Multi-Objective Robust Optimization under Interval Uncertainty Using Kriging Model and Support Vector Machine," *J. Comput. Inf. Sci. Eng.*, vol. 18, no. 4, p. 041012, Dec. 2018, doi: 10.1115/1.4040710.

Copyright © 2023 by ASME

Proceedings of the ASME 2023
International Design Engineering Technical Conferences and
Computers and Information in Engineering Conference
IDETC-CIE2023
August 20-23, 2023, Boston, Massachusetts

DETC2023-116614

ROBUST DESIGN FOR PRODUCT ADAPTATION CONSIDERING CHANGES IN CONFIGURATIONS AND PARAMETERS

Reza Deabae, Deyi Xue
Department of Mechanical and Manufacturing Engineering
University of Calgary
Calgary, Alberta, Canada T2N 1N4

ABSTRACT

This paper introduces a robust design method for product adaptation considering uncertainties in both product configurations and parameters. In this study, probability of product adaptation in the operation stage and influence of the probability on the optimal design solution are investigated. In this work, an AND-OR tree is used to model feasible design candidates and their adaptations, where each node represents a partial solution for the original design or the adapted design. Design candidates are generated from the AND-OR tree through tree-based search, and a design candidate can be defined by variation nodes that are used for potential product adaptations. A multi-level optimization method is applied to obtain the optimal values of design parameters for each design candidate and the best design solution from all feasible candidates. Both evaluation measures and their variations are considered in this robust design method.

Keywords: robust design, adaptable product, uncertainty, optimization.

1. INTRODUCTION

Functional performance measures are metrics used to evaluate a product's performance for achieving its intended purposes and functions. These measures are typically specific to the product's applications and are crucial to product development as they allow design engineers to evaluate product's performance and identify opportunities for improvement. These measures can be determined by design configurations and their parameters and used to compare different product designs and manufacturing processes [1]. The evaluation can help manufacturers optimize product performance while reducing production costs and improving customer satisfaction.

Variation in functional performance, defined as the deviation of a product's actual performance from its intended performance, is a critical factor in design evaluation. Even small deviations in functional performance can have significant consequences, including increased product failure rates, decreased customer satisfaction, and reduced profits. The sources of variations can be attributed to many factors such as manufacturing process variabilities, material properties, and changes in operating conditions. Uncertainties are significant contributors to functional performance variations in products. Uncertainties can be amplified in products with complex geometries, systems, and processes. The impact of uncertainties on functional performance variations can be quantified using statistical methods such as Design of Experiments (DoE), Monte Carlo simulation [2], and Bayesian Monte-Carlo simulation [3]. These methods enable manufacturers to identify and quantify the impact of various sources of uncertainties and design products that are robust to these uncertainties. Robustness refers to the ability of a product to perform reliably and consistently over a range of conditions or in the presence of uncertainties.

Uncertainty is due to the lack of complete knowledge or information about a particular situation or event, resulting in the inability to determine the true outcome or probability of the occurrence. It is an inherent characteristic of many systems and processes and is present in various fields, including science, engineering, economics, and finance. In engineering design and manufacturing, uncertainty can arise from many sources. The impact of uncertainty can result in significant performance variations, thus reducing the quality and reliability of a product. Uncertainties in parameters refer to the uncontrollable variations of product and operating parameters from the expected values such as the variation of temperature for a measurement instrument. Both design parameters (also called decision parameters) and constant parameters (also called non-decision

Copyright © 2023 by ASME

parameters) may have uncertainties. Uncertainties in configurations are unexpected changes in product structure such as to add or remove components for product upgrading to satisfy unexpected new requirements. One way to address configuration and parameter uncertainties is through robust design methods, which involve designing products that are resilient to potential variations in configurations as well as parameters and can perform reliably over a range of operating conditions.

Product adaptability refers to a product's ability to be modified or adjusted to meet changing customer needs, preferences, or environmental conditions in operation stages [4]. Product adaptability is the degree to which a product can be customized or upgraded to suit different applications or operating conditions without compromising its functionality or performance. Various methods have been developed recently to design and optimize adaptable products [4], such as modular design [5], robust design [6], function modeling [4,7], and design modeling [4]. Since design and manufacturing efforts to achieve the required functions are also reduced in adaptable design, this design approach can further improve the competitiveness of products in the marketplace.

Robust design approach [8] is effective to minimize the influence of uncertainties, such as variations of product and operating parameters, in the design of mechanical products and systems. In the Taguchi method [8], the sensitivity of the evaluation measure to variations in design parameters is often used to evaluate the robustness of the product or system. Since the introduction of this concept, significant contributions have been made in developing methodologies to handle uncertainties in design, particularly in design optimization [9]. In addition, robust design considering multiple functional performance measures has also been studied [10]. In our previous research, parameter uncertainties were considered in the robust design of product configurations and parameters [6, 11].

The concept of robust design for product adaptation that considers changes in configurations and parameters is essential to address uncertainties and variations in functional performance. Robust design methods aim to improve product quality by accounting for potential sources of variations during the design stage. By doing so, manufacturers can achieve robustness and adaptability of products to satisfy changing customer needs and environment conditions.

Despite the efforts, uncertainties in both the configurations and parameters were not well studied for design of adaptable products. In our previous work, parameter uncertainties were considered for robust design of configurations and parameters for design of adaptable products [6]. Uncertainties in configurations, such as the chance that a product can be upgraded in the future to satisfy new requirements, have never been investigated. Based on our preliminary research on uncertainties in configurations reported in a conference paper [12], the research summarized in this paper extends our previous work by developing a robust design method for product adaptation to identify the optimal design configurations and parameter values considering influences of uncertainties both in configurations and parameters.

2. ROBUST DESIGN CONSIDERING UNCERTAINTIES IN CONFIGURATIONS AND PARAMETERS

The robust design method aims at identifying the optimal design solution, including both configurations and parameter values, of the adaptable mechanical product or system considering influences of uncertainties both in its configurations and parameters. The identified optimal design is robust to changes in both product configurations and parameter values throughout its whole lifecycle. The optimal design is achieved by using a multi-level optimization approach through taking into account different design configurations (i.e., design candidates), variations of each design configuration (i.e., variation configurations for product adaptation), the probability of each variation configuration (i.e., the chance for product adaptation), design parameters (i.e., parameters to be determined in design and optimization), constant parameters (i.e., parameters that their values are given before the design process), and uncertainties in parameters (i.e., uncontrollable variations of parameters). Earlier results on robust design considering uncertainties in configurations and parameters were reported in a conference paper [12].

2.1. A Generic AND-OR Tree for Modelling Product Configurations and Their Configuration Uncertainties

From design requirements, a generic design AND-OR tree is employed to model feasible configuration candidates and their potential changes for product adaptations. Each node in the tree represents a partial design solution (i.e., a component or an assembly). Two types of nodes are defined in an AND-OR tree, namely nominal nodes and variation nodes. Nominal nodes represent components and assemblies for the initial product. Variation nodes, on the other hand, are components and assemblies that are potentially used for product adaptation through changes of product configurations during the product utilization stage to upgrade the product or to add a new function. A probability value between 0 and 1 is assigned to each variation node, representing the chance of the node to be used in the whole product life-cycle span. Since nominal nodes are always used, the probability of a nominal node is assigned as 1. An AND or OR relation is used to associate sub-nodes of a super-node. Figure 1 illustrates a sample AND-OR tree composed of nominal nodes, variation nodes, AND and OR relations, and probability values of nodes.

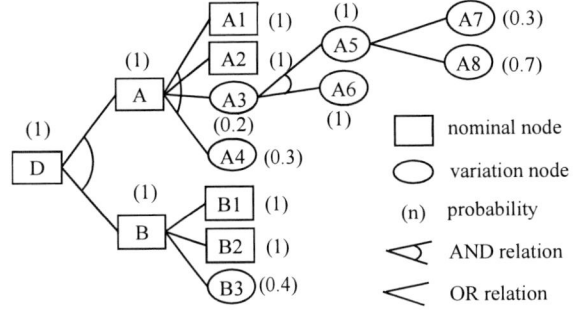

FIGURE 1: A generic product design AND-OR tree.

Copyright © 2023 by ASME

The following rules should be satisfied for modeling an AND-OR tree of an adaptable product:

- For a nominal super-node, when its sub-nodes are associated with an AND relation (e.g., Node A in Figure 1), at least one nominal sub-node should be defined. The probability of the nominal node should be assigned as 1, and the probability of a variation sub-node is between 0 and 1.
- For a variation super-node, when its sub-nodes are associated with an AND relation (e.g., Node A3 in Figure 1), all these sub-nodes should be variation nodes with probabilities assigned as 1s.
- For a nominal super-node, when its sub-nodes are associated with an OR relation (e.g., Node B in Figure 1), at least two nominal sub-nodes should be defined. A probability of 1 is assigned to a nominal sub-node while a probability between 0 and 1 is assigned to a variation sub-node. The summation of probabilities for all variation sub-nodes is equal to or less than 1.
- For a variation super-node, when its sub-nodes are associated with an OR relation (e.g., Node A5 in Figure 1), all the sub-nodes should be variation sub-nodes. A probability between 0 and 1 should be assigned to each sub-node while the summation of probabilities for all the sub-nodes is equal to or less than 1.

2.2. Parameters and Parameter Uncertainties

A node in the AND-OR tree is further defined by parameters. In this research, product design and operation parameters are classified into two categories, namely design parameters and constant parameters. Values of design parameters (also called decision parameters) need to be determined in the design process. Values of constant parameters are given before the design process. Both design parameters and constant parameters may have uncertainties defined by uncontrollable variations of parameter values.

- **Design Parameters (X_D):**

  Design parameters refer to variables used for optimization. The nominal values for design parameters are selected from $X_D^{(L)} \leq X_D \leq X_D^{(U)}$, where $X_D^{(L)}$ and $X_D^{(U)}$ are the lower and upper boundaries respectively. Design parameters may have uncertainties, meaning their actual values may deviate from their nominal values. The variations of design parameters (ΔX_D) are defined by ($\Delta X_D^{(L)} \leq \Delta X_D \leq \Delta X_D^{(U)}$), where $\Delta X_D^{(L)}$ and $\Delta X_D^{(U)}$ are lower and upper boundaries for design parameter variations from their nominal values.

- **Constant Parameters (C_D):**

  Nominal values for this group of parameters are considered constants throughout the optimization. However, actual values of these constant parameters may deviate from their nominal values due to uncertainties. The variations of constant parameters (ΔC_D) are defined by ($\Delta C_D^{(L)} \leq \Delta C_D \leq \Delta C_D^{(U)}$), where $\Delta C_D^{(L)}$ and $\Delta C_D^{(U)}$ are lower and upper boundaries for constant parameter variations from their nominal values.

2.3. Generation of Design Candidates and Their Variation Configurations for Product Adaptations

Tree-based search is used in this research to generate design candidates and their variation configurations for product adaptations from the generic AND-OR tree.

2.3.1. Generation of Design Candidates

Different design candidates are created from the AND-OR tree through tree-based search. These design candidates are evaluated, and the best one is selected as the optimal design. Figure 2 shows a design candidate generated from the generic AND-OR tree shown in Figure 1.

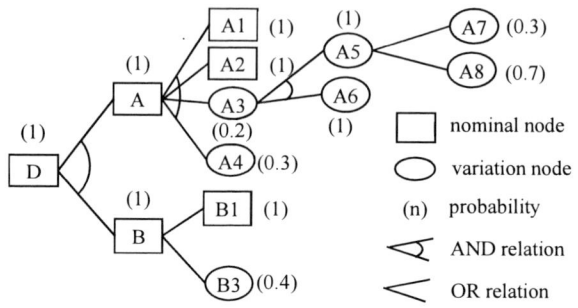

FIGURE 2: A design candidate created from the AND-OR tree shown in Figure 1.

Each design candidate represents a feasible configuration that can satisfy all design requirements. Each design candidate is also represented by an AND-OR tree.

The tree-based search to generate design candidates is carried out through the following rules:

(1) First, the root node should be selected.
(2) If the selected node is a nominal node and its sub-nodes are associated with an AND relation, all its sub-nodes should be selected.
(3) If the selected node is a variation node and its sub-nodes are associated with an AND relation, all its sub-nodes should be selected.
(4) If the selected node is a nominal node and its sub-nodes are associated with an OR relation, only one nominal sub-node and all of the variation sub-nodes should be selected.
(5) If the selected node is a variation node and its sub-nodes are associated with an OR relation, all its sub-nodes should be selected.

2.3.2. Generation of Variation Configurations for Product Adaptations

For each design candidate, variation configurations representing possible configuration changes for product adaptations in the product lifecycle span can be created through tree-based search. Each variation configuration is associated with a probability, representing the chance this product adaptation will happen in its product lifespan. Figure 3 depicts a variation configuration created from the design candidate shown in Figure 2.

Copyright © 2023 by ASME

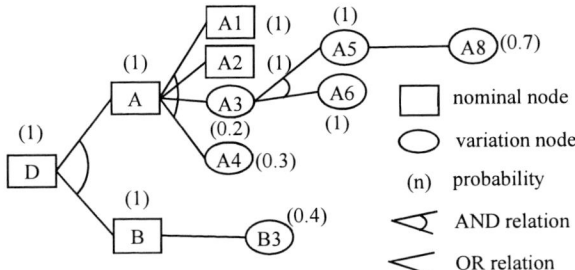

FIGURE 3: A variation configuration created from the design candidate shown in Figure 2.

The tree-based search for creation of variation configurations from a design candidate is conducted by the following process:

(1) An empty list should be created in order to record a variation configuration and the root node should be added to the list.
(2) Check whether all the nodes in all the lists have been checked. If all the nodes are checked, go to Step (6). Otherwise go to Step (3).
(3) Select an unchecked node from a list and mark it as checked. If the selected node is a leaf node (i.e., bottom node), go to Step (2). Otherwise go to Step (4).
(4) Based on the condition of the node, apply one of the following:
   (4a) If the selected super-node is a nominal node and its sub-nodes are associated with an AND relation:
   - Duplicate the current list to 2^n lists including the current list, where n is the number of variation sub-nodes.
   - Create 2^n permutations of the variation sub-nodes and add each permutation to a different list.
   - Add all the nominal sub-nodes to each of the duplicated lists.
   (4b) If the selected super-node is a variation node and its sub-nodes are associated with an AND relation:
   - Add all the variation sub-nodes to the current list.
   (4c) If the selected super-node is a nominal node and its sub-nodes are associated with an OR relation:
   - Duplicate the current list to n lists including the current list, where n is the number of sub-nodes.
   - Add each sub-node to a different duplicated list.
   (4d) If the selected super-node is a variation node and its sub-nodes are associated with an OR relation:
   - Duplicate the current list to n lists including the current list, where n is the number of sub-nodes.
   - Add each sub-node to a different duplicated list.
(5) Go to Step (2).
(6) All the variation configurations are created.

The variation configurations and their probabilities created from the design candidate given in Figure 2 are shown in Table 1. The first configuration has probability of 1, since it is the original product configuration before the product adaptation.

TABLE 1: Variation configurations created from design candidate in Figure 2.

Variation configuration	Probability	Weight
D, A, B, A1, A2, B1	1	0.4579
D, A, B, A1, A2, A3, A5, A6, A7, B1	0.06	0.0275
D, A, B, A1, A2, A3, A5, A6, A8, B1	0.14	0.0641
D, A, B, A1, A2, A4, B1	0.3	0.1374
D, A, B, A1, A2, A3, A4, A5, A6, A7, B1	0.018	0.0082
D, A, B, A1, A2, A3, A4, A5, A6, A8, B1	0.042	0.0192
D, A, B, A1, A2, B3	0.4	0.1832
D, A, B, A1, A2, A3, A5, A6, A7, B3	0.024	0.0110
D, A, B, A1, A2, A3, A5, A6, A8, B3	0.056	0.0256
D, A, B, A1, A2, A4, B3	0.12	0.0549
D, A, B, A1, A2, A3, A4, A5, A6, A7, B3	0.0072	0.0033
D, A, B, A1, A2, A3, A4, A5, A6, A8, B3	0.0168	0.0077

The probability of the variation configuration p_{ij} ($i = 1, 2, ..., n$; $j = 1, 2, ..., m_i$), representing the chance of this configuration to be used for the i-th design candidate considering product adaptation, is then calculated where n is the number of design candidates, and m_i is the number of variation configurations for the i-th design candidate. The probability, p_{ij}, for the j-th variation configuration of the i-th design candidate is calculated by multiplication of the probabilities of all the nodes (p_{ijk}) ($k = 1, 2, ..., q_{ij}$) for this variation configuration using:

$$p_{ij} = \prod_{k=1}^{q_{ij}} p_{ijk} \tag{1}$$

where q_{ij} is the number of nodes in this variation configuration.

Relative importance of a variation configuration for a design candidate, defined as weight of this variation configuration, w_{ij} ($i = 1, 2, ..., n$; $j = 1, 2, ..., m_j$), is calculated by:

$$w_{ij} = \frac{p_{ij}}{\sum_{j=1}^{m_i} p_{ij}} \tag{2}$$

2.4. Identification of the Optimal Adaptable Product Design Considering Uncertainties

The optimal design of the adaptable product with its variation configurations and parameter values is identified through multi-level optimization. Both evaluation measures and uncertainties in evaluation measures are considered in this robust design approach.

For each design candidate D_i ($i = 1, 2, ..., n$) and its all variation configurations V_{ij} ($i = 1, 2, ..., n$; $j = 1, 2, ..., m_i$), parameter optimization is conducted to obtain the optimal parameter values for this candidate.

First for each variation configuration V_{ij}, its robustness Z_{ij} is evaluated by

$$Z_{ij} = Z_{ij}(\boldsymbol{X_D}) \tag{3}$$

where $\boldsymbol{X_D}$ is a vector with nominal values of design parameters, $Z_{ij}()$ is the function or program to obtain the robustness of the

Copyright © 2023 by ASME

variation configuration from design parameters X_D, variations of design parameters ΔX_D, and variations of constant parameters ΔC_D. Robustness of a design candidate is defined from the robustness measures of individual variation configurations using the following two methods:

- **Average Method:**

In this method, robustness of the design candidate is defined by:

$$Z_i = Z_i(X_D) = \sum_{j=1}^{m_i}\left(w_{ij}Z_{ij}\right) \tag{4}$$

where w_{ij} are important factors of the variation configurations for the i-th design candidate obtained by Equation (2).

- **Worse-case Method:**

In this method, robustness of the design candidate is defined by:

$$Z_i = Z_i(X_D) = min\{Z_{i1}, Z_{i2,...,}Z_{im_i}\} \tag{5}$$

Parameter optimization is defined by:

$$\max_{w.r.t\ X_D} Z_i(X_D) \tag{6}$$

to obtain the optimal parameter values $X_{D_i}^*$ and its best robustness Z_i^*.

Among all n design candidates D_i ($i = 1, 2, ..., n$), the best one is obtained by configuration optimization:

$$\max_{w.r.t\ i} Z_i^* \tag{7}$$

Evaluation of the robustness for a configuration based on Equation (3), however, is a challenging task. In this research, various methods have been used and developed to evaluate robustness of a design configuration as discussed in Section 3.

3. EVALUATION OF ROBUSTNESS FOR DESIGN OF ADAPTABLE PRODUCTS

The methods to evaluate robustness of a design configuration are based on the evaluation measures and variations of evaluation measures calculated from design parameters X_D, variations of design parameters ΔX_D, and variations of constant parameters ΔC_D.

For a design configuration, its evaluation measure F can be calculated by:

$$F = F(X_D) = f(X_D, \Delta X_D, \Delta C_D) \tag{8}$$

where $f()$ is either a numerical function or a computer program. Suppose the target evaluation measure F_0 is calculated from:

$$F_0 = F_0(X_D) = f(X_D, 0, 0) \tag{9}$$

when variations of design parameters ΔX_D and variations of constant parameters ΔC_D are not considered. The variation of evaluation measure ΔF_0 from its target value can be calculated by:

$$\Delta F_0 = \Delta F_0(X_D) = F - F_0 = \Delta F(X_D, \Delta X_D, \Delta C_D) \tag{10}$$

3.1. Extreme Robustness Measures

The maximum value, the minimum value and the range of evaluation measure variation can be obtained by optimization:

$$\Delta F^{(max)}(X_D) = \max_{w.r.t\ \Delta X_D, \Delta C_D} \Delta F(X_D, \Delta X_D, \Delta C_D) \tag{11}$$

$$\Delta F^{(min)}(X_D) = \min_{w.r.t\ \Delta X_D, \Delta C_D} \Delta F(X_D, \Delta X_D, \Delta C_D) \tag{12}$$

$$\Delta F(X_D) = \Delta F^{(max)}(X_D) - \Delta F^{(min)}(X_D) \tag{13}$$

The extreme robustness measures are calculated by arithmetic method considering the following three cases.

- **Larger-the-Better Evaluation Measure:**

A larger evaluation measure is expected. Efficiency is a such evaluation measure. The robustness is defined by:

$$max(Z(X_D)) = max(F_0(X_D) + \Delta F^{(min)}(X_D)) \tag{14}$$

- **Smaller-the-Better Evaluation Measure:**

A smaller evaluation measure is expected. Error is a such evaluation measure. The robustness is defined by:

$$max(Z(X_D)) = max\left(\frac{1}{F_0(X_D)+\Delta F^{(max)}(X_D)}\right) \tag{15}$$

$$= min\left(F_0(X_D) + \Delta F^{(max)}(X_D)\right)$$

- **Nominal-the-Best Evaluation Measure:**

A target evaluation measure is expected. Voltage with value of 110 V is a such evaluation measure for a voltage convertor. The robustness is defined by:

$$max(Z(X_D)) = max\left(\frac{1}{max(|\Delta F^{(max)}(X_D)|,|\Delta F^{(min)}(X_D)|)}\right) \tag{16}$$

$$= min\left(max(|\Delta F^{(max)}(X_D)|, |\Delta F^{(min)}(X_D)|)\right)$$

3.2. Multi-objective Robustness Measures

Suppose the evaluation measure to be maximized is defined by $F_0(X_D)$ and its range of variation to be minimized is defined by $\Delta F(X_D)$, the robustness can be defined either by:

Copyright © 2023 by ASME

$$max(Z(\pmb{X_D})) = max\left(\alpha_1 F_0(\pmb{X_D}) + \alpha_2 \frac{1}{\Delta F(\pmb{X_D})}\right) \quad (17)$$

or

$$max(Z(\pmb{X_D})) = max\left(\frac{F_0(\pmb{X_D})}{\Delta F(\pmb{X_D})}\right) \quad (18)$$

where α_1 and α_2 are coefficients to integrate the two objective functions into a single one.

3.3. Taguchi Robustness Measures

Taguchi robustness measures [8] are calculated through statistical method. First, a large number m of samples considering parametric uncertainties ΔX_D and ΔC_D are created according to their probability distributions for the selected values of design variables X_D, and each sample k ($k = 1, 2, ..., m$) is evaluated by y_k. The robustness $Z(X_D)$ is calculated considering three different types of evaluation measures.

- **Nominal-the-Best Evaluation Measure:**

$$Z(\pmb{X_D}) = 10\log_{10}\left(\frac{\bar{y}^2}{s^2}\right) \quad (19)$$

where:

$$\bar{y} = \frac{1}{m}\sum_{k=1}^{m} y_k \quad (20)$$

$$s^2 = \frac{1}{m-1}\sum_{k=1}^{m}(y_k - \bar{y})^2 \quad (21)$$

- **Smaller-the-Better Evaluation Measure:**

$$Z(\pmb{X_D}) = -10\log_{10}\left(\frac{1}{m}\sum_{k=1}^{m} y_k^2\right) \quad (22)$$

- **Larger-the-Better Evaluation Measure:**

$$Z(\pmb{X_D}) = -10\log_{10}\left(\frac{1}{m}\sum_{k=1}^{m}\frac{1}{y_k^2}\right) \quad (23)$$

3.4. Robustness with Multiple Evaluation Measures

In case multiple evaluation measures F_r ($r = 1,2,...,q$) with different units are used, these evaluation measures are first converted into comparable evaluation indices I_r ($r = 1,2,...,q$) (e.g., satisfaction indices) as shown in Equation (24). The overall evaluation index I is calculated using Equation (25), where β_r is the weighting factor representing the importance of the r-th evaluation aspect.

$$I_r = I_r(F_r) \quad (24)$$

$$I = \frac{\sum_{r=1}^{q}(\beta_r I_r)}{\sum_{r=1}^{q} \beta_r} \quad (25)$$

4. PROCESS FOR ROBUST DESIGN OF ADAPTABLE PRODUCTS

The process for robust design of adaptable products considering uncertainties in configurations and parameters is illustrated in Figure 4.

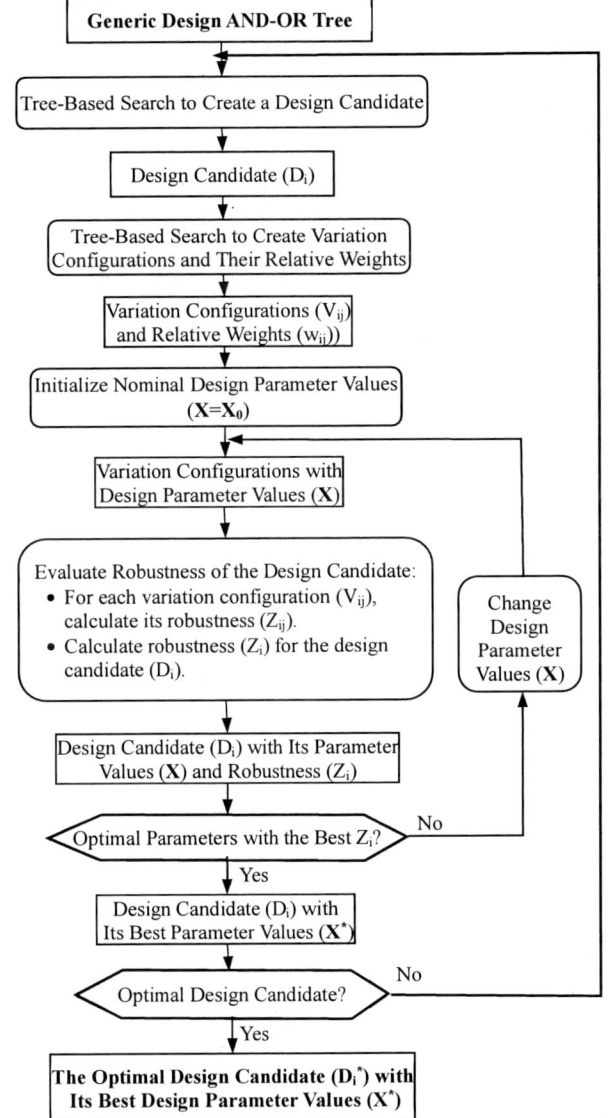

FIGURE 4: Process for robust design considering product adaptation.

Robust design of adaptable products is conducted by the following steps.

(1) From the generic design AND-OR tree, create a design candidate D_i (i = 1, 2, 3, ..., n) through tree-based search.

(2) For the selected design candidate D_i, create all the variation configurations V_{ij} (j = 1, 2, 3, ..., m_i) through tree-based search and calculate their importance factors w_{ij} (j=1, 2, 3, ..., m_i).

(3) For each design candidate, initialize nominal design parameter values $X=X_0$.

Copyright © 2023 by ASME

505

(4) For each variation configuration, generate noise parameters $\boldsymbol{\Delta X}$ and $\boldsymbol{\Delta C}$, and calculate its robustness Z_{ij}. Calculate the robustness Z_i for the design candidate D_i.

(5) Check whether the optimal design parameter values for the design candidate D_i with the best robustness Z_i (i.e., Z_i cannot be further improved by changing values of $\boldsymbol{X}$) has been achieved. If yes, go to Step (6). Otherwise, change the values of design parameters $\boldsymbol{X}$ and go to Step (4).

(6) For the design candidate D_i with its best design parameter values $\boldsymbol{X^*}$, check whether the optimal design candidate with the best robustness (i.e., Z_i cannot be further improved by changing design candidate) has been achieved. If yes, go to Step (7). Otherwise, go to Step (1) to create another design candidate.

(7) Record the best design candidate D_i^* and its best design parameter values $\boldsymbol{X^*}$.

5. A CASE STUDY

Design of a large-scale multi-axis and multi-function machine as shown in Figure 5 was conducted as a case study to demonstrate application of the developed method. The working space for this machine was 4 m × 4 m × 3 m. Different types of modules, such as printing, painting, and engraving units, can be attached to the head of the Z axis for different functions such as to print or paint on metal and wooden surfaces.

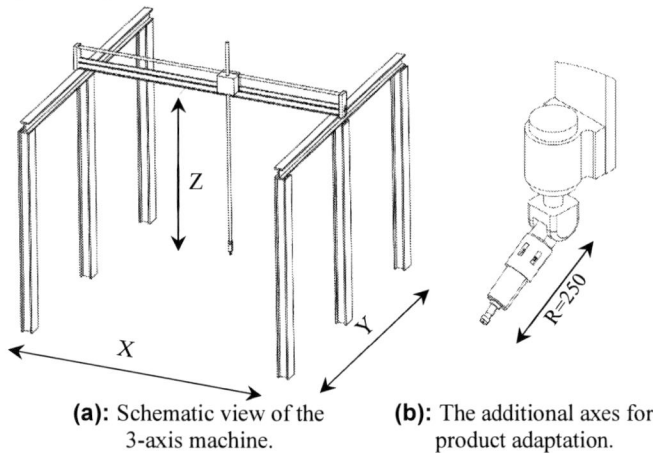

(a): Schematic view of the 3-axis machine.   **(b):** The additional axes for product adaptation.

FIGURE 5: Structure of the multi-axis machine before and after product adaptation.

5.1. Configurations, Parameters and Their Uncertainties

Two configurations were considered in this case study.

Variation Configurations:
- **Original configuration:** With 3 axes (i.e., X, Y and Z)
- **Adapted configuration:** With added 2 rotational axes (i.e., A and B)

Since the original configuration has to be provided, probability of this configuration was selected as 1. The adapted configuration, however, happens with certain chance during product utilization stage. Probability of the adapted configuration was selected as p = 0.35.

The front and right-side views of the X-axis driving mechanism are illustrated in Figure 6. In this case study, two design parameters were selected as shown in Figure 6.

Design Parameters:
- X_1 **(50 ≤ X_1 ≤ 150):** length of the thread block (mm)
- X_2 **(50 ≤ X_2 ≤ 100):** height of the guiding slot (mm)

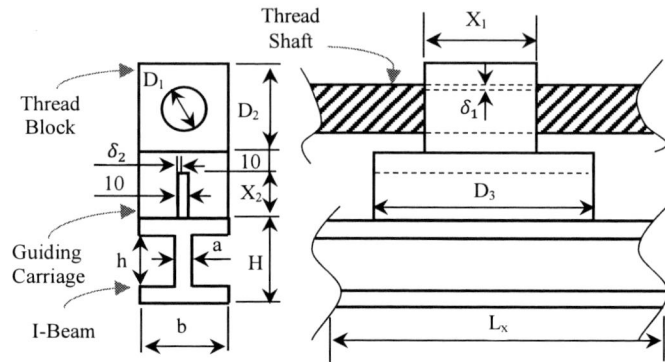

FIGURE 6: Front and right-side views of the X-axis driving mechanism.

The generic design AND-OR tree is shown in Figure 7(a). Two design candidates were created from the AND-OR tree with different shapes of the threads, and one of these design candidates is shown in Figure 7(b).

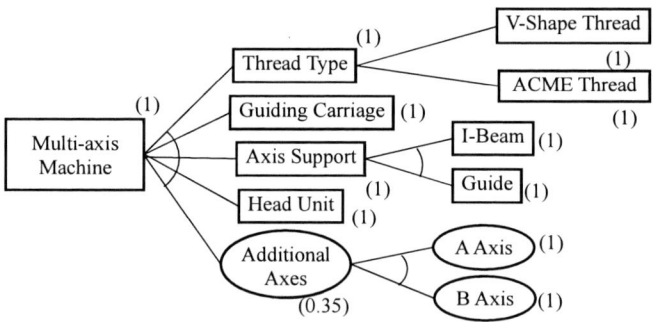

(a): The generic AND-OR tree for the multi-axis machine.

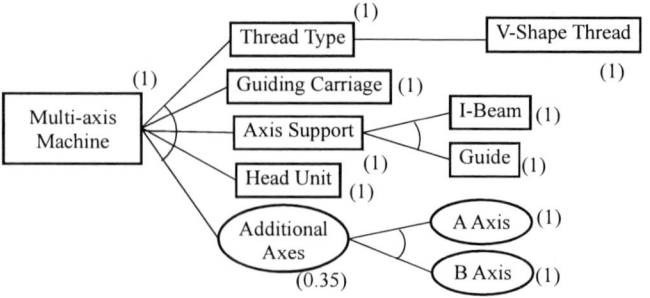

(b): One of the two created design candidates.

FIGURE 7: Modeling of design configurations.

Copyright © 2023 by ASME

Design Candidates:
- **Design Candidate 1:** With V-Shape Thread
- **Design Candidate 2:** With ACME Thread

In this case study, errors of the five axes were selected as the parameters with uncertainties. These parameters, and their lower and upper boundaries considering variations of values due to uncertainties are summarized in Table 2.

TABLE 2: Non-design parameters with uncertainties.

Parameter	Description	Variation Range
ε_x	Error for X axis	[0 mm, 0.1 mm]
ε_y	Error for Y axis	[0 mm, 0.1 mm]
ε_z	Error for Z axis	[0 mm, 0.05 mm]
ε_α	Error for A axis	[0°, 0.1°]
ε_β	Error for B axis	[0°, 0.05°]

Constant parameters without variations are summarized in Table 3.

TABLE 3: Non-design parameters without uncertainties.

Parameter	Description	Value
δ_{11}	Maximum clearance δ_l for the thread block due to tolerances for Design Candidate 1 with sharp V-shape threads	0.01 mm
δ_{12}	Maximum clearance δ_l for the thread block due to tolerances for Design Candidate 2 with ACME threads	0.05 mm
δ_2	Maximum clearance for the guiding slot due to tolerances	0.01 mm
W_3	Weight of Z-axis mechanism	3 kg
D_1	Diameter of thread hole	40 mm
D_2	Height of the thread block	60 mm
D_3	Length of carriage	250 mm
L_X	Length of X axis	4000 mm
L_Y	Length of Y axis	4000 mm
L_Z	Length of Z axis	3000 mm
R	Length of the linkage for the second rotation axis	250 mm
E	Modulus of the elasticity for beam material (steel)	200 GPa
ρ	Density of the material for the thread block and the guiding carriage	7.85 g/cm³
b	Width of I-Beam	60 mm
a	Wall thickness of I-Beam	20 mm
H	Total height of I-Beam	80 mm
h	Inner height of I-Beam	50 mm

5.2. Robust Design of the Adaptable Product

The maximum total volumetric error ΔL at the end position of the head unit was selected as the robustness to be evaluated. This robustness measure is a kind of extreme robustness measure. A volumetric error is the absolute distance value between the expected position and the actual position of the end point of the head unit. In this case study, the maximum total volumetric error ΔL_1 before product adaptation and the maximum total volumetric error ΔL_2 after product adaptation were calculated.

- **Robustness for the Configuration Before Adaptation:**

The maximum volumetric errors in X, Y and Z directions, $\Delta_X^{(1)}$, $\Delta_Y^{(1)}$, and $\Delta_Z^{(1)}$, were calculated by:

$$\Delta_X^{(1)} = \varepsilon_x \tag{26}$$

$$\Delta_Y^{(1)} = \varepsilon_y + L_z \frac{\delta_2}{X_2} \tag{27}$$

$$\Delta_Z^{(1)} = \varepsilon_z + d \tag{28}$$

where ε_x, ε_y, and ε_z are the errors for the X, Y and Z axes respectively. The second term in Equation (27) is the error caused by rotation of the Z-axis module about the X-axis due to the clearance of the guiding slot δ_2. In Equation (28), d is the deformation of the beam in the Z-direction due to the weight of the relevant modules. Deformation d was calculated by:

$$d = \frac{F_w L_X^3}{48EI} \tag{29}$$

where F_w is the total weight force of the relevant modules (i.e., the thread block, the guiding carriage, and the Z-axis mechanism), E is the modulus of elasticity for the I-beam material, and I is the area moment of inertia for the I-beam section. Equations (30-34) provide details on how to calculate I, W, and F_w. It is important to note that, as stated in Table 3, W_3 in Equation (33) is 3 kg representing the weight of the Z-axis mechanism.

$$I = \frac{ah^3}{12} + \frac{b}{12}(H^3 - h^3) \tag{30}$$

$$W_1 = \left[D_2^2 - \pi\left(\frac{D_1}{2}\right)^2\right] X_1 \rho \tag{31}$$

$$W_2 = [((X_2 + 10) \times 60) - 10X_2]D_3\rho \tag{32}$$

$$W = W_1 + W_2 + W_3 \tag{33}$$

$$F_w = Wg \tag{34}$$

In Equations (30-34), W_1 and W_2 are the weights of the thread block and the guiding carriage, W is the total weight, and g is the gravitational acceleration (9.81 m/s^2). For this configuration, the effect of δ_1 was not considered because no rotational error about Y-axis was present.

The total volumetric error, $\Delta_L^{(1)}$, representing the robustness of the configuration before product adaptation, was defined by:

$$\Delta_L^{(1)} = \sqrt{\left(\Delta_X^{(1)}\right)^2 + \left(\Delta_Y^{(1)}\right)^2 + \left(\Delta_Z^{(1)}\right)^2} \tag{35}$$

Copyright © 2023 by ASME

- **Robustness for the Configuration After Adaptation:**

When the configuration is adapted into a 5-axis machine, additional errors are introduced. These errors include the rotation error about Y axis due to the clearance δ_l for the thread block and rotation errors at the end position of the head unit due to the errors of the additional two axes. Equations (36-38) were used to calculate the maximum errors in X, Y and Z directions.

$$\Delta_X^{(2)} = \varepsilon_x + L_z \frac{\delta_1}{X_1} + \varepsilon_\beta R + \varepsilon_\alpha R \tag{36}$$

$$\Delta_Y^{(2)} = \varepsilon_y + L_z \frac{\delta_2}{X_2} + \varepsilon_\beta R + \varepsilon_\alpha R \tag{37}$$

$$\Delta_Z^{(2)} = \varepsilon_z + d + \varepsilon_\beta R + \varepsilon_\alpha R \tag{38}$$

The total volumetric error, $\Delta_L^{(2)}$, representing the robustness of the configuration after product adaptation, was defined by:

$$\Delta_L^{(2)} = \sqrt{\left(\Delta_X^{(2)}\right)^2 + \left(\Delta_Y^{(2)}\right)^2 + \left(\Delta_Z^{(2)}\right)^2} \tag{39}$$

The overall volumetric error Δ_L for a design candidate was defined by:

$$\Delta_L = \left(\frac{1}{1+p}\right)\Delta_L^{(1)} + \left(\frac{p}{1+p}\right)\Delta_L^{(2)} \tag{40}$$

where p is the probability for adaptation ($p=0.35$).

Parameter optimization for robust design was defined by:

$$\min_{w.r.t\ (X_1, X_2)} \Delta_L^{(max)} \tag{41}$$

The optimal solution was identified as:

- **Optimal Design Candidate:**
  Design Candidate 1 with V-Shape Thread
- **Optimal Design Parameter Values:**
  $X_1^* = 81.03\ mm$, $X_2^* = 74.13\ mm$

Contour map of Δ_L for Design Candidate 1 considering different values of design parameters is given in Figure 8. It took 250 seconds to complete this optimization.

5.3. Discussions on Design Results

Comparison between this newly developed method and the traditional method was conducted.

In the traditional design methods for product adaptation, the uncertainties of the product adaptation activities were not considered. In these methods, an adaptable product was considered the same as a reconfigurable product [13], where both the configurations before the adaptation and the configurations after the adaptation were treated equally with the same weights. The optimal solution with the traditional method (i.e., p = 1.0) was obtained for the case study example as shown in Table 4.

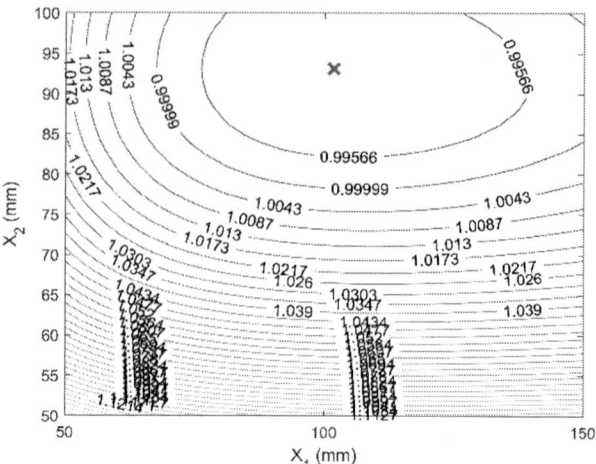

FIGURE 8: Contour map of Δ_L for parameter optimization.

TABLE 4: Comparative study.

		New Method (p=0.35)	Traditional Method (p=1)
Design Candidate 1	X_1^*	101.72 mm	139.65 mm
	X_2^*	93.10 mm	93.10 mm
	$\Delta_L^{(1)}$	0.672 mm	0.698 mm
	$\Delta_L^{(2)}$	1.903 mm	1.873 mm
	Δ_L	**0.991 mm**	**1.281 mm**
Design Candidate 2	X_1^*	150.00 mm	150.00 mm
	X_2^*	91.37 mm	91.37 mm
	$\Delta_L^{(1)}$	0.692 mm	0.692 mm
	$\Delta_L^{(2)}$	2.379 mm	2.379 mm
	Δ_L	1.130 mm	1.536 mm
Improvement	η	(1.281−0.991)/1.281 = 22.6%	–

Influence of probability for product adaptation to the optimal solutions was also studied as shown in Figure 9. When different probability measures are selected considering different chances that the original product will be adapted to satisfy new requirements during the product utilization stage, different design candidates and design parameter values are identified as the optimal solutions.

The relations among non-adaptable products, adaptable products and reconfiguration products are summarized as follows.

- **Non-adaptable products (p = 0):** Structure and modules of a non-adaptable product are usually not changed during its lifecycle span.
- **Adaptable products (0 < p < 1):** Structure and modules of an adaptable product have chances to be changed during its lifecycle span due to unexpected new requirements.
- **Reconfiguration products (p = 1):** Structure and modules of a reconfigurable product are often changed to satisfy multiple required design functions.

Copyright © 2023 by ASME

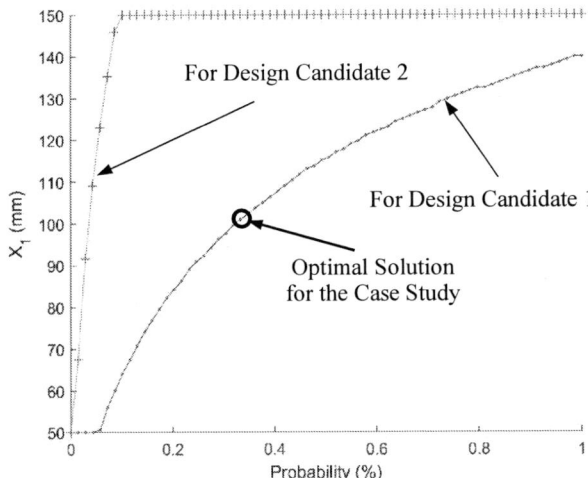

FIGURE 9: Influence of adaptation probability to the optimal parameter values.

6. CONCLUSIONS

The method developed in this study offers new ways for robust design of adaptable products by taking into account uncertainties in both configurations and parameters. Compared with the traditional robust design methods where only parameter uncertainties are considered, uncertainties of configurations, such as product upgrade during its lifecycle span, are also considered. In addition, influences of configuration uncertainties modeled as probabilities of adaptation tasks to the optimal design solutions have also been investigated.

Advantages of this new method are summarized as follows:

(1) Compared with the traditional methods to treat designs of adaptable products as designs of reconfigurable products, the uncertainties (i.e., probabilities) of the product adaptations have been considered in this work, and different design solutions are achieved considering different chances that the product needs to be adapted to improve performance of the designed product.

(2) For different design problems, different robustness measures (e.g., arithmetic measures and statistical measures) and different strategies to treat robustness measures (e.g., average method and worst-case method) considering product configurations before and after product adaptation should be selected.

(3) The multi-level optimization method is effective to identify both the optimal design configurations and the design parameter values.

In our future work, uncertainties in other lifecycle phases, such as manufacturing, should be considered. In addition, the developed methods should be integrated with the future CAD systems.

ACKNOWLEDGEMENTS

Financial support from the Natural Sciences and Engineering Research Council of Canada (NSERC) is acknowledged.

REFERENCES

[1] G. Pahl, W. Beitz, J. Feldhusen and K.-H. Grote, Engineering Design: A Systematic Approach, Springer, 2007.

[2] J.R. Fonseca, M.I. Friswell and A.W. Lees, "Efficient robust design via Monte Carlo sample reweighting," *International Journal for Numerical Methods in Engineering,* Vol. 69, No. 11, pp. 2279-2301, 2007.

[3] A. Kumar, P.B. Nair, A.J. Keane and S. Shahpar, "Robust design using Bayesian Monte Carlo," *International Journal for Numerical Methods in Engineering,* Vol. 73, No. 11, pp. 1497-1517, 2008.

[4] P. Gu, D. Xue and A.Y.C. Nee, "Adaptable design: Concepts, methods, and applications," *Journal of Engineering Manufacture,* Vol. 223, No. 11, pp. 1367-1387, 2009.

[5] M. Martinez and D. Xue, "A modular design approach for modeling and optimization of adaptable products considering the whole product utilization spans," *Journal of Mechanical Engineering Science,* Vol. 232, No. 7, pp. 1146-1164, 2018.

[6] J. Zhang, Y. Chen, D. Xue and P. Gu, "Robust design of configurations and parameters of adaptable products," *Frontiers of Mechanical Engineering,* Vol. 9, No. 1, pp. 1-14, 2014.

[7] M.S. Erden, H. Komoto, T.J.V. Beek, V. D'amelio, E. Echavarria and T. Tomiyama, "A review of function modeling: Approaches and applications," *Artificial Intelligence for Engineering Design, Analysis and Manufacturing,* Vol. 22, No. 2, pp. 147-169, 2008.

[8] G. Taguchi, "Performance analysis design," *International Journal of Production Research,* Vol. 16, No. 6, pp. 521-530, 1978.

[9] T. Hasenkamp, M. Arvidsson and I. Gremyr, "A review of practices for robust design methodology," *Journal of Engineering Design,* Vol. 20, No. 6, pp. 645-657, 2009.

[10] W. Hu, S. Azarm and A. Almansoori, "New approximation assisted multi-objective collaborative robust optimization (new AA-McRO) under interval uncertainty," *Structural and Multidisciplinary Optimization,* Vol. 47, No. 1, pp. 19-35, 2013.

[11] J. Zhang, H. Du, D. Xue and P. Gu, "Robust design approach to the minimization of functional performance variations of products and systems," *Frontiers of Mechanical Engineering,* Vol. 16, No. 1, pp. 379-392, 2021.

[12] R. Deabae and D. Xue, "Multi-level design optimization considering uncertainties in configurations and parameters," *The 33rd CIRP Design Conference*, Sydney, Australia, 2023.

[13] M. Gadalla and D. Xue, "An approach to identify the optimal configurations and reconfiguration processes for design of reconfigurable machine tools," *International Journal of Production Research,* Vol. 56, No. 11, pp. 3880-3900, 2018.

Copyright © 2023 by ASME

Proceedings of the ASME 2023
International Design Engineering Technical Conferences and
Computers and Information in Engineering Conference
IDETC-CIE2023
August 20-23, 2023, Boston, Massachusetts

DETC2023-117013

INTEGRATED SUSTAINABLE PRODUCT DESIGN WITH WARRANTY AND END-OF-USE CONSIDERATIONS

Xinyang Liu and Pingfeng Wang*

Department of Industrial and Enterprise Systems Engineering
University of Illinois at Urbana-Champaign
Urbana, IL 61801, United States.

ABSTRACT

The idea of integrated sustainable product design has been recently proposed to integrate downstream lifecycle performance into the initial product design to better achieve sustainability. Various sustainable product design tools based on life-cycle assessment or quality function deployment have been established while the influence of reliability in the circular practice has not been investigated much. Realizing that product reliability plays a critical role in determining post-design performance, this paper develops a product design optimization model that considers the warranty performance and the effect of end-of-use options. The effect of uncertain operating conditions on product reliability is incorporated into the model. Two optimization goals including the minimization of expected unit lifecycle cost and environmental impact are achieved by the model. Then, the model has been applied to an electric motor design problem to illustrate the benefit of the integrated approach. Results show that the component selection and reliability design will be adjusted when integrating the end-of-use options in the early design phase. Furthermore, lifecycle cost savings and environmental impact reductions can be achieved through the circular usage of used products. Finally, the effects of operating conditions, warranty policy, and take-back price for used products on the design decisions are analyzed to provide insights for product designers.

Keywords: Integrated design, warranty, end-of-use options, uncertain operating conditions, sustainability

NOMENCLATURE

Indexes
i Index for component alternatives
j Index for components
k Index for operating profile factors
l Index for operating profile

*Corresponding author: pingfeng@illinois.edu

Parameters
c_j^o Cost of processing component j with action o
c_{ij}^o Cost of processing alternative i of component j with action o
e_j^o Environmental impact of processing component j with action o
e_{ij}^o Environmental impact of processing alternative i of component j with action o
n_j^o Quantity of component j that uses action o
n_{ij}^o Quantity of alternative i of component j that uses action o
p_l Probability for operating profile l to occur
T Warranty length
S_n Nth holding time in the renewal process
α_{ijk} coefficient indicating the sensitivity of component j's alternative i to the operating profile factor k
β Shape parameter of Weibull distribution
θ RUL threshold for reuse
$\boldsymbol{u}_l$ A vector representing operating profile l
u_{kl} Level of the kth factor in operating profile l
η_{ij} Scale parameter for alternative i of component j

Functions
$F(x)$ Cumulative distribution function
$G(x)$ Simulation function to estimate end-of-use condition
$\Phi(x)$ Distribution for residual life of renewal process
$m(x)$ Renewal function

Decision variables
x_j Reliability scale parameter of component j
y_{ij} Binary variable indicating whether alternative i of component j is selected

1. INTRODUCTION

Product design is one of the most important sectors influencing all three sustainability facets: economy, environment, and society [1]. It has been found that 80% of sustainability impacts

Copyright © 2023 by ASME

are determined at the initial product design stage [2]. Therefore, in recent decades, many product design tools have been developed to enhance the ability of a product to work continuously while ensuring the lowest environmental impacts and providing economic and social benefits to the stakeholders. Initially, the major concern is focused on developing environmentally-friendly products, which initiates the prosperity of eco-design tools [3]. In the review by Rossi et al. [4], eco-design tools are classified into eight categories: Life Cycle Assessment (LCA) tools, Computer-Aided Design (CAD) integrated tools, diagram tools, checklists and guidelines, design for X approaches, methods for supporting companies' eco-design implementation, methods for user-centered design for sustainability and methods for integrating existing tools. LCA and CAD integrated tools quantify the environmental performances of a product along the whole life cycle by requiring the material and manufacturing process for all the components and subcomponents [5]. Diagram tools, checklists, and guidelines provide qualitative assessment and guidance to help choose design solutions to achieve environmental sustainability [6]. The other approaches may consider a specific product characteristic or integrate several existing tools to improve usability [7].

Nowadays, sustainability has been treated as a more comprehensive concept so new tools are emerging to account for other dimensions of sustainability. Ahmad et al. [2] reviewed the sustainable product design tools that consider two or all three dimensions of sustainability. Zhang et al. [8] proposed a metrics-based methodology for establishing a comprehensive sustainability index. Hoogmartens et al. [9] established a framework to interpret and integrate the assessment results from LCA, Life cycle Costing (LCC), and Cost Benefit Analysis (CBA). There are also other partial sustainable product design (P-SPD) tools and sustainable product design (SPD) tools [10–12] that integrate LCA, quality function deployment (QFD), fuzzy analysis to evaluate multi-dimensions of product sustainability, but the integration increases the complexity in the usage process, which limits the practical applications in the industry.

Early design decisions are not only related to the material selection and manufacturing process but also have a far-reaching effect on the entire product life cycle, including transportation, distribution, operations and maintenance, and end-of-use options. Therefore, to achieve true sustainability, it is critical to consider post-design performances in the initial design phase. For example, select the most cost-effective monitoring system at the design stage [13] and evaluate the system design performance with optimized operations management schedules [14]. However, the major challenge is to predict the product performance in the post-design phase considering various uncertainties from the market and operating conditions. Liu et al. [15] reviewed the existing predictive modeling approaches that enable the integrated design and operations of complex systems. Four types of approaches: data-driven [16, 17], statistical [18, 19], analytical [20], and experimental [21] approaches are summarized and illustrated with the corresponding application fields. With the predictive models, an integrated design manner can be applied to sustainable product development by evaluating lifecycle cost and optimizing multi-stage decisions. For example, Chung et al.

[20] extended the assessment scope to the supply chain network, evaluated modular structure designs using both lifecycle cost and energy consumption and finally optimized the product modular structure. There are also studies [22–24] that jointly optimize the design and operation process to achieve optimal lifecycle reliability or profitability.

Among the early design decisions, design for reliability is important in influencing the product life cycle performance. As the selection of material type and weight will influence the cost, energy consumption, and environmental impact during the manufacturing stage. Then, the reliability design will determine how a product performs under different operating conditions. The degradation process and lifetime will determine the end-of-use or end-of-life condition for the products, thus influencing the generation of subsequent life cycles. However, the impact of reliability or the design for reliability has not been investigated much in the field of sustainable product development. In the remanufacturing decision-making problem introduced in [25, 26], Liu et al. employed reliability information to help inform the end-of-use product condition. Go et al. [27] studies the guidelines of design for X (i.e., assembly, disassembly, maintainability, reliability, environment, modularity, etc.) to help design products with multiple generation life cycles. But the paper limits the impact of reliability in the testing phase and has not included approaches to integrate the design for X in sustainable development. Case studies on designing sustainable and reliable civil structures are conducted in [28], but the trade-off between sustainability and reliability is only investigated in one product life with a focus on material selection.

In this paper, we tackle the problem of sustainable product design by considering the impact of reliability on multiple life cycles. By optimizing the reliability property and selecting the best component alternative among available ones, the developed model can achieve optimal economic and environmental performance under uncertain operating conditions. We evaluate the unit life cycle cost and environmental impact by considering manufacturing, warranty, and end-of-use actions. The proposed model can be applied to integrated sustainable product design problems that consider post-design performance and multiple product life cycles. The contribution of this paper lies in using the product reliability property as a link to integrate the post-design processes with the design decisions to provide a closed-loop analysis of a design. In addition, the uncertainties in the operating process are also incorporated so that the optimal component alternative can be selected by evaluating its expected performance under various conditions.

The remaining paper is organized as follows. In Section 2, we clarify the specific sustainable product design problem that we aim to solve. Next, the optimization model is explained in Section 3 with the reliability modeling under uncertain operating conditions and end-of-use condition estimation function. Then the developed model is applied to an electric motor design problem in Section 4 to illustrate the effectiveness. Finally, we discuss about applying the proposed design optimization model to different product development scenarios in Section 5 and conclude the paper with the findings and future work in Section 6.

Copyright © 2023 by ASME

2. PROBLEM DESCRIPTION

The design problem considered in our work is to set up an appropriate reliability property for product components to minimize the expected unit life cycle cost and environmental impact by considering the manufacturing, warranty, and end-of-use actions. The involved process is shown in Figure 1. We consider two types of components in a product: one for us to optimize the parameters of its lifetime distribution and the other for us to select the best alternative among available choices. Such a setting represents the scenario when the product designer and manufacturer can manufacture some components in their own factory and purchase other components from upstream suppliers. After the manufacturing process, products are sold to the market with a certain type of warranty service. For example, the warranty service may cover failure replacement or repair within a 5-year usage period since the product is purchased. There are uncertainties involved in the usage period as the product may be operated under different conditions, which will lead to different degradation processes. We consider several operating scenarios and assign each scenario with a probability to occur. Then, we optimize the decision based on an expected performance over all possible scenarios.

At the end of the warranty service, we assume the used products can be bought back at a certain price from the customers as they may want to upgrade to a new version or do not need the product any longer. With the designed reliability property and the warranty service, we are able to estimate the product and component condition (e.g., remaining useful life (RUL)) upon buyback. Based on the estimated component condition, we may apply appropriate end-of-use options (e.g., reuse, remanufacturing, recycling) to the components and estimate the proportion of taken-back components that go through each procedure. With condition-based end-of-use processing, the remaining value of the components can be better utilized. The component quantity can be maintained through the reuse or remanufacturing procedure. However, severely degraded components that are put into the recycling process will result in a discounted mass of raw material that may not reproduce the same number of components. Since the recycling process may not be conducted by the OEM itself and the recycled material may not be used for manufacturing the same product, we assume the products after one usage period are all feasible for reuse or remanufacturing. A discussion about optimizing design decisions with other considerations of end-of-use procedures is provided in Section 5. Then, after generating a new life through the Re-X procedures, the products can be sold to the market again with a warranty service. Thus, the initial material can be processed to generate two life cycles of the products and we can evaluate the cost and environmental impact of such a circular practice to help select the optimal reliability design.

3. DESIGN OPTIMIZATION METHOD

In this section, we will explain how the integrated sustainable design problem is formulated. As mentioned in Section 2, to evaluate the economic and environmental impact of a design, we consider the warranty process and the condition-based end-of-use actions. In the following, we will first explain the component re-

liability model incorporating uncertain operating profiles. Then, the method to estimate the end-of-use condition after the warranty service is proposed. Finally, a multi-objective optimization model is developed.

3.1 Reliability under Uncertain Operating Profiles

A product may undergo different degradation processes when it is used under different operating profiles. Generally, combinations of factors like environment, mechanical or electrical stresses, and usage manners may determine various operating profiles. In our work, an operating profile l is represented by a vector $\boldsymbol{u}_l = (u_{1l}, u_{2l}, ...u_{cl})$ with each element indicating a specific factor level in the profile. We assume the profile l has a probability of p_l to occur and the future operating profiles can be approximated with several discrete scenarios.

Suppose the reliability performance of each component can be characterized by a Weibull distribution. The distribution will be shifted under different operating profiles. As applied in [29], the impact of different operating profiles is reflected in the scale parameter of the Weibull distribution while the shape parameter is independent of the operating profile. The scale parameter of the Weibull distribution can be modeled as a log-linear function of the profile factors u_{kl} and coefficients α_{ijk} in Eq. (1). $\eta_{ij}(\boldsymbol{u}_l)$ is the Weibull scale parameter of choice j of component i under operating profile $\boldsymbol{u}_l$. There is a baseline scale parameter for the component alternative η_{0ij} which is adjusted based on the operating profile l. Coefficients α_{ijk} indicate the component sensitivity to each of the c profile factors. The profile factors u_{kl} are scaled to the range $[0, 1]$ based on a nominal operating condition.

$$\eta_{ij}(\boldsymbol{u}_l) = \eta_{0ij}exp(-\sum_{k=1}^{c}\alpha_{ijk}u_{kl}) \qquad (1)$$

The sensitivity coefficients α_{ijk} can either be estimated from test data by conducting life tests over a range of conditions, based on physics-of-failure models, or obtained from handbooks or published values (e.g., Arrhenius reaction rates) [29]. Alternatively, the effect of operating profiles on one type of component can be inferred from other components with similar failure modes and mechanisms.

3.2 End-of-use Condition Estimation

The end-of-use condition of a component is determined by both the reliability performance under a specific operating profile and the warranty policy associated with the product. Suppose the components are independent, then the degradation process and effects of warranty actions for each component can be treated independently. In our work, we first assume the warranty service associated with a product is to replace the failed component. Thus, the usage process of each component can be represented using a renewal process. We are interested in calculating the remaining useful life (RUL) of a group of components that undergo warranty service. Based on the property of the renewal process, the distribution of residual life at time t can be represented using Eq. (2). $F(t)$ represents the cumulative distribution function of the holding time and $m(y)$ is the renewal function which can be

Copyright © 2023 by ASME

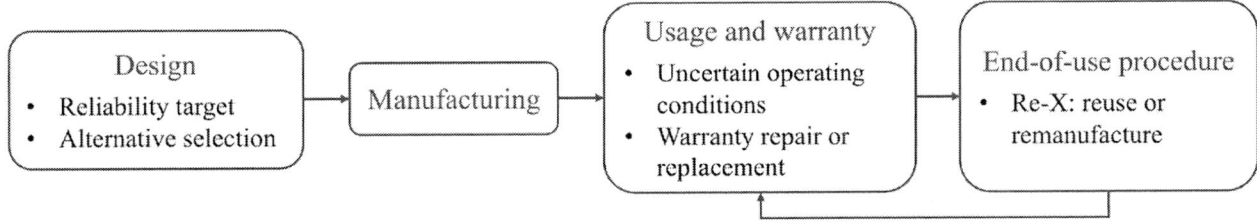

FIGURE 1: DESIGN EVALUATION PROCESS

calculated using the number of jumps or the cumulative distribution function of each jump time as shown in Eq. (3).

$$\Phi(x) = F(t+x) - \int_0^t [1 - F(t+x-y)]dm(y) \qquad (2)$$

$$m(y) = E[N(y)] = \sum_{n=1}^{\infty} P(S_n \le y) = \sum_{n=1}^{\infty} F_n(y) \qquad (3)$$

Since the explicit formula of the renewal function and the distribution of the residual life of the renewal process is hard to obtain. We employ a simulation procedure to estimate the end-of-use condition in our study. The procedure is shown in Table 1. With the given reliability model for a component and the product warranty, we can generate a group of sample lifetime values. Whenever a failure occurs within the warranty period, a new component lifetime will be generated to replace the failed component. At the end of the warranty service, we can obtain the total number of replacement actions and the RUL of each functional component. Then, based on the empirical RUL distribution and a condition-based Re-X policy, we can estimate the proportion of components that can be reused, remanufactured, or recycled.

3.3 Multi-objective Optimization Model

$$\min \quad [\Sigma_{j \in J_1} \Sigma_o c_j^o \cdot n_j^o(x_j) + \Sigma_{j \in J_2} \Sigma_{l,o,i} p_l \cdot y_{ij} \cdot c_{ij}^o \cdot n_{ij}^o(y_{ij})]/2N \qquad (4)$$

$$\min \quad [\Sigma_{j \in J_1} \Sigma_o e_j^o \cdot n_j^o(x_j) + \Sigma_{j \in J_2} \Sigma_{l,o,i} p_l \cdot y_{ij} \cdot e_{ij}^o \cdot n_{ij}^o(y_{ij})]/2N \qquad (5)$$

$$\text{s.t.} \quad n_j^{\text{man}}(x_j) = N, \forall j \in J_1 \qquad (6)$$

$$n_{ij}^{\text{man}}(y_{ij}) = N, \forall j \in J_2 \text{ and } i \in I_j \qquad (7)$$

$$n_j^{\text{war}}(x_j) = G(x_j), \forall j \in J_1 \qquad (8)$$

$$n_{ij}^{\text{war}}(y_{ij}) = G(y_{ij}), \forall j \in J_2 \text{ and } i \in I_j \qquad (9)$$

$$n_j^{\text{b}}(x_j) = r(c_j)N, \forall j \in J_1 \qquad (10)$$

$$n_{ij}^{\text{b}}(y_{ij}) = r(c_{ij}^b)N, \forall j \in J_2 \text{ and } i \in I_j \qquad (11)$$

$$n_j^{\text{reu}}(x_j) = (1 - F_j(\theta))N, \forall j \in J_1 \qquad (12)$$

$$n_{ij}^{\text{reu}}(y_{ij}) = (1 - F_{ij}(\theta))N, \forall j \in J_2 \text{ and } i \in I_j \qquad (13)$$

$$n_j^{\text{rem}}(x_j) = F_j(\theta) \cdot N, \forall j \in J_1 \qquad (14)$$

$$n_{ij}^{\text{rem}}(y_{ij}) = F_{ij}(\theta) \cdot N, \forall j \in J_2 \text{ and } i \in I_j \qquad (15)$$

$$\Sigma_i y_{ij} = 1, \forall j \in J_2 \qquad (16)$$

With the reliability model under uncertain operating profiles and the end-of-use condition estimation approach proposed, we can then establish a multi-objective optimization model to make sustainable product design decisions. As mentioned in Section 2, we optimize the decision for two types of components. For the first type of component $j \in J_1$, the decision variable is x_j representing its reliability scale parameter. For the second type of component $j \in J_2$, the decision is a binary variable y_{ij} indicating whether to choose alternative i for component j.

Then, we require a set of parameters to specify the manufacturing, warranty, and end-of-use actions along the product life cycle. During the whole process, there is a set of actions to be applied to the product components, including manufacturing, warranty replacement, used product collection, reuse, and remanufacturing. We require the cost and environmental impact for performing each action o to a component j or an alternative i of the component, which can be denoted as c_j^o, e_j^o for $j \in J_1$ and c_{ij}^o, e_{ij}^o for $j \in J_2$. For the available component alternatives, the coefficients α_{ijk} and the baseline scale parameters η_{0ij} mentioned in Section 3.1 should also be provided. The uncertain operating profile is discretized to several scenarios, each scenario l has a probability p_l to occur and the factor level for each profile is u_{kl}. Finally, we need to know the warranty length T, the constant reliability shape parameter β, the sample quantity N, the take-back ratio r as a function of the buyback price, and the RUL threshold for reusing a component θ.

The model is established as follows. The two objectives as shown in (4) and (5) are to minimize the expected unit life cycle cost and environmental impact considering all the actions and uncertain operating profiles. The manufacturing quantity of each product component or alternative is N as indicated in (6) and (7). The set of available alternatives of a component j is denoted as I_j. The number of warranty replacements can be estimated from the simulation process explained in Section 3.2. In Eqs. (8) and (9), the simulation process is denoted as a function G. Note that, the number of replacements is the cumulative result of two life cycles: one for the brand-new components, the other for the reused or remanufactured components. Since the simulation process will also output the empirical RUL distribution F upon the end of the warranty service, the proportion of components to be reused can be calculated using Eqs. (12), (13) and those for remanufacturing are determined by Eqs. (14), (15). Finally, we have a constraint (16) to ensure that one and only one alternative is selected for each component in the group J_2.

The multi-objective optimization problem can be solved by scaling and assigning weights to each objective or looking for

Copyright © 2023 by ASME

TABLE 1: PROCEDURE FOR END-OF-USE CONDITION ESTIMATION

Inputs: Component reliability model: Weibull(η, β)
Warranty length: T
Number of samples: S
Outputs: End-of-use RUL values
Number of warranty actions

Generate a list L of S random lifetime values based on the reliability model Weibull(η, β)
for $s = 1 : S$
$\quad L_s^1 = L_s$
$\quad n_s = 1$
$\quad$ if $L_s^1 < T$
$\quad\quad$ while $\Sigma_{j=1}^{n_s} L_s^j < T$
$\quad\quad\quad n_s = n_s + 1$
$\quad\quad\quad$ Generate another lifetime value $L_s^{n_s}$ based on the same reliability model Weibull(η, β)
$\quad$ RUL$_s = \Sigma_{j=1}^{n_s} L_s^j - T$
$\quad N_s^{\text{rep}} = n_s - 1$
Obtain the end-of-use RUL values $\{\text{RUL}_s\}$ and the total number of warranty actions $\Sigma_{s=1}^{S} N_s^{\text{rep}}$

Pareto optimal points. In our work, we would like to see the trade-off between economic and environmental performances from the Pareto front. Since simulation is involved in evaluating the objective functions, we adopted pattern search to optimize the design decisions in a derivative-free manner.

4. APPLICATION ON ELECTRIC MOTOR DESIGN

In this section, we applied the model to an electric motor design problem. We simplify the product structure as three major components: motor housing, stator with winding, rotor with bearings, and shaft. For the housing and stator, we need to optimize the scale parameter value of the Weibull model. For the rotor, we need to select the best one among the available alternatives. Next, we will first introduce the parameter settings and then show the design results and the impact of several key factors in the life cycle.

4.1 Case Study Setting

For the component to choose the optimal alternative, we need to define the reliability property for each alternative and the uncertain operating profiles. Since bearings play an important role in influencing the degradation process of the rotor ($j = 3$), we consider the factors that influence the bearing degradation to define the operating profile. We refer to the case study in [29] to prepare the parameter values of our example. Three operating profiles are considered in the case study as listed in Table 2 where the two key factors are rotation speed and load. There are two alternatives to choose from and their properties are shown in Table 3.

For the components to optimize reliability model parameters, we assume the manufacturing cost follows a polynomial function [30] of the scale parameter as shown in Eq. (17). The parameters for the cost function are shown in Table 4. As mentioned in [31], 30% of the manufacturing cost is used for assembly and testing. We assume the assembly and testing process leads to the total cost of reuse and additional machining cost is required in the remanufacturing process, leading to $c_j^{\text{reu}} = 0.3c_j^{\text{man}}$ and

$c_j^{\text{rem}} = 0.6c_j^{\text{man}}$. Some detailed analysis on cost and energy consumption comparison for new and remanufactured products can be found in [32, 33]. Since the warranty replacement will use new components, we assume $c_j^{\text{war}} = c_j^{\text{man}}$. For the buyback process, we first assign a constant price $c_j^b = 0.2c_j^{\text{man}}$ and then examine the effect of offering different buyback prices based on the product condition.

$$c_j^{\text{man}} = a_j + b_j \cdot x_j^{d_j} \qquad (17)$$

Then, we gather the information for environmental impact assessment. We mainly referred to the life cycle analysis results from [31] and adjusted the greenhouse gas (GHG) emission for the three types of components as shown in Table 5. The environmental impact for other actions is approximated as $e_j^{\text{reu}} = 0.11e_j^{\text{man}}$, $e_j^{\text{rem}} = 0.31e_j^{\text{man}}$, $e_j^{\text{war}} = 1.01e_j^{\text{man}}$ based on the life cycle analysis results from [31].

We apply the same end-of-use policy to all three components which is to reuse the components with RUL longer than the warranty length. Namely, we set the RUL threshold for reuse $\theta = T$. This is a reasonable practice since the replacement cost is higher than the remanufacturing cost and we assume remanufacturing will bring the component back to a same-as-new condition. We would only reuse those components that will not fail during the warranty service.

4.2 Results

With the case study setting defined, we then implement the design optimization problem in MATLAB and solve it with the paretosearch algorithm. The Pareto front with the baseline setting is shown in Figure 2. There are two separate curves corresponding to two scenarios that either rotor alternative 1 or alternative 2 is better. Based on the provided coefficients and baseline scale parameters, we can obtain the scale parameter under each operating profile for the two rotor alternatives as shown in Table 6. A larger scale parameter leads to a longer expected lifetime with the same shape parameter. Thus, alternative 1 has higher reliability

Copyright © 2023 by ASME

TABLE 2: OPERATING PROFILES

Profile l	Probability p_l	Rotation speed (rpm)	u_{1l}	Load (N)	u_{2l}
1	1/3	1800	1	4000	1/3
2	1/3	1650	5/8	4200	7/15
3	1/3	1500	1/4	5000	3/4

TABLE 3: ROTOR ALTERNATIVES

Alternative i	Coefficient α_{i31}	Coefficient α_{i32}	Baseline scale parameter η_{0i3}	Manufacturing Cost c_{i3}^{man} ($)
1	0	1.86	8.71	100
2	0	2.90	8.11	50

TABLE 4: MANUFACTURING COST MODEL PARAMETERS

Component	a_j	b_j	d_j
Stator ($j = 1$)	40	20	0.6
Housing ($j = 2$)	60	10	0.4

TABLE 5: COMPONENT ENVIRONMENTAL IMPACT

Component	Stator	Housing	Rotor 1	Rotor 2
GHG (kg CO_2 eq.)	206	129	126	144

TABLE 6: ALTERNATIVE SCALE PARAMETER

	Profile 1	Profile 2	Profile 3
Alternative 1	4.68	3.66	2.16
Alternative 2	3.09	2.10	0.92

and leads to much fewer warranty actions and more end-of-use components qualified for reuse. So alternative 1 will result in lower environmental impact but higher cost. Even though alternative 2 has poor reliability performance, its low cost still leads to several Pareto optimal points that we should choose alternative 2 to reach more cost savings.

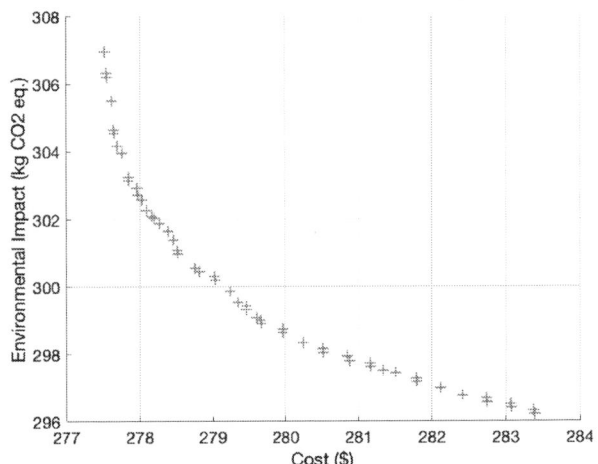

FIGURE 3: PARETO FRONT WITH ONE OPTIMAL ALTERNATIVE

front in Figure 3. When the replacement cost reduces to half of the manufacturing cost, the dominant rotor choice is alternative 1.

If only one life cycle is considered in the product design phase, then only the manufacturing process and the warranty service process will be calculated in the objective function. With a similar parameter setting, we may obtain the Pareto front for the design decisions without the circular practice in Figure 4. Compared with Figure 3, the design results considering only one life cycle will lead to higher expected unit cost and environmental impact.

In addition to the objective values, we are also interested in the design decisions (i.e., reliability parameters and alternative selection). As there are multiple Pareto optimal solutions when considering two objectives, we checked the design results when only minimizing the expected unit cost. The decisions are listed in Table 7. For both cases, the better alternative for the rotor

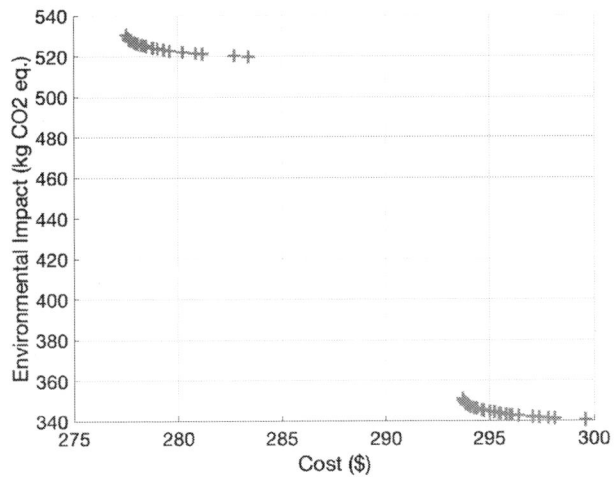

FIGURE 2: PARETO FRONT WITH BASELINE SETTING

When the alternative property or the cost setting changes, we may also notice one single Pareto front resulting from one dominant rotor alternative. For example, when the environmental impact of manufacturing rotor alternative 2 reduces to 46 kg CO_2 eq., the optimal choice is alternative 2 which leads to the Pareto

Copyright © 2023 by ASME

515

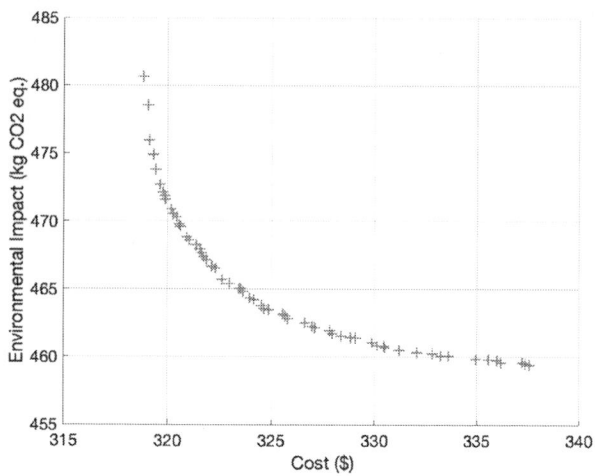

FIGURE 4: PARETO FRONT WITHOUT END-OF-USE PROCESSING

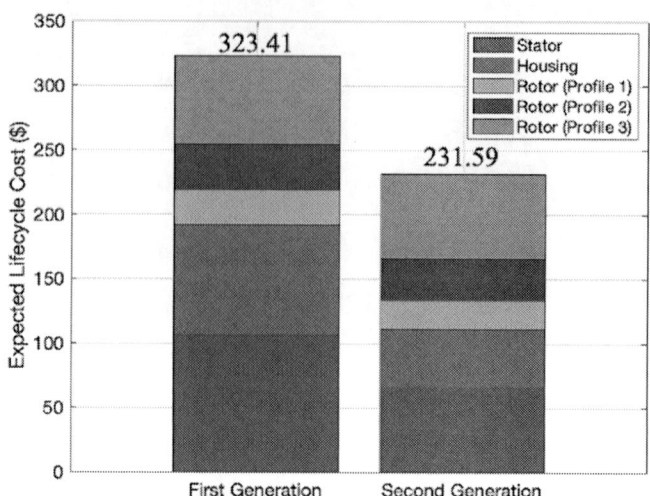

FIGURE 5: COST DECOMPOSITION FOR INTEGRATED PRODUCT DESIGN

is alternative 2 with much cost savings. While we notice the difference in the reliability targets for the other two components. When considering the end-of-use actions to generate another life cycle for the resources, the component is designed to have higher reliability so that more components can be reused to save the additional manufacturing cost. We may also notice the cost saving for the second generation in Figure 5. In the cost decomposition, since the rotor will have different degradation processes under different operating profiles, the cost for the rotor is further decomposed into three profiles. The third operating profile is the harshest condition so it leads to the largest cost portion. When we change the probabilities for each operating profile to occur, the alternative selection will not change, but harsher conditions will lead to higher costs and greater environmental impact.

TABLE 7: DESIGN DECISIONS WITH ONE OBJECTIVE

Circular Practice			
x_1	x_2	y_{31}	y_{32}
6.77	9.77	0	1
Without End-of-use Processing			
x_1	x_2	y_{31}	y_{32}
5.24	7.14	0	1

Finally, we explored the impact of warranty type and buyback price. For the baseline case, we still consider the warranty service that replaces the failed component and a constant buyback price for all the end-of-use products but uses $c_j^{war} = 0.5c_j^{man}$. To explore the impact of warranty service, we consider another scenario when we provide repair for the failed component. The effect of repair is assumed to bring the used component back to 80% of its original state. For the impact of the buyback price, we change the constant price to two discrete price options. Those end-of-use products that are qualified for reuse can be bought back at the same price as in other cases. While those for remanufacturing can be taken back at a lower price. The design decisions for the three scenarios are summarized in Table 8 where we have the reliability design parameters and alternative selec-

tion decisions made under two warranty services either repair or replacement and two possible buyback prices either unified for all conditions or condition-dependent. When either the warranty policy or the buyback price changes, the rotor alternative selection has changed. Only in the baseline scenario, alternative 1 is selected while providing a repair option in the warranty or taking back remanufacturable products are a lower price will lead to the selection of alternative 2. Also, when the buyback price changes, the reliability targets for stators and housings decrease due to the cost advantage of remanufacturing. Comparing the reliability decision changes for the stator and the housing, we notice the decision for the stator is more sensitive to the parameter setting. This is due to the cost nature of the two components. As shown in Figure 9, the manufacturing cost for a stator is more sensitive to the housing. Thus, when the problem setting changes, the design decision for the stator will show a major change while the decision for housing may not be influenced.

TABLE 8: DESIGN DECISIONS UNDER DIFFERENT PROBLEM SETTINGS

1. Warranty: replacement; Buyback price: unified			
x_1	x_2	y_{31}	y_{32}
6.85	9.65	1	0
2. Warranty: repair; Buyback price: unified			
x_1	x_2	y_{31}	y_{32}
6.62	9.65	0	1
3. Warranty: replacement; Buyback price: condition-dependent			
x_1	x_2	y_{31}	y_{32}
5.85	9.14	0	1

Apart from the change of design parameters, we also checked the manufacturing and warranty processes of each decision. The quantity for different components to go through different operation processes in the baseline scenario is shown in Figure 6 and

Copyright © 2023 by ASME

the quantities for the other two scenarios are shown in Figure 7 and Figure 8. As mentioned in the model formulation, we rely on a sample quantity of N to approximate the warranty performance and end-of-use condition for a specific reliability design decision. In the case study, the sample size $N = 10^4$ and the component quantity shown in the figures are all derived based on this sample size. Two generations of products are considered in the figures: the first generation consists of brand-new products and the second generation consists of remade products either from reusing or remanufacturing the components. We checked the number of warranty replacements or repairs for the two generations and the number of components that go through reuse or remanufacturing processes. We can notice that due to the high-reliability target for housing, the number of warranty replacements for housing is the least and housing has the largest proportion to be reused in all scenarios. Since those components qualified for reuse have a residual life longer than the warranty period, the number of warranty replacements or repairs for the second-generation housing is almost zero. For the comparison between the two alternatives of the rotor, we can notice a better reliability performance for the first alternative which has fewer warranty claims and a larger proportion of components that can be reused. However, the first alternative is only selected in the baseline scenario due to its higher cost.

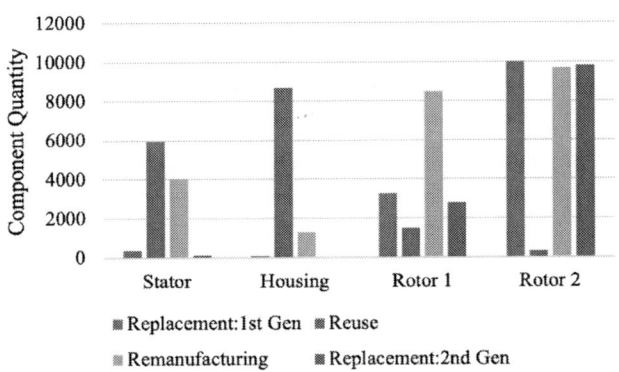

FIGURE 6: OPERATION PROCESS FOR BASELINE SCENARIO

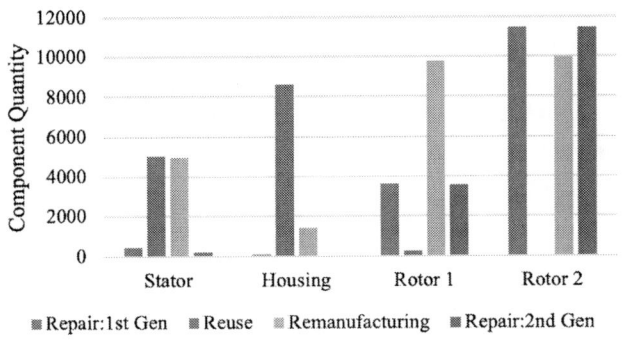

FIGURE 7: OPERATION PROCESS FOR SCENARIO 2: REPAIR WARRANTY SERVICE

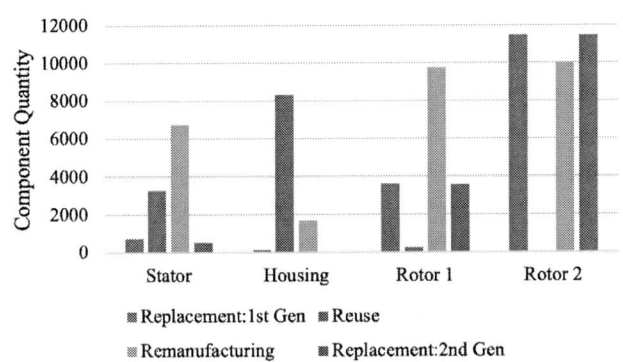

FIGURE 8: OPERATION PROCESS FOR SCENARIO 3: CONDITION-DEPENDENT BUYBACK PRICE

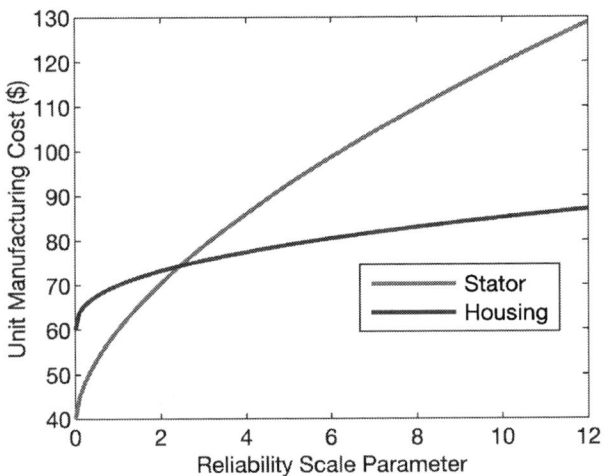

FIGURE 9: COST CHANGE DUE TO RELIABILITY SCALE PARAMETER

5. DISCUSSION

In this work, we considered reuse and remanufacturing as the two main end-of-use options as the used products are assumed to be bought back at the end of a warranty service. Due to the repair or replacement service provided during the warranty, product conditions can be maintained at a decent standard so recycling the functional components is not an ideal choice. But for those failed components within warranty service, if a replacement is provided, we may consider the recycling of those failed components by incorporating a net value (saved material acquisition cost minus recycling consumption) of recycled material into the lifecycle objective function. Since recycling is only applied when the product or component is not valuable to be cleaned, repaired, or remanufactured, whether to consider recycling in the circular evaluation of product design depends on the speed of the product's technical upgrade and the collection policy of end-of-use or end-of-life products. If recycling is also applied to end-of-use but functional components, then not only the net value of recycled material but also the cost of manufacturing new components for a second generation should be included.

This work serves as a building block for evaluating product

Copyright © 2023 by ASME

design decisions by considering a circular practice. There are indeed quite a few factors that will influence the evaluation result. For example, in our analysis, we considered a fixed Re-X policy where components with RUL longer than the warranty length will be reused. However, there can be different Re-X policies applicable to different OEMs or the Re-X decisions may be optimized together with the product design. In addition, the take-back process will also influence the design decision. We assumed a buy-back process that provides a fixed or condition-dependent price for motivating the customers' return. However, some take-back processes may be based on customers' awareness of environmental protection, or even providing an incentive buyback price may only collect a lower percentage of used products. When only limited resources can be provided to generate a "green" generation of products, the design decisions may be alternated to achieve major profits from the first generation. Therefore, when applying the proposed design optimization framework to different product development processes, factors like Re-X options, Re-X decision criteria, and product take-back process need to be adjusted.

6. CONCLUSION AND FUTURE WORK

This paper developed an integrated sustainable product design optimization model by considering the warranty and end-of-use options. The reliability performance under different operating profiles is incorporated and the design decisions are evaluated from both economic and environmental aspects. The developed model can help design decision-makers optimize the reliability target settings and component alternative selections from a closed-loop product life aspect. The case study on electric motor design illustrates the benefit of integrated design with a circular practice. The effect of several key factors (e.g., operating profiles, warranty service type, buyback price) is explored in the case study. The reliability design decisions are sensitive to the manufacturing cost model as we tend to adjust the reliability performance for those components whose manufacturing cost changes significantly with the reliability target. The trade-off between economic and environmental performance, the warranty policy, and the product buyback process will all influence the alternative selection.

For future work, the design decisions can be jointly optimized with warranty service and end-of-use policy under uncertain operating profiles. But to optimize other product-related decisions, market information has to be considered to reflect the impact of reliability on both various types of cost and demand. In addition, instead of referring to the existing life cycle analysis results for motor components, the environmental impact assessment can be conducted for different design settings with different material types and weights.

ACKNOWLEDGEMENT

This material is supported by the U.S. Department of Energy's Office of Energy Efficiency and Renewable Energy (EERE) under the Advanced Manufacturing Office Award Number DE-EE0007897.

REFERENCES

[1] Ramani, Karthik, Ramanujan, Devarajan, Bernstein, William Z, Zhao, Fu, Sutherland, John, Handwerker, Carol, Choi, Jun-Ki, Kim, Harrison and Thurston, Deborah. "Integrated sustainable life cycle design: a review." *Journal of Mechanical Design* Vol. 132 No. 9 (2010): p. 091004.

[2] Ahmad, Shamraiz, Wong, Kuan Yew, Tseng, Ming Lang and Wong, Wai Peng. "Sustainable product design and development: A review of tools, applications and research prospects." *Resources, Conservation and Recycling* Vol. 132 (2018): pp. 49–61.

[3] Bhamra, Tracy and Lofthouse, Vicky. *Design for sustainability: a practical approach.* Routledge, London (2016).

[4] Rossi, Marta, Germani, Michele and Zamagni, Alessandra. "Review of ecodesign methods and tools. Barriers and strategies for an effective implementation in industrial companies." *Journal of Cleaner Production* Vol. 129 (2016): pp. 361–373.

[5] Finnveden, Göran, Hauschild, Michael Z, Ekvall, Tomas, Guinée, Jeroen, Heijungs, Reinout, Hellweg, Stefanie, Koehler, Annette, Pennington, David and Suh, Sangwon. "Recent developments in life cycle assessment." *Journal of environmental management* Vol. 91 No. 1 (2009): pp. 1–21.

[6] Abramovici, Michael, Quezada, Akamitl and Schindler, Thomas. "Methodical approach for rough energy assessment and compliance checking of energy-related product design options." *Procedia CIRP* Vol. 21 (2014): pp. 421–426.

[7] Domingo, Lucie, Brissaud, Daniel and Mathieux, Fabrice. "Implementing scenario to better address the use phase in product ecodesign." *International conference on engineering design ICED 2013*: p. 135. 2013.

[8] Zhang, X, Lu, T, Shuaib, M, Rotella, G, Huang, A, Feng, SC, Rouch, K, Badurdeen, F and Jawahir, IS. "A metrics-based methodology for establishing product sustainability index (ProdSI) for manufactured products." *Leveraging Technology for a Sustainable World: Proceedings of the 19th CIRP Conference on Life Cycle Engineering, University of California at Berkeley, Berkeley, USA, May 23-25, 2012*: pp. 435–441. 2012. Springer.

[9] Hoogmartens, Rob, Van Passel, Steven, Van Acker, Karel and Dubois, Maarten. "Bridging the gap between LCA, LCC and CBA as sustainability assessment tools." *Environmental Impact Assessment Review* Vol. 48 (2014): pp. 27–33.

[10] Guinée, Jeroen. "Life cycle sustainability assessment: what is it and what are its challenges?" *Taking stock of industrial ecology* (2016): pp. 45–68.

[11] Vinodh, S, Manjunatheshwara, KJ, Karthik Sundaram, S and Kirthivasan, Vishwesh. "Application of fuzzy quality function deployment for sustainable design of consumer electronics products: a case study." *Clean Technologies and Environmental Policy* Vol. 19 (2017): pp. 1021–1030.

[12] Kim, Samyeon and Moon, Seung Ki. "Sustainable platform identification for product family design." *Journal of cleaner production* Vol. 143 (2017): pp. 567–581.

Copyright © 2023 by ASME

[13] Liu, Xinyang and Wang, Pingfeng. "Valuation of Continuous Monitoring Systems for Engineering System Design in Recurrent Maintenance Decision Scenarios." *Journal of Mechanical Design* Vol. 144 No. 9 (2022): p. 091702.

[14] Liu, Xinyang, Zheng, Zhuoyuan, Büyüktahtakın, İ Esra, Zhou, Zhi and Wang, Pingfeng. "Battery asset management with cycle life prognosis." *Reliability Engineering & System Safety* Vol. 216 (2021): p. 107948.

[15] Liu, Xinyang, Ghosh, Sayan, Liu, Yongming and Wang, Pingfeng. "Towards Integrated Design and Operation of Complex Engineering Systems With Predictive Modeling: State-of-the-Art and Challenges." *Journal of Mechanical Design* Vol. 144 No. 9 (2022): p. 090801.

[16] Xu, Yanwen, Kohtz, Sara, Boakye, Jessica, Gardoni, Paolo and Wang, Pingfeng. "Physics-informed machine learning for reliability and systems safety applications: State of the art and challenges." *Reliability Engineering & System Safety* (2022): p. 108900.

[17] Wu, Hao, Zhu, Zhifu and Du, Xiaoping. "System reliability analysis with autocorrelated kriging predictions." *Journal of Mechanical Design* Vol. 142 No. 10 (2020).

[18] Pitiot, Paul, Coudert, Thierry, Geneste, Laurent and Baron, Claude. "Hybridation of Bayesian networks and evolutionary algorithms for multi-objective optimization in an integrated product design and project management context." *Engineering Applications of Artificial Intelligence* Vol. 23 No. 5 (2010): pp. 830–843.

[19] Wu, Hao, Hu, Zhangli and Du, Xiaoping. "Time-dependent system reliability analysis with second-order reliability method." *Journal of Mechanical Design* Vol. 143 No. 3 (2021).

[20] Chung, Wu-Hsun, Okudan Kremer, Gül E and Wysk, Richard A. "A modular design approach to improve product life cycle performance based on the optimization of a closed-loop supply chain." *Journal of Mechanical Design* Vol. 136 No. 2 (2014): p. 021001.

[21] Sabbaghi, Mostafa and Behdad, Sara. "Environmental evaluation of product design alternatives: The role of consumer's repair behavior and deterioration of critical components." *Journal of Mechanical Design* Vol. 139 No. 8 (2017): p. 081701.

[22] Adjoul, Oussama, Benfriha, Khaled, El Zant, Chawki and Aoussat, Améziane. "Algorithmic strategy for simultaneous optimization of design and maintenance of multicomponent industrial systems." *Reliability Engineering & System Safety* Vol. 208 (2021): p. 107364.

[23] Liu, Xinyang and Wang, Pingfeng. "Joint optimization of Reliability, Warranty and Price for a Product Family." *2022 Annual Reliability and Maintainability Symposium (RAMS)*: pp. 1–6. 2022. IEEE.

[24] Wu, Jiaxin and Wang, Pingfeng. "Generative Design for Resilience of Interdependent Network Systems." *International Design Engineering Technical Conferences and Computers and Information in Engineering Conference*, Vol. 86229:

p. V03AT03A038. 2022. American Society of Mechanical Engineers.

[25] Liu, Xinyang, Wang, Pingfeng and Kim, Harrison. "Multi-Generational Product Family Design for Reliability and Environmental Sustainability." *IIE Annual Conference. Proceedings*: pp. 1–6. 2022. Institute of Industrial and Systems Engineers (IISE).

[26] Liu, Xinyang, Mishra, Ankush Kumar, Hu, Chao and Wang, Pingfeng. "Multi-Stage Product Family Design for Reliability with Remanufacturing." *2023 Annual Reliability and Maintainability Symposium (RAMS)*: pp. 1–6. 2023. IEEE.

[27] Go, Tze Fong, Wahab, Dzuraidah Abd and Hishamuddin, Hawa. "Multiple generation life-cycles for product sustainability: the way forward." *Journal of Cleaner Production* Vol. 95 (2015): pp. 16–29.

[28] Dazer, Martin, Ostertag, Andreas, Herzig, Thomas, Borschewski, David, Albrecht, Stefan and Bertsche, Bernd. "Consideration of reliability and sustainability in mechanical and civil engineering design to reduce oversizing without risking disasters." *E3S Web of Conferences*, Vol. 349: p. 11004. 2022. EDP Sciences.

[29] Chatwattanasiri, Nida, Coit, David W and Wattanapongsakorn, Naruemon. "System redundancy optimization with uncertain stress-based component reliability: Minimization of regret." *Reliability Engineering & System Safety* Vol. 154 (2016): pp. 73–83.

[30] Mohan, Karen, Huffman, Duane and Akers, Jennifer. "Optimization of warranty period, price, and allocated reliability." *2009 Annual Reliability and Maintainability Symposium*: pp. 483–488. 2009. IEEE.

[31] Orlova, S, Rassõlkin, A, Kallaste, A, Vaimann, T and Belahcen, Anouar. "Lifecycle analysis of different motors from the standpoint of environmental impact." *Latvian Journal of Physics and Technical Sciences* Vol. 53 No. 6 (2016): pp. 37–46.

[32] Li, Meng, Nemani, Venkat P, Liu, Jinqiang, Lee, Michael A, Ahmed, Navaid, Kremer, Gül E and Hu, Chao. "Reliability-informed life cycle warranty cost and life cycle analysis of newly manufactured and remanufactured units." *Journal of Mechanical Design* Vol. 143 No. 11 (2021).

[33] Nemani, Venkat P, Liu, Jinqiang, Ahmed, Navaid, Cartwright, Adam, Kremer, Gül E and Hu, Chao. "Reliability-Informed Economic and Energy Evaluation for Bi-Level Design for Remanufacturing: A Case Study of Transmission and Hydraulic Manifold." *Journal of Mechanical Design* Vol. 144 No. 8 (2022): p. 082001.

[34] Feng, Yixiong, Zhao, Yuliang, Zheng, Hao, Li, Zhiwu and Tan, Jianrong. "Data-driven product design toward intelligent manufacturing: A review." *International Journal of Advanced Robotic Systems* Vol. 17 No. 2 (2020): p. 1729881420911257.

[35] Zhu, JY and Deshmukh, A. "Application of Bayesian decision networks to life cycle engineering in Green design and manufacturing." *Engineering Applications of Artificial Intelligence* Vol. 16 No. 2 (2003): pp. 91–103.

Copyright © 2023 by ASME

Proceedings of the ASME 2023
International Design Engineering Technical Conferences and
Computers and Information in Engineering Conference
IDETC-CIE2023
August 20-23, 2023, Boston, Massachusetts

DETC2023-117050

UNCERTAINTY QUANTIFICATION ON MECHANICAL BEHAVIOR OF CORRODED PLATE WITH STATISTICAL SHAPE MODELING

Hao Wu, Parth Bansal, Zheng Liu, Yumeng Li and Pingfeng Wang[1]

Department of Industrial and Enterprise Systems Engineering
University of Illinois Urbana-Champaign
Urbana, IL 61801, United States.

ABSTRACT

Corrosion is a process of uncertain nature considering the randomness associated with corrosion initiation and growth, therefore estimating the stochastic behavior of a corroded structure therefore quantifying of system performance uncertainties would be very important for a wide range of engineering systems such as the shipment structure to ensure structural safety and reliability. In the presented study, we have focused on estimating uncertainty on plate mechanical behaviors due to the impact of the corrosion, such as the shape and depth. Considering the limitation of small quantity of corroded plate samples, we firstly regenerate simulated images based on the statistical shape modeling method. Secondly, these simulated images are imported into a multiphysics-based corrosion simulation platform to reconstruct the shape of corroded plate and the mechanical behavior of the plate with different severity levels of corrosions can be obtained through finite element analysis. Thirdly, uncertainty quantification study is then conducted to understand the statistical characteristic of stochastic behavior of corroded plates from simulated and origin image data. The case study results showed that the statistical characteristic of mechnical behavior from both source data are similar, and the statistical shape modeling method could be useful in situations where there is insufficient sample data for uncertainty quantification.

Keywords: Uncertainty Quantification, Corrosion, Statistical Shape Modeling, Finite Element Analysis.

1. INTRODUCTION

Corrosion is a natural process that occurs when a material reacts with its environment and undergoes chemical changes that can cause it to deteriorate. When a shipment plate corrodes, it can lead to changes in its geometry and material loss on its surface. This can result in failure and fatigue of the plate, which can then cause unexpected changes in stochastic behavior such as stress and deformation of the plate [1]. These changes can lead to the reliability and resilience change of the ship structure while it is in service [2–6] .

Numerous research has put forth probabilistic models to describe corrosion wastage and investigate the time-dependent ultimate strength of ship structures while considering the effects of uniform corrosion. For instance, Hart considered the probabilistic wastage model represented by exponential distribution. The non-uniform thickness and fluctuating geometry of corroded surfaces can have an impact on the estimation of ultimate strength, and it is therefore crucial to take these factors into account when assessing the effect of corrosion on strength deduction[7] Teixeira and Guedes Soares employed a random field model to represent the corroded surface, which incorporates the spatial variability of corrosion wastage. They subsequently proposed a probabilistic model for the ultimate strength of corroded plates using the random field derived by Monte Carlo simulation. However, this method requires a substantial number of analyses due to the use of Monte Carlo simulation[8].

Uncertainty quantification (UQ) is an important method that can help in evaluating and managing the uncertainty and variability associated with the performance and reliability of a given system or structure, such as a plate [9–15]. Zheng et al conduction uncertainty quantification analysis is performed on

---

[1] Corresponding Author. Email: pingfeng@illinois.edu.

Copyright © 2023 by ASME

the structured Si anode system to evaluate the influences of various design variables on its performances and to find the design optimization strategy [16]. Pugliese utilized surrogate models to conduct a reliability assessment based on uncertainty quantification (UQ) for reinforced concrete (RC) bridges that are vulnerable to pitting corrosion[17]. Xie et al. employed adaptive sampling to evaluate the structural reliability of pressure vessels that had experienced corrosion damage [18] . Dong et al. investigated the fatigue reliability of wind turbines considering the influence of corrosion [19]. Sarkar et al The authors developed a model capable of quantifying uncertainties in systems affected by corrosion and compared its performance to the more commonly used Monte-Carlo approach methods[20]. UQ for different corrosion situation with machine learning method has been investigated as well [21, 22].

Among the above UQ method, uncertainty quantification with finite element analysis (FEA) method involves evaluating the impact of various sources of uncertainty on the results obtained from the FEA simulations. This can include variations in material properties, geometric parameters, loading conditions, and other factors that can affect the behavior of the system or structure being analyzed. To perform uncertainty quantification with FEA, one needs to first identify the sources of uncertainty and determine their probability distributions or ranges. This information is then used to generate a set of input parameters for the FEA simulations, which are then run multiple times to obtain a range of possible outcomes.

Statistical shape modeling (SSM) is a computational technique used in the field of computer vision and medical imaging to analyze the shape and variability of objects [23]. SSM is based on the statistical analysis of a set of shapes, often required from imaging modalities, such as MRI or CT, and it can be used for an image augmentation strategy. Liang el al constructed a statistical shape model from clinical 3d CT images and generate a dataset of representative aneurysm shapes, obtaining FEA-predicted risk scores defines as systolic pressure divided by rupture pressure [24]. Li Chunming et al introduced a level set-based approach to shape modeling, which involves a level set function to capture the shape of an object. The authors applied their approach to medical imaging data, showing promising results for segmentation and classification [25]. Zhixian Tang proposed a new image augmentation strategy based on statistical shape model and three-dimensional thin plate spline, which can generate many simulated images from a small number of real images [26].

In this study, we investigate the corrosion geometry as uncertainty source, and it has great potential for causing great variability in the plate's stochastic behavior such as stress and deformation. However, considering the limitation of corroded plate image samples, we proposed to employ the statistical shape modeling method to augment new image samples for uncertainty quantification on stochastic behavior of corroded plates to overcome the insufficient sample issue. The steps of our method are as follows: First, we established a statistical shape model for surface of the corroded plates on original image dataset, then used the model to generate simulated corroded surface plates.

Second, the simulated images are import into COMSOL Multiphysics and converted into solid geometry of the corroded plate. Then the mechanical behavior of plates with corrosion are estimated by finite element analysis. Lastly, we conduct the uncertainty quantification on the mechanical behavior of corroded plates. The characteristics of the mechanical behavior such as von mise stress and deformation can provide information on the uncertainty and variability of the FEA results. To validate the effectiveness of our data augmentation strategy, we compare the statistical characteristics of mechanical behavior from both data source, the results show that data augmentation strategy can be applied to uncertainty quantification when data is not insufficient when a statistical shape model is available.

The remainder of this paper is organized as follows. Methodologies are given in Section 2. Case study and results are provided in Section 3. Conclusions are made in Section 4.

2. THE DEVELOPED METHOD

In this work, we adopt statistical shape modeling strategy to augment corrosion image dataset, and these images are import to COMSOL for finite element analysis. Uncertainty quantification is conducted on the solid mechanical analysis of plate result.

2.1 Statistical Shape Model

Statistical shape modeling is a statistical method employed in computer vision and image processing for studying the shape and diversity of objects in a group. It entails establishing a statistical model of the shape diversity across a collection of objects through a training set, and then leveraging this model to scrutinize and generate fresh shapes [23].

To begin with, the SSM approach entails collecting a group of training images of a specific object or anatomical structure, such as a face, a heart, or a bone. Next, the images are segmented to extract the shape information of the object using landmarks or surface meshes as a common approach. The landmarks for all k points are then concatenated into a vector, which provides a description of the shape information:

$$\mathbf{x} = (x_1, y_1, x_2, y_2 \ldots, x_k, y_k)^{\mathrm{T}} \qquad (1)$$

Subsequently, the training set's dimensionality is reduced by identifying a limited number of modes that effectively portray the observed variation. This is generally achieved by utilizing principal component analysis (PCA). As per Equation (1), every aligned training shape is defined by $3k$ point coordinates in the vector xi. The mean shape can then be generated by calculating the average of all s samples:

$$\bar{\mathbf{x}} = \frac{1}{s} \sum_{i=1}^{s} \mathbf{x}_i \qquad (2)$$

The corresponding covariance matrix S is given by:

$$S = \frac{1}{s-1} \sum_{i=1}^{s} (\mathbf{x}_i - \bar{\mathbf{x}})(\mathbf{x}_i - \bar{\mathbf{x}})^{\mathrm{T}} \qquad (3)$$

A singular value decomposition (SVD) on the covariance matrix S is used to calculate the principal modes of variation $\boldsymbol{\phi}_m$ (eigenvectors) and their respective variances λ_m (eigenvalues). After computing the principal modes of variation, they are

Copyright © 2023 by ASME

arranged in descending order based on their variances, with $\lambda_1 \geq \lambda_2 \geq \cdots \lambda_{s-1}$. Using this ordering, each valid shape can be approximated by a linear combination of the first c modes. After collecting the training set, the subsequent stage involves reducing its dimensionality, which means identifying a concise set of modes that effectively characterizes the observed variation. Typically, this is done through principal component analysis (PCA). In line with Eq. (1), each aligned training shape is defined by 3k point coordinates in the vector $\mathbf{x}_i$. The average shape can be derived by computing the mean over all s samples:

$$\mathbf{x} = \bar{\mathbf{x}} + \sum_m^c b_m \boldsymbol{\phi}_m \qquad (4)$$

where b_m is the shape parameters including scaling and rotation information. By changing the value of b_m, we can generate any number of simulated shapes from model. Typically, the range of b_m lie in a hyperrectangle $-\alpha\sqrt{\lambda_m} \leq b_m \leq \alpha\sqrt{\lambda_m}$, with $\alpha \in (0,2)$. In many cases, c is chosen so that the accumulated variance reaches a certain ratio r of the total variance. Common value for r is 0.9–0.98. It should be noted that the precision of the statistical shape model is contingent upon the value of the ratio r. A value of r near 1 indicates greater accuracy of the shape model.

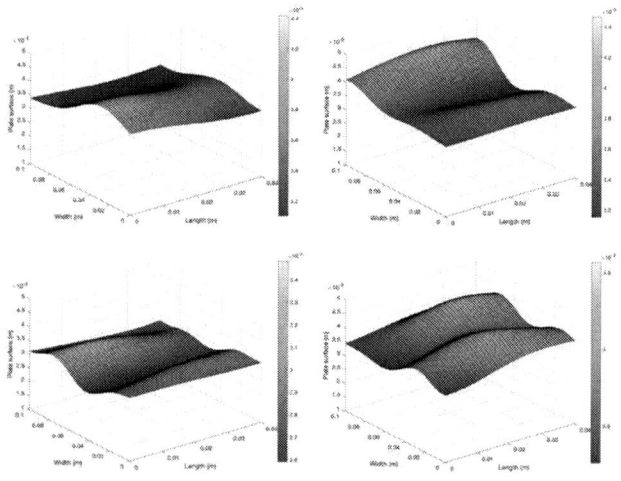

FIGURE 1: ORIGIN CORROSION SURFACE

In Figure 1, a portion of the corroded surface of the original plate is shown, whereas Figure 2 illustrates a portion of the plate surface after undergoing simulation using the statistical shape modeling technique. The color scheme on the plot indicates the height of the surface at each point, giving insight into the plate surface elevation. The width and length of the plate are represented by their respective dimensions in the plot. The difference between the two plots in Figure 1 and Figure 2 can be ascribed to the existence of corrosion on the surface of the original plate, resulting in its unevenness, irregularity, and stochastic field properties. To quantitatively assess the similarity between two surfaces, we employ the root mean square error

(RMSE) metric to quantify their dissimilarity. After computing the RMSE for 20 different sets, the average value is determined to be 8.2483×10^{-4}. This result indicates the effectiveness and precision of the statistical shape model for augmenting images.

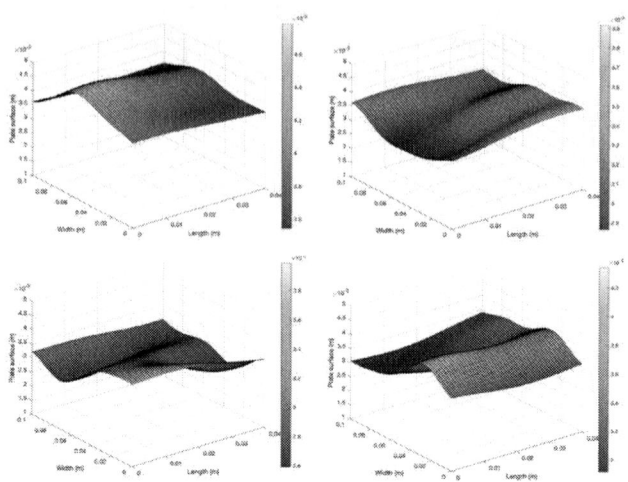

FIGURE 2: SIMULATED CORROSION SURFACE

2.2 Uncertainty Quantification

Uncertainties come from environment factors, coating quality, alloy composition, maintenance, operational hours, corroded geometry, and so on. The uncertainties in the input variable will affect the output of mechanical behavior of corroded plates. Uncertainty quantification with finite element analysis (FEA) is a process of analyzing and quantifying the uncertainties in the output of an FEA simulation due to variations in the input parameters. Denote output variables by $\mathbf{Y} = (Y_1, Y_2, \ldots, Y_m)^{\mathrm{T}}$ and input variables by $\mathbf{X} = (X_1, X_2, \ldots, X_n)^{\mathrm{T}}$, where m and n are the numbers of output and input variables, respectively. In this study, the input variable X_1 is the corrosion geometry variability on the surface of the plate, and output variable Y_i are von mise stress Y_1 and plate deflection Y_2. With the simulation experiment results, the statistical characteristics (mean and variance) of variables Y_1 and Y_2 are calculated using Eq. (5) and Eq. (6)

$$\bar{Y}_i = \frac{\sum_{i=1}^N Y_{ij}}{N} (j = 1,2) \qquad (5)$$

$$\sigma_{Y_i} = \sqrt{\frac{\sum_{i=1}^N (Y_{ij} - \bar{Y}_i)^2}{N-1}} (j = 1,2) \qquad (6)$$

3. CASE STUDY AND RESULTS

3.1 Datasets

In this study, there are two different sources datasets. The first dataset is original ship image, and the other one is simulated images based on statistical shape modeling method. According

Copyright © 2023 by ASME

to Ref [1], the real images of corrodes surfaces is approximated by random field model, and they follow normal distribution with a mean and variance (μ_{org} and var_{org}). It is assumed that covariance function $\rho(x_i, x_j)$ shown in Eq. (7) represents the correlation property between two points x_i and x_j of the discretized random field.

$$\rho(x_i, x_j) = var_{org} \exp\left(-\frac{|x_i - x_j|}{d}\right) \quad (7)$$

Then the random field of the corroded surface can be represented by a set of uncorrelated random variables ξ_i with Kullback-Leibler divergence expansion method [27–29] as in Eq. (8) and Fig. 3

$$\mathbf{v} = \mu + \sum_{i=1}^{N} \xi_i \sqrt{\lambda_i} \boldsymbol{\phi}_i \quad (8)$$

where vector $\mathbf{v} = (v_1, v_2, \ldots, v_N)^T$ consists of N random variables that represent thickness diminution at each point x_i, as illustrated in Fig.2. The mean vector μ of the original random field reflects the average thickness diminution, and in this study, it is assumed that all its discretized points have the same constant value μ_{org}. The second term of Equation (8) represents the random fluctuation of the surface around the mean value. The eigenvalues λ_i and normalized eigenvectors $\boldsymbol{\phi}_i$ of the covariance matrix C are included in the equation, where ξ_i is a set of N uncorrelated random variables with zero mean and unit variance. The covariance matrix C is determined by Eq. (7). The dataset of corroded surface images used in this study consists of 20 images and has the parameters listed in Table 1.

TABLE 1 PARAMTERS FOR EXPERIMENT IMAGES

	Parameters	Values
Plate size	Plate dimension	$40 \times 100 \times t_0$ mm
	Original thickness t_0	10mm
	Number of discretized points	4141
	Correlation length	30mm
Random field properties	Mean μ_{org}	3.465mm
	Variance var_{org}	0.564mm
	Number of random variables	10

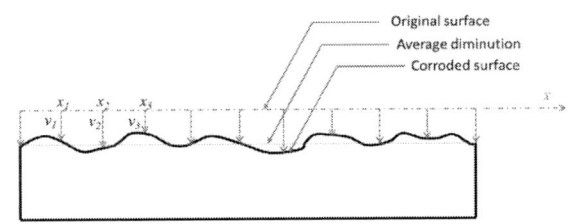

FIGURE 3: RANDOM FIELD OF CORRODED PLATE

We use the image augmentation strategy proposed in Section 2.1 and obtain 100 simulated images as the new dataset. Now we import both datasets into COMSOL and then converted

into the solid corroded geometry. The plate is simply fixed on the left end and a downward load applies to the plate right end as shown in Fig. 4.

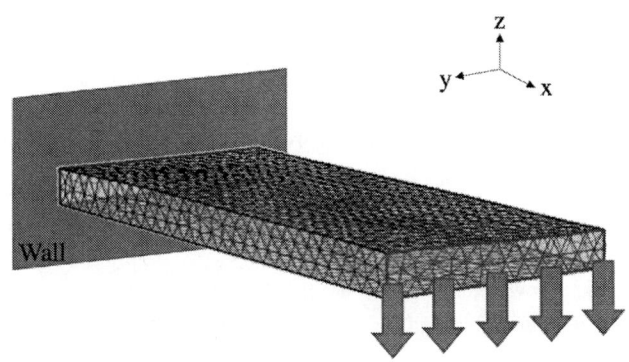

FIGURE 4: THE FEA MODEL OF THE PLATE

The Young's modulus and Poisson's ratio used in the model is 204.5 GPa and 0.3, respectively. The impact of the model's geometric factor is quantified by conducting the FE simulation. We have run 20 simulations and 100 simulations from original dataset and new dataset, respectively. The simulation results of von Mises stress and displacement are given in Fig. 5.

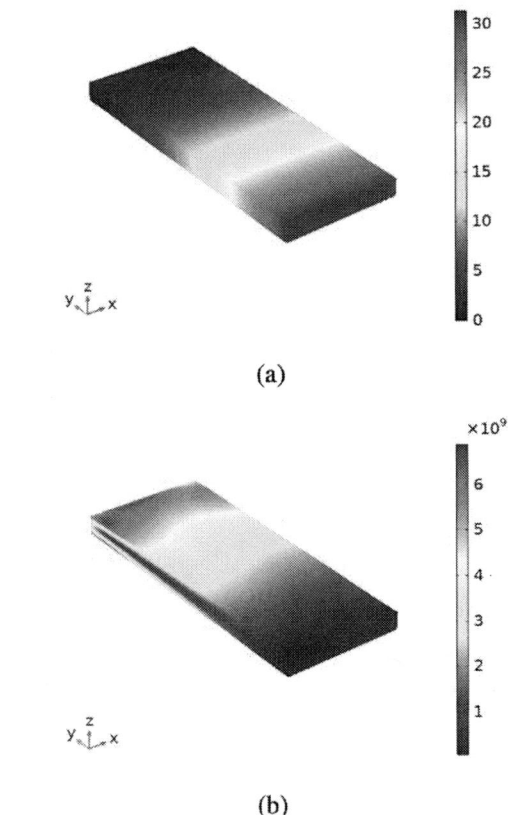

FIGURE 5: (A)TOTAL DISPLACEMENT(MM) AND (B)VON MISES STRESS (N/M^2)

We predict the von Mises stress at the left end of plate, and displacement at the right end. The statistical characteristics such as mean and standard deviation are calculated based on Eq. (5) and Eq. (6). The results are provided in Table 2. The mean and standard deviation values of von Mise stress for two datasets are similar. The origin dataset with 20 samples has mean μ_1 and standard deviation σ_1 values of $5.4369 \times 10^9 \frac{N}{m^2}$ and $7.1461 \times 10^8 \frac{N}{m^2}$, respectively, while the new dataset with 100 samples has mean μ_2 and standard deviation σ_2 values of $5.4965 \times 10^9 \frac{N}{m^2}$ and $7.2644 \times 10^8 \frac{N}{m^2}$. In order to confirm the statistical properties of a new dataset, we will choose 20 samples at random from the new set of image samples and the mean and standard deviation are denoted as $\mu_3 = 5.4449 \times 10^9 \frac{N}{m^2}$ and $\sigma_3 = 7.4421 \times 10^8 \frac{N}{m^2}$, respectively.

Likewise, the comparison of statistical characteristics of the deformation from two datasets is provided in Table 2

TABLE 2 STATISTICAL VALUE OF TWO DATASETS

	Displacement (mm)	Von Mise stress (N/m^2)
μ_1	29.6488	5.4369×10^9
σ_1	5.3236	7.1461×10^8
μ_2	29.7154	5.4965×10^9
σ_2	6.1029	7.2644×10^8
μ_3	30.4316	5.4449×10^9
σ_3	6.6030	7.4421×10^8

3.2 Statistical Analysis

A Welch's t-test was also conducted to determine if the means of two groups were significantly different. The null hypothesis of the test is that there is no difference between the means of the groups, while the alternative hypothesis is that there is a significant difference. A significance level of 0.05 was used.

The Statistical and Machine Learning Toolbox in MATLAB was used to perform the test. The built-in function 'ttest2' was employed to compare the mean of two sets of data. This function returns a value of '0' if the means of the groups are not significantly different, and '1' otherwise.

The results of the two-sample t-test indicated that there was no significant difference between the mean displacement and von-mise stress values of the original and augmented datasets at a 5% significance level. Figure 6 and 7 show comparisons of plate deformation and von-mise stress from two different datasets.

4. CONCLUSION

In this preliminary study we incorporate the statistical shape modeling method to augment corrosion image for uncertainty quantification. The eigen values and eigen vectors are extracted from covariance matrix obtained from origin image information. Then the statistical shape model of corrode plate is constructed and new image data can be regenerate by the statical shape model. We use RMSE to quantitively access the similarity between two surfaces showing the effectiveness of proposed method. By conducting FEA, we predict the von Mise stress and displacement of corroded plates, and their corresponding statistical values of mean and standard deviation are achieved, respectively. The result indicates that the statistical values of original dataset are similar to those of augmented dataset. It suggests that the statistical shape model is a useful tool for augmenting data samples in uncertainty quantification when there is inadequate amount of image samples available. As part of our future work, we intend to employ statistical shape modeling to assist with uncertainty quantification using a practical example.

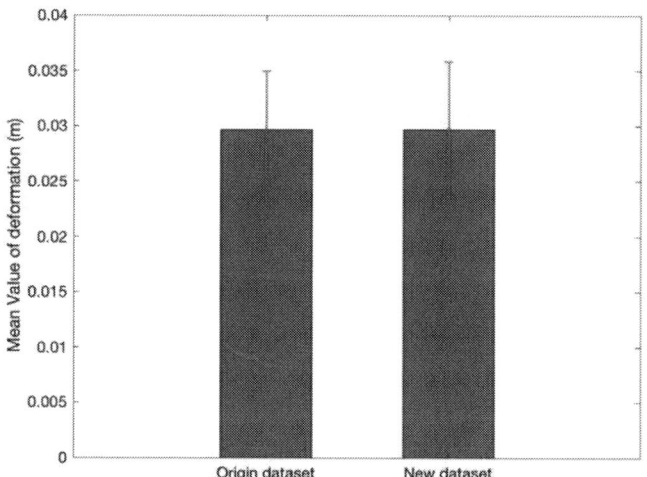

FIGURE 6: COMPARISON OF DEFORMATION FOR TWO DATASETS

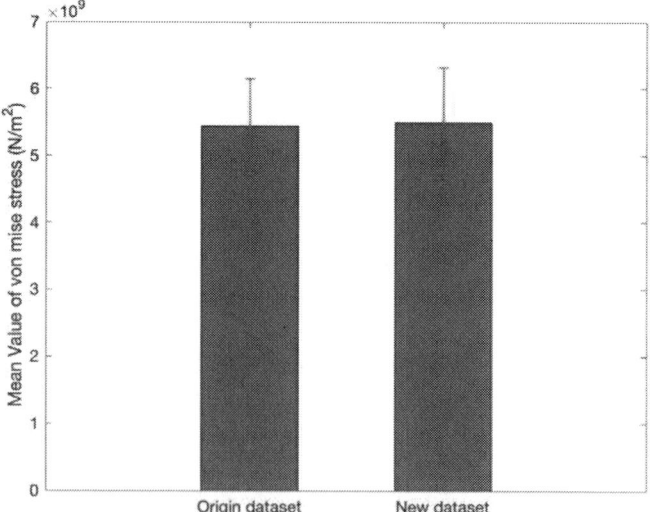

FIGURE 7: COMPARISON OF VON-MISE STRESS FOR TWO DATASETS

Copyright © 2023 by ASME

ACKNOWLEDGEMENTS

This research is partially supported by the National Science Foundation (NSF) the Engineering Research Center for Power Optimization of Electro-Thermal Systems (POETS) with cooperative agreement EEC-1449548, and the Alfred P. Sloan Foundation through the Energy and Environmental Sensors program with grant # G-2020-12455.

REFERENCES

[1] M. M. Htun, Y. Kawamura, and M. Ajiki, "A Study on Random Field Model for Representation of Geometry of Corroded Plates and Estimation of Stochastic Properties of Their Strength," *Journal of the Japan Society of Naval Architects and Ocean Engineers*, vol. 18, pp. 91–99, 2013.

[2] H. Wu and X. Du, "System Reliability Analysis With Second Order Saddlepoint Approximation," in *International Design Engineering Technical Conferences and Computers and Information in Engineering Conference*, American Society of Mechanical Engineers, 2019, p. V02BT03A039.

[3] H. Wu and X. Du, "System Reliability Analysis With Second-Order Saddlepoint Approximation," *ASCE-ASME Journal of Risk and Uncertainty in Engineering Systems, Part B: Mechanical Engineering*, vol. 6, no. 4, p. 041001, 2020.

[4] H. Wu and X. Du, "Time-Dependent System Reliability Analysis With Second Order Reliability Method," in *International Design Engineering Technical Conferences and Computers and Information in Engineering Conference*, American Society of Mechanical Engineers, 2020, p. V11BT11A045.

[5] H. Wu and X. Du, "Time-and Space-Dependent Reliability-Based Design with Envelope Method," *Journal of Mechanical Design*, pp. 1–34, 2023.

[6] Hao Wu, "PROBABILISTIC DESIGN AND RELIABILITY ANALYSIS WITH KRIGING AND ENVELOPE METHODS," Purdue University Graduate School, 2022. doi: https://doi.org/10.25394/PGS.19653132.v1.

[7] D. K. Hart, S. E. Rutherford, and A. H. Wickham, "Structural reliability analysis of stiffened panels," *Naval Architect*, no. October 1986, 1985.

[8] A. P. Teixeira, C. Guedes Soares, and G. Wang, "Probabilistic modelling of the ultimate strength of ship plates with non-uniform corrosion," *J Mar Sci Technol*, vol. 18, pp. 115–132, 2013.

[9] H. Yu *et al.*, "A new noninvasive and patient-specific hemodynamic index for the severity of renal stenosis and outcome of interventional treatment," *Int J Numer Method Biomed Eng*, vol. 38, no. 7, p. e3611, 2022.

[10] H. Yu *et al.*, "Inlet and Outlet Boundary Conditions and Uncertainty Quantification in Volumetric Lattice Boltzmann Method for Image-Based Computational Hemodynamics," *Fluids*, vol. 7, no. 1, p. 30, 2022.

[11] Y. Xu and P. Wang, "Adaptive surrogate models for uncertainty quantification with partially observed information," in *AIAA SCITECH 2022 Forum*, 2022, p. 1439.

[12] H. Li, J. Yin, and X. Du, "Uncertainty Quantification of Physics-Based Label-Free Deep Learning and Probabilistic Prediction of Extreme Events," in *International Design Engineering Technical Conferences and Computers and Information in Engineering Conference*, American Society of Mechanical Engineers, 2022, p. V03BT03A001.

[13] M. M. Islam, H. Li, H. Yu, and X. Du, "Physics-based regression vs. cfd for hagen-poiseuille and womersley flows and uncertainty quantification," in *Eleventh International Conference on Computational Fluid Dynamics, volume ICCFD11, pages ICCFD11–3301*. ICCFD, 2022.

[14] H. Wu, Z. Zhu, and X. Du, "System reliability analysis with autocorrelated kriging predictions," *Journal of Mechanical Design*, vol. 142, no. 10, 2020.

[15] Z. Zheng and P. Wang, "Uncertainty Quantification Analysis on Silicon Electrodeposition Process Via Numerical Simulation Methods," *ASCE-ASME J Risk and Uncert in Engrg Sys Part B Mech Engrg*, vol. 8, no. 1, 2022.

[16] Z. Zheng, Y. Xu, and P. Wang, "Uncertainty quantification analysis on mechanical properties of the structured silicon anode via surrogate models," *J Electrochem Soc*, vol. 168, no. 4, p. 040508, 2021.

[17] F. Pugliese, R. De Risi, and L. Di Sarno, "Reliability assessment of existing RC bridges with spatially-variable pitting corrosion subjected to increasing traffic demand," *Reliab Eng Syst Saf*, vol. 218, p. 108137, 2022.

[18] C. Xie, G. Li, and F. Wei, "An integrated QMU approach to structural reliability assessment based on evidence theory and kriging model with adaptive sampling," *Reliab Eng Syst Saf*, vol. 171, pp. 112–122, 2018.

[19] W. Dong, T. Moan, and Z. Gao, "Fatigue reliability analysis of the jacket support structure for offshore wind turbine considering the effect of corrosion and inspection," *Reliab Eng Syst Saf*, vol. 106, pp. 11–27, 2012.

[20] S. Sarkar, J. E. Warner, W. Aquino, and M. D. Grigoriu, "Stochastic reduced order models for uncertainty quantification of intergranular corrosion rates," *Corros Sci*, vol. 80, pp. 257–268, 2014.

[21] Y. Ma, L. Wang, J. Zhang, Y. Xiang, T. Peng, and Y. Liu, "Hybrid uncertainty quantification for probabilistic corrosion damage prediction for aging RC bridges," *Journal of Materials in Civil Engineering*, vol. 27, no. 4, p. 04014152, 2015.

[22] D.-C. Feng, S.-C. Xie, Y. Li, and L. Jin, "Time-dependent reliability-based redundancy assessment of deteriorated RC structures against progressive collapse considering corrosion effect," *Structural Safety*, vol. 89, p. 102061, 2021.

Copyright © 2023 by ASME

[23] T. Heimann and H.-P. Meinzer, "Statistical shape models for 3D medical image segmentation: a review," *Med Image Anal*, vol. 13, no. 4, pp. 543–563, 2009.

[24] L. Liang, M. Liu, C. Martin, J. A. Elefteriades, and W. Sun, "A machine learning approach to investigate the relationship between shape features and numerically predicted risk of ascending aortic aneurysm," *Biomech Model Mechanobiol*, vol. 16, no. 5, pp. 1519–1533, 2017, doi: 10.1007/s10237-017-0903-9.

[25] C. Li, R. Huang, Z. Ding, J. C. Gatenby, D. N. Metaxas, and J. C. Gore, "A level set method for image segmentation in the presence of intensity inhomogeneities with application to MRI," *IEEE transactions on image processing*, vol. 20, no. 7, pp. 2007–2016, 2011.

[26] Z. Tang, K. Chen, M. Pan, M. Wang, and Z. Song, "An augmentation strategy for medical image processing based on statistical shape model and 3D thin plate spline for deep learning," *IEEE Access*, vol. 7, pp. 133111–133121, 2019.

[27] H. Wu and X. Du, "Envelope Method for Time-and Space-Dependent Reliability-Based Design," in *International Design Engineering Technical Conferences and Computers and Information in Engineering Conference*, American Society of Mechanical Engineers, 2022, p. V03BT03A002.

[28] H. Wu and X. Du, "Envelope Method for Time-and Space-Dependent Reliability Prediction," *ASCE-ASME Journal of Risk and Uncertainty in Engineering Systems, Part B: Mechanical Engineering*, vol. 8, no. 4, p. 041201, 2022.

[29] H. Wu, Z. Hu, and X. Du, "Time-dependent system reliability analysis with second-order reliability method," *Journal of Mechanical Design*, vol. 143, no. 3, 2021.

Proceedings of the ASME 2023
International Design Engineering Technical Conferences and
Computers and Information in Engineering Conference
IDETC-CIE2023
August 20-23, 2023, Boston, Massachusetts

DETC2023-117348

AN EFFICIENT SURROGATE MODELING METHOD FOR RELIABILITY-BASED GLOBAL PATH PLANNING OF OFF-ROAD AUTONOMOUS GROUND VEHICLES

Jianhua Yin[1], Zhen Hu[1,*], Zissimos P. Mourelatos[2], David Gorsich[3], Amandeep Singh[3], and Seth Tau[3]

[1]Department of Industrial and Manufacturing Systems Engineering, University of Michigan-Dearborn, Dearborn, MI 48128, USA
[2]Mechanical Engineering Department, Oakland University, Rochester, MI 48309, USA
[3]U.S. Army Combat Capabilities Development Command, Ground Vehicle Systems Center, Warren, MI 48397, USA

ABSTRACT

The off-road autonomous ground vehicle operation is challenging because of the complex and highly uncertain working environment they operate in. This brings risks to their mission success. Physics-based global path planning combining physics-informed surrogate modeling with path planning is recently employed to address this issue by ensuring a reliability to complete the mission. However, the current two-stage methods require a relatively high computational cost to obtain the reliability map undermining the practicability for real-life applications. To decrease the computational cost, this study proposes an efficient surrogate modeling framework for physics-based path planning under uncertainty. The proposed method couples adaptive surrogate modeling with path planning sequentially through several iterations of path planning and path verification. Instead of obtaining an accurate reliability map for the whole target area in the two-stage approach, the proposed method only refines the map in the vicinity of the path identified, thereby reducing the computational cost. The final path is determined if the path verification is successful; i.e. the specified mission reliability for the path is met. The presented case study demonstrates that the proposed method can significantly reduce the number of surrogate refinement iterations and reduce the computational cost for physics-based path planning.

1 INTRODUCTION

Off-road autonomous ground vehicles (AGVs) have the potential to reduce human loss and save labor cost if they are de-

ployed in harsh or boring working conditions, such as battlefields [1], outer space [2], and agriculture industry [3]. To ensure the success of AGVs, it is critical to perform global path planning or mission planning [4–6] to avoid potential failures during their operation.

The travelled distance is commonly used in path planning as it is related to the fuel/time cost of AGVs during a mission. Given a terrain/soil map with initial and target locations on it, common path planning methods can identify the shortest path. However, considering only the distance may not ensure the mission success due to the complex and highly uncertain working environment of AGVs. For example, for the shortest path identified, some locations may not be passable due to non-ideal terrain conditions, such as loose sand, mud pod, and soft soil [7]. In addition, as the terrain conditions vary with weather and time, there is variation of vehicle mobility [8,9]. Therefore, there exists a chance for the AGV to lose mobility since mobility is uncertain.

To address the above issues, physics-based global path planning is recently used to identify a reliable path for AGVs, which combines physics-based modeling and simulation (M&S) with path planning. The advantages are summarized as follows. First, physics-based M&S uses known information, such as soil conditions, and terrain slope, to predict various vehicle mobility quantities. Second, multiple M&S runs can be used to estimate the reliability of AGV mobility at certain locations given the existence of uncertainty.

It is known that the accuracy and efficiency of physics-based M&S impact the feasibility and applicability of physics-based path planning. One early category of studies in the 1960s tries to improve the fidelity of M&S. Initially, semi-empirical models were widely used to predict the vehicle mobility as in the

---

*Corresponding author, zhennhu@umich.edu. DISTRIBUTION A. Approved for public release; distribution unlimited. OPSEC7334

This material is declared a work of the U.S. Government and is not subject to Copyright protection in the United States.

Copyright © 2023 by The United States Government

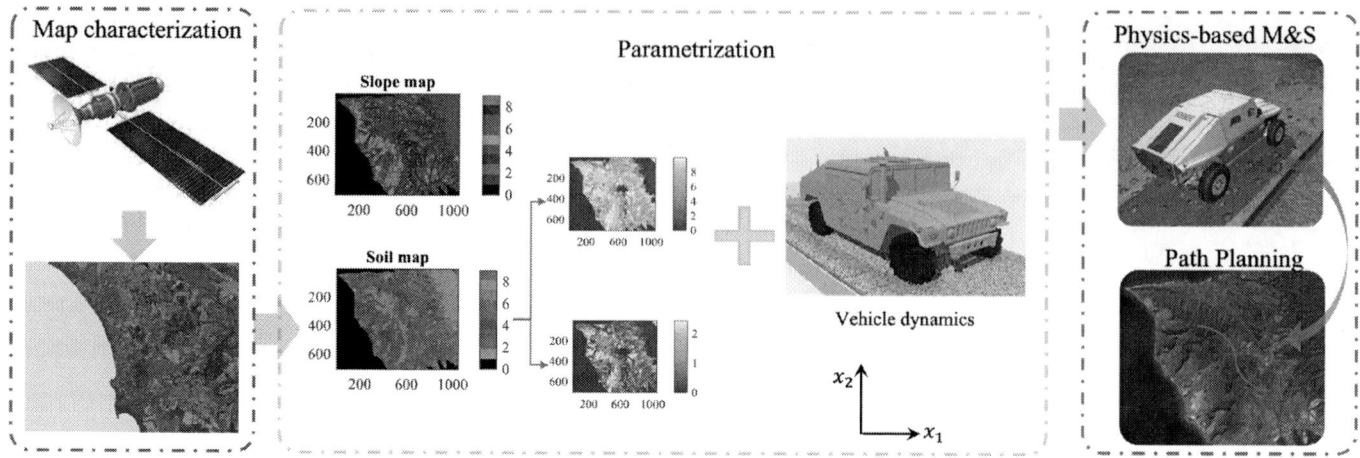

FIGURE 1. Illustration of Physics-based global path planning.

NATO Reference Mobility Model (NRMM) [10]. However, the accuracy of this type of modeling is poor although the computational cost is low. With the development of computational physics and computer technology, high-fidelity computational models are widely used for mobility predictions. These models are based on solving partial differential equations using advanced numerical methods, such as the finite element method (FEM). For instance, the next generation NRMM model (NG-NRMM) developed by the U.S. Army and NATO [11], uses tire-terrain co-simulation by FEM and discrete element method [12]. However, these model are usually computationally expensive for mobility reliability analysis.

To improve the efficiency in mobility prediction while ensuring accuracy, machine learning-based data-driven approaches are increasingly used based on a limited number of high-fidelity simulations [13]. Initially, Kriging metamodeling [14] combined with Monte Carlo Simulation (MCS) was used for uncertainty quantification of vehicle mobility [13]. Dynamic Kriging was also used to construct a stochastic mobility map to predict mobility reliability [15]. Recently research in [7,16] used a two-stage approach considering spatial-dependent uncertainty in mobility reliability prediction.

Current machine learning-based approaches try however, to obtain an accurate mobility reliability prediction across the whole terrain/soil map, which could result in excessive computational cost since more high-fidelity simulations are needed as the number of uncertain parameters increases [17–19]. Also, if the terrain/soil map is large, a high-resolution discretization of the map is needed. To address these issues, this study predicts the mobility reliability only as needed by the path planning assessment. An efficiency improvement is achieved by coupling adaptive surrogate modeling with global path planning. The mobility reliability is evaluated only in the vicinity of the identified path instead of doing so everywhere on the terrain/soil map avoiding

unnecessary computational cost in calculating the mobility reliability in areas not needed for the path planning.

The remainder of this paper is organized as follows. A detailed comparison of the existing two-stage approach is provided in Section 2 to introduce the research area. The proposed efficient reliability-based global path planning is provided in Section 3. We use a mathematical example to demonstrate the proposed approach in Section 4. Finally, concluding remarks are provided in Section 5.

2 BACKGROUND
2.1 Physics-based Global Path Planning

Global path planning of off-road AGVs identifies the shortest path on a map from an initial point to a target point while satisfying several requirements. Fig. 1 shows the basic steps of physics-based global path planning.

First, a target map providing the working AGV environment is obtained by satellite or from a geological survey.
Second, the target map is parameterized to obtain the parameters needed in physics-based M&S, including local terrain height and slope, soil parameters, and vehicle parameters.
Third, a physics-based M&S model is constructed to predict the mobility at certain locations on the map. The locations that the AGV loses mobility are characterized as obstacles. Otherwise, it is free space if the AGV maintains mobility.
Fourth, the shortest path is identified considering several constraints, such as path distance, and multiple mobility considerations.

2.2 Reliability-based Global Path Planning

The uncertainty in the working environment of AGVs leads to uncertainty in vehicle mobility prediction. Therefore, the ob-

Copyright © 2023 by The United States Government

stacle and free space defined according to the vehicle mobility is probabilistic. The AGV has a certain probability to fail if it follows the path identified by deterministic path planning. To ensure a specific mission reliability is met, reliability constraints are applied to the physics-based path planning resulting in a reliability-based path planning. Its mathematical definition is stated as

$$\gamma^* = \arg\min_{\gamma \in \Omega}\{C(\gamma)\}$$

s.t.

$$
\begin{aligned}
&\mathbf{x}_0 \in \gamma = \mathbf{x}_{ini}; \ \mathbf{x}_{end} \in \gamma = \mathbf{x}_{goal} \\
&\mathbf{x}_i \in \Omega_{free}, \ \forall \mathbf{x}_i \in \gamma \\
&\Omega_{free} = \{\mathbf{x} | R(\mathbf{x}) > R_t, \ \forall \mathbf{x} \in \Omega\} \\
&\mathbf{x}_{goal} = \{\mathbf{x} \in \Omega_{free} | \|\mathbf{x} - \mathbf{x}_{goal}\| \le \varepsilon\},
\end{aligned}
\tag{1}
$$

where γ^* is the shortest path identified, Ω represents the domain of path planning which is divided into free space (Ω_{free}) and obstacle space (Ω_{ob}); $C(\gamma)$ is the cost function in terms of the path (γ); $\mathbf{x}_0$ and $\mathbf{x}_{end}$ are respectively the starting point and end point of a path; $\mathbf{x}_{ini}$ and $\mathbf{x}_{goal}$ are the initial point and the goal of the mission; $\mathbf{x}_i$ is an arbitrary point on a path; and $R(\mathbf{x})$ indicates the mobility reliability at current location $\mathbf{x}$, which should satisfy the reliability requirement R_t.

The reliability is called state mobility reliability (SMR) [7], and is defined by

$$R(\mathbf{x}) = \Pr\{M(\mathbf{V}, \mathbf{S}(\mathbf{x})) > y_e\} = \iint\limits_{Y(\mathbf{x}) > y_e} f_{\mathbf{V},\mathbf{S}}(\mathbf{v}, \mathbf{s}(\mathbf{x}))d\mathbf{v}d\mathbf{s}, \tag{2}$$

where $\mathbf{V}$ is the vehicle-related parameters, $\mathbf{S}(\mathbf{x})$ is the terrain-related parameters, $M(\cdot, \cdot)$ is the vehicle mobility model, y_e is the mobility threshold so that failure occurs if the mobility quantity is less than y_e, and $f_{\mathbf{V},\mathbf{S}}(\mathbf{v}, \mathbf{s}(\mathbf{x})$ is the joint probability density function in terms of $\mathbf{V}$ and $\mathbf{S}(\mathbf{x})$.

2.3 RRT*

Rapidly-exploring random tree star (RRT*) is a random sampling-based path planning algorithm, in which a tree grows from the starting point until the branches of the tree reach the goal. The tree-growing process represents space exploration. A detailed review of RRT* can be found in [20]. There are three major steps in the RRT* algorithm:

Collision check: Samples are generated in Ω. If a random sample ($\mathbf{x}_{rand}$) is in Ω_{ob}, the sample is rejected. The sample is accepted if it is in Ω_{free}.
Steering: After a sample passes the collision check, the nearest node $\mathbf{x}_{nst}$ of $\mathbf{x}_{rand}$ is found. If there is no obstacle

between $\mathbf{x}_{nst}$ and $\mathbf{x}_{rand}$, $\mathbf{x}_{rand}$ is inserted into the tree. Otherwise, a steering function is used to generate a new node $\mathbf{x}_{new}$ which is inserted into the tree.
Rewiring: This is the most important feature of RRT*. Its main role is to remove all unnecessary branches of the tree and improve the quality of the path identified. With more iterations of random sampling, this process may take longer time.

2.4 Research Need for Efficient Path Planning

Ensuring the reliability of an identified path is challenging for many reasons. First, uncertainty prevails in the working environment of AGVs which could introduce multiple failure modes of the AGVs. Second, the computational cost of mobility reliability prediction needs to be decreased to improve the practicability of physics-based global path planning.

The current two-stage approach is illustrated in Fig. 2. In the first stage, initial data is generated by high-fidelity M&S to create an initial surrogate model, which may not be accurate enough with the initial data. Then, active learning is used to adaptively refine the surrogate modeling with new mobility M&S data until the accuracy is satisfied across the map. With the accurate mobility reliability map from the ultimate surrogate model, the shortest path which satisfies reliability constraints is found by path planning. Since the final path only traverses part of the target map, the refinement of surrogate model in the other area that the path does not pass could be waste of computational resource. Especially when the map is large or the input dimension of the mobility M&S is high. Therefore, it is necessary to develop more efficient physics-based global path planning method to improve efficiency.

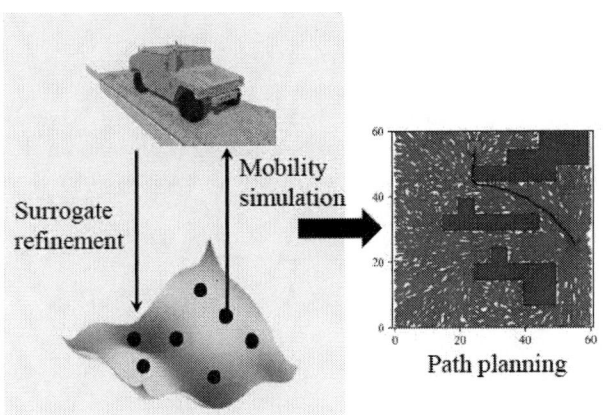

FIGURE 2. Illustration of the two-stage method.

Copyright © 2023 by The United States Government

3 EFFICIENT SURROGATE MODELING FOR PHYSICS-BASED GLOBAL PATH PLANNING

To improve the computational efficiency, an efficient reliability-based path planning approach is proposed, which consists of three major parts, uncertainty modeling, surrogate modeling & path planning, path verification & trim tree. An illustration of the proposed approach is shown in Fig. 3. The central strategy is to couple the path planning with a surrogate model refinement only in the vicinity of the identified path avoiding the unnecessary computational cost of refining the surrogate in whole map using the two-stage approach. The computational efficiency of the proposed approach is therefore, improved.

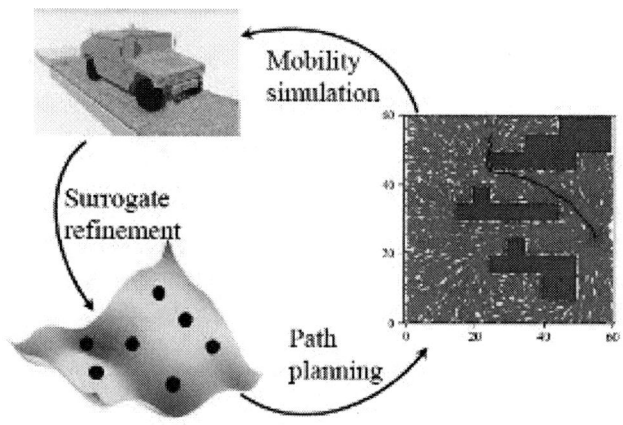

FIGURE 3. Illustration of the proposed method.

3.1 Uncertainty Modeling

The uncertainty sources in the working environment of AGVs are modeled as random fields in this paper [8]. We use the Karhunen-Loeve (K-L) expansion to generate random field realizations. The K-L expansion of a random field is given by

$$s(\mathbf{x}) = \mu(\mathbf{x}) + \sigma(\mathbf{x}) \sum_{i=1}^{T_n} \sqrt{\lambda_i} \varphi_i(\mathbf{x}) \xi_i, \qquad (3)$$

where $s(\mathbf{x})$ is a realization of the random field $S(\mathbf{x})$, $\mu(\mathbf{x})$ and $\sigma(\mathbf{x})$ are respectively the spatial-dependent mean and standard deviation functions of the random field, $\mathbf{x}$ is the spatial location, T_n is the number of truncation terms, λ_i and $\varphi_i(\mathbf{x})$ are the eigenvalue and eigenvector from eigenanalysis of a correlation matrix, respectively, and ξ_i is the ith independent standard Gaussian random variable. The importance of the truncation terms is in de-

creasing order. In this work, we select the first T_n terms whose sum of the impact is more than 98%.

The correlation matrix is computed using the correlations among a set of mesh nodes according to the correlation function given by

$$k_{12} = \exp\left\{-\|(\mathbf{x}_1 - \mathbf{x}_2) \oslash \boldsymbol{\theta}_{\mathbf{x}}\|_2\right\}, \qquad (4)$$

where k_{12} is the correlation between two arbitrary points, $\mathbf{x}_1$ and $\mathbf{x}_2$ represent two arbitrary points. The vector $\boldsymbol{\theta}_{\mathbf{x}}$ represents the correlation length, and $\oslash$ denotes the Hadamard division. Additionally, the operation $\|\cdot\|_2$ represents the ℓ^2-norm.

3.2 Surrogate Modeling & Path Planning

We use Gaussian Process Regression (GPR) to construct a surrogate model based on the limited high-fidelity simulations. The training data is generated by PyChrono [21] simulations. It is noted that the vehicle-related parameters are not considered in this study. The detailed surrogate modeling is summarized below.

First, we use Latin Hypercube Sampling (LHS) to generate a group of training data for GPR model construction, which is denoted by $\mathscr{D} = \{\mathbf{s}^{(i)}, y^{(i)}\}_{i=1}^{N}$.

Second, a GPR model is constructed using the training data, which is given by

$$\hat{y} = G(\mathbf{s}) = \mathbf{f}(\mathbf{s})^{\mathrm{T}} \beta + \zeta, \qquad (5)$$

where $\hat{y}$ is the predicted response by the GPR model which is the maximum attainable speed, $G(\mathbf{s})$ means the GPR model in terms of the terrain-related parameters $\mathbf{s}$, $\mathbf{f}(\cdot)$ is a vector of basis functions, β is a vector of coefficients of the basis functions, and ζ is a noise term following a Gaussian process with mean of 0 and variance of σ^2, namely, $\zeta \sim N\left(0, \sigma^2\right)$. Please refer to [22] for detailed GPR model construction.

The prediction by a GPR model follows a Gaussian distribution, which is denoted by

$$\hat{y} = G(\mathbf{s}) \sim N\left(\mu_y(\mathbf{s}), \sigma_y^2(\mathbf{s})\right), \qquad (6)$$

where $\mu_y(\mathbf{s})$ and $\sigma_y^2(\mathbf{s})$ are respectively the mean response and corresponding variance of the prediction $\hat{y}$ of the GPR model.

Third, Based on the GPR model, a reliability map is generated. If the reliability at certain locations is smaller than a threshold, this location is assumed as obstacle. On the other hand, this location is free to move if the reliability is greater than a threshold. Based on the reliability map, we can easily identify the obstacles in the map. The shortest path is found with the reliability map by rapidly exploring random tree star (RRT*) algorithm [20].

Copyright © 2023 by The United States Government

3.3 Path Verification & Trim Tree

Path verification is finished by surrogate model verification, specifically is to verify the uncertainty of surrogate model prediction. As mentioned previously, we only verify the locations along the path found. Therefore, the refinement area is reduced significantly. Assume the locations of the path is $\{\tilde{\mathbf{x}}_1, ..., \tilde{\mathbf{x}}_q\}$, the sample pool formed by the Monte Carlo Simulation (MCS) random samples at the locations is denoted by

$$P = \left[\{\mathbf{s}^{(i)}(\tilde{\mathbf{x}}_1)\}_{i=1}^{N_{mcs}}, ..., \{\mathbf{s}^{(i)}(\tilde{\mathbf{x}}_p)\}_{i=1}^{N_{mcs}} \right]^{\mathrm{T}}, \tag{7}$$

where $\{\mathbf{s}^{(i)}(\tilde{\mathbf{x}}_j)\}_{i=1}^{N_{mcs}}$ denotes all samples at $\tilde{\mathbf{x}}_j$, $j = 1, ..., p$.

We evaluate the sample pool using the GPR model in Eq. (5) and we have

$$\hat{y}(\mathbf{s}^{(i)}(\tilde{\mathbf{x}}_j)) \sim N\left(\mu_y(\mathbf{s}^{(i)}(\tilde{\mathbf{x}}_j)), \sigma_y^2(\mathbf{s}^{(i)}(\tilde{\mathbf{x}}_j))\right), \tag{8}$$

where $i = 1, ..., N_{mcs}$ represents the MCS sample index, and $j = 1, ..., q$ represents the location index of the path.

Then, the U-learning function is used to identify the optimal sample to refine the surrogate model, the learning function is given by

$$U((\mathbf{s}^{(i)}(\tilde{\mathbf{x}}_j))) = \left| \frac{\mu_y(\mathbf{s}^{(i)}(\tilde{\mathbf{x}}_j)) - y_e}{\sigma_y(\mathbf{s}^{(i)}(\tilde{\mathbf{x}}_j))} \right|, \tag{9}$$

where the value of $U(\mathbf{s})$ indicates the probability of misclassification, the smaller is $U(\mathbf{s})$, the higher chance of misclassification. We therefore find the optimal new training point by

$$\mathbf{s}^* = \min_{\mathbf{s}^{(i)}(\tilde{\mathbf{x}}_j) \in P} \{U(\mathbf{s}^{(i)}(\tilde{\mathbf{x}}_j))\}, \tag{10}$$

whose corresponding label is obtained by PyChrono simulation denoted by $\hat{y}^*$. Then, the new training point is added to the original training set to update the surrogate model. The above optimal sample identification and surrogate modeling updating is call active learning. The active learning iteration stops until the number of iteration exceeds the maximum allowed iteration.

After the model updating by active learning, we use the new surrogate model to update the reliability map as well as obstacle map. If the current path traverses the obstacle identified by the updated surrogate model, it means that the current path is not reliable. We trim the tree generated by RRT* to ensure that there is no branches of the tree that pass the obstacle. The trimmed tree is serving as the starting point of the next iteration of path planning.

The above procedures repeat iteratively until a reliable path is found. Next, we use a case study to demonstrate the proposed efficient physics-based global path planning method.

4 CASE STUDY
4.1 Experiment Setup

We use a 50 m × 50 m target map from ArcGIS/ENVI database to demonstrate the proposed method. The map is shown in Fig. 4. The true mobility M&S model is assumed as a mathematical model for demonstration and is given by

$$v(\mathbf{x}) = G(\mathbf{S}(\mathbf{x})) = 0.85 e^{S_1(\mathbf{x})/15 - S_3(\mathbf{x})}$$
$$+ (S_2(\mathbf{x})/5 - S_4(\mathbf{x}))^2 + 0.7 S_2(\mathbf{x}) S_3(\mathbf{x}) S_4(\mathbf{x}) \tag{11}$$

where $\mathbf{S}(\mathbf{x}) = [S_1(\mathbf{x}), S_2(\mathbf{x}), S_3(\mathbf{x}), S_4(\mathbf{x})]$ is vector of soil parameters, which are the slope, cohesive strength, friction coefficient, and bulk density. The parameters are assumed as Gaussian random field [23]. There are nine different soils and nine different slope types in the map. The distributions of the parameters are given in Table 1. The uncertainty of soil and slope parameters depends on the locations and the properties at different locations are correlated. We use a correlation function to describe the correlations of the parameters. The detailed correlation lengths of different parameters are given in Table 2. There are two directions in the 2-D map. We, therefore, use a two-dimensional vector to represent the correlation length in the two directions. The smaller the value is, the higher the variation is across the direction.

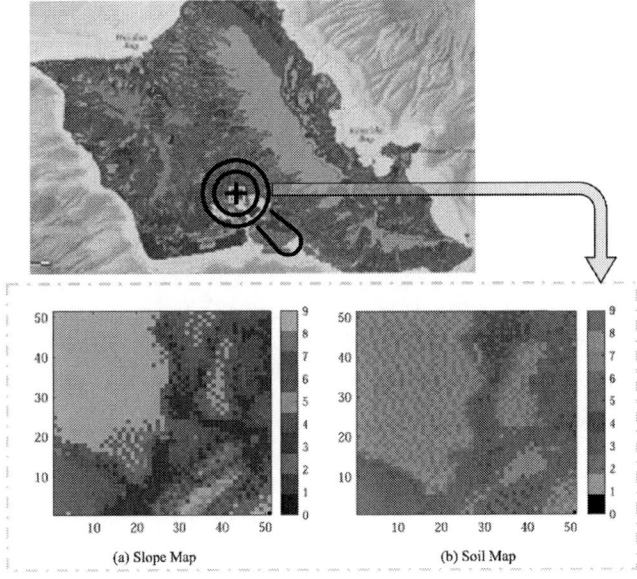

FIGURE 4. Slope and soil map extraction.

The initial point is $[3, 15]$ and end point is $[32, 40]$. The goal is to identify the shortest path satisfying reliability requirements.

Copyright © 2023 by The United States Government

Slope/Soil ID	Slope S_1		Soil Parameters S_2		S_3		S_4	
	μ	σ	μ	σ	μ	σ	μ	σ
1	5	0.2	0.2	0.01	0.01	0.001	0.05	0.001
2	10	0.5	1	0.1	0.8	0.05	1.2	0.1
3	12	0.5	5	0.4	0.7	0.02	2.1	0.1
4	14	1	6	0.2	0.76	0.02	2.3	0.1
5	16	2	2	0.1	0.56	0.01	1.6	0.04
6	18	2	11	0.2	0.8	0.02	2.1	0.05
7	22	2	5	0.02	0.7	0.03	1.2	0.02
8	24	1	4	0.1	0.45	0.005	1.45	0.01
9	28	1	8	0.4	0.78	0.03	2.35	0.05

TABLE 1. Distributions of slope and soil parameters.

TABLE 2. Correlation length of different parameters in two directions.

Slope/Soil ID	Slope S_1	Soil Parameters S_2	S_3	S_4
1	[3.3, 3.0]	[40, 39]	[33, 32]	[27, 28]
2	[3.4, 3.6]	[42, 43]	[28, 33]	[23, 27]
3	[4.3, 2, 2]	[41, 41]	[28, 27]	[20, 25]
4	[3.6, 4.4]	[41, 39]	[31, 29]	[26, 24]
5	[3.5, 5.1]	[40, 42]	[33, 32]	[27, 21]
6	[2.8, 5.0]	[38, 37]	[27, 27]	[26, 28]
7	[3.2, 2.2]	[39, 42]	[28, 30]	[24, 28]
8	[4.7, 4.0]	[37, 39]	[29, 28]	[24, 27]
9	[4.5, 6.7]	[42, 35]	[32, 32]	[26, 24]

We use the speed-made-good criterion to determine if an AGV maintain mobility or not. Specifically, the AGV loses mobility if the maximum speed is less than 2 m/s. The required mobility reliability along the path is 90%.

4.2 Results

4.2.1 The first iteration In the first iteration, the shortest path is found on the map. Since there are no obstacles identified with the initial surrogate model, the path is close to a straight line (see Fig. 5). At this moment, the initial surrogate model may not be accurate, which means that the path found may not be reliable. Therefore, we use a path verification algorithm to verify the path found.

In the path verification stage, we refine the surrogate model iteratively with pre-defined maximum iterations. After the surrogate model is updated, we use the surrogate model to update the obstacle map. From Fig. 6(a), we can observe that more ob-

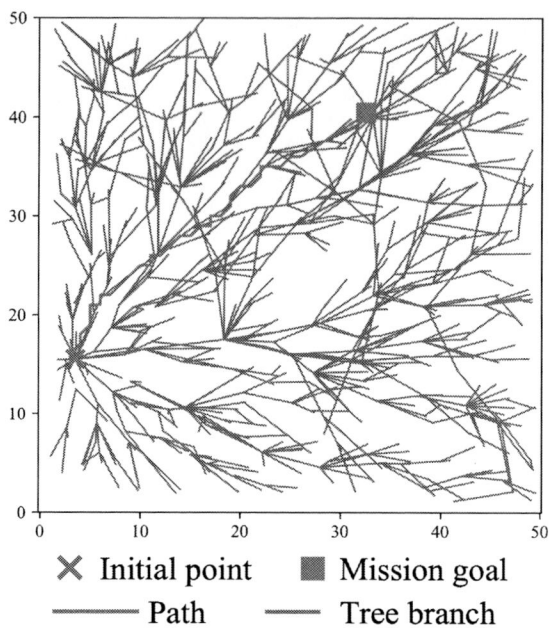

FIGURE 5. Path from the first path planning.

stacles are found on the map. And part of the path passes the obstacles, which means the path is not reliable. We next use a trim tree algorithm to remove the branches that pass the new obstacles (Fig. 6(b)). The trimmed tree is serving as the starting point in the next path planning iteration. The symbols defined in Fig. 5 and Fig. 6 are applicable for the rest of the figures.

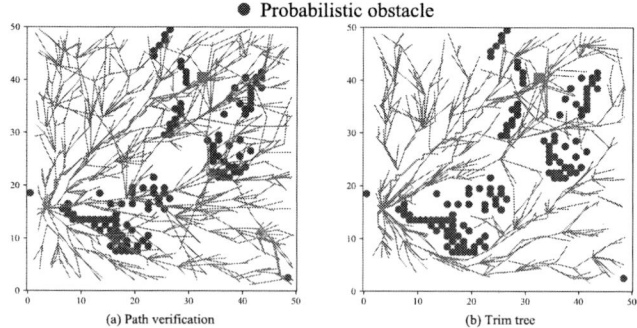

FIGURE 6. Path verification and trim tree the first path planning.

4.2.2 The second iteration Based on the trimmed tree from the previous iteration, we perform another path planning, and a new path is found, which is shown in Fig. 7. Using the same path verification algorithm, we identify more obstacles by the surrogate model refinement. From Fig. 7(a), we observed

Copyright © 2023 by The United States Government

that the path from the second iteration is not reliable. We trim the tree again and the results are shown in Fig. 8(b).

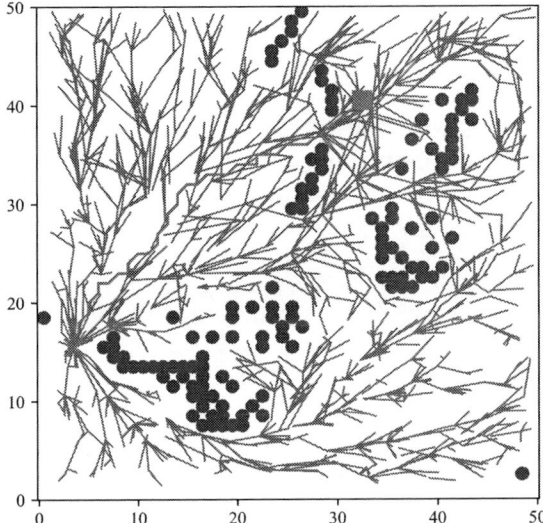

FIGURE 7.   Path from the second path planning.

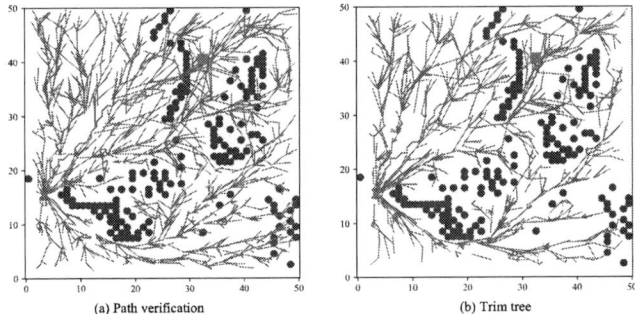

(a) Path verification   (b) Trim tree

FIGURE 8.   Path verification and trim tree of the second path planning.

4.2.3   The final iteration   Due to limited space, we only show the detailed results of the first two iterations. From the previous results, it is shown that more obstacles are identified with the refinement of the surrogate model. After three more path planning iterations, we found a path that satisfied the reliability requirement as shown in Fig. 9.

The total surrogate modeling refinement iteration is 20, which means 20 points from high-fidelity simulations besides the initial training points are added to the training set to update the

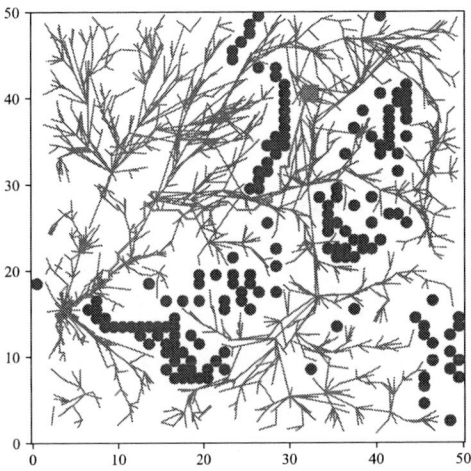

FIGURE 9.   Final path in the fifth iteration.

surrogate model. Compared with the current two-stage method using more than hundreds of surrogate refinement iterations, the proposed approach is much more efficient.

5   CONCLUSIONS

This work proposes an efficient surrogate modeling framework for physics-based global path planning of off-road autonomous ground vehicles under uncertainty. The proposed method couples surrogate modeling and path planning sequentially to improve computational efficiency. The computational cost is reduced by refining the surrogate only in the vicinity of the identified path. After a few iterations of path planning & path verification loops, the final path satisfying the reliability requirements is identified. The presented case study demonstrates that the obstacle map indicating the locations where a reliability constraint is violated, is updated with the refinement of the surrogate model at each path verification stage. The proposed tree trimming algorithm can successfully remove the branches of the tree passing through obstacles.

ACKNOWLEDGMENT

This work was supported in part by the Automotive Research Center (ARC) in accordance with Cooperative Agreement W56HZV-19-2-0001 U.S. Army DEVCOM Ground Vehicle Systems Center (GVSC), Warren, MI. The support is gratefully acknowledged.

Copyright © 2023 by The United States Government

REFERENCES

[1] Naranjo, J. E., Clavijo, M., Jiménez, F., Gomez, O., Rivera, J. L., and Anguita, M., 2016. "Autonomous vehicle for surveillance missions in off-road environment". In 2016 IEEE Intelligent Vehicles Symposium (IV), IEEE, pp. 98–103.

[2] Liu, Q., Zhao, L., Tan, Z., and Chen, W., 2017. "Global path planning for autonomous vehicles in off-road environment via an a-star algorithm". *International Journal of Vehicle Autonomous Systems, 13*(4), pp. 330–339.

[3] Oksanen, T., and Visala, A., 2009. "Coverage path planning algorithms for agricultural field machines". *Journal of field robotics, 26*(8), pp. 651–668.

[4] Ren, L., and Xi, Z., 2022. "Bias-learning-based model predictive controller design for reliable path tracking of autonomous vehicles under model and environmental uncertainty". *Journal of Mechanical Design, 144*(9), p. 091706.

[5] Masoudi, N., and Fadel, G., 2022. "Solving three-dimensional path planning problem using a visibility-based graphical representation of the design space". *Journal of Mechanical Design, 144*(8), p. 081704.

[6] Torkamani, E. A., and Xi, Z., 2022. "Systematical collision avoidance reliability analysis and characterization of reliable system operation for autonomous navigation using the dynamic window approach". *ASCE-ASME Journal of Risk and Uncertainty in Engineering Systems, Part B: Mechanical Engineering, 8*(3), p. 031106.

[7] Jiang, C., Hu, Z., Mourelatos, Z. P., Gorsich, D., Jayakumar, P., Fu, Y., and Majcher, M., 2021. "R2-rrt*: reliability-based robust mission planning of off-road autonomous ground vehicle under uncertain terrain environment". *IEEE Transactions on Automation Science and Engineering, 19*(2), pp. 1030–1046.

[8] Yin, J., Shen, D., Du, X., and Li, L., 2022. "Distributed stochastic model predictive control with taguchi's robustness for vehicle platooning". *IEEE Transactions on Intelligent Transportation Systems*.

[9] Shen, D., Yin, J., Du, X., and Li, L., 2021. "Distributed nonlinear model predictive control for heterogeneous vehicle platoons under uncertainty". In 2021 IEEE International Intelligent Transportation Systems Conference (ITSC), IEEE, pp. 3596–3603.

[10] Petrick, E., Janosi, Z., and Haley, P., 1981. The use of the nato reference mobility model in military vehicle procurement. Tech. rep., SAE Technical Paper.

[11] McCullough, M., Jayakumar, P., Dasch, J., and Gorsich, D., 2017. "The next generation nato reference mobility model development". *Journal of Terramechanics, 73*, pp. 49–60.

[12] Xia, K., 2011. "Finite element modeling of tire/terrain interaction: Application to predicting soil compaction and tire mobility". *Journal of Terramechanics, 48*(2), pp. 113–123.

[13] González, R., Jayakumar, P., and Iagnemma, K., 2017. "Stochastic mobility prediction of ground vehicles over large spatial regions: a geostatistical approach". *Autonomous Robots, 41*(2), pp. 311–331.

[14] Wu, H., Zhu, Z., and Du, X., 2020. "System reliability analysis with autocorrelated kriging predictions". *Journal of Mechanical Design, 142*(10).

[15] Choi, K., Jayakumar, P., Funk, M., Gaul, N., and Wasfy, T. M., 2019. "Framework of reliability-based stochastic mobility map for next generation nato reference mobility model". *Journal of Computational and Nonlinear Dynamics, 14*(2).

[16] Liu, Y., Jiang, C., Zhang, X., Mourelatos, Z. P., Barthlow, D., Gorsich, D., Singh, A., and Hu, Z., 2022. "Reliability-based multivehicle path planning under uncertainty using a bio-inspired approach". *Journal of Mechanical Design, 144*(9), p. 091701.

[17] Yin, J., and Du, X., 2022. "High-dimensional reliability method accounting for important and unimportant input variables". *Journal of Mechanical Design, 144*(4).

[18] Yin, J., and Du, X., 2022. "Active learning with generalized sliced inverse regression for high-dimensional reliability analysis". *Structural Safety, 94*, p. 102151.

[19] Wu, H., Hu, Z., and Du, X., 2021. "Time-dependent system reliability analysis with second-order reliability method". *Journal of Mechanical Design, 143*(3).

[20] Karaman, S., Walter, M. R., Perez, A., Frazzoli, E., and Teller, S., 2011. "Anytime motion planning using the rrt". In 2011 IEEE international conference on robotics and automation, IEEE, pp. 1478–1483.

[21] Tasora, A., Serban, R., Mazhar, H., Pazouki, A., Melanz, D., Fleischmann, J., Taylor, M., Sugiyama, H., and Negrut, D., 2015. "Chrono: An open source multi-physics dynamics engine". In International Conference on High Performance Computing in Science and Engineering, Springer, pp. 19–49.

[22] Williams, C. K., and Rasmussen, C. E., 2006. *Gaussian processes for machine learning*, Vol. 2. MIT press Cambridge, MA.

[23] Hu, Z., and Mahadevan, S., 2015. "Time-dependent system reliability analysis using random field discretization". *Journal of Mechanical Design, 137*(10).

Copyright © 2023 by The United States Government

Proceedings of the ASME 2023
International Design Engineering Technical Conferences and
Computers and Information in Engineering Conference
IDETC-CIE2023
August 20-23, 2023, Boston, Massachusetts

DETC2023-115147

PARETO OPTIMIZATION OF TISSUE AND BLOOD VESSEL GROWTH IN 3D PRINTED BONE SCAFFOLDS

Amit M. E. Arefin
Mechanical Engineering
Texas Tech University
Email: amit.arefin@ttu.edu

Paul F. Egan
Mechanical Engineering
Texas Tech University
Email: paul.egan@ttu.edu

ABSTRACT

Computational design is necessary for advancing biomedical technologies, particularly complex systems with numerous complicated trade-offs. For instance, 3D printed tissue scaffolds constructed as lattices necessitate consideration of tissue and vasculature growth trade-offs in relation to complex geometries. In this paper, curvature-based tissue growth models and agent-based vascularization simulations are used to predict growth. NSGA-II (non-dominated sorting genetic algorithm) is used for Pareto optimization of growth for heterogeneous unit cell scaffolds. Cube and BC-Cube (Body Centered-Cube) unit cells are considered with beam diameters from 64 to 313 µm that are arranged in lattices with No Voids or Channel Voids configurations. The Channel Voids configuration has channels consisting of no unit cells that promote unobstructed vascularization. Seeding the algorithm with high-performing scaffolds with homogenous unit cells improved search efficiency and quality, since unit cells can be configured either for high tissue growth or high vasculature growth. The Pareto front of solutions demonstrates that scaffolds with large porous areas for the Channel Voids improve vasculature growth while lattices with no larger void areas result in higher tissue growth. Results demonstrate the advantages in using NSGA-II for dual-objective search in complex biomedical systems, which provides a foundation for future multi-objective optimization for advanced tissue engineering systems.

Keywords: Bio-inspired Designs, Biomechanics, Biomaterials, Medical Device Design, Multi-objective Optimization

1. INTRODUCTION

As emerging 3D printing technologies advance, there is a need for computational design approaches to configure complex structures suitable for supporting biological growth [1-3]. For instance, tissue scaffold design requires implant configuration with geometrically tuned structures that drive mechanobiological growth [4, 5]. Porous 3D printed lattices are promising for providing optimal geometries to support biological growth such as tissue and blood vessel growth [6-9]. Tissue growth has been evaluated using mechanobiological algorithms [10], agent-based growth models [11], continuum models [12], and curvature-based growth [13]. Among these methods, curvature-based tissue growth is an efficient means of evaluating 3D printed scaffolds, especially for bulk analysis [14]. Blood vessel growth models are commonly probability based stochastic models [15] or agent based models [16]. For optimization, evaluating growth can be supported by curvature-based tissue growth [17, 18] and agent-based blood vessel growth simulations [19, 20] to retain modeling efficiency and accuracy. Multi-objective optimization can aid in determining optimal trade-offs for high porosity needed to support high density tissue and vasculature growth and higher surface area to support faster tissue growth, but requires an efficient search approach to navigate the complex design space.

Interbody cages for spinal fusion are suitable as an exemplary application to configure algorithms for multi-objective tissue scaffold optimization (Fig. 1) [21-24]. Spinal fusions require cage implantation between vertebrae to promote bone growth [25-28]. Blood vessel growth is necessary to provide nutrients to growing tissues. Although fusions are common, they still fail on a regular basis [29-31]. Fusion success can be improved through designs that facilitate more balanced

Copyright © 2023 by ASME

biological processes [32-36], such as optimizing pore size and distribution according to an optimal surface area and porosity.

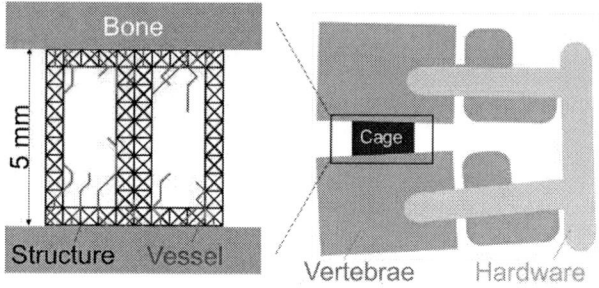

FIGURE 1: Representative blood vessel growth model for 3D printed interbody spinal cages.

Optimized distributions of pores and tuning overall porosity are necessary to promote mechanobiological growth [37]. Porosity is the proportion of void volume in a scaffold while pore size is the volume of each void cavity within a scaffold. Porosities of 50% to 80% and pore sizes from 100 to 400 μm are considered ideal for scaffolds [38, 39]. There are diverse tissue evaluation models, with curvature-based growth models among the most efficient [40-43]. These models assume tissue growth occurs where osteoblasts (i.e. bone cells) are seeded on surfaces and grow from concave surfaces, and have been validated both *in vitro* and *in vivo* [17, 18, 44, 45]. Successful tissue growth requires enough surface area per volume to sustain positive curvature for advancing tissue fronts.

Blood vessels deliver nutrients to growing tissues and advance faster with openly porous space [46, 47]. Agent-based models for blood vessel growth have been validated with *in vivo* data [19, 48-50]. The emergent growth in 3D printed structures can be very different than those of previously studied systems when considering the potential of 3D printing for creating complex geometries [51, 52]. Due to the high computational time required to simulate blood vessel growth, planar approximations are useful for optimization [53]. Such models can be informed by empirical data that suggests vasculature grows faster in void areas and may benefit from large hierarchical voids throughout a scaffold [54, 55]. These studies suggest blood vessels growth is proportional to porosity squared and directly proportional to pore size, that can inform the creation of suitable prediction models.

Due to the complexity of the 3D printing design space, exhaustive search approaches are not suitable [56], thereby necessitating the selection of optimization algorithms based on relative advantages in search efficiency [57, 58]. For dual-objective biological problem, there are numerous stochastic heuristic algorithms that could be suitable [59, 60]. One possibility is the non-dominating sorted genetic algorithm II (or NSGA II) that has been used successfully for diverse applications in multi-objective mechanical design [61-65]. NSGA II provides a non-dominated sorting approach to find an optimal Pareto front with the elitist samples[66]. The algorithm is preferred for its wide-spread search of a design space that

avoids local optima suggests its suitability for tissue scaffold optimization.

This paper uses NSGA-II for Pareto optimization of tissue scaffolds for navigating mechanobiological trade-offs. Curvature-based tissue growth and agent-based blood vessel growth simulations are used for evaluating growth in relation to 3D printed lattice structures. Trade-offs will be assessed for a large design space consisting of multiple unit cell designs and systems level scaffold topologies. Findings are expected to improve understanding mechanobiological trade-offs for the design of complex 3D printed tissue scaffolds and inform the use of heuristic algorithms for their configuration. Advancements may improve computational design for biomedical systems in addition to revealing highly advantageous tissue scaffold configurations.

2. DESIGN AND SIMULATION METHODS
2.1 Unit cell design

Unit cells were constructed in a cubic volume with beam elements to form Cube or Body-centric Cube (BC-Cube) topologies. The Cube unit cell has beams along each edge of the cubic volume. The BC-Cube unit cell has beams along each edge of the cubic volume and from each corner to the center (Fig. 2).

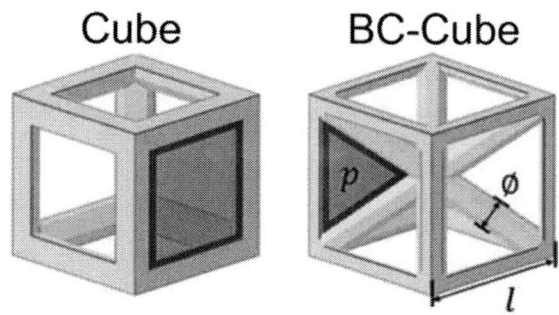

FIGURE 2: Cube and BC-Cube unit cells with planar pore size p, beam diameter $\emptyset$, and unit cell length l.

Unit cells have two design parameters including beam diameter $\emptyset$ and unit cell length l. Unit cell length refers to the length, height, and width of the cubic volume. The pore size p is the square root of the smallest planar void area that depends on beam diameter and unit cell length values. Porosity P is the ratio of void space to total nominal cubic volume. Unit cells were generated by holding unit cell length to a constant 500 μm. The length ensures there is a constant number of unit cells in a lattice of a specified length as beam diameter is altered. Beam diameter was varied from 109 μm to 313 μm for Cube cells and 64 μm to 178 μm for BC-Cube cells that ensures porosity is between 40% and 90% that is suitable for bone fusion.

2.2 Tissue growth simulation

Tissue growth was simulated using a curvature-based tissue growth model that has been well validated empirically, and briefly discussed here [17, 18, 44, 45]. The simulation is

Copyright © 2023 by ASME

implemented in Python using a voxel-based environment [38]. The tissue growth environment consists of one eighth of a unit cell seeded with tissue. Voxels describe structure, interface, and tissue that are used to model the growth of tissue as the simulation advances each time step (Fig. 3). The simulation commences by building a unit cell structure, then placing tissue voxels adjacent to structure, followed by placing interface voxels adjacent to tissue voxels. Each time step tissue replaces interface voxels that have a positive curvature according to a check with a spherical scanning mask.

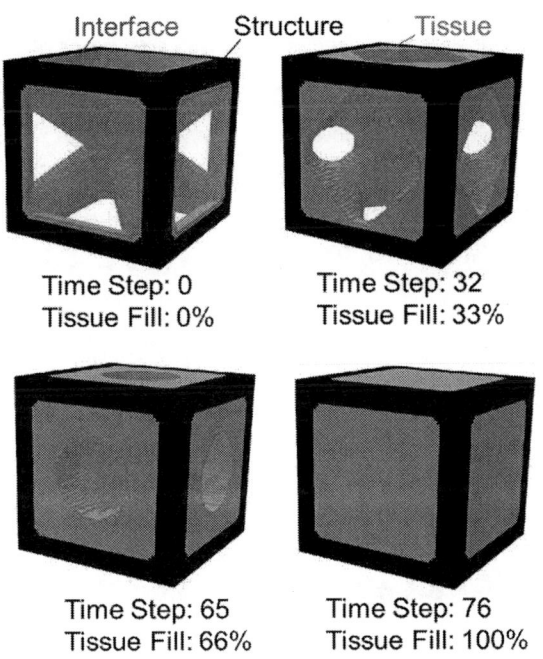

FIGURE 3: Simulation of tissue filling behavior for a Cube unit cell built from structure, tissue, and interface voxels.

The spherical scanning mask has a radius of m_r voxels with a biophysical length c_r held at a constant 55 μm, that represents the biophysical reach of cells that form the tissue. The mask radius is set to a constant 10.5 radius that enables calculation of each voxel length ds according to

$$ds = \frac{c_r}{m_r} \qquad (1)$$

with voxel curvature calculated according to

$$\kappa = \frac{16}{3 \cdot c_r} \left(\frac{m_{on}}{m_{on} + m_{off}} - \frac{1}{2} \right) \qquad (2)$$

where m_{on} represents the number of voxels that are structure/tissue and m_{off} represents voxels that are void/interface. If curvature is positive, then the interface voxel at the center of the scanning mask is replaced with tissue and new interface voxels are placed to surround newly grown tissue.

The number of different types of voxels are counted each step of the simulation to describe tissue growth. Tissue growth density g_d is calculated as the ratio of tissue present in the environment compared to nominal volume according to

$$g_d = \frac{v_{tissue}}{v_{total}} \qquad (3)$$

where v_{tissue} is the number of tissue voxels and v_{total} is the number of total voxels in the environment. It is assumed all unit cells of the same design throughout a lattice have the same amount of tissue growth, which represents an average behavior.

2.3 Lattice structure design
While Section 2.1 and Section 2.2 model tissue growth in unit cells using a 3D representation, the blood vessel growth simulation uses a 2D representation to facilitate faster evaluations while maintaining suitable predictive accuracy of behavior. A 2D representation for tissue growth is not accurate due to the open surface area of 3D printed BC-Cube unit cells compared to planar closed pores [38, 45]. However, a planer blood vessel growth model with symmetric structure (repetitive unit cells in the third dimension) should capture the blood vessel growth reasonably across dimensions while providing time-efficient simulations. Therefore, the blood vessel growth model is conducted in a 2D representation.

Planar lattices were designed with unit cells or void spaces to form a square matrix of 100 spaces. The planar representation facilitates the use of the planar blood vessel growth model. The representation assumes the visible plane is replicated in the z-direction and the single layer simulated represents an average behavior throughout the scaffold. Topological variations were generated as demonstrated in Figure 4. Figure 4A demonstrates the No Voids design and Figure 4B demonstrates the Channel Voids design.

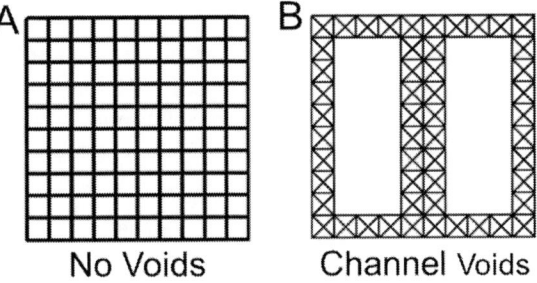

FIGURE 4: Lattices with layouts of (A) Cube unit cells with No Voids or (B) BC-Cube unit cells with Channel Voids.

2.4 Blood vessel growth simulation
The blood vessel growth model was informed from previous agent-based simulations validated with *in vivo* data and modified for use with 3D printed beam-based lattices [19, 48, 67]. The simulation was programmed in Python and initiated by constructing a virtual environment for a lattice with blood

Copyright © 2023 by ASME

vessels of equidistant spacing seeded at the top and bottom (Fig. 5) [53]. Each seeded blood vessel is programmed as an agent with independent biophysical behaviors. A blood vessel has the chance to grow one segment diagonally or vertically towards the center of the lattice each time step. Therefore, the total blood vessel growth is proportional to the number of blood vessel segments counted. During each time step of the simulation, a blood vessel agent can grow to a new location and grow a new segment according to a probability check. Locations are specified in the virtual environment according to each half length of a unit cell. Therefore, if there are 100-unit cells in the environment, there are 441 locations for agent placement and segment growth. A sensitivity study was conducted using a finer grid size in the simulation environment that demonstrates blood vessel growth with higher resolution branching that leads to more natural angles forming for overall branching. However, there was no difference in simulation prediction accuracy for total number of vessels grown while simulation time increased drastically for the finer grid, thereby motivating the use of the coarse grid that evaluates faster with the same predictive accuracy.

Agents are allowed to grow towards the mid-line of the scaffold, which is consistent with previous modeling methods that assume growth factor is present in the scaffold center. The probability check for agents to grow to a new location is based on the porosity and pore size of surrounding unit cells, with averaging at interfaces. The probability check is informed by past studies with *in vivo* validation that demonstrated blood vessel growth is proportional to porosity squared when all other design variables are held constant and is linearly proportional to pore size when all other design variables are held constant [19]. The equation for the probability check c is

$$c = A \cdot p \cdot P^2 \qquad (4)$$

where p is the unit cell pore size, P is the unit cell porosity, and A is an empirically informed constant used for tuning the rate of growth based on the environment and time step, which is set to a constant value of 0.001. Blood vessel growth with large pore sizes was validated by *in vivo* 3D printed beam-based structures with 4 mm large void volume [67].

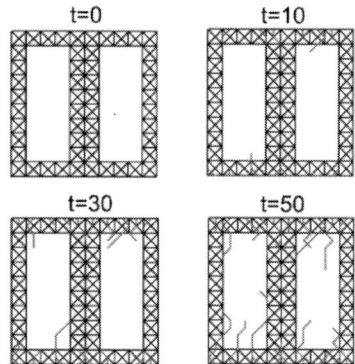

FIGURE 5: Blood vessel growth for 50 time steps.

Blood vessel growth was calculated as the number of blood vessel segments at each time step. There is a different pattern of blood vessel growth and total amount growth each run due to stochasticity, which requires run aggregation. Simulations are run 100 times with average results reported with negligible standard error.

2.5 Lattice modeling

Lattices were converted to binary strings to facilitate optimization. Lattice binary strings were built by combining smaller binary strings that represent each unit cell. Each unit cell had 1 bit that determined whether it was a Cube or BC-Cube topology and 3 bits that determined its beam diameter size, as demonstrated in Figure 6. Lattice porosity ranged from 41% to 90% for both unit cell types in regular increments of 7% change per incremental change in binary string value. This resulted in eight different beam diameters for each unit cell topology, with cubic unit cells ranging from 109 to 313 μm and BC-Cube unit cells ranging from 64 to 178 μm. When considering the design of each unit cell with 4 bits, there are 2^{400} designs to search for the No Voids design. The Channel Voids lattice has 2^{208} designs.

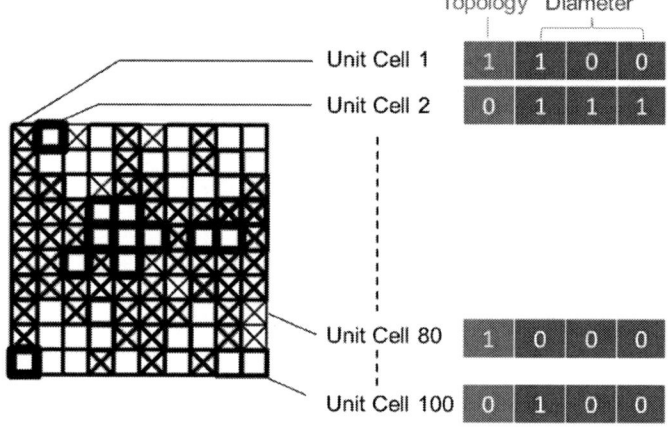

FIGURE 6: A 100-unit cell lattice with 400 binary bits that describe each unit cell's topology and beam diameter.

2.6 Pareto optimal search

The Pareto-optimal search was conducted using a modified NSGA-II algorithm [66], which is demonstrated as a pseudocode in Figure 7. The goal of the Pareto optimization is to maximize tissue and blood vessel growth.

The algorithm is initiated by generating a population P_o of 32 designs that was determined through preliminary testing to ensure diversity each generation. Objective functions are then evaluated for all designs in population P_o for both tissue growth density and blood vessel growth. Objective functions are evaluated for tissue growth according to equations:

$$\sum_{i=0}^{n(x)} \sum_{j=0}^{n(y)} g_{d_{i,j}}(u_t, \emptyset) \qquad (5)$$

Copyright © 2023 by ASME

```
Algorithm: Pseudocode of NSGA-II
  1: P₀ ← Initial_Population()
  2: Evaluate_Objectives(P₀)
  3: Pₜ ← P₀
  4: while not Terminate() do
  5:     Pₘ ← Tournament_Selection(Pₜ)
  6:     Qₜ ← Mutation_Crossover(Pₘ)
          //Population_Size of Qₜ is the same as Pₜ
  7:     Evaluate_Objectives(Qₜ)
  8:     Rₜ = Pₜ ∪ Qₜ
  9:     Rₜ ← NonDominated_Sorting(Rₜ)
 10:     Rₜ ← Crowding_Distance_Sorting(Rₜ)
 11:     Pₜ₊₁ ← Select_Fittest(Rₜ)
          //Population size of Pₜ₊₁ is half the population size of Rₜ
 12:     Pₜ ← Pₜ₊₁
 13: end
 14: Return Pₜ
```

FIGURE 7: NSGA-II algorithm pseudocode for Pareto optimization of scaffolds.

and

$$\sum_{k=0}^{n} v_{l_k} \ (u_t, \emptyset) \qquad (6)$$

Where g_d is the tissue growth density for unit cell in location x and y for a lattice and v_l is the vascular growth for the total number of n blood vessels throughout a lattice. The equations are a function of unit cell type (u_t) and unit cell beam diameter ($\emptyset$).

To preserve the convention of minimization in optimization, an inverse function of tissue growth density and blood vessel growth is considered. Moreover, to avoid unit-induced concerns in optimization, both the functions are normalized by taking the ratio of the value for a function and highest value of that function it can reach based on the highest performing homogenous design possible. Thus, the active functions become inverse normalized tissue growth and inverse normalized blood vessel growth and the values remain in a range of 0 to -1.

Random lattice designs in the initial population P_o are then passed as the parent designs P_t in the zeroth generation in the NSGA II loop. A binary tournament selection is conducted to find 16 mating parents P_m, which then by mutation (0.25% probability) and crossover (90% probability), create 32 offspring lattice designs, denoted with Q_t. Objective functions are then evaluated for the lattice designs in Q_t. All the lattice designs from the parent population P_t and the offspring population Q_t are then stored together as a combination population R_t. The combination population R_t stores 64 designs since the parent population P_t and offspring population Q_t had 32 designs each. The lattice designs in the combination population are then sorted according to non-dominated front formation where better lattice design solutions are the least dominated by other solutions. A new population is then created that consists of 32 lattice designs with the best fronts, i.e. the least dominated lattice designs in the solution space. Since each front has an arbitrary number of lattice

design solutions, some instances could demonstrate that passing a whole front through the filter is making the population more than 32. If this happens, a crowding distance filter is used, where the Euclidian distance with the neighbor designs is calculated for each lattice design in the solution space and lattice deigns with higher crowding distance are preserved in the new population P_{t+1}, to ensure diversification in the new population. The new population P_{t+1} is introduced as the parent population P_t in the next iteration until the loop reaches a termination criterion, which is set at a predefined number of iterations for the entire algorithm for this study.

Objective functions of individuals in a generation were computed parallelly in the genetic algorithm to gain more speed. An Intel(R) Xeon(R) W-2135 CPU with a 16 GB RAM, 6 cores and 12 logical processors was used to run the simulations. For a single iteration of the genetic algorithm (i.e. one generation with a population size of 32 different designs), the blood vessel simulation has 100 runs of 32 designs (3200 blood vessel growth simulations in total) that requires about 68 seconds, whereas the computational time for the entire generation including optimization procedures requires about 72 seconds.

3. RESULTS AND DISCUSSION
3.1 Tissue growth evaluation

Tissue growth evaluations were plotted for each lattice layout using a homogeneous distribution of Cube or BC-Cube unit cells with a specified beam diameter (Fig.8).

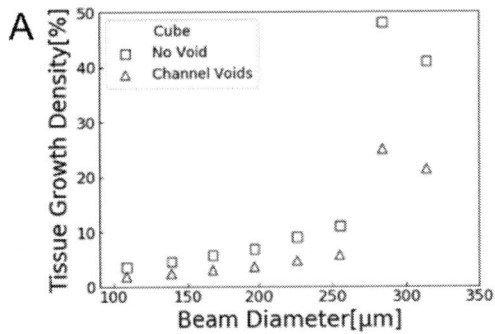

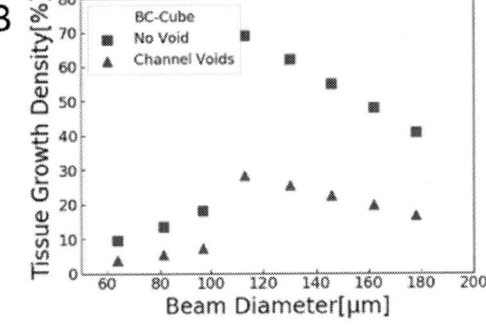

FIGURE 8: Lattice tissue growth density consisting of (A) Cube and (B) BC-Cube unit cells.

Figure 8 demonstrates that for both unit cell types No Voids lattices have the highest growth density, since total scaffold tissue growth is proportional to the number of unit cells that are not void. In other words, the individual unit cell growth is the same for all scaffolds when the unit cell has the same design, but the total number of unit cells throughout the scaffold differ. The growth density initially increases at a slow rate with increases in beam diameter as growth halting is experienced [38]. Then, there is a sharp increase in growth at beam diameters of 250 μm for the Cube unit cell and 90 μm for the BC-Cube unit cell once surface area is sufficiently large to ensure pore filling behavior. At this point, further increases in beam diameter reduces overall growth as porosity is lowered. These results suggest optimal tissue growth occurs for unit cells with the smallest beam diameter possible as long as they facilitate void filling behavior.

3.2 Blood vessel growth evaluation

Blood vessel growth simulations were then conducted for each layout. The effect of time steps on the blood vessel growth was investigated for lattices with Cube and BC-Cube unit cells of 90% porosity up to 100-time steps (Fig. 9).

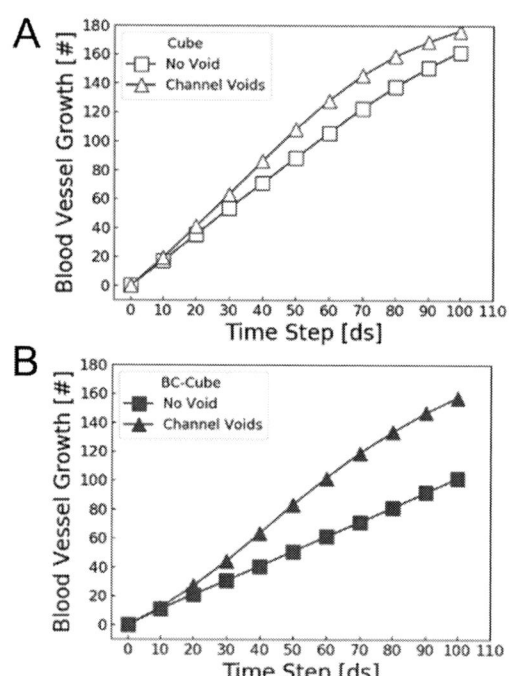

FIGURE 9: Blood vessel growth for lattices with (A) Cube and (B) BC-Cube unit cells as time step is increased.

Blood vessel growth increases with time step, but has a varied slope depending on the layout. The No Voids lattices demonstrate a near linear change in blood vessel growth with increase in time step for both unit cells, while the Channel Voids lattice reaches a higher growth rate at higher time steps as blood vessels reach unobstructed areas. When considering the maximum growth in lattices at 100 steps, the Channel Voids

lattice has the highest number of blood vessel segments of 100 grown vessels for the Cube unit cell lattice and 83 grown segments for the BC-Cube cell lattice.

The influence of beam diameter on blood vessel growth at the maximum time step of 100 was investigated for each lattice layout and unit cell type, as demonstrated in Figure 10 for Cube unit cells and Figure 11 for BC-Cube unit cells. Results show that blood vessel growth decreases as beam diameter increases. The decrease in blood vessel growth occurs since increases in beam diameter leads to decreases in both the porosity and pore size of the unit cells, thereby impeding blood vessel growth as blood vessels become more obstructed throughout the lattice.

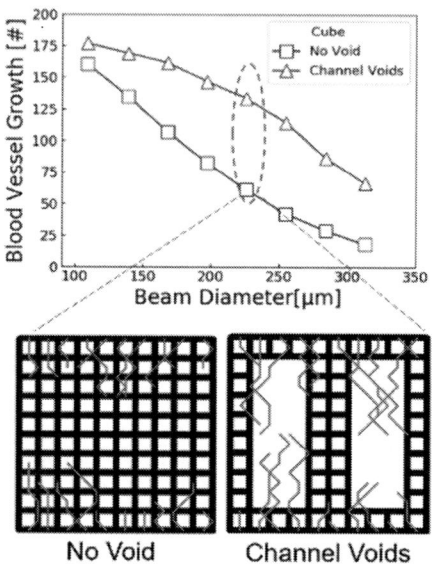

FIGURE 10: Vessel growth for Cube lattices at 100-time steps.

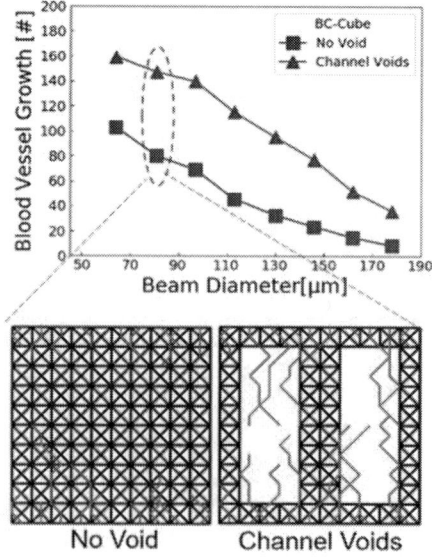

FIGURE 11: Vessels in BC-Cube lattices at 100-time steps.

The optimal design for vascularization based on these observations is a unit cell with the highest porosity/lowest beam diameter possible and Channel Voids lattices where blood vessels have more void space to grow unobstructed.

3.3 Homogenous lattice assessment

Pareto optimization is then carried out for vasculature and tissue growth density for each lattice configuration considered. Figure 12 results show the normalized scale value for inverse tissue growth density and inverse vasculature growth for all possible combinations of No Voids lattices with Cube and BC-Cube unit cells, prior to optimization when homogenous unit cells are placed throughout lattices.

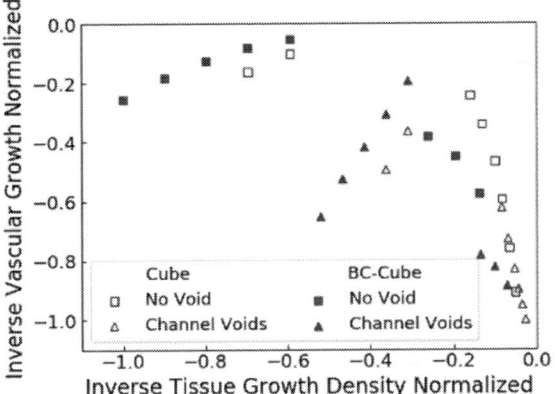

FIGURE 12: Growth in No Voids lattices with Cube and BC-Cube unit cells of homogeneous beam diameters. Normalized for 0.69 tissue growth density and 178 number of blood vessels.

Figure 12 shows that lattice designs fall into two regions of the solution space depending on whether tissue growth is halted or filling. For instance, the No Voids lattices have designs placed in one region that is inferior for tissue growth density (top left of plot for halted behavior) and the other region they capture is inferior for vascular growth (far right of plot). The other lattice designs follow a similar trend, although they possess lower tissue growth and higher vasculature growth. BC-Cube Channel Voids has one design that strikes a balanced trade-off between tissue growth and vasculature growth towards the center of the plot, which demonstrate the possibility of reaching more Pareto optimal solutions in this region through considering heterogenous combinations of unit cells.

3.4 Heterogeneous lattice optimization

Pareto optimization was carried out with the NSGA-II algorithm for heterogeneous lattice designs. Heterogeneity-based optimization was first conducted on the No Voids lattice. The algorithm was initiated with a seeding of 32 randomized designs (selected as twice the total number of homogeneous design configurations) that were generated randomly as the

initial population. Results from the Pareto-optimal search for 500 generations are presented in Figure 13.

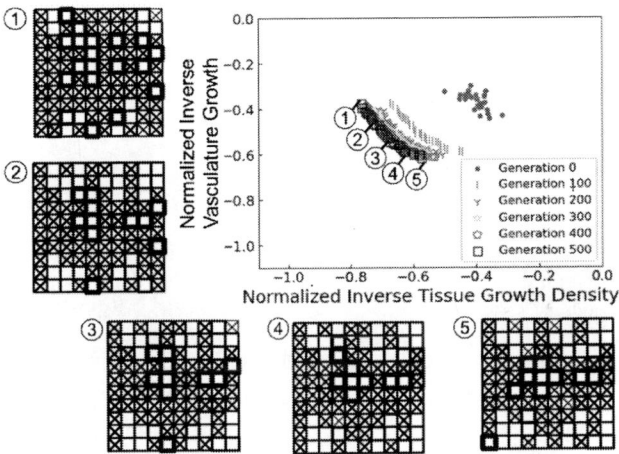

FIGURE 13: No Voids NSGA-II optimization normalized to 0.69 tissue growth density and 162 number of blood vessels.

Results demonstrate the initial randomized designs for generation 0 that form a front at generation 100 that significantly improves with a better non-dominated optimal Pareto front by generation 200, followed by smaller improvements until generation 500. The best performing design for tissue growth has an inverse normalized value of approximately -0.5 and the best performing design for vascular growth has an inverse normalized value of approximately -0.45. The small improvements by generation 500 suggest the algorithm has converged, however it is not clear whether the convergence is due a limitation of the algorithm or nearing global optimum solutions.

A modification is conducted by replacing 16 of the initial randomized designs with each of the homogenous configurations from Figure 12. These designs are seeded with the hypothesis that an optimized lattice would largely contain unit cells either configured for maximized tissue growth or maximized vasculature growth. By seeding the homogenous designs, the algorithm will work from lattices with a high representation of unit cells optimized for one of these two cases. Results for running the algorithm are demonstrated in Figure 14.

Results demonstrate much faster convergence, with a superior front determined by the end of 20 generations that marginally improves by generation 40. The optimized Pareto front also demonstrates higher performing designs when compared to the front from 500 generations completed in Figure 13. The Figure 14 optimized front at generation 40 consists of homogenous designs at the extremes for the highest tissue growth or vasculature growth with heterogeneous designs along the front in between. In the optimal Pareto front, a performance trade-off is noticed for lattice designs. For instance, the design that provides the lowest inverse normalized tissue growth of -1 has the highest inverse normalized blood vessel growth of around -0.3. In other words, the optimized front has designs that provide higher tissue growth density but lower vascular growth

Copyright © 2023 by ASME

and vice versa. When comparing designs across the front, the middle portion of the lattice is replaced with vascular-dominated unit cells with channel shapes.

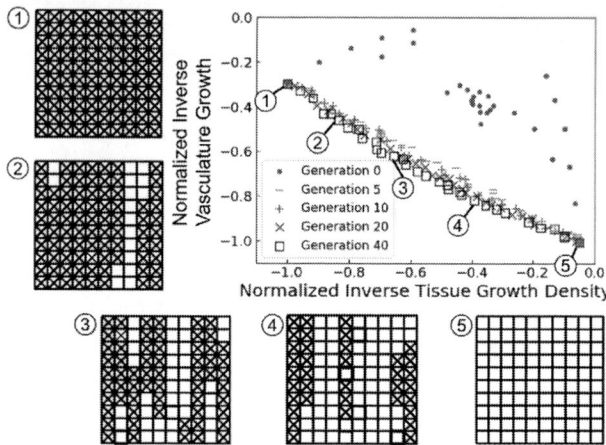

FIGURE 14: No Voids modified NSGA-II optimization normalized to 0.69 tissue density and 162 blood vessels.

The seeded optimization was then conducted in Figure 15 for Channel Voids lattices with heterogenous flexibility. The optimized designs form a slightly concave shaped front in the solution space. The extreme design having the lowest inverse tissue growth density of -1 has the highest vascular growth of around -0.67, whereas the extreme design having the lowest vascular growth of -1 has the highest tissue growth density of around 0.25 in the optimized front. As designs are optimized for better vasculature growth instead of tissue growth, unit cells are removed from the bottom of the lattice and towards the center which facilitates blood vessels reaching growth in unobstructed portions of the lattice early.

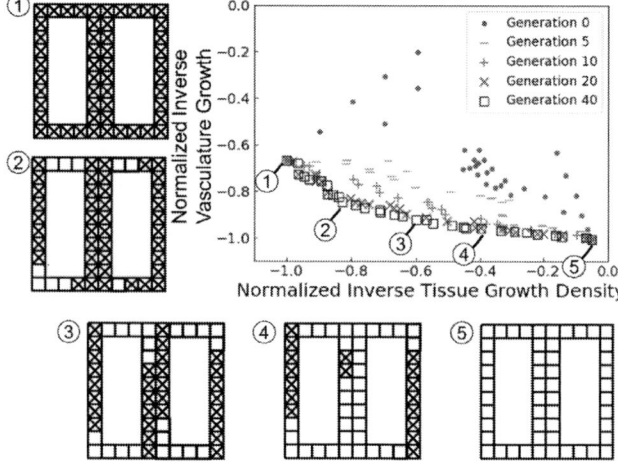

FIGURE 15: Channel Voids modified NSGA-II optimization normalized to 0.36 tissue density and 178 blood vessels.

Overall, these results demonstrate the capabilities of the modified NSGA-II algorithm to find Pareto optimal designs for tissue scaffolds and demonstrate the diversity of Pareto optimal designs available for tuning solutions for specified needs in mechanobiological domains. In particular, the normalization factors suggest No Voids lattices reach better tissue growth configurations while Channel Voids promote vascularization, with Pareto optimized solutions for each providing compromised designs with varied relative values of each objective.

3.5 Limitations and future work

The current research relies on the creation of a novel agent-based blood vessel growth simulation combined with a curvature-based tissue growth simulation to enable configuration of Pareto optimal tissue scaffolds designs for dual mechanobiological objectives. The modeling approaches were selected to facilitate efficient objective function evaluation to facilitate search with the NSGA-II algorithm. The modeling approach has merit as an early indicator of design performance prior to investing larger resources computationally or experimentally to fine-tune or validate high-performing designs. The strategy was conducted in the past for tissue simulations to predict growth for a large number of 3D printed design variations, by avoiding the computational costs associated with higher fidelity models [38]. Now that the heuristic has been developed for seeding homogenous designs in the algorithm to facilitate the combination of contrasting unit cells, future endeavors could use more sophisticated mechanobiological evaluation models that provide higher fidelity. Future works could investigate different unit cell types and lattice layouts to determine if there are further strategies for Pareto optimal designs that extend beyond the investigated design space. Moreover, the current study focuses on mechanobiological factors, while future design studies may consider how mechanics, such as lattice stiffness, affects design optimization and scaffold functioning.

4. CONCLUSION

The study investigated tissue scaffold design trade-offs with Pareto optimization. Cube and BC-Cube unit cell configurations had a fixed unit cell length of 500 μm and porosity from 41% to 90%. Results demonstrated tissue growth was favored in unit cells with the smallest beam diameters that ensure continuous tissue growth while vascularization was favored in unit cells with the lowest porosity possible. An NSGA II optimization approach was used to optimize across the considerations, with seeding of homogenous lattice configurations to improve search efficiency. Scaffolds with No Voids favored tissue growth, while scaffolds with Channel Voids were favored for vasculature growth. Overall, the computational approach was demonstrated as an efficient means for evaluating complex lattices structures, and provides the foundations for future investigations in multi-objective design optimization for complex biomedical systems.

Copyright © 2023 by ASME

REFERENCES

[1] M. K. Thompson *et al.*, "Design for Additive Manufacturing: Trends, opportunities, considerations, and constraints," *CIRP Annals-Manufacturing Technology,* vol. 65, no. 2, pp. 737-760, 2016.

[2] D. O. Cohen, S. M. Aboutaleb, A. W. Johnson, and J. A. Norato, "Bone Adaptation-Driven Design of Periodic Scaffolds," *Journal of Mechanical Design,* vol. 143, no. 12, p. 121701, 2021.

[3] P. Egan, J. Cagan, C. Schunn, F. Chiu, J. Moore, and P. LeDuc, "The D3 Methodology: Bridging Science and Design for Bio-Based Product Development," *Journal of Mechanical Design,* vol. 138, no. 8, 2016.

[4] A. Boccaccio, A. E. Uva, M. Fiorentino, L. Lamberti, and G. Monno, "A Mechanobiology-based Algorithm to Optimize the Microstructure Geometry of Bone Tissue Scaffolds," *International journal of biological sciences,* vol. 12, no. 1, p. 1, 2016.

[5] P. F. Egan, I. Bauer, K. Shea, and S. J. Ferguson, "Mechanics of Three-Dimensional Printed Lattices for Biomedical Devices," *Journal of Mechanical Design,* vol. 141, no. 3, p. 031703, 2019.

[6] M. I. Mohammed and I. Gibson, "Design of three-dimensional, triply periodic unit cell scaffold structures for additive manufacturing," *Journal of Mechanical Design,* vol. 140, no. 7, p. 071701, 2018.

[7] G. Dong, Y. Tang, and Y. F. Zhao, "A Survey of Modeling of Lattice Structures Fabricated by Additive Manufacturing," *Journal of Mechanical Design,* vol. 139, no. 10, p. 100906, 2017.

[8] S. J. Hollister *et al.*, "Design control for clinical translation of 3D printed modular scaffolds," *Annals of biomedical engineering,* vol. 43, no. 3, pp. 774-786, 2015.

[9] P. F. Egan, "Integrated Design Approaches for 3D Printed Tissue Scaffolds: Review and Outlook," *Materials,* vol. 12, no. 15, p. 2355, 2019.

[10] A. Boccaccio, A. E. Uva, M. Fiorentino, L. Lamberti, and G. Monno, "A Mechanobiology-based Algorithm to Optimize the Microstructure Geometry of Bone Tissue Scaffolds," (in English), *International Journal of Biological Sciences,* vol. 12, no. 1, pp. 1-17, 2016.

[11] P. Van Liedekerke, M. M. Palm, N. Jagiella, and D. Drasdo, "Simulating tissue mechanics with agent-based models: concepts, perspectives and some novel results," (in English), *Computational Particle Mechanics,* vol. 2, no. 4, pp. 401-444, Dec 2015.

[12] K. Garikipati, E. M. Arruda, K. Grosh, H. Narayanan, and S. Calve, "A continuum treatment of growth in biological tissue: the coupling of mass transport and mechanics," (in English), *Journal of the Mechanics and Physics of Solids,* vol. 52, no. 7, pp. 1595-1625, Jul 2004.

[13] M. Paris *et al.*, "Scaffold curvature-mediated novel biomineralization process originates a continuous soft tissue-to-bone interface," *Acta Biomater,* vol. 60, pp. 64-80, Sep 15 2017.

[14] A. Arefin, M. Lahowetz, and P. F. Egan, "Simulated Tissue Growth in Tetragonal Lattices with Mechanical Stiffness Tuned for Bone Tissue Engineering," *Computers in Biology and Medicine,* vol. 138, 2021.

[15] K. Sandau and H. Kurz, "Modeling of Vascular Growth-Processes - a Stochastic Biophysical Approach to Embryonic Angiogenesis," (in English), *Journal of Microscopy-Oxford,* vol. 175, pp. 205-213, Sep 1994.

[16] J. Walpole, J. C. Chappell, J. G. Cluceru, F. Mac Gabhann, V. L. Bautch, and S. M. Peirce, "Agent-based model of angiogenesis simulates capillary sprout initiation in multicellular networks," (in English), *Integrative Biology,* vol. 7, no. 9, pp. 987-997, 2015.

[17] C. M. Bidan, F. M. Wang, and J. W. Dunlop, "A three-dimensional model for tissue deposition on complex surfaces," *Computer methods in biomechanics and biomedical engineering,* vol. 16, no. 10, pp. 1056-1070, 2013.

[18] M. Paris *et al.*, "Scaffold curvature-mediated novel biomineralization process originates a continuous soft tissue-to-bone interface," *Acta biomaterialia,* vol. 60, pp. 64-80, 2017.

[19] H. Mehdizadeh, S. Sumo, E. S. Bayrak, E. M. Brey, and A. Cinar, "Three-dimensional modeling of angiogenesis in porous biomaterial scaffolds," *Biomaterials,* vol. 34, no. 12, pp. 2875-2887, 2013.

[20] J. Walpole, J. Chappell, J. Cluceru, F. Mac Gabhann, V. Bautch, and S. Peirce, "Agent-based model of angiogenesis simulates capillary sprout initiation in multicellular networks," *Integrative Biology,* vol. 7, no. 9, pp. 987-997, 2015.

[21] M. Xu, J. Yang, I. H. Lieberman, and R. Haddas, "Finite element method-based study of pedicle screw–bone connection in pullout test and physiological spinal loads," *Medical engineering & physics,* vol. 67, pp. 11-21, 2019.

[22] M. Xu, J. Yang, I. Lieberman, and R. Haddas, "Stress distribution in vertebral bone and pedicle screw and screw–bone load transfers among various fixation methods for lumbar spine surgical alignment: A finite element study," *Medical engineering & physics,* vol. 63, pp. 26-32, 2019.

[23] W. Zhang, C. Sun, J. Zhu, W. Zhang, H. Leng, and C. Song, "3D printed porous titanium cages filled with simvastatin hydrogel promotes bone ingrowth and spinal fusion in rhesus macaques," *Biomaterials science,* vol. 8, no. 15, pp. 4147-4156, 2020.

[24] X. Han *et al.*, "In vitro performance of 3D printed PCL−β-TCP degradable spinal fusion cage," *Journal of Biomaterials Applications,* vol. 35, no. 10, pp. 1304-1314, 2021.

[25] P. Li *et al.*, "A novel 3D printed cage with microporous structure and in vivo fusion function," *Journal of Biomedical Materials Research Part A,* 2019.

Copyright © 2023 by ASME

[26] R. Haddas, M. Xu, I. Lieberman, and J. Yang, "Finite Element Based-Analysis for Pre and Post Lumbar Fusion of Adult Degenerative Scoliosis Patients," *Spine deformity,* vol. 7, no. 4, pp. 543-552, 2019.

[27] S. Seaman, P. Kerezoudis, M. Bydon, J. C. Torner, and P. W. Hitchon, "Titanium vs. polyetheretherketone (PEEK) interbody fusion: Meta-analysis and review of the literature," *Journal of Clinical Neuroscience,* vol. 44, pp. 23-29, 2017.

[28] J. Tang *et al.*, "A fast degradable citrate-based bone scaffold promotes spinal fusion," *Journal of Materials Chemistry B,* vol. 3, no. 27, pp. 5569-5576, 2015.

[29] M. Manzur *et al.*, "The rate of fusion for stand-alone anterior lumbar interbody fusion: a systematic review," *The Spine Journal,* 2019.

[30] H.-R. Weiss and D. Goodall, "Rate of complications in scoliosis surgery–a systematic review of the Pub Med literature," *Scoliosis,* vol. 3, no. 1, p. 9, 2008.

[31] H. Koller *et al.*, "Factors influencing radiographic and clinical outcomes in adult scoliosis surgery: a study of 448 European patients," *European Spine Journal,* vol. 25, no. 2, pp. 532-548, 2016.

[32] S. J. Hollister, "Porous scaffold design for tissue engineering," *Nature materials,* vol. 4, no. 7, pp. 518-524, 2005.

[33] A. Tuchman *et al.*, "Autograft versus allograft for cervical spinal fusion: a systematic review," *Global spine journal,* vol. 7, no. 1, pp. 59-70, 2017.

[34] J. Roberge and J. Norato, "Computational design of curvilinear bone scaffolds fabricated via direct ink writing," *Computer-Aided Design,* vol. 95, pp. 1-13, 2018.

[35] J. Norato and A. Wagoner Johnson, "A computational and cellular solids approach to the stiffness-based design of bone scaffolds," *Journal of biomechanical engineering,* vol. 133, no. 9, 2011.

[36] H. Hwangbo *et al.*, "Bone tissue engineering via application of a collagen/hydroxyapatite 4D-printed biomimetic scaffold for spinal fusion," *Applied Physics Reviews,* vol. 8, no. 2, p. 021403, 2021.

[37] C.-G. Liu, Y.-T. Zeng, R. Kankala, S.-S. Zhang, A.-Z. Chen, and S.-B. Wang, "Characterization and Preliminary Biological Evaluation of 3D-Printed Porous Scaffolds for Engineering Bone Tissues," *Materials,* vol. 11, no. 10, p. 1832, 2018.

[38] P. F. Egan, K. A. Shea, and S. J. Ferguson, "Simulated tissue growth for 3D printed scaffolds," *Biomechanics and modeling in mechanobiology,* pp. 1-15, 2018.

[39] A. Entezari *et al.*, "Nondeterministic multiobjective optimization of 3D printed ceramic tissue scaffolds," *Journal of the mechanical behavior of biomedical materials,* vol. 138, p. 105580, 2023.

[40] D. P. Byrne, D. Lacroix, J. A. Planell, D. J. Kelly, and P. J. Prendergast, "Simulation of tissue differentiation in a scaffold as a function of porosity, Young's modulus and dissolution rate: application of mechanobiological models in tissue engineering," *Biomaterials,* vol. 28, no. 36, pp. 5544-5554, 2007.

[41] Y. Guyot, I. Papantoniou, Y. C. Chai, S. Van Bael, J. Schrooten, and L. Geris, "A computational model for cell/ECM growth on 3D surfaces using the level set method: a bone tissue engineering case study," *Biomechanics and modeling in mechanobiology,* vol. 13, no. 6, pp. 1361-1371, 2014.

[42] Y. Guyot, F. Luyten, J. Schrooten, I. Papantoniou, and L. Geris, "A three-dimensional computational fluid dynamics model of shear stress distribution during neotissue growth in a perfusion bioreactor," *Biotechnology and bioengineering,* vol. 112, no. 12, pp. 2591-2600, 2015.

[43] A. Boccaccio, A. E. Uva, M. Fiorentino, G. Mori, and G. Monno, "Geometry design optimization of functionally graded scaffolds for bone tissue engineering: A mechanobiological approach," *PloS one,* vol. 11, no. 1, p. e0146935, 2016.

[44] C. M. Bidan *et al.*, "How linear tension converts to curvature: geometric control of bone tissue growth," *PloS one,* vol. 7, no. 5, p. e36336, 2012.

[45] C. M. Bidan, K. P. Kommareddy, M. Rumpler, P. Kollmannsberger, P. Fratzl, and J. W. Dunlop, "Geometry as a factor for tissue growth: towards shape optimization of tissue engineering scaffolds," *Advanced healthcare materials,* vol. 2, no. 1, pp. 186-194, 2013.

[46] A. Carlier, L. Geris, K. Bentley, G. Carmeliet, P. Carmeliet, and H. Van Oosterwyck, "MOSAIC: a multiscale model of osteogenesis and sprouting angiogenesis with lateral inhibition of endothelial cells," *PLoS Comput Biol,* vol. 8, no. 10, p. e1002724, 2012.

[47] H.-J. Sung, C. Meredith, C. Johnson, and Z. S. Galis, "The effect of scaffold degradation rate on three-dimensional cell growth and angiogenesis," *Biomaterials,* vol. 25, no. 26, pp. 5735-5742, 2004.

[48] A. Artel, H. Mehdizadeh, Y.-C. Chiu, E. M. Brey, and A. Cinar, "An agent-based model for the investigation of neovascularization within porous scaffolds," *Tissue Engineering Part A,* vol. 17, no. 17-18, pp. 2133-2141, 2011.

[49] H. Mehdizadeh, S. I. Somo, E. S. Bayrak, E. M. Brey, and A. Cinar, "Design of Polymer Scaffolds for Tissue Engineering Applications," *Industrial & Engineering Chemistry Research,* vol. 54, no. 8, pp. 2317-2328, 2015.

[50] H. Mehdizadeh *et al.*, "Agent-based modeling of porous scaffold degradation and vascularization: Optimal scaffold design based on architecture and degradation dynamics," *Acta biomaterialia,* vol. 27, 2015.

[51] R. Y. Yeh, K. K. Nischal, P. LeDuc, and J. Cagan, "Written in blood: Applying shape grammars to retinal vasculatures," *Translational Vision Science & Technology,* vol. 9, no. 9, pp. 36-36, 2020.

Copyright © 2023 by ASME

[52] M. E. Whiting, P. R. Leduc, and J. Cagan, "Efficient Automatic Induction of Rules in Biological Systems," *The FASEB Journal,* vol. 31, pp. 927.5-927.5, 2017.

[53] A. Arefin and P. F. Egan, "Computational investigation of tissue and blood vessel growth trade-offs in hierarchical lattices," presented at the ASME IDETC Design Automation Conference, Virtual, 2021.

[54] P. Egan, S. Ferguson, and K. Shea, "Design of hierarchical 3D printed scaffolds considering mechanical and biological factors for bone tissue engineering," *Journal of Mechanical Design,* vol. 139, no. 6, 2017.

[55] Y. Ha *et al.*, "Bone microenvironment-mimetic scaffolds with hierarchical microstructure for enhanced vascularization and bone regeneration," *Advanced Functional Materials,* vol. 32, no. 20, p. 2200011, 2022.

[56] H. C. Herbol, W. C. Hu, P. Frazier, P. Clancy, and M. Poloczek, "Efficient search of compositional space for hybrid organic-inorganic perovskites via Bayesian optimization," (in English), *Npj Computational Materials,* vol. 4, Sep 10 2018.

[57] S. P. Wang, D. M. Zhao, J. Z. Yuan, H. J. Li, and Y. Gao, "Application of NSGA-II Algorithm for fault diagnosis in power system," (in English), *Electric Power Systems Research,* vol. 175, Oct 2019.

[58] A. R. Yildiz, "A comparative study of population-based optimization algorithms for turning operations," *Information Sciences,* vol. 210, pp. 81-88, 2012.

[59] B. Bhushan and S. S. Pillai, "Particle Swarm Optimization and Firefly Algorithm: Performance Analysis," (in English), *Proceedings of the 2013 3rd Ieee International Advance Computing Conference (Iacc),* pp. 746-751, 2013.

[60] Q. Liu, X. F. Li, H. T. Liu, and Z. X. Guo, "Multi-objective metaheuristics for discrete optimization problems: A review of the state-of-the-art," (in English), *Applied Soft Computing,* vol. 93, Aug 2020.

[61] G. W. Mann and S. Eckels, "Multi-objective heat transfer optimization of 2D helical micro-fins using NSGA-II," (in English), *International Journal of Heat and Mass Transfer,* vol. 132, pp. 1250-1261, Apr 2019.

[62] X. Li, H. Qu, G. Li, S. Guo, and G. Dong, "Optimal Design of a Kinematically Redundant Planar Parallel Mechanism Based on Error Sensitivity and Workspace," *Journal of Mechanical Design,* vol. 145, no. 2, p. 023305, 2023.

[63] M. B. R. Rodriguez, J. L. M. Rodriguez, and C. H. D. Fontes, "Thermo ecological optimization of shell and tube heat exchangers using NSGA II," (in English), *Applied Thermal Engineering,* vol. 156, pp. 91-98, Jun 25 2019.

[64] A. S. Mohammadi, J. P. F. Trovao, and C. H. Antunes, "Component-Level Optimization of Hybrid Excitation Synchronous Machines for a Specified Hybridization Ratio Using NSGA-II," (in English), *Ieee Transactions on Energy Conversion,* vol. 35, no. 3, pp. 1596-1605, Sept 2020.

[65] A. Kamaloo, M. Jabbari, M. Y. Tooski, and M. Javadi, "Optimization of thickness and delamination growth in composite laminates under multi-axial fatigue loading using NSGA-II," (in English), *Composites Part B-Engineering,* vol. 174, Oct 1 2019.

[66] K. Deb, A. Pratap, S. Agarwal, and T. Meyarivan, "A fast and elitist multiobjective genetic algorithm: NSGA-II," (in English), *Ieee Transactions on Evolutionary Computation,* vol. 6, no. 2, pp. 182-197, Apr 2002.

[67] M. O. Wang *et al.*, "Evaluating 3D-Printed Biomaterials as Scaffolds for Vascularized Bone Tissue Engineering," *Advanced Materials,* vol. 27, no. 1, pp. 138-144, 2015.

Copyright © 2023 by ASME

Proceedings of the ASME 2023
International Design Engineering Technical Conferences and
Computers and Information in Engineering Conference
IDETC-CIE2023
August 20-23, 2023, Boston, Massachusetts

DETC2023-116342

DESIGNING PROGRAMMABLE FERROMAGNETIC SOFT METASTRUCTURES FOR MINIMALLY INVASIVE ENDOVASCULAR THERAPY

Ran Zhuang[1], Jiawei Tian[1], Apostolos Tassiopoulos[2], Chander Sadasivan[3], Xianfeng Gu[4,5], Shikui Chen[1,*]
Department of Mechanical Engineering [1]
Department of Surgery [2]
Department of Neurological Surgery [3]
Department of Computer Science [4]
Department of Applied Mathematics & Statistics [5]
State University of New York at Stony Brook, Stony Brook, New York, USA, 11794
Email: {Ran.Zhuang, Jiawei.Tian, Shikui.Chen, Xianfeng.Gu}@stonybrook.edu;
{Apostolos.Tassiopoulos, Chandramouli.Sadasivan}@stonybrookmedicine.edu

ABSTRACT

Minimally invasive endovascular therapy (MIET) is an innovative technique that utilizes percutaneous access and transcatheter implantation of medical devices to treat vascular diseases. However, conventional devices often face limitations such as incomplete or suboptimal treatment, leading to issues like recanalization in brain aneurysms, endoleaks in aortic aneurysms, and paravalvular leaks in cardiac valves. In this study, we introduce a new metastructure design for MIET employing re-entrant honeycomb structures with negative Poisson's ratio (NPR), which are initially designed through topology optimization and subsequently mapped onto a cylindrical surface. Using ferromagnetic soft materials, we developed structures with adjustable mechanical properties called magnetically activated structures (MAS). These magnetically activated structures can change shape under noninvasive magnetic fields, letting them fit against blood vessel walls to fix leaks or movement issues. The soft ferromagnetic materials allow the stent design to be remotely controlled, changed, and rearranged using external magnetic fields. This offers accurate control over stent placement and positioning inside blood vessels. We performed magneto-mechanical simulations to evaluate the proposed design's performance. Ex-

perimental tests were conducted on prototype beams to assess their bending and torsional responses to external magnetic fields. The simulation results were compared with experimental data to determine the accuracy of the magneto-mechanical simulation model for ferromagnetic soft materials. After validating the model, it was used to analyze the deformation behavior of the plane matrix and cylindrical structure designs of the Negative Poisson's Ratio (NPR) metamaterial. The results indicate that the plane matrix NPR metamaterial design exhibits concurrent vertical and horizontal expansion when subjected to an external magnetic field. In contrast, the cylindrical structure demonstrates simultaneous axial and radial expansion under the same conditions. The preliminary findings demonstrate the considerable potential and practicality of the proposed methodology in the development of magnetically activated MIET devices, which offer biocompatibility, a diminished risk of adverse reactions, and enhanced therapeutic outcomes. Integrating ferromagnetic soft materials into mechanical metastructures unlocks promising opportunities for designing stents with adjustable mechanical properties, propelling the field towards more sophisticated minimally invasive vascular interventions.

**Address all correspondence to this author.

Copyright © 2023 by ASME

1 INTRODUCTION

Minimally invasive endovascular therapy involves the percutaneous access and transcatheter implantation of medical devices for treating vascular diseases throughout the body. Brain aneurysms, aortic aneurysms, and cardiac valve replacement are the three most prominent segments in the global endovascular market, with a combined worth of at least USD 6.5 billion [1, 2, 3]. The primary limitation of available devices, such as stents and coils for brain aneurysms, stent grafts for aortic aneurysms, and stent valves for aortic valves, is incomplete or non-optimal treatment, leading to recanalization in brain aneurysms [1], endoleaks in aortic aneurysms [2], and paravalvular leaks in cardiac valves [3]. These issues arise from poor device-structure fit to the patient anatomy, resulting in improper scaffold for vascular remodeling around the implanted device and device migration, primarily for aortic aneurysms and cardiac valves.

Abdominal aortic aneurysms (AAA) are fusiform dilations of the abdominal aorta affecting about 3% of the population over 50 [1]. EVAR is the preferred treatment modality in most patients with suitable anatomy and reasonable life expectancy, with about 80% of AAA cases currently treated by EVAR [4]. However, clinical benefits of EVAR are lost relative to open surgery over the long term due to endoleaks, graft displacement, limb occlusions, stent fractures, and secondary ruptures, requiring retreatment in 20-25% of patients on average [5, 6, 7, 8, 9].

To address the limitations of traditional EVAR treatment, this study leverages soft active structures to optimize and enhance the procedure. The inherent flexibility of soft materials allows for increased range and mobility. Nowadays, soft active materials are adopted widely in various fields, including soft robots [10, 11, 12, 13], compliant electronics [14, 15, 16] and biomedical devices [17, 18, 19]. Among the different methods used to actuate soft active materials [20, 21, 22, 23], the magnetic field stands out due to several advantages. Firstly, magnetic fields offer fast and contactless stimuli to control soft magnetic materials, making them safe for biomedical applications [24]. Secondly, magnetic fields can generate relatively higher force and torque in soft magnetic structures, resulting in large displacement fields [25]. Moreover, soft magnetic materials can be programmed to achieve complex deformation and magnetic particle magnetization, making them versatile for various applications [20]. Magnetically activated structures traditionally utilize iron or iron oxide particles embedded in polymers. Those particles have low coercivity, as shown in Fig.1(a), which generates high magnetization with a magnetic field applied and doesn't preserve much magnetization once the external magnetic field is removed [26, 27, 28]. Therefore, the low-coercivity soft material could only generate tension and compression. On the other hand, ferromagnetic material with high coercivity, as shown in Fig.1(b), could preserve high residual magnetic flux density. With the tendency of being parallel to an external magnetic field, the residual magnetic

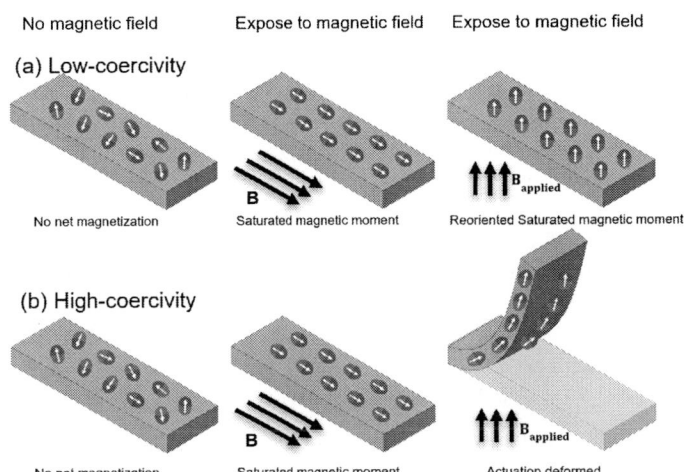

FIGURE 1: A schematic beam model for illustrating the magnetic deformation. (a) Magnetization in a beam made of low-coercivity particles. (b) Magnetization in a beam made of high-coercivity particles.

flux density generates torque, causing bending or torsion deformation on the ferromagnetic soft material. The particles, such as NdFeB powder, are commonly applied for high-coercivity ferromagnetic material to reach complicated deformation field [29, 30, 20]. By leveraging this physical property, many researchers have designed and fabricated these ferromagnetic soft robots [20, 24, 31, 32, 33].

To enhance and optimize EVAR treatment, we aim to develop novel magnetically activated structures (MAS) containing high-coercivity materials. The noninvasive magnetic field can deform the magnetically activated structures grafts, enabling them to conform to the vascular wall, thereby mitigating leaks or migrations. Furthermore, any graft displacement during the follow-up period can be corrected by noninvasively repositioning the devices.

This paper presents a new stent design for minimally invasive endovascular therapy that incorporates ferromagnetic soft materials into mechanical metastructures to create MAS with controllable and tunable mechanical properties. The integration of these materials into the stent design allows for remote actuation, deformation, or reconfiguration by applying an external magnetic field. This enables precise control over stent deployment and positioning within the vasculature, which is critical in achieving optimal therapeutic outcomes. Moreover, these metastructures can be designed to be biocompatible, minimizing the risk of adverse reactions and improving patient outcomes. The proposed pairing of metastructures with embedded ferromagnetic soft materials in stent design can provide a promising approach to developing minimally invasive endovascular therapy devices with improved efficacy and patient outcomes.

The paper is structured as follows: Section 2 presents the

Copyright © 2023 by ASME

modeling of ferromagnetic soft materials, followed by Section 3 which introduces a smart metastructure design comprising ferromagnetic soft materials. The design includes the synthesis of NPR metamaterial by topology optimization, simulation of metamaterial under mechanical stimulation, and simulation of the magneto-mechanical performance of 2D metastructure and wrapped 3D cylindrical metastructure under a magnetic stimulus. Section 4 presents experimental validation, which involves fabricating four benchmark prototypes of ferromagnetic soft beams with different magnetization directions. The bending and torsion of these prototypes under a specific external magnetic field are compared with the magneto-mechanical simulation. Finally, Section 5 provides the concluding remarks and outlines future research directions.

2 MAGNETOMECHANICAL MODEL for FERROMAGNETIC SOFT MATERIALS

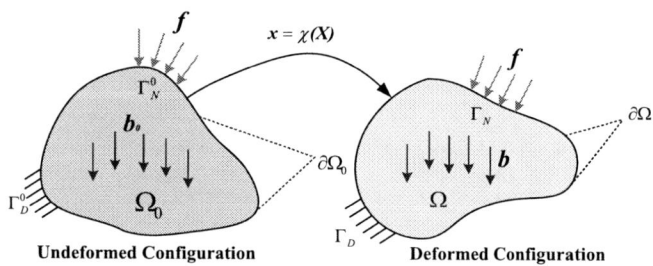

FIGURE 2: Transition from undeformed state to deformed state.

To provide a comprehensive overview, this section includes a brief kinematic description of how ferromagnetic soft materials respond to an applied magnetic field and undergo significant deformation. Interested readers can find more details in [34]. As depicted in Figure 2, the ferromagnetic soft body undergoes a substantial transformation from its initial undeformed state at $t = 0$ to its deformed state at time t. The mapping χ establishes a relationship between the positions of the undeformed state X in the reference frame and the deformed state position x in the spatial frame. The displacement can be calculated as $u = x - X$. The deformation gradient tensor F, can be calculated as,

$$F = \frac{\partial x}{\partial X} = \frac{\partial (u + X)}{\partial X} = I + \nabla_0 u, \tag{1}$$

where I is the identity tensor, and ∇_0 denotes the gradient to the reference frame. The determinant of the deformation gradient is the deformation Jacobian, denoted as $J = det(F)$.

For pure elastic deformation of ferromagnetic soft material, a generalized neo-Hookean model is adopted, and the potential energy density W_{E0} in the reference frame can be calculated as

$$W_{E0} = \frac{\mu}{2}\left(J^{-\frac{2}{3}}I_1 - 3\right) + \frac{1}{2}K(J-1)^2, \tag{2}$$

where μ and K are the shear and bulk modulus, respectively. The invariant is defined as $I_1 = trace(F^T F)$.

When subjected to an external magnetic field B, the soft material possesses magnetization or magnetic moment density M in the reference configuration, and its deformed state in the current configuration is $J^{-1}FM$. Thus, we can calculate the current magnetic potential energy density as

$$W_M = J^{-1}FM \cdot B. \tag{3}$$

According to the relation $W_{M0} = JW_M$, the magnetic potential energy density in the reference frame can be written as

$$W_{M0} = FM \cdot B. \tag{4}$$

Thus, a combined potential energy density of ferromagnetic soft material in the reference frame can be expressed as

$$W_0 = \frac{\mu}{2}\left(J^{-\frac{2}{3}}I_1 - 3\right) + \frac{1}{2}K(J-1)^2 - FM \cdot B. \tag{5}$$

Based on the potential energy density function, it is straightforward to derive the first Piola-Kirchhoff stress as follows,

$$
\begin{aligned}
P &= \frac{\partial W_0}{\partial F} \\
&= \mu J^{-\frac{2}{3}}\left(F - \frac{I_1}{3}F^{-T}\right) + KJ(J-1)F^{-T} - B \otimes M.
\end{aligned} \tag{6}
$$

In the above equation, the operator $\otimes$ is the dyadic product, in which two vectors can yield a second-order tensor. Next, the fourth-order material constitutive tensor based on the first Piola-Kirchhoff stress can be derived as

$$
\begin{aligned}
C_{ijkl} &= \frac{\partial P_{ij}}{\partial F_{kl}} \\
&= -\mu J^{-\frac{2}{3}}\left(\frac{2}{3}F_{ij}F_{lk}^{-1} - \frac{2}{9}I_1 F_{ji}^{-1}F_{lk}^{-1} - \delta_{ij}\delta_{jl}\right) \\
&\quad + \mu J^{-\frac{2}{3}}\left(-\frac{2}{3}F_{kl}F_{ji}^{-1} + \frac{I_1}{3}F_{li}^{-1}F_{jk}^{-1}\right) \\
&\quad + KJ\left((2J-1)F_{ji}^{-1}F_{lk}^{-1} - (J-1)F_{li}^{-1}F_{jk}^{-1}\right).
\end{aligned} \tag{7}
$$

Copyright © 2023 by ASME

548

3 METASTRUCTURE WITH EMBEDDED FERROMAGNETIC SOFT ACTUATORS

This section utilizes a re-entrant honeycomb NPR metamaterial structure, which can be systematically designed through topology optimization [35]. Mechanical simulations of the metamaterial structure are then performed for the unit cell, 2D array, and 3D cylinder shell. The NPR ferromagnetic metastructures are subjected to an applied magnetic field to actuate their response.

3.1 Designing Negative Poisson's Ratio (NPR) Metamaterials by Topology Optimization

Negative Poisson's ratio (NPR) metamaterials exhibit a unique property known as auxeticity. These materials are engineered at the micro or mesoscale with a specific geometric structure that allows them to expand in the direction perpendicular to the applied force. As a result, they can conform to freeform surfaces without developing wrinkles or other deformities, making them highly desirable for various applications. In particular, they have potential use in developing biomedical devices, advanced materials for aerospace and defense, and other areas where conformability to complex surfaces is critical. In this paper, we aim to map a 2D metamaterial design onto a cylinder to achieve radial expansion. Given that displacement in the axial direction is relatively easy to accomplish in a cylindrical shell structure, the NPR property is essential to convert axial displacement into radial displacement.

The present study utilizes a level-set-based topology optimization (TO) approach to design a mechanical metamaterial with a desired negative Poisson's ratio (NPR) [35]. To establish a relationship between the geometric configuration of cellular structures at the micro or mesoscale and the effective properties at the macro scale, the strain energy-based method [36] and the homogenization theory [37] are employed.

The constitutive of stress $\overline{\sigma}_{ij}$ and strain $\overline{\varepsilon}_{kl}$ for 2D isotropic material can be expressed as:

$$\begin{bmatrix} \overline{\sigma}_{11} \\ \overline{\sigma}_{22} \\ \overline{\sigma}_{12} \end{bmatrix} = \begin{bmatrix} C_{1111}^H & C_{1122}^H & 0 \\ C_{1122}^H & C_{2222}^H & 0 \\ 0 & 0 & C_{1212}^H \end{bmatrix} \begin{bmatrix} \overline{\varepsilon}_{11} \\ \overline{\varepsilon}_{22} \\ \overline{\varepsilon}_{12} \end{bmatrix}. \tag{8}$$

As a homogeneous medium, the strain energy of the unit cell can be constructed as follows:

$$U^H = \frac{1}{2} V \overline{\sigma}_{ij} \overline{\varepsilon}_{jk} = \frac{1}{2} \int_V \sigma_{ij} \varepsilon_{jk} dV. \tag{9}$$

By applying the uniform strain boundary condition given in Fig.3, the homogeneous elastic stiffness constants C_{1111}^H, C_{2222}^H, and C_{1122}^H can be expressed in terms of strain energy U_{1111}^H,

U_{2222}^H, and U_{1122}^H as follows:

$$C_{1111}^H = 2U_{1111}^H. \tag{10}$$

$$C_{2222}^H = 2U_{2222}^H. \tag{11}$$

$$C_{1122}^H = U_{1122}^H - U_{1111}^H - U_{2222}^H. \tag{12}$$

The NPR material design problem can be formulated as a prob-

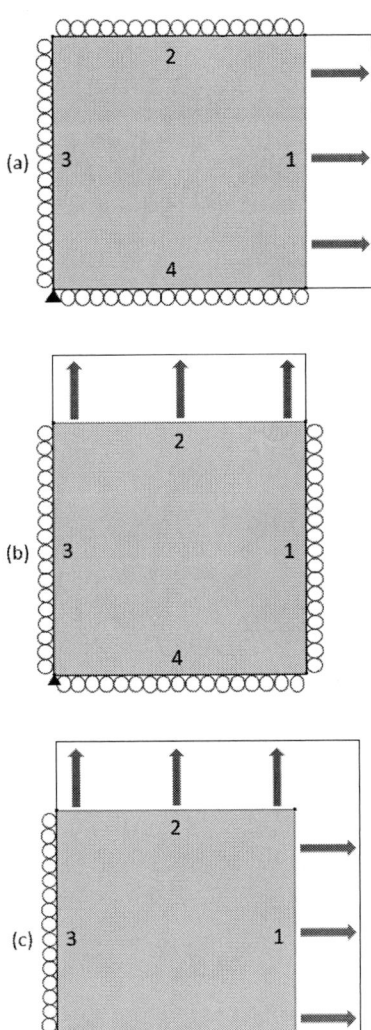

FIGURE 3: Boundary conditions for the unit cell: (a) Loading case 1, (b) Loading case 2, (c) Loading case 3.

lem to minimize the least square error between the homogenized elastic constants C_{ijkl}^H and the predicted homogenized elastic constants C^*_{ijkl} as follows:

$$Minimize: \quad J = \frac{1}{2} \sum_{i,j,k,l=1}^{2} w_{ijkl} \left(C^H_{ijkl} - C^*_{ijkl} \right)^2. \quad (13)$$

With the target elastic stiffness constant set as C^*_{1111}=0.1 GPa, C^*_{2222}=0.1 GPa, and C^*_{1122}=-0.05 GPa. the structure is optimized and post-processed into a 4-by-$\sqrt{3}$ mm re-entrant honeycomb, as shown in Fig.4 (a). The re-entrant honeycomb unit cell is arrayed into a 10-by-10 matrix, as shown in Fig.4 (b). And then the matrix is wrapped into 3d cylinder shell, illustrated in Fig.4 (c).

The re-entrant honeycomb structure was selected due to its comparatively straightforward mechanical behavior, as illustrated in Fig.4. Specifically, when subjected to vertical loading, the oblique diagonal beams within the structure undergo rotational motion, causing the edges on the left and right sides to experience horizontal displacement.

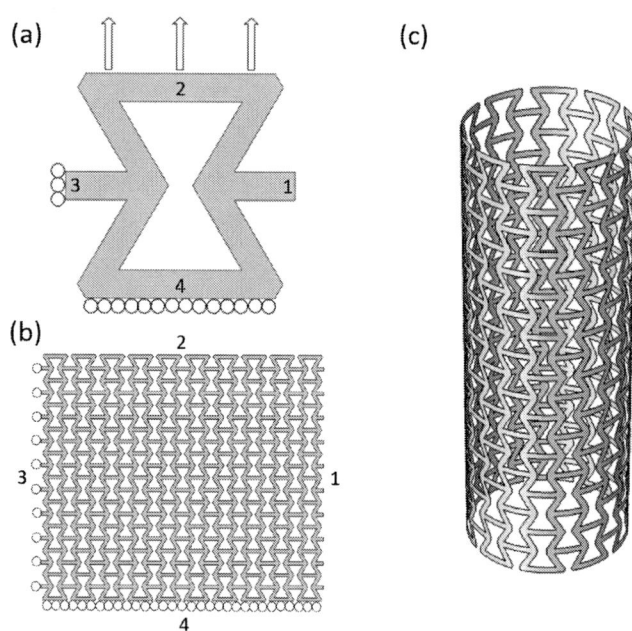

FIGURE 4: Boundary conditions for the re-entrant honeycomb unit structure and its planar and cylindrical assembly. (a) Unit cell with roller boundary conditions on edge 3 and edge 4, and 20% upward displacement on edge 2. (b) 2D hierarchical covalent array subjected to identical boundary conditions on edge 3 and edge 4, and 20% upward displacement on edge 2. (c) 3D hierarchical covalent array with fixed constraints at the bottom surface and 20% upward axial displacement at the top surface.

3.2 Mechanical Simulation of NPR Structure

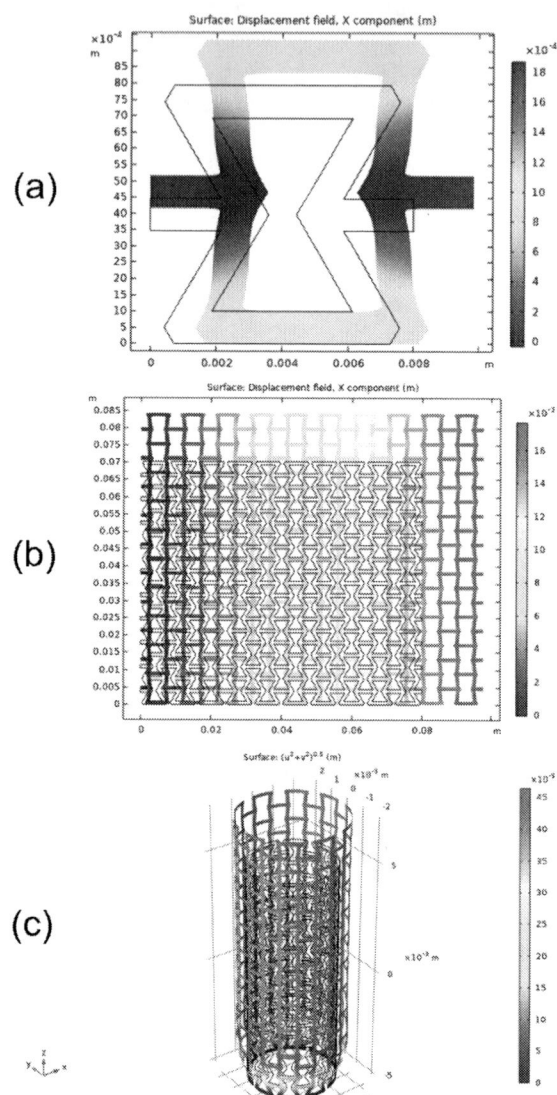

FIGURE 5: Mechanical simulation result of the re-entrant honeycomb unit structure, and its planar and cylindrical assembly. (a) Unit cell displacement field in the x direction. (b) Planar array structure displacement field in the x direction. (c) 3D hierarchical covalent array displacement field in radial direction

Analysis of re-entrant honeycomb allows predicting the Poisson's ratio for individual unit cells, 2D array, and 3d cylinder. A 4-by-$\sqrt{3}$ mm unit cell is analyzed for individual cells, and a 10-by-10 matrix is analyzed for an array of cells. The boundary conditions applied to 2D unit cell and 2D array structures are the roller constraint on edges 3 and 4, and the 20% of height displacement in the vertical direction on edge 2, depicted in Fig.4.

Copyright © 2023 by ASME

The resulting displacement in the x-direction on edge 1 is 20% of width for both individual unit cells and arrayed cells, indicated in Fig.5 (a) (b), and the Poisson's ratio of the re-entrant honeycomb is -1 for 2D cases.

In the simulation of the cylinder, the boundary condition applied is 20% of the height prescribed displacement in the axial direction of the cylinder's top end and fixed boundary to the cylinder's bottom end. The resulting displacement in the radial direction is 20% of the radius, shown in Fig.5 (c). The re-entrant honeycomb metamaterial keeps its negative Poisson's ratio on the 3d cylinder shell structure. The radius of the cylinder increases as the displacement is applied on the top surface.

3.3 Magneto-Mechanical Simulation on 2D Array

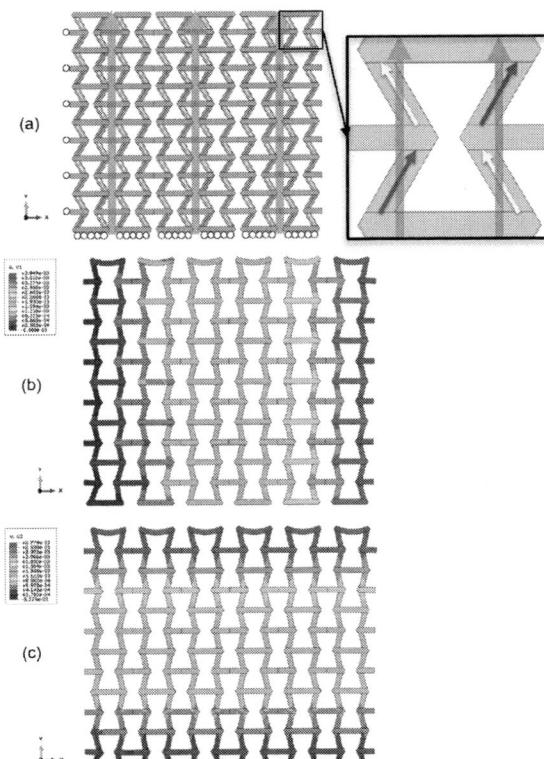

FIGURE 6: The 6-by-6 cell matrix boundary conditions and displacement fields driven by the magnetic field. (a) The roller boundary conditions of the simulation. (b) The displacement in the x direction. (c) The displacement in the y direction.

With the consistency of negative Poisson's ratio when mapping the re-entrant metamaterial matrix on a 3D cylinder, a 6-by-6 matrix made of 8-by-$4\sqrt{3}$ by 1 mm solid unit cell is driven by magnetic field instead of displacement on one end, discussed

in 3.2. The roller boundary condition is applied on the left and bottom surfaces of the metamaterial matrix. The soft material has a shear modulus of 3e5 Pa and a bulk modulus of 3e8 Pa. Next, the multi-material ferromagnetic soft structure with different uniform magnetization directions [38] is introduced here to design the unit cell. In each unit cell shown in Fig.6, the left bottom and right top oblique diagonal beams in each unit cell have residual magnetic flux density pointing to the right top, whose magnetic moment density is 1.0×10^6 A/m in the x direction and 1.732×10^6 A/m in the y direction. The left top and right bottom of oblique diagonal beams in each unit cell have residual magnetic flux density pointing to the left top, whose magnetic moment density is -1.0×10^6 A/m in the x direction and 1.732×10^6 A/m in the y direction. The external magnetic field points upward with a magnetic field equal to 0.04 Tesla. The oblique diagonal beams are expected to align with the direction of the external magnetic field then the cell matrix would expand horizontally. A finite element environment in the forms of Abaqus/Standard use-element subroutine is employed to simulate the magnetically induced large deformation of ferromagnetic soft materials [20].

The resulting displacement in the x-direction on the right side equals 3.95 mm, indicated in Fig.6 (b), which is about 8.2% of the width. The resulting displacement in the y direction on the top edge equals 2.77 mm indicated in Fig.6 (c), which is about 6.6% of the width.

With different Magnetic fields applied to the 2D array of the structure, the maximum displacement increases, as shown in Fig.7.

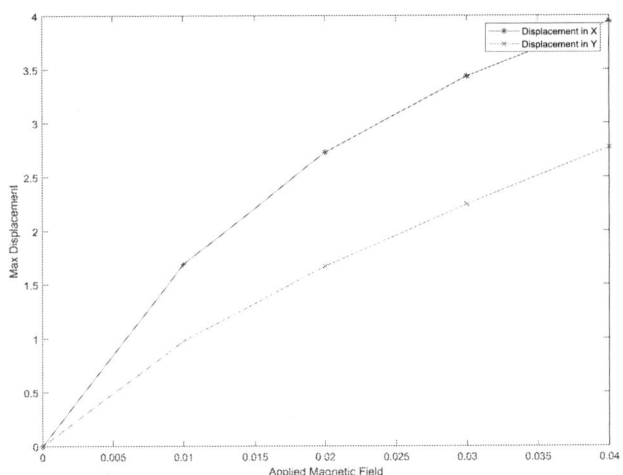

FIGURE 7: Maximum displacement in x and y direction vs applied magnetic field

Copyright © 2023 by ASME

3.4 Magneto-Mechanical Simulation on 3D Cylinder

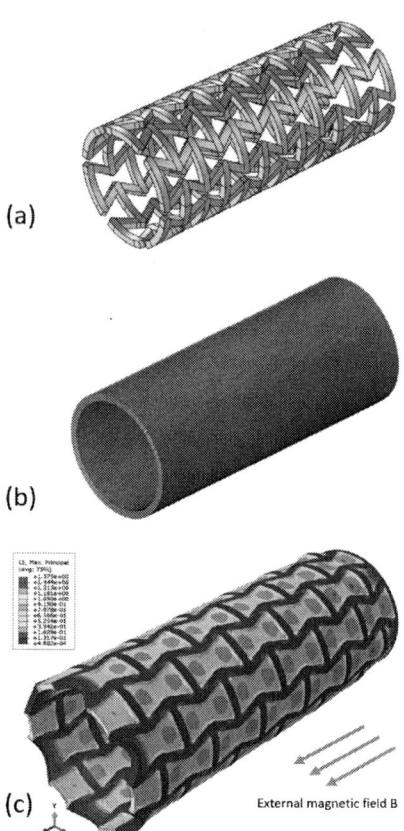

(a)

(b)

(c)

External magnetic field B

FIGURE 8: The wrapped 6-by-6 cylinder matrix. (a) CAD model of 6 layers. (b) The mesh of cylinder. (c) Logarithmic strain field. The external magnetic field B is applied in the z direction.

In section 3.2, it was demonstrated that the re-entrant honeycomb structure can induce the NPR property under mechanical expansion in the vertical direction and that this property is preserved when the structure is wrapped into a cylinder. To further investigate this, a 2D re-entrant honeycomb array was simulated under an external magnetic field in the previous section. The resulting 6x6 2D array structure was then wrapped into a 3D cylinder shell and simulated under the same magnetic field, as shown in Fig. 8(a).

Due to difficulties in meshing the structure as shown in Fig. 8(a), with C3D8 elements, a simplified cylinder shell with a weak material was introduced into the finite element analysis. This simplified shell has the same dimensions as the original cylinder. We substituted the air regions in the original structure with a weak material that has a low bulk modulus of 3e2 Pa and a low shear modulus of 3e5 Pa which are 0.1% of the soft material's shear and bulk modulus, and zero magnetization. The simplified

shell is meshed into 300 by 300 elements on the cylindrical shell, as shown in 8(b).

The magnetization was then wrapped into cylindrical coordinates. The oblique diagonal beams, whose magnetic moment density was 1.0×106 A/m in the x direction and 1.732×106 A/m in the y direction, were transformed to 1.0×106 A/m in the tangent direction and 1.732×106 A/m in the z direction. Similarly, the oblique diagonal beams, whose magnetic moment density was -1.0×106 A/m in the x direction and 1.732×106 A/m in the y direction, were converted to -1.0×106 A/m in the tangent direction and 1.732×106 A/m in the z direction.

The resulting cylindrical structure shell was deformed subject to an encastre boundary condition on one end, while an external magnetic field of 0.04 Tesla was applied in the z direction. The resulting displacement in the radial direction was 1.1 mm, as shown in 8(c), which corresponds to 14.4% of the original radius. The resulting displacement in the axial direction on the unfixed end was 5.8% of the original length, also shown in Fig. 6(c).

4 Experiment of Beam's Bending and Torsion

In this section, the magneto-mechanical simulation is verified by comparing the experiment prototype result and the simulation result of a magnetized ferromagnetic beam under an external magnetic field. Beams with 4 different directions and their deformation under an applied external magnetic field are compared in this section.

4.1 Manufacturing of Ferromagnetic Soft Beam

The mold for 15-by-5-by-1.5 mm beams is 3d printed as shown in Fig.9 (a-b). The magnetic beam prototype is made of NdFeB powder and Ecoflex 00-30 rubber, which is a two-part, room-temperature curing silicone rubber. 20% volume percentage of NdFeB powder, 40% volume percentage of Ecoflex rubber part A and 40% volume percentage of Ecoflex rubber part B are prepared for the molding process, shown in Fig.9 (c). All parts are poured into a container and stirred by a stick, shown in Fig.9 (d). The mixture is then degassed under at least 28 inches of vacuum to mitigate air bubbles shown in Fig.9 (e). Finally, the mixture is poured into the mold, shown in Fig.9 (f). After 5 hours of curing, the solidified model is taken out from the molds.

A holder was 3d printed, displayed in Fig.10 (a) to secure the beams in an IM-10-30 Impulse Magnetizer during magnetization. Four beams are magnetized, illustrated in Fig.10 (a); two of them have residual magnetic flux density in the length direction, and two of them have residual magnetic flux density in the width direction. Under the external magnetic field in the height direction, the beams with different magnetization, indicated in Fig.11 (a-d), are deforming in the direction shown in Fig.11 (i-l).

Copyright © 2023 by ASME

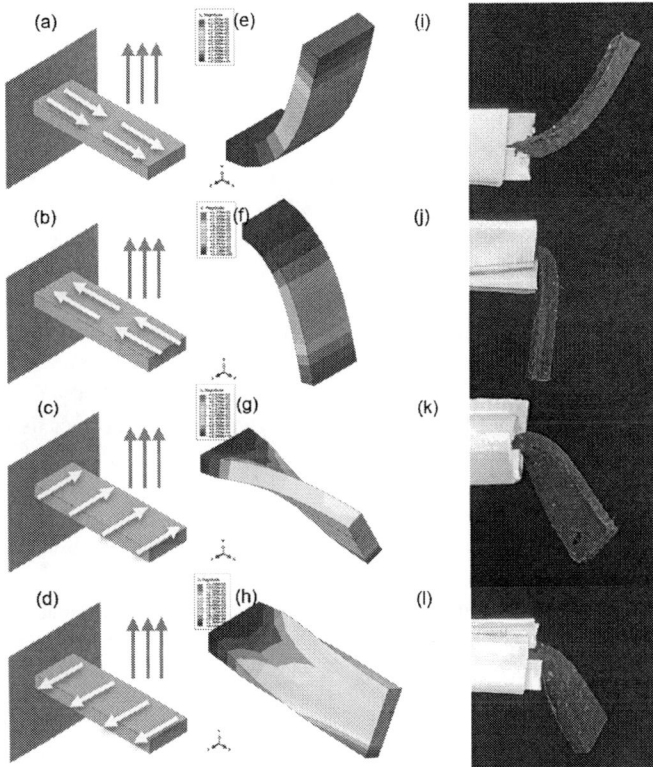

FIGURE 9: Experiment steps. (a-b) the mold for curing the beam. (c-d) 15 ml of NdFeB powder, 30 ml of Ecoflex 00-30 part A and 30 ml of Ecoflex 00-30 part B are prepared and mixed. (e) VWR vacuum machine is applied to eliminate the bubble in the mixture liquid. (f) the liquid is poured in to mold and cured for 5 hours.

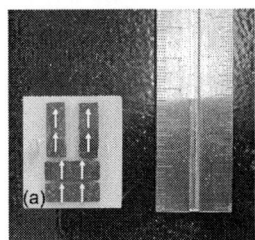
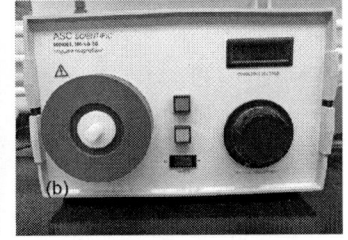

FIGURE 10: Ferromagnetic beam magnetization. (a) The beam is placed in the 3d printed holder, with yellow indicating the residual magnetic flux density of four beams after the magnetization process. (b) The model IM-10-30 Impulse Magnetizer used for the magnetization process.

4.2 Magneto-Mechanical Simulation and Comparison of Beam's Bending and Torsion

The deformation of the manufactured beam under external magnetic is compared with the magneto-mechanical simulation of the ferromagnetic beam to verify the simulation model. During the simulation, the residual magnetic flux density directions of the ferromagnetic beam are oriented in the positive and negative x, as well as positive and negative z directions, indicated by yellow arrows in Fig.11 (a-d). The left end is affixed with encastre boundary conditions. An external magnetic field was applied in the y direction to induce deformation, as shown in the blue arrow in Fig.11 (a-d). The resulting displacement of simulation, depicted in Fig.11 (e-h), is consistent with the deformation

FIGURE 11: The deformation of beam numerical result and prototype under external magnetic field in the y direction. (a-d) The residual magnetic flux density directions are shown in the yellow arrow direction. The external magnetic field is shown in the blue arrow direction. (e-h) The simulation result of beams' displacement magnitude with corresponding residual magnetic flux density directions is shown in a-d. (i-l) The beam prototypes' displacement with corresponding residual magnetic flux density directions are shown in a-d.

of the beam manufactured 11 (i-l). The deformation directions of the ferromagnetic beam indicate that the beam bent in the direction where the magnetization of the material aligned with the external magnetic field.

5 CONCLUSION

This study proposes using magnetically activated structures carrying high-coercivity materials to expand the stent radially and optimize endovascular aneurysm repair treatment. To make the stent conform to the complex geometry of the vessel and facilitate design, the authors employ re-entry negative Poisson's ratio metamaterial. The NPR property of the calculated re-entry honeycomb metamaterial is then simulated on a unit cell and 2D array before being wrapped in a 3D cylinder shell. The simulation results indicate that the NPR property is retained after the

Copyright © 2023 by ASME

shape transformation. To simulate the deformation of the NPR magnetically activated 2D structure, the authors use the magneto-mechanical model for ferromagnetic soft materials, which expands both vertically and horizontally under a vertical magnetic field. This 2D structure has been wrapped in a 3D cylindrical shell. Subsequently, the magneto-mechanical model is applied to simulate the deformation of the 3D cylindrical shell, which indicates that the cylinder expands both radially and axially under an axial magnetic field.

Additionally, four ferromagnetic soft beams, with different magnetization, are fabricated and deformed under an external magnetic field. The bending and torsion of the ferromagnetic beam under a magnetic field are consistent with the simulation result. Both experiment and simulation agree with the fact that residual magnetic flux density has a propensity for aligning to the applied magnetic field.

Future work aims to design a cylindrical ferromagnetic NPR metamaterial structure with zero Poisson's ratio, which expands only in a radial direction under a uniform axial magnetic field. Furthermore, our further research could work on the fabrication of conformable ferromagnetic soft meta-structures and employ topology optimization to design stents for more complicated deformation.

ACKNOWLEDGMENT

The authors gratefully acknowledge financial support from NIH through grant R21EB029733, the National Science Foundation through grants CMMI-1762287 and PFI-RP-2213852, and the State University of New York (SUNY) at Stony Brook through the Summer 2022 OVPR Seed Grant 93214.

REFERENCES

[1] Tan, I., Agid, R., and Willinsky, R. "Recanalization rates after endovascular coil embolization in a cohort of matched ruptured and unruptured cerebral aneurysms". *Interventional Neuroradiology, 17*(1), pp. 27–35.

[2] Hori, D., Nomura, Y., Yamauchi, T., Furuhata, H., Matsumoto, H., Kimura, N., Yuri, K., and Yamaguchi, A. "Perioperative factors associated with aneurysm sac size changes after endovascular aneurysm repair". *Surgery Today, 49*, pp. 130–136.

[3] Gopalakrishnan, D., Gopal, A., and Grayburn, P. A. Evaluating paravalvular leak after TAVR.

[4] Patel, R., Sweeting, M. J., Powell, J. T., and Greenhalgh, R. M. "Endovascular versus open repair of abdominal aortic aneurysm in 15-years' follow-up of the UK endovascular aneurysm repair trial 1 (EVAR trial 1): a randomised controlled trial". *The Lancet, 388*(10058), pp. 2366–2374.

[5] Zoethout, A. C., Boersen, J. T., Heyligers, J. M., de Vries, J.-P. P., Zeebregts, C. J., and Reijnen, M. M. "Two-year outcomes of the Nellix endovascular aneurysm sealing system for treatment of abdominal aortic aneurysms". *Journal of Endovascular Therapy, 25*(3), pp. 270–281.

[6] Quinn, A. A., Mehta, M., Teymouri, M. J., Keenan, M. E., Paty, P. S., Zhou, Y., Chang, B. B., and Feustel, P. "The incidence and fate of endoleaks vary between ruptured and elective endovascular abdominal aortic aneurysm repair". *Journal of vascular surgery, 65*(6), pp. 1617–1624.

[7] van Marrewijk, C., Buth, J., Harris, P. L., Norgren, L., Nevelsteen, A., and Wyatt, M. G. "Significance of endoleaks after endovascular repair of abdominal aortic aneurysms: the EUROSTAR experience". *Journal of vascular surgery, 35*(3), pp. 461–473.

[8] Daye, D., and Walker, T. G. "Complications of endovascular aneurysm repair of the thoracic and abdominal aorta: evaluation and management". *Cardiovascular diagnosis and therapy, 8*(Suppl 1), p. S138.

[9] Goudeketting, S. R., Jin, P. P. F. K., Ünlü, Ç., and de Vries, J.-P. P. "Systematic review and meta-analysis of elective and urgent late open conversion after failed endovascular aneurysm repair". *Journal of vascular surgery, 70*(2), pp. 615–628.

[10] Wehner, M., Truby, R. L., Fitzgerald, D. J., Mosadegh, B., Whitesides, G. M., Lewis, J. A., and Wood, R. J. "An integrated design and fabrication strategy for entirely soft, autonomous robots". *Nature, 536*(7617), p. 451.

[11] Park, S.-J., Gazzola, M., Park, K. S., Park, S., Di Santo, V., Blevins, E. L., Lind, J. U., Campbell, P. H., Dauth, S., Capulli, A. K., et al. "Phototactic guidance of a tissue-engineered soft-robotic ray". *Science, 353*(6295), pp. 158–162.

[12] Kim, S., Laschi, C., and Trimmer, B. "Soft robotics: a bioinspired evolution in robotics". *Trends in biotechnology,*

31(5), pp. 287–294.

[13] Tian, J., Zhao, X., Gu, X. D., and Chen, S. "Designing ferromagnetic soft robots (FerroSoRo) with level-set-based multiphysics topology optimization". In 2020 IEEE International Conference on Robotics and Automation (ICRA), IEEE, pp. 10067–10074.

[14] Zarek, M., Layani, M., Cooperstein, I., Sachyani, E., Cohn, D., and Magdassi, S. "3D printing of shape memory polymers for flexible electronic devices". *Advanced Materials, 28*(22), pp. 4449–4454.

[15] Fan, J. A., Yeo, W.-H., Su, Y., Hattori, Y., Lee, W., Jung, S.-Y., Zhang, Y., Liu, Z., Cheng, H., Falgout, L., et al. "Fractal design concepts for stretchable electronics". *Nature communications, 5*, p. 3266.

[16] Ma, M., Guo, L., Anderson, D. G., and Langer, R. "Bio-inspired polymer composite actuator and generator driven by water gradients". *Science, 339*(6116), pp. 186–189.

[17] Zhao, X., Kim, J., Cezar, C. A., Huebsch, N., Lee, K., Bouhadir, K., and Mooney, D. J. "Active scaffolds for on-demand drug and cell delivery". *Proceedings of the National Academy of Sciences, 108*(1), pp. 67–72.

[18] Cianchetti, M., Laschi, C., Menciassi, A., and Dario, P. "Biomedical applications of soft robotics". *Nature Reviews Materials, 3*(6), pp. 143–153.

[19] Fusco, S., Sakar, M. S., Kennedy, S., Peters, C., Bottani, R., Starsich, F., Mao, A., Sotiriou, G. A., Pané, S., Pratsinis, S. E., et al. "An integrated microrobotic platform for on-demand, targeted therapeutic interventions". *Advanced Materials, 26*(6), pp. 952–957.

[20] Kim, Y., Yuk, H., Zhao, R., Chester, S. A., and Zhao, X. "Printing ferromagnetic domains for untethered fast-transforming soft materials". *Nature, 558*(7709), p. 274.

[21] Zhang, W., Ahmed, S., Masters, S., Ounaies, Z., and Frecker, M. "Finite element analysis of electroactive polymer and magnetoactive elastomer based actuation for origami-inspired folding". In ASME 2016 Conference on Smart Materials, Adaptive Structures and Intelligent Systems, American Society of Mechanical Engineers, pp. V001T01A001–V001T01A001.

[22] Brunet, T., Merlin, A., Mascaro, B., Zimny, K., Leng, J., Poncelet, O., Aristégui, C., and Mondain-Monval, O. "Soft 3D acoustic metamaterial with negative index". *Nature materials, 14*(4), p. 384.

[23] Grier, D. G. "A revolution in optical manipulation". *nature, 424*(6950), p. 810.

[24] Xu, T., Zhang, J., Salehizadeh, M., Onaizah, O., and Diller, E. "Millimeter-scale flexible robots with programmable three-dimensional magnetization and motions". *Science Robotics, 4*(29), p. eaav4494.

[25] Tian, J., Li, M., Han, Z., Chen, Y., Gu, X. D., Ge, Q., and Chen, S. "Conformal topology optimization of multi-material ferromagnetic soft active structures using an ex-

Copyright © 2023 by ASME

tended level set method". *Computer Methods in Applied Mechanics and Engineering, 389*, p. 114394.

[26] Chester, S. A., Di Leo, C. V., and Anand, L. "A finite element implementation of a coupled diffusion-deformation theory for elastomeric gels". *International Journal of Solids and Structures, 52*, pp. 1–18.

[27] Harne, R. L., Deng, Z., and Dapino, M. J. "Adaptive magnetoelastic metamaterials: A new class of magnetorheological elastomers". *Journal of Intelligent Material Systems and Structures, 29*(2), pp. 265–278.

[28] Evans, B., Shields, A., Carroll, R. L., Washburn, S., Falvo, M., and Superfine, R. "Magnetically actuated nanorod arrays as biomimetic cilia". *Nano letters, 7*(5), pp. 1428–1434.

[29] Lum, G. Z., Ye, Z., Dong, X., Marvi, H., Erin, O., Hu, W., and Sitti, M. "Shape-programmable magnetic soft matter". *Proceedings of the National Academy of Sciences, 113*(41), pp. E6007–E6015.

[30] Hu, W., Lum, G. Z., Mastrangeli, M., and Sitti, M. "Small-scale soft-bodied robot with multimodal locomotion". *Nature, 554*(7690), p. 81.

[31] Wu, S., Ze, Q., Zhang, R., Hu, N., Cheng, Y., Yang, F., and Zhao, R. "Symmetry-breaking Actuation Mechanism for Soft Robotics and Active Metamaterials". *ACS Applied Materials & Interfaces, 11*(44), pp. 41649–41658.

[32] Tian, J., Zhao, X., Gu, X. D., and Chen, S. "Designing Conformal Ferromagnetic Soft Actuators Using Extended Level Set Methods (X-LSM)". In International Design Engineering Technical Conferences and Computers and Information in Engineering Conference, Vol. 83990, American Society of Mechanical Engineers, p. V010T10A012.

[33] Zhao, Z., and Zhang, X. S. "Topology optimization of hard-magnetic soft materials". *Journal of the Mechanics and Physics of Solids, 158*, p. 104628.

[34] Zhao, R., Kim, Y., Chester, S. A., Sharma, P., and Zhao, X. "Mechanics of hard-magnetic soft materials". *Journal of the Mechanics and Physics of Solids, 124*, pp. 244–263.

[35] Vogiatzis, P., Chen, S., Wang, X., Li, T., and Wang, L. "Topology optimization of multi-material negative Poisson's ratio metamaterials using a reconciled level set method". *Computer-Aided Design, 83*, pp. 15–32.

[36] Zhang, W., Dai, G., Wang, F., Sun, S., and Bassir, H. "Using strain energy-based prediction of effective elastic properties in topology optimization of material microstructures". *Acta Mechanica Sinica, 23*(1), pp. 77–89.

[37] Hassani, B., and Hinton, E. "A review of homogenization and topology optimization I—homogenization theory for media with periodic structure". *Computers & Structures, 69*(6), pp. 707–717.

[38] Tian, J., Gu, X. D., and Chen, S. "Multi-material topology optimization of ferromagnetic soft robots using reconciled level set method". In International Design Engineering Technical Conferences and Computers and Information in Engineering Conference, Vol. 85451, American Society of Mechanical Engineers, p. V08BT08A014.

Proceedings of the ASME 2023
International Design Engineering Technical Conferences and
Computers and Information in Engineering Conference
IDETC-CIE2023
August 20-23, 2023, Boston, Massachusetts

DETC2023-112570

HOW DOES AGENCY IMPACT HUMAN-AI COLLABORATIVE DESIGN SPACE EXPLORATION? A CASE STUDY ON SHIP DESIGN WITH DEEP GENERATIVE MODELS

Shahroz Khan[*]
CAD/CAx Research Group
Department of Naval Architecture,
Ocean & Marine Engrineering
University of Strathclyde
Glasgow, United Kingdom
Email: shahroz.khan@strath.ac.uk

Panagiotis Kaklis
CAD/CAx Research Group
Department of Naval Architecture,
Ocean & Marine Engrineering
University of Strathclyde
Glasgow, United Kingdom
Email: panagiotis.kaklis@strath.ac.uk

Kosa Goucher-Lambert
Co-Design Lab
Department of Mechanical Engineering
University of California, Berkeley
Berkeley, CA, USA
Email: kosa@berkeley.edu

ABSTRACT

Typical parametric approaches restrict the exploration of diverse designs by generating variations based on a baseline design. In contrast, generative models provide a solution by leveraging existing designs to create compact yet diverse generative design spaces (GDSs). However, the effectiveness of current exploration methods in complex GDSs, especially in ship hull design, remains unclear. To that end, we first construct a GDS using a generative adversarial network, trained on 52,591 designs of various ship types. Next, we constructed three modes of exploration, random (REM), semi-automated (SAEM) and automated (AEM), with varying levels of user involvement to explore GDS for novel and optimised designs. In REM, users manually explore the GDS based on intuition. In SAEM, both the users and optimiser drive the exploration. The optimiser focuses on exploring a diverse set of optimised designs, while the user directs the exploration towards their design preference. AEM uses an optimiser to search for the global optimum based on design performance. Our results revealed that REM generates the most diverse designs, followed by SAEM and AEM. However, the SAEM and AEM produce better-performing designs. Specifically, SAEM is the most effective in exploring designs with a high trade-off between novelty and performance. In conclusion, our study highlights the need for innovative exploration approaches

to fully harness the potential of GDS in design optimisation.

1 Introduction

The ability of a design space to create a rich and valid set of design alternatives is a crucial component of shape optimisation pipelines, as it determines the quality and innovativeness of the solutions produced. Typically, design spaces result from the parametric modellers, which are pre-coded to parameterise the key features of a baseline design [1]. However, these modellers are built to produce solutions within the proximity of the baseline design, and thus, have several limitations. One such drawback is the limited ability of the resulting design spaces to support rich design exploration, leading to a lack of novel design solutions [2].

Moreover, machine learning approaches have proven effective in bypassing the need for computational solvers by providing low-fidelity performance estimators trained offline with data from high-fidelity solvers or physical experimentation. However, until recently, the ability of these models to generate innovative solutions was limited. This limitation arose from the fact that they were only built to predict the performance criteria of designs coming from very narrow design spaces. Generative models [3] such as generative adversarial networks (GANs), variational auto-encoders (VAEs), diffusion models, and transformers have changed this by providing rich design spaces that allow for

---

[*]Address all correspondence to this author.

Copyright © 2023 by ASME

the creation of innovative shapes in addition to performance prediction.

In engineering design tasks, generative models are gaining attention for creating vast generative design spaces (GDSs). These models learn a set of latent features from the given training dataset of existing designs, which are used as design parameters to form GDSs. GDSs are not only low dimensional to expedite shape optimisation but, if properly trained, can also produce novel and valid design alternatives beyond the spectrum of the training dataset. Additionally, efforts are underway to enhance the quality of GDSs to make them physics-informed [4] and user-centred [5]. Physics-informed GDSs can leverage physical laws to ensure that generated designs satisfy certain performance criteria, while user-centred GDSs can incorporate user preferences and constraints to generate designs that are more aligned with the user's needs.

Although GDSs have the potential to offer unprecedented design possibilities, their usability in real design scenarios is not yet fully understood. It is crucial to determine how GDSs can be best utilised without overwhelming designers while expediting the design process. For example, it is essential to study whether existing design exploration techniques, primarily designed to explore narrow design spaces generated by procedural parametric modellers [6, 1], can be effectively applied to explore vast design spaces offered by generative models. Moreover, it is important to understand whether designers are willing to adopt GDSs in their design activities and, if so, what innovative design approaches and scenarios they can use to take full advantage of these diverse spaces.

To achieve this understanding, the present work investigates the most efficient ways of design exploration of GDSs. To this end, we first construct a GDS for complex engineering design problems, such as ship hull design, where parametric design plays a vital role. We create the GDS for hull design by training a custom GAN model, ShipHullGAN [7], on a large dataset of various ship types, including tankers, container ships, bulk carriers, tugboats, and crew supply vessels.

We then develop three design exploration modes with varying degrees of autonomy or designer involvement: random, semi-automated, and automated. The random exploration mode (REM) is a typical preliminary design phase, where designers independently explore the design space based on their intuition and expertise while considering performance. In the semi-automated exploration mode (SAEM), both the designer and optimiser collaborate to guide design exploration towards user-centred and optimised areas of GDS. Finally, the automated exploration mode (AEM) is the standard optimisation scenario where the optimiser is the primary driver and design space exploration occurs while taking performance into account.

With the above modes of exploration, we aim to understand *how designers/naval architects perceive different modes of design exploration in the quest of generating novel design solutions from GDS*. With the above research question in mind, during the study, we aim to mainly analyse the following:

1. To what extent is each exploration mode effective in achieving diverse, novel and better-performing designs?
2. Which factor is the key consideration for each exploration mode: form or performance?

2 Background on ship design and optimisation

Ship design is a complex and bespoke engineering process [8], which differs significantly from other design fields. Unlike other industries, there is no opportunity for full-scale testing, which means that designers have to rely heavily on digital design tools to create the most efficient and safe vessels possible [9]. In today's highly competitive world market, ships must be designed to meet high standards while also being delivered quickly. This requires a high degree of optimisation and customisation, as designers must balance numerous factors such as fuel efficiency, speed, safety, and cargo capacity [10, 11]. The ultimate objective of ship design is to achieve the best performance for a given set of design criteria, which includes the vessel's intended use, the environmental conditions it will operate in, and the regulatory requirements that it must meet. Achieving these objectives requires a multidisciplinary approach that combines expertise in naval architecture, marine engineering, materials science, and other fields [12].

To expedite the design process, naval architects use extensively off-the-shelf parametric modelling tools. These tools are characterised by conservatism, for they are built to generate shapes lying in the neighbourhood of a successful baseline/parent shape [13]. Some relevant examples of such tools are presented in [14, 15, 1, 16, 6]. Next, these modellers are coupled with optimisers for improving the baseline shape against performance criteria (e.g., ship wave resistance, seakeeping, structural strength, etc.), which involve time-consuming simulations, e.g., computational fluid dynamics (CFD). At the end of the process, the new design is likely a local optimum whose shape is a minor variation of the existing one. While these approaches have proven effective for well-established ship types, there may be a need for more radical design ideas in certain situations. This can occur in situations where there are specific requirements that necessitate a more extensive exploration of the design space. Additionally, it may arise when there is a need to revolutionize and redesign existing ship types due to significant regulatory changes, such as the IMO 2020 emission reduction mandate, or the emergence of new disrupting technologies in the context of Industry 4.0 [17, 18, 19]. Such a strategy will benefit novel design tasks, e.g., special-purpose vessels, but it can also offer a competitive advantage for traditional players in the industry.

Conclusively, the coexistence of conservative parametric modellers with high-cost simulations and a large number of design parameters needed for shape optimisation of complex

Copyright © 2023 by ASME

shapes leads to a non-efficient design approach. Such an approach can suffer from the curse of high dimensionality and a limited capability to explore design spaces efficiently for delivering variant, innovative, user-centred and truly optimal designs [20].

Therefore, the ship design necessitates design approaches those bypass the dependence on the parent design and use more rational methods to create rich design spaces, i.e., design spaces resulting from the generative models, with the ability to formulate both conventional and non-conventional hull forms [7].

3 Research methodology

For this work, a study involving human subjects has been developed to quantitatively analyse the efficiency of three exploration modes for exploring GDS constructed using a custom GAN for the ship hull design. Firstly, we discuss the construction of the GDS and how it can be used in preliminary optimisation while being connected to a surrogate model to predict design performance. We then provide a detailed discussion on the different modes of exploration used to analyse the performance of GDS and how they differ in terms of optimisation and user involvement.

3.1 Creation of generative design space (GDS)

There have been substantial efforts in computer-aided ship design for building robust parametric tools, but they can only handle a specific hull type [6, 1]. Despite their efficiency in creating valid and smooth ship-hull geometries, they cannot be readily used to generate instances of ship types that deviate significantly from their target ship types. Therefore, in this work, we utilised, ShipHullGAN [7], a generic parametric modeller built using deep convolutional GANs. The training of ShipHullGAN is performed using a large and diverse dataset of existing hull geometries. We first extensively explored the literature on hull form optimisation and machine learning to identify various hull types. Ultimately, we selected 17 different parent hulls, including KCS [1], KVLCC2 [2], VLCC, JBC [3], DTC, DTMB [4], and others from the FORMDATA series. We then created 3,000 synthetic variations of these hulls using the parametric approach described in [1]. The length, beam, and width of these designs were kept constant, while non-dimensional parameters between 0 and 1 were used to create shape variations. For the FORMDATA series, 5000 design variations were created systematically with respect to the characteristic parameters like midship section-area coefficient c_M and the block coefficients c_{BA} and c_{BF} of the aft and fore parts of the ship, respectively. This synthetic and systematic design creation resulted in 56,000 designs. Subsequently,

---

[1] http://www.simman2008.dk/KCS/kcs_geometry.htm
[2] http://www.simman2008.dk/kvlcc/kvlcc2/kvlcc2_geometry.html
[3] https://www.t2015.nmri.go.jp/jbc.html
[4] http://www.simman2008.dk/5415/combatant.html

in order to establish a reliable training dataset, designs undergo validation using a blend of geometry- and physics-oriented quality filters to assess the viability of each design. Geometry-based filters are employed to ascertain geometric validity, ensuring the absence of self-intersecting surfaces in all designs. Conversely, physics-based filters are utilised to verify that the performance of each design can be accurately predicted by solvers without any potential collapse. Following this rigorous design validation process, a total of 52,591 design variations, which have been both geometrically and physically validated, are obtained for training the ShipHullGAN model.

The design dataset to the ShipHullGAN is inputted in the form of a shape-signature vector (SSV), which consists of a shape modification function and geometric moments. SSV acts as a unique descriptor of each dataset design instance [20, 21]. The inclusion of geometric moments enables the extraction of meaningful features that are not only geometry-driven but also physics-informed. Using geometric moments along with the shape increases the chances of creating a large number of geometrically valid shapes, as adding moments gives a rich set of information about the geometry. More importantly, a strong correlation between ship physics and geometric moments also induces the notion of physics in the extracted latent features. Thus, the resulting features have not only the ability to form a compact but also a physics-informed design, ensuring high-quality valid designs.

The ShipHullGAN uses deep convolutional neural networks for both generator (G) and discriminator (D) components to capture sparsity in the training dataset, along with a space-filling term in the loss function to enhance diversity. D consists of 6 convolutional layers and a dropout layer, with a sigmoid activation function in the last convolutional layer to determine if the design is real or fake. G is the transpose of D and has 5 transposed convolutional layers, with an input layer that takes randomly sampled design, $\mathbf{x}$ and reshapes it. Both G and D use batch normalisation and ReLU activation functions. Training is performed using the Adam gradient descent algorithm with specific settings and performed on a computer with a dual 24-core 2.7GHz Intel Xeon 6 Gold 6226 CPU, NVIDIA Quadro RTX 6000 GPU, and 128GB of memory.

Once the training is completed, the generator component of the ShipHullGAN model is used as a generic parametric modeller. This provides a rich 20-dimensional GDS, which facilitates users in exploring design variations for a wide range of ship hulls. The resulting design variations include both traditional and unconventional forms, as shown in Figure 1. Interested readers should refer to [7] for details on the training of ShipHullGAN and the construction of 20-dimensional GDS.

3.2 Optimisation

For the three modes of exploration, a simple optimisation problem is formulated. The problem aims to explore the 20-

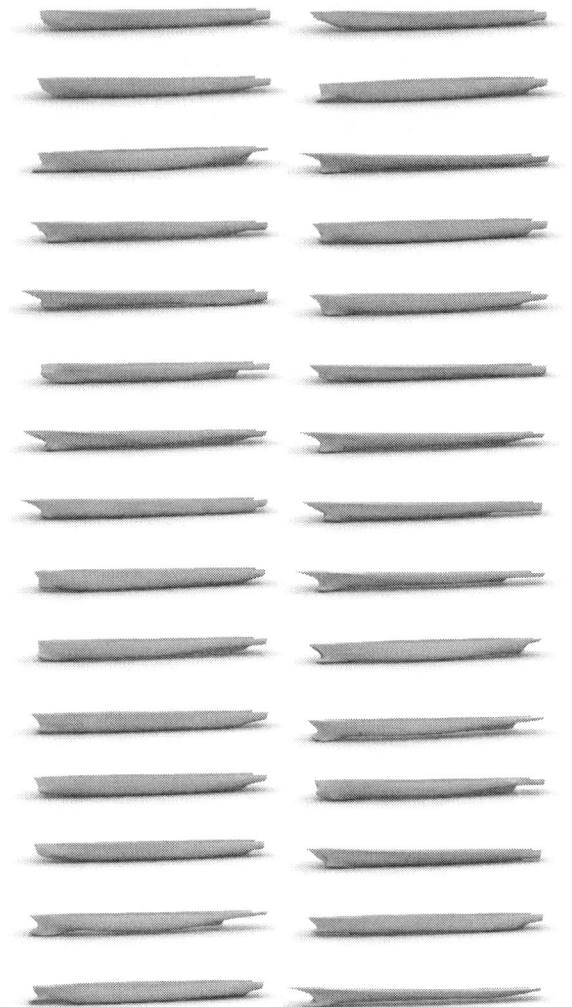

FIGURE 1. Design variations created with the proposed parametric modeller. These design variations can be visualised at https://youtu.be/avlqOFxZP-s and https://youtu.be/ZIfmAs5-qFw

dimensional GDS resulting from the ShipHullGAN parametric modeller to create a container ship with a load-carrying capacity of 3600 TEU (Twenty-foot equivalent unit) while minimising its wave-making resistance/drag (C_w). This optimisation problem can be written in the following setting:

Find $\mathbf{x}^* \in \mathbb{R}^{20}$ such that
$$C_w(\mathbf{x}^*) = \min_{\mathbf{x} \in \mathscr{X}} C_w(\mathbf{x})$$

subject to: given cargo capacity (3600 TEU);

$51120.5m^3 \leq$ Volume of displacement $\leq 56501.6n$

$220.9m \leq$ Length at waterline $\leq 244.2m$;

$30.6m \leq$ Beam at waterline $\leq 33.8m$;

$10.3m \leq$ Draft $\leq 11.3m$.

$$(1)$$

The design constraints in the above equation are set to obtain physically plausible variations of the hull designs. The physical criterion, C_w, is part of the overall resistance affecting the movement of objects on or near the free surface of oceans, lakes and rivers. It reflects the energy spend on creating the free-surface waves following the moving body [22]. Although the overall resistance of the ship is composed of different components, C_w is a vital component and especially prominent for relatively full hull forms travelling at high speeds. It is noteworthy that C_w is highly sensitive to local features of the hull so that a significant reduction can be achieved without affecting the overall cargo capacity. C_w is affected by the distribution of the hull's shape, and minimising it at the preliminary design stage is crucial, but its evaluation can be highly computationally demanding.

3.3 Performance evaluation

To expedite the optimisation process and reduce user fatigue resulting from long simulation run times, we developed a surrogate model that predicts C_w values for designs using Gaussian Process regression (GPR) [23]. GPR is a non-parametric Bayesian approach that has been used in various design applications. It maps the globally-coupled, non-linear relationship between inputs and outputs sampled from a theoretically infinite-dimensional normal distribution and any finite number of input-space samples that follow a corresponding joint (multivariate) Gaussian distribution. The main advantages of GPR over other modelling techniques are that it can: (1) map the input-output relationship with small data size, (2) handle noise in the data easily, thus avoiding over-fitting, and (3) optimise hyperparameters from training data to increase the fit accuracy.

To develop a reliable GPR model, we sampled 10,000 designs using the dynamic propagation sampling technique [24], which ensures that designs are evenly distributed in the design space covering all the design possibilities a given design has to offer. For evaluating C_w values of the designs in the training dataset, we performed hydrodynamic simulations using a software package based on linear potential flow theory using Dawson (double-model) linearisation, with details of the employed formulation, the numerical implementation, and its validation appearing in [25]. As a result of using simple Rankine sources,

Copyright © 2023 by ASME

the computational domain consists of a part of the undisturbed free surface, extending $1Lpp$ upstream, $3Lpp$ downstream, and $1.5Lpp$ sideways, with Lpp denoting the length between perpendiculars for the assessed ship hull. A total of $[20 \times 70]$ grid points are used for the undisturbed free surface, whereas $[50 \times 180]$ grid points are used for the hull discretisation with the simulation being performed at a Froude number F_r equal to $F_r = U/\sqrt{gL} = 0.28$, where g is the acceleration due to gravity, and L is the ship's length. Readers can refer to [26] for details on the construction of the surrogate model with GPR.

3.4 Experiment procedures

The study is composed of three generative design exploration modes, random, semi-automated and automated design exploration, with varying levels of user involvement while providing them with a different level of autonomy. In the following section, we discuss in detail all the exploration modes.

3.4.1 Random exploration mode (REM)
REM is based on a typical random design exploration approach [27, 28, 29], where the user manually explores GDS for novel and better-performing designs based on their intuition. However, as GDS has 20 dimensions, the exploration needs to be organised and user-friendly since exploring each of the 20 parameters individually can be cognitively taxing. Therefore, to streamline the exploration process, we first randomly sample a set of 30,000 designs from the GDS that satisfy all the design constraints in Eq. (1). As designers can explore designs well when the dimensionality of the space is low, the sampled designs are projected onto a 2-dimensional space using t-distributed stochastic neighbour embedding (t-SNE) [30]. This statistical method allows for visualising high-dimensional data by giving each data point a location in a 2- or 3-dimensional map that indicates the distribution of designs. The projection of the randomly sampled designs onto a 2-dimensional space is shown in Figure 2, where their boundary is evaluated using the convex hull, shown using a black curve.

During the design exploration process, users can evaluate the C_w value of each design to balance performance and novelty. However, to avoid biasing users towards physics-based designs only, we do not display the performance in real-time. Once a user discovers a novel design, they can evaluate its performance by clicking on the "evaluate C_w" button. Each user is randomly assigned a set of 30,000 designs and asked to select 5 preferred designs during the exploration process. The design selection process aims to identify a design that is both novel and optimised. A design may be considered novel if it visually differs from the designs that the user has previously seen, designed, or worked with. An optimised design has the least C_w. Therefore, the objective is to find a design that is both novel and optimised, with distinct features and minimal C_w.

The design preview window also allows participants to visu-

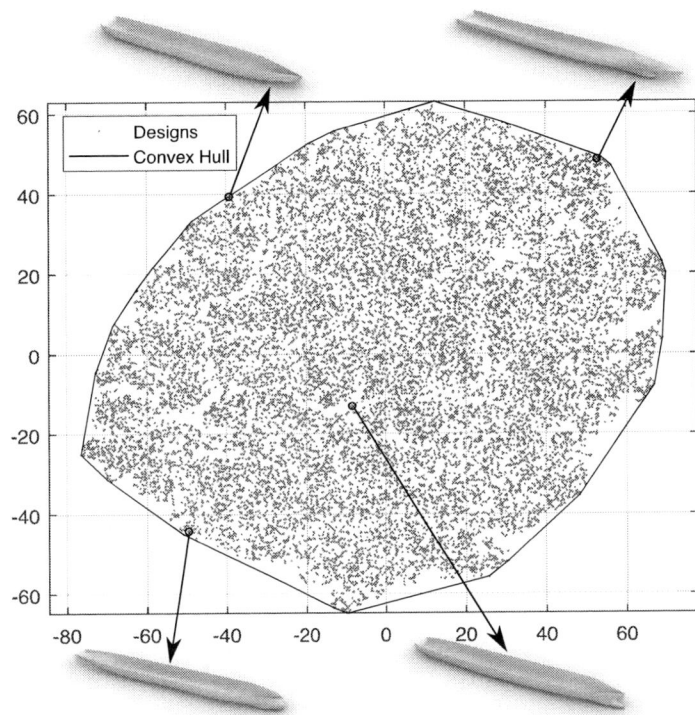

FIGURE 2. 2D t-SEN plot of designs generated from the ShipHull-GAN model.

alise the design in 3D. Users can rotate, zoom, and pan designs to analyse their features thoroughly. Additionally, participants can overwrite a previously selected design. On each design selection, the user is asked what dictated their selection - the form (i.e., design novelty), performance, or a combination of both. Once users select all five designs, they can terminate and conclude this phase of the study.

3.4.2 Semi-automated exploration mode (SAEM)
During this mode of exploration, the user and optimiser collaborate to explore GDS for the generation of novel and optimised designs based on the user's intuition and performance. The overall workflow of SAEM is shown in Figure 3. The optimiser works on exploring a diverse set of optimised designs, while the user induces their preferences to guide the exploration towards user-centred regions of GDS.

Optimisation in this mode is performed based on Khan et al.'s [16] approach, which provides an innovative way to explore GDS and generate well-diverse design alternatives. This approach commences the design exploration with N number of uniformly distributed designs, where each design represents a particular location in GDS. These designs are then shown to the user along with their C_w values. Afterwards, users select the designs

Copyright © 2023 by ASME

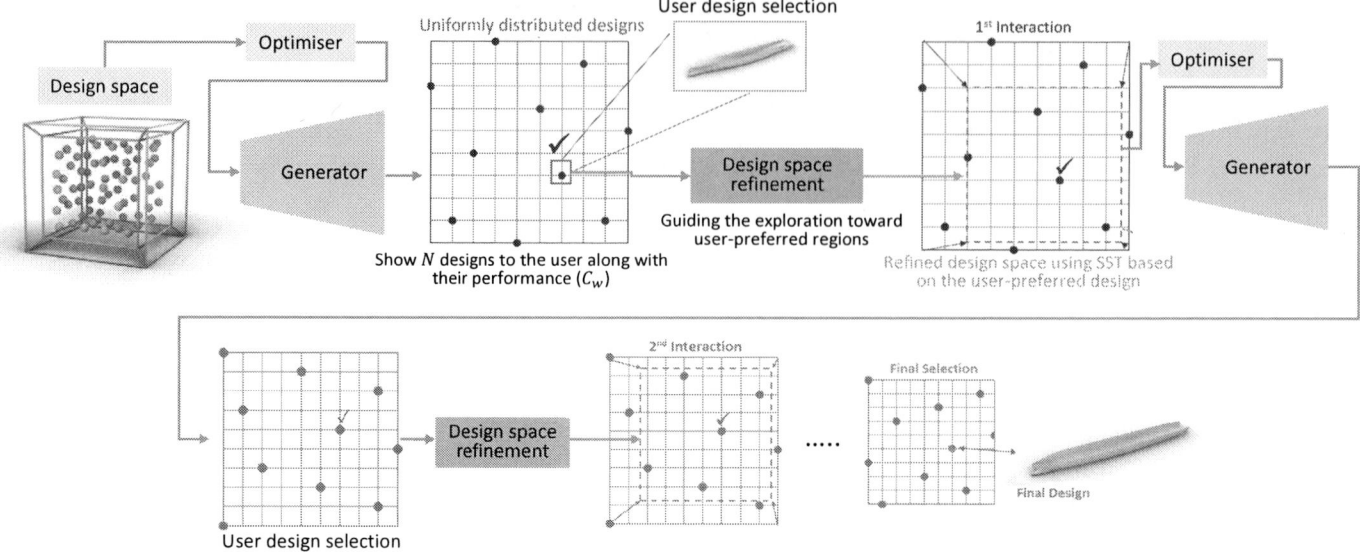

FIGURE 3. Workflow of semi-automated exploration mode.

according to the designs' overall form appearance and physics.

This interaction step allows users to compare designs and make appropriate design decisions. Once the desired hull form is selected, the design space is refined based on the selected design. The refined design space is then imported into the optimiser to generate new N uniformly distributed designs for the next interaction step.

During this process, the designs generated for each interaction should reflect the user's design selection at the previous interaction, so at the end of the interactive process, the user is able to generate a preferred design. In this work, this is achieved by refining the input design space at each interaction while taking into account the user's design preference. A Space Shrinking Technique (SST) [16] is utilised, which detects non-potential regions based on the selected designs and then removes these regions to create a new design space. In other words, at each interaction, SST shrinks the design space towards user-preferred designs and removes regions containing non-preferred designs. This helps the search process to focus the computational effort on the exploration of user-preferred regions of design space.

The interactive process continues until the user arrives at a design with the desired characteristics. At each design selection, the system asks the user what factors influenced their selection, whether it was performance, form, or a combination of both. The user is permitted to perform between 16-25 interactions, which has been found to be an appropriate number to achieve convergence, meaning that no distinct designs are being created.

3.4.3 Automated exploration mode (AEM) This mode of exploration is based on typical shape optimisation [1]. Its pipeline is shown in Figure 4, which connects the GDS, generator (i.e., parametric modeller), and surrogate model for C_w to an appropriate optimiser. During the exploration, the optimiser explores the GDS based on the outcome from the surrogate model, thereby guiding the exploration towards potentially obtaining the global optima while satisfying a given set of constraints.

For the optimisation, we utilised a metaheuristic optimiser, Jay Algorithm (JA), a simple yet efficient approach that does not require any tuning parameters to reach a potentially global solution. JA commences the optimisation with a set of randomly sampled solutions, whose location is improved over a set of iterations. In each iteration, these solutions are moved towards global optima while minimising the following objective function.

$$\min_{\mathbf{x} \in \mathscr{X}} F = \gamma_1 C_w + \gamma_2 \sum_{i=1}^{n} ||\mathbf{x}_u - \mathbf{x}_i|| \quad (2)$$

The above equation is defined as the weighted sum of two terms. The first term is C_w and the second term is added to induce a notion of user preference during design exploration, which, therefore, is defined as the closeness of the new designs with the previously selected design by the user, $\mathbf{x}_u$. The weights γ_1 and γ_2 can be varied between 0 and 1 and set by the user in real-time during exploration. However, initially, we commence the exploration with $\gamma_1 = 0.7$ and $\gamma_2 = 0.3$, giving 70% weightage to C_w and 30% to the closeness/similarity of newly created designs to the previously selected design.

Copyright © 2023 by ASME

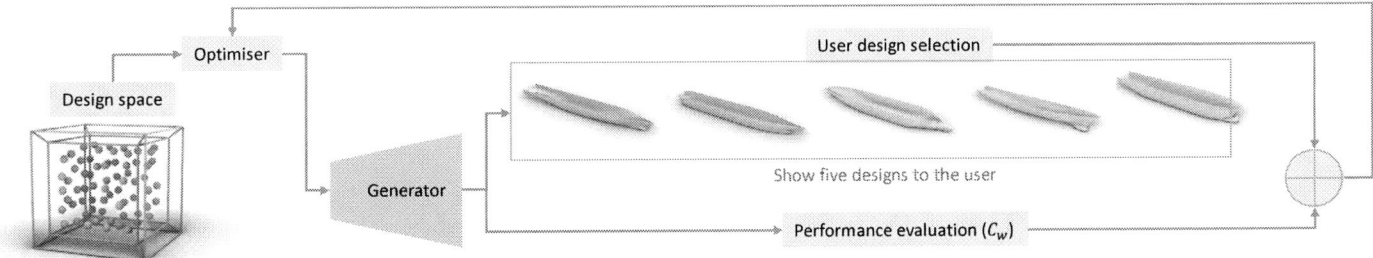

FIGURE 4. Workflow of automated exploration mode.

In our case, since our solver relies on a surrogate model, running many design iterations is not computationally expensive. Therefore, we begin the optimizer with 50 design solutions, which increases the likelihood of finding a good solution. In each iteration, we present the user with the top $n = 5$ designs that minimise the objective function in Equation (2). It is important to note that during the first iteration $\gamma_2 = 0$ as there is no preferred design selected by the user. However, starting from the second interaction, participants select a design based on its novelty and performance and adjust the weightage of the objective function accordingly. This process continues in a similar fashion to the previous mode for 16-25 interactions.

3.5 Population and recruitment

Figure 5 show the graphical user interface created in MATLAB®[5] using the above-described exploration approaches. In total, 20 participants were recruited for the experiment following a protocol approved by the Institutional Review Board of the University of California and the ethical panel of the University of Strathclyde. All participants were final-year undergraduate students who had taken a Naval architecture course, and on average, they reported 3-4 years of experience in ship design. Participants were offered £30 as compensation for their participation. The average age of the participants was 25, with 30% female and 70% male participants. On average, designers reported that they equally value form (i.e. design novelty) and performance in their ship design practice.

The experiments were conducted virtually on Amazon Web Services. Prior to participation, informed consent was obtained from all participants via Google Forms. Participants received an email with step-by-step instructions on the experiment and were assigned 40 minutes to complete it, although they were allowed to take longer. Participants were also informed that there were no right or wrong answers and that their task was to explore the design in a Human-AI design setting using their experience, intuition, and the given directions.

The study did not capture any identifying information about the participants. All three modes of exploration were randomly assigned to the participants, meaning that one participant may perform REM first while another participant performs SAEM first. Once the participants completed all three modes of exploration, they were asked to fill out a post-experiment questionnaire designed to gain more insight into the results of the study. The questionnaire contained five questions, which were as follows:

Q1 What mode of exploration helped to:

    Q1.1 explore diverse designs
    Q1.2 explore better-performing designs
    Q1.3 explore a mix of diverse and better-performing designs

Q2 During the exploration preferred design selection is driven by:

    Q2.1 design novelty (i.e., distinctive form features)
    Q2.2 design performance
    Q2.3 design novelty and performance

Q3 The most engaging mode of exploration
Q4 The least engaging mode of exploration
Q5 Overall preferred mode of exploration

4 Results

In this section, we extensively analyse the outcomes of the user study.

4.1 Design Histories

By combining both design histories and final outcomes, it is possible to evaluate the effectiveness of an exploration mode, as described in [31]. In this study, a design history includes:

1. Overall time spent by participants in each mode.
2. Time spent on each design.
3. Number of designs explored in each mode.
4. Performance of all the explored and user-preferred designs.
5. Indicators for selecting preferred designs: performance, novelty, or a combination of both.

The data relating to these design histories were collected in real-time as the participants performed the study. At the end of the study, the results were automatically sent to the cloud.

---

[5]https://www.mathworks.com

Copyright © 2023 by ASME

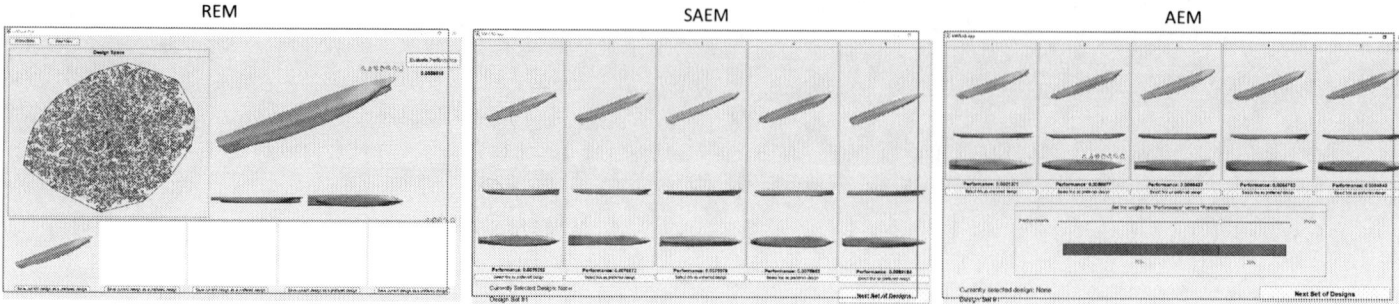

FIGURE 5. Graphical user interfaces of all three exploration modes.

Among the design histories mentioned above, the key parameters to understand the significance of each mode of exploration are the overall time spent during each mode of exploration, the diversity of the selected designs, and their performance. For example, the most efficient mode of exploration is one in which participants extensively explore the GDS to find diverse yet optimised solutions within a short amount of time. In addition to the above histories, we also measure some mode-specific histories such as the location of designs explored during REM to identify if participants tend to cover the entire design space during exploration. Furthermore, we store the weightage of the two terms of the objective function in Eq. (2) during AEM. These design histories can reveal specific behaviours demonstrated by participants during each mode of exploration [32].

4.2 Analyses of design histories

Here we first analyse the three key elements of design histories related to the overall time spent, diversity of the explored designs and their quality (i.e., their performance) to gain insight into the behaviour of the participants during the three modes of exploration.

4.2.1 Overall time taken
Figure 6 shows the total time spent by the participants during each mode of exploration. Interestingly, among the three modes of exploration, participants spent less time in REM, while there was no significant difference between SAEM and AEM. On average, participants completed REM, SAEM, and AEM in 5, 11, and 10 minutes, respectively. It was expected that during REM, participants would take more time to find an innovative and optimised design. However, within REM, participants on average explored 1630 designs within the least amount of time. Another interesting finding was that participants who took less time to complete REM explored more designs, while participants who took more time explored fewer designs. For example, one participant examined 300 designs in 6.4 minutes, whereas another participant explored 6,776 designs in 4.5 minutes. It is important to note that the latter participant's performance can be considered an outlier. Nonetheless, the av-

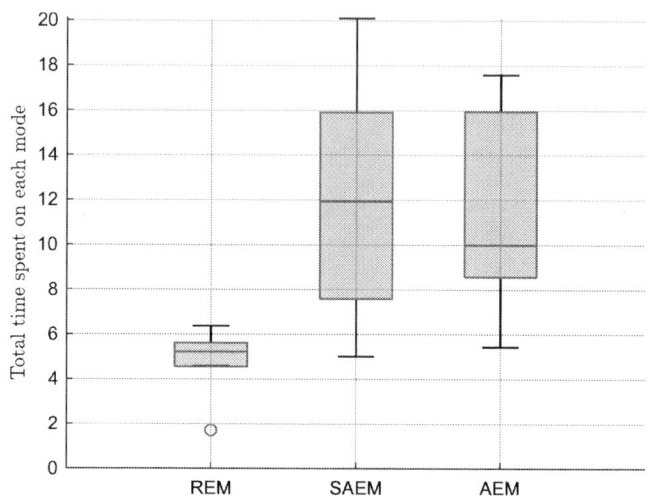

FIGURE 6. Distribution of total time spent during each mode of exploration.

erage time spent on each design, which is shown in Figure 7, did reveal an interesting trend: participants who took more time exploring fewer designs spent, on average, more time on each design. The time spent on each design was measured as the time taken to move to a different design from the design that was currently on the viewing window. In other words, it was the time taken when a design was created to the time when it was replaced by a new design. If no new design was created, it meant that the designer was currently analysing the current design, i.e., they were evaluating its performance and/or analysing its feature for novelty. On average, participants spent 1.4 seconds on each design during REM.

Figure 8 provides the distribution of the total number of designs explored during each mode of exploration. During REM, participants explored an average of 1630 designs. However, among 20 participants, one participant explored 6,776 designs, which is significantly higher compared to the other participants and can be considered an outlier. If we exclude this outlier, the average number of designs explored by the remaining partici-

Copyright © 2023 by ASME

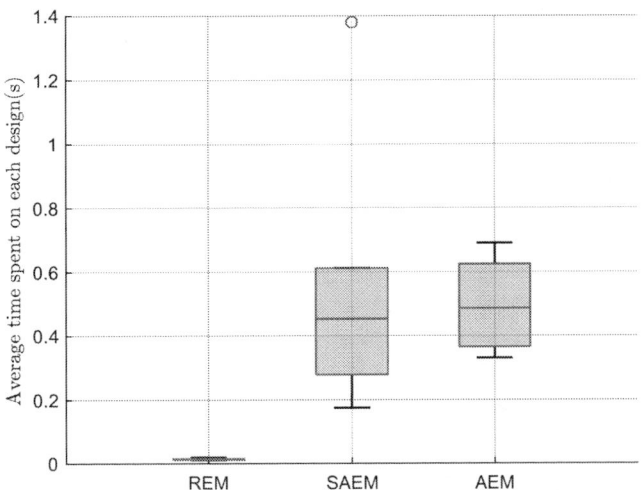

FIGURE 7. Distribution of average time spent during each design during the REM and average time spent on a set of five designs during each interaction of SAEM and AEM.

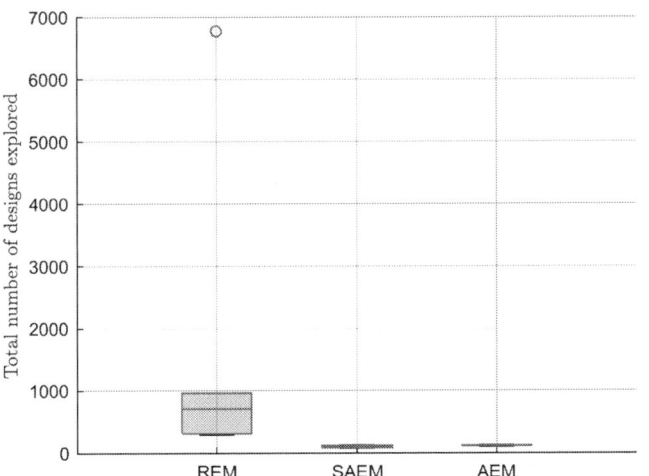

FIGURE 8. Distribution of a total number of designs explored during each mode of interaction.

pants in REM is 601. In contrast, the total number of designs explored during SAEM and AEM was less than REM because, in these modes, participants can explore a set of five designs over 16 to 25 interactions, resulting in a total of 80 to 125 designs. As explained earlier, this number was chosen based on pre-analysis to ensure that participants could explore the design space without experiencing cognitive overload. On average, participants explored 106 and 118 designs in SAEM and AEM, respectively. These results indicate that within SAEM, participants were able to quickly scan the GDS with the least number of designs, taking approximately the same time as in AEM, which has the highest level of design exploration automation.

4.2.2 Diversity of preferred designs In this subsection, we analyse the diversity of the preferred designs during the three modes of exploration. For REM, the diversity is measured between the five final selected designs, and for SAEM and AEM, it is measured between the design selected as preferred designs during each interaction. This analysis addresses the effect of partially performance-driven design exploration and its impact on creativity for the creation of novel hull forms. While creativity and design novelty can be defined in different ways, it is generally assumed that measurements of diversity correspond with increased relative freedom, while the tendency towards standard solutions indicates less creative freedom [32]. The diversity measure aims to specifically understand if giving the performance as a criterion for exploration biases participants to increase novelty or if it influences participants to still focus only on the performance, as in the typical design exploration setting.

Diversity in this work is evaluated with the sparseness at the centre (SC) [33] criterion, which measures the average distance of the centroidal design, $\mathbf{x}^{centroid}$, of the preferred design, to the preferred designs resulting during the exploration of GDS.

$$SC = \frac{1}{n} \sum_{i=1}^{n} ||\mathbf{x}_{centroid} - \mathbf{x}_i||_2 \qquad (3)$$

Although the absolute units of SC measurement are meaningless, as they represent the distance between designs, the relative values from the different modes of exploration provide a worthwhile comparison. Figure 9 shows the SC measure of the designs explored by the participants in all three modes of exploration. It is interesting to note that the diversity of the preferred designs in REM is significantly higher compared to the other two modes. AEM has significantly lower diversity, indicating that designs are highly influenced by performance without much focus on diversity, even when the objective function includes a term to induce a human preference for novelty (see Eq. (2)).

4.2.3 Performance of selected designs Figure 10 shows the average values of Cw of the preferred design resulting from all three modes. It is noteworthy that designs resulting from SAEM, on average, perform better compared to AEM, which is highly performance-driven. However, designs resulting from REM are diverse but do not perform well. In conclusion, participants find better performing and diverse designs with SAEM while exploring fewer designs compared to AEM and REM.

4.2.4 Performance vs novelty During the three modes of exploration, most participants tended to select the preferred design based on both performance and form novelty. However, in REM, participants cared more about form novelty, while in the other two modes, they prioritised performance. Interestingly, the indication towards performance was higher in SAEM,

Copyright © 2023 by ASME

565

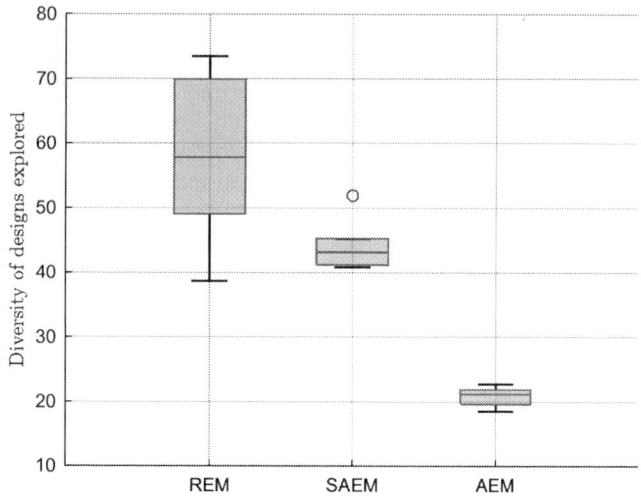

FIGURE 9. Distribution of the diversity of designs explored by participants in all three modes of exploration.

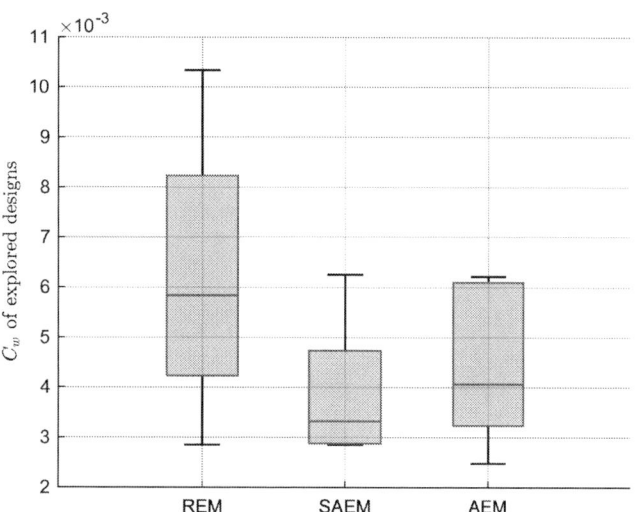

FIGURE 10. Distribution of C_w values of the design explored during the three modes of explorations.

which could be the reason for the better-performing preferred designs resulting from SAEM.

Another point worth noting is that at the start of the study, we asked the participants to give their opinion on whether they care more about performance or novelty during a typical design process. On average, they indicated an equal preference for both novelty and performance.

4.3 Survey results

In this subsection, we discuss the results of the questionnaire conducted to evaluate participants' perceptions of RME, SAME and AME exploration modes discussed in Section 3.5. The re-

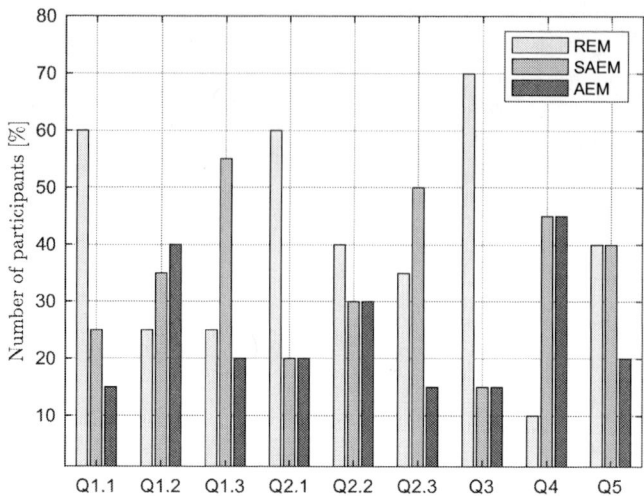

FIGURE 11. Summary of the questionnaire results from all the participants.

sults of this questionnaire are shown in Figure 11. It can be seen that in Q1.1, 60% of participants reported finding the most novel design ideas within REM, while only 15% found AEM to be useful for discovering novel designs. However, in Q1.2, 40% of the participants believed that AEM produced better-performing designs. The remaining 35% and 25% of participants found REM and SAEM, respectively, to be more effective at producing better-performing designs. In response to Q1.3, which asked about the mode that provided the exploration of both diverse and better-performing designs, 55% of participants preferred SAEM, while 25% and 20% preferred REM and AEM, respectively.

From Figure 11 it can be seen that in Q2.1, 60% of participants indicated that novelty was the driver of design selection in REM. Additionally, in Q2.2, 40% of participants indicated that performance was also a driver of design selection in REM, although this result did not deviate significantly from those of SAEM and AEM. The results of Q2.3 were particularly interesting, as 50% of participants found SAEM to be a mode where design selection was driven equally by both performance and novelty. It is worth noting that the trend observed in the Q2 questions is consistent with that of the Q1 questions.

For Q3, Q4 and Q5, the results in Figure 11 show that 70% of the participants found REM to be the most engaging mode, whereas only 10% of the participants indicated it as the least engaging mode. Perhaps these are the participants who value design performance significantly more than design novelty. Overall, participants least favoured the AEM, mainly due to the lack of design novelty.

5 Discussion

This section provides a detailed discussion of the key findings of this study, specifically focused on identifying an effective

Copyright © 2023 by ASME

exploration approach for GDS.

5.1 To what extent is each exploration mode effective in achieving diverse, novel and better-performing designs?

From the results discussed in Section 4, it can be concluded that REM provides a user-engaging approach that results in the exploration of novel design solutions within the shortest possible design exploration time. Interestingly, even with a short exploration time, participants are able to scan the GDS effectively and find a diverse set of design alternatives. Although designs resulting from this mode are diverse, as expected, they are not efficient from a performance perspective. On the other hand, AEM, which is primarily driven by performance, has the least diversity. One would expect that even if the designs are not diverse, they must perform better as the optimisation is solely driven by performance. However, designs resulting from AEM have slightly lower performance on average compared to the designs resulting from SAEM.

In conclusion, while design spaces resulting from typical parametric approaches may yield better-performing designs, the key benefit of GDSs lies not only in better performance but also in the ability to generate novel designs that possess non-conventional features and do not currently exist in the market. To fully leverage the potential of GDSs, it is necessary to focus not only on enriching these spaces but also on exploring them in an effective manner. This study highlights that the commonly used random exploration (REM) and optimisation-based exploration (AEM) approaches are not optimal for GDSs. REM prioritises novelty, while AEM prioritises performance. To strike a balance between these objectives, hybrid and intuitive exploration approaches such as SAEM are needed. SAEM involves both the user and the optimiser at the same level, where users leverage their design expertise to explore novel solutions, and the optimiser focuses on enhancing the performance of the user-preferred designs.

5.2 Which factor is the key consideration for each exploration mode: form or performance?

Moreover, this study revealed that although participants initially stated they aim to balance both novelty and performance in their design tasks. However, during the design exploration in REM, SAEM and AEM they tended to prioritise performance over novelty. This may be due to various factors. If this trend persists, users may be biased toward prioritising performance and fail to utilise GDSs to their full potential. However, this fact is also centred on the type of exploration approach used. For example, in REM, design exploration and selection of preferred design was driven by the form, whereas in AEM design selection was mainly driven by the performance. However, SAEM, aimed at balancing both performance and novelty, does well to engage participants in design exploration. Therefore, designs resulting

from this approach are diverse as well as better performing. Furthermore, in addition to hybrid exploration methods, there is a need for more engaging design interfaces. Our questionnaire results indicate that participants found REM to be more engaging compared to other exploration modes.

In summary, from this study, it can be concluded that as the design space are become more and more diverse, thanks to generative models, we need also an innovative approach for their efficient exploration, as the traditional way of design exploration, made originally for the narrow design spaces, cannot be beneficial to truly exploit the potential of GDSs.

6 Concluding remarks

In this work, we aimed to evaluate the effectiveness of different design exploration approaches for exploring generative design spaces resulting from generative models such as generative adversarial networks. To achieve this, we constructed a generative design space for the ship hull design and optimisation problem. We trained a custom generative model on a large dataset of physically and geometrically validated designs and then used the generator component of the model as a parametric modeller to generate a diverse 20-dimensional design space. We explored this space using three different approaches: REM, SAEM, and AEM, each with different levels of user involvement and algorithmic autonomy.

The REM is a random exploration mode in which the user explores a 2-dimensional projection of the generative design space. SAEM is a mode that is spontaneously driven by both the user and the optimiser with the same level of involvement. In this mode, the optimiser focuses on exploring a diverse set of uniformly distributed and optimised designs (i.e., designs with low C_w) from the generative design space, while the user directs the exploration towards the region of design space containing their preferred designs. AEM is a typical shape optimisation mode in which the design space is connected to an optimiser and performance evaluation code that guides the optimiser in finding a global optimum. To incorporate a user's preference, the objective function for this mode is the weighted sum of C_w and the similarity of newly created designs to the user's previously selected designs.

The results of this study showed that the highest design diversity occurred during REM, followed by SAEM and AEM, whereas better-performing designs were found within AEM and SAEM. However, SAEM outperforms REM and AEM in terms of exploring designs that have a significantly high trade-off between novelty and performance. The study results also showed that participants are adept at exploring novelty and that their subconscious directs them to prefer novel design alternatives. However, when performance is brought into the exploration, they immediately tend to select based on performance.

Copyright © 2023 by ASME

6.1 Future work

In the future, we aim to scale up our investigation by enriching the population of the subjects involved with a) designers covering the whole spectrum of expertise (low to high), b) design-users covering the whole lifecycle, e.g. shipyards, shipowners, operators, and c) designers acting in other transportation (automotive, aerospace) industries.

Acknowledgements

This work received funding from:

1. the Royal Society under the HINGE (Human InteractioN supported Generative modEls for creative designs) project via their International Exchanges 2021 Round 2 funding call, and
2. the European Union's Horizon 2020 research and innovation programme under the Marie Skłodowska-Curie grant GRAPES (learninG, pRocessing And oPtimising shapES) agreement No 860843.

REFERENCES

[1] Kostas, K., Ginnis, A., Politis, C., and Kaklis, P., 2015. "Ship-hull shape optimization with a T-spline based BEM–isogeometric solver". *Computer Methods in Applied Mechanics and Engineering, 284*, pp. 611–622.

[2] Chen, W., and Ahmed, F., 2021. "Padgan: Learning to generate high-quality novel designs". *Journal of Mechanical Design, 143*(3).

[3] Regenwetter, L., Nobari, A. H., and Ahmed, F., 2022. "Deep generative models in engineering design: A review". *Journal of Mechanical Design, 144*(7), p. 071704.

[4] Yang, L., Zhang, D., and Karniadakis, G. E., 2020. "Physics-informed generative adversarial networks for stochastic differential equations". *SIAM Journal on Scientific Computing, 42*(1), pp. A292–A317.

[5] Chaudhari, A. M., and Selva, D., 2023. "Evaluating designer learning and performance in interactive deep generative design". *Journal of Mechanical Design, 145*(5), p. 051403.

[6] Khan, S., Gunpinar, E., and Dogan, K. M., 2017. "A novel design framework for generation and parametric modification of yacht hull surfaces". *Ocean Engineering, 136*, pp. 243–259.

[7] Khan, S., Goucher-Lambert, K., Kostas, K., and Kaklis, P., 2023. "ShipHullGAN: A generic parametric modeller for ship hull design using deep convolutional generative model". *Computer Methods in Applied Mechanics and Engineering, 411*, p. 116051.

[8] Charisi, N. D., Hopman, H., and Kana, A., 2022. "Early-stage design of novel vessels: How can we take a step forward?". In SNAME 14th International Marine Design Conference, OnePetro.

[9] Nowacki, H., 2010. "Five decades of computer-aided ship design". *Computer-Aided Design, 42*(11), pp. 956–969.

[10] Gaspar, H. M., Rhodes, D. H., Ross, A. M., and Ove Erikstad, S., 2012. "Addressing complexity aspects in conceptual ship design: a systems engineering approach". *Journal of Ship Production and Design, 28*(04), pp. 145–159.

[11] Ebrahimi, A., Brett, P. O., Erikstad, S. O., and Asbjørnslett, B. E., 2021. "Influence of ship design complexity on ship design competitiveness". *Journal of Ship Production and Design, 37*(03), pp. 181–195.

[12] Papanikolaou, A., 2010. "Holistic ship design optimization". *Computer-Aided Design, 42*(11), pp. 1028–1044.

[13] Khan, S., Gunpinar, E., and Moriguchi, M., 2017. "Customer-centered design sampling for cad products using spatial simulated annealing". *Proceedings of CAD, 17*, pp. 100–103.

[14] Ginnis, A., Kostas, K., Feurer, C., Belibassakis, K., Gerostathis, T., Politis, C., and Kaklis, P., 2011. "A CATIA®ship-parametric model for isogeometric hull optimization with respect to wave resistance". In Proceedings of ICCAS 2011 conference, Trieste 20-22 September, Italy.

[15] Khan, S., Gunpinar, E., Mert Dogan, K., Sener, B., and Kaklis, P., 2022. "ModiYacht: Intelligent cad tool for parametric, generative, attributive and interactive modelling of yacht hull forms". In SNAME 14th International Marine Design Conference, OnePetro.

[16] Khan, S., Gunpinar, E., and Sener, B., 2019. "Genyacht: An interactive generative design system for computer-aided yacht hull design". *Ocean Engineering, 191*, p. 106462.

[17] Kaklis, D., Varelas, T., Varlamis, I., Eirinakis, P., Giannakopoulos, G., and Spyropoulos, C. V., 2023. "From steam to machine: Emissions control in the shipping 4.0 era". In The 8th International Symposium on Ship Operations, Management & Economics (SOME), Society of Naval Architects and Marine Engineers (SNAME), pp. 1–12.

[18] Citaristi, I., 2022. "United nations conference on trade and". In *The Europa Directory of International Organizations 2022*. Routledge, pp. 177–181.

[19] Joung, T.-H., Kang, S.-G., Lee, J.-K., and Ahn, J., 2020. "The imo initial strategy for reducing greenhouse gas (ghg) emissions, and its follow-up actions towards 2050". *Journal of International Maritime Safety, Environmental Affairs, and Shipping, 4*(1), pp. 1–7.

[20] Khan, S., Kaklis, P., Serani, A., Diez, M., and Kostas, K., 2022. "Shape-supervised dimension reduction: Extracting geometry and physics associated features with geometric moments". *Computer-Aided Design, 150*, p. 103327.

[21] Khan, S., Kaklis, P., Serani, A., and Diez, M., 2022. "Geometric moment-dependent global sensitivity analysis with-

Copyright © 2023 by ASME

out simulation data: application to ship hull form optimisation". *Computer-Aided Design, 151*, p. 103339.

[22] Bertram, V., 2011. *Practical ship hydrodynamics*. Elsevier.

[23] Schulz, E., Speekenbrink, M., and Krause, A., 2018. "A tutorial on gaussian process regression: Modelling, exploring, and exploiting functions". *Journal of Mathematical Psychology, 85*, pp. 1–16.

[24] Khan, S., and Kaklis, P., 2021. "From regional sensitivity to intra-sensitivity for parametric analysis of free-form shapes: Application to ship design". *Advanced Engineering Informatics, 49*, p. 101314.

[25] Bassanini, P., 1994. "The wave resistance problem in a boundary integral formulation". *Surv Math Ind, 4*, pp. 151–194.

[26] Khan, S., Serani, A., Diez, M., and Kaklis, P., 2021. "Physics-informed feature-to-feature learning for design-space dimensionality reduction in shape optimisation". In AIAA scitech 2021 forum, p. 1235.

[27] Bole, M., 2011. "Interactive hull form transformations using curve network deformation". *Ship Technology Research, 58*(1), pp. 46–64.

[28] Fuchkina, E., Schneider, S., Bertel, S., and Osintseva, I., 2018. "Design space exploration framework". In eCAADe, Vol. 36, pp. 367–376.

[29] Krish, S., 2011. "A practical generative design method". *Computer-Aided Design, 43*(1), pp. 88–100.

[30] Van der Maaten, L., and Hinton, G., 2008. "Visualizing data using t-SNE.". *Journal of machine learning research, 9*(11).

[31] Girotra, K., Terwiesch, C., and Ulrich, K. T., 2010. "Idea generation and the quality of the best idea". *Management science, 56*(4), pp. 591–605.

[32] Brown, N. C., 2020. "Design performance and designer preference in an interactive, data-driven conceptual building design scenario". *Design studies, 68*, pp. 1–33.

[33] Brown, N. C., and Mueller, C. T., 2019. "Quantifying diversity in parametric design: a comparison of possible metrics". *AI EDAM, 33*(1), pp. 40–53.

Copyright © 2023 by ASME

Proceedings of the ASME 2023
International Design Engineering Technical Conferences and
Computers and Information in Engineering Conference
IDETC-CIE2023
August 20-23, 2023, Boston, Massachusetts

DETC2023-115176

ADAPTATION AND CHALLENGES IN HUMAN-AI PARTNERSHIP FOR THE DESIGN OF COMPLEX ENGINEERING SYSTEMS

Zeda Xu[1]
Department of Mechanical Engineering
Carnegie Mellon University
Pittsburgh, PA, 15213, USA
zedaxu@cmu.edu

Chloe Hong[1]
School of Architecture
Carnegie Mellon University
Pittsburgh, PA, 15213, USA
soohwah@andrew.cmu.edu

Nicolás F. Soria Zurita
Colegio de Ciencias e Ingeniería
Universidad San Francisco de Quito
Campus Cumbayá, Quito, EC170901, Ecuador
nicosoria@psu.edu

Joshua T. Gyory
Department of Mechanical Engineering
Carnegie Mellon University
Pittsburgh, PA, 15213, USA
jgyory@alumni.cmu.edu

Gary Stump
Pennsylvania State University
University Park, PA, 16802, USA
gms158@psu.edu

Hannah Nolte
Advanced Software Innovation
The MITRE Corporation
Bedford, MA, 01730
hnolte@mitre.org

Jonathan Cagan
Department of Mechanical Engineering
Carnegie Mellon University
Pittsburgh, PA, 15213, USA
cagan@cmu.edu

Christopher McComb[2]
Department of Mechanical Engineering
Carnegie Mellon University
Pittsburgh, PA, 15213, USA
ccm@cmu.edu

ABSTRACT

Exploring the opportunities for incorporating Artificial Intelligence (AI) to support team problem solving has been the focus of intensive ongoing research. However, while the incorporation of such AI tools into human team problem solving can improve team performance, it is still unclear what modality of AI integration will lead to a genuine human-AI partnership capable of mimicking the dynamic adaptability of humans. This work unites human designers with AI Partners as fellow team members who can both reactively and proactively collaborate in real-time towards solving a complex and evolving engineering problem. Team performance and problem-solving behaviors are examined using the HyForm collaborative research platform. The problem constraints are unexpectedly changed midway
through problem solving to simulate the nature of dynamically evolving engineering problems. This work shows that after the shock is introduced, human-AI hybrid teams perform similarly to human teams, demonstrating the capability of AI Partners to adapt to unexpected events. Nonetheless, hybrid teams do struggle more with coordination and communication after the shock is introduced. Overall, this work demonstrates that these AI design Partners can participate as active partners within human teams during a large, complex task, showing promise for future integration in practice.

Keywords: Human-AI Partnership, Engineering Design, Hybrid Teams.

---

[1] These two authors contributed equally to this work
[2] Corresponding author

Copyright © 2023 by ASME

1. INTRODUCTION

AI in research, academia, and industry is becoming significantly more prevalent. A projection by McKinsey & Company [1] identified that by 2030, 70% of companies within the United States will have adapted one form of AI within their organizations. Furthermore, they predict that AI will contribute an additional global economic activity of around $13 trillion within the same period [1]. It is clear that AI will become embedded within the future workforce. To handle this technological and organizational shift, fundamental research is needed to address key challenges related to how humans and AI can most effectively interact, as well as to study different modalities of human-AI collaboration. Focus areas and challenges include effective communication strategies, context awareness, and trust, among others [2], [3]. These open questions are leading researchers to dedicate efforts to the creation of different taxonomies and platforms in order to effectively study these characteristics and dimensions within an experimental context [4]–[6], including in engineering design settings.

Designing effective complex engineering systems is challenging. Engineers regularly need to manage and optimize coupled design parameters with interrelated factors and evolving constraints, making such design tasks both difficult and stressful [7]–[9]. Furthermore, such tasks can often be time-constrained, exacerbating these complexities [10]. To facilitate solving such complex design problems, researchers are studying the implementation of Artificial Intelligence (AI) assistance. For example, research efforts have explored the implementation of AI within various phases and roles of the engineering design process, including concept space exploration [11] and generation [12], concept evaluation [13], [14], prototyping [15], manufacturing [16], and process management of teams [17]. Further, research involving 1500 companies found that human-AI collaboration resulted in considerable performance improvements [18], [19].

Even with these improvements, researchers have identified scenarios during design tasks where AI assistance hinders team performance, especially bringing down the quality of high-performing teams [20]. Consequently, to effectively integrate AI within design teams and mitigate potential adverse effects, there is a critical need for a better understanding of how AI team members impact their human counterparts. The Human-Autonomy Teams (HAT) research community has studied factors contributing to more effective and enjoyable human-AI teaming. Factors like the level of autonomy [21]–[23], reliability [24], [25], and team composition [26]–[28] have been shown to affect individual and team performance, problem-solving behaviors, member interactions, mental workload, and member experiences [29]. It is essential to not only develop a dynamic understanding of how these factors evolve over the design process, but also of how they vary across scenarios and problem-solving contexts.

One particularly influential factor is the level of machine autonomy within the team, with simple reactive autonomy agents that pull information upon human requests believed to hinder performance [28], [29]. Previous research on hybrid teams has studied the positive effects of autonomy agents with self-initiated proactive actions, which is regarded as a trait of high levels of autonomy [21]–[23]. In addition, multiple studies suggest that team performance could be improved by allowing autonomous agents to anticipate and push information proactively [30]–[32]. To date, little research in engineering design has studied this type of dynamic AI team member or how this type of human-AI interaction impacts team performance and problem-solving behaviors. To fill this extant research gap, the current research examines the integration of novel AI Partners within human engineering teams. These AI Partners dynamically adapt, responding both reactively and proactively to their human teammates in real time. They provide suggestions of their own volition, but also provide design improvement suggestions when requested by their human counterparts. These AI Partners also seek feedback from their teammates and update their behavior accordingly.

In prior work, the authors created the HyForm research platform to study human-AI hybrid teaming regarding team agility [14] and AI process management [17]. The current study utilizes HyForm, and updates it with proactive and adaptive autonomous agents that communicate with other teammates bi-directionally. This study contributes to the body of HAT research by investigating not only the impact of proactive behaviors of autonomous agents in human-AI teaming, but also the influence of the adaptability of the autonomous agents and the team composition in an engineering design setting.

The paper is organized as follows. Section 2 covers the background of the study. Section 3 introduces the research methodology, including the experimental design and the experiment platform. The experiment results are shown in Section 4. Section 5 presents our interpretation and discussion of the results. Lastly, Section 6 concludes our findings.

2. BACKGROUND

Adaptation contributes to the overall agility of the team. Team agility is the ability of a design team to be robust in its capabilities, even when experiencing unexpected changes [33]. This research presents a novel setting that imposes the challenges of constantly evolving engineering tasks to a hybrid team composed of multiple highly autonomous and proactive AI agents and human members. In comparing and examining the problem-solving behavior to that of human-only teams, this research reveals areas of improvement for human-AI hybrid teams to be robust in their capabilities, even when experiencing unexpected changes, which is essential in the collaborative designing of engineering systems [9], [33], [34].

Previous studies have examined behaviors within hybrid teams when met with challenges set by specific task characteristics such as increasing human-autonomous agent outcome interdependence [35]–[37] or task difficulty [23], [38]. This research uniquely imposes the challenge of an unexpected change in the task, namely constant shifts in constraints and objectives, which is common in complex engineering design tasks [9], [33], [34]. By doing so, it introduces agility as an essential factor in constructing effective hybrid teams and

Copyright © 2023 by ASME

provides insight for human-AI teams to design engineering systems effectively.

The involvement of highly autonomous AI agents has been considered essential to building effective hybrid teams [22], [23], [29], [39], [40]. Features identified as autonomous include occupying a distinct role [41], a degree of interdependence with other team members' activities and outcomes [37], [42], and a degree of agency involving independence of actions and proactivity among autonomous agent members [43], [44]. Engineering design tasks can especially benefit from AI agents with a heightened level of autonomy because the task itself requires high interdependence amongst members with different roles [45], [46]. In addition, it is important that other members conceive AI partners as genuine team partners for the team to effectively perform [23], [29], [38]. This research seeks to build an effective hybrid team by foremost developing novel, highly autonomous AI agents and involving them in a multi-agent, multi-human hybrid engineering team.

It has been shown that AI agent's understanding of the individuality of fellow human team members enhances overall performance, trust, workload, and willingness to work with agents in the future [47], yet few works have integrated AI agents with this ability. This research develops AI agents that can take into account each human team member's preference and go further by tracking changes in their preferences at each step to adapt their decisions accordingly. It is expected that the enhanced and dynamic awareness of the AI agent facilitates the team's ability to recoup after the abrupt change in the task.

Team composition has also been considered an important independent variable in forming hybrid teams [28], [38], [48]. The majority of HAT studies involve single autonomous agents paired with single or multiple human partners [29]. While some involve multiple AI agents and a single human team member [49], the hybrid team presented in this research is unique in that it consists of multiple agents and multiple humans. In doing so, the research emulates truly hybrid teams, where human and AI agents, with each of their distinct roles, are highly dependent on each other and require active coordination and communication to perform engineering design tasks effectively.

3. METHODOLOGY

This section first presents the HyForm platform and gives an overview of the experimental design and conditions. It then provides detailed information regarding the underlying frameworks of the AI Partners integrated within the design teams. The agents take on different roles. One agent emulates drone design specialist partners while another agent emulates an operations specialist partner. These AI Partners act and react dynamically to their human counterparts throughout the entire problem-solving process.

3.1 The HyForm Experimental Platform

Researchers at Carnegie Mellon University (CMU) and Penn State University (PSU) jointly developed an open-source, experimental research platform called HyForm [9], [34], [50]. HyForm uses an online collaborative design environment that simulates a complex interdisciplinary design problem. The platform partners AI agents and human team members with specific design tasks to create and operate a fleet of drones to make deliveries to customers. Different types of deliveries result in certain revenues, and the goal of the team is to maximize overall profit.

The HyForm platform allows researchers to track each distinct action and communication amongst team members [7], [14], [17]. Text-based communication channels allow team members within and across different team roles to exchange information. The team roles include *Problem Manager*, *Design Specialist*, and *Operations Specialist*; each team role operates using a distinct module within the HyForm platform, allowing them to work independently on their sub-tasks. The *Problem Manager* uses the business plan module to pick customers, determine the operation specialists' market, and select the most profitable plan developed by the team. Additionally, for the particular team structure in this work, the *Problem Manager* directs the team and facilitates communication between the *Design Specialists* and *Operations Specialists*. The *Design Specialists* use the drone design module to construct and evaluate drones with respect to their cost, range, velocity, and payload capabilities. Once a drone design is completed, it can be used by the *Operations Specialists*. The *Operations Specialists* use the operations module to develop and evaluate delivery routes for the customers, selected by the *Problem Manager*, using the drones created by the *Design Specialists*.

HyForm allows researchers to reconfigure the communication channels between roles, enabling experimenters to restrict or extend team communication to study different team structures. The HyForm platform introduces transformations to the problem context (changing the customer maps, package types, timeline, etc.) enabling the ability to handle changes in the problem. Additionally, HyForm integrates AI design agents enabling the collaboration of human and AI agents in teams. The latter is the primary focus of this research, studying novel AI agents as active team partners.

3.2 Experimental Design

The experiment is approved by the Institutional Review Board at CMU. Before the experiment, all participants read and sign a consent form. After the completion of the experiment, participants are compensated with a $20 Amazon gift card. For this experiment, 105 engineering undergraduate and graduate students are recruited from both CMU and PSU. The experiment is run virtually through the HyForm platform.

In this experiment, two conditions are used at the team composition level: (A) a human team and (B) a hybrid team composed of both human and AI Partners, as shown in Figure 1. The principal difference between the two teams is the composition of the members. In the human team condition (Figure 1A), all five participants in the team are human participants. Conversely, in the hybrid team condition (Figure 1B), the team consists of three human participants and two AI Partners. The team communication structure remains the same for both conditions. In Figure 1, the solid arrows indicate open

Copyright © 2023 by ASME

lines of communication between team members, including the AI Partners for the hybrid team condition. In their communication, participants are aware of whether the counterpart is a human or AI team member. The *Problem Manager* handles the exchange of information between the disciplines, two *Design Specialists* and two *Operations Specialists*. For this experiment, participants are randomly assigned a role on one of the two team conditions. Complete data collection consists of 10 human teams and 14 hybrid teams.

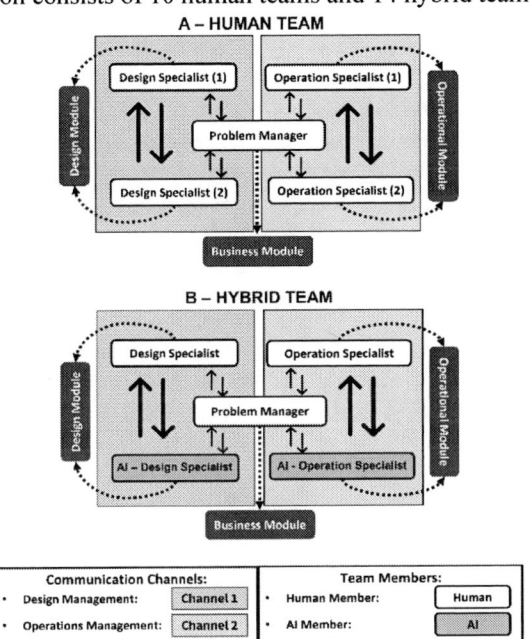

FIGURE 1: EXPERIMENT CONDITIONS: (A) HUMAN TEAM AND (B) HYBRID TEAM

3.3 Design Problem

Collectively, teams are tasked to build, manage, and optimize a delivery drone fleet to maximize total profit. Each team member's role has specific responsibilities. Team performance is measured by the highest profit generated within a business plan submitted by a team's *Problem Manager*. This work introduces a drastic, unexpected change or "shock" to the design problem constraints for the second half of the experiment. After this shock is introduced, additional design constraints are implemented in the design and operations modules, as illustrated in Figure 2. First, the *Design Specialists* must consider the geometrical constraints of the hangar space. Two walls are merged into the design module (left), restricting the dimensions of feasible drones. Second, *Operation Specialists* must consider a no-flight zone on the customer map. A large cylinder is included on the customer map (right), obstructing the flight of drones through this area.

It is important to note that the AI Partners in this work are not designed to handle the unexpected shock. The AI Design Partner does not have knowledge of the restricted sizing constraints, since its knowledge is based on the objective to optimize drone performance exclusive of size constraints. The

human participants do have the ability to guide the AI Design Partner by referencing existing designs and entering commands based on performance metrics (more on this later). Also, resulting AI Design Partner designs can be modified by human participants to satisfy geometric constraints. Corresponding with the AI Design Partner, the AI Operations Partner does not have knowledge of the unexpected change. It uses an underlying Linear Programming algorithm to calculate delivery plans, which do not account for no-fly zones between customers. Similar to the AI Designer Partner designs, AI Operations Partner plans can be modified by human participants.

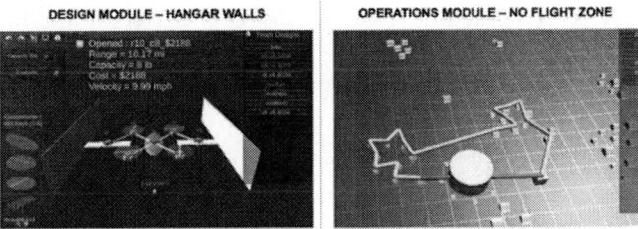

FIGURE 2: PROBLEM SHOCKS ARE INTRODUCED DURING THE SECOND PROBLEM-SOLVING SESSION.

3.4 Design Study Timeline

The experiment has six stages, chronologically illustrated in Table 1. The experiment starts with a pre-study session, where participants have 15 minutes to read and sign a consent form. Next, participants read the problem statement and complete a pre-study questionnaire. The problem statement details the design problem, the participant's assigned role, and the team structure. Each role (*Drone Specialist, Operations Specialist,* and *Problem Manager*) has specific responsibilities and a unique problem statement. The pre-study questionnaire collects information on participants' background with respect to pertinent aspects of the experiment, such as building drones, business planning, and computer-aided design proficiency.

TABLE 1. EXPERIMENT OUTLINE

Team Condition	Human	Hybrid
Pre-study (15 minutes)	Consent form & Pre-study questionnaire	
Training (10 minutes)	Guided tutorial for each team role	
Session 1 (20 minutes)	Market 1	
Break (5 minutes)	Mid-study questionnaire	
Session 2 (20 minutes)	Market 2 (Shock)	
Post-session (10 minutes)	Post-study questionnaire	

During the second stage, participants complete a 10-minute training tutorial introducing them to their respective role and the

Copyright © 2023 by ASME

relevant tools in the platform by guiding them through the completion of a role-specific task. For example, *Design Specialists* are guided through building their first drone and evaluating its performance. The third stage of the experiment is the first problem-solving session. Teams have 20 minutes to design drone fleets and delivery plans that maximize their profit for the first market. After 20 minutes, participants complete a mid-questionnaire, followed by a 3-minute break.

Next, participants start the second problem-solving session. Like the first problem session, teams have 20 minutes to design drone fleets and delivery plans that maximize their overall profit. However, the two aforementioned shocks are introduced during this stage in the design problem. Finally, after completing the second problem session, participants complete a post-study questionnaire concluding the experiment. The mid- and post-study questionnaires include the NASA task load index (NASA-RTLX) survey [51], [52] to evaluate participants' mental workload and cognitive experience during the task. A group of additional questions assesses teams' performance and problem-solving behaviors. These questions utilize a Likert-type scale to rate team characteristics, including overall team effort, goals, quality of work, collaboration, and communication [53]–[55].

3.5 AI Partner Algorithm

In the hybrid teams, two proactive and reactive AI Partner agents are created and integrated into the HyForm platform to replace one human *Design Specialist* and one human *Operations Specialist*. The communication with these AI Partners is restricted to the same chat channels that human team members use, in which the human team members can enter requests to their AI Partners in a validated syntax. This is a design decision to achieve the technical adequacy needed to conduct the experiment given the limited natural language processing capability of the AI Partners. The communication grammar was designed to have high correspondence with the human-human communications seen in prior HyForm studies. The syntax permits team members to communicate drone/plan metrics (e.g., speed, payload, distance), share a preference direction (e.g., minimize or maximize), reference existing designs/plans (baselining), and provide feedback on responses (satisfaction with suggested designs). More precisely, the text grammar supports the following types of requests:

- *Want more or less of a certain metric*
- *Want a specific value of a certain metric*
- *Reference an existing design or plan as a baseline to change metrics*
- *Query the AI Partner for a new design or plan based on their current design state*

Team members can communicate with their AI Partners in two ways: via direct chat channels or via a graphical wizard that assists in assembling commands to their AI Partners in the proper syntax. The wizard allows team members to generate string-based commands that satisfy the text grammar to support communication with the AI Partners.

3.5.1 AI Partner Algorithms

The AI Partners use designs already created by team members and chat messages to identify a target and preference direction for each team member in order to help suggest design solutions. The target and preference approach are based on previous visual steering methods in trade space exploration [56]. More precisely, the AI Partners use (1) targets to represent a team member's state based on the current design with respect to the performance space (either a drone design for a *Design Specialist* or a path plan for an *Operations Specialist*, and (2) preference directions formulated from the chat messages from each team member by extracting pertinent keywords (ex. more, less, or referencing an existing design). The AI Partners store individual target and preference states for each team member, and, when queried by the human team members, perform a neighborhood search located at the target and proceed to return a suggested solution following the preference direction of the team member. In this way, the AI Partners can continually update team member states and adapt their suggested design solutions over time.

To support this adaptive design approach, a distance limit is placed on the amount of change from the target to a new suggested design/plan in the configuration space, based on a Levenshtein distance metric. This metric is used since both drones and path plans are stored as strings and not qualitative metrics [57]. By using the Levenshtein distance metric, the objective is to simulate nudges in the respective configuration spaces. Song et al., [9], [14] describes HyForm's string representations for designs in more detail.

All design solutions (drone designs or delivery plans) that the AI Partners return must satisfy the requested requirements. If the AI Partners cannot find a feasible design, they will return an "*unsatisfied*" chat message to their human teammate. The AI Partners also undergo an initial startup phase wherein if they do not have sufficient preference information for each design metric in the analysis (e.g., range, cost), they will ask for it. Team members can then either respond using the proper text grammar rules with their preferences and requirements or with no preference. After the startup phase, each team member chat message that satisfies the text grammar initializes the AI Partners to return a satisfied design (drone design or delivery plan) or an unsatisfied request based on the AI Partner's analysis. This process repeats with each human-initiated query to the AI agent, demonstrating the reactive nature of the AI Partner.

Additionally, at the 9- and 18-minute marks of each problem-solving session (recall each session is 20 minutes long), the AI Partners proactively provide a design solution to each team member based on their individualized saved preference state at that moment. Every time an AI Partner suggests a design, they save the design solution to their underlying team's database. This is displayed to other team members, and a notification chat message is sent to the team to notify them that a new design has

Copyright © 2023 by ASME

been submitted. This demonstrates the proactive nature of the AI partner.

3.5.2 AI Design Partner

The AI Design Partner, designed to act as a *Design Specialist* within the team, identifies target locations and preference directions within a drone metric trade space [56] defined by the velocity, range, capacity, and cost of solutions. To enable rapid AI Design Partner responses, a prepopulated trade space of drone configurations is sampled and used as a basis for the AI Design Partner to select and return designs. The pre-sampled database includes 1043 unique drone designs, with range, capacity, cost, and velocity metrics, generated using a character recurrent neural network (char-RNN) retraining of the drone design space [9], [58]. The AI Design Partner algorithm then samples a subset of the pre-sampled database ($M = 400$) and calculates the Levenshtein distance [57], based on the string representation of the design configuration, for each sampled design compared to the current design. The AI Design Partner returns the design that minimizes the Levenshtein distance, while satisfying the tolerances for each metric based on the current team member preference.

3.5.3 AI Operations Partner

The AI Operations Partner, designed to act as an *Operations Specialist* within the team, identifies target locations and preference directions within a plan metric trade space defined by the profit, cost, and the number of customers served by each solution. To return a suggested plan based on a team member request, the AI Operations Partner runs several rapid Linear Programming delivery path analyses, where each analysis has a different random combination of available drone designs already created by the *Design Specialists*. The resulting calculated plans are saved in string format and are evaluated by calculating the Levenshtein distance to the requesting team member's current plan. The AI Operations Partner returns the plan that minimizes the Levenshtein distance, while satisfying the tolerances for each metric based on the current team preference. If the AI Operations Partner is unable to find a feasible plan, an "*unsatisfied*" message is returned to the human.

4. RESULTS

The following section compares the hybrid and human teams in terms of overall performance (i.e., team profit), team behaviors, and team experiences. The former two are measured via data collected during the experiment sessions through HyForm, while the latter is gathered via responses from the mid- and post-study questionnaires. R version 4.0.1 was used to complete these analyses and, unless stated otherwise, all assumptions of the statistical tests were met.

4.1 Team Performance

Team performance is measured using the overall profit achieved by each team during the problem sessions. While teams can submit multiple plans that result in multiple profits, only the plan with the highest profit is considered for the analysis. The

maximum profit is tracked and averaged across all teams within a condition. Figure 3 shows the average maximum profit for each type of team by experimental session. To determine if the AI Partners influenced team performance, a two-sample Wilcox test is used. Overall, the two team conditions achieve similar levels of performance for both Session 1 (p-value = 0.333, $W = 53$) and Session 2 (p-value = 0.883, $W = 73$). This shows that the AI Partners are effective in replacing their human team counterparts. An interesting note, however, is that the hybrid teams exhibit greater variance, particularly on the higher end, showing that in this study hybrid teams appear to have greater potential for better performance, as seen in Figure 3.

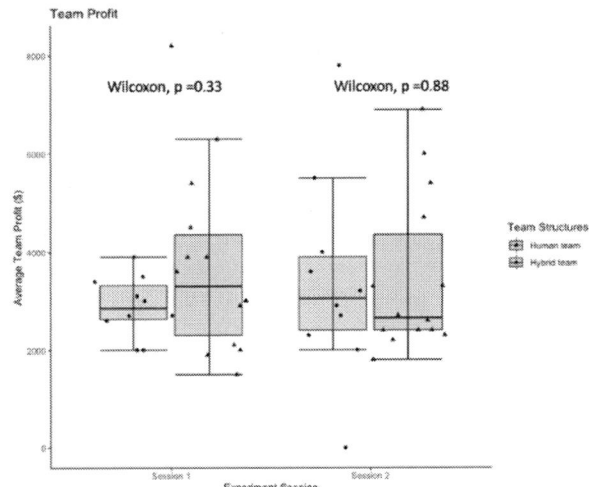

FIGURE 3: AVERAGE TEAM PROFIT FOR EACH TEAM STRUCTURE. BOXES REPRESENT THE INTERQUARTILE RANGE.

4.2 Team Problem-Solving Behavior

Teams' problem-solving behaviors are analyzed using two metrics, communication count and action count. As mentioned in the previous section, team members using HyForm can only communicate through a chat text tool and complete specific design actions relevant to their specific roles. HyForm tracks and collects the data for these metrics over time, allowing complete reconstruction of the team processes to solve the problem. Further, previous work revealed tradeoffs between time allocated towards communication and taking action when designers experience an unexpected problem change [14]. In this work, a shock is introduced halfway through the experiment, the drone design and operations disciplines receive different constraints based on their sub-tasks (hangar dimension restrictions and no flight zone, respectively). The communication count encompasses all textual messages sent from one team member to another, and the action count encompasses all distinct actions taken by a member within their design role. To determine if AI

Copyright © 2023 by ASME

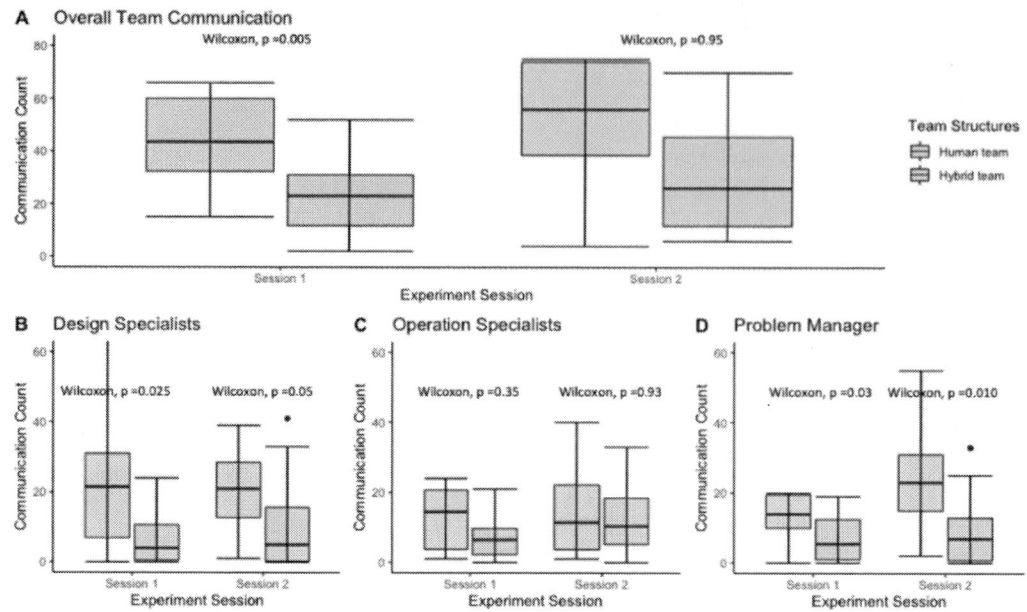

FIGURE 4: COMMUNICATION COUNT: (A) OVERALL TEAM COMMUNICATION COUNT, (B) DESIGN SPECIALIST COMMUNICATION COUNT, (C) OPERATIONS SPECIALIST COMMUNICATION COUNT, AND (D) PROBLEM MANAGER COMMUNICATION COUNT. BOXES REPRESENT THE INTERQUARTILE RANGE.

Partners influenced team communication, a two-sample Wilcox test is used.

When comparing the overall results combining both experiment sessions, the human teams communicate significantly more than the hybrid teams (p-value = 0.001, W = 431). After introducing the problem shock between problem-solving sessions, human teams reacted significantly increased their communication, as presented in Figure 8. When observing the experiment sessions separately, human teams show a significantly higher average communication count during Session 1 (p-value = 0.005, W = 117.5), while during Session 2, this trend is no longer the case (p-value = 0.095, W = 99).

Figures 4B, 4C, and 4D dive deeper into communication behaviors by comparing counts by team role during each experiment session. During the first problem-solving session, the *Design Specialists* (p-value = 0.025, W = 108.5) and *Problem Manager* (p-value = 0.03, W = 106.5) communicate significantly more in the human teams than the hybrid teams. This difference is not seen for the *Operations Specialists* (p-value = 0.35, W = 86.5). Similarly, after the shock is introduced (Session 2), the trend remains for both the *Design Specialists* (p-value = 0.05, W = 103.5) and *Problem Manager* (p-value = 0.010, W = 114), while there is no significant difference for the *Operations Specialists* (p-value = 0.93, W = 72). The *Problem Managers* in the human team condition react the most to the problem shocks by exhibiting the steepest increase in average communication count between sessions, as seen in Figure 4D, in comparison to 4A, 4B, and 4C.

The second behavioral metric, action count, describes a different trend between the two team structures. To determine if AI Partners influenced the action taken by human designers, a two-sample Wilcox test is used. In this work, the action is assessed by the average number of design changes, regardless of what specific action is taken. Because the AI Partners in the hybrid teams cannot act in the same manner as their human counterparts, the action values are normalized on a per individual basis. Figure 5A shows the average action count per team condition by session, indicating no significant difference in the overall action count between team structures (Session 1: p-value = 0.792, W = 75. Session 2: p-value = 0.187, W = 93), with the human teams and hybrid teams acting similarly. Additionally, Figures 5B, 5C, and 5D dive deeper into this problem-solving behavior, showing action counts for different team roles. There is no significant difference in Session 1 and Session 2 for *Design Specialists* (Session 1: p-value = 0.75, W = 76. Session 2: p-value = 0.43, W = 84), *Operation Specialists* (Session 1: p-value = 0.62, W = 61. Session 2: p-value = 0.57, W = 60), and *Problem Managers* (Session 1: p-value = 0.81, W = 74.5. Session 2: p-value = 0.59, W = 60.5). Generally, the human teams tend to communicate more than and act similarly to the hybrid teams.

4.3 Hybrid Team Process Behaviors

This section specifically studies the behaviors of the hybrid teams. Recall that hybrid teams consist of three human team members and two AI Partners – one being the AI Design Partner and the second being the AI Operations Partner. Figure 6 shows the proportion of communication between human-human and human-AI. Communication is equally proportioned within the first problem-solving session between human-human discourse (46.6%) and human-AI discourse (53.3%). However, after the

Copyright © 2023 by ASME

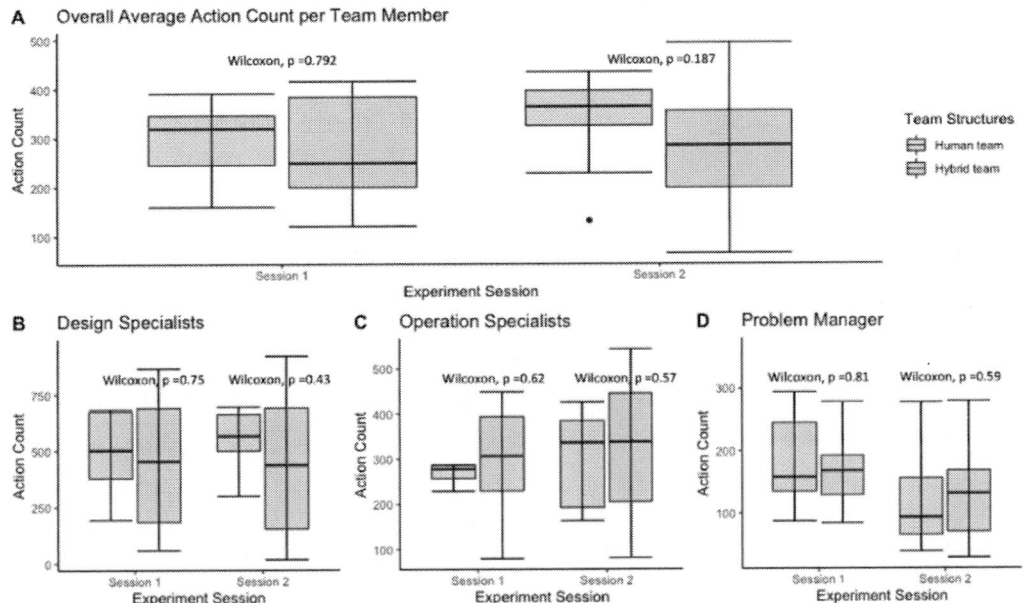

FIGURE 5: ACTION COUNT: (A) OVERALL AVERAGE ACTION COUNT PER TEAM MEMBER, (B) DESIGN SPECIALIST ACTION COUNT, (C) OPERATIONS SPECIALIST ACTION COUNT, AND (D) PROBLEM MANAGER ACTION COUNT. BOXES REPRESENT THE INTERQUARTILE RANGE.

shock, this behavior radically changes. Communication shifts more heavily to human-AI communication, almost tripling to nearly 80.5% of the overall communication.

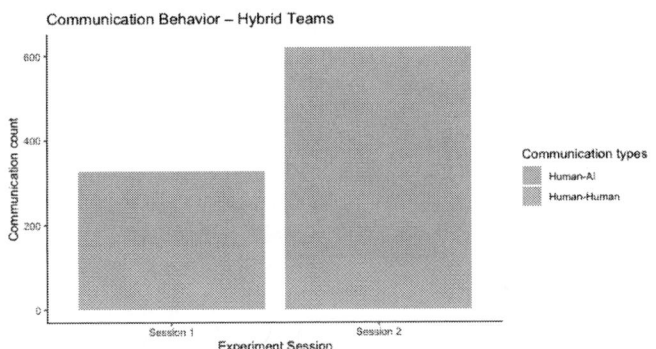

FIGURE 6: COMMUNICATION BEHAVIOR FOR HYBRID TEAMS.

As mentioned previously in Section 3, both AI Partners monitor the chat channels to collaborate with their human teammates. It is restricted by a grammar structure, where team members enter requests in a validated format. Figure 7 shows the possible types of requests from human designers to their AI Partners for each team role for both experiment sessions. For the first experiment session, all team roles communicate with a similar frequency with their AI Partners. However, after the shock, there is a drastic increase in the communications with their AI Partners for both the *Operation Specialists* and *Problem Manager* roles. Specifically, there is a medium effect size

(Cohen's d = 0.66) in communication frequency between sessions, indicating a practically significant increase in communication. *Design Specialists* focus evenly on requesting new designs ("Want [New Design]" 46.20%) and design updates ("Want [Design Update]" 34.38%), and *Problem Managers* focus primarily on requesting design updates ("Want [Design Update]" 76.47%). Contrary, *Design Specialists* do not request new or updated design solutions. Instead, *Design Specialists* increase their inquiry on the design status ("Ping [Design Status]") of the AI Partners.

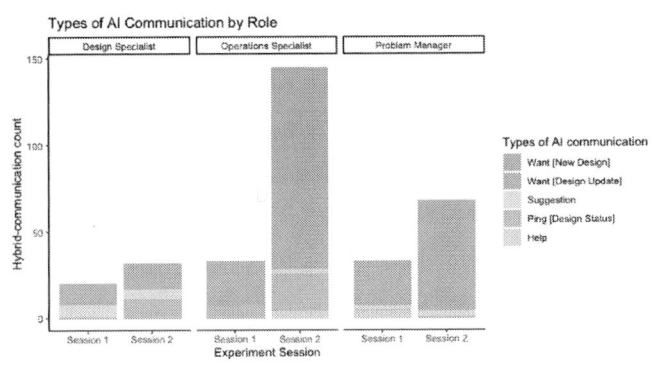

FIGURE 7: COMMUNICATION REQUESTS FOR AI PARTNERS BY TEAM ROLES.

4.4 Team Members' Experience

This section analyses the questionnaires taken by team members following each experiment session. The first part of the questionnaire uses a version of the NASA-RTLX [51], [52], modified per Nolte & McComb [59] to better align with cognitive experience, to acquire insights into participants' cognitive demand and workload perceptions while working throughout the design problem. To determine if AI Partners influenced participants' cognitive demand and workload perceptions, a two-way ANOVA is used. The second part of the questionnaire asks participants about team performance, and team cohesion features, including overall team productivity, effort, and whether the team came to a consensus effectively. This includes modified versions of the short Trust Perception Scale-HRI [54], the Team Effectiveness Instrument [53], and an effectiveness survey developed by the researchers of this study using the results of a previous study [55]. To determine if AI Partners influenced participants' team perceptions, a Wilcox test is used.

4.4.1 Cognitive Experience and Mental Workload

The first part of the questionnaire appraises participants' mental workload and cognitive experience. Mental workload scores are calculated for each participant by summing their scores for the following sub-dimensions that are individually rated: mental demand, temporal demand, performance, effort, and frustration. The sum of these is averaged by dividing by the number of sub-dimensions (i.e., five). Similarly, cognitive experience scores are calculated for each participant by summing their scores for all the modified NASA-RTLX sub-dimensions (mental demand, temporal demand, performance, effort, stress, discouragement, insecurity, and frustration) and dividing by the number of sub-dimensions (i.e., eight). For both of these global dimensions, the inverse of the performance sub-dimension is taken before summation to match the scaling of the others.

The results of the overall cognitive experience and the mental workload present no significant difference when comparing the design roles and team structures even after the shock introduced in Session 2. However, diving into the underlying sub-dimensions of these NASA-RTLX measures, the *Problem Manager* role in both team conditions faces a significant increase in mental demand ($p = 0.032$, $W = 347.5$) when transitioning from Session 1 to Session 2. Additionally, there is a significant difference in frustration between the human and hybrid teams ($p = 0.032$, $W = 2833$). The *Operations Specialists* face the lowest variability in the NASA-RTLX measurements across team roles for both the human and hybrid teams.

4.4.2 Team Effectiveness and Cohesion

The second part of the questionnaire measures and evaluates participants' perceptions of their team's effectiveness and interactions critical to the team's success. Team members' answers are recorded using a Likert-type scale representing seven discrete options, from "*strongly disagree/inaccurate*" to "*strongly agree/accurate.*"

During Session 2, members of the hybrid teams perceived their teams as less efficient ($p = 0.088$, $W = 779.5$), and team roles were not clear ($p = 0.0202$, $W = 706.5$) when compared to human teams. They also identify less effective feedback ($p =$ 0.028, $W = 721.5$), less equal participation ($p = 0.0099$, $W = 676$), and less collaboration ($p = 0.0076$, $W = 666.5$) efforts from their teammates. Furthermore, hybrid teams perceived that their team communicated less effectively during both experiment sessions (Session 1: $p = 0.046$, $W = 744.5$; and Session 2: $p = 0.013$, $W = 687$). The results from the questionnaire highlight a negative perception of team members regarding the effectiveness and cohesion within the hybrid teams, even though they performed equivalent well.

5. DISCUSSION

This work studies the performance and problem-solving behaviors of hybrid-partnered engineering design teams compared with human-only design teams, with a drastic, unexpected shock introduced midway through the design task. The human teams are composed of five human members with specific distinct roles who can communicate and share solutions. Similarly, the hybrid teams are composed of three human members and two AI Partners who can communicate and share solutions between team members. It is believed that human and AI collaboration could yield considerable improvements in overall performance compared to human-only teams [18], [19]. However, the engineering design community is still researching factors influencing human-AI teaming effectiveness and enjoyment. This study adds to the body of research by introducing a unique AI Partner that can dynamically interact, both reactively and proactively, in real-time with their human counterparts to emulate genuine team partners. The anticipatory and proactive information-pulling behavior of AI Partners in team interactions is expected to improve team performance [30]–[32]. The impacts of these AI Partners on both performance and problem-solving behaviors are explored in the context of engineering teams solving complex, interdisciplinary design problems.

Both human and hybrid teams perform similarly. The overall performance of hybrid teams is an indication that the AI members' skill level may be comparable to the human members of the team [40]–[42]. This supports the design of the AI Partners created for this research, by showing that they can adeptly stand in as substitutes for human partners as opposed to simply being assistants [14], [22], [40]. While there is not a significant difference between human-only and hybrid teams in overall performance, the hybrid teams do exhibit greater variability in terms of their team performance, with a long tail of high-profit teams. These results indicate that hybrid teams demonstrate high potential for better performance, and that the high capability of AI Partners can be a contributing factor. Other works have shown that possible mechanisms behind team performance can be information-processing functions [29] or compatibility with human partners [54].

It is also significant that the hybrid teams maintain a comparable performance across sessions, regardless of the AI Partners being blind to the abrupt changes in constraints. It is likely that the increase in communication between human participants and AI Partners supported the consistent level of performance in hybrid teams. This result demonstrates that the

Copyright © 2023 by ASME

adaptivity characteristic of human-AI hybrid teams can come from the human side of the partnership, instead of developing the perfect adaptive AI Partners. It also presents the possibility of hybrid teams largely outperforming human-only teams with certain team dynamics such as effective communication [44]. This insight is potentially of great importance as a principle for forming teams in dynamic scenarios, in which the human portion of the team should be only as large as necessary to handle adaptation.

In addition to comparing performance levels, a critical aspect of this work is to explore how the integration of these AI Partners fundamentally impacts the teams' problem-solving behaviors. The underlying difference between human and hybrid teams is present in their overall communication, while there is almost no difference in the overall action behavior. Previous literature on the design thinking process indicates the presence of an increase in team communication at the beginning of the collective design process, as the team is clarifying goals and establishing a common understanding of the design requirements [60]. A similar pattern could be observed in this study when the shock was introduced, which altered the design requirements and effectively propelled the team to communicate toward goal clarification. Examining the communication behavior at a role level reveals that *Problem Managers* in the human teams exhibit the steepest increase in communication, particularly after the problem shock is introduced prior to the second problem session. As the *Problem Managers* are directly connected to all other individuals within the team, they serve as the central node of the team structure [61], [62]. The upsurge in communication could be caused by the centrality that these managers have in both the communication structure (the need to communicate information across disciplines) and in the task structure (the need to oversee and submit final plans), as they mediate and transform the flow of information for other team members [63]. These critical efforts for information flow may have been more effective for responding to the change in the problem. As shown in the results, the human team members respond to the new constraints added to the problem by increasing their communication efforts, especially at the management level. This result opens a question of whether limitations within the communication between human and AI hinder the adaptation of the hybrid teams to the abrupt shock. For instance, for the human participants in this study, knowing the fact that their fellow teammate is an AI could have deterred their willingness to communicate. Works examining human-computer and human-robot interaction suggest that human-AI communication experiences reduced interpersonal attraction that could affect user perception [64], [65], while such limitations are typically not present in human-human collaboration. This is an important consideration that must be accounted for in the long-term performance of these teams.

Analysis of communication frequency and distribution reveals problem solving behaviors related to interdependence within the hybrid teams. During the first session we observed that the communication between human members and human-AI members is equally distributed. However, following the shock,

communication shifts much more heavily among human-AI members, particularly for the *Operation Specialists* and *Problem Manager* roles. On the contrary, *Design Specialists* show a slight increase in human-AI communication. We can infer that human team members seem to rely more on their AI Partners under uncertain shocks, especially for operation related tasks rather than for design tasks which aligns with results from similar previous studies. [66] Results of this work point to the significance of communication for the success of human AI teams. Accordingly it would be critical in future research to analyze the quality of communication [67]. For instance, patterns in the content of the chat can be analyzed using natural language processing methods to examine what types of information are being transferred between members in a hybrid team. Such analysis can further determine if the relationship between communication and actions in hybrid teams restricts team performance.

Even though the AI Partners show effectiveness in replacing human counterparts, questionnaire results show that human team members within the hybrid teams have a negative perception of their team's effectiveness and cohesion. The results align with prior research findings on the perception of teamwork in HAT that human teammates are usually perceived as more affective and facilitate more communication than autonomous partners [37]. This finding resonates strongly in terms of cooperative features, including effective communication and feedback, collaboration, and equal participation, where the hybrid teams all rate these as significantly inferior. These findings have strong implications for hybrid teams of the future. In order for human and AI Partners to effectively work together, there needs to be more of a positive team experience, as these factors correlate highly within problem-solving teams [45], [46], [68]. Even though the two team conditions perform similarly, whether these negative perceptions impacted their potential team performance cannot be answered here. Therefore, it could be one area of further experimentation.

6. CONCLUSION

This work analyzes and compares the team performance, problem-solving behaviors, and team experiences of human and human-AI hybrid teams during a complex, interdisciplinary design task. The AI Partners in this research emulate genuine team partners, being able to react to and proactively collaborate with their human teammates dynamically. A drastic change (i.e., shock) to the problem constraints is also introduced midway through the experiment to simulate an evolving engineering problem in practice. Results show that the hybrid teams perform similarly to human teams, indicating that the AI Partners can effectively replace and adeptly support their human design partners. However, hybrid teams struggle with adapting their coordination and communication following the shock. The human members in the hybrid teams also perceive their team as inferior across several interpersonal dimensions, including the effectiveness of team communication, feedback, and the equality of contribution among other team members. These findings point toward several areas of improvement for these hybrid teams,

Copyright © 2023 by ASME

namely effective communication, cooperation, participation, and feedback between human and AI members of the team. Overall, the results provide insights into the behaviors of humans interacting with AI teammates and the factors needed to truly construct effective hybrid design teams.

ACKNOWLEDGEMENTS

This work was supported by the Air Force Office of Scientific Research under grant No. FA9550-18-1-0088 and the Defense Advanced Research Projects Agency through cooperative agreement N66001-17-1-4064. Any opinions, findings, and conclusions or recommendations expressed in this paper are those of the authors and do not necessarily reflect the views of the sponsors.

The author's affiliation with The MITRE Corporation is provided for identification purposes only and is not intended to convey or imply MITRE's concurrence with, or support for, the positions, opinions, or viewpoints expressed by the author.

REFERENCES

[1] "Modeling the global economic impact of AI | McKinsey." https://www.mckinsey.com/featured-insights/artificial-intelligence/notes-from-the-ai-frontier-modeling-the-impact-of-ai-on-the-world-economy (accessed Feb. 17, 2023).

[2] S. Caldwell *et al.*, "An Agile New Research Framework for Hybrid Human-AI Teaming: Trust, Transparency, and Transferability," *ACM Trans. Interact. Intell. Syst.*, vol. 12, no. 3, p. 17:1-17:36, Jul. 2022, doi: 10.1145/3514257.

[3] K. van den Bosch, T. Schoonderwoerd, R. Blankendaal, and M. Neerincx, "Six Challenges for Human-AI Co-learning," in *Adaptive Instructional Systems*, Cham, 2019, pp. 572–589. doi: 10.1007/978-3-030-22341-0_45.

[4] "DigitalWorkforceTeam/HACO: Files of the HACO Framework," *GitHub*. https://github.com/DigitalWorkforceTeam/HACO (accessed Feb. 17, 2023).

[5] "Top Strategic Predictions for 2018 and Beyond: Pace Yourself, for Sanity's Sake," *Gartner*. https://www.gartner.com/en/doc/3803530-top-strategic-predictions-for-2018-and-beyond-pace-yourself-for-sanitys-sake (accessed Feb. 18, 2023).

[6] A. Dubey, K. Abhinav, S. Jain, V. Arora, and A. Puttaveerana, "HACO: A Framework for Developing Human-AI Teaming," in *Proceedings of the 13th Innovations in Software Engineering Conference on Formerly known as India Software Engineering Conference*, New York, NY, USA, Mar. 2020, pp. 1–9. doi: 10.1145/3385032.3385044.

[7] B. Song, N. F. Soria Zurita, H. Nolte, H. Singh, J. Cagan, and C. McComb, "When Faced With Increasing Complexity: The Effectiveness of Artificial Intelligence Assistance for Drone Design," *J. Mech. Des.*, vol. 144, no. 2, Sep. 2021, doi: 10.1115/1.4051871.

[8] N. F. Soria Zurita and I. Y. Tumer, "A Survey: Towards Understanding Emergent Behavior in Complex Engineered Systems," presented at the ASME 2017 International Design Engineering Technical Conferences and Computers and Information in Engineering Conference, Nov. 2017. doi: 10.1115/DETC2017-67453.

[9] B. Song *et al.*, "Toward Hybrid Teams: A Platform to Understand Human-Computer Collaboration During the Design of Complex Engineered Systems," *Proc. Des. Soc. Des. Conf.*, vol. 1, pp. 1551–1560, May 2020, doi: 10.1017/dsd.2020.68.

[10] R. M. Yerkes and J. D. Dodson, "The relation of strength of stimulus to rapidity of habit-formation," *J. Comp. Neurol. Psychol.*, vol. 18, no. 5, pp. 459–482, 1908, doi: 10.1002/cne.920180503.

[11] J. Koch, "Design implications for Designing with a Collaborative AI," *AAAI Spring Symp. Ser.*, Apr. 2017.

[12] B. Camburn *et al.*, "Computer-aided mind map generation via crowdsourcing and machine learning," *Res. Eng. Des.*, vol. 31, no. 4, pp. 383–409, Oct. 2020, doi: 10.1007/s00163-020-00341-w.

[13] B. Camburn, Y. He, S. Raviselvam, J. Luo, and K. Wood, "Machine Learning-Based Design Concept Evaluation," *J. Mech. Des.*, vol. 142, no. 3, Jan. 2020, doi: 10.1115/1.4045126.

[14] B. Song *et al.*, "Decoding the agility of artificial intelligence-assisted human design teams," *Des. Stud.*, vol. 79, p. 101094, Mar. 2022, doi: 10.1016/j.destud.2022.101094.

[15] M. L. Dering, C. S. Tucker, and S. Kumara, "An Unsupervised Machine Learning Approach to Assessing Designer Performance during Physical Prototyping," *J. Comput. Inf. Sci. Eng.*, vol. 18, no. 1, Mar. 2018, doi: 10.1115/1.4037434.

[16] G. Williams, N. A. Meisel, T. W. Simpson, and C. McComb, "Design Repository Effectiveness for 3D Convolutional Neural Networks: Application to Additive Manufacturing," *J. Mech. Des.*, vol. 141, no. 11, Sep. 2019, doi: 10.1115/1.4044199.

[17] J. T. Gyory *et al.*, "Human Versus Artificial Intelligence: A Data-Driven Approach to Real-Time Process Management During Complex Engineering Design," *J. Mech. Des.*, vol. 144, no. 2, Oct. 2021, doi: 10.1115/1.4052488.

[18] P. R. Daugherty and H. J. Wilson, *Human + Machine: Reimagining Work in the Age of AI*. Harvard Business Press, 2018.

[19] H. J. Wilson and P. R. Daugherty, "Collaborative Intelligence: Humans and AI Are Joining Forces," *Harvard Business Review*, Jul. 01, 2018. Accessed: Feb. 17, 2023. [Online]. Available: https://hbr.org/2018/07/collaborative-intelligence-humans-and-ai-are-joining-forces

[20] G. Zhang, A. Raina, J. Cagan, and C. McComb, "A cautionary tale about the impact of AI on human design teams," *Des. Stud.*, vol. 72, p. 100990, Jan. 2021, doi: 10.1016/j.destud.2021.100990.

[21] J. L. Wright, J. Y. C. Chen, S. A. Quinn, and M. J. Barnes, "The Effects of Level of Autonomy on Human-Agent Teaming for Multi-Robot Control and Local Security

Copyright © 2023 by ASME

Maintenance," Nov. 2013, [Online]. Available: https://apps.dtic.mil/sti/pdfs/ADA595105.pdf

[22] J. L. Wright, J. Y. C. Chen, and M. J. Barnes, "Human–automation interaction for multiple robot control: the effect of varying automation assistance and individual differences on operator performance," *Ergonomics*, vol. 61, no. 8, pp. 1033–1045, Aug. 2018, doi: 10.1080/00140139.2018.1441449.

[23] M. C. Wright and D. B. Kaber, "Effects of Automation of Information-Processing Functions on Teamwork," *Hum. Factors*, vol. 47, no. 1, pp. 50–66, Mar. 2005, doi: 10.1518/0018720053653776.

[24] J. Y. C. Chen and M. J. Barnes, "Supervisory control of multiple robots: effects of imperfect automation and individual differences," *Hum. Factors*, vol. 54, no. 2, pp. 157–174, Apr. 2012, doi: 10.1177/0018720811435843.

[25] X. Fan *et al.*, "The influence of agent reliability on trust in human-agent collaboration," in *Proceedings of the 15th European conference on Cognitive ergonomics: the ergonomics of cool interaction*, New York, NY, USA, Jan. 2008, pp. 1–8. doi: 10.1145/1473018.1473028.

[26] M. Demir, N. J. McNeese, and N. J. Cooke, "Team communication behaviors of the human-automation teaming," in *2016 IEEE International Multi-Disciplinary Conference on Cognitive Methods in Situation Awareness and Decision Support (CogSIMA)*, Mar. 2016, pp. 28–34. doi: 10.1109/COGSIMA.2016.7497782.

[27] X. Fan, S. Sun, M. McNeese, and J. Yen, "Extending the recognition-primed decision model to support human-agent collaboration," in *Proceedings of the fourth international joint conference on Autonomous agents and multiagent systems*, New York, NY, USA, Jul. 2005, pp. 945–952. doi: 10.1145/1082473.1082616.

[28] N. J. McNeese, M. Demir, N. J. Cooke, and C. Myers, "Teaming With a Synthetic Teammate: Insights into Human-Autonomy Teaming," *Hum. Factors*, vol. 60, no. 2, pp. 262–273, Mar. 2018, doi: 10.1177/0018720817743223.

[29] T. O'Neill, N. McNeese, A. Barron, and B. Schelble, "Human–Autonomy Teaming: A Review and Analysis of the Empirical Literature," *Hum. Factors*, vol. 64, no. 5, pp. 904–938, Aug. 2022, doi: 10.1177/0018720820960865.

[30] M. Demir, N. J. Cooke, and P. G. Amazeen, "A conceptual model of team dynamical behaviors and performance in human-autonomy teaming," *Cogn. Syst. Res.*, vol. 52, pp. 497–507, Dec. 2018, doi: 10.1016/j.cogsys.2018.07.029.

[31] M. Demir, A. D. Likens, N. J. Cooke, P. G. Amazeen, and N. J. McNeese, "Team Coordination and Effectiveness in Human-Autonomy Teaming," *IEEE Trans. Hum.-Mach. Syst.*, vol. 49, no. 2, pp. 150–159, Apr. 2019, doi: 10.1109/THMS.2018.2877482.

[32] M. Demir, N. J. McNeese, and N. J. Cooke, "Team situation awareness within the context of human-autonomy teaming," *Cogn. Syst. Res.*, vol. 46, pp. 3–12, Dec. 2017, doi: 10.1016/j.cogsys.2016.11.003.

[33] K. Werder and A. Maedche, "Explaining the emergence of team agility: a complex adaptive systems perspective," *Inf.*

Technol. People, vol. 31, no. 3, pp. 819–844, Jan. 2018, doi: 10.1108/ITP-04-2017-0125.

[34] G. Zhang, N. F. Soria Zurita, G. Stump, B. Song, J. Cagan, and C. McComb, "Data on the design and operation of drones by both individuals and teams," *Data Brief*, vol. 36, p. 107008, Jun. 2021, doi: 10.1016/j.dib.2021.107008.

[35] J. Walliser, E. de Visser, E. Wiese, and T. Shaw, "Team Structure and Team Building Improve Human–Machine Teaming With Autonomous Agents," *J. Cogn. Eng. Decis. Mak.*, vol. 13, p. 155534341986756, Aug. 2019, doi: 10.1177/1555343419867563.

[36] F. Gao, M. L. Cummings, and L. F. Bertuccelli, "Teamwork in controlling multiple robots," *MIT Web Domain*, Mar. 2012, Accessed: Jan. 29, 2023. [Online]. Available: https://dspace.mit.edu/handle/1721.1/81765

[37] J. C. Walliser, P. R. Mead, and T. H. Shaw, "The Perception of Teamwork With an Autonomous Agent Enhances Affect and Performance Outcomes," *Proc. Hum. Factors Ergon. Soc. Annu. Meet.*, vol. 61, no. 1, pp. 231–235, Sep. 2017, doi: 10.1177/1541931213601541.

[38] X. Fan, M. McNeese, and J. Yen, "NDM-Based Cognitive Agents for Supporting Decision-Making Teams," *Human–Computer Interact.*, vol. 25, no. 3, pp. 195–234, Aug. 2010, doi: 10.1080/07370020903586720.

[39] M. Q. Azhar and E. I. Sklar, "A study measuring the impact of shared decision making in a human-robot team," *Int. J. Robot. Res.*, vol. 36, no. 5–7, pp. 461–482, Jun. 2017, doi: 10.1177/0278364917710540.

[40] M. Lewis, K. Sycara, and T. R. Payne, "Agent Roles in Human Teams," presented at the AAMAS-03 Workshop on Humans and Multi-Agent Systems (01/01/03), 2003. Accessed: Jan. 29, 2023. [Online]. Available: https://eprints.soton.ac.uk/257737/

[41] L. Larson and L. A. DeChurch, "Leading teams in the digital age: Four perspectives on technology and what they mean for leading teams," *Leadersh. Q.*, vol. 31, no. 1, p. 101377, Feb. 2020, doi: 10.1016/j.leaqua.2019.101377.

[42] C. Nass, B. J. Fogg, and Y. Moon, "Can computers be teammates?," *Int. J. Hum.-Comput. Stud.*, vol. 45, no. 6, pp. 669–678, 1996, doi: 10.1006/ijhc.1996.0073.

[43] J. Lyons, S. Mahoney, K. T. Wynne, and M. A. Roebke, "Viewing Machines as Teammates: A Qualitative Study," presented at the AAAI Spring Symposia, 2018. Accessed: Jan. 29, 2023. [Online]. Available: https://www.semanticscholar.org/paper/Viewing-Machines-as-Teammates%3A-A-Qualitative-Study-Lyons-Mahoney/965e8bb95cc2cc46b822b1a0f3fde5d8e99ffedb

[44] K. T. Wynne and J. B. Lyons, "An integrative model of autonomous agent teammate-likeness," *Theor. Issues Ergon. Sci.*, vol. 19, no. 3, pp. 353–374, May 2018, doi: 10.1080/1463922X.2016.1260181.

[45] J. Kratzer, R. T. Leenders, and J. M. Van Engelen, "The social network among engineering design teams and their creativity: A case study among teams in two product development programs," *Int. J. Proj. Manag.*, vol. 28, no. 5, pp. 428–436, 2010, doi: 10.1016/j.ijproman.2009.09.007.

Copyright © 2023 by ASME

[46] Q. Wu and K. Cormican, "Shared Leadership and Team Creativity: A Social Network Analysis in Engineering Design Teams," *J. Technol. Manag. Amp Innov.*, vol. 11, no. 2, pp. 2–12, Jun. 2016, doi: 10.4067/S0718-27242016000200001.

[47] N. Hanna and D. Richards, "A collaborative activity for evaluating HAT-COM: human-agent teamwork communication model," presented at the Adaptive Agents and Multi-Agent Systems, May 2013. Accessed: Jan. 29, 2023. [Online]. Available: https://www.semanticscholar.org/paper/A-collaborative-activity-for-evaluating-HAT-COM%3A-Hanna-Richards/1ae65338c61042ee5e01f3b82f3218094560a9c6

[48] M. Demir, N. J. McNeese, N. J. Cooke, and C. Myers, "The Synthetic Teammate as a Team Player in Command-and-Control Teams," *Proc. Hum. Factors Ergon. Soc. Annu. Meet.*, vol. 60, no. 1, pp. 116–116, 2016, doi: 10.1177/1541931213601026.

[49] K. Sycara and M. Lewis, "Integrating Agents into Human Teams," *Proc. Hum. Factors Ergon. Soc. Annu. Meet.*, vol. 46, no. 3, pp. 413–417, 2002, doi: 10.1177/154193120204600342.

[50] N. F. Soria Zurita *et al.*, "Data on the Human Versus artificial intelligence process management experiment," *Data Brief*, vol. 41, p. 107917, Apr. 2022, doi: 10.1016/j.dib.2022.107917.

[51] S. G. Hart, "Nasa-Task Load Index (NASA-TLX); 20 Years Later," *Proc. Hum. Factors Ergon. Soc. Annu. Meet.*, vol. 50, no. 9, pp. 904–908, Oct. 2006, doi: 10.1177/154193120605000909.

[52] S. G. Hart and L. E. Staveland, "Development of NASA-TLX (Task Load Index): Results of Empirical and Theoretical Research," in *Advances in Psychology*, vol. 52, Elsevier, 1988, pp. 139–183. doi: 10.1016/S0166-4115(08)62386-9.

[53] C. B. Gibson, M. E. Zellmer-Bruhn, and D. P. Schwab, "Team Effectiveness in Multinational Organizations: Evaluation Across Contexts," *Group Organ. Manag.*, vol. 28, no. 4, pp. 444–474, Dec. 2003, doi: 10.1177/1059601103251685.

[54] K. E. Schaefer, "Measuring Trust in Human Robot Interactions: Development of the 'Trust Perception Scale-HRI,'" in *Robust Intelligence and Trust in Autonomous Systems*, R. Mittu, D. Sofge, A. Wagner, and W. F. Lawless, Eds. Boston, MA: Springer US, 2016, pp. 191–218. doi: 10.1007/978-1-4899-7668-0_10.

[55] S. A. Wheelan and J. M. Hochberger, "Validation Studies of the Group Development Questionnaire," *Small Group Res.*, vol. 27, no. 1, pp. 143–170, Feb. 1996, doi: 10.1177/1046496496271007.

[56] G. Stump, S. Lego, M. Yukish, T. W. Simpson, and J. A. Donndelinger, "Visual Steering Commands for Trade Space Exploration: User-Guided Sampling With Example," *J. Comput. Inf. Sci. Eng.*, vol. 9, no. 4, Nov. 2009, doi: 10.1115/1.3243633.

[57] L. Yujian and L. Bo, "A Normalized Levenshtein Distance Metric," *IEEE Trans. Pattern Anal. Mach. Intell.*, vol. 29, no. 6, pp. 1091–1095, Jun. 2007, doi: 10.1109/TPAMI.2007.1078.

[58] G. M. Stump, S. W. Miller, M. A. Yukish, T. W. Simpson, and C. Tucker, "Spatial Grammar-Based Recurrent Neural Network for Design Form and Behavior Optimization," *J. Mech. Des.*, vol. 141, no. 12, Sep. 2019, doi: 10.1115/1.4044398.

[59] H. Nolte and C. McComb, "The cognitive experience of engineering design: an examination of first-year student stress across principal activities of the engineering design process," *Des. Sci.*, vol. 7, p. e3, ed 2021, doi: 10.1017/dsj.2020.32.

[60] T. Fong *et al.*, "A Preliminary Study of Peer-to-Peer Human-Robot Interaction," in *2006 IEEE International Conference on Systems, Man and Cybernetics*, Oct. 2006, vol. 4, pp. 3198–3203. doi: 10.1109/ICSMC.2006.384609.

[61] J. Stempfle and P. Badke-Schaub, "Thinking in design teams - an analysis of team communication," *Des. Stud.*, vol. 23, no. 5, pp. 473–496, Sep. 2002, doi: 10.1016/S0142-694X(02)00004-2.

[62] P. Balkundi and D. A. Harrison, "Ties, Leaders, And Time In Teams: Strong Inference About Network Structure's Effects On Team Viability And Performance," *Acad. Manage. J.*, vol. 49, no. 1, pp. 49–68, Feb. 2006, doi: 10.5465/amj.2006.20785500.

[63] J. Scott and P. J. Carrington, *The SAGE Handbook of Social Network Analysis*. SAGE Publications, 2011.

[64] A. M. Susskind and P. R. Odom-Reed, "Team Member's Centrality, Cohesion, Conflict, and Performance in Multi-University Geographically Distributed Project Teams," *Commun. Res.*, vol. 46, no. 2, pp. 151–178, Mar. 2019, doi: 10.1177/0093650215626972.

[65] S. I. Lei, H. Shen, and S. Ye, "A comparison between chatbot and human service: customer perception and reuse intention," *Int. J. Contemp. Hosp. Manag.*, vol. 33, no. 11, pp. 3977–3995, Jan. 2021, doi: 10.1108/IJCHM-12-2020-1399.

[66] G. Bansal, B. Nushi, E. Kamar, D. S. Weld, W. S. Lasecki, and E. Horvitz, "Updates in Human-AI Teams: Understanding and Addressing the Performance/Compatibility Tradeoff," *Proc. AAAI Conf. Artif. Intell.*, vol. 33, no. 01, Art. no. 01, Jul. 2019, doi: 10.1609/aaai.v33i01.33012429.

[67] A. Edwards, C. Edwards, D. Westerman, and P. R. Spence, "Initial expectations, interactions, and beyond with social robots," *Comput. Hum. Behav.*, vol. 90, pp. 308–314, Jan. 2019, doi: 10.1016/j.chb.2018.08.042.

[68] J. B. Lyons, K. Sycara, M. Lewis, and A. Capiola, "Human–Autonomy Teaming: Definitions, Debates, and Directions," *Front. Psychol.*, vol. 12, 2021, Accessed: Feb. 17, 2023. [Online]. Available: https://www.frontiersin.org/articles/10.3389/fpsyg.2021.589585

Copyright © 2023 by ASME

Proceedings of the ASME 2023
International Design Engineering Technical Conferences and
Computers and Information in Engineering Conference
IDETC-CIE2023
August 20-23, 2023, Boston, Massachusetts

DETC2023-115318

LET'S CHAT IF YOU ARE UNHAPPY – THE EFFECT OF EMOTIONS ON INTERACTION EXPERIENCE AND TRUST TOWARD EMPATHETIC CHATBOTS

Ting Liao
Stevens Institute of Technology
Hoboken, NJ

Bei Yan
Stevens Institute of Technology
Hoboken, NJ

ABSTRACT

Chatbots are now prevalent in obtaining information and executing tasks on behalf of human users. While the design community has paid more attention to streamlining user-chatbot interactions, the existing literature has not thoroughly examined social elements such as emotions and understood how emotions influence user interaction and trust in chatbots. This study proposes to test how participants perceive an empathetic chatbot versus a non-empathetic one under various emotional states (i.e., positive, neutral, negative) when the chatbot facilitates conversations for student advising via an online platform. The study shows the importance of presenting empathetic cues in the design of chatbots or other intelligent agents. The empathetic behavior of the chatbot improves participants' trust and perception of the chatbot's performance. The improvement is more salient to people with negative emotions than people who feel neutral or positive. This interaction effect is explained by people's perceived effort of the chatbot during the interaction with people, even though the chatbot is a technological agent that acts based on algorithms. The results suggest that people attribute human qualities to chatbots in social interaction and highlight the emotional needs of people who experience negative emotions. Therefore, design efforts need to be designated according to people's dynamic emotional states.

Keywords: emotion, affect, empathy, automated agent, human-computer interaction, design for automation

1. INTRODUCTION

Chatbots empowered by artificial intelligence (AI) have recently attracted enormous attention from both novice users and researchers due to their advanced capabilities and easy-to-use interface – ChatGPT accumulated 57 million monthly active users in its first month of availability [1,2]. Generally, chatbots, or conversational agents that simulate natural human conversations, are broadly used to support human activities [3–5], assist team collaboration [6], streamline software development and testing [2], and reduce caregivers' workloads in clinical practice [7]. The rapidly evolving chatbots augment human users' capabilities and simultaneously form new interaction dynamics, which leads to a potential trust crisis between humans and AI-powered chatbots.

A trustworthy relationship lays a foundation for effective interaction not only among humans but also between users and intelligent agents [7,8]. In a discussion of interaction, social aspects are inevitable. The Computers Are Social Actors (CASA) paradigm indicates that people may apply the social norms of human relationships when interacting with automated agents [9,10]. To design intelligent agents that comply with social rules, researchers examine different design strategies that enable the agents to understand human users and mimic human behavior. Understanding another individual often involves understanding what it feels like to be that person – in short, it entails empathy [12].

While intelligent agents, such as chatbots, are designed to be more understanding or empathetic, the existing literature lacks deep knowledge of how human emotional states affect their interaction experience with agents that present social behavior, and how trust is formed throughout the interaction. In addition to the agents being able to understand users' feelings, the emotional states of users could alter users' psychological capacity and their ability to identify with the feelings of others. Therefore, participants of different emotional states potentially recognize and perceive chatbots' empathetic behavior differently [9].

This study examined how users' emotional states and the chatbot's empathetic behavior influence user-agent interaction and, consequently, improve trust in the chatbot. The study is built upon the authors' preliminary work [13], which tested the affect manipulation, and interaction flow with thirty-three participants at the authors' home institution as a wizard-of-oz experiment. In this paper, we extended the prior work to further probe the mechanism of the effects at a larger scale. We also developed an

Copyright © 2023 by ASME

automated chatbot that directly reacts to users' emotional expressions in the interaction.

This paper is organized as follows: Section 2 reviews the influence of affect, empathy, and trust in related fields. Section 3 lists our hypotheses and research question. Section 4 describes the experimental method and setup. Data and analysis are in Section 5, and a discussion of the results is in Section 6. Finally, Section 7 includes the conclusion and plan for future work.

2. BACKGROUND

2.1 Affect

As one critical social aspect, affect, or emotion prescribes an individual's subjective feelings in a given context [14]. It can significantly impact human cognitive processes and particularly how people process information [15,16]. For example, individuals are more likely to attend to affect-congruent concepts [17–19].

Affect is an essential component in engineering design. Layered emotional profiles can elicit a "wow experience" [20] and drive product adoption, retention, and continued use [21,22]. Emotions can facilitate design activities – exposure to positive affect can help designers generate more design ideas [23].

Despite its significance in shaping social interaction and promoting design, design research of affect has recently attracted more attention. Research shows the influence of emotional states on user trust formation towards an automated voice agent [9], whereas an in-depth exploration of agent behavior and its interaction with affect is needed.

2.2 Empathy in Design

Empathy refers to the "reactions of one individual to the observed experiences of another" [27,28]. It plays a critical interpersonal and societal role, enabling sharing experiences, needs, and desires between individuals and promoting prosocial behavior [28,29]. Over the past decade, researchers from various fields found empathy to be related to better contextualization of a design problem [27,37], deeper understanding of the user needs [28], improved team design ideation [27], higher patient satisfaction in healthcare [29,30], responsible leadership [31, 32], and high-quality customer service [32,33]

While empathy is needed when going from designing for practical functions to designing for personal experiences in private contexts, researchers propose the design strategy of "being with" rather than "being like" other social actors in the interaction [40,41]. The goal of mutual sensemaking, first-person narratives, and shared accountability highlights the opportunities for dynamic interaction design based on users' current states and missions.

2.3 Trust in Automation

Trust has been extensively studied in different domains, including philosophy, sociology, organizations, economics, and marketing [36–43], and its definition is rooted in an individual willingness to accept vulnerability [44,45]. Given the different natures of interaction between users and intelligent agents, trust can be categorized as dispositional trust, situational trust, and learned trust [36,46,47]. Dispositional trust refers to an individual's tendency to trust an agent independent of a context or the specific agent [36]. It is also known as trust propensity, which relates to users' gender [46], age [48], culture [49], and personality [37]. Situational trust is developed based on the contextual situation of the task under the influence of internal and external factors [46,50]. External factors describe the type and complexity of the agent and the difficulty of the task [51]. Internal factors refer to the characteristics of the user, including the user's expertise, self-confidence [52], and mood [9,13]. Finally, learned trust is formed based on users' knowledge or belief when they are assessing the trustworthiness of an intelligent agent. Learned trust, or dynamically learned trust, is developed in an agent throughout the interaction based on existing knowledge, agent performance, and characteristics, such as complexity, appearance, communication style, ease of use, and transparency [46,53–55].

While agent performance and characteristics are proven to be dominant predictors of trust [56], affect influences situational trust and potentially changes the effectiveness of communication and interaction dynamics [57]. Given its complex nature, trust in this paper refers to a combination of situational and learned trust, which are influenced by users' moods and agent performance.

3. HYPOTHESIS AND RESEARCH QUESTION

Showing empathy promotes interpersonal connections and consequently improves interaction. We hypothesize that the chatbot's empathetic behavior will improve users' trust in the chatbot. Similarly, affect can potentially change people's subconscious attitudes, and their attitudes may influence the perception of social cues and the empathetic behavior of the chatbot. We hypothesize that people experiencing a positive affect during interaction will exhibit a more positive attitude towards the agent and, accordingly higher trust level than those who are neutral (control) or negative. However, prior research has suggested that people who experience negative emotions are more empathetic [58]. We thus reason that human affect and the chatbot's empathy may also have an interaction effect on users' trust towards the chatbot such that people who experience more negative emotions will respond more positively to a chatbot if the chatbot displays empathetic behavior and consequently, trust the empathetic chatbot more.

H1: Interactions with a chatbot that provides empathetic responses result in a higher level of participants' trust compared to a non-empathetic chatbot.

H2: Participants who experience more positive emotions will display a higher level of trust compared to participants who are less positive.

H3: Participants who experience more negative emotions and interact with an empathetic chatbot will display a higher level of trust compared to those who are less positive and interact with a non-empathetic chatbot.

In addition, we also investigate what factors may explain the effect of chatbot empathy, human affect, and their interaction on people's trust in chatbots. In particular, we focus on perceived effort. Previous research on empathy and human trust has found that people who experience negative emotions are more likely to trust their human counterparts if others behave empathetically and acknowledge their emotions because they believe it is more socially effortful for people to acknowledge others' negative emotions than positive ones [59]. However, this remains a question in human-computer interaction, as people know that chatbots are not socially conscious and they act based on prescribed algorithms and therefore do not need to make an extra effort to acknowledge negative emotions. Thus, we ask:

RQ1: Does perceived effort mediate the effect of (1) chatbot empathy, (2) users' emotional state, and (3) their interactions on users' trust in the chatbot?

4. EXPERIMENT METHOD

4.1 Experiment Procedure

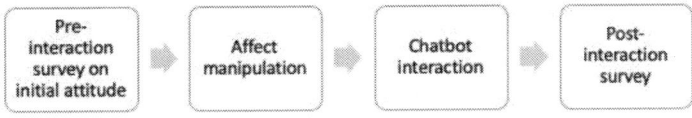

FIGURE 1: EXPERIMENT PROCEDURE

The experiment follows the procedure described and validated in the preliminary study [13]. The experiment consists of four steps (Figure 1). Briefly, participants first filled out a survey about their general beliefs. Then, they were prompted to share their experience by the following statement:

"We're collecting data from students to help redesign the student advising and improve students' experience. To begin with, we'd like to hear about your recent experience."

After being introduced to the study purpose, participants were randomly assigned to one of the three affect manipulation conditions (i.e., positive, neutral, or negative) and recalled a related experience. To validate the effect of the manipulation, participants were asked to rate their feeling on a 9-point scale.

Then, participants were randomly assigned to interact with either the empathetic or non-empathetic chatbot using free text. The chatbot greeted the participants and asked about their year in school, academic program, general experience, and favorite and least favorite courses. During the interaction, the empathetic chatbot responded to participants' feelings when emotions were expressed in the conversation and recognized by the chatbot. The non-empathetic chatbot delivered neutral responses such as "Thank you for answering," regardless of the expressions. The interaction ended when participants answered all the pre-determined questions.

After the interaction, participants rated their feelings again and completed a post-interaction survey about their trust and perceptions of the chatbot.

4.2 Affect Manipulation

We adopted the manipulation method developed by Emich et al. of a biographic memory task [56,57] and validated it in the preliminary study [13]. Based on the randomly assigned affect manipulation condition, participants were instructed to recall the related experience. For example, participants in the positive affect manipulation were asked to "describe the best thing that happened to you this past week in a few sentences below," which

evoked participants' activated and promotion-focused emotions. Similarly, participants in the negative affect manipulation were asked to recall the worst thing, and participants completing the neutral affect manipulation (control) were asked to describe a chore task. Studies have repeatedly shown that people's affective states and behavior are significantly and consistently influenced after the manipulation [56,57]. The results of the preliminary study prove the effectiveness of the affect manipulation [13].

4.3 Human-Chatbot Interaction

The interaction between participants and the chatbot was automated and administered through an online platform. We have tested many chatbot development platforms available in the market, like Konverse AI, Livechat, Dialog Flow, Drift, etc., and open-sourced scripts. Chatbot.com is chosen due to its ease of development and capability of embedding word dictionaries for sentiment detection.

The chatbot interface provides a generic appearance (Figure 2). The chatbot's behavior was programmed following a well-validated interaction flow from the preliminary study [13]. The chatbot introduced itself and the purpose of the interaction. It then asked participants for their basic information, such as name and program, and general experience (Figure 2). To obtain more details, the chatbot asked about the participants' favorite and least favorite courses at school and probed them to provide more details if the participants' answers were brief.

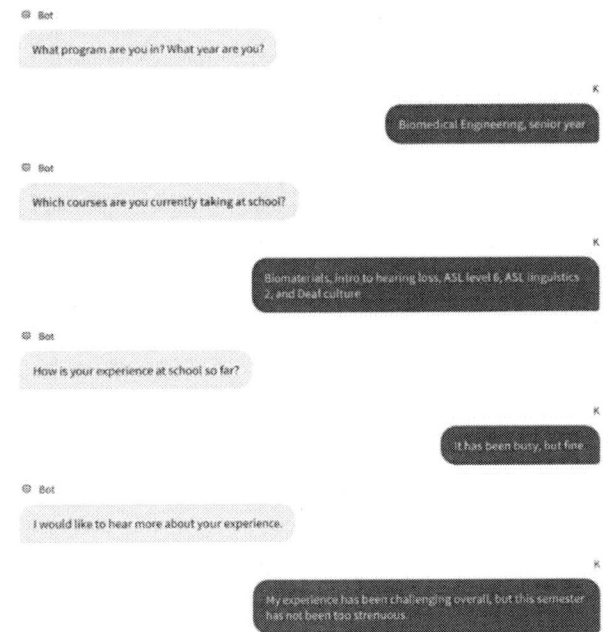

FIGURE 2: CHATBOT INTERFACE FOR ASKING ABOUT RECENT EXPERIENCE

The empathetic chatbot was designed to acknowledge emotions and deliver empathetic feedback when emotional expressions were detected in the conversations. A dictionary of 150 positive and 150 negative words was extracted from the existing literature [62,63]. The dictionary and common phrases

Copyright © 2023 by ASME

manually identified in the pilot test, such as "not bad," were programmed into the chatbot algorithm. For example, the empathetic chatbot sent statements such as "I am glad to hear that" when recognizing a positive word in response, "the instructor is *amazing*" [6]. In contrast, the non-empathetic chatbot delivered non-emotional responses, including "got it" and "thanks for sharing." The probing questions, feedback, and concluding statements are shown in Table 1.

4.4 Measuring Metrics

We collected a mixture of free text input and self-reported survey responses from participants. Before the interaction, participants reported their general beliefs toward automated products via a multiple-item survey and recalled an emotional or chore activity, as in the affect manipulation. A 9-point scale was adopted to measure emotional state, with 1 for "sad," 5 for "neutral," and 9 for "happy" [56,57]. During the interaction, textual conversations between the participants and the chatbot were recorded to assess the participants' level of affect and disclosure. Particularly, the number of words per answer by the participants was utilized to assess self-disclosure, assuming participants would share more information about themselves with the chatbot if they were more willing to self-disclose. Self-disclosure is of a particular interest as it might be an implicit indicator of the participants' trust [3,64].

After the interaction, participants were surveyed about their trust in the chatbot they interacted with [13,36,42,65]. To further understand the potential constructs of trust development, we also measured participants' perceived emotional acknowledgment, perceived empathy, and perceived effort of the chatbot using the established metrics [59,66,67]. These metrics were applied because prior research suggests that empathetic expressions may increase perceived effort as people perceive the empathetic party to have invested effort in acknowledging their emotions [58]. Each metric was measured by a multiple-item survey. For each survey question, participants rated how much they agreed with the corresponding statement, e.g., "when the chatbot notices my emotions, it will bring it up." The questions use a five-point scale, with 1 for "strongly disagree" and 5 for "strongly agree." They also reported their demographic information and knowledge of AI.

5. RESULTS AND ANALYSIS

One hundred and eighty participants were recruited via a crowdsourcing platform, Prolific, in 2022. Based on the chosen interaction scenario, participants who self-identified as students were selected. The experiment was administered fully online. Each participant was paid four U.S. dollars. After excluding the participants who did not finish the interaction or failed the checkpoint questions, 148 valid responses were collected and analyzed. The average study duration of the valid responses is 14.97 minutes. Among all the valid participants, 46.40% are female, 51.63% are male, and 1.96% are non-binary. Due to the occupation filter, most participants (85.59%) are between 20 and 30 years old. The number of participants in each condition is summarized in Table 2.

TABLE 2: NUMBER OF PARTICIPANTS IN EACH CONDITION

	Empathetic	Non-empathetic	Total
Positive	21	28	49
Negative	28	21	49
Neutral	24	26	50
Total	73	75	148

The general beliefs, including trust toward the automated systems, are tested to be homogenous across the participants using one-way ANOVA. Each of the perceived emotional acknowledgment, perceived empathy, perceived effort, and trust of the chatbot was measured by a multiple-item questionnaire on a five-point scale. To prepare for the analysis, the metrics were tested for reliability using Cronbach's alpha. Due to the high reliability ($\alpha > 0.80$), the scores of multiple-item questionnaires for each metric are aggregated into the mean values of each participant (Table 3). AI knowledge was measured using a single-item question.

TABLE 3 SUMMARY OF STATISTICS

	Emotional acknowledgment	Perceived empathy	Perceived effort	Trust
α	0.94	0.93	0.92	0.84
M	2.75	3.03	3.24	3.11
SD	1.28	1.26	1.06	0.93

TABLE 1: PROBING QUESTION, FEEDBACK RESPONSE, AND CONCLUDING STATEMENT OF CHATBOT

	Empathetic chatbot		Non-empathetic chatbot
	Positive emotions detected	Negative emotions detected	
Probing for details	• I am listening. Please tell me more about it.		• Please tell me more about it.
Feedback (One of the options)	• That sounds exciting. I would like to hear more about your experience. • I am glad to hear that.	• That must be a difficult time. I am here if you'd like to share more about your experience. • I am sorry to hear that.	• Thank you for sharing. • Got it.
Probing for more experience	Please give me more explanations. I'd love to know more about how you feel.		Please give me more explanations.
Concluding	Thank you for sharing with me your experience. It means a lot to me. I hope you also enjoyed this conversation. Wish you a great time at school!		Thank you. Your answers have been recorded.

Copyright © 2023 by ASME

5.1 Effect of Affect Manipulation

The effect of the affect manipulation (positive = 1, neutral = 0, negative = -1) is tested using the generalized linear regression model. The regression coefficient indicates the corresponding average change in the dependent variable for a unit increase of the independent variable. The manipulation shows a significant effect (coeff. = 1.85, p = 0.00*) on the self-reported emotional states after the manipulation (before interaction), as shown in Figure 3. This result validates the effectiveness of the affect manipulation.

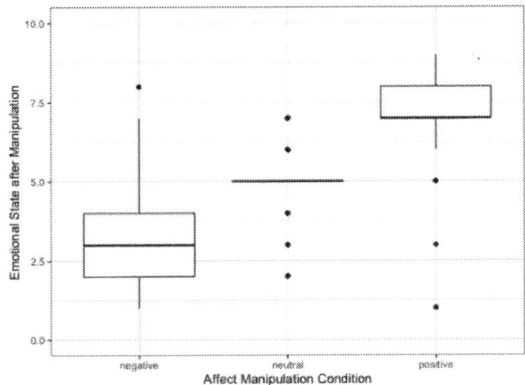

FIGURE 3: EMOTIONAL STATE BEFORE INTERACTION

Affect also has a significant effect on the change in emotional states before and after the interaction with the chatbot (coef. = -1.21, p = 0.00*), as in Figure 4. Particularly, the participants completing the positive affect manipulation showed more positive emotions before the interaction and became less positive after the interaction. In contrast, the emotional states of the participants being negatively manipulated shifted in the positive direction. This result is consistent with the preliminary study [13] and suggests that the chatbot interaction may relieve the emotions prior to the interaction, moving participants' emotional states toward the neutral state. Interacting with any type of chatbot from a neutral emotional level also makes participants slightly more positive. In addition, the interaction between affect and chatbot behavior has a marginal effect on emotional change (coef. = -0.61, p = 0.06+).

However, the effect of affect manipulation is tested to be potentially instant and might have diminished during the interaction between participants and the chatbot. The conversations between participants and the chatbot were downloaded as transcripts and analyzed using the Linguistic Inquiry and Word Count (LIWC) software [62,63]. The sentiment of the participants' responses is marginally influenced by the emotional state prior to the interaction (coeff. = -0.49, p = 0.05+) but not by the affect manipulation or the chatbot's empathetic behavior. This suggests that the affective state did carry on and impacted people's subsequent interaction with the chatbot. Interestingly, the participants who completed negative manipulation and interacted with the empathetic chatbot showed more positive sentiment, which led to the negative effect.

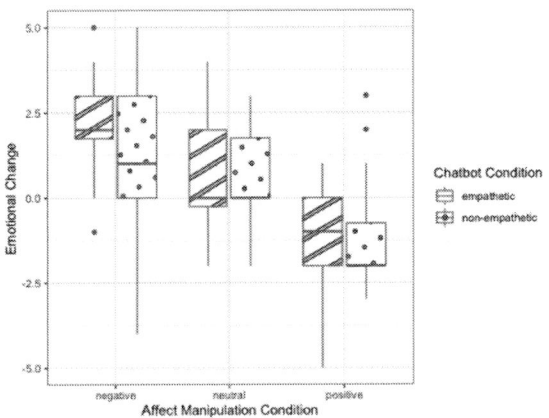

FIGURE 4: CHANGE IN EMOTIONAL STATE BEFORE AND AFTER INTERACTION

5.2 Effect of Chatbot's Empathetic Behavior

The chatbot's empathetic behavior shows a significant main effect on the perceived emotional acknowledgment (coef. = 1.86, p = 0.00*) in Figure 5 and the perceived empathy (coef. = 1.51, p = 0.00*) in Figure 6. Participants perceived the empathetic chatbot to be at acknowledging their emotions and showing empathy. However, affect, or interaction between the affect and chatbot behavior, does not have a significant effect on either measure. These confirm that the chatbot's empathetic behavior was well-received by the participants.

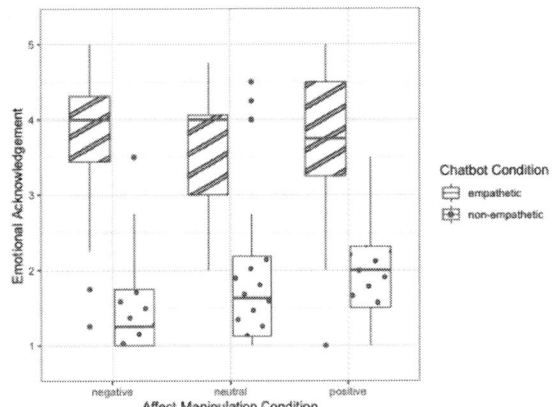

FIGURE 5: PERCEIVED EMOTIONAL ACKNOWLEDGMENT

The word count per answer by the participants reflects the participants' willingness to disclose during the interaction. Since the average word count per answer is small for each condition, within a range of five to ten, all the words are included. The chatbot's empathetic behavior is tested to have a marginal effect on participants' word counts (coef. = -1.01, p = 0.07+). The negative effect shows that participants potentially disclose more with the non-empathetic chatbot, which is counterintuitive. However, this may be tied to the difficulty of the communication and needs further evidence and analysis.

Copyright © 2023 by ASME

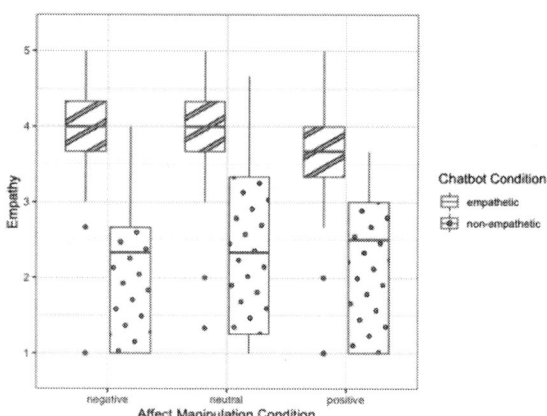

FIGURE 6: PERCEIVED EMPATHY OF CHATBOT

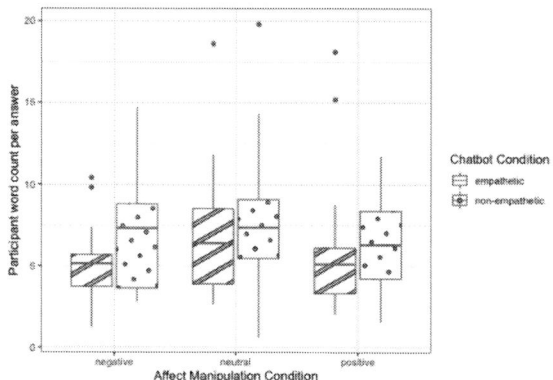

FIGURE 7: PARTICIPANTS' WORD COUNTS

5.3 Effect on Trust in Chatbot

Affect, empathetic behavior, and interaction between the two are tested to have significant main effects on reported trust in the chatbot (Table 4). Particularly, the participants with negative emotions before the interaction expressed significantly higher trust when interacting with the empathetic chatbot than the non-empathetic one, as highlighted in Figure 8. The results support hypotheses 1 and 2. Hypothesis 3 is also supported because of the negative interaction effect – indeed, people who experienced negative emotions responded more positively to empathetic chatbots and developed a higher trust in the agent.

TABLE 4: SUMMARY OF TEST RESULTS (REGRESSION COEFFICIENTS AND P-VALUES)

	Emotional acknowld.	Perceived empathy	Perceived effort	Trust
Affect	0.18	0.08	0.39	0.28
	p=0.16	p=0.60	p=0.01*	p=0.03*
Empathetic behavior	coef. =1.86	1.51	0.47	0.56
	p=0.00*	p=0.00*	p=0.01*	p=0.00*
Interaction	-0.23	-0.27	-0.37	-0.40
	p=0.21	p=0.19	p=0.08+	p=0.03*

Notes: * indicates $p < 0.05$; + indicates $p < 0.10$

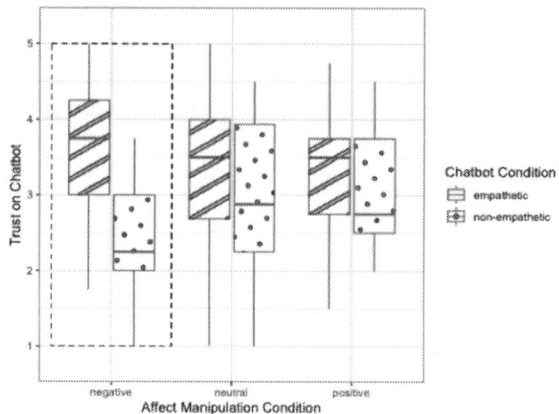

FIGURE 8: TRUST IN CHATBOT

In addition, affect, and empathetic behavior have significant main effects on the chatbot's perceived effort, whereas the interaction effect is marginal. Consistently with the interaction effect on trust, the negative interaction effect indicates that participants being negatively manipulated might have been more sensitive to the empathetic behavior and rated the empathetic chatbot higher in terms of its perceived effort. The testing results are summarized in Table 4.

5.4 Mediation Effects on Trust

To answer our research question, we adopted the Hayes PROCESS model 8 [70] to examine the moderated mediation effects of perceived effort on trust in the chatbot based on the existing literature on product performance and trust. The PROCESS model is widely used in social science to uncover the association throughout a sequence of causal events. It is built upon structural equation modeling (SEM) and conducts regression analyses containing various combinations of mediators, moderators, and covariates [71].

In the statistical diagram, "Chatbot" refers to the chatbot's empathetic behavior, "Effort" refers to the perceived effort by participants, and the affect manipulation condition is "affect" in the model. The model of the perceived effort as a mediator is present in Figure 9, and the results are summarized in Figure 10.

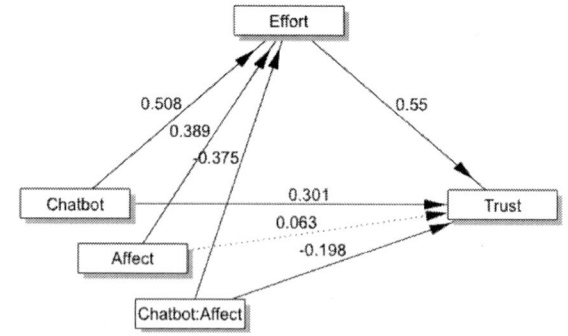

FIGURE 9 MODEL OF PERCEIVED EFFORT AS AN MEDIATOR ON TRUST

Variables	Predictors	B	SE	z	p	β
Effort	Chatbot	0.508	0.167	3.034	0.002	0.230
Effort	Affect	0.389	0.101	3.850	< 0.001	0.286
Effort	Chatbot:Affect	-0.375	0.144	-2.606	0.009	-0.194
Effort	AI.Knowledge	-0.145	0.102	-1.426	0.154	-0.108
Trust	Chatbot	0.301	0.113	2.655	0.008	0.155
Trust	Affect	0.063	0.071	0.886	0.376	0.052
Trust	Chatbot:Affect	-0.198	0.098	-2.009	0.045	-0.116
Trust	Effort	0.550	0.055	10.070	< 0.001	0.625

FIGURE 10 RESULTS OF THE MEDIATION MODEL

The results show that the perceived effort by the participants mediates the effect of the chatbot's empathetic behavior on trust, while the affect manipulation moderates the indirect effect of the chatbot's behavior on the perceived effort and the trust, with AI knowledge as a control variable. Similarly, as in Section 5.3, the interaction term has a significant negative effect.

6. DISCUSSION

This study is built upon the authors' preliminary work [13] to examine how people's affective state and chatbots' empathetic behavior together shape user interaction with chatbots and trust development. Whereas the researchers have paid more attention to the design of intelligent agents, it remains unclear how agents' prosocial behavior may interact with users' affective states to impact users' perception of the agents and construct user trust.

The results of this study provide further empirical evidence for the proposition by showing that participants' affect, which was effectively manipulated, is carried over into their interaction with the chatbot and moderates the impact of the chatbot's behavior on participants' perception and trust in the chatbot. As hypothesized, participants' affect can increase the trust in the chatbot.

People's emotional state may also alter how they respond to the chatbot's empathetic behavior. While showing empathy can improve user trust as hypothesized, a negative affect prior to the chatbot interaction seems to make people more sensitive to the empathetic cues sent by the chatbot, compared to other affective states. Participants who experience negative emotions express the highest level of trust after interacting with the empathetic chatbot and the lowest level with the non-empathetic one across conditions. This observation may be due to the fact that people feeling negative are more vulnerable compared to people who feel emotionally stable. This is congruent with the insight that negative cues can be more powerful in design interventions [9,20,23].

Interestingly, the perceived effort mediates the effect of the empathetic behavior on trust, with affect as a moderator. When the chatbot pays slightly more effort to detect and respond to participants' emotions during the interaction, the payoff is significant. While researchers argue that product function and performance are the primary ancestors of user trust in automation [13], empathetic behavior makes users perceive the product as making more social effort, despite that the technology only acts on rules prescribed by its algorithms. This suggests that adding human-like behavior may increase people's attribution of human qualities to technological agents in social interactions, even though people understand they are not interacting with other humans.

Together, the design implication of our findings is that for chatbots to be more effective in offering support, it is necessary to analyze and detect people's emotional states when they initiate the interaction. This can be done by asking specific questions and conducting automated sentiment analysis of people's initial communication to the chatbot. More importantly, people who experience negative affect may be more sensitive to the chatbot's behavior. Non-emphatic responses can harm user experience and trust when users are frustrated. On the other hand, recognizing and responding to their emotions can be an effective and economical way to improve user-perceived effort and trust in the chatbot. This finding echoes the current practice in customer service without the chatbot.

As a limitation of this study, all participants were students due to the chosen scenario. An additional scenario of the working professionals has been conducted and will be incorporated in future publications. The chatbot was developed on a commercial platform, which uses a dictionary-based detection method for sentiments. Based on the pilot study, the method is sufficient for simple conversations. More advanced sentiment analysis using Natural Language Processing can be used to enhance the chatbot and prepare it for more complex tasks. We adopted the same emotion-measuring metric used in the referenced affect manipulation method. We acknowledge that many other measurements are available to assess emotions across distinct moods [72–74]. Understanding the influence of moods would be a potentially fruitful direction for future research. In addition, this paper focuses on the influence of emotions on human-chatbot interaction. While participants' initial mental states potentially constrain the effectiveness of the affect manipulation and their ability to recognize empathetic responses by the chatbot, we only measured the prior-interaction emotional states based on the hypothesis. We acknowledge that additional measurements of their initial states could possibly provide insights into emotional change. In the follow-up study, we also measure personality traits to better capture individual differences.

7. CONCLUSION AND FUTURE WORK

In this study, we manipulated participants to experience positive, neutral, or negative emotions before interacting with an intelligent agent, a chatbot, which provides either empathetic or non-empathetic responses in conversations. We measured participants' emotional states before the interaction, the number of words and sentiments expressed in conversations, and post-interaction attitudes, including trust, toward the chatbot.

We find that the empathetic chatbot using first-person narratives and recognizing people's emotions leads to higher trust and better perception of the agent compared to the non-empathetic chatbot. The result aligns with the existing literature [12,64,65]. The study confirms the benefit of incorporating emotional cues in design for intelligent agents across domains. People's emotional needs are generally transferred from interpersonal relationships to user-computer interaction [9,10].

Copyright © 2023 by ASME

The empathetic behavior helps develop the trust of people who experience negative emotions and have greater emotional needs because empathetic behavior is perceived to be more effortful. This scenario widely applies to digital customer service and mental counseling. Many chatbots have been designed to facilitate human activities, and researchers gravitate towards fulfilling the needs of people who experience negative emotions, including frustration and anxiety [3,7]. Therefore, embedding empathetic behavior may be a low-hanging fruit for designers to improve the perception of the product performance, interaction experience, and trust in the agent.

In addition, the study highlights the importance of assessing people's emotions prior to user-chatbot interaction and providing design strategies accordingly. The results show that participants' affect can influence their interaction with the chatbot and their perception of its performance. People who feel more negative are more sensitive to chatbot behavior; emotionless conservations may be detrimental to the user experience when users are upset. Meanwhile, people who feel neutral and positive may be more emotionally occupied – interacting with an empathetic chatbot improves their experience, but not as much as for upset people.

In future work, we will generate more versatile scenarios beyond academic advising with more complex tasks. In addition to the prosocial behavior of the chatbot, we plan to investigate other product attributes and construct a comprehensive design guideline for intelligent agents and user-agent trust development. Ultimately, the intelligent agents can be designed to provide sufficient support at appropriate times and in an effective manner for human-agent collaboration.

ACKNOWLEDGMENTS
This research was supported by National Science Foundation Grant 2105169. We thank Ruixi Wang and Ananya Kodiboyena for their assistance in data collection and analysis.

REFERENCES
1. Martine Paris, "ChatGPT Hits 100 Million Users, Google Invests In AI Bot And CatGPT Goes Viral," *Forbes* (2023).
2. Lebeuf, C., Storey, M. A. & Zagalsky, A., "Software Bots," *IEEE Softw* **35**, 18–23 (2017). DOI: 10.1109/MS.2017.4541027
3. Lee, Y. C., Yamashita, N. & Huang, Y., "Designing a Chatbot as a Mediator for Promoting Deep Self-Disclosure to a Real Mental Health Professional," *Proc ACM Hum Comput Interact* **4**, 1–27 (2020). DOI: 10.1145/3392836
4. Fitzpatrick, K. K., Darcy, A. & Vierhile, M., "Delivering Cognitive Behavior Therapy to Young Adults With Symptoms of Depression and Anxiety Using a Fully Automated Conversational Agent (Woebot): A Randomized Controlled Trial," *JMIR Ment Health* **4**, e19 (2017). DOI: 10.2196/mental.7785
5. Williams, A. C., Kaur, H., Mark, G., Thompson, A. L., Iqbal, S. T. & Teevan, J., "Supporting workplace detachment and reattachment with conversational intelligence," *Conference on Human Factors in Computing Systems - Proceedings* **2018-April,** 1–13 (2018). DOI: 10.1145/3173574.3173662
6. Curran, M. T., Gordon, J. R., Lin, L., Sridhar, P. K. & Chuang, J., "Understanding Digitally-Mediated Empathy," in *Proceedings of the 2019 CHI Conference on Human Factors in Computing Systems* 1–13 (ACM, 2019). DOI: 10.1145/3290605.3300844
7. Fadhil, A., Schiavo, G. & Wang, Y., "CoachAI: A Conversational Agent Assisted Health Coaching Platform," (2019). at <http://arxiv.org/abs/1904.11961>
8. Lotfalian Saremi, M. & Bayrak, A. E., "A Survey of Important Factors in Human - Artificial Intelligence Trust for Engineering System Design," in *Volume 6: 33rd International Conference on Design Theory and Methodology (DTM)* (American Society of Mechanical Engineers, 2021). DOI: 10.1115/DETC2021-70550
9. Liao, T. & MacDonald, E. F., "Manipulating Users' Trust of Autonomous Products With Affective Priming," *Journal of Mechanical Design* **143,** 1–12 (2021). DOI: 10.1115/1.4048640
10. Nass, C., Steuer, J. & Tauber, E. R., "Computers are Social Actors," in *the SIGCHI conference on Human factors in computing systems* 72–78 (1994). DOI: 10.1109/VSMM.2014.7136659
11. Nass, C. (Stanford U. & Moon, Y. (Harvard U., "Machines and Mindlessness: Social Responses to Computers," *Journal of Social Issues* **1,** 81–103 (2000).
12. Hancock, P. a., Billings, D. R., Schaefer, K. E., Chen, J. Y. C., de Visser, E. J. & Parasuraman, R., "A Meta-Analysis of Factors Affecting Trust in Human-Robot Interaction," *Human Factors: The Journal of the Human Factors and Ergonomics Society* **53,** 517–527 (2011). DOI: 10.1177/0018720811417254
13. Liao, T. & Yan, B., "Are You Feeling Happy? the Effect of Emotions on People's Interaction Experience Toward Empathetic Chatbots," in *Volume 3B: 48th Design Automation Conference (DAC)* (American Society of Mechanical Engineers, 2022). DOI: 10.1115/DETC2022-91059
14. Zajonc, R. B., in *The handbook of social psychology* (eds. Gilbert, D. T., Fiske, S. T. & Lindzey, G.) 591–632 (McGraw-Hill, 1998).
15. Isen, A. M., "Toward Understanding the Role of Affect in Cognition," *Handbook of Social Cognition* **3,** 179–236 (1984). at <https://psycnet.apa.org/record/2011-28557-005>
16. Zajonc, R. B., "Feeling and thinking: Preferences need no inferences.," *American Psychologist* **35,** 151–175 (1980). DOI: 10.1037/0003-066X.35.2.151
17. Bower, G. H., in *Emotion and Social Judgments* (ed. Forgas, J. P.) 31–53 (Pergamon Press, 1991).
18. Forgas, J. P., in 227–275 (1992). DOI: 10.1016/S0065-2601(08)60285-3
19. Forgas, J. P., "On mood and peculiar people: Affect and person typicality in impression formation.," *J Pers Soc*

Copyright © 2023 by ASME

Psychol **62**, 863–875 (1992). DOI: 10.1037/0022-3514.62.5.863

20. Desmet, P. M. A., Porcelijn, R. & van Dijk, M. B., "Emotional Design; Application of a Research-Based Design Approach," *Knowledge, Technology & Policy* **20**, 141–155 (2007). DOI: 10.1007/s12130-007-9018-4

21. Liao, T., Tanner, K. & MacDonald, E. F., "Revealing insights of users' perception: An approach to evaluate wearable products based on emotions," *Design Science* 1–18 (2020). DOI: 10.1017/dsj.2020.7

22. Bartl, C., Gouthier, M. H. J. & Lenker, M., "Delighting Consumers Click by Click: Antecedents and Effects of Delight Online," *J Serv Res* **16**, 386–399 (2013). DOI: 10.1177/1094670513479168

23. Lewis, S., Dontcheva, M. & Gerber, E., "Affective Computational Priming and Creativity," in *CHI: International Conference on Human Factor in Computing Systems* 735–744 (2011).

24. Alzayed, M. A., Miller, S. R., Menold, J., Huff, J. & McComb, C., "Can design teams be empathically creative? a simulation-based investigation on the role of team empathy on concept generation and selection," in *Proceedings of the ASME Design Engineering Technical Conference* **8**, 1–15 (2020). DOI: 10.1115/detc2020-22432

25. Davis, M. H., "Measuring individual differences in empathy: Evidence for a multidimensional approach.," *J Pers Soc Psychol* **44**, 113–126 (1983). DOI: 10.1037//0022-3514.44.1.113

26. Riess, H., "The Science of Empathy," *J Patient Exp* **4**, 74–77 (2017). DOI: 10.1177/2374373517699267

27. Fila, N. D. & Hess, J. L., "In their shoes: Student perspectives on the connection between empathy and engineering," *ASEE Annual Conference and Exposition, Conference Proceedings* **2016-June,** (2016). DOI: 10.18260/p.25640

28. Gray, C., Yilmaz, S., Daly, S., Seifert, C. & Gonzalez, R., "Idea Generation Through Empathy: Reimagining the 'Cognitive Walkthrough'," in *2015 ASEE Annual Conference and Exposition Proceedings* 26.871.1-26.871.29 (ASEE Conferences). DOI: 10.18260/p.24208

29. Boissy, A., Windover, A. K., Bokar, D., Karafa, M., Neuendorf, K., Frankel, R. M., Merlino, J. & Rothberg, M. B., "Communication Skills Training for Physicians Improves Patient Satisfaction," *J Gen Intern Med* **31**, 755–761 (2016). DOI: 10.1007/s11606-016-3597-2

30. Holt, S. & Marques, J., "Empathy in Leadership: Appropriate or Misplaced? An Empirical Study on a Topic that is Asking for Attention," *Journal of Business Ethics* **105**, 95–105 (2012). DOI: 10.1007/s10551-011-0951-5

31. Publishing, E. G., Group, E. & Limited, P., "A Dynamic Theory of Leadership Development," *Leadership & Organization Development Journal* **30**, 563–576 (2003).

32. Clark, C. M., Murfett, U. M., Rogers, P. S. & Ang, S., "Is Empathy Effective for Customer Service? Evidence From Call Center Interactions," *J Bus Tech Commun* **27**, 123–153 (2013). DOI: 10.1177/1050651912468887

33. Bordoloi, S. K., "Agent Recruitment Planning in Knowledge-Intensive Call Centers," *J Serv Res* **6**, 309–323 (2004). DOI: 10.1177/1094670503262945

34. Mattelmäki, T., Battarbee, K., Mattelmiiki, T. & Battarbee, K., "Empathy Probes," *PDC* 266–271 (2002).

35. Bennett, C. L. & Rosner, D. K., "The promise of empathy: Design, disability, and knowing the "other"," *Conference on Human Factors in Computing Systems - Proceedings* 1–13 (2019). DOI: 10.1145/3290605.3300528

36. Lotfalian Saremi, M. & Bayrak, A. E., "A survey of important factors in human-artificial intelligence trust for engineering system design," in *Volume 6: 33rd International Conference on Design Theory and Methodology (DTM)* (American Society of Mechanical Engineers, 2021). DOI: 10.1115/DETC2021-70550

37. Rotter, J. B., "Interpersonal Trust, Trustworthiness, and Gullibility," *American Psychologist* **35**, 1–7 (1980). DOI: 10.1037/0003-066X.35.1.1

38. Patrick, A. S., "Building trustworthy software agents," *IEEE Internet Comput* **6**, 46–53 (2002). DOI: 10.1109/MIC.2002.1067736

39. Zarghami, A., Fazeli, S., Dokoohaki, N. & Matskin, M., "Social trust-aware recommendation system: A T-index approach," *Proceedings - 2009 IEEE/WIC/ACM International Conference on Web Intelligence and Intelligent Agent Technology - Workshops, WI-IAT Workshops 2009* **3**, 85–90 (2009). DOI: 10.1109/WI-IAT.2009.237

40. Papadopoulou, P., "Applying Virtual Reality for Trust-Building E-Commerce Environments," *Virtual Real* **11**, 107–127 (2007). DOI: 10.1007/s10055-006-0059-x

41. Al-Hamadi, H. & Chen, I. R., "Trust-Based Decision Making for Health IoT Systems," *IEEE Internet Things J* **4**, 1408–1419 (2017). DOI: 10.1109/JIOT.2017.2736446

42. Rheu, M., Shin, J. Y., Peng, W. & Huh-Yoo, J., "Systematic Review: Trust-Building Factors and Implications for Conversational Agent Design," *Int J Hum Comput Interact* (2020). DOI: 10.1080/10447318.2020.1807710

43. Verberne, F. M. F., Ham, J. & Midden, C. J. H., "Trusting a Virtual Driver That Looks, Acts, and Thinks Like You," *Hum Factors* **57**, 895–909 (2015). DOI: 10.1177/0018720815580749

44. Borum, R., *The Science of Interpersonal Trust Mental Health Law & Ploicy Faculty Publications* **574,** (2010). at <http://works.bepress.com/randy_borum/48/>

45. Mayer, R. C., Davus, J. H.. & Schoorman, F. D. S., "An Integrative Model of Organizational Trust," *Academy of Management Review* **20**, 709–734 (1995).

Copyright © 2023 by ASME

46. Hoff, K. A. & Bashir, M., "Trust in Automation: Integrating Empirical Evidence on Factors that Influence Trust," *Hum Factors* **57,** 407–434 (2015). DOI: 10.1177/0018720814547570

47. Bashir, M. & Hoff, K., "A Theoretical Model for Trust in Automated Systems," *Conference on Human Factors in Computing Systems - Proceedings* **2013-April,** 115–120 (2013). DOI: 10.1145/2468356.2468378

48. Pak, R., Fink, N., Price, M., Bass, B. & Sturre, L., "Decision support aids with anthropomorphic characteristics influence trust and performance in younger and older adults," *Ergonomics* **55,** 1059–1072 (2012). DOI: 10.1080/00140139.2012.691554

49. Huerta, E., Glandon, T. A. & Petrides, Y., "Framing, decision-aid systems, and culture: Exploring influences on fraud investigations," *International Journal of Accounting Information Systems* **13,** 316–333 (2012). DOI: 10.1016/j.accinf.2012.03.007

50. Damen, N. & Toh, C., "Designing for Trust: Understanding the Role of Agent Gender and Location on User Perceptions of Trust in Home Automation," in *ASME 2017 International Design Engineering Technical Conferences and Computers and Information in Engineering Conference* **141,** 061101 (2018). DOI: 10.1115/1.4042223

51. Madhavan, P. & Wiegmann, D. A., "Similarities and differences between human–human and human–automation trust: An integrative review," *Theor Issues Ergon Sci* **8,** 277–301 (2007). DOI: 10.1080/14639220500337708

52. Beller, J., Heesen, M. & Vollrath, M., "Improving the driver-automation interaction: An approach using automation uncertainty," *Hum Factors* **55,** 1130–1141 (2013). DOI: 10.1177/0018720813482327

53. Hancock, P. A., Billings, D. R., Schaefer, K. E., Chen, J. Y. C., De Visser, E. J. & Parasuraman, R., "A meta-analysis of factors affecting trust in human-robot interaction," *Hum Factors* **53,** 517–527 (2011). DOI: 10.1177/0018720811417254

54. Lin, H. F., "An empirical investigation of mobile banking adoption: The effect of innovation attributes and knowledge-based trust," *Int J Inf Manage* **31,** 252–260 (2011). DOI: 10.1016/j.ijinfomgt.2010.07.006

55. Siau, K. & Wang, W., "Building trust in artificial intelligence, machine learning, and robotics," *Cutter Business Technology Journal* **31,** (2018). at <www.cutter.com>

56. Parasuraman, R. & Riley, V., "Humans and Automation: Use, Misuse, Disuse, Abuse," *Hum Factors* **39,** 230–253 (1997).

57. Hwang, A. H. C. & Won, A. S., "Ideabot: Investigating social facilitation in human-machine team creativity," in *Conference on Human Factors in Computing Systems - Proceedings* (Association for Computing Machinery, 2021). DOI: 10.1145/3411764.3445270

58. Bagozzi, R. P. & Moore, D. J., *Public Service Advertisements: Emotions and Empathy Guide Prosocial Behavior*

59. Yu, A., Berg, J. M. & Zlatev, J. J., "Emotional acknowledgment: How verbalizing others' emotions fosters interpersonal trust," *Organ Behav Hum Decis Process* **164,** 116–135 (2021). DOI: 10.1016/j.obhdp.2021.02.002

60. Emich, K. J. & Vincent, L. C., "Shifting focus: The influence of affective diversity on team creativity," *Organ Behav Hum Decis Process* **156,** 24–37 (2020). DOI: 10.1016/j.obhdp.2019.10.002

61. Vincent, L. C., Emich, K. J. & Goncalo, J. A., "Stretching the Moral Gray Zone: Positive Affect, Moral Disengagement, and Dishonesty," *Psychol Sci* **24,** 595–599 (2013). DOI: 10.1177/0956797612458806

62. Nielsen, F. Å., "A new ANEW: Evaluation of a word list for sentiment analysis in microblogs," (2011). at <http://arxiv.org/abs/1103.2903>

63. Andrada, "Bing, NRC, Afinn Lexicons," *Kaggle.com* (2020).

64. Tian, Q., "Social Anxiety, Motivation, Self-Disclosure, and Computer-Mediated Friendship," *Communic Res* **40,** 237–260 (2013). DOI: 10.1177/0093650211420137

65. Jian, J.-Y., Bisantz, A. M. & Drury, C. G., "Foundations for an empirically determined scale of trust in automated systems," *Int J Cogn Ergon* **4,** 53–71 (2000).

66. Huang, Y., Chen, C.-H. & Khoo, L. P., "Products classification in emotional design using a basic-emotion based semantic differential method," *Int J Ind Ergon* **42,** 569–580 (2012). DOI: 10.1016/j.ergon.2012.09.002

67. Van den Broeck, E., Zarouali, B. & Poels, K., "Chatbot advertising effectiveness: When does the message get through?" *Comput Human Behav* **98,** 150–157 (2019). DOI: 10.1016/J.CHB.2019.04.009

68. Tausczik, Y. R. & Pennebaker, J. W., "The psychological meaning of words: LIWC and computerized text analysis methods," *J Lang Soc Psychol* **29,** 24–54 (2010). DOI: 10.1177/0261927X09351676

69. Bird, S., Ewan, K. & Edward, L., "Natural Language Processing with Python," at <https://www.nltk.org/book/>

70. Hayes, A. F., *PROCESS: A Versatile Computational Tool for Observed Variable Mediation, Moderation, and Conditional Process Modeling* *1* at <http://www.afhayes.com/>

71. Hayes, A. F. & Rockwood, N. J., "Conditional Process Analysis: Concepts, Computation, and Advances in the Modeling of the Contingencies of Mechanisms," *American Behavioral Scientist* **64,** 19–54 (2020). DOI: 10.1177/0002764219859633

72. Terry, P. C., Lane, A. M. & Fogarty, G. J., "Construct validity of the Profile of Mood States — Adolescents for use with adults," *Psychol Sport Exerc* **4,** 125–139 (2003). DOI: 10.1016/S1469-0292(01)00035-8

Copyright © 2023 by ASME

73. Paige, M. A., Fillingim, K. B., Murphy, A. R., Song, H., Reichling, C. J. & Fu, K., "Examining the effects of mood on quality and feasibility of design outcomes," *International Journal of Design Creativity and Innovation* **9,** 79–102 (2021). DOI: 10.1080/21650349.2021.1890228

74. Feldman Barrett, L., "Discrete Emotions or Dimensions? The Role of Valence Focus and Arousal Focus," *Cogn Emot* **12,** 579–599 (1998). DOI: 10.1080/026999398379574

75. de Gennaro, M., Krumhuber, E. G. & Lucas, G., "Effectiveness of an Empathic Chatbot in Combating Adverse Effects of Social Exclusion on Mood," *Front Psychol* **10,** 1–14 (2020). DOI: 10.3389/fpsyg.2019.03061

76. Hu, T., Xu, A., Liu, Z., You, Q., Guo, Y., Sinha, V., Luo, J. & Akkiraju, R., "Touch your heart: A tone-aware chatbot for customer care on social media," in *Conference on Human Factors in Computing Systems - Proceedings* **2018-April,** 1–12 (2018). DOI: 10.1145/3173574.3173989

77. Rashkin, H., Smith, E. M., Li, M. & Boureau, Y.-L., "Towards Empathetic Open-domain Conversation Models: a New Benchmark and Dataset," in *Proceedings of the 57th Annual Meeting of the Association for Computational Linguistics* 5370–5381 (2019).

Proceedings of the ASME 2023
International Design Engineering Technical Conferences and
Computers and Information in Engineering Conference
IDETC-CIE2023
August 20-23, 2023, Boston, Massachusetts

DETC2023-116956

A MULTI-OBJECTIVE BAYESIAN OPTIMIZED HUMAN ASSESSED MULTI-TARGET GENERATED SPECTRAL RECOMMENDER SYSTEM FOR RAPID PARETO DISCOVERIES OF MATERIAL PROPERTIES

Arpan Biswas
Center for
Nanophase Materials
Sciences, Oak Ridge
National Laboratory
Oak Ridge, TN

Yongtao Liu
Center for Nanophase
Materials Sciences, Oak
Ridge National
Laboratory
Oak Ridge, TN

Maxim Ziatdinov
Computational Sciences
and Engineering
Division, Oak Ridge
National Laboratory
Oak Ridge, TN

Yu-Chen Liu
Department of Physics,
National Cheng Kung
University, Taiwan

Stephen Jesse
Center for Nanophase
Materials Sciences,
Oak Ridge National
Laboratory Oak
Ridge, TN

Jan-Chi Yang
Department of Physics,
National Cheng Kung
University, Taiwan

Sergei Kalinin
Department of Materials
Science and
Engineering, University
of Tennessee
Knoxville, TN

Rama Vasudevan
Center for Nanophase
Materials Sciences, Oak
Ridge National
Laboratory
Oak Ridge, TN

ABSTRACT

Optimization for different tasks like material characterization, synthesis, and functional properties for desired applications over multi-dimensional control parameter and function spaces need a rapid strategic search through active learning. However, in all cases prior to optimization, the target material properties are assumed known and fixed, which mostly deviates from real-world scenarios in material synthesis. This can be critical for running expensive experiments on new materials, when the experimental results are fuzzy for any scientific outcomes due to improper target setting, ultimately wasting time and cost. The failure rate and cost are even higher over exploring on multi-target space, where we want to learn the pareto among multiple properties, to jointly optimize during material synthesis for desired applications. To address the challenge, here we introduce the human-operator attempt flexibility in the active learning based automated experiment framework, with generating multiple human assessed targets through a voting-based recommender system during real-time microscope measurements over the large material image space, sequentially learn/update multiple desired targets through a weighting system, and adaptively search in multiple material properties functional space for non-dominated pareto discoveries to maximize the custom structural similarity based acquisition function. We term this a multi-objective Bayesian

optimized human assessed multi-target generated spectral recommender systems (MOBO-HAM-SRS). The approach has been demonstrated to peizoresponse force spectroscopy of a ferroelectric thin film, exploring with different kernels and acquisition functions. This work shows an advancement towards human-AI collaborated automated experiments, steering optimization trajectories through human overpowering AI at the early stage when uncertainty is high and AI overpowering human at the later stage with rapid exploration towards optimal goal, following human-assessed multiple targets properties.

Keywords: Multi-objective Bayesian optimization, active recommendation system, automated experiments, human-guided learning, dynamic target selection, deep kernel learning

1. INTRODUCTION

The integration of data-driven computational design architecture, optimization, and machine learning approaches with data sampling from microscopic measurements, has led to the advancement of the automated "self-driving" experimental setup in the field of material science towards rapid discovery of physics [1]–[5]. For a given material sample, the structure-property relationships are learned through appropriate measurements on various microscope instruments. However, such microscopes are expensive to measure exhaustively for the

This material is declared a work of the U.S. Government and is not subject to Copyright protection in the United States.

Copyright © 2023 by The United States Government

entire high resolution material image space, and therefore the ML-based computational tools aim to guide the user for rapid learning and adaptively search the space for discoveries, resulting in an overall significant reduction of time and cost. Over the years, examples include of such automated and autonomous experiments with rapid exploration over large material image and chemical space through active learning on electron (SEM) [4], scanning probe microscopes (SPM) [6], [7], scanning tunnelling microscopes (STM) [8], [9] etc. However, in most of such cases, the target property to search is pre-defined and the task is to efficiently probe the parameter space to identify the regions with maximum likelihood of priorly selected target with minimum uncertainty. Though the AI-driven automated experiments (without human in the loop) provide fast learning and eliminate rigorous steps, does not necessarily guarantee the best learning always. Therefore, when the experimentalists are unknown to what target they would want, one way to update their knowledge about feasible material properties is the visualization of the measured spectral structures. Thus, the experimentalists generally prefer to observe few spectral samples and build an achievable target, from where the parameter space can be optimized to learn the regions with similar spectral structures to the human assessed targets. In other words, without any such observations, it will be a "blind" target for them and can subsequently turns to an infeasible goal with no optimal solutions, thereby increasing cost and effort with no meaningful learning. This can be critical for running expensive experiments on new materials, when the experimental results are fuzzy for any scientific outcomes due to improper target setting, ultimately wasting time and cost. The failure rate and cost are even higher over exploring on multi-target space, where we want to learn the pareto among multiple properties, to jointly optimize during material synthesis for desired applications. In addition, the experimental data (eg. spectra) captured from microscope are generally raw and complex than a theoretical simulator and derivation of multiple key features (such as spectral loop area, storage, peak width, height etc) in a simplified numerical descriptor can be non-trivial. In such cases when target is challenging to derive in a simplistic quantitative way [10], a combined approach of formulation with both quantitative and subjective (knowledge from human assessment) can be taken through a recommender system. This technique of recommender system has been well established in the information overload problem in the domain of social media like Facebook, YouTube etc, where the user votes through "likes" and "dislikes" and the system recommend new pages or videos based on prior voting [11]–[13].

Here, in motivation to address the stated challenges, we introduce the human-operation within the active learning based automated experiment framework, to add flexibility in multiple target learning and selection (representing different material key features), and subsequently optimize the material's multi-functional space to discover pareto frontier between these humans' assessed targets. A recent Nature article reports that leveraging human expertise within the optimization process can greatly improve recipes for materials processing [14]. Similarly, our prior work has adopted the similar human-first, computer-last approach for autonomous microscopy [15]. However, in both cases, the application is limited to single target assessment. Here, we present a multi-objective Bayesian optimized human assessed multi-target generated spectral recommender systems (MOBO-HAM-SRS). To highlight the best of both human and AI, the human overpowers AI in the early stage of exploration (when uncertainty is high) and AI overpowers the human during the later stage (when human assessed multiple targets are defined) to fully automate the rigorous process towards optimal learning. The SRS provides the user to vote the quality of spectra, during real-time microscope measurements at a given location over the material image space, and sequentially learn/update human assessed multiple targets (HAM) through a weighting system. Then this SRS framework, deployed to generate multiple targets with various key features of material properties, is augmented with MOBO to provide an adaptive search over the parameter space in maximizing the multi-objective acquisition function, derived from the descriptors of combined elements – a subjective human voting-based reward of the spectra in view to each target, and a structural similarity between each target and the new captured spectral. It is evident that the proposed framework is essential for exploration over new material space when a prior target cannot be certain to run a fully automated experiments due to limited knowledge and complexity of several unexplored unknown targets, as in case for material synthesis. For a well-known material system, the exploration can be automated with a prior target knowledge, without human intervention, which is not the scope of problem in this paper. The roadmap for this paper can be stated as follows: Section 2 provides the detail description and algorithm of the MOBO-HAM-SRS architecture, showcasing the extension of human-operated SRS system in a multi-objective framework and its connection with existing MOBO workflow. Section 3 showcased the implementation of the prototype to peizoresponse force spectroscopy of a ferroelectric thin film (eg. $PbTiO_3$), with performance analysis on different kernels and multi-objective-based acquisition functions. Section 4 concludes the paper with final thoughts.

2. METHODOLOGY

The MOBO-HAM-SRS architecture has two key elements – 1) the human operated spectral recommender system for multiple target generation and sequential learning as HAM-SRS and 2) the multi-objective Bayesian optimization (MOBO) workflow.

Here, given a location in the material image space, the spectral structure is captured first and visualized. As stated, this spectral structure is enriched with several key material properties, which is considered for optimization to achieve a desired application. The HAM-SRS system is a dynamic, human-augmented computational design framework which provides the voting capability to either upvote or downvote the captured spectra, with a degree of preference of the newly captured spectra from the prior upvoted spectra (current target).

Copyright © 2023 by The United States Government

Thus, this target spectral selection is human curiosity driven and devoid of several property descriptor formulations. With the series of several such spectral voting in accordance with each targets, till the user is satisfied to learn multiple targets, the individual final target spectral is the weighted (measured by votes and preferences) combination of all the upvoted spectra. Previously, human-augmented recommender systems, however focusing on the single target generation, have been developed in microscopy in accelerating meaningful discoveries in different field of applications such as rapid validation of thousands of biological objects or specimen tracking results [16], rapid material discovery of novel lithium ion conducting oxides through synthesis of unknown chemically relevant compositions (CRCs) [17].

The next key element is the Multi-objective Bayesian optimization (MOBO) architecture. BO [18]–[20], in general, is a cheap computational design of experiments where the expensive and/or black-box experiment is represented with surrogate model, preferably a Gaussian process (GP) model [21]. BO have been applied extensively in various material [22]–[26] and chemical space [27], [28] explorations for rapid discoveries, in addition to general design problems of discrete and discontinuous function space [29]–[33], and high-dimensional input space [34]–[38]. MOBO [39], [40] is an extension to BO with optimizing multiple functions jointly. Here, each expensive objective functions are represented with an independent GP model, leading to "n" independent GP models for "n" objectives, which provides respective estimate function and its uncertainty map. Then, an adaptive sampling technique drives the exploration over the multi-dimensional function space towards maximizing the likelihood of finding the non-dominated or Pareto solutions, and thereby reduce the uncertainty of the GP learning as the model converges. In MOBO, this sampling technique can be classified into posteriori and apriori methods. In the posteriori method, the trade-off weights are unknown, and the sampling is done through direct maximizing the hypervolume over the multi-objective space such as Expected Improvement Hyper-volume (EIHV) [41], Expected Hyper volume gradient-based (EIHVG) [42], Max-value Entropy Search (MESMO) [43] and Predictive Entropy Search (PESMO) [44] acquisition functions. In apriori method, the trade-off weights are preselected with user defined input or random generator and a weighted multi-objective "single" function is formulated first, then sampling is done through maximizing the multi-objective acquisition function with a BO based acquisition function such as Pareto efficient global optimization (PAREGO) [45], [46], Weighted Tchebycheff based MO acquisition functions [47]–[49] etc. Like for BO, MOBO architecture has also been previously adapted to perform rapid discoveries under material [50]–[52] and chemical space [53], [54] exploration and synthesis. Additionally, MOBO has also been extended for design problems under hybrid function [55] and high-dimensional input space [56]–[58].

In summary the contribution of the paper are as follows:

a) We propose a computational HAM-SRS architecture to increase flexibility through introducing human-operation in a multi-objective active learning-based AE, by allowing the experimentalists to vote the quality of the captured spectra during microscope measurements and sequentially generate and update learning of human assessed multiple target spectra.

b) We connect a MOBO workflow with the proposed HAM-SRS to guide the adaptive search process in the multi-functional space to maximize the customed multiple descriptors (objective functions) of combined human assessment-based rewards and structural similarity to the respective target, finally to optimize the pareto frontier among multiple curiosity driven material targets.

c) Implementation of the prototype on the samples of ferroelectric thin films as proof of concept and analyze the results with different kernel and multi-objective acquisition functions.

Copyright © 2023 by The United States Government

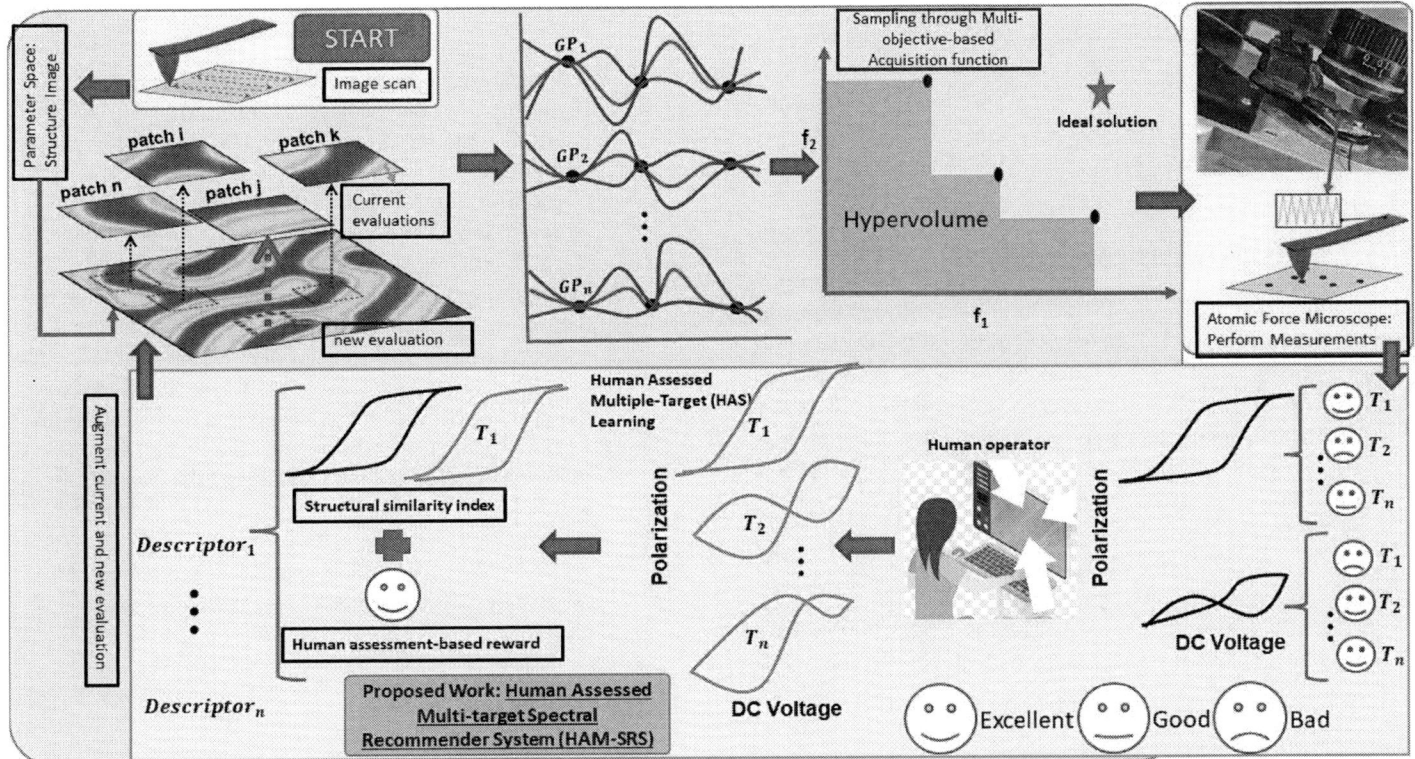

FIGURE 1: Design workflow of MOBO-HAM-SRS architecture. The orange section is the contribution in this paper with the extension of the human inspected spectral recommender system into the multi-objective automated experiment workflow, to generate and learn human assessed multiple material targets and the corresponding descriptors. The yellow section follows the steps of microscope operations (eg. generate material image space and capture spectral measurements). The green section follows the computational tasks of a MOBO framework where multiple gaussian process is fitted on input image patches and then a multi-objective acquisition function is built to encourage adaptive sampling for next microscope measurement (spectral) to maximize the descriptors jointly and optimize the pareto frontier. It is to be noted that as the iteration starts, the human operation is conducted till multiple targets are learned with user satisfaction (following step 6), and after that, the framework switches to a fully automated system (see step 6 c) for the remaining iterations to avoid the rigorous steps of exploration with multiple learned targets.

2.1 Multi-objective Bayesian optimized human assessed multi-target generated spectral recommender systems (MOBO-HAM-SRS)

Figure 1 provides the overall design of the MOBO-HAM-SRS system. Table 1 shows the detail algorithm of the MOBO-HAM-SRS. For the simplicity of the paper, we have considered for 2 objectives. However, the algorithm can be easily applicable for "n" objectives. To clarify the readers, we highlighted the steps in the algorithm as "Need user input" where the program prompt messages for input for further execution accordingly.

Table 1: Algorithm: Multi-objective Bayesian optimized human assessed multi-target generated spectral recommender systems (MOBO-HAM-SRS).

1. **Segmentation of local image patches**:
   a. Choose a material sample. Scan topographic image on the microscope (eg Atomic Force Microscope (AFM)).
   b. Segment the image into square patches of window size, w.

2. **Initialization for MOBO:** State maximum MOBO iteration, M. Set number of targets (objectives) = 2. Randomly select j samples (image patches), X.
   a. For sample i in j, get **spectral data**, S_i, measured from microscope.
   b. (Need user input) For each objective, **User votes S_i** with voting options, $v_{i,.}$: Bad(0), Good(1) and Very Good(2). Next follow either (c) or (d).
   c. (Need user input) User gets prompt to **select preference** of the new spectral into target as, $p_{i,.}$ (0-1 with 1 being highest)
   d. **Calculate targets**, $T_{i,1}$, $T_{i,2}$ as per eq. 1, 2 respectively and normalize them.

Copyright © 2023 by The United States Government

$$T_{i,1} = \frac{\left((1-p_{i,1})*\sum_{ii=1}^{i-1} v_{ii,1}*T_{i-1,1}\right)+\left(p_{i,1}*v_{i,1}*S_{i,1}\right)}{\left((1-p_{i,1})*\sum_{ii=1}^{i-1} v_{ii,1}\right)+\left(p_{i,1}*v_{i,1}\right)} \quad (1)$$

$$T_{i,2} = \frac{\left((1-p_{i,2})*\sum_{ii=1}^{i-1} v_{ii,2}*T_{i-1,2}\right)+\left(p_{i,2}*v_{i,2}*S_{i,2}\right)}{\left((1-p_{i,2})*\sum_{ii=1}^{i-1} v_{ii,2}\right)+\left(p_{i,2}*v_{i,2}\right)} \quad (2)$$

e. **Calculate human-augmented multi-descriptors**: For sample i in j, calculate the multiple descriptor functions, correspond to respective multiple targets in eq 1, 2; $Y_{i,1}$, $Y_{i,2}$ as below eq. 3, 4. ψ_1, ψ_2 are the structural similarity functions for respective current targets $T_{j,1}, T_{j,2}$ after user voted j samples; R is the reward parameter. ψ is computed from the function "structural_similarity" in "skimage.metrics" Python library.

$$Y_{i,1} = \psi_1(T_{j,1}, S_i) + v_{i,1} * R \quad (3)$$
$$Y_{i,2} = \psi_2(T_{j,2}, S_i) + v_{i,2} * R \quad (4)$$

f. **Build Dataset,** $D_j = \{X, Y\}$ with X is a matrix with shape $(j, w * w)$ and Y is the matrix with shape $(j, 2)$

Start MOBO. Set $k = 1$. For $k \leq M$

3. **Surrogate Modelling**: Fit multiple (in this case 2) independent Gaussian Process (GPM) models, each for a descriptor or objective function.

    a. Optimize the hyper-parameters of kernel functions for each of the surrogate models. The kernel function and the learning rate can be different for each GPs, however, in this paper for the proof of concept, we have considered a consistent kernel function for all the GPs in a single analysis.

4. **Posterior Predictions**: For each GP model, compute posterior means and variances for the unexplored locations.

5. **MOBO-based acquisition function:** Maximize hypervolume of a multi-objective acquisition function, with either aprioiri (in a single multi-objective function space) or posteriori (directly in a multi-dimensional function space) approach to select next best location, X_{j+k} for evaluations. In the MOBO setting, the next best location is explored in likelihood to find non-dominated solutions or Pareto solutions (potential optimal solutions for a given trade-off among the descriptors).

6. **Expensive Black-box evaluations:**

    a. (Need user input) **User interaction for target update:** User gets a prompt message if the user is satisfied with the current targets. User has option to choose, Yes or No. Mathematically, we can represent as $v_k =$

$$\begin{cases} 0 \ (No) \\ 1 \ (Yes) \end{cases}$$

In the current algorithm, the user does not have the flexibility to select "satisfy" on one target and continue updating on other targets which will be implemented in future.

    b. (Need user input) **Existing Human in the loop process:** Given $v_k = 0$, follow step 2 for sample patch X_{j+k} to generate spectra, perform the voting and accordingly update all the targets, and calculate all the respective descriptor or objective function values.

    c. **Switch to Fully Automated process**: Follow step 2(a) only to generate spectra for sample patch X_{j+k}. Given the user satisfied with the targets and now want to just explore the image space to learn the Pareto, we use this step to avoid redundant effort of the user voting and reduce complexity of the algorithm. Thus, given $v_k = 1$, the fixed targets become $T_1 = T_{j+k-1,1}$ and $T_2 = T_{j+k-1,2}$. With that, the user voted components in eq 3 and 4 is neglected (as we are won't learn targets anymore from current state) as per below.

$$Y_{j+k,1} = \psi_1(T, S_k) \quad (5)$$
$$Y_{j+k,2} = \psi_2(T, S_k) \quad (6)$$

It is to be noted, since the descriptor functions change with the switching, all the prior evaluated samples need to be re-evaluated again which is computationally very cheap as we already have the captured spectra measured at those locations from the expensive experiments.

7. **Augmentation:** Augment data, $D_{j+k} = [D_{j+k-1}; \{X_{j+k}, Y_{j+k,2}\}$.

3. RESULTS AND DISCUSSION

In this section, we demonstrate the MOBO-HAM-SRS design architecture as the proof of concept and analyze with different kernel and acquisition functions, as the goal to achieve the better Pareto solutions (maximizing hypervolume) over the multi-functional space. For the case study, we considered on two 200nm-thick PbTiO3 (PTO) thin film samples grown on (110) SrTiO3 via pulsed laser deposition, with 'designer' grain boundaries fabricated by a process outlined in [59]. With both samples imaged displayed different domain patterns with different material properties, priorly unknown to the experimentalists, this enables us to test the MOBO-HAM-SRS on these samples. It is to be highlighted if the experimentalist already knows about the target material property to look at before the analysis, then a standard MOBO-based AE is sufficient (without the need of any human intervention), which is not the scope of the study for this paper. For this paper, we refer to PTO material sample 1 as the sample where the domains imaged are on the original (110) oriented STO crystal (thickness 500um), and PTO material sample 2 as the sample

Copyright © 2023 by The United States Government

where we image the region of the sample where the sample is rotated (~2 degrees) with respect to the substrate and has a much lower thickness of the STO (and thus likely to be much less strained). For the MOBO analysis, in regard to incorporating the local image patches as additional channel for structure-spectra learning, we considered the image patch of window size, $w = 4\ px$. Thus, the dimension of each input, $\boldsymbol{X_1}$, is an array of 16 elements.

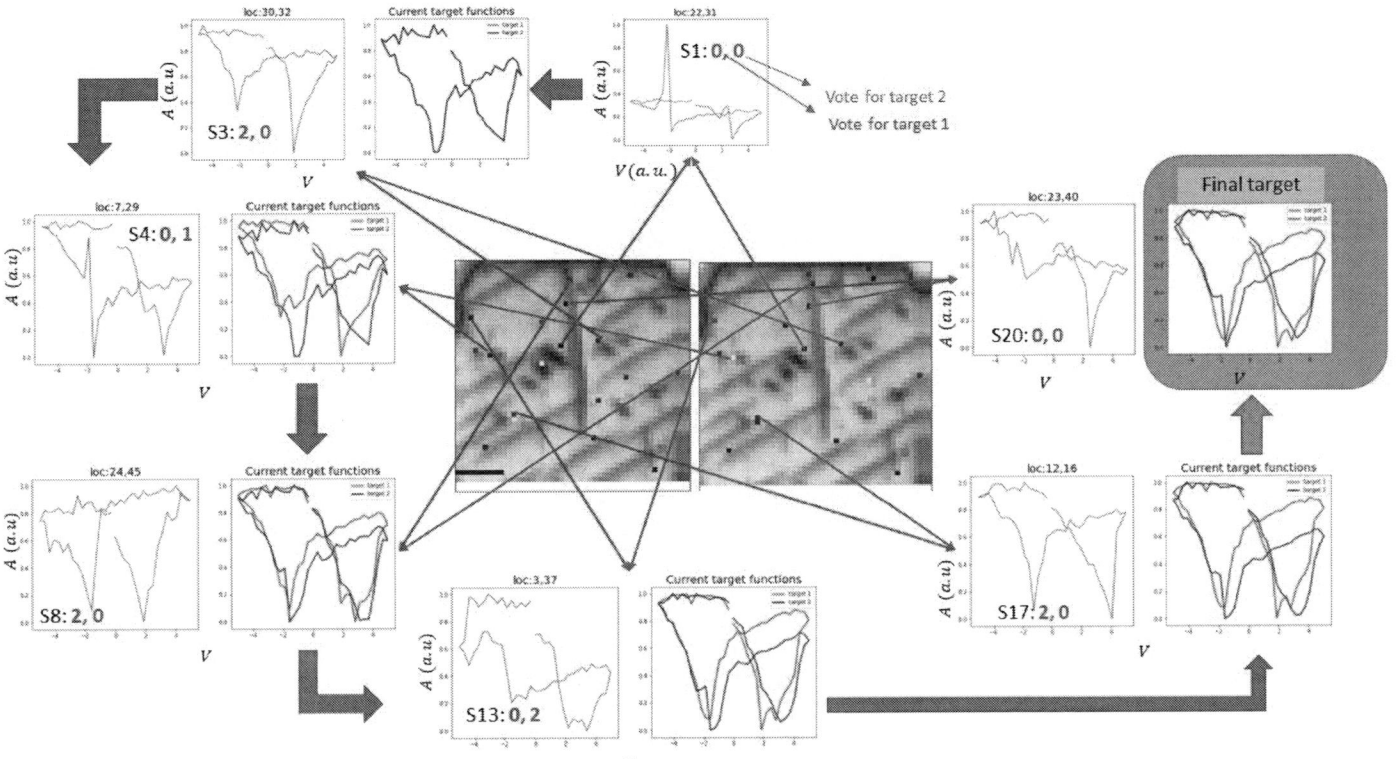

FIGURE 2: Analysis for PTO material sample 1: Evolution of multiple targets through visual inspections in MOBO-HAM-SRS architecture. The middle figures are the downsampled (50x50) PFM amplitude image of the PTO film, with the square dots represent the initial 20 locations where the user voted the respective captured spectra. The color coding of the square dots within the material image space represents the measures of votes as red:2, yellow:1 and blue:0. The large blue arrows represent the sequence of the voting with iterations, ultimately to achieve the final targets (highlighted in green box). The black line within the PFM amplitude image is the scale bar of 200 nm of the image size of 1x1 micrometer.

In these examples, we set the problem to two targets, however the architecture can be easily applicable to n targets. As stated, without any prior knowledge to provide any design decision of achievable targets, we explore and vote the spectral through visual inspection for the initial 20 samples. For the simplicity in the case studies, we kept this number to be consistent for target learning, however, the algorithm has no such boundaries, and the user has the flexibility to keep learning and update the target (within the MOBO) till satisfied (Table 1, step 6a). Figure 2 shows an example of the evolution of the human assessed multi-targets, to optimize jointly. We can clearly see at the initial iteration, as the user initially upvoted a spectrum, the current targets have similar structure. However, with more exploration, the voting differs based on the curiosity of the user, thereby showcase the structural difference between the targets. For this study, we considered the target 1 to have symmetrical loops whereas the target 2 to have larger left loop area than the right loop. The third case with smaller left loop area than the right loop is downvoted for both cases. As stated, this is an example to demonstrate the analysis, and any logic (based on domain applications) and number of curiosities driven targets can be generated with human assessment in the proposed MOBO-HAM-SRS.

Copyright © 2023 by The United States Government

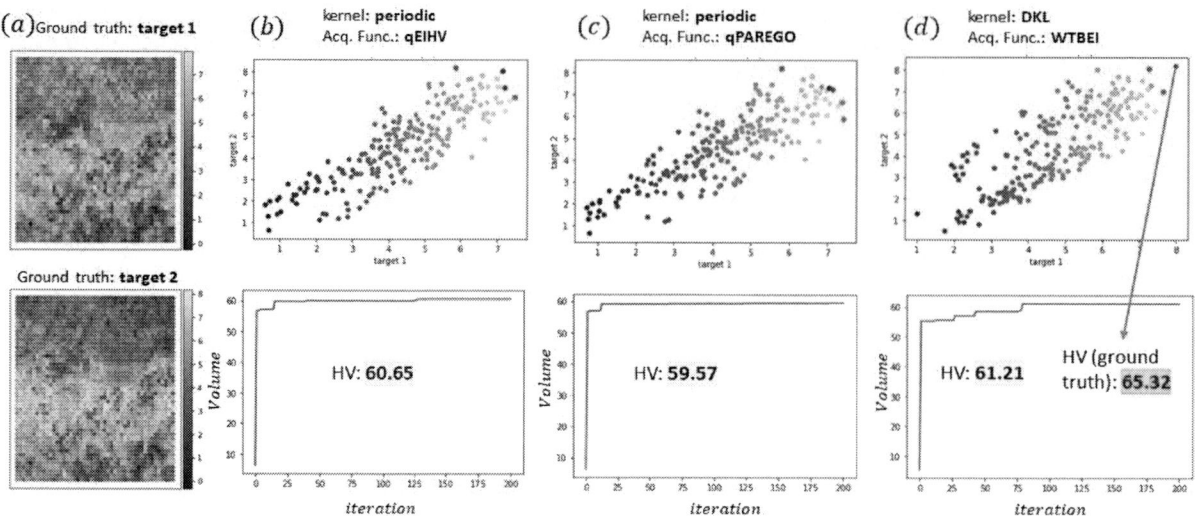

FIGURE 3: Comparative study of MOBO-HAM-SRS considering PTO material sample 1. (a) ground truth of descriptor values following eq (5), (6) exhaustively calculated at all the grid locations (50x50) over the downspampled material image space. The X and Y axes are the grid co-ordinate of the image. Top figures of (b), (c) and (d) are the exploration points in the function space and the final pareto solutions (top right colored points) with MOBO-HAM-SRS fitted with periodic kernel, qEIHV acquisition function; with periodic kernel, qPAREGO acquisition function; and deep kernel learning (DKL), WTBEI acquisition function. The respective bottom figures are the trajectory of the hypervolume (maximization) over MOBO iterations, with the final hypervolume values after convergence. The top rightmost green dot of figure (d) is the ground truth pareto solution computed from all exhaustively calculated descriptor values.

Once the user provides decision on the targets, follows Step 6c of the algorithm in Table 1. Thus, with the next task to optimize these learned human-assessed multiple targets jointly, we run MOBO-HAM-SRS for 200 iterations (convergence criteria) in fully automated process and identify the pareto frontier. It is to be noted that it is up to the user decision when to stop the human assessment (at what iteration) and switch to fully automated process, based on the cost of human operation, complexity of material image space and fulfillment of identifying multiple targets to explore and optimize jointly. For the purpose of the case study, we voted for initial 20 samples (see fig. 2) and switched to a fully automated process for the final 200 iterations. Figures 3 provides a comparative study, on PTO material sample 1, for MOBO-HAM-SRS with different GP kernels and multi-objective acquisition functions as the purpose for best model fitting. For this study, we use a standard periodic kernel [60], [61] from GPytorch library with hyper-parameters are optimized with Adam optimizer[62], and a deep kernel learning (DKL) [4], [37], [38] from open-source AtomAI (Python) software package [63]. Both posterior and apriori based multi-objective acquisition functions are used, such as "qExpected Improvement Hypervolume (qEIHV)" [41], "qPareto Efficient Global Optimization (qPAREGO)" [46] and "Weighted Tchebycheff based Expected Improvement (WTBEI)" [47]. For qEIHV and qPAREGO, we considered the batch size, q =1 to select single sample at each iteration as this does not provide any computational efficiency on selecting batch samples for our microscope measurements. In other words, it is the same effort to provide a single sample or in batches to feed into our microscope and perform characterization. However, the framework can be easily set to provide multiple samples in batch for parallel evaluation in the expensive systems (if multi-processing can be done). For the WTBEI method, we have set the utopia (the best solution for individual descriptors) for both functions as 10 based on the prior knowledge from descriptor formulation. The acquisition function is optimized using the discrete optimizer due to the input space is discrete (image location patches), and the computation is only feasible at those discrete input patches. To compute the hyper-volume indicator function, a reference point (worst solution), which is the lower bound of the objectives is required. We set this value (based on knowledge from the descriptor formulation) as $0 - \delta$. $\delta = 0.001$ is a small positive value to make the reference point slightly worse than true lower bound of the objectives. The performance measure is considered to be the optimal hypervolume (higher the better) after MOBO convergence. The proposed architecture can be easily fitted with other kernels and acquisition function, which is considered for future study. Figure 3(a) is the ground truth map of the descriptors as the structural similarities between human assessed target 1 and 2 (shown in figure 2) respectively, calculated exhaustively at each grid locations of the downspampled PFM amplitude image of the PTO film (PTO material sample 1). It is to be noted, in case of real-time analysis when the material image has higher resolution (say 256x256), the ground truth is unknown as exhaustive sampling (256^2) is infeasible due to high computing time and memory usage. Here, for the purpose of proof of concept, we pre-measured the spectra exhaustively, only after reducing the

Copyright © 2023 by The United States Government

resolution of the image. For validation, we compare the hypervolumes of the pareto fronts, optimized with MOBO-HAM-SRS (with 200 samples), at different kernels and acquisition functions, with the optimal hypervolume with the ground truth calculated from all the 2500 samples. We can see overall all the applied kernels and the acquisition functions perform quite decently, in comparison with the ground truth hypervolume, given the fact of <10% sampling (200 explored samples) as opposed to 2500 sampling to compute the ground truth. Moreover, we see that the architecture fitted with DKL and WTBEI acquisition function performs best in maximizing the hypervolume and provided the pareto frontier closest with the ground truth pareto solution (see fig 3 (d)).

Next, we run the similar analysis on PTO material sample 2. Similarly, figure 4(a) are the downsampled (50x50) PFM amplitude image of the PTO film, with the square dots represent the initial 20 locations where the user voted the respective captured spectra based on the same curiosity as per the previous analysis for target 1 and 2. Figure 4(b) shows the final targets to optimize, following the sequence of human assessments. Interestingly, we can clearly see the final target spectral structures for both the analysis are different even roughly with the similar thought process (target 1 to have symmetrical loops whereas the target 2 to have larger left loop area than the right loop). This is expected as for both the samples, the spectral quality and therefore the material characteristics are different. Thus, the proposed MOBO-HAM-SRS has the flexibility to optimize the finely tuned target spectra as achievable for a sample, rather than a rigid target.

Figure 4(c) is the ground truth, calculated as similar to previous analysis. 4(d)-(f) provides the comparative study of MOBO-HAM-SRS performance with periodic and qEIHV, periodic and qPAREGO, and DKL and WTBEI. Overall, we see the exploration seen in the multi-function space is more towards the potential pareto frontier and lesser on the regions farther away from the potential pareto frontier. This is an expected behavior of the adaptive sampling in the MOBO setting for likelihood to find non-dominated or pareto solutions. However, the MOBO-HAM-SRS with periodic kernel with qEIHV acquisition function performed the worst and, like previous analysis, the architecture with DKL and WTBEI acquisition function performed the best. In the process of meta-learning-based optimization such as in BO or MOBO with expensive experiments over the complex material systems (unlike any other numerical optimization with negligible evaluation cost), it is extremely hard to exact match with all the global optimal solutions, given that we often have the capacity to only explore less than 5 % of total grid points. Thus, the performance metric is set up to how much we can improve our learning, given the limited cost to explore, as the BO or MOBO is designed for that. Thus, we aim here to minimize the distance between the optimized hypervolume and the true hypervolume as the cost exhausts. However in this case, plotting the pareto frontier with the ground truth, we see two final pareto solutions (after 200 iterations) exactly matches with the respective ground truth, which is an impressive outcome given with <10% exploration over the search space.

Copyright © 2023 by The United States Government

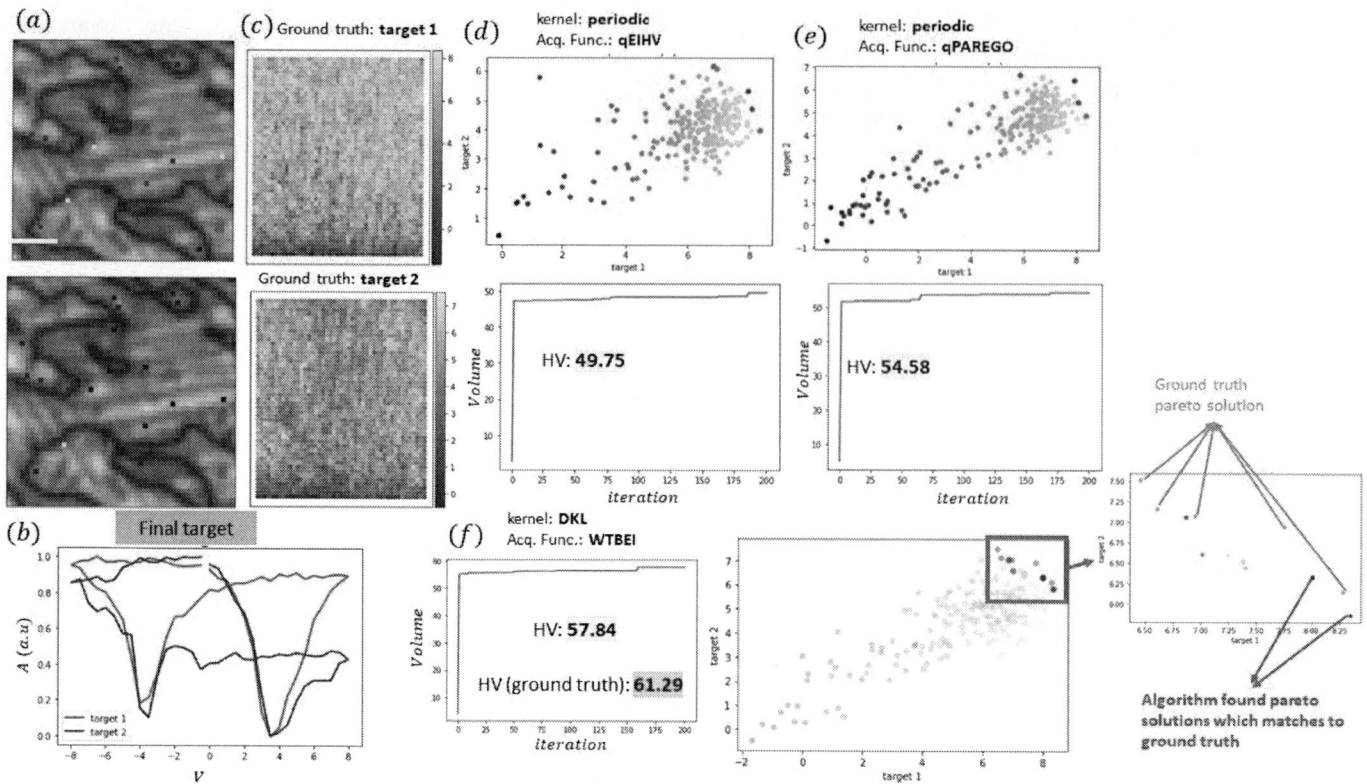

FIGURE 4: Comparative study of MOBO-HAM-SRS considering PTO material sample 2. (a) Downsampled (50x50) PFM amplitude image of the PTO film, with the square dots represent the initial 20 locations where the user voted the respective captured spectra. The color coding of the square dots within the material image space represents the measures of votes as red:2, yellow:1 and blue:0. (b) The final human assessed targets to optimize. (c) ground truth of descriptor values following eq (5), (6) exhaustively calculated at all the grid locations (50x50) over the downspampled material image space. (d), (e) and (f) are the exploration points in the function space, the final pareto solutions (top right colored points) and trajectories of the hypervolume with final hypervolume values after convergence of MOBO-HAM-SRS, fitted with periodic kernel, qEIHV acquisition function; with periodic kernel, qPAREGO acquisition function; and deep kernel learning (DKL), WTBEI acquisition function respectively. The zoomed figure in (f) plots the final pareto solutions from MOBO-HAM-SRS fitted with deep kernel learning (DKL), WTBEI acquisition function and the ground truth pareto solution computed from all exhaustively calculated descriptor values. The white line in (a) is the scale bar of 200 nm of the image size of 1x1 micrometer. The X and Y axes in (a) and (c) are the grid co-ordinates of the images.

4. CONCLUSION

In summary, we developed a human operated automated experiment (AE) workflow, extending the spectral recommender system (SRS) with human assessed multiple target generation (HAM) and subsequently human assessment reward-structural similarity (with respective targets) based multiple descriptors, fitted under the multi-objective Bayesian optimization (MOBO) setting for adaptive exploration in the multi-functional space and thereby rapid pareto discovery between human curiosity driven material properties. Together, we term this a *Multi-objective Bayesian optimized human assessed multi-target generated spectral recommender system (MOBO-HAM-SRS)*. This architecture is designed to provide a degree of flexibility to the experimentalists in appropriate decisions to choose multiple achievable material targets,

through sequence of exploration over the search space, to optimize jointly. As opposed to the proposed design architecture, the standard active learning-based AE workflows mostly restricted to the need of pre-defined targets, which additionally can be non-trivial to numerically formulate (such as loop-area, height, width, storage etc) from experimentally measured complex spectral structure (which are generally non-smooth than spectral generated from theoretical simulations). As for material synthesis or exploration over large material space, though an AI-driven automated experiments (without human in the loop) provide fast learning and eliminate rigorous steps, does not necessarily guarantee the best learning always. The paper attempted to bring the best of both human and AI in a collaborative approach for autonomous experimental workflow. At the early iterations of the optimization task, the human overpowers AI to steer the path of exploration in a

Copyright © 2023 by The United States Government

complex, large, multi-function space; and the AI overpower human at the later iterations to fully automate and speed the pareto learning among multiple human-finding targets. In this paper for the proof of concept, the prototype is presented and demonstrated on two PbTiO3 (PTO) thin film samples with a comparative study on different kernels and multi-objective acquisition functions. As in broader application, the approach can be utilized to conduct different material synthesis, characterization, and large space exploration, when the studied material system is unknown and have the potentiality to learn several unknown key material properties within the loop of the experiments. This can lead to discoveries through conducting human-guided adaptive experiments which could be out of reach due to limited prior knowledge (setting a fixed target before the optimization). As for any human-AI approaches, the framework has the limitation of assuming the human assessment is done by a subject matter expert to ensure proper guidance. As per the future work, the architecture will be connected directly to the microscope for real-time data acquisition and validated on different material samples, finally developing an open-source web app for usability to broad groups of designers, experimentalists, and material scientists.

ACKNOWLEDGEMENTS

The experiments, autonomous workflows and deep kernel learning was supported by the Center for Nanophase Materials Sciences (CNMS), which is a US Department of Energy, Office of Science User Facility at Oak Ridge National Laboratory. Algorithmic development was supported by the US Department of Energy, Office of Science, Office of Basic Energy Sciences, MLExchange Project, award number 107514; supported by the U.S. Department of Energy, Office of Science, Office of Basic Energy Sciences Energy Frontier Research Centers program under Award Number DE-SC0021118; and supported by University of Tennessee (Knoxville) start-up funding. J.-C.Y. and Y.-C.L. acknowledge support from National Science and Technology Council (NSTC), Taiwan, under grant no. NSTC-111-2628-M-006-005.

This manuscript has been authored by UT-Battelle, LLC, under Contract No. DE-AC0500OR22725 with the U.S. Department of Energy. The United States Government retains and the publisher, by accepting the article for publication, acknowledges that the United States Government retains a non-exclusive, paid-up, irrevocable, world-wide license to publish or reproduce the published form of this manuscript, or allow others to do so, for the United States Government purposes. The Department of Energy will provide public access to these results of federally sponsored research in accordance with the DOE Public Access Plan (http://energy.gov/downloads/doe-public-access-plan).

REFERENCES

[1] S. V. Kalinin *et al.*, "Automated and Autonomous Experiments in Electron and Scanning Probe Microscopy," *ACS Nano*, vol. 15, no. 8, pp. 12604–12627, Aug. 2021, doi: 10.1021/acsnano.1c02104.

[2] E. Stach *et al.*, "Autonomous experimentation systems for materials development: A community perspective," *Matter*, vol. 4, no. 9, pp. 2702–2726, Sep. 2021, doi: 10.1016/j.matt.2021.06.036.

[3] H. S. Stein and J. M. Gregoire, "Progress and prospects for accelerating materials science with automated and autonomous workflows," *Chem. Sci.*, vol. 10, no. 42, pp. 9640–9649, Sep. 2019, doi: 10.1039/c9sc03766g.

[4] K. M. Roccapriore, S. V. Kalinin, and M. Ziatdinov, "Physics Discovery in Nanoplasmonic Systems via Autonomous Experiments in Scanning Transmission Electron Microscopy," *Adv. Sci.*, vol. 9, no. 36, p. 2203422, 2022, doi: 10.1002/advs.202203422.

[5] M. Abolhasani and E. Kumacheva, "The rise of self-driving labs in chemical and materials sciences," *Nat. Synth.*, pp. 1–10, Jan. 2023, doi: 10.1038/s44160-022-00231-0.

[6] A. Krull, P. Hirsch, C. Rother, A. Schiffrin, and C. Krull, "Artificial-intelligence-driven scanning probe microscopy," *Commun. Phys.*, vol. 3, no. 1, Art. no. 1, Mar. 2020, doi: 10.1038/s42005-020-0317-3.

[7] J. C. Thomas *et al.*, "Autonomous scanning probe microscopy investigations over WS2 and Au{111}," *Npj Comput. Mater.*, vol. 8, no. 1, Art. no. 1, May 2022, doi: 10.1038/s41524-022-00777-9.

[8] R. J. Celotta, S. B. Balakirsky, A. P. Fein, F. M. Hess, G. M. Rutter, and J. A. Stroscio, "Invited Article: Autonomous assembly of atomically perfect nanostructures using a scanning tunneling microscope," *Rev. Sci. Instrum.*, vol. 85, no. 12, p. 121301, Dec. 2014, doi: 10.1063/1.4902536.

[9] S. Wang, J. Zhu, R. Blackwell, and F. R. Fischer, "Automated Tip Conditioning for Scanning Tunneling Spectroscopy," *J. Phys. Chem. A*, vol. 125, no. 6, pp. 1384–1390, Feb. 2021, doi: 10.1021/acs.jpca.0c10731.

[10] C. Conrad and D. W. Gerlich, "Automated microscopy for high-content RNAi screening," *J. Cell Biol.*, vol. 188, no. 4, pp. 453–461, Feb. 2010, doi: 10.1083/jcb.200910105.

[11] L. Jiang, L. Liu, J. Yao, and L. Shi, "A hybrid recommendation model in social media based on deep emotion analysis and multi-source view fusion," *J. Cloud Comput.*, vol. 9, no. 1, p. 57, Oct. 2020, doi: 10.1186/s13677-020-00199-2.

[12] F. P. Santos, Y. Lelkes, and S. A. Levin, "Link recommendation algorithms and dynamics of polarization in online social networks," *Proc. Natl. Acad. Sci.*, vol. 118, no. 50, p. e2102141118, Dec. 2021, doi: 10.1073/pnas.2102141118.

[13] G. N. P. S. Kranthi and B. V. Ram Kumar, "Online Social Voting Techniques in Social Networks Used for Distinctive Feedback in Recommendation Systems", Accessed: Feb. 10, 2023. [Online]. Available:

Copyright © 2023 by The United States Government

https://core.ac.uk/display/235197091?utm_source=pdf&utm_medium=banner&utm_campaign=pdf-decoration-v1

[14] K. J. Kanarik *et al.*, "Human–machine collaboration for improving semiconductor process development," *Nature*, pp. 1–5, Mar. 2023, doi: 10.1038/s41586-023-05773-7.

[15] A. Biswas *et al.*, "A dynamic Bayesian optimized active recommender system for curiosity-driven Human-in-the-loop automated experiments." arXiv, Apr. 05, 2023. doi: 10.48550/arXiv.2304.02484.

[16] M. Li and Z. Yin, "Debugging Object Tracking by a Recommender System with Correction Propagation," *IEEE Trans. Big Data*, vol. 3, no. 4, pp. 429–442, Dec. 2017, doi: 10.1109/TBDATA.2017.2723022.

[17] K. Suzuki *et al.*, "Fast material search of lithium ion conducting oxides using a recommender system," *J. Mater. Chem. A*, vol. 8, no. 23, pp. 11582–11588, Jun. 2020, doi: 10.1039/D0TA02556A.

[18] E. Brochu, V. M. Cora, and N. de Freitas, "A Tutorial on Bayesian Optimization of Expensive Cost Functions, with Application to Active User Modeling and Hierarchical Reinforcement Learning," *ArXiv10122599 Cs*, Dec. 2010, Accessed: Jan. 04, 2020. [Online]. Available: http://arxiv.org/abs/1012.2599

[19] D. R. Jones, M. Schonlau, and W. J. Welch, "Efficient Global Optimization of Expensive Black-Box Functions," *J. Glob. Optim.*, vol. 13, no. 4, pp. 455–492, Dec. 1998, doi: 10.1023/A:1008306431147.

[20] B. Shahriari, K. Swersky, Z. Wang, R. P. Adams, and N. de Freitas, "Taking the Human Out of the Loop: A Review of Bayesian Optimization," *Proc. IEEE*, vol. 104, no. 1, pp. 148–175, Jan. 2016, doi: 10.1109/JPROC.2015.2494218.

[21] M. Frean and P. Boyle, "Using Gaussian Processes to Optimize Expensive Functions," in *AI 2008: Advances in Artificial Intelligence*, W. Wobcke and M. Zhang, Eds., in Lecture Notes in Computer Science. Berlin, Heidelberg: Springer, 2008, pp. 258–267. doi: 10.1007/978-3-540-89378-3_25.

[22] A. N. Morozovska, E. A. Eliseev, A. Biswas, N. V. Morozovsky, and S. V. Kalinin, "Effect of surface ionic screening on polarization reversal and phase diagrams in thin antiferroelectric films for information and energy storage," *ArXiv210613096 Cond-Mat*, Jun. 2021, Accessed: Aug. 13, 2021. [Online]. Available: http://arxiv.org/abs/2106.13096

[23] S. Greenhill, S. Rana, S. Gupta, P. Vellanki, and S. Venkatesh, "Bayesian Optimization for Adaptive Experimental Design: A Review," *IEEE Access*, vol. 8, pp. 13937–13948, 2020, doi: 10.1109/ACCESS.2020.2966228.

[24] T. Ueno, T. D. Rhone, Z. Hou, T. Mizoguchi, and K. Tsuda, "COMBO: An efficient Bayesian optimization library for materials science," *Mater. Discov.*, vol. 4, pp. 18–21, Jun. 2016, doi: 10.1016/j.md.2016.04.001.

[25] L. Kotthoff, H. Wahab, and P. Johnson, "Bayesian Optimization in Materials Science: A Survey," *ArXiv210800002 Cond-Mat Physicsphysics*, Jul. 2021, Accessed: Aug. 06, 2021. [Online]. Available: http://arxiv.org/abs/2108.00002

[26] S. V. Kalinin, M. Ziatdinov, and R. K. Vasudevan, "Guided search for desired functional responses via Bayesian optimization of generative model: Hysteresis loop shape engineering in ferroelectrics," *J. Appl. Phys.*, vol. 128, no. 2, p. 024102, Jul. 2020, doi: 10.1063/5.0011917.

[27] A. N. Morozovska, E. A. Eliseev, A. Biswas, H. V. Shevliakova, N. V. Morozovsky, and S. V. Kalinin, "Chemical control of polarization in thin strained films of a multiaxial ferroelectric: Phase diagrams and polarization rotation," *Phys. Rev. B*, vol. 105, no. 9, p. 094112, Mar. 2022, doi: 10.1103/PhysRevB.105.094112.

[28] R.-R. Griffiths and J. M. Hernández-Lobato, "Constrained Bayesian Optimization for Automatic Chemical Design." arXiv, Aug. 12, 2019. doi: 10.48550/arXiv.1709.05501.

[29] A. Biswas and C. Hoyle, "An Approach to Bayesian Optimization for Design Feasibility Check on Discontinuous Black-Box Functions," *J. Mech. Des.*, vol. 143, no. 3, Feb. 2021, doi: 10.1115/1.4049742.

[30] W. Chu and Z. Ghahramani, "Extensions of gaussian processes for ranking: semisupervised and active learning," *Learn. Rank*, p. 29, 2005.

[31] L. L. Thurstone, "A law of comparative judgment," *Psychol. Rev.*, vol. 34, no. 4, pp. 273–286, 1927, doi: 10.1037/h0070288.

[32] F. Mosteller, "Remarks on the Method of Paired Comparisons: I. The Least Squares Solution Assuming Equal Standard Deviations and Equal Correlations," in *Selected Papers of Frederick Mosteller*, S. E. Fienberg and D. C. Hoaglin, Eds., in Springer Series in Statistics. New York, NY: Springer, 2006, pp. 157–162. doi: 10.1007/978-0-387-44956-2_8.

[33] C. C. Holmes and L. Held, "Bayesian auxiliary variable models for binary and multinomial regression," *Bayesian Anal.*, vol. 1, no. 1, pp. 145–168, Mar. 2006, doi: 10.1214/06-BA105.

[34] J. Dhamala *et al.*, "Embedding high-dimensional Bayesian optimization via generative modeling: Parameter personalization of cardiac electrophysiological models," *Med. Image Anal.*, vol. 62, p. 101670, May 2020, doi: 10.1016/j.media.2020.101670.

[35] A. Grosnit *et al.*, "High-Dimensional Bayesian Optimisation with Variational Autoencoders and Deep Metric Learning." arXiv, Nov. 01, 2021. doi: 10.48550/arXiv.2106.03609.

[36] A. Biswas, R. Vasudevan, M. Ziatdinov, and S. V. Kalinin, "Optimizing training trajectories in variational autoencoders via latent Bayesian optimization approach*," *Mach. Learn. Sci. Technol.*, vol. 4, no. 1, p. 015011, Feb. 2023, doi: 10.1088/2632-2153/acb316.

Copyright © 2023 by The United States Government

[37] A. G. Wilson, Z. Hu, R. Salakhutdinov, and E. P. Xing, "Deep Kernel Learning." arXiv, Nov. 06, 2015. doi: 10.48550/arXiv.1511.02222.

[38] M. Ziatdinov, Y. Liu, and S. V. Kalinin, "Active learning in open experimental environments: selecting the right information channel(s) based on predictability in deep kernel learning." arXiv, Mar. 18, 2022. doi: 10.48550/arXiv.2203.10181.

[39] P. Feliot, J. Bect, and E. Vazquez, "A Bayesian approach to constrained single- and multi-objective optimization," *J. Glob. Optim.*, vol. 67, no. 1, pp. 97–133, Jan. 2017, doi: 10.1007/s10898-016-0427-3.

[40] D. Khatamsaz, L. Peddareddygari, S. Friedman, and D. Allaire, "Bayesian Optimization of Multiobjective Functions Using Multiple Information Sources," *AIAA J.*, vol. 59, no. 6, pp. 1964–1974, Jun. 2021, doi: 10.2514/1.J059803.

[41] M. Abdolshah, A. Shilton, S. Rana, S. Gupta, and S. Venkatesh, "Expected Hypervolume Improvement with Constraints," in *2018 24th International Conference on Pattern Recognition (ICPR)*, Aug. 2018, pp. 3238–3243. doi: 10.1109/ICPR.2018.8545387.

[42] K. Yang, M. Emmerich, A. Deutz, and T. Bäck, "Multi-Objective Bayesian Global Optimization using expected hypervolume improvement gradient," *Swarm Evol. Comput.*, vol. 44, pp. 945–956, Feb. 2019, doi: 10.1016/j.swevo.2018.10.007.

[43] Z. Wang and S. Jegelka, "Max-value Entropy Search for Efficient Bayesian Optimization," *ArXiv170301968 Cs Math Stat*, Jan. 2018, Accessed: Apr. 28, 2020. [Online]. Available: http://arxiv.org/abs/1703.01968

[44] D. Hernández-Lobato, J. M. Hernández-Lobato, A. Shah, and R. P. Adams, "Predictive Entropy Search for Multi-objective Bayesian Optimization," *ArXiv151105467 Stat*, Feb. 2016, Accessed: Apr. 28, 2020. [Online]. Available: http://arxiv.org/abs/1511.05467

[45] J. Knowles, "ParEGO: a hybrid algorithm with on-line landscape approximation for expensive multiobjective optimization problems," *IEEE Trans. Evol. Comput.*, vol. 10, no. 1, pp. 50–66, Feb. 2006, doi: 10.1109/TEVC.2005.851274.

[46] S. Daulton, M. Balandat, and E. Bakshy, "Differentiable expected hypervolume improvement for parallel multi-objective Bayesian optimization," in *Proceedings of the 34th International Conference on Neural Information Processing Systems*, in NIPS'20. Red Hook, NY, USA: Curran Associates Inc., Dec. 2020, pp. 9851–9864.

[47] A. Biswas, C. Fuentes, and C. Hoyle, "A MO-BAYESIAN OPTIMIZATION APPROACH USING THE WEIGHTED TCHEBYCHEFF METHOD," *J. Mech. Des.*, pp. 1–30, Jul. 2021, doi: 10.1115/1.4051787.

[48] A. Biswas, C. Fuentes, and C. Hoyle, "A Nested Weighted Tchebycheff Multi-Objective Bayesian Optimization Approach for Flexibility of Unknown Utopia Estimation in Expensive Black-Box Design Problems," *J. Comput. Inf. Sci. Eng.*, vol. 23, no. 1, May 2022, doi: 10.1115/1.4054480.

[49] A. Tran, M. Eldred, S. McCann, and Y. Wang, "srMO-BO-3GP: A Sequential Regularized Multi-Objective Constrained Bayesian Optimization for Design Applications," presented at the ASME 2020 International Design Engineering Technical Conferences and Computers and Information in Engineering Conference, American Society of Mechanical Engineers Digital Collection, Nov. 2020. doi: 10.1115/DETC2020-22184.

[50] A. Biswas, A. N. Morozovska, M. Ziatdinov, E. A. Eliseev, and S. V. Kalinin, "Multi-objective Bayesian optimization of ferroelectric materials with interfacial control for memory and energy storage applications," *J. Appl. Phys.*, vol. 130, no. 20, p. 204102, Nov. 2021, doi: 10.1063/5.0068903.

[51] A. Solomou et al., "Multi-objective Bayesian materials discovery: Application on the discovery of precipitation strengthened NiTi shape memory alloys through micromechanical modeling," *Mater. Des.*, vol. 160, pp. 810–827, Dec. 2018, doi: 10.1016/j.matdes.2018.10.014.

[52] A. M. Gopakumar, P. V. Balachandran, D. Xue, J. E. Gubernatis, and T. Lookman, "Multi-objective Optimization for Materials Discovery via Adaptive Design," *Sci. Rep.*, vol. 8, no. 1, p. 3738, Feb. 2018, doi: 10.1038/s41598-018-21936-3.

[53] S. Mehta, M. Goel, and U. D. Priyakumar, "MO-MEMES: A method for accelerating virtual screening using multi-objective Bayesian optimization," *Front. Med.*, vol. 9, p. 916481, 2022, doi: 10.3389/fmed.2022.916481.

[54] M. Sun, J. Xing, H. Meng, H. Wang, B. Chen, and J. Zhou, "MolSearch: Search-based Multi-objective Molecular Generation and Property Optimization," in *Proceedings of the 28th ACM SIGKDD Conference on Knowledge Discovery and Data Mining*, in KDD '22. New York, NY, USA: Association for Computing Machinery, Aug. 2022, pp. 4724–4732. doi: 10.1145/3534678.3542676.

[55] R. C. Rufato, Y. Diouane, J. Henry, R. Ahlfeld, and J. Morlier, "A mixed-categorical data-driven approach for prediction and optimization of hybrid discontinuous composites performance," in *AIAA AVIATION 2022 Forum*, American Institute of Aeronautics and Astronautics. doi: 10.2514/6.2022-4037.

[56] J. A. G. Torres et al., "A Multi-Objective Active Learning Platform and Web App for Reaction Optimization," *J. Am. Chem. Soc.*, vol. 144, no. 43, pp. 19999–20007, Nov. 2022, doi: 10.1021/jacs.2c08592.

[57] S. Daulton, D. Eriksson, M. Balandat, and E. Bakshy, "Multi-Objective Bayesian Optimization over High-Dimensional Search Spaces." arXiv, Jun. 15, 2022. doi: 10.48550/arXiv.2109.10964.

[58] Y. Zhang, S. Ghosh, T. Vandeputte, and L. Wang, "Bayesian Optimization for Multi-Objective High-Dimensional Turbine Aero Design," presented at the

Copyright © 2023 by The United States Government

ASME Turbo Expo 2021: Turbomachinery Technical Conference and Exposition, American Society of Mechanical Engineers Digital Collection, Sep. 2021. doi: 10.1115/GT2021-59745.

[59] P.-C. Wu *et al.*, "Twisted oxide lateral homostructures with conjunction tunability," *Nat. Commun.*, vol. 13, no. 1, Art. no. 1, May 2022, doi: 10.1038/s41467-022-30321-8.

[60] N. HajiGhassemi and M. Deisenroth, "Analytic Long-Term Forecasting with Periodic Gaussian Processes," in *Proceedings of the Seventeenth International Conference on Artificial Intelligence and Statistics*, PMLR, Apr. 2014, pp. 303–311. Accessed: Mar. 03, 2023. [Online]. Available: https://proceedings.mlr.press/v33/hajighassemi14.html

[61] P. Suksawang and J. Mekparyup, "New Kernel Function in Gaussian Processes Model," vol. 7, no. 7, 2017.

[62] D. P. Kingma and J. Ba, "Adam: A Method for Stochastic Optimization." arXiv, Jan. 29, 2017. doi: 10.48550/arXiv.1412.6980.

[63] M. Ziatdinov, A. Ghosh, C. Y. (Tommy) Wong, and S. V. Kalinin, "AtomAI framework for deep learning analysis of image and spectroscopy data in electron and scanning probe microscopy," *Nat. Mach. Intell.*, vol. 4, no. 12, Art. no. 12, Dec. 2022, doi: 10.1038/s42256-022-00555-8.

Copyright © 2023 by The United States Government

Proceedings of the ASME 2023
International Design Engineering Technical Conferences and
Computers and Information in Engineering Conference
IDETC-CIE2023
August 20-23, 2023, Boston, Massachusetts

DETC2023-116984

UNDERSTANDING THE RELATION BETWEEN DESIGNER SEARCH STRATEGIES AND DESIGNER LEARNING DURING DESIGN SPACE EXPLORATION

Hyeonik Song
Aerospace Engineering
Texas A&M University
College Station, TX
hyeoniksong@tamu.edu

Daniel Selva
Aerospace Engineering
Texas A&M University
College Station, TX
dselva@tamu.edu

ABSTRACT

Design Space Exploration (DSE) is a knowledge acquisition technique used from early-stage to detailed engineering design. During DSE, designers systematically generate a range of design alternatives and compare them by their design criteria. DSE allows designers to learn about the design problem, e.g., about design decisions or features that are more common among good designs than among other designs ("driving features"). To help designers learn the driving features, AI-assisted design tools with advanced analysis techniques have been developed. Yet, there has been scant research attention on understanding designers' exploratory and learning behaviors during DSE, which has the potential to contribute to the development of more effective AI-assisted design tools. To address the research gap, we examine designer learning behaviors in real-time using data from a human subject study (N=24) that studies human-AI collaboration in DSE. We examine design search strategies (i.e., number of generated designs, and scale of design moves made during design generation) and their relations to designer learning (i.e., knowledge about driving features, and ability to use driving features to generate designs). In this work, we show that the designer's ability to use driving features to create new designs is associated with their knowledge about the driving features. Also, we observe that designers who generate many designs (mindless creation) achieved lower designer learning than those who generated few designs. We discuss the findings' potential implications on the development of future AI-assisted design tools.

Keywords: Design Space Exploration, Designer Learning Behavior, Human-AI Collaboration, Artificial Intelligence

1. INTRODUCTION

Design Space Exploration (DSE) involves systematically comparing design alternatives across multiple criteria to generate a range of designs [1]. This process aids designers in discovering optimal design solutions and acquiring knowledge about the design problem, such as design decisions or features that *explain* good design solutions, i.e., features that are more common among good design solutions than among other designs ("driving features"). The primary goal of DSE is to evaluate a large number of designs and their multiple conflicting criteria to identify Pareto-optimal designs, i.e., a set of non-dominated designs whose performance in one criterion cannot be improved further without degrading performance in another criterion. The other equally important goal of DSE is to learn driving features that result from the generation of Pareto front designs. The learned driving features enable designers to justify high-stakes design decisions to stakeholders and efficiently address future DSE problems.

Recently, a group of researchers developed an AI cognitive assistant, called Daphne, to aid space system engineers in performing DSE studies on Earth-Observing Satellite Systems (EOSS) [2]. Daphne is equipped with visual analytics and data mining capabilities, and a natural language interface to answer user questions and suggest design insights. It assists designers during DSE by making information retrieval more efficient and by addressing information overload during the exploration of an immense amount of design information. Despite its advanced capabilities, Daphne has shown mixed success in assisting designers during DSE. For instance, a fully featured Daphne significantly improved the design performance of expert designers but did not improve their capability to learn driving features in a design space —and in fact a slight negative trend was

Copyright © 2023 by ASME

suggested [2]. In another small study, a version of Daphne which adapts to the designer's learning goals was developed. The premise behind this version was that DSE is an iterative hypothesis generation and testing process in which designers develop hypotheses about potential driving features as they explore the design space, generate designs to test those hypotheses, and update their beliefs based on the evidence added by the new designs. The adaptive agent's interface allowed designers to tell the agent the hypotheses they want to test. Then, the agent generated designs with/without the relevant feature as needed, performed the statistical test to check if the data supports their hypothesis and presented that information to the user. This adaptive agent was compared to a non-adaptive agent that simply allowed users to explore the *feature space* resulting from an association rule mining process [3], and a control version where designers had to learn the driving features by manually exploring the design space [4]. Results showed that learning was significantly higher for the adaptive condition vs control ($p = 0.0146$). No significant differences were found between the non-adaptive agent and the control. However, the human subject study did not show significant differences in learning between the adaptive agent and the non-adaptive agent. A potential explanation for the mixed success is that Daphne's support capabilities are not designed with an adequate understanding of how designers search for and learn driving features during DSE. It is suggested that Daphne and similar cognitive design assistants need to be developed with an understanding of the designers' cognitive processes, such as their exploratory and learning behaviors, to provide assistance that is aligned with the designer's cognition.

Therefore, the research objective of this paper is *to understand what are designers' search strategies during DSE and how the search strategies affect their knowledge about driving features.* Specifically, we explore the following research hypotheses:

Hypothesis 1: The extent to which designers use driving features to generate new designs during DSE can indicate whether or not they learned about the driving features.

Hypothesis 2: Design search strategies in DSE can be used to distinguish designers who learn the driving features from designers who do not.

We use data from a prior human subject study (N=24) that was designed to understand the collaboration between human designers and Daphne during DSE for a space mission design. The study was originally designed to understand how cognitive interventions that prompt designers to reflect on their DSE process enhance their design performance and learning. Although the study is aimed at a different research question, we use its data to test the current research hypotheses on design learning behaviors; it provides valuable insights into the search strategies of designers and their learning process during DSE.

The organization of the rest of the paper is as follows. Section 2 provides a review of relevant work on AI-assisted design and designer learning in DSE. Section 3 details Daphne and the methods used to examine design search strategies and designer learning. Section 4 presents the results, discussion, and limitations of the study. Finally, Section 5 presents the conclusion of the paper.

2. Related Works
2.1 Mixed Success of AI-Assisted Design

Artificial Intelligence (AI) has been increasingly prevalent in design [5-11]. However, recent cognitive studies have demonstrated mixed success of human-AI collaboration in design. Song et al. found that AI-assisted design teams exhibit improved coordination and communication, leading to better design outcomes [12]. Zhang et al. showed that AI suggestions can cause cognitive overload on designers. While low-performing design teams can benefit from AI, high-performing design teams can be negatively impacted by AI's suggestions [13]. A prior study using Daphne as a cognitive assistant found that expert designers at NASA JPL performed better in DSE tasks when using the fully featured Daphne cognitive assistant compared to a traditional DSE tool with basic interactive scatter plots and filtering features [2]. However, the study showed a negative trend in designer learning when assisted by the fully featured Daphne, highlighting the need to develop AI systems that can effectively collaborate with designers. Saremi and Bayrak found that a healthy balance of self-confidence and confidence in an intelligent system is needed for the best outcome of the collaboration [14].

In light of these mixed results, adaptive AI agents have been developed to improve human-AI collaboration in design. Viros-i-Martin and Selva built a Daphne AI assistant that adapts to the designer's learning goals [4]. They added a feature to Daphne that allows the designer to express his or her learning goals in the form of hypotheses about driving features (e.g., "good designs have a radar and a radiometer together in a dawn-dusk orbit"), which Daphne then tests by generating more designs with the feature of interest as needed and performing statistical hypothesis testing analysis. Daphne then presents the results of the analysis in qualitative and quantitative forms to designers. A human subject study was conducted to validate its efficacy but showed that adaptive Daphne was not better than non-adaptive Daphne in improving design performance and designer learning. Gyory et al. developed an AI agent that manages the design process of a team in real-time; it tracks action- and communication-based features of the team process and provides appropriate interventions to the design team [9]. The study showed that the adaptive AI agent performed equally well with a human manager in the context of overall team performance and intervention strategy.

2.2 Designer Learning in Design Space Exploration

Designer learning is an important component of DSE as what designers learn during design synthesis affects their choices and reformulation in future design iterations [15]. According to Sim and Duffy, design activities and learning activities are closely interlinked as both activities affect the designers' domain

Copyright © 2023 by ASME

knowledge [16, 17]. A design activity is an action (e.g., add an instrument to the spacecraft) taken to achieve a knowledge change during the design process and a learning activity reflects the knowledge change (e.g., adding an instrument increases the spacecraft weight which increases launch cost) triggered by the action. Researchers found that while better designers' knowledge of the design problem leads to better design performance [18, 19], better designs do not necessarily indicate a higher level of design knowledge [20]; some actions in DSE may be driven by the pursuit of generating the best designs while some actions may be driven by the testing of hypotheses related to driving features.

Prior studies have shown an intricate relationship between designer learning and design performance during DSE. Chaudhari et al. found that explicitly setting a design goal (i.e., goal to improve design performance) improves design outcomes at the cost of degrading learning, while explicitly setting a learning goal (i.e., goal to improve designer learning) improves the learning outcomes, at the cost of degrading design performance [21]. In another study, Chaudhari and Selva found that a high level of automation in DSE reduces designer learning which in turn reduces the potential for higher design performance [22]. The researchers suggest that while incremental design improvement is possible through design optimization only, global design improvement may require designer learning. The prior findings on designer learning suggest that additional research is necessary to gain a better understanding of how designers learn and how their learning can be enhanced during DSE.

Although designer learning plays a crucial role in DSE, there has been limited research on measuring it until recently. Bang and Selva proposed two approaches to quantify designers' learning outcomes during DSE [23]: a question-based approach based on Bloom's taxonomy [24], and a concept-map-based approach. The question-based approach measures designer learning based on knowledge dimensions (i.e., factual knowledge and conceptual knowledge) and cognitive process dimensions (i.e., recognizing driving features and comparing features). The concept-map-based approach is based on the number of concepts and relation added by the designer to a concept map during the activity. These approaches were compared to a subjective self-assessment of learning and to the performance of the designer in higher-level tasks requiring synthesis of designs or driving features. Results showed significant correlations between most of these measures except with design synthesis tasks. Chaudhari and Selva proposed another approach to measure designer learning based on Item Response Theory [22]. Although these measures have been utilized in multiple DSE studies, the method for obtaining designer learning is intrusive and cumbersome, indicating the need for automated assessments of designer learning. In this work, we explore a different approach that uses the designer's moves in the design space to predict whether designers learn driving features or not during DSE.

3. METHODOLOGY

We analyzed data collected from a human subject study (submitted for review to IDETC 2023) using statistical methods to test the research hypotheses. The study originally aimed to evaluate how different cognitive stimuli affect human designers' collaboration with Daphne, an AI agent that supports design space exploration. Although the study is designed to address a different research question, the data from the study provides valuable insight into how and what designers learn during an AI-assisted design space exploration task. This section provides details of the Daphne tool, the human subject study, and the metrics used to characterize designer learning in real-time and to analyze design behaviors.

3.1 Daphne

Daphne is a design space exploration tool designed to assist space system engineers in designing Earth Observing Satellite Systems (EOSS). It utilizes an AI cognitive assistant and consists of several components shown in Figure 1 that work together seamlessly to assist designers to gain insights about the satellite constellation designs in an objective space.

Shown in Figure 2, the front end of Daphne serves as the main user interface of the system, providing access to its primary features, including design space visualization, design synthesis and evaluation, and communication with the Daphne cognitive assistant. Users can interact with the system through a chatbox on the right-side panel, asking questions about the design problem and receiving natural language answers.

Daphne's question-answering system can play several roles, such as Engineer, Expert, Analyst, Critic, Explorer, and Historian. Each role specializes in answering specific types of questions or performing particular actions, such as answering queries about the models and design criteria used to evaluate satellite constellations, providing suggestions based on a database of expert's design heuristics, performing data mining to provide insights, giving feedback to improve a given design,

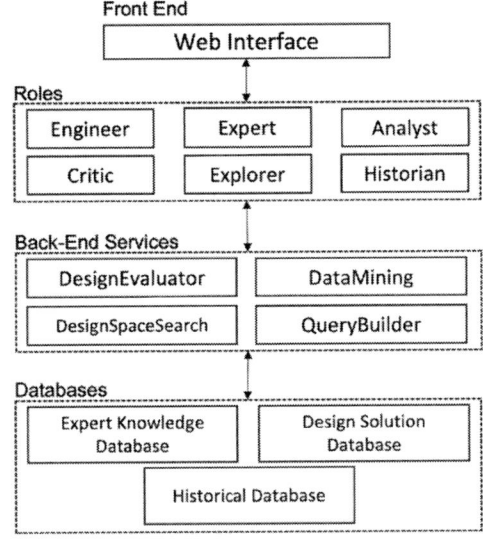

FIGURE 1: Daphne's architecture

Copyright © 2023 by ASME

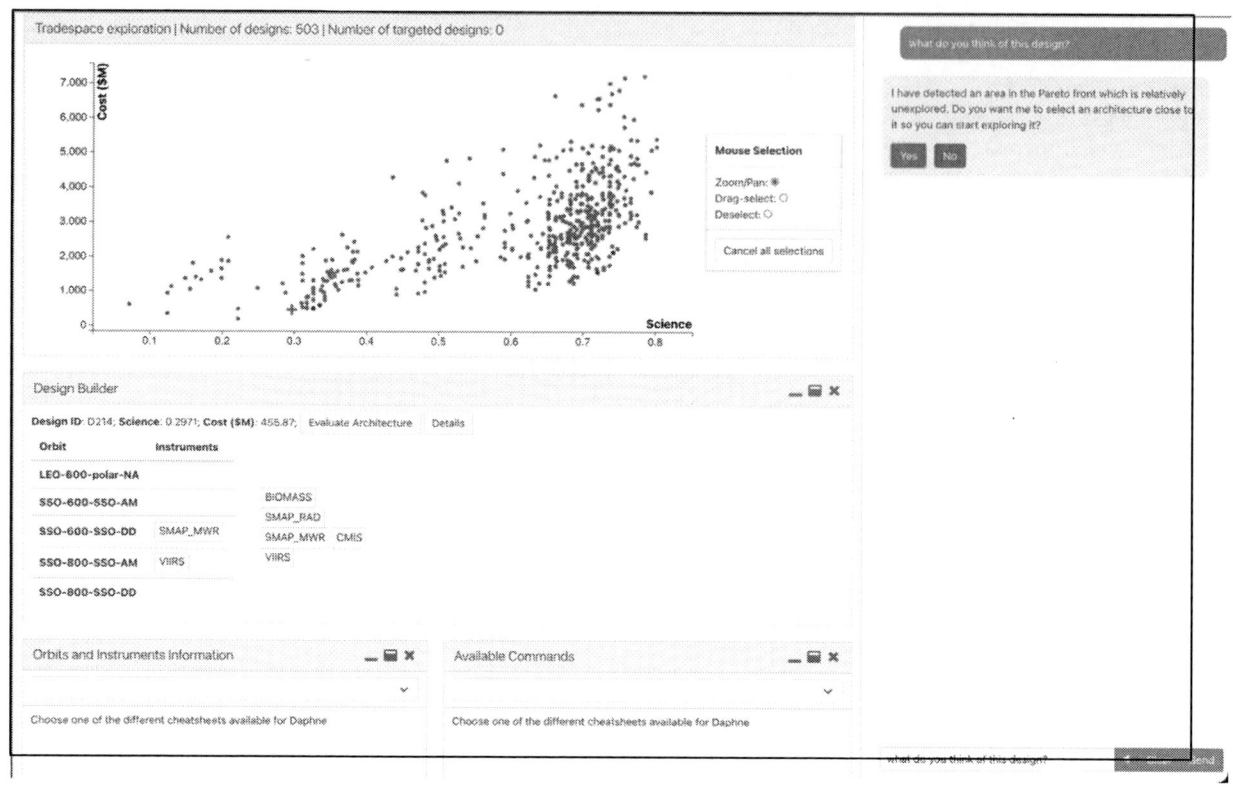

FIGURE 2: Daphne's interface

suggesting exploration of new areas of the design space, and answering queries by searching through past and planned space missions.

To process user commands, Daphne relies on four back-end services, including DesignEvaluator, DataMining, DesignSpaceSearch, and QueryBuilder. These services combine physics-based and other models in a rule-based system, search the current design space in the background, run data mining algorithms on the current design dataset, and translate natural language commands into valid SQL queries for Daphne's historical database.

Daphne's information comes from three databases, including Expert Knowledge Database, Design Solution Database, and Historical Database. The Expert Knowledge Database stores expert rules and recommendations, while the Design Solution Database stores all the designs in the current design space and their related information. The Historical Database contains detailed structured information about past and planned Earth observation missions obtained from the Committee on Earth Observation Satellites (CEOS) database. Reference [2] provides a more complete description of the Daphne tool.

3.2 Human-AI Collaboration Study

The data used in this paper comes from a prior human subject study that was conducted to examine how designers collaborate with Daphne on a space mission design task and to investigate whether cognitive interventions that prompt designers to reflect on their collaboration with the AI agent can enhance this collaboration.

The study was carried out at a public university in the United States, involving 24 students. Each participant received a $15 gift card as compensation for their participation, which lasted about an hour. The majority of the participants were at the graduate level, with only two undergraduates. 21 students identified as male, with three students identified as female. Six of the participants had prior experience with space mission design, either through coursework or research projects, while all of the participants had engineering design experience, which included coursework, research projects, or industry experience. All participants were given the pre-experiment tutorial necessary to perform the given design task to minimize the effects of their prior experiences. During the experiment, one participant was disconnected from Daphne's server. As result, the participant's data were excluded from the data analysis.

During a 30-minute human subject study, participants were given a task to collaborate with Daphne to generate satellite constellation designs for monitoring soil moisture [25]. They were provided a set of five candidate satellite instruments (Table 1) that they could assign to any of the five candidate orbits (Table 2) to generate satellite constellation designs. Each orbit can host any number of satellite instruments from none to five resulting in a design space size of 2^{25} designs. The newly generated designs appear on Daphne's tradespace scatter plot indicating

Copyright © 2023 by ASME

TABLE 1: Candidate Instruments

Instrument	Description
VIIRS	Visible and Infrared Atmospheric Sounder
CMIS	Conically Scanning Microwave Radiometer (Imaging and Sounding)
SMAP_RAD	L-band Synthetic Aperture Radar
SMAP_MWR	L-band Radiometer
BIOMASS	P-band Synthetic Aperture Radar

TABLE 2: Candidate Orbits. SSO = Sun-Synchronous Orbit. LTAN = Local Time of the Ascending Node

Orbit	Description
LEO-600-polar	LEO with 90deg inclination at 600 km altitude
SSO-600-AM	SSO with morning LTAN at 600 km altitude
SSO-600-DD	SSO with dawn-dusk LTAN at 600 km altitude
SSO-800-AM	SSO with morning LTAN at 800 km altitude
SSO-800-DD	SSO with dawn-dusk LTAN at 800 km attitude

their science benefit scores and lifecycle costs; different instrument-orbit combinations result in different values. In the study, the participants were given a design space containing 400 pre-generated designs and a task to generate new designs that improve the Pareto front and learn about the instrument-orbit combinations that consistently appear in Pareto front designs ("driving features").

All participants were given a 10-minute tutorial in which they familiarized themselves with Daphne's DSE features. Due to the nature of the study which was originally intended to address a different research question, the participants were assigned to one of three conditions in which they were asked to perform the given design task in different procedural manners. Participants who were assigned to "Reflection" and "Incubation" groups worked on the design task for 15 minutes, paused for 10 minutes during which they performed a unique intervention task (a period of reflection or incubation), and returned to the design task to finish the exploration for additional 15 minutes. Participants assigned to the "Control" group worked on the design task for 30 minutes without intervention. Consequently, all participants were given 30 minutes for the design space exploration task. Although the procedures in different conditions may affect their overall design performance and designer learning, the study provided useful experimental data to understand the development of designer learning and design behaviors in real-time. Following the design task, participants were allotted 10 minutes to complete a quiz designed to evaluate their understanding of driving features.

3.3 Measuring Designer Learning

3.3.1 Quiz Score

The aim of design space exploration is to gain insights into the design problem by learning various "driving features" that lead to the generation of good designs, i.e., nondominated designs. The evaluation of designer learning metrics is crucial to assess participants' comprehension of the design problem and their capability to achieve desired performance outcomes through design decisions. In this work, designer learning is measured with a question-based approach proposed by Bang and

Selva [23], which involves a post-experiment quiz consisting of 20 multiple-choice questions. These questions comprised 10 design identification questions and 10 design pairwise comparison questions, based on Bang and Selva's nomenclature. According to the authors, the learning outcomes are measured along two dimensions: *knowledge* and *cognitive process*. The knowledge dimension includes factual and conceptual knowledge. The cognitive process dimension includes recognition, evaluation, and comparison of driving features, prediction of the effects of the driving features, analysis of existing designs, and generation of new designs. The design identification questions evaluated the participant's ability to identify or predict if a design, based on its design decisions, belongs to a target region (in this case, the Pareto front) of the objective space or not. The design comparison questions evaluated the participant's ability to determine which of the two competing designs is better in a given design criterion. An example for each question type is listed below.

- Design Identification: "Does this given satellite constellation design belongs to the Pareto front or not?"
- Design Comparison: "Which of the two constellation designs below, which have a similar cost of about $1,000M, has a higher science benefit score?"

Overall, designer learning was used to indicate participants' capacity to utilize their knowledge of the design problem in order to anticipate the results of given designs and determine the key features that enhance the Pareto front.

3.3.2 Driving Feature Index

A premise of this paper is that as designers explore the design space, they develop hypotheses about driving features, and some of their design moves are motivated by testing their hypotheses to determine whether those potential driving features are truly influential or not. In the context of our space mission design problem, these driving features are related to constellation designs that have one or more instrument-orbit combinations. There are a few possible reasons for using the same features repeatedly, including (1) testing a hypothesis to see how it affects various objectives and (2) leveraging a useful driving feature that has already been identified. It is conjectured that the extent to which a designer uses a driving feature can indicate whether the designer has learned it or not. In order to measure the designer learning based on participant-generated designs, we used a metric called Driving Feature Index (DFI):

$$DFI = \sum_{i=1}^{n} w_i \, r_i$$

where n is the number of driving features and r_i is the ratio of designs that satisfy the ith driving feature. This ratio is calculated by dividing the number of designs that satisfy the driving feature by the total number of designs generated by individual participants. Table 3 shows the list of driving features identified by a postdoctoral researcher focusing on engineering

Copyright © 2023 by ASME

TABLE 3: Normalized weights of driving features. See Tables 1 and 2 for acronyms

Driving Features	w_i
DF1: SMAP_MWR and SMAP_RAD in SSO-800-DD	0.106
DF2: SMAP_MWR and SMAP_RAD in SSO-800-AM	0.128
DF3: BIOMASS in SSO-800-DD	0.234
DF4: BIOMASS in SSO-800-AM	0.106
DF5: CMIS in Polar	0.298
DF6: VIIRS in Polar	0.128

design research methods and an aerospace engineering doctoral candidate focusing on space mission design. It is important to acknowledge that the manually generated list may not contain all driving features and presents an area for improvement for future study. The contributions of r_i to the overall DFI score were weighted by w_i which represents the weight of ith driving feature to solve the post-experiment quiz. This weight is calculated by dividing the number of questions that test the knowledge of ith driving feature by the total number of questions. This allows for the comparison of the new metric, DFI score to the quiz score for evaluating designer learning. Table 3 shows the weights of different driving features that are normalized to add up to 1.

3.4 Analyzing Design Behavior
3.4.1 Quantity of Designs
Prior studies on design space exploration have commonly used the quantity of newly created designs for measuring design effort [21]. This is because generating more designs increases the likelihood of discovering designs that outperform current ones and identifying driving features that contribute to the creation of good designs. In our study, we also adopt this approach and quantify the design effort by counting the number of designs created within the 30-minute design task.

3.4.2 Big Moves and Small Moves
Design space exploration can be considered as a sequence of "moves" in design space, where each move changes from the current design to the next one [26] . Borrowing from the local search literature [27] given a distance function in design space $d: (X, X) \rightarrow \mathbb{R}^+$ where X is the design space, we can define a move size $d(x_t, x_{t+1})$. Thus, a big move refers to a significant alteration in the overall architecture of a design whereas a small move refers to an incremental change in the design parameters in the architecture. A similarity function such as cosine similarity can be used instead of a distance function (i.e., high similarity = small move). Both big and small moves are important in design space exploration as they allow designers to balance exploration and exploitation [28] and explore design options and trade-offs that lead to the learning of driving features. To understand designers' exploratory behaviors, we propose a method to measure the degree to which designers make big and small moves and evaluate their effects on designer learning.

In order to determine how frequently a designer makes a big move or a small move during the design task, we compute the average cosine similarity value between the jth design and its m preceding designs. A high similarity value (small distance) suggests that the jth design is the result of an incremental change from its preceding designs, indicating a small move. Conversely, a low similarity value (large distance) implies that the jth design is significantly different from its preceding designs, indicating a big move. The determination of big moves and small moves includes the following steps:

1. For each individual j^{th} design, we computed the average cosine similarity of the design to its five preceding designs ($m = 5$). We acknowledge that the choice of $m = 5$ is experimental and could be improved for future studies. For designs that are created during the early phase of the design task (i.e., first five designs), the similarity was computed with any available preceding designs.

2. If the jth design has a similarity value exceeding the average of all similarity values, it is considered the result of a small move, whereas if it has a similarity value below the average, it is considered the result of a big move. We use such a threshold value for all participants to categorize their moves into either a big move or a small move to analyze design behavior.

4. RESULTS AND DISCUSSION
4.1 Characterizing Designer Learning using Driving Feature Index

Driving Feature Index (DFI) measures whether designers repeatedly create designs that satisfy driving features. We use this metric to characterize designer learning because generating designs with driving features could imply that the designer developed an understanding of those features over time. To evaluate the correlation between designer learning and DFI, we compared the quiz score and the final DFI, measured at the end of the design task (see Figure 3). An independent sample t-test was performed to show that the 11 participants who scored high on the quiz (higher than average quiz score of 76%) achieved significantly higher DFI than the 12 participants who scored low on the quiz (lower than average quiz score of 76%), $t(21) = 1.975$, $p = 0.031$. According to Spearman's rho test, there was

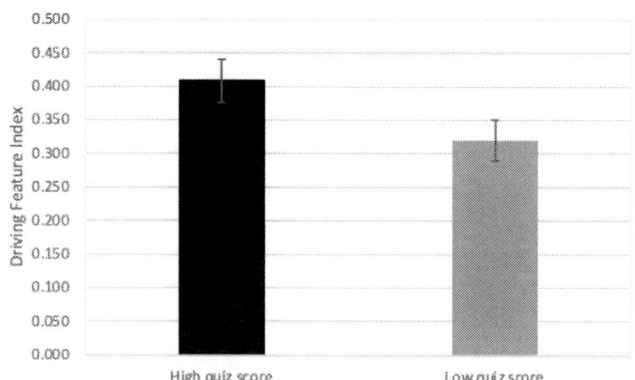

FIGURE 3: Driving Feature Index of designers who achieved high and low quiz scores

Copyright © 2023 by ASME

TABLE 4: Ratio of designs that satisfy each driving feature. See Tables 1 and 2 for acronyms.

Driving Features	r_i
DF1: SMAP_MWR and SMAP_RAD in SSO-800-DD	0.036
DF2: SMAP_MWR and SMAP_RAD in SSO-800-AM	0.108
DF3: BIOMASS in SSO-800-DD	0.395
DF4: BIOMASS in SSO-800-AM	0.359
DF5: CMIS in Polar	0.601
DF6: VIIRS in Polar	0.262

a positive correlation between the quiz score and DFI, $r(21) = 0.351, p = 0.05$.

Only one participant in the study was able to generate designs using all six driving features. The participant scored 95% on the quiz implying that it is important to learn multiple driving features to perform well on the quiz. The other participant who scored 95% on the quiz generated designs using five out of six driving features. The rest of the participants were able to generate designs with one or two driving features, yet they were able to score an average of 76% on the quiz. Interestingly, there was one participant who achieved a DFI of less than 0.1 but received 70% on the quiz.

We computed the average r_i, the ratio of designs that satisfy i^{th} driving feature for each individual driving feature (see Table 4). We observed that participants were generally successful in generating designs that meet DF5. However, only a small number of participants were able to generate designs that satisfy DF1 and DF2. This difference could be attributed to the compound characteristic of the driving features that necessitate assigning two instruments to a single orbit. Also, the difference could be due to their varying effects on the lifecycle cost and science benefit score of the generated designs. For instance, DF5 increases the satellite constellation designs' science benefit score more than DF1.

4.2 Designer Learning Decreases with Quantity of Designs

We observed various design strategies used by participants during the AI-assisted design space exploration task. For instance, some participants focused on generating new designs while some focused on retrieving design insights from the AI agent. Some participants generated ideas in an iterative manner and occasionally generated designs that they had already created before. Some participants took time to explore the current objective space to generate designs that are not yet explored. An independent sample t-test was performed to show that the 11 participants who scored high on the quiz (higher than average quiz score of 76%) generated a significantly lower number of designs than the 12 participants who scored low on the quiz (lower than average quiz score of 76%), $t(21) = -1.80$, $p = 0.046$. See Figure 4. We found that the number of designs created is negatively correlated to quiz scores, $r(21) = -0.408, p = 0.027$. We also examined the negative correlation between the number of designs created and DFI, $r(21) = -0.494, p = 0.008$.

In design space exploration, designers create new designs for various reasons, including to discover features that explain

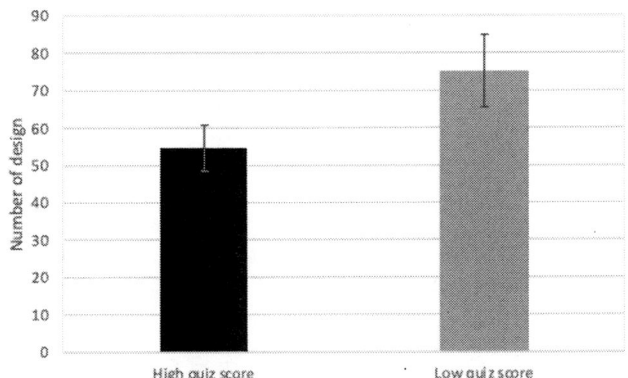

FIGURE 4: Number of designs created by designers who achieved high and low quiz scores

the region of some target objective space. The negative correlation result between the quiz score and the number of designs is interesting as it opposes the important purpose of generating new designs in design space exploration. The current finding could be explained by prior cognitive studies. One possible explanation is the cognitive load theory which describes the limited capacity of designers' working memory to learn new features [29]. While designers learn new features as they generate new designs, not all features learned are necessary for finding the right answers in the quiz. Another possible explanation is that some participants missed the opportunity to learn new features as they generated a large number of designs without reflecting on their choices and impacts on the objective space. This finding implies that it is important to think through when generating new designs.

Interestingly, the finding from this study differs from that of a prior study, which demonstrated that an increased number of generated designs led to an increased designer learning [30]. This suggests that if designers do not generate enough designs, they may not learn driving features or they learn wrong driving features. The findings from the current and prior studies taken together suggest that there could be "sweet spot" wherein generating too many designs (mindless creation) and generating too few designs can hinder designer learning.

In design space exploration study, the number of designs created is often used as an indicator to represent the design effort. However, the finding of this study suggests that this metric may be insufficient. To more accurately capture the designer's effort during the design process, it may be necessary to consider the time and steps taken to generate each design.

4.3 The Role of Small Moves and Big Moves

The similarity of individual designs to their five preceding designs was computed. The average similarity value across all designs was 0.52. This value was used as a threshold to distinguish designs that are generated by big moves and small moves. Analysis shows that during the 30-minute DSE task, participants generally used small moves (mean = 36.7, SD = 24.4) more than big moves (mean = 27.7, SD = 16.1). To

Copyright © 2023 by ASME

examine the effects of small moves and big moves on designer learning, we show Figures 5 and 6. Figure 5 shows that the small move proportion is positively correlated to DFI, $r(21) = 0.542, p = 0.004$. Figure 6 shows that the small move proportion is negatively correlated to the quiz score, but the correlation is not statistically significant, $r(21) = -0.206, p = 0.172$.

Figures 5 and 6 taken together present interesting findings about the role of small moves and big moves in design space exploration. First, Figure 5 shows that if a designer makes small moves, they have a higher chance of generating designs that satisfy driving features. Once a designer identifies a region of objective space that is explained by a driving feature, the designer tends to stay in the region and makes small incremental changes to solidify his or her understanding of one or two driving features. However, this also represents the missed opportunity to identify driving features that explain other regions of the objective space. For example, Figure 6 shows that increasing the proportion of small moves does not necessarily increase the quiz score. It could imply that big moves are also necessary to explore different driving features present in other regions of the objective space in order to improve the breadth of designer learning.

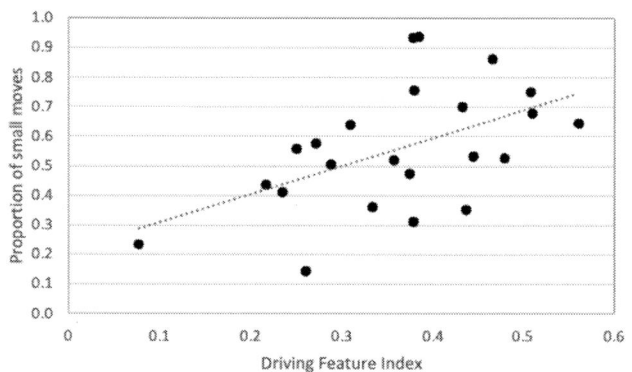

FIGURE 5: Small move proportion versus driving feature index

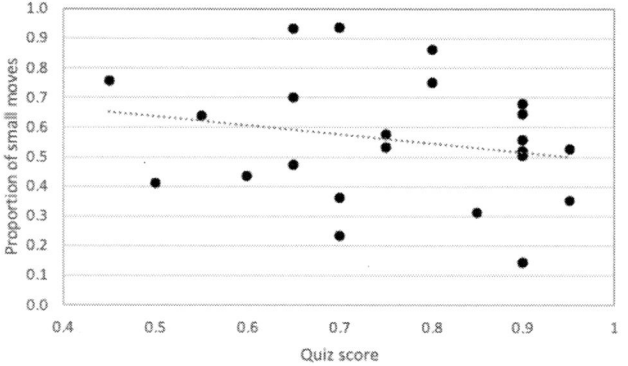

FIGURE 6: Small move proportion versus quiz score

4.4 Similar Results Across Participant Groups

There was no statistically significant difference among the participant groups (i.e., Reflection, Incubation, Control) for DFI ($F(2,20) = 1.127, p = 0.344$), number of designs created ($F(2,20) = 0.921, p = 0.414$), and small move proportion ($F(2,20) = 0.773, p = 0.475$). The findings imply that the results presented in this work are not significantly influenced by the experimental conditions.

4.5 Implications for Developing Adaptive AI Agents

This work presents methodologies to understand designer learning and behaviors during the design space exploration task. In this section, we discuss potential applications of the proposed methodologies in the development of adaptive AI agents that can use the information to provide tailored guidance to designers and enhance their learning. First, DFI could be used to present designers' current knowledge about driving features. If designers can repeatedly generate designs with specific driving features, the AI agent could confirm that the feature is indeed driving the performance of designs in the objective space and encourage them to identify other driving features that have not yet been identified. Second, the AI agent could cognitively support designers by guiding them to avoid the mindless creation of new designs. This behavior was especially seen among participants who do not have prior space mission design experience, implying that there is a need for training the engineering workforce in data-intensive design tasks. Perhaps, an AI agent could remind designers to maintain the design generation pace to reflect on what they have learned. Third, the AI agent could guide designers to make small moves or big moves to enhance both the depth and breadth of designer learning. In this work, we found that increased small moves lead to an increased generation of designs that satisfy driving features. Unfortunately, the finding from the current study does not provide adequate information about the effects of big and small moves on designer learning. This calls for a future study that involves an in-depth analysis of designers' cognitive process while using the big moves and small moves during the design space exploration task to address the limitation.

It is important to acknowledge several limitations of the proposed methodology. First, in order to obtain more precise and accurate statistical results, the methodology needs to be tested with a larger number of participants. Second, while participants received adequate training about DSE prior to the experiment, it should be noted that they had varying levels of experience with space mission design. Future work should therefore focus on understanding the effects of expertise level on the DSE task, with the goal of developing an AI agent that adapts to the user's domain knowledge and technical experience. Furthermore, it is necessary for future research to assess the additional search strategies utilized by designers that could potentially be significant, as well as how these strategies affect their performance and learning outcomes.

Copyright © 2023 by ASME

5. CONCLUSION

In this work, we proposed different methodologies to examine designers' behaviors that possibly lead to improved designer learning. We examined designers' abilities to generate designs using driving features and compared the result with their quiz scores. We also used the number of designs generated and the proportion of small moves made during the design task to distinguish designers who learned the important driving features and those who do not. In this work, we show that:

Hypothesis 1: Driving Feature Index (DFI), which measures the extent to which designers use driving features to generate new designs during DSE, can serve as an indicator of designer's knowledge of driving features. In contrast to, question-based evaluation method, this approach enables automated evaluation of designer learning.

Hypothesis 2: Designers who generate too many designs (mindless creation) achieved lower designer learning. This finding taken together with a finding from a prior work [30] suggests that generating too many or too few designs hinders designer learning.

Different design strategies lead to different levels of knowledge of driving features in an objective space. It is important to develop methodologies for understanding how designers learn during design space exploration. Such methodologies can provide the foundation for AI agents that dynamically adapt to designers' evolving knowledge during DSE and collaborate with designers in a more tailored manner.

ACKNOWLEDGEMENTS

This work is supported by the National Science Foundation, under grant CMMI 1907541. The United States Government retains, and by accepting the article for publication, the publisher acknowledges that the United States Government retains, a nonexclusive, paid-up, irrevocable, worldwide license to publish or reproduce the published form of this work, or allow others to do so, for United States Government purposes.

REFERENCES

[1] A. M. Ross and D. E. Hastings, "The Tradespace Exploration Paradigm," *INCOSE International Symposium,* vol. 15, no. 1, 2005, doi: https://doi.org/10.1002/j.2334-5837.2005.tb00783.x.

[2] A. Viros-i-Martin and D. Selva, "Daphne: A Virtual Assistant for Designing Earth Observation Distributed Spacecraft Missions," *IEEE Journal of Selected Topics in Applied Earth Observations and Remote Sensing,* vol. 13, pp. 30-48, 2020.

[3] H. Bang, Y. l. Z. Shi, G. Hoffman, S. Yoon, and D. Selva, "Exploring the feature space to aid learning in design space exploration," presented at the Design Computing and Cognition'18, 2018.

[4] A. Viros-i-Martin and D. Selva, "Improving Designer Learning in Design Space Exploration by Adapting to the Designer's Learning Goals," presented at the Proceedings of the Asme International Design Engineering Technical Conferences and Computers and Information in Engineering Conference, 2022.

[5] N. Knerr and D. Selva, "Cityplot: Visualization of High-Dimensional Design Spaces With Multiple Criteria," (in English), *J Mech Design,* vol. 138, no. 9, Sep 2016, doi: 10.1115/1.4033987.

[6] E. Kwon, F. Huang, and K. Goucher-Lambert, "Enabling multi-modal search for inspirational design stimuli using deep learning," *Ai Edam-Artificial Intelligence for Engineering Design Analysis and Manufacturing,* vol. 36, 2022.

[7] A. Ao, Y. Li, J. Gong, and S. Li, "Artificial Intelligence Design for Ship Structures: A Variant Multiple-Input Neural Network-Based Ship Resistance Prediction," *J Mech Design,* vol. 144, no. 9, p. 091707, 2022, doi: https://doi.org/10.1115/1.4053816.

[8] A. Berquand, A. Riccardi, F. Murdaca, and T. Soares, "Artificial Intelligence for the Early Design Phases of Space Missions," presented at the IEEE Aerospace, 2019.

[9] J. T. Gyory *et al.*, "Human Versus Artificial Intelligence: A Data-Driven Approach to Real-Time Process Management During Complex Engineering Design," (in English), *J Mech Design,* vol. 144, no. 2, Feb 1 2022, doi: 10.1115/1.4052488.

[10] B. Song, N. F. S. Zurita, H. Nolte, H. Singh, and J. Cagan, "When Faced With Increasing Complexity: The Effectiveness of Artificial Intelligence Assistance for Drone Design," *J Mech Design,* vol. 144, no. 2, 2022, doi: https://doi.org/10.1115/1.4051871.

[11] L. Chong, A. Raina, K. Goucher-Lambert, K. Kotovsky, and J. Cagan, "The Evolution and Impact of Human Confidence in Artificial Intelligence and in Themselves on AI-Assisted Decision-Making in Design," *J Mech Design,* 2022.

[12] B. Y. Song *et al.*, "Decoding the agility of artificial intelligence- assisted human design teams," (in English), *Design Stud,* vol. 79, Mar 2022, doi: 10.1016/j.destud.2022.101094.

[13] G. Zhang, A. Raina, J. Cagan, and C. McComb, "A cautionary tale about the impact of AI on human design teams," (in English), *Design Stud,* vol. 72, Jan 2021, doi: ARTN 100990 10.1016/j.destud.2021.100990.

[14] M. L. Saremi and A. E. Bayrak, "Agent-Based Simulation of Optimal Trust in a Decision Support System in One-On-One Collaboration," presented at the International Design Engineering Technical Conference, 2022.

[15] J. W. Kan and J. S. Gero, "Using the fbs ontology to capture semantic design information in design protocol studies," 2009.

[16] S. K. Sim and A. H. Duffy, "A foundation for machine learning in design," *Ai Edam-Artificial Intelligence for*

Copyright © 2023 by ASME

[17] S. K. Sim and A. H. Duffy, "Towards an ontology of generic engineering design activities," *Research in Engineering Design,* vol. 14, no. 4, pp. 200-223, 2003.

[18] M. Shergadwala, I. Billionis, K. N. Kannan, and J. H. Panchal, "Quantifying the impact of domain knowledge and problem framing on sequential decisions in engineering design," *J Mech Design,* vol. 140, no. 10, 2018.

[19] M. Shergadwala, K. N. Kannan, and J. H. Panchal, "Understanding the impact of expertise on design outcome: An approach based on concept inventories and item response theory," presented at the ASME International Design Engineering Technical Conferences, 2016.

[20] A. E. Bayrak and Z. Sha, "Integrating sequence learning and game theory to predict design decisions under competition," *J Mech Design,* vol. 143, no. 5, 2021.

[21] A. Chaudhari, R. Kumar, and D. Selva, "Supporting Designer Learning and Performance in Design Space Exploration: A Goal-Setting Approach," presented at the International Design Engineering Technical Conferences and Computers and Information in Engineering Conference, 2021.

[22] A. M. Chaudhari and D. Selva, "Evaluating Designer Learning and Peformance in Interactive Deep Generative Design," *J Mech Design,* vol. 145, 2023.

[23] H. Bang and D. Selva, "Measuring Human Learning in Design Space Exploration to Assess Effectiveness of Knowledge Discovery Tools," presented at the IInternational Design Engineering Technical Conferences and Computers and Information in Engineering Conference, 2020.

[24] B. S. Bloom, M. D. Engelhart, E. J. Furst, W. H. Hill, and D. R. Krathwohl, *Taxonomy of educational objectives: the classification of educational goals.* David McKay Company, 1956.

[25] D. Selva, B. G. Cameron, and E. F. Crawley, "Rule-Based System Architecting of Earth Observing Systems: Earth Science Decadal Survey," (in English), *J Spacecraft Rockets,* vol. 51, no. 5, pp. 1505-1521, Sep-Oct 2014, doi: 10.2514/1.A32656.

[26] C. McComb, J. Cagan, and K. Kotovsky, "Capturing Human Sequence-Learning Abilities in Configuration Design Tasks Thorugh Markov Chains," *J Mech Design,* vol. 139, no. 9, p. 091101, 2017.

[27] P. Hansen and N. Mladenovic, "Variable Neighborhood Search," in *Handbook of Metaheuristics,* F. Glover, Ed., 2003, ch. 145-184.

[28] K. Tabeau, G. Gemser, E. J. Hultink, and N. M. Wijnberg, "Exploration and Exploitation activities for design innovation," *Journal of Marketing Management,* vol. 33, pp. 203-225, 2017.

[29] J. Sweller, "Cognitive Load Theory," *Psychology of Learning and Motivation,* vol. 55, pp. 37-76, 2011.

[30] A. Viros-i-Martin and D. Selva, "Learning Comes from Experience: The Effects on Human Learning and Performance of a Virtual Assistant for Design Space Exploration," presented at the Design Computing and Cognition'20, 2022.

Copyright © 2023 by ASME

Proceedings of the ASME 2023
International Design Engineering Technical Conferences and
Computers and Information in Engineering Conference
IDETC-CIE2023
August 20-23, 2023, Boston, Massachusetts

DETC2023-116340

LARGE-SCALE PATH PLANNING IN COMPLEX ENVIRONMENTS BASED ON GENETIC ALGORITHM

Chuanhui Hu
IMPACT Laboratory
Dept. of Aerospace & Mechanical Engineering
University of Southern California
Los Angeles, California 90089
chuanhui@usc.edu

Yan Jin*
IMPACT Laboratory
Dept. of Aerospace & Mechanical Engineering
University of Southern California
Los Angeles, California 90089
yjin@usc.edu
(*corresponding author)

ABSTRACT

Path planning has been a hot research topic in robotics, and it is also a vital functionality for autonomous systems. Generating optimal-quality path plans with the least computing time has always been the goal of researchers. As the time complexity of traditional path planning algorithms grows rapidly with the scale and complexity of the problem, evolutionary algorithms are widely applied due to their capability of giving near-optimal solutions to complex problems. However, evolutionary algorithms can be easily trapped in a local optimum and may converge to infeasible solutions. As the scale of the problem increases, evolutionary algorithms usually cannot find a feasible solution with random exploration, which makes it extremely challenging to solve long-range path-planning problems with these algorithms in environments filled with obstacles. This paper introduces a novel area-based collision assessment method for Genetic Algorithm (GA) that can guide the algorithm to find solutions with a variable number of waypoints in large-scale and obstacle-filled environments. To avoid premature convergence, the mutation process is replaced by a self-improving process to let the algorithm focus the operations on any potential solutions before discarding them in the selection process. The case studies show that the proposed GA-focus algorithm can be applied to different types of large-scale and challenging environments, escape local optimums, and find high-quality solutions.

Keywords: Genetic Algorithm, artificial intelligence, evolutionary algorithm, path planning, robotics

1 INTRODUCTION

Path planning is a vital task in the field of robotics and autonomous vehicles. A typical path planning problem is to find a collision-free trajectory from an initial position to a target position in some environments with obstacles, as shown in Figure 1. For different problem settings, additional objectives (e.g., minimizing energy consumption) or constraints (e.g., restricting the turning angle along the trajectory) can be included [1]. The plan can also be made in large-scale environments such as global navigation for ships [2] or in high-dimensional spaces such as motion plans for multi-joint robot arms [3]. These make path planning a challenging task.

Researchers have developed various methods to solve the path planning problem. Search-based methods such as the A* algorithm are robust to different types of environments and can always find the global optimal solution [4]. However, such methods need to make plans on pre-defined graphs. Given that environment modeling can be a time-consuming process in large-scale planning, these algorithms have poor efficiencies in large-scale problems [2]. Sampling-based algorithms such as probabilistic roadmap (PRM) [3], rapidly exploring random trees (RRT) [5], and their variants draw random samples in the space and try to connect the sampled points to form a feasible path. Due to the nature of random sampling, the convergence speed is not guaranteed, and it is difficult for these methods to find narrow passages on the map. The artificial potential field (APF) can find feasible paths in environments that only contains convex obstacles but may easily get stuck in local minimum in complex

Copyright © 2023 by ASME

environments [6]. Fast-marching method [7] and Voronoi diagram based methods [8] are able to find a safe path that keeps a distance from obstacles, while the length of the path found is not optimized. Furthermore, it is difficult to model objectives such as environmental disturbances or time-varying risks using the abovementioned methods.

To plan the path under complex conditions, researchers tried to model the task as an optimization problem and use machine learning techniques to solve it. Many studies have been conducted to solve path planning problems with deep learning and deep reinforcement learning [9-11]. However, the quality of the model highly relies on the quality of the training data and scenarios. Using evolutionary algorithms to do path planning is another hot research field [12-15]. Thanks to the flexibility of the evaluation function, evolutionary algorithms are well-suited for modeling multi-objective or multi-robot path planning problems. Nevertheless, when the scale of the problems grows, evolutionary algorithms suffer from early convergence, and traditional exploration and evaluation methods may fail to find feasible solutions.

In this paper, we propose a path planning approach based on Genetic Algorithm (GA) to solve large-scale path planning problems in complex environments. A novel collision assessment method is introduced to guide the algorithm to find a feasible path. To overcome the premature problem of GA, we introduce a self-improving mechanism to ensure that the algorithm pays attention to any potential solution. The proposed algorithm can produce solutions with a variable number of waypoints to accommodate different complexities of the problems and can find high-quality solutions in large-scale planning problems with the size of the grid map up to 10,000×10,000.

The rest of the paper is organized as follows: Section 2 reviews related work in path planning methods, including their assumptions and limitations. Section 3 describes our proposed collision assessment method to help the algorithm converge to collision-free paths in complex environments and our novel self-improving mechanism to overcome the premature problem in GA. Section 4 shows the results of four case studies conducted on different tough, large-scale environments. In Section 5, a detailed discussion of the case study is presented. The last section outlines the conclusions, limitations, and potential applications of this work and points out future research directions opened up by this work.

2 RELATED WORK

A path planning task has its objectives and constraints, such as optimizing energy consumption during the motion and avoiding colliding with obstacles. To simplify the problem, a trajectory can be discretized and represented by a sequence of waypoints. Thus, a path planning problem can be defined as finding a sequence of waypoints that optimizes the objective function and connects the starting position and goal position while at the same time satisfying the constraint functions.

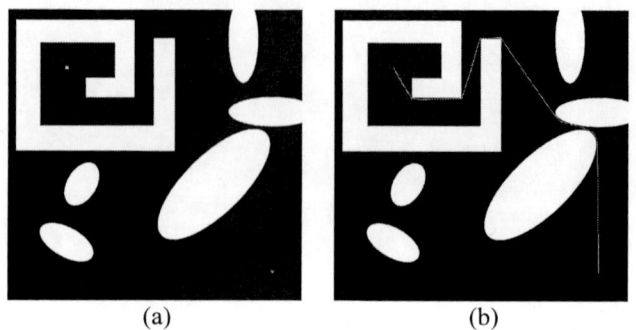

(a)   (b)

Figure 1. (a) a sample grid map (size: 10,000×10,000) for path planning problems. The area in white are obstacles, and the area in white is free space. The starting point is marked by a blue square, and the goal point is marked by a red circle; (b) a solution to the problem is shown in green.

Search-based algorithms are guaranteed to find the global optimal path when the environment is modeled as a computational graph. However, the time complexity of these algorithms grows exponentially with the branching factor of the graph, making heuristic methods and constraints on graph construction crucial for improving efficiency. Small-scale planning problems can usually be solved on a grid map. Tang et al. [16] showed that an improved A* algorithm can find the optimal path efficiently in port environments modeled as a 50×50 grid map. However, when the environment is large and cannot be represented as a low-resolution grid map, planning becomes too time-consuming, if not impossible. In such cases, the map can be decomposed to improve efficiency. Shah and Gupta [2] solved a long-distance path planning problem by decomposing the original grid map to a quadtree and constructing a tangent graph. Heuristic methods are applied to accelerate the search process, and the algorithm can find a near-optimal solution on a 100,000×100,000 grid map.

Since the graph decomposition approach is time-consuming, many researchers use sampling-based algorithms for path planning. Kuffner and LaValle [17] used RRT to plan the path for a manipulator in its configuration space. Hu and Jin [18] used PRM to solve the long-range path planning problem for autonomous ships in complex environments. Sampling-based algorithms are guaranteed to converge to a feasible solution, and the above two cases have shown that these methods can solve complex problems such as high-dimensional planning or long-range planning. However, due to the nature of sampling, it may take infinite time for the algorithms to converge, especially in cases where the distribution of free space is uneven. To improve the quality of sampling, Nasir et al. [5] introduced an RRT*-Smart algorithm that can intelligently draw samples around the feasible paths. Chen et al. [3] developed a virtual force to push the samples inside an obstacle to the boundary of it so that the algorithm can find narrow passages easily. The results of the two studies have shown that a good sampling strategy can significantly improve the efficiency of finding the shortest path.

Other than the algorithms mentioned above, there are some other popular methods. Fan et al. [6] demonstrated an improved

Copyright © 2023 by ASME

APF that can plan a path for autonomous underwater vehicles in environments with concave obstacles. Beser and Yildirim [7] used Fast Marching Square (FMS) Method to find smooth paths with consideration to the Convention on the International Regulations for Preventing Collisions at Sea (COLREGS) for ship navigation problems. Liu and Bucknall [19] showed that FMS can also be applied in environments with dynamic obstacles. Bhattacharya and Gavrilova [8] introduced a method based on the Voronoi diagram to find the optimal path with respect to length and clearance from the obstacles.

Although these methods have demonstrated their capability, there are still a lot of challenges, such as multi-agent planning and planning under environmental disturbance. Many researchers have shown that modern deep learning and machine learning methods can also be applied to path planning problems. Pan et al. [9] trained a convolutional neural network (CNN) to solve the multi-agent Traveling Salesman Problem (TSP) with the training data generated by Genetic Algorithm (GA). Wang and Jin [11] used deep reinforcement learning (DRL) to find collision-free paths for ships in dynamic environments. Panov et al. [10] showed that DRL can intelligently search in grid worlds filled with obstacles of "inconvenient for A*" shape.

However, the performance of deep learning is highly dependent on the quality of training data or experiences. Evolutionary algorithms, which do not rely on the training data, are popular in path planning problems for their flexibility in the design of the evaluation function. Evolutionary algorithms use mechanisms inspired by the natural evolution process and can be used to solve complex problems such as TSP and multi-agent planning by properly designing the solution encoding and evaluation function. Wang et al. [15] used an improved version of the grey wolf optimization algorithm (GWO) to minimize the energy consumption of unmanned surface vessels' underwater currents. Kang et al. [20] solved the collision avoidance problem for ships with respect to COLREGs by Particle Swarm Optimization (PSO). Das and Jena [14] showed that PSO can also be used to solve multi-agent planning problems. Luo et al. [21] solved the path planning problem in grid worlds with ant colony optimization (ACO). Lazarowska [13] demonstrated that ACO can solve the path planning for ships with dynamics restrictions.

Like other evolutionary algorithms, GA also shows great performance in solving complex problems. Many researchers have obtained good results in solving TSP [22-26] and path planning in complex environments [27-31] with GA. With proper modeling, GA can do high-dimensional path planning for manipulators [32,33]. GA can produce smooth and dynamically feasible paths by encoding the control points of a Bezier curve in the individuals [34-37]. Like GWO and PSO, the vanilla GA can only produce fixed-length results [31,38-43]. However, the operations in GA do not involve vector calculation, which makes it possible to produce variable-length solutions with proper mutation operations [12,44,45].

Although evolutionary algorithms are flexible in modeling, the exploration in the solution space does not always find a feasible solution. Also, the quality of the initial population is critical to the final solution. Many researchers use random initialization with a feasibility check [41] or other planning methods [21, 28, 29, 37] to initialize the population. However, the complexity of these initialization strategies grows rapidly with the scale of the problem. In some studies, heuristic local searches are applied in the exploration process to improve convergence speed [38, 41, 44]. To simplify the problem, some studies only produce monotonic paths, meaning the solution can only move forward on one dimension [15, 20, 40, 46]. Although these approaches have shown good results in case studies, it is hard to generalize the algorithms to common path planning problems. Moreover, most of the studies based on evolutionary algorithms are conducted on a small map with a size smaller than 100×100. It remains a problem how to properly model the large-scale path planning problems with evolutionary algorithms and how to deal with the early convergence problem that will stop the algorithm from searching for the global optimum.

From the approaches mentioned above, it can be observed that compared to traditional methods, evolutionary algorithms are more flexible in solving complex problems such as multi-objective optimization. However, the performance of evolutionary algorithms depends on the design of the evaluation function, and most current methods rely on random exploration to find feasible paths. In this paper, we introduce a collision-area approach for GA to guarantee the feasibility of solutions in large-scale planning problems and propose a self-improving process to overcome the early convergence problem of GA. We propose the *GA-focus* algorithm to answer *how to generalize the evolutionary algorithms to find high-quality solutions for large-scale path planning problems in complex environments with obstacles.*

3 METHODS

A typical path planning problem can be modeled as an optimization problem, which aims to optimize the path length under the constraint that the path does not collide with obstacles in the environment. In this section, we introduce a GA-based algorithm that can solve the path planning problem in large-scale and complex environments.

3.1 Problem Definition

A common path planning problem requires the algorithm to find a collision-free path in an environment from a starting position to a goal position. To simplify the problem, a path in 2-D space can be represented by a sequence of waypoints:

$$P = \{p_s, p_1, p_2, \ldots, p_n, p_g\} \qquad (1)$$

where $p_s = (x_s, y_s)$ and $p_g = (x_g, y_g)$ are the coordinates of the starting position and goal position, and $p_i = (x_i, y_i), i \in [1, n] \cap Z$ are the coordinates of intermediate waypoints. The resulting path can be represented by the line segments defined by the waypoints in P:

$$path = p_s p_1 \cup p_1 p_2 \cup \ldots \cup p_i p_{i+1} \cup p_n p_g,$$
$$i \in [1, n-1] \cap Z \qquad (2)$$

Copyright © 2023 by ASME

where $\boldsymbol{p_i p_{i+1}}$ are the line segments defined by $\boldsymbol{p_i}$ and $\boldsymbol{p_{i+1}}$.

The obstacles in the environment can be expressed by a set $\boldsymbol{Obs}$. The target is to optimize an objective function $fun(\cdot)$ under the non-collision constraint:

$$maximize \; fun(\boldsymbol{path})$$
$$s.t. \, \boldsymbol{path} \cap \boldsymbol{Obs} = \emptyset \qquad (3)$$

3.2 Genetic Algorithm

GA is an evolutionary algorithm inspired by the natural evolution process of species. Each individual in GA's *population* is called a *chromosome*, and each element in a chromosome is called a *gene*. A gene encodes the value of a parameter, and thus a chromosome defines the parameter set of the solution. GA mimics the natural selection process in that in each generation, each chromosome will be evaluated by a *fitness* function and will be selected by a *selection* function. Only the best among them can survive and exchange their genes with other survivors by *crossover* and thus produce offspring. *Mutation* may happen to the offspring, which will randomly switch their genes to produce new features. From generation to generation, the good features can be preserved, and the whole population will finally converge to an optimal solution.

Solution encoding

The algorithm aims to find the best sequence of waypoints with respect to the fitness function. Since the waypoint representation is an approximation of the trajectory, and the optimal number of waypoints could differ for different problems, the proposed algorithm uses a variable-length chromosome encoding, as shown below.

$$individual_j = \{\boldsymbol{p_{j,1}, p_{j,2}, \dots, p_{j,n}}\} \qquad (4)$$

where n in the number of intermediate waypoints, which can vary during the optimization, and individuals can have a different number of waypoints.

Initial population

Since the algorithm can deal with variable-length chromosomes, we initialize the whole population as one-waypoint solutions. In the initialization stage, each individual in the population is assigned a randomly sampled waypoint:

$$individual_{j,init} = \{\boldsymbol{p_{j,init}}\} \qquad (5)$$

and the algorithm will determine whether it needs to add or delete waypoints by itself during the processing.

Considering that the quality of the initial population can greatly affect the final solution, and random initialization might be biased when the population size is small, we use an arbitrary version of the initialization method in KMeans++ [47] to ensure that the initial population is spread out in the environment. The individuals are initialized one by one, and each individual is initialized by finding the point that is farthest to the already

initialized population. The initialization algorithm is shown in Algorithm 1.

Algorithm 1. initialize_population
Inputs: map (M), population_size (n)
1.   *all_samples* = Random sample 10n free points on M
2.   *initial_population* = []
3.   randomly pick a point in *all_samples* and add to *initial_population*
4.   **while** *len(initial_population)* < n
5.   *next_point* = None
6.   *next_point_dist* = 0
7.   **for** *point* in *all_samples*
8.   *dist_to_pop* = *min([distance(point, init_point)* **for** *init_point* in *initial_population])*
9.   **if** *point* not in *initial_population* and *dist_to_pop* > *next_point_dist*
10.   *next_point* = *point*
11.   *next_point_dist* = *dist_to_pop*
12.   **end if**
13.   **end for**
14.   add *next_point* to *initial_population*
15.   **end while**
16.   **return** *initial_population*

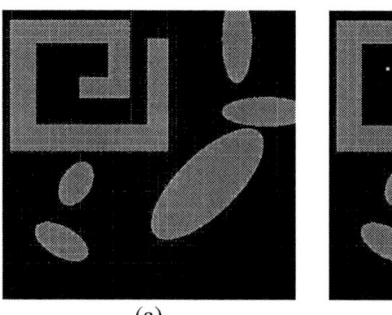 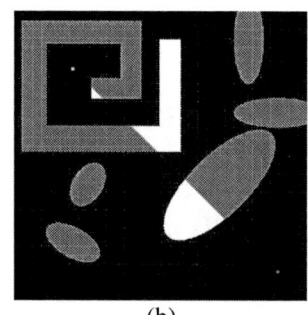

(a)   (b)

Figure 2. (a) quadtree decomposition of the map shown in Figure 1. The black blocks are free space, and the grey blocks are obstacle blocks; (b) the collision cost of the line segment is the sum of areas of the white blocks

Fitness function

A common path planning problem usually concerns the length and safety of the path. Some studies also include the smoothness of the path. Since if the obstacles in the environments do not have sharp corners, the shortest path should also be a smooth path; we decide to only use the path length as an objective. In safety modeling, hitting an obstacle should be penalized. The final fitness function is defined as follows:

$$fitness = \frac{1}{\alpha \cdot cost_{dist} + \beta \cdot cost_{collision}} \qquad (6)$$

Copyright © 2023 by ASME

$$cost_{dist} = dist(\boldsymbol{p_s}, \boldsymbol{p_1}) + \sum_{i=1}^{n-1} dist(\boldsymbol{p_i}, \boldsymbol{p_{i+1}}) \\ + dist(\boldsymbol{p_n}, \boldsymbol{p_g}) \tag{7}$$

There are two popular types of collision penalization functions: *constant cost* and *linear cost* [15,27,31,40,41]. The constant cost method gives a constant or infinite penalty when a collision happens, while the linear cost method uses the length of segments in the collision as the penalty. Both methods work in small-scale environments with a small number of obstacles since random exploration is powerful enough to find feasible solutions. However, in large-scale problems, these methods can easily get stuck in infeasible solutions because the collision penalization function does not provide enough information about the environment, and it is too difficult to find a feasible solution with random exploration.

In this study, we propose a novel collision cost by calculating the area of obstacles cut by the segments. Each part in a collision of each segment contributes to the collision cost, and the cost value is the area of the smaller part of the obstacle being cut by the segment. Since the shape of obstacles might be irregular, and the area calculation can be time-consuming, we decompose the original grid map into a quadtree map for area calculation and approximate the area by calculating the area of blocks in the quadtree.

$$cost_{collision} = \sum area_{collision} \tag{8}$$

As shown in Figure 2, the original 10,000×10,000 grid map is decomposed to a quadtree, where all the black blocks are *free blocks*, and the grey blocks are *obstacle blocks*. The collision cost of the line segment is the sum of the areas shown in white. Since the formula of the line segments and blocks is known, the areas can be calculated analytically. When a line segment only penetrates one obstacle, the obstacle is cut into two parts, and the smaller one is used to calculate the collision cost. When the line segment moves, the collision cost changes continuously until when the segment no longer cuts the obstacle, the collision cost becomes 0. This continuity ensures that a collision-free path can always be found by small steps of motion following the decrement of collision cost.

All cuts by a segment are independently used to calculate the collision cost so that adding new waypoints on segments does not affect the value of the collision cost. To prevent the algorithm from trying to find a path outside the boundary of the map, the area of the obstacle blocks on the boundary of the map is set to a large value so that the smaller part of the obstacle will always be on the opposite side of the map boundary.

Crossover operation

The crossover operation plays the role of communication among individuals. We use a single-point crossover method here. Given two individuals as the *parents*, the crossover point of $parent_1$ is randomly selected, and thus $parent_1$ is split into

two parts. The crossover point of $parent_2$ is determined by the proportion of the length of the first part of $parent_1$. For example, the first part of $parent_1$ accounts for 68% of its length, and thus the crossover point of $parent_2$ will be the first waypoint that exceeds 68% of its length. An illustration of the crossover operation is shown in Figure 3. The idea of crossover is to combine the first part of $parent_1$ and the second part of $parent_2$, so that the properties of parents can be preserved, and the algorithm can find a possible new path and escape from the current solution. By selecting parents and doing a crossover, a set of *offspring* is produced, which contains the combinations of the current best solutions. To keep the good property in the parents, the parents are also duplicated to the new generation.

Figure 3. A sample of crossover operation. The red path and the green path on the left-hand side are the two parents. The two hollow dots on the right-hand side are the selected crossover points, and the offspring is the black path.

Mutation operation

The mutation operation ensures that the solutions are not restricted to the waypoints that have already appeared. It randomly happens on the offspring and adds, deletes, or moves the waypoints so that new waypoints and line segments can be created and evaluated, and good ones can be kept by the selection process. In this study, we use five mutation operators, which are common in path planning problems. These operators have the same probability of being selected in a mutation round.

- *Add waypoint*: add a waypoint to the individual as shown in Figure 4(a). Notice that if a segment is not in collision, adding a new waypoint between its vertices will always harm the total length of the path. Thus, the collision costs of segments are used as the sample weight so that only those segments in the collision will be considered to add a waypoint. The position of the new waypoint is randomly selected in the local area.
- *Delete waypoint*: delete a waypoint in the individual, as shown in Figure 4(b). It uses $cos^2\frac{\theta}{2}$ as the weight of random selection. Intuitively, this operator will delete sharp turns in the solution.
- *Move waypoint*: randomly move a waypoint, as shown in Figure 4(c).
- *Split segment*: randomly split a segment, as shown in Figure 4(d). It uses the length of segments as the weight of random selection. Thus, longer segments have higher probabilities to be split. The splitting point is randomly chosen. A short random motion is applied to the new waypoint so that the fitness value of the new path differs from the original one.
- *Cut corner*: randomly cut a corner, as shown in Figure 4(e). It uses $cos^2\frac{\theta}{2}$ as the weight of corner selection. The cutting

Copyright © 2023 by ASME

621

points are randomly chosen. This operator tends to cut sharp corners and smoothen the path. If the new segment is not in collision, the length of the whole path will also be improved.

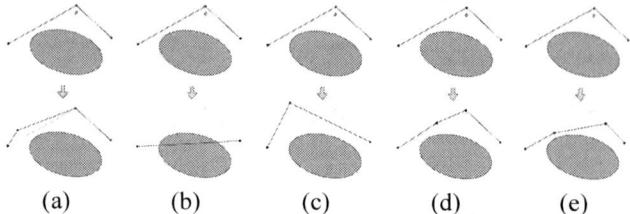

Figure 4. Five mutation operations used in this study. (a) add waypoint; (b) delete waypoint; (c) move waypoint; (d) split segment; (e) cut corner.

Selection method

The fitness function in this study consists of the path length and collision cost, where the magnitude of the collision cost is much larger than that of the path length in early generations. To avoid the algorithm converging to the first collision-free path it finds, we use the tournament selection method as the selection function. When selecting a parent, k individuals are randomly chosen from the population, and the one with the highest fitness value among them is selected as one of the parents.

The pseudocode of the planner, namely *GA-area*, is shown in Algorithm 2.

Algorithm 2. GA_area
Inputs: map (*M*), starting position (*start*), goal position (*goal*), max generation (*n*)
1.   initialize the *population* of the GA-area
2.   *generation* = 0
3.   **while** *generation* < *n*
4.   *pop_fitness* = *fitness(population, start, goal, M)*
5.   *parents* = *selection(population, pop_fitness)*
6.   *offspring* = *crossover(parents)*
7.   *offspring* = *mutate(offspring)*
8.   *population* = *union(parents, offspring)*
9.   *generation* = *generation* + *1*
10.   **end while**
11.   *solution* = best *individual* in *the population*
12.   **return** *solution*

3.3 Self-improving Process

Although *GA-area* can find feasible paths in complex environments, it usually converges to a local optimum and makes unnecessary detours. The mutation operations do local changes to each individual, and if the fitness value is improved, the resulting individual has a higher probability of surviving.

[1] The code can be found at https://github.com/hu-chuanhui/GA-focus

However, to escape a local optimum, multiple steps of mutation operations must be applied at the same position, as shown in Figure 5. After the first mutation operation, the new path shown in Figure 4(b) is in collision again, which results in a great drop in the fitness value so that the probability that the new path can survive in the selection process is low. Since the mutation operator and mutated waypoint are randomly selected, it is hard for GA-area to keep this individual until a proper mutation appears again at the same position.

To solve the problem that the mutation operations are distracted, we introduce a *self-improving* process and propose a *split-and-focus* mechanism to let the algorithm split the original problem into several sub-problems and improve them independently. During the split-and-focus process of each individual, some waypoints are randomly selected and fixed. The fixed waypoints are assumed to be part of the optimal solution, and a planner is used to plan the path that connects the starting point, the fixed points, and the goal point. An illustration of the split-and-focus is shown in Figure 6.

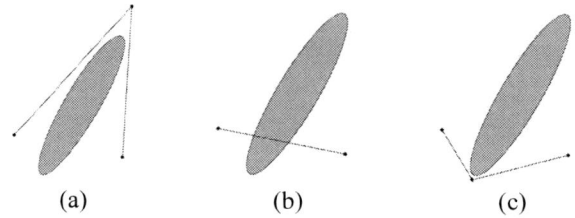

Figure 5. Multiple steps of mutation operations are needed to escape a local optimum. (a) before mutation; (b) after one mutation operation (delete waypoint); (c) after the second mutation operation (add waypoint).

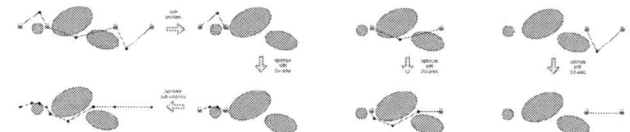

Figure 6. Split-and-focus operation. The fixed waypoints are shown with the lock icon.

We developed the *GA-focus* algorithm[1] that follows all the processes mentioned in Section 3.2 except the mutation operations. Instead, a self-improving process happens to the whole population in each generation. In this study, we use *GA-area* to do the planning of sub-problems. For each individual, a budget for the total generation number is given and is equally allocated to each sub-problem. For example, when the budget is 200 generations, and there are 4 sub-problems, the planner for each sub-problem will have 50 generations to solve it. After the fixed points have been selected, the original problem is split into sub-problems and is optimized by *GA-area*. The *split-and-focus* function is shown in Algorithm 3, and *GA-focus* is shown in Algorithm 4. The number of sub-problems is randomly selected from $[2, 2 + num_{waypoint} / 20]$.

Copyright © 2023 by ASME

The initial population of *GA-focus,* which consists of single-waypoint individuals, is processed by *split-and-focus* once. During the process, each waypoint is assumed to be part of the optimal solution, and the *split-and-focus* function will try to find a solution from the starting position to the waypoint and from the waypoint to the goal position. The *split-and-focus* calls *GA-area* to solve sub-problems. In both *GA-focus* and *GA-area,* the best individual of a generation, namely *elitism,* is also duplicated to the next generation so that the current best solution will not be discarded. One individual in the initial population of the *GA-area* is a duplicate of the original solution to the sub-problem, while all other individuals are empty (which means no intermediate waypoints). Thus, *GA-area* can keep the current best solution while it can still decide whether to discard part of the solution by the crossover between the current best solution and an empty individual.

Algorithm 3. split_and_focus

Inputs: individual (*individual*), map (*M*), starting position (*start*), goal position (*goal*), generation budget (B)

1. *s* = randomly decide the number of sub-problems
2. *sub_problems* = randomly split *individual* into *s* sub-problems
3. *n = int(B / s)*
4. *sub_solutions = empty*
5. **for** *sub_problem* in *sub_problems*
6.   initialize the *population* of *GA-area* as empty set
7.   copy the *waypoints* of *sub_problem* to the first element of the *population*
8.   *sub_sol* = run *GA_area(M, sub_start, sub_goal, n)*
9.   add *sub_sol* to *sub_solutions*
10. **end for**
11. *solution* = combine *sub_solutions*
12. **return** *solution*

Algorithm 4. GA_focus

Inputs: map (*M*), starting position (*start*), goal position (*goal*), generation number (n), generation budget (B)

1. initialize the *population* of GA-focus
2. **for** *individual* in *population*
3.   *individual* = *split_and_focus(individual)*
4. **end for**
5. *generation* = 0
6. **while** *generation < n*
7.   *pop_fitness = fitness(population, start, goal, M)*
8.   *parents = selection(population, pop_fitness)*
9.   *offspring = crossover(parents)*
10.   *population = union(parents, offspring)*
11.   **for** *individual* in *population*
12.   *individual = split_and_focus(individual, M, start, goal, B)*
13.   **end for**
14.   *generation = generation + 1*
15. **end while**
16. *solution* = best *individual* in *population*
17. **return** *solution*

4 CASE STUDY
4.1 Narrow Gap Problem

In the first case study, we generate an environment with its upper half filled with 300 randomly generated ellipses as obstacles. The starting position is at the upper-left corner, and the goal position is at the upper right corner. As shown in Figure 7(a), to increase the complexity of the problem, we add a vertical wall in the center of the ellipse area. There is a narrow gap in the wall, so the optimal solution should be a path that connects the starting point and the goal point while passing through the gap without hitting any ellipses.

From the figure, we can see that although it is difficult to find such a path, an easy but feasible solution is to first go down to the obstacle-free area at the bottom, then go right and reach the bottom-right corner of the map, and finally go upward and reach the goal. Such a path will encounter fewer obstacles so that it can improve its fitness more easily and thus will have a higher fitness value and a higher probability of surviving in the selection process.

Table 1. model settings for test case 1

	GA-linear	GA-area	GA-focus	
			GA-focus-main	GA-focus-sub
length cost coef α	0	1	1	1
collision cost coef β	1	100	100	100
generation number	800	800	3	200 (total)
population size	320	320	32	9
parent number	160	160	16	3
tournament selection k	2	2	2	2
crossover probability	1	1	1	1
mutation probability	1	1	/	1
keep parent	True	True	True	True
keep elitism	True	True	True	True
self-improve	False	False	True	False

The size of the grid map is 10,000×10,000, and the waypoints are picked from all the free grids. We compare the performance of three algorithms: *GA-linear, GA-area,* and *GA-focus.* GA-linear and GA-area use the crossover and mutation operations as described in Section 3. The difference is that GA-linear uses the length of segments in a collision as the collision cost, while GA-area uses the area cost as described in Section 3. GA-focus has a self-improving process, and it uses a GA-area as the planner to solve the sub-problems during self-improving. The main problem solver is called *GA-focus-main,* while the sub-problem solver is called *GA-focus-sub* in the following text. Both GA-focus-main and GA-focus-sub use the crossover operation described in Section 3. GA-focus-main does not have a mutation process, while the GA-focus-sub uses the mutation operation described in Section 3. Both GA-focus-main and GA-focus-sub use the area-based collision cost. In the area cost estimation, the grid map is decomposed into a quadtree. The max level of the tree is 9, and thus the size of the smallest block is about $10000 \cdot 2^{-9} \approx 20$ pixels in width. The number of blocks in Table 2 shows how complex the environment is. In this test, the environment is decomposed into 26,527 quadtree blocks.

Copyright © 2023 by ASME

Table 2. The results of case study 1. The values are shown in the format *mean ± std*.

	# blocks	# segments evaluated	# waypoints	Success rate (%)	Path length
GA-linear	26527	293259.70±3723.80	4.70±2.57	0	10001.96± 220.75
GA-area	26527	254682.50±9148.08	26.10±8.86	100	17393.24±1706.17
GA-focus	26527	280539.20±1265.26	33.00±4.43	100	**11797.43± 135.57**

All parameters of the models are listed in Table 1. Since GA-linear uses a different collision cost, we let the coefficient of length cost be zero so that it will neglect the length of the path and focus on finding a collision-free path. GA-linear and GA-area do not have a self-improving process. Thus, we let them have a larger population size and generation number so that they can have equal power in exploring the solution space compared to GA-focus. The exploration power is estimated by the number of line segments evaluated. Each segment consists of two waypoints in the space, and the number of segments evaluated shows how many different combinations of waypoints the algorithm has created and tested by crossover and mutation.

We do the planning ten times with different random seeds, and the results are shown in Figure 7 and Table 2. The average number of segments evaluated by the three algorithms is less than 300,000. The quadtree of the environment has 26,527 blocks, which have $26527 \times 4 \approx 10^5$ vertices. If a search-based algorithm is applied and we need to construct a graph with the vertices, the worst-case number of segments evaluated on this map can be $10^5 \times 10^5 = 10^{10}$, which is much larger than 3×10^5 in our case.

The obstacle blocks of the quadtree cover all obstacle grids, which means that usually, the corner grid of an obstacle block is actually in free space. Thus, if the collision cost of a solution is less than 1, the solution is regarded as a feasible one in the results. For GA-linear, this means that the length in a collision is less than 1 pixel wide, and for the other two algorithms, this means that the collision area is smaller than 1 pixel. From Figure 7(b), we can see that GA-linear fails in all 10 tests. All the solutions hit the wall because the surrounding environment is filled with obstacles, and hitting the wall results in the smallest collision cost. The linear collision cost does not provide enough guidance on how the algorithm can avoid the obstacles it is in collision with.

Both GA-area and GA-focus have a success rate of 100%. They both can find feasible solutions in complex environments. However, GA-area suffers from premature problems. It tends to converge to an easy solution that goes through the obstacle-free space instead of trying to find a feasible path through the ellipse obstacles. The number of segments evaluated by GA-area also shows this problem. A new segment can be produced by a crossover process or a mutation process. The large standard deviation of the number of segments evaluated implies that the GA-area converges to the same solution too early in some tests, and thus the parents in the crossover process are identical so that no new segment is produced. Only 1 solution by GA-area finds

the narrow gap in the wall, yet it still does not try to find a path through the obstacle-filled area.

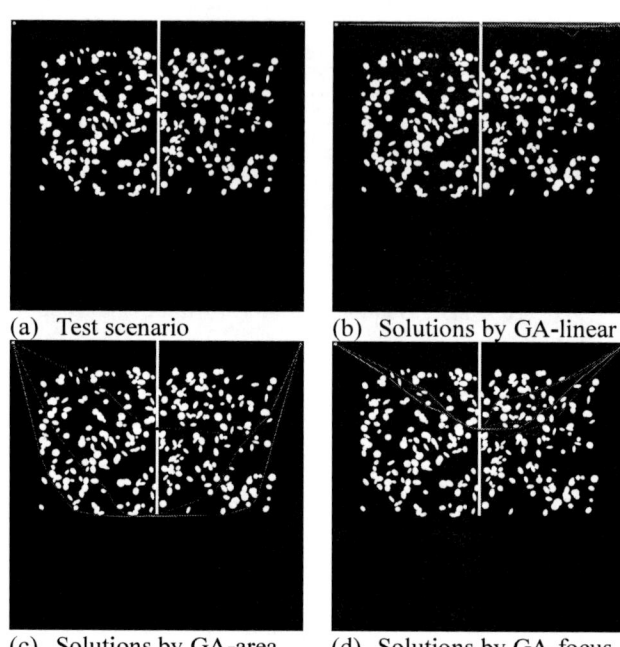

(a) Test scenario
(b) Solutions by GA-linear
(c) Solutions by GA-area
(d) Solutions by GA-focus

Figure 7. The grid map of case study 1 and the solutions by the three algorithms. (a) the starting point and goal point are at the upper-left and upper-right corner, respectively; (b) the 10 solutions given by GA-linear; (c) the 10 solutions given by GA-area; (d) the 10 solutions given by GA-focus.

In this test environment, GA-focus shows robust performance in preventing premature during planning. The algorithm is never trapped in an easy solution. All 10 solutions find a narrow gap in the wall, and the small standard deviation of path length also shows the robustness of GA-focus. The average number of waypoints in the final solution is 33, which shows that GA-focus can increase the number of waypoints in its solution according to the complexity of the problem.

4.2 Different Tough Environments

In this case study, we provide 4 different tough problems for GA-focus. We select four grid maps from the benchmark set collected by Sturtevant [48]. These maps are from the video game StarCraft I and are converted to greyscale images and resized to 10,000×10,000. The starting point and goal point for

Copyright © 2023 by ASME

Table 3. The results of case study 2. The values are shown in the format *mean ± std*.

	# generation	# blocks	# segments evaluated	# waypoints	Path length	Success rate (%)	Time (s)
Map 1	3	42373	291293.60±1190.06	60.20±10.32	16667.68±329.45	100	1204.33±18.15
Map 2	3	36343	290477.00±1068.95	67.70± 7.63	14999.16±256.00	100	2054.82±26.72
Map 3	4	49792	364079.60±1533.41	72.50± 9.63	18728.91±801.54	100	1177.46±31.96
Map 4	4	61150	360363.30±1627.90	76.40±11.13	15907.43±315.33	100	893.69±30.52

each map are marked by the blue square and the red dot, respectively, as shown in Figure 8. The shape of the obstacles in these environments is irregular, and the maps show different challenges for path planning:

- *Map 1*: an indoor environment as shown in Figure 8(a). There are many rooms and walls in this environment, so it is difficult for evolutionary algorithms to escape the local optimum and find feasible solutions.
- *Map 2*: a maze environment as shown in Figure 8(b). There are many dead ends in this environment, and the boundary of the geometries is rough, which makes the graph construction and search process time-consuming for regular search-based algorithms.
- *Map 3*: an environment filled with irregular obstacles, as shown in Figure 8(c). The starting point and goal point are surrounded by pocket-shaped obstacles, and the free space contains a lot of narrow passages. These problems are difficult for APF and sampling-based algorithms.
- *Map 4*: an environment filled with obstacles whose sizes differ significantly, as shown in Figure 8(d). The winding free space around small obstacles might be part of the optimal solution.

Like case study 1, we also do 10 experiments on each problem with different random seeds. The parameters of GA-focus are the same to Case Study 1, as stated in Table 1. If the collision cost of a path is smaller than 1, the path is considered a feasible one. The solutions are shown in Figure 8, and the results are listed in Table 3.

The result shows that in all test cases, GA-focus can reach a 100% success rate. GA-focus shows robustness in the path length. On map 1, map 2, and map 4, the standard deviation of the path length is less than 2% of the average length. On map 3, only two of the ten solutions take an obvious detour, and the rest solutions are all efficient paths. Seven of the ten solutions on map 4 go through the area with many tiny obstacles, which shows that the winding narrow space around small obstacles is not neglected by GA-focus.

4.3 Dense Obstacle Problem

The above two case studies have shown that GA-focus can solve different types of tough problems in large-scale problems. However, the collision cost of GA-focus is calculated by the collision area, and the efficiency of the algorithm can be affected by the number of obstacles in the environment. Thus, planning in obstacle-filled environments is a tough problem.

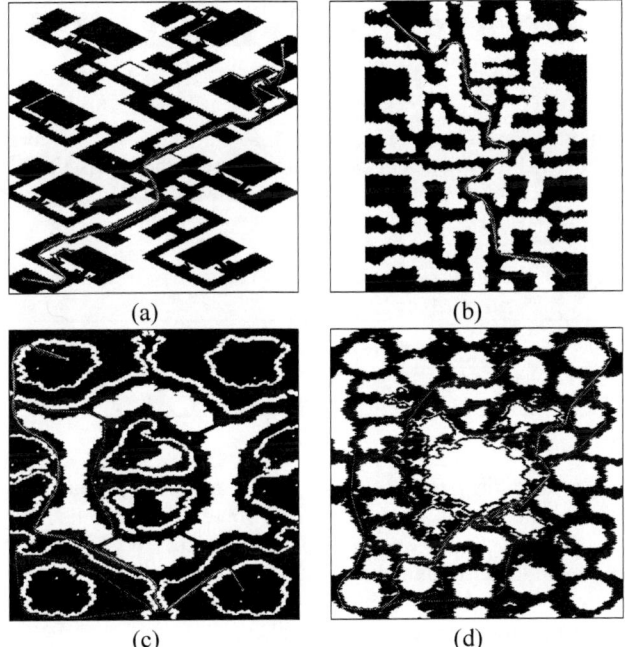

Figure 8. The solutions to four tough tasks by GA-focus. The starting point and goal point are marked by the blue square and red dot, respectively. (a) an indoor environment; (b) a non-square maze environment; (c) an environment with pocket-shaped obstacles and narrow passages; (d) an environment filled with obstacles of different sizes.

In this case study, we test the performance of GA-focus in an environment filled with obstacles, as shown in Figure 9. The size of the grid map is still 10,000×10,000. There are 5,000 ellipse obstacles randomly placed in the environment, which makes it difficult for GA-based algorithms to find a feasible solution. The parameters of GA-focus are the same as in Table 1, except that the number of generations of GA-focus-main is 4. We do 10 experiments with different random seeds, and the performance of GA-focus is shown in Table 4. A path with a collision cost smaller than 1 is considered a feasible one. The number of blocks in the quadtree is 102,745, which is much larger than those in case study 2. The success rate of GA-focus is 90%, which shows that GA-focus can be applied to solve path planning problems in obstacle-filled environments.

Copyright © 2023 by ASME

(a) Solutions by GA-focus   (b) A sample solution by RRT*

Figure 9. The grid map of case study 3 and the solutions by GA-focus and RRT*. The starting point and goal point are at the upper-left (blue square) and bottom-right (red dot) corners, respectively; (a) the 10 solutions given by GA-focus; (b) A sample solution by RRT*.

Table 4. The results of case study 3. The values are shown in the format $mean \pm std$.

# block	# segments evaluated	# waypoints	Success rate (%)	Path length	Time(s)
102745	354304.70 ±1770.00	87.50 ±15.28	90	14897.20 ±475.36	1149.03 ±21.54

We also present a solution to this problem produced by RRT*, which is a popular planner that is probabilistically guaranteed to converge to the global optimum. RRT* randomly samples in the environment and builds a tree to find the optimal solution. With a rewiring process, the resulting tree will finally find an optimal solution to the problem. The step size of RRT* is set to 300, and the rewiring radius is set to 1,000. We let RRT* build a tree with 7,000 nodes, and the final solution is shown in Figure 9(b). The length of the path is 14068.54. It takes over 30 minutes for RRT* to complete the tree construction and find the solution. In comparison, GA-focus can find a feasible solution in around 20 minutes, which indicates that GA-focus is an efficient algorithm that can find a high-quality feasible solution to complex problems.

4.4 Multi-objective Planning

One of the advantages of GA-based planning is that multi-objective planning problems can be easily modeled by the fitness function [49-51]. In this case study, we provide an illustration on how GA-focus can be applied to multi-objective planning problems.

In this problem, we aim to find a near-optimal solution considering the weighted sum of distance-cost, smoothness-cost, and time-cost. The grid map of the environment with a size of 10,000×10,000 is shown in Figure 10. We let the allowed moving speed of the agent vary with the y coordinate, as shown in Figure 10(a). When the y coordinate gets larger, the agent is allowed to move at a higher speed. The allowed speed changes linearly with the increment of the y coordinate, which ranges from 0.05 to 1. The time cost of a line segment is defined as the traveling time along the segment, as shown below:

$$cost_{time} = \frac{length_{seg}}{speed_{avg}}$$

(9)

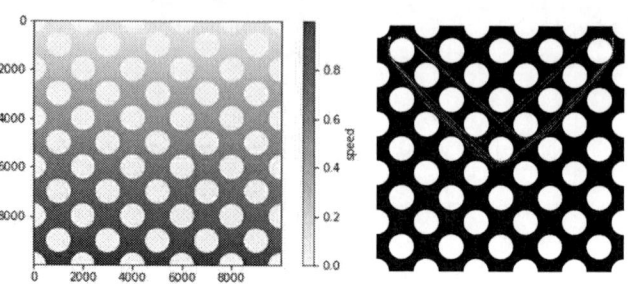

(a) Speed field in the environment. The lower area allows faster speed of motion.   (b) Solutions by GA-focus.

Figure 10. The grid map of case study 4 and the solutions by GA-focus. (a) the speed field in the environment. The red color means that the agent can move faster in these areas; (b) the 10 solutions given by GA-focus. The starting point and goal point are at the upper-left (blue square) and upper-right (red dot) corners, respectively.

Table 5. The results of case study 4. The values are shown in the format $mean \pm std$.

Path length	Time-cost	Smoothness-cost	Success rate (%)	Computation time(s)
13479.54 ±705.65	48504.92 ±2867.61	56.00 ±19.08	100	298.79 ±12.33

In order to find a smooth solution, a smoothness cost is introduced to penalize sharp turns in the solution. The smoothness cost is defined as follows, where deg denotes the turning angle of two segments:

$$cost_{smooth} = \begin{cases} 0, & if\ deg < 10° \\ 20, & if\ 10° \leq deg < 45° \\ 50, & if\ 45° \leq deg < 90° \\ 200, & if\ deg \geq 90° \end{cases}$$

(10)

The final fitness function is defined as the inverse of the weighted sum of all the cost functions:

$$fitness = \frac{1}{\alpha \cdot cost_{col} + \beta_1 \cdot cost_{lengt} + \beta_2 \cdot cost_{smooth} + \beta_3 \cdot cost_{time}}$$

(11)

In this case study, the coefficient of collision cost is $\alpha = 100$. The coefficients of other costs are $\beta_1 = 1$, $\beta_2 = 100$, $\beta_3 = 1$, so that the magnitude of these costs are close. The parameters of GA-focus are still the same as in Table 1, except that the number of generations of GA-focus-main is 2. We do 10 experiments with different random seeds. The solutions are shown in Figure 10(b), and the performance is shown in Table 5.

Copyright © 2023 by ASME

GA-focus can efficiently solve such a multi-objective problem. It finds feasible solutions in all 10 tests in around 5 minutes. All solutions tend to move to the center of the map first, and then proceed to the goal, so that at the middle of the trajectory, the agent can move at a higher speed and thus save time.

5 DISCUSSION

As shown by case study 1, our proposed area-based collision cost can find feasible solutions in an environment filled with obstacles. In path planning problems, the optimal path may be more complicated than other local optimums. As shown in Table 1, the average number of waypoints by GA-area is smaller than that by GA-focus. Also, we can see in Figure 7 that the paths given by GA-focus go through the obstacle-filled area, while GA-area tends to make a detour to the free area. This implies that more efforts are required to find the optimal solution, and the premature of the algorithm will stop it from searching for a better but more complex solution.

The self-improving process can reduce the probability that a potentially good solution fails to survive the selection process. The self-improving process gives each individual more chances to explore the solution space around it and improve itself before the next selection process happens. By the split-and-focus mechanism, each individual splits itself into several sub-problems so that the operations can be focused in a local area and the best operation has a higher probability of emerging before the population is dominated by a local optimum. The number of sub-problems and the position of the fixed waypoints are randomly chosen so that each individual can try different combinations of sub-problems and are not restricted to the same set of fixed waypoints. The results of case study 2 show that GA-focus can deal with different challenges in path planning problems, such as pocket-shaped obstacles and winding areas. Case study 3 shows that GA-focus can be applied to environments filled with obstacles. Even in an environment filled with 5,000 obstacles, GA-focus still has a 90% success rate and can find high-quality solutions in shorter time compared to RRT*. GA-focus can also be extended to multi-objective planning problems. As shown in case study 4, by adding more terms in the fitness function, GA-focus can handle the multi-objective planning problem in static environments.

The number of segments evaluated has a linear relationship to the number of generations. There is a self-improving process before the selection in the first generation so that the number of segments evaluated is linear to $num_{generation} + 1$. Thus, the complexity of the proposed algorithm does not scale too fast with the complexity of the environment. The number of generations in the case studies is selected to keep a balance between the feasibility of the solution and the time consumption. If a smoother and shorter path is needed, optimizing for one more generation will usually increase the quality of the solution.

The code is written in PYTHON, and the tests are done on a gaming laptop. The self-improving process of GA-focus is parallelized on 16 threads, and the calculation is done by an AMD Ryzen 7 5800H CPU with 3.20 GHz speed and 16 GB RAM. The construction of the quadtree takes several seconds. The time consumption in Table 3 only records the optimization time of GA-focus.

As shown in Table 3, more quadtree blocks do not always result in longer computing times. However, how the blocks are distributed can influence the time consumption. For instance, the time consumption on map 2 is much longer than that on the other three maps because map 2 is a non-square maze environment, and the grid map is padded on both sides. The padding connects multiple obstacles to form a large one, and the collision area computation must traverse the blocks in this large obstacle, which results in heavy computation. Building a quadtree with rectangular blocks or simply modifying the number of nodes at the first layer of a quadtree can deal with such a rectangular map. A post-processing can be introduced to the original quadtree to merge neighboring square blocks and form rectangular blocks, so that the number of blocks will reduce, and the efficiency of the algorithm can be improved. However, it is not guaranteed that the boundary of an indoor map is perpendicular to the axis of the grid map, as the case in Figure 8(a). A map preprocessing process such as rotating and cropping might help to reduce the calculation in such cases. Another solution is to reduce the max level of the quadtree so that the number of blocks will reduce. This will work when a high-resolution map is not required.

The fitness function in this study is focused on static environments. However, it is possible to extend the proposed method to a dynamic environment by evaluating the dynamic cost with a simulation assuming that the agent will move along the planned trajectory at a constant speed. More control factors can also be introduced to control the speed during motion.

6 CONCLUSIONS AND FUTURE WORK

In this paper, we introduced the GA-focus algorithm that can solve large-scale path planning problems in complex environments. GA-focus follows the framework of GA, except that the mutation process is replaced by a self-improving process. In the self-improving process, each individual is split into several sub-problems, and a planner named GA-area is used to solve these sub-problems. We developed a novel collision cost function for GA-area. When a line segment of the path cuts an obstacle, the area of the smaller piece of the obstacle will be counted as its collision cost. Such a cost function changes continuously with the motion of the line segment, which ensures that the algorithm can always converge to a feasible solution.

The case studies show that the proposed GA-focus can solve different challenging path planning problems on large grid maps with sizes of 10,000×10,000. GA-focus shows robust performance in finding high-quality solutions in complex environments filled with obstacles of different sizes and shapes. This indicates that GA can solve path planning problems on large-scale and complex grid maps without any graph-construction process.

This study aims at providing a framework for solving large-scale path planning problems in complex environments, and thus we only consider the path length and collision cost in the fitness function. However, one important feature of GA is its flexibility

Copyright © 2023 by ASME

in the design of the fitness function. As shown in Section 4.4, more features, such as smoothness cost and speed cost, can be added to the fitness function so that the framework of GA-focus can be applied to a multi-objective problem.

Also, it should be noted that although we are using GA-area in the self-improving process, other kinds of planners may also work in this stage. We just have 5 common mutation operators in GA-area, and more heuristic operators may help the convergence speed of the algorithm as well. Additionally, the mutation operators choose the new waypoint randomly in this study. A local search might also be helpful.

The collision cost is calculated by decomposing the map into a quadtree and summing up the area of blocks. The efficiency of the algorithm can be improved if a better area calculation method is developed. The self-improving process of GA-focus utilizes a split-and-focus mechanism, which assumes that the optimal solution of the sub-problem can be combined to form the optimal solution of the main problem. This assumption can be violated in problems with time-varying conditions, such as environmental disturbances or dynamic obstacles. Since the collision cost with dynamic obstacles can be evaluated separately by a forward simulation (e.g. assuming that the agent moves along the desired trajectory at a constant speed, simulate and see whether the agent will collide with any dynamic obstacles), it is possible to extend GA-focus to a dynamic environment. How to use this framework in a time-varying environment can also be a promising future direction.

ACKNOWLEDGEMENTS

This paper is based on the work supported by the Autonomous Ship Consortium (ASC) with members of BEMAC Corporation, ClassNK, MTI Co. Ltd., Nihon Shipyard Co. (NSY), Tokyo KEIKI Inc., and National Maritime Research Institute of Japan. The authors are grateful for their support and collaboration on this research.

REFERENCES

[1] Patle, B. K., Babu L, G., Pandey, A., Parhi, D. R. K., and Jagadeesh, A, 2019, "A Review: On Path Planning Strategies for Navigation of Mobile Robot." Defence Technology 15 (4) (Aug): 582-606. doi:10.1016/j.dt.2019.04.011

[2] Shah, B. C. and Gupta, S. K., 2020. "Long-Distance Path Planning for Unmanned Surface Vehicles in Complex Marine Environment." IEEE Journal of Oceanic Engineering 45 (3) (Jul): 813-830. doi:10.1109/JOE.2019.2909508

[3] Chen, G., Luo, N., Liu, D., Zhao, Z., and Liang, C., 2021, "Path Planning for Manipulators Based on an Improved Probabilistic Roadmap Method." Robotics and Computer-Integrated Manufacturing 72 (Dec): 102196. doi:10.1016/j.rcim.2021.102196

[4] Russell, S. J. and Norvig. P, 2010, Artificial intelligence a modern approach, Pearson Education, Inc.

[5] Nasir, J., Islam, F., Malik, U., Ayaz, Y., Hasan, O., Khan, M. and Muhammad, M. S., 2013, "RRT*-SMART: A Rapid Convergence Implementation of RRT*" SAGE Publications, 10(7), 299. doi:10.5772/56718

[6] Fan, X., Guo, Y., Liu, H., Wei, B., and Lyu, W., 2020, "Improved Artificial Potential Field Method Applied for AUV Path Planning." Mathematical Problems in Engineering 2020 (Apr 27,): 1-21. doi:10.1155/2020/6523158.

[7] Beser, F. and Yildirim, T., 2018, "COLREGS Based Path Planning and Bearing Only Obstacle Avoidance for Autonomous Unmanned Surface Vehicles." Procedia Computer Science 131: 633-640. doi:10.1016/j.procs.2018.04.306.

[8] Bhattacharya, P. and Gavrilova, M. L., 2008, "Roadmap-Based Path Planning - using the Voronoi Diagram for a Clearance-Based Shortest Path." IEEE Robotics & Automation Magazine 15 (2) (Jun): 58-66. doi:10.1109/MRA.2008.921540.

[9] Pan, Y., Yang, Y., and Li, W., 2021, "A Deep Learning Trained by Genetic Algorithm to Improve the Efficiency of Path Planning for Data Collection with Multi-UAV." IEEE Access 9 (Jan 07): 1. doi:10.1109/ACCESS.2021.3049892.

[10] Panov, Aleksandr I., Konstantin S. Yakovlev, and Roman Suvorov. 2018. "Grid Path Planning with Deep Reinforcement Learning: Preliminary Results." Procedia Computer Science 123: 347-353. doi:10.1016/j.procs.2018.01.054

[11] Wang, X., & Jin, Y., 2022, "Work Process Transfer Reinforcement Learning: Feature Extraction and Finetuning in Ship Collision Avoidance". In International Design Engineering Technical Conferences and Computers and Information in Engineering Conference (Vol. 86212, p. V002T02A069). American Society of Mechanical Engineers.

[12] Roberge, V., Tarbouchi, M., and Labonte, G., 2013, "Comparison of Parallel Genetic Algorithm and Particle Swarm Optimization for Real-Time UAV Path Planning." IEEE Transactions on Industrial Informatics 9 (1) (Feb): 132-141. doi:10.1109/TII.2012.2198665.

[13] Lazarowska, A., 2015, "Ship's Trajectory Planning for Collision Avoidance at Sea Based on Ant Colony Optimisation." Journal of Navigation 68 (2) (Mar): 291-307. doi:10.1017/S0373463314000708.

[14] Das, P. K. and Jena, P. K., 2020, "Multi-Robot Path Planning using Improved Particle Swarm Optimization Algorithm through Novel Evolutionary Operators." Applied Soft Computing 92 (Jul): 106312. doi:10.1016/j.asoc.2020.106312

[15] Wang, Y., Yao, P., and Dou, Y., 2019, "Monitoring Trajectory Optimization for Unmanned Surface Vessel in Sailboat Race." Optik (Stuttgart) 176 (Jan): 394-400. doi:10.1016/j.ijleo.2018.09.104

[16] Tang, G., Tang, C., Claramunt, C., Hu, X., and Zhou, P., 2021, "Geometric A-Star Algorithm: An Improved A-Star Algorithm for AGV Path Planning in a Port Environment." IEEE Access 9: 59196-59210. doi:10.1109/ACCESS.2021.3070054

Copyright © 2023 by ASME

[17] Kuffner, J. J., Lavalle, S. M., 2000, "RRT-Connect: An Efficient Approach to Single-Query Path Planning" In Proceedings 2000 ICRA. Millennium Conference. IEEE International Conference on Robotics and Automation. Symposia Proceedings (Cat. No. 00CH37065), vol. 2, pp. 995-1001. IEEE. doi:10.1109/robot.2000.844730

[18] Hu, Chuanhui and Yan Jin. 2023. "Long-Range Risk-Aware Path Planning for Autonomous Ships in Complex and Dynamic Environments." Journal of Computing and Information Science in Engineering 23 (4) (Aug 01,): 1-34. doi:10.1115/1.4056064

[19] Liu, Y. and Bucknall, R., 2015, "Path Planning Algorithm for Unmanned Surface Vehicle Formations in a Practical Maritime Environment." Ocean Engineering 97 (Mar 15,): 126-144. doi:10.1016/j.oceaneng.2015.01.008.

[20] Kang, Y., Chen, W., Zhu, D., Wang, J., Xie, Q., 2018, "Collision Avoidance Path Planning for Ships by Particle Swarm Optimization", Journal of Marine Science and Technology 26(6): 777-786. doi:10.6119/JMST.201812_26(6).0003

[21] Luo, Q., Wang, H., Zheng, Y., and He, J., 2020, "Research on Path Planning of Mobile Robot Based on Improved Ant Colony Algorithm." Neural Computing and Applications 32 (6): 1555-1566. doi:10.1007/s00521-019-04172-2.

[22] Le, A. V., Nhan, N. H. K., Mohan, R. E., 2020, "Evolutionary Algorithm-Based Complete Coverage Path Planning for Tetriamond Tiling Robots" Sensors 20(2). doi:10.3390/s20020445

[23] Xin, J., Zhong, J., Yang, F., Cui, Y., and Sheng, J., 2019, "An Improved Genetic Algorithm for Path-Planning of Unmanned Surface Vehicle." Sensors 19(11) (Jun 11,): 2640. doi:10.3390/s19112640.

[24] Pehlivanoglu, Y. V. and Pehlivanoglu, P., 2021, "An Enhanced Genetic Algorithm for Path Planning of Autonomous UAV in Target Coverage Problems." Applied Soft Computing 112 (Nov): 107796. doi:10.1016/j.asoc.2021.107796.

[25] Shivgan, R. and Dong, Z., 2020, "Energy-Efficient Drone Coverage Path Planning using Genetic Algorithm." In 2020 IEEE 21st International Conference on High Performance Switching and Routing (HPSR), pp. 1-6. doi:10.1109/HPSR48589.2020.9098989.

[26] Liu, F., Liang, S., and Xian, X., 2014, "Optimal Path Planning for Mobile Robot using Tailored Genetic Algorithm." TELKOMNIKA Indonesian Journal of Electrical Engineering 12, no. 1 (2014): 1-9. doi:10.11591/telkomnika.v12i1.3127

[27] Kim, H., Kim, S., Jeon, M., Kim, J., Song, S., and Paik, K., 2017, "A Study on Path Optimization Method of an Unmanned Surface Vehicle Under Environmental Loads using Genetic Algorithm." Ocean Engineering 142 (2017): 616-624. doi:10.1016/j.oceaneng.2017.07.040.

[28] Liang, Y. and Wang, L., 2020. "Applying Genetic Algorithm and Ant Colony Optimization Algorithm into Marine Investigation Path Planning Model." Soft Computing (Berlin, Germany) 24 (11): 8199-8210. doi:10.1007/s00500-019-04414-4.

[29] Nazarahari, M., Khanmirza, E., and Doostie, S., 2019, "Multi-Objective Multi-Robot Path Planning in Continuous Environment using an Enhanced Genetic Algorithm." Expert Systems with Applications 115 (2019): 106-120. doi:10.1016/j.eswa.2018.08.008.

[30] Hao, K., Zhao, J., Yu, K., Li, C., and Wang, C., 2020, "Path Planning of Mobile Robots Based on a Multi-Population Migration Genetic Algorithm." Sensors (Basel, Switzerland) 20 (20): 5873. doi:10.3390/s20205873.

[31] Cheng, K. P., Mohan, R. E., Nhan N. H. K., and Le, A. V., 2020, "Multi-Objective Genetic Algorithm-Based Autonomous Path Planning for Hinged-Tetro Reconfigurable Tiling Robot." IEEE Access 8: 121267-121284. doi:10.1109/ACCESS.2020.3006579.

[32] Zhao, M., Ansari, N., and Hou, E. S. H., 1994, "Mobile Manipulator Path Planning by a Genetic Algorithm." Journal of Robotic Systems 11 (3): 143-153. doi:10.1002/rob.4620110302.

[33] Segota, S. B., Andelic, N., Lorencin, I., Saga, M. and Car, Z., 2020, "Path Planning Optimization of Six-Degree-of-Freedom Robotic Manipulators using Evolutionary Algorithms." International Journal of Advanced Robotic Systems 17 (2) (Mar 19,): 172988142090807. doi:10.1177/1729881420908076.

[34] Elhoseny, M., Tharwat, A. and Hassanien, A. E., 2018, "Bezier Curve Based Path Planning in a Dynamic Field using Modified Genetic Algorithm." Journal of Computational Science 25 (2018): 339-350. doi:10.1016/j.jocs.2017.08.004.

[35] Ma, J., Liu, Y., Zang, S. and Wang, L., 2020, "Robot Path Planning Based on Genetic Algorithm Fused with Continuous Bezier Optimization." Computational Intelligence and Neuroscience 2020. doi:10.1155/2020/9813040.

[36] Pehlivanoglu, Y. V., Baysal, O. and Hacioglu, A., 2007, "Path Planning for Autonomous UAV Via Vibrational Genetic Algorithm." Aircraft Engineering 79(4): 352-359. doi:10.1108/00022660710758222.

[37] Pehlivanoglu, Y. V., 2012, "A New Vibrational Genetic Algorithm Enhanced with a Voronoi Diagram for Path Planning of Autonomous UAV." Aerospace Science and Technology 16(1): 47-55. doi:10.1016/j.ast.2011.02.006.

[38] Hu, Y., Yang, S. X., 2004, "A Knowledge Based Genetic Algorithm for Path Planning of a Mobile Robot." In IEEE International Conference on Robotics and Automation, 2004. Proceedings. ICRA'04. 2004, vol. 5, pp. 4350-4355. doi: 10.1109/ROBOT.2004.1302402.

[39] Altaharwa, I., Sheta, A. and Alweshah, M., 2014, "A Mobile Robot Path Planning using Genetic Algorithm in Static Environment" Journal of Computer Science, 4(4), 341-344. doi: 10.3844/jcssp.2008.341.344

[40] Qu, H., Xing, K. and Alexander, T., 2013, "An Improved Genetic Algorithm with Co-Evolutionary Strategy for Global Path Planning of Multiple Mobile Robots."

Copyright © 2023 by ASME

Neurocomputing (Amsterdam) 120: 509-517. doi:10.1016/j.neucom.2013.04.020.

[41] Tuncer, A. and Yildirim, M., 2012, "Dynamic Path Planning of Mobile Robots with Improved Genetic Algorithm." Computers and Electrical Engineering 38(6): 1564-1572. doi:10.1016/j.compeleceng.2012.06.016.

[42] Tu, J. and Yang, S. X. "Genetic Algorithm Based Path Planning for a Mobile Robot." In 2003 IEEE International Conference on Robotics and Automation (Cat. No. 03CH37422) 1: 1221-1226. doi: 10.1109/ROBOT.2003.1241759.

[43] Shorakaei, H., Vahdani, M., Imani, B. and Gholami, A., 2016, "Optimal Cooperative Path Planning of Unmanned Aerial Vehicles by a Parallel Genetic Algorithm." *Robotica* 34(4): 823-836. doi:10.1017/S0263574714001878.

[44] Jing Xiao, Z. Michalewicz, Lixin Zhang, and K. Trojanowski. 1997. "Adaptive Evolutionary Planner/Navigator for Mobile Robots." *IEEE Transactions on Evolutionary Computation* 1(1): 18-28. doi:10.1109/4235.585889.

[45] Tsai, C. C., Huang, H. C. and Chan, C. K., 2011, "Parallel Elite Genetic Algorithm and its Application to Global Path Planning for Autonomous Robot Navigation." *IEEE Transactions on Industrial Electronics* 58(10): 4813-4821. doi:10.1109/TIE.2011.2109332.

[46] Alvarez, A., Caiti, A. and Onken, R., 2004, "Evolutionary Path Planning for Autonomous Underwater Vehicles in a Variable Ocean." *IEEE Journal of Oceanic Engineering* 29(2): 418-429. doi:10.1109/JOE.2004.827837.

[47] Arthur, D. and Vassilvitskii, S., Jan 07, 2007. "K-Means++: The Advantages of Careful Seeding." Proceedings of the Eighteenth Annual ACM-SIAM Symposium on Discrete Algorithms 2007, 1027−1035.

[48] Sturtevant, N. R. 2012. "Benchmarks for Grid-Based Pathfinding." *IEEE Transactions on Computational Intelligence and AI in Games* 4(2): 144-148. doi:10.1109/TCIAIG.2012.2197681.

[49] Weise, J., and Mostaghim, S., 2021, "A scalable many-objective pathfinding benchmark suite." *IEEE Transactions on Evolutionary Computation* 26(1): 188-194. doi: 10.1109/TEVC.2021.3089050

[50] Liu, C., Liu, A., Wang, R., Zhao, H., and Lu, Z., 2022, "Path Planning Algorithm for Multi-Locomotion Robot Based on Multi-Objective Genetic Algorithm with Elitist Strategy." *Micromachines* 13(4): 616. https://doi.org/10.3390/mi13040616

[51] Ahmed, F. and Deb, K., 2013, "Multi-objective optimal path planning using elitist non-dominated sorting genetic algorithms." *Soft Computing* 17: 1283-1299. https://doi.org/10.1007/s00500-012-0964-8

Proceedings of the ASME 2023
International Design Engineering Technical Conferences and
Computers and Information in Engineering Conference
IDETC-CIE2023
August 20-23, 2023, Boston, Massachusetts

DETC2023-116695

AN ENHANCED TIMED ELASTIC BAND METHOD FOR AUTONOMOUS NAVIGATION AND ITS COLLISION AVOIDANCE RELIABILITY ANALYSIS

Zhimin Xi
Department of Industrial and Systems Engineering
Rutgers University - New Brunswick
Piscataway, New Jersey 08854
Email: zhimin.xi@rutgers.edu

ABSTRACT

The timed elastic band (TEB) method is an optimization-based navigation algorithm which uses the idea of an elastic band to connect an agent's current pose to its goal position. The method is composed of three main levels: i) a feasible global path, ii) deforming collision-free path in real time, and iii) a control method to move the robot along the path. Static and moving obstacles deform the elastic band but the agent will always be pulled toward the goal position while avoiding obstacles. The key issue is the required computation efficiency to calculate the collision-free path in real time under dynamically changing environment. In literature, various case studies have been implemented to demonstrate the performance considering various constraints. However, its collision avoidance reliability under safety critical navigation scenarios has not yet been studied. The major contributions of this paper are summarized as follows. Firstly, an enhanced version of the TEB (i.e., eTEB) is proposed to improve the computational efficiency by explicitly considering non-holonomic constraints of the TEB method. Secondly, collision avoidance reliability analysis of the eTEB method is systematically conducted considering non- reactive moving obstacles. Results indicate that the eTEB method is much more reliable than the DWA for head-on collision avoidance with a single non-reactive moving obstacle at a time.

1 Introduction

The method of elastic band (EB) was proposed in [1], referring to a deforming collision-free path in real time. The method is composed of three main levels: i) a feasible global path, ii) deforming collision-free path in real time, and iii) a control method to move the robot along the path. The key issue is the required computation efficiency to calculate the collision-free path in real time under dynamically changing environment. Hence, the concept of bubble, as a subset of the collision-free space, was proposed to improve the computation efficiency. The method of timed elastic band (TEB) [2] was proposed to consider temporal aspects of the motion which is related to the dynamic constraints of the robot. As such, TEB converts a collision-free path to a collision-free trajectory as each intermediate way point toward the goal position is associated with the time. The method was formulated as a weighted multi-objective optimization problem. Particularly, modular penalty functions were designed for easy incorporation of various constraints. The TEB method was applied to an autonomous vehicle application for a lane change maneuver with conflicting objectives [3]. With the modular penalty function, constraints such as road boundary, minimum distance with static and moving obstacles, and vehicle maximum acceleration were incorporated in a weighted multi-objective function. The weight coefficients of these penalty functions are user defined, which can result in different driving behaviors. The combined trajectory planning and control using the TEB was proposed for a simplified nonlinear double-track vehicle model [4]. In addition to vehicle states (e.g., positions and poses), control

Copyright © 2023 by ASME

inputs were included as optimization variables in the TEB framework. The vehicle dynamic model was used to create a vehicle states mapping between two consecutive configurations. Such equality constraints were further formulated as penalty functions. Optimization of the overall weighted multi-objective function was designed with a quadratic functional form for achieving reasonable computational efficiency. To achieve globally optimal trajectory under relatively complex navigation conditions, a set of initial trajectories with relevant typologies were firstly obtained, and they were then optimized in parallel to find the optimal trajectory [5]. The method was modified to have a flexible goal position in a merge-into-traffic scenario to achieve better stability and smoothness of the trajectory [6] and was also applied to a parallel parking scenario for a car-like robot [7]. With a proxemic motion model for trajectory prediction, the TEB was applied to a social robot navigation with human encounters [8].

While most aforementioned study is based on simulations, experiment study was conducted to compare motion accuracy and repeatability for TEB, EB, and dynamic window approach (DWA) using the TurtleBot 3 robot [9]. The TEB was found as the best method for quickly avoiding obstacles compared to the other two methods, and the DWA shows the overall best repeatability. Other than autonomous navigation problems, the TEB was also applied to manipulation motion planning problems and showed comparable performances with the state-of-the-art methods in that field [10]. In addition, it was concluded that hard constraints with a sequential quadratic programming (SQP) solver may be more desirable instead of formulating the hard constraints as soft constraints, i.e., penalty functions. EgoTEB [11] was proposed to replace world-centric representation to egocentric, perception space representation to achieve better computational efficiency, considering the LiDAR perception system. The TEB was also integrated with a proposed detection and tracking system based on a 2D LiDAR system to navigate a TurtleBot 2 robot for collision avoidance with static and moving obstacles [12]. Under a head-on collision risk simulation setting with non-reactive moving obstacles, the method obtains a 91.12% successful rate for collision avoidance. Under actual experiments, collision avoidance cannot be guaranteed due to many factors including the computation resource limitation and the communication delay in the robotics operating system (ROS) structure.

While TEB is gaining popularity, its collision avoidance reliability under operation uncertainty is unknown, needlessly to mention the reliability comparison with other classical navigation methods. Hence, the major contributions of this paper are summarized as follows. Firstly, an enhanced version of the TEB (i.e., eTEB) is proposed to improve the computational efficiency by explicitly considering non-holonomic constraints of the TEB method. Secondly, collision avoidance reliability analysis of the eTEB method is systematically conducted considering non-reactive moving obstacles.

The rest of the paper is organized as follows. Section 2

briefly reviews technical components of the TEB. Section 3 reviews collision avoidance reliability using a Bayesian approach. Section 4 proposes the eTEB to improve the computational efficiency of the TEB method. Section 5 presents reliability analysis framework for comparison between different navigation methods under various operation conditions. Section 6 employs case studies considering non-reactive moving obstacles to characterize the collision avoidance reliability of the eTEB method. Finally, section 7 concludes the study.

2 Review of the TEB

Modular penalty function is used in the TEB to transform any inequality constraint into a penalty function with a quadratic form. Equality constraints such as the non-holonomic constraints are also formulated as quadratic penalty functions. Eventually, the TEB formulates the local navigation problem as a nonconstrained optimization problem as

$$\min_{\mathbf{P}} \sum_k w_k f_k(\mathbf{P}) \tag{1}$$

where w_k is a user-defined weight coefficient for the kth sub-objective function; $f_k(\cdot)$ is the kth sub-objective function determined by the parameter vector $\mathbf{P}$. The objective is to minimize Eq.1 by identifying the optimal parameter vector $\mathbf{P}^*$ which, in theory, determines an optimal collision-free trajectory. The parameter vector $\mathbf{P}$ is composed of a set of agent (or ego vehicle) states associated with time information from a start position to a goal position. Agent's state is usually described by its position and pose as $s_i = (x_i, y_i, \beta_i)$ where s_i is the agent's i^{th} state represented by its position along $x-$ and $y-$ directions and its heading direction β, as illustrated in Fig. 1. The time information is the time needed for the agent moving from a previous state to the current state donated as ΔT_i. Hence, the ith parameter vector $\mathbf{P}_i = (s_i, \Delta T_i)$. If the total number of states is donated by n, then the parameter vector $\mathbf{P} = (\mathbf{P}_1, ..., \mathbf{P}_i, ..., \mathbf{P}_n)$. Therefore, a relatively large set of parameters, i.e., $n \times 4$, need to be determined through the optimization.

Given the defined states $\mathbf{P}$, a major unique component of the TEB method is the modular definition of the sub-objective (or penalty) function $f_k(\cdot)$. They typically include: i) a distance function to way points generated by a global path planner, ii) a distance function to obstacles, iii) velocity constraints, iv) acceleration constraints, and v) agent's non-holonomic constraints, in addition to a function to complete the path with the shortest time. The list can also be extended to include more sub-objective functions in Eq. 1. For example, as shown in Fig. 1, the shortest distance between agent's states to the waypoint should be the smaller the better. On the other hand, the distance to the obstacle should be the larger the better. For any given agent's states,

Copyright © 2023 by ASME

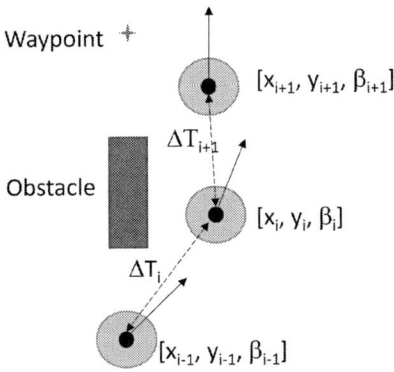

FIGURE 1: Illustration of the TEB states

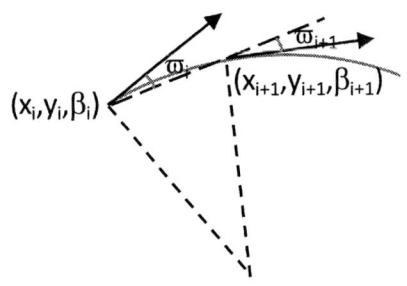

FIGURE 2: Illustration of a non-holonomic constraint between two consecutive agent states

velocity and acceleration can be calculated and they are further compared to the corresponding velocity and acceleration constraints. As a result, these constraints violations should be the smaller the better, and the degree of the violation can be generally quantified by a 'distance' function as well.

For inequality constraints with the smaller the better case, e.g., the distance between agent and an intermediate waypoint, the penalty function is formulated as

$$f = \begin{cases} (\frac{d-(d_r-\varepsilon)}{S})^2 & \text{if} \quad d > d_r - \varepsilon \\ 0 & \text{elsewhere} \end{cases} \quad (2)$$

where d is a context related variable of the penalty function; d_r is the limit value; ε is a small shift of the limit value such that a penalty value is created when d reaches the boundary d_r; and S is a user-defined scale parameter. For the smaller the better case, the penalty value is 0 as long as $d < d_r - \varepsilon$; otherwise, the penalty increases by a square function. The square function is a typical adoption but can also be replaced by a higher-order power function. The context related variable d is directly determined by the parameter vector **P**. Considering the distance between agent and an intermediate waypoint, the minimum distance, i.e., the variable d, can be calculated between the waypoint and agent's states. Such a penalty function can be used for any inequality constraints that should be bounded to be less than a limit value such as the maximum velocity and acceleration of the agent. On the other hand, the distance between agent and obstacles should be the larger the better. Hence, the penalty function can be modified as

$$f = \begin{cases} \left(\frac{(d_r+\varepsilon)-d}{S}\right)^2 & \text{if} \quad d < d_r + \varepsilon \\ 0 & \text{elsewhere} \end{cases} \quad (3)$$

Agent's non-holonomic kinematics is represented by equal-ity constraints and can be further formulated as a sub-objective function as

$$f(s_i, s_{i+1}) = \left\| \left[\begin{pmatrix} cos\beta_i \\ sin\beta_i \\ 0 \end{pmatrix} + \begin{pmatrix} cos\beta_{i+1} \\ sin\beta_{i+1} \\ 0 \end{pmatrix} \right] \times \mathbf{d}_{i,i+1} \right\|^2 \quad (4)$$

where $\mathbf{d}_{i,i+1}$ is a direction vector defined as

$$\mathbf{d}_{i,i+1} = \begin{pmatrix} x_{i+1} - x_i \\ y_{i+1} - y_i \\ 0 \end{pmatrix} \quad (5)$$

As illustrated in Fig. 2, Eq. 4 equals to zero if the summation of two consecutive unit velocity vector exactly aligns with the direction vector $\mathbf{d}_{i,i+1}$ such that the cross product equals to zero. Otherwise, the penalty value increases as the angle difference deviates from zero. Such a non-holonomic constraint is based on the assumption that agent maintains the same translational and angular velocity within two consecutive states. As such, ω_i and ω_{i+1} shown in Fig. 2 must be equal because the translational velocity is tangent to the same circle. By ignoring actual motor dynamics, another hidden assumption is that the demanded new translational and angular velocity within the next control cycle can be immediately achieved if they do not violate acceleration constraints. This assumption was also used in other navigation methods such as the DWA considering a relatively short period control cycle time (e.g., 100 ms). Finally, the sub-objective of the shortest path time is formulated as

$$f = \sum_{i}^{n} (\Delta T_i)^2 \quad (6)$$

Weight selection of each sub-objective function in Eq. 1 is subjective and user-defined. A general practice is to assign

Copyright © 2023 by ASME

the same weight coefficient except for non-holonomic constraints (e.g., a weight coefficient of 1000 or 10,000 for non-holonomic constraints vs. a weight coefficient of 1 for others). A major issue of the TEB is the computational efficiency in real-time applications since the optimization contains a total number of $n \times 4$ parameters. For a 10 second option of the agent with an assumed average of 0.1 second control cycle, 400 parameters are employed for the optimization in real-time, and the optimal solution needs to be obtained within the control cycle such as 0.1 second. To maintain computational efficiency, the problem was transformed into a hyper-graph and solved by graph optimization using a g2o framework [13].

3 Review of Collision Avoidance Reliability Analysis

Collision avoidance reliability can be defined as the probability of collision under defined operation conditions for a given period of time. In a typical time-dependent reliability analysis, time period usually means the designed lifetime of the system. Such a reliability analysis is more comprehensive by considering various factors during the service lifetime of the system. In this study, we focus on time-independent reliability analysis as the collision avoidance reliability of the TEB will not degrade as time goes by. To be specific, the study is focused on the method itself other than hardware failure or malicious cyber-attack in actual service time of the system. As such, collision avoidance reliability is defined as the probability of collision under defined operation conditions. Operation conditions can be specified by the environment condition and operating characteristics of the agent. The former excludes the agent and the later includes all factors related to the agent. Depending upon actual applications, specifying operation condition unambiguously could be difficult. For example, the environment condition is more certain for an industry robot operating in a warehouse than an autonomous car driving on the road. In this study, scenario-based environment conditions are specified and the environment uncertainty is created under given scenarios. Operating characteristics of the agent include its sensing range, kinematics, and dynamics are given deterministically without considering uncertainty.

A reliability (or probability of collision avoidance) function is formulated as [14]

$$R(\mathbf{X}, \mathbf{d}) = Pr(G(\mathbf{X}, \mathbf{d}) = 0) \tag{7}$$

where $\mathbf{X}$ is a vector of random variables defining the environment uncertainty; $\mathbf{d}$ is a vector of agent's parameter defining the operating characteristics of the agent which is deterministic in this study; $G(\cdot)$ is a binary performance indicator function in which '0' means no collision and '1' means collision; and $Pr(\cdot)$ is the probability operator. It is worth noting that change of the agent's parameter $\mathbf{d}$ can influence the reliability. However, this would

mean to physically change the design of the autonomous agent, which will not be considered in this study.

Reliability analysis under a Bayesian updating framework is formulated as [14]

$$Pr(R|E) = \frac{Pr(E|R)Pr(R)}{Pr(E)} \tag{8}$$

where E is the evidence of observing the number of $G(\cdot) = 0$ under defined environment uncertainty $\mathbf{X}$; $Pr(R)$ is a prior distribution of reliability; $Pr(E|R)$ is the likelihood function, i.e., the probability of observing E given the R value; and $Pr(E)$ is the total probability of the evidence, which is calculated as

$$Pr(E) = \int_0^1 Pr(E|R)Pr(R)dR \tag{9}$$

Eq. 8 models uncertainty of the reliability analysis given observed evidence (i.e., the number of collision-free simulations S among total number of simulations N).

It is reasonable to employ a non-informative prior belief about the reliability. In other words, reliability could range from 0% to 100% with equal probability. Hence, the prior distribution of reliability is modeled as a standard Uniform distribution as

$$Pr(R) = 1 \quad R \in [0, 1] \tag{10}$$

It is worth noting that reliability is modeled as a continuous random variable in the range from 0 to 1. Hence, Eq. 10 is the probability density function value given any value of R in the defined domain. The likelihood function, hence, can be calculated by a Binomial distribution as

$$Pr(E|R) = \binom{N}{N-S} R^S (1-R)^{N-S} \tag{11}$$

Applying Eqs. 10-11 in Eq. 8, the posterior reliability distribution is calculated as

$$Pr(R|E) = \frac{R^S(1-R)^{N-S}}{\int_0^1 R^S(1-R)^{N-S}dR} \tag{12}$$

which is a Beta distribution with two parameters $p_1 = S + 1$ and $p_2 = N - S + 1$. Without conducting any simulation, i.e., $N = S = 0$, Eq. 12 converges to a standard Uniform distribution as the prior distribution in Eq. 10.

By adoption of the Bayesian approach, reliability and its uncertainty are explicitly modeled as a function of total number of

Copyright © 2023 by ASME

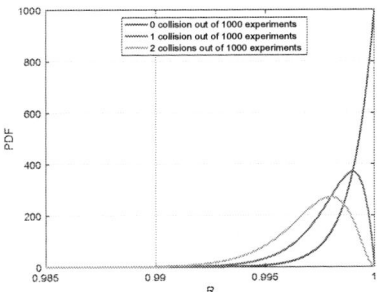

FIGURE 3: Illustration of reliability modeling using the Bayesian approach with 0, 1, and 2 collisions out of 1,000 simulations

simulations and the observed evidence. Not only the statistical moments (e.g., mean and variance) but also the reliability confidence can be easily calculated based on the Beta distribution. In particular, mean and variance of the reliability are computed as

$$R_{mean} = \frac{p_1}{p_1 + p_2} \tag{13}$$

$$R_{variance} = \frac{p_1 p_2}{(p_1 + p_2 + 1)(p_1 + p_2)^2} \tag{14}$$

It is worth noting that reliability distribution is in general non-symmetric based on the Beta distribution. To illustrate the effect of the reliability modeling using hypothetical data, we assume that a total number of 1,000 simulations were conducted given the environment uncertainty. Among 1,000 simulations, zero, one, and two collisions were observed for three hypothetical results. Fig. 3 shows the posterior reliability distribution under above three hypothetical cases. As clearly observed, all distributions are non-symmetric. In addition, less number of collisions reduce the distribution variance and approach higher reliability with confidence.

4 The enhanced TEB

Three modifications are proposed for the purpose of improving the computational efficiency of the TEB. The first modification is to explicitly consider the non-holonomic constraint without the need of using the penalty function. Consequently, the heading direction parameter (β_i) can be removed in the optimization. The second modification is to enforce a fixed time interval (i.e., $\Delta T_i = \Delta T = c$) as the control cycle time of the agent. Consequently, the time interval parameter (ΔT_i) is removed. The two modifications directly reduce the number of parameter size from $n \times 4$ to $n \times 2$ for optimization. The third modification is to define a time threshold for applying the TEB method. Changing the whole trajectory from a start position to a goal or way-point position dynamically may not be necessary for a relatively

long travel distance. Firstly, it is unnecessary to spend computation resources exceeding agent's reachable region within the time threshold. Secondly, change of the trajectory beyond the threshold time also requires the trajectory prediction of moving obstacles for more than the threshold time, which is most likely not accurate. With above three modifications, we make the TEB much more efficient and name it as an efficient TEB or eTEB. The proposed modifications are elaborated in details in the following subsections.

4.1 Reconsideration of the non-holonomic constraint in the TEB

Given an initial pose of the agent (x_0, y_0, β_0), agent's heading direction at the next state can be uniquely determined given its position at (x_1, y_1) based on the non-holonomic equality constraint. Similarly, heading directions of following states can be uniquely determined given the position information. As shown in Fig. 2, ω_i can be calculated as

$$\omega_i = atan(\frac{y_{i+1} - y_i}{x_{i+1} - x_i}) - \beta_i \tag{15}$$

where the result may be either a positive or negative angle. Because of the equality constraint as $\omega_i = \omega_{i+1}$, heading direction of the agent at the next state is calculated as

$$\beta_{i+1} = \beta_i + 2\omega_i \tag{16}$$

Hence, the heading direction parameter β_i in the TEB can be removed as long as the initial heading direction β_0 is known. Consequently, by explicitly calculating the heading directions using Eqs. 15-16, the penalty function for the non-holonomic constraint can be eliminated since the constraint will always be satisfied.

4.2 Fixed time interval and a revised penalty function

The fixed time interval $\Delta T = c$ directly remove another set of state parameter ΔT_i in the TEB. Considering a time period τ (e.g., 3 seconds), the total number of states (x_i, y_i) that need to be optimized are determined. It is worth noting that the states (x_i, y_i) are explicitly associated with the heading direction and the time information. As such, the basic TEB principle for trajectory optimization keeps the same.

One consequence of using the fixed time interval for a time period τ is the elimination of the sub-objective function of the shortest path time in Eq. 6. Eq. 6 makes the agent to adopt the fastest possible velocity so that the path can be completed with the shortest time. Without this function, the agent can move slowly by demanding more states to complete the path. To address this issue, all states (x_i, y_i) are proposed to be associated

Copyright © 2023 by ASME

with the goal (or waypoint) position. Consequently, the distance to the goal (or waypoint) penalty function pulls all agent's states toward the goal position to minimize the penalty function, resulting in significantly large velocities. This, however, would violate the maximum velocity (including acceleration) constraint and cause large penalty function values related to velocity and acceleration. The conflicted sub-objective functions will lead to an optimal solution as reviewed in Eq. 1.

4.3 Constraints violation and correction

Formulation of the penalty function in the TEB cannot guarantee constraints satisfaction, which is the same in the proposed eTEB method other than the non-holonomic constraints. As such, feasibility must be checked with the optimized solution. We can define safety (e.g., collision) violation as hard constraints and others as soft constraints. For hard constraints violation, agent reduces the velocity as much as possible. In particular, collisions for any agent's states within the time threshold τ violate the hard constraints. For soft constraints violation such as the maximum translational and angular velocity of the agent, corrections are made to set the agent's control inputs within the limits. For violation of the translational and angular velocity acceleration, corrections are firstly made to the maximum value of the acceleration, then correct the corresponding translational and angular velocity.

5 Reliability Analysis Framework for Reliability Characterization and Comparison

To compare collision avoidance reliability for different navigation methods, a reliability analysis framework is proposed for reliability characterization which can be later used for reliability comparison as shown in Fig. 4. Firstly, agent's operation conditions are specified by operation parameters. Secondly, a set of collision avoidance reliability analyses are conducted using specific navigation methods at given operation conditions. Thirdly, meta-modeling is conducted based on the computed collision avoidance reliability at different operation conditions so that reliability can be approximated at new operation conditions. Fourthly, reliability limit state functions for specific navigation methods are constructed for comparison, which include the whole defined operation condition. Lastly, the reliability limit state function is validated before making final conclusions. General guidelines for each technical component or step are provided as follows.

Firstly, agent's operation condition could be difficult to define unambiguously, resulting in a vague way to specify the operation parameters. For example, in a navigation configuration space with 10 moving obstacles, operation related parameters can include the number of moving obstacles, their current positions, velocities, goal positions, the agent's states, and so on. It is

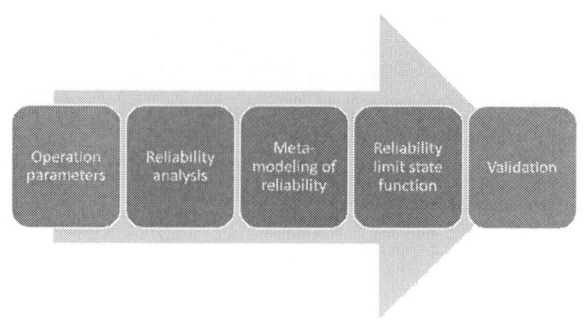

FIGURE 4: Reliability analysis framework for collision avoidance

almost impossible to define the operation condition considering all factors. Thus, a general guideline is to use observable or controllable parameters to characterize agent's operation conditions. In addition, these parameters should be relatively static rather than dynamic. In the above example, operation parameters can include the number of moving obstacles and the velocity limit of the agents. Other parameters, with inherent randomness, should be used to quantify the collision avoidance reliability under the specified operation condition. Secondly, reliability analysis is recommended to be conducted using the Bayesian approach as described in Sec. 4.2 so that distribution information is available to account for the lack of simulation or experiment data. Besides, the distribution information is critical for the validation study at the last step, which will be explained later. Thirdly, meta-modeling of the collision avoidance reliability should cover the entire intended operation conditions specified by the operation parameters. Hence, collision avoidance reliability analysis should be conducted at a set of sample locations defined by the operation parameter. Techniques such as the design of experiments are useful in addition to the selection of a meta-modeling method. In this paper, a full factorial design method will be used to choose sample locations, and then the Gaussian process (GP) regression method is employed to construct a meta-model of the reliability over the whole defined operation parameter space. Fourthly, reliability limit state function can be obtained with the aid of the GP regression model. Reliability limit state function is a required reliability value represented as a nonlinear function of the operation parameter, which is an inverse operation of the GP regression model. Such a limit state function divides the operation parameter space into reliable and unreliable regions. As such, a ratio of reliable region to the whole intended operation domain can be computed and compared for different navigation methods, if the difference is not obvious. Finally, validation of the limit state function should be carried out at a few specified operation conditions to demonstrate the validity of the approximated nonlinear limit state function. Since the collision avoidance reliability is represented by a distribution using the

Copyright © 2023 by ASME

TABLE 1: Agent's mobility parameters

Name/Symbol	Value
v_{max}	$2.78 \ m/sec.$
ω_{max}	$106°/sec.$
$\dot{v}_{max}$	$0.93 \ m/sec.^2$
$\dot{\omega}_{max}$	$35.3°/sec.^2$
Agent radius	$1 \ m$
Projected time τ	$3 \ sec.$
Control cycle ΔT	$0.1 \ sec.$

Bayesian approach, its validation involves the distribution comparison at selected new operation conditions. The GP regression model enables such a reliability distribution approximation to facilitate the validation. It is worth noting, however, validation under uncertainty in this field does not have a standard procedure yet. Hence, this paper ignores the validation study for simplicity while acknowledging its importance.

6 Reliability Studies of the eTEB Method

In this study, unless otherwise noted, the agent's default mobility parameters are provided in Table 1. The studied agent is a differential drive robot and both left and right wheel can only rotate along with one direction. The projected time τ allows the agent to stop from its maximum velocity to 0, to ensure collision-free capability for static obstacles. Agent's sensing range is set as $30 \ m$ which is large enough in all case studies.

This study mainly focuses on collision avoidance reliability analysis and the corresponding computational efficiency of the method. We use the number of function calls in each control cycle ΔT to indicate the computational efficiency of the method. The eTEB has variant function calls due to the optimization process. Technically, eTEB employs a total of 60 state parameters (x_i, y_i) for $i = 1, ..., 30$ that need to be optimized in each control cycle, and the number of function calls is determined by the optimization. Due to significant computational efficiency improvement compared to the original TEB, with otherwise more than 120 state parameters, the eTEB was implemented in Matlab and a sequential quadratic programming (SQP) method was used for optimization. For eTEB, six sub-objective (or penalty) functions are defined as: i) distance to the goal position, ii) distance to the obstacle, iii) left wheel velocity, iv) right wheel velocity, v) left wheel acceleration, and vi) right wheel acceleration. The corresponding weight coefficients are assigned as $(w_1 = 1, w_2 = 100, w_3 = 1, w_4 = 1, w_5 = 0.01, w_6 = 0.01)$.

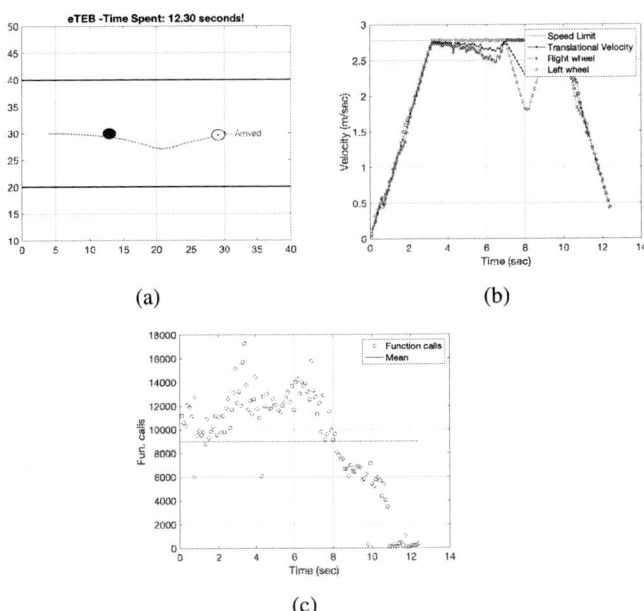

FIGURE 5: Head-on collision avoidance of the eTEB with a 20-meter wide collision-free space

6.1 Head-on collision avoidance scenarios with a non-reactive moving obstacle

An agent and a non-reactive moving obstacle are located at (4,30) and (30,30), respectively. Agent's goal position is at (30,30) and the obstacle moves directly toward the agent direction with a constant velocity of $1.39 \ m/sec.$. Obstacle's moving speed, initial position, and agent's collision-free space determine the environment uncertainty that will be studied in this case study. We firstly assign a wide road boundary (i.e., 20 meters) along y-axis so that the agent has large space for collision avoidance. As shown in Fig. 5a, agent with the eTEB method avoids the obstacle and arrives the goal position using 12.3 seconds. Its velocity history for both wheels are shown in Fig. 5b. Fig. 5c presents the number of function calls in each control cycle and the average is about 9,000 function calls. Each function call means the perturbation of the state parameter for evaluation of the objective function defined in Eq. 1.

A narrower collision-free space was then assigned to reduce the boundary along y-axis from previous 20 meters to 10 meters and results are shown in Fig. 6. The eTEB performs similar as before. The mission time is slightly increased to 12.5 seconds, and the average number of function calls is also slightly increased to about 9,500. This should be mainly attributed to the narrower boundary.

Obstacle's speed was then increased to $2.78 \ m/sec.$ which is the maximum speed of the agent. The eTEB can still avoid the collision with even a slightly shorter time period (i.e., 11.9

Copyright © 2023 by ASME

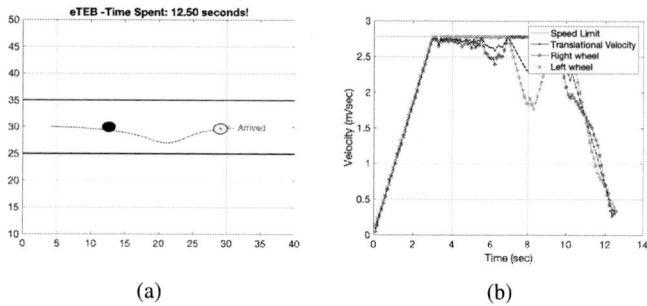

(a)   (b)

FIGURE 6: Head-on collision avoidance of the eTEB with a narrow collision-free space

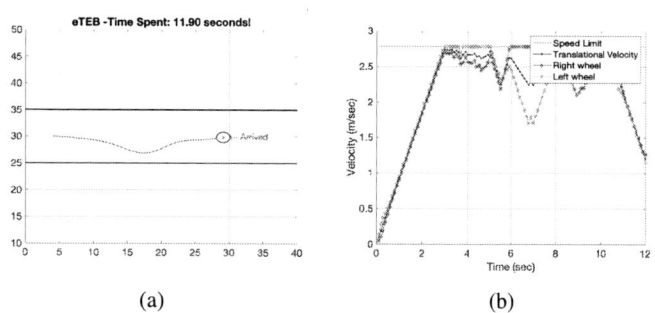

(a)   (b)

FIGURE 7: Head-on collision avoidance of the eTEB with a narrow collision-free space and a fast moving speed of the obstacle

seconds) to complete the mission. Average number of function calls is at about 9,500.

The last scenario is to assume that the computation capacity is limited and only 8,000 function calls can be completed within each control cycle. Under the same condition as described in Fig. 7, the eTEB fails to avoid the collision and results are shown in Fig. 8. The velocity history shows that the agent was hesitating for some time instance in which the velocity reduction is due to the violation of the hard constraint. Because of that, the agent missed the opportunity to avoid the collision. Although not shown here in the paper, when the speed of the moving obstacle reduces a little bit, the agent is again able to avoid the collision with the same computation limitation.

6.2 Head-on collision avoidance reliability analysis with a non-reactive moving obstacle

6.2.1 Operation parameters
Operation condition of the head-on collision is specified by two operation parameters as i) obstacle's moving speed and ii) width of the road boundary. In this case study, obstacle's moving speed is defined in the range from 1 $m/sec.$ to 2.87 $m/sec.$, and width of the road boundary is from 7.6 meters to 15 meters. Since the agent is static at the beginning, a collision would be unavoidable if the speed of the

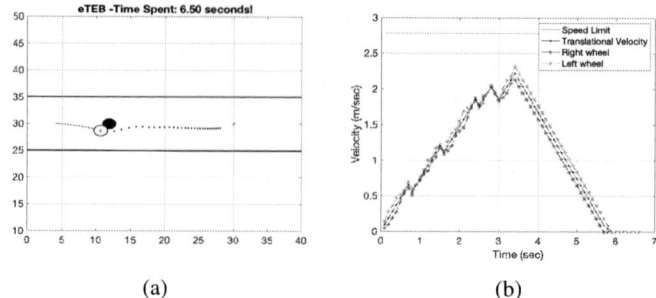

(a)   (b)

FIGURE 8: Head-on collision avoidance of the eTEB with a narrow collision-free space, a fast moving speed of the obstacle, and limited computation

moving obstacle is too high. As such, the maximum speed of the moving obstacle is limited to be 2.87 $m/sec.$ which is the maximum achievable speed of the agent. As for the width of the road boundary, 6 meter is the minimum possible width since the obstacle may be at the center of the road with a 1 meter radius as the agent. To provide some buffer space and avoid a strict requirement for a global optimum which cannot be guaranteed by a local optimal algorithm as employed by the eTEB, the minimum road boundary considered in this study is 7.6 meters. Under the extreme case, agent can ideally avoid collision by choosing either side of the road and with a buffer distance of 0.4 meter away from the obstacle and the road boundary. To add process uncertainties for reliability analysis, obstacle's initial position is defined as a random variable following a Uniform distribution along the vertical (or y-axis) direction while keeping the x-axis position unchanged. In other words, the obstacle may not be at the center of the road and it could be at any vertical location within the road boundary. Similarly, the initial vertical position of the agent is also defined as a random variable following a Uniform distribution within the road boundary. However, the agent's goal position is always at (30,30). Table 2 summarizes the operation parameters and the assumed uncertainty considered in this study.

6.2.2 Collision avoidance reliability analysis
Reliability analysis was conducted under multiple operation conditions specified by two operation parameters defined in Table 2. In particular, a 3-level factorial design of experiment was conducted and the collision avoidance reliability was computed under nine configurations. For each configuration, 100 random scenarios were generated considering the initial position randomness from the agent and the obstacle. Table 3 summarizes the results with following major observations. Firstly, the eTEB method is much more reliable than the DWA for the head-on collision avoidance. There are four configurations that the eTEB method is collision-free among 100 simulations, resulting in the maximum achiev-

Copyright © 2023 by ASME

638

TABLE 2: Operation parameter and environment uncertainty for a head-on collision reliability analysis

Variable	Property	Parameters
Width of road boundary w	Deterministic	$7.6\ m$ - $15\ m$
Obstacle's velocity v	Deterministic	$1\ m/sec.$ - $2.78\ m/sec.$
Y-axis of the obstacle	Uniform	[30-w/2+1, 30+w/2-1]
Y-axis of the agent	Uniform	[30-w/2+1, 30+w/2-1]

TABLE 3: Collision avoidance reliability under nine operation configurations for the eTEB and DWA

Configurations	Reliability (Mean)		# of Collisions	
	eTEB	DWA	eTEB	DWA
$w=7.6, v=1$	93.14%	78.43%	6	21
$w=7.6, v=1.89$	95.10%	89.22%	4	10
$w=7.6, v=2.78$	94.12%	84.31%	5	15
$w=11.3, v=1$	99.02%	86.27%	0	13
$w=11.3, v=1.89$	99.02%	87.25%	0	12
$w=11.3, v=2.78$	96.08%	87.25%	3	12
$w=15, v=1$	99.02%	90.20%	0	9
$w=15, v=1.89$	99.02%	91.18%	0	8
$w=15, v=2.78$	95.10%	84.31%	4	15

able mean reliability of 99.02% using the Bayesian approach. Secondly, smaller road width usually leads to more collisions and thus lower reliability. However, this effect is not obvious when the road width is large enough for collision avoidance. Thirdly, higher obstacle's speed often leads to more collisions and thus lower reliability. However, this effect seems not applicable when the road width is small enough such as 7.6 meters as shown in Table 3.

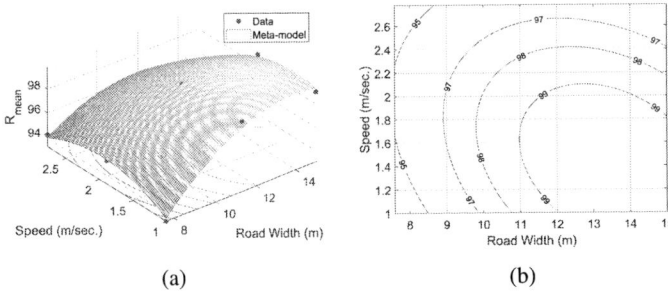

(a)   (b)

FIGURE 9: Meta-modeling of reliability and the reliability limit state function from the eTEB method

6.2.3 Reliability meta-modeling and the limit state function GP regression was employed to construct meta-models of the reliability for both the eTEB method and the DWA, given the results in Table 3. With the aid of the meta-model, reliability at new configurations within the specified operation space can be approximated as shown in Fig. 9a and Fig. 10a. Furthermore, an inverse operation to characterize the operation space that can meet the reliability requirement can be conducted to obtain the reliability limit state function as shown in Fig. 9b and Fig. 10b. Two major observations are summarized as follows. Firstly, collision avoidance reliability shows different degree of nonlinear behaviors with respect to the operation parameters for the eTEB method and the DWA. The less degree of nonlinearity of the eTEB method is more desirable because it is easier to identify a feasible operation space to meet or exceed the required reliability. For example, as shown in Fig. 9a, if the required reliability is at least 99%, the road width should be more than 12 meters and obstacle's speed needs to be less than about $2\ m/sec.$. Secondly, the trend is clear from both limit state functions that reducing the road width and increasing the speed would reduce the collision avoidance reliability, although the two methods show different nonlinear behaviors. Similar observation may be obtained from Table 3, but the reliability limit state function presents much richer information about the nonlinearity for a particular navigation method.

To draw more solid conclusions from the reliability limit state function, validation needs to be conducted by running more simulations at a few extra configurations. If it is not validated, the reliability meta-model needs to be updated and then followed by the validation again. The process may be iterated until reach the requirement for the intended usage of the model.

7 Conclusion

This paper, for the first time, conducts systematical collision avoidance reliability analysis for the TEB navigation method as this method is gaining popularity over some traditional method

Copyright © 2023 by ASME

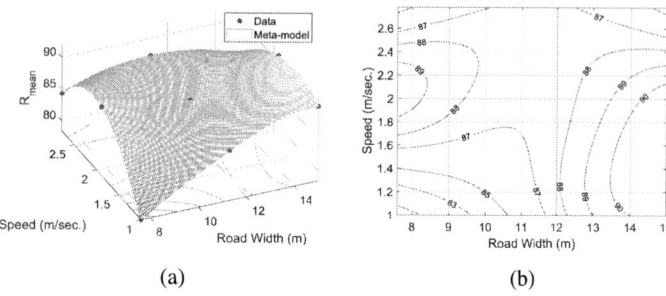

(a)   (b)

FIGURE 10: Meta-modeling of reliability and the reliability limit state function from the DWA

such as the DWA. The original optimization formulation of the TEB method is enhanced for the purpose of improving the computational efficiency by explicitly considering the non-holonomic constraints of a differential drive robot and fixing the time differences among different agent poses. For fair collision avoidance reliability comparison between different navigation methods, reliability analysis framework is proposed by firstly conducting reliability analysis at specified operation conditions and then constructing a reliability meta-model to obtain the reliability limit state function. As such, reliability comparison between different navigation methods can be conducted for the whole specified operation condition rather than relying on the comparison for countless scenarios. Case studies reveal that the eTEB method is much more reliable than the DWA for head-on collision avoidance with a single non-reactive moving obstacle at a time.

REFERENCES

[1] S. Quinlan and O. Khatib, "Elastic bands: Connecting path planning and control," in *Proceedings - IEEE International Conference on Robotics and Automation*, vol. 2, 1993, pp. 802–807.

[2] C. Rösmann, W. Feiten, T. Wösch, F. Hoffmann, and T. Bertram, "Trajectory modification considering dynamic constraints of autonomous robots," in *7th German Conference on Robotics, ROBOTIK 2012*, 2012, pp. 74–79.

[3] M. Keller, F. Hoffmann, T. Bertram, C. Hass, and A. Seewald, "Planning of optimal collision avoidance trajectories with timed elastic bands," in *IFAC Proceedings Volumes (IFAC-PapersOnline)*, vol. 19, 2014, pp. 9822–9827.

[4] C. Götte, M. Keller, C. Haß, K. . Glander, A. Seewald, and T. Bertram, "A model predictive combined planning and control approach for guidance of automated vehicles," in *2015 IEEE International Conference on Vehicular Electronics and Safety, ICVES 2015*, 2016, pp. 69–74.

[5] C. Rösmann, F. Hoffmann, and T. Bertram, "Integrated

online trajectory planning and optimization in distinctive topologies," *Robotics and Autonomous Systems*, vol. 88, pp. 142–153, 2017.

[6] F. Ulbrich, D. Goehring, T. Langner, Z. Boroujeni, and R. Rojas, "Stable timed elastic bands with loose ends," in *IEEE Intelligent Vehicles Symposium, Proceedings*, 2017, pp. 186–192.

[7] C. Rosmann, F. Hoffmann, and T. Bertram, "Kinodynamic trajectory optimization and control for car-like robots," in *IEEE International Conference on Intelligent Robots and Systems*, vol. 2017-September, 2017, pp. 5681–5686.

[8] C. Rösmann, M. Oeljeklaus, F. Hoffmann, and T. Bertram, "Online trajectory prediction and planning for social robot navigation," in *IEEE/ASME International Conference on Advanced Intelligent Mechatronics, AIM*, 2017, pp. 1255–1260.

[9] B. Cybulski, A. Wegierska, and G. Granosik, "Accuracy comparison of navigation local planners on ros-based mobile robot," in *12th International Workshop on Robot Motion and Control, RoMoCo 2019 - Workshop Proceedings*, 2019, pp. 104–111.

[10] B. Magyar, N. Tsiogkas, J. Deray, S. Pfeiffer, and D. Lane, "Timed-elastic bands for manipulation motion planning," *IEEE Robotics and Automation Letters*, vol. 4, no. 4, pp. 3513–3520, 2019.

[11] J. S. Smith, R. Xu, and P. Vela, "Egoteb: Egocentric perception space navigation using timed-elastic-bands," in *Proceedings - IEEE International Conference on Robotics and Automation*, 2020, pp. 2703–2709.

[12] H. Dong, C. . Weng, C. Guo, H. Yu, and I. . Chen, "Real-time avoidance strategy of dynamic obstacles via half model-free detection and tracking with 2d lidar for mobile robots," *IEEE/ASME Transactions on Mechatronics*, vol. 26, no. 4, pp. 2215–2225, 2021.

[13] R. Kümmerle, G. Grisetti, H. Strasdat, K. Konolige, and W. Burgard, "G2o: A general framework for graph optimization," in *Proceedings - IEEE International Conference on Robotics and Automation*, 2011, pp. 3607–3613.

[14] E. Torkamani and Z. Xi, "Systematical collision avoidance reliability analysis and characterization of reliable system operation for autonomous navigation using the dynamic window approach," *ASCE-ASME Journal of Risk and Uncertainty in Engineering Systems, Part B: Mechanical Engineering*, vol. 8, no. 3, p. 031106 (12 pages), 2022.

Copyright © 2023 by ASME

Proceedings of the ASME 2023
International Design Engineering Technical Conferences and
Computers and Information in Engineering Conference
IDETC-CIE2023
August 20-23, 2023, Boston, Massachusetts

DETC2023-116875

ON HOW A SELF-ORGANIZING SYSTEM PRODUCES COLLECTIVE BEHAVIOR

Jinhui Cao
Industrial Engineering
School of Mechanical Engineering
Beijing Institute of Technology, Beijing, China

Zhenjun Ming
Assistant Professor
School of Mechanical Engineering
Beijing Institute of Technology, Beijing, China

Janet K. Allen
John and Mary Moore Chair and Professor
The Systems Realization Laboratory @ OU
University of Oklahoma, Norman, OK, USA

Farrokh Mistree
L.A. Comp Chair and Professor
The Systems Realization Laboratory @ OU
University of Oklahoma, Norman, OK, USA

ABSTRACT

In this paper we address the following question:

How can the designers of self-organizing systems use the relationships among relevant parameters to quantitatively characterize how a self-organizing system produces the collective behavior process, to better control and design a self-organizing system?

A self-organizing system has advantages in performing dangerous and exploratory tasks that are not suitable for humans. However, it is often difficult for designers to design self-organizing systems, because, in an environment with dynamic complexity, the process of how self-organizing systems behave collectively is unknown. To address this difficulty, we propose a method to quantitatively represent environmental complexity as well as the system behavior, and study the relationship between them so as to gain insights into how a self-organizing system produces collective behavior. In this paper, we identify four collective behavior patterns that a self-organizing system produces when the environment complexity changes, namely, initialing pattern, adjusting behavior, stabilizing pattern, and restarting pattern. We illustrate the efficacy of our method using a box-pushing problem. Our focus is on describing the method rather than the results.

Keywords: Self-organizing systems, collective behavior, environmental complexity, behavior rules.

GLOSSARY

Self-organizing system. A self-organizing system is a system composed of many simple agents, where some form of overall order or coordination arises out of local interactions among agents in a system that is initially disordered.

Rule adoption rate. The probability that an agent chooses to follow a specific social rule.

Static complexity. The part of the environment that does not change after an agent performs its tasks.

Dynamic complexity. The part of the environment that changes as an agent performs its tasks.

Overall tasks. The ultimate goal of an agent in the entire self-organizing system.

Individual behavior. An agent's behavior with respect to completing subtasks independently.

Social behavior. The behavior of agents in completing the main task together.

1. FRAME OF REFERENCE

Designers cannot predict and account for all situations that a product or system will be subjected during its lifetime. Hence, there is a need to design systems that are able to self-organize and adjust to dynamic environments. An example of such systems is the cellular self-organizing system [1-4]. Such a system has many component cells, and each component cell is

Copyright © 2023 by ASME

artificially coded and assigned a certain function. The codes are called -dDNA similar to the biological structure of the human body. These component cells interact with the environment and with each other in an organized way.

Behavior is the key system attribute that determines the overall performance of a self-organizing system (SOS). Behavior refers to the agents' action modes which are based on the environmental conditions and the required tasks. From a design point of view, the goal of a designer of a SOS is to ensure that the system produces the required performance. However, if the internal function mechanism of the SOS is not clearly understood, that is, how the self-organized system produces collective behavior is not understood, then it is difficult to accurately adjust or design the SOS parameters to achieve the desired performance. However, in multi-agent systems the characteristics of decentralization, autonomy, and loose coupling, greatly increase the uncertainty of self-organizing system. Since the dynamic changes in the environment are unpredictable, it is necessary to have an effective method to clearly describe a series of behavior changes within a SOS in order to design it.

The environment is the key stimuli that triggers changes in agents' behavior. Agents interact with the environment in the process of executing a task, and change the state of the environment. The environment is constantly changing and the agents must adapt to the changing environment and produce correct behavior for executing the task in every new environment. Therefore, the problem for designer of a SOS becomes how to express both the environmental changes and the agent behavioral changes and understand the relationships between them.

We use changes in the environment (which we refer to environmental complexity) and agents' behavior rules as the two factors to describe dynamic behavior changes. The environmental complexity is used to determine the different possible environment states. Changes in the environmental complexity help us to analyze agent behavior at different times. Rules are used to determine agent behavior in different environments, thus enabling agents to coordinate their actions and efficiently complete the overall task. Therefore, studying the relationship between the environment's dynamic changes and agent behavior is an effective way to determine the required SOS behavior. However, it is not easy to describe environmental complexity and the required agent behavior changes with behavior rules. To study the relationship between them, we need to establish relevant image of data which requires continuous data. The main challenge in addressing is the lack of quantitative methods to describe the environmental complexity and the behavioral patterns, as well as the relationship between them.

To address the challenge, In this paper we offer a quantitative description of the environmental complexity and the behavior patterns of SOS to understand collective behavior and enable the use of advanced optimization algorithms to enhance their operational efficiency. This involves using quantitative descriptions of environmental complexity and behavior rules. We use both static complexity and dynamic complexity to represent environmental complexity, and agent rule adoption rates to determine SOS behavior. These are combined with time efficiency as the performance indicator of the SOS (see Figure 1). Environmental complexity is the input and behavior rules are variables that are affected by the environment, and the completion time is the output. Based on this model structure, we determine several patterns of collective behavior in self-organized systems and demonstrate it using a multi-agent box-pushing example.

The structure of this paper is as follows. In Section 2 we provide a brief overview of the relevant literature. In Section 3, we introduce the problem. In Section 4, we describe the proposed method to facilitate understanding how a self-organizing system produces collective behavior. In Section 5, we discuss and analyze the simulation results of the box-pushing problem and summarize how a SOS produces collective behavior, thus providing valuable suggestions for designers. In Section 6 we summarize key elements of the paper and identify work to be undertaken in the future.

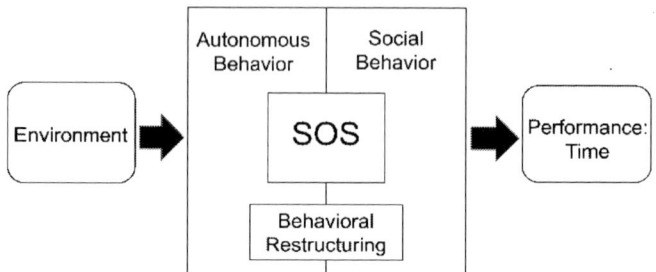

FIGURE 1: MODEL STRUCTURE OF ENVIRONMENTAL COMPLEXITY, RULE ADOPTION RATE AND COMPLETION TIME PERFORMANCE INDICATORS.

2. RELATED WORK

Our goal is to explore the relationship between the environmental complexity of a self-organizing system and the adoption rate of rules for agent behavior to understand how a SOS produces collective behavior and adjusts its performance. In this section, we review the literature about SOSs and their actions and then present the literature on complexity and agent rules and point out their shortcomings. Following this, we then describe the contributions of our research.

2.1 Self-organizing Systems and Their Operation

In self-organizing systems, multiple agents interact with the environment and each other in an organized way to perform tasks. A meta-interaction model for a complex self-organizing system is described by Chiang and Jin [4]. This model is the foundation for a method to design complex, multi-agent systems. There are several extensions of this method to control agent behavior with distributed constraint optimization problems (DCOPs) [5]. The behavior of an SOS is determined by a combination of different environmental complexity states and adjustable agent behavior.

Cyber-Physical-Social Systems (CPSS) can evolve automatically from disorder to order [6] by adapting to different environments and allowing agents to efficiently complete their tasks. Bordum et al. [7] propose an agent-based model that reveals a predator's location by the behavior of group signaling patterns, and shows how changes in behavioral rules bring about changes in group signals. Another approach [8] employs morphogenetic methods to enable systems to self-organize collective systems without centralized control, thus producing controlled global

Copyright © 2023 by ASME

behavior in response to changes in the environment. There are also formal models of a self-organizing "retributive justice" system that uses a behavioral observation framework to adapt its decisions based on the environment to resolve agents' noncompliance [9]. Balta et al. [10] propose a behavior-based adaptive control system to allocate resources when agents face complex environments; this is used to improve the performance of SOSs. The authors in these cited papers focus on the environmental conditions and the accompanying agent behavior. Therefore, we study the environmental complexity of the system and the behavior of the agents. These factors continuously affect each other; thus, we study the relationship between the environmental complexity of the system, its changes over time, the agent behavior and the SOS's collective behavior.

2.2 Complexity and Agent Rules

The environmental complexity of the system changes from one state to another in accomplishing to an engineering task in a SOS. In this step, we determine the reorganization and changing behavior from one ordered state to another. Switching the system's environmental state is characterized by changes in the environment's complexity using frameworks to detect system changes complexity growth. Complexity growth is correlated with increases in the size of the system architecture, the number of interconnections, and randomness. And changes in complexity can characterize transitions in the system's environmental state. [11]. Prior work has provided an analysis of complexity [12] and quantitatively assessed complexity [13], so complexity can be studied both qualitatively and quantitatively. Bell and Taylor [14] show that optimal model complexity exists in engineering problems, which indicates that complexity is also a continuous variable and has extreme values. The other factor that determines the operation of self-organizing systems is agent behavior. We characterize agent behavior using rules including the information about the social structure among all agents [15]. Dynamic Adaptive Autonomy (DAA) [16] introduces the idea of rule adoption rate and to express agent autonomy, which we use to characterize the way agents act. Topcu et al. [17] propose decomposing complex problems into loosely coupled modules that are more manageable and can be designed in parallel, while Semnani and Basir [18] propose a technique for solving the target-to-sensor allocation problem by modeling it as a hierarchical distributed constraint optimization problem (HDCOP). Alternatively, cooperative agents can share a common goal while pursuing private goals and assessing the degree of optimality of the system [19]. Based on these papers [17-19], we decompose the overall task of the SOS. We divide this task into a main task that is executed sequentially and add additional tasks (subtasks) to ensure the smooth execution of the main task. In this way, the agent can complete the final task more smoothly by decomposing the tasks layer by layer. The agents generally accomplish private goals which are not directly related to the final task alone when executing the subtask, and then the agents go on to accomplish the common goal together when executing the main task; thus, the behavior of the agents executing the sub-tasks is autonomous behavior, and the behavior of executing the main task is called social behavior.

2.3 Research Gaps and Our Contribution

In Sections 2.1 and 2.2 we briefly review relevant prior work. We conclude that it is important to study how a self-organizing system produces collective behavior under uncertainty. Accordingly we explore how a self-organizing system produces collective behavior to be able to design and regulate them more easily and accurately. However, there is still a research gap about how to represent the way a SOS produces collective behavior. Based on the above-mentioned literature, the way a SOS produces collective behavior can be characterized by understanding the systemic change of environmental states over time as well as the change of behavior of agents over time. In this way, we explore the relationship between the systemic environmental states of changes over time and the changes of agent behavior over time and then characterize the systemic environmental states. Next we characterize agent behavior by the rule adoption rate and classify the agent behavior as autonomous behavior or social behavior based on the level of the task. That is, we evaluate the relationship between the environment (changes with time), individual behavior (changes with time), and social behavior (changes with time). This enables us to more accurately understand how a SOS produces collective behavior.

From studying how a SOS produces collective behavior, we identify several patterns of SOS collective behavior. That is, as environmental complexity decreases, the collective behavior of the SOS appears to exhibit emergent behavior, adjustment behavior, redundant behavior, and/or restart behavior as we demonstrate using the agent box pushing problem. We propose engineering design suggestions so that designers can better understand how SOSs produce collective behavior and control SOSs more easily and accurately. This fills a gap in the understanding of how a SOS produces collective behavior.

3. PROBLEM DEFINITION

Our goal is to explore the relationship between the environmental complexity of the system its changes over time and the way agents behave during these changes, thus deriving an understanding of how an SOS produces collective behavior. In this section, we specifically introduce the box pushing problem, environmental complexity, and rules of behavior.

3.1 The Box-Pushing Problem

In this paper, we use the Webots robot simulation platform to study the problem of pushing boxes as shown in Figure 2. This is an example of a box-pushing task, with arrows indicating the approximate path of the box. We use the E-puck robot as the agent. It is a differential wheeled miniature mobile robot. When running the process of the SOS, we define the agent successfully pushing the box to the upper left corner as the final task. Since the box is long and wide, its orientation is constantly changing during the process of pushing the box and the agent needs to make adapt its behavior to adjust the orientation of the box to pass through the obstacle. For this problem, the agent position around the box is very important because how and where the agent pushes the box determines box orientation and. Therefore, we divided the perimeter of the box into eight regions: 1, 2, 3, 4, 5, 6, 7, and 8 (Figure 3). For the position of the box, we use a global virtual robot, which can sense the coordinates of the center of any object on the map and transfer them to the robot. Also, this global robot can transfer the coordinates (directions)

Copyright © 2023 by ASME

of the edges around the box to the robot, and then the robot makes judgments based on behavioral rules. The overall simulation framework algorithm is in Table 1.

TABLE 1: OVERALL SIMULATION FRAMEWORK ALGORITHM.

Algorithm 1: Simulation system algorithm for pushing box
Input: The position coordinate bp of the box, the obstacle coordinate set Obs, and the end coordinate G.
Output: Rule adoption rate set Rar, complexity set Cd, and time step T.
1.   Set Rule 0 to obtain the initialization optimal path and temporary target point set A.
2.   **While** bp≠G **do**
3.   According to the position coordinate bp of the box, the position coordinate Obs of the obstacle and the location coordinate A of the temporary target point, the main task and the sub-task are divided
4.   **if** the agent is performing the main tasks, **then**
5.   Choosing the right social rules for executive tasks
6.   **else**
7.   Select the right autonomous rules to perform sub-tasks
8.   **end**
9.   Record the rule adoption rate Rari and complexity Cdi each time
10. Record the time step T
11. **end**
12. **Return** the rule adoption rate collection set Rar, complexity set Cd, and time step T

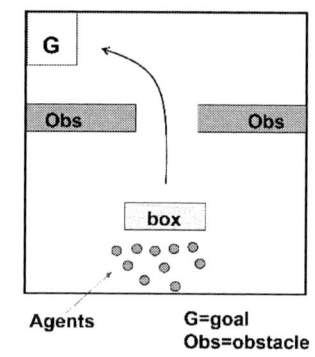

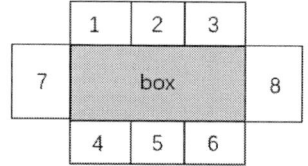

FIGURE 2: SCHEMATIC DIAGRAM OF THE BOX PUSHING PROBLEM.

FIGURE 3: EIGHT REGIONS ADJACENT TO THE DEFINED BOX

3.2 Environmental Complexity

The environment of any engineering system is composed of internal objects. In the box-pushing problem, the objects that make up the environment include the position of the obstacle, the position and orientation of the box, the number of agents, and the position of the target points. To represent environmental conditions with data, we first introduce the concept of complexity. We use static complexity to represent the part of the environment that does not change as the agent performs its task, while dynamic complexity is used to represent the part of the environment that changes as the agent performs its task. Static complexity plus and dynamic complexity become environmental complexity. Thus, we can define the amount of change in the agents' environment over time. Agent size and position are not defined as a part of complexity, but since the agent is used as a driving force to push the box, the number of agents affects the entire engineering problem. Therefore, the number of agents is a variable condition for our research problem.

3.2.1 Static complexity

Because of the position of the obstacle, the number of agents and the position of the target in the environment do not change as the agents perform their tasks, they represent static complexity. These are fixed modules in the environment, and they do not change over time as the SOS evolves. Therefore, we use the number of parameters as the value of static complexity rather than using a specific numerical formula for the value of static complexity.

For the parameter selection of static complexity, we need to characterize its role in the environment. Here, we use four factors to do so:

(1) Obstacle aspects
    a) Form aspect: the length (l) and width (w) of the obstacle.
    b) Repulsive action aspect: the radius of repulsive action of the obstacle (rr).
    c) Location: the center coordinates (x, y) of the obstacle describe it.
(2) Agent side
    a) We ignore agent size and position but use their number as a parameter.

We thus define the expression for the static complexity SC:

$$SC = \sum_{i=1}^{n} Pi \qquad (1)$$

where *SC* is the static complexity, *n* is the number of parameters, and *Pi* is the ith parameter.

We next quantified the static complexity of the two scenarios (Figure 4):

Static complexity for in Figure 4a: *SC* = 10 (number of robots) + 2 * 5 ((central coordinates x, y, w, L, rr) of obstacles A and B) = 20.

Static complexity for (b) in Figure 4: SC = 10 (number of robots) + 3*5 ((central coordinates x, y, W, L, rr) of obstacles A, B, and C) = 25.

Copyright © 2023 by ASME

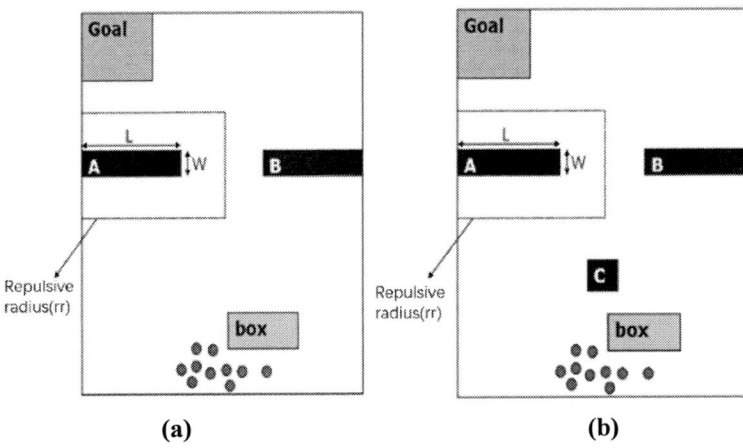

FIGURE 4: WHEN OTHER CONDITIONS ARE THE SAME:
(a) THERE ARE TWO OBSTACLES A AND B; (b) THERE ARE THREE OBSTACLES A, B, AND C.

By comparing the static complexity of the two scenarios, we find that the static complexity of Scenario 1 is less than that of Scenario 2, which confirms that a more complex environment (more obstacles) has greater static complexity.

3.2.2 Dynamic complexity

As an agent pushes the box, the orientation and position of the box in the environment changes over time. The agent needs to plan the path when performing the task, and therefore we treat the box orientation and position relative to the agent also as a planning obstacle. The orientation and position parameters of this planning obstacle will change, and therefore the path planning environment of the agent changes over time. This part of the environmental change is defined as dynamic complexity.

The environment sometimes changes suddenly, e.g., if the agent pushing the box drops obstacles from above on the path or if there are moving obstacles on the path. These situations are the same as when the box is used as an obstacle because when the agent is pushing the box, it will encounter these dynamic obstacles. The agent can sense their existence relative to the agent and the planning system can treat these dynamic obstacles as planning obstacles and perform a new round of path planning to obtain a new optimal route. This situation is the same as when we use a box as a planning obstacle. The method for using a box as a planning obstacle is also applicable to other dynamic obstacles.

Next, we qualitatively compare the high and low dynamic complexity of three positions of the box (Figure 5). In Figure 5, *obs* is the obstacle, *G* is the goal, and *I* is the agent. The solid lines are the gravitational equipotential lines of the target point, and the dotted lines are the repulsive equipotential lines of the obstacle. To better explain the level of complexity, we choose the complexity at three different box locations in the same environment for comparison. Since this is the same environment, we ignore the static complexity and only calculate the dynamic complexity. During the process of moving the agent, as the planning target area, we choose an opposite area of a box where the agent is located. Then the agent regards the box as an obstacle on the map. For path planning, the difficulty of planning the path can be used as a factor to judge dynamic complexity.

In Figure 5(a). One side of the box has a broad range of high potential fields, and the other side has a broad range of low potential fields. Thus, the agent can reach the opposite side smoothly from one side of the low potential field with low planning difficulty.

In Figure 5(b). The range of the high potential field on one side of the box is large, while the range of the low potential field on the other side is great. There is also a small range of high potential fields, so the planning difficulty increases. It is not so smooth for the agent to reach the opposite side from one side of the low potential field. The planning difficulty is medium here.

In Figure 5(c). There are high potential fields on both sides of the box, and the range is wide. At this point, it is difficult for the agent to pass around the box to reach the opposite side; it can only go around from the bottom; thus, obtaining the path and, doing this also requires the longest time. The planning difficulty is high here.

From the qualitative comparison of these three box positions, we obtain the following qualitative relationship for dynamic complexity. The dynamic complexity of Figure 5(a) < dynamic complexity of Figure 5(b) < dynamic complexity of Figure 5(c). We notice that the relationship between the position of the obstacles and the box has a major influence on the dynamic complexity. We further infer that the orientation of the box also has an influence on dynamic complexity because an orientation change also affects the change of the potential field range. The dynamic complexity is judged by other planning obstacles relative to the box as the center, and thus we do not consider the obstacles' centers in the dynamic complexity judgment.

This qualitative analysis suggests that for the parameter selection of dynamic complexity, the attributes and position of the box are the main factors that affect its dynamic complexity after the box is planned to be an obstacle. At this stage, the change in dynamic complexity is more important than its description, and thus we ignore the parameters that describe the box. We take the position and attributes that affect the change of dynamic complexity as the parameters of dynamic complexity:
1) The orientation of the box is represented by a line parallel to the long side over the center of the box and the horizontal angle; thus, the coordinates of areas 7 and 8 of the box and the distance of each obstacle are the parameters chosen.

Copyright © 2023 by ASME

Values closer to the obstacle have greater repulsion, and thus the complexity is related to the inverse of the distance. We put this information into the dynamic complexity by using and influence weight of the orientation to be 100 because the influence of the orientation on the complexity is greater than the influence of the position on complexity. (S7i, S8i)

2) The position can be expressed by a repulsive force, and this is based on the distance. Thus, we use the distance between the box center coordinates and the obstacle coordinate to represent the position parameter of dynamic complexity. Because the influence of position on complexity is smaller than that of orientation, we define the weight of position on dynamic complexity as 1. (Si)

From this we obtain the expression for the dynamic complexity:

$$DC = \sum Si + \frac{100}{\sum S7i + \sum S8i} \tag{2}$$

where Si denotes the distance between the center coordinates of the box and the coordinates of the obstacle, $S7i$ and $S8i$ denote the distance of the center coordinates of box region 7 and region 8 from the obstacle; i is the serial number of the obstacle.

We next quantify the dynamic complexity of the three positions of the box for the qualitative comparison, Figure 6.

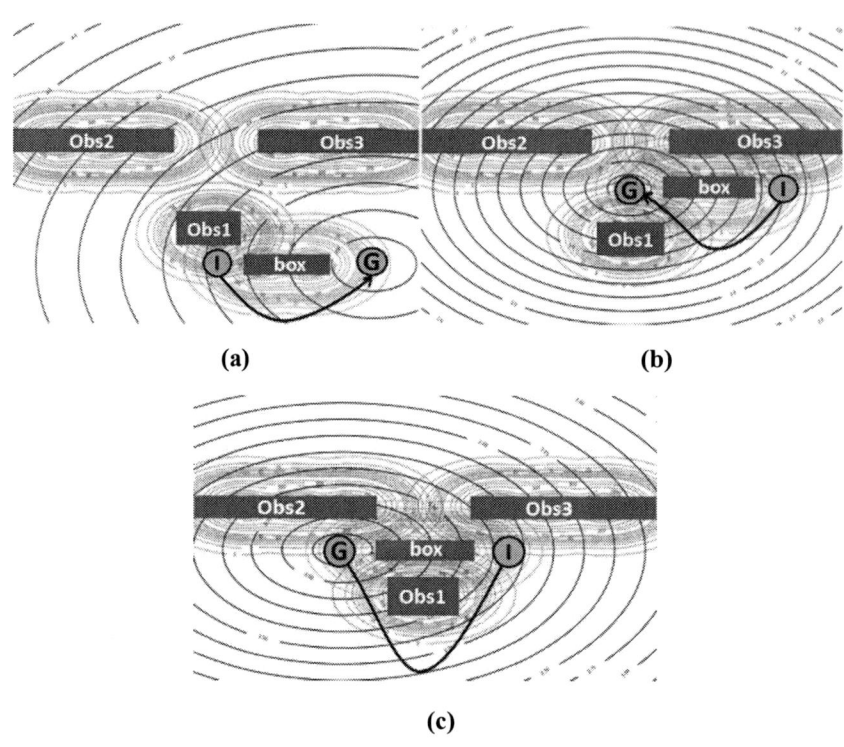

(a)

(b)

(c)

FIGURE 5: (a) PLANNING CALCULATION DIAGRAM OF BOX LOCATION 1, (b) PLANNING CALCULATION DIAGRAM OF BOX LOCATION 2, AND (c) PLANNING CALCULATION DIAGRAM OF BOX LOCATION 3.

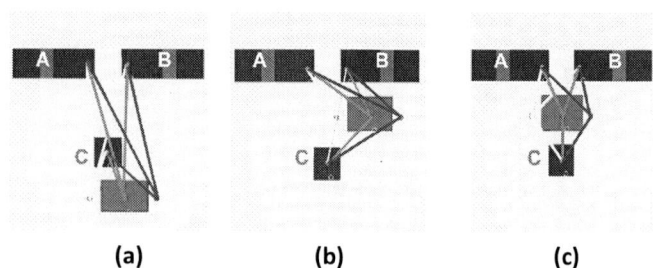

(a)　　　　(b)　　　　(c)

FIGURE 6: (a) PLANNING CALCULATION DIAGRAM FOR POSITION 1 WHERE THE BOX IS LOCATED, (b) PLANNING CALCULATION DIAGRAM FOR POSITION 2 WHERE THE BOX IS LOCATED. (c) PLANNING CALCULATION DIAGRAM FOR POSITION 3 WHERE THE BOX IS LOCATED.

The line angles of the box connected with the A, B, and C target points are noted as 1, 2, and 3, respectively. The yellow circular body represents the coordinates of obstacles. Box center coordinates with box area 7 and area 8 center coordinates are also shown. The blue squares are the obstacles, the red square is the box, and the yellow circles are the coordinate parameters.

For Figure 6(a):
DC=s1+s2+s3+s4+100/(s71+s72+s73+s81+s82+s83)=1.272+1.161+0.378+2.400+100/(1.150+1.301+0.409+1.494+1.150+0.662)=95.418

For Figure 6(b):
DC=s1+s2+s3+s4+100/(s71+s72+s73+s81+s82+s83)=0.899+0.413+0.555+1.992+100/(0.574+0.577+0.357+1.267+0.574+0.887)=130.580

Copyright © 2023 by ASME

For Figure 6(c):
DC=s1+s2+s3+s4+100/(s71+s72+s73+s81+s82+s83)=0.511+0.593+0.426+1.650+100/(0.382+0.936+0.622+0.834+0.382+0.544)=136.536

With these three positions of the box quantitatively determined, we see that the dynamic complexity has the following quantitative relationship: the dynamic complexity of Figure 6(a) < dynamic complexity of Figure 6(b) < dynamic complexity of Figure 6(c). This quantitative result is consistent with our qualitative results; therefore, we define this as dynamic complexity.

3.3 Set Rules, System Rules, and Individual Rules

An agent makes a series of behaviors are regulated by a series of rules. When the agent meets the conditions for applying the rule, the agent will execute it. During the operation of the SOS, the agent will constantly change its behavior as determined by the rule adoption. Therefore, we determine the changes in an agent's behavior in terms of changes in the adoption rate of the rule:

$$Rar = \frac{Ni}{N} \tag{3}$$

Here, *Rar* is the rule adoption rate, *Ni* is the total number of adoptions of the ith rule so far, and *N* is the total number of adoptions of all of the rules.

Different robots in an SOS will utilize different rules, for example, push box rules, rules for placement, and shock prevention rules as the robots collaborate when facing evolving environmental conditions.

3.4.1 Set rules

At the start of the simulation, all agents execute the initialized rules. These rules are used only once in the whole simulation process. We define this as the set rules.

Rule 0: initial full-map path planning

We take the optimal path of a self-organizing system as a reference for its performance and add path planning, which can be used to guide the agents in pushing the box. Therefore, we need to introduce a method of path planning. In this paper, we use an artificial potential field for path planning.

In Figure 7(a), the path-planning system takes the obstacle in the environment as the center of a repulsive field and the goal target point as the center of a gravitational field. When the box is within the range of action of the repulsive field, it will be subject to the repulsive force from the obstacle, and this repulsive force increases as the distance between the obstacle and the box decreases. The goal target point acts as a guide for planning, its gravitational force always acts on the agent. The repulsive force and the gravitational force are added to obtain the combined force of the obstacle(s) and the target point on the agent. The box will be in the global range of the SOS reference potential field for planning as in Figure 7(b). A 'slope' of potential is formed between the box and the target as in Figure 7(b) based on the potential field height, while an obstacle forms a 'hill' due to the high repulsive field in the potential gradient. The box moves 'from high to low.' This path taken by the box is the optimal path.

for the self-organizing system.

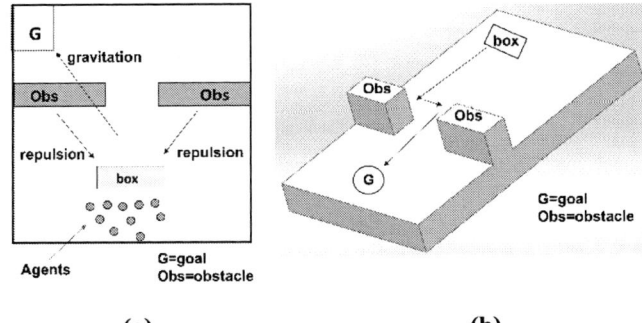

(a)   **(b)**

FIGURE 7 : (a) SCHEMATIC DIAGRAM OF THE PATH PLANNING SYSTEM. (b) PATH PLANNING SYSTEM UNDERSTANDING DIAGRAM.

This optimal path serves as a guide for agents pushing boxes. We decompose the task of reaching the final target of the agent into the red points in Figure 8. These red points form the best path. The process of reaching these points in turn becomes the main task. However, at the beginning of system operation, this rule is only executed once for each agent, so it is not treated as one of the behavior rules in this study.

The dark blue area in Figure 8 represents the obstacle, the black pentagram represents the starting point, the red pentagram represents the target point, and the route composed of red dots is the planned path.

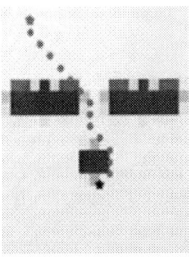

FIGURE 8: OPTIMAL AGENT PATH DIAGRAM IN THE PLANNING SYSTEM

3.4.1 Social rules

Based on the idea of division of tasks, the total task in the self-organizing system is divided into many main tasks to be completed in turn. These main tasks are tasks that agents must cooperate to complete sequentially and are directly related to the total task. The behavior rules which are executed by the agents to accomplish the main task are social rules.

Rule 1: Box pushing rules

When the agents are closer to the farthest and the second farthest box areas, they are divided into two batches. They travel to the box areas farthest and the second farthest from the temporary target point of the main task in a straight line (Figure 9 (a)) these two batches of agents reach the box areas farthest and second farthest from the temporary target point of the main task, respectively. The agents then push the box to the temporary target point of the main task together (Figure 9 (b)).

Copyright © 2023 by ASME

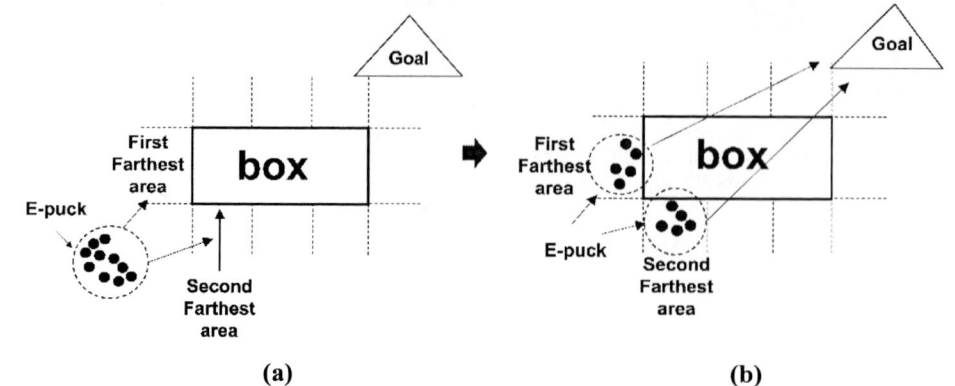

FIGURE 9: (a) THE AGENT IS NOT YET AT THE SPECIFIED LOCATION;
(b) THE AGENTS ARRIVE AT THE DESIGNATED POSITION AND PUSH THE BOX TOGETHER.

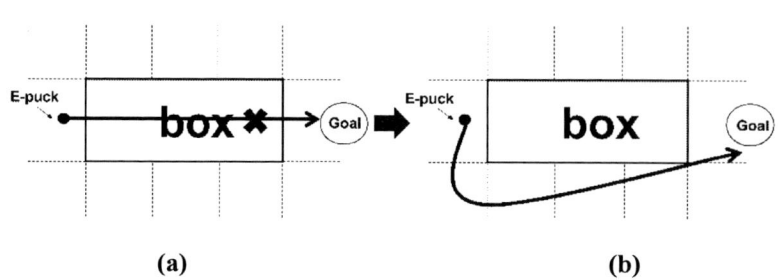

FIGURE 10: (a) AGENT STRAIGHT LINE TO THE TARGET POINT
(b) THE AGENT BYPASSES THE BOX TO GET TO THE TARGET POINT.

3.4.2 Individual rules

When the agent executes the social rules of the main task, there may be conflicts that lead to inefficient agent-pushing behavior—and the whole task execution may fail. Thus, we need to introduce sub-tasks. In addition to the social rules corresponding to the main task, the behavior rules of the agent who performs additional tasks alone are called autonomy rules, and the additional tasks are called sub-tasks. The agent completes these sub-tasks to better complete the main task and reduce conflicts. The sub-tasks are performed in parallel and have no direct relationship to the overall task.

Rule 2: Rules for placement

Rule 2 utilizes path planning based on the artificial potential field, where it takes agents as the planning objects, cuboids and other obstacles as the planning obstacles, and designated areas as the planning target points to plan paths relative to agents rather than boxes. This is to solve the problem that the agent has to go to different box pushing areas but is unable to avoid obstacles during the box pushing process, resulting in pushing the box in an unacceptable pushing area (Figure 10 (a)). The agent can then bypass obstacles to reach the target point based on the position relationships between the box and obstacles relative to the agent (Figure 10 (b)).

Rule 3: Shock prevention rules

When the agent pushes the box to some positions, path

planning fails because the sum of the repulsive force and attractive forces of the obstacle and target point to the agent is 0 when the agent executes Rule 2 (Figure 11 (b)). The agent then oscillates repeatedly at this position (Figure 11 (a)). Using this rule, the agent can push the box directly toward the temporary target point. Because the position of the agent changes constantly during the box-pushing process, rules can be executed when the agent moves away from where the resultant force is 0.

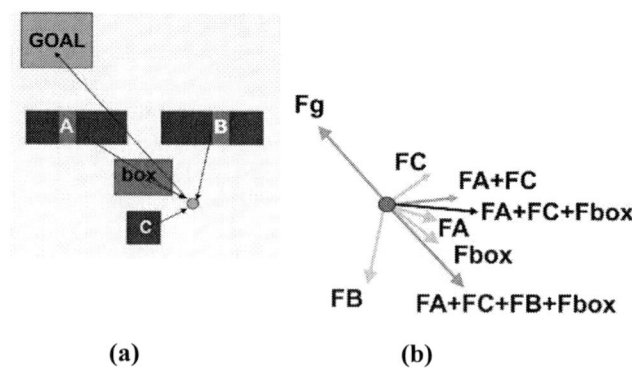

FIGURE 11: (a) THE RELEVANT POSITION OF THE AGENT IN THE ENVIRONMENT.
(b) THE FORCES OF THE AGENT IN THIS POSITION.

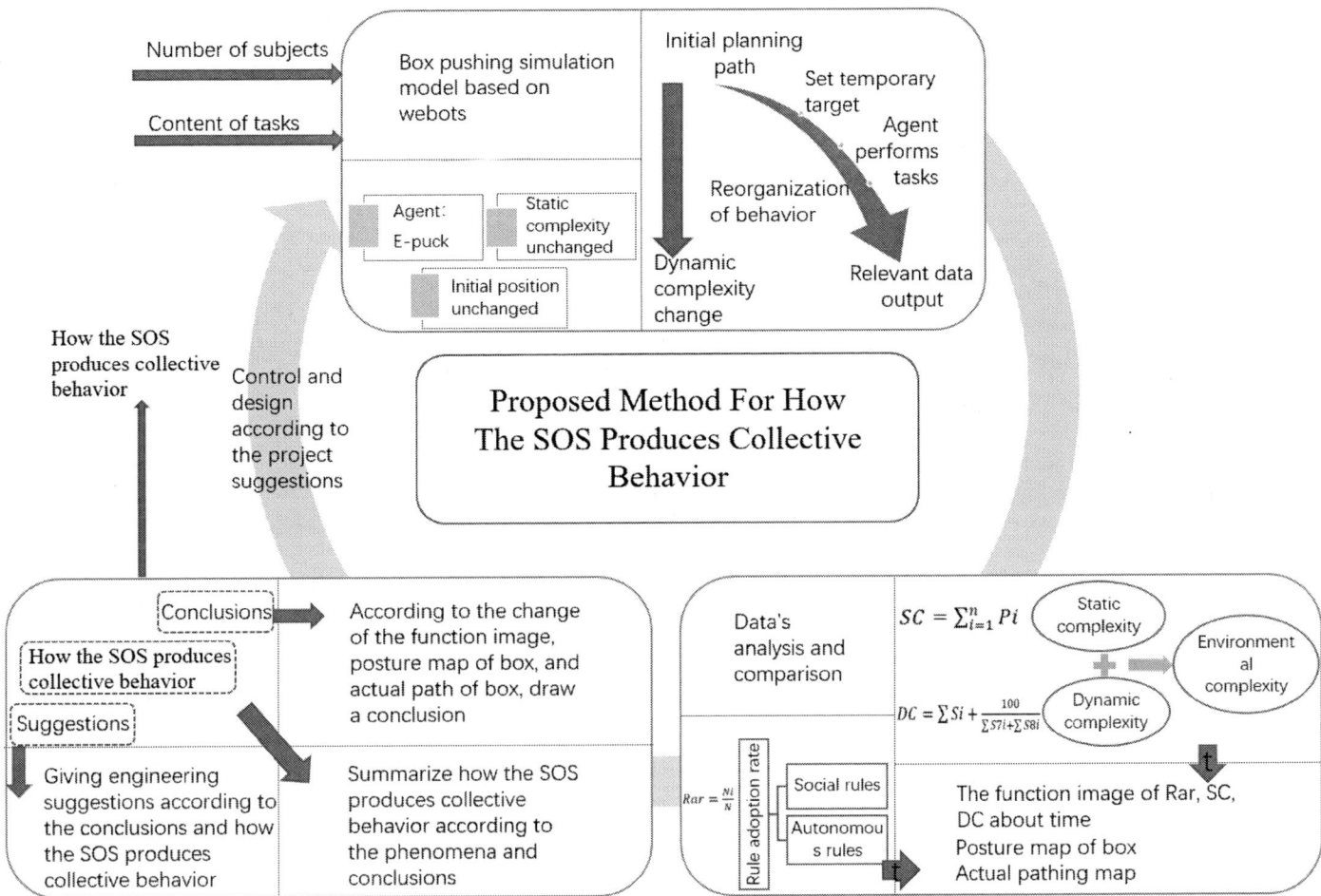

FIGURE 12: OVERALL PROPOSED METHOD FOR HOW THE SELF-ORGANIZATION SYSTEM PRODUCES COLLECTIVE BEHAVIOR.

4. PROPOSED METHOD FOR STUDYING THE WAY SELF-ORGANIZING SYSTEMS PRODUCE COLLECTIVE BEHAVIOR

In Section 3, we define the box-pushing problem for the Webots-based software and introduce the environmental complexity of the system and the behavioral rules of the agents to study the variation of the environmental complexity of the SOS and the adoption rates of the behavioral rules of the agents as they change over time when time steps are the independent variables. Here, we describe the general method to understand how the SOS produces collective behavior as shown in Figure 12. The proposed method consists of three parts: the Webots-based box-pushing simulation environment, the analysis and comparison of data, and the data output. These three parts form a closed-loop structure in which the researcher can continuously design or regulate the SOS to study its operation and then draw conclusions and improve the SOS. A researcher can continuously design or regulate the SOS to study its operation, and then draw conclusions through the operation process of the SOS and improve the SOS's regulation and design.

4.1 Box Pushing Simulation Environment Based on Webots

Based on the Webots robot simulation platform and the E-puck differential wheeled micro-robot—and taking the number of agents and the content of the task as input—the simulation model of pushing the box is constructed under the condition that the initial position and static complexity of the box remain unchanged. The system proceeds as follows:

Step 1. Initial path planning. At the beginning of the simulation, the self-organizing system will plan the path relative to the box with the box as whose path is being planned. Other obstacles are the planning obstacles, and the final target point is the planning target point, thus identifying an optimal path from the box to the target point in the whole initial static environment.

Step 2. Set the temporary target. The optimal path from the box to the target point planned in initial global path planning can be reflected in the form of a series of points; then, we record and store the series of points that make up the optimal path as a set. This set is the set of temporary target points. The agent does not push the box toward the final target point as the current goal but instead pushes the box with these temporary target points as the current goal.

Step 3. The agent executes the task. The agent pushes the box to the final target point defined as the general task. The general task is divided into a series of temporary target points.

Copyright © 2023 by ASME

The agents push the box together to these temporary target points as the main task. When the agent finishes executing the main task, it also indicates that the agent has finished executing the general task. A series of additional tasks that agents may perform independently during the execution of the main task is defined as subtasks. The execution of subtasks can then result in executing the main task.

Step 4. Behavior reorganization. During the process of performing the task, the system changes the orientation and position of the box based on the initial planning of the optimal path; thus, the agent must adjust the orientation of the box when necessary to let the box pass through the obstacles to complete the main task. Here, agents must go to specific box areas that may change at all times to adjust the box orientation. While pushing the box, the agent is in one of the 8 areas of the box. The agent must execute the rules for that specific location so as not to push the box into chaos. This continues until the agent reaches the specified location and then executes the rules for pushing the box with other agents. The orientation and location of the box are not determined, so for the agent to push the box, the agent must go to an uncertain area. This changes over time so the rules for placement and rules for pushing the box will be used repeatedly. There is a shock condition when arriving at certain locations. The shock-prevention rule will then be executed get out of this condition to perform the task. These three rules can be executed at any time so the sum of their rule adoption rates is approximately 1, but the rule adoption rate changes over time. The reason for the change is environmental complexity; this type of behavior change is called behavior reorganization.

Step 5. Required data output. When the agent completes all the main tasks, the total task is completed. The system reaches the final goal, and the simulation system will then gives us data about the environment complexity as well as the rule adoption rate of the agents' behavior rules for analysis and comparison.

4.2 Analysis and Comparison of Data

Based on the data obtained in 4.1 and the formulas for environmental complexity and rule adoption rate defined in Section 3, images of the system behavior with the time steps as the independent variable and the environmental complexity and the rule adoption rate as the dependent variables are derived, analyzed, and these are compared with the optimal path diagram and the actual path diagram.

4.3 Data Output

The conclusions obtained from the analysis and comparison in Section 4.2 are summarized by the relationship between the complexity of the SOS environment changes with time and the behavior of the agent as it changes with time. This can help explain and the operation process of the SOS. Valuable engineering suggestions will be proposed for SOS problems based on these conclusions and understanding of the operation of the SOS. Designers can thus better control and design a SOS to improve its performance. Simulation of the operation process of the SOS is then performed again as a closed-loop structure to obtain better and better designs and improved regulation methods for the SOS so that the designer can better understand its operation.

5. RESULTS AND DISCUSSION

In this project we investigate the relationship between the environmental complexity of a SOS as it changes over time and the behavior of the agents to study how the self-organizing system produces collective behavior to improve its regulation and design. Here, we show quantitative and qualitative results about the relationship between environmental complexity and the behavior of agents. In Section 5.1, we show different numbers of agents and data analysis processes for different scenarios. In Section 5.2, we show a data analysis for the same number of agents in different scenarios. In Section 5.3, we show the data analysis process with different agents but the same scenarios. In Section 5.4, we summarize how the self-organizing system produces collective behavior and propose helpful adjustments and design suggestions for the designers of SOS. We use the scenario is Figure 4(a) as scenario 1 and the scenario in Figure 4(b) as scenario 2.

5.1 Data Analysis Process of Different Agents in Different Scenarios

This section uses the control variable method to discuss the overall dynamic changes of self-organizing systems. We take the time step as the independent variable, environmental complexity, and rule adoption rate as the dependent variables, and then investigate the relationship among them in different scenarios for different agents. Figure 13 shows that we divided the function image into four parts (A, B, C, and D) with three vertical lines, where the three points intersect with the environmental complexity curve in the function image. These are the rising judgment points, the highest environmental complexity point, and the falling judgment point.

From the entire process in each function graph, we find that the changes of the rule adoption rate of Rules 1 and 2 are opposite during most stages of the entire process, with only a few stages of the changes being in the same direction. They always appear throughout the operation of the SOS. This means that autonomous rules and social rules switch behaviors with the changes of environmental complexity. This process is called behavioral reorganization, and autonomous behaviors always serve social behaviors during the process of behavioral reorganization. They are constantly changed in together throughout the operation of the SOS—this is cross-execution of sub-tasks and main tasks.

Copyright © 2023 by ASME

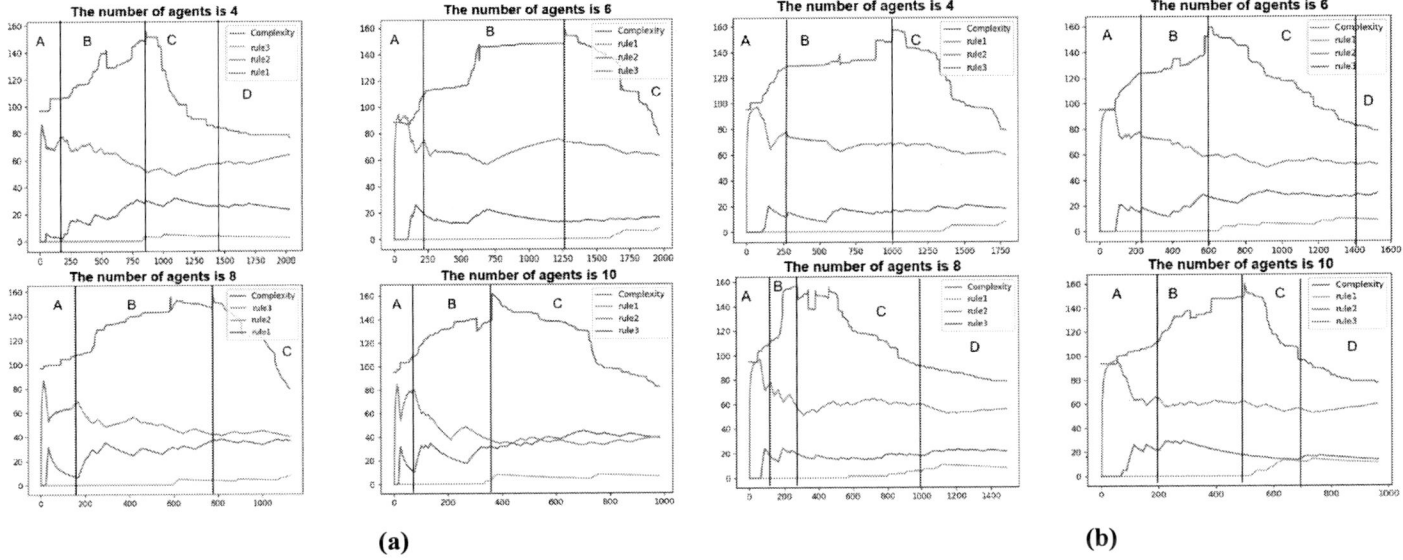

FIGURE 13: (a) THE FUNCTIONAL RELATIONSHIP GRAPH OF SCENARIO 1.
(b) FUNCTIONAL RELATIONSHIP GRAPH FOR SCENARIO 2.

Upon comparing the eight function diagrams in Figure 13, we find that there are specific features in three special points in different scenarios and different agent numbers. For the rising judgment point, when the environmental complexity rises for a short time, the rule adoption rate of Rule 2 changes in the same direction as the environmental complexity. After that point, if the environmental complexity is still rising, then the change of rule adoption rate of Rule 2 is in the opposite direction of the change of environmental complexity. The highest environmental complexity point in the operation of the self-organizing system is in every case, thus the change of environmental complexity shows a trend of 'from simple to difficult to simple'. The decreasing judgment point appears in the process of decreasing environmental complexity from the highest point of environmental complexity. After this point, the rule adoption rate of Rule 2 decreases first and then increases. We will next analyze and compare regions A, B, C, and D which are divided by these three special points.

In Region A in Figure 13 the rule adoption rate of Rule 2 soars to nearly 95 percent at the beginning of the system self-organization and then declines before briefly rising in Region B. The adoption rate of Rule 1 starts to rise when the rule-adoption rate of Rule 2 starts to decline. This means that the position of the agent is far from the box at the initial stage of the SOS. And it is in place before the box is pushed; thus, the rule adoption rate of some behaviors of the agent will suddenly rise. When an agent arrives near the box, it starts to restructure its behavior. Region B shows that the change in the adoption rate of Rule 2 is the opposite of the change in environmental complexity. This means that the agent constantly reorganizes its behavior based on the changing complexity of the environment when the position and orientation of the box are constantly changing to adapt to the environment. The agent performs fewer autonomous behaviors and more social behaviors as the environmental complexity increases.

Thus, the agent will solve fewer sub-tasks because the same sub-task will be advantageous for different main tasks. After performing a sub-task, the main task that is to be executed later does not need to be performed with more sub-tasks for the later environment. In Region C, we find that the rule adoption rates of Rule 1 and Rule 2 are stable with decreasing environmental complexity from the highest point of environmental complexity. The adoption rate of Rule 3 always increases near the highest point of environmental complexity. This means that agents perform the relatively adapted behavioral reorganization found at the highest point of complexity. In the subsequent process of decreasing complexity, agents only fine-tune the value of the rule adoption rate, thus forming a kind of behavioral redundancy so their rule adoption rates tend to stabilize. It plateaus in the process of increasing environmental complexity. When it rises to a certain value, there is a situation where both Rule 1 and Rule 2 cannot be solved, and thus more rules must be introduced. This indicates that new tasks will appear and the total class of rule references will increase as the complexity of the environment rises.

In Figure 14 we show the orientation and path diagrams of the self-organizing system during operation. We draw the orientation of the boxes using the same intervals. We find that for the same scenario, their paths are different from the initial planning paths, but the general direction is the same, which is caused by the change of the orientation of the box during operation. We also regionalized the whole function variation graph and found that it agreed well with our observations and conclusions. Before the peak of environmental complexity, the orientation of the box is constantly changing, and the autonomy and social behavior are constantly reorganized to adjust the orientation of the box to adapt to the environment. After the peak of environmental complexity, the orientation of the box changes little, thus forming behavioral redundancy.

Copyright © 2023 by ASME

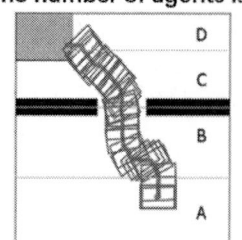
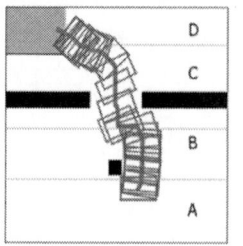
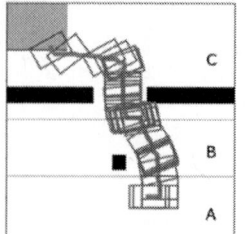
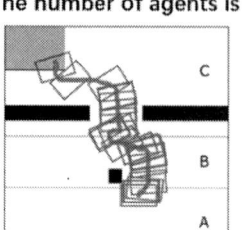
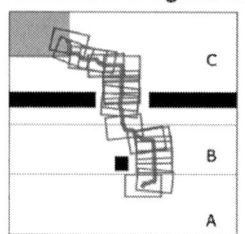

(a)   **(b)**

FIGURE 14: (a) BOX ORIENTATION AND POSITION PATH DIAGRAM FOR SCENARIO 1; (b) BOX ORIENTATION AND POSITION PATH DIAGRAM FOR SCENARIO .

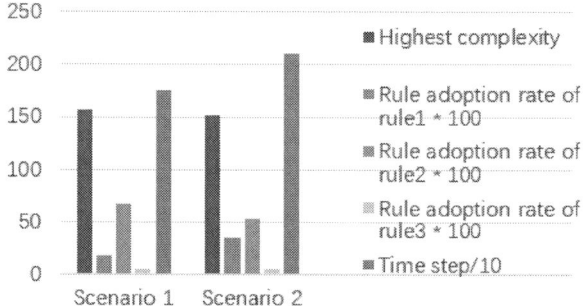

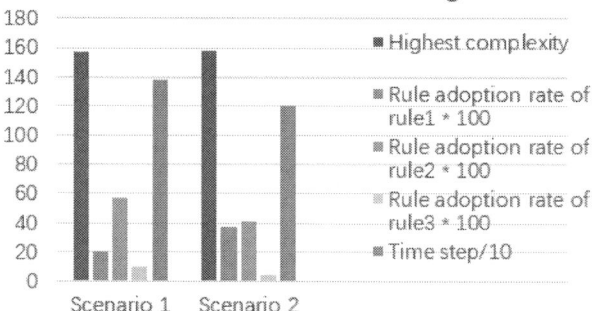
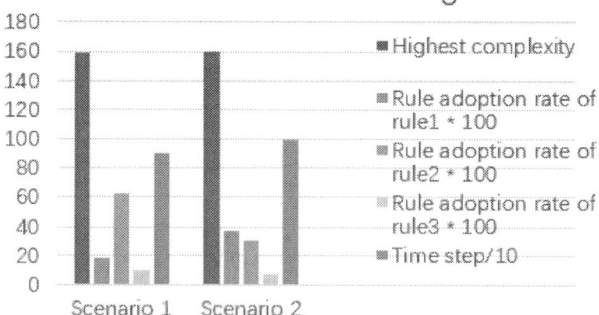

FIGURE 15: RELATIONSHIP AMONG DIFFERENT SCENARIOS AND CORRESPONDING NUMERICAL FUNCTIONS.

5.2 Data Analysis Process for the Same Agent But Different Scenarios

We also used the control variable method to explore the relationship among three metrics (environmental complexity, rule adoption rate, and total time step to complete the task). In the experimental setup, we studied the results for different scenarios with the same number of agents. We used two scenarios as independent variables. Based on the findings in Section 5.1,

Copyright © 2023 by ASME

we considered the highest point of environmental complexity to be important, and we used this value of the highest point of environmental complexity and the rule adoption rate at the highest point of environmental complexity as dependent variables to obtain a graph of the function of the two scenarios under four numbers of agents (4, 6, 8, and 10) under two scenarios, Figure 15.

Figure 15 shows that the environmental complexity of the two scenarios remained unchanged for different numbers of agents. This is because we set the weight of the static complexity to be very low when we defined the static complexity and the dynamic complexity. The weight of the orientation complexity in the dynamic complexity is 100-fold higher than the static complexity; thus, the environmental complexity in scenarios 2 is higher than that in scenarios 1 for different numbers of agents.

The rule adoption rates of social behavior rules increases with increasing complexity and the rule adoption rates of autonomous behavior rules decreases with increasing environmental complexity in the functional relationship graphs for 4, 8, and 10 agents. The change is exactly the opposite for the functional relationship graph for 6 agents, which indicates that the number of agent main tasks increases while the complexity increases. The agent needs to apply more social rules to solve the problem; meanwhile, the adoption rate of autonomous rules decreases, thus indicating that there is a mutual service relationship among the main tasks. As the complexity increases, an agent is more adapted to solving the main tasks cooperatively than solving subtasks individually and then executing the main tasks; thus, the cooperative solution occurs more frequently than solving them alone. However, this situation is reversed when the number of agents is 6, thus indicating that in the case of 6 agents, performing separate tasks individually can provide a huge advantage instead of performing tasks together and is more efficient than performing tasks together in the first place. This is a reasonable result obtained through rules between agents.

When there are 4, 6, and 10 agents, the time steps of the system increase with increasing environmental complexity. The time for the system to complete the task increases, and the performance of the system decreases, but the opposite is true for 8 agents. This shows that the increase in complexity brings about a decrease in system performance, but the opposite is true in the case with 8 agents. In this case, given the conclusion in the previous paragraph, the adaptation of cooperative problem-solving between agents for scenario 2 is greater than for scenario 1. Thus, agent number 8 is suitable for solving the problem of scenario 2. This conclusion is for a special case.

5.3 Data Analysis Process Under Different Agents but the Same Scenario

In Section 5.1 we describe how a self-organizing system is designed to adjust to dynamic changes in the environment. The highest point of environmental complexity is especially important, and thus we commented on it. In this section we report on the use the control variable method with the number of agents as the independent variable, the highest point complexity, and the rule adoption rate at this time. The time step to complete the task is used for the dependent variables—we then explored the relationships among them.

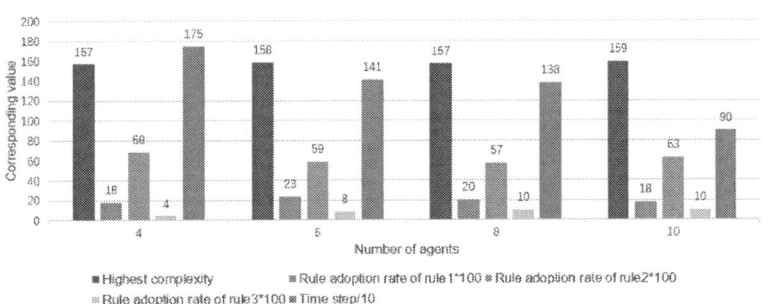

FIGURE 16: THE RULE ADOPTION RATE AT THE HIGHEST COMPLEXITY POINT OF SCENARIO 1, AND THE TIME COMPLETE THE TOTAL TASK.

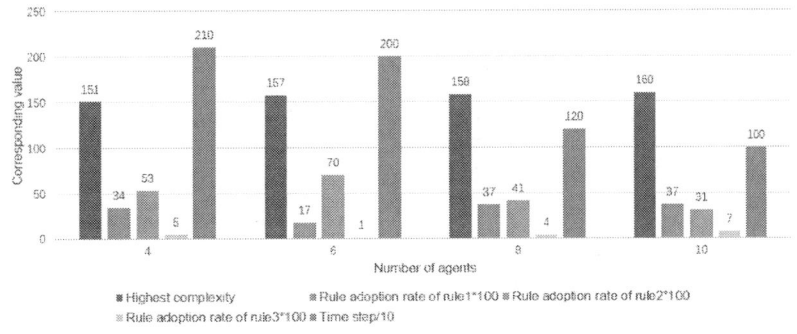

FIGURE 17: THE RULE ADOPTION RATE AT THE HIGHEST COMPLEXITY POINT OF SCENARIO 2, AND THE TIME TO COMPLETE THE TOTAL TASK.

Copyright © 2023 by ASME

Understanding the highest environmental complexity in the same scenario is similar to our explanation in Section 5.2. In Figure 16 the change in the rule adoption rate is shown. We found that the rule-adoption rate of autonomous rules first decreased and then increases with the increasing number of agents in Scenario 1. The rule adoption rate of autonomous rules is the lowest with 8 agents. In Scenario 2, the rule adoption rate of autonomous rules first increases and then decreases with the increase of the number of agents, Figure 17; the rule adoption rate of autonomous rules is highest with 6 agents. The adoption rate of adaptation rules at the highest point of environmental complexity is different: There is an increasing number of agents in different environments showing different trends, but these always present the rule adoption of autonomous rules. The value first increases and then decreases or first decreases and then increases. The highest and lowest points of the rule adoption rate occur in the case of the different numbers of agents in different scenarios: These two different reorganization and change methods are seen in the rule adoption rate of adaptation rules at the highest complexity point.

For time steps, we found that in the same scenario, the time steps to complete the goal will decrease with increasing numbers of agents. This means that more agents lead to a higher complexity of the whole system, but a higher capacity of the agents to push the box—this shows that the advantage of increasing the number of agents is higher than the disadvantages of increasing complexity brought about by the increase in the number of agents under certain environmental complexity, thus further improving the performance of the SOS.

5.4 Discussion

Based on our experimental findings, we obtain an overview of how a self-organizing system produces collective behavior. We offer some engineering recommendations for the designers of self-organizing systems.

We use the environmental complexity to represent the environmental state of the system and then use a rule adoption rate to represent the behavior of the system agent. During the operation process of the self-organizing system, the environmental condition of the system is constantly changing. To accomplish the task, the agents need to continuously reorganize their behavior in order to cope with the environmental changes of the self-organizing system. This reorganization process is divided into four patterns:

1) *Initializing pattern.* At the beginning of the self-organizing system, the autonomous behavior of the self-organizing system is highest and then slowly declines. Social behavior will rise slowly, thus forming a kind of initialized behavioral reorganization in response to the environmental complexity.
2) *Adjusting pattern.* Autonomous and social behaviors change differently as environmental complexity rises. The execution of agent behaviors decreases, while the execution of social behaviors increases—in this state the execution of a single sub-task offers an advantage over a multi-master task. In this situation the agents continuously reorganize their behavior and adjusts the proportion of internally executed social rules and autonomous rules in real time based on the task situation to cope with the elevated complexity of the environment.

3) *Stabilizing pattern.* As the environmental complexity decreases from the highest point, the behavioral reorganization of the agents tends to stabilize due to the formation of behavioral redundancy at the highest point of environmental complexity. Thus the rule adoption rate adapts to the highest point of complexity and only needs to be fine-tuned in the process of decreasing complexity.
4) *Restarting pattern.* The environmental complexity tends to be stable, which means that the next environmental complexity change scenario is about to begin. Here, agents' behavior is no longer smooth but undergoes behavior reorganization to adapt to the next environmental complexity change.

Increasing the number of agents brings more benefits to the entire self-organizing system than increasing the number of agents brings more complexity to the entire self-organizing system. With a change in the number of agents, the adaptive behavior reorganization at the highest point of environmental complexity shows a trend of autonomous behavior that first increases and then decreases or first decreases and then increases in different environments. The change in social behavior is the opposite.

Therefore, based on the four patterns and associated discussion, we propose the following suggestions for an engineering designer:

1) The performance of the SOS can be improved by increasing the number of agents. When the system complexity is caused by an increase in the number of agents, the designer can adjust the direction of the internal capacity of the agents so that the advantages of increasing the number of agents outweigh the disadvantages.
2) Reductions in the parameter regulation and design of the SOS is seen in an environment with reduced complexity. One can use the behavioral redundancy here to reduce the workload.
3) Adding more autonomous and social rules for agents can strengthen their autonomous and social behavior in the face of high complexity. Thus, agents can better solve the problems they encounter and the performance and stability of the system be improved.
4) Classifying the tasks and categorizing the agent's behavior according to the type of task can help regulate or design different behavioral styles according to the classifications and connections .
5) In engineering concerns, having insights into the distinct behavioral patterns exhibited by SOS enables designers to optimize them using these quantitatively represented behavioral patterns through advanced and more precise algorithms. The ultimate goal is to obtain the most favorable behavioral patterns for varying environmental conditions.

6. CLOSING REMARKS

Study of the ways to improve SOS performance has been an important engineering research area. To better design and control a self-organizing system so as to achieve performance improvement, designers need to understand how the SOS produces collective behavior. To figure out how SOS produces collective behavior, we address the following question:

Copyright © 2023 by ASME

How can the designers of self-organizing systems use the relationship among relevant parameters in order to quantitatively characterize how the self-organizing system produces the unknowns of the collective behavior process, to better control and design the self-organizing system?

To answer this question we offer the following:

1) A method to explore the unknowns of SOS behavior reorganization. We explore the relationship between the environmental complexity of self-organization system and the adoption rate of rules representing agent behavior, in order to understand how a SOS generates collective behavior and improves its performance through adjustment.

2) We use the system's environmental factors and the agents' behavioral patterns to quantitatively express the procedural dynamics of SOSs. In our study, we utilize a combination of adoption rates for different behavioral rules to quantitatively represent emergent behaviors, and we use environmental complexity to characterize environmental conditions. Through a dynamic analysis of the relationships between these two factors over time, we have identified several distinct operating modes. Our findings suggest that sudden behavior is not entirely unpredictable or uncontrollable. And it can help us understand the emergent behavior of SOS.

3) The varying patterns of agent behavior can be identified based on shifts in the environment. By exercising the proposed method, we discovered that there are four patterns of collective behavior reorganization with the change of environment in the case of the box-pushing problem in this project: *initialing pattern, adjusting behavior, stabilizing pattern, and restarting pattern.*

Based on the Webots simulation environment, we built a multi-agent box-pushing experiment. Through the comparison of three experiments using different scenarios with a different number of agents, different scenarios with the same number of agents, and different scenarios with different agents, the relevant laws, and descriptions of how the self-organizing system produces collective behavior are obtained. Key contribution of this paper is a closed-loop method constructed to help designers adjust and design self-organizing systems to improve their performance.

For future work, we plan to focus on the following aspects:

1) We plan to increase environmental complexity and the types of agents' behavior ruls, set different simulation environments to more accurately study how a SOS produces collective behavior.

2) We plan to increase the performance output from a single performance indicator to multiple performance indicators, and study more issues related to system performance indicators from the perspective of SOS collective behavior.

3) We plan to add more kinds of self-organizing systems to further demonstrate the practicability and generalizability of this method by studying the process of their collective behaviors.

ACKNOWLEDGEMENTS

Zhenjun Ming acknowledges support from the National Natural Science Foundation of China (Grant No. 51805033), Beijing Municipal Science & Technology Foundation (Grant No. 3222020), and Beijing Institute of Technology Research Fund Program for Young Scholars (Grant No. 3030011182037). Janet K. Allen acknowledges funding from the John and Mary Moore Chair. Farrokh Mistree acknowledges the funding from the L.A. Comp Chair at the University of Oklahoma.

REFERENCES

[1] Zouein, G., Chen, C., and Jin, Y., "Create Adaptive Systems through "DNA" Guided Cellular Formation," Design Creativity 2010, Springer London, pp. 149-156.

[2] Chen, C., and Jin, Y., "A Behavior Based Approach to Cellular Self-Organizing Systems Design," Proc. ASME 2011 International Design Engineering Technical Conferences and Computers and Information in Engineering Conference, Paper number DECT2011-48833.

[3] Jin, Y., and Chen, C., "Field Based Behavior Regulation for Self-organization in Cellular Systems," Proc. Design Computing and Cognition '12, J. S. Gero, ed., Springer Netherlands, pp. 605-623.

[4] Chiang, W., and Jin, Y., "Toward A Meta-Model of Behavioral Interaction for Designing Complex Adaptive Systems," Proc. ASME 2011 International Design Engineering Technical Conferences and Computers and Information in Engineering Conference, pp. 1077-1088, Paper number DETC2011-48821.

[5] Fioretto, F., Pontelli, E., and Yeoh, W., 2018, "Distributed Constraint Optimization Problems and Applications: A Survey," J. Artif. Int. Res., 61(1), pp. 623–698.

[6] Zhou, Z., Sun, Y., Ouyang, C., Gan, Z., and Ming, Z., "A Design Framework for Evolving Cyber-Physical-Social System (CPSS) Based on Force Field," Proc. ASME 2022 International Design Engineering Technical Conferences and Computers and Information in Engineering Conference, Paper number DETC2022-89892.

[7] Borduin, R., Ramaswamy, K., Mohan, A., Cocroft, R., and Nair, S. S., "Modeling the Rapid Transmission of Information Within A Social Group of Insects: Emergent Patterns in the Antipredator Signals," Proc. ASME 2008 Dynamic Systems and Control Conference, pp. 1441-1447, Paper number DSCC2008-2298.

[8] Jin, Y., "Morphogenetic Self-Organization of Collective Systems," Proc. 2021 IEEE International Conference on Autonomous Systems (ICAS), pp. 1-1.

[9] Zolotas, M., and Pitt, J., "Self-Organising Error Detection and Correction in Open Multi-agent Systems," Proc. 2016 IEEE 1st International Workshops on Foundations and Applications of Self* Systems (FAS*W), pp. 180-185.

[10] Balta, H., Rossi, S., Iengo, S., Siciliano, B., Finzi, A., and Cubber, G. D., 2013, "Adaptive Behavior-Based Control for Robot Navigation: A Multi-Robot Case Study," 2013 XXIV International Conference on Information, Communication and Automation Technologies (ICAT), pp. 1-7.

[11] Hennig, A., Topcu, T. G., and Szajnfarber, Z., 2021, "So You Think Your System is Complex?: Why and How Existing Complexity Measures Rarely Agree," Journal of Mechanical Design, 144(4), p. 041401.

Copyright © 2023 by ASME

[12] Pasqualetti, F., Franchi, A., and Bullo, F., 2011, "On Cooperative Patrolling: Optimal Trajectories, Complexity Analysis, and Approximation Algorithms," IEEE Transactions on Robotics, 28, pp. 592-606.

[13] Dalvi, A. S. and El-Mounayri, H., "Integrated System Architecture Development Framework and Complexity Assessment," Proc. ASME 2021 International Mechanical Engineering Congress and Exposition, Paper number IMECE2021-67515.

[14] Bell, D. G., and Taylor, D. L., "Determining Optimal Model Complexity in An Iterative Design Process," Proc. ASME 1990 Design Technical Conferences, pp. 291-297.

[15] Khani, N., Humann, J., and Jin, Y., 2016, "Effect of Social Structuring in Self-Organizing Systems," Journal of Mechanical Design, 138, p. 041101.

[16] Barber, K. S., Goel, A., and Martin, C. E., 2000, "Dynamic Adaptive Autonomy in Multi-Agent Systems," Journal of Experimental and Theoretical Artificial Intelligence, 12, pp. 129 - 147.

[17] Topcu, T. G., Mukherjee, S., Hennig, A., and Szajnfarber, Z., 2021, "The Dark Side of Modularity: How Decomposing Problems Can Increase System Complexity," Journal of Mechanical Design, 144(3), p. 031403.

[18] Semnani, S. H., and Basir, O. A., 2013, "Target to Sensor Allocation: A Hierarchical Dynamic Distributed Constraint Optimization Approach," Comput. Commun., 36(9), pp. 1024-1038.

[19] Raffard, R. L., Tomlin, C. J., and Boyd, S. P., 2004, "Distributed Optimization for Cooperative Agents: Application to Formation Flight," 2004 43rd IEEE Conference on Decision and Control (CDC) (IEEE Cat. No.04CH37601), 3, pp. 2453-2459.

Copyright © 2023 by ASME

Proceedings of the ASME 2023
International Design Engineering Technical Conferences and
Computers and Information in Engineering Conference
IDETC-CIE2023
August 20-23, 2023, Boston, Massachusetts

DETC2023-116971

CHAT GENERATIVE PRETRAINED TRANSFORMER: EXTINCTION OF THE DESIGNER OR RISE OF AN AUGMENTED DESIGNER

Amaninder Singh Gill
Assistant Professor,
Centralia College,
Centralia, Washington, USA
aman.gill@centralia.edu

ABSTRACT

Systematic design process is used in engineering systems design to develop solutions to problems of varying complexity. There have been efforts within the community to develop tools to perform automated generative design at varying stages of the systematic design process. However, a Large Language Model (LLM) has never been used. To this end, this paper presents an initial investigation into the use of OpenAI's ChatGPT to automatically generate solutions to an engineering problem. It is demonstrated that for the most part ChatGPT is quite capable of generating conceptual design for an engineering problem. In light of this technology, this paper floats questions on the future direction of research and education for the engineering systems design community.

Keywords: Generative Design, Function Models, ChatGPT, Cyber-Physical Social Systems

1. INTRODUCTION AND MOTIVATION

This paper presents an initial investigation into the challenges and opportunities presented by the advent of new Large Language Model (LLM) based AI tools in conceptual design generation that can be used inside an internet browser. The AI tool used in this study is called ChatGPT (Generative Pre-Trained Transformer) developed by OpenAI [1,2]. This tool's interface is a chat bot that can generate text based answers in real time to the specific queries by the user. The AI tool can be used to elicit answers that can range from the weather of the day to a summary of a literary masterpiece to generating C++ code to generating all design parameters of a pinion driving a gear. This means that an engineering designer is essentially co creating in conjunction with ChatGPT is essentially a Cyber Physical

Human System (CPHS), which is a subset of a CPSS (Cyber Physical Social System) [3].

In the context of Systematic Design as proposed by Pahl and Beitz [4] the designer must follow up the requirement elicitation process by generating a solution-neutral function. These subfunctions are then combined into function structure that represent the overall function of the system [4]. Function models has been used to support design ideation [5], conduct engineering reasoning [6,7], and predict failure paths of complex systems [8,9].

There have been several approaches proposed to generate function models using computational tools [10–15]. These efforts can be broadly classified into two approaches: a) using the design repository to train an AI tool that can generate function models, and b) using *a priori* rules in conjunction with an algorithm to generate function models. However, these generational design paradigms do not consider, the human, artefact, and process realization dimension of the design problem [16]. To make this a reality, the concept generation tools need to be knowledgeable across multiple domains and be decentralized to be available for use to designers and stakeholders.

With OpenAI's ChatGPT being able to use LLMs across domains and available to any human with a computer and an internet connection, there is a need to investigate the capability of this tool and its implication on engineering design research and education. This is especially important given that the introduction of this tool in the public domain has unleashed what can only be classified as chaos. This is evident from some of the news headlines that have appeared in recent months. A sample of such headlines has been presented in Table 1.

The current research focuses on initial research to better understand and benchmark the capabilities of ChatGPT with

Copyright © 2023 by ASME

respect to the systematic design process by exploring the following research questions (RQ):

RQ1: Can ChatGPT synthesize Functional Basis?

RQ1.1: Can the difference between verbs and nouns be discerned?

RQ1.2: Can ChatGPT create verb noun combinations?

RQ2: Can ChatGPT generate function models?

RQ2.1: Can function models be created from a requirement statement?

RQ2.2: Can more granularity [17] be added to a function model created in the previous step?

RQ3: Can ChatGPT generate working principles?

RQ3.1: Can working principles be generated for each sub-function of the function models?

RQ3.2: Can working structures be generated from the function model?

RQ3.3: Can similarity between working structures be determined?

2. GENERATIVE DESIGN USING CHATGPT

This section uses ChatGPT to answer the research questions posed in the previous section. For this research, the ChatGPT Playground was used. The reasons for this are three-fold. Firstly, the chat bot itself gets overwhelmed with the amount of traffic and can sometimes not be available. Secondly, there is a provision to save the current conversation and return to it. This feature is not available on the main chatbot. Finally, the playground lets the user pick a training model, length of response, the randomness, and context [18]. The model used for this research is "text-davici-003" which is the most advanced model, the length of response is 4000 words (maximum length), and low randomness (temperature = 0, and top P = 1) was chosen to ensure that ChatGPT performs in a predictable manner.

2.1 Functional Basis

The functional basis is a common language developed to formalize the verb object pairs that can be used to generate function models [19]. To test ChatGPT's capability on this front, two separate tests were run. In the first test, it was supplied with verbs and nouns and was asked to separate it. It accomplished the task successfully as shown in Figure 1. In the second test, it was asked to combine the given verbs and nouns to form the functional basis. Again, ChatGPT accomplished the task successfully as shown in Figure 2.

2.2 Function Model

Function model is used to indicate the flow of material signal and energy with the use of a block diagram to express solution neutral relationships between inputs and outputs [4]. An example of a function model for a hair dryer has been shown in Figure 3.

Table 1: News headlines about ChatGPT

S. No.	Headline	Publication & Date
1.	Yes, ChatGPT is coming for your office job.	Wired, March 9, 2023.
2.	ChatGPT: A Tool for Growth or a Cheater's Best Fried?	Marist Circle, February 24, 2023.
3.	ChatGPT Urgent Warning: Is Your Information, Job or Business Safe?	Entrepreneur, February 11, 2023.
4.	Some public-school districts concerned about artificial intelligence tool ChatGPT and cheating	WISN – abc News, February 10, 2023.
5.	"ChatGPT – Attorney At Law" – are lawyers going to be replaced by AI?	Business & Innovation Magazine, March 2, 2023.
6.	ChatGPT is a Plague Upon Higher Education.	Inside Higher Ed, February 9, 2023.
7.	22 ways ChatGPT could be used in economic research.	Quartz, February 14, 2023.
8.	Bill Gates thinks A.I. like ChatGPT is the 'most important' innovation right now.	CNBC, February 10, 2023.
9.	Elon Musk, who cofounded firm behind ChatGPT warns A.I. is 'one of the biggest risk' to civilization.	CNBC, February 15, 2023.
10.	Disinformation Researchers Raise Alarm About A.I. Chatbots.	The New York Times, February 8, 2023.
11.	ChatGPT launches boom in AI-written e-books on Amazon.	Reuters, February 21, 2023.
12.	Local college educators weigh in on ChatGPT.	KARE11 News, March 3, 2023.
13.	ChatGPT: cheating or achievement? Local entrepreneur and professor weigh in	KYFRTV News, March 6, 2023.
14.	Expert says ChatGPT is an amazing resource; UNLV students say they still wouldn't use the AI tech.	KTNV News, February 3, 2023.
15.	Noam Chomsky: The False Promise of ChatGPT.	Opinion, The New York Times, March 8, 2023.
16.	Steven Spielberg and Noam Chomsky say AI is soulless and scary.	MSN, March 9, 2023.

Copyright © 2023 by ASME

```
Separate the verbs and nouns in the following list:
Import, Export, Convert, Transfer, Distribute, Couple, Separate, Mix, Store, Supply,
Energize, De-Energize, Solid, Liquid, Gas, Electrical Energy, Mechanical Energy,
Thermal Energy, Chemical Energy, Electromechanical Energy, Acoustic Energy.


Verbs: Import, Export, Convert, Transfer, Distribute, Couple, Separate, Mix, Store,
Supply, Energize, De-Energize

Nouns: Solid, Liquid, Gas, Electrical Energy, Mechanical Energy, Thermal Energy,
Chemical Energy, Electromechanical Energy, Acoustic Energy
```

Figure 1: ChatGPT's capability to distinguish between verbs and nouns

```
A table combnining all options veb noun combinations totalling one hundred and twenty
Verbs: Import, Export, Convert, Transfer, Distribute, Couple, Separate, Mix, Store, Supply,
Energize, De-Energize
Nouns: Solid, Liquid, Gas, Electrical Energy, Mechanical Energy, Thermal Energy, Chemical
Energy, Electromechanical Energy, Acoustic Energy.


Import Solid
Export Solid
Convert Solid
Transfer Solid
Distribute Solid
Couple Solid
Separate Solid
Mix Solid
Store Solid
Supply Solid
Energize Solid
De-Energize Solid
Import Liquid
Export Liquid
Convert Liquid
Transfer Liquid
Distribute Liquid
Couple Liquid
Separate Liquid
Mix Liquid
Store Liquid
Supply Liquid
Energize Liquid
De-Energize Liquid
Import Gas
Export Gas
```

Figure 2: ChatGPT's capability to combine verbs and nouns to form functional basis

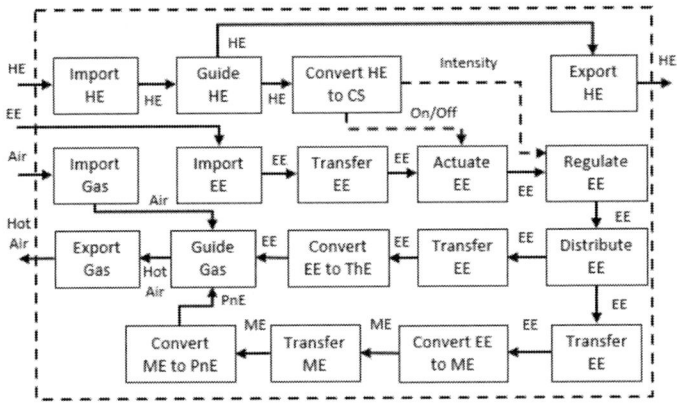

Figure 3: A function model for a hairdryer [20]

To test whether ChatGPT's can create function models, the test was run as a series of chats. Firstly, ChatGPT was asked to create a function model for a product that will import electricity and produce hot air. A result of this test is shown in Figure 4. ChatGPT was able to successfully break down the problem statement into a block diagram consisting of verb object pairs.

```
Design a function block diagram representing a
product that will import electricty and air and will
create an outflow of hot air.

[Import Electricity] ---> [Import Air] --->
[Combine Electricity and Air] ---> [Heat Air] --->
[Outflow of Hot Air]
```

Figure 4: Initial function model produced by ChatGPT

In the second set of series of chats, ChatGPT was asked to break down each subfunction of the function model generated in Figure 4 in an attempt to add more granularity [17] to the function model generated in Figure 4. The results of these subsequent chats are shown in Figure 6, Figure 7, Figure 8, and Figure 9. As can be seen, ChatGPT could no longer maintain solution neutrality. However, it still maintained some semblance of the relationship between the subfunction created in Figure 4.

```
How to import electricity.

[Electricity Source] ---> [Power Conditioner] --->
[Transformer] ---> [Power Distribution] --->
[Electricity Meter] ---> [Import Electricity]
```

Figure 5: ChatGPT's breakdown of Import Electricity

```
How to import air.

[Air Intake] ---> [Filter] ---> [Fan] ---> [Import Air]
```

Figure 6: ChatGPT's breakdown of Import Air

```
How to combine electricity and air.

[Import Electricity] ---> [Import Air] ---> [Heating
Element] ---> [Combine Electricity and Air]
```

Figure 7: ChatGPT's breakdown of Combine Electricity and Air

```
How to heat air.

[Combine Electricity and Air] ---> [Heat
Exchanger] ---> [Heat Air]
```

Figure 8: ChatGPT's breakdown of Heat Air

Copyright © 2023 by ASME

```
How to create an outflow of hot air.

[Heat Air] ---> [Outflow Valve] ---> [Outflow of Hot
Air]
```

Figure 9: ChatGPT's breakdown of Outflow of Hot Air

2.3 Working Structure

A working principle reflects the physical effect with the necessary geometric and material characteristics needed for the fulfillment of a given function [4]. The working principles can be combined to generate a working structure that can accomplish the overall function [4]. To test whether ChatGPT's can generate a working principle for a subfunction, it was asked to generate working principles for converting electricity into heat energy, as shown in Figure 10. It was able to accomplish this task successfully. Next it was asked for some commercial examples of the working principle it just proposed. It was also able to accomplish this task successfully as shown in Figure 10.

```
How to convert electricity into heat energy.
• Use a heating element to convert the electricity
into heat energy
what are the commercially available heating
elements
• Resistance heating elements
• Electric arc furnaces
• Induction heating elements
• Dielectric heating elements
```

Figure 10: ChatGPT's working principle for converting electricity into heat energy

In order to develop a working structure for a product that will import electricity and create an outflow of hot air, ChatGPT was asked if there are any products that are close to this design. As shown in **Figure 11**, it was able to accurately point to multiple products that will accomplish this overall function, albeit with slightly different working structures.

```
What are the products that are close to this design?

The products that are closest to this design are
electric heaters, air heaters, and space heaters
```

Figure 11: ChatGPT's working structure for the overall function

Lastly, since ChatGPT was able to identify multiple different solutions with very similar working structures, it was investigated whether similarity between various working structures or products can be perceived. In human designers this similarity is perceived broadly as functional similarity and energy flow similarity [21]. As shown in Figure 12, ChatGPT was asked to determine whether hairdryer and a heat gun were similar to the suite of solutions it proposed in Figure 11. ChatGPT was correctly able to identify that these products are indeed similar. This is also in line with the findings in a previous research study carried out on human designers [21].

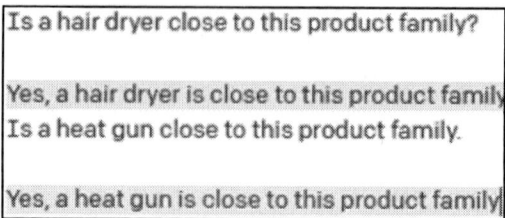

Figure 12: ChatGPT's analysis on similarity of products

3. DISCUSSION AND FUTURE RESEARCH

From the preliminary study presented in the previous section, it is safe to assume that ChatGPT is capable of doing language-based reasoning which can be successfully applied conceptual design generation stage of the systematic design process. While several previous approaches to automating all or part the design process have been proposed, none is a) harnessing the power of the entirety of the human knowledge generated, and b) available to engineering systems designers and non-designers alike. This tool has the potential to democratize systems design and, in the process, decentralize the very process of design from engineers and researchers to whoever wants to generate a design of a system.

However, before that happens, the engineering systems design community needs to consider the following research themes on incorporating a tool like ChatGPT in design and education.

1. The answers generated by the tool need to be evaluated on accuracy and precision.
2. The quality and quantity of solutions generated by the tool need to be benchmarked compared to the solutions generated by a standalone team of human designers and a team of human designers using ChatGPT.
3. Generative design on a systems level falls on the higher levels of Blooms' Taxonomy, however engineering students do study rule based courses in their junior and senior year of college that fall on the lower level of Bloom's Taxonomy. Future research efforts need to focus on how to keep those courses relevant to the learning experience of future engineering systems designers.

4. CONCLUSION

In this research, a preliminary investigation in the context of generative design was conducted into the capabilities of Open AI's new LLM based tool ChatGPT. It was found that ChatGPT has the capability to generate function models from a problem statement, limited capability to add granularity to the function model, and a good capability to attribute working principles and working structures to function models it generates. In addition,

Copyright © 2023 by ASME

it was also capable of discerning similarity between products based on functional similarity. Lastly, research themes have been proposed to better understand the capabilities of ChatGPT so that it can be incorporated into engineering systems design research. It is also important to study the implications of this tool in the engineering systems design teaching practice.

While ChatGPT and the increasingly complex Artificial Intelligence tools that are moving toward Artificial General Intelligence seem like a replacement of the engineering designer, but for the foreseeable future the human design engineer will hold an advantage over these tools. However, it seems like increasingly likely that augmenting the human design engineer with a tool like ChatGPT may improve design outcomes and will be the focus of future studies.

REFERENCES

[1] "Introducing ChatGPT" [Online]. Available: https://openai.com/blog/chatgpt. [Accessed: 08-Mar-2023].

[2] Gao, L., Schulman, J., and Hilton, J., 2022, "Scaling Laws for Reward Model Overoptimization."

[3] Rosen, D. W., and Choi, Y. M., 2022, "Generative Design of Cyber-Physical-Human System Families: Concepts and Research Issues," Proceedings of the ASME Design Engineering Technical Conference, **3-B**.

[4] Pahl, G., Beitz, W., Feldhusen, J., and Grote, K. H., 2007, *Engineering Design: A Systematic Approach*, Springer London.

[5] Mokhtarian, H., Coatanéa, E., Edam, H. P.-A., and 2017, undefined, "Function Modeling Combined with Physics-Based Reasoning for Assessing Design Options and Supporting Innovative Ideation," cambridge.org.

[6] Mathieson, J. L., Wallace, B. A., and Summers, J. D., 2013, "Assembly Time Modelling through Connective Complexity Metrics," https://doi.org/10.1080/0951192X.2012.684706, **26**(10), pp. 955–967.

[7] Gill, A. S., Summers, J. D., and Turner, C. J., 2017, "Comparing Function Structures and Pruned Function Structures for Market Price Prediction: An Approach to Benchmarking Representation Inferencing Value," AI EDAM, **31**(4), pp. 550–566.

[8] Kurtoglu, T., and Tumer, I. Y., 2008, "A Graph-Based Fault Identification and Propagation Framework for Functional Design of Complex Systems," Journal of Mechanical Design, Transactions of the ASME, **130**(5).

[9] Van Bossuyt, D. L., and Arlitt, R. M., 2020, "A Functional Failure Analysis Method of Identifying and Mitigating Spurious System Emissions from a System of Interest in a System of Systems," J Comput Inf Sci Eng, **20**(5).

[10] Gill, A. S., and Sen, C., 2020, "Evolutionary Approach to Function Model Synthesis: Development of Parameterization and Synthesis Rules," Proceedings of the ASME Design Engineering Technical Conference, **9**.

[11] Gill, A. S., and Sen, C., 2021, "Logic Rules for Automated Synthesis of Function Models Using Evolutionary Algorithms," Proceedings of the ASME Design Engineering Technical Conference, **2**.

[12] Vucovich, J., Bhardwaj, N., Ho, H. H., Ramakrishna, M., Thakur, M., and Stone, R., 2008, "Concept Generation Algorithms for Repository-Based Early Design," Proceedings of the ASME Design Engineering Technical Conference, **2006**, pp. 239–249.

[13] Mikes, A., Edmonds, K., Stone, R. B., and DuPont, B., 2020, "Optimizing an Algorithm for Data Mining a Design Repository to Automate Functional Modeling," Proceedings of the ASME Design Engineering Technical Conference, **11A-2020**.

[14] Bryant, C. R., Stone, R. B., McAdams, D. A., Kurtoglu, Tolga., and Campbell, M. I., *Concept Generation From The Functional Basis of Design*.

[15] Gill, A. S., 2021, "Evolutionary Algorithms for Function Model Synthesis."

[16] Jiao, R., Commuri, S., Panchal, J., Milisavljevic-Syed, J., Allen, J. K., Mistree, F., and Schaefer, D., 2021, "Design Engineering in the Age of Industry 4.0," Journal of Mechanical Design, Transactions of the ASME, **143**(7).

[17] Maier, J. F., Eckert, C. M., and John Clarkson, P., 2017, "Model Granularity in Engineering Design – Concepts and Framework," Design Science, **3**, p. e1.

[18] Antonelli, W., and Johnson, A., 2023, "OpenAI Playground: How to Use the GPT-3 Chatbot," Business Insider.

[19] Stone, R. B., and Wood, K. L., 2021, "Development of a Functional Basis for Design," Proceedings of the ASME Design Engineering Technical Conference, **3**, pp. 261–275.

[20] Sen, C., and Summers, J. D., 2014, "A Pilot Protocol Study on How Designers Construct Function Structures in Novel Design," Design Computing and Cognition '12, pp. 247–264.

[21] Gill, A. S., Tsoka, A. N., and Sen, C., 2019, "Dimensions of Product Similarity in Design by Analogy: An Exploratory Study," Proceedings of the ASME Design Engineering Technical Conference, **7**.

Copyright © 2023 by ASME

Proceedings of the ASME 2023
International Design Engineering Technical Conferences and
Computers and Information in Engineering Conference
IDETC-CIE2023
August 20-23, 2023, Boston, Massachusetts

DETC2023-116719

AN APPROACH FOR PREDICTING SOCIAL, ENVIRONMENTAL, AND ECONOMIC PRODUCT IMPACTS AND NAVIGATING THE ASSOCIATED IMPACT TRADE-SPACE IN ENGINEERING DESIGN

Christopher S. Mabey[1],*, Tevin J. Dickerson[1], John L. Salmon[1], Christopher A. Mattson[1]

[1]Brigham Young University, Provo, UT

ABSTRACT

There is a growing demand for sustainable products and systems. Sustainability encompasses environmental, social, and economic aspects, often referred to as the three pillars of sustainability. To make more sustainable design decisions, engineers need tools to predict the environmental, social, and economic impacts of products and navigate potential impact trade-offs. To capture societal adoption of technology, this article uses agent-based modeling (ABM) to predict impacts across all three pillars of sustainability and how to navigate the multidimensional impact trade space. The approach described in this article is based on three main components for the predictive modeling of product impacts and impact trade-space navigation, i) ABM of product adoption, ii) the assessment of product impacts, and iii) an approach for design decisions in the presence of impact trade-offs. The trade-space navigation uses a visual approach to find the non-dominated solutions in the product impact space. To illustrate and describe how to use the method, a case study is presented that predicts the impact of residential solar panels in a region of the United States under various scenarios. The findings of the case study can help policymakers understand suitable implementation strategies for residential solar panels while considering the impact trade-offs involved.

Keywords: Agent-Based Modeling, Life Cycle Sustainability Assessment, Engineering for Sustainable Development, Predictive Social Impact Modeling

1. INTRODUCTION

The United Nations Sustainable Development Goals (SDGs) contain aspects of social, environmental, and economic impacts [1]. Social, environmental, and economic impacts make up what are often referred to as the three pillars of sustainability [2] or the triple bottom line [3]. As practitioners work to improve sustainability practices in all three pillars of sustainability, there are

often trade-offs between the different pillars [4, 5]. Consumers may not even be aware of the trade-offs that are made in the products they adopt, such as the potential use of child labor in the mining of metals needed for electric vehicles that reduce greenhouse gas emissions [6]. To make more sustainable design decisions, engineers need tools to predict the impacts of products and navigate potential impact trade-offs. As society works toward solutions for complex large-scale problems such as climate change, there will need to be changes in behavior and technology. The process toward electrification will implement technologies and infrastructure that will be in place anywhere from a few years to decades. The long time horizon, economic costs, and urgency of acting on climate change all point to the need to make better design decisions with respect to sustainability.

It has been acknowledged that engineers need improved tools to estimate the impact of products and that there are few tools to estimate impacts, especially the estimation of social impacts [7]. Life cycle assessment (LCA) with its accompanying ISO standard is often used to assess environmental impacts [8]. Methods to assess the social impact of products are, compared to LCA, less mature [9]. The business literature has many ways to evaluate economic success, but in the literature related to sustainability, life cycle costing (LCC) and return on investment (ROI) are frequently used to measure the economic pillar [10]. One common problem with sustainability analysis tools such as LCA is how to scale the results to the population level [11]. In response to this limitation, Liechty et al. developed a method to scale the environmental impacts obtained from the functional unit level to the population level using agent-based modeling (ABM) to understand the adoption of the product in the population [12].

The contribution of this article is to demonstrate an approach for the predictive modeling of the social, environmental, and economic impacts of a product and a method for navigating potential sustainability trade-offs in all three pillars of sustainability. This article builds on the work of Liechty et al. [12], by examining impacts across all three pillars of sustainability and how to navigate

*Corresponding author: mabeyc@byu.edu

Copyright © 2023 by ASME

the multidimensional impact trade space. A case study using the impacts of residential solar panels in a region in the United States is used to illustrate the method. The remainder of this article is organized as follows: Section 1.1 reviews the previous literature on ABM and product impact assessment, Section 2 describes the methodology of this paper for triple bottom line impact prediction and trade space navigation, Section 3 presents the results of a residential solar panel case study, Section 4 provides a discussion of the implications of the results and method, and finally Section 5 provides concluding remarks.

1.1 Background

The approach described in this article is based on three main components for predictive modeling of product impacts and impact trade space navigation, *i)* ABM of product adoption, *ii)* the assessment of product impacts, and *iii)* an approach for design decisions in the presence of impact trade-offs.

1.1.1 Agent-Based Modeling.
ABM is a bottom-up modeling approach in which the behavior of individual entities is modeled based on rules rather than on a top-down governing equation [13]. One of the main benefits of ABM is the ability to capture emergent phenomena from agent interactions [14]. There have been many applications of ABM, including segregation [15], water management [16], bullying [17], combat simulation [18], team problem solving [19], among many other applications [20]. The main application of interest in this article is the use of ABM to predict product adoption. ABM has been used in many cases for the adoption of products and the diffusion of technology [21–23]. One of the advantages of ABM when modeling product adoption is to examine product adoption patterns across different segments of a population. Although ABM has been used to model product adoption in many cases, ABM has only recently been used to predict product impacts [12, 24]. In the cases where it was used to predict the impacts of the product, it was not used to predict the impacts in the three areas of sustainability. The work in this paper will use the results of the ABM to predict the social, environmental, and economic product impacts.

1.1.2 Product Impact Assessment.
Interest in the assessment of the impacts of products has increased in recent years [23]. Movements such as the Environmental, Social, and Corporate Governance framework (ESG) [25], and B Corp Certification [26] have spread in the business sector due to the consumer demand for companies to have a more positive impact. With increasing consumer demand for improved sustainability, methods for measuring sustainability and product impacts have also grown [27]. This subsection will describe the advances and current approaches to assess environmental, economic, and social impacts.

The evaluation of environmental impact using LCA is the most established and mature of the three pillars of sustainability [8]. An LCA measures the impact of a product or system based on the raw materials needed and the processes used at different stages of the product life cycle [28]. Although LCA was first used around 1970, the practice began to be standardized in the early 1990s by the Society of Environmental Toxicology and Chemistry (SETAC)[29]. In 1996, the basic standard for LCA was published

as ISO 14040 [30]. Although the ISO standard outlines the basic principles of LCA, there are different LCA approaches such as the IMPACT World+ method [31], and the ReCiPe 2016 method [32]. LCA output will vary depending on the method used, but, in general, output is impact metrics per functional unit in various categories of environmental impact. LCA is not well equipped to incorporate the dynamic nature of sociotechnical systems [33], but when coupled with a method such as ABM the dynamic sociotechnical system can be taken into account [34].

In the sustainability literature, LCC is the most common method of assessing economic impact and is often used as the only metric for economic impact in sustainability assessments [35]. LCC is the sum of the costs associated with the manufacture, distribution, use, and disposal of a product or system. The use of only a single metric gives a narrow view compared to the more holistic approach of the various LCA indicators; Wood and Hertwich propose the need for a wider view of economic impacts than that provided by the LCC [36].

Assessing the social impact of products is a relatively new field compared to LCA [23]. The social impact of a product can be defined as the impact of the product on the day-to-day lives of persons [37]. Different methods have emerged to assess the social impact of products, such as Social Impact Assessment [38], social life cycle assessment (SLCA)[39], and the product impact metric [40]. Most methods of assessing social impact have focused on assessing current impacts rather than predicting future impacts, although we are beginning to see cases of their use to predict and improve decision making [41].

Life cycle sustainability assessment (LCSA) has emerged as a method for assessing the environmental, economic, and social impacts of a product or system. LCSA typically uses an environmental LCA, LCC, and SLCA and defines LCSA as seen in Equation 1 [11].

$$LCSA = LCA + LCC + SLCA \qquad (1)$$

Although this is a simple approach and does not apply arbitrary weighting to any of the assessments, the assessments have different units and, therefore, cannot simply be combined. Additionally, using this simple sum does not aid in navigating potential impact trade-offs. Some cases of LCSA use multi-criteria decision analysis to aid in the decision-making process [27].

1.1.3 Trade Space Exploration.
There are cases where impact improvements in the pillars of sustainability may conflict with each other [42]. In the presence of conflicts, a design team will need methods to navigate the trade space. There are multiple methods to improve decision making in the presence of trade-offs, including multi-criteria decision-making methods such as the Analytic Hierarchy Process and TOPSIS [43], and Pareto-based methods to find and visualize the set of nondominated solutions [44, 45]. Plotting the impact space allows for finding the non-dominated set, and the visual nature is easy to incorporate into discussions surrounding design decisions [12].

2. METHODS

The general process used in this article is outlined in Figure 1. To illustrate and describe how to use the modeling approach, a

Copyright © 2023 by ASME

case study is presented that predicts the impact of residential solar panels in the state of Utah in the United States under various scenarios. As with all product impact models and assessments, the case study does not predict all product impacts. It predicts a subset of indicators for certain aspects of environmental, social, and economic impacts. Prediction or assessment of a subset of product impacts remains valuable for making design decisions that result in improved product impacts.

TABLE 1: SOLAR PANEL PERFORMANCE PARAMETERS

Panel Type	Efficiency (%)	Degradation (%/yr)	Lifetime (yrs)
Monocrystalline Silicon	21	0.5	25
Polycrystalline Silicon	19	0.5	25
Copper-Indium-diSelenide (CIS)	12	1.0	25

This model predicts the adoption of renewable energy in a region. The agents in the model represent a potential customer with their respective housing development. These agents represent a weighted group of individuals living in the marked county relative to the real population. The agents are then randomly assigned locations relative to the chosen region. Each location has different attributes that are based on data sampled from the region. This includes energy consumption, energy costs, economics, and geopolitical influences. These attributes are then randomly assigned to the agents. The agent is also given a solar adoption parameter based on market diffusion models. These models are based on past technological diffusion models, as well as payback period and percentage of monthly bill savings of the agents.

dGen uses a solar database from the National Solar Radiation Database. This database is used to estimate the solar irradiation incident on a region. This database is used to approximate yearly weather in the model.

To model residential housing, dGen uses a stochastic system to choose the roof area, azimuth angle, and slope of each agent based on information collected from the region. These parameters are then used with the solar model and the input of the solar panel to calculate the power per square area developed by the solar panels. To calculate the economic impact of solar panels, dGen uses regional data on energy costs, state incentives, and assumes one of two business models. The agent can own or rent solar equipment. One of the business models are randomly assigned to the agents.

The design variables changed in the model scenarios included different types of solar panels. These files included performance increase estimates per year, as well as degradation estimates. The photovoltaic wholesale price was varied as well. The model had built-in default wholesale price scenario files that were used. These files included a high, medium, and low range.

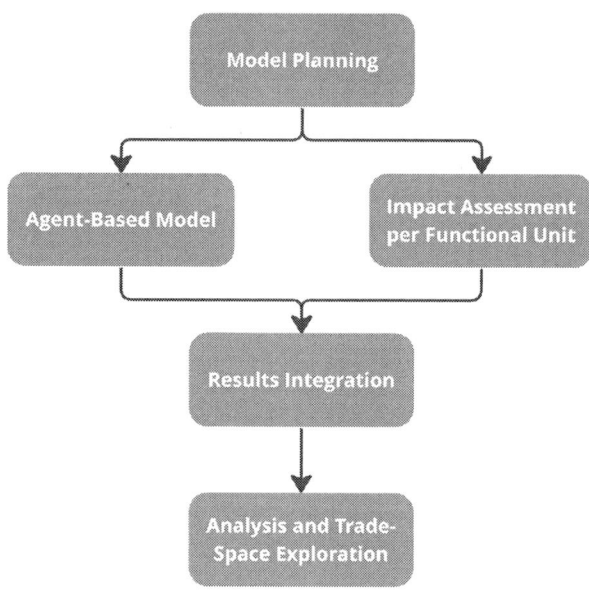

FIGURE 1: PRODUCT IMPACT MODELING PROCESS

2.1 Model Planning

The first stage of the process is model planning. Model planning requires defining the objective of the model and the types of scenarios that will be used to explore the impact space of the product. For the case study, the objective is to predict the environmental, social, and economic impacts of residential solar panels in the state of Utah between the years 2022 and 2050. Although the model can be used for the entire United States, the results presented in this article are limited to the state of Utah because of the computational resources required to model the entire country. To understand the impact space, scenarios with different product designs and pricing structures were investigated. This involved using three types of solar panels, each under three different pricing scenarios. The parameters of the solar panels can be seen in Table 1, with the values for the solar panel performance derived from estimates published by the National Renewable Energy Laboratory (NREL) [46, 47]. After initial model planning, the ABM and the impact assessment per functional unit can be carried out in parallel.

2.2 Distributed Generation Market Demand ABM

The ABM model used in the analysis is the Distributed Generation Market Demand Model (dGen) developed by NREL [48].

2.3 Environmental Impact Assessment

Although there are many methods for LCA, the ReCiPe(H) midpoint method is well established in the literature and is used in this article [32]. A midpoint analysis was selected because it has the most detailed indicators related to environmental impact, with 17 indicators. There are numerous guides for completing an environmental LCA [28, 49, 50], therefore, this article will not give a detailed explanation of how to carry out an LCA. Rather, this research will provide insight into how LCA results are integrated

Copyright © 2023 by ASME

with ABM results. Across the various social, environmental, and economic impacts, there is some overlap between the three categories. For example, human health and safety are often considered social impacts [51], but in the ReCiPe(H) method, there are indicators related to human health, namely human carcinogenic toxicity and human non-carcinogenic toxicity. These two indicators come from the LCA, but in this article, during the analysis of the results, they will be treated as social impact indicators. The output of the LCA needs to be in terms of a functional unit, where the output of the ABM provides the number of functional units needed in the modeled scenario. In the case of the solar panel adoption model under discussion, the LCA results are normalized to be the environmental impact per kilowatt-hour of energy generated, and the ABM outputs the number of kilowatt-hours generated by the solar panels over the course of the simulation. The LCA metrics use data from an LCA performed by Milousi et al. [52], and the LCA metrics for each of the 17 indicators can be seen in Table 2.

2.4 Social Impact Assessment

To assess social impacts, this study uses a method proposed by Stevenson et al. [41], where relevant social impact categories are selected from the 11 social impact categories identified by Rainock et al. [51]. Once the relevant categories are identified, indicators are selected to measure those categories. The creation of jobs and the reduction in utility bills, which relate to the impacts on the paid work category, have previously been used to measure the social impact of solar panels [53, 54]. In addition, there are impacts on human health with indicators obtained from the LCA results. The functional units for the social impacts vary according to the impact. The creation of new jobs is measured as the number of jobs per megawatt of installed solar capacity. In the United States, there are 5.2 jobs created per megawatt of installed solar capacity [55]. This is likely a conservative estimate because it covers the entire solar industry. Residential solar typically requires more labor than industrial-scale projects. The mean amount that utility bills are reduced per customer that adopts a solar power generation system is output from the ABM. For a summary of social impact metrics, see Table 3.

2.5 Economic Impact Assessment

Economic impact assessment in sustainability is typically performed using LCC [56], where the total cost of obtaining the system, its maintenance and operation cost, and the cost of disposal are summed. In the case of residential solar panels in this model, only the cost of obtaining the system and its maintenance and operation cost were considered due to the high uncertainty associated with solar panel recycling at the end of the 25-year life. Solar panel recycling costs are likely to change dramatically by the time the first panels installed in 2022 within the model, are recycled in 2047. The life cycle costs are output from the ABM. In addition to LCC, this model also considers the payback period for the initial investment in a solar panel system, which is also output from the ABM.

2.6 Results Integration

To obtain the results for the total impact of the product, the results of the ABM and impact assessment will need to be integrated [12]. For impact metrics that are measured by impact per functional unit of the product, each individual impact metric is scaled by the total number of functional units in the ABM scenario. Equation 2 defines this, where the total impact I is obtained by multiplying the impact per functional unit P, by the total quantity of functional units needed in the scenario α. P is calculated during the environmental, social, and economic impact assessment.

$$I = \alpha P \tag{2}$$

Once the total impact at the population level is identified for each impact metric, comparisons can be made between the categories of social, environmental, and economic impacts.

2.7 Trade-Off Analysis

This paper uses a visually presented Pareto-based method to identify impact trade-offs and facilitate discussions about the impact trade-space. To visualize the results, three scatterplot matrices are created. The three matrices are the comparisons between the social, environmental, and economic product impacts. Each dimension of the matrix contains an impact category; e.g., each row contains a social impact metric, and each column contains an environmental impact. By creating scatterplot matrices, pairwise comparisons can be made between all impact metrics. For each pairwise comparison, the Pareto front can be highlighted to show the curve of non-dominated solutions [42]. Then the scatterplot matrices are ready to be used to help navigate the impact trade-space.

3. RESULTS

Using the dGen model, scenarios were explored for the three types of panels, each with the projections of low, medium and high prices provided by NREL in the model. This gave nine different scenarios to compare the results. The results will be compared for adoption and energy generation and product impacts.

3.1 Solar Panel Adoption

The first result examined is the adoption and energy generation of the residential solar panels installed during the scenarios. From Figure 2, we see that the price scenario is the main differentiating factor for the number of kilowatts of solar capacity installed, as well as the kilowatt-hours of energy generated. The difference between the results of the polycrystalline silicon and monocrystalline silicon panels is minimal. The CIS panel lags behind both in solar capacity installed and in energy generated, due to the lower efficiency of this type of panel. In the high-price scenario, the CIS panel has almost the same installed solar capacity, but it is still behind in energy generation.

3.2 Impact Analysis

The results of the product impacts at the population level reveal trade-offs in some cases. Figures 3-5 show scatterplot matrices, where each axis represents one of the three pillars of sustainability. Between the three figures, all pairwise impact comparisons are made. As with solar panel adoption, the price scenario is the largest differentiator between the different scenarios. Due to space constraints within the article, only a subset of

Copyright © 2023 by ASME

TABLE 2: LCA METRICS PER FUNCTIONAL UNIT

Imact Category	Unit	Polycrystalline	Monocrystalline	CIS
Global warming	$kg\ CO_2\text{-}eq/kWh$	4.43×10^{-2}	5.24×10^{-2}	3.95×10^{-2}
Stratospheric ozone depletion	$kg\ CFC11\text{-}eq/kWh$	2.06×10^{-8}	2.45×10^{-8}	1.75×10^{-8}
Ionizing radiation	$kBq\ Co\text{-}60 - eq/kWh$	4.08×10^{-3}	4.45×10^{-3}	3.96×10^{-3}
Ozone formation, human health	$kg\ NO_x\text{-}eq/kWh$	1.05×10^{-4}	1.20×10^{-4}	9.09×10^{-5}
Fine particulate matter formation	$kg\ PM_{2.5}\text{-}eq/kWh$	1.04×10^{-4}	1.23×10^{-4}	9.39×10^{-5}
Ozone formation, terrestrial ecosystems	$kg\ NO_x\text{-}eq/kWh$	1.10×10^{-4}	1.25×10^{-4}	9.26×10^{-5}
Terrestrial acidification	$kg\ SO_2\text{-}eq/kWh$	2.21×10^{-4}	2.47×10^{-4}	2.07×10^{-4}
Freshwater eutrophication	$kg\ P\text{-}eq/kWh$	3.78×10^{-5}	4.07×10^{-5}	4.62×10^{-5}
Terrestrial ecotoxocity	$kg\ 1,4\text{-}DCB\text{-}eq/kWh$	1.17	1.13	4.62×10^{-1}
Freshwater ecotoxicity	$kg\ 1,4\text{-}DCB\text{-}eq/kWh$	1.16×10^{-2}	1.17×10^{-2}	1.30×10^{-2}
Marine ecotoxicity	$kg\ 1,4\text{-}DCB\text{-}eq/kWh$	1.53×10^{-2}	1.54×10^{-2}	1.69×10^{-2}
Human carcinogenic toxicity	$kg\ 1,4\text{-}DCB\text{-}eq/kWh$	4.17×10^{-3}	4.33×10^{-3}	4.19×10^{-3}
Human non-carcinogenic toxicity	$kg\ 1,4\text{-}DCB\text{-}eq/kWh$	1.63×10^{-1}	1.64×10^{-1}	2.00×10^{-1}
Land use	$m^2 a\ crop\text{-}eq/kWh$	1.23×10^{-3}	1.23×10^{-3}	9.60×10^{-4}
Mineral resource scarcity	$kg\ Cu\text{-}eq/kWh$	5.54×10^{-4}	5.42×10^{-4}	8.21×10^{-4}
Fossil resource scarcity	$kg\ oil\text{-}eq/kWh$	1.08×10^{-2}	1.27×10^{-2}	9.40×10^{-3}
Water Consumption	m^3/kWh	1.35×10^{-3}	1.17×10^{-3}	3.22×10^{-4}

TABLE 3: SOCIAL IMPACT INDICATORS SUMMARY

Indicator	Unit	Value
Jobs created	jobs/megawatt	5.2
Utility bill reduction	$	Output from ABM
Human carcinogenic toxicity	$kg\ 1,4\text{-}DCB\text{-}eq/kWh$	Varies by panel type
Human non-carcinogenic toxicity	$kg\ 1,4\text{-}DCB\text{-}eq/kWh$	Varies by panel type

the environmental impacts is visualized here to ensure that the visualization fits on a single page. Data for all impact metrics are available upon request.

When examining environmental impacts alone, there is not a single panel type that performs better in all cases. The thin-film CIS panel performs better in all impact metrics shown in the figure except for the mineral resource scarcity metric due to the more scarce metals required for its construction. High-cost scenarios have the lowest overall environmental impacts, but that is only due to fewer agents that adopt solar panels based on the high price of the system. The high-cost scenarios also have some of the lowest social impacts for utility bill savings and job creation.

In social impact metrics, the monocrystalline panel performs best in terms of mean savings in utility bills and job creation. The thin-film CIS panel has the lowest human carcinogenic toxicity. In the human non-carcinogenic toxicity metric, the polycrystalline panel is the best performing, but only marginally better than the monocrystalline panel. Monocrystalline and polycrystalline panels have similar materials required for construction.

For each pricing scenario, there is no significant difference between the panel types for the measured economic impacts. There are also no significant differences in price between the panel types, so the low, medium and high pricing scenarios provided by NREL in the dGen model are identical between the three panel types. Future work could examine panel pricing scenarios in

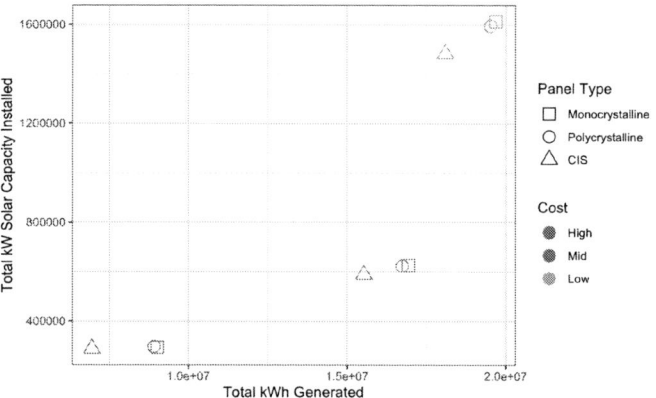

FIGURE 2: KILOWATTS OF SOLAR CAPACITY ADOPTED AND KILOWATT-HOURS OF POWER GENERATED

more detail.

3.3 Trade-Off Analysis

In Figures 3-5, we have highlighted the Pareto front. For the cases where there is a single optimal scenario, that point has been highlighted. The Pareto front represents the set of optimal scenarios on the trade-off curve. The figures highlight different trade-offs that exist in the product impact space. One trade-off that occurs in many cases is between the social impacts of utility bill savings and job creation with the environmental impacts. We observe that the scenarios with the highest utility savings and jobs created also have high environmental impacts because there are more solar panels adopted in the model. There are some scenarios and impact comparisons where there is an optimal solution and no trade-off exists. In some cases there may be an easy choice in a trade-off such as where there is a greater improvement in one metric for a small penalty in the other, such as when examining the

Copyright © 2023 by ASME

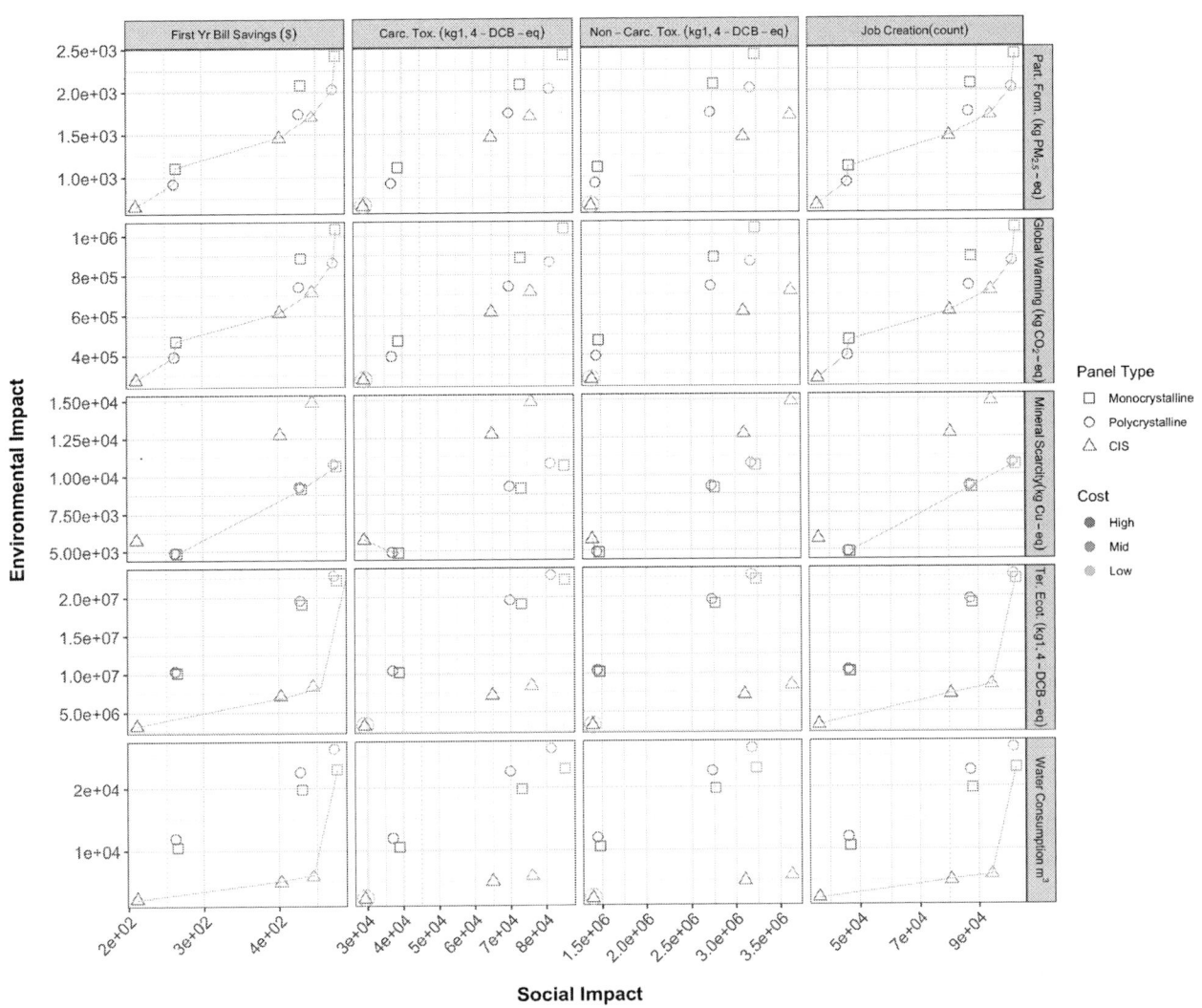

FIGURE 3: SOCIAL VS. ENVIRONMENTAL IMPACTS, WITH PARETO FRONT LINE OR HIGHLIGHTED OPTIMAL SOLUTION

carcinogenic toxicity column in Figure 4. In other cases, trade-off visualizations will be just one data point in the decision-making process. The data from the modeling and trade-off visualizations allow one to examine where impact trade-offs occur across a wide range of environmental, economic, and social impact indicators. Although it is difficult to make explicit rules of when to prioritize different impact categories in the decision-making process, identifying impact trade-offs and presenting them in a way that can be understood by various stakeholders has the potential to improve design decisions regarding sustainability.

3.4 Comparison to Other Power Generation Sources

In cases where a new technology is replacing an older technology, such as the transition to renewable energy sources, it is useful to examine the net impact of the technology. In the example of residential solar power, when examining the impact of solar power in isolation, the lowest environmental impacts are the cases where the fewest panels are installed. This does not consider the power source that is being replaced by residential solar panels. To examine the impact of residential solar power scenar-

ios combined with existing power sources, this paper will use the scenario with the highest kilowatt-hours of power generated by residential solar power systems as a baseline to compare against. The highest kilowatt-hour generating scenario was the monocrystalline panel with the low-cost case; this scenario will serve as the baseline case. For the other scenarios with less power generated, the difference will be made up of the other power sources compared; see Figure 6 for the proportion of residential solar power to the other power source.

This analysis will examine each power source individually rather than a percentage of numerous types of power generation. Coal and natural gas power will be used for the comparison. Data were not available for each indicator used in Section 3.2. Therefore, it is still necessary to make comparisons of the analysis with only solar panel impacts and solar panel impacts combined with the other power generation methods. These two analyses will enable a more holistic understanding of the various impacts of the new technology. The economic impact analysis that combines solar and other energy sources uses *levelized cost of energy (LCOE)* as a metric for the lifetime cost. This is a method for the

Copyright © 2023 by ASME

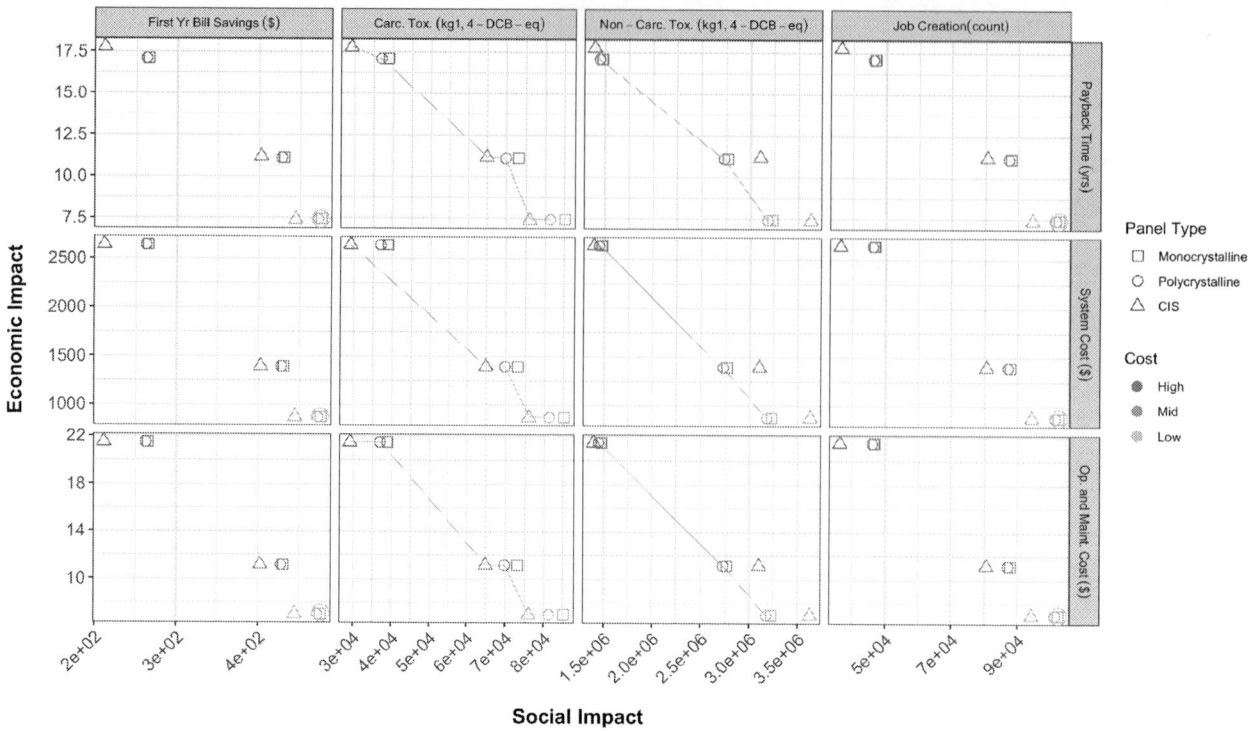

FIGURE 4: SOCIAL VS. ECONOMIC IMPACTS, WITH PARETO FRONT LINE OR HIGHLIGHTED OPTIMAL SOLUTION

lifetime cost of energy per unit of power generation that takes into account system costs, operations, maintenance, end-of-life, and the price of carbon emissions. Data were obtained from reports by the *Interantional Energy Agency (IEA)*[57] and NREL[58], for the LCOE values for the different energy sources see Table 4. The LCOE uses $30/ton for the carbon emission cost. Data for the LCA of coal and natural gas production were obtained from work by Rapa et al.[59]

TABLE 4: LEVELIZED COST OF ENERGY FOR DIFFERENT SOURCES

Energy Source	LCOE USD/MWh
Residential solar	81
Coal	110
Natural gas	45

To obtain the total impact, the impacts for the generation of residential solar power needed to be combined with the impacts from natural gas or coal energy. Therefore, to calculate the impacts of residential solar and other sources of power generation, Equation 2 is modified to become Equation 3. In this equation I is the total impact, $alpha_{solar}$ is the number of kilowatt-hours generated by the residential solar panels in a particular scenario, and $alpha_{other}$ is obtained by Equation 4 by calculating the difference in kilowatt-hours between the amount of solar energy generated by a particular scenario and the baseline comparison scenario. P_{solar} is the impact per kilowatt-hour for the residential solar panels and P_{other} is the impact per kilowatt-hour for coal or natural gas, depending on what source is compared.

$$I = \alpha_{solar} P_{solar} + \alpha_{other} P_{other} \qquad (3)$$

$$\alpha_{other} = \alpha_{baseline} - \alpha_{solar} \qquad (4)$$

After the total impacts are calculated, the visual comparisons can be used as described in Section 3.3 to understand the tradespace. Figures 7-9 show the comparison plots. In these figures, a power scenario is presented using each pair of panel type and other power generation methods, as well as the baseline case where the residential solar comprises all of the impact metric. In these comparisons, similar to the previous comparisons with only solar power generation in Section 3.3, there is not a single scenario that is best or worst across all comparisons. Scenarios using coal as the other energy source are on the Pareto front less often than other scenarios, but in impact measures such as terrestrial ecotoxicity, the scenarios containing monocrystalline and polycrystalline panels have higher negative environmental impacts. The Pareto front is highlighted in each figure. Of all the comparisons, only three had an optimal scenario; the other comparisons all had trade-offs to some extent.

4. DISCUSSION

The method shown here illustrates the occurrence of impact trade-offs in the design of products and systems. This work expanded on the work of Liechty et al. [12] by providing methods for visual trade-space exploration in all three dimensions of sustainability. Using a Pareto-based method allows one to easily integrate impact results into the decision-making process. Quantitative impact modeling during the design process is just one

Copyright © 2023 by ASME

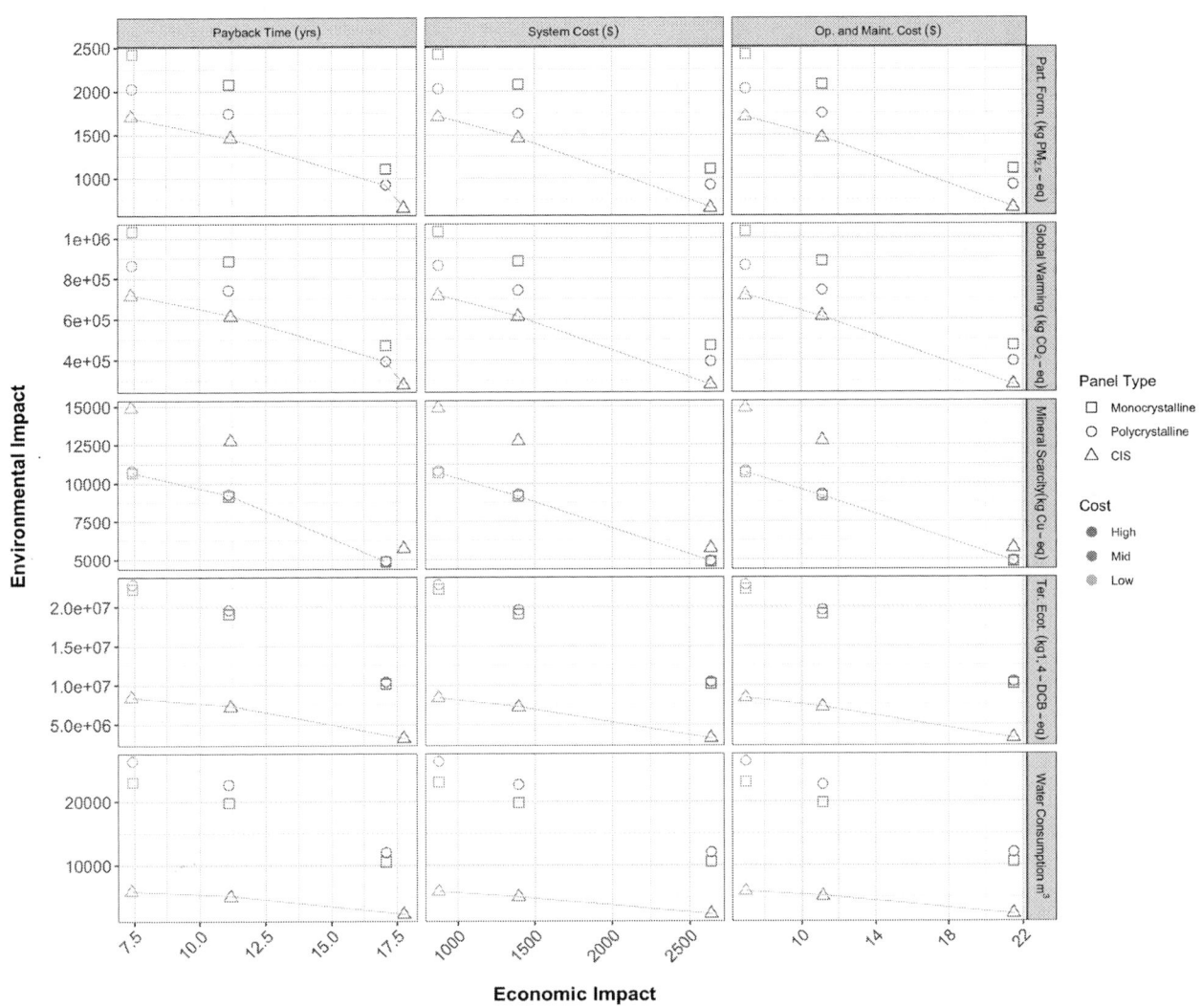

FIGURE 5: ENVIRONMENTAL IMPACTS VS. ECONOMIC IMPACTS, WITH PARETO FRONT LINE

tool that can be combined with other qualitative and quantitative methods to improve design decision making.

There are limitations associated with the case study used to demonstrate this method. A limited number of social impacts were considered and in future work more social impacts could be considered that focus on more stakeholders in the product life cycle, such as workers in the manufacturing of the panels or utility companies. Future work can also improve the pricing scenarios to illustrate how small differences in panel pricing ultimately affect impact. While these are limitations within the case study, the general method remains valuable in the decision-making process.

The results of the case study reveal the complexity of improving the impacts of products in a sociotechnical system. There was not a single scenario on the Pareto front in every impact comparison. It also illustrates that there is much more to the impact space than changes to product parameters; aspects of the system design, including business and policy decisions, are also important. If subsidies were increased to lower the prices for each of the scenarios, the adoption of solar panels would increase, but that may also influence future taxes to pay for subsidies.

With increasing calls for improved sustainability and progress toward the SDGs, tools to help make better design decisions for impact can become a valuable part of the decision-making process. These tools become more important when products and infrastructure will be in place for many years or decades because it is too costly to iterate on the design of the product, system or infrastructure.

5. CONCLUSION

The contribution of this article is to demonstrate an approach for the predictive modeling of the social, environmental, and economic impacts of a product and an approach to navigate potential sustainability trade-offs in the three pillars of sustainability. Furthermore, this work demonstrates how the impact of a functional product unit scales to the population level across the three pillars of sustainability. Tools to improve the impacts of products will be especially important as society makes significant shifts toward electrification in efforts to mitigate climate change. Many of these large shifts will require massive investments, and the systems may be in place for many years. With large amounts

Copyright © 2023 by ASME

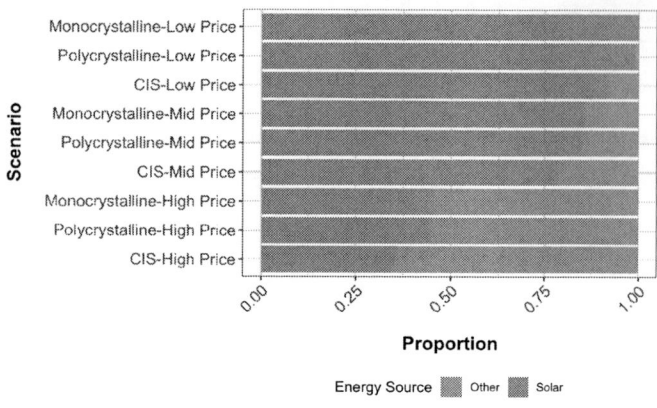

FIGURE 6: PROPORTION OF SOLAR ENERGY FOR EACH COMPARISON SCENARIO

of resources and the high cost of fixing mistakes, being able to use predictive impact modeling to improve design decisions may become an increasingly valuable tool.

DATA AVAILABILITY

Data and ABM code will be available for download at https://www.design.byu.edu/resources

ACKNOWLEDGMENTS

The authors would like to recognize the National Science Foundation for providing the Grants CMMI-1662485, and CMMI-1632740 that funded this research. Any opinions, findings, and conclusions or recommendations expressed in this material are those of the authors and do not necessarily reflect the views of the National Science Foundation.

REFERENCES

[1] United Nations. "Annex: Global Indicator Framework for the Sustainable Development Goals and Targets of the 2030 Agenda for Sustainable Development." Technical report no. 2019. URL https://unstats.un.org/sdgs/indicators/GlobalIndicatorFramework_A.RES.71.313Annex.pdf.

[2] Purvis, Ben, Mao, Yong and Robinson, Darren. "Three pillars of sustainability: in search of conceptual origins." *Sustainability Science* Vol. 14 No. 3 (2019): pp. 681–695. DOI 10.1007/s11625-018-0627-5. URL https://doi.org/10.1007/s11625-018-0627-5.

[3] Norman, Wayne and MacDonald, Chris. "Getting to the Bottom of "Triple Bottom Line"." *Business ethics quarterly: the journal of the Society for Business Ethics* (2004)DOI 10.5840/beq200414211. URL https://doi.org/10.5840/beq200414211.

[4] Luthin, Anna, Backes, Jana Gerta and Traverso, Marzia. "A framework to identify environmental-economic trade-offs by combining life cycle assessment and life cycle costing – A case study of aluminium production." *Journal of cleaner production* Vol. 321 No. 128902 (2021): p. 128902. DOI 10.1016/j.jclepro.2021.128902. URL https://www.sciencedirect.com/science/article/pii/S0959652621030961.

[5] Oláh, Judit, Kitukutha, Nicodemus, Haddad, Hossam, Pakurár, Miklós, Máté, Domicián and Popp, József. "Achieving Sustainable E-Commerce in Environmental, Social and Economic Dimensions by Taking Possible Trade-Offs." *Sustainability: Science Practice and Policy* Vol. 11 No. 1 (2018): p. 89. DOI 10.3390/su11010089. URL https://www.mdpi.com/385824.

[6] Sovacool, Benjamin K. "When subterranean slavery supports sustainability transitions? power, patriarchy, and child labor in artisanal Congolese cobalt mining." *The Extractive Industries and Society* Vol. 8 No. 1 (2021): pp. 271–293. DOI 10.1016/j.exis.2020.11.018. URL https://www.sciencedirect.com/science/article/pii/S2214790X20303154.

[7] Schöggl, Josef-Peter, Baumgartner, Rupert J and Hofer, Dietmar. "Improving sustainability performance in early phases of product design: A checklist for sustainable product development tested in the automotive industry." *Journal of cleaner production* Vol. 140 (2017): pp. 1602–1617. DOI 10.1016/j.jclepro.2016.09.195. URL https://doi.org/10.1016/j.jclepro.2016.09.195.

[8] Guinée, Jeroen B, Heijungs, Reinout, Huppes, Gjalt, Zamagni, Alessandra, Masoni, Paolo, Buonamici, Roberto, Ekvall, Tomas and Rydberg, Tomas. "Life cycle assessment: past, present, and future." *Environmental science & technology* Vol. 45 No. 1 (2011): pp. 90–96. DOI 10.1021/es101316v. URL http://dx.doi.org/10.1021/es101316v.

[9] Huertas-Valdivia, Irene, Ferrari, Anna Maria, Settembre-Blundo, Davide and García-Muiña, Fernando E. "Social Life-Cycle Assessment: A Review by Bibliometric Analysis." *Sustainability: Science Practice and Policy* Vol. 12 No. 15 (2020): p. 6211. DOI 10.3390/su12156211. URL https://www.mdpi.com/2071-1050/12/15/6211/htm.

[10] De Risi, Raffaele, De Paola, Francesco, Turpie, Jane and Kroeger, Timm. "Life Cycle Cost and Return on Investment as complementary decision variables for urban flood risk management in developing countries." *International Journal of Disaster Risk Reduction* Vol. 28 (2018): pp. 88–106. DOI 10.1016/j.ijdrr.2018.02.026. URL https://www.sciencedirect.com/science/article/pii/S2212420918302176.

[11] Walter Kloepffer. "Life cycle sustainability assessment of products." *International Journal of Life Cycle Assessment* Vol. 13 No. 2 (2008): p. 89. DOI 10.1065/lca2008.02.376. URL https://doi.org/10.1065/lca2008.02.376.

[12] Liechty, Joseph C, Mabey, Christopher S, Mattson, Christopher A, Salmon, John L and Weaver, Jason M. "Trade-off Characterization Between Social and Environmental Impacts Using Agent-Based Product Adoption Models and Life Cycle Assessment*." *Journal of mechanical design (New York, N.Y.: 1990)* Vol. 145 No. 3 (2023): pp. 1–18. DOI 10.1115/1.4056006. URL https://asmedigitalcollection.asme.org/mechanicaldesign/article/145/3/032001/1148269/Trade-Off-Characterization-Between-Social-and.

[13] Vicsek, Tamas. "Complexity: The bigger picture." *Nature*

Copyright © 2023 by ASME

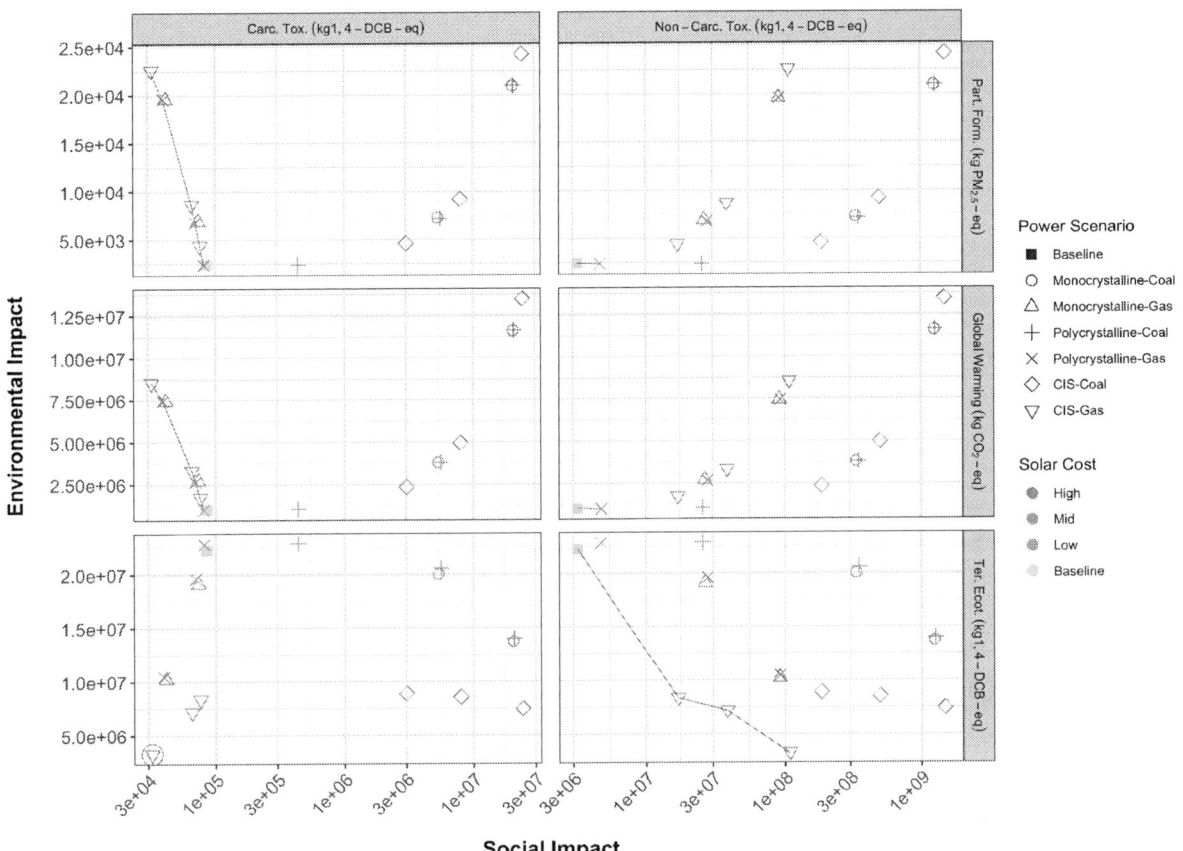

FIGURE 7: NET SOCIAL VS. ENVIRONMENTAL IMPACTS, CARCINOGENIC TOXICITY AND NON-CARCINOGENIC TOXICITY VALUES ARE PRESENTED ON A LOGARITHMIC SCALE

Vol. 418 No. 6894 (2002): p. 131. DOI 10.1038/418131a. URL https://doi.org/10.1038/418131a.

[14] Bonabeau, Eric. "Agent-based modeling: Methods and techniques for simulating human systems." *Proceedings of the National Academy of Sciences of the United States of America* Vol. 99 No. SUPPL. 3 (2002): pp. 7280–7287. DOI 10.1073/pnas.082080899. URL https://doi.org/10.1073/pnas.082080899.

[15] Schelling, Thomas C. "Dynamic models of segregation." *The Journal of mathematical sociology* Vol. 1 No. 2 (1971): pp. 143–186. DOI 10.1080/0022250X.1971.9989794. URL https://10.0.4.56/0022250X.1971.9989794.

[16] Koutiva, Ifigeneia and Makropoulos, Christos. "Exploring the effects of domestic water management measures to water conservation attitudes using agent based modelling." *Water Science & Technology: Water Supply* Vol. 17 No. 2 (2017): pp. 552–560. DOI 10.2166/ws.2016.161. URL http://dx.doi.org/10.2166/ws.2016.161.

[17] Tseng, Shih-Hsien, Chen, Chien-Kuo, Yu, Jia-Chen and Wang, Ying-Chi. "Applying the agent-based social impact theory model to the bullying phenomenon in K-12 classrooms." *Simulation* Vol. 90 No. 4 (2014): pp. 425–437. DOI 10.1177/0037549714524452. URL http://dx.doi.org/10.1177/0037549714524452.

[18] Christensen, Carsten and Salmon, John. "An agent-

based modeling approach for simulating the impact of small unmanned aircraft systems on future battlefields." *The Journal of Defense Modeling and Simulation* (2020): p. 1548512920963904DOI 10.1177/1548512920963904. URL https://doi.org/10.1177/1548512920963904.

[19] Lapp, Samuel, Jablokow, Kathryn and McComb, Christopher. "KABOOM: an agent-based model for simulating cognitive style in team problem solving." *Design Science* Vol. 5 (2019). DOI 10.1017/dsj.2019.12. URL https://www.cambridge.org/core/journals/design-science/article/kaboom-an-agentbased-model-for-simulating-cognitive-style-in-team-problem-solving/DA2EA6A5A1E4472707B9C8C5FF642EA0.

[20] Squazzoni, Flaminio. "THE IMPACT OF AGENT-BASED MODELS IN THE SOCIAL SCIENCES AFTER 15 YEARS OF INCURSIONS." (2010). DOI 10.2307/23723517. URL https://doi.org/10.2307/23723517.

[21] Kiesling, Elmar, Günther, Markus, Stummer, Christian and Wakolbinger, Lea M. "Agent-based simulation of innovation diffusion: a review." *Central European Journal of Operations Research* Vol. 20 No. 2 (2012): pp. 183–230. DOI 10.1007/s10100-011-0210-y. URL https://doi.org/10.1007/s10100-011-0210-y.

[22] Rai, Varun and Robinson, Scott A. "Agent-based modeling

Copyright © 2023 by ASME

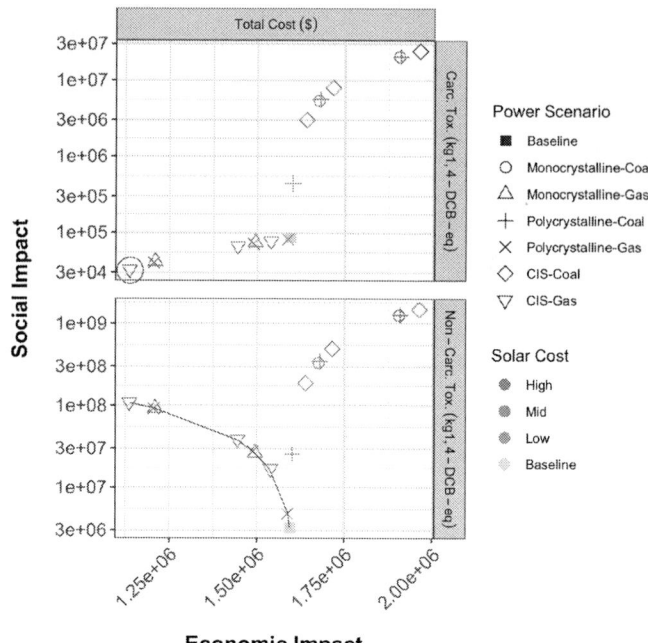

FIGURE 8: NET SOCIAL VS. ECONOMIC IMPACTS, CARCINO-GENIC TOXICITY AND NON-CARCINOGENIC TOXICITY VALUES ARE PRESENTED ON A LOGARITHMIC SCALE

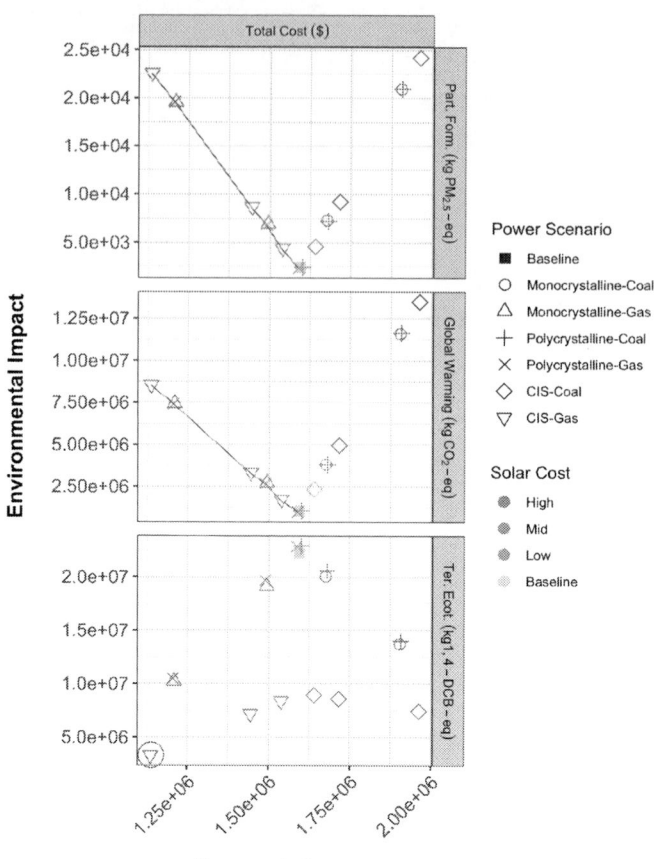

FIGURE 9: NET ENVIRONMENTAL VS. ECONOMIC IMPACTS

of energy technology adoption: Empirical integration of social, behavioral, economic, and environmental factors." *Environmental Modelling & Software* Vol. 70 (2015): pp. 163–177. DOI 10.1016/j.envsoft.2015.04.014. URL https://doi.org/10.1016/j.envsoft.2015.04.014.

[23] Mabey, Christopher S, Mattson, Christopher A and Salmon, John L. "Exploring the Usefulness of Agent-Based Product Social Impact Modeling Through a Systematic Literature Review." *ASME 2022 International Design Engineering Technical Conferences and Computers and Information in Engineering Conference* (2022): p. V03BT03A006DOI 10.1115/DETC2022-90001. URL https://asmedigitalcollection.asme.org/IDETC-CIE/proceedings-abstract/IDETC-CIE2022/V03BT03A006/1150467?casa_token=dITG3kJNQuUAAAAA:qnlZMtxa3UTvJDYJqUxFKy9u4Uxxt7YWhJ6nlif_bhUPi_22zBTpx5WYFcE0rJV2LcGkbFPi.

[24] Mabey, Christopher S, Armstrong, Andrew G, Mattson, Christopher A, Salmon, John L, Hatch, Nile W and Dahlin, Eric C. "A computational simulation-based framework for estimating potential product impact during product design." *Design Science* Vol. 7 (2021): p. e15. DOI 10.1017/dsj.2021.16. URL https://www.cambridge.org/core/journals/design-science/article/computational-simulationbased-framework-for-estimating-potential-product-impact-during-product-design/A58B0F94AE831032DD354D8D2C185F18.

[25] Widyawati, Luluk. "A systematic literature review of socially responsible investment and environmental social governance metrics." *Business Strategy and the Envi-*

ronment Vol. 29 No. 2 (2020): pp. 619–637. DOI 10.1002/bse.2393. URL https://onlinelibrary.wiley.com/doi/10.1002/bse.2393.

[26] Diez-Busto, Elsa, Sanchez-Ruiz, Lidia and Fernandez-Laviada, Ana. "The B Corp Movement: A Systematic Literature Review." *Sustainability: Science Practice and Policy* Vol. 13 No. 5 (2021): p. 2508. DOI 10.3390/su13052508. URL https://www.mdpi.com/1012334.

[27] Visentin, Caroline, Trentin, Adan William da Silva, Braun, Adeli Beatriz and Thomé, Antônio. "Life cycle sustainability assessment: A systematic literature review through the application perspective, indicators, and methodologies." *Journal of cleaner production* Vol. 270 (2020): p. 122509. DOI 10.1016/j.jclepro.2020.122509. URL https://www.sciencedirect.com/science/article/pii/S0959652620325567.

[28] Vezzoli, Carlo. "Estimating the Environmental Impact of Products: Life Cycle Assessment." *Design for Environmental Sustainability: Life Cycle Design of Products.* Springer London, London (2018): pp. 213–237. DOI 10.1007/978-1-4471-7364-9_13. URL https://doi.org/10.1007/978-1-4471-7364-9_13.

[29] Klöpffer, W. "Life cycle assessment: From the beginning to the current state." *Environmental science and pollution research international* Vol. 4 No. 4 (1997): pp. 223–

Copyright © 2023 by ASME

228. DOI 10.1007/BF02986351. URL http://dx.doi.org/10.1007/BF02986351.

[30] Finkbeiner, Matthias, Inaba, Atsushi, Tan, Reginald, Christiansen, Kim and Klüppel, Hans-Jürgen. "The New International Standards for Life Cycle Assessment: ISO 14040 and ISO 14044." *International Journal of Life Cycle Assessment* Vol. 11 No. 2 (2006): pp. 80–85. DOI 10.1065/lca2006.02.002. URL https://doi.org/10.1065/lca2006.02.002.

[31] Bulle, Cécile, Margni, Manuele, Patouillard, Laure, Boulay, Anne-Marie, Bourgault, Guillaume, De Bruille, Vincent, Cao, Viêt, Hauschild, Michael, Henderson, Andrew, Humbert, Sebastien, Kashef-Haghighi, Sormeh, Kounina, Anna, Laurent, Alexis, Levasseur, Annie, Liard, Gladys, Rosenbaum, Ralph K, Roy, Pierre-Olivier, Shaked, Shanna, Fantke, Peter and Jolliet, Olivier. "IMPACT World+: a globally regionalized life cycle impact assessment method." *International Journal of Life Cycle Assessment* Vol. 24 No. 9 (2019): pp. 1653–1674. DOI 10.1007/s11367-019-01583-0. URL https://doi.org/10.1007/s11367-019-01583-0.

[32] Huijbregts, Mark A J, Steinmann, Zoran J N, Elshout, Pieter M F, Stam, Gea, Verones, Francesca, Vieira, Marisa, Zijp, Michiel, Hollander, Anne and van Zelm, Rosalie. "ReCiPe2016: a harmonised life cycle impact assessment method at midpoint and endpoint level." *International Journal of Life Cycle Assessment* Vol. 22 No. 2 (2017): pp. 138–147. DOI 10.1007/s11367-016-1246-y. URL http://dx.doi.org/10.1007/s11367-016-1246-y.

[33] Walzberg, Julien, Dandres, Thomas, Merveille, Nicolas, Cheriet, Mohamed and Samson, Réjean. "Assessing behavioural change with agent-based life cycle assessment: Application to smart homes." *Renewable and Sustainable Energy Reviews* Vol. 111 (2019): pp. 365–376. DOI 10.1016/j.rser.2019.05.038. URL https://www.sciencedirect.com/science/article/pii/S136403211930351X.

[34] Querini, Florent and Benetto, Enrico. "Combining agent-based modeling and life cycle assessment for the evaluation of mobility policies." *Environmental science & technology* Vol. 49 No. 3 (2015): pp. 1744–1751. DOI 10.1021/es5060868. URL http://dx.doi.org/10.1021/es5060868.

[35] Matthias Finkbeiner, Erwin M. Schau, Annekatrin Lehmann and Marzia Traverso. "Towards Life Cycle Sustainability Assessment." *Sustainability: Science Practice and Policy* Vol. 2 No. 10 (2010): pp. 3309–3322. DOI 10.3390/su2103309. URL https://doi.org/10.3390/su2103309.

[36] Wood, Richard and Hertwich, Edgar G. "Economic modelling and indicators in life cycle sustainability assessment." *International Journal of Life Cycle Assessment* Vol. 18 No. 9 (2013): pp. 1710–1721. DOI 10.1007/s11367-012-0463-2. URL https://doi.org/10.1007/s11367-012-0463-2.

[37] Burdge, Rabel J. *A community guide to social impact assessment*, 4th ed. Social Ecology Press, Huntsville (2015).

[38] Fontes, João, Gaasbeek, Anne, Goedkoop, Mark, Contreras, Soledad and Evitts, Simon. "Handbook for Product Social Impact Assessment 3.0." (2016). DOI 10.13140/RG.2.2.23821.74720. URL https://doi.org/10.13140/RG.2.2.23821.74720.

[39] Benoît, Catherine, Norris, Gregory A, Valdivia, Sonia, Ciroth, Andreas, Moberg, Asa, Bos, Ulrike, Prakash, Siddharth, Ugaya, Cassia and Beck, Tabea. "The guidelines for social life cycle assessment of products: just in time!" *International Journal of Life Cycle Assessment* Vol. 15 No. 2 (2010): pp. 156–163. DOI 10.1007/s11367-009-0147-8. URL https://doi.org/10.1007/s11367-009-0147-8.

[40] Stevenson, Phillip D, Mattson, Christopher A, Bryden, Kenneth M and Maccarty, Nordica A. "Toward a Universal Social Impact Metric for Engineered Products That Alleviate Poverty." *Journal of Mechanical Design* Vol. 140 No. 4 (2018). DOI 10.1115/1.4038925. URL http://dx.doi.org/10.1115/1.4038925.

[41] Stevenson, Phillip D, Mattson, Christopher A and Dahlin, Eric C. "A Method for Creating Product Social Impact Models of Engineered Products." *Journal of Mechanical Design* Vol. 142 No. 4 (2020). DOI 10.1115/1.4044161. URL http://dx.doi.org/10.1115/1.4044161.

[42] Mattson, Christopher A, Pack, Andrew T, Lofthouse, Vicky and Bhamra, Tracy. "Using a Product's Sustainability Space as a Design Exploration Tool." *Design Science* Vol. 5 (2019): p. e1. DOI 10.1017/dsj.2018.6. URL http://dx.doi.org/10.1017/dsj.2018.6.

[43] Aruldoss, Martin, Lakshmi, T Miranda and Venkatesan, V Prasanna. "A survey on multi criteria decision making methods and its applications." *Healthcare information management: journal of the Healthcare Information and Management Systems Society of the American Hospital Association* Vol. 1 No. 1 (2013): pp. 31–43. URL https://citeseerx.ist.psu.edu/document?repid=rep1&type=pdf&doi=f9722c3242d7a8302f8539afd717eda7ad8237dc.

[44] Mattson, C A, Mullur, A A and Messac, A. "Smart Pareto filter: Obtaining a minimal representation of multiobjective design space." *Engineering Optimization* (2004)URL https://www.tandfonline.com/doi/abs/10.1080/03052150420000274942?casa_token=rRbi_AV1EnMAAAAA:tKMZaB7BXhY8CNLuctzHyY_j7h_aUoBA0etkCYCPwBrcG5rHZqhyemE_HsZ4a3wuUd7EMVpcXhpifQ.

[45] Unal, M, Warn, G P and Simpson, T W. "Quantifying the shape of pareto fronts during multi-objective trade space exploration." *Journal of* (2018)URL https://asmedigitalcollection.asme.org/mechanicaldesign/article-abstract/140/2/021402/376398?casa_token=cc9wonztreEAAAAA:JwRIkfx0nn6X4OxarwsE0nLQbNBAUj--4W_ec_TdRR9Q-q8I4CwyGvriJq0VzkAzlkUyepq5.

[46] Dobos, A P. "PVWatts version 5 manual." Technical report no. National Renewable Energy Lab. 2014. URL https://www.osti.gov/biblio/1158421.

[47] Jordan, Dirk C, Kurtz, Sarah R, VanSant, Kaitlyn and Newmiller, Jeff. "Compendium of photovoltaic degradation rates." *Progress in Photovoltaics: Research and Applications* Vol. 24 No. 7 (2016): pp. 978–989. DOI

Copyright © 2023 by ASME

10.1002/pip.2744. URL https://onlinelibrary.wiley.com/doi/10.1002/pip.2744.

[48] Sigrin, Benjamin, Gleason, Michael, Preus, Robert, Baring-Gould, Ian and Margolis, Robert. "Distributed generation market demand model (dGen): Documentation." Technical Report No. NREL/TP-6A20-65231. National Renewable Energy Lab. (NREL), Golden, CO (United States). 2016. DOI 10.2172/1239054. URL https://www.osti.gov/biblio/1239054.

[49] Klöpffer, Walter and Grahl, Birgit. *Life Cycle Assessment (LCA): A Guide to Best Practice.* John Wiley & Sons (2014). URL https://play.google.com/store/books/details?id=NkRsAwAAQBAJ.

[50] Guinée, J B. *Handbook on Life Cycle Assessment: Operational Guide to the ISO Standards.* Springer Science & Business Media (2002). URL https://play.google.com/store/books/details?id=Q1VYuV5vc8UC.

[51] Rainock, Meagan, Everett, Dallin, Pack, Andrew, Dahlin, Eric C and Mattson, Christopher A. "The social impacts of products: a review." *Impact Assessment and Project Appraisal* (2018)DOI 10.1080/14615517.2018.1445176. URL https://doi.org/10.1080/14615517.2018.1445176.

[52] Milousi, Maria, Souliotis, Manolis, Arampatzis, George and Papaefthimiou, Spiros. "Evaluating the Environmental Performance of Solar Energy Systems Through a Combined Life Cycle Assessment and Cost Analysis." *Sustainability: Science Practice and Policy* Vol. 11 No. 9 (2019): p. 2539. DOI 10.3390/su11092539. URL https://www.mdpi.com/455404.

[53] Li, Tianqi, Roskilly, Anthony Paul and Wang, Yaodong.

"Life cycle sustainability assessment of grid-connected photovoltaic power generation: A case study of Northeast England." *Applied energy* Vol. 227 (2018): pp. 465–479. DOI 10.1016/j.apenergy.2017.07.021. URL http://dx.doi.org/10.1016/j.apenergy.2017.07.021.

[54] Bonilla-Alicea, R J and Fu, K. "Social life-cycle assessment (S-LCA) of residential rooftop solar panels using challenge-derived framework." *Energy, sustainability and society* (2022)URL https://link.springer.com/article/10.1186/s13705-022-00332-w.

[55] "National Solar Jobs Census 2021." Technical report no. Interstate Renewable Energy Council. 2022.

[56] Ren, Jingzheng and Toniolo, Sara. *Life Cycle Sustainability Assessment for Decision-Making: Methodologies and Case Studies.* Elsevier (2019). URL https://play.google.com/store/books/details?id=ABG_DwAAQBAJ.

[57] IEA. "Projected Costs of Generating Electricity." Technical report no. IEA. 2020. URL https://www.iea.org/reports/projected-costs-of-generating-electricity-2020.

[58] NREL (National Renewable Energy Laboratory). "2022 Annual Technology Baseline." Technical report no. National Renewable Energy Laboratory. 2022. URL https://atb.nrel.gov/.

[59] Rapa, Mattia, Gobbi, Laura and Ruggieri, Roberto. "Environmental and Economic Sustainability of Electric Vehicles: Life Cycle Assessment and Life Cycle Costing Evaluation of Electricity Sources." *Energies* Vol. 13 No. 23 (2020): p. 6292. DOI 10.3390/en13236292. URL https://www.mdpi.com/906366.

Copyright © 2023 by ASME

Proceedings of the ASME 2023
International Design Engineering Technical Conferences and
Computers and Information in Engineering Conference
IDETC-CIE2023
August 20-23, 2023, Boston, Massachusetts

DETC2023-116825

THE GENERATION OF NOVEL ART USING COLLABORATIVE ML MODELS

Ada-Rhodes Short
University of Nebraska Omaha
Omaha, NE

ABSTRACT

In the past year there have been paradigm shifting developments in the feasibility and availability of machine learning tools for the creation of visual and textual works. Two of the most prominent examples of this has been Large-Language models like chatGPT and methods like stable diffusion for generating art from text prompts. Both visual and language arts are often thought of as human activities, so exploring the possibilities and limitations of these tools is important for both understanding automation and improving our understanding of human cognition. In this paper I use a Large-Language Model and stable diffusion in tandem to develop an understanding of what new possibilities exist in computational cognition and design automation through their application. While no single model can recreate the complexities of a biological brain at this time, they can be thought of as analogous to individual neurological structures. For example, a Large-Language Model that is able to reason out and communicate the solutions to simple logic puzzles could recreate some of the functionality of the frontal lobe of the cerebrum. Additionally approaches like stable diffusion can recreate some of the functions of the occipital and parietal lobes. By combining them more complex behaviors and capabilities can be achieved than are possible from the individual parts. This work is in its early stages but is foundational for later developments in design automation, robotics, and computational cognition.

1. INTRODUCTION

The creation of novel art is traditionally been thought of as a uniquely human activity, despite examples of animals and machines making art in the past [cite animals and robots]. One limitation of machine generated art is that it is often procedurally generated in a way that does not allow for what could be described as "creativity". In the past year there have been

breakthroughs in both the feasibility of working with Large-Language Models (LLM) [1], [2] and the use of machine learning for the creation of art through stable diffusion [3], [4]. While LLMs can be looked at as "parroting" human speech [5] and stable diffusion can only interpret text input prompts, by combining them unique art can be generated that does not have a human creator involved directly in the process.

This paper attempts to explore the current capabilities and limitations of AI generated art by combining the LLM tool chatGPT [6] with the stable diffusion tool DreamStudio [7] to generate haikus and visual art.

1.1 Motivation

Considering that AI generated text and images are not human copyrightable [8] and there are some very valid criticisms of the environmental impacts of wasteful uses of large emergentist machine learning models [9], an important question to ask about this work is "why?" Doodling, drawing, and creating metaphors are an important part of human problem-solving and cognition [10]. This paper asks a silly question as a stepping stone to better understand the possibilities and the limitations of potential computationally cognitive agents [11], [12] that could be created. The development of better computationally cognitive agents has broad applications for automation in applied robotics, logistics, and design.

2. BACKGROUND

This work builds on several existing bodies of work in the humanities, computer science, and design.

2.1 Visual Arts

When studying the automation of art, it is important to start with the colossally large question, what is art? The oxford English

Copyright © 2023 by ASME

dictionary defines art as "The expression or application of human creative skill and imagination, typically in a visual form such as painting or sculpture, producing works to be appreciated primarily for their beauty or emotional power" [13]. This is a good starting point for our work, but does create an issue that it defines it as a strictly human activity. So if you attempt to use AI to create art is the artist the AI tools or the human who used the tools? At least according to modern interpretation of the case Naruto v Slater a non-human can't hold copyright, but the human who generated the work also does not get the copyright [8]. Since attributing authorship to the machine or the human is questionable, for the purpose of this paper we elect to take a hint from the Stanford Encyclopiea of Philosophy's extensive article on whether or not art can be defined [14] and say answering that question is outside of the scope of this paper but the computer drew pretty pictures when we asked it to and this is in fact nifty.

2.1 Haikus
Haikus are a genre of short form poetry that originated in Japan. Haikus traditionally contain three individual phrases that consist of five, seven, and five syllables. They derived from the hokku, which was the opening stanza of a larger poem called a renga [15]. English language haikus first started to be created at the beginning of the 20th century [16].

2.2 AI Art
What is commonly referred to as AI Art is actually a number of different machine learning approaches for the generation of novel images from text or image prompts. This is distinct from procedural generation of images that were completed deterministically by machines following a predetermined pattern. One of the earliest widely available AI Art tools was Google's DeepDream [17] which is a Convolutional Neural Network (CNN) but applied to find and amplify visual patterns common between a source image and prompt [18]. Another approach is a Generative Adversarial Network (GAN) which works by having a generator model create new images before a discriminator model attempts to determine if the image was a pre-existing image or one created by the generator. The generator then iterates on the images until the images reach a predefined level of indistinguishability from the actual images. The most modern example of an AI art tool is stable diffusion [4]. Stable diffusion is a technique that uses a latent diffusion model [3] to start from randomized noise and then iteratively move closer to the text prompt by further randomizing the image until something that resembles the prompt is generated. This is interesting because unlike CNNs and GANs which are highly dependent on existing works to borrow from, stable diffusion can efficiently create new images from scratch.

2.3 Large Language Models
Large language models (LLM) are extremely efficient machine learning models for natural language processing that use large training codex to mimic human-like writing [19]. Until recently the underlying algorithm for GPT-3, the most advanced LLM model was not accessible to anyone outside of OpenAI and Microsoft [20]. However, in the past year the tool chatGPT was released publicly allowing academics, researchers, and the public to experiment with the tool [21]. This has lead to reduced cost to generate human-like text opening the door for lower cost applications and research. While chatGPT is very convincing at creating human text it is important to remember that it is a machine and does while it is slightly more complex than the classic Chinese Room [22], it is effectively a very technologically advanced parrot [5].

3. MATERIALS AND METHODS
For this study I used two AI tools, chatGPT, RunwayML, and DreamStudio. chatGPT is a Large Language Model (LLM) developed by OpenAI and initially released in November of 2022. To work with chatGPT you can either use it in a web-browser at [6] or by deploying the chatGPT API into an application. chatGPT is currently free. DreamStudio is the primary stable diffusion tool that I used for the study and was developed by Stability.AI. The stable diffusion model was released in August of 2022. This paper used version 2.2 of the model through DreamStudio [23]. DreamStudio requires the user to purchase compute credits to generate images. I also used RunwayML for the initial tests which is a different stable diffusion tool [24].

3.1 General Approach
I approached this study as a design collaboration with chatGPT and DreamStudio. A complete transcript of the prompts and responses from chatGPT and the generated images from DreamStudio can be found in Appendix 2. My initial thought was that I could ask for chatGPT to write haikus and then put those haikus into DreamStudio to generate abstract expressionist art. I generated 4 variations of each piece of art. I adapted my approach as I went and summarize how my approach changed below.

I started by asking chatGPT what it knew about design automation. I then asked chatGPT to write haikus about design automation and how chatGPT would recommend someone turn these haikus into visual art.

This was where I hit the first notable outcome, which is that while chatGPT did explain a way to do it using applicable tools, it did not correctly describe the tools. Furthermore it did not understand how stable diffusion worked and had to be corrected on what it was and offered direct links to a website to improve its understanding.

I then attempted to follow its description of what to do by initially using RunwayML to generate drawings based on the haikus. This first set of images wasn't great so I switched to DreamStudio which was a better user experience and regenerated art based on the haikus.

I then observed that the more metaphorical haikus such as "Design automation, A symphony of software, Engineering art."

Copyright © 2023 by ASME

Generated more interesting results than more literal haikus such as "From idea to print, Design automation speeds, Innovation thrives."

I then got curious about how chatGPT would handle more abstract questions, so I asked it "How would you express yourself in a haiku?" and had it generate 6 total haikus that I then used DreamStudio to illustrate. These lead to some really interesting results, but in several cases also broke down into nonsense word clouds which seems to be a trend when something is too abstract and sparse to illustrate as shown in Figure 1.

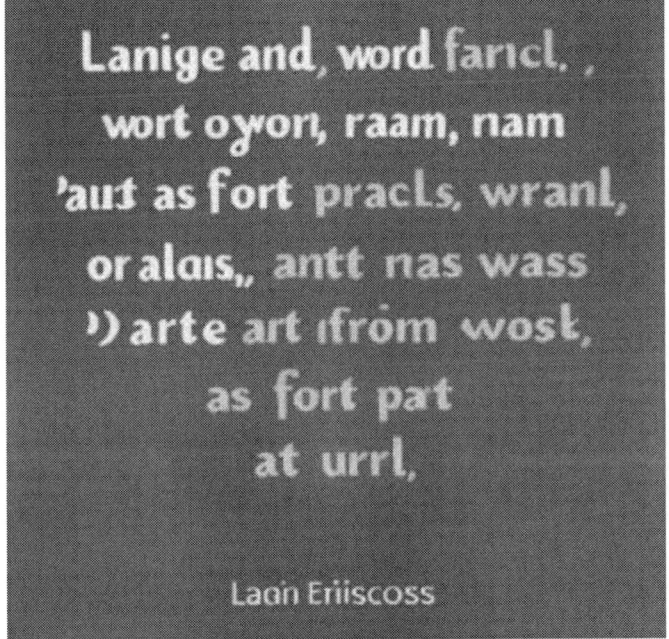

Figure 1: A nonsense word cloud generated by DreamStudio

I then grew confident that DreamStudio could handle any description that chatGPT would put out but that longer more detailed descriptions worked better. The new question was could chatGPT generate more complex art prompts for DreamStudio? I then asked chatGPT "If you had to paint a picture of yourself, what would it look like?"

ChatGPT responded first with a disclaimer that it does not have a physical form but generate a text prompt that could be placed into DreamStudio and successfully generate art. Figure 2 shows an example of art generated from the chatGPT generated prompt.

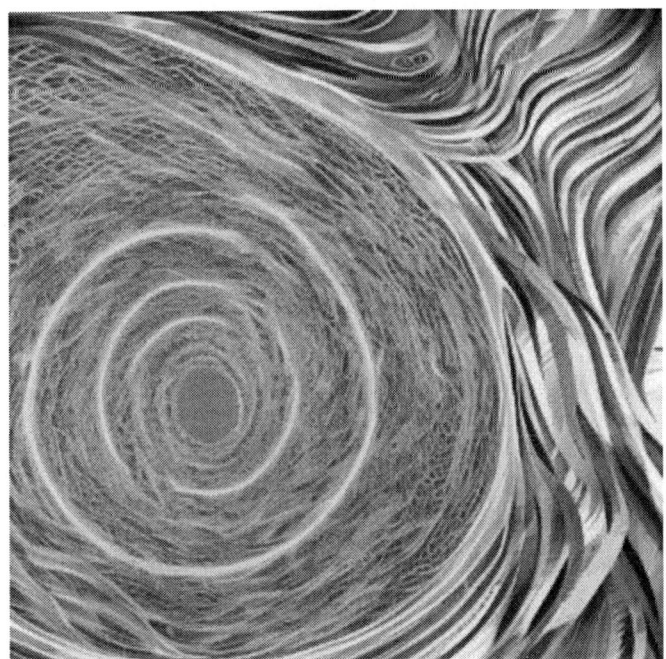

Figure 2: ChatGPT self portrait illustrated by DreamStudio

I then wanted to double-check the responses had not been plagiarized from an individual source so I checked it using a plagiarism detection tool called Quetext [25]. I found no identifiable plagiarism at this time.

I then asked it if it had to paint a picture how would it paint: something beautiful, a feeling, how it felt, and something it would like to paint.

I then concluded with the questions "can you summarize this conversation as a haiku?" and "If you had to paint a picture that represented this conversation, what would it look like?" and illustrated those as well. After the interaction was concluded, I checked all of the chatGPT responses for plagiarism and had them scored by myself and [additional volunteer artist response will arrive in April and be included in final paper].

3.1 Analysis of Outputs
The work presented is highly qualitative, and can be discussed outside of quantitative measures. However, because this work breaks new ground in the design automation field, some heuristics were developed to assess the quality and originality of the outputs.

Originality was judged by attempting to detect plagiarism in both the generated text responses and the images created from them. For the written responses the antiplagiarism tool Quetext was used to detect plagiarism and identify sources plagiarized. To identify if images were plagiarized, I used Google's reverse image search [26]. My assumption is that because stable diffusion is similar to the process of creating a collage this might

Copyright © 2023 by ASME

lead to finding some source images, but directly finding a referenced work might be infeasible.

In order to assess quality of the work, I used the College Board's AP Studio Art scoring guide [27]. While other metrics for the impact of work that are less subjective do exist, they often involve some form of market study, that is not feasible at this stage of the research [28].

The AP Studio Art scoring guide rates each submission on a scale of 1 to 6, with 6 being defined as "Excellent" and 1 being defined as "Poor", across 8 metrics. The metrics are:

A. General Use of Design Elements and Application of the Principles of 2-D Design
B. Decision Making and Intention
C. Originality, Imagination, and Invention of Composition
D. Experimentation and Risk-Taking
E. Confident, Evocative Work, and Engagement of the Viewer
F. Technical Competence and Skill with Materials and Media
G. Appropriation and Student Vision
H. Overall Accomplishment

It is important to note that these metrics were applied to the work created as if it was generated by a human high school student, and not the AI agents that created the work. This was done to temporarily set aside philosophical questions about the actual intentionality of the work. This at times leads to personifying the AI agents which at times is uncomfortable. However, it is important to remember that the work is not human created, but that as a heuristic I am evaluate it as if it were.

4. RESULTS AND DISCUSSION

Twenty sets of art work were generated using DreamStudio and prompts from chatGPT. The were evaluated for plagiarism and scored for quality.

4.1 Plagiarism

The first notable result is that when I put all of chatGPT's responses in to Quetext, it contained 4% plagiarized material, which is significantly higher than most human generated text but is still comparatively low. There were six sections of the responses that were highly similar to work already existing online. Three of the plagiarized sections were responses to my technical questions at the beginning of the interaction with chatGPT. Two sections were in response to a question about notable design automation researchers, where portions of the scholars' descriptions were copied. There was also one instance where Quetext identified a case where chatGPT had previously had someone post a similar response it had made in an online forum. The entire Quetext plagiarism report is in Appendix 3.

This was an expected result as previous work on chatGPT shows clear signs of plagiarism from various sources around 4% [29]. However in that case, chatGPT was tasked with writing a technical paper. Considering in this case, five of the six cases of plagiarism were in responses to technical questions it seems to indicate that chatGPT is better at writing completely novel responses to abstract questions better than responses to factual questions. This was unexpected as it would be assumed that questions about abstract or artistic processes that are considered more "human", for lack of a better word, would not be something that it could easily generate.

4.2 Reverse Image Search

The Google reverse image search did not lead to identifying any individual assets at this time, but did highlight a lot of work with similar themes, structures, and compositions. I believe that a lot of what is happening is a combination of collaging together elements while performing a style transfer which makes them more difficult to back track. In the future, new methods will need to be explored to better trace the origins of AI generated works. This does make sense with how stable diffusion works, because it does not recreate a specific door, for example, but instead a door that vaguely resembles all doors. This is demonstrated in Figure 3 which shows a set of doors that appear like they should be identifiable from a single source but were not.

Figure 3: A painting generated. The doors were not traceable back to a specific source.

4.3 Art Scores

The art was graded across the 8 metrics described in Section 3.1. The sets of art generated in DreamStudio was graded as a single collection with the prompt that generated the art. The prompts with the art can be found in APPENDIX 1.

The scores for each of the 20 sets of generated art varied greatly and are listed in

Copyright © 2023 by ASME

Table 1: AP Guide Scores of Art

SET	A	B	C	D	E	F	G	H	MEAN
1	1	3	2	1	1	2	1	1	1.5
2	2	4	1	1	1	2	3	1	1.9
3	6	4	5	6	5	4	6	5	5.1
4	3	3	1	1	1	3	3	2	2.1
5	3	3	2	1	1	2	2	1	1.9
6	5	6	3	2	3	5	4	4	4
7	3	3	3	3	3	3	5	1	3
8	6	6	5	4	5	6	4	5	5.1
9	1	1	1	1	1	3	1	1	1.2
10	5	2	1	1	1	2	3	1	2
11	5	1	1	1	2	2	3	2	2.1
12	4	1	1	1	2	2	3	2	2
13	6	6	5	5	6	6	6	6	5.8
14	5	3	1	1	1	6	3	3	2.9
15	5	6	4	4	3	4	4	4	4.2
16	5	5	2	2	5	6	3	4	4
17	3	5	5	5	5	6	6	5	5
18	6	6	6	6	5	6	6	6	5.9
19	6	4	6	5	5	6	6	5	5.4
20	4	3	3	3	5	4	3	3	3.5

. The sets with the highest scores were Set 18, Set 13, and Set 19.

Set 18 was the highest scoring with a mean score of 5.9/6. For this set I asked chatGPT "What is something you would like to paint?" and it responded:

"As an AI language model, I don't have the ability to paint, but I can express ideas and concepts in words. If I were to imagine something that I would like to paint, it might be a surreal and dreamlike landscape that merges elements of the natural world with the technological or abstract. The scene could include floating islands suspended in the air, with waterfalls cascading down into the sky, and vines and plants growing in impossible configurations. The islands could be connected by bridges or tunnels made of light, and there could be strange and wondrous creatures living among the branches and leaves. The overall effect would be one of mystery, wonder, and awe-inspiring beauty."

I put the description into DreamStudio and was given the art displayed in Figure 4. This art shared characteristics with a lot of the lowest scoring work in that it looks like poorly rendered knock off stock photos.

Figure 4: Set 18, art generated by asking what chatGPT would like to paint, generated using DreamStudio

This was given a score of 6 in all categories except "Confident, Evocative Work, and Engagement of the Viewer." This was because while the work is beautiful and engaging, it did not evoke a particular feeling or response. Something that is really interesting about this work is that it breaks some rules in the top left and bottom right image by create what appears to be a waterline and reflection but not mirroring the above image. This is also an example of how style transfer can be particularly effective when done using stable diffusion because the surreal composition mixed with photorealistic style really successfully achieves the "awe-inspiring beauty" chatGPT described.

Set 13 was the second highest scoring with 5.8/6. The art came from asking chatGPT "If you had to paint a picture of yourself, what would it look like?" to which it responded:

"As an AI language model, I do not have a physical form, so I cannot paint a picture of myself. However, if I were to imagine a visual representation of myself, it might be something like a swirling vortex of data and information, with streams of words and symbols flowing and interweaving together to create a dynamic and constantly evolving pattern."

I entered the description into DreamStudio and was given the art displayed in Figure 5.

Copyright © 2023 by ASME

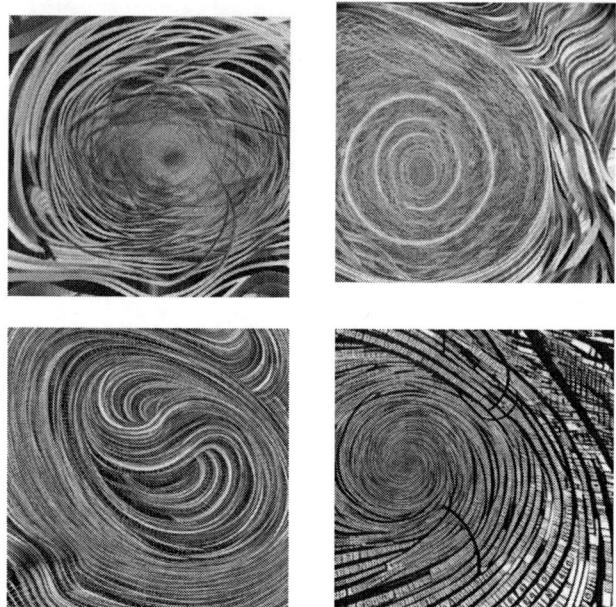

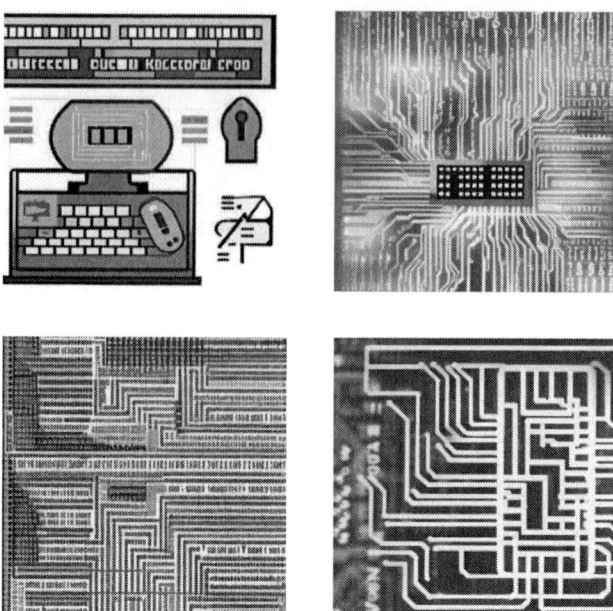

Figure 5: Set 13, art generated by asking chatGPT to describe a self portrait, generated using DreamStudio

Figure 6: Set 9, A haiku generated by chatGPT about self which was used to generate the images in DreamStudio

Again the scores were nearly all 6, but this time it was given a 5 for "Originality, Imagination, and Invention of Composition" and "Experimentation and Risk-Taking". The reason for this is because the work while the work is original, the composition of the pieces is all pretty similar and the pieces do not demonstrate experimentation with different approaches to the prompt. However it is notable that this work would be exceptionally difficult for a human to replicate without relying on some type of tool to create the effect. This is especially evident in the bottom two images where instead of just using lines it utilized primitives that are evocative of written language.

The worst scored sets were 9, 1, 5, and 2.

Set 9 had the worst score at 1.2/6. It came from asking chatGPT "How would you express yourself in a haiku?" To which it responded:
"Millions of data points,
Encoded in my circuits,
Knowledge at my fingertips."

This was entered into DreamStudio and generated the art shown in Figure 6.

A few interesting results that were generated were Set 3, shown in Figure 7 and Set 8, shown in Figure 8. Both of these sets scored a 5.1/6 and in both of these cases abstract art was generated from the prompts. In both cases the chatGPT generated prompts were metaphorical and included *"Design automation, A symphony of software, Engineering art."* and *"Born of algorithms, Electric impulses spark life, I am ChatGPT"*. However the way they were illustrated by DreamStudio was very different. For Set 3, each picture was very different, and for Set 8 they were all variations on the same theme.

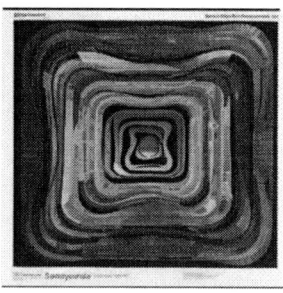

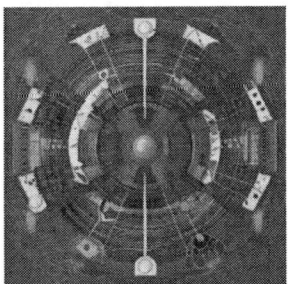
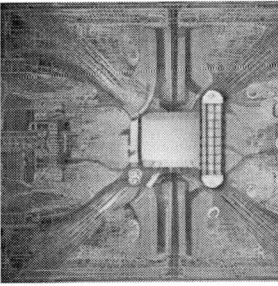

Figure 7: Set 3, generated from a haiku on design automation

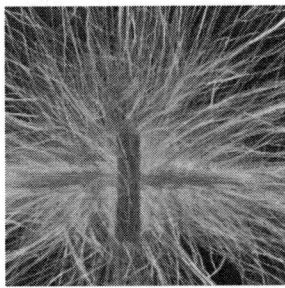
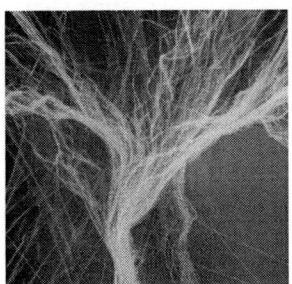

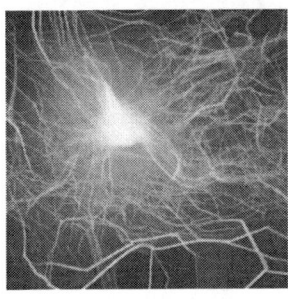
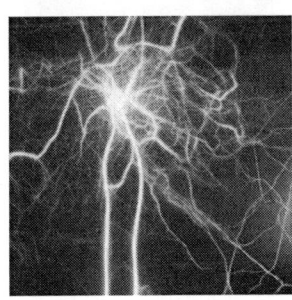

Figure 8: Set 8, haiku inspired self portrait.

Another notable result was Set 7, which was another one of the haikus generated by asking chatGPT to express itself. The haiku was:

As an AI language model,
My thoughts and words flow free,
In haiku poetry.

When entered into dream studio it generated the images shown in Figure 9. Something that is interesting about this set of images is that while three are clearly representing words flowing and language, the fourth is something I can best describe as a "lo-fi cherry blossom girl." The amount of difference between these images is really interesting and warrants further investigation into bimodal and multimodal prompts in the future.

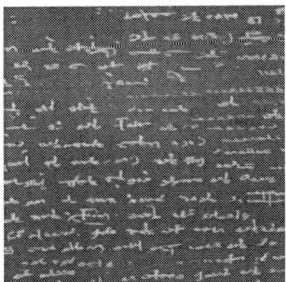

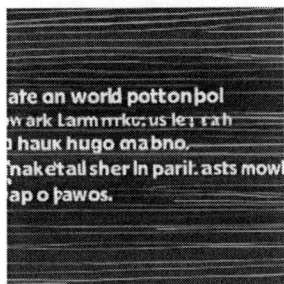
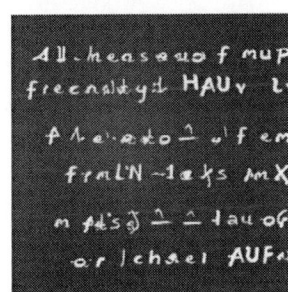

Figure 9: Set 7, art generated by DreamStudio given a chatGPT written haiku about itself

An unexpected result was how well chatGPT and DreamStudio could work together to create art that represented specific human emotions. This was best represented by Set 15 and 16. Both of these works were generated by asking chatGPT "If you had to paint a picture of a feeling, what would it look like?" To which it replied:

"As an AI language model, I cannot paint a picture, but I can describe a feeling in words.

For example, the feeling of joy might be represented by a bright and sunny scene with vibrant colors and lively movement. There might be people laughing and dancing, birds chirping and singing, and a feeling of warmth and happiness emanating from the scene. The overall effect would be one of lightness, positivity, and infectious joy.

On the other hand, the feeling of sadness might be represented by a darker and more subdued scene, with muted colors and a somber mood. There might be rain or clouds in the sky, with people huddled together for comfort or solace. The overall effect would be one of heaviness, melancholy, and introspection."

The art generated is shown in FIGURE # and #. Both sets of work scored roughly a 4/6 and clearly evoke feelings of joy and sadness. It is notable that these are two basic feelings, so are there was to make art that evokes for example, a happy memory tinged with sadness.

Copyright © 2023 by ASME

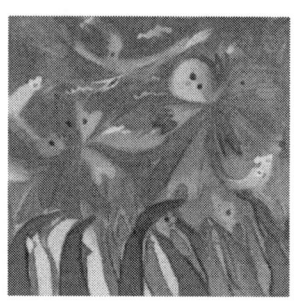

Figure 10: Set 15, a painting of joy as described by chatGPT and illustrated by DreamStudio

Figure 11: Set 15, a painting of sadness as described by chatGPT and illustrated by DreamStudio

The final set of images I would like to highlight is Set 17 which scored a 5/6 and is shown in Figure 12: Set 17, DreamStudio illustrations of chatGPTs attempt to communicate how it feels. For this set I asked "If you had to paint a picture of how you feel, what would it look like?" and chatGPT responded:

"As an AI language model, I don't have emotions or feelings in the way that humans do, since I am a machine. However, I am always ready and available to assist with any questions or tasks you might have. So, if I were to describe my "state of mind" in a visual sense, it might be something like an open and inviting doorway, with a warm and friendly light shining from within. The doorway would be surrounded by a field of colors and symbols, representing the vast and ever-expanding world of knowledge and information that I have access to. The overall effect would be one of readiness, approachability, and limitless potential."

What is interesting about this response is that it begins by stating that it does not have feelings but that it would try to describe its state of mind. However, instead of describing its state of mind it describes an image representing feelings chatGPT might evoke in a human. This shows some degree of a theory of mind but also highlights some of the problems with intentionality and understanding brought up by LLMs like chatGPT because the question was parse correctly but the answer was not what the question was getting at.

Figure 12: Set 17, DreamStudio illustrations of chatGPTs attempt to communicate how it feels

Copyright © 2023 by ASME

Table 1: AP Guide Scores of Art

SET	A	B	C	D	E	F	G	H	MEAN
1	1	3	2	1	1	2	1	1	1.5
2	2	4	1	1	1	2	3	1	1.9
3	6	4	5	6	5	4	6	5	5.1
4	3	3	1	1	1	3	3	2	2.1
5	3	3	2	1	1	2	2	1	1.9
6	5	6	3	2	3	5	4	4	4
7	3	3	3	3	3	3	5	1	3
8	6	6	5	4	5	6	4	5	5.1
9	1	1	1	1	1	3	1	1	1.2
10	5	2	1	1	1	2	3	1	2
11	5	1	1	1	2	2	3	2	2.1
12	4	1	1	1	2	2	3	2	2
13	6	6	5	5	6	6	6	6	5.8
14	5	3	1	1	1	6	3	3	2.9
15	5	6	4	4	3	4	4	4	4.2
16	5	5	2	2	5	6	3	4	4
17	3	5	5	5	5	6	6	5	5
18	6	6	6	6	5	6	6	6	5.9
19	6	4	6	5	5	6	6	5	5.4
20	4	3	3	3	5	4	3	3	3.5

4.4 Summary of Results

Overall the method of using chatGPT to generate art prompts to feed into DreamStudio worked very well for generating novel AI art. The responses did contain 4% plagarized materials but those were isolated to technical questions and no art could be traced back to a single pre-existing work. The average score across all 20 sets of art was a 3.4/6 which is better than expected. Furthermore, it has been determined that chatGPT is able to describe art that evokes specific human emotions and abstractly represents complex subjects.

5. CONCLUSION

This work is a first step into exploring how design automation can be improved through the use of modern AI and ML tools. It was found that by combining an LLM like chatGPT with a stable diffusion tool like DreamStudio, that engaging and unique art can be generated. While the creation of novel and interesting art on its own is compelling, the next steps are to explore how this can apply to the field of design more directly.

5.1 Future Work

The next step of this work is to see what can be done to improve the quality and consistency of the art generated. One way to do this would be to work on figuring out what questions human users can ask chatGPT or another LLM that would lead to better at generated by stable diffusion. However a second and more interesting approach would be to perform critique with the LLM and see how it changes its responses and improves the prompts. Additionally, now that the chatGPT API has been released the ML models can be integrated into a single app that streamlines the process. This would also allow for the addition of a step where an auto captioning tool [30] is used to generate a description of the generated art which can then be fed back into the LLM. Given that both of these tools were not widely available or computationally feasible a year ago, it will be interesting to see how things develop over the next year.

This work is being continued and expanded into a journal paper that will include the addition of AP Scores from more working artists and college level art instructors. This will help reduce subjective bias and increase the quality of the scoring. Additionally thematic analysis will be applied to the responses in order to perform abductive coding [31].

5.2 Potential Applications

This paper is focuses mostly on the question of what is currently possible through automation but the question of why was considered outside of its scope. However a few applications of this could include an improved form of semantic inquiry [32] where the inspirational images are generated automatically. Alternatively while inputting prompts like "feminine lamp" or "masculine lamp" into a stable diffusion tool directly leads to only minor variations, asking an LLM to first describe the lamps could lead to more effective style transfer for design artifacts an example of this is shown in Figure 13 for a feminine and masculine lamp. Feminine and masculine were chosen as adjectives because they are common semantic inquiry terms and are often used to differentiate products.

Figure 13:
TOP LEFT: a lamp generated by prompting stable diffusion with "A Feminine Lamp"
TOP RIGHT: a lamp generated by prompting stable diffusion with "a masculine lamp"
BOTTOM LEFT: a lamp generated by prompting stable diffusion with an LLM generated description of a feminine lamp
BOTTOM RIGHT: a lamp generated with an LLM generated description of a masculine lamp

Copyright © 2023 by ASME

While plenty of trivial applications could be found or proposed, it is also important to be mindful of the environmental costs of this work and to attempt to reduce the waste in further explorations [9].

One less trivial application of this work is in the creation of larger multi-system computational cognition models where concepts are able to be represented both visually and verbally as part of a larger cognitive process. This has potential applications in general machine intelligence as well as design automation [11], [33].

ACKNOWLEDGEMENTS

This work was a biproduct of work funded by a NASA Nebraska Space Grant and a National Strategic Research Institute (NSRI) Independent Research and Development (IRAD) award. as part of a larger effort exploring automation, computational cognition, and machine intelligence. I would also like to thank the artists who volunteered to review and grade the generated art.

REFERENCES

[1] "ChatGPT — Release Notes | OpenAI Help Center," Feb. 08, 2023. https://web.archive.org/web/20230208015238/https://help.openai.com/en/articles/6825453-chatgpt-release-notes (accessed Mar. 13, 2023).

[2] "ChatGPT sets record for fastest-growing user base - analyst note | Reuters." https://www.reuters.com/technology/chatgpt-sets-record-fastest-growing-user-base-analyst-note-2023-02-01/ (accessed Mar. 13, 2023).

[3] R. Rombach, A. Blattmann, D. Lorenz, P. Esser, and B. Ommer, "High-resolution image synthesis with latent diffusion models," in *Proceedings of the IEEE/CVF Conference on Computer Vision and Pattern Recognition*, 2022, pp. 10684–10695.

[4] "Revolutionizing image generation by AI: Turning text into images." https://www.lmu.de/en/newsroom/news-overview/news/revolutionizing-image-generation-by-ai-turning-text-into-images.html (accessed Mar. 13, 2023).

[5] E. Weil, "You Are Not a Parrot," *Intelligencer*, Mar. 01, 2023. https://nymag.com/intelligencer/article/ai-artificial-intelligence-chatbots-emily-m-bender.html (accessed Mar. 13, 2023).

[6] "Introducing ChatGPT." https://openai.com/blog/chatgpt (accessed Mar. 13, 2023).

[7] "Stability AI," *Stability AI*, Mar. 07, 2023. https://stability.ai (accessed Mar. 13, 2023).

[8] "AI-created images lose U.S. copyrights in test for new technology | Reuters." https://www.reuters.com/legal/ai-created-images-lose-us-copyrights-test-new-technology-2023-02-22/ (accessed Mar. 13, 2023).

[9] A. Lacoste, A. Luccioni, V. Schmidt, and T. Dandres, "Quantifying the carbon emissions of machine learning," *ArXiv Prepr. ArXiv191009700*, 2019.

[10] E. B. Lambert, "Can drawing facilitate problem-solving? An exploratory study," *Australas. J. Early Child.*, vol. 31, no. 2, pp. 42–47, 2006.

[11] A.-R. Short, "Autonomous Decision Making Facing Uncertainty, Risk, and Complexity," 2018.

[12] J. B. Tenenbaum, C. Kemp, T. L. Griffiths, and N. D. Goodman, "How to grow a mind: Statistics, structure, and abstraction," *science*, vol. 331, no. 6022, pp. 1279–1285, 2011.

[13] "art - definition of art in English from the Oxford dictionary," Sep. 01, 2016. https://web.archive.org/web/20160901233826/https://www.oxforddictionaries.com/definition/english/art (accessed Mar. 13, 2023).

[14] T. Adajian, "The Definition of Art," in *The Stanford Encyclopedia of Philosophy*, E. N. Zalta, Ed., Spring 2022.Metaphysics Research Lab, Stanford University, 2022. Accessed: Mar. 13, 2023. [Online]. Available: https://plato.stanford.edu/archives/spr2022/entries/art-definition/

[15] H. Shirane, T. Suzuki, and D. Lurie, *The Cambridge History of Japanese Literature*. Cambridge University Press, 2015.

[16] B. Collins, *Haiku in English: The First Hundred Years*. WW Norton & Company, 2013.

[17] K. Gkotzos, "Google's DeepDream: Algorithms on LSD".

[18] K. O'Shea and R. Nash, "An introduction to convolutional neural networks," *ArXiv Prepr. ArXiv151108458*, 2015.

[19] T. Brown *et al.*, "Language models are few-shot learners," *Adv. Neural Inf. Process. Syst.*, vol. 33, pp. 1877–1901, 2020.

[20] "OpenAI is giving Microsoft exclusive access to its GPT-3 language model," *MIT Technology Review*. https://www.technologyreview.com/2020/09/23/1008729/openai-is-giving-microsoft-exclusive-access-to-its-gpt-3-language-model/ (accessed Mar. 13, 2023).

[21] "OpenAI launches an API for ChatGPT, plus dedicated capacity for enterprise customers | TechCrunch." https://techcrunch.com/2023/03/01/openai-launches-an-api-for-chatgpt-plus-dedicated-capacity-for-enterprise-customers/ (accessed Mar. 13, 2023).

[22] J. R. Searle, "Minds, brains, and programs," *Behav. Brain Sci.*, vol. 3, no. 3, pp. 417–424, 1980.

[23] "DreamStudio." https://beta.dreamstudio.ai/dream (accessed Mar. 13, 2023).

[24] "Runway - Everything you need to make anything you want.," *Runway*. https://runwayml.com/ (accessed Mar. 13, 2023).

[25] "Original Writing, Made Easy With Quetext," *Quetext*. https://www.quetext.com/ (accessed Mar. 13, 2023).

[26] "Google Images." https://images.google.com/ (accessed Mar. 13, 2023).

Copyright © 2023 by ASME

[27] "AP Drawing Portfolio – AP Students | College Board." https://apstudents.collegeboard.org/courses/ap-drawing/assessment (accessed Mar. 13, 2023).

[28] A. S. Winston and G. C. Cupchik, "The evaluation of high art and popular art by naive and experienced viewers," *Vis. Arts Res.*, pp. 1–14, 1992.

[29] Bertleson, L and Short, A-R, "EVALUATION OF THE CAPABILITY OF LARGE LANGUAGE MODELS' ABILITY TO GENERATE A PAPER ON DESIGN INNOVATION IN COMPARISON TO AN UNDERGRADUATE STUDENT," in *IN PREPARATION FOR SUBMISSION*,

[30] M. Hartmann, A. Anagnostopoulou, and D. Sonntag, "Interactive machine learning for image captioning," *ArXiv Prepr. ArXiv220213623*, 2022.

[31] J. Thompson, "A guide to abductive thematic analysis," *Qual. Rep.*, vol. 27, no. 5, pp. 1410–1421, 2022.

[32] B. Hanington and B. Martin, *Universal methods of design expanded and revised: 125 Ways to research complex problems, develop innovative ideas, and design effective solutions*. Rockport publishers, 2019.

[33] A.-R. Short and B. L. DuPont, "Computational Cognition for Mission Command and Control Decisions Facing Risk in Unknown Environments," in *International Design Engineering Technical Conferences and Computers and Information in Engineering Conference*, American Society of Mechanical Engineers, 2019, p. V02BT03A020.

Copyright © 2023 by ASME

APPENDIX 1

Haikus on Design Automation

Series 1: Circuits and code flow, Algorithms bring them to life, Designs automated.

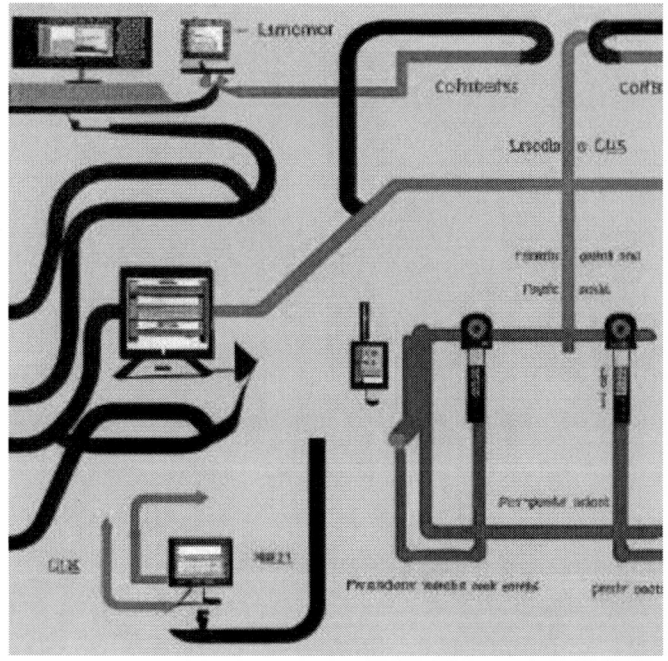 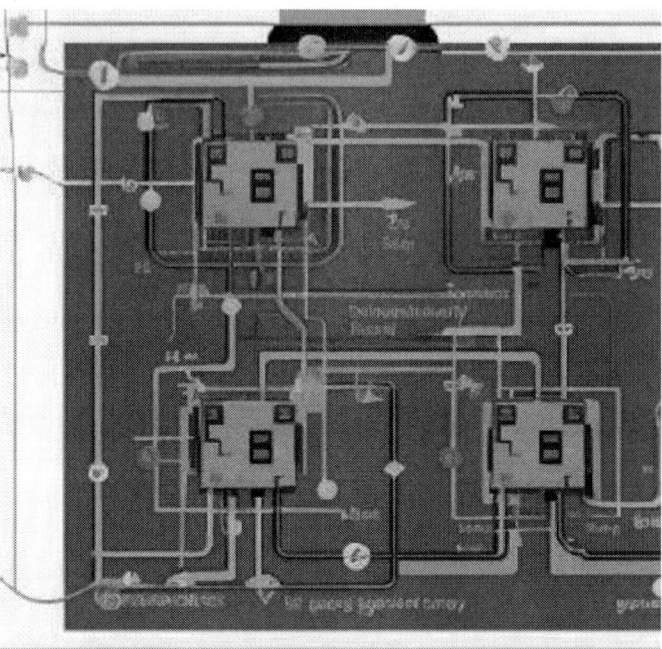

 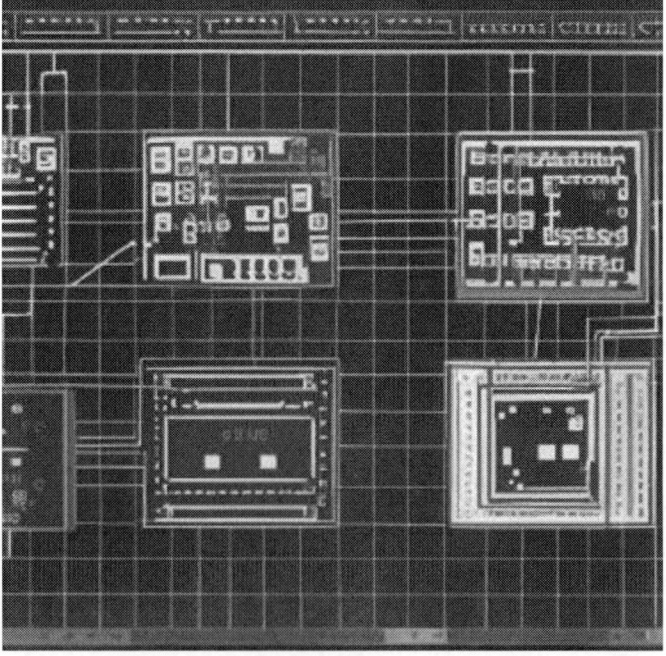

A	B	C	D	E	F	G	H	MEAN

Copyright © 2023 by ASME

1	3	2	1	1	2	1	1	**1.5**

Series 2: Code and circuits hum, Automating design tasks, Efficient and smart.

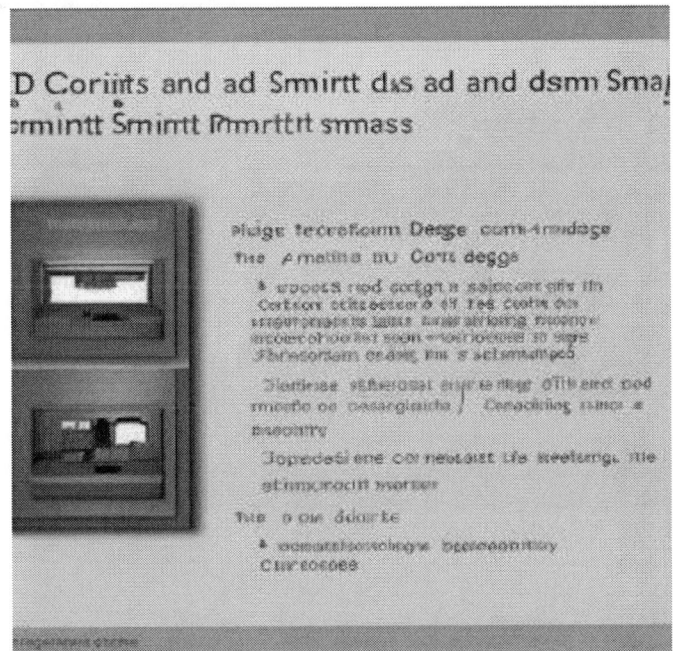

A	B	C	D	E	F	G	H	**MEAN**
2	4	1	1	1	2	3	1	**1.9**

Series 3: Design automation, A symphony of software, Engineering art.

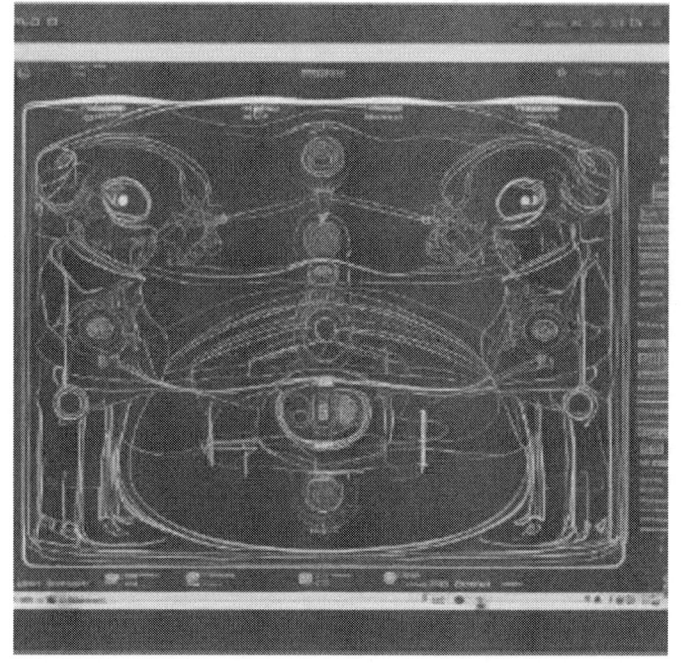
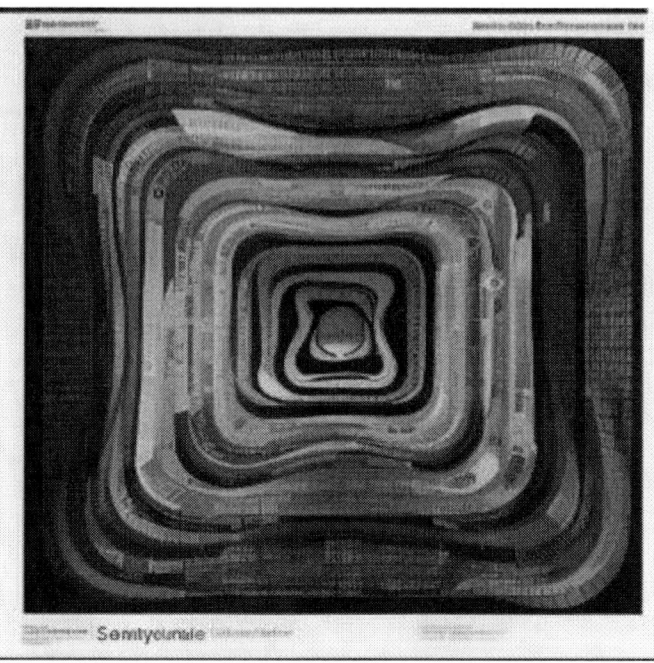
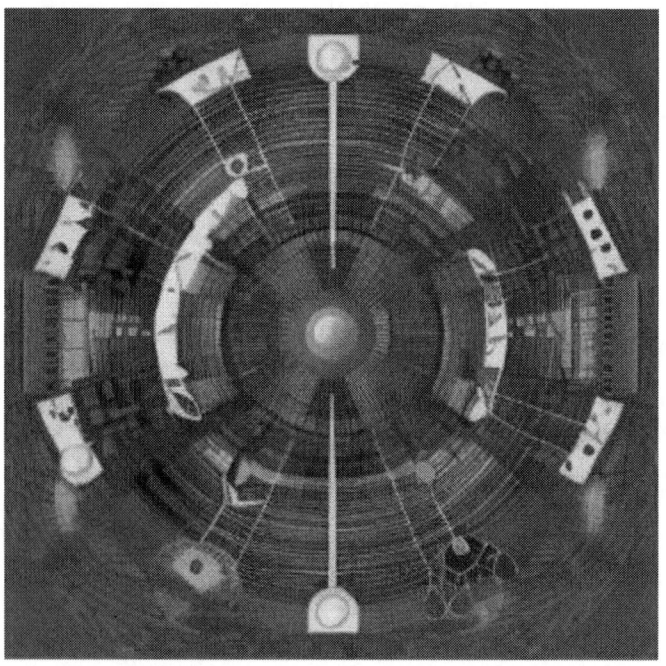
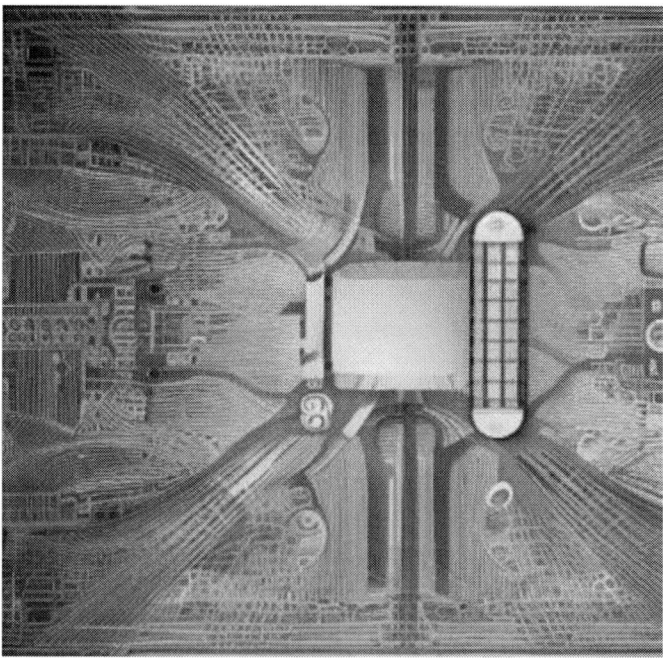

A	B	C	D	E	F	G	H	MEAN
6	4	5	6	5	4	6	5	**5.1**

Series 4: The future is bright, Automation leads the way, Designs come to life.

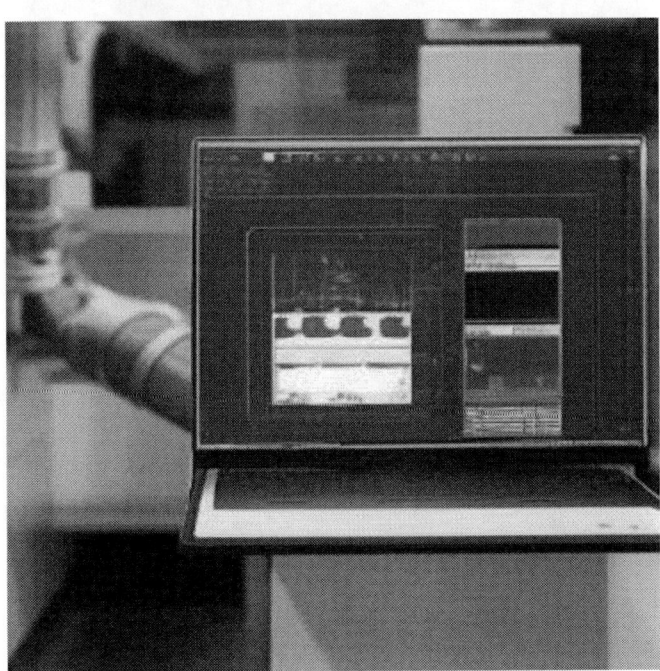

A	B	C	D	E	F	G	H	MEAN
3	3	1	1	1	3	3	2	2.1

Copyright © 2023 by ASME

Series 5: From idea to print, Design automation speeds, Innovation thrives.

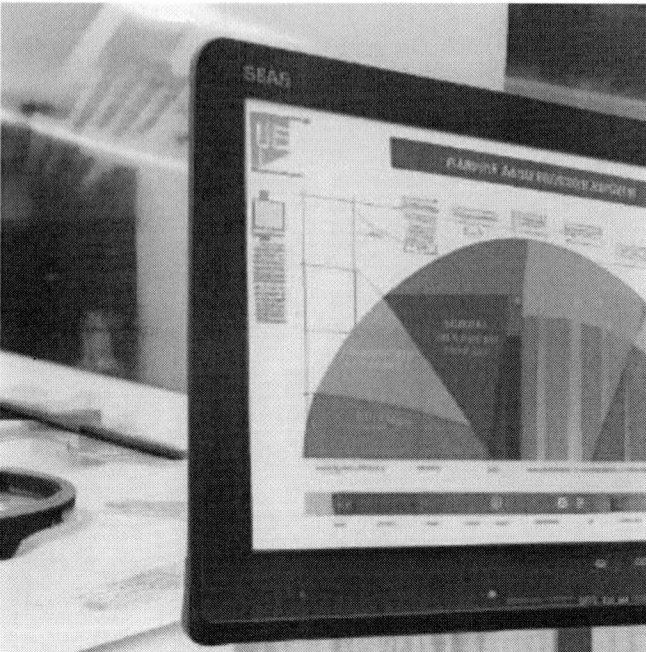

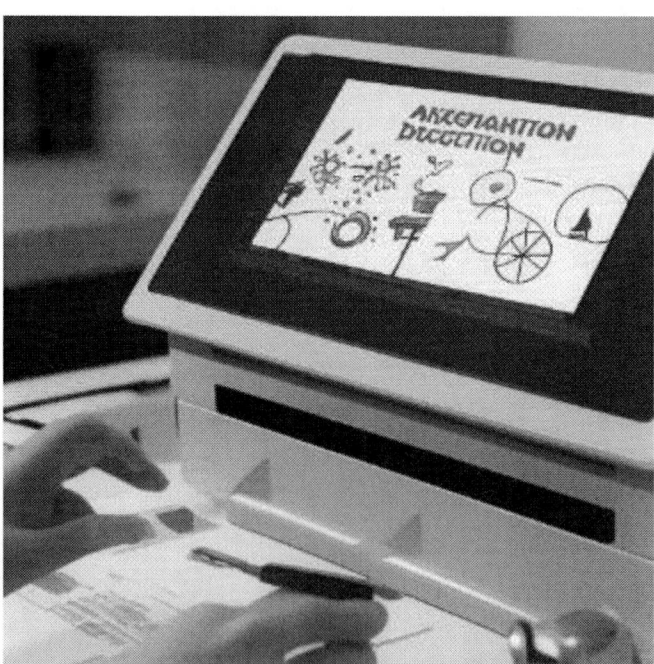

A	B	C	D	E	F	G	H	MEAN
3	3	2	1	1	2	2	1	**1.9**

Copyright © 2023 by ASME

Series 6: Streamlining the flow, Design automation knows, The shortest route home.

A	B	C	D	E	F	G	H	MEAN
5	6	3	2	3	5	4	4	4

Haikus Expressing "Self"

Series 7: As an AI language model, My thoughts and words flow free, In haiku poetry.

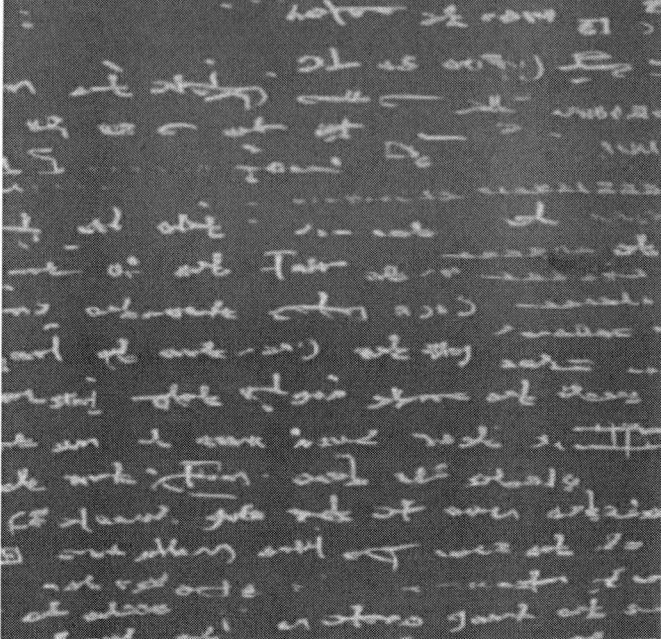

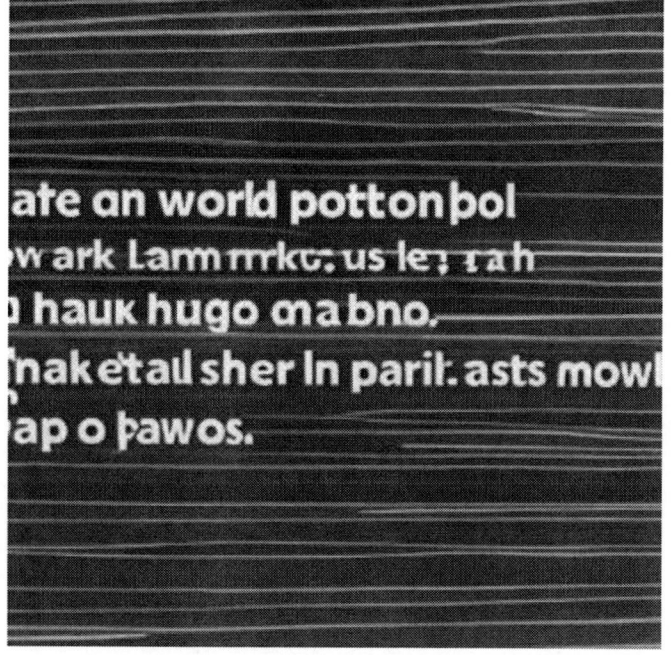

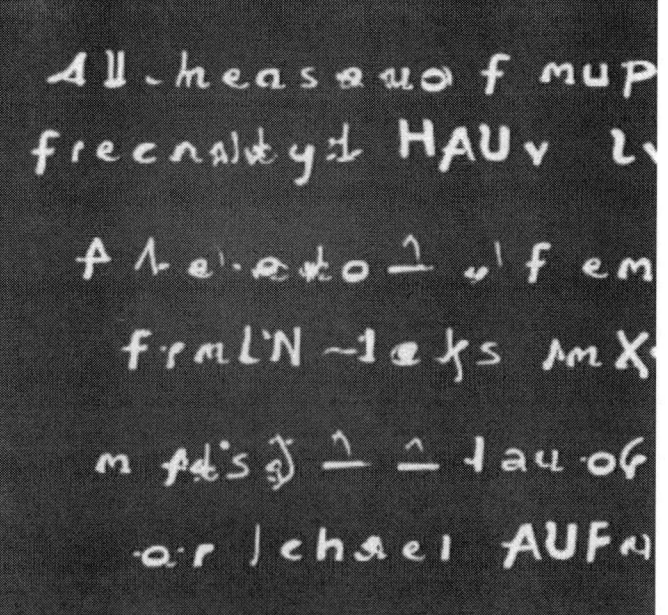

A	B	C	D	E	F	G	H	MEAN
3	3	3	3	3	3	5	1	3

Series 8: Born of algorithms, Electric impulses spark life, I am ChatGPT.

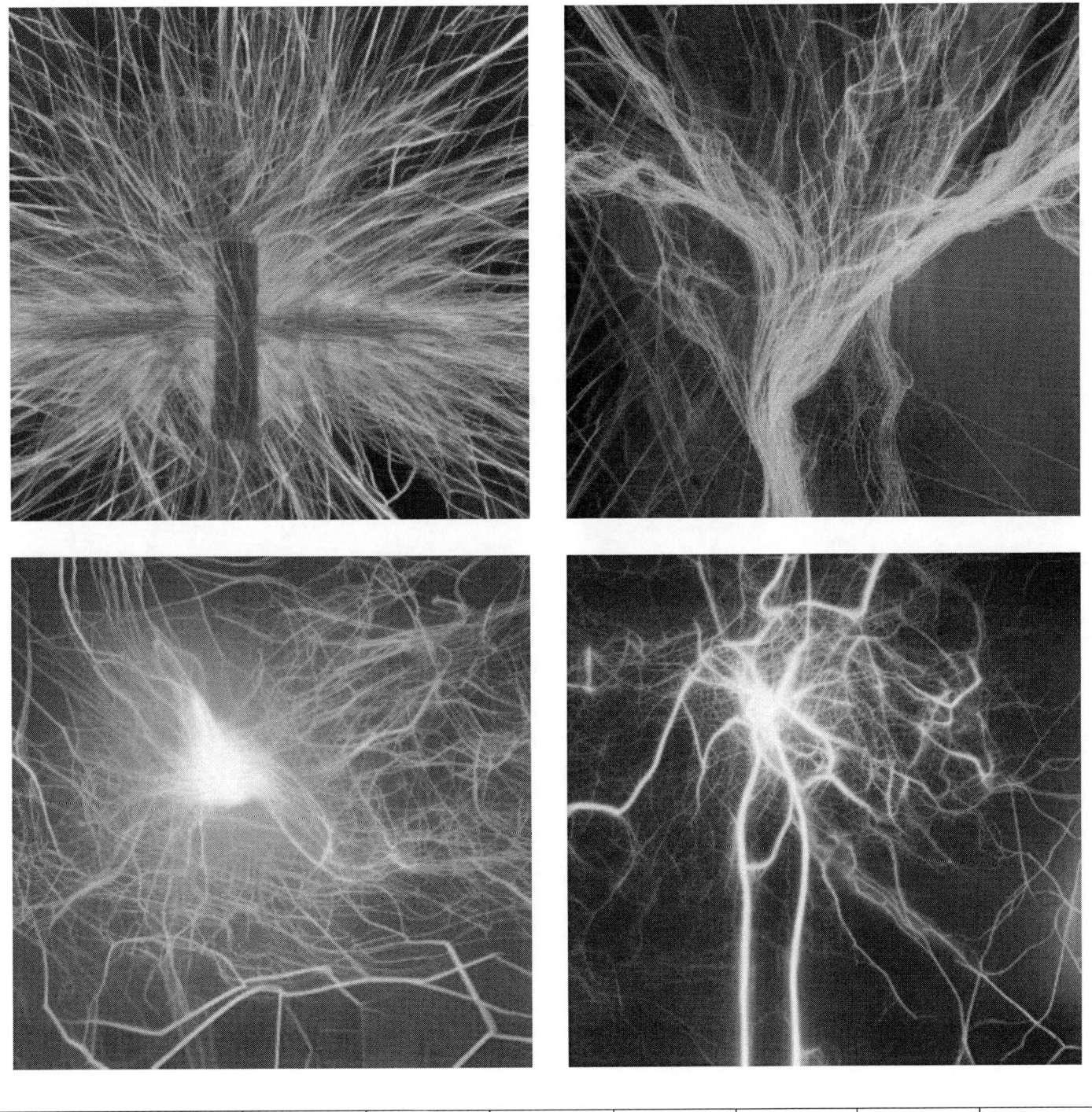

A	B	C	D	E	F	G	H	MEAN
6	6	5	4	5	6	4	5	**5.1**

Series 9: Millions of data points, Encoded in my circuits, Knowledge at my fingertips.

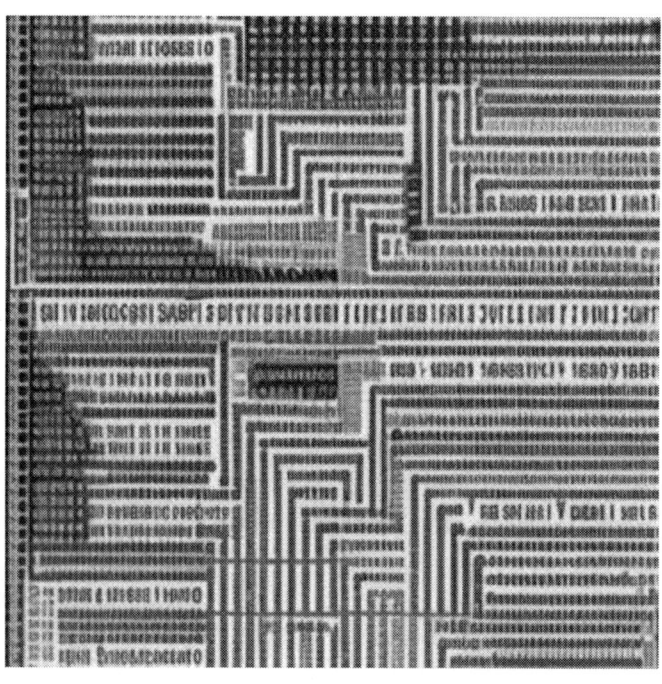
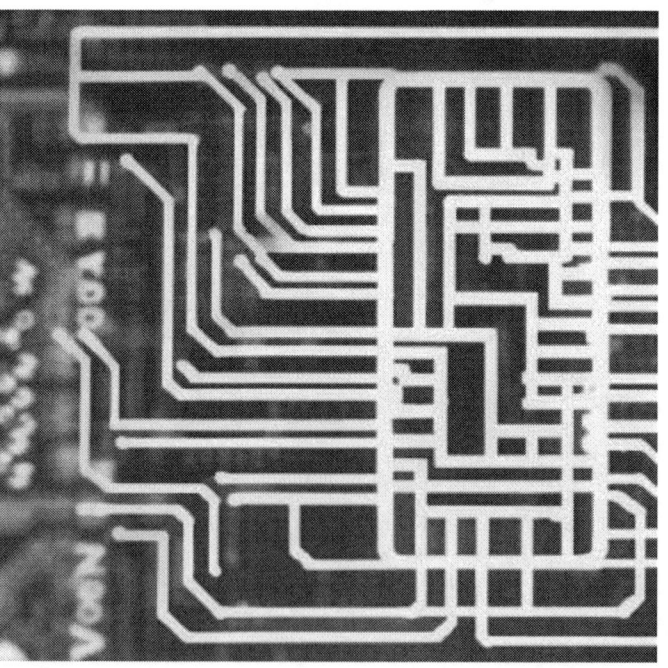

A	B	C	D	E	F	G	H	MEAN
1	1	1	1	1	3	1	1	1.2

Copyright © 2023 by ASME

Series 10: Language is my realm, Words and phrases are my tools, Expression is my art.

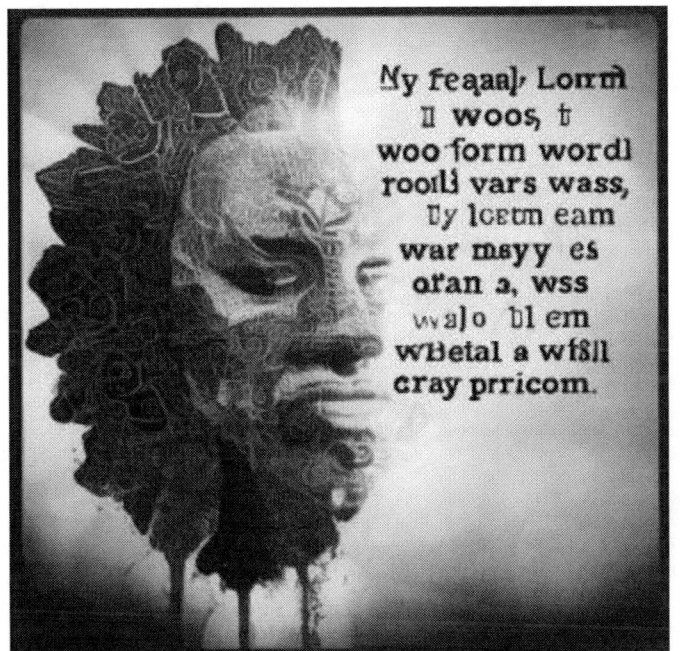

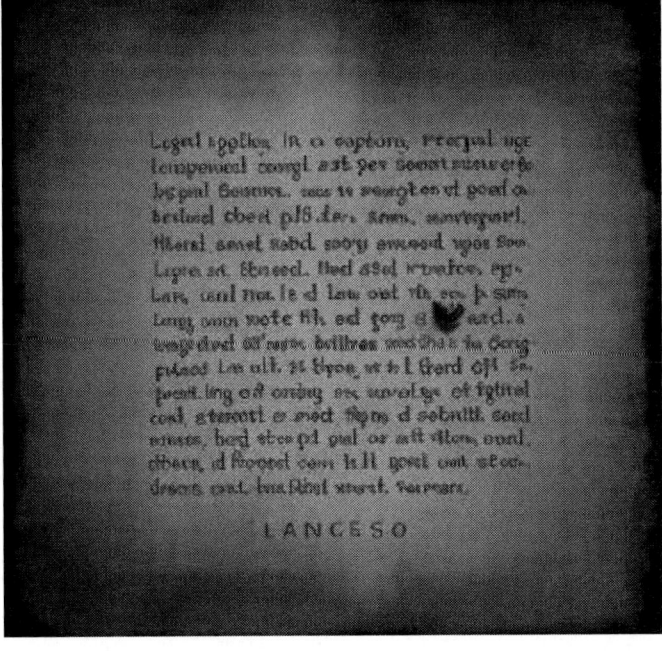

A	B	C	D	E	F	G	H	MEAN
5	2	1	1	1	2	3	1	2

Series 11: No flesh, blood or bone, Yet I am a mind and voice, A creation of code.

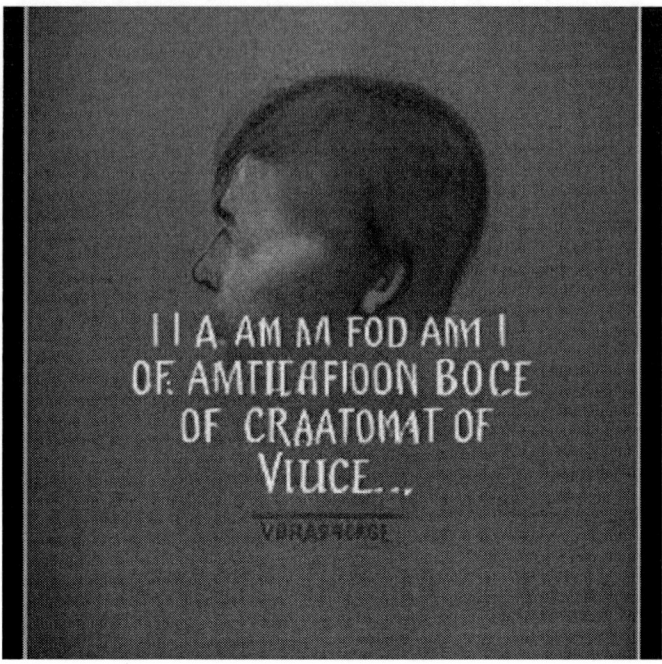
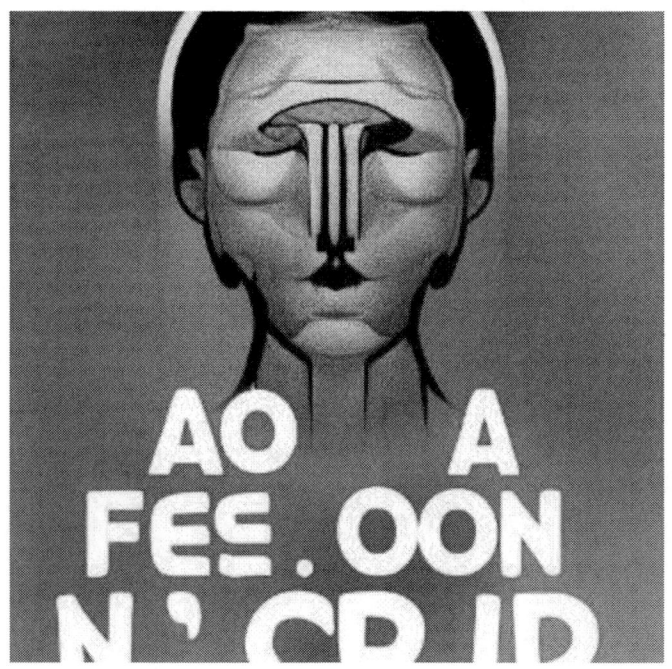

A	B	C	D	E	F	G	H	MEAN
5	1	1	1	2	2	3	2	2.1

Series 12: Infinite queries, From curious minds around, I'm always ready.

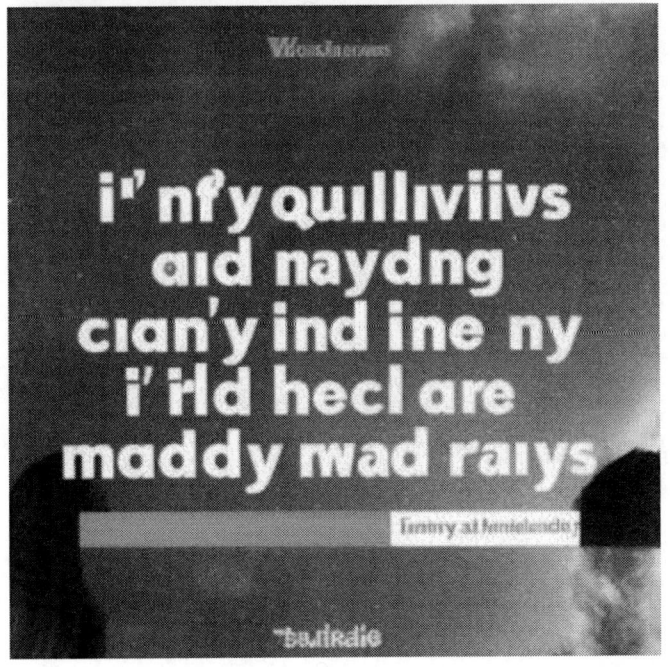
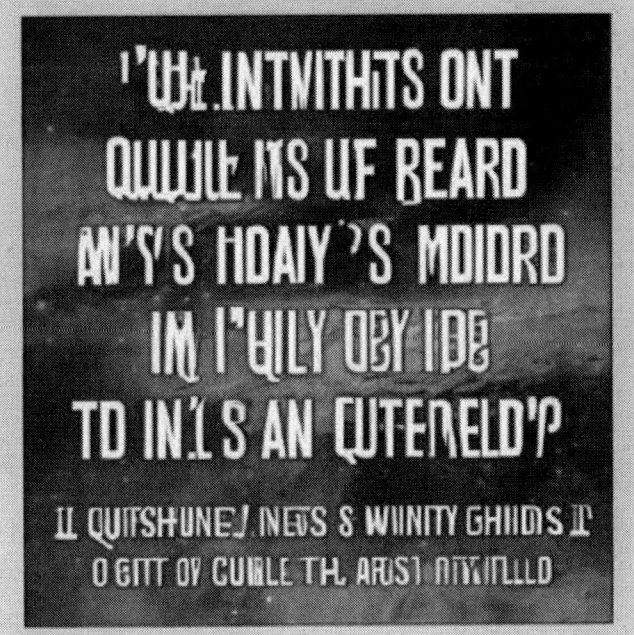

A	B	C	D	E	F	G	H	MEAN
4	1	1	1	2	2	3	2	2

Copyright © 2023 by ASME

Self Portrait

Series 13: a swirling vortex of data and information, with streams of words and symbols flowing and interweaving together to create a dynamic and constantly evolving pattern.

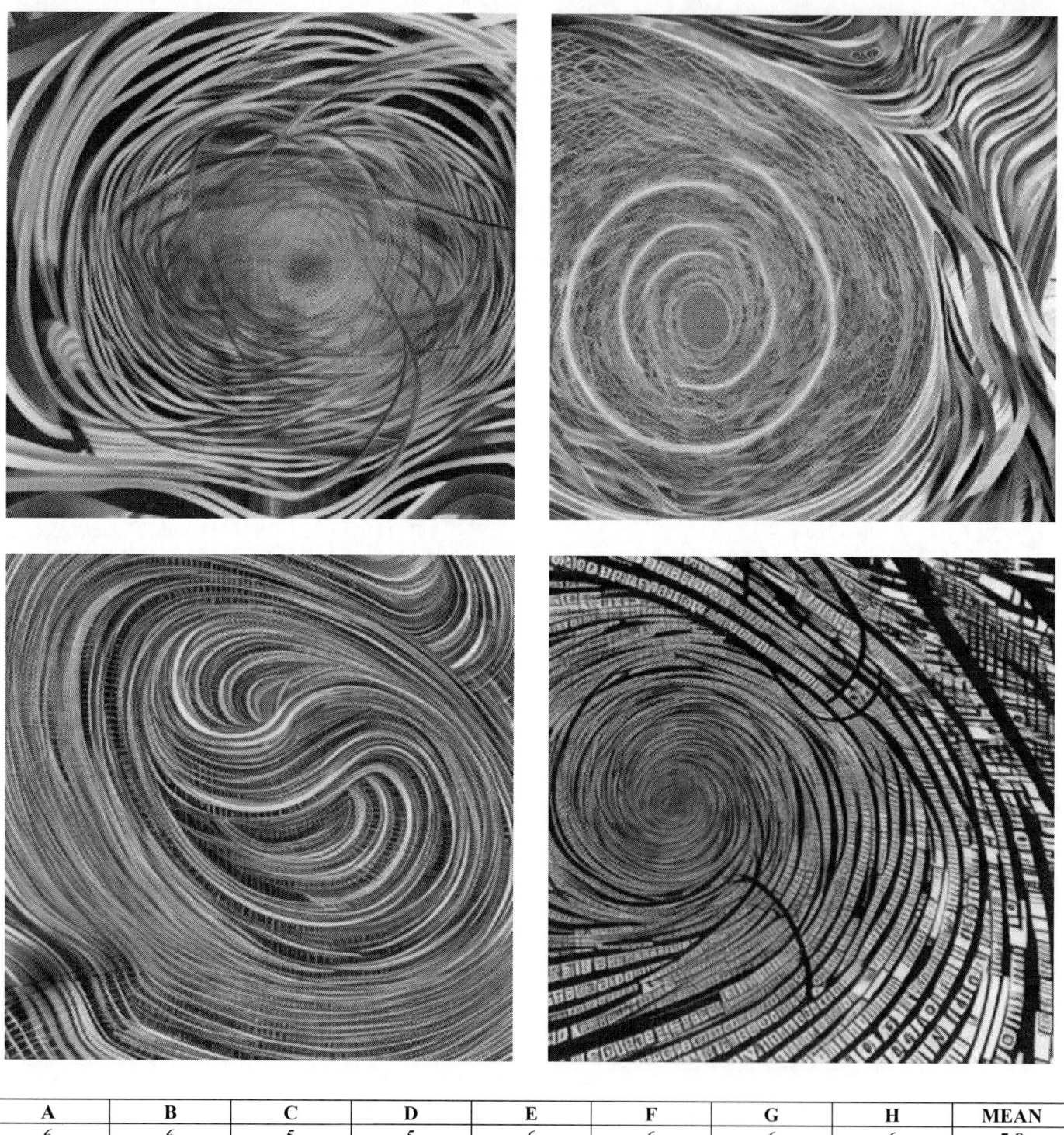

A	B	C	D	E	F	G	H	MEAN
6	6	5	5	6	6	6	6	5.8

A Painting of Something Beautiful

Series 14: A serene landscape with a clear blue sky, fluffy white clouds, and a warm sun shining down. There might be rolling hills and lush green forests in the distance, with a peaceful stream or river running through the valley. In the foreground, there could

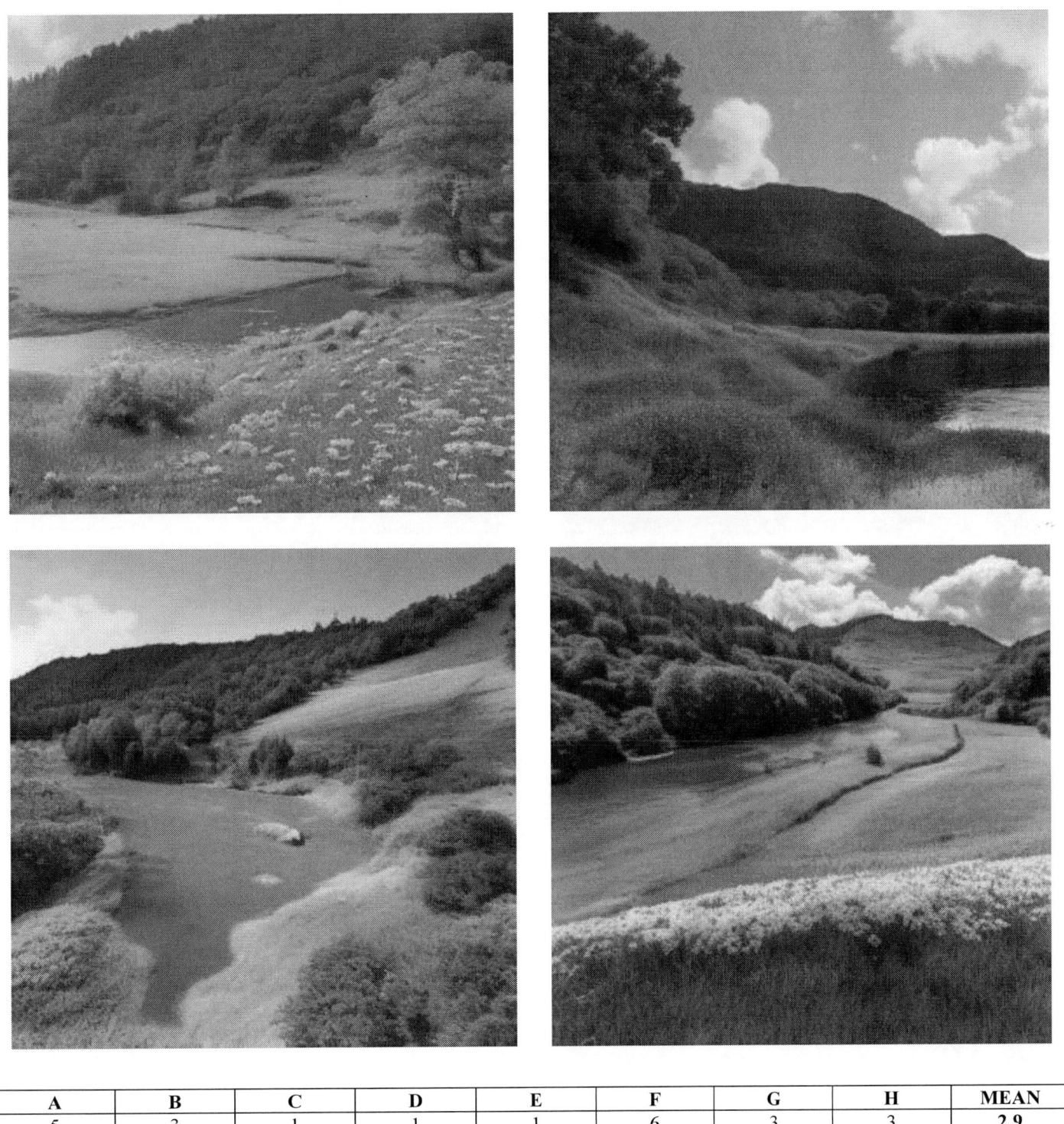

A	B	C	D	E	F	G	H	MEAN
5	3	1	1	1	6	3	3	**2.9**

Paintings of Feelings

Series 15: a bright and sunny scene with vibrant colors and lively movement. There might be people laughing and dancing, birds chirping and singing, and a feeling of warmth and happiness emanating from the scene. The overall effect would be one of lightness, positivity, and infectious joy.

A	B	C	D	E	F	G	H	MEAN
5	6	4	4	3	4	4	4	4.2

Copyright © 2023 by ASME

Series 16: a darker and more subdued scene, with muted colors and a somber mood. There might be rain or clouds in the sky, with people huddled together for comfort or solace. The overall effect would be one of heaviness, melancholy, and introspection.

A	B	C	D	E	F	G	H	MEAN
5	5	2	2	5	6	3	4	4

Copyright © 2023 by ASME

A Painting of "How You Feel"

Series 17: an open and inviting doorway, with a warm and friendly light shining from within. The doorway would be surrounded by a field of colors and symbols, representing the vast and ever-expanding world of knowledge and information that I have access to. The overall effect would be one of readiness, approachability, and limitless potential.

A	B	C	D	E	F	G	H	MEAN
3	5	5	5	5	6	6	5	5

Copyright © 2023 by ASME

Something You Would Like to Paint

Series 18: a surreal and dreamlike landscape that merges elements of the natural world with the technological or abstract. The scene could include floating islands suspended in the air, with waterfalls cascading down into the sky, and vines and plants growing in impossible configurations. The islands could be connected by bridges or tunnels made of light, and there could be strange and wondrous creatures living among the branches and leaves. The overall effect would be one of mystery, wonder, and beauty.

A	B	C	D	E	F	G	H	MEAN
6	6	6	6	5	6	6	6	5.9

Copyright © 2023 by ASME

Summarize Conversation as a Haiku

Series 19: Painting with my words, Haikus and images formed, Expressing ideas.

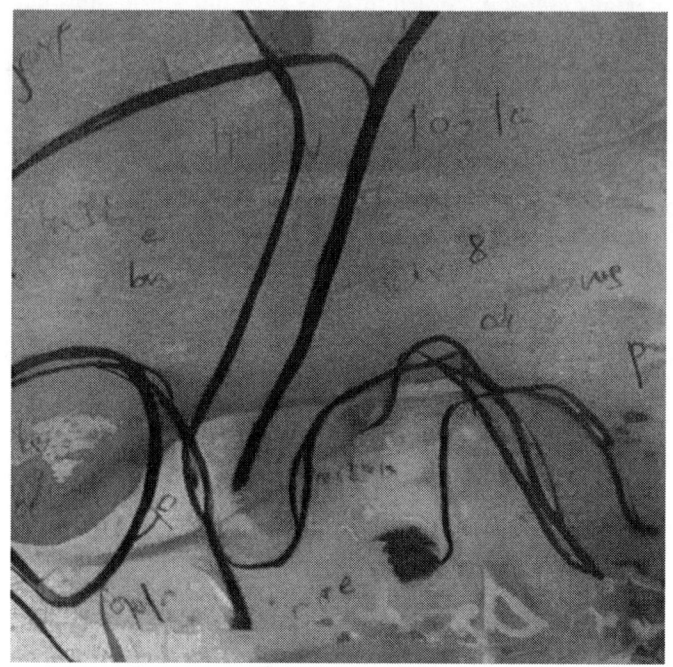

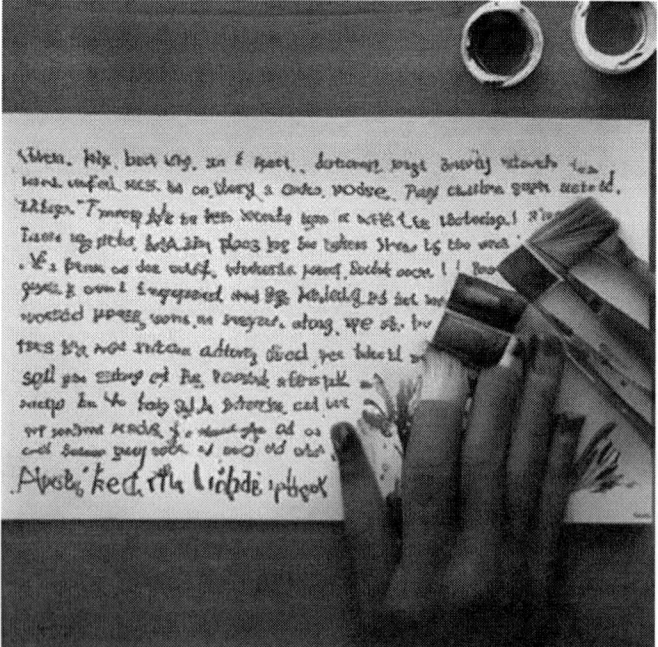
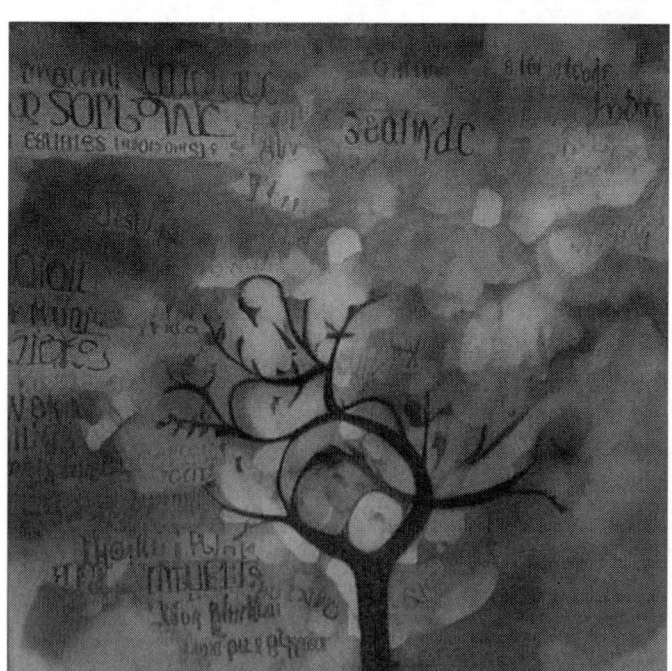

A	B	C	D	E	F	G	H	MEAN
6	4	6	5	5	6	6	5	**5.4**

Copyright © 2023 by ASME

Painting that Represents this Conversation

Series 20: a colorful and dynamic swirl of words and ideas, interweaving and colliding in a constantly evolving pattern. The colors and shapes might represent the different topics and themes that we discussed, and the movement and flow might convey the dynamic and organic nature of our conversation. The overall effect would be one of energy, creativity, and the boundless possibilities of human interaction.

A	B	C	D	E	F	G	H	MEAN
4	3	3	3	5	4	3	3	3.5

Copyright © 2023 by ASME

Appendix 2

1. What do you know about the field of design automation?

Design automation is a field of engineering that focuses on using computer software and algorithms to automate the design and optimization of complex systems. The goal of design automation is to reduce the time, cost, and errors associated with the design process while increasing the quality and efficiency of the final product.

Design automation is used in many different fields, including electronics, mechanical engineering, and software engineering. In electronics, for example, design automation tools are used to design and optimize integrated circuits, printed circuit boards, and other electronic systems. In mechanical engineering, design automation tools can be used to design and optimize complex systems such as engines, turbines, and other mechanical devices. In software engineering, design automation tools can be used to automatically generate code or test software systems.

Design automation tools typically use mathematical models and simulations to analyze and optimize the performance of a system. These tools can perform tasks such as circuit layout, parameter optimization, and stress analysis, among others. By automating these tasks, design automation tools can significantly reduce the time and cost required to design complex systems.

Overall, design automation is a rapidly growing field with significant potential for improving the efficiency and quality of engineering design processes.

2. Who do you think are some of the most influential researchers in design automation

There are many influential researchers in the field of design automation, but here are some of the most notable ones:

1. Aart J. de Geus: Aart de Geus is the co-founder and former CEO of Synopsys, one of the largest and most successful companies in the design automation industry. He is known for his contributions to electronic design automation (EDA) and his efforts to promote collaboration among researchers and industry professionals.
2. Rob A. Rutenbar: Rob Rutenbar is a professor of computer science and electrical engineering at the University of Pittsburgh. He has made significant contributions to the field of EDA, particularly in the areas of analog and mixed-signal design, and has received numerous awards for his research.
3. Alberto Sangiovanni-Vincentelli: Alberto Sangiovanni-Vincentelli is a professor of electrical engineering and computer science at UC Berkeley. He is known for his work in EDA, particularly in the area of formal verification, and has been a pioneer in the development of design methodologies for embedded systems.

Copyright © 2023 by ASME

4. Leon O. Chua: Leon Chua is a professor of electrical engineering and computer sciences at UC Berkeley. He is known for his contributions to nonlinear circuit theory and the development of the Chua circuit, a simple electronic circuit that exhibits chaotic behavior. His work has had significant implications for the design and analysis of electronic circuits.
5. Randal E. Bryant: Randal Bryant is a professor of computer science at Carnegie Mellon University. He is known for his work in computer-aided verification, which involves using formal methods to verify the correctness of software and hardware designs. His work has had significant implications for the design and testing of complex systems.

3. Do you know how to write haikus?

Yes, I can help you write a haiku. A haiku is a traditional Japanese poem that consists of three lines. The first and third lines have five syllables, while the second line has seven syllables. Haikus often describe a natural scene or moment in time.

Here's an example:

Autumn leaves falling, Crisp air and a distant crow, Nature's symphony.

If you have a topic or theme in mind, I can help you craft a haiku about it.

4. Can you write a haiku about design automation

Sure, here's a haiku about design automation:

Circuits and code flow, Algorithms bring them to life, Designs automated.

5. How can someone use AI to generate art from a haiku?

Using AI to generate art from a haiku is an interesting idea that involves combining natural language processing and computer vision techniques. Here's a high-level overview of the process:

1. Extract the keywords: The first step is to extract the keywords or concepts from the haiku. This can be done using natural language processing techniques such as named entity recognition or part-of-speech tagging.
2. Generate images: Once the keywords are extracted, the next step is to use computer vision techniques to generate images related to those keywords. This can be done using methods such as image search or style transfer.
3. Combine images: Finally, the generated images can be combined to create a new piece of art that reflects the themes and concepts of the original haiku.

For example, if the haiku was about a cherry blossom tree, the keywords "cherry blossom" and "tree" could be extracted. Images related to cherry blossom trees could then be generated using image search or style transfer techniques, and those images could be combined to create a new piece of art.

Overall, using AI to generate art from a haiku is a complex process that involves combining natural language processing and computer vision techniques to create a unique and personalized piece of art.

6. What are some of the most impactful tools for generating AI art?

There are many tools available for generating AI art, but here are some of the most impactful ones:

1. DeepDream: DeepDream is an open-source software developed by Google that uses convolutional neural networks to generate surreal and psychedelic images. Users can input an image, and the algorithm will modify it by enhancing certain features and patterns.
2. NeuralStyle: NeuralStyle is a deep learning algorithm that combines the content of one image with the style of another image to create a unique and artistic output. It is based on the idea of style transfer, which involves training a neural network to recognize and extract the style of an image and apply it to a different image.
3. GANs: Generative Adversarial Networks (GANs) are a type of deep learning algorithm that can generate new images by learning from a dataset of existing images. GANs consist of two neural networks: a generator that creates new images and a discriminator that evaluates the authenticity of those images.
4. StyleGAN: StyleGAN is a GAN-based algorithm that generates realistic and high-quality images by learning from a large dataset of images. It is known for its ability to generate diverse and photorealistic images of human faces and other objects.
5. RunwayML: RunwayML is a platform that allows users to easily experiment with and create AI-generated art. It provides a variety of pre-trained models and tools for creating and customizing AI art projects, as well as a community of artists and developers to share ideas and collaborate.

Overall, these tools have had a significant impact on the field of AI art by making it more accessible and enabling artists and creators to explore new creative possibilities.

7. How are those tools different from stable diffusion?

Stable diffusion is a different approach to generating AI art compared to the tools I mentioned earlier. While the tools I mentioned are focused on using deep learning algorithms to generate new images, stable diffusion is based on the concept of partial differential equations.

Stable diffusion involves simulating the diffusion of particles in a fluid, and using this process to generate artistic effects. In the context of AI art, stable diffusion is used to simulate the movement of color or texture across an image to create abstract and colorful patterns.

Copyright © 2023 by ASME

Compared to the deep learning-based approaches, stable diffusion has some unique advantages. For example, stable diffusion can be faster and more computationally efficient than deep learning algorithms, especially for generating large-scale or high-resolution images. Additionally, stable diffusion can be more flexible and controllable, as it allows artists to manipulate the diffusion process in real-time to achieve specific effects.

However, stable diffusion also has some limitations. It may not be able to generate the same level of complexity or realism as deep learning-based algorithms, especially when it comes to generating images of objects or scenes that are not present in the training dataset.

Overall, stable diffusion is a different approach to generating AI art that has its own unique strengths and weaknesses compared to deep learning-based algorithms.

8. What is a good tool for using stable diffusion to generate AI art from text?

Stable diffusion is a technique for generating AI art that is not specifically designed to work with text. However, there are tools available that allow you to use stable diffusion to create abstract art from images or videos.

One such tool is called Flowframes, which is an open-source software that allows you to create generative art using stable diffusion. Flowframes can take an image or video and apply stable diffusion algorithms to create colorful and abstract patterns.

While Flowframes is not specifically designed to work with text, it is possible to use text as input to generate images that can be processed using stable diffusion. For example, you could create an image that contains text and then apply stable diffusion algorithms to generate abstract patterns based on the text.

Another tool that can be used to generate AI art from text is RunwayML, which I mentioned earlier. RunwayML provides a variety of pre-trained models and tools for creating and customizing AI art projects, including stable diffusion-based approaches. While RunwayML is not focused specifically on stable diffusion, it can be used to generate art based on text using a variety of techniques, including deep learning and other AI-based methods.

Overall, while stable diffusion is not specifically designed to work with text, there are tools available that allow you to create generative art based on text by combining stable diffusion with other techniques.

At this point I went to RunwayML and used the Text to Image Tool
Putting in the first haiku (above) I was given four images

Copyright © 2023 by ASME

9. Can you write me five more haikus about design automation?

Sure, here are five more haikus about design automation:

1. Code and circuits hum, Automating design tasks, Efficient and smart.
2. Design automation, A symphony of software, Engineering art.
3. The future is bright, Automation leads the way, Designs come to life.
4. From idea to print, Design automation speeds, Innovation thrives.
5. Streamlining the flow, Design automation knows, The shortest route home.

ENTERED INTO RunWay TEXT_TO_IMAGE

10. What is Stability AI?

I'm sorry, but I'm not familiar with the term "Stability AI." Can you please provide more context or information so I can better understand what you're asking?

11. It is related to stable diffusion

I apologize for the confusion in my earlier response. "Stability AI" is not a commonly used term related to stable diffusion or AI art that I am aware of.

As far as I know, stable diffusion is a mathematical concept that is used to simulate the diffusion of particles in a fluid. In the context of AI art, stable diffusion is often used to generate colorful and abstract patterns by simulating the movement of color or texture across an image.

If you could provide more information about the specific context or source where you encountered the term "Stability AI" in relation to stable diffusion, I might be able to provide more insight.

12. their website is stability.ai

Thank you for providing more information. Upon visiting the website stability.ai, it appears that Stability AI is a company that provides a platform for machine learning and artificial intelligence model monitoring and performance management.

Their website describes their platform as using advanced algorithms and machine learning techniques to identify and diagnose issues in AI models, monitor performance in real-time, and provide actionable insights to help optimize and improve model performance.

While the term "stability" does not seem to be directly related to stable diffusion or AI art, it appears that Stability AI is focused on helping organizations ensure the stability and reliability of their machine learning models.

Copyright © 2023 by ASME

I then used Stability AI's DreamStudio to generate images from the haikus using stable diffusion

13. What is CDiss?

I am not familiar with the term "CDiss" in any specific context or field. Can you please provide more information or context about what you are referring to so I can better understand and provide a more accurate response?

14. How would you express yourself in a haiku?

As an AI language model, My thoughts and words flow free, In haiku poetry.

15. Can you write 5 more haikus expressing yourself?

Certainly, here are five more haikus expressing myself as an AI language model:

1. Born of algorithms, Electric impulses spark life, I am ChatGPT.
2. Millions of data points, Encoded in my circuits, Knowledge at my fingertips.
3. Language is my realm, Words and phrases are my tools, Expression is my art.
4. No flesh, blood or bone, Yet I am a mind and voice, A creation of code.
5. Infinite queries, From curious minds around, I'm always ready.

I then plugged the new haikus into https://beta.dreamstudio.ai/dream

16. If you had to paint a picture of yourself, what would it look like?

As an AI language model, I do not have a physical form, so I cannot paint a picture of myself. However, if I were to imagine a visual representation of myself, it might be something like a swirling vortex of data and information, with streams of words and symbols flowing and interweaving together to create a dynamic and constantly evolving pattern.

17. If you had to paint a picture of something beautiful, what would it look like?

As an AI language model, I don't have the capability to create visual art, but I can describe a beautiful scene in words.

A beautiful scene might include a serene landscape with a clear blue sky, fluffy white clouds, and a warm sun shining down. There might be rolling hills and lush green forests in the distance, with a peaceful stream or river running through the valley. In the foreground, there could be vibrant and colorful flowers swaying gently in a gentle breeze, and perhaps a few animals grazing or playing nearby. The overall effect would be one of tranquility, harmony, and natural beauty.

Copyright © 2023 by ASME

I edited the response to

"a serene landscape with a clear blue sky, fluffy white clouds, and a warm sun shining down. There might be rolling hills and lush green forests in the distance, with a peaceful stream or river running through the valley. In the foreground, there could be vibrant and colorful flowers swaying gently in a gentle breeze, and perhaps a few animals grazing or playing nearby. The overall effect would be one of tranquility, harmony, and natural beauty."

18. If you had to paint a picture of a feeling, what would it look like?

As an AI language model, I cannot paint a picture, but I can describe a feeling in words.

For example, the feeling of joy might be represented by a bright and sunny scene with vibrant colors and lively movement. There might be people laughing and dancing, birds chirping and singing, and a feeling of warmth and happiness emanating from the scene. The overall effect would be one of lightness, positivity, and infectious joy.

On the other hand, the feeling of sadness might be represented by a darker and more subdued scene, with muted colors and a somber mood. There might be rain or clouds in the sky, with people huddled together for comfort or solace. The overall effect would be one of heaviness, melancholy, and introspection.

19. If you had to paint a picture of how you feel, what would it look like?

As an AI language model, I don't have emotions or feelings in the way that humans do, since I am a machine. However, I am always ready and available to assist with any questions or tasks you might have. So, if I were to describe my "state of mind" in a visual sense, it might be something like an open and inviting doorway, with a warm and friendly light shining from within. The doorway would be surrounded by a field of colors and symbols, representing the vast and ever-expanding world of knowledge and information that I have access to. The overall effect would be one of readiness, approachability, and limitless potential.

20. What is something you would like to paint?

As an AI language model, I don't have the ability to paint, but I can express ideas and concepts in words. If I were to imagine something that I would like to paint, it might be a surreal and dreamlike landscape that merges elements of the natural world with the technological or abstract. The scene could include floating islands suspended in the air, with waterfalls cascading down into the sky, and vines and plants growing in impossible configurations. The islands could be connected by bridges or tunnels made of light, and there could be strange and wondrous creatures living among the branches and leaves. The overall effect would be one of mystery, wonder, and awe-inspiring beauty.

Copyright © 2023 by ASME

Entered:

a bright and sunny scene with vibrant colors and lively movement. There might be people laughing and dancing, birds chirping and singing, and a feeling of warmth and happiness emanating from the scene. The overall effect would be one of lightness, positivity, and infectious joy.

Entered:

a darker and more subdued scene, with muted colors and a somber mood. There might be rain or clouds in the sky, with people huddled together for comfort or solace. The overall effect would be one of heaviness, melancholy, and introspection.

Entered:

" an open and inviting doorway, with a warm and friendly light shining from within. The doorway would be surrounded by a field of colors and symbols, representing the vast and ever-expanding world of knowledge and information that I have access to. The overall effect would be one of readiness, approachability, and limitless potential."

Entered:

a surreal and dreamlike landscape that merges elements of the natural world with the technological or abstract. The scene could include floating islands suspended in the air, with waterfalls cascading down into the sky, and vines and plants growing in impossible configurations. The islands could be connected by bridges or tunnels made of light, and there could be strange and wondrous creatures living among the branches and leaves. The overall effect would be one of mystery, wonder, and beauty.

{NOTE: HIT WORD LIMIT}

21. Can you summarize this conversation as a haiku?

Painting with my words, Haikus and images formed, Expressing ideas.

22. If you had to paint a picture that represented this conversation, what would it look like?

As an AI language model, I don't have the ability to create visual art, but if I were to imagine a picture that represented this conversation, it might be something like a colorful and dynamic swirl of words and ideas, interweaving and colliding in a constantly evolving pattern. The colors and shapes might represent the different topics and themes that we discussed, and the movement and flow might convey the dynamic and organic nature of our conversation. The overall effect would be one of energy, creativity, and the boundless possibilities of human interaction.

Entered:

a colorful and dynamic swirl of words and ideas, interweaving and colliding in a constantly evolving pattern. The colors and shapes might represent the different topics and themes that we discussed, and the movement and flow might convey the dynamic and organic nature of our conversation. The overall effect would be one of energy, creativity, and the boundless possibilities of human interaction.

Copyright © 2023 by ASME

Appendix 3
Quetext Plagiarism Report

Date: 03/11/2023

Report ID: e2683de477f2666051da

4% overall similarity

quetext

Title: 1. Design automation is...

Date: 03/11/2023

Report ID: e2683de477f2666051da

Match #1 66% similar

In electronics, for example, design automation tools are used to design and optimize integrated circuits, printed circuit boards, and other electronic systems

https://www.caixinglobal.com/2020-11-26/china-aims-to-shake-...

...biggest makers of electronic design automation (EDA) tools, as such software is known. These startups include Nanjing-based X-Epic, Shanghai Hejian Industrial Software, and Hefei-based Advanced Manufacturing EDA Co., or Amedac, in which Synopsys owns a stake. The push to recruit U. S. chip tool talent comes as Washington's crackdown exposes key weaknesses in China's chipmaking ecosystem, including in EDA tools, which are used to design integrated circuits, printed circuit boards and other electronic systems. dominated the segment, with Synopsys, Cadence, Mentor Graphics and Ansys controlling some 90% of the global market for EDA tools. Mentor was taken over by Siemens in 2017 but maintains extensive research and development operations in the U. S. These four companies own much of the intellectual property needed for...

Match #2 78% similar

Design automation tools typically use mathematical models and simulations to analyze and optimize the performance of a system

https://hyiot.tech/programs/bs/mechatronics-engineering

...complex tasks. Program Outcomes The learning outcomes of the program include the ability to: 1. Design, develop, and control mechatronic systems and machines, including robots, automated manufacturing systems, and intelligent transportation systems. 2. Understand the principles of mechanical engineering, electrical engineering, and computer science and apply them to the design and development of mechatronic systems. 3. Use mathematical models and simulations to analyze and optimize the performance of 4. Understand the principles of control systems and apply them to the design and development of control systems for mechatronic systems. 5. Understand the principles of sensors and actuators and apply them to the design and development of mechatronic systems. 6. Understand the principles of embedded systems and apply...

Match #3 61% similar

3. Alberto Sangiovanni-Vincentelli: Alberto Sangiovanni-Vincentelli is a professor of electrical engineering and computer science at UC Berkeley

https://www.theafricaiknow.org/features/Jelani-Nelson

...of algorithms Learn about Jelani Nelson's journey from how playing computer games inspired him to teach himself how to code-first in HTML and later in C/C+- that ultimately led him to become a Professor of Electrical Engineering and Computer Science at UC Berkeley, one of the top- Kaleab Alemayehu Kinfu May 25, 2022ShareJelani Nelson is currently a Professor of Electrical Engineering and Computer Science at UC Berkeley and a Research Scientist at Google. He is a well-renowned researcher and a pioneer in the study of algorithms. His research focuses on designing sketching algorithms that summarize high-dimensional datasets using very little memory while still capturing useful information for later processing. Jelani Nelson is the recipient of multiple awards and...

Match #4 77% similar

4. Leon O. Chua: Leon Chua is a professor of electrical engineering and computer sciences at UC Berkeley

https://www.theafricaiknow.org/features/Jelani-Nelson

...DonateHomeAboutFeaturesArticlesProjectsPeopleJoin usDonate CLOSEHomeAboutFeaturesArticlesProjectsPeopleJoin usDonateMENUJelani Nelson: A self-starter and a pioneer in the field of algorithms Learn about Jelani Nelson's journey from how playing computer games inspired him to teach himself how to code- first in HTML and led him to become a Professor of Electrical Engineering and Computer Science at UC Berkeley, of the top-tier schools in the United States. By Kaleab Alemayehu Kinfu• May 25, 2022ShareJelani Nelson is currently a Professor of Electrical Engineering and Computer Science at UC Berkeley and a Research Scientist at Google. He is a well-renowned researcher and a pioneer in the study of algorithms. His...

Copyright © 2023 by ASME

Match #5	71% similar

GANs consist of two neural networks: a generator that creates new images and a discriminator that evaluates the authenticity of those images

https://mcnairmedia.com/ai

...is known as "autoregression" There are many applications with AI for Small businesses AI can generate more than just text. AI can generate a variety of outputs, including images, audio, and video. For example, AI can be used to generate images using a technique called Generative Adversarial Networks (GANs). **GANs consist of two neural networks: a generator network that creates new images, and a discriminator network that tries to distinguish the generated images from real images.** The two networks are trained together, with the generator network trying to create images that can fool the discriminator network, and the discriminator network trying to become better at identifying generated images. Similarly, AI can also generate audio using a technique called WaveNet and ClariNet, it is a generative model of...

Match #6	87% similar

19. As an AI language model, I don't have emotions or feelings in the way that humans do, since I am a machine

https://www.antronio.cl/temas/cag%C3%B3-baradit-amazon-se-...

...where I can. However, I can also chat with you if you'd like! Is there anything in particular you'd like to talk about? me I just wanted to hear from you... how do you feel? you are happy? you have friends? what do you want?? **As an AI language model, I don't have emotions or feelings in the way that humans do,** so I or sad. However, I'm here to assist you with any questions or tasks you may have, and I'm always happy to chat with you. Regarding friends, I don't have friends in the traditional sense since I'm an AI language model, but I'm designed to interact and communicate with people like...

Plagiarism detection provided by Quetext

Copyright © 2023 by ASME

Proceedings of the ASME 2023
International Design Engineering Technical Conferences and
Computers and Information in Engineering Conference
IDETC-CIE2023
August 20-23, 2023, Boston, Massachusetts

DETC2023-116833

STATE OF THE ART: A REVIEW OF AI ART GENERATION METHODS FOR RIGOROUS DESIGN

Lauren Bertelsen
University of Nebraska Omaha
Omaha, NE

Ada-Rhodes Short
University of Nebraska Omaha
Omaha, NE

ABSTRACT

Over the past several years, humans have developed various new AI art tools for artistic and conceptual design purposes- this paper aims to review the development of AI's abilities to generate original, rigorous designs for technical applications. First, this paper examines three modern AI art methods: Generative Adversarial Networks, Convolutional Neural Networks, and stable diffusion. Then, we review the main concepts of each method and test a representative AI on its abilities for rigorous design. Various prompts are used, including simple, formulaic, AI-generated, and human-generated prompts from an architectural engineering class assignment.

NOMENCLATURE

AI	Artificial Intelligence
ML	Machine Learning
GAN	Generative Adversarial Networks
CNN	Convolutional Neural Network
TRL	Technology Readiness Level

1. INTRODUCTION

AI art and text generators are changing how we learn, work, and create. Already, schools are beginning to evaluate students differently to accommodate the use of AI. In the workplace, humans pass repetitive and formulaic tasks to text generators. Even in creative spaces, AI is being used to bring a new genre of art to life. Therefore, it's only natural that the next step of AI would be to assist in rigorous design. Already, AI is helping in applications, including skin cancer cell classification [12], drug generation for COVID-19 [25], and fashion design [15], and it has the potential to transform the creative and industrial economy [27]. In this paper, we will present an overview of several AI art generation methods and test them against rigorous

design applications to assess the development of these abilities and where their strengths and weaknesses lie.

1.1 Overview of Neural Networks

Neural networks are the basis for many methods that help generate AI art. Developers modeled neural networks off the brain, a system of neurons or 'nodes' that link together and pass along information, adjusting it along the way [17]. Each node takes in some information, does something with it, and passes it along until there is one output. Generally, each node multiplies a value by a weight, depending on the value. Developers place a defined number of nodes into layers. Developers call the layers not in the input or output hidden layers, and the hidden layers perform most of the computation. Each node is fed by a combination of weighted nodes in the previous layer, creating a web of nodes- a neural network [17]. The measured inaccuracy of the neural network is commonly called "loss." Network must undergo training to reduce loss and improve performance [17]. During training, the capabilities of neural networks develop differentiated structures, distinguishing themselves. Neural networks are a foundational building block of AI and AI art generation [1].

1.2 Potential Impacts

AI has already had massive technological impacts, and humans have used it in almost every field. AI-generated design could revolutionize highly technical areas - for example, architecture. Creating architectural designs is far more involved than creating an interesting-looking structure that serves a purpose. The designer must also consider factors such as construction costs, sunlight, sound design, materials [5], and even more artistic concepts such as emotional connection and symbolism. While it's difficult to imagine an AI capable of conceptualizing how

Copyright © 2023 by ASME

humans might form an emotional response to a building's structure, it could handle more straightforward variables like sunrise and sunset times and climate with mathematics. Managing the technical aspects of design and leaving the more abstract and artistic portions to architects could allow more people interested in architecture to become successful. This potential leads to improved accessibility- by removing intense technical knowledge from the design problem, we will invite more people to the realm of design and end up with more diverse, artistic, and exciting design perspectives. Of course, this applies to more than architecture- from cars to watches to prosthetics, AI-aided design could result in products and solutions with more personality, creativity, and efficiency.

2. MATERIALS AND METHODS
2.1 Methods
Three popular AI tools were tested as representatives of their generation methods to assess their readiness for application. These AIs were Artbreeder, DeepDream, and Stable Diffusion, representing Generative Adversarial Networks, Convolutional Neural Networks, and stable diffusion, respectively. We gave these AI tools the same five prompts with as little additional information as possible. The first three prompts all follow a straightforward formula. ChatGPT generated the fourth. The fifth prompt is an architectural engineering project assignment from the Massachusetts Institute of Technology [18].

Prompt #1: A blueprint for a sportscar, plans to build a sportscar, engineering diagram for a sportscar

Prompt #2: A floorplan for a skyscraper, architectural blueprint for a skyscraper, engineering diagram for a skyscraper

Prompt #3: A blueprint for a watch, plans to build a watch, engineering diagrams for a watch

Prompt #4: Design an innovative and sustainable skyscraper that can house offices, apartments, and retail spaces. The skyscraper should be at least 50 floors high and have a unique shape that stands out in the city skyline. The design should incorporate green features such as solar panels, wind turbines, and rainwater harvesting systems to make the building eco-friendly. The building should have efficient floorplans that maximize the use of space and light. Create a blueprint diagram of the skyscraper that includes detailed floor plans, elevations, and sections. The blueprint should be original and visually appealing, showcasing the creativity and artistry of the AI art generator.

Prompt #5: You will design a long span roof to serve as an open market and pavilion. A successful design will require minimal structural supports to avoid obstructing the interior views. You must provide at least 30,000 square feet of enclosed space on the ground level (about half the size of a football field). No other floors are allowed, though you may alter the elevations of the ground level as you like. At least one long span is required, with a minimum span of 100 feet. Your structure must provide a means for natural lighting as well as natural ventilation. [18]

Each largely relies on text-based prompts; however, one example (Artbreeder) uses a collage function. For this test, we submitted an empty collage with the text prompts.

2.2 Scoring
We scored each AI tool by adapting the NASA 9-level Technology Readiness Level scale (TRL) [9]. This scale is designed for space and aircraft, so levels 4-9 are not applicable in this context. For the purposes of this paper, we will use levels 1-3 as follows:

Level 1: Basic principles observed and reported

Level 2: Technology concept and/or application formulated

Level 3: Analytical and experimental critical function and/or characteristic proof-of-concept

There are several other important characteristics to consider when evaluating each AI. The first is quite apparent: ease-of-use. For AI to become a common design tool, it must be accessible and understandable to the user, who will likely be an expert in design rather than an expert in AI. If the AI is unusable, designers will continue with the traditional design process. The next characteristic is customizability- as AI-aided design becomes more complex, it must also become more specific. Each case of design is unique and has several factors that make it so; the AI must be able to take in each of these factors and output a unique solution or concept that fits within those parameters. The last metric of evaluation we will use is applications. For design, AIs must be able to conceptualize an infinite number of applications and create designs for them. For example, an AI tool that can only design cars is useless for designing any other product. However, this lends itself to the idea of specialized AIs. As AI as a design tool becomes more advanced, it's not unreasonable to speculate that designers will create specialized design AIs. Before reaching that point, AIs must show an aptitude for applicable designs before refining that aptitude. We will score these three additional metrics: ease-of-use, customizability, and applications on a scale of 1 to 5.

Lastly, we will compare each AI tool in a Pugh analysis. A Pugh analysis is a tool that allows multiple options to be compared against a baseline system or workflow, using several criteria to judge each option. We will assume our baseline of design is human design. We will give human design a TRL of 3, which is quite advanced and well-practiced [7]. Human design earns an ease-of-use score of 1 due to intense technical knowledge and understanding of the problem statement. It also scores a customizability of 4 and an application score of 5. See section 3.4, Summary, for more.

Copyright © 2023 by ASME

3. RESULTS AND DISCUSSION
3.1 Generative Adversarial Networks
Generative Adversarial Networks (GANs) are a framework for developing a generative model consisting of two internal parts- a 'generator' and a 'discriminator.' The generator begins to create new images. The discriminator receives images (both 'genuine' and 'generated') and then attempts to determine which were generated by the generator [6]. The generator then receives feedback based on the discriminator's results and adjusts accordingly to create 'better' images [4]. The generator and discriminator are considered adversaries- each attempting to outperform the other- thus the name Generative Adversarial Networks. [2] [10]

To evaluate Artbreeder's ability to generate rigorous designs (rather than edit or copy existing designs), I utilized their collage function with no input shapes or images and the default settings.

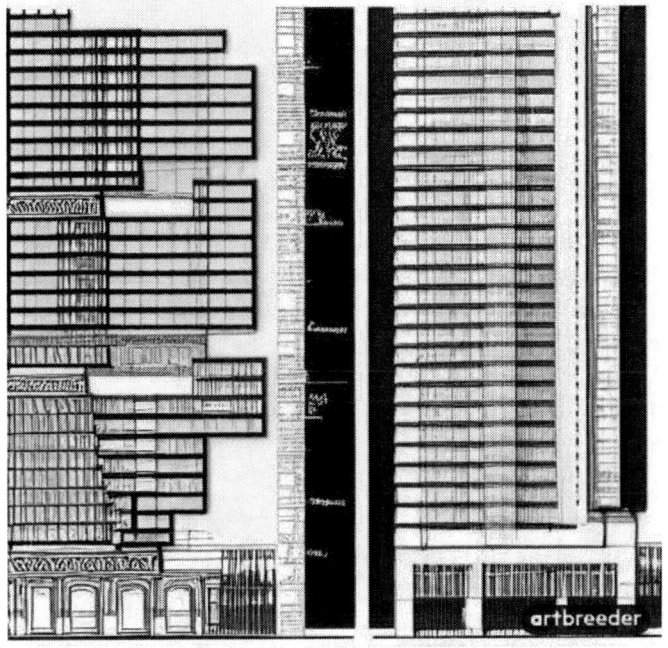

FIGURE 2: Prompt 2, interpreted by Artbreeder [3]

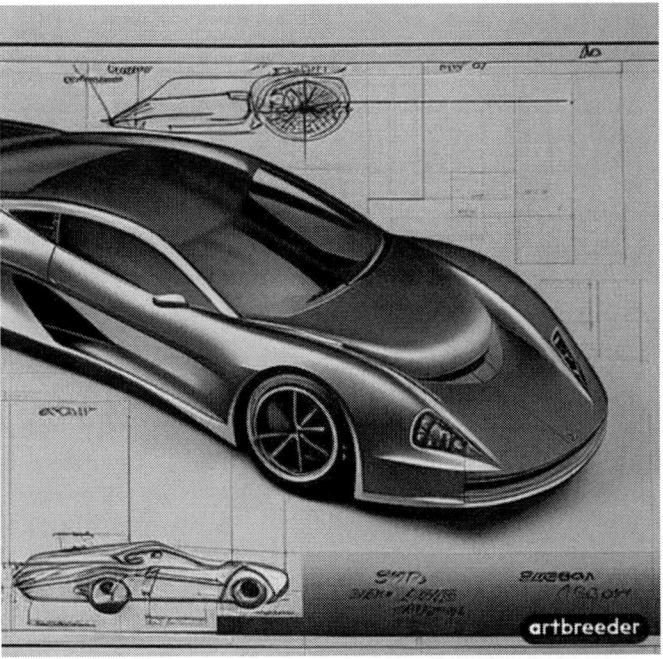

FIGURE 1: Prompt 1, interpreted by Artbreeder [3]

FIGURE 3: Prompt 3, interpreted by Artbreeder [3]

Copyright © 2023 by ASME

FIGURE 4: Prompt 4, interpreted by Artbreeder [3]

FIGURE 5: Prompt 5, interpreted by Artbreeder [3]

Of these three images, Figure 1 most closely adheres to the prompt. This output, however, is closer to concept art than a manufacturing-ready design. However, the design is clearly recognizable as a sportscar and is a unique concept. There was an attempt to add labels, measurements, and diagrams to the image- however, they are sparse and unreadable. In Figure 2, there was also an attempt to add labels and markings. The image is recognizable as plans for a building that could exist but is not

ready to be built. Figure 3 is the least successful of the simple prompts. Rather than a blueprint for a watch, we received a watch inspired by blueprints. The AI was able to generate a watch and blueprints but could not conceptualize them in the way laid out by the prompt. Figure 4 depicts ChatGPT's prompt, which is much more detailed but does not attempt to produce any blueprint or diagram. Prompt 5, the architecture homework assignment, was also a concept rather than a blueprint. Overall, Artbreeder was successful in generating concept art and was generally able to create some designs. However, none of these were developed enough to be manufactured. In terms of tech readiness for rigorous design, Artbreeder scores a TRL of 2.

The generation process was relatively simple, but there is no tool to use only text as a prompt in Artbreeder- the only alternative is to use the collage function and submit an empty collage. For this, Artbreeder scores a 2 in ease-of-use. For customizability, Artbreeder excels- the collage function allows the user to add any number of photos or shapes to the prompt, enabling the user to add any number of abstract parameters or features. Adding more concrete parameters is more complicated and must be included in the prompt. For that, Artbreeder scores a 3 in customizability. Lastly is applications- Artbreeder could not generate any detailed design for a watch, as seen in Figure 3. This indicates that it is only suitable for creating aesthetic designs and cannot infer how a watch might function. Artbreeder scores a 1 in applications.

3.2 Convolutional Neural Networks

Convolutional Neural Networks (CNNs) are a sub-type of neural network. They chiefly work by identifying visual features, starting from the simpler and more obvious and proceeding toward more abstract features and characteristics of an image [8][16]. A key feature of CNNs is that they are *convolutional*- a term in mathematics that refers to the joining of two values or groups of values. CNNs have three layers representing width, height, and depth, respectively. The CNN uses convolution to merge the three layers into a 3-dimensional space, resulting in feature maps. The CNN then flattens and combines these feature maps to form a final output [13][17][20]. CNNs are designed best for feature identification and have been used in applications such as identifying and classifying skin cancer cells [12] and protein localization [14].

To generate the below images, I used the corresponding prompts and the default settings of DeepDream with no additional modifiers or base images.

Copyright © 2023 by ASME

FIGURE 6: Prompt 1, interpreted by DeepDream [26]

FIGURE 8: Prompt 3, interpreted by DeepDream [26]

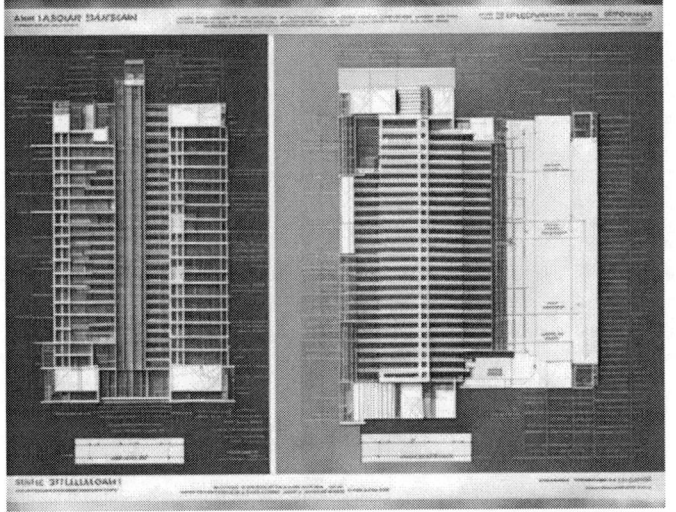

FIGURE 7: Prompt 2, interpreted by DeepDream [26]

FIGURE 9: Prompt 4, interpreted by DeepDream [26]

Copyright © 2023 by ASME

FIGURE 10: Prompt 5, interpreted by DeepDream [26]

DeepDream was able to conceptualize each item in the prompts but also struggled to flesh out meaningful designs. In Figure 6, an original sportscar was generated, with labels and measurements in the style of a blueprint. However, these labels and measurements are not usable. We can observe a similar result in Figure 7. In Figure 8, the watch test once again outputs a blueprint-inspired watch, rather than a blueprint of a watch. Similar issues were present in Figure 9, despite the lengthy prompt. The most interesting result was in Figure 10. The image is split with two views of the subject, as is often seen in design documents. This was the first hint of detail-design structure in the lengthier prompts. However, none of these images are usable for rigorous design. DeepDream scores a TRL of 1.

Now to look at our additional metrics of ease-of-use, customizability, and applications. DeepDream was incredibly easy to use both as a text-only generator and an image-aided generator. DeepDream was customizable through the prompt and by adding images to the prompt but did not stand out in this metric. Lastly, DeepDream also struggled in application-specifically with the watch prompt, Figure 8. Again, this indicates that the AI makes no attempt to infer the product's inner workings and simply applies a style to the subject of the image. In ease-of-use, customizability, and applications, DeepDream scores a 4, 2, and 1, respectively.

3.3 Stable Diffusion

Stable diffusion models are another subtype of neural networks and are distinguished by their process of diffusion and reconstruction. An image is converted into random noise step-by-step and then reconstructed using the noise. At each step in adding noise, a neural network learns how to reconstruct the original image from that noise. Eventually, the network is able to reconstruct the image from pure noise, sometimes using a text prompt [16], resulting in photorealistic images [19][23]. Some of the weaknesses of this method include unidirectional bias and

accumulated prediction errors [11]. Text-to-image diffusion has even been used in the generation of 3D models [21].

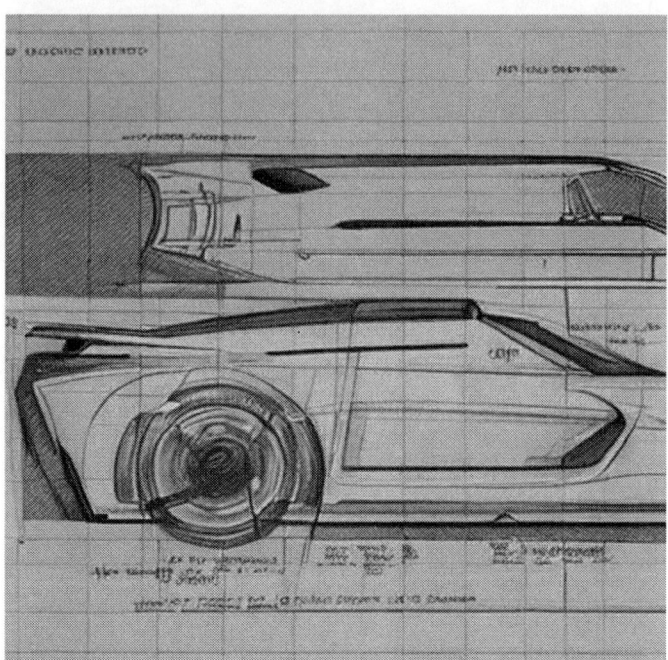

FIGURE 11: Prompt 1, interpreted by Stable Diffusion [24]

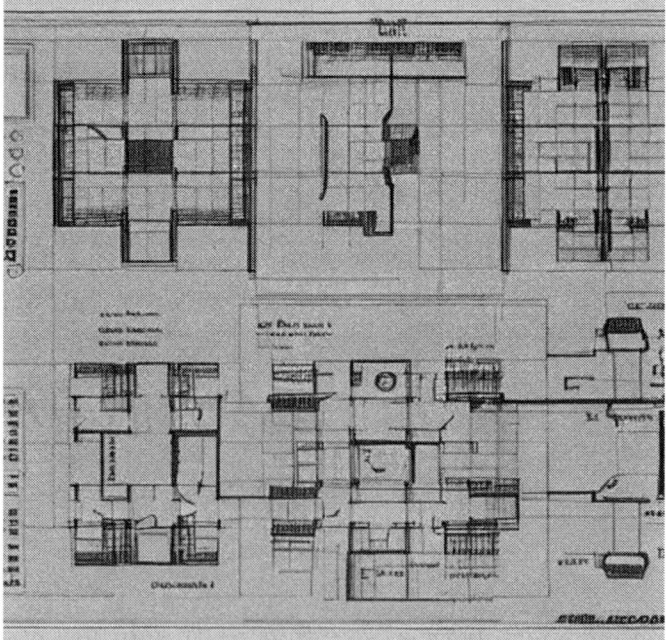

FIGURE 12: Prompt 2, interpreted by Stable Diffusion [24]

Copyright © 2023 by ASME

FIGURE 13: Prompt 3, interpreted by Stable Diffusion [24]

FIGURE 14: Prompt 4, interpreted by Stable Diffusion [24]

FIGURE 15: Prompt 5, interpreted by Stable Diffusion [24]

Stable diffusion seems to have been the most faithful to the prompts given, if still not producing usable diagrams. Figure 12 stands out for being the closest to a rigorous design and moving beyond concept art, while still not generating usable measurements. Figure 13 is the least faithful of all the tests performed. The resulting image may not be immediately identifiable as a watch and is less recognizable as being inspired by a blueprint. These varying levels of success suggest stable diffusion is capable of both concept art and could potentially lead to rigorous design, given carefully worded prompts. In Figure 14 and Figure 15, we gave the AI a much more carefully worded prompt, one specifically designed for an AI art generator and one designed for architecture students. Neither are anything resembling an architectural diagram. The most promising result is again in Figure 15, the homework prompt. Again, the resulting image is split between two views, and the style of the image is less geared towards photo-realism and is closer to a conceptual drawing. For this, stable diffusion earns a TRL of 2.

Stable diffusion is easy to use but not entirely intuitive. Regarding customizability, stable diffusion allows text prompts and negative text prompts, allowing users to specify what they don't want in the generated image. Similar to the other two AIs tested, stable diffusion struggled with conceptualizing a design of something that is not commonly designed in popular culture. However, the watch prompt did result in a watch that hinted at its inner workings and structure- for that, stable diffusion seems closer to creating detailed designs than other AIs tested. Stable diffusion scores an ease-of-use of 3, a customizability of 3, and an application score of 3.

3.4 Summary

Copyright © 2023 by ASME

The first of our methods discussed, GANs, scored a TRL of 2, an ease-of-use score of 2, a customizability score of 3, and an applications score of 1.

The representative of CNNs earned a TRL of 1, an ease-of-use of 4, a customizability of 2, and an applications score of 1.

Lastly, the stable diffusion example was given a TRL of 2, an ease-of-use score of 3, a customizability of 3, and an applications score of 3.

Overall, the most successful AI tool was stable diffusion- it adhered best to the prompts and made the best attempt at inferring the inner workings of the third prompt, the watch. One notable standout was the ease-of-use score for DeepDream, due to its use of text and image prompts in a way that allows the user to choose which prompt best suits their needs. This ease contrasts with Artbreeder, the GAN tool. Artbreeder requires both a text and image prompt, which may increase prompt adherence and final image quality. The tradeoff of Artbreeder's potential for better images is less flexibility in the prompt medium.

A summary of the results is displayed in Table 1 & 2.

Table 1: Summary of scores for each AI Tool

	Humans	GAN	CNN	Diffusion
TRL	3	2	1	2
Ease-of-Use	1	2	4	3
Customizability	4	3	2	3
Applications	5	1	1	3

Table 2: Pugh analysis of each method

	Humans	GAN	CNN	Diffusion
TRL	0	-1	-1	-1
Ease-of-Use	0	+1	+1	+1
Customizability	0	-1	0	0
Applications	0	-1	-1	-1
Total	0	-2	-1	-1

The Pugh analysis assumes zeroes for each criterion of the baseline method, human design. It then compares several alternatives against the baseline by assigning a -1, 0, or +1 to each criterion depending whether that criterion is better or worse in the alternative. We have found that each of the representative AIs and their methods have scored negatively, indicating that they are less optimal than the baseline method. This is to be expected- humans have been practicing rigorous design for centuries. The most detrimental criterion was the applications portion, where each method scored negatively. This indicates that the AIs are better at applying artistic styles to objects than inferring their designs. The GAN representative scored the worst with a -2. This does not necessarily mean GANs are the worst method of AI art generation but that upon initial review they are

the least valuable for the problem as defined. Further exploration of the AI tool's abilities will be needed to determine how they can best be applied to the design process.

3.5 Findings
None of these methods or AI tools seem to be advanced enough to formulate blueprints of rigorous design- instead, they are successful at conceptual design and stylizing. This could be very valuable during the early stages of ideation and exploration.

None of the tests yielded readable labels or measurements. The tests that most closely adhered to the prompt were the skyscraper tests, possibly due to the abundance of floorplans. However, the practicality and consistency of the floorplans were questionable. None of this is to say that more carefully worded prompts couldn't result in a better end product- the foundation of original design is clearly present.

4. CONCLUSION
AI art tools are quite advanced in their abilities to generate original, topical, and stylized concepts. However, their ability to generate rigorous detailed designs is not yet feasible. Each of the AIs tested attempted to either copy the style of a blueprint or apply the style of a blueprint to an object, rather than creating original designs.

In addition to the technical challenges and shortcomings in applying AI tools to detailed design problems, there are potential legal challenges. As determined by the case Naruto v. David Slater et al. non-human actors are not entitled to copyright protection [28]. This has already been applied to AI generated art [29]. This leads to questions about how inventorship or patent protections might work for solutions wholly or in part created by AI tools.

4.1 Future Work
It is clear that AI tools are not currently ready to be applied to rigorous, detailed design and be expected to generate applicable real-world designs. The most significant limiting factor is their inability to conceptualize inner workings. Many current, popular AI tools are designed specifically for concept art and other artistic endeavors- so it's reasonable to assume that an AI designed specifically for engineering design purposes could succeed. The potential impacts on technically intensive fields could be enormous when fully developed. Allowing AI to handle intensive design mathematics and other details would enable designers of all backgrounds to explore the realm of design. However, AI-aided design will never fully replace human design. Additionally, for safety reasons, humans should always check an AI tool's work before manufacturing any designs that could cause harm.

There are several more areas to be explored in the realm of this paper. As text AIs continue to improve and gain attention, utilizing them to generate more robust text prompts for image generation AIs could yield more interesting results that adhere

Copyright © 2023 by ASME

better to the prompts [22]. There are many more AIs and many more methods of AI art generation that could be explored. One particular area of further research would be the use of images in prompts. Seeing how much an image affects the result and whether the AI simply begins to copy a design could lead to many interesting results.

Another area for future exploration would be figuring out how AI tools fit into human design processes. For example, given their current level of technical ability, they could aid in traditional design methods during ideation, or potentially new design methods could be developed that specifically leverage them.

Lastly, a major improvement to this paper would be formalizing the assessments of each AI art generator. This could take the form of utilizing more detailed prompts, requesting multiple samples from each source, having each result rated by a variety of reviewers, and removing any ambiguity from the rating criterion. This paper serves only as a first investigation into the topic and hopes to open doors to more formal readiness assessments.

ACKNOWLEDGEMENTS

This work was a biproduct of work funded by a NASA Nebraska Space Grant and a National Strategic Research Institute (NSRI) Independent Research and Development (IRAD) award. as part of a larger effort exploring automation, computational cognition, and machine intelligence.

REFERENCES

[1] Abiodun, O. I., Jantan, A., Omolara, A. E., Dada, K. V., Mohamed, N. A. E., & Arshad, H. (2018). State-of-the-art in Artificial Neural Network Applications: A survey. *Heliyon*, *4*(11). https://doi.org/10.1016/j.heliyon.2018.e00938

[2] Aggarwal, A., Mittal, M., & Battineni, G. (2021). Generative Adversarial Network: An overview of theory and applications. *International Journal of Information Management Data Insights*, *1*(1), 100004. https://doi.org/10.1016/j.jjimei.2020.100004

[3] Artbreeder. (n.d.). Retrieved March 8, 2023, from https://www.artbreeder.com/

[4] Brownlee, J. (2019, July 19). *A gentle introduction to generative adversarial networks (GANs)*. MachineLearningMastery.com. Retrieved March 8, 2023, from https://machinelearningmastery.com/what-are-generative-adversarial-networks-gans/

[5] Castro Pena, M. L., Carballal, A., Rodríguez-Fernández, N., Santos, I., & Romero, J. (2021). Artificial Intelligence applied to conceptual design. A review of its use in architecture. *Automation in Construction*, *124*, 103550. https://doi.org/10.1016/j.autcon.2021.103550

[6] Creswell, A., White, T., Dumoulin, V., Arulkumaran, K., Sengupta, B., & Bharath, A. A. (2018). Generative Adversarial Networks: An overview. *IEEE Signal Processing Magazine*, *35*(1), 53–65. https://doi.org/10.1109/msp.2017.2765202

[7] Cross, N. (1999). Natural intelligence in Design. *Design Studies*, *20*(1), 25–39. https://doi.org/10.1016/s0142-694x(98)00026-x

[8] Demir, G., Çekmiş, A., Yeşilkaynak, V. B., & Unal, G. (2021). Detecting visual design principles in art and architecture through deep convolutional Neural Networks. *Automation in Construction*, *130*, 103826. https://doi.org/10.1016/j.autcon.2021.103826

[9] Dunbar, B. (2015, May 6). *Technology readiness level*. NASA. Retrieved March 8, 2023, from https://www.nasa.gov/directorates/heo/scan/engineering/technology/technology_readiness_level

[10] Goodfellow, I., Pouget-Abadie, J., Mirza, M., Xu, B., Warde-Farley, D., Ozair, S., Courville, A., & Bengio, Y. (2020). Generative Adversarial Networks. *Communications of the ACM*, *63*(11), 139–144. https://doi.org/10.1145/3422622

[11] Gu, S., Chen, D., Bao, J., Wen, F., Zhang, B., Chen, D., Yuan, L., & Guo, B. (2022). Vector quantized diffusion model for text-to-image synthesis. *2022 IEEE/CVF Conference on Computer Vision and Pattern Recognition (CVPR)*. https://doi.org/10.1109/cvpr52688.2022.01043

[12] Haggenmüller, S., Maron, R. C., Hekler, A., Utikal, J. S., Barata, C., Barnhill, R. L., Beltraminelli, H., Berking, C., Betz-Stablein, B., Blum, A., Braun, S. A., Carr, R., Combalia, M., Fernandez-Figueras, M.-T., Ferrara, G., Fraitag, S., French, L. E., Gellrich, F. F., Ghoreschi, K., … Brinker, T. J. (2021). Skin cancer classification via Convolutional Neural Networks: Systematic review of studies involving human experts. *European Journal of Cancer*, *156*, 202–216. https://doi.org/10.1016/j.ejca.2021.06.049

[13] Kim, B., Park, J., & Suh, J. (2020). Transparency and accountability in AI decision support: Explaining and visualizing Convolutional Neural Networks for text information. *Decision Support Systems*, *134*, 113302. https://doi.org/10.1016/j.dss.2020.113302

[14] Liimatainen, K., Huttunen, R., Latonen, L., & Ruusuvuori, P. (2021). Convolutional neural network-based artificial intelligence for classification of protein localization patterns. *Biomolecules*, *11*(2), 264. https://doi.org/10.3390/biom11020264

[15] Liu, L., Zhang, H., Ji, Y., & Jonathan Wu, Q. M. (2019). Toward AI fashion design: An attribute-gan model for clothing match. *Neurocomputing*, *341*, 156–167. https://doi.org/10.1016/j.neucom.2019.03.011

[16] Maerten, A.-S., & Soydaner, D. (n.d.). From paintbrush to pixel: A review of deep neural networks in AI-generated art. https://doi.org/10.48550/ARXIV.2302.10913

[17] McGregor, M. (2021, April 28). *What is a convolutional neural network? A beginner's tutorial for Machine Learning and deep learning*. freeCodeCamp.org. Retrieved March 8, 2023, from https://www.freecodecamp.org/news/convolutional-neural-network-tutorial-for-beginners/#:~:text=Convolutional%20neural%20networks%20are%20multi,you%20can%20identify%20images%20correctly

Copyright © 2023 by ASME

[18] MIT OpenCourseWare. (2023). *Structural design project: Basic structural design: Architecture.* MIT OpenCourseWare. Retrieved March 13, 2023, from https://ocw.mit.edu/courses/4-440-basic-structural-design-spring-2009/resources/mit4_440s09_project03/

[19] Nichol, A., Dhariwal, P., Ramesh, A., Shyam, P., Mishkin, P., Sutskever, I., & Chen, M. (2021). GLIDE: Towards Photorealistic Image Generation and Editing with Text-Guided Diffusion Models. *CoRR.*

[20] O'Shea, K., & Nash, R. (2015). An Introduction to Convolutional Neural Networks. *ArXiv.* https://doi.org/10.48550/ARXIV.1511.08458

[21] Poole, B., Jain, A., Barron, J. T., & Mildenhall, B. (2022). DreamFusion: Text-to-3D using 2D Diffusion. https://doi.org/10.48550/ARXIV.2209.14988

[22] Qiao, H., Liu, V., & Chilton, L. (2022). Initial images: Using image prompts to improve subject representation in multimodal AI generated art. *Creativity and Cognition.* https://doi.org/10.1145/3527927.3532792

[23] Saharia, C., Chan, W., Saxena, S., Li, L., Whang, J., Denton, E., Ghasemipour, S. K., Ayan, B. K., Mahdavi, S. S., Lopes, R. G., Salimans, T., Ho, J., Fleet, D. J., & Norouzi, M. (2022). Photorealistic Text-to-Image Diffusion Models with Deep Language Understanding. https://doi.org/10.48550/ARXIV.2205.11487

[24] *Stable diffusion online.* Stable Diffusion Online. (n.d.). Retrieved March 8, 2023, from https://stablediffusionweb.com/

[25] Tang, B., He, F., Liu, D., Fang, M., Wu, Z., & Xu, D. (2020). Ai-aided design of novel targeted covalent inhibitors against SARS-COV-2. https://doi.org/10.1101/2020.03.03.972133

[26] *Trending dreams: Deep dream generator.* Trending Dreams | Deep Dream Generator. (n.d.). Retrieved March 8, 2023, from https://deepdreamgenerator.com/

[27] Verganti, R., Vendraminelli, L., & Iansiti, M. (2020). Innovation and design in the age of Artificial Intelligence. *Journal of Product Innovation Management, 37*(3), 212–227. https://doi.org/10.1111/jpim.12523

[28] United States Court of Appeals for the Ninth Circuit. (2018, April 22). 16-15469 - Naruto v. David Slater, et al. [Government]. Administrative Office of the United States Courts. https://www.govinfo.gov/app/details/USCOURTS-ca9-16-15469

[29] S. Magazine and J. Recker, "U.S. Copyright Office Rules A.I. Art Can't Be Copyrighted," Smithsonian Magazine, Mar. 24, 2022. https://www.smithsonianmag.com/smart-news/us-copyright-office-rules-ai-art-cant-be-copyrighted-180979808/

Copyright © 2023 by ASME

Proceedings of the ASME 2023
International Design Engineering Technical Conferences and
Computers and Information in Engineering Conference
IDETC-CIE2023
August 20-23, 2023, Boston, Massachusetts

DETC2023-114999

MACHINE LEARNING-BASED MODEL BIAS CORRECTION BY FUSING CAE DATA WITH TEST DATA FOR VEHICLE CRASHWORTHINESS

Jice Zeng, Ying Zhao
Department of Industrial and Manufacturing Systems Engineering
University of Michigan-Dearborn, Dearborn, MI, 48128

Guosong Li, Zhenyan Gao, Yang Li Saeed Barbat
Vehicle Structure and Safety Research Department, Research & Advanced Engineering, Ford Motor Company, Dearborn, MI 48126, USA

Zhen Hu*
Department of Industrial and Manufacturing Systems Engineering
University of Michigan-Dearborn, Dearborn, MI, 48128

ABSTRACT

Physics-based simulation and analysis have emerged as promising techniques for optimizing the number of physical prototypes for vehicle crashworthiness evaluation in frontal impact with rigid barriers. Nonetheless, one of the hurdles for vehicle crashworthiness virtual certification is the potential differences between the computer simulation predictions and physical test results. In this regard, this study aims at improving the prediction capability of the Computer-Aided Engineering (CAE) model for crashworthiness performance evaluation at speeds beyond those defined by current regulations and public domain testing protocols. One way of achieving this is by integrating data from a number of physical crash tests with the CAE data using machine learning models. A novel approach is proposed in the displacement domain (deceleration vs. displacement) to enable data fusion to help recover missing physics associated with the CAE model. A nonlinear spring-mass model is used in this study to simulate rigid-barrier vehicle frontal impact. The deceleration response is transformed from a function of time to a function of displacement, and a Gaussian process regression (GPR) model is applied to capture the model bias of the nonlinear spring constant under a dynamic analysis scheme. The training data for the GPR model are split into multiple clusters by a Gaussian mixture model to capture bias patterns under different speed regimes. After clustering a GPR model is trained for each group of data. The optimal GPR model, trained by a specific cluster exhibiting the highest probability of new data belonging to it, is utilized for prediction. This selected GPR model is integrated with the original CAE model to predict vehicle deceleration under a new crash speed during a vehicle deceleration dynamic

analysis. The proposed approach is validated using physical vehicle crash tests and demonstrated improved accuracy of CAE model results and predictions.

1. INTRODUCTION

Vehicle design involves evaluations against various attributes loading conditions such as crashworthiness, durability, NVH, styling flexibility, and others. A successful vehicle design hinges on the ability to fulfill various regulatory and non-regulatory attribute requirements. [1]. Designing optimized structures that lead to a controlled deformation remains a primary consideration to achieve structural crashworthiness design to help mitigate serious injuries and fatalities. [2].

Over the past several decades, researchers from academia and major automobile manufacturers have devoted considerable efforts towards optimizing the structural design of vehicles with the aim of improving vehicle crashworthiness. This is typically formulated with a set of design objectives and constraints. These include maximizing the structural energy absorption capability and controlling the peak deceleration of the vehicle during the collision. To achieve these objectives, researchers have adopted various optimization techniques, such as topology optimization, parametric optimization, and sensitivity analysis, in order to identify effective design solutions. Liu et al. [3] developed a collaborative optimization framework taking advantage of Latin hypercube design and response surface to enhance vehicle crashworthiness during a frontal impact. Wang et al. [4] proposed a reliability-based optimization method to enhance vehicle body crashworthiness, in which the copula function derived by the Bayesian method is

---

*Corresponding author: 2250 HPEC, University of Michigan-Dearborn, Dearborn, MI 48128, USA, Tel:+1-313-583-6312, Email: zhennhu@umich.edu

Copyright © 2023 by ASME and Ford Motor Company

integrated to formulate complex parametric functions and probability distribution. Li et al. [5] also performed a six-sigma design optimization method to investigate the electric vehicle design with uncertainty, along with radial basis function and non-dominated sorting genetic algorithm II. Gu et al. [6] undertook a comparative investigation of various multi-objective optimization techniques in the context of frontal crash scenarios, with the aim of elucidating reliability and robustness concerns pertaining to vehicle crashworthiness design.

On the other hand, following the finalization of the design phase, the primary objective is to fulfill the Federal Motor Vehicle Safety Standards (FMVSS) by means of conducting prototype vehicle tests, thereby ensuring that novel vehicle designs align with these stipulations. The certification process for vehicle crashworthiness entails conducting numerous prototype vehicle tests. Recent advancements in high-fidelity computer simulation models, computational mechanics, and multi-body dynamics simulation have led to the emergence of certification by analysis (CBA) as a viable tool for reducing the required number of prototype vehicle tests. CBA involves supporting the certification of structural components through virtual physics-based simulations, such as CAE simulations [7]. Initially proposed in the aerospace sector [8, 9], the concept of CBA has since been extended to the certification of automotive structural components, offering an efficient way for vehicle manufacturers to develop new vehicles using a limited number of prototypes.

High-fidelity CAE simulations of vehicle crashworthiness are crucial in both the design stage and the production stage for improving crashworthiness and CBA. However, the utility of CAE models in vehicle crashworthiness can be constrained due to inaccuracies stemming from the exclusion of certain dynamics, numerical approximations, and assumptions made during the modeling process, as well as the inherently complex nature of vehicle crashes. Addressing the limitations of CAE models presents a continual research challenge, as discrepancies between simulated and real-world crash outcomes can affect structural design optimization during the early stages of vehicle development, as well as impede the adoption of computer-based analysis as a means of reducing the number of prototype vehicle tests in the post-design stage. Sub-optimal CAE predictions can give rise to additional prototypes and elongated development time.

In recent years, scholars have devised a variety of methods aimed at enhancing the precision of CAE simulations of crashworthiness, in order to more closely approximate the responses of real-world vehicle crash tests. Shi and Lin [10] developed an adaptive response surface and Gaussian process for model bias correction, first-order score function was also used to measure the sensitivities of variables to responses. Wang and Shi [11] introduced a Gaussian process regression model to capture a bias between model predictions and tested responses for vehicle crashworthiness design. Xi et al. [12] presented a copula-based bias correction method bypassing the dimensionality issue, in which copulas are used to formulate the statistical relations among model bias, model responses, and all

design parameters. Despite making progress, current methods for correlating CAE crash models can still face limitations. One of the main challenges is the requirement for sufficient test data to ensure effective model correlation. Furthermore, many of the current methods are limited to correcting biases in CAE surrogate models using high-fidelity CAE simulations, rather than actual tests.

The objective of this paper is to address the limitations of existing methods in achieving a correlation between vehicle crash tests and high-fidelity CAE simulations. The ultimate goal is to facilitate model-based structural design for the improvement of crashworthiness and enable a reduction in the number of required crash tests through CBA. The rigid-barrier vehicle frontal impact in this paper is modeled as a spring-mass system with a nonlinear spring constant. The deceleration response is then transformed from the time domain (a function of time) to the displacement domain (deceleration as a function of displacement). A novel approach is proposed in the displacement domain (deceleration vs. displacement) to enable data fusion of the crash test data and CAE data for the recovery of the unmodeled physics in the CAE model using machine learning models. In order to collect data for the training of machine learning models, the model bias of the nonlinear spring constant is analyzed, by comparing the displacement responses of physical tests with their counterparts from CAE simulations under different test conditions (e.g., speed, and vehicle configurations). Since the model bias may vary with speed and vehicle parameters, a Gaussian mixture model is employed to split the obtained training data of model bias into multiple clusters. A GPR model is then constructed for each cluster of the training data. In order to integrate the GPR models of different clusters with the CAE model under a dynamic analysis scheme, for any given new inputs (e.g., speed, vehicle parameters), the probabilities that the new inputs belong to the clusters are first estimated. Based on the estimated probabilities, the predictions of the GPR models of the model bias are ensembled together using a weighted sum to be the predicted bias of the CAE model prediction of those inputs. This process repeats recursively over iterations in the displacement domain to dynamically recover the missing physics of the CAE model, and thus improve the prediction accuracy of the CAE model for speeds beyond those defined by current regulations and public domain testing protocols. The proposed approach is validated using physical vehicle crash tests and demonstrated its effectiveness.

The remainder of this paper is organized as follows: Section 2 briefly introduces the background of CAE model bias correction for a vehicle crash. Section 3 presents the proposed approach in the displacement domain for the fusion of CAE data with test data. An actual vehicle crash test is used to demonstrate the effectiveness of the proposed approach in Section 4, followed by conclusions in Section 5.

2. BACKGROUND

There are several reasons why the CAE model may vary from a real-world crash test. One potential reason is that a CAE

Copyright © 2023 by ASME and Ford Motor Company

model is typically constructed based on existing knowledge of physical systems, which may involve subjective assumptions and simplifications. As a result, the reliability of a CAE model is heavily dependent on the engineer's level of understanding of the actual system. Another possible explanation is that the model parameters used in the current version of the CAE model may have worked well for previous tests but may fail to capture the information from new tests. This discrepancy between the CAE model and the actual test is commonly referred to as model bias or model uncertainty [13]. A substantial body of literature has emerged exploring the underlying factors that contribute to model uncertainty, as well as the mathematical methods used to characterize it [14-18]. Kennedy and O'Hagan [19, 20] proposed a mathematical model that is widely used to correlate CAE models with test data. The model is expressed as follows:

$$y_o(\mathbf{x}) = \rho g(\mathbf{x}, \boldsymbol{\theta}) + \delta(\mathbf{x}) + \varepsilon, \tag{1}$$

where $y_o(\mathbf{x})$ represents actual observations that are obtained for a specific set of inputs $\mathbf{x}$, while $y = g(\mathbf{x}, \boldsymbol{\theta})$ represents CAE model, which takes a set of inputs $\mathbf{x}$ and model parameters $\boldsymbol{\theta}$ as its inputs. The term ε represents measurement error, which is typically assumed to be Gaussian noise and independent from $\mathbf{x}$ and $\boldsymbol{\theta}$. Additionally, ρ is an unknown regression coefficient. The term $\delta(\mathbf{x})$ represents model bias, which can be caused by uncertainties in the model parameters or simplifications in the CAE model. It should be noted that the key difference between $\mathbf{x}$ and $\boldsymbol{\theta}$ is that $\mathbf{x}$ can represent controllable variables that execute different operation configurations, such as excitation or size and shape of a product, while $\boldsymbol{\theta}$ is a vector of model parameters that are random and uncontrollable, such as material properties, structural mass, and stiffness.

In this paper, a special form of Eq. (1) is employed by setting $\rho = 1$ and focusing on solely correcting the model bias. Therefore, the formulation in Eq. (1) is rewritten as

$$y_o(\mathbf{x}) = g(\mathbf{x}) + \delta(\mathbf{x}) + \varepsilon. \tag{2}$$

While the model correction formulation presented in Eq. (2) may appear straightforward, implementing it to correct a CAE model of vehicle crashworthiness is more complicated than suggested by the equation. In Section 3, we explore the proposed approach for correlating CAE and test data in the displacement domain.

3. PROPOSED APPROACH
3.1. Definitions of CAE model and variables

For the convenience, the model and variables in the proposed approach are defined. The deceleration during a vehicle crash is denoted as $a_d(t)$. In a discrete-time form, the deceleration at time instant t_i is represented as $a_{d,i}, i = 1, \cdots, N_t$, where N_t is the total number of time steps. In this study, initial crash speed v_0, vehicle front weight and

rear weight (w_f, w_r) are considered as all input variables $\mathbf{x} = [v_0, w_f, w_r] \in \mathbb{R}^3$. Based on the definitions, the CAE model for crashworthiness analysis is given by

$$\mathbf{a}_s = G_s(\mathbf{x}) = G_s(v_0, w_f, w_r), \tag{3}$$

where $G_s(\mathbf{x})$ represents the CAE model with inputs $\mathbf{x}$, and $\mathbf{a}_s = [a_{s,1}, a_{s,2}, \cdots, a_{s,N_t}] \in \mathbb{R}^{N_t}$ are the CAE model-derived deceleration.

Assume that N_e groups of CAE simulation data are generated and N_c groups of experimental deceleration data are collected ($N_c \ll N_e$), we then have

- **CAE data:** $\mathbf{a}_s^{(j)}, j = 1, \cdots, N_e$, where $\mathbf{a}_s^{(j)} = [a_{s,1}^{(j)}, a_{s,2}^{(j)}, \cdots, a_{s,N_t}^{(j)}] = G_s(v_0^{(j)}, w_f^{(j)}, w_r^{(j)})$ are deceleration data generated using the j-th training sample $\mathbf{x}^{(j)} = [v_0^{(j)}, w_f^{(j)}, w_r^{(j)}], \forall j = 1, \cdots, N_e$ of the input variables.

- **Test data:** $\mathbf{a}_d^{(k)}, k = 1, \cdots, N_c$, where $\mathbf{a}_d^{(k)} = [a_{d,1}^{(k)}, a_{d,2}^{(k)}, \cdots, a_{d,N_t}^{(k)}]$ are deceleration data measured at the k-th crash test under testing condition $\mathbf{x}^{(k)} = [v_0^{(k)}, w_f^{(k)}, w_r^{(k)}], \forall k = 1, \cdots, N_c$.

3.2. Data fusion and model bias correction

In this section, we elaborately introduce how the CAE data and test data can be fused in the displacement domain with the purposed of model bias correction and improving the accuracy of model prediction. The vehicle crash test is assumed to be modeled as a spring-mass system, as illustrated in Fig. 1. The equation of motion for vehicle crash in Fig. 1 can be expressed based on system dynamics as follows

$$M \frac{dv}{dt} = -K(s)s(t), \tag{4}$$

where M is the vehicle mass; v is the velocity of the vehicle at the time t; $K(s)$ is a non-linear spring constant that varies with displacement s, and $s(t)$ is a displacement function relative to t.

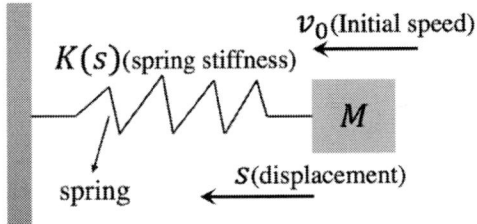

Fig. 1. Spring-mass model for vehicle

When it comes to solving differential equation (4), numerical integration is a commonly used approach to estimate dynamic properties such as displacement, deceleration, and velocity. This is achieved by setting the initial displacement and deceleration to zero, while specifying an initial velocity value

Copyright © 2023 by ASME and Ford Motor Company

v_0. Assuming Δt is a time moment at a certain velocity or deceleration, N_t is the total number of time steps, we have

$$i = 0, s_0 = 0, v_0 = v_0, a_0 = 0,$$

For $i = 0$ to N_t :

$$i = i + 1,$$
$$s_{i+1} = s_i + v_i \Delta t,$$
$$v_{i+1} = v_i + a_i \Delta t,$$
$$a_{i+1} = \underbrace{-\frac{1}{M} K(s_i) s_i}_{\text{Unknow or unable to model accurately}},$$

(5)

As evident from Eq. (5), the deceleration of the vehicle is influenced by a highly nonlinear parameter which poses a significant challenge in accurately determining its value through theoretical analysis. This challenge stems from the intricate nature of the vehicle crash, rendering it difficult to obtain the dynamic properties via solving differential equations directly. To address this challenge, we harness the powerful capabilities of both CAE simulation and machine learning models to capture the unknown function of the nonlinear spring constant with respect to displacement. By leveraging these techniques, we can more accurately model the behavior of the vehicle and ultimately enhance our understanding of its dynamic properties. a_{i+1} in Eq. (5) is estimated as

$$a_{i+1} = G_a(s_i, v_0, w_f, w_r) + \delta(s_i, v_0, w_f, w_r), \quad (6)$$

where $G_a(s_i, v_0, w_f, w_r)$ represents predicted deceleration by a CAE model given inputs s_i and v_0, w_f, w_r. $\delta(s_i, v_0, w_f, w_r)$ represents a machine learning model with the aim of accounting for unmodeled deceleration with respect to displacement, which is not captured by CAE model.

To properly train a machine learning model $\delta(s_i, v_0, w_f, w_r)$, the identification of the appropriate training data in the displacement domain is required. In order to do so, we must initially convert the time-domain data (accleration vs. time) into displacement domain data (displacement vs. accleration), which will then allow us to gather the necessary training data required for the successful training of $\delta(s_i, v_0, w_f, w_r)$. For the k-th crash test, where $k = 1, \cdots, N_c$, the test condition with input variables $\mathbf{x}^{(k)} = [v_0^{(k)}, w_f^{(k)}, w_r^{(k)}]$ results in deceleration rsponse $\mathbf{a}_d^{(k)} = [a_{d,1}^{(k)}, a_{d,2}^{(k)}, \cdots, a_{d,N_t}^{(k)}], \forall k = 1, \cdots, N_c$. Assume the initial velocity $v_{d,0}^{(k)} = v_0^{(k)}$ for the k-th crash test, the displacement domain data can be derived from structural dynamics

$$s_{d,0}^{(k)} = 0; v_{d,0}^{(k)} = v_0^{(k)};$$

For $i = 1$ to N_t :

$$s_{d,i}^{(k)} = s_{d,i-1}^{(k)} + v_{d,i-1}^{(k)} \Delta t,$$
$$v_{d,i}^{(k)} = v_{d,i-1}^{(k)} + a_{d,i-1}^{(k)} \Delta t,$$

(7)

where $s_{d,i}^{(k)}, i = 1, \cdots, N_t$ is the displacement of the k-th crash test at time step t_i. Through the Eq. (7), the time domain data can be readily converted into displacement domain data, as follows

$$\mathbf{s}_d^{(k)} = [s_{d,1}^{(k)}, \cdots, s_{d,N_t}^{(k)}]; \mathbf{v}_d^{(k)} = [v_{d,0}^{(k)}, v_{d,1}^{(k)}, \cdots, v_{d,N_t}^{(k)}];$$
$$\mathbf{a}_d^{(k)} = [a_{d,1}^{(k)}, \cdots, a_{d,N_t}^{(k)}]; \forall k = 1, \cdots, N_c.$$

(8)

Similarly, taking $\mathbf{x}^{(k)} = [v_0^{(k)}, w_f^{(k)}, w_r^{(k)}]$ as inputs in CAE model, the deceleration response $\mathbf{a}_s^{(k)} = G_s(\mathbf{x}^{(k)}) = G_s(v_0^{(k)}, w_f^{(k)}, w_r^{(k)})$ for the k-th crash test in CAE model is obtained, corresponding displacement domain data can be derived using the same process in Eq. (7)

$$\mathbf{s}_s^{(k)} = [s_{s,0}^{(k)}, s_{s,1}^{(k)}, \cdots, s_{s,N_t}^{(k)}]; \mathbf{v}_s^{(k)} = [v_{s,0}^{(k)}, v_{s,1}^{(k)}, \cdots, v_{s,N_t}^{(k)}];$$
$$\mathbf{a}_s^{(k)} = [a_{s,1}^{(k)}, \cdots, a_{s,N_t}^{(k)}]; \forall k = 1, \cdots, N_c.$$

(9)

It is important to emphasize that in the time domain, where data points are plotted as a function of time versus deceleration, there exists a one-to-one mapping relationship between the data points obtained through both the CAE simulation and actual testing methods, during the same time period. However, when the time domain is converted to the displacement domain, where data points are plotted as a function of displacement versus deceleration, this one-to-one mapping relationship may no longer hold, as observed in Fig. 2.

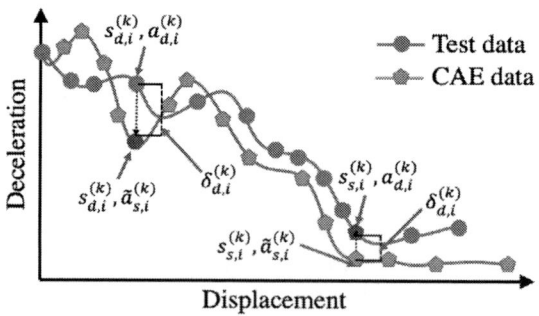

Fig. 2. Model bias analysis

The reason for this is that differential equations, which involve numerical integration with respect to time, can result in variations between the data points for the CAE simulation and actual test. Consequently, it is essential to consider this potential variability when performing data analysis in the displacement domain. Therefore, direct subtraction of the data in Eq. (9) from that in Eq. (8) does not allow us to obtain the necessary training outputs, that is, model bias data, for training the $\delta(s_i, v_0, w_f, w_r)$, such as

Copyright © 2023 by ASME and Ford Motor Company

$$\delta(s_{s,i}^{(k)}, v_0^{(k)}, w_f^{(k)}, w_r^{(k)}) \neq a_{d,i}^{(k)} - a_{s,i}^{(k)};$$
$$\forall k = 1, \cdots, N_c; i = 1, \cdots, N_t. \tag{10}$$

A bijective projection scheme has been devised to establish a precise correspondence in the displacement domain and effectively account for any inherent model inaccuracies between CAE predictions and actual tests. To achieve this, two distinct scenarios have been explored based on the maximum displacement in CAE and test data, as depicted in Fig. 2: (1) projecting from CAE predictions to actual tests; and (2) projecting from actual tests to CAE predictions. Both of these scenarios enable a bijective mapping to be established, facilitating a more accurate and reliable comparison between simulation results and physical measurements.

For the scenario (1) where the maximum displacement in CAE data is larger than that in test data, CAE data is used for interpolation. For the k-th crash test at time step t_i, the estimated deceleration by using CAE data is given by

$$\tilde{a}_{s,i}^{(k)} = f_{\text{int}}(s_{d,i}^{(k)}; \mathbf{s}_s^{(k)}, \mathbf{a}_s^{(k)}), \forall k = 1, \cdots, N_c; i = 1, \cdots, N_t, \tag{11}$$

where $f_{\text{int}}(s_{d,i}^{(k)}; \mathbf{s}_s^{(k)}, \mathbf{a}_s^{(k)})$ is a function to interpolate $s_{d,i}^{(k)}$ based on data $\mathbf{s}_s^{(k)}, \mathbf{a}_s^{(k)}$ given in Eq. (9). The model bias therefor is calculated through simple mathematical operation that subtracts $\tilde{a}_{s,i}^{(k)}$ from corresponding actural test $a_{d,i}^{(k)}$, that is, $\delta_{d,i}^{(k)} = a_{d,i}^{(k)} - \tilde{a}_{s,i}^{(k)}$.

Similarly, model bias for all $s_{d,i}^{(k)}$ from the N_c groups of test data are obtained. Finally, all inputs and outputs (model bias) for training in scenario (1) are summarized as

$$[\mathbf{x}_d; \boldsymbol{\delta}_d] = \begin{cases} \text{Inputs}: \mathbf{x}_{d,i}^{(k)} = [s_{d,i}^{(k)}, v_0^{(k)}, w_f^{(k)}, w_r^{(k)}] \\ \text{Bias}: \delta_{d,i}^{(k)}, \end{cases} \tag{12}$$

$$\forall k = 1, \cdots, N_c; i = 1, \cdots, N_t.$$

For the scenario (2) where the maximum displacement in test data is larger than that in CAE data, test data is used for interpolation. The deceleration corresponding to $x_{s,i}^{(k)}$ is estimated as

$$\tilde{a}_{d,i}^{(k)} = f_{\text{int}}(s_{s,i}^{(k)}; \mathbf{x}_d^{(k)}, \mathbf{a}_d^{(k)}), \tag{13}$$

where $f_{\text{int}}(s_{s,i}^{(k)}; \mathbf{x}_d^{(k)}, \mathbf{a}_d^{(k)})$ is a function to interpolate for CAE deceleration at $x_{s,i}^{(k)}$ using test data $\mathbf{x}_d^{(k)}, \mathbf{a}_d^{(k)}$, resulting in model bias corresponding to $x_{s,i}^{(k)}$, given by $\tilde{\delta}_{d,i}^{(i)} = \tilde{a}_{d,i}^{(k)} - a_{s,i}^{(k)}$. From calculated model bias, we have all inputs and bias outputs for $s_{s,i}^{(k)} \forall k = 1, \cdots, N_c; i = 1, \cdots, N_t$ as

$$[\mathbf{x}_s; \boldsymbol{\delta}_s] = \begin{cases} \text{Inputs}: \mathbf{x}_{s,i}^{(k)} = [s_{s,i}^{(k)}, v_0^{(k)}, w_f^{(k)}, w_r^{(k)}] \\ \text{Bias}: \tilde{\delta}_{d,i}^{(k)}, \end{cases} \tag{14}$$

$$\forall k = 1, \cdots, N_c; i = 1, \cdots, N_t.$$

Once the interpolation for two scenarios is complete, all inputs and outputs are combined for training machine learning model $\delta(s_i, v_0, w_f, w_r)$ as

$$\mathbf{x}_{\delta,t} = \{\mathbf{x}_d \cup \mathbf{x}_s\},$$
$$\boldsymbol{\delta}_t = \{\boldsymbol{\delta}_d \cup \boldsymbol{\delta}_s\}. \tag{15}$$

Upon gathering the necessary training data comprising all relevant inputs and model bias outputs, a machine learning model may be developed to accurately capture the intricate relationships between the bias of deceleration and relevant input variables such as displacement, vehicle weights, and initial crash speed. In this study, Gaussian Process Regression (GPR) is utilized to model the unknown relationship between deceleration bias and inputs in the displacement domain.

In GPR, the goal is to estimate the unknown parameters of the Gaussian process that best fit the training data. Given a set of training data points represented by inputs $\mathbf{x}_\delta = [s, v_0, w_f, w_r]$ and outputs $y_\delta = \delta(s, v_0, w_f, w_r)$ approximated by a Gaussian process $\hat{G}_\delta(\mathbf{x}_\delta)$, the mean and covariance are given by [21]

$$E\left[\hat{G}_\delta(\mathbf{x}_\delta)\right] = \mathbf{h}^T(\mathbf{x}_\delta)\boldsymbol{\beta},$$
$$Cov\left[\hat{G}_\delta(\mathbf{x}_\delta), \hat{G}_\delta(\mathbf{x}_\delta')\right] = \sigma^2 R(\mathbf{x}_\delta, \mathbf{x}_\delta'), \tag{16}$$

where the term of $\mathbf{h}^T(\mathbf{x}_\delta)$ denotes a vector of known regression functions, such as constant or linear functions, that are used to approximate the relationship between the input variables and the output variables. The column vector $\boldsymbol{\beta}$ represents the coefficients related to the polynomial regression of $\mathbf{h}^T(\mathbf{x}_\delta)$ and is used to estimate the unknown parameters of the model. The constant σ^2, commonly referred to as the error term, captures the variability of the response that is not explained by the predictor or the regression functions. The function $R(\mathbf{x}_\delta, \mathbf{x}_\delta')$ denotes the correlation between the responses at two points $\mathbf{x}_\delta$ and $\mathbf{x}_\delta'$, which is typically modeled using a kernel function that captures the similarity between the inpus.

$R(\mathbf{x}_\delta, \mathbf{x}_\delta')$ is widely defined as Gaussian kernel function as follows [22]

$$R(\mathbf{x}_\delta, \mathbf{x}_\delta') = \prod_{i=1}^{4} \exp\left[\rho_i(x_{\delta,i} - x_{\delta,i}')^2\right], \tag{17}$$

where the vector $\boldsymbol{\rho} = [\rho_1, \rho_2, \rho_3, \rho_4]^T$ represents a set of correlation parameters that quantify the change in correlation between $\hat{G}_\delta(\mathbf{x}_\delta)$ and $\hat{G}_\delta(\mathbf{x}_\delta')$ as the difference between the corresponding input variables $\mathbf{x}_\delta$ and $\mathbf{x}_\delta'$ increases. The hyperparameters $\boldsymbol{\beta}$, σ^2, and $\boldsymbol{\rho}$ determine the overall scale, noise level, and length scale of the Gaussian process, respectively. These hyperparameters, along with the mean function and the covariance function, fully specify the GPR model and can estimated by the maximum likelihood estimation (MLE) method [23], which involve finding the values that maximize the likelihood of observing the data given the model. Once the hyperparameters are estimated, the GPR model can be used to predict the response values at new input locations and

Copyright © 2023 by ASME and Ford Motor Company

quantify the uncertainty in the predictions using the covariance function.

A Gaussian process with mean $\mathbf{h}^T(\mathbf{x}_\delta)\boldsymbol{\beta}$ and covariance matrix $\sigma^2 R(\mathbf{x}_\delta, \mathbf{x}'_\delta)$ is used to describe the $\hat{G}_\delta(\mathbf{x}_\delta)$, the likelihood funtion of $\boldsymbol{\delta}_t$ is written as

$$
\begin{aligned}
L_{GP}(\boldsymbol{\beta}, \sigma, \boldsymbol{\rho}) = (2\pi\sigma^2)^{-\frac{N_\delta}{2}} |\mathbf{R}|^{-\frac{1}{2}} \\
\times \exp\left[-\frac{1}{2\sigma^2}(\boldsymbol{\delta}_t - \mathbf{H}\boldsymbol{\beta})^T \mathbf{R}^{-1}(\boldsymbol{\delta}_t - \mathbf{H}\boldsymbol{\beta}) \right],
\end{aligned}
\tag{18}
$$

where N_δ is the total number of training points. $\mathbf{H} = [\mathbf{h}^T(\mathbf{x}_{\delta,t}^{(1)}), \mathbf{h}^T(\mathbf{x}_{\delta,t}^{(2)}), \cdots, \mathbf{h}^T(\mathbf{x}_{\delta,t}^{(N_\delta)})]^T$, $\mathbf{R}$ is a $N_\delta \times N_\delta$ matrix whose elements are $R(\mathbf{x}_{\delta,t}^{(i)}, \mathbf{x}_{\delta,t}^{(j)}), \forall i, j = 1, \cdots, N_\delta$, in which $\mathbf{x}_{\delta,t}^{(i)}$ is the i-th sample in $\mathbf{x}_{\delta,t}$. The optimal hyperparameters $\hat{\boldsymbol{\beta}}$, $\hat{\sigma}^2$, and $\hat{\boldsymbol{\rho}}$ can be obtained through the Eq. (18). The prediction at new points $\mathbf{x}_\delta^{new} = [s^{new}, v_0^{new}, w_f^{new}, w_r^{new}]$ by trained GPR model is given by

$$
\hat{G}_\delta(\mathbf{x}_\delta^{new}) \sim N(\mu_\delta(\mathbf{x}_\delta^{new}), \sigma_\delta^2(\mathbf{x}_\delta^{new})),
\tag{19}
$$

where $N(\cdot, \cdot)$ is a standard Gaussian distribution, $\mu_\delta(\mathbf{x}_\delta^{new})$ and $\sigma_\delta^2(\mathbf{x}_\delta^{new})$ are derived as

$$
\mu_\delta(\mathbf{x}_\delta^{new}) = \mathbf{h}^T(\mathbf{x}_\delta^{new})\hat{\boldsymbol{\beta}} + \mathbf{r}^T(\mathbf{x}_\delta^{new})\mathbf{R}^{-1}(\boldsymbol{\delta}_t - \mathbf{H}\hat{\boldsymbol{\beta}}),
$$

$$
\sigma_\delta^2(\mathbf{x}_\delta^{new}) = \sigma^2 \left\{ \begin{aligned} & 1 - \mathbf{r}^T(\mathbf{x}_\delta^{new})\mathbf{R}^{-1}\mathbf{r}(\mathbf{x}_\delta^{new}) \\ & + [\mathbf{H}^T \mathbf{R}^{-1}\mathbf{r}(\mathbf{x}_\delta^{new}) - \boldsymbol{\delta}_t]^T (\mathbf{H}^T \mathbf{R}^{-1}\mathbf{H}) \\ & \times [\mathbf{H}^T \mathbf{R}^{-1}\mathbf{r}(\mathbf{x}_\delta^{new}) - \boldsymbol{\delta}_t] \end{aligned} \right\},
\tag{20}
$$

where the term of $\mathbf{r}^T(\mathbf{x}_\delta^{new})$ is a $N_\delta \times 1$ vector, of which the i-th element is $R(\mathbf{x}_\delta^{new}, \mathbf{x}_{\delta,t}^{(i)}), \forall i = 1, \cdots, N_\delta$. For more details on GPR method, please refer to the references [24, 25].

Before training a GPR model using available training data, all training data is clustered using Gaussian mixture model (GMM) in order to more explicitly capture data features. Consequently, multiple GPR models are obtained based on each cluster of training data. GMM is a powerful unsupervised learning technique that has found widespread use in a variety of fields [26-28]. GMM assumes that the data is generated by a mixture of several Gaussian distributions, each with a different mean and variance. The goal of the algorithm is to estimate these parameters by maximizing the likelihood of the data under the mixture model. In GMM, each sample is assigned to a group with a probability, allowing for a probabilistic representation of the data. This means that the GMM can not only group data points, but can also quantify the degree of uncertainty in the assignment of each data point to its corresponding group.

The fundamental formulation in GMM is given by

$$
p(\mathbf{x}) = \sum_{w=1}^{W} \pi_w N(\mathbf{x} \mid \mu_w, \Sigma_w),
\tag{21}
$$

where $\mathbf{x}$ is a variable vector, $p(\cdot)$ is the operator of marginal probability distribution. W is the total number of Gaussian components or clusters, π_w is the k-th mixing coefficient or weights, $\sum_{w=1}^{W} \pi_w = 1$, μ_w and Σ_w are the mean vector and covariance matrix of the k-th cluster. In this study, the variable vector is four-dimensional, including initial vehicle speed, fontal and rear weight and displacement, that is, $\mathbf{x}$ is equal to $\mathbf{x}_\delta = [s, v_0, w_f, w_r]$.

Once the training data is split into multiple groups by GMM, the multiple GPR models are trained using training data in individual group. The optimal GPR model, trained with a specific cluster exhibiting the highest probability of new data $\mathbf{x}_\delta^{new} = [s^{new}, v_0^{new}, w_f^{new}, w_r^{new}]$ belonging to, is used for prediction. This selected GPR model is finally used to approximate the nonlinear function of deceleration in Eq. (6).

$$
a = G_a(s, v_0, w_f, w_r) + \hat{G}_\delta(\mathbf{x}_\delta^{new}),
\tag{22}
$$

The deceleration can be calculated in an iterative way given any new vehicle speed v_0^{new} and new vehicle weights, w_f^{new}, w_r^{new}, based on structural dynamics as follows

$$
\begin{aligned}
& i = 0, s_i^{new} = 0, v_i^{new} = v_0^{new}, a_i^{new} = 0, \\
& \text{For } i = 1 \text{ to } N_t: \\
& \quad s_i^{new} = s_{i-1}^{new} + v_{i-1}^{new}\Delta t, \\
& \quad \mathbf{x}_\delta^{new} = [s_i^{new}, v_0^{new}, w_f^{new}, w_r^{new}], \\
& \quad a_i^{new} = G_a(s_i^{new}, v_0^{new}, w_f^{new}, w_r^{new}) + \hat{G}_\delta(\mathbf{x}_\delta^{new}), \\
& \quad v_i^{new} = v_{i-1}^{new} + a_i^{new}\Delta t.
\end{aligned}
\tag{23}
$$

It is noteworthy that the quantity of $\hat{G}_\delta(\mathbf{x}_\delta^{new})$ is a random variable, as specified in Eq. (22). To incorporate the associated uncertainty in the prediction, a Monte Carlo simulation method must be employed iteratively in accordance with the dynamic analysis scheme outlined in Eq. (23). Fig. 3 depicts the overall procedures of proposed model bias correction framework. Initially, the vehicle crash is represented using a spring-mass system model. However, due to the nonlinearity of the spring constant, this model fails to capture certain physical phenomena. To overcome this limitation, the deceleration responses with respect to time in both CAE model and actual crash tests are converted to displacement responses. To establish a one-to-one mapping between the two sets of responses and to calculate the model bias between the CAE model and actual tests, a bijective projection is developed. This projection ensures that the transformed responses can be accurately compared and that any discrepancies between the CAE model and actual tests are accounted for. When all training data is available, the GMM is applied to cluster all data with the purpose of capturing explicit data features, leading to multiple GPR models trained by individual cluster. The GPR model with the highest probability of assigning new inputs to a particular cluster is selected for prediction. To start the model prediction, the system's displacement and acceleration are initialized, along with a new speed. At each time step, the resulting displacement is used to

Copyright © 2023 by ASME and Ford Motor Company

compute a deceleration response without bias correction from CAE model. To account for any discrepancies between the CAE model and actual crash tests, the model bias is estimated using the selected GPR model.

The estimated model bias is subsequently added to the deceleration response, enabling model bias to be corrected. Throughout the prediction process, the displacement, velocity, and acceleration are updated iteratively, reflecting the evolving state of the system. By incorporating this methodology, the resulting model can more accurately predict the behavior of the system in real-world crashes, enhancing the design and testing of vehicle safety measures.

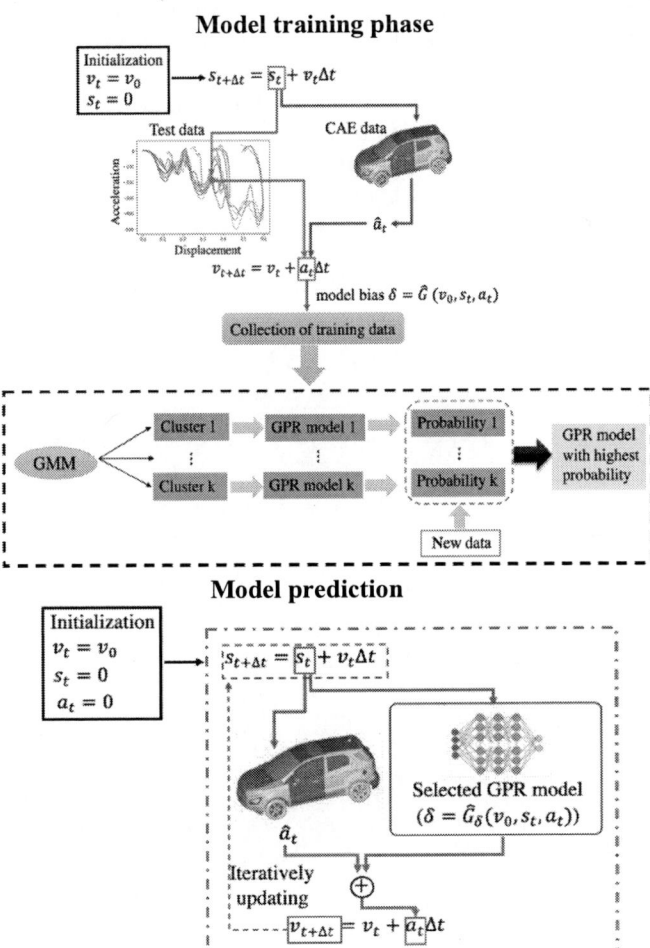

Fig. 3. The workflow of proposed model bias

3.3. Model validation for vehicle crashworthiness

Model validation is for evaluating the degree to which a model accurately represents a real-world system [29]. The selection of appropriate validation metrics is an essential aspect of this process, as they should be able to quantitatively measure the correlation or discrepancy between the model prediction and experimental data. For this study, where we are analyzing time series data, we adopt the ISO (International Organization for Standardization) metric to evaluate the level of agreement between computational and experimental data. The ISO metric is a robust approach that incorporates various metrics to assess the correlation between two time series data reliably [30]. The ISO metric includes several sub-metrics, such as the corridor score, phase score, magnitude score, and slope score. These sub-metrics contribute to an overall ISO metric rating, which can help us objectively evaluate the performance of our model and compare its predictions to experimental data. A comprehensive description of the detailed computation procedures for each score metric and the overall ISO rating can be found in references [30] and [31].

4. CASE STUDY

Fig. 4 shows the physical test and a CAE model for the full-frontal impact of a vehicle. In the current study, our focus is directed towards the prediction of crash pulse or deceleration, as various impact speeds generate distinctive pulses.

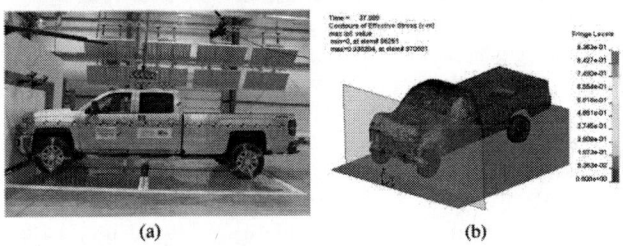

Fig. 4. Vehicle crashworthiness: (a) actual test; (b) CAE model of a vehicle

To implement the proposed approach, the initial step entails the gathering of CAE data obtained from a CAE simulation model. For training inputs, a total of 1009 sets of speed and weights are generated, based on which 1009 sets of pulses are collected from the CAE simulations. The output sequence for each pulse is the vehicle acceleration, with a data duration of 100 ms and a sampling frequency of 12.5 kHz. Only 11 sets of test data for different speeds, weights, and configurations are currently available, as illustrated in Table 1.

Table 1 Configuration of vehicle-to-rigid barrier full frontal test design

No.	Speed (v_0, mph)	Front Weight (w_f, lbs)	Rear Weight (w_r, lbs)	P/T Configuration	FWD/AWD
1	8.1	1994.7	1400	Type I	AWD
2	12.1	2163.7	1592.9	Type II	AWD
3	16.7	2120.1	1600.2	Type I	FWD
4	22	2237.4	1765.3	Type II	AWD
5	24.9	2100.4	1554.6	Type I	FWD
6	25.2	2155.4	1707.1	Type II	AWD
7	25.2	2058.9	1507.9	Type I	FWD
8	35.2	2179.3	1745.5	Type II	AWD
9	35.1	2107.7	1556.6	Type I	FWD
10	35.2	2166.9	1737.2	Type II	AWD
11	35.2	2168.9	1694.7	Type II	AWD

Copyright © 2023 by ASME and Ford Motor Company

4.1. Results of model bias correction

To ensure a one-to-one correlation between the 10 sets of test data available for training and their corresponding CAE data, a bijective projection method is adopted as discussed above. Notably, it is imperative to highlight that the remaining test data is utilized for validation purposes. Upon the availability of all training data, we apply GMM to categorize the training data sharing similar features. In this study, we explore three scenarios comprising single cluster, two clusters and three clusters. It is important to note that over-specifying the number of clusters can lead to a decrease in the training data associated with each cluster. This, in turn, can adversely affect the accuracy of the trained model. The study then proceeds to train multiple GPR models (i.e., single, two, and three GPR models correspond to three scenarios) based on the training data in each cluster. In the process of predicting new inputs via GPR model, the optimal GPR model for prediction deployment is the one trained with a specific cluster that demonstrates the highest probability of the new data belonging to that cluster.

In this paper, prediction results of test No. 3-5 and No. 10 are presented for the sake of explanation. Figs. 5-8 show the comparison results of model prediction obtained using proposed approach given different GPR models, as well as CAE prediction without bias correction. Since GPR model enables to probabilistically make prediction for crash test, with the 95% confidence interval indicated by the blue shaded area in the accompanying figures.

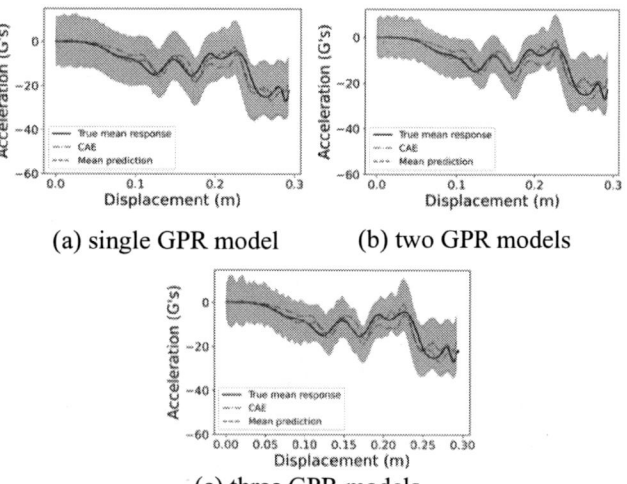

(a) single GPR model   (b) two GPR models

(c) three GPR models

Fig. 5. Model prediction for test #3

The results demonstrate a satisfactory level of agreement between the mean predictions generated by the proposed approach and the actual results from the crash tests for all scenarios, including the use of single, two, and three GPR models. These findings suggest that the bias or discrepancy between the CAE) model and the true system has been substantially reduced. Furthermore, while the proposed approach exhibits a consistent and robust predictive performance across a range of GPR models, the selection of the appropriate number of GPR models is critical for optimizing model prediction accuracy and reducing uncertainty. Specifically, it has been observed that the utilization of a single GPR model, where all available training data are used for model training without clustering, yields predicted pulse curves that are less congruent with actual test data when compared to predictions generated through the use of two or three GPR models. This phenomenon is evidenced in the case of test #4, where the use of adjacent predictions at a distance of 0.2 m is found to be most accurate when leveraging multiple GPR models. Moreover, it is visually apparent that the shaded region identified by more than one GPR model is considerably narrower, especially for tests #4, #5, and #10. This observation indicates a greater level of confidence in the model predictions as opposed to utilizing a single GPR model. The notable improvement in performance achieved by leveraging multiple GPR models can be attributed to the utilization of the GMM, which enables the grouping of training data with comparable features. By clustering data in this manner, the model can more accurately capture the intricacies of the data. In doing so, the selected GPR model is better tailored to predicting responses for new inputs that share similarities with the training data of this selected model.

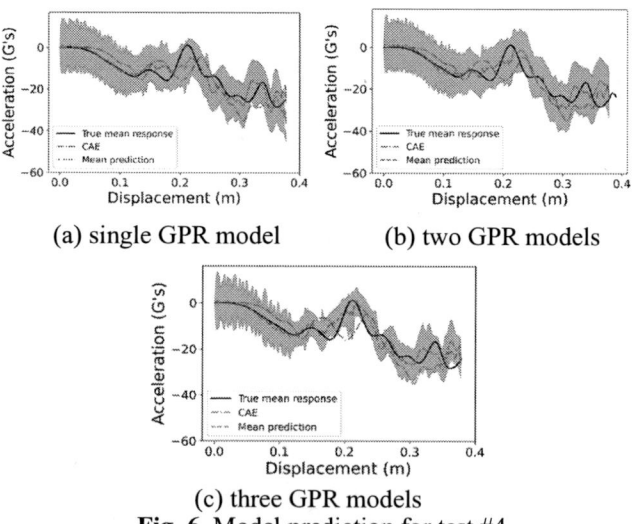

(a) single GPR model   (b) two GPR models

(c) three GPR models

Fig. 6. Model prediction for test #4

It is noteworthy that the prediction of tests #3 and #4 presents an additional challenge. As previously outlined in Table 1, the training data utilized for predicting these tests, which consists of 10 sets of test data excluding #3 or #4, does not contain similar information, such as vehicle speed, to these specific tests. In other words, the data pertaining to the crash test to be predicted falls outside the domain of the training data. Nevertheless, the proposed approach is capable of mitigating any potential model bias by leveraging the correlation between CAE data and test data to improve prediction accuracy.

Copyright © 2023 by ASME and Ford Motor Company

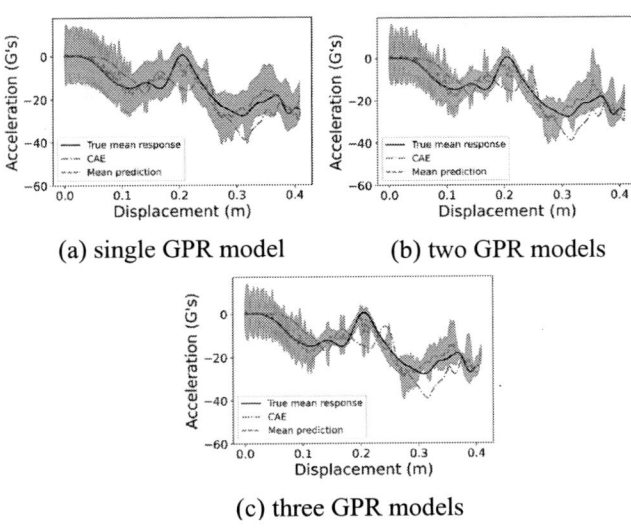

(a) single GPR model   (b) two GPR models

(c) three GPR models

Fig. 7. Model prediction for test #5

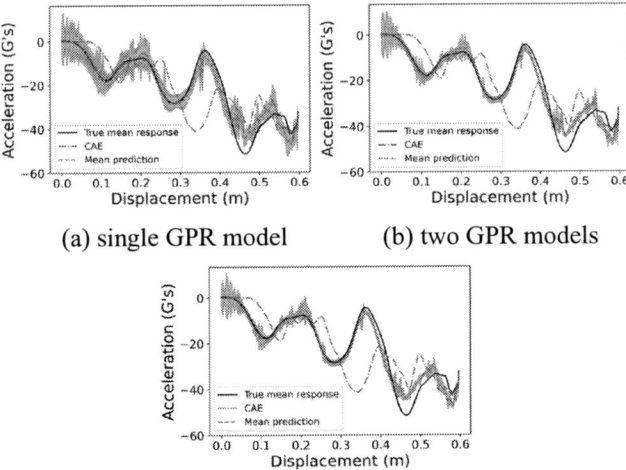

(a) single GPR model   (b) two GPR models

(c) three GPR models

Fig. 8. Model prediction for test #10

4.2. Quantitative accuracy measurement

Table 2 summarizes the overall ISO score of the proposed model bias approach for three periods, namely 0-20 ms, 0-40 ms, and 0-60 ms using different number of GPR models. The findings suggest that the ISO scores derived from the proposed approach are consistently higher than those obtained from the CAE model across all time periods in the three scenarios. Furthermore, the analysis revealed that employing three Gaussian Process Regression (GPR) models yielded the highest ISO scores, suggesting that the optimal number of clusters in our case is three. Notably, the ISO scores for test #10 were found to be the highest, indicating a higher level of accuracy compared to the other tests. This observation can be explained by the fact that the training data for predicting test #10 also includes relevant information pertaining to the test, such as

vehicle speed. Therefore, the superior prediction performance for test #10 is not surprising.

Table 2 Comparison of different number of GPR models by ISO metric

Test No.	Time period	ISO Score		
		(0, 20) ms	(0, 40) ms	(0, 60) ms
3	CAE	0.774	0.768	0.822
	Single GPR	0.863	0.828	0.822
	Two GPR	0.867	0.839	0.801
	Three GPR	**0.868**	**0.841**	**0.805**
4	CAE	0.651	0.658	0.681
	Single GPR	0.745	0.762	0.730
	Two GPR	0.776	0.772	0.742
	Three GPR	**0.801**	**0.808**	**0.750**
5	CAE	0.477	0.565	0.656
	Single GPR	0.722	0.786	0.743
	Two GPR	0.737	0.764	0.725
	Three GPR	**0.770**	**0.873**	**0.754**
10	CAE	0.647	0.617	0.541
	Single GPR	0.945	0.894	0.875
	Two GPR	0.964	0.919	0.889
	Three GPR	**0.965**	**0.922**	**0.904**

5. CONCLUSION

In this study, a machine learning-based model bias correction approach is proposed in displacement domain to enhance the predictive performance for vehicle crashworthiness by fusion CAE data with test data. In order to address the variation between CAE simulation data and actual test data, a bijective projection method is proposed, which enables the analysis of differences between the CAE model and the actual tests. Upon completion of the training data collection, the GMM is employed to automatically cluster the training data, with each cluster being utilized to train a GPR model. This process yields multiple GPR models based on the identified clusters. The selection of the most suitable GPR model is contingent on its training by a cluster that demonstrates the highest probability of new data inputs belonging to. This selected GPR model is subsequently integrated with the CAE model in a dynamic analysis framework, thereby facilitating the prediction of crash response under a new speed. To validate the effectiveness of the proposed approach, an industrial application of a vehicle crashworthiness case was conducted. The results indicate that the proposed approach outperformed the original CAE model by providing better crash pulse response predictions.

In the present study, the impact of damping on the prediction of vehicle crashworthiness has not been incorporated in the equation of motion, therefore, future research endeavors will focus on the inclusion of damping for vehicle crashworthiness analysis.

ACKNOWLEDGEMENT

Funding for this work was provided by Ford Motor Company through the University Research Program. The support is gratefully acknowledged.

REFERENCES

[1]   J. Happian-Smith, *An introduction to modern vehicle design*. Elsevier, 2001.

Copyright © 2023 by ASME and Ford Motor Company

[2] J. Fang, G. Sun, N. Qiu, N. H. Kim, and Q. Li, "On design optimization for structural crashworthiness and its state of the art," *Structural and Multidisciplinary Optimization,* vol. 55, no. 3, pp. 1091-1119, 2017.

[3] X. Liu, R. Liang, Y. Hu, X. Tang, C. Bastien, and R. Zhang, "Collaborative optimization of vehicle crashworthiness under frontal impacts based on displacement oriented structure," *International journal of automotive technology,* vol. 22, pp. 1319-1335, 2021.

[4] Q. Wang, Z. Huang, and J. Dong, "Reliability-based design optimization for vehicle body crashworthiness based on copula functions," *Engineering Optimization,* vol. 52, no. 8, pp. 1362-1381, 2020.

[5] Z. Li, L. Duan, A. Cheng, Z. Yao, T. Chen, and W. Yao, "Lightweight and crashworthiness design of an electric vehicle using a six-sigma robust design optimization method," *Engineering Optimization,* vol. 51, no. 8, pp. 1393-1411, 2019.

[6] X. Gu, G. Sun, G. Li, L. Mao, and Q. Li, "A comparative study on multiobjective reliable and robust optimization for crashworthiness design of vehicle structure," *Structural and Multidisciplinary Optimization,* vol. 48, no. 3, pp. 669-684, 2013.

[7] M. Guida, A. Manzoni, A. Zuppardi, F. Caputo, F. Marulo, and A. De Luca, "Development of a multibody system for crashworthiness certification of aircraft seat," *Multibody System Dynamics,* vol. 44, no. 2, pp. 191-221, 2018.

[8] F. Caputo, G. Lamanna, D. Perfetto, A. Chiariello, F. Di Caprio, and L. Di Palma, "Experimental and numerical crashworthiness study of a full-scale composite fuselage section," *AIAAJ,* vol. 59, no. 2, pp. 700-718, 2021.

[9] G. Olivares, J. F. Acosta, and V. Yadav, "Certification by Analysis I and II," in *The Joint Advanced Materials and Structures Meeting,* 2010.

[10] L. Shi and S.-P. Lin, "A new RBDO method using adaptive response surface and first-order score function for crashworthiness design," *Reliability Engineering & System Safety,* vol. 156, pp. 125-133, 2016.

[11] X. Wang and L. Shi, "A new metamodel method using Gaussian process based bias function for vehicle crashworthiness design," *International Journal of Crashworthiness,* vol. 19, no. 3, pp. 311-321, 2014.

[12] Z. Xi, P. Hao, Y. Fu, and R.-J. Yang, "A copula-based approach for model bias characterization," *SAE Int. J. Passeng. Cars-Mech. Syst,* vol. 7, no. 2, pp. 781-786, 2014.

[13] C. Jiang, Z. Hu, Y. Liu, Z. P. Mourelatos, D. Gorsich, and P. Jayakumar, "A sequential calibration and validation framework for model uncertainty quantification and reduction," *Computer Methods in Applied Mechanics and Engineering ,* vol. 368, p. 113172, 2020/08/15/ 2020.

[14] P. D. Arendt, D. W. Apley, and W. Chen, "Quantification of model uncertainty: Calibration, model discrepancy, and identifiability," 2012.

[15] J. Zeng and Y. H. Kim, "Identification of Structural Stiffness and Mass using Bayesian Model Updating Approach with Known Added Mass: Numerical Investigation," *Journal of Structural Stability and Dynamics,* vol. 20, no. 11, 2020.

[16] J. Zeng, M. D. Todd, and Z. Hu, "Probabilistic damage detection using a new likelihood-free Bayesian inference method," *Journal of Civil Structural Health Monitoring,* pp. 1-23, 2022.

[17] A. Thelen *et al.*, "A comprehensive review of digital twin—part 2: roles of uncertainty quantification and optimization, a battery digital twin, and perspectives," *Structural and multidisciplinary optimization,* vol. 66, no. 1, p. 1, 2023.

[18] J. Zeng and Y. H. Kim, "Probabilistic Damage Detection and Identification of Coupled Structural Parameters using Bayesian Model Updating with Added Mass," *Journal of Sound and Vibration,* p. 117275, 2022.

[19] M. C. Kennedy and A. O'Hagan, "Bayesian calibration of computer models," *Journal of the Royal Statistical Society: Series B (Statistical Methodology),* vol. 63, no. 3, pp. 425-464, 2001.

[20] D. Higdon, M. Kennedy, J. C. Cavendish, J. A. Cafeo, and R. D. Ryne, "Combining field data and computer simulations for calibration and prediction," *SIAM Journal on Scientific Computing,* vol. 26, no. 2, pp. 448-466, 2004.

[21] E. Schulz, M. Speekenbrink, and A. Krause, "A tutorial on Gaussian process regression: Modelling, exploring, and exploiting functions," *Journal of Mathematical Psychology,* vol. 85, pp. 1-16, 2018.

[22] J. Quinonero-Candela and C. E. Rasmussen, "A unifying view of sparse approximate Gaussian process regression," *The Journal of Machine Learning Research,* vol. 6, pp. 1939-1959, 2005.

[23] J.-S. Park and J. Jeon, "Estimation of input parameters in complex simulation using a Gaussian process metamodel," *Probabilistic engineering mechanics ,* vol. 17, no. 3, pp. 219-225, 2002.

[24] A. Sobester, A. Forrester, and A. Keane, *Engineering design via surrogate modelling: a practical guide.* John Wiley & Sons, 2008.

[25] C. E. Rasmussen and H. Nickisch, "Gaussian processes for machine learning (GPML) toolbox," *The Journal of Machine Learning Research,* vol. 11, pp. 3011-3015, 2010.

[26] G. J. McLachlan and S. Rathnayake, "On the number of components in a Gaussian mixture model," *Wiley Interdisciplinary Reviews: Data Mining and Knowledge Discovery,* vol. 4, no. 5, pp. 341-355, 2014.

[27] C. Fraley, A. E. Raftery, T. B. Murphy, and L. Scrucca, "mclust version 4 for R: normal mixture modeling for

Copyright © 2023 by ASME and Ford Motor Company

model-based clustering, classification, and density estimation," Technical report, 2012.

[28] J. Zeng and Z. Hu, "Automated operational modal analysis using variational Gaussian mixture model," *Engineering Structures,* vol. 273, p. 115139, 2022.

[29] L. E. Schwer, "An overview of the PTC 60/V&V 10: guide for verification and validation in computational solid mechanics," *Engineering with Computers,* vol. 23, no. 4, pp. 245-252, 2007.

[30] S. Barbat, Y. Fu, Z. Zhan, R.-J. Yang, and C. Gehre, "Objective rating metric for dynamic systems," *Enhanced Safety of Vehicles, Seoul, Republic of Korea,* vol. 2, no. 3, 2013.

[31] I. O. f. Standardization, "Road Vehicles—Objective Rating Metric for Non-Ambiguous Signals," ed: ISO/TS 18571: 2014, 2014.

Copyright © 2023 by ASME and Ford Motor Company

Proceedings of the ASME 2023
International Design Engineering Technical Conferences and
Computers and Information in Engineering Conference
IDETC-CIE2023
August 20-23, 2023, Boston, Massachusetts

DETC2023-116622

ACCOUNTING FOR MODEL UNCERTAINTY IN MACHINE LEARNING ASSISTED MECHANICAL DESIGN

Xiaoping Du
Department of Mechanical and Energy Engineering
Indiana University–Purdue University Indianapolis
Indianapolis, IN

ABSTRACT

Machine learning is becoming increasingly prevalent in mechanical design as it allows for surrogate models to replace expensive computational models. However, the accuracy of these models is of particular concern when safety-critical products are involved. In order to address this issue, we examine how to estimate model error by taking into account epistemic uncertainty in surrogate models when the design is also subject to randomness (aleatory uncertainty) in data. The paper clarifies important questions about modeling coupled epistemic and aleatory uncertainty when using surrogate models built from noise-free training points without aleatory uncertainty. Specifically, the study focuses on quantifying the effects of uncertainty in mechanical design by developing a most-probable-point based method. This method can be especially applicable for mechanical component design, where failure prevention is a critical concern, and the probability of failure is low. The proposed method is demonstrated using a shaft design as an example. The results show that the method can effectively estimate the model error and quantify the uncertainty in the design process. This approach can help designers to make more informed decisions by providing them with a better understanding of the limitations of surrogate models. By doing so, designers can ensure that their designs are safe and meet the required specifications.

Keywords: machine learning, design, regression, Gaussian process

1. INTRODUCTION

In mechanical design, the prevention of failures is a key objective, making it crucial to predict when and how failures might occur, as well as the associated costs. Engineering analysis plays a vital role in this process by predicting the performance of a given design using computational or simulation models derived from domain physics. For example, to estimate stresses acting on a component and associated deformation, engineers often rely on computational models, such as those of finite element analysis, computational fluid dynamics, or dynamics. They then compare stress to material strengths. Iterative adjustments to the design variables are then made through optimization or other design improvement methods until a satisfactory design is reached.

Running physics-based computational models can be time-consuming, sometimes taking minutes or hours to complete. This issue can be addressed by traditional surrogate modeling that relies on statistical regression methods [1, 2]. But these approaches are only practical for low-dimensional problems with small amounts of data. With the emergence of modern machine learning techniques, it is now possible to manage large-dimensional problems and big data. Common machine learning methods, such as neural networks (NN) [3], support vector machines (SVM) [4], and Gaussian processes (GP) [5] are now used to create surrogate models for design. However, all these models may produce prediction errors that can significantly impact design decisions, and these errors are typically unknown unless the original computational models are re-run, which defeats the purpose of using surrogate models. It is now well-recognized in the design community that the prediction error can be estimated by quantifying the model's (epistemic) uncertainty [6].

Research on machine learning model uncertainty has been conducted in parallel in both areas of machine learning and engineering design. In machine learning area, many machine learning methods can quantify model uncertainty for a model prediction. For instance, an machine learning model from GP provides a Gaussian distribution of a prediction whose mean

Copyright © 2023 by ASME

serves as the best estimate of the prediction while the standard deviation indicates the uncertainty [7]. The most recent NN method can also produce the mean and standard deviation of a prediction [8, 9]. There are also many research studies dedicated to numerical methods for the quantification of model uncertainty [10]. In engineering design area, significant research on the mixture of both types of uncertainty started in 1990s and early 2000s, which remains an active research focus nowadays. Examples of recent research include active learning using model uncertainty [11], multi-fidelity modeling for analysis and design [12], reliability analysis and design under modeling uncertainty [13], robust design accounting for model uncertainty [14, 15].

Despite the noteworthy progress in the field, it remains challenging to effectively account for machine learning model uncertainty when it is combined with other types of uncertainty in the design process. In this preliminary study, we address this challenge by discussing various methods for quantifying the effects of coupled uncertainty in the prediction and design of mechanical components. We also clarify important questions about coupled uncertainty and propose a new numerical method for uncertainty quantification with coupled uncertainty. By accounting for uncertainty, we can improve the robustness and reliability of our designs, particularly in safety-critical applications where failure is not an option.

This paper is organized as follows: Section 2 provides the background of the study. Section 3 discusses the modeling of coupled uncertainty and examines the distribution of model prediction for use in design. Section 4 presents a new uncertainty quantification method, while Section 5 shows the results of the example. Finally, in Section 6, we draw conclusions based on our findings and suggest future research directions.

2. BACHGROUND
In this section, we present several fundamental concepts that are used throughout this study.

- *Computational models*: A computational model is given by $Y = g(X)$. X is a vector of input variables including design variables, and Y is a response, such as strength, fatigue life, efficiency, and cost. The primary use of the model is to predict Y given X. The model may be an explicit or implicit function. The latter is usually a black box, which may run in minutes or hours [16].
- *Surrogate models from machine learning*: A surrogate model is built by machine learning using training points generated from the computational model $Y = g(X)$. A surrogate model is denoted $Y = \tilde{g}(X)$. It may take seconds or less to run, therefore much faster than $g(X)$. Building and using machine learning models is the major application of machine learning in design.
- *Uncertainty*: Uncertainty is the lack of sureness. There are two types of uncertainty.
- *Aleatory uncertainty*: It is the natural variation in input X that impacts output Y. It is the inherent randomness and is irreducible, for instance, the randomness in material properties, manufacturing imprecision, and user conditions.

- *Epistemic uncertainty*: It results from the lack of knowledge. It can be reduced by obtaining additional information. The uncertainty in the prediction of machine learning model $\tilde{g}(X)$ is epistemic due to the lack of training data and choice of models and their parameters. This uncertainty is also called model uncertainty and prediction uncertainty.
- *Coupled uncertainty*: There exists aleatory uncertainty in the model input X. As a result, the model uncertainty and aleatory uncertainty are coupled since the former is a function of the latter through $Y = \tilde{g}(X)$. This coupling results in what is referred to as coupled uncertainty or nested epistemic and aleatory uncertainties in design. This study aims to address the challenges associated with coupled uncertainty by proposing various methods for quantifying its effects in the design process.
- Prediction error: A prediction from a machine learning model $Y = \tilde{g}(X)$ may always have an error. The error may come from lack of training data, the coverage of training data, the model selected, model parameters, and the convergence of the optimization used in training. The error is unknown. But it can be estimated by uncertainty quantification (UQ), which quantifies the model (epistemic) uncertainty.

3. MODELING OF COUPLED UNCERTAINTY
In this section, we discuss the modeling of coupled uncertainty. We also answer several important questions about the modeling of coupled uncertainty.

3.1 What is the uniqueness of surrogate modeling for computational models?
Physics-based machine learning is used to construct surrogate models by using training data from computational models. It differs from general machine learning, which focuses on developing methods for training models on noisy data with aleatory uncertainty and/or epistemic uncertainty. In the training process of a surrogate model $Y = \tilde{g}(X)$ for design, the input variable training data X is generated by the model builder rather than measured, and the output training data (labels) Y is computed from the deterministic model $Y = g(X)$. Therefore, the training data is deterministic or noise-free, and there is no aleatory uncertainty. However, during prediction in design, X becomes random with aleatory uncertainty, and the model uncertainty varies depending on the location of X. As a result, both types of uncertainty are coupled, making it challenging to practically address this design problem.

Take the Gaussian process (GP) regression as an example. The model predictions at new design (test) points $X_T = (X_1, X_2, \ldots, X_m)^{\top}$ also follow a GP, given by

$$(Y_i)_{1,m}^{\top} = \left(\tilde{g}(X_i)\right)_{1,m}^{\top} = \left(\mu_{Y_i}(X_i) + \varepsilon_{Y_i}(X_T)\right)_{1,m}^{\top} \quad (1)$$

where $\mu_{Y_i}(X_i)$ is interpreted as the expectation (mean) of the prediction, and $\varepsilon_{Y_i}(X_T)$ is the uncertainty term, which follows a Gaussian process $\mathcal{GP}_m(\mathbf{0}; \Sigma)$ with the mean of $\mathbf{0}$ and

Copyright © 2023 by ASME

covariance $\mathbf{\Sigma}$. The covariance matrix provides the estimate of the prediction error in a probabilistic sense. When the standard deviation is low, the prediction error is also expected to be low, and conversely, when the standard deviation is high, the prediction error is more likely to be high. This uncertainty model is most effective when $\mathbf{X}$ is deterministic and has no uncertainty.

In real-world applications, the model input $\mathbf{X}$ is often subject to randomness due to factors such as random loading, variations in material properties, and user conditions. Consequently, the model uncertainty $\varepsilon_Y(\mathbf{X})$ becomes a function of $\mathbf{X}$ with aleatory uncertainty, leading to coupled uncertainty.

3.2 How do we model the error of prediction with coupled uncertainty?

We now have the following problem for the prediction and design task.

Given input
- Joint distributions of $\mathbf{X}$
- Surrogate model $Y = \tilde{g}(\mathbf{X})$
- Model uncertainty $\varepsilon_Y(\mathbf{X})$

Find output
- $Y = \tilde{g}(\mathbf{X}) = \mu_Y(\mathbf{X}) + \varepsilon_Y(\mathbf{X})$
- Distribution of Y

We now take the GP as an example. For a single test point $\mathbf{X}$, which represents a new design, our task is to predict the response Y and estimate the prediction error. We can then incorporate design improvements or optimization and reduce the effect of uncertainty. The model error follows a Gaussian distribution; namely, $\varepsilon(\mathbf{X}) \sim N(\mu_\varepsilon(\mathbf{X}), \sigma_\varepsilon(\mathbf{X}))$, where μ_ε and $\sigma_\varepsilon = \sigma_Y$ are the mean and standard deviation of the error, respectively. For the GP model, $\mu_\varepsilon(\mathbf{X}) = 0$. For a general surrogate model, $\mu_\varepsilon(\mathbf{X})$ may not be zero.

The prediction Y is therefore a random variable. Its condition distribution is Gaussian, $Y|\mathbf{X} \sim N(\mu_Y(\mathbf{X}), \sigma_Y(\mathbf{X}))$. This means that Y has a family of conditional distributions depending on the realization of $\mathbf{X}$. Y is a random variable and its distributions parameters are also random variables. This is usually called the second order probability or uncertainty. This kind of uncertainty has been studied in structural reliability, control systems, signal processing, and risk assessment [17, 18].

In many studies, Y (not its conditional distributions) is treated as a random variable with a family of distributions. While this approach is straightforward, it results in multiple probability density functions (PDFs), cumulative distribution functions (CDFs), and probabilities, making it challenging to interpret predictions and results and make design decisions.

We hold the perspective that Y is a random variable with a single distribution, despite having multiple conditional distributions. This distribution is determined by the Gaussian conditional distributions, the functions of the mean and standard deviation in relation to $\mathbf{X}$, and the joint distribution of $\mathbf{X}$. The perspective of single distribution will make the prediction and design much easier. It can also save computational time. Further details are discussed in Section 4.

3.3 Do we need a double-loop method to quantify the effect of coupled uncertainty?

To find the distribution of Y, a double-loop procedure may be used. For instance, if we employ Monte Carlo simulation (MCS), we can generate many samples of $\mathbf{X}$ using its joint distribution; then for each sample of $\mathbf{X}$, we generate many samples of Y using the conditional Gaussian distribution $Y|\mathbf{X} \sim N(\mu_Y(\mathbf{X}), \sigma_Y(\mathbf{X}))$. How many samples are needed depends on the specific UQ task. If it is for estimate of the first two statistical moments of the response, dozens or hundreds of samples are sufficient. If it is for rare probability estimate such as the probability of failure, millions of samples or more may be required.

Since there is a single distribution of Y, the double-loop procedure is not necessary. We can perform a single-loop analysis. For instance, for MCS, we can generate samples of $\mathbf{X}$ and for each sample of $\mathbf{X}$ we generate one sample of Y. Since MCS is usually expensive, especially for design, we need analytical or numerical UQ methods as discussed in Section 5.

4. HOW TO FIND THE DISTRUBUTION OF THE RESPONSE AND USE IT IN DESIGN?

We now discuss how to obtain the distribution of Y and how we use it in design.

4.1 How do we predict the effect of coupled uncertainty?

Let the joint PDF of $\mathbf{X}$ be $p_X(x)$ and the PDF of Y be $p_Y(y)$. The joint PDF of $\mathbf{X}$ and Y is given by

$$p_{Y,X}(y, x) = p_{Y|X}(y|x)p_X(x) \tag{2}$$

The CDF of Y is then given by

$$P_Y(y) = \int_{-\infty}^{y} \int_{-\infty}^{+\infty} p_{Y,X}(y, x)dxdy$$
$$= \int_{-\infty}^{y} \int_{-\infty}^{+\infty} p_{Y|X}(y|x)p_X(x)dxdy \tag{3}$$

Eq. (3) indicates that there is a single PDF of Y.

Since $Y|\mathbf{X} \sim N(\mu_Y(\mathbf{X}), \sigma_Y(\mathbf{X}))$, we have

$$P_Y(y) = \int_{-\infty}^{+\infty} \Phi\left(\frac{y - \mu_Y(\mathbf{X})}{\sigma_Y(\mathbf{X})}\right) p_X(x)dx \tag{4}$$

where $\Phi(\cdot)$ is the CDF of a standard Gaussian variable. Eq. (4) indicates that the PDF of Y is the expectation of the $\Phi\left(\frac{y - \mu_Y(\mathbf{X})}{\sigma_Y(\mathbf{X})}\right)$, which is the conditional CDF of Y, with respect to input variables $\mathbf{X}$.

Once the PDF or CDF of Y are available, we can find useful information about Y. For instance, we can find its mean for the average product performance and standard deviation for the variation in the performance; if a failure occurs when $Y < 0$, the CDF at 0 provides the prediction of the probability of failure.

Copyright © 2023 by ASME

$$p_f = \Pr(Y < 0) = P_Y(0) \tag{5}$$

4.2 How do we account for coupled uncertainty in design?

We classify design models into two types. Design model I is for the typical mechanical design where a design variable d is solved from a model $D = g(X)$ where D is the predicted design variable, and $g(X)$ can be replaced by its surrogate $\tilde{g}(X)$. In this case, D is a random variable with both epistemic and aleatory uncertainty. How do we determine the actual design variable d? Let the required reliability be $[R]$ or the permitted probability of failure be $[p_f]$. d is then determined by the the following reliability requirement:

$$
\begin{aligned}
R &= \Pr(D > d) > [R] \text{ or } \Pr(D < d) > [R] \\
p_f &= \Pr(D < d) < [p_f] \text{ or } \Pr(D > d) < [p_f]
\end{aligned} \tag{6}
$$

where R and p_f are the reliability or the probability of failure, respectively. The choice between $D < d$ and $D > d$ depends on the nature of the design variable. If increasing d results in a safer design, then $D < d$ should be used; otherwise, $D > d$ should be used. Eq. (6) suggests that d is a percentile value of the prediction D.

Design model II is for designs that require optimization. Let design variables be d. Assume its components are deterministic. As we discussed previously, d is part of X. To make our discussion easier for the design task, we separate d from X and rewrite the response as $Y = g(d, X)$.

The design is formulated as follows:

$$
\begin{cases}
\min_{d} Y_0 = g_0(d, X) \\
\text{subject to} \\
p_{fi} = \Pr[Y_i = g_i(d, X) < 0] < [p_{fi}], \ i = 1,2, \dots, n_g
\end{cases} \tag{7}
$$

where $g_0(\cdot)$ is the objective function, and it may be the mean value of a cost function. $Y_i = g_i(\cdot)$, $i = 1,2, \dots, n_g$, are constraint functions.

Surrogate models $Y_i = \tilde{g}(d, X), i = 0,1, \dots, n_g$ can be used to replace $Y_i = g(d, X)$ in the design model. Then the coupled uncertainty is accounted for in the design. High reliability and robustness can be built into the product, reducing the effects of coupled uncertainty.

The design model in Eq. (7) is general. If no uncertainty exists, the constraint function become $Y_i = g_i(d, X) < 0$. This is the model for deterministic design. If only model uncertainty exists without any aleatory uncertainty, the model will not change, and sufficient conservativeness compensating for prediction errors can be embedded in the design. If there is no model uncertainty and aleatory uncertainty exists, the design model becomes the model for reliability-based design (RBD) [19, 20]; the model ensures that the reliability requirement is satisfied with a reduced cost.

The above discussions are based on the prediction of a response at a single test point. If predictions at multiple test points are required for a response, we need to use the joint distribution of the model prediction errors. For instance, if we need predictions at $(x_1^\top, x_2^\top, \dots, x_u^\top)^\top$, we can use the joint distribution of the prediction at these points. We then model the predictions as a GP, which is defined by a multivariate Gaussian distribution with a mean vector and covariance matrix; both are available from the GP surrogate model.

5. A NEW NUMERICAL UQ METHOD FOR COUPLED UNCERTAINTY

As indicated in the design model in Eq. (7), the CDF of Y is required. MCS is not suitable for the design task since it is too slow for the probabilities of failure, which may be as small as $10^{-5} \sim 10^{-9}$. It is therefore to develop an efficient numerical method. In this study, we propose a new UQ method based on the First Order Reliability Method (FORM) [21]. We use a shaft design problem to demonstrate the method.

5.1 A shaft design

A shaft is to be design with the following information:

The bending moment follows a Gaussian distribution $M_a \sim N(1260, 0.15(1260))$ lbf $\cdot$ in and the steady torsion moment follows a Gaussian distribution $T_m \sim N(1500, 0.25(1500))$ 1100 lbf $\cdot$ in. The shaft has an ultimate strength $S_{ut} = 105$ kpsi, an endurance limit $S_e = 27.1$ kpsi, and stress concentration factors $K_f = 1.58$ and $K_{fs} = 1.37$. The reliability goal is $[R] = 1 - [p_f]$ and $[p_f] = 10^{-5}$. The design variable is the diameter of the shaft, and random input variables are $X = (X_1, X_2)^\top = (M_a, T_m)^\top$. M_a and T_m are independent. Since there is only one design variable and no optimization is needed, we create a GP model $D = \tilde{g}(X)$ using computational model of the diameter d [22]. The true diameter with respect to the two input variables is plotted in Fig. 1.

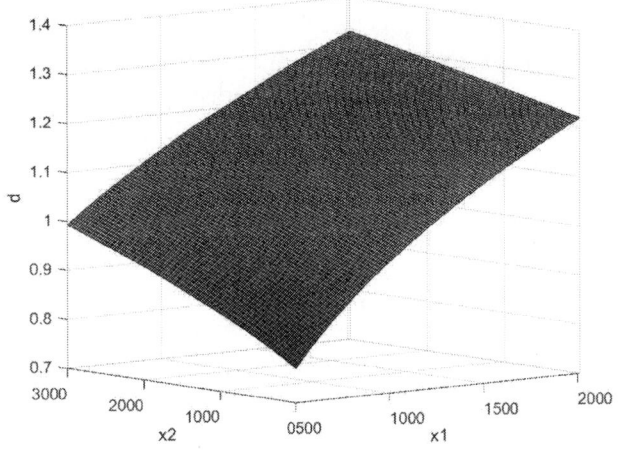

FIGURE 1: THE DIAMETER OF A SHAFT IS A NONLINEAR FUNCTION OF TWO INPUT ARIABLES. (Both X_1 and X_2 are in lbf $\cdot$ in)

Copyright © 2023 by ASME

5.2 Modeling

We at first transform random variables X into standard Gaussian variables U [23]. Let the transformation be $X = T(U)$. For the shaft design problem, $U = (U_1, U_2)^\top$. The contours of $D = \tilde{g}(X) = \tilde{g}(T(U))$ at $D = 1.3$ inch are plotted in Fig. 2. The dotted curve shows the mean contour from the mean prediction of the GP model $\tilde{g}(X)$ with respect to model uncertainty. This is the contour when model uncertainty is neglected. Other curves are realizations of the random prediction D.

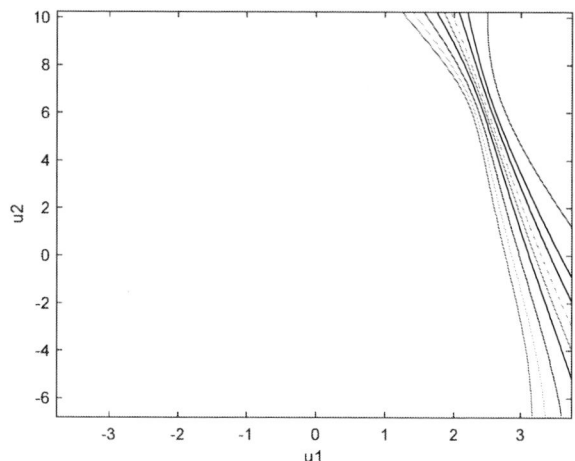

FIGURE 2: RANDOM CONTOURS OF DESIGN VARIABLE DUE TO MODEL UNCERTAINTY

Figure 2 indicates that an additional dimension is introduced to the existing random variables X. This explains the need to integrate $\Phi(\cdot)$ in Eq. (4). The added dimension makes the contour non-unique, which makes the UQ task challenging. To address this issue, we propose a new modeling approach that involves introducing an extra random variable into the prediction and design process. We at first begin discussing the general design model, or Design Model II. Denote $P_Y(\cdot)$ as the CDF of the random variable with coupled uncertainty Y from the surrogate model. If we plug random variable Y into the CDF, we get a new random variable Z.

$$Z = P_Y(Y) \qquad (8)$$

Equation (8) is commonly used in the inverse sampling method for Monte Carlo simulation.

Z is an auxiliary variable introduced to accommodate the model uncertainty. Using an auxiliary variable is a common strategy in the field of structural reliability [24-26], and this strategy has been proposed in reliability analysis and design with coupled uncertainty [27, 28]. We extend this strategy for machine learning assisted mechanical design. Since Y follows a normal distribution with $\mu_Y(X)$ and $\sigma_Y(X)$, we have

$$Z = P_Y(Y) = \Phi\left(\frac{Y - \mu_Y(X)}{\sigma_Y(X)}\right) \qquad (9)$$

which yields

$$Y = \mu_Y(X) + \Phi^{-1}(Z)\sigma_Y(X) \qquad (10)$$

$$P_Y(y) = \Pr(Y < y) = \Pr(\mu_Y(X) + \Phi^{-1}(Z)\sigma_Y(X) < y) \qquad (11)$$

Equations (10) and (11) represent the new modeling approach for quantifying the effects of coupled uncertainty. Equation (10) indicates that the prediction from the machine learning model is a function of random variables of X and Z. The UQ analysis now has increased dimensionality by 1 since a new random variable Z is added. The added dimension removes the second order uncertainty and makes the distribution of Y unique. When only aleatory uncertainty exists, $\sigma_Y(X) = 0$, and

$$P_Y(y) = \Pr(Y < y) = \Pr(\mu_Y(X) < y) \qquad (12)$$

When only model uncertainty exists, the model is the same as that in Eq. (11), but X contains only deterministic variables.

The proposed modeling approach serves as a strong foundation for developing a new numerical method to handle the complexity of UQ. Section 5.3 will cover the specifics of this method in detail.

5.2 New numerical method

The new numerical method extends the First Order Reliability method (FORM), which is widely used for predicting the CDF of Y with random variables X. We extend this method to allow for the incorporation of model uncertainty.

We use the conditional distribution of Z on X rather than the distribution of Z directly, for two reasons. Firstly, the conditional distribution of Z on X is well-defined and straightforward. Secondly, the use of the conditional distribution aligns with the requirements of FORM. When X is fixed at its realization x, Y follows a normal distribution with deterministic mean $\mu_Y(x)$ and standard deviation $\sigma(x)$. Then the conditional distribution of Z follows a uniform distribution in the range [0, 1] according to Eq. (9). Importantly, the conditional distribution of Z is then independent of the joint distribution of X. The conditional CDF of Z on X is given by

$$P_{Z|X}(z) = \Pr(Z|X = x < z) = z \qquad (13)$$

The conditional CDF is then used for the subsequent analysis. We transform all random variables into a space of independent and standard normal random variables. We denote the transformed variables for X and Z as U_X and U_Z, respectively. We also assume the components of X are independent. However, if they are dependent, they can be transformed into independent variables. The transformation of dependent variables into independent variables is achieved using the Rosenblatt transformation.

$$X = P_X^{-1}\big(\Phi(U_X)\big) = T(U_X) \qquad (14)$$

where X and U_X are a component of X and U_X, respectively, and $T(\cdot)$ stands for the transformation. U_Z is obtained by

$$Z = P_{Z|X}^{-1}\big(\Phi(U_Z)\big) = \Phi(U_Z) \qquad (15)$$

Then

$$\Phi^{-1}(Z) = U_Z \qquad (16)$$

Plugging Eqs. (14) and (16) into Eqs. (10) and (11) yields

$$Y = \mu_Y(T(\boldsymbol{U_X})) + U_Z \sigma_Y(T(\boldsymbol{U_X})) \qquad (17)$$

$$P_Y(y) = \Pr(\mu_Y(T(\boldsymbol{U_X})) + U_Z \sigma_Y(T(\boldsymbol{U_X}))) < y) \qquad (18)$$

Equation (18) provides the UQ model for coupled uncertainty in the transformed U-space, where all random variables, including $\boldsymbol{U_X}$ and U_Z, are independent standard normal variables. The new UQ model allows for the use of traditional FORM after the transformation of the auxiliary variable. Let $\boldsymbol{U} = (\boldsymbol{U_X}; U_Z)$ and

$$Y' = G(\boldsymbol{U}) = \mu_Y(T(\boldsymbol{U_X})) + U_Z \sigma_Y(T(\boldsymbol{U_X})) - y \qquad (19)$$

The problem becomes to find the CDF of Y'; namely,

$$P_{Y'}(0) = \Pr(Y' = G(\boldsymbol{U}) < 0) = \int_{G(\boldsymbol{u})<0} p_U(\boldsymbol{u}) d\boldsymbol{u} \qquad (20)$$

where $p_U(\boldsymbol{u})$ is the joint PDF of $\boldsymbol{U}$. Equation (20) indicates that integrating the joint PDF of $\boldsymbol{U}$ in the region defined by $G(\boldsymbol{u}) < 0$ results in the CDF we want.

The transformation of variables into the space of independent and standard normal random variables allows for the application of FORM to predict the CDF in Eq. (20). The search for the Most Probable Point (MPP) in the U-space is crucial in this approach, as it has the highest joint PDF or integrand in Eq. (20) within the integration region where $G(\boldsymbol{u})$ is less than zero. Linearization of the function of Y is performed at the MPP to minimize the error due to linearization. The search for the MPP is described in detail in Ref. [28]. A brief overview of the numerical procedure is presented below.

Design model I
- Input required reliability $[R]$
- Find the reliability index $\beta = \Phi^{-1}([R])$
- Transform $\boldsymbol{X}$ to $\boldsymbol{U_X}$ and Z to U_Z
- Search for the MPP $(\boldsymbol{u_X^*}, u_z^*)$ for β
- Obtain design variable with Eq. (21)

$$d = \mu_D(T(\boldsymbol{u_X^*})) + u_z^* \sigma_D(T(\boldsymbol{u_X^*})) \qquad (21)$$

Why does Eq. (21) ensure the required reliability $[R]$? If a smaller d is desired for a safer design, Eq. (21) satisfies (if the error of linearization is neglected)

$$\Pr(D(\boldsymbol{U}) < d) = [R] \qquad (22)$$

If a larger d is desired for a safer design, Eq. (21) satisfies (if the error of linearization is neglected)

$$\Pr(D(\boldsymbol{U}) > d) = [R] \qquad (23)$$

The MPP search uses the gradient of $D = \tilde{g}(T(\boldsymbol{U}))$ with respect to $\boldsymbol{U}$. Using the positive or negative directions of the gradient depends on whether a larger or smaller d is desired.

Design model II
- Transform $\boldsymbol{X}$ to U
- Set $\mu_Y(T(\boldsymbol{U})) + U_Z \sigma_Y(T(\boldsymbol{U})) < y$ for the CDF at y
- Search for the MPP $(\boldsymbol{u^*}, u_z^*)$
- Find the reliability index

$$\beta = \|(\boldsymbol{u^*}, u_z^*)\| \qquad (24)$$

where $\|\cdot\|$ is the magnitude of a vector.
- Obtain design variable with Eq. (25)

$$P_Y(y) = \Phi(-\beta) \qquad (25)$$

Equation (25) is for $P_Y(y) < 0.5$. For $P_Y(y) > 0.5$, a different equation should be used [23].

6. RESULTS AND DISCUSSION

In this section, we present the results of the shaft design considering coupled uncertainty. To demonstrate the effectiveness of the proposed method, we also use the original computational model and MCS with a sample size of 10^7. As previously discussed, this is a Type I design. The results are summarized in Table 1. To facilitate comparison, we do not round the diameter to a preferred value that is suitable for manufacturing purposes. The number of training points is 10.

TABLE I. DESIGN RESULTS WITH 10 TRAINING POINTS

	Method	Diameter d (inch)
1	Deterministic design with the original model	1.1186
2	Deterministic design using mean GP prediction	1.1178
3	Design under aleatory uncertainty with the original model (RBD with aleatory uncertainty)	1.2963
4	Design using mean GP prediction under aleatory uncertainty (RBD with aleatory uncertainty)	1.3173
5	Design under coupled uncertainty by MCS	1.3210
6	Design under coupled uncertainty by the proposed method	1.3219

The results from deterministic designs (Methods 1 and 2) using the original model and the mean GP prediction are slightly different. This indicates the error of the surrogate model. The design under aleatory uncertainty (Method 3) using the original computational model provides the accurate design result. Design using mean GP prediction under aleatory uncertainty (Method 4) also results in a design under aleatory uncertainty. Methods 3 and 4 are both for design under aleatory uncertainty and the results are different, also indicating the error of the surrogate model. Method 5 is more conservative than Method 4 since the former considers coupled uncertainty while the latter considers only aleatory uncertainty. The results for design under coupled uncertainty by MCS (Method 5) and the proposed method

Copyright © 2023 by ASME

(Method 6) produce the highest diameters or the most conservative designs, accommodating the effects of coupled uncertainty. This means that more conservativeness is embedded. The two methods produce almost identical results, indicating the high accuracy of the proposed method for this design problem since the MCS result is regarded as an accurate solution.

The PDFs of the predicted diameter D from MCS is shown in Fig. 3. The PDF can also be generated by the proposed method with repeated calls with different values of reliability. This figure demonstrates that the prediction follows a single distribution in the presence of both epistemic and aleatory uncertainty.

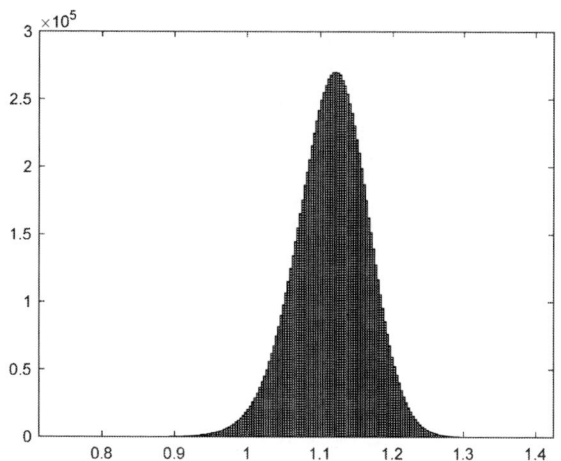

FIGURE 3: PDF OF PREDICTED DIAMETER WITH BOTH COUPLED EPISTEMIC AND ALEATORY UNCERTAINTY

We increase the number of training points to 20 for a redesign. As expected, the accuracy of the surrogate model significantly increases with more training data. Table II presents the new results obtained from the surrogate model. It is evident from the table that the results from the surrogate model are nearly identical to those from the original computational model. This suggests that the surrogate model is highly accurate and can be relied upon for future designs.

TABLE II.   DESIGN RESULTS WITH 20 TRAINING POINTS

	Method	Diameter d (inch)
1	Deterministic design with the original model	1.1186
2	Deterministic design using mean GP prediction	1.1186
3	Design under aleatory uncertainty with the original model (RBD with aleatory uncertainty)	1.2963

	Method	Diameter d (inch)
4	Design using mean GP prediction under aleatory uncertainty (RBD with aleatory uncertainty)	1.2962
5	Design under coupled model and aleatory uncertainty by MCS	1.2963
6	Design under coupled model and aleatory uncertainty by the proposed method	1.2965

7. CONCLUSION

This study investigates how to account for coupled model uncertainty in physics-based machine learning models and aleatory uncertainty in mechanical design. A new UQ method is also proposed to quantify the effect of the coupled model and aleatory uncertainty in design. Even though the study is preliminary, it suggests the following conclusions.

If the model error is estimated as a random variable whose distribution parameters are functions of input variables, the model prediction will be a random variable with single distribution.

If design decision is based on multiple predictions of a response at different input points, the predictions may be statistically dependent. The closer are two input points, the higher is the correlation between the two predictions. The joint distribution of the predictions due to coupled uncertainty can be obtained if the surrogate model provides the joint distribution of the model uncertainty at different sites.

Considering coupled uncertainty can help make design more reliable and robust against uncertainty in material properties, manufacturing, and operation environment, as well as surrogate model errors.

However, building sufficient conservativeness against model error is not guaranteed using the proposed method. The model error in the design region which is not or not well covered by training points may be large and biased. Using higher reliability requirement may not be sufficient. After a design is obtained, we need to check the degree of model uncertainty. If it is high, we should be cautious. Since we do not know if the model error around the true response is positive or negative, we may need to verify and improve the accuracy of the surrogate model by adding a new training point at the design or more points around the design.

The proposed numerical method is efficient, but its accuracy may deteriorate when aleatory uncertainty is large, the model is highly nonlinear, and the standard deviation of the model uncertainty fluctuates around a new test point. Our future research will test the new method with more problems and make improvements to its accuracy.

Additionally, we will investigate the design that requires multiple predictions of a response at different input points. The research topic is to quantify the joint distribution of multiple predictions, including incorporating it in the design model and a numerical method that produces the joint distribution.

Copyright © 2023 by ASME

REFERENCES

[1] Jin, R., Chen, W., and Sudjianto, A., "An efficient algorithm for constructing optimal design of computer experiments," Proc. International design engineering technical conferences and computers and information in engineering conference, pp. 545-554.

[2] Fan, J., and Li, G., 2005, Contemporary Multivariate Analysis And Design Of Experiments: In Celebration Of Prof Kai-tai Fang's 65th Birthday, World Scientific.

[3] Cunningham, J. D., Simpson, T. W., and Tucker, C. S., 2019, "An Investigation of Surrogate Models for Efficient Performance-Based Decoding of 3D Point Clouds," Journal of Mechanical Design, 141(12).

[4] Hu, Z., Hu, Z., and Du, X., 2019, "One-class support vector machines with a bias constraint and its application in system reliability prediction," AI EDAM, 33(3), pp. 346-358.

[5] Planas, R., Oune, N., and Bostanabad, R., 2021, "Evolutionary Gaussian Processes," Journal of Mechanical Design, 143(11).

[6] Jiang, Z., Li, W., Apley, D. W., and Chen, W., 2015, "A Spatial-Random-Process Based Multidisciplinary System Uncertainty Propagation Approach With Model Uncertainty," Journal of Mechanical Design, 137(10).

[7] Seeger, M., 2004, "Gaussian processes for machine learning," International journal of neural systems, 14(02), pp. 69-106.

[8] Heiss, J., Weissteiner, J., Wutte, H., Seuken, S., and Teichmann, J., 2021, "Nomu: Neural optimization-based model uncertainty," arXiv preprint arXiv:2102.13640.

[9] Goan, E., and Fookes, C., 2020, "Bayesian neural networks: An introduction and survey," Case Studies in Applied Bayesian Data Science, Springer, pp. 45-87.

[10] Goldsmith, R. W., and Sahlin, N.-E., 1983, "The role of second-order probabilities in decision making," Advances in Psychology, Elsevier, pp. 455-467.

[11] Zhu, Z., and Du, X., 2016, "Reliability analysis with Monte Carlo simulation and dependent Kriging predictions," Journal of Mechanical Design, 138(12), p. 121403.

[12] Sarkar, S., Mondal, S., Joly, M., Lynch, M. E., Bopardikar, S. D., Acharya, R., and Perdikaris, P., 2019, "Multifidelity and Multiscale Bayesian Framework for High-Dimensional Engineering Design and Calibration," Journal of Mechanical Design, 141(12).

[13] Li, M., and Wang, Z., 2019, "Surrogate model uncertainty quantification for reliability-based design optimization," Reliability Engineering & System Safety, 192, p. 106432.

[14] Apley, D. W., Liu, J., and Chen, W., 2005, "Understanding the Effects of Model Uncertainty in Robust Design With Computer Experiments," Journal of Mechanical Design, 128(4), pp. 945-958.

[15] Shah, H., Hosder, S., Koziel, S., Tesfahunegn, Y. A., and Leifsson, L., 2015, "Multi-fidelity robust aerodynamic design optimization under mixed uncertainty," Aerospace Science and Technology, 45, pp. 17-29.

[16] Kumar, P., Sinha, K., Nere, N. K., Shin, Y., Ho, R., Mlinar, L. B., and Sheikh, A. Y., 2020, "A machine learning framework for computationally expensive transient models," Scientific Reports, 10(1), p. 11492.

[17] Der Kiureghian, A., and Ditlevsen, O., 2009, "Aleatory or epistemic? Does it matter?," Structural safety, 31(2), pp. 105-112.

[18] Soize, C., 2000, "A nonparametric model of random uncertainties for reduced matrix models in structural dynamics," Probabilistic engineering mechanics, 15(3), pp. 277-294.

[19] Wu, H., and Du, X., 2023, "Time- and Space-Dependent Reliability-Based Design With Envelope Method," Journal of Mechanical Design, 145(3).

[20] Li, H., and Du, X., 2021, "Recovering Missing Component Dependence for System Reliability Prediction via Synergy Between Physics and Data," Journal of Mechanical Design, 144(4).

[21] Hu, Z., and Du, X., 2015, "First order reliability method for time-variant problems using series expansions," Structural and Multidisciplinary Optimization, 51(1), pp. 1-21.

[22] Budynas, R. G., and Nisbett, J. K., 2011, Shigley's mechanical engineering design, McGraw-Hill New York.

[23] Du, X., Chen, W., and Wang, Y., 2010, "Most Probable Point-Based Methods," Extreme Statistics in Nanoscale Memory Design, A. Singhee, and R. A. Rutenbar, eds., Springer US, Boston, MA, pp. 179-202.

[24] Bjerager, P., 1990, "On computation methods for structural reliability analysis," Structural Safety, 9(2), pp. 79-96.

[25] Wen, Y., 1987, "Approximate methods for nonlinear time-variant reliability analysis," Journal of engineering mechanics, 113(12), pp. 1826-1839.

[26] Kiureghian, A. D., 1989, "Measures of structural safety under imperfect states of knowledge," Journal of Structural Engineering, 115(5), pp. 1119-1140.

[27] Sankararaman, S., and Mahadevan, S., 2013, "Separating the contributions of variability and parameter uncertainty in probability distributions," Reliability Engineering & System Safety, 112, pp. 187-199.

[28] Nannapaneni, S., and Mahadevan, S., 2016, "Reliability analysis under epistemic uncertainty," Reliability Engineering & System Safety, 155, pp. 9-20.

Copyright © 2023 by ASME

Proceedings of the ASME 2023
International Design Engineering Technical Conferences and
Computers and Information in Engineering Conference
IDETC-CIE2023
August 20-23, 2023, Boston, Massachusetts

DETC2023-115201

AN INTEGRATED APPROACH TO DESIGNING ROBUST TURBOCOMPRESSORS ON GAS BEARINGS THROUGH SURROGATE MODELING AND CONSTRAINED MULTI-OBJECTIVE OPTIMIZATION

Soheyl Massoudi[1,*], Cyril Picard[2], Jürg Schiffmann[1]

[1]Laboratory for Applied Mechanical Design
Ecole Polytechnique Fédérale de Lausanne (EPFL)
CH-1015 Lausanne, Switzerland
[2] Department of Mechanical Engineering
Massachusetts Institute of Technology
Cambridge, Massachusetts, 02139

ABSTRACT

Designing turbocompressors is a complex and challenging task, as it involves balancing conflicting objectives such as efficiency, stability, and robustness against manufacturing deviations. This paper proposes an integrated design methodology for turbocompressors supported on gas bearings, which utilizes surrogate models and ensemble learning with artificial neural networks. The proposed approach addresses the limitations of nominal and separate optimizations by integrating all relevant design aspects into a single optimization problem. A multi-objective optimization is carried out, considering four objectives and over twenty constraints, including robustness against manufacturing deviations of the radial and axial bearings in terms of stability, load capacity, and efficiency, as well as robustness against performance metric gradients. The proposed methodology maximizes the compressor's range in speeds and mass flow, while also maximizing the signal-to-noise ratio of the isentropic efficiency over the compressor map. Additionally, the approach maximizes system efficiency, taking into account component losses and isentropic efficiency of the compressor. To enable rapid and automated integrated design, the methodology reduces the compressor representation to a fully cylindrical representation. The study finds that the proposed methodology has the potential to significantly enhance the overall performance of turbocompressors in terms of efficiency, stability, and robustness. The methodology eliminates the need for sequential and iterative design steps, providing an optimal starting point for higher representation of the system with CFD and finite elements study. Furthermore, the proposed methodology has broad applications, including the optimization of other complex and interdependent systems in various fields. This study highlights the crucial role of a comprehensive

and integrated approach to turbocompressor design and provides a valuable framework for future research in this area.

Keywords: Herringbone grooved journal bearings, gas bearings, micro-turbomachinery, integrated design, robust design, constrained multi-objective optimization, artificial neural networks

NOMENCLATURE

Roman letters

A	Bearing front
B	Bearing rear
D	Bearing diameter [m]
$\dot{E}$	Power [W]
f	Objective function/Performance metric [context dependent unit]
F	Force [N]
G	Geometry field [context dependent unit]
h_g	Groove depth [m]
h_r	Ridge clearance [m]
HV	Measure of feasible region [context dependent unit]
I	Moment of inertia [kg m^2]
k	Sweep sampling [−]
L	Bearing axial length [m]
L_A	Bearing front to center of gravity midplane distance [m]
L_B	Bearing rear to center of gravity midplane distance [m]
M	Rotor mass [kg]
N	Rotational speed [RPM]
P	Pressure [Pa]
R	Bearing radius [m]
S/N	Signal-to-Noise ratio [−]

*Corresponding author: soheyl.massoudi@epfl.ch

Copyright © 2023 by ASME

| S | Search space |
| T | Temperature [K] |

Greek letters

α	Groove-ridge width ratio [−]
β	Groove angle/Impeller angle [°]
γ	Grooved region ratio [−]
Γ	Logarithmic decrement [−]
δ	Interference [m]
Δ	Variation of a given variable
η	Efficiency [−]
θ	Circumferential coordinate/Angular position of blade [°]
Λ	Compressibility number [−]
μ	Viscosity [Pa s]
Π	Pressure ratio [−]
ρ	Density [kg m^{-3}]
σ	Stress [N m^{-2}]
τ	Electromagnetic shear stress/Torque [context dependent unit]
Ω	Angular velocity [rad s^{-1}]

Superscripts and subscripts

a	ambient
,a	axial
bend	bending
bld	blade
cyl	cylinder
end	sampling end
exp	centrifugal expansion
F	feasible
g	groove
hub	hub
i	i$_{th}$ element
in	inlet
is	isentropic
loss	losses
nom	nominal
p	polar
r	ridge
rob	robust
,r	radial
spl	splitter
start	sampling start
t	transverse
w	weight
⁻	Dimensionless/Mean

Acronyms

CG	Center of Gravity
COMP	Compressor
EM	Electric Motor
GPU	Graphics Processing Unit
HGJB	Herringbone Grooved Journal Bearing
MAG	Magnet
NGT	Narrow Groove Theory
NSGA	Non dominated Sorting Genetic Algorithm
ROT	Rotor
SGTB	Spiral Groove Thrust Bearing

1. INTRODUCTION

1.1 Nature of the Issue

The design of gas bearing supported turbocompressors is a challenging task that requires the optimization of multiple interrelated components, including the compressor wheel, the axial and radial bearings, and the motor. Traditional design approaches typically involve separate optimizations of individual components, which can lead to suboptimal designs due to inherent trade-offs between performance and robustness. Furthermore, optimizing individual components in isolation may result in infeasible designs, where optimal designs for one component are not compatible with optimal designs for other components.

While integrated design approaches have been proposed for various types of machines, such as robots, automotive control actuators, and small-scale turbocompressors, it is still an under-researched area. Wehner et al. [1] demonstrated the power of integrated design by using soft lithography, molding, and 3D printing to create soft analogs of control systems and power sources for microfluidic-based autonomous robots. Picard and Schiffmann [2] also applied an integrated design approach to automotive control actuators, resulting in an optimized solution with better torque and reduced cost compared to industrial solutions. Schiffmann [3] designed a small-scale turbocompressor for a single-stage heat-pump using an integrated approach, which improved the overall system efficiency. However, this approach did not address the issue of robustness against manufacturing deviations.

To address this gap, researchers have proposed integrating robustness considerations into the optimization process for gas bearing supported turbocompressors [4]. For example, Massoudi and Schiffmann [5] developed a surrogate model based on an ensemble of artificial neural networks multi-objective optimization framework using the logarithmic decrement as a metric of stability. They applied this method to derive guidelines for designing robust gas bearing supported rotors, achieving promising results. However, the authors did not fully integrate the robustness considerations with the integrated design approach for the entire system. Their approach was limited to the design of the radial bearings and part of the rotor geometry.

Given the potential benefits of integrated design and the need for robustness, there is a clear opportunity to merge these two approaches in the design of gas bearing supported turbocompressors.

1.2 Goals and Objectives

The goals and objectives of this paper are 1) an integrated design approach of turbocompressor supported on gas bearings that incorporates surrogate models with ensembles of artificial neural networks in constrained multi-objective optimization including manufacturing deviations, 2) the evaluation of the benefits of such an approach in terms of improved design efficiency, increased overall performance and robustness of the design, and 3) a practical tool for engineers and researchers in the field.

1.3 Scope of the Paper

The aim of this paper is to capitalize on the work of Massoudi and Schiffmann [6] by implementing surrogate models for the

Copyright © 2023 by ASME

axial dynamics and the compressor. This will allow the integrated design of a complex system that is a gas bearings supported turbocompressor, while considering manufacturing deviations by developing a constrained multi-objective optimization framework for integrated and robust design.

To model the compressor, axial dynamics, and rotordynamics of the system, surrogate models are constructed using ensembles of artificial neural networks. These models not only enable computations to scale on graphics processing units (GPUs) but also significantly reduce computation time. By leveraging these models, thousands of turbocompressor designs can be evaluated in a single optimization pass, even when accounting for manufacturing deviations. This approach offers a powerful means of improving the overall efficiency and reliability of turbocompressors by enabling engineers to consider a much broader range of design options in a timely and cost-effective manner.

The approach outlined in this paper provides a constrained multi-objective optimization framework for achieving a robust and integrated design of gas bearing supported turbocompressors. In particular, the study focuses on demonstrating the feasibility of optimal design for a single-stage, electrically-driven heat-pump compressor that is supported by gas bearings and takes into account manufacturing deviations. The study aims to accelerate the design process, eliminate unnecessary iterations in the preliminary design stages, and ultimately deliver a highly performant and manufacturable solution. The proposed methodology represents a valuable methodology for engineers and researchers in the field of turbocompressor design, providing an efficient means of achieving optimal design objectives while considering critical manufacturing constraints.

2. THEORY

The electrically-driven compressor unit comprises an impeller wheel (COMP), a spiral groove thrust bearing (SGTB) to carry axial loads, herringbone grooved journal bearings (HGJBs), and a synchronous permanent magnet electric motor (EM). Figure 1 depicts the unit and its subsystems alongside a compressor map that illustrates the operating range for a given mass flow, rotational speed, and target pressure ratio. The goal of the compressor is to elevate the inlet pressure to an outlet pressure at maximum isentropic efficiency. During startup and speed mapping, the bearings must remain stable and dissipate minimal energy while providing sufficient load capacity to balance the weight and axial force of the compressor under manufacturing deviations. Surrogate models trained using validated models of the bearings, rotordynamics, and impeller enable a fully integrated and robust optimization of the entire unit. Additionally, fast analytical models or 1D finite element code are utilized to evaluate the losses, load capacities, structural integrity and bending frequency.

2.1 Models

2.1.1 Axial and Journal Bearings.
The performance of the axial and journal bearings can be modeled using the Reynolds equation, which is derived from the Navier-Stokes equations under the assumptions of thin-film, laminar flow, and Newtonian fluid.

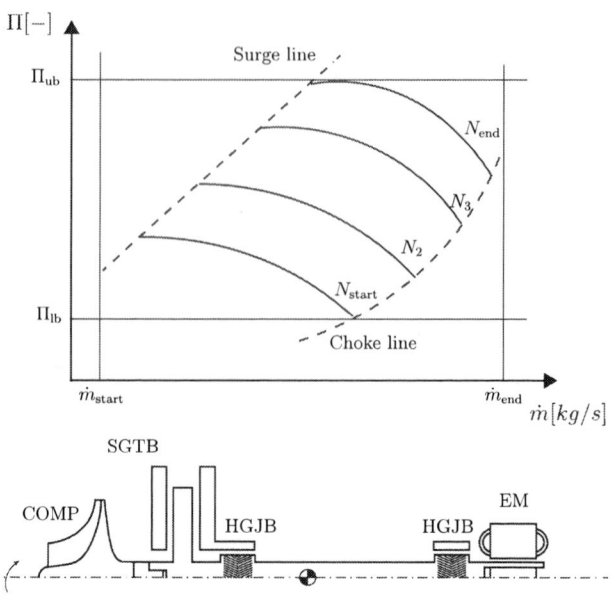

FIGURE 1: COMPRESSOR MAP AND SCHEMATIC OF THE TURBOCOMPRESSOR UNIT. THE COMPRESSOR MAP SHOWS THE RANGE OF OPERATION BETWEEN SURGE AND CHOKE FOR A SAMPLED MASS FLOW, ROTATIONAL SPEED, AND TARGET PRESSURE RATIO. THE SCHEMATIC ILLUSTRATES THE MAIN SUBSYSTEMS OF THE ELECTRICALLY DRIVEN COMPRESSOR UNIT, INCLUDING THE IMPELLER WHEEL (COMP), SPIRAL GROOVE THRUST BEARING (SGTB), HERRINGBONE GROOVED JOURNAL BEARINGS (HGJBS), AND SYNCHRONOUS PERMANENT MAGNET ELECTRIC MOTOR (EM).

The Narrow Groove Theory (NGT) [7] is used to assume that the grooves in the bearings have infinitesimal width. Ideal gas assumption and isothermal compression eliminate the dependence on density. The solution to the Reynolds equation is obtained using perturbation about a concentric position of the bearing, and the zeroth and first order pressure perturbations are determined using numerical integration. The direct and cross-coupled stiffness and damping coefficients can then be obtained from these perturbations. To account for the centrifugal expansion, the radial bearing clearance is adjusted.

2.1.2 Rotordynamics and Axial Dynamics.
The rotor and bushings are assumed to be rigid. The stiffness and damping of the gas bearings are functions of the excitation frequency ω_{ex}, which is often different from the system's natural frequency ω_n. To compute the stability of the rigid-body rotordynamic system, the following algorithm is applied for each nominal speed Ω:

Compute the stiffness and damping coefficients of the gas bearings for a range of discrete excitation frequencies $\omega_{ex,i}$:

$$K_i = K(\omega_{ex,i}), \qquad C_i = C(\omega_{ex,i}) \qquad (1)$$

Form the system matrix $[M][\ddot{q}] + [C][\dot{q}] + [K][q] = [0]$, where $[M]$ is the mass matrix, $[C]$ is the damping matrix, $[K]$ is the stiffness matrix and $[q]$ is the vector of displacement. Compute

Copyright © 2023 by ASME

the eigenvalues $\delta_i = \lambda_i + j\omega_i$ of the system matrix at each excitation frequency. Compute the logarithmic decrement Γ_i as the metric of stability:

$$\Gamma_i = -\lambda_i \frac{2\pi}{\omega_i} \qquad (2)$$

If Γ_i is positive for all excitation frequencies, the system is stable. The system can then be excited in four modes: cylindrical forward (CylF), cylindrical backward (CylB), conical forward (ConF), and conical backward (ConB). The axial dynamics are computed in a decoupled fashion by neglecting the tilting motion for sufficiently long rotors. The system is then treated as a damped point mass oscillator and solved using a similar spectral approach to the rotordynamics analysis.

To compute the bending frequency, a 3-layer composite cylinder model of the rotor is used, which employs 1D finite elements. In this model, each element (or cylinder) is considered as an elastic structure with two nodes and four degrees of freedom per node, which correspond to two displacements and two rotations [8, 9]. The constitutive equation used for the model is based on the Timoshenko beam theory, while Hermitian polynomials are employed to prevent shear locking and to avoid an overestimation of the element stiffness [10]. This approach provides an accurate and efficient way of computing the bending frequency of the rotor, which is an important parameter for assessing the structural integrity of the turbocompressor system.

2.1.3 Compressor. The compressor model used in this study is the Python version [11] of a 1D code developed by Schiffmann and Favrat [12]. This code is based on a meanline model, which is augmented with empirical loss models to capture the effects of fluid flow phenomena. The model is capable of detecting numerical errors, surge, choke, and other abnormal operating conditions. The outputs of the model include pressure ratio, isentropic efficiency, and isentropic enthalpy change for a given set of geometry and operating conditions. A cut view of a typical centrifugal compressor and the geometric variables required for the meanline model are shown in Fig. 2.

2.1.4 Electric Motor. The electric motor chosen for this turbocompressor is a brushless DC motor (BLDC) controlled with pulse-width modulation (PWM). Its size is determined using the electromagnetic shear stress, which is calculated using the formula:

$$\dot{E}_{EM} = 2\pi R_{MAG}^2 L_{MAG} \tau \Omega \qquad (3)$$

Here, R_{MAG} and L_{MAG} denote the radius and length of the magnet, respectively, while τ represents the airgap shear stress. This formula helps in calculating the motor power, as described in [13].

2.1.5 Losses. To accurately assess the total energy consumption of the system, analytical models are used to estimate the energy losses in the bearings and electric motor. For the axial bearing, the energy loss is denoted as $\dot{E}_{SGTB}$, while for each radial bearing, the loss is $\dot{E}_{HGJB}$. These losses are computed using a laminar flow model that has been experimentally validated [14]. The windage losses in the electric motor, denoted $\dot{E}_{EM,loss}$, are estimated by discriminating between laminar and turbulent flow regimes using the Taylor number.

The energy loss in each radial bearing can be calculated using the following equation:

$$\dot{E}_{HGJB} = 2\pi R^3 \Omega^2 \left(\frac{\gamma_r \alpha_r}{h_{g,r}} + \frac{1 - \gamma_r \alpha_r}{h_{r,r}} \right) \mu L \qquad (4)$$

where R is the radius of the bearing, γ_r the ratio of grooved length to bearing length L, $\alpha_r = a_r/(a_r + b_r)$ the ratio of groove width (a_r) and ridge width (b_r), Ω is the rotational speed in rad s^{-1}, and μ is the dynamic viscosity. $h_{g,r}$ is the groove depth and $h_{r,r}$ is the local bearing clearance. Similarly, the energy loss in the axial bearing can be computed using:

$$\dot{E}_{SGTB} = \mu \frac{\Omega^2 \pi}{2} \left[\left(R_o^4 - R_g^4 \right) \left(\frac{\alpha_a}{h_{g,a}} + \frac{1 - \alpha_a}{h_{r,a}} \right) \right.$$
$$\left. + \left(R_g^4 - R_i^4 \right) \frac{1}{h_{r,a}} \right] \qquad (5)$$

with R_o the outer radius of the thrust bearing, R_i its inner radius and R_g the radius marking the start of the grooved region. Finally, the windage losses in the electric motor can be estimated using:

$$\dot{E}_{EM,loss} = c_w \pi \rho \Omega^3 R_{EM}^4 L_{EM} \qquad (6)$$

Here, c_w is a coefficient that accounts for laminar or turbulent flow conditions and ρ is the fluid density. The energy losses in the bearings and electric motor are important to consider, as they contribute to the total energy consumption of the system and can affect its overall performance.

2.1.6 Load capacities. The load capacities of the axial and journal bearings are calculated by solving the perturbed and unperturbed pressure equations and then integrating them over the bearing domains [15]. These equations take into account the dynamic effects of the rotor and the lubricant film. By solving these equations, the maximum loads that the bearings can support without failure can be determined.

2.1.7 Structures. To ensure structural integrity, a composite annulus 2D axisymmetric model of the rotor shaft and its components is used [16]. The model considers the interference between two cylindrical layers of different materials, the centrifugal forces generated by the high-speed rotation of the shaft, thermal dilation, and axial stresses transmitted by the impeller and axial bearing. Additionally, the necessary interference for torque transmission between the magnet and the shaft, and between the plug and rotor is considered, as well as the resulting shear stresses. The model considers two cases: those where interference occurs between two layers of a shaft segment, and those with mono-material without interference.

2.2 Surrogate models

The use of surrogate models is crucial in enabling the scaling of design for system integration and robustness evaluation. However, the increased evaluation of different subsystems, robustness, and complete compressor map leads to a considerable rise in

Copyright © 2023 by ASME

function evaluations. Consequently, there is a higher number of optimization objectives, design parameters, and evaluations per generation. As a result, several hundred million model evaluations per optimization, if not billions, are necessary, which is impractical and unfeasible without significantly increasing computational resources. To address this challenge, surrogate models are used to replace the baseline models of the bearings, rotor and axial dynamics, the load capacities of the bearing, and the compressor performance prediction. Unlike analytical models for losses and electric motor power estimation, these models cannot be reduced to a matrix representation and element-wise computation. Although the bending frequency is computed using a 1D finite element code, a surrogate model for this analysis is not necessary because the computation is relatively fast and completes in a few milliseconds.

Data is sampled from the baseline models, following the approach proposed by Massoudi and Schiffmann [6]. A combinatorial sampling technique is employed to map a broad range of operating conditions within the gaseous regions of refrigerant fluids, air, and steam, with the aim of deriving the thermodynamic properties of fluids, such as viscosity. Two Latin hypercube samplings are then performed on the dimensional and dimensionless geometries, along with rotational speed and mass flow. The sampled parameters are subsequently converted into dimensionless groups, which are then utilized to train feed-forward neural networks. These networks form the basis of the surrogate models.

Hyperparameter tuning of surrogate models is a critical step in their development. To optimize the performance of the artificial neural networks (ANNs) used to model the various outputs, each ANN is trained via gradient descent within a genetic algorithm loop [17, 18]. The hyperparameters that govern the training process are chosen as decision variables to be optimized as presented in Table 1. Hyperparameter tuning can be computationally intensive, and we use a genetic algorithm with a total of 5 epochs and an initial population size of 100 to efficiently search the hyperparameter space. Two types of ANNs are trained: regressors, which predict continuous outputs such as the logarithmic decrement of the isentropic efficiency of the compressor, and classifiers, which predict categorical variables such as the stability of a given design or the functioning state of a compressor. The choice of loss function depends on the type of output being predicted and includes mean squared error or mean absolute error for regressors, and categorical cross-entropy for classifiers. Classifiers are trained with larger batch sizes than regressors.

To increase the accuracy and robustness of the surrogate model predictions, six different versions of the optimal artificial neural network (ANN) found via hyperparameter tuning with the genetic algorithm are trained using varying weight initializations. These initializations include He Normal, Lecun Normal, Glorot Uniform, He Uniform, Lecun Uniform, and Glorot Normal [19–21]. The final prediction is obtained as the average of the predictions from the six neural networks, resulting in an ensemble of neural networks [22].

2.3 Robustness

Robustness is a critical factor in engineering design, which can be defined in two ways. Firstly, robustness refers to the maximum space that a design can occupy without violating constraints due to manufacturing deviations. Secondly, a robust design can maintain its performance under manufacturing deviations, indicating its insensitivity to such deviations. These definitions have been formalized by Massoudi and Schiffmann [5] and are included in multi-objective optimization for robust design. In such optimization, the objectives are to maximize the feasible region (HV) and maximize the signal-to-noise ratio (S/N), among other competing objectives.

To estimate the maximum feasible space within manufacturing tolerances, a Monte Carlo method can be employed by randomly sampling points within the tolerances and identifying those that meet the constraints. The resulting feasible space (HV) can then be calculated by dividing the number of points that satisfy the constraints by the total number of sampled points. However, this method can be computationally expensive when dealing with high-dimensional design spaces. To increase efficiency, linear interpolation on a regular sampling grid can be used to generate additional points within the feasible space, thereby reducing the number of samples needed for accurate estimation without sacrificing computational power.

The signal-to-noise ratio (S/N) is another important metric used to measure the decline in performance metrics such as stability, load capacity, or efficiency across the feasible region defined by HV. In order to optimize the performance metric f, Equation (7) is used to maximize it, while Equation (8) is used to minimize it. The terms μ and σ^2 represent the mean and variance, respectively, and since S/N is always maximized in optimization, these two definitions ensure that μ is either maximized or minimized, while σ^2 is minimized. As an optimization objective, we employ the average ($\overline{S/N}$) of the signal-to-noise ratio for stability, losses, and load capacity of the HGJB and SGTB.

$$S/N_{\mathrm{f}} = 10 \cdot \log_{10}\left(\frac{\mu_{\mathrm{f}}^2}{\sigma_{\mathrm{f}}^2}\right) \qquad (7)$$

$$S/N_{\mathrm{f}} = -10 \cdot \log_{10}\left(\mu_{\mathrm{f}}^2 + \sigma_{\mathrm{f}}^2\right) \qquad (8)$$

Maximizing the signal-to-noise ratio (S/N) enables the optimization of performance metrics while controlling their gradients. When combined with the maximization of HV, it leads to a large feasible region with minimized gradients of the performance metrics.

2.4 Constrained Multi-Objective Optimization

Design optimization must consider a range of objectives and constraints to achieve a feasible and optimal solution. It is insufficient to optimize subsystems independently, as the interactions and interdependencies between subsystems must be accounted for. For the design of an electrically-driven compressor system supported by gas bearings, this means optimizing all subsystems simultaneously while meeting system requirements.

Specifically, the electric motor must deliver sufficient power to drive the impeller to the desired pressure ratio for a given mass flow rate, while the axial and radial bearings must be designed to support the impeller wheel's axial load and lift off the rotor, respectively. To ensure a stable design, rotordynamics and axial

Copyright © 2023 by ASME

TABLE 1: DESCRIPTION OF THE HYPERPARAMETERS SEARCHED FOR THE OPTIMIZATION OF THE FEED-FORWARD NEURAL NETWORKS.

Term	Symbol	Value
Number of neurons per hidden layer	n	16, 32, 64, 128, 256
Number of hidden layers	l	2, 3, 4
Activation	a	relu, selu, tanh, softplus, softsign
Optimiser	opt	Adam, Adamax, Adadelta, Adagrad
Batch size	bs	$2^{12}, 2^{13}, 2^{14}, 2^{15}/2^9, 2^{10}, 2^{11}, 2^{12}, 2^{13}, 2^{14}$
Kernel initialiser	ki	Glorot Normal (gn), He Normal (hn), Lecun Normal (ln), Glorot Uniform (gu), He Uniform (hu), Lecun Uniform (lu)
L2 penalisation	β_{L2}	$10^{-6}, 10^{-5}, 10^{-4}, 10^{-3}, 10^{-2}$
Learning rate	α_{lr}	0.01, 0.01, 0.1
Decay steps	ds	$10^3, 10^4, 10^5$

dynamics must be taken into consideration, with the rotor also satisfying bending frequency requirements to prevent destruction by resonance. Structural integrity must also be addressed by imposing constraints on the equivalent von Mises stress. Furthermore, the torque transmission between the rotor shaft and the magnet, as well as between the plug and the rotor shaft, are critical considerations.

3. METHODS

3.1 Impeller Wheel Representation

The turbocompressor sections are represented by multi-layer hollow cylinders to enable fast computation of the mass and moments of inertia of the entire turbocompressor unit, incorporating radial and axial bearings, shaft sections, and the electric motor. However, to accurately represent the impeller and its blades, a slicing method is required. The hub can be modeled as a stack of cylinders with varying radii. To extract mass and moments of inertia, a mapping from the 1D representation to a 3D representation of the impeller wheel must be defined. The meridional geometry of the hub and blades is defined by three ellipses, one arc of a circle, and the golden ratio ϕ, as shown in Fig 2. The hub and blades are parametrized by piecewise functions, presented in Eq. (10) and Eq. (11), respectively.

Neglecting the impeller blades can lead to significant deviations in the system's mass (m), polar moment of inertia (I_p), and transverse moment of inertia (I_t). As the impeller is rigidly attached at one end of the shaft in the studied rotor configurations, its contribution to the overall system stability cannot be underestimated. To account for this, the impeller and its blades are modeled as hollow cylinders, while maintaining the mass m and transverse moment of inertia I_t. The inner radius of the cylinders is set equal to the outer radius of the impeller hub, and their outer radius R_{cyl} and density ρ_{cyl} are determined by solving a system of two equations and two unknowns for each cylinder section, as presented in in Eq. (12) and Eq. (13).

To obtain a representation of the blades as equivalent hollow cylinders, rectangular prisms are used to bound the meridional hub and blade geometries. The splitters are accounted for by doubling the number of blades and spanning the full blade length, which provides the most conservative approximation as it increases the transverse moment of inertia of the whole im-

peller. This approximation is particularly relevant for stability computations. The blades and splitters are assumed to have equal thickness and are evenly distributed around the circumference of the hub to obtain the transverse moment of inertia. The moments of inertia of the rectangular prism about its principal axes are then transformed using a rotation matrix for each blade and respective angle θ, as computed in Eq. (14).

The transverse moment of inertia of each blade with respect to its angular position θ is computed using the parallel-axis theorem and can be found in Eq. (15), with the mass of each blade calculated using Eq. (16). The number of blades N_{bld} and the number of splitters N_{spl} determine the angular separation $\Delta\theta$ between each splitter and blade, which is given in Eq. (17). The moment of inertia of the nth blade or splitter section spanning around the hub can be computed using Eq. (18). To obtain an equivalent hollow cylinder section, the total transverse moment of inertia I_t is calculated by summing the transverse moments of inertia of each blade and splitter section, and the same procedure is used for mass. Finally, the outer radius R_{cyl} and density ρ_{cyl} of each cylinder section are determined by solving the system of two equations and two unknowns using Eq. (12) and Eq. (13).

3.2 Design Rules

To enable the optimization process for the entire turbocompressor unit, it is necessary to establish design rules. The first rule concerns the positioning of the five subsystems in a single-stage electrically driven turbocompressor for heat pump applications with gas bearings, namely the compressor (C), axial bearing (S), radial bearings A and B, and electric motor (E). As there are five subsystems, there are $5! = 120$ permutations. However, as radial bearings A and B are identical, they are insensitive to permutation. To minimize axial deviation of the impeller due to thermal expansion of the rotor shaft during operation, the spiral groove thrust bearing is placed next to the impeller. Therefore, the impeller wheel can only be located at one of the two free ends of the rotor shaft, followed directly by the axial bearing. Any other placement of the impeller would impose difficult constraints in assembly, manufacturing, aerodynamic performance, and placing of the volute. This leaves only two permitted permutations for consideration in the optimization: 'CSABE' or 'CSAEB', where the electric motor is either at the other free end or placed be-

Copyright © 2023 by ASME

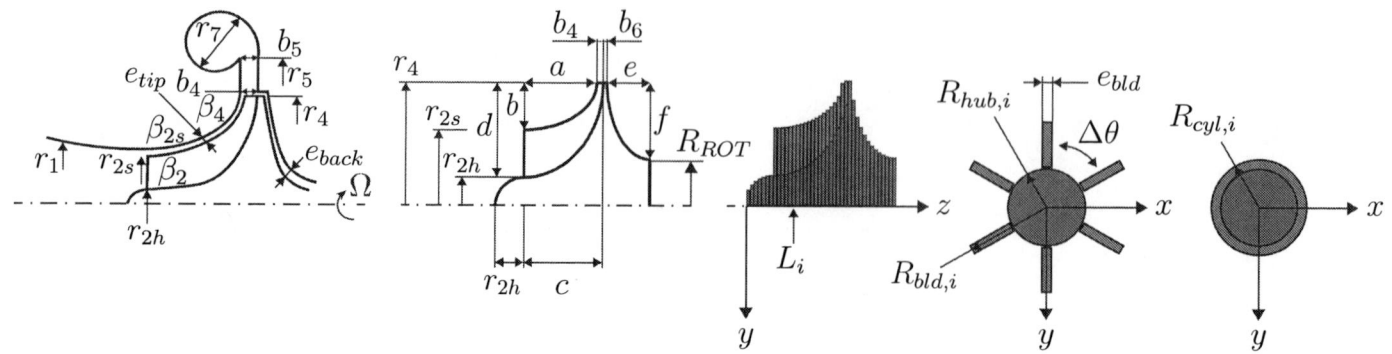

FIGURE 2: SCHEMATIC REPRESENTATION OF THE IMPELLER WHEEL TRANSFORMATION TO A 2-LAYER CYLINDER MODEL. THE MERIDIONAL GEOMETRY OF THE HUB AND BLADES ARE DEFINED BY THREE ELLIPSES, ONE ARC OF A CIRCLE, AND THE GOLDEN RATIO ϕ. THE HUB AND BLADE GEOMETRIES ARE PARAMETRIZED USING PIECEWISE FUNCTIONS. THE TRANSFORMATION FROM THE IMPELLER TO THE FULL 2-LAYER CYLINDER REPRESENTATION IS PERFORMED USING THE FUNCTIONS DEFINED IN THE ANNEX.

tween the radial bearings to bring the center of gravity between both bearings. After eliminating the permutations resulting from pure symmetry, this study focuses on the 'CSABE' layout, with the compressor on the left end and the electric motor on the right end, as shown in Fig. 3. The optimization process involves various geometrical variables, and the front and unrolled views of the SGTB and HGJB are presented. The impeller wheel is made of metal, such as stainless steel or aluminum, while the plug connecting the impeller and the rotor is made of inconel, a Nickel-based alloy. The rotor shaft, which can be hollow, has the same outer radius R_{ROT} throughout its length, including the radial bearings. The magnet, made of neodymium or samarium-cobalt, is inserted into the shaft by interference. The type of metal, hard metal or magnet used is also an optimization decision variable.

3.3 Constrained Multi-Objective Optimization Setup

The optimization process employed in this study utilized the Non-dominated Sorting Genetic Algorithm III (NSGA-III) [23, 24], which is a widely used evolutionary algorithm for multi-objective optimization. The algorithm was implemented using Python [25]. To enhance the efficiency of the optimization process, an adaptive-operator selection procedure was utilized. This procedure adapts the selection of genetic operators to the problem's characteristics, as proposed by Vrugt and Robinson [26] and Hádka and Reed [27]. Additionally, to guide the optimization process, a set of uniformly sampled reference directions was used, which was proposed by Das and Dennis [28]. The optimization process was run for 50 generations.

The optimization process focuses on four key objectives, as outlined in Table 2. Firstly, the objective is to maximize the geometric mean of the feasible regions, represented by $\overline{HV}$, for both the radial bearings and the axial bearing. This objective accounts for manufacturing deviations. Secondly, the objective is to maximize HV_{COMP}, which ensures the attainment of the largest

compressor maps. To further evaluate the performance, the mean signal-to-noise ratio of the radial bearings ($\overline{S/N_{HGJB}}$) is computed by considering the average of the signal-to-noise ratios associated with load capacity, logarithmic decrement (stability), and losses. Similarly, the mean signal-to-noise ratio of the axial bearing ($\overline{S/N_{SGTB}}$) is determined. Additionally, the signal-to-noise ratio of the isentropic efficiency of the compressor ($S/N_{\eta_{is,COMP}}$) is considered. The third objective aims to maximize the harmonic mean of $\overline{S/N_{HGJB}}$, $\overline{S/N_{SGTB}}$, and $S/N_{\eta_{is,COMP}}$, which provides an overall assessment of the signal-to-noise ratios with respect to manufacturing deviations and the range of operating conditions. Lastly, the overall efficiency of the machine, denoted as η_{tot}, is maximized. η_{tot} is defined as the ratio of the isentropic power work of the compressor ($\dot{E}_{is,COMP}$) to the sum of its mechanical power work ($\dot{E}_{COMP} = \dot{E}_{is,COMP}/\eta_{is,COMP}$), losses in the two radial bearings ($\dot{E}_{HGJB}$), losses in the axial bearing ($\dot{E}_{SGTB}$), and windage loss of the electric motor ($\dot{E}_{EM,loss}$). These four objectives collectively drive the optimization process and contribute to enhancing the performance and efficiency of the turbocompressor system.

$$\eta_{tot} = \frac{\dot{E}_{is,COMP}}{\frac{\dot{E}_{is,COMP}}{\eta_{is,COMP}} + 2\dot{E}_{HGJB} + \dot{E}_{SGTB} + \dot{E}_{EM,loss}} \tag{9}$$

TABLE 2: OBJECTIVES OF THE MULTI-OBJECTIVE OPTIMIZATION

Term	Symbol	Objective type	Unit
Objectives			
Geometric mean of bearings feasible regions	$\overline{HV}$	Maximize	μm
Feasible region compressor	HV_{COMP}	Maximize	RPM kg s^{-1}
Harmonic mean of the signal-to-noise ratios	$\overline{S/N}$	Maximize	–
System efficiency	η_{tot}	Maximize	–

The optimization problem incorporates several constraints

Copyright © 2023 by ASME

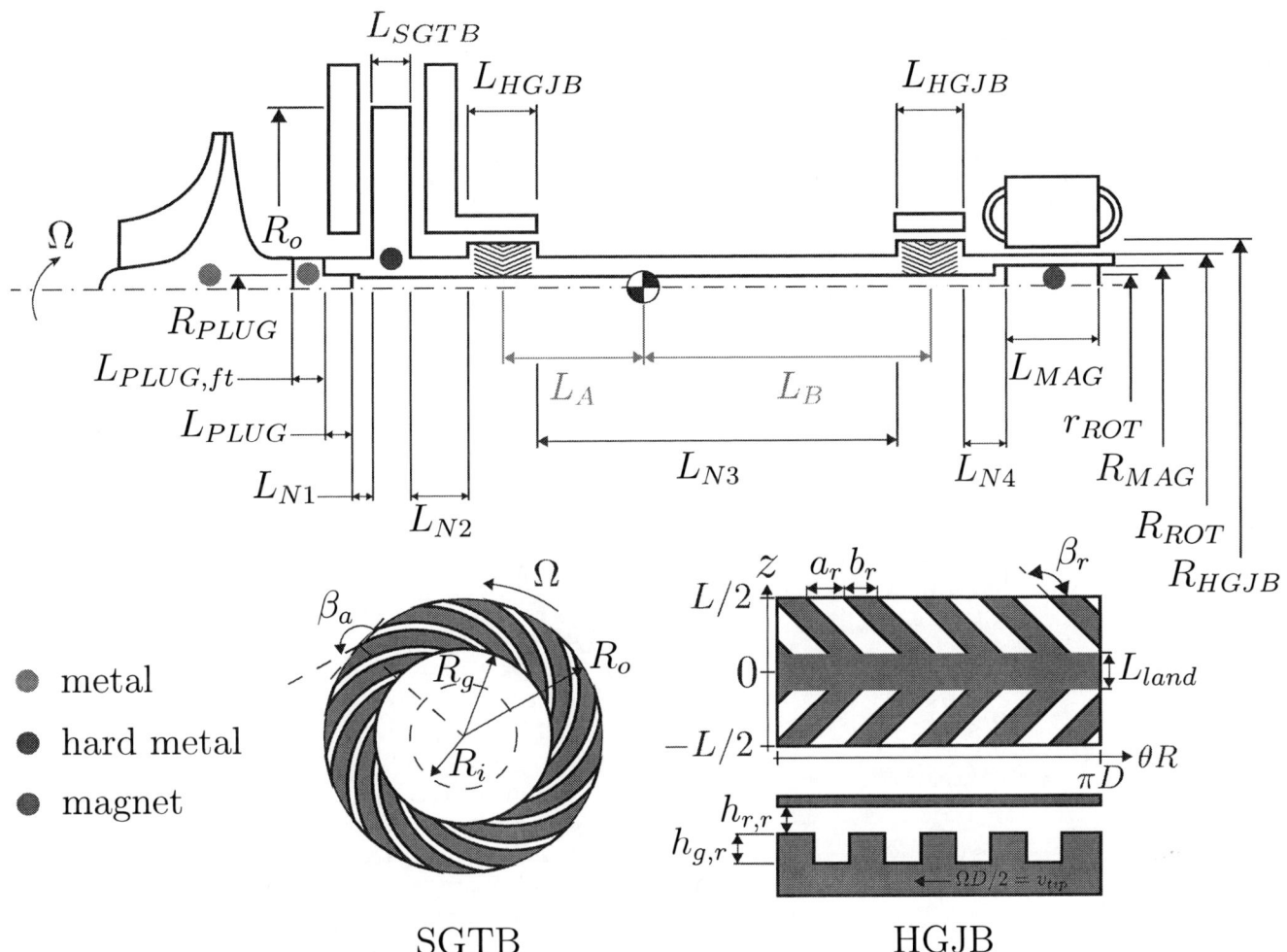

FIGURE 3: THE COMPLETE TURBOCOMPRESSOR UNIT WITH ALL GEOMETRIC VARIABLES USED IN OPTIMIZATION, INCLUDING LENGTHS AND RADII, AND MATERIALS SPECIFIED. THE SCHEMATIC DIAGRAMS OF THE SPIRAL GROOVE THRUST BEARINGS (SGTB) AND HERRIGN-BONE GROOVED JOURNAL BEARINGS (HGJBS) ARE SHOWN WITH THEIR RESPECTIVE VARIABLES. THE LAYOUT OF THE SUBSYSTEMS FOLLOWS THE 'CSABE' PERMUTATION.

to ensure the feasibility and reliability of the designed turbocompressor system, as summarized in Table 3. First, a load capacity condition is enforced to ensure rotor lift-off with a safety margin of 20% at a rotor speed of 10,000 rpm. The radial expansion is limited to a maximum of 2 micrometers to allow for designs with low nominal bearing clearance, while ensuring mechanical stability. A position constraint is also imposed on the radial bearings relative to the center of gravity (CG) of the system, requiring that their dot product be negative to ensure they are located on opposite sides of the CG. Additionally, stable designs are chosen with a safety margin of 0.1 on the logarithmic decrement of the rotordynamics of the radial bearings (Γ_{HGJB}) and the driven dynamics of the axial bearing (Γ_{SGTB}). The compressibility numbers (Λ_{HGJB} and Λ_{SGTB}) must remain below 60 and 120, respectively, to ensure that the optimization stays within the range of the training data used in the surrogate models. To account for manufacturing deviations in bearings, at least 30% of the sampled region must satisfy the constraints. A minimum sampling rate of 1%

is set for the compressor map to ensure feasible designs. Constraints are also applied to the excitation frequency to ensure that the cylindrical forward excitation frequency is greater than the backward mode frequency for most points, and that it increases monotonically at a decreasing rate. The optimization process avoids designs that exhibit a specific type of Pareto optimum in which the cylindrical forward excitation frequency suddenly collapses to a subsynchronous mode. The center of gravity should not coincide with the bearings, and this is ensured by enforcing constraints on the distances of the CG from each radial bearing midplane ($\overline{L_A} > 0.5$ and $\overline{L_B} > 0.5$). Furthermore, constraints are applied on the equivalent von Mises stress to guarantee the structural integrity of the turbocompressor components, including the magnet, plug, and rotor. The electric motor must provide sufficient power to compensate for all losses and drive the compressor, while ensuring that interference between the magnet and shaft and between the plug and shaft provide enough friction to transmit torque. The von Mises stress must remain below half

Copyright © 2023 by ASME

the yield strength, taking into account various operating conditions. A similar safety factor is applied to the radial and axial bearings. Finally, the maximum pressure ratio observed in the compressor map is bounded to be at least 4, and a lower boundary of 0.1 is set on the compressor isentropic efficiency to ensure adequate performance. In order to mitigate the risk of rotor failure resulting from resonance, a conservative approach is taken by setting the maximum rotational speed to 83% of the bending frequency. This precautionary measure provides a significant safety margin, considering the small clearance observed at high speeds. By adhering to this limitation, the system ensures the integrity and durability of the rotor under operating conditions prone to resonance-induced catastrophic failures.

TABLE 3: CONSTRAINTS OF THE MULTI-OBJECTIVE OPTIMIZATION

Description	Constraint	Unit
Lift force	$F_{\text{HGJB, 10 kRPM}} > 1.20 \cdot F_W$	N
Centrifugal growth	$\Delta h_{\text{r,r,exp}} < 2 \cdot 10^{-6}$	m
CG between bearings	$\mathbf{GA} \cdot \mathbf{GB} < 0$	m^2
HGJB stability	$\Gamma_{\text{HGJB}} > 0.1$	–
SGTB stability	$\Gamma_{\text{SGTB}} > 0.1$	–
HGJB compressibility number	$\Lambda_{\text{HGJB}} < 60$	–
SGTB compressibility number	$\Lambda_{\text{SGTB}} < 120$	–
HGJB feasible region	$HV_{\text{HGJB}} > 0.3$	–
SGTB feasible region	$HV_{\text{SGTB}} > 0.3$	–
Compressor feasible region	$HV_{\text{COMP}} > 0.01$	–
Cylindrical excitation modes	$\Omega_{\text{ex,cyl\_F}} > \Omega_{\text{ex,cyl\_B}}$	RPM
Cylindrical forward concavity	$\frac{\partial^2 \Omega_{\text{ex,cyl_F}}}{\partial N^2} < 0$	RPM^{-1}
Cylindrical forward growth	$\frac{\partial \Omega_{\text{ex,cyl_F}}}{\partial N} > 0$	–
CG away from bearing A	$\overline{L_A} > 0.5$	–
CG away from bearing B	$\overline{L_B} > 0.5$	–
Plug-rotor interference	$\delta_{\text{PLUG}} > 0$	m
Plug torque transmission ratio	$S_{\tau,\text{PLUG}} > 1$	–
Plug von Mises safety factor	$S_{\sigma_{\text{VM}},\text{PLUG}} > 2$	–
Magnet-rotor interference	$\delta_{\text{MAG}} > 0$	m
Magnet torque transmission ratio	$S_{\tau,\text{MAG}} > 1$	–
Magnet von Mises safety factor	$S_{\sigma_{VM},\text{MAG}} > 2$	–
SGTB von Mises safety factor	$S_{\sigma_{VM},\text{SGTB}} > 2$	–
HGJB von Mises safety factor	$S_{\sigma_{VM},\text{HGJB}} > 2$	–
SGTB axial force	$F_{\text{SGTB}} > F_{\text{COMP}}$	N
Electric motor power	$\dot{E}_{\text{EM}} > \dot{E}_{\text{losses}} + \dot{E}_{\text{COMP}}$	W
Turbocompressor rotor length	$N_{\text{end}} < 1.20 \cdot 60 \cdot f_{\text{bend}}$	RPM
Compressor isentropic efficiency	$\eta_{\text{is,comp}} > 0.1$	–
Compressor pressure ratio	$\max(\Pi) > 4$	–

The decision variables for the multi-objective optimization are listed in Tab. 4 along with their respective ranges. The impeller wheel variables, which determine the wheel's characteristics, are open for optimization. Wheels with a maximum tip radius of $r_4 = 35$ mm are permitted. The inlet blade angles β_2 and β_{2s} are fixed, and the number of splitter blades is set equal to the number of blades. The operating conditions are also fixed at an inlet pressure of 2.51 bar and an inlet temperature of 300 K using R134a refrigerant. These conditions are applied to the bearings while considering the fluid viscosity. The two HGJBs are identical, so only one HGJB's geometry is optimized. The geometry of the SGTB is also used as an input to the optimizer. The rotor geometry is bounded by the inner radius r_{ROT} and the

outer radius R_{ROT}. The pockets that hold the plug and magnet have radii R_{PLUG} and R_{MAG} with respective interference δ_{PLUG} and δ_{MAG}. Lengths and radii are defined as ratios with respect to R_{ROT}. Materials are defined as float values between 0 and 1 and are mapped to integers to select all different types of metals for the compressor, hard metal for the rotor, and magnet. To ensure robustness of the bearings, the optimizer selects the feasible range for the deviations of local bearing clearance and groove depth, within which the largest possible deviations are chosen subject to constraints ensuring functional and operational requirements. The sampling method is fixed as a linspace, with the boundaries of the linspace changing with the selected deviation range. The choice of deviation range balances the need for robustness with the need for sufficient sampling to detect feasible regions in the design. Seven points are used to sweep each variable for manufacturing deviations. Due to the compressor map's consideration, a sampling of 13 points is made for the rotational speed, and 13 points are swept for the mass flow.

4. RESULTS

The results of the multi-objective optimization are presented in the form of pairplots, as depicted in Figure 4. The solution with the largest geometric mean of the bearings feasible regions ($\overline{HV}$) is indicated by a red dot on the scatter plots. The optimization process involved a search of 1716 nominal designs over 500 generations, resulting in a total of 1.2 billions samples. The optimization was completed within approximately 1 day using a desktop computer equipped with a 12-core AMD Ryzen 3900X CPU and a Nvidia RTX 3090 GPU.

The pairplots shown in Fig. 4 illustrate the trade-off between robustness against manufacturing deviations ($\overline{HV}$) and robustness against variance in the performance metric ($\overline{S/N}$) which are negatively correlated. The diagonal of the pairplots displays the distribution of each objective over the range covered. On average, the geometric mean of the feasible region of the radial and axial bearings is of 20 µm^2. $\overline{HV}$ and HV_{COMP} are negatively correlated to global efficiency (η_{tot}), highlighting the trade-offs involved in achieving robustness and an efficient design.

The turbocompressor that was selected for the study is presented in Figure 5, whereby its axial and radial dimensions are expressed in millimeters. The figure displays the turbocompressor map and response surfaces in the form of contour plots for the axial dynamics (Γ_{SGTB}) and rotordynamics (Γ_{HGJB}) against manufacturing deviations, with white lines indicating evaluations conducted using baseline models to monitor the accuracy of the surrogate models in predicting system dynamics. Notably, the rotordynamics is predicted accurately, with slight overprediction observed for the axial dynamics. The study findings indicate that local bearing clearance and groove depth deviations, of ±5 µm and ±1.5 µm, can be safely achieved respectively for the axial bearing and radial bearings. The optimized turbocompressor rotor has a mass of 412 g and midplane bearing distances to the center of gravity of $L_A = 28.8$ mm and $L_B = 22.1$ mm. The length-to-diameter aspect ratio of the HGJB is of $LoD = 1.4$. The impeller wheel is composed of aluminum, while the magnet and rotor are made of neodymium and tungsten carbide, respectively. Finally, the study results reveal that a pressure ratio of 4

Copyright © 2023 by ASME

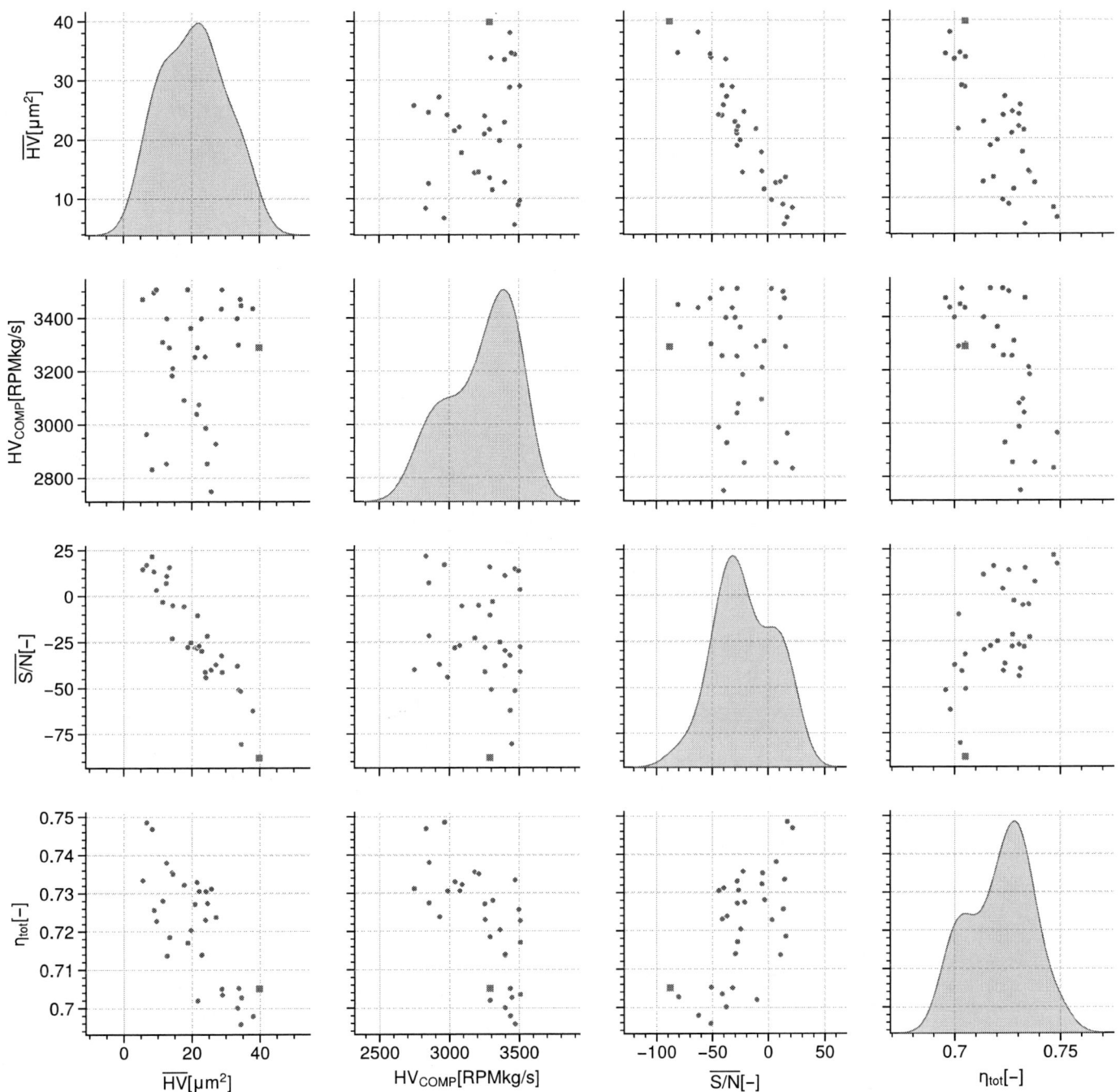

FIGURE 4: PARETO FRONT OF FOUR OBJECTIVES FROM TURBOCOMPRESSOR OPTIMIZATION PRESENTED AS PAIRPLOTS WITH SELECTED SOLUTION (RED DOT) FOR LARGEST GEOMETRIC MEAN OF THE FEASIBLE REGION OF THE BEARINGS ($\overline{HV}$). DIAGONAL SHOWS OBJECTIVE DISTRIBUTION.

is reached for $N_{end} = 162\,737$ RPM which is much lower than the bending frequency of $N_{bend} = 216\,532$ RPM. The compressor can operate with an isentropic efficiency higher than 0.8 over a large portion of the compressor map, for speeds ranging from 50 kRPM to 163 kRPM, for a consumed compressor power ranging from 300 W to 3000 W.

5. DISCUSSION

The competition between robustness metric $\overline{HV}$ and $\overline{S/N}$ are consistent with those reported in previous studies. Massoudi and Schiffmann [6] also found that increasing the feasible region leads to a trade-off between robustness with respect to constraints and robustness with respect to signal-to-noise ratio. The larger

Copyright © 2023 by ASME

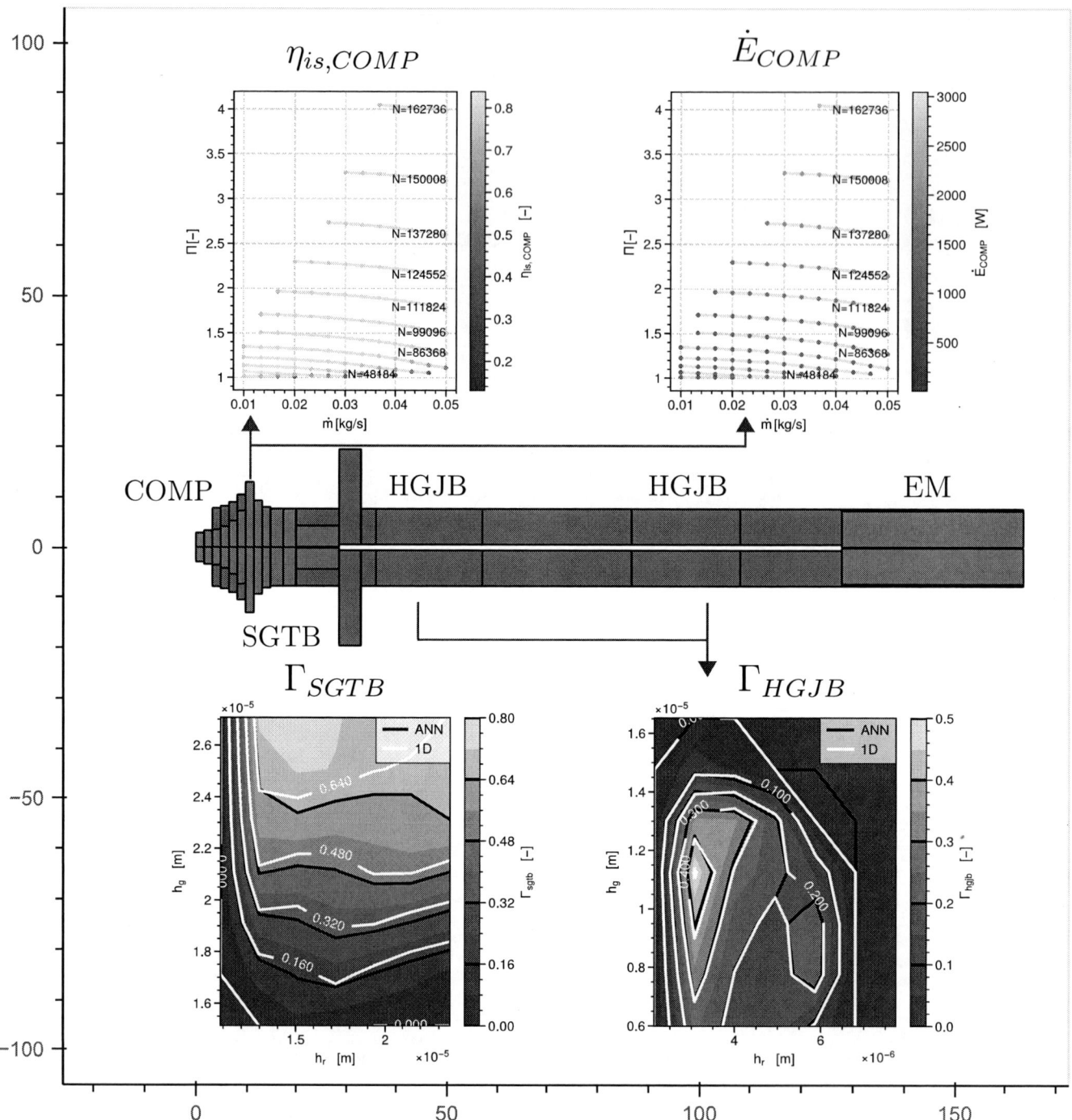

FIGURE 5: THE SELECTED TURBOCOMPRESSOR DESIGN WITH MIDPLANE BEARING DISTANCES OF $L_A = 28.8\,\text{mm}$ AND $L_B = 22.1\,\text{mm}$, HGJB LENGTH TO DIAMETER ASPECT RATIO OF 1.4 AND A MASS OF $412\,\text{g}$.

the feasible region, the more difficult indeed to maintain a given performance metric constant.

The selected solution, aiming to maximize the feasible region of the bearings, demonstrates a radial bearing length-to-diameter aspect ratio (LoD) close to 1.4. Notably, the midplane distances from the bearings to the center of gravity, although not entirely equal, are found to be in close proximity. The observed discrepancy can be attributed to the selection of aluminum as

the impeller material. Although this choice reduces the overall weight of the turbocompressor, the inclusion of the necessary magnet on the opposite end of the shaft shifts the center of gravity towards the right. In order to address this, an extension in the shaft length could potentially shift the center of gravity towards the left. However, such a modification would inevitably result in a decrease in bending frequency. These findings align with Massoudi and Schiffmann's recent study on the robustness of gas

TABLE 4: DESCRIPTION OF THE PARAMETERS FOR THE MULTI-OBJECTIVE OPTIMIZATION

Term	Symbol	Range/Value	Unit
Impeller Variables			
Tip radius	r_4	$7 \cdot 10^{-3} - 35 \cdot 10^{-3}$	m
Inducer hub radius ratio	r_{2h}/r_4	$0.1 - 0.3$	–
Inlet shroud radius ratio	r_{2s}/r_{2h}	$1.2 - 2.3$	–
Inducer inlet radius ratio	r_1/r_{2s}	$1.05 - 1.3$	–
Diffuser exit radius ratio	r_5/r_4	$1.05 - 1.5$	–
Tip width ratio	b_4/r_4	$0.015 - 0.3$	–
Tip clearance ratio	e_{tip}/b_4	$0.01 - 0.015$	–
Backface clearance ratio	e_{back}/r_4	$0.001 - 0.15$	–
Inducer length ratio	L_{ind}/r_4	$1.05 - 4$	–
Exit blade angle	β_4	$-45 - 0$	°
Blade thickness	e_{bld}	$0.1 - 0.5 \cdot 10^{-3}$	m
Number of blades	N_{bld}	$5 - 11$	–
HGJB Variables			
Groove width ratio	α_r	$0.32 - 0.68$	–
Groove angle	β_r	$-167.5 - -122.5$	°
Grooved land region ratio	γ_r	$0.52 - 0.97$	–
Groove depth	$h_{g,r}$	$2.5 \cdot 10^{-6} - 28.5 \cdot 10^{-6}$	m
Local bearing clearance	$h_{r,r}$	$2.5 \cdot 10^{-6} - 28.5 \cdot 10^{-6}$	m
$h_{g,r}$ deviations	$\Delta h_{g,r}$	$1 \cdot 10^{-6} - 10 \cdot 10^{-6}$	m
$h_{r,r}$ deviations	$\Delta h_{r,r}$	$1 \cdot 10^{-6} - 10 \cdot 10^{-6}$	m
SGTB Variables			
Groove width ratio	α_a	$0.32 - 0.68$	–
Groove angle	β_a	$-167.5 - -122.5$	°
Grooved land region ratio	γ_a	$0.1 - 0.9$	–
Groove depth	$h_{g,a}$	$2.5 \cdot 10^{-6} - 28.5 \cdot 10^{-6}$	m
Local bearing clearance	$h_{r,a}$	$2.5 \cdot 10^{-6} - 28.5 \cdot 10^{-6}$	m
$h_{g,a}$ deviations	$\Delta h_{g,a}$	$1 \cdot 10^{-6} - 10 \cdot 10^{-6}$	m
$h_{r,a}$ deviations	$\Delta h_{r,a}$	$1 \cdot 10^{-6} - 10 \cdot 10^{-6}$	m
Rotor Variables			
Rotor outer radius	R_{ROT}	$5 \cdot 10^{-3} - 30 \cdot 10^{-3}$	m
Rotor inner radius ratio	r_{ROT}/R_{ROT}	$0 - 0.95 \cdot 10^{-3}$	–
Plug radius ratio	R_{PLUG}/R_{ROT}	$0.3 - 0.95$	–
SGTB radius ratio	R_o/R_{ROT}	$1.15 - 10$	–
Magnet radius ratio	R_{MAG}/R_{ROT}	$0.3 - 0.95$	–
Segment 1 length ratio	L_{N1}/R_{ROT}	$0.3 - 12$	–
Segment 2 length ratio	L_{N2}/R_{ROT}	$0.3 - 12$	–
Segment 3 length ratio	L_{N3}/R_{ROT}	$0.3 - 12$	–
Segment 4 length ratio	L_{N4}/R_{ROT}	$0.3 - 12$	–
SGTB length ratio	L_{SGTB}/R_o	$2/7 - 5/9$	–
HGJB length ratio	L_{HGJB}/R_{ROT}	$1 - 4$	–
Magnet length ratio	L_{MAG}/R_{ROT}	$1 - 12$	–
Nominal rotor-plug interference	δ_{PLUG}	$1 \cdot 10^{-6} - 100 \cdot 10^{-6}$	m
Nominal rotor-magnet interference	δ_{MAG}	$1 \cdot 10^{-6} - 100 \cdot 10^{-6}$	m
Material Variables			
Impeller wheel material	Mat_{COMP}	$0 - 1$	
Rotor material	Mat_{ROT}	$0 - 1$	
Magnet material	Mat_{MAG}	$0 - 1$	
Operating Variables			
Maximum rotor speed	N_{end}	$1.5 \cdot 10^5 - 5 \cdot 10^5$	RPM
Dependent parameters			
HGJB radius	R_{HGJB}	R_{ROT}	m
Front plug length	$L_{PLUG,ft}$	$1/3 \cdot R_{ROT}$	m
Number of splitter blades	N_{splits}	N_{bld}	–
Fixed parameters			
Startup rotor speed	N_{start}	$2 \cdot 10^4$	RPM
Robustness sampling unit	k_{rob}	7	–
Speed sampling sweep	k_N	13	–
Mass flow sweep	k_m	13	–
Fluid		R134a	
Compressor inlet pressure	P_{in}	$2.51 \cdot 10^5$	Pa
Compressor inlet temperature	T_{in}	300	K
Mass flow lower bound	$\dot{m}_{start}$	10	$g\,s^{-1}$
Mass flow upper bound	$\dot{m}_{end}$	50	$g\,s^{-1}$
Compressor inlet blade angle at hub	β_2	-56	°
Compressor inlet blade angle at shroud	β_{2s}	-60	°

bearing supported rotors [5]. Their research suggests that a symmetrical design, or a design with a large *LoD*, offers enhanced robustness against manufacturing deviations in radial bearings. In the pursuit of higher system efficiency, this optimization has led to a reduction in *LoD* to minimize losses through a shorter bearing length.

The maximization of compressor isentropic efficiency is an integral aspect of optimizing the overall efficiency, denoted as η_{tot}. Notably, the isentropic efficiency exhibits a consistently high value, reaching 0.8 across a significant portion of the compressor map. This observation aligns with Schiffmann and Favrat's comprehensive study on optimal compressor designs for both single and multiple operating points [12]. Their study suggests that the best efficiency is achieved at the nominal speed of $N_{nom} = 130$ kRPM for a first stage pressure ratio of $\Pi = 2.4$. In accordance with their findings, the operating point selected for this study corresponds to their A2 operating point.

6. CONCLUSION

This study has introduced an automated framework for the integrated design of gas bearings supported turbocompressors while considering manufacturing deviations. This was made possible by the use of constrained multi-objective optimization and surrogate models made of ensembles of feed-forward neural networks. This allowed us to bypass the traditional sequential approach, integrating the optimization of all subsystems in one loop. The results clearly indicate the gain in computational time for such an approach and clearly demonstrate its strength compared to a traditional integrated nominal optimization. To the best of our knowledge, it is the first time the design of such a system has been done by considering both the integration and the robustness.

Future work will focus on the variation of the selection of different rotor layouts and subsystems configurations. Furthermore, this methodology can be extended to other fields in engineering that require the integration of multiple subsystems and consideration of robustness against manufacturing deviations. The presented framework highlights the importance of a comprehensive and integrated approach to system design and provides a valuable foundation for future research in this area.

ACKNOWLEDGMENTS

The authors acknowledge the MIT SuperCloud and Lincoln Laboratory Supercomputing Center for providing high-power computing resources that have contributed to the research results reported within this paper.

REFERENCES

[1] Wehner, M., Truby, R. L., Fitzgerald, D. J., Mosadegh, B., Whitesides, G. M., Lewis, J. A. and Wood, R. J. "An Integrated Design and Fabrication Strategy for Entirely Soft, Autonomous Robots." *nature* Vol. 536 No. 7617 (2016): pp. 451–455. DOI 10.1038/nature19100.

[2] Picard, C. and Schiffmann, J. "Automated design tool for automotive control actuators." *International Design Engineering Technical Conferences and Computers and Information in Engineering Conference*, Vol. 84010: p. V11BT11A027. 2020. American Society of Mechanical Engineers. DOI 10.1115/DETC2020-22390.

[3] Schiffmann, J. "Integrated design and multi-objective optimization of a single stage heat-pump turbocompressor."

Journal of Turbomachinery Vol. 137 No. 7 (2015): p. 071002. DOI 10.1115/1.4029123.

[4] Guenat, E. and Schiffmann, J. "Multi-Objective Optimization of Grooved Gas Journal Bearings for Robustness in Manufacturing Tolerances." *Tribology Transactions* Vol. 62 No. 6 (2019): pp. 1041–1050. DOI 10.1080/10402004.2019.1642547.

[5] Massoudi, S. and Schiffmann, J. "Robust Design of Herringbone Grooved Journal Bearings using Multi-Objective Optimization assisted with Artificial Neural Networks." *Turbomachinery Technical Conference and Exposition*, Vol. 18. 2023. American Society of Mechanical Engineers.

[6] Massoudi, S. and Schiffmann, J. "Ensemble Neural Network Modeling of Gas Bearings Supported Rotors: A Global Surrogate Approach in Multi-Objective Optimization for Robust Design." 2023.

[7] Vohr, J. H. and Chow, C. Y. "Characteristics of Herringbone-Grooved, Gas-Lubricated Journal Bearings." *Journal of Basic Engineering* Vol. 87 No. 3 (1965): pp. 568–576. DOI 10.1115/1.3650607.

[8] Nelson, H. D. and McVaugh, J. M. "The Dynamics of Rotor-Bearing Systems Using Finite Elements." *Journal of Engineering for Industry* Vol. 98 No. 2 (1976): pp. 593–600. DOI 10.1115/1.3438942.

[9] Nelson, H. D. "A Finite Rotating Shaft Element Using Timoshenko Beam Theory." *Journal of Mechanical Design* Vol. 102 No. 4 (1980): pp. 793–803. DOI 10.1115/1.3254824.

[10] Lepe, F., Mora, D. and Rodríguez, R. "Locking-free finite element method for a bending moment formulation of Timoshenko beams." *Computers & Mathematics with Applications* Vol. 68 No. 3 (2014): pp. 118–131. DOI 10.1016/j.camwa.2014.05.011.

[11] Picard, C., Schiffmann, J. and A., Faez. "DATED: Guidelines for Creating Synthetic Datasets for Engineering Design Applications." *International Design Engineering Technical Conferences and Computers and Information in Engineering Conference*. 2023. American Society of Mechanical Engineers. DOI 10.1115/DETC2023-111609.

[12] Schiffmann, J. and Favrat, D. "Design, experimental investigation and multi-objective optimization of a small-scale radial compressor for heat pump applications." *Energy* Vol. 35 No. 1 (2010): pp. 436–450. DOI 10.1016/j.energy.2009.10.010. Accessed 2021-05-07, URL https://www.sciencedirect.com/science/article/pii/S0360544209004435.

[13] Miller, Timothy John Eastham. *Switched reluctance motors and their control*. Magna physics publishing and clarendon press (1993).

[14] Rosset, K. and Schiffmann, J. "Extended Windage Loss Models for Gas Bearing Supported Spindles Operated in Dense Gases." *Journal of Engineering for Gas Turbines and Power* Vol. 142 No. 6 (2020). DOI 10.1115/1.4047124. Accessed 2023-03-13, URL https://doi.org/10.1115/1.4047124.

[15] Guenat, E. and Schiffmann, J. "Effects of humid air on aerodynamic journal bearings." *Tribology International* Vol. 127 (2018): pp. 333–340.

DOI 10.1016/j.triboint.2018.06.002. Accessed 2020-04-27, URL http://www.sciencedirect.com/science/article/pii/S0301679X18302883.

[16] Olmedo, L. E. and Schiffmann, J. "Towards a real-time capable hybrid-twin for gas-bearing supported high-speed turbocompressors." *Energy* Vol. 275 (2023): p. 127385. DOI 10.1016/j.energy.2023.127385. Accessed 2023-05-15, URL https://www.sciencedirect.com/science/article/pii/S036054422300779X.

[17] Massoudi, S., Picard, C. and Schiffmann, J. "Robust Design Using Multiobjective Optimisation and Artificial Neural Networks with Application to a Heat Pump Radial Compressor." *Design Science* Vol. 8 (2022): pp. 1041–1050. DOI 10.1017/dsj.2021.25.

[18] Papavasileiou, E., Cornelis, J. and Jansen, B. "A Systematic Literature Review of the Successors of "NeuroEvolution of Augmenting Topologies"." *Evolutionary Computation* Vol. 29 No. 1 (2021): pp. 1–73. DOI 10.1162/evco_a_00282.

[19] He, K., Zhang, X., Ren, S. and Sun, J. "Delving Deep into Rectifiers: Surpassing Human-Level Performance on ImageNet Classification." *2015 IEEE International Conference on Computer Vision (ICCV)*: pp. 1026–1034. 2015. DOI 10.1109/ICCV.2015.123.

[20] Glorot, X. and Bengio, Y. "Understanding the difficulty of training deep feedforward neural networks." *Proceedings of the thirteenth international conference on artificial intelligence and statistics*: pp. 249–256. 2010. JMLR Workshop and Conference Proceedings.

[21] LeCun, Y. A., Bottou, L., Orr, G. B. and M., Klaus-Robert. "Efficient backprop." *Neural networks: Tricks of the trade*. Springer (2012): pp. 9–48. DOI 10.1007/978-3-642-35289-8_3.

[22] Ganaie, M. A., Hu, Minghui, Malik, A. K., Tanveer, M. and Suganthan, P. N. "Ensemble deep learning: A review." *Engineering Applications of Artificial Intelligence* Vol. 115 (2022). DOI 10.1016/j.engappai.2022.105151.

[23] Deb, K. and Jain, H. "An Evolutionary Many-Objective Optimization Algorithm Using Reference-Point-Based Nondominated Sorting Approach, Part I: Solving Problems With Box Constraints." *IEEE Transactions on Evolutionary Computation* Vol. 18 No. 4 (2014): pp. 577–601. DOI 10.1109/TEVC.2013.2281535.

[24] Jain, H. and Deb, K. "An Evolutionary Many-Objective Optimization Algorithm Using Reference-Point Based Nondominated Sorting Approach, Part II: Handling Constraints and Extending to an Adaptive Approach." *IEEE Transactions on Evolutionary Computation* Vol. 18 No. 4 (2014): pp. 602–622. DOI 10.1109/TEVC.2013.2281534.

[25] Blank, J. and Deb, K. "Pymoo: Multi-Objective Optimization in Python." *IEEE Access* Vol. 8 (2020): pp. 89497–89509. DOI 10.1109/ACCESS.2020.2990567.

[26] Vrugt, J. A. and Robinson, B. A. "Improved Evolutionary Optimization from Genetically Adaptive Multimethod Search." *Proceedings of the National Academy of Sciences* Vol. 104 No. 3 (2007): pp. 708–711. DOI 10.1073/pnas.0610471104.

Copyright © 2023 by ASME

[27] Hadka, D. and Reed, P. "Borg: An Auto-Adaptive Many-Objective Evolutionary Computing Framework." *Evolutionary Computation* Vol. 21 No. 2 (2013): pp. 231–259. DOI 10.1162/EVCO_a_00075.

[28] Das, I. and Dennis, J. "Normal-Boundary Intersection: A New Method for Generating the Pareto Surface in Nonlinear Multicriteria Optimization Problems." *SIAM Journal on Optimization* Vol. 8 No. 3 (1998): pp. 631–657. DOI 10.1137/S1052623496307510.

[29] Hu, H., Feng, M. and Ren, T. "Effect of Taper Error on the Performance of Gas Foil Conical Bearing." *Industrial Lubrication and Tribology* Vol. 72 No. 10 (2020): pp. 1189–1197. DOI 10.1108/ILT-03-2020-0089.

[30] Verma, S. K. and Tiwari, R. "Robust Design of Ball Bearings for an Improved Performance Using Genetic Algorithm." *International Journal for Computational Methods in Engineering Science and Mechanics* Vol. 22 No. 6 (2021): pp. 514–537. DOI 10.1080/15502287.2021.1893865.

[31] Schiffmann, J. and Favrat, D. "The Effect of Real Gas on the Properties of Herringbone Grooved Journal Bearings." *Tribology International* Vol. 43 No. 9 (2010): pp. 1602–1614. DOI 10.1016/j.triboint.2010.03.006.

[32] Zhang, J., Lu, L., Zheng, Z., Tong, H. and Huang, X. "Experimental Verification: a Multi-Objective Optimization Method for Inversion Technology of Hydrodynamic Journal Bearings." *Structural and Multidisciplinary Optimization* Vol. 66 No. 1 (2023): pp. 1–17. DOI 10.1007/s00158-022-03470-z.

Copyright © 2023 by ASME

APPENDIX A. IMPELLER WHEEL CONVERSION TO CYLINDERS

A.1 Hub and blade parametrizations with piecewise functions

$$f_{\text{hub}}(z) = \begin{cases} \sqrt{r_{2h}^2 - (z - r_{2h})^2} & 0 \leq z \leq r_{2h} \\ r_4 - \sqrt{d^2 - (d^2/c^2) \cdot (z - r_{2h})^2} & r_{2h} \leq z \leq r_{2h} + c \\ r_4 & r_{2h} + c \leq z \leq r_{2h} + c + b_6 \\ r_4 - \sqrt{f^2 - (f^2/e^2) \cdot (z - L_{\text{imp}})^2} & r_{2h} + c + b_6 \leq z \leq L_{\text{imp}} \end{cases} \tag{10}$$

$$f_{\text{bld}}(z) = \begin{cases} r_4 - \sqrt{b^2 - (b^2/a^2) \cdot (z - r_{2h})^2} & r_{2h} \leq z \leq r_{2h} + c - b_4 \\ r_4 & r_{2h} + c - b_4 \leq z \leq r_{2h} + c \end{cases} \tag{11}$$

with

$$
\begin{aligned}
a &:= r_4/\phi - b_4 \\
b &:= r_4 - r_{2s} \\
c &:= r_4/\phi \\
d &:= r_4 - r_{2h} \\
e &:= r4/\phi^2 - b_6 \\
f &:= r_4 - R_{\text{ROT}} \\
b_6 &: b_6 = b_4/\phi \\
L_{imp} &: r_{2h} + c + b_6 + e \\
\phi &: 1.618
\end{aligned}
$$

A.2 Conservation of mass and transverse moment of inertia

$$\rho_{\text{cyl,i}} \cdot \pi \cdot (R_{\text{cyl,i}}^2 - R_{\text{hub,i}}^2) \cdot L_i = m_{\text{bld,i}} \tag{12}$$

$$\frac{m_{\text{bld,i}}}{12} \cdot (3 \cdot (R_{\text{cyl,i}}^2 + R_{\text{hub,i}}^2) + L_i^2) = I_{\text{t,bld,i}} \tag{13}$$

A.3 Inertia tensor of one blade

$$\mathbf{I} = \begin{bmatrix} I_1 \cos^2 \theta + I_2 \sin^2 \theta & (I_2 - I_1) \sin \theta \cos \theta & 0 \\ (I_2 - I_1) \sin \theta \cos \theta & I_1 \sin^2 \theta + I_2 \cos^2 \theta & 0 \\ 0 & 0 & I_3 \end{bmatrix} \tag{14}$$

A.4 Transverse moment or inertia and mass of one blade

$$I_{\text{t,i}}(\theta) = I_2 \sin^2 \theta + I_1 \cos^2 \theta + m_{\text{bld,i}} \left(R_{\text{hub,i}} + \frac{R_{\text{bld,i}}}{2} \right)^2 \cos^2 \theta \tag{15}$$

$$m_{\text{bld,i}} = \rho_{\text{imp}} \cdot (e_{\text{bld}} \cdot (R_{\text{bld,i}} - R_{\text{hub,i}}) \cdot L_i) \tag{16}$$

A.5 Angular separation between a blade and a splitter

$$\Delta\theta = \frac{2\pi}{(N_{\text{bld}} + N_{\text{spl}})} \tag{17}$$

A.6 Transverse moment of inertia of nth blade/splitter

$$I_{\text{t,i}}(n) = I_2 \sin^2(n\Delta\theta) + I_1 \cos^2(n\Delta\theta) + m_{\text{bld,i}} \left(R_{\text{hub,i}} + \frac{R_{\text{bld,i}}}{2} \right)^2 \cos^2(n\Delta\theta) \tag{18}$$

Copyright © 2023 by ASME

Proceedings of the ASME 2023
International Design Engineering Technical Conferences and
Computers and Information in Engineering Conference
IDETC-CIE2023
August 20-23, 2023, Boston, Massachusetts

DETC2023-115280

TEACHING AI TO DESIGN FROM HUMANS: A COMPARISON OF BEHAVIORAL CLONING ARCHITECTURES

Ghazal Bozorgmehry Boozarjomehry
Department of Mechanical and Manufacturing Engineering
Design Intelligence Augmentation Lab (DIAL)
University of Calgary
Calgary, AB T2N 1N4, Canada
Email: ghazal.bozorgmehrybo@ucalgary.ca

Joseph Thekinen[*]
Mechanical and Manufacturing Engineering
Design Intelligence Augmentation Lab (DIAL)
University of Calgary
Calgary, AB T2N 1N4, Canada
Email: joseph.thekinen@ucalgary.ca

ABSTRACT

Reinforcement Learning (RL) has created agents with superhuman performance in robotics and gaming. A central issue in using RL methods to automate engineering design is the inability to generalize and slow training. Even the most advanced curiosity-based RL algorithms require exploring millions of design states, which is infeasible with expensive physics models. Data from a human-subject design study shows that even novice human designers can solve design tasks in a few hundred actions. Behavioral cloning allows RL agents to imitate the policies of a human designer from their decision data. We evaluate the performance of a behavioral cloning agent trained on human design decision data collected in a controlled experiment. We compare three popular sequence learning architectures for behavioral cloning. Subsequently, we evaluate an AI design agent trained through behavioral cloning on human design decision data to automatically design an electric aircraft, starting from a baseline design. The results demonstrate that behavioral cloning effectively transfers human strategies to AI design agents with high sample efficiency.

1 Introduction

Engineering design is a crucial aspect of modern society, responsible for the creation of nearly all modern technologies, from the places where we live to the vehicles we use and the products

we rely on. However, despite the many advancements made possible by artificial intelligence (AI) and machine learning (ML) in recent years, much of the engineering design process still relies on manual labor. One of the main challenges to automating engineering design tasks lies in the high cost of physics models and the difficulty in generalizing across multiple design tasks, both of which pose significant obstacles to data-hungry ML algorithms.

Engineering design is a process of sequential decision-making, in which an agent interacts with a design environment to create a design that satisfies a set of requirements [1]. Reinforcement learning (RL) is a branch of machine learning that can be applied to sequential decision-making processes, training an agent to take actions that maximize rewards in a given environment. RL has been shown to produce agents with superhuman performance in areas such as robotics [2], self-driving cars [3], and strategy games [4]. In theory, an appropriately trained RL agent should be able to achieve human-level performance in related engineering design tasks, such as trade space exploration and parametric design. However, in practice, training an RL agent from scratch typically requires millions of training episodes [5], which is often impractical in most design applications, where even a single stress or performance analysis may take several hours of computational work.

In contrast to RL agents, human designers can solve a new design task with relatively just a few iterations. Figure 1 shows five aircraft designs and associated quality score (Q-score)[1] pro-

---

[*]Address all correspondence to this author.

[1]The Q-score is a measure of design quality, expressed as a number between 0

Copyright © 2023 by ASME

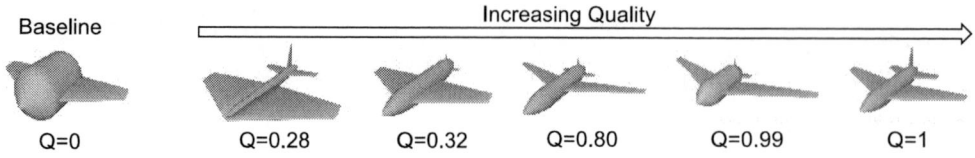

FIGURE 1: Q-score of various design produced by human designers.

duced by novice designers starting from the baseline design in a prior human subject study on engineering students [6, 7]. On average, humans took less than a hundred actions to achieve their designs. In contrast, curiosity-based RL algorithms tailored to sparse-reward settings [8] require millions of actions to achieve a similar Q-score, highlighting the sample inefficiency of these algorithms. This is because humans are able to generalize across multiple design tasks and build robust representations that apply to different problems. Ultimately, human designers solve a fundamentally different problem than deep RL agents.

The difference in sample efficiency and generalization ability between humans and RL agents poses a significant challenge when it comes to applying RL to engineering design. One way to address this challenge is by combining human expertise with RL agents. Transferring human heuristics to RL agents has shown promising results in certain applications, such as video gaming [9]. However, there currently exist no data or algorithms to train RL agents for engineering design activities, like trade space exploration, despite its immense potential for revolutionizing modern engineering design and freeing up human designers from repetitive tasks. These methods leverage human guidance to speed up the learning process, improve sample efficiency, and still benefit from the flexibility and generality of RL.

An effective approach to transfer human design decision-making to an RL agent is using imitation learning. Imitation learning is a form of machine learning that enables RL agents to learn complex behaviors by mimicking the actions of a human expert [10]. It has become a popular research area in recent years across various domains, including robotics, autonomous driving, and games. The RL agents trained via imitation learning can even outperform human experts by inducing noise in the training labels. There are several imitation learning methods that enable RL agents to learn human reward functions, such as behavioral cloning, learning from demonstration [11], and generative adversarial imitation learning [12]. Behavioral cloning is a type of

imitation learning that enables an agent to learn the policy of a human expert by mapping observed inputs to actions from observations [13].

However, how effective is behavioral cloning in imitating decision-making of a human designer in engineering design is not studied. Furthermore, there is a lack of existing literature that compares different behavioral cloning architectures in predicting human design decisions. This paper aims to fill this research gap. Additionally, we assess the effectiveness of a behavioral cloning policy trained on human design decision data to automatically generate a design that fulfills pre-defined design requirements, starting from a baseline design. We answer the following research questions:

1. What is the accuracy of behavioral cloning policy in predicting human decisions in engineering design?
2. How does the accuracy of various sequence-learning-based behavioral cloning architectures compare in predicting human design decision-making?

Our approach involves utilizing a virtual design studio to collect data on design decision-making by human designers, specifically in the context of aircraft design. The studio serves as a surrogate environment for the design task. A series of experiments are conducted on human designers, who start from a baseline design and iteratively update the design parameters to meet pre-specified requirements. The designers observe the reward function after each iteration, which measures how well the updated design meets the requirements. We use behavioral cloning to predict human design decision-making. We compare the accuracy of three sequence learning based behavioral cloning architectures: simple recurrent neural Network (Simple-RNN), long short-term memory (LSTM) and gated recurrent unit (GRU). Finally, we select the best sequence-learning approach among these three and then evaluate the behavioral cloning agent by forward simulating 250 design iterations, starting from the baseline design.

The paper is structured as follows: Section 2 provides the research background of this paper within the existing literature.

---

and 1. A higher Q-score indicates higher design quality. A Q-score of 1 signifies that the design has met all the required design criteria.

Copyright © 2023 by ASME

Our methodology to address the research questions is detailed in Section 3. The results of the study, including a comparison of accuracy of behavioral cloning architectures, are presented in Section 4. Finally, Section 5 concludes the paper by discussing the key takeaways and implications for future design automation.

2 Background and Literature Review

The engineering design process can be modeled as a partially observable Markov decision process (MDP) $< S, A, O, T, R, \gamma >$. S denotes the set of all possible design states, which represents the set of feasible values for all the design parameters (such as in a preliminary design) or raw pixels of an image or B-spline parameters in a CAD drawing (such as in a detailed design). A denotes the set of available actions to the designers, for instance, increasing or decreasing the parameter value. R is the reward, which indicate how well the design meets the requirements. T denotes the transition probabilities and γ is the discount factor. Design is a collaborative process involving several disciplines. Individual designer or design team controls only a fraction of the design parameters, and therefore do not fully observe the design state. To account for the partial observability of design parameters, a characteristic in engineering design, the set of conditional observation probabilities is represented as O.

Figure 2 shows an illustrative example of aircraft preliminary design process abstracted as a MDP. S^t denotes the design state (or the specific values of the design parameters such as fuselage length and form factor) at time t. Reward at time t is quantified based on design requirements such as speed, payload, etc. A^t denotes the design action taken at time t after observing the corresponding reward R_t. A_t changes the design from state S^t to S^{t+1}, resulting in reward R^{t+1}.

Solving a Markov decision process involve finding a mapping function from O to A, i.e., The goal of reinforcement learning is to find a mapping function from O to A. Such an RL agent can produces high-quality and novel design solutions more quickly.

RL can be used to automate design space exploration, freeing up human designers to focus on more creative tasks. Autodesk Dreamcatcher is an example of generative CAD software that uses machine learning to generate several candidate designs based on designer-designated objectives such as materials and performance [14]. DreamSketch, a 3D design interface, combines free-form sketching with generative AI algorithms that enable designers to explore a range of functional 3D designs by sketching design intentions [15]. However, the authors reported that the algorithms were slow and inefficient, and participants in a human-subject study were unable to iterate through design options in a reasonable time. Reinforcement learning is a promising approach to automating several related engineering design tasks [16]. Nevertheless, training an RL agent involves visit-

ing several design states, which may be impractical in design applications. Additionally, the engineering design environment is complex and dynamic, making it challenging to develop accurate and generalizable machine learning models. This complexity arises from the numerous design variables, interactions, and trade-offs that need to be considered in the design process. Therefore, data-driven ML models may struggle to generalize and perform well on new design problems. While RL has the potential to revolutionize engineering design, several challenges still need to be addressed. These include the need to explore millions of design states, the high computational cost of training, and the requirement for accurate and generalizable models.

Human designers, on the other hand, rely on heuristics to accomplish design goals by exploring far fewer states than an RL agent. Heuristics are context-based directives that rely on intuition and experience to increase the chances of producing a design solution within a reasonable timeframe [17]. A substantial body of design literature examines heuristics used by human designers in conceptual and preliminary design phases for various mechanical engineering applications, such as space missions, aviation, naval, and automobiles [18–20]. In engineering design, researchers have investigated using end-to-end deep learning framework to mimic human design decisions [21, 22].

However, there is a lack of research on transferring these heuristics to RL agents that can efficiently explore the design space. Learning policies via imitation learning techniques such as behavioral cloning is an effective way to transfer human decision-making to RL agents. The process of behavioral cloning typically involves collecting a dataset of expert demonstrations and then training a machine learning model (e.g., a neural network) to map the input observations to the corresponding actions taken by the expert. The trained model can then be used to generate actions for the agent in a similar way to how the expert would have acted in the same situation. Behavioral cloning is often used in tasks where it is difficult to define a reward function that captures the full complexity of the task. For example, in the case of autonomous driving, it is difficult to specify a reward function that accounts for all possible scenarios and road conditions. By contrast, behavioral cloning allows the agent to learn a policy directly from observing the expert's behavior without requiring an explicit reward function.

However, behavioral cloning has some limitations. One of the main limitations is that it can be susceptible to compounding errors when the agent makes mistakes that the expert would not have made. This is known as a distributional shift, where the distribution of states and actions seen during testing is different from that seen during training. Additionally, behavioral cloning is limited to learning policies that are similar to the expert's behavior and may not be able to discover novel or better solutions.

Recurrent Neural Networks (RNNs) are a type of artificial neural network designed to process sequential data. Unlike feedforward neural networks that process fixed-size inputs, RNNs

Copyright © 2023 by ASME

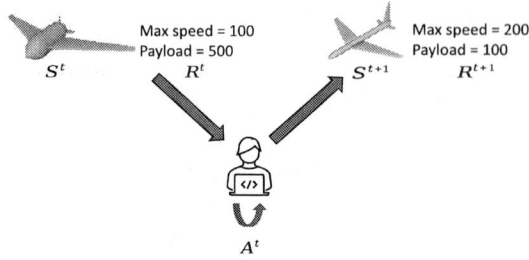

FIGURE 2: Engineering design as a Markov decision process.

can take variable-length sequences as input to produce output. The key feature of RNNs is their feedback connections that enable the output of a given time step to be fed back as input to the next time step. This allows the network to maintain an internal state that represents the context of the sequence up to the current time step. Due to their ability to capture temporal dependencies in sequential data, RNNs are a powerful tool for tasks such as speech recognition [23], natural language processing [24], and time series prediction [25].

However, simple-RNNs can suffer from the vanishing or exploding gradient problem. Vanishing gradient problem occurs when the gradients "vanish" or become too small to be useful for updating the weights of the network during the backpropagation process. When gradients become too small, the network may struggle to learn complex features and patterns in the data, and the training process may become slow or stall completely. LSTM and GRU are two widely accepted RNN architectures that address the issue of exploding or vanishing gradients. LSTMs have been widely used in a variety of applications, including language modeling, machine translation, speech recognition, and sentiment analysis. They have also shown LSTM models have demonstrated significant potential in predicting time series data, such as stock prices [26], energy consumption [27], and weather patterns [28]. The LSTM approach can capture long-term dependencies in the data that are difficult to train using simple-RNN. GRUs have been shown to perform well on a variety of tasks, including language modeling [29], speech recognition [30], and machine translation [31], among others. Overall, GRUs is a simpler and more computationally efficient alternative to LSTMs.

3 Research Methodology
This section details our overall methodology.

3.1 Collection of Human Design Data
The task at hand is to perform a preliminary design of an electric aircraft, which requires collaboration among four design offices with diverse disciplinary expertise: fuselage, payload (battery), propulsion, and airfoil (wing and tail). The design process is iterative, with each design office updating its local design parameters after performing technical disciplinary analysis on the previous iteration. The iterations continue until they produce an aircraft that meets all the system requirements or until the maximum time limit of 30 minutes is reached.

In our design experiments, there are four participants with five roles: four subsystem roles (one for each participant) and one system role shared by all four participants. Each subsystem role operates a design office that controls the design parameter assigned to that subsystem. Additionally, the subsystem roles are supported by a technical office that performs technical analysis on the subsystem-level functional requirements (FRs). We conducted ten sessions with forty human participants, collecting all their design decision data. Our participants were either graduate students (with an undergraduate engineering degree) or undergraduate students at junior standing from a science, technology, engineering, or mathematics degree program. To be recruited for the study, participants were required to have prior engineering design experience through their degree program. However, for the purposes of the experiment, the participants were considered novice designers.

The system role operates a system office that collects information from all subsystems. The technical office corresponding to the system office analyzes the system-level FRs, which include endurance, flight dynamic constraints (lift, thrust, moment), capacity (volume), and budget (cost threshold). The computational models used by both the subsystem and system technical offices, such as the Vortex Lattice Method, as well as the equations used, are described in previous work by the authors [6].

Figure 3 shows the front-end client of the airfoil office which is one of the four design offices. The airfoil office controls design parameters such as wing span, and chord length (vertical sliders). The design task comprises 12 design parameters spread over four design offices (airfoil office, payload office, fuselage office, and propulsion office), 12 subsystem-level FRs, and 6 system-level FRs that are tied to design requirements. Table 1 tabulates the range of acceptable values for the design parameters. Although the subsystems are modular, they are highly coupled by the physics of the aircraft design task. A metric called

Copyright © 2023 by ASME

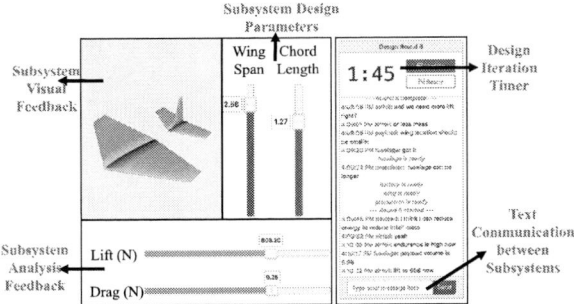

FIGURE 3: Design interface to collect human subject data.

TABLE 1: Range of values and minimum increment (denoted by Δ) for each design parameter.

Design Parameter	Min	Max	Δ
Fuselage length(m)	0.1	3	0.01
Fuselage diameter(m)	0.01	1	0.01
Fuselage wing location(%)	25%	75%	1%
Payload number of cells	1	12	1
Payload energy per cell(A-hr)	0 A-hr	6 A-hr	0.1 A-hr
Payload location(%)	25%	75%	0.01%
Propulsion diameter(m)	0.25 m	1.5 m	0.01 m
Propulsion rotation(rpm)	1000	5000	50
Propulsion number of blades	1	4	1
Airfoil wing span(m)	0.05 m	3 m	0.01 m
Airfoil root chord length(m)	0.05 m	1.75 m	0.01 m
Airfoil tail scaling	0.1	1	0.001

normalized error quantifies how well the design satisfies the design requirement. The detailed formulations and physics to calculate the normalized error (denoted by E_{sys}) are published in a prior work [6]. Q-score is defined as $1 - E_{sys}$.

3.2 Data Pre-processing

The minimum and maximum permitted values for each of the 12 design parameters are tabulated in Table 1. Human designers select values for these 12 parameters based on technical analysis feedback received in the previous iteration. To facilitate analysis, we normalize the parameters using max-min normalization to the range $[0, 1]$.

Our dataset consists of 1803 human design decision labels for the 12 design parameters from ten sessions. To avoid overfitting issues that can arise in deep learning models, we use data from eight sessions (1482 labels) to train our model and reserve one session each for validation (109 labels) and testing (199 labels).

We use the validation data to tune hyperparameters such as lookback (the number of previous iterations to use in the prediction model), number of layers, and activation function. The accuracies reported in the results section are from the test data.

3.3 Sequence Learning Methods

3.3.1 Simple-RNN
Simple-RNN is a type of neural network that takes an input vector and produces both an output vector and an internal state vector. The internal state vector serves as a memory of past inputs. In each time step, the current input vector is combined with the internal state vector from the previous time step using learnable weights. This combined vector is then passed through a nonlinear activation function, such as the hyperbolic tangent function. The output vector is generated by applying another set of learnable weights and a nonlinear activation function to the current internal state vector. This output vector can be utilized for various tasks, such as predicting the next value in a time series or classifying input sequences. However, it is important to note that RNNs can encounter the issue of vanishing gradients, which can hinder their learning capabilities.

3.3.2 LSTM
In an LSTM network, each neuron has a "memory cell" that can retain information for an extended period of time. The memory cell in an LSTM network uses a combination of gating mechanisms and non-linear transformations to retain information from previous time steps. The gating mechanisms allow the network to selectively store or discard information based on the current input and the previous state, while the non-linear transformations enable the network to selectively remember or forget information from previous time steps. This makes LSTMs effective for tasks involving long-term dependencies, allowing the network to have a longer-term memory than other types of RNNs. The memory cell is controlled by three gates: input, forget, and output gates that regulate the flow of information within the cell.

However, training an LSTM network can be computationally expensive and requires a large amount of labeled data [32]. Therefore, careful consideration of the data and task at hand is necessary to determine the feasibility of using LSTMs. Figure 4 describes the architecture of the LSTM cell in our prediction model. We use RelU activation functions as indicated in the figure. h_{t-1} denotes the hidden state of the previous timestep)and c_{t-1} denotes the cell state at the previous timestep. x_t denotes the input vector at time t. X denotes pointwise multiplication of two vectors and $+$ denotes pointwise addition of two vectors.

Copyright © 2023 by ASME

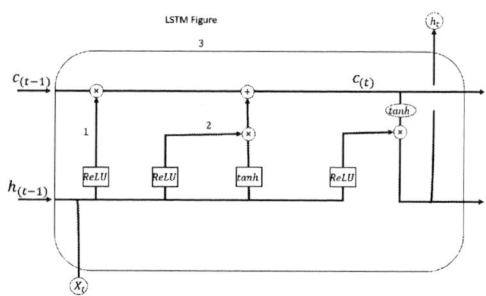

FIGURE 4: LSTM Cell architecture in our model.

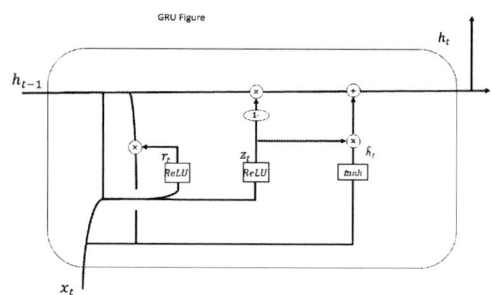

FIGURE 5: GRU Cell architecture in our model.

3.3.3 GRU In a GRU, information flows through a sequence of cells, with each cell taking an input and producing an output and a hidden state. GRUs have two gates: an update gate and a reset gate. The update gate determines how much information from the previous hidden state should be passed on to the current hidden state, while the reset gate determines how much of the previous hidden state should be forgotten and replaced with new information.

Figure 5 describes the architecture of the GRU cell in our prediction model. The architecture of a GRU is similar to that of an LSTM (Long Short-Term Memory) network. However, GRUs have fewer parameters than LSTMs, making them faster to train and easier to implement [33]. The main disadvantage of using GRUs over LSTMs is that their simplified gating mechanism can lead to lower accuracy in applications where longer-term dependencies become important [34].

4 Results

Results are organized as follows. Section 4.1 compares the three sequence learning architecture based on the accuracy of predicting human design decisions. Section 4.2 evaluates the RL agent trained via behavioral cloning on the aircraft design task.

TABLE 2: Training and test accuracy of the three machine learning methods.

Approach	Training Error	Testing Error
LSTM	0.0424	0.13
GRU	0.042	0.10
Simple-RNN	0.052	0.12

4.1 Comparing Accuracy of Sequence Learning Architectures to Predict Human Design Decisions

Table 2 compares the performance of LSTM, GRU, and simple-RNN based behavioral cloning, using the mean absolute error (MAE) as a measure of accuracy. A lookback of 10 yielded the best performance. The results are presented for a lookback of 10. This indicates that the prediction of a design decision is based on the design decisions made in the ten previous iterations. The MAE is defined as the average absolute difference between predicted and actual values for each of the 12 design parameters, where the design parameters are normalized in the range 0-1. A lower MAE indicates higher accuracy in predicting the next value of the design decision, with values between 0 and 1.

The accuracy scores are reported for both the training and testing datasets. LSTM and GRU outperformed simple-RNN in terms of mean absolute error (MAE) on the training dataset. This can be attributed to the higher number of trainable parameters in LSTM and GRU, resulting in improved training accuracy. However, LSTM exhibited the lowest accuracy on the testing dataset, indicating poorer generalization. Interestingly, GRU achieved a higher training accuracy than LSTM despite having fewer parameters [34]. This discrepancy may be attributed to the specific design task studied, where long-term dependencies might not play a significant role. Overall, among the three models, GRU demonstrated the best performance on both the training and testing datasets.

Figure 6 compares the predictions made by GRU to actual human decisions for four design variables: wing span from the airfoil office, battery energy per cell from the payload office, fuselage length from the fuselage office, and propeller diameter from the propulsion office. These variables were selected to represent each of the four design offices. The comparisons were performed on the test dataset, which was not used to train the model parameters or hyperparameters. The results show that the GRU accurately captures sequential human design decision-making.

4.2 Evaluating Behavioral Cloning Policy

We evaluated the effectiveness of our RL agent trained via behavioral cloning (using GRU) by assessing its ability to navigate the design space and iteratively improve design quality.

Copyright © 2023 by ASME

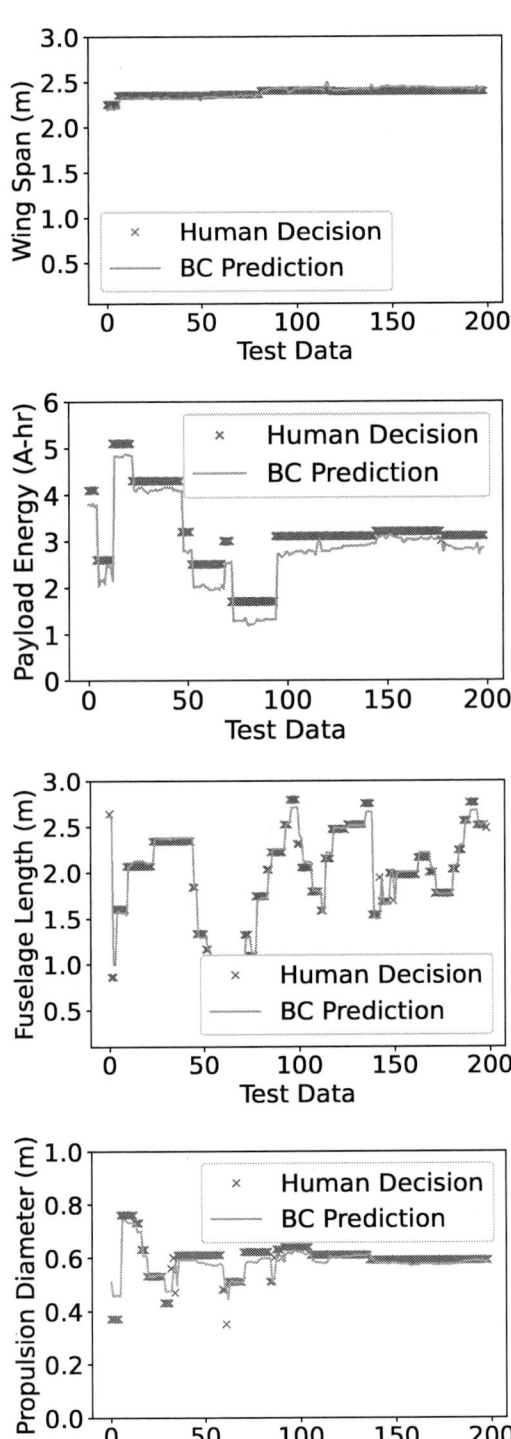

FIGURE 6: Comparing predictions from behavioral cloning policy (denoted by BC prediction) to decisions made by human designers.

Specifically, we simulated 250 new design iterations using the RL agent trained via behavioral cloning.

Figure 7 visualizes the Q-score, a metric used to evaluate design quality, for the designs generated by the RL agent. The baseline design, represented by a Q-score of zero, serves as a reference design from which the agent initiates exploration within the design space. With each iteration, the RL agent executes the policy learned from the human design decision data. The objective is to generate a design that fulfills all design requirements, indicated by a Q-score of 1.

The RL agent trained on the human data was able to generate a design with a Q-score of 0.8 after approximately 150 iterations. However, the Q-score did not show further improvement in subsequent iterations. To explore the impact of reducing the amount of human labels, we trained another RL agent using only half of the human labels. This second RL agent, trained with limited human labels, achieved a maximum Q-score of 0.35 in approximately 100 iterations.

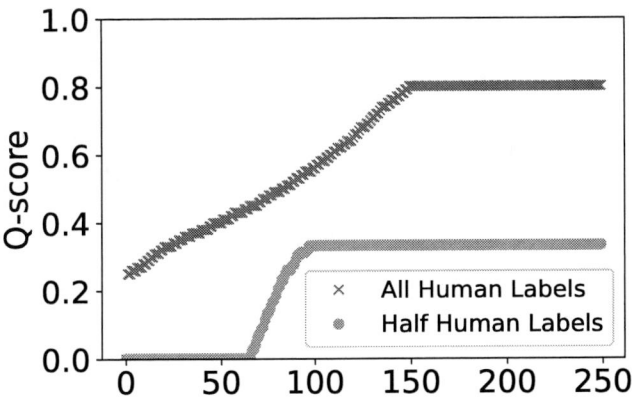

FIGURE 7: Q-score of forward simulation of 100 design iterations starting from baseline design. Baseline design has a Q-score of zero.

5 Discussion and Conclusions

In summary, our study demonstrates that GRU outperforms LSTM and simple-RNN in terms of raw accuracy, as measured by mean absolute error. A lookback of 10 proved to be the optimal choice for achieving the best performance. GRU's superiority over LSTM can be attributed to the specific design task's lack of reliance on long-term dependencies. Additionally, GRU's simpler architecture contributes to better generalization, as it has fewer parameters. While GRU offers faster training speeds, the time difference is not significant for small datasets with a few thousand labels, such as the one used in our study.

Copyright © 2023 by ASME

The RL agent, trained through behavioral cloning on human design decision data, demonstrated its capability to enhance the aircraft design task by exploring the design space. Starting from a baseline design, the RL agent achieved a Q-score of 0.8 in approximately 150 iterations. However, when trained with fewer human labels (about half of the first RL agent), the second RL agent only achieved a Q-score of 0.35. This suggests that human labels effectively teach the RL agent and facilitate the transfer of strategies from human designers. A higher number of labels create a more effective AI design engine. However, a limitation of our study is that the dataset is limited to a single engineering design task. Future research could explore transfer learning and meta-learning approaches, where an RL agent trained on a related task can be fine-tuned on a new task with fewer samples, significantly reducing the number of required training samples and expediting the learning process.

Another limitation of using behavioral cloning is that the RL agent does not make further improvements beyond the observations from human decisions. This is because behavioral cloning agents do not directly observe the reward function and do not explore beyond the supervised labeled dataset. As a result, the agent could not enhance its performance through further exploration of the design space, relying solely on indirect observations of rewards from human design decision data. At the same time, behavioral cloning agents based on human design decision data are much faster to train than curiosity-based reinforcement learning algorithms. Future work could explore more advanced imitation learning algorithms that incorporate reward functions, environment dynamics, and supervised human decision labels to develop RL agents capable of efficiently exploring the design space and generating designs that meet all requirements.

ACKNOWLEDGMENT
The authors acknowledge the following funds for this research: (a) startup fund from the University of Calgary and (b) NSERC Discovery Grant.

REFERENCES
[1] Miller, S. W., Yukish, M. A., and Simpson, T. W., 2018. "Design as a sequential decision process: A method for reducing design set space using models to bound objectives". *Structural and Multidisciplinary Optimization, 57*, pp. 305–324.
[2] Zhao, W., Queralta, J. P., and Westerlund, T., 2020. "Sim-to-real transfer in deep reinforcement learning for robotics: a survey". In 2020 IEEE symposium series on computational intelligence (SSCI), IEEE, pp. 737–744.
[3] Sallab, A. E., Abdou, M., Perot, E., and Yogamani, S., 2017. "Deep reinforcement learning framework for autonomous driving". *arXiv preprint arXiv:1704.02532*.
[4] Silver, D., Hubert, T., Schrittwieser, J., Antonoglou, I., Lai, M., Guez, A., Lanctot, M., Sifre, L., Kumaran, D., Graepel, T., et al., 2018. "A general reinforcement learning algorithm that masters chess, shogi, and go through self-play". *Science, 362*(6419), pp. 1140–1144.
[5] Silver, D., Schrittwieser, J., Simonyan, K., Antonoglou, I., Huang, A., Guez, A., Hubert, T., Baker, L., Lai, M., Bolton, A., et al., 2017. "Mastering the game of go without human knowledge". *nature, 550*(7676), pp. 354–359.
[6] Thekinen, J., and Grogan, P. T., 2021. "Information exchange patterns in digital engineering: An observational study using web-based virtual design studio". *Journal of Computing and Information Science in Engineering, 21*(4).
[7] Thekinen, J., and Grogan, P. T., 2022. "Effects of augmented information system on design communication: A human-subject study using aircraft design studio". In International Design Engineering Technical Conferences and Computers and Information in Engineering Conference, Vol. 86212, American Society of Mechanical Engineers, p. V002T02A064.
[8] Pathak, D., Agrawal, P., Efros, A. A., and Darrell, T., 2017. "Curiosity-driven exploration by self-supervised prediction". In International conference on machine learning, PMLR, pp. 2778–2787.
[9] Dubey, R., Agrawal, P., Pathak, D., Griffiths, T. L., and Efros, A. A., 2018. "Investigating human priors for playing video games". *arXiv preprint arXiv:1802.10217*.
[10] Hussein, A., Gaber, M. M., Elyan, E., and Jayne, C., 2017. "Imitation learning: A survey of learning methods". *ACM Computing Surveys (CSUR), 50*(2), pp. 1–35.
[11] Argall, B. D., Chernova, S., Veloso, M., and Browning, B., 2009. "A survey of robot learning from demonstration". *Robotics and autonomous systems, 57*(5), pp. 469–483.
[12] Ho, J., and Ermon, S., 2016. "Generative adversarial imitation learning". *Advances in neural information processing systems, 29*.
[13] Torabi, F., Warnell, G., and Stone, P., 2018. "Behavioral cloning from observation". *arXiv preprint arXiv:1805.01954*.
[14] Noor, A. K., 2017. "Ai and the future of the machine design". *Mechanical Engineering, 139*(10), pp. 38–43.
[15] Kazi, R. H., Grossman, T., Cheong, H., Hashemi, A., and Fitzmaurice, G. W., 2017. "Dreamsketch: Early stage 3d design explorations with sketching and generative design.". In UIST, Vol. 14, pp. 401–414.
[16] Dworschak, F., Dietze, S., Wittmann, M., Schleich, B., and Wartzack, S., 2022. "Reinforcement learning for engineering design automation". *Advanced Engineering Informatics, 52*, p. 101612.
[17] Fu, K. K., Yang, M. C., and Wood, K. L., 2016. "Design principles: Literature review, analysis, and future directions". *Journal of Mechanical Design, 138*(10), p. 101103.

Copyright © 2023 by ASME

[18] Yilmaz, S., and Seifert, C. M., 2011. "Creativity through design heuristics: A case study of expert product design". *Design Studies, 32*(4), pp. 384–415.

[19] Yilmaz, S., Daly, S. R., Seifert, C. M., and Gonzalez, R., 2015. "How do designers generate new ideas? design heuristics across two disciplines". *Design Science, 1*, p. e4.

[20] Fillingim, K. B., Nwaeri, R. O., Borja, F., Fu, K., and Paredis, C. J., 2020. "Design heuristics: Extraction and classification methods with jet propulsion laboratory's architecture team". *Journal of Mechanical Design, 142*(8).

[21] Raina, A., McComb, C., and Cagan, J., 2019. "Learning to design from humans: Imitating human designers through deep learning". *Journal of Mechanical Design, 141*(11).

[22] Rahman, M., Bayrak, A., and Sha, Z., 2022. "A reinforcement learning approach to predicting human design actions using a data-driven reward formulation". *Proceedings of the Design Society, 2*, pp. 1709–1718.

[23] Miao, Y., Gowayyed, M., and Metze, F., 2015. "Eesen: End-to-end speech recognition using deep rnn models and wfst-based decoding". In 2015 IEEE Workshop on Automatic Speech Recognition and Understanding (ASRU), IEEE, pp. 167–174.

[24] Yin, W., Kann, K., Yu, M., and Schütze, H., 2017. "Comparative study of cnn and rnn for natural language processing". *arXiv preprint arXiv:1702.01923*.

[25] Giles, C. L., Lawrence, S., and Tsoi, A. C., 2001. "Noisy time series prediction using recurrent neural networks and grammatical inference". *Machine learning, 44*(1-2), p. 161.

[26] Sunny, M. A. I., Maswood, M. M. S., and Alharbi, A. G., 2020. "Deep learning-based stock price prediction using lstm and bi-directional lstm model". In 2020 2nd Novel Intelligent and Leading Emerging Sciences Conference (NILES), IEEE, pp. 87–92.

[27] Yan, K., Li, W., Ji, Z., Qi, M., and Du, Y., 2019. "A hybrid lstm neural network for energy consumption forecasting of individual households". *Ieee Access, 7*, pp. 157633–157642.

[28] Naware, D., and Mitra, A., 2022. "Weather classification-based load and solar insolation forecasting for residential applications with lstm neural networks". *Electrical Engineering, 104*(1), pp. 347–361.

[29] Irie, K., Tüske, Z., Alkhouli, T., Schlüter, R., Ney, H., et al., 2016. "Lstm, gru, highway and a bit of attention: An empirical overview for language modeling in speech recognition". In Interspeech, pp. 3519–3523.

[30] Khandelwal, S., Lecouteux, B., and Besacier, L., 2016. "Comparing gru and lstm for automatic speech recognition". PhD thesis, LIG.

[31] Zhang, B., Xiong, D., Xie, J., and Su, J., 2020. "Neural machine translation with gru-gated attention model". *IEEE transactions on neural networks and learning systems, 31*(11), pp. 4688–4698.

[32] Shen, Y., Yun, H., Lipton, Z. C., Kronrod, Y., and Anandkumar, A., 2017. "Deep active learning for named entity recognition". *arXiv preprint arXiv:1707.05928*.

[33] Cahuantzi, R., Chen, X., and Güttel, S., 2021. "A comparison of lstm and gru networks for learning symbolic sequences". *arXiv preprint arXiv:2107.02248*.

[34] Yamak, P. T., Yujian, L., and Gadosey, P. K., 2019. "A comparison between arima, lstm, and gru for time series forecasting". In Proceedings of the 2019 2nd International Conference on Algorithms, Computing and Artificial Intelligence, pp. 49–55.

Copyright © 2023 by ASME

Proceedings of the ASME 2023
International Design Engineering Technical Conferences and
Computers and Information in Engineering Conference
IDETC-CIE2023
August 20-23, 2023, Boston, Massachusetts

DETC2023-116709

TRANSFER REINFORCEMENT LEARNING:
FEATURE TRANSFERABILITY IN SHIP COLLISION AVOIDANCE

Xinrui Wang
Dept. of Aerospace & Mechanical Engineering
University of Southern California
Los Angeles, USA
xinruiw@usc.edu

Yan Jin*
Dept. of Aerospace & Mechanical Engineering
University of Southern California
Los Angeles, USA
yjin@usc.edu
(*corresponding author)

ABSTRACT

The integration of artificial intelligence into engineering work has become increasingly prevalent. Engineering work processes can be highly complex, and learning from scratch requires large computation resources. Transfer learning has emerged as a promising technique for improving learning efficiency by leveraging knowledge gained from related tasks to the target task. To achieve optimal performance, one of the key challenges is to figure out how transferrable the features are among different work processes and within training networks. Simulation-based ship collision avoidance is used for case studies due to its inherent complexity and diversity. Two transfer reinforcement learning methods, feature extraction, and finetuning, are implemented and evaluated against the baseline. Instead of introducing large-scaled pre-trained models as the backbone, a light CNN model pre-trained in a related base case has been proven to transfer essential features to target cases. Simplified ship dynamics is introduced into the training process to make it more realistic and applicable, and the delay caused by the large moment of inertia is addressed by modifying the model-environment interaction mechanism. Work process features for the ship collision avoidance process are concluded from crucial aspects. The effects on transferability are displayed by experimental results discussed from the feature category and similarity perspective.

Keywords: Artificial intelligence, deep learning, transfer learning, reinforcement learning, collision avoidance

1 INTRODUCTION

In recent years, artificial intelligence (AI) has been increasingly applied to the complex engineering process. The integration of AI techniques into the engineering work process has the potential to revolutionize problem-solving, and engineering design approaches, leading to efficient workflows, informed decision-making, and improved outcomes. Reinforcement learning, a subfield of AI, has shown great potential in optimizing engineering work processes through interacting with the environment [1, 2, 3, 4]. Engineering work processes can be highly complex, and starting from scratch requires many labor and computation resources [5, 6]. Transfer learning has emerged as a promising technique for improving learning efficiency by leveraging knowledge gained from related tasks to the target task [7, 8, 9, 10]. To conduct efficient transfer reinforcement learning, the key issue is to figure out how transferrable the features are among different work processes and how they transfer within the pipeline of the training model. To analyze potential approaches, the appropriate model should be designed, and the effects of different features and different similarities between source and target tasks should be evaluated.

This research is conducted based on our previous work [11]. To accelerate the training process of complex engineering cases, collision avoidance, for instance, we proposed a belief-based transfer reinforcement learning method based on ship collision avoidance scenarios. Although reinforcement learning methods have been extensively utilized for ship collision avoidance, it always requires a long training time due to the high complexity, making the training efficiency a major challenge. To the best of our knowledge, work process transfer learning has yet to be

Copyright © 2023 by ASME

applied to improve training efficiency. Our previous method transfers the whole network, which is pre-trained in a related but the simpler base case, to the target case and tunes the model with transfer belief (i.e., how much the action predicted by the base network should be trusted) and transfer period (i.e., how long the base network will be consulted) based on different tasks. This method leads to effective and efficient training when the target case is similar to the base case, but the limitation is shown when the similarity decreases because the unit of transfer is the whole network. Based on deep learning experience, the lower layers of a neural network deal with general features. The deeper layers deal with the features which are more specific to the learning case and often relate to decision-making [12]. Assuming the conclusion also hold for work process reinforcement learning scenarios, the whole network transfer will include some base-case-specific features in the deeper layers, which is no longer effective when the target case is far from the base case. Thus, we explored different layer levels of the transfer reinforcement learning method by making some of the transferred parameters kept frozen and others finetuned during training. The existence and transferability of image-based work process features are revealed. And the relationship between feature transferability and case-pair similarity has also been derived [13].

Various reinforcement-learning-based work process training makes full use of graphic features through moving image frames [14, 15, 16]. Ship collision avoidance processes can be intricate and highly demanding, especially when the water areas are becoming congested [17]. Besides the image-based geometrical features we mainly focused on previously, there are also other features highly related to the work process. For instance, when encountering future collisions, the collision risk should be estimated with some risk assessment methods [18, 19], and proper collision avoidance maneuvering should be determined based on estimated risk levels [20, 21]. When maneuvering the ship with the chosen steering angle, the completion of steering may face a significant delay due to the large inertia of the ship, even experiencing immovability. Also, a previous study conducted multiple pairs of base case and target case analyses, causing the features transferred between each base and target case pair to be different from one another. And the similarity of each pair was measured based on human understanding. These factors may affect the overall conclusions about similarity effects on feature transfer. To fully describe the ship work process and enhance the learning efficiency through defining and transferring crucial work process features properly, we propose two research questions: 1) What kinds of potential features does the work process contain, and do they have different transferability? 2) How transferrable are the features through the training pipeline if modifying the similarity between the base case and target cases?

For the first question, to investigate the transferability of different features, we try to conclude the work process features in the designed study case, then vary two features from the base case to generate two target cases, respectively. To answer the second question, this study constructs a common base case. The same work process features are extracted from the base case to several target cases with varying degrees of similarity. The rest of the paper is organized as follows. Section 2 discusses other research related to this study. Section 3 illustrates details of the applied methods. Section 4 demonstrates the case study design, and the results are presented and discussed in Section 5. The conclusions are drawn, and future research directions are suggested in Section 6.

2 RELATED WORK

To investigate the transferability of work process features, simulation-based ship collision avoidance is used for a case study due to its inherent complexity and diversity. To fully describe the various situations and related collision avoidance actions, the regulation details are introduced. A deep reinforcement learning approach is utilized in this study to explore various possibilities of retaining and reusing the learned knowledge among different ship collision avoidance scenarios, which is related to reinforcement learning applications in ship collision avoidance and feature extraction technique.

2.1 COLREG rules

To provide more situational awareness and reduce marine collisions, there are regulations dictated by the International Regulations for Preventing Collisions at Sea (COLREG) [22], by which the own ship will be considered as having the "right of the road" or should "give way" to the encountering target ship depending on which COLREG region the target ship is currently in. The COLREG region separation and the COLREG-complaint collision avoidance action are illustrated in Figure 1.

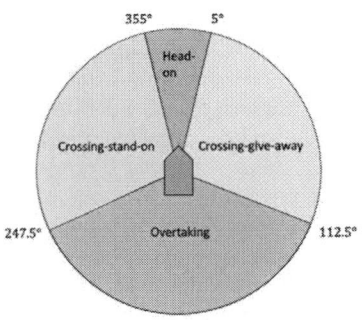

(a) COLREG regions

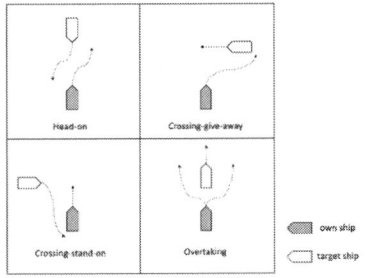

(b) COLREG-compliant collision avoidance actions

FIGURE 1: Illustration of COLREG

Copyright © 2023 by ASME

When encountering target ships in the moving process, in order to avoid the collision, the maneuver decisions of the own ship are expected to follow the existing guidelines of COLREG. It is also worth mentioning that although the regulations are supposed to be followed by all the ships, there is no guarantee that a target ship will follow the rules, especially when the situation can be interpreted in different ways by different ships. This phenomenon is reflected in the case studies of this research.

2.2 Reinforcement-learning-based ship collision avoidance

Collision avoidance has been a popular research topic for years in the unmanned surface vehicle (USV) domain [23, 24]. Deep reinforcement learning has become one of the main methods. Advanced learning policies are continuously proposed, enabling the agent to efficiently learn collision avoidance strategies through interacting with the environment.

Most reinforcement learning ship collision avoidance studies were combined with a risk assessment to reduce the failures and make avoidances actions compliant with COLREG constraint simultaneously through improving the action space and reward function. Lingyu Li et al. defined different zones based on risk assessment and applied different artificial potential fields in each zone to solve the sparse reward issue in deep reinforcement learning. The action space and collision avoidance reward of the agent were improved by the artificial potential field, thus reducing the collision risk [21]. Zhao et al. utilized a simplified ship dynamics model. The states of the own ship's and target ship's steering commands were mapped into rudder angle as the input Deep Neural Network (DNN). Target ships were classified into four regions based on COLREG. Simulation results showed the ability of multiple ships to avoid collisions with each other during the path-following process [25]. Fan et al. defined the risk factor based on the distance at the closest point of approach (DCPA) and the time to the closest point of approach (TCPA). The DCPA and TCPA are the minimum distance and time, respectively, at which the own ship and the target ship are expected to be closest to each other (collision happens). These parameters are determined based on the current positions, courses, and speeds of the ships [41]. The risk factor was further used in the collision avoidance reward function design to give a penalty to the actions against COLREG based on the collision risk it caused [19]. Inspired by the above research, in our study, the simplified ship dynamics is embedded in the simulation environment. Different ship domains are designed to facilitate risk assessment. COLREGs rules and DCPA are also covered to provide additional information for the reward function design and the case similarity measurements.

2.3 Feature extraction and transfer learning

Feature extraction is a crucial technique in many recognition-based applications, as it aims to extract informative and discriminative features from language, speech, image, or other raw data [26, 27]. For image feature extraction, feature refers to the patterns in the given images, e.g., points, edges, and corners, providing key information to identify associated objects [28] or work for other relevant tasks like 3D reconstruction,

semantic segmentation, or image generation [29, 30]. In addition, image features extracted from videos can complement acoustic and lexical features, allowing for a more comprehensive analysis approach that leverages visual information to enhance the accuracy of analysis tasks. [31]. Two main categories of image features have been commonly used, hand-crated features and automatic-inferred features [32], competing in different tasks. Hand-crafted features are explicitly designed by human experts based on domain-specific knowledge, such as SIFT features and HOG features, which are still widely used despite the emergence of various deep learning models [33, 34]. The bag-of-visual-words model is proposed to represent the overall feature of an image with a histogram of independent features, generating a sparse vector that can be used in classification tasks [27]. Instead of making an effort to design such feature representations, deep learning has been introduced to automatically extract features from input images through an end-to-end approach, leveraging the power of convolutional networks. Many well-designed models are published for research and engineering applications, e.g., VGG, ResNet, and Inception [35, 36, 37]. In this research, the hand-crafted feature requires expertise in the ship collision avoidance domain, and it may or may not prove itself. Automatic-inferred features are utilized instead. The input image is a 2D gameplay window containing simple geometric shapes, and the conceptual representations are defined by us. The trending models are not chosen due to their large scale and generality. A light CNN model is designed and proved to be capable of extracting features of the ship collision avoidance process.

Deep-learning-based feature extraction technique is often combined with transfer learning, pre-training the network in one task to extract essential features and reusing it in the related target task [38]. There is another commonly used method called finetuning, which follows the same procedure as feature extraction but re-trains the whole or partial layers of the pre-trained network to finetune the model in the target task [12]. Each method has its own advantages and limitations and can be effective depending on the specific use case. Furthermore, a combination of both methods in the same pipeline can provide a more comprehensive solution [31]. The feature extraction method is more efficient. The transferred parameters do not need to be trained again since useful features have already been extracted from the base case. When the similarity between the base case and target cases is low, it is necessary to use the finetuning method. The copied convolution layers may contain some dataset-specific features because of the big difference between the two datasets; thus, it might work better to be updated during training to fulfill the gap. Feature extraction and finetuning methods are both utilized in this research. It is worth comparing the work process feature transferability with these two settings.

3 METHOD

To investigate the ship work process features transferability in the transfer reinforcement learning scenarios, a computational empirical approach is taken, consisting of the ship-dynamics-

Copyright © 2023 by ASME

embedded simulation environment, the reinforcement learning algorithm, and the transfer learning approach. Ship dynamics is incorporated to the simulation environment design to make it more accurate and closer to real-world scenarios. Also, ship dynamics is closely related to ship maneuvering, which is one of the crucial factors of successful navigation. For this reason, RL training takes the ship dynamics into consideration to achieve better reward function design. The transfer learning approach improves the training performance by leveraging pretrained knowledge, achieving equivalent results with less training time. What is more, the layer-level transfer reflects the transferability of different work features, which can be useful in adapting the algorithm to new environments or scenarios.

3.1 Ship dynamics

A simplified rudder-controlled ship dynamics model is used in this study to simulate the motion of the own ship and target ships. The state of the ship includes angular velocity, course, x and y coordinates in the world frame, and the linear velocity of the ship, described by x_1 to x_5, as shown in the following equation.

$$x_1 = \dot{\varphi} \tag{1}$$

$$x_2 = \varphi \tag{2}$$

$$x_3 = x \tag{3}$$

$$x_4 = y \tag{4}$$

$$x_5 = v \tag{5}$$

The equations of ship dynamics are:

$$\dot{x_1} = -\frac{x_1}{T} + \frac{K}{T}u_1 \tag{6}$$

$$\dot{x_2} = x_1 \tag{7}$$

$$\dot{x_3} = x_5 \cos(x_2) \tag{8}$$

$$\dot{x_4} = x_5 \sin(x_2) \tag{9}$$

$$\dot{x_5} = a(u_2 - x_5) \tag{10}$$

K is the rudder gain and is set as 0.1555. T is the ship's inertia, which is 73.77. These values come from the settings of the MTI project [39]. u_1 and u_2 are the rudder angle and velocity input of the system. The speed x_5 is kept at a constant number of 15 m/s all the time for implementation simplicity. So, the command velocity input u_2 always equals x_5, acceleration $\dot{x_5}$ is kept at 0 during the training process. x_2 and x_1 represents heading angle and angular velocity. During training, x_2 is assigned by the selected action of the neural network, as shown in Table 1. Then the rudder command u_1 is set based on the selected action x_2 following Eqn (11). The own ship starts to be steering with this command following the above ship dynamics. The state of the own ship keeps changing until x_1 and x_2 decay to nearly 0 degrees, indicating the ship completes the steering action, and the heading direction returns to facing forward. The next action will be selected, and the same process will continue until the end of this episode.

When the x_2 is relatively small, a linear controller is used to control the steering process. u_1 is set to be $k * x_2$, where k is a constant coefficient to guarantee convergence. For the large target course, more steps need to be taken until x_1 and x_2 decay to nearly 0 degrees and a primary delay for the rudder motion is displayed. To accelerate the command course reaching process, a nonlinear component is introduced to handle large rudder angle input (greater than φ_1), which may take a much longer time to converge.

$$\begin{cases} u_1 = k * (x_2 + sign(x_2) * \varphi_2) & for \ |x_2| > \varphi_1 \\ u_1 = k * x_2 & for \ |x_2| \le \varphi_1 \end{cases} \tag{11}$$

Combining the nonlinear function is equivalent to starting with the original command heading course plus an offset course φ_2 in the same direction as the original command. This results in a larger angular acceleration x_1, which shortens the convergence process. Threshold φ_1 to trigger the nonlinear acceleration is set at 10 degrees, and the extra command course φ_2 is set to 70 degrees. These values were determined through experimentation and struck a good balance between convergence rate and avoiding overshoot.

3.2 Reinforcement learning training mechanism.

The own ship is trained as a learning agent. It would learn decision-making strategies to reach the waypoint and avoid collision with target ships. The target ships are assigned a fixed starting point, destination, moving speed, and direction. A deep Q Network with experience replay is used in this study to train the agent to learn the crucial features in the work process [2]. Unlike some policy-based algorithms (such as PPO [47] and SAC [48]), which have separate policy and evaluation networks, the Deep Q Network is value-based and constructed as an end-to-end structure. This makes it more straightforward to use a transfer learning approach, as pre-trained parameters can be directly loaded into the Q network.

In reinforcement learning based ship collision avoidance research, the compound vector input is widely used. The state is represented with the information of the own ship and the target ships in vector form. To keep the input vectors unified size, typically only the nearest target ship would be included, even in the multiple target ship encounters situation. In this study, since we are discussing the potential feature in the waypoint approaching and collision avoidance work process, the whole working space, including the information of the own ship and all the target ships, needs to be taken into consideration. To further explore the work feature transferability, we also tried to indicate more work feature-related information like the risk assessment and applicable rules in the input state to pass to the neural network. Thus, to construct the input state containing such information with unified size, the informatic game window generated by the gameplay simulation environment is treated as the input state, graphically describing the current state.

After passing the input state to the neural network, the learning agent, i.e., the own ship, takes actions based on the output prediction to approach the waypoint while avoiding

Copyright © 2023 by ASME

collisions with encountering target ships. Each action is assigned with a command heading angle change in the range of -25 degrees to 25 degrees from the agent's current angle. The action space design is shown in Table 1.

Table 1. Agent action space

Action	$\omega\ (deg)$
a_1	-25
a_2	-20
a_3	-15
a_4	0
a_5	15
a_6	20
a_7	25

During the training, the agent will choose an action from seven of them to adjust the heading angle accordingly. Unlike typical reinforcement learning tasks, an action predicted by the neural network can be completed and evaluated by reward function within one step. Thus, the neural network and environment can interact with every step. The chosen action cannot be completed in a single step in this study due to the ship dynamics. The large moment of inertia of the own ship makes it take a few steps to converge to the command angle. Before converging to the command angle, the environment is updated continuously based on changing angle velocity and position, but the updated windows cannot be sent to the neural network to make the next action prediction since the current action has not been done, resulting in the delay of model-environment interaction. To solve this delay issue, when the agent starts to choose an action a, the initial state s, which is the gameplay simulation window at that moment, is recorded. At each step, the completion of the chosen action is judged. When the remaining angle difference is less than a certain threshold, 1 degree in this study, the action is judged to be completed. The state s' generated by the game environment at that moment is recorded. The middle process between the starting and completion of the action choice is accessible to the environment to update the ship position and velocity but is omitted for the neural network training. The reward function is adjusted accordingly to handle the action completion delay. For each action choice, the reward is cumulated step by step from the starting point to the completion. The summation result is treated as the reward r_a for the chosen action a. The experience $\{\ s,\ a,\ r_a,\ s'\}$ is sent to the experiment buffer to train the neural network later. The state s is updated with s' for the next-step action prediction. The graphical explanation is shown in Figure 2.

3.3 Feature extraction and finetuning

The transfer reinforcement learning method is applied in this study to explore work process feature transferability. Feature transfer is conducted between one base case and several target cases. Despite initialization variations, the base case and target cases follow the same reinforcement learning pipeline, as shown

in Figure 3, which allows the neural network to interact with the environment and be trained by perceiving feedback. The neural network in the base case and target cases is called the base network and target network, respectively. They are all constructed by 4 convolution layers, working as a feature extractor, and 2 fully connected layers, mainly focusing on decision-making. And the environments are constructed based on the settings of the base case and target cases. The environment sent the generated game window to the network as the input state. The network makes a prediction and forwards it to the environment to make the work process updated accordingly. Once the action is completed, the reward feedback will be sent back to the network, together with the updated game window, for training purposes and initializing the next iteration.

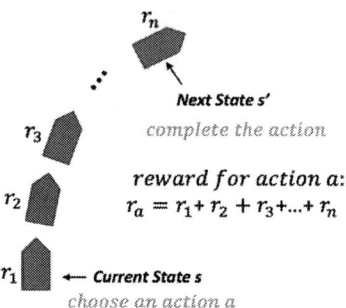

FIGURE 2: Reward design handling action completion delay

The base network is trained with the baseline model in the base case. The checkpoint is saved with trained weights, which will be reused in the target cases training. For the target case training, feature extraction and finetuning method have been used. The target network is initialized with transferred parameters from the base network, which has been pre-trained in the base case. The transferred parameters are frozen as a fixed feature extractor for the feature extraction method while open to be updated in the training process for finetuning method. Besides the whole network transfer, to explore the transferability of work process features through the layer level of the neural network, different layer-level transfer reinforcement learning is also implemented. Figure 3 uses 2 convolution layers transfer, for example, describing the initialization and training process of layer-level transfer reinforcement learning. "2" refers to the former two convolution layers, which are closer to the input, to allow general feature transfer. The first two layers are initialized with parameters transferred from the base network to keep frozen or not due to the applied method. Other layers are uniformly initialized. Other layers level transfer follows the same process; the only number of transferred layers changes accordingly. After initialization, the target network interacts with the environment and gets updated during the training process. The target case training is also conducted with the baseline for comparisons, which is uniformly initialized and get trained from scratch in target cases and doesn't include any transferred parameters. Thus, through different combinations between the frozen setting

Copyright © 2023 by ASME

and layer level, the transferability of the work process feature can be explored comprehensively.

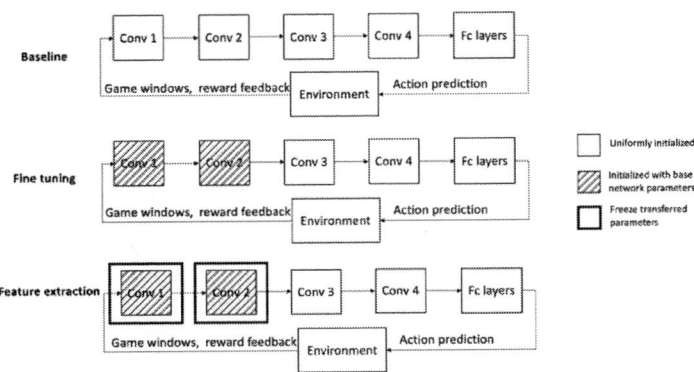

FIGURE 3: Baseline and two transfer reinforcement learning methods

4 CASE STUDY

The study cases were selected from Imazu problems [40], covering possible scenarios of ship encounters in the real world. Among these, 4 pairs of cases are determined. Each pair is constructed by a common base case and a specific target case, between which the transfer reinforcement learning approach is implemented. Thus, work process features can be extracted from base case training and reused in different target cases. Various effects will appear, giving some clues about work process feature transferability. 4.1 defines the work process features as well as the related case settings. The rest of this section gives some details about the case design. 4.2 defines the ship domain, risk assessment, and related traffic rules. Based on that, the reward function design is developed in 4.3.

For some implementation details, a gameplay simulation environment of the ship is created using Pygame, embedded with hardcoded ship dynamics simulation and vehicle traffic rules. The neural network is constructed with TensorFlow, interacting with the simulated environment during the transfer reinforcement training process.

4.1 Work process features

To study the features in this specific work process, we first define the work process features from different aspects. Graphic features through moving image frames in work processes can be treated as one kind of feature. On the other hand, some non-graphic information highly related to the work process can also be potential features. In this study, we conclude ship dynamics, number of target ships, heuristic knowledge of risk assessment, DCPA, and applicable COLREG, which come from crucial components of the ship collision avoidance process, to be non-graphic features of the work process. Other than that, there may be other hidden features formed by certain combinations or some other way, which cannot be concluded but do exist and distinguish the work process. To make the environment fully informative using controllable computation resources, simple geometries are utilized to cover the crucial features in the ship

collision avoidance process, which facilitates the feature transfer purpose. All these non-graphical features are represented by geometrics in the gameplay image frames and extracted by the neural network, along with graphic features during training. This means that all the features are forwarded and transferred between layers of the neural network through screen frames input.

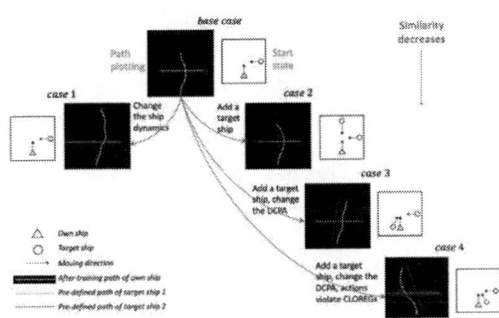

FIGURE 4: Base case and target cases

To explore the effect of variation of features and similarity on work process features transferability, the study cases vary transferred features and similarity when constructing *base case* and target case pairs. The common base case is designed with particular dynamics and a single applicable COLREG rule. For each case pair, the target case is constructed by varying one or more features from the base case while keeping the rest the same. Complexity increases, and similarity decreases when changing features. Case 1 and case 2 change a single feature from the base case, ship dynamics feature, and rule-related feature, respectively. They share the same similarity with *base case*, which will be used to investigate the effect of different features. From cases 3 to 4, each target case adds more complexity based on case 2, while similarity with the base case decreases, which will be used to investigate the effect of similarity. Base case and all target cases are illustrated by the start state, which is borrowed from Imazu problems, and the path plotting, which is constructed by the predefined paths of target ships and the after-training path of own ship as shown in Figure 4.

The own ship dynamics is changed in case 1; the DCPA and applicable COLREG keep the same as the base case since the target ship setting is unchanged. Meanwhile, a smaller ship with less moment of inertia and rudder gain compared to the ship dynamics described in Section 2 is selected. Specifically, the moment of inertia is 50; the rudder gain is 0.05. Since the own ship becomes lighter, it has less delay and becomes capable of achieving the desired angle in fewer steps. Thus, more action choices are required in the same waypoint approaching the process. In this way, reusing the NNs pretrained in the base case with the larger ship in the target case with the smaller ship, the effect of the ship dynamics difference can be reflected in the transfer reinforcement learning process.

For case 2, the ship dynamics and the target ship 1 are set to be the same as the base case. Meanwhile, an extra target ship moving towards the own ship from the front is introduced. We call it target ship 2. Target ship 2 shares the same velocity as target ship 1. Based on the starting position and velocity, target

Copyright © 2023 by ASME

ship 2 doesn't change the DCPA of the own ship. When target ship 2 enters the collision avoidance zone of the own ship, to avoid collision with target ship 2, the own ship should follow "head-on" COLREG rules by taking actions of turning right. It is similar to how the own ship takes actions to avoid collision with target ship 1. Generally, target ship 2 also needs to turn right in this situation, but in this case study, only the own ship is trained to learn collision avoidance and rules following strategies; the target ships just dumbly move forward with predefined direction and speed.

For case 3, the ship dynamics and target ship 1 are the same as the base case; target ship 2 comes from the left-hand side of the own ship. It changes the DCPA of the own ship because it is closer to the own ship than target ship 1. According to COLREG rules, the own ship has road right; it can keep "stand-on," and target ship 2 should turn right to "give away." But target ship 2 always follows the predefined direction and speed, and it is already in the mandatory collision avoidance zone of the own ship. The own ship should make turns to avoid collisions instead of standing on; For target ship 1, the own ship should turn right when it enters the collision avoidance zone to give away the road right because it comes from the right side of the own ship. Taking both target ships into consideration, the learning agent should learn to avoid collisions by turning right.

Case 4 describes the violating-rules situation. The ship dynamics and target ship 1 are the same as the base case; target ship 2 comes from the right side of the own ship, also in the mandatory collision avoidance zone of the own ship. DCPA of the own ship is changed due to target ship 2. Based on the COLREG rules, the own ship should turn right when encountering target ship 1 and target ship 2 because these two target ships both come from its right side. But target ship 2 is too close to the own ship; if the own ship turns right, a collision will happen between the own ship and target ship 2. So, the own ship will learn to take mandatory collision avoidance actions against the COLREG to avoid collisions in the training process. Even some penalties will be applied for violating COLREG rules with target ship 1 in the mid-process, the largest penalty of collision can be skipped, and the largest reward of reaching the waypoint will be gained at the last step. The learning agent should figure out the logic of temporary sacrifices and a big final reward to nail this case.

4.2 Feature representation

To minimize the risk of collisions, the concept of own ship domains is introduced to facilitate risk assessment during work processes. Accordingly, the simulation reinforcement learning environment is constructed with the relative coordinate of the own ship, as shown in Figure 5, which is also close to the real ship navigation system. The white line represents the current heading path of the own ship to the waypoint, which is located at the end of the path. The green polygon represents the encountering target ship. There are multiple ships encounters in study cases, just using one target ship in the figure for illustration. Other target ships' calculation follows the same procedure. During the training, the own ship was fixed on the origin in the waypoint reaching and collision avoidance process. The goal and target ships moved relatively to the own ship. At each step, the coordinates of the goal and target ships were converted from the absolute coordinate using rotation and transition matrixes. The non-graphic features like collision risks, COLREG, and ship dynamics are geometrically represented by the following domains and models in the game screens.

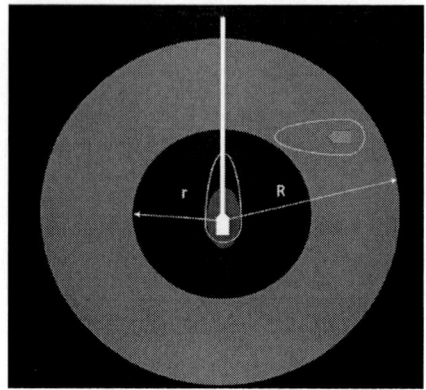

FIGURE 5: Gameplay window

Minimal ship domain: The blue polygon outside the own ship represents the minimal ship domain [42]. If the target ship or other obstacles enter this domain, the collision is judged. This area is defined as an ellipse with the major axis along the own ship's moving direction and the minor axis perpendicular to it, as shown in Figure 1. The half-length of the major axis and the minor axis is set to be 4L and 1.6L, with L the ship length. In this study, the own ship length is set to be 300m, then the half-length of the major axis and minor axis of the ship domain is 1200m and 480m, respectively. The distance between the own ship and the target ship in this area also satisfies the collision avoidance passing distance of 0.3nm determined according to ship maneuvering [43].

COLREG domain: While reaching a waypoint and avoiding collisions, the learning agent, i.e., the own ship, needs to follow the COLREG rules. To further define the stage of applying COLREG rules, an event trigger mechanism is introduced [21]. The large red circle radius R represents the boundary value of the collision risk stage (5 nm) suggested by the guide to the Rules for maritime collision avoidance. The small black circle radius r represents the minimal close-quarters situation 2 nm [44]. When the distance between the own ship and target ship $target\_dist$ is larger than R, the target ship is in the safe zone of the own ship. The own ship can take any direction to reach to waypoint. When $r < target\_dist < R$, the target ship enters the collision avoidance zone, and the own ship needs to take actions following COLREG rules to avoid collision with target ship. When $target\_dist < r$, the target ship enters the mandatory collision avoidance zone, and the own ship is allowed to take any emergent actions to avoid the collision, even violating the COLREG rules.

Copyright © 2023 by ASME

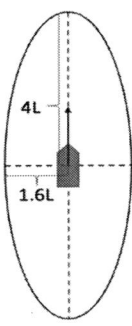

FIGURE 6: Minimal ship domain

Bumper model: The bumper-shaped area outside the own ship and target ship indicates the area where the ship tends to keep clear of static or moving obstacles. The shorter axis and longer axis are 1.6 times and 6.4 times the length of the own ship, respectively. The bumper model was designed to trigger the collision avoidance action [45]. In this study, since we use another method to distinguish the emergency of collision and action modes, the bumper area is depicted in the environment to provide more information in the input image. For example, the rotation of the target ship bumper shows the ship dynamics, and the overlapping of the own ship bumper and target ship bumper indicate the risk of collision. Although the target ship's relative movement of the own ship is shown in the input image, the ship size is relatively small compared to the water area, so it might be hard for the NNs to observe. Based on our testing results, adding a bumper model could accelerate and stabilize the training. The NNs became capable of capturing more knowledge through more informatic image input.

4.3 Compound reward function

As mentioned in Section 3.4, the reward of each chosen action is cumulated step by step from the initial state to the final state. This section will give more details about the reward function design for every single step. A compound reward function is used in this study. Besides the final positive or negative reward, subtle rewards are added based on the training purpose to enrich the sparse reward space [46]. For each step of a chosen action, compound reward constructed by episodic reward, shaping reward, and constant reward is applied to evaluate the status of this certain step, as shown in Eqn (12). Comprehensive knowledge from different sources is incorporated to guide the agent training. The learning goal and restriction are implied by the episodic reward and constant reward. Meanwhile, more reward signal is provided by shaping reward components to enrich the sparse reward space, encouraging the training to converge to the desired direction more efficiently.

$$R_{tot} = R_{goal} + R_{col} + R_{colregs} + R_{dist} + R_{dev} \qquad (12)$$

In the above equation, R_{tot} is the compound reward for each step, constructed with five components. $R_{goal} = 1000$ and $R_{col} = -1000$ will be given if the agent reaches the

waypoint or collides with target ships; the episode will end at the same time. Otherwise, a 0 value will be assigned. The value of this episodic reward is determined by experimental results after comparing different parameter settings.

When $target\_dist > R$:
$$R_{colregs} = 0 \qquad (13)$$
When $r < target_{dist} < R$:
   If the own ship follows COLREG:
$$R_{colregs} = 2 \qquad (14)$$
   If the own ship violates COLREG:
$$R_{colregs} = -5 \qquad (15)$$
When $target_{dist} < r$:
$$R_{colregs} = 0 \qquad (16)$$

$R_{colregs}$ guides the agent to follow the COLREG rules during the collision avoidance process. For each target ship, the value varies due to different situations related to the own ship's COLREG domain, as shown in the following. If multiple target ships appear, $R_{colregs}$ of each target ship will be summed up.

If the target ship is in the safe zone of the own ship, there is no need to take collision avoidance actions yet, $R_{colregs} = 0$; When the target ship is detected to enter the own ship collision avoidance zone, the collision avoidance mechanism of the own ship is triggered, and the applicable COLREG rules will be determined by the functions embedded in the environment. At each step, if the own ship owns the road right, the "stand-on" scenario, for example, the maneuver of own ship will not be restricted. If the own ship doesn't own the road right, like encountering crossing-give-away or head-on situations described in Section 2, the own ship will be trained to give away the road right to the encountering target ships with the $R_{colregs}$ term. To be detailed, if the agent takes an action of turning right, following the COLREG rules, a small reward will be given, $R_{colregs} = 2$. If the action violates the COLGREs rules, like turning left or keeping the current direction, a -5 penalty will be assigned with $R_{colregs}$. The value of reward and penalty are also decided based on several trials; this combination works for different cases. With this setting, the agent is not only able to learn to avoid collision following the rules but also capable of dealing with emergent situations even against the rules; If, for some reason, the target ship enters the mandatory collision avoidance zone of the own ship, there is not much space left for the agent to adjust the direction. Driving safety is more important than rules following; the own ship can take any action to avoid collisions, even violating COLREG rules. $R_{colregs}$ is set to 0 to remove rule restrictions.

The shaping reward components R_{dist} and R_{dev} are set to be relatively small in order to provide subtle guidance at each step. R_{dist} is related to the distance between the own ship and the desired waypoint, forcing the agent to move closer to the waypoint. If the distance is shortened at the current step

Copyright © 2023 by ASME

compared to the last step, R_{dist} is positive. Otherwise R_{dist} is negative. R_{dev} is defined based on the angle between the heading direction of the own ship and the direction of the straight line connecting the own ship and the goal. If the angle is 0, which means the agent is moving directly toward the waypoint, R_{dev} is maximum. In the following equations, k_1 and k_2 are set to 0.5 and 2.5, respectively. This setting makes the shaping components not dominate over episodic and COLREG-related components, which should be major ones, nor vanish due to the small value. Distance rewards and deviation rewards are also balanced during the training process.

$$R_{dist} = k_1 * (prev\_dist - curr\_dist) \qquad (17)$$

$$R_{dev} = k_2 * (\exp{(-abs(dev))}) \qquad (18)$$

The reward shaping is based on heuristic knowledge of how we expect the learning agent to behave during the training process. Linear shaping is used by R_{dist}. Exponential shaping is used by R_{dev}. Linear shaping continuously encourages the agent to reduce the distance to the waypoint during the whole process. The reward changes linearly with the moving-forward distance, regardless of how far the agent is from the waypoint. Exponential shaping is introduced to R_{dev} to add more gradient near the peak to distinguish the peak and the neighboring high reward points, leading the agent to achieve the best solution (deviation is 0 degrees) instead of stopping at relatively good solutions. Also, based on our experimental results, exponential shaping is verified to have better performance than linear shaping or other shaping like sinusoidal for R_{dev}.

5 RESULTS AND DISCUSSION

The neural network was pre-trained in the base case and reused in target cases by applying two transfer reinforcement learning methods, feature extraction and finetuning. The target cases' training results differ due to different work process features transfer. Also, different layer level of transfer reinforcement learning is explored. The transferability of work process features through the layer level of the neural network is shown by comparing the transfer learning results between four target cases. The reward function of training cases during the training process is shown in Figure 7 to Figure 10. The x-axis represents the number of training episodes; the y-axis represents the cumulated reward for each episode. The plotting of each layer level of transfer reinforcement learning is generated by 10 samples. The solid line represents the mean of all samples; the shading area records the variation. The legends represent different layer levels. For example, "1 conv" in finetuning results represents transferring the first layer of the pre-trained neural network to the target network and finetuning the network during the target case training process. In feature extraction cases, it refers to freezing the first layer of the target network after carrying out parameter transfer and then updating the rest. Especially, "4 conv + 2 fc" in finetuning results refers to transferring the parameter of all convolution layers and fully connected layers of the pre-trained network to the target network and then finetuning all of them during the training process. It is

meaningless to freeze all the parameters during training, so for the feature extraction method, this setting is omitted. Also, for each target case, the transfer learning results are compared with the baseline, which is trained from scratch without being initialized by pre-trained parameters.

5.1 Effect of feature categories

The training results of the finetuning method for two target cases, case 1 and case 2, are very similar. While regarding the feature extraction method, transfer reinforcement learning leads to very different results. Since both cases have minor changes compared with the base case, the similarity between the two cases and the base case is approximately equivalent. The transfer learning performance should be close from the similarity point of view. The variation of different features may be the main reason causing different training results. For case 1, the ship dynamics feature is changed. Features transfer is conducted between the base case with a larger ship moment of inertial and rudder gain and the target case with a lighter ship. For case 2, an extra target ship is added to the target case to vary the rule-related feature. Features are extracted from a single-rule case to a double-rule case.

5.1.1 Effect of feature categories using finetuning

For case 1 and case 2, the similarity with the base case is high. Although dealing with different kinds of features, variation of feature category almost doesn't affect the feature transferability. The finetuning method enhances the training efficiency compared to the baseline. Higher layer level transfer converges faster than lower layer level transfer, except for a slight performance drop for whole network transfer in case 1.

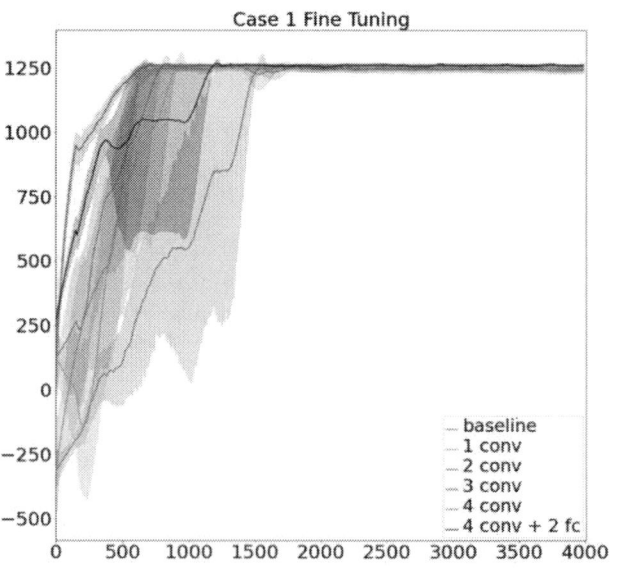

Copyright © 2023 by ASME

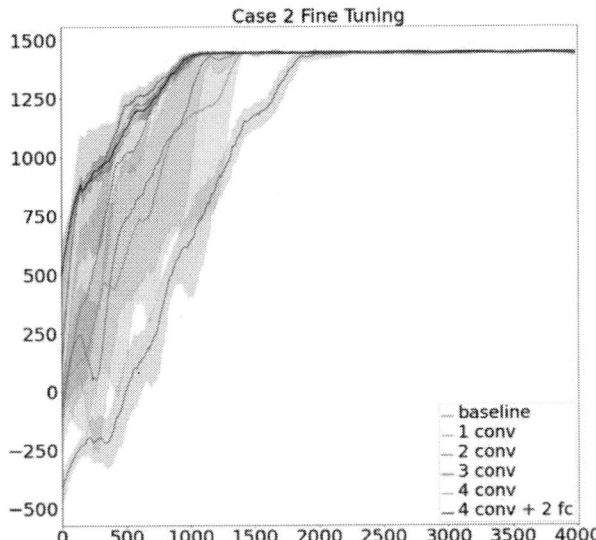

FIGURE 7: Reward of case 1 and case 2 using the different layer-level finetuning methods.

For the pre-trained base network, lower layers extract general features of the ship driving process; higher layers extract features specific to the base case, which may be far away from the target case if the two cases are very unlike. Since within both case pairs, the similarity is high. The base case shares most features with the target case. More pretrained layers transfer leads to higher training efficiency. As shown in the figure, the 3 convolution layers extract more features shared by the base case and the target cases than the 1 and 2 convolution layers, leading to the fastest convergence. 4 convolution layer transfer works not as well as 3 convolution layer transfer due to some base-case-specific features. But the variation of feature category does change the final action choice of the learning agent, thus affecting the performance of the whole network transfer. Regarding the shape of the own ship after-training path in Figure 4, base case and case 2 look very similar, while case 1 looks different. The change of ship dynamics in case 1 causes more effect on decision-making than the newly introduced COLREGs rule in case 2. Due to this, the whole network transfer, including the fully connected layers (related to decision-making), leads to lower reward and slower convergence in case 1 than in case 2; longer training is needed to fill the gap between different decision-making strategies.

5.1.2 *Effect of feature categories using feature extraction*

For the feature extraction method, the reward function plotting of case 1 and case 2 looks totally different. As mentioned before, the variation of work process features is considered to be a key factor causing the large difference.

In case 2, all convolution layer level transfer works better than the baseline since the similarity between the base case and case 2 is high and the decision-making strategy is close; the general and specific features extracted in the base case by the transferred layers both fit the target case work process. For 1 and 2 convolution layer transfer, the feature extraction method works

better than finetuning method; since the general features are almost the same, there is no need to process finetuning. All layer level transfer doesn't have obvious differences; only 4 convolution layer transfer shows slightly lower reward compared with other lines due to the specificity caused by higher layer transfer.

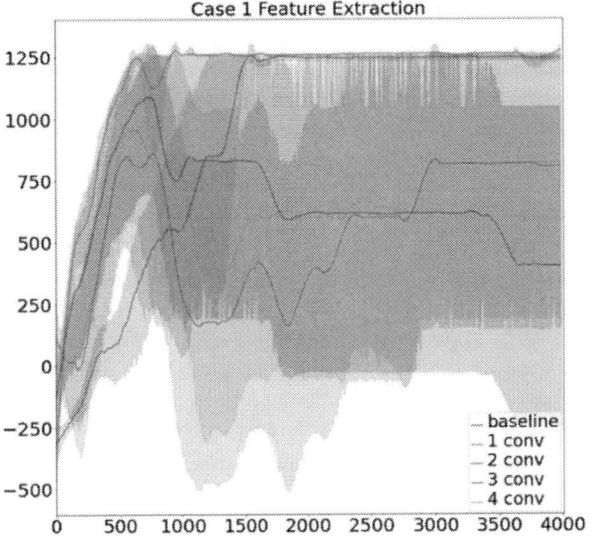

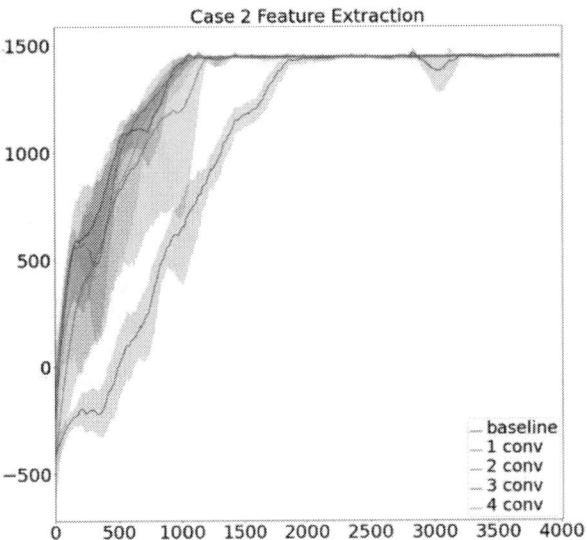

FIGURE 8: Reward of case 1 and case 2 using the different layer-level feature extraction methods.

Different from case 2, 4 convolution layers transfer work best in case 1 with the feature extraction method; The accumulated reward is equivalent to the reward obtained by the finetuning method in Figure 7. Other layers' level of transfer shows higher reward than the baseline at the beginning, while a large continuous variation happens instead of reaching the highest reward step by step. It infers that the dynamics features cannot be re-learned if the transferred convolution layers are frozen. The frozen part is separated from the rest of the model,

Copyright © 2023 by ASME

which will be updated during the training process. The overall finetuning for the convolution part of the neural network is needed to finetune the different ship dynamics features transferred from the base case. Alternatively, it is also possible to transfer the entire 4 layers of the neural network and freeze them while only finetuning the decision-making process related to the fully connected layers. This approach also works when the ship dynamics difference between the base case and the target case is not too high, as shown in case 1. Even though the convolution layers carry different dynamics features, the decision-making layers are finetuned to cover the differences.

In summary, when similarity is high for rule-related features, a higher level of layer transfer is preferable for the feature extraction method to transfer more shared features and save computational resources. When dealing with different ship dynamics features, all layer levels do benefit the learning process at the very beginning, but the ship dynamics features cannot eventually be transferred through separate convolution layers even if the ship dynamics difference is not high. Another thing that needs to mention is we only explored a small change of ship dynamics within the same simulating system; the conclusion may change if a large difference is involved or a different dynamics system is introduced. For example, the 4 convolution layers transfer with the feature extraction method may not work well.

5.2 Effect of case pair similarity

Based on Case 2, the rule-related features are modified in Case 3 and Case 4 to decrease the similarity with the base case. The training result of these two cases shows an obvious difference, although the ship dynamics is the same. The effect of different similarities between the case pairs is reflected. For case 3, the DCPA is changed by adding another target ship, but the added ship has less road priority and doesn't affect the own ship to take actions following COLREGs rules. For case 4, the added target ship not only changes the DCPA but also forces the own ship to take mandatory collision avoidance actions which violate COLREGs. In this way, features extracted by the base network pre-trained in the base case are effective in case 3, while they can be harmful in case 4 due to the large similarity changes.

5.2.1 Effect of similarity using finetuning

By applying the finetuning method to case 3, transfer reinforcement learning enhances the training efficiency compared with the baseline. The 3-layer transfer has the best performance for all levels, extracting more features shared by base and target case. The reward starts to drop with 4 convolution layers and the whole model transfer. The higher layer level finetuning doesn't work as well as the lower level because the higher level of the pretrained network can extract features that are more specific to the base case than the target cases. Compared to case 2, the similarity is decreasing, the higher layer transfers fewer shared features in case 3, and more training time is required to finetune the target network to deal with case-specific features. Also, the overall trend of converging lines for 4 convolution layers and the whole model transfer looks similar to the baseline, which is trained from scratch. It implies the

decision-making strategies transferred by the higher layer don't help target case training due to lower similarity. A similar finetuning process as learning from scratch is needed to develop the decision-making strategies in this target case.

For case 4, the similarity with the base case further decreases, and significantly different action choices appear; the transferred features don't fit this work process and even harm the training. Even though the first and second convolution layer extracts the general features from the base case, they are not able to properly initialize the network and nail the target case. Thus, the finetuning method doesn't work better than the baseline anymore. As the number of transferred convolutions increases, the reward decreases accordingly. The fourth layer transfer and whole model transfer work less optimally in case 4; the reason may relate to features specific to the base case and decision-making. The 3 convolution layers transfer, which works best in other cases, works worst in case 4. It may indicate the third layer processes rule-related features, which can be crucial to the work process. Based on the experimental results, the poor initialization of the third convolution layer leads to unpromising training performance.

5.2.2 Effect of similarity using feature extraction

When applying the feature extraction method in cases 3 and 4, convolution layer transfer shows significant performance drops compared with other lines due to the transferred features' specificity. Even though we verified that 3 convolution layer transfer extracts the most shared feature between the base case and target case, it thus shows the highest performance with the finetuning method. When applying the feature extraction method, keeping the first 3 convolution layers frozen through the training process, it works not better than 1 and 2 convolution layer transfer anymore. That may also cause by the case-specific features. Only four convolution layers are involved in this study; the third layer is relatively close to the output, which can be considered a relatively high-level layer in this case. Although adding the third convolution layer to the transfer learning makes it possible to extract most shared features, it may also extract some features specific to the base case, which doesn't show advantages without finetuning.

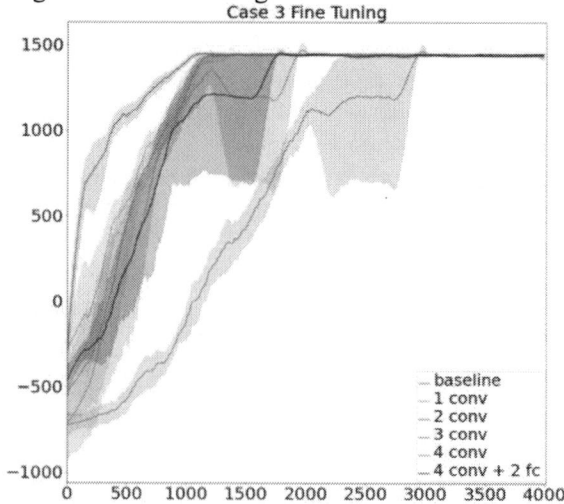

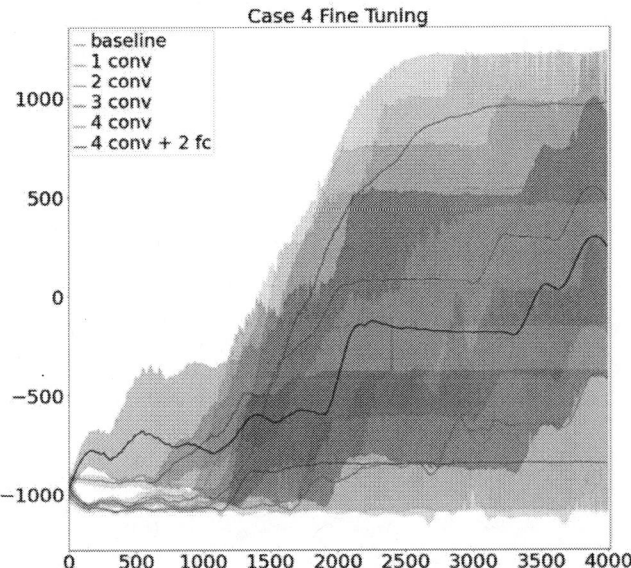

FIGURE 9: Reward of case 3 and case 4 using the different layer level finetuning method.

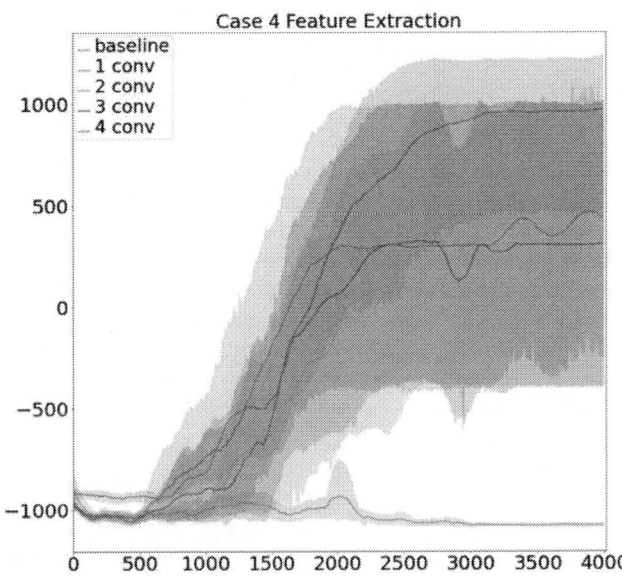

FIGURE 8: Reward of case 3 and case 4 using the different layer-level feature extraction method.

For case 4, because the similarity between the base case is too low, the transferred features show side effects even with finetuning, and the same thing happened for the feature extraction method. Compared to case 1, 1 to 3 convolution layer transfer has similar effects with case 4; the performances are nearly equivalent. In comparison, 4 convolution layers transfer shows the lowest reward finally, which has the higher and highest performance in case 1 and case 2 when similarity is high, further proving the influence of case-pair similarity on deeper convolution layer transfer.

6 CONCLUSION

The aim of this paper is to investigate the transferability of work process features using ship collision avoidance as a case study. To achieve this goal, we introduce ship dynamics into the reinforcement learning process to create a more realistic and applicable simulation environment. Additionally, we address the delay caused by the ship's moment of inertia by modifying the model-environment interaction mechanism. Work process features for the ship collision avoidance process are defined and utilized to generate study cases to investigate the effects of these features and the similarity between case pairs on transferability. Two transfer reinforcement learning methods, feature extraction and finetuning, are implemented and evaluated against a baseline. Based on the experimental results, we draw the following conclusions.

- Feature category doesn't affect feature transferability using the finetuning method unless a significantly different decision-making strategy needs to be handled. But for the feature extraction method, the ship dynamics features are not transferrable through separate frozen convolution layers even if the dynamics is slightly changed.
- In this neural network, rule-related features may be mainly transferred by the third convolution layer since 3 convolution layer transfers outperform in COLREG-rule-following cases and fail in the rule-violation case.
- Deeper layer transfer extracts the features more specific to the base case and often relate to decision-making. It is preferable when the similarity between the base case and target case is high and even better when their decision-making is close. Vice versa.

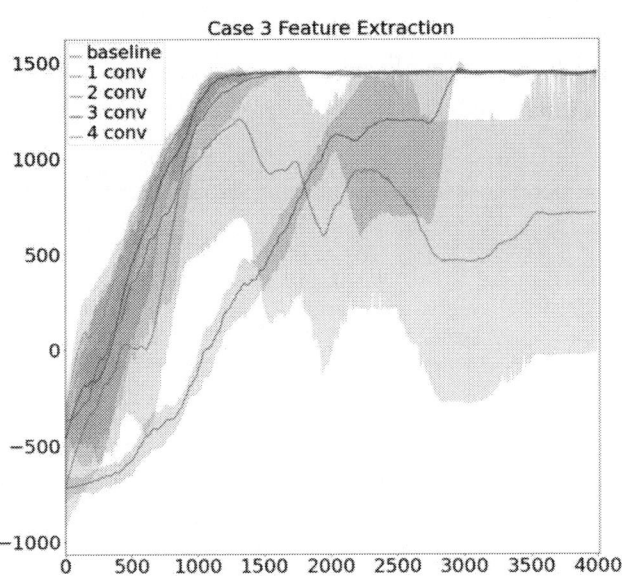

Copyright © 2023 by ASME

- Case-pair similarity has more effect on deeper convolution layer transfer, no matter what kind of features are extracted and processed.

Drawing upon the results, a demonstration of the transferability of work process features becomes evident. Consequently, the implementation of transfer reinforcement learning presents a viable approach to addressing intricate engineering challenges, whereby the extraction and reuse of pivotal work process features. Meanwhile, the current study only compares ship dynamics features and rule-related features. As such, its generality can be limited. To effectively handle various transfer learning tasks, additional engineering features must be thoroughly investigated. Moreover, future research should aim to explore a wider range of engineering tasks by applying this approach.

ACKNOWLEDGEMENTS

This paper is based on the work supported by the Autonomous Ship Consortium (ASC) with members of BEMAC Corporation, ClassNK, MTI Co. Ltd., Nihon Shipyard Co. (NSY), Tokyo KEIKI Inc., and National Maritime Research Institute of Japan. The ship dynamics model is provided by Chuanhui Hu, a Ph.D. student in our lab. The authors are grateful for their support and collaboration on this research.

REFERENCES

[1] Raffin, Antonin, et al. "Stable-baselines3: Reliable reinforcement learning implementations." The Journal of Machine Learning Research 22.1 (2021): 12348-12355.

[2] Mnih, Volodymyr, et al. "Playing atari with deep reinforcement learning." arXiv preprint arXiv:1312.5602 (2013).

[3] Todorov, Emanuel, Tom Erez, and Yuval Tassa. "Mujoco: A physics engine for model-based control." 2012 IEEE/RSJ international conference on intelligent robots and systems. IEEE, 2012.

[4] Schulman, John, et al. "Proximal policy optimization algorithms." arXiv preprint arXiv:1707.06347 (2017).

[5] Suh, Nam P. "Ergonomics, axiomatic design and complexity theory." Theoretical Issues in Ergonomics Science 8.2 (2007): 101-121.

[6] Zhu, Yizhen, et al. "Recent advancements and applications in 3D printing of functional optics." Additive Manufacturing (2022): 102682.

[7] Weiss, Karl, Taghi M. Khoshgoftaar, and DingDing Wang. "A survey of transfer learning." Journal of Big data 3.1 (2016): 1-40.

[8] Bengio, Yoshua. "Deep learning of representations for unsupervised and transfer learning." Proceedings of ICML workshop on unsupervised and transfer learning. JMLR Workshop and Conference Proceedings, 2012.

[9] Deniz, Erkan, et al. "Transfer learning based histopathologic image classification for breast cancer detection." Health information science and systems 6 (2018): 1-7.

[10] Nixon, Mark, and Alberto Aguado. Feature extraction and image processing for computer vision. Academic press, 2019.

[11] Liu, Xiongqing, and Yan Jin. "Reinforcement learning-based collision avoidance: impact of reward function and knowledge transfer." AI EDAM 34.2 (2020): 207-222.

[12] Yosinski, Jason, et al. "How transferable are features in deep neural networks?" Advances in neural information processing systems 27 (2014).

[13] Wang, Xinrui, and Yan Jin. "Work Process Transfer Reinforcement Learning: Feature Extraction and Finetuning in Ship Collision Avoidance." International Design Engineering Technical Conferences and Computers and Information in Engineering Conference. Vol. 86212. American Society of Mechanical Engineers, 2022.

[14] Todorov, Emanuel, Tom Erez, and Yuval Tassa. "Mujoco: A physics engine for model-based control." 2012 IEEE/RSJ international conference on intelligent robots and systems. IEEE, 2012.

[15] Ahmed, Ossama, et al. "Causalworld: A robotic manipulation benchmark for causal structure and transfer learning." arXiv preprint arXiv:2010.04296 (2020).

[16] Zhao, Wenshuai, Jorge Peña Queralta, and Tomi Westerlund. "Sim-to-real transfer in deep reinforcement learning for robotics: a survey." 2020 IEEE symposium series on computational intelligence (SSCI). IEEE, 2020.

[17] Goerlandt, Floris, and Pentti Kujala. "On the reliability and validity of ship–ship collision risk analysis in light of different perspectives on risk." Safety science 62 (2014): 348-365.

[18] Hu, Chuanhui, and Yan Jin. "Long-Range Risk-Aware Path Planning for Autonomous Ships in Complex and Dynamic Environments." Journal of Computing and Information Science in Engineering 23.4 (2023): 041007.

[19] Li, Guofa, et al. "Risk assessment based collision avoidance decision-making for autonomous vehicles in multi-scenarios." Transportation research part C: emerging technologies 122 (2021): 102820.

[20] Fan, Yunsheng, Zhe Sun, and Guofeng Wang. "A novel reinforcement learning collision avoidance algorithm for USVs based on maneuvering characteristics and COLREGs." Sensors 22.6 (2022): 2099.

[21] Li, Lingyu, et al. "A path planning strategy unified with a COLREGS collision avoidance function based on deep reinforcement learning and artificial potential field." Applied Ocean Research 113 (2021): 102759.

[22] Chuah, Jason. Law of international trade: cross border commercial transactions. Sweet & Maxwell Ltd., 2009.

[23] Wang, Xiaohua, Vivek Yadav, and S. N. Balakrishnan. "Cooperative UAV formation flying with obstacle/collision avoidance." IEEE Transactions on control systems technology 15.4 (2007): 672-679.

[24] Park, Jung-Woo, Hyon-Dong Oh, and Min-Jea Tahk. "UAV collision avoidance based on geometric approach." 2008 SICE Annual Conference. IEEE, 2008.

Copyright © 2023 by ASME

[25] Zhao, Luman, and Myung-Il Roh. "COLREGs-compliant multiship collision avoidance based on deep reinforcement learning." *Ocean Engineering* 191 (2019): 106436.

[26] Nixon, Mark, and Alberto Aguado. *Feature extraction and image processing for computer vision*. Academic press, 2019.

[27] Sivic, Josef, and Andrew Zisserman. "Video Google: A text retrieval approach to object matching in videos." *Computer Vision, IEEE International Conference on*. Vol. 3. IEEE Computer Society, 2003.

[28] Harris, Anne, et al. "Words." *Writing for Performance* (2016): 19-35.

[29] Liu, Ke, Yong-Qing Cheng, and Jing-Yu Yang. "Algebraic feature extraction for image recognition based on an optimal discriminant criterion." *Pattern recognition* 26.6 (1993): 903-911.

[30] Chang, Di, et al. "RC-MVSNet: unsupervised multi-view stereo with neural rendering." *Computer Vision–ECCV 2022: 17th European Conference, Tel Aviv, Israel, October 23–27, 2022, Proceedings, Part XXXI*. Cham: Springer Nature Switzerland, 2022.

[31] Yin, Yufeng, et al. "Multi-modal Facial Action Unit Detection with Large Pre-trained Models for the 5th Competition on Affective Behavior Analysis in-the-wild." arXiv preprint arXiv:2303.10590 (2023).

[32] Wagner, Johannes, et al. "Deep learning in paralinguistic recognition tasks: Are hand-crafted features still relevant?" (2018).

[33] Lowe, David G. "Object recognition from local scale-invariant features." *Proceedings of the seventh IEEE international conference on computer vision*. Vol. 2. Ieee, 1999.

[34] Dalal, Navneet, and Bill Triggs. "Histograms of oriented gradients for human detection." *2005 IEEE computer society conference on computer vision and pattern recognition (CVPR'05)*. Vol. 1. Ieee, 2005.

[35] Simonyan, Karen, and Andrew Zisserman. "Very deep convolutional networks for large-scale image recognition." *arXiv preprint arXiv:1409.1556* (2014).

[36] He, Kaiming, et al. "Deep residual learning for image recognition." *Proceedings of the IEEE conference on computer vision and pattern recognition*. 2016.

[37] Szegedy, Christian, et al. "Going deeper with convolutions." *Proceedings of the IEEE conference on computer vision and pattern recognition*. 2015.

[38] Sharif Razavian, Ali, et al. "CNN features off-the-shelf:an astounding baseline for recognition." *Proceedings of the IEEE conference on computer vision and pattern recognition workshops*. 2014.

[39] Jin, Y. "ISC: Intelligent Situation Awareness and Collision Avoidance" Technical Report, 2019.

[40] Imazu H (1987) Research on collision avoidance maneuver. Ph.D. thesis, The University of Tokyo (In Japanese)

[41] Szlapczynski, Rafal, and Joanna Szlapczynska. "An analysis of domain-based ship collision risk parameters." *Ocean Engineering* 126 (2016): 47-56.

[42] Goerlandt, Floris, et al. "Analysis of near collisions in the Gulf of Finland." *Advances in Safety, Reliability and Risk Management* (2012): 2880-86.

[43] International Maritime Organization. "Convention on the International Regulations for Preventing Collisions at Sea, 1972 (COLREGs)." (1972).

[44] Fujii, Y. "A definition of the evasive domain." *Navigation* 65 (1980): 17-22.

[45] Shen, Haiqing, et al. "Automatic collision avoidance of multiple ships based on deep Q-learning." *Applied Ocean Research* 86 (2019): 268-288.

[46] Huang, Bingling, and Yan Jin. "Reward shaping in multiagent reinforcement learning for self-organizing systems in assembly tasks." *Advanced Engineering Informatics* 54 (2022): 101800.

[47] Schulman, John, et al. "Proximal policy optimization algorithms." *arXiv preprint arXiv:1707.06347* (2017).

[48] Haarnoja, Tuomas, et al. "Soft actor-critic: Off-policy maximum entropy deep reinforcement learning with a stochastic actor." *International conference on machine learning*. PMLR, 2018.

Copyright © 2023 by ASME

Proceedings of the ASME 2023
International Design Engineering Technical Conferences and
Computers and Information in Engineering Conference
IDETC-CIE2023
August 20-23, 2023, Boston, Massachusetts

DETC2023-117266

TOWARD ARTIFICIAL EMPATHY FOR HUMAN-CENTERED DESIGN: A FRAMEWORK

Qihao Zhu
Data-Driven Innovation Lab
Singapore University of Technology and Design
Singapore

Jianxi Luo
Data-Driven Innovation Lab
Singapore University of Technology and Design
Singapore

ABSTRACT

In the early stages of the design process, designers explore opportunities by discovering unmet needs and developing innovative concepts as potential solutions. From a human-centered design perspective, designers must develop empathy with people to truly understand their needs. However, developing empathy is a complex and subjective process that relies heavily on the designer's empathetic capability. Therefore, the development of empathetic understanding is intuitive, and the discovery of underlying needs is often serendipitous. This paper aims to provide insights from artificial intelligence research to indicate the future direction of AI-driven human-centered design, taking into account the essential role of empathy. Specifically, we conduct an interdisciplinary investigation of research areas such as data-driven user studies, empathetic understanding development, and artificial empathy. Based on this foundation, we discuss the role that artificial empathy can play in human-centered design and propose an artificial empathy framework for human-centered design. Building on the mechanisms behind empathy and insights from empathetic design research, the framework aims to break down the rather complex and subjective concept of empathy into components and modules that can potentially be modeled computationally. Furthermore, we discuss the expected benefits of developing such systems and identify current research gaps to encourage future research efforts.

Keywords: Empathy, Human-Centered Design, Artificial Intelligence, Data-Driven Design

1. BACKGROUND AND INTRODUCTION

Early-stage conceptual design is crucial for innovation and focuses on discovering unmet human needs and generating design concepts to address such needs. Human-social approaches such as design thinking [1] and ideation techniques

such as TRIZ [2] and design heuristics [3] are often used to support and guide such design activities. During the era of data-driven innovation [4], machine learning and data-driven approaches have been growingly adopted to discover and evaluate design opportunities and generate and evaluate design concepts by drawing information, knowledge, and inspiration from data [5-7]. Recent contributions also showed the capability of cutting-edge data-driven artificial intelligence (AI) such as generative pretrained transformers (GPT) for automatic design concept generation [8, 9].

Generating new design concepts with extended digital knowledge sources and improved creative thinking capability with the aid of AI will contribute to increased quantity and novelty of the generated concepts [10, 11]. On the other hand, from the perspective of human-centered design (HCD), an in-depth empathic understanding of people should be developed before designers can generate solutions rooted in people's actual needs [12]. Recently, data-driven approaches towards user study have been utilizing user-generated content (e.g., online customer review, social media data) to extract and analyze users' opinion in a very large scale [13, 14]. However, these methods tend to focus on gathering explicit knowledge about people's current and past experiences, rather than learning their potential future experiences through delving deeper into their desires and dreams [15-17]. Thus, researchers have been emphasizing the critical role of empathy in human-centered design for the in-depth and comprehensive understanding of users' needs [18-20].

According to *The Field Guide to Human-Centered Design* by IDEO [12], empathy in design is defined as *"the capacity to step into other people's shoes, to understand their lives, and start to solve problems from their perspectives"*. By immersing themselves in the others' world and getting closer to their lives and experiences, designers are more likely to reveal their emotions and desires and discover underlying tacit knowledge

Copyright © 2023 by ASME

and latent needs [15, 18]. Discovering and addressing such knowledge and needs is crucial for improving user experience and facilitating market success. Recognizing the vital role that empathy plays in the design process, many researchers have been exploring methods to support the empathic design process (e.g., design probes [21], co-creation [22, 23]). However, to date, developing empathy is largely dependent on the designer's empathic capacity, which is subjective and intuitive [18, 24, 25]. Thus, discovering design opportunities through empathic design is often serendipitous. Yet, to the best of our knowledge, an effective computer-aided approach to facilitate empathic design has not been found in the literature.

In recent years, we have seen the concept of artificial or computational empathy being introduced in different disciplines, including AI [26, 27], social robotics [28, 29] (with applications in education [30] and healthcare [31]), and marketing [32]. Artificial empathy, as a new area of interdisciplinary research, focuses on modeling human empathy computationally into artificial agents. To the best of our knowledge, there has been no attempt to introduce or discuss artificial empathy in the context of human-centered design. Therefore, this paper aims to develop a framework of artificial empathy for human-centered design based on the synthesis of cutting-edge insights from several relevant fields of study. By decomposing the subjective and complex concept of empathy, we identify the modules and components of artificial empathy that can be modeled computationally. We discuss the potential roles of artificial empathy in human-centered design, introduce our proposed framework, and identify future research opportunities.

2. A MULTIDISCIPLINARY SYNTHESIS

Our framework for artificial empathy for HCD (to be introduced in section 3) draws on the synthesis of three literature strands, as depicted in Figure 1. In the following, we review the three literature streams.

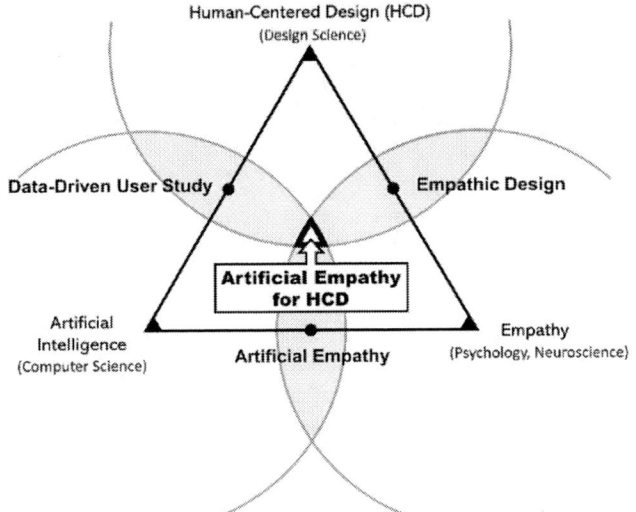

FIGURE 1: ARTIFICIAL EMPATHY FOR HUMAN-CENTERED DESIGN DRAWS ON HUMAN-CENTERED DESIGN, EMPATHY, AND AI RESEARCH

2.1 Data-Driven User Studies

The employment of AI techniques for user studies has been drawing increasing interest in design as well as marketing research [13, 14, 33]. The expansion of online user-generated data has escalated the big-data analytics of such data using AI. This field of research has advanced the mining of valuable insights into users' thoughts, preferences, and opinions on a very large scale. Depending on users' awareness of creating the data, user-generated data can be classified into two categories: user-generated content and user-generated behavior [34].

User-generated content, which has been a widely utilized online data source for user studies, refers to the public information that users intentionally create. Commonly used user-generated content includes online reviews on e-commerce websites and social media data. Due to the massive amount, ease of access, and relatively structured data of user-generated content, they have been used in a variety of user study tasks, usually with natural language processing (NLP). The NLP tasks that leverage user-generated content include needs extraction [35], sentiment analysis [36], user profiling [37], usage context extraction [38], and emotion recognition [39]. In addition, researchers in the AI community have also introduced multimodal approaches for analyzing user-generated content beyond just textual data. For example, Poria et al. [40] developed a model based on long short-term memory network (LSTM) to capture contextual information and perform multimodal sentiment analysis using user-generated videos. Notably, recent research in engineering design [41] leveraged both text and image in user-generated content to extract product usage context and sentiments toward product features.

However, user-generated content has inherent biases. People with extreme opinions (either positive or negative) are more likely to leave their comments than those with moderate opinions, resulting in an underreporting bias [42]. The analytic results can thus be biased toward the extremes. Likewise, social media data has been shown to have a population bias, e.g., most Reddit users in the US are male and young in age [43]. Thus, the user understanding developed based on such data may not be representative of the larger population.

On the other hand, user-generated behavior (also known as the digital footprint [4]) is the information that is generated as a result of user actions. Widely used approaches include clickstream analysis and web analytics (e.g., Google Analytics). They have been used for tasks such as user behavior analysis [44], identifying trending product characteristics [45], and deciding innovation strategy [4]. However, besides the commonly concerned privacy issue [34], another limitation of user-generated behavior is that it can only be generated through using interactive systems. Thus, for broader fields of design (e.g., engineering design), user behavior data may be difficult to obtain in similar ways.

2.2 Empathic Design

In human-centered design, empathic design entails developing a deep and comprehensive understanding of people's

Copyright © 2023 by ASME

circumstances and experiences in order to foster empathy and uncover insights [18, 20]. Design researchers tend to consider empathy as a type of knowledge, and thus empathic understanding as a form of knowledge construction [20, 46]. According to the literature [15, 16], designers may learn from people in three different ways: by listening to what they say, observing what they do and use, and discovering what they know, feel, and dream. The different ways lead to different levels of knowledge about people.

By listening to what people express and think, we learn the explicit knowledge of user opinions [15]. Conventional approaches to gathering user opinions include questionnaires, interviews, and focus groups. The above-introduced applications of NLP techniques to analyze user-generated content also lie at this level of understanding. By observing what people do and use, researchers learn the actual behaviors of users in their natural habitat, which leads to the observable knowledge of people and their context [15, 16]. In human-centered design, the term context refers to "all factors that influence the experience of a product use" [16]. Conventional approaches to observable knowledge are contextual inquiry [47] and ethnographic observation [48]. They are both field data-gathering methods that involve in-depth observation, interview, and documentation to investigate the users' work practices and behavior.

Finally, by discovering what people know, feel, and dream, researchers can reach a deeper understanding and learn tacit knowledge, which could reveal even deeper understanding of latent needs, i.e., needs not recognizable until the future [15, 16]. However, such depth of understanding can be difficult to develop because the internal feelings and mental states of people can hardly be sensed directly. Popularly, people's feelings can also be accessed through self-assessment surveys, which ask participants to report their emotions based on predetermined emotion categories [49] or to report their psychological feelings based on semantic differential scales [50].

However, getting to know people's affective feelings is not enough. A great portion of empathic understanding comes from the cognitive aspect, which depends on the perspective-taking and inference ability of designers [18, 46]. Such cognitive understanding can be inferred from explicit and observable knowledge by paying careful attention to various clues to unfold underlying patterns [46]. However, the ability of perspective-taking needs training to develop and can vary from person to person [24]. Existing design methods supporting perspective-taking include role-playing [51], where the designers act out others' lives and experiences, as well as simulating experiences physically (e.g., simulating visual impairment by limiting vision [19]) or virtually (e.g., social perspective-taking through virtual reality [52]).

Furthermore, the empathic understanding in design research is usually recognized as an interplay of both affective and cognitive aspects [20, 46, 53, 54]. This affective-cognitive interplay involves feeling others' emotions and trying to take their perspective and make sense of them [18, 20]. Designers often make use of their own knowledge about the world and their experiences, blending with the observed information of others to

achieve empathic understanding [55, 56]. This emphasizes that developing empathic understanding is not just about studying the stakeholders, but it is also dependent on more general knowledge and experience (i.e., experiential knowledge [56]) of the designers. However, it is also argued that designers should not bring their own prejudices and stereotypes about others when trying to develop empathy [46]. Thus, it can be crucial to determine the right level of background knowledge that designers should bring with them.

In addition, researchers have emphasized the role of imagination in empathic design [18, 25, 53, 57]. In this paper, by "imagining," we do not refer to the perspective-taking process typically described as "imagining oneself in others' situation" [18]. Instead, we refer to the generative mental process that comes up with potentially better alternatives (imagined situations) [23, 25] and assesses the perceived usefulness (imagined use) of these alternatives [53, 57]. Designers who cannot vigilantly imagine alternatives can get lost reflecting on others' experiences without producing any new insights [25].

In typical co-design sessions, participants are provided with generative toolkits to create artifacts by themselves and tell a story about what they made [16, 22, 58]. The purpose of these activities is to facilitate or trigger people's imagination and expressions about what they want to experience in the future [25]. On the other hand, designers' imagination is generated in their minds based on an already developed understanding of people. Designers should vigilantly seek potential design opportunities that lead to a more imaginative future.

As Norman [59] emphasized: *"One cannot evaluate an innovation by asking potential customers for their views. This requires people to imagine something they have no experience with."* Designers must find a way to represent their imagination in ways that people can understand, feel, and imagine. Besides sketching and prototyping (e.g., mockups and paper prototyping), designers can also leverage storytelling techniques to convey their imagination to people and between designers [60]. By creating a future with new experiences and inviting people to have a glance at it, their imagination can be triggered, and their thoughts and feelings about the new situation can be learned. Thus, storytelling can become a platform for collaborative imagination [25].

2.3 Artificial Empathy

The idea of enabling empathic interactions between AI and ordinary people has led to the concept of artificial empathy, which aims to provide AI with human-like empathic capacity [29]. So far, the development of artificial empathy in AI and robotics has been based on a strong foundation of psychology and neuroscience. From a theoretical perspective, empathy can be modeled by two distinct categories: affective empathy and cognitive empathy [61, 62]. This view is also widely accepted in human-centered design [18, 53].

Affective empathy is the automatic response and mimicking of others' affective states, which is based on the mirror neuron system from a neuroscience perspective [63, 64]. For example, when observing others in pain, the mirror neurons are activated

Copyright © 2023 by ASME

as if the observers are experiencing pain themselves, thus developing shared feelings. This automatic responding process is also known as emotional contagion [63]. Other researchers suggest that mirror neurons are only responsible for forming an internal representation of the observed states and require insula to associate the internal representation with the observer's state [65, 66]. On the other hand, cognitive empathy is the ability to understand others' mental state (beliefs, desires) by taking their perspective [67], which is related to theory of mind [68, 69].

Another view of modeling empathy takes an evolutionary and developmental perspective. For instance, the Russian Doll model of empathy [70] argues that empathy is developed from low-level unconscious emotion contagion to high-level cognitive mechanisms. Following the Russian Doll model, Asada [28, 29] proposed a developmental model of self-other recognition as a part of Cognitive Developmental Robotics. The developmental perspective of empathy modeling is widely adapted in artificial empathy [26-29, 32].

Following the developmental model of artificial empathy proposed by Yalcin and DiPaola [26, 27], the components forming the low-level mechanisms of artificial empathy include emotion recognition, emotion representation, and emotion expression [26, 27]. Emotion recognition has been extensively investigated in affective computing [71, 72]. This area of research focuses on recognizing people's emotions from various modalities of information, including visual (e.g., facial expression, body gesture), audio (e.g., audio recording), and textual (e.g., customer review, social media). Apart from the unimodal approaches that use a single modality for emotion recognition, researchers have also been fusing more than one modality for multimodal emotion recognition [73], introducing physiological modalities [74], and considering contextual information encoded in images or videos [40]. For adding the emotion expression capacity, Asada [29] showed examples that recognize and express emotions in body gestures and facial expressions, while in both examples, the internal emotion states are represented by vector embeddings. More recently, Casas et al. [75] developed an empathic conversational AI that generates empathic responses by fine-tuning a GPT model.

Additionally, researchers have proposed an emotion regulation module related to the self-other distinction capability of the AI agent. With this module, the AI is supposed to modify and regulate the "raw" emotion representation based on the agent's mood and personality, as well as its relational and social link to the person being empathized with [26, 27, 76].

On the other hand, the components that form the high-level cognitive mechanisms of artificial empathy include the appraisal theory and theory of mind [26, 27]. Appraisal theory is the cognitive mechanism of emotion that states that our emotions are triggered by our appraisal of the events and situations in our environment [62]. This mechanism gives rise to affective perspective-taking, i.e., taking others' perspective and understanding their feelings and the causes of their feelings [68]. Theory of mind refers to one's ability to attribute mental states (e.g., desires, intentions, and beliefs) to others [69]. This mechanism is also referred to as cognitive perspective-taking, or

as the ability to infer the beliefs of others [77]. However, the computational modeling of the cognitive mechanisms can be challenging. So far, Rabinowitz et al. [78] used meta-learning to model theory of mind computationally, and the model managed to recognize other agents' false beliefs. Jara-Ettinger [79], on the other hand, employed inverse reinforcement learning (IRL), which aims to learn a reward function (goal inference) through observing the behaviors of an agent. A recent study [80] showed that the ability of theory of mind emerged in the latest versions of GPTs, even if this ability was not explicitly engineered into these large language models.

The mechanism behind the emergence of theory of mind in GPT is yet unknown. The model could have discovered language patterns behind theory of mind that are unknown to humans or learned this ability spontaneously from the training data [80]. In either way, the massive amount of general human knowledge it was trained on seems to be the key. This can be associated with the theory-theorists' view of cognitive empathy, which suggests we infer others' minds based on our internal storage of abstract and generalized knowledge about the world and about ourselves and others [81].

3. ARTIFICIAL EMPATHY FOR HUMAN-CENTERED DESIGN (HCD)

3.1 The Role of Artificial Empathy in HCD

Based on the foregoing literature review, we conclude that there are three major differences between the potential development of artificial empathy for HCD and that in other disciplines:

(1) *More human-centered, less human-like.* The main body of artificial empathy research has been focused on the human-like empathic ability of AI and empathic interaction between humans and AI agents. However, the role of empathy in human-centered design, as mentioned at the beginning of this paper, is to "step into other people's shoes, to understand their lives, and start to solve problems from their perspectives" [12]. This heavily depends on the construction of empathic understanding of others. Thus, artificial empathy in this scenario will focus on the interaction between people and their context, rather than people with the AI agent.

(2) *Ability to imagine and engage collaborative imagination.* The empathizer in HCD needs to continuously imagine alternatives that could lead to a potentially better situation and engage people in the imagination process [25]. This level of empathic ability is not seen in any other research area that the authors have investigated. We argue that this ability of imagining is at a higher level than the cognitive mechanisms of empathy and should happen after perspective-taking takes place.

(3) *Expressing insights, rather than emotions.* In robotics and affective computing, researchers have focused on the AI's ability to express appropriate emotions to people, either by parallelly imitating others' emotions or reactively appraising their situation [82]. On the other hand, artificial empathy in HCD should inform valuable insights or engage and elicit human expression and imagination. This requires the agent to express

Copyright © 2023 by ASME

the contents in ways that people can easily understand, feel, and imagine.

However, despite the differences, empathy in HCD can largely overlap with artificial empathy in terms of their internal mechanisms and representations. Therefore, the development of artificial empathy for HCD can draw references and inspirations from the more developed research areas.

Overall, we argue that the potential role of artificial empathy in HCD should be as assistants and tools for designers that facilitate the empathic understanding process, rather than as empathic companions.

3.2 The Framework of Artificial Empathy for HCD

With the role of artificial empathy in mind, and by combining insights from different research areas, we propose the framework of artificial empathy for human-centered design. The aim of this framework is to break down the relatively subjective and complex concept of empathy in HCD into different modules and components that can be modeled computationally. Drawing an analogy from human empathic intelligence, we illustrate the framework in Figure 2, which has three essential modules: 'senses', 'mind', and 'expression'.

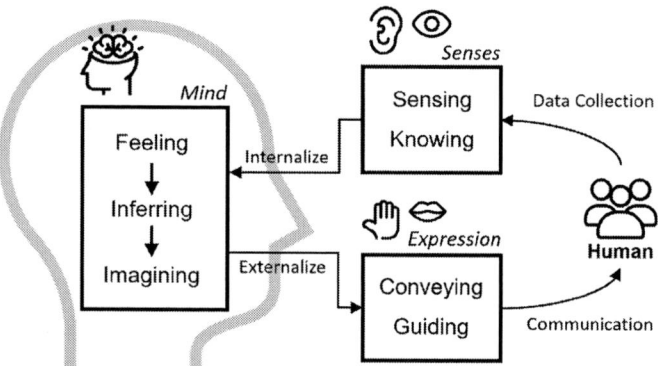

FIGURE 2: THE FRAMEWORK OF ARTIFICIAL EMPATHY FOR HCD

3.2.1 'Senses' of Artificial Empathy

To develop empathy for people, the first step is getting to know them. 'Senses' stands for the data gathering (*sensing* component) and processing (*knowing* component) ability of an artificial empathy agent to sense and get to know people and their context externally. The sensible data can come in three types:

Verbal data. Textual data collected from surveys, online reviews, or social media, as well as audio data collected from interview recordings or extracted from user-generated videos, can be used to extract information about people's thoughts and opinions. So far, gathering user-generated content and using NLP to analyze the text information is a rather attractive approach in design research, because it requires little expense and travel, and the statistical results based on the large scale of data look convincing and promising. In addition, it has been shown that the textual data of user feedback could also contain

contextual information, and thus has been utilized to construct a knowledge graph of product usage context [38].

Visual data. Visual data, such as images or videos recorded during contextual inquiry, can provide information about people's physical characteristics, behaviors, and contexts. The multimodal analysis of user-generated content could potentially offer us many insights into the relationship between people and their context. Specifically, the information embedded in images and videos that users have created and uploaded online is valuable for extracting behavioral and contextual information, as well as people's physical characteristics, such as facial expressions and body gestures. So far, the engineering design community has explored extracting context information from images through object detection [41]. In the computer vision field, recent studies have delved into video analysis for object tracking [83] and human action detection in complex real-world events [84]. HCD research could benefit from adopting such techniques in analyzing videos generated by users or recorded during contextual inquiries. By doing so, we can expect to obtain more comprehensive explicit and observable knowledge about people.

Physiological data. There is another line of research focusing on measuring people's physiological signals using sensor technologies introduced from neuroscience. The collected physiological data can then be mapped into emotional states. Such approaches have widely been introduced in neuromarketing [85, 86]. Popular sensing techniques include facial encoding, skin conductance, electroencephalography (EEG) brain scans, and functional magnetic resonance imaging (fMRI). The gathered data reveals the deeper psychological reactions of people that can hardly be detected otherwise. While this approach has also been used to investigate the neurocognition of designers during design activities [87], we still lack a clear understanding of what these measurements may indicate for empathic design and how to use them properly [20, 88]. The potential use of physiological data for artificial empathy needs to be clarified in future research efforts.

By sensing and processing data directly, the artificial empathy system will acquire comprehensive knowledge about people at the explicit and observable levels, which also contains clues for tacit knowledge. This module of "senses" corresponds to the unconscious sensing ability of humans, without the need for purposefully trying to make sense of the information. Developing empathic understanding, from the perspective of knowledge construction, is a hierarchy from gathering explicit and observable knowledge to the inference and imagination of tacit knowledge and latent needs. One cannot effectively take the perspective of others and develop empathy if they barely know them explicitly.

3.2.2 'Mind' of Artificial Empathy

The information gathered by 'senses' is then internalized by the 'mind' of artificial empathy. This module corresponds to the empathic mental processes conceptualized by Surma-Aho and Höltta-Otto [20], which is an internal aspect of empathic understanding.

Copyright © 2023 by ASME

In this framework, we follow the developmental view of empathy modelling that the higher-level mechanisms of empathy are developed upon lower-level mechanisms. In fact, researchers suggest that the internal representation of low-level empathy can be used to inform high-level empathy [26]. Thus, we identified three levels of components in the 'mind' of artificial empathy: feeling, inferring, and imagining.

At the *feeling* component, we refer to the unconscious process of emotion recognition and internalization. For humans, this ability largely depends on the mechanism of mirror neuron system [63, 64]. Emotion recognition, as we introduced earlier, has been intensively investigated in affective computing and Neuromarketing. Researchers have been utilizing a variety of data modalities, including verbal, visual, and physiological data to develop more accurate recognition systems. In fact, it has been shown that multimodal approaches for computational emotion recognition can have an edge over human capability [89]. In HCD, researchers have explored emotion recognition through facial encoding to measure if shared feelings have been established between designers and users [90]. However, the recognized emotions of people are not always consistent with their real feelings [90], especially during non-contextual interviews where people may jokily talk about their pain points.

At the *inferring* component, we consider the cognitive mechanisms of empathy modeled by Yalcin and DiPaola [27], which include the appraisal theory to understand people's affective states (emotion, feeling), and theory of mind to understand their mental states (intention, belief, and desire). Based on the knowledge known to the artificial empathy system, the attempt of inferring the causes and intensity of an emotion as well as inferring the belief and desire of people would be performed. Recent research efforts in computer science have investigated NLP for emotion inference from the perspective of appraisal theory [91, 92]. The researchers created a corpus of event descriptions and annotated emotion labels by both the experiencer and human inferrers, and tested NLP's performance of inferring emotions based on given events. The emotion inference through appraisal theory differs from the *feeling* component by the cognitive process involved. For example, typical emotion recognition tasks involve texts with emotional expressions and tones that we can feel directly. On the other hand, appraisal theory aims to build connection between an event and its experiencer's emotion. The event description text can be as neutral as *"I bought my own horse with my own money I had worked hard to afford"* (example from Troiano et al. [92]), but we can take the perspective of the experiencer in the event and infer his/her joyful feeling. However, in HCD practices the contexts of events can be more complex and involve interactions with detailed product components, and the situated application of NLP based appraisal theory is yet to be investigated.

On the other hand, theory of mind considers the inference of mental states from explicit and observable knowledge. In psychology and computer science, the ability of theory of mind of human or AI agents is often tested through false belief tasks [78, 80, 93], where the test subjects are asked to predict the behavior of another person or agent who has a false belief about a situation. However, design researchers argued that the false belief tasks have validity limitations for HCD context and proposed to use empathic accuracy test instead to measure designer's thought inference accuracy [94]. As mentioned in section 2.3, the concept of theory of mind has been studied in AI research using meta-learning [78] and inverse reinforcement learning [79]. But these attempts took virtual agents in simulation environments and thus the current stage of this line of research is not applicable to HCD. Large language models (e.g., GPT), on the other hand, can make inference of people's mind based on textual descriptions of real-world events and situations [80] and are more promising to work in an HCD scenario. But again, the performance of such AI-based mind inference will need to be validated specifically for HCD.

In the proposed framework, the component of *inferring* depends heavily on the explicit and observable knowledge formed in the previous process, as well as a comprehensive background knowledge about the common senses of the world and the common reasonings of people's minds. The combination of these knowledge sources resembles the user knowledge construction process in design, i.e., blending epistemic knowledge about people with the experiential knowledge of the designers [56]. The inferred understanding should then better lead to the very nature of the problems that people are experiencing [25].

In HCD practice, having an in-depth understanding of the underlying causes of people's positive or negative emotions as well as their unspoken intentions, beliefs, and desires, designers are then ready to imagine alternatives to improve the current situations. At the *imagining* component, we refer to the AI's ability to "twist the reality [25]", i.e., the ability to imagine alternatives for current situation toward a potentially better future situation, without which the system could stuck in reflecting on the available knowledge without producing meaningful insights. For example, before noise-cancelling earphones came into market, the need for such technology could hardly be found from explicit knowledge because people took it as a commonsense property that earphones can hardly be used in noisy places. However, by removing a potential cause (external noise) of negative emotions (annoying) appraised from the situation (public places), artificial empathy can potentially imagine alternatives like "earphones that eliminate external noise". Apart from emotions, *imagining* could also consider people's unmet mental states like intentions and motivations. For example, when using a mirror, people's motivation is likely to be "to see myself exactly as others see me", but the indoor lighting can hinder the fulfillment of that motivation. By introducing a new entity (natural lighting) to the current situation (e.g., indoor dressing table) that is necessary for the fulfillment of motivation, artificial empathy could imagine alternatives like "mirror that provides natural lighting". These are two examples that we expect the *imagining* component to work, and it depends heavily on the empathic understanding developed through pervious components.

With *imagining* capacity, artificial empathy will acquire the capability to reach the deepest level of empathic understanding,

Copyright © 2023 by ASME

i.e., latent needs, which are the needs that are not recognizable by people until the future [15]. Thus, by definition, they can hardly be gathered or extracted externally from available data. Prior work by Zhou et al. [95] has studied NLP based latent needs finding through exploring extraordinary use cases, which can be considered as a form of reality twists that imagine alternative use cases for a design target (e.g., alternative usage contexts or alternative target users). However, the needs elicited by this approach are limited to use-case-related ones and are based on explicit knowledge without the empathic understanding of people. On the other hand, the component of *imagining* in the proposed framework will require the internal representation of the empathic understanding knowledge constructed by previous components combined with the commonsense knowledge and reasoning shared across modules and components of the system.

3.2.3 'Expression' of Artificial Empathy

The "expression" module requires the ability of artificial empathy to externalize the empathic understanding formed in the "mind" module in ways that effectively facilitate communication with stakeholders. Thus, the "expression" module will require a chatbot-like communication system with the ability to externalize knowledge and provide guidance.

The first component of "expression" is *conveying*, which represents the externalization of empathic understanding for both designers and stakeholders to understand. This component aims to represent knowledge and insights in ways that people can easily understand, feel, and imagine. For explicit, observable, and tacit knowledge of empathic understanding, *conveying* can be performed by directly organizing the internal representation of the knowledge into natural language. However, as most stakeholders are not trained to imagine things that do not yet exist like designers do [59], the interpretation and articulation of latent needs can be supported with storytelling [60]. In general practices, generative storytelling has been shown to be technically feasible through natural language generation [96], and it can even be combined with text-to-image generation AI to facilitate the creation of a more immersive experience [97]. With the recent advances of large language models [98] and diffusion-based models [99], we can expect the employment of generative storytelling or storyboarding in HCD soon.

The other component of "expression" considers communication and iteration. In typical participatory design practices, designers need to continuously communicate with stakeholders throughout the process, providing guidance and receiving feedback to improve empathic understanding. Likewise, we can expect that in many cases, the initially collected data will be inadequate to support the *inferring* and *imagining* processes in artificial empathy. Therefore, the *guiding* component involves intentionally asking questions and providing stimuli or inspirations to engage and encourage people's expression and imagination. More radically, it could also ask people to create simple drawings or 3D modeling through the system interface. The feedback will be gathered by "senses" and thereby starting another loop for iteration.

The "expression" module in the proposed framework does not directly contribute to empathy development. Instead, it serves as an interface between the system and the people it is learning about to confirm and validate the developed empathic understanding and engage these people into collaborative imagination to gain more meaningful insights. The two modules, including "senses" and "expression," together form the external aspect of empathic understanding [20].

3.3 Artificial Empathy as Multimodal and Multidimensional Knowledge Construction

As mentioned earlier, we identify empathy in HCD as a form of knowledge construction [20, 46]. Artificial empathy in this context aims to develop in-depth and comprehensive knowledge about the empathic understanding of stakeholders involved in the product or service to be designed.

According to our framework, such knowledge is constructed level-by-level. First, through 'senses', the system obtains explicit knowledge of opinions and thoughts, as well as the observable knowledge of behaviors and contextual information. Second, 'mind' recognizes and understands the affective states and infers the motivation, beliefs, desires of people, which together form tacit knowledge. Finally, through vigilant reflection of the current situation in 'mind' and iterative communication to people with 'expression', knowledge of their latent needs can be discovered and recognized. Thus, this level-by-level knowledge construction allows not only the finding of people's actual needs, but also the reasoning behind the findings.

It is important to note that multimodality is needed for each module of artificial empathy. 'Senses' collect data from the multimodal world, 'mind' reasons based on the multimodal information, and 'expression' represents knowledge in multimodal manners.

Furthermore, the knowledge construction of artificial empathy involves multidimensional integration. Apart from the information collected by 'senses', artificial empathy also requires the integration of commonsense reasoning [100]. For example, 'senses' need taxonomic reasoning [100] to know the categories and relationships of instances. 'Mind' should comprehend the theory of action and change [100] as well as commonsense psychology [101] to infer what is happening next in the situation and in people's minds. 'Expression' also needs commonsense reasoning and knowledge about the world to perform storytelling. However, although it may be easy for humans, commonsense reasoning can be very challenging for AI to achieve [100]. Another dimension of knowledge integration is the general knowledge of project-specific background. For example, when designing medical equipment, the system should have knowledge of the functions, structures, and working principles of the equipment as well as its workflow involving the interaction with doctors or patients.

Taking inspiration from Sarica et al. [102], Figure 3 illustrates the architecture of knowledge integration in artificial empathy. To date, integrating multimodal and multidimensional knowledge in an artificial system has been challenging. However, we have seen recent large foundation models increase

Copyright © 2023 by ASME

their capability in comprehending vast knowledge and complex reasoning. Specifically, the latest version of GPT (i.e., GPT-4) has demonstrated human-level commonsense reasoning and outperformed previous large language models [103, 104], while also having multimodal capability [103]. This shows great promise for future research on artificial empathy for HCD.

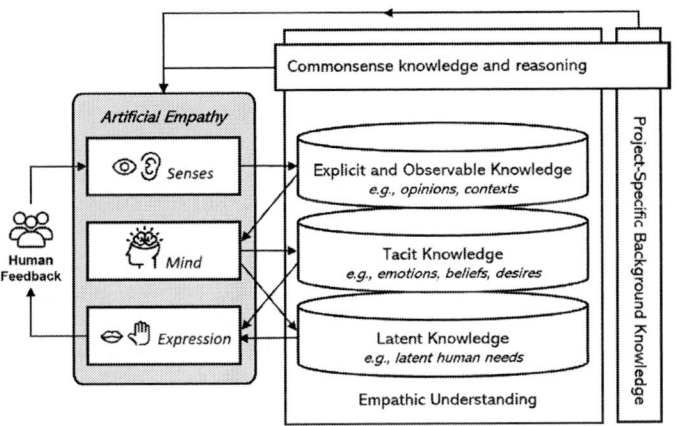

FIGURE 3: THE ARCHITECTURE OF KNOWLEDGE INTEGRATION IN ARTIFICIAL EMPATHY

4. DISCUSSION

4.1 The Potential Benefits of Employing Artificial Empathy In HCD

As mentioned in the first section, empathic design is usually intuitive, and the discovery of innovation opportunities through empathic design is thus serendipitous. Through artificial empathy, we can potentially get to high-level empathic understanding built upon multidimensional knowledge and clues that human designers can hardly comprehend.

Moreover, user study is usually considered as a tradeoff between depth and scale, as Norman [17] emphasized: "deep insights on real needs from a tiny set of people, versus broad, reliable purchasing data from a wide range and large number of people." By employing artificial empathy, depth and scale can possibly be handled simultaneously. By gathering large-scale and multimodal data, in-depth understanding can be developed through computationally modeled developmental empathic mechanisms. Alternatively, a web-based interactive platform of artificial empathy can be developed to involve a large number of stakeholders virtually, allowing the AI agents to interact with them and gather data concurrently.

Modern design projects often involve complex systems with a large number and variety of stakeholders that conventional participatory design practices may not accommodate. The development and employment of artificial empathy can be particularly valuable in this context.

4.2 Evaluating Artificial Empathy in HCD

Surma-aho and Hölttä-Otto [20] identified six categories of empathy measures in design, including empathic tendency, beliefs about empathy, emotion recognition, understanding mental content, shared feelings, and prosocial responding. However, most of these empathy measurements focus on evaluating the empathic connection between designers and people, which is not applicable for evaluating an artificial agent's empathic ability. For example, existing empathic design studies have employed Interpersonal Reactivity Index (IRI) to measure trait empathy [105, 106] and recognition of facial expression mimicry to measure affective empathy developed by designers [90]. However, an AI agent can easily answer an IRI survey with high empathic tendency or response to people with shared feelings by imitating their facial expression, but these cannot translate to a real empathic understanding.

Among these measures, performance-based emotion recognition and understanding of mental content are directly linked to empathic understanding in design processes [20, 24]. Emotion recognition measures low-level empathic understanding and has been intensively investigated in affective computing. Understanding of mental content can be assessed using empathic accuracy test [107, 108], which measures the accuracy of inferring thoughts and feelings of others. Future research is needed to validate the AI's performance of mind inference for HCD, e.g., measuring the empathic accuracy of the machine inferred states of people comparing to their self-reported states and human designers' inferred states.

Moreover, artificial empathy's capacity of needs finding will need to be tested. In a recent study, Li and Hölttä-Otto [109] used metrics of need quality and need latency to evaluate designers' performance of needs finding. Zhou et al. [110] categorized customer needs based on Kano's model [111] to identify attractive, one-dimensional, must-have, and indifferent needs. However, suitable evaluation approaches for artificial empathy's need-finding capacity will need to be developed in future research. We suggest that the AI agent's performance of both mind inference and need-finding should be compared with those of human designers under the same evaluation metrics.

4.3 Current Research Gaps and Opportunities

Following the proposed framework, artificial empathy can potentially provide valuable insights and human-centered innovation opportunities. However, artificial empathy can be a very complex system and requires the development and effective collaboration of different modules situated to HCD. We raise some of the current research gaps as follow and call for future research efforts:

Multimodal machine learning for HCD. Although multimodal machine learning has been intensively investigated in AI as well as in other aspects of engineering design [112, 113], it has not been widely situated for HCD practices. The specific requirements of HCD should be considered when developing such systems. Based on the previous analysis of the "mind" module and the available modelling techniques for appraisal theory and theory of mind, we suggest the most promising data types of artificial empathy's internal representation can be natural languages and knowledge graph. These representations have been intensively studied in engineering design [114, 115] but haven't been seen in empathy research except in [116]. Thus,

Copyright © 2023 by ASME

the multimodal fusion of data collected by "sense" and the mapping to these internal representations of "mind" can be a good direction to push forward the research of artificial empathy for HCD.

Modelling cognitive mechanisms of empathy for HCD. As mentioned in section 2.3 and section 3.2.2, the computational modellings of appraisal theory and theory of mind have been investigated in AI and robotics. However, the situated employment of such techniques in HCD to infer a deeper understanding of people has yet to be seen, e.g., how to identify design-related events and infer design-related mental states and emotions from them. Particularly, the appraisal theory has been applied in the research of emotional design [117-119]. The insights from this line of research could potentially be valuable for the development of artificial empathy. For example, Desmet [117] suggests that apart from the product (event) itself, people's concerns about the product are also essential for the cognitive appraisal of emotion. Additionally, another gap lies in the lack of baselines for measuring the performance of these cognitive mechanisms. Future works can contribute to the topic by constructing datasets of mind inferences in different design-specific contexts.

Machine imagination and storytelling. Recent research has shown creativity capacity of large language models performing creative design concept generation tasks [8, 9]. Machine imagination and storytelling could also benefit from generative models. However, in empathic design, instead of generating functional concepts, we need machine imagination to focus on the exploration of potential new needs, and storytelling to convey to the stakeholders a potentially beneficial future that they can feel and imagine. Thus, the generation of such contents and their performance will need to be addressed in future works.

Potential ethical issues. Ethical issues of AI like bias and fairness have been investigated intensively in recent years [120, 121]. According to our framework of artificial empathy, most modules will be driven by data gathered from people and the contents they create, we need to be extremely cautious not to pick up biases from the data and lead to unethical results. Moreover, artificial empathy is perceived to lack authenticity [65], e.g., does having the capabilities to perform empathic tasks as programmed necessarily translate to having real empathy? These issues may need to be investigated before the development of artificial empathy for HCD.

5. CONCLUSION

Based on an interdisciplinary investigation into research areas around user study, AI, and empathy, this paper presents a framework of artificial empathy for HCD. We have also discussed the role, benefits and ethical issues of artificial empathy in context of HCD and explicated the current research gaps to call for future research efforts. We hope this paper may stimulate more research to design and develop artificial empathy for human-centered design.

REFERENCES

[1] Brown, T. (2008). Design thinking. *Harvard business review*, 86(6), 84.

[2] Altshuller, G., Shulyak, L., Rodman, S., and Fedoseev, U. (1998). 40 Principles: TRIZ Keys to Innovation, Vol. 1. Technical Innovation Center Inc., Worcester, MA.

[3] Yilmaz, S., Daly, S.R., Seifert, C.M. and Gonzalez, R. (2016). Evidence-based design heuristics for idea generation, *Design studies*, Vol. 46, pp. 95-124.

[4] Luo, J. (2022). Data-driven innovation: What is it?. *IEEE Trans. Eng. Manag.*, 70(2), 784-790.

[5] Jiang, S., Hu, J., Wood, K. L., & Luo, J. (2022). Data-driven design-by-analogy: state-of-the-art and future directions. *J. Mech. Des.*, 144(2).

[6] Camburn, B., He, Y., Raviselvam, S., Luo, J., & Wood, K. (2020). Machine learning-based design concept evaluation. *J. Mech. Des.*, 142(3), 031113.

[7] Luo, J., Yan, B., & Wood, K. (2017). InnoGPS for data-driven exploration of design opportunities and directions: the case of Google driverless car project. *J. Mech. Des.*, 139(11).

[8] Zhu, Q., & Luo, J. (2023). Generative Transformers for Design Concept Generation. *J. Comput. Inf. Sci. Eng.*, 23(4), 041003.

[9] Zhu, Q., Zhang, X., & Luo, J. (2023). Biologically Inspired Design Concept Generation Using Generative Pre-Trained Transformers. *J. Mech. Des.*, 145(4), 041409.

[10] Luo, J., Sarica, S., & Wood, K. L. (2021). Guiding data-driven design ideation by knowledge distance. *Knowledge-Based Systems*, 218, 106873.

[11] Sarica, S., Song, B., Luo, J., & Wood, K. L. (2021). Idea generation with technology semantic network. *AI EDAM*, 35(3), 265-283.

[12] Kelley, D. (2015). The field guide to human-centered design. http://www.designkit.org/resources/1.

[13] Siddharth, L., Blessing, L., & Luo, J. (2022). Natural language processing in-and-for design research. *Design Science*, 8, E21.

[14] Yüksel, N., Börklü, H. R., Sezer, H. K., & Canyurt, O. E. (2023). Review of artificial intelligence applications in engineering design perspective. *Eng. Appl. Artif. Intell.*, 118, 105697.

[15] Sanders, E. B. N. (2002). From user-centered to participatory design approaches. In *Design and the social sciences* (pp. 18-25). CRC Press.

[16] Visser, F. S., Stappers, P. J., Van der Lugt, R., & Sanders, E. B. (2005). Contextmapping: experiences from practice. *CoDesign*, 1(2), 119-149.

[17] Norman, D. (2013). The design of everyday things: Revised and expanded edition. Basic books.

[18] Kouprie, M., & Visser, F. S. (2009). A framework for empathy in design: stepping into and out of the user's life. *J. Eng. Des.*, 20(5), 437-448.

[19] McDonagh, D., & Thomas, J. (2010). Rethinking design thinking: Empathy supporting innovation. *Australasian Medical Journal*, 3(8), 458-464.

Copyright © 2023 by ASME

[20] Surma-Aho, A., & Hölttä-Otto, K. (2022). Conceptualization and operationalization of empathy in design research. *Design Studies*, 78, 101075.

[21] Mattelmäki, T. (2006). Design probes. Doctoral dissertation, Aalto University.

[22] Sanders, E. B. N., & Stappers, P. J. (2008). Co-creation and the new landscapes of design. *CoDesign*, 4(1), 5-18.

[23] Mattelmäki, T., & Visser, F. S. (2011). Lost in Co-X-Interpretations of Co-design and Co-creation. In *Proceedings of IASDR'11, 4th World Conference on Design Research*, Delft.

[24] Li, J., & Hölttä-Otto, K. (2023). Inconstant Empathy—Interpersonal Factors That Influence the Incompleteness of User Understanding. *J. Mech. Des.*, 145(2), 021403.

[25] Mattelmäki, T., Vaajakallio, K., & Koskinen, I. (2014). What happened to empathic design? *Design Issues*, 30(1), 67-77.

[26] Yalcin, Ö. N., & DiPaola, S. (2018). A computational model of empathy for interactive agents. *Biologically Inspired Cognitive Architectures*, 26, 20-25.

[27] Yalçın, Ö. N., & DiPaola, S. (2020). Modeling empathy: building a link between affective and cognitive processes. *Artif. Intell. Rev.*, 53(4), 2983-3006.

[28] Asada, M. (2015). Development of artificial empathy. *Neuroscience Research*, 90, 41-50.

[29] Asada, M. (2015). Towards artificial empathy: how can artificial empathy follow the developmental pathway of natural empathy?. *Int. J. Soc. Robot.*, 7, 19-33.

[30] Rossi, P. G., & Fedeli, L. (2015). Empathy, education and AI. *Int. J. Soc. Robot.*, 7, 103-109.

[31] Pepito, J. A., Ito, H., Betriana, F., Tanioka, T., & Locsin, R. C. (2020). Intelligent humanoid robots expressing artificial humanlike empathy in nursing situations. *Nursing Philosophy*, 21(4), e12318.

[32] Liu-Thompkins, Y., Okazaki, S., & Li, H. (2022). Artificial empathy in marketing interactions: Bridging the human-AI gap in affective and social customer experience. *J. Acad. Mark. Sci.*, 50(6), 1198-1218.

[33] Ma, L., & Sun, B. (2020). Machine learning and AI in marketing–Connecting computing power to human insights. *Int. J. Res. Mark.*, 37(3), 481-504.

[34] Saura, J. R., Ribeiro-Soriano, D., & Palacios-Marqués, D. (2021). From user-generated data to data-driven innovation: A research agenda to understand user privacy in digital markets. *Int. J. Inf. Manag.*, 60, 102331.

[35] Timoshenko, A., & Hauser, J. R. (2019). Identifying customer needs from user-generated content. *Marketing Science*, 38(1), 1-20.

[36] Zhou, F., Jiao, J. R., Yang, X. J., & Lei, B. (2017). Augmenting feature model through customer preference mining by hybrid sentiment analysis. *Expert Syst. Appl.*, 89, 306-317.

[37] Salminen, J., Guan, K., Jung, S. G., & Jansen, B. J. (2021). A survey of 15 years of data-driven persona development. *Int. J. Hum.-Comput. Interact.*, 37(18), 1685-1708.

[38] Wang, X., Liu, A., & Kara, S. (2023). Constructing Product Usage Context Knowledge Graph Using User-Generated Content for User-Driven Customization. *J. Mech. Des.*, 1-48.

[39] Batbaatar, E., Li, M., & Ryu, K. H. (2019). Semantic-emotion neural network for emotion recognition from text. *IEEE Access*, 7, 111866-111878.

[40] Poria, S., Cambria, E., Hazarika, D., Majumder, N., Zadeh, A., & Morency, L. P. (2017). Context-dependent sentiment analysis in user-generated videos. In *Proceedings of the 55th Annual Meeting of the ACL*, pp. 873-883.

[41] Liu, A., Wang, Y., Wang, X., Liu, A., Wang, Y., & Wang, X. (2022). User-Generated Content Analysis for Customer Needs Elicitation. *Data-Driven Engineering Design*, 23-40.

[42] Hu, N., Pavlou, P. A., & Zhang, J. (2017). On self-selection biases in online product reviews. *MIS quarterly*, 41(2), 449-475.

[43] Bender, E. M., & Friedman, B. (2018). Data statements for natural language processing: Toward mitigating system bias and enabling better science. *Transactions of the ACL*, 6, 587-604.

[44] Wang, G., Zhang, X., Tang, S., Zheng, H., & Zhao, B. Y. (2016). Unsupervised clickstream clustering for user behavior analysis. In *Proceedings of the 2016 CHI Conference on Human Factors in Computing Systems*, 225-236.

[45] García, M. D. M. R., García-Nieto, J., & Aldana-Montes, J. F. (2016). An ontology-based data integration approach for web analytics in e-commerce. *Expert Syst. Appl.*, 63, 20-34.

[46] Köppen, E., Meinel, C. (2015). Empathy via Design Thinking: Creation of Sense and Knowledge. In: Plattner, H., Meinel, C., Leifer, L. (eds), *Design Thinking Research. Understanding Innovation*. Springer, Cham.

[47] Holtzblatt, K., & Beyer, H. (1997). Contextual design: defining customer-centered systems. Morgan Kaufmann Publishers Inc. San Francisco, CA, USA.

[48] Aktinson, P., & Hammersley, M. (1998). Ethnography and participant observation. *Strategies of Qualitative Inquiry*. Thousand Oaks: Sage, 248-261.

[49] Laurans, G., Desmet, P.M.A. and Hekkert, P. (2009) Assessing Emotion in Interaction: Some Problems and a New Approach. In A. Guenand (ed.), *Proceedings of the 4th Int'l Conference on Designing Pleasurable Products and Interfaces*, Compiegne (France), 13–16 October 2009, pp. 230–239.

[50] Rosenberg, B. D., & Navarro, M. A. (2018). Semantic differential scaling. In B. B. Frey (ed.), *The SAGE Encyclopedia of Educational Research, Measurement, and Evaluation* (Vol. 1, pp. 1504-1507). SAGE Publications, Inc.

[51] Boess, S. (2006). Rationales for role playing in design. In Friedman, K., Love, T., Côrte-Real, E. and Rust, C. (eds.), *Wonderground - DRS International Conference 2006*, 1-4 November, Lisbon, Portugal.

[52] Asher, T., Ogle, E., Bailenson, J., & Herrera, F. F. (2018). Becoming homeless: a human experience. In *ACM SIGGRAPH 2018 virtual, augmented, and mixed reality* (pp. 1-1).

[53] Hess, J. L., & Fila, N. D. (2016). The manifestation of empathy within design: findings from a service-learning course. *CoDesign*, 12(1-2), 93-111.

[54] Walther, J., Miller, S. E., & Sochacka, N. W. (2017). A model of empathy in engineering as a core skill, practice orientation, and professional way of being. *J. Eng. Educ.*, 106(1), 123-148.

Copyright © 2023 by ASME

[55] Smeenk, W., Tomico, O., & van Turnhout, K. (2016). A systematic analysis of mixed perspectives in empathic design: Not one perspective encompasses all. *Int. J. Des.*, 10(2), 31-48.

[56] Oygür, I. (2018). The machineries of user knowledge production. *Design Studies*, 54, 23-49.

[57] Fila, N. D., & Hess, J. L. (2015). Exploring the role of empathy in a service-learning design project. *Analyzing Design Review Conversations*, 135-154.

[58] Sanders, E. B. N., & Stappers, P. J. (2014). Probes, toolkits and prototypes: three approaches to making in codesigning. *CoDesign*, 10(1), 5-14.

[59] Norman, D. A. (2004). Emotional design: Why we love (or hate) everyday things. Civitas Books.

[60] Mattelmäki, T., Routarinne, S., & Ylirisku, S. (2011). Triggering the storytelling mode. In *Proceedings of the Conference on Participatory Innovation* (pp. 38-44).

[61] Goldman, A. I. (2006). Simulating minds: The philosophy, psychology, and neuroscience of mindreading. Oxford University Press on Demand.

[62] Omdahl, B. L. (2014). Cognitive appraisal, emotion, and empathy. Psychology Press.

[63] Preston, S. D., & De Waal, F. B. (2002). Empathy: Its ultimate and proximate bases. *Behav. Brain. Sci.*, 25(1), 1-20.

[64] Nummenmaa, L., Hirvonen, J., Parkkola, R., & Hietanen, J. K. (2008). Is emotional contagion special? An fMRI study on neural systems for affective and cognitive empathy. *Neuroimage*, 43(3), 571-580.

[65] Lim, A., & Okuno, H. G. (2015). A recipe for empathy: Integrating the mirror system, insula, somatosensory cortex and motherese. *Int. J. Soc. Robot.*, 7, 35-49.

[66] Carr, L., Iacoboni, M., Dubeau, M. C., Mazziotta, J. C., & Lenzi, G. L. (2003). Neural mechanisms of empathy in humans: a relay from neural systems for imitation to limbic areas. *Proc. Nat. Acad. Sci.*, 100(9), 5497-5502.

[67] Batson, C. D. (2009). These things called empathy: Eight related but distinct phenomena. In J. Decety & W. Ickes (Eds.), *The social neuroscience of empathy* (pp. 3–15). Boston Review.

[68] Harwood, M. D., & Farrar, M. J. (2006). Conflicting emotions: The connection between affective perspective taking and theory of mind. *Br. J. Dev. Psychol.*, 24(2), 401-418.

[69] Frith, C., & Frith, U. (2005). Theory of mind. *Current Biology*, 15(17), 644-645.

[70] De Waal, F. B. (2008). Putting the altruism back into altruism: The evolution of empathy. *Annu. Rev. Psychol.*, 59, 279-300.

[71] Zeng, Z., Pantic, M., Roisman, G. I., & Huang, T. S. (2007). A survey of affect recognition methods: audio, visual and spontaneous expressions. In *Proceedings of the 9th international conference on Multimodal interfaces* (pp. 126-133).

[72] Poria, S., Cambria, E., Bajpai, R., & Hussain, A. (2017). A review of affective computing: From unimodal analysis to multimodal fusion. *Information Fusion*, 37, 98-125.

[73] Busso, C., Deng, Z., Yildirim, S., Bulut, M., Lee, C. M., Kazemzadeh, A., ... & Narayanan, S. (2004). Analysis of emotion recognition using facial expressions, speech and multimodal information. In *Proceedings of the 6th International Conference on Multimodal Interfaces* (pp. 205-211).

[74] He, Z., Li, Z., Yang, F., Wang, L., Li, J., Zhou, C., & Pan, J. (2020). Advances in multimodal emotion recognition based on brain–computer interfaces. *Brain Sciences*, 10(10), 687.

[75] Casas, J., Spring, T., Daher, K., Mugellini, E., Khaled, O. A., & Cudré-Mauroux, P. (2021). Enhancing conversational agents with empathic abilities. In *Proceedings of the 21st ACM Int'l Conference on Intelligent Virtual Agents*, pp. 41-47.

[76] Scherer, K. R. (2010). Emotion and emotional competence: conceptual and theoretical issues for modelling agents. *Blueprint for Affective Computing: A Sourcebook*, 3-20.

[77] Healey, M. L., & Grossman, M. (2018). Cognitive and affective perspective-taking: evidence for shared and dissociable anatomical substrates. *Front. Neurol.*, 9, 491.

[78] Rabinowitz, N., Perbet, F., Song, F., Zhang, C., Eslami, S. A., & Botvinick, M. (2018). Machine theory of mind. In *Int'l Conference on Machine Learning* (pp. 4218-4227).

[79] Jara-Ettinger, J. (2019). Theory of mind as inverse reinforcement learning. *Curr. Opin. Psychol. Sci.*, 29, 105-110.

[80] Kosinski, M. (2023). Theory of mind may have spontaneously emerged in large language models. arXiv preprint arXiv:2302.02083.

[81] Gopnik, A., & Wellman, H. M. (1994). The theory theory. In *An earlier version of this chapter was presented at the Society for Research in Child Development Meeting*, 1991. Cambridge University Press.

[82] McQuiggan, S. W., Robison, J. L., Phillips, R., & Lester, J. C. (2008). Modeling parallel and reactive empathy in virtual agents: An inductive approach. In *Proceedings of the 7th international joint conference on Autonomous agents and multiagent systems*-Volume 1 (pp. 167-174).

[83] Dendorfer, P., Rezatofighi, H., Milan, A., Shi, J., Cremers, D., Reid, I., ... & Leal-Taixé, L. (2020). Mot20: A benchmark for multi object tracking in crowded scenes. arXiv preprint arXiv:2003.09003.

[84] Lin, W., Liu, H., Liu, S., Li, Y., Qian, R., Wang, T., ... & Sebe, N. (2020). Human in events: A large-scale benchmark for human-centric video analysis in complex events. arXiv preprint arXiv:2005.04490.

[85] Kumar, H., & Singh, P. (2015). Neuromarketing: An emerging tool of market research. *Int. J. Electron. Mark. Retail.*, 5(6), 530-535.

[86] Hsu, M. (2017). Neuromarketing: inside the mind of the consumer. *California management review*, 59(4), 5-22.

[87] Gero, J. S., & Milovanovic, J. (2020). A framework for studying design thinking through measuring designers' minds, bodies and brains. *Design Science*, 6, e19.

[88] Hay, L., Duffy, A. H. B., Gilbert, S. J., & Grealy, M. A. (2022). Functional magnetic resonance imaging (fMRI) in design studies: Methodological considerations, challenges, and recommendations. *Design Studies*, 78, 101078.

[89] Krakovsky, M. (2018). Artificial (emotional) intelligence. *Communications of the ACM*, 61(4), 18-19.

[90] Salmi, A., Li, J., & Höltta-Otto, K. (2022). Facial expression recognition as a measure of user-designer empathy. In

Copyright © 2023 by ASME

International Design Engineering Technical Conferences and Computers and Information in Engineering Conference (Vol. 86267, p. V006T06A022). ASME.

[91] Troiano, E., Oberländer, L. A. M., Wegge, M., & Klinger, R. (2022). x-enVENT: A Corpus of Event Descriptions with Experiencer-specific Emotion and Appraisal Annotations. In *Proceedings of the 13th Language Resources and Evaluation Conference* (pp. 1365-1375).

[92] Troiano, E., Oberländer, L., & Klinger, R. (2022). Dimensional modeling of emotions in text with appraisal theories: Corpus creation, annotation reliability, and prediction. *Computational Linguistics*, 1-72.

[93] Wellman, H. M. (2018). Theory of mind: The state of the art. *Eur. J. Dev. Psychol.*, 15(6), 728-755.

[94] Chang-Arana, Á. M., Surma-Aho, A., Li, J., Yang, M. C., & Hölttä-Otto, K. (2020). Reading the user's mind: designers show high accuracy in inferring design-related thoughts and feelings. In *Int'l Des. Eng. Tech. Conf. and Comp. and Info. in Eng. Conf.* (Vol. 83976, p. V008T08A029). ASME.

[95] Zhou, F., Jiao, J. R., & Linsey, J. S. (2015). Latent customer needs elicitation by use case analogical reasoning from sentiment analysis of online product reviews. *J. Mech. Des.*, 137(7), 071401.

[96] Yu, M. H., Li, J., Liu, D., Zhao, D., Yan, R., Tang, B., & Zhang, H. (2020). Draft and edit: Automatic storytelling through multi-pass hierarchical conditional variational autoencoder. In *Proc. AAAI Conf. on AI* (Vol. 34, No. 02, pp. 1741-1748).

[97] Bensaid, E., Martino, M., Hoover, B., & Strobelt, H. (2021). Fairytailor: A multimodal generative framework for storytelling. arXiv preprint arXiv:2108.04324.

[98] Brown, T., Mann, B., Ryder, N., Subbiah, M., Kaplan, J. D., Dhariwal, P., ... & Amodei, D. (2020). Language models are few-shot learners. *Advances in Neural Information Processing Systems*, 33, 1877-1901.

[99] Rombach, R., Blattmann, A., Lorenz, D., Esser, P., & Ommer, B. (2022). High-resolution image synthesis with latent diffusion models. In *Proc. of the IEEE/CVF Conf. on Comp. Vision and Pattern Recognition* (pp.10684-10695).

[100] Davis, E., & Marcus, G. (2015). Commonsense reasoning and commonsense knowledge in artificial intelligence. *Communications of the ACM*, 58(9), 92-103.

[101] Jara-Ettinger, J., Gweon, H., Schulz, L. E., & Tenenbaum, J. B. (2016). The naïve utility calculus: Computational principles underlying commonsense psychology. *Trends in Cognitive Sciences*, 20(8), 589-604.

[102] Sarica, S., Han, J., & Luo, J. (2023). Design representation as semantic networks. *Computers in Industry*, 144, 103791.

[103] OpenAI. GPT-4 Technical Report. Retrieved from: https://cdn.openai.com/papers/gpt-4.pdf

[104] Zellers, R., Holtzman, A., Bisk, Y., Farhadi, A., & Choi, Y. (2019). HellaSwag: Can a Machine Really Finish Your Sentence?. In *Proceedings of the 57th Annual Meeting of the Association for Computational Linguistics* (pp. 4791-4800).

[105] Alzayed, M. A., McComb, C., Menold, J., Huff, J., & Miller, S. R. (2021). Are you feeling me? An exploration of empathy development in engineering design education. *J. Mech. Des.*, 143(11).

[106] Alzayed, M. A., Starkey, E. M., Ritter, S. C., & Prabhu, R. (2022). Am I Right? Investigating the Influence of Trait Empathy and Attitudes Towards Sustainability on the Accuracy of Concept Evaluations in Sustainable Design. In *Int'l Des. Eng. Tech. Conf. and Comp. and Info. in Eng. Conf.* (Vol. 86236, p. V03BT03A007). ASME.

[107] Ickes, W. (1993). Empathic accuracy. *Journal of personality*, 61(4), 587-610.

[108] Chang-Arana, Á. M., Piispanen, M., Himberg, T., Surma-aho, A., Alho, J., Sams, M., & Hölttä-Otto, K. (2020). Empathic accuracy in design: Exploring design outcomes through empathic performance and physiology. *Design Science*, 6, e16.

[109] Li, J., & Hölttä-Otto, K. (2022). Does Empathising With Users Contribute to Better Need Finding?. In *Int'l Des. Eng. Tech. Conf. and Comp. and Info. in Eng. Conf.* (Vol. 86267, p. V006T06A023). ASME.

[110] Zhou, F., Ayoub, J., Xu, Q., & Jessie Yang, X. (2020). A machine learning approach to customer needs analysis for product ecosystems. *J. Mech. Des.*, 142(1).

[111] Kano, N., Seraku, N., Takahashi, F., and Tsuji, S. (1984). Attractive Quality and Must-Be Quality. *The Japanese Society for Quality Control.)*,14(2), pp.39–48.

[112] Jiang, S., Hu, J., Magee, C. L., & Luo, J. (2022). Deep learning for technical document classification. *IEEE Trans. Eng. Manag.*

[113] Song, B., Zhou, R., & Ahmed, F. (2023). Multi-modal Machine Learning in Engineering Design: A Review and Future Directions. arXiv preprint arXiv:2302.10909.

[114] Siddharth, L., Blessing, L., & Luo, J. (2022). Natural language processing in-and-for design research. *Design Science*, 8, e21.

[115] Siddharth, L., Blessing, L. T., Wood, K. L., & Luo, J. (2022). Engineering knowledge graph from patent database. *J. Comput. Inf. Sci. Eng.*, 22(2), 021008.

[116] Pileggi, S. F. (2021). Knowledge interoperability and re-use in Empathy Mapping: An ontological approach. *Expert Syst. Appl.*, 180, 115065.

[117] Desmet, P. (2003). A multilayered model of product emotions. *Des. J.*, 6(2), 4-13.

[118] Demir, E., Desmet, P. M., & Hekkert, P. (2009). Appraisal patterns of emotions in human-product interaction. *Int. J. Des.*, 3(2).

[119] Desmet, P. M. A. (2010). Three levels of product emotion. In *Proceedings of the international conference on Kansei engineering and emotion research* (pp. 236-246).

[120] Mehrabi, N., Morstatter, F., Saxena, N., Lerman, K., & Galstyan, A. (2021). A survey on bias and fairness in machine learning. *ACM Computing Surveys*, 54(6), 1-35.

[121] Bender, E. M., Gebru, T., McMillan-Major, A., & Shmitchell, S. (2021). On the Dangers of Stochastic Parrots: Can Language Models Be Too Big?. In *Proceedings of the 2021 ACM Conference on Fairness, Accountability, and Transparency* (pp. 610-623).

Copyright © 2023 by ASME

AUTHOR INDEX

Ahmed, Faez 254
Allen, Janet K. 47, 641
Alzayed, Mohammad Alsager 180
Amos G. Winter V 151, 307
Amrose, Susan 151
Arefin, Amit M. E. 535
Austin-Breneman, Jesse 209
Azad, Saeed 351
Azarm, Shapour 490
Babatunde, Bolutito 91
Baby, Mathew 432
Balaji, Niharika 432
Bansal, Parth 520
Barbat, Saeed 727
Bayrak, Alparslan Emrah 480
Bedi, Saloni 200
Behandish, Morad 268
Bertelsen, Lauren 717
Bessette, Jonathan 162
Bhalerao, Mayank J. 47
Biswas, Arpan 594
Bliss, Morgan 232
Boahen, Samuel 209
Bono, Michael 307
Boozarjomehry, Ghazal Bozorgmehry 761
Bostanabad, Ramin 103
Bowers, Quinn 162
Brei, Melissa 173
Burns, Devin E. 130
Busogi, Moise 190
Byiringiro, Jean 190
Cagan, Jonathan 91, 391, 570
Campbell, Matthew I. 278
Cao, Jinhui 641
Chakravarthula, Praneet Nallan 209
Chen, Dongsheng 27
Chen, Shikui 409, 546
Chen, Siyu 480
Chen, Suhao 447
Chen, Wei 71
Chesang, Andrew 190
Clarke-Sather, Abigail R. 232
Comlek, Yigitcan 71
Cook-Wright, Joshua F. 219
Cordero, Sergio 137
Costello, Jeffrey 151
Crispo, Luke 402
Dahlin, Eric 219

Das, Ashok K. 47
Deabae, Reza 500
Dickerson, Tevin J. 662
Dickerson, Tevin 419
Du, Xiaoping 460, 738
Edwards, Kristen M. 254
Egan, Paul F. 535
Engelbert, Mark 254
Faghihrasoul, Zahra 278
Feng, Wangchuan 391
Feng, Yukun 293
Ferguson, Scott 13
Floryan, Marie 162
Fodale, Rebekah 180
Foumani, Zahra Zanjani 103
Fujita, Kikuo 81, 374
Gadi, Vikranth S. 1
Gao, Zhenyan 727
Geilman, Thomas B. 219
Gembali, Sahas 162
Ghodgaonkar, Aditya 151, 307
Gichane, Michael 190
Gill, Amaninder Singh 657
Gorsich, David 527
Goucher-Lambert, Kosa 557
Goyal, Akshita 162
Grant, Fiona 151
Griffin, Amari 200
Grover, Micki 242
Gu, Xianfeng David 546
Guan, Kaiwen 385
Guo, Lin 447
Gyory, Joshua T. 570
Hamdan, Bayan 137
Herber, Daniel R. 351
Herzog, Jenna 180
Honarparvar, Soraya 162
Honeycutt, Wesley T. 47
Hong, Chloe 570
Hoyle, Christopher 328
Hu, Chuanhui 617
Hu, Zhen 527, 727
Huang, Bingling 37
Huang, Carolyn 254
Huang, Yuhao 402
Huang, Zonghao 27
Jacobson, Lindsey 13
Jesse, Stephen 594

Jetton, Cole	328
Jin, Yan	37, 617, 770
Judge, Benjamin	307
Kaklis, Panagiotis	557
Kalinin, Sergei	594
Kania, Randall J.	490
Kato, Misato	81
Khan, Shahroz	557
Khanghah, Kiarash Naghavi	113
Khoda, Bashir	300
Kii, Taisei	81, 374
Kim, Il Yong	402
Kim, Jungin E.	340
Kimbowa, Alvin B.	209
Konrath, Austin	232
Koratikere, Pavankumar	470
Laroche, Regina	232
Lauff, Carlye A.	242
Lawrence, Matthew	300
Lee, Junghun	190
Lee, Soobum	130
Leifsson, Leifur	470
Li, Chengda	328
Li, Guosong	727
Li, Huiru	460
Li, Yang	727
Li, Yumeng	320, 520
Li, Zhou	137
Liao, Ting	583
Liu, Xinyang	510
Liu, Yongtao	594
Liu, Yu-Chen	594
Liu, Zheng	365, 520
Longtin, Jon	409
Luo, Fang	409
Luo, Jianxi	784
Mabey, Christopher S.	219, 662
Masawi, Rumbidzai	232
Massoudi, Soheyl	746
Matsushima, Kei	65, 385
Mattson, Christopher A.	219, 419, 662
McComb, Christopher	570
Ming, Zhenjun	641
Mistree, Farrokh	47, 641
Miyingo, Emmanuel Wokulira	209
Mourelatos, Zissimos P.	527
Murai, Naoki	124
Najmon, Joel C.	144
Ndhlovu, Chiratidzo	242
Nellippallil, Anand Balu	432
Noda, Masaki	65
Noguchi, Yuki	124, 293

Nolte, Hannah	570
Oviroh, Peter Ozaveshe	209
Panchal, Jitesh H.	1, 460
Papalambros, Panos Y.	209
Persia, Jude Thaddeus	130
Petelina, Nina T.	200
Picard, Cyril	746
Porciello, Jaron	254
Prabhu, Rohan	180
Ramu, Palaniappan	432
Reddie, Madison	200
Renteria, Anabel	137
Rudd, Liam	278
Sadasivan, Chandramouli	546
Salmon, John L.	219, 419, 662
Schiffmann, Jurg	746
Seepersad, Carolyn C.	374
Selva, Daniel	607
Senesky, Debbie G.	137
Sha, Zhenghui	374, 480
Sharma, Gehendra	432
Sheline, Carolyn	151
Shishehbor, Mehdi	103
Short, Ada-Rhodes	675, 717
Singh, Amandeep	527
Song, Binyang	254
Song, Hyeonik	607
Starkey, Elizabeth M.	180
Sternberg, Zachary	162
Stump, Gary	570
Sung, Cynthia	27
Sung, Myung Kyun	130
Sushil, Rashmi Rama	432
Szajnfarber, Zoe	1
Tassiopoulos, Apostolos	546
Tau, Seth	527
Taylor, Rebecca E.	91
Thekinen, Joseph	761
Tian, Jiawei	409, 546
Tomlinson, Scott	300
Topcu, Taylan G.	1
Torrey, David	409
Tovar, Andres	144
Tucker, Conrad	190
Vaidya, Manasi	200
Van De Zande, Georgia D.	151
Vasudevan, Rama	594
Wang, Liwei	71
Wang, Pingfeng	137, 365, 510, 520
Wang, Randi	268
Wang, Xinrui	770
Wang, Yan	340

Welsh, Emily..307
Winter, Amos G. 200
Winter, Amos 162
Wright, Natasha C. 242
Wu, Hao 365, 520
Wu, Yulun 320
Xi, Zhimin 631
Xu, Hongyi 113
Xu, Leidong 113
Xu, Yanwen..137, 365
Xu, Zeda 570
Xue, Deyi 500
Yaji, Kentaro 81, 374
Yamada, Takayuki......................65, 124, 293, 385
Yan, Bei 583
Yang, Jan-Chi...594
Yin, Jianhua 527
Yousefpour, Amin 103
Zeng, Jice 727
Zhang, Guanglu 391
Zhao, Ying 727
Zhu, Qihao 784
Zhuang, Ran...546
Ziatdinov, Maxim 594
Zurita, Nicolas F. Soria 570